EDA for IC System Design, Verification, and Testing

Electronic Design Automation for Integrated Circuits Handbook

Edited by
Louis Scheffer, Luciano Lavagno, and Grant Martin

EDA for IC System Design, Verification, and Testing

EDA for IC Implementation, Circuit Design, and Process Technology

EDA for IC System Design, Verification, and Testing

Edited by

Louis Scheffer
Cadence Design Systems
San Jose, California, U.S.A.

Luciano Lavagno
Cadence Berkeley Laboratories
Berkeley, California, U.S.A.

Grant Martin
Tensilica Inc.
Santa Clara, California, U.S.A.

Taylor & Francis
Taylor & Francis Group
Boca Raton London New York

A CRC title, part of the Taylor & Francis imprint, a member of the Taylor & Francis Group, the academic division of T&F Informa plc.

Published in 2006 by
CRC Press
Taylor & Francis Group
6000 Broken Sound Parkway NW, Suite 300
Boca Raton, FL 33487-2742

Printed in the United States of America on acid-free paper
10 9 8 7 6 5 4 3 2 1

International Standard Book Number-10: 0-8493-7923-7 (Hardcover)
International Standard Book Number-13: 978-0-8493-7923-9 (Hardcover)
Library of Congress Card Number 2005052924

Library of Congress Cataloging-in-Publication Data

EDA for IC system design, verification, and testing / editors, Louis Scheffer, Luciano Lavagno, Grant Martin.
p. cm. -- (Electronic design and automation for integrated circuits handbook)
Includes bibliographical references and index.
ISBN 0-8493-7923-7
1. Integrated circuits--Computer-aided design. 2. Integrated circuits--Verification--Data processing. I. Title: Electronic design automation for integrated circuit system design, verification, and testing. II. Scheffer, Louis. III. Lavagno, Luciano, 1959- IV. Martin, Grant (Grant Edmund) V. Series.

TK7874.E26 2005
621.3815--dc22 2005052924

informa
Taylor & Francis Group
is the Academic Division of Informa plc.

Visit the Taylor & Francis Web site at
http://www.taylorandfrancis.com

and the CRC Press Web site at
http://www.crcpress.com

Acknowledgments and Dedication for the EDA Handbook

The editors would like to acknowledge the unsung heroes of EDA, those who have worked to advance the field, in addition to advancing their own personal, corporate, or academic agendas. These are the men and women who have played a variety of key roles — they run the smaller conferences, they edit technical journals, and they serve on standards committees, just to name a few. These largely volunteer jobs do not make anyone rich or famous despite the time and effort that goes into them, but they do contribute mightily to the remarkable and sustained advancement of EDA. Our kudos to these folks, who do not get the credit they deserve.

On a more personal note, Louis Scheffer would like to acknowledge the love, support, encouragement, and help of his wife Lynde, his daughter Lucynda, and his son Loukos. Without them this project would not have been possible.

Luciano Lavagno would like to thank his wife Paola and his daughter Alessandra Chiara for making his life so wonderful.

Grant Martin would like to acknowledge, as always, the love and support of his wife, Margaret Steele, and his two daughters, Jennifer and Fiona.

Preface

Preface for Volume 1

Electronic Design Automation (EDA) is a spectacular success in the art of engineering. Over the last quarter of a century, improved tools have raised designers' productivity by a factor of more than a thousand. Without EDA, Moore's law would remain a useless curiosity. Not a single billion-transistor chip could be designed or debugged without these sophisticated tools — without EDA we would have no laptops, cell phones, video games, or any of the other electronic devices we take for granted.

Spurred on by the ability to build bigger chips, EDA developers have largely kept pace, and these enormous chips can still be designed, debugged, and tested, even with decreasing time-to-market.

The story of EDA is much more complex than the progression of integrated circuit (IC) manufacturing, which is based on simple physical scaling of critical dimensions. EDA, on the other hand, evolves by a series of paradigm shifts. Every chapter in this book, all 49 of them, was just a gleam in some expert's eye just a few decades ago. Then it became a research topic, then an academic tool, and then the focus of a start-up or two. Within a few years, it was supported by large commercial EDA vendors, and is now part of the conventional wisdom. Although users always complain that today's tools are not quite adequate for today's designs, the overall improvements in productivity have been remarkable. After all, in which other field do people complain of *only* a 21% compound annual growth in productivity, sustained over three decades, as did the International Technology Roadmap for Semiconductors in 1999?

And what is the future of EDA tools? As we look at the state of electronics and IC design in 2005–2006, we see that we may soon enter a major period of change in the discipline. The classical scaling approach to ICs, spanning multiple orders of magnitude in the size of devices over the last 40+ years, looks set to last only a few more generations or process nodes (though this has been argued many times in the past, and has invariably been proved to be too pessimistic a projection). Conventional transistors and wiring may well be replaced by new nano- and biologically based technologies that we are currently only beginning to experiment with. This profound change will surely have a considerable impact on the tools and methodologies used to design ICs. Should we be spending our efforts looking at Computer Aided Design (CAD) for these future technologies, or continue to improve the tools we currently use?

Upon further consideration, it is clear that the current EDA approaches have a lot of life left in them. With at least a decade remaining in the evolution of current design approaches, and hundreds of thousands or millions of designs left that must either craft new ICs or use programmable versions of them, it is far too soon to forget about today's EDA approaches. And even if the technology changes to radically new forms and structures, many of today's EDA concepts will be reused and built upon for design of technologies well beyond the current scope and thinking.

The field of EDA for ICs has grown well beyond the point where any single individual can master it all, or even be aware of the progress on all fronts. Therefore, there is a pressing need to create a snapshot of this extremely broad and diverse subject. Students need a way of learning about the many disciplines and topics involved in the design tools in widespread use today. As design grows multi-disciplinary, electronics designers and EDA tool developers need to broaden their scope. The methods used in one subtopic may well have applicability to new topics as they arise. All of electronics design can utilize a comprehensive reference work in this field.

With this in mind, we invited many experts from across all the disciplines involved in EDA to contribute chapters summarizing and giving a comprehensive overview of their particular topic or field. As might be appreciated, such chapters represent a snapshot of the state of the art in 2004–2005. However, as surveys and overviews, they retain a lasting educational and reference value that will be useful to students and practitioners for many years to come.

With a large number of topics to cover, we decided to split the Handbook into two volumes. Volume One covers system-level design, micro-architectural design, and verification and test. Volume Two covers the classical "RTL to GDS II" design flow, incorporating synthesis, placement and routing, along with related topics; analog and mixed-signal design, physical verification, analysis and extraction, and technology CAD topics for IC design. These roughly correspond to the classical "front-end/back-end" split in IC design, where the front-end (or logical design) focuses on making sure that the design does the right thing, assuming it can be implemented, and the back-end (or physical design) concentrates on generating the detailed tooling required, while taking the logical function as given. Despite limitations, this split has persisted through the years — a complete and correct logical design, independent of implementation, remains an excellent handoff point between the two major portions of an IC design flow. Since IC designers and EDA developers often concentrate on one side of this logical/physical split, this seemed to be a good place to divide the book as well.

In particular, Volume One starts with a general introduction to the topic, and an overview of IC design and EDA. System-level design incorporates many aspects — application-specific tools and methods, special specification and modeling languages, integration concepts including the use of Intellectual Property (IP), and performance evaluation methods; the modeling and choice of embedded processors and ways to model software running on those processors; and high-level synthesis approaches. ICs that start at the system level need to be refined into micro-architectural specifications, incorporating cycle-accurate modeling, power estimation methods, and design planning. As designs are specified and refined, verification plays a key role — and the handbook covers languages, simulation essentials, and special verification topics such as transaction-level modeling, assertion-based verification, and the use of hardware acceleration and emulation as well as emerging formal methods. Finally, making IC designs testable and thus cost-effective to manufacture and package relies on a host of test methods and tools, both for digital and analog and mixed-signal designs.

This handbook with its two constituent volumes is a valuable learning and reference work for everyone involved and interested in learning about electronic design and its associated tools and methods. We hope that all readers will find it of interest and that it will become a well-thumbed resource.

Louis Scheffer
Luciano Lavagno
Grant Martin

Editors

Louis Scheffer

Louis Scheffer received the B.S. and M.S. degrees from Caltech in 1974 and 1975, and a Ph.D. from Stanford in 1984. He worked at Hewlett Packard from 1975 to 1981 as a chip designer and CAD tool developer. In 1981, he joined Valid Logic Systems, where he did hardware design, developed a schematic editor, and built an IC layout, routing, and verification system. In 1991, Valid merged with Cadence, and since then he has been working on place and route, floorplanning systems, and signal integrity issues.

His main interests are floorplanning and deep submicron effects. He has written many technical papers, tutorials, invited talks, and panels, and has served the DAC, ICCAD, ISPD, SLIP, and TAU conferences as a technical committee member. He is currently the general chair of TAU and ISPD, on the steering committee of SLIP, and an associate editor of IEEE Transactions on CAD. He holds five patents in the field of EDA, and has taught courses on CAD for electronics at Berkeley and Stanford. He is also interested in SETI, and serves on the technical advisory board for the Allen Telescope Array at the SETI institute, and is a co-author of the book SETI-2020, in addition to several technical articles in the field.

Luciano Lavagno

Luciano Lavagno received his Ph.D. in EECS from U.C. Berkeley in 1992 and from Politecnico di Torino in 1993. He is a co-author of two books on asynchronous circuit design, of a book on hardware/software co-design of embedded systems, and of over 160 scientific papers.

Between 1993 and 2000, he was the architect of the POLIS project, a cooperation between U.C. Berkeley, Cadence Design Systems, Magneti Marelli and Politecnico di Torino, which developed a complete hardware/software co-design environment for control-dominated embedded systems.

He is currently an Associate Professor with Politecnico di Torino, Italy and a research scientist with Cadence Berkeley Laboratories. He serves on the technical committees of several international conferences in his field (e.g., DAC, DATE, ICCAD, ICCD) and of various workshops and symposia. He has been the technical program and tutorial chair of DAC, and the technical program and general chair of CODES. He has been associate and guest editor of IEEE Transactions on CAD, IEEE Transactions on VLSI and ACM Transactions on Embedded Computing Systems.

His research interests include the synthesis of asynchronous and low-power circuits, the concurrent design of mixed hardware and software embedded systems, as well as compilation tools and architectural design of dynamically reconfigurable processors.

Grant Martin

Grant Martin is a Chief Scientist at Tensilica, Inc. in Santa Clara, California. Before that, Grant worked for Burroughs in Scotland for 6 years; Nortel/BNR in Canada for 10 years; and Cadence Design Systems for 9 years, eventually becoming a Cadence Fellow in their Labs. He received his Bachelors and Masters degrees in Mathematics (Combinatorics and Optimization) from the University of Waterloo, Canada, in 1977 and 1978.

Grant is a co-author of *Surviving the SOC Revolution: A Guide to Platform-Based Design*, 1999, and *System Design with SystemC*, 2002, and a co-editor of the books *Winning the SoC Revolution: Experiences in Real Design*, and *UML for Real: Design of Embedded Real-Time Systems*, June 2003, all published by Springer (originally by Kluwer). In 2004, he co-wrote with Vladimir Nemudrov the first book on SoC design published in Russian by Technosphera, Moscow. Recently, he co-edited *Taxonomies for the Development and Verification of Digital Systems* (Springer, 2005), and *UML for SoC Design* (Springer, 2005).

He has also presented many papers, talks and tutorials, and participated in panels, at a number of major conferences. He co-chaired the VSI Alliance Embedded Systems study group in the summer of 2001, and is currently co-chair of the DAC Technical Programme Committee for Methods for 2005 and 2006. His particular areas of interest include system-level design, IP-based design of system-on-chip, platform-based design, and embedded software. He is a senior member of the IEEE.

Contributors

Iuliana Bacivarov
SLS Group, TIMA Laboratory
Grenoble, France

Mike Bershteyn
Cadence Design Systems, Inc.
Cupertino, California

Shuvra Bhattacharyya
University of Maryland
College Park, Maryland

Joseph T. Buck
Synopsys, Inc.
Mountain View, California

Raul Camposano
Synopsys Inc.
Mountain View, California

Naehyuck Chang
Seoul National University
Seoul, South Korea

Kwang-Ting (Tim) Cheng
University of California
Santa Barbara, California

Alain Clouard
STMicroelectronics
Crolles, France

Marcello Coppola
STMicroelectronics
Grenoble, France

Robert Damiano
Synopsys Inc.
Hillsboro, Oregon

Marco Di Natale
Scuola Superiore S. Anna
Pisa, Italy

Nikil Dutt
Donald Bren School of Information and
Computer Sciences,
University of California, Irvine
Irvine, California

Stephen A. Edwards
Columbia University
New York, New York

Limor Fix
Design Technology, Intel
Pittsburgh, Pennsylvania

Harry Foster
Jasper Design Automation
Mountain View, California

Frank Ghenassia
STMicroelectronics
Crolles, France

Miltos D. Grammatikakis
ISD S.A.
Athens, Greece

Rajesh Gupta
University of California, San Diego
San Diego, California

Sumit Gupta
Tensilica Inc.
Santa Clara, California

Ahmed Jerraya
SLS Group, TIMA Laboratory, INPG
Grenoble, France

Bozena Kaminska
Simon Fraser University and
Pultronics Incorporated
Burnaby, British Colombia,
Canada

Bernd Koenemann
Mentor Graphics, Inc.
San Jose, California

Luciano Lavagno
Cadence Berkeley Laboratories
Berkeley, California

Steve Leibson
Tensilica, Inc.
Santa Clara, California

Enrico Macii
Politecnico di Torino
Torino, Italy

Laurent Maillet-Contoz
STMicroelectronics
Crolles, France

Erich Marschner
Cadence Design Systems
Berkeley, California

Grant Martin
Tensilica Inc.
Santa Clara, California

Ken McMillan
Cadence Berkeley Laboratories
Berkeley, California

Renu Mehra
Synopsys, Inc.
Mountain View, California

Prabhat Mishra
University of Florida
Gainesville, Florida

Ralph H.J.M. Otten
Eindhoven University of Technology
Eindhoven, Netherlands

Massimo Poncino
Politecnico di Torino
Torino, Italy

John Sanguinetti
Forte Design Systems, Inc.
San Jose, California

Louis Scheffer
Cadence Design Systems
San Jose, California

Sandeep Shukla
Virginia Tech
Blacksburg, Virginia

Gaurav Singh
Virginia Tech
Blacksburg, Virginia

Jean-Philippe Strassen
STMicroelectronics
Crolles, France

Vivek Tiwari
Intel Corp.
Santa Clara, California

Ray Turner
Cadence Design Systems
San Jose, California

Li-C. Wang
University of California
Santa Barbara, California

John Wilson
Mentor Graphics
Berkshire, United Kingdom

Wayne Wolf
Princeton University
Princeton, New Jersey

Contents

SECTION III Micro-Architecture Design

SECTION IV Logical Verification

SECTION V Test

SECTION I

INTRODUCTION

1 Overview

Luciano Lavagno
Cadence Berkeley Laboratories
Berkeley, California

Grant Martin
Tensilica Inc.
Santa Clara, California

Louis Scheffer
Cadence Design Systems
San Jose, California

Introduction to Electronic Design Automation for Integrated Circuits

Modern integrated circuits (ICs) are enormously complicated, often containing many millions of devices. Design of these ICs would not be humanly possible without software (SW) assistance at every stage of the process. The tools used for this task are collectively called electronic design automation (EDA).

EDA tools span a very wide range, from purely logical tools that implement and verify functionality, to purely physical tools that create the manufacturing data and verify that the design can be manufactured. The next chapter, *The IC Design Process and EDA*, by Robert Damiano and Raul Camposano, discusses the IC design process, its major stages and design flow, and how EDA tools fit into these processes and flows. It particularly looks at interfaces between the major IC design stages and the kind of information — abstractions upwards, and detailed design and verification information downwards — that must flow between these stages.

A Brief History of Electronic Design Automation

This section contains a *very* brief summary of the origin and history of EDA for ICs. For each topic, the title of the relevant chapter(s) is mentioned in *italics*.

The need for tools became clear very soon after ICs were invented. Unlike a breadboard, ICs cannot be modified easily after fabrication, so testing even a simple change involves weeks of delay (for new masks and a new fabrication run) and considerable expense. Furthermore, the internal nodes of an IC are difficult to probe because they are physically small and may be covered by other layers of the IC. Even if these problems can be worked around, the internal nodes often have very high impedances and hence are difficult to measure without dramatically changing the performance. Therefore circuit simulators were crucial to IC design almost as soon as ICs came into existence. These programs are covered in the chapter *Analog Simulation: Circuit Level and Behavioral Level*, and appeared in the 1960s.

Next, as the circuits grew bigger, clerical help was required in producing the masks. At first there were digitizing programs, where the designer still drew with colored pencils but the coordinates were transferred to the computer, written to magnetic tape, and then transferred to the mask making machines. Soon, these early programs were enhanced into full-fledged layout editors. These programs were first developed in the late 1960s and early 1970s. Analog designs in the modern era are still largely laid out manually, with some tool assistance, as *Layout Tools for Analog ICs and Mixed-Signal SoCs: A Survey* will attest, although some developments in more automated optimization have been occurring, along with many experiments in more automated layout techniques.

As the circuits grew larger, getting the logic design correct became difficult, and *Digital Simulation* (i.e., logic simulation) was introduced into the IC design flow. Also, testing of the completed chip proved to be difficult, since unlike circuit boards, internal nodes could not be observed or controlled through a "bed of nails" fixture. Therefore automatic test pattern generation (ATPG) programs were developed that generate test vectors that only refer to the visible pins. Other programs that modified designs to make them more controllable, observable, and testable were not far behind. These programs, covered in *Design-for-Test* and *Automatic Test Pattern Generation*, were first available in the mid-1970s. Specialized *Analog and Mixed-Signal Test* needs were met by special testers and tools.

As the number of design rules, number of layers, and chip sizes all continued to increase, it became increasingly difficult to verify by hand that a layout met all the manufacturing rules, and to estimate the parasitics of the circuit. Therefore *Design Rule Checking*, and *Layout Extraction* programs were developed, starting in the mid-1970s. As the processes became more complex, with more layers of interconnect, the original analytic approximations to R, C, and L values became inadequate, and *High-Accuracy Parasitic Extraction* programs were required to determine more accurate values, or at least calibrate the parameter extractors.

The next bottleneck was doing the detailed designing of each polygon. Placement and routing programs allowed the user to specify only the gate-level netlist — the computer would then decide on the location of the gates and the wires connecting them. Although some silicon efficiency was lost, productivity was greatly improved, and IC design opened up to a wider audience of logic designers. The chapters *Digital Layout — Placement* and *Routing* cover these programs, which became popular in the mid-1980s.

Even just the gate-level netlist soon proved to be of too much detail, and synthesis tools were developed to create such a netlist from a higher level specification, usually expressed in a hardware description language (HDL). This is called *Logic Synthesis* and became available in the mid-1980s. In the last decade, *Power Analysis and Optimization from Circuit to Register Transfer Levels* has become a major area of concern and is becoming the number one optimization criterion for many designs, especially portable and battery powered ones.

Around this time, the large collection of tools that need to be used to complete a single design became a serious problem. Electronic design automation *Design Databases* were introduced to cope with this problem. In addition, *Design Flows* began to become more and more elaborate in order to hook tools together, as well as to develop and support both methodologies and use models for specific design groups, companies, and application areas.

In the late 1990s, as the circuits continued to shrink, noise became a serious problem. Programs that analyzed power and ground networks, cross-talk, and substrate noise in systematic ways became commercially available. The chapters *Design and Analysis of Power Supply Networks*, *Mixed-Signal Noise Coupling in System-on-Chip Design: Modeling, Analysis and Validation*, and *Noise Considerations in Digital ICs* cover these topics.

Gradually through the 1990s and early 2000s, chips and processes became sufficiently complex that the designs that optimize yield were no longer only a minimization of size. *Design for Manufacturability in the Nanometer Era,* otherwise known as "Design for Yield", became a field of its own. Also in this time frame, the size of the features on the chip became comparable to, or less than, the wavelength of the light used to create them. To compensate for this as much as possible, the masks were no longer a direct copy of what the designer intended. The creation of these more complex masks is covered in *Resolution Enhancement Techniques and Mask Data Preparation.*

On a parallel track, developing the process itself was also a difficult problem. *Process Simulation* tools were developed to predict the effects of changing various process parameters. The output from these programs, such as doping profiles, was useful to process engineers but too detailed for electrical analysis. Another suite of tools (see *Device Modeling — from Physics to Electrical Parameter Extraction*) that predict device performance from a physical description of devices was needed and developed. These models were particularly useful when developing a new process.

One of the areas that developed very early in the design of electronic systems, at least in part, but which is the least industrialized as a standard process, is that of system-level design. As the chapter on *Using Performance Metrics to Select Microprocessor Cores for IC Designs* points out, one of the first instruction set simulators appeared soon after the first digital computers did. However, until the present day, system-level design has consisted mainly of a varying collection of tricks, techniques, and *ad hoc* modeling tools.

The logic simulation and synthesis processes introduced in the 1970s and 1980s, respectively, are, as was discussed earlier, much more standardized. The front–end IC design flow would not have been possible to standardize without the introduction of standard HDLs. Out of a huge variety of HDLs introduced from the 1960s to the 1980s, Verilog and VHDL have become the major *Design and Verification Languages.* For a long time — till the mid to late 1990s, verification of digital design seemed stuck at standard digital simulation — although at least since the 1980s, a variety of *Hardware Acceleration and Emulation* solutions have been available to designers. However, advances in verification languages and the growth in design complexity have triggered interest in more advanced verification methods, and the last decade has seen considerable interest in *Using Transactional Level Models in a SoC Design Flow, Assertion-based Verification,* and *Formal Property Verification. Equivalence Checking* has been the formal technique most tightly integrated into design flows, since it allows designs to be compared before and after various optimizations and back-end-related modifications, such as scan insertion.

For many years, specific systems design domains have fostered their own application-specific *Tools and Methodologies for System-Level Design* — especially in the areas of algorithm design from the late 1980s through to this day. The late 1990s saw the emergence of and competition between a number of C/C++-based *System-Level Specification and Modeling Languages.* With the possibility of now incorporating the major functional units of a design (processors, memories, digital and mixed-signal HW blocks, peripheral interfaces, and complex hierarchical buses) all onto a single silicon substrate, the mid-1990s to the present day have also seen the rise of the System-on-chip (SoC). It is thus that the area of *SoC Block-Based Design and IP Assembly* has grown, in which the complexity possible with advanced semiconductor processes is ameliorated to some extent via reuse of blocks of design. Concomitant with the SoC approach has been the development, during the last decade, of *Performance Evaluation Methods for MPSoC Design,* development of embedded processors through specialized *Processor Modelling and Design Tools,* and gradual and still-forming links to *Embedded Software Modelling and Design.* The desire to raise HW design productivity to higher levels has spawned considerable interest in (*Parallelizing*) *High Level Synthesis* over the years. It is now seeing something of a resurgence driven by C/C++/SystemC as opposed to the first-generation high-level synthesis (HLS) tools driven by HDLs in the mid-1990s.

After the system level of design, architects need to descend one level of abstraction to the micro-architectural level. Here, a variety of tools allow one to look at the three main performance criteria: timing or delay (*Cycle-accurate System-Level Modeling and Performance Evaluation*), power (*Micro-Architectural Power Estimation and Optimization*), and physical *Design Planning.* Micro-architects need to make trade-offs between the timing, power, and cost/area attributes of complex ICs at this level.

The last several years have seen a considerable infilling of the design flow with a variety of complementary tools and methods. Formal verification of function is only possible if one is assured that the timing is correct, and by keeping a lid on the amount of dynamic simulation required, especially at the postsynthesis and postlayout gate levels, good *Static Timing Analysis* tools provide the assurance that timing constraints are being met. It is also an underpinning to timing optimization of circuits and for the design of newer mechanisms for manufacturing and yield. Standard cell-based placement and routing are not appropriate for *Structured Digital Design* of elements such as memories and register files, leading to specialized tools. As design groups began to rely on foundries and application specific integrated circuit (ASIC) vendors and as the IC design and manufacturing industry began to "de-verticalize", design libraries, covered in *Exploring Challenges of Libraries for Electronic Design*, became a domain for special design flows and tools. It ensured the availability of a variety of high performance and low power libraries for optimal design choices and allowed some portability of design across processes and foundries. *Tools for Chip-Package Codesign* began to link more closely the design of IOs on chip, the packages they fit into, and the boards on which they would be placed. For implementation "fabrics" such as field-programmable gate arrays (FPGAs), specialized *FPGA Synthesis and Physical Design Tools* are necessary to ensure good results. And a renewed emphasis on *Design Closure* allows a more holistic focus on the simultaneous optimization of design timing, power, cost, reliability, and yield in the design process.

Another area of growing but specialized interest in the analog design domain is the use of new and higher level modeling methods and languages, which are covered in *Simulation and Modeling for Analog and Mixed-Signal Integrated Circuits.*

A much more detailed overview of the history of EDA can be found in [1]. A historical survey of many of the important papers from the International Conference on Computer-Aided Design (ICCAD) can be found in [2].

Major Industry Conferences and Publications

The EDA community, formed in the early 1960s from tool developers working for major electronics design companies such as IBM, AT&T Bell Labs, Burroughs, Honeywell, and others, has long valued workshops, conferences, and symposia, in which practitioners, designers, and later, academic researchers, could exchange ideas and practically demonstrate the techniques. The Design Automation Conference (DAC) grew out of workshops, which started in the early 1960s, and although held in a number of U.S. locations, has in recent years tended to stay on the west coast of the United States or a bit inland. It is the largest combined EDA trade show and technical conference held annually anywhere in the world. In Europe, a number of country-specific conferences held sporadically through the 1980s, and two competing ones held in the early 1990s, led to the creation of the consolidated Design Automation and Test in Europe (DATE) conference, which started in the mid-1990s and has grown consistently in strength ever since. Finally, the Asia-South Pacific DAC (ASP-DAC) started in the mid to late 1990s and completes the trio of major EDA conferences spanning the most important electronics design communities in the world.

Complementing the larger trade show/technical conferences has been ICCAD, which for over 20 years has been held in San Jose, and has provided a more technical conference setting for the latest algorithmic advances in EDA to be presented, attracting several hundred attendees. Various domain areas of EDA knowledge have sparked a number of other workshops, symposia, and smaller conferences over the last 15 years, including the International Symposium on Physical Design (ISPD), International Symposium on Quality in Electronic Design (ISQED), Forum on Design Languages in Europe (FDL), HDL and Design and Verification conferences (HDLCon, DVCon), High-level Design, Verification and Test (HLDVT), International Conference on Hardware–Software Codesign and System Synthesis (CODES+ISSS), and many other gatherings. Of course, the area of Test has its own long-standing International Test Conference (ITC); similarly, there are specialized conferences for FPGA design (e.g., Forum on Programmable Logic [FPL]) and a variety of conferences focusing on the most advanced IC designs such as the International Solid–State Circuits Conference (ISSCC) and its European counterpart (ESSCC).

There are several technical societies with strong representation of design automation: one is the Institute of Electrical and Electronics Engineers (IEEE, pronounced "eye-triple-ee"), and the other is the Association for Computing Machinery (ACM).

Various IEEE and ACM transactions contain major work on algorithms and design techniques in print — a more archival-oriented format than conference proceedings. Among these, the IEEE Transactions on computer-aided design (CAD), the IEEE Transactions on VLSI systems, and the ACM Transactions on Design Automation of Electronic Systems are notable. A more general readership magazine devoted to Design and Test and EDA topics is IEEE Design and Test.

As might be expected, the EDA community has a strong online presence. All the conferences mentioned above have web pages describing locations, dates, manuscript submission and registration procedures, and often detailed descriptions of previous conferences. The journals above offer online submission, refereeing, and publication. Online, the IEEE (http://ieee.org), ACM (http://acm.org), and *CiteSeer* (http://citeseer.ist.psu.edu) offer extensive digital libraries, which allow searches through titles, abstracts, and full text. Both conference proceedings and journals are available. Most of the references found in this volume, at least those published after 1988, can be found in at least one of these libraries.

Structure of the Book

In the simplest case of digital design, EDA can be divided into system-level design, micro-architecture design, logical verification, test, synthesis-place-and-route, and physical verification. System-level design is the task of determining which components (bought and built, HW and SW) should comprise a system that can do what one wants. Micro-architecture design fills out the descriptions of each of the blocks, and sets the main parameters for their implementation. Logical verification verifies that the design does what is intended. Test ensures that functional and nonfunctional chips can be told apart reliably, and inserts testing circuitry if needed to ensure that this is the case. Synthesis, place, and route take the logical description, and map it into increasingly detailed physical descriptions, until the design is in a form that can be built with a given process. Physical verification checks that the design is manufacturable and will be reliable. In general, each of these stages works with an increasingly detailed description of the design, and may fail due to problems unforeseen at earlier stages. This makes the flow, or sequence of steps that the users follow to finish their design, a crucial part of any EDA methodology.

Of course not all, or even most chips, are fully digital. Analog chips and chips with a mixture of analog and digital signals (commonly called mixed-signal chips) require their own specialized tool sets.

All these tools must work on circuits and designs that are quite large, and do so in a reasonable amount of time. In general, this cannot be done without models, or simplified descriptions of the behavior of various chip elements. Creating these models is the province of Technology CAD (TCAD), which in general treats relatively small problems in great physical detail, starting from very basic physics and building the more efficient models needed by the tools that must handle higher data volumes.

The division of EDA into these sections is somewhat arbitrary, and below a brief description of each of the chapters of the book is given.

System Level Design

Tools and Methodologies for System-Level Design by Bhattacharyya and Wolf

This chapter covers very high level system-level design approaches and associated tools such as Ptolemy, the Mathworks tools, and many others, and uses video applications as a specific example illustrating how these can be used.

System-Level Specification and Modeling Languages by Buck

This chapter discusses the major approaches to specify and model systems, and the languages and tools in this domain. It includes issues of heterogeneous specifications, models of computation and linking multidomain models, requirements on languages, and specialized tools and flows in this area.

SoC Block-Based Design and IP Assembly by Wilson

This chapter approaches system design with particular emphasis on SoC design via IP-based reuse and block-based design. Methods of assembly and compositional design of systems are covered. Issues of IP reuse as they are reflected in system-level design tools are also discussed.

Performance Evaluation Methods for Multiprocessor Systems-on-Chip Design by Bacivarov and Jerraya

This chapter surveys the broad field of performance evaluation and sets it in the context of multi-processor systems-on-chip (MPSoC). Techniques for various types of blocks — HW, CPU, SW, and interconnect — are included. A taxonomy of performance evaluation approaches is used to assess various tools and methodologies.

System-Level Power Management by Chang, Macii, Poncino and Tiwari

This chapter discusses dynamic power management approaches, aimed at selectively stopping or slowing down resources, whenever this is possible while still achieving the required level of system performance. The techniques can be applied both to reduce power consumption, which has an impact on power dissipation and power supply, and energy consumption, which improves battery life. They are generally driven by the software layer, since it has the most precise picture about both the required quality of service and the global state of the system.

Processor Modeling and Design Tools by Mishra and Dutt

This chapter covers state-of-the-art specification languages, tools, and methodologies for processor development used in academia and industry. It includes specialized architecture description languages and the tools that use them, with a number of examples.

Embedded Software Modeling and Design by di Natale

This chapter covers models and tools for embedded SW, including the relevant models of computation. Practical approaches with languages such as unified modeling language (UML) and specification and description language (SDL) are introduced and how these might link into design flows is discussed.

Using Performance Metrics to Select Microprocessor Cores for IC Designs by Leibson

This chapter discusses the use of standard benchmarks, and instruction set simulators, to evaluate processor cores. These might be useful in nonembedded applications, but are especially relevant to the design of embedded SoC devices where the processor cores may not yet be available in HW, or be based on user-specified processor configuration and extension. Benchmarks drawn from relevant application domains have become essential to core evaluation and their advantages greatly exceed that of the general-purpose ones used in the past.

Parallelizing High-Level Synthesis: A Code Transformational Approach to High-Level Synthesis by Singh, Gupta, Shukla, and Gupta

This chapter surveys a number of approaches, algorithms, and tools for HLS from algorithmic or behavioral descriptions, and focuses on some of the most recent developments in HLS. These include the use of techniques drawn from the parallel compiler community.

Micro-Architecture Design

Cycle-Accurate System-Level Modeling and Performance Evaluation by Coppola and Grammatikakis

This chapter discusses how to use system-level modeling approaches at the cycle-accurate micro-architectural level to do final design architecture iterations and ensure conformance to timing and performance specifications.

Micro-Architectural Power Estimation and Optimization by Macii, Mehra, and Poncino

This chapter discusses the state of the art in estimating power at the micro-architectural level, consisting of major design blocks such as data paths, memories, and interconnect. *Ad hoc* solutions for optimizing both specific components and the whole design are surveyed.

Design Planning by Otten

This chapter discusses the topics of physical floor planning and its evolution over the years, from dealing with rectangular blocks in slicing structures to more general mathematical techniques for optimizing physical layout while meeting a variety of criteria, especially timing and other constraints.

Logical Verification

Design and Verification Languages by Edwards

This chapter discusses the two main HDLs in use — VHDL and Verilog, and how they meet the requirements for design and verification flows. More recent evolutions in languages, such as SystemC, System Verilog, and verification languages such as OpenVera, e, and PSL are also described.

Digital Simulation by Sanguinetti

This chapter discusses logic simulation algorithms and tools, as these are still the primary tools used to verify the logical or functional correctness of a design.

Using Transactional Level Models in a SoC Design Flow by Clouard, Ghenassia, Maillet-Contoz, and Strassen

This chapter discusses a real design flow at a real IC design company to illustrate the building, deployment, and use of transactional-level models to simulate systems at a higher level of abstraction, with much greater performance than at register transfer level (RTL), and to verify functional correctness and validate system performance characteristics.

Assertion-Based Verification by Foster and Marschner

This chapter introduces the relatively new topic of assertion-based verification, which is useful for capturing design intent and reusing it in both dynamic and static verification methods. Assertion libraries such as OVL and languages such as PSL and System Verilog assertions are used for illustrating the concepts.

Hardware Acceleration and Emulation by Bershteyn and Turner

This chapter discusses HW-based systems including FPGA, processor based accelerators/emulators, and FPGA prototypes for accelerated verification. It compares the characteristics of each type of system and typical use models.

Formal Property Verification by Fix and McMillan

This chapter discusses the concepts and theory behind formal property checking, including an overview of property specification and a discussion of formal verification technologies and engines.

Test

Design-for-Test by Koenemann

This chapter discusses the wide variety of methods, techniques, and tools available to solve design-for-test (DFT) problems. This is a huge area with a huge variety of techniques, many of which are implemented in tools that dovetail with the capabilities of the physical test equipment. The chapter surveys the specialized techniques required for effective DFT with special blocks such as memories as well as general logic cores.

Automatic Test Pattern Generation by Wang and Cheng

This chapter starts with the fundamentals of fault modeling and combinational ATPG concepts. It moves on to gate-level sequential ATPG, and discusses satisfiability (SAT) methods for circuits. Moving on beyond traditional fault modeling, it covers ATPG for cross talk faults, power supply noise, and applications beyond manufacturing test.

Analog and Mixed-Signal Test by Kaminska

This chapter first overviews the concepts behind analog testing, which include many characteristics of circuits that must be examined. The nature of analog faults is discussed and a variety of analog test equipment and measurement techniques surveyed. The concepts behind analog built-in-self-test (BIST) are reviewed and compared with the digital test.

RTL to GDS-II, or Synthesis, Place, and Route

Design Flows by Hathaway, Stok, Chinnery, and Keutzer

The RTL to GDSII flow has evolved considerably over the years, from point tools hooked loosely together, to a more integrated set of tools for design closure. This chapter addresses the design flow challenges based on the rising interconnect delays and new challenges to achieve closure.

Logic Synthesis by Khatri and Shenoy

This chapter provides an overview and survey of logic synthesis, which has since the early 1980s, grown to be the vital center of the RTL to GDSII design flow for digital design.

Power Analysis and Optimization from Circuit to Register Transfer Levels by Monteiro, Patel, and Tiwari

Power has become one of the major challenges in modern IC design. This chapter provides an overview of the most significant CAD techniques for low power, at several levels of abstraction.

Equivalence Checking by Kuehlmann and Somenzi

Equivalence checking can formally verify whether two design specifications are functionally equivalent. The chapter defines the equivalence-checking problem, discusses the foundation for the technology, and then discusses the algorithms for combinational and sequential equivalence checking.

Digital Layout — Placement by Reda and Kahng

Placement is one of the fundamental problems in automating digital IC layout. This chapter reviews the history of placement algorithms, the criteria used to evaluate quality of results, many of the detailed algorithms and approaches, and recent advances in the field.

Static Timing Analysis by Sapatnekar

This chapter overviews the most prominent techniques for static timing analysis. It then outlines issues relating to statistical timing analysis, which is becoming increasingly important to handle process variations in advanced IC technologies.

Structured Digital Design by Mo and Brayton

This chapter covers the techniques for designing regular structures, including data paths, programable logic arrays, and memories. It extends the discussion to include regular chip architectures such as gate arrays and structured ASICs.

Routing by Scheffer

Routing continues from automatic placement as a key step in IC design. Routing creates all the wires necessary to connect all the placed components while obeying the process design rules. This chapter discusses various types of routers and the key algorithms.

Exploring Challenges of Libraries for Electronic Design by Hogan and Becker

This chapter discusses the factors that are most important and relevant for the design of libraries and IP, including standard cell libraries, cores, both hard and soft, and the design and user requirements for the same. It also places these factors in the overall design chain context.

Design Closure by Cohn and Osler

This chapter describes the common constraints in VLSI design, and how they are enforced through the steps of a design flow that emphasizes design closure. A reference flow for ASIC is used and illustrated. Future design closure issues are also discussed.

Tools for Chip-Package Codesign by Franzon

Chip-package co-design refers to design scenarios, in which the design of the chip impacts the package design or vice versa. This chapter discusses the drivers for new tools, the major issues, including mixed-signal needs, and the major design and modeling approaches.

Design Databases by Bales

The design database is at the core of any EDA system. While it is possible to build a bad EDA tool or flow on *any* database, it is impossible to build a good EDA tool or flow on a bad database. This chapter describes the place of a design database in an integrated design system. It discusses databases used in the past, those currently in use as well as emerging future databases.

FPGA Synthesis and Physical Design by Betz and Hutton

Programmable logic devices, both complex programmable logic devices (CPLDs) and FPGAs, have evolved from implementing small glue-logic designs to large complete systems. The increased use of such devices — they now are the majority of design starts — has resulted in significant research in CAD algorithms and tools targeting programmable logic. This chapter gives an overview of relevant architectures, CAD flows, and research.

Analog and Mixed-Signal Design

Analog Simulation: Circuit Level and Behavioral Level by Mantooth and Roychowdhury

Circuit simulation has always been a crucial component of analog system design and is becoming even more so today. In this chapter, we provide a quick tour of modern circuit simulation. This includes starting on the ground floor with circuit equations, device models, circuit analysis, more advanced analysis techniques motivated by RF circuits, new advances in circuit simulation using multitime techniques, and statistical noise analysis.

Simulation and Modeling for Analog and Mixed-Signal Integrated Circuits by Gielen and Philips

This chapter provides an overview of the modeling and simulation methods that are needed to design and embed analog and RF blocks in mixed-signal integrated systems (ASICs, SoCs, and SiPs). The role of behavioral models and mixed-signal methods involving models at multiple hierarchical levels is covered. The generation of performance models for analog circuit synthesis is also discussed.

Layout Tools for Analog ICs and Mixed-Signal SoCs: A Survey by Rutenbar and Cohn

Layout for analog circuits has historically been a time-consuming, manual, trial-and-error task. In this chapter, we cover the basic problems faced by those who need to create analog and mixed-signal layout, and survey the evolution of design tools and geometric/electrical optimization algorithms that have been directed at these problems.

Physical Verification

Design Rule Checking by Todd, Grodd, and Fetty

After the physical mask layout is created for a circuit for a specific design process, the layout is measured by a set of geometric constraints or rules for that process. The main objective of design rule checking is to achieve high overall yield and reliability. This chapter gives an overview of design rule checking (DRC) concepts and then discusses the basic verification algorithms and approaches.

Resolution Enhancement Techniques and Mask Data Preparation by Schellenberg

With more advanced IC fabrication processes, new physical effects, which could be ignored in the past, are being found to have a strong impact on the formation of features on the actual silicon wafer. It is now essential to transform the final layout via new tools in order to allow the manufacturing equipment to deliver the new devices with sufficient yield and reliability to be cost-effective. This chapter discusses the compensation schemes and mask data conversion technologies now available to accomplish the new design for manufacturability (DFM) goals.

Design for Manufacturability in the Nanometer Era by Dragone, Guardiani, and Strojwas

Achieving high yielding designs in state-of-the-art IC process technology has become an extremely challenging task. Design for manufacturability includes many techniques to modify the design of ICs in order to improve functional and parametric yield and reliability. This chapter discusses yield loss mechanisms and fundamental yield modeling approaches. It then discusses techniques for functional yield maximization and parametric yield optimization. Finally, DFM-aware design flows and the outlook for future DFM techniques are discussed.

Design and Analysis of Power Supply Networks by Blaauw, Pant, Chaudhry, and Panda

This chapter covers design methods, algorithms, tools for designing on-chip power grids, and networks. It includes the analysis and optimization of effects such as voltage drop and electro-migration.

Noise Considerations in Digital ICs by Kariat

On-chip noise issues and impact on signal integrity and reliability are becoming a major source of problems for deep submicron ICs. Thus the methods and tools for analyzing and coping with them, which are discussed in this chapter, have been gaining importance in recent years.

Layout Extraction by Kao, Lo, Basel, Singh, Spink, and Scheffer

Layout extraction is the translation of the topological layout back into the electrical circuit it is intended to represent. This chapter discusses the distinction between designed and parasitic devices, and discusses the three major parts of extraction: designed device extraction, interconnect extraction, and parasitic device extraction.

Mixed-Signal Noise Coupling in System-on-Chip Design: Modeling, Analysis, and Validation by Vergese and Nagata

This chapter describes the impact of noise coupling in mixed-signal ICs, and reviews techniques to model, analyze, and validate it. Different modeling approaches and computer simulation methods are presented, along with measurement techniques. Finally, the chapter reviews the application of substrate noise analysis to placement and power distribution synthesis.

Technology Computer-Aided Design

Process Simulation by Johnson

Process simulation is the modeling of the fabrication of semiconductor devices such as transistors. The ultimate goal is an accurate prediction of the active dopant distribution, the stress distribution, and the device geometry. This chapter discusses the history, requirements, and development of process simulators.

Device Modeling — from Physics to Electrical Parameter Extraction by Dutton, Choi, and Kan

Technology files and design rules are essential building blocks of the IC design process. Development of these files and rules involves an iterative process that crosses the boundaries of technology and device development, product design, and quality assurance. This chapter starts with the physical description of IC devices and describes the evolution of TCAD tools.

High-Accuracy Parasitic Extraction by Kamon and Iverson

This chapter describes high-accuracy parasitic extraction methods using fast integral equation and random walk-based approaches.

References

[1] A. Sangiovanni-Vincentelli, The tides of EDA, *IEEE Des. Test Comput.*, 20, 59–75, 2003.
[2] A. Kuelhmann, Ed., *20 Years of ICCAD*, Kluwer Academic Publishers (now Springer), Dordrecht, 2002.

2

The Integrated Circuit Design Process and Electronic Design Automation

Robert Damiano
Synopsys Inc.
Hillsboro, Oregon

Raul Camposano
Synopsys Inc.
Mountain View, California

2.1 Introduction

In this chapter, we describe the design process, its major stages, and how electronic design automation (EDA) tools fit into these processes. We also examine the interfaces between the major integrated circuit (IC) design stages as well as the kind of information — both abstractions upwards, and detailed design and verification information downward — that must flow between stages. We assume Complementary Metal Oxide Semiconductor (CMOS) is the basis for all technologies.

We will illustrate with a continuing example. A company wishes to create a new system on chip (SoC). The company assembles a product team, consisting of a project director, system architects, system verification engineers, circuit designers (both digital and analog), circuit verification engineers, layout engineers, and manufacturing process engineers. The product team determines the target technology and geometry as well as the fabrication facility or foundry. The system architects initially describe the system-level design (SLD) through a transaction-level specification in a language such as C++, SystemC, or Esterel. The system verification engineers determine the functional correctness of the SLD through simulation. The engineers validate the transaction processing through simulation vectors. They monitor the results for errors. Eventually, these same engineers would simulate the process with an identical set of vectors through the system implementation to see if the specification and the implementation match. There is some ongoing research to check this equivalence formally.

The product team partitions the SLD into functional units and hands these units to the circuit design teams. The circuit designers describe the functional intent through a high-level design language (HDL). The most popular HDLs are Verilog and VHDL. SystemVerilog is a new language, adopted by the IEEE, which contains design, testbench, and assertion syntax. These languages allow the circuit designers to express the behavior of their design using high-level functions such as addition and multiplication. These languages allow expression of the logic at the register transfer level (RTL), in the sense that an assignment of registers expresses functionality. For the analog and analog mixed signal (AMS) parts of the design, there are also high-level design languages such as Verilog-AMS and VHDL-AMS. Most commonly, circuit designers use Simulation Program with Integrated Circuit Emphasis (SPICE) transistor models and netlists to describe analog components. However, high-level languages provide an easier interface between analog and digital segments of the design and they allow writing higher-level behavior of the analog parts. Although the high-level approaches are useful as simulation model interfaces, there remains no clear method of synthesizing transistors from them. Therefore, transistor circuit designers usually depend on schematic capture tools to enter their data.

The design team must consider functional correctness, implementation closure (reaching the prioritized goals of the design), design cost, and manufacturability of a design. The product team takes into account risks and time to market as well as choosing the methodology. Anticipated sales volume can reflect directly on methodology; whether it is better to create a fully custom design, semicustom design, use standard cells , gate arrays, or a field programmable gate array (FPGA). Higher volume mitigates the higher cost of fully custom or semicustom design, while time to market might suggest using an FPGA methodology. If implementation closure for power and speed is tantamount, then an FPGA methodology might be a poor choice. Semicustom designs, depending on the required volume, can range from microprocessor central processor units (CPUs), digital signal processors (DSPs), application-specific standard parts (ASSP) or application-specific integrated circuits (ASIC). In addition, for semicustom designs, the company needs to decide whether to allow the foundry to implement the layout, or whether the design team should use customer owned tools (COT). We will assume that our product team chooses semicustom COT designs. We will mention FPGA and fully custom methodologies only in comparison.

In order to reduce cost, the product team may decide that the design warrants reuse of intellectual property (IP). Intellectual property reuse directly addresses the increasing complexity of design as opposed to feature geometry size. Reuse also focuses on attaining the goals of functional correctness. One analysis estimates that it takes 2000 engineering years and 1 trillion simulation vectors to verify 25 million lines of RTL code. Therefore, verified IP reuse reduces cost and time to market. Moreover, IP blocks themselves have become larger and more complex. For example, the 1176JZ-S ARM core is 24 times larger than the older 7TDI-S ARM core. The USB 2.0 Host is 23 times larger than the Universal Serial Bus (USB) 1.1 Device. PCI Express is 7.5 times larger than PCI v 1.1.

Another important trend is that SoC-embedded memories are an increasingly large part of the SoC real estate. While in 1999, 20% of a 180-nm SoC was embedded memory, roadmaps project that by 2005, embedded memory will consume 71% of a 90-nm SoC. These same roadmaps indicate that by 2014, embedded memory will grow to 94% of a 35-nm SoC.

Systems on chips typically contain one or more CPUs or DSPs (or both), cache, a large amount of embedded memory and many off-the-shelf components such as USB, Universal Asynchronous Receiver-Transmitter (UART), Serial Advanced Technology Attachment (SATA), and Ethernet (cf. Figure 2.1). The differentiating part of the SoC contains the new designed circuits in the product.

The traditional semicustom IC design flow typically comprises up to 50 steps. On the digital side of design, the main steps are functional verification, logical synthesis, design planning, physical implementation which includes clock-tree synthesis, placement and routing, extraction, design rules checking (DRC) and layout versus schematic checking (LVS), static timing analysis, insertion of test structures, and test pattern generation. For analog designs, the major steps are as follows: schematic entry, SPICE simulation, layout, layout extraction, DRC, and LVS. SPICE simulations can include DC, AC, and transient analysis, as well as noise, sensitivity, and distortion analysis. Analysis and implementation of corrective procedures for the manufacturing process such as mask synthesis and yield analysis, are critical at smaller geometries. In order to verify an SoC system where many components reuse IP, the IP provider may supply verification IP, monitors, and checkers needed by system verification.

There are three basic areas where EDA tools assist the design team. Given a design, the first is verification of functional correctness. The second deals with implementation of the design. The last area deals with analysis and corrective procedures so that the design meets all manufacturability specifications. Verification, layout, and process engineers on the circuit design team essentially own these three steps.

SPICE reportedly is an acronym for Simulation Program with Integrated Circuit Emphasis

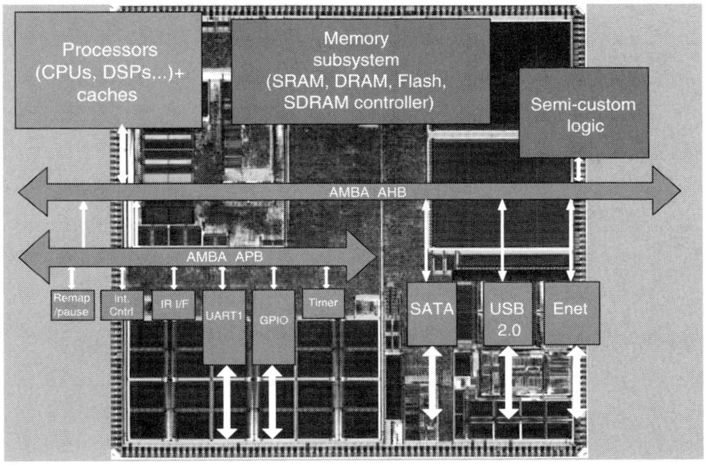

FIGURE 2.1. SoC with IP.

2.2 Verification

The design team attempts to verify that the design under test (DUT) functions correctly. For RTL designs, verification engineers rely highly on simulation at the cycle level. After layout, EDA tools, such as equivalence checking, can determine whether the implementation matches the RTL functionality. After layout, the design team must check that there are no problem delay paths. A static timing analysis tool can facilitate this. The team also needs to examine the circuit for noise and delay due to parasitics. In addition, the design must obey physical rules for wire spacing, width, and enclosure as well as various electrical rules. Finally, the design team needs to simulate and check the average and transient power. For transistor circuits, the design team uses SPICE circuit simulation or fast SPICE to determine correct functionality, noise, and power.

We first look at digital verification (cf. Figure 2.2). RTL simulation verifies that the DUT behavior meets the design intent. The verification engineers apply a set of vectors, called a testbench, to the design through an event-driven simulator, and compare the results to a set of expected outputs. The quality of the verification depends on the quality of the testbench. Many design teams create their testbench by supplying a list of the vectors, a technique called directed test. For a directed test to be effective, the design team must know beforehand what vectors might uncover bugs. This is extremely difficult since complex sequences of vectors are necessary to find some corner case errors. Therefore, many verification engineers create testbenches that supply stimulus through random vectors with biased inputs, such as the clock or reset signal. The biasing increases or decreases the probability of a signal going high or low. While a purely random testbench is easy to create, it suffers from the fact that vectors may be illegal as stimulus. For better precision and wider coverage, the verification engineer may choose to write a constrained random testbench. Here, the design team supplies random input vectors that obey a set of constraints.

The verification engineer checks that the simulated behavior does not have any discrepancies from the expected behavior. If the engineer discovers a discrepancy, then the circuit designer modifies the HDL and the verification engineer resimulates the DUT. Since exhaustive simulation is usually impossible, the design team needs a metric to determine quality. One such metric is coverage. Coverage analysis considers how well the test cases stimulate the design. The design team might measure coverage in terms of number of lines of RTL code exercised, whether the test cases take each leg of each decision, or how many "reachable" states encountered.

Another important technique is for the circuit designer to add assertions within the HDL. These assertions monitor whether internal behavior of the circuit is acting properly. Some designers embed tens of thousands of assertions into their HDL. Languages like SystemVerilog have extensive assertion syntax based

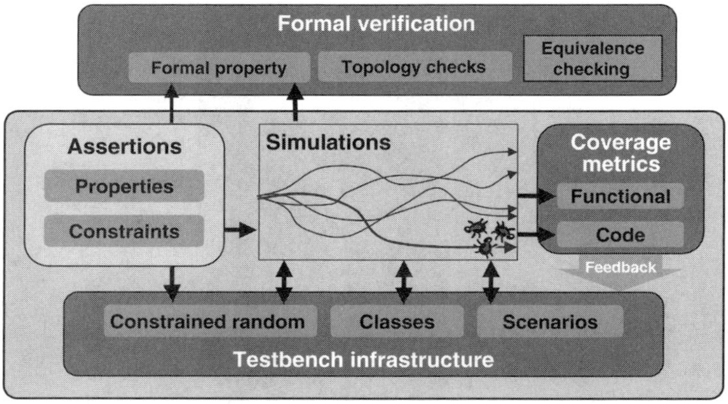

FIGURE 2.2. Digital Simulation/Formal Verification.

on linear temporal logic. Even for languages without the benefit of assertion syntax, tool-providers supply an application program interface (API), which allows the design team to build and attach its own monitors.

The verification engineer needs to run a large amount of simulation, which would be impractical if not for compute farms. Here, the company may deploy thousands of machines, 24/7, to enable the designer to get billions of cycles a day; sometimes the machines may run as many as 200 billion cycles a day. Best design practices typically create a highly productive computing environment. One way to increase throughput is to run a cycle simulation by taking a subset of the chosen verification language which is both synchronous and has a set of registers with clear clock cycles. This type of simulation assumes a uniformity of events and typically uses a time wheel with gates scheduled in a breadth first manner.

Another way to tackle the large number of simulation vectors during system verification is through emulation or hardware acceleration. These techniques use specially configured hardware to run the simulation. In the case of hardware acceleration, the company can purchase special-purpose hardware, while in the case of emulation the verification engineer uses specially configured FPGA technology. In both cases, the system verification engineer must synthesize the design and testbench down to a gate-level model. Tools are available to synthesize and schedule gates for the hardware accelerator. In the case of an FPGA emulation system, tools can map and partition the gates for the hardware.

Of course, since simulation uses vectors, it is usually a less than exhaustive approach. The verification engineer can make the process complete by using assertions and formal property checking. Here, the engineer tries to prove that an assertion is true or to produce a counterexample. The trade-off is simple. Simulation is fast but by definition incomplete, while formal property checking is complete but may be very slow. Usually, the verification engineer runs constrained random simulation to unearth errors early in the verification process. The engineer applies property checking to corner case situations that can be extremely hard for the testbench to find. The combination of simulation and formal property checking is very powerful. The two can even be intermixed, by allowing simulation to proceed for a set number of cycles and then exhaustively looking for an error for a different number of cycles. In a recent design, by using this hybrid approach , a verification engineer found an error 21,000 clock cycles from an initial state. Typically, formal verification works well on specific functional units of the design. Between the units, the system engineers use an "assume/guarantee" methodology to establish block pre- and postconditions for system correctness.

During the implementation flow, the verification engineer applies equivalence checking to determine whether the DUT preserves functional behavior. Note that functional behavior is different from functional intent. The verification engineer needs RTL verification to compare functional behavior with functional intent. Equivalence checking is usually very fast and is a formal verification technology, which is exhaustive in its analysis. Formal methods do not use vectors.

For transistor-level circuits, such as analog, memory, and radio frequency (RF), the event-driven verification techniques suggested above do not suffice (cf. Figure 2.3). The design team needs to compute signals

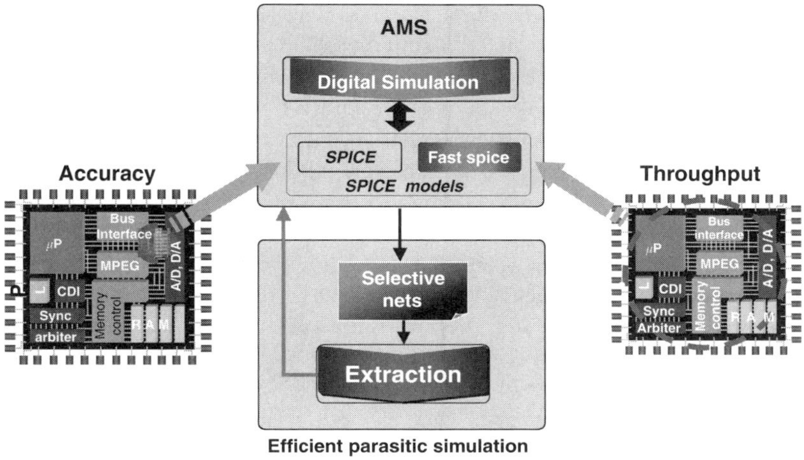

FIGURE 2.3. Transistor simulation with parasitics.

accurately through SPICE circuit simulation. SPICE simulation is very time consuming because the algorithm solves a system of differential equations. One way to get around this cost is to select only a subset of transistors, perform an extraction of the parasitics, and then simulate the subset with SPICE. This reduction gives very accurate results for the subset, but even so, the throughput is still rather low. Another approach is to perform a fast SPICE simulation. This last SPICE approach trades some accuracy for a significant increase in throughput. The design team can also perform design space exploration by simulating various constraint values on key goals such as gain or phase margin to find relatively optimal design parameters. The team analyzes the multiple-circuit solutions and considers the cost trade-offs. A new generation of tools performs this "design exploration" in an automatic manner. Mixed-level simulation typically combines RTL, gate and transistor parts of the design and uses a communication back-plane to run the various simulations and share input and output values.

Finally, for many SoCs, both hardware and software comprise the real system. System verification engineers may run a hardware–software co-simulation before handing the design to a foundry. All simulation system components mentioned can be part of this co-simulation. In early design stages, when the hardware is not ready, the software can simulate ("execute") an instruction set model (ISM), a virtual prototype (model), or an early hardware prototype typically implemented in FPGAs.

2.3 Implementation

This brings us to the next stage of the design process, the implementation and layout of the digital design. Circuit designers implement analog designs by hand. Field programmable gate array technologies usually have a single basic combinational cell, which can form a variety of functions by constraining inputs. Layout and process tools are usually proprietary to the FPGA family and manufacturer. For semicustom design, the manufacturer supplies a precharacterized cell library, either standard cell or gate array. In fact, for a given technology, the foundry may supply several libraries, differing in power, timing, or yield. The company decides on one or more of these as the target technology. One twist on the semicustom methodology is structured ASIC. Here, a foundry supplies preplaced memories, pad-rings and power grids as well as sometimes preplaced gate array logic, similar to the methodology employed by FPGA families. The company can use semicustom techniques for the remaining combinational and sequential logic. The goal is to reduce nonrecurring expenses by limiting the number of mask-sets needed and by simplifying physical design.

By way of contrast, in a fully custom methodology, one tries to gain performance and limit power consumption by designing much of the circuit as transistors. The circuit designers keep a corresponding RTL design. The verification engineer simulates the RTL and extracts a netlist from the transistor description. Equivalence checking compares the extracted netlist to the RTL. The circuit designer manually places and routes the transistor-level designs. Complex high-speed designs, such as microprocessors, sometimes use full custom methodology, but the design costs are very high. The company assumes that the high volume will amortize the increased cost. Fully custom designs consider implementation closure for power and speed as most important. At the other end of the spectrum, FPGA designs focus on design cost and time to market. Semicustom methodology tries to balance the goals of timing and power closure with design cost (cf. Figure 2.4).

In the semicustom implementation flow, one first attempts to synthesize the RTL design into a mapped netlist. The circuit designers supply their RTL circuit along with timing constraints. The timing constraints consist of signal arrival and slew (transition) times at the inputs, and required times and loads (capacitances) at the outputs. The circuit designer identifies clocks as well as any false or multiple-cycle paths. The technology library is usually a file that contains a description of the function of each cell along with delay, power, and area information. Either the cell description contains the pin-to-pin delay represented as look-up table functions of input slew, output load, and other physical parameters such as voltage and temperature, or as polynomial functions that best fit the parameter data. For example, foundries provide cell libraries in Liberty or OLA (Open Library Application Programming Interface) formats. The foundry also provides a wire delay model, derived statistically from previous designs. The wire delay model correlates the number of sinks of a net to capacitance and delay.

Several substages comprise the operation of a synthesis tool. First, the synthesis tool compiles the RTL into technology-independent cells and then optimizes the netlist for area, power, and delay. The tool maps the netlist into a technology. Sometimes, synthesis finds complex functions such as multipliers and adders in parameterized (area/timing) reuse libraries. For example, the tool might select a Booth multiplier from the reuse library to improve timing. For semicustom designs, the foundry provides a standard cell or gate array library, which describes each functional member. In contrast, the FPGA supplier describes a basic combinational cell from which the technology mapping matches functional behavior of subsections of the design. To provide correct functionality, the tool may set several pins on the complex gates to constants. A post-process might combine these functions for timing, power, or area.

A final substage tries to analyze the circuit and performs local optimizations that help the design meet its timing, area and power goals. Note that due to finite number of power levels of any one cell, there are limits to the amount of capacitance that functional cell types can drive without the use of buffers. Similar restrictions apply to input slew (transition delay). The layout engineer can direct the synthesis tool by enhancing or omitting any of these stages through scripted commands. Of course, the output must be a mapped netlist.

To get better timing results, foundries continue to increase the number of power variations for some cell types. One limitation to timing analysis early in the flow is that the wire delay models are statistical estimates of the real design. Frequently, these wire delays can differ significantly from those found after routing. One interesting approach to synthesis is to extend each cell of the technology library so

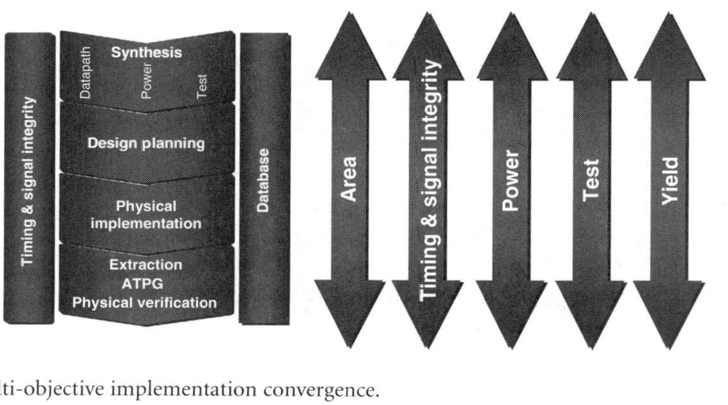

FIGURE 2.4. Multi-objective implementation convergence.

that it has an infinite or continuous variation of power. This approach, called gain-based synthesis, attempts to minimize the issue of inaccurate wire delay by assuming cells can drive any wire capacitance through appropriate power level selection. In theory, there is minimal perturbation to the natural delay (or gain) of the cell. This technique makes assumptions such as that the delay of a signal is a function of capacitance. This is not true for long wires where resistance of the signal becomes a factor. In addition, the basic approach needs to include modifications for slew (transition delay).

To allow detection of manufacturing faults, the design team may add extra test generation circuitry. Design for test (DFT) is the name given to the process of adding this extra logic (cf. Figure 2.5). Sometimes, the foundry supplies special registers, called logic-sensitive scan devices. At other times, the test tool adds extra logic called Joint Test Action Group (JTAG) boundary scan logic that feeds the registers. Later in the implementation process, the design team will generate data called scan vectors that test equipment uses to detect manufacturing faults. Subsequently, tools will transfer these data to automatic test equipment (ATE), which perform the chip tests.

As designs have become larger, so has the amount of test data. The economics of the scan vector production with minimal cost and design impact leads to data compression techniques. One of most widely used techniques is deterministic logic built in self-test(BIST). Here, a test tool adds extra logic on top of the DFT to generate scan vectors dynamically.

Before continuing the layout, the engineer needs new sets of rules, dealing with the legal placement and routing of the netlist. These libraries, in various exchange formats, e.g., LEF for logic, DEF for design and PDEF for physical design, provide the layout engineer physical directions and constraints. Unlike the technology rules for synthesis, these rules are typically model-dependent. For example, there may be information supplied by the circuit designer about the placement of macros such as memories. The routing tool views these macros as blockages. The rules also contain information from the foundry.

Even if the synthesis tool preserved the original hierarchy of the design, the next stages of implementation need to view the design as flat. The design-planning step first flattens the logic and then partitions the flat netlist as to assist placement and routing;—in fact, in the past, design planning was sometimes known as floor planning. A commonly used technique is for the design team to provide a utilization ratio to the design planner. The utilization ratio is the percentage of chip area used by the cells as opposed to the nets. If the estimate is too high, then routing congestion may become a problem. If the estimate is too low, then the layout could waste area. The design-planning tool takes the locations of hard macros into account. These macros are hard in the sense that they are rectangular with a fixed length, fixed width, and sometimes a fixed location on the chip. The design-planning tool also tries to use the logical hierarchy of the design as a guide to the partitioning. The tool creates, places and routes a set of macros that have fixed lengths, widths, and locations. The tool calculates timing constraints for each macro and routes the power

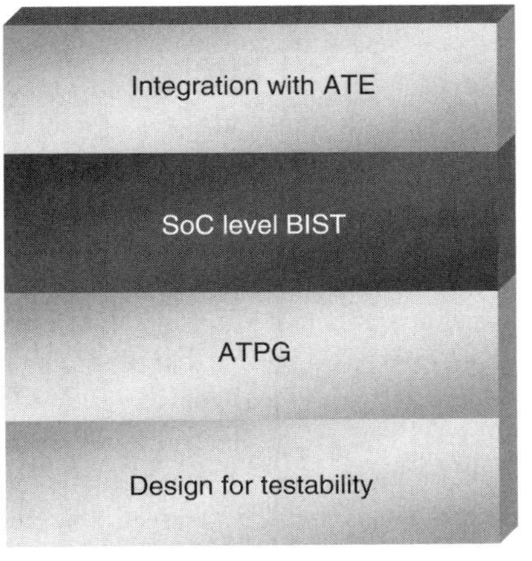

FIGURE 2.5. Design for test.

and ground grids. The power and ground grids are usually on the chip's top levels of metal and then distributed to the lower levels. The design team can override these defaults and indicate which metal layers should contain these grids. Sometimes design planning precedes synthesis. In these cases, the tool partitions the RTL design and automatically characterizes each of the macros with timing constraints.

After design planning, the layout engineer runs the physical implementation tools on each macro. First, the placer assigns physical locations to each gate of the macro. The placer typically moves gates while minimizing some cost, e.g., wire length or timing. Legalization follows the coarse placement to make sure the placed objects fit physical design rules. At the end of placement, the layout engineer may run some more synthesis, like re-sizing of gates. One of the major improvements to placement over the last decade is the emergence of physical synthesis. In physical synthesis, the tool interleaves synthesis and placement. Recall that previously, logic synthesis used statistical wire capacitance. Once the tool places the gates, it can perform a global route and get capacitances that are more accurate for the wires, based on actual placed locations. The physical synthesis tool iterates this step and provides better timing and power estimates.

Next, the layout engineer runs a tool that buffers and routes the clock tree. Clock-tree synthesis attempts to minimize the delay while assuring that skew, that is the variation in signal transport time from the clock to its corresponding registers, is close to zero.

Routing the remaining nets comes after clock-tree synthesis. Routing starts with a global analysis called global route. Global route creates coarse routes for each signal and its outputs. Using the global routes as a guide, a detailed routing scheme, such as a maze channel or switchbox, performs the actual routing. As with the placement, the tool performs a final legalization to assure that the design obeys physical rules. One of the major obstacles to routing is signal congestion. Congestion occurs when there are too many wires competing for a limited amount of chip wire resource. Remember that the design team gave the design planner a utilization ratio in the hope of avoiding this problem.

Both global routing and detailed routing take the multilayers of the chip into consideration. For example, the router assumes that the gates are on the polysilicon layer, while the wires connect the gates through vias on 3–8 layers of metal. Horizontal or vertical line segments comprise the routes, but some recent work allows 45° lines for some foundries. As with placement, there may be some resynthesis, such as gate resizing, at the end of the detailed routing stage.

Once the router finishes, an extraction tool derives the capacitances, resistances, and inductances. In a two-dimensional (2-D) parasitic extraction, the extraction tool ignores 3-D details and assumes that each chip level is uniform in one direction. This produces only approximate results. In the case of the much slower 3-D parasitic extraction, the tool uses 3-D field solvers to derive very accurate results. A 2½-D extraction tool compromises between speed and accuracy. By using multiple passes, it can access some of the 3-D features. The extraction tool places its results in a standard parasitic exchange format file (SPEF).

During the implementation process, the verification engineer continues to monitor behavioral consistency through equivalence checking and using LVS comparison. The layout engineer analyzes timing and signal integrity issues through timing analysis tools, and uses their results to drive implementation decisions. At the end of the layout, the design team has accurate resistances, capacitances, and inductances for the layout. The system engineer uses a sign-off timing analysis tool to determine if the layout meets timing goals. The layout engineer needs to run a DRC on the layout to check for violations.

Both the Graphic Data System II (GDSII) and the Open Artwork System Interchange Standard (OASIS) are databases for shape information to store a layout. While the older GDSII was the database of choice for shape information, there is a clear movement to replace it by the newer, more efficient OASIS database. The LVS tool checks for any inconsistencies in this translation.

What makes the implementation process so difficult is that multiple objectives need consideration. For example, area, timing, power, reliability, test, and yield goals might and usually cause conflict with each other. The product team must prioritize these objectives and check for implementation closure.

Timing closure—that is meeting all timing requirements—by itself is becoming increasingly difficult and offers some profound challenges. As process geometry decrease, the significant delay shifts from the cells to the wires. Since a synthesis tool needs timing analysis as a guide and routing of the wires does not occur until after synthesis, we have a chicken and egg problem. In addition, the thresholds for noise sensitivity also shrink with smaller geometries. This along with increased coupling capacitances, increased current densities and sensitivity to inductance, make problems like crosstalk and voltage (IR) drop increasingly familiar.

Since most timing analysis deals with worst-case behavior, statistical variation and its effect on yield add to the puzzle. Typically timing analysis computes its cell delay as function of input slew (transition delay) and output load (output capacitance or RC). If we add the effects of voltage and temperature variations as well as circuit metal densities, timing analysis gets to be very complex. Moreover, worst-case behavior may not correlate well with what occurs empirically when the foundry produces the chips. To get a better predictor of parametric yield, some layout engineers use statistical timing analysis. Here, rather than use single numbers (worst case, best case, corner case, nominal) for the delay-equation inputs, the timing analysis tool selects probability distributions representing input slew, output load, temperature, and voltage among others. The delay itself becomes a probability distribution. The goal is to compute the timing more accurately in order to create circuits with smaller area and lower power but with similar timing yield.

Reliability is also an important issue with smaller geometries. Signal integrity deals with analyzing what were secondary effects in larger geometries. These effects can produce erratic behavior for chips manufactured in smaller geometries. Issues such as crosstalk, IR drop, and electromigration are factors that the design team must consider in order to produce circuits that perform correctly.

Crosstalk noise can occur when two wires are close to each other (cf. Figure 2.6). One wire, the aggressor, switches while the victim signal is in a quiet state or making an opposite transition. In this case, the aggressor can force the victim to glitch. This can cause a functional failure or can simply consume additional power. Gate switching draws current from the power and ground grids. That current, together with the wire resistance in the grids, can cause significant fluctuations in the power and ground voltages supplied to gates. This problem, called IR drop, can lead to unpredictable functional errors. Very high frequencies can produce high current densities in signals and power lines, which can lead to the migration of metal ions. This power electromigration can lead to open or shorted circuits and subsequent signal failure.

Power considerations are equally complex. As the size of designs grow and geometries shrink, power increases. This can cause problems for batteries in wireless and hand-held devices, and thermal management in microprocessor, graphic and networking applications. Power consumption falls into two areas: dynamic power (cf. Figure 2.7), the power consumed when devices switch value; and leakage power (cf. Figure 2.8), the power leaked through the transistor. Dynamic power consumption grows directly with increased capacitance and voltage. Therefore, as designs become larger, dynamic power increases. One easy way to reduce dynamic power is to decrease voltage. However, decreased voltage leads to smaller noise margins and less speed.

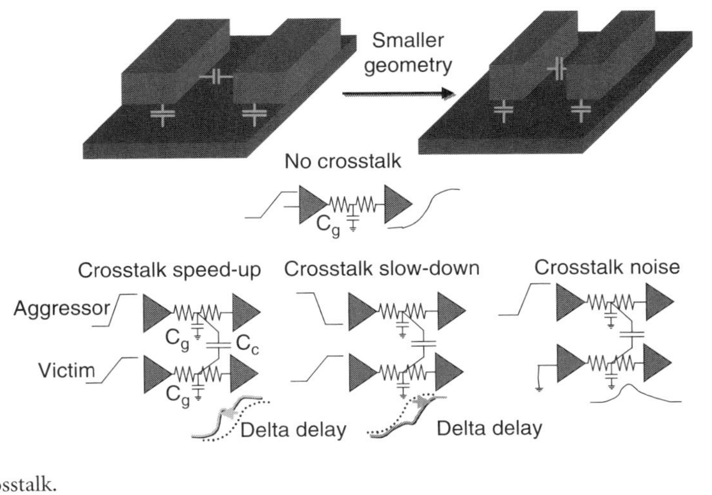

FIGURE 2.6. Crosstalk.

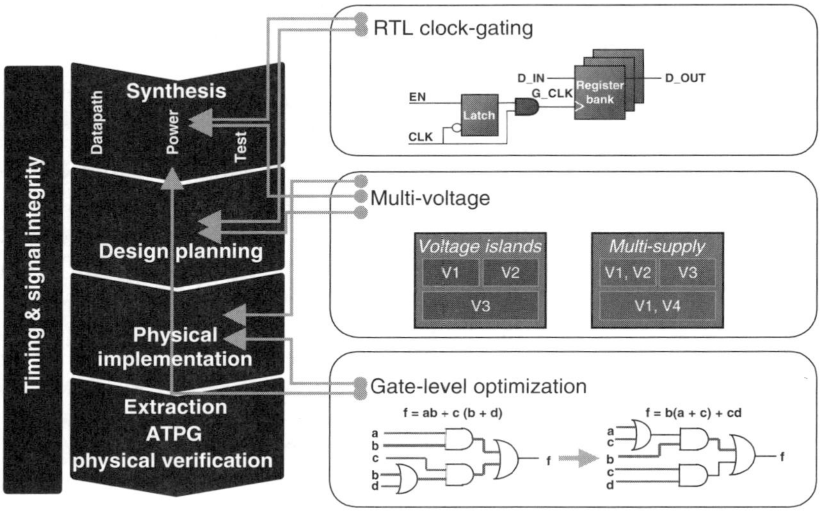

FIGURE 2.7. Dynamic power management.

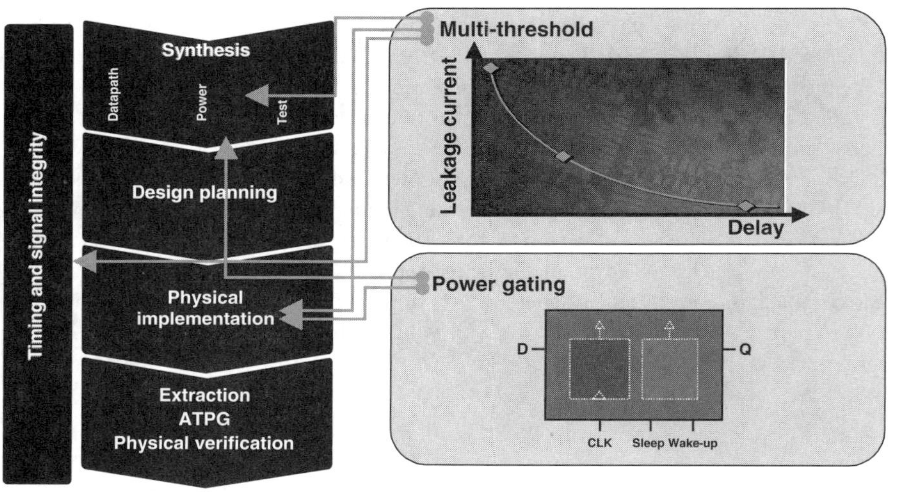

FIGURE 2.8. Static power management (leakage).

A series of novel design and transistor innovations can reduce the power consumption. These include operand isolation, clock gating, and voltage-islands. Timing and power considerations are very often in conflict with each other, so the design team must employ these remedies carefully.

A design can have part of its logic clock-gated by using logic to enable the bank of registers. The logic driven by the registers is quiescent until the clock-gated logic enables the registers. Latches at the input can isolate parts of a design that implement operations (e.g. an arithmetic logic unit (ALU)), when results are unnecessary for correct functionality, thus preventing unnecessary switching. Voltage-islands help resolve the timing vs. power conflicts. If part of a design is timing critical, a higher voltage can reduce the delay. By partitioning the design into voltage-islands, one can use lower voltage in all but the most timing-critical parts of the design. An interesting further development is dynamic voltage/frequency scaling, which consists of scaling the supply voltage and the speed during operation to save power or increase performance temporarily.

The automatic generation of manufacturing fault detection tests was one of the first EDA tools. When a chip fails, the foundry wants to know why. Test tools produce scan vectors that can identify various manufacturing faults within the hardware. The design team translates the test vectors to standard test data format and the foundry can inject these inputs into the failed chip through automated test equipment (ATE). Remember that the design team added extra logic to the netlist before design planning, so that test equipment could quickly insert the scan vectors, including set values for registers, into the chip. The most common check is for stuck at 0 or stuck at 1 faults where the circuit has an open or short at a particular cell. It is not surprising that smaller geometries call for more fault detection tests. An integration of static timing analysis with transition/path delay fault automatic test pattern generation (ATPG) can help, for example, to detect contact defects; while extraction information and bridging fault ATPG can detect metal defects.

Finally, the design team should consider yield goals. Manufacturing becomes more difficult as geometries shrink. For example, thermal stress may create voids in vias. One technique to get around this problem is to minimize the vias inserted during routing, and for those inserted, to create redundant vias. Via doubling, which converts a single via into multiple vias, can reduce resistance and produce better yield. Yield analysis can also suggest wire spreading during routing to reduce cross talk and increase yield. Manufacturers also add a variety of manufacturing process rules needed to guarantee good yield. These rules involve antenna checking and repair through diode insertion as well as metal fill needed to produce uniform metal densities necessary for copper wiring chemical–mechanical polishing (CMP). Antenna repair has little to do with what we typically view as antennas. During the ion-etching process, charge collects on the wires connected to the polysilicon gates. These charges can damage the gates. The layout tool can connect small diodes to the interconnect wires as a discharge path.

Even with all the available commercial tools, there are times when layout engineers want to create their own tool for analysis or small implementation changes. This is analogous to the need for an API in verification. Scripting language and C-language-based APIs for design databases such as MilkyWay and OpenAccess are available. These databases supply the user with an avenue to both the design and rules. The engineer can directly change and analyze the layout.

2.4 Design for Manufacturing

One of the newest areas for EDA tools is design for manufacturing. As in other areas, the driving force of the complexity is the shrinking of geometries. After the design team translates their design to shapes, the foundry must transfer those shapes to a set of masks. Electron beam (laser) equipment then creates the physical masks for each layer of the chip from the mask information. For each layer of the chip, the foundry applies photoresistive material, and then transfers the mask structures by the stepper optical equipment onto the chip. Finally, the foundry etches the correct shapes by removing the excess photoresist material.

Since the stepper uses light for printing, it is important that the wavelength is small enough to transcribe the features accurately. When the chip's feature size was 250 nm, we could use lithography equipment that produced light at a wavelength of 248 nm. New lithography equipment that produces light of lower wavelength needs significant innovation and can be very expensive. When the feature geometry gets significantly smaller than the wavelength, the detail of the reticles (fine lines and wires), transferred to the chip from the mask can be lost. Electronic design automation tools can analyze and correct this transfer operation without new equipment, by modifying the shapes data— a process known as mask, synthesis (cf. Figure 2.9). This process uses resolution enhancement techniques and methods to provide dimensional accuracy.

One mask synthesis technique is optimal proximity correction (OPC). This process takes the reticles in the GDSII or OASIS databases and modifies them by adding new lines and wires, so that even if the geometry is smaller than the wavelength, optical equipment adequately preserves the details. This technique successfully transfers geometric features of down to one-half of the wavelength of the light used. Of course given a fixed wavelength, there are limits beyond which the geometric feature size is too small for even these tricks.

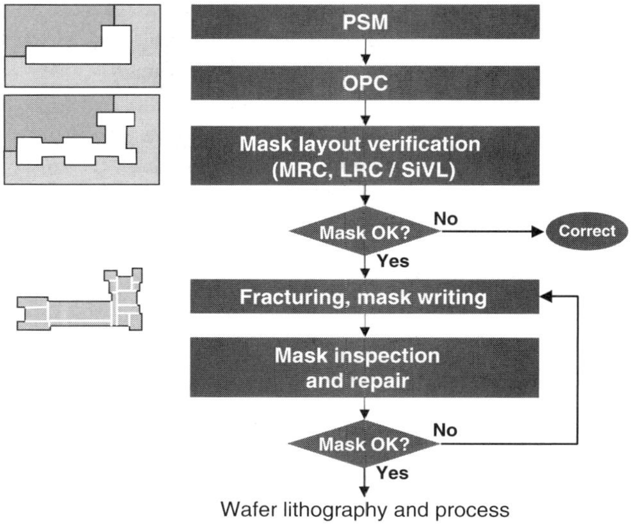

FIGURE 2.9. Subwavelength: from layout to masks.

For geometries of 90 nm and below, the lithography EDA tools combine OPC with other mask synthesis approaches such as phase shift mask (PSM), off-axis illumination and assist features (AF). For example, PSM is a technique where the optical equipment images dark features at critical dimensions with 0° illumination on one side and 180° illumination on the other side. There are additional manufacturing process rules needed such as minimal spacing and cyclic conflict avoidance, to avoid situations where the tool cannot map the phase.

In summary, lithography tools proceed through PSM, OPC, and AF to enhance resolution and make the mask more resistive to process variations. The process engineer can perform a verification of silicon vs. layout and a check of lithography rule compliance. If either fails, the engineer must investigate and correct, sometimes manually. If both succeed, another EDA tool "fractures" the design, subdividing the shapes into rectangles (trapezoids), which can be fed to the mask writing equipment. The engineer can then transfer the final shapes file to a database, such as the manufacturing-electron-beam-exposure system (MEBES). Foundry equipment uses the MEBES database (or other proprietary formats) to create the physical masks. The process engineer can also run a "virtual" stepper tool to pre-analyze the various stages of the stepper operation. After the foundry manufactures the masks, a mask inspection and repair step ensures that they conform to manufacturing standards.

Another area of design for manufacturing analysis is prediction of yield (cf. Figure 2.10). The design team would like to correlate some of the activities during route with actual yield. Problems with CMP, via voids and cross talk can cause chips to unexpectedly fail. EDA routing tools offer some solutions in the form of metal fill, via doubling and wire spacing. Library providers are starting to develop libraries for higher yields that take into account several yield failure mechanisms. There are tools that attempt to correlate these solutions with yield. Statistical timing analysis can correlate timing constraints to parametric circuit yield.

Finally, the process engineer can use tools to predict the behavior of transistor devices or processes. Technology computer aided design (TCAD) deals with the modeling and simulation of physical manufacturing process and devices. Engineers can model and simulate individual steps in the fabrication process. Likewise, the engineer can model and simulate devices, parasitics or electrical/thermal properties, therefore providing insights into their electrical, magnetic or optical properties.

For example, because of packing density, foundries may switch isolation technology for an IC from the local oxidation of silicon model toward the shallow trench isolation (STI) model. Under this model, the

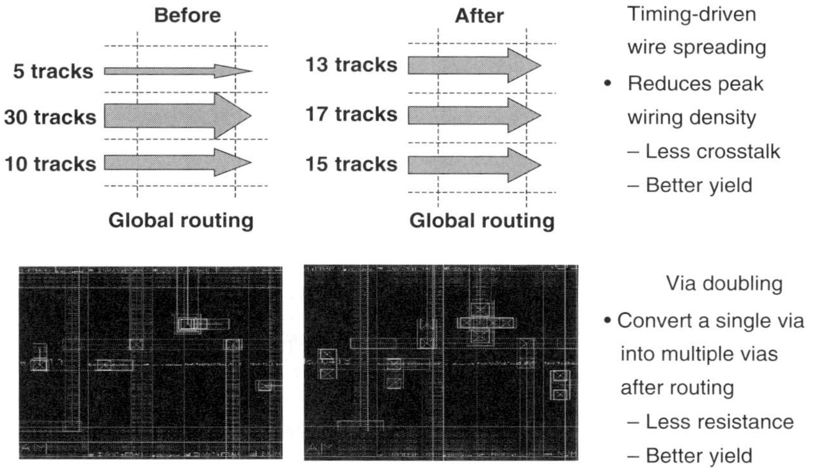

FIGURE 2.10. Yield enhancement features in routing.

process engineer can analyze breakdown stress, electrical behavior such as leakage, or material vs. process dependencies. Technology computer aided design tools can simulate STI effects, extract interconnect parasitics, such as diffusion distance, and determine SPICE parameters.

References

[1] M. Smith, *Application Specific Integrated Circuits*, Addison-Wesley, Reading, MA, 1997.

[2] A. Kuehlmann, *The Best of ICCAD*, Kluwer Academic Publishers, Dordrecht, 2003.

[3] D. Thomas and P. Moorby, *The Verilog Hardware Description Language*, Kluwer Academic Publishers, Dordrecht, 1996.

[4] D. Pellerin and D. Taylor, *VHDL Made Easy*, Pearson Education, Upper Saddle River, N.J., 1996.

[5] S. Sutherland, S. Davidson, and P. Flake, *SystemVerilog For Design: A Guide to Using SystemVerilog for Hardware Design and Modeling*, Kluwer Academic Publishers, Dordrecht, 2004.

[6] T. Groetker, S. Liao, G. Martin, and S. Swan, *System Design with SystemC*, Kluwer Academic Publishers, Dordrecht, 2002.

[7] G. Peterson, P. Ashenden, and D. Teegarden, *The System Designer's Guide to VHDL-AMS*, Morgan Kaufman Publishers, San Francisco, CA, 2002.

[8] K. Kundert and O. Zinke, *The Designer's Guide to Verilog-AMS*, Kluwer Academic Publishers, Dordrecht, 2004.

[9] M. Keating and P. Bricaud, *Reuse Methodology Manual for System-on-a-Chip Designs*, Kluwer Academic Publishers, Dordrecht, 1998.

[10] J. Bergeron, *Writing Testbenches*, Kluwer Academic Publishers, Dordrecht, 2003.

[11] E. Clarke, O. Grumberg, and D. Peled, *Model Checking*, The MIT Press, Cambridge, MA, 1999.

[12] S. Huang and K. Cheng, *Formal Equivalence Checking and Design Debugging*, Kluwer Academic Publishers, Dordrecht, 1998.

[13] R. Baker, H. Li, and D Boyce, *CMOS Circuit Design, Layout, and Simulation*, Series on Microelectronic Systems, IEEE Press, New York, 1998.

[14] L. Pillage, R. Rohrer, and C. Visweswariah, *Electronic Circuit and System Simulation Methods*, McGraw-Hill, New York, 1995.

[15] J. Elliott, *Understanding Behavioral Synthesis: A Practical Guide to High-Level Design*, Kluwer Academic Publishers, Dordrecht, 2000.

[16] S. Devadas, A. Ghosh, and K. Keutzer, *Logic Synthesis*, McGraw-Hill, New York, 1994.

[17] G. DeMicheli, *Synthesis and Optimization of Digital Circuits*, McGraw-Hill, New York, 1994.

[18] I. Sutherland, R. Sproull, and D. Harris, *Logical Effort: Defining Fast CMOS Circuits*, Academic Press, New York, 1999.

[19] N. Sherwani, *Algorithms for VLSI Physical Design Automation*, Kluwer Academic Publishers, Dordrecht, 1999.
[20] F. Nekoogar, *Timing Verification of Application-Specific Integrated Circuits (ASICs)*, Prentice-Hall PTR, Englewood Cliffs, NJ, 1999.
[21] K. Roy and S. Prasad, *Low Power CMOS VLSI: Circuit Design*, Wiley, New York, 2000.
[22] C-K.Cheng, J. Lillis, S. Lin, and N. Chang, *Interconnect Analysis and Synthesis*, Wiley, New York, 2000.
[23] W. Dally and J. Poulton, *Digital Systems Engineering*, Cambridge University Press, Cambridge, 1998.
[24] M. Abramovici, M. Breuer, and A. Friedman, *Digital Systems Testing and Testable Design*, Wiley, New York, 1995.
[25] A. Wong, *Resolution Enhancement Techniques in Optical Lithography*, SPIE Press, Bellingham, WA, 2001.
[26] International Technology Roadmap for Semiconductors (ITRS), 2004, URL: http://public.itrs.net/.

SECTION II

SYSTEM LEVEL DESIGN

3

Tools and Methodologies for System-Level Design

Shuvra Bhattacharyya
University of Maryland
College Park, Maryland

Wayne Wolf
Princeton University
Princeton, New Jersey

3.1 Introduction

System-level design has long been the province of board designers, but levels of integration have increased to the point that chip designers must concern themselves about system-level design issues. Because chip design is a less forgiving design medium — design cycles are longer and mistakes are harder to correct — system-on-chip (SoC) designers need a more extensive tool suite than may be used by board designers.

System-level design is less amenable to synthesis than are logic or physical design. As a result, system-level tools concentrate on modeling, simulation, design space exploration, and design verification. The goal of modeling is to correctly capture the system's operational semantics, which helps with both implementation and verification. The study of models of computation provides a framework for the description of digital systems. Not only do we need to understand a particular style of computation such as dataflow, but we also need to understand how different models of communication can reliably communicate with each other. Design space exploration tools, such as hardware/software codesign, develop candidate designs to understand trade-offs. Simulation can be used not only to verify functional correctness but also to supply performance and power/energy information for design analysis.

We will use video applications as examples in this chapter. Video is a leading-edge application that illustrates many important aspects of system-level design. Although some of this information is clearly specific to video, many of the lessons translate to other domains.

The next two sections briefly introduce video applications and some SoC architecture that may be the targets of system-level design tools. We will then study models of computation and languages for system-level modeling. Following this, we will survey simulation technique. We will close with a discussion of hardware/software codesign.

3.2 Characteristics of Video Applications

The primary use of SoCs for multimedia today is for video encoding — both compression and decompression. In this section, we review the basic characteristics of video compression algorithms and the implications for video SoC design.

Video compression standards enable video devices to inter-operate. The two major lines of video compression standards are MPEG and H.26x. The MPEG standards concentrate on broadcast applications, which allow for a more expensive compressor on the transmitter side in exchange for a simpler receiver. The H.26x standards were developed with videoconferencing in mind, in which both sides must encode and decode. The advanced video codec (AVC) standard, also known as H.264, was formed by the confluence of the H.26x and MPEG efforts.

Modern video compression systems combine lossy and lossless encoding methods to reduce the size of a video stream. Lossy methods throw away information as a result of which the uncompressed video stream is not a perfect reconstruction of the original; lossless methods do allow the information provided to them to be perfectly reconstructed. Most modern standards use three major mechanisms:

- The discrete cosine transform (DCT) together with quantization
- Motion estimation and compensation
- Huffman-style encoding

The first two are lossy while the third is lossless. These three methods leverage different aspects of the video stream's characteristics to encode it more efficiently.

The combination of DCT and quantization was originally developed for still images and is used in video to compress single frames. The DCT is a frequency transform that turns a set of pixels into a set of coefficients for the spatial frequencies that form the components of the image represented by the pixels. The DCT is preferred over other transforms because a two-dimensional (2D) DCT can be computed using two one-dimemsional (1D) DCTs, making it more efficient. In most standards, the DCT is performed on an 8×8 block of pixels. The DCT does not by itself lossily compress the image; rather, the quantization phase can more easily pick out information to acknowledge the structure of the DCT. Quantization throws out fine details in the block of pixels, which correspond to the high-frequency coefficients in the DCT. The number of coefficients set to zero is determined by the level of compression desired.

Motion estimation and compensation exploit the relationships between frames provided by moving objects. A reference frame is used to encode later frames through a motion vector, which describes the motion of a macroblock of pixels (16×16 in many standards). The block is copied from the reference frame into the new position described by the motion vector. The motion vector is much smaller than the block it represents. Two-dimensional correlation is used to determine the position of the macroblock's position in the new frame; several positions in a search area are tested using 2D correlation. An error signal encodes the difference between the predicted and the actual frames; the receiver uses that signal to improve the predicted picture.

MPEG distinguishes several types of frames: I (inter) frames, which are not motion-compensated; P (predicted) frames, which have been predicted from earlier frames; and B (bidirectional) frames, which have been predicted from both earlier and later frames.

The results of these lossy compression phases are assembled into a bit stream and compressed by using lossless compression such as Huffman encoding. This process reduces the size of the representation without further compromising image quality.

It should be clear that video compression systems are actually heterogeneous collections of algorithms. We believe that this is true of other applications of SoCs as well. A video platform must run several algorithms; those algorithms perform very different types of operations, imposing very different requirements on the architecture.

This has two implications for tools: first, we need a wide variety of tools to support the design of these applications; second, the various models of computation and algorithmic styles used in different parts of an application must at some point be made to communicate to create the complete system.

Several studies of multimedia performance on programmable processors have remarked on the significant number of branches in multimedia code. These observations contradict the popular notion of video as regular operations on streaming data. Fritts and Wolf [1] measured the characteristics of the MediaBench benchmarks.

They used path ratio to measure the percentage of instructions in a loop body that were actually executed. They found that the average path ratio of the MediaBench suite was 78%, which indicates that a significant number of loops exercise data-dependent behavior. Talla et al. [2] found that most of the available parallelism in multimedia benchmarks came from inter-iteration parallelism.

3.3 Other Application Domains

Video and multimedia are not the only application domains for SoCs. Communications and networking are the other areas in which SoCs provide cost/performance benefits. In all these domains, the SoC must be able to handle multiple simultaneous processes. However, the characteristics of those processes do vary. Networking, for example, requires a lot of packet-independent operations. While some networking tasks do require correlating multiple packets, the basic work is packet independent. The large extent of parallelism in packet-level processing can be exploited in the micro-architecture. In the communications world, SoCs are used today primarily for baseband processing, but we should expect SoCs to take over more traditional high-frequency radio functions over time. Since radio functions can operate at very high frequencies, the platform must be carefully designed to support these high rates while providing adequate programmability of radio functions. We should expect highly heterogeneous architectures for high-frequency radio operations.

3.4 Platform Characteristics

Many SoCs are heterogeneous multiprocessors and the architectures designed for multimedia applications are no exceptions. In this section, we review several SoCs, including some general-purpose SoC architectures as well as several designed specifically for multimedia applications.

Two very different types of hardware platforms have emerged for large-scale applications. On the one hand, many custom SoCs have been designed for various applications. These custom SoCs are customized by loading software onto them for execution. On the other hand, platform field-programmable gate arrays (FPGAs) provide FPGA fabrics along with CPUs and other components; the design can be customized by programming the FPGA as well as the processor(s). These two styles of architecture represent different approaches for SoC architecture and they require very different sorts of tools: custom SoCs require large-scale software support, while platform FPGAs are well suited to hardware/software codesign.

3.4.1 Custom System-on-Chip Architectures

The Viper chip [3], shown in Figure 3.1, was designed for high-definition video decoding and set-top box support. Viper is an instance of the Philips Nexperia™ architecture, which is a platform for multimedia applications.

The Viper includes two CPUs: a MIPS32 processor and a Trimedia processor. The MIPS32 is an RISC architecture. The Trimedia is a 5-issue VLIW designed to support video applications. The Trimedia handles all video tasks. It also schedules operations since it is faster than the MIPS32. The MIPS32 runs the operating system and the application stack provided by the service provider (i.e., non-Philips-provided applications). It also performs graphics operations and handles access requests (Figure 3.1).

Each processor has its own bus. A third large, point-to-point bus connects to the off-chip bulk memory used for video memory. Several bridges connect the three high-speed busses. The Viper includes many on-chip devices. The peripheral system provides various general-purpose types of I/Os such as GPIO and IEEE 1394. The audio subsystem includes I/O units that support the SPDIF standard. The video system includes a number of units that video perform processing as well as I/O: picture input processing, a video scaler, an advanced image composition processor, a processor for the MPEG system layer, etc. The infrastructure system includes a memory management unit and the busses.

The Texas Instruments OMAP processor (http://www.omap.com) is designed for mobile multimedia. As shown in Figure 3.2, it includes two processors, an ARM9 CPU and a TI C5x-series digital signal, image and video processing (DSP), connected by a shared memory. OMAP implementations include a wide variety of peripherals. Some peripherals are shared between the two processors, including UARTs and mailboxes for interprocessor communication. Many peripherals, however, are owned by one processor. The DSP owns serial ports, DMA, timers, etc. The ARM9 owns the majority of the I/O devices, including USB, LCD controller, camera interface, etc.

The ST Microelectronics Nomadik is also designed for mobile multimedia applications. (Both the TI OMAP and Nomadik implement the OMAPI software standard.) As shown in Figure 3.3, it can be viewed as a hierarchical multiprocessor. The overall architecture is a heterogeneous multiprocessor organized around a bus, with an ARM processor attached to audio and video processors. But the audio and video processors are each heterogeneous multiprocessors in their own right. The video processor provides an extensive set of hardwired accelerators for video operations organized around two busses; a small processor known as an MMDSP+ provides programmability. The audio processor makes heavier use of an MMDSP+ processor to implement audio-rate functions; it also includes two busses.

All the above SoCs have been architected for particular markets. The ARM multiprocessor core takes a more general-purpose approach. Its MPCore™ is a core designed to be used in SoCs. It can implement

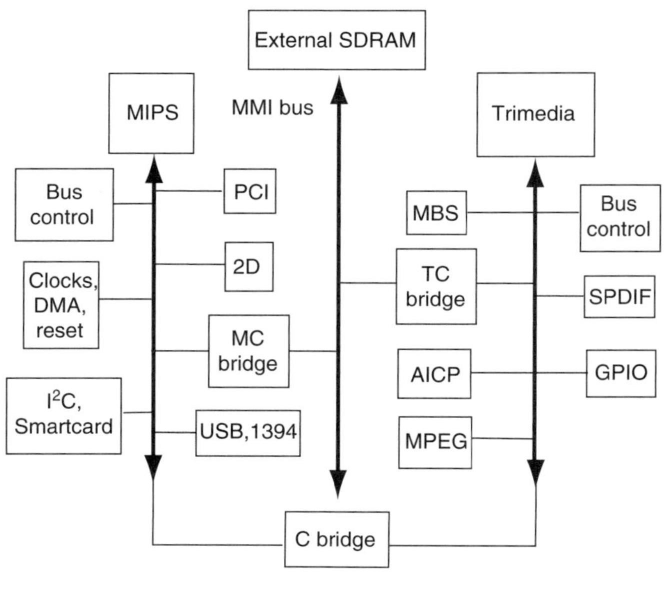

FIGURE 3.1 Hardware architecture of the Viper.

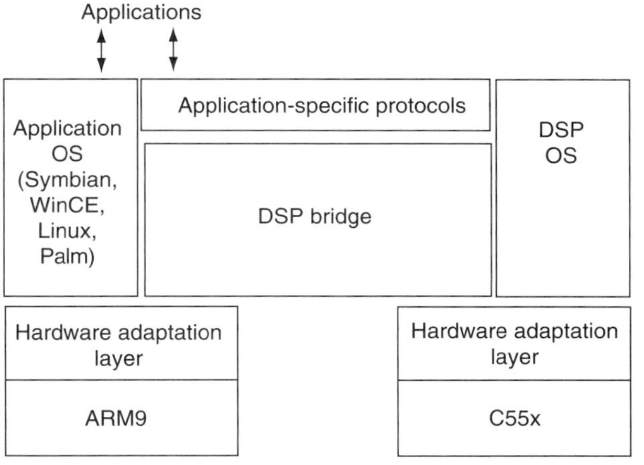

FIGURE 3.2 Hardware and software architecture of the Texas Instruments OMAP.

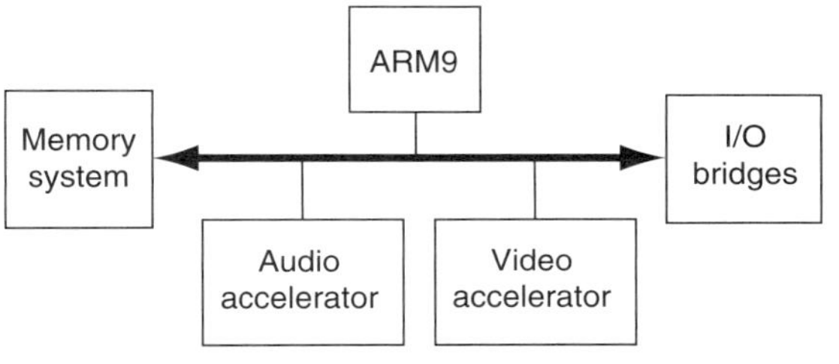

FIGURE 3.3 Architecture of the ST Micro Nomadik.

up to four ARM cores. Those cores can be arranged in any combination of symmetric and asymmetric multiprocessing. Symmetric multiprocessing requires cache coherence mechanisms, which may slow down execution to some extent and cost additional energy, but simplify software design. Asymmetric multiprocessing does not rely on shared memory.

3.4.2 Platform Field-Programmable Gate Arrays

Field-programmable gate arrays [4] have been used for many years to implement logic designs. The FPGA provides a more general structure than programmable logic devices, allowing denser designs. They are less energy-efficient than custom ASICs but do not require the long ASIC design cycle.

Field-programmable gate array logic has now become dense enough that manufacturers provide platform FPGAs. While this term has been used in more than one way, most platform FPGAs provide one or more CPUs in addition to a programmable FPGA fabric. Platform FPGAs provide a very different sort of heterogeneous platform than custom SoCs. The FPGA fabric allows the system designer to implement new hardware functions. While they generally do not allow the CPU itself to be modified, the FPGA logic is a closely coupled device with high throughput to the CPU and to memory. The CPU can also be programmed using standard tools to provide functions that are not well suited to FPGA implementation.

The ATMEL AT94 family (http://www.atmel.com) provides an RISC CPU along with an FPGA core. The AVR CPU core is relatively simple, without a memory management unit. The FPGA core is reprogrammable and configured at power-up.

The Xilinx Virtex II Pro (http://www.xilinx.com) is a high-performance platform FPGA. Various configurations include one to four PowerPCs. The CPUs are connected to the FPGA fabric through the PowerPC bus. The large FPGA fabric can be configured to provide combinations of logic and memory functions.

3.5 Models of Computation and Tools for Model-Based Design

Based on our discussion of application characteristics and hardware platforms, we can now consider tools for multimedia system design. Increasingly, developers of hardware and software for embedded computer systems are viewing aspects of the design and implementation processes in terms of domain-specific models of computation. A domain-specific model of computation is designed to represent applications in a particular functional domain such as DSP; control system design; communication protocols or more general classes of discrete, control-flow intensive decision-making processes; graphics; and device drivers. For discussions of some representative languages and tools that are specialized for these application domains, see [5–13] .

Chapter 4 includes a broad, integrated review of domain-specific languages and programming models for embedded systems. This section discusses in more detail particular modeling concepts and tools that are particularly relevant for multimedia systems. Since DSP techniques constitute the "computational core" of many multimedia applications, a significant part of this section focuses on effective modeling of DSP functionality.

As described in Chapter 4, domain-specific computational models specialize the types of functional building blocks (components) that are available to the programmer and the interactions between components to characterize more intuitively or explicitly relevant characteristics of the targeted class of applications compared to general-purpose programming models. In doing so, domain-specific models of computation often exchange some of the *Turing-complete* expressive power (ability to represent arbitrary computations) of general-purpose models and often also a large body of legacy infrastructure for advantages such as increased intuitive appeal, support for formal verification, and more thorough optimization with respect to relevant implementation metrics.

3.5.1 Dataflow Models

For most DSP applications, a significant part of the computational structure is well suited to modeling in a dataflow model of computation. In the context of programming models, dataflow refers to a modeling methodology where computations are represented as directed graphs in which vertices (actors) represent functional components and edges between actors represent first-in-first-out (FIFO) channels that buffer data values (tokens) as they pass from an output of one actor to an input of another. Dataflow actors can represent computations of arbitrary complexity; typically in DSP design environments, they are specified using conventional languages such as C or assembly language, and their associated tasks range from simple, "fine-grained" functions such as addition and multiplication to "coarse-grain" DSP kernels or subsystems such as FFT units and adaptive filters. The development of application modeling and analysis techniques based on dataflow graphs was inspired significantly by the computation graphs of Karp and Miller [14], and the process networks of Kahn [15]. A unified formulation of dataflow modeling principles, as they apply to DSP design environment, is provided by the dataflow process networks model of computation of Lee and Parks [16].

A dataflow actor is enabled for execution whenever it has sufficient data on its incoming edges (i.e., in the associated FIFO channels) to perform its specified computation. An actor can execute whenever it is enabled (data-driven execution). In general, the execution of an actor results in some number of tokens being removed (consumed) from each incoming edge, and some number being placed (produced) on each outgoing edge. This production activity in general leads to the enabling of other actors.

The order in which actors execute, called the schedule, is not part of a dataflow specification, and is constrained only by the simple principle of data-driven execution defined above. This is in contrast to many alternative computational models, such as those that underlie procedural languages, in which execution order is overspecified by the programmer [17]. The schedule for a dataflow specification may be determined at compile time (if sufficient static information is available), at run-time, or using a mixture of compile- and run-time techniques. A particularly powerful class of scheduling techniques, referred to as quasi-static scheduling (see, e.g., [18]), involves most, but not all of the scheduling decisions being made at compile time.

Figure 3.4 shows an illustration of a video processing subsystem that is modeled using dataflow semantics. This is a design, developed using the Ptolemy II tool for model-based embedded system design [19]

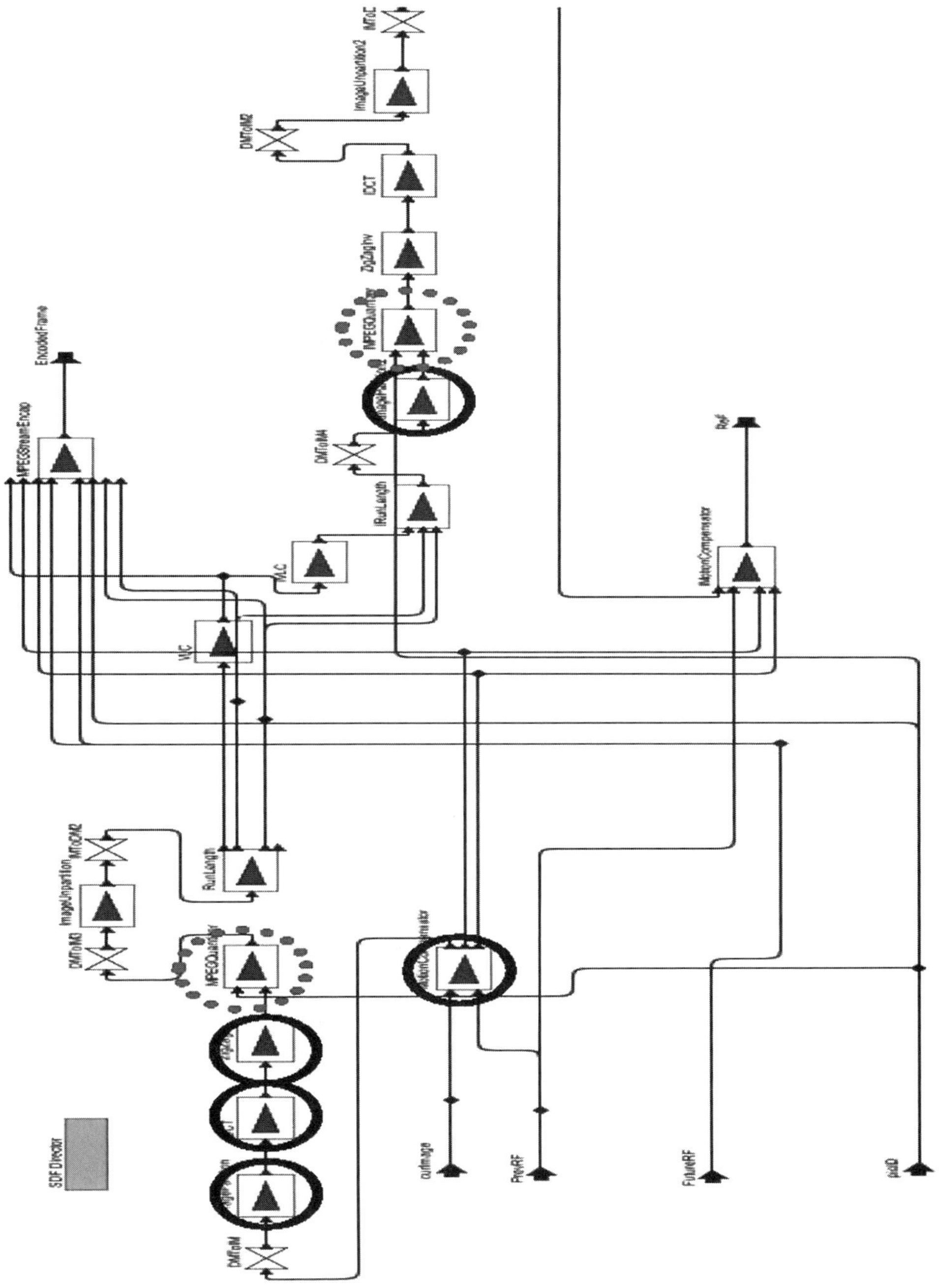

FIGURE 3.4 A video processing subsystem modeled in dataflow.

of an MPEG2 subsystem for encoding the P frames that are processed by an enclosing MPEG2 encoder system. A thorough discussion of this MPEG2 system and its comparison to a variety of other modeling representations is presented in [20]. The components in the design of Figure 3.4 include actors for the DCT, zig-zag scanning, quantization, motion compensation and run length coding. The arrows in the illustration correspond to the edges in the underlying dataflow graph. The actors and their interactions all conform to the semantics of *synchronous dataflow* (SDF), which is a restricted form of dataflow that is efficient for describing a broad class of DSP applications and has particularly strong formal properties and optimization advantages [21,22]. Specifically, SDF imposes the restriction that the number of tokens produced and consumed by each actor on each incident edge is constant. Many commercial DSP design tools have been developed that employ semantics that are equivalent to or closely related to SDF. Examples of such tools include Agilent's ADS, Cadence's SPW (now developed by CoWare), and MCCI's Autocoding Toolset.

3.5.2 Dataflow Modeling for Video Processing

In the context of video processing, SDF permits accurate representation of many useful subsystems, such as the P-frame encoder shown in Figure 3.4. However, such modeling is often restricted to a highly coarse level of granularity, where actors process individual frames or groups of successive frames on each execution. Modeling at such a coarse granularity can provide compact, top-level design representations, but greatly limits the benefits offered by the dataflow representation since most of the computation is subsumed by the general-purpose, intra-actor program representation. For example, the degree of parallel processing and memory management optimizations exposed to a dataflow-based synthesis tool becomes highly limited at such coarse levels of actor granularity.

A number of alternative dataflow modeling methods have been introduced to address this limitation of SDF modeling for video processing and more generally, multidimensional signal processing applications. For example, the multidimensional synchronous dataflow (MD-SDF) model extends SDF semantics to allow constant-sized, n-dimensional vectors of data to be transferred across graph edges, and provides support for arbitrary sampling lattices and lattice-changing operations [23]. The computer vision synchronous dataflow (CV-SDF) model is designed specifically for computer vision applications, and provides a notion of *structured buffers* for decomposing video frames along graph edges; accessing neighborhoods of image data from within actors, in addition to the conventional production and consumption semantics of dataflow, and allowing actors to access efficiently previous frames of image data [24,25]. Blocked dataflow (BLDF) is a metamodeling technique for efficiently incorporating hierarchical, block-based processing of multidimensional data into a variety of dataflow modeling styles, including SDF and MD-SDF [20].

Multidimensional synchronous dataflow, CV-SDF, and BLDF are still at experimental stages of development and to our knowledge, none of them has yet been incorporated into a commercial design tool, although practical image and video processing systems of significant complexity have been demonstrated using these techniques. Integrating effective dataflow models of computation for image and video processing into commercial DSP design tools is an important area for further exploration.

3.5.3 Control Flow

As described previously, modern video processing applications are characterized by some degree of control flow processing for carrying out data-dependent configuration of application tasks and changes across multiple application modes. For example, in MPEG2 video encoding, significantly different processing is required for I, P, and B frames. Although the processing for each particular type of frame (I, P, or B) conforms to the SDF model, as illustrated for P-frame processing in Figure 3.4, a layer of control flow processing is needed to integrate efficiently these three types of processing methods into a complete MPEG2 encoder design. The SDF model is not well suited for performing this type of control flow processing and, more generally, for any functionality that requires dynamic communication patterns or activation/deactivation across actors.

A variety of alternative models of computation have been developed to address this limitation, and integrate flexible control flow capability with the advantages of dataflow modeling. In Buck's Boolean dataflow model [26], and the subsequent generalization as integer-controlled dataflow [27], provisions for such flexible processing were incorporated without departing from the framework of dataflow, and in a manner that facilitates construction of efficient hybrid compile-/run-time schedules. In Boolean dataflow, the number of tokens produced or consumed on an edge is either fixed or is a two-valued function of a control token present on a control terminal of the same actor. It is possible to extend important SDF analysis techniques to Boolean dataflow graphs by employing symbolic variables. In particular, in constructing a schedule for Boolean dataflow actors, Buck's techniques attempt to derive a quasi-static schedule, where each conditional actor execution is annotated with the run-time condition under which the execution should occur. Boolean dataflow is a powerful modeling technique that can express arbitrary control flow structures. However, as a result, key formal verification properties of SDF, such as bounded memory and deadlock detection, are lost in the context of general Boolean dataflow specifications.

In recent years, several modeling techniques have also been proposed that enhance expressive power by providing precise semantics for integrating dataflow or dataflow-like representations with finite-state machine (FSM) models. These include El Greco [28], which has evolved into the Synopsys Cocentric System Studio, and that provides facilities for "control models" to configure dynamically specification parameters; *charts (pronounced starcharts) with heterochronous dataflow as the concurrency model [29]; the FunState intermediate representation [30]; the DF* framework developed at K. U. Leuven [31]; and the control flow provisions in bounded dynamic dataflow [32].

Figure 3.5 shows an illustration of a specification of a complete MPEG2 video encoder system that builds on the P-frame-processing subsystem of Figure 3.4, and employs multiple dataflow graphs nested within an FSM representation. Details on this specification can be found in [20].

Cocentric System Studio is a commercial tool for system design that employs integrated representations using FSMs and dataflow graphs. We will discuss the Cocentric tool in more detail later in this chapter.

3.5.4 Ptolemy

The Ptolemy project at U.C. Berkeley has had considerable influence on models of computation for DSP and multimedia, and also on the general trend toward viewing embedded systems design in terms of models of computation. The first origins of this project are in the BLOSIM [33] tool, which developed block diagram simulation capability for signal processing systems. Work on BLOSIM led to the development of the SDF formalism [21] and the Gabriel design environment [10], which provided simulation, static scheduling, and single- and multiprocessor software synthesis capability for SDF-based design.

Ptolemy (now known as Ptolemy Classic) is the third-generation tool that succeeded Gabriel [34]. The design of Ptolemy Classic emphasized efficient modeling and simulation of embedded systems based on the interaction of heterogeneous models of computation. A key motivation was to allow designers to represent each subsystem of a design in the most natural model of computation associated with that subsystem, and allow subsystems expressed in different models of computation to be integrated seamlessly into an overall system design. A key constraint imposed by the Ptolemy Classic approach to heterogeneous modeling is the concept of *hierarchical heterogeneity*. It is widely understood that in hierarchical modeling, a system specification is decomposed into a set C of subsystems in which each subsystem can contain one or more hierarchical components, each of which represents another subsystem in C. Under hierarchical heterogeneity, each subsystem in C must be described using a uniform model of computation, but the nested subsystem associated with a hierarchical component H can be expressed in a model of computation that is different from the model of computation that expresses the subsystem containing H.

Thus, under hierarchical heterogeneity, the integration of different models of computation must be achieved entirely through the hierarchical embedding of heterogeneous models. A key consequence is that whenever a subsystem S_1 is embedded in a subsystem S_2 that is expressed in a different model of computation, the subsystem S_1 must be abstracted by a hierarchical component in S_2 that conforms to the model of computation associated with S_2. This provides precise constraints for interfacing different models

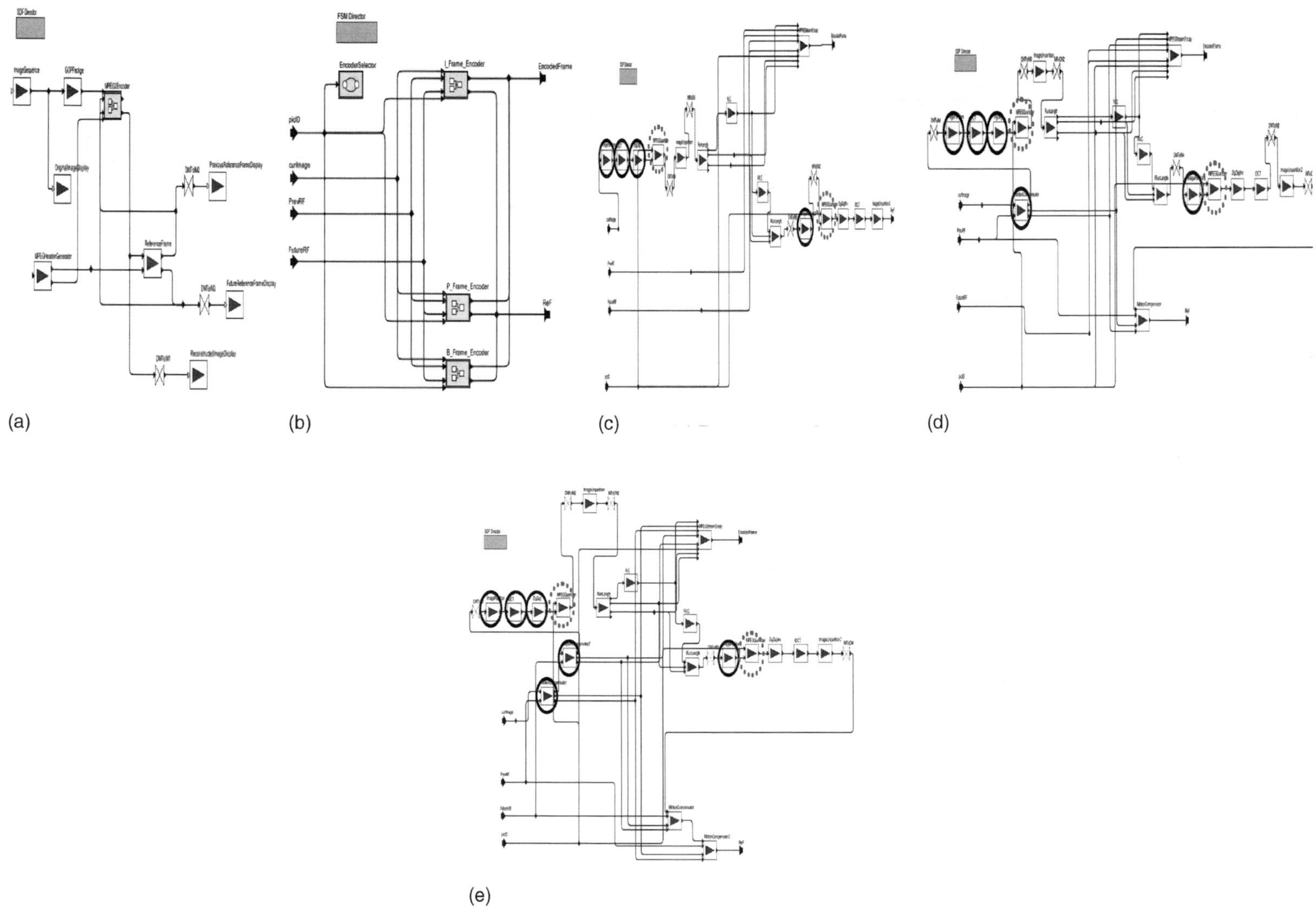

FIGURE 3.5 An MPEG2 video encoder specification. (a) MPEG2 encoder (top); (b) inside the FSM; (c) I-frame encoder; (d) P-frame encoder; and (e) B-frame encoder.

of computation. Although these constraints may not always be easy to conform to, they provide a general and unambiguous convention for heterogeneous integration and perhaps even more importantly, the associated interfacing methodology allows each subsystem to be analyzed using the techniques and tools available for the associated model of computation.

Ptolemy Classic was developed through a highly flexible, extensible, and robust software design, and this has facilitated experimentation with the underlying modeling capabilities in various aspects of embedded systems design by many research groups. Other major areas of contribution associated with development of Ptolemy Classic include hardware/software codesign, as well as further contributions in dataflow-based modeling and synthesis (see, e.g., [23,35,36]).

The current incarnation of the Ptolemy project, called Ptolemy II, is a Java-based tool that furthers the application of model-based design and hierarchical heterogeneity [19], and provides an even more malleable software infrastructure for experimentation with new techniques involving models of computation.

An important theme in Ptolemy II is the reuse of actors across multiple models and computational models. Through an emphasis in Ptolemy II on support for *domain polymorphism*, the same actor definition can in general be applicable across a variety of models of computation. In practice, domain polymorphism greatly increases reuse of actor code. Techniques based on interface automata [37] have been developed to characterize systematically the interactions between actors and models of computation, and reason about their compatibility (i.e., whether or not it makes sense to instantiate the actor in specifications that are based on the model) [38].

3.5.5 Compaan

MATLAB® is one of the most popular programming languages for algorithm development and high-level functional simulation for DSP applications. In the Compaan project developed at Leiden University, systematic techniques have been developed for synthesizing embedded software and FPGA-based hardware implementations from a restricted class of MATLAB programs known as parameterized, static nested loop programs [39]. In Compaan, an input MATLAB specification is first translated into an intermediate representation based on the Kahn process network model of computation [15]. The Kahn process network model is a general model of data-driven computation that subsumes as a special case as in the dataflow process networks mentioned earlier in this chapter. Like dataflow process networks, Kahn process networks also consist of concurrent functional modules that are connected by FIFO buffers with non-blocking writes and blocking reads; however, unlike in the dataflow process network model, modules in Kahn process networks do not necessarily have their execution decomposed *a priori* into well-defined, discrete units of execution [16].

Through its aggressive dependence analysis capabilities, Compaan combines the widespread appeal of MATLAB at the algorithm development level, with the guaranteed determinacy, compact representation, simple synchronization, and distributed control features of Kahn process networks for efficient hardware/software implementation.

Technically, the Kahn process networks derived in Compaan can be described as equivalent cyclo-static dataflow graphs [40], which we discuss in more detail later in this chapter, and therefore fall under the category of dataflow process networks. However, these equivalent cyclo-static dataflow graphs can be very large and unwieldy to work with and, therefore, analysis in terms of the Kahn process network model is often more efficient and intuitive.

Development of the capability for translation of MATLAB to Kahn process networks was originally developed by Kienhuis et al. [41], and this capability has since then evolved into an elaborate suite of tools for mapping Kahn process networks into optimized implementations on heterogeneous hardware/software platforms consisting of embedded processors and FPGAs [39]. Among the most interesting optimizations in the Compaan tool suite are dependence analysis mechanisms that determine the most specialized form of buffer implementation, with respect to reordering and multiplicity of buffered values, for implementing interprocess communication in Kahn process networks [42].

3.5.6 CoWare

CoWare, developed originally by IMEC and now by CoWare, Inc., is a tool for system-level design, simulation, and hardware/software synthesis of DSP systems [43]. Integration of dataflow and control flow modeling is achieved in CoWare through networks of processes that communicate through remote procedure call semantics. Processes can have one or more flows of control, which are called threads, and through a remote procedure call, a process can invoke a thread in another process that is connected to it. A thread in a process can also execute autonomously, independent of any remote procedure call, in what is effectively an infinite loop that is initiated at system start-up. During run time, the overhead of switching between threads can be significant, and furthermore, such points of switching present barriers for important compiler optimizations. Therefore, remote procedure calls are inlined during synthesis, thereby merging groups of compatible, communicating processes into monolithic processes.

Buffered communication of data can also be carried out between processes. A variety of protocols can be used for this purpose, including support for the FIFO buffering associated with dataflow-style communication. Thus CoWare provides distinct mechanisms for interprocess control flow and dataflow that can be flexibly combined within the same system or subsystem specification. This is in contrast to control flow/dataflow integration techniques associated with methods such as Boolean dataflow and general dataflow process networks, where all intermodule control flow is mapped into a more general dataflow framework, and also with techniques based on hierarchical heterogeneity, where control flow and dataflow styles are combined only through hierarchical embedding.

3.5.7 Cocentric System Studio

At Synopsys, a design tool for hardware/software cosimulation and synthesis has been developed based on principles of hierarchical heterogeneity, in particular, hierarchical combinations of FSMs and dataflow graphs. This tool, originally called El Greco [28], along with its predecessor COSSAP [44], which was based on a close variant of SDF semantics [45], evolved into Synopsys's Cocentric System Studio.

Cocentric incorporates two forms of dataflow modeling — cyclo-static dataflow modeling for purely deterministic dataflow relationships, and a general form of dataflow modeling for expressing dynamic dataflow constructs. Cyclo-static dataflow developed at Katholieke Universiteit Leuven, is a useful extension of SDF, in which the numbers of tokens produced and consumed by an actor are allowed to vary at run time as long as the variation takes the form of a fixed, periodic pattern [40]. At the expense of some simplicity in defining actors, cyclo-static dataflow has been shown to have many advantages over SDF, including more economical buffering of data, more flexible support for hierarchical specifications, improved ability to expose opportunities for behavioral optimization, and facilitation of more compact design representations (see, e.g., [40,46,47]).

Control flow models in Cocentric can be integrated hierarchically with dataflow models. Three types of hierarchical control flow models are available: an *or-model* represents conditional transitions, and associated actions, among a collection of mutually exclusive states; an *and-model* represents a group of subsystems that execute in parallel with broadcast communication across the subsystems; and a *gated model* is used to conditionally switch models between states of activity and suspension. To increase the potential for compile-time analysis and efficient hardware implementation, a precise correspondence is maintained between the control semantics of Cocentric and the synchronous language Esterel [48].

3.5.8 Handel-C

One important direction in which innovative models of computation are being applied to embedded systems is in developing variants of conventional procedural languages, especially variants of the C language, for the design of hardware and for hardware/software codesign. Such languages are attractive because they are based on a familiar syntax, and are thus easier to learn and to adapt to existing designs. A prominent example of one such language is Handel-C, which is based on embedding concepts of the abstract Handel language [49] into a subset of ANSI C. Handel in turn is based on the communicating sequential processes (CSPs) model of computation [50]. CSP provides a simple and intuitive form of support to coordinate parallel execution of multiple hardware subsystems through its mechanism of synchronized, point-to-point communication channels for interprocess communication. Handel was developed at Oxford University and prototyped originally by embedding its semantics in the SML language [51]; subsequently, the concepts of Handel were ported to ANSI C and commercialized through Celoxica Limited as Handel-C (see, e.g., [52]).

In Handel-C, parallelism is specified explicitly by the programmer using a *par* statement, and communication across parallel subsystems through CSP semantics is specified using an associated *chan* (short for "channel") construct.

Handel-C has been targeted to platform FPGAs, such as Altera Excalibur and Xilinx Virtex II Pro.

3.5.9 Simulink

At present, Simulink [53], developed by the The MathWorks, is perhaps the most widely used commercial tool for model-based design of DSP hardware and software. Simulink provides a block-diagram interface and extensive libraries of predefined blocks; highly expressive modeling semantics with support for continuous time, discrete time, and mixed-signal modeling; support for fixed-point data types; and capabilities for incorporating code in a variety of procedural languages, including Matlab and C. Simulink can be augmented with various capabilities to enhance greatly its utility in DSP and multimedia system implementation. For example, various add-on libraries provide rich collections of blocks geared toward signal processing, communications, and image/video processing. The Real-time Workshop provides code generation capabilities to translate automatically Simulink models into ANSI C. Stateflow provides the capability to augment Simulink with sophisticated control flow modeling features. The Xilinx System Generator is a plug-in tool for Simulink, which generates synthesizable hardware description language code targeted to Xilinx devices. Supported devices include the Virtex-II Pro platform FPGA, which was described earlier in this chapter. The Texas Instruments Embedded Target links Simulink and the Real-time Workshop with the Texas Instruments Code Composer Studio to provide a model-based environment for design and code generation that is targeted to fixed- and floating-point programmable digital signal processors in the Texas Instruments C6000 series.

3.5.10 Prospects for Future Development of Tools

The long list (over 300 at present) of other third-party products and services that work with Simulink demonstrates the significance of this tool in the embedded systems industry as well as the growing adoption of model-based design techniques. Although model-based design is used widely for signal processing, which is an important building block for multimedia applications, model-based techniques are not yet so extensively used for the overall multimedia systems design process. This section has discussed key challenges in developing effective design tools that are centered around domain-specific computational models for multimedia. These include efficient, integrated representation of multidimensional data streams, high-level control flow, and reconfigurable application behavior and closer interaction, such as that facilitated by the relationship between Cocentric and Esterel, of models of computation with design representations that are used in the back end of the hardware/software synthesis process. As these challenges are studied further, and more experience is gained in the application of models of computation to multimedia hardware and software implementation, we expect increasing deployment of model-based tools in the multimedia domain.

3.6 Simulation

Simulation is very important in SoC design. Simulation is not limited to functional verification, as with logic design. System-on-chip designers use simulation to measure the performance and power consumption of their SoC designs. This is due in part to the fact that much of the functionality is implemented in software, which must be measured relative to the processors on which it runs. It is also due to the fact that the complex input patterns inherent in many SoC applications do not lend themselves to closed-form analysis.

SystemC (http://www.systemc.org) is a simulation language that is widely used to model SoCs. SystemC leverages the C++ programming language to build a simulation environment. SystemC classes allow designers to describe a digital system using a combination of structural and functional techniques. SystemC supports simulation at several levels of abstraction. Register-transfer level simulations, for example, can be performed with the appropriate SystemC model. SystemC is most often used for more abstract models. A common type of model built in SystemC is a transaction-level model. This style of modeling describes the SoC as a network of communicating machines, with explicit connections between the models and functional descriptions for each model. The transaction-level model describes how data are moved between the models.

Hardware/software cosimulators are multimode simulators that simultaneously simulate different parts of the system at different levels of detail. For example, some modules may be simulated in register-transfer mode while software running on a CPU is simulated functionally. Cosimulation is particularly useful for debugging the hardware/software interface, such as debugging driver software. The Seamless cosimulator from Mentor Graphics is a well-known example of a hardware/software cosimulator. The VaST Systems CoMET simulator is designed to simulate networks of processors and hardware devices.

Functional validation, performance analysis, and power analysis of SoCs require simulating large number of vectors. Video and other SoC applications allow complex input sequences. Even relatively compact tests can take up tens of millions of bytes. These long input sequences are necessary to run the SoC through a reasonable number of the states implemented in the system. The large amounts of memory that can be integrated into today's systems, whether they are on-chip or off-chip, allow the creation of SoCs with huge numbers of states that require long simulation runs.

Simulators for software running on processors have been developed over the past several decades. Both computer architects and SoC designers need fast simulators to run the large benchmarks required to evaluate architectures. As a result, a number of simulation techniques covering a broad range of accuracy and performance have been developed.

A simple method of analyzing CPU is to sample the program counter (PC) during program execution. The Unix *prof* command is an example of a PC-sampling analysis tool. Program counter sampling is subject to the same limitations on sampling rate as any other sampling process, but sampling rate is usually not a major concern in this case. A more serious limitation is that PC sampling gives us relative performance but not absolute performance. A sampled trace of the PC tells us where the program spent its time during execution, which gives us valuable information about the relative execution time of program modules that can be used to optimize the program. But it does not give us the execution time on a particular platform — especially if the target platform is different from the platform on which the trace is taken — and so we must use other methods to determine the real-time performance of programs.

Some simulators concentrate on the behavior of the cache, given the major role of the cache in determining overall system performance. The *dinero* simulator is a well-known example of a cache simulator. These simulators generally work from a trace generated from the execution of a program. The program is to be analyzed with additional code that records the execution behavior of the program. The *dinero* simulator then reconstructs the cache behavior from the program trace. The architect can view the cache in various states or calculate cache statistics.

Some simulation systems model the behavior of the processor itself. A functional CPU simulator models instruction execution and maintains the state of the programming model, that is, the set of registers visible to the programmer. The functional simulator does not, however, model the performance or energy consumption of the program's execution.

A cycle-accurate simulator of a CPU is designed to predict accurately the number of clock cycles required to execute every instruction, taking into account pipeline and memory system effects. The CPU model must therefore represent the internal structure of the CPU accurately enough to show how resources in the processor are used. The SimpleScalar simulation tool [54] is a well-known toolkit for building cycle-accurate simulators. SimpleScalar allows a variety of processor models to be built by a

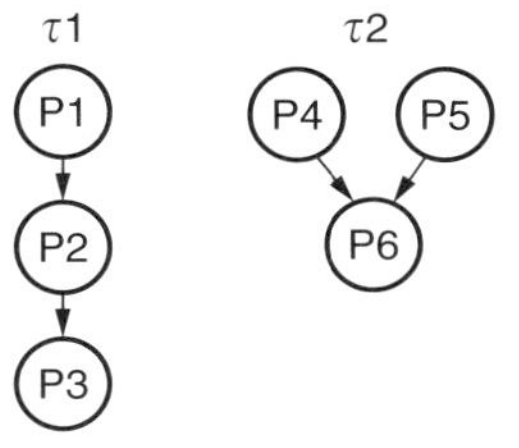

FIGURE 3.6 A task graph.

combination of parameterization of existing models and linking new simulation modules into the framework.

Power simulators are related to cycle-accurate simulators. Accurate power estimation requires models of the CPU micro-architecture, which are at least as detailed as those used for performance evaluation. A power simulator must model all the important wires in the architecture since capacitance is a major source of power consumption. Wattch [55] and SimplePower [56] are the two best-known CPU power simulators.

3.7 Hardware/Software Cosynthesis

Hardware/software cosynthesis tools allow system designers to explore architectural trade-offs. These tools take a description of a desired behavior that is relatively undifferentiated between hardware and software. They produce a heterogeneous hardware architecture and the architecture for the software to run on that platform. The software architecture includes the allocation of software tasks to the processing elements of the platform, and the scheduling of computation and communication.

The functional description of an application may take several forms. The most basic is a task graph, as shown in Figure 3.6. A task graph is a simplified form of dataflow graph in which each graph component runs at its own rate; however, the rates of the tasks need not be integrally related. The graph describes data dependencies between a set of processes. Each component of the graph (i.e., each set of connected nodes) forms a task. Each task runs periodically and every task can run at a different rate. The task graph model generally does not concern itself with the details of operations within a process. The process is characterized by its execution time. Several variations of task graphs that include control information have been developed. In these models, the output of a process may enable one of several different processes.

An alternative representation for behavior is a programming language. Several different codesign languages have been developed and languages such as SystemC have been used for cosynthesis as well. These languages may make use of constructs to describe parallelism that were originally developed for parallel programming languages. Such constructs are often used to capture operator-level concurrency. The subroutine structure of the program can be used to describe task-level parallelism.

The most basic form of hardware/software cosynthesis is hardware/software partitioning. As shown in Figure 3.7, this method maps the design into an architectural template. The basic system architecture is bus-based, with a CPU and one or more custom hardware processing elements attached to the bus. The type of CPU is determined in advance, which allows the tool to estimate accurately software performance. The tool must decide what functions go into the custom processing elements; it must also schedule all the operations, whether implemented in hardware or software. This approach is known as hardware/software partitioning because the bus divides the architecture into two partitions, and partitioning algorithms can be used to explore the design space.

Two important approaches to searching the design space during partitioning were introduced by early tools. The Vulcan system [57] starts with all processes in the custom processing elements and iteratively moves selected processes to the CPU to reduce the system cost. The COSYMA system [58] starts with all operations running on the CPU and moves selected operations from loop nests into the custom processing element to increase performance.

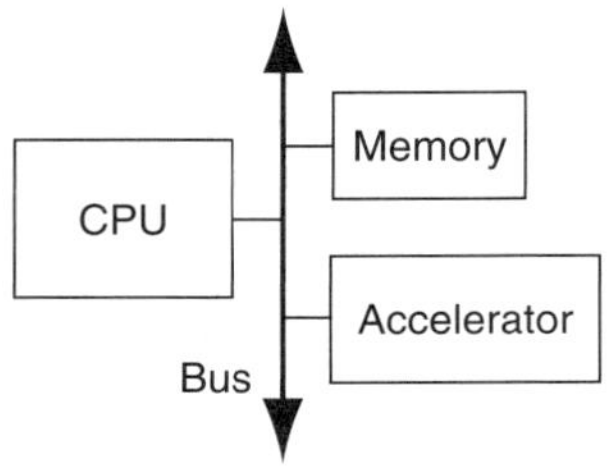

FIGURE 3.7 A template for hardware/software partitioning.

Hardware/software partitioning is ideally suited to platform FPGAs, which implement the bus-partitioned structure and use FPGA fabrics for the custom processing elements. However, the cost metric is somewhat different than in custom designs. Because the FPGA fabric is of a fixed size, using more or less of the fabric may not be important so long as the design fits into the amount of logic available.

Other cosynthesis algorithms have been developed that do not rely on an architectural template. Kalavade and Lee [59] alternately optimize for performance and cost to generate a heterogeneous architecture. Wolf [60] alternated cost reduction and load balancing while maintaining a performance-feasible design. Dick and Jha [61] used genetic algorithms to search the design space.

Scheduling is an important task during cosynthesis. A complete system schedule must ultimately be constructed; an important aspect of scheduling is the scheduling of multiple processes on a single CPU. The study of real-time scheduling for uniprocessors was initiated by Liu and Layland [62], who developed rate-monotonic scheduling (RMS) and earliest-deadline-first (EDF) scheduling. Rate-monotonic scheduling and EDF are priority-based schedulers, which use priorities to determine which process to run next. Many cosynthesis systems use custom, state-based schedulers that determine the process to be executed based upon the state of the system.

Design estimation is an important aspect of cosynthesis. While some software characteristics may be determined by simulation, hardware characteristics are often estimated using high-level synthesis. Henkel and Ernst [63] used forms of high-level synthesis algorithms to synthesize quickly a hardware accelerator unit and estimate its performance and size. Fornaciari et al. [64] used high-level synthesis and libraries to estimate power consumption.

Software properties may be estimated in a variety of ways, depending on the level of abstraction. For instance, Li and Wolf [65] built a process-level model of multiple processes interacting in the cache to provide an estimate of the performance penalty due to caches in a multitasking system. Tiwari et al. [66] used measurements to build models of the power consumption of instructions executing on processors.

3.8 Summary

System-level design is challenging because it is heterogeneous. The applications that we want to implement are heterogeneous — we may mix data and control, synchronous and asynchronous, etc. The architectures on which we implement these applications are also heterogeneous combinations of custom hardware, processors and memory. As a result, system-level tools help designers manage and understand complex, heterogeneous systems. Models of computation help designers cast their problem in a way that can be clearly understood by both humans and tools. Simulation helps designers gather important design characteristics. Hardware/software cosynthesis helps explore design spaces. As applications become more complex, we should expect to see tools continue to reach into the application space to aid with the transition from algorithm to architecture.

Over the next few years, we should expect to see simulation tools improve. Commercial simulation tools are well suited to networks of embedded processors or to medium-size SoCs, but new tools may be needed for large heterogeneous single-chip multiprocessors. Most multiprocessor simulators are designed for symmetric multiprocessors, but the large number of heterogeneous multiprocessor SoCs being designed assures that more general simulators will be developed.

Design modeling languages continue to evolve and proliferate. As CAD companies introduce new languages, it becomes harder to support consistent design flows. We hope that the design community will settle on a small number of languages that can be supported by a flexible set of interoperable tools.

References

[1] J. Fritts and W. Wolf, Evaluation of static and dynamic scheduling for media processors, *Proceedings, MICRO-33 MP-DSP2 Workshop*, ACM, Monterey, CA, December 2000.

[2] D. Talla, L. John, V. Lapinskii, and B.L. Evans, Evaluating signal processing and multimedia applications on SIMD, VLIW and superscalar architectures, *Proceedings of the IEEE International Conference on Computer Design*, Austin, TX, 2000.

[3] S. Dutta et al., Viper: a multiprocessor SOC for advanced set-top box and digital TV systems, *IEEE Design Test Comput.*, 18, 21–31, 2001.

[4] W. Wolf, *FPGA-Based System Design*, PTR Prentice Hall, New York, 2004.

[5] A. Basu, M. Hayden, G. Morrisett, and T. von Eicken, A language-based approach to protocol construction, *ACM SIGPLAN Workshop on Domain-Specific Languages*, Paris, France, January 1997.

[6] C.L. Conway and S.A. Edwards, NDL: a domain-specific language for device drivers, *Proceedings of the Workshop on Languages Compilers and Tools for Embedded Systems*, Washington, D.C., June 2004.

[7] S.A. Edwards, *Languages for Digital Embedded Systems*, Kluwer Academic Publishers, Dordrecht, 2000.

[8] K. Konstantinides and J.R. Rasure, The Khoros software-development environment for image-processing and signal-processing, *IEEE Trans. Image Process.*, 3, 243–252, 1994.

[9] R. Lauwereins, M. Engels, M. Ade, and J.A. Peperstraete, Grape-II: a system-level prototyping environment for DSP applications, *IEEE Comput. Mag.*, 28, 35–43, 1995.

[10] E.A. Lee, W.H. Ho, E. Goei, J. Bier, and S.S. Bhattacharyya, Gabriel: a design environment for DSP, *IEEE Trans. Acoust., Speech, Signal Process.*, 37, 1751–1762, 1989.

[11] V. Manikonda, P.S. Krishnaprasad, and J. Hendler, Languages, behaviors, hybrid architectures and motion control, in *Essays in Mathematical Control Theory (in Honor of the 60th Birthday of Roger Brockett)*, J. Baillieul and J.C. Willems, Eds., Springer, Heidelberg, 1998, pp. 199–226.

[12] K. Proudfoot, W.R. Mark, S. Tzvetkov, and P. Hanrahan, A real-time procedural shading system for programmable graphics hardware, *Proceedings of SIGGRAPH*, Los Angeles, CA, 2001.

[13] S.A. Thibault, R. Marlet, and C. Consel, Domain-specific languages: from design to implementation application to video device drivers generation, *IEEE Trans. Software Eng.*, 25, 363–377, 1999.

[14] R.M. Karp and R.E. Miller, Properties of a model for parallel computations: determinacy, termination, queuing, *SIAM J. Appl. Math.*, 14, 1966.

[15] G. Kahn, The semantics of a simple language for parallel programming, *Proceedings of the IFIP Congress*, Stockholm, Sweden, 1974.

[16] E.A. Lee and T.M. Parks, Dataflow process networks, *Proceedings of the IEEE*, May 1995, pp. 773–799.

[17] A.L. Ambler, M.M. Burnett, and B.A. Zimmerman, Operational versus definitional: a perspective on programming paradigms, *IEEE Comput. Mag.*, 25, 28–43, 1992.

[18] S. Ha and E.A. Lee, Compile-time scheduling and assignment of data-flow program graphs with data-dependent iteration, *IEEE Trans. Comput.*, 40, 1225–1238, 1991.

[19] J. Eker, J.W. Janneck, E.A. Lee, J. Liu, X. Liu, J. Ludvig, S. Neuendorffer, S. Sachs, and Y. Xiong, Taming heterogeneity — the Ptolemy approach, *Proceedings of the IEEE*, January 2003.

[20] D. Ko and S.S. Bhattacharyya, Modeling of block-based DSP systems, *J. VLSI Signal Process. Syst. Signal, Image, and Video Technol.*, 40, 289–299, 2005.

[21] E.A. Lee and D.G. Messerschmitt, Synchronous dataflow, *Proc. IEEE*, 75, 1235–1245, 1987.

[22] S.S. Bhattacharyya, P.K. Murthy, and E.A. Lee, *Software Synthesis from Dataflow Graphs*, Kluwer Academic Publishers, Dordrecht, 1996.

[23] P.K. Murthy and E.A. Lee, Multidimensional synchronous dataflow, *IEEE Trans. Signal Process.*, 50, 2064–2079, 2002.

[24] D. Stichling and B. Kleinjohann, CV-SDF — a model for real-time computer vision applications, *Proceedings of the IEEE Workshop on Application of Computer Vision*, Orlando, FL, December 2002.

[25] D. Stichling and B. Kleinjohann, CV-SDF — a synchronous data flow model for real-time computer vision applications, *Proceedings of the International Workshop on Systems, Signals and Image Processing*, Manchester, UK, November 2002.

[26] J.T. Buck and E.A. Lee, Scheduling dynamic dataflow graphs using the token flow model, *Proceedings of the International Conference on Acoustics, Speech, and Signal Processing*, Minneapolis, MN, April 1993.

[27] J.T. Buck, Static scheduling and code generation from dynamic dataflow graphs with integer-valued control systems, *Proceedings of the IEEE Asilomar Conference on Signals, Systems, and Computers*, Pacific Grove, CA, October 1994, pp. 508–513.

[28] J. Buck and R. Vaidyanathan, Heterogeneous modeling and simulation of embedded systems in El Greco, *Proceedings of the International Workshop on Hardware/Software Co-Design*, San Diego, CA, May 2000.

[29] A. Girault, B. Lee, and E.A. Lee, Hierarchical finite state machines with multiple concurrency models, *IEEE Trans. Comput.-Aid. Design Integrated Circuits Syst.*, 18, 742–760, 1999.

[30] L. Thiele, K. Strehl, D. Ziegenbein, R. Ernst, and J. Teich, FunState — an internal representation for codesign, *Proceedings of the International Conference on Computer-Aided Design*, San Jose, CA, November 1999.

[31] N. Cossement, R. Lauwereins, and F. Catthoor, DF*: an extension of synchronous dataflow with data dependency and non-determinism, *Proceedings of the Forum on Design Languages*, Tubingen, Germany, September 2000.

[32] M. Pankert, O. Mauss, S. Ritz, and H. Meyr, Dynamic data flow and control flow in high level DSP code synthesis, *Proceedings of the International Conference on Acoustics, Speech, and Signal Processing*, Adeliade, Australia, 1994.

[33] D.G. Messerschmitt, Structured interconnection of signal processing programs, *Proceedings of the IEEE Global Telecommunications Conference*, Atlanta, GA, 1984.

[34] J.T. Buck, S. Ha, E.A. Lee, and D.G. Messerschmitt, Ptolemy: a framework for simulating and prototyping heterogeneous systems, *Int. J. Comput. Simul.*, 4, 155–182, 1994.

[35] S.S. Bhattacharyya, J.T. Buck, S. Ha, and E.A. Lee, Generating compact code from dataflow specifications of multirate signal processing algorithms, *IEEE Trans. Circuits Syst. — I: Fundam. Theory Appl.*, 42, 138–150, 1995.

[36] A. Kalavade and E.A. Lee, A hardware/software codesign methodology for DSP applications, *IEEE Design Test Comput. Mag.*, 10, 16–28, 1993.

[37] L. de Alfaro and T. Henzinger, Interface automata, *Proceedings of the Joint European Software Engineering Conference and ACM SIGSOFT International Symposium on the Foundations of Software Engineering*, Vienna, Austria, 2001.

[38] E.A. Lee and Y. Xiong, System-level types for component-based design, *Proceedings of the International Workshop on Embedded Software*, Tahoe City, CA, October 2001, pp. 148–165.

[39] T. Stefanov, C. Zissulescu, A. Turjan, B. Kienhuis, and E. Deprettere, System design using Kahn process networks: the Compaan/Laura approach, *Proceedings of the Design, Automation and Test in Europe Conference and Exhibition*, Paris, France, February 2004.

[40] G. Bilsen, M. Engels, R. Lauwereins, and J.A. Peperstraete, Cyclo-static dataflow, *IEEE Trans. Signal Process.*, 44, 397–408, 1996.

[41] B. Kienhuis, E. Rijpkema, and E. Deprettere, Compaan: deriving process networks from Matlab for embedded signal processing architectures, *Proceedings of the International Workshop on Hardware/Software Co-Design*, San Diego, CA, May 2000.

[42] A. Turjan, B. Kienhuis, and E. Deprettere, Approach to classify inter-process communication in process networks at compile time, *Proceedings of the International Workshop on Software and Compilers for Embedded Processors*, Amsterdam, The Netherlands, September 2004.

[43] D. Verkest, K. Van Rompaey, I. Bolsens, and H. De Man, CoWare — a design environment for heterogeneous hardware/software systems, *Readings in Hardware/Software Co-design*, Kluwer Academic Publishers, Dordrecht, 2001.

[44] S. Ritz, M. Pankert, and H. Meyr, High level software synthesis for signal processing systems, *Proceedings of the International Conference on Application Specific Array Processors*, Berkeley, CA, August 1992.

[45] S. Ritz, M. Pankert, and H. Meyr, Optimum vectorization of scalable synchronous dataflow graphs, *Proceedings of the International Conference on Application Specific Array Processors*, Venice, Italy, October 1993.

[46] T.M. Parks, J.L. Pino, and E.A. Lee, A comparison of synchronous and cyclo-static dataflow, *Proceedings of the IEEE Asilomar Conference on Signals, Systems, and Computers*, Pacific Grove, CA, November 1995.

[47] S.S. Bhattacharyya, Hardware/software co-synthesis of DSP systems, in *Programmable Digital Signal Processors: Architecture, Programming, and Applications*, Y.H. Hu, Ed., Marcel Dekker, New York, 2002, pp. 333–378.

[48] G. Berry and G. Gonthier, The Esterel synchronous programming language: design, semantics, implementation, *Sci. Comput. Programming*, 19, 87–152, 1992.

[49] I. Page, Constructing hardware/software systems from a single description, *J. VLSI Signal Process.*, 12, 87–107, 1996.

[50] C.A.R. Hoare, *Communicating Sequential Processes*, Prentice-Hall, New York, 1985.

[51] L.C. Paulson, *ML for the Working Programmer*, Cambridge University Press, London, 1996.

[52] S. Chappell and C. Sullivan, Handel-C for Co-processing & Co-design of Field Programmable System on Chip, Technical report, Celoxica Limited, September 2002.

[53] J.B. Dabney and T.L. Harman, *Mastering Simulink*, Prentice Hall, New York, 2003.

[54] D.C. Burger and T.M. Austin, The SimpleScalar Tool Set, Version 2.0, U W Madison Computer Sciences Technical Report #1342, June, 1997.

[55] D. Brooks, V. Tiwari, and M. Martonosi, Wattch: a framework for architectural-level power analysis and optimizations, *Proceedings of the 27th International Symposium on Computer Architecture*, Vancouver, Canada, June 2000.

[56] N. Vijaykrishnan, M. Kandemir, M.J. Irwin, H.Y. Kim, and W. Ye, Energy-driven integrated hardware-software optimizations using SimplePower, *Proceedings of the International Symposium on Computer Architecture*, Vancouver, Canada, June 2000.

[57] R.K. Gupta and G. De Micheli, Hardware-software cosynthesis for digital systems, *IEEE Design Test Comput.*, 10, 29–41, 1993.

[58] R. Ernst, J. Henkel, and T. Benner, Hardware-software cosynthesis for microcontrollers, *IEEE Design Test Comput.*, 10, 64–75, 1993.

[59] A. Kalavade and E.A. Lee, The extended partitioning problem: hardware/software mapping, scheduling, and implementation-bin selection, *Design Autom. Embedded Syst.*, 2, 125–163, 1997.

[60] W. Wolf, An architectural co-synthesis algorithm for distributed, embedded computing systems, *IEEE Trans. VLSI Syst.*, 5, 218–229, 1997.

[61] R.P. Dick and N.K. Jha, MOGAC: a multiobjective genetic algorithm for the hardware-software co-synthesis of distributed embedded systems, *IEEE Trans. Comput.-Aid. Design*, 17, 920–935, 1998.

[62] C.L. Liu and J.W. Layland, Scheduling algorithms for multiprogramming in a hard- real-time environment, *J. ACM*, 20, 46–61, 1973.

[63] J. Henkel and R. Ernst, A path-based estimation technique for estimating hardware runtime in HW/SW-cosynthesis, *Proceedings of the 8th IEEE International Symposium on System Level Synthesis*, Cannes, France, 1995, pp. 116–121.

[64] W. Fornaciari, P. Gubian, D. Sciuto, and C. Silvano, Power estimation of embedded systems: a hardware/software codesign approach, *IEEE Trans. VLSI Syst.*, 6, 266–275, 1998.

[65] Y. Li and W. Wolf, Hardware/software co-synthesis with memory hierarchies, *IEEE Trans. CAD*, 18, 1405–1417, 1999.

[66] V. Tiwari, S. Malik, and A. Wolfe, Power analysis of embedded software: a first step towards software power minimization, *IEEE Trans. VLSI Syst.*, 2, 437–445, 1994.

4

System-Level Specification and Modeling Languages

Joseph T. Buck
Synopsys, Inc.
Mountain View, California

4.1 Introduction

This chapter is an overview of high-level, abstract approaches to the specification and modeling of systems, and the languages and tools in this domain. Many of the most effective approaches are domain-specific, and derive their power from the effective exploitation of a *model of computation* that is effective for expressing the problem at hand. The chapter first surveys some of the most used models of computation, and introduces the specialized tools and techniques used for each. Heterogeneous approaches that allow for the combination of more than one model of computation are then discussed.

It is hardly necessary to explain the reader that because of Moore's law, the size and complexity of electronic circuit design is increasing exponentially. For a number of years, most advanced chip designs have included at least one programmable processor, meaning that what is being designed truly qualifies for the name "system-on-chip" (SoC), and success requires that both hardware and embedded software be designed, integrated, and verified. It was common in the late 1990s to see designs that included a general-purpose microprocessor core, a programmable digital signal processor, as well as application-specific synthesized digital logic circuitry. Leading-edge designs now include many processors as well as on-chip networks. As a consequence, the task of the SoC architect has become not merely quantitatively but also qualitatively more complex.

To keep up with the increasing design complexity, there are two principal means of attacking the problem. The first tactic is to increase design reuse, i.e., assemble new designs from previously designed and verified components. The second tactic is to raise the level of abstraction, so that larger designs can be kept manageable. The purpose of the languages, approaches and tools described in this chapter is to raise the level of abstraction. Given suitable higher-level models for previously designed components, these approaches can also help with design reuse.

However, there is a price to be paid; some of the most powerful system-level language approaches are also domain-specific. Data flow-oriented approaches (Kahn process networks [7]) have great utility for digital signal processing intensive subsystems (including digital audio, image and video processing as well as digital communication), but are cumbersome to the point of unsuitability if complex control must be implemented. As a consequence, there has been great interest in hybrid approaches.

It has long been common practice for designers to prototype components of systems in a traditional higher level language, most commonly C or C++. A high level language raises the level of abstraction to some extent, but for problems involving signal processing, many designers find Matlab [1] preferable. However, since these are sequential languages, they do not provide systematic facilities for modeling concurrency, reactivity, interprocess communication, or synchronization.

When the system designer begins to think about concurrency and communication, it quickly becomes apparent that suitable models for explaining/understanding how this will work are problem-specific. A designer of synchronous logic can consider a decomposition of a problem into a network of extended finite-state machines (FSMs), operating off a common clock. A designer of a video processing application, which will run on a network of communicating processors and sharing a common bus, will require a very different model.

The organizing concept of this chapter is the notion of a *model of computation*. Edward Lee, who is largely responsible for drawing attention to the concept, describes a model of computation as "the laws of physics that govern component interactions" [2]. In [3], Berry takes the analogy further, comparing the instantaneous communication in the synchronous-reactive model to Newtonian gravitation and the bounded delays of the discrete-event (DE) model to the theory of vibration.

Models of computation are related to languages, but the relationship is not exact. For example, while Esterel [4] is specifically designed to support the synchronous-reactive model of computation, it is also common to implement support for a model of computation with a C++ class library (e.g., SystemC [5]) or by using a subset of a language. Using the synthesizable subset of VHDL or Verilog, and working with synthesis semantics, is to use a very different model of computation than to use the full languages and to work with the simulation behavior.

4.2 A Survey of Domain-Specific Languages and Methods

Any designer manages complexity by subdividing a design into manageable components or modules, and block diagrams are also common in many disciplines. To a hardware designer, the components or blocks have a direct physical significance, as logic blocks to be instantiated or synthesized, and they communicate by means of wires. To a designer of a digital communication system, the blocks might be processes communicating by means of streams of data and connections between components might be first-in first-out (FIFO) queues. Alternatively, the blocks might be tasks or system modes, and the connections might be precedence constraints or even state transitions. The model of computation in the sense we use it here, forms the "rules of the game" and describes how the components execute and communicate.

This chapter will first discuss those models of computation that have proved effective in hardware and embedded software design and modeling, and will then discuss hybrid approaches, where more than one model of computation is used together. The Ptolemy project [6] was, to the best of the author's knowledge, the first systematic attempt to allow designers to combine many models of computation in a systematic way, but there are also many other important hybrid approaches.

4.2.1 Kahn Process Networks and Dataflow

A Kahn process network (KPN) [7] is a network of processes that communicate by means of unbounded FIFO queues. Each process has zero or more input FIFOs, and zero or more output FIFOs. Each FIFO is connected to one input process and one output process. When a process writes data to an output FIFO, the write always succeeds; the FIFO grows as needed to accommodate the data written. Processes may read their inputs only by means of blocking reads; if there are insufficient data to satisfy a request, the reading process blocks until the writing process for the FIFO provides more. The data written by a process depend only on the data read by that process and the initial state of the process. Under these conditions, Kahn [7] showed that the trace of data written on each FIFO is independent of process scheduling order. As a result, KPNs are a popular paradigm for the description and implementation of systems for the parallel processing of streaming data.

Because any implementation that preserves the semantics will compute the same data streams, KPN representations or special cases of them, are a useful starting point for the system architect, and are often used as executable specifications.

Dataflow process networks are a special case of KPNs, in which the behavior of each process (often called an *actor* in the literature) can be divided into a sequence of execution steps called *firings* by Lee and Parks [8]. A firing consists of zero or more read operations from input FIFOs, followed by a computation, and then by zero or more write operations to output queues (and possibly a modification to the process's internal state). This model is widely used in both commercial and academic software tools such as SPW [9], COSSAP [10], System Studio [11], and Ptolemy. The subdivision into firings, which are treated as indivisible quanta of computation, can greatly reduce the context switching overhead in simulation, and can enable synthesis of software and hardware. In some cases (e.g., Yapi [12]), the tools permit processes to be written as if they were separate threads, and then split the threads into individual firings by means of analysis. The thread representation allows read and write directives to occur anywhere, while the firing representation can make it easier to understand the data rates involved, which is important for producing consistent designs.

In an important special case of dataflow process networks, the number of values read and written by each firing of each process is fixed, and does not depend on the data. This model of computation was originally called synchronous dataflow (SDF) [13], which is now widely considered an unfortunate choice of terminology because of confusion with synchronous languages, and because SDF is an untimed model. In fact, the term was originally used for the very different LUSTRE language [14]; LUSTRE is synchronous but not dataflow; SDF is dataflow but not synchronous. The term "static dataflow" is now considered preferable; Lee himself uses it in [8]. Fortunately, the widely used acronym "SDF" still applies.

Figure 4.1 shows a simple SDF graph. In the diagram, the numbers adjacent to the inputs and outputs of the actors indicate how many data values are written to, or read from, the attached FIFO queue on each actor execution. While not shown in this example, it is possible for an edge of an SDF graph to have initial logical delays which can be thought of as initial values in the queues. If the graph contains cycles, initial values must be present to avoid a deadlock.

The first task in the analysis of an SDF graph is to derive an equation from each FIFO that constrains the number of actor firings the schedule must contain: the number of values written to the FIFO must equal the number of values read from the FIFO. Let n_{ij} be the number of values written by actor i to the edge (FIFO queue) j. If actor i reads, rather than writes from edge j, then a negative number is stored in n_{ij} (the negative of the number of values read). Now, let r_i be the number of times actor i is "fired" in one iteration of the schedule. It is required that the number of values in each queue be the same after a completion of the schedule, so we obtain a series of *balance equations:*

$$\sum_{i=1}^{K} n_{ij} r_i, \; j=1,\ldots,L$$

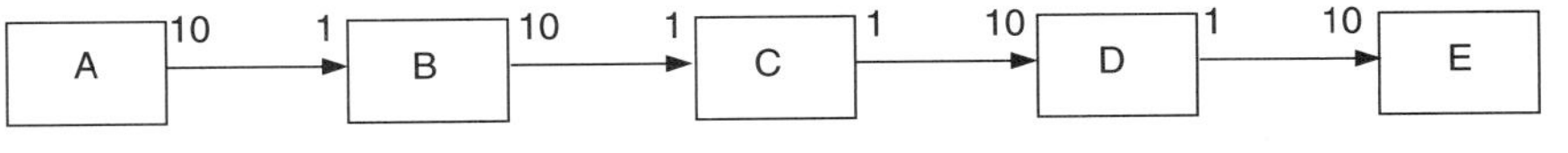

FIGURE 4.1 A simple static dataflow graph. Numbers indicate the number of values read or written on each firing of the adjacent dataflow actor.

where K is the number of actors and L is the number of edges. Alternatively, treating N as the matrix formed by the n_{ij} terms, and $\hat{r}$ as the vector of rates, we have

$$N\hat{r} = \hat{0}$$

If the matrix N is nonsingular, only the trivial solution that sets all rates to zero exists, and we say that the graph is inconsistent. In Figure 4.2, graph (a) is an example of an inconsistent graph. Such a graph can be executed by a dynamic dataflow simulator, but the number of values in some queues must increase without bound. In [13] it is shown that, for a connected graph, if the matrix is nonsingular the null space has rank 1, so that all solutions for $\hat{r}$ are proportional. We are interested in the least integral solution; as shown in [13], there is a simple algorithm to find the minimal integer solution in $O(K + L)$ time. Note that even if a graph has consistent rates, if there are cycles with insufficient delay, we have a deadlock condition and a schedule is not possible. Figure 4.2(b) is an example of such a graph.

In the case of Figure 4.1, a minimal schedule must execute A and E once, B and D 10 times, and C 100 times. Linear schedules containing 100 invocations of C are clearly not efficient, so looped schedules are of great interest for implementations, especially for sequential software implementations. Looped schedules, in which the invocation of each actor appears exactly once — the so-called *single appearance schedules* — are treated in detail in [15]. One possible single appearance schedule for our example graph can be written as A,10(B,10(C),D),E.

For static dataflow networks, efficient static schedules are easily produced, and bounds can be determined for all of the FIFO buffers, whether for a single programmable processor, multiple processors, or hardware implementations. An excellent overview of the analysis of SDF designs as well as the synthesis of software for a single processor from such designs, can be found in [16].

For multiple processors, one obvious alternative is to form a task dependence graph from the actor firings that make up one iteration of the SDF system, and apply standard task scheduling techniques such as list scheduling to the result. Even for uniform-rate graphs where there is no looping, the scheduling problem is NP-hard. However, because it is likely that implicit loops are present, a linear schedule is likely to be too large to be handled successfully. Efficient multiprocessor solutions usually require preservation of the hierarchy introduced by looping — see [17] for one approach to do just that.

Engels et al. [18] proposed an extension of static dataflow that allows the input–output pattern of an actor to vary in a periodic manner; this model is called cyclo-static dataflow. Complete static schedules can still be obtained, but in most cases the interconnecting FIFO queues can be made much shorter, which is particularly advantageous for hardware implementation.

The Gabriel system [19] was one of the earliest examples of a design environment that supported the SDF model of computation for both simulation and code generation for DSPs. Gabriel's successor, Ptolemy, extended and improved Gabriel's dataflow simulation and implementation capabilities [20]. Another successful early SDF-based code generation system was Descartes [21], which was later commercialized as part of COSSAP by Cadis (later acquired by Synopsys). The GRAPE-II system [22] supported implementation using the cyclo-static dataflow model.

The term "dynamic dataflow" is often used to describe dataflow systems that include data-dependent firing and therefore are not static. COSSAP [10] was apparently the first true dynamic dataflow simulator;

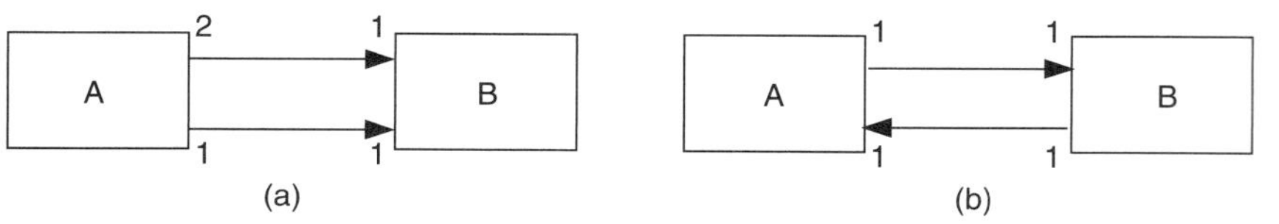

Figure 4.2 Inconsistent graphs. (a) has inconsistent rates; (b) is rate-consistent but deadlocked.

Messerschmitt's Blosim [23], while older, required the user to specify sizes of all FIFO buffers and writes to full FIFOs blocked (while COSSAP buffers grow as needed), so it was not a true dataflow (or KPN) simulator. However, the method used by COSSAP for dynamic dataflow execution is suitable for simulation only, and not for embedded systems implementation.

While there have been many graphical, block diagram-based systems supporting simulation as well as software and hardware implementation, there are also textual languages whose model of computation corresponds to SDF. The first of these was Silage [24]. Silage is a declarative language, in which all loops are bounded; it was designed to allow DSP algorithms to be efficiently implemented in software or hardware. The DFL language [25] was derived from Silage and a set of DFL-based implementation tools was commercialized by Mentor Graphics as DSPstation.

While there are many algorithmic problems or subproblems that can be modeled as SDF, at least some dynamic behavior is required in most cases. Hence there has long been a interest in providing for at least some data-dependent execution of actors in tools, without paying for the cost of full dynamic dataflow.

The original SPW tool from Comdisco (later Cadence, now CoWare) [9] used a dataflow-like model of computation that was restricted in a different way: each actor had a hold signal. If connected, the actor reads a value from the hold signal. If a "true" value is read, the actor does not execute; otherwise the actor reads one value from each input, does a computation, and writes one value to each output. This model is more cumbersome than SDF for static multirate operation, but can express dynamic behaviors that SDF cannot express, and the one place buffers simplified the generation of hardware implementations. It is a special case of dynamic dataflow (although limited to one place buffers). Later versions of SPW added full SDF and dynamic dataflow support.

Boolean-controlled dataflow (BDF) [26], later extended to allow for integer control streams (IDF) [27], was an attempt to extend SDF analysis and scheduling techniques to a subset of dynamic dataflow. While in SDF, the number of values read or written by each I/O port is fixed, in BDF the number of values read by any port can depend on the value of a Boolean data value read by, or written by some other port, called a control port. In the BDF model, as in SDF, each port of each actor is annotated with the number of values transferred (read or written) during one firing. However, in the case of BDF, instead of a compile-time constant, the number of values transferred can be an expression containing Boolean-valued variables. These variables are the data values that are arrived at or are written by a *control port*, a port of the BDF actor that must transfer one value per firing. The SPW model, then, is a special case of BDF where there is one control port, which controls all other ports, and the number of values transferred must be 1 or 0. Integer control streams allow control streams to be integers.

While BDF is still restricted compared to general dataflow, it is sufficiently expressive to be Turing-equivalent. Unfortunately, this means that a number of analysis problems, including the important question of whether buffer sizes can be bounded, are undecidable in general (as shown in [26]). Nevertheless, clustering techniques can be used to convert an IDF graph into a reduced graph consisting of clusters; each individual cluster has a static or quasi-static schedule, and only a subset of the buffers connecting clusters can potentially grow to an unbounded size. This approach was taken in the dataflow portion of Synopsys's System Studio [11], for example.

The BDF model has also been used for hardware synthesis by Zepter [28]. Zepter's approach can be thought of as a form of interface synthesis: given a set of predesigned components that read and write data periodically, perhaps with different periods and perhaps controlled by enable signals, together with a behavioral description of each component as a BDF model, Zepter's Aden tool synthesized the required control logic and registers to correctly interface the components.

There is a close correspondence between the cases that can be handled by Aden and in the cases handled by the LUSTRE language (see Section 4.2.5), because the Boolean conditions that control the dataflow become gated clocks in the implementation.

Later work based on Zepter's concept [29] relaxed the requirement of periodic component behavior and provided more efficient solutions, but handled only the SDF case.

4.2.2 Dataflow with Bounded Queues

Many implementers that start from a general dataflow description of the design choose an implementation with fixed bounds on the size of the queues. Given a fixed queue size, a write to a full queue can get blocked, but if analysis has already shown that the sizes of the queues suffice, this does not matter.

One of the most commonly used fixed-sized queue implementations at the prototyping level is the sc_fifo<T> channels in SystemC [5]. An example of implementation-oriented approach that uses fixed-size queues is the task transfer level (TTL) approach [30], in which a design implemented in terms of fixed-sized queues is obtained by refinement from an executable specification written in KPN form in an environment called Yapi [12].

4.2.3 Matlab

The most widely used domain-specific tool for digital signal processing intensive applications is Matlab [1], from the MathWorks, Inc. Matlab is both a high-level language and an interactive environment. The language (sometimes called M-code, to contrast it with C-code) treats vectors and arrays as first-class objects, and includes a large set of built-in functions as well as libraries for the manipulation of arrays, signal processing, linear algebra, and numerical integration; the integration features can be used effectively in continuous-time modeling. The environment provides extensive data visualization capabilities. Matlab is an interpreted language, but for code that manipulates vectors and matrices, the interpreter overhead is small.

An algorithm written in Matlab is sequential. Other tools build on Matlab by allowing functional blocks to be written in Matlab or C. The MathWorks provides Simulink [31], a block diagram dataflow tool that allows functional blocks to be written in C or in M-code. A number of other tools in this category (including SPW, COSSAP, System Studio, and others) all provide some form of interface to Matlab.

4.2.4 Statecharts and its Variants

A very influential methodology for representing hierarchical control structures is Statecharts, invented by Harel [32], and commercialized by the company i-Logix, which Harel cofounded, with the name Statemate.

Statecharts representations are hierarchical extended finite state machines (FSMs). Two types of models are supported, called *or-states* and *and-states.* An or-state Statecharts model looks like an FSM diagram, with guard conditions and actions associated with each transition between states. The constituent states of a Statecharts model can be atomic states, or can be Statecharts models themselves. The other form of Statecharts model is called an and-state, and consists of two or more Statecharts models that execute in parallel. In Harel's original version of Statecharts, transition arrows can cross levels, in either

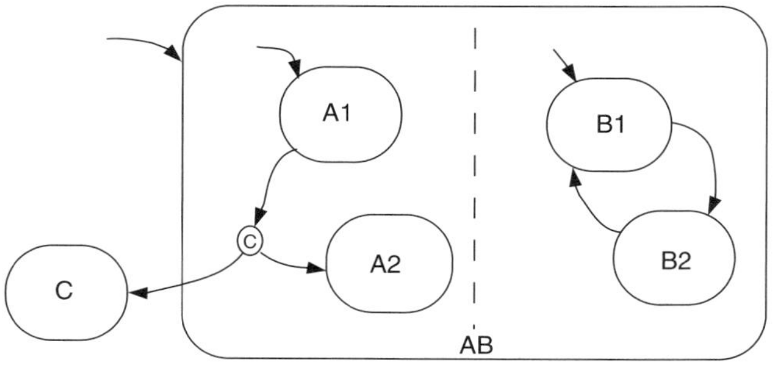

FIGURE 4.3 A simple Statechart. The graph as a whole is an or-state (either in AB or in C); AB is an and-state (we enter both A1 and B1 at the same instant), which is composed of two or-states (A1 and A2; B1 and B2). Transition arcs are annotated with conditions and actions (not shown here).

direction, which provides for great expressive power, but at the price of modularity. Figure 4.3 is an example Statechart.

Statecharts can be used for generation of C or hardware description language (HDL) code for implementation purposes; however the representation seems to be used most commonly today as a specification language rather than as an implementation tool [33]; Harel's Statecharts were combined with a variety of other specification methodologies to form Unified Modeling Language (UML) [34].

The form of Statecharts has been widely accepted, but there has been a great deal of controversy about the semantics, and as a result many variants of Statecharts have been created. In [35], von der Beek identifies 20 Statecharts variants in the literature and proposes a 21st !

The main points of controversy are:

- *Modularity.* Many researchers, troubled by Harel's level-crossing transitions, eliminated them and came up with alternative approaches to make them unnecessary. In many cases, signals are used to communicate between levels.
- *Microsteps.* Harel's formulation handles cascading transitions, which occur based on one transition of the primary inputs by using delta cycles, as is done in VHDL and Verilog. Others, such as Argos [36] and SyncCharts [37], have favored synchronous-reactive semantics and find a fixpoint, so that what requires a series of microsteps in Harel's statecharts becomes a single atomic transition. If a unique fixpoint does not exist, then formulations of Statecharts that require a fixpoint reject the specification as ill formed.
- *Strong preemption vs. weak preemption.* When a hierarchical state and an interior state both have an outgoing transition that is triggered by the same event, in a sense we have a race. With strong preemption, the outer transition "wins"; the inner transition is completely preempted. With weak preemption, both transitions take place (meaning that the action associated with the inner transition is performed), with the inner action taking place first (this order is required because the outer transition normally causes the inner state to terminate). Strong preemption can create causality violations, since the action on an inner transition can cause an outer transition that would preempt the inner transition. Many Statecharts variants reject specifications with this kind of causality violation as ill formed. Some variants permit strong or weak preemption to be specified separately for each transition.
- *History and suspension.* When a hierarchical state is exited and reentered, does it "remember" its previous state? If it does, is the current state remembered at all levels of hierarchy (deep history), or only at the top level (shallow history)? In Harel's Statecharts, shallow history or deep history can be specified as an attribute of a hierarchical state. In some other formulations, a suspension mechanism is used instead, providing the equivalent of deep history by "freezing" a state (which might correspond to the gating of a clock in a hardware implementation). Figure 4.4 shows an oversimplified traffic light state machine with a "pause mode", implemented first with a Harel-style history mechanism, and then with a SyncCharts-like suspension mechanism.

Most of these issues are also relevant for control-oriented synchronous languages, and will be discussed in the next section.

4.2.5 Synchronous/Reactive Languages

In the synchronous-reactive model of computation, we consider reactive systems, meaning systems that respond to events from the environment. A reactive system described with a synchronous model instantaneously computes its reaction, modifies its state, and produces outputs. Of course, physical realizations cannot react instantaneously but as for synchronous circuits, what is really required is that the reaction can be computed before the deadline arrives for the processing of the next event (typically the next clock cycle boundary, minus any setup and hold times for a synchronous circuit). The concepts used in synchronous languages originate in the synchronous process calculi of Milner [38]. The best known textual synchronous languages are Esterel [4], LUSTRE [14], and SIGNAL [39]. There are also several variants of

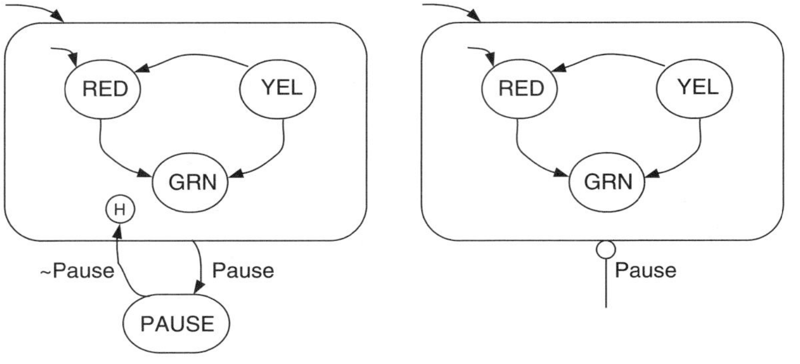

FIGURE 4.4 History vs. suspension. The traffic light controller freezes when the pause signal is true, and resumes when it is false. On the left, this is done using a history mechanism; on the right, it is done with suspension, which works like gating a clock.

Statecharts that have synchronous-reactive semantics:* Argos [36] and SyncCharts [37]. An excellent survey of the synchronous model and principal languages can be found in [40].

The synchronous languages can be subdivided into two groups: those that support hierarchical control, and those that define time series by means of constraints. These two groups of languages appear very different in form, but it turns out that they share an underlying structure.

Of the languages that we have mentioned, Esterel, Argos, and SyncCharts fall into the first category; in fact, SyncCharts can be thought of as a successful effort to extend Argos into a graphical form of the Esterel language.

Here is a simple example of an Esterel program from [3]. In this case, a controller has two Boolean input signals, A and B, as well as a reset signal, R. The task is to wait for an event on both A and B; the events can be received in either order or simultaneously. When both events are seen, an output event is to be produced on O and then the controller should restart. However, if a reset event is received on R, the controller should restart without producing an event on O. The code looks like

```
module ABO:
input A, B, R;
output O; loop
     [ await A | await B ];
     emit O
each R end
module
```

In this example, we have synchronous parallelism (for the two awaited statements) as well as strong preemption (the loop ... each construct aborts and restarts the loop on each receipt of the reset event).

The corresponding representation of this example in SyncCharts is shown in Figure 4.5; the graphical form has the same semantics as the textual form.

Esterel's well-behaved hierarchical control structure makes it a good language to write model checkers for; XEVE [41] is one such checker. Esterel and XEVE have been successfully used to achieve verifiably high state coverage for the control-dominated parts of a commercial DSP [42], by writing the reference model in a mixture of C (for datapath) and Esterel (for control).

The first implementation of Esterel required that the entire program be flattened into a state transition graph (essentially a flat FSM). While this approach yielded fast software implementations, the code size tended to explode for large program sizes. For many years, the implementation of choice was to translate the Esterel program into a circuit representation that preserves the control hierarchy of the original

*Harel's original Statecharts has discrete event semantics, including microsteps that resemble those of VHDL and Verilog.

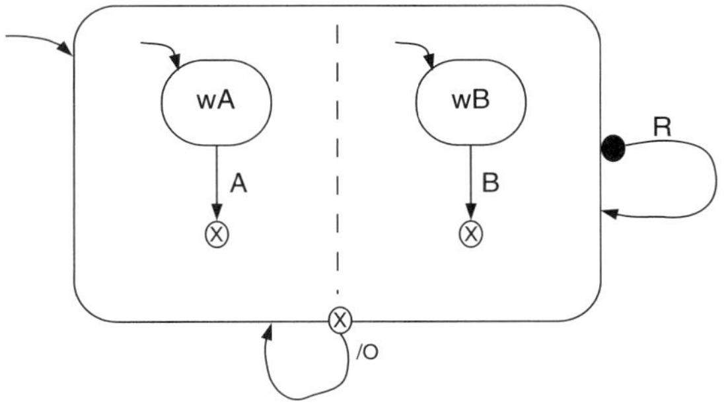

FIGURE 4.5 Syncharts version of "ABO". The x-in a-circle denotes normal termination; if on the edge of a subsystem, it indicates the action to be taken on termination. The circle on the edge triggered by R indicates strong preemption; strong preemption edges take priority over termination edges.

program; the two implementation approaches are described and compared in [43]. The software implementation produced by the popular "version 5" Esterel software release was in effect a simulator for this hierarchical control circuit, which had the disadvantage of being slower than necessary due to the frequent update of circuit signals that had no effect. There has been recent interest in more efficient software implementations of Esterel, particularly due to the work of Edwards [44].

An approach with close connections to Esterel is Seawright's production-based specification system, Clairvoyant [45]. The "productions" correspond to regular expressions, and actions (written in VHDL or Verilog) are associated with these expressions. The specification style closely resembles that of the Unix tool *lex* [46]. Seawright et al. [47] later extended Clairvoyant, designed a graphical interface, and produced a commercial implementation called Protocol Compiler. While not a commercial success, the tool was effectively used to synthesize and verify complex controllers for SDH/Sonet applications [48]. The circuit synthesis techniques used were quite similar to those used for Esterel in [43].

LUSTRE and SIGNAL are a very different style of synchronous language, in that their variables are streams; specifically, they are time series. With each time series there is a clock associated; two streams that are defined at the same points in time are said to have the same clock. While clocks impose a global ordering, there is neither a mechanism nor a need to define time scales in more detail.

It is possible to sub-sample a stream, based on a Boolean-valued stream, to produce a stream with a slower clock; this is analogous to a down-sampling operation in digital signal processing.

In LUSTRE, variables correspond to streams, and unary and binary arithmetic and logical operations are extended to operate pointwise on the elements of those streams. LUSTRE is a declarative language, meaning that streams are defined once in terms of other streams. The following is a simplified example from one in [14]: here

```
node COUNTER(reset: bool) returns (n: int)
let
      n = 0 -> if reset then 0 else pre(n) + 1;
tel.
```

defines a counter with reset, that increments with every basic clock cycle and that resets when the reset input is true. Every node (corresponding to what might be called a module or an entity in other languages) has a basic clock, corresponding to the fastest clock that is input to the node. Note that both n and reset are streams.

The keyword produces a sub-sampled stream, and the current keyword interpolates a stream (using a sample-and-hold mechanism) to match the current basic clock. Definitions can be cyclic as long as each cycle contains at least one preoperator.

The LUSTRE language does not allow basic clock time intervals to be split into smaller ones; we can sub-sample a stream or interpolate it to match the clock of a faster stream. The SIGNAL language (which we will not discuss further) allows this kind of subdivision. This distinction makes it much easier to map

LUSTRE programs into digital circuits than SIGNAL; pre makes a register, when gates the clock, current is effectively a latch.

There is a strong connection between the stream-based synchronous languages and certain dataflow languages such as Silage and SDF-based systems. Even the name "synchronous dataflow" suggests the connection. The important distinction between the two approaches is in the concept of time. In LUSTRE, every pair of events in the system has a defined global ordering, and the location of each event can be correlated to a particular tick of the highest-rate clock in the system. In Silage or SDF, there is only data, and the specification gives only a partial ordering of data production events (all required inputs must be computed before an output is produced). Edward Lee has long argued (e.g., in [8]) that the specification of global ordering where it is not warranted can over-constrain the solution space.

However, this distinction is perhaps less important that it appears as a practical matter, in that an implementer could start with a LUSTRE specification and ignore timing issues. Similarly, in the Ptolemy system [6], it is necessary to assign time values to stream elements produced by an SDF "master" system to interface it to a DE subsystem, and in DSP applications assigning equally spaced times to sample points is often the most useful choice, which makes the SDF system look much like Esterel.

4.2.6 Communicating Sequential Processes

Communicating sequential processes (CSP) with rendezvous is a model of computation first proposed by Hoare [49]. Like dataflow, CSP is an untimed model of computation, in which processes synchronize based on the availability of data. Unlike KPNs, however, there is no storage element at all between the connected processes (so for example, a hardware implementation might have only wires, and no registers, at the point of connection).

Given a pair of communicating processes, the read request and matching write request form a rendezvous point; whichever process performs the I/O request first, waits for its peer. The Occam language [50] was designed around CSP.

The use of remote procedure calls (RPC) as a form of inter-process communication is a form of CSP. This is because the communication between the client and server is completely synchronized; there are no buffers to store outstanding requests (though protocols built on top of RPC often implement buffering at a higher level).

4.2.7 Polis and Related Models

The Polis system [51] uses a locally synchronous and globally asynchronous model of computation. Each process in a design is a "codesign finite state machine" (CFSM), a form of extended FSM. The interaction between the CFSM tasks differs from the standard model of interacting concurrent FSMs in that an unbounded (and nonzero) delay is added to each interaction between CFSMs. The intent of this design style is to avoid introducing a bias toward hardware or software implementation: a synchronous hardware design might introduce one cycle of delay for communication, while a software implementation or a partition of the design that requires tasks to communicate over a network, might require a much larger time for a communication. The processes communicate with each other asynchronously using one place buffers; this means that data loss is a possibility.

The unconstrained model, where there are no time deadlines or protection against overwritten buffers, is only intended as a starting point. The designer is expected to add constraints sufficient to assure that deadlines are met. It is not strictly necessary to guard against overwritten communication buffers, as there are some cases where the data loss is not a problem or is even desirable. For example, in an automotive application, a speedometer task might repeatedly compute a current speed estimate, while a digital display of current speed only needs to sample these estimates occasionally.

Cadence produced a commercial tool, VCC, that was based on the Polis model.

Polis permits the CFSM tasks to be written in any of several forms that can be translated into FSMs, including Esterel, graphical FSMs, or HDL subsets. Many of the analysis tools in the Polis framework require that the FSM tasks be completely flattened and that individual states be enumerated, and as a result Polis has been most effective on designs with many relatively simple communicating tasks. Automotive applications have been a principal driver for Polis and VCC and seem most suitable for this design style; for example, Ref. [52], which describes the design of an engine management system in VCC.

Specification and description language (SDL) [53] is a standard specification language used widely in telephony, particularly in Europe. Like Polis, SDL represents tasks as communicating FSMs, and communication between tasks can introduce an unbounded delay. There is more detail about SDL in Chapter 9.

4.2.8 Discrete Events and Transaction-Level Modeling

In the discrete event (DE) model of computation, no doubt most familiar to the reader as the underlying model for hardware description languages (HDLs) such as Verilog and VHDL, each event has associated with it a specific time. However, unlike the synchronous-reactive model, reactions to events are not instantaneous, instead simulated time elapses. Even for reactions that conceptually take zero time, there is still an event ordering that takes place, often in the form of microsteps or delta cycles. SystemC also has discrete event semantics.

We will not deal further with Verilog or VHDL in this chapter; the languages are discussed in detail in Chapter 15. In this chapter, we are interested in DE modeling at higher levels of abstraction.

Transaction-level modeling (TLM) has become an extremely popular term in recent years. Grötker et al., [5] define TLM as "a high-level approach to modeling digital systems where details of communication among modules are separated from the details of the implementation of the functional units or of the communication architecture." The term "transaction-level modeling" was coined by the original SystemC development team; an alternative, "transaction-based modeling", was also considered and might have been a preferable choice, as TLM does not correspond to a particular level of abstraction in the same sense that, for example, register transfer level (RTL) does [54]. However, distinctions between TLM approaches and register-transfer level can clearly be made: while in an RTL model of a digital system, the detailed operation of the protocol, address, and data signals on a bus are represented in the model, with TLM a client of a bus-based interface might simply issue a call to high-level read() or write () functions. Chapter 8 of Grötker et al.[5] gives a simple but detailed example of a TLM approach to the modeling of a bus-based system with multiple masters, slaves, and arbiters.

While the SystemC project coined the TLM term, it clearly did not invent the concept. The SpecC language [55], through its channel feature, permits the details of communication to be abstracted away in much the same manner that SystemC supports. Furthermore, the SystemVerilog language [56] supports TLM as well, through a new type of port that is declared with the interface keyword. It is possible to use the separation of functional units and communication simply as a cleaner way to organize one's SystemVerilog code, and still represent the full RTL representation of the design in all of its detail, and Sutherland et al. [56] recommend just this.

With TLM as it is commonly used, the model of computation is still DE simulation, and it is common for a design that is in the process of being refined to mix levels, with detailed RTL representation of some components and TLM representations of others. It is possible, however, to disregard time in a very high-level TLM representation, resulting in a model of computation that more closely resembles communicating sequential processes with rendezvous, or even dataflow with bounded queues.

At present, methodology for TLM in system design is not particularly mature, and there is disagreement about the number of abstraction levels that need to be represented in the design flow. One example of an attempt to make TLM and refinement more rigorous is given by Cai and Gajski [57]. In their model, the designer starts with an untimed specification model, and separately refines the model of communication and of computation. Many intermediate points are identified, for example, their bus-functional models have accurate timing for communication and approximate timing for computation.

4.3 Heterogeneous Platforms and Methodologies

The first systematic approach that accepted the value of domain-specific tools and models of computation, yet sought to allow designers to combine more than one model in the same design, was Ptolemy [6]. The approach to heterogeneity followed in the original Ptolemy was to consistently use block diagrams for designs, but to assign different semantics to a design based on its "domain", roughly corresponding to a model of computation. Primitive actors were designed to function only in particular domains, and hierarchical designs were simulated based on the rules of the current domain. To achieve heterogeneity, Ptolemy allowed a hierarchical design belonging to one domain (e.g., SDF) to appear as an atomic actor following the rules of another domain (e.g., DE simulation). For this to work correctly, mechanisms had to be developed to synchronize schedulers operating in different domains.

The original Ptolemy, now called Ptolemy Classic, was written in C++ as an extensible class library; its successor, Ptolemy II [58] was thoroughly rearchitected and written in Java.

Ptolemy Classic classed domains as either untimed (e.g., dataflow) or timed (e.g., DE). From the perspective of a timed domain, actions in an untimed domain will appear to be instantaneous. In the case of a mixture of SDF and DE for example, one might represent a computation with SDF components and the associated delay in the DE domain, thus separating the computation from the delay involved in a particular implementation. When two distinct timed domains are interfaced, a global time is maintained, and the schedulers for the two domains are kept synchronized. The details of implementation of scheduler synchronization across domain boundaries are described in [6].

Ptolemy Classic was successful as a heterogeneous simulation tool, but it possessed a path to implementation (in the form of generated software or HDL code) only for dataflow domains (SDF and BDF). Furthermore, all of its domains shared the characteristic that atomic blocks represented processes and connections represented data signals.

One of the more interesting features added by Ptolemy II was its approach to hierarchical control [59]. The concept, flowing logically out of the Ptolemy idea, was to extend Statecharts to allow for complete nesting of data-oriented domains (e.g., SDF and DE) as well as synchronous-reactive domains, working out the semantic details as required. Ptolemy II calls designs that are represented as state diagrams, modal designs. If the state symbols represent atomic states, we simply have an extended FSM (extended because, as in Statecharts, the conditions and actions on the transition arcs are not restricted to Boolean signals). However, the states can also represent arbitrary Ptolemy subsystems. When a state is entered, the subsystem contained in the state begins execution. When an outgoing transition occurs, the subsystem halts its execution. The so-called time-based signals that are presented to the modal model propagate downward to the subsystems that are "inside" the states.

Girault et al. [59] claim that the semantic issues with Statecharts variants identified by von der Beek [35] can be solved by orthogonality: nesting FSMs together with domains providing the required semantics, thereby obtaining, for example, either synchronous-reactive or DE behavior. In the author's view, this only partially solves the problem, because there are choices to be made about the semantics of FSMs nested inside of other FSMs.

A similar project to allow for full nesting of hierarchical FSMs and dataflow with a somewhat different design, was part of Synopsys's System Studio [11] (originally code-named "El Greco"). System Studio's functional modeling combines dataflow models with Statecharts-like control models that have semantics very close to those of SyncCharts [37], as both approaches started with Esterel semantics. Like Ptolemy II, System Studio permits any kind of model to be nested inside of any other, and state transitions cause interior subsystems to start and stop. One unique feature of System Studio is that parameter values, which in Ptolemy are set at the start of a simulation run or are compiled in when code generation is performed, can be reset to different values each time a subsystem inside a state transition diagram is started.

There have been several efforts to make the comparison and the combination of models of computation more theoretically rigorous. Lee and Sangiovanni-Vincentelli [60] introduced a meta-model that represents signals as sets of events. Each event is a pair consisting of a value and a tag, where the tags can come from either a totally ordered or a partially ordered set. Synchronous events share the same tag. The model can represent important features of a wide variety of models of computation, including most of those discussed in this chapter; however, in itself it does not lead to any new results. It can be thought of, perhaps, as an educational tool.

Metropolis [61] is a new design environment for heterogeneous systems. Like Ptolemy, it supports more than one model of computation. Unlike Ptolemy, Metropolis attempts to treat functionality and architecture as orthogonal, to make it easier to create mappings between functional and architectural representations, and to make it possible to represent explicitly constraints as well as to verify properties with model checkers [62].

4.4 Conclusions

This chapter has described a wide variety of approaches, all of them deserving of more depth than could be presented here. Some approaches have been more successful than others.

It should not be surprising that there is resistance to learning new languages. It has long been argued that system-level languages are most successful when the user does not recognize that what is being proposed is, in fact, a new language. Accordingly there has sometimes been less resistance to graphical approaches (especially when the text used in such an approach is from a familiar programming language such as C++ or an HDL), and to class libraries that extend C++ or Java. Lee [2] makes a strong case that such approaches are really languages, but acceptance is sometimes improved if users are not told about this. A strong case can be made that approaches based on KPNs, dataflow, and Matlab have been highly successful in a variety of application areas that require digital signal processing. These include wireless; audio, image and video processing; radar; 3-D graphics, and many others. Hierarchical control tools, such as those based on Statecharts and Esterel, have also been successful, though their use is not quite as widespread. Most of the remaining tools described here have been found useful in smaller niches, though some of these are important. However, it is the author's belief that higher-level tools are underused, and that as the complexity of systems to be implemented continues to increase, designers who exploit domain-specific system-level approaches will benefit by doing so.

References

[1] The MathWorks Inc., *MATLAB Reference Guide*, Natick, MA, USA, 1993.

[2] E.A. Lee, Embedded software, in M. Zelowitz, Ed., *Advances in Computers*, Vol. 56, Academic Press, New York, 2002.

[3] G. Berry, The foundations of Esterel, in G. Plotkin, C. Stirling, and M. Tofte, Eds., *Proof, Language and Interaction: Essays in Honor of Robin Milner*, MIT Press, Cambridge, MA, 2000, pp. 425–454.

[4] G. Berry and G. Gonthier, The Esterel synchronous programming language: design, semantics, implementation, *Sci. Comput. Progr.*, 19, 87–152, 1992.

[5] T. Grotker, S. Liao, G. Martin, and S. Swan. *System Design with SystemC*, Kluwer Academic Publishers, Dordrecht, 2002.

[6] J.T. Buck, S. Ha, E. A. Lee, and D.G. Messerschmitt, Ptolemy: a framework for simulating and prototyping heterogeneous systems, *Int. J. Comput. Simul.*, 4, 155–182, 1994.

[7] G. Kahn, The semantics of a simple language for parallel programming, *Information Processing 74:Proceedings of IFIP Congress 74*, Stockholm, Sweden,1974, pp. 471–475.

[8] E.A. Lee and T.M. Parks, Dataflow process networks, *Proc. IEEE*, 83, 773–801, 1995.

[9] Cadence Design Systems, *Signal Processing WorkSystem (SPW)*, 1994.

[10] J. Kunkel, COSSAP: a stream driven simulator, *IEEE Int. Workshop Microelectrion. Commun.*, 1991.

[11] J.T. Buck and R. Vadyanathan, Heterogeneous modeling and simulation of embedded systems in El greco, *Proceedings of the Eighth International Workshop on Hardware/Software Codesign (CODES)*, 2000.

[12] E.A. de Kock, G. Essink, W.J.M. Smits, P. van der Wolf, J.Y. Brunel, W.M. Kruijtzer, P. Lieverse, and K.A. Vissers, Yapi: application modeling for signal processing systems, *Proceedings of the 37th Design Automation Conference*, Los Angeles, CA, 2000.

[13] E.A. Lee and D.G. Messerschmitt, Synchronous data flow, *Proc. IEEE*, 75(9), 1235–1245, 1987.

[14] N. Halbwachs, P. Caspi, P. Raymond, and D. Pilaud, The synchronous data flow programming language LUSTRE, *Proc. IEEE*, 79, 1305–1320, 1991.

[15] S.S. Bhattacharyya and E.A. Lee, Looped schedules for dataflow descriptions of multirate signal processing algorithms, *Form. Method. Sys. Design*, 5, 183–205, 1994.

[16] S.S. Bhattacharyya, P.K. Murthy, and E.A. Lee, *Software Synthesis from Dataflow Graphs*, Kluwer Academic Publishers, Dordrecht, 1996.

[17] J.L. Pino, S.S. Bhattacharyya, and E.A. Lee, A hierarchical multiprocessor scheduling system for DSP applications, *Proceedings of IEEE Asilomar Conference on Signals, Systems, and Computers*, Pacific Grove, CA, 1995.

[18] M. Engels, G. Bilsen, R. Lauwereins, and J. Peperstraete, Cyclo-static data flow: model and implementation, *Proceedings of the 28th Asilomar Conference on Signals, Systems, and Computers*, Pacific Grove, CA, 1994, pp. 503–507.

[19] J. Bier, E. Goei, W. Ho, P. Lapsley, M. O'Reilly, G. Sih, and E.A. Lee, Gabriel: a design environment for DSP, *IEEE Micro Mag.*, 10, 28–45, 1990.

[20] J.L. Pino, S. Ha, E.A. Lee, and J.T. Buck. Software synthesis for DSP using Ptolemy, *J. VLSI Signal Process.*, 9, 7–21, 1995.

[21] S. Ritz, M. Pankert, and H. Meyr, High level software synthesis for signal processing systems, *International Conference on Application Specific Array Processors*, 1992, pp. 679–693.

[22] R. Lauwereins, M. Engels, M. Ade, and J.A. Perperstraete, GRAPE-II: a tool for the rapid prototyping of multi-rate asynchronous DSP applications on heterogeneous multiprocessors, *IEEE Comput.*, 28, 35–43, 1995.

[23] D.G. Messerschmitt, A tool for structured functional simulation, *IEEE J. Selec. Area. Comm.*, SAC-2, 137–147, 1984.

[24] P.N. Hilfinger, J. Rabaey, D. Genin, C. Scheers, and H. De Man, DSP specification using the Silage language, *Proceedings of the International Conference on Acoustics, Speech, and Signal Processing*, Albuquerque, NM, USA, 1990, pp. 1057–1060.

[25] P. Willekens, D. Devisch, M. Ven Canneyt, P. Conflitti, and D. Genin, Algorithm specification in DSP station using data flow language, *DSP Appl.*, 3, 8–16, 1994.

[26] J.T. Buck, Scheduling Dynamic Dataflow Graphs with Bounded Memory Using the Token Flow Mode,. Ph.D. thesis, University of California at Berkeley, 1993.

[27] J.T. Buck, Static scheduling and code generation from dynamic dataflow graphs with integer-valued control systems, *Proceedings of the IEEE Asilomar Conference on Signals, Systems, and Computers*, Pacific Grove, CA, 1994.

[28] P. Zepter and T. Grotker, Generating synchronous timed descriptions of digital receivers from dynamic data flow system level configurations, *Proceedings of European Design And Test Conference*, Paris, France, 1994.

[29] J. Horstmannshoff and H. Meyr, Efficient building block based RTL/HDL code generation from synchronous data flow graphs, *Proceedings of the 37th Design Automation Conference*, Los Angeles, CA, 2000.

[30] P. van der Wolf, E. de Kock, T. Henriksson, W.M. Kruijtzer, and G. Essink, Design and programming of embedded multiprocessors: an interface-centric approach, *Proceedings of CODES+ISSS 2004*, 2004, pp. 206–216.

[31] The MathWorks Inc., *Simulink: dynamic system simulation for MATLAB*, Natick, MA, USA, 1997.

[32] D. Harel, Statecharts: a visual approach to complex systems, *Sci. Comput. Progr.*, 8, 231–275, 1987.

[33] D. Harel and E. Gery, Executable object modeling with statecharts, *Computer*, 30, 31–42, 1997.

[34] J. Rumbaugh, I. Jacobson, and G. Booch, *Unified Modelling Language Reference Manual*, Addison Wesley Longman, Inc., Reading, MA, USA, 1998.

[35] M. von der Beek, A comparison of statechart variants, in L. de Roever and J. Vytopil, Eds., *Formal Techniques in Real-Time and Fault Tolerant Systems*, Springer, Berlin, 1994, pp. 128–148.

[36] F. Maraninchi, The Argos language: graphical representation of automata and description of reactive systems, *IEEE International Conference on Visual Languages*, Kobe, Japan, 1991.

[37] C. Andre, Representation and analysis of reactive behaviors: a synchronous approach, *Proceedings of CESA 1996*, 1996.

[38] R. Milner, Calculi for synchrony and asynchrony, *Theor. Comput. Sci.*, 25 ,267–310, 1983.

[39] P. Le Guernic and T. Gautier, Data-flow to von Neumann: the SIGNAL approach, in L. Bic and J.-L. Gaudiot, Eds., *Advanced Topics in Dataflow Computing*, Prentice-Hall, New York, 1991.

[40] N. Halbwachs, *Synchronous Programming of Reactive Systems*, Kluwer Academic Publishers, Dordrecht, 1993.

[41] A. Bouali, XEVE, an Esterel verification environment, *Proceedings of the International Conference on Computer-Aided Verification (CAV)*, 1998, pp. 500–504.

[42] L. Arditi, A. Bouali, H. Boufaied, G. Clave, M. Hadj-Chaib, and R. de Simone, Using Esterel and formal methods to increase the confidence in the functional validation of a commercial DSP, *Proceedings of ERCIM Workshop on Formal Methods for Industrial Critical Systems*, 1999.

[43] G. Berry, A hardware implementation of pure Esterel, *Sadhana-Acad. P. Eng. S.*, 17, 95–139, 1992. (Special issue on real time systems.)

[44] S.A. Edwards, Compiling Esterel into sequential code, *Proceedings of the 37th Design Automation Conference*, 2000.

[45] A. Seawright and F. Brewer, Clairvoyant: a synthesis system for production-based specification, *Proc. IEEE Trans. VLSI Syst.*, 2, 172–185, 1994.

[46] M.E. Lesk, LEX: A Lexical Analyzer Generator, Technical Report Computing Science Technical Report 39, AT&T Bell Laboratories, 1975.

[47] A. Seawright, U. Holtmann, W. Meyer, B. Pangrle, R. Verbrugghe, and J.T. Buck, A system for compiling and debugging structured data processing controllers, *Proceedings of EuroDAC 96*, 1996.

[48] W. Meyer, A. Seawright, and F. Tada, Design and synthesis of array structured telecommunication processing applications, *Proceedings of the 34th Design Automation Conference*, Anaheim, CA, 1997, pp. 486–491.

[49] C.A.R. Hoare, CSP — Communicating Sequential Processes, *International Series in Computer Science*, Prentice-Hall, New York, 1985.

[50] INMOS Limited, *Occam Programming Manual*, Prentice-Hall, New York, 1984.

[51] F. Balarin, E. Sentovich, M. Chiodo, P. Giusto, H. Hsieh, B. Tabbara, A. Jurecska, L. Lavagno, C. Passerone, K. Suzuki, and A. Sangiovanni-Vincentelli, *Hardware-Software Co-design of Embedded Systems — The POLIS Approach*, Kluwer Academic Publishers, Dordrecht, 1997.

[52] M. Baleani, A. Ferrari, A. Sangiovanni-Vincentelli, and C. Turchetti, Hw/sw codesign of an engine management system, *Proceedings of DATE 2000*, 2000.

[53] CCITT, *Specification and Description Language, CCITT Z.100*, International Consultative Committee on Telegraphy and Telephony, 1992.

[54] T. Grotker, private communication, 2004.

[55] D. Gajski, J. Zhu, R. Dömer, A. Gerstlauer, and S. Zhao, *SpecC: Specification Language and Methodology*, Kluwer Academic Publishers, Dordrecht, 2000.

[56] S. Sutherland, S. Davidmann, and P. Flake, *SystemVerilog For Design: A Guide to Using SystemVerilog for Hardware Design and Modeling*, Kluwer Academic Publishers, Dordrecht, 2004.

[57] L. Cai and D. Gajski, Transaction level modeling: an overview, *First International Conference on HW/SW Codesign and System Synthesis (CODES+ISSS 2003)*, Newport Beach, CA, 2003.

[58] J. Davis, R. Galicia, M. Goel, C. Hylands, E.A. Lee, J. Liu, X. Liu, L. Muliadi, S. Neuendorffer, J. Reekie, N. Smyth, J. Tsay, and Y. Xiong, Heterogeneous Concurrent Modeling and Design in Java, Technical Report UCB/ERL M98/72, EECS, University of California, Berkeley, 1998.

[59] A. Girault, B. Lee, and E.A. Lee, Hierarchical finite state machines with multiple concurrency models, *IEEE T. Comput. Aid. D.*, 18, 742–760, 1999.

[60] E.A. Lee and A. Sangiovanni-Vincentelli, Comparing models of computation, *Proceedings of the IEEE/ACM International Conference on Computer-aided Design*, 1996, pp. 234–241.

[61] F. Balarin, H. Hsieh, L. Lavagno, C. Passerone, A. Sangiovanni-Vincentelli, and Y. Watanabe, Metropolis: an integrated environment for electronic system design, *IEEE Comput.*, 36(4), 45–52, 2003.

[62] X. Chen, F. Chen, H. Hsieh, F. Balarin, and Y. Watanabe, Formal verification of embedded system designs at multiple levels of abstraction, *Proceedings of the International Workshop on High Level Design, Validation, and Test*, Cannes, France, 2002, pp. 125–130.

5

SoC Block-Based Design and IP Assembly

John Wilson
Mentor Graphics
Berkshire, United Kingdom

Block-based design strategies, where existing IP is reused in new designs, are reaching prominence as designers start to examine how best to take advantage of the capacity now available on silicon.

Traditional design processes start with an initial concept, and are followed by a process of refinement and detailing that eventually results in an implementation of the original idea. This craftsman's approach, where each module in the implementation is created specifically for the design in which it ultimately exists, results in a cohesive design, but at some cost. If an IP module that implements a required function already exists, it may be more costly to train a designer to acquire enough knowledge to reproduce that functionality in-house [1]. With first time deployment of a design module, there is always an increased risk of encountering design bugs.

A block-based approach offers the possibility of reusing existing design modules in a new design by trading off the design optimization benefits of custom design work against the time and cost savings offered by reusing existing components. In making the "design" or "reuse" decision, designers have to weigh a large number of factors, and many of these factors are economic rather than design driven.

Processor vendors such as ARM and MIPS are prominent amongst the top IP revenue generators [2]. Processors are relatively small IP modules (in terms of gate count) and are not especially hard to design. So there must be some characteristics that make processor IP compelling to buy rather than build.

The first factor is the development effort required to create the software tools like assemblers, compilers, linkers, debuggers, and OS ports. Without this support infrastructure, the processor IP may be easy to deploy in designs, but impossible to utilize effectively.

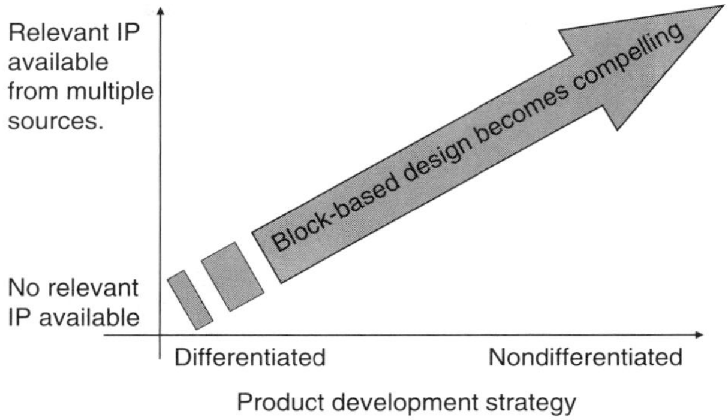

FIGURE 5.1 When block-based design becomes compelling.

The second factor is the design maturity. In creating home-grown IP, designers are deploying prototype modules with the inherent risk of initial deployment. There may be significant cost and reliability advantages in deploying IP that is already mature and improved through iteration [3].

As more IP begins to exhibit the characteristics of processors (extended infrastructure, maturity acquired over multiple projects), deploying reusable IP becomes a viable design choice.

In reality, designers should not be choosing between the block-based and traditional approaches, but examining where best to deploy those strategies on each project. Design modules that embody the design group's key knowledge and competencies — the differentiated part of a design — are the key areas on which to focus project resources and effort. Nondifferentiated parts of a design are ideal candidates for buying and configuring external IP, instead of using up valuable design team resources on designing something that already exists and offers no competitive advantage to the product (Figure 5.1).

There are some significant challenges in reusing IP in a new design, and these should not be dismissed lightly. However, a number of different industry trends are converging to ease these problems.

Design teams need to pay careful attention to these industry trends because they have the potential to offer significant competitive advantage and, increasingly teams will be expected to deliver their designs as reusable modules for use by other design teams within and outside their organizations.

5.1 The Economics of Reusable IP and Block-Based Design

For both IP creators and designers, IP becomes compellingly reusable as it increases in complexity, and standards emerge to enable that IP to be deployed rapidly into designs. A rough rule of thumb is that IP complexity (as measured by gate count and functionality) increases by five times over 3 years. Figure 5.2 illustrates the complexity growth by comparing the gate count of the highest performance members of two commonly used IP families over a 4-year period. It is unlikely that designer productivity can match this complexity growth without aggressive adoption of reuse strategies.

Standards are important for IP providers because that increases the potential market for their products. Lack of standards means that requests to customize IP for each customer are very likely. Customizing IP increases design, verification, testing, and support costs. Requirements to customize can increase design costs by such an extent that the design project is more appropriately priced as a consulting engagement. At this point, all the advantages of reusing IPs are lost. To make reuse a reality, it is incumbent on the IP providers to build enough flexibility and customization options into their designs to allow the IP to be used in an SoC, and for designers to create their designs according to standards to maximize the chances of being able to deploy reusable IP without requiring customization.

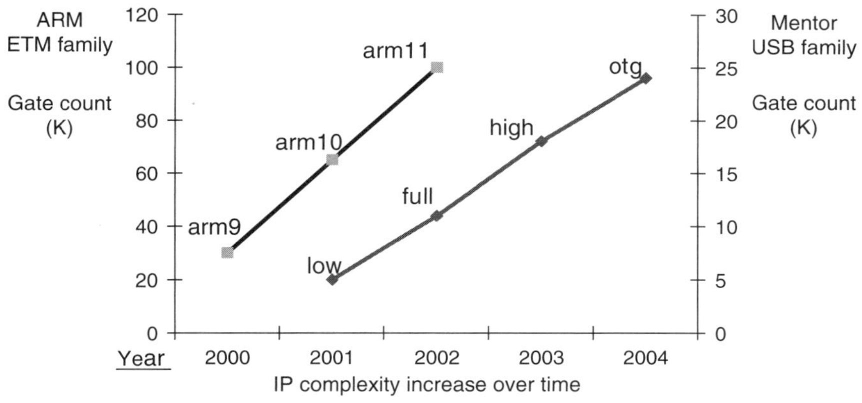

FIGURE 5.2 Complexity (as measured by gate count) increase in two IP product families over successive generations.

Many IP providers were caught up in the HDL language wars where customers were demanding Verilog or VHDL models. Many IP providers had to deliver the same IP in multiple model formats, doubling the modeling and verification effort and the costs.

The emergence of multilanguage HDL simulators and synthesis tools has made the choice of language almost totally irrelevant. It is straightforward to mix and match IP modules written in different languages. Most IP companies can be confident that supplying IP in one language is not a barrier to adoption by potential customers using the other.

In fact, some commercial IP is now supplied as mixed VHDL/Verilog design modules. An example of this is the Mentor Graphics Ethernet controller, where one of the major functional blocks was designed as a Verilog module by an external organization, and was subsequently integrated with other blocks written in VHDL, to create the overall IP module.

Up to now, IP reuse has focused on individual IP modules and how they should be structured [4] but much less emphasis was placed on how the IP modules could be deployed in a design.

Many large organizations use proprietary IP formats and demand that external IP supplied to the organization meet those format requirements [5],[6]. Reformatting and validating IP packages to meet those proprietary standards is a significant burden for IP providers.

Under the auspices of VSIA (www.vsia.org), an IP infrastructure (IPI) standard is being developed to enable a standard IP packaging format to be followed, and to allow the creation of IP readers that can reformat the IP package to meet each organization's IP structure requirements. This offers the possibility that IP deliverables can be packaged in a generic format by the IP creators, and unpackaged according to the format required by the recipient organization. Many large organizations with internal IP standards spend considerable amounts of time qualifying bought-in IP packages. In many cases, qualifying means rearranging the contents of the IP package to match the internal company standards, and that is just another customization cost.

Each standard that emerges makes block-based design more compelling for both IP creators and designers who want to buy and deploy that IP. Creating IP that conforms to recognized standards gives IP providers a bigger, more viable market for their products, and the power to resist costly requests for customization [7].

5.2 Standard Bus Interfaces

The first point of contact for most designers integrating IP into systems is the bus interfaces and protocols. Nonstandard and custom interfaces make integration a time-consuming task. On-chip bus standards such as AMBA [8], OCP [9], and CoreConnect [10] have become established, and most IP will be expect to conform to one of those interfaces unless there is a very good reason to deviate from that standard.

The benefits of using IP with standard interfaces are overwhelming.

Designers who have integrated an IP module with a standard bus interface into a system are able to integrate a different IP module using the same interface with ease.

Because this task has moved from a custom to a repeatable operation, specialist designers who understand how to craft optimally standard interconnect between IP modules have emerged. In turn, their understanding has resulted in tools that can automatically generate the required interconnect infrastructure for each bus [11]. Some of these tools are extremely sophisticated, taking into account bandwidth and power requirements to generate bus infrastructure IP that may be significantly more complex than the IP to which it is connected. A typical On-The-Go USB IP configuration might use 30K gates, while a fully configured adapter for the Sonics [11]On Chip bus can exceed 100K gates.

Verification also becomes easier because the same test modules used to validate a bus interface on one IP module can be reused on another. Again, the emergence of standard interfaces has stimulated new markets — there are many companies now supplying verification IP [12] to validate standard interface protocols. Using verification IP from one source to check IP modules from another is a very robust check.

So far, most of the focus has been on system bus for interconnecting CPUs, memory, and peripheral devices. But standard interfaces are emerging for other IP functions: for instance, many USB providers now have a standard interface to the PHY, enabling the designer to choose the best mix of USB and PHY for their design [13].

Designers may think interfaces only refer to hardware but consistent software interfaces are equally important. There have been attempts to create standards around hardware-dependent software [14], but these have not gained widespread acceptance, partly because of concerns over debug capabilities and performance (driver functions are called frequently and can have a disproportionate effect on system performance). However, some proprietary formats (for example, Xilinx's MLD) have been interfaced successfully with multiple commercial operating systems without any significant loss of performance. So it seems that some level of software driver standardization can be expected to gain acceptance in the near future.

The use of standard interfaces, wherever possible, is a key strategy for IP reuse and deployment, and enables the automation of block-based design.

5.3 Use of Assertion-Based Verification

IP creators spend significant time and effort creating testbenches to exercise and validate their "standalone" IP over a wide range of design scenarios. Testbenches are delivered to customers as part of the IP package to help designers build up confidence that the IP is functional.

There is one fundamental limitation in this approach: testbenches cannot be easily reused when the IP is embedded in a design. The decision to buy in the IP indicates the lack of a certain level of design knowledge and sophistication in the design team, so most designers do not have the specialized knowledge to develop their own design-dependent functional tests for that IP module. The testbenches supplied with the IP are not modular, so when designers integrate the IP into their designs, they have to start from scratch to develop testbenches that are appropriate for the new IP design context.

This has caused a "chicken-and-egg" problem for the IP industry. It is difficult to convince designers to use IP if it cannot be demonstrated to be successful, but how can the IP be shown to be successful if it does not get deployed in designs? Currently, the gold standard indicator that an IP is reusable is the length of the list of working designs in which that IP has been successfully used.

One technology that has emerged to ease this problem is assertion-based verification (ABV). When a design is created, designers can also write down a series of assertions about how the IP module should react to various events. When the IP is simulated as part of a design, the assertions can be used to check that the IP is not encountering previously unknown scenarios that are untested or cause malfunctions.

```
 -- Generate parity serially, clearing down after every
character TX
  Process (CLOCK, RESETN)
  begin
    if (RESETN = '0') then
      DP <= '0';
    elsif (CLOCK' event and CLOCK = '1') then
      if (TxState = C_TX_START) and (TxClkEnab = '1') then
         DP <=  '0' --initialise parity = 0
      else
         if ((HStep and TxClkEnab) = '1') then
           DP <= (TXMD xor DP); --EXOR DP and current data bit
         end if;
      end if;
    end if;
  end process;

-- Generate parity, if required of the right polarity
process (EPS, SP, DP)
begin
PTY <= not (EPS xor (not (SP or not DP)));
end process;


-- Assert 0: check illegal Transmit State Machine states
-- psl     property TXE_FSM_illegal is never
--        ((TxState = "1101") or (TxState = "1110") or
--         (TxState = "1111") @ rose(CLOCK));
-- psl     assert TXE_FSM_illegal;
```

In the same way that code-coverage tools can be used to indicate areas of HDL that have not been adequately tested, designers can also use code coverage to check that the assertions associated with the IP module have been activated. Unexecuted assertions would indicate that the original IP designer's intent has not been checked for the design in which the IP is included. More tests are needed to prove that the IP is executing correctly (or not, as the case may be) in the design.

There are a number of different assertion languages, some proprietary and some standards-based. PSL [15] and System Verilog Assertions (SVA)[16] are emerging as the languages most likely to be supported in standard HDL simulators.

5.4 Use of IP Configurators and Generators

IP reuse becomes compelling for the IP creator if there are opportunities to reuse the IP in a wide range of designs. A common strategy is to make the IP highly configurable, giving designers the best chance of setting up an optimal configuration that will work in the target design.

The first generation of configurable IP was statically configurable. That is, parameterization features of the underlying HDL were used to allow designers to select and set particular options.

However, many of the IP configurations are so complex and interdependent that this is well outside the scope of HDL configuration capabilities. A single configuration choice may result in complete sets of interfaces being added or deleted from the IP. Increasingly, IP is provided with their own configuration scripts and programs enabling designers to customize the IP and the supporting files delivered with that IP. This is dynamically configurable IP. A simple example of dynamically configurable IP is the ARM MultiLayer AMBA switched bus infrastructure IP. The version included in the Arm Design Kit v1.1 did not include the IP directly, but did include a program to be run by the designer to generate the HDL for the IP that matched the designer's specific requirements.

For some highly complex IP, the delivery package contains no IP at all, but only the generator program. Running the generator program creates the IP and support files, customized according to each designer's requirements.

The effort required to create generator programs is far greater than creating just one instance of the IP, but the economics can be compelling if the potential market for the IP increases commensurately, and will become even more compelling as design environments that support heterogeneous IP integration emerge.

One interesting aspect of IP configuration is how processor vendors choose to handle extensions to the core processor architecture. The traditional approach is to create a standard co-processor interface onto which designers can add additional functions that are tightly bound into the processor operation to augment the standard processor capabilities. This approach generally lacks support from the software tools associated with the processor, which can make it difficult for programmers to take full advantage of the co-processor.

An alternative approach pioneered by Tensilica and ARC is to create generators that create custom instances of a processor, including the ability to add custom instructions. The software tools designed for these processors also follow the configuration, so that the software tools match the processor capabilities.

Both approaches have merit. Configurable processors come more into their own if significant additions are required to processor capabilities. The co-processor approach has been boosted by the emergence of tools that can synthesize software system functionality into hardware co-processors using standard interfaces to accelerate overall system performance. It is still too early to say how much utility these tools will have: software underperformance is often not diagnosed until the hardware prototype is complete, at which stage it may be too late to change the hardware design.

Some companies offer IP already integrated into larger processor-based design modules. The IP in the design is often configurable but cannot be deleted, and standard bus interfaces are exposed for designers to connect on additional IP [17].

There are significant advantages providing highly integrated modules. Additional levels of high-value IP integration can be offered — the module test structures, and OSs can be pre-configured to run on such systems. This is platform-based design (PBD).

The first generation of PBD IP was very generic in nature. Often consisting of a processor, essential core IP such as interrupt and memory controllers, and a set of standard peripherals such as UARTs, timers and GPIO modules, it was also supplied with parameterizable boot code to enable the user to start executing the code quickly. The push behind the creation of this PBD IP was processor companies who were finding that it was very difficult for designers to understand how to create the support infrastructure required to make the processors execute the code efficiently.

Second-generation PBD IP tends to be strongly themed for particular classes of target applications. Alongside the standard CPU subsystems, specialist IP and software are integrated for applications such as mobile phones [18] or video applications [19], and significant effort is applied to provide customized configuration tools for the second-generation PBD IP [20].

Second-generation PBD IP providers have taken nondifferentiated processor subsystems and added their specialist knowledge and IP to create a value-added differentiated design.

Platform-based design will dominate nondifferentiated design strategies. Many designers will start by searching for the platform IP that is closest to their requirements and build, from that point, IP connected using standard buses and interfaces.

5.5 The Design Assembly and Verification Challenge

Designers creating designs using block-based design techniques face an interesting dichotomy. The design choice to use external IP is almost certainly driven by the lack of resources or expertise to create that IP on the project.

However, when the IP is integrated into the design, the designer is still faced with the task of understanding the IP enough to create appropriate system-level verification tests, and this is a significant and often underestimated task.

Where IP with standard interfaces are deployed, there are often standard protocol tests available as verification IP.

ABV-enabled IP, in addition to enabling in-system checks, provide some new possibilities when used in conjunction with code coverage tools. Assertions state designer intent. Measuring the extent of activation of these assertions gives designers a good indication of how well the IP functionality has been covered.

Building system-level verification environments involves many of the same characteristics as automating the design build. The same tools that automate the design-build processes would be expected to generate appropriate system-level tests.

As an example, it is interesting to think about a simple processor-based SoC design that incorporates some external IP such as a USB and Bluetooth module connected on a standard system bus.

If the system bus infrastructure is generated ("stitched"), then it is reasonable to expect monitors to be automatically enabled in the design environment to identify any deviations from the bus standard. The monitor may be used to check specific protocols at recognized interfaces on the bus, or examine the overall block operation to help gather and evaluate performance and other data.

Providing stimulus to exercise automatically the design is more challenging. It is quite easy to build automatically software programs to run on the processor to provide "ping" tests for the IP, which can be used to verify that the processor-to-peripheral functionality is correct.

It is also reasonable to provide "traffic generator" IP to stimulate external standard interfaces. If a design exports (makes visible externally) an interface, it is not too difficult to automatically construct a testbench that attaches a traffic generator module to that interface. However, when more than one traffic generator is used in a design, there are major problems in automating the intent and synchronization of the activities of different traffic generators to create desired test scenarios.

For processor-based designs, one common technique is to construct a "loopback" style of testbench where a component in the design with a processor and communications interface (for instance, a UART, USB, or Bluetooth controller) is connected to a matching component in the testbench. The processor interfaces on the testbench components are connected back to the processor in the design (Figure 5.3).

This allows test programs to be created that can be run on the processor, enabling both the design under test to be set up and activated, and cause events via the testbench components which can be used to verify if the design components are operational.

This methodology meshes neatly with the techniques used by software programmers to test their programs outside the host-system environment.

While this methodology is not generally applicable, associating test modules with IP modules is the first step to partially automating testbench creation in parallel with design creation.

There are limits to what can be automated. It would be very difficult for a designer to create a system with Bluetooth and USB modules under the control of an RTOS running on a processor, and expect automatic generation of tests to insert a message into the Bluetooth module and see the message emerge from the USB port. Automation tools cannot reasonably be expected to understand design intent for complex test scenarios.

Ideally, a designer would get involved in verification test development for higher level system functionality, and leave the tools to automate the mechanical parts of the testbench construction, analogous to the way that the designer selects the IP for inclusion in a design but uses automated tools for the detailed integration.

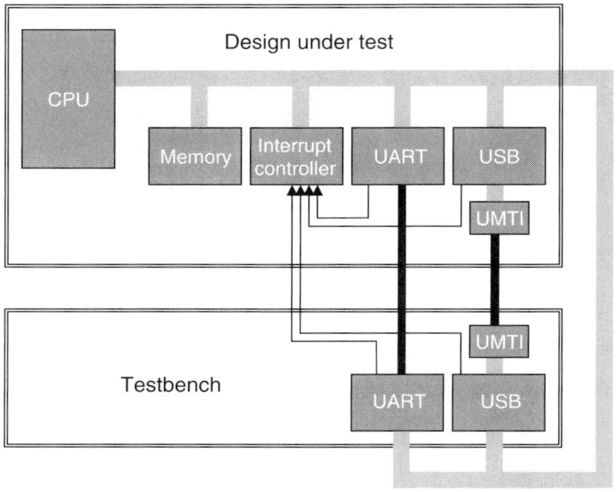

FIGURE 5.3 "Loopback" style of testbench.

However, for the moment, it seems that the lack of agreed standards in this area restricts even this level of automation to proprietary systems where organizations have full control over the test interfaces and methodology.

5.6 The SPIRIT XML Databook Initiative

Abstraction level paradigm shifts are common when designers need an order of magnitude increase in design productivity. Working with logic gates instead of transistors, and with MSI modules instead of gates were consistent trends in previous decades.

Each of these abstractions is consistent within its own reference framework: there is no logic gate that could not be connected to another logic gate, for instance. But block-based design is different.

Links to many disparate sources of information are required to make sensible design choices. For instance, if an IP module is supplied with an OS driver stack, then it may be easy to connect the module to the processor from an HW viewpoint, but this does not mean that the IP is functional. The driver software may be supplied in a form that is incompatible with the processor or the operating system running on that processor.

There have been many credible attempts to create a unified design flow [21] but most put a significant burden on IP creators to model IP in a specific way, and do not deal with enough of the myriad disparate design aspects to be immediately compelling. It is very difficult for the IP creators to justify the effort to support such tools within a nascent market environment and this, rather than technical considerations, have tended to restrict promising initiatives.

The SPIRIT initiative (www.spiritconsortium.com) changes the emphasis from requiring IP to be supplied in a fixed format, to one of documenting whatever aspects of the IP are available in XML [22]. Typically, this would include, but not be restricted to, information about the configurable IP parameters, bus interface information, and pointers to the various resources — like ABV files and software driver modules — associated with that IP module [23].

The power of SPIRIT [24] is that the information enables a consistent HW, SW, and system view of a design to be maintained. The same information can be used to create HW views and software programmer's views for a design.

Design environments capable of reading this XML information can make intelligent design configuration and integration choices [25][24]. From an HW designer's point of view, that could be an IP "stitcher" creating detailed HDL descriptions. The same information can be used to build SW programs to run on the design, design documentation, and setup verification environments, etc.

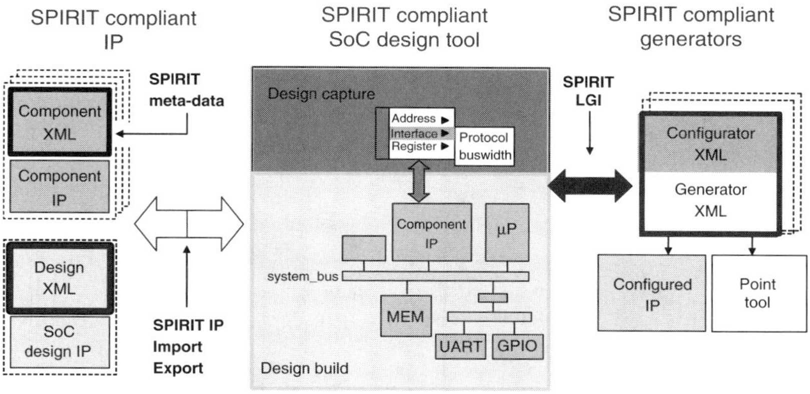

FIGURE 5.4 The SPIRIT design environment architecture. (From Mentor Graphics Platform Express, www.-mentor.com/products/embedded_software/platform_baseddesign/index.cfm.)

SPIRIT enables the deployment of disparate IP generators within design environments so many of the IP configurators and generators already in existence can be reused. In fact, their use is enhanced because instead of designers being required to guess at the parameters to be supplied to configure the IP appropriately, the XML-based design database can be used to extract and calculate the required settings automatically (Figure 5.4).

In turn, this enables the creation of new classes of "design-wide" generators, which assess the information available to individual IP modules and build aspects of the design, e.g., HDL, software programs to run on the design or design documentation. Nobody expects all IP to be delivered with every conceivable view: but where information is missing (from the design generator's point of view), the generators are expected to have a strategy to handle that missing information.

An example of a tool that can process SPIRIT formatted information is Mentor Graphic's Platform Express [25]. SPIRIT XML component and design information is read into the tool, along with other XML information, to build the design database. Generators extract information from this design database, sometimes creating design modules such as HDL netlists, software programs, and documentation; sometimes creating tool modules such as configuration files for external tools used in the design process; or creating new information to add to the design database for use by other generators later in the design creation process.

There is also a realization in SPIRIT that it is impossible to create a single global unifying standard that can encompass all aspects of design. Instead, the approach has been to agree on core information, and provide mechanisms for extending and customizing the data. This allows arbitrary file types to be associated with IP. It enables individual organizations to add their own custom data where specialist data processing is required (this is a common requirement for highly configurable IP).

Most importantly, it enables *de facto* data standards for areas not directly covered by SPIRIT to be developed by expert interested groups and used in conjunction with SPIRIT.

An organic, evolutionary approach to evolve data standards may seem a strange path to follow, but where the information domain is characterized by high complexity and unstructured relationships, enabling different data models to be supported and watching for winners to emerge is a powerful strategy. The moderating factor that will prevent an explosion of incompatible or overlapping standards is the availability of generators and features inherent in XML. IP providers will provide data in the format that can be utilized by the most functionally useful generators, and as more IP support the data, more generators will emerge to use that data. For data provided in different formats, XSL [26] (a standard within the XML standards family) can be used to extract and reformat that data for use with existing generators.

Ultimately, the goal of SPIRIT is to enable designers to mix and match IP and design generators from many different sources, and leverage the emerging block-based design standards to build and verify high-quality designs rapidly and reliably.

5.7 Conclusions

No single strategy enables block-based design strategies, but a number of important initiatives are converging to make block-based design compelling. For example, IP supplied with standard SoC bus interfaces ready for integration into a design; configurators to help designers to choose valid options; ABV modules to catch deployment errors; and XML databook and catalog information will be easier to deploy in many different designs.

SPIRIT does not mandate specific IP standards, but is likely to accelerate the emergence of various *de facto* standards because generators will search out compatible information supplied in the IP package to build various aspects of designs. IP blocks which do not supply that information can be handled, but designers are likely to choose IP that the generators can help deploy most effectively, and ignore the IP that requires additional manual design effort.

The real challenge for design teams will be deciding on their core design competencies on which to focus their expertise and resources, and being ruthless in deciding that compromising on bought-in IP modules is a more efficient design choice than requiring custom IP design modifications for noncritical system components.

References

[1] K. Pavitt, What we know about strategic management of technology, in *Implementing New Technologies: Innovation and the Management of Technology*, 2nd ed., E. Rhodes and D. Wield, Eds., NCC Blackwell, Oxford, UK,1994, pp. 176–178.

[2] G. Moretti, Third Party IP: A Shaky Foundation for SoC Design, *EDN*, February 3, magazine (online), 2005.

[3] C.M. Christensen, *The Innovator's Dilemma*, Harvard Business School Press, Cambridge, Mass, USA, 1997, p. 9.

[4] M. Keating and P. Bricaud, *Reuse Methodology Manual for System-on-a-Chip Designs,* Kluwer Academic Publishers, Boston, 1998, Chap. 9.

[5] Repositories are Better for Reuse, *Electronics Times* (U.K.), March 27, 2000.

[6] Philips Execs Find IP Unusable, but Necessary, *Electronics Times* (U.K.), March 22, 2000.

[7] M. E. Porter, How Competitive Forces Shape Strategy, *Harvard Business Review*, March 1, 1979.

[8] AMBA Specification, www.arm.com/products/solutions/AMBAHomePage.html

[9] OCP Specification, www.ocpip.org/getspec

[10] CoreConnect Specification, www-03.ibm.com/chips/products/coreconnect

[11] Sonics, www.sonicsinc.com

[12] For example, Synopsys (www.synopsys.com/products/designware/dwverificationlibrary.html), Cadence (www.cadence.com/products/ip) are just two of many companies. The Design and Reuse catalogue gives many examples — see www.us.design-reuse.com/verificationip.

[13] S. McGowan, *USB 2.0 Transceiver Macrocell Interface (UTMI)*, Intel, March 29, 2001. www.intel.com/technology/usb/download/2_0_Xcvr_Macrocell_1_05.pdf

[14] Hardware-dependent software, in *Taxonomies for the Development and Verification of Digital Systems*, B. Bailey, G. Martin and T. Anderson, Eds., Springer, Heidelberg, 2005, pp. 135–168.

[15] Accellera, Property Specification Language Reference Manual, June 9, 2004, www.eda.org/vfv/docs/PSL-v1.1.pdf

[16] System Verilog Assertions — see the System Verilog LRM, www.eda.org/sv/SystemVerilog_3.1a.pdf

[17] H. Chang, L. Cooke, M. Hunt, G. Martin, A. McNelly, and L. Todd, *Surviving the SoC Revolution,* Kluwer Academic Publishers, Dordrecht, 1999, pp. 155–182.

[18] P. Cumming, The TI OMAP™ platform approach to SoC, in *Winning the SoC Revolution*, G. Martin and H. Chang, Eds., Kluwer Academic Publishers, Boston, 2003, Chap. 5, pp. 97–118.

[19] J. Augusto de Oliveira and H. Van Antwerpen, The Philips Nexperia digital video platform, in *Winning the SoC Revolution*, G. Martin and H. Chang, Eds., Kluwer Academic Publishers, Boston, 2003, Chap. 4, pp. 67–96.

[20] G. Mole, Philips Semiconductors Next Generation Architectural IP ReUse Developments for SoC Integration, *IP/SOC 2004*, Grenoble, France, December 2004.

[21] S. Krolikoski, F. Schirrmeister, B. Salefski, J. Rowson, and G. Martin, Methodology and Technology for Virtual Component Driven Hardware/Software Co-Design on the System Level, paper 94.1, *ISCAS 1999*, Orlando, FL, 1999.
[22] W3C Consortium, XML Standard, www.w3.org/XML
[23] SPIRIT Consortium, SPIRIT — Examples [presentation], www.spiritconsortium.com, December 8, 2004.
[24] SPIRIT Consortium, "SPIRIT — The Dream" [presentation], www.spiritconsortium.com, December 8, 2004.
[25] Mentor Graphics Platform Express, www.mentor.com/products/embedded_software/platform_baseddesign/index.cfm
[26] W3C Consortium, XSL Standard, www.w3.org/Style/XSL

6

Performance Evaluation Methods for Multiprocessor System-on-Chip Design

Ahmed Jerraya
SLS Group, TIMA Laboratory, INPG
Grenoble, France

Iuliana Bacivarov
SLS Group, TIMA Laboratory
Grenoble, France

6.1 Introduction

Multi-processor systems-on-chip (MPSoCs) require the integration of heterogeneous components (e.g., micro-processors, DSP, ASIC, memories, buses, etc.) on a single chip. The design of MPSoC architectures requires the exploration of a huge space of architectural parameters for each component. The challenge of building high-performance MPSoCs is closely related to the availability of fast and accurate performance evaluation methodologies. This chapter provides an overview of the performance evaluation methods developed for specific subsystems. It then proposes to combine subsystem performance evaluation methods to deal with MPSoC.

Performance evaluation is the process that analyzes the capabilities of a system in a particular context, i.e., a given behavior, a specific load, or a specific set of inputs. Generally, performance evaluation is used to validate design choices before implementation or to enable architecture exploration and optimization from very early design phases.

A plethora of performance evaluation tools have been reported in the literature for various subsystems. Research groups have approached various types of subsystems i.e., software (SW), hardware (HW), or interconnect, differently, by employing different description models, abstraction levels, performance metrics, or technology parameters. Consequently, there is currently a broad range of methods and tools for performance evaluation, addressing virtually any kind of design and level of hierarchy, from very specific subsystems to generic, global systems.

Multi-processor system-on-chip (MPSoC) is a concept that aims at integrating multiple subsystems on a single chip. Systems that put together complex HW and SW subsystems are difficult to analyze. Additionally, in this case, the design space exploration and the parameter optimization can quickly become intractable. Therefore, the challenge of building high-performance MPSoCs is closely related to the availability of fast and accurate performance evaluation methodologies.

Existing performance evaluation methods have been developed for specific subsystems. However, MPSoCs require new methods for evaluating their performance. Therefore the purpose of this study is to explore different methodologies used for different subsystems in order to propose a general framework that tackles the problem of performance evaluation for heterogeneous MPSoC. The long-term goal of this work is to build a global MPSoC performance evaluation by composing different tools. This kind of evaluation will be referred to as holistic performance evaluation.

The chapter is structured as follows: Section 6.2 defines the key characteristics of performance evaluation environments. It details the analyzed subsystems, their description models and environments, and the associated performance evaluation tools and methods. Section 6.3 is dedicated to the study of MPSoC performance evaluation. Section 6.4 proposes several trends that could guide future research toward building efficient MPSoC performance evaluation environments.

6.2 Overview of Performance Evaluation in the Context of System Design Flow

This section defines typical terms and concepts used for performance evaluation. First, the performance-evaluation process is positioned within a generic design flow. Three major axes define existing performance evaluation tools: the subsystem under analysis, the performance model, and the performance evaluation methodology. They are detailed in this section. An overview of different types of subsystems is provided, focusing on their abstraction levels, performance metrics, and technology parameters. In the end, the main performance evaluation approaches are introduced.

6.2.1 Major Steps in Performance Evaluation

This section analyzes the application domain of the performance evaluation process within the systems design flow. A designed system is evaluated by a suitable performance evaluation tool where it is represented by a performance model. Next, this section presents how evaluation results may influence decisions during system design.

A design flow may include one or more performance evaluation tools. These evaluation tools could be used for different purposes, e.g., to verify if a system meets the constraints imposed or runs properly and to help making design choices.

Figure 6.1 presents a generic design flow. The initial specification is split into a functional and a nonfunctional part of the subsystem to be analyzed. The functional part contains the behavior of the subsystem under analysis, described it as an executable program or as a formal model (e.g., equation). However, the set of evaluation constraints or quality criteria selected from the initial specification constitute the nonfunctional part.

The performance model cannot be separated from the evaluation methodology, because it provides the technology characteristics used to compute performance results prior to the real implementation. Moreover, it selects the critical characteristics to be analyzed, such as the model performance metrics: timing, power, and area. Eventually it chooses the measurement strategy (e.g., an aggregation approach). Both performance metrics and technology parameters may be built into the evaluation tool or given as external libraries.

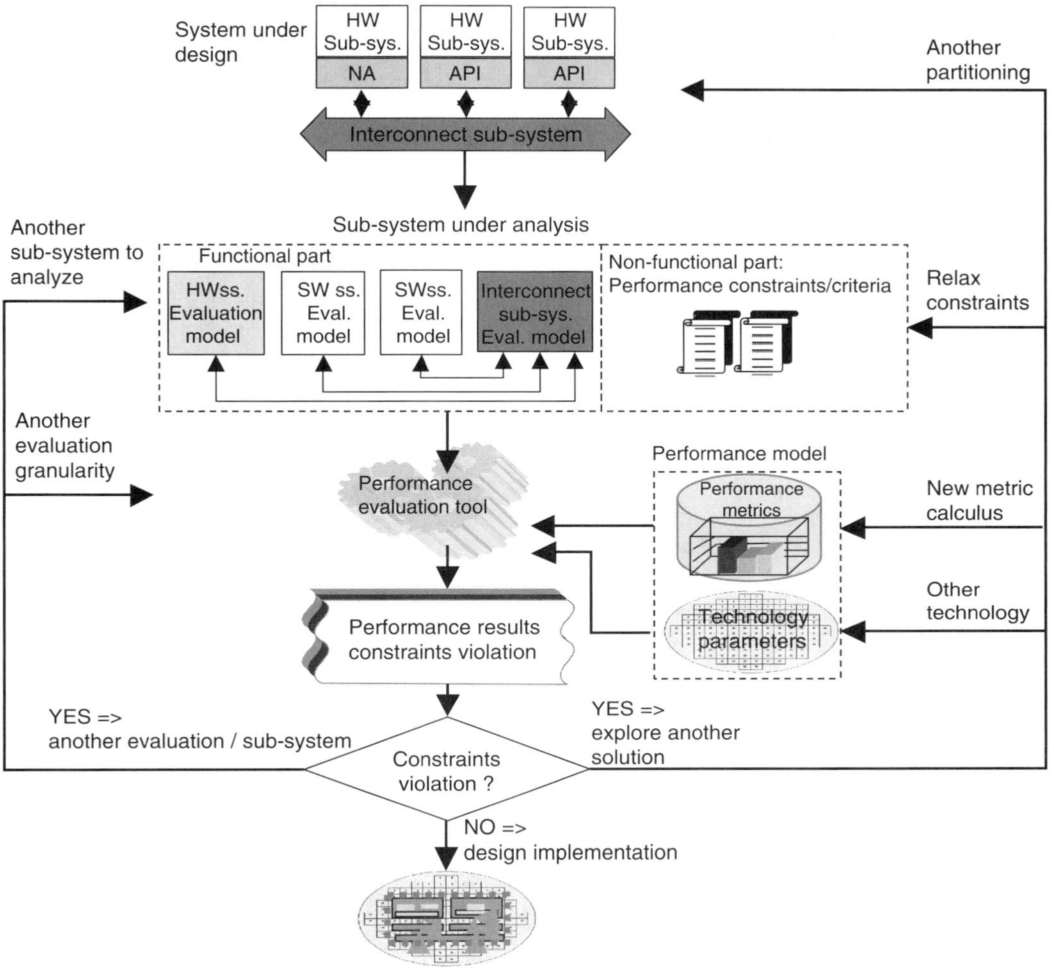

FIGURE 6.1 General performance evaluation and design optimization flow.

The design process may be made of several iterations when the system needs to be tuned or partially redesigned. Several design loops may modify the performance model. Each time a new calculation of the metrics (e.g., the chip area can be enlarged in order to meet timing deadlines) or technological parameters (e.g., a new underlying technology or increased clock frequency) is initiated.

6.2.2 The Key Characteristics of Performance Evaluation

Figure 6.2 details on three axes, the three main characteristics of the performance evaluation process: the subsystem under analysis, its abstraction level, and the performance evaluation methodology. Analysis of five kinds of subsystems will be considered: HW, SW, CPUs, interconnect subsystems, and MPSoCs. Each basic subsystem has specific design methods and evaluation tools. They may be designed at different abstraction levels, of which we will consider only three. Also, three performance evaluation methodologies will be considered, with metrics and technology parameters specific to different subsystems.

A subsystem under analysis is characterized by its type (e.g., HW, SW, etc.), and its abstraction level. The performance evaluation may be applied to different kinds of subsystems varying from simple devices to

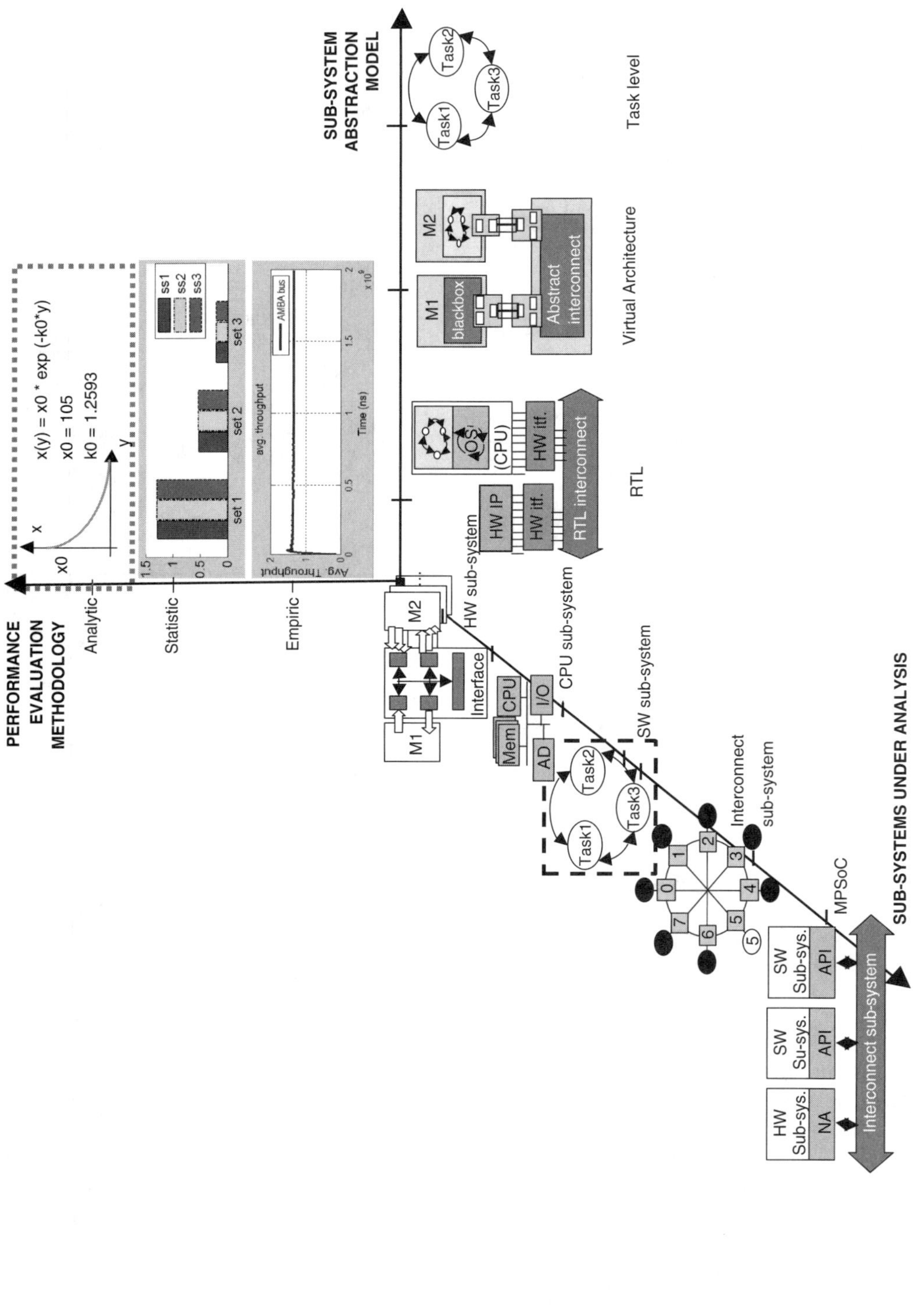

FIGURE 6.2 Performance evaluation environment characteristics.

sophisticated modules. We will consider five main kinds of subsystems: HW subsystems, CPU subsystems, SW subsystems, interconnect subsystems, and constituent parts of MPSoC. Traditionally they are studied by five different research communities.

Classically, the HW community [1] considers HW subsystems as HDL models. They are designed with electronic design automation (EDA) tools that include specific performance analysis [2–5]. The computer architecture community e.g., [6] considers CPU subsystems as complex microarchitectures. Consequently, specialized methodologies for CPU design and performance evaluation have been developed [7]. The SW community [8,9] considers SW subsystems as programs running parallel tasks. They are designed with computer-aided SW engineering (CASE) tools and evaluated with specific methods [10–22]. The networking community [23–30] considers interconnect subsystems as a way to connect diverse HW or SW components. The network performance determines the overall system efficiency, and consequently it is an intensively explored domain.

Each of these communities uses different *abstraction levels* to represent their subsystem. Without any loss of generality, in this study only three levels will be considered: register transfer level (RTL), virtual-architecture level, and task level. These may be adapted for different kinds of subsystems.

Performance evaluation uses a specific methodology and a system performance model. The methodology may be simulation-based, analytic (i.e., using a mathematical description), or statistical. The system performance model takes into account the performance metrics and technology parameters.

The performance metrics are used for assessing the system under analysis. They may be physical metrics related to real system functioning or implementation (e.g., execution timings, occupied area, or consumed power), or quality metrics that are related to nonfunctional properties (e.g., latency, bandwidth, throughput, jitter, or errors).

The technology parameters are required to fit the performance model to an appropriate analysis domain or to customize given design constraints. The technology parameters may include architectural features of higher level (e.g., the implemented parallelism, the network topology), or lower level (e.g., the silicon technology, the voltage supply).

6.2.3 Performance Evaluation Approaches

The two main classes of performance evaluation reported in literature are: statistical approaches and deterministic approaches. For statistical approaches, the performance is a random variable characterized by several parameters such as a probability distribution function, average, standard deviation, and other statistical properties. Deterministic approaches are divided into empirical and analytical approaches. In this case, the performance cost function is defined as a deterministic variable, a function of critical parameters. Each of these approaches is defined as follows:

The *statistical approach* [17,19] proceeds in two phases. The first phase finds the most suitable model to express the system performance. Usually parameters are calibrated by running random benchmarks. The second phase makes use of the statistical model previously found to predict the performance of new applications. In most cases, this second phase provides a feedback for updating the initial model.

The *empirical* approach can be accomplished either by measurement or simulation. *Measurement* is based on the real measurement of an already built or prototyped system. It generally provides extremely accurate results. Because this approach can be applied only late in the design cycle when a prototype can be made available, we do not include it in this study. The *simulation approach* [3,16,21,24–28,31] relies on the execution of the complete system using input scenarios or representative benchmarks. It may provide very good accuracy. Its accuracy and speed depend on the abstraction level used to describe the simulated system.

The *analytical approach* [2,10–12,14,15,32] formally investigates system capabilities. The subsystem under analysis is generally described at a high level of abstraction by means of algebraic equations. Mathematical theories applied to performance evaluation make possible a complete analysis of the full system performance at an early design stage. Moreover, such approaches provide fast evaluation because they replace time-consuming system compilation and execution. Building an analytical model could be very complex. The dynamic behavior (e.g., program context switch and wait times due to contentions or collisions) and refinement steps (e.g., compiler optimizations) are hard to model. However, this approach may be useful for worst-case analysis or to find corner cases that are hard to cover with simulation.

6.2.4 Hardware Subsystems

6.2.4.1 Definition

An HW subsystem is a cluster of functional units with a low programmability level like FPGA or ASIC devices. It can be specified by finite state machines (FSMs) or logic functions. In this chapter, the HW concept excludes any modules that are either CPUs or interconnection like. We also restrain the study to digital HW.

6.2.4.2 Abstraction Levels

HW abstraction is related to system timing, of which we consider three levels: high-level language (HLL), bus cycle-accurate (BCA), and RTL. At HLL, the behavior and communication may hide clock cycles, by using abstract channels and high-level communication of primitives, e.g., a system described by untimed computation and transaction-level communication. At BCA level, only the communication of the subsystem is detailed to the clock cycle level, while the computation may still be untimed. At RTL, both the computation and communication of the system are detailed to clock cycle level. A typical example is a set of registers and some combinatorial functions.

6.2.4.3 Performance Metrics

Typical performance metrics are power, execution time, or size, which could accurately be extracted during low-level estimations and used in higher abstraction models.

6.2.4.4 Technology Parameters

Technology parameters abstract implementation details of the real HW platform, depending on the abstraction level. At HLL, as physical signals and behavior are abstracted, the technology parameters denote high-level partitioning of processes with granularity of functions (e.g., C function) and with reference to the amount of exchanged transactions. At BCA level, the technology parameters concern data formats (e.g., size, coding, etc.), or behavioral data processing (e.g., number of bytes transferred, throughputs, occupancies, and latencies). At RTL, the HW subsystem model is complete. It requires parameters denoting structural and timing properties (e.g., for the memory or communication subsystems) and implementation details (e.g., the FPGA mapping or ASIC implementation).

There are different performance evaluation tools for HW subsystems, which make use of different performance evaluation approaches: simulation-based approaches [3], analytical approaches [2], mixed analytical and statistical approaches [18], mixed simulation and statistical approaches [5], and mixed analytical, simulation, and statistical approaches [4].

6.2.5 CPU Modules

6.2.5.1 Definition

A CPU module is a hardware module executing a specific instruction set. It is defined by an instruction set architecture (ISA) detailing the implementation and interconnection of the various functional units, the set of instructions, register utilization, and memory addressing.

6.2.5.2 Abstraction Levels

For CPU modules, three abstraction levels can be considered: RTL, also known as the micro-architecture level, the cycle-accurate ISA level, and the assembler (ASM) ISA level. The RTL (or micro-architecture level) offers the most detailed view of a CPU. It contains the complete detailed description of each module, taking into account the internal data, control, and memory hierarchy. The cycle-accurate ISA level details the execution of instructions with clock accuracy. It exploits the real instruction set model and internal resources, but in an abstract CPU representation. The ASM ISA level increases the abstraction, executing programs on a virtual CPU representation, with abstract interconnections and parameters, e.g., an instruction set simulator.

6.2.5.3 Performance Metrics

The main performance metrics for CPU subsystems are related to timing behavior. We can mention among these the throughput that expresses the number of instructions executed by CPU per time unit, the utilization that represents the time ratio spent on executing tasks, and the time dedicated to the execution of a program or to respond to a peripheral. Other performance evaluation metrics are power consumption and memory size.

6.2.5.4 Technology Parameters

Technology parameters abstract the CPU implementation details, depending on the abstraction level. At RTL, only the technology for the CPU physical implementation is abstract. The ISA level abstracts the control and data path implementation, but it still details the execution with clock-cycle accuracy using the real instruction set (load/store, floating point, or memory management instructions), the internal register set, and internal resources. And finally, the ASM level abstracts the micro-architecture (e.g., pipeline and cache memory), providing only the instruction set to program it.

Different performance evaluation tools for CPU subsystems exist, making use of different performance evaluation approaches: simulation-based approaches [31,33] analytical approaches [32,10], statistical approaches [19], mixed analytical and statistical approaches [34], and mixed analytical, simulation, and statistical approaches [7].

Chapter 9 of this book ("Using Performance Metrics to Select Microprocessor Cores for IC Designs") has as its main objective, the measurement of CPU performance. For a thorough discussion on CPU performance evaluation utilizing benchmarking techniques, we refer the reader to Chapter 10.

6.2.6 Software Modules

6.2.6.1 Definition

A software module is defined by the set of programs to be executed on a CPU. They may have different representations (procedural, functional, object-oriented, etc.), different execution models (single- or multi-threaded), different degrees of responsiveness (real time, nonreal time), or different abstraction levels (from HLL down to ISA level).

6.2.6.2 Abstraction Levels

We will consider three abstraction levels for SW modules. At HLL, parallel programs run independently on the underlying architecture, interacting by means of abstract communication models. At transaction-level modeling (TLM) level, parallel programs are mapped and executed on generic CPU subsystems. They communicate explicitly but their synchronization remains implicit. And finally, at ISA level, the code is targeted at a specific CPU and it targets explicit interconnects.

6.2.6.3 Performance Metrics

The metrics most used for SW performance evaluation are run time, power consumption, and occupied memory (footprint). Additionally, SW performance may consider concurrency, heterogeneity, and abstraction at different levels [35].

6.2.6.4 Technology Parameters

For SW performance evaluation, technology parameters abstract the execution platform. At HLL, technology parameters abstract the way different programs communicate using, for example, coarse-grain send()/receive() primitives. At TLM level, technology parameters hide the technique or resources used for synchronization, such as using a specific Operating System (OS), application program interfaces (APIs) and mutex_lock()/unlock()-like primitives. At ISA level, technology parameters abstract the data transfer scheme, the memory mapping, and the addressing mode.

Different performance evaluation tools for SW models exist, making use of different performance evaluation approaches: simulation-based [21], analytical [12], and statistical [17].

6.2.6.5 Software Subsystems

6.2.6.5.1 Definition.

When dealing with system-on-chip design, the SW is executed on a CPU subsystem, made of a CPU and a set of peripherals. In this way, the CPU and the executed SW program are generally combined in an SW subsystem.

6.2.6.5.2 Abstraction Levels.

The literature presents several classifications for the abstraction of SW subsystems, among which we will consider three abstraction levels: The HLL, the OS level, and the hardware abstraction layer (HAL) level. At the HLL, the application is composed of a set of tasks communicating through abstract HLL primitives provided by the programming languages (e.g., send()/receive()). The architecture, the interconnections, and the synchronization are abstract. At the OS level, the SW model relies on specific OS APIs, while the interconnections and the architecture still remain abstract. Finally, at the HAL level, the SW is bound to use a specific CPU insturtion set and may run on an RTL model of the CPU. In this case, the interconnections are described as an HW model, and the synchronization is explicit.

6.2.6.5.3 Performance Metrics.

The main performance metrics are the timing behavior, the power consumption, and the occupied area. They are computed by varying the parameters related to the SW program and to the underlying CPU architecture.

6.2.6.5.4 Technology Parameters.

In SW subsystem performance evaluation, technology parameters abstract the execution platform, and characterize the SW program. At HLL, technology parameters mostly refer to application behavioral features, abstracting the communication details. At the OS level, technology parameters include OS features (e.g., interrupts, scheduling, and context switching delays), but their implementation remains abstract. At HLL, technology parameters abstract only the technology of implementation, while all the other details, such as the data transfer scheme, the memory mapping or the addressing mode are explicitly referred to.

Different performance evaluation tools for SW subsystems exist, making use of different performance evaluation approaches: simulation-based approaches [16,20], analytical approaches [11,13], statistical approaches [19], and mixed analytical, simulation, and statistical approaches [22].

6.2.7 Interconnect Subsystems

6.2.7.1 Definition

The interconnect subsystem provides the media and the necessary protocols for commusnication between different subsystems.

6.2.7.2 Abstraction Levels

In this study, we will consider RTL, transactional, and service or HLL models. At the HLL, different modules communicate by requiring services using an abstract protocol, via abstract network topologies. The transactional level still uses abstract communication protocols (e.g., send/receive) but it fixes the communication topology. The RTL communication is achieved by means of explicit interconnects like physical wires or buses, driving explicit data.

6.2.7.3 Performance Metrics

The performance evaluation of interconnect subsystem focuses on the traffic, interconnection topology (e.g., network topology, path routing, and packet loss within switches), interconnection technology (e.g., total wire length and the amount of packet switch logic), and application demands (e.g., delay, throughput, and bandwidth).

6.2.7.4 Technology Parameters

A large variety of technology parameters emerge at different levels. Thus, at HLL, parameters are the throughput or latency. At TLM level, the parameters are the transaction times and the arbitration strategy. At the RTL, the wires and pin-level protocols allow system delays to be measured accurately.

Simulation is the performance evaluation strategy most used for interconnect subsystems at different abstraction levels: behavioral [24], cycle-accurate level [25], and TLM level [27]. Interconnect simulation models can be combined with HW/SW co-simulation at different abstraction levels for the evaluation of full MPSoC [26,28].

6.2.8 Multiprocessor Systems-on-Chip Models

6.2.8.1 Definition

An MPSoC is a heterogeneous system built of several different subsystems like HW, SW, and interconnect, and it takes advantage of their synergetic collaboration.

6.2.8.2 Abstraction levels

MPSoCs are made of subsystems that may have different abstraction levels. For example, in the same system, RTL HW components can be coupled with HLL SW components, and they can communicate at the RTL or by using transactions. In this study, we will consider the interconnections, synchronization, and interfaces between the different subsystems. The abstraction levels considered are the functional level, the virtual architecture model level, and the level that combines RTL models of the hardware with instruition set architecture models of the CPU. At the functional level (like message passing interface (MPI)) [36], the HW/SW interfaces, the interconnections, and synchronization are abstracted and the subsystems interact through high-level primitives (send, receive). For the virtual architecture level, the interconnections and synchronization are explicit but the HW/SW interfaces are still abstract. The lowest level considered deals with an RTL architecture for the HW-related sections coupled with the ISA level for the SW. This kind of architecture explicitly presents the interconnections, synchronization, and interfaces. In order to master the complexity, most existing methods used to assess heterogeneous MPSoC systems are applied at a high abstraction level.

6.2.8.3 Performance Metrics

MPSoC performance metrics can be viewed as the union of SW, HW and interconnect subsystems performance metrics, for instance, execution time, memory size, and power consumption.

6.2.8.4 Technology Parameters

A large variety of possible technology parameters emerge for each of the subsystems involved, mostly at different levels and describing multiple implementation alternatives. They are considered during system analysis, and exploited for subsystem performance optimization.

Different performance-evaluation tools for MPSoC exist. They are developed for specific subsystems [37–41], or for the complete MPSoC [42–44]. Section 6.3 deals with performance-evaluation environments for MPSoCs.

6.3 MPSoC Performance Evaluation

As has been shown in previous sections, many performance evaluation methods and tools are available for different subsystems: HW, SW, interconnect and even for MPSoCs. They include a large variety of measurement strategies, abstraction levels, evaluated metrics, and techniques. However, there is still a considerable gap between particular evaluation tools that consider only isolated components and performance evaluation for a full MPSoC.

The evaluation of a full MPSoC design containing a mixture of HW, SW and interconnect subsystems, needs to cover the evaluation of all the subsystems, at different abstraction levels.

Few MPSoC evaluation tools are reported in the literature [37–39,41,42,44,45]. The key restriction with existing approaches is the use of a homogeneous model to represent the overall system, or the use of slow evaluation methods that cannot allow the exploration of the architecture by evaluating a massive number of solutions.

For example, the SymTA/S approach [45] makes use of a standard event model to represent communication and computation for complex heterogeneous MPSoC. The model allows taking into account complex behavior such as interrupt control and data dependant computation. The approach allows accurate performance analysis; however, it requires a specific model of the MPSoC to operate.

The ChronoSym approach [41] makes use of a time-annotated native execution model to evaluate SW execution times. The timed simulation model is integrated into an HW/SW co-simulation framework to consider complex behaviors such as interactions with the HW resources and OS performance. The approach allows fast and accurate performances analysis of the SW subsystem. However, for the entire MPSoC evaluation, it needs to be combined with other approaches for the evaluation of interconnect and HW subsystems.

Co-simulation approaches are also well suited for the performance evaluation of heterogeneous systems. The co-simulation offers flexibility and modularity to couple various subsystem executions at different abstraction levels and even specified in different languages. The accuracy of performance evaluation by co-simulation depends on the chosen subsystem model and on their global synchronization.

A complete approach aiming at analyzing the power consumption for the entire MPSoC by using several performance models is presented in [42]. It is based on interconnecting different simulations (e.g., SystemC simulation and ISS execution) and different power models for different components, in a complete system simulation platform named MPARM. A related work is [43], where the focus is on MPSoC communication-performance analysis.

The co-estimation approach in [44] is based on the concurrent and synchronized execution of multiple power estimators for HW/SW system-on-chip-power analysis. Various power estimators can be plugged into the co-estimation framework, possibly operating at different levels of abstraction. The approach is integrated in the POLIS system design environment and PTOLEMY simulation platform. The power co-estimation framework drives system design trade-offs, e.g., HW/SW partitioning, component, or parameter selection. A similar approach is represented in [46,47]. The tool named ROSES allows different subsystems that may be described at different abstraction levels and in different languages to be co-simulated.

However, when applied to low abstraction levels, evaluation approaches based on co-simulation [43,44,46] appear to be slow. They cannot explore large solution spaces. An alternative would be to combine co-simulation with analytical methods in order to achieve faster evaluation. This is similar to methods used for CPU architecture exploration [31,33].

Figure 6.3 shows such a scheme for MPSoC. The key idea is to use co-simulation for computing extensive data for one architectural solution. The results of co-simulation will be further used to parameterize an analytical model. A massive number of new design solutions can be evaluated faster using the newly designed analytic model.

The left branch of Figure 6.3 represents the performance evaluation of the full system by co-simulation. This can be made by using any existing co-simulation approach [43,44,46]. The input of the evaluation process is the specification of the MPSoC architecture to be analyzed. The specification defines each subsystem, the communication model, and the interconnection interfaces.

The right branch of Figure 6.3 describes the evaluation of the full system using an analytical model. The figure represents the analytical model construction with dotted lines. This is done through component characterizations and parameter extraction from the base co-simulation model.

After the construction phase, the analytical branch can be decoupled from the co-simulation. The stand-alone analytical performance evaluation provides quick and still accurate evaluations for new designs. The two branches of Figure 6.3, i.e., the co-simulation and the analytical approach, lead to similar performance results, but they are different in terms of evaluation speed and accuracy.

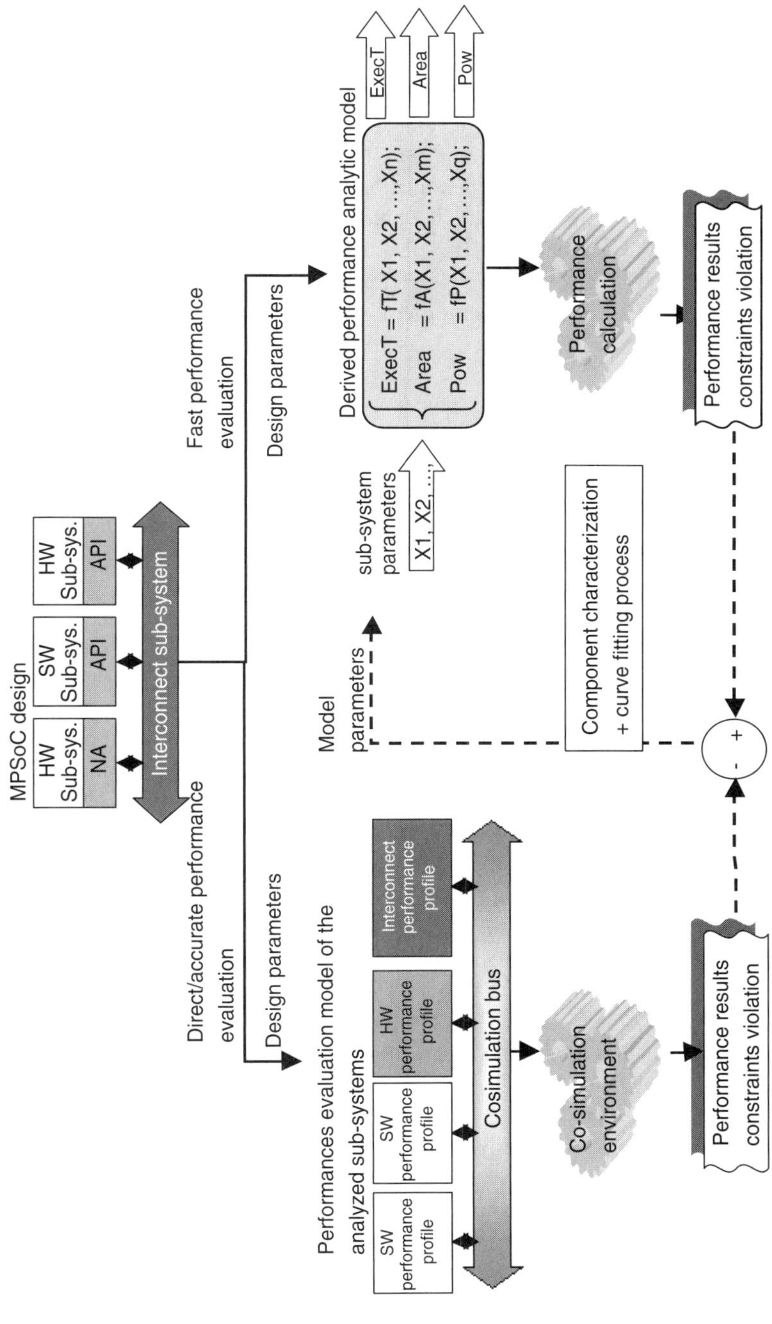

FIGURE 6.3 Global heterogeneous MPSoC evaluation approach.

The proposed strategy is based on the composition of different evaluation models, for different MPSoC subsystems. It combines simulation and analytical models for fast and accurate evaluation of novel MPSoC designs. The further objective is to develop a generic framework for design space exploration and optimization, where different simulation based or analytical evaluation methods could be applied to different subsystems and at different levels of abstraction.

6.4 Conclusion

MPSoC is an emerging community trying to integrate multiple subsystems on a single chip and consequently requiring new methods for performance evaluation [48]. Therefore, the aim of this study was to explore different methodologies for the different subsystems that may compose the MPSoC.

We defined a general taxonomy to handle the heterogeneity and diversity of performance-evaluation solutions. This taxonomy introduced the attributes of an evaluation tool: abstraction levels, modeling techniques, measured metrics, and technology parameters. Finally, we proposed an evaluation framework based on the composition of different methods in which simulation and analytical methods could be combined in an efficient manner, to guide the design space exploration and optimization.

References

[1] G. de Micheli, R. Ernst, and W. Wolf, *Readings in Hardware/Software Co-Design*, Morgan Kaufmann, San Francisco, CA, 1st ed., June 1, 2001, ISBN 1558607021.

[2] S. Dey and S. Bommu, Performance analysis of a system of communicating processes, *International Conference on Computer-Aided Design (ICCAD 1997)*, ACM and IEEE Computer Society, San Jose, CA, 1997, pp. 590–597.

[3] V. Agarwal, M.S. Hrishikesh, S.W. Keckler, and D. Burger, Clock rate versus IPC: The end of the road for conventional microarchitectures, *Proceedings of the 27th International Symposium on Computer Architecture*, Vancouver, Canada, June 2000, pp. 248–259.

[4] H. Yoshida, K. De, and V. Boppana, Accurate pre-layout estimation of standard cell characteristics, *Proceedings of ACM/IEEE Design Automation Conference (DAC)*, San Diego, California, United States, June 2004, pp. 208–211.

[5] C. Brandolese, W. Fornaciari, and F. Salice, An area estimation methodology for FPGA based designs at SystemC-level, *Proceedings of ACM/IEEE Design Automation Conference (DAC)*, San Diego, California, USA, June 2004, pp. 129–132.

[6] A. D. Patterson and L. J. Hennessy, *Computer Organization and Design, the Hardware/SW Interface*, 2nd ed., Morgan-Kaufmann, San Francisco, CA, 1998, ISBN 155860 -491-X.

[7] D. Ofelt and J.L. Hennessy, Efficient performance prediction for modern microprocessors, *SIGMETRICS*, 2000, pp. 229–239.

[8] B. Selic, An efficient object-oriented variation of the statecharts formalism for distributed real-time systems, CHDL 1993, *IFIP Conference on Hardware Description Languages and their Applications*, Ottawa, Canada, 1993, pp. 28–28.

[9] B. Selic and J. Rumbaugh, *Using UML for Modeling Complex Real-Time Systems*, Whitepaper, Rational Software Corp., 1998, http://www.rational.com/media/whitepapers/umlrt.pdf

[10] J. Walrath and R. Vemuri, A performance modeling and analysis environment for reconfigurable computers, IPPS/SPDP Workshops, 1998, pp. 19–24.

[11] B. Spitznagel and D. Garlan, Architecture-based performance analysis, *Proceedings of the 1998 Conference on Software Engineering and Knowledge Engineering*, San Francisco, California, 1998.

[12] P. King and R. Pooley, Derivation of petri net performance models from UML specifications of communications SW, *Proceedings of Computer Performance Evaluation Modelling Techniques and Tools: 11th International Conference, TOOLS 2000*, Schaumburg, IL, 2000.

[13] F. Balarin, STARS of MPEG decoder: a case study in worst-case analysis of discrete event systems, *Proceedings of the International Workshop on HW/SW Codesign*, Copenhagen, Denmark, April 2001, pp. 104–108.

[14] T. Schuele and K. Schneider, Abstraction of assembler programs for symbolic worst case execution time analysis, *Proceedings of ACM/IEEE Design Automation Conference (DAC)*, San Diego, California, USA, June 2004, pp. 107–112.

[15] C Lu, J.A. Stankovic, T.F. Abdelzaher, G. Tao, S.H. Son, and M. Marley, Performance specifications and metrics for adaptive real-time systems, *IEEE Real-Time Systems Symposium (RTSS 2000)*, Orlando, FL, 2000.

[16] M. Lajolo, M. Lazarescu, and A. Sangiovanni-Vincentelli, A compilation-based SW estimation scheme for hardware/SW cosimulation, *Proceedings of the 7th IEEE International Workshop on Hardware/SW Codesign*, Rome, Italy, 3–5, 1999, pp. 85–89.

[17] E.J. Weyuker and A. Avritzer, A metric for predicting the performance of an application under a growing workload, *IBM Syst. J.*, 41, 45–54, 2002.

[18] V.D. Agrawal and S.T. Chakradhar, Performance estimation in a massively parallel system, *SC*, 1990, pp. 306–313.

[19] E.M. Eskenazi, A.V. Fioukov, and D.K. Hammer, Performance prediction for industrial software with the APPEAR method, *Proceedings of STW PROGRESS Workshop*, Utrecht, The Netherlands, October 2003.

[20] K. Suzuki, and A.L. Sangiovanni-Vincentelli, Efficient software performance estimation methods for hardware/software codesign, *Proceedings ACM/IEEE Design Automation Conference (DAC)*, Los Vegas, Nevada, United States, 1996, ISBN:0-89791-779-0, pp.605–610.

[21] Y. Liu, I. Gorton, A. Liu, N. Jiang, and S.P. Chen, Designing a test suite for empirically-based middleware performance prediction, *The Proceedings of TOOLS Pacific 2002*, Sydney, Australia, 2002.

[22] A. Muttreja, A. Raghunathan, S. Ravi, and N.K. Jha, Automated energy/performance macromodeling of embedded software, *Proceedings ACM/IEEE Design Automation Conference (DAC)*, San Diego, CA, USA, June 2004, ISBN:1-58113-828-8, pp. 99–102.

[23] K. Lahiri, A. Raghunathan, and S. Dey, Fast performance analysis of bus-based system-on-chip communication architectures, *Proceedings ACM/IEEE Design Automation Conference (DAC)*, San Jose, CA, United States, June 1999, ISBN:0-7803-5862-5, pp. 566–572.

[24] M. Lajolo, A. Raghunathan, S. Dey, L. Lavagno, and A. Sangiovanni-Vincentelli, A case study on modeling shared memory access effects during performance analysis of HW/SW systems, *Proceedings of the 6th IEEE International Workshop on Hardware/SW Codesign*, Seattle, WA, 15–18, 1998, pp. 117–121.

[25] J.A. Rowson and A.L. Sangiovanni-Vincentelli, Interface-based design, *Proceedings of the 34th Conference on Design Automation*, Anaheim Convention Center, ACM Press, Anaheim, CA, ISBN 0-89791-920-3, 9–13, 1997, pp. 178–183.

[26] K. Hines and G. Borriello, Optimizing communication in embedded system co-simulation, *Proceedings of the 5th International Workshop on Hardware/Software Co-Design*, Braunschweig, Germany, March 1997, ISBN:0-8186-7895-X, p. 121.

[27] S. G. Pestana, E. Rijpkema, A. Radulescu, K.G.W. Goossens, and O.P. Gangwal, Cost-performance trade-offs in networks on chip: a simulation-based approach, *DATE*, 2004, 764–769.

[28] F. Poletti, D. Bertozzi, L. Benini, and A. Bogliolo, Performance analysis of arbitration policies for SoC communication architectures, *Kluwer J. Des. Autom. Embed. Syst.*, 8, 189–210, 2003.

[29] L. Benini and G.D. Micheli, Networks on chips: a new SoC paradigm, *IEEE Comp.*, 35, 70–78, 2002.

[30] E. Bolotin, I. Cidon, R. Ginosar, and A. Kolodny, QNoC: QoS architecture and design process for network on chip, *J. Syst. Archit.*, 49, 2003. Special issue on Networks on Chip.

[31] K. Chen, S. Malik, and D.I. August, Retargetable static timing analysis for embedded SW, *International Symposium on System Synthesis ISSS*, 2001, 39–44.

[32] A. Hergenhan and W. Rosenstiel, Static timing analysis of embedded SW on advanced processor architectures, *Proceedings of Design, Automation and Test in Europe*, Paris, 2000, pp. 552–559.

[33] V. Tiwari, S. Malik, and A. Wolfe, Power analysis of embedded SW: a first step towards SW power minimization, *IEEE T. VLSI Syst.*, 2, 437–445, 1994.

[34] P.E. McKenney, Practical performance estimation on shared-memory multiprocessors, *Parall. Distr. Comput. Syst.*, 1999, pp. 125–134.

[35] M.K. Nethi and J.H. Aylor, Mixed level modelling and simulation of large scale HW/SW systems, *High Performance Scientific and Engineering computing*: Hardware/Software Support, Kluwer Academic Publishers, Norwell, MA, USA, 2004, ISBN:1-4020-7580-4. pp. 157–166.

[36] The MPI Standard, http://www-unix.mcs.anl.gov/mpi/standard.html.

[37] V. Gupta, P. Sharma, M. Balakrishnan, and S. Malik, Processor evaluation in an embedded systems design environment, *13th International Conference on VLSI Design (VLSI-2000)*, Calcutta, India, 2000, pp. 98–103.

[38] A. Maxiaguine, S. Künzli, S. Chakraborty, and L. Thiele, Rate analysis for streaming applications with on-chip buffer constraints, *ASP-DAC*, Yokohama, Japan, 2004.

[39] R. Marculescu, A. Nandi, L. Lavagno, and A. Sangiovanni-Vincentelli, System-level power/performance analysis of portable multimedia systems communicating over wireless channels, *Proceedings of IEEE/ACM International Conference on Computer Aided Design*, San Jose, CA, 2001.

[40] Y. Li and W. Wolf, A task-level hierarchical memory model for system synthesis of multiprocessors, *Proceedings of the ACM/IEEE Design Automation Conference (DAC)*, Anaheim, CA, United States, June 1997, pp. 153–156.

[41] I. Bacivarov, A. Bouchhima, S. Yoo, and A.A. Jerraya, ChronoSym — a new approach for fast and accurate SoC cosimulation, *Int. J. Embed. Syst.*, Interscience Publishers, ISSN (Print): 1741-1068, in press.

[42] M. Loghi, M. Poncino, and L. Benini, Cycle-accurate power analysis for multiprocessor systems-on-a-chip, *ACM Great Lakes Symposium on VLSI*, 2004, 410–406.

[43] M. Loghi, F. Angiolini, D. Bertozzi, L. Benini, and R. Zafalon, Analyzing on-chip communication in a MPSoC environment, *Proceedings of the Design, Automation and Test in Europe (DATE)*, Vol. 2, Paris, France, February 2004, pp. 752–757.

[44] M. Lajolo, A. Raghunathan, S. Dey, and L. Lavagno, Efficient power co-estimation techniques for systems-on-chip design, *Proceedings of Design Automation and Test in Europe*, Paris, 2000.

[45] R. Henia, A. Hamann, M. Jersak, R. Racu, K. Richter, and R. Ernst, System level performance analysis — the SymTA/S approach, *IEE Proceedings Comput. Dig. Tech.*, 152, 148–166, 2005.

[46] W. Cesario, A. Baghdadi, L. Gauthier, D. Lyonnard, G. Nicolescu, Y. Paviot, S. Yoo, A.A. Jerraya, and M. Diaz-Nava, Component-based design approach for multicore SoCs, *Proceedings of the ACM/IEEE Design Automation Conference (DAC)*, New Orleans, LA, USA, June 2002, ISBN~ISBN:0738-100X, 1-58113-461-4, pp. 789–794.

[47] A. Baghdadi, D. Lyonnard, N-E. Zergainoh, and A.A. Jerraya, An efficient architecture model for systematic design of application-specific multiprocessor SoC, *Proceedings of the Conference on Design, Automation and Test in Europe (DATE)*, Munich, Germany, March 2001, ISBN:0-7695-0993-2, pp. 55–63.

[48] *4th International Seminar on Application-Specific Multi-Processor SoC Proceedings*, 2004, Saint-Maximin la Sainte Baume, France, available at http://tima.imag.fr/mpsoc/2004/index.html.

7

System-Level Power Management

Naehyuck Chang
Seoul National University
Seoul, South Korea

Enrico Macii
Politecnico di Torino
Torino, Italy

Massimo Poncino
Politecnico di Torino
Torino, Italy

Vivek Tiwari
Intel Corp.
Santa Clara, California

7.1 Introduction

Power consumption can be drastically reduced if it is considered from the very early stages of the design flow. Power-aware system design has thus become one of the most important areas of investigation in the recent past, although only a few techniques that have been proposed for addressing the problem have already undergone the automation process. One of the approaches that has received a lot of attention, both from the conceptual and the implementation sides is certainly the so-called dynamic power management (DPM). The idea behind this technique, which is very broad and thus may come in very many different flavors, is that of selectively stopping or under-utilizing for some time the system resources that for some periods are not executing useful computation or that do not have to provide results at maximum speed. The landscape of DPM techniques is wide, and exhaustively surveying it is a hard task, which goes beyond the scope of this handbook (the interested reader may refer to the excellent literature on the subject, for instance [1–4]). However, some of the solutions that have been proposed so far have shown to be particularly effective, and are thus finding their way in commercial products. This chapter will review such techniques in some detail.

In general terms, a system is a collection of components whose combined operations provide a service to a user. In the specific case of electronic systems, the components are processors, digital signal processors (DSPs), memories, buses, and macro-cells. Power efficiency in an electronic system can be achieved by optimizing: (1) the architecture and the implementation of the components, (2) the communication between the components, and (3) the usage of the components.

In this chapter, we will restrict our attention to point (3), that is, to the issues of reducing the power consumed by an electronic system by means of properly managing its resources during the execution of the tasks the system is designed for.

The underlying assumption for all the solutions we will discuss is that the activity of the system components is event-driven; for example, the activity of display servers, communication interfaces, and user-interface functions is triggered by external events and it is often interleaved with long periods of quiescence. An intuitive way of reducing the average power dissipated by the whole system consists of shutting down (or reducing the performance of) the components during their periods of inactivity (or under-utilization). In other words, one can adopt a dynamic power management (DPM) policy that dictates how and when the various components should be powered (or slowed) down in order to minimize the total system power budget under some performance/throughput constraints.

A component of an event-driven system can be modeled through a finite state machine that, in the simplest case, has two states: *active* and *idle*. When the component is idle, it is desirable to shut it down by lowering its power supply or by turning off its clock; in this way, its power dissipation can be drastically reduced. Modern components are able to operate in a much wider range of modes, making the issue of modeling the power behavior of an electronic system a much more complex task. We will discuss a modeling approach based on the concept of power and energy state machine (ESM) at the beginning of the next section.

The simplest dynamic power management policies that have been devised are time-out-based: a component is put in its power-down mode only T time units after its finite state machine model has entered the idle state. This is because it is assumed that there will be a very high chance for the component to be idle for a much longer time if it has been in the idle state for at least T time units.

Time-out policies can be inefficient for three reasons: first, the assumption that if the component is idle for more than T time units, then it will remain idle for much longer, may not be true in many cases. Second, whenever the component enters the *idle* state, it stays powered for at least T time units, wasting a considerable amount of power in that period. Third, speed and power degradations due to shut downs performed at inappropriate times are not taken into account; in fact, it should be kept in mind that the transition from power-down to fully functional mode has an overhead: it takes some time to bring the system up to speed, and it may also take more power than the average, steady-state power.

To overcome the limitations of time-out-based policies, more complex strategies have been developed. Predictive and stochastic techniques are at the basis of effective policies; they rely upon complex mathematical models and they all try to exploit the past history of the active and idle intervals to predict the length of the next idle period, and thus decide whether it may be convenient to turn off or power-down a system component.

Beside component idleness, also component under-utilization can be successfully used to reduce power consumption by a fair amount. For example, if the time a processing unit requires to complete a given task is shorter than the actual performance constraint, execution time can be increased by making the processor run slower; this can be achieved by either reducing the clock frequency, or by lowering the supply voltage, or both, provided that the hardware and the software enable this kind of operation. In all cases, substantial power savings are guaranteed basically with no performance penalty. Policies for dynamic clock and voltage control are of increasing popularity in electronic products that feature dynamic power management capabilities.

The core of the chapter focuses on DPM policies, including those targeting dynamic voltage scaling (DVS). Some attention, in this context, will be devoted to battery-driven DPM and, more in general, to the problem of optimizing the usage of batteries, a key issue for portable systems. Finally, the software perspective on the DPM problem, which entails the availability of some capabilities for software-level power estimation, is briefly touched upon in the last section of the chapter.

7.2 Dynamic Power Management

7.2.1 Power Modeling for DPM: Power and Energy State Machines

The formalism of finite state machines can be extended to model power consumption of an electronic system. A power state machine (PSM) illustrates power consumption variations as the state of the system changes. PSMs associate the workload values with a set of resources relevant to its power behavior [5]. At any given point of time, each resource is in a specific power state, which consumes a discrete amount of power.

Figure 7.1 shows the example of the PSM for the Intel StrongARM SA-1100 processor. Each state denotes the power value or performance of the system in a specific state. The transition of the state machine often denotes transition overhead or transition probability. Once we come up with a PSM of all the components in a system, we can easily track the current power values of all resources as the state of the system changes. This approach enables us to have the average or peak power dissipation by looking at the power state transition over time for a given set of environmental conditions. The power consumption in each state may be a designer's estimation, a value obtained from simulation, or a manufacturer's specification.

The PSM formalism can be enhanced to account for the power cost (in addition to the time cost) due to speed changes, such as clock frequency changes, if a device consumes a nontrivial amount of dynamic power. In fact, dynamic power in a state depends on frequency, and it increases as the clock frequency of the device decreases.

Energy state machines (ESMs) separately denote the dynamic energy and leakage power, while PSMs do not distinguish them [6]. To eliminate time dependency, the dynamic portion is represented by energy, while the leakage portion is represented by power. Figure 7.2 shows the concept of ESM. Dynamic energy consumption, ξ_i, is associated with a transition, and leakage power consumption, ϕ_i , is associated with a state. Each state change clearly requires a different amount of dynamic energy. If we slow down the clock frequency, only the tenure time of the states becomes longer.

The ESM represents the actual behavior of a system and its energy consumption. However, measuring leakage power and dynamic energy consumption separately is far from trivial, without a very elaborate power estimation framework. Special measurement and analysis tools that separately handle dynamic energy and leakage power in system level are often required to annotate the ESM [7,8].

7.2.2 Requirements and Implementation of Dynamic Power Management

In addition to models for power-managed systems, as those discussed in the previous section, several other issues must be considered when implementing DPM in practice. These include the choice of an implementation style for the power manager (PM) that must guarantee accuracy in measuring inter-arrival times and service times for the system components, flexibility in monitoring multiple types of components, low perturbation, and marginal impact on component usage. Also key is the choice of the style for monitoring component activity; options here are off-line (traces of events are dumped for later analysis) and online (traces of events are analyzed while the system is running, and statistics related to component usage are constantly updated) monitoring. Finally, of utmost importance is the definition of an appropriate power management policy, which is essential for achieving good power/performance

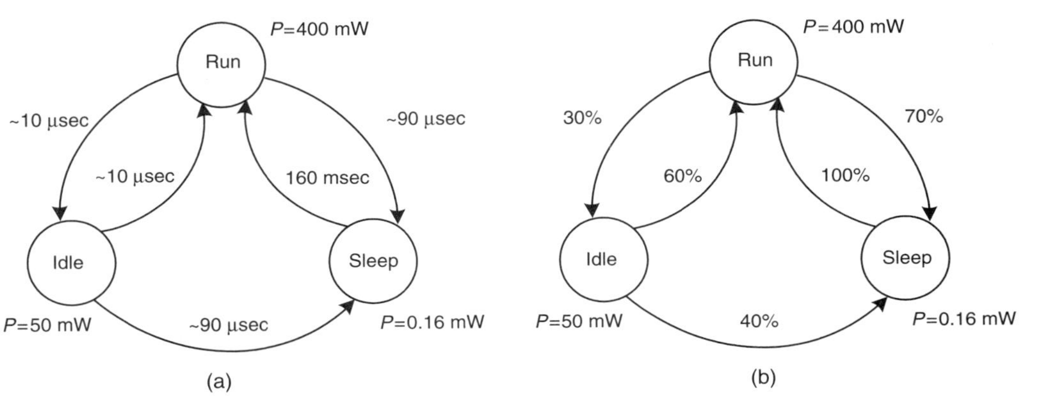

FIGURE 7.1 Power state machine for StrongARM SA-1100: (a) with transition overhead; (b) with transition probability.

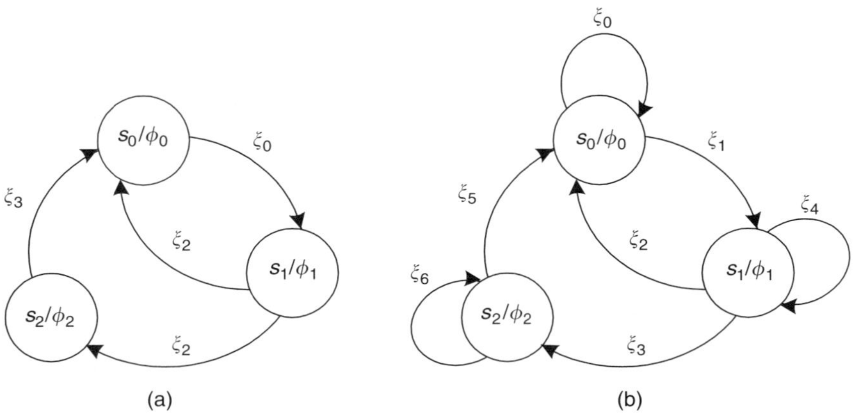

FIGURE 7.2 Energy state machine: (a) asynchronous ESM; (b)synchronous ESM.

trade-offs. The remainder of this section is devoted to the discussion of various options for DPM policies, including those regarding DVS.

7.2.3 Dynamic Power Management Policies

Dynamic power management is the process of dynamically reconfiguring the operating mode of the resources of a system. Its goal is to activate only a minimum number of components or a minimum load on such components, while providing the requested service and performance levels.

DPM encompasses a set of policies that achieve energy-efficient computation by selectively turning off or reducing the performance of system components when they are idle or partially unexploited. The fundamental premise for the applicability of DPM is that systems and their components experience nonuniform workloads during operation time. Such an assumption is generally valid for most battery-operated systems. The second assumption of DPM is that it is possible to predict, with a certain degree of confidence, the fluctuations of workloads. Obviously, workload observation and prediction should not consume significant energy in order to minimize the overhead.

A successful implementation of DPM relies on two key assumptions: (1) the availability of power manageable components that support multiple operational states, which express a trade-off between power and performance; and (2) the existence of a *power manager* that drives component transitions with the correct protocol and implements a power management policy (an algorithm that decides when to shut down idle components and to wake up sleeping components). This section aims at covering and relating different approaches to system-level DPM. We will highlight benefits and pitfalls of different power management policies. We classify power management approaches into two major classes: *predictive* and *stochastic* control techniques. Within each class, we introduce approaches being applied to system design and described in the literature.

7.2.3.1 Predictive Policies

The rationale in all the predictive policies is that of exploiting the correlation that may exist between the past history of the workload and its near future in order to make reliable predictions about future events. For the purpose of DPM, we are interested in *predicting idle periods,* which are long enough to go to the low-power state. This can be expressed as $T_{\text{idle}} > T_{\text{breakeven}}$, where T_{idle} and $T_{\text{breakeven}}$ are the idle time and the transition overhead to and from the idle mode, respectively. Good predictors should minimize mispredictions. We define overprediction (underprediction) as a predicted idle period longer (shorter) than the actual one. Overpredictions result in a performance penalty, while underpredictions imply a waste of power without incurring a performance penalty.

The quality of a predictor is measured by two figures: *safety* and *efficiency*. Safety is the complement of the risk of making overpredictions, and efficiency is the complement of the risk of making underpredictions. We call a predictor with maximum safety and efficiency an *ideal predictor*. Unfortunately, predictors of practical interest are neither safe nor efficient, thus resulting in suboptimum controls.

There are a lot of power management policies with predictive behavior, the simplest one being the *fixed time-out policy*. The fundamental assumption in the fixed time-out policy is that the probability of T_{idle} being longer than $T_{breakeven} + T_{timeout}$, given that $T_{idle} > T_{timeout}$, is close to 1:

$$P(T_{idle} > T_{breakeven} + T_{timeout} \mid T_{idle} \geq T_{timeout}) \approx 1 \tag{7.1}$$

Thus, the critical design parameter is the choice of the time-out value, $T_{timeout}$. The time-out policy has two main advantages: first, it is general; second, its safety can be improved by simply increasing the time-out value. Unfortunately, large timeout values compromise the efficiency, because of the trade-off between safety and efficiency. In addition, as mentioned in the introduction, timeout policies have two more disadvantages. First, they tend to waste a sizeable amount of power while waiting for the timeout to expire. Second, they suffer from a performance penalty upon wake-up. These two disadvantages can be addressed by using a predictive shut-down or wake-up policy, respectively.

The *predictive shut-down policy* forces the PM to make a decision about whether or not to transit the system to the low-power state as soon as a new idle period starts. The decision is based on the observation of past idle/busy periods, and it eliminates the waste of power due to the waiting for the time-out to expire.

Two kinds of predictive shut-down policies were proposed in [9]. In the first policy, a nonlinear regression equation is obtained from the previous "on" and "off" time, and the next "turn-on" time is estimated. If the predicted idle period (T_{pred}) is longer than $T_{breakeven}$, the system is immediately shut down as soon as it becomes idle. This policy, however, has two disadvantages. First, there is no automatic way to decide the type of regression equation. Second, it requires one to perform on-line data collection and analysis for constructing and fitting the regression model.

In the second predictive shut-down policy, the idle period is predicted based on a threshold. The duration of the busy period preceding the current idle period is observed. If the previous busy period is longer than the threshold, the current idle period is assumed to be longer than $T_{breakeven}$, and the system is shut down. The rationale of this policy is that short busy periods are often followed by long idle periods. Similar to the time-out policy, the critical design decision parameter is the choice of the threshold value $T_{threshold}$.

The predictive wake-up policy is proposed in [10], and it addresses the second limitation of the fixed time-out policy, namely the performance penalty that is always paid upon wake-up. To reduce this cost, the power manager performs predictive wake-up when the predicted idle time expires, even if no new request has arrived. This policy may increase power consumption if the idle time has been under-predicted, but it decreases the response time of the first incoming request after an idle period. As usual, in this context, power is traded for performance.

7.2.3.1.1 Adaptive Predictive Policies.

The aforementioned static predictive policies are all ineffective when the workload is either unknown *a priori* or it is nonstationary, since the optimality of DPM strategies depends on the workload statistics. Several adaptive techniques have been proposed to deal with nonstationary workloads.

One option is to maintain a set of time-out values, each of which is associated with an index indicating how successful it would have been [11]. The policy chooses, at each idle time, the time-out value that would have performed best among the set of available ones. In alternative, a weighted list of time-out values is kept, where the weight is determined by relative performance of past requests with respect to the optimal strategy [12], and the actual time-out is calculated as a weighted average of all the time-out values in the list [13]. The time-out value can be increased when it results in too many shut-downs, and it can otherwise be decreased when more shut-downs are desirable.

Another adaptive shut-down policy has been proposed in [10]. The idle time prediction is calculated as a weighted sum of the last idle period and the last prediction (exponential average):

$$T^{n}_{\text{pred}} = \alpha T^{\text{n-1}}_{\text{idle}} + (1 - \alpha) T^{\text{n-1}}_{\text{pred}} \tag{7.2}$$

The last underprediction is mitigated by reevaluating T_{pred} periodically, if the system is idle and it has not been shut down. The last overprediction is reduced by imposing a saturation constraint, $C_{\max}$, on predictions:

$$T^{\text{n}}_{\text{pred}} \leq C_{\max}\, T^{\text{n}-1}_{\text{pred}} \tag{7.3}$$

7.2.3.2 Stochastic Policies

Although all the predictive techniques address workload uncertainty, they assume deterministic response time and transition time of a system. However, the system model for policy optimization is very abstract, and abstraction introduces uncertainty. Hence, it is safer and more general to assume a stochastic model for the system as well as the workload. Moreover, real-life systems support multiple power states, which cannot be handled by simple predictive techniques. In fact, the latter are based on heuristic algorithms, and their optimality can be gauged only through comparative simulations. Finally, predictive techniques are only geared toward the power minimization, and cannot finely control performance penalty.

Stochastic control techniques formulate policy optimization as an optimization problem under the uncertainty of the system as well as of the workload. They assume that both the system and the workload can be modeled as Markov chains, and offer significant improvement over previous power management techniques in terms of theoretical foundations and of robustness of the system model. Using stochastic techniques allows one to (1) model the uncertainty in the system power consumption and the state-transition time; (2) model complex systems with multiple power states; and (3) compute globally optimum power management policies, which minimize the energy consumption under performance constraints or maximize the performance under power constraints.

A typical Markov model of a system consists of the following entities [14]:

- A service requester (SR) that models the arrival of service requests.
- A service provider (SP) that models the operation states of the system.
- A PM that observes the state of the system and the workload, makes a decision, and issues a command to control the next state of the system.
- Some cost metrics that associate power and performance values to each command.

7.2.3.2.1 Static Stochastic Policies.

Static stochastic control techniques [14] perform policy optimization based on the fixed Markov chains of the SR and of the SP. Finding a globally power-optimal policy that meets given performance constraints can be formulated as a linear program (LP), which can be solved in polynomial time in the number of variables. Thus, policy optimization for Markov processes is exact and computationally efficient. However, several important points must be understood: (1) The performance and power obtained by a policy are expected values, and there is no guarantee of the optimality for a specific workload instance. (2) We cannot assume that we always know the SR model beforehand. (3) The Markov model for the SR or SP is just an approximation of a much more complex stochastic process, and thus the power-optimal policy is also just an approximate solution.

The Markov model in [14] assumes a finite set of states, a finite set of commands, and discrete time. Hence, this approach has some shortcomings: (1) The discrete-time Markov model limits its applicability since the power-managed system should be modeled in the discrete-time domain. (2) The PM needs to send control signals to the components in every time-slice, which results in heavy signal traffic and heavy load on the system resources (therefore more power dissipation). Continuous-time Markov models [15] overcome these shortcomings by introducing the following characteristics: (1) A system model based on continuous-time Markov decision process is closer to the scenarios encountered in practice.

(2) The resulting power management policy is asynchronous, which is more suitable for implementation as a part of the operating system (OS).

*7.2.3.2.2 **Adaptive Stochastic Policies.***
An adaptive extension of the stochastic control techniques is proposed to overcome a limitation of the static approach. As it is not possible to have the complete knowledge of the system (SP) and its workload (SR) *a priori*, even though it is possible to construct a model for the SP once and for all, the system workload is generally much harder to characterize in advance. Furthermore, workloads are often nonstationary. Adaptation consists of three main phases: policy precharacterization, parameter learning, and policy interpolation [16]. Policy precharacterization constructs an n-dimensional table addressed by n parameters for the Markov model of the workload. The table contains the optimal policy for the system under different workloads. Parameter learning is performed online during system operation by short-term averaging techniques. The parameter values are then used for addressing the table and for obtaining the power management policy. If the estimated parameter values are not in accordance with the exact values, which are used for addressing the table, interpolation may obtain an appropriate policy as a combination of the policies in the table. Experimental results reported in [16] indicate that the adaptive technique performs nearly as well as the ideal policy computed off-line, assuming perfect knowledge of the workload parameters over time.

7.2.4 Dynamic Voltage Scaling

Supply voltage scaling is one of the most effective techniques in power minimization of CMOS circuits because the dynamic energy consumption of CMOS devices is quadratically related to the supply voltage. Unfortunately, the supply voltage has a strong relation to the circuit delay: the lower the supply voltage, the larger the circuit delay and the smaller the maximum operating frequency, which may degrade the performance of the target system.

Dynamic voltage scaling is the power management technique that controls the supply voltage according to the current workload at run-time to minimize the energy consumption without having an adverse effect on system performance. Dynamic voltage scaling can be viewed as a variant of DPM in which DPM is applied not just to *idle* components but also to those resources that are *noncritical* in terms of performance, running the resource at different power/speed points. In other words, DVS introduces the notion of *multiple active states,* besides multiple idle states exploited by traditional DPM. Moreover, since in DVS power/speed trade-off points are defined by different supply voltage levels, DVS is traditionally applied to CPUs, rather than to other components, and it is thus exploited at the task granularity.

The DVS technique typically utilizes the slack time of tasks to avoid performance degradation. For example, when the current task has less remaining time than the expected execution time at the maximum frequency, the voltage scheduler lowers the supply voltage, and extends the execution time of this task up to the arrival time of the next task. To apply DVS to real systems, hardware support for voltage scaling is required [17], and the software support that monitors the task execution and gives the voltage control command to the DVS hardware is needed as well.

The issues related to the implementation of DVS have been investigated in many studies. A majority of them developed an energy-efficient scheduling method for a system with real-time requirements. Each work suggests different run-time slack estimation and distribution schemes [18] that are trying to achieve the theoretical lower bound of energy consumption that is calculated statically for a given workload [19].

7.2.4.1 Task Scheduling Schemes for Dynamic Voltage Scaling

The various voltage scheduling methods in DVS-enabled systems differ by when to adjust frequency and voltage, how to estimate slacks, and how to distribute these to waiting tasks. Several DVS scheduling schemes are summarized and evaluated in [18]. In this section, DVS scheduling algorithms for hard-real time systems are classified by the *granularity* of the voltage adjustments. Also, the slack estimation methods for some DVS scheduling algorithms are introduced.

7.2.4.1.1 *Voltage Scheduling Granularity.*

DVS scheduling schemes are classified by voltage scheduling granularity and fall into two categories: inter-task DVS algorithms and intra-task DVS algorithms. In the intra-task DVS algorithms, a task is partitioned into multiple pieces such as time slots [20] or basic blocks [21], and a frequency and consequent voltage assignment is applied during the task execution. The actual execution time variation is estimated at the boundary of each time slot or each basic block and used for the input of adaptive operation frequency and voltage assignment.

In the inter-task DVS algorithms, voltage assignment is executed at the task's boundaries. After a task is completed, a new frequency and consequent voltage setting are applied by static or dynamic slack time estimation. The slack time estimation method has to be aggressive for the system energy reduction. At the same time, it must be conservative so that every task is successfully scheduled within its deadline. These two rules are conflicting with each other, making it difficult to develop an effective slack estimation algorithm.

7.2.4.1.2 *Slack Time Estimation.*

DVS techniques for hard-real-time systems enhance the traditional earliest deadline first (EDF) or rate monotonic (RM) scheduling to exploit slack time, which is used to adjust voltage and frequency of voltage-scalable components. Therefore, the primary objective is to estimate the slack time accurately for more energy reduction. Various kinds of static and dynamic slack estimation methods have been proposed to exploit most of the slack time without violating the hard-real-time constraints.

Many approaches for inter-task DVS algorithms deal with static slack estimation methods [22–26]. These methods typically aim at finding the lowest possible operating frequency at which all the tasks meet their deadlines. These methods rely on the worst-case execution time (WCET) to guarantee hard-real-time demands. Therefore, the operating frequency can be lowered to the extent that each task's WCET does not exceed the deadline. The decision of each task's frequency can be done statically because it is a function of WCET and deadline, which are not changed during run-time.

In general, the actual execution time is quite shorter than WCET. Therefore, WCET-based static slack time estimation cannot fully exploit actual slacks. To overcome this limitation, various dynamic slack estimation methods have been proposed.

The cycle-conserving RT-DVS technique utilizes the extra slack time to run other remaining tasks at a lower clock frequency [24]. In this approach, operating frequency is scaled by the CPU utilization factor. The utilization factor is updated when any task is released or completed. When any task is released, the utilization factor is calculated according to the task's WCET. After a task is completed, the utilization factor is updated by using the actual execution time. The operation frequency may be lowered until the next arrival time of that task.

The next release time of a task can be used to calculate the slack budget [23]. This approach maintains two queues: the run queue and the delay queue. The former holds tasks that are waiting by their priority order, while the latter holds tasks that are waiting for next periods, ordered by their release schedule. When the active queue is empty and the required execution time of an active task is less than its allowable time frame, the operation frequency is lowered using that slack time. According to Shin et al. [23], the allowable time frame is defined by

$$\text{min(active\_task.deadline, delay\_queue.head.release\_time)} - \text{current\_time} \tag{7.4}$$

A path-based slack estimation method for intra-task DVS algorithms is presented in [21]. The control flow graph (CFG) is used for slack time estimation during the execution of the given hard-real-time program. Each node of the CFG is a basic block of the target program, and each edge indicates the control dependency between basic blocks. When the thread of execution control branches to the next basic block, the expected execution time is updated. If the expected execution time is smaller than the tasks's WCET, the operating frequency can be lowered.

7.2.4.2 Practical Considerations in Dynamic Voltage Scaling

Many DVS studies, especially those focusing on task scheduling, have assumed a target system: (1) consisting of all voltage-scalable components whose supply voltage can be set to any value within a given range of supply voltage;(2) in which only dynamic energy is considered; and (3) in which the speed settings of the tasks do not affect other components of the system. Although these assumptions simplify the calculation of energy consumption and development of a scheduling scheme, the derived scheduling may not perform well because the assumption does not reflect a realistic setting. In fact, recent advancement in CMOS technology makes leakage power consumption significant. Recent studies have addressed the impact of these nonidealities on DVS.

7.2.4.2.1 Discretely Variable Voltage Levels.
Unlike assumption (1) above, most commercial microprocessors supporting supply voltage scaling (e.g., Intel XScale, Transmeta Crusoe) can select only a small number of predefined voltages as supply voltage. To get a more practical DVS schedule, some scheduling techniques are proposed for discretely variable supply voltage levels instead of continuously variable ones.

An optimal voltage allocation technique for a single task with discontinuously variable voltages is proposed in [27] using integer linear programming. In case only a small number of discrete voltages can be used, they show that the schedule with at most two voltages for each task minimizes the energy consumption under a timing constraint. Another work deals with the static voltage allocation problem for circuits with multiple supply voltages [28]. This scheduling problem for discretely variable voltage levels is extended to get an energy-optimal schedule of multiple tasks for dynamically voltage-variable microprocessors controlled by software [29].

7.2.4.2.2 Leakage-Aware Dynamic Voltage Scaling.
As the supply voltage of CMOS devices becomes lower, the threshold voltage should also be reduced, which results in dramatic increase of the subthreshold leakage current. Therefore, the static power including the leakage current as well as the dynamic power are major contributors to the total power dissipation.

If static energy consumption is considered, the total energy consumption is no longer a monotonically increasing function of the supply voltage. Since the performance degradation due to the reduction of the supply voltage may increase the execution (or wake-up) time, this may result an increase in the static energy consumption. Consequently, if the supply voltage is reduced below a certain limit, the energy consumption becomes larger again. Inspired by this convex energy curve, which is no longer monotonically increasing with the supply voltage, a leakage-aware DVS scheduling scheme is proposed in [30]. It finds the voltage that minimizes the total energy consumption including leakage current, and avoids supply voltage scaling below the limit, even though there is still some slack time available.

7.2.4.2.3 Memory-Aware Dynamic Voltage Scaling.
As the complexity of modern systems increases, other components beside microprocessors, e.g., memory devices, contribute more to system power consumption. Thus, their power consumption must be considered when applying a power management technique. Unfortunately, many offchip components do not allow supply voltage scaling. Since they are controlled by the microprocessor, their active periods may become longer when we slow down the microprocessor using DVS. The delay increase due to lower supply voltage of a microprocessor may increase the power consumption of devices that do not support DVS. In these cases, the energy gain achieved from DVS on the processor must be traded for the energy increase of other devices.

There are some related studies about DVS in systems including devices without scalable supply voltages, especially focusing on memories. In both [31] and [32], it is shown that aggressive reduction of supply voltage of a microprocessor results in an increase of total energy consumption, because the static energy consumption of memory devices becomes larger as the execution time gets longer, and thus chances of actuating some power-down decreases. In [32], an analytic method to obtain an energy-optimal frequency assignment of memories not supporting DVS and CPUs supporting DVS is proposed as a viable improvement in this context.

7.2.4.3 Nonuniform Switching Load Capacitances

Even if multiple tasks are scheduled, most studies characterize these tasks only by the timing constraints. This assumes that all tasks with the same voltage assignment and the same period will consume the same amount of energy irrespective of their operation. This means that a uniform switching load capacitance in the energy consumption equation is assumed for all the tasks.

However, in practice, different tasks may utilize different data-path units that produce different energy consumption even for the tasks with the same period. A scheduling method for multiple tasks with nonuniform switching load capacitances has been proposed in [29]. This approach modifies the algorithm of Yao et al. [19] so as to obtain an energy-optimal schedule for nonuniform switching load capacitances, which helps in better matching the real power behavior of the overall system.

7.3 Battery-Aware Dynamic Power Management

All the dynamic power management techniques described in some detail in the previous section do implicitly assume an ideal power supply. While this simplifying assumption may be considered very reasonable for electronic systems connected to a fixed power supply, it is simply not correct in the case of battery-operated systems.

7.3.1 Battery Properties

Batteries are nonideal charge storage units, as pointed out in any battery handbook [33]. From a designer's standpoint, there are two main nonidealities of real-life battery cells that need to be considered:

- *The capacity of a battery depends on the discharge current.* At higher currents, a battery is less efficient in converting its chemically stored energy into available electrical energy. This fact is shown on the top panel of Figure 7.3, where the capacity of the battery is plotted as a function of the average-current load. We observe that, for increasing load currents, the battery capacity progressively deviates from the nominal value (broken line). Moreover, battery capacity is also affected by the discharge rate: At higher rates, the cell is less efficient at converting its stored energy into available electrical energy. This fact implies that battery lifetime will be negatively correlated to the *variance* of the current load; for a given average current value, a constant load will result in the longest battery lifetime of all load profiles.
- *Batteries have some (limited) recovery capacity when they are discharged at high current loads.* A battery can recover some of its deliverable charge if the periods of discharge are interleaved with rest periods, i.e., periods in which no current is drawn. The bottom panel of Figure 7.3 shows how an intermittent current load (broken line) results in a longer battery lifetime than a constant current load (solid line), for identical discharge rate. The x-axis represents the actual elapsed time of discharge, that is, it does not include the time during which the battery has been disconnected from the load.

Accounting for the aforementioned nonidealities is essential, since it can be shown that power management techniques (both with and without DVS) that neglect these issues may actually result in an increase in energy [34,35].

7.3.2 Battery-Aware Dynamic Power Management

The most intuitive solution consists of incorporating *battery-driven* policies into the DPM framework, either implicitly (i.e., using a battery-driven metric for a conventional policy) [36] or explicitly (i.e., a truly battery-driven policy) [37,38]. A simple example of the latter type could be a policy whose decision rules used to control the system operation state are based on the observation of a battery's output voltage, which is (nonlinearly) related to the state of charge.

More generally, it is possible to directly address the above-mentioned nonidealities to shape the load current profile so as to increase as much as possible the effective capacity of the battery.

The issue of load-dependent capacity can be faced along two dimensions. The dependency on the average current can be tackled by shaping the current profile in such a way that high-current-demanding operations are executed first (i.e., with fully charged battery), and low-current-demanding ones are executed later (i.e., with a reduced-capacity battery) [39].

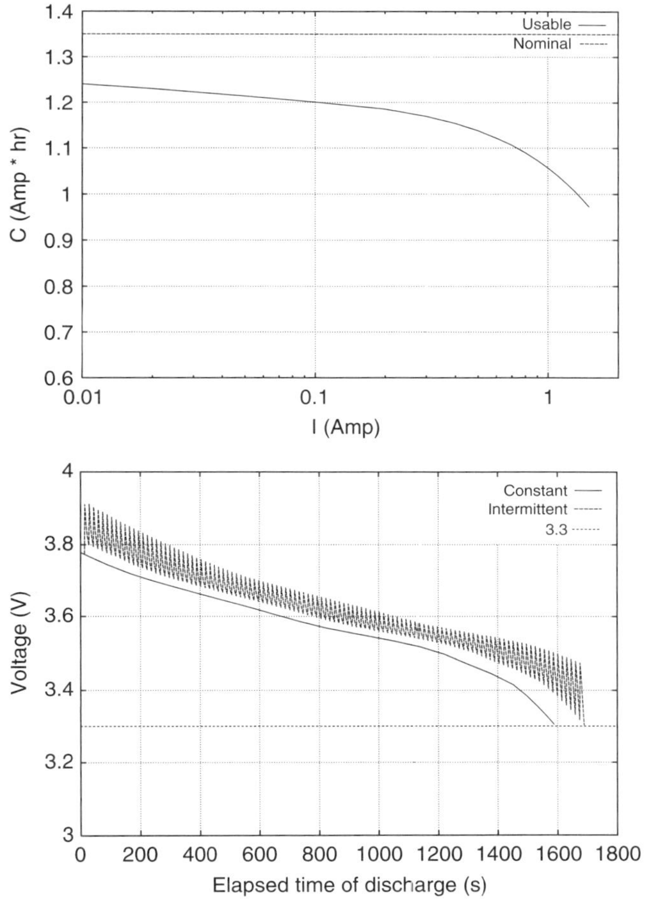

FIGURE 7.3 Capacity variation as a function of load and charge recovery in intermittent discharge.

This principle fits well at the task granularity, where the shape of the profile corresponds to task scheduling. Intuitively, the solution that maximizes battery efficiency and thus its lifetime is the one in which tasks are scheduled in nondecreasing order of their average current demand (Figure 7.4), compatibly with possible deadline or response time constraints [40].

The issue of charge recovery can be taken into account by properly arranging the idle times in the current profile. In particular, idle periods can be *inserted* between the execution of tasks. Notice that this is different from typical current profiles, where active and idle intervals alternate in relatively long bursts (Figure 7.5a). Inserting idle slots between task execution will allow the battery to recover some of the charge so that lifetime can be increased (Figure 7.5b). In the example, it may happen that after execution of T_2 the battery is almost exhausted and execution of T_3 will not complete; conversely, the insertion of an idle period will allow the battery to recover part of the charge so that execution of T_3 becomes possible. Idle time insertion can be combined with the ordering of tasks to exploit both properties and achieve longer lifetimes [41].

7.3.3 Battery-Aware Dynamic Voltage Scaling

The possibility of scaling supply voltages at the task level adds further degrees of freedom in choosing the best shaping of the current loads. Voltage scaling can be in fact viewed as another opportunity for reducing the current demand of a task, at the price of increased execution time. Under a simple, first-order

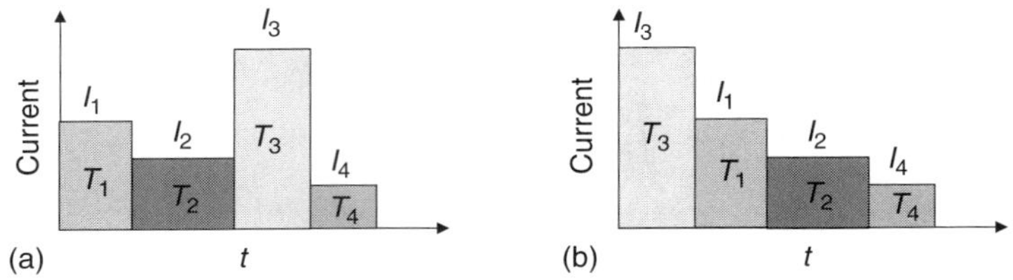

FIGURE 7.4 A set of tasks (a) and its optimal sequencing (b).

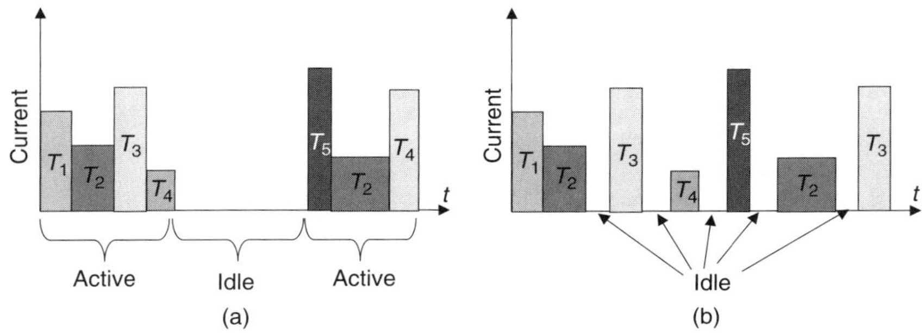

FIGURE 7.5 Active/idle bursts (a) and idle time insertion (b).

approximation, the drawn current I is proportional to V^3, while delay increases proportionally with V. Therefore, the trade-off is between a decrease (increase) in the the discharge current, and an increase (decrease) in the duration of the stress [41].

Battery-aware DVS exploits the above-mentioned dependency of battery capacity on the load, since it can be proved that scaling voltage is always more efficient than inserting idle periods [40]. Therefore, anytime a slack is available, it should be filled by scaling the voltage (Figure 7.6), compatibly with performance constraints. This is equivalent to stating that the impact on lifetime of the rate-dependent behavior of batteries dominates that due to the charge recovery effect.

Solutions proposed in the literature typically start from a nondecreasing current profile and reclaim the available slack from the tail of the schedule by slowing down tasks according to the specified constraints [40,42,43].

7.3.4 Battery Scheduling

In the case of supply systems consisting of multiple batteries, the load-dependent capacity of the batteries has deeper implications, which open additional opportunities for optimization. In fact, since at a given point in time the load is connected to only one battery, the other ones are idle. Dynamic Power Management with multiple batteries amounts thus to the problem of assigning (i.e., scheduling) battery usage over time; hence this problem is often called *battery scheduling*.

The default battery scheduling policy in use in most devices is a nonpreemptive one that sequentially discharges one battery after another, in some order. Because of the above-mentioned nonidealities, this is clearly a suboptimal policy [44].

Similar to task scheduling, battery scheduling can be either independent or dependent from the workload. In the former case, batteries are attached to the load in a round-robin fashion for a fixed amount of time. Unlike task scheduling, the choice of this quantity is dictated by the physical property of batteries. It can be shown, in fact, that the smaller this interval, the higher is the equivalent capacity of the battery sets [45]. This is because by rapidly switching between full load and no load, each battery perceives an

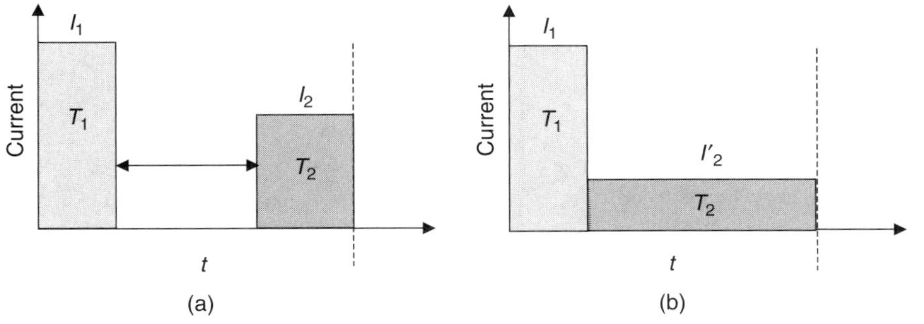

FIGURE 7.6 Idle period insertion (a) vs. voltage scaling (b).

effective averaged discharge current proportional to the fraction of time it is connected to the load. In other words, if a battery is connected to the load current *I* for a fraction $\alpha < 1$ of the switching period, it will perceive a load current αI. This is formally equivalent to connecting the two batteries in parallel, without incurring into the problem of mutually discharging the batteries.

In the latter case, we assign a battery to the load depending on its characteristics. More precisely, one should select which battery to connect to the load based on run-time measurement of the current drawn, in an effort to match a load current to the battery that better responds to it [37,44,45]. A more sophisticated, workload-dependent scheme consists of adapting the round-robin approach to heterogeneous multibattery supplies (i.e., having different nominal capacities and discharge curves). In these cases, the current load should be split nonuniformly over all the cells in the power supply. Therefore, the round-robin policy can be modified so that the time slice has different durations. For instance, in a two-battery system, this is equivalent to connecting the batteries to the load following a square wave with unbalanced duty cycle [46].

7.4 Software-Level Dynamic Power Management

7.4.1 Software Power Analysis

Software constitutes a major component of systems where power is a constraint. Its presence is very visible in a mobile computer, in the form of the system software and application programs running on the main processor. But software also plays an even greater role in general digital applications. An ever-growing fraction of these applications are now being implemented as embedded systems. In these systems the functionality is partitioned between a hardware and a software component.

The software component usually consists of application-specific software running on a dedicated processor. Relating the power consumption in the processor to the instructions that execute on it provides a direct way of analyzing the impact of the processor on the system power consumption. Software impacts system power at various levels. At the highest level, this is determined by the way functionality is partitioned between hardware and software. The choice of the algorithm and other higher level decisions about the design of the software component can affect system power consumption significantly. The design of the system software, the actual application source code, and the process of translation into machine instructions — all of these determine the power cost of the software component.

In order to systematically analyze and quantify this cost, however, it is important to start at the most fundamental level, that is, at the level of the individual instructions executing on the processor. Just as logic gates are the fundamental units of computation in digital hardware circuits, instructions can be thought of as the fundamental unit of software. Accurate modeling and analysis at this level is therefore essential. Instruction level models can then be used to quantify the power costs of the higher constructs of software (application programs, system software, algorithm, etc.).

It would be helpful to define the terms "power" and "energy," as they relate to software. The average power consumed by a processor while running a given program is given by

$$P = IV_{dd}$$

where P is the average power, I the average current, and V_{dd} the supply voltage. Power is also defined as the rate at which energy is consumed. Therefore, the energy consumed by a program is given by

$$E = P\,T$$

where T is the execution time of the program. This, in turn, is given by $T = N\tau$, where N is the number of clock cycles taken by the program and τ the clock period. Energy is thus given by

$$E = I\,V_{dd}\,N\,\tau$$

Note that if the processor supports dynamic voltage and frequency switching, then V_{dd} and τ can vary over the execution of the program. It is then best to consider the periods of code execution with different (V_{dd},τ) combinations as separate components of the power/energy cost. As it can be seen from the above discussion, the ability to obtain an estimate of the current drawn by the processor during the execution of the program is essential for evaluating the power/energy cost of software. These estimates can either be obtained through simulations or through direct measurements.

7.4.1.1 Software Power Estimation through Simulation

The most commonly used method for power analysis of circuits is through specialized power analysis tools that operate on abstract models of the given circuits. These tools can be used for software power evaluation too. A model of the given processor and a suitable power analysis tool are required. The idea is to simulate the execution of the given program on the model. During simulation the power analysis tool estimates the power (current) drawn by the circuit using predefined power estimation formulas, macro-models, heuristics, or algorithms.

However, this method has some drawbacks. It requires the availability of models that capture the internal implementation details of the processor. This is proprietary information, which most software designers do not have access to. Even if the models are available, there is an accuracy vs. efficiency trade-off. The most accurate power analysis tools work at the lower levels of the design — switch- or circuit-level. These tools are slow and impractical for analyzing the total power consumption of a processor as it executes entire programs. More efficient tools work at the higher levels — register transfer or architectural. However, these are limited in the accuracy of their estimates.

7.4.1.2 Measurement-Based Instruction-Level Power Modeling

The above problems can be overcome if the current being drawn by the processor during the execution of a program is physically measured. A practical approach to current measurement as applied to the problem of instruction-level power analysis has been proposed [47]. Using this approach, empirical instruction-level power models were developed for three commercial microprocessors. Other researchers have subsequently applied these concepts to other microprocessors. The basic idea is to assign power costs to individual instructions (or instruction pairs to account for inter-instruction switching effects) and to various inter-instruction effects like pipeline stalls and cache misses. These power costs are obtained through experiments that involve the creation of specific programs and measurement of the current drawn during their execution. These costs are the basic parameters that define the instruction-level power models. These models form the basis of estimating the energy cost of entire programs. For example, for the processors studied in [47], for a given program, the overall energy cost is given by

$$E = \sum_{i} (B_i \cdot N_i) + \sum_{i,j}(O_{i,j} \cdot N_{i,j}) + \sum_{k} E_k$$

The base power cost, B_i, of each instruction, i, weighted by the number of times it will be executed, N_i, is added up to give the base cost of the program. To this the circuit state switching overhead, $O_{i,j}$, for each pair of consecutive instructions, i,j , weighted by the number of times the pair is executed, $N_{i,j}$, is added. The energy contribution, E_k, of the other inter-instruction effects, k (stalls and cache misses) that would occur during the execution of the program, is finally added.

The base costs and overhead values are empirically derived through measurements. The other parameters in the above formula vary from program to program. The execution counts N_i and $N_{i,j}$ depend on the execution path of the program. This is dynamic, run-time information that has to be obtained using software performance analysis techniques. In certain cases, it can be determined statically, but in general it is best obtained from a program profiler. For estimating E_k, the number of times the pipeline stalls and cache misses occur have to be determined. This is again dynamic information that can be statically predicted only in certain cases. In general, it can be obtained from a program profiler and a cache simulator, which are able to track the dynamic nature of program execution.

The processors whose power models have been published using the ideas above, have so far been in-order machines with relatively simple pipelines. Out-of-order superscalar machines present a number of challenges to an instruction oriented modeling approach and provide good opportunities for future research.

7.4.1.3 Idle Time Evaluation

The above discussion is for the case when the processor is active and is constantly executing instructions. However, a processor may not always be performing useful work during program execution. For example, during the execution of a word processing program, the processor may simply be waiting for keyboard input from the user and may go into a low-power state during such idle periods. To account for these low-power periods, the average power cost of a program is thus given by

$$P = P_{\text{active}}\, T_{\text{active}} + P_{\text{idle}}\, T_{\text{idle}}$$

where P_{active} represents the average power consumption when the microprocessor is active, T_{active} the fraction of the time the microprocessor is active, and P_{idle} and T_{idle} the corresponding parameters for when the microprocessor is idle and has been put in a low-power state. T_{active} and T_{idle} need to be determined using dynamic performance analysis techniques. In modern microprocessors, a hierarchy of low-power states is typically available, and the average power and time spent for each state would need to be determined.

7.4.2 Software-Controlled Power Management

For systems in which part of the functionality is implemented in software, it is natural to expect that there is potential for power reduction through modification of the software. Software power analysis (whether achieved through physical current measurements or through simulation of models of the processors) as described in Section 7.4.1, helps in identifying the reasons for the variation in power from one program to another. These differences can then be exploited in order to search for best low-power alternatives for each program.

The information provided by the instruction-level analysis can guide higher-level design decisions like hardware–software partitioning and choice of algorithm. It can also be directly used by automated tools like compilers, code generators, and code schedulers for generating code targeted toward low power. Several ideas in this regard have been published, starting with the work summarized in [47]. Some of these ideas are based on specific architectural features of the subject processors and memory systems. The most important conclusion though is that in modern general-purpose CPUs, software energy and performance track each other, i.e., for a given task, a faster program implementation will also have lower energy. Specifically, it is observed that the difference in average current for instruction sequences that perform the same function is not large enough to compensate for any difference in the number of execution cycles.

Thus, given a function, the least energy implementation for it is the one with the faster running time. The reason for this is that CPU power consumption is dominated by a large common cost factor (power consumption due to clocks, caches, instruction fetch hardware, etc.) that for the most part is independent of instruction functionality and does not vary much from one cycle to the other. This implies that the large body of work devoted to software performance optimization provides direct benefits for energy as well. Power management techniques, such as increased use of clock gating and multiple on-chip voltages, indicate that future CPUs may show greater variation in power consumption from cycle to cycle. However, CPU design and power consumption trends do suggest that the relationship between software energy and power that was observed before will continue to hold. In any case, it is important to realize that software directly impacts energy/power consumption, and thus it should be designed to be efficient with respect to these metrics. A classic example of inefficient software is "busy wait loops." Consider an application such as a spreadsheet that requires frequent user input. During the times when the spreadsheet is recalculating values, high CPU activity is desired in order to complete the recalculation in a short time. In contrast, when the application is waiting for the user to type in values, the CPU should be inactive and in a low-power state. However, a busy wait loop will prevent this from happening, and will keep the CPU in a high-power state. The power wastage is significant. Converting the busy wait loops into an instruction or system call that puts the CPU into a low-power state from which it can be woken up on an I/O interrupt will eliminate this wasted power.

7.4.2.1 OS-Directed Dynamic Power Management

An important trend in modern CPUs is the ability for software to control the operating voltage and frequency of the processor. These different voltage/frequency operating points, which represent varying levels of power consumption, can be switched dynamically under software control. This opens up additional opportunities for energy-efficient software design, since the CPU can be made to run at the lowest-power state that still provides enough performance to meet the task at hand. In practice, the DPM and DVS techniques described in Section 7.2 from a hardware-centric perspective (because they managed multiple power states of *hardware components)* can be also viewed from the software perspective.

In this section, we describe implementation issues of power management techniques. In general, the OS is the best software layer where a DPM policy can be implemented. OS-directed power management (OSPM) has several advantages: (1) The power/performance dynamic control is performed by the software layer that manages computational, storage, and I/O tasks of the system. (2) Power management algorithms are unified in the OS, yielding much better integration between the OS and the hardware, (3) Moving the power management functionality into the OS makes it available on every machine on which the OS is installed. Implementation of OSPM is a hardware/software co-design problem because the hardware resources need to interface with OS-directed software PM, and because both hardware resources and the software applications need to be designed so that they could cooperate with OSPM.

The advanced configuration and power interface (ACPI) specification [48] was developed to establish industry common interfaces enabling a robust OSPM implementation for both devices and entire systems. Currently, it is standardized by Hewlett-Packard, Intel, Microsoft, Phoenix, and Toshiba. It is the key element in OSPM, since it facilitates and accelerates the co-design of OSPM by providing a standard interface to control system components. From a power management perspective, OSPM/ACPI promotes the concept that systems should conserve energy by transitioning unused devices into lower power states, including placing the entire system in a low-power state (sleep state) when possible. ACPI-defined interfaces are intended for wide adoption to encourage hardware and software designers to build ACPI-compatible (and, thus, OSPM-compatible) implementations. Therefore, ACPI promotes the availability of power-manageable components that support multiple operational states. It is important to notice that ACPI specifies neither how to implement hardware devices nor how to realize power management in the OS. No constraints are imposed on implementation styles for hardware and on power management policies. The implementation of ACPI-compliant hardware can leverage any technology or architectural optimization as long as the power-managed device is controllable by the standard interface specified by ACPI. The power management module of the OS can be implemented using any kind of DPM policy.

Experiments were carried out in [49,50] to measure the effectiveness of different DPM policies on ACPI-compliant computers.

7.5 Conclusions

The design of energy-efficient systems goes through the optimization of the architecture of the individual components, the communication between them, and their utilization. Out of these three dimensions, the latter offers a lot of opportunities from the system perspective, as it enables the exploitation of very abstract models of the components and it does not usually require detailed information about their implementation.

In this chapter, we have analyzed the problem of optimizing the usage of components of an electronic system, and we have discussed a number of solutions belonging to the wide class of DPM techniques. DPM aims at dynamically choosing the operating states of the components that best fit the required performance levels, thus reducing the power wasted by idle or under-utilized components.

In its simplest form, DPM entails the decision between keeping a given component active or turning it off, where the "off" state may consist of different power/performance trade-off values. When combined with the possibility of dynamically varying the supply voltage, DPM generalizes to the more general problem of DVS, in which multiple active states are possible. Different policies for actuating DPM/DVS solutions have been studied in the literature, accounting for different aspects of the problem, including, for instance, nonidealities such as the use of a discrete set of voltage levels, the impact of leakage power, and the presence of nonideal supply sources (i.e., batteries).

As mentioned earlier in this chapter, run-time power management is only one facet of the problem of optimizing power/energy consumption of electronic systems. Component and communication design are equally important in the global context, but the way they can be addressed differs substantially from the case of DPM/DVS, as the effectiveness of the latter is mainly dependent on the system workload.

References

[1] L. Benini and G. De Micheli, *Dynamic Power Management: Design Techniques and CAD Tools*, Kluwer Academic Publishers, Dordrecht, 1998.

[2] L. Benini and G. De Micheli, System level power optimization: techniques and tools, *ACM Trans. Des. Autom. Electron. Syst.*, 5, 115–192, 2000.

[3] L. Benini, A. Bogliolo, and G. De Micheli, A survey of design techniques for system-level dynamic power management, *IEEE Trans. VLSI Syst.*, 8, 299–316, 2000.

[4] E. Macii, Ed., *IEEE Design Test Comput.*, Special Issue on Dynamic Power Management, 18, 2001.

[5] L. Benini, R. Hodgson, and P. Siegel, System-level power estimation and optimization, *ISLPED-98: International Symposium on Low Power Electronics and Design*, August 1998, pp. 173–178.

[6] H. Shim, Y. Joo, Y. Choi, H. G. Lee, K. Kim, and N. Chang, Low-energy off-chip SDRAM memory systems for embedded applications, *ACM Trans. Embed. Comput. Syst.*, Special Issue on Memory Systems, 2, 98–130, 2003.

[7] N. Chang, K. Kim, and H. G. Lee, Cycle-accurate energy consumption measurement and analysis: case study of ARM7TDMI, *IEEE Trans. VLSI Syst.*, 10, 146–154, 2002.

[8] I. Lee, Y. Choi, Y. Cho, Y. Joo, H. Lim, H. G. Lee, H. Shim, and N. Chang, A web-based energy exploration tool for embedded systems, *IEEE Des. Test Comput.*, 21, 572–586, 2004.

[9] M. Srivastava, A. Chandrakasan, and R. Brodersen, Predictive system shutdown and other architectural techniques for energy efficient programmable computation, *IEEE Trans. VLSI Syst.*, 4, 42–55, 1996.

[10] C.-H. Hwang and A. Wu, A predictive system shutdown method for energy saving of event-driven computation, *ICCAD-97: International Conference on Computer-Aided Design*, November 1997, pp. 28–32.

[11] P. Krishnan, P. Long, and J. Vitter, Adaptive Disk Spin-Down via Optimal Rent-to-Buy in Probabilistic Environments, *International Conference on Machine Learning*, July 1995, pp. 322–330.

[12] D. Helmbold, D. Long, and E. Sherrod, Dynamic disk spin-down technique for mobile computing, *Proceedings of the International Conference on Mobile Computing,* November 1996, pp. 130–142.

[13] F. Douglis, P. Krishnan, and B. Bershad, Adaptive disk spin-down policies for mobile computers, *Proceedings of the 2nd USENIX Symposium on Mobile and Location-Independent Computing,* April 1995, pp. 121–137.

[14] L. Benini, G. Paleologo, A. Bogliolo, and G. D. Micheli, Policy optimization for dynamic power management, *IEEE Trans. Comput.-Aid. Des.,* 18, 813–833, 1999.

[15] Q. Qiu and M. Pedram, Dynamic power management based on continuous-time Markov decision processes, *DAC-36: Design Automation Conference,* June 1999, pp. 555–561.

[16] E. Chung, L. Benini, A. Bogliolo, and G. D. Micheli, Dynamic power management for non-stationary service requests, *DATE-99: Design Automation and Test in Europe,* March 1999, pp. 77–81.

[17] T. Burd and R. Brodersen, Design issues for dynamic voltage scaling, *ISLPED-00: International Symposium on Low Power Electronics and Design,* July 2000, pp. 9–14.

[18] W. Kim, D. Shin, H. S. Yun, J. Kim, and S. L. Min, Performance comparison of dynamic voltage scaling algorithms for hard real-time systems, *Real-Time and Embedded Technology and Applications Symposium,* September 2002, pp. 219–228.

[19] F. Yao, A. Demers, and A. Shenker, A scheduling model for reduced CPU energy, *FOCS-95: Foundations of Computer Science,* October 1995, pp. 374–382.

[20] S. Lee and T. Sakurai, Run-time voltage hopping for low-power real-time systems, *DAC-37: Design Automation Conference,* June 2000, pp. 806–809.

[21] D. Shin, J. Kim, and S. Lee, Low-energy intra-task voltage scheduling using static timing analysis, *DAC-38: Design Automation Conference,* June 2001, pp. 438–443.

[22] Y.-H. Lee and C. M. Krishna, Voltage-clock scaling for low energy consumption in real-time embedded systems, *6th International Workshop on Real-Time Computing Systems and Applications Symposium,* December 1999, pp. 272–279.

[23] Y. Shin, K. Choi, and T. Sakurai, Power optimization of real-time embedded systems on variable speed processors, *ICCAD-00: International Conference on Computer-Aided Design,* November 2000, pp. 365–368.

[24] P. Pillai and K. G. Shin, Real-time dynamic voltage scaling for low-power embedded operating systems, *Proceedings of the 18th ACM Symposium on Operating Systems Principles,* October 2001, pp. 89–102.

[25] F. Gruian, Hard real-rime scheduling using stochastic data and DVS processors, *ISLPED-01: International Symposium on Low Power Electronics and Design,* August 2001, pp. 46–51.

[26] G. Quan and X. S. Hu, Energy efficient fixed-priority scheduling for real-time systems on variable voltage processors, *DAC-38: Design Automation Conference,* June 2001, pp. 828–833.

[27] T. Ishihara and H. Yasuura, Voltage scheduling problem for dynamically variable voltage processors, *ISLPED-98: International Symposium on Low Power Electronics and Design,* August 1998, pp. 197–202.

[28] Y. Lin, C. Hwang, and A. Wu, Scheduling techniques for variable voltage low power designs, *ACM Trans. Des. Autom. Electron. Syst.,* 2, 81–97, 1997.

[29] W. Kwon and T. Kim, Optimal voltage allocation techniques for dynamically variable voltage processors, *DAC-40: Design Automation Conference,* June 2003, pp. 125–130.

[30] R. Jejurikar, C. Pereira, and R. Gupta, Leakage aware dynamic voltage scaling for real-time embedded systems, *DAC-41: Design Automation Conference,* June 2004, pp. 275–280.

[31] X. Fan, C. Ellis, and A. Lebeck, The synergy between power-aware memory systems and processor voltage, *Workshop on Power-Aware Computer Systems,* December 2003, pp. 130–140.

[32] Y. Cho and N. Chang, Memory-aware energy-optimal frequency assignment for dynamic supply voltage scaling, *ISLPED-04: International Symposium on Low Power Electronics and Design,* August 2004, pp. 387–392.

[33] D. Linden, *Handbook of Batteries,* 2nd ed., McGraw-Hill, Hightstown, NJ, 1995.

[34] T. Martin and D. Sewiorek, Non-ideal battery and main memory effects on CPU speed-setting for low power, *IEEE Trans. VLSI Syst.,* 9, 29–34, 2001.

[35] L. Benini, G. Castelli, A. Macii, E. Macii, M. Poncino, and R. Scarsi, Discrete-time battery models for system-level low-power design, *IEEE Trans. VLSI Syst.,* 9, 630–640, 2001.

[36] M. Pedram and Q. Wu. Design considerations for battery-powered electronics, *IEEE Trans. VLSI Syst.,* 10, 601–607, 2002.

[37] L. Benini, G. Castelli, A. Macii, and R. Scarsi, Battery-driven dynamic power management, *IEEE Des. Test Comput.*, 18, 53–60, 2001.

[38] P. Rong and M. Pedram, Extending the lifetime of a network of battery-powered mobile devices by remote processing: a Markovian decision-based approach, *DAC-40: Design Automation Conference*, June 2003, pp. 906–911.

[39] A. Macii, E. Macii, and M. Poncino, Current-controlled battery management policies for lifetime extension of portable systems, *ST J. Syst. Res.*, 3, 92–99, 2002.

[40] P. Chowdhury and C. Chakrabarti, Static task-scheduling algorithms for battery-powered DVS systems, *IEEE Trans. VLSI Syst.*, 13, 226–237, 2005.

[41] D. Rakhmatov and S. Vrudhula, Energy management for battery-powered embedded systems, *ACM Trans. Embed. Comput. Syst.*, 2, 277–324, 2003.

[42] J. Luo and N. Jha, Battery-aware static scheduling for distributed real-time embedded systems, *DAC-38: Design Automation Conference*, June 2001, pp. 444–449.

[43] D. Rakhmatov, S. Vrudhula, and C. Chakrabarti, Battery-conscious task sequencing for portable devices including voltage/clock scaling, *DAC-39: Design Automation Conference*, June 2002, pp. 211–217.

[44] Q. Wu, Q. Qiu, and M. Pedram, An interleaved dual-battery power supply for battery-operated electronics, *ASPDAC-00: Asia and South Pacific Design Automation Conference*, January 2000, pp. 387–390.

[45] L. Benini, G. Castelli, A. Macii, E. Macii, M. Poncino, and R. Scarsi, Scheduling battery usage in mobile systems, *IEEE Trans. VLSI Syst.*, 11, 1136–1143, 2003.

[46] L. Benini, D. Bruni, A. Macii, E. Macii, and M. Poncino, Extending lifetime of multi-battery mobile systems by discharge current steering, *IEEE Trans. Comput.*, 53, 985–995, 2003.

[47] V. Tiwari, S. Malik, A. Wolfe, and T. C. Lee, Instruction level power analysis and optimization software, *J. VLSI Signal Process.*, 13, 1–18, 1996.

[48] Hewlett-Packard Corporation, Intel Corporation, Microsoft Corporation, Phoenix Technologies Ltd., and Toshiba Corporation, *Advanced Configuration and Power Interface Specification Revision 3.0*, September 2004.

[49] Y. Lu, T. Simunic, and G. D. Micheli, Software controlled power management, *CODES-99: International Workshop on Hardware-Software Codesign*, May 1999, pp. 151–161.

[50] Y. Lu, E. Y. Chung, T. Simunic, L. Benini, and G. D. Micheli, Quantitative comparison of power management algorithms, *DATE-00: Design Automation and Test in Europe* , March 2000, pp. 20–26.

8

Processor Modeling and Design Tools

Prabhat Mishra
University of Florida
Gainesville, Florida

Nikil Dutt
Donald Bren School of Information and Computer Sciences,
University of California, Irvine
Irvine, California

8.1 Introduction

This chapter covers state-of-the-art specification languages, tools, and methodologies for processor development in academia as well as industry. Time-to-market pressure coupled with short product lifetimes creates a critical need for design automation in processor development. The processor is modeled using a specification language such as Architecture description language (ADL). The ADL specification is used to generate various tools (e.g., simulators, compilers, and debuggers) to enable exploration and validation of candidate architectures. The goal is to find the best possible processor architecture for the given set of application programs under various design constraints such as cost, area, power, and performance. The ADL specification is also used to perform various design automation tasks, including hardware generation and functional verification of processors.

Computing is an integral part of daily life. We encounter two types of computing devices everyday: desktop-based computing devices and embedded computer systems. Desktop-based computing systems encompass traditional "computers", including personal computers, notebook computers, workstations, and servers. Embedded computer systems are ubiquitous — they run the computing devices hidden inside a vast array of everyday products and appliances such as cell phones, toys, handheld PDAs, cameras, and microwave ovens. Both types of computing devices use programmable components such as processors, coprocessors, and memories to execute the application programs. These programmable components are also referred as *programmable architectures.* Figure 8.1 shows an example embedded system with programmable architectures. Depending on the application domain, the embedded system can have application-specific hardwares, interfaces, controllers, and peripherals. The complexity of the programmable architectures is increasing at an exponential rate due to technological advances as well as demand for realization of ever more complex applications in communication,

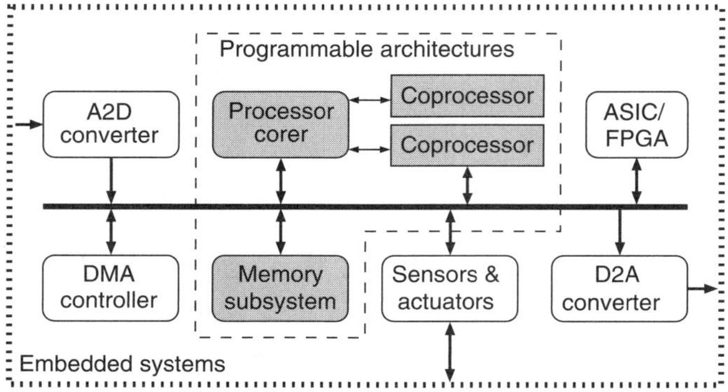

FIGURE 8.1 An example of an embedded system.

multimedia, networking, and entertainment. Shrinking time-to-market coupled with short product lifetimes create a critical need for design automation of increasingly sophisticated and complex programmable architectures.

Modeling plays a central role in design automation of processors. It is necessary to develop a specification language that can model complex processors at a higher level of abstraction and also enable automatic analysis and generation of efficient prototypes. The language should be powerful enough to capture high-level description of the programmable architectures. On the other hand, the language should be simple enough to allow correlation of the information between the specification and the architecture manual. Specifications widely in use today are still written informally in natural language like English. Since natural language specifications are not amenable to automated analysis, there are possibilities of ambiguity, incompleteness, and contradiction: all problems that can lead to different interpretations of the specification. Clearly, formal specification languages are suitable for analysis and verification. Some have become popular because they are input languages for powerful verification tools such as a model checker. Such specifications are popular among verification engineers with expertise in formal languages. However, these specifications are not acceptable by designers and other tool developers. Therefore, the ideal specification language should have formal (unambiguous) semantics as well as easy correlation with the architecture manual.

Architecture description languages (ADLs) have been successfully used as a specification language for processor development. The ADL specification is used to perform early exploration, synthesis, test generation, and validation of processor-based designs as shown in Figure 8.2. The ADL specification can also be used for generating hardware prototypes [1,2]. Several researches have shown the usefulness of ADL-driven generation of functional test programs [3] and JTAG interface [4]. The specification can also be used to generate device drivers for real-time operating systems (RTOS)[5]. The rest of the chapter is organized as follows: Section 2 describes processor modeling using ADLs. Section 3 presents ADL-driven methodologies for software toolkit generation, hardware synthesis, exploration, and validation of programmable architectures. Finally, Section 4 concludes the chapter.

8.2 Processor Modeling Using ADLs

The phrase ADL has been used in the context of designing both software and hardware architectures. Software ADLs are used for representing and analyzing software architectures [6,7]. They capture the behavioral specifications of the components and their interactions that comprise the software architecture. However, hardware ADLs capture the structure (hardware components and their connectivity), and the behavior (instruction-set) of processor architectures. The concept of using machine description languages for specification of architectures has been around for a long time. Early ADLs such as ISPS [8] were used for simulation, evaluation, and synthesis of computers and other digital systems. This section describes contemporary hardware ADLs. First, it tries to answer why ADLs (not other languages) are used

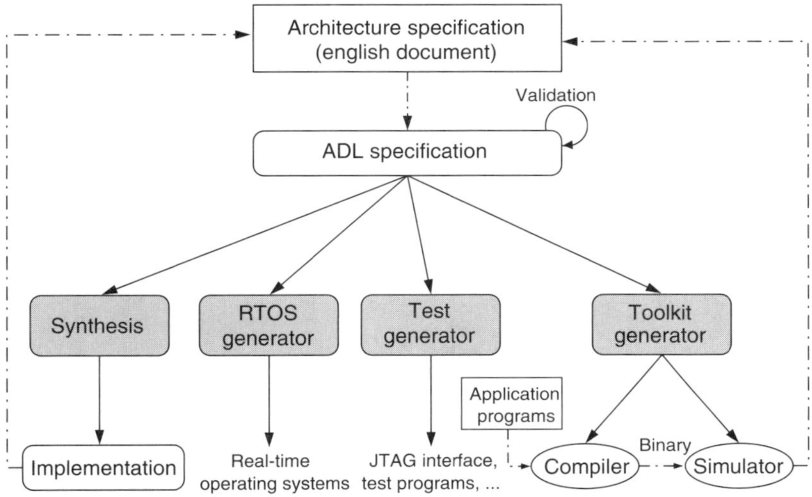

FIGURE 8.2 Architecture description languages-driven exploration, synthesis, and validation of programmable architectures.

for modeling and specification. Next, it surveys contemporary ADLs to compare their relative strengths and weaknesses in the context of processor modeling and ADL-driven design automation.

8.2.1 ADLs and other Languages

How do ADLs differ from programming languages, hardware description languages (HDLs), modeling languages, and the like? This section attempts to answer this question. However, it is not always possible to answer the following question: Given a language for describing an architecture, what are the criteria for deciding whether it is an ADL or not?

In principle, ADLs differ from programming languages because the latter bind all architectural abstractions to specific point solutions whereas ADLs intentionally suppress or vary such binding. In practice, architecture is embodied and recoverable from code by reverse engineering methods. For example, it might be possible to analyze a piece of code written in C language and figure out whether it corresponds to *Fetch* unit or not. Many languages provide architecture level views of the system. For example, C++ language offers the ability to describe the structure of a processor by instantiating objects for the components of the architecture. However, C++ offers little or no architecture-level analytical capabilities. Therefore, it is difficult to describe architecture at a level of abstraction suitable for early analysis and exploration. More importantly, traditional programming languages are not natural choice for describing architectures due to their inability in capturing hardware features such as parallelism and synchronization.

ADLs differ from modeling languages (such as UML) because the latter are more concerned with the behaviors of the whole rather than the parts, whereas ADLs concentrate on the representation of components. In practice, many modeling languages allow the representation of cooperating components and can represent architectures reasonably well. However, the lack of an abstraction would make it harder to describe the instruction set (IS) of the architecture. Traditional HDLs, such as VHDL and Verilog, do not have sufficient abstraction to describe architectures and explore them at the system level. It is possible to perform reverse-engineering to extract the structure of the architecture from the HDL description. However, it is hard to extract the instruction set behavior of the architecture. In practice, some variants of HDLs work reasonably well as ADLs for specific classes of programmable architectures.

There is no clear line between ADLs and non-ADLs. In principle, programming languages, modeling languages, and hardware description languages have aspects in common with ADLs, as shown in Figure 8.3. Languages can, however, be discriminated from one another according to how much architectural

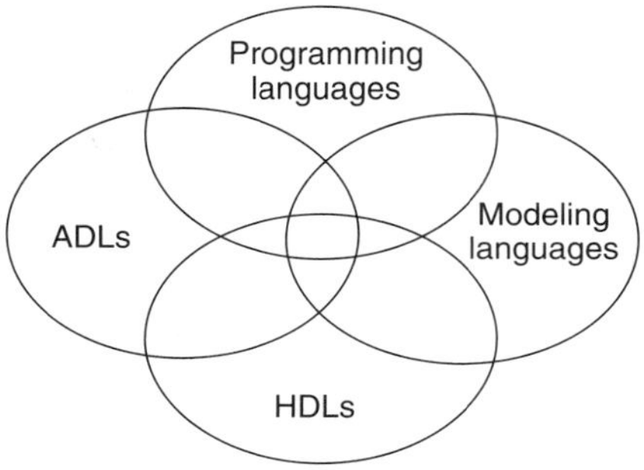

FIGURE 8.3 ADLs vs. non-ADLs.

information they can capture and analyze. Languages that were born as ADLs show a clear advantage in this area over languages built for some other purpose and later co-opted to represent architectures.

8.2.2 Contemporary Architecture Description Languages

This section briefly surveys some of the contemporary ADLs in the context of processor modeling and design automation. There are many comprehensive ADL surveys available in the literature including ADLs for retargetable compilation [9], programmable embedded systems [10], and system-on-chip (SOC) design [11]. Figure 8.4 shows the classification of ADLs based on two aspects: *content* and *objective.* The content-oriented classification is based on the nature of the information an ADL can capture, whereas the objective-oriented classification is based on the purpose of an ADL. Contemporary ADLs can be classified into six categories based on the following objectives: simulation, synthesis, test generation, compilation, validation, and operating system (OS) generation.

ADLs can be classified into four categories based on the nature of the information: structural, behavioral, mixed, and partial. The structural ADLs capture the structure in terms of architectural components and their connectivity. The behavioral ADLs capture the instruction-set behavior of the processor architecture. The mixed ADLs capture both structure and behavior of the architecture. These ADLs capture complete description of the structure or behavior or both. However, the partial ADLs capture specific information about the architecture for the intended task. For example, an ADL intended for interface synthesis does not require internal structure or behavior of the processor. Traditionally, structural ADLs are suitable for synthesis and test-generation. Similarly, behavioral ADLs are suitable for simulation and compilation. It is not always possible to establish a one-to-one correspondence between content- and objective-based classifications. For example, depending on the nature and amount of information captured, partial ADLs can represent any one or more classes of the objective-based ADLs. This section presents the survey using content-based classification of ADLs.

8.2.2.1 Structural ADLs

ADL designers consider two important aspects: the level of abstraction vs. generality. It is very difficult to find an abstraction to capture the features of different types of processors. A common way to obtain generality is to lower the abstraction level. Register transfer level (RTL) is a popular abstraction level — low enough for detailed behavior modeling of digital systems, and high enough to hide gate-level implementation details. Early ADLs are based on RTL descriptions. This section briefly describes a structural ADL: MIMOLA [12].

8.2.2.1.1 MIMOLA

MIMOLA [12] is a structure-centric ADL developed at the University of Dortmund, Germany. It was originally proposed for micro-architecture design. One of the major advantages of MIMOLA is that the same description can be used for synthesis, simulation, test generation, and compilation. A tool chain including the MSSH hardware synthesizer, the MSSQ code generator, the MSST self-test program compiler, the MSSB functional simulator, and the MSSU RTL simulator were developed based on the MIMOLA language [12].

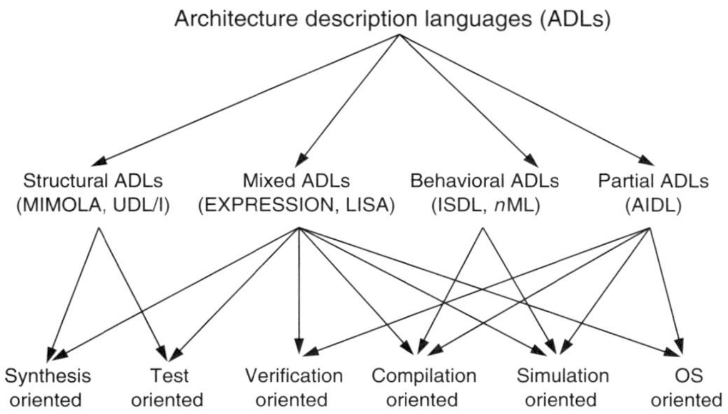

FIGURE 8.4 Taxonomy of ADLs.

MIMOLA has also been used by the RECORD [12] compiler. MIMOLA description contains three parts: the algorithm to be compiled, the target processor model, and additional linkage and transformation rules. The software part (algorithm description) describes application programs in a PASCAL-like syntax. The processor model describes micro-architecture in the form of a component netlist. The linkage information is used by the compiler in order to locate important modules such as program counter and instruction memory. The following code segment specifies the program counter and instruction memory locations [12]:

```
LOCATION_FOR_PROGRAMCOUNTER  PCReg;
LOCATION_FOR_INSTRUCTIONS  IM[0..1023];
```

The algorithmic part of MIMOLA is an extension of PASCAL. Unlike other high-level languages, it allows references to physical registers and memories. It also allows the use of hardware components using procedure calls. For example, if the processor description contains a component named MAC, programmers can write the following code segment to use the multiply-accumulate operation performed by MAC:

```
res  := MAC(x, y, z);
```

The processor is modeled as a netlist of component modules. MIMOLA permits modeling of arbitrary (programmable or nonprogrammable) hardware structures. Similar to VHDL, a number of predefined, primitive operators exist. The basic entities of MIMOLA hardware models are modules and connections. Each module is specified by its port interface and its behavior. The following example shows the description of a multi-functional ALU module [12]:

```
MODUL ALU
  (IN inpl inp2: 31:0);
  OUT outp (31:0)  ;
  IN ctrl  ;
  )
CONBEGIN
   outp <—    CASE ctrl OF
      0: inpl + inp2  ;
      1: inpl - inp2  ;
      END;

CONEND;
```

The CONBEGIN/CONEND construct includes a set of concurrent assignments. In the example a conditional assignment to output port *outp* is specified, which depends on the two-bit control input *ctrl*. The netlist structure is formed by connecting ports of module instances. For example, the following MIMOLA description connects two modules: *ALU* and accumulator *ACC*.

```
CONNECTIONS ALU.outp → ACC.inp
            ACC.outp → ALU.inp
```

The MSSQ code generator extracts instruction-set information from the module netlist. It uses two internal data structures: connection operation graph (COG) and instruction tree (I-tree). It is a very difficult task to extract the COG and I-trees even in the presence of linkage information due to the flexibility of an RTL structural description. Extra constraints need to be imposed in order for the MSSQ code generator to work properly. The constraints limit the architecture scope of MSSQ to micro-programmable controllers, in which all control signals originate directly from the instruction word. The lack of explicit description of processor pipelines or resource conflicts may result in poor code quality for some classes of VLIW or deeply pipelined processors.

8.2.2.2 Behavioral ADLs

The difficulty of instruction-set extraction can be avoided by abstracting behavioral information from the structural details. Behavioral ADLs explicitly specify the instruction semantics and ignore detailed hardware structures. Typically, there is a one-to-one correspondence between behavioral ADLs and instruction-set reference manual. This section briefly describes two behavioral ADLs: nML [13] and Instruction set description language [14].

8.2.2.2.1 nML

nML is an instruction-set-oriented ADL proposed at the Technical University of Berlin, Germany. nML has been used by code generators CBC [15] and CHESS [16], and instruction-set simulators (ISSs) Sigh/Sim [17] and CHECKERS. Currently, CHESS/CHECKERS environment is used for automatic and efficient software compilation and instruction-set simulation [18]. nML developers recognized the fact that several instructions share common properties. The final nML description would be compact and simple if the common properties are exploited. Consequently, nML designers used a hierarchical scheme to describe instruction-sets. The instructions are the topmost elements in the hierarchy. The intermediate elements of the hierarchy are partial instructions (PI). The relationship between elements can be established using two composition rules: AND- and OR-rules. The AND-rule groups several PIs into a larger PI and the OR-rule enumerates a set of alternatives for one PI. Therefore, instruction definitions in nML can be in the form of an and–or tree. Each possible derivation of the tree corresponds to an actual instruction.

To achieve the goal of sharing instruction descriptions, the instruction-set is enumerated by an attributed grammar [19]. Each element in the hierarchy has a few attributes. A nonleaf element's attribute values can be computed based on its children's attribute values. Attribute grammar is also adopted by other ADLs such as ISDL [14] and TDL [20]. The following nML description shows an example of instruction specification [13]:

```
op numeric_instruction(a:num_action, src:SRC, dst:DST)
action {
  temp_src = src;
  temp_dst = dst;
  a.action;
  dst = temp_dst;
}
op num_action = add|sub
op add()
action = {
  temp_dst = temp_dst + temp_src
}
```

The definition of *numeric_instruction* combines three PI with the AND-rule: *num_action,* SRC, and DST. The first PI, *num_action,* uses OR-rule to describe the valid options for actions: *add* or *sub.* The number of all possible derivations of *numeric_instruction* is the product of the size of *num_action, SRC* and *DST.* The common behavior of all these options is defined in the *action* attribute of *numeric_instruction.* Each option for *num_action* should have its own action attribute defined as its specific behavior, which is referred by the *a.action* line. For example, the above code segment has action description for *add* operation. Object-code image and assembly syntax can also be specified in the same hierarchical manner.

nML also captures the structural information used by instruction-set architecture. For example, storage units should be declared since they are visible to the instruction-set. nML supports three types of storages: RAM, register, and transitory storage. Transitory storage refers to machine states that are retained only for a limited number of cycles, e.g., values on buses and latches. Computations have no delay in nML timing model — only storage units have delay. Instruction delay slots are modeled by introducing storage units as pipeline registers. The result of the computation is propagated through the registers in the behavior specification. nML models constraints between operations by enumerating all valid combinations. The enumeration of valid cases can make nML descriptions lengthy. More complicated constraints, which often appear in DSPs with irregular instruction-level parallelism (ILP) constraints or VLIW processors with multiple issue slots, are hard to model with nML. For example, nML cannot model the constraint that operation *I1* cannot directly follow operation *I0.* nML explicitly supports several addressing modes. However, it implicitly assumes an architecture model, which restricts its generality. As a result it is hard to model multicycle or pipelined units and multiword instructions explicitly. A good critique of nML is given in [61].

8.2.2.2.2 ISDL

Instruction Set Description Language (ISDL) was developed at MIT and used by the Aviv compiler [21] and GENSIM simulator generator [22]. The problem of constraint modeling is avoided by ISDL with explicit specification. ISDL language is mainly targeted toward VLIW processors. Similar to nML, ISDL primarily describes the instruction-set of processor architectures. ISDL consists of mainly five sections: instruction word format, global definitions, storage resources, assembly syntax, and constraints. It also contains an optimization information section that can be used to provide certain architecture specific hints for the compiler to make better machine-dependent code optimizations. The section on instruction word format defines fields of the instruction word. The instruction word is separated into multiple fields each containing one or more subfields. The global definition section describes four main types: tokens, nonterminals, split functions, and macro definitions. Tokens are the primitive operands of instructions. For each token, assembly format and binary encoding information must be defined. An example token definition of a binary operand is

```
Token X[0..1]   X_R ival {yylval.ival = yytext[1] – '0';}
```

In this example, following the keyword *Token* is the assembly format of the operand. *X_R* is the symbolic name of the token used for reference. The *ival* is used to describe the value returned by the token. Finally, the last field describes the computation of the value. In this example, the assembly syntax allowed for the token *XR* is *X0* or *X1,* and the values returned are 0 or 1, respectively. The value (last) field is used for behavioral definition and binary encoding assignment by nonterminals or instructions. Nonterminal is a mechanism provided to exploit commonalities among operations. The following code segment describes a nonterminal named *XYSRC:*

```
Non_Terminal ival XYSRC:   X_D {$$ = 0;}       |
                           Y_D {$$ = Y_D + 1;};
```

The definition of *XYSRC* consists of the keyword *Nonterminal,* the type of the returned value, a symbolic name as it appears in the assembly, and an action that describes the possible token or nonterminal combinations, and the return value associated with each of them. In this example, *XYSRC* refers to tokens *X_D* and *Y_D* as its two options. The second field *(ival)* describes the returned value type. It returns 0 for *X_D* or incremented value for *Y_D.* Similar to nML, storage resources are the only structural information modeled by ISDL. The storage section lists all the storage resources visible to the programmer. It lists the names and sizes of the memory, register files, and special registers. This information is used by the compiler to determine the available resources and how they should be used.

The assembly syntax section is divided into fields corresponding to the separate operations that can be performed in parallel within a single instruction. For each field, a list of alternative operations can be described. Each operation description consists of a name, a list of tokens or nonterminals as parameters, a set of commands that manipulate the bitfields, RTL description, timing details, and costs. RTL description captures the effect of the operation on the storage resources. Multiple costs are allowed including operation execution time, code size, and costs due to resource conflicts. The timing model of ISDL describes when the various effects of the operation take place (e.g., because of pipelining). In contrast to nML, which enumerates all valid combinations, ISDL defines invalid combinations in the form of Boolean expressions. This often leads to a simple constraint specification. It also enables ISDL to capture irregular instruction sequencing constraints. The following example shows how to describe the constraint that instruction I1 cannot directly follow instruction I0. The "[1]" indicates a time shift of one instruction fetch for the I0 instruction. The ~ is a symbol for NOT and '&' is for logical AND.

```
~(I1 *) & ([1] I0 *, *)
```

ISDL provides the means for compact and hierarchical instruction-set specification. However, it may not be possible to describe instruction-sets with multiple encoding formats using the simple tree-like instruction structure of ISDL.

8.2.2.3 Mixed ADLs

Mixed ADLs capture both structural and behavioral details of the architecture. This section briefly describes three mixed ADLs: HMDES, EXPRESSION, and LISA.

8.2.2.3.1 HMDES

Machine description language HMDES was developed at the University of Illinois at Urbana-Champaign for the IMPACT research compiler [23]. C-like preprocessing capabilities such as file inclusion, macro expansion, and conditional inclusion are supported in HMDES. An HMDES description is the input to the MDES machine description system of the Trimaran compiler infrastructure, which contains IMPACT as well as the Elcor research compiler from HP Labs. The description is first pre-processed, then optimized and translated to a low-level representation file. A machine database reads the low-level files and supplies information for the compiler backend through a predefined query interface.

MDES captures both structure and behavior of target processors. Information is broken down into sections such as format, resource usage, latency, operation, and register. For example, the following code segment describes register and register file. It describes 64 registers. The register file describes the width of each register and other optional fields such as generic register type (virtual field), speculative, static, and rotating registers. The value '1' implies speculative and '0' implies nonspeculative.

```
SECTION Register {
        R0();R1(); ... R63 ();
  'R[0]'()... 'R[63]'();
...
}

SECTION Register_File {
RF_i(width(32) virtual(i) speculative (1)
  static(R0 ...R63 rotating ('R[0]' ... 'R[63]'));
  ...
}
```

MDES allows only a restricted retargetability of the cycle-accurate simulator to the HPL-PD processor family [24]. MDES permits description of memory systems, but limited to the traditional hierarchy, i.e., register files, caches, and main memory.

8.2.2.3.2 EXPRESSION

Traditional mixed ADLs require explicit description of Reservation Tables (RT). Processors that contain complex pipelines, large amounts of parallelism, and complex storage sub-systems, typically contain a large number of operations and resources (and hence RTs). Manual specification of RTs on a per-operation basis thus becomes cumbersome and error-prone. The manual specification of RTs (for each configuration) becomes impractical during rapid architectural exploration. The EXPRESSION ADL [25] describes a processor as a netlist of units and storages to generate automatically RTs based on the netlist [26]. Unlike MIMOLA, the netlist representation of EXPRESSION is coarse grain. It uses a higher level of abstraction similar to block-diagram level description in architecture manual.

EXPRESSION ADL was developed at the University of California, Irvine. The ADL has been used by the retargetable compiler (EXPRESS [27]) and simulator (SIMPRESS [28]) generation framework. The framework also supports a graphical user interface (GUI) and can be used for design space exploration of programmable architectures consisting of processor cores, coprocessors, and memories [29]. An EXPRESSION description is composed of two main sections: behavior and structure. The behavior section has three subsections: operations, instruction, and operation mappings. Similarly, the structure section consists of three subsections: components, pipeline/data-transfer paths, and memory subsystem.

The operation subsection describes the instruction-set of the processor. Each operation of the processor is described in terms of its opcode and operands. The types and possible destinations of each operand are also specified. A useful feature of EXPRESSION is operation group that groups similar operations together for the ease of later reference. For example, the following code segment shows an operation group *(alu_ops)* containing two ALU operations: *add* and *sub.*

```
(OP_GROUP alu_ops
  (OPCODE add
    (OPERANDS (SRC1 reg)(SRC2  reg/imm) (DEST  reg))
    (BEHAVIOR DEST = SRC1 + SRC2)
  ...)
(OPCODE sub
    (OPERANDS (SRC1 reg)  (SRC2 reg/imm) (DEST reg))
    (BEHAVIOR DEST = SRC1 - SRC2)
  ...)
)
```

The instruction subsection captures the parallelism available in the architecture. Each instruction contains a list of slots (to be filled with operations), with each slot corresponding to a functional unit. The operation mapping subsection is used to specify the information needed by instruction selection and architecture-specific optimizations of the compiler. For example, it contains mapping between generic and target instructions. The component subsection describes each RTL component in the architecture. The components can be pipeline units, functional units, storage elements, ports, and connections. For multicycle or pipelined units, the timing behavior is also specified. The subsection on pipeline/data-transfer path describes the netlist of the processor. The *pipeline path description* provides a mechanism to specify the units, which comprise the pipeline stages, while the *data-transfer path description* provides a mechanism for specifying the valid data transfers. This information is used to both retarget the simulator, and to generate reservation tables needed by the scheduler [26]. An example path declaration for the DLX architecture [30] (Figure 8.5) is shown below. It describes that the processor has five pipeline stages. It also describes that the *Execute* stage has four parallel paths. Finally, it describes each path, e.g., it describes that the *FADD* path has four pipeline stages.

```
(PIPELINE Fetch Decode Execute MEM WriteBack)
(Execute (ALTERNATE IALU MULT FADD DIV))
(MULT (PIPLINE MUL1 MUL2...MUL7))
(FADD(PIPELINE FADD1 FADD2 FADD3 FADD4))
```

The memory subsection describes the types and attributes of various storage components (such as register files, SRAMs, DRAMs, and caches). The memory netlist information can be used to generate memory aware compilers and simulators [31]. Memory aware compilers can exploit the detailed information to hide the latency of the lengthy memory operations [32]. EXPRESSION captures the data path information in the processor. The control path is not explicitly modeled. The instruction model requires extension to capture inter-operation constraints such as sharing of common fields. Such constraints can be modeled by ISDL through cross-field encoding assignment.

8.2.2.3.3 LISA

Language for Instruction Set Architecture (LISA) [33] was developed at Aachen University of Technology, Germany, with a simulator centric view. The language has been used to produce production quality simulators [34]. An important aspect of LISA is its ability to capture control path explicitly. Explicit modeling of both data path and control is necessary for cycle-accurate simulation. LISA has also been used to generate retargetable C compilers [35,36]. Its descriptions are composed of two types of declarations: resource and operation. The resource declarations cover hardware resources such as registers, pipelines, and memories. The pipeline model defines all possible pipeline paths that operations can go through. An example pipeline description for the architecture shown in Figure 8.5 is as follows:

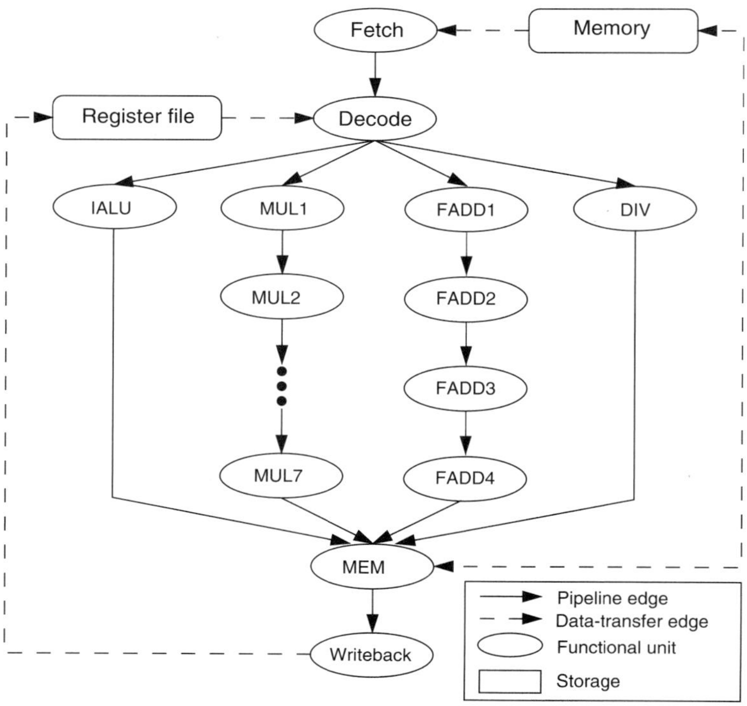

FIGURE 8.5 The DLX architecture.

```
PIPELINE int    = {Fetch; Decode; IALU; MEM; WriteBack}
PIPELINE flt    = {Fetch; Decode; FADD1; FADD2;
                   FADD3; FADD4; MEM; WriteBack}
PIPELINE mul    = {Fetch; Decode; MUL1; MUL2; MUL3; MUL4;
                   MUL5; MUL6; MUL7; MEM; WriteBack)
PIPELINE div    = {Fetch; Decode; DIV; MEM; WriteBack}
```

Operations are the basic objects in LISA. They represent the designer's view of the behavior, the structure, and the instruction-set of the programmable architecture. Operation definitions capture the description of different properties of the system such as operation behavior, instruction-set information, and timing. These operation attributes are defined in several sections. LISA exploits the commonality of similar operations by grouping them into one. The following code segment describes the decoding behavior of two immediate-type (i-type) operations (ADDI and SUBI) in the DLX *Decode* stage. The complete behavior of an operation can be obtained by combining its behavior definitions in all the pipeline stages.

```
OPERATION i-type IN pipe_int.Decode {
  DECLARE {
    GROUP opcode = {ADD|| SUBI}
    GROUP rsi, rd = {fix_register};
  }
  CODING {opcode rsi rd immediate}
  SYNTAX {opcode rd " , " rsi " , " immediate}
  BEHAVIOUR {reg_a = rsi; imm = immediate; cond = 0;
  }
  ACTIVATION {opcode, writeback}
  }
```

A language similar to LISA is RADL. RADL [37] was developed at Rockwell, Inc. USA as an extension of the LISA approach that focuses on explicit support of detailed pipeline behavior to enable generation of production quality cycle- and phase-accurate simulators.

8.2.2.4 Partial ADLs

The ADLs discussed so far capture complete description of the processor's structure, behavior, or both. There are many description languages such as AIDL [38] that capture partial information of the architecture needed to perform a specific task. AIDL is an ADL developed at the University of Tsukuba for design of high-performance superscalar processors [38]. It seems that AIDL does not aim at data path optimization but aims at the validation of the pipeline behavior such as data-forwarding and out-of-order completion. In AIDL, the timing behavior of the pipeline is described using interval temporal logic. AIDL does not support software toolkit generation. However, AIDL descriptions can be simulated using the AIDL simulator.

8.3 ADL-Driven Methodologies

This section describes the ADL-driven methodologies used for processor development. It presents the following three methodologies that are used in academic research as well as industry:

- software toolkit generation and exploration;
- generation of hardware implementation;
- top–down validation.

8.3.1 Software Toolkit Generation and Exploration

Embedded systems present a tremendous opportunity to customize designs by exploiting the application behavior. Rapid exploration and evaluation of candidate architectures are necessary due to time-to-market pressure and short product lifetimes. ADLs are used to specify processor and memory architectures and generate software toolkit including compiler, simulator, assembler, profiler, and debugger. Figure 8.6 shows a traditional ADL-based design space exploration flow. The application programs are compiled and simulated, and the feedback is used to modify the ADL specification with the goal of finding the best possible architecture for the given set of application programs under various design constraints such as area, power, and performance. An extensive body of recent work addresses ADL-driven software toolkit generation and design space exploration of processor-based embedded systems, in both academia: ISDL [14], Valen-C [39], MIMOLA [12], LISA [33], nML [13], Sim-nML [40], EXPRESSION [25], and industry: ARC [41], Axys [42], RADL [37], Target [18], Tensilica [43], MDES [24].

One of the main purposes of an ADL is to support the automatic generation of a high-quality software toolkit including at least an ILP compiler and a cycle-accurate simulator. However, such tools require detailed information about the processor, typically in a form that is not concise and easily specifiable. Therefore, it becomes necessary to develop procedures to generate automatically such tool-specific information from the ADL specification. For example, RTs are used in many ILP compilers to describe resource conflicts. However, manual description of RTs on a per-instruction basis is cumbersome and error-prone. Instead, it is easier to specify the pipeline and datapath resources in an abstract manner, and generate RTs on a per-instruction basis [26]. This section describes some of the challenges in automatic generation of software tools (focusing on compilers and simulators) and survey some of the approaches adopted by current tools.

8.3.1.1 Compilers

Traditionally, the software for embedded systems was hand-tuned in assembly. With increasing complexity of embedded systems, it is no longer practical to develop software in assembly language or to optimize it manually except for the critical sections of the code. Compilers which produce optimized machine-specific code from a program specified in a high-level language (HLL) such as C/C++ and Java are necessary in order to produce efficient software within the time budget. Compilers for embedded systems have been the focus of several research efforts recently [44].

The compilation process can be broadly broken into two steps: analysis and synthesis. During analysis, the program (in HLL) is converted into an intermediate representation (IR) that contains all the desired information such as control and data dependences. During synthesis, the IR is transformed and

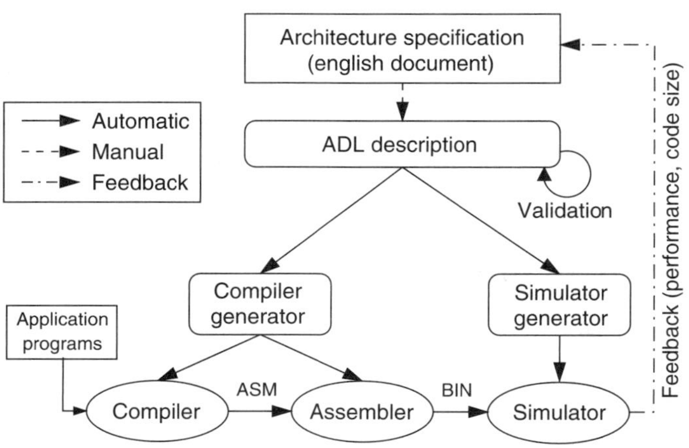

FIGURE 8.6 ADL-driven design space exploration.

optimized in order to generate efficient target-specific code. The synthesis step is more complex and typically includes the following phases: instruction selection, scheduling, resource allocation, code optimizations/transformations, and code generation. The effectiveness of each phase depends on the algorithms chosen and the target architecture. A further problem during the synthesis step is that the optimal ordering between these phases is highly dependent on the target architecture and the application program. As a result, traditionally, compilers have been painstakingly hand-tuned to a particular architecture (or architecture class) and application domain(s). However, stringent time-to-market constraints for SOC designs no longer make it feasible to generate manually compilers tuned to particular architectures. Automatic generation of an efficient compiler from an abstract description of the processor model becomes essential.

A promising approach to automatic compiler generation is the "retargetable compiler" approach. A compiler is classified as retargetable if it can be adapted to generate code for different target processors with significant reuse of the compiler source code. Retargetability is typically achieved by providing target machine information (in an ADL) as input to the compiler along with the program corresponding to the application. The complexity in retargeting the compiler depends on the range of target processors it supports and also on its optimizing capability. Due to the growing amount of ILP features in modern processor architectures, the difference in quality of code generated by a naive code conversion process and an optimizing ILP compiler can be enormous. Recent approaches on retargetable compilation have focused on developing optimizations/transformations that are "retargetable" and capturing the machine-specific information needed by such optimizations in the ADL. The retargetable compilers can be classified into three broad categories, based on the type of the machine model accepted as input.

8.3.1.1.1 Architecture Template Based

Such compilers assume a limited architecture template, which is parameterizable for customization. The most common parameters include operation latencies, number of functional units, number of registers, etc. Architecture template-based compilers have the advantage that both optimizations and the phase ordering between them can be manually tuned to produce highly efficient code for the limited architecture space. Examples of such compilers include the Valen-C compiler [39] and the GNU-based C/C++ compiler from Tensilica Inc. [43]. The Tensilica GNU-based C/C++ compiler is geared toward the Xtensa parameterizable processor architecture. One important feature of this system is the ability to add new instructions (described through an instruction extension language) and automatically generate software tools tuned to the new instruction-set.

8.3.1.1.2 Explicit Behavioral Information Based

Most compilers require a specification of the behavior in order to retarget their transformations (e.g., instruction selection requires a description of the semantics of each operation). Explicit behavioral information-based retargetable compilers require full information about the instruction-set as well as explicit resource conflict information. Examples include the AVIV [21] compiler using ISDL, CHESS [16] using nML, and Elcor [24] using MDES. The AVIV retargetable code generator produces machine code, optimized for minimal size, for target processors with different instruction-set. It solves the phase-ordering problem by performing a heuristic branch-and-bound step that performs resource allocation/assignment, operation grouping, and scheduling concurrently. CHESS is a retargetable code generation environment for fixed-point DSP processors. CHESS performs instruction selection, register allocation, and scheduling as separate phases (in that order). Elcor is a retargetable compilation environment for VLIW architectures with speculative execution. It implements a software pipelining algorithm (modulo scheduling) and register allocation for static and rotating register files.

8.3.1.1.3 Behavioral Information Generation Based

Recognizing that the architecture information needed by the compiler is not always in a form that may be well suited for other tools (such as synthesis) or does not permit concise specification, some research has focused on extraction of such information from a more amenable specification. Examples include the MSSQ and RECORD compiler using MIMOLA [12], retargetable C compiler based on LISA [35], and the

EXPRESS compiler using EXPRESSION [25]. MSSQ translates Pascal-like HLL into microcode for micro-programmable controllers, while RECORD translates code written in a DSP-specific programming language, called data flow language, into machine code for the target DSP. The retargetable C compiler generation using LISA is based on the reuse of a powerful C compiler platform with many built-in code optimizations and generation of mapping rules for code selection using the instruction semantics information [35]. The EXPRESS compiler tries to bridge the gap between explicit specification of all information (e.g., AVIV) and implicit specification requiring extraction of instruction-set (e.g., RECORD) by having a mixed behavioral/structural view of the processor.

8.3.1.2 Simulators

Simulators are critical components of the exploration and software design toolkit for the system designer. They can be used to perform diverse tasks such as verifying the functionality and/or timing behavior of the system (including hardware and software), and generating quantitative measurements (e.g., power consumption), which can be used to aid the design process. Simulation of the processor system can be performed at various abstraction levels. At the highest level of abstraction, a functional simulation of the processor can be performed by modeling only the instruction-set. Such simulators are termed instruction-set simulators (ISS) or instruction-level simulators (ILS). At lower levels of abstraction are the cycle- and phase-accurate simulation models that yield more detailed timing information. Simulators can be further classified based on whether they provide bit-accurate models, pin-accurate models, exact pipeline models, and structural models of the processor.

Typically, simulators at higher levels of abstraction (e.g., ISS and ILS) are faster but gather less information as compared to those at lower levels of abstraction (e.g., cycle- and phase-accurate). Retargetability (i.e., ability to simulate a wide variety of target processors) is especially important in the arena of embedded system design with emphasis on the exploration and co-development of hardware and software. Simulators with limited retargetability are very fast but may not be useful in all aspects of the design process. Such simulators typically incorporate a fixed architecture template and allow only limited retargetability in the form of parameters such as the number of registers and ALUs. Examples of such simulators are numerous in the industry and include the HPL-PD [24] simulator using the MDES ADL. The model of simulation adopted has significant impact on the simulation speed and flexibility of the simulator. On the basis of the simulation model, simulators can be classified into three types: interpretive, compiled, and mixed.

8.3.1.2.1 Interpretation-Based

Such simulators are based on an interpretive model of the processors instruction-set. Interpretive simulators store the state of the target processor in host memory. It then follows a fetch, decode, and execute model: instructions are fetched from memory, decoded and then executed in serial order. Advantages of this model include ease of implementation, flexibility, and the ability to collect varied processor state information. However, it suffers from significant performance degradation as compared to other approaches primarily due to the tremendous overhead in fetching, decoding, and dispatching instructions. Almost all the commercially available simulators are interpretive. Examples of research interpretive retargetable simulators include SIMPRESS [28] using EXPRESSION, and GENSIM/XSIM [22] using ISDL.

8.3.1.2.2 Compilation-Based

Compilation-based approaches reduce the runtime overhead by translating each target instruction into a series of host machine instructions, which manipulate the simulated machine state. Such translation can be done either at compile time (static compiled simulation) where the fetch-decode-dispatch overhead is completely eliminated, or at load time (dynamic compiled simulation), which amortizes the overhead over repeated execution of code. Simulators based on the static compilation model are presented by Zhu et al. [45] and Pees et al. [34]. Examples of dynamic compiled code simulators include the Shade simulator [46], and the Embra simulator [47].

8.3.1.2.3 Interpretive + Compiled

Traditional interpretive simulation is flexible but slow. Instruction decoding is a time-consuming process in a software simulation. Compiled simulation performs compile time decoding of application programs to improve the simulation performance. However, all compiled simulators rely on the assumption that the complete program code is known before the simulation starts and is furthermore run-time static. Due to the restrictiveness of the compiled technique, interpretive simulators are typically used in embedded systems design flow. Two recently proposed simulation techniques (JIT-CCS [48] and IS-CS [49]) combine the flexibility of interpretive simulation with the speed of the compiled simulation. The JIT-CCS technique compiles an instruction during runtime, *just-in-time* before the instruction is going to be executed. Subsequently, the extracted information is stored in a simulation cache for direct reuse in a repeated execution of the program address. The simulator recognizes if the program code of a previously executed address has changed and initiates a re-compilation. The IS-CS technique performs time-consuming instruction decoding during compile time. In case, an instruction is modified at run-time, the instruction is re-decoded prior to execution. The IS-CS technique also uses an *instruction abstraction* technique to generate aggressively optimized decoded instructions that further improves simulation performance [49,50].

8.3.2 Generation of Hardware Implementation

Recent approaches on ADL-based software toolkit generation enable performance-driven exploration. The simulator produces profiling data and thus may answer questions regarding the instruction-set, the performance of an algorithm and the required size of memory and registers. However, the required silicon area, clock frequency, and power consumption can only be determined in conjunction with a synthesizable HDL model. There are two major approaches in the literature for synthesizable HDL generation. The first one is a parameterized processor core-based approach. These cores are bound to a single processor template, whose architecture and tools can be modified to a certain degree. The second approach is based on processor specification languages.

8.3.2.1 Processor Template Based

Examples of processor template-based approaches are Xtensa [43], Jazz [51], and PEAS [52]. Xtensa [43] is a scalable RISC processor core. Configuration options include the width of the register set, caches, and memories. New functional units and instructions can be added using the Tensilica Instruction Extension (TIE) language. A synthesizable hardware model along with software toolkit can be generated for this class of architectures. Improv's Jazz [51] processor is supported by a flexible design methodology to customize the computational resources and instruction-set of the processor. It allows modifications of data width, number of registers, depth of hardware task queue, and addition of custom functionality in Verilog. PEAS [52] is a GUI-based hardware/software codesign framework. It generates HDL code along with software toolkit. It has support for several architecture types and a library of configurable resources.

8.3.2.2 Specification Language Based

Figure 8.7 shows a typical framework of processor description language-driven HDL generation. Structure-centric ADLs such as MIMOLA are suitable for hardware generation. Some of the behavioral languages (such as ISDL and nML) are also used for hardware generation. For example, the HDL generator HGEN [22] uses ISDL description, and the synthesis tool GO [18] is based on nML. Itoh et al. [53] have proposed a microoperation description based synthesizable HDL generation. It can handle simple processor models with no hardware interlock mechanism or multicycle operations. Mixed languages such as LISA and EXPRESSION capture both structure and behavior of the processor. The synthesizable HDL generation approach based on LISA language [2] produces an HDL model of the architecture. The designer has the choice to generate a VHDL, Verilog, or SystemC representation of the target architecture [2]. The HDL generation methodology presented by Mishra et al. [1] combines the advantages of the processor template-based environments and the language-based specifications using EXPRESSION ADL.

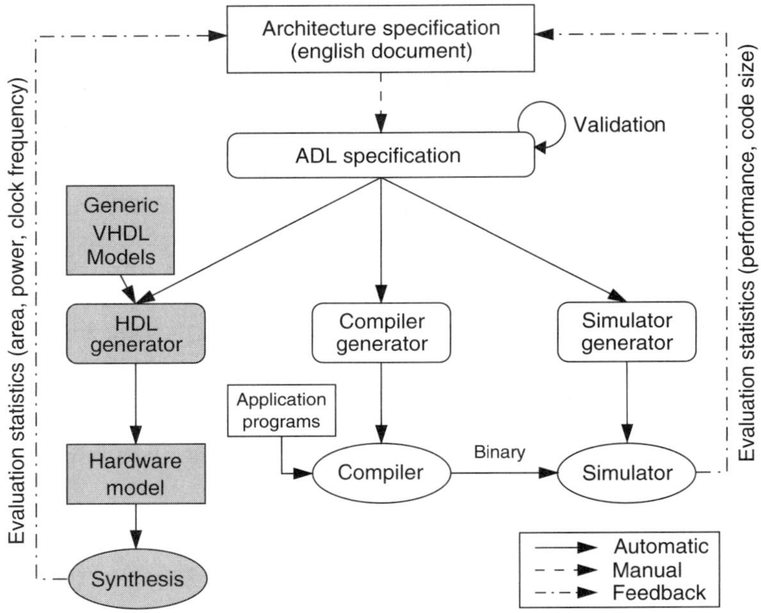

FIGURE 8.7 ADL-driven implementation generation and exploration.

8.3.3 Top-Down Validation

The validation of microprocessors is one of the most complex and important tasks in the current SOC design methodology. Figure 8.8 shows a traditional architecture validation flow. The architect prepares an informal specification of the microprocessor in the form of an English document. The logic designer implements the modules in the RTL. The *RTL design* is validated using a combination of simulation techniques and formal methods. One of the most important problems in today's processor design validation is the lack of a golden reference model that can be used for verifying the design at different levels of abstraction. Thus, many existing validation techniques employ a *bottom–up approach* to pipeline verification, where the functionality of an existing pipelined processor is, in essence, reverse-engineered from its RTL implementation.

Mishra et al. [54] have presented an ADL-driven validation technique that is complementary to these bottom–up approaches. It leverages the system architects knowledge about the behavior of the programmable architectures through ADL constructs, thereby allowing a powerful *top-down approach* to microprocessor validation. Figure 8.9 shows an ADL-driven top–down validation methodology. This methodology has two important steps: validation of ADL specification and specification-driven validation of programmable architectures.

8.3.3.1 Validation of ADL Specification

It is important to verify the ADL specification to ensure the correctness of the architecture specified and the generated software toolkit. Both static and dynamic behavior need to be verified to ensure that the specified architecture is well formed. The static behavior can be validated by analyzing several static properties, such as connectedness, false pipeline and data-transfer paths, and completeness using a graph-based model of the pipelined architecture [55]. The dynamic behavior can be validated by analyzing the instruction flow in the pipeline using a finite state machine (FSM)-based model to verify several important architectural properties such as determinism and in-order execution in the presence of hazards and multiple exceptions [56].

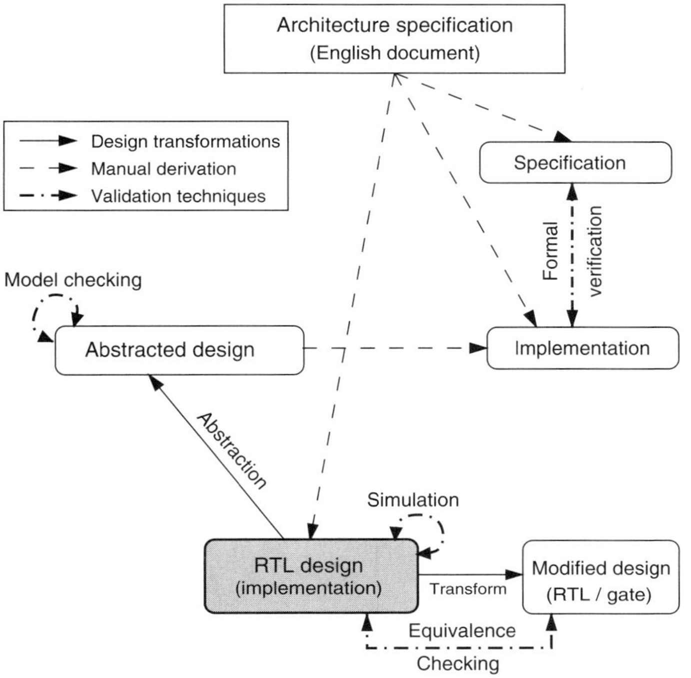

FIGURE 8.8 Traditional bottom–up validation flow.

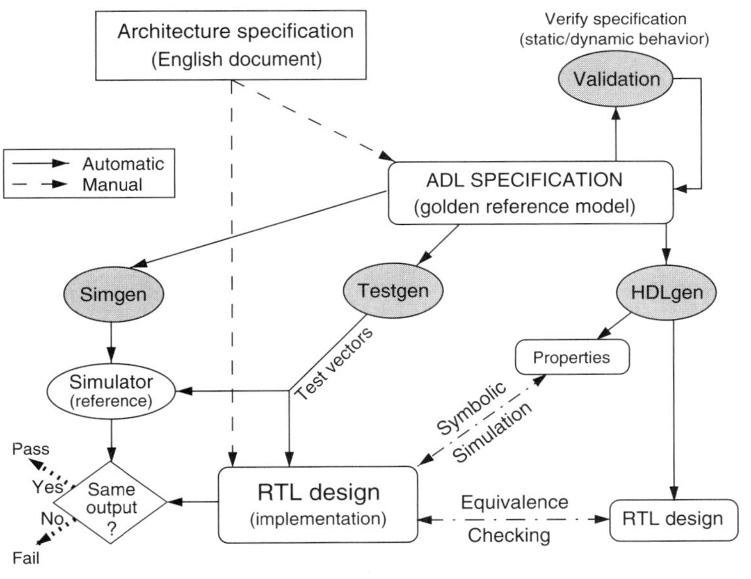

FIGURE 8.9 Top–down validation flow.

8.3.3.2 Specification-Driven Validation

The validated ADL specification can be used as a golden reference model for top–down validation of programmable architectures. The top–down validation approach has been demonstrated in two directions: functional test program generation, and design validation using a combination of equivalence checking and symbolic simulation. Test generation for functional validation of processors has been demonstrated using MIMOLA [12], EXPRESSION [57], and nML [18]. A model checking based approach is used to generate automatically functional test programs from the processor specification using EXPRESSION ADL [57]. It generates graph model of the pipelined processor from the ADL specification. The functional test programs are generated based on the coverage of the pipeline behavior. ADL-driven design validation using equivalence checking has been demonstrated using EXPRESSION ADL [58]. This approach combines ADL-driven hardware generation and validation. The generated hardware model is used as a reference model to verify the hand-written implementation *(RTL design)* of the processor. To verify that the implementation satisfies certain properties, the framework generates the intended properties. These properties are applied using symbolic simulation [58].

8.4 Conclusions

Design and verification of today's complex processors require the use of automated tools and techniques. ADLs have been successfully used in academic research as well as in industry for processor development. The processor is modeled using an ADL. The ADL specification is used to generate software tools including compiler and simulator to enable early design space exploration. The ADL specification is also used to perform other design automation tasks including hardware generation and functional verification.

The early ADLs were either structure-oriented (MIMOLA, UDL/I) or behavior-oriented (nML, ISDL). As a result, each class of ADLs is suitable for specific tasks. For example, structure-oriented ADLs are suitable for hardware synthesis, and unfit for compiler generation. Similarly, behavior-oriented ADLs are appropriate for generating compiler and simulator for instruction-set architectures, and unsuited for generating cycle-accurate simulator or hardware implementation of the architecture. The later ADLs (LISA and EXPRESSION) adopted the mixed approach where the language captures both structure and behavior of the architecture. The existing ADLs are getting modified with the new features and methodologies to perform software toolkit generation, hardware generation, instruction-set synthesis, and test generation for validation of architectures. For example, nML is extended by Target Compiler Technologies to perform hardware synthesis and test generation [18]. Similarly, LISA has been used for hardware generation [2], instruction encoding synthesis [59], and JTAG interface generation [4]. Likewise, EXPRESSION has been used for hardware generation [1], instruction-set synthesis [60], test generation [3], and specification validation [54].

ADLs designed for a specific domain (such as DSP or VLIW) or for a specific purpose (such as simulation or compilation) can be compact and it is possible to automatically generate efficient (in terms of area, power, and performance) tools and hardwares. However, it is difficult to design an ADL for a wide variety of architectures to perform different tasks using the same specification. Generic ADLs require the support of powerful methodologies to generate high-quality results compared to domain or task-specific ADLs. In the future, the existing ADLs will go through changes in two dimensions. First, ADLs will specify not only processor, memory, and co-processor architectures but also other components of the system-on-chip architectures including peripherals and external interfaces. Second, ADLs will be used for software toolkit generation, hardware synthesis, test generation, instruction-set synthesis, and validation of microprocessors. Furthermore, multiprocessor SOCs will be captured and various attendant tasks will be addressed. The tasks include support for formal analysis, generation of RTOS, exploration of communication architectures, and support for interface synthesis. The emerging ADLs will have these features.

References

[1] P. Mishra, A. Kejariwal, and N. Dutt, Synthesis-driven exploration of pipelined embedded processors, *Proceedings of International Conference on VLSI Design*, 2004, pp. 921–926.

[2] O. Schliebusch, A. Chattopadhyay, M. Steinert, G. Braun, A. Nohl, R. Leupers, G. Ascheid, and H. Meyr, RTL processor synthesis for architecture exploration and implementation, *Proceedings of Design Automation and Test in Europe (DATE)*, 2004, pp. 156–160.

[3] P. Mishra and N. Dutt, Functional coverage driven test generation for validation of pipelined processors, *Proceedings of Design Automation and Test in Europe (DATE)*, 2005, pp. 678–683.

[4] O. Schliebusch, D. Kammler, A. Chattopadhyay, R. Leupers, G. Ascheid,and H. Meyr, Automatic generation of JTAG interface and debug mechanism for ASIPs, *GSPx*, 2004.

[5] S. Wang and S. Malik, Synthesizing operating system based device drivers in embedded systems, *Proceedings of International Symposium on Hardware/Software Codesign and System Synthesis (CODES+ISSS)*, 2003, pp. 37–44.

[6] N. Medvidovic and R. Taylor, A framework for classifying and comparing architecture description languages, in *Proceedings of European Software Engineering Conference (ESEC)*, M. J. and H. Schauer, Ed., Springer, Berlin, 1997, pp. 60–76.

[7] Paul C. Clements, A survey of architecture description languages, *Proceedings of International Workshop on Software Specification and Design (IWSSD)*, 1996, pp. 16–25.

[8] M. R. Barbacci, Instruction set processor specifications (ISPS): The notation and its applications. *IEEE Trans. Comp.*, C-30, 24–40, 1981.

[9] W. Qin and S. Malik, *Architecture Description Languages for Retargetable Compilation, in The Compiler Design Handbook: Optimizations & Machine Code Generation*, CRC Press, Boca Raton, FL, 2002.

[10] P. Mishra and N. Dutt, Architecture description languages for programmable embedded systems. *IEE Proceedings on Computers and Digital Techniques*, 2005.

[11] H. Tomiyama, A. Halambi, P. Grun, N. Dutt, and A. Nicolau, Architecture description languages for systems-on-chip design, *Proceedings of Asia Pacific Conference on Chip Design Language*, 1999, pp. 109–116.

[12] R. Leupers and P. Marwedel, Retargetable code generation based on structural processor descriptions, *Design Automat. Embedded Syst.*, 3, 75–108, 1998.

[13] M. Freericks, The nML Machine Description Formalism, Technical Report TR SM-IMP/DIST/08, TU Berlin, Germany, 1993.

[14] G. Hadjiyiannis, S. Hanono, and S. Devadas, ISDL: an instruction set description language for retargetability, *Proceedings of Design Automation Conference (DAC)* , 1997, pp. 299–302.

[15] A. Fauth and A. Knoll, Automatic generation of DSP program development tools, *Proceedings of International Conference Acoustics, Speech and Signal Processing (ICASSP)*, 1993, pp. 457–460.

[16] D. Lanneer, J. Praet, A. Kifli, K. Schoofs, W. Geurts, F. Thoen, and G. Goossens, CHESS: Retargetable code generation for embedded DSP processors, in *Code Generation for Embedded Processors*, Kluwer Academic Publishers, Dordrecht, 1995, pp. 85–102.

[17] F. Lohr, A. Fauth, and M. Freericks, Sigh/sim: An Environment for Retargetable Instruction Set Simulation,. Technical Report 1993/43, Berlin, Germany, 1993.

[18] http://www.retarget.com. *Target Compiler Technologies.*

[19] J. Paakki, Attribute grammar paradigms — a high level methodology in language implementation, *ACM Comp. Surv.*, 27(2):196-256, 1995.

[20] D. Kastner, TDL: A Hardware and Assembly Description Languages, Technical Report TDL 1.4, Saarland University, Germany, 2000.

[21] S. Hanono and S. Devadas, Instruction selection, resource allocation, and scheduling in the AVIV retargetable code generator, *Proceedings of Design Automation Conference (DAC)*, 1998, pp. 510–515.

[22] G. Hadjiyiannis, P. Russo, and S. Devadas, A methodology for accurate performance evaluation in architecture exploration, *Proceedings of Design Automation Conference (DAC)*, 1999, pp. 927–932.

[23] J. Gyllenhaal, B. Rau, and W. Hwu,. HMDES Version 2.0 Specification, Technical Report IMPACT-96-3, IMPACT Research Group, University of Illinois, Urbana, IL, 1996.

[24] The MDES User Manual, http://www.trimaran.org, 1997.

[25] A. Halambi, P. Grun, V. Ganesh, A. Khare, N. Dutt, and A. Nicolau, EXPRESSION: a language for architecture exploration through compiler/simulator retargetability, *Proceedings of Design Automation and Test in Europe (DATE)*, 1999, pp. 485–490.

[26] P. Grun, A. Halambi, N. Dutt, A. Nicolau, RTGEN: an algorithm for automatic generation of reservation tables from architectural descriptions, *Proceedings of International Symposium on System Synthesis (ISSS)*, 1999, pp. 44–50.

[27] A. Halambi, A. Shrivastava, N. Dutt, and A. Nicolau, A customizable compiler framework for embedded systems, *Proceedings of Software and Compilers for Embedded Systems (SCOPES)*, 2001.

[28] A. Khare, N. Savoiu, A. Halambi, P. Grun, N. Dutt, and A. Nicolau, V-SAT: a visual specification and analysis tool for system-on-chip exploration, *Proceedings of EUROMICRO Conference*, 1999, pp. 1196–1203.

[29] http://www.ics.uci.edu/~express. *Exploration framework using EXPRESSION.*

[30] J. Hennessy and D. Patterson, *Computer Architecture: A Quantitative Approach.* Morgan Kaufmann Publishers Inc, San Mateo, CA, 1990.

[31] P. Mishra, M. Mamidipaka, and N. Dutt, Processor-memory co-exploration using an architecture description language, *ACM Trans. Embedded Comput. Syst. (TECS)*, 3, 140–162, 2004.

[32] P. Grun, N. Dutt, and A. Nicolau, Memory aware compilation through accurate timing extraction, *Proceedings of Design Automation Conference (DAC)*, 2000, pp. 316–321.

[33] V. Zivojnovic, S. Pees, and H. Meyr, LISA — machine description language and generic machine model for HW/SW co-design, *IEEE Workshop on VLSI Signal Processing*, 1996, pp. 127–136.

[34] S. Pees, A. Hoffmann, H. Meyr, Retargetable compiled simulation of embedded processors using a machine description language. *ACM Trans. Design Automat. of Electron. Syst.*, 5, 815–834, 2000.

[35] M. Hohenauer, H. Scharwaechter, K. Karuri, O. Wahlen, T. Kogel, R. Leupers, G. Ascheid, H. Meyr, G. Braun, and H. Someren, A methodology and tool suite for c compiler generation from ADL processor models, *Proceedings of Design Automation and Test in Europe (DATE)*, 2004, pp. 1276–1283.

[36] O. Wahlen, M. Hohenauer, R. Leupers, and H. Meyr, Instruction scheduler generation for retargetable compilation. *IEEE Design Test Comput.*, 20, 34–41, 2003.

[37] C. Siska, A processor description language supporting retargetable multi-pipeline DSP program development tools, *Proceedings of International Symposium on System Synthesis (ISSS)*, 1998, pp. 31–36.

[38] T. Morimoto, K. Yamazaki, H. Nakamura, T. Boku, and K. Nakazawa, Superscalar processor design with hardware description language aidl, *Proceedings of Asia Pacific Conference on Hardware Description Languages (APCHDL)*, 1994.

[39] A. Inoue, H. Tomiyama, F. Eko, H. Kanbara, and H. Yasuura, A programming language for processor based embedded systems, *Proceedings of Asia Pacific Conference on Hardware Description Languages (APCHDL)*, 1998, pp. 89–94.

[40] V. Rajesh and R. Moona, Processor modeling for hardware software codesign, *Proceedings of International Conference on VLSI Design*, 1999, pp. 132–137.

[41] ARC Cores. *http://www.arccores.com.*

[42] http://www.axysdesign.com. *Axys Design Automation.*

[43] Tensilica Inc. *http://www.tensilica.com.*

[44] P. Marwedel and G. Goossens, *Code Generation for Embedded Processors*, Kluwer Academic Publishers, Dordrecht, 1995.

[45] J. Zhu and D. Gajski, A retargetable, ultra-fast, instruction set simulator, *Proceedings of Design Automation and Test in Europe (DATE)*, 1999.

[46] R. Cmelik and D. Keppel, Shade: a fast instruction-set simulator for execution profiling. *ACM SIGMETRICS Performance Evaluation Rev.*, 22, 128–137, 1994.

[47] E. Witchel, and M. Rosenblum, Embra: fast and flexible machine simulation, *Measurement and Modeling of Computer Systems*, 1996, pp. 68–79.

[48] A. Nohl, G. Braun, O. Schliebusch, R. Leupers, H. Meyr, and A. Hoffmann, A universal technique for fast and flexible instruction-set architecture simulation, *Proceedings of Design Automation Conference (DAC)*, 2002, pp. 22–27.

[49] M. Reshadi, P. Mishra, and N. Dutt, Instruction set compiled simulation: a technique for fast and flexible instruction set simulation, *Proceedings of Design Automation Conference (DAC)*, 2003, pp. 758–763.

[50] M. Reshadi, N. Bansal, P. Mishra, N. Dutt, An efficient retargetable framework for instruction-set simulation, *Proceedings of International Symposium on Hardware/Software Codesign and System Synthesis (CODES+ISSS)*, 2003, pp. 13–18.

[51] http://www.improvsys.com. *Improv Inc.*

[52] M. Itoh, S. Higaki, Y. Takeuchi, A. Kitajima, M. Imai, J. Sato, and A. Shiomi, PEAS-III: an ASIP design environment, in *Proceedings of International Conference on Computer Design (ICCD)*, 2000.

[53] M. Itoh, Y. Takeuchi, M. Imai, A. Shiomi, Synthesizable HDL generation for pipelined processors from a micro-operation description, *IEICE Trans. Fundamentals*, E00-A(3), March 2000.

[54] P. Mishra, Specification-Driven Validation of Programmable Embedded Systems, Ph.D. thesis, University of California, Irvine, 2004.

[55] P. Mishra and N. Dutt, Automatic modeling and validation of pipeline specifications, *ACM Transactions on Embedded Computing Systems (TECS)*, 3, 114–139, 2004.

[56] P. Mishra, N. Dutt, and H. Tomiyama, Towards automatic validation of dynamic behavior in pipelined processor specifications, *Kluwer Design Automat.Embedded Syst. (DAES)*, 8, 249–265, 2003.

[57] P. Mishra, and N. Dutt, Graph-based functional test program generation for pipelined processors, *Proceedings of Design Automation and Test in Europe (DATE)*, 2004, pp. 182–187.

[58] P. Mishra, and N. Dutt, N. Krishnamurthy, and M. Abadir, A top-down methodology for validation of microprocessors, *IEEE Design Test Comput.*, 21, 122–131, 2004.

[59] A. Nohl, V. Greive, G. Braun, A. Hoffmann, R. Leupers, O. Schliebusch, and H. Meyr, Instruction encoding synthesis for architecture exploration using hierarchical processor models, *Proceedings of Design Automation Conference (DAC)*, 2003, pp. 262–267.

[60] P. Biswas, and N. Dutt, Reducing code size for heterogeneous-connectivity-based VLIW DSPs through synthesis of instruction set extensions, *Proceedings of Compilers, Architectures, Synthesis for Embedded Systems (CASES)*, 2003, pp. 104–112.

[61] M. Hartoog, J. Rowson, P. Reddy, S. Desai, D. Dunlop, E. Harcourt, and N. Khullar, Generation of software tools from processor descriptions for hardware/software codesign, *Proceedings of Design Automation Conference (DAC)*, 1997, pp. 303–306.

9

Embedded Software Modeling and Design

Marco Di Natale
Scuola Superiore S. Anna
Pisa, Italy

Abstract

The development of correct complex software for reactive embedded systems requires automated support for the verification of functional and nonfunctional properties by formal analysis or by simulation and testing. Reduced time-to-market and increased complexity also demand the reuse of components, which also brings the issue of composability of software artifacts. Currently, a language (or a design methodology) that can provide all these desirable features without incurring excessive inefficiency is not available, and a strong semantic characterization of the modeling language or separation of the concerns between the functional and the architecture-level design is the solution advocated by many. This chapter provides an overview of existing models and tools for embedded software, starting from an introduction to the fundamental concepts and the basic theory of existing models of computation, both synchronous and asynchronous. The chapter also features an introduction to commercial languages and tools such as Unified Modeling Language (UML) and Specification and Description Language (SDL), and a quick peek at research work in software models and tools, to give a firm understanding of future trends and currently unsolved issues.

9.1 Introduction

The increasing cost necessary for the design and fabrication of ASICs, together with the need for the reuse of functionality, adaptability, and flexibility, are among the causes for an increasing share of software-implemented functions in embedded projects. Figure 9.1 represents a typical architectural framework for embedded systems, where application software runs on top of a real-time operating system (RTOS) (and possibly a middleware layer) which abstracts from the hardware and provides a common API for the reuse of functionality (such as the OSEK standard in the automotive domain).

Unfortunately, mechanisms for improving the reuse of software at the level of programming code, such as RTOS- or middleware-level APIs, are still not sufficient to achieve the desired levels of productivity, and the error rate of software programs is exceedingly high. Today, model-based design of software bears the promise of a much-needed step-up in productivity and reuse.

The use of abstract software models at the highest possible level in the development process may significantly increase the chances that the design and its implementation are correct. Correctness can be achieved in many ways. Ideally, it should be mathematically proved by formal reasoning upon the model of the system and its desired properties (provided the model is built on solid mathematical foundations and its properties are expressed by some logic function(s)). Unfortunately, in many cases, formal model checking is not possible or simply impractical. In this case, the modeling language should at least provide abstractions for the specification of reusable components, so that software artifacts can be clearly identified with the provided functions or services.

Also, when exhaustive proof of correctness cannot be achieved, a secondary goal of the modeling language should be providing support for simulation and testing. In this view, formal methods can also be used to guide the generation of the test suite and to guarantee coverage. Finally, modeling languages and tools should ensure that the model of the software, after being checked by means of formal proof or by simulation, is correctly implemented in a programming language executed on the target hardware (this requirement is usually satisfied by automatic code generation tools).

Industry and research groups have been working now for decades in the software engineering area looking for models, methodologies, and tools to improve the correctness and increase the reusability of software components. Traditionally, software models and formal specifications have had their focus on behavioral properties and have been increasingly successful in the verification of functional correctness. However, embedded software is characterized by concurrency and resource constraints and by nonfunctional properties, such as deadlines or other timing constraints, which ultimately depend upon the computation platform.

This chapter attempts at providing an overview of (visual and textual) languages and tools for embedded software modeling and design. The subject is so wide, evolving so rapidly, and encompassing so many different issues that only a short survey is possible in the limited space allocated to this chapter. As of today, despite all efforts, existing methodologies and languages fall short in achieving most of the desirable goals and yet they are continuously being extended in order to allow for the verification of at least a subset of the properties of interest.

The objective is to provide the reader with an understanding of what are the principles for functional and nonfunctional modeling and verification, what are the languages and tools available in the market and what can be possibly achieved with respect to practical designs. The description of (some) commercial languages, models, and tools is supplemented with a survey of the main research trends and the

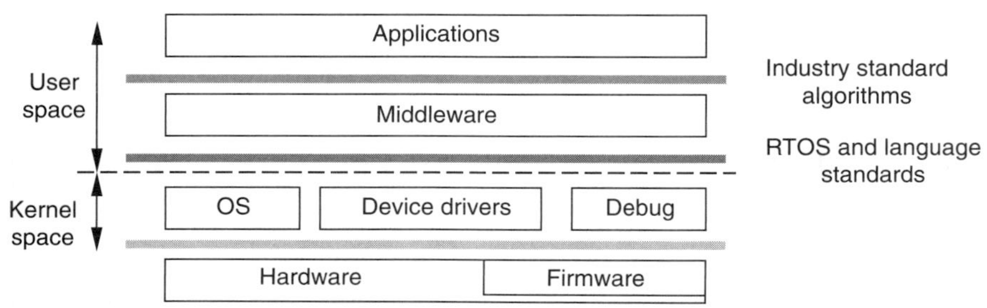

FIGURE 9.1 Common architecture for embedded software.

results that may open new possibilities in the future. The reader is invited to refer to the cited papers in the bibliography section for wider discussion on each issue.

The organization of the chapter is as follows: the introduction section defines a reference framework for the discussion of the software-modeling problem and provides a short review of abstract models for functional and temporal (schedulability) analysis. The second section provides a quick glance at the two main categories of available languages and models purely synchronous as opposed to general asynchronous models. Then, an introduction to the commercial modeling languages Unified Modeling Language (UML) and Specification and Description Language (SDL) is provided. The discussion of what can be achieved with both, with respect to formal analysis of functional properties, schedulability analysis, simulation and testing follows. This chapter also discusses the recent extensions of existing methodologies to achieve the desirable goal of component-based design. Finally, a quick glance at the research work in the area of embedded software design, methods, and tools, closes the chapter.

9.1.1 Challenges in the Development of Embedded Software

According to a typical development process (represented in Figure 9.2), an embedded system is the result of possibly multiple refinement stages encompassing several levels of abstraction, from user requirements, to system testing, and sign-off. At each stage, the system is described using an adequate formalism, starting from abstract models for the user domain entities at the requirements level. Lower-level models, typically developed in the later stages, provide an implementation of the abstract model by means of design entities representing hardware and software components. The implementation process is a sequence of steps that constrain the generic specification by exploiting the possible options (non-determinism) available from higher levels.

The designer's problem is making sure that the models of the system developed at the different stages do satisfy the properties required from the system, and that low-level descriptions of the system are correct implementations of higher-level specifications. This task can be considerably easier if the models of the system at different abstraction levels are homogeneous, i.e., if the computational models on which they are based share a common semantics and, possibly, a common notation.

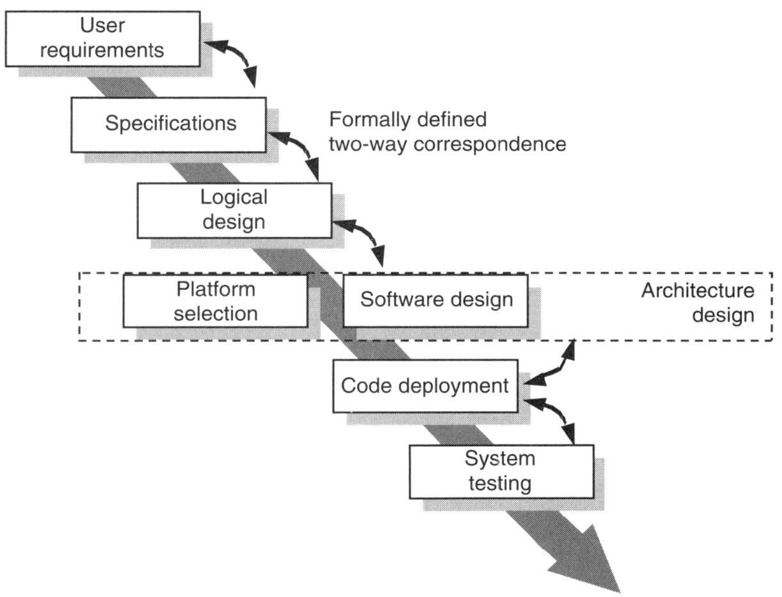

FIGURE 9.2 Typical embedded software development process.

The problem of correct mapping from a high-level specification, employing an abstract model of the system, to a particular software and hardware architecture or *platform* is one of the key aspects of the design of embedded systems.

The separation of the two main concerns of functional and architectural specification and the mapping of functions to architecture elements are among the founding principles of many design methodologies, such as the platform-based design [1] and tools like the Ptolemy and Metropolis frameworks [2,3], as well as of emerging standards and recommendations, such as the UML profile for schedulability, performance, and time (SPT) from the Object Management Group (OMG) [4] and industry-based practices, such as the V-cycle of software development common in the automotive industry [5]. A keyhole view of the corresponding stages is represented in Figure 9.3.

The main design activities taking place at this stage and the corresponding challenges can be summarized as follows:

- *The specification of functionality* is concerned with the development of logically correct system abstractions. If the specification is defined using a formal model, *formal verification* allows checking that the functional behavior satisfies a given set of properties.
- The *system software and hardware* platform components are defined in the *architecture design* level.
- After the definition of logical and physical resources available for the execution of the functional model and the definition of the *mapping of functional model elements onto the platform (architecture elements)* executing them, *formal verification of nonfunctional properties*, such as timing properties, and schedulability analysis usually take place.

Complementing the above two steps, *implementation verification* is the process of checking that the behavior of the logical model, when mapped onto the architecture model, correctly implements the high-level formal specifications.

9.1.2 A Short Introduction to Formal Models and Languages and to Schedulability Analysis

A short review of the most common models of computation (formal languages) proposed by academia and possibly supported by the industry with the objective of formal or simulation-based verification is fundamental for understanding commercial models and languages, and it is also important for understanding present and future challenges (refer to [6,7] for more details).

9.1.2.1 Formal Models

Formal models are formal languages that specify the semantics of computation and communication (also defined as *model of computation* [MOC] [6]). Models of computations may be expressed, for example, by means of a language or automaton formalisms.

A system-level MOC is used to describe the system as a (possibly hierarchical) collection of design entities (blocks, actors, tasks, processes) performing units of computations represented as transitions or

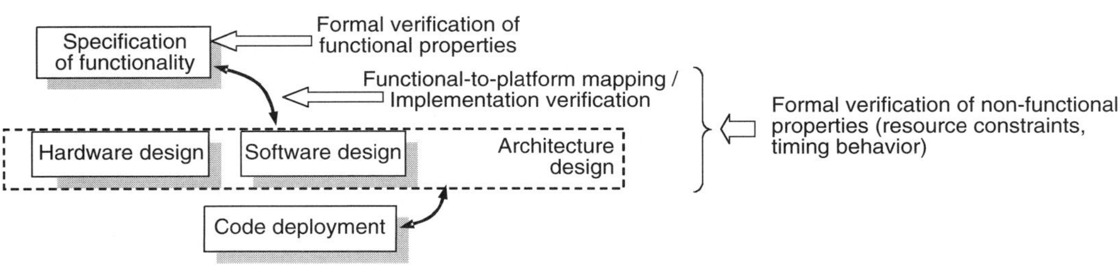

FIGURE 9.3 Mapping formal specifications to a hardware/software platform.

actions, characterized by a state and communicating by means of events (tokens) carried out by signals. Composition and communication rules, concurrency models, and time representation are among the most important characteristics of an MOC.

Once the system specifications are given according to a formal MOC, *formal methods* can be used to achieve design-time verification of *properties* and *implementation* as in Figure 9.3. In general, properties of interest go under the two general categories of *ordered* and *timed executions*:

- Ordered execution relates to verification of event and state ordering. Properties such as *safety, liveness, absence of deadlock, fairness,* or *reachability,* belong to this category.
- Timed execution relates to event enumeration, such as checking that no more that *n* events (including time events) occur between any two events in the system, for example, *timeliness* and some notions of *fairness.*

Verification of desirable system properties may be quite hard or even impossible to achieve by logical reasoning on formal models. Formal models are usually classified according to the *decidability* of properties. Decidability of properties in timed and untimed models depends on many factors, such as the type of logic (propositional or first-order) for conditions on transitions and states, the real-time semantics, including the definition of the time domain (discrete or dense) and the linear or branching time logic that is used for expressing properties (interested readers may refer to [7] for a survey on the subject). In practice [6], decidability should be carefully evaluated. In some cases, even if it is decidable, the problem cannot be practically solved since the required run time may be prohibitive and, in other instances, even if undecidability applies to the general case, it may happen that the problem at hand admits a solution.

Verification of model properties can take many forms. In the *deductive* approach, the system and the property are represented by statements (clauses) written in some logic (for example, expressed in the linear time temporal logic (LTL) [8] or in the branching time Computation Tree Logic (CTL) [9]) and a theorem proving tool (usually under the direction of a designer or some expert) applies deduction rules until (hopefully) the desired property reduces to a set of axioms, or a counterexample is found. In model checking, the system and possibly the desirable properties are expressed using an automaton or some other kind of executable formalism. The verification tool ensures that no executable transition or reachable system state violates the property. To do so, it can generate all the potential (finite) states of the system (exhaustive analysis). When the property is violated, the tool usually produces the (set of) counterexample(s).

The first model checkers worked by constructing the whole structure of states prior to property checking, but modern tools are able to perform verification as the states are produced. This means that the method does not necessarily require the construction of the whole state graph. On the fly model checking and the SPIN [10] toolset provide an instance and an implementation of this approach, respectively.

To give some examples (Figure 9.4), checking a system implementation I against a specification of a property P in case both are expressed in terms of automata (*homogeneous verification*) requires the following steps. The implementation automaton A_I is composed with the complementary automaton A_P expressing the negation of the desired property. The implementation I violates the specification property if the product automaton $A_I \| A_P$ has some possible run and it is verified if the composition has no runs. *Checking by observers* can be considered as a particular instance of this method, very popular for synchronous models.

In the very common case in which the property specification consists of a logical formula and the implementation of the system is given by an automaton, the verification problem can be solved algorithmically or deductively by transforming it into an instance of the previous cases, for example, by transforming the negation of a specification formula f_S into the corresponding automaton, and by using the same techniques as in homogeneous verification.

Verification of implementation is usually obtained by exploiting simulation and bisimulation properties.

Following is a very short survey of formal system models, starting with finite-state machines (FSMs), probably the most popular and the basis for many extensions.

In FSMs process behavior is specified by enumerating the (finite) set of possible system states and the transitions among them. Each transition connects two states and it is labeled with the subset of input

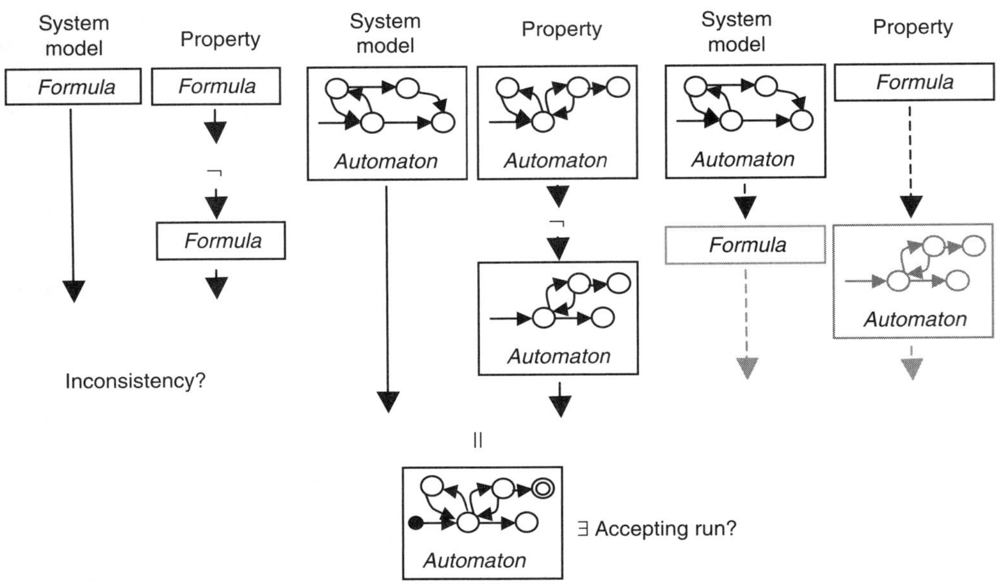

FIGURE 9.4 Checking the system model against some property.

variables (and possibly the guard condition upon their values) that triggers its execution. Furthermore, each transition can produce output variables. In Mealy FSMs, outputs depend on both state and input variables, while in the Moore model, outputs only depend on the process state. Guard conditions can be expressed according to different logics, such as propositional logic, first-order logic, or even Turing-complete programming code.

In the *Synchronous* FSM model, signal propagation is assumed to be instantaneous. Transitions and the evaluation of the next state happen for *all the system components* at the same time. Synchronous languages such as Esterel and Lustre are based on this model. In the *Asynchronous* model, two asynchronous FSMs never execute a transition at the same time except when rendezvous is explicitly specified (a pair of transitions of the communicating FSMs occur simultaneously). The SDL process behavior is an instance of this general model.

Composition of FSMs is obtained by the construction of a product transition system, i.e., a single FSM where the set of states is the cartesian product of the sets of the states of the component machines. The difference between synchronous and asynchronous execution semantics is quite clear when compositional behaviors are compared. Figure 9.5 portrays an example showing the differences in the synchronous and asynchronous composition of two FSMs.

When there is a cyclic dependency among variables in interconnected synchronous FSMs, the Mealy model like any other model where outputs are instantaneously produced based on the input values, may result in a fixed-point problem and possibly inconsistency (Figure 9.6 shows a simple functional dependency). The existence of a unique fixed-point solution (and its evaluation) is not the only problem resulting from the composition of FSMs.

In large, complex systems, composition may easily result in a huge number of states. This phenomenon is known as *state explosion.*

In the *Statecharts* extension to FSMs [11], Harel proposed three mechanisms to reduce the size of an FSM for modeling practical systems: state hierarchy, simultaneous activity, and nondeterminism. In *Statecharts,* a state can possibly represent an enclosed state machine. In this case, the machine is in one of the states enclosed by the super state(s) and concurrency is achieved by enabling two or more state machines to be active simultaneously (and-states, such as lap and off in Figure 9.7).

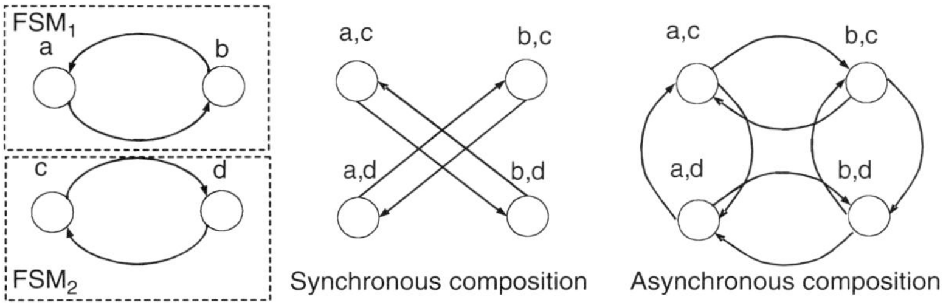

FIGURE 9.5 Composition of synchronous and asynchronous FSMs.

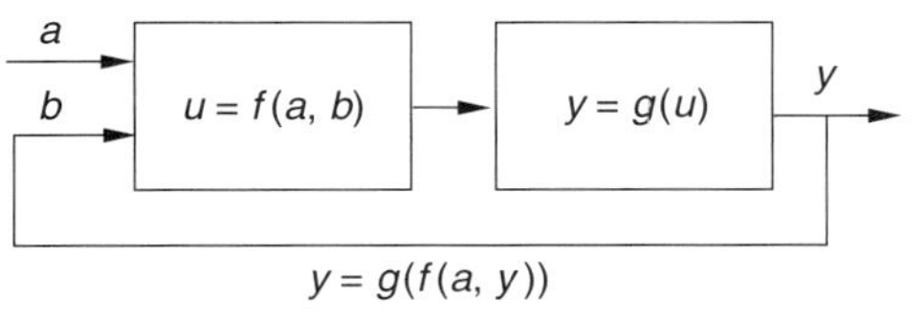

FIGURE 9.6 A fixed-point problem arising from the composition and cyclic dependencies.

In *Petri net* (PN) models, the system is represented by a graph of places connected by transitions. The places represent unbounded channels that carry *tokens,* and the state of the system is represented at any given time by the number of tokens existing in a given subset of places. Transitions represent the elementary reactions of the system. A transition can be executed (fired) when it has a fixed, prespecified number of tokens in its input places. When fired, it consumes the input tokens and produces a fixed number of tokens on its output places. Since more than one transition may originate from the same place, one transition can execute while disabling another one by removing the tokens from the shared input places. Hence, the model allows for non-determinism and provides a natural representation of concurrency by allowing simultaneous execution of multiple transitions (Figure 9.8, left).

The FSM and PN models have originally been developed with no reference to time or time constraints, but the capability of expressing and verifying timing requirements is key in many design domains (including embedded systems). Hence, both have been extended in order to allow time-related specifications. Time extensions differ according to the time model that is assumed. Models that represent time with a discrete-time base are said to belong to the family of discrete-time models, while the others are based on continuous (dense) time. Furthermore, proposals of extensions differ in how time references should be used in the system, whether a global clock or local clocks should be used, and how time should be used in guards on transitions or states, inputs and outputs.

When building the set of reachable states, the time value adds another dimension, further contributing to the state explosion problem. In general, discrete-time models are easier to analyze if compared with dense time models, but synchronization of signals and transitions results in fixed-point evaluation problems whenever the system model contains cycles without delays.

Note that discrete-time systems are naturally prone to an implementation based on the *time-triggered* paradigm, where all actions are bound to happen at multiples of a time reference (usually implemented by means of a response to a timer interrupt), and continuous-time (asynchronous systems) conventionally corresponds to implementations based on the event-based design paradigm, where system actions can happen at any time instant. This does not imply a correspondence between time-triggered systems and synchronous systems. The latter are characterized by the additional constraints that all system components must perform an action synchronously (at the same time) at each tick in a periodic time base.

Many models have been proposed in research literature for time-related extensions. Among those, TPN [12,13] and timed automata (TA) [14] are probably the best known.

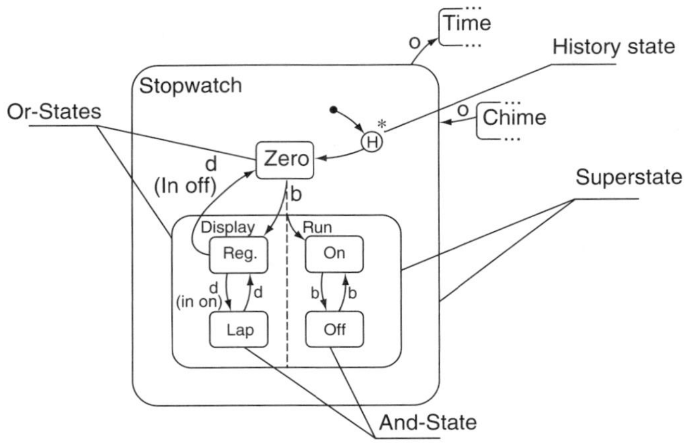

FIGURE 9.7 An example of statechart.

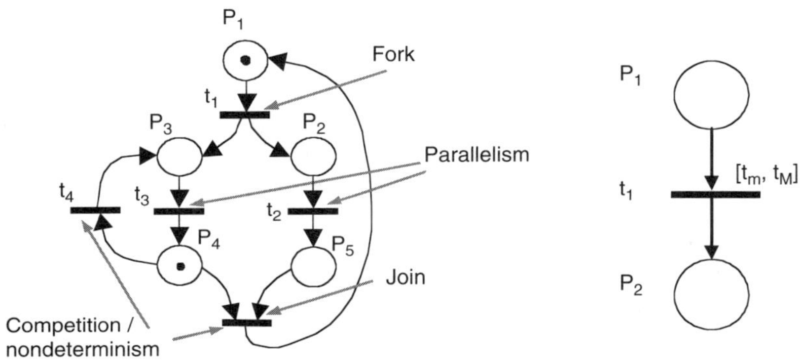

FIGURE 9.8 A sample PN showing examples of concurrent behavior and non-determinism (left) and notations for the time Petri net (TPN) model (right).

Timed automata (an example is given in Figure 9.9) operate with a finite set of locations (states) and a finite set of real-valued clocks. All clocks proceed at the same rate and measure the amount of time that passed since they were started (reset). Each transition may reset some of the clocks and each defines a restriction on the value of the symbols as well as on the clock values required for it to happen. A state may be reached only if the values of the clocks satisfy the constraints and the proposition clause defined on the symbols evaluates to true.

Timed Petri Nets (TDPN) [15] and Time Petri Nets (TPN) are two different, albeit similarly named, models that extend the Petri net formalism allowing for the expression of time-related constraints. The two differ in the way time advances. In timed Petri nets time advances in transitions, thus violating the instantaneous nature of transitions (which makes the model much less prone to analysis). In the TPN model, time advances while token(s) are in places (Figure 9.10). Enabling and deadline times can be associated to transitions, the enabling time being the time a transition must be enabled before firing, and the deadline being the time instant by which the transition must be taken (Figure 9.8, right).

The additional notion of stochastic time allows the definition of the (generalized) stochastic PNs [16,17] used for the purpose of performance evaluation.

Many further extensions have been proposed for both TAs and TPNs. The task of comparing the two models for expressiveness should take into account all the possible variants and is probably not particularly

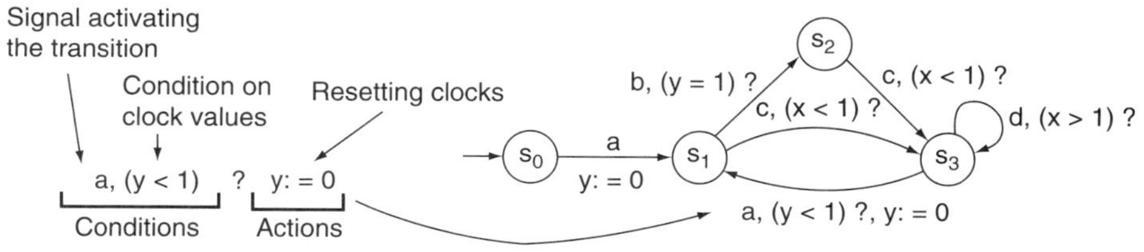

FIGURE 9.9 An example of TA.

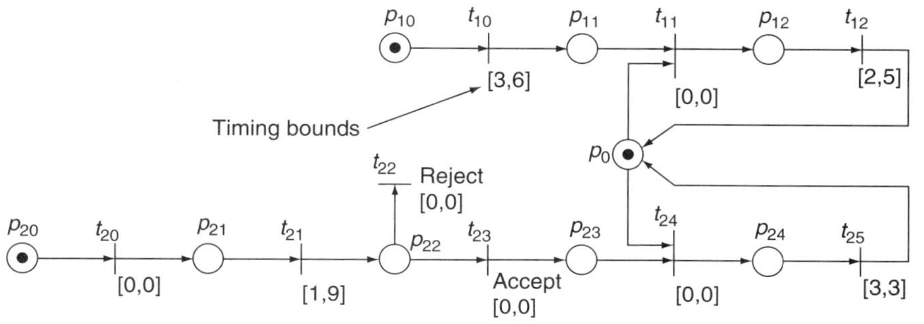

FIGURE 9.10 A sample time Petri net.

interesting in itself. For most problems of practical interest, however, both models are essentially equivalent when it comes to expressive power and analysis capability [18].

A few tools based on the TA paradigm have been developed and are very popular. Among those, we cite Kronos [19] and Uppaal [20]. The Uppaal tool allows modeling, simulation, and verification of real-time systems modeled as a collection of nondeterministic processes with finite control structure and real-valued clocks, communicating through channels or shared variables [20,21]. The tool is free for non-profit and academic institutions.

TAs and TPNs allow the formal expression of the requirements for logical-level resources, timing constraints, and timing assumptions, but timing analysis only deals with abstract specification entities, typically assuming infinite availability of physical resources (such as memory or CPU speed). If the system includes an RTOS with the associated scheduler, the model needs to account for preemption, resource sharing and the non-determinism resulting from them. Dealing with these issues requires further evolution of the models.

For example, in TA, we may want to use clock variables for representing the execution time of each action. In this case, however, only the clock associated with the action scheduled on the CPU should advance, with all the others being stopped.

The hybrid automata model [22] combines discrete transition graphs with continuous dynamical systems. The value of system variables may change according to a discrete transition or it may change continuously in system states according to a trajectory defined by a system of differential equations. Hybrid automata have been developed for the purpose of modeling digital systems interacting with (physical) analog environments, but the capability of stopping the evolution of clock variables in states (first derivative equal to 0) makes the formalism suitable for the modeling of systems with preemption.

Time Petri Nets and TA can also be extended to cope with the problem of modeling finite computing resources and preemption. In the case of TA, the extension consists of the Stopwatch Automata model, which handles suspension of the computation due to the release of the CPU (because of real-time scheduling), implemented in the HyTech [23] for linear hybrid automata. Alternatively, the scheduler is modeled with an extension to the TA model, allowing for clock updates by subtraction inside transitions (besides normal clock resetting). This extension, available in the Uppaal tool, avoids the undecidability of the model where clocks associated with the actions not scheduled on the CPU are stopped.

Likewise, TPNs can be extended to the preemptive TPN model [24], as supported by the ORIS tool [25]. A tentative correspondence between the two models is traced in [26]. Unfortunately, in all these cases, the complexity of the verification procedure caused by the state explosion poses severe limitations upon the size of the analyzable systems.

Before moving on to the discussion of formal techniques for the analysis of time-related properties at the architecture level (schedulability), the interested reader is invited to refer to [27] for a survey on formal methods, including references to industrial examples.

9.1.2.2 Schedulability Analysis

If specification of functionality aims at producing a logically correct representation of system behavior, architecture-level design is where physical concurrency and schedulability requirements are expressed. At this level, the units of computation are the *processes* or *threads* (the distinction between these two operating systems concepts is not relevant for the purpose of this chapter and in the following text, the generic term *task* will be optionally used for both), executing concurrently in response to environment stimuli or prompted by an internal clock. Threads cooperate by exchanging data and synchronization or activation signals and contend for the use of the execution resource(s) (the processor) as well as for the other resources in the system. The physical architecture level is also the place, where the concurrent entities are mapped onto target hardware. This activity entails the selection of an appropriate scheduling policy (for example, offered by an RTOS), and possibly support by timing or schedulability analysis tools.

Formal models, exhaustive analysis techniques, and model checking are now evolving toward the representation and verification of time and resource constraints together with the functional behavior. However, applicability of these models is strongly limited by state explosion. In this case, exhaustive analysis and joint verification of functional and nonfunctional behavior can be sacrificed for the lesser goal of analyzing only the *worst-case timing* behavior of coarse-grain design entities representing concurrently executing threads.

Software models for time and schedulability analysis deal with preemption, physical and logical resource requirements, and resource management policies and are typically limited to a somewhat simplified view of functional (logical) behavior, mainly limited to synchronization and activation signals.

To give an example if, for the sake of simplicity, we limit discussion to single processor systems, the scheduler assigns the execution engine (CPU) to threads (tasks) and the main objective of the real-time scheduling policies is to formally guarantee the timing constraints (deadlines) on the thread response to external events.

In this case, the software architecture can be represented as a set of concurrent tasks (threads). Each task τ_i executes periodically or according to a sporadic pattern and it is typically represented by a simple set of attributes, such as the tuple $(C_i, \theta_i, p_i, D_i)$ representing the worst-case computation time, the period (for periodic threads) or the minimum inter-arrival time (for sporadic threads), the priority, and the relative (to the release time r_i) deadline of each thread instance.

Fixed priority scheduling and rate monotonic analysis (RMA) [28,29] are by far the most common real-time scheduling and analysis methodologies. Rate monotonic analysis provides a very simple procedure for assigning static priorities to a set of *independent periodic tasks* together with a formula for checking schedulability against deadlines.

The highest priority is assigned to the task having the highest rate, and schedulability is guaranteed by checking the worst-case scenario that can possibly happen. If the set of tasks is schedulable in that condition, then it is schedulable under all circumstances. For RMA the critical condition happens when all the tasks are released at the same time instant, initiating the largest busy period (continuous time interval when the processor is busy executing tasks of a given priority level).

By analyzing the busy period (from $t = 0$), it is possible to derive the worst-case completion time W_i for each task τ_i. If the task can be proven to complete before or at the deadline ($W_i \leq D_i$), then it can be guaranteed. The iterative formula for computing W_i (in case $\theta_i \leq D_i$) is

$$W_i = C_i + \sum_{\forall j \in he(i)} \frac{W_i}{\theta_j} C_j$$

where *he(i)* are the indices of those tasks having a priority higher than or equal to p_i.

Rate monotonic scheduling was developed starting from a very simple model, where all tasks are periodic and independent. In reality, tasks require access to shared resources (apart from the processor) that can only be used in an exclusive way, such as, for example, communication buffers shared among asynchronous threads.

In this case, it is possible that one task is blocked because another task holds a lock on the shared resources. When the blocked task has a priority higher than the blocking task, *priority inversion* occurs and finding the optimal priority assignment becomes an NP-hard problem. Real-time scheduling theory settles at finding resource assignment policies that provide at least a worst-case bound upon the blocking time. The priority inheritance (PI) and the (immediate) priority ceiling (PC) protocols [30] belong to this category.

The essence of the PC protocol (which has been included in the RTOS OSEK standard issued by the automotive industry) consists in raising the priority of a thread entering a critical section to the highest among the priorities of all the threads that may possibly request access to the same critical section. The thread returns to its nominal priority as soon as it leaves the critical section. The PC protocol ensures that each thread can be blocked at most once, and bounds the duration of the blocking time to the largest critical section shared between itself or higher priority threads and lower priority threads.

When the blocking time due to priority inversion is bound for each task and its worst-case value is B_i, the evaluation of the worst-case completion time in the schedulability test becomes

$$W_i = C_i + \sum_{\forall j \in he(i)} \frac{W_i}{\theta_j} C_j + B_i$$

9.1.2.3 Mapping the Functional Model into the Architectural Model

The mapping of the actions defined in the functional model onto architectural model entities is the critical design activity where the two views are reconciled. In practice, the actions or transitions defined in the functional part must be executed in the context of one or more system threads. The definition of the architecture model (number and attributes of threads) and the selection of resource management policies, the mapping of the functional model into the corresponding architecture model and the validation of the mapped model against functional and nonfunctional constraints is probably one of the major challenges in software engineering.

Single-thread implementations are quite common and are probably the only choice that allows for (practical) verification of implementation and schedulability analysis, meaning that there exist CASE tools that can provide both according to a given MOC. The entire functional specification is executed in the context of a single thread performing a never-ending cycle where it serves events in a noninterruptable fashion according to the run-to-completion paradigm. The thread waits for an event (either external, like an interrupt from an I/O interface, or internal, like a call or signal from one object or FSM to another), fetches the event and the associated parameters, and, finally, executes the corresponding code.

All the actions defined in the functional part need be scheduled (statically or dynamically) for execution inside the thread. The schedule is usually driven by the partial order in the execution of the actions, as defined by the MOC semantics. Commercial implementations of this model range from the code produced by the Esterel compiler [31] to single-thread implementations by Embedded Coder toolset from Mathworks and TargetLink from DSpace (of Simulink models) [32,33] or the single-thread code generated by Rational Rose Technical Developer [34] for the execution of UML models.

The scheduling problem is much simpler than it is in the multithreaded case, since there is no need to account for thread scheduling and preemption and resource sharing usually result in trivial problems.

On the other extreme, one could define *one thread for every functional block* or every possible action. Each thread can be assigned its own priority, depending on the criticality and on the deadline of the corresponding action. At run time, the operating system scheduler properly synchronizes and sequentializes the tasks so that the order of execution respects the functional specification.

Both approaches may easily prove inefficient. The single-thread implementation suffers from large priority inversion due to the need for completing the processing of each event before fetching the next event in the global queue. The one-to-one mapping of the functions or actions to threads suffers from excessive scheduler overhead caused by the need for a context switch at each action. Considering that the action specified in a functional block can be very short and that the number of functional blocks is usually quite high (in many applications it is of the order of hundreds), the overhead of the operating system could easily prove unbearable.

The designer essentially tries to achieve a compromise between these two extremes, balancing responsiveness with schedulability, flexibility of the implementation, and performance overhead.

9.1.3 Paradigms for Reuse: Component-Based Design

One more dimension can be added to the complexity of the software design problem if the need for maintenance and reuse is considered. To this purpose, component-based and object-oriented (OO) techniques have been developed for constructing and maintaining large and complex systems.

A component is a product of the analysis, design, or implementation phases of the lifecycle and represents a prefabricated solution that can be reused to meet (sub)system requirement(s). A component is commonly used as a vehicle for the reuse of two basic design properties:

- *Functionality.* The functional syntax and semantics of the solution that the component represents.
- *Structure.* The structural abstraction that the component represents. These can range from "small grain" to architectural features, at the subsystem or system level.

The generic requirement for "reusability" maps into a number of issues. Probably the most relevant property that components should exhibit is *abstraction* meaning the capability of hiding implementation details and describing relevant properties only. Components should also be easily adaptable to meet the changing processing requirements and environmental constraints through controlled modification techniques (like *inheritance* and *genericity*). In addition, *composition* rules must be used to build higher-level components from existing ones. Hence, an ideal component-based modeling language should ensure that properties of components (functional properties, such as liveness, reachability, and deadlock avoidance, or nonfunctional properties such as timeliness and schedulability) are preserved or at least decidable after composition. Additional (practical) issues include support for implementation, separate compilations, and imports.

Unfortunately, reconciling the standard issues of software components, such as context independence, understandability, adaptability, and composability, with the possibly conflicting requirements of timeliness, concurrency, and distribution, typical of hard real-time system development is not an easy task and is still an open problem.

Object-oriented design of systems has traditionally embodied the (far from perfect) solution to some of these problems. While most (if not all) OO methodologies, including the UML, offer support for inheritance and genericity, adequate abstraction mechanisms, and especially composability of properties, are still subject of research.

With its latest release, UML has reconciled the abstract interface abstraction mechanism with the common box-port-wire design paradigm. The lack of an explicit declaration of the required interface and the absence of a language feature for structured classes were among the main deficiencies of classes and objects, if seen as components. In UML 2.0, ports allow a formal definition of a required as well as a provided interface. The association of protocol declaration with ports further improves the clarification of the semantics of interaction with the component. In addition, the concept of a structured class allows a much better definition of a component.

Of course, port interfaces and the associated protocol declarations are not sufficient for specifying the semantics of the component. In UML 2.0, OCL can also be used to define behavioral specifications in the form of invariants, preconditions and post-conditions, in the style of the contract-based design methodology (implemented in Eiffel [35]).

9.2 Synchronous vs. Asynchronous Models

The verification of the functional and nonfunctional properties of software demands for formal semantics and a strong mathematical foundation of the models. Many argue that a fully analyzable model cannot be constructed unless shedding generality and restricting the behavioral model to simple and analyzable semantics.

Among the possible choices, the *synchronous reactive* model enforces determinism and provides a sound methodology for checking functional and nonfunctional properties at the price of expensive implementation and performance limitations. Moreover, the synchronous model is built on assumptions (computation times neglectable with respect to the environmental dynamics and synchronous execution) that do not always apply to the controlled environment and to the architecture of the system.

Asynchronous or general models typically allow for (controlled) non-determinism and more expressiveness, at the cost of strong limitations in the extent of the functional and nonfunctional verification that can be performed.

Some modeling languages, such as UML, are deliberately general enough, so that it is possible to use them for specifying a system according to a generic asynchronous or synchronous paradigm, provided that a suitable set of extensions (semantics restrictions) are defined.

By the end of this chapter, it will hopefully be clear how neither of the two design paradigms (synchronous or asynchronous) is currently capable of facing all the implementation challenges of complex systems. The requirements of the synchronous assumption (on the environment and the execution platform) are difficult to meet, and component-based design is very difficult (if not impossible). The asynchronous paradigm on the other hand, results in implementations which are very difficult to analyze for logical and time behavior.

9.3 Synchronous Models

In the *synchronous reactive model*, time advances at discrete instants and the program progresses according to the successive *atomic reactions* (sets of synchronously executed actions), which are performed instantaneously (zero computation time), meaning that the reaction is fast enough with respect to the environment. The resulting discrete-time model is quite natural to many domains, such as control engineering and (hardware) synchronous digital logic design (VHDL).

The composition of system blocks implies product combination of the states and the conjunction of the reactions for each component. In general, this results in a fixed-point problem and the composition of the function blocks is a relation, not a function, as outlined in Section 9.1.

The French synchronous languages *Signal*, *Esterel*, and *Lustre* are probably the best representatives of the synchronous modeling paradigm.

Lustre [36,37] is a declarative language based on the dataflow model where nodes are the main building blocks. In Lustre, each flow or stream of values is represented by a variable, with a distinct value for each tick in the discrete-time base. A node is a function of flows: it takes a number of typed input flows and defines a number of output flows by means of a system of equations.

A Lustre node (an example is given in Figure 9.11) is a pure functional unit except for the `pre` and initialization (−>) expressions, which allow referencing the previous element of a given stream or forcing an initial value for a stream. Lustre allows streams at different rates, but in order to avoid nondeterminism it forbids syntactically cyclic definitions.

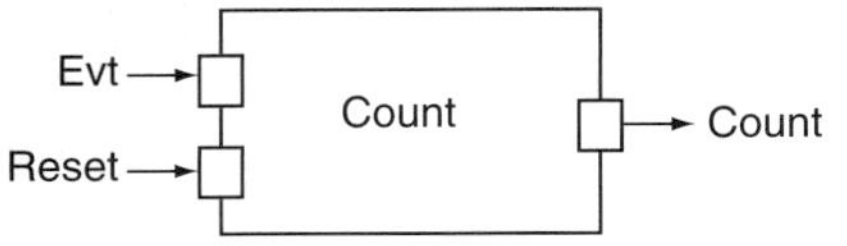


```
Node Count(evt, reset: bool)
     returns(count: int);
Let
     Count = if (true → reset) then 0
               else if evt then pre(count)+1
               else pre(count)
Tel
```

```
Mod60 = Count(second, pre(mod60 = 59));
Minute = (mod60 = 0);
```

FIGURE 9.11 An example of Lustre node and its program.

Esterel [38] is an imperative language, more suited for the description of control. An Esterel program consists of a collection of nested, concurrently running threads. Execution is synchronized to a single, global clock. At the beginning of each reaction, each thread resumes its execution from where it paused (e.g., at a pause statement) in the last reaction, executes imperative code (e.g., assigning the value of expressions to variables and making control decisions), and finally either terminates or pauses, waiting for the next reaction.

Esterel threads communicate exclusively through signals representing globally broadcast events. A signal does not persist across reactions and it is present in a reaction if and only if it is emitted by the program or by the environment.

Esterel allows cyclic dependencies and treats each reaction as a fix-point equation, but the only legal programs are those that behave functionally in every possible reaction. The solution of this problem is provided by *constructive causality* [39], which amounts to checking if, regardless of the existence of cycles, the output of the program (the binary circuit implementing it) can be formally proven to be causally dependent from the inputs for all possible input assignments.

The language allows for conceptually sequential (operator ;) or concurrent (operator ||) execution of reactions, defined by language expressions handling signal identifiers (as in the example of Figure 9.12). All constructs take zero time except `await` and `loop ... each ...`, which explicitly produce a program pause. Esterel includes the concept of preemption, embodied by the `loop ... each R` statement in the example of Figure 9.12 or the `abort` *`action`* `when` *`signal`* statement. The reaction contained in the body of the loop is preempted (and restarted) when the signal R is set. In case of an abort statement, the reaction is preempted and the statement terminates.

Formal verification was among the original objectives of Esterel. In synchronous languages, the verification of properties typically requires the definition of a special program called the *observer*, which observes the variables or signals of interest and at each step decides if the property is fulfilled. A program satisfies the property if and only if the observer never complains during any execution.

The verification tool takes the program implementing the system, an observer of the desired property, and another program modeling the assumptions on the environment. The three programs are combined in a synchronous product, as in Figure 9.13, and the tool explores the set of reachable states. If the observer never reaches a state where the system property is not valid before reaching a state where the assumption observer declares violation of the environment assumptions, then the system is correct. The process is described in detail in [40].

Finally, the commercial package *Simulink* by Mathworks [41] allows modeling and simulation of control systems according to a synchronous reactive model of computation, although its semantics is neither formally nor completely defined. Rules for translating a Simulink model into Lustre have been outlined in [42], while [43] discusses the very important problem of how to map a zero-execution time Simulink semantics into a software implementation of concurrent threads where each computation necessarily requires a finite execution time is discussed.

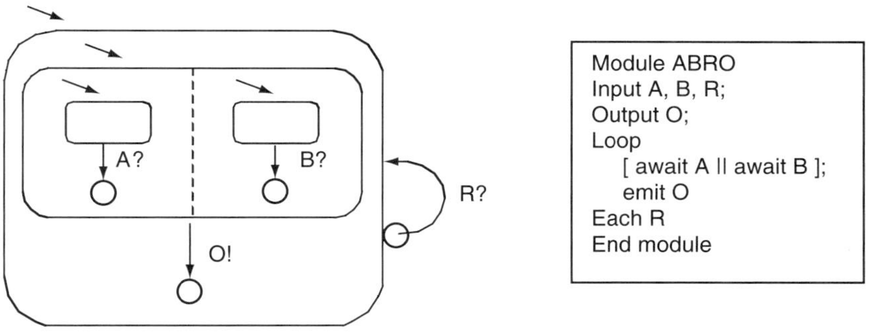

FIGURE 9.12 An example showing features of the Esterel language with an equivalent statechart-like visual formalization.

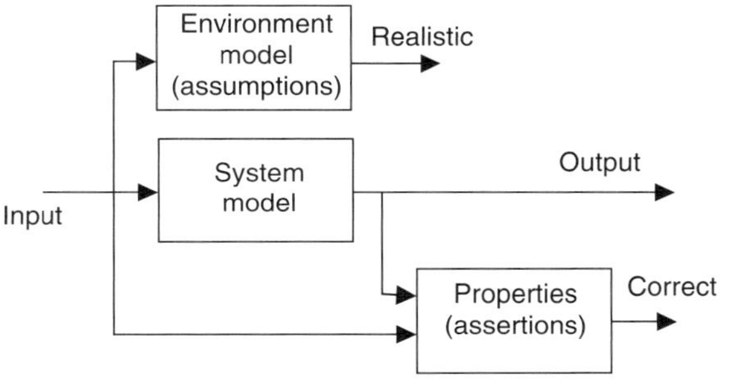

FIGURE 9.13 Verification by observers.

9.3.1 Architecture Deployment and Timing Analysis

Synchronous models are typically implemented as a single task that executes according to an event server model. Reactions decompose into atomic actions that are partially ordered by the causality analysis of the program. The scheduling is generated at compile time, trying to exploit the partial causality order of the functions in order to make the best possible use of hardware and shared resources. The main concern is checking that the synchrony assumption holds, i.e., ensuring that the longest chain of reactions ensuing from any internal or external event is completed within the step duration. Static scheduling means that critical applications are deployed without the need for any operating system (and the corresponding overhead). This reduces system complexity and increases predictability, avoiding preemption, dynamic contention over resources and other non-deterministic operating systems functions.

9.3.2 Tools and Commercial Implementations

Lustre is implemented by the commercial toolset Scade, offering an editor that manipulates both graphical and textual descriptions; two code generators, one of which is accepted by certification authorities for qualified software production; a simulator; and an interface to verification tools such as the plug-in from Prover [44].

The early *Esterel* compilers were developed by Gerard Berry's group at INRIA/CMA and freely distributed in binary form. The commercial version of Esterel was first marketed in 1998 and it is now available from Esterel Technologies, which later acquired the Scade environment. Scade has been used in many industrial projects, including integrated nuclear protection systems (Schneider Electric), flight control software (Airbus A340–600), and track control systems (CS Transport). Dassault Aviation was one of the earliest supporters of the Esterel project, and has long been one of its major users.

Several verification tools use the synchronous observer technique for checking Esterel programs [40]. It is also possible to verify the implementation of Esterel programs with tools exploiting explicit state space reduction and bi-simulation minimization (FC2Tools) and finally, tools can also be used to automatically generate test sequences with guaranteed state/transition coverage.

The very popular *Simulink* tool by Mathworks was developed with the purpose of simulating control algorithms, and has since its inception been extended with a set of additional tools and plug-ins, such as, the Stateflow plug-in for the definition of the FSM behavior of a control block, allowing modeling of hybrid systems, and a number of automatic code generation tools, such as the Realtime Workshop and Embedded Coder by Mathworks and TargetLink by DSpace.

9.3.3 Challenges

The main challenges and limitations that the Esterel language must face when applied to complex systems are the following:

- Despite improvements, the space and time efficiency of the compilers is still not satisfactory.
- Embedded applications can be deployed on architectures or control environments that do not comply with the synchronous reactive model.
- Designers are familiar with other dominant methods and notations. Porting the development process to the synchronous paradigm and languages is not easy.

The efficiency limitations are mainly due to the formal compilation process and the need to check for constructive causality. The first three Esterel compilers used automata-based techniques and produced efficient code for small programs, but they did not scale to large-scale systems because of state explosion. Versions 4 and 5 are based on translation into digital logic and generate smaller executables at the price of slow execution. (The program generated by these compilers wastes time evaluating each gate in every clock cycle.) This inefficiency can produce code 100 times slower than that from the previous compilers [45].

Version 5 of the compiler allows cyclic dependencies by exploiting Esterel's constructive semantics. Unfortunately, this requires evaluating all the reachable states by symbolic state-space traversal [46], which makes it extremely slow.

As for the difficulty in matching the basic paradigm of synchrony with system architectures, the main reasons of concern are:

- the bus and communication lines, if not specified according to a synchronous (time triggered) protocol, and the interfaces with the analog world of sensors and actuators.
- the dynamics of the environment, which can possibly invalidate the instantaneous execution semantics.

The former has been discussed at length in a number of papers (such as [47,48]), giving conditions for providing a synchronous implementation in distributed systems.

Finally, in order to integrate synchronous languages with mainstream commercial methodologies and languages, translation and import tools are required. For example, it is possible from Scade to import discrete-time Simulink diagrams and Sildex allows importing Simulink/Stateflow discrete-time diagrams. Another example is UML with the attempt at an integration between Esterel Studio and Rational Rose, and the proposal for an Esterel/UML coupling drafted by Dassault [49] and adopted by commercial Esterel tools.

9.4 Asynchronous Models

UML and SDL are languages developed in the context of general purpose computing and (large) telecommunication systems respectively. The Unified Modeling Language is the merger of many OO design methodologies aimed at the definition of generic software systems. Its semantics is not completely specified and intentionally retains many variation points in order to adapt to different application domains.

For example, in order to be practically applicable to the design of embedded systems, further characterization (a specialized *profile* in UML terminology) is required. In the 2.0 revision of the language, the system is represented by a (transitional) model where active and passive components, communicating by means of connections through port interfaces, cooperate in the implementation of the system behavior. Each reaction to an internal or external event results in the transition of a *Statechart* automaton describing the object behavior.

The Specification and Description Language (SDL) has a more formal background since it was developed in the context of software for telecommunication systems, for the purpose of easing the implementation of verifiable communication protocols. An SDL design consists of blocks cooperating by means of asynchronous signals. The behavior of each block is represented by one or more (conceptually concurrent) processes. Each process in turn implements an extended FSM.

Until the recent development of the UML profile for Schedulability, Performance and Time (SPT), standard UML did not provide any formal means for specifying time or time-related constraints, or for specifying resources and resource management policies. The deployment diagrams were the only (inadequate) means for describing the mapping of software onto the hardware platform and tool vendors had tried to fill the gap by proposing nonstandard extensions.

The situation with SDL is not very different, although SDL offers at least the notion of global and external time. Global time is made available by means of a special expression and can be stored in variables or sent in messages.

The implementation of asynchronous languages typically (but not necessarily) relies on an operating system. The latter is responsible for scheduling, which is necessarily based on static (design time) priorities if a commercial product is used for this purpose. Unfortunately, as it will be clear in the following, real-time schedulability techniques are only applicable to very simple models and are extremely difficult to generalize to most models of practical interest or even to the implementation model assumed by most (if not all) commercial tools.

9.4.1 Unified Modeling Language

The UML represents a collection of engineering practices that have proven successful in the modeling of large and complex systems, and has emerged as the software industry's dominant OO-modeling language.

Born at Rational in 1994, UML was taken over in 1997 at version 1.1 by the OMG Revision Task Force (RTF), which became responsible for its maintenance. The RTF released UML version 1.4 in September 2001 and a major revision, UML 2.0, which also aims to address the embedded or real-time dimension, has recently been adopted (late 2003) and it is posted on the OMG's website as "UML 2.0 Final Adopted Specification" [4].

The Unified Modeling Language has been designed as a wide-ranging, general-purpose modeling language for specifying, visualizing, constructing, and documenting the artifacts of software systems. It has successfully been applied to a wide range of domains, ranging from health and finance to aerospace and e-commerce, and its domains go even beyond software, given recent initiatives in areas such as systems engineering, testing, and hardware design. A joint initiative between OMG and INCOSE (International Council on Systems Engineering) is working on a profile for Systems Engineering, and the SysML consortium has been established to create a systems modeling language based on UML. At the time of this writing, over 60 UML CASE tools can be listed from the OMG resource page (http://www.omg.org).

After revision 2.0, the UML specification consists of four parts:

- UML 2.0 infrastructure, defining the foundational language constructs and the language semantics in a more formal way than they were in the past.
- UML 2.0 superstructure, which defines the user level constructs.
- OCL 2.0 object constraint language (OCL), which is used to describe expressions (constraints) on UML models.
- UML 2.0 diagram interchange, including the definition of the XML-based XMI format, for model interchange among tools.

The Unified Modeling Language comprises of a metamodel definition and a graphical representation of the formal language, but it intentionally refrains from including any design process. The UML in its general form is deliberately semiformal and even its state diagrams (a variant of statecharts) retain sufficient semantics variation points in order to ease adaptability and customization.

The designers of UML realized that complex systems cannot be represented by a single design artifact. According to UML, a system model is seen under different views, representing different aspects. Each view corresponds to one or more diagrams, which taken together, represent a unique model. Consistency of this multiview representation is ensured by the UML metamodel definition. The diagram types included in the UML 2.0 specification are represented in Figure 9.14, as organized in the two main categories that relate to *structure* and *behavior.*

When domain-specific requirements arise, more specific (more semantically characterized) concepts and notations can be provided as a set of stereotypes and constraints, and packaged in the context of a *profile.*

Structure diagrams show the static structure of the system, i.e., specifications that are valid irrespective of time. Behavior diagrams show the dynamic behavior of the system. The main diagrams are:

- *use case diagram* a high-level (user requirements-level) description of the interaction of the system with external agents;
- *class diagram* representing the static structure of the software system, including the OO description of the entities composing the system, and of their static properties and relationships;
- *behavior diagrams* (including *sequence diagrams* and *state diagrams* as variants of Message Sequence Charts and Statecharts) providing a description of the dynamic properties of the entities composing the system, using various notations;
- *architecture diagrams*, (including *composite* and *component diagrams*, portraying a description of reusable components) a description of the internal structure of classes and objects, and a better characterization of the communication superstructure, including communication paths and interfaces.
- *implementation diagrams*, containing a description of the physical structure of the software and hardware components.

The class diagram is typically the core of a UML specification, as it shows the logical structure of the system. The concept of classifiers (class) is central to the OO design methodology. Classes can be defined as user-defined types consisting of a set of attributes defining the internal state, and a set of operations (signature) that can be possibly invoked on the class objects resulting in an internal transition. As units

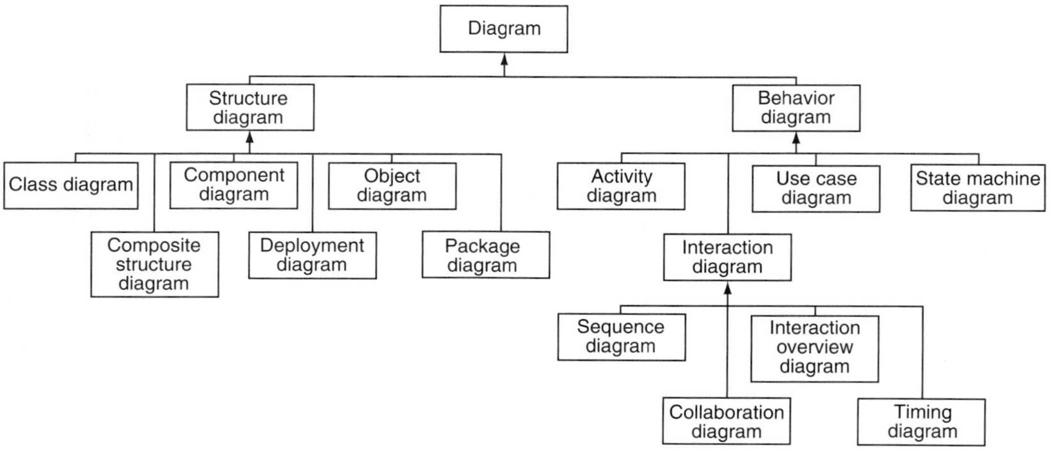

FIGURE 9.14 A taxonomy of UML 2.0 diagrams.

of reuse, classes embody the concepts of encapsulation (or information) hiding and abstraction. The signature of the class abstracts the internal state and behavior, and restricts possible interactions with the environment. Relationships exist among classes and relevant relationships are given special names and notations, such as aggregation and composition, use and dependency. The generalization (or refinement) relationship allows controlled extension of the model by allowing a derived class specification inherit all the characteristics of the parent class (attributes and operations, but also, selectively, relationships) while providing new ones (or redefining the existing).

Objects are instances of the type defined by the corresponding class (or classifier). As such, they embody all of the classifier attributes, operations, and relationships. Several books [50,51] have been dedicated to the explanation of the full set of concepts in OO design. The interested reader is invited to refer to literature on the subject for a more detailed discussion.

All diagram elements can be annotated with constraints, expressed in OCL or in any other formalism that the designer sees as appropriate. A typical class diagram showing dependency, aggregation, and generalization associations is shown in Figure 9.15.

UML 2.0 finally acknowledged the need for a more formal characterization of the language semantics and for better support for component specifications. In particular, it became clear that simple classes provide a poor match for the definition of a reusable component (as outlined in previous sections).

As a result, necessary concepts, such as the means to clearly identify *interfaces* which are provided and (especially) those which are required, have been added by means of the port construct. An interface is an abstract class declaring a set of functions with their associated signature. Furthermore, structured classes and objects allow the designer to specify formally the internal communication structure of a component configuration.

UML 2.0 classes, structured classes, and components are now encapsulated units that model active system components and can be decomposed into contained classes communicating by signals exchanged over a set of *ports*, which models communication terminals (Figure 9.16). A port carries both structural information on the connection between classes or components, and *protocol* information that specifies what messages can be exchanged across the connection. A state machine or a UML Sequence Diagram may be associated to a protocol to express the allowable message exchanges. Two components can interact if there is a connection between any two ports that they own and that support the same protocol in

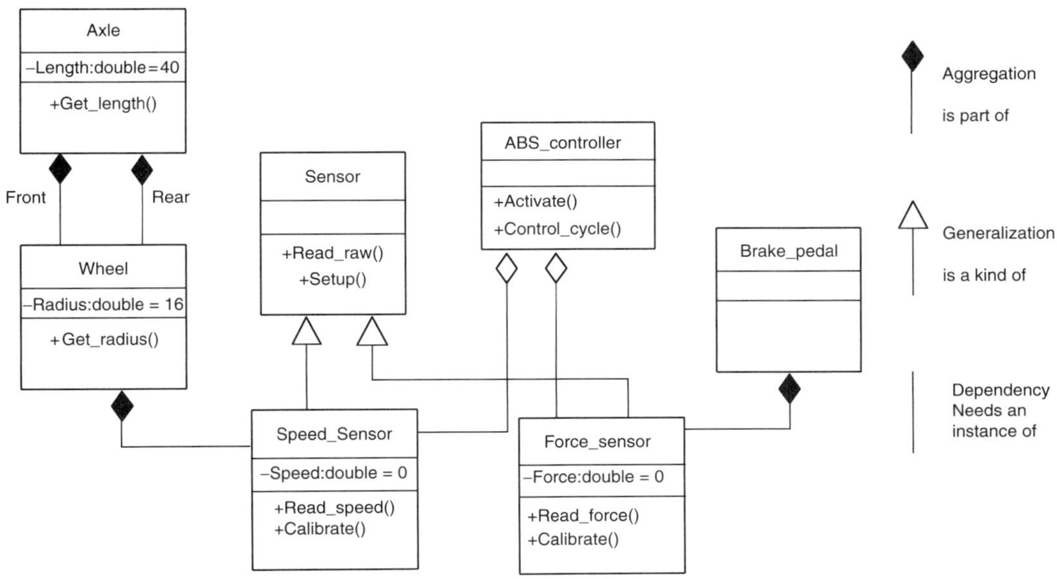

FIGURE 9.15 A sample class diagram.

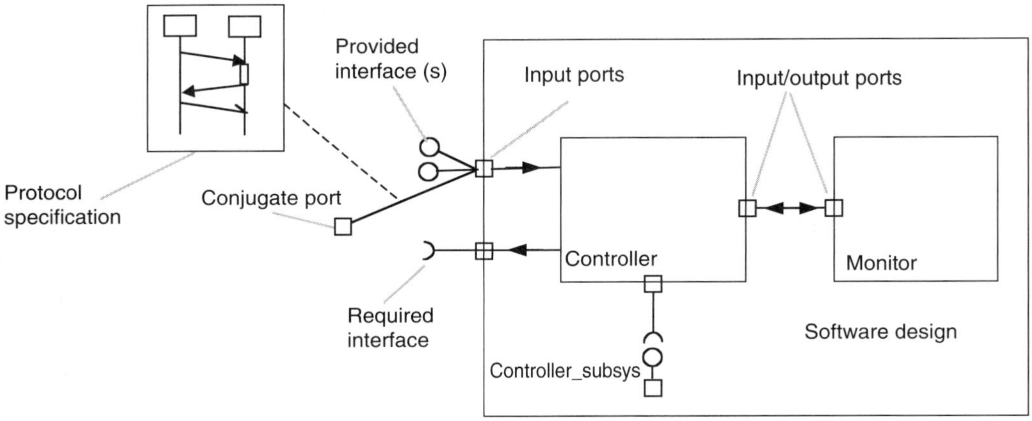

FIGURE 9.16 Ports and components in UML 2.0.

complementary (or *conjugated*) roles. The behavior or reaction of a component to an incoming message or signal is typically specified by means of one or more statechart diagrams.

Behavior diagrams comprise *statechart diagrams*, *sequence diagrams* and *collaboration diagrams*.

Statecharts [11] describe the evolution in time of an object or an interaction between objects by means of a hierarchical state machine. UML statecharts are an extension of Harel's statecharts, with the possibility of defining actions upon entering or exiting a state as well as actions to be executed when a transition is made. Actions can be simple expressions or calls to methods of the attached object (class) or entire programs. Unfortunately, not only does the Turing completeness of actions prevents decidability of properties in the general model, but UML does not even clarify most of the semantics variations left open by the standard statecharts formalism.

Furthermore, the UML specification explicitly gives actions a run-to-completion execution semantics, which makes them nonpreemptable and makes the specification (and analysis) of typical RTOS mechanisms such as interrupts and preemption impossible.

To give an example of UML statecharts, Figure 9.17 shows a sample diagram where, upon entry of the composite state (the outermost rectangle), the subsystem enters three concurrent (and-type) states, named `Idle`, `WaitForUpdate`, and `Display_all`, respectively. Upon entry into the `WaitForUpdate` state, the variable `count` is also incremented. In the same portion of the diagram, reception of message `msg1` triggers the exit action setting the variable `flag` and the (unconditioned) transition with the associated *call action* `update()`. The `count` variable is finally incremented upon reentry in the state `WaitForUpdate`.

Statechart diagrams provide the description of the state evolution of a single object or class, but are not meant to represent the emergent behavior deriving from the cooperation of more objects; nor are they appropriate for the representation of timing constraints. *Sequence diagrams* partly fill this gap. *Sequence diagrams* show the possible message exchanges among objects, ordered along a time axis. The time points corresponding to message-related events can be labeled and referred to in constraint annotations. Each sequence diagram focuses on one particular scenario of execution and provides an alternative to temporal logic for expressing timing constraints in a visual form (Figure 9.18).

Collaboration diagrams also show message exchanges among objects, but they emphasize structural relationships among objects (i.e., "who talks with whom") rather than time sequences of messages (Figure 9.19).

Collaboration diagrams are also the most appropriate way for representing logical resource sharing among objects. Labeling of messages exchanged across links defines the sequencing of actions in a similar (but less effective) way to what can be specified with sequence diagrams.

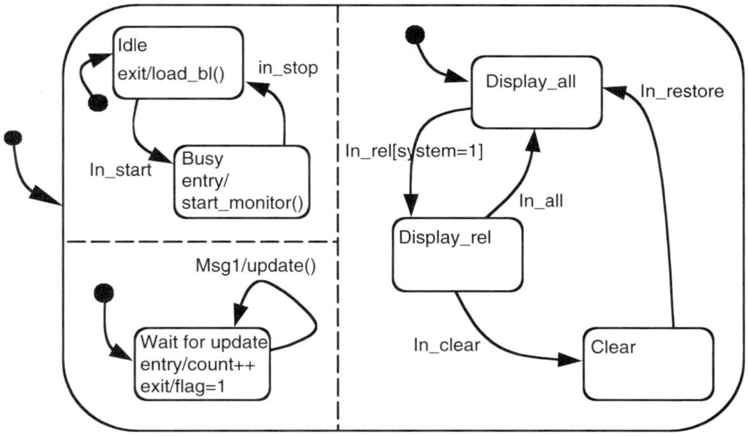

FIGURE 9.17 An example of UML statechart.

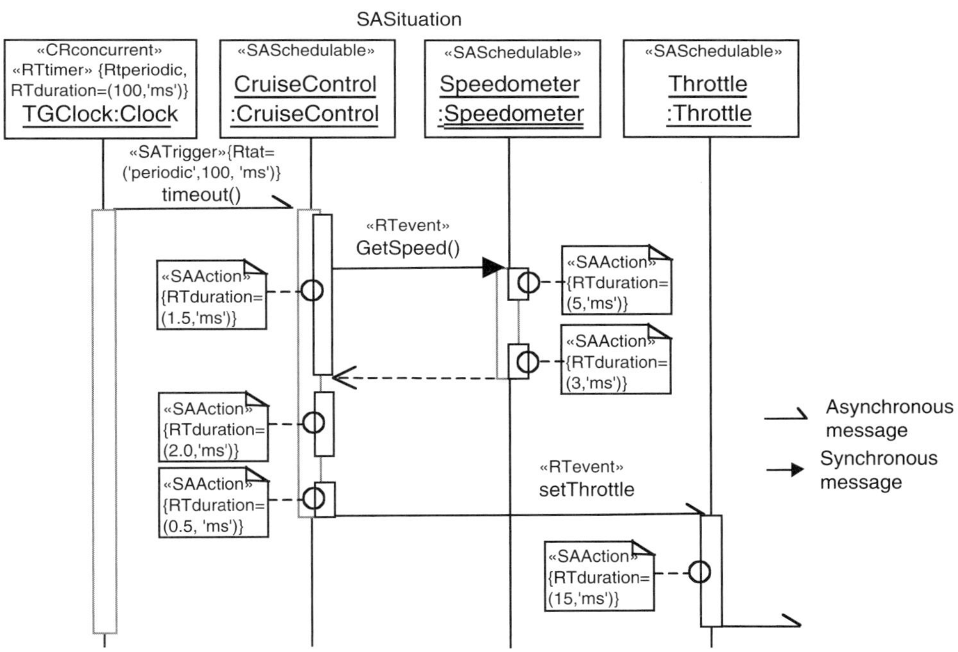

FIGURE 9.18 A sample sequence diagram with annotations showing timing constraints.

Despite the availability of multiple diagram types (or may be because of it), the UML metamodel is quite weak when it comes to the specification of dynamic behavior. The UML metamodel concentrates on providing structural consistency among the different diagrams and provides sufficient definition for the static semantics, but the dynamic semantics is never adequately addressed, up to the point that a major revision of the UML action semantics has become necessary. UML is currently headed in a direction where it will eventually become an executable modeling language, which would for example, allow early verification of system functionality. Within the OMG, a standardization action has been purposely defined with the goal of providing a new and more precise definition of actions. This activity goes under the name of *action semantics for the UML*. Until UML actions are given a more precise semantics, a

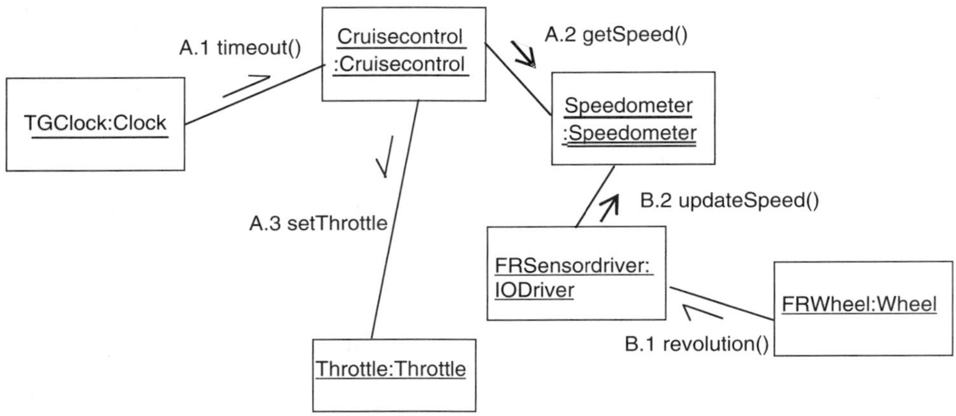

FIGURE 9.19 Defining message sequences in a Collaboration diagram.

faithful model, obtained by combining the information provided by the different diagrams is virtually impossible. Of course, this also nullifies the chances for formal verification of functional properties on a *standard UML model*.

However, simulation or verification of (at least) some behavioral properties and (especially) automatic production of code are features that tool vendors cannot ignore if UML is not to be relegated to the role of simply documenting software artifacts. Hence, CASE tools provide an interpretation of the variation points. This means that validation, code generation, and automatic generation of test cases are tool-specific and depend upon the semantics choices of each vendor.

Concerning formal verification of properties, it is important to point out that UML does not provide any clear means for specifying the properties that the system (or components) are expected to satisfy, nor any means for specifying assumptions on the environment. The proposed use of OCL in an explicit contract section to specify assertions and assumptions acting upon the component and its environment (its users) can hopefully fill this gap in the future.

As of today, research groups are working on the definition of a formal semantic restriction of UML behavior (especially by means of the statecharts formalism) in order to allow for formal verification of system properties [52,53]. After the definition of such restrictions, UML models can be translated into the format of existing validation tools for timed message sequence charts (MSC) or TA.

Finally, the last type of UML diagrams are *Implementation diagrams*, which can be either component diagrams or deployment diagrams. Component diagrams describe the physical structure of the software in terms of software components (modules) related to each other by dependency and containment relationships. Deployment diagrams describe the hardware architecture in terms of processing or data storage nodes connected by communication associations, and show the placement of software components onto the hardware nodes.

The need to express, in UML, timeliness-related properties and constraints, and the pattern of hardware and software resource utilization as well as resource allocation policies and scheduling algorithms found a (partial) response only in 2001 with the OMG issuing a standard SPT profile. The specification of timing attributes and constraints in UML designs will be discussed in Section 9.4.3. Finally, work is currently being conducted in the OMG to develop a test profile for UML 2.0. With the aid of this profile, it will be possible to derive and validate test specifications from a formal UML model.

9.4.1.1 Object Constraint Language

The OCL [54] is a formal language used to describe (constraint) expressions on UML models. An OCL expression is typically used to specify invariants or other types of constraint conditions that must hold for the system. Object constraint language expressions refer to the *contextual instance*, which is the model element to which the expression applies, such as classifiers, like types, classes, interfaces, associations (acting as types), and data types. Also all attributes, association-ends, methods, and operations without side-effects that are defined for these types can be used.

Object constraint language can be used to specify *invariants* associated with a classifier. In this case, it returns a boolean type and its evaluation must be true for each instance of the classifier at any moment in time (except when an instance is executing an operation).

Pre- and *Post-conditions* are other types of OCL constraints that can be possibly linked to an operation of a classifier, and their purpose is to specify the conditions or contract under which the operation executes. If the caller fulfills the precondition before the operation is called, then the called object ensures that the post-condition holds after execution of the operation, but of course only for the instance that executes the operation (Figure 9. 20).

9.4.2 Specification and Description Language

The SDL is an International Telecommunications Union (ITU-T) standard promoted by the SDL Forum Society for the specification and description of systems [55].

Since its inception, a formal semantics has been part of the SDL standard (Z.100), including visual and textual constructs for the specification of both the architecture and the behavior of a system. The behavior of (active) SDL objects is described in terms of concurrently operating and asynchronously communicating abstract state machines (ASMs). SDL provides the formal behavior semantics that UML is currently missing: it is a language based on communicating state machines enabling tool support for simulation, verification, validation, testing, and code generation.

In SDL, systems are decomposed into a hierarchy of *block agents* communicating via (unbounded) channels that carry typed signals. Agents may be used for structuring the design and can in turn be decomposed into sub-agents until leaf blocks are decomposed into *process agents*. Block and process agents differ since blocks allow internal concurrency (subagents) while process agents only have an interleaving behavior.

The behavior of process agents is specified by means of extended finite and communicating state machines (SDL services) represented by a connected graph consisting of states and transitions. Transitions are triggered by external stimuli (signals, remote procedure calls) or conditions on variables. During a transition, a sequence of actions may be performed, including the use and manipulation of data stored in local variables or asynchronous interaction with other agents or the environment via signals that are placed into and consumed from channel queues.

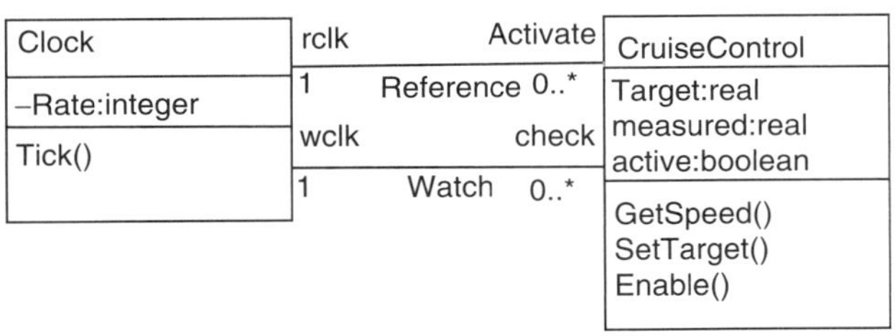


```
Context: CruiseControl
inv: not active or
     abs(target-measured) < 10
context: Clock
inv: activate->size() <= 3
pre:active
```

```
context: Clock
inv: activate <> check

context: CruiseControl::GetSpeed()
pre: active
post: result=measured
```

FIGURE 9.20 Examples of OCL adding constraints or defining behavior of operations in a class diagram.

Channels are asynchronous (as opposed to synchronous or rendezvous) FIFO queues (one for each process) and provide a reliable, zero- or finite-delay transmission of communication elements from a sender to a receiver agent.

Signals sent to an agent will be delivered to the input port of the agent. Signals are consumed in the order of their arrival either as a trigger of a transition or by being discarded in case there is no transition defined for the signal in the current state. Actions executed in response to the reception of input signals include expressions involving local variables or calls to procedures.

In summary, an agent definition consists of

- a *behavior specification* given by the agent extended FSM;
- a *structure and interaction diagram* detailing the hierarchy of system agents with their internal communication infrastructure;
- the variables (*attributes*) under the control of each agent;
- the black box or *component view* of the agent defining the interaction points (ports) with the provided and required interfaces.

Figure 9.21 shows an example of SDL notation with five interacting blocks and one leaf block (C2) decomposed into processes. The behavior of the process is specified in terms of its states and reactions as represented in the diagram on the right, where rectangles with concave points represent messages received, and rectangles with convex points denote messages sent. The SDL notation is a representation of the behavior of the equivalent extended state machine (an example is shown on the right side of Figure 9.22).

Figure 9.22 shows a process behavior and a matching representation by means of an extended FSM (right). The figure shows the use of a variable (var1) as a *continuous input* signal for the state S3. A continuous input is an SDL Boolean expression in terms of one or more variables. If the expression evaluates to true and no transition on regular signals is available, the corresponding branch is taken. Note that continuous signals are evaluated after all other signals are checked (i.e., they have the lowest priority).

The state with asterisk reached from the continuous input signal transition is a placeholder, representing the starting state of the transition (S3) itself.

Agents combine behavior and structure in a single entity and fit nicely into the concept of active objects in OO languages (active objects are objects that possess a thread of control and may initiate a reaction).

The latest SDL version, SDL 2000, extends the language by including typical features of OO and component-based modeling techniques, including encapsulation of services by ports (gates) and interfaces, classifiers and associations, and specialization (refinement) of virtual class structure and behavior. In SDL 2000, the user can define classifiers (types in SDL) for blocks, processes, and services, and use instances

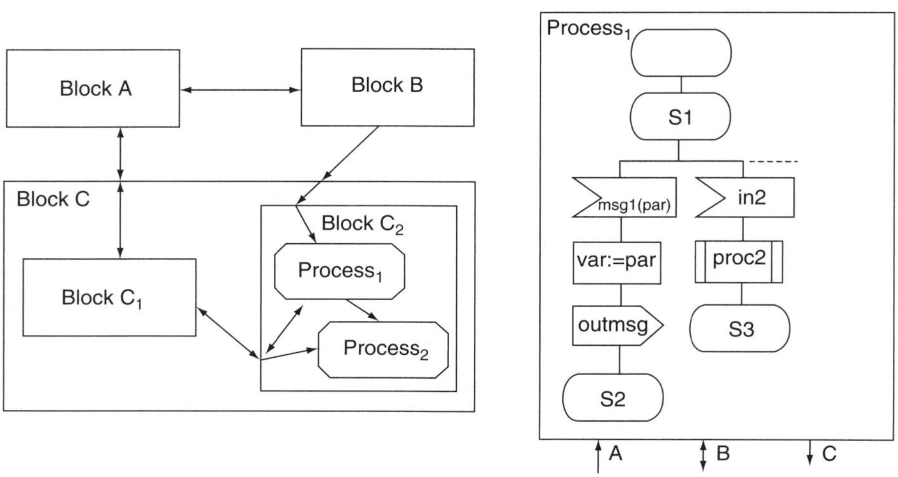

FIGURE 9.21 Interacting SDL blocks, block decomposition, and the description of the behavior of a process.

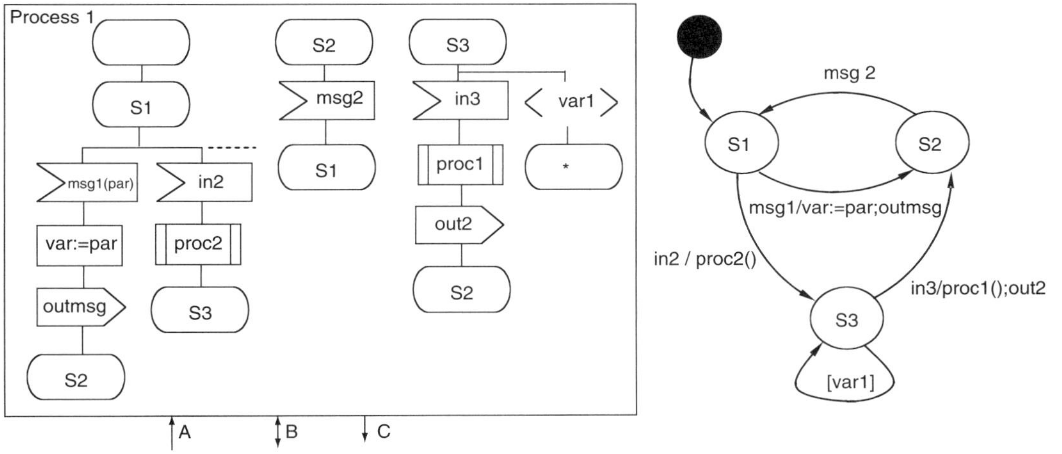

FIGURE 9.22 A process behavior and the corresponding extended FSM.

of types for block, process, service, procedure, and signal description. The specification of behavior for automatic generation of implementations is further improved by the support for exception handling and state machine decomposition.

In SDL 2000, two communicating agents must provide gates with matching interfaces (required or supported) and a connection between the gates must exist.

In the example given in Figure 9.23, an SDL 2000 block exposing two gates (`Bin` of input/output type and the output gate `Bout`) is shown. The block defines two internal signals msg and ack exchanged on the channel connecting the internal processes—`Sender` and `Receiver`.

SDL types can be specialized, leading to subtypes by exploiting the inheritance mechanism. Virtual objects or transitions can be redefined and additional objects or actions can be added to the specification of virtual objects. A redefined transition or object is subject to further specialization (if virtual) unless it is marked as finalized.

SDL offers native mechanisms for representing external (global) time. Time is available by means of the predefined variable *now*, the `now()` primitive and timer constructs. Process actions can set timers, that is the specification of a signal at a predefined point in time, wait and eventually receive a timer expiry signal (as in Figure 9.24). SDL timer timeouts are always received in the form of asynchronous messages and timer values are only meant to be minimal bounds, meaning that any timer signal may remain enqueued for an unbounded amount of time.

In SDL, processes inside a block are meant to be executed concurrently, and no specification for a sequential implementation by means of a scheduler, necessary when truly concurrent hardware is not available, is given. Activities implementing the processes behavior (transitions between states) are executed in a run-to-completion fashion. From an implementation point of view, this raises the same concerns that hold for the implementation of UML models. In SDL, however, the exception mechanism is available to model preemption and interrupt handling.

Other language characteristics make verification of time-related properties impossible: the Z.100 SDL semantics says that external time progresses in both states and actions. However, each action may take an unbounded amount of time to execute, and each process can remain in a state for an indeterminate amount of time before taking the first available transition.

Furthermore, for timing constraints specification, SDL does not include the explicit notion of event, therefore it is impossible to define a time tagging of most events of interest such as transmission and reception of signals, although Message Sequence Charts (MSCs, similar in nature to UML Sequence Diagrams and defined in part Z.120 of the standard [56]) are typically used to fill this gap since they allow the expression of constraints on time elapsing between events.

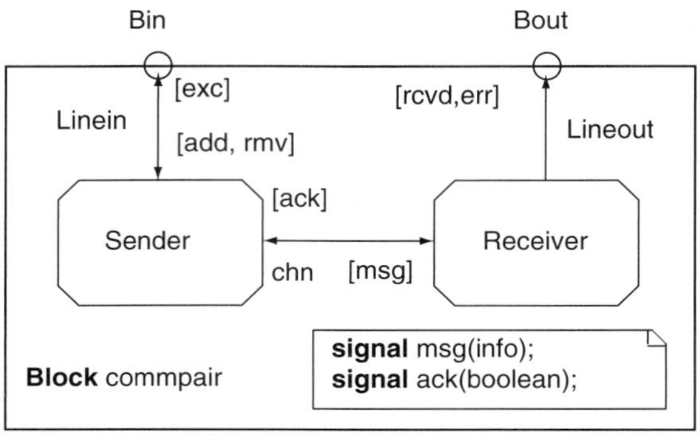

FIGURE 9.23 Specification and description language blocks communicating by ports implementing a signal interface.

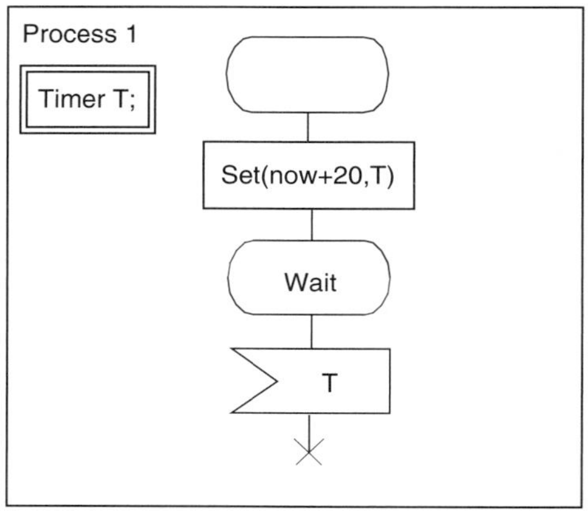

FIGURE 9.24 Example of timer use.

Incomplete specification of time events and constraints prevents timed analysis of SDL diagrams, but the situation is not much better for untimed models. The properties of interest are in general undecidable because of infiniteness of data domains, unbounded message buffers, and the semi-formal modeling style, where SDL is mixed with code fragments inside conditions and actions (the formal semantics of SDL is aimed at code generation rather than at simulation and verification). The untimed verification of properties is only possible in restricted models and seldom performed by commercial or research tools dealing with SDL code. The SDL description, (often restricted to a subset of the language) is provided as an input to transformation tools, which convert it to other formalisms, such as Petri Nets (analyzed by [25, 57]) or TA (for analysis by the Kronos/IF [19,58,59] or Uppaal [20] toolsets).

Research on the verification of SDL designs has been done in academic and industrial laboratories and centers. For example, Regensburger and Barnard [60] describes the verification of a layer of the GSM protocol (6 processes and 1013 reachable states) performed at Siemens by using BDD-based model checking and Bozga et al. [61] report on several case studies related to telecommunication protocols.

9.4.3 Architecture Deployment, Timing Specification, and Analysis

Both UML and SDL have been developed outside the context of embedded systems design and it is clear from previous sections that they are poorly suited to modeling resource allocation and sharing, or to the minimum requirements for timing analysis. In fact, almost nothing exists in standard UML and SDL for modeling (or analyzing) nonfunctional aspects, nor can scheduling or placement of software components on hardware resources be specified and analyzed.

Several proposals exist, most notably the OMG profile for SPT for UML, and others for SDL, which enhance the standard language defining timed models of systems, including time assumptions on the environment and platform-dependent aspects like resource availability and scheduling. Such model extensions allow formal or simulation-based validation of the timing behavior of the software.

The RT-UML profile is currently implemented by commercial tools, and OMG reports that proof of feasibility has been obtained by the interaction with schedulability analysis tools such as TimeWiz by TimeSys and Pert by TriPacific.

9.4.3.1 Unified Modeling Language

The OMG Profile for SPT, voted for adoption in November 2001, aims at substituting a number of proposals for time-related extensions that have appeared in recent years. For example, in order to support better the mapping of active objects into concurrent threads, many research and commercial systems introduced additional nonstandard diagrams. One such example is Artisan's Real-Time Studio, which features the Concurrency diagram, capturing the multitasking and concurrency aspects of the system, and a System Architecture diagram.

The SPT profile defines a comprehensive conceptual framework that extends the UML meta-model with a much broader scope than any other real-time extension and applies to all diagrams. Currently, it consists mostly of a notation framework or vocabulary, with the purpose of providing the necessary concepts for schedulability and performance analysis of (timed) behavioral diagrams or scenarios. However, the SPT inherits from UML the deficiencies related to its incomplete semantics and, at least as of today, it lacks a sufficiently established practice. A major revision of the profile is now being issued by OMG following the request for a profile proposal for real-time embedded systems (MARTES).

The current version of the profile is based on extensions (stereotyped model elements, tagged values and constraints) belonging to three main framework packages, further divided into subpackages (as in Figure 9.25).

Of the three frameworks, the *General Resource Modeling* framework contains the fundamental definitions for modeling *time* in the RTtimeModeling subpackage, generic *resource* specification and usage patterns in the RTresourceModeling profile, and extensions for modeling *concurrency* in the RTconcurrencyModeling profile. The *analysis models* package contains specialized concepts for modeling *schedulability* (RAprofile) and *performance* (PAprofile) analysis. Finally, the *infrastructure models* package deals with real-time middleware specification (RealTimeCORBAModel).

In the SPT, the *time model* provides both continuous and discrete-time models as well as global and local clocks, including drift and offset specifications. The profile allows referencing to time instances (associated to events) of *Time* type, and to the time interval between any two instances of time of *Duration* type in attributes or constraints specifications inside any UML diagram.

The time package not only contains definitions for a formal model of time, but also stereotyped definitions for the two basic mechanisms of timer and clock. Timers can be periodic (as in SDL), they can be set or reset, paused or restarted and, when expired, they send timeout signals. Clocks are specialized periodic timers capable of generating Tick events.

Time values are typically associated to Events, defined in UML as a "specification of a type of observable occurrence" (change of state). A pairing of an event with the associated time instance (time tag) is defined in the SPT profile as a TimedEvent.

The *resource model* package defines schedulability analysis as a quality of service (QoS) matching problem between a client and the resource it uses. The profile provides two interpretations: in the *peer interpretation* the client and the resource coexist at the same level; in the *layered interpretation*, the resources are used to implement the client. This is an attempt at capturing the distinctive aspect of schedulability analysis as a match between the timing (or in general QoS) requirements of *logical entities* in the *logical layer* and the corresponding QoS offers of an implementation or *engineering layer*.

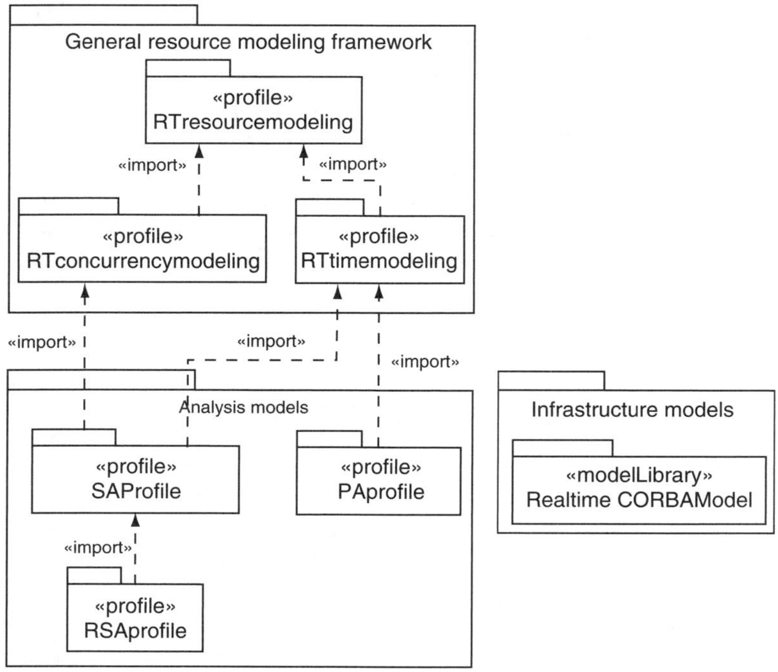

FIGURE 9.25 The framework packages and the subpackages in the OMG SPT.

In the OMG profile, the mapping between the logical entities and the physical architecture supporting their execution is a form of *realization layering* (synonymous of *deployment*). The semantics of the mapping provides a further distinction between the "deploys" mapping, indicating that instances of the supplier are located on the client, an example being the relation between a package and the HW node on which it executes, and the "requires" mapping, which is a specialization indicating that the client provides a minimum deployment environment as required by the supplier, typically in the context of a QoS contract (Figure 9.26).

Although a visual representation of both is possible, it is not clear in which of the existing diagrams it should appear. Hence, the SPT recommendation for expressing the mappings consists of a table of relationships expressing the deploy and requires associations among logical and engineering components.

The QoS match is very simplified view of the schedulability problem with very few practical instances. To give one possible example (Figure 9.26, right) consider the sufficient schedulability condition for rate monotonic, based on processor utilization [28]

$$U = \sum_{i=1,n} \frac{C_i}{\theta_i} \leq U_{\text{lub}} = n(2^{(1/n)} - 1)$$

where

$$U_i = \frac{C_i}{\theta_i}$$

The QoS demand of each thread consists of a demand for utilization U_i to the operating system/scheduling policy and a demand for a target worst-case computation time C_i to the hardware. The QoS offer from the RM scheduling policy is the guaranteed least upper bound on the utilization $U_{\text{lub.}}$ and the QoS matching rule easily follows.

The *schedulability analysis* model is based on stereotyped scenarios (scheduling "situations" in the SPT language). Each Scheduling situation is in practice a sequence diagram, or a collaboration diagram (better

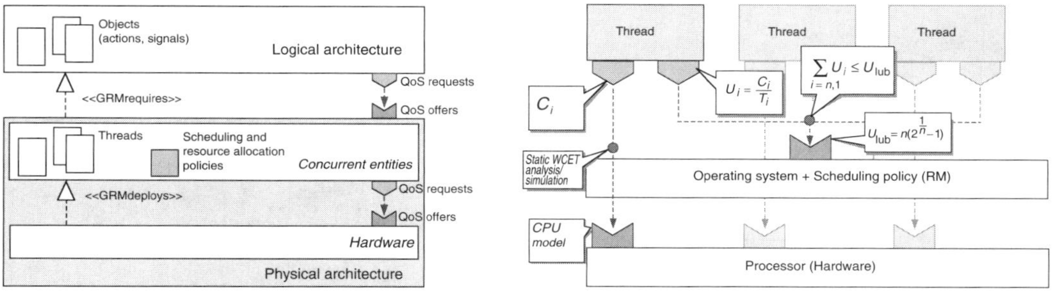

FIGURE 9.26 Schedulability analysis requires creating a correspondence between logical and physical architecture

suited for representing resource sharing), where one or more trigger events result in the actions to be scheduled within the deadline associated with the trigger.

Rate monotonic analysis is the method of choice for analyzing simple models with a restricted semantics, conforming to the so-called *task-centric* design paradigm. This requires updating the current definition of UML actions in order to allow for preemption (which is a necessary prerequisite of RMA). Such a recommendation has been explicitly formulated in the SPT profile and is among the recommended changes in the agenda of the OMG action semantics group.

In this task-centric approach, the behavior of each active object or task consists of a combination of reading input signals, performing computation, and producing output signals (Figure 9.28, left). Each active object can request the execution of actions of other passive objects in a synchronous or asynchronous fashion.

Each task is logically concurrent; it is activated periodically, and handles a single event. Active objects cooperate only by means of pure asynchronous messages, possibly implemented by means of memory mailboxes, a kind of protected (shared resource) object (Figure 9.27).

In the SPT, the TimedEvent that triggers the execution of the task action is identified with the stereotype "SATrigger" and it is the model element to which the deadline constraint (for the execution of the action) and the schedulability response are associated (Figure 9.27).

Unfortunately, the task-centric model, even if simple and effective in some contexts (see also other design methodologies [62–64], where the analysis of UML models is made possible by restricting the semantics of the analyzable models), only allows analysis of simple models where active objects do not interact with each other. In general UML models, each action is part of multiple end-to-end computations (with the associated timing constraints) and the system consists of a complex network of cooperating active objects, implementing state machines, and exchanging asynchronous messages.

In their work [65–67], Saksena, Karvelas, and Wang present an integrated methodology that allows dealing with more general OO models where multiple events can be provided as input to a single thread. According to their model, each thread has an incoming queue of events possibly representing real-time transactions and the associated timing constraints (Figure 12.10).

Consequently, scheduling priority (usually a measure of time urgency or criticality) is attached to events rather than threads. This design and analysis paradigm is called *event-centric design*. Each event has an associated priority (for example, deadline monotonic priority assignment can be used). Event queues are ordered by priority (i.e., threads process events based on their priority) and threads inherit the priority of the events they are currently processing. This model entails a two-level scheduling: the events enqueued as input to a thread need to be scheduled to find their processing order. At the system level, the threads are scheduled by the underlying RTOS (a preemptive priority-based scheduler is assumed).

Schedulability analysis of the general case is derived from the analysis methodology for generic deadlines (possibly greater than the activation rates of the triggering events) by computing response times of actions relative to the arrival of the external event that triggers the end-to-end reaction (transaction) T

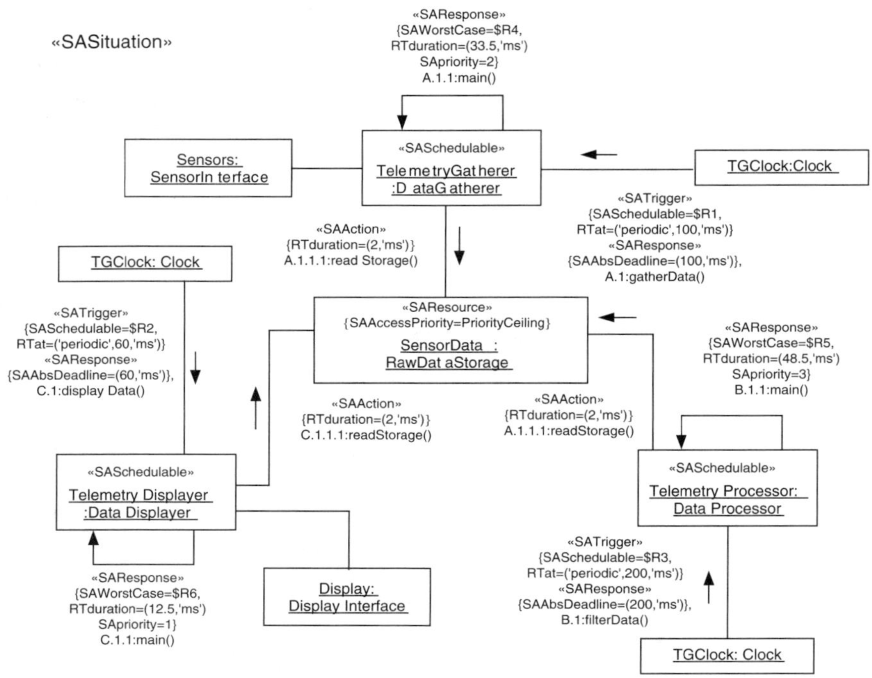

FIGURE 9.27 A sample collaboration instance representing a situation suitable for RMA. (Adapted from the OMG profile document.)

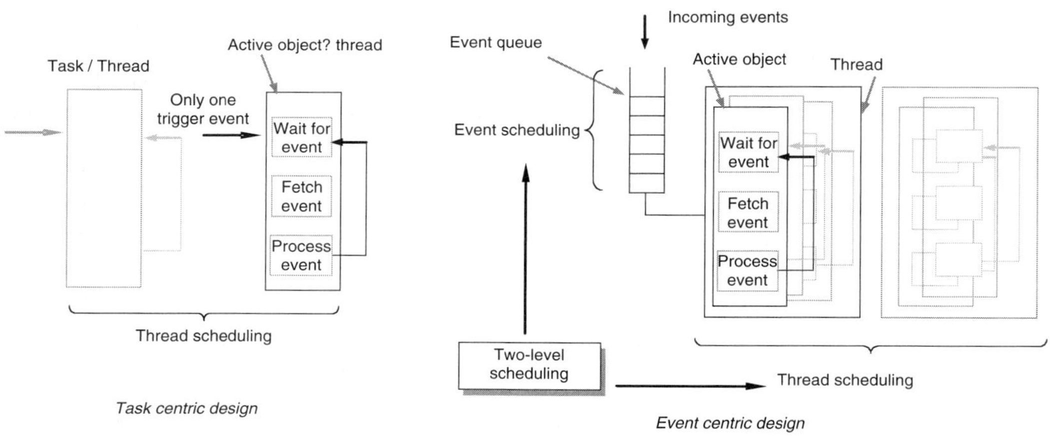

FIGURE 9.28 The dual-scheduling problem in the event-centric analysis.

containing the action. The analysis is based on the standard concepts of critical instant and busy period for task instances with generic deadlines (adapted to the transaction model.)

In [66], many implementation options for the event-centric model are discussed, for example single-thread implementation, multithread implementation with fixed priority or dynamic priority multi-threaded implementation and a formula or procedure for schedulability analysis is provided or at least discussed for each of these models. The interested reader should refer to [66] for details.

From a time analysis point of view, it makes sense to define the set of (nested) actions that are called synchronously in response to an event. This set is called *synchronous set.*

In a *single-thread implementation*, the only application thread processes pending events in priority order. Since there is only one thread, there is only one level of scheduling. Actions inherit their priority from the priority of the triggering event. Also, a synchronously triggered action inherits its priority from its caller. In single-threaded implementations, any synchronous set that starts executing is completed without interruption. Hence, the worst-case blocking time of an action starting at $t = 0$ is bound by the longest synchronous set of any lower priority action that started prior to $t = 0$ and interference can be computed as the sum of the interference terms from other transactions and the interference from actions in the same transaction.

The single-thread model can be analyzed for schedulability (details in [66]) and it is also conformant to the nonpreemptable semantics of UML actions. Most of the existing CASE tools support a single-threaded implementation and some of them (such as Rational Rose RT) support the priority-based enqueuing of activation events (messages).

In *multithreaded implementations*, each event represents a request for an end-to-end sequence of actions, executed in response to its arrival. Each action is executed in the context of a thread. Conceptually, we can reverse the usual meaning of thread scheduling, considering events (and the end-to-end chain of actions) as the main scheduling item and the threads required for the execution of actions as special mutual exclusion (mutex) resources, because of the impossibility of having a thread preempt itself. This insight allows using threads and thread priorities in a way that facilitates response time analysis.

If threads behave as mutexes, then it makes sense to associate with each thread a ceiling priority as the priority at which the highest priority event is served. As prescribed by PI or PCP, *threads inherit the priority of the highest priority event* in their waiting queue and this allows bounding priority inversion. In this case, the worst-case blocking time is restricted to the processing of a lower priority event. Furthermore, before processing an event, a thread locks the active object within which the event is to be processed (this is necessary when multiple threads may handle events that are forwarded to the same object), and a ceiling priority and a priority inheritance rule must be defined for active objects as well as for threads.

For the multithreaded implementation, a schedulability analysis formula is currently available for the aforementioned case in which threads inheriting the priority of the events. The schedulability analysis formula generally results in a reduced priority inversion with respect to the single-thread implementation, but its usefulness is hindered by the fact that (to the author's knowledge) there is no CASE tool supporting generation of code with the possibility of assigning priorities to events and run-time support for executing threads inheriting the priority of the event they are currently handling.

Assigning *static priorities to threads* is tempting (static priority scheduling is supported by most RTOSs) and it is quite a common choice for multithreaded implementations (Rational Rose RT and other tools use this method). Unfortunately, when threads are required to handle multiple events it is not clear what priority should be assigned to them and the computation of the worst-case response time is in general very hard because of the two-level scheduling and because of the difficulty of constructing an appropriate critical instant. The analysis needs to account for multiple priority inversion that arises from all the lower priority events handled by a higher priority thread. However, it may still be possible to use static thread priorities in very simple cases when it is easy to estimate the amount of priority inversion.

The discussion of the real-time analysis problem bypasses some of the truly fundamental problems in designing a schedulable software implementation of a functional (UML) model. The three main degrees of freedom in a real-time UML design subject to schedulability analysis are:

- assigning priorities to events
- defining the optimal number of tasks, and especially
- defining a mapping of methods (or entire objects) to tasks for their execution.

Timing analysis (as discussed here) is only the last stage, after the mapping decisions have been taken.

What is truly required is a set of design rules, or even better, an automatic synthesis procedure that helps in the definition of the set of threads and especially in the mapping of the logical entities to threads and to the physical platform.

The problem of synthesizing a schedulable implementation is actually a complex combinatorial optimization problem. Saksena et al. [67], propose an automatic procedure for synthesizing the three main degrees of freedom in a real-time UML design. The synthesis procedure uses a heuristic strategy based on a decomposition approach, where priorities to events/actions are assigned in a first stage and mapping is performed in a separate stage.

This problem, probably the most important in the definition of a software implementation of real-time UML models, is still an open problem.

9.4.3.2 Specification and Description Language

SDL contains primitives for dealing with global time, but nevertheless, when the specification of time behavior and resource usage is an issue, it is impossible to capture most time specifications and constraints.

For example, SDL does not offer the possibility of defining local time, timers cannot be periodic and it is difficult to express clocks, where periodic ticks are instantaneously consumed or lost. The main reason for these deficiencies is that the SDL notion of time was aimed at code generation, rather than at simulation and verification.

As stated in the previous section, SDL timer values only represent minimal bounds. This means that the SDL semantics provides no means to specify constraints on the time passed in states or actions, whereas the verification or simulation of a property often depends on finite lower and upper bounds. Also, the delays in the transmission of signals are unbounded.

The use of MSC and annotations to express the timing constraints that standard SDL misses is commonplace in commercial tools and research proposals. In particular, some tools allow the expression of duration constraints on actions or other simple constraints in the form of special comments (with the problem that each tool defines its own syntax and semantics.) Other proposals allow the definition of upper and lower time bounds for the delays applied to messages conveyed by a channel, or the definition of the delays probabilities.

Concerning the expressivity problem, a general real-time framework for SDL is presented in [68,69], (the first one is aimed at hardware software codesign). The extension of SDL with more powerful timers and clocks as well as local clocks has been proposed by many (see e.g., [70]). Other proposals for expression of timing requirements can be found in [71,72].

Another problem of SDL is that it provides neither deployment and resource usage information, nor the notion of preemptable or nonpreemptable actions that is necessary for schedulability analysis. One proposal [73] consists in defining a list of resources with an attribute defining their preemptability. Block, Process, and Task definitions need to be complemented with the specification of the resource on which they are executed. Scheduling policies are defined by keywords, such as RMA or Earliest Deadline First (EDF), or by priority rules attached to resources or agents. Priorities may be static or dynamic and in the latter case, they may be specified based on preconditions, or by means of an observer, which explicitly updates priorities depending on the observed states and events.

Some tools, for example time-enhanced versions of ObjectGeode [74] as presented in [75] and Tau [76,77] and the tool and methods based on Queuing SDL [78] propose to attach explicit timing information to SDL constructs such as tasks, and to provide some minimal deployment information. The ObjectGeode simulator, for example, allows associating an execution time (interval) with each action. Schedulability of SDL models is also discussed in [79].

Deficiencies in the specification of time also clearly affect the simulation of SDL models with time constraints. Any rigorous attempt to construct the simulation graph of an SDL system (the starting point for simulation and verification) must account for all possible combinations of execution times, timer expirations, and consumptions. Since no control over time progress is possible, many undesirable executions might be obtained during this exhaustive simulation.

In practice, existing simulation and verification tools make simplifying assumptions on execution and idle times. The usual convention is that statements or entire transactions are atomic; that actions take zero time to execute, and that the system executes immediately whatever it can execute (time passes only where explicitly required by waiting conditions). This option is justified by the fact that it generates the highest degree of determinism, thus reducing the state space by an important factor (making the system practically analyzable).

In [80], an interesting integration of SDL model simulation with resource scheduling is proposed: annotations related to timing constraints and resource usage are translated into scheduler and resource manager models that can be used to simulate the time behavior of the model with respect to schedulability constraints on a standard SDL tool. Performance evaluation of SDL models is the subject of other research works [75, 81, 82]

The verification of time-related properties of SDL models can often be done on a finite control abstraction, which is a system with finite data domains, bounded message buffers, and bounded agent creation. A number of interesting verification problems are decidable on such a finite control abstraction under the condition that in the timed automata obtained by translation, clocks can only be reset to zero, stopped, and restarted, and that the only allowed tests are comparisons of clock values or differences of clock values with constants, which, in practice allows transformation of the model into the standard semantics of TA.

Some SDL tools provide the facility to define observers as a means for defining constraints and properties, but SDL does not provide any standard notation for them.

9.4.4 Tools and Commercial Implementations

There is a large number of specialized CASE tools for UML modeling of embedded systems. Among those, Rhapsody of I-Logix [83], Real-Time Studio of Artisan [84], TAU Generation 2 from Telelogic [76] and Rational Rose RT (now Rational Rose Technical Developer [34], which is in essence a profile implementing ROOM in the UML framework) of IBM (Rational) are probably the most common. In contrast with general-purpose UML tools, all these try to answer the designer's need for automated support for model simulation, verification, and testing. In order to do so, they all provide an interpretation of the dynamic semantics of UML. Common features include interactive simulation of models and automatic generation of code. Nondeterminism is forbidden or eliminated by means of semantics restrictions. Some timing features and design of architecture level models are usually provided, although in nonstandard form if compared to the specifications of the SPT profile. Furthermore, the support for OCL and user-defined profiles is often quite weak.

Formal validation of untimed and timed models is typically not provided, since commercial vendors focus on providing support for simulation, automatic code generation, and (partly) automated testing. Third-party tools inter-operating by means of the standard XMI format provide schedulability analysis and research work on restricted UML semantics models demonstrates that formal verification techniques can be applied to UML behavioral models. This is usually done by transformation of UML models into a formal MOC and subsequent formal verification by existing tools (for both untimed and timed systems). A number of tools, including those previously cited, allow the definition of UML models including real-time constraints, and later possible verification of schedulability properties.

SDL is supported by powerful development environments integrating advanced facilities (like simulation, model checking, test generation, and auto-coding). Tools like ObjectGeode or the Telelogic Tau2 series support many phases of software development, ranging from some restricted form of analysis to implementation (code generation) and on-target deployment. Properties are expressed by MSC or by means of a tool-specific language allowing the definition of observers. No schedulability analysis is currently available from commercial vendors.

The most recent versions of some of these tools attempt to combine various modelling approaches in order to unite the best of both worlds. One example is SDL-UML and the Telelogic TAU series of products [76], where the TTCN language (set to become one of the UML test profile recommended notations) is used to describe formally the test cases and sequences [85].

9.5 Research on Models for Embedded Software

The models and tools described in the previous sections are representatives of a larger number of methods, models, languages, and tools (at different levels of maturity) that are currently being developed to face the challenges posed by embedded software design.

This section provides an insight on other (often quite recent) proposals for solving advanced problems related to the modeling, simulation, and verification of functional and nonfunctional properties, including support for compositionality and possibly for integration of heterogeneous models, where heterogeneity applies to both the execution model and the semantics of the component communications or, in general, interactions.

In particular, the request for heterogeneity arises from the observation that there is no clear winner among the different models and tools and that different parts of complex embedded systems may be better suited to different modeling and analysis paradigms (such as dataflow models for data-processing blocks and FSMs for control blocks). The first objective in this case is to reconcile by unification the synchronous and asynchronous execution paradigms.

The need to handle timing and resource constraints (hence preemption) together with the modeling of system functionality is another major requirement for modern methodologies.

Performance analysis by simulation of timed (control) systems with scheduling and resource constraints is, for example, the main goal of the TrueTime Toolset from the LTH Technical University in Lund [86].

The TrueTime toolset is one example of modeling paradigm where scheduling and resource management policies can be handled as separate modules to be plugged in together with the blocks expressing the functional behavior. TrueTime consists of a set of Simulink blocks that simulate real-time kernel policies as found in commercial RTOSs. The system model is obtained by connecting kernel, network, and ordinary Simulink blocks representing the functional behavior of the control application. The toolbox is capable of simulating the system behavior of the real-time system with interrupts, preemption and scheduling of resources.

The Times tool [87] is another research framework (built on top of Uppaal and available at http://www.timestool.org) attempting an integration of the TA formalism with methods and algorithms for task and resource scheduling. In Times, a timed automaton can be used to define an arbitrary complex activation model for a set of deadline-constrained tasks, to be scheduled by fixed or dynamic (Earliest Deadline First [28]) priorities. Model checking techniques are used to verify the schedulability of the task set, with the additional advantage (if compared with standard worst-case analysis) of providing the possible runs that fail the schedulability test.

Other design methodologies bring further distinction among (at least) three different design dimensions representing:

- The *(functional) behavior* of the system and the timing constraints imposed by the environment and/or resulting from design choices
- The *communication network* and the *semantics of communication* defining interactions on each link
- The *platform* onto which the system is mapped and the timing properties of the system blocks resulting from the binding

The Ptolemy and the Metropolis environments are probably the best-known examples of design methodologies founded on the previous principles.

Ptolemy (http://www.ptolemy.eecs.berkeley.edu/) is a simulation and rapid prototyping framework for heterogeneous systems developed at the Center for Hybrid and Embedded Software Systems (CHESS) in the Department of Electrical Engineering and Computer Sciences of the University of California at Berkeley [2]. Ptolemy (now at version II) targets complex heterogeneous systems encompassing different operations such as signal processing, feedback control, sequential decision-making, and possibly even user interfaces.

Prior to UML 2.0, Ptolemy II had already introduced the concept of actor-oriented design, by complementing traditional object-orientation with concurrency and abstractions of communication between components.

In Ptolemy II, the model of the application is built as a hierarchy of interconnected actors (most of which are domain polymorphic). Actors are units of encapsulation (components) that can be composed hierarchically, producing the design tree of the system. Actors have an interface abstracting their internals and providing bidirectional access to functions. The interface includes ports that represent points of communication, and parameters, which are used to configure the actor operations.

Communication channels pass data from one port to another according to some communication semantics. The abstract syntax of actor-oriented design can be represented concretely in several ways, one example being the graphical representation provided by the Ptolemy II Vergil GUI (Figure 9.29).

Ptolemy is not built on a single, uniform, model of computation, but rather, it provides a finite library of *Directors* implementing different models of computations. A Director, when placed inside a Ptolemy II actor, defines its abstract execution semantics (MOC) and the execution semantics of the interactions among its component actors (its *domain*).

A component that has a well-defined behavior under different models of computation is called a domain polymorphic component, which means that its behavior is polymorphic with respect to the domain or MOC that is specified at each node of the design tree.

The available MOCs in Ptolemy II include:

- Communicating Sequential Processes (CSP, with synchronous rendezvous)
- continuous time (CT, where behavior is specified by a set of algebraic or differential equations)
- discrete events, (DE, consisting of a value and a time stamp)
- Finite State Machines (FSM)
- (Kahn) Process networks (PN)
- Synchronous dataflows (SDF)
- Synchronous/reactive (SR) models and Time-triggered synchronous execution (the Giotto framework [88])

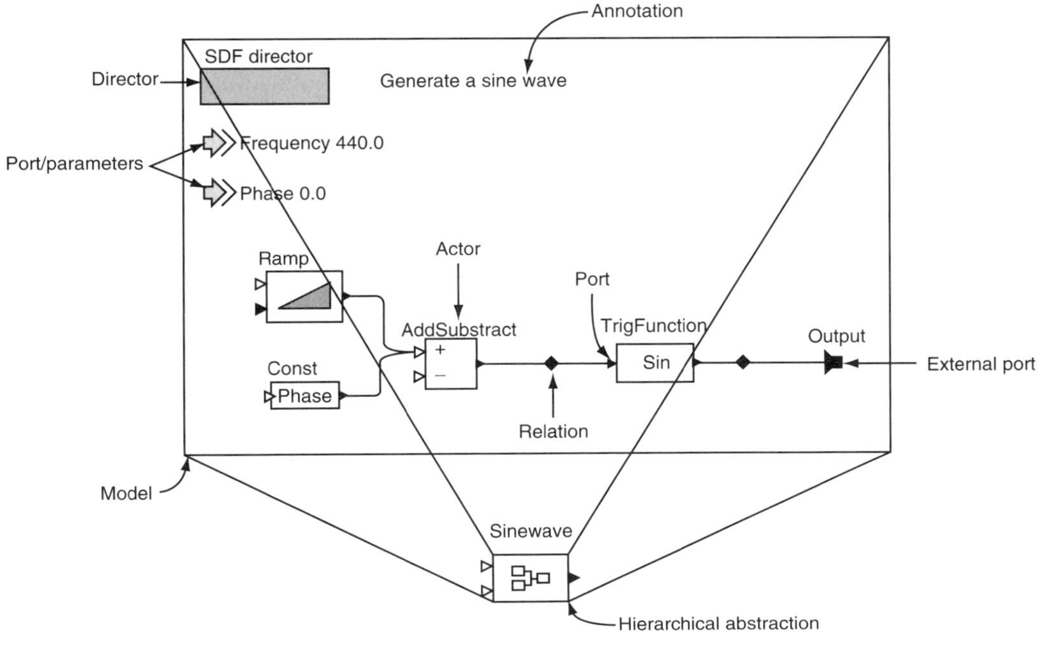

FIGURE 9.29 A Ptolemy actor and its black-box (higher level) representation.

The CT and SR domains have fixed-point semantics, meaning that in each iteration the domain may repeatedly fire the components (execute the available active transitions inside them) until a fixed point is found.

Of course, key to the implementation of polymorphism is proving that an aggregation of components under the control of a domain in turn defines a polymorphic component. This has been proven possible for a large number of combinations of MOCs [89].

In Ptolemy II, the implementation language is Java, and an experimental module for automatic Java code generation from a design is now available at http://ptolemy.eecs.berkeley.edu/. The source code of the Ptolemy framework is available for free from the same web site. Currently, the software has hundreds of active users from industry and academia at various sites worldwide.

The Metropolis environment embodying the platform-based design methodology [1] for design representation, analysis, and synthesis, is under development at the University of California at Berkeley, under the sponsorship of the MARCO Gigascale System Research Center.

In the Metropolis toolset [90], system design is seen as the result of a progressive refinement of high-level specifications into lower-level models, *possibly heterogeneous* in nature. The environment deals with all phases of the design from conception to final implementation.

Metropolis is designed as a very general infrastructure based on a meta-model with precise semantics that is general enough to support existing computation models and accommodate new ones. The metamodel supports not only functionality capture and analysis through simulation and formal verification, but also architecture description and the mapping of functionality to architectural elements (Figure 9.30).

The Metropolis Metamodel (MMM) language provides basic building blocks that are used to describe communication and computation semantics in a uniform way. These components represent *computation*, *communication*, and *coordination* or *synchronization* as follows:

- *Processes* for describing computation
- *Media* for describing communication between processes
- *Quantity Managers* for enforcing a scheduling policy of processes
- *Netlists* for describing interconnection of objects and for instantiating and interconnecting quantity managers and media

A process is an active object (it possesses its own thread of control) that executes a sequence of actions (instructions, subinstructions, function calls, and awaits [91]) and generates a sequence of events. Each process in a system evolves by executing one action after the other. At each step (which is formally

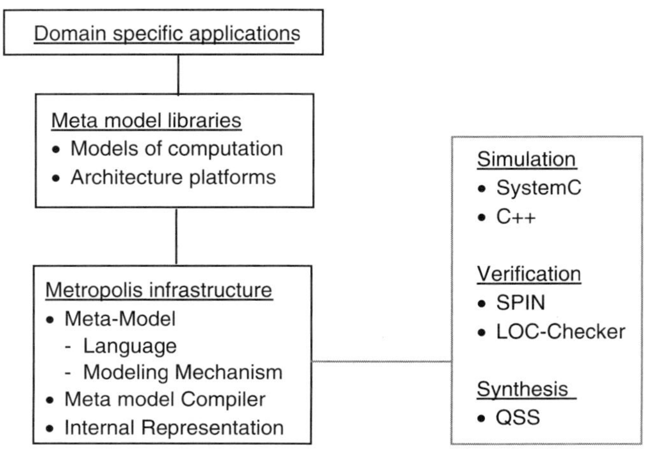

FIGURE 9.30 The Metropolis framework.

described in [91] in terms of a global execution index), each process in the system executes one action and generates the corresponding event. This synchronous execution semantics is relaxed to asynchronicity by means of a special event called NOP that can be freely interleaved in the execution of a process (providing non-determinism).

Each *process* defines its *Ports* as the possible interaction points. *Ports* are typed objects associated with an interface, which declares the services that can be called by the process or called from external processes (bi-directional communication) in a way similar to what is done in UML 2.0.

Processes cannot connect directly to other processes but the interconnection has to go through a *Medium*, which has to define (i.e., implement) the services declared by the interface associated with the ports. The separation of the three orthogonal concepts of Process, Port, and Medium provides maximum flexibility and reusability of behaviors, that is, the meaning of the communication can be changed and refined without changing the computation description that resides in the processes (Figure 9.31).

A Metropolis program can be seen as a variant of a system-encompassing TA where the transition between states is labeled by event vectors (one event per process). At each step, there is a set of event vectors that could be executed to make a transition from the current state to the next state. Unless a suitably defined Quantity Manager restricts the set of possible executions by means of scheduling constraints, the choice among all possible transitions is performed nondeterministically.

Quantity Managers define the scheduling of actions by assigning tags to events. A tag is an abstract quantity from a partially order set (such as, for example, time or execution order). Multiple requests for action execution can be issued to a quantity manager that has to resolve them and schedule the processes in order to satisfy the ordering relation on the set of tags. The tagged-signal model (Lee and Sangiovanni-Vincentelli [92]) is the formalism that defines the unified semantic framework of signals and processes that stands at the foundation of the MMM.

By defining different communication primitives and different ways of resolving concurrency, the user can, in effect, specify different MoCs. For instance, in a synchronous model of computation, all events in an event vector have the same tag.

The mapping of a functional model onto a platform is performed by enforcing a synchronization of function and architecture. Each action in the function side is correlated with an action on the architecture side using synchronization constraints.

The precise semantics of Metropolis allows for system simulation, synthesis and formal analysis. The metamodel includes constraints that represent in abstract form requirements not yet implemented or assumed to be satisfied by the rest of the system and its environment.

Metropolis uses a logic language to capture nonfunctional constraints (time or energy constraints, for example). Constraints are declarative formulas of two kinds: linear temporal logic (LTL) and logic of constraints (LOC).

Although two verification tools are currently provided to check that constraints are satisfied at each level of abstraction, the choice of the analysis and synthesis method or design tool depends on the application domain and the design phase. Metropolis clearly cannot possibly provide algorithms and tools for all possible design configurations and phases, but it provides mechanisms to compactly store and communicate all design information, including a parser that reads metamodel designs and a standard API

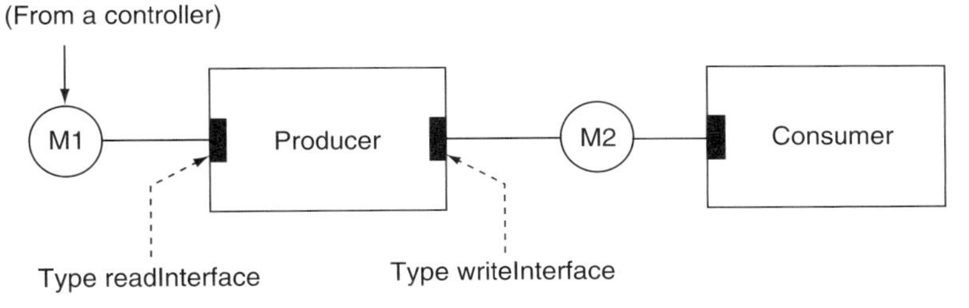

FIGURE 9.31 Processes, Ports, and Media in Metropolis.

that allows developers to browse and modify design information so that they can plug in the required algorithms for a given application domain. This mechanism has been exploited to integrate the Spin software verification tool [10].

Finally, as the last example, separation of behavior, communication or interaction, and execution models are among the founding principles of the component-based design and verification procedure outlined in [93,94]. The methodology is based on the assumption that the system model is a transition system with dynamic priorities restricting non-determinism in the execution of system actions. Priorities define a strict partial order among actions ($a_1 \prec a_2$). A priority rule is a set of pairs $\{(C^j, \prec^j)\}_{j \in J}$ such that $\prec^j$ is a priority order and C^j is a state constraint specifying when the rule applies.

In [95], a component is defined as the superposition of the three models defining its behavior, interactions, and execution model. The transition-based formalism is used to define the behavior, and interaction specification consists of a set of connectors, defining a maximal set of interacting actions and specification of actions that must occur simultaneously in the context of an interaction. The set of actions that are allowed to occur in the system consists of the *complete actions* (Figure 9.32, triangles). Incomplete actions (Figure 9.32, circles) can only occur in the context of an interaction defining a higher-level complete action (as in the definition of $IC[K_1]^+$ in Figure 9.32, which defines the complete action $a_5|a_9$, meaning that incomplete action a_9 can only occur synchronously with a_5). This definition of interaction rules allows for a general model of asynchronous components, which can be possibly synchronized upon a subset of their actions if and when needed. Finally, the execution model consists of a set of dynamic priority rules.

Composing a set of components means applying a set of rules at each of the three levels. The rules defined in [95] define an associative and commutative composition operator. Further, the composition

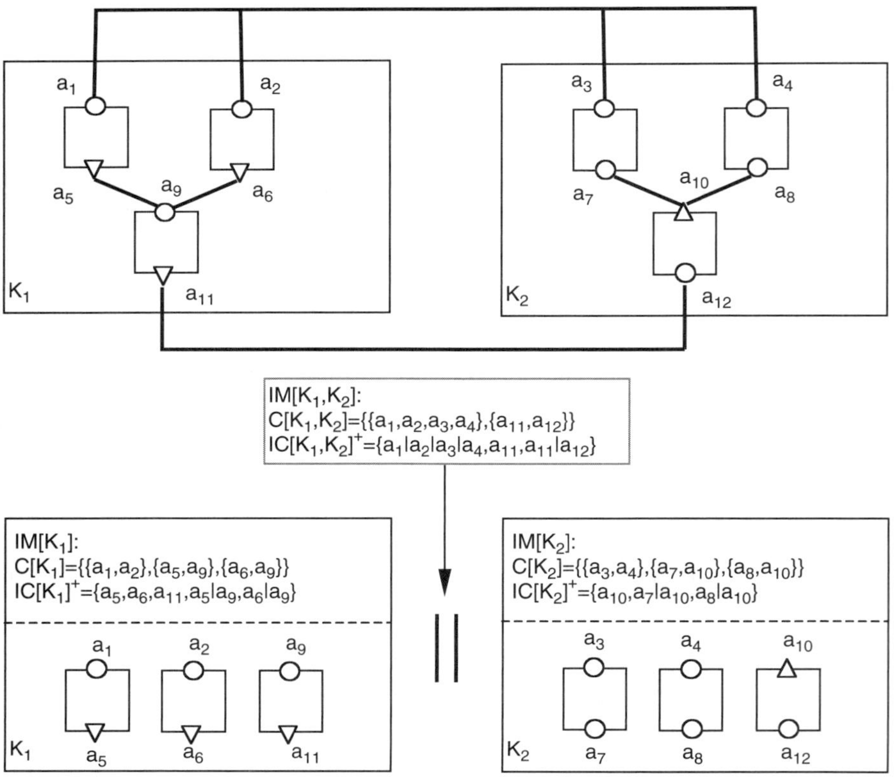

FIGURE 9.32 Interaction model among system components. (From Gossler, G. and Sifakis, J., Composition for component-based modeling. With permission.)

model allows for *composability* (properties of components are preserved after composition) and compositionality (properties of the compound can be inferred from the properties of the components) with respect to deadlock freedom (liveness).

Gossler and Sifakis [95] propose a methodology for analysis of timed systems (and a framework for the composition rules for component-based real-time models) based on the framework represented in Figure 9.33. According to the proposed design methodology, the system architecture consists of a number of layers, each capturing a different set of properties.

At the lowest level, threads (processes) are modeled according to the TA formalism, capturing the functional behavior and the timing properties and constraints.

At the preemption level, system resources and preemptability are accounted for by adding one or more preemption transitions (right-hand side of Figure 9.34), one for each preemptable resource, and mutual exclusion rules are explicitly formulated as a set of constraints acting upon the transitions of the processes.

Finally, the resource management and scheduling policies are represented as additional constraints $K_{pol}=K_{adm}\wedge K_{res}$ where K_{adm} are the admission control constraints and K_{res} are the constraints specifying how resource conflicts are to be resolved.

Once scheduling policies and schedulability requirements are in place, getting a correctly scheduled system amounts to finding a nonempty control invariant K such that $K \Rightarrow K_{sched}\wedge K_{pol}$.

Finally, the last example of research framework for embedded software modeling is the generic modeling environment (GME), developed at Vanderbilt University [96], which is a configurable toolkit offering a meta–meta modeling facility for creating domain-specific models and program synthesis environments. The configuration of a domain-specific metamodel can be achieved by defining the syntactic, semantic, and presentation information regarding the domain. This implies defining the main

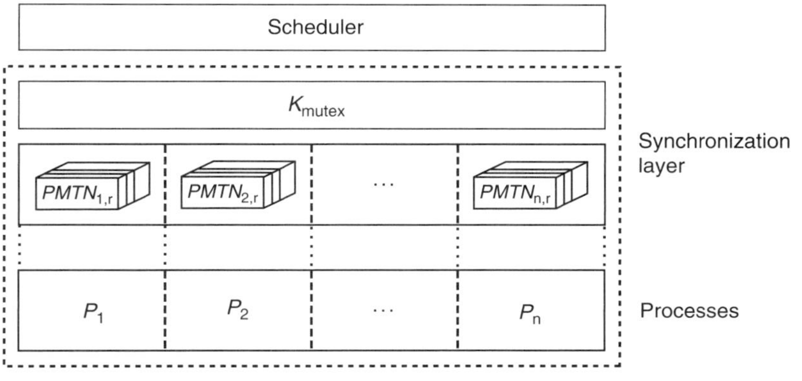

FIGURE 9.33 The architecture of the system model as defined. (From Gossler, G. and Sifakis, J., Composition for component-based modeling. With permission.)

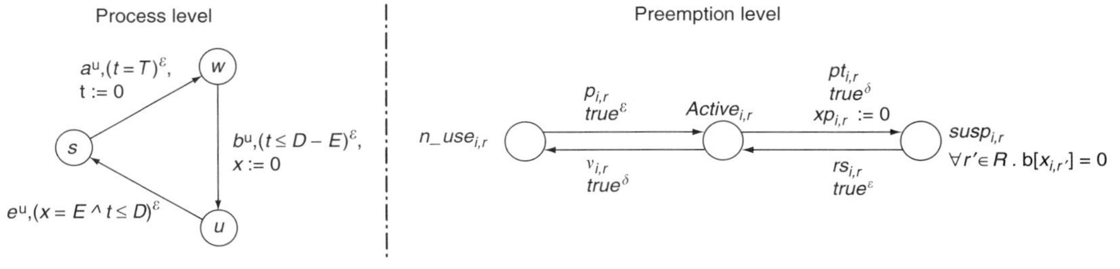

FIGURE 9.34 Process and preemption modeling. (From Gossler, G. and Sifakis, J., Composition for component-based modeling. With permission.)

domain concepts, the possible relationships and the constraints restricting the possible system configurations as well as the visibility rules of object properties.

The vocabulary of the domain-specific languages implemented by different GME configurations is based on a set of generic concepts built into GME itself. These concepts include hierarchy, multiple aspects, sets, references, and constraints. Models, atoms, references, connections, and sets are first-class objects.

Models are compound objects that can have parts and inner structure. Each part in a container is characterized by a role. The modeling instance determines what parts are allowed and in which roles. Models can be organized in a hierarchy, starting with the root module. The aspects provide visibility control. The relationships can be (directed or undirected) connections, further characterized by attributes. The model specification can define several kinds of connections, which objects can participate in a connection and further explicit constraints. The connection only appears between two objects in the same model entity. References help establish connections to external objects as well.

9.6 Conclusions

This chapter discussed the use of software models for the design and verification of embedded software systems. It attempted to classify and a survey existing formal models of computation, following the classical divide between synchronous and asynchronous models and between models for functionality as opposed to models for software architecture specification. Problems like formal verification of system properties, both timed and untimed and schedulability analysis are discussed. This chapter also provided an overview of commercially relevant modeling languages UML and SDL, and discusses recent extensions to both these standards. The discussion of each topic was supplemented with an indication of the available tools that implemented the available methodologies and analysis algorithms. Finally, the chapter contained a short survey of the results of recent researches and a discussion of open issues and future trends.

References

[1] Sangiovanni Vincentelli, A., Defining platform-based design, *EEDesign of EETimes*, February 2002.

[2] Lee, E.A., Overview of the Ptolemy Project, Technical Memorandum UCB/ERL M03/25, July 2, 2003, University of California, Berkeley, CA.

[3] Balarin, F., Hsieh, H., Lavagno, L., Passerone, C., Sangiovanni-Vincentelli, A., and Watanabe, Y., Metropolis: an integrated environment for electronic system design, *IEEE Computer*, April 2003.

[4] *UML Profile for Schedulability, Performance and Time Specification.* OMG Adopted Specification, July 1, 2002, http://www.omg.org.

[5] Beck, T., Current trends in the design of automotive electronic systems, *Proceedings of the Design Automation and Test in Europe Conference*, 2001.

[6] Edwards, S., Lavagno, L., Lee, E.A., and Sangiovanni-Vincentelli, A., Design of embedded systems: formal models, validation and synthesis, *Proceedings of the IEEE*, March 1997.

[7] Alur, R. and Henzinger, T.A., Logics and models of real time: a survey, *Real-Time: Theory in Practice, REX Workshop*, Lecture Notes in Computer Science, Vol. 600, Springer, Berlin, 1991, pp. 74–106.

[8] Pnueli, A., The temporal logic of programs, *Proceedings of the 18th Annual Symposium on the Foundations of Computer Science, IEEE*, November 1977, pp. 46–57.

[9] Emerson, E.A., Temporal and modal logics, in *Handbook of Theoretical Computer Science*, Vol. B, van Leeuwen, J., Ed., Elsevier, Amsterdam, 1990, pp. 995–1072.

[10] Holzmann, G.J., *Design and Validation of Computer Protocols*, Prentice-Hall, Eaglewood cliffs, NJ, 1991.

[11] Harel, D., Statecharts: a visual approach to complex systems, *Sci. Comp. Prog.*, 8, 231–275, 1987.

[12] Merlin, P.M. and Farber, D.J., Recoverability of communication protocols, *IEEE Trans. Comm.*, 24, 36–103, 1976.

[13] Sathaye, A.S. and Krogh, B.H. Synthesis of real-time supervisors for controlled time Petri nets, *Proceedings of 32nd IEEE Conference on Decision and Control*, Vol. 1, San Antonio, 1993, pp. 235–238.
[14] Alur, R. and Dill, D.L., A theory of timed automata, *Theoret. Comput. Sci.*, 126, 183–235, 1994.
[15] Ramchandani, C., Analysis of Asynchronous Concurrent Systems by Timed Petri Nets, Ph. D. thesis, MIT, Cambridge, MA, 1974.
[16] Molloy, M.K., Performance analysis using stochastic Petri nets, *IEEE Trans. Comp.*, 31, 913–917, 1982.
[17] Ajmone Marsan, M., Conte, G., and Balbo, G., A class of generalized stochastic Petri nets for the performance evaluation of multiprocessor systems, *ACM Trans. Comp. Sys.*, 2, 93–122, 1984.
[18] Haar, S., Kaiser, L., Simonot-Lion, F., Toussaint, J., On Equivalence Between Timed State Machines and Time Petri Nets, Rapport de recherche de l'INRIA – Lorraine, 2000.
[19] Yovine, S., Kronos: a verification tool for real-time systems, *Springer Int. J. Software Tools Technol. Transfer*, 1, 123–133, 1997.
[20] Larsen, K.G., Pettersson, P., and Yi, W. Uppaal in a nutshell, *Springer Int. J. Software Tools Technol. Transfer* 1, 134–152, 1997.
[21] Yi, W., Pettersson, P., and Daniels, M., Automatic verification of real-time communicating systems by constraint solving, *Proceedings of the 7th International Conference on Formal Description Techniques*, Berne, Switzerland, 4–7 October 1994.
[22] Henzinger, T.A., The theory of hybrid automata, *Proceedings of the 11th Annual Symposium on Logic in Computer Science (LICS)*, IEEE Computer Society Press, Washington, DC, 1996, pp. 278–292.
[23] Henzinger, T.A., Ho, P.-H., and Wong-Toi, H., HyTech: a model checker for hybrid systems, *Software Tools Technol. Trans.*, 1, 110–122, 1997.
[24] Vicario, E., Static analysis and dynamic steering of time-dependent systems using time Petri nets, *IEEE Trans. Software Eng.*, 27, 728–748, 2001.
[25] The ORIS tool web page: http://www.dsi.unifi.it/~vicario/Research/ORIS/oris.html.
[26] Lime, D. and Roux, O.H., A translation based method for the timed analysis of scheduling extended time Petri nets, *The 25th IEEE International Real-Time Systems Symposium*, December 5–8, Lisbon, Portugal, 2004.
[27] Clarke, E.M. and Wing, J.M., Formal Methods: State of the Art and Future Directions, *Technical Report CMU-CS-96-178*, Carnegie Mellon University (CMU), Sept. 1996.
[28] Liu, C. and Layland, J., Scheduling algorithm for multiprogramming in a hard real-time environment, *J. ACM*, 20, 46–61, 1973.
[29] Klein, M.H. et al., *A practitioner's Handbook for Real-Time Analysis: Guide to Rate Monotonic Analysis for Real-Time Systems*, Kluwer Academic Publishers, Hingham, MA, 1993.
[30] Rajkumar, R., *Synchronization in Multiple Processor Systems, Synchronization in Real-Time Systems: A Priority Inheritance Approach*, Kluwer Academic Publishers, Dordrecht, 1991.
[31] Benveniste, A., Caspi, P., Edwards, S.A., Halbwachs, N., Le Guernic, P., and de Simone, R., The synchronous languages 12 years later, *Proc. IEEE*, 91–1, 64–83, 2003.
[32] Real-Time Workshop Embedded Coder 4.2, web page: http://www.mathworks.com/products/rtwembedded/.
[33] dSPACE Produkte: Production Code Generation Software, web page: www.dspace.de/ww/de/pub/products/sw/targetli.htm.
[34] Rational Rose Technical Developer, web page: http://www-306.ibm.com/software/awdtools/developer/technical/.
[35] Meyer, B., An overview of Eiffel, in *The Handbook of Programming Languages, Vol. 1, Object-Oriented Languages*, Salus, P.H., Ed., Macmillan Technical Publishing, New York, 1998.
[36] Caspi, P., Pilaud, D., Halbwachs, N., and Plaice, J.A., LUSTRE: a declarative language for programming synchronous systems, *ACM Symposium Principles of Programmimg Language (POPL)*, Munich, Germany, 1987, pp. 178–188.
[37] Halbwachs, N., Caspi, P., Raymond, P., and Pilaud, D., The synchronous data flow programming language LUSTRE, *Proc. IEEE*, 79, 1305–1320, 1991.
[38] Boussinot, F. and de Simone, R., The Esterel language, *Proc. IEEE*, 79, 1293–1304, 1991.
[39] Berry, G., The constructive semantics of pure Esterel, *5th Algebraic Methodology and Software Technology Conference*, Munich, Germany, 1996, pp. 225–232.

[40] Westhead, M. and Nadjm-Tehrani, S., Verification of embedded systems using synchronous observers, in Lecture Notes in Computer Science, Formal Techniques in Real-Time and Fault-Tolerant Systems, Vol. 1135, Springer-Verlag, Berlin, 1996.

[41] The Mathworks Simulink and StateFlow. web page: http://www.mathworks.com.

[42] Scaife, N., Sofronis, C., Caspi, P., Tripakis, S., and Maraninchi, F., Defining and translating a "safe" subset of Simulink/Stateflow into Lustre, *Proceedings of 2004 Conference on Embedded Software, EMSOFT'04*, Pisa, Italy, September 2004, Springer, Berlin.

[43] Scaife, N. and Caspi, P., Integrating model-based design and preemptive scheduling in mixed time- and event-triggered systems, *16th Euromicro Conference on Real-Time Systems (ECRTS'04)*, Catania, Italy, 2004, pp. 119–126.

[44] Prover Technology, http://www.prover.com/.

[45] Edwards, S.A., An Esterel compiler for large control-dominated systems, *IEEE Trans. Comput. -Aided Des. Integr. Circuits Syst.*, 21, 2002, 169–183.

[46] Shiple, T. R., Berry, G., and Touati, H., Constructive analysis of cyclic circuits, *European Design and Test Conference*, 1996.

[47] Benveniste, A., Caspi, P., Le Guernic, P., Marchand, H., Talpin, J.-P., and Tripakis, S., A protocol for loosely time-triggered architectures, *Proceedings of 2002 Conference on Embedded Software, EMSOFT'02*, Sifakis, J. and Sangiovanni-Vincentelli, A., Eds., Lecture Notes in Computer Science, Vol. 2491 , Springer, Berlin, pp. 252–265.

[48] Benveniste, A., Caillaud, B., Carloni, L., Caspi, P., and Sangiovanni-Vincentelli, A., Heterogeneous reactive systems modeling: capturing causality and the correctness of loosely time-triggered architectures (LTTA), *Proceedings of the Conference on Embedded Software, EMSOFT'04*, Buttazzo, G. and Edwards, S., Eds., September 2004, pp. 27–29.

[49] Biannic, Y. L., Nassor, E., Ledinot, E., and Dissoubray, S., UML object specification for real-time software. *RTS Show*, 2000.

[50] Selic, B., Gullekson, G., and Ward, P.T., *Real-Time Object-Oriented Modeling*, Wiley, New York, 1994.

[51] Douglass, B. P., *Doing Hard Time: Developing Real-Time Systems with Objects, Frameworks, and Patterns*, Addison-Wesley, Reading, MA, 1999.

[52] Latella, D., Majzik, I., and Massink, M., Automatic verification of a behavioural subset of UML statechart diagrams using the SPIN modelchecker, *Formal Aspects Comput.* 11, 637–664, 1999.

[53] del Mar Gallardo, M., Merino, P., and Pimentel, E., Debugging UML designs with model checking, *J. Object Technol.*, 1, 2002.

[54] UML 2.0 OCL Final Adopted specification, available at http://www.omg.org/cgi-bin/doc?ptc/2003-10-14.

[55] ITU-T. Recommendation Z.100. Specification and Description Language (SDL). Z-100, Int. Telecom. Union Standard. Sect., 2000.

[56] ITU-T. Recommendation Z.120. Message Sequence Charts. Z-120, Int. Telecom. Union Standard. Sect, 2000.

[57] The PEP tool (Programming Environment based on Petri Nets), Documentation and user guide: http://parsys.informatik.uni-oldenburg.de/~pep/Paper/PEP1.8_doc.ps.gz.

[58] Bozga, M., Ghirvu, L., Graf, S., and Mounier, L., IF: a validation environment for timed asynchronous systems, *Comp. Aided Verification, CAV*, Lecture Notes in Computer Science, Vol. 1855, Springer, Berlin, 2000.

[59] Bozga, M., Graf, S., and Mounier, L., IF-2.0: a validation environment for component-based real-time systems. *Comp. Aided Verification, CAV*, Lecture Notes in Computer Science, Vol. 2404, Springer, Berlin, 2002, pp. 343–348.

[60] Regensburger, F and Barnard, A., Formal verification of SDL systems at the Siemens mobile phone department, *Tools and Algorithms for the Construction and Analysis of Systems. 4th International Conference, TACAS'98*, Lecture Notes in Computer Science, Vol. 1384, Springer, Berlin, 1998, pp. 439–455.

[61] Bozga, M., Graf, S., and Mounier, L., Automated validation of distributed software using the IF environment, *Workshop on Software Model Checking, Electronic Notes in Theoretical Computer Science*, Vol. 55, Elsevier, Amsterdam, 2001.

[62] Gomaa, H. *Software Design Methods for Concurrent and Real-Time Systems*, Addison-Wesley, Reading, MA, 1993.

[63] Burns A. and Wellings, A.J., HRT-HOOD: a design method for hard real-time, *J. Real-Time Sys.*, 6, 73–114, 1994.
[64] Awad, M., Kuusela, J., and Ziegler J., *Object-Oriented Technology for Real-Time Systems: A Practical Approach Using OMT and Fusion*, Prentice-Hall, Eaglewood Cliffs, NJ, 1996.
[65] Saksena, M., Freedman, P., and Rodziewicz, P., Guidelines for automated implementation of executable object oriented models for real-time embedded control systems, *Proceedings IEEE Real-Time Systems Symposium*, December 1997, pp. 240–251.
[66] Saksena, M. and Karvelas, P., Designing for schedulability: integrating schedulability analysis with object-oriented design, *Proceedings of the Euromicro Conference on Real-Time Systems*, Stockholm, June 2000.
[67] Saksena, M., Karvelas, P., and Wang, Y., Automatic synthesis of multi-tasking implementations from real-time object-oriented models, *Proceedings IEEE International Symposium on Object-Oriented Real-Time Distributed Computing*, March 2000.
[68] Slomka, F., Dörfel, M., Münzenberger, R., and Hofmann, R., Hardware/software codesign and rapid-prototyping of embedded systems. *IEEE Design Test Computers, (Special issue: Design Tools for Embedded Systems)*, Vol. 17, 28–38, 2000.
[69] Bozga, M., Graf, S., Mounier, L., Ober, I., Roux, J.-L., and Vincent, D., Timed extensions for SDL, *Proceedings of the SDL Forum 2001*, Lecture Notes in Computer Science, Vol. 2078, Springer, Berlin, June 2001.
[70] Münzenberger, R. Slomka, F. Dörfel, M., and Hofmann, R., A general approach for the specification of real-time systems with SDL, in *Proceedings of the 10th International SDL Forum*, Reed, R. and Reed, J., Eds., Lecture Notes in Computer Science, Vol. 2078, Springer, Berlin, 2001.
[71] Algayres, B., Lejeune, Y., and Hugonnet, F., GOAL: observing SDL behaviors with Geode, *Proceedings of SDL Forum 95.*
[72] Dörfel, M., Dulz, W., Hofmann, R., and Münzenberger, R., SDL and Non-Functional Requirement, Internal Report IMMD7 05/01, University of Erlangen, August 20, 2001.
[73] Mitschele-Thiel, S and Muller-Clostermann, B., Performance engineering of sdl/msc systems, *Computer Networks*, 31,1801–1815, 1999.
[74] Telelogic ObjectGeode, web page: http://www.telelogic.com/products/additional/objectgeode/index.cfm.
[75] Roux, J.L., SDL Performance analysis with ObjectGeode, *Workshop on performance and time in SDL*, 1998.
[76] Telelogic TAU Generation2, web page: http://www.telelogic.com/products/tau/tg2.cfm.
[77] Bjorkander, M., Real-time systems in UML (and SDL), *Embedded System Engineering*, October/November 2000, published by EDA.
[78] Diefenbruch, M., Heck, E., Hintelmann, J., and Müller-Clostermann, B., Performance evaluation of SDL systems adjunct by queuing models, *Proceedings of the SDL-Forum'95*, 1995.
[79] Alvarez, J.M., Diaz, M., Llopis, L.M., Pimentel, E., and Troya, J.M., Deriving hard-real time embedded systems implementations directly from SDL specifications, *International Symposium on Hardware/Software Codesign CODES*, 2001.
[80] Bucci, G., Fedeli, A., and Vicario, E., Specification and simulation of real time concurrent systems using standard SDL Tools, *11th SDL Forum*, Stuttgart, July 2003.
[81] Spitz, S; Slomka, F., and Dörfel, M., SDL* — an annotated specification language for engineering multimedia communication systems, *Workshop on High Speed Networks*, Stuttgart, October 1999.
[82] Malek, M., PerfSDL: interface to protocol performance analysis by means of simulation, *Proceedings of the SDL Forum* 99.
[83] I-Logix Rhapsody, web page: http://www.ilogix.com/rhapsody/rhapsody.cfm.
[84] Artisan Real-Time Studio, web page: http://www.artisansw.com/products/professional_overview.asp.
[85] Telelogic TAU TTCN Suite, web page: http://www.telelogic.com/products/tau/ttcn/index.cfm.
[86] Henriksson, D., Cervin A. and Årzén, K.-E., Truetime: simulation of control loops under shared computer resources, *Proceedings of the 15th IFAC World Congress on Automatic Control*, Barcelona, Spain, July 2002.
[87] Amnell, T., et al., Times — a tool for modelling and implementation of embedded systems, in *Proceedings of 8th International Conference, TACAS* 2002, Grenoble, France, April 8–12, 2002.

[88] Henzinger, T.A., Giotto: a time-triggered language for embedded programming, in *Proceedings of 1st International Workshop on Embedded Software (EMSOFT'2001)*, Tahoe City, CA, USA, Oct. 2001, Vol. 2211, Lecture Notes in Computer Science, Springer, Berlin, 2001, pp. 166–184.

[89] Lee, E.A. and Xiong, Y., System-level types for component-based design, *Proceedings 1st International Workshop on Embedded Software (EMSOFT'2001)*, Tahoe City, CA, October 2001, Lecture Notes in Computer Science, Vol. 2211, Springer, Berlin, 2001, pp. 237–253.

[90] Balarin, F. Lavagno, L. Passerone, C., and Watanabe, Y., Processes, interfaces and platforms. embedded software modeling in Metropolis, *Proceedings of the EMSOFT Conference*, Grenoble, France, 2002, 407–416.

[91] Balarin, F., Lavagno, L., Passerone, C., Sangiovanni Vincentelli, A., Sgroi M., and Watanabe, Y., Modeling and design of heterogeneous systems, Lecture Notes in Computer Science, Vol. 2549, Springer, Berlin, 2002, pp. 228–273.

[92] Lee, E.A. and Sangiovanni-Vincentelli, A., Comparing models of computation, *Proceedings of the IEEE/ACM International Conference on Computer-aided Design*, 1996, pp. 234–241.

[93] Gossler, G. and Sifakis. J., Composition for component-based modeling, *Proceedings of FMCO'02*, November 2002, Leiden, The Netherlands, Lecture Notes in Computer Science, Vol. 2852, Springer, Berlin, pp. 443–466.

[94] Altisen, K., Goessler, G., and Sifakis, J. Scheduler modeling based on the controller synthesis paradigm J.Real-Time Sys. (special issue on Control Approaches to Real-Time Computing), 23, 55–84, 2002.

[95] Gossler, G. and Sifakis, J., Composition for component-based modeling, *Proceedings of the FMCO Conference*, Leiden, The Netherlands, November 5–8, 2002.

[96] Ledeczi, A., Maroti, M., Bakay, A., Karsai, G., Garrett, J., Thomason, IV, C., Nordstrom, G., Sprinkle, J., and Volgyesi, P., The generic modeling environment, *Workshop on Intelligent Signal Processing*, Budapest, Hungary, May 17, 2001.

[97] Jantsch, A., *Modeling Embedded Systems and SoCs-Concurrency and Time in Models of Computation*, Morgan Kaufman, San Fransisco, CA, June 2003, ISBN 1558609253.

10

Using Performance Metrics to Select Microprocessor Cores for IC Designs

Steve Leibson
Tensilica, Inc.
Santa Clara, California

10.1 Introduction

Just as they have become essential building blocks for board-level designs, microprocessors are now absolutely essential components in integrated circuit (IC) design. The reason is simple. Microprocessor cores are the most reusable on-chip components a design team can use because of their easy software programmability and because of the extensive development-tool environment that surrounds any good microprocessor or microprocessor core. Although it is easy to envision using microprocessor cores to implement many tasks on an IC, it is difficult to select an appropriate processor from the many cores now available because contemporary microprocessor cores are complex, multi-function elements. Consequently, there is no single, simple, meaningful way to evaluate the suitability of a microprocessor core for specific embedded tasks on an IC. Many factors must be considered including processor performance, gate count, power dissipation, the availability of suitable interfaces, and the processor's software-development support. This chapter deals with the objective measurement of processor-core performance for the purpose of selection of microprocessor cores for IC design.

Designers measure microprocessor and microprocessor-core performance through benchmarking. Processor chips already realized in silicon are considerably easier to benchmark than processor cores. All major microprocessor vendors offer their processors in evaluation kits, which include complete evaluation boards and software-development tool suites (which include compilers, assemblers, linkers, loaders, instruction-set simulators [ISSs], and debuggers). Consequently, benchmarking chip-level microprocessor performance consists of compiling selected benchmarks for the target processor, downloading and running those benchmarks on the evaluation board, and recording the results.

Microprocessor cores are incorporeal and they are not usually available as stand-alone chips. In fact if these cores were available as chips, they would be far less useful to IC designers because on-chip processors can have vastly greater I/O capabilities than processors realized as individual chips. Numerous wide and fast buses are the norm for microprocessor cores to accommodate instruction and data caches, local memories, and private I/O ports.This is not true for microprocessor chips because these additional buses would greatly increase pin count, which would drive component cost beyond a tolerable level for microprocessor chips. However, extra I/O pins for microprocessor cores are not costly, so they usually have many buses and ports.

Benchmarking microprocessor cores has become increasingly important in the IC-design flow because any significant 21st century IC design incorporates more than one processor core — at least two, often several, and occasionally hundreds (see Figure 10.1). As the number of microprocessor cores used on a chip increases, and as these processor cores perform more tasks that were previously implemented with blocks of logic that were manually designed and coded by hand, measuring processor-core performance becomes an increasingly important task in the overall design of the IC.

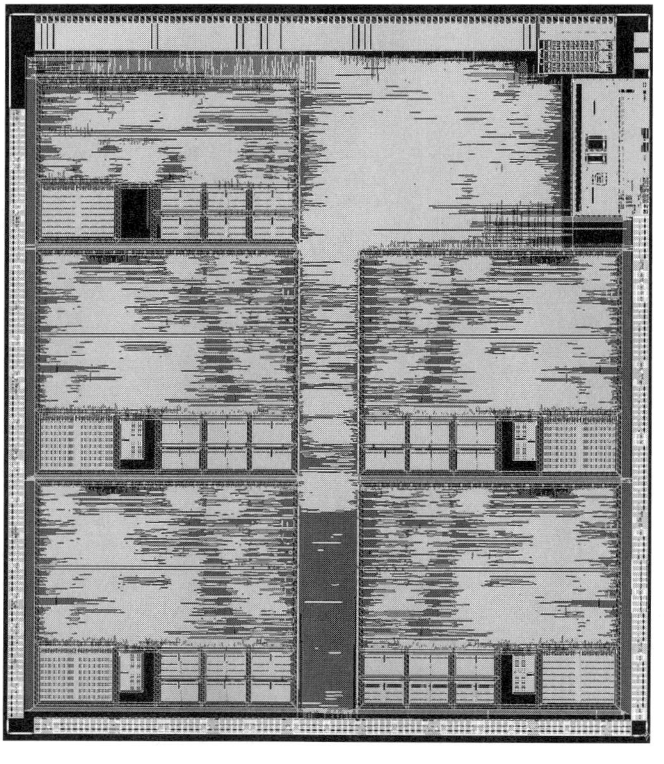

FIGURE 10.1 A chip from the MediaWorks family of configurable media processor products which is a DVD-resolution MPEG video and audio encoder/decoder system on a chip and is intended for use in solid-state camcorders and portable video products. The chip contains five loosely coupled heterogeneous processors. Reproduced with permission.

Measuring the performance of microprocessor cores is a bit more complex than benchmarking microprocessor chips: another piece of software, an ISS, must stand in for a physical processor chip. Each microprocessor and each member of a microprocessor family requires its own specific ISS because the ISS must provide a cycle-accurate model of the processor's architecture to produce accurate benchmark results. Note that it is also possible to benchmark a processor core using gate-level simulation, but this approach is three orders of magnitude slower (because gate-level simulation is much slower than instruction-set simulation) and is therefore used infrequently [1]. The selected benchmarks are compiled and run on the ISS to produce the benchmark results. All that remains is to determine where the ISS and the benchmarks are to be obtained, how they are to be used, and how the results will be interpreted. These are the subjects of this chapter.

10.2 The ISS as Benchmarking Platform

Instruction-set simulators serve as benchmarking platforms because processor cores as realized on an IC rarely exist as a chip. Instruction-set simulators simulate the software-visible state of a microprocessor without employing a gate-level model of the processor so that they run quickly, often 1000 times faster than gate-level processor simulations running on HDL simulators. The earliest ISS was created for the EDSAC (Electronic Delay Storage Automatic Calculator), which was developed by a team led by Maurice V. Wilkes at the University of Cambridge's Computer Laboratory (shown in Figure 10.2). The room-sized EDSAC I was the world's first fully operational, stored-program computer (the first von Neumann machine) and went online in 11949. EDSAC II became operational in 1958.

The EDSAC ISS was first described in a paper on debugging EDSAC programs, which was written and published in 1951 by S. Gill, one of Wilkes' research students [2]. The paper describes the operation of a "tracing simulator", which operates by fetching the simulated instruction, decoding the instruction to save trace information, updating the simulated program counter if the instruction is a branch, or else placing the nonbranch instruction in the middle of the simulator loop and executing it directly, and then

FIGURE 10.2 The EDSAC I, developed at the Computer Laboratory, University of Cambridge, UK, which became the first fully functional stored-program computer in 1948. This computer also had the first instruction-set simulator, which was described in an article published in 1951. Copyright Computer Laboratory, University of Cambridge. Reproduced with permission.

returning to the top of the simulator loop. Thus the first operational processor ISS predates the introduction of the first commercial microprocessor (Intel's 4004) by some 20 years.

Cycle-accurate ISSs, the most useful simulator class for processor benchmarking, compute exact instruction timing by accurately modeling the processor's pipeline. All commercial microprocessor cores have at least one corresponding ISS. They are obtained in different ways. Some core vendors rely on third parties to provide an ISS for their microprocessors. Other vendors offer an ISS as part of their tool suite. Some ISSs must be purchased and some are available in evaluation packages from the processor-core vendor.

To serve as an effective benchmarking tool, an ISS must not only simulate the operation of the processor core, but must also provide the instrumentation needed to provide the critical statistics of the benchmark run. These statistics include cycle counts for the various functions and routines executed, and main- and cache-memory usage. The better the ISS instrumentation, the more comparative information the benchmark will produce.

10.3 Ideal Versus Practical Processor Benchmarks

Integrated-circuit designers use benchmarks to help them pick the "best" microprocessor core for a given task to be performed on the IC. The original definition of a benchmark was literally a mark on a workbench that provided some measurement standard. Eventually, the first benchmarks (carved into the workbench) were replaced with standard measuring tools such as yardsticks. Processor benchmarks provide yardsticks for measuring processor performance. The ideal yardstick would be one that could measure any processor for any task. The ideal processor benchmark would produce results that are relevant, reliable, objective, comparable, and applicable. Unfortunately, no such processor benchmark exists.

In one sense, the ideal processor benchmark would be the actual application code that the processor will run. No other piece of code can possibly be as representative of the actual task to be performed as the actual code that executes that task. No other piece of code can possibly replicate the instruction-use distribution, register and memory use, or data-movement patterns of the actual application code. In many ways however, the actual application code is less than ideal as a benchmark.

First and foremost, the actual application code may not exist when candidate processors are benchmarked, because benchmarking and processor selection must occur early in the project. A benchmark that does not exist is worthless.

Next, the actual application code serves as an extremely specific benchmark. It will indeed give a very accurate prediction of processor performance for a specific task, *and for no other task*. In other words, the downside of a highly specific benchmark is that the benchmark will give a less-than-ideal indication of processor performance for other tasks. Because on-chip processor cores are often used for a variety of tasks, the ideal benchmark may well be a suite of application programs and not just one program.

Yet another problem with application-code benchmarks is their lack of instrumentation. The actual application code has almost always been written to execute the task, not to measure a processor core's performance. Appropriate measurements may require modification of the application code. This modification consumes time and resources, which may not be readily available. Even so, with all of these issues that make the application-code benchmark less than ideal, the application code (if it exists) provides invaluable information on processor-core performance and should be used to help make a processor-core selection whenever possible.

10.4 Standard Benchmark Types

Given that the "ideal" processor benchmark proves less than ideal, the industry has sought "standard" benchmarks that it can use when the target application code is either not available or not appropriate. There are four types of standard benchmarks: full-application or "real-world" benchmarks, synthetic or small-kernel benchmarks, hybrid or derived benchmarks that mix and match aspects of the full-application and synthetic benchmarks, and micro-benchmarks (not to be confused with microprocessor benchmarks).

Full-application benchmarks and benchmark suites employ existing system- or application-level code drawn from real applications, although probably not the specific application of interest to any given IC-design team. These benchmarks may incorporate many thousands of lines of code, have large instruction-memory footprints, and consume large amounts of data memory.

Synthetic benchmarks tend to be smaller than full-application benchmarks. They consist of smaller code sections representing commonly used algorithms. They may be extracted from working code or they may be written specifically as a benchmark. Writers of synthetic benchmarks try to approximate instruction mixes of real-world applications without replicating the entire application.

Hybrid benchmarks mix and match large application programs and smaller blocks of synthetic code to create a sort of torture track (with long straight sections and tight curves) for exercising microprocessors. The hybrid-benchmark code is augmented with test data sets taken from real-world applications. A microprocessor core's performance around this torture track can give a good indication of the processor's abilities over a wide range of situations, although probably not the specific use to which the processor will be put on an IC.

Micro-benchmarks are very small code snippets designed to exercise some particular processor feature or to characterize a particular machine primitive in isolation from all other processor features and primitives. Micro-benchmark results can define a processor's peak capabilities and reveal potential architectural bottlenecks, but peak performance is not a very good indicator of overall application performance. Nevertheless, a suite of micro-benchmarks may approximate the ideal benchmark for certain applications, if the application is a common one with many standardized, well-defined functions to perform.

10.5 Prehistoric Performance Ratings: MIPS, MOPS, and MFLOPS

Lord Kelvin could have been predicting processor performance measurements when he said: "When you can measure what you are speaking about, and express it in numbers, you know something about it; but when you cannot measure it, when you cannot express it in numbers, your knowledge is of a meager and unsatisfactory kind. It may be the beginning of knowledge, but you have scarcely, in your thoughts, advanced to the stage of science". [3]

The need to rate processor performance is so great that at first, microprocessor vendors grabbed any and all numbers at hand to rate performance. These prehistoric ratings are measures of processor performance that lack code to standardize the ratings. Consequently, these performance ratings are not benchmarks. Just as an engine's RPM (revolutions per minute) reading does not give one sufficient information to measure the performance of an internal combustion engine (you need engine torque plus transmission gearing, differential gearing, and tire diameter to compute load), the prehistoric, clock-related processor ratings of MIPS, MOPS, MFLOPS, and VAX MIPS are akin to clock rate: they tell one almost nothing about the true performance potential of a processor.

Before it became the name of a microprocessor vendor, the term "MIPS" was an acronym for "millions of instructions per second." If all processors had the same instruction-set architecture (ISA), then a MIPS rating could possibly be used as a performance measure. However, all processors do not have the same ISA. In fact, microprocessors and microprocessor cores have very different ISAs. Consequently, some processors can do more with one instruction than other processors, as large automobile engines can do more than smaller ones at the same RPM.

This problem was already bad in the days when only CISC (complex-instruction-set computers) processors roamed the Earth. The problem went from bad to worse when RISC (reduced-instruction-set computers) processors arrived on the scene. One CISC instruction would often do the work of several RISC instructions (by design), so that a CISC MIPS rating did not correlate well with a RISC MIPS rating because of the work differential between RISC's simple instructions and CISC's more complex instructions.

The next step in creating a usable processor performance rating was to switch from MIPS to VAX MIPS, which was accomplished by setting the extremely successful VAX 11/780 minicomputer — introduced in 1977 by the now defunct Digital Equipment Corporation (DEC) — as the standard against which all other processors are measured. So, if a processor ran twice as fast as a VAX 11/780 on a set of programs, it was rated at 2 VAX MIPS. The original term "MIPS" then became "native MIPS," so as not to confuse the original ratings with VAX MIPS. DEC referred to VAX MIPS as VUPs (VAX units of performance) to keep things interesting.

Both native and VAX MIPS are inadequate measures of processor performance because they are usually provided without specifying the software (or even the programming language) used to make the measurement. Because different programs have different mixes of instructions, different memory-usage patterns, and different data-movement patterns, the same processor may earn one MIPS rating on one set of programs and quite another rating on a different set. Because MIPS ratings are not linked to a specific benchmark program suite, the "MIPS" acronym has come to stand for "meaningless indication of performance."

A further problem with the VAX MIPS measure of processor performance is that the concept of using a VAX 11/780 minicomputer as the gold performance standard is an idea that is more than a bit long in the tooth in the 21st century. There are no longer many (or any) VAX 11/780s available for running benchmark code and DEC effectively disappeared when Compaq Computer Corp. purchased what was left of the company in January, 1998 following the decline of the minicomputer market. Hewlett-Packard absorbed Compaq in May 2002, submerging DEC's identity even further.

Even more tenuous than the MIPS performance rating is the concept of MOPS, an acronym that stands for "millions of operations per second." Every algorithmic task requires the completion of a certain number of fundamental operations, which may or may not have a one-to-one correspondence with machine instructions. Count these fundamental operations in the millions and they become MOPS. If they are floating-point operations, one gets MFLOPS (millions of floating-point operations per second). Both MOPS and MFLOPS ratings suffer from the same drawback as the MIPS rating: there is no standard software to serve as the benchmark that produces the ratings. In addition, the "conversion factor" for computing how many operations a processor performs per clock (or how many processor instructions constitute one operation) is somewhat fluid as well, which means that the processor vendor is free to develop a conversion factor on its own. Consequently, MOPS and MFLOPS performance ratings exist for various processor cores but they really do not help an IC-design team pick a processor core because they are not true benchmarks.

10.6 Classic Processor Benchmarks (The Stone Age)

Like ISSs, standardized processor performance benchmarks predate the 1971 introduction of Intel's 4004 microprocessor, but barely. The first benchmark suite to attain *de facto* "standard" status was a set of programs known as the Livermore Kernels (also popularly called the Livermore Loops).

10.6.1 The LFK Benchmark

The Livermore Kernels were first developed in 1970 and consist of 14 numerically intensive application kernels written in FORTRAN. Ten more kernels were added in the 1980s and the final suite of benchmarks was discussed in a paper published in 1986 by F. H. McMahon of the Lawrence Livermore National Laboratory (LLNL), Livermore, CA [4]. The Livermore Kernels actually constitute a supercomputer benchmark, measuring a processor's floating-point computational performance in terms of MFLOPS. Because of the somewhat frequent occurrence of floating-point errors in many computers, the Livermore Kernels test both the processor's speed and the system's computational accuracy. Today, the Livermore Kernels are called the Livermore FORTRAN Kernels (LFK).

The LFK are real samples of floating-point computations taken from a diverse workload of scientific applications extracted from operational program code used at LLNL. The kernels were extracted from programs in use at LLNL because those programs were generally far too large to serve as useful benchmark programs; they included hardware-specific subroutines for performing functions such as I/O, memory management, and graphics that were not appropriate for benchmark testing of floating-point performance; and they were largely classified, due to the nature of the work done at LLNL. Some kernels represent widely used, generic computations such as dot and matrix (SAXPY) products, polynomials, first sum and differences, first-order recurrences, matrix solvers, and array searches. Some kernels typify often-used FORTRAN constructs while others contain constructs that are difficult to compile into efficient machine code. These kernels were selected to represent both the best and worst cases of common FORTRAN programming practice to produce results that measure a realistic floating-point performance range by challenging the FORTRAN compiler's ability to produce optimized machine code. Table 10.1 lists the 24 LFK .

A complete LFK run produces 72 timed results, by timing the execution of the 24 LFK three times using different DO-loop lengths.

The LFK are a mixture of vectorizable and nonvectorizable loops, and test the computational capabilities of the microprocessor hardware and the ability of the software tools to create efficient code. The LFK benchmark also tests a processor's vector abilities and the associated software tools' abilities to vectorize code.

At first glance, the LFK benchmark appears to be nearly useless for the benchmarking of embedded microprocessor cores in ICs. It is a floating-point benchmark written in FORTRAN that looks for good vector abilities. FORTRAN compilers for embedded microprocessor cores are also quite rare and unusual. Today, very few real-world applications run tasks like those appearing in the 24 LFK , which are far more suited to research on the effects of very high speed nuclear reactions than the development of commercial, industrial, or consumer products. It is unlikely that the processors in a mobile phone handset or an MP3 music player will ever need to perform 2-D hydrodynamic calculations. Embedded-software developers working in FORTRAN are also rare. Consequently, microprocessor-core vendors are quite unlikely to tout LFK benchmark results for their cores. The LFK benchmarks are far more suited to testing supercomputers such as a Cray XD1 or Intel's ASCI Option Red. However, as on-chip gate counts grow, as microprocessor cores gain SIMD (single-instruction, multiple-data) and floating-point execution units, and as IC designs increasingly tackle tougher number-crunching applications including the implementation of audio and video codecs, the LFK benchmark could become valuable for testing more than just supercomputers.

TABLE 10.1 The 24 Kernels in the Livermore FORTRAN Kernels

LFK Number	Kernel Description
1	An excerpt from a hydrodynamic application
2	An excerpt from an Incomplete Cholesky-Conjugate Gradient program
3	The standard inner-product function from linear algebra
4	An excerpt from a banded linear equations routine
5	An excerpt from a tridiagonal elimination routine
6	An example of a general linear recurrence equation
7	Equation of state code fragment (as used in nuclear weapons research)
8	An excerpt of an alternating direction, implicit integration program
9	An integrate predictor program
10	A difference predictor program
11	A first sum
12	A first difference
13	An excerpt from a 2-D particle-in-cell program
14	An excerpt from a 1-D particle-in-cell program
15	A sample of casually written FORTRAN to produce sub-optimal machine code
16	A search loop from a Monte–Carlo program
17	An example of an implicit conditional computation
18	An excerpt from a 2-D explicit hydrodynamic program
19	A general linear recurrence equation
20	An excerpt from a discrete ordinate transport program
21	A matrix product calculation
22	A Planckian distribution procedure
23	An excerpt from a 2-D implicit hydrodynamic program
24	Finds the location of the first minimum in X

10.6.2 LINPACK

LINPACK is a collection of FORTRAN subroutines that analyze and solve linear equations and linear least-squares problems. Jack Dongarra assembled the LINPACK collection of linear algebra routines at the Argonne National Laboratory, Argonne, IL. The first versions of LINPACK existed in 1976 but the first users' guide was published in 1977 [5,6]. The package solves linear systems whose matrices are general, banded, symmetric-indefinite, symmetric positive definite, triangular, and tridiagonal square. In addition, the package computes the QR and singular value decompositions of rectangular matrices and applies them to least-squares problems. The LINPACK routines are not strictly speaking a benchmark, but they exercise a computer's floating-point capabilities and, as of 2005, Dongarra maintains an online list of computer systems and their LINPACK performance results at www.netlib.org.

Originally, LINPACK benchmarks performed computations on a 100×100 matrix, but with the rapid increase in computing performance that has occurred, LINPACK performance numbers for 1000×1000 arrays now appear on Dongarra's site as well. The original versions of LINPACK were written in FORTRAN but there are now versions written in C and Java as well. Performance is reported in single- and double-precision MFLOPS, which reflects the benchmark's (and the national laboratories') focus on "big iron" machines (supercomputers). The LINPACK benchmarks are not commonly used to measure microprocessor or microprocessor-core performance.

10.6.3 The Whetstone Benchmark

The Whetstone benchmark was written by Harold Curnow of the now defunct CCTA (Central Computer and Telecommunications Agency, the British government agency tasked with computer procurement) to test the performance of a proposed computer. It is the first program to appear in print that was designed as a synthetic benchmark to test processor performance, although it was specifically developed to test the performance of only one computer: the hypothetical Whetstone machine.

The Whetstone benchmark is based on application-program statistics gathered by Brian A. Wichmann at the National Physical Laboratory in England. Wichmann was using an Algol-60 compilation system that compiled Algol statements into instructions for the hypothetical Whetstone computer system, which was named after the small town of Whetstone located just outside the city of Leicester, England, where the compilation system was developed. Wichmann developed statistics on instruction usage for a wide range of numerical computation programs then in use.

An Algol-60 version of the Whetstone benchmark was released in November 1972 and single- and double-precision FORTRAN versions appeared in April 1973. The FORTRAN versions became the first widely used, general-purpose, synthetic computer performance benchmarks. Information about the Whetstone benchmark was first published in 1976 [7].

The Whetstone benchmark suite consists of several small loops that perform integer and floating-point arithmetic, array indexing, procedure calls, conditional jumps, and elementary function evaluations. These loops reside in three subprograms (called p3, p0, and pa) called from a main program. The subprograms call trigonometric (sine, cosine, and arctangent) and other math-library functions (exponentiation, log, and square root). The benchmark authors empirically managed the Whetstone's instruction mix by manipulating loop counters within the modules to match Wichmann's instruction-mix statistics. Empirical Whetstone loop weights range from 12 to 899.

The Whetstone benchmark produces speed ratings in terms of thousands of Whetstone instructions per second (KWIPS), thus using the hypothetical Whetstone computer as the gold standard for this benchmark. In 1978, self-timing versions (written by Roy Longbottom, also of CCTA) produced speed ratings in MOPS and MFLOPS and overall rating in MWIPS.

As with the LFK and LINPACK benchmarks, the Whetstone benchmark focuses on floating-point performance. Consequently, it is not commonly applied to embedded processor cores that are destined to execute integer-oriented and control tasks on an IC because such a benchmark really will not tell the design team much they need to know about the processor(s) in question. However, the Whetstone benchmark's appearance spurred the creation of a plethora of "stone," benchmarks and one of those, the Dhrystone, became the first widely used benchmark to rate microprocessor performance.

10.6.4 The Dhrystone Benchmark

Reinhold P. Weicker, working at Siemens-Nixdorf Information Systems, wrote and published the first version of the Dhrystone benchmark in 1984 [8]. The benchmark's name, "Dhrystone," is a pun derived from the Whetstone benchmark. The Dhrystone benchmark is a synthetic benchmark that consists of 12 procedures in one measurement loop. It produces performance ratings in Dhrystones per second. Originally, the Dhrystone benchmark was written in a "Pascal subset of Ada." Subsequently, versions in Pascal and C have appeared, and the C version is the one most used today.

The Dhrystone benchmark differs significantly from the Whetstone and these differences make the Dhrystone far more suitable as a microprocessor benchmark. First, the Dhrystone is strictly an integer program. It does not test a processor's floating-point abilities because most microprocessors in the early 1980s (and even in the early 21st century) have no native floating-point computational abilities. The Dhrystone does devote a lot of time executing string functions (copy and compare), which microprocessors often execute in real applications.

The original version of the Dhrystone benchmark quickly became successful. One indication of the Dhrystone's success were the attempts by microprocessor vendors to inflate their Dhrystone ratings by "gaming" the benchmark (cheating). Weicker had made no attempt to thwart compiler optimizations when he wrote the Dhrystone benchmark because he reasonably viewed those optimizations as typical of real-world programming. However, version 1 of the Dhrystone benchmark did not print or use the results of its computations allowing "properly" written optimizing compilers using dead-code-removal algorithms to optimize away most of the benchmark by removing the benchmark program's computations that produced never-used results. Clearly, such optimizations go far beyond the spirit of the benchmark. Weicker corrected this problem by publishing version 2.1 of the Dhrystone benchmark in 1988 [9]. At the same time, he acknowledged some of the benchmark's limitations and published his criteria for using the Dhrystone benchmarks to compare processors [10].

Curiously, DEC's VAX also plays a role in the saga of the Dhrystone benchmarks. The VAX 11/780 minicomputer could run 1757 version 2.1 Dhrystones/sec. Because the VAX 11/780 was (erroneously) considered a 1-MIPS computer, it became the Dhrystone standard machine, which resulted in the emergence of the DMIPS or D-MIPS (Dhrystone MIPS) rating. By dividing a processor's Dhrystone 2.1 performance rating by 1757, processor vendors could produce an official-looking DMIPS rating for their products. Thus, DMIPS fuses two questionable rating systems (Dhrystones and VAX MIPS) to create a third, derivative, equally questionable microprocessor-rating system.

The Dhrystone benchmark's early success as a marketing tool encouraged abuse, and this abuse became rampant. For example, vendors quickly (though unevenly) added specialized string routines written in machine code to some of their compiler libraries because accelerating these heavily used library routines boosted Dhrystone ratings, even though the actual performance of the processor did not change at all. Some compilers started in-lining these machine-coded string routines for even better ratings. As the technical marketing teams at the microprocessor vendors continued to study the benchmark, they found increasingly better ways of improving their products' ratings by introducing compiler optimizations that only applied to the benchmark. Some compilers even had pattern recognizers that could recognize the Dhrystone benchmark source code or a "Dhrystone command-line switch" that would cause the compiler to "generate" the entire program using a previously prewritten, precompiled, hand-optimized version of the benchmark program.

It is not even necessary to alter a compiler to produce wildly varying Dhrystone results using the same processors. Using different compiler optimization settings can drastically alter the outcome of a benchmark test, even if the compiler has not been "Dhrystone optimized." For example, Tensilica's Xtensa LX processor core produces different Dhrystone benchmark results that differ by almost 2:1 depending on the setting of the "in-lining" switch of its compiler (shown in Table 10.2).

TABLE 10.2 Performance Difference in Dhrystone Benchmark between In-Lined and Non-In-Lined Code

Xtensa LX Compiler Setting	DMIPS rating (at 150 MHz)	DMIPS/MHz rating
No in-line code	183.187	1.221
In-line code	322.154	2.148

In-lining code is not permissible under Weicker's published rules, but as there is no organization to enforce Weicker's rules, there is no way to ensure that processor vendors benchmark fairly, and no way to force vendors to fully disclose the conditions that produced their benchmark results. Weaknesses associated with conducting Dhrystone benchmark tests and reporting the results highlight a problem: if the benchmarks are not conducted by objective third parties under controlled conditions that are disclosed with the benchmark results, the benchmark results must be suspect even if there are published rules as to a benchmark's use, because there is nobody to enforce the rules and not everyone follows rules when the sales of a new microprocessor are at stake.

Over time, other technical weaknesses have appeared in the Dhrystone benchmark. For example, the actual Dhrystone code footprint is quite small and as microprocessor instruction caches grew, some processors were able to run the Dhrystone benchmark entirely from their caches, which boosted their performance on the Dhrystone benchmark but did not represent real performance gains because most application programs will not fit entirely into a processor's cache. Processors with wider data buses and wide-load and -store instructions that move many string bytes per bus transaction also earn better Dhrystone performance ratings. Although this behavior also benefits application code, it can mislead the unwary who think that the benchmark is comparing the relative computational performance of processors.

One of the worst abuses of the Dhrystone benchmark occurred when processor vendors ran benchmarks on their own evaluation boards and on competitors' evaluation boards, and then published the results. The vendor's own board would have fast memory and the competitors' boards would have slow memory but this difference would not be revealed when the scores were published. Differences in memory speed do affect Dhrystone results, so publishing the Dhrystone results of competing processors without disclosing the use of dissimilar memory systems to compare the processors is clearly dishonest, or at least mean-spirited. Other sorts of shenanigans were used to distort Dhrystone comparisons as well. One of the Dhrystone benchmark's biggest weaknesses is therefore the absence of a governing body or an objective third party that can review, approve, and publish Dhrystone benchmark results.

Despite all of these weaknesses, microprocessor and microprocessor-core vendors continue to publish Dhrystone benchmark results — or at least they have Dhrystone results handily available — because Dhrystone ratings are relatively easy to generate and the benchmark code is very well understood after two decades of use. Vendors also keep Dhrystone ratings handy because the lack of such numbers might raise a red flag for some prospective customers. The Dhrystone's explosive popularity as a tool for making processor comparisons, coupled with its technical weaknesses and the abuses heaped upon the benchmark program, helped to spur the further evolution of processor benchmarks in the 1990s.

10.6.5 The *EDN* Microprocessor Benchmarks

EDN magazine, a trade publication for the electronics industry, was a very early advocate of microprocessor use for general electronic system design, and extensively covered the introduction of new microprocessors starting in the early 1970s. In 1981, just after the first wave of commercial 16-bit microprocessors had been introduced, *EDN* published the first comprehensive article on this subject. The article used a set of microprocessor benchmarks to compare the four leading 16-bit microprocessors of the day: DEC's LSI-11/23, Intel's 8086, Motorola's 68000, and Zilog's Z8000 [11]. The Intel, Motorola, and Zilog processors often competed for system design wins in the early 1980s, and this article provided design engineers with some of the first microprocessor benchmark ratings to be published by an objective third party.

This article was written by Jack E. Hemenway, a consulting editor for *EDN*, and Robert D. Grappel of MIT's Lincoln Laboratory. Hemenway and Grappel summed up the reason that the industry needs objective benchmark results succinctly in their article:

Why the need for a benchmark study at all? One sure way to start an argument among computer users is to compare each one's favorite machine with the others. Each machine has strong points and drawbacks, advantages and liabilities, but programmers can get used to one machine and see all the rest as inferior. Manufacturers sometimes don't help: Advertising and press releases often imply that each new machine is the ultimate in computer technology. Therefore, only a careful, complete and unbiased comparison brings order out of the chaos.

The *EDN* article continues with an excellent description of the difficulties associated with microprocessor benchmarking:

Benchmarking anything as complex as a 16-bit processor is a very difficult task to perform fairly. The choice of benchmark programs can strongly affect the comparisons' outcome so the benchmarker must choose the test cases with care.

Hemenway and Grappel used a set of benchmark programs created in 1976 by a research group at Carnegie-Mellon University (CMU). These benchmarks were first published as a paper, which was presented by the CMU group in 1977 at the National Computer Conference [12]. The tests in these benchmark programs — interrupt handlers, string searches, bit manipulation, and sorting — are very representative of types of tasks microprocessors and processor cores must perform in most embedded applications. The *EDN* authors excluded the CMU benchmarks that tested floating-point computational performance because none of the processors in the *EDN* study had native floating-point resources, something which is true even of most contemporary microprocessor cores.

The seven programs in the 1981 *EDN* benchmark study appear in Table 10.3.

Significantly, the authors of this article allowed programmers from each microprocessor vendor to code each benchmark for their company's processor. Programmers were required to use assembly language to code the benchmarks, which removed the associated compiler from the equation and allowed the processor hardware architectures to compete directly. Considering the state of microprocessor compilers in 1981, this was probably an excellent decision.

The authors refereed the tests by reviewing each program to ensure that no corners were cut and that the programs faithfully executed each benchmark algorithm. The authors did not force the programmers to code in a specific way and allowed the use of special instructions (such as instructions designed to perform string manipulation) because use of these instructions fairly represented actual processor use.

To ensure fairness, the study's authors ran the programs and recorded the benchmark results. They published the results and included both the program size and execution speed. Publishing the program size recognized that memory was costly and limited, a situation that still holds true in IC design. The authors also published the scores of each of the seven benchmark tests separately, acknowledging that a combined score would tend to mask information about a processor's specific abilities.

TABLE 10.3 Programs in the 1981 *EDN* Benchmark Article by Hemenway and Grappel

EDN 1981 Microprocessor Benchmark Component	Benchmark Description
Benchmark A	Simple priority interrupt handler
Benchmark B	FIFO interrupt handler
Benchmark E	Text string search
Benchmark F	Primitive bit manipulation
Benchmark H	Linked-list insertion (tests addressing modes and 32-bit operations)
Benchmark I	Quicksort algorithm (tests stack manipulation and addressing modes)
Benchmark K	Matrix transposition (tests bit manipulation and looping)

In 1988, *EDN* published a similar benchmarking article covering digital signal processors (DSPs) [13]. The article was written by David Shear, one of *EDN*'s regional editors. Shear benchmarked 18 DSPs using 12 benchmark programs selected from 15 candidate programs. The DSP vendors participated in selecting the DSP benchmark programs that were used from the 15 candidate programs. The final set of DSP benchmark programs included six DSP filters, three math benchmarks (a simple dot product and two matrix multiplications), and three fast Fourier transforms (FFTs). Shear used benchmark programs to compare DSPs that are substantially different from the programs used by Grappel and Hemenway to compare general purpose microprocessors because, as a class, DSPs are applied quite differently from general-purpose processors. The 12 DSP benchmark programs of Shear's benchmark set appear in Table 10.4.

Notably, Shear elected to present the performance results of the benchmarks (both execution speed and performance) as bar charts rather than in tabular form, noting that the bar charts emphasized large performance differences while masking differences of a few percent, which he dismissed as "not important."

This article also recognized the potential for cheating by operationalizing the "Three Not-So-Golden Rules of Benchmarking" outlined by *EDN* editor Walt Patstone in 1981, as a follow-up to the article by Hemenway and Grappel [14]:

> Rule 1 — All's fair in love, war, and benchmarking.
> Rule 2 — Good code is the fastest possible code.
> Rule 3 — Conditions, cautions, relevant discussion, and even actual code never make it to the bottom line when results are summarized.

These *EDN* microprocessor benchmark articles established and demonstrated all of the characteristics of an effective, modern microprocessor benchmark:

- Conduct a series of benchmark tests that exercise salient processor features for a class of tasks
- Use benchmark programs that are appropriate to the processor class being studied
- Allow experts to code the benchmark programs
- Have an objective third party check and run the benchmark code to ensure that the experts do not cheat
- Publish benchmark results that maximize the amount of available information about the tested processor
- Publish both execution speed and memory use for each processor on each benchmark program because there is always a tradeoff between a processor's execution speed and its memory consumption

These characteristics shaped the benchmarking organizations and their benchmark programs in the 1990s.

TABLE 10.4 The 12 Programs in EDN's 1988 DSP Benchmark Article

EDN 1988 DSP Benchmark Program	Benchmark Description
Benchmark 1	20-tap FIR filter
Benchmark 2	64-tap FIR filter
Benchmark 3	67-tap FIR filter
Benchmark 4	8-pole canonic IIR filter (4×) (direct form II)
Benchmark 5	8-pole canonic IIR filter (5×) (direct form II)
Benchmark 6	8-pole transpose IIR filter (direct form I)
Benchmark 7	Dot product
Benchmark 8	Matrix multiplication, 2×2 times 2×2
Benchmark 9	Matrix multiplication, 3×3 times 3×1
Benchmark 10	Complex 64-point FFT (radix-2)
Benchmark 11	Complex 256-point FFT (radix-2)
Benchmark 12	Complex 1024-point FFT (radix-2)

10.7 Modern Processor Performance Benchmarks

As the number of available microprocessors mushroomed in the 1980s, the need for good benchmarking standards became increasingly apparent. The Dhrystone experiences showed how useful a standard benchmark could be, and these experiences also demonstrated the lengths (both fair and unfair) to which processor vendors would go to earn top benchmark scores. The *EDN* articles had shown that an objective third party could bring order out of benchmarking chaos. As a result, both private companies and industry consortia stepped forward with the goal of producing "fair and balanced" processor benchmarking standards.

10.7.1 SPEC: The Standard Performance Evaluation Corporation

One of the first organizations to tackle seriously the need for good microprocessor benchmarking standards was SPEC (originally the System Performance Evaluation Cooperative, and now the Standard Performance Evaluation Corporation). A group of workstation vendors including Apollo, Hewlett-Packard, MIPS Computer Systems, and Sun Microsystems working in conjunction with the trade publication *Electronic Engineering Times* founded SPEC in 1988. The consortium produced its first processor benchmark, SPEC89, the following year. SPEC89 provided a standardized measure of compute-intensive microprocessor performance with the express purpose of replacing the existing, vague MIPS and MFLOPS rating systems. Because high-performance microprocessors were primarily used in high-end workstations at the time, and because SPEC was formed as a cooperative by workstation vendors, the SPEC89 benchmark consisted of source code that was to be compiled for UNIX.

The Dhrystone benchmark had already demonstrated that benchmark code quickly rots over time due to rapid advances in processor architecture and compiler technology. (It is no small coincidence that Reinhold Weicker, Dhrystone's creator and one of the people most familiar with processor benchmarking, became a key member of SPEC and has been heavily involved with the ongoing creation of SPEC benchmarks.) The likelihood of "benchmark rot" was no different for SPEC89, so the SPEC organization has regularly improved and expanded its benchmarks, producing SPEC92, SPEC95 (with separate integer and floating-point components called CINT95 and CFP95), and finally SPEC CPU2000 (consisting of CINT2000 and CFP2000).

Table 10.5 and Table 10.6 list the SPEC CINT2000 and CFP2000 benchmark component programs, respectively.

As Table 10.5 and Table 10.6 show, SPEC benchmarks are application-based benchmarks, not synthetic benchmarks. The SPEC benchmark programs are excellent as workstation/server benchmarks because they use the actual applications that are likely to be assigned to these machines. SPEC publishes benchmark performance results for various computer systems on its web site (www.spec.org) and sells its benchmark code.

TABLE 10.5 SPEC CINT2000 Benchmark Component Programs

CINT2000 Benchmark Component	Language	Category
164.gzip	C	Compression
175.vpr	C	FPGA circuit placement and routing
176.gcc	C	C programming language compiler
181.mcf	C	Combinatorial optimization
186.crafty	C	Game playing: chess
197.parser	C	Word processing
252.eon	C++	Computer visualization
253.perlbmk	C	PERL programming language
254.gap	C	Group theory, interpreter
255.vortex	C	Object-oriented database
256.bzip2	C	Compression
300.twolf	C	Place and route simulator

TABLE 10.6 SPEC CFP2000 Benchmark Component Programs

CFP2000 Benchmark Component	Language	Category
168.wupwise	Fortran 77	Physics/quantum chromodynamics
171.swim	Fortran 77	Shallow water modeling
172.mgrid	Fortran 77	Multi-grid solver: 3-D potential field
173.applu	Fortran 77	Parabolic/elliptic partial differential equations
177.mesa	C	3-D graphics library
178.galgel	Fortran 90	Computational fluid dynamics
179.art	C	Image recognition/neural networks
183.equake	C	Seismic wave propagation simulation
187.facerec	Fortran 90	Image processing: face recognition
188.ammp	C	Computational chemistry
189.lucas	Fortran 90	Number theory/primality testing
191.fma3d	Fortran 90	Finite-element crash simulation
200.sixtrack	Fortran 77	High energy nuclear physics accelerator design

As of 2004, The CPU2000 benchmark suite costs $500 ($125 for educational use). A price list for the CPU2000 benchmarks as well as a wide range of other computer benchmarks appears on the SPEC web site.

Because the high-performance microprocessors used in workstations and servers are sometimes used as embedded processors and some of them are available as microprocessor cores for use on SOCs, microprocessor and microprocessor-core vendors sometimes quote SPEC benchmark scores for their products. Use these performance ratings with caution because the SPEC benchmarks are not necessarily measuring performance that is meaningful to embedded applications. For example, no mobile phone is likely to be required to simulate seismic-wave propagation, except for the possible exception of handsets sold in California.

A paper written by Jakob Engblom of Uppsala University compares the static properties of the SPECint95 benchmark programs with code from 13 embedded applications consisting of 334,600 lines of C source code [15]. Engblom's static analysis discovered several significant differences between the static properties of the SPECint95 benchmark and the 13 embedded programs. Significant differences include:

- Variable sizes. Embedded programs carefully control variable size to minimize memory usage. Workstation-oriented software like SPECint95 does not limit variable size nearly as much because workstation memory is relatively plentiful.
- Unsigned data are more common in embedded code.
- Logical (as opposed to arithmetic) operations occur more frequently in embedded code.
- Many embedded functions only perform side effects (such as flipping an I/O bit). They do not return values.
- Embedded code employs global data more frequently for variables and for large constant data.
- Embedded programs rarely use dynamic memory allocation.

Although Engblom's intent in writing this paper was to show that SPECint95 code was not suitable for benchmarking embedded development tools such as compilers, his observations about the characteristic differences between SPECint95 code and typical embedded application code are also significant for embedded processors and processor cores. Engblom's observations underscore the maxim that the best benchmark code is always the actual target application code.

10.7.2 BDTI: Berkeley Design Technology, Inc.

Berkeley Design Technology, Inc. (BDTI) is a technical services company that has focused exclusively on the applications of DSPs since 1991. BDTI helps companies select, develop, and use DSP technology by providing expert advice and consulting, technology analysis, and highly optimized software development services. As part of those services, BDTI has spent more than a decade developing and applying DSP benchmarking tools. As such, BDTI serves as a private third party that develops and administers DSP benchmarks.

BDTI introduced its core suite of DSP benchmarks, formally called the "BDTI Benchmarks," in 1994. The BDTI Benchmarks consist of a suite of 12 algorithm kernels, which represent key DSP operations used in common DSP applications. BDTI revised, expanded, and published information about the 12 DSP algorithm kernels in its BDTI Benchmark in 1999 [16]. The 12 DSP kernels in the BDTI Benchmark appear in Table 10.7.

BDTI's DSP benchmark suite is an example of a specialized processor benchmark. DSPs are not general-purpose processors and therefore require specialized benchmarks. BDTI does not use full applications as does SPEC. The BDTI Benchmark is a hybrid benchmark that uses code (kernels) extracted from actual DSP applications.

BDTI calls its benchmarking methodological approach "algorithm kernel benchmarking and application profiling." The algorithm kernels used in the BDTI benchmarks are functions that constitute the building blocks used by most signal-processing applications. They have been extracted by BDTI from examples of DSP application code. These kernels are the most computationally intensive portions of the donor DSP applications. The resulting DSP kernels are all written in assembly code and must be hand-ported to each new DSP architecture. BDTI believes that the extracted kernel code is more rigorously defined than the large DSP applications such as a V.90 modem or a Dolby AC-3 audio codec, which perform many functions in addition to the core DSP algorithms.

The "application-profiling" portion of the BDTI Benchmarks methodology refers to a set of techniques used by BDTI to either measure or estimate the amount of time, memory, and other resources that an application spends executing various code subsections, including subsections that correspond to the DSP kernels in the BDTI Benchmark. BDTI develops kernel-usage estimates from a variety of information sources including application-code inspections, instrumented code-run simulations, and flow-diagram analysis.

TABLE 10.7 BDTI Benchmark Kernels

BDTI Benchmark Kernel Function	Function Description	Example Applications
Real block FIR	Finite impulse response filter that operates on a block of real (not complex) data	Speech processing (e.g., G.728 speech coding)
Complex block FIR	FIR filter that operates on a block of complex data	Modem channel equalization
Real single-sample FIR	FIR filter that operates on a single sample of real data	Speech processing, general filtering
LMS adaptive FIR	Least-mean-squares adaptive filter, operates on a single sample of real data	Channel equalization, servo control, linear predictive coding
IIR	Infinite impulse response filter that operates on a single sample of real data	Audio processing, general filtering
Vector dot product	Sum of the point-wise multiplication of two vectors	Convolution, correlation, matrix multiplication, multi-dimensional signal processing
Vector add	Pointwise addition of two vectors, producing a third vector	Graphics, combining audio signals or images
Vector maximum	Finding the value and location of the maximum value in a vector	Error control coding, algorithms using block floating-point
Viterbi decoder	Decoding a block of bits that have been convolutionally encoded	Error control coding
Control	A sequence of control operations (test, branch, push, pop, and bit manipulation)	Virtually all DSP applications include some control code
256-point, in-place FFT	Fast Fourier transform converts a time-domain signal into the frequency domain	Radar, sonar, MPEG audio compression, spectral analysis
Bit unpack	Unpacks variable-length data from a bit stream	Audio decompression, protocol handling

BDTI originally targeted its benchmark methodology at realized-in-silicon DSP processors, not processor cores. The key portion of the benchmark methodology that limited its use to processors and not cores was the need to conduct the benchmark tests at actual clock speeds. BDTI did not want processor manufacturers to test this year's processors and then extrapolate the results to next year's clock speeds. However, SOC designers must evaluate and compare processor-core performance long before a chip has been fabricated. Processor-core clock speeds will depend on the IC fabrication technology and cell libraries used to create the SOC. BDTI uses results from a cycle-accurate simulator and worst-case projected clock speeds (based on gate-level processor-core simulation and BDTI's evaluation of the realism of that projection) to obtain benchmark results from processor cores.

BDTI does not release the BDTI Benchmark code. Instead, it works with processor vendors (for a fee) to port the benchmark code. BDTI then acts as the objective third party; conducts the benchmark tests; and measures execution time, memory usage, and energy consumption for each benchmark kernel. BDTI also sells some of the results of its benchmark tests in the form of reports on specific DSP processors and in the form of a book called the *Buyers Guide to DSP Processors* (the 2004 edition of BDTI's DSP buyers guide sells for $2695).

To satisfy people who use one number to make a rough cut of processor candidates, BDTI rolls a processor's execution times on the 12 DSP kernels into one composite number that it has dubbed the BDTImark2000. To prevent the mixing of processor benchmark scores verified with hardware and simulated processor-core scores, BDTI publishes the results of simulated benchmark scores under a different name: the BDTIsimMark2000. Both the BDTImark2000 and the BDTIsimMark2000 scores are available without charge on BDTI's website (www.bdti.com). Figure 10.3 shows BDTImark2000™ and BDTIsimMark2000™ scores for ARM's ARM7 and ARM9E, CEVA CEVA-X, LAS LSI Logic ZSP400 and

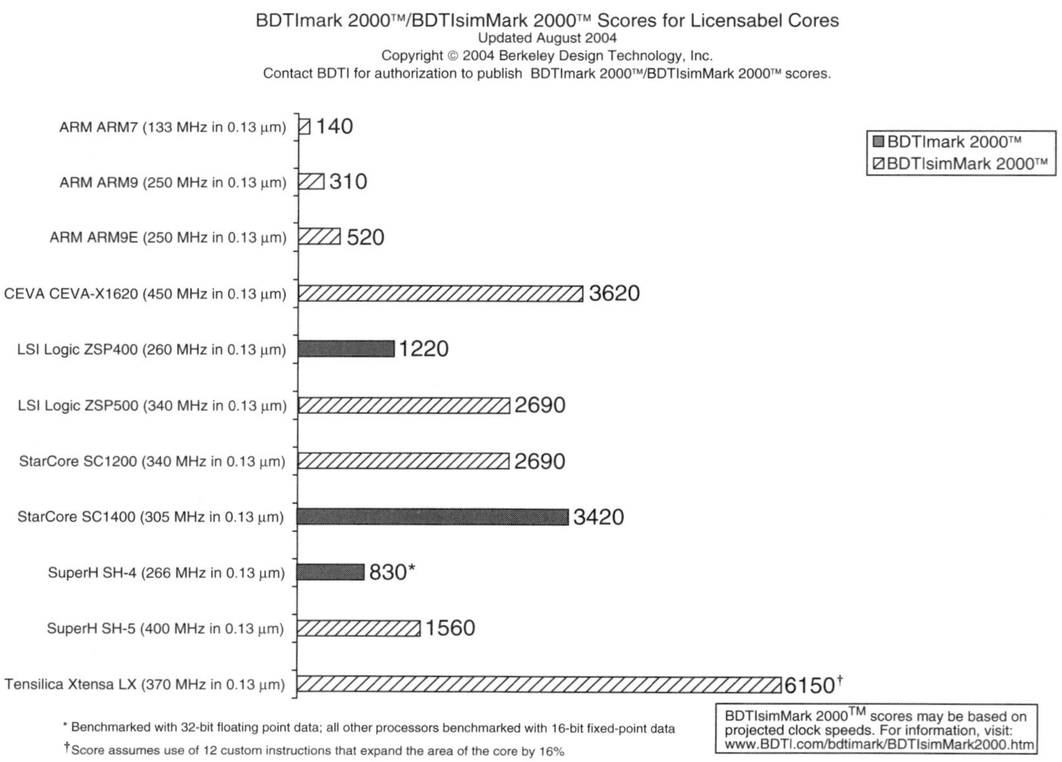

FIGURE 10.3 The BDTImark2000™/BDTIsimMark2000™ provides a summary measure of signal processing speed. For more information and scores see www.BDTI.com. Scores copyright 2004 BDTI.

ZSP500, StarCore SC1200 and SC1400, SuperH SH-4 and SH-5, and Tensilica Xtensa LX microprocessor and DSP cores.

10.7.3 EEMBC (Embedded Microprocessor Benchmark Consortium)

EDN magazine's legacy of microprocessor benchmark articles grew into full-blown realization when *EDN* editor Markus Levy founded the nonprofit, embedded-benchmarking organization called EEMBC (*EDN* Embedded Benchmark Consortium, later dropping the "*EDN*" from its name but not the corresponding "E" from its abbreviation) in 1997. EEMBC's stated goal was to produce accurate and reliable metrics based on real-world embedded applications for evaluating embedded-processor performance. Levy drew remarkably broad industry support from microprocessor and DSP vendors for his concept of a benchmarking consortium which had picked up 21 founding corporate members by the end of 1998: Advanced Micro Devices, Analog Devices, ARC, ARM, Hitachi, IBM, IDT, Lucent Technologies, Matsushita, MIPS, Mitsubishi Electric, Motorola, National Semiconductor, NEC, Philips, QED, Siemens, STMicroelectronics, Sun Microelectronics, Texas Instruments, and Toshiba [17]. EEMBC (pronounced "embassy") spent nearly three years working on a suite of benchmarks for testing embedded microprocessors and introduced its first benchmark suite at the Embedded Processor Forum in 1999. EEMBC released its first certified scores in 2000 and, in the same year, announced that it would start to certify benchmarks run on simulators so that processor cores could be benchmarked. As of 2005, EEMBC has more than 50 corporate members.

The EEMBC benchmarks are contained in six suites loosely grouped according to application. The six suites are: automotive/industrial, consumer, Java GrinderBench, networking, office automation, and telecom. Each of the suites contains several benchmark programs that are based on small, derived kernels extracted from application code. All EEMBC benchmarks are written in C, except for the benchmark programs in the Java benchmark suite, which are written in Java. The six benchmark suites and descriptions of the programs in each suite appear in the Tables 10.8 through 10.13.

EEMBC's benchmark suites with their individual benchmark programs allow designers to select the benchmarks that are relevant to a specific design, rather than lumping all of the benchmark results into one number. EEMBC's benchmark suites are developed by separate subcommittees, each working on one application segment. Each subcommittee selects candidate applications that represent the application segment and dissects each application for the key kernel code that performs the important work. This kernel code coupled with a test harness becomes the benchmark. Each benchmark has published guidelines in an attempt to force the processor vendors to play fair.

However, the industry's Dhrystone experience proved conclusively that some processor vendors will not play fair without a fair and impartial referee, so EEMBC created one: the EEMBC Certification Laboratories, LLC (ECL). Just as major sports leagues hire referee organizations to conduct games and enforce the rules of play, EEMBC employs ECL to conduct benchmark tests and enforce the rules of EEMBC benchmarking. Although a processor vendor ports the EEMBC benchmark code and test harnesses to its own processor and runs the initial benchmark tests, only ECL can certify the results. ECL takes the vendor-supplied code, test harness, and a description of the test procedures used, and then certifies the benchmark results by inspecting the supplied code and rerunning the tests. In addition to code inspection, ECL makes changes to the benchmark code (such as changing variable names) to counteract "over optimized" compilers. Vendors cannot publish EEMBC benchmark scores without ECL certification.

There is another compelling reason for an EEMBC benchmark referee: EEMBC rules allow for two levels of benchmarking "play" (the major league and the minors, to continue the sports analogy). The lower level of EEMBC "play" produces "out-of-the-box" scores. Out-of-the-box EEMBC benchmark tests can use any compiler (in practice, the compiler selected has changed the performance results by as much as 40%) and any selection of compiler switches, but cannot modify the benchmark source code. The "out-of-the-box" results therefore give a fair representation of the abilities of the processor/compiler combination without adding in programmer creativity as a wild card. The higher level of EEMBC play is called "full-fury." Processor vendors seeking to improve their full-fury EEMBC scores (posted as "optimized"

TABLE 10.8 EEMBC Automotive/Industrial Benchmark Programs

EEMBC Automotive/Industrial Benchmark Name	Benchmark Description	Example Application
Angle to time conversion	Compute engine speed and crankshaft angle	Automotive engine control
Basic integer and floating point	Calculate arctangent from telescoping series	General purpose
Bit manipulation	Character and pixel manipulation	Display control
Cache "Buster"	Long sections of control algorithms with pointer manipulation	General purpose
CAN Remote Data Request	Controller Area Network (CAN) communications	Automotive networking
Fast Fourier transform (FFT)	Radix-2 decimation in frequency, power spectrum calculation	DSP
Finite impulse response (FIR) filter	High- and low-pass FIR filters	DSP
Inverse discrete cosine transform (iDCT)	iDCT using 64-bit integer arithmetic	Digital video, graphics, image recognition
Inverse fast Fourier transform (iFFT)	Inverse FFT performed on real and imaginary values	DSP
Infinite impulse response (IIR) filter	Direct-for II, N-cascaded, second-order IIR filter	DSP
Matrix arithmetic	LU decomposition of N×N input matrices, determinant computation, cross product	General purpose
Pointer chasing	Search of a large, doubly linked list	General purpose
Pulse width modulation (PWM)	Generate PWM signal for stepper-motor control	Automotive actuator control
Road speed calculation	Determine road speed from successive timer/counter values	Automotive cruise control
Table lookup and interpolation	Derive function values from sparse 2-D and 3-D tables	Automotive engine control, antilock brakes
Tooth to spark	Fuel-injection and ignition-timing calculations	Automotive engine control

TABLE 10.9 EEMBC Consumer Benchmark Programs

EEMBC Consumer Benchmark Name	Benchmark Description	Example Applications
High pass grey-scale filter	2-D array manipulation and matrix arithmetic	CCD and CMOS sensor signal processing
JPEG	JPEG image compression and decompression	Still-image processing
RGB to CMYK conversion	Color-space conversion at 8 bits/pixel	Color printing
RGB to YIQ conversion	Color-space conversion at 8 bits/pixel	NTSC video encoding

TABLE 10.10 EEMBC GrinderBench Java 2 Micro Edition (J2ME) Benchmark Programs

EEMBC GrinderBench for Micro Java 2 Edition (J2ME) Benchmark Description	Benchmark Name	Example Applications
Chess	Machine -vs.machine chess matches, 3 games, 10 moves/game	Cell phones, PDAs
Cryptography	DES, DESede, IDEA, Blowfish, Twofish encryption and decryption	Data security
kXML	XML parsing, DOM tree manipulation	Document processing
ParallelBench	Multi-threaded operations with mergesort and matrix multiplication	General purpose
PNG decoding	PNG image decoding	Graphics, web browsing
Regular expression	Pattern matching and file I/O	General purpose

TABLE 10.11 EEMBC Networking Benchmark Programs

EEMBC Networking 2.0 Benchmark Name	Benchmark Description	Example Applications
IP Packet Check	IP header validation, checksum calculation, logical comparisons	Network router, switch
IP Network Address Translator (NAT)	Network-to-network address translation	Network router, switch
OSPF version 2	Open shortest path first/Djikstra shortest path first algorithm	Network routing
QoS	Quality of service network bandwidth management	Network traffic flow control

TABLE 10.12 EEMBC Office Automation Benchmark Programs

EEMBC Office Automation Benchmark Name	Benchmark Description	Example Applications
Dithering	Grayscale to binary image conversion	Color and monochrome printing
Image rotation	90-degree image rotation	Color and monochrome printing
Text processing	Parsing of an interpretive printing control language	Color and monochrome printing

TABLE 10.13 EEMBC Telecom Benchmark Programs

EEMBC Telecom Benchmark Name	Benchmark Description	Example Applications
Autocorrelation	Fixed-point autocorrelation of a finite-length input sequence	Speech compression and recognition, channel and sequence estimation
Bit allocation	Bit-allocation algorithm for DSL modems using DMT	DSL modem
Convolutional encoder	Generic convolutional coding algorithm	Forward error correction
Fast Fourier transform (FFT)	Decimation in time, 256-point FFT using Butterfly technique	Mobile phone
Viterbi decoder	IS-136 channel decoding using Viterbi algorithm	Mobile phone

scores on the EEMBC web site) can use hand-tuned code, assembly-language subroutines, special libraries, special CPU instructions, co-processors, and other hardware accelerators. Full-fury scores tend to be much better than out-of-the-box scores, just as application-optimized production code generally runs much faster than code that has merely been run through a compiler. The free-for-all nature of EEMBC full-fury benchmarking rules underscores the need for a benchmark referee.

EEMBC does not publish its benchmark code. Processor vendors gain access to the EEMBC benchmark source code only by joining EEMBC, so EEMBC benchmark results are available for microprocessors and microprocessor cores supplied only by companies that have joined the consortium. With permission from the processor vendor, EEMBC will publish certified EEMBC benchmark-suite scores on the EEMBC website (www.eembc.org). Many processor vendors do allow EEMBC to publish their processors' scores and the EEMBC site already lists more than 300 certified benchmark-suite scores for a wide range of microprocessors and microprocessor cores.

Vendors need not submit their processors for testing with every EEMBC benchmark suite and some vendors do not allow their products' scores to be published at all, although they may share the certified results privately with prospective customers. Vendors tend to publish information about the processors that do well and allow poorer-performing processors to remain shrouded in anonymity.

EEMBC's work is ongoing. To prevent benchmark rot, EEMBC's subcommittees constantly evaluate revisions to the benchmark suites. The Networking suite is already on version 2.0 and the Consumer suite is undergoing revision as of the writing of this chapter. To date, EEMBC has created benchmarks that focus on measuring execution speed but the organization is developing power- and energy-oriented benchmarks because power efficiency and low-power operation have become important processor characteristics.

10.7.4 Modern Processor Benchmarks from Academic Sources

The EEMBC benchmarks have become the *de facto* industry standard for measuring benchmark performance, but two aspects of the EEMBC benchmarks make them less than ideal for all purposes. The EEMBC benchmark code is proprietary and secret. Therefore it is not open to evaluation, analysis, and criticism by industry analysts, journalists, and independent third parties. It is also not available to designers who want to perform their own microprocessor evaluations. Such independent evaluations would be unnecessary if all of the EEMBC corporate members tested their processors and published their results. However, fewer than half of the EEMBC corporate members have published any benchmark results for their processors and few have tested all of their processors [18].

10.7.4.1 UCLA's MediaBench 1.0

There are two sets of university-developed microprocessor benchmarks that can somewhat fill the gaps left by EEMBC. The first is MediaBench, developed by the Computer Science and Electrical Engineering Departments at the University of California at Los Angeles (UCLA) [19]. The MediaBench 1.0 benchmark suite was created to explore compilers' abilities to exploit instruction-level parallelism (ILP) in processors with VLIW (very long instruction word) and SIMD structures. It is targeted at microprocessors and compilers that target new-media applications and consists of several applications culled from image-processing, communications, and DSP applications and a set of input test data files to be used with the benchmark code. (All of the MediaBench files are located at http://cares.icsl.ucla.edu/MediaBench.) Applications used in MediaBench 1.0 appear in Table 10.14.

10.7.4.2 MiBench from the University of Michigan

A more comprehensive benchmark suite called MiBench was presented by its developers from the University of Michigan at the IEEE's 4th Annual Workshop on Workload Characterization in December 2001 [20]. MiBench intentionally mimics EEMBC's benchmark suite. It includes a set of 35 embedded applications in six application-specific categories: automotive and industrial, consumer devices, office automation, networking, security, and telecommunications. Table 10.15 lists the 35 benchmark programs (plus two repeated benchmark tests) in the six categories. All of the benchmark programs in the MiBench suite are available in standard C source code at http://www.eecs.umich.edu/mibench.

TABLE 10.14 MediaBench 1.0 Benchmark Programs

MediaBench 1.0 Application	Application Description	Example Applications
JPEG	Standardized compression method for full-color and gray-scale images	Digital cameras, printers
MPEG	MPEG2 codec for high-quality digital video transmission	Digital television, DVD
GSM	European GSM 06.10 full-rate speech codec	Mobile phone
G.721 Voice Compression	Reference implementations of CCITT G.711, G.721, and G.723 voice compression	Mobile phone
PGP	"Pretty Good Privacy" cryptographic algorithm	Secure data communications
PEGWIT	Publish key encryption and authentication	Secure data communications
Ghostscript	PostScript language interpreter	Graphical display, printing
Mesa	OpenGL 3-D graphics library	Workstation graphics
RASTA	Speech-recognition algorithm	Advanced user interfaces
EPIC	Experimental image-compression algorithm	Digital cameras, printers
ADPCM	Adaptive differential pulse code modulation (audio codec)	Digital telephony

TABLE 10.15 MiBench Benchmark Programs

Auto/Industrial	Consumer Devices	Office Automation	Networking	Security	Telecom
basicmath	JPEG	Ghostscript	Dijkstra	Blowfish encoder	CRC32
bitcount	lame	ispell	patricia	Blowfish decoder	FFT
qsort	mad	rsynth	(CRC32)	PGP sign	IFFT
susan (edges)	tiff2bw	sphinx	(sha)	PGP verify	ADPCM encode
susan (corners)	tiff2rgba	stringsearch	(Blowfish)	Rijndael encoder	ADPCM decoder
susan (smoothing)	tiffdither			Rijndael decoder	GSM encoder
	tiffmedian typeset			sha	GSM decoder

The MiBench automotive and industrial benchmark group tests a processor's control abilities. These routines include:

- *basicmath*: cubic function solving, integer square root, angle conversions
- *bitcount*: counts the number of bits in an integer array
- *qsort*: quicksort algorithm applied to a large string array
- *susan*: image recognition program developed to analyze MRI brain images

The MiBench applications in the consumer devices category include:

- *JPEG encode/decode*: still-image codec
- *tiff2bw*: converts a color tiff image into a black-and-white image
- *tiff2rgba*: converts a color image into one formatted into red, green, and blue
- *tiffdither*: dithers a black-and-white image to reduce its size and resolution
- *tiffmedian*: reduces an image's color palette by taking several color-palette medians
- *lame*: the Lame MP3 music encoder
- *mad*: a high-quality audio decoder for the MPEG1 and MPEG2 video formats
- *typeset*: a front-end processor for typesetting HTML files

Applications in the MiBench office-automation category include:

- *Ghostscript*: a Postscript language interpreter minus a graphical user interface
- *stringsearch*: searches a string for a specific word or phrase
- *ispell*: a fast spelling checker
- *rsynth*: a text-to-speech synthesis program
- *sphinx*: a speech decoder

The set of MiBench networking applications includes:

- *Dijkstra*: computes the shortest path between nodes
- *Patricia*: a routing-table algorithm based on Patricia tries

The networking group also reuses the CRC32, sha, and Blowfish applications from the MiBench suite's security applications group.

The security group of MiBench applications includes:

- *Blowfish encryption/decryption*: a symmetric block cipher with a variable-length key
- *sha*: a secure hashing algorithm
- *Rijndale encryption/decryption*: the cryptographic algorithm used by the AES encryption standard
- *PGP sign/verify*: a public-key cryptographic system called "Pretty Good Privacy"

MiBench's telecommunications application suite includes the following applications:

- *FFT/IFFT*: fast Fourier transform and the inverse transform
- *GSM encode/decode*: European voice codec for mobile telephony

- *ADPCM encode/decode*: a speech-compression algorithm
- *CRC32*: a 32-bit cyclic redundancy check

With its 35 component benchmark applications, MiBench certainly provides a thorough workout for any processor/compiler combination. Both MediaBench and MiBench solve the problem of proprietary code for anyone who wants to conduct a private set of benchmark tests by providing a standardized set of benchmark tests at no cost and with no use restrictions. Industry analysts, technical journalists, and researchers can publish results of independent tests conducted with these benchmarks although none seem to have done so, to date. The tradeoff made with these processor benchmarks from academia is the absence of an official body that enforces benchmarking rules and provides result certification. For someone conducting their own benchmark tests, self-certification may be sufficient. For anyone else, the entity publishing the results must be scrutinized for fairness in the conducting of tests, for bias in test comparisons among competing processors, and for bias in any conclusions.

10.8 Configurable Processors and the Future of Processor-Core Benchmarks

For the past 20 years, microprocessor benchmarks have attempted to show how well specific microprocessor architectures work. As their history demonstrates, the benchmark programs that have been developed have had mixed success in achieving this goal. One thing that has been constant over the years is the use of benchmarks to compare microprocessors and microprocessor cores with fixed ISAs. Nearly all microprocessors realized in silicon have fixed architectures and the microprocessor cores available for use in ASICs and SOCs also had fixed architectures. However, the transmutable silicon of ASICs and SOCs makes it feasible to employ microprocessor cores with configurable architectures instead of fixed ones.

Configurable architectures allow designers to add new instructions and registers to the processor's architecture that boost the processor's performance for specific, targeted applications. Use of configurable processor cores has a profound impact on the execution speed of any program, including benchmark programs. Processor vendors are taking two fundamental approaches to making configurable-ISA processors available to IC designers. The first approach provides tools that allow the designer to modify an existing base processor architecture to boost performance on a target application or set of applications. Companies taking this approach include ARC International, MIPS Technologies, and Tensilica, Inc. The other approach employs tools that compile entirely new processor architectures for each application. Companies taking this approach include ARM, CoWare/LISATek, Silicon Hive, and Target Compilers.

The ability to tailor a processor's ISA to a target application (which includes benchmarks) can drastically increase execution speed of that program. For example, Figure 10.4 shows EEMBC Consumer benchmark suite scores for Tensilica's configurable Xtensa V microprocessor core (labeled Xtensa T1050 in Figure 10.4). The right-hand column of Figure 10.4 shows benchmark scores for a standard processor configuration. The center column shows the improved benchmark scores for a version of the Xtensa V core that has been tailored by a design engineer specifically to run the EEMBC Consumer benchmark programs. Table 10.16 summarizes the performance results of the benchmarks for the stock and tailored processor configurations.

Note that the stock Xtensa V processor core performs on par with, or slightly faster than, other 32-bit RISC processor cores, and that tailoring the Xtensa V processor core for the individual tasks in the EEMBC Consumer benchmark suite by adding new instructions that are matched to the needs of the benchmark computations boosts the processor's performance on the individual benchmark programs from 4.6 × to more than 100 ×. The programmer modifies the benchmark or target application code by adding C intrinsics that use these processor extensions. The resulting performance improvement is similar to results that designers might expect to achieve when tailoring a processor for their target application code, so in a sense, the EEMBC benchmarks are still performing exactly as intended. The optimized Xtensa V EEMBC Consumer benchmark scores reflect the effects of focusing 3 months worth of an engineering graduate student's "full fury" to improve the Xtensa V processor's benchmark performance, optimizing both the processor and the benchmark code.

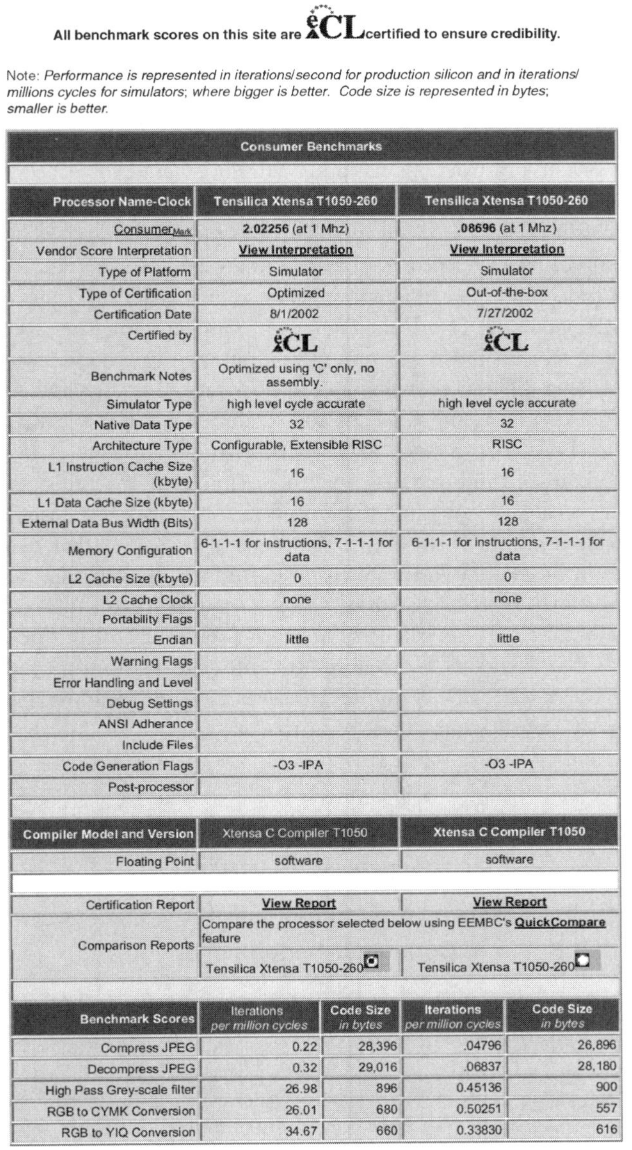

All benchmark scores on this site are ECL certified to ensure credibility.

Note: *Performance is represented in iterations/second for production silicon and in iterations/ millions cycles for simulators; where bigger is better. Code size is represented in bytes; smaller is better.*

Consumer Benchmarks		
Processor Name-Clock	Tensilica Xtensa T1050-260	Tensilica Xtensa T1050-260
ConsumerMark	2.02256 (at 1 Mhz)	.08696 (at 1 Mhz)
Vendor Score Interpretation	View Interpretation	View Interpretation
Type of Platform	Simulator	Simulator
Type of Certification	Optimized	Out-of-the-box
Certification Date	8/1/2002	7/27/2002
Certified by	ECL	ECL
Benchmark Notes	Optimized using 'C' only, no assembly.	
Simulator Type	high level cycle accurate	high level cycle accurate
Native Data Type	32	32
Architecture Type	Configurable, Extensible RISC	RISC
L1 Instruction Cache Size (kbyte)	16	16
L1 Data Cache Size (kbyte)	16	16
External Data Bus Width (Bits)	128	128
Memory Configuration	6-1-1-1 for instructions, 7-1-1-1 for data	6-1-1-1 for instructions, 7-1-1-1 for data
L2 Cache Size (kbyte)	0	0
L2 Cache Clock	none	none
Portability Flags		
Endian	little	little
Warning Flags		
Error Handling and Level		
Debug Settings		
ANSI Adherance		
Include Files		
Code Generation Flags	-O3 -IPA	-O3 -IPA
Post-processor		
Compiler Model and Version	Xtensa C Compiler T1050	Xtensa C Compiler T1050
Floating Point	software	software
Certification Report	View Report	View Report
Comparison Reports	Compare the processor selected below using EEMBC's QuickCompare feature Tensilica Xtensa T1050-260	Tensilica Xtensa T1050-260

Benchmark Scores	Iterations per million cycles	Code Size in bytes	Iterations per million cycles	Code Size in bytes
Compress JPEG	0.22	28,396	.04796	26,896
Decompress JPEG	0.32	29,016	.06837	28,180
High Pass Grey-scale filter	26.98	896	0.45136	900
RGB to CYMK Conversion	26.01	680	0.50251	557
RGB to YIQ Conversion	34.67	660	0.33830	616

FIGURE 10.4 Tailoring of a processor core's architecture for a specific set of programs that can result in significant performance gains. Copyright EEMBC. Reproduced with permission.

TABLE 10.16 Optimized Versus Out-of-the-Box EEMBC Consumer Benchmark Scores

EEMBC Consumer Benchmark Name	Optimized Xtensa T1050 Score (Iterations/M cycles)	Out-of-the-Box Xtensa T1050 Score (Iterations/M cycles)	Performance Improvement
Compress JPEG	0.22	0.04796	4.59 ×
Decompress JPEG	0.32	0.06837	4.68 ×
High pass grey-scale filter	26.98	0.45136	59.77 ×
RGB to CMYK conversion	26.01	0.50251	51.76 ×
RGB to YIQ conversion	34.67	0.33830	102.48 ×

Evolution in configurable-processor technology is allowing the same sort of effects to be achieved on EEMBC's out-of-the-box benchmark scores. In 2004, Tensilica introduced the XPRES Compiler, a processor-development tool that analyzes C/C++ code and automatically generates performance-enhancing processor extensions from its code analysis. Tensilica's Xtensa Processor Generator accepts the output from the XPRES Compiler and automatically produces the specified processor and a compiler that recognizes the processor's architectural extensions as native instructions and automatically generates object code that uses these extensions. Under EEMBC's out-of-the-box benchmarking rules, results from code compiled with such a compiler and run on such a processor qualify as out-of-the-box (not optimized) results. The results of an EEMBC Consumer benchmark run using such a processor appear in Figure 10.5. This figure compares the performance of a stock Xtensa V processor-core configuration with that of an Xtensa LX processor core that has been extended using the XPRES Compiler. (Note that Figure 10.5 compares the Xtensa V and Xtensa LX microprocessor cores running at 260 and 330 MHz, respectively, but the scores are reported on a per-MHz basis, canceling out any performance differences attributable to clock rate. Also, the standard versions of the Xtensa V and Xtensa LX processor cores have the same ISA and produce the same benchmark results.)

The XPRES Compiler boosted the stock Xtensa processor's performance by 1.17 × to 18.7 ×, as reported under EEMBC's out-of-the-box rules. The time required to effect this performance boost was 1 hour. These performance results demonstrate the sort of performance gains designers might expect from automatic processor configuration. However, these results also suggest that benchmark performance comparisons become even more complicated with the introduction of configurable-processor technology and that the designers making comparisons of processors using such benchmarks must be especially careful to understand how the benchmarks tests for each processor are conducted. Other commercial offerings of configurable-processor technology include ARM's OptimoDE, Target Compilers' Chess/Checkers, and CoWare's LISATek compiler. However as far as could be determined, there are no published benchmark results available for these products.

Benchmark score iterations comparison

Consumer benchmarks

Processor name-clock	Tensilica Xtensa T1050 -260	Tensilica Xtensa LX 330 MHz
Consumermark™	0.08696	0.51997
Compress JPEG	**0.04796**	0.056
Decompress JPEG	**0.06837**	0.083
High pass grey-scale filter	**0.45136**	8.661
RGB to CYMK conversion	**0.50251**	7.572
RGB to YIQ conversion	**0.33830**	6.310

Processor name-clock	Tensilica Xtensa T1050 -260	Tensilica Xtensa LX 330 MHz
Consumermark™	1.00	5.98
Compress JPEG	**1.00**	1.17
Decompress JPEG	**1.00**	1.21
High pass grey-scale filter	**1.00**	19.19
RGB to CYMK conversion	**1.00**	15.07
RGB to YIQ conversion	**1.00**	18.65

FIGURE 10.5 Tensilica's XPRES Compiler that can produce a tailored processor core with enhanced performance for a target application program in very little time. Copyright EEMBC. Reproduced with permission.

10.9 Conclusion

As a growing number of processor cores are employed to implement a variety of on-chip tasks, ASIC and SOC designers increasingly rely on benchmarking tools to evaluate processor-core performance. The tool of first recourse is always the actual application code for the target task but this particular tool is not always conveniently at hand. Standard processor benchmark programs run by objective parties must often stand in as replacement tools for processor evaluation. These tools have undergone more than two decades of evolution and their use (and abuse) is now well understood.

Benchmark evolution has not ended. Power-oriented benchmarks are in the wings. The transition to sub-100 nm lithography has placed a new emphasis on the amount of power and energy required to execute on-chip tasks. Benchmarking organizations are developing new processor benchmarks to measure power dissipation on standardized tasks. These new benchmarks will arrive in the next couple of years and will make processor performance benchmarking tools even more valuable to ASIC and SOC design teams.

References

[1] J. Rowson, Hardware/Software Co-Simulation, *Proceedings of the 31st Design Automation Conference (DAC'94)*, San Diego, 1994.

[2] S. Gill, The diagnosis of mistakes in programmes on the EDSAC, *Proceedings of the Royal Society Series A Mathematical and Physical Sciences*, 22, 1951, Cambridge University Press, London, pp. 538–554.

[3] Lord Kelvin (William Thomson), *Popular Lectures and Addresses*, Vol. 1, p. 73 (originally: Lecture to the Institution of Civil Engineers, 3, 1883).

[4] F. McMahon, The Livermore FORTRAN Kernels: A Computer Test of the Numerical Performance Range, Technical Report, Lawrence Livermore National Laboratory, 1986.

[5] J.J. Dongarra, LINPACK Working Note #3, FORTRAN BLAS Timing, Argonne National Laboratory, Argonne, IL, 1976.

[6] J.J. Dongarra, J.R. Bunch, C.M. Moler, and G.W. Stewart, LINPACK Working Note #9, Preliminary LINPACK User's Guide, ANL TM-313, Argonne National Laboratory, Argonne, IL, 1977.

[7] H.J. Curnow and B.A. Wichmann, A synthetic benchmark, *Comput. J.*, 19, 43–49, 1976.

[8] R.P. Weicker, Dhrystone: a synthetic systems programming benchmark, *Commun. ACM*, 27, 1013–1030, 1984.

[9] R.P. Weicker, Dhrystone benchmark: rationale for version 2 and measurement rules, *SIGPLAN Notices*, 1988, pp. 49–62.

[10] R.P. Weicker, Understanding variations in Dhrystone performance, *Microprocessor Report*, 1989, pp. 16–17.

[11] R.D. Grappel and J.E. Hemenway, A tale of four µPs: benchmarks quantify performance, *EDN*, 1, 1981, pp.179–232.

[12] S.H. Fuller, P. Shaman, D. Lamb, and W. Burr, Evaluation of computer architectures via test programs, *AFIPS Conference Proceedings*, Vol. 46, 1977, pp. 147–160.

[13] D. Shear, EDN's DSP Benchmarks, *EDN*, 29, 1988, pp. 126–148.

[14] W. Patstone, 16-bit µP benchmarks — an update with explanations, *EDN*, September 16, 1981, p. 169.

[15] J. Engblom, Why SpecInt95 Should Not Be Used to Benchmark Embedded Systems Tools, *Proceedings of the ACM SIGPLAN 1999 Workshop on Languages, Compilers, and Tools for Embedded Systems (LCTES'99)*, Atlanta, GA, 5, 1999.

[16] *Evaluating DSP Processor Performance*, Berkeley Design Technology, Inc. (BDTI), 1997–2000.

[17] M. Levy, At last: benchmarks you can believe, *EDN*, 5, 1988, pp. 99–108.

[18] T. Halfhill, Benchmarking the Benchmarks, *Microprocessor Report*, 30, 2004.

[19] C. Lee, M. Potkonjak et al., MediaBench: A Tool for Evaluating and Synthesizing Multimedia and Communications Systems, MICRO 30, *Proceedings of the Thirtieth Annual IEEE/ACM International Symposium on Microarchitecture*, 1–3, 1997, Research Triangle Park, NC.

[20] M. Guthaus, J. Ringenberg et al., MiBench: A Free, Commercially Representative Embedded Benchmark Suite, *IEEE 4th Annual Workshop on Workload Characterization*, Austin, TX, 2001.

11

Parallelizing High-Level Synthesis: A Code Transformational Approach to High-Level Synthesis

Gaurav Singh
Virginia Tech
Blacksburg, Virginia

Sumit Gupta
Tensilica Inc.
Santa Clara, California

Sandeep Shukla
Virginia Tech
Blacksburg, Virginia

Rajesh Gupta
University of California, San Diego,
San Diego, California

Abstract

Synthesis is the process of generating circuit implementations from descriptions of what a component does. Synthesis in part consists of refinement, elaboration as well as transformations and optimizations at multiple levels to generate circuits that compare favorably with manual designs. High-level synthesis (HLS) or behavioral synthesis specifically refers to circuit synthesis from algorithmic (or behavioral) descriptions. In this chapter, we focus on the developments in code transformation techniques for HLS over the past two decades, and describe recent progress in coordinated compiler and HLS transformations that seeks to combine effectively advances in parallelizing compiler techniques. We describe how coordinated compiler and HLS techniques can yield efficient circuits through examples of a class of source-level and dynamic transformations. We also describe recent developments in system-level modeling techniques and languages that attempt to raise the level of abstraction in the design process.

11.1 Introduction

The semiconductor industry has seen the widespread acceptance and use of language-level modeling of digital designs over the past decade. Digital designs are modeled using program-like descriptions, often using languages described in previous chapters, instead of schematic or block diagram descriptions of circuit blocks. The typical design process starts with design entry in a hardware description language at the register-transfer level (RTL), followed by logic synthesis.

The increasing willingness of designers to code circuits in a hardware description language and, in limited cases, yield control of exactly where the clock boundaries (i.e., storage elements) are placed in order to meet a given performance target has led system architects to explore at even higher levels of abstractions that allow algorithmic specification of the system. Using higher levels of abstraction reduces the size of the model, thus improving the efficiency of simulation and verification tasks. System architects can explore different algorithmic alternatives and different implementation choices and even hardware–software partitions. Transformations and optimizations applied at higher levels of abstraction have a larger impact on the quality of results of the final synthesized design as compared to applying optimizations at the register-transfer or circuit level.

One key enabling technology for specifying designs and systems in higher levels of abstraction is the ability to generate automatically or synthesize circuits that implement these designs. This process is generally referred to as *high-level synthesis* (HLS) or behavioral synthesis. HLS enables designers to shorten the design times and reduce the possibility of introducing errors or bugs during the process of refining system descriptions to RTL code.

However, current HLS efforts have several limitations: language-level optimizations are few and their impact on final circuit area and performance cannot always be predicted or controlled. The quality of synthesis results is poor, particularly for designs with moderately complex control flow (containing conditionals and loops).

Poor synthesis results can be attributed to several factors. Most of the optimizing transformations that have been proposed over the years are *operation-level* transformations. That is, these transformations typically operate on three-operand computational expressions or instructions [1]. In contrast, *language-* or *source-level* optimizations refer to transformations that require structural and hierarchical information available in the source code to operate. For example, loop transformations such as loop unrolling, loop fusion, and loop-invariant code motion (LICM), use information about the loop structure and the loop index variables and their increments. However, few language-level optimizations have been explored in the context of HLS and their effects on final circuit area and performance are not well understood. Often the effectiveness of transformations has been demonstrated in isolation from other transformations and results are often presented in terms of scheduling results, with little or no analysis of control and hardware costs. Furthermore, designs used as HLS benchmarks are often small, synthetic designs. It is thus difficult to judge if an optimization has a positive impact beyond scheduling results, on large moderately complex designs.

Recently, Gupta et al. [2,3] proposed a new approach to HLS that they termed as *parallelizing high-level synthesis* (PHLS). Parallelizing high-level synthesis seeks to create efficient circuit implementations through a combined strategy of compiler and synthesis transformations. It uses aggressive code parallelizing and code motion techniques to discover circuit optimization opportunities beyond what is possible with traditional HLS. The chief strength of these heuristics is the ability to select the code transformations so as to improve the overall synthesis results.

Parallelizing high-level synthesis addresses the limitations of current HLS approaches by employing techniques and tools that achieve the best compiler optimizations and synthesis results, irrespective of the programming style used in the high-level descriptions. In the past, several compiler techniques have been tried for HLS, albeit with mixed success. This is partly because direct application of compiler techniques does not necessarily optimize hardware synthesis results. The PHLS methodology has shown that for HLS, compiler techniques such as code transformations have to be selected and guided based on their effects on the control, interconnect, and area costs, and not based solely on their impact on instruction schedules. For instance, speculatively executing operations *may* lead to more complex control logic, which in turn can increase the longest combinational path through the design, and thus, increase the clock frequency of the synthesized circuit. Thus, we must examine the overall impact of applying speculation on the synthesized circuit, beyond just looking at scheduling results.

The rest of this chapter is organized as follows: in Section 11.2, we give an overview of the technical background of prior research work in HLS and compilers, and briefly discuss some commercial HLS tool offerings. We also discuss the challenges faced by HLS. We discuss the role of PHLS approach in Section 11.3 and the various transformations used in this approach. The PHLS approach has been validated by implementing a range of PHLS techniques in the *SPARK* software framework, presented in Section 11.4. We conclude this chapter in Section 11.5 with a summary and an outlook for the future of system-level design.

11.2 Background and Survey of the State of the Art

High-level synthesis and compilers are two separate technical streams that have recently merged, leading to the evolution of PHLS methodology. In this section, we provide an overview of prior research work in the HLS and compiler communities.

11.2.1 Early Work in HLS

High-level synthesis techniques have been investigated for two decades now. Over the years, HLS techniques have been chronicled in several books [4–9]. At its core, HLS consists of the following five main tasks:

- *Resource allocation.* This task consists of determining the number of resources that have been allocated or allotted to synthesize the hardware circuit. Typically, designers can specify an allocation in terms of the number of resources of each resource type. Resources consist not only of functional units (like adders and multipliers), but may also include registers and interconnection components (such as multiplexers and buses).
- *Scheduling.* The scheduling problem is to determine the time step or clock cycle in which each operation in the design executes. The ordering between the "scheduled" operations is constrained by the data (and possibly control) dependencies between the operations. Scheduling is often done under constraints on the number of resources (as specified in the resource allocation).
- *Module Selection.* Module selection is the task of determining the resource type from the resource library that an operation executes on. The need for this task arises because of the presence of several resources of different types (and different area and timing) that an operation may execute on. For example, an addition may execute on an adder, an ALU, or a multiply-accumulate (MAC) unit. There are area, performance, and power trade-offs in choosing between these components. Module selection must make a judicious choice between different resources such that a metric like area or timing is minimized.
- *Binding.* Binding determines the mapping between the operations, variables, and data (and control) transfers in the design and the specific resources in the resource allocation. Hence, operations are mapped to specific functional units, variables to registers, and data/control transfers to interconnection components.
- *Control synthesis and optimization.* Control synthesis generates a control unit that implements the schedule. This control unit generates control signals that control the flow of data through the data path (i.e., through the multiplexers). Control optimization tries to minimize the size of the control unit and hence, improve metrics such as area and power.

Each of these tasks are NP-complete and often dependent on each other. To delve deeper into each of these tasks, we refer the reader to [4–6,8,9].

A number of the early work in HLS was focused on scheduling heuristics for data-flow designs. The simplest scheduling approach is to schedule all the operations *as soon as possible* (ASAP) [10–13]. The converse approach is to schedule the operations *as late as possible* (ALAP). Scheduling heuristics such as *urgency scheduling* [14] and *force-directed scheduling* [15] schedule operations based on their urgency and mobility, respectively. Urgency of an operation is the minimum number of control steps from the last step in the schedule that the operation can be scheduled in before a timing constraint is violated. Mobility of an operation is the difference between the ASAP and ALAP start times of the operation.

Another category of heuristics uses a *list scheduling approach* whereby operations are scheduled after they are ordered based on control and data dependencies [16–18]. There are several other types of scheduling approaches that either iteratively reschedule the design [19] or schedule operations along the critical path through the behavioral description [20]. In the MAHA system [20], operations on the critical path are scheduled first and then the critical path is divided into several pipeline stages. Finally, the operations on the off-critical paths are scheduled based on a notion of freedom. This notion of freedom of an operation is similar to the mobility of an operation as defined above.

Resource allocation and binding techniques ranging from clique partitioning [11] to knowledge-based systems [10] have been explored in the past. The optimization goals for resource binding vary from reducing registers and functional units to reducing wire delays and interconnect costs [4,6,10]. Tseng and Siewiorek [11] use clique partitioning heuristics to find a clique cover for a module allocation graph. Paulin and Knight [21] perform exhaustive weight-directed clique partitioning of a register-compatibility graph to find the solution with the lowest combined register and interconnect costs. Stok and Philipsen [22] use a network flow formulation for minimum module allocation while minimizing interconnect. Mujumdar et al. [23] consider operations and registers in each time-step one at a time and use a network flow formulation to bind them.

Since the subtasks of HLS are highly interrelated, there have been several works that have attempted to perform these tasks simultaneously. Approaches using *integer linear programming* (ILP) formulations frequently fall into this category [24–27]. Gebotys and Elmasry [25] present an integer-programming model for simultaneous scheduling and allocation that minimizes interconnect. A 0/1 integer-programming model is proposed for simultaneous scheduling, allocation and binding in the OSCAR system [26]. Wilson et al. [27] presented a generalized ILP approach that gives an integrated solution to the various HLS subtasks. Safir and Zavidovique [28] and Kollig and Al-Hashimi [29] use simulated annealing to search the design space while performing simultaneous scheduling, allocation, and binding.

Pipelining has been the primary technique to improve performance for data-flow design [15,18,30–32]. The Sehwa system [30] automatically pipelines designs into a fixed number of stages to achieve the maximum performance. In the HAL system [15], designs are pipelined, with the help of user-specified pipeline boundaries, so that the resources are distributed uniformly across concurrent control steps. In the Maha system [20], the critical path is repeatedly divided into pipeline stages till the user-specified timing bound is satisfied.

Several optimization techniques such as algebraic transformations, retiming, and code motions across multiplexers have been proposed for improved synthesis results [33–35]. Walker and Thomas [34] proposed a set of transformations that were implemented within the *System architect's workbench* [36]. In this system, the input description is captured in a variant of data-flow graphs known as a *value trace* [37]. The system first applied coarse-level transformations such as function inlining, inverse function inlining, and removal of uncalled functions (or dead code elimination of whole functions). Then the system applies transformations that move operations into and out of select operations (or multiplexers). These code-motion transformations are similar to the speculative code motions used in designs with control flow. Another commonly used flow graph transformation is *tree height reduction* [38,39]. In this technique, the height of an expression tree or number of operations that execute in a chain is reduced by employing the commutative and distributive properties of the operators.

11.2.2 Recent Developments in HLS for Complex Control Flow

The HLS of designs with control flow (conditional branches and loops) has been studied by several groups over the last decade. Path-based scheduling (PBS) [40] is an exact approach that creates a as-fast-as-possible (AFAP) schedule by scheduling each control path independently. However, the order of operations on each control path is fixed and scheduling is formulated as a clique-covering problem on an interval graph; this means that speculative code motions are not supported. An extension of this technique that allowed operation reordering within basic blocks was presented in [41]. On the other hand, tree-based scheduling [42] supports speculative code motion and removes all the join nodes from the design, thus leaving the control-data flow graph (CDFG) looking like a tree. Both these approaches deal with loops by removing the loop's feedback path and scheduling just one iteration of the loop. The PUBSS approach [43] captures scheduling information in the behavioral finite state machine (BFSM) model. The input description (in VHDL) is captured as a network of communicating BFSMs and the schedule is generated using constraint-solving algorithms.

In the CVLS approach from NEC [44–46], a condition vector is generated for each operation based on the conditions under which the operation executes. Operations whose condition vectors are complementary are considered mutually exclusive and are free to be scheduled on the same resource. Speculation and operation duplication across a merge node are supported as well.

Radivojevic and Brewer [47] present an exact symbolic formulation that schedules each control path (or trace) independently and then creates an ensemble schedule of valid, scheduled traces. The traces are validated by using a backtracking procedure that can be computationally expensive. Haynal [48] uses an automata-based approach for symbolic scheduling of cyclic behaviors under sequential timing and protocol constraints. This is an exact approach, but can grow exponentially in terms of internal representation size. The *Waveschedule* approach [49] incorporates speculative execution into HLS to achieve its objective of minimizing the expected number of cycles. Kim et al. [50] transform a CDFG into a dataflow graph (DFG) by exploring various conditional resource sharing possibilities. The DFG is scheduled and then transformed back into a CDFG.

Santos et al. [51–53] and Rim et al. [54] support generalized code motions during scheduling in synthesis systems whereby operations can be moved globally irrespective of their position in the input. Santos creates and evaluates the cost of a set of solutions for the given design and then picks the solution with the lowest cost. This approach supports all types of code motions and employs code motion pruning techniques to reduce the search space and hence the run time of the approach.

Paulin et al. [55] present a hardware–software co-design methodology that uses the Amical tool [56] for synthesizing the control-intensive parts of the design and the Cathedral-2/3 tool [17] for DSP-oriented synthesis. The Amical tool enables the use of large function components in the hardware resource library. The Cathedral-2/3 tool is specifically targeted toward the synthesis of high-throughput application specific units.

The Olympus system [57] is a complete set of HLS, logic-level synthesis and simulation tools. The HLS tool, *Hercules*, accepts an input description in a variant of C called *HardwareC* [58], which introduces the notion of concurrency and timing into C. *Relative scheduling* [59] is used to schedule a design with operations that have unbounded delay.

Prior work on presynthesis transformations has focused on altering the control flow or extracting the maximal set of mutually exclusive operations [60,61]. Li and Gupta [62] restructure the control flow and attempt to extract common sets of operations within conditionals to improve synthesis results. Kountouris and Wolinski [63] perform operation movement and duplication before scheduling and also attempt to detect and eliminate false control paths in the design. Function inlining, where a function call is replaced by the contents of the function, was presented in the context of HLS by Walker and Thomas [34].

11.2.3 Related Work in Compiler and Parallelizing Compiler Approaches

Compiler transformations such as common subexpression elimination (CSE) and copy propagation predate HLS and are standard in most software compilers [1,64]. These transformations are applied as passes on the input program code and for cleanup at the end of scheduling before code generation. Compiler transformations were developed for improving code efficiency for sequential program execution. Their use in digital circuit synthesis has been limited. For instance, CSE has been used for throughput improvement [35], for optimizing multiple constant multiplications [65,66] and as an algebraic transformation for operation cost minimization [67,68].

A converse of CSE, namely, *common subexpression replication* has been proposed to aid scheduling by adding redundant operations [69,70]. *Partial redundancy elimination* [71] inserts copies of operations present in only one conditional branch into the other conditional branch, so as to eliminate common subexpressions in subsequent operations. Janssen et al. [67] and Gupta et al. [72] propose doing CSE at the source level to reduce the effects of the factorization of expressions and control flow on the results of CSE.

In the context of parallelizing compilers, *mutation scheduling* [73] also performs local optimizations such as CSE during scheduling in an opportunistic manner. A range of parallelizing code transformation techniques has also been previously developed for high-level language software compilers (especially parallelizing compilers) [74,75].

Although the basic transformations (e.g., dead code elimination, copy propagation) can be used in synthesis as well, other transformations need to be reinstrumented for synthesis by incorporating ideas of mutual exclusivity of operations, resource sharing, and hardware cost models. Cost models of operations and resources in compilers versus synthesis tools are particularly very different. For example, in compilers, there is generally a uniform push toward executing operations ASAP by speculative code motions since this always lead to improved execution in sequential (and pipelined) processors. Indeed, the optimality claims in *percolation* [76] and *trace scheduling* [74] are based entirely upon maximum movement of operations out of conditional branches. In the context of HLS, such notions of optimality have little relevance because of the interdependency of scheduling and resource utilization in a parallel hardware execution model. In circuit synthesis, code transformations that lead to increased resource utilization may also lead to higher hardware costs in terms of steering logic and associated control circuits. Some of these costs can be mitigated by interconnect minimizing resource binding techniques [77].

The use of code motions as a transformation technique has been explored at length in the context of parallelizing compilers for VLIWs [74,76,78–82]. Trace scheduling [74] uses profiling information to aggregate sequences of basic blocks (or control paths) that execute frequently into *traces*. These traces in the design are scheduled one at a time starting with the most time critical one. Hence, code motions are allowed only within traces and not across traces; in effect, code motions that duplicate or eliminate operations across multiple traces are not supported. Compensation code must be added in a bookkeeping pass applied after scheduling.

Percolation scheduling [76,83,84] applies a set of atomic semantic-preserving transformations to move or *percolate* operations in the design and create a resource-constrained schedule. Given an infinite number of resources, percolation scheduling has been shown to produce optimal schedules. Percolation scheduling suffers from two main limitations: (1) operations are moved linearly by visiting each node on the control path being traversed, and (2) this may lead to duplication of operations in multiple control paths. Trailblazing [80] overcomes the limitations of percolation scheduling by moving operations across large pieces of code with the help of a hierarchically structured intermediate representation [85].

Loop transformations such as loop unrolling and loop pipelining have received some attention in previous HLS works. Early work, such as the Flamel system [13], applied loop unrolling for designs with only data-flow to increase the parallelism of the design. Potasman et al. [86] demonstrated the improvements that can be obtained by applying the parallelizing compiler technique of perfect loop pipelining to data-flow designs. Holtmann and Ernst [87] apply loop pipelining for designs with control flow. In this approach, loops on the program path that have the highest predicted probability of being taken are pipelined, thereby deferring operations belonging to other paths.

11.2.4 High-Level Synthesis Tools in the Industry

Given the focus on HLS for over two decades, several tools and systems for HLS have been built in industry. Below, we list some of them. While technical details about these tools are not always available, our goal is to describe them in sufficient detail to make meaningful comparisons.

(1) *Cynthesizer* from *Forte Design Systems* [88] delivers an implementation path from SystemC to RTL. As shown in Figure 11.1, it includes a combination of behavioral synthesis functions: operation

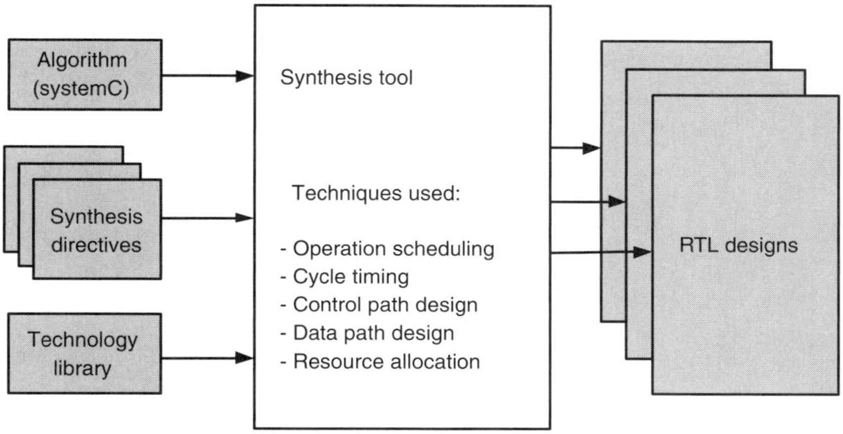

FIGURE 11.1 Forte's HLS tool flow.

scheduling, cycle timing, control and data path design, resource allocation, and RTL generation. Starting from untimed high-level SystemC models, Forte's synthesis tool builds timed RTL hardware implementations based on an external set of directives or constraints specified by the designer, such as clock speed, latency, pipelining, and loop unrolling. As an example, a block of code can be constrained to execute in no more than ten cycles by inserting a latency directive in the SystemC code itself. The block of code can be a straight line code enclosed within braces ({ }), a switch or if/else branch or a loop body.

(2) *PICO Express* from *Synfora* [89] is a tool that synthesizes ANSI-C into RTL. In an ASIC design flow, it takes in the C program code, a data set, and system requirements such as latency, bandwidth, clocking, and area. The other input is Synfora's library (currently characterized for TSMC's 0.13 μm process technology), which enables the tool to calculate area and performance. This tool is based on the PICO-NPA system developed at HP Labs [90]. The PICO-NPA system automatically synthesizes nonprogrammable accelerators (NPAs) to be used as coprocessors for functions expressed as loop nests in C. The NPAs it generates consist of a synchronous array of one or more customized processor datapaths, their controller, local memory, and interfaces. The tool is particularly powerful at finding loop-level parallelism and applying optimizations such as software pipelining, besides instruction-level parallelism. The tool determines the required hardware resources by examining the designer-specified performance requirements. Finally, it schedules operations, after which it allocates resources, and maps the components in Synfora's library.

(3) *Catapult C* from *Mentor Graphics* [91] synthesizes algorithmic descriptions in C++ and provides a graphical user interface to analyze alternative microarchitecture implementations. In addition to providing feedback (in terms of area estimates, latency, and throughput values) for each architectural modification, the micro-architecture solutions can be saved for reexploration at a later time. The tool allows each solution to be reused, thus allowing a single source code representation to be used to derive multiple implementations. The designer can apply constraints on steps such as loop unrolling, pipelining, and merging, array mapping, resource allocation, etc. The design flow through Catapult is as shown in Figure 11.2.

(4) The *Agility compiler* from *Celoxica* [92] provides a synthesis path from SystemC code and outputs electronic design interchange format (EDIF) netlists for FPGA logic and RTL HDL (hardware description language) code for ASIC synthesis. The implemented optimizations include automatic tree balancing, fine-grained logic sharing, and retiming.

(5) *Cyber* from *NEC* [46] is a HLS system used internally within NEC. *Cyber* accepts behavioral descriptions in C or behavioral VHDL and generates RT descriptions in C(BDL), Verilog, VHDL, and NEC's original RT language. C(BDL) is a subset of C with some extensions added to describe

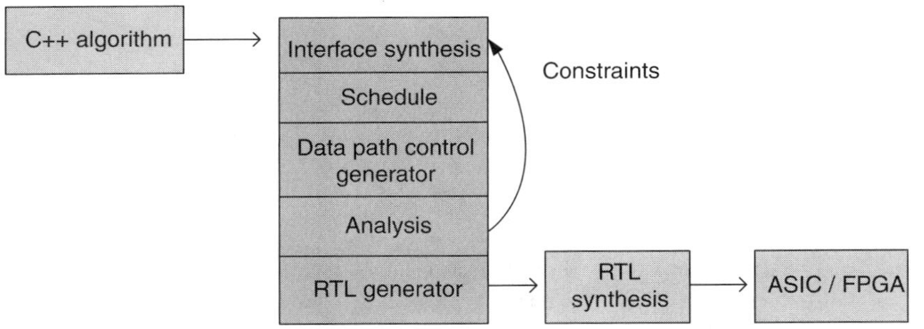

FIGURE 11.2 Mentor graphics's catapult C synthesis flowchart.

hardware. These extensions are provided as specialized comments or pragmas for the Cyber synthesizer. *Cyber* generates bit and cycle-accurate C code at the RT level, which can be compiled with off-the-shelf C compilers. *Cyber* generates a condition vector for each operation based on the conditions under which the operation executes; thus, condition vectors capture control-flow information. Operations whose condition vectors are complementary are considered mutually exclusive and are free to be scheduled on the same resource. Operation can be reordered using speculation and by duplicating operations across merge nodes in the control-flow graph (CFG).

(6) *BlueSpec Compiler* (*BSC*) from *BlueSpec* [93] synthesizes a hardware-centric input description specified in SystemVerilog and generates both the RTL and cycle-accurate C model that can be used for system-level functional verification. The compiler uses the notion of *operation-centric* descriptions [94], wherein the behavior of a system is described as a collection of guarded atomic actions or rules, which operate on the state of the system. Each rule specifies the condition or guard under which it is enabled, and a function that determines the consequent allowable state transition. The atomicity requirement simplifies the task of hardware description by permitting the designer to formulate each rule as if the rest of the system is static. Using this methodology, any legitimate behavior of the system can be expressed as a series of atomic actions on the state.

In the Bluespec methodology, the designer first explicitly instantiates the state elements of the system (registers, FIFOs, memories, etc.). Next, the designer describes the behavior of the system using a collection of rules, which operate on the state of the system. There is a straightforward translation from rules into hardware as shown in Figure 11.3. Assuming all state is accessible (no port contention), each guard and transition function can be implemented easily as combinational logic. A hardware scheduler and control circuit is added so that in every cycle the scheduler dynamically picks one function whose corresponding condition is satisfied and the control circuit updates the state of the system with the result of the selected function. The cycle time in such a synthesis is determined by the slowest guard and the slowest functions. Thus, the challenge in generating efficient hardware from sets of atomic actions is to generate a scheduler, which in every cycle picks a maximal set of rules that can be executed simultaneously. In the BlueSpec compiler, this synthesis is done using term rewriting systems [94].

Early results from BlueSpec show that the method of HLS from the guarded atomic actions is particularly useful for micro-architectural exploration in the design of complex ASICs.

(7) *Behavioral compiler* (BC) from *Synopsys* [8] (now discontinued) is the oldest commercial HLS offering. *BC* accepts the input description in a subset of behavioral VHDL. The VHDL is further restricted by coding guidelines and *pragma* constructs. *BC* uses technology libraries for resource allocation and for mapping functional units to standard cells. *BC* takes the unscheduled input description and generates a schedule for the operations. However, *BC* requires the user to insert some scheduling structure in the code in terms of VHDL *wait* statements. *Wait* statements are used for tasks such as partitioning two sections of the behavioral code that must execute in different clock cycles, or for two consecutive writes to the same variable, or for ensuring that the true and

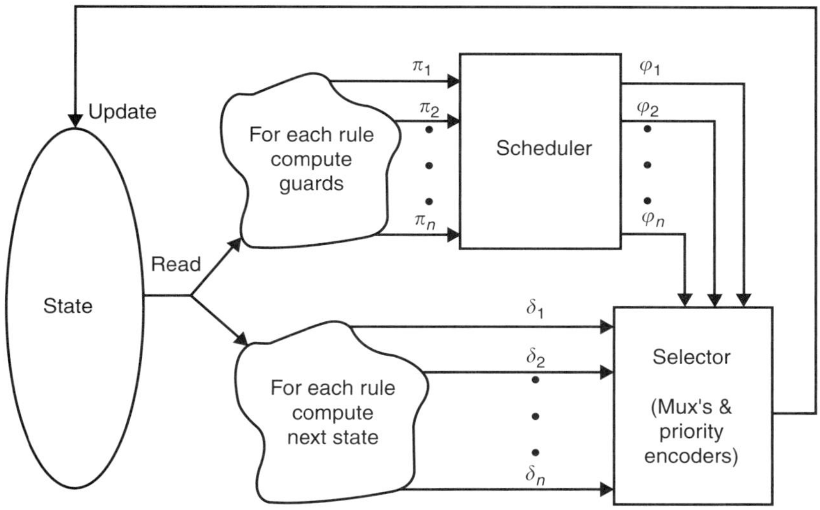

FIGURE 11.3 Synthesis from guarded atomic actions.

false branches of a conditional construct execute in the same number of cycles. The scheduler in *BC* has two different scheduling modes, *cyclefixed* and *superstatefixed*:

- *Cyclefixed mode.* In *cyclefixed* mode, *BC* preserves the cycle-to-cycle behavior of the I/O as defined by the behavioral description. It is used when the designer wants to describe a cycle-accurate description of the design.
- *Superstatefixed mode.* The section of code between two wait statements in the behavioral description is denoted as a *superstate.* In *superstatefixed* mode, *BC* can insert additional clock cycles in a superstate if required. Designers can use the superstatefixed mode to explore the latency versus area trade-offs in their designs.

Apart from these scheduling modes, *BC* can read three other schedule-constraining commands. These commands — *setcycles, setmincycles,* and *setmaxcycles* — can be used to constrain the exact, minimum, and maximum number of cycles between two operations or for a given loop. For resource allocation, *BC* tries to minimize latency by maximizing resource utilization. There is an *extendlatency* scheduling mode in which *BC* minimizes resource allocation at the cost of increased latency. Intermediate designs can be explored using *setcycles* command.

(8) The *XPRES compiler* from *Tensilica* [95,96] creates tailored instructions for the Tensilica Xtensa LX processor from algorithmic specifications in C/C++ code. Thus, it is not a behavioral synthesis tool in its classical sense. However, we mention this tool here since the tool applies parallelizing and vectorizing code transformations on the C/C++ code and generates customized hardware functional units and vector register files and adds them to the base Xtensa processor. The C/C++ code is then reinstrumented using intrinsics to use special instructions that are mapped to these customized functional units. Tensilica has demonstrated the efficacy of this tool through a range of benchmarks [96].

The XPRES compiler uses three techniques for accelerating the application: operator fusion, SIMD vectorization, and using flexible length instruction extensions (FLIX) instructions.

- *Operator fusion* combines a frequently occurring sequence of simple operations into a single instruction. A functional unit is generated corresponding to this new instruction and the code is instrumented to use this instruction. This reduces the number of instructions executed within, say a loop, and also reduces the number of instructions that must be fetched from memory.
- *SIMD vectorization* refers to vectorizing of loops by creating an instruction with multiple identical execution units that operate on multiple data items in parallel. For example, XPRES may

generate a two-operation MAC SIMD instruction that operates on two 16-byte data items simultaneously. Hence, this doubles performance over executing a MAC instruction for each of the two data items sequentially. Note that the rest of the Tensilica software tool will generate vector (or SIMD) functional units and vector register files to implement these instructions.

- The third technique used by XPRES is to create *FLIX instructions* to accelerate code by executing multiple *independent* operations in the same instruction. This is analogous to VLIW instructions in which several operations execute in parallel. These new "FLIX" instructions again execute more operations in fewer cycles, hence, accelerating the application. Note that each of these techniques leads to generation of extra hardware functional units, data path connections, register files, etc.

Thus, XPRES extends a base instruction set architecture with VLIW instructions, vector operations, fusion operations, and vector register files to create an application-specific instruction-set processor optimized for a particular application. Because XPRES is integrated with Tensilicas configurable processor tool chain, the new instructions, operations, and register files are automatically recognized and exploited by the entire software tool chain, including the C/C++ compiler.

Table 11.1 presents the input and output description languages for each of the HLS tools discussed above.

11.2.5 Continuing Challenges in HLS

Despite extensive research in HLS techniques, practical adoption of HLS has suffered from two challenges:

(1) Manual implementations are often superior to HLS solutions due to the ability of designers to apply a large range of RTL and circuit-level optimizations. Commercial HLS tools are often touted for digital signal processing designs with straight-line code. The presence of control flow (conditionals and loops) often leads to poor synthesis results.

(2) The flexibility of high-level and procedural languages such as C, C++, and their variants gives designers the freedom to specify the same algorithm using widely different programming styles and control constructs. This is particularly problematic for behavioral descriptions with complex and nested conditionals and loops, and leads to unpredictable quality of synthesis results. Current HLS tools address this issue by either limiting the input language or by asking designers to conform to preferred coding styles.

(3) During HLS, designers are often given minimal controllability over the transformations that affect synthesis. Designers are, thus, unable to use their design intuition to control and guide the synthesis process.

Speculative code motions have been proposed for improving synthesis results in designs with complex control flow. Whereas these transformations have been shown to be useful for improving schedule lengths, it is not clear if a transformation has a positive impact beyond scheduling, particularly on the control and area costs. Industry experience shows that, often, critical paths in control-intensive designs

TABLE 11.1
High-Level Synthesis Tools

Tools	Input (Behavioral)	Output (RTL)
Forte's cynthesizer	SystemC	SystemC/Verilog
Mentor's catapult	C++/C	Verilog, VHDL
Synfora's PICO express	C	Verilog
Celoxica's agility compiler	SystemC	EDIF netlists/HDL
NEC's cyber	C/VHDL	C(BDL)/Verilog/VHDL/NEC's RTL
BlueSpec's BSC	SystemVerilog	Verilog
Synopsys's behavioral compiler	VHDL	VHDL
Tensilica's XPRES	C++/C	Verilog

pass through the control unit and steering logic. To this end, Rim et al. [54] use an analytical model to estimate the cost of additional interconnect and control caused by code duplication during code motions. Bergamaschi [97] proposes the behavioral network graph to bridge the gap between high-level and logic-level synthesis and aid in estimating the effects of one on the other.

Several approaches use intermediate representations such as CDFGs for design entry. They often use restricted and stylized input descriptions that reduce the appeal of using high-level languages for design entry. Furthermore, language-level optimizations are few and their effects on final circuit area and speed are not well understood.

All these factors continue to limit the acceptance of HLS tools among designers. As mentioned earlier, manual implementations are often superior to HLS solutions because of the range of design optimizations that a system designer has at his/her disposal. While HLS tools have been catching up, just as they have already done so at lower levels of abstraction, a significant gap remains. *Parallelizing high-level synthesis* has been shown to address these factors and has demonstrated through experimental results that it can produce designs that are competitive with manual designs. This can make HLS part of the microelectronic design flow, thus, bringing about the much needed productivity gain required to design increasingly complex circuits.

11.3 Parallelizing HLS

Recent HLS approaches have employed beyond-basic-block code motions such as speculation — derived from the compiler domain — to increase resource utilization. The PHLS methodology systematically incorporates several similar techniques that attempt to increase the scope of code motions and increase available parallelism to improve resource utilization. The transformations in PHLS are structured into two groups: *presynthesis* and *dynamic* transformations. Before discussing these transformations, we briefly describe the intermediate model used to carry out these transformations.

11.3.1 Design Description Modeling

One-key component of the PHLS methodology proposed in the *SPARK* HLS system [98] (described in the next section) is modeling the design description using a novel *three-layered graph representation.* Control and data flow in the input description is captured using a CFG and a DFG, respectively. Additionally, the program structure and hierarchy of the control flow are captured using a hierarchical intermediate representation called *hierarchical task graphs* (HTGs) [85]. The nodes of these — the CFGs, DFGs, and HTGs — form a three-layered graph such that there is a relation between the nodes of each successive layer. Layered graphs are defined as follows:

Definition 11.3.1

A *k-layered graph* is a connected graph in which the vertices are partitioned into k sets $L = l_1, \ldots, l_k$ and edges run between the vertices of successive layers, l_i and l_{i-1}.

The top-level layer consists of the nodes from the HTG, the next layer has nodes from the CFG and the lowest level layer has nodes from the DFG. Data-flow graphs and CFGs have been used extensively in past work: DFGs capture the read-after-write data dependencies between operations and CFG capture the flow of control between basic blocks in the design [4,5]. We present HTGs in the next section.

11.3.1.1 Hierarchical Task Graphs: A Model for Designs with Complex Control Flow

Traditionally, CDFGs have been the most popular intermediate representation for HLS [5]. Control-data flow graphs consist of operation and control nodes with edges for both data flow and control flow. However, to apply the range of coarse- and fine-grain parallelizing compiler transformations proposed in the PHLS methodology, we also require structural information from the source code. Thus, PHLS complements CDFGs with an intermediate representation that also maintains the hierarchical structuring of the design, such as if-then-else blocks and for- and while-loops. This intermediate representation consists of basic blocks encapsulated in HTGs [85,80].

A *HTG* is a directed acyclic graph with unique *Start* and *Stop* nodes such that there exists a path from the *Start* node to every node in the HTG and a path from every node in the HTG to the *Stop* node. Edges in a HTG represent control flow between HTG nodes. Each node in an HTG can be one of the following three types:

1. *Single nodes* are nodes that have no subnodes and are used to encapsulate basic blocks.
2. *Compound nodes* are hierarchical in nature, i.e., they contain other HTG nodes. They are used to represent structures like if-then-else blocks, switch-case blocks or a series of HTGs.
3. *Loop nodes* are used to represent the various types of loops (for, while-do, do-while). Loop nodes consist of a loop head and a loop tail that are single nodes, and a loop body that is a compound node.

The *Start* and *Stop* nodes for all compound and loop HTG nodes are always single nodes. Note that the *Start* and *Stop* nodes of a single node are the node itself.

An example "C" description and its corresponding HTG representation is shown in Figures 11.4(a) and (b). This representation consists of a root HTG (*Htg0*) that has several subnodes. One of them is a If-HTG (compound) node, *Htg2*. This If-HTG has four subnodes of itself. In Figure 11.4(c), we show how the control and DFGs can be overlaid onto the HTG graph. Each basic block is encapsulated in a single node.

Since HTGs maintain a hierarchy of nodes, they are able to retain coarse, high-level information about program structure in addition to operation-level and basic block-level information. This aids in coarse-grain code restructuring (such as that done by loop transformations) and also, in operation movement by reducing the amount of compensation code required. Furthermore, nonincremental moves of operations across large blocks of code are possible without visiting each intermediate node [80].

The layering of the HTG, CFG, and DFG representations of the same example C description is shown in Figure 11.5. Each operation in the DFG has a one-to-one mapping with a basic block in the CFG and each basic block in the CFG is encapsulated in a single HTG node in the HTG representation.

Note that HTGs capture information about the control flow between hierarchical nodes and CFGs capture information about the control flow between basic blocks. Hence, we maintain both graphs since HTGs

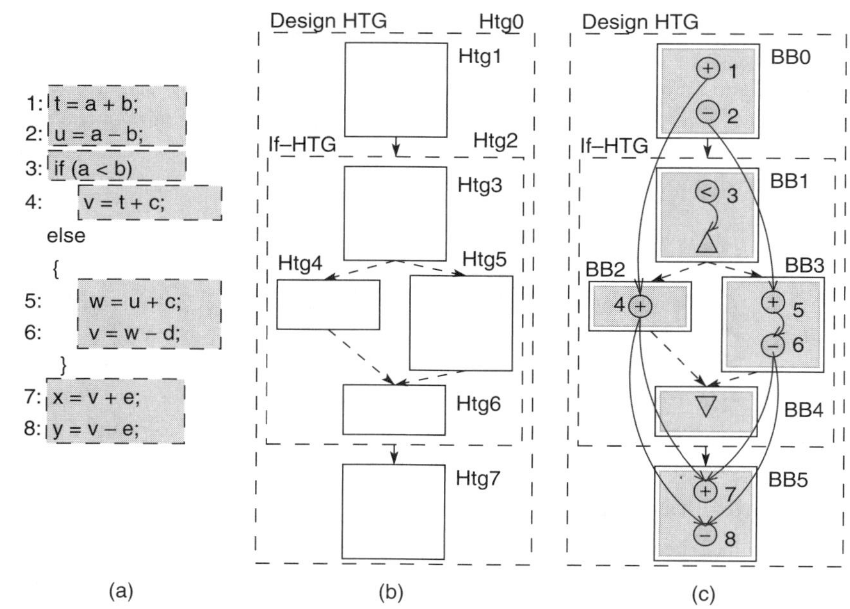

FIGURE 11.4 (a) Example C description; (b) corresponding HTG representation; and (c) HTG representation with the control and DFGs overlaid on top. Each basic block is encapsulated in a single HTG node. Basic block BB4 is added to the CFG as a join node for the If-HTG node.

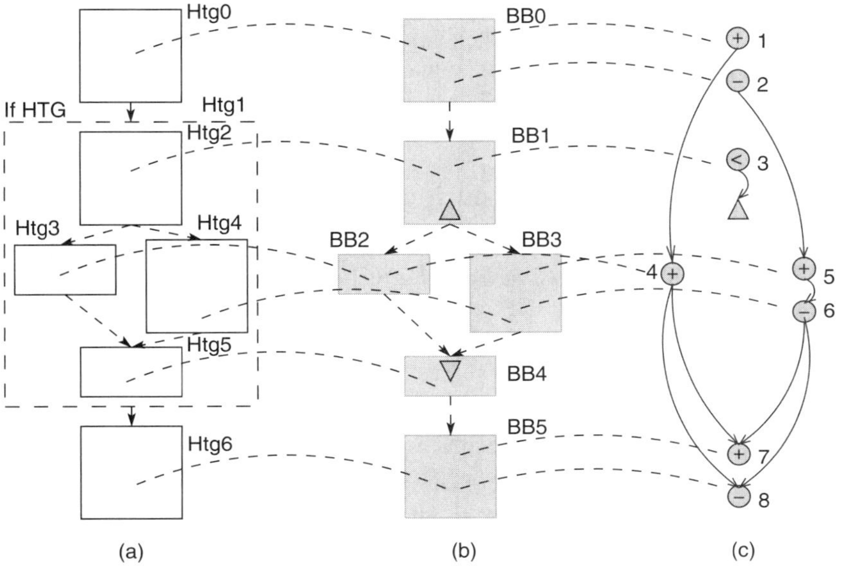

FIGURE 11.5 (a) Hierarchical task graph representation of the example C description; (b) CFG for the example; and (c) DFG for the example. Together these form a three-layered design graph, VQ. The dashed lines denote the edges between the different layers of the design graph.

are an efficient way to traverse the hierarchy of the design (useful for code transformation algorithms), whereas CFGs are efficient for traversing the basic blocks in the design (useful for scheduling algorithms).

11.3.2 Presynthesis Optimizations

Presynthesis optimizations are applied prior to elaboration of the source code into operation-level descriptions. These transformations include: (1) coarse-level code restructuring by function inlining and loop transformations (loop unrolling, loop fusion, etc.), (2) transformations that remove unnecessary and redundant operations such as CSE, copy propagation, and dead code elimination, (3) transformations such as LICM, induction variable analysis and operation strength reduction, that reduce the number of operations within loops and replace expensive operations (multiplications and divisions) with simpler operations (shifts, additions, and subtractions).

Loop transformations such as loop unrolling and loop fusion restructure the code, thereby, exposing many more opportunities for optimizations later on during scheduling. These have the maximum impact on the quality of results and have to be carefully guided, often with designer intervention. The *SPARK* HLS framework described in the next section uses synthesis scripts to guide these transformations [2,3].

Common subexpression elimination is a well-known basic compiler transformation that attempts to detect repeating subexpressions in a piece of code, compute them once and reuse the stored result wherever the subexpression occurs subsequently [1]. Often compilers apply CSE on three-operand instructions. However, Gupta et al. [3,72] have shown that applying CSE on the source code can often expose opportunities for optimizations not otherwise available. In addition to this, CSE can also be applied dynamically during scheduling as explained later in Section 11.3.3.2. *Loop-invariant* operations are computations within a loop body that produce the same results each time the loop is executed [1,64]. These computations can be moved outside the loop body, without changing the results of the code. In this way, these computations will execute only once *before* the loop, instead of for *each* iteration of the loop body.

The PHLS methodology uses a number of compiler transformations such as CSE, LICM, dead code elimination, etc., during the presynthesis stage to remove redundant and expensive operations. Although these transformations have been used extensively in compilers [1,64,99], HLS tools have to take into account the additional control and area (in terms of interconnect) costs of these transformations. Gupta et al. [3,72] have shown that the common subexpression elimination and LICM transformations are particularly effective in improving the quality of HLS results for multimedia applications. These applications are characterized by array accesses within nested loops. The index variable calculation for these array accesses often contains complex arithmetic expressions that can be optimized significantly by CSE and LICM.

11.3.3 Transformations Applied during Scheduling

A range of classical HLS and parallelizing compiler transformations are applied during the scheduling phase of the PHLS methodology. In this chapter, in the interest of space, we only briefly describe the speculative code motions and dynamic transformations. We refer the reader to [3] for details of these techniques and descriptions of other techniques such as chaining of operations across conditional boundaries, multicycle operation scheduling, etc.

11.3.3.1 Speculative Code Motions

The PHLS methodology proposes a set of speculative code motion transformations to alleviate the problem of poor synthesis results in the presence of complex control flow in designs. Effectively, these code motions reorder operations to minimize the effects of the choice of control flow (conditionals and loops) in the input description. These beyond-basic-block code motion transformations attempt to extract the inherent parallelism in designs and increase resource utilization.

Generally, speculation refers to the unconditional execution of operations that were originally supposed to have executed conditionally. However, in the context of PHLS, speculation comes into four forms: (1) speculation, (2) reverse speculation, (3) conditional speculation, and (4) early condition execution. Frequently there are situations when there is a need to move operations *into* conditionals [77,100]. This may be done by *reverse speculation,* where operations before conditionals are moved into *subsequent* conditional blocks and executed conditionally, or this may be done by *conditional speculation,* wherein an operation from after the conditional block is duplicated *up* into *preceding* conditional branches and executed conditionally. Reverse speculation can be cupled with another novel transformation, namely, *early condition execution.* This transformation evaluates conditional checks ASAP. This removes the control dependency on the operations in the conditional branches and thereby makes them available for scheduling.

The movement of operations due to the various speculative code motions is shown in Figure 11.6 using solid arrows. Also shown is the movement of operations across entire hierarchical blocks, such as if-then-else blocks or loops. Since these code motions reorder, speculate, and duplicate operations, they

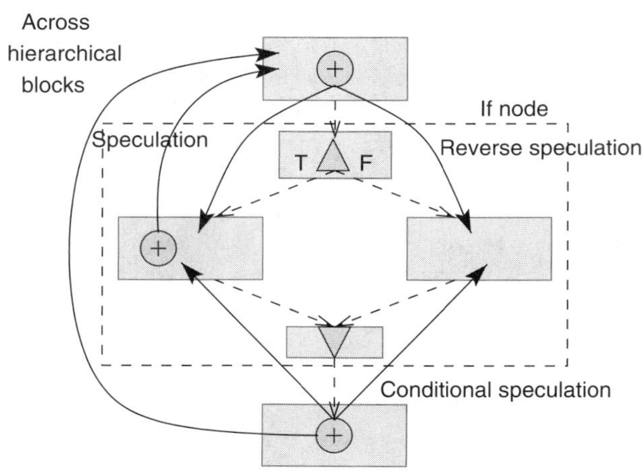

FIGURE 11.6 Various speculative code motions: operations may be speculated, reverse speculated, conditionally speculated or moved across entire conditional blocks.

often create new opportunities for dynamically applying transformations such as CSE and copy propagation, during scheduling.

11.3.3.2 Dynamic Transformations

Dynamic transformations are transformations that are deployed during various HLS tasks, most prominently during operation scheduling and resource allocation. Compiler and synthesis transformations suffer from the classical phase ordering problem, whereby the order in which transformations are applied impacts the effectiveness of the transformations. Dynamic transformations offer a way around this problem by creating opportunities for and applying optimizations during the scheduling and resource allocation phases. Dynamic transformations are particularly effective when combined with speculative code motions.

We have earlier proposed two instances of dynamic transformations in the PHLS methodology: dynamic branch balancing and dynamic CSE. Both these transformations operate during scheduling and are closely interlinked with the speculative code motions. *Dynamic branch balancing* adds scheduling steps in the shorter branches of conditional branch constructs, so that the longest path through the conditional branch is not affected [101]. The additional scheduling steps can then be used to schedule speculatively operations that would otherwise have been scheduled in later time steps. *Dynamic CSE* opportunistically looks for common subexpressions that can be eliminated during scheduling [102]. These are additional opportunities for CSE that are created as a result of code motions and code restructuring, which were unavailable before scheduling.

11.4 The *SPARK* PHLS Framework

The PHLS methodology, along with the various transformations presented in the previous sections has been implemented in the *SPARK* PHLS framework [103]. An overview of this framework is shown in Figure 11.7. *SPARK* provides the ability to control and thus, experiment with the various code transformations using user-defined scripts and command-line options. *SPARK* consists of 100,000+ lines of C++ code and uses a commercial C/C++ front-end from Edison design group (EDG) [104].

One of the main features of *SPARK* is the distinction between code transformation algorithms and the use of these techniques in a given design. This allows the designer to control the transformations applied to the design description by way of synthesis scripts [2]. The scripting ability is used to experiment with new transformations and develop heuristics to guide these transformations, which enables one to find the combination of techniques that leads to the best synthesis results. Various code motions, branch balancing algorithms, methods of calculating priority and other required transformations can be enabled using synthesis scripts. A typical *SPARK* script file has three main sections: the scheduler functions, the list of allowed code motions (code motion rules) and the cost of code motions.

Experiments performed using *SPARK* on moderately complex designs show significant improvements over classical (nonparallelizing) synthesis approaches. Experiments were conducted on a range of multimedia (MPEG-1, MPEG-2, and image processing applications (GIMP, Susan)). The results demonstrated that the various code transformations led to improvements in the performance and controller size of all the designs. The synthesis results for most of the designs follow a similar pattern. By far, the speculative code motions and presynthesis optimizations have the maximum impact on the synthesis results. The overall improvements obtained when all the transformations are applied in the *SPARK* toolkit compared to when only nonspeculative code motions are applied are up to a maximum of 80% and an average of 60% in the total delay through the circuit, with either no increase or a modest increase of 5% in circuit area.

The RTL VHDL generated by *SPARK* for these benchmarks was synthesized using the Synopsys *Design Compiler* logic synthesis tool [105]. The logic synthesis results demonstrated that the improvements in performance seen earlier translate over to the synthesis results of the final netlist. Hence, the total delay through the circuit reduces by up to 50% as the code motions are enabled with negligible impact on critical path length (clock period). However, these aggressive code motions do lead to an increase in area by about 20% when all the code motions are enabled. This area increase is due to the additional steering and control logic caused by increased resource sharing.

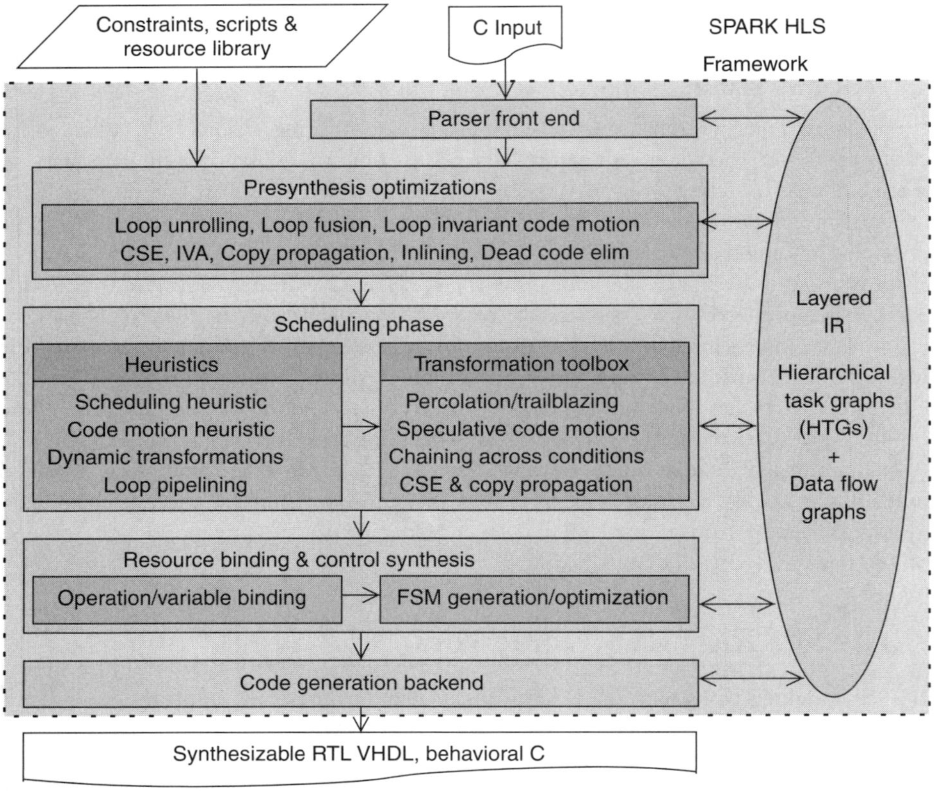

FIGURE 11.7 The *SPARK* PHLS framework.

Dynamic CSE leads to lower area due to fewer operations and the corresponding reductions in interconnect. This ends up counterbalancing the increase in area seen due to the code motions. Thus, enabling all the code motions along with CSE and dynamic CSE, leads to overall reductions in the total delay through the circuit by nearly 70% as compared to only within basic block code motions, while the design area remains almost constant (and sometimes even decreases).

11.5 Summary

The semiconductor industry's needs are shifting from pure ASIC synthesis to hardware–software co-design and direct support for synthesis on reprogrammable (FPGA) or customizable platforms (such as those offered by Tensilica and Arc). System architects are gravitating toward *system-level modeling and design methodologies** that enable them to first specify applications without notions of hardware or software, and then explore different hardware–software partitions and different hardware, communication, and memory architectures. Ideally, software compilers and synthesis tools should then be able to generate the hardware, software, and communication interfaces from these system-level specifications.

The push toward system-level design and synthesis methodologies has led to a renewed interest in HLS, as indicated by a number of startup companies in this domain. In this chapter, we described recent developments in HLS and also reviewed the considerable work done over the past two decades. We focused on a new code transformational, parallelizing approach to HLS that attempts to address the increasing size and

* See chapters on "System-level specification and modeling languages" by J. Buck and "Design and Verification Languages" by S. Edwards in this book for more on system-level design methodologies and languages.

complexity of contemporary designs. The PHLS methodology represents an important step forward in the automated synthesis of hardware components from algorithmic descriptions, so that the quality of results is not compromised by coding style and the use of control constructs. To do so, this methodology applies a coordinated set of coarse- and fine-grain parallelizing compiler and classical HLS transformations. The PHLS methodology has been validated by implementation in the *SPARK* software framework and early results demonstrate the effectiveness of this methodology in producing high-quality synthesized designs.

We believe that code transformational approaches such as the one proposed in the PHLS methodology are key in moving the level of abstraction from RTL hardware design to system-level modeling, design space exploration, refinement, and synthesis, thereby enhancing designer productivity.

References

[1] A. Aho, R. Sethi, and J. Ullman, *Compilers: Principles and Techniques and Tools*, Addison-Wesley, Reading, 1986.
[2] S. Gupta, Coordinated Coarse-Grain and Fine-Grain Optimizations for High-Level Synthesis, PhD thesis, University of California, Irvine, 2003.
[3] S. Gupta, R.K. Gupta, N.D. Dutt, and A. Nicolau, *SPARK: A Parallelizing Approach to the High-Level Synthesis of Digital Circuits*, Kluwer, Dordrecht, 2004.
[4] G. De Micheli, *Synthesis and Optimization of Digital Circuits*, McGraw-Hill, New York, 1994.
[5] D.D. Gajski, N.D. Dutt, A.C.-H. Wu, and S. Y.-L. Lin, *High-Level Synthesis: Introduction to Chip and System Design*, Kluwer, Dordrecht, 1992.
[6] R. Camposano and W. Wolf, *High Level VLSI Synthesis*, Kluwer, Dordrecht, 1991.
[7] D.C. Ku and G. De Micheli, *High Level Synthesis of ASICs Under Timing and Synchronization Constraints*, Kluwer, Dordrecht, 1992.
[8] D.W. Knapp, *Behavioral Synthesis: Digital System Design using the Synopsys Behavioral Compiler*, Prentice-Hall, Englewood Cliff, NJ, 1996.
[9] A.A. Jerraya, H. Ding, P. Kission, and M. Rahmouni, *Behavioral Synthesis and Component Reuse with VHDL*, Kluwer, Dordrecht, 1997.
[10] T.J. Kowalski and D.E. Thomas, The VLSI design automation assistant: what's in a knowledge base, *Design Automation Conference*, 1985.
[11] C.J. Tseng and D.P. Siewiorek, Automated synthesis of data paths in digital systems, *IEEE Trans. Comput-Aided Des., Integrated Circuits and Syst.*, 5, 379–395, 1986.
[12] P. Marwedel, A new synthesis for the MIMOLA software system, *Design Automation Conference*, 1986.
[13] H. Trickey, Flamel: a high-level hardware compiler, *IEEE Trans. Comput-Aided Des., Integrated Circuits and Syst.*, 6, 259–269, 1987.
[14] E. Girczyc, Automatic Generation of Micro-sequenced Data Paths to Realize ADA Circuit Descriptions, PhD thesis, Carleton University, 1984.
[15] P.G. Paulin and J.P. Knight, Force-directed scheduling for the behavioral synthesis of ASIC's, *IEEE Trans. Comput. Aided Des.*, 8, 661–678, 1989.
[16] C.Y. Hitchcock and D.E. Thomas, A method of automatic data path synthesis, *Design Automation Conference*, 1983.
[17] H. De Man, J. Rabaey, P. Six, and L. Claesen, Cathedral-II: A silicon compiler for digital signal processing, *IEEE Des. Test Mag.*, 13, 13–25, 1986.
[18] B.M. Pangrle and D.D. Gajski, Slicer: A state synthesizer for intelligent silicon compilation, ICCAD'86, Santaclara, CA, 1986.
[19] I.-C. Park and C.-M. Kyung, Fast and near optimal scheduling in automatic data path synthesis, *Design Automation Conference*, 1991.
[20] A.C. Parker, J. Pizarro, and M. Mlinar, MAHA: A program for datapath synthesis, *Design Automation Conference*, 1986.
[21] P.G. Paulin and J.P. Knight, Scheduling and binding algorithms for high-level synthesis, *Proceedings of the 26th ACM/IEEE Conference on Design automation*, 1989, pp. 1–6.
[22] L. Stok and W.J.M. Philipsen, Module allocation and comparability graphs, *IEEE International Symposium on Circuits and Systems*, 5, June 1991, pp 2862–2865.
[23] A. Mujumdar, R. Jain, and K. Saluja, Incorporating performance and testability constraints during binding in high-level synthesis, *IEEE Trans. Comp. Aided Des.*, 15, 1212–1225, 1996.

[24] C.T. Hwang, T.H. Lee, and Y.C. Hsu, A formal approach to the scheduling problem in high level synthesis, *IEEE Trans. Comput. Aided Des.*, 10, 464–475, 1991.

[25] C.H. Gebotys and M.I. Elmasry, Optimal synthesis of high-performance architectures, *IEEE J. Solid State Circuits*, 27, 389–397, 1992.

[26] B. Landwehr, P. Marwedel, and R. Doemer, Oscar: Optimum simultaneous scheduling, allocation and resource binding based on integer programming, *European Design Automation Conference*, 1994.

[27] T.C. Wilson, N. Mukherjee, M.K. Garg, and D. K. Banerji, An ILP solution for optimum scheduling, module and register allocation, and operation binding in datapath synthesis, *VLSI Des.*, 3, No.1, 21–36, 1995.

[28] A. Safir and B. Zavidovique, Towards a global solution to high level synthesis problems, *European Design Automation Conference*, March 1990, pp. 283–288.

[29] P. Kollig and B.M. Al-Hashimi, Simultaneous scheduling, allocation and binding in high level synthesis, *Electron. Lett.*, 33, 1516–1518, 1997.

[30] N. Park and A. Parker, Sehwa: A software package for synthesis of pipelines from behavioral specifications, *IEEE Trans. Comput. Aided Des.*, 7, 356–370, 1988.

[31] E. Girczyc, Loop winding — a data flow approach to functional pipelining, *International Symposium of Circuits and Systems*, 1987.

[32] L.-F. Chao, A.S. LaPaugh, and E.H.-M. Sha, Rotation scheduling: A loop pipelining algorithm, *Design Automation Conference*, 1993.

[33] M. Potkonjak and J. Rabaey, Optimizing resource utlization using tranformations, *IEEE Trans. Comput. Aided Des.*, 13, 277–293, 1994.

[34] R. Walker and D. Thomas, Behavioral transformation for algorithmic level IC design, *IEEE Trans. Comput. Aided Des.*, 8, 1115–1128, 1989.

[35] Z. Iqbal, M. Potkonjak, S. Dey, and A. Parker, Critical path optimization using retiming and algebraic speed-up, *Design Automation Conference*, 1993.

[36] D.E. Thomas, E.M. Dirkes, R.A. Walker, J.V. Rajan, J.A. Nestor, and R.L. Blackburn, The system architect's workbench, *Design Automation Conference*, 1988.

[37] M.C. McFarland, The Value Trace: A Database for Automated Digital Design, Technical Report, Carnegie-Mellon University, 1978.

[38] R. Hartley and A.E. Casavant, Tree-height minimization in pipelined architectures, *International Conference on Computer-Aided Design*, November 1989, pp. 112–115.

[39] A. Nicolau and R. Potasman, Incremental tree height reduction for high level synthesis, *Design Automation Conference*, June 1991, pp. 770–774.

[40] R. Camposano, Path-based scheduling for synthesis, *IEEE Trans. Comput. Aided Des.*, 10, 85–93, 1991.

[41] R.A. Bergamaschi, S. Raje, and L. Trevillyan, Control-flow versus data-flow-based scheduling: combining both approaches in an adaptive scheduling system, *IEEE Trans. VLSI Syst.*, 5, 82–100, 1997.

[42] S.H. Huang, Y.L. Jeang, C.T. Hwang, Y.C. Hsu, and J.F. Wang, A tree-based scheduling algorithm for control dominated circuits, *Proceedings of the 30th international Conference on Design Automation*, 1993, pp. 578–582.

[43] W. Wolf, A. Takach, C.-Y. Huang, R. Manno, and E. Wu, The Princeton University behavioral synthesis system, *Design Automation Conference*, 1992, pp. 182–187.

[44] K. Wakabayashi and T. Yoshimura, A resource sharing and control synthesis method for conditional branches, *ICCAD-89*, Nov 1989, pp. 62–65.

[45] K. Wakabayashi and H. Tanaka, Global scheduling independent of control dependencies based on condition vectors, *Design Automation Conference*, 1992, pp. 112–115.

[46] K. Wakabayashi, C-based synthesis experiences with a behavior synthesizer, "Cyber", *Design, Automation and Test in Europe*, 1999, pp. 390–393.

[47] I. Radivojevic and F. Brewer, A new symbolic technique for control-dependent scheduling, *IEEE Trans. Comput. Aided Des.*, 15, 45–57, 1996.

[48] S. Haynal, Automata-Based Symbolic Scheduling, PhD thesis, University of California, Santa Barbara, 2000.

[49] G. Lakshminarayana, A. Raghunathan, and N.K. Jha, Incorporating speculative execution into scheduling of control-flow intensive behavioral descriptions, *Design Automation Conference*, 1998, pp. 108–113.

[50] T. Kim, N. Yonezawa, J.W.S. Liu, and C.L. Liu, A scheduling algorithm for conditional resource sharing — a hierarchical reduction approach, *IEEE Trans. Comput. Aided Des.*, 13, 425–438, 1994.

[51] L.C.V. dos Santos and J.A.G. Jess, A reordering technique for efficient code motion, *Design Automation Conference*, 1999, pp. 296–299.

[52] L.C.V. dos Santos, A method to control compensation code during global scheduling, *Workshop on Circuits, Systems and Signal Processing*, 1997.

[53] L.C.V. dos Santos, Exploiting Instruction-Level Parallelism: A Constructive Approach, PhD thesis, Eindhoven University of Technology, 1998.

[54] M. Rim, Y. Fann, and R. Jain, Global scheduling with code-motions for high-level synthesis applications, *IEEE Trans. VLSI Syst.*, 3, 379–392, 1995.

[55] P. Paulin, J. Frehel, M. Harrand, E. Berrebi, C. Liem, F Nacabal, and J.-C. Herluison, High-level synthesis and codesign methods: an application to a videophone codec, *European Design Automation Conference with EURO-VHDL*, 1995, pp. 444–451.

[56] P. Kission, H. Ding, and A.A. Jerraya, Structured design methodology for high-level design, *Design Automation Conference*, 1994.

[57] G. De Micheli, D.C. Ku, F. Mailhot, and T. Truong, The Olympus synthesis system, *IEEE Des. Test Comput.*, 7, 37–53, 1990.

[58] D. Ku and G. De Micheli, HardwareC — A Language for Hardware Design, Technical Report CSL-TR-90-419, Stanford University, 1988.

[59] D. Ku and G. De Micheli, Relative scheduling under timing constraints, *Design Automation Conference*, 1990, pp. 59–64.

[60] J. Li and R.K. Gupta, HDL optimizations using timed decision tables, *Design Automation Conference*, 1996.

[61] O. Penalba, J.M. Mendias, and R. Hermida, Maximizing conditional reuse by pre-synthesis transformations, *Design, Automation and Test in Europe*, 2002, p. 1097.

[62] J. Li and R.K. Gupta, Decomposition of timed decision tables and its use in presynthesis optimizations, *International Conference on Computer Aided Design*, 1997.

[63] A. Kountouris and C. Wolinski, High level pre-synthesis optimization steps using hierarchical conditional dependency graphs, *Euromicro Conference*, Vol. 1, 1999, pp. 290–294.

[64] S.S. Muchnick, *Advanced Compiler Design and Implementation*, Morgan Kaufmann, San Francisco, 1997.

[65] M. Potkonjak, M.B. Srivastava, and A. Chandrakasan, Multiple constant multiplications: Efficient and versatile framework and algorithms for exploring common subexpression elimination, *IEEE Trans. Comput Aided Des.*, 15, 151–165, 1996.

[66] R. Pasko, P. Schaumont, V. Derudder, S. Vernalde, and D. Durackova, A new algorithm for elimination of common subexpressions, *IEEE Trans. Comput. Aided Des.*, 18, 58–68, 1999.

[67] M. Janssen, F. Catthoor, and H. De Man, A specification invariant technique for operation cost minimisation in flow-graphs, *International Symposium on High-level Synthesis*, 1994, pp. 146–151.

[68] M. Miranda, F. Catthoor, M. Janssen, and H. De Man, High-level address optimisation and synthesis techniques for data-transfer intensive applications, *IEEE Trans. VLSI Syst.*, 6, 677–686, 1998.

[69] D.A. Lobo and B.M. Pangrle, Redundant operator creation: A scheduling optimization technique, *Design Automation Conference*, 1991.

[70] M. Potkonjak and J. Rabaey, Maximally fast and arbitrarily fast implementation of linear computations, *International Conference on CAD*, 1992, pp. 304–308.

[71] R. Kennedy, S. Chan, S.-M. Liu, R. Io, P. Tu, and F. Chow, Partial redundancy elimination in SSA form, *ACM Trans. Progr. Languages Syst.*, 1999.

[72] S. Gupta, M. Miranda, F. Catthoor, and R. Gupta, Analysis of high-level address code transformations for programmable processors, *Design, Automation and Test in Europe*, 2000, pp. 9–13.

[73] S. Novack and A. Nicolau, Mutation scheduling: A unified approach to compiling for fine-grain parallelism, in *Languages and Compilers for Parallel Computing*, Springer-Verlag, London, UK, 1994, 16–30.

[74] J. Fisher, Trace scheduling: A technique for global microcode compaction, *IEEE Trans. Comput.*, 1981.

[75] K. Ebcioglu and A. Nicolau, A global resource-constrained parallelization technique, *3rd International Conference on Supercomputing*, 1989.

[76] A. Nicolau, A Development Environment For Scientific Parallel Programs, Technical Report, Cornell University, 1985.

[77] S. Gupta, N. Savoiu, N.D. Dutt, R.K. Gupta, and A. Nicolau, Conditional speculation and its effects on performance and area for high-level synthesis, *International Symposium on System Synthesis*, 2001, pp. 171–176.

[78] W.W. Hwu et al., The superblock: An effective technique for VLIW and superscalar compilation, *J. Supercomput.*, 7, 1993, pp. 229–248.

[79] R. Gupta and M.L. Soffa, Region scheduling: An approach for detecting and redistributing parallelism, *IEEE Trans. Software Engineering*, 421–431, 1990.

[80] A. Nicolau and S. Novack, Trailblazing: A hierarchical approach to Percolation Scheduling, *International Conference on Parallel Processing*, 1993.

[81] D.W. Wall, Limits of instruction-level parallelism, *International Conference on Architectural Support for Programming Languages and Operating System (ASPLOS)*, 1991.

[82] M.S. Lam and R.P. Wilson, Limits of control flow on parallelism, *International Symposium on Computer Architecture*, 1992, pp. 46–57.

[83] A. Aiken and A. Nicolau, A development environment for horizontal microcode, *IEEE Trans. Software Eng.*, 14, 584–594, 1988.

[84] A. Nicolau, Uniform parallelism exploitation in ordinary programs, *International Conference on Parallel Processing*, 1985.

[85] M. Girkar and C.D. Polychronopoulos, Automatic extraction of functional parallelism from ordinary programs, *IEEE Trans. Parallel Distribut. Syst.*, 3, 166–178, 1992.

[86] R. Potasman, J. Lis, A. Nicolau, and D. Gajski, Percolation based synthesis, *Design Automation Conference*, 1990, pp. 444–449.

[87] U. Holtmann and R. Ernst, Combining MBP-speculative computation and loop pipelining in high-level synthesis, *European Design and Test Conference*, 1995, pp. 550–556.

[88] Forte Design Systems — http://www.forteds.com, Behavioral Design Suite.

[89] Synfora Inc. — http://www.synfora.com/, PICO Express.

[90] R. Schreiber, S. Aditya, S. Mahlke, V. Kathail, B.R. Rau, D. Cronquist, and M. Sivaraman, Pico-npa: High-Level Synthesis of Nonprogrammable Hardware Accelerators, Technical Report HPL-2001-249, HP Labs, October 2001.

[91] Mentor Graphics — http://www.mentor.com/, Catapult C Synthesis.

[92] Celoxica Limited — http://www.celoxica.com, Agility Compiler — Advanced Synthesis Technology For SystemC.

[93] BlueSpec Inc. — http://www.bluespec.com, BlueSpec Compiler (BSC).

[94] J.C. Hoe and A. Arvind, Synthesis of operation-centric hardware descriptions, *IEEE/ACM Intl. Conf. on Computer Aided Design (ICCAD)*, 511–518, 2000.

[95] Tensilica Inc. — http://www.tensilica.com, XPRES Compiler: Xtensa PRocessor Extension Synthesis.

[96] D. Goodwin and D. Petkov, Automatic generation of application specific processors, *International Conference on Compilers, Architectures and Synthesis for Embedded Systems (CASES)*, 2003.

[97] R.A. Bergamaschi, Behavioral network graph unifying the domains of high-level and logic synthesis, *Design Automation Conference*, 1999, pp. 213–218.

[98] SPARK parallelizing high-level synthesis framework website, http://mesl.ucsd.edu/ spark.

[99] D.F. Bacon, S.L. Graham, and O.J. Sharp, Compiler transformations for high-performance computing, *ACM Comput. Surv.*, 26, 345–420, 1994.

[100] S. Gupta, N. Savoiu, S. Kim, N.D. Dutt, R.K. Gupta, and A. Nicolau, Speculation techniques for high level synthesis of control intensive designs, Design Automation Conference, 2001.

[101] S. Gupta, N.D. Dutt, R.K. Gupta, and A. Nicolau, Dynamically increasing the scope of code motions during the high-level synthesis of digital circuits, *Invited Paper in Special Issue of IEE Proc.-Comput. and Digital Tech.: Best of DATE 2003*, 150, 330–337, 2003.

[102] S. Gupta, M. Reshadi, N. Savoiu, N.D. Dutt, R.K. Gupta, and A. Nicolau, Dynamic common subexpression elimination during scheduling in high-level synthesis, *International Symposium on System Synthesis*, 2002, pp. 261–266.

[103] S. Gupta, R.K. Gupta, N.D. Dutt, and A. Nicolau, Coordinated parallelizing compiler optimizations and high-level synthesis, *ACM Trans. Design Autom. Electron. Syst.*, 9, 441–470, 2004.

[104] Edison Design Group (EDG) compiler frontends, http://www.edg.com.

[105] S. Gupta, N.D. Dutt, R.K. Gupta, and A. Nicolau, SPARK: A high-level synthesis framework for applying parallelizing compiler transformations, *International Conference on VLSI Design*, 2003, pp. 461–466.

SECTION III

MICRO-ARCHITECTURE DESIGN

12

Cycle-Accurate System-Level Modeling and Performance Evaluation

Marcello Coppola
STMicroelectronics
Grenoble, France

Miltos D. Grammatikakis
ISD S.A.
Athens, Greece

12.1 Introduction

The number of transistors on a chip grows exponentially, pushing technology toward highly integrated systems-on-a-chip (SoC) [1]. Existing design tools fail to exploit all the possibilities offered by this technological leap, and shorter than ever time-to-market trends drive the need for innovative design methodology and tools. It is expected that a productivity leap can only be achieved by focusing on higher levels of abstraction, enabling optimization of the top part of the design where important algorithmic and architectural decisions are made, and massive reuse of predesigned system and block components.

Detailed VHDL or Verilog models are inadequate for system-level description due to poor simulation performance. Advanced system-level modeling may lead from a high-level system model derived from initial specifications, through successive functional decomposition and refinement, to implementing an optimized, functionally correct, unambiguous protocol, i.e., without deadlock conditions or race hazards. A popular open-source C++ system-level modeling and simulation library is OSCI's SystemC [2]. SystemC consists of a collection of C++ classes describing mainly hardware concepts and a simulation kernel implementing the runtime semantics. It provides all basic concepts used by HDLs, such as fixed-point data types, modules, ports, signals, time, and abstract concepts such as interfaces, communication channels, and events. SystemC 2.0 allows the development and exchange of fast system models, providing seamless integration of tools from a variety of vendors [3]. A detailed system-level modeling framework can rely on a SystemC-based C++ Intellectual Property (IP) modeling library, a powerful simulation engine, a runtime and test environment, and refinement methodology. Two commercial tools for domain-independent SystemC-based design have been designed previously.

- In September 1998, the Felix initiative of the Alta Group of Cadence Design Systems delivered the Cierto Virtual Component Co-design tool suite (VCC 2.0) [4] VCC supports both top-down design based on communication refinement, and bottom-up reuse-oriented design. When a design is fully refined and its performance meets system requirements, the designer may export an implementation. Hardware elements are ready for HDL simulation, floor planning, and logic synthesis, while software models are mapped to processor-specific code for linking to a selected real-time operating system (RTOS) configured with the appropriate interrupt handlers, counter/timer initialization, and static schedulers. Cadence claims that only VCC permits a consistent level of design abstraction, even after detailed refinement. In addition, VCC generates complete test benches for verifying implementations using popular coverification tools. These tools combine a simulation of the hardware description, with software running on a processor simulator. Today, Cadence has withdrawn VCC from the market.
- In June 2001, Synopsys formally announced the Cocentric System Studio [5] and Cocentric SystemC Compiler [6]. Today, the first one is called just System Studio and the second has been withdrawn by Synopsys from the market. The tool is a design entry and modeling environment that utilizes block and state diagramming techniques using graphical abstractions to represent concurrency, reactive control, and dynamic switching concepts. System Studio allows the user to execute models either on a time- or an event-driven simulation engine. The Cocentric SystemC Compiler integrates behavioral and RTL synthesis engines to transform SystemC descriptions into gate-level netlists, including support for resource sharing, multicycle/pipelined operations, FSM generation, and a "MemWrap" graphical interface for mapping to specific types of memories. In addition, "Bcview" provides a graphical cross-linking capability to source code.

Despite current system design efforts, there is not yet a complete and efficient SystemC-based development environment capable of:

- providing parameterized architecture libraries, synthesis, and compilation tools for fast user-defined creation and integration of concise, precise, and consistent SystemC models;
- enabling IP reuse at the system and block level, which results in significant savings in project costs, time scale, and design risk [7,8]; and
- supporting efficient design space exploration.

In this chapter, we focus on general methodologies for cycle-accurate system-level modeling and performance evaluation. More specifically, in Section 12.2, we consider system modeling, design flow, and design space exploration. We first introduce abstraction levels, and outline general system-level design methodology, including *top-down and bottom-up protocol refinement, hierarchical modeling* (i.e., decomposing system functionality into subordinate modules [blocks and subblocks]), *orthogonalization of concerns* (i.e., separating module specification from architecture, and behavior from communication), and *communication protocol layering*. Then, we examine design space exploration, a key process to enhancing design quality.

In Section 12.3, we outline SystemC-based system-level modeling objects, such as module, clock, intramodule memory, intramodule synchronization, and inter-module communication channels. Then we discuss, with examples, back-annotation issues, especially back-annotating time for transactional clock-accurate modeling. We also examine a more complex example of back-annotating time on a user-defined SystemC-based hierarchical finite-state machine (`HFsm`). This object provides two descriptions of model behavior; using *states, conditions, actions, and delays* as in traditional (flat) finite-state machines, or *states and events*.

In Section 12.4, we focus on system-level modeling methodologies for collecting performance statistics from modeling objects, including automatic extraction of statistical properties. Our description focuses on the efficient, open-source on-chip communication network (OCCN) framework that includes an object-oriented C++ library built on top of SystemC [2]. Advanced high-level performance modeling environments may be based on advanced system-level monitoring activities. Thus, we examine the design of integrated system-level tools, *generating, processing, presenting, and disseminating* system monitoring information.

In Section 12.5, we discuss open system-level modeling issues related to asynchronous modeling, parallel and distributed system-level simulation, and interoperability with other design tools. By resolving these issues, SystemC-based system-level modeling could achieve a higher degree of productivity in contrast to other similar libraries currently available in the market.

12.2 System Modeling and Design Methodology

We examine fundamental concepts in system-level design, including abstraction level, system-level modeling methodology, and design exploration.

12.2.1 Abstraction Levels

A fundamental issue in system design is model creation. A model is a concrete representation of IP functionality. In contrast to component IP models, a *virtual platform prototype* refers to system modeling. Virtual platform enables integration, simulation, and validation of system functionality, reuse at various levels of abstraction, and design space exploration for various implementations and appropriate hardware/software partitioning. A virtual prototype consists of:

- *Hardware modules,* including peripheral IP block model (e.g., I/O, timers, Audio, Video code, or DMA), processor emulator via instruction set simulator (ISS) (e.g., ARM V4, PowerPC, ST20, or Stanford DLX), and communication network with internal or external memory (e.g., bus, crossbar, or network-on-chip).
- *System software,* including hardware dependent software, models of RTOS, device drivers, and middleware.
- *Environment simulation,* including application software, benchmarks, and stochastic models.

Notice that a virtual prototype may hide, modify, or omit system properties. As shown in Figure 12.1, abstraction levels in system modeling span multiple levels of accuracy ranging from functional to transactional cycle- and bit-accurate to gate level. Each level introduces new model details [9]. We now provide an intuitive description for the most used abstraction levels from the most abstract to the most specific.

- *Functional models* have no notion of resource sharing or time, i.e., functionality is executed instantaneously and the model may or may not be bit-accurate. This layer is suitable for system concept validation and functional partitioning between control and data. This includes definition of abstract data types, specification of hardware or software (possibly RTOS) computation, communication, and synchronization mechanisms, and algorithm integration to high-level simulation using Ada, Matlab, ODAE Solver, OPNET, SDL, SIMSCRIPT, SLAM, SystemC, or UML technology.
- *Transactional behavioral models* (denoted simply as transactional) are functional models mapped to a discrete-time domain. Transactions are atomic operations with their duration stochastically determined, i.e., a number of clock cycles in a synchronous model. Although general transactions on buses cannot be modeled, transactional models are fit for modeling pipelining, RTOS introduction, basic communication protocols, test-bench realization, and preliminary performance estimation.
- *Transactional clock-accurate models* (denoted transactional CA) map transactions to a clock. Thus, synchronous protocols, wire delays, and device access times can be accurately modeled. Using

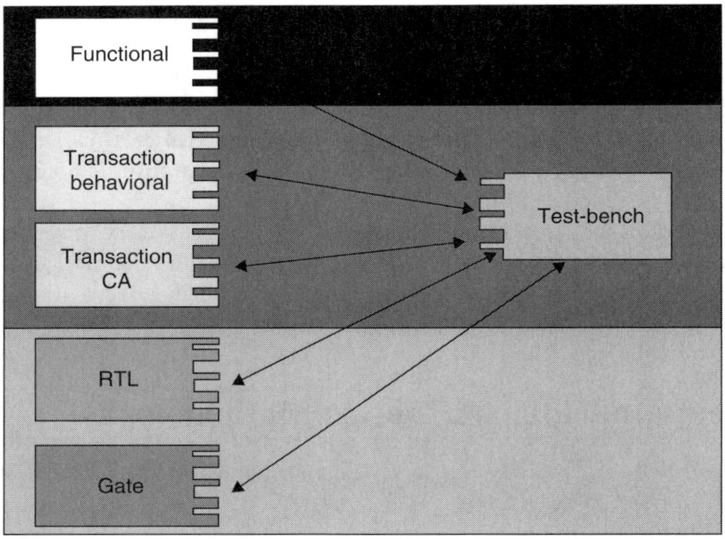

FIGURE 12.1 IP modeling in various abstraction levels.

discrete-event systems, this layer allows for simple, generic, and efficient cycle-accurate performance modeling of abstract processor core wrappers (called bus functional models), bus protocols, signal interfaces, peripheral IP blocks, instruction set simulators, and test benches. Time delays are usually back-annotated from RTL models, since transactional CA models are not always synthesizable.

- *RTL models* correspond to the abstraction level from which synthesis tools can generate gate-level descriptions (or netlists). Register-transfer-level systems are usually visualized as having two components: data and control. The data part consists of registers, operators, and data paths, while the control part provides the time sequence of signals that evoke activities in the data part. Data types are bit-accurate, interfaces are pin-accurate, and register transfer is time-accurate. Propagation delay is usually back-annotated from gate models.
- *Gate models* are described in terms of primitives, such as boolean logic with timing data and layout configuration. For simulation reasons, gate models may be internally mapped to a continuous-time domain, including currents, voltages, noise, clock rise, and fall times. Storage and operators are broken down into logic implementing the digital functions corresponding to these operators, while timing for individual signal paths can be obtained. Thus, according to the above, an embedded physical SRAM memory model may be defined as:
  - a functional model in a high-level programming language such as Ada, C, C++, or Java;
  - a transactional CA model, allowing validation of its integration with other components;
  - implementation-independent RTL logic described in VHDL or Verilog;
  - a vendor gate library described using NAND, flip-flop schematics; or,
  - a detailed and fully characterized mask layout at the physical level, depicting rectangles on chip layers and geometrical arrangement of I/O and power locations.

12.2.2 General System Modeling Methodology

Optimal system modeling methodology is a combination of stepwise protocol refinement, hierarchical modeling, orthogonalization of concerns, and communication layering techniques. Stepwise protocol refinement is achieved through a top-down and bottom-up approach.

- In *bottom-up refinement*, IP-reuse-oriented integration with optimal evaluation, composition, and deployment of prefabricated or predesigned IP block and system components drive the process.

- In *top-down stepwise refinement,* emphasis is placed on specifying unambiguous semantics, capturing desired system requirements, optimal partitioning of system behavior into simpler behaviors, and gradually refining the abstraction level down to a concrete low-level architecture by adding details and constraints in a narrower context, while preserving desired properties. Top-down refinement allows the designer to explore modeling at different level of abstractions, thus trading model accuracy with simulation speed. This continuous re-arrangement of existing IP in ever-new composites is a key process to new marketing ideas. It also allows for extensive and systematic reuse of design knowledge and application of formal correctness techniques. The concrete architecture obtained through formal refinement must satisfy:
- *relative correctness*, i.e., it must logically imply the abstract architecture, and
- *faithfulness*, i.e., no new properties could be derived.

While formal refinement is hard to achieve since there are no general proof techniques, relative refinement is based on *refinement patterns* consisting of a pair of architectural schemas that are relatively correct with respect to a given mapping. By applying refinement patterns, e.g., a general optimized state transformation [10], we can systematically and incrementally transform an abstract architecture to an equivalent lower-level form. Notice that formal architectural transformation is related to providing an augmented calculus with annotations for properties such as correctness, reliability, and performance [11]. If architectural components eventually become explicit formal semantic entities, then architectural compatibility can be checked in a similar way to type-checking in a high-level programming language [12].

Hierarchy is a fundamental relation for modeling conceptual and physical processes in complex systems of organized complexity based either on a top-down analytical approach, i.e., the divide and conquer paradigm, or a bottom-up synthesis approach, i.e., the design-reuse paradigm. Hierarchical modeling based on simpler, subordinate models controlled by high-level system models enables a better encapsulation of the design's unique properties, resulting in efficient system design with simple, clean, pure, self-contained, and efficient modular interfaces. For example, hierarchical modeling of control, memory, network, and test bench enables the system model to become transparent to changes in sub-block behavior and communication components. A systematic framework supporting hierarchical modeling may be based on the SystemC module object, and the inheritance and composition features of the C++ language [13,14]. Thus, SystemC modules consist of SystemC submodules in nested structures. Moreover, SystemC facilitates the description of hierarchical systems by supporting module class hierarchy very efficiently since simulation of large, complex hierarchical models imposes *no* impact on performance over the corresponding nonhierarchical models.

Stepwise transformation of a high-level behavioral model of an embedded system into actual implementation can be based on the concept of *orthogonalization of concerns* [15]:

- Separation of function specification from final architecture, i.e.,, *what* are the basic system functions for system-level modeling vs. how the system organizes hardware and software resources in order to implement these functions. This concept enables system partitioning and hardware/software codesign by focusing at progressively lower levels of abstraction.
- Separation of communication from computation (called behavior) enables plug-and-play system design using communication blocks that encapsulate and protect IP cores. This concept reduces ambiguity among designers, and enables efficient design space exploration.

Layering simplifies individual component design by defining functional entities at various abstraction levels and implementing protocols to perform each entity's task. Advanced inter-module communication refinement may be based on establishing distinct communication layers,thus, greatly simplifying the design and maintenance of communication systems [16,17]. We usually adopt two layers, namely:

- The *communication layer* which provides a generic message API, abstracting away the fact that there may be a point-to-point channel, a bus, or an arbitrarily complex network-on-chip.
- The *driver layer* which builds the necessary communication channel adaptation for describing high-level protocol functionality, including compatible protocol syntax (packet structure) and semantics (temporal ordering).

12.2.3 Design Space Exploration

In order to evaluate the vast number of complex architectural and technological alternatives, the architect must be equipped with a highly parameterized, user-friendly, and flexible design space exploration methodology. This methodology is used to construct an initial implementation from system requirements, mapping modules to appropriate system resources. This solution is subsequently refined through an iterative-improvement strategy based on reliability, power, performance, and resource contention metrics obtained through domain- or application-specific performance evaluation using *stochastic analysis models and tools,* and *benchmarks* for common applications, e.g., commercial networking or multimedia. The proposed solution provides values for all system parameters, including configuration options for sophisticated multiprocessing, multithreading, pre-fetching, and cache hierarchy components.

After generating an optimized mapping of behavior onto architecture, the designer may either manually decompose hardware components to the RTL of abstraction, or load system-level configuration parameters onto available behavioral synthesis tools, such as the Synopsys coreBuilder and coreConsultant. These tools integrate a pre-loaded library of configurable high-level (soft) IPs, e.g., STMicroelectronics' `STbus` [18,19], VSIA's `VCI` [20], OCP [9], and generic interface prototypes. The target is to parameterize IP in order to generate the most appropriate synthesis strategy automatically. Now, design flow can proceed normally with routing, placement, and optimization by interacting with tools, such as Physical Compiler, Chip Architect, and PrimeTime.

12.3 Back-Annotation of System-Level Modeling Objects

In this section, we consider back-annotation of system-level models, focusing especially on hardware modeling objects.

12.3.1 System-Level Modeling Objects

In this section, we examine a high-level modeling environment consisting of a SystemC-based C++ modeling library and associated simulation kernel. A proprietary runtime and test environment may be provided externally [21]. Thus, our high-level modeling environment provides user-defined C++ building blocks (macro functions) that simplify system-level modeling, enable concurrent design flow (including synthesis), and provide the means for efficient design space exploration and hardware/software partitioning.

High-level SystemC-based modeling involves hardware objects as well as system and application software components. For models of complex safety critical systems, software components may include calls to an RTOS model [22]. An RTOS model abstracts a real-time operating system by providing generic system calls for operating system management (RTOS kernel and multitask scheduling initialization), task management (fork, join, create, sleep, activate), event handling (wait, signal, notify), and real-time modeling. SystemC 3.0 is expected to include user-defined scheduling constructs on top of the core SystemC scheduler, providing RTOS features such as thread creation, interrupt, and abort. During software synthesis, an RTOS model may be replaced with a commercial RTOS.

Focusing on hardware components and on-chip communication components, we will describe system-level modeling objects (components) common in SystemC models. Most objects fall into two main categories: *active* and *passive modeling objects.* While active objects include at least one thread of execution, initiating actions that control system activity during the course of simulation, passive objects do not have this ability and only perform standard actions required by active objects. Notice that communication objects are neither active nor passive. Each hardware block is instantiated within a SystemC class called `sc_module`. This class is a container of a user-selected collection of SystemC clock domains, and active, passive, and communication modeling objects including:

- *SystemC and user-defined data types* providing low-level access, such as `Bits` and `Bytes`, allowing for bit-accurate, platform-independent modeling.

- *User-defined passive memory objects* such as `Register`, `FIFO`, `LIFO`, `circular FIFO`, `Memory`, `Cache` as well as user-defined collections of externally addressable hierarchical memory objects.
- *SystemC and user-defined intramodule communication and synchronization objects* based on message passing, such as `Monitor`, or concurrent shared memory such as `Mutex`, `Semaphore`, `Conditional Variables`, `Event Flag`, and `Mailbox`. This includes `Timer` and `Watchdog Timer` for generating periodic and nonperiodic time-out events.
- *Active control flow objects* such as SystemC processes (`SC_Method`, asynchronous `Thread`, and clocked `Thread`), and `HFsm`. Intramodule interaction including computation, communication, and synchronization is performed either by local processes or by remote control flow objects defined in other modules.
- *inter-module communication object* usually based on message passing, e.g., Message [23].
- *user-defined inter-module point-to-point and multi-point communication channels and corresponding interfaces*, e.g., peripheral, basic, and advanced VCI [20], Amba bus (`AHB`, `APB`) [24], ST Microelectronics' proprietary Request Acknowledge (`RA`), Request Valid (`RV`), and `Stbus` (type1, type2 and type3) [18,19]. These objects are built on top of SystemC in a straightforward manner. For example, OCCN provides a library of objects with the necessary semantics for modeling on-chip communication infrastructure [23].

12.3.2 Back-Annotating Time Delay in System Models

When high-level views and increased simulation speed are desirable, systems may be modeled at the transactional level using an appropriate abstract data type, e.g., processor core, video line, or network communication protocol-specific data structure. With transactional modeling, the designer is able to focus on IP functionality rather than on details of the physical interface, including data flows, FIFO sizes, or time constraints. For efficient design space exploration, transactional models (including application benchmarks) must be annotated with a number of architectural parameters.

- Realizability parameters refer to system design issues that control system concurrency such as peripheral, processor, memory, network interface, router, and their interactions, e.g., packetization, arbitration, packet transfer, or instruction execution delays; RTOS models, e.g., context switch or operation delays; and VLSI layout, e.g., time-area tradeoffs, clock speed, bisection bandwidth, power consumption models, pin count, or signal delay models.
- Serviceability parameters refer to reliability, availability, performability, and fault-recovery models for transient, intermittent, and permanent hardware/software errors. When repairs are feasible, fault recovery is usually based on detection (through checkpoints and diagnostics), isolation, rollback, and reconfiguration. While reliability, availability and fault-recovery processes are based on two-state component characterization (faulty or good), performability metrics evaluate degraded system operation in the presence of faults, e.g., increased latency due to packet congestion when there is limited loss of network connectivity.

Time delay is only one of many possible annotations. In order to provide accurate system performance measurements, e.g., power consumption, throughput rates, packet loss, latency statistics, or QoS requirements, all transactional models involving computation, communication, and synchronization components must be back-annotated with an abstract notion of time. Thus, timing characteristics, obtained either through hardware synthesis of cycle-accurate hardware models, or by profiling software models on an RTOS-based embedded processor with cycle-accurate delays associated to each operation in its instruction set, can be back-annotated to the initial TLM model for transactional cycle-accurate (CA) modeling. The cycle-accurate system architecture model can be used for system-level performance evaluation, design space exploration, and also as an executable specification for the hardware and software implementation.

However, architectural delays are not always cycle-accurate but cycle-approximate. For example, for computational components mapped to a particular RTOS, or large communication transactions mapped to a particular shared bus, it is difficult to estimate accurately thread delays, which depend on precise system configuration and load. Similarly, for deep submicron technology, wire delays that dominate protocol timings cannot be determined until layout time. Thus, in order to avoid system revalidation for deadlock and data race hazards, and ensure correct behavior independent of computation, communication, and synchronization delays, one must include all necessary synchronization points and interface logic in the transactional models. Analysis using parameter sweeps helps estimate sensitivity of system-level design due to perturbations in the architecture, and thus examine the possibility of adding new features in derivative products [1,25].

For most SystemC-based hardware objects, time annotation is performed statically, during instantiation time. Notice that time delays also occur due to the underlying protocol semantics. As an example, we now consider memory and communication channel objects; `HFsms` are discussed in the next section:

- Time is annotated during instantiation of memory objects by providing the number of clock cycles (or absolute time) for respective read/write memory operations. The timing semantics of the memory access protocol implemented within an active control flow object may contribute additional dynamic delays. For example, a *congestion delay* may be related to allowing a single-memory operation to proceed within a particular time interval, or a *large packet delay* may be due to transferring only a small amount of information per unit time.
- Time is annotated during instantiation of the processes connected to the communication channels through timing constraints implemented within an active control flow object, blocking calls to intramodule communication and synchronization objects, and SystemC `wait` or `wait_until` statements. Furthermore, blocking calls within inter-module communication objects instantiated within the SystemC channel object or adherence to specific protocol semantics may imply additional dynamic delays, e.g., compatibility in terms of the size of communicated tokens; this parameter may be annotated during communication channel instantiation.

12.3.3 An Example in Back-Annotation: Hierarchical Finite-State Machine (HFsm)

Finite-state machines are very important for modeling dynamic aspects of hardware systems. Most hardware systems are highly state-dependent, since their actions depend not only on their inputs, but also on what has previously happened. Thus, states represent a well-defined system status over an interval of time, e.g., a clock cycle. In cycle-accurate simulation, we assume that system events occur at the beginning (or the end) of a clock cycle. In this way, we can capture the right behavior semantics, and we can back-annotate hardware objects in a very detailed way.

Hierarchical finite-state models define complex, control-dominated systems as a set of smaller subsystems, by allowing the model developer to consider each subsystem in a modular way. Scoping is used for types, variable declaration, and procedure calls. Hierarchy is usually disguised into two forms, namely:

- Structural hierarchy which enables the realization of new components based on existing ones. A system is a set of interconnected components, existing at various abstraction levels.
- *Behavioral hierarchy* which enables system modeling using either *sequential*, e.g., procedure, or recursion, or *concurrent computation*, e.g., parallel or pipelined decomposition.

As asynchronous and clocked threads defined in SystemC, an `HFsm` modeled in C++/SystemC is a primary and powerful unit for capturing description of dynamic behavior within a SystemC model. `HFsm` may be used to capture formally the dynamic behavior of a hardware module, e.g., a circuit representing an embedded system controller or a network protocol. For simple control tasks, asynchronous or clocked threads in SystemC and `HFsm` implementations can be made equivalent. Although clocked threads are usually more efficient from the simulation time point of view, an `HFsm` implementation is more general, and improves clarity and efficiency when specifying complex control tasks. It also allows one to back-annotate system-level models in a more detailed way, thus making cycle accuracy easier to obtain. Since outer `HFsm` states are typically behavioral generalizations of inner ones, `HFsm` facilitates the implementation of policy mechanisms such as signaling exceptions to a higher scope and pre-emptive events. It also avoids pitfalls in implementations of traditional (flat) finite-state machine implementations, such as:

- Exponential explosion in the number of states or transitions when composing sub-states, hence the system-level model is characterized using less redundant timing information.
- Difficulty in modeling complex hierarchical control flow, with transitions spanning different levels of the behavioral hierarchy.
- Difficulty in handling hierarchical group transitions, or global control signals specified for a set of states, e.g., reset and halt.
- No support for concurrent, multithreaded computation.

HFsm are based on Harel statecharts [26] and at least 20 variants [27].

12.3.3.1 Flat `HFsm` Description Using States, Conditions, Actions, and Delays

With a simple, commonly used user–interface description, a flat `HFsm` is represented using states, conditions, delays, and actions. States are used to identify `HFsm` status, conditions are used to drive actions, and delay times are used to postpone actions. Composite transitions are defined as a quadruple formed by the next state, a condition, an action, and a delay. A flat `HFsm` represented graphically is shown in Figure 12.2.

An equivalent representation can be given in tabular form.

- State `S1`={{`C1, A1, t1, S2`}, {NULL}};
- State `S2`={{`C3, A3, t3, S3`}, {NULL}};
- State `S3`={{`C4, A4, t4, S1`}, {`C3, A3, t5, S3`}, {NULL}};

Let us consider the state `S3`. In this state we have two transitions. In the first one, `S1` is the next state, `C4` is the condition, `A4` is the action, and `t4` is a constant delay. The semantics of this transition are as follows. If the `HFsm` is in state `S3` and condition `C4` is true, then `HFsm` performs action *`A4`* and transfers control to state `S1` after a delay of *`t4`* clock cycles. Thus, `HFsm` designer could re-define this order, e.g., condition-action-delay.

Furthermore, if `HFsm` conditions were *not* allowed to make blocking function calls, then the delay associated in performing the corresponding action would always be *deterministic* and could possibly be estimated by the model designer. *Dynamic delays*, e.g., delays due to concurrent access to shared resources, could be handled at the beginning of the action routine by placing special blocking function calls to intramodule communication and synchronization objects.

12.3.3.2 `HFsm` Description using States and Events

For more complex examples, hierarchical states may have "parents" or "children", placed at different levels. This approach is common with machines providing various operating modes, i.e., machine features

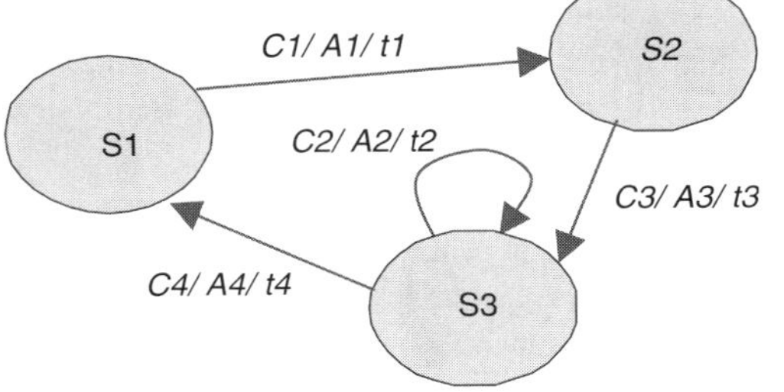

FIGURE 12.2 Graphical illustration of a flat, clocked finite-state machine.

that define multiple, alternative configurations. Then, `Hfsm` is described optimally using discrete-event systems, i.e., a finite set of states, a finite set of event categories, start-up and regular state transitions mapping state-event pairs to other states, and actions associated with transitions. This description is natural, abstract, powerful, and efficient.

- *Initial start-up transitions* originate during initial transitions from a `top` superstate. These transitions invoke entry and start-up actions (e.g., setting state parameters) for the state associated with the transition as well as for intermediate superstates, if a transition crosses two or more levels of hierarchy.
- *Regular transitions* occur at the current state during normal `Hfsm` processing. They are triggered by events selected either randomly or through external stimula from a user-defined enumerated event list. These transitions cause entry (and exit) actions upon entering (respectively, exiting) an `HFsm` state. Depending on the position in the state hierarchy of the new vs. the current state, entry and exit actions are executed in order, either from the least deeply nested to the most deeply nested state, or *vice versa*. Entry and exit actions always precede processing actions possibly associated with the newly entered hierarchical state.

With this description, a SystemC `wait` or `wait_until` statement can be associated with each state event. Moreover, blocking calls to intramodule communication and synchronization objects may account for additional dynamic delays.

12.4 Automatic Extraction of Statistical Features

System-level modeling is an essential ingredient of system design flow. Data and control flow abstraction of the system hardware and software components express not only functionality, but also performance characteristics that are necessary to identify system bottlenecks. For example, virtual channels are used to avoid both network and protocol deadlock, and also to improve performance, and provide quality of service. While for software components it is usually the responsibility of the user to provide appropriate performance measurements, for hardware components and interfaces it is necessary to provide a statistical package that hides internal access to the modeling objects. This statistical package can be used for evaluating system performance characteristics. The statistical data may be analyzed using visualization software, e.g., the open source Grace tool or dumped to a file for subsequent data processing, e.g., via an electronic spreadsheet or a specialized text editor.

12.4.1 Statistical Classes for System-Level Modeling

Considering the hardware system modeling objects previously proposed, we observe that dynamic performance characteristics, e.g., latency, throughput, packet loss, resource utilization, and possibly power consumption (switching activity) are definitely required from:

- *inter-module communication and synchronization objects* (`Message`) reflecting communication channel performance metrics for throughput, latency, or packet loss; and
- *intramodule passive memory objects* (`Register`, `FIFO`, `LIFO`, circular `FIFO`, `Memory`, `Cache`) reflecting memory performance metrics for throughput, latency, packet loss, buffer size, and hit ratio. Although similar metrics for certain joint intramodule communication and synchronization objects, e.g., `Mailboxes` are possible, these objects may be implemented in hardware in various ways, e.g., using control logic and static memory.

Assuming a common 2-D graphic representation, a statistical API for the above metrics can be based on a function `enable_stat(args)` enabling the data-capturing activity. Its arguments specify a distinct name of the modeling object, the absolute start and end time for statistics collection, the title and legends for the x and y axes, the time window for windowed statistics, i.e., the number of consecutive points averaged for generating a single statistical point, and a boolean flag for stopping or restarting statistics during simulation.

Since complex systems involve both time- (instant) and event-driven (duration) statistics, we may provide two general monitoring classes, collecting instant and duration measurements from system components with the following functionality.

In *time-driven simulation*, signals usually have instantaneous values. During simulation, these values can be captured at a specific time by calling the following function:

- `stat_write (double time, double value).`

In *event-driven simulation*, recorded statistics for events must include arrival and departure time (or duration). Since the departure time is known later, the interface must be based on two functions `stat_event_start` and `stat_event_end`. Thus, first the user invokes an operation

- `a = int stat_event_start(double arrival_time)`

to record the arrival time and save the unique location of the event within the internal table of values in a local variable `a`. Then, when the event's departure time is known, this time is recorded within the internal table of values at the correct location, by calling the `stat_event_end` function with the appropriate departure time.

- `void stat_event_end(double departure_time, a).`

The `stat_event_start/end` operations may take into account memory addresses in order to compute duration statistics for consecutive operations, e.g., consecutive read/write accesses (or enqueue/dequeue) operations corresponding to the same memory block (or packet) for a `Register`, `FIFO`, `Memory`, `Cache`, or inter-module communication object, i.e., `Message`. Thus, a modified `StatDurationLib` class performs the basic `stat_event_start/end` operations using (transparently to the user) internal library object pointers. Then, upon an event arrival, we invoke the command

- `void stat_event_start(long int MAddr, double arrival_time)`

in order to record the following in the next available location within the internal table:

- the current memory address (`MAddr`),
- the arrival time (`arrival_time`), and
- an undefined (-1) departure time.

Then, upon an event departure, we invoke the command

- `void stat_event_end(long int MAddr, double departure_time)`

to search for the current `MAddr` in the internal table of values, and update the entry corresponding to the formerly undefined, but now defined departure time.

For duration statistics in 2-D graphic form, the y-axis point may indicate time of occurrence of a read operation performed on the same address as a previous cache write operation whose time of occurrence is shown at the corresponding point on the x-axis. Event duration, e.g., latency for data access from the same memory block can be obtained by subtracting these two values. For example, using `Grace`™ this computation can be performed very efficiently using the `Edit data sets: create_new (using Formula)` option. Furthermore, notice that the write-read mapping is one to one, i.e., it is first written and then it is read, while a *reverse access*, i.e., read before write, causes an error.

As explained before, statistics collection may either be performed by the user, or directly by the hardware modeling objects, e.g., bus channels, such as `APB`, `AHB`, or `StBus`, using library-internal object pointers. In the latter case, software probes are inserted into the *source code of library routines*, either manually by setting sensors and actuators, or more efficiently through the use of a monitoring segment which automatically compiles the necessary probes. Software probes share resources with the system model, thus offering low cost, simplicity, flexibility, portability, and precise application performance measurement in a timely, frictionless manner.

Furthermore, based on the previous statistical functions, we can derive application-specific statistics for the following system modeling objects (similar to instant and duration classes, `enable_stat(args)` functions are provided for initializing parameters):

- *average throughput over a specified time window* of `Register`, `FIFO` objects, `Memory`, `Cache`, and inter-module communication objects (`Message`),
- *cumulative average throughput over consecutive time windows* of `Register`, `FIFO` objects, `Memory`, `Cache`, and inter-module communication objects (`Message`),
- *instant value of counter-based objects* such as `FIFO`, `LIFO`, and circular `FIFO` objects; this class can also be used to compute the instant `Cache` hit ratio (for read/write accesses) and cell loss probability with binary instantaneous counter values,
- *average instant value over a specified time window of counter-based objects;* this class can also be used to compute the average `Cache` hit ratio and cell loss probability,
- *cumulative average value over consecutive time windows* of counter-based objects; this class can also be used to compute the cumulative average `Cache` hit ratio and cell loss probability,
- *latency* statistics for consecutive read/write (or enqueue/dequeue) accesses on `Register`, `FIFO`, `Memory`, `Cache`, and inter-module communication objects (`Message`).

In addition to the previously described classes which cover all basic cases, it is sometimes necessary to combine statistical data from different modeling objects, e.g., from hierarchical memory units, in order to compare average read vs. write access times, or for computing cell loss in ATM layer communication. For this reason, we need new user-defined *joint or merged statistics* built on top of time- and event-driven statistics. Special parameters, e.g., boolean flags for these joint detailed statistic classes can lead to more detailed statistics.

We have discussed automatic extraction of statistical properties from hardware modeling objects. This statistical approach may also be extended to software modeling. As an example, the `enable_stat_` construct for obtaining write throughput statistics (read throughput is similar) from a FIFO object `buffer` is listed below. This function call will generate Figure 12.3.

```
//Enable statistics in [0,500], with time window = 1 sample
enable_stat_throughput_read("buffer", 0, 500, 1,
      "Simulation Time", "Average Throughput for Write Access");
```

For more details on the implementation of statistical classes, and the use of online or off-line `Grace`-based statistical graphs, the reader is referred to the OCCN user manual and the statistical test benches that accompany the OCCN library [23].These test benches are compatible with both Solaris and Linux operating systems.

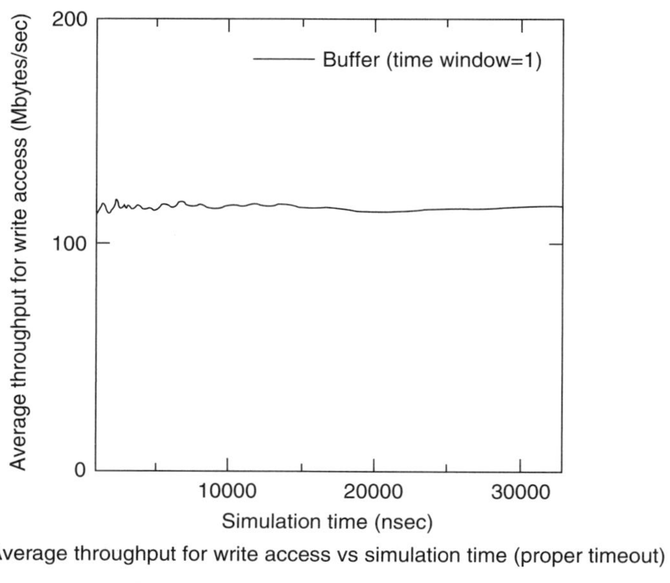

FIGURE 12.3 Performance results using Grace for a transport layer protocol.

12.4.2 Advanced System Monitoring

Conventional text output and source-code debugging are inadequate for monitoring and debugging complex and inherently parallel system models. Current tools, such as the Synopsys System Studio generate `vcd` files for signal tracing, or build relational databases in the form of tables for data recording, visualization, and analysis. Advanced high-level performance modeling environments may be based on advanced system-level monitoring activities. Although these activities may correspond to distinct monitoring phases occurring in time sequence, potentially there is a partial overlap between them.

- *Generation* refers to detecting events, and providing *status and event reports* containing monitoring traces (or histories) of system activity.
- *Processing* refers to functions that deal with monitoring information such as filtering, merging of traces, correlation, analysis, validation, and model updating. These functions convert low-level monitoring data to the required format and level of detail.
- *Presentation* refers to displaying monitoring information in an appropriate form.
- *Dissemination* concerns the distribution of selected monitoring reports to system-level developers and external processing entities.

In addition to the above main activities, implementation issues relating to intrusiveness and synchronization are crucial to the design and evaluation of monitoring activities.

12.4.2.1 Generation of Monitoring Reports: Status and Event Reports

In order to describe the dynamic behavior of an object (or group of objects) over a period of time, status and event reports are recorded in time order as *monitoring traces*. A *complete monitoring trace* contains all monitoring reports generated by the system since the start of the monitoring session, while a *segmented trace* is a sequence of reports collected during a limited period of time, e.g., due to overflow of a trace buffer or deliberate halting of trace generation. A monitoring trace may also be used to generate nonindependent traces based on various logical views of objects or system activity. For each trace we need to identify the reporting entity, the monitored object, the type of the report, as well as user-defined parameters, e.g., start and end time, time-window, priority, and size. We also provide browsing or querying facilities (by name or content) or runtime adjustments, e.g., examining the order of event occurrences or readjusting the interval between event occurrences.

A *status report* contains a subset of system state information, including object properties, such as time stamp, status, and identity. Appropriate criteria define the sampling rate, the reporting scheme, and the contents of each report. For example, the report may either be generated periodically, i.e., based on a predetermined finite-state machine or Thread schedule, or on demand, i.e., upon receiving a request for solicited reporting. The request may itself be periodic, i.e., via polling or on a random basis.

System events may be detected immediately upon occurrence, or after a delay. For example, signals on an internal system bus may be monitored in real time, while alternatively, status reports may be generated, stored, and processed in order to detect events at a later time. Event detection may be internal to the monitored object, i.e., typically performed as a function of the object itself, or external, i.e., performed by an external agent who receives status reports and detects changes in the state of the object. Once the occurrence of an event is detected, an *event report* is generated. In general, an event report contains a variable number of attributes such as reporting entity, monitored object, event identifier, type, priority, time of occurrence, state of the object immediately before and after event occurrence, application-specific state variables, time stamps, text messages, and possibly pointers to detailed information.

12.4.2.2 Processing of Monitoring Information

A system-level model may generate large amounts of monitoring information. Design exploration is successful, if the data can be used to identify design problems and provide corrective measures. Thus, the data collection phase is split into four different phases.

Validation of monitoring information provides data consistency checks, possibly specified in a formal language, ensuring that the system is monitored correctly. This includes:

- *sanity tests* based on the validity of individual monitoring traces, e.g., by checking for correct token values in event fields, such as an identity or time stamp, and
- *validation of monitoring reports against each other*, e.g., by checking against known system properties, including temporal ordering.

Filtering reduces the amount of monitoring data to a suitable rate and level of detail. For example, filter mechanisms reduce the complexity of displayed process communications by:

- incorporating a variable report structure
- displaying processes down to a specified level in the module hierarchy
- masking communication signals and data using filter dialogs and
- providing advanced filter functionality for displaying only tokens with pre-determined values

Analysis processes monitoring traces based on special user-selected criteria. Since analysis is application-dependent, it relies on sophisticated stochastic models involving combinatorics, probability theory, and Markov chain theory. *Analysis* techniques enable:

- simple data processing, e.g., average, maxima, variance statistics of particular state variables,
- statistical trend analysis for forecasting using data analysis packages such as Maple, Grace, SAS, or S from AT & T,[28,29] and *correlation* (merging, combination, or composition) of monitoring traces which raises the abstraction level.

Together with filtering, it prevents the users from being overwhelmed by an immense amount of detailed information, and helps identify system bottlenecks. Thus, for example, events generated by sensors or probes may be combined using `and`, `or`, and `not` operators to provide appropriate high-level reliability metrics. Since correlation of system-level monitoring information is a very challenging task, a relational database, such as mini-SQL, that includes selection, projection, and join operators is sometimes useful.

12.4.2.3 Presentation of Monitoring Information

Various visualization techniques such as simple textual representation, time-process diagrams, and animation may be provided.

Textual representation (ascii) increases its expressive power by providing appropriate indentation, color, and highlighting to distinguish information at different abstraction levels. Events may be displayed in a causal rather than temporal order by including parameters such as event type, name of process initiating the event, name of process(es) handling the event, and contents of transmitted messages.

A *time-process diagram* is a 2-D diagram illustrating the current system state and the sequence of events leading to that state. The horizontal axis represents events corresponding to various processes, while the vertical one represents time. In synchronous systems, the unit of time corresponds to actual time, while in asynchronous systems, it corresponds to the occurrence of an event. In the latter case, the diagram is called a *concurrency map*, with time dependencies between events shown as arrows. An important advantage of time-process diagrams is that monitoring information may be presented on a simple text screen or a graphical one.

An *animation* captures a snapshot of the current system state. Both textual and graphical event representations e.g., input, output, and processing events, can be arranged in a 2-D display window. Graphical representations use various formats such as icons, boxes, Kiviat diagrams, histograms, bar charts, dials, *X*–*Y* plots, matrix views, curves, pie charts, and performance meters. Subsequent changes in the display occur in single step or continuous fashion, and provide an animated view of system evolution.For online animation, the effective rates at which monitoring information is produced and presented to the display must be matched. For each abstraction level, animation parameters include enable/disable event monitoring or visualization, clock precision, monitoring interval, or level of detail, and viewing/printing statistics.

12.4.2.4 Dissemination of Monitoring Information

Monitoring reports would have to reach designers, managers, or processing entities. Thus, dissemination schemes range from very simple and fixed, to very complex and specialized. Selection criteria contained within the subscription request are used by the dissemination system to determine which reports (and with what contents, format, and frequency) should be delivered.

12.4.2.5 Other Important Monitoring Design Issues

Specifications for the actual graphical user interface of the monitoring tool depend on the designer's imagination, experience, and time. We present here desirable features of a system-level monitoring interface:

- *Visualization at different abstraction levels* which enables the user to observe behavior at various abstraction levels on a per object, per block, or per system basis. Thus, the user may start observation at a coarse level and progressively (or simultaneously) focus on lower levels. At each level, application performance and platform-specific metrics may be presented in appropriate easy-to-read charts and graphs. The communication volume (reliability or power consumption) may be visualized by adjusting the width or color of the lines interconnecting (respectively) the boxes representing the corresponding modules,
- A *history function* which visualizes inherent system parallelism by permitting the user to
  - scroll the display of events forwards or backwards in time by effectively changing a simulated system clock, and
  - control the speed at which system behavior is observed using special functions, e.g., to start, stop, pause, or restart an event display, perform single-step or continuous animation, and allow for real-time or slow-motion animation.
- *Visibility of interactions* which enables the user to visualize dynamically contents of a particular communication message, object data structure, module or driver configuration, or general system data such as log files or filtering results.
- *Placement of monitoring information* which greatly enhances visibility and aids human comprehension. Placement may either be automatic, i.e., using computational geometry algorithms, or manual by providing interface functions, e.g., for moving or resizing boxes representing objects, system events, or coupled links representing process communication.
- *Multiple views* use multiple windows representing system activities from various points of view, thus providing a more comprehensive picture of system behavior.
- *Scalable monitoring* focuses on monitoring large-scale models, with tens or hundreds of thousands of objects. Large-scale monitoring could benefit from efficient message queues, or nonblocking and blocking concurrent queues [30,31] that achieve a high degree of concurrency at low implementation cost compared to other general methods [32,33].

  In addition, the monitoring system designer should focus on the issues of intrusiveness and synchronization occurring in distributed system design.
- Intrusiveness refers to the effect that monitoring may have on the monitored system due to sharing common resources, e.g., processing power, communication channels, and storage space. Intrusive monitors may lead not only to system performance degradation, e.g., due to increased memory access, but also to possible deadlock conditions and data races, e.g., when evaluating symmetric conditions which result in globally inconsistent actions.
- Distributed simulation can support time- and space-efficient modeling of large modular systems at various abstraction levels, with performance orders of magnitude better than existing commercial simulation tools [34]. Distributed systems are more difficult to monitor due to increased parallelism among processes or processors, random and nonnegligible communication delays, possible process failures, and unavailable global synchronized time. These features cause interleavings of monitoring events that might result in different output data from repeated runs of deterministic distributed algorithms, or different views of execution events from various objects

[35–38]. However, these problems do not occur with SystemC scheduling (and early SystemC-based parallel efforts), since current versions of the kernel are sequential, offering simulated parallelism and limited event interleaving.

12.5 Open System-Level Modeling Issues

Cycle-accurate transactional system-level modeling is performed at an abstraction level higher than traditional hardware description languages, thus improving design flow and enhancing design quality by offering an increased design space, efficient simulation, and block- and system-level reuse. We have described efforts toward a general, rich, and efficient system-level performance evaluation methodology that allows multiple levels of abstraction, enables hardware/software codesign, and provides interoperability between several state-of-the-art tools. Activities that could enhance this methodology include

- System reliability and fault-tolerance models [39].
- Power-consumption models [40–42].
- Asynchronous modeling [43], e.g., by modifying the SystemC kernel, and providing waves, concurrency map data structures, and system snapshots.
- Graphical visualization [44,45,46], including appropriate GUIs for:
  - *interactive model design* based on importing ready-to-use library modules, e.g., `HFsms`, or reusable IP block or system components,
  - *simulation control*, such as saving into files, starting, pausing, and restarting simulation as well as displaying waveforms by linking to graphical libraries, and dumping or changing simulation and model parameters during initialization or runtime [47],
  - *advanced monitoring features*, based on generation, processing, presentation, and dissemination activities, and
  - *platform-specific performance metrics*, such as simulation efficiency, and computation, and communication load, which help improve simulation performance in parallel platforms, e.g., through automatic data allocation, latency hiding, or dynamic load balancing.
- Executable model specification by customizing existing tools, such as Doxygen or Doc++.
- Interoperability between libraries and tools, e.g., ISS, system and hardware description languages, and simulation interfaces which can provide long-term reuse opportunities. This could be achieved with middleware, e.g., by inventing an appropriate interface definition language.
- System validation and verification — existing block- and system-level verification tools are usually external to the system-level modeling library and tightly linked to actual implementation and this causes a complete rebuilt of the verification environment, even if only very few architectural parameters change.

References

[1] Carloni, L.P. and Sangiovanni-Vincentelli, A.L., Coping with latency in SoC design,. *IEEE Micro* (Special Issue on Systems on Chip), 22–25, 24–35, 2002.

[2] SystemC, http://www.systemc.org

[3] SystemC 2.0, *User's Guide*, 2001, see http://www.systemc.org and http://www-ti.informatik.uni-tuebingen.de/~systemc

[4] Krolikoski, S., Schirrmeister, F., Salefski, B., Rowson, J., and Martin, G., Methodology and Technology for Virtual Component Driven Hardware/Software Co-Design on the System Level, Paper 94.1, ISCAS 99, Orlando, FL, May 30–June 2, 1999.

[5] SystemC Modeling with the Synopsys Cocentric SystemC Studio, Synopsys, 2002. See http://www.synopsys.com/products/cocentric_studio/cocentric_studio.html

[6] SystemC Synthesis with the Synopsys Cocentric SystemC Compiler, Synopsys, 2002, see http://www.synopsys.com/products/cocentric_systemC/cocentric_systemC.html

[7] Ferrari, A. and Sangiovanni-Vincentelli, A., System design: traditional concepts and new paradigms, *Proceedings of the International Conference on Computer Design*, 1999, pp. 2–13.

[8] Rowson, J.A. and Sangiovanni-Vincentelli, A.L., Interface-based design, *Proceedings of the Design Automation Conference*, 1997, pp. 178–183.

[9] Haverinen, A., Leclercq, M., Weyrich, N., and Wingard, D., SystemC-based SoC communication modeling for the OCP protocol, white paper submitted to SystemC, 2002. Also see http://www.ocpip.org/home

[10] Moriconi, M., Qian, X., and Riemenschneider, R.A., Correct architecture refinement, *IEEE Trans. Software Eng.*, SE-21, 356–372, 1995.

[11] Medvidovic, N. and Taylor, R.N., A framework for classifying and comparing architecture description languages, *Proceedings of the Software Engineering Conference*, Lecture Notes in Computer Science, Vol. 1301, Spinger, Berlin, 1997, pp. 60–76.

[12] Allen, R. and Garlan, D., A formal basis for architectural connection, *ACM Trans. Software Eng. Methodol.*, 6, 213–249, 1997.

[13] Charest, L., Aboulhamid, E.M., and Tsikhanovich, A., Designing with SystemC: multi-paradigm modeling and simulation performance evaluation, *Proceedings of the HDL Conference*, 2001, pp. 33–45.

[14] Virtanen, S., Truscan, D., and Lilius, J., SystemC based object oriented system design, *Proceedings* of the *Forum on Design Languages*, 2001, pp. 1–4.

[15] Gajski, D.D., Zhu, J., Doemer, A., Gerstlauer, S., and Zhao, S., *SpecC: Specification Language and Methodology*, Kluwer, Dordrecht, 2000. Also see, http://www.specc.org

[16] Brunel, J.-Y., Kruijtzer, W.M., Kenter, H.J.H.N., Petrot, F., Pasquier, L., de Kock, E.A, and Smits, W.J.M., Cosy communication IP's, *Proceedings of the Design Automation Conference*, 2000, pp. 406–409.

[17] Coppola, M., Curaba, S., Grammatikakis, M.D., Locatelli, R., Marrucia, G., and Papariello, F., OCCN: a NoC modeling framework for design exploration, *J. Syst. Arch. — Special Issue on Network-on-Chip*, Vol. 50, Elsevier, North Holland, 2004, pp. 129–163.

[18] Carey, J., *STbus Superhighway: Type 3*, internal document, ST Microelectronics, March 2001.

[19] Scandurra, A., *STbus C++ Class*, internal document, ST Microelectronics, December 2000.

[20] VSI Alliance, http://www.vsi.org/

[21] Lyoyd, A., *Runtests Lite User Guide*, internal document, ST Microelectronics, June 2000.

[22] Melkonian, M., Get by without an RTOS. Embedded Systems, 13 (10), September 2000, available from http://www.embedded.com/2000/0009/0009feat4.htm

[23] Coppola, M., Curaba, S., Grammatikakis, M., Maruccia, G., and Papariello, F., *The OCCN user manual*, unpublished report, 2004. Available from http://occn.sourceforge.net

[24] Amba Bus, Arm, http://www.arm.com

[25] Carloni, L.P., McMillan, K.L., and Sangiovanni-Vincentelli, A.L., Theory of latency-insensitive design, *IEEE Trans. Computer-Aided Design of Integrated Circuits & Syst.*, 20–29, 1059–1076, 2001.

[26] Harel, D., Statecharts: a visual formalism for complex systems, *Sci. Comput. Program.*, 8, 231–274, 1987.

[27] von der Beeck, M., A comparison of statecharts variants, *Proceedings of the Formal Techniques in Real Time and Fault Tolerant Systems*, Lecture Notes in Computer Science, Vol. 863, Spinger, Berlin, 1994, pp. 128–148.

[28] Zimmerman, C., *The Quantitative Macroeconomics and Real Business Cycle*, for statistical software see: http://dge.repec.org/software.html

[29] Grace see http://plasma-gate.weizmann.ac.il/Grace/

[30] Hunt, G.C., Michael, M., Parthasarathy, S., and Scott, M.L., An efficient algorithm for concurrent priority queue heaps, *Inf. Proc. Lett.*, 60, 151–157, 1996.

[31] Michael, M. and Scott, M.L., Simple, fast and practical non-blocking and blocking concurrent queue algorithms, *Proceedings of the ACM Symposium on Princ. Distrib. Comput.*, 1996 , pp. 267–275.

[32] Prakash, S., Yann-Hang, L., and Johnson, T., A non-blocking algorithm for shared queues using compare-and-swap, *IEEE Trans. Comput.*, C-43, 548–559, 1994.

[33] Turek, J., Shasha, D., and Prakash, S., Locking without blocking: making lock-based concurrent data structure algorithms nonblocking, *Proceedings of the ACM Symposium on Princ. Database Systems*, 1992, pp. 212–222.

[34] Grammatikakis, M.D. and Liesche, S., Priority queues and sorting for parallel simulation, *IEEE Trans. Soft. Eng.*, SE-26, 401–422, 2000.
[35] Chandy, K. M. and Lamport, L., Distributed snapshots: determining global states of distributed systems, *ACM Trans. Comp. Syst.*, 3, 63–75, 1985.
[36] Christian, F., Probabilistic clock synchronization, *Distr. Comput.*, 3, 146–158, 1989.
[37] Fidge, C.J. Partial orders for parallel debugging, *Proceedings of the ACM Workshop Parallel Distr. Debug.*, 1988, pp. 183–194.
[38] Lamport, L., Time, clocks and the ordering of events in distributed systems, *Comm. ACM*, 21, 558–564, 1978.
[39] Grammatikakis, M.D., Hsu, D.F., and Kraetzl, M., *Parallel System Interconnections and Communications*, CRC Press, Boca Raton, FL, 2000.
[40] Benini, L. and De Micheli, G., System-level power optimization: techniques and tools, *ACM Trans. Autom. Electr. Syst.*, 5, 115–192, 2000.
[41] Benini, L., Ye, T.T., and De Micheli, G., Packetized on-chip interconnect communication analysis for mpsoc, *Proceedings of the Design Automation & Test in Europ Conf.*, 2003.
[42] Ye, T.T., Benini, L., and De Micheli, G., Analysis of power consumption on switch fabrics in network routers, *Proceedings of the Design Automation Conf.*, 2002.
[43] Mueller, W., Ruf, J., Hoffmann, D., Gerlach, J., Kropf, T., and Rosenstiehl, W., The simulation semantics of SystemC, *Proceedings of the Design, Automation & Test in Europe Conf.*, 2001, pp. 64–71.
[44] Koutsofios, E. and North, S.C., *Dot User Manual*, AT&T Bell Labs, Murray Hill, NJ, 1993. See http://www.cs.brown.edu/cgc/papers/dglpttvv-ddges-97.ps.gz
[45] Reid, M., Charest, L., Tsikhanovich, A., Aboulhamid, E.M., and Bois, F., Implementing a graphical user interface for SystemC, *Proceedings of the HDL Conf.*, 2001, pp. 224—231.
[46] Sinha, V., Doucet, F., Siska, C., Gupta, R., Liao, S., and Ghosh, A., YAML: a tool for hardware design, visualization, and capture, *Proceedings* of the *International Symp. Syst. Synthesis*, 2000, pp. 9–17.
[47] Charest, L., Reid, M., Aboulhamid, E.M., and Bois, G., A methodology for interfacing open source SystemC with a third party software, *Proceedings of the Design, Automation & Test in Europe Conference*, 2001, pp. 16–20.

13

Micro-Architectural Power Estimation and Optimization

Enrico Macii
Politecnico di Torino
Torino, Italy

Renu Mehra
Synopsys, Inc.
Mountain View, California

Massimo Poncino
Politecnico di Torino
Torino, Italy

Abstract

As power consumption has become one of the major limiting factors in current electronic systems, this chapter introduces innovative methodologies for successfully dealing with power estimation and optmization during the early stages of the design process. In particular, the presentation offers an insight into state-of-the-art techniques for power estimation at the micro-architectural level, describing how power consumption of components like data-path macros, glue and steering logic, memory macros, buses, interconnect, and clock wires can be efficiently modeled for fast and accurate power estimation. Then, the focus shifts to power optimization, covering the most popular classes of techniques, such as those based on clock-gating, on exploitation of common-case computation, and on threshold and supply voltage management. Ad-hoc optimization solutions for specific components, such as on-chip memories and global buses are also briefly discussed for the sake of completeness.

Most of the aforementioned approaches to micro-architectural power estimation and optimization have now reached a significant level of maturity, and are thus finding their way into commercial CAD tools that are currently hitting the EDA market. Strengths and limitations of the design technology that is at the basis of such tools will be discussed in detail throughout this chapter.

13.1 Introduction

Low-power consumption has become one of the most important features of modern electronic devices, due to several factors, ranging from the increased market share of mobile and portable systems for telecom and computing to the increased cost of implementing, packaging and manufacturing circuits working at high speed and temperature. As a consequence, techniques and tools that enable tight power consumption control during design are required.

Pioneering work on low-power design techniques has focused on gate and transistor levels where, due to the available information on the structure and the macroscopic parameters of the devices, accurate power estimates are expected and satisfactory methods for both estimation and optimization are available.

The increased complexity of modern designs, facilitated by the advent of aggressively scaled technologies, and the augmented pressure of time-to-market constraints, called for modifications to the way ICs are designed. This resulted in a substantial shift upward in the design paradigm and in the supporting tool flow. Today, electronic systems are specified and then developed starting from the micro-architectural (also called register-transfer, RT) level, in which the basic design entities are no longer elementary objects such as transistors or logic gates, but rather complex blocks capable of performing nontrivial functions. Typical components at the micro-architectural level include data-path macros, such as adders and multipliers, storage elements, such as registers and memory banks, communication resources (e.g., buses) and steering elements (e.g., multiplexers, codecs). Along this line, which defines the RT level as the common entry point for digital design, technologies, methodologies, and tools for low-power design applicable at the RTL have come up in the last decade, now providing the industrial design community with integrated EDA solutions for power analysis, estimation, and optimization starting from this level of abstraction.

As electronic technology keeps evolving, so does EDA technology; higher levels of abstraction are currently being proposed as possible starting points for new design development. For instance, in the system-on-chip (SoC) domain, hardware–software combined specification, design and synthesis are becoming common practice, and new methodologies and tools are being investigated. It is fair to say, though, that fully automated solutions are still not commercially available, especially that concerning power-oriented design. As of today, the RTL is thus the highest level of abstraction for which extensive EDA support is guaranteed.

The objective of this chapter is to provide a comprehensive overview of the most advanced, yet well-established EDA solutions for power estimation and optimization. For the reasons highlighted above, the focus of the discussion will be restricted to the RT level. The chapter will start by providing some background information concerning power consumption in CMOS circuits; it will continue with a brief overview of the architectural template assumed by most of the power modeling, estimation, and optimization approaches, which constitute the core of the chapter and which are the subjects of Sections 13.4 and 13.5.

13.2 Background

CMOS is, by far, the most common technology used for manufacturing digital circuits. There are three major sources of power dissipation in a CMOS circuit [1]:

$$P = P_{\mathrm{Sw}} + P_{\mathrm{SC}} + P_{\mathrm{L}} \tag{13.1}$$

P_{Sw}, called *switching* or *dynamic power*, is due to charging and discharging capacitors driven by the gates in the circuit. P_{SC}, called *short-circuit power*, is caused by the short-circuit currents that arise when pairs of PMOS/NMOS transistors are conducting simultaneously. Finally, P_{L}, called *leakage* or *static* or *stand-by power*, originates from subthreshold currents caused by transistors with low threshold voltages and from gate currents caused by reduced thickness of the gate oxide.

For older technologies (e.g., 0.25-μm), P_{Sw} was dominant. For deep-submicron processes, P_{L} becomes more important. For instance, in ASIC designs, leakage power accounts for around 5–10% of the total power budget at 180 nm, and this fraction grows to 20–25% at 130 nm and to 35–50% at 90 nm. Therefore, leakage power minimization must be addressed from the design standpoint, and not just at the technology or process level, as was done in the past.

Design methods for leakage power control are currently the subject of intensive investigation; approaches based on variable-threshold (VTCMOS), dual-threshold (DTCMOS), and multi-threshold CMOS devices (MTCMOS); the insertion of (possibly distributed) sleep transistors; the adoption of multi-voltage gates (MVCMOS); and the application of reverse and forward body biasing are some examples of promising solutions for reducing the impact of leakage in nanometer circuits. Yet, most of the methods and tools for low-power design in use today are still primarily targeting the minimization of the dynamic component of the power; this is because research in this domain has been carried out for a much longer time, and the available solutions have now reached a significant degree of maturity. As a consequence, a significant part of this chapter will be focused on techniques addressing switching power modeling, estimation, and optimization.

Switching power for a CMOS gate working in a synchronous environment is modeled by the following equation:

$$P_{\mathrm{Sw}} = \frac{1}{2} C_{\mathrm{L}} V_{\mathrm{dd}}^2 f_{\mathrm{Ck}} E_{\mathrm{Sw}} \tag{13.2}$$

where C_{L} is the output load of the gate, V_{dd} the supply voltage, f_{Ck} the clock frequency, and E_{Sw} the *switching activity* of the gate, defined as the probability of the gate's output of making a logic transition during one clock cycle.

Reductions of P_{Sw} are achievable by combining the minimization of the four parameters in the equation above.

Historically, supply voltage scaling has been the most used approach to power optimization, since it normally yields considerable savings owing to the quadratic dependence of P_{Sw} on V_{dd}. The major shortcoming of this solution, however, is that lowering the supply voltage affects circuit speed. As a consequence, both design and technological solutions must be applied in order to compensate the decrease in circuit performance introduced by reduced voltage. In other words, speed optimization is applied first, followed by supply voltage scaling, which brings the design back to its original timing but with a lower power requirement.

A similar problem, i.e., performance decrease, is encountered when power optimization is obtained through frequency scaling. Techniques that rely on reductions of the clock frequency to lower power consumption are thus usable under the constraint that some performance slack does exist. Although this may seldom occur for designs considered in their entirety, it happens quite often that some specific units in a larger architecture do not require peak performance for some clock/machine cycles. Selective frequency scaling (as well as voltage scaling) on such units may thus be applied, at no penalty in the overall system speed.

Optimization approaches that have a lower impact on performance, yet allowing significant power savings, are those targeting the minimization of the *switched capacitance* (i.e., the product of the capacitive load and the switching activity). Static solutions (i.e., applicable at design time) handle switched capacitance minimization through area optimization (that corresponds to a decrease in the capacitive load) and switching activity reduction via exploitation of different kinds of signal correlations (temporal, spatial, and spatio-temporal). Dynamic techniques, on the other hand, aim at eliminating power wastes that may be originated by the application of certain system workloads (i.e., the data being processed).

Static and dynamic optimizations can be achieved at different levels of design abstraction. Actually, addressing the power problem from the very early stages of design development offers enhanced opportunities to obtain significant reductions of the power budget and to avoid costly redesign steps. Power-conscious design flows must then be adopted; these require, at each level of the design hierarchy, the exploration of different alternatives, as well as the availability of power estimation tools that could provide accurate feedback on the quality of each design choice.

As discussed in the introduction, in this chapter, we will devote our attention to the RT level, as it represents the actual front-end for automated design today. More specifically, although RTL descriptions may contain components of different kinds (see the next section for more details on the architectural template), we will primarily concentrate on estimation and optimization of the portion of the architecture that is normally designed with the help of automatic tools (e.g., data-path, controller, interconnect and clock tree). Modeling, estimation and optimization techniques for components with peculiar characterisitics, such as memories and buses will also be surveyed in order to make the picture as complete as possible.

13.3 Architectural Template

The definition of a "micro-architecture" goes through the assumption of an *architectural template,* which serves as a reference for all micro-architectural descriptions. There are two main advantages in assuming such a template; first, it allows one to infer a fine-grain partition of the entire design into *objects* of manageable size, for which customized models and specific optimizations can be devised. Second, by assuming that all micro-architectural descriptions will map onto this template, the chance of reusing models and optimization techniques increases.

A traditional micro-architectural template views a design as the interaction of a *data-path* and a *controller,* which fits well to the so-called Finite State Machine with Data-path FSMD model [2]. Variants to the base FSMD model concern the structure of the controller (e.g., sparse logic implementation vs. wired logic), the structure of the interconnect (e.g., number, type, and size of the buses), or the supported arithmetic operations (e.g., number and type of the available data-path units).

Figure 13.1 shows a possible architecture of a typical FSMD that exposes its basic building blocks: the controller (shown on the left) and the data-path (right), consisting of a register file (or equivalently, a set of sparse registers), a memory, various interconnection buses, and some data-path units (integer and/or floating point). Besides the blocks themselves, the interconnection between them (not explicitly depicted in the figure) is also a source of power consumption; these wires can be global or local, depending on whether they span the entire architecture or not. Examples of the former classes are the clock signal and the global buses; examples of the latter are the signals connecting the units to the global wires. The most important of all wires is the clock that constitutes a significant source of consumption, because of its large load and high activity. While most of the local wires are usually hidden inside the primitive blocks, global wires need a special treatment because they cannot be captured otherwise.

The architectural template introduced above is sometimes defined as *structural RTL,* with the purpose of emphasizing the explicit notion of structure it contains, as well as the RT level nature of the operations that take place during execution.

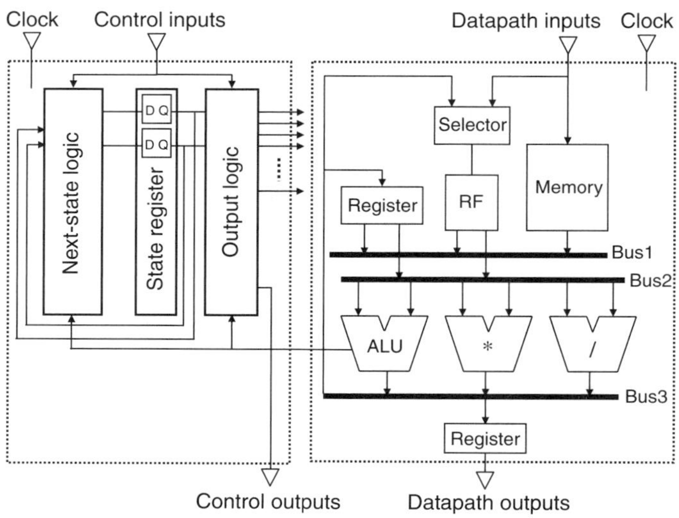

FIGURE 13.1 Architectural template for RTL power estimation.

The key issue is that fitting a micro-architectural description to this template allows us to restrict the granularity of the power models and of the optimizations to that of the following main building blocks: *controller, data-path units, memory, buses,* and *wires.* Registers may fall either in the class of memory devices (for a register-file implementation style) or in that of data-path units (for "sparse" style). This approach is followed by most micro-architectural power estimation and optimization frameworks that have been proposed in the recent literature [3–10].

The FSMD template of Figure 13.1 also nicely fits micro-architectural descriptions specified using hardware description languages (e.g., VHDL or Verilog). They are in fact described as a state machine (the controller), whose states consist of a series of micro-architectural operations (e.g., assignments, arithmetic, or logic operations) corresponding to data-path operations. It is worth emphasizing, however, that such descriptions, after parsing by standard HDL compilers, tend to lose some of their structural semantics and are partly flattened into a netlist of finer-grain primitives such as abstract logic gates [11], in which only memory macros are preserved.

13.4 Micro-Architectural Power Modeling and Estimation

The problem of power estimation at the RTL amounts to building a *power model* that relates the power consumption of the target design to suitable quantities. In formula,

$$P = P(X_1, \dots, X_n)$$

where X_i, $i = 1, \dots, n$, represent the *n model parameters.* The construction of the model shown above implies addressing several issues that call for the development of various modeling alternatives, and are discussed in detail in the sequel.

13.4.1 Modeling Issues

13.4.1.1 Model Granularity

This issue has already been addressed in Section 13.3, where the definition of an architectural template has been shown to determine the granularity of the power models; power estimation of a design requires thus the construction of power models for the following classes of building blocks: *controller, data-path units, memory, buses, and wires.*

13.4.1.2 Model Parameters

The choice of the parameters to be included in the model is constrained by the fact that they must be quantities that are observable at the abstraction level at which they are used. Under the architectural model of Section 13.3, the abstract model of switching power as shown in Equation (13.2) translates into the following high-level expression:

$$P_{\text{total}} = k \sum_{\forall \text{component } i} A_i C_i \tag{13.3}$$

where A_i and C_i denote the *switching activity* and the *physical capacitance* of component *i*, respectively; *k* lumps the fV_{dd}^2 terms, which can be considered as scaling factors at the RTL.

Equation (13.3) decouples the problem of building a power model into that of building a model for activity and a model for capacitance, for each type of component. In general, activity and capacitance models will depend on different parameters because they are affected by different physical quantities.

Activity is easier to model, because at the RTL it is a well-defined quantity. Activity models rely on *activity parameters,* such as bit-wise (referred to specific input or output signals of a component) or word-wise (regarding input or output values of a component) transition and static probabilities. Most activity models proposed in the literature use these probability measures as parameters. Other choices for activity parameters may include variants of transition probabilities such as *transition density* [12], defined as an average (over time) switching rate or various *correlation measures.*

Modeling physical capacitance is a more problematic task than modeling activity. The term "physical" suggests the difficulty in linking capacitance to quantities observable at the RTL. In spite of that, RTL capacitance models can be derived with a reasonable degree of accuracy. They all rely on the intuitive observation that capacitance will be roughly related to the number of "objects" (gates or similar lower-level primitives) of the target component. In other words, physical capacitance at the RTL is approximated by *complexity,* and we thus speak of *complexity models,* based on *complexity parameters.*

Complexity parameters that are available at the RTL are restricted to the *width* of a component (i.e., its number of inputs and outputs) or the *number of states* (which is relevant only to the controller, since notion of state is explicit.) Any complexity parameter different from the above would require some additional information derived from back-annotation of physical information of previous implementations.

13.4.1.3 Model Semantics

Models can be distinguished in terms of how the values that they return can be interpreted. The most intuitive option is to assume that models express *average power,* which is commonly used as a metric to track battery lifetime or average heat dissipation. In this case, the semantics of the model is that of a single figure to represent the consumption of the target description. Average power models are called *cumulative power models* [13].

However, the notion of *cycle* intrinsic to RTL descriptions allows us to obtain a power model with a richer semantic by simply changing the way we collect statistics. The first step in this direction consists of expanding the model of Equation (13.3) into

$$P_{\text{total}} = k \sum_{\forall \text{cycle } j} P_j = k \sum_{\forall \text{cycle } j} \sum_{\forall \text{component } i} A_{ij} C_{ij} \tag{13.4}$$

where P_j denotes the power consumption at cycle *j*, which can be obtained by summing the power consumption for each component (as in Equation [13.3]), this time using activities and capacitances of component *i* at each cycle *j*. The semantics of the model of Equation (13.4) is *cycle-accurate,* because it allows to track cycle-by-cycle (total) power.

The use of a cycle-accurate model affects the choice of the parameters. For example, transition or static probabilities are not suitable quantities anymore, since they are intrinsically "average." Conversely, cycle accurate models should use cycle-based activity measures, such as the *number of bit toggles* between consecutive patterns (i.e., the *Hamming distance)* [14,15] or *the values* of consecutive input patterns [13,16].

A cycle-accurate model provides several advantages over a cumulative one. First, it allows one to go beyond the bare evaluation of average power and can be used to perform sophisticated analysis of power consumption over time, such as reliability, noise, or IR drop analysis. In addition, a cycle-accurate model is more accurate than a cumulative one, but not just because it provides a set of power values rather than a single one. In fact, the relation between input statistics and power is nonlinear: average consumption is usually different from the consumption associated to average input statistics, especially when power consumption varies significantly over time. Therefore, even when average power is the objective, averaging the series of cycle-by-cycle values will yield a more accurate estimate than a model of average power. On the negative side, cycle-accurate models require significantly larger storage than cumulative ones.

13.4.1.4 Model Construction

Concerning model construction, there are two options to build an RTL power model: a *top-down* (or *analytical)* approach, and a *bottom-down* (or *empirical)* one [3].

Top-down approaches relate activity and capacitance of an RTL component to the model parameters through a closed formula. The term "top-down" refers to the fact that the model is derived directly from the micro-architectural description, and is not based on lower-level information. For this reason, such formula usually has a *physical interpretation*. Analytical models are particularly useful either when dealing with a newly designed circuit for which no information of previous implementations is available, or when the implementation of the circuit follows some predictable template, which can be exploited to force some specific relation between the model parameters. A memory is a typical example of an entity for which an analytical model is suitable: its internal organization is quite fixed (cell array, bit and wordlines, decoders, MUXes, and sense amps), thus allowing accurate modeling based on "internal" parameters [17,18].

If we exclude these special cases, however, top-down models are not very accurate, since their links to the implementation (e.g., technology and synthesis constraints) are quite weak.

Bottom-up approaches, conversely, are based on "measuring" the power consumption of existing implementations, from which the actual power model is derived. Typically, the template of the power model (i.e., the parameters and a set of coefficients used to weigh the parameters) is defined up front; statistical techniques are then used to fit the model template to the measure of power values. This approach is known as *macromodeling*, and has proved to be a very accurate and robust methodology for RTL power estimation, and can be considered the state-of-the-art solution. Section 13.4.2 will be devoted to the detailed description of the macromodeling flow.

13.4.1.5 Model Storage

The issue of model storage is concerned with the *shape* of the model. Since models express a mathematical relation between power and a set of parameters, the problem amounts to that of representing such a relation. The two options are to store it as an equation or as a look-up table, corresponding to the choice of representing a relation as a continuous function (equation-based models), or a discrete function approximated by points (look-up table-based models). These two types of models differ in their storage requirements and *robustness*, that is, the sensitivity of the model to the conditions (i.e., the experiments) used for its construction. In that sense, robustness is an issue only for empirical models. Concerning storage requirements, equation-based models are much more compact than table-based ones. In general, an equation will only require the storage of the coefficients of the model, as opposed to a full table. In addition, the accuracy of a table-based model is proportional to its size (the denser the table, the higher the accuracy), whereas the accuracy of an equation-based model is independent of the model size.

13.4.2 RTL Power Models

In this section we will review the most relevant results on power modeling of data-path units, controller, and wires. On the other hand, as already mentioned in Section 13.2, we will not consider components such as memories and communication buses.

13.4.2.1 Power Modeling of Data-Path Units

There are three main reasons for which achieving accurate power estimates for data-path units must resort to empirical models. First, because of the variety of available data-path units, a distinct analytical model for each type of component would be required. Second, analytical models, because of their functional meaning, would not be able to capture differences due to various implementation styles of a given component (e.g., a ripple-carry vs. a carry-lookahead adder). Third, accuracy of analytical models would be too low for units that have poorly regular structures (e.g., dividers or floating-point operators).

Besides being able to cope with these issues, empirical models allow one to devise a single-model template that can be used for all types of data-path units. Empirical power models are commonly called *macromodels*. The term has been borrowed from the statistical domain to denote the fact that such models have a "coarse" level of detail, and are used to relate quantities pertaining to different abstraction levels, such as RTL parameters, to the actual measured power. We thus talk of "macromodeling" when defining the process of building an empirical power model for a generic component.

Before reviewing the most important results in data-path macromodeling, it is worth summarizing the flow that is normally used for building such macromodels.

13.4.2.1.1 Macromodeling Flow

Macromodeling involves four major steps, as depicted in Figure 13.2:

1. *Choice of model parameters.* Although this step applies to non-empirical models also, it is especially important for macromodels, since it defines the parameters space and it affects the complexity of the following steps.
2. *Design of the training set.* The training set is a representative subset of the set of all possible input vector pairs that will be used to determine the model. Key points in the choice of the training set are its size (i.e., the total number of vector pairs) and its statistical distribution. The former is responsible for the simulation time, while the latter impacts accuracy; a bad distribution of the training set may offset the advantage of a large number of vector pairs. What defines a "good" distribution depends on the chosen parameters. A requirement for the training set is that it should span the domain of all the model parameters as much as possible. When one or more domains are not sufficiently covered by the training set, we say that the model is *insufficiently trained.* For instance, if the parameter of the model is the switching activity, the choice of random patterns as a training set would not be a good one, since only a very small portion of the activity domain (the one around a switching activity of 0.5) would be exercised.
3. *Characterization.* This step consists of using the training set to generate a set of points in the *(power, parameters)* space. For each element in the training set (a vector pair), a corresponding value of power is obtained by means of a low-level power simulator (gate or transistor-level).
4. *Model extraction.* This step consists of deriving the model from the set of measurements obtained in the previous step. The actual calculation depends on how the model is stored. For equation-based models, a least-mean-square (LMS) regression engine is typically applied to the sample measurements; if a look-up table is used, extraction consists of collecting the power values for each of the discrete points of the parameters space.

13.4.2.1.2 Macromodels for Data-Path Units

The literature on power macromodeling for data-path units is quite rich, and it includes solutions that cover all points of the power model space discussed in Section 13.4.1, with different accuracy/effort trade-offs. In spite of the vast amount of available material, it is quite easy to identify the key results in the power macromodeling domain.

The technique described by Powell and Chau [19] can be considered as the first proposal of a power macromodel for data-path units; it is actually just a capacitance macromodel, because it assumes fixed activity, that is, no activity parameters are included. Landman and Rabaey [20] recognized that modeling activity is essential, by observing that even "random" data do not exhibit true randomness on all bits,

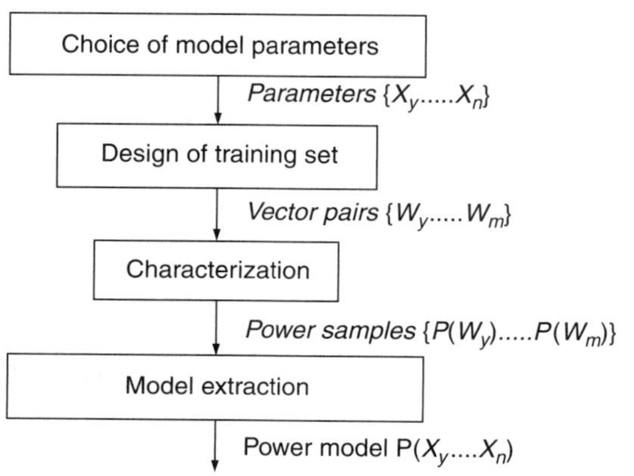

FIGURE 13.2 Macromodeling flow.

because of sign extension bits due to reduced dynamic ranges of typical values. Their model, called *dual-byte type* (DBT), separates data bits in two regions (sign and random), with distinct activity parameters. Since each unit may have different combinations of input and output regions depending on their functionalities, each unit must have a distinct model.

Gupta and Najm [21] proposed a power macromodel that can be considered as the state of the art; it consists of a three-dimensional look-up table of power coefficients, whose dimensions are P_{in}, the average input probability, D_{in}, the average input *transition density*, and D_{out}, the average output transition density. Characterization is based on quantizing each dimension into equally spaced intervals.

Figure 13.3 shows the three dimensions of the look-up table in the case of a quantization interval of 0.2. Notice how the relation between D_{in} and P_{in} constrains the number of feasible points in the (D_{in}, P_{in}) parameters' space.

Under this model, characterization involves the generation of a number of input streams for each feasible value of parameters D_{in} and P_{in}. Values of parameter D_{out}, instead, are extracted from the resulting output stream (obtained through simulation), and are discretized using a binning technique.

The three-dimensional look-up table model provides two main advantages with respect to the DBT model (or its variants). First, the use of a look-up table with discretized dimensions significantly improves the robustness of the estimates. Second, since parameters are normalized values, the same model template can apply to any data-path unit, without the need of (1) customizing it to its functionality, and (2) including an explicit "capacitance" (i.e., complexity) parameter (thus, without the need of being parameterized).

Many macromodels published in the literature build around this three-dimensional table-based model, by either improving some specific aspects that help to increase accuracy (e.g., interpolation schemes and local accuracy improvements) [22–25], or by generalizing it into a parameterized model (i.e., where the bit-width of the operators becomes a parameter, as in the DBT model) [26].

13.4.2.2 Power Modeling of Controllers

Control logic typically has a smaller impact than the data-path; however, its dissipation is not negligible for several reasons. First, some designs are control-intensive, and may contain many interacting controllers whose impact on power may be sizable. Second, data-path units are often hand-crafted while controllers are synthesized; then, they use silicon area less efficiently than data-path. Third, controllers are usually active for a large fraction of the operation time, while the data-path can be partially or completely idle.

What makes controller power modeling more challenging than data-path units is the fact that controllers are usually specified in a very abstract fashion even at the RT level and they are specific to each

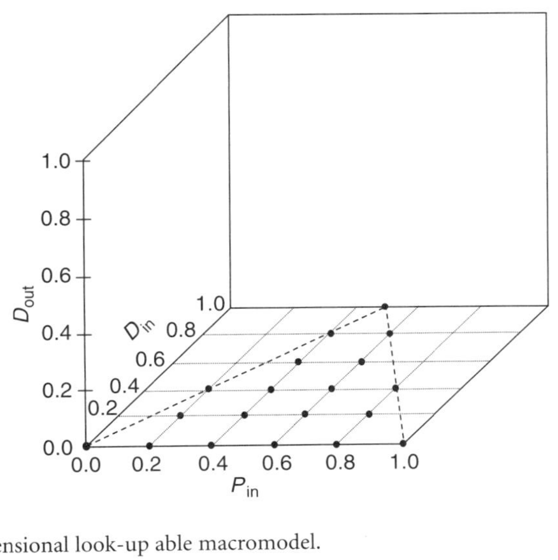

FIGURE 13.3 Three-dimensional look-up able macromodel.

design. In contrast, data-path units are usually instantiated from a library that can be precharacterized once and for all. Some controller implementation styles employ regular structures, like PLAs or ROMs, that can be precharacterized with good accuracy. On the other hand, the most common choice for controllers is to synthesize them as sparse logic. Because of its irregular structure, this makes power modeling considerably harder.

One additional difficulty lies in the fact that the description of controllers is more abstract than that of data-path components because states are usually specified symbolically. Thus, controller synthesis goes through two steps, namely state assignment and logic synthesis.

Power modeling of controllers at the micro-architectural level can be classified into *pre-state assignment* [27,28] and *post-state assignment* [4,6,29] depending on the target description of the controller. The former is generally less accurate, and can be used to provide upper and lower bounds to power consumption. The latter tries to give actual estimates of average power. Strictly speaking however, RTL descriptions of controllers typically fall in the former category.

Due to the poor link between an abstract, state-based description and the actual implementation, power models for controllers that have some practical value are empirical. The analytical model presented in [27] provides in fact theoretical bounds for the switching in FSMs, but the correlation to actual power figures is intrinsically very weak. For controllers, macromodeling is thus more than just an option. The difference with respect to data-path components, however, is that the model template is not so intuitive, because it is difficult to estimate the correlation between parameters and the actual power consumption. Therefore, the choice of the parameters is even more critical than in the case of data-path units. Possible parameters for controllers include:

- *Static behavioral* parameters (i.e., parameters that can be obtained from the behavioral specification): the number of inputs, the number of outputs and the number of states.
- *Dynamic behavioral* parameters (i.e., parameters that can be obtained from functional simulation): the average input and output signal and transition probabilities.
- *Static structural* parameters (i.e., parameters that are available after state encoding): the number of state variables.
- *Dynamic structural* parameters (i.e., parameters that can be obtained from RTL simulation): the average transition probability of the state variables.

A framework for the exploration of all possible models of order M with up to N terms is described in [30]; it defines a third-order model with four terms that yields an average relative error of about 30% and relative standard deviation of the error of about the same magnitude. The model uses only a subset of the above parameters, but from all of the four above-described categories. This model has been shown to be far more accurate and robust than other models, including the "intuitive" model, that is, the one that uses a linear combination of switching activities (for inputs, outputs, and state variables) weighted by their corresponding cardinalities (number of inputs, outputs, and state variables).

13.4.2.3 Power Modeling of Memories

As for any other component, memory power can be fitted into the template of Equation (13.2):

$$P_{\text{mem}} = \frac{1}{2} C_{\text{mem}} V_{\text{dd}}^2 f_{Ck} E_{\text{Sw}}$$

where C_{mem} denotes the capacitance that is switched on a memory access. Since the capacitance for a read access is, in general, different from that for a write access, C_{mem} should be considered as an average capacitance.

For this reason, it is more accurate to resort to a cycle-accurate model, and express average power as follows:

$$P_{\text{mem}} = \frac{1}{2} \frac{1}{N_{\text{cycles}}} (N_{\text{read}} C_{\text{read}} + N_{\text{write}} C_{\text{write}}) f_{Ck} V_{\text{dd}}^2 \quad (13.5)$$

where N_{cycles} is the total number of cycles, N_{read} (N_{write}) the number of read (write) accesses, and C_{read} (C_{write}) the corresponding capacitances.

Memories, and in particular SRAM arrays, unlike generic RTL blocks, lend themselves to a relatively easy modeling of capacitance, owing to their well-defined internal structure which is characterized by high regularity. In other words, it is feasible to adopt an analytical model that expresses the total memory power as the sum of the various components: read/write circuitry, decoders, cells, bitlines, wordlines, MUXes, and sense amplifiers [17,31,32]. These models are very accurate, but suffer from being strongly dependent on technology; in fact, they require the knowledge of parameters such as the capacitance of the bitline or the wordline, or the capacitance of a minimum-sized transistor, which cannot always be easily accessible.

A simpler, yet effective model for micro-architectural estimation is an empirical one, where *all* model parameters are chosen in order to be easily available at a higher level of abstraction (e.g., the number of words) and the relation to the technology is established in a characterization run. For example, a typical capacitance model can be expressed as

$$C_{\text{mem}} = a + bW + cN + dW\,N$$

where W is the number of rows and N the number of columns of the memory array [4]. A similar model that also included the number of words as parameters was used in [33]. The characterization proceeds very similarly to the conventional macromodeling flow, in which the values of a, b, c, and d are determined by means of LMS regression. If we stick to the template of Equation (13.5), separate characterizations for read and write capacitances are required.

13.4.2.4 Power Modeling of Wires

If we assume that the power consumption of data-path units and controller is based on empirical macromodels, their power estimates include the contribution of (internal) wires. There are then two categories of wires that need a custom power model: *Global* wires, such as reset and clock, and *inter-component* wires, such as those connecting RTL blocks.

For both types of wires similar considerations apply: in fact, power is consumed by wires when charging and discharging the corresponding capacitances. Therefore, the model of Equation (13.3) where A_i and C_i are the switching probability and the parasitic capacitance of the ith net, respectively well fits wires also. Since factor A_i is available for both global wires and inter-component wires, the problem of modeling wiring power reduces to that of modeling wiring capacitance, or, equivalently, wiring length. Wiring capacitances are unknown at the RTL, but realistic estimates can be obtained based on structural information, area estimates, and low-level wiring models. The problem of estimating the total wire length at high levels of abstraction is quite well understood; it is based on variants of Rent's rule [34–36] which relate the length of the interconnect to macroscopic parameters that can be easily inferred from a high-level specification, such as the number of I/O pins.

The main difference between the power models for global wires and inter-component wires lies in their "context." Area-based estimates work reasonably well for global wires, since they span the entire design. As a matter of fact, typical power models for global wires directly relate wire length to a power k of design area (often $k = 0.5$), and the latter to a power q of the number of I/O pins [20].

For inter-component wires, conversely, their length is more weakly correlated to the entire design area. The solution adopted is very similar, yet on a smaller scale—that is, rather than referring to the entire design, each wire refers to the components it connects. In this case, wire length is made proportional to a power of the component's area, which is related to a power of the number of I/O pins of the component [37].

Figure 13.4 shows the conceptual topology of an inter-component wire. The output capacitance C_{out} of a component is implicitly accounted for by its power model. Conversely, the input capacitance C_{in} is not included in the power model, since power estimates provided by low-level simulation represent the power drawn from the supply net, while input capacitors are directly charged by the primary input lines. The total wire capacitance should thus take into account the fanout of a wire (known at the RTL), both externally

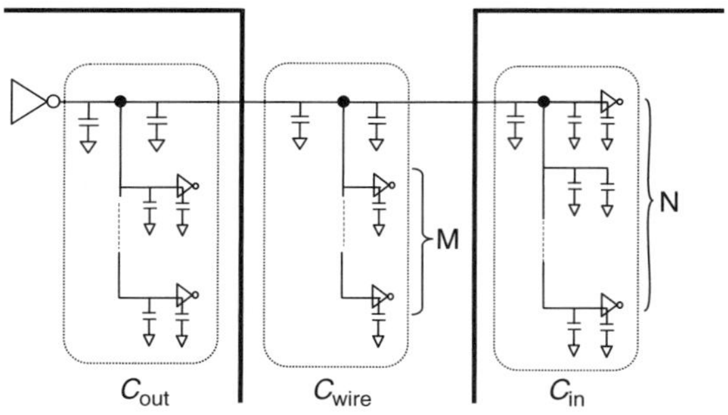

FIGURE 13.4 Hierarchical topology of an inter-component wire.

(C_{wire}) and internally (C_{in}) by splitting each contribution into a fixed term (the "stem" capacitance) and a quantity proportional to the number of fanouts (the "branch" capacitance).

It is worth emphasizing that in ultra-deep-submicron technologies, where wires dominate chip area, the estimation of wire length requires some early floorplanning information in order to be reliable. Such information can then be back-annotated into the RTL description to guide wire length estimation based on the empirical models discussed above.

13.4.2.5 Power Modeling of Buses

At the micro-architectural level, buses are relatively straightforward entities consisting of a set of wires. However, we believe they deserve a separate treatment from the generic wires we have considered in Section 13.4.2.4, since they are usually seen as a single-interconnection resource that will eventually be routed together as *a set* of wires; this physical "grouping" of bus wires is also reflected by typical HDL descriptions, where buses are represented as arrays of signals.

Buses may be shared (that is, they may not just be point-to-point connections) and, mostly, are single-master buses relying on a simple protocol that defines the physical (i.e., the signaling scheme) and the "data-link" (i.e., the binary representation of the transmitted "values") layers. More complex on-chip bus architectures that include features such as support for multi-master and complex protocols (i.e., those required by modern multi-processor SoC architectures) tend to be categorized as hardware blocks rather than as a wiring infrastructure, because of the high amount of hardware control they require [38]. This trend is witnessed by the wide availability of synthesizable IP blocks that can be modularly used by RTL designers and automatically synthesized (e.g., Synopsys' DesignWare IPs for AMBA buses [39]). In this section, we will focus on single-master buses that are typical in non-core-based designs.

Specifically, with reference to the template of Figure 13.1, buses are the resources that are used to connect blocks of two types (possibly in a shared manner): computational units and memories.

A conceptual model of the average power consumed by a bus consisting of n lines can be obtained by simply summing up the contribution of each wire, according to the model of Equation (13.2):

$$P_{bus} = \frac{1}{2}\sum_{i=1}^{n}(C_L V_{dd}^2 f_{Ck} E_{Sw}^i)$$

where C_L is, in this case, the capacitance of a single bus line, and E_{Sw}^i the switching activity of the ith bus line. The equation is based on the assumption that all lines have roughly the same capacitance. This model would be reasonably accurate if the bus lines were routed independently of each other, possibly on different metal layers. Conversely, the "grouped" nature of buses makes the model unrealistic, because it considers each wire as if it were isolated from the others, thus completely ignoring the contribution due to *coupling capacitances*, that is, the mutual capacitive effect of two neighboring wires.

Figure 13.5 shows a simplified view of the capacitances switched between two adjacent wires. C_L is the ground capacitance considered in the model (also called *self*-capacitance), while C_C denotes the coupling

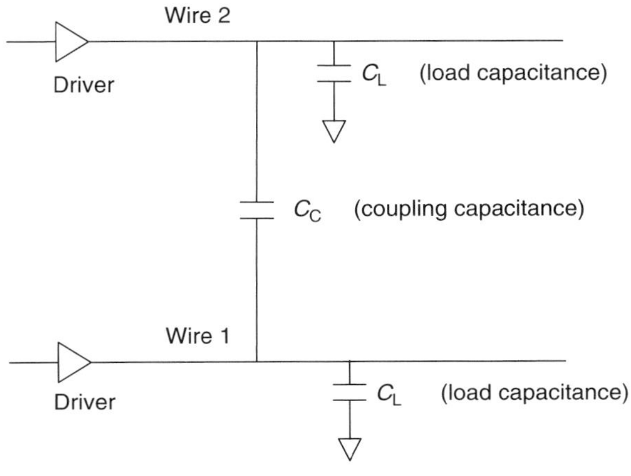

FIGURE 13.5 Capacitances involved in bus line switching.

capacitance. In deep-submicron technologies, the magnitude of C_C far exceeds the self-capacitance: Electrical-level simulations for 0.13 μm technologies have shown that C_C is more than three times the value of C_L [40], and that this factor will further increase in future technologies.

In view of the discussion above, accounting for the contribution of C_C in on-chip buses is mandatory; this can be done by augmenting the basic power model as follows [41]:

$$P_{bus} = \frac{1}{2}\left(\sum_{i=1}^{n}(E^i_{Sw}C_L + E^i_{Sw}C_C)f_{Ck}V^2_{dd}\right) \tag{13.6}$$

E^i_{Sw} denotes the *coupling switching activity*, i.e., a quantity related to the simultaneous switching of two adjacent lines. In fact, when the transitions on two adjacent lines, *a* and *b*, are aligned in time, there are only two transition pairs that cause C_C to switch: (1) when both *a* and *b* switch to different final values; (2) when one of the two lines switches, while the other one does not, and their final values are different.

Table 13.1 shows the normalized switched capacitance for a two-line bus, when all capacitive effects are considered. λ represents the ratio C_C/C_L. In the table, only $0 \rightarrow 1$ transitions are counted as power-dissipating transitions on C_L.

The table clearly shows that increasingly larger values of λ will tend to emphasize the importance of the energy due to switching of the coupling capacitance.

13.4.3 Power Estimation

The modeling technology discussed in the previous sections can be successfully used to enhance state-of-the-art RTL-to-physical design flows with power estimation capabilities.

Assuming that the design to be estimated is described by an FSMD, as defined in Section 13.3, the estimation procedure consists of three basic steps [8], as shown in Figure 13.6.

The first operation implies identifying and separating the data-path components from each other and from the FSM that represents the control. This is needed in order to enable the power estimator to generate the power models for each component in the FSMD.

The FSMD then needs to be simulated. To this purpose, an RTL simulator is used to trace all the internal signals that define the boundaries between the various components in the FSMD. This is of fundamental importance for the evaluation of the power models that is needed to complete the estimation procedure.

Finally, the actual power estimation must take place. The design hierarchy is traversed at the top level, from the inputs toward the outputs, and for each component, model construction and model evaluation

TABLE 13.1 Normalized Switched Capacitance on Two Adjacent Bus Lines

		(b^t, b^{t+1})			
		$0 \to 0$	$0 \to 1$	$1 \to 0$	$1 \to 1$
	$0 \to 0$	0	$1+\lambda$	0	0
(a^t, a^{t+1})	$0 \to 1$	$1+\lambda$	2	$1+2\lambda$	1
	$1 \to 0$	0	$1+2\lambda$	0	λ
	$1 \to 1$	0	1	λ	0

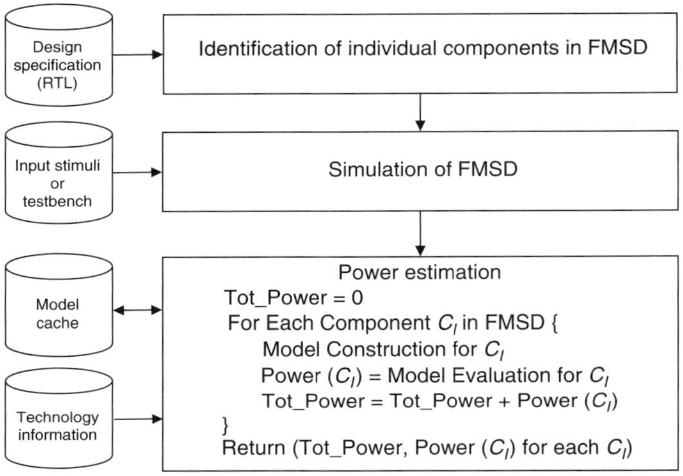

FIGURE 13.6 Micro-architectural power estimation flow.

are carried out. Model construction entails building the proper model according to the type of component being considered. A caching strategy may be used to limit the number of times models are built; more specifically, after the model for a given component is built, it is stored in a cache, so that it can possibly be reused at later times. Model evaluation, on the other hand, requires that the parameter values obtained during the RTL simulation phase are plugged into each model to get the actual power values.

The total power information for the design is then obtained by summing up the contribution of the model of each component; both a total power budget and a power breakdown can thus be reported to the user.

13.5 Micro-Architectural Power Optimization

This section presents some of the most popular micro-architectural power optimization techniques used in designs today. We first deal with data-path components and controllers; in this case, we classify the existing solutions into three categories:

- Those based on clock-gating, whose objective is to stop the clock for some cycles of operation in order to achieve a reduction of the switching component of the power.
- Those based on exploitation of common-case computation, whose objective is to optimize switching power consumption for the most common operation conditions.
- Those based on dynamic management of threshold and supply voltages, whose goal is that of reducing either leakage or switching power, or both by operating the logic in a multi-V_{th}/multi-V_{dd} regime.

For each class of approaches we discuss the basic idea as well as the algorithms used and challenges faced in applying these techniques in automated flows. We then move to techniques applicable to memories and buses. Owing to the peculiarty of these components, their optimization is treated separately, toward the end of this section.

13.5.1 Clock-Gating

Clock-gating dynamically shuts off the clock to portions of a design that are "idle" for some cycles. The theory of this technique has been investigated extensively in the past [42–44], and clock-gating is now considered the most successful and widely adopted solution for switching power reduction in real designs [45–48].

In some cases, it may be possible to shut off the clock to an entire block in the design, thereby saving large amounts of power when the block is not functioning. Perfect cases for this are when a block is used only for a specific mode of operation, e.g., the receiver and transmitter parts in a transceiver may not be active at the same time, and the receiver can be shut off during transmit stages, or vice versa.

It is also possible to gate the clock of a single register or set of registers. For instance, synchronous, load-enabled registers are usually implemented using a clocked D-type flip-flop and a recirculating multiplexor, with the flip-flop being clocked at every cycle as shown in Figure 13.7a.

In the gated-clock version of Figure 13.7b, the register does not get the clock signal in the cycles when no new data are loaded, therefore reducing switching power. Savings are further enhanced by the removal of the multiplexor. Gating a single-bit register, however, has the associated penalty of power consumption in the clock-gating logic. The key is then to amortize such a penalty over a large number of registers, by saving the flip-flop clocking power and the multiplexor power of all of them using a single clock-gating circuit.

The transparent latch is used to guarantee that spurious glitches in the enable signal occurring when the clock is high are not propagated to the clock input of the register. The latch freezes the output at the rising edge of the clock, and ensures that the new enable signal, EN1, at the AND gate is stable when the clock is high. Meanwhile, the enable signal can time-borrow from the latch, so that it has the entire clock period avaliable to propagate.

We cover advanced methods for detecting clock-gating opportunities in Section 13.5.1.1; next, we present issues associated with implementing clock-gating in a typical automatic flow, along with possible solutions in Sections 13.5.1.2–13.5.1.7.

13.5.1.1 Advanced Clock-Gating

The clock-gating conditions that depend on the enable signal of the register bank, which can be identified by means of topological inspection and analysis of the RTL description, can be extended by considering the functional behavior of the circuit. In particular, it is possible to augment the opportunities for stopping the clock that feeds a given register anytime the output of such a register is not observable, that is,

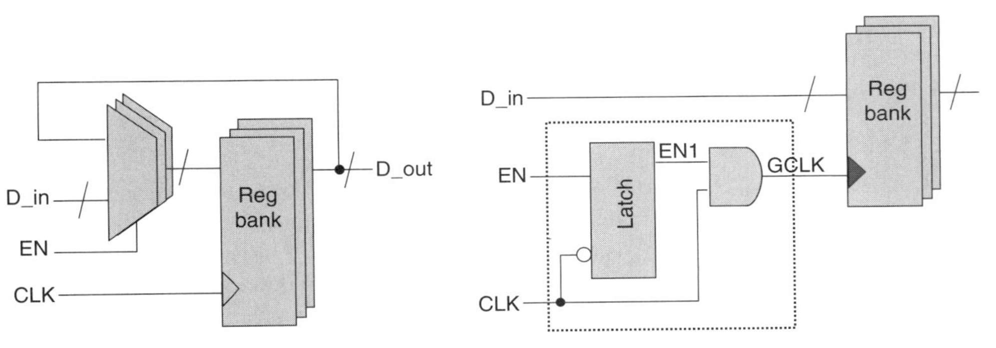

FIGURE 13.7 Clock-gating: (a) load-enabled register implementation; (b) gated-clock implementation.

the value of the output is not used by the gates in the register's fanout cone. Calculating these additional options for clock-gating entails finding the *observability don't care* (ODC) conditions for the output of the register; Babighian et al. [49] propose to extract idle conditions from a RTL netlist, by focusing on control signals that drive steering modules (e.g., multiplexers, tristate buffers, and load-enabled registers). Referring, for instance, to a multiplexer, all but one data inputs are unobservable at any given time depending on the value(s) that are carried out by the control signal(s), so that the logic values of the unobservable branches become irrelevant to the correct operation of the circuit. Hence, the clock signals of the registers in the fanin cone of an unobservable multiplexer branch can be gated without compromising the functionality.

The goal of the approach of Babighian et al. [49] is to improve the effectiveness of clock-gating by creating an activation function that can stop the clock of a set of registers for a significant fraction of cycles when the register outputs are unobservable. This is done in two steps. First, by performing ODC computation, which is based on a backward traversal of the data-path in decreasing topological order, considering only ODCs created by steering modules. Second, by generating the activation function, which implies the synthesis of the ODC function and the addition of clock-gating logic to the RTL netlist. Even though ODC computation is simple for a single-steering module, ODC expressions can become quite large if the netlist has many levels of steering modules and many fanout points. Therefore, traversal of only a limited number of levels in the netlist is allowed.

Once the ODC expression is computed, the corresponding logic must be instantiated in order to drive properly the clock-gating logic. The main difficulty in this step is due to the fact that ODC conditions masking register in clock cycle T may be used to gate their clock in cycle T-1. In other words, the clock-gating logic may need to be active in the clock cycle immediately before the register becomes unobservable. Unfortunately, the control signals at the inputs of the ODC functions are generated one clock cycle too late.

If the control signals are available directly as outputs of the registers, the instantiation of the clock-gating logic is relatively straightforward. Logic gates implementing the ODC expressions are inserted and their inputs are connected to the inputs of the registers. In real-life designs, however, the control inputs of the steering modules seldom come directly from the registers; instead, they are often generated by additional logic. In this case, the entire cone of logic between registers and control signals should be duplicated and connected at the inputs of the registers, and ODC computation gates should then be connected at the outputs of the duplicated cones. Clearly, the addition of this extra logic may represent a nonnegligible overhead.

Most of today's commercial synthesis tools deal with this issue by restricting the type of activation function used to gate the clock. In practice, they just detect ODCs generated locally to the registers by assuming the register's outputs to be always observable by the environment. This corresponds to considering, as ODC function, the complement of the enable signal, which feeds the clock-gating logic.

By detecting clock-gating conditions only when enable signals are present, no precomputation of clock-gating conditions is required and thus no duplication is needed. Notice that ODC-based clock-gating subsumes traditional automatic clock-gating as a very special limit case (i.e., backward traversal is completely avoided). Thus, the clock-gating conditions computed by the ODC-based approach are guaranteed to be a superset of those targeted by tools that introduce clock-gating only for load-enabled registers (see Figure 13.7).

13.5.1.2 Clock-Skew Issues

A problem in latch-based architectures comes from the fact that clock skew between the latch and the AND gate can result in glitches at the gated-clock output. This is explained in Figure 13.8. In particular, Figure 13.8a shows the case when the clock arrives much earlier at the AND gate than at the latch. Here, the clock skew between the latch and the AND gate should be less than the clock-to-output delay of the latch for the circuit to function properly. Figure 13.8b depicts the situation when the clock arrives earlier at the latch. In this case, the clock skew between the AND gate and latch should be less than the sum of the setup time of the latch and the input-to-output delay of the latch to function properly. Therefore, the

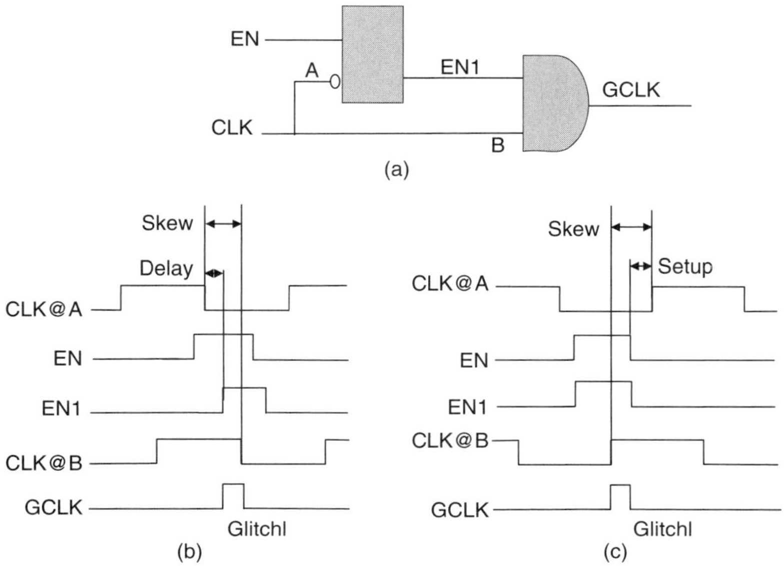

FIGURE 13.8 Clock skew within clock-gating logic: (a) clock-gating logic; (b) Glitches due to positive clock skew between latch and AND gate; (c) glitches due to negative clock skew between latch and AND gate.

clock-skew between the latch and AND gate, C_s, should be carefully controlled according to the following equation:

$$-(s + d_{in}) < C_s < d_{clk}$$

where s is the setup time of the latch, d_{in} the input-to-output delay of the latch, C_s the difference in clock arrival time between the latch and the AND gate (the clock arrival time at the AND gate minus the clock arrival time at the latch), and d_{clk} the clock-to-output delay of the latch.

Depending on the relative placement of the latch and the AND gate, these requirements may pose very stringent constraints on the clock-tree synthesis (CTS) tool.

The best way to control the relative timing of the two clock signals is to keep the entire structure in a single cell, called the *integrated clock-gating* (ICG) cell. The cell should be designed specifically for clock-gating, with the explicit requirements discussed above. Most technology libraries today do include the ICG cell as part of their primitives. Another way to address this issue is to ensure that the latch and the AND gate are close to each other during the placement phase of the design, by placing hard constraints on the distance between them. This makes it simpler for CTS tools to reduce the clock skew between them during the clock routing phase.

13.5.1.3 Clock Latency Issues

In order to maximize the power savings, a single clock-gating cell may be used to gate several registers, if the activation function is common to all of them. But the gated-clock may not have enough strength to drive all these registers, calling for a clock tree at its output. If a clock tree is introduced between the ICG cell and the registers it controls, the clock signal at the gating logic arrives much before the clock signal at the registers, and the activation signal must be ready before the clock arrives at the gating logic. This applies strict timing constraints on the activation signal, which must be addressed during synthesis.

13.5.1.4 Clock-Tree Synthesis

In the presence of massively gated-clocks, CTS tools must automatically address the presence of gated-clocks, both combinational and sequential, and also the integrated clock-gating cell on the clock line.

Clock tree synthesis tools must support different relative latency requirements at different points in the clock tree, since the clock latency at the gated-clocks can be very different from the latency at the registers. If integrated clock-gating cell is not used, the CTS tool would need to provide stringent control of clock skew between the latch and the AND gate.

13.5.1.5 Physical Clock-Gating

Physical clock-gating simultaneously takes into account the factors mentioned in Sections 13.5.1.2–13.5.1.4, namely skew, latency and CTS issues. There exists a spectrum of clock-gating approaches with regard to the placement of clock-gating cells into a clock tree. Designers often opt to place the clock-gating cells as close as possible to the final placement of their corresponding registers, as shown in Figure 13.9a. This placement can be enforced during physical synthesis by specifying a bound for the proximity of the clock-gating cells to the registers. Some advantages of this approach are that it makes it easier to estimate the latency from the clock-gating cell and it also increases the amount of available slack for the arrival of the activation signal. The impact on the clock tree is fairly minimal since the clock-gating cells are placed close to the registers and can eliminate the need for buffer insertion after clock-gating cell insertion.

A disadvantage to this approach is that it leaves most of the clock-tree switching even when branches are leading to registers that will have the clock blocked by a gated-clock. In order to save as much power as possible, it is desirable to gate as many buffers on the clock tree as possible. This is difficult for the designer to do without knowledge of the actual physically induced timing constraints.

In a physically aware clock-gating system, CTS works in conjunction with placement and clock-gating to determine an optimized placement and insertion of the clock-gating cells into the clock tree. This information is used to balance the delay on the activation signal with the amount of potential power saved by placing the clock-gating cell closer to the root of the clock tree, as shown in Figure 13.9b.

13.5.1.6 Clock Tree Planning

Cell placement, clock-gating cell insertion and CTS can also be decoupled, provided that the process of clock tree planning is started after cell placement and it does not interfere with clock routing. The objective of the approach in [50,51] is to build a power-optimal gated-clock tree structure fully compatible with state-of-the-art physical design tools to perform detailed clock routing and buffering. Thus, the output of the proposed clock tree planning methodology is not a completely routed clock tree; instead, it is a clock netlist (including clock-gating cells and related control logic) and constraints that, provided as input to CTS tools, lead to a low-power gated-clock tree, while still accounting for all nonpower-related requirements (e.g., controlled skew and low crosstalk-induced noise).

The methodology consists of three steps:

- Calculation of the clock-gating activation functions
- Generation of the clock tree logical topology
- Instantiation and propagation of the clock-gating cells and related logic

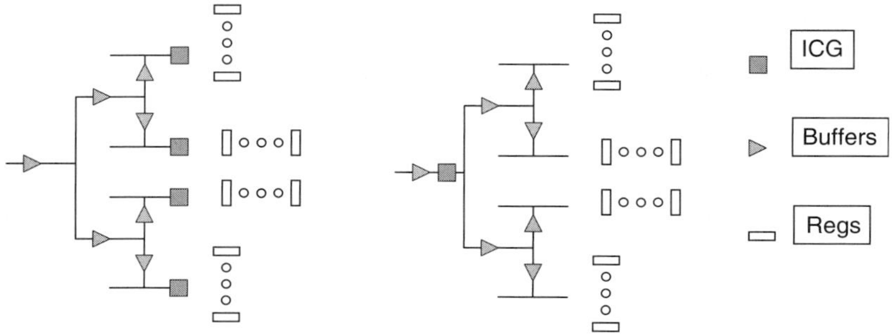

FIGURE 13.9 Integrated clock-gating cell placement: (a) close to registers; (b) buffering after ICG cell insertion.

The gated-clock activation functions for all the RTL modules are computed first. Next, according to the activation functions and the physical position of the registers, the logical topology of the clock tree is planned. This entails balancing the reduction in clock-switching activity against clock and activation function capacitive loads. Clock-gating cells are then inserted into the clock-tree topology and propagated upward in the tree whenever this is convenient, thus balancing the clock power consumption against the power of the gated-clock subtree. The information about the gated-clock tree is finally passed to the back-end portion of the flow, which will take care of clock-tree routing and buffering.

13.5.1.7 Testability Issues

Clock-gating reduces test coverage of the circuit since gated-clock registers are not clocked unless the activation signal is high. During test or scan modes, test vectors need to be loaded into the registers, and hence they must be clocked irrespective of the value of the activation signal. One way to address this, investigated from the theoretical standpoint in [52], is to include a control point or control gate at the activation signal, as shown in Figure 13.10a. This allows clock-gating signal, EN, to be overridden during the scanning in or out of vectors by the test-mode signal. In this way, during the test clock cycles, the clock signal is not gated by the activation signal, EN, and the register can be tested to see if it holds the correct state. Further, the test-mode signal is held at logic "1" during test mode, making any stuck-at fault on the activation signal unobservable. If full observability is required, this signal must be explicitly made observable by tapping it into an observability exclusive or (XOR) tree, as shown in Figure 13.10b.

A growing concern around scan-based testing is the power consumed during the scanning in and out of test vectors [53,54]. The changing register values during scanning can create activity levels that are much higher than those experienced during "normal" operation and lead to "good" chips failing during testing.

13.5.2 Exploitation of Common-Case Computation

It is well known that in complex digital architectures, some functionalities are far more exercised than others; this is due, mainly, to the fact that the data to be processed (i.e., the workload) may not have equiprobable distribution. Optimizing the common-case computation has thus become an established practice in high-throughput design [55], where variable-latency units replace fixed-latency ones to improve the overall system performance by adapting the latency of the data path to the length of the computation to be executed. Architectural retiming [56], speculative completion [57], and telescopic units [58,59] are examples of application, even in an automatic fashion, of the optimization paradigm based on exploitation of common-case computation.

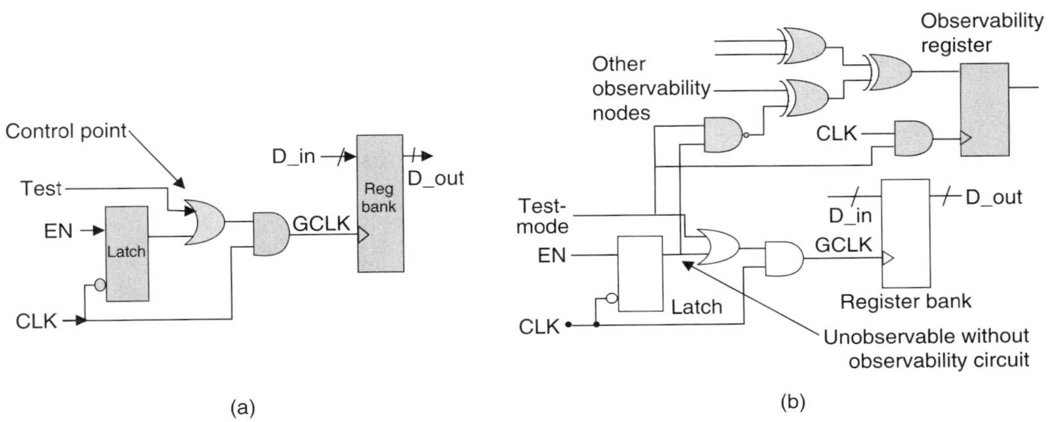

FIGURE 13.10 Testability issues with clock-gating: (a) adding controllability; (b) improving observability.

This idea has been extended recently to the case of power minimization; more specifically, as switching power depends on what the units of a design are which are activated by the input data as well as on the type of data to be processed, it may be possible to come up with variable-power architectural solutions which guarantee minimum power demand for the most probable execution conditions [60].

Figure 13.11 shows a possible architecture implementing the common-case optimization approach.

Block *A* supports the full functionality of the design while block *B* only covers the most common subset of it. *A* and *B* work in mutual exclusion, owing to the latch-enabled registers placed on the primary inputs of the two blocks. Based on the next datum to be processed, block *SEL* is in charge of selecting whether *A* or *B* should be activated in the next clock cycle. As *B* is much smaller than *A*, any time *B* is active, power is reduced. On the other hand, when *A* has to compute, there may be a penalty in power, as the whole design is larger (thus there are more gates that have to switch) if compared to block *A* alone. Let q be the probability of block *B* to be active, let P_A and P_B be the average power consumed by blocks *A* and *B*, respectively, when they are computing, and let P_{Ov} be the power overhead due to block *SEL* and the additional logic needed to make the architecture working. Actual power savings are achieved if

$$(1 - q)P_A + qP_B + P_{Ov} < P_A \tag{13.7}$$

Equation (13.7) tells us that the architecture exploiting the common-case computation obtains power reductions over block *A* alone for high values of probability q. Clearly, high q usually implies a larger *SEL* and a larger *B* (intuitively, what happens is that as q increases, some of the functionality of *A* is incorporated also into *B*). The challenge is thus that of designing the smallest possible *SEL* and *B* blocks that maximize the value of q and limit the impact on area and delay penalty.

In the sequel we review some solutions proposed in the recent literature that can be considered as practical actuation, with a slightly different flavor, of the design framework discussed above.

13.5.2.1 Operand Isolation

The concept at the basis of *operand isolation* [61] is illustrated in Figure 13.12.

In Figure 13.12a, we observe that the output of the multiplier is only used when the control signals to the multiplexors, SEL_0 and SEL_1, are both high. In cycles when either of the control signals is low, if the multiplier inputs change, the multiplier performs computation but its result is not used. The wasted power may be substantial if these idle cycles occur for long periods of time. Figure 13.12b shows operand isolation applied to the multiplier. First, the activation signal AS is created to detect the idle cycles of the multiplier. AS is high in the "active" clock cycles when the multiplier output is being used, and otherwise low. This signal is used to isolate the multiplier by freezing its inputs during idle cycles using a set of gates, called *isolation logic.* In Figure 13.12b, the isolation logic consists of AND gates, but OR gates or latches may also be used. Using AND/OR gates avoids the introduction of new sequential elements and reduces the impact on the rest of the flow. Also, AND/OR gates are cheaper and tend to give better power savings overall.

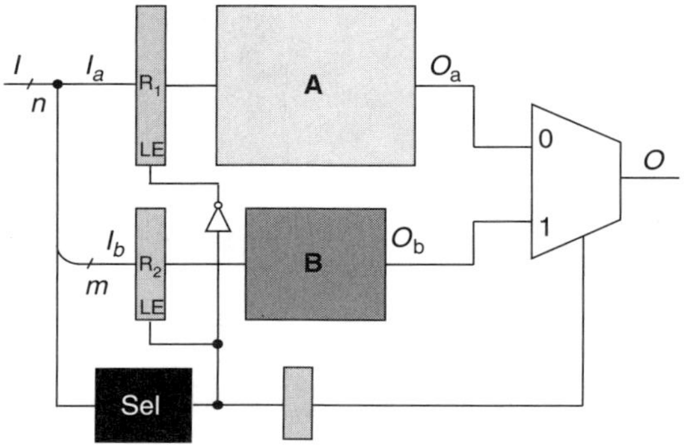

FIGURE 13.11 Exploitation of common-case computation.

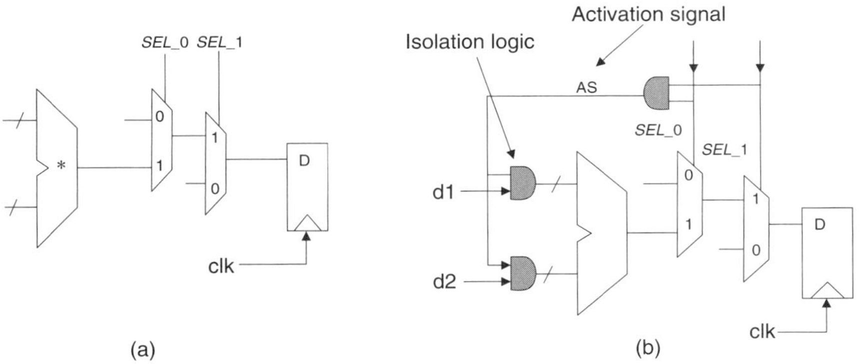

FIGURE 13.12 Operand isolation: (a) original circuit; (b) after operand isolation.

Operand isolation saves power by reducing switching in the operator being isolated, but it also introduces timing, area, and power overhead from the additional circuitry for the activation signal and the isolation logic. This overhead must be carefully evaluated against the power savings obtained to ensure a net power saving without too much delay or area penalty.

13.5.2.2 Precomputation

Precomputation [62,63] relies on duplication of part of the logic with the purpose of precomputing the circuit output values one clock cycle before they are required, and then using these values to reduce the total amount of switching in the circuit during the next clock cycle. Knowing the output values one clock cycle in advance allows the original logic to be turned off during the next time frame, thus eliminating any charging/discharging of internal capacitances.

The size of the logic that precalculates the output values must obviously be kept under control, since its contribution to the total power balance may offset the savings achieved by blocking the switching inside the original circuit. Several variants to the basic architecture can be adopted to take care of this problem. In particular, it may sometimes be convenient to resort to partial, rather than global, shutdown, i.e., to select for precomputation only a (possibly small) subset of the circuit inputs.

As an example, consider Figure 13.13a; the combinational block, *A*, implements an *N*-input, single-output Boolean function, *f*, and it has the I/O pins connected to registers R_1 and R_2. A possible precomputation architecture is depicted in Figure 13.13b.

Key elements of the precomputation architecture are the two N-input, single-output predictor functions, called g_1 and g_0, whose behavior is required to satisfy the following constraints:

$$g_1 = 1 \Rightarrow f = 1$$

$$g_0 = 1 \Rightarrow f = 0$$

The consequence is that if at the present clock cycle either g_1 or g_0 evaluates to 1, the load-enable (LE) signal goes to 0, and the inputs to block *A* at the next clock cycle are forced to retain the current values. Hence, no gate output transitions inside block *A* occur, while the correct output value for the next time frame is provided by the two registers located on the outputs of g_1 and g_0.

As mentioned earlier, the choice of the predictor functions is a difficult task. Perfect prediction requires $g_1 \equiv f$ and $g_0 \equiv f'$. However, this solution would not give any advantage in terms of power consumption over the original circuit, since it would entail the triplication of block *A*, and thus it would cause the same number of switchings as before, but with an area three times as large as the original network. Consequently, the objective to be reached is the realization of two functions for which the probability of their logical sum (i.e., $g_1 + g_0$) being 1 is as high as possible, but for which the area penalty due to their implementations is very limited. Also, the delay of the implementation of g_1 and g_0 should be given some attention, since the prediction circuitry may be on the critical path and, therefore, it may impact the performance of the optimized design.

One way of guaranteeing functions g_1 and g_2 to be much less complex than function f, thus implying a marginal area overhead in addition to a remarkable power savings, consists of making the two predictor functions depend on a limited number of inputs as compared to f.

Precomputation-based power optimization has been shown to be effective in the case of designs with pipelined structure. On the contrary, it seems to be hardly applicable to the case of sequential circuits with feedback. The reason for this is that the precomputation functions never attempt to stop the present-state inputs, which represent the majority of the inputs to the combinational logic for sequential circuits with a realistic number of memory elements (i.e., flip-flops).

13.5.2.3 Computational Kernel Extraction

It is known that, when in their steady state, complex sequential circuits tend to run through a limited set of states. Once such a set, called *computational kernel* [64], is extracted from a given circuit specification, it can be successfully used for various types of optimization, including power minimization.

Given a sequential circuit with the traditional topology shown in Figure 13.14a, the paradigm proposed in [64] for improving its power dissipation is based on the architecture depicted in Figure 13.14b.

The essential elements of the architecture are the following:

- The combinational portion of the original circuit (block C)
- The computational kernel (block K)
- The selector function (block SEL)

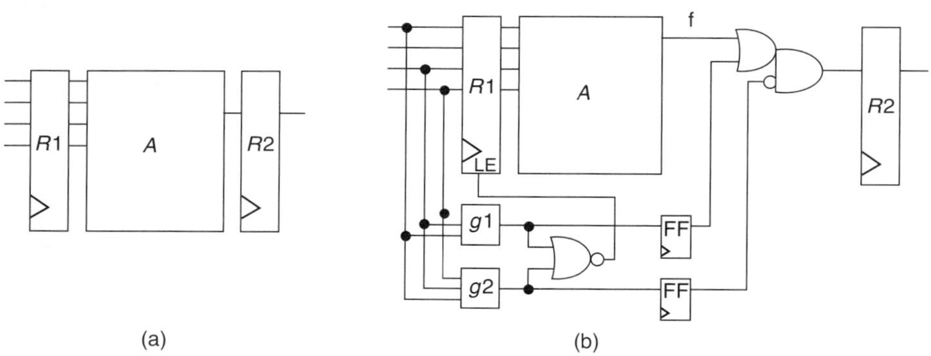

FIGURE 13.13 Precomputation: (a) original circuit; (b) precomputed architecture.

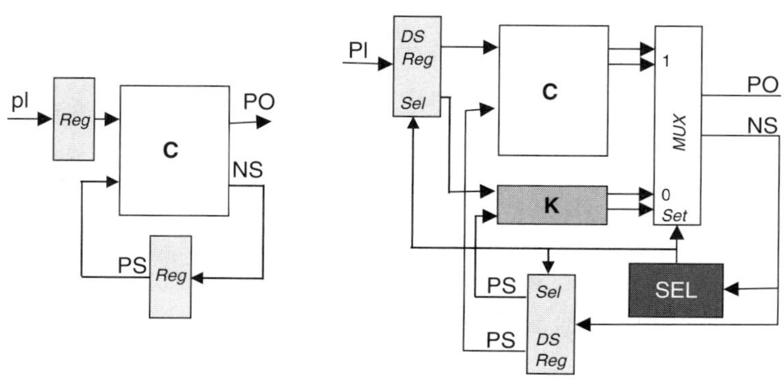

FIGURE 13.14 Computational kernel extraction: (a) original circuit; (b) kernel-based architecture.

- The dual-state registers (block *DS Reg*)
- The output multiplexor (block *MUX*)

The computational kernel can be seen as a "dense" implementation of the circuit it has been extracted from. In other terms, *K* implements the core functions of the original circuit, and because of its reduced complexity, it usually implements such functions in a faster and less power consuming way.

The purpose of the selector function *SEL* is that of deciding what logic block, between *C* and *K*, will provide the output value and the next state in the following clock cycle. To take a decision, *SEL* examines the values of the next-state outputs at clock cycle n. If the output and next-state values in clock cycle $n + 1$ can be computed by the kernel *K*, then *SEL* generates the value 1. Otherwise, it generates the value 0. The output of block *SEL* is connected to the multiplexor that selects which block produces the output and the next-state, as well as to the control input of the dual-state registers, which steer the primary and state inputs to the appropriate combinational block (either *C* or *K*). The optimized implementation is functionally equivalent to the original one.

The scheme in Figure 13.14b is just one among several possible architectures. The peculiar feature of this solution concerns the topology of the selection logic. In particular, the choice of having a selection function that only depends on the next-state outputs is dictated by the need of obtaining a small implementation. Reducing the size of the support of *SEL*, i.e., not including the primary inputs, is one way of pursuing this objective.

Fundamental for a successful application of the kernel-based optimization paradigm is the procedure adopted for kernel extraction. If the circuit is described by means of its state transition graph, the kernel can be determined exactly through symbolic (i.e., binary decision diagram (BDD-based)) procedures similar to those employed for FSM reachability analysis. On the other hand, when the state transition graph is too large to be managed, there are two options. If a gate-level description of the circuit is available, then block *K* can be iteratively synthesized by means of implication analysis followed by redundancy removal. Otherwise (i.e., if only a functional description of the circuit is available), kernel identification and synthesis can be performed by resorting to simulation of typical input traces and by then running probabilistic analysis and resynthesis of the state transition graph based on the results of the simulation. Obviously, in case approximate kernel extraction is adopted, the savings that can be obtained are usually more limited.

13.5.3 Managing Voltages

As discussed in Section 13.2, power consumption in CMOS designs heavily depends both on the operating supply voltage (switching power) and on the transistors' threshold voltage (leakage power).

The two variables are not independent, and switching speed constitutes the link. Starting from the 0.5 μm technology node, supply voltage levels have been scaling at approximately 1 V/0.1 μm. As we are now below the 100 nm feature size, the reduced operating voltages are forcing threshold voltages down to 0.25 V and below, in order to preserve speed. This has had a major impact on the leakage current of the transistors built into these technologies. As Equation (13.8) shows, the subthreshold leakage current grows exponentially as the threshold voltage decreases:

$$I_{sub} = I_0(e^{(-V_{th}/S)}(1 - e^{-qV_{ds}/kT})) \quad (\text{at } V_{gs} = 0) \tag{13.8}$$

For approximately every 65–85 mV decrease in threshold voltage (depending on temperature), there is an order of magnitude increase in subthreshold leakage current.

13.5.3.1 Managing Threshold Voltages

Silicon foundries have started to offer multiple threshold devices at the same process node to address the need to control leakage current and enabling designers to trade leakage and performance [65]. Along with the standard V_{th}, a low and high V_{th} transistor may be offered where the low-V_{th} device may have order of magnitude higher leakage than the standard V_{th} device, and the high-V_{th} device may have an order of magnitude lower leakage than the standard. This reduction in leakage is not free though, and it comes at the expense of the speed of the device — there could be a 20% to 2*x* delay penalty between the standard and the high-V_{th} devices — and an increase in the cost of the fabrication process.

Synthesis algorithms can be used to optimize leakage by using high-threshold voltages while still meeting the timing requirements using low V_{th} devices on the more critical paths. The threshold voltage of a transistor can be dynamically changed by varying its back bias. The change in V_{th} is roughly proportional to the square root of the back-bias voltage. As threshold voltages drop below 0.25 V, variable back biasing may gain more appeal. A distinct advantage of this approach is that during periods when heavy processing is needed, V_{th} can be reduced, thus speeding up the cells. When the cells are in a slower drowsy or idle mode, the threshold voltage can be raised, thus lowering the leakage. One significant impact of using variable back-biasing is that there are two new terminals for each cell that now need to be routed. A common ASIC design practice is to create cells that tie the *n-well* regions to V_{dd}, and the *p-well* regions to ground. In the physical implementation, these are simply pre-defined contacts designed into the cell that are connected as part of the power and ground routes. To enable back-biasing, new voltage lines are routed to control the bias. These can be to individual cells or, more likely, to regions that contain multiple cells sharing the same *well* and a common tie-cell to control the *well bias* [66,67].

13.5.3.2 Managing Supply Voltages

13.5.3.2.1 Power Gating

For applications like cell phones that have long shut-down periods, the power consumption is dominated by the leakage power consumption during off periods. *Power gating* addresses this issue by shutting off the power supply to blocks that are not in use. This not only eliminates the switching power dissipation in the block, but also hugely saves the leakage power dissipation when the block is shut down. In the simplest case of power gating, the voltage level across the chip is the same, but different power supply grids are used in different parts of the chip. Besides both leakage and switching power savings, this helps to control the IR drop on the power grids. These design techniques create interesting switching on the chip. When one block is shut down, the voltage on its output ports may drift to undefined values causing large leakage and unexpected functionality in the gates that it drives. Therefore, it is necessary to isolate all the output ports of a block that is shut down. Further, if a block is shut down, it might be important for it to *remember* some of its previous state values. The registers used to store these values need to be specially designed to be powered by a secondary supply that will allow them to retain their values during shut down and restore it when the primary supply is up [68,69].

Figure 13.15 shows a typical floorplan of a power-gated chip with isolation cells and retention registers.

EDA tools need to insert automatically isolation cells and retention registers for powered-down blocks. To complete the entire flow, it is important to be able to represent all this information in the source description (the RTL), and infer both isolation and retention logic during RTL synthesis. Since the power-down behavior directly impacts the functionality of the chip, it is important to be able to verify this functionality both with simulation tools and formal verification tools.

To enable power gating in a complete automated design flow, place, route and back-end optimization tools need to be aware of the different power supply regions. Cells powered by each supply need to be placed in separate areas, and signals should be routed so that they do not traverse areas serviced by a different power supply, since subsequent repeater/buffer insertion on these signals will not be legal.

With complex power-gating schemes, the role of power and power-network analysis tools becomes more important. Turning the voltage ON or OFF to a block can cause large transients on the power grid, affecting other blocks on the chip. These effects should be accounted for by the power analysis tools.

13.5.3.2.2 Multi-V_{dd} Design

Over the last few years, more and more chips are using multivoltage design in one form or the other, and EDA tools need to support the various special needs of these design styles [70,71]. For instance, the power-gating technique discussed earlier can be thought of as the simplest version of multivoltage design, where the ON voltage level is the same for all the blocks, but can be changed to 0 V during shut down.

A more complex methodology uses different voltages for different blocks on the chip, which are powered with separate power supplies that can be independently shut down [72].

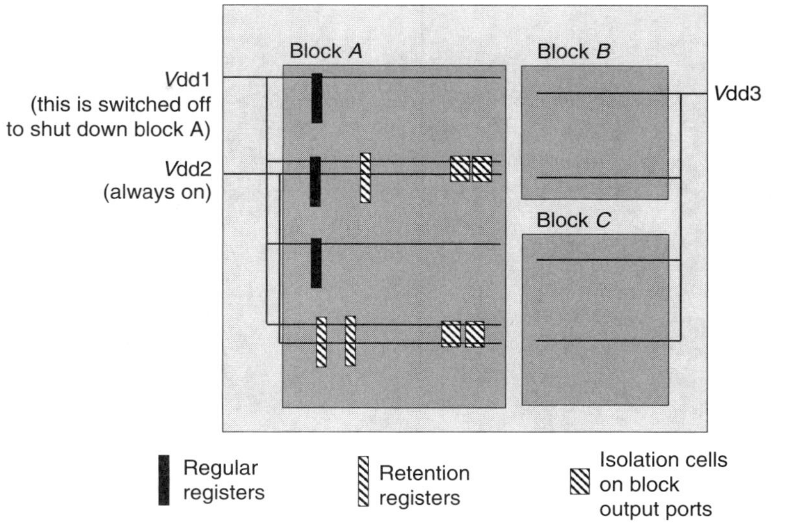

FIGURE 13.15 Applying power-gating.

An example of a multivoltage chip is shown in Figure 13.16.

Variations of multivoltage design styles include dynamic V_{dd} scaling, where on-chip controls are used to switch the voltage on the different parts of the chip to different predefined levels. Adaptive scaling takes this a step further, where high-level control feedback loops can be used to change the V_{dd} values of different blocks based on the speed/power requirements for the chip at any given time.

Along with the issues associated with power gating, multivoltage design has other requirements from EDA tools. First, tools need to understand the impact of voltage levels on timing and power of library cells (hence, on the design) and need to provide both analysis and optimization capabilities that accurately account for the different V_{dd} levels. Secondly, the implementation tools should be be able to insert automatically level shifters to be able to adapt signals from one level to another. Instantiation and optimization of level shifter cells is key to multivoltage design [73]. The tools have to manage cells that have more than one supply rail, and circuitry that can vary or completely shut down the supply voltage to a block.

Clock tree generators need to account for buffers that operate at different voltages to provide clock signals to each block, and the router needs to account for buffer placement in the context of different voltage regions on the chip. Routing a feed-through signal via a region may now require the insertion of level shifters in order to drive the signal adequately. Analysis tools need to understand these different situations — tracking all of these new voltage-based modes — and provide useful feedback to the designer.

An additional major impact on the design flow is the need to treat the supply line as another variable. For most previous mainstream designs, logical netlists only specified the input and output connections between gates. V_{dd} pins for all the cells (as also the V_{ss} pins) were attached to a single power (ground) network after the place and route step was done. In a multivoltage design, different cells are connected to different power (ground) networks and this information needs to be managed in the entire design flow starting from the RT level, to be able to simulate correctly and verify the design right from the start.

A final design implication that optimization and analysis tools must account for is the signal integrity impact of driving some lines at higher voltages than others. The higher-voltage lines can cause larger spikes in neighboring low-voltage lines than other lower-voltage aggressors, which impacts timing analysis, power, and the routing of lines on the chip.

13.5.4 Memory Power Optimization

The power model of Equation (13.5) exposes the two quantities that can be targeted to reduce memory power, namely, the number of (read or write) accesses and the (read or write) capacitance. The former are

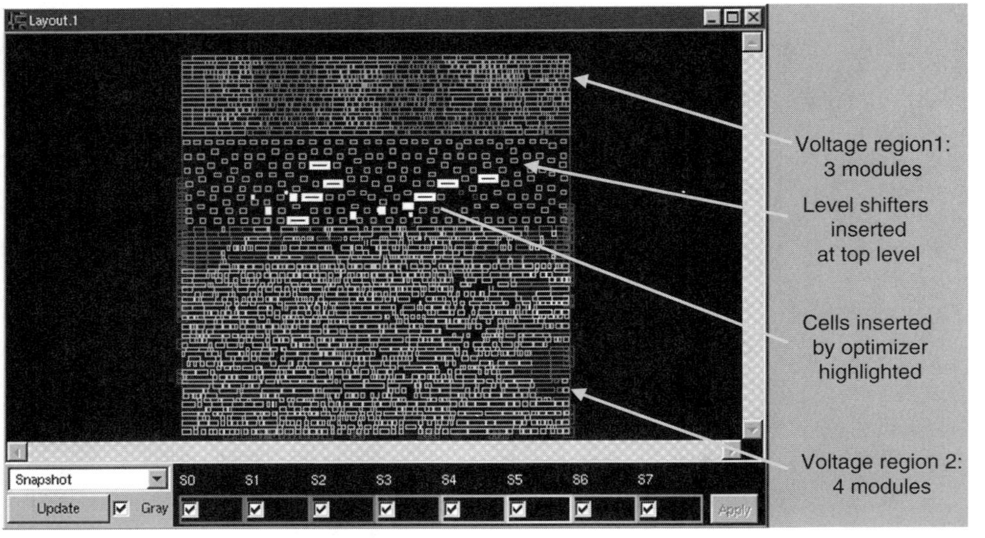

FIGURE 13.16 Example of multivoltage chip.

actually dictated by the scheduling of the operations resulting from register-transfer level design or high-level synthesis and should be considered as fixed. Attention should then focus on capacitance minimization.

Apparently, the read or write capacitances are related to physical parameters and are also uniquely defined for a given technology. In practice, C_{read} and C_{write} should be regarded as *average* read or write capacitances, that is, they should represent the cost of an *average* read or write access. This subtle difference opens the way to a class of optimization techniques that modify the *organization* of the memory, which can be seen as the RTL counterpart of techniques that, in a more general macro-architectural context, aim at optimizing the *memory hierarchy* of a system. The most well-known example is the use of a cache between processor and memory, whose net effect is that of reducing the *average cost* of accessing memory.

Such a reduction of average cost (i.e., capacitance) can be achieved by either augmenting the hierarchy vertically (i.e., adding levels to the hierarchy), or by growing it horizontally (i.e., adding support structures in parallel to existing hierarchy levels), or both [74,75]. In an RTL context, the cycle-accurate timing information contained in the description rules out the vertical transformations, since they affect the average cycle time of a memory. We can reasonably assume that, in a micro-architecture, only one level of memory hierarchy does (typically) exist. The latter option is therefore the only viable one that allows a relatively seamless integration in a micro-architecture. Horizontal modification of a memory can be restated in terms of the usual common-case paradigm discussed in Section 13.5.2.

In this specific situation, the common case is represented by the memory cells that are accessed most. The idea is that of "isolating" this common case from the average case; since both are identified by memory cells, this amounts to instantiating a memory block "in parallel" to the main memory. This is shown in the example in Figure 13.17.

The left side of the figure depicts the original memory, consisting of $W = 1000$ words, while on the right-hand side, the W memory locations are mapped onto two different memory blocks of sizes $W_1 = 200$ and $W_2 = 800$. Assume that for a given workload the memory cells in the monolithic block Mem are accessed a total of $N = 10000$ times. For the partitioned architecture, the common case, i.e., the most frequently accessed memory cells are placed into the smaller block Mem_1 (with $N_1 = 8000$ accesses) and the least frequently accessed locations are mapped to the larger block Mem_2 (with $N_2 = 2000$ accesses).

We realistically assume that the power cost for accessing a cell in a memory block is directly related to the size of the block itself (the larger the block, the higher the power cost). Specifically, for our example, we assume the power access cost for cells in Mem to be $C_{\text{Mem}} = 0.5$, the cost for cells in Mem_1 to be $C_{\text{Mem1}} = 0.1$, and the cost for cells in Mem_2 to be $C_{\text{Mem2}} = 0.4$.

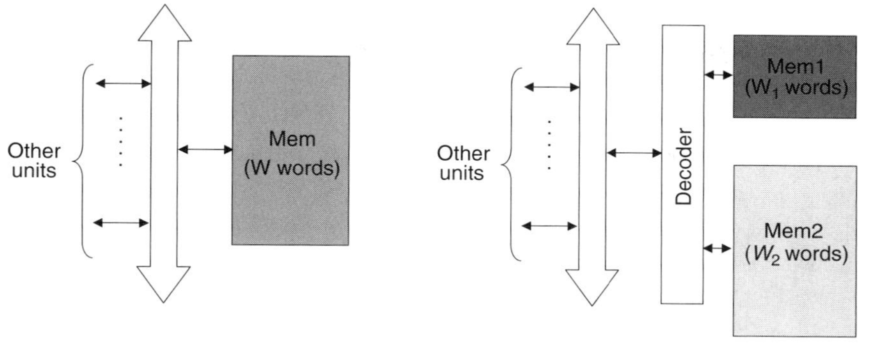

FIGURE 13.17 Common-case optimization applied to memory.

The average power for the monolithic memory is $P_{\text{mono}} = P_{\text{Mem}} = \dfrac{NC_{\text{Mem}}}{N} = 0.5$. For the partitioned memory, neglecting the decoder overhead, the power is $P_{\text{part}} = P_{\text{Mem1}} + P_{\text{Mem2}} = \dfrac{N_1 C_{\text{Mem1}} + N_2 C_{\text{Mem2}}}{N} = \dfrac{8000 \cdot 0.1 + 2000 \cdot 0.4}{10000} = 0.16$. The partitioned scheme is more convenient as N_1 gets larger w.r.t. N_2, meaning that the most frequently accessed locations should be mapped as much as possible onto smaller blocks.

The generic template illustrated in Figure 13.17 lends itself to numerous variants, depending for instance on whether the subblocks are overlapping or not, on the choice of the number and of the size of the subblocks, on the possibility of having noncontiguous partitions (i.e., addresses could be *relocated* within blocks), on the architecture and implementation style of the decoder, on the organization and routing of the address and data buses, on the placement of the subblocks, etc.

The common-case principle applied to memories requires proper extra logic to drive the accesses to the correct memory bank (the block generically denoted as "encoder" in the figure) whose power and performance overhead constrains the type of partitioning scheme allowed. Some memory partitioning variants, such as those proposed in [76], provide significant power reductions with a very limited hardware overhead that can be easily tolerated into an RTL design.

It is important to mention that, although the delay of the decoder is on the critical path, it usually does not affect the overall cycle time, since the access times of the subblocks are smaller than that of the monolithic memory; it suffices that $(t_{\text{decoder}} + \max_i(t_{\text{block}i})) \leq t_{\text{mem,mono}}$ in order to guarantee a seamless integration of the partitioned scheme ($t_{\text{mem,mono}}$ is the access time of the monolithic memory, and $t_{\text{block}i}$ is the access time of the generic subblock after partitioning).

13.5.5 Bus Power Optimization

Among the various parameters that appear in the model of Equation (13.6), only switching activity can be exploited at the micro-architectural level in order to reduce power. While supply voltage and frequency are somehow assigned up-front, capacitance values (both switching and coupling) can only be reduced during physical design through proper wire sizing (for self-capacitance), spacing, or shielding (for coupling capacitance).

Reducing the activity factors amounts to modifying the binary values that are transmitted to the bus, in other terms, *encoding* the values. For correct operations, clearly, both encoding and decoding are required at each end of the bus.

Since we are dealing with hardware components, encoding/decoding must be implemented through specialized hardware blocks. This obvious requirement poses serious limitations on the encoding that can be applied on an on-chip bus: codecs will in fact consume power, which if not kept under control may easily offset the power gain achieved from the reduction of switching and coupling activities. Moreover, the codec also adds delay to the corresponding paths, which must be limited since the cycle time is fixed up-front at the RTL. There is therefore a clear tradeoff between codec complexity (power, but also delay) and the reduction in the number of transitions. Since the latter are weighted by the switched capacitance, this trade-off can be cleanly expressed as a minimum value (switching and coupling) of capacitance that represents the break-even point of the power cost function; beyond this minimum value, the encoding scheme becomes advantageous.

This trade-off drastically limits the type of encoding to be chosen. The existing literature on low-power bus encoding is vast and it consists, in most of the cases, of quite complex encoding schemes that are more suitable for off-chip buses or for long global on-chip buses rather than for short data-path buses. Moreover, technology scaling plays against the application of encoding to on-chip buses: as wire capacitance progressively increases with respect to that of the cells, the average length of the wires must be decreased, thus making it more and more difficult to amortize the cost of the codec.

13.5.5.1 Bus-Encoding Schemes

Bus encoding can be applied at the physical level *(signal encoding)* or at the data-link level *(data encoding)*. The former consists of modifying (in time or space) the way the binary 0's and 1's are represented. For instance, a 1 could be encoded as a $0 \rightarrow 1$ transition. Data encoding, instead, consists of modifying the way the binary patterns are represented. For example, a word could be represented by adding one parity bit. The application of the two types of encoding is not mutually exclusive. However, signal encoding may have an impact on the technology used to implement the design. For instance, a typical signal encoding technique consists of reducing the voltage swing on (possibly some) of the bus wires. To make this possible, the technology on which the system will be implemented must be able to support that voltage level. For this reason, signal-encoding techniques are more related to the physical level of abstraction, and will not be analyzed further. Conversely, data encoding is more general, since the only underlying assumption is that of tolerating the insertion of additional hardware to perform the encoding.

The problem of defining a code that minimizes the number of (self and/or coupling) transitions enables a theoretical formulation that has led to solutions that mix results from information and probability theory. Although surveying these techniques is out of the scope of this chapter, a rough classification may help in understanding the problem better. Bus-encoding techniques can be categorized based on two dimensions:

- *The amount of redundancy allowed.* Some encoding schemes rely on spatial or temporal redundancy. Spatial redundancy implies the addition of extra bus lines, whereas temporal redundancy implies the addition of extra cycles to the bus transfers.
- *The amount of knowledge on the statistics of the transmitted data.* Some schemes assume *a priori* knowledge of the statistical properties of the information transmitted on the bus, which can be exploited to customize the encoding functions to the most typical behavior. An example is that of address buses, in which, although the specific patterns sent on the bus are not known, there exists a high degree of correlation between them, because of the sequential execution of a program.

Redundant codes are very popular in literature. One of the most referenced ones is the *bus-invert* (BI) scheme [77], in which the transmitter computes the Hamming distance between the word to be sent and the previously transmitted one. If the distance is larger than half of the bus width, the word to be transmitted is inverted, i.e., complemented. The information about inversion is carried by an additional wire that is used at the receiver end to restore the original data.

The BI scheme has some interesting properties. First, the worst-case number of transitions of an n-bit bus is $n/2$ at each cycle. Second, if we assume that data are uniformly randomly distributed, it is possible to show that the average number of transitions with this code is lower than that of any other encoding scheme with just one redundant line. Moreover, the basic 1-bit redundant BI code has the property that the average number of transitions per line increases as the bus gets wider, and asymptotically converges to 0.5, which is also the average switching per line of an unencoded bus, and is already close to this value for 32-bit buses. This shortcoming has spun a number of variants of the basic BI scheme, based on the partitioning of the bus into smaller blocks and on the use of bus inversion on each block independently. Since the trivial application of this partitioned variant on an m-block bus would require m control lines, these methods have tried to reduce this additional complexity (see, for example, the work in [78,79]).

Addition of redundancy is not very desirable at the RTL. Temporal redundancy obviously alters the timing of the operations, thus giving rise to performance issues, while spatial redundancy may require the modification of the bus interface to support the extra connections, which may not be feasible when connecting synthesizable IP blocks with predefined I/O.

Discarding redundancy drastically limits the spectrum of applicability of bus encoding. In particular, if no assumption on the statistical properties of the transmitted data can be made, results from information theory show that it is not possible to reduce the number of transitions [80]. Some irredundant codes proposed in the literature bypass this theoretical limitation by building statistical information online over a given timing window [81,82]. These adaptive schemes require however a significant hardware overhead that is not generally affordable at the RTL.

When some knowledge of the statistical properties is available *a priori*, more effective encodings can be devised. The most realistic option at the RTL is the case of address buses, for which there exists a high degree of correlation between consecutive addresses; in particular, addresses generated by processors typically exhibit a high degree of *sequentiality;* this is particularly true for data-dominated applications, where the few control structures only occasionally break the sequentiality of the address stream.

Some authors have suggested the adoption of Gray coding [83] for address buses. This code achieves its asymptotic best performance of a single transition per emitted address when infinite streams of consecutive addresses are considered. This average can be lowered to asymptotic zero transitions, at the price of adding some spatial redundancy [84]. Alternatively, the asymptotic zero-transition behavior can be achieved without any redundancy by exploiting the decorrelating characteristics of the XOR function, when applied to consecutive bus patterns. In this way, the values on the bus are encoded using a transition signaling scheme [85]. Notice that the decorrelation implies an operation of a signal with its previous copy and may not be feasible at the RTL for timing reasons. Conversely, Gray encoding is a fully combinational transformation and it is potentially feasible to insert Gray codecs into an RTL description, should the cycle time constraints allow it.

So far, we have discussed techniques that aim at reducing switching activity. When addressing coupling activity, things are even more critical, because the encoding involves pairs of wires, thus making the hardware overhead required by the codec more complex by construction. In addition, solutions based on encoding the data to reduce coupling activity tend to ignore the most important implication of coupling, that is, its impact on timing due to *crosstalk*. Minimizing the number of simultaneous transitions on adjacent wires may reduce power consumption of coupling capacitance, yet it does not reduce crosstalk by itself. Crosstalk mainly affects signal integrity, and even a significant reduction of crosstalk-induced power is of little interest for designers, if it does not guarantee the proper functionality of the circuit. Therefore, a solution to the problem must be consistent with performance-oriented crosstalk reduction techniques. In other terms, since crosstalk is mainly a capacitive effect, the only way to reduce it is by reducing the capacitance that causes it, and let power reduction come as a by-product. This makes the power optimization an issue to be dealt with during the physical design step, with the application of techniques ranging from (static or dynamic) wire permutation [86–88] to nonuniform wire spacing [89], from the insertion of shielding lines [90–92] to different combinations of the approaches above [93,94]. Solutions based on encoding of the data, however, may still be applied on top of capacitance reduction techniques, should the timing constraint allow it.

13.6 Conclusions

This chapter has reviewed the basic principles of power modeling, estimation, and optimization for digital CMOS circuits described at the micro-architectural level. The most common and successful modeling solutions for various types of micro-architectural components, such as data-path macros, controllers, memories, wires and buses have been discussed in detail, offering a comprehensive overview of the state of the art in this domain. The flow for power estimation that makes use of the various models has then been illustrated.

The focus has then been shifted to power optimization technologies; here, solutions based on clock-gating, exploitation of common-case computation and dynamic V_{th}/V_{dd} management, memory partitioning, and bus encoding have been explored, pointing out the main context in which they can be more suitably applied. Emphasis has been put on design automation aspects of most of the techniques considered, with the objective of making this chapter of practical use not only to IC designers and architects, but also to EDA engineers.

References

[1] A.P. Chandrakasan, S. Sheng, and R.W. Brodersen, Low-power CMOS digital design, *IEEE J. Solid State Circuits,* 27, 473–484, 1992.

[2] D.D. Gajski, N.D. Dutt, A.C.-H. Wu, and S.Y.-L. Lin, *High-Level Synthesis: Introduction to Chip and System Design,* Kluwer Academic Publishers, Dordrecht 1992.

[3] P. Landman, High-level power estimation, *ISLPED-96: ACM/IEEE International Symposium on Low Power Electronics and Design,* Monterey, CA, August 1996, pp. 29–35.

[4] P. Landman and J. Rabaey, Activity-sensitive architectural power analysis, *IEEE Trans. Comput. Aided Des. Integr. Circuits Syst,* 15, 571–587, 1996.

[5] P. Landman, R. Mehra, and J. Rabaey, An integrated CAD environment for low-power design, *IEEE Des. Test Comput.,* 13, 72–82, 1996.

[6] A. Raghunathan, S. Dey, and N. Jha, Register-Transfer Level Estimation Techniques for Switching Activity and Power Consumption, *ICCAD-96: IEEE/ACM International Conference in Computer-Aided Design,* San Jose, CA, November 1996, pp. 158–165.

[7] S. Katkoori and R. Vemuri, Architectural power estimation based on behavioral profiling, *J. VLSI Des.,* 7, 255–270, 1998.

[8] A. Bogliolo, I. Colonescu, R. Corgnati, E. Macii, and M. Poncino, An RTL power estimation tool with on-line model building capabilities, *PATMOS-01: IEEE International Workshop on Power and Timing Modeling, Optimization and Simulation,* Yverdon-les-Bains, Switzerland, September 2001, pp. 2.3.1–2.3.10.

[9] S. Ravi, A. Raghunathan, and S. Chakradhar, Efficient RTL power estimation for large designs, *IEEE International Conference on VLSI Design,* New Delhi, India, January 2003, pp. 431–439.

[10] D. Helms, E. Schmidt, A. Schulz, A. Stammermann, and W. Nebel, An improved power macro-model for arithmetic datapath components, *PATMOS-02: IEEE International Workshop on Power and Timing Modeling, Optimization and Simulation,* Sevilla, Spain, September 2002, pp. 16–24.

[11] M. Bruno, A. Macii, and M. Poncino, A statistical power model for non-synthetic RTL operators, *PATMOS-03: IEEE International Workshop on Power and Timing Modeling, Optimization and Simulation,* Torino, Italy, September 2003, pp. 208–218.

[12] F. Najm, Transition density: a new measure of activity in digital circuits, *IEEE Trans. Comput. Aided Des. Integr. Circuits Syst.,* 12, 310–323, 1993.

[13] Q. Qiu, Q. Wu, C.-S. Ding, and M. Pedram, Cycle-accurate macro-models for RT-level power analysis, *IEEE Trans. VLSI Syst.,* 6, 520–528, 1998.

[14] H. Mehta, R.M. Owens, and M.J. Irwin, Energy characterization based on clustering, *DAC-33: ACM/IEEE Design Automation Conference,* Las Vegas, NV, June 1996, pp. 702–707.

[15] S. Gupta and F. Najm, Energy-per-cycle estimation at RTL, *ISLPED-99: ACM/IEEE International Symposium on Low-Power Electronics and Design,* San Diego, CA, August 1999, pp. 16–17.

[16] L. Benini, A. Bogliolo, M. Favalli, and G. De Micheli, Regression models for behavioral power estimation, *PATMOS-96: IEEE International Workshop on Power and Timing Modeling, Optimization and Simulation,* Bologna, Italy, October 1996, pp. 125–130.

[17] D. Liu and C. Svensson, Power consumption estimation in CMOS VLSI chips, *IEEE J. Solid State Circ.,* 29, 663–671, 1994.

[18] E. Schmidt, G. Jochens, L. Kruse, F. Theeuwen, and W. Nebel, Memory power models for multilevel power estimation and optimization, *IEEE Trans. VLSI Syst.*, 10, 106–109, 2002.
[19] S. Powell and P. Chau, Estimating power dissipation of VLSI signal processing chips: the PFA technique, *VLSI Signal Processing*, 4, 250–259, 1990.
[20] P. Landman and J. Rabaey, Architectural power analysis: the dual-bit type model, *IEEE Trans. VLSI Syst.*, 3, 173–187, 1995.
[21] S. Gupta and F. Najm, Power macromodeling for high level power estimation, *DAC-34: ACM/IEEE Design Automation Conference*, Anaheim, CA, June 1997, pp. 365–370.
[22] M. Barocci, L. Benini, A. Bogliolo, B. Ricco, and G. De Micheli, Look-up table power macromodels for behavioral library components, *IEEE Alessandro Volta Memorial Workshop on Low-Power Design*, Como, Italy, March 1999, pp. 173–181.
[23] R. Corgnati, E. Macii, and M. Poncino, Clustered table-based macromodels for RTL power estimation, *GLS-VLSI-99: IEEE/ACM Great Lakes Symposium on VLSI*, Ann Arbor, MI, March 1999, pp. 354–357.
[24] A. Bogliolo, E. Macii, V. Mihailovici, and M. Poncino, Combinational characterization-based power macro-models for sequential macros, *PATMOS-99: IEEE International Workshop on Power and Timing Modeling, Optimization and Simulation*, Kos, Greece, October 1999, pp. 293–302.
[25] M. Anton, I. Colonescu, E. Macii, and M. Poncino, Fast characterization of RTL power macromodels, *ICECS-01: IEEE International Conference on Electronics, Circuits and Systems*, La Valletta, Malta, September 2001, pp. 1591–1594.
[26] A. Bogliolo, R. Corgnati, E. Macii, and M. Poncino, Parameterized RTL power models for combinational soft macros, *IEEE Trans. VLSI Syst.*, 9, 880–887, 2001.
[27] A. Tyagi, Entropic bounds on FSM switching, *ISLPED-96: ACM/IEEE International Symposium on Low Power Electronics and Design*, Monterey, CA, August 1996, pp. 323–327.
[28] D. Marculescu, R. Marculescu, and M. Pedram, Theoretical bounds for switching activity analysis in finite-state machines, *ISLPED-98: ACM/IEEE International Symposium on Low Power Electronics and Design*, Monterey, CA, August 1998, pp. 36–41.
[29] S. Katkoori and R. Vemuri, Simulation-based architectural power estimation for PLA-based controllers, *ISLPED-96: ACM/IEEE International Symposium on Low Power Electronics and Design*, Monterey, CA, August 1996, pp. 121–124.
[30] L. Benini, A. Bogliolo, E. Macii, M. Poncino, and M. Surmei, Regression-based RTL power models for controllers, *GLS-VLSI-00: ACM/IEEE Great Lakes Symposium on VLSI*, Evanston, IL, March 2000, pp. 147–152.
[31] U. Ko and P. T. Balsara, Characterization and design of a low-power, high-performance cache architecture, *IEEE International Symposium on VLSI Technology, Systems and Applications*, May 1995, pp. 235–238.
[32] M.B. Kamble and K. Ghose, Analytical energy dissipation models for low power caches, *ISLPED-97: ACM/IEEE International Symposium on Low Power Design*, Monterey, CA, August 1997, pp. 143–148.
[33] S.L. Coumeri and D.E. Thomas, Memory modeling for system synthesis, *IEEE Trans. VLSI Syst.*, 8, 327–334, 2000.
[34] B.S. Landman and R.L. Russo, On a pin vs. block relationship for partitions of logic graphs, *IEEE Trans. Comput.*, 20, 1469–1479, 1971.
[35] S. Sastry and A.C. Parker, Stochastic models for wireability of analysis of gate arrays, *IEEE Trans. Comput. Aided Des.Integr. Circuits Syst.*, 5, 52–65, 1986.
[36] F. J. Kurdahi and A.C. Parker, Techniques for area estimation of VLSI layouts, *IEEE Trans. Comput.-Aided Des. Integr. Circuits Syst.*, 8, 81–92, 1989.
[37] C. Anton, A. Bogliolo, P. Civera, I. Colonescu, E. Macii, and M. Poncino, RTL estimation of steering logic power, *PATMOS-00: IEEE International Workshop on Power and Timing Modeling, Optimization and Simulation*, Gottingen, Germany, September 2000, pp. 36–45.
[38] L. Benini and G. De Micheli, Networks on chips: a new SoC paradigm, *IEEE Comput.*, 35, 70–78, 2002.

[39] Synopsys Design Ware Library, *http://www.synopsys.com.*
[40] R. Ho, K.W. Mai, and M.A. Horowitz, The future of wires, *Proceedings of the IEEE,* Vol. 89, No. 4, April 2001, pp. 490–504.
[41] P. Sotiriadis and A. Chandrakasan, A bus energy model for deep submicron technology, *IEEE Trans. VLSI Syst.,* 10, 341–350, 2002.
[42] L. Benini, P. Siegel, and G. De Micheli, Automatic Synthesis of Gated Clocks for Power Reduction in Sequential Circuits, *IEEE Des. Test Comput.,* 11, 32–40, 1994.
[43] L. Benini and G. De Micheli, Transformation and synthesis of FSMs for low power gated clock implementation, *IEEE Trans. Comput. Aided Des. Integr. Circuits Syst.,* 15, 630–643, 1996.
[44] L. Benini, G. De Micheli, E. Macii, M. Poncino, and R. Scarsi, Symbolic synthesis of clock-gating logic for power optimization of synchronous controllers, *ACM Trans. Des. Automat. Electron. Syst.,* 4, 351–375, 1999.
[45] M. Gowan, L.L. Biro, and D.B. Jackson, Power considerations in the design of the Alpha 21264 microprocessor, *DAC-35: ACM/IEEE Design Automation Conference,* San Francisco, CA, June 1998, pp. 726–731.
[46] A. Correale, Overview of the power minimization techniques employed in IBM powerPC 4xx embedded controllers, *ISLPD-95: ACM International Symposium on Low Power Design,* Dana Point, CA, August 1995, pp. 75–80.
[47] V. Tiwari, D. Singh, S. Rajgopal, G. Mehta, R. Patel, and F. Baez, Reducing power in high-performance microprocessors, *DAC-35: ACM/IEEE Design Automation Conference,* San Francisco, CA, June 1998, pp. 732–737.
[48] Z. Khan, and G. Mehta, Automatic clock gating for power reduction, *SNUG-99: Synopsys Users Group,* San Jose, CA, March 1999.
[49] P. Babighian, L. Benini, and E. Macii, A scalable algorithm for RTL insertion of gated clocks based on observability don't cares computation, *IEEE Trans. Comput. Aided Des. Integr. Circuits Syst.,* 24, 29–42, 2005.
[50] L. Benini, M. Donno, A. Ivaldi, and E. Macii, Clock-tree power optimization based on RTL clock-gating, *DAC-40: ACM/IEEE Design Automation Conference,* Anaheim, CA, June 2003, pp. 622–627.
[51] M. Donno, E. Macii, and L. Mazzoni, Power-aware clock tree planning, *ISPD-04: ACM/IEEE International Symposium on Physical Design,* Phoenix, Arizona, April 2004, pp. 138–147.
[52] L. Benini, M. Favalli, and G. De Micheli, Design for testability of gated-clock FSMs, *EDTC-96: IEEE European Design and Test Conference,* Paris, France, March 1996, pp. 589–596.
[53] B. Pouya and A. Crouch, Optimization for vector volume and test power, *ITC-00: IEEE International Test Conference,* Atlantic City, NJ, October 2000, pp. 873–881.
[54] J. Saxena, K. Butler, and L. Whetsel, An analysis of power reduction techniques in scan testing, *ITC-01: IEEE International Test Conference,* Baltimore, MD, October 2001, pp. 670–677.
[55] J.L. Hennessy and D.A. Patterson, *Computer Architecture: A Quantitative Approach,* Morgan Kaufmann Publishers, San Francisco, CA, 1996.
[56] S. Hassoun and C. Ebeling, Architectural retiming: pipelining latency-constrained circuits, *DAC-33: ACM/IEEE Design Automation Conference,* Las Vegas, NV, June 1996, pp. 708–713.
[57] S. Nowick, Design of a low-latency asynchronous adder using speculative completion, *IEE Proc.–Comput. Digit. Tech.,* 143, 301–307, 1996.
[58] L. Benini, G. De Micheli, E. Macii, and M. Poncino, Telescopic units: a new paradigm for performance optimization of VLSI designs, *IEEE Trans. Comput. Aided Des. Integr. Circuits Syst.,* 17, 220–232, 1998.
[59] L. Benini, G. De Micheli, A. Lioy, E. Macii, G. Odasso, and M. Poncino, Automatic synthesis of large telescopic units based on near-minimum timed supersetting, *IEEE Trans. Comput.,* 48, 769–779, 1999.
[60] G. Lakshminarayana, A. Raghunathan, K. S. Khouri, N. K. Jha, and S. Dey, Common case computation: a new paradigm for energy and performance optimization, *IEEE Trans. Comput. Aided Des. Integr. Circuits Syst.,* 23, 33–49, 2004.

[61] M. Munch, B. Wurth, R. Mehra, J. Sproch, and N. Wehn, Automating RT-level operand isolation to minimize power consumption in datapaths, *DATE-00: IEEE Design Automation and Test in Europe,* Paris, France, March 2000, pp. 624–631.

[62] M. Alidina, J. Monteiro, S. Devadas, A. Ghosh, and M. Papaefthymiou, Precomputation-based sequential logic optimization for low power, *IEEE Trans. VLSI Syst.,* 2, 426–436, 1994.

[63] J. Monteiro, S. Devadas, and A. Ghosh, Sequential logic optimization for low power using input-disabling precomputation architectures, *IEEE Trans. Comput. Aided Des. Integr. Circuits Syst.,* 17, 279–284, 1998.

[64] L. Benini, G. De Micheli, A. Lioy, E. Macii, G. Odasso, and M. Poncino, Synthesis of power-managed sequential components based on computational kernel extraction, *IEEE Trans. Comput. Aided Des. Integr. Circuits Syst.,* 20, 1118–1131, 2001.

[65] S. Svilan, J.B. Burr, and G.L. Tyler, Effects of elevated temperature on tunable near-zero threshold CMOS, *ISLPED-01: ACM/IEEE International Symposium on Low Power Electronics and Design,* Huntington Beach, CA, August 2001, pp. 255–258.

[66] K. Flautner, D. Flynn, and M. Rives, A combined hardware–software approach for low-power SoCs: applying adaptive voltage scaling and the vertigo performance-setting algorithms, *DesignCon,* Paper SA2-3, Santa Clara, CA, January 2003.

[67] S. Martin, K. Flautner, T. Mudge, and D. Blaauw, Combined dynamic voltage scaling and adaptive body biasing for lower power microprocessors under dynamic workloads, *ICCAD-02: ACM/IEEE International Conference on Computer Aided Design,* San Jose, CA, November 2002, pp. 721–725.

[68] S. Shigematsu, S. Mutoh, Y. Matsuya, Y. Tanabe, and J. Yamada, A 1-V high-speed MTCMOS circuit scheme for power-down application circuits, *IEEE J. Solid State Circuits,* 32, 861–869, 1997.

[69] V. Zyuban and S. Kosonocky, low power integrated scan-retention mechanism, *ISLPED-02: ACM/IEEE International Symposium on Low Power Electronics and Design,* Monterey, CA, August 2002, pp. 98–102.

[70] K. Usami, M. Igarashi, F. Minami, T. Ishikawa, M. Kawakawa, M. Ichida, and K. Nogami, Automated low-power technique exploiting multiple supply voltages applied to media processor, *IEEE J. Solid State Circuits,* 33, 463–472, 1998.

[71] L. Wei, K. Roy, and V. De, Low power low voltage CMOS design techniques for deep submicron ICs, *VLSI-00: International Conference on VLSI Design,* January 2000, pp. 24–29.

[72] D.E. Lackey, S. Gould, T.R. Bednar, J. Cohn, and P.S. Zuchowski, Managing power and performance for system-on-chip designs using voltage islands, *ICCAD-02: ACM/IEEE International Conference on Computer Aided Design,* San Jose, CA, November 2002, pp. 195–202.

[73] F. Ishihara, F. Sheikh, and B. Nikolic, Level conversion for dual supply systems, *ISLPED-03: ACM/IEEE International Symposium on Low Power Electronics and Design,* Seoul, Korea, August 2003, pp. 164–167.

[74] A. Macii, L. Benini, and M. Poncino, *Memory Design Techniques for Low-Energy Embedded Systems,* Kluwer Academic Publishers, Dordrecht, 2002.

[75] P. Panda and N. Dutt, *Memory Issues in Embedded Systems-on-Chip Optimization and Exploration,* Kluwer Academic Publishers, Dordrecht, 1999.

[76] L. Benini, L. Macchiarulo, A. Macii, and M. Poncino, Layout-driven memory synthesis for embedded systems-on-chip, *IEEE Trans. VLSI Syst.,* 10, 96–105, 2002.

[77] M. Stan, and W.P. Burleson, Bus-invert coding for low-power I/O, *IEEE Trans. VLSI Syst.* 3, 49–58, 1995.

[78] Y. Shin, S.-I. Chae, and K. Choi, Partial bus-invert coding for power optimization of application-specific systems, *IEEE Trans. VLSI Syst.* 9, 377–383, 2001.

[79] U. Narayanan, K.-S. Chung, and K. Taewhan, Enhanced bus invert encodings for low-power, *ISCAS-02: IEEE International Symposium on Circuits and Systems,* Vol. 5, Scottsdale, AZ, May 2002, pp. 25–28.

[80] S. Ramprasad, N. Shanbhag, and I. Hajj, Signal coding for low power: fundamental limits and practical realizations, *IEEE Trans. Circuits Syst. II: Analog Digit. Signal Process.,* 46, 923–929, 1999.

[81] L. Benini, A. Macii, E. Macii, M. Poncino, and R. Scarsi, Architectures and synthesis algorithms for power-efficient bus interfaces, *IEEE Trans. Comput. Aided Des. Integr. Circuits Syst.*, 19, 969–980, 2000.

[82] M. Mamidipaka, D. Hirschberg, and N. Dutt, Low power address encoding using self-organizing lists, *ISLPED-01: ACM/IEEE International Symposium on Low Power Electronics and Design*, Huntington Beach, CA, August 2001, pp. 188–193.

[83] H. Mehta, R. M. Owens, and M. J. Irwin, Some issues in gray code addressing, *GLS-VLSI-96: IEEE/ACM Great Lakes Symposium on VLSI*, Ames, IA, March 1996, pp. 178–180.

[84] L. Benini, G. De Micheli, E. Macii, D. Sciuto, and C. Silvano, Asymptotic zero-transition activity encoding for address busses in low-power microprocessor-based systems, *GLS-VLSI-97: IEEE/ACM Great Lakes Symposium on VLSI*, Urbana-Champaign, IL, March 1997, pp. 77–82.

[85] Y. Aghaghiri, F. Fallah, and M. Pedram, Irredundant address bus encoding for low power, *ISLPED-01: ACM/IEEE International Symposium on Low Power Electronics and Design*, Huntington Beach, CA, August 2001, pp. 182–187.

[86] Y. Shin and T. Sakurai, Coupling-driven bus design for low-power application-specific systems, *DAC-38: ACM/IEEE Design Automation Conference*, Las Vegas, NV, June 2001, pp. 750–753.

[87] L. Macchiarulo, E. Macii, and M. Poncino, Low-energy encoding for deep-submicron address buses, *ISLPED-01: ACM/IEEE International Symposium on Low Power Electronics and Design*, Huntington Beach, CA, August 2001, pp. 176–181.

[88] J. Henkel and H. Lekatsas, A2BC: adaptive address bus coding for low-power deep sub-micron designs, *DAC-38: ACM/IEEE Design Automation Conference*, Las Vegas, NV, June 2001, pp. 744–749.

[89] L. Macchiarulo, E. Macii, and M. Poncino, Wire placement for crosstalk energy minimization in address buses, *DATE-02: IEEE Design Automation and Test in Europe*, Paris, France, March 2002, pp. 158–162.

[90] A. Vittal and M. Sadowska, Crosstalk reduction for VLSI, *IEEE Trans. Comput. Aided Des. Integr. Circuits Syst.*, 16, 290–298, 1997.

[91] H. Kaul, D. Sylvester, and D. Blauuw, Active shields: a new approach to shielding global wires, *GLS-VLSI-02: ACM/IEEE Great Lakes Symposium on VLSI*, New York, NY, March 2002, pp. 112–117.

[92] S. Salerno, E. Macii, and M. Poncino, Crosstalk energy reduction by temporal shielding, *ISCAS-04: IEEE International Symposium on Circuits and Systems*, Vol. 2, Vancouver, BC, Canada, May 2004, pp. 749–752.

[93] R. Arunachalam, E. Acar, and S. Nassif, Optimal shielding/spacing metrics for low power design, *ISVLSI-03: IEEE Annual Symposium on VLSI*, Tampa, FL, February 2003, pp. 167–172.

[94] S. Salerno, E. Macii, and M. Poncino, Combining wire swapping and spacing for low-power deep-submicron buses, *GLS-VLSI-03: IEEE/ACM Great Lakes Symposium on VLSI*, Washington, DC, April 2003, pp. 198–202.

14

Design Planning

Ralph H.J.M. Otten
Eindhoven University of Technology
Eindhoven, Netherlands

14.1 Introduction

In the past decade, *design automation* faced a sequence of what were called *closure problems.* "How to achieve wireability in placement of components or modules on a chip," and "how to allocate resources in order to optimize schedules" are among the early ones. In the 1990s timing closure was a dominant challenge; that is "how to ensure timing convergence with minimal size." Toward the end of that decade so many additional characteristics and constraints, such as (static and active) power, signal integrity, and electromagnetic compliance, had to be considered that industry started to speak of the ever elusive *design closure* (For an analysis of these characteristics and constraints, see the chapter "Design Closure", by John Cohn, and Peter Osler in volume II of this book). Yet, that is today's challenge of design automation: "how to specify a function to be implemented on a chip, feed it to an *EDA* tool, and get, without further interaction, a design that meets all requirements of functionality, speed, size, power, yield and other costs."

14.1.1 Wiring Closure

Around 1980, designers realized that the complexity of chips in the decade to come forced them to use more than just a router and occasionally a placer to find a starting point. At the same time, it was obvious that the two tasks were heavily dependent on each other. Routing without a placement was inconceivable, while at the same time a placement might render any routing infeasible. This "phase problem" between placement and routing was tackled with error-prone spacing repairs and wasteful reservations, which turned out to be inadequate. Thus the quest for *wiring closure* was born.

The answer came with the introduction of *floorplans,* data structures that capture relative positions of objects rather than co-ordinates. Once such a "topology" was generated and stored, optimizations could convert it into geometry, often a dissection of the rectangular chip into rectangular regions for the modules. The restriction to rectangular shapes was convenient because the optimization could be organized as trading off dimensions in order to reserve an adequate shape for each module while achieving an optimal overall contour. Under a mild structural constraint this required only polynomial time.

In a sense, floorplanning is simply a generalization of placement. Whereas placement was the manipulation of geometrically fixed objects to arrange them in a configuration without overlaps, floorplanning handled objects for which the shape did not have to be predetermined. It lent itself much better to top-down approaches where the shape and sometimes even the precise size of the modules were unknown. Certain restraints enabled *stepwise refinement* in that the tree representation of the floorplan was forced to be a refinement of a given hierarchy tree. These features caused a strong association of the floorplan concept with hierarchical design.

The salient feature of floorplans is that they allow designers to perform early analysis on their design decisions so that performance can be improved without resorting to lengthy iterations (in its original context floorplanning was presented as an iteration-free approach to chip design!). It thus enables *silicon virtual prototyping*, where sophisticated assessment procedures are integrated with hierarchical floorplan design to adapt modules in a timely manner to their (preliminary) environments, while respecting intellectual property reuse.

Today it can safely be stated that placement has been replaced by floorplanning, followed by a legalization step. Many modern back-end tools start with generating point configurations (with or without size-dependent spacing) or overlapping geometrical objects, mostly by analytic or stochastic optimizers that can handle complex but flat incidence structures with library elements of fixed shape. Once the optimum for this configuration is established, legalizers ensure that all overlap is removed, and that routability is enhanced. Ironically, this was exactly what early fully automatic floorplanners did [1]. Other back-end tools start by partitioning a netlist while giving relative positions to the blocks. Details about the algorithms can be found in the chapter "Digital Layout — Placement" by Andrew Kahng and Sherief Reda in volume II of this book.

In Section 14.2, we work from the definition of a floorplan to the many models and configurations that have served to capture relative positions in one way or another. We emphasize the importance of meaningful and easy to handle metrics for evaluating floorplans, although the in-depth treatment and comparison is left to other chapters in this book.

14.1.2 Timing Closure

With wiring as the main closure problem of the 1980s, timing closure became the target of the next decade. Two timing closure approaches emerged in the middle of the 1990s, both founded on the observation that gate delay can be kept constant under load variation by sizing [2]. Applying this rule to the whole design flat is one of them, and it has been commercialized successfully. There is however a flaw in this approach: as interconnects get longer, wire resistance cannot be neglected anymore. Unfortunately, keeping the delay constant by sizing is a model valid only when there is no resistance between the driving gate and the capacitive load(s). Line buffering is a patch that often helps, but cannot rescue the method. The other approach is by *wire planning*. It allocates delay to *global* interconnect first, then assigns budgets to modules, and applies the constant delay method to each of these modules. The assumptions of this method are that interconnect delay is linear with its length, and the size of the individual models is small compared to the whole design. The latter is used in two ways: interconnect delay within the modules cannot be reduced by buffer insertion, and modules can be treated as points during the global delay allocation.

If interconnect delay is proportional to interconnect length, path delay is insensitive to module position as long as detours are avoided. Interconnect linear with length is achieved by optimal segmentation of wires with buffers. To achieve minimal delay, buffers should be optimized in size. This fixes line impedance, thus giving up degrees of freedom to keep module delay within the budget. Adapting line impedance changes the delay per unit length. Fortunately, delay change by size variation around the optimum is small, and by minimizing area during the budget assignment, slack is created that can be used to compensate for additional interconnect delay.

What remains is to implement an algorithm for time budget assignment. This implies trading module size for speed: the smaller the module, the slower the signal propagation through it. If we know the relation between the two performance characteristics for every module, a minimization of the total area under timing constraints is easily formulated. Every path imposes a constraint on the budget assignment, and every budget assignment to a module implies a certain amount of area. The mathematical program obtained in this way can be efficiently solved if the trade-off function is convex, i.e., there is an algorithm with runtime polynomial in the size of the tableau.

There may however be exponentially many paths in a graph, and the size of the tableau could be exponential in the size of the graph. What is needed is a formulation that is polynomial in the size of the graph. Such a formulation will be presented in Section 14.3.3. Before that, we will review constant delay synthesis, and summarize the assumptions of wire planning. We conclude with a Section that shows how mild violations of these assumptions can be accommodated.

14.1.3 Design Closure

Up until 2000, the EDA industry has concentrated on tools and techniques for achieving closure with respect to one target design parameter. Moreover, these parameters were mainly related to that part of the trajectory that synthesized mask level data from an *rtl* description or even lower levels. After wiring and timing, power is likely to become the closure problem of this decade.

Today's chip synthesis requires tools and methodologies for manipulating designs at higher levels of abstraction to meet a large variety of performance constraints. What is needed is a formal system to handle trade-offs between performance characteristics over many levels of abstraction. Such a system will be indispensable to achieve *design closure,* in the sense alluded to at the beginning of this chapter. It should free design trajectories from the need for iterations and interactions to which users resort when their tools fail in one or more aspects to achieve closure, not knowing whether the process will converge or not, and in the latter case not even being sure whether a solution within specifications exists.

The approach hinges on a generalization of the concept of a performance characteristic, and a set of operations over such characteristics that extend and reduce performance spaces without ever losing candidates for optimal solutions with respect to any monotonic cost function. Of course, the key is the effectiveness in keeping the search space small, and the efficiency of the algorithms which do that.

After analyzing successes of the past in achieving closure, and pointing to the efforts of today, we formulate a number of observations that are the basis for the construction of such a formal system, an algebra for trade-offs, the topic of Section 14.4.

14.2 Floorplans

Formally, a floorplan is a data structure that captures the relative positions of objects which get their final shape by optimizing some objective without violating constraints.

This definition does not imply any underlying hierarchical structure. It simply generalizes the notion of placement: the manipulation of objects with fixed geometrical features in order to allocate them without any overlap in an enclosing plane region. The generalization refers to both the uncertainty or flexibility of the shape of some or all of the objects, and the possibility of overlap when these objects have a shape or area associated with them.

It is evident that hierarchical approaches that want to find regions for allocating subsets of modules almost always must resort to floorplanning in some way, because it is neither always wise nor often feasible to predetermine the shape of all modules in a hierarchy. This obvious need for floorplanning in approaches that want to preserve, completely or partly, properties of earlier partitioning, caused an almost complete identification of floorplanning with hierarchical layout design. The elegant formulation of stepwise refinement for layout synthesis, where preserving the hierarchical structure of a functional design occurs automatically by adoptation of a structural restraint that expanded, ordered and labeled the original hierarchy tree, planted this idea firmly in the minds of back-end tool developers of the late 1980s.

This is easily understood if we see that a supermodule's floorplan (that is the relative positions of its modules) was derived from:

- a preliminary shape of the supermodule
- a preliminary environment (e.g., pin positions)

- shape information for the modules (e.g., shape constraints)
- an incidence structure (i.e., a netlist)
  and possibly
- timing information of external signals
- module delay information
- path delay information.

The result was subsequently used in

- creating a preliminary environment for the supermodule's modules
- assessing wire space requirements
- estimating the delays.

All these are controlled by the (given or derived) hierarchy.
Figure 14.1 illustrates this, and essentially invites the recursive application of the central floorplan function. The terminology, the creation, and the use of module environments, and the explicit mention of hierarchy in the last item, seems to bind floorplanning exclusively to hierarchical design. However, a flat design with a single supermodule and a single level of leaf cells (undivided modules) is perfectly consistent, and was always the most important practical application of the floorplanning concept. It has rendered classical constructive placement obsolete, because of its flexibility and relative efficiency.

14.2.1 Models

In this section, we look at three classes of models for floorplans, not necessarily in chronological order of origin, but fairly representative of what has been conceived over three decades for capturing relative positions of modules.

14.2.1.1 Graphs

The oldest floorplans are graphs. Drawing a rectangle dissection already produces a graph, where line segments are interpreted as edges and the points where these come together (mostly *T-junctions*, sometimes *crossings*) as vertices. It should be noted that different dissections may lead to the same graph. Special properties that such a graph should have is that its edges can be divided into two classes, *h*- and *v-edges*, and it should possess a plane representation where these edges are represented by orthogonal line segments, one orientation per class. This graph is called the *floorplan graph* (Figure 14.2a).

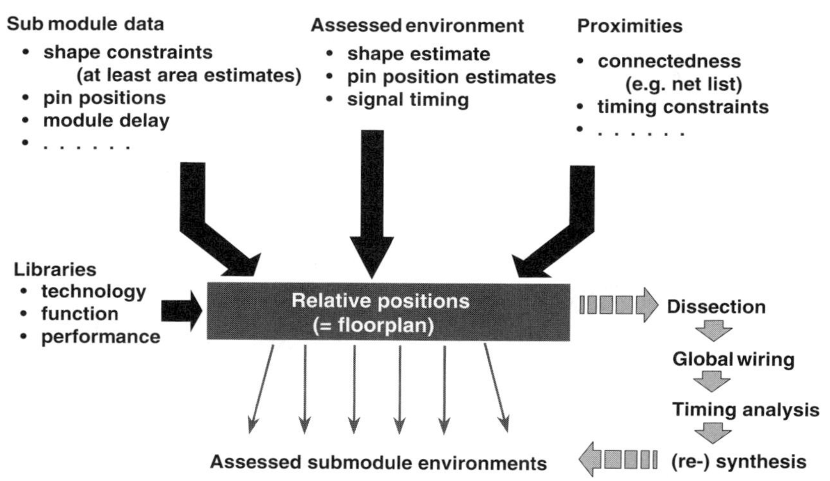

FIGURE 14.1 Floorplan design.

The first formal floorplanning procedure started with a graph in which the edges represented desired adjacencies. The vertices represent the objects, that is modules, but in the first applications, *rooms.* Obviously, not every such graph can be converted into a rectangle dissection. They should be planar, but that is not enough. In addition, they should have a dual that has a plane representation with the two orthogonal sets of line segments, as described above. Graphs with such a dual are called *Grason graphs* in honor of the first scientist to publish a floorplanning procedure [3]. This pioneer also gave a characterization of Grason graphs, but efficient algorithms for recognizing Grason graphs and constructing an associated rectangle dissection came some 15 years later [4–7].

On the basis of the division into h- and v-edges, we can also divide the edges of the Grason graph, yielding two graphs: the *Grason-h-graph* and the *Grason-v-graph.* By directing these graphs consistently with the relative positions of the rectangles of their vertices, we obtain the *Grason-h-digraph* (Figure 14.2b) and the *Grason-v-digraph.*

Equivalent to this digraph pair (or the colored Grason graph), and also to the colored floorplan graph (i.e., with explicitly given h- and v-edgesets) is the *channel digraph* (Figure 14.2d) where each channel (i.e., maximal line segment) has a vertex and each T-junction has an arc pointing to the "bar of the T" [8]. This graph has rarely been used though it has some characteristic properties and retains generality, where the much more popular *polar digraphs* are not capable of capturing all adjacencies.

Although known already in the first half of the last century, these polar graphs were introduced [9] into layout synthesis in 1970 by Tatsuo Ohtsuki. The *polar h-digraph* (Figure 14.2c) takes the maximal horizontal

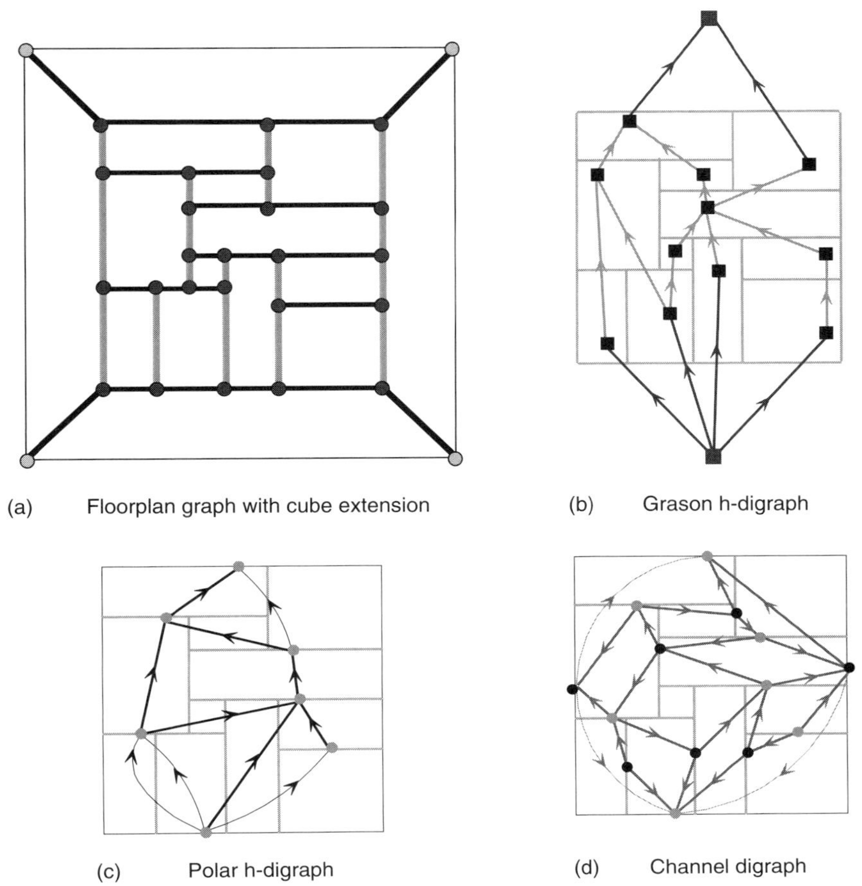

FIGURE 14.2 Graphs.

line segments as its vertices, and has an arc for every rectangle. It does capture incidence of rectangles to line segments, but adjacency between rectangles is not exactly covered. Consequently, the *Grason h-digraph* is a subgraph of the line graph of the *polar h-digraph.*

The relations between the graphs acting as floorplans for the same rectangle dissection are brought together in the table of Figure 14.3, where "D" stands for dualization and "L" for forming the line digraph. The upward arrow toward the polar digraphs and superset signs indicate the loss of information when going to polar graphs. Polar graphs are nevertheless the most popular graph models because they can be transformed into sets of linear equalities that must be satisfied by the rectangle dimensions of any compatible dissection*. This forms a good starting point for sizing using mathematical programming.

14.2.1.2 Point Configurations

Plane point sets are by now the most commonly used floorplans. If the only information that is used is the sequence of the points (representing the modules) after projecting them on an axis, then it is better to speak of a sequence pair, because all distance information is left unused.

Methods that try to embed a distance space into a euclidean space, and finally into a two-dimensional euclidean space often use eigenvalues to find the plane with maximum spread and to lose as little distance information as possible when projecting on that plane. Depending on the distance metric certain statements about optimality can be made. However, the euclidean distance is not really a useful metric and projection may force modules apart that should and can be kept together. Eigenvalue methods are good mainly for recovering structure in low-dimensional configurations.

With the rise of annealing as a competitive method for improving complex configurations, it was soon considered for floorplan design [10]. The question was how to represent relative positions in such a way that total wiring length was calculated very quickly. The first answer was a pair of sequences, which meant that the moves were transpositions of two elements in one such sequence, but the points were then spaced according to their area (not the square root of area!).

Sequence pairs obtained a more specific meaning when the two permutations of modules implied the following meaning for relative positions:

- A module m_i is to the left of module m_j if it precedes that module in both sequences.
- A module m_i is above a module m_j if it precedes that module in the first sequence, and if it follows that module in the second sequence.

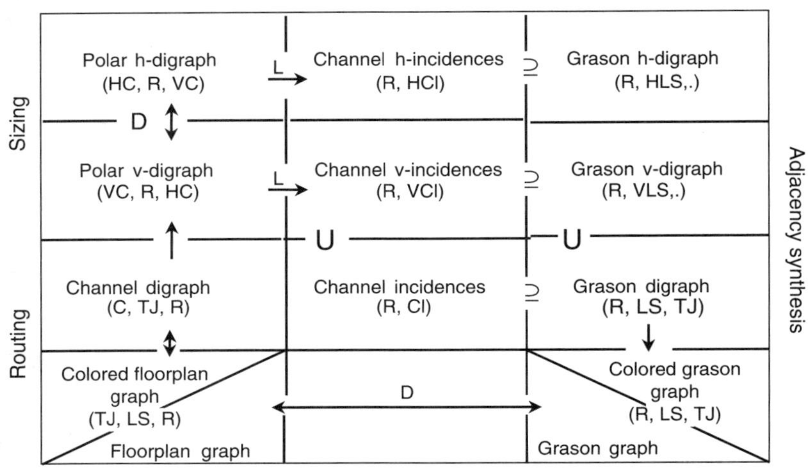

FIGURE 14.3 Relation table.

*In fact, these equalities are the Kirchhoff equations from basic electric circuit theory, with e.g., the widths as current intensities, and the heights as voltage differences between the nodes.

Till now sequence pairs are only generated by stochastic algorithms such as annealing. Evaluations are time consuming, although polynomial or even linear, which causes long runtimes, because these evaluations are placed in the inner loop. This makes such algorithms impractical. In favor of sequence pairs, it should be noted that they can easily handle empty rooms without explicitly encoding them in the sequences. This is not possible, for example, with polar graphs!

14.2.1.3 Trees

Trees for representing rectangle configurations have become popular recently, as shown by the appearance of *O-*, *bi-*, *star* and *stair trees*. These representations compete in their efficiency to solve the packing problem for their compatible rectangle configurations. These packing problems are easy to solve for rectangles with fixed shape and orientation. An example is the *labeled ordered tree* (Figure 14.4) from which a packing can be computed in linear time [11]. Among all possible trees with the adequate number of vertices there is one that produces the smallest possible packing. There are however very many possible trees though their number is bounded by the *Catalan number* times the number of permutations, and this number is asymptotically smaller than $(n!)^2$, the number of sequence pairs.

14.2.2 Design

In many ways, one can argue that floorplan design has not been solved adequately. The various algorithms for generating floorplans modeled by graphs are either confined to subspaces that do not have to contain optimum or even near-optimum plans, or produce in essence a number of configurations that cannot be bounded by a polynomial in the number of modules.

14.2.2.1 Adjacency-Based Floorplan Design

The first thoughts on floorplan design were directed toward realizing the required or desired adjacencies. These adjacencies form a graph, and if this graph is a Grason graph, a rectangle dissection with these adjacencies exists. Linear-time algorithms† for testing graphs for this property exist. In general, these adjacencies do not form a Grason graph, and seldom a subgraph of a Grason graph. The adjacency requirements therefore have to be pruned to such a subgraph and subsequently extended to a suitable Grason graph, either explicitly or implicitly. A generic procedure is given in Figure 14.5.

Graphs with rectangular duals (even graphs with duals, when combinatorially defined) have to be planar. Procedures that try to realize adjacencies therefore often start with heuristics for planarizing the

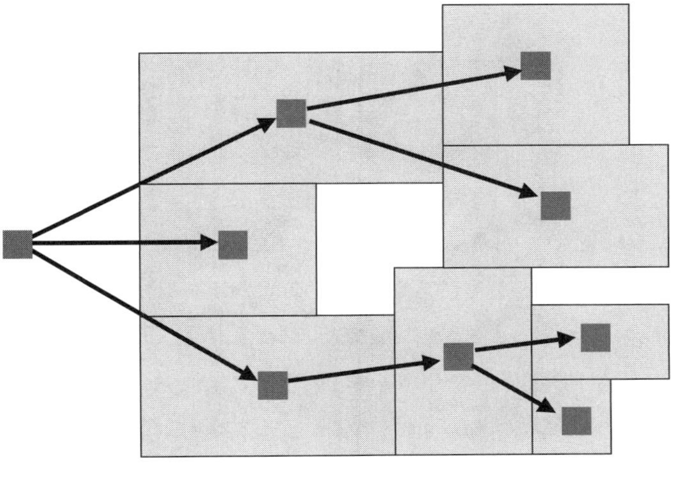

FIGURE 14.4 Labeled ordered tree.

†Do not trust the textbooks on this! Consult the original articles.

corresponding graph. A straightforward way for doing that is by removing "weak" adjacencies. Removing enough edges will planarize a graph sooner or later. But adding nodes also can have this effect. These "phony" rectangles (because nodes are supposed to represent rectangles) are called wiring nodes in [12]. Other heuristics can be found, but after the first stage in Figure 14.5 the graph is supposed to be planar, so it has a dual and of course a plane representation. But does it have a rectangular dual? Not for a representation with *complex triangles* - that is, representations with circuits of three edges with other edges inside. If that is the case there are a number of options:

- modify the representation
- remove the complex triangles
- triangulate avoiding complex triangles.

In [13], all options are exploited although in the reverse sequence. The remaining freedom can be used for assigning the corners.

The resulting graph has a rectangular dual, and therefore a corresponding rectangle dissection can be found. A constructive procedure for generating such a dissection in an efficient way was published in [5]. An elegant method, based on the observation that construction is in essence assigning T-junctions, was *complete matching* in [14].

A flow such as in Figure 14.5 does not consider shape requirements (not even size requirements) in its floorplan generation. Shape constraints are taken into account only afterward in a floorplan optimization step, mostly a tableau of inequalities holding the floorplan (e.g., by the equations derived from polar graphs) and the shape constraints. This often leads to unacceptable results, the main reason why pure adjacency-based floor-planning was never successful. The approach in the next section is much better equipped to take shapes into account.

14.2.2.2 Proximity-Based Floorplan Design

Point configurations can be generated efficiently depending on the objective. Eigenvalue methods are very fast but are bound to use a euclidean metric. Besides, they usually use projection to lower the

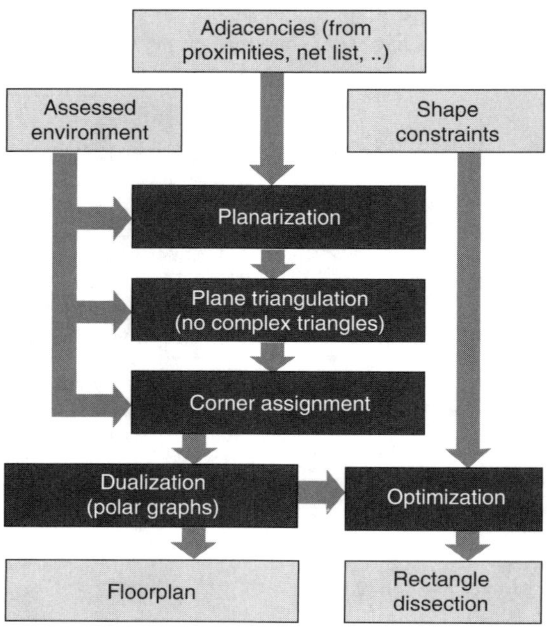

FIGURE 14.5 Adjacency-based floorplan design.

dimensionality of the solution, which may reduce distances, leaving other modules unnecessarily far apart. Repeated probing based on eigenvalues makes the generation slow.

Probabilistic techniques, such as annealing, genetic algorithms, and evolution, need very fast evaluations, which makes them resort to crude measures. For fixed rectangles, techniques such as in Section 14.2.1.3 can be of advantage, but they cannot handle more flexible shape constraints.

Currently the best results seem to be obtained with analytic optimizers of the quadratically convergent Newton-type, followed by legalizers. They are usually applied to flat designs, that is designs without any hierarchical structure, regardless of the number of objects to be treated. Since these analytic optimizers do not produce nonoverlapping configurations of the objects to be allocated, they are classical examples of floorplanners, although the literature often calls them (global) placement procedures, which is only justified when combined with a valid legalizer.

In any way, critical for floorplanners is the metric by which the results are evaluated. In the chapter "Digital Layout - Placement", by Andrew Kahng and Sherief Reda, this issue will be addressed.

14.2.3 Constrained Floorplans

Much more can be said about floorplanning, especially about the misconceptions and misinterpretations surrounding it. A particular topic that attracts many scientists (but few tool developers) is the avoidance of special constraints. Early in its history the restraint of preserving sliceability was recommended [1]. It was exactly that constraint that made floorplanning fit as a glove with stepwise refinement from a given hierarchy. Floorplans with that property could be represented as a tree, but there were other and more important properties. The fact that slicing structures have the minimum number of "channels" may have had its day as well as the conflict-free routing sequence of those channels. But the fact that many algorithms that are at least *NP*-hard for general floorplan structures become tractable for slicing structures, remains of paramount importance. Among these algorithms is of course the floorplan optimization problem [15]. Later it was proven by Stockmeyer [16] that even the simple special case of fixed module geometries and free orientations was *NP*-hard! Given a slicing tree and the more general shape constraints defined in [17] makes floorplan optimization efficiently solvable. Also, labeling a given tree to find the optimum slicing tree with that topology can be done in polynomial time [18]. And last but not the least, given a point configuration (but actually a pair of sequences), the optimal compatible slicing structure can be found in polynomial time. "Compatible" means that every slice in the floorplan corresponds exactly with a rectangular set in the point configuration [19].

Considering the place and the nature of floorplanning capable of handling uncertainties, flexibility, and preliminary positions, there is not much sense in paying any price for also including nonslicing floorplan topologies [20, 21].

14.3 Wireplans

A wireplan is an incidence structure of modules and global wires. A global wire is one that can be sped up by buffer insertion. The latter implies that its (minimum) length and layer are known. For the modules, a trade-off between speed and size should be available. If timing constraints are given for the inputs and outputs, time budgets can be assigned to the modules so that the total size is minimal. The modules can then be designed using the concept of constant delay so that they do not exceed their budget.

14.3.1 Constant Delay

14.3.1.1 Delay Models

In [2], the observation was made that sizing a gate proportional to its load keeps the gate delay constant. This confirms the delay model introduced in [22] that reads as

$$\tau = g\,\frac{C_L}{C_{in}} + p$$

where p is the inherent parasitic delay and g the computational effort. The latter depends on the function, the gate topology, and the relative dimensioning of the transistors. Both p and g are size-independent, i.e., increasing the gate size with an arbitrary factor does not affect these parameters. The quotient C_L/C_{in} is called the *restoring effort* of the gate.

14.3.1.2 Sizing formulation

Constant delay synthesis starts from a network of gates where each gate has a fixed delay assigned to it. In order to realize this delay, its *restoring effort* has to have the appropriate value. Satisfying these requirements for all gates in the network simultaneously leads to a set of linear equations, as can be seen from Figure 14.6.

Using the notations in that figure the set can be written as

$$\mathbf{c} = \mathbf{q} + \mathbf{N}\mathbf{f}^{D}\mathbf{c} \quad \text{or} \quad (\mathbf{I} - \mathbf{N}\mathbf{f}^{D})\,\mathbf{c} = \mathbf{q} \tag{14.1}$$

The algebraic and numerical aspects of this set have been analyzed in [23] where a detailed derivation can be found as well.

The sizing procedure is summarized in Figure 14.7. It shows that if synthesis produces, beside a netlist, also a vector **f** of reciprocal values of restoring efforts (thereby fixing the delay of each gate), solving set Equation (14.1) yields the vector **c** of input capacitances (and implicitly the gate sizes). Placement of these sized gates may reveal changes in the wire capacitances, and consequently in the vector **q**. Iterating the calculation converges to a set of sizes and imposed capacitances consistent with the placement.

14.3.1.3 Limits

The gate model of Section 14.3.1.1, which underlies the procedure in Section 14.3.1.2, assumes negligible resistance between the gate and its load. For relatively short wires in the network this assumption is certainly acceptable. Modules of up to 100,000 gates can be safely dimensioned in this way. For larger modules wires may get so long that wire resistance is no longer negligible. In today's chips, it is even said that interconnect dominates performance. It is not possible to derive an equally simple and numerically robust iteration method when resistance has to be accounted for. Moreover, improving speed by sizing does not work when there is line resistance. Line segmentation and buffering is effective up to a certain level. This limits the validity of constant delay synthesis (Figure 14.8).

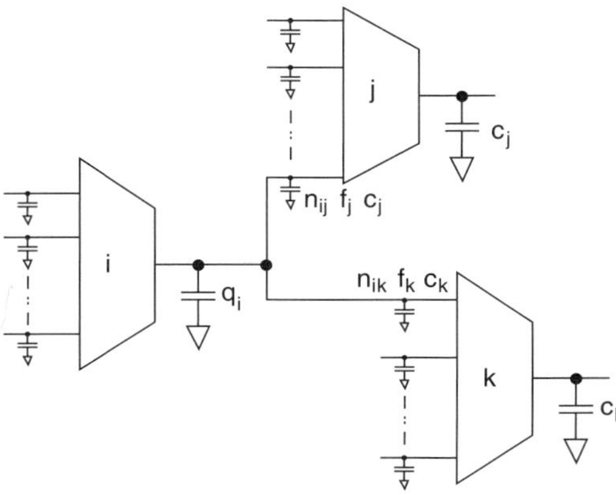

FIGURE 14.6 Composing the load capacitance of gate i : it consists of an "imposed" capacitance q_i, and the input capacitances in its fanout, which are proportional to the respective load capacitances.

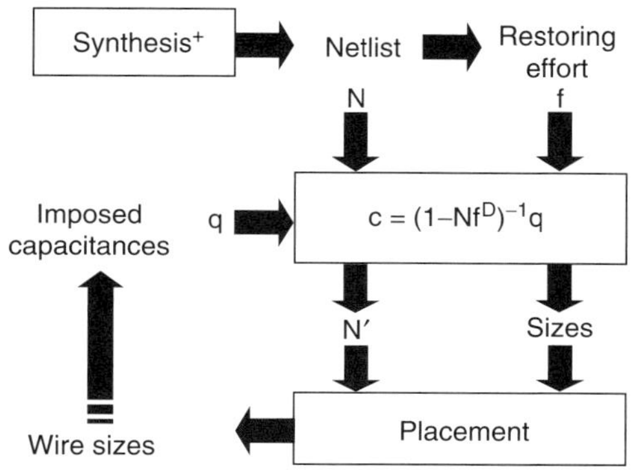

FIGURE 14.7 Computational procedure for sizing the gates.

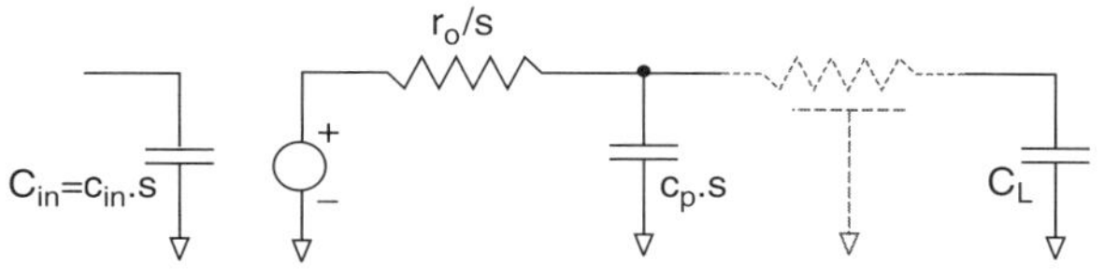

FIGURE 14.8 Wire resistance, here de-emphasized, is neglected in the model underlying the constant delay synthesis.

14.3.2 Wire Planning

14.3.2.1 Problem Statement

Larger networks will have wires with nonnegligible wire resistance; if some wires have to connect widely separated points this can only be avoided in very regular designs by only-neighbor connections. To extend the methodology of the previous section to these designs, such wires have to be treated appropriately. Not uncommon in complex designs is dividing the whole into parts that can be handled by the sizing procedure. Such a partitioning often comes naturally in functional design. The time plan can then identify connections that allow for (or have to have) delays that exceed a critical value. If this value is taken to be the delay of a segment in an optimally buffered interconnection, the thus identified interconnect can be characterized as wires with a length that enables delay reduction by segmentation and buffering. It has been shown that the critical value only depends on the chosen technology and not, for example, on the layer to which the wire is going to be assigned [23][‡].

The approach is to allocate the delay of such identified interconnect first, and then divide the remaining time budget over the modules. These modules do not contain such interconnect and can be sized with the procedure described in Section 14.3.1.2, and thus be held within the budget assigned to that module. The problem to be solved is therefore: given a network of modules connected by *global* interconnect and timing requirements on paths in that network, assign size and delay to each module so that a network can be produced that satisfies the timing requirements. Of course, the relation between size and delay of each module has to be known. This relation will be discussed in Section 14.3.2.3. But first, possible properties of global interconnect will be exploited.

14.3.2.2 Wire Delay

Optimally segmented wires have a delay linear in the length [23]. And if wire delay is linear with its length, its contribution to the path delay between two points on a chip does not depend on the position

[‡]The delay of a segment in an optimally buffered line is indeed a process constant while the length of a segment-depends on the layer. That length does not depend on the buffer sizes!

of the modules on this path. Even stronger, if the path is detour-free, the wire delay on a path is fixed: it only depends on the coordinates of the two endpoints.

Of course, it may in general not be possible to realize the paths of a network in a detour-free manner when the coordinates of these endpoints are given.[§] An easy net-by-net criterion has been formulated in [23], answering the question whether for a given endpoint placement, a detour-free realization exists. Even a point placement can be derived straightforwardly if that criterion is tested (and satisfied) for all nets. The given endpoint placement is called *monotonic* in that case. Also it has been demonstrated there that every acyclic functional network has a realization that makes every endpoint placement monotonic. For logic networks a synthesis procedure has been implemented that preserves monotonicity [25].

Here we assume that an acyclic network is given with the coordinates of its endpoints and that it can be realized without path detours. In Section 14.3.4, we discuss what can be done if detours cannot be avoided and extra wire delay is incurred. The endpoints can be the location of connections to the outside, entries and exits of a supermodule containing all modules in the network, the positions of preplaced registers, etc.

14.3.2.3 Module Size and Delay

The relation between the size and delay of a module is essentially a trade-off: speeding up a module is usually paid for with additional area. Plotted in an area versus delay quadrant, the realization values will be in a region bounded below for both performance characteristics by a monotonically decreasing function with the points of interest on that function. In practice, that function can be well approximated by a convex curve. In [26] this curve is approximated by finding a best fit for $(delay - c)\,(area + b) = a$ through the identified *Pareto* points among the design points.

14.3.3 Time Budgetting

14.3.3.1 Problem Formulation

Assume that a wireplan is given with a monotonic placement of the endpoints (in the sense of Section 14.3.2.2). This means that there exists a point placement for its modules enabling simultaneous detour-free routing of all paths between inputs and outputs. The Manhattan or L_1 length of any input–output path ($L_{i,o}$) can therefore be made equal to the L_1 distance between the input and output terminal pins of that path. The wire delay on such a "monotonic" path is proportional to this length, regardless of the position of the modules on this path, because the delay of optimally buffered wires is linear with length, and the summed delay of the wire segments between the modules is the summed length of the segments multiplied by some constant. The main conclusion is that we can calculate the wire delay of a monotonic path directly from the input and output pin positions. Because module positions are not needed to do that delay calculation, it is possible to do time budgeting even before relative module positions are known.

To obtain the total path delay, the delay of its modules has to be added to the wire delay. That delay depends on the implementation of the modules. Let P^{jk} be the set of all paths from input j to output k; in a given wire plan and p a single path in this set. Then the constraints imposed by the timing requirements read as follows:

$$\forall_{j,k}\forall_{p\in P_{jk}}\left[D_{W_{jk}} + \sum_{m\in p} D_m(A_m) \leq \mathrm{T}_{\mathrm{req}_{jk}}\right]$$

These constraints simply state that for each single path between two pins, the wire delay plus the summed module delay should not exceed a given timing constraint. $T_{\mathrm{req}_{jk}}$ is an upper bound on the delay between pins *j* and *k*. $D_{W_{jk}}$ is the wire delay on the path, and thus equal to L_1 distance between both pins multiplied by a (technology/layer dependent) constant. The minimum (Pareto) delay of a module *m* is approximated by $D_m(A_m)$, a function of its size (area) A_m.

[§]In practice, the overwhelming majority of two-point routes are without detours as experiments in [24] show: only 1.37% were detoured!

The first question that can be posed is whether the wire plan can be realized in a footprint of a given size, while satisfying the timing requirements. It can be answered by minimizing total area, ΣA_i, subject to the given constraints. If the obtained minimal area is larger than the area of the footprint, the wire plan does not fit. Otherwise, not only is the answer to the question obtained, but also the available "slack" area that can be used to adapt modules to the wire characteristics.

The method to use for solving the above optimization problem depends on the delay model, that is on the module delay as a function of module area. Linearizing the D_m curves, and using linear program solvers yields a method which can be polynomial in the size of the program, but the linearizations need to be sufficiently coarse. Other time budgeting techniques have appeared in the literature. Among the more popular techniques, we have to count the *zero-slack algorithm* of Nair et al. [27] and its variants, the *knapsack of approach* of Karkowski [28] and the convex optimization of Sapatnekar, et al. [29].

This optimization problem can be solved by *geometric program* solvers if both object function and constraints have the form $f(t) = \Sigma c_j \Pi t_i^{aij}$. Choosing for the module delay

$$D(s) = d/s + r.$$

achieves that. The function captures the essence of common area-delay trade-offs. D is the delay as a function of a size factor s, making the area of a module $A(s) = a_0 s$. For each module d, r and a_0 are given. Now, the resulting program for n modules is

$$\text{minimize} \sum_{i=1}^{n} a_{0_i} s_i$$

with constraints of the following form:

$$\sum_{j \in p} \left[d_j s_j^{-1} + r_j \right] + D_{W_{jk}} \leq T_{\text{req}_{jk}}$$

A certain class of geometric programs (including the example above) has *posynomial* functions for both object function and constraints $C_j > 0$. By applying the transformation $t_i = e^{x_i}$, both object function and constraints of the posynomial formulation become convex, and can be solved in polynomial time. Further, general convex solvers can be used. The number of constraints depends on the number of paths. Unfortunately, the latter is exponential in the number of edges, resulting in a program of unmanageable size.

14.3.3.2 Complexity Reduction

The number of *constraints* is determined by the total number of input–output paths. The number of terms in a constraint is the number of modules in the associated path. Note that the number of terms is far greater than the number of variables (module sizes), since modules may be on several paths. The number of terms is of interest here, and renders the straightforward formulation impractical already for relatively small designs.

It is easy to see that the number of paths in a graph can be exponentional with regard to the number of edges, but there may be some hope that in practical designs the number of paths is small. This is not the case, and an approach based on considering all paths is out of the question.

However, all the necessary information can be captured by maintaining a number of variables at each node. The delay from a primary input i to an internal node n depends on the *maximum* delay of the paths from i to n. If s of those subpaths exists, s constraints appear for each subpath from n to an output. Since only maximum delays count, only the delay of the slowest (sub)path is of interest. To use that observation, we introduce at each node, for each primary input in its fanin cone, a variable representing the maximum module delay on the subpath from the input to this node. Using the fact that we have a directed, acyclic graph, this value can easily be calculated as the sum of the delay of this node and the *maximum* of the values of its predecessors.

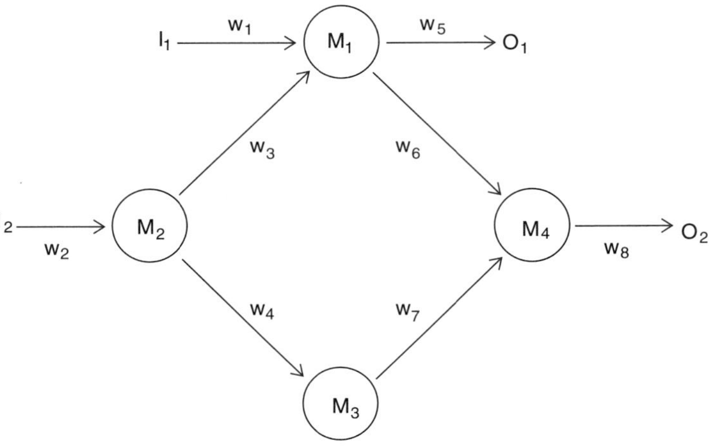

FIGURE 14.9 A circuit of modules.

For example, in Figure 14.9, there are two paths between I_2 and O_2. With w_x denoting the wire delay of segment x, and M_y the module delay of module y, we find (in the straight forward formulation) as constraints for these path

$$\omega_2 + M_2 + \omega_3 + M_1 + \omega_6 + M_4 + \omega_8 \leq T_{\text{req},I2O2}$$
$$\omega_2 + M_2 + \omega_4 + M_3 + \omega_7 + M_4 + \omega_8 \leq T_{\text{req},I2O2}$$

Using monotonicity and linear wire delay

$$\omega_2 + \omega_3 + \omega_6 + \omega_8 = \omega_2 + \omega_4 + \omega_7 + \omega_8 = Dw_{I2,O2}$$

we obtain the form of section 14.3.3.1

$$M_2 + M_1 + M_4 + Dw_{I2,O2} \leq T_{req,I2O2}$$
$$M_2 + M_3 + M_4 + Dw_{I2,O2} \leq T_{req,I2O2}$$

In the formulation of this section, introducing the variable AT_n^I to denote the maximum module delay from input I to node n, we find

$$AT_{M_4}^{I_2} + Dw_{I2,O2} \leq T_{\text{req}}I_2O_2$$

$$AT_{M_4}^{I_2} = M_4 + max\,(AT_{M_4}^{I_2}, AT_{M_3}^{I_2})$$

$$AT_{M_3}^{I_2} = M_3 + max\,(AT_{M_2}^{I_2})$$

$$\vdots$$

$$AT_{M_2}^{I_2} = M_2 + 0$$

Although the savings in efficiency are not noticeable in such a small example, it is obvious that they are tremendous in large examples. In fact, the number of constraints becomes equal to the number of nodes plus the product of the number of edges and the number of primary inputs. Each node has for each primary input in the fanin cone a variable representing the module delay up to here. The number of these variables is bound by the number of primary inputs. For each edge, a constraint is generated for each of those variables at the source node, linking the delay of a node with the delay of its successor. For each node, an additional constraint representing its own delay is generated. Both kinds of constraints are simple and have only two terms.

This shift from paths to nodes makes the *timing graph* from [30] a more convenient model. It is in essence the line digraph of the original: the arrows become nodes, and two arrows that have a head–tail connection at a module node yield an arrow in the timing graph. The conversion is illustrated in Figure 14.10.

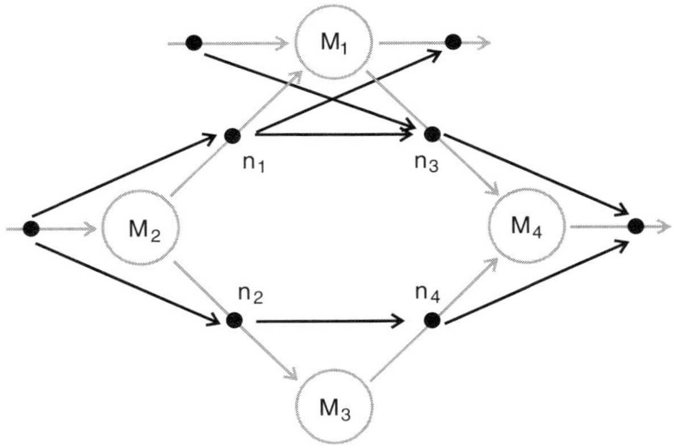

FIGURE 14.10 The line graph of the digraph in Figure 14.9.

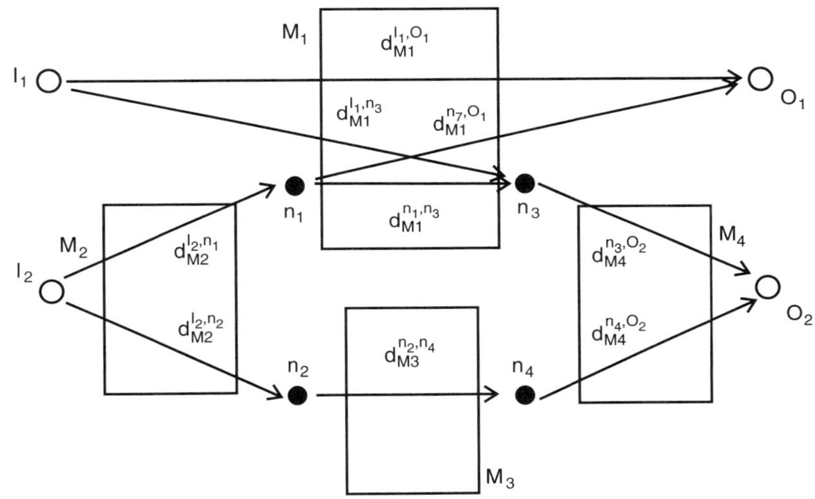

FIGURE 14.11 The timing graph of the digraph in Figure 14.9.

In this model, we can distinguish different delays between different module pin pairs and associate them with the corresponding arrow. Figure 14.11 shows the timing graph that corresponds with the wire plan of Figure 14.9.

As an alternative to [30], we have for each node as many variables AT as there are primary inputs in its input cone. In Figure 14.11, node n_3 has two variables associated, $AT_{n_3}^{I_2}$ and $AT_{n_3}^{I_2}$. In this way, all input–output delays can be taken into account. The complexity of the problem is now, as is the case with static timing analysis, polynomial in the size of the graph.

The problem formulation after this complexity reduction becomes

$$\text{minimize} \sum_M A_M$$

subject to

$$\forall I \in C(m)\ \forall(m,n) \in E\left[AT_n^I \geq AT_m^I + d_M^{mn}(A_M)\right] \quad \text{(E)} \tag{14.2}$$

$$\forall I \in PI\ \forall o \in Po\ \left[AT_n^I \leq T_{req,IO} - Dw_{IO}\right]$$

where *PI* and *PO* are the sets of primary inputs and outputs, respectively, *E* the set of edges in the timing graph, and *C(m)* the set of primary inputs in the input cone of *m*.

Moving from the straightforward path-based formulation of Section 14.3.3.1 to the nodebased approach of this section enables an efficient area optimization under timing constraints. But there are some more techniques to reduce the problem even further and save on runtime.

A simple reduction in the number of constraints is obtained by not restricting the formulation to variables derived from looking at maximum delays from the inputs. The first set of constraints in (E) contains

$$\sum_{m} \gamma(m) |C(m)|$$

constraints, g(m) being the outdegree of node m. If the generation of constraints is done with reference to the primary outputs, the total number of constraints is determined by output cones and indegrees. This may yield a quite different and sometimes considerably lower number of constraints. Although a graph may be analyzed for this comparison in a straightforward manner, the tableau of constraints may be generated for both cases after which the smaller one is used. The computation time for generating such a tableau is small compared to the time taken by the actual computation.

A more intricate improvement in efficiency is obtained by using a technique called *pruning* [31]. It reduces problem size, degeneracy, and redundancy without sacrificing accuracy. As illustrated by Figure.14.12, the basic pruning operation is a graph transformation that replaces a node with touching arcs, and replaces it with other arcs. The variables on the arcs are such that the associated optimization problem is equivalent to the original one.

Generally, solver performance depends on the number of constraints, the number of variables, and the total number of terms in the constraints. Pruning affects these numbers. It is possible to assign to each node a *gain,* a measure of the benefit in case this node was pruned, taking these effects into account. If, for example, two constraints with four variables are replaced by one constraint with six variables, this may or may not be beneficial, depending on the used solver. This is reflected by a positive or negative gain. Nodes are greedily pruned until no nodes with positive gains exist anymore.

In [31], only the numbers of variables and constraints are taken into account, and given equal weights, but one can also consider the number of terms, and tailor the associated weights to the solver used. In the original formulation, only one *AT* variable resides at each node, while one for each primary input may also exist. Therefore, the pruning procedure has to be adjusted to calculate gains for *AT variables* rather than nodes. In this way, an *AT* at node *n* may be pruned for primary input I_1, but not for primary input I_2. Another way of looking at this is that *for each primary input,* a timing graph is being constructed, consisting of the input and its fanout cone, and to which the original pruning procedure can be applied. Then, the resulting optimization program is simply the sum of object functions and the concatenation of the associated constraints. The pruning procedure requires only one graph traversal, and results not only in a dramatically more compact formulation, but also one that is numerically better conditioned.

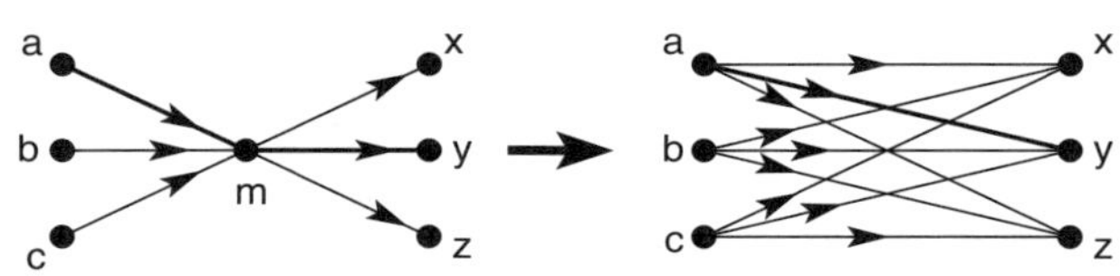

$$\left.\begin{aligned} AT^I_m &\ge AT^I_a + d^{am} \\ AT^I_y &\ge AT^I_m + d^{my} \end{aligned}\right\} \Rightarrow AT^I_y \ge AT^I_a + d^{am} + d^{my}$$

FIGURE 14.12 The basic pruning operation with constraints for the bold path.

14.3.4 Robustness

The time budgets are based on assumptions concerning path lengths, monotonicity, and trade-offs. In real life, however, some of these assumptions may be hard to meet later on: one may, for instance, sometimes need to deviate from monotonicity. Another uncertainty is the accuracy of the trade-off models. More generally, one can say that in a noniterative design flow, one needs a certain amount of "slack separation"[¶] in order to obtain "robustness" with respect to uncertainties later on. Fortunately, slack on the majority of paths can be introduced at very low cost.

First, the formulation itself ensures a certain amount of slack: wire delay is calculated as the path length times some constant. When the modules are realized, however, they will occupy space, effectively reducing wire length, and thus delay. Second, the formulation ensures that all input–output pairs will have a critical path: a path whose delay equals the constraint. This implies that every module is on a critical path. It does not however imply that every wire segment is. This offers the possibility of detouring these wires. Third, if simple bounds on area are used, this may also result in slack. Finally, wire delay is calculated for ideally buffered wires. If logic is pulled out of the module, and spread over the wires, buffers may be replaced by "useful" gates. Therefore, wire delay estimation is kept conservative.

If the slack inherent in the formulation is still not sufficient, Westra et al. [26] introduces a way to provide additional slack at little area cost. It also has the advantage that *truely critical paths* that limit circuit performance most are revealed, making it clear where the main thrust of design effort should focus.

14.4 A Formal System For Trade-Offs

It is interesting to look back on how wiring closure was achieved, though keeping in mind that in those days wiring was realized in two or three layers and allocated to channels between modules. Critical was the generalization of placement, concerned with the arrangement of geometrically fixed objects without overlap in the plane (possible bounded or preferably small) into floorplan design, where only relative positions of objects are being fixed. Accepting slicing as a structural restraint then guaranteed sufficient wiring space that could be treated uniformly with the modules during sizing, and yielded a conflict-free sequence of channel routing problems. In addition, it rendered several optimizations tractable, including sizing which was in essence a trade-off between the dimensions of height and width. These dimension pairs are partially ordered by a *dominance relation:* one pair dominates another if neither of its dimensions is larger than the corresponding dimension of the other pair.

Similar observations can be made with respect to timing closure. The essential shift there is from sizing with speed as an arbitrary outcome to timing constraints with size beyond control. For smaller modules the concept of constant delay enabled closure. However, the underlying assumption that the delay of a gate can be kept constant by varying its size linearly with its load is only valid when resistance between the gate and its load can be neglected. This required planning of all wires for which resistance could not be neglected. They take a considerable but constant part of the timing budget leaving the remainder to be divided over the modules. Again this comes down to trade-offs, now between size and speed, and a corresponding *dominance relation.*

After timing, the primary concern for many of today's *SoC* devices is power consumption, which if appropriate paradigm shifts can be found, will lead to similar trade-offs between speed and power. The problem however is that performance characteristics can no longer be dealt with by considering them in isolated pairs or triples. What is needed is a formal system that allows us to manipulate a design without losing potentially optimal final solutions. Such a system is the topic of this section, but before introducing it we observe the following.

[¶] *Negative slack* means constraints are violated, *(positive) slack* means there is room to tighten a constraint, and *slack separation* is the difference between the most critical path and other paths.

Closure has up to now been achieved by Pareto-style trade-offs, where every realization implies a combination of values of performance characteristics. These values are linearly ordered, though the realization points are partially ordered by dominance. Only points that are not dominated by others can be optimal under monotone objectives. At the end, the best one with respect to such an objective is chosen. The two examples above enable hierarchical application of the methods.

Today, it is of paramount importance to avoid premature commitment while being unaware of the consequences, because of the many relevant performance characteristics that interact with different levels of design. Choosing the best one at a certain level may preclude optimal ones in later stages. Yet, multidimensional Pareto analysis may be, and mostly is, very complex.

14.4.1 Realization Spaces

A performance characteristic is a set of realization values that indicate a quality aspect of a product or design step. Performance characteristics that quickly come to mind in chip design are *size, delay,* and *power.* Considered by themselves they are of the type "the smaller the better." Without loss of generality we will assume all characteristics to be of that type. Another property of these classic performance characteristics is that they are totally ordered.[‖] That is, for every pair of realizations we can say whether one is better than the other with respect to such a parameter. The performance characteristics measurable in the final product often share this property. Although performance characteristics without that property do not easily come to mind, we will not adopt such a constraint for performance characteristics. We assume that they are ordered, i.e., there is a reflexive, transitive, and anti-symmetric relation associated with any performance characteristic, but not all of its values are comparable. Any trade-off between totally ordered performance characteristics may produce such a relation. Power and delay, for example, form pairs that are so ordered by dominance: such a pair dominates another pair if it is at least as good in both aspects. Obviously, two pairs each of which improves upon the other in some aspect are not comparable. If we consider a number of realizations of a given product specification and compare them on the basis of power consumption and speed, we are in fact looking at a number of pairs that are *partially ordered.* We call such a partially ordered set of realizations with respect to some performance characteristic a *realization space,* and its ordering relation *dominance,* i.e., a realization *dominates* another realization if it is at least as good with respect to every concerned performance characteristic. Abstractly, a *realization space* is simply a set Q with a partial order $\preccurlyeq_Q$, and we say $q \in Q$ *dominates* $q' \in Q$ whenever $q \preccurlyeq_Q q'$.

Let us consider all performance characteristics that might be relevant for a design. Some of these characteristics will be measurable in the final result. Others might be absorbed during the design process after they have played a role in certain stages of that process. It may even be so that certain characteristics have not been encountered at all upon completion because early decisions precluded them. That is, at each stage of the design process, realization spaces concerned with subsets of the performance characteristics are being manipulated and transformed into other realization spaces. It is important to note here that when realization spaces are combined to form a new space they induce a new partial ordering. In other words, the result is also a realization space. We no longer distinguish spaces concerned with basic performance characteristics and results that are associated with partial orders!

Each element (or point) in a realization space S represents a (partial) realization of a system. For convenience, we write them as vectors, $\bar{c} = (c_1, c_2, \ldots, c_n)$, where each c_i is a realization value of a characteristic Q_i. Then $\bar{a}$ in S *dominates* $\bar{b}$ in S ($\bar{a} \preccurlyeq_S \bar{b}$) if $a_i \preccurlyeq_{Q_i} b_i$ for all i. Dominance still expresses the fact that the underlying realization is in no aspect worse than the other. It can be raised to subsets of a realization space to say that any point in the dominated subset can be matched or improved upon by some point in the dominating subset: A subset A of a realization space dominates another subset B if for every element $\bar{b} \in B$ there is an element $\bar{a} \in A$ such that $\bar{a} \preccurlyeq_C \bar{b}$; this is written as $A \preccurlyeq B$. If A and B dominate each other, they are called *Pareto-equivalent* which is denoted as $A \equiv B$. Pareto-equivalence between two subsets means that neither subset contains a point that cannot be matched by a point in the other subset.

‖ A performance characteristic whose ordering is total (or linear) is called a basic performance characteristic here.

Realization points that are strictly dominated by other points in the same set can be removed without losing interesting realization options. In general, we do not want to maintain larger sets than necessary. This brings about the notion of minimality: a subset C of a realization space is *Pareto-minimal* if the removal of any of its elements yields a subset that is not Pareto-equivalent to C. That is, no element dominates another element in a Pareto-minimal subset. Or in other words, the Pareto-minimal subsets are *anti-chains* of the partial order.

Note that two Pareto-equivalent subsets need not be the same. However, when they are Pareto-minimal they have to be identical! More disturbing is the fact that not all realization spaces have a unique Pareto-equivalent subset that is Pareto-minimal. Whenever it exists for a realization space S we denote it by $\min(S)$. A sufficient requirement for such a *Pareto-frontier* is that every chain in the space has a smallest element (i.e., the space is *well-ordered)*. This implies that *every finite realization space has a Pareto-equivalent subset that is Pareto-minimal.* In the present context, we mostly have only discrete points obtained by profiling in the realization spaces. From that perspective, the implication is certainly satisfactory. We assume in the rest of this section that all realization spaces are well-ordered.

14.4.2 Operators

Of course, the unbridled combination of the performance characteristics yields immense, unmanageable spaces. Essential, therefore, is an efficient algorithm that takes a realization space and returns a so-called *Pareto frontier.* This potentially reduces realization spaces, and might, if applied intelligently, keep a design flow manageable. Other operations may be defined provided that no potentially optimal candidate realizations are lost. Operations should preferably *preserve minimality,* i.e., applying the operation on minimized sets produces a minimal resulting set. However, a number of obviously necessary operations do not have that property. In such cases, we have to settle for *optimum preserving,* meaning that minimizing the result of such an operation on minimized operands, is the same as minimizing the result of that same operation on the unminimized operands.

14.4.2.1 Minimization

We denote by min(C) the unique Pareto-minimal set of a realization spaces C. It contains all realizations from that space that we would be interested in practice because all other realizations are dominated by some realization in min(C).

We call the operation to obtain that subset *minimization.* It is the key operation in the system. It would be nice if other operations when applied to minimized realization spaces would not require minimization to obtain the Pareto-minimal equivalent space. More precisely, if we have an operator Ψ that takes n spaces as its operands, we would like

$$\min(\Psi\,(C_1, \dots C_n)) = \Psi\,(\min(C_1), \dots, \min(C_n)) \tag{14.3}$$

We say then that Ψ *preserves minimality.* In that case minimization after applying operator Ψ is unnecessary. The result is always minimal when the operands have been minimized, which saves time in computing the results of the operation.

Unfortunately, some indispensable operations will not preserve minimality. That does not mean that it is not advantageous to apply the operation on minimized spaces. The computational effort spent on these generally smaller spaces is almost always considerably less than when the unreduced spaces are the arguments. If only a Pareto-equivalent subset is obtained, that is

$$\Psi\,(C_1, \dots C_n) \equiv \Psi\,(\min\,(C_1), \dots, \min C_n)) \tag{14.4}$$

Subsequent minimization then yields the same realization space as when the operation was applied to $C_1, \dots, C_n$. So by minimizing operands, which are often intermediate results in the design process, before applying Ψ, no optimal realizations are lost. Every practical operator should *support minimization* in that sense, for otherwise all realizations have to be analyzed to identify the optimal ones.

If Pareto-equivalence is a congruence with respect to the operator Ψ, that is

$$\forall_{1\leq i\leq n}[C_i' \equiv C_i] \rightarrow [\Psi\ (C_i', \dots, C_n') \equiv \Psi\ (C_1, \dots, C_n)]$$

then the operation necessarily supports minimization. This means that to prove that minimization support it suffices to show that the operator preserves dominance:

$$\forall_{1\leq i\leq n}[C_i' \preccurlyeq C_i] \rightarrow [\Psi\ (C_1', \dots, C_n') \preccurlyeq \Psi\ (C_1, \dots, C_n)]$$

14.4.2.2 Free Product

When designing subsystems independently with their own realization spaces and tradeoffs, they have to be put together later, which means that their realization spaces have to be combined. The corresponding operator is called *free product.*

Let C_1 and C_2 be realization spaces each with their own performance characteristics. The free product $C_1 \bullet C_2$ consists of the points $\bar{c}_1 \cdot \bar{c}_2$, the concatenation of the "vectors" of each point $\bar{c}_1 \in C_1$ and each point $\bar{c}_2 \in C_2$

Obviously, not only is Pareto-equivalence a congruence with respect to free product, the operator even preserves minimality, which was to be expected when its purpose is considered.

14.4.2.3 Union

To enable a choice between realizations resulting from different design paths, the operator *union* is introduced. It can be thought of as the realization points of both spaces being brought into one and the same space spanned by the same set of performance characteristics as the two individual spaces. The notation is $C_1 \sqcup C_2$.

The operator does not preserve minimality. Obviously, a design trajectory may produce a realization that is dominated by one resulting from the other design path. So, minimization may result in a subset strictly included in its operand. But minimization is supported, because it can be seen as easily that $C_1 \sqcup C_2 \equiv \min (C_1) \sqcup \min (C_2)$. It is even true that Pareto-equivalence is a congruence with respect to the union operator.

14.4.2.4 Constraints

When we have a set of realizations, some of which are invalid because of additional constraints, then we can apply a constraint to filter out only those realizations that satisfy the constraint. Such constraints can be formally expressed as a set D of acceptable realizations or equivalently as a proposition on the realization space, identifying the acceptable realizations. Application of the constraint D to a realization space C leaves those realizations of C satisfying the constraint. We denote the new space as $C \sqcap D$.

When defined in that general way, constraints present a problem. It may easily happen that a constraint filters out a realization that dominated one that was not filtered out. Minimization cannot be supported in this way, because that operator may have removed such dominated realization points, while they may be part of the minimized result of the constraint applied to the unminimized space. The constraints in our algebra — we will call them *safe constraints* — will never have a dominating realization passing while the dominating realization fails the test. More precisely, a constraint D is called a safe constraint when for every $\bar{a}$ and $\bar{b}$ with $\bar{a} \preccurlyeq \bar{b}$ $\bar{b} \in D$ implies that $\bar{a} \in D$. The notation for applying a safe constraint D on a realization space C is $C \breve{\sqcap} D$.

14.4.2.5 Projection

In certain design stages, certain performance characteristics may no longer be relevant for future decisions. Projection can then be used to remove such a dimension of the realization space. In general, minimality is not preserved by projection, but dominance is. Therefore congruence and hence minimization support follow.

14.4.3 Cost Functions

The selection of a final realization point often happens with a cost function. Such a cost function should be one which never selects a point that is dominated under the space's partial order by another point. If that were possible, the whole formal system constructed here would be pointless, because the purpose was to manipulate and form realization spaces without losing any candidates for optimal realizations. In the process, strictly dominated realizations are being removed when reducing a space to the Pareto frontier. To ensure that cost functions are in concord with the motivation for building this formal system, we require *monotonicity*, i.e., whenever $\bar{c}_1 \preccurlyeq \bar{c}_2$ the cost function f has to satisfy $f(\bar{c}_1) \leq f(\bar{c}_2)$.

Thus defined, it requires cost functions to assign real numbers to every point in the final realization space. This is perfectly in order when the real objective is like that, as for example the chip size (or area). Often the selection is not so unambiguous, and designers resort to weighted combinations. Setting these weights is more often than not a trial-and-error affair. Yet the outcome may depend heavily on the exact setting. In essence, it is a consequence of restricting the range of cost functions to a basic performance characteristic. This is not necessary. If we allow the range to be any performance characteristic, that is any partially ordered set, then a simple adaptation of the definition of monotonicity preserves the essential property cost functions should have.

Monotone functions from a realization space C into a performance characteristic are then defined to be functions satisfying

$$\forall_{C_1, C_2 \in C}[\bar{c}_1 \preccurlyeq \bar{c}_2 \rightarrow \bar{c}_2 f(\bar{c}_1 \preccurlyeq f(\bar{c}_2)]$$

Thus defined they still can only select realizations in the Pareto-minimal set of the space as optimal. Moreover, any member of that set can be selected by such a function. That is, for every point in the Pareto-minimal set of a realization space, there is a monotone cost function that will select that point as minimal. This means that further reduction of spaces is not possible without giving up the guarantee that no potentially optimal candidate realizations are lost, if the cost function is not known. With the generalized cost function however a partial selection of candidates is possible, leaving fewer realizations to consider in subsequent steps.

14.4.4 Concluding Remarks

This section describes the algebra (C, min, $\bullet$, $\sqcup, \downarrow, \overset{\cup}{\sqcap}$), where C is the set of all possible realization spaces that can be encountered during the design of a chip. The algebra was introduced in [32] where it mainly served as a formal system for runtime reconfiguration. The manipulations and calculations then have to take place on mostly resource-constrained devices. Here it is used for design-time exploration with a wide choice in computer power and lesser constraints on compute time. Not surprisingly, different algorithms may be chosen in these two application areas.

In the latter application, the algebra is meant for supporting the design while it forms realization spaces by combining subsystems and their realization spaces. The number of realizations, also when restricted to Pareto-minimal sets, can grow very fast, and it is therefore crucial to keep the space sizes small. Reducing them to their Pareto-minimal spaces is the most important means here, but pruning the space may still be necessary. The challenge is then to find the best approximation of realization space by one with a limited number of points [33, 34].

To date the most efficient minimization algorithm has time complexity $O(N\ (\log N)^d)$ [35], where d is the number of dimensions of the realization space and N the number of realizations. The number of points for which the algorithm is faster than simpler algorithms such as the *simple cull* in [36], having a complexity of $O(N^2)$, is quite high. This makes the former algorithm the choice only for large spaces, while *simple cull* and hybrid algorithms [36] are recommendable for smaller numbers of realizations. A similar effect is observed for data structures for storing sets of realizations. The quad-tree data structure [37, 38] is shown to be more efficient than linear lists in [38], but here also, the gain is achieved for large numbers of data points. Yukish[36] shows that when keeping points lexicographically sorted, normalized linear tables can be advantageous for computing the union and intersection of sets.

References

[1] R.H.J.M. Otten, Automatic floorplan design, *Proceedings of the 19th Design Automation Conference*, Las Vegas, USA, 1982, pp. 261–267.

[2] J. Grodstein, E. Lehman, H. Harkness, B. Grundmann, and Y. Watanabe, A delay model for logic synthesis of continuously-sized networks, *Proceedings of ICCAD*, November 1995.

[3] J. Grason, *A Dual Linear Graph Representation for Space-filling Location Problems of the Floor-Planning Type*, MIT Press, Cambridge, Massachusetts, USA, 1970.

[4] K. Kozminsky and E. Kinnen, Rectangular duals of planar graphs, *Networks*, 15, 145–157, 1985.

[5] J. Bhasker and S. Sahni, Representation and generation of rectangular dissections, *Proceedings of the 23rd Design Automation Conference*, Las Vegas, USA , 1986, pp. 108–114.

[6] J. Bhasker and S. Sahni, A linear time algorithm to check for the existence of a rectangular dual of a planar triangulated graph , *Networks*, 17, 307-317, 1987.

[7] R. Schmid, Synthese von Floorplans auf der Basis von Dualgraphen mit rechtwinkliger Einbettung, Ph.D. Thesis, Universität Karlsruhe, West Germany, 1987.

[8] U. Flemming, Representation and generation of rectangular dissections, *Proceedings of the 15th Design Automation Conference*, Las Vegas, USA, 1978, 138–144.

[9] T. Ohtsuki, N. Sugiyama, and H. Kawanishi, An optimization technique for integrated circuit lay-out design *Proceedings of ICCST*, Kyoto, Japan, 1970, pp. 67–68.

[10] R.H.J.M. Otten, L.P.P.P. van Ginneken, Floorplan design using annealing, *Proceedings of ICCAD*, SantaClara, 1984.

[11] T. Takahashi, A new encoding scheme for rectangle packing problem, *Proceedings ASP-DAC 2000*, Yokohama, Japan, 2000, pp. 175-178.

[12] W.R.Heller, G. Sorkin, and K. Maling, The planar package planner for system designers, *Proceedings of the 19th Design Automation Conference*, Las Vegas, USA, 1982, pp. 663–670.

[13] S. Tsukiyama, K. Koike, and I. Shirakawa, An algorithm to eliminate alle complex triangles in a maximal planar graph for use in VLSI floorplan, *Proceedings ISCAS*, Philadelphia, PA, USA, 1986, pp. 321-324.

[14] S.M. Leinwand and Yen-Tai Lai, An algorithm for building rectangular floor-plans, *Proceedings of the 21st Design Automation Conference*, Albuquerque, USA, 1984, pp. 663-664.

[15] R.H.J.M. Otten and P.S.Wolfe, Algorithm for wiring space assignment in slicing floorplans, Technical Report Y0882-0682, IBM Thomas J. Watson Research Center, Yorktown Heights, USA, 1982.

[16] L. Stockmeyer, Optimal orientations of cells in slicing floorplan designs, *Inform. Control*, 57, 91-101, 1983.

[17] R.H.J.M. Otten, Efficient floorplan optimization, *Proceedings of ICCD*, Port Chester, USA, 1983, pp. 499–502.

[18] W. Keller, F. Schreiner, B. Schurman, E. Siepmann, and G. Zimmerman, Hierarchisches Top Down Chip Planning , Report of the University of Kaiserslautern, West Germany, 1987.

[19] L.P.P.P. van Ginneken and R.H.J.M. Otten, Optimal slicing of plane point placements, *Proceeding of the European Design Automation Conference*, Glasgow, 1990, pp. 322–326.

[20] Young, F.Y. and D. F. Wong, How good are slicing floorplans?, *INTEGRATION: VLSI J.* 23, 61-73, 1997.

[21] Lai, M. and D.F. Wong, Slicing Tree is a complete floorplan representation, *Proceedings of the Design, Automation and Test in Europe Conference (DATE)*, March 2001.

[22] I. Sutherland and R. Sproull, The theory of logical effort: designing for speed on the back of an envelope, in *Advanced Research in VLSI*, C. Séquin, Ed., MIT Press, Cambridge, 1991.

[23] R.H.J.M. Otten, A design flow for performance planning, in *Architecture Design and Validation Methods*, E. Borger, Ed., Springer, Berlin, 2000.

[24] J. Westra, C. Bartels, and P.R. Groeneveld, Probabilistic congestion prediction, *Proceedings of the International Symposium on Physical Design*, 2004, pp. 204-209.

[25] W. Gosti, A. Narayan, R. K. Brayton, and A. L. Sangiovanni-Vincentelli, Wireplanning in Logic Synthesis, *Proceedings of ICCAD*, 1998.

[26] J. Westra, D.-J. Jongeneel, R.H.J.M. Otten, and C.Visweswariah, Time budgeting in a wire planning context, *Proceedings of DATE,* 2003.

[27] R. Nair, C.L. Berman, P.S. Hauge, and E.J. Yoffa, Generation of performance constraints for layout, *IEEE Transactions on CAD,* Vol. 8, August 1989.

[28] I. Karkowski, Circuit delay optimization as a multiple choice linear knapsack problem, *Proceedings of the European Design Automation Conference (EDAC),* March 1993, pp. 419-423.

[29] S.S. Sapatnekar, V.B. Rao, P.M. Vaidya, and S.M. Kang, An exact Solution to the transistor sizing problem for CMOS circuits using convex optimization, *IEEE Trans. CAD,* 12, 1621–1634, 1993.

[30] A.R. Conn, I.M. Elfadel, W.W. Molzen, Jr., P.R. O'Brien, P.N. Strenski, C. Visweswariah, and C.B. Whan, Gradient-based optimization of custom circuits using a static-timing Formulation, *Proceedings of the 36th Design Automation Conference,* June 1999, p. 452–459, 1999.

[31] C. Visweswariah and A.R. Conn, Formulation of static circuit optimization with reduced size, degeneracy and redundancy by timing graph manipulation, *Proceedings of the International Conference on Computer-Aided Design (ICCAD),* 1999, p.244.

[32] M. Geilen, A.A. Basten, B. Theelen, and R.H.J.M. Otten, An algebra of Pareto points, pp. 88–97, *Proceedings of ASCD 2005.*

[33] C. Mattson, A. Mullur, and A. Messac, Minimal representations of multiobjective design space using a smart Pareto filter, *Proceedings of the 9th AIAA/ISSMO Symposium on Multidisciplinary Analysis and Optimization,* 2002.

[34] E. Zitzler, L.Thiele, M. Laumanns, C. Fonseca, and V.G.D.Fonseca, Performance assessment of multiobjective optimizers: an analysis and a review, *IEEE Trans. Evol. Comput.,* 7,117-132, 2003.

[35] J. Bentley, Multidimensional command-and-conquer, *Commun. ACM,* 23, 214–229, 1980.

[36] M. Yukish, Algorithms to Identify Pareto Points in Multidimensional Data Sets, Ph. D. Thesis, Pennsylvania State University, August 2004.

[37] R. Finkel and J. Bentley, Quad trees: a data structure for retrieval on composite keys, *Acta Inform.,* 4, 1–9, 1974.

[38] M. Sun and R.E.Steuer, Quad trees and linear lists for identifying nondominated criterion vectors, *ORSA J. Comput.,* 8, 367–375, 1996.

[39] L. Cheng, L. Deng, and M.D.F. Wong, Floorplan Design for 3-D VLSI design, *Proceedings of the IEEE Asia South Pacific Design Automation Conference (ASP-DAC),* January 2005.

SECTION IV

LOGICAL VERIFICATION

15

Design and Verification Languages

Stephen A. Edwards
Columbia University
New York, New York

Abstract

After a few decades of research and experimentation, register–transfer dialects of two standard languages — Verilog and VHDL — have emerged as the industry standard starting point for automatic large-scale digital integrated circuit synthesis. Writing register-transfer-level descriptions of hardware remains a largely human process and hence the clarity, precision, and ease with which such descriptions can be coded correctly has a profound impact on the quality of the final product and the speed with which the design can be created.

While the efficiency of a design (e.g., the speed at which it can run or the power it consumes) is obviously important, its correctness is usually the paramount issue, consuming the majority of the time (and hence money) spent during the design process. In response to this challenge, a number of the so-called verification languages have arisen. These have been designed to assist in a simulation-based or formal verification process by providing mechanisms for checking temporal properties, generating pseudorandom test cases, and for checking how much of a design's behavior has been exercised by the test cases.

Through examples and discussion, this chapter describes the two main design languages — VHDL and Verilog — as well as SystemC, a language currently used to build large simulation models; System Verilog, a substantial extension of Verilog; and OpenVera, e, and PSL, the three leading contenders for becoming the main verification language.

15.1 Introduction

Hardware description languages (HDLs) are now the preferred way to enter the design of an integrated circuit, having supplanted graphical schematic capture programs in the early 1990s. A typical design methodology in 2005 starts with some back-of-the-envelope calculations that lead to a rough architectural design.

This architecture is refined and tested for functional correctness by implementing it as a large simulator written in C or C++ for speed. Once this high-level model is satisfactory, it is passed on to designers who recode it in a register–transfer-level (RTL) dialect of VHSIC (Very High-Speed Integrated Circuit) Hardware Description Language (VHDL) or Verilog — the two industry-dominant HDLs. This new model is often simulated to compare it to the high-level reference model, then fed to a logic synthesis system such as Synopsys' Design Compiler, which translates the RTL into an efficient gate-level netlist. Finally, this netlist is given to a place-and-route system that ultimately generates the list of polygons that will become wires and transistors on the chip.

None of these steps, of course, is as simple as it might sound. Translating a C model of a system into RTL requires adding many details, ranging from protocols to cycle-level scheduling. Despite many years of research, this step remains stubbornly manual in most designs. Synthesizing a netlist from an RTL dialect of an HDL has been automated, but is the result of many years of university and industrial research, as are all the automated steps after it.

Verifying that the C or RTL models are functionally correct presents an even more serious challenge. At the moment, simulation remains the dominant way of raising confidence in the correctness of these models, but has many drawbacks. One of the more serious is the need for simulation to be driven by appropriate test cases. These need to exercise the design, preferably the difficult cases that expose bugs, and be both comprehensive and relatively short since simulation takes time.

Knowing when simulation has exposed a bug and estimating how complete a set of test cases actually is, are two other major issues in a simulation-based functional verification methodology. The verification languages discussed here attempt to address these problems by providing facilities for generating biased, constrained random test cases, checking temporal properties, and checking functional coverage.

15.2 History

Many credit Reed [1] with the first HDL. His formalism, simply a list of Boolean functions that define the inputs to a block of flip-flops driven by a single clock (i.e., a synchronous digital system), captures the essence of an HDL: a semi-formal way of modeling systems at a higher level of abstraction. Reed's formalism does not mention the wires and vacuum tubes that would actually implement his systems, yet it makes clear how these components should be assembled.

In the five decades since Reed, both the number and the need for HDLs has increased. In 1973, Omohundro and Tracey [2] could list nine languages and dozens more have been proposed since.

The main focus of HDLs has shifted as the cost of digital hardware has dropped. In the 1950s and 1960s, the cost of digital hardware remained high and was used primarily for general-purpose computers. Chu's CDL [3] is representative of languages of this era: it uses a programming-language-like syntax; has a heavy bias toward processor design; and includes the notions of arithmetic, registers and register transfer, conditionals, concurrency, and even microprograms. Bell and Newell's influential ISP (described in their 1971 book [4]) was also biased toward processor design.

The 1970s saw the rise of many more design languages [5, 6]. One of the more successful was ISP'. Developed by Charles Rose and his student Paul Drongowski at Case Western Reserve in 1975 to 1976, ISP' was based on Bell and Newell's ISP and was used in a design environment for multiprocessor systems called N.mPc [7]. Commercialized in 1980 (and since owned by a variety of companies), it enjoyed some success, but starting in 1985, the Verilog simulator (and accompanying language) began to dominate the market.

The 1980s brought Verilog and VHDL, which remain the dominant HDLs to this day (2004). Initially successful because of its superior gate-level simulation speed, Verilog started life in 1984 as a proprietary language in a commercial product, while VHDL was designed at the behest of the US Department of Defense as a unifying representation for electronic design [8].

While the 1980s was the decade of the widespread commercial use of HDLs for simulation, the 1990s brought them an additional role as input languages for logic synthesis. While the idea of automatically synthesizing logic from an HDL dates back to the 1960s, it was only the development of multilevel logic synthesis in the 1980s [9] that made them practical for specifying hardware, much as compilers for software require optimization to produce competitive results. Synopsys was one of the first to release a commercially successful logic synthesis system that could generate efficient hardware from RTL Verilog specifications and by the end of the 1990s, virtually every large integrated circuit was designed this way.

Hardware description languages continue to be important for providing inputs for synthesis and modeling for simulation, but their importance as aids to validation has recently grown substantially. Long an important part of the design process, the use of simulation to check the correctness of a design has become absolutely critical, and languages have evolved to help perform simulation quickly, correctly, and judiciously.

Clearly articulated in features recently added to System Verilog, it is now common to generate automatically simulation test cases using biased random variables (e.g., to generate random input sequences in which reset occurs very infrequently), check that these cases thoroughly exercise the design (e.g., by checking whether certain values or transitions have been overlooked), and check whether invariants have been violated during the simulation process (e.g., making sure that each request is followed by an acknowledgment). Hardware description languages are expanding to accommodate such methodologies.

15.3 Design Languages

15.3.1 Verilog

The Verilog hardware description language [10–12] was designed and implemented by Phil Moorby at Gateway Design Automation in 1983 to 1984 (see Moorby's history of the language for more details [6]). The Verilog product was very successful, buoyed largely by the speed of its "XL" gate-level simulation algorithm. Cadence bought Gateway in 1989 and largely because of the pressure from the competing, open VHDL language, made the language public in 1990. Open Verilog International (OVI) was formed shortly thereafter to maintain and promote the standard. IEEE adopted it in 1995, and ANSI in 1996.

The first Verilog simulator was event-driven and very efficient for gate-level circuits, the fashion of the time, but the opening of the Verilog language in the early 1990s paved the way for other companies to develop more efficient compiled simulators, which traded up-front compilation time for simulation speed.

Like tree rings, the syntax and semantics of the Verilog language embodies a history of simulation technologies and design methodologies. At its conception, gate- and switch-level simulations were in fashion, and Verilog contains extensive support for these modeling styles that are now little used. (Moorby had worked with others on this problem before designing Verilog [13].)

Like many HDLs, Verilog supports hierarchy for structural modeling, but was originally designed assuming modules would have at most tens of connections. Hundreds or thousands of connections are now common, and Verilog-2001 [11] added a more succinct connection syntax to address this problem.

Procedural or behavioral modeling, once intended mainly for specifying testbenches, was pressed into service first for RTL specifications, and later for the so-called behavioral specifications. Again, Verilog-2001 added some facilities to enable this (e.g., always @* to model combinational logic procedurally) and SystemVerilog has added additional support (e.g., always_comb, always_ff).

The syntax and semantics of Verilog have always been a compromise between modeling clarity and simulation efficiency. A "reg" in Verilog, the variable storage class for behavioral modeling, is exactly a shared variable. This means it simulates very efficiently (e.g., writing to a reg is just an assignment to memory), but also means that it can be misused (e.g., when written to by two concurrently running processes) and misinterpreted (e.g., its name suggests a memory element such as a flip-flop, but it often represents purely combinational logic).

The book by Thomas and Moorby [14] has long been the standard text on the language (Moorby was the main designer), and the language reference manual [10], since it was adopted from the original Verilog simulator user manual, is surprisingly readable. Other references include Palnitkar [15] for an overall description of the language, and Mittra [16] and Sutherland [17] for the programming language interface (PLI). Smith [18] compares Verilog and VHDL. French et al. [19] present a clever way of compiling Verilog simulations and also discuss more traditional ways.

15.3.1.1 Coding in Verilog

A Verilog description is a list of modules. Each module has a name; an interface consisting of a list of named ports, each with a type, such as a 32-bit vector, and a direction; a list of local nets and regs; and a body that can contain instances of primitive gates such as ANDs and ORs, instances of other modules (allowing hierarchical structural modeling), continuous assignment statements, which can be used to model combinational datapaths, and concurrent processes written in an imperative style.

Figure 15.1 shows the various modeling styles supported in Verilog. The two-input multiplexer circuit in Figure 15.1(a) can be represented in Verilog using primitive gates (Figure 15.1(b)), a continuous assignment (Figure 15.1(c)), a user-defined primitive (a truth table, Figure 15.1(d)), and a concurrent process (Figure 15.1(e)). All of these models exhibit roughly the same behavior (minor differences occur when some inputs are undefined) and can be mixed freely within a design.

One of Verilog's strengths is its ability to also represent testbenches within the model being tested. Figure 15.1(f) illustrates a testbench for this simple mux, which applies a sequence of inputs over time and prints a report of the observed behavior.

Communication within and among Verilog processes takes place through two distinct types of variables: *nets* and *regs*. Nets model wires and must be driven either by gates or by continuous assignments. Regs are exactly shared memory locations and can be used to model memory elements. Regs can be assigned only by imperative assignment statements that appear in *initial* and *always* blocks. Both nets and regs can be single bits or bit vectors, and regs can also be arrays of bit vectors to model memories. Verilog also has limited support for integers and floating-point numbers. Figure 15.2 shows a variety of declarations.

The distinction between regs and nets in Verilog is pragmatic: nets have slightly more complicated semantics (e.g., they can be assigned a capacitance to model charge storage and they can be connected to multiple tri-state drivers to model buses), but regs behave exactly like memory locations and are therefore easier to simulate quickly. Unfortunately, the semantics of regs makes it easy to inadvertently introduce nondeterminism in the language (e.g., when two processes simultaneously attempt to write to the same reg, the result is undefined). This will be discussed in more detail in the next section.

Figure 15.3 illustrates the syntax for defining and instantiating models. Each module has a name and a list of named ports, each of which has a direction and a width. Instantiating such a module consists of giving the instance a name and listing the signals or expressions to which it connected. Connections can be made positionally or by port name, the latter being preferred for modules with many (perhaps ten or more) connections.

Continuous assignments are a simple way to model both Boolean and arithmetic datapaths. A continuous assignment uses Verilog's comprehensive expression syntax to define the function to be computed, and its semantics are such that the value of the expression on the right of a continuous expression is always copied to the net on the left (regs are not allowed on the left of a continuous assignment). Practically, Verilog simulators implement this by recomputing the expression on the right whenever any variable it references changes. Figure 15.4 illustrates some continuous assignments.

Behavioral modeling in Verilog uses imperative code enclosed in *initial* and *always* blocks that write to reg variables to maintain state. Each block effectively introduces a concurrent process that is awakened by an event and runs until it hits a delay or a wait statement. The example in Figure 15.5 illustrates basic behavioral usage.

Figure 15.6 shows a more complicated behavioral model, in this case a simple state machine. This example is written in a common style where the combinational and sequential parts of a state machine are written as two separate processes. The first process is purely combinational. The @(a or b or state) directive triggers its execution when signals a, b, or state change. The code consists of a multiway choice — a case statement — and performs procedural assignments to the o and nextState registers. This illustrates one of the odder aspects

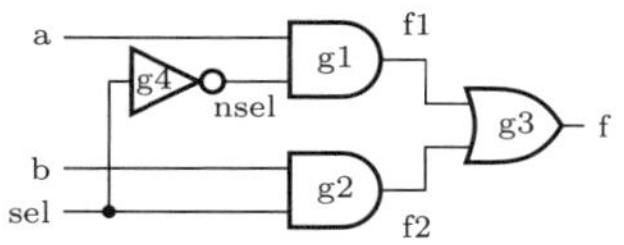

(a)

```
module mux(f,a,b,sel);
output f;
input a, b, sel;

and g1(f1, a, nsel),
    g2(f2, b, sel);
or  g3(f, f1, f2);
not g4(nsel, sel);

endmodule
```

(b)

```
module mux(f,a,b,sel);
output f;
input a, b, sel;

assign f = sel ? a : b;

endmodule
```

(c)

```
primitive
   mux(f,a,b,sel);
output f;
input a, b, sel;

table
  1?0 : 1;
  0?0 : 0;
  ?11 : 1;
  ?01 : 0;
  11? : 1;
  00? : 0;
endtable
endprimitive
```

(d)

```
module mux(f,a,b,sel);
output f;
input a, b, sel;
reg f;

always @(a or b or sel)
  if (sel) f = a;
  else f = b;

endmodule
```

(e)

```
module testbench;
  reg a, b, sel;
  wire f;

  mux dut(f, a, b, sel);

  initial begin
    $display("a,b,sel -> f");
    $monitor($time,,
             "%b%b%b -> ",
             a, b, sel, f);
    a = 0; b = 0 ; sel = 0;
    #10 a = 1;
    #10 sel = 1;
    #10 b = 1;
    #10 sel = 0;
  end
endmodule
```

(f)

FIGURE 15.1 Verilog examples: (a) a multiplexer circuit; (b) the multiplexer described as a Verilog structural model; (c) the multiplexer described using a continuous assignment; (d) a user-defined primitive for the multiplexer; (e) the multiplexer described with imperative code; (f) a testbench for the multiplexer.

of Verilog modeling: both are declared reg, yet neither corresponds to the output of a latch or flip-flop. This reflects the Verilog requirement that procedural assignment can only be performed on regs.

The second process models a pair of flip-flops that holds the state between cycles. The @(posedge clk or reset) directive makes the process sensitive to the rising edge of the clock or a change in the reset signal. At the positive edge of the clock, the process captures the value of the nextState variable and copies it to state.

The example in Figure 15.6 illustrates the two types of behavioral assignments. The assignments used in the first process are the so-called blocking assignments, written =, and take effect immediately. Nonblocking assignments are written <= and have somewhat subtle semantics. Instead of taking effect

```
wire a;                     // Simple wire
tri [15:0] dbus;            // 16-bit tristate bus
tri #(5,4,8) b;             // Wire with delay
reg [-1:4] vec;             // Six-bit register
trireg (small) q;           // Wire stores a small charge
integer imem[0:1023];       // Array of 1024 integers
reg [31:0] dcache[0:63]; // A 32-bit memory
```

FIGURE 15.2 A sampling of Verilog *net* and *reg* definitions.

```
module mymod(out1, out2, in1, in2);
output out1;       // Outputs first by convention
output [3:0] out2; // four-bit vector
input in1;
input [2:0] in2;

// Module body: instances,
// continuous assignments,
// initial and always blocks

endmodule

module usemymod;
reg a;
reg [2:0] b;
wire c, e, g;
wire [3:0] d, f, h;

mymod m1(c, d, a, b);           // simple instance
mymod m2(e, f, c, d[2:0]),      // instance with part-select input
      m3(.in1(e), .in2(f[2:0]), // connect-by-name
         .out1(g), .out2(h));

endmodule
```

FIGURE 15.3 Verilog structure: an example of a module definition and another module containing three instances of it.

immediately, the right-hand sides of nonblocking assignments are evaluated when they are executed, but the assignment itself does not take place until the end of the current time instant. Such behavior effectively isolates the effect of nonblocking assignments to the next clock cycle, much like the output of a flip-flop is only visible after a clock edge. In general, nonblocking assignments are preferred when writing to state-holding elements for exactly this reason. See Figure 15.7 and the next section for a more extensive discussion of blocking vs. nonblocking assignments.

15.3.1.2 Verilog Shortcomings

Compared to VHDL, Verilog does a poor job at protecting users from themselves. Verilog's regs are exactly shared variables and the language permits all the standard pitfalls associated with them, such as races and nondeterministic behavior. Most users avoid such behavior by following certain rules (e.g., by restricting assignments to a shared variable to a single concurrent process), but Verilog allows more dangerous usage. Tellingly, a number of EDA companies exist solely to provide lint-like tools for Verilog that report such poor coding practices. Gordon [20] provides a more detailed discussion of the semantic challenges of Verilog.

Nonblocking assignments are one way to ameliorate most problems with nondeterminism caused by shared variables, but they too can lead to bizarre behavior. To illustrate the use of shared variables, consider a three-stage shift register. The implementation in Figure 15.7(a) appears to be correct, but in fact may not behave as expected because the language says the simulator is free to execute the three *always* blocks in any

```
module add8(sum, a, b, carryin);
output [8:0] sum;
input [7:0] a, b;
input carryin;

// unsigned arithmetic
assign sum = a + b + carryin;
endmodule

module datapath(addr_2_0, icu_hit, psr_bm8, hit);
output [2:0]  addr_2_0
output        icu_hit
input         psr_bm8;
input         hit;
wire   [31:0] addr_qw_align;
wire   [3:0]  addr_qw_align_int;
wire   [31:0] addr_d1;
wire          powerdown;
wire          pwdn_d1;

// part select, vector concatenation is {}
assign addr_qw_align = { addr_d1[31:4], addr_qw_align_int[3:0] };

// if-then-else operator
assign addr_offset = psr_bm8 ? addr_2_0[1:0] : 2'b00;

// Boolean operators
assign icu_hit = hit & !powerdown & !pwdn_d1;

// ...

endmodule
```

FIGURE 15.4 Verilog modules illustrating continuous assignment. The first is a simple eight-bit full adder producing a nine-bit result. The second is an excerpt from a processor datapath.

```
module behavioral;
reg [1:0] a, b;

initial begin
  a = 'b1;
  b = 'b0;
end

always begin
  #50 a = ~a;  // Toggle a every 50 time units
end

always begin
  #100 b = ~b;  // Toggle b every 100 time units
end

endmodule
```

FIGURE 15.5 A simple Verilog behavioral model. The code in the *initial* block runs once at the beginning of simulation to initialize the two registers. The code in the two *always* blocks runs periodically: once every 50 and 100 time units.

order when they are triggered. If the processes execute top-to-bottom, the module becomes a one-stage shift register, but if they execute bottom-to-top, the behavior is as intended.

Figure 15.7(b) shows a correct implementation of the shift register that uses nonblocking assignments to avoid this problem. The semantics of these assignments is such that the value on the right-hand side

```
module FSM(o, a, b, reset);
output o;
reg o;  // declared reg only because o is assigned procedurally
input a, b, reset;
reg [1:0] state, nextState; // only "state" holds state

// Combinational logic block: sensitive to changes on all inputs and
// outputs o and nextState always assigned

always @(a or b or state)
 case (state)
    2'b00: begin
      o = a & b;
      nextState = a ? 2'b00 : 2'b01;
    end
    2'b01: begin
      o = 0; nextState = 2'b10;
    end
    default: begin
      o = 0; nextState = 2'b00;
    end
 endcase

// Sequential block: sensitive to clock edge

always @(posedge clk or reset)
  if (reset)
    state <= 2'b00;
  else
    state <= nextState;

endmodule
```

FIGURE 15.6 A Verilog behavioral model for a state machine illustrating the common practice of separating combinational and sequential blocks.

of the assignment is captured when the statement runs, but the actual assignment of values is only done at the "end" of each instant in time, i.e., after all three blocks have finished executing. As a result, the order in which the three assignments are executed does not matter and therefore the code always behaves like a three-stage shift register.

However, the use of nonblocking assignments can also be deceptive. In most languages, the effect of an assignment can be felt by the instruction immediately following it, but the delayed-assignment semantics of a nonblocking assignment violates this rule. Consider the erroneous decimal counter in Figure 15.8(a). Without knowing the subtle semantics of Verilog nonblocking assignments, the counter would appear to count from 0 to 9, but in fact it counts to 10 before being reset because test of o by the *if* statement gets the value of o from the previous clock cycle, not the results of the o $<=$ o + 1 statement. A corrected version is shown in Figure 15.8(b), which uses a local variable count to maintain the count, blocking assignments to touch it, and finally a nonblocking assignment to o to send the count outside the module.

Coupled with the rules for register inference, the circuit implied by the counter in Figure 15.8(b) is actually just fine. While both o and count are marked as reg, only the count variable will actually become a state-holding element.

15.3.2 VHDL

Although VHDL and Verilog have grown to be nearly interchangeable, they could not have had more different histories. Unlike Verilog, VHDL was deliberately designed to be a standard HDL after a lengthy, deliberate process. As Dewey explains [8], VHDL was created at the behest of the U. S. Department of Defense in response to the desire to incorporate integrated circuits (specifically very high-speed integrated

```
module bad_sr(o, i, clk);
output o;
input i, clk;
reg a, b, o;

always @(posedge clk) a = i;
always @(posedge clk) b = a;
always @(posedge clk) o = b;

endmodule
```
(a)

```
module good_sr(o, i, clk);
output o;
input i, clk;
reg a, b, o;

always @(posedge clk) a <= i;
always @(posedge clk) b <= a;
always @(posedge clk) o <= b;

endmodule
```
(b)

FIGURE 15.7 Verilog examples illustrating the difference between blocking and nonblocking assignments. (a) An erroneous implementation of a three-stage shift register that may or may not work depending on the order in which the simulator chooses to execute the three *always* blocks. (b) A correct implementation using nonblocking assignments, which make the variables take on their new values after all three blocks are done for the instant.

```
module bad_counter(o, clk);
output o;
input clk;
reg [3:0] o;

always @(posedge clk) begin
  o <= o + 1;
  if (o == 10)
    o <= 0;
end

endmodule
```
(a)

```
module good_counter(o, clk);
output o;
input clk;
reg [3:0] o;
reg [3:0] count;

always @(posedge clk) begin
  count = count + 1;
  if (count == 10)
    count = 0;
  o <= count;
end

endmodule
```
(b)

FIGURE 15.8 Verilog examples illustrating a pitfall with nonblocking assignments: (a) an erroneous implementation of a counter, which counts to 10, not 9; (b) A correct implementation using a combination of blocking and nonblocking assignments.

circuits (VHSICs)) in military hardware, hence the name of the program from which VHDL evolved. Starting with a summer study at Woods Hole, Massachusetts in 1981, requirements for and scope of the language were first established; then, after a bidding process, a contract to develop the language was awarded in 1983 to three companies: Intermetrics, which was the prime contractor for Ada, the software programming language developed for the U. S. military in the early 1980s; Texas Instruments; and IBM. Dewey and de Geus [21] describe this history in more detail.

The VHDL language was created in 1983 and 1984, essentially concurrent with Verilog, and first released publicly in 1985. Interest in an IEEE standard HDL was high at the time, and VHDL was eventually adapted and adopted as IEEE standard 1076 in 1987 [22] and revised in 1993 [23]. Verilog, meanwhile, remained proprietary until 1990. The standardization and growing popularity of VHDL was certainly instrumental in Cadence's decision to make Verilog public.

The original objectives of VHDL [8] were to provide a means of documenting hardware (i.e., as an alternative to imprecise English descriptions) and of verifying it through simulation. As such, a VHDL simulator was developed along with the language.

VHDL is vast, complicated, and has a verbose syntax obviously derived from Ada. While its popularity as a means of formal documentation is questionable, it has succeeded as a modeling language for hardware simulation, and, like Verilog, more recently as a specification language for RTL logic synthesis.

Although there are many features absent in Verilog that are unique to VHDL (and vice versa), in practice most designers use a nearly identical subset because this is what the synthesis tools accept. Thus although the syntax of the two languages differs greatly, and the semantics appear different, their usage has converged to a common core.

Unlike Verilog, VHDL has spawned a plethora of books discussing its proper usage. Basic texts include Lipsett et al. [24] (one of the earliest), Dewey [25], Bhasker [26], Perry [27], and Ashenden [28, 29]. More advanced is Cohen [30], who suggests preferred idioms in VHDL, and Harr and Stancluescu [31], who discuss using VHDL for a variety of modeling tasks, not just RTL.

15.3.2.1 Coding in VHDL

Like Verilog, VHDL describes designs as a collection of hierarchical modules. But unlike Verilog, VHDL splits them into interfaces — called entities — and their implementations — called architectures. In addition to named input and output ports, entities also define compile-time parameters (generics), types, constants, attributes, use directives, and others.

Figure 15.9 shows code for the same two-input multiplexer roughly equivalent to Verilog examples in Figure 15.1. Figure 15.9(a) is the entity declaration for the multiplexer, which defines its input and output ports. Figure 15.9(b) is a purely structural description of the multiplexer; it defines internal signals, the interface to the Inverter, AndGate, and OrGate components, and instantiates four of these gates. The name of the architecture, "structural" is arbitrary; it is used to distinguish among various architectures. The Verilog equivalent in Figure 15.1(b) did not define the internal signals, relying instead on Verilog's rule of automatically considering undeclared signals to be single-bit wires. Furthermore, the Verilog example used the built-in gate-level primitives; VHDL itself does not know about logic gates, but can be taught about them.

Figure 15.9(c) illustrates a dataflow model for the multiplexer with each logic gate made explicit. VHDL does have built-in logical operators. Figure 15.9(d) shows an even more succinct implementation, which uses the multi-way *when* conditional operator.

Finally, Figure 15.9(e) shows a behavioral implementation of the mux. It defines a concurrently running process sensitive to the three mux inputs (a, b, and c) and uses an if-then-else statement (VHDL provides most of the usual control-flow statements) to select between copying the a and the b signal to the output d.

One of the design philosophies behind the VHDL was to maximize its flexibility by making most things user-definable. As a result, unlike Verilog, it has only the most rudimentary built-in types of (e.g., Boolean variables, but nothing to model four-valued logic), but has a much more powerful type system that allows such types to be defined. The Bit used in the examples in Figure 15.9 is actually a predefined part of the standard environment, i.e.,

```
type BIT is ('0', '1');
```

which is a character enumeration type whose two values are the characters 0 and 1 (VHDL is case insensitive; Bit and BIT are equivalent.)

There are advantages and disadvantages to this approach. While it allows more things to be added to the language later, it also makes for a few infelicities. For example, the predicate in the *if* statement in Figure 15.9(e) must be written "c = '1'" instead of just "c" because the argument must be of type Boolean, not Bit. While not a serious issue, it is yet another thing that contributes to VHDL's verbosity.

Figure 15.10 is a more elaborate example showing an implementation of the classic traffic light controller from Mead and Conway [32]. This is written in a synthesizeable dialect, using the common practice of separating the output and next-state logic from the state-holding element. Specifically, the first process is sensitive only to the clock signal. The *if* statement in the first process checks for an event on the clock (VHDL signals have a variety of attributes; *event* is true whenever the value has changed) and the clock being high, i.e., the rising edge of the clock. The second process is sensitive only to the inputs and present state of the machine, not the clock, and is meant to model

```
entity mux2 is
  port (a, b, c : in Bit; d : out Bit);
end;
```

(a)

```
architecture structural of mux2 is

  signal cbar, ai, bi : Bit;       -- internal signals

  component Inverter               -- component interfaces
    port (a:in Bit; y: out Bit);
  end component;
  component AndGate
    port (a1, a2:in Bit; y: out Bit);
  end component;
  component OrGate
    port (a1, a2:in Bit; y: out Bit);
  end component;

begin
  I1: Inverter port map(a => c, y => cbar); -- connect-by-name
  A1: AndGate  port map(a, c, ai);          -- connect-by-position
  A2: AndGate  port map(a1 => b, a2 => cbar, y => bi);
  O1: OrGate   port map(a1 => ai, a2 => bi, y => d);
end;
```

(b)

```
architecture dataflow1 of mux2 is
  signal cbar, ai, bi : Bit;
begin
  cbar <= not c;
  ai   <= a and c;
  bi   <= b and cbar;
  d    <= ai or bi;
end;
```

(c)

```
architecture dataflow2 of mux2 is
begin
  d <= a when c = '1' else
       b;
end;
```

(d)

```
architecture behavioral of mux2 is
begin
  process(a, b, c)  -- define sensitivity list
  begin
    if c = '1' then
      d <= a;
    else
      d <= b;
    end if;
  end process;
end;
```

(e)

FIGURE 15.9 VHSIC Hardware Development Language (VHDL) code for a two-input multiplexer: (a) the entity definition for the multiplexer; (b) a structural implementation instantiating primitive gates; (c) a dataflow implementation with an expression for each gate; (d) a direct dataflow implementation. (e) a behavioral implementation.

combinational logic. It illustrates the multi-way conditional *case* statement, constants, and bit vectors. It employs types (i.e., std_ulogic and std_ulogic_vector) and operators from the ieee.std_logic_1164 library, an IEEE standard library [33] for modeling logic that can represent unknown values ("X") as well as 0s and 1s.

```
library ieee;
use ieee.std_logic_1164.all;

entity tlc is
  port (
    clk            : in  std_ulogic;
    reset          : in  std_ulogic;
    cars           : in  std_ulogic;
    short          : in  std_ulogic;
    long           : in  std_ulogic;
    highway_yellow : out std_ulogic;
    highway_red    : out std_ulogic;
    farm_yellow    : out std_ulogic;
    farm_red       : out std_ulogic;
    start_timer    : out std_ulogic);
end tlc;

architecture imp of tlc is
signal current_state, next_state : std_ulogic_vector(1 downto 0);
constant HG : std_ulogic_vector := "00";
constant HY : std_ulogic_vector := "01";
constant FY : std_ulogic_vector := "10";
constant FG : std_ulogic_vector := "11";
begin

P1: process (clk)    -- Sequential process
begin
  if (clk'event and clk = '1') then
    current_state <= next_state;
  end if;
end process P1;

-- Combinational process: sensitive to input changes
P2: process (current_state, reset, cars, short, long)
begin
  if (reset = '1') then
    next_state <= HG;
    start_timer <= '1';
  else
    case current_state is
      when HG =>
        highway_yellow <= '0'; highway_red <= '0';
        farm_yellow    <= '0'; farm_red    <= '1';
        if (cars = '1' and long = '1') then
          next_state <= HY; start_timer <= '1';
        else
          next_state <= HG; start_timer <= '0';
        end if;
      when HY =>
        highway_yellow <= '1'; highway_red <= '0';
        farm_yellow    <= '0'; farm_red    <= '1';
        if (short = '1') then
          next_state <= FG; start_timer <= '1';
        else
          next_state <= HY; start_timer <= '0';
        end if;
      when FG =>
        highway_yellow <= '0'; highway_red <= '1';
        farm_yellow    <= '0'; farm_red    <= '0';
        if (cars = '0' or long = '1') then
          next_state <= FY; start_timer <= '1';
        else
          next_state <= FG; start_timer <= '0';
        end if;
      when FY =>
        highway_yellow <= '0'; highway_red <= '1';
        farm_yellow    <= '1'; farm_red    <= '0';
        if (short = '1') then
          next_state <= HG; start_timer <= '1';
        else
          next_state <= FY; start_timer <= '0';
        end if;
      when others =>
        next_state     <= "XX"; start_timer <= 'X';
        highway_yellow <= 'X'; highway_red    <= 'X';
        farm_yellow    <= 'X'; farm_red       <= 'X';
    end case;
  end if;
end process P2;
end imp;
```

FIGURE 15.10 The traffic-light controller from Mead and Conway [32] implemented in VHDL, illustrating the common practice of separating combinational and state-holding processes.

15.3.2.2 VHSIC Hardware Description Language (VHDL) Shortcomings

One shortcoming of VHDL is its obvious verbosity: the use of begin/end pairs instead of braces, the need to separate entities and their architectures, the need to spell out things like ports, its lengthy names for standard logic types (e.g., std_ulogic_vector), and its requirement of enclosing Boolean values and vectors in quotes. Many of these are artifacts of its roots in the Ada language, another fairly verbose language commissioned by the U. S. Department of Defense, but others are due to questionable design decisions. Consider the separation of entity/architecture pairs. While separating these concepts is a boon to abstraction and simplifies the construction of simulations of the same system in different configurations (e.g., to run a simulation using a gate-level architecture in place of a behavioral one for more precise timing estimation), in practice most designers only ever write a single architecture for a given entity and such pairs are usually written together.

The flexibility of VHDL has advantages and disadvantages. Its type system is much more flexible than Verilog's, providing things such as aggregate types and overloaded functions and operators, but this flexibility comes with a need for standardization and also tends to increase the verbosity of the language. Some of the need for standardization was recognized early, resulting in libraries such as the widely supported IEEE 1164 library for multivalued logic modeling. However, a standard for signed and unsigned arithmetic on logic vectors was slower in coming (it was eventually standardized in 1997 [34]), prompting both Synopsys and Mentor to each introduce similar but incompatible and incomplete versions of such a library.

Fundamentally, many of the problems stem from a desire to make the language too general. Aspects of the type system suffer from this as well. While the ability to define new enumerated types for multi-valued logic modeling is powerful, it seems a little odd to require virtually every VHDL program (since its main use has long been specification for RTL synthesis) to include one or more standard library. This also leads to the need to be constantly comparing signals to the literal '1' instead of just using a signal's value directly, and requires a user to carefully watch the types of subexpressions.

15.3.3 SystemC

SystemC is a relative latecomer to the HDL wars. Developed at Synopsys in the late 1990s, primarily by Stan Liao, SystemC was originally called Scenic [35] and was intended to replace Verilog and VHDL as the main system description language for synthesis (see Arnout [36] for some of the arguments for SystemC). SystemC is not so much a language as a C++ library along with a set of coding rules, but this is exactly its strength. It evolved from the common practice of first writing a high-level simulation model in C or C++, refining it, and finally recoding it in RTL Verilog or VHDL. SystemC was intended to smooth the refinement process by removing the need for a separate HDL.

SystemC, especially in its original incarnation, can be thought of as a dialect of C++ for modeling digital hardware. Like Verilog and VHDL, it supports hierarchical models whose blocks consist of input/output ports, internal signals, concurrently running imperative processes, and instances of other blocks. The SystemC libraries make two main contributions: an inexpensive mechanism for running many processes concurrently (based on a lightweight thread package; see Liao et al. [35]), and an extensive set of types for modeling hardware systems, including bit vectors and fixed-point numbers. A SystemC model consists of a series of class definitions, each of which define a block.

Methods defined for such a class become concurrently running processes, and the constructor for each class starts these processes running by passing them to the simulation kernel. Simulating a SystemC model starts by calling the constructors for all blocks in the design, then invoking the scheduler which is responsible for executing each of the concurrent processes as needed.

The computational model behind earlier versions of SystemC was cycle-based instead of the event-driven model of Verilog and VHDL. This meant that the simulation was driven by a collection of potentially asynchronous, but periodic clocks. Later versions (SystemC 2.0 and higher) adopted an event-driven model much like VHDLs. Also added were many constructs useful for higher-level modeling, such as abstract channels and interfaces, which has pushed the use of SystemC toward system-level modeling and away from its origins as hardware description.

SystemC books have only appeared recently. Grötker et al. [37] provide a good introduction to SystemC 2.0. Bhasker [38] is also an introduction. The volume edited by Müller et al. [39] surveys more advanced SystemC modeling techniques.

15.3.3.1 Coding in SystemC

Figure 15.11 shows a small SystemC model for a 0 to 99 counter driving a pair of seven segment displays. It defines two modules (the decoder and counter *structs)* and an sc_main function that defines some internal signals, instantiates two decoders and a counter, and runs the simulation while printing out what it does.

The two modules in Figure 15.11 illustrate two of the three types of processes possible in SystemC. The decoder module is the simpler one: it models a purely combinational process by defining a method (called, arbitrarily, "compute") that will be invoked by the scheduler every time the number input changes, as indicated by the "sensitive << number;" statement beneath the definition of compute as an SC_METHOD.

In the second module, counter is an SC_CTHREAD process: a method (here called "tick") that is invoked in response to a clock edge (here, the positive edge of the clk input, as defined by the SC_CTHREAD(tick, clk.pos()); statement) and can suspend itself with the wait() statement. Specifically, the scheduler resumes the method when the clock edge occurs, and the method runs until it encounters a wait() statement, at which point its state is saved and control passes back to the scheduler.

This example illustrates only a very small fraction of the SystemC type libraries. It uses unsigned integers (sc_uint), bit vectors (sc_bv), and a clock (sc_clock). The nonclock types are wrapped in sc_signals, which behave like VHDL signals. In particular, when an SC_CTHREAD method assigns a value to a signal, the effect of this assignment is felt only after all the processes triggered by the same clock edge have been run. Thus, such assignments behave like blocking assignments in Verilog to ensure that the nondeterministic order in which such processes are invoked (the scheduler is allowed to invoke them in any order) does not affect the ultimate outcome of simulating the system.

15.3.3.2 SystemC Shortcomings

Like many languages, the most common use of SystemC has diverged from its designers' original intentions i.e., as an input for hardware synthesis in the case of SystemC 1.0. A number of synthesis tools for the language do exist, but since version 2.0, SystemC has been primarily (and quite successfully) used for system modeling. This does mean, however, that it does not solve the "separate language for synthesis problem."

A big disadvantage of SystemC is that C++ was never intended for modeling digital hardware and as a result, is even more lax about enforcing rules than Verilog. The syntax, similarly, is somewhat awkward and relies on some very tricky macro preprocessor definitions. On detailed models, the simulation speed of a good compiled-code Verilog or VHDL simulator may be better, although SystemC is much faster for higher-level models. For such systems which consist of complex processes, SystemC should be superior since the simulation becomes nearly a normal C++ program. However, the context-switching cost in SystemC is higher than that of a good Verilog or VHDL simulator when running a more detailed model, so a system with many small processes would not simulate as quickly.

Another issue is the ease with which a SystemC model can inadvertently be made nondeterministic. Although carefully following a discipline of only communicating among processes through signals will ensure the simulation is nondeterministic, any slight deviation from this will cause problems. For example, library functions that use a hidden global variable may cause nondeterminism if called from different processes. Accidentally holding state in an SC_METHOD process (easily done if class variables are assigned) can also cause problems since such methods are invoked in an undefined order.

Many argue that nondeterministic behavior in a language can be desirable for modeling nondeterministic systems, which certainly exist and need to be modeled. However, the sort of nondeterminism in a language such as SystemC or Verilog creeps in unexpectedly and is difficult to use as a modeling tool.

```
#include "systemc.h"
#include <stdio.h>

struct decoder : sc_module {
  sc_in<sc_uint<4> > number;
  sc_out<sc_bv<7> > segments;

  void compute() {
    static sc_bv<7> codes[10] = {
      0x7e, 0x30, 0x6d, 0x79, 0x33,
      0x5b, 0x5f, 0x70, 0x7f, 0x7b };
    if (number.read() < 10)
      segments = codes[number.read()];
  }

  SC_CTOR(decoder) {
    SC_METHOD(compute);
    sensitive << number;
  }
};

struct counter : sc_module {
  sc_out<sc_uint<4> > tens;
  sc_out<sc_uint<4> > ones;
  sc_in_clk clk;

  void tick() {
    int one = 0, ten = 0;
    for (;;) {
      if (++one == 10) {
        one = 0;
        if (++ten == 10) ten = 0;
      }
      ones = one;
      tens = ten;
      wait();
    }
  }

  SC_CTOR(counter) {
    SC_CTHREAD(tick, clk.pos());
  }
};

int sc_main(int argc, char *argv[])
{
  sc_signal<sc_uint<4> > ones, tens;
  sc_signal<sc_bv<7> > ones_segments, tens_segments;
  sc_clock clk;

  decoder decoder1("decoder1");
  decoder1(ones, ones_segments);
  decoder decoder2("decoder2");
  decoder2(tens, tens_segments);

  counter counter1("counter1");
  counter1(tens, ones, clk);

  for (int i = 0 ; i < 12 ; i++) {
    sc_start(clk, 1);
    printf("%d %d %x %x\n",
           (int)tens.read(), (int)ones.read(),
           (int)(sc_uint<7>)tens_segments.read(),
           (int)(sc_uint<7>)ones_segments.read());
  }

}
```

FIGURE 15.11 A SystemC model for a two-digit decimal counter driving two seven-segment displays.

For the simulation of a nondeterministic model to be interesting, there needs to be some way of seeing the different possible behaviors, yet a nondeterministic artifact such as an SC_METHOD process that holds state provides no mechanism for ensuring that it is not, in fact, predictable. As a result, a designer has a hard time answering whether a model of a nondeterministic system can exhibit undesired behavior, even through a careful selection of test cases.

15.4 Verification Languages

Thanks to dramatic improvements in integrated circuit fabrication technology, it is now possible to build very large, complex integrated circuits. Hardware description languages, logic synthesis, and automated place-and-route technology have similarly made it possible to design such complicated systems. Unfortunately, technology to validate these designs, i.e., to identify design errors, has had a hard time keeping pace with these trends.

The time required to validate a design now greatly outstrips the time required to design or fabricate it. Although many novel techniques have been proposed to address the validation problem, simulation remains the preferred method. So-called formal verification techniques, which amount to efficient exhaustive simulation, have been gaining ground, but suffer from capacity problems.

Simulation applies a stimulus to a model of a design to predict the behavior of the fabricated system. Naturally, there is a trade-off between highly detailed models that can predict many attributes, say, logical values, timing, and power consumption, and simplified models that can only predict logical behavior but execute much faster.

Because the size of the typical design has grown exponentially over time, functional simulation, which only predicts the logical behavior of a synchronous circuit at clock-cycle boundaries, has become the preferred form of simulation because of its superior speed. Furthermore, designers have shied away from more timing-sensitive circuitry such as gated clocks and transparent latches because they require more detailed simulation models and are therefore more costly to simulate.

Simulation-based validation raises three important questions: what the stimulus should be, whether it exposes any design errors, and whether the stimulus is long and varied enough to have exposed "all" design errors. Historically, these three questions have been answered manually, i.e., by having a test engineer write test cases, check the results of simulation, and make some informed guesses about how comprehensive the test suite actually is.

A manual approach has many shortcomings. Writing test cases is tedious and the number necessary for "complete" verification grows faster than the size of the system description. Manually checking the output of simulation is similarly tedious and subtle errors can be easily overlooked. Finally, it is difficult to judge quantitatively how much of a design has really been tested.

More automated methodologies, and ultimately languages, have evolved to address some of these challenges, although the verification problem remains one of the most difficult. Biased random test case generation has become standard practice, although it has only supplemented manual test case generation, not completely supplanted it. Designer-inserted assertions, long standard practice for software development, have also become standard for hardware, although the sort of assertions needed in hardware are more complicated than the typical "the argument must be nonzero" sort of checking that works well in software. Finally, automated "coverage" checking, which attempts to quantify how much of a design's behavior has been exercised in simulation, has also become standard.

All of these techniques, while an improvement, are not a panacea. While biased random test case generation can quickly generate many interesting tests, it provides no guarantees of completeness, meaning bugs may go unnoticed. Because they must often check a temporal property (e.g., "acknowledge arrives within three cycles of every request"), good assertions in hardware systems are much more difficult to write than those for software (which most often check data structure consistency) and again, there is no way to know when enough assertions have been added, and it is possible that the assertions themselves have flaws (e.g., they let bugs by). Finally, test cases that achieve 100% coverage can also let bugs by because the criteria for coverage are necessarily weak. Coverage typically checks what states particular variables have been in, but it cannot consider all combinations because they grow exponentially quickly with design size. As a result, certain important combinations may not be checked even though coverage checks report "complete coverage."

While the utility of biased random test generation and coverage metrics is mostly limited to simulation, assertion specification techniques are useful for, and have been heavily influenced by, formal verification. Pure formal techniques consider all possible behaviors by definition and therefore do not require explicit test cases (implicitly, they consider all possible test cases) and also do not need to consider coverage. But knowing what behavior is unwanted is crucial for formal techniques, whose purpose is to either expose unwanted behavior or formally prove it cannot occur.

Recently, a sort of renaissance has been occurring in verification languages. Temporal logics, specifically linear temporal logic (LTL) and computation tree logic (CTL), form the mathematical basis for most assertion checking, but their traditional mathematical syntax is awkward for hardware designers. Instead, a number of more traditional computer languages, which combine a more human-readable syntax for the bare logic with a lot of "syntactic sugar" for more naturally expressing common properties, have been proposed for expressing properties in these logics. Two industrial efforts from Intel (ForSpec) and IBM (Sugar) have emerged as the most complete ones.

Meanwhile, some EDA companies have produced languages designed for writing testbenches and checking simulation coverage. Vera, originally designed by Systems Science and since acquired by Synopsys, and e, designed and sold by Verisity, acquired in 2005 by Cadence, have been the two most commercially successful. Bergeron [40] discusses how to use these two languages.

All four of these languages have recently undergone extensive crossbreeding. Vera has been made public, rechristened Open Vera, had Intel's ForSpec assertions grafted onto it, and added almost in its entirety to SystemVerilog. Sugar, meanwhile, has been adopted by the Accelera standards committee, rechristened the Property Specification Language (PSL), and has also been added in part to SystemVerilog. Verisity's e has changed the least, having only recently been made public, but is being standardized by the IEEE in 2005 (P1647).

There are obvious advantages in having a single industry-standard language for assertions, so it seems likely that most of these languages will eventually disappear, but as of this writing (2004), there is no obvious winner.

The sections that follow describe languages that are currently public. As mentioned above, most started as proprietary in-house or commercial.

15.4.1 Open Vera

OpenVera began life around 1995 as Vera, a proprietary language implemented by Systems Science, Inc. mainly for creating testbenches for Verilog simulations. As such, its syntax was heavily influenced by Verilog. VHDL support was added later. Synopsys bought the company in 1998, rechristened it OpenVera, and released the language to the public in 2001.

OpenVera is a concurrent, imperative language designed for writing testbenches. It executes in concert with a Verilog or VHDL simulator and can both provide stimulus to the simulator and observe the results. In addition to the usual high-level imperative language constructs, such as conditionals, loops, functions, strings, and associative arrays, it provides extensive facilities for generating biased random patterns (designed to be applied to the hardware design under test) and monitoring what values particular variables take on during the simulation (for checking coverage).

In the process of making OpenVera public, Synopsys added the assertion specification capabilities of Intel's ForSpec language, making it easy to check whether certain behaviors ever appear during the simulation.

In a further nod to crossbreeding, around 2003 much of Vera was incorporated into System Verilog 3.1. In particular, its style of generating biased random variables and checking for behavior coverage have been adopted more or less verbatim.

The following is a quick overview of OpenVera 1.3, the latest version as of this writing (September 2004). References for the language include the OpenVera language reference manual, available from the OpenVera website, and Haque et al. [41].

OpenVera has three main facilities that separate it from more traditional programming languages: biased random variable generation subject to constraints, monitoring facilities for reporting coverage of state variables, and the ability to specify and check temporal assertions.

Figure 15.12 shows a simple OpenVera program that demonstrates the language's ability to generate biased random variables. First, a Java/C++-like class is defined containing the 16-bit field addr and the 32-bit field data. These are marked rand, meaning their values will be set by a call to the randomize() method implicitly defined for every class.

Following the definition of the fields, a constraint (named arbitrarily word_align) is defined for objects of this class. Such constraints restrict the possible values the randomize() method may assign to various fields. This particular constraint simply restricts the two lowest bits of addr to be zero, i.e., aligned on a four-byte boundary.

The "demo" program creates a new Bus object then, for 50 times, invokes randomize() to generate a new set of "random" values (in fact, they are taken from a pseudorandom sequence guaranteed to be the same each time the program runs) for the two fields in the bus object that conform to the given constraint.

Overly constrained or inconsistent constraints may lead randomize() to fail. It signals this with a non-OK return value (OK and FAIL are reserved words in Vera), here assigned to the local integer variable result. For this simple example, randomization will always succeed. The constraint solver guarantees that it will only fail if there is no consistent solution to the supplied constraints.

This example of course only illustrates a fraction of OpenVera's facilities for biased random variable generation. It can also add constraints on-the-fly, impose set membership constraints, follow user-defined distributions, impose conditional constraints, impose constraints between variables, selectively randomize and disable randomization of user-specified variables, and generate random sequences from a language specified by a grammar.

OpenVera has extensive support for so-called functional coverage checking, which can monitor state variables and state transitions (unlike software coverage, which usually monitors which statements and branches have been executed). It uses a bin model: each bin represents a particular state or transition and when a matching activity occurs, the counter for that bin is incremented. The number of bins that remain empty after simulation therefore gives a rough idea of what behavior has yet to be exercised.

The coverage_group construct defines a type of monitor. Each specifies a set of variables to monitor and an event that triggers a check of the variables, typically the positive edge of a clock. Like a class, these constructs must be explicitly instantiated and there may be multiple copies of each, especially useful when a coverage group has parameters. Figure 15.13 illustrates a simple coverage group.

Like the earlier example, Figure 15.13 shows only the most basic coverage functionality. Open Vera can also monitor cross coverage, i.e., the combinations of values taken by two or more variables; selectively

```
class Bus {
  rand bit[15:0] addr;
  rand bit[31:0] data;

  constraint world_align { addr[1:0] == '2b0; }
}

program demo {
  Bus bus = new();
  repeat (50) {
    integer result = bus.randomize();
    if (result == OK)
      printf("addr = %16h data = %32h\n", bus.addr, bus.data);
    else
      printf("randomization failed: overconstrained.\n");
  }
}
```

FIGURE 15.12 A simple Vera program illustrating its ability to generate biased random variables and its mix of imperative and object-oriented styles. From the OpenVera 1.3 LRM.

```
bit clk;
bit [15:0] addr;
bit [7:0] data;

coverage_group MyChecker {
  sample_event = @(posedge clk);
  sample addr, data;
}

program demo_coverage {
  MyChecker checker = new();
  ...
}
```

FIGURE 15.13 A simple Vera program containing a single coverage group that monitors which values appear on the addr and data state variables. From the OpenVera 1.3 LRM.

disable coverage checking, e.g., during system reset; allow the user to specify bins explicitly and the values mapped to them; and monitor transition coverage, i.e., combinations of values taken by the same variable in successive cycles.

OpenVera's assertions, which were adapted from Intel's ForSpec language, provide a way to check temporal properties such as "an acknowledge signal must occur within three clock cycles after any request." Because of its source, the syntax for assertions is a little unusual.

Figure 15.14 illustrates the assertion language. It defines a checker called "mychecker" that takes a single integer parameter *p*, an enable signal, a clock, and an eight-bit bus, and looks for the sequence *p* (4), 6, 9, 3 on the bus. The clock construct defines a collection of events (patterns) synchronized to the positive edge of the clock signal. Events e0 through e3 are very simple properties: they look for particular patterns on the result signal. The myseq pattern is the interesting one: it looks for the appearance of the four events separated by a single clock cycle (a one-cycle delay is written 2 #1). The assert directive means to check for the myseq event and report an error otherwise.

Finally, the bind directive indicates where to instantiate the checker, names the particular instance of it myinst, passes the parameter 4, and connects the checker to the enable, clock, and result signals.

Once more, the example in Figure 15.14 barely scratches the surface. Sequences can also contain consecutive and nonconsecutive repetition, explicit delays, simultaneous and disjoint sequences, and sequence containment.

15.4.2 The e Language

The e language was developed by Verisity as part of its Specman product as a tool for efficiently writing testbenches. Like Vera, it is an imperative object-oriented language with concurrency, the ability to generate constrained random values, mechanisms for checking functional (variable value) coverage, and a way to check temporal properties (assertions). Books on e include Palnitkar [42] and Iman and Joshi [43].

The syntax of e is a little unusual. First, all code must be enclosed in <' and '>' symbols, otherwise it is considered a comment. Unlike C, e declarations are written "name : type." The syntax for fields in compound types (e.g., structs) includes particles such as % and !, which indicate when a field is to be driven on the device-under-test and not randomly computed, respectively.

Figure 15.15 shows a fragment of an e program that defines an abstract test strategy for a very simple microprocessor, specifically how to generate instructions for it. It illustrates the type system of the language as well as the utility of constraints. It defines two enumerated scalar types, opcode and reg, and defines the width of each. The struct instr defines a new compound type (instr) that represents a single instruction. First is the op field, which is one of the opcodes defined earlier. It, the op1, and the op2 fields are marked with %, indicating that they should be considered by the pack built-in procedure, which marshals data to send to the simulation.

The kind field is also an enumerated scalar, but is used here as a type tag. It is not marked with %, which means that its value will not be included when the structure is packed and sent to the simulation.

```
/* Checks for the sequence p, 6, 9, 3 */
unit mychecker
  #(parameter integer p = 0)
  (logic en, logic clk, logic [7:0] result);

  clock posedge (clk) {
    event e0: (result == p);
    event e1: (result == 6);
    event e2: (result == 9);
    event e3: (result == 3);
    event myseq: if (en) then (e0 #1 e1 #1 e2 #1 e3);
  }

  assert myassert: check(myseq, "Missed a step.");
endunit

/* Watch for the sequence 4, 6, 9, 3 on the outp bus */
bind instances cnt_top.dut:
  mychecker myinst #(4) (enable, clk, outp);
```

FIGURE 15.14 A sample of OpenVera's assertion language. It defines a checker that watches for the sequence *p*, 6, 9, 3 and creates an instance of it that checks for the sequence 4, 6, 9, 3 on the outp bus. From the OpenVera 1.3 LRM.

The two when directives define two subtypes, i.e., "reg instr" and "imm instr." Such subtypes are similar to derived classes in object-oriented programming languages. Here, the value of the kind field, which can be either imm or reg, determines the subtype.

The two keep directives impose constraints between the kind field and the opcode, ensuring, e.g., that ADD and SUB instructions are of the reg type. Although these constraints are simple, e is able to impose much more complicated constraints on the values of fields in a struct.

The final when directive further constrains the JMP and CALL instructions, i.e., by restricting what values the op2 field may take for these instructions.

The extend sys directive adds a field named instrs to the sys built-in structure, which is the basic environment. The leading ! makes the system create an empty list of instructions, which will be filled in later.

Figure 15.16 illustrates how the definition of Figure 15.15 can be used to generate tests that exercise the ADD and ADDI instructions. It first adds constraints to the instr class (the template for instructions defined in Figure 15.15) that restrict the opcodes to either ADD or ADDI, then imposes a constraint on the top level (sys) that makes it generate exactly ten instructions. Running the source code of Figure 15.15 and Figure 15.16 together makes the system generate a sequence of ten pseudorandom instructions.

15.4.3 Property Specification Language

The property specification language, PSL, evolved from the proprietary Sugar language developed at IBM. Its focus is narrower than either OpenVera or e, since its goal is purely to specify temporal properties to be checked in hardware designs, but is more disciplined and has more formal semantics.

Beer et al. [44] provide a good introduction to an earlier version of the language, which they explain evolved over many years. It has been used within IBM in the RuleBase formal verification system since 1994 and was also pressed into service as a checker generator for simulators in 1997. Accelera, an EDA standards group, adopted it as their formal property language in 2002. Cohen [45] provides instruction on the language.

PSL is based on CTL [46], a powerful but rather cryptic temporal logic for specifying properties of finite-state systems. It is able to specify both safety properties ("this bad thing will never happen") as well as liveness properties ("this good thing will eventually happen"). Liveness properties can only be checked formally because they make a statement about all the possible behaviors of a system while safety properties can also be tested in simulation. Linear Temporal Logic (LTL), a subset of CTL, expresses only safety properties and can therefore be turned into checking automata meant to be run in concert with a simulation

```
Instruction encoding for a very simple processor
<'
type opcode : [ ADD, SUB, ADDI, JMP, CALL ] (bits: 4);
type reg : [ REG0, REG1, REG2, REG3 ] (bits: 2);

struct instr {
  %op  : opcode;      // Four-bit opcode
  %op1 : reg;         // Two-bit operand
  kind : [imm, reg]; // Whether instruction is immediate or register

  when reg instr { %op2 : reg; }  // Second operand register
  when imm instr { %op2 : byte; } // Second operand and immediate byte

  // Constrain certain instructions to be register, immediate
  keep op in [ ADD, SUB ] => kind == reg;
  keep op in [ ADDI, JMP, CALL ] => kind == imm;

  // Constrain the second operand for JMP and CALL instructions
  when imm instr {
    keep opcode in [JMP, CALL] => op2 < 16;
  }
};

extend sys {
  !instrs : list of instr; // Add a non-generated field called instrs
};
'>
```

FIGURE 15.15 e Code defining instruction encoding for a simple eight-bit microprocessor. After an example in the Specman tutorial.

to look for unwanted behavior. PSL carefully defines which subset of its properties are purely LTL and are therefore candidates for use in simulation-based checking.

PSL is divided into four layers. The lowest, Boolean, consists of instantaneous Boolean expressions on the signals in the design under test. The syntax of this layer follows that of the HDL to which PSL is being applied, and can be Verilog, SystemVerilog, VHDL, or GDL. For example, a[0:3] & b[0:3] and a(0 to 3) and b(0 to 3) represent the bit-wise *and* of the four most significant bits of vectors a and b in the Verilog and VHDL flavors, respectively.

The second layer, temporal, is where PSL gets interesting. It allows a designer to state properties that hold across multiple clock cycles. The always operator, which states that a Boolean expression holds in every clock cycle, is one of the most basic. For example, always !(ena & enb) states that the signals ena and enb will never be true simultaneously in any clock cycle.

More interesting are operators that specify delays. The next operator is the simplest one. The property always (req -> next ack) states that in every cycle that the req signal is true, the ack signal is true in the next cycle. The -> symbol denotes implication, i.e., if the expression to the left is true, that on the right must also be true. The next operator can also take an argument, e.g., always req -> next [2] ack means that ack must be true two cycles after each cycle in which req is true.

PSL provides an extended form of regular expressions convenient for specifying more complex behaviors. Although it would be possible to write always (req -> next (ack -> next !cancel)) to indicate that ack must be true after any cycle in which req is true, and cancel must be false in the cycle after that, it is much easier instead to write always {req ; ack ; !cancel}. This illustrates a basic principle of PSL: most operators are actually just "syntactic sugar;" the set of fundamental operators is quite small.

PSL draws a clear distinction between "weak" operators that can be checked in simulation (i.e., safety properties) and "strong" operators, which express liveness properties and can only be checked formally. Strong operators are marked with a trailing exclamation point (!), and some operators come in both strong and weak varieties.

```
<'
extend instr {
  keep opcode in [ADD, ADDI];
  keep op1 == REG0;
  when reg instr { keep op2 == REG1; }
  when imm instr { keep op2 == 0x3; }
};

extend sys {
   keep instrs.size() == 10;
};
'>
```

FIGURE 15.16 e Code that uses the instruction encoding of Figure 15.15 to randomly generate ten instructions. After an example in the Specman tutorial.

The eventually! operator illustrates the meaning of strong operators. The property always (req -> eventually! ack) states that after req is asserted, ack will always be asserted eventually. This is not something that can be checked in simulation: if a particular simulation saw req but did not see ack, it would be incorrect to report that this property failed because running that particular simulation longer might have produced ack. This is the fundamental difference between safety and liveness properties. Safety states something bad never happens; liveness states something good eventually happens.

Another subtlety is that it is possible to express properties in which time moves backward through a property. A simple example is always ((a && next [3] (b)) -> c, which states that when a is true and b is true three clock cycles later, c is true in the first cycle, i.e., when a is true. While it is possible to check this in simulation (for each cycle in which a is true, remember whether c is true and look three clock cycles later for b), it is more difficult to build automata that check such properties in simulation.

The third layer of PSL, the verification layer, instructs a verification tool what tests to perform on a particular design. It amounts to a binding between properties defined with expressions from the Boolean and temporal layer, and modules in the design under test. The following simple example:

```
vunit ex1a(top_block.i1.i2) {
  A1: assert never (ena && enb);
}
```

declares a "verification unit" called ex1a, binds it to the instance named top_block. i1. i2 in the design under test, and declares (the assertion named A1) that the signals ena and enb in that instance are never true simultaneously.

In addition to assert, verification units may also include assume directives, which state the tool may assume a given property; assume_guarantee, which both assumes and tests a particular property; restrict, which constrains the tool to only consider those behaviors that have the given property; cover, which asks the tool to check whether a certain property was ever observed; and fairness, which instructs the tool to only consider paths in which the given property occurs infinitely often, e.g., only when the system does not wait indefinitely.

The fourth, modeling layer of PSL essentially allows Verilog, System Verilog, VHDL, or GDL code to be included inline in a PSL specification. The intention here is to allow additional details about the system under test to be included in the PSL source file.

15.4.4 System Verilog

Recently, many aspects of the Vera, Sugar, and ForSpec verification languages have been merged into Verilog, along with the higher-level programming constructs of Superlog [47] (which were taken nearly verbatim from C and C++) to produce SystemVerilog [12]. As a result, Verilog has become the English of the HDL world: voraciously assimilating parts of other languages and making them its own.

```
class Bus;
  rand bit[15:0] addr;
  rand bit[31:0] data;

  constraint world_align { addr[1:0] = 2'b0; }
endclass

initial begin
  Bus bus = new;

  repeat (50) begin
    if (bus.randomize() == 1)
      $display("addr = %16h data = %h\n",
                 bus.addr, bus.data);
    else
      $display("overconstrained: no satisfying values exist\n");
  end
end

typdef enum { low, mid, high } AddrType;

class MyBus extends Bus;
  rand AddrType atype; // Additional random variable

  // Additional constraint on address: still word-aligned}
  constraint addr_range {
    (atype == low ) -> addr inside { [0:15] };
    (atype == mid ) -> addr inside { [16:127] };
    (atype == high) -> addr inside { [128:255] };
  }
endclass

task exercise_bus;
  int res;

  // Restrict to low addresses
  res = bus.randomize() with { atype == low; };

  // Restrict to particular address range
  res = bus.randomize()
        with { 10 <= addr && addr <= 20 };

  // Restrict data to powers of two
  res = bus.randomize() with { data & (data - 1) == 0 };

  // Disable word alignment
  bus.word_align.constraint_mode(0);

  res = bus.randomize with { addr[0] || addr[1] };

  // Re-enable word alignment
  bus.word_align.constraint_mode(1);
endtask
```

FIGURE 15.17 Constrained random variable constructs in System Verilog. The example starts with a simple definition of a Bus class that constrains the two least-significant bits of the address to be zero, then invokes the randomize method to randomly generate address/data pairs and print the result. Next is a refined version of the Bus class that adds a field taken from an enumerated type that further constrains the address depending on its value. The example ends with a task that illustrates various ways to control the constraints. After examples in the SystemVerilog LRM [12].

The C- and C++-like features added to SystemVerilog read like the list of features in those languages. SystemVerilog adds enumerated types, record types *(structs)*, *typedefs*, type casting, a variety of operators such as +=, operator overloading, control-flow statements such as break and continue, as well as object-oriented programming constructs such as classes, inheritance, and dynamic object creation and deletion.

```
enum { red, green, blue } color;

bit [3:0] adr, offset;

covergroup g2 @(posedge clk);
  Hue:    coverpoint color;
  Offset: coverpoint offset;

  // Consider (color, adr) pairs, e.g.,
  // (red, 3'b000), (red, 3'b001), ..., (blue, 3'b111)
  AxC:    cross color, adr;

  // Consider (color, adr, offset) triplets
  // Creates 3 * 16 * 16 = 768 bins
  all:    cross color, adr, Offset;
endgroup

g2 g2_inst = new; // Create a watcher

bit [9:0] a; // Takes values 0--1023

covergroup cg @(posedge clk);

  coverpoint a {
    // place values 0--63 and 65 in bin a
    bins a = { [0:63], 65 };

    // create 65 bins, one for 127, 128, ..., 191
    bins b[] = { [127:150], [148:191] };

    // create three bins: 200, 201, and 202
    bins c[] = { 200, 201, 202 };

    // place values 1000--1023 in bin d
    bins d = { [1000:$] };

    // place all other values (e.g., 64, 66, .., 126, 192, ...) in their own bin
    bins others[] = default;
  }

endgroup

bit [3:0] a;

covergroup cg @(posedge clk);
  coverpoint a {
    // Place any of the sequences 4 -> 5 -> 6, 7 -> 11, 8 -> 11, 9 -> 11, 10 ->11,
    // 7 -> 12, 8 -> 12, 9 -> 12, and 10 -> 12 into bin sa.
    bins sa = (4 => 5 => 6), ([7:9],10 => 11,12);

    // Create separate bins for 4 -> 5 -> 6, 7 -> 10, 8 -> 10, and 9 -> 10
    bins sb[] = (4 => 5 => 6), ([7:9] => 10);

    // Look for the sequence 3 -> 3 -> 3 -> 3
    bins sc = 3 [* 4];

    // Look for any of the sequences 5 -> 5, 5 -> 5 -> 5, or 5 -> 5 -> 5 -> 5
    bins sd = 5 [* 2:4];

    // Look for any sequence of the form 6 -> ... -> 6 -> ... -> 6
    // where "..." represents any sequence that excludes 6
    bins se = 6 [-> 3];
  }
endgroup
```

FIGURE 15.18 SystemVerilog covers constructs. The example begins with a definition of a "covergroup" that considers the values taken by the color and offset variables as well as combinations. Next is a covergroup illustrating the variety of ways "bins" may be defined to classify values for coverage. The final covergroup illustrates SystemVerilog's ability to look for and classify sequences of values, not just simple values. After examples in the System Verilog LRM [12].

```
// Make sure req1 or req2 is true if we are in the REQ state}
always @(posedge clk)
  if (state == REQ)
    assert (req1 || req2);

// Same, but report the error ourselves
always @(posedge clk)
  if (state == REQ)
    assert (req1 || req2)
    else
      $error("In REQ; req1 || req2 failed (\%0t)", $time);

property req_ack;
  @(posedge clk) // Sample req, ack at rising clock edge
    // After req is true, between one and three cycles later,
    // ack must have risen.
    req ##[1:3] $rose(ack);
endproperty

// Assert that this property holds, i.e., create a checker
as_req_ack: assert property (req_ack);

// The own_bus signal goes high in 1 to 5 cycles,
// then the breq signal goes low one cycle later.
sequence own_then_release_breq;
  ##[1:5] own_bus ##1 !breq
endsequence

property legal_breq_handshake;
  @(posedge clk)      // On every clock,
  disable iff (reset) // unless reset is true,
  // once breq has risen, own_bus should rise and breq should fall.
  $rose(breq) |-> own_then_release_breq;
endproperty

assert property (legal_breq_handshake);
```

FIGURE 15.19 SystemVerilog assertions. The first two *always* blocks check simple safety properties, i.e., that req1 and req2 are never true at the positive edge of the clock. The next property checks a temporal property: that ack must rise between one and three cycles after each time req is true. The final example shows a more complex property: when reset is not true, a rising breq signal must be followed by own_bus rising between one and five cycles later and breq falling.

At the highest level, it also adds strings, associative arrays, concurrent process control (e.g., fork/join), semaphores, and mailboxes, giving it features only found in concurrent programming languages such as Java.

Perhaps the most interesting features added to SystemVerilog are those directly related to verification. Specifically, SystemVerilog includes constrained, biased random variable generation, user-defined functional coverage checking, and temporal assertions, much like those in Vera, e, PSL, and ForSpec.

Figure 15.17 illustrates some of the random test-generation constructs in SystemVerilog, which were largely taken from Vera. Compare this with Figure 15.12.

Figure 15.18 illustrates some of the coverage constructs in SystemVerilog. In general, one defines "covergroups," which are collections of bins that sample values on a given event, typically a clock edge. Each covergroup defines the sorts of values it will be observing (e.g., values of a single variable, combinations of multiple variables, and sequences of values on a single variable) and rules that define the "bins" each of these values will be placed in. In the end, the simulator reports which bins were empty, indicating that none of the matching behavior was observed. Again, much of this machinery was taken from Vera (cf. Figure 15.13).

Figure 15.19 shows some of SystemVerilog's assertion constructs. In addition to signaling an error when an "instantaneous" condition does not hold (e.g., a set of variables are taking on mutually incompatible values), SystemVerilog has the ability to describe temporal sequences such as "ack must rise between one and five cycles after req rises" and check whether they appear during simulation. Much of the syntax comes from PSL/Sugar.

15.5 Conclusions

VHDL and Verilog remain the dominant HDLs and are likely to be with us for a long time, although perhaps they will become what assembly language has become to programming: a part of the compilation chain, but not generally written manually. Both have deep flaws, but these can be largely avoided by adhering to coding conventions, and in practice are fairly practical design entry vehicles.

The future of the verification languages discussed in this chapter is less certain. Clearly, there is a need to automate the validation process as much as possible, and these languages do provide useful assistance in the form of biased constrained pseudorandom test case generation, temporal property assertions, and coverage estimates. However, none has clearly proven itself essential to modern IC design, and the plethora of variants and derivatives suggests that their evolution is not complete. An even more serious question is whether the complexity of these languages, especially in their specification of temporal properties, creates more problems than it solves.

The fundamental burdens of specifying digital hardware and verifying its correctness will continue to fall on design and verification languages. Even if those in the future bear little resemblance to those described here, the current crop forms a strong foundation on which to build.

References

[1] I.S. Reed, Symbolic synthesis of digital computers, in *Proceedings of the ACM National Meeting*, Toronto, Canada, September 1952, pp. 90–94.

[2] W.E. Omohundro and J. H. Tracey, Flowware — a flow charting procedure to describe digital networks, in *Proceedings of the First International Conference on Computer Architecture (ISCA)*, Gainesville, FL, December 1973, pp. 91–97.

[3] Y. Chu, An ALGOL-like computer design language, *Commn. ACM*, 8, 607–615, 1965.

[4] C. Gordon Bell and A. Newell, *Computer Structures: Readings and Examples*, McGraw-Hill, New York, 1971.

[5] Y. Chu, D.L. Dietmeyer, J.R. Duley, F.J. Hill, M. R. Barbacci, C.W. Rose, G. Ordy, B. Johnson, and M. Roberts, Three decades of HDLs: Part I, CDL through TI-HDL. *IEEE Design Test Comput.*, 9, 69–81, 1992.

[6] D. Borrione, R. Piloty, D. Hill, K.J. Lieberherr, and P. Moorby, 2nd Three decades of HDLs: Part II, Colan through Verilog, *IEEE Design Test Comput.*, 9, 54–63, 1992.

[7] F.I. Parke, An introduction to the N.mPc design environment, in *Proceedings of the 16th Design Automation Conference*, San Diego, CA, June 1979, pp. 513–519.

[8] Al Dewey, VHSIC hardware description (VHDL) development program, *Proceedings of the 20th Design Automation Conference*, Miami Beach, FL, 1983, pp. 625–628.

[9] R.K. Brayton, G.D. Hachtel, and A.L. Sangiovanni-Vincentelli, Multilevel logic synthesis, *Proc. IEEE*, 78, 264–300, 1990.

[10] *IEEE Standard Hardware Description Language Based on the Verilog Hardware Description Language (1364–1995)*, IEEE Computer Society, New York, NY, 1996.

[11] *IEEE Standard Verilog Hardware Description Language (1364–2001)*, IEEE Computer Society, New York, NY, September 2001.

[12] Accelera, 1370 Trancas Street #163, Napa, CA 94558. *SystemVerilog 3.1a Language Reference Manual: Accellera's Extensions to Verilog*, May 2004.

[13] P.L. Flake, P.R. Moorby, and G. Musgrave, An algebra for logic strength simulation, *Proceedings of the 20th Design Automation Conference*, Miami Beach, FL, June 1983, pp. 615–618.

[14] D.E. Thomas and P.R. Moorby, *The Verilog Hardware Description Language*, (5th ed.), Kluwer, Boston, MA, 2002.

[15] S. Palnitkar, *Verilog HDL: A Guide to Digital Design and Synthesis,* Prentice-Hall, Upper Saddle River, NJ, 1996.
[16] S. Mittra, *Principles of Verilog PLI,* Kluwer, Boston, MA, 1999.
[17] S. Sutherland, *The Verilog PLI Handbook,* Kluwer, Boston, MA, 1999.
[18] D.J. Smith, VHDL & Verilog compared & constrasted — plus modeled examples written in VHDL, Verilog, and C, in *Proceedings of the 33rd Design Automation Conference,* Las Vegas, Nevada, June 1996, pp. 771–776.
[19] R.S. French, M.S. Lam, J.R. Levitt, and K. Olukotun, A general method for compiling event-driven simulations, in *Proceedings of the 32nd Design Automation Conference,* San Francisco, CA, June 1995, pp. 151–156.
[20] M.J.C. Gordon, The semantic challenge of Verilog HDL, in *Proceedings of the Tenth Annual IEEE Symposium on Logic in Computer Science (LICS),* San Diego, CA, June 1995.
[21] A. Dewey and A.J. de Geus, VHDL: toward a unified view of design, *IEEE Design Test Comput.,* 9, 8–17, 1992.
[22] *IEEE Standard VHDL Reference Manual (1076–1987),* The Institute of Electrical and Electronics Engineers (IEEE), New York, NY, 1988.
[23] *IEEE Standard VHDL Language Reference Manual (1076–1993),* IEEE Computer Society, New York, NY, 1994.
[24] R. Lipsett, C.F. Schaefer, and C. Ussery, *VHDL: Hardware Description and Design,* Kluwer, Boston, MA, 1989.
[25] A.M. Dewey, *Analysis and Design of Digital Systems with VHDL,* Brooks/Cole Publishing (Formerly PWS), Pacific Grove, CA, 1997.
[26] J. Bhasker, *A VHDL Synthesis Primer.* 2nd ed., Star Galaxy Publishing, Allentown, PA, 1998.
[27] D.L. Perry, *VHDL,* 3rd ed., McGraw-Hill, New York, 1998.
[28] P.J. Ashenden, *The Designer's Guide to VHDL,* Morgan Kaufmann, San Francisco, CA, 1996.
[29] P.J. Ashenden, *The Student's Guide to VHDL,* Morgan Kaufmann, San Francisco, CA, 1998.
[30] B. Cohen, *VHDL Coding Styles and Methodologies,* , 2nd ed., Kluwer, Boston, MA, 1999.
[31] R.E. Harr and A.G. Stanculescu, Eds., *Applications of VHDL to Circuit Design,* Kluwer, Boston, MA, 1991.
[32] C. Mead and L. Conway, *Introduction to VLSI Systems,* Addison-Wesley, Reading, MA, 1980.
[33] *IEEE Standard Multivalue Logic System for VHDL Model Interoperability (Std_logic_1164),* The Institute of Electrical and Electronics Engineers (IEEE), New York, NY, 1993.
[34] *IEEE Standard VHDL Synthesis Packages (1076.3–1997),* The Institute of Electrical and Electronics Engineers (IEEE), New York, NY, 1997.
[35] S. Liao, S. Tjiang, and R. Gupta, An efficient implementation of reactivity for modeling hardware in the Scenic design environment, in *Proceedings of the 34th Design Automation Conference,* Anaheim, CA, June 1997, pp. 70–75.
[36] G. Arnout, SystemC standard, *Proceedings of the Asia South Pacific Design Automation Conference (ASP-DAC),* Yokohama, Japan, January 2000, pp. 573–578.
[37] T. Grötker, S. Liao, G.Martin, and S. Swan, *System Design with SystemC,* Kluwer, Boston, MA, 2002.
[38] J. Bhasker, *A SystemC Primer,* 2nd ed., Star Galaxy Publishing, Allentown, PA, 2004.
[39] W. Müller, W. Rosenstiel, and J. Ruf, Eds., *SystemC: Methodologies and Applications,* Kluwer, Boston, MA, 2003.
[40] J. Bergeron, *Writing Testbenches: Function Verification of HDL Models,* 2nd ed., Kluwer, Boston, MA, 2003.
[41] F. Haque, J. Michelson, and K. Khan, *The Art of Verification with Vera,* Verification Central, www.verificationcentral.com, 2001.
[42] S. Palnitkar, *Design Verification with e,* Prentice-Hall, Upper Saddle River, NJ, 2003.
[43] S. Iman and S. Joshi, *The e Hardware Verification Langauge,* Kluwer, Boston, MA, 2004.
[44] I. Beer, S. Ben-David, C. Eisner, D. Fisman, A. Gringauze, and Y. Rodeh, The temporal logic Sugar, in *Proceedings of the 13th International Conference on Computer-Aided Verification (CAV), Lecture Notes in Computer Science,* Paris, France, Springer, Berlin, 2001, Vol. 2102, pp. 363–367.
[45] B. Cohen, S. Venkataramanan, and A. Kumari, *Using PSL/Sugar for Formal Verification,* VhdlCohen Publishing, Palos Verdes Peninsula, CA, 2004.

[46] E.M. Clarke and E.A. Emerson, Design and synthesis of synchronization skeletons using branching time temporal logic, in *Proceedings of the Workshop on Logic of Programs, Lecture Notes in Computer Science,* Yorktown Heights, New York, May 1981. Springer, Berlin, Vol. 131, pp. 52–71.

[47] P.L. Flake and S.J. Davidmann, Superlog, a unified design language for system-on-chip, *Proceedings of the Asia South Pacific Design Automation Conference (ASP-DAC),* Yokohama, Japan, January 2000, pp. 583–586.

16

Digital Simulation

John Sanguinetti
Forte Design Systems, Inc.
San Jose, California

16.1 Introduction

Logic simulation is the primary tool used for verifying the logical correctness of a hardware design. In many cases logic simulation is the first activity performed in the process of taking a hardware design from concept to realization. Modern hardware description languages are both simulatable and synthesizable. Designing hardware today is really writing a program in the hardware description language. Performing a simulation is just running that program. When the program (or model) runs correctly, then one can be reasonably assured that the logic of the design is correct, *for the cases that have been tested in the simulation.*

Simulation is the key activity in the design verification process. That is not to say that it is an ideal process. It has some very positive attributes:

1. It is a natural way for the designer to get feedback about his design. Because it is just running a program — the design itself — the designer interacts with it using the vocabulary and abstractions of the design. There is no layer of translation to obscure the behavior of the design.
2. The level of effort required to debug and then verify the design is proportional to the maturity of the design. That is, early in the design's life, bugs and incorrect behavior are usually found more quickly. As the design matures, it takes longer to find the errors.
3. Simulation is completely general. Any hardware design can be simulated. The only limits are time and computer resources.

On the negative side, simulation has two drawbacks, one of which is glaring:

1. There is (usually) no way to know when you are done. It is not feasible to completely test, via simulation, all possible states and inputs of any nontrivial system.
2. Simulation can take an inordinately large amount of computing resources, since typically it uses a single processor to reproduce the behavior of many (perhaps millions of) parallel hardware processes.

Every design project must answer the question "have we simulated enough to find all the bugs?" and every project manager has taped out his design knowing that the truthful answer to that question is either "no" or "I don't know." It is this fundamental problem with simulation that has caused so much effort to be spent looking for both tools to help answer the question, and formal alternatives to simulation.

Code coverage, functional coverage, and logic coverage tools have all been developed to help gauge the completeness of simulation testing. None are complete solutions, though they all help. Formal alternatives have been less successful. Just as in the general software world where proving programs correct has proven intractable, formal methods for verifying hardware designs have still not proven general enough to replace simulation. That is not surprising, since it is the same problem.

The second drawback motivates much of the material in this chapter. That is, simulation is always orders of magnitude slower than the system being simulated. If a hardware system runs at 1 GHz, a simulation of that system might run at 10–1000 Hz, depending on the level of the simulation and the size of the system. That is a slowdown from 10^6 to 10^8! Consequently, many people have spent considerable time and effort finding ways to speed up logic simulation.

Considering both the advantages and disadvantages of logic simulation, it is really quite a good tool for verifying the correctness of a hardware design. Despite its drawbacks, simulation remains the first choice for demonstrating correctness of a design before fabrication, and its value has been well established.

16.1.1 Levels of Abstraction

Because simulation is a general technique, a hardware design can be simulated at different levels of abstraction. Often it is useful to simulate a model at several levels of abstraction in the same simulation run. The commonly used levels of abstraction are gate level, register transfer level (RTL), and behavioral (or algorithmic) level. However, it is possible to incorporate lower levels like transistor level or even lower physical levels as well as higher levels such as transaction level or domain-specific levels. For this discussion, we will restrict our attention to behavioral, RTL, and gate levels, with the understanding that other levels are completely compatible with the techniques described here.

16.1.2 Discrete Event Simulation

> "we decided to focus on simulation ..., because that's the only really interesting thing to do with a computer" (Alan Kay, Second West Coast Computer Faire, 1978)

Logic simulation is a special case of the more general discrete event simulation methods, which were initially developed in the 1960s [5]. Discrete event simulation is a method of representing the behavior of a system over time, using a computer. The system may be either real or hypothetical, but its state is assumed to change over time due to some combination of external stimulus and internal state. The fundamental idea is that the behavior of any system can be decomposed into a set of discrete instants of time at which things happen. Those instants are called events, and the "things that happen" are state changes. This is very analogous to the way we digitize continuous physical phenomena, like audio sampling. In essence, we digitize a time-based process by dividing it up into discrete events. It is easy to see that with a fine enough granularity, one can get an adequately accurate representation for just about any purpose.

The basic operation of a discrete event simulation is given in Figure 16.1.

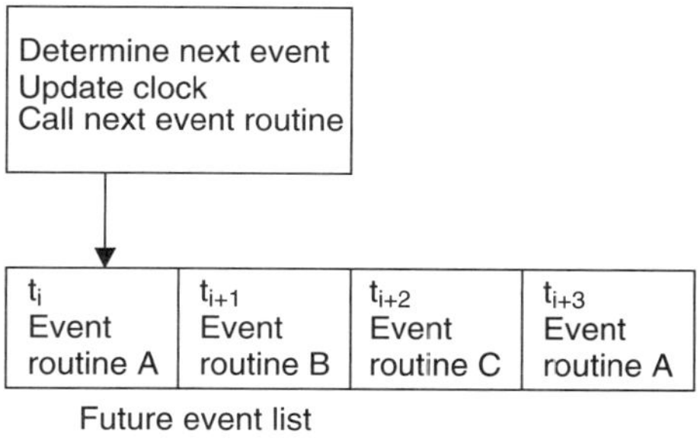

FIGURE 16.1 Discrete event simulation basic operation.

Each event in the system is represented by an event routine. An event routine is some code to be executed to represent the action at that event, which usually amounts to a state change and a determination of the occurrence of one or more future events. The set of possible events is usually (though not always) fixed, but the number of occurrences of each type of event can be variable.

The list of events shown in Figure 16.1 is a list of those events that are scheduled to happen at known times in the future (t_{i+1}, t_{i+2},...). The entries on this list contain the time at which the event will occur, and the type of event it is; that is, what event routine should be called to realize that event's behavior. It is common to have more than one instance of any particular type of event scheduled to occur at various times on this list. This list is called the future event list.

16.2 Event- vs. Process-Oriented Simulation

The basic discrete event simulation paradigm was in use in the early 1960s (and probably earlier), and several subroutine libraries were developed for FORTRAN and Algol to support it. It was soon realized that there were more natural ways to write simulation models of many systems, and that led to the development of specialized simulation languages, particularly GPSS, SIMSCRIPT, and Simula [3]. Of these, GPSS is still used today to model systems as extensive as the U.S. air traffic control system. Simula was the first object-oriented programming language and introduced the concepts of classes and objects.

Both GPSS and Simula took a process-oriented view of modeling. That is, instead of writing a separate routine for each event, one would write a routine that represented a process which might include several events within it taking place over a nonzero period of simulated time. The routine would have wait statements in it to indicate that some time would pass between one event and the next. This did not affect the underlying simulation mechanism, but it did change the way the simulation model was written. A process orientation is particularly useful for hardware simulation at a behavioral level, while an event orientation works for RTL and gate levels. The most popular hardware description languages, Verilog [6,7], VHDL [2], and SystemC [4], are process-oriented simulation languages which are indirectly descended from Simula.

16.3 Logic Simulation Methods and Algorithms

Logic simulation is essentially a process of computing a state trajectory of the system over time. The system's state is defined by the state variables, which are the storage elements in the circuit. We usually think of these as the registers and latches in the design. How the state, taken as a whole, changes over time is called the state trajectory.

A simple state trajectory as a function of time can be written as follows:

$$f(m, n, t + 1) = m_t + n_t$$

where m is a state variable and n is an input.

It is obvious how to compute this function:

$$
\begin{aligned}
&\text{Do } (i = 1, \text{endtime}) \\
&\quad t = t + 1 \\
&\quad m = m + n \\
&\quad n = \text{new input} \\
&\text{enddo}
\end{aligned}
$$

This example illustrates how a function is simulated over time. There is a loop, which consists of updating the system time or clock, evaluating the logic components, and updating the state variables. Compare this loop to the process of Figure 16.1. There is only one event routine, and it gets executed repeatedly at a regular time interval.

Now we can expand this example to two functions,

$$f(m, n, t + 1) = m_t + n_t$$
$$g(x, n, t + 1) = x_t \wedge n_t$$

Evaluating these two functions together would look like this:

$$
\begin{aligned}
&\text{Do } (i = 1, \text{endtime}) \\
&\quad t = t + 1 \\
&\quad m = m + n \\
&\quad x = x \wedge n \\
&\quad n = \text{new input} \\
&\text{enddo}
\end{aligned}
$$

This set of state functions when taken together, make up a system with a state variable, which is a duple (m, x). Running this simulation computes the trajectory of the state vector (m, x) over the simulation interval.

Notice that the way we have defined the functions f and g, the value of n used in the computation is the value of n at time t, even though the time of the evaluation is $t + 1$. This leads to a further refinement of the simulation process. In real hardware, state variables (registers or latches) do not get updated instantaneously. That is, the computation even if it is a simple assignment, takes some nonzero amount of time. So the simulation loop could be rewritten as in Figure 16.2.

```
do (i=1, endtime)
   t = t + 1
   // evaluate the functions
   t_m = m + n
   t_x = x ^ n
   t_n = new input
   // update the variable values
   m = t_m
   x = t_x
   n = t_n
enddo
```

FIGURE 16.2 Basic simulation loop.

While we have not represented the actual delays involved in the computation, we have abstracted them into a behavior that takes into account their effect. We now have the typical event-processing loop in a digital logic simulation: advance the clock, evaluate all the logic functions, and update the variable values.

Note that this has incorporated an abstraction of physical behavior over time, since we are moving time from one discrete moment to another, and we are assuming that nothing interesting happens in between those two instants. We also assumed that m, x, and n do not change their values instantaneously, but they do change before the next time instant $(t + 1)$. Note that they all change together after all the evaluation has been done. This is usually called a "delta cycle," meaning it is a set of events that happen at time t, but after all the evaluation events. That is, the new-value assignments happen at $t + \delta t$.

A further complication arises when time delays must be introduced into a computation. That is, the function being computed may look like this:

$$f(m, n, t) = g(m_{t-k}, n_{t-k})$$

That is, the new value does not get updated into the state variable until k time units after the evaluation. To handle this, the simulation loop would look like Figure 16.3.

It is common practice to replace the array *t_m*() in Figure 16.3 with a linked list, which is ordered by the index t. In order to make this work, the delay is usually incorporated into the new-value computation, and the new value is put on a future update list at time $t + k$. The future update list is analogous to the future event list, and in fact does not need to be separate from it. The simulation loop would then look like Figure 16.4.

Here we have assumed that the future update list has values only for the variable m. Generally, there may be many different state variables whose new values may be saved on the future update list, so each entry

```
do (i=1, endtime)
   t = t + 1
   // evaluate the functions
   t_m(t+k) = m + n
   t_n = new input
   // update the variable values
   m = t_m(t)
   n = t_n
enddo
```

FIGURE 16.3 Simulation loop with update step.

```
do (i=1, endtime)
   t = t + 1
   // evaluate the functions
   t_m= m + n
   put_on_future_update_list(t_m, t+k)
   t_n = new input
   // update the variable values
   m = take_off_future_update_list(t)
   n = t_n
enddo
```

FIGURE 16.4 Simulation loop with update list.

on the list must identify the state variable that the value is associated with. Note also that the new value is always at the front of the list, if the list is ordered by increasing time t. That is, at any given time t, there are no values on the list with time $<t$ (those would be values in the past), so all the variable updates can be found at the front of the list. Of course, there may be no values on the list associated with time t, that is, the first entry on the list may have time $> t$, in which case no variable would be updated at time t.

16.3.1 Synchronous and Asynchronous Logic

The example above shows synchronous, or sequential, logic. That is, each variable's new value is computed at a regular interval ($t, t + 1, \ldots$). There is also asynchronous, or combinational, logic that might be present in a system to be simulated. The salient characteristic of asynchronous logic is that the function is computed continuously. That is, when the inputs change, the output changes immediately, at least as compared to the granularity of the simulation time. For example, in an and gate followed by an inverter, the two inputs are *and*ed into the inverter whose output is created with no (apparent) delay (Figure 16.5).

This would be represented by $d = \sim(c \,\&\, b)$. In terms of our simulation loop, asynchronous logic functions have their outputs updated during the evaluation phase, not delayed until the update phase. That is, there is no temporary variable created to hold the new value.

It now is apparent that we need some way of distinguishing between the kind of assignment that is immediate and the previous, synchronous, kind, which is delayed. A simulation language typically has the ability to represent instantaneous change as well as delayed change.

16.3.2 Propagation

Combinational logic leads to a further complication of our simulation loop, namely, where do we do the evaluation of a combinational expression? Combinational logic must be evaluated whenever one of its constituent inputs changes. Sequential logic can be evaluated regularly, at time intervals corresponding to a clock signal. That is why it was natural to write the simulation loop with the evaluations happening immediately after updating the time variable. But where do we put combinational logic update?

The answer can be found by looking at what would cause a combinational expression to be evaluated. That is, when would its inputs change? If we look at a typical circuit that has both combinational and sequential logic, it might look like Figure 16.6.

Looking at this, we see that the input to a combinational expression is either the value of a state variable (a register) or the output of another combinational expression. Going back to the simulation loop, the logical place to put the evaluation of combinational expressions is after the update of state variables. If our combinational logic consisted of

$$c = m \mid a$$
$$d = \sim(c \,\&\, b)$$

Then the loop would look like Figure 16.7.

The simulation loop as it is described in Figure 16.7 is now sufficient to simulate circuits that have the form of Figure 16.6, but with an important caveat. That caveat is that the combinational expressions must be ordered such that each one is evaluated only after all of its inputs have been updated. If a circuit does not contain any combinational feedback loops, then it is possible to satisfy this ordering requirement. Ordering the expressions is called *levelizing*.

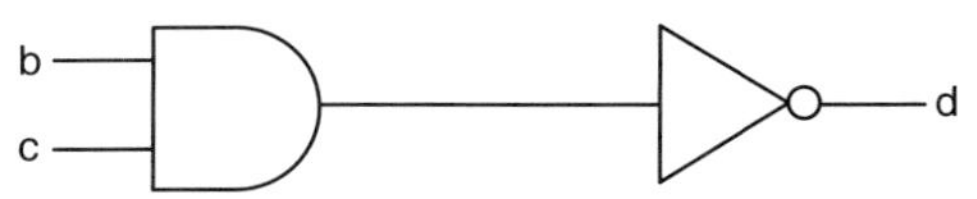

FIGURE 16.5 Example asynchronous logic.

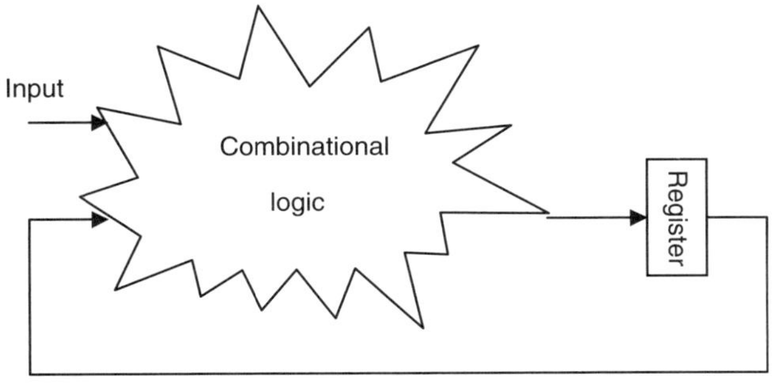

FIGURE 16.6 Example with synchronous and asynchronous logic.

```
do (i=1, endtime)
   t = t + 1
   // evaluate the state variables
   t_m = m + n
   put_on_future_update_list(t_m, t)
   t_n = new input
   // update the variable values
   m = take_off_future_update_list(t)
   n = t_n
   // update all combinational expressions
   c = m | a
   d = ~(c & b)
enddo
```

FIGURE 16.7 Simulation loop with asynchronous logic.

If the delays for the sequential variables are all zero, which is a common case, then this simulation loop is quite efficient, and in fact many logic simulators have been created with just this simulation loop. The drawback of this approach however, is that all of the expressions in the circuit must be evaluated on every loop iteration, i.e., at each time instant. When very much of the circuit remains unchanging for long periods, this can be quite wasteful, as expressions that do not change get continually reevaluated.

To improve the simulation efficiency in this common case, we can return to the basic event scheduling idea of discrete event simulation, and only evaluate an expression when one or more of its inputs has changed. This is called *propagation* of a value from an output to an input, or in language terms, from a left-hand side of an assignment to a right-hand side. The simulation loop now would look like Figure 16.8.

Just as we introduced a future update list, we can also create a propagation list, where the entry on the list is the expression whose inputs have changed. Rewriting the simulation loop using that, we get Figure 16.9.

Notice here that the propagate list contains a pointer or other identifier of the expression that needs to be evaluated. Notice also that the expressions that depend on any variable are given by the function $c(v)$. The function $c(v)$ is static, since in logic simulation, the expressions that depend on any variable are known at compile time. Finally, notice that it is a simple efficiency improvement to only propagate those values that have changed. That is, if the new value of a variable is the same as the old value, there is no need to reevaluate any dependent expressions. With that enhancement, we have a complete simulation loop for a logic simulator.

```
do (i=1, endtime)
  t = t + 1
  // evaluate the state variables
  t_m = m + n
  put_on_future_update_list(t_m, t)
  t_n = new input
  // update the variable values
  m = take_off_update_event_list(t)
  n = t_n
  // propagate m and n to combinational expressions
  c = m | a
  // propagate c to combinational expressions
  b = ~(c & d)
 // propagate b to combinational expressions
  ...
enddo
```

FIGURE 16.8 Simulation loop with propagation.

```
do (i=1, endtime)
  t = t + 1
  for (all state variables m)
    t_m= f(.)   // evaluate the state variables
    put_on_future_update_list(m, t_m, t)
  endfor
  // update the variable values
  while (u = take_off_future_update_list(t) != empty)
    u.v = u.t_v
     put_on_propagate_list(c(u.v), t+u.d)
  endwhile
  // propagate changed values to combinational expressions
  while (c = take_off_propagate_list(t) != empty)
    c.v = eval(c.expr)
    put_on_propagate_list(c(c.v), t+c.d)
  endwhile
enddo
```

FIGURE 16.9 Simulation loop with propagation list.

The only remaining necessary generalization is time. We have dealt with time as simply an ordered set of instants, t_1, t_2, … This would be sufficient if each time instant was able to be mapped onto a real time line, for example, if each t_i corresponded to clock cycle i. However, in the general case, there may be nonuniform time intervals at which events happen. Just as there could be a delay in updating the state variables with new values, there could be delays updating combinational variables with new values (this would represent gate delays). So both the update event list and the propagate list would be ordered by time, and each event on either list would include a time at which it was to occur. Now, when all the events at the current time t are finished, the main loop would be iterated upon and time would be advanced to the nearest time in the future from either of the two lists (Figure 16.10).

The entry which is put on the future event list is the sequential expression, which will be evaluated at a given time in the future (often a clock-cycle boundary). This now has all the elements of a logic simulation loop, sometimes also called the scheduling loop or the simulation kernel. People often talk about a "simulation engine" in an attempt to make simulators sound more sophisticated than they really are. The scheduler is about the only thing that could be called a "simulation engine," and as we have seen, conceptually it is pretty simple.

16.3.3 Processes

We have described a logic simulation model as a collection of synchronous state variables (registers) and asynchronous combinational expressions. This corresponds to a typical RTL-style description of a logic circuit. We might write a description of such a circuit as follows:

at rising edge of clock:

$\quad reg_i = newval_i$

$\quad reg_j = newval_j$

$\quad \ldots$

asynchronous:

$\quad newval_i = f(reg_i, reg_j, \ldots)$

$\quad newval_j = g(reg_i, reg_j, \ldots)$

$\quad \ldots$

This maps directly onto the simulation loop described in Figure 16.10. However, this is a pretty low level of description, and it would be laborious to write circuit models this way routinely. By introducing the

```
do (while t < endtime)
  t = min(future_event_list, future_update_list, propagate_list)
  while (m = take_off_future_event_list(t) != empty)
    t_m= eval(m.expr)   // evaluate the state variable in this event
    put_on_future_update_list(m.v, t_m, t+m.d)
    put_on_future_event_list(m.v, m.expr, t+m.next)
  endwhile
  // update the variable values
  while (u = take_off_future_update_list(t) != empty)
    u.v = u.t_v
    put_on_propagate_list(c(u.v), t+u.d)
  endwhile
  // propagate changed values to combinational expressions
  while (c = take_off_propagate_list(t) != empty)
    c.v = eval(c.expr)
    put_on_propagate_list(c(c.v), t+c.d)
  endwhile
enddo
```

FIGURE 16.10 Simulation loop with future event, update, and propagate lists.

idea of a process, we can gain expressive power and still use the same simulation mechanics. A process would look like this:

process
  wait (rising edge of clock)
  $reg_i <= f(reg_i, reg_j, \ldots)$
  $reg_j <= g(reg_i, reg_j, \ldots)$
  ...
endprocess

Here we have combined the combinational logic with the sequential logic. Note that we have used a special symbol $<=$, to indicate that the assignment is a delayed assignment. This is not strictly necessary, as we could infer that any assignment to a state variable should be a delayed assignment. Verilog uses a special assignment symbol, VHDL and SystemC do not.

We could extend the process by allowing time to pass during the execution of the process, as in Figure 16.11.

This process represents a communication protocol with another process whereby data are exchanged using a ready/valid handshake, and the output happens ten clock-cycles after the input. It is straightforward to map this process onto our simulation loop by decomposing the process into events that get put onto the future event list. Each of the events corresponds to a wait in the process. This is left as an exercise for the reader.

The notion of a process becomes more powerful when a system has multiple processes, which can operate on common variables. Processes are inherently independent. That is, the events in two processes may or may not have any ordering relationship between them. Typically, all processes are assumed to begin at the beginning of simulation (time 0), but the events they wait on may be different and unrelated. Events within processes may be synchronized by means of common events or variables, but they do not have to be.

It is easy to see how arbitrarily complicated behavior can be described with multiple processes, and it is straightforward to map the events in each process into our simulation structure. Indeed, it is fairly easy to construct models that cannot be physically realized using these structures. While it may not be apparent what the value of describing unrealizable systems is, this power can be very useful when modeling the environment that the system is subjected to. This is usually called the *testbench*, while the target system is called the *design under test*. It is very useful to be able to describe both in the same simulation environment.

```
process
    while (forever)
        while (!valid)
            wait (rising edge of clock)
        d <= data.in
        ready <= 0
        wait (rising edge of clock)
        for (i=0; i<10; i++)
            wait (rising edge of clock)
        dataout <= f(d)
        ready <= 1
    endwhile
endprocess
```

FIGURE 16.11 An example process.

16.3.4 Race Conditions

A race condition can occur in a concurrent system when the behavior of the system depends on the order of execution of two events that are logically unordered. The most common cause of this is when one process modifies a variable and another reads the same variable at the same simulated time. This will not happen with state variables when delayed assignment is used, but it can happen with combinational variables, or with state variables if delayed assignment is not used. VHDL took the approach of making all assignments to state variables delayed, while Verilog did not. Thus, it is easier to write a model with race conditions in Verilog than in VHDL. There is an efficiency cost to delayed assignment of course, which is one of the reasons that VHDL simulators are typically slower than Verilog simulators.

16.4 Impact of Languages on Logic Simulation

The three major hardware description languages, Verilog, VHDL, and SystemC, are the primary languages used for hardware simulation today. All of them have the richness required to represent the vast majority of hardware designs. Verilog has more low-level capabilities than VHDL or SystemC, while SystemC has more high-level capabilities than VHDL or Verilog. They have common features that enable efficient simulation of hardware constructs, primarily hardware data types, hardware-oriented hierarchy, and hardware-oriented timing and synchronization.

16.4.1 Data Types

One of the ways that hardware description languages differ from other programming languages is the data types they offer. Logic signals are either 1 or 0, or in some cases neither (i.e., undriven or floating). But real hardware can be more complicated, so it is often convenient to be able to model signals as a range of strengths. That is, a strong signal can override a weak signal. It is also convenient for simulation purposes to include an unknown value (x), which indicates that a value is either uninitialized, or driven by conflicting values. This range of possible values is most useful when modeling at low levels, which is why Verilog has the richest set of signal values and SystemC has the least.

All hardware description languages have data types that allow the explicit specification of bit widths. That is, it can be specified that a variable is n bits wide, where the maximum value of n is usually some large number. This is useful when describing buses and collections of signals that are to be treated as a single variable. There are also operations to go along with these data types, like concatenation and subset selection. Operations on these data types cause the simulator to do more work than would be done in a normal C program, since the underlying computer must use several instructions, to accomplish them, rather than a single native instruction.

16.4.2 Variables

Verilog, VHDL, and SystemC make a distinction between variables which are state variables and those which are combinational variables. This makes the simulator's job easier because the classification is made by the programmer. In fact, a simulator could determine which variables are state variables and which are not by context. Verilog goes even further and requires the programmer to indicate which assignments are delayed assignments and which are not.

16.4.3 Hierarchy

The organization of a model in a hardware description language consists of a collection of modules in a tree structure. This corresponds with the way hardware is built. Signals are passed between modules in the tree by means of input and output ports. As far as the simulation goes, the port connections between parent and child are the same as combinational logic assignments. That is, for an input port, the left-hand side of the assignment is the port variable in the child module, and the right-hand side is the port variable in the parent module. By including these constructs in the language, the user does not have to write so much code, but the simulator still has to do the same amount of work.

16.4.4 Time Control

Hardware description languages offer a clock-based model of synchronization and event control. This makes synchronous logic circuits easy to write. Again, it does not make the simulator do any less work. The main characteristic is the definition of events in relation to one or more common signals, which are usually interpreted as a system clock signal, or a set of clock signals. Thus, a process can wait on the rising or falling edge of a signal, or either edge. The common way of indicating that two processes execute at the same time is to have both wait on the same edge of a common signal. At the RTL, where registers are explicitly instantiated, it is common to have hundreds or thousands of registers all triggered by the rising edge of a clock. This can be taken advantage of by the simulator to reduce the overhead of these events. Note however, that there is nothing inherently different about a clock signal from any other signal. The only difference is how it is used.

A Verilog description of a clock signal is shown in Figure 16.12a, a register using it is shown in Figure 16.12b, and a process using it is shown in Figure 16.12c.

16.4.5 Combinational Logic

A distinctive feature of hardware description languages from a simulation point of view, is the inclusion of a separate construct to describe combinational or continuous logic. As previously described, combinational logic is composed of assignments to variables that are done reactively. That is, the assignment is evaluated and performed whenever any of its constituent variables changes. We saw earlier how propagation is handled in the simulator. Combinational expressions are evaluated only as a result of propagation. Their use is a natural way to represent hardware at the RTL, but their use imposes complications for the simulator, since a poor choice of when to evaluate them can have dramatic consequences on the running time of the simulation. Figure 16.13a shows a Verilog continuous assignment, and Figure 16.13b shows the same variable assignment written as a process. Note that in both cases, the expression evaluation and assignment will be performed whenever one of the right-hand side variables changes.

```
reg clk;
clk = 0;
always #10
  clk = ~clk;
```
(a)

```
reg [7:0]  state, newstate;
always @(posedge clk)
  state <= newstate;
```
(b)

```
always begin
  while ( go == 0)
    @(posedge clk) ;   // wait for go
  count = 0;
  while (count < 10)   // wait for 10 cycles
    @(posedge clk) count = count + 1;
  done <= 1;           // raise flag
  @(posedge clk)
    done <= 0;   // drop flag after one cycle
end
```
(c)

FIGURE 16.12 Description and use of a clock signal in verilog.

```
assign var = a + (b ^ c);        always @(a or b or c)
                                   var = a + (b ^ c);
```

(a) (b)

FIGURE 16.13 Logic written as a) Continuous assignment; b) process.

16.5 Logic Simulation Techniques

Simulation speed is defined as the ratio of simulation time to simulated time. Simulation time is the real time required to execute the simulation model, while simulated time is the time represented in the model. Because an essentially unbounded amount of simulation is required to verify the correct behavior of a complex digital design, simulation speed is very important. It is interesting to note that in logic simulation, the simulation speed ratio is often called the slow-down, since the simulation takes longer than the real system, usually by orders of magnitude. However, in general system simulation, the simulation speed ratio is often called a speedup, since the simulated time scale may be very large.

The underlying mechanics of logic simulation are as described above. Except in the case of cycle-based simulation as described below, all simulators use a scheduler that functions pretty much the same. The scheduler is responsible for selecting the next event to execute and transferring control to it. There are well-known algorithms for implementing schedulers, and all mainstream logic simulators have reasonably well-optimized schedulers.

Nevertheless, different simulators can have vastly different performances on the same model written in the same language (or the same model written in a different language). The difference in execution efficiency between simulators is due to how the event routines are executed, how operations are executed, and especially how events are scheduled. It is not uncommon for two different simulators to execute a different number of events — differing by a factor of 2 or 3 — for the same model, and still yield the same results.

16.5.1 Interpreted Simulation

Most early logic simulators were interpreters. That is, the simulator read in the model source, built some internal data structures, and encoded the event routine operations in a unique instruction set. Then, as the simulation was run, the event routines were interpreted by a special piece of code, the interpreter, which "executed" those custom instructions. This is a technique that has been used for many years to translate and execute programs in a variety of languages. In general, interpreted execution offers a good opportunity for debugging the program, since the interpreter can relate any errors directly to the source of the program. However, there is an efficiency cost. Interpreted execution is slow, because the interpreter has to do a lot of work for each instruction. As a result, there are few modern logic simulators that are interpreters.

16.5.2 Compiled Code Simulation

Logic simulators have been categorized as "compiled code" simulators to distinguish them from interpreters. A better description would be simply "compiler." [1] A compiled code simulator is nothing more than a compiler for the simulation language it implements. Fundamentally, it differs from an interpreter in how the event routines are executed. While an interpreter executes the event routines by executing a sequence of operations represented as higher-level instructions, a compiler prepares the event routines so that they can be executed directly by the host machine. That is, an event routine is compiled into machine code so that it can be called as a subroutine. Essentially, the compiler has to do more work initially, but it produces event routines that execute much faster. It is common for compiled code to execute one to two orders of magnitude faster than interpreted code. That difference in efficiency has been observed across many different programming languages and many different underlying machine architectures. It is no surprise that compiled logic simulators typically run between 10 and 20 times faster than interpreted logic simulators on the same models.

It is common for compiled logic simulators to emit event routine code in C and then use the host machine's C compiler to produce machine code. The C code emitted is nothing more than an intermediate form of the compiled program, with the C compiler serving as the last phase of the compiling process. In general, there is little difference in efficiency between compiled simulators that use the host C compiler to produce machine instructions and those that emit machine instructions directly.

It is worth noting that the relative efficiency of interpreters and compilers is dependent on the level of abstraction of the model being simulated. The lower the level, the smaller the difference. The fastest gate-level simulator, even today, is still an interpreted Verilog simulator.

16.5.3 Cycle-Based Simulation

In the general case, updating a variable value may have a time delay which results in an event being scheduled and then executed by the scheduling loop. In a typical synchronous logic model, every clock cycle could have many intermediate events within the cycle when different variables are updated. In most cases however, the same state behavior will be observed at the clock boundaries if those intermediate events are collapsed to two events, one at the clock edge and one immediately after. This is generally called cycle-based simulation. Since the state behavior is all that is important when verifying logical correctness, this simplification is appealing. The question then becomes, how much faster can you simulate a model using this abstraction?

A number of different cycle-based techniques have been tried. The obvious way to do it was mentioned previously, where every expression in the model is ordered and executed once in every cycle. This is sometimes called ubiquitous cycle-based simulation. The big advantage this has is that there is no overhead of putting events on the future event list or taking them off. In an RTL logic simulation, this overhead can amount to as much as 40% of the execution time. For some designs this static scheduling has proven effective, but the big drawback it has is that if the model has few state variables changing in each cycle, there is a lot of wasted work done. This technique is seldom used for commercial logic simulators now.

However, the ubiquitous cycle-based technique is amenable to hardware acceleration. The scheduling loop is simple enough in that it is relatively easy to program an FPGA-based hardware device to perform the simulation. Products that do this are called emulators, and they can perform logic simulations several orders of magnitude faster than software simulators.

16.5.4 Level of Abstraction and Speed

As noted above, while not limited to these levels, the primary levels of abstraction that logic simulation is concerned with are gate level, RTL, and behavioral level. From a simulation perspective, the main difference between these levels is the number of events that are executed. Between RTL and gate level, the primary abstraction is the width of operands and results. At the RTL, multiple-bit variables, or *vectors* in Verilog terminology, are operated on as a unit. At gate level, vectors are typically split into their individual bits. Thus, at gate level, every logical operation is performed by a hardware element, which requires an event to produce its output. If the timing of the design is important for the simulation, then simulating each physical component is required, since each component may have a different delay.

At the RTL, operations are typically performed as aggregates on their input variables, producing vectors as results. It is easy to see that *and*ing two 16-bit vectors to produce a 16-bit result can be done with one event at RTL, while it will take 16 events at gate level. Since the event routine itself takes about the same amount of time to *and* two 1-bit inputs as two 16-bit inputs, the gate-level simulation will run 16 times slower than the RTL simulation. It is common for the same design to simulate an order of magnitude faster at RTL than at gate level.

Moving up to behavioral level, the primary abstraction is the reduction of clocked events. That is, in behavioral code, variable values are computed without regard to the mechanics of the computation. A computation that may take several cycles in the ultimate hardware will be done in just one event. Synchronization in behavioral code is typically done via communication signals rather than fixed-cycle counts. Consequently, the simulation can be done with significantly fewer events at behavioral level than at RTL. Again, the difference between simulation time for a behavioral model and the same design at the RTL can be an order of magnitude.

16.5.5 Co-Simulation Methods

There are a variety of hardware simulation languages that have been invented and used over the years. In addition to the primary HDLs, Verilog, VHDL, and SystemC, other languages have been created to write testbench code in, most prominently e and Vera. In addition, code written in other more general-purpose languages (C, PERL, etc.) is often incorporated into a simulation model. While these other languages are mainly used for modeling the parts of the system that make up the environment for the hardware design, there is no difference between them and the hardware design as far as the simulator is concerned. An event is an event, whether in the hardware model or the environment model.

Integrating this multiplicity of languages into a single simulation model can be a problem. Integrating a general-purpose programming language with a simulation language is not especially hard, since there is no concept of time in general-purpose languages. Thus, all one really needs to do is allow an escape mechanism so that code written in the general-purpose language can be called in an event routine. Providing an API so that the general-purpose language code can schedule events and get called as an event routine is pretty straightforward. All mainstream logic simulators provide this capability.

Integrating a simulator with another logic simulator is a more difficult problem. Ideally, a logic simulator would be able to understand all the required simulation languages and compile a model written in all of them into a single model. Practically, logic simulators are created to handle one language, and integration with other languages is done at a coarser level of granularity.

The simplest way to integrate models written in two or more languages is to compile the components with separate compilers, and then put them together by coordinating their schedulers. This is called *co-simulation*. That is, each simulator runs as it normally would, keeping its own scheduler and future event list as well as other local data, and synchronizing through the schedulers. This can be done as shown in Figure 16.14. The points marked as <==n==> are points at which the schedulers must synchronize with each other. All schedulers must have finished the preceding section before any can proceed to the next section.

The only remaining difficulty is that propagation may be required between parts of the model that reside in different simulators. That is, a Verilog module could have a variable whose value depends on a variable in

```
do (forever)                                          do (forever)
  t = next t                           <==1==>          t = next t
  while (m = future_event_list(t))                      while (m = future_event_list(t))
    execute event routine m                               execute event routine m
  endwhile                             <==2==>          endwhile
  // update the variable values                         // update the variable values
  while (u = update_list(t))                            while (u = update_list(t))
    u.v = u.t_v                                           u.v = u.t_v
    put_on_propagate_list(c(u.v), t+u.d)                  put_on_propagate_list(c(u.v), t+u.d)
  endwhile                             <==3==>          endwhile
  // propagate changed values                           // propagate changed values
  while (c = propagate_list(t))                         while (c = propagate_list(t))
    c.v = eval(c.expr)                                    c.v = eval(c.expr)
    put_on_propagate_list(c(c.v), t+c.d)                  put_on_propagate_list(c(c.v), t+c.d)
  endwhile                             <==4==>          endwhile
enddo                                                 enddo
```

FIGURE 16.14 Synchronization points in co-simulation.

a VHDL module. It is easy to see how to do this — just provide a way for the VHDL model to propagate a value onto the Verilog model's propagate list. Actually doing it in a concise and automatic way is more problematic, and this problem has been solved in a variety of ways through a combination of language features (to identify a variable as external) and an API. Sometimes, this ability is simply restricted to module ports.

16.5.6 Single-Kernel Simulators

Co-simulation with multiple schedulers works, but the coarse granularity of the synchronization can impose a substantial overhead if there are many propagation events across the simulator boundary. In general, the API calls between simulators involve a fair amount of overhead as each one must establish its environment on every call. It would be more efficient if the model components written in all the languages could use the same scheduler. This is the approach taken by the so-called single-kernel simulators.

In essence, simulators that handle multiple languages with a single scheduler have separate compilers for each language but they all use the same scheduler. That avoids the need for synchronization between different schedulers, and can save a substantial amount of overhead. There are commercial single-kernel simulators for Verilog and VHDL, and Verilog, VHDL, and SystemC.

16.6 Impact of HVLs on Simulation

Hardware verification languages (HVLs) came into vogue as special-purpose languages for writing testbenches. They provide convenient means of generating stimulus for a hardware model, and also provide useful abstractions for modeling the hardware model's environment. For instance, it is generally easier to write a communication protocol that provides correct, variable input in an HVL than it is to write the same thing in an HDL. From the simulator's point of view, the mechanics of dealing with an HVL are the same as co-simulation. Indeed, an HVL is just another flavor of simulation language. Simulators that integrate an HVL and an HDL into a single scheduler are becoming more common.

16.7 Summary

In this chapter, we have covered many of the details of digital logic simulation. Logic simulation is simply a special case of discrete event simulation, which has a long history in general system modeling. The speed of simulation is proportional to the level of detail in the simulation model, which in turn is determined by the level of abstraction at which the model is expressed. In logic simulation, the three common levels of abstraction are gate level, RTL, and behavioral level. There are techniques like cycle simulation used to speed up simulation, as well as techniques like co-simulation used to improve the hardware modeling capability. Ultimately, logic simulation is the most general technique available to verify that a hardware design does what it is intended to do.

References

[1] A. Aho, R. Sethi, and J. Ullman, *Compilers Principles, Techniques, and Tools*, Addison-Wesley, Reading, 1986.
[2] P. Ashenden, *The Designer's Guide to VHDL*, Morgan Kaufmann Publishers, San Francisco, CA, 1995.
[3] G.M. Birtwistle, O.-J. Dahl, B. Myhrhaug, and K. Nygaard, *Simula Begin*, Auerbach Publishers, Philadelphia, PA, 1973.
[4] T. Grötker, S. Liao, G. Martin, and S. Swan, *System Design with SystemC*, Kluwer Academic Publishers, Dordrecht, 2002.
[5] M.H. McDougall, Computer system simulation: an introduction. *Comput. Surv.*, 2, 1970, pp. 191–209.
[6] S. Palnitkar, *Verilog HDL, A Guide to Digital Design and Synthesis*, 2nd ed., Sunsoft Press, Mountain View, CA, 2003.
[7] D.E. Thomas and P. Moorby, *The Verilog Hardware Description Language*, Kluwer Academic Publishers, Dordrecht, 1991.

17

Using Transactional-Level Models in an SoC Design Flow

Alain Clouard
STMicroelectronics
Crolles, France

Frank Ghenassia
STMicroelectronics.
Crolles, France

Laurent Maillet-Contoz
STMicroelectronics.
Crolles, France

Jean-Philippe Strassen
STMicroelectronics
Crolles, France

17.1 Introduction

Multimillion gate circuits currently under design with the latest CMOS technologies include not only hardwired functionalities but also embedded software, running most often on more than a single processor. This complexity is driving up the need for extensions to the traditional register-transfer level (RTL) to layout design and verification flow. Indeed, these circuits are complete systems: system-on-chip (SoC).

Systems-on-chip, as the name implies, are complete systems composed of processors, busses, hardware accelerators, I/O peripherals, analog/RF devices, memories, and embedded software. Less than a decade ago, these components were assembled on boards; nowadays, they can be embedded into a single circuit.

This skyrocketing complexity results in two major consequences: (1) mandatory reuse of many existing Intellectual Property (IPs) to avoid redesigning the entire chip from scratch for every new generation; and (2) employment of embedded software to provide major parts of the expected functionality of the chip. Allowing software development to start very early in the development cycle is therefore of paramount importance in reducing time-to-market. Meanwhile, real-time requirements are key parameters of the specifications, especially in the application domains targeted by STMicroelectronics (e.g., automotive, multimedia, telecom). It is therefore equally important to be able to analyze the expected real-time behavior of a defined SoC architecture. Another crucial issue is the functional verification of IPs that compose the system, as well as verification of their integration. The design flow must support an efficient verification process to reduce development time and to avoid silicon re-spins that could jeopardize the return on investment of the product under design. We will also describe how the proposed approach strengthens the verification process.

At STMicroelectronics, one approach that addresses these two issues is to extending the CAD solution made available to product divisions, Unicad, beyond the RTL entry point; this extension is referred to as the system-to-RTL flow. As the current ASIC flow mainly relies on three implementation views of a design: layout, gate, and RTL levels; the extended flow adds two new views: transaction-level model (TLM) and algorithmic.

In the remainder of this chapter, we first introduce the system-to-RTL design and verification flow with its objectives and benefits. We then focus on the TLM view with a description of the modeling style and also discuss the application of these models to support critical activities of platform-based design: early embedded software development, early architecture analysis, and providing a reference model for the functional verification. Finally, we present how we used this approach in the design of a multimedia multiprocessor platform.

17.2 Related Work

Transaction-level model has become a topic of much greater interest in the industry for the last couple of years, since the delivery of SystemC 2.0 [1]. It has rapidly raised hopes of delivering the long-awaited benefits of a system-design approach for industrial designs.

There are various possible "system-level" abstraction levels of a hardware and software SoC, ranging from paper specification of the functionality before hardware/software partitioning, down to synthesizable RTL of the complete hardware and source code of embedded software (in C typically) [2,3]. Transaction-level model is an intermediate level between these two extremes, and is most often based on C++ — especially using SystemC as base set of C++ classes [4].

Even the most complex SoC platforms can be modeled using a TLM approach [5,6]. Actually there have been various abstraction levels referred to as TLM. We will describe them, following an ordering from less abstract to most abstract. Some tool frameworks encompass several of these levels for mixed-level simulation [7]. The following is a bottom-up description for easy understanding by hardware-oriented readers. However, a real project flow would be essentially top-down, including some reuse.

17.2.1 BCA or CA-TLM

Transaction-based modeling is sometimes understood as only using a C++ variable to contain the value represented by a bus signal for a word with a given bus width, at a given bus cycle. Such modeling of transfers between IPs in an SoC may be called cycle-accurate TLM (CA-TLM) but is more suitably called bus cycle-accurate (BCA) modeling, and is very close to RTL modeling [8]. It requires the IP behaviors to interface with the bus through a bus interface that includes protocol finite-state machines (FSMs) that are difficult to code, debug, and maintain consent with an evolving design of synthesizable VHDL/Verilog RTL.

17.2.2 Protocol-Phase-Aware TLM

A slightly higher level of abstraction, yet sometimes referred to as TLM, is modeling only the phase, but not the cycles of a bus protocol [9]. A transaction is composed of several subtransactions, e.g., bus request, bus grant, transfer single word, or burst transfer. However, the data transferred are structured according to bus width and bus protocol. In such modeling, the modeling engineer must, as in the BCA approach, select a given protocol, understand it, and manage interfaces between his IP behaviors and the bus protocol.

17.2.3 Our TLM and PV

Our definition of TLM, as described later in the chapter, raises the level of abstraction compared to BCA or protocol-phase-aware TLM modeling. Yet it is a model of the SoC hardware architecture after hardware/software partitioning. In our TLM, the data of a transaction can be any structure of words and bits (e.g., as little granularity as a bus half-word transferred between two peripheral registers, or much larger granularity — a full VGA image transferred between two memory buffers) that is meaningful as a transfer between two IPs, at a time determined by an event from the circuit (system) specification. IP models can be pure C/C++ along with register-accurate interfaces accessed using a simple TLM transport application programming interface (API). This level of TLM modeling has proven to be extremely useful to create reference models for functional verification of hardware, and also for SoC simulation platforms for embedded software developers.

Optionally, the transaction definition of our TLM definition can be refined into a structure that is bus-protocol aware (bus width, burst capability, lock, etc.), but this is optional and useful only for SoC architects for fine analysis of arbitrations and other characteristics of the SoC interconnect.

Our TLM definition is slightly different from the one described in [10] because we define an abstraction that models an SoC architecture with *no requirement for* timing [11]. To our knowledge, this abstraction level is unique, and is the foundation for the numerous benefits of our methodology. Timing remains, however, a vital part of the system specification; and a timing-complementary view of TLM is widely used for timing-behavior analysis as described in [12–15]. We also describe how timing can be managed in our methodology without compromising the untimed TLM models.

17.2.4 Higher-Level Models

Our definition of TLM, being register-accurate for the IP interfaces but not necessarily protocol-accurate for the transfers between IPs, is an intermediate between the protocol-phase aware TLM on the one hand, and more abstract models on the other.

More abstract models, e.g., untimed functional models [16], have some high-level architectural features such as queues but no registers and no address map, preventing execution of the embedded software. There exist variants such as timed functional models [16,17], abstract virtual component (VC) models [18], or timing extensions of multithread graphs [19], which may be useful for architects but are not ready for use by embedded software developers. Even higher abstraction models, including for example, communicating sequential processes that only have links between behaviors without any hardware architectural concept [20], may be useful.

17.2.5 TLM/PV — At the Heart of System-to-RTL Flow

From our experience, the TLM/PV level is the most useful one in which to invest modeling effort. This is because it can be created either by hardware-dependent software developers, SoC architects, SOC RTL integration engineers, or hardware functional verification engineers with C++ knowledge. Then TLM/PV models can be used by any or all of these categories of hardware and software engineers, with or without the timed extension to the base TLM model.

For instance, let us describe how TLM/PV addresses functional verification. Most efforts in this domain, as described in [21–23], have been focused on the measurement of test coverage. Our TLM practice provides a complementary benefit enabling the automatic generation of expected results that can then be compared with the output generated by the design under test. A similar approach is in embedding an HDL cycle-accurate (CA) reference model into the test-bench so that outputs can be compared cycle by cycle. Our approach is different because we use an untimed reference model that leads to great savings in the model development and maintenance. For functional verification, a similar technique using an untimed model is sometimes applied for processor verification [24]. We generalize this approach for any digital hardware IP.

17.2.6 TLM Standardization and Tools

Our TLM methodology and associated SystemC-based classes have been submitted in a joint proposal in May 2003 to the TLM working group of the Open SystemC Initiative (OSCI, www.systemc.org), and the OSCI standardization work has been using it intensively. The OSCI standardization process with various partners in the OSCI TLM WG also benefited from inputs by the OCP-IP TLM WG [25]. Various papers reflect the work done in the OSCI TLM WG [2,9].

Currently, various CAD vendors have started offering TLM support in their tools and libraries, either with an untimed, system event-based approach , which we promote as our main modeling style , or complementing it with protocol-aware APIs [9].

17.2.7 Conclusion on Related Work

In this chapter, we uniquely address the rationalization of an industrial ASIC/SoC design flow around the real deployment of a new higher abstraction level, i.e., TLM. This abstraction enables the systematic practice of three key activities of an SoC design: embedded software development, architecture analysis, and functional verification.

17.3 Overview of the System-to-RTL Design Flow

Our system-to-RTL flow depends on three abstraction levels (see Figure 17.1):

- The *SoC functional view* models the expected behavior of the circuit without taking into account how it is implemented.
- The *SoC architecture view* (SoC-A or SoC TLM platform) captures all information required to program embedded software on the processors in the circuit.
- The *SoC microarchitecture view* (SoC-MA or SoC RTL platform) captures all information that enables CA simulation of the SoC. Most often, it is modeled at the RTL using VHDL or Verilog. Such models are almost always available because they are used as input for logic synthesis.

The three views have complementary objectives that avoid the conflict between the need for accurate descriptions, which leads to undesirably slow simulation speed and the need for simulating real-time embedded application software.

The *SoC functional view*, as implied by its name, specifies the expected behavior of the circuit perceived by the user. It is an executable specification of the SoC function that is normally composed of algorithmic software. It contains neither architecture nor address mapping information. Performance figures are

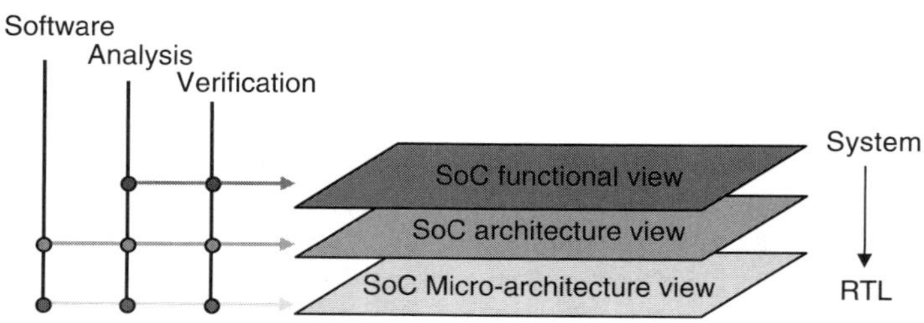

FIGURE 17.1 Overview of the design flow.

specified separately, very often as a paper specification.

Farther down in the flow is the *SoC architecture view* that focuses on:

- *Early firmware development.* It allows the hardware-dependent software to be validated before running on a prototype.
- *SoC architecture exploration.* In the early phase of the design cycle, architects have to define the architecture that will best fit the customer requirements. The early availability of the TLM platform is a means to obtain the quantitative figures needed for determining the appropriate architectural decisions.
- *Functional verification.* It provides a reference model for the verification engineer. This "golden" model is used to generate tests that can be applied to the implementation model in order to verify whether the functionality matches the expected behavior.

The *SoC microarchitecture view* at the end of the system-to-RTL design flow is intended for:

- *Low-level embedded software debug and validation in the real hardware simulation environment.* Our goal is to enable the debugging of device drivers, and their integration into the target operating system before the first breadboard or even the hardware emulator is available.
- *SoC functional verification.* We aim at verifying that the IPs, once integrated, still provide the expected functionality, and that parallel block execution and communication do not corrupt the IP behavior. Certain blocks are processors that run software. The verification activity must take this fact into account and validate all possible interactions between the hardware and software components.
- *SoC microarchitecture validation.* In order to sustain the real-time requirements of the application, the embedded software is usually optimized and the hardware is configured accordingly. In case of insufficient performance, the SoC architecture may be upgraded to match the requirements. Any part requiring cycle accuracy will count on an RTL view.

A discussion of these activities is available [26].

The three abstract views can be integrated gracefully into the SoC development process as explained below. The functionality is independent of the architecture so that development of its corresponding view can be initiated as soon as there is a need for a product specification.

While software and architecture teams are using the same reference—SoC TLM—for firmware development and coarse-grain architecture analysis, the RTL development takes place in parallel, leading to an SoC RTL platform. Hence, concurrent hardware/software engineering is achieved as depicted in Figure 17.2. By the time the SoC RTL platform is available, some tasks closely related to the hardware implementation are ready to start: fine-grain architecture analysis, hardware verification, and low-level software integration with hardware. These tasks are conducted concurrently with the emulation setup, the synthesis, and back-end implementation of the standard ASIC design flow.

Finally, as the first breadboard is ready, the software (drivers, firmware, and a simplified application) will have been completed with good confidence and the SoC will have gone through a thorough verification process. This will certainly increase the probability of achieving first-time silicon and system success.

Note that we might have to mix SoC architecture and microarchitecture models in a single simulation model. Such a mixture of abstraction levels requires a standard mechanism that provides a methodology with an automated assembly tool, i.e., design flow automation (see Section 17.7). This standard automation will not only ensure the compatibility of the different abstraction models within the existing tools, but it also gives us a benefit with a gain in time.

Currently, one difficulty with the proposed approach is that in the absence of synthesis tools that can generate these views from a reference model, coherence with the other two views must be maintained throughout the circuit development. To address this issue, we reuse the functional subset of "system" test vectors across all views, to ensure their conformity with the expected functionality. Alternatively, we could run the same embedded software (without modification) on different simulation environments. Identical behavior proves the compatibility of the underlying models. Obviously, the coherence is ensured up to

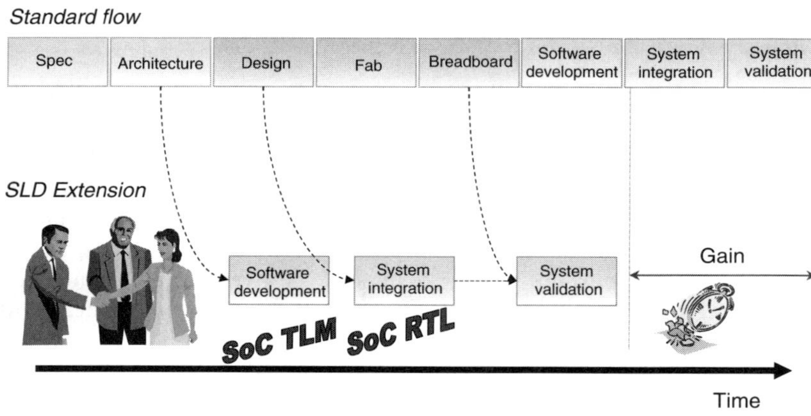

FIGURE 17.2 Concurrent hardware/software engineering.

the level of functional coverage that the tests provide. Also, as mentioned earlier, TLM models may be used as executable specifications of an SoC for RTL verification.

17.4 TLM — A Complementary View for the Design Flow

17.4.1 TLM Modeling Concepts

When addressing the need for early embedded software development and architecture analysis, engineers list the following requirements as barriers for this activity:

- *Speed.* Models used for the activities above involve millions of simulated cycles in certain cases; it is thus unacceptable to wait for even a day to complete a simulation run, especially in the case of interactive sessions.
- *Accuracy.* Certain analysis requires full-CA models to obtain reliable results. The minimum requirement is that the model should be detailed enough to run the related embedded software.
- *Lightweight modeling.* Any modeling activity besides the RTL coding (which is anyway mandatory for silicon) must be simple enough so that the modeling cost is not higher than the resulting benefits to the SoC project management.

Complex designs usually start with the development of an algorithmic model (e.g., based on a dataflow simulation engine for DSP-oriented design). The advantage of an algorithmic model is that it simulates extremely quickly, since it captures the algorithm free of the implementation details. However, such a model has no notion of the hardware and software components; it does not model registers or system synchronizations related to the SoC architecture. Thus, this model cannot fulfill our need of enabling embedded software execution.

With the emerging C-based dialects that support hardware concepts, it is often believed that writing CA models in a C-based environment could meet all the stated requirements. This approach has several major drawbacks [8,25,27–29]:

- Much information captured in such a model is not available in the IP documentation, but is maintained only in the designer's mind ... and in the RTL source code. As a result, the RTL designer has to spend considerable time to update the CA model development, or the modeling engineer is forced to reverse-engineer the RTL code. Either way ends up being tedious and time-consuming, without solving the issue.

- The CA model is merely an order of magnitude faster than the equivalent RTL simulation … very close to the speed of a cycle-based VHDL/Verilog model.

Not only is the speed too low to run a significant amount of embedded software in a reasonable time frame, the development cost is also very high compared to the benefits.

Meanwhile, we have noticed that architects and software engineers do not require cycle accuracy for all of their activities. For instance, software development may not require cycle accuracy until engineers are ready to work on software optimization; earlier in the process, they do not need high simulation speed coupled with cycle accuracy, e.g., verification of cycle-dependent behavior.

Based on this experience and new understanding, we have adopted the "divide and conquer" approach by providing two complementary environments:

- An SoC TLM platform for early use with a relatively lightweight development cost. This abstraction level sits between the CA bit-true model and the untimed algorithmic model. This is, in our opinion, an adequate trade-off between speed and accuracy with a complementary SoC RTL platform (at least 1000 times faster than RTL simulation in our experience, as depicted in Figure 17.3).
- An SoC RTL platform for fine-grained CA simulations at the cost of slower simulation speed and later availability.

Let us now define some terms in order to understand the TLM modeling approach. A system is a mixture of hardware and software that provides functionality with adequate performance to its user. An SoC is a particular system in which both hardware and software are embedded in a single circuit. A TLM system model is composed of a set of *modules* that:

- Execute part of the expected system behavior through one or more concurrent *threads.*
- Communicate data to each other (and, potentially, to a test-bench) to perform the system functionality. Each set of data exchanges is a *transaction.* Two consecutive transactions may transfer data sets of various sizes. The size corresponds to the amount of data being exchanged between two system synchronizations.
- Synchronize with each other (and potentially with the test-bench). System *synchronization* is an explicit action between at least two modules that need to coordinate some behavior distributed over the modules that depends on this synchronization. Such coordination is essential to assure predictable system behavior. A synchronization example is an interrupt raised by a DMA to notify a transfer completion.

The simulation speedup achieved by a TLM model vs. the equivalent RTL model correlates with the ratio of the average number of cycles between two system synchronizations. An explicit system synchronization is compulsory for the TLM model to behave properly because it is the only mechanism available for synchronizing the different processes in the system. Our experience shows that all systems comply with this rule. Figure 17.3 provides approximate ratios in terms of simulation speedup and modeling efforts.

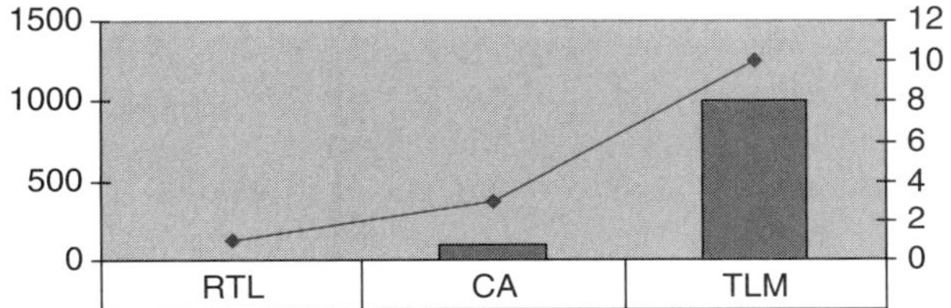

	RTL	CA	TLM
Simulation speed-up vs. RTL	1	100	1000
Modeling speed-up vs. RTL	1	3	10

FIGURE 17.3 Efficiency of modeling strategies.

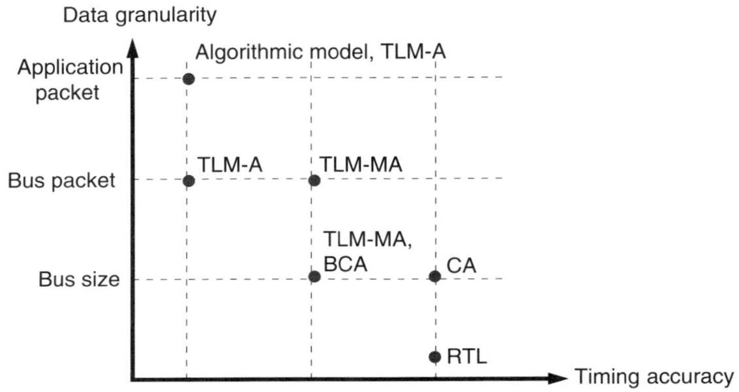

FIGURE 17.4 Modeling accuracy.

Figure 17.4 gives a better picture for the different models that we are going to quote throughout this chapter according to their timing accuracy and data granularity.

All of the models included in Figure 17.4 have bit-true behavior except the algorithmic specification model. TLM-A is the TLM architecture model that gives an untimed SoC architectural view as presented earlier. For some particular purposes, it may be necessary to refine a TLM model to include some information related to microarchitecture. We label this refined, timed model as TLM-MA, i.e., TLM microarchitecture model. To ensure satisfactory efficiency, we provide a method to compile and build the TLM IP source code, without any modification, as a TLM-A or TLM-MA model. In this chapter, unless explicitly stated, the term TLM refers to TLM-A. Further down in abstraction is the BCA model, a timing-accurate model for communication structure and interface. Following that is the CA model. Finally, we have the RTL model, the most precise timed model for a system module in the range shown.

17.4.2 Embedded Software Development

Software development (at least debugging and validation) starts only if it can be executed on the target platform. The traditional approach considers the physical prototype, e.g., an emulator or FPGA board prototype, as the starting point for software development. The advantage is earlier availability before the chip has been fabricated, whereas the drawback is the fact that hardware development is almost complete. As a result, if the software execution uncovers any need to modify the hardware at this stage, it will be be very costly.

Hardware/software coverification could start executing software on the target-simulated hardware earlier. Any small hardware modification is still possible if this technique is employed with the first top-level VHDL/Verilog netlist. The downside is the dependence on VHDL/Verilog models for simulation; this results in a slow execution speed running only a reduced part of the embedded software. Any major hardware modifications will be too costly because the design is way too advanced.

An SoC TLM platform can be delivered shortly after the architecture is specified. As a result, it is available for software development very early in the hardware design phase. A true hardware/software codevelopment is therefore attained. Hardware and software teams cooperate in line with the "contract" between them — the TLM platform. The software developers regard the TLM model as the reference standard on which to run their code while the hardware designers use it as the golden model for their design.

17.4.3 Functional Verification

17.4.3.1 Block-Level Verification

Functional verification assures that the IP implementation is in compliance with the functional specification. Using the paper specification, a verification engineer defines test scenarios that will be applied to the RTL IP model. Frequently, the engineer needs to determine "manually" the expected result of each scenario.

The TLM model is an actual functional specification of an IP. It is also referred to as the executable specification of the IP since it only captures the intended behavior perceived by the user, but excludes any internal implementation details. In other words, it models the architecture of the IP but not its microarchitecture.

Therefore, a TLM model can replace the manual process undertaken by the verification engineer to generate expected results. As demonstrated in Figure 17.5, the test scenario AVP (architecture verification pattern) is applied to the TLM model to generate automatically the expected output (i.e., reference output). A common question is: how can it be ensured that the executable specification, the TLM, is compliant with the written specification? And thus, what is the level of confidence that its resulted output can be used as a reference? There is no answer to this question. Also, the same question can be asked of the output results manually generated by the verification engineer. No warranty exists that can grant that the engineer has not misunderstood the specification or that he has not made a mistake while generating the reference output.

Note that an implicit assumption for TLM IP verification is the existence of a TLM IP model prior to the availability of an RTL model. Legacy IP may not comply with this criterion since it is normally delivered as an RTL model only. In this case, a TLM model needs to be developed based on the RTL reference model. The test scenarios and vectors are extracted and recorded from the RTL test-bench. A TLM test-bench will then load the transaction traces and signal values to drive the TLM IP model developed later. Once the TLM view is validated against its RTL counterpart, it is ready to be integrated into an SoC platform for software development and architecture analysis.

17.4.3.2 SoC System-Level Verification

In a typical design flow, any independently verified IPs must be verified in an SoC environment as well. Functional integration tests are developed to verify the following features:

- *Memory map.* To prevent registers from being mapped at the same addresses and; guarantee that hardware and software teams use the same mapping.
- *Data consistency.* To ensure that data generated by an IP matches the input data (in terms of format and content) of another IP that will reuse the generated data.
- *Concurrency.* To ensure that concurrent execution of all IPs is controlled for deterministic expected behavior.

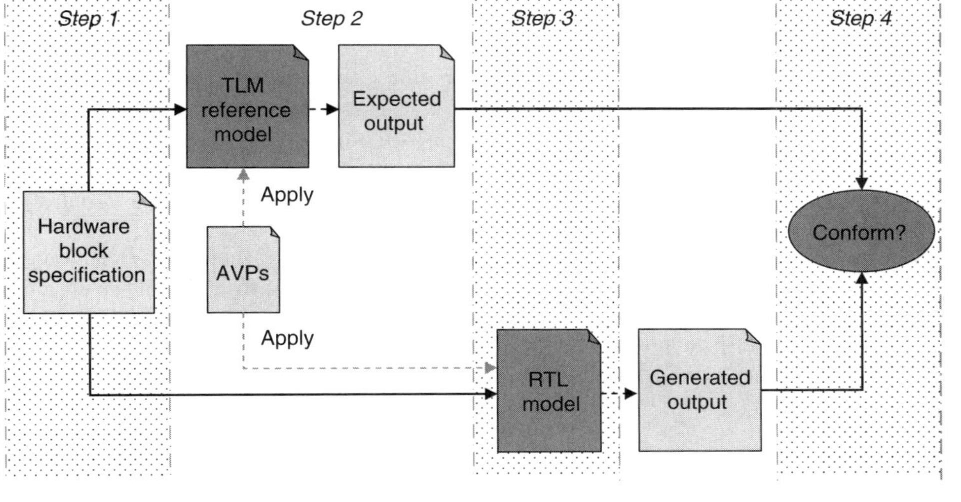

FIGURE 17.5 Functional verification using a TLM model.

The same approach as for block-level verification holds here. The SoC TLM platform is used to generate expected system reference output for a given scenario.

17.4.4 Early System Integration

System integration is the assembly of different system components to create a properly functioning system. A stepwise process with successive partial integrations throughout the development of the hardware and software components is adopted. TLM models participate in such an integration process in the following manner:

- Software is executed on the reference model of the hardware IP for a common understanding of the hardware specification.
- Hardware RTL implementation is checked against the reference model used by software developers.
- Earlier integration of hardware components can take place by mixing TLM and RTL views of IPs instead of waiting for the release of all RTL blocks.
- Software can be executed on a partial RTL platform by using TLM models for IPs whose RTL code is not available yet. It enables earlier hardware/software integration.

Abstraction-level adaptors are required for interfacing TLM and RTL models in this sort of mixed simulation. Thus, reusable bus functional models are developed for different communication protocols such as STBus or AMBA.

The process of system integration could consume considerable effort since it involves diverse parts from different points of a given design. Front-end design flow automation could be very helpful in reducing this effort (see Section 17.7).

17.4.5 Performance Analysis and Verification

A system must be carefully controlled such that its circuit respects the real-time constraints stated in the initial system definition and specification. Every design flow step is verified against this condition. A system performance analysis and verification will achieve this purpose. The system performance can be studied once the timing information is included. The following two cases may arise:

- *Timed TLM-A model.* It may be sufficient to add delays to execution paths of a TLM-A model to mimic its timing behavior. The delay computation may vary from a fixed value to a complex function whose returned value varies over time and depends on information collected at run-time. This technique is illustrated in Section 17.6.
- *Timed TLM-MA model.* A simple time annotation on a TLM-A model might result in unacceptable inaccuracies, particularly for complex IPs. It is therefore indispensable to refine the model into a TLM-MA model and then sum up the timing annotations. It is often necessary to refine an architectural IP model if it represents a shared resource. An arbitration policy must replace the simple first-in first-out policy* of the TLM-A model. Typical models of this category are buses and memory controllers.

Both cases model the timing delays with calls to the SystemC "wait" statement. Depending on the design phase, timing information is either obtained from the architect or from the analysis of the equivalent RTL simulation. For a true top–down flow, the architect, based on some preliminary benchmarking experiments, will initially insert the timing information. Once the RTL is available, timing information will be extracted from the RTL simulation for annotation into the equivalent TLM model, i.e., back-annotation [30].

* Such a policy is optimized for simulation speed because it removes the need for a delta cycle.

17.5 TLM Modeling Application Programming Interface

17.5.1 Modeling Environment

As described in depth in the literature [31], our modeling environment should support modularity, concurrency, hierarchy, separation of behavior from communication, and hardware and algorithmic data types. After evaluation of C/POSIX and synchronous language environments, which had some drawbacks with regard to our specific TLM goal, we evaluated and selected SystemC [1,4] — a language that complies with most of these requirements. In addition, as it is becoming a de facto industry standard, SystemC is supported by most CAD tools, hence facilitating a smooth transition to successive steps of the design flow (see Section 17.8).

Each module of an SoC TLM platform is characterized by:

- Bit-true behavior of the block
- Register-accurate block interface
- System synchronizations managed by the block

For the integration of modules into a platform, SystemC does not offer the inherent ability to perform transactions among several modules connected to a single channel based on a destination identifier (e.g., address map in case of bus-based architectures). To address this, we created a user-defined interface along with the associated bus channels (see the next section).

The computation model derived from the previous TLM definition implies that all system synchronizations are potentially rescheduling points in a simulation environment for accurate concurrent modeling. System synchronizations could be modeled by fixed means (e.g., event, signal) or data exchanges (e.g., polling). The simulated system will behave correctly only if all these potential system synchronizations cause a call to the simulation kernel, thus enabling its scheduler to activate other modules. SystemC events and signals comply with this requirement. The simulator is however unable to distinguish data exchanges of transactional models from implicit synchronizations. A conservative approach must therefore be adopted to allow the kernel to reschedule on each transaction completion.

Once the TLM platform is integrated, embedded software can be executed either as a native compilation on the simulation host for higher simulation speed, or as a cross-compilation on the embedded processor architecture for an instruction set simulator (ISS) execution.

17.5.2 Modeling Application Programming Interface

This section discusses the definition of an API for IP models to support data exchange among the different IPs. The general characteristics follow:

- *IP-A model reuse.* Reuse an IP-A model in an SoC-A model for simulation without any code modification, regardless of its microarchitectural implementation.
- *Port.* Enable emission and reception of data to/from communication structures.
- *Communication/behavior separation.* Separate IP behavior from the communication mechanism such that various communication schemes could be tried with the same IP-A model.
- *Information locality.* Information should be captured locally by the model that owns it.
- *Simulation efficiency.* The API should not result in significant simulation degradation.
- *Intuitive API.* Minimize learning curve and potential misuse through an intuitive API.
- *Abstraction and protocol adaptors.* Minimize the number of these models that must be developed.
- *Automation.* Should automate activities such as monitoring, adaptor insertion, code generation, and consistency check at minimum tool development cost.

There are several important differences between IP-A and IP-MA models. First, the granularity of the data sent and received by an IP-A model corresponds to the model function, while IP-MA models exchange data as managed by the communication layer. Second, an IP-A model only expects to read and write data in the communication control layer whereas an IP-MA model expects to control the handshaking mechanism required by a data transfer. Third, no timing information is necessary for an IP-A model since deterministic execution relies on a defined order of system event occurrences. In contrast, an IP-MA model captures details of data exchange and the internal IP structure to provide an accurate timing annotation.

17.5.2.1 TLM API: TLM Transport

The TLM API enables the transport of transactions among the IPs for information exchange. The information conveyed by a transaction includes the initiator dataset, target dataset, initiator access port, and target access port. A corresponding transaction structure follows:

```
class tlm_transaction        {
      tlm_port *m_initiator_port;       //Initiator port pointer
      tlm_port *m_target_port;          //Target port pointer
      tlm_transaction_data *m_data;     //Transaction data pointer
      tlm_metadata *m_metadata;         //Metadata pointer
      ...                          };
```

Transactions are exchanged among the different IPs via the ports. Such a transaction is atomic, and the IP will resume its execution when the transaction is completed. The TLM interface is defined as a subclass of `sc_interface`. Then, this TLM interface defines a function named `transport` that will be called to serve a transaction. A C++ function prototype looks like:

```
tlm_status transport(tlm_transaction &transaction);
```

where the parameter `transaction` is a reference to the transaction structure, and the return value provides transaction status information.

This API is considered as a generic transport layer for TLM. The actual content of the transaction is defined as another layer on top of it. In other words, a protocol is defined based on this generic API wherein the actual data type being transferred is specified. Such specific data are derived from the empty class, `tlm_transaction_data`. Subsequently, IP models can be developed based on this protocol. This idea is demonstrated in Figure 17.6.

17.5.2.2 TLM API: TLM Protocol

Different protocols could be defined based on the transport API to reflect the rules required for data exchange. Structures corresponding to the following TLM protocols are available as examples:

- *TAC* for supporting the interconnection of IP-A models. A transaction is either a read or a write operation.
- *AMBA* for support of IP-MA models based on the AMBA protocol. Single transfers, bursts, and locked transfers are supported.
- *STBus* for support of IP-MA models based on the STBus protocol. Request/acknowledge and separation between requests and responses are supported.
- *OCP* for support of IP-MA models based on the OCP protocol. Features are very similar to the STBus.

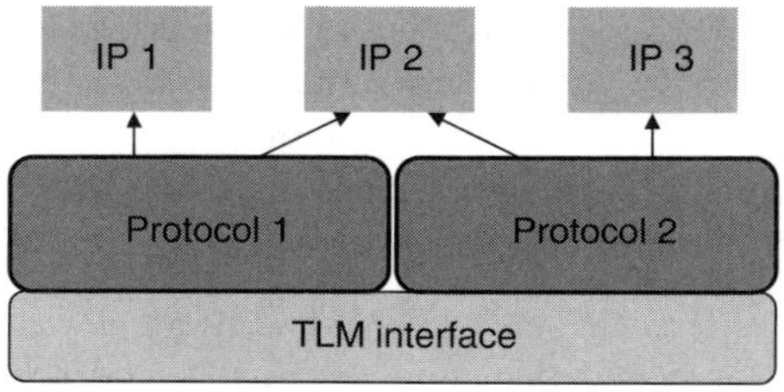

FIGURE 17.6 TLM layers.

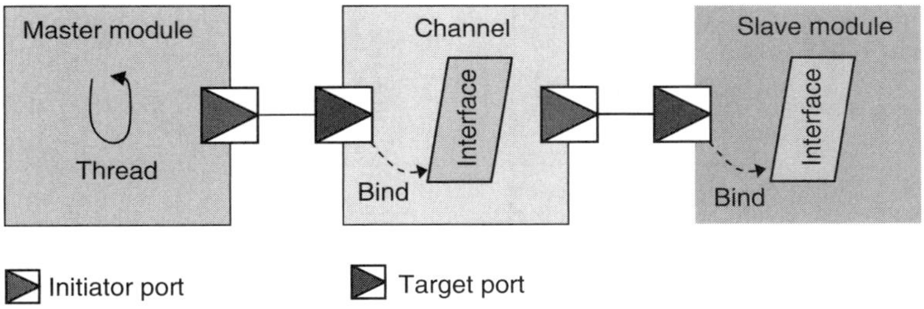

FIGURE 17.7 TLM modeling API.

Figure 17.7 illustrates the foundation of our TLM modeling API. Communication protocols that require more than one exchange to transfer data can be decomposed into multiway transfers. The request is initiated when the port is accessed, and acknowledged when the `transport` function returns on the initiator side. An identical scheme holds for the response. Such a decoupling between requests and responses enables separate pipelining of requests and responses.

17.6 Example of a Multimedia Platform

We present here a model of an MPEG4 coder–decoder (CODEC) based on our TLM TAC protocol. An SoC architecture model is indispensable for a CODEC because an efficient simulation model is required for embedded software development and interactive debugging.

17.6.1 System Model

The MPEG4 platform employs a distributed multiprocessing architecture with two internal buses and four application-specific processors (ASIP), on which are distributed the embedded software, a general-purpose host processor managing the application-level control, and seven hardware blocks. Consequently, there is much software parallelism with explicit synchronization elements in the software. Each processing block is dedicated to a particular part of the CODEC algorithm.

The hardware and software are partitioned according to the complexity and flexibility needed by the CODEC algorithm. All operators work at the macroblock level (the video unit for the CODEC). Descriptions of some major system blocks follow (see Figure 17.8 for the platform block diagram).

Multisequencer (MSQ). Multisequencer is an RISC processor for video pipeline management. It is mainly firmware with a hardware scheduler for fast context switches and process management.

Multichannel controller (MCC). Multichannel controller is a hardware with microprogrammed DMA. It arbitrates all the present requests from the operators and performs I/O with external memory. Scheduling is done for each request where a request is a memory burst with variable size. An I/O request is considered as a single transaction even if it may take several cycles to execute.

VLIW Image predictor (VIP). VLIW image predictor is a mix of hardware and firmware for performing motion compensation. Control is done by firmware while processing is done by hardware with a special VLIW instruction set. The instruction-set is modeled.

Encoder (COD). Encoder is a pure hardware. It computes the difference between the acquired and predicted image, DCT, zigzag, quantization and run/level coding.

In terms of system behavior, the external processor posts commands to the mailbox, HIF. The MSQ (programmed in C) takes the command into consideration and is in charge of the internal control of the CODEC. Consequently, the MSQ activates the different internal hardware or programmable blocks to achieve the coding or decoding of the video flow by reading status registers and writing command registers for each block. All the operators are pipelined and communicate with the system memory through a

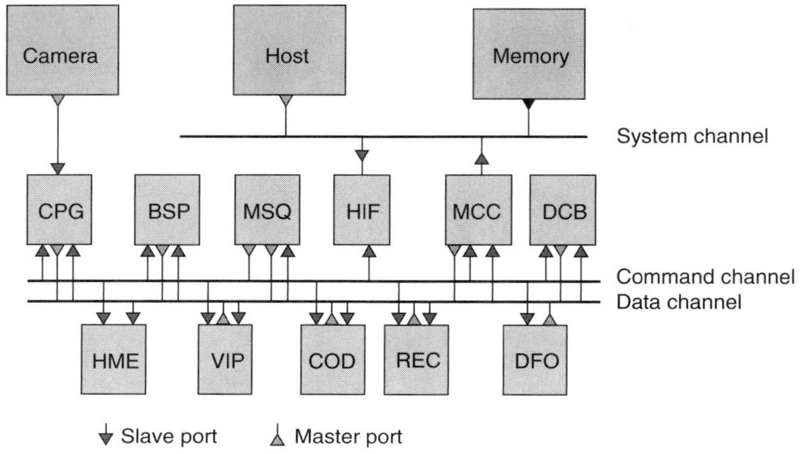

FIGURE 17.8 The multimedia platform block diagram.

memory controller. This memory controller receives requests from the internal operators and generates transactions on the top-level bus to access the system memory. It communicates with the internal operators through well-identified input/output registers that contain the values to be stored into and loaded from the memory.

The two internal channels on the platform, command and data, serve different purposes. For those modules that need to give access to their control register, their ports are bound to the command channel, whereas for those modules that handle data communication, their ports are bound to the data channel. The MSQ is a particular case since it has a double role: it generates read/write operations to control the system behavior, and hence its command master port is connected to the command channel; it also initiates transactions to the memory, and hence its communication port is bound to the data channel.

17.6.2 Design Choices

The requirement of being able to simulate the embedded software prevents us from developing a pure functional and sequential circuit model. Each computing block is modeled as a SystemC module with its own processes and associated synchronization elements.

17.6.2.1 Programmable Modules

This multiprocessor design is composed of C-programmed modules— pure hardware blocks as well as mixed blocks that encompass hardware and programmable operators. There are three categories with specific modeling strategies as stated below.

Software block. We rely on the sequential aspect of the codes for MSQ, BSP, HME, and VIP. All firmware is written in C. This enables the code to be natively compiled on the workstation and to communicate with the TLM model through an I/O library (via simple stubs to call C++ from C). The specific built-in instructions of the processors are modeled as C functions. One part of the C model is used directly for ROM code generation using retargetable C cross-compilers. An ISS is integrated in the environment to run the cross-compiled application software on the SoC host. Once the software is written, it remains unchanged despite its environment: TLM or RTL simulations, emulation, or application board.

Hardware block. A high-level model is written in SystemC. It is a functional, bit-true model with the memory transactional representation. We do not characterize the internal block structure. Instead, we model the input/output of the block, the synchronization, and the internal computation at the functional level. The complete SystemC model is used as a reference for RTL validation.

Let us consider an example. The coder block (COD) is a hardware pipeline with five operations: Delta, DCT, Zigzag, Quantization, and Run/Level. A FIFO is inserted among all these operations; the computation, controlled by an FSM, is pipelined; and a DMA manages the inputs and outputs. The RTL model is fully representative of this architecture. Developing such a model requires at least 3-person-months effort for a senior designer. In contrast, the corresponding transactional model consists of the input data acquisition (acquired and predicted blocks in our example), computes results with a C function and the resulting output. The gain is twofold: (1) easier and faster modeling; and (2) greater simulation speed. The corresponding effort is only about 1-person-week.

Mixed hardware/Software block. Mixed hardware/software blocks are modeled as a mix of the two previous categories.

17.6.3 System Integration Strategy

Some features of the modeling environment applied to the MPEG4 platform are illustrated in this section. We will focus our attention only on the fundamental aspects of our approach.

System synchronization in the platform is twofold. First, the MSQ is in charge of the global control of the platform. It is responsible for activating the coding/decoding tasks according to the current system status. A task will then be executed only if the relevant data are available; otherwise it will be suspended. Several tasks may be activated at the same time due to the internal operation pipeline. Synchronization is achieved by writing or reading the command and status registers of the different operators. Second, data exchanges between the operators and the memory are blocking operations. The platform synchronization scheme ensures that an operator will resume its computation only when the previous transaction is completed from the system point of view.

We model the data exchange with arbitrary sizes in the platform with respect to the system semantics of the exchange. For instance, transactions between the camera and grabber are line-per-line image transfer while transactions between MSQ and other operators are 32-bits wide.

17.6.4 Experimental Results — Simulation Figures

We compare the performance figures obtained for the MPEG4 platform in terms of code size and execution speed on different complementary environments as shown in Figure 17.9.

The modeling choices of the IP bring a significant gain in terms of model size: TLM models are 10 times smaller than RTL. As a result, they are easier to write and faster to simulate. The TLM simulation speed is 1400 times faster than an RTL model on a SUN Ultra 10 workstation of 330 MHz, 256 MB memory.

The RTL model simulates a coded image in an hour while the TLM model in SystemC only requires 2.5 sec per coded image. Before the Accellera SCE-MI interface [32] is available on the Mentor Celaro hardware emulator, an ad-hoc co-emulation transactional interface has been implemented through the C API; it provides both controllability and observability of the emulated design (clock control, memory and register load/dump, and programmable built-in logic analyzer). On the hardware side, synthesizable models have been developed for the camera, the memory, and several RTL transactors [32], that translate a data packet from a transaction into a BCA data exchange. On the software side, the host TLM model has been extended for HW debug and performance evaluation purposes. The coemulation requires 35 sec per coded image, i.e., a system clock at about 40–60 kHz. This speed (more than 30 times faster than a cycle-based coemulation) enables us to run the software developed onto the SoC TLM platform without any external or synthesizable CPU core being required for the host modeling.

The performances obtained for the TLM environment are well suited for embedded software development, since it runs a significant test-bench of 50 images in a couple of minutes on an Ultra 10 with full source level debugging facilities.

17.7 Design Flow Automation

Systems-on-chips are composed of a large number of IP blocks. They are interconnected with complex communication structures such as layered buses and crossbars. In addition, as we have discussed earlier,

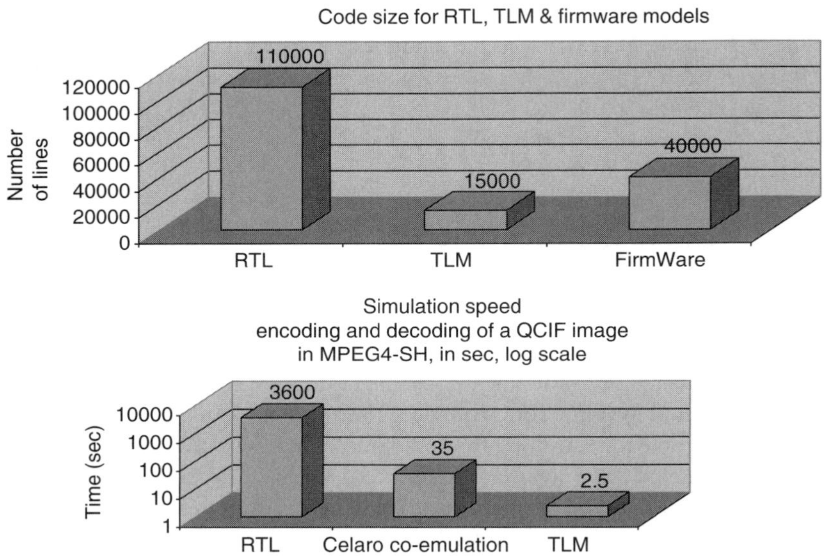

FIGURE 17.9 Performance figures of the MPEG4 platform.

different abstraction models are frequently mixed in a single simulation platform. The creation of such hardware platforms is a tedious and error-prone procedure with many repetitive tasks. In addition, low-level embedded software is extremely dependent on the hardware configuration.

Because of this complexity, design flow automation is a vital approach to reduce the time-to-market of an SoC product. We propose a platform automation tool that automates all the repetitive tasks in creating an SoC platform while leaving room for users to intervene manually in the procedure wherever necessary. This section discusses the idea of such a tool that aims to support system-to-RTL activities, and relies on the emerging SPIRIT† standard [33].

17.7.1 Overview of Design Flow Cockpits

The design and verification flow for an SoC encompasses the traditional ASIC flow as well as other activities related to the embedded software development and (micro-) architecture definition.

The design flow is split into two main categories:

- *Front-end cockpit.* Technology-independent tools and methods such as embedded software, architecture definition, and functional verification.
- *Back-end cockpit.* Technology-dependent tools and methods requiring information on the target technology. This includes logic synthesis as well as place-and-route activities.

The distinction made above does not, however, suggest that technology-independent tools would not rely on any implementation-dependent information (thus being technology-dependent). It simply means that there are far fewer dependencies. These dependencies must be identified and the related information must be exchanged between the front-end and back-end cockpits.

An obvious example is the clock frequency. Front-end activities depend on a target clock frequency to run CA (RTL or BCA) simulations. This frequency will remain an assumption until the design has gone through the physical implementation. Front-end activities that rely on this number should then be validated with the final number provided by tools of the back-end cockpit. Figure 17.10 illustrates the front-end and back-end cockpits for an SoC design flow.

† Structure for Packaging, Integrating and Re-using IP within Tool-flows.

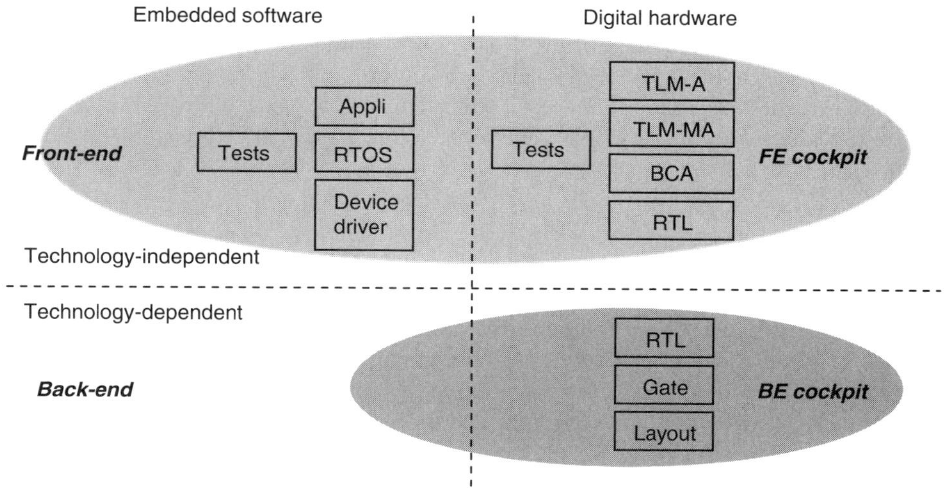

FIGURE 17.10 Design flow cockpits.

17.7.2 Front-End Cockpit

The front-end cockpit automates activities related to:

- embedded software development
- architecture definition and validation
- IP and SoC functional verification

The automation process relies on the availability of reusable IPs, for both their hardware, and associated software such as drivers. The meta-description of these IPs based on the SPIRIT standard will enable the automation tool to automate their usage in the design flow.

The main expected features of the automation tool are:

- Generation of SoC models at various levels of abstraction, namely TLM, BCA, and RTL, out of a single description of an SoC architecture. An editor, for instance a schematic editor, can be provided to edit the architecture.
- Assembly of all low-level embedded software that is necessary to boot the platform and initialize the peripherals.
- Assembly of specified software. For example, diagnostic routines can be assembled and executed to verify whether the generated platform behaves as expected.

Getting an IP meta-description ready for such an automation tool entails a repetititive effort. An industry standard is thus being developed through the SPIRIT Consortium to define a standard mechanism for describing and handling IP. This standard enables automated IP configuration and integration.

Within the SPIRIT-defined Design Environment, an IP meta-description or meta-data is written as an XML file. It could then be imported or exported by any SPIRIT-compliant CAD tool — that is, the design flow automation tool. The automation tool will configure an IP based on its XML file and subsequently generate a design by assembling the different IPs. This tool also has the ability to invoke any potential plug-in required by the configuration and generation of the IPs.

17.8 Conclusion

As highlighted in this chapter, we have introduced a new abstraction level in our SoC design flow to address the needs of early embedded software development, architecture analysis, and functional verification.

This level captures the architecture of the IP (or SoC), and complements the RTL level that models the microarchitecture. The combined usage of the two views avoids the usual conflict between the need for accurate descriptions and the request to simulate the real-time embedded application software at high speed. This flow extension has successfully been applied to the SoC design and verification activities in an industrial context where there are very large teams with limited manpower dedicated to the introduction of new techniques.

Acknowledgments

This chapter is an overview of the system-level design flow developed at STMicroelectronics. We earnestly appreciate the contributions of all our colleagues to the developments described in this chapter. Our genuine gratitude goes to Philippe Magarshack for his continuous support and encouragement throughout the development of the flow. We thank our many colleagues who have contributed to transaction-level model (TLM) deployment in STMicroelectronics, and especially Kshitiz Jain and Rohit Jindal for their specific additions to the methodology. We also gratefully acknowledge the professional contribution of Jim-Moi Wong for her active review and improvement of the chapter.

References

[1] Open SystemC Initiative, SystemC 2.0.1 Language Reference Manual, www.system.org

[2] A. Donlin, Transaction level modeling: flows and use models, *CODES+ISSS 2004*, Stockholm, Sweden, September 2004. pp. 75–80.

[3] A. Donlin, A. Braun, and A. Rose, SystemC for the design and modeling of programmable systems, *FPL 2004*, Antwerpen, Belgium, August 2004.

[4] SystemC 2.0 Functional Specification, Open SystemC Initiative 2002, http://www.systemc.org

[5] P.G. Paulin, C. Pilkington, and E. Bensoudane, StepNP: a system-level exploration platform for network processors, *IEEE Design Test Comput.*, 19, No. 6, Nov.–Dec. 2002, pp. 17–26.

[6] P.G. Paulin, C. Pilkington, M. Langevin, E. Bensoudane, and G. Nicolescu, Parallel programming models for a multi-processor SoC platform applied to high-speed traffic management, *Proceedings of CODES 2004*, Stockholm, Sweden, September 2004, pp. 48–53.

[7] G. Nicolescu, S. Yoo, and A. Jerraya, Mixed-Level cosimulation for fine gradual refinement of communication in SoC design, *Proceedings of the Design, Automation and Test in Europe (DATE) Conference*, Munich, Germany, March 2001, pp. 754–759.

[8] T. Groetker, Reducing the SoC Design Cycle: Transaction-level SoC Verification with SystemC , 2002, http://www.synopsys.com

[9] B. Vanthournout, S. Goossens, and T. Kogel, Developing Transaction-Level Models in SystemC, Coware Inc. White Paper, September 2004.

[10] L. Cai and D. Gajski, Transaction level modeling: an overview, *CODES+ISSS'03*, Newport Beach, California, 19–24, 2003.

[11] F. Ghenassia et al., Using transaction-level models in a SoC design flow, in *SystemC: Methodologies and Applications*, W. Müller, W. Rosenstiel, and J. Ruf, Eds., Kluwer Academic Publishers, Dordrecht, 2003, pp. 29–63.

[12] F. Carbognani, C.K. Lennard, C. Norris Ip, A. Cochrane, and P. Bates, Qualifying precision of abstract SystemC models using the SystemC verification standard, *Proceedings of the Design, Automation and Test in Europe Conference and Exhibition (DATE'03)*, Munich, Germany, 2003, pp. 88–94.

[13] P. Martinelli, A. Wellig, and J. Zory, Transaction-level prototyping of a UMTS outer-modem for system-on-chip validation and architecture exploration, *Proceedings of the 15th IEEE International Workshop on Rapid System Prototyping (RSP'04)*, Geneva, Switzerland, 2004, pp. 193–200.

[14] P. Hardee, Transaction-level Modeling and the ConvergenSC™ Products, Version 0.1, http://www.coware.com

[15] S. Pasricha, N. Dutt, and M. Ben-Romdhane, Extending the TLM approach for fast communication architecture exploration, *Proceedings of Design Automation Conference (DAC'04)*, San Diego, California, 2004, pp. 113–118.

[16] W. Rosenstiel, and J. Ruf, Y. Vanderperren, M. Pauwels, W. Dehaene, A. Berna, and F. Ozdemir, A SystemC based system on chip modelling, and Design methodology. Chapter 1, *SystemC: Methodologies and Applications*, W. Müller, W. Rosenstiel, and J. Ruf, Eds., Kluwer Academic Publishers, Dordrecht, 2003.

[17] M. Baleani, A. Ferrari, A. Sangiovanni-Vincentelli, and C. Turchetti. HW/SW codesign of an engine management system, *Proceedings of Design, Automation and Test in Europe conference*, Paris, France, March 2000, pp. 263–267.

[18] G. Martin, H. Chang, L. Cooke, M. Hunt, A. McNelly, and L. Todd, *Surviving the SOC Revolution — A Guide to Platform-Based Design*, Kluwer Academic Publishers, Dordrecht, 1999, chap. 4.

[19] P. Thoen and F. Catthoor, *Modeling, Verification and Exploration of Task-Level Concurrency in Real-Time Embedded Systems*, Kluwer Academic Publishers, Dordrecht, 2000.

[20] C. Hoare, *Communicating Sequential Processes (CSP)*, International Series in Computer Science, Prentice-Hall, New York, 1985.

[21] M. Abrahams and J. Barkley, RTL verification strategies, *IEEE WESCON'98*, Anaheim, California, 1998, pp. 130–134.

[22] S. Asaf, E. Marcus, and A. Ziv, Defining coverage views to improve functional coverage analysis, *Proceedings of the 41st Annual Conference on Design Automation*, San Diego, California, June 2004, pp. 41–44.

[23] TransEDA web site: http://www.transeda.com

[24] F. Casaubieilh, A. McIsaac, M. Benjamin, M. Bartley, F. Pogodalla, F. Rocheteau, M. Belhadj, J. Eggleton, G. Mas, G. Barrett, and C. Berthet, Functional verification methodology of chameleon processor, *Proceedings of the 33rd Annual Conference on Design Automation*, Las Vegas, Nevada, 1996, pp. 421–426.

[25] A. Haverinen, M. Leclercq, N. Weyrich, and D. Wingard, SystemC-based SoC Communication Modeling for the OCP Protocol, 2002, http://www.ocpip.org

[26] C. Chevallaz, N. Mareau, and A. Gonier, Advanced methods for SoC concurrent engineering, *Proceedings of Design, Automation and Test in Europe Conference (DATE'02)*, 2002, Paris, France, pp. 59–63.

[27] J. Gerlach and W. Rosenstiel, System level design using the SystemC modeling platform, *Proceedings of the Forum on Specification & Design Languages (FDL'00)*, Tüübingen, Gernamy, 2000.

[28] L. Semeria and A. Ghosh, Methodology for hardware/software co-verification in C/C++, *Proceedings of the High-Level Design Validation and Test Workshop (HLDVT '99)*, San Diego, California, 1999.

[29] A. Fin, F. Fummi, M. Martignano, and M. Signoretto, SystemC: a homogenous environment to test embedded systems, *Proceedings of the IEEE International Symposium on Hardware/Software Codesign (CODES'01)*, Copenhagen, Denmark, 2001, pp. 17–22.

[30] A. Clouard, G. Mastrorocco, F. Carbognani, A. Perrin, and F. Ghenassia, Towards bridging the precision gap between SoC transactional and cycle-accurate, *Proceedings of Design, Automation and Test in Europe Conference (DATE'02)*, Paris, France, 2002.

[31] D. Gajski, J. Zhu, R. Domer, A. Gerstlauer, and S. Zhao, *SpecC: Specification Language and Methodology*, Kluwer Academic Publishers, Boston/Dordrecht/London, 2000.

[32] Accellera Interface Technical Committee, 2002, http://www.eda.org/itc

[33] SPIRIT Consortium web site: http://www.spiritconsortium.org

18
Assertion-Based Verification

Erich Marschner
Cadence Design Systems
Berkeley, California

Harry Foster
Jasper Design Automation
Mountain View, California

18.1 Introduction

Functional verification is a process of confirming that the *intent* of the design was preserved during its implementation. Hence, this process requires two key components: a *specification of design intent* and a *design implementation.* Yet historically, describing design intent in a fashion useful to the verification process has been problematic. For example, typical forms of specification are based on natural languages, which certainly do not lend themselves to any form of automation during the verification process. Furthermore, ambiguities in the specification often lead to misinterpretation in the design and verification environments. The problem is compounded when a verification environment cannot be shared across multiple verification processes (that is, the lack of interoperability between the specification and the various verification environments for simulation, acceleration, emulation, or formal verification).

In this chapter, we introduce a new approach to addressing the functional verification challenge, known as *assertion-based verification,* which provides a unified methodology for unambiguously specifying design intent across multiple verification processes using assertions. Informally, an assertion is a statement of design intent that can be used to specify design behavior [1]. Assertions may specify internal design behaviors (such as a specific FIFO structure) or external design behavior (such as protocol rules for a design's interfaces or even higher-level requirements that span across design blocks). One key characteristic of assertions is that they allow the engineer to specify *what* the design is supposed to do at a high level of abstraction, without having to describe the details of *how* the design intent is to be implemented. Thus, this abstract view of the design intent is ideal for the verification process — whether we are specifying high-level requirements or lower-level internal design behavior.

18.1.1 Controllability and Observability

Fundamental to the discussion of assertion-based verification is an understanding of the concepts of *controllability* and *observability.* Controllability refers to the ability to influence an embedded finite state-machine,

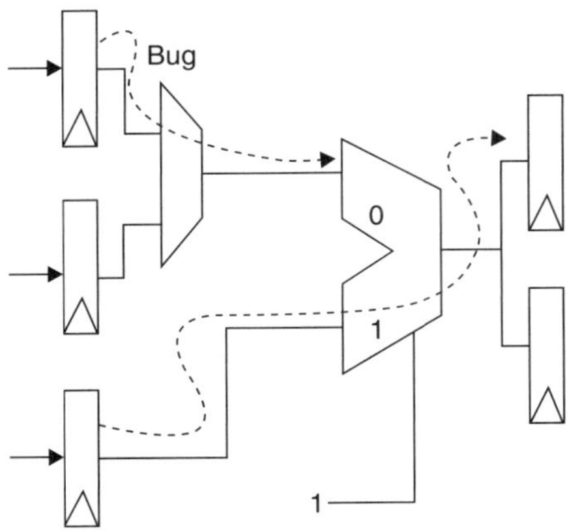

FIGURE 18.1 Poor observability misses bugs.

structure, or specific line of code within the design by stimulating various input ports. Note that while in theory, a simulation testbench has high controllability of the design model's input ports, during verification it can have very low controllability of an internal structure within the model. Observability, in contrast, refers to the ability to observe the effects of a specific internal finite state-machine, structure, or stimulated line of code. Thus, a testbench generally has limited observability if it only observes the external ports of the design model (because the internal signals and structures are often hidden from the testbench).

To identify a design error using the testbench approach, the following conditions must hold (i.e., evaluate to True):

1. The testbench must generate proper input stimulus to activate (i.e., sensitize) a bug.
2. The testbench must generate proper input stimulus to propagate all effects resulting from the bug to an output port.

It is possible, however, to set up a condition where the input stimulus activates a design error that does not propagate to an observable output port (Figure 18.1). In these cases, the first condition cited above applies; however, the second condition is absent.

Embedding assertions in the design model increases observability. In this way, the verification environment no longer depends on generating proper input stimulus to propagate a bug to an observable port. Thus, any improper or unexpected behavior can be caught closer to the source of the bug, in terms of both time and location in the design intent.

While assertions help solve the observability challenge in simulation, they do not help with the controllability challenge. However, by adopting an assertion-based, constraint-driven simulation environment, or applying formal property checking techniques to the design assertions, we are able to address the controllability challenge.

18.2 History

In this section, we discuss different approaches to solving the functional specification challenge. We begin by briefly introducing propositional temporal logic, which forms the basis for modern property specification languages. Building on this foundation, we then present a historical perspective for various forms of assertion specification.

18.2.1 Reasoning about Behavior

Logic, whose origins can be traced back to ancient Greek philosophers, is a branch of philosophy (and mathematics) concerned with reasoning about a model. In a classical logic system, we state a proposition, and then deduce (or infer) if a given model satisfies our proposition, as illustrated in Figure 18.2.

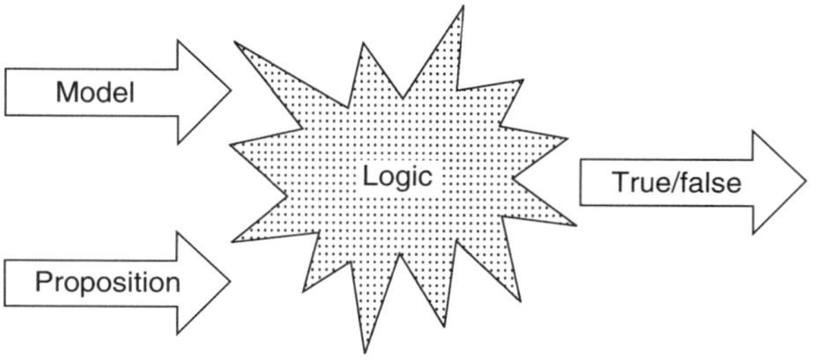

FIGURE 18.2 Classical logic system.

For example, consider the following set of propositions:

- *The moon is a satellite of the earth.*
- *The moon is rising (now).*

If we let the universe be our model, then using classical logic we can check whether our set of propositions holds for the given model.

Classical logic is good for describing *static* situations. However, it is unsuitable for describing dynamic behavior (i.e., situations involving time). Returning to the previous example, it would not be possible to express the following proposition in classical logic, since it involves time:

- *The moon will rise again and again.*

Note that our interest in functional verification of hardware systems requires that we use a logic that is expressive enough to describe properties of reactive systems. For a reactive system, components of the system concurrently maintain ongoing interaction with their environment, as well as other concurrent components of the system. Hence, in the next section, we discuss a more expressive logic that involves time.

18.2.1.1 Propositional Temporal Logic

In this section, we build a foundation for understanding propositional temporal logic by introducing a few basic concepts. The advantage of using temporal logic to specify properties of reactive systems is that it enables us to reason about these systems in a very simple way. That is, temporal logic eliminates the need to explicitly specify the time relationships between system events which are represented as Boolean formulas that describe states of the design. For example, instead of writing the property expression involving time explicitly, such as

```
∀t.!(grant1(t) & grant2(t))
```

which specifies for all values of time **t**, `grant1` and `grant2` are mutually exclusive, we simply write in a temporal property language, such as the Accellera Property Specification Language (PSL):

```
always !(grant1 & grant2)
```

which states that `grant1` and `grant2` should not hold at the same time.

In temporal logic, we define a computational path π as an infinite sequence of states:

$$\pi = s_0, s_1, s_2, s_3, \ \ldots$$

that represents a forward progression of time and a succession of states, s_i. When proving a temporal assertion, we may assume that a point in time or given state along the path has a unique future or successor state (for example, $s_0 \rightarrow s_1$), in which case the assertion is checked on a given path or trace of execution (for example, a single simulation trace). Thus, each possible computational path of a system is

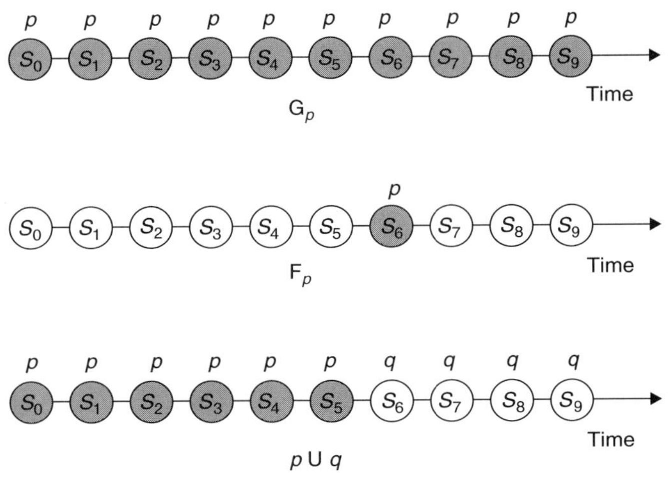

FIGURE 18. 3 Temporal formula path operator semantics.

considered separately. We refer to this form of logic as *linear-time temporal logic*, and Linear Temporal Logic (LTL) is one example [2]. Alternatively, we may assume that each point in time or given state along a path may split into multiple futures or successor states (for example, $s_0 \rightarrow s_1$ and $s_0 \rightarrow s_5$). In that case, all computational paths are considered concurrently, and are usually represented as a tree of infinite computational paths. We refer to this form of logic as *branching-time temporal logic*, and Computation Tree Logic (CTL) is one example [3].

CTL and LTL play essential roles in formal hardware verification for branching-time temporal logic and linear-time temporal logic, respectively. To show how these two types of logic differ, we first introduce CTL* [4]. CTL* contains operators for reasoning about paths of computation, such as the *path formula* operators **G** (always), **F** (eventually), **U** (until) and **X** (next), and operators for reasoning on branching paths of execution, that is, the *state formula* operators **A** (for all paths of execution) and **E** (for some path of execution). In addition to these quantifiers, any Boolean compositions of CTL* formulas are CTL* formulas as well.

For any temporal formulas p and q, as illustrated in Figure 18.3, the temporal formula **G** p specifies that the temporal formula p holds for all states of a path π (or simply, p always holds). The temporal formula **F** p specifies that p holds for some future state of a path π. The temporal formula p **U** q specifies that the temporal formula p holds in all states of a path π until q holds in some future state of π.

The temporal formula **A**f specifies that for all paths π starting from the current state, f holds. The temporal formula **E**f means that there is a path π starting from the current state for which f holds.

As seen in the previous paragraphs, CTL* can be separated into *state formulas* (involving **A** and **E**) and *path formulas* (involving **G**, **F**, **X**, and **U**). Any atomic proposition p (or Boolean expression) over *state formulas* is by definition a *state formula*. In addition, existential quantification over path formulas (for example, **E**f, where **E** is the existential quantifier and f a *path formula*) is also by definition a *state formula*.

Any *state formula* is a *path formula*, as are Boolean compositions of *path formulas*. In addition, path formulas can be composed using the temporal operators **X** f and f_1 **U** f_2.

Note that the **F** operator can be thought of as an alias for the unary form of the *until* operator (for example, **F**p = *true* **U** p), and that **G** and **F** are dual (that is, **G**p is equivalent to $\neg$**F**$\neg p$). The rationale behind the first alias is that *eventually* (**F**) is equivalent to waiting vacuously until p is valid; and the rationale behind the second formula is that saying "*p always valid*" is equivalent to saying that "*it is not the case that ¬p will be valid in the future.*" Similarly, it is not difficult to show that **E** and **A** are dual.

Note that in CTL* we do not make any restriction on the order in which temporal and branching operators appear in a valid formula. As a result, **FG** and **AG** are valid CTL* formulas. The first formula (**FG**) states that *eventually our proposition will be valid forever.* The second formula (**AG**) states that *for all paths starting from the current state, our proposition will always be valid.*

Now that we have presented CTL*, we can restrict this logic to CTL and LTL.

- *CTL:* A CTL formula is a CTL* formula beginning with a branch quantifier (**A** and **E**), restricting that temporal operators (**F**, **G**, **U**, and **X**) be proceeded by branch quantifiers. For example, the formula **AG** p is a valid CTL formula, but **FG** p is not.
- *LTL:* An LTL formula is the subset of CTL* formulas obtained by simply restricting the valid formulas to path formulas. For example, the formula **FG** p is a valid LTL formula, but **EG** p is not.

If we consider an implicit universal quantifier for all paths (that is, **A**) in front of an LTL formula, we can see that certain behaviors, such as **A**(**GF** p), cannot be represented by CTL, although it is a valid LTL formula. Similarly **AG**(**EF** p) is a valid CTL formula, but not a valid LTL formula. While the first formula says that for all states of all paths, eventually p holds (a fairness constraint), the second formula says that for all branches of all states, at least in one of the paths, eventually p holds.

Note that neither LTL nor CTL can express a property that involves counting, such as

- *p is asserted in every even cycle.*

In the next section, we introduce extended regular expressions, which allow us to express behavior that involves counting.

We have not thoroughly discussed the semantics of CTL*, CTL, and LTL in this chapter, but we have given a short introduction on this subject to subsidize the remainder of this chapter. We refer the reader to [5] for a more complete definition of these logics' semantics and complexities.

18.2.1.2 Extended Regular Expressions

Regular expressions provide a convenient way to define sets of computations (that is, a temporal pattern used to match various *sequences* of states). Extended regular expressions are regular expressions extended with conjunctions and negation. Hence, the computational path defined by an extended regular expression can be combined to form the building blocks used to specify hardware design assertions. Regular expressions can express counting *modulo n* type behavior through the use of the * operator. Hence, regular expressions allow us to specify properties that cannot be described by either CTL or LTL, such as `{`true; p}`[*], which states that *p is asserted in every even cycle.*

Note that not all properties can be expressed with extended regular expressions. For example, the property *eventually p holds forever* cannot be expressed using extended regular expressions, yet this property can be expressed in LTL (that is, **FG** p).

18.2.2 Assertion Languages

Assertions may be expressed either declaratively or procedurally. A declarative assertion is always active, and is evaluated concurrently with other components in the design. A procedural assertion, on the other hand, is a statement within procedural code, and is executed sequentially in its turn within the procedural description. Hence, declarative properties are natural for specifying block-level interfaces, as well as systems. Similarly, a procedural assertion is convenient for expressing implementation-level properties that must hold in the context of procedural code. A key difference between the declarative assertion and the procedural assertion is that the declarative assertion concurrently monitors the assertion expression, while the procedural assertion only validates the assertion expression during sequential visits through the procedural code.

In the following sections, we discuss various languages and techniques in use today for expressing assertions, including: the VHSIC Hardware Description Language (*VHDL*), the *Open Verification Library* (OVL), and *SystemVerilog Assertions* (SVA). A comprehensive discussion of Accellera PSL follows in Section 18.3.

18.2.2.1 VHSIC Hardware Description Language (VHDL) Assertions

The concept of assertions was introduced quite early in the development of VHDL. VHDL was originally developed in the early 1980s, by a team consisting of Intermetrics, Inc., IBM, and Texas Instruments, under a DoD-sponsored effort to create a design and documentation language for VHSIC-class electronic designs. The initial requirements for this language were defined in June 1981 during a workshop in Woods Hole, Massachusetts. The requirements published in the proceedings of that workshop [6] included a requirement to support exception handling, but they did not explicitly mention assertions. However, by January 1983, when these requirements were included in the Request for Proposal (RFP) for the VHDL effort [7], they had been extended to include many of the so-called "Steelman" requirements developed earlier for the Ada programming language [8], including a requirement for assertions that was almost identical to the assertion requirement for Ada (section 3.8.6 Assertions):

> It shall be possible to include assertions in programs. If an assertion is false when encountered during simulation, it shall raise an exception. It shall also be possible to include assertions, such as the expected frequency for selection of a conditional path, that cannot be verified. [Note that assertions can be used to aid optimization and maintenance.]

Ironically, the original Ada requirement did not lead to an explicit assertion construct in Ada, yet the derivative requirement for VHDL resulted in the creation of not just one, but two kinds of assertions in VHDL: a sequential assertion statement, which can appear within sequential (i.e., procedural) code in a process or subprogram; and a concurrent assertion statement, which can appear in a block, and acts like an independent process.

VHDL assertions involve only combinational expressions; no temporal operators are available in VHDL, other than the very limited capability provided by certain signal-valued attributes such as S'delayed. Even so, VHDL assertions quickly proved to be a very useful mechanism for expressing invariants that are expected to hold within a design, such as a relationship among variables that is expected to hold at a given point in the execution of algorithmic code (expressed with a sequential assertion statement) or a relationship among signals that is expected to hold at all times (expressed with a concurrent assertion statement).

The simulation semantics for VHDL assertions, which are defined in the VHDL Language Reference Manual [9], ensure that an assertion will issue an error message any time it is executed and the asserted expression evaluates to False. However, error detection during simulation is not the only value of such assertions. They also can play a significant role in documentation of code and therefore maintenance of the code over time. Furthermore, as formal equivalence checking developed as an alternative to simulation, VHDL assertions were adapted to that verification method as well. For example, in the early to mid-1990s, VHDL assertions were used in one formal verification tool [10] to specify both "axioms" or assumptions about the environment, and assertions about the design that needed to be verified.

At the time of this writing, work is in progress to extend VHDL assertions further by incorporating PSL v1.1 (see Section 18.3) into the next version of the VHDL standard. This extension will enable PSL temporal operators, properties, sequences, and directives to be used alongside the existing VHDL concurrent assertion statements to express much more complex behavioral requirements.

18.2.2.2 Open Verification Library

One of the challenges in creating an assertion-based methodology is ensuring consistency within a design project. Any inconsistencies between multiple stakeholders involved in the design and verification process can become so problematic that the benefits the assertions provide during the verification process are overshadowed by an unmanageable methodology. For example, an assertion-based methodology needs to provide a consistent manner of:

- controlling assertions (for example, enabling and disabling assertions)
- reporting assertion violations
- specifying reset

Aside from methodology consistency, another challenge related to assertion adoption has been the effort required to teach engineers new languages that specify assertions.

OVL was developed as a means to overcome these challenges within an assertion-based methodology [11]. Actually, OVL is not an assertion language, and lacks the expressiveness of languages such as PSL or SystemVerilog Assertions (SVA). On the contrary, OVL is a library of simulation monitors written in both Verilog and VHDL. Hence, OVL could be classified as a declarative form of assertion specification. During the RTL development process, engineers select and then instantiate OVL monitors into the RTL model. Each assertion monitor is then used to check for a specific Boolean or temporal violation during the verification process. The following is an example of a Verilog instantiated OVL monitor that checks for the case when `grant1` and `grant2` are not mutually exclusive:

```
assert_always mutex (clk, reset_n, !(grant1 & grant2));
```

OVL monitors address assertion-based methodology considerations by encapsulating a unified and systematic method of reporting, and a common mechanism for enabling and disabling assertions during the verification process. The reporting and enable/disable features use a consistent process, which provides uniformity and predictability within an assertion-based methodology.

OVL offers a wide set of monitors, enabling the engineer to capture a large class of assertions such as the simple `assert_always` Boolean invariant shown in the previous example, one-hot checking, as well as multicycle temporal checks such as `assert_next` and `assert_change`.

18.2.2.3 SystemVerilog Assertions

Unlike VHDL, the IEEE-1364 Verilog language never contained a Boolean assertion construct. It was not uncommon for designers to capture assertions in their Verilog RTL in an ad hoc fashion using the Verilog `$display()` system task. The problem with this approach is that not all `$display()` calls could be treated or identified as assertions, which means that any assertions specified using a `$display()` could not be verified by formal verification tools. In addition, this approach of specifying assertions required a significant amount of extra modeling to express sequences or other complex temporal behaviors.

SystemVerilog Assertions was developed to provide engineers the means to describe complex behaviors about their designs in a clear and concise manner. SVA is based on LTL built over sublanguages of regular expressions. Although SVA lacks the full expressiveness of LTL (that is, it only supports *globally* (G) in the form of the SystemVerilog `always` or an implicit *globally* of a concurrent assertion), most engineers will find SVA sufficient to express most common assertions required for hardware design.

SVA supports two types of assertions: *procedural* (or immediate) and *declarative* (or concurrent). Both the types of assertion are intended to convey the intent of the design engineer and to identify the source of a problem as quickly and directly as possible. In addition, engineers can use SVA to define temporal correctness properties and coverage events.

The following is an example of a SystemVerilog concurrent assertion used to check for mutual exclusion of grant signals:

```
property mutex;
  @(posedge clk) disable iff (!reset_n) (!(grant1 & grant2));
endproperty

assert property (mutex);
```

At the time of this writing, the Accellera 3.1a SystemVerilog LRM [12] has moved to the IEEE for standardization under the auspices of the IEEE's Corporate Standards Group (CAG) as IEEE P1800. In 2004, great effort was taken within Accellera to ensure semantics alignment between SVA with PSL. However, there are a few syntactical differences between SVA and PSL due to the divergent objectives of these two languages. Most notably, SVA was designed to provide an embedded assertion capability directly within the SystemVerilog language, while PSL was designed as a standalone assertion language that harmoniously works across multiple hardware description languages (HDLs), such as SystemVerilog, Verilog, and VHDL.

18.3 State of the Art

The following section introduces principles and concepts behind Accellera PSL [13].

18.3.1 Property Specification Language Principles

PSL is a comprehensive language that includes both LTL and CTL constructs. This enables PSL to support various kinds of formal verification flows, including event- and cycle-driven simulation and various algorithms for formal verification (model checking).

Most tools are based on a single algorithm and therefore cannot support all of PSL. In particular, most simulation tools are characterized by a single trace and monotonically advancing time, which preclude support for many CTL-based properties and make it difficult to support some LTL-based properties efficiently. The subset of PSL that can be supported efficiently in both simulation and formal verification is known as the "Simple Subset" and is defined by a short list of simple syntactic restrictions.

The temporal semantics of PSL are formally defined. This formal definition ensures precision; it is possible to understand precisely what a given PSL construct means, and does not mean. The formal definition also enables careful reasoning about the interaction of PSL constructs, or about their equivalence or difference. The formal definition has enabled validation of the PSL semantic definition, either manually or through the use of automated reasoning [14].

For practical application, it is imperative that PSL statements be grounded in the domain to which they apply. This is accomplished in PSL by building on expressions in the HDL used to describe the design that is the domain of interest. At the bottom level, PSL deals with Boolean conditions that represent states of the design. By using HDL expressions to represent those Boolean conditions, PSL temporal semantics are connected to, and smoothly extend, the semantics of the underlying HDL.

18.3.2 Basic PSL Concepts

PSL is defined in *layers* (Figure 18.4). The Boolean layer is the foundation; it supports the temporal layer that uses Boolean expressions to specify behavior over time. The verification layer consists of directives that specify how tools should use temporal layer specifications to verify functionality of a design. Finally, the modeling layer defines the environment within which verification is to occur.

PSL is also defined in various *flavors* (Figure 18.4). Each flavor corresponds to a particular underlying HDL. Currently defined flavors include VHDL, Verilog, and SystemVerilog. A SystemC flavor has been added in the recently completed IEEE 1850–2005 standard [14a].

Since PSL works with various underlying HDLs, it must adapt to the various forms of expression allowed in each language, and the various notions of datatype that exist in each language. To that end,

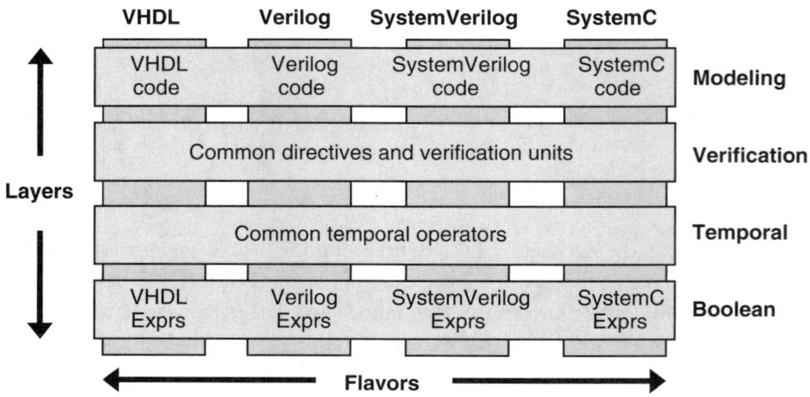

FIGURE 18.4 PSL layers and flavors.

PSL defines a number of expression type classes — Bit, Boolean, BitVector, Numeric, and String — and specifies how native and user-defined datatypes in VHDL, Verilog, SystemVerilog (and now, SystemC) all map to those type classes.

18.3.2.1 Boolean Layer

The Boolean layer consists of expressions that belong to the Boolean type class, i.e., those whose values are, or map to, True or False. These expressions represent conditions within the design, e.g., the state of the design, or the values of inputs, or a relationship among control signals. Typically, this includes any expression that is allowed as the condition in an 'if' statement in a given HDL. PSL extends this set of expressions with a collection of built-in functions that check for rising or falling signals, signal stability, and other useful utilities.

18.3.2.2 Temporal Layer

The temporal layer involves time, and Boolean expressions that *hold* (i.e., are True) at various points in time. This includes expressions that hold pseudo-continuously (i.e., at every point in time considered by the verification tool) as well as expressions that hold at selected points in time, such as those points at which a clock edge occurs, or an enabling condition is True. Temporal operators enable the specification of complex behaviors that involve multiple conditions over time.

One class of temporal operators is used to construct sequential extended regular expressions (SEREs) or *sequences*. A sequence describes behavior in the form of a series of conditions that hold in succession. A sequence may specify that a given condition repeats for a minimum, maximum, or even unbounded number of times, and it may specify that two or more subordinate sequences overlap or hold in parallel. If the behavior described by a sequence matches the behavior of the design starting at a given time, then the sequence holds at that time. PSL also supports *endpoints*, which are identical to sequences except that they hold at the time in which the behavior completes.

For example, consider the following sequence declarations, which describe a simple handshake-based bus protocol. In this case, a default clock declaration specifies the times at which successive expressions are evaluated:

```
default clock = (posedge clk);
sequence GetBus = { req[+]; (req && ack) };
sequence HoldBus = { (req && ack)[*] };
sequence RlsBus = { (!req && ack); !ack };
sequence ReadOp = { rwb && ardy; !drdy[*0:2]; drdy };
sequence ReadT = { GetBus; {HoldBus && ReadOp}; RlsBus };
```

The first three define the series of control conditions involved in arbitrating for control of the bus. Sequence GetBus describes behavior in which a request signal is high for one or more cycles, followed by a cycle in which the request signal is still high and an acknowledge signal goes high. Sequence HoldBus describes behavior in which both request and acknowledge signals stay high for an indefinite length of time. Sequence RlsBus describes behavior in which the request signal drops, and in the next cycle the acknowledge signal drops. Similarly, sequence ReadOp describes the control signals involved in a read operation on the bus: in the first cycle, a read operation is indicated and "address ready" signal is asserted; following that, "data ready" is asserted after a delay of 0, 1, or 2 cycles. Sequence ReadT combines all of these to represent a read transaction in which the GetBus sequence occurs first, then the ReadOp sequence occurs in parallel with the HoldBus sequence, and finally the RlsBus sequence occurs.

A second class of temporal operators provides a means of expressing conditional behavior using implication. This includes Boolean implication (a −> b) as well as suffix implication, in which the antecedent may be a sequence rather than a simple Boolean. In the latter case, the consequent holds at the end of the antecedent sequence. A third class of temporal operators consists of a set of English keywords that describe temporal relationships. These include **always**, **never**, **next**, **eventually!**, **until**, **before**, and slight variations thereof. Temporal operators may be combined with Booleans and sequences to describe behaviors or *properties* of a design.

For example, consider the following property declarations, which describe certain characteristics of the same simple bus protocol:

property BusArbitrationCompletes =
    always GetBus |=> {HoldBus; RlsBus};

property ReqSteady =
    always rose(req) −> **next** req **until** ack;

property AckSteady =
    always rose(ack) −> **next** !req **before** !ack;

property AckWithin (constant N) =
    always rose(req) −> **next** {{ack} **within** [*N]};

The first declaration defines a property BusArbitrationCompletes: if sequence GetBus occurs, then it will always be followed by sequence HoldBus and then sequence RlsBus. The second declaration defines property ReqSteady: if signal req rises, then it will stay high until signal ack is high. Similarly, the third declaration defines property AckSteady: if signal ack rises, then signal req will go low before ack goes low. The final declaration defines parameterized property AckWithin: after req rises, ack will occur within *N* cycles, where *N* is to be provided later.

18.3.2.3 Verification Layer

The verification layer specifies how sequences and properties are expected to apply to the design, and therefore how verification tools should attempt to verify the design using those sequences and properties. PSL *directives* indicate what to do with a given property or sequence. The **assert** directive says that a given property is expected to hold for the design, and that this should be checked during verification. The **assume** directive says that a given property should be taken for granted — it can be used in verifying other properties, but it need not be checked. The **restrict** directive is similar; it says that that a given sequence should be taken for granted. Typically, **assert** directives apply to the design being verified, while **assume** and **restrict** directives are used to constrain the environment within which the design is being verified. Additional directives (in particular, the **cover** directive, described below) provide additional capabilities.

The Verification layer also includes *verification units*, which enable packaging a collection of related PSL declarations and directives so they can be applied as a group to a particular part of the design. A verification unit may be bound to a particular instance within the design, or to a module used within the design. This causes the PSL directives to apply to that instance, or to all instances of that module, respectively. Unbound verification units may be defined to create reusable packages of PSL definitions. One verification unit may *inherit* another verification unit, which makes the definitions of the latter available for use in the former.

18.3.2.4 The Modeling Layer

The Modeling layer consists of HDL code used to model the environment of the design and/or to build auxiliary state machines that simplify the construction of PSL sequences or properties. Modeling the environment of the design under verification is primarily of interest for formal verification (see Chapter 20 *Formal Property Verification*). Building auxiliary state machines applies to both formal verification and simulation, and is addressed here.

Counters are one class of auxiliary state machines that are often necessary in specifying behavior. The modeling layer of PSL allows the underlying HDL to be used for constructing such a counter within a verification unit, so that it is kept separate from the design itself, yet is still available for use in a PSL assertion. For example, consider a byte-serial bus protocol that specifies the number of bytes being transferred as the first byte following the packet header. Assuming the Verilog flavor of PSL, the following verification unit might be written to specify the behavior of this protocol:

```
vunit SerialProtocol (serial_interface_module) {
   integer count, crc;

   always @(posedge clk)
     if (header)
       begin
       count ⇐ currentbyte;
       crc ⇐ 0;
       end
   else if (count > 0)
       begin
       crc ⇐ compute_crc(crc, currentbyte);
       count ⇐ count - 1;
       end

endpoint header = {currentbyte == `PKT_HDR; 1};
sequence data = {(count>0)[*]; crc == currentbyte};

assert always header -> next data;

}
```

In this example, verification unit SerialProtocol contains Verilog code (in Courier font) that acts as an auxiliary state machine supporting the assertion at the end of the verification unit. The Verilog code maintains a counter, loaded from the first byte after the header, and decrements the counter as successive data bytes appear. In parallel, it computes a cyclic redundancy check (CRC) for the packet, to compare against the CRC embedded at the end of the packet.

This example illustrates reference to a PSL endpoint within the HDL modeling layer code. Endpoint "header" is a sequence consisting of the header byte of a packet followed by the packet size. When this sequence has been recognized, it triggers the loading of the packet size into the counter in the Verilog code. The assertion also uses this endpoint as the indication that a packet has started, which implies that the data must follow (and that the CRC computed by the Verilog code must match the byte following the data in the packet).

Although PSL is defined as a separate language, it is also possible to embed PSL directly into HDL code. One approach involves a convention adopted by many PSL tools in which PSL code is embedded in HDL comments, following the comment delimiter and the keyword "psl." Another approach involves incorporation of PSL directly into the underlying HDL, as is planned for the next version of IEEE 1076 VHDL. Embedded assertions are of most use for designers who want to add assertions as they go, to document their assumptions. External assertions (in verification units) are often of more use to verification engineers, who want to verify blocks without modifying the source code of those blocks.

18.3.3 Future Direction

The current interest in assertion-based verification is focused on the use of assertions in simulation. Assertions are usually added after the fact, to increase observability of events within the design so that bugs can be detected more closely to their source. But this approach is only beginning to take advantage of the power of assertion languages. Over time, we can expect to see increased use of the coverage capabilities of assertion languages to enable coverage-driven verification. At the same time, taking advantage of the formal semantics of assertion languages such as PSL, formal verification of designs will become more common. Eventually, we can expect to see assertion languages used to thoroughly specify the behavior of design IP blocks and their interfaces [15], with such specifications developed before, rather than after, the design is done. This may even lead to use of assertion languages as a vehicle for design, if appropriate methods for assertion-based design synthesis or component selection can be developed.

In parallel with increasing application of assertion languages, development of those languages will continue to increase their capabilities and applicability within a variety of design languages and environments. Much of this development is occurring within IEEE standards activities. The IEEE 1800 SystemVerilog working group is pursuing integration of IEEE 1364 Verilog and Accellera 3.1a SystemVerilog, and in the process has refined SystemVerilog Assertions further. The IEEE 1850 PSL working group has completed the definition of IEEE Std 1850-2005 Property Specification Language (PSL), which refines and enhances Accellera PSL v1.1.

References

[1] H. Foster, A. Krolnik, and D. Lacey, *Assertion-Based Design*, 2nd ed., Kluwer Academic Press, Dordrecht, 2004.
[2] A. Pnueli, The temporal logic of programs, *Proceedings of 18th IEEE Symposium on Foundations of Computer Science*, 1977, pp. 46–57.
[3] E.M. Clarke and E.A. Emerson, Design and synthesis of synchronization skeletons using branching time temporal logic, *Proceedings of Workshop on Logic of Programs*, Lecture Notes in Computer Sciences, Springer, Berlin, Vol. 131, 1981, pp. 52–71.
[4] E.A. Emerson and J.Y. Halpern, "Sometimes" and "not never" revisited: on branching vs. linear time, *Proceedings of 10th ACM Symposium on Principles of Programming Languages*, 1983, pp. 127–140.
[5] T. Kropf, *Introduction to Formal Hardware Verification*, Springer, Berlin,1998.
[6] G.W. Preston, Report of IDA Summer Study on Hardware Description Language (IDA Paper P-1595), Institute of Defense Analyses, October 1981.
[7] Department of Defense Requirements for Hardware Description Languages, January 1983.
[8] Department of Defense Requirements for High Order Computer Programming Languages ("Steelman"), June 1978, at http://www.adahome.com/History/Steelman.
[9] IEEE Standard 1076-2002, *IEEE Standard VHDL Language Reference Manual*, IEEE, 17 May 2002.
[10] VFormal User Guide, COMPASS Design Automation, August 1995.
[11] H. Foster and C. Coelho, Assertions targeting a diverse set of verification tools, *Proceedings of 11th Annual International HDL Conference*, March 2001.
[12] Accellera, 1370 Trancas Street, #163, Napa, CA 94558. *SystemVerilog Language Reference Manual*, 2004. http://www.eda.org/sv/SystemVerilog_3.1a.pdf.
[13] Accellera, 1370 Trancas Street, #163, Napa, CA 94558. *Property Specification Language, Reference Manual*, 2004. www.eda.org/vfv/docs/PSL-v1.1.pdf.
[14] M.J.C. Gordon, Validating the PSL/Sugar semantics using automated reasoning, in *Formal Aspects of Computing*, Vol. 15, Springer, London, 2003, pp. 406–421.
[14a] IEEE Std 1850-2005, *IEEE Standard for Property Specification Lanuage (PSL)*, IEEE, New York, September 2005.
[15] E. Marschner, B. Deadman and G. Martin, IP reuse hardening via embedded sugar assertions, in *Proceedings of the IP-Based System-on-Chip Design Workshop*, Grenoble, France, October 30–31, 2002.

19 Hardware Acceleration and Emulation

Ray Turner
Cadence Design Systems, San Jose, California

Mike Bershteyn
Cadence Design Systems, Inc. Cupertino, California

19.1 Introduction

According to a Collett International North American survey, 82% of silicon respins are due, at least in part, to functional errors. Thus, comprehensive functional verification is key to reducing developmental costs and delivering a product on time. Functional verification of a design is most often performed using logic simulation and prototyping. There are advantages and disadvantages to each, and often both methods are used. Logic simulation is easy, accurate, flexible, and low cost. But it is not fast enough for large designs and it is much too slow to run application software against the hardware design. Field-programmable gate array (FPGA)-based prototypes are fast and inexpensive. But the time required to implement a large design into several FPGAs can be very long and it is error-prone. Changes to fix design flaws also take a long time to implement and may require board wiring changes. Since FPGA prototypes have little debugging capability, probing signals inside the FPGAs in real time is very difficult, if not impossible, and recompiling FPGAs to move probes takes too long. Clearly, simulation should be used earlier in the verification process when bugs and fixes are frequent, and prototyping should be used at the end of the development cycle when the design is basically complete, and speed is needed to get sufficient testing to uncover any remaining system-level bugs. Prototyping is also popular for testing software.

Simulation acceleration can address the performance shortcomings of simulation to an extent. Here, the design is mapped into a hardware accelerator to run faster, and the testbench (and any behavioral design code) continues to run on the simulator on the workstation. A high bandwidth, low latency channel connects the workstation to the accelerator to exchange signal data between testbench and design. By Amdahl's law, the slowest device in the chain will determine the speed achievable. Normally, this is the testbench in the simulator. With a very efficient testbench (written in C or transaction-based), the channel may become the bottleneck.

In-circuit emulation (ICE) greatly reduces the long time required to implement and change designs typically seen with FPGA prototyping, and provides a comprehensive, efficient debugging capability. While it takes weeks or months to implement an FPGA prototype, it takes only days to implement emulation. And design changes take a few hours or less. Emulation does this at the expense of running speed and cost compared to FPGA prototypes. Looking at emulation from the other direction, it improves on acceleration's performance by substituting "live" stimulus for the simulated testbench. This stimulus can come from a target system (the product being developed) or from test equipment. At 10,000 to 1,00,000 times the speed of simulation, emulation alone delivers the speed necessary to test application software while still providing a comprehensive hardware debug environment. Later we will discuss creation of emulation environments.

It is worth noting that simulation and prototyping involve two different styles of execution. Simulation executes the register transfer level (RTL) code serially while a prototype executes fully in parallel. This leads to differences in debugging. In simulation, you set a breakpoint and stop simulation to inspect the design state, interact with the design, and resume simulation. You can stop execution "midcycle" as it were, with only part of the code executed. You can see any signal in the design and the contents of any memory location at any time. You can even back up (if you have saved a checkpoint) and rerun. With a prototype, you use a logic analyzer for visibility, so you can see only a limited number of signals, which you determine ahead of time (by clipping on probes). The target does not stop when the logic analyzer triggers, so each time you change the probes or trigger condition, you have to reset the environment and start again from the beginning. Acceleration and emulation are more like prototyping and silicon in terms of RTL execution and debugging, since the entire design executes simultaneously as it will in the silicon. Since the same hardware is often used to provide both simulation acceleration and ICE, these systems provide a blend of these two very different debugging styles.

Another difference between simulation, and acceleration and emulation is a consequence of accelerators using hardware for implementation — they are fundamentally two-state machines — acting the way the silicon will when fabricated. Thus, they are not useful for analyzing X-state initialization or strength resolution. Accelerators also do not model precise circuit timing, hence they will probably not find any race conditions or set-up and hold time violations. These tasks are properly carried out during simulation or with a static timing analysis tool. A key distinction between an emulator and an FPGA prototyping system is that the emulator provides a rich debug environment, while a prototyping system has little to no debug capability and is primarily used after the design is debugged, to create multiple copies for system analysis and software development.

Logic Simulation	Acceleration	In-Circuit Emulation	FPGA Prototype
Accurate	Faster	Fast enough for embedded software	Inexpensive
Flexible	Handles large designs	Real world stimulus	Fast enough for embedded software
Not fast enough for large designs	Debugs via simulator UI	Short implementation time from RT level	Long implementation time
Not fast enough to test embedded software	Not fast enough for embedded software	Full debugging capabilities	Little, if any debug capability Error-prone
	Setup time for porting your testbench	Slower than FPGA prototype	Hard to change design
		Higher cost than FPGA prototype	

19.1.1 Hardware-Accelerated Verification Systems

In the 1980s, IBM developed the Engineering Verification Engine (EVE) family of simulation acceleration hardware for internal use. In 1988, Quickturn delivered the first commercial emulator when FPGA technology had reached sufficient capacity to allow their implementation. In 1997, Quickturn (now Cadence) delivered a processor-based emulator based on custom silicon.

19.1.1.1 FPGA-Based Systems

FPGAs would seem to be an obvious choice for implementing an emulator. And, if the design would fit into 2 to 5 FPGAs, they would be. But, the number of gates an FPGA can hold drops dramatically as a design is partitioned into multiple FPGAs. Literally, hundreds of FPGAs are required to implement a general-purpose accelerator/emulator that has comprehensive debugging capability. When a design is partitioned into so many FPGAs, the FPGA I/O pin count rather than the vendor-rated gate count determines the capacity of the FPGA. Rent's Rule may be used to calculate this [10–12].

FPGA-Based Emulation Systems

Mentor Celaro
Mentor (formerly Ikos and VMW) VStation
Cadence (formerly Verisity and Axis) Xtreme
Cadence (formerly Quickturn) RPM, Mars, Enterprise, and System Realizer
Cadence Mercury and Mercury^{Plus}

19.1.1.2 Processor-Based Systems

Processor-based emulators consist of a massive array of Boolean processors able to share data with one another, running at very high speed. The software technology consists of partitioning a design among the processors and scheduling individual Boolean operations in the correct time sequence and in an optimal way. Initially, performance did not match FPGA-based emulators, but compile times of less than an hour, and the elimination of timing problems that plagued FPGA-based systems made the new technology appealing for many use models, especially simulation acceleration. Future generations of this technology eventually surpassed FPGA systems in emulation speed while retaining the huge advantage in compilation times — and without a farm of a hundred PCs for compilation. Advances in software technology extended the application of processor-based emulators to handle asynchronous designs with any number of clocks. Other extensions supported 100% visibility of *all* signals in the design, visibility of all signals at *any* time from the beginning of the emulation run, and dynamic setting of logic analyzer trigger events without recompilation. At the same time that the emulation speed of FPGA-based systems was decreasing (due to heavily multiplexing pins), new generations of processor-based systems not only increased emulation speed, but also proved scalable in capacity to hundreds of millions of gates.

Processor-Based Emulation Systems

Synopsys Arkos Tharas Hammer Cadence (formerly Quickturn) CoBALT, and CoBALT[Plus] Cadence Palladium and Palladium II

Comparison of General Characteristics of FPGA-Based and Processor-Based Emulators

	FPGA-Based Emulators	Processor-Based Emulators
Technology	Uses commercial off-the-shelf components	Requires custom silicon
Compile speed from RTL	Takes many hours on many CPUs	10–30 million gates/h on 1 CPU
Emulation speed	150–800 kHz	400–1500 kHz
Predictable, reliable compiles	Fair	Excellent
Maximum capacity	30–120 million gates	256 million gates
Ability to handle asynchronous designs	Good	Good
Power requirements	50–150 W/million gates	100–250 W/million gates
Partitioning problems	Not found until individual FPGA compiles fail	Found early in process
Timing issues in emulation model	Potential for setup and hold issues and subtle, hard to find timing problems	Avoided completely
Size	0.1–0.3 ft^3/million gates	0.25–1 ft^3/million gates
Ability to make small changes quickly	Must recompile entire FPGA(s)	Can compile and download incremental changes to processor

19.2 Emulator Architecture Overview

19.2.1 Small-Scale Emulation and Logic Prototyping with FPGA

Field-programmable gate arrays provide a convenient and inexpensive mechanism for prototyping low complexity logic designs. In fact, recently the logic gate and memory capacity of the FPGA has grown so much that the high-end FPGA parts are roughly equivalent in capacity to a median size application specific integrated circuit (ASIC), that is, about 2 to 5 million gates (usable capacities of FPGAs is typically ¼ to $^1/_{10}$ the rated capacity). Similarly, modern FPGA can operate at speeds up to hundreds of megahertz which is also close to an average ASIC clock rate. These factors have resulted in the growth in popularity of FPGA prototyping for small- to medium-size logic designs.

In the majority of cases, the designers of such prototypes create and customize them independently, based on the requirements and characteristics of the systems in which their logic designs are intended to operate. The external inputs and outputs are connected either to a production board or to a prototype board specifically designed for that purpose. A minimal set of tools and infrastructure required for prototyping is supplied by FPGA vendors for whom such prototyping activities represent a significant market. Such tools typically consist of the compilation software that converts the logic design into an FPGA programming bitstream, and the means for debugging the design once compiled into an FPGA.

For example, Xilinx (www.xilinx.com) offers a comprehensive ISE design environment that covers logic synthesis, placement, routing, and timing closure for all the FPGA parts available from the supplier. After the design is mapped and downloaded into a part, the *ChipScope*™ debugger can be used to observe the internal signals under specified triggering conditions. This debugger acts as an embedded logic analyzer and consists of the logic core embedded into the user's design and the software to operate the debugger and display the results.

Similarly, Altera (www.altera.com) offers the *QuartusII*™ software kit for design compilation and the *SignalTap*™ embedded logic analyzer that is used with this company's FPGA devices.

As much as FPGA vendors attempt to simplify prototyping, there are still some difficulties that the developers of such a prototype must overcome. For example, FPGA devices are volatile. Their configurations must be stored in nonvolatile memory and downloaded at powerup. The facilities for that purpose need to be created. In many cases, the verification environment for the design is implemented in software. For that, the FPGA needs to be interfaced with the workstation that runs the software. These issues create a business opportunity for a number of vendors of small FPGA-based rapid prototyping systems that are typically implemented on boards, which contain the target FPGA and the facilities for design download and workstation interfaces. The use of such systems shortens the time required for prototype bringup in comparison with the completely custom-designed setups. FPGA prototyping system companies include Eve, Simpod, Hardi, and ExpressIP.

Things rapidly become more complicated when the design size grows beyond the capacity of one FPGA. Although multiple FPGA prototyping systems are also available, they all suffer from difficult problems of partitioning, creating signal connectivity, and multi-FPGA timing closure.

19.2.2 Large-Scale Emulation with FPGA Arrays

19.2.2.1 FPGA Arrays with Direct Connection Pattern

Over the last 15 years, FPGA density has increased much faster than the density of an average very Large Scale Integrated Circuit (VLSI). Because of that, early emulation systems faced the problem of modeling the target design with multiple FPGA parts. Many more FPGA devices were necessary to implement a logic design intended for an average ASIC than would be needed today.

The systems of the early 1990s had already identified the major difficulties of large-scale FPGA array emulation: partitioning, timing closure, and interconnect schemes. One distinct approach used for FPGA interconnect was a regular pattern of direct FPGA to FPGA connections. Although moderately successful, this scheme significantly magnified the difficulties of timing closure and partitioning because many signals had to visit several FPGA parts in order to reach their destinations.

19.2.2.2 FPGA Arrays with Partial Crossbar

This architecture, highly successful in the 1990s, implements the interconnection of FPGA parts using a number of special crossbar devices [1,15]. Each such device is a fully programmable crossbar with N terminals. Assuming that the FPGA array consists of K parts each having M external inputs and outputs, every crossbar device will connect to at most N/K inputs or outputs of each FPGA. This ratio, commonly called the "richness" of a partial crossbar, determines the routability of the interconnection scheme. The number of pins of a single crossbar device defines the limit to which a one-level partial crossbar can be scaled. Further capacity increase is accomplished using the multilevel schemes, where the crossbar devices are again connected in a partial crossbar pattern. It has been shown that with clever use of the internal symmetry of FPGA pins, even low richness values allow most practical logic designs to be routed successfully. Hybrid schemes that combine direct connections with partial crossbar have also been implemented [14].

The problem of multi-FPGA timing closure has also been addressed. This problem in FPGA-based emulators primarily manifested itself with hold time violations. Unlike the insufficient setup time which can be corrected simply by reducing the operating speed of an emulator, hold time violations affect the functionality of the design and must be removed. A number of approaches have been proposed, which utilize partitioning constraints and clock signal duplications [2] or special circuit modification [3,4,15].

The problem that presented the FPGA array emulators with their biggest challenge was that of multi-FPGA partitioning. As commercial FPGA devices were increasingly architected to contain functionally complete designs or modules, they did not maintain a pin/gate ratio that would result from automatic partitioning of the large logic design into many (often hundreds of) parts. As a consequence, the usable gate utilization was typically way below the level that a single FPGA could offer. As FPGA capacitycontinued to grow much faster than their pin count, the problem was aggravated with every new FPGA generation.

19.2.2.3 FPGA Arrays with Time-Multiplexed Connections

As an answer to the challenge of insufficient FPGA pin count, several time division multiplexing (TDM) schemes have been proposed, which divide the bandwidth available at an FPGA input/output pin among several signals that need to be transmitted between the parts.

A virtual wire scheme [5,13,16], though defined in general terms, has been primarily applied to direct FPGA connectivity architecture. It schedules the transmission of signals between FPGA parts as well as the storage of signal values at design flip/flops, in terms of the cycles T_V of a single clock signal which is distributed throughout the emulator. After each logical signal that crosses an FPGA partition boundary is assigned to both the physical pin and a T_V cycle, the total number of the T_V cycles necessary to advance the circuit to the next state is determined. This number essentially constitutes a TDM factor M for this design, which will be equal to the ratio of the design clock cycle (in a simplifying assumption that it was a single-clock synchronous design to start with) to T_V. Circuits are later built in the FPGA programmable logic that implement and control TDM and signal evaluation schedules. It is important to note that virtual wire technology assumes that the design can be statically scheduled (although exceptions are made for asynchronous signals which are not to be multiplexed at all) and that the execution of exactly M T_V cycles assures its correct transition to the next state. This methodology effectively transforms the original circuit to a synchronous equivalent, which operates with a clock period T_V. It has to assure that the signal setup time requirements for such synchronous circuit are satisfied and therefore no combinatorial path exists with the propagation time larger than T_V.

A different scheme [6], though again defined in general terms, has been primarily used with the partial crossbar architecture. This scheme is different in that it makes no attempt to transform the circuit to its synchronous equivalent, and it makes no assumptions about the signal propagation times inside the FPGA. All it does is the high-speed multiplexing of the FPGA inputs and outputs with the maximal speed permitted by the partial crossbar, printed circuit board, etc. The multiplexing period T_V does not depend at all on the design being emulated and is in fact constant, as is the multiplexing factor M. On the other hand, there are no guarantees that exactly M T_V cycles would be sufficient for advancing the design state. Signal transitions are propagated freely until they reach the FPGA boundary where they are synchronized to the TDM clock. As in the virtual wires technique, some signals can be excluded from multiplexing (and the clocks always are) so that the asynchronous behavior can be emulated more closely. The crossbar parts have the ability to demultiplex signals before switching them. Therefore, the fact that some signals share a physical wire does not impose any constraints on the way these signals should be routed.

For each solution, it has been argued that it has advantages over the other in the areas of emulation speed, support for asynchronous design styles, as well as the overall cost and the ease of adoption. However, it is important to point out the similar difficulties encountered by both. The virtual wires technology allows the increase of effective bandwidth of the FPGA pins only to the extent that T_V is less than the design state evaluation time. In order to keep the emulation speed from going down, it is necessary therefore to keep T_V low. However, to satisfy setup time requirements, T_V must be larger than any combinatorial path delay. Controlling these delays in an FPGA, while keeping compilation time low, is notoriously difficult. Besides, the delays are known only after the FPGA compilation is finished while the signal schedule has to be determined before this compilation starts. As a result, pin bandwidths cannot be increased so that they keep up with the ever-growing gate/pin ratio. Although the technique by Sample et al. [6] does not explicitly require controlling setup time, in the end it suffers from similar limitations.

There is an intriguing observation one can make while considering the multi-FPGA systems with TDM. A configurable logic structure (typically, a lookup table) that produces a signal which is subject to time multiplexing is severely underused. While its output needs to be valid only during one T_V period, this output is in fact being produced continuously. Thus on average, only one out of

M configurable logic structures could be in actual use at any given time. Conversely, we could use the configurable logic much more efficiently if it could compute different signals during different T_V cycles. Had this been the case however, it would effectively turn a lookup table into a processor that executes instructions over time.

19.2.3 Processor Arrays for Emulation

A processor-based hardware emulator [7] consists of bit-wide processors which execute a different Boolean function of N variables every T_V cycle. The input values on which the Boolean functions operate are supplied by sets of multiplexers that switch every T_V cycle as well. The Boolean functions and the control codes for the input multiplexers (addresses) together constitute instructions that are stored in the instruction memory. The depth of such memory determines the TDM factor M that applies both to the processors and to the interconnect structures. In this approach, every processor effectively performs the work that M FPGA lookup tables formally had.

In the semiconductor implementations of a processor-based emulator (Figure 19.1), routing structures are significantly simplified compared with FPGA. In place of a large variety of wire segments that could be arbitrarily combined with programmable switches, processor-based emulation chips contain a two-level routing structure [8] that recognizes processor cluster-level routing in addition to the uniform chip-level routing. Because of that, signal propagation delays are uniform. Processor-based emulators offer tighter control over the signal timing than an FPGA-based emulator's virtual logic can allow, and corresponding low duration of T_V cycle and high values of M.

To better understand how a processor-based emulator works, it is useful to briefly review how a logic simulator works. Recall that a computer's arithmetic-logic unit (ALU) can perform basic

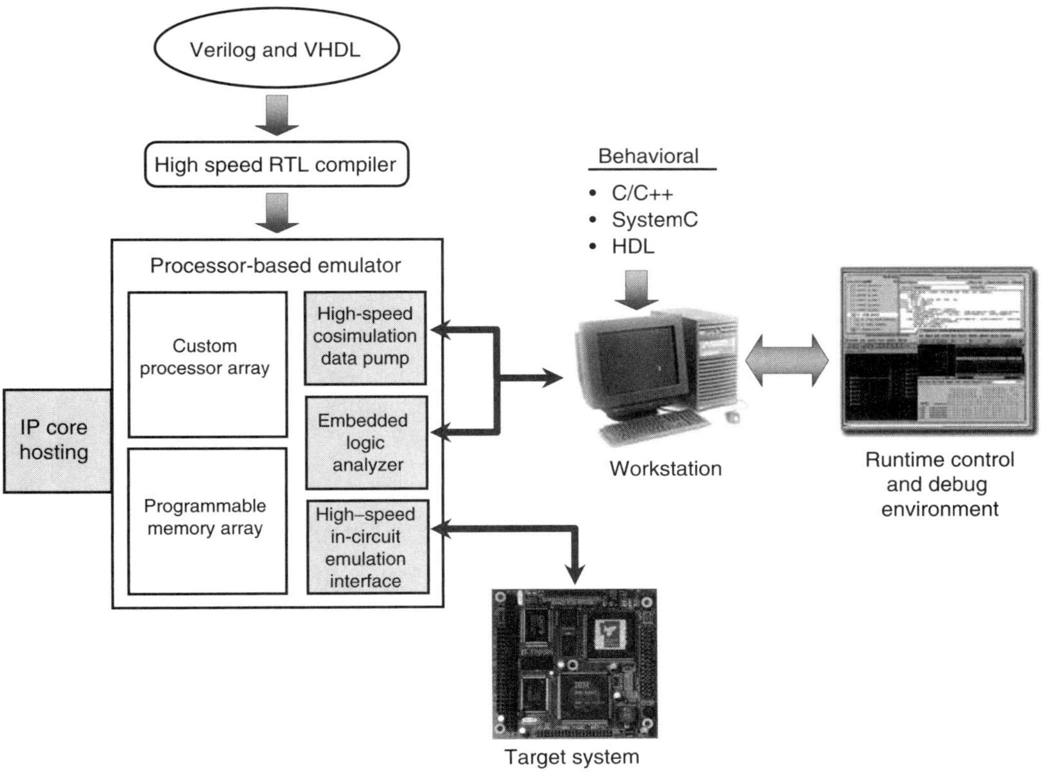

FIGURE 19.1 Processor-based emulator architecture.

Boolean operations on variables, e.g. AND, OR, NOT and that a language construct such as "always @ (posedge Clock) Q = D" forms the basis of a flip-flop. In the case of gates (and transparent latches), simulation order is important. Signals race through a gate chain schematically "left-to-right" so to speak, or "top-to-bottom" in RTL source code. Flip-flops (registers) break up the gate chain for ordering purposes.

One type of simulator, a levelized compiled logic simulator, performs the Boolean equations one-at-a-time in the correct order. (Time delays are not relevant for *functional* logic simulation.) If two ALUs were available, you can imagine breaking the design up into two independent logic chains and assigning each chain to an ALU, thus parallelizing the process and reducing the time required, perhaps to one half. A processor-based emulator has from tens of thousands to hundreds of thousands of ALUs, which are efficiently scheduled to perform all the Boolean equations in the design in the correct sequence. The following series of drawings (starting with Figure 19.2 and Figure 19.3) illustrate this process.

Step 1: Reduce Boolean Logic to Four-Input Functions

The following sequencing constraint set applies:

- The flip-flops must be evaluated first
- S must be calculated before M
- M must be calculated before P
- Primary inputs B, E, and F, must be sampled before S is calculated
- Primary inputs G, H, and J, must be sampled before M is calculated
- Primary inputs K, L, and N, must be sampled before P is calculated

Note that primary input A can be sampled at any time after the flip-flops.

Step 2: Scheduling Logic Operations among Processors and Time Steps

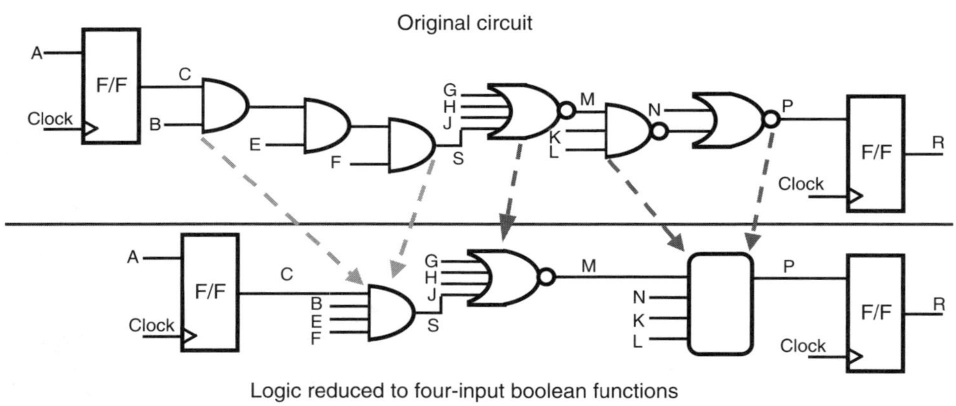

FIGURE 19.2 Result of reducing logic to four-input functions.

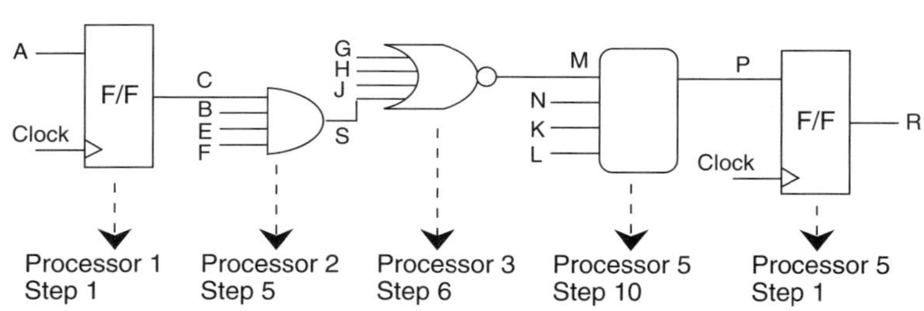

FIGURE 19.3 Result of scheduling logic.

Time Step	Processor 1	Processor 2	Processor 3	Processor 4	Processor 5
1	Calculate C	Sample B	Sample G		Calculate R
2		Sample E	Sample H		Sample N
3		Sample F	Sample J		
4					Sample K
5		Calculate S			Sample L
6			Calculate M		
7					Receive M
8					
9					
10	Sample A				Calculate P

One possible scheduling is shown above. An *emulation cycle* consists of running all the steps for a complete modeling of the design . Large designs take typically from 60 to 300 steps to schedule, depending on whether the compiler optimizes for speed or capacity.

In addition to the Boolean equations representing the design logic, the emulator must also efficiently implement memories, encrypted "soft" intellectual property (IP), and support physical IP "bonded-out" cores. For simulation acceleration it must also have a fast, low-latency channel to connect the emulator to a simulator on the workstation, since behavioral code cannot be synthesized into the emulator. For in-circuit operation, the emulator needs to support connections to a target system and test equipment. Finally, to provide visibility into the design for debugging, the emulator contains a logic analyzer. The logic analyzer will be described later.

The balance of this chapter will focus primarily on processor-based emulators as they represent the state-of-the-art currently, with comparative information where FPGA-based emulator capabilities are significantly different.

Key Characteristics of Processor-Based Accelerator/Emulator

- Maps any design style; automatically maps multiple clock domains, gated clocks, asynchronous loops
- Fast compilation without needing dozens of workstations
- High running speed
- Complete debug capabilities without recompiling design
- Architecture scales to 200 million gates and beyond
- Several hundred megabytes of design memory per million gates
- Ten thousand I/O pins or more for in-circuit applications
- Many simultaneous users
- No performance or capacity penalty for 100% debug visibility
- Built-in logic analyzer with complex triggering capability
- Flexible licensing
- Hard and soft IP support

19.3 Design Modeling

As was demonstrated above, the further we move along the line from a simple FPGA-based prototype to a complicated, highly scalable emulation system, the less the actual model execution mechanism resembles the target design implementation in silicon. In other words, for emulation, the user's design needs to be modeled. Below, we shall cover the major reasons and methods of design modeling.

19.3.1 Tri-State Bus Modeling

Tri-state busses are modeled with combinatorial logic. When none of the enables are on, emulators give the user a choice of "pull-up," "pull-down," or "retain-state." In the latter case, a latch is inserted into the design to hold the state of the bus when no drivers are enabled. In case multiple enables are on, for pull-up and retain-state logic, 0 will "win," and for pull-down logic, 1 will win. (Note that this is a good place to use assertions.)

19.3.2 Asynchronous Loop Breaking

Emulators do not model the precise gate-level timing of silicon. FPGA-based emulators have random "gate" delays and processor-based emulators have delays of 0. Because of the 0 delay characteristic of processor-based emulators, asynchronous loops are broken automatically by a delay flip-flop during compilation. However, by allowing the user to specify where loop breaks should occur, emulator performance may be enhanced, since the performance is related to the length of long combinatorial paths.

19.3.3 Clock Handling in Processor-Based Emulators

As described earlier, clocking was one of the prime sources of unreliability in FPGA-based emulators. Processor-based emulators completely avoid this problem. They can generate all the clocks necessary for a design or they can accept externally generated clocks. It is much more convenient to have the emulator generate all the design clocks needed — and it runs faster. To provide the maximum possible emulation speed while retaining the asynchronous accuracy required, processor-based emulators provide two methods of handling asynchronous design clocks: *aligned edge* and *independent edge* (Figure 19.4).

19.3.4 Independent Edge Clocking

Since emulators do not model design timing, but rather are *functional* equivalents, the exact timing between asynchronous clock edges is not relevant. It is only necessary that clock edges that are not simultaneous in "real life" are emulated independently. With independent edge clocking (Figure 19.5 and Figure 19.6), an emulation cycle is scheduled for every edge of every clock unless an edge is naturally coincident with an already scheduled edge. This is very similar to an event-driven simulator.

19.3.5 Aligned Edge Clocking

Aligned edge clocking is based on the fact that although many clocks in a design *happen* to have noncoincident edges because of their frequencies, proper circuit operation does not *depend* on the edges being independent (Figure 19.7). In this case, while proper frequency relationships are maintained, clock edges are aligned to the highest frequency clock, thereby reducing the number of emulation cycles required (compared to independent edge clocking), and increasing emulation speed.

In aligned edge clocking, an emulation cycle is first scheduled for each edge of the fastest clock in the design. Then, all other clocks are scheduled relative to this clock, with the slower clock edges "aligned" to the next scheduled emulation cycle. Note that no additional emulation cycles have been added for the second and third (slower) clocks (Figure 19.8 and Figure 19.9). Thus emulation speed is maintained. Also note that while edges are moved to following fastest clock edge, frequency relationships are maintained which could be essential for proper circuit operation. Giving the user the flexibility to switch between

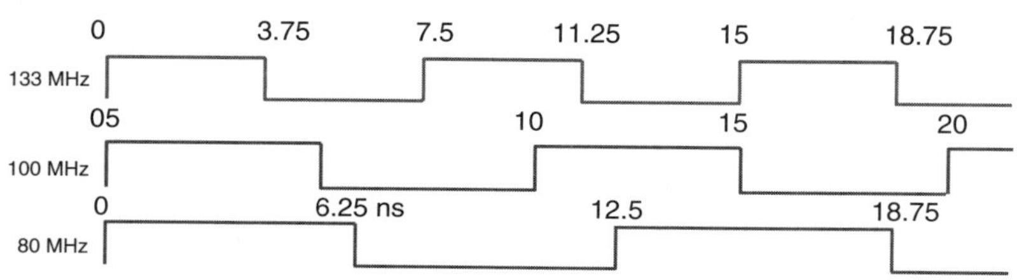

FIGURE 19.4 Clocking example — three asynchronous clocks.

aligned edge clocking and independent edge clocking in processor-based emulators provides both high asynchronous accuracy and the fastest possible emulation speed for a wide variety of design styles.

Comparison of Clocking Techniques in Processor-Based Emulator

	Independent Edge Clocking	Aligned Edge Clocking
Preserves frequency relationships	Yes	Yes
Matches simulation	Yes	May not
Maintains speed with more clocks	No	Yes
Maintains edge independence	Yes	No

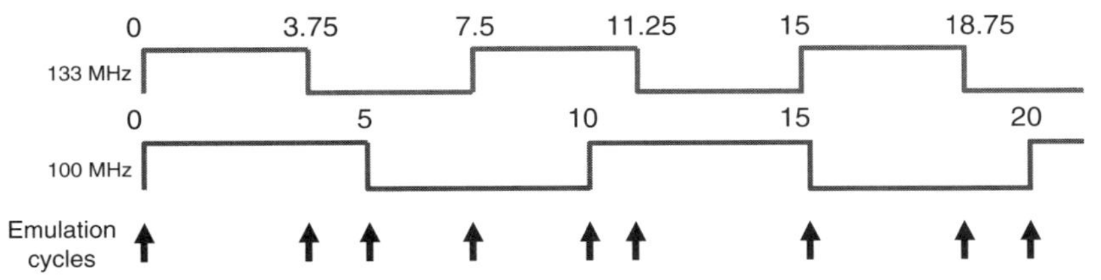

FIGURE 19.5 Independent edge clocking of two asynchronous clocks.

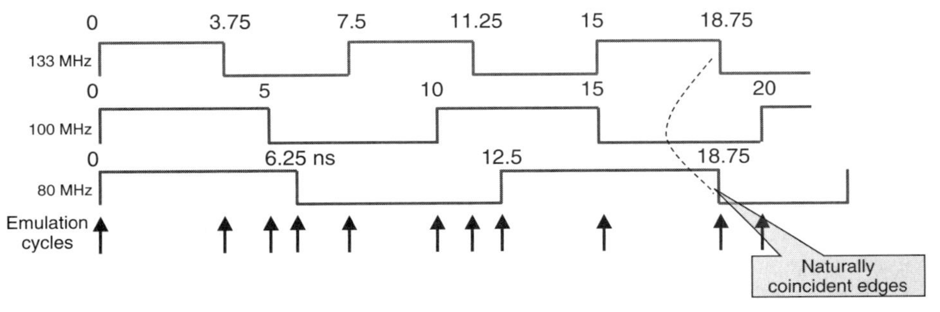

FIGURE 19.6 Adding a third asynchronous clock with independent edge clocking.

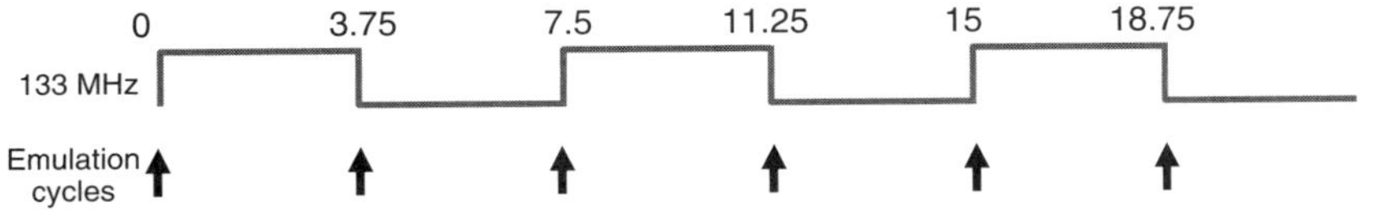

FIGURE 19.7 Aligned edge clocking starts by assigning an emulation cycle for each edge of the fastest clock.

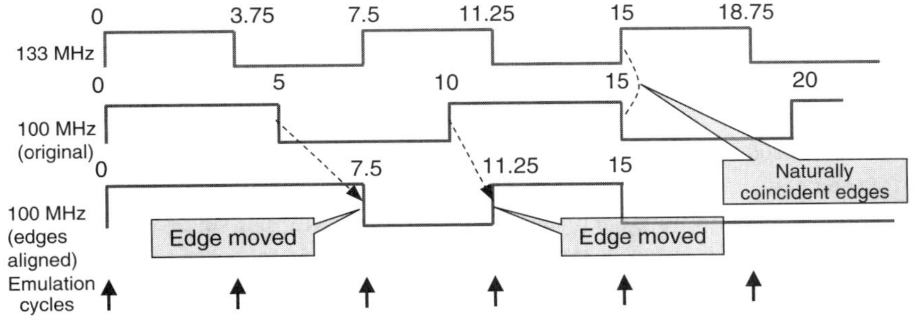

FIGURE 19.8 Add second clock aligning edges to fastest clock.

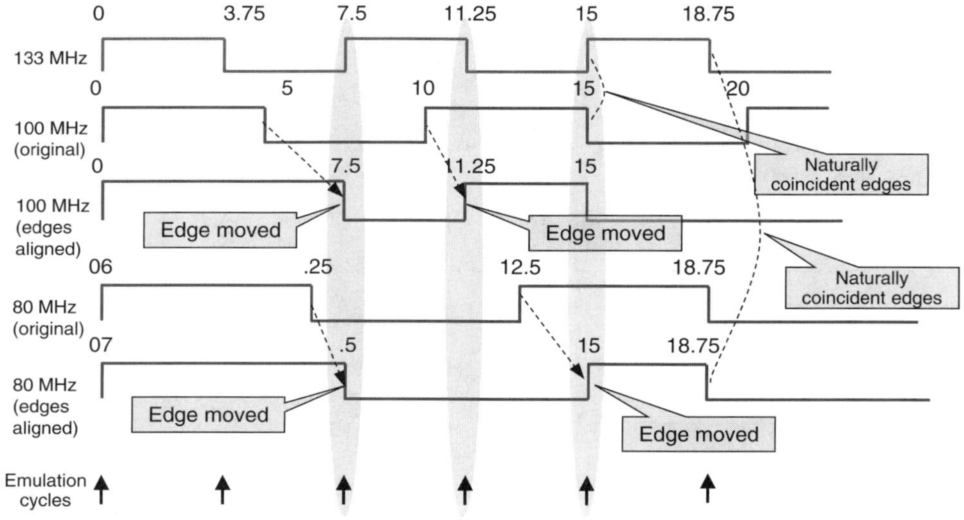

FIGURE 19.9 Add additional clocks by aligning edges to fastest clock.

19.3.6 Timing Control on Output

When interfacing to the real world, it is sometimes necessary to control the relative timing of output signals. A dynamic random access memory (DRAM) memory interface is one such example — all the address lines must be stable before asserting Write-Enable. Since processor-based emulators schedule logic operations to occur in sequence, it is easy to add a constraint on the timing within the emulation cycle on individual (or groups of) output signals, to control the timing to a very high resolution relative to other output signals (Figure 19.10). The compiler then schedules this output calculation at the appropriate point in the emulation cycle. This is not possible with FPGA-based emulators since they have no control over timing within a design clock.

19.3.7 Timing Control on Input

In a similar way, and for similar reasons, timing on input signals (sampling) may be controllable by the user to meet specific situations. Again, processor-based emulators can simply schedule specific input pins to be sampled before or after others within the emulation cycle. With FPGA-based emulators the user must "tweak" timing by adding delays to certain signals — including a large guard-band, because FPGA-based emulators cannot control absolute timing along different logic paths. For FPGA-based emulators, this is a hit-or-miss proposition and may vary from compile to compile.

19.3.8 Generating High-Speed Clocks

Occasionally, it may be necessary to generate a clock at a multiple of the highest design clock. An example is a chip with an internal clock divider. It would reduce emulation performance to run the entire emulation at this higher speed when only a couple of flip-flops are running at this rate. Since there are usually over a hundred steps in an emulation cycle, it is easy for processor-based emulators to generate a higher frequency clock by scheduling multiple changes on an output within a single emulation cycle (Figure 19.11). As shown below, the user may exercise control over the timing of all the changes and create fairly complex clock signals, if needed. FPGA-based emulators cannot do this because they have no control over timing within the emulation cycle.

19.3.9 Handling IP

Intellectual property can be an important part of a design. For acceleration and emulation design, IP may be supplied in two basic forms: (1) as physical silicon — a "bonded-out core," or (2) as a synthesizable RTL source file. Note that if IP is supplied in behavioral form, accelerated cosimulation must be used, with the behavioral code running on the workstation, which will reduce verification performance.

19.3.9.1 Physical IP

Most emulators and accelerators provide a mechanism for incorporating physical IP into the design model. The IP component will be mounted on an IP block of some sort and inserted into the emulator. Some emulation vendors also supply turn-key IP blocks with various components mounted, e.g., a large, programmable FPGA with a flash card reader on the front panel, or an ARM IP block which mates with the ARM-supplied IP tile and accepts ARM programming files (Figure 19.12).

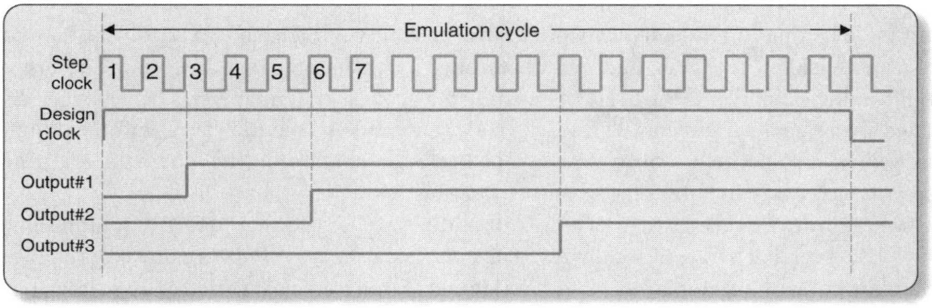

FIGURE 19.10 Processor-based emulators can adjust output timing with high precision.

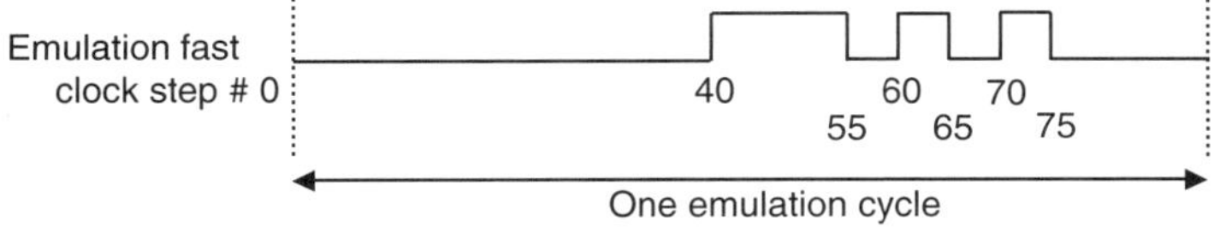

FIGURE 19.11 Special, high frequency clocks may be generated by processor-based emulators.

FIGURE 19.12 Physical IP prototype block with TriMedia core mounted.

The user may also mount a physical IP on their target system, but it must be noted that these connections may be treated differently when running vector regressions.

19.3.9.2 Source Code IP

RTL IP source files may be supplied and are included along with the user's design files in the compilation flow. RTL files for IP may be encrypted by the IP vendor for confidentiality reasons. In such a case, the user cannot inspect the encrypted block or any signals inside the block However, the IP vendor may provide access to defined signals the user might reasonably need in order to debug designs using the IP. An example is a microprocessor register set, which may be needed for debugging purposes. In other respects, the IP block is handled as part of the design.

19.4 Debugging

19.4.1 Logic Analyzer and Signal Trace Display

Most of the time spent verifying designs is on debugging problems. Thus, productivity in debugging is paramount in delivering time-to-market. Core to this capability is being able to rapidly display any signal in the design, and being able to trace back rapidly through a circuit to find the cause of the observed anomaly. The built-in logic analyzer is the heart of an emulator's debugging capability. The debugger displays in an integrated way, signal waveforms, RTL source code, generated schematic views, and the design hierarchy. Signal values at a specific point in time can be back-annotated onto the schematic and RTL source code displays to accelerate the user's grasp of the situation.

To give some context, consider that an emulator with a very large design with 20 million signals, running at 500 kHz would generate 60 Terabytes of signal trace data *per minute*! In any logic analyzer, there is a finite amount of memory for signal trace data storage. Emulators deal with this by presenting the user with trade-offs between width (number of signals) and depth (number of clock cycles). Emulators also use several techniques to "extend" the logic analyzer trace capacity greatly. Some of these techniques are covered under various patents granted or applications pending.

Emulators only need to capture design inputs and the outputs of flip-flops and memories. The values of other signals are Boolean combinations of these and can be quickly calculated, *when needed*. This capability would be necessary in any event since combinatorial logic in the design is optimized into n-input Boolean functions and thus some combinatorial signals in the design do not exist individually in the emulator.

Oftentimes, the engineer doing the debugging may not know ahead of time which signals will need to be probed to debug a problem. This is especially true when verification is performed by a separate team of engineers. In the course of debugging in such situations, the engineer may need to see *any* signal in the design to trace the problem back to the root cause. Emulators often allow the user to trade-off display speed, logic analyzer depth, and the support of dynamic target systems to get the best debugging capabilities for their particular needs.

In contrast to 100% signal visibility, *Dynamic Probing* provides a very deep logic analyzer with a smaller number of probes. Users must specify prior to running — and also prior to compilation with FPGA-based emulators — the signals they wish to see. Much as with simulation, adding a signal to the list allows display of that signal going forward in time, but not backwards in time. ("Forward" refers to clock cycles that have not yet been emulated, "backward" refers to clock cycles that have already been emulated.) Processor-based emulators accomplish this rapid change of signals to be probed (<1 sec to change 1000 signal probes) by utilizing otherwise "idle" time steps in the processor array to route signals to available logic analyzer memory channels. The compiler reserves a small percentage of the processor array bandwidth for this purpose when using Dynamic Probing, allowing new signals to be probed without changing the "program" for emulating the design. Note that in FPGA-based emulators, when not using 100% visibility, recompiling the entire design with the new probes (or at least a few FPGAs) is necessary — which changes the timing of the design model, sometimes resulting in different functional behavior or reduced operating speed.

For any of the above vision choices, emulators often also provide the ability to extend the depth of signal tracing (number of samples) to infinity — from the beginning of emulation, showing "*all the signals, all the time*" — but with some trade-offs. To support "infinite" trace data, the emulator is stopped periodically (as the physical logic analyzer becomes full) to upload the necessary data to the workstation. Then emulation continues. This approach also requires a use model, which tolerates periodic stopping of the clocks to the target system (typically a static target system). The advantage is that the logic analyzer now has infinite depth, and is able to display signal trace data from the beginning of the emulation run for any or *all* signals in the design. The trade-off is that stopping the clocks to upload the logic analyzer data results in reduced net emulation speed — usually 2 to 3 times slower. However, the user can turn on and off the infinite trace recording at run time so that the reduced speed is only experienced during times of interest. For example, one could boot a PC at full speed and then turn on infinite tracing to study a particular problem. Thus, in spite of having a finite, practical amount of logic analyzer trace memory, emulators can provide a rich set of vision choices from which the user can select a suitable set of trade-offs for their use model.

19.4.2 Definition of Trigger Condition

Emulators allow the user to specify complex trigger conditions for the logic analyzer. In the course of debugging a problem, the need to change the trigger condition is often encountered. Processor-based emulators support changing the trigger condition instantly, without needing to recompile the entire design. FPGA-based emulators may require recompiling one or more FPGAs in order to change trigger conditions when the change involves new signals.

19.4.3 Design Experiments with Set, Force, and Release Commands

Design Experiments allow the user to make certain design changes instantly, without recompilation, and see the result. This can be useful

- To disable sections of the circuit that are interfering with the section you want to inspect
- To force signals on or off to enable/disable signal effects
- To inject changes into the circuit, including complex time sequences, that duplicate the proposed fix and observe their effect

This allows the user to experiment with ideas without having to recompile the design. Design experiments are accomplished using the Set, Force, and Release commands, possibly included in a script for time sequencing. *Set* instantly sets a flip-flop to a state of 1 or 0, but then allows normal circuit operation to change it. *Force* and *Release* are complimentary commands; *Force* will freeze a signal to 1 or 0 until a *Release* command is given for that signal.

19.5 Use Models

Emulators and accelerators can be used in a number of ways to accelerate functional verification.

19.5.1 Accelerated, Signal-Based CoSimulation

Using a logic simulator such as Cadence Incisive, in this use model achieves faster simulation. The synthesizable RTL is moved into the emulator, where it runs several orders of magnitude faster than on the workstation. Behavioral code (e.g., testbench and memory models) continues to be run on the workstation by the simulator. Overall performance is limited by the communication channel between emulator and workstation and also by the testbench execution on the simulator. Most users in this mode prefer to debug in the familiar simulation environment, providing visibility into the signals in the RTL code in the emulator.

19.5.2 C/C++ Testbench

Similar to cosimulation, this use model has a C testbench (and possibly models) instead of a hardware description language (HDL) testbench. The C code is interfaced to the emulator through an application programming interface (API). This mode can provide higher performance than HDL cosimulation because C code executes faster than HDL in a simulator. C testbenches are becoming more popular because C is often used for system-level modeling before designing the ASIC and some of the test routines may be reusable for the ASIC testbench. Debugging in this mode involves using the emulator debug environment (which the user prefers to look like the simulator) as well as a C debugger.

19.5.3 Transaction-Based Acceleration

Transaction-Based Acceleration (TBA) is a form of cosimulation in which *transactions* are sent to the emulator rather than bit-by-bit signal exchanges. This reduces the traffic between workstation and emulator, allowing higher performances to be achieved (where the channel is the limiting factor). Transactors are designed in synthesizable RTL and put into the emulator with the design. These transactors map the transactions from the workstation into bit-by-bit operations for the design in the emulator. TBA has the additional advantage of raising the level of debugging of the design from individual logic levels to transactions which are more meaningful to the designer. There is a significant issue of acquiring a rich library of transaction IP that can be (or has been) synthesized. The Accellera committee (www.accellera.org) has standardized an interface between simulators and emulators as the Standard Co-Emulation Modeling Interface (SCE-MI) which, if supported, allows mix-and-match use of emulators/accelerators with simulators for transaction-based testbenches.

19.5.4 Vector Regression

This use model applies large sets of vectors for regression testing small changes in the design. The vectors may come from simulation, testers, or other sources. Typically, they are used close to tape-out when small changes are made to the logic and the user must ensure that nothing was "broken" by the change. Comparison of resulting vectors against known good results may be done either in the emulator or the workstation (depending on emulator capabilities). Emphasis is on fast execution and pass/fail results.

19.5.5 Embedded Testbench

In this use model, the testbench (and the entire design) is written in synthesizable RTL code and the design and testbench are run in the emulator at full emulation speeds. Few users are willing to write a testbench in synthesizable RTL because of the extra effort and the existing investment in behavioral testbench code. This use model is most often applied to processor chips, because performance is valued so highly as to make the extra testbench effort worthwhile. When the ASIC contains a processor core, embedded testbench mode can be applied using a simple synthesizable testbench and processor object code executing out of emulator memory.

19.5.6 In-Circuit Emulation

This model of operation connects the emulator with a user's *target system* — a prototype of the system the user is designing. The emulator typically replaces the ASIC(s) being designed for the target system, allowing system-level and software testing prior to silicon availability. Because the emulator runs much slower than the ASIC being designed, slow-down solutions must be implemented to match the real-world interfaces to the emulator. Clocks may be derived in the target system or in the emulator. In some simple environments, the target system may be a piece of test equipment that generates and verifies test data — network testers are a typical example of this.

19.6 The Value of In-Circuit Emulation

In-circuit emulation provides the highest performance and delivers the following value for verification:

1. Tests the design with "live" data in real world environment
   a. Finding bugs you did not imagine test cases for
   b. Reducing the number of test cases you have to write
   c. Reducing your risk of silicon respin
2. Allows software testing to begin before first silicon, even before tape-out
   a. Hardware/software interaction uncovers interface bugs
   b. Supports software verification with live data
   c. Enables optimal hardware/software trade-offs
   d. Saves months off software delivery cycle
3. Delivers products to market sooner
4. Reduces risk of shipping products with bugs requiring a product recall
5. Avoids performance reduction as design size grows compared to simulation

19.7 Considerations for Successful Emulation

Creating a successful ICE environment involves a few additional considerations relative to simulation or acceleration.

19.7.1 Creating an In-Circuit Emulation Environment

Emulation generally runs at 200 kHz to 1 MHz while the actual system you are designing will run much faster. When the emulator must interface with real world, full speed devices, an emulation environment which bridges the speed difference between the real-world speeds and emulation speeds must be created. There are a variety of ways this might be accomplished. Sometimes it is as simple as using a tester that can be slowed down. For example, for Ethernet interfaces, usually the PHY is bypassed. A Spirent *SmartBits* tester can slow down the Ethernet bit stream to emulation speed (both transmit and receive).

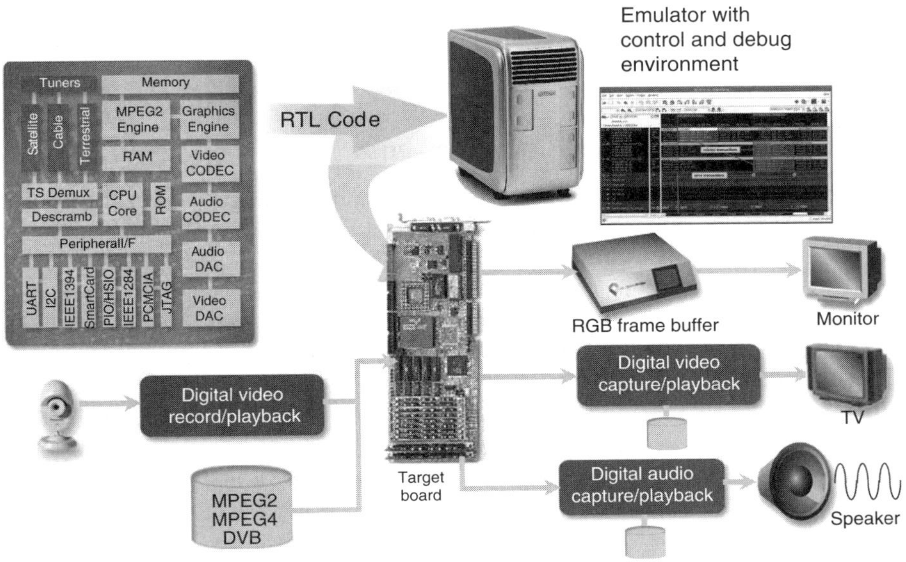

FIGURE 19.13 Emulation environment for set-top box design.

Alternatively, you can use an Ethernet bidirectional packet buffer, such as a Cadence SpeedBridge™, between multiple real-world connections and slowed-down emulation speed Ethernet ports.

Another emulation environment example is the verification of an ASIC for a video graphics card. The red-green-blue (RGB) data will be produced at emulation speeds, but the monitor cannot sync the image at these speeds. You can use a video frame buffer, such as the Cadence RGB SpeedBridge, which accepts the RGB data at emulation speeds and puts the pixel data into a frame buffer. The frame buffer is then read out at normal scan rates and drives a monitor. In this way, the user views the image produced by the emulated design but the frame update rate is reduced. Such an environment is common as it allows the user to "see" the results of their design.

A set-top box emulation environment is shown in Figure 19.13. The RF circuitry is dropped — emulation is for doing functional verification of the digital portion of the design. The rest of the ASIC is compiled into the emulator and the emulator plugged into the ASIC socket on the PC board. A digital video recorder/player is used to deliver the digital data stream representing the demodulated RF. This data stream could come from some standard data sets for motion picture expert group (MPEG) compression testing or it could be data that the user recorded from a video camera and microphone. The video frame buffer is used to display (nearly) still images. A second digital video recorder/player captures video data at emulation speed, records it to hard disk, and then plays it back at full speed. Being able to view motion is a critically important factor in MPEG decoder work, since the MPEG process does not reproduce the exact bit stream put through it (termed "lossy"). Hence, decoder algorithms are evaluated subjectively. In a similar way, audio is produced via a digital audio recorder/player which buffers and replays audio at full speed so that you can hear the quality of reproduction.

19.7.2 Debug Issues with Dynamic Targets

Some emulators temporarily suspend the clocks during operation, either to capture or to calculate and display signal data. While the emulator is doing this, it is not available to emulate with the target system. As long as the target system is static, there is no problem. Emulation resumes normally when the clocks are restarted. Dynamic targets however, may not recover automatically from having their clocks stopped for a few seconds. With dynamic target systems it may be necessary to select a different choice for capturing and/ displaying signal data to avoid this problem — if one is available. Note that some SpeedBridges solve this problem by keeping the target clocks running even when the emulator is not supplying clocks to it. If you expect to have any dynamic target systems, this would be a key issue in selection of an emulator; some emulators do not offer alternate vision choices and hence cannot be used at all with dynamic targets.

19.7.3 Considerations for Software Testing

Most electronic products today have software content — and many of them have extensive software content. The traditional approach to software verification is to wait for (mostly) working silicon to begin software debugging. This makes the hardware and software debugging tasks largely sequential and increases the product's development time. It also means that a serious system problem may not be found until after first silicon, requiring a costly respin and delaying the project for 2 to 3 months. The objective of hardware/software coverification is to make the hardware and software debugging tasks as concurrent as possible. At a minimum, this means starting software debug as soon as the IC is "taped-out," rather than waiting for good silicon. But even greater concurrency is possible. Software debugging with the actual design can begin as soon as the hardware design achieves some level of correct functionality. Starting software debugging early can save from 2 to 6 months of product development time. Early software testing can also uncover a bug in the ASIC early enough to fix it in the ASIC, thereby avoiding a software patch that may degrade usability or performance of product. The increased level of software testing available, with the speed of ICE, also reduces the risk of a software bug going undetected and getting released with the product, and the expensive upgrades, recalls, or lost business that might result.

There are additional benefits to be obtained by starting software verification prior to freezing the hardware design. If system problems or performance issues are found, designers can make intelligent tradeoffs in deciding whether to change the hardware or software, possibly avoiding a degradation in product functionality, reduced performance, or an increase in product cost.

Early software integration on "real" hardware provides enormous value

- Software is developed on "real" hardware before implementation is complete
- Software verification can be done with "live" data (in-circuit)
- Early software development platforms can run faster than with other tools
- There's an opportunity to fix the chip design before tape-out rather than patching the software, allowing an optimal hardware or software solution
- System-level verification allows hardware/software tradeoffs

Whether a microprocessor in-circuit emulator (MP-ICE) or RTOS debugger is used, the software debugging and hardware debugging environments can be synchronized so that hardware/software interface issues can be debugged conveniently. The breakpoint/trigger systems of the emulator and MP-ICE are crossconnected such that the emulator's logic analyzer trigger is one of the MP-ICE breakpoint conditions and the MP-ICE breakpoint trap signal is set as an emulator logic analyzer trigger condition. Therefore, if a software breakpoint is reached, the emulator captures the condition of the ASIC at the same moment. If an ASIC event occurs that triggers the logic analyzer, the software is stopped at that moment. This allows inspection of the hardware events that led to a software breakpoint, or of the ASIC operation resulting from executing a set of software instructions. This kind of coordinated debugging is extremely valuable for understanding subtle problems that occur at the hardware/software interface.

There are a variety of ways to interface to a microprocessor for software testing. JTAG, RS-232, or Ethernet connections can control a resident debug monitor. The OS in the processor may have its own debug environment, or a MP-ICE may be used. The microprocessor itself could be a packaged part or a bonded-out core mounted on the target system, an RTL model which gets mapped into the emulator with the rest of the design, or an instruction-set simulator (ISS) model running on the workstation. Of these, the ISS will suffer a considerable performance penalty compared to the others. Detailed information for these various approaches may be found in [9]. Another approach in which a silicon vendor supports high-performance software development for embedded processors prior to silicon availability is described in [17].

The most common configuration for high-performance software testing is a physical model connected via a JTAG port to the software debugger. The JTAG interface spec does not allow for the JTAG clock to be stopped midstream. This also can cause difficulty with some of the vision choices in emulators. Some emulators do not offer alternate vision choices and, hence cannot be used for software testing with systems using a JTAG port for connection (without suffering huge performance penalties by using an ISS). Other emulators have a single 100% visibility capability which does not allow for operating a JTAG port to the software debugger except by reverting to compiled (and constantly recompiled) probes. As software testing becomes more and more important in achieving time-to-market demands, it is critical to know what software development environments an emulator can support with high performance.

19.7.4 Multi-User Access

Some emulators allow their total capacity to be dynamically partitioned among a number of simultaneous users. This increases the usage considerably, especially early in the design cycle for block-level acceleration and in environments where multiple projects are sharing the emulator. As the design progresses and full-chip and system-level testing is required, a larger percentage of the emulator capacity (perhaps all of it) is used to emulate it.

19.8 Summary

Hardware accelerators and emulators provide much higher verification performance than logic simulators, but require some additional effort to deploy. ICE provides the highest performance — often 10,000 — 1,000,000 times faster than a simulator, but requires an emulation environment to be built around it with speed buffering devices. Accelerators and emulators require the user to be aware of the differences between simulation and silicon (i.e., emulators and chips)

1. Simulation has 12 states or more, silicon has but two states.
2. Simulation generally executes RTL statements sequentially, silicon "executes" RTL concurrently.
3. Simulation is highly interactive, silicon less so.

FPGA-based emulators use commercial FPGAs and are smaller and consume less power, while processor-based emulators require custom silicon designs and consume more power but they compile designs much faster and can deliver higher performance. The key benefit of using emulation for design verification is shorter time-to-market and higher product quality.

References

[1] M.R. Butts et al., U.S. Patent 5,036,473.
[2] N.-P. Chen et al., U.S. Patent 5,475,830.
[3] R.A. Price et al., U.S. Patent 5,259,006.
[4] W. Dai et al., U.S. Patent 5,452,253,
[5] A. Agarwal et al., U.S. Patent 5,596,742.
[6] S.P. Sample et al., U.S. 5,960,191.
[7] W.F. Beausoleil et al., U.S. Patent 5,551,013.
[8] W.F. Beausoleil et al., U.S. Patent 6,035,117.
[9] R. Turner, Accelerated hardware/software co-verification speeds first silicon and first software, *Cadence Des. Sys.*, available on www.cadence.com.
[10] P. Christie and D. Stroobandt, The interpretation and application of Rent's rule, *IEEE Trans.VLSI Syst.*, 8, pp. 639–648, December 2000.
[11] B. Landman and R. Russo, On a pin versus block relationship for partitioning of logic graphs, *IEEE Trans. Comput*, C20, pp. 1469–1479, 1971.
[12] http://www.ecs.umass.edu/ece/tessier/courses/697c/ps3/rents_rule.pdf
[13] J. Babb, R. Tessier, and A. Agarwal. Virtual wires: overcoming pin limitations in FPGA-based logic emulators, Proceedings of IEEE Workshop on FPGA Based Custom Computing Machines, April 1993, pp. 142–151.
[14] K. Kuusilinna, et al., Designing BEE: a hardware emulation engine for signal processing in low-power wireless applications, *EURASIP J. Appl. Signal Process.*, 2003, pp. 502–513, 2003.
[15] J. Varghese, M. Butts, and J. Batcheller, An efficient logic emulation system, *IEEE Trans. VLSI Syst.*, 1, 171–174, 1993.
[16] J. Babb et al., Logic emulation with virtual wires, *IEEE Trans. CAD*, 16, pp. 609–626, 1997.
[17] G. A. Dada, Reducing time to market: parallel software development with emulation and simulation tools for MXC architectures, Motorola, http://www.freescale.com/files/wireless_comm/doc/brochure/MXCSWDEVWP.pdf.

20
Formal Property Verification

Limor Fix
Design Technology, Intel
Pittsburgh, Pennsylvania

Ken McMillan
Cadence Berkeley Laboratories
Berkeley, California

20.1 Introduction

In 1994, Intel spent 500 million dollars to recall the Pentium CPU due to a functional bug in the floating-point division result which was off by 61/10,00,000. The growing usage of hardware and software in our world and their insufficient quality has already cost an enormous price in terms of human life, time, and money. Both hardware and software designs are experiencing a verification crisis since simulation-based logic verification is becoming less and less effective due to the growing complexity of the systems. To fight the low coverage in simulation, large systems are often decomposed into blocks and each of these blocks is simulated separately in its own environment. Only when the block-level simulations are stable, will integration and full-chip/system simulation be carried out. This methodology caused a dramatic increase in the effort invested in simulation-based verification since a complete simulation environment, including tests, checkers, coverage monitors, and environment models, is developed and maintained for each block. Industrial reports show that half of the total design effort is devoted to verification, the number of bugs is growing exponentially, and hundreds of billions of simulation cycles are consumed [1]. Despite the huge investment in verification, only a negligible percentage of all possible executions of the design is actually being tested.

Formal property verification has major advantages over simulation. It offers exhaustive verification technology, which is orders of magnitude more efficient as compared to scalar exhaustive simulation. In formal verification, all possible executions of the design are analysed and no tests need to be developed. The main deficiency of formal verification is its limited capacity compared with simulation. The state-of-the-art formal property verification engines can handle models in the order of 10 K memory elements and with complementary automatic model reduction techniques properties are verified on hardware designs and software programs with 100 to 500 K memory elements.

To give the reader some initial intuition into the formal verification approach, let us consider the simple program:

$$Int\ x;$$

$$read\ (x); \quad if\ x < 0\ then\ x = -x; \quad print(x);$$

Assume that x is a 32-bit integer with a range of 2^{32} different values. To verify fully that for every legal input the program prints a nonnegative number, simulation needs to cover the entire input space, i.e., 2^{32} simulation runs. In a formal verification approach, a single symbolic simulation run provides full coverage of the input space and thus exhaustively verifies the program. In this example, the input to the symbolic simulation is a set of n Boolean variables representing the input bits and the result of the symbolic simulation is a Boolean expression that reflects the program computation over the input variables. The resulting Boolean expression is then checked to be nonnegative.

A *property* is a specification of some aspect of a system's behavior that is considered necessary, but perhaps not sufficient for correct operation. Properties can range from simple "sanity checks" to more complex functional requirements. For example, we might specify that a mutually exclusive pair of control signals is never asserted simultaneously, or we might specify that a packet router chip connected to the PCI bus always conforms to the rules for correct signalling on the bus. In principle, we could even go so far as to specify that packets are delivered correctly, but full functional specifications of this sort are rare, because of the difficulty in verifying them. More often, formal verification is restricted to a collection of properties considered to be of crucial importance, or difficult to check adequately by simulation.

In hardware design, formal property verification is used throughout the design cycle.

As Table 20.1 illustrates, a hardware design project consists of five major simultaneous activities, called here (rows): RTL, Timing, Circuit, Layout, and Post-silicon. Over the duration of the project (columns), each activity has several consecutive subactivities. For example, the RTL activity starts with uArch development, which is followed by RTL development and it ends with RTL validation. At the early stages of the project when the architecture and the microarchitecture are designed, formal models of new ideas and protocols are developed and formally verified against a set of requirements. For example, the formal verification of a new cache coherence protocol may be established at this stage. This verification is usually carried out by the validation team working closely with the architects. An important benefit from this activity is the intimate familiarity of the validation team, early at the design stage, with the new microarchitecture. Later, while the RTL is being developed, the formal property verification

TABLE 20.1 Formal Property Verification in Hardware Design Flow

Project phases	Design	Stabilization	Implementation	Convergence	Debug
RTL	uArch development	RTL development	RTL validation		
Timing	Timing specification at block level	Full timing specification	Converging timing to project goals	Timing converged	
Circuit	Schematic entry	25% schematic	100% schematic	Only bug fix in schematic	
Layout			Layout clean at block level	Layout assembled and clean	
Post-silicon				A0 tape-out	Functional silicon
FPV	Verify micro architectural properties on FV models				
		Verify RTL properties			
			Verify micro architectural properties on the RTL		

activity shifts to RTL verification. The properties to be verified at this stage are either derived from the high-level microarchitectural specification, for example, the IEEE specification of the arithmetic operations, or from the implementation constraints that need to be verified on the RTL, for example verifying that all timing paths are within the required range. Constraints consumed by synthesis, static timing analysis, and power analysis are formally verified to be valid and thus formal verification is offering not only better functional correctness but also debugging of implementation problems. Formal property verification of the RTL has been mostly carried out by validation teams over the last 10 years. However, recently, with the increased capacity of the formal verification technology and increased ease of use, RTL and circuit designers are starting to use formal property verification products. Toward the end of the project and even after tape-out, the focus of formal property verification moves back to the verification of the high-level microarchitecture protocols — but this time, these properties are verified against the RTL and not against artificial formal models. It is still a big effort to verify high-level microarchitecture protocols against the RTL — only a few high-risk areas in the design are selected to be verified and the verification is carried out by the validation teams.

In the hardware electronic industry, tools for formal property verification have been developed since the early 1990s in internal CAD groups of big VLSI companies like Intel, IBM, Motorola, Siemens [2,3], and Bell-labs [4]. In the late 1990s, several start-up companies, like @HDL, 0-in, Real Intent, Verplex, and others, have begun offering similar tools. The large EDA companies are also offering products for formal property verification today.

In advanced electronic systems, software and hardware are becoming more and more interchangeable. One common usage of software is in high-level modeling of complex systems. These systems are designed using high-level software languages like C and SystemC [5] and only later in the design cycle, parts of the system (or all of it) are designated to be implemented in hardware. The other common usage of software is in the design of Embedded Systems. These systems usually consist of embedded software that executes on top of an embedded microprocessor, or increasingly, processors. Modern CPUs may also include an embedded software called *microcode* that is used to simplify the instruction set implemented in hardware.

Simulation is currently the main verification tool for software. However, the increased frequency of fatal and catastrophic software errors is driving the electronic design community to look for better, more exhaustive verification solutions. In the last 5 years, intensive efforts have been directed to finding new and fully automatic ways to apply formal methods to software verification.

Two applications of formal verification are currently under research and development. One is *property verification of software*, which is aimed at establishing the functional correctness of the software with respect to properties developed by the programmer and/or automatically extracted sanity properties that check, for example, array bounds [6–11]. The second application of formal verification to software is *translation validation*, which is aimed at verifying the correct translation of the high-level software model into the corresponding hardware description, e.g., Verilog, or into a lower-level software program [12–15].

Formal property verification of software is just starting to move into the industry — most promising results were reported by the research team in Microsoft, verifying Windows NT drivers. As Bill Gates said in at a Keynote address at WinHec, April 2002: "*Things like even software verification, this has been the Holy Grail of computer science for many decades but now in some very key areas, for example, driver verification we're building tools that can do actual proof about the software and how it works in order to guarantee the reliability.*" [16]

Formal automatic translation validation of software is also at the initial transfer stage in the electronics industry, for example Intel has developed a formal tool for verifying correspondence between microcode programs [15]. A few EDA startups, like Calypto, are offering C to Verilog translation validation. Results reported for academic systems demonstrate ability to verify 10,000 lines of source code in C against the compiler results [17] and being able to verify an academic microprocessor with 550 C lines against its RTL implementation with 1200 latches [14].

20.2 Formal Property Verification Methods and Technologies

20.2.1 Formal Property Specification

Hardware designs and embedded software generally fall into a class known as *reactive systems*. As defined by Pnueli [18], these are systems that interact continuously with their environment, receiving input and producing output. This is in contrast to a computer program, such as a compiler, that receives input once, then executes to termination, producing output once. To specify a reactive system, we need to be able to specify the valid input/output *sequences*, not just the correct function from input to output. For this purpose, Pnueli proposed the use of a formalism known as *temporal logic* that had previously been used to give a logical account of how temporal relationships are expressed in natural language. Temporal logic provides operators that allow us to assert the truth of a proposition at certain times relative to the present time. For example, the formula *Fp* states that p is true at *some* time in the future, while Gp says that p is true at *all* times in the future. These operators allow us to express succinctly a variety of commonly occurring requirements of reactive systems. We can state, for example, that it is never the case that signals `grant1` and `grant2` are asserted at the same time, or that if `req1` is asserted now, that eventually (at some time in the future), `grant1` is asserted. The former is an example of a *safety* property. It says that some bad condition never occurs. The latter property is an example of a *liveness* property. It says that some good condition must eventually occur. To reason about liveness properties, we must consider infinite executions of a system, since the only way to violate the property is to execute infinitely without occurrence of the good condition.

Since Pnueli introduced the use of temporal logic for specification, a number of notations have been developed to make specifications either more expressive or more succinct. For example, we may want to specify that a condition must occur not just eventually, but within a given amount of time. In a real-time temporal logic [19], the formula $F_{(<5)}p$ might be used to specify that p must become true within five time units. Moreover, because hardware systems are typically pipelined, it is common to refer to the conditions occurring in consecutive clock cycles. For this purpose, it is more convenient to use a notation similar to regular expressions. For example, using the PSL language, a standard in the EDA industry, we can say "if `req1` is followed by `grant1`, then eventually `dav` is asserted" like this:

```
(req1; gnt1) |-> F dav
```

The semicolon represents sequencing. Such notations also allow us to express some "counting" properties that have no expression in ordinary temporal logic. For example, we can state that p holds in every other clock cycle like this: (p; true)*.

Modern property specification languages used in EDA provide some additional features that are convenient for hardware specifications. For example, one can specify a particular signal as the "clock" by which time is measured in a given property, and one can specify a "reset" condition that effectively cancels the requirements of the property. The latter is useful for dealing with reset and exception conditions.

20.2.1.1 History of Temporal Specification Languages

Pnueli suggested the use of temporal logic in the late 1970s [20]. Starting in the early 1980s, a variety of logics and notations have been used in research on automatic verification, including linear temporal logic (LTL), computational tree logic (CTL), CCS, dynamic logic, and temporal logic of actions (TLA) [20,21–29]. Some of these notations, notably CCS and TLA, are used for both modelling systems and specifying properties. In the 1990s, property languages specialized to EDA were developed. These include ForSpec, developed at Intel [30], Sugar from IBM [31], temporal "e" from Verisity language, and CBV from Motorola. These languages were donated to the Accellera standards body, and have been the basis for defining a new IEEE standard proposal called PSL [32].

20.2.2 Hardware Formal Verification Technologies/Engines

Having specified properties of a system in a suitable notation, we would then like to be able to verify these properties formally, in as automated a manner as possible. Fully automated verification of temporal properties is generally referred to as *model checking*, in reference to the first such technique developed by

Clarke and Emerson [22]. Model checking can not only verify temporal properties, but also provide *counter-examples* for properties that are false. Counterexamples are traces of incorrect system behavior that are valuable in diagnosing errors.

Since hardware systems are finite state, they can in principle always be verified by model checking. Advances in model checking algorithms have allowed designs of significant size to be checked.

20.2.2.1 The Automata-Theoretic Approach

We can think of a property as specifying a set of acceptable input/output traces of a system. If all possible traces of a given system fall into this set, then the system satisfies the property. Most algorithms for model checking work by translating the property into a *finite automaton* that accepts exactly the set of traces accepted by the property [33]. The automaton is a state graph, whose edges are labelled with input/output pairs. The *language* of the automaton is the set of input/output traces observed along accepting paths in this graph. By defining "accepting path" in various ways, we can define different kinds of languages. For example, if we specify the legal initial and final states of an accepting path, we obtain a set of finite traces. To obtain a language of infinite traces, we can specify the initial states, and the sets of states that may occur "infinitely often" on an accepting infinite path. This allows us to represent liveness properties with automata. Once the property has been translated into an equivalent automaton, we need no longer concern ourselves with the property language.

To check the property on a system, we also represent the system with an equivalent automaton. We now have only to check that every trace of the system automaton is a trace of the property automaton. This can be done by combing the automaton for the complement of the property with the system automaton to produce what is called a "product automaton". If this automaton accepts any traces, then the system does not satisfy the property. This test can be made, in the simplest case, by searching the state graph of the product automaton for an accepting path. For finite traces, this can be accomplished by a depth- or breadth-first search, starting form the initial states. This kind of search-based model checking (referred to as "explicit-state") is exemplified by systems such as COSPAN [34], SPIN [6] and Murphi [7], which are quite effective at verifying software-based protocols, cache coherence protocols, and other systems with relatively small state spaces.

20.2.2.2 In Search of Greater Capacity

The main practical difficulty with the above approach is that the state graphs of the automata can be prohibitively large. The number of states of the system automaton is, in the worst case, the number of possible configurations of the registers (or other state-holding elements) in the system, which is exponential. For most property languages, the number of states in the automaton corresponding to a property is also exponential in the size of the property text. Thus, the brute-force approach described above is hopeless for large hardware designs. Instead, a number of heuristic methods have been developed that avoid explicit construction of the state graph. Here, we discuss a few of these methods that are currently used in EDA.

20.2.2.3 Binary Decision Diagram-Based Symbolic Model Checking

The symbolic model checking (SMC) approach was introduced in the late 1980s [35, 75] and was first implemented in a tool called SMV [36]. In this approach, automata are not explicitly constructed. Instead, a logical formula is used to characterize implicitly the set of possible transitions between states. In this formula, variable x represents the state of register x in the current state while x' represents the state of register x at the next time. The possible transitions of a sequential machine are easily characterized in this way, as a set of Boolean equations (see Figure 20.1).

For most property languages, the property automaton can also be characterized in this way, without an exponential expansion. A breadth-first search can then be accomplished purely by operations on Boolean formulas. To do this, we use a formula to stand for the set of states that satisfies it. The set of successors P' of a set of states P can then be obtained by the following equation, where V is the set of state variables, R the transition formula, and ∃ the existential quantifier over Boolean variables:

$$P'(V') = \exists V.P(V) \wedge R(V,V')$$

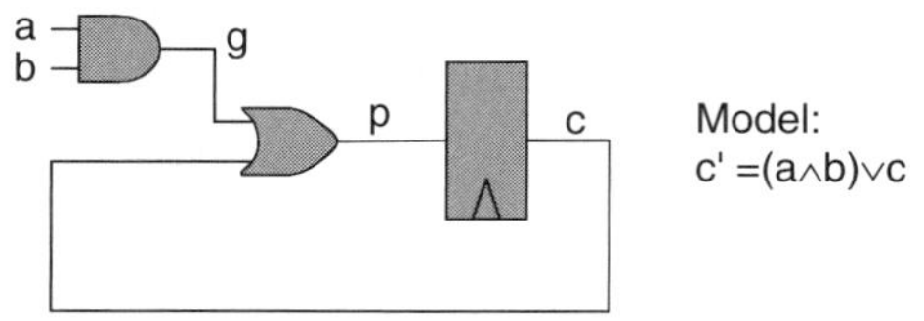

FIGURE 20.1 Characterizing transitions of a circuit.

A symbolic breadth-first search starts with a formula characterizing the initial states, and iterates the above "image" operation until a fixed point occurs (i.e., no new states are obtained).

To use this idea effectively, we need a compact representation for Boolean formulas, on which Boolean operations and existential quantification can be efficiently applied. For this purpose, binary decision diagrams (BDDs) are commonly used. A BDD is a decision graph in which variables occur in the same order along all paths, and common subgraphs are combined. In 1986, Bryant gave an efficient algorithm for Boolean operations on this structure [37]. Symbolic model checking using BDDs provides a means of property checking for circuits of modest size (typically up to a few hundred registers) in cases where the state graph is far too large to be constructed explicitly. This method was used successfully to find errors in the cache coherence protocols of a commercial multiprocessor [38], and has since been used in a number of commercial EDA tools, with many improvements introduced in the 1990s.

20.2.2.4 Satisfiability Problem-Based Bounded and Unbounded Model Checking

The Boolean satisfiability problem (SAT) is to determine whether there exists a truth assignment that makes a given Boolean formula true (or equivalently, whether there exists an input pattern that makes the output of a given Boolean circuit one). The first algorithm to solve this classic NP-complete problem was described in 1960 [76], and since then many improvements have been made. The success of recent SAT solvers such as Chaff [39] and Grasp [40] in solving very large problem instances prompted interest in using SAT solvers for model checking. In 1999, the notion of *bounded model checking* (BMC) using a SAT solver was introduced [41]. As in SMC, the transitions of the system and property automata are characterized by Boolean formulas. By *unfolding* this representation (i.e., making k consecutive copies, corresponding to consecutive time frames) the question of existence of a counter-example of k steps can be posed as a SAT problem. This made it possible to exploit recent advances in SAT solvers to find counterexamples to properties for systems much larger than can be handled by SMC (typically up to a few thousand registers).

20.2.2.5 Unbounded Model Checking Using a SAT Solver

The limitation of BMC is that it can only find counterexamples. It does not provide a practical way of proving that a property holds for all possible behaviors of a system. However, the success of BMC in handling large designs led to new methods for complete model checking that exploit SAT. One method, called k-induction [42], uses a SAT solver to test whether there is a path of k distinct good states leading to a bad state. If not, and if there is no path of k steps from an initial state to a bad state, then by induction the property holds at all times. This method can prove properties using just a SAT solver, though in practice the required k value can be quite large, making the BMC step prohibitive.

Another approach, called *localization*, uses a SAT solver only to decide which components of a system are relevant to proving a particular property. This is done by using the SAT solver to refute either a partial counter-example or all counter-examples of a given length [43,44]. The components of the system used in this refutation are then used to attempt to verify the property using BDD-based SMC or other methods. Since modern SAT solvers are quite effective at ignoring irrelevant facts, this process can produce a significant reduction in the size of the system to be checked. Another approach uses a technique called *interpolation* to derive an approximation of the "successor states" of a given state set from a bounded model-checking run [45].

This allows SMC using just a SAT solver. These techniques allow designs with up to a few thousand latches to be fully verified, at least in cases where only a small part of the system is relevant to the property, the domain in which SAT-based BMC is effective.

20.2.2.6 Symbolic Simulation

In 1990, the technique of *symbolic simulation* was introduced for hardware verification [46]. This technique resembles ordinary simulation in the sense that the user provides an input sequence to drive the design. However, in symbolic simulation, the input values can be symbolic variables as well as numeric zeros and ones. Thus, a single simulation run may in effect represent many runs of the design. The outputs of the simulation are Boolean formulas over the symbolic variables instead of numeric values. The outputs can then be compared to the desired functions. This comparison can be done using BDDs or a SAT solver. The complexity of this check can be reduced by introducing logical "unknown" (X) values at inputs that are considered irrelevant to the case being verified. By itself, symbolic simulation does not prove properties because, like BMC, it simulates only bounded runs of the system. However, complete property verification is possible using the *symbolic trajectory evaluation* (STE) methodology, based on symbolic simulation [47]. This method requires the user to provide part of the proof, but has the advantage that it can be applied to fairly large designs.

20.2.2.7 Theorem Proving Methods

From the1950s into the 1980s, a great deal of progress was made in mechanizing mathematics and logical deduction. Classic papers in the field include [48–55]. Beginning in the 1970s with the Boyer–Moore theorem prover NQTHM [56] and LCF [57], many practical tools have been developed that assist the user in formalizing and proving mathematical theorems. Later systems include ACL2 [68], PVS [59], and HOL [60]. These systems usually provide some form of mechanized *proof search* to aid in the construction of proofs. The proof search mechanism may be fully automatic, as in NQTHM, or customizable by writing *tactics*, as in LCF and its descendants.

To use a theorem prover for formal property verification, we translate both the system model and the desired property into the logic of the prover. This can be done, for example, by automatically translating the program into a logical assertion that characterizes its transition behavior (a *shallow embedding*), or by treating the program itself as an object in the formalism and writing an interpreter for the programming language in the logic (a *deep embedding*). In either case, verifying the property reduces to proving an appropriate theorem in the logic.

To prove properties of complex systems with a theorem prover, a considerable amount of user guidance is required, in the form of lemmas, tactics, or manual guidance of the deduction process. The most common proof approach is to construct an *inductive invariant* of the system (or of the system augmented with auxiliary structures). An inductive invariant is a fact that is true initially, and implies itself at the next time, thus is true always. Once a suitable invariant is obtained, the proof that it is inductive can be aided by the use of a *ground decision procedure* [61]. This approach greatly reduces the required amount of proof guidance. However, manually constructing an inductive invariant is still a time-consuming and intellectually taxing process.

For this reason, a number of tools that combine a theorem prover and a model checker have been developed. In such systems, the overall proof of a property can be reduced to lemmas that can be discharged by a model checker. For example, the HOL-VOSS system [62] and its successor Forte [63] use a theorem prover to reduce the proof to lemmas that can be checked by symbolic simulation. The Cadence SMV system [64] can be used to reduce the proof to verification of temporal logic properties by SMC. Such systems have the potential to shift a substantial amount of the proof effort from the user onto automated tools. To make this possible, however, we require an overall proof strategy that reduces the problem to a scale that can be handled by model checking.

20.2.3 Modularization and Abstraction

The most common proof approaches for dealing with complex systems are *modularization* and *abstraction*. By modularization, we mean breaking a property to be proved (a *goal*) into two or more properties (or *sub-goals*) that are more easily verified automatically, usually because they depend on fewer system components. We must show, in turn, that the subgoals imply the goal. In a strictly modular approach, such as the *assume-guarantee paradigm* [65], we verify a system consisting of two modules A and B, by proving properties of A and B in isolation, then combining these to deduce a property of the system. Strictly modular proofs can be difficult to obtain, however, because of the need to capture in the subgoals all properties of one module that are required by the other. In practice, one can take a more relaxed approach. It is necessary only that the sub-goals be provable in a coarser *abstraction* of the system than the original goal. The abstraction may be obtained by localization (i.e., considering only a subset of system components), or by various methods of abstracting data (including *predicate abstraction*, which is described below).

In fact, the key to all known methods for verifying large, complex systems is *abstraction*. This means replacing a complex system with a simpler system that captures the required properties. An abstract system can be constructed manually, in which case we must prove that the abstract system in fact preserves the properties of interest. Alternatively, an abstraction can be constructed automatically, perhaps according to parameters given by the user. A general framework for this approach is provided by *abstract interpretation* [66]. In this framework, the user chooses a suitable representation for facts about data in the system, and an automatic analysis computes the strongest inductive invariant of the system that can be expressed in this representation. As an example, we might use a representation that can express all linear affine relations between program variables. An important instance of abstract interpretation is *predicate abstraction* [67]. Here, the user provides a set of simple predicates, like $x < y$. The analysis constructs the strongest inductive invariant that can be expressed as a Boolean combination of these predicates.

Of course, the invariant constructed by abstract interpretation may or may not be strong enough to prove the desired property. The key to this approach is to choose an abstraction appropriate to the given property. Thus, in predicate abstraction, the predicates must be chosen carefully, so that they are sufficient to prove the property, without being so numerous as to make the analysis intractable. Recently, considerable attention has been given to automating the selection of predicates [10,68]. This has proved to be especially effective in software model checking as we will see. Abstract interpretation has also been effectively used to prove the subgoals arising in a modularization approach [64]. Here, the key is to choose subgoals that can be proved using a coarse abstraction. By this approach, much of the work of constructing invariants can be done automatically, thus simplifying the manually constructed part of the proof.

20.3 Software Formal Verification

A big challenge in applying formal verification to software is the need to translate the original code into a formal representation that can be reasoned about. In particular difficulties are caused by dynamic memory or process allocations, dynamic loops, dynamic jump-targets, aliases, and pointers. The state explosion problem is another big challenge in formal verification of software — as software programs often have a very large state space. Below is a brief survey of automatic software verification techniques.

20.3.1 Explicit Model Checking of Protocols

In 1991 to 1992, protocols were verified using dedicated modelling languages like Promela [6] and Murphi [7] and explicit model checkers. This approach best fits concurrent asynchronous software systems. The explicit model-checking algorithm performs a search on the state space of the system, and on-the-fly verifies for each newly visited state that the property holds for it. The algorithm stores all the states it encounters in a large hash table. When a state is generated that is already in the hash table, the search does not expand to its successor states. To allow efficient search of the state space, several optimization techniques have been developed. The most dominating one is the partial order reduction. Not all possible successors of a given state are generated and included in the search. However, states and flows of the program which are not visited in the search are proven to be redundant for establishing the correctness of the property. For example, if the program has two processes *P1:: a ; b;* (in process *P1*, statement *a* is performed first and then statement *b*) and *P2:: c ; d ;* then there are six possible execution flows (orders) *abcd, acbd, acdb, cdab, cabd,* or *cadb.* However, if the correctness of the property to be checked is only influenced by the order of execution of the statements *b* and *d,* the search can be limited to check only two flows: *acbd* and *acdb.* This technique has been proven to be successful for protocols that have high degree of concurrency and thus partial reduction can be very beneficial. In particular, industrial cache coherence protocols, cryptographic, and security-related protocols were successfully verified. Known academic and industrial model checkers based on this family of techniques are SPIN [6], Murphi [7] and TLA+ [69]. The main drawback of this approach is that the verification is not applied to the "golden model" of the design, that is, to the actual C and Java programs.

20.3.2 Explicit and Symbolic Model Checking of Java and C Programs

In 1999 to 2001, Java programs were abstracted using abstract interpretation [66] and translated into finite-state models to be verified using explicit and symbolic model checkers [8,9,70,71]. Similar technology has been proposed for translating C programs, using *predicate abstraction* [67] into Boolean programs, and submitting them to a symbolic model checker [10]. The extracted Boolean program has the same control-flow structure as the original program; however, it has only Boolean variables, each representing a Boolean predicate over the variables of the original program. For example, if x and y are complex data structures with the same type, e.g., structures, then a Boolean variable in the extracted program may represent the predicate $x=y$, t, thus resulting in a much smaller state space in the Boolean program.

To give the reader some intuition into predicate abstraction, let x be an integer variable in the original C program and let the Boolean program have only two predicates that refer to x, "the predicate *b1* is defined as $x > 2$" and "the predicate *b2* is defined as $x=5$".

Original program	Boolean program
pc1: x=5;	*pc1: b1=true; b2=true;*
pc2: ..	*pc2: ..*

In this case, it is easy to see that if x is assigned 5 in the original program then both the predicates *b1* and *b2* should become *true* in the corresponding program location, *pc1,* in the Boolean program. A more interesting case is

Original program	Boolean program
pc1: x=x + 1;	*pc1: b1=?; b2=?;*
pc2: ..	*pc2: ..*

In this case, since *b1* represents $x>2$, *b1* should be assigned *true* if $x+1>2$, i.e., if at program location *pc1*, $x>1$ holds. The condition $x>1$ is called the Weakest Precondition that would guarantee that $x>2$ holds after the assignment $x=x+1$. Similarly, since *b2* represents $x=5$, *b2* should be assigned *true* if $x+1=5$, that is, if at program location *pc1*, $x=4$ holds. However, at location *pc1* in the Boolean program, we only have access to the truth value of *b1* and *b2* and not to the values of the predicates $x>1$ and $x=4$ which are not included in the set of predicates used for the generation of the Boolean program. In this case, an automated decision procedure is invoked to strengthen these predicates into an expression over the set of the chosen predicates. In our example, the predicate $x>1$ will be strengthen to $x>2$ and the predicate $x=5$ will strengthen into *false.* Thus, we get:

Original program	Boolean program
pc1: x=x+1;	*pc1: b1=if b1 then true else false;* *b2=false;*
pc2: ..	*pc2: ..*

The predicate abstraction technology has been extended to handle pointers, procedures and procedure calls [10]. For example, let the predicate *b1* be **p>5,*

Original program	Boolean program
pc1: x=3;	*pc1: b1=?;*
pc2: ..	*pc2: ..*

if *x* and **p* are aliases, then *b1* should be assigned *false* since **p* becomes *3*, else *b1* should retain its previous value. Thus, we get:

Original program	Boolean program
pc1: x=3;	*pc1: b1=(if &x=p then false else retain);*
pc2: ..	*pc2: ..*

The SMC algorithm is applied to the Boolean program to check the correctness of the property. If successful, we can conclude that the property holds also for the original program. If failed, the Boolean program is refined to more precisely represent the original program. Additional Boolean predicates are automatically identified from the counter-example and a new Boolean program is constructed. This iterative approach is called iterative abstraction refinement [77]. Using this technique has been proven to be successful for checking safety properties of Windows NT device drivers and for discovering invariants regarding array bounds. Academic and industrial model checkers based on this family of techniques include SLAM by Microsoft [10], Bandera [70], Java Pathfinder [9], TVLA [72], Feaver [73], Java-2-SAL [74], and Blast [68].

20.3.3 Bounded Model Checking of C Programs

Most recently, in 2003 to 2004, C and SpecC programs were translated into propositional formulae and then formally verified using a bounded model checker [14,11]. This approach can handle the entire ANSI-C language consisting of pointers, dynamic memory allocations and dynamic loops, i.e., loops with conditions that cannot be evaluated statically. These programs are verified against user defined safety properties and also against automatically generated properties about pointer safety and array bounds.

The C program is first translated into an equivalent program that uses only *while, if, goto* and *assignment* statements. Then each loop of the form *while (e) inst* is translated into *if (e) inst; if (e) inst; …; if (e) inst; {assertion !e}*, where *if (e) inst* is repeated *n* times. The assertion *!e* is later formally verified and if it does not hold, *n* is increased until the assertion holds. The resulting program that has no loops, and the properties to be checked are then translated into a propositional formula that represents the model after unwinding it *k* times. The resulting formula is submitted to a SAT solver and if a satisfying assignment is found, it represents an error trace. During the *k*-steps unwinding of the program, pointers are handled(punctuation needed?). For example,

```
int a, b, p;
 if (x) p=&a; else p=&b;
*p=5;
```

The above program is first translated into:

```
p1=(x ? &a: p0)∧p2=(x? p1: &b)
*p2=5;
```

That is, three copies of the variable *p* are created *p0, p1* and *p2.* If *x* is true then *p1=&a* and *p2=p1*, and if *x* is *false p1=p0* and *p2=&b*. Then **p2=5* is replaced by

**(x? p1; &b)=5*	which is replaced by
*(x ? *p1; *&b)=5*	which is replaced by
*(x? *p1; b)=5*	which is replaced by
*(x? *(x? &a :p0); b)=5*	which is replaced by
*(x? *&a ; b)=5*	which is replaced by
(x? a ; b)=5	

After all these automatic transformation the pointers have been eliminated and the resulting statement is *(x? a; b)=5.* For details on dynamic memory allocations and dynamic loops see [11].

20.3.4 Translation Validation of Software

Two technological directions are currently pursued for formally verifying the correct translation of software programs. One [12,17], which automatically establishes an abstraction mapping between the source program and the object code, offers an alternative to the verification of synthesizers and compilers. The other direction [13,14], which automatically translates the two programs given in C and Verilog into a BMC formula and submits it to a SAT solver. The value of each Verilog signal at every clock cycle is visible to the C program. Thus, the user can specify and formally verify the desired relation between the C variables and the Verilog signals. Both the C and Verilog programs are unwound for *k* steps as we have described in the previous section.

20.4 Summary

Formal property verification provides a means of ensuring that a hardware or software system satisfies certain key properties, regardless of the input presented. Automated methods developed over the last two decades have made it possible, in certain cases, to verify properties of large, complex systems with minimal user interaction, and to find errors in these systems. These methods can be used as an adjunct to simulation and testing in the design verification process, to enable the design of more robust and reliable systems.

References

[1] B. Bentley, Validating the Intel Pentium 4 Microprocessor, *Proceedings of the 38th Design Automation Conference (DAC)*, ACM Press, New York, 2001, pp. 244–248.

[2] G. Kamhi, O. Weissberg, L. Fix, Z. Binyamini, and Z. Shtadler, Automatic datapath extraction for efficient usage of HDD, *Proceedings of the 9th International Conference Computer Aided Verification (CAV)*, Lecture Notes in Computer Science, Vol. 1254, Springer, Berlin, 1997, pp. 95–106.

[3] D. Geist and I. Beer, Efficient model checking by automated ordering of transition relation partitions, *Proceedings of Computer-Aided Verification (CAV)*, Springer-Verlag, Berlin,1994.

[4] R.P. Kurshan, Formal verification on a commercial setting, *Proceedings of Design Automation Conference (DAC)*, 1997, pp. 258–262.

[5] T. Grotker, S. Liao, G. Martin, and S. Swan, *System Design with SystemC*, Kluwer Academic Publishers, Dordrecht, 2002.

[6] G.J. Holzmann, *Design and Validation of Computer Protocols*, Prentice-Hall, Englewood Cliffs, NJ, 1991.

[7] D.L. Dill, A.J. Drexler, A.J. Hu, and C. Han Yang, Protocol verification as a hardware design aid, *IEEE International Conference on Computer Design: VLSI in Computers and Processors*, IEEE Computer Society, Washington, DC,, 1992, pp. 522–525.

[8] C. Demartini, R. Iosif, and R. Sisto, A deadlock detection tool for concurrent Java programs. *Software — Practice and Experience*, 29, 577–603, 1999.

[9] K. Havelund and T. Pressburger,. Model checking Java programs using Java PathFinder, *Int. J. Software Tools Technol. Trans.*, 2, 366–381, 2000.
[10] T. Ball, R. Majumdar, T. Millstein, and S.K. Rajamani, Automatic Predicate Abstraction of C Programs, *PLDI 2001, SIGPLAN Notices*, 36, 203–213, 2001.
[11] E. Clarke, D. Kroening, and F. Lerda, A tool for checking ANSI-C program, in *Proceedings of 10th International Conference, on Tools and Algorithms for Construction and Analysis of Systems (TACAS)*, Lecture Notes in Computer Science, Vol. 2988, Springer, Berlin, 2004.
[12] A. Pnueli, M. Siegel, and O. Shtrichman, The code validation tool (CVT)-automatic verification of a compilation process, *Int. J. Software Tools Technol. Trans.* (STTT), 2, 192–201, 1998.
[13] E. Clarke and D. Kroening, Hardware verification using ANSI-C programs as a reference, *Proceedings of ASP-DAC 2003*, IEEE Computer Society Press, Washington, DC, 2003, pp. 308–311.
[14] E. Clarke, D. Kroening, and K. Yorav, Behavioral consistency of C and Verilog programs using bounded model checking, *Proceedings of the 40th Design Automation Conference (DAC)*, ACM Press, New York, 2003, pp. 368–371.
[15] L. Fix, M. Mishaeli, E. Singerman, and A. Tiemeyer, MicroFormal: FV for Microcode. Intel Internal Report, 2004.
[16] Bill Gates, Keynote address at WinHec 2002. www.microsoft.com/billgates/speeches/2002/04-18winhec.asp
[17] L. Zuck, A. Pnueli, Y. Fang, and B. Goldberg, Voc: a translation validator for optimizing compilers, *J. Univ. Comput. Sci.*, 3, 223–247, 2003. Preliminary version in ENTCS, 65, 2002.
[18] A. Pnueli, Specification and development of reactive systems, in *Information Processing 86*, Kugler, H.-J., Ed., IFIP, North-Holland, Amsterdam, 1986, pp. 845–858.
[19] R. Alur and T.A. Henzinger, Logics and models of real time: a survey, in *Real Time: Theory in Practice,* de Bakker, J.W., Huizing, K., de Roever, W.-P., and Rozenberg, G., Eds., *Lecture Notes in Computer Science*, Vol. 600, Springer, Berlin, 1992, pp. 74–106.
[20] A. Pnueli, The temporal logic of programs, *Proceedings of the 18th Symposium on Foundations of Computer Science*, 1977, pp. 46–57.
[21] M. Ben-ari, Z. Manna, and A. Pnueli, The temporal logic of branching time, *Acta Inform.*, 20, 207–226, 1983.
[22] E.M. Clarke and E.A. Emerson, Design and synthesis of synchronization skeletons using branching time temporal logic, *Proceedings of Workshop on Logic of Programs*, Lecture Notes in Computer Science, Vol. 131, Springer, Berlin, 1981, pp. 52–71.
[23] M.J. Fischer and R.E. Lander, Propositional dynamic logic of regular programs, *J. Comput. Syst. Sci.*, 18, 194–211, 1979.
[24] L. Lamport, Sometimes is sometimes "not never" — on the temporal logic of programs, *Proceedings ofthe 7th ACM Symposium on Principles of Programming Languages*, January 1980, pp. 174–185.
[25] B. Misra and K.M. Chandy, Proofs of networks of processes, *IEEE Trans. Software Eng.*, 7, 417–426, 1981.
[26] R. Milner, A calculus of communicating systems, Lecture Notes in Computer Science, Vol. 92. Springer, Berlin, 1980.
[27] A. Pnueli, The temporal semantics of concurrent programs, *Theor. Comput. Sci.*, 13, 45–60, 1981.
[28] J.P. Queille and J. Sifakis, Specification and verification of concurrent systems in Cesar. *Proceedings of 5th International Symposium On Programming*, Lecture Notes in Computer Science, Vol. 137, Springer, Berlin, 1981, pp. 337–351.
[29] P. Wolper, Synthesis of Communicating Processes from Temporal Logic Specifications, Ph.D. thesis, Stanford University, 1982.
[30] R. Armoni, L. Fix, A. Flaisher, R. Gerth, B. Ginsburg, T. Kanza, A. Landver, S. Mador-Haim, E. Singerman, A. Tiemeyer, M. Vardi, and Y. Zbar, The ForSpec temporal logic: a new temporal property specification language, *Proceedings of TACAS*, 2002, pp. 296–311.
[31] I. Beer, S. Ben-David, C. Eisner, D. Fisman, A. Gringauze, and Y. Rodeh, The temporal logic sugar, *Proceedings of Conference on Computer-Aided Verification (CAV'01)*, Lecture Notes in Computer Science, Vol. 2102, Springer, Berlin, 2001, pp. 363–367.
[32] Property Specification Language Reference Manual, Accellera, 2003.
[33] M.Y. Vardi and P. Wolper, An automata-theoretic approach to automatic program verification, *Proceedings of IEEE Symposium on Logic in Computer Science*, Computer Society Press, Cambridge, 1986, pp. 322–331.

[34] R. P. Kurshan, *Computer-Aided Verification of Coordinating Processes*, Princeton Univ. Press, Princeton, 1994.
[35] O. Coudert, C. Berthet, and J.C. Madre, Verification of synchronous sequential machines based on symbolic execution, *Proceedingsof International Workshop on Automatic Verification Methods for Finite State Systems*, Lecture Notes in Computer Science, Vol. 407, Springer, Berlin, 1989.
[36] K.L. McMillan, Symbolic Model Checking: An Approach to the State Explosion Problem, Ph.D. thesis, CMU CS-929131, 1992.
[37] R.E. Bryant, Graph-based algorithms for Boolean function manipulation, *Proceedings of IEEE Transactions on Computers*, Vol. C-35, 1986, pp. 677–691.
[38] K. L. McMillan and J. Schwalbe, Formal verification of the gigamax cache consistency protocol, *International Symposium on Shared Memory Multiprocessing*, 1991, pp. 242–251.
[39] M.W. Moskewicz, C. F. Madigan, Y. Zhao, L. Zhang, and S. Malik, Chaff: engineering an efficient SAT solver, *Proceedings of the 38th Design Automation Conference (DAC'01)*, June 2001.
[40] J.P. Marques-Silva and K.A. Sakallah, GRASP: a new search algorithm for satisfiability, *IEEE Trans. Comput.*, 48, 506–521, 1999.
[41] A. Biere, A. Comatti, E. Clarke, and Y. Zhu, Symbolic model checking without BDDs, *Proceedings of the Workshop on Tools and Algorithms for the Construction and Analysis of Systems (TACAS'99)*, Lecture Notes in Computer Science Springer, Berlin, 1999.
[42] M. Sheeran, S. Singh, and G. Stalmarck, Checking safety properties using induction and a sat-solver, *Proceedings of International Conference on Formal Methods in Computer Aided Design (FMCAD 2000)*, W.A. Hunt and S.D. Johnson, Eds., 2000.
[43] P. Chauhan, E. M. Clarke, J. H. Kukula, S. Sapra, H. Veith, and D. Wang, Automated abstraction refinement for model checking large state spaces using SAT-based conflict analysis, *Proceedings of FMCAD 2002*, Lecture Notes inComputer Science, Vol. 2517, Springer, Berlin, 2002, pp. 33–51.
[44] K.L. McMillan and N. Amla, Automatic abstraction without counterexamples, *Proceedings of Tools and Algorithms for the Construction and Analysis of Systems (TACAS'03)*, Lecture Notes in Computer Science, Vol. 2619, Springer, Berlin, 2003, pp. 2–17.
[45] K.L. McMillan, Interpolation and SAT-based model checking, *Proceedings of CAV 2003*, Lecture Notes in Computer Science, Vol. 2725, Springer, Berlin, 2003, pp. 1–13.
[46] R.E. Bryant and C-J. Seger, Formal verification of digital circuits using symbolic ternary system models, *Workshop on Computer Aided Verification. 2nd International Conference, CAV'90*, Clarke, E.M. and Kurshan, R.P., Eds., Lecture Notes in Computer Science, Vol. 531, Springer, Berlin, 1990.
[47] D.L. Beatty, R.E. Bryant, and C-J. Seger, Formal hardware verification by symbolic ternary trajectory evaluation, *Proceedings of the 28th ACM/IEEE Design Automation Conference.* IEEE Computer Society Press, Washington, DC, 1991.
[48] S.C. Kleene, *Introduction to Metamathematics.* Van Nostrand, Princeton, NJ, 1952.
[49] J. McCarthy, Computer programs for checking mathematical proofs, *Proceedings of the Symposia in Pure Mathematics, Vol. V, Recursive Function Theory.* American Mathematical Society, Providence, RI, 1962.
[50] J.A. Robinson,. A machine-oriented logic based on the resolution principle. *J. ACM*, 12, 23–41, 1965.
[51] D. Scott, Constructive validity, *Symposium on Automatic Demonstration*, Lecture Notes in Mathematics, Vol. 125, Springer, New York, 1970, pp. 237-275.
[52] C.A.R. Hoare, Notes on data structuring, in *Structural Programming*, O.J. Dahl, E.W. Dijkstra, and C.A.R. Hoare, Academic Press, New York, 1972, pp. 83–174.
[53] R. Milner and R. Weyhrauch, Proving computer correctness in mechanized logic, in *Machine Intelligence*, Vol. 7, B. Meltzer and D. Michie, Eds., Edinburgh University Press, Edinburgh, Scotland, 1972, pp. 51–70.
[54] G.P. Huet, A unification algorithm for types lambda-calculus, *Theor. Comput. Sci.*, 1, 27–58, 1975.
[55] D.W. Loveland, *Automated Theorem Proving: A Logical Basis*, North-Holland, Amsterdam, 1978.
[56] R. S. Boyer and J S. Moore, *A Computational Logic*, Academic Press, New York, 1979.
[57] M. Gordon, R. Milner, and C. P. Wadsworth, Edinburgh LCF: a mechanised logic of computation, Lecture Notes in Computer Science, Vol. 78, Springer-Verlag, Berlin, 1979.
[58] M. Kaufmann and J. Moore, An industrial strength theorem prover for a logic based on common lisp, *IEEE Trans. Software Eng.*, 23, April 1997, pp. 203–213.

[59] S. Owre, J. Rushby, and N. Shankar, PVS: a prototype verification system, in *Proceedings of 11th International Conference on Automated Deduction*, D. Kapur, Ed., Lecture Notes in Artificial Intelligence, Vol. 607, Springer, Berlin, 1992, pp. 748–752.

[60] M. J. C. Gordon and T. F. Melham, Eds., *Introduction to HOL: A Theorem Proving Environment for Higher Order Logic*, Cambridge University Press, London, 1993.

[61] G. Nelson and D. C. Oppen, Simplification by cooperating decision procedures, *ACM Trans. Programming Lang. Syst.*, 1, 245—257, 1979.

[62] J. Joyce and C.-J. H. Seger, The HOL-Voss system: model-checking inside a general-purpose theorem-prover, *Proceedings of the 6th International Workshop on Higher Order Logic Theorem Proving and its Applications*, 1993, pp. 185–198.

[63] M. Aagaard, R. B. Jones, C.-Johan H. Seger: Lifted-FL: a pragmatic implementation of combined model checking and theorem proving, in *Theorem Proving in Higher Order Logics (TPHOLs)*, Y. Bertot, G. Dowek, and A. Hirschowitz, Eds., Springer, Berlin, 1999, pp. 323–340.

[64] K.L. McMillan, A methodology for hardware verification using compositional model checking, *Sci. Comput. Programming*, 37, 279–309, 2000.

[65] A. Pnueli, In transition from global to modular temporal reasoning about programs, in *Logics and Models of Concurrent Systems, sub-series F: Computer and System Science*, K.R. Apt, Ed., Springer-Verlag, Berlin, 1985, pp. 123–144.

[66] P. Cousot and R. Cousot, Abstract interpretation: a unified lattice model for static analysis of programs by construction and approximation of fixpoints, in *Proceedings of 4th ACM Symposium on Principles of Programming Languages*, 1977, pp. 238–252.

[67] S. Graf and H. Saidi, Construction of abstract state graphs with PVS, *CAV97: Computer-Aided Verification*, Lecture Notes in Computer Science, Vol. 1254, Springer, Berlin, 1997, pp. 72–83.

[68] T. A. Henzinger, R. Jhala, R. Majumdar, and G. Sutre, Lazy abstraction, *Proceedings of the 29th Annual Symposium on Principles of Programming Languages (POPL)*, ACM Press, 2002, pp. 58–70.

[69] Y. Yu, P. Manolios, and L. Lamport, Model checking TLA+ specifications, *Proceedings of Correct Hardware Design and Verification Methods.*

[70] J. Corbett, M. Dwyer, J. Hatcliff, S. Laubach, C. Pasareanu, Robby, and H. Zheng, Bandera: extraction finite-state models from Java source code, *Proceedings of 22nd International Conference on Software Engineering (ICSE2000)*, June 2000.

[71] M. Dwyer, J. Hatcliff, R. Joehanes, S. Laubach, C. Pasareanu, Robby, W. Visser, and H. Zheng, Tool supported program abstraction for finite-state verification, in *ICSE 01: Software Engineering*, 2001.

[72] T. Lev-Ami and M. Sagiv, TVLA: a framework for Kleene-based static analysis, *Proceedings of the 7th International Static Analysis Symposium*, 2000.

[73] G. Holzmann, Logic verification of ANSI-C code with SPIN, *Proceedings of 7th International SPIN Workshop*, K. Havelund Ed., Lecture Notes in Computer Science, Vol. 1885, Springer, Berlin, 2000, pp. 131–147.

[74] D.Y.W. Park, U. Stern, J.U. Skakkebaek, and D.L. Dill, Java model checking, *Proceedings of the First International Workshop on Automated Program Analysis, Testing and Verification*, June 2000.

[75] J.R. Burch, E.M. Clarke, K.L. McMillan, D.L. Dill, and L.J. Hwang, Symbolic model checking: 10^{20} states and beyond, *Proceedings of the Fifth Annual IEEE Symposium on Logic in Computer Science, IEEE Computer Society Press*, Washington, D.C., 1990. pp. 1–33.

[76] M. Davis and H. Putnam, A computing procedure for quantification theory, *J. ACM*, 7, 201–215, 1960.

[77] E.M. Clarke, O. Grumberg, S. Jha, Y. Lu, and H. Veith, Counterexample-guided abstraction refinement, in *Computer Aided Verification, LNCS 1855, 12th International Conference, CAV 2000*, E.A. Emerson and A.P. Sistla, Eds., Springer, Chicago, IL, USA, July 15–19, 2000, pp. 154–169.

SECTION V

TEST

21

Design-For-Test

Bernd Koenemann
Mentor Graphics, Inc.
San Jose, California

21.1 Introduction

Design-for-test or design-for-testability (DFT) is a name for design techniques that add certain testability features to a microelectronic hardware product design. The premise of the added features is that they make it easier to develop and apply manufacturing tests for the designed hardware. The purpose of manufacturing tests is to validate that the product hardware contains no defects that could adversely affect the product's correct functioning.

Tests are applied at several steps in the hardware manufacturing flow and, for certain products, may also be used for hardware maintenance in the customer's environment. The tests generally are driven by test programs that execute in automatic test equipment (ATE) or, in the case of system maintenance, inside the assembled system itself. In addition to finding and indicating the presence of defects (i.e., the test fails), tests may be able to log diagnostic information about the nature of the encountered test fails. The diagnostic information can be used to locate the source of the failure.

Design-for-test plays an important role in the development of test programs and as an interface for test application and diagnostics.

Historically speaking, DFT techniques have been used at least since the early days of electric/electronic data-processing equipment. Early examples from the 1940s and 1950s are the switches and instruments that allowed an engineer to "scan" (i.e., selectively) probe the voltage/current at some internal nodes in an analog computer (analog scan). Design-for-test is often associated with design modifications that provide improved access to internal circuit elements such that the local internal state can be controlled (controllability) or observed (observability) more easily. The design modifications can be strictly physical in nature (e.g., adding a physical probe point to a net) or add active circuit elements to facilitate controllability/observability (e.g., inserting a multiplexer into a net). While controllability and observability improvements for internal circuit elements definitely are important for the test, they are not the only type of DFT. Other guidelines, for example, deal with the electromechanical characteristics of the interface between the product under test and the test equipment, e.g., guidelines for the size, shape, and spacing of probe points, or the suggestion to add a high-impedance state to drivers attached to probed nets such that the risk of damage from backdriving is mitigated.

Over the years, the industry has developed and used a large variety of more or less detailed and more or less formal guidelines for desired and/or mandatory DFT circuit modifications. The common understanding of DFT in the context of electronic design automation (EDA) for modern microelectronics is shaped to a large extent by the capabilities of commercial DFT software tools as well as by the expertise and experience of a professional community of DFT engineers researching, developing, and using such tools. Much of the related body of DFT knowledge focuses on digital circuits while DFT for analog/mixed-signal circuits takes something of a backseat. The following text follows this scheme by allocating most of the space to digital techniques.

21.2 The Objectives of Design-For-Test for Microelectronics Products

Design-for-test affects and depends on the methods used for test development, test application, and diagnostics. The objectives, hence, can only be formulated in the context of some understanding of these three key test-related activities.

21.2.1 Test Generation

Most tool-supported DFT practiced in the industry today, at least for digital circuits, is predicated on a "structural" test paradigm. "Functional" testing attempts to validate the circuit under test functions according to its functional specification. For example, does the adder really add? Structural testing, by contrast, makes no direct attempt to ascertain the intended functionality of the circuit under test. Instead, it tries to make sure that the circuit has been assembled correctly from some low-level building blocks as specified in a structural netlist. For example, are all logic gates there that are supposed to be there and are they connected correctly? The stipulation is that if the netlist is correct (e.g., somehow it has been fully verified against the functional specification) and structural testing has confirmed the correct assembly of the structural circuit elements, then the circuit should be functioning correctly.

One benefit of the structural paradigm is that the test generation can focus on testing a limited number of relatively simple circuit elements rather than having to deal with an exponentially exploding multiplicity of functional states and state transitions. While the task of testing a single logic gate at a time sounds simple, there is an obstacle to overcome. For today's highly complex designs, most gates are deeply embedded whereas the test equipment is only connected to the primary I/Os and/or some physical test points. The embedded gates, hence, must be manipulated through intervening layers of logic. If the intervening logic contains state elements, then the issue of an exponentially exploding state space and state transition sequencing creates an unsolvable problem for test generation. To simplify test generation, DFT addresses the accessibility problem by removing the need for complicated state transition sequences when trying to control or observe what is happening at some internal circuit element.

Depending on the DFT choices made during circuit design/implementation, the generation of structural tests for complex logic circuits can be more or less automated. One key objective of DFT methodologies, hence, is to allow designers to make trade-offs between the amount and type of DFT and the cost/benefit (time, effort, quality) of the test generation task.

21.2.1.1 Test Application

Complex microelectronic products typically are tested multiple times. Chips, for example may be tested on the wafer before the wafer is diced into individual chips (wafer probe/sort), and again after being packaged (final test). More testing is due after the packaged chips have been assembled into a higher level package such as a printed circuit board (PCB) or a multi-chip module (MCM). For products with special reliability needs, additional intermediate steps such as burn-in may be involved which, depending on the flow details, may include even more testing (e.g., pre- and postburn-in test, or *in situ* test during burn-in). The cost of test in many cases is dominated by the test equipment cost, which in turn depends on the number of I/Os that need to be contacted, the performance characteristics of the tester I/O (i.e., channel) electronics, and the depth/speed of pattern memory behind each tester channel. In addition to the cost of the tester frame itself, interfacing hardware (e.g., wafer probes and prober stations for wafer sort, automated pick-and-place handlers for final test, burn-in-boards (BIBs) for burn-in, etc.) is needed that connects the tester channels to the circuit under test.

One challenge for the industry is keeping up with the rapid advances in chip technology (I/O count/size/placement/spacing, I/O speed, internal circuit count/speed/power, thermal control, etc.) without being forced to upgrade continually the test equipment. Modern DFT techniques, hence, have to offer options that allow next generation chips and assemblies to be tested on existing test equipment and/or reduce the requirements/cost for new test equipment. At the same time, DFT has to make sure that test times stay within certain bounds dictated by the cost target for the products under test.

21.2.2 Diagnostics

Especially for advanced semiconductor technologies, it is expected that some of the chips on each manufactured wafer will contain defects that render them nonfunctional. The primary objective of testing is to find and separate those nonfunctional chips from the fully functional ones, meaning that one or more responses captured by the tester from a nonfunctional chip under test differ from the expected response. The percentage of chips that fail the test, hence, should be closely related to the expected functional yield for that chip type. In reality, however, it is not uncommon that all chips of a new chip type arriving at the test floor for the first time fail (so-called zero-yield situation). In that case, the chips have to go through a debug process that tries to identify the reason for the zero-yield situation. In other cases, the test fall out (percentage of test fails) may be higher than expected/acceptable or fluctuate suddenly. Again, the chips have to be subjected to an analysis process to identify the reason for the excessive test fall out.

In both cases, vital information about the nature of the underlying problem may be hidden in the way the chips fail during test. To facilitate better analysis, additional fail information beyond a simple pass/fail is collected into a fail log. The fail log typically contains information about when (e.g., tester cycle), where (e.g., at what tester channel), and how (e.g., logic value) the test failed. Diagnostics attempt to derive from the fail log the logical/physical location inside the chip at which the problem most likely started. This location provides a starting point for further detailed failure analysis (FA) to determine the actual root cause. Failure analysis, in particular physical FA (PFA), can be very time consuming and costly, since it typically involves a variety of highly specialized equipment and an equally specialized FA engineering team. The throughput of the FA labs is very limited, especially if the initial problem localization from diagnostics is poor. That adversely affects the problem turnaround time and how many problem cases can be analyzed. Additional inefficiency arises if the cases handed over to the FA lab are not relevant for the tester fall out rate.

In some cases (e.g., PCBs, MCMs, embedded, or stand-alone memories), it may be possible to repair a failing circuit under test. For that purpose, diagnostics must quickly find the failing unit and create a work-order for repairing/replacing the failing unit. For PCBs/MCMs, the replaceable/repairable units are the chips and/or the package wiring. Repairable memories offer spare rows/columns and some switching logic that can substitute a spare for a failing row/column. The diagnostic resolution must match the granularity of replacement/repair. Speed of diagnostics for replacement is another issue. For example, cost reasons may dictate that repairable memories must be tested, diagnosed, repaired, and retested in a single test insertion. In that scenario, the fail data collection and diagnostics must be more or less real-time as the test is applied. Even if diagnostics are to be performed offline, failure data collection on expensive production test equipment must be efficient and fast or it will be too expensive.

Design-for-test approaches can be more or less diagnostics-friendly. The related objectives of DFT are to facilitate/simplify fail data collection and diagnostics to an extent that can enable intelligent FA sample selection, as well as improve the cost, accuracy, speed, and throughput of diagnostics and FA.

21.2.3 Product Life-Cycle Considerations

Test requirements from other stages of a chip's product life cycle (e.g., Burn-In, PCB/MCM test, [sub]system test, etc.) can benefit from additional DFT features beyond what is needed for the chip manufacturing test proper. Many of these additional DFT features are best implemented at the chip level and affect the chip design. Hence, it is useful to summarize some of these additional requirements, even though the handbook primarily focuses on EDA for IC design.

21.2.3.1 Burn-In

Burn-in exposes the chips to some period of elevated ambient temperature to accelerate and weed out early life fails prior to shipping the chips. Typically, burn-in is applied to packaged chips. Chips designated for direct chip attach assembly may have to be packaged into a temporary chip carrier for burn-in and subsequently be removed again from the carrier. The packaged chips are put on BIBs, and several BIBs at a time are put into a burn-in oven. In static burn-in the chips are simply exposed to an elevated temperature, then removed from the oven and retested. Burn-in is more effective if the circuit elements on the chips are subjected to local electric fields during burn-in. Consequently, at a minimum, some chip power pads must be connected to a power supply grid on the BiBs. The so-called dynamic burn-in further requires some switching activity, typically meaning that some chip inputs must be connected on the BiBs and be wired out of the burn-in oven to some form of test equipment. The most effective form of burn-in, called *in situ* burn-in, further requires that some chip responses can be monitored for failures while in the oven. For both dynamic and *in situ* burn-in, the number of signals that must be wired out of the oven is of concern because it drives the complexity/cost of the BiBs and test equipment.

Burn-in friendly chip DFT makes it possible to establish a chip burn-in mode with minimal I/O footprint and data bandwidth needs.

21.2.3.2 Printed Circuit Board/Multi-Chip Module Test

The rigors (handling, placing, heating, etc.) of assembling multiple chips into a higher-level package can create new assembly-related defects associated with the chip attach (e.g., poor solder connection) and interchip wiring (e.g., short caused by solder splash). In some cases, the chip internals may also be affected (e.g., bare chips for direct chip attach are more vulnerable than packaged chips). The basic PCB/MCM test approaches concentrate largely on assembly-related defects and at best use very simple tests to validate that the chips are still "alive."

Although functional testing of PCBs/MCMs from the edge connectors is sometimes possible and used, the approach tends to make diagnostics very difficult. In-circuit testing is a widely practiced alternative or complementary method. In-circuit testing historically has used so-called bed-of-nails interfaces to contact physical probe points connected to the interchip wiring nets. If every net connected to the chip is contacted by a nail, then the tester can essentially test the chip as if stand-alone. However, it often is difficult to prevent some other chip driving the same net that the tester needs to control for testing the currently selected chip. To overcome this problem, the in-circuit circuit tester drivers are strong enough to override (backdrive) other chip drivers. Backdriving is considered a possible danger, and reliability problem for some types of chip drivers and some manufacturers may discourage backdriving. Densely packed, double-sided PCBs or other miniaturized packages may not leave room for landing pads on enough nets, and the number and density of nets to be probed may make the bed-of-nail fixtures too unwieldy.

Design-for-test techniques implemented at the chip level can remove the need for backdriving from a physical bed-of-nails fixture or use electronic alternatives to reduce the need for complete physical in-circuit access.

21.2.3.3 (Sub)System Support

Early proto-type bring-up, and in the case of problems, debug, pose a substantial challenge in the development of complex microelectronics systems. It often is very difficult to distinguish between hardware, design, and software problems. Debug is further complicated by the fact that valuable information about the detailed circuit states that could shed light on the problem may be hidden deep inside the chips in the assembly hierarchy. Moreover, the existence of one problem (e.g., a timing problem) can prevent the system from reaching a state needed for other parts of system bring-up, verification, and debug.

System manufacturing, just like PCB/MCM assembly, can introduce new defects and possibly damage the components. The same may apply to hardware maintenance/repair events (e.g., hot-plugging a new memory board).

Operating the system at the final customer's site can create additional requirements, especially if the system must meet stringent availability or safety criteria.

Design-for-test techniques implemented at the chip level can help enable a structural hardware integrity test that quickly and easily validates the physical assembly hierarchy (e.g., chip to board to backplane, etc.), and that the system's components (e.g., chips) are operating. DFT can also increase the observability of internal circuit state information for debug, or the controllability of internal states and certain operating conditions to continue debug in the presence of problems.

21.3 Overview of Chip-Level Design-For-Test Techniques

Design-for-test has a long history with a large supporting body of theoretical work as well as industrial application. Only a relatively small and narrow subset of the full body of DFT technology has found its way into the current EDA industry.

21.3.1 Brief Historical Commentary

Much of the DFT technology available in today's commercial DFT tools has its roots in the electronic data-processing industry. Data-processing systems have been complex composites made up of logic, memory, I/O, analog, human interface, and mechanical components long before the semiconductor industry invented the system-on-chip (SoC) moniker. Traditional DFT as practiced by the large data-processing system companies since at least the 1960s represents highly sophisticated architectures of engineering utilities that simultaneously address the needs of manufacturing, product engineering, maintenance/service, availability, and customer support. The first commercial DFT tools for IC design fielded by the EDA industry, by contrast, were primitive scan insertion tools that only addressed the needs of automatic test pattern generation (ATPG) for random logic. Tools have become more sophisticated and comprehensive, offering more options for logic test (e.g., built-in self-test [BIST]), support for some nonrandom logic design elements (e.g., BIST for embedded memories), and support for higher-level package testing (e.g., boundary scan for PCB/MCM testing). However, with few exceptions, there still is a lack of comprehensive DFT architectures for integrating the bits and pieces, and a lack of consideration for applications besides manufacturing test (e.g., support for nondestructive memory read for debug purposes is not a common offering by the tool vendors).

21.3.2 About Design-For-Test Tools

There are essentially three types of DFT-related tools:

- *Design-for-test synthesis (DFTS)*. Design-for-test involves design modification/edit steps (e.g., substituting one flip-flop type with another one) akin to simple logic transformation or synthesis. DFTS performs the circuit modification/edit task.
- *Design rules checking (DRC)*. Chip-level DFT is mostly used to prepare the circuit for some ATPG tool or to enable the use of some type of manufacturing test equipment. The ATPG tools and test equipment generally impose constraints on the design under test. DRC checks the augmented design for compliance with those constraints. Note that this DRC should not be confused with physical verification design rule checking (also called DRC) of ICs, as is discussed in the chapter on 'Design Rule Checking' in this handbook.
- *Design-for-test intellectual property (DFT IP) creation, configuration, and assembly*. In addition to relatively simple design modifications, DFT may add test-specific function blocks to the design.

Some of these DFT blocks can be quite sophisticated and may rival some third-party IP in complexity. And, like other IP blocks, the DFT blocks often must be configured for a particular design and then be assembled into the design.

21.3.3 Chip Design Elements and Element-Specific Test Methods

Modern chips can contain different types of circuitry with vastly different type-specific testing needs. Tests for random logic and tests for analog macros are very different, for example. DFT has to address the specific needs of each such circuit type and also facilitate the integration of the resulting type-specific tests into an efficient, high-quality composite test program for all pieces of the chip. System-on-chip is an industry moniker for chips made up of logic, memory, analog/mixed-signal, and I/O components. The main categories of DFT methods needed, and to a reasonable extent commercially available, for today's IC manufacturing test purposes can be introduced in the context of a hypothetical SoC chip.

Systems-on-chip are multiterrain devices consisting of predesigned and custom design elements:

- Digital logic, synthesized (e.g., cell-based) or custom (e.g., transistor-level)
- Embedded digital cores (e.g., processors)
- Embedded memories (SRAM, eDRAM, ROM, CAM, Flash, with or without embedded redundancy)
- Embedded register files (large number, single port, and multiport)
- Embedded Field Programmable Gate Array (eFPGA)
- Embedded Analog/Mixed-Signal (PLL/DLL, DAC, ADC)
- High-Speed I/Os (e.g., SerDes)
- Conventional I/Os (large number, different types, some differential)

The following overview will introduce some key type-specific DFT features for each type of component, and then address chip-level DFT techniques that facilitate the integration of the components into a top-level design.

21.3.3.1 Digital Logic

The most common DFT strategies for digital logic help prepare the design for ATPG tools. ATPG tools typically have difficulties with hard-to-control or hard-to-observe nets/pins, sequential depth, and loops.

21.3.3.1.1 Control/Observe Points.

The job of an ATPG tool is to locally set up suitable input conditions that *excite* a fault (i.e., trigger an incorrect logic response according to the fault definition; for example, to trigger a stuck-at-1 fault at a particular logic gate input, that input must receive a logic 0 from the preceding gates), and that *propagate* the incorrect value to an observable point (i.e., side-inputs of gates along the way must be set to their noncontrolling values). The run-time and success rate of test generation depend not least on the search space the algorithm has to explore to establish the required excitation and propagation conditions.

Control points provide an alternative means for the ATPG tool to more easily achieve a particular logic value. In addition to providing enhanced controllability for test, it must be possible to disable the additional logic such that the original circuit function is retained for normal system operation. In other words, DFT often means the implementation of multiple distinct modes of operation, for example, a *test mode* and a *normal mode.*

The second control point type is mostly used to override unknown/unpredictable signal sources, in particular for signal types that impact the sequential behavior, e.g., clocks. In addition to the two types of control points, there are other types for improved 1-controllability (e.g., using an OR gate) and for "randomization" (e.g., using an XOR gate). The latter type, for example, is useful in conjunction with pseudo-random test methods that will be introduced later. As can be seen from the examples, the implementation of control points tends to add cost due to one or more additional logic levels that affect the path delay and require additional area/power for transistors and wiring. The additional cost for implementing DFT is generally referred to as "overhead," and over the years there have been many, sometimes heated, debates juxtaposing the overhead against the benefits of DFT.

Observe points are somewhat "cheaper" in that they generally do not require additional logic in the system paths. The delay impact, hence, is reduced to the additional load posed by the fan-out and (optional) buffer used to build an observe.

21.3.3.1.2 Scan Design.

Scan design is the most common DFT method associated with synthesized logic. The concept of scan goes back to the very early days of the electronics industry, and it refers to certain means for controlling or observing otherwise hidden internal circuit states. Examples are manual dials to connect measurement instruments to probe points in analog computers, the switches, and lights on the control panel of early digital computers (and futuristic computers in Sci-Fi flicks), and to more automated electronic mechanisms to accomplish the objective, for example the use of machine instructions to write or read internal machine registers. Beginning with the late 1960s or so, scan has been implemented as a dedicated, hardware-based operation that is independent of, and does not rely on, specific intelligence in the intended circuit function.

21.3.3.1.2.1 Implementing the Scan Function. Among the key characteristics of scan architectures are the choice of which circuit states to control/observe, and the choice of an external data interface (I/Os and protocol) for the control/observe information. In basic scan methods, all (full scan) or some (partial scan) internal sequential state elements (latches or flip-flops) are made controllable and observable via a serial interface to minimize the I/O footprint required for the control/observe data. The most common implementation strategy is to replace the functional state elements with dual-purpose state elements (scan cells) that can operate as originally intended for functional purposes and as a serial shift register for scan. The most commonly used type of scan cell consists of an edge-triggered flip-flop with two-way multiplexer (scan mux) for the data input (mux-scan flip-flop).

The scan mux is typically controlled by a single control signal called scan_enable that selects between a scan-data and a system-data input port. The transport of control/observe data from/to the test equipment is achieved by a serial shift operation. To that effect, the scan cells are connected into serial shift register strings called scan chains. The scan-in port of each cell is either connected to an external input (scan-in) for the first cell in the scan chain or to the output of a single predecessor cell in the scan chain. The output from the last scan cell in the scan chain must be connected to an external output (scan-out). The data input port of the scan mux is connected to the functional logic as needed for the intended circuit function.

There are several commercial and proprietary DFTS tools available that largely automate the scan-chain construction process. These tools operate on register transfer level (RTL) and/or gate-level netlists of the design. The tools typically are driven by some rules on how to substitute nonscan storage elements in the prescan design with an appropriate scan cell, and how to connect the scan cells into one or more scan chains. In addition to connecting the scan and data input ports of the scan cells correctly, attention must be given to the clock input ports of the scan cells. To make the shift-registers operable without interference from the functional logic, a particular circuit state (*scan state*), established by asserting designated scan state values at certain primary inputs and/or by executing a designated initialization sequence, must exist that:

1. switches all scan muxes to the scan side (i.e., the local scan_enable signals are forced to the correct value);
2. assures that each scan cell clock port is controlled from one designated external clock input (i.e., any intervening clock gating or other clock manipulation logic is overridden/disabled);
3. all other scan cell control inputs like set/reset are disabled (i.e., the local control inputs at the scan cell are forced to their inactive state);
4. all scan data inputs are sensitized to the output of the respective predecessor scan cell or the respective scan_in port for the first scan cell in the chain (i.e., side-inputs of logic gates along the path are forced to a nondominating value);
5. the output of the last scan_cell in the scan chain is sensitized to its corresponding scan_out port or the side-inputs of logic gates along the path are forced to a nondominating value, such that pulsing the designated external clock (or clocks) once results in shifting the data in the scan chains by exactly one-bit position.

This language may sound pedantic, but the DFTS and DFT DRC tools tend to use even more detailed definitions for what constitutes a valid scan chain and scan state. Only a crisp definition allows the tools to validate the design thoroughly and, if problems are detected, write error messages with enough diagnostic information to help a user find and fix the design error that caused the problem.

21.3.3.1.2.2 About Scan Architectures. Very few modern chips contain just a single scan chain. In fact, it is fairly common to have several selectable scan-chain configurations. The reason is that scan can be used for a number of different purposes. Facilitating manufacturing test for synthesized logic is one purpose. In that case, the scan cells act as serially accessible control and observe points for logic test, and test application essentially follows a protocol such as:

1 Establishing the manufacturing test-scan configuration and associated scan state, and serially load the scan chains with new test input conditions
2 Applying any other primary input conditions required for the test
3 Waiting until the circuit stabilizes, and measure/compare the external test responses at primary outputs;
4 Capturing the internal test responses into the scan cells (typically done by switching scan_enable to the system side of the scan mux and pulsing a clock)
5 Reestablishing the manufacturing test-scan configuration and associated scan state, and serially unload the test responses from the scan chains into the tester for comparison

Steps 1 and 5 in many cases can be overlapped, meaning that while the responses from one test are unloaded through the scan-out pins, new test input data are simultaneously shifted in from the scan-in pins. The serial load/unload operation requires as many clock cycles as there are scan cells in the longest scan chain. Manufacturing test time, and consequently test cost, for scan-based logic tests are typically dominated by the time used for scan load/unload. Hence, to minimize test times and cost, it is preferable to implement as many short, parallel scan chains as possible. The limiting factors are the availability of chip I/Os for scan-in/-out or the availability of test equipment channels suitable for scan. Modern DFT tools can help optimize the number of scan chains and balance their length according to the requirements and constraints of chip-level manufacturing test. Today's scan-insertion flows also tend to include a post-placement *scan reordering* step to reduce the wiring overhead for connecting the scan cells. The currently practiced state of the art generally limits reordering to occur within a scan chain. Research projects have indicated that further improvements are possible by allowing the exchange of scan cells between scan chains. All practical tools tend to give the user some control over partial ordering, keeping subchains untouched, and placing certain scan cells at predetermined offsets in the chains.

In addition to building the scan chains proper, modern DFT tools also can insert and validate *pin-sharing* logic that makes it possible to use functional I/Os as scan-in/-out or scan control pins, thus avoiding the need for additional chip I/Os dedicated to the test. In many practical cases, a single dedicated pin is sufficient to select between normal mode and test mode. All other test control and interface signals are mapped onto the functional I/Os by inserting the appropriate pin-sharing logic.

Besides chip manufacturing test, scan chains often are also used for access to internal circuit states for higher-level assembly (e.g., board-level) testing. In this scenario, it generally is not possible or economically feasible to wire all scan-in/-out pins used for chip testing out to the board connectors. Board-level wiring and connectors are very limited and relatively "expensive." Hence, the I/O footprint dedicated to scan must be kept at a minimum and it is customary to implement another scan configuration in the chips, wherein all scan cells can be loaded/unloaded from a single pair of scan-in/-out pins. This can be done by concatenating the short scan chains used for chip manufacturing test into a single, long scan chain, or by providing some addressing mechanism for selectively connecting one shorter scan chain at a time to the scan-in/-out pair. In either case, a *scan switching network* and associated control signals are required to facilitate the reconfiguration of the scan interface.

In many practical cases, there are more than two scan configurations to support additional engineering applications beyond chip and board-level manufacturing test, for example, debug or system configuration.

The scan architectures originally developed for large data-processing systems in the 1960s and 1970s, for example, were designed to facilitate comprehensive engineering access to all hardware elements for testing, bring-up, maintenance, and diagnostics. The value of comprehensive scan architectures is only now being rediscovered for the complex system-level chips possible with nanometer technologies.

21.3.3.1.3 Timing Considerations and At-Speed Testing.
Timing issues can affect and plague both the scan infrastructure as well as the application of scan-based logic test.

21.3.3.1.3.1 Scan-Chain Timing. The frequently used mux-scan methodology uses edge-triggered flip-flops as storage elements in the scan cells. And the edge-clock is used for both the scan operation and for capturing test responses into the scan cells, making both susceptible to hold-time errors due to clock skew. Clock skew not only exists between multiple *clock domains* but also within each clock domain. The latter tend to be more subtle and easier to overlook. To deal with interdomain issues, the DFT tools have to be aware of the clock domains and clock-domain boundaries. The general rule-of-thumb for scan-chain construction is that each chain should only contain flip-flops from the same clock domain. Also, leading-edge and falling-edge flip-flops should be kept in separate scan chains even if driven from the same clock source. These strict rules of "division" can be relaxed somewhat if the amount of clock skew is small enough to be reliably overcome by inserting a *lock-up latch or flip-flop* between the scan cells. The susceptibility of the scan operation to hold-time problems can further be reduced by increasing the delay between scan cells, for example, by adding buffers to the scan connection between adjacent scan cells.

In practice, it is not at all unusual for the scan operation to fail for newly designed chips. To avoid the likelihood of running into these problems, it is vitally important to perform a very thorough timing verification on the scan mode. In the newer nanometer technologies, signal integrity issues such as static/dynamic IR-drop have to be taken into account in addition to process and circuit variability. A more "radical" approach is to replace the edge-triggered scan clocking with a level-sensitive multiphase clocking approach as, for example, in level-sensitive scan design (LSSD). In this case, the master and slave latches in the scan cells are controlled from two separate clock sources that are pulsed alternately. By increasing the nonoverlap period between the clock phases it is possible to overcome any hold-time problems during scan without the need for lock-up latches or additional intercell delay.

21.3.3.1.3.2 Scan-Based Logic Test Timing Considerations. Clock skew and hold-time issues also affect the reliable data transmission across interdomain boundaries. For example, if the clocks of two interconnected domains are pulsed together for capture, then it may be impossible to predict whether old or new data are captured. If the data-change and clock edge get too close together, the receiving flip-flop could even be forced into metastability. ATPG tools traditionally try to avoid these problems by using a "capture-by-domain" policy in which only one clock domain is allowed to be captured in a test. This is only possible if DFT makes sure that it is indeed possible to issue a capture clock to each clock domain separately (e.g., using a separate test clock input for each domain or by de-gating the clocks of other domains). The "capture-by-domain" policy can adversely affect test time and data volume for designs with many clock domains, by limiting fault detection for each test to a single domain. Some ATPG tools nowadays offer sophisticated multiclock compaction techniques that overcome the "capture-by-domain" limitation (e.g., if clock-domain analysis shows that there is no connection between certain domains, then capture clocks can be sent to all of those domains without creating potential hold-time issues; if two domains are connected, then their capture clocks can be issued sequentially with enough pulse separation to assure predictability of the interface states). Special treatment of the boundary flip-flops between domains in DFT is an alternative method.

21.3.3.1.3.3 Scan-Based At-Speed Testing. In static, fully complementary CMOS logic there is no direct path from power to ground except when switching. If a circuit is allowed to stabilize and settle down from all transitions, a very low power should be seen in the stable (quiescent) state. That expectation is the basis of IDDq (quiescent power) testing. Certain defects, for example shorts, can create power-ground paths and therefore be detectable by an abnormal amount of quiescent power. For many years, low-speed stuck-at testing combined with IDDq testing have been sufficient to achieve reasonable quality levels for many CMOS designs. The normal quiescent background current unfortunately increases with each new technology generation, which reduces the signal-to-noise ratio of IDDq measurements. Furthermore, modern process technologies are increasingly susceptible to interconnect opens and resistive problems that are less easily detectable with IDDq to begin with. Many of these defects cause additional circuit delays and cannot be tested with low-speed stuck-at tests. Consequently, there is an increasing demand for at-speed delay fault testing.

In scan-based delay testing, the circuit is first initialized by a scan operation. Then a rapid sequence of successive input events is applied at tight timings to create transitions at flip-flop outputs in the circuit, have them propagate through the logic, and capture the responses into receiving flip-flops. The responses finally are unloaded by another scan operation for comparison. Signal transitions at flip-flop outputs are obtained by loading the initial value into the flip-flop, placing the opposite final value at the flip-flop's data input, and pulsing the clock. In the case of mux-scan, the final value for the transition can come from the scan side (*release from scan*) or the system side (*release from capture*) of the scan mux, depending on the state of the scan-enable signal. The functional logic typically is connected to the system side of the scan mux. The transitions will, hence, generally arrive at the system side of the receiving flip-flop's scan mux, such that the scan-enable must select the system side to enable capture. The release from scan method, therefore, requires switching the scan-enable from the scan side to the system side early enough to meet setup time at the receiving flip-flop, but late enough to avoid hold-time issues at the releasing flip-flop. In other words, the scan-enable signal is subject to a two-sided timing constraint and accordingly must be treated as a timing-sensitive signal for synthesis, placement, and wiring. Moreover, it must be possible to synchronize the scan-enable appropriately to the clocks of each clock domain. To overcome latency and synchronization issues with high fan-out scan-enables in high-speed logic, the scan-enables sometimes are pipelined. The DFT scan insertion tools must be able to construct viable pipelined or non-pipelined scan-enable trees and generate the appropriate timing constraints and assertions for physical synthesis and timing verification.

Most ATPG tools do not have access to timing information and, in order to generate predictable results, tend to assume that the offset between release and capture clocks is sufficient to avoid completely setup time violations at the receiving flip-flops. That means the minimal offset between the release and capture clock pulses are dominated by the longest signal propagation path between the releasing and receiving flip-flops. If slow maintenance paths, multicycle paths, or paths from other slower clock domains get mixed in with the "normal" paths of a particular target clock domain, then it may be impossible to test the target domain paths at their native speed. To overcome this problem, some design projects disallow multicycle paths and insist that all paths (including maintenance paths) that can be active during the test of a target domain, must fit into the single-cycle timing window of that target domain. Another approach is to add enough timing capabilities to the ATPG software to identify all potential setup and hold-time violations at a desired test timing. The ATPG tool can then avoid sending transitions through problem paths (e.g., holding the path inputs stable) or set the state of all problem flip-flops to "unknown". Yet other approaches use DFT techniques to separate multicycle paths out into what looks like another clock domain running at a lower frequency.

21.3.3.1.3.4 Power and Noise Considerations. Most scan-based at-speed test methods perform the scan-load/-unload operations at a relatively low frequency, not the least to reduce power consumption during shift. ATPG patterns tend to have close to 50% switching probability at the flip-flop outputs during scan, which in some cases can be ten times more than what is expected during normal operation. Such abnormally high switching activity can cause thermal problems, excessive static IR drop, or exceed the tester power-supply capabilities when the scan chains are shifted at full system speed. The reality of pin electronic capabilities in affordable test equipment sets another practical limit to the data rate at which the scan interface can be operated. One advantage of lower scan speeds for design is that it is not necessary to design the scan chains and scan-chain interface logic for full system speed. That can help reduce the placement, wiring, and timing constraints for the scan logic.

Slowing down the scan rate helps reduce average power, but does little for dynamic power (d*i*/d*t*). In circuits with a large number of flip-flops, in particular when combined with tightly controlled low-skew clocks, simultaneous switching of flip-flop outputs can result in unexpected power/noise spikes and dynamic IR drops, leading to possible scan-chain malfunctions. These effects may need to be considered when allocating hold-time margins during scan-chain construction and verification.

21.3.3.1.3.5 On-Chip Clock Generation. The effectiveness of at-speed tests not only depends on the construction of proper test event sequences but may also critically depend on being able to deliver these sequences at higher speed with higher accuracy than supported by the test equipment. Many modern chips use on-chip clock frequency multiplication (e.g., using phase-locked loops [PLLs]) and phase alignment, and there is an increasing interest in taking advantage of this on-chip clocking infrastructure for testing. To that effect, clock system designers add programmable test waveform generation features to the clock circuitry. Programmable in this context tends to mean the provision of a serially loadable control register that determines the details of which and how many clock phases are generated. The actual sequence generation is triggered by a (possibly asynchronous) start signal that can be issued from the tester (or some other internal controller source) after the scan load operation has completed. The clock generator will then produce a deterministic sequence of edges that are synchronized to the PLL output. Some high-performance designs may include additional features (e.g., programmable delay lines) for manipulating the relative edge positions over and above what is possible by simply changing the frequency of the PLL input clock.

The ATPG tools generally cannot deal with the complex clock generation circuitry. Therefore, the on-product clock generation (OPCG) logic is combined into an OPCG macro and separated from the rest of the chip by cut points. The cut points look like external clock pins to the ATPG tool. It is the responsibility of the user to specify a list of available OPCG programming codes and the resulting test sequences to the ATPG tool, which in turn is constrained to use only the thus specified event sequences. For verification, the OPCG macro is simulated for all specified programming codes, and the simulated sequences appearing at the cut points are compared to the input sequences specified to the ATPG tool.

Some circuit elements may require finer-grained timing that requires a different approach. One example is clock jitter measurement, another memory access time measurement. Test equipment uses analog or digitally controlled, tightly calibrated delay lines (timing verniers), and it is possible to integrate similar features into the chip. A different method for measuring arbitrary delays is to switch the to-be-measured delay path into an inverting recirculating loop and measure the oscillation frequency (e.g., counting oscillations against a timing reference). Small delays that would result in extremely high frequencies are made easier to test by switching them into and out of a longer delay path and comparing the resulting frequencies. Oscillation techniques can also be used for delay calibration to counteract performance variations due to process variability.

21.3.3.1.4 Custom Logic.

Custom transistor-level logic, often used for the most performance-sensitive parts of a design, poses a number of unique challenges to the DFT flow and successful test generation. Cell-based designs tend to use more conservative design practices and the libraries for cell-based designs generally come with premade and preverified gate-level test generation models. For transistor-level custom designs, by contrast, the gate-level test generation models must somehow be generated and verified "after-the-fact" from the transistor-level schematics. Although the commercial ATPG tools may have some limited transistor-level modeling capabilities, their wholesale use generally leads to severe tool run-time problems, and hence is strongly discouraged. The construction of suitable gate-level models is complicated by the fact that custom logic often uses dynamic logic or other performance/area/power-driven unique design styles, and it is not always easy to determine what should be explicitly modeled and what should be implied in the model (e.g., the precharge clocks and precharge circuits for dynamic logic). Another issue for defect coverage as well as diagnostics is to decide which circuit-level nets to explicitly keep in the logic model vs. simplifying the logic model for model size and tool performance.

The often-extreme area, delay, and power sensitivity of custom-designed structures are often met with a partial scan approach, even if the synthesized logic modules have full scan. The custom design team has to make a trade-off between the negative impact on test coverage and tool run-time vs. keeping the circuit overhead small. Typical "rules-of-thumb" will limit the number of nonscannable levels and discourage feed-back loops between nonscannable storage elements (e.g., feed-forward pipeline stages in data paths are often good candidates for partial scan). Area, delay, and power considerations also lead to a more widespread use of pass-gates and other three-state logic (e.g., three-state buses) in custom-design circuitry. The control inputs of pass-gate structures, such as the select lines of multiplexers, and enables of three-state bus drivers tend to require specific decodes for control (e.g., "one-hot") to avoid three-state contention (i.e., establishing a direct path from power to ground) that could result in circuit damage due to burn-out. To avoid potential burn-out, DFT and ATPG must cooperate to assure that safe control states are maintained during scan and during test. If the control flip-flops are included in the scan chains, then DFT hardware may be needed to protect the circuits (e.g., all bus-drivers are disabled during scan).

Other areas of difficulty are complex memory substructures with limited scan and possibly unusual or pipelined decodes that are not easily modeled with the built-in memory primitives available in the DFT/ATPG tools.

21.3.3.1.5 ***Logic Built-In Self-Test.***

Chip manufacturing test (e.g., from ATPG) typically assumes full access to the chip I/Os plus certain test equipment features for successful test application. That makes it virtually impossible to port chip manufacturing tests to higher-level assemblies and into the field. Large data-processing systems historically stored test data specifically generated for in-system testing on disk and applied them to the main processor complex through a serial maintenance interface from a dedicated service processor that was delivered as part of the system. Over the years it became too cumbersome to store and manage vast amounts of test data for all possible system configurations and ECO (engineering change order) levels, and the serial maintenance interface became too slow for efficient data transfer. Hence, alternatives were pursued that avoid the large data volume and data transfer bottleneck associated with traditional ATPG tests.

21.3.3.1.5.1 Pseudo-Random Pattern Testing. The most widely used alternative today is logic built-in self-test (logic BIST) using pseudo-random patterns. It is known from coding theory that pseudo-random patterns can be generated easily and efficiently in hardware, typically using a so-called pseudo-random pattern generator (PRPG) macro utilizing a linear feedback shift register (LFSR). The PRPG is initialized to a starting state called PRPG seed and in response to subsequent clock pulses, produces a state sequence that meets certain tests of randomness. However, the sequence is not truly random. In particular, it is predictable and repeatable if started from the same seed. The resulting pseudo-random logic states are loaded into scan chains for testing in lieu of ATPG data. The PRPGs nowadays are generally built into the chips. That makes it possible to use a large number of relatively short on-chip scan chains because the test data do not need to be brought in from the outside through a narrow maintenance interface or a limited number of chip I/Os. Simple LFSR-based PRPGs connected to multiple parallel scan chains result in undesirable, strong value correlations (*structural dependencies*) between scan cells in adjacent scan chains. The correlations can reduce the achievable test coverage. To overcome this potential problem, some PRPG implementations are based on *cellular automata* rather than LFSRs. LFSR-based implementations use a *phase-shifting network* constructed out of exclusive-OR (XOR) to eliminate the structural dependencies. The phase-shifting network can be extended into a *spreading network* that makes it possible to drive a larger number of scan-chain inputs from a relatively compact LFSR.

More scan chains generally mean shorter scan chains that require fewer clock cycles for load/unload, thus speeding up test application (each scan test requires at least one scan load/unload), assuming that the DFTS tool used for scan insertion succeeds in reducing and balancing the length of the scan chains. Modern DFTS tools are capable of building scan architectures with multiple separately balanced selectable chain configurations to support different test methods on the same chip. It should be noted that the achievable chain length reduction can be limited by the length of preconnected scan-chain segments in hard macros. Hence, for logic BIST it is important to assure that large hard macros are preconfigured with several shorter chain segments rather than a single long segment. As a rule-of-thumb, the maximum chain length for logic BIST should not exceed 500 to 1000 scan cells. That means a large chip with 1M scan cells requires 1K scan chains or more. The overhead for the PRPG hardware is essentially proportional to the number of chains (a relatively constant number of gates per scan chain for the LFSR/CA, phase shifting/spreading network, and scan-switching network).

The flip side of the coin for pseudo-random logic BIST is that pseudo-random patterns are less efficient for fault testing than ATPG-generated, compacted test sets. Hence, ten times as many or more pseudo-random patterns are needed for equivalent nominal fault coverage, offsetting the advantage of shorter chains. Moreover, not all faults are easily tested with pseudo-random patterns. Practical experience with large-scale data-processing systems has indicated that a stuck-at coverage of around up to 95% can be achievable with a reasonable (as dictated by test time) number of pseudo-random test patterns. Going beyond 95% requires too much test application time to be practical. Coverage of 95% can be sufficient for burn-in, higher-level assembly, system, and field testing, but may be unacceptable for chip manufacturing test. Consequently, it is not unusual to see ATPG-based patterns used for chip manufacturing tests and pseudo-random logic BIST for all subsequent tests.

Higher test coverage, approaching that of ATPG, can be achieved with pseudo-random patterns only by making the logic more testable for such patterns. The 50–50 pseudo-random signal probability at the outputs of the scan cells gets modified by the logic gates in the combinational logic between the scan cells. Some internal signal probabilities can be skewed so strongly to 0 or 1 that the effective controllability or observability of downstream logic becomes severely impaired. Moreover, certain faults require many more specific signal values for fault excitation or propagation than achievable with a limited set of pseudo-random patterns (it is like rolling dice and trying to get 100+ sixes in a row). Wide comparators and large counters are typical architectural elements afflicted with that problem. Modern DFT synthesis and analysis tools offer so-called *pseudo-random testability analysis* and *automatic test-point insertion features*. The testability analysis tools use *testability measures* or signal local characteristics captured during good machine fault simulation to identify nets with low pseudo-random controllability or observability for pseudo-random patterns. The test point insertion tools generate a suggested list of control or observe points that should be added to the netlist to improve test coverage. Users generally can control how many and what type of test points are acceptable (control points tend to be more "expensive" in circuit area and delay impact) for manual or automatic insertion. It is not unusual for test points to be "attracted" to timing-critical paths. If the timing-critical nets are known ahead of time, they optionally can be excluded from modification. As a rule-of-thumb, one test point is needed per 1K gates to achieve the 99%+ coverage objective often targeted for chip manufacturing test. Each test point consumes roughly ten gates, meaning that 1% additional logic is required for the test points.

It should be noted that some hard-to-test architectural constructs such as comparators and counters can be identified at the presynthesis RTL, creating a basis for RTL analysis and test point insertion tools.

21.3.3.1.5.2 Test Response Compression. Using pseudo-random patterns to eliminate the need for storing test input data from ATPG solves only part of the data volume problem. The expected test responses for ATPG-based tests must equally be stored in the test equipment for comparison with the actual test responses. As it turns out, LFSR-based hardware macros very similar to a PRPG macro can be used to implement error detecting code (EDC) generators. The most widely used EDC macro implementation for logic BIST is called multiple-input signature register (MISR), which can sample all scan-chain outputs in parallel. As the test responses are unloaded from the scan chains they are simultaneously clocked into the MISR where they are accumulated. The final MISR state after a specified number of scan tests have been applied is called the *signature*, and this signature is compared to an expected signature that has been precalculated by simulation for the corresponding set of test patterns.

The MISR uses only shifting and XOR logic for data accumulation, meaning that each signature bit is the XOR sum of some subset of the accumulated test response bit values. One property of XOR-sums is that even a single unknown or unpredictable summand makes the sum itself unknown or unpredictable. If the signature of a defect-free product under test is unknown or unpredictable, then the signature is useless for testing. Hence, there is a "Golden Rule" for signature-based testing that no unknown or unpredictable circuit state can be allowed to propagate to the MISR. This rule creates additional design requirements over and above what is required for scan-based ATPG. The potential impact is further amplified by the fact that pseudo-random patterns, unlike ATPG patterns, offer little to no ability for intelligently manipulating the test stimulus data. Hence, for logic BIST, the propagation of unknown/unpredictable circuit states (also known as x states) must generally be stopped by hardware means. For example, microprocessor designs tend to contain tens of thousands of three-state nets, and modern ATPG tools have been adapted to that challenge by constructively avoiding three-state contention and floating nets in the generated test patterns. For logic BIST, either test hardware must be added to prevent contention or floating nets, or the outputs of the associated logic must be de-gated (rendering it untestable) so that they cannot affect the signature. Some processor design projects forbid the use of three-state logic, enabling them to use logic BIST. Other design teams find that so radical an approach is entirely unacceptable.

Three-state nets are not the only source of x states. Other sources include unmodeled or incompletely modeled circuit elements, uninitialized storage elements, multiport storage element write conflicts, and set-up/hold-time timing violations. Design-for-test synthesis and DRC tools for logic BIST must analyze the netlist for potential x-state sources and add DFT structures that remove the x-state generation potential (e.g., adding exclusive gating logic to prevent multiport conflicts, or by assuring the proper initialization of storage elements), or add de-gating logic that prevents x-state propagation to the MISR (e.g., at the outputs of uninitialized or insufficiently modeled circuit elements). Likewise, hardware solutions may be required to deal with potential setup/hold-time problems (e.g., clock-domain boundaries, multicycle paths, nonfunctional paths, etc.). High-coverage logic BIST, overall, is considered to be significantly more design-intrusive than ATPG methods where many of the issues can be dealt within pattern generation rather than through design modifications.

21.3.3.1.5.3 At-Speed Testing and Enhanced Defect Coverage with Logic BIST. At-speed testing with logic BIST essentially follows the same scheme as at-speed testing with ATPG patterns. (Historical note: contrary to frequent assertions by logic BIST advocates that pseudo-random pattern logic BIST is needed to enable at-speed testing, ATPG-based at-speed test has been practiced long before logic BIST became popular and is still being used very successfully today.) Most approaches use slow scan (to limit power consumption, among other things) followed by the rapid application of a short burst of at-speed edge events. Just as with ATPG methods, the scan and at-speed edge events can be controlled directly by test equipment or from an OPCG macro. The advantage of using slow scan is that the PRPG/MISR and other scan-switching and interface logic need not be designed for high speed. That simplifies timing closure and gives the placement and wiring tools more flexibility. Placement/wiring consideration may still favor using several smaller, distributed PRPG/MISR macros. Modern DFT synthesis, DRC, fault grading, and signature-simulation tools for logic BIST generally allow for distributed macros.

Certain logic BIST approaches, however, may depend on performing all or some of the scan cycles at full system speed. With this approach, the PRPG/MISR macros and other scan interface logic must be designed to run at full speed. Moreover, scan chains in clock domains with different clock frequencies may be shifted at different frequencies. That affects scan-chain balancing, because scan chains operating at half frequency in this case should only be half as long as chains operating at full frequency. Otherwise they would require twice the time for scan load/unload.

The fact that pseudo-random logic BIST applies ten times (or more) as many tests than ATPG for the same nominal fault coverage can be advantageous for the detection of unmodeled faults and of defects that escape detection by the more compact ATPG test set. This was demonstrated empirically in early pseudo-random test experiments using industrial production chips. However, it is not easy to extrapolate these early results to the much larger chips of today. Today's chips can require 10K or more ATPG patterns. Even ATPG vectors are mostly pseudo-random, meaning that applying the ATPG vectors is essentially equivalent to applying 10K+ logic BIST patterns, which is much more than what was used in the old hardware experiments.

Only recently has the debate over accidental fault/defect coverage been refreshed. The theoretical background for the debate is the different *n-detect* profiles for ATPG and logic BIST. ATPG test sets are optimized to the extent that some faults are only detected by one test (1-detect) in the set. More faults are tested by a few tests and only the remainder of the fault population gets detected ten times (10-detect) or more. For logic BIST tests, by contrast, almost all faults are detected many times. Static bridging fault detection, for example, requires coincidence of a stuck-at fault test for the victim net with the aggressor net being at the fault value. If the stuck-at fault at the victim net is detected only once, then the probability of detecting the bridging fault is determined by the probability of the aggressor net being at the faulty value, e.g., 50% for pseudo-random values. A 2-detect test set would raise the probability to 75% and so on. Hence, multidetection of stuck-at faults increases the likelihood of detecting bridging faults. The trend can be verified by running bridging-fault simulation for stuck-at test sets with different *n*-detect profiles and comparing the results. Recent hardware experiments confirmed the trend for production chips.

It must be noted, however, that modern ATPG tools can and have been adapted to optionally generate test sets with improved *n*-detect profiles. The hardware experiments cited by the BIST advocates in fact were performed with ATPG-generated *n*-detect test sets, not with logic BIST tests. It should also be noted that if the probability of the aggressor net being at the faulty value is low, then even multiple detects may not do enough. ATPG experiments that try to constructively enhance the signal probability distribution have shown some success in that area. Finally, a new generation of tools is emerging that extract realistic bridging faults from the circuit design and layout. ATPG tools can and will generate explicit tests for the extracted faults, and it has been shown that low *n*-detect test sets with explicit clean-up tests for bridging faults can produce very compact test sets with equally high or higher bridging fault coverage than high *n*-detect test sets.

21.3.3.1.5.4 Advanced Logic BIST Techniques. Experience with logic BIST on high-performance designs reveals that test points may be helpful to improve nominal test coverage, but can have some side-effects for characterization and performance screening. One reported example shows a particular defect in dynamic logic implementing a wide comparator that can only be tested with certain patterns. Modifying the counter with test points as required for stuck-at and transition fault coverage creates artificial nonfunctional short paths. The logic BIST patterns use artificial paths for coverage and never sensitize the actual critical path. Knowing that wide comparators were used in the design, some simple weighting logic had been added to the PRPG macro to create the almost-all-1s or almost-all-0s patterns suited for testing the comparators. The defect was indeed only detected by the weighted tests and escaped the normal BIST tests. Overall, the simple weighting scheme measurably increased the achievable test coverage without test points.

If logic BIST is intended to be used for characterization and performance screening, it may also be necessary to enable memory access in logic BIST. It is not uncommon that the performance-limiting paths in high-speed design traverse embedded memories. The general recommendation for logic BIST is to fence embedded memories off with boundary scan during BIST. That again creates artificial paths that may not be truly representative of the actual performance-limiting paths. Hardware experience with high-performance processors shows that enabling memory access for some portion of the logic BIST tests does indeed capture unique fails. Experiments with (ATPG-generated) scan tests for processor's performance binning have similarly shown that testing paths through embedded memories is required for better correlation with functional tests.

As a general rule-of-thumb, if BIST is to be used for characterization, then the BIST logic may have to be designed to operate at higher speeds than the functional logic it is trying to characterize.

21.3.3.1.5.5 Logic BIST Diagnostics. Automated diagnosis of production test fails has received considerable attention recently and is considered a key technology for nanometer semiconductor technologies. In traditional stored pattern ATPG testing, each response bit collected from the device under test is immediately compared to an expected value, and most test equipment in the case of a test fail (i.e., mismatch between actual and expected response values) allows for optionally logging the detailed fail information (e.g., tester cycle, tester channel, fail value) into a so-called fail set for the device under test. The fail sets can then be postprocessed by automated logic diagnostic software tools to determine the most likely root cause locations.

In logic BIST, the test equipment normally does not see the detailed bit-level responses because these are intercepted and accumulated into a signature by the on-chip MISR. The test equipment only sees and compares highly compressed information contained in the accumulated signatures. Any difference between an actual signature and the expected signature indicates that the test response must contain some erroneous bits. It generally is impossible to reconstruct bit-level fail sets from the highly compressed information in the signatures. The automated diagnostic software tools, however, need the bit-level fail sets. The diagnosis of logic BIST fails, hence, require an entirely different analysis approach or some means for extracting a bit-level fail set from the device under test. Despite research efforts aimed at finding alternative methods, practitioners tend to depend on the second approach. To that effect, the logic BIST tests are structured such that signatures are compared after each group of n tests, where n can be 1, 32, 256, or some other number. The tests further must be structured such that each group of n tests is independent. In that case, it can be assumed that a signature mismatch can only be caused by bit-level errors in the associated group of n tests. For fail-set extraction, the n tests in the failing group are repeated and this time the responses are directly scanned out to the test equipment without being intercepted by the MISR (*scan dump operation*). Existing production test equipment generally can only log a limited number of failing bits and not the raw responses. Hence, the test equipment must have stored expect data available for comparison. Conceptually, that could be some "fake" expect vector like all 0s, but then even correct response tests would result in failing bits that would quickly exceed the very limited fail buffers on the testers, meaning that the actual expect data must be used. However, the number of logic BIST tests tends to be so high that it is impractical to store all expect vectors in the production test equipment. Bringing the data in from some offline medium would be too slow. The issue can be overcome if it is only desired to diagnose a small sample of failing devices, for example, prior to sending them to the FA lab. In that case, the failing chips can be sent to a nonproduction tester for retesting and fail-set logging.

Emerging, very powerful, statistical yield analysis and yield management methods require the ability to log large numbers of fail sets during production testing, which means fail-set logging must be possible with minimal impact to the production test throughput. That generally is possible and fairly straightforward for ATPG tests (as long as the test data fit into the test equipment in the first place), but may require some logistical ingenuity for logic BIST and other signature-based test methods.

21.3.3.1.6 Test Data Compression.

Scan-based logic tests consume significant amounts of storage and test time on the ATE used for chip manufacturing test. The data volume in first order is roughly proportional to the number of logic gates on the chip and the same holds for the number of scan cells. Practical considerations and test equipment specifications oftentimes limit the number of pins available for scan-in/-out and the maximum scan frequency. Consequently, the scan chains for more complex chips tend to be longer and it takes commensurately longer to load/unload the scan chains. There is a strong desire to keep existing test equipment and minimize expensive upgrades or replacements. Existing equipment on many manufacturing test floors tend to have insufficient vector memory for the newer chip generations. Memory re-load is very time-consuming and should be avoided if possible. The purpose of test data compression is to reduce the memory footprint of the scan-based logic tests such that they comfortably fit into the vector memory of the test equipment.

21.3.3.1.6.1 Overview. Test equipment tends to have at least two types of memory for the data contained in the test program. One type is the vector memory that essentially holds the logic levels for the test inputs and for the expected responses. The memory allocation for each input bit could include additional space for waveform formats (the actual edge timings are kept in a separate time-set memory space) over and above the logic level. On the output side, there typically are at least two bits of storage for each expected response bit. One bit is a mask value that determines if the response value should be compared or ignored (for example, unknown/unpredictable responses are masked) and the other bit defines the logic level to compare with. The other memory type contains nonvector program information like program op-codes for the real-time processing engine in the test equipment's pin electronics.

Some memory optimization for scan-in data is possible by taking advantage of the fact that the input data for all cycles of the scan load/unload operation use the same format. The format can be defined once upfront and only a single bit is necessary to define the logic level for each scan cycle. A test with a single-scan load/unload operation may thus consume three bits of vector memory in the test equipment, and possibly only two bits of no response masking is needed.

Scan-based logic test programs tend to be simple in structure, with one common loop for the scan/load unload operation and short bursts of other test events in between. Consequently, scan-based test programs tend to consume only very little op-code memory and the memory limitation for large complex chips is only in the vector memory.

It is worth noticing that most production test equipment offers programming features such as branching, looping, logic operations, etc., for functional testing. The scan-based ATPG programs, however, do not typically take advantage of these features for two reasons. First, some of the available features are equipment-specific. Second, much of the data volume is for the expected responses. While the ATPG tools may have control over constructing the test input data, it is the product under test, not the ATPG software that shapes the responses. Hence, taking full advantage of programming features for really significant data reduction would first require some method for removing the dependency on storing the full amount of expected responses.

21.3.3.1.6.2 Input Data Compression. Input data compression, in general, works by replacing the bit-for-bit storage of each logic level for each scan cell with some means for algorithmically constructing multiple input values on-the-fly from some compact source data. The algorithms can be implemented in software or hardware on the tester or in software or hardware inside the chips under test. To understand the nature of the algorithms, it is necessary to review some properties of the tests generated by ATPG tools.

Automatic test pattern generation proceeds by selecting one yet untested fault and generating a test for that one fault. To that effect, the ATPG algorithm will determine sufficient input conditions to excite and propagate the selected fault. That generally requires that specific logic levels must be asserted at some scan cells and primary inputs. The remaining scan cells remain unspecified at this step in the algorithm. The thus constructed, partially specified, vector is called a *test cube.* It has become customary to refer to the specified bits in the test cube as "care bits" and to the unspecified bits as "don't care bits." All ATPG tools used in practice perform vector *compaction,* which means that they try to combine the test cubes for as many faults as possible into a single test vector. The methods for performing vector compaction vary but the result generally is the same. Even after compaction, for almost all tests, there are many more don't care bits than care bits. In other words, scan-based tests generated by ATPG tools are characterized by a low *care bit density* (percentage of specified bits in the compacted multifault test cube). After compaction, the remaining unspecified bits will be filled by some arbitrary fill algorithm. Pseudo-random pattern generation is the most common type of algorithm used for fill. That is noteworthy because it means that the majority of bit values in the tests are generated by the ATPG software algorithmically from a very compact seed value. However, neither the seed nor the algorithm is generally included in the final test data. And, without that knowledge, it tends to be impossible to recompact the data after the fact, with any commonly known compression algorithm (e.g., zip, lzw, etc.). The most commonly practiced test data compression methods in use today focus on using simple software/hardware schemes to regenerate algorithmically fill data in the test equipment or in the chip under test. Ideally, very little memory is needed to store the seed value for the fill data and only the care bits must be stored explicitly.

The "cheapest" method for input fill data compression is to utilize common software features in existing test equipment without modifications of the chip under test. Run length encoding (RLE) is one software method used successfully in practice. In this approach, the pseudo-random fill algorithm in ATPG is replaced with a repeat option that simply repeats the last value until a specified bit with the opposite value is encountered. The test program generation software is modified to search for repeating input patterns in the test data from ATPG. If a repeating pattern of sufficient length is found, it will be translated into a single pattern in vector memory, and a repeat op-code in op-code memory, rather than storing everything explicitly in vector memory. Practical experience with RLE shows that care bit density is so low that repeats are possible and the combined op-code plus vector memory can be 10× less than storing everything in vector memory. As a side-effect, tests with repeat fill create less switching activity during scan and, hence, are less power-hungry than their pseudo-random brethren.

Run length encoding can be effective for reducing the memory footprint of scan-based logic tests in test equipment. However, the fully expanded test vectors comprised of care and don't care bits are created in the test equipment and these are the expanded vectors that are sent to the chip(s) under test. The chips have the same number of scan chains and the same scan-chain length as for normal scan testing without RLE. Test vector sets with repeat fill are slightly less compact than sets with pseudo-random fill. Hence, test time suffers slightly with RLE, meaning that RLE is not the right choice if test time reduction is as important as data volume reduction.

Simultaneous data volume and test time reduction is possible with on-chip decompression techniques. In that scenario, a hardware macro for test input data compression is inserted between the scan-in pins of the chip and the inputs of a larger number of shorter scan chains (i.e., there are more scan chains than scan-in pins). The decompression macro can be combinational or sequential in nature, but in most practical implementations it tends to be linear in nature. Linear in this context means that the value loaded into each scan cell is a predictable linear combination (i.e., XOR sum and optional inversion) of some of the compressed input values supplied to the scan-in pins. In the extreme case, each linear combination contains only a single term, which can, for example, be achieved by replication or shifting.

Broadcast scan, where each scan-in pin simply fans-out to several scan chains without further logic transformation, is a particularly simple decompression macro using replication of input values. In broadcast scan, all scan chains connected to the same scan-in pin receive the same value. That creates strong correlation (replication) between the values loaded into the scan cells of those chains. Most ATPG software implementations have the ability to deal directly with correlated scan cell values and will automatically imply the appropriate values in the correlated scan cells. The only new software needed is for DFT synthesis to create automatically the scan fan-out and DFT DRC for analyzing the scan fan-out network and setting up the appropriate scan cell correlation tables for ATPG. The hard value correlations created by broadcast scan can make some faults hard to test and make test compaction more difficult because the correlated values create care bits even if the values are not required for testing the target fault. It must therefore be expected that ATPG with broadcast scan has longer run-times, creates slightly less compact tests, and achieves slightly lower test coverage than ATPG with normal scan. With a scan fan-out ratio of 1:32 (i.e., each scan-in fans out to 32 scan chains), it is possible to achieve an effective data volume and test time reduction of 20× or so. It is assumed that the scan chains can be reasonably well balanced and that there are no hard macros with preconnected scan segments that are too long.

The more sophisticated decompression macros contain XOR gates and the values loaded into the scan cells are a linear combination of input values with more than one term. Industrial ATPG does understand hard correlations between scan cells but not Boolean relationships like linear combinations with more than one term. Hence, the ATPG flow has to be enhanced to deal with the more sophisticated decompression techniques. Each care bit in a test cube creates a linear equation with the care bit value on the one side and an XOR sum of some input values on the other side. These specific XOR sums for each scan cell can be determined upfront by symbolic simulation of the scan load operation. Since a test cube typically has more than one care bit, a system of linear equations is formed, and a linear equation solver is needed to find a solution for the system of equations. In some cases the system of equations has no solution; for example, if the total number of care bits exceeds the number of input values supplied from the tester, which can happen when too many test cubes are compacted into a single test. In summary, similar to broadcast scan, ATPG for the more sophisticated schemes also adds CPU time, possibly reduces test coverage, and increases the number of tests in the test set. The sophisticated techniques require more hardware per scan chain (e.g., one flip-flop plus some other gates) than the simple fan-out in broadcast scan. However, the more sophisticated methods tend to offer more flexibility and should make more optimal compression results possible.

Although the so-called weighted random pattern (WRP) test data compression approach is proprietary and not generally available, it is worth a brief description. The classical WRP does not exploit the low care bit density that makes the other compression methods possible, but a totally different property of scan-based logic tests generated by ATPG. The property is sometimes called *test cube clustering*, meaning that the specified bit values in the test cubes for groups of multiple faults are mostly identical and only very few care bit values are different (i.e., the test cubes in a cluster have a small Hamming distance from each other). That makes it possible to find more compact data representations for describing all tests in a cluster (e.g., a common base vector and a compact difference vector for each cluster member). Many years of practical experience with WRP confirms that input data volume reductions in excess of 10× are possible by appropriately encoding the cluster information.

It has been suggested that two-level compression should be possible by taking advantage of both the cluster effect and the low care bit density of scan tests. Several combined schemes have been proposed, but they are still in the research stage.

21.3.3.1.6.3 Response Data Compression/Compaction. Since the amount of memory needed for expected responses can be more than for test input data, any efficient data compression architecture must include test response data compression (sometimes also called compaction). WRP, the oldest data compression method with heavy-duty practical production use in chip manufacturing test, for example, borrows the signature approach from logic BIST for response compaction. Both the WRP decompression logic and the signature generation logic were provided in proprietary test equipment. The chips themselves were designed for normal scan and the interface between the chips and the test equipment had to accommodate the expanded test input and test response data. The interface for a given piece of test equipment tends to have a relatively fixed width and bandwidth, meaning that the test time will grow with the gate count of the chip under test. The only way to reduce test time in this scenario is to reduce the amount of data that has to go through the interface bottleneck. That can be achieved by inserting a response compression/compaction macro between the scan chains and the scan-out pins of the chip so that only compressed/compacted response data has to cross the interface.

How that can be done with EDCs using MISR macros was already known from and proven in practice by logic BIST, and the only "new" development required was to make the on-chip MISR approach work with ATPG tests with and without on-chip test input data decompression. From a DFT tool perspective, that means adding DFT synthesis and DFT DRC features for adding and checking the on-chip test response compression macro, and adding a fast signature simulation feature to the ATPG tool. The introduction of signatures instead of response data also necessitated the introduction of new data types in the test data from ATPG and corresponding new capabilities in the test program generation software, for example, to quickly reorder tests on the test equipment without having to resimulate the signatures.

One unique feature of the MISR-based response compression method is that it is not necessary to monitor the scan-out pins during the scan load/unload operation (the MISR accumulates the responses on-chip and the signature can be compared after one or more scan load/unload operations are completed). The test equipment channels normally used to monitor the scan-outs can be reallocated for scan-in, meaning that the number of scan-ins and scan chains can be doubled, which reduces test time by a factor of 2 if the scan chains can be rebalanced to be half as long as before. Furthermore, no expected values are needed in the test equipment vector memory for scan, thus reducing the data volume by a factor of 2 or more (the latter comes into play if the test equipment uses a two-bit representation for the expected data and the input data can be reformatted to a one-bit representation).

Mathematically speaking, the data manipulations in an MISR are very similar to those in a linear input data decompression macro. Each bit of the resulting signature is a linear combination (XOR sum) of a subset of test response bit values that were accumulated into the signature. Instead of using a sequential state machine like an MISR to generate the linear combinations, it is also possible to use a combinational XOR network to map the responses from a large number of scan-chain outputs to a smaller number of scan-out pins. Without memory, the combinational network cannot accumulate responses on-chip. Hence, the scan-out pins must be monitored and compared to expected responses for each scan cycle. The data reduction factor is given by the number of internal scan-chain outputs per scan-out pin.

Selective compare uses multiplexing to connect one out of several scan-chain outputs to a single scan-out pin. The selection can be controlled from the test equipment directly (i.e., the select lines are connected to chip input pins) or through some intermediate decoding scheme. Selective compare is unique in that the response value appearing at the currently selected scan-chain output is directly sent to the test equipment without being combined with other response values, meaning that any mismatches between actual and expected responses can be directly logged for analysis. The flip side of the coin is that responses in the currently de-selected scan chains are ignored, which reduces the overall observability of the logic feeding into the de-selected scan cells and could impair the detection of unforeseen defects. It also should be noted that in addition to monitoring the scan-out pins, some input bandwidth is consumed for controlling the selection.

Mapping a larger number of scan-chain outputs onto a smaller number of scan-out pins through a combinational network is sometimes referred to as *space compaction.* A MISR or similar sequential state machine that can accumulate responses over many clock-cycles performs *time compaction.* Of course, it is possible to combine space and time compaction into a single architecture.

21.3.3.1.6.4 X-State Handling. Signatures are known for not being able to handle unknown/unpredictable responses (x states). In signature-based test methods, e.g., logic BIST, it is therefore customary to insist that x-state sources should be avoided or disabled, or if that is not possible then the associated x states must under no circumstances propagate into the signature generation macro.

x-state avoidance can be achieved by DFT circuit modifications (e.g., using logic gates instead of pass-gate structures) or by adjusting the test data such that the responses become predictable (e.g., asserting enable signals such that no three-state conflicts are created and no unterminated nets are left floating). The latter is in many cases possible with ATPG patterns; however, that typically increases ATPG run-time and can adversely affect test coverage. Design modifications may be considered too intrusive, especially for high-performance designs, which has hampered the widespread acceptance of logic BIST.

Disabling x-state propagation can be achieved by local circuit modifications near the x-state sources or by implementing a general-purpose *response-masking* scheme. Response masking typically consists of a small amount of logic that is added to the inputs of the signature generation macro (e.g., MISR). This logic makes it possible to selectively de-gate (i.e., force a known value) onto the MISR input(s) that could carry x states. A relatively simple implementation, for example, consists of a serially loadable *mask vector* register with one mask bit for each MISR input. The value loaded into the mask bit determines whether the response data from the associated scan-chain output are passed through to the MISR or are de-gated. The mask vector is serially preloaded prior to scan load/unload and could potentially be changed by reloading during scan load/unload. A single control signal directly controlled from the test equipment can optionally activate or de-activate the effect of the masking on a scan cycle by scan cycle basis. More sophisticated implementations could offer more than one dynamically selectable mask bit per scan-chain or decoding schemes to dynamically update the mask vector. The dynamic control signals as well as the need to preload or modify mask vectors do consume input data bandwidth from the tester and add to the input data volume. This impact generally is very minimal if a single mask vector can be preloaded and used for one or more than one full scan load/unload without modification. The flip side of the coin in this scenario is some loss of observability due to the fact that a number of predictable responses that could carry defect information may be masked in addition to the x states.

If a combinational space compaction network is used and the test equipment monitors the outputs, then it becomes possible to let x states pass through and mask them in the test equipment. Flexibility could, for example, be utilized to reduce the amount of control data needed for a selective compare approach (e.g., if a "must-see" response is followed by an x state in a scan chain, then it is possible to leave that chain selected and to mask the x state in the tester rather than expending input bandwidth to change the selection).

If an XOR network is used for space compaction, then an x state in a response bit will render all XOR-sums containing that particular bit value equally unknown/unpredictable, masking any potential fault detection in the other associated response bits. The "art" of constructing x-tolerant XOR networks for space compaction is to minimize the danger of masking "must-see" response bit values. It has been shown that suitable and practical networks can indeed be constructed if the number of potential x states appearing at the scan-chain outputs in any scan cycle is limited.

Newer research adds some memory to the XOR-network approach. Unlike in MISR, there is no feedback loop in the memory. The feedback in the MISR amplifies and "perpetuates" x states (each x state, once captured, is continually fed back into the MISR and will eventually contaminate all signature bits). The x-tolerant structures with memory, by contrast, are designed such that x states are flushed out of the memory fairly quickly (e.g., using a shift-register arrangement without feedback).

It should be noted that many of the x-tolerant schemes are intended for designs with a relatively limited number of x states. However, certain test methodologies can introduce a large and variable number of x states. One example is delay-test with aggressive strobe timings that target short paths but cause setup time violations on longer paths.

21.3.3.1.6.5 Logic Diagnostics. Automated logic diagnostics software generally is set up to work from a bit-level fail set collected during test. The fail set identifies which tests failed and were data-logged, and within each data-logged failing test, which scan cells or output pins encountered a mismatch between the expected and the actual response value. Assuming binary values, the failing test can be expressed as the bitwise XOR sum of the correct response vector and an *error vector*. The diagnosis software can use a *precalculated fault dictionary or posttest simulation*. In both cases, fault simulation is used to associate individual faults from a list of model faults with error-simulated vectors. Model faults like stuck-at or transition faults are generally attached to gate-level pins in the netlist of the design under test. Hence the fault model carries gate-level locality information with it (the pin the fault is attached to). The diagnosis algorithms try to find a match between the error vectors from the data-logged fail sets and simulated error vectors that are in the dictionary or are created on the fly by fault simulation. If a match is found, then the associated fault will be added to the so-called *call-out* list of faults that partially or completely match the observed fail behavior.

Response compression/compaction reduces the bitwise response information to a much smaller amount of data. In general, the reduction is lossy in nature, meaning that it may be difficult or impossible to unambiguously re-construct the bit-level error vectors for diagnosis. That leaves essentially two options for diagnosing fails with compressed/compacted responses. The first option is to first detect the presence of fails in a test, then reapply the same test and extract the bit-level responses without compression/compaction. This approach has already been discussed in the section on logic BIST. Quick identification of the failing tests in signature-based methods can be facilitated by adding a reset capability to the MISR macro and comparing signatures at least once for each test. The reset is applied in between tests and returns the MISR to a fixed starting state even if the previous test failed and produced an incorrect signature. Without the reset, the errors would remain in the MISR, causing incorrect signatures in subsequent tests even if there are no further mismatches. With the reset and signature compared at least once per test, it is easier to determine the number of failing tests and schedule all or some of them for retest and data logging.

As already explained in the section on logic BIST, bit-level data logging from retest may make it necessary to have the expected responses in the test equipment's vector memory. To what extent that is feasible during production test is a matter of careful data management and sampling logistics. Hence, there is considerable interest in enabling meaningful *direct diagnostics* that use the compacted responses without the need for retest and bit-level data logging.

Most response compression/compaction approaches used in practice are linear in nature, meaning that the *difference vector* between a nonfailing compacted response and the failing compacted response is only a function of the bit-level error vector. In other words, the difference vector contains reduced information about the error vector. The key question for direct diagnostics is to what extent it is possible to associate this reduced information with a small enough number of "matching" faults to produce a meaningful call-out list. The answer depends on what type of and how much fault-distinguishing information is preserved in the mapping from the un-compacted error vector to the compacted difference vector, and how that information can be accessed in the compacted data. Also important is what assumptions can be realistically made about the error distributions and what type of encoding is used for data reduction. For example, if it is assumed that for some grouping of scan chains, at most one error is most likely to occur per scan cycle in a group, then using an error-correcting code (ECC) for compaction would permit complete reconstruction of the bit-level error vector. Another issue is what matching criteria are used for deciding whether to include a fault in the call-out. A complete match between simulated and actual error vector generally also entails a complete match between simulated and observed compacted difference vectors. That is, the mapping preserves the matching criterion (albeit the fact that the information reduction leads to higher ambiguity). On the other hand, partial matching success based on proximity in terms of Hamming distance may not be preserved. For example, MISR-based signatures essentially are hash codes that do not preserve proximity. Nor should we forget the performance considerations. The bit-level error information is used in posttest simulation methods not only to determine matching but also to greatly reduce the search space by only including faults from the back-trace cones feeding into the failing scan cells or output pins.

Selective compare at first blush appears to be a good choice for diagnostics, because it allows for easy mapping of differences in the compacted responses back to bit-level errors. However, only a small subset of responses is visible in the compacted data and it is very conceivable that the ignored data contains important fault-distinguishing information.

To make a long story short, direct diagnostics from compacted responses are an area for potentially fruitful research and development. Some increasingly encouraging successes of direct diagnosis for certain linear compression/compaction schemes have been reported recently.

21.3.3.1.6.6 Scan-Chain Diagnostics. Especially for complex chips in new technologies it must be expected that defects or design issues affect the correct functioning of the scan load/unload operation. For many design projects, scan is not only important for test but also for debug. Not having fully working scan chains can be a serious problem.

Normal logic diagnostics assume fully working scan chains and are not immediately useful for localizing scan chain problems. Having a scan-chain problem means that scan cells downstream from the problem location cannot be controlled and scan cells upstream from the problem location cannot be observed by scan. The presence of such problems can mostly be detected by running scan-chain integrity tests, but it generally is difficult or impossible to derive the problem location from the results. For example, a hold-time problem that results in race-through makes the scan chain look too short and it may be possible to deduce that from the integrity test results. However, knowing that one or more scan cells were skipped does not necessarily indicate which cells were skipped.

Given that the scan load/unload cannot be reliably used for control and observation, scan-chain diagnostics tend to rely on alternative means of controlling/observing the scan cells in the broken scan chains. If mux-scan is used, for example, most scan cells have a second data input (namely the system data input) other than the scan data input and it may be possible to control the scan cell from that input (e.g., utilizing scan cells from working scan chains). Forcing known values into scan cells through nonscan inputs for the purpose of scan-chain diagnostics is sometimes referred to as *lateral insertion.* And most scan cell outputs not only feed to the scan data input of the next cell in the chain, but also feed functional logic that in turn may be observable. It also should be noted that the scan cells downstream from the problem location are observable by scan.

Design-for-test can help with diagnostics using lateral insertion techniques. A drastic approach is to insist that all scan cells have a directly controllable set and clear to force a known state into the cells. Other scan architectures use only the clear in combination with inversion between all scan cells to help localize scan-chain problems. Having many short scan chains can also help as long as the problem is localized and affects only a few (ideally one) chain. In that scenario the vast majority of scan chains are still working and can be used for lateral insertion and observation. In this context, it is very useful to design the scan architecture such that the short scan chains can be individually scanned out to quickly determine which chains are working and which not.

21.3.3.2 Embedded Memories

The typical DFT Synthesis tools convert only flip-flops or latches into scan cells. The storage cells in embedded memories generally are not automatically converted. Instead, special memory-specific DFT is used to test the memories themselves as well as the logic surrounding the memories.

21.3.3.2.1 Types of Embedded Memories.

Embedded memories come in many different flavors, varying in functionality, usage, and in the way they are implemented physically. From a test perspective, it is useful to grossly distinguish between register files and dense custom memories.

Register files are often implemented using design rules and cells similar to logic cells. The sensitivity to defects and failure modes is similar to that of logic, making them suitable for testing with typical logic tests. They tend to be relatively shallow (small address space) but can be wide (many bits per word), and have multiple ports (to enable read or write from/to several addresses simultaneously). Complex chips can contain tens or even many hundreds of embedded register files.

Dense custom memories, by contrast, tend to be hand-optimized and may use special design rules to improve the density of the storage cell array. Some memory types, for example embedded DRAM (eDRAM), use additional processing steps. Because of these special properties, dense custom memories are considered to be subject to special and unique failure modes that require special testing (e.g., retention time testing, and pattern sensitivities). On the other hand, the regular repetitive structure of the memory cell arrays, unlike "random" logic, lends itself to algorithmic testing. Overall, memory testing for stand-alone as well as embedded memories has evolved in a different direction than logic testing.

In addition to "normally" addressed random access memories (RAMs) including register files, static RAMs (SRAMs), and dynamic RAMs (DRAMs), there are read-only memories (ROMs), content addressable memories (CAMs), and other special memories to be considered.

21.3.3.2.2 Embedded Memories and Logic Test.

Logic testing with scan design benefits from the internal controllability and observability that comes from converting internal storage elements into scan cells. With embedded memories, the question arises as to whether and to what extent the storage cells inside the memories should be likewise converted into scan cells. The answer depends on the type of memory and how the scan function is implemented. Turning a memory element into a scan cell typically entails adding a data port for scan data, creating master–slave latch pairs or flip-flops, and connecting the scan cells into scan chains. The master–slave latch pair and scan-chain configuration can be fixed or dynamic in nature.

In the fixed configuration, each memory storage cell is (part of) one scan cell with a fixed, dedicated scan interconnection between the scan cells. This approach requires modification of the storage cell array in the register file, meaning that it only can be done by the designer of the register file. The overhead, among other things, depends on what type of cell is used for normal operation of the register file. If the cells are already flip-flops, then the implementation is relatively straightforward. If the cells are latches, then data port and a single-port scan-only latch can be added to each cell to create a master–slave latch pair for shifting. An alternative is to only add a data port to each latch and combine cell pairs into the master–slave configuration for shifting. The latter implementation tends to consume less area and power overhead, but only half of the words can be controlled or observed simultaneously (pulsing the master or slave clock overwrites the data in the corresponding latch types). Hence, this type of scan implementation is not entirely useful for debug where a nondestructive read is preferred, and such a function may have to be added externally if desired.

In the dynamic approach, a shared set of master or slave latches is temporarily associated with the latches making up one word in the register file cell array, to establish a master–slave configuration for shifting. Thus, a serial shift register through one word of the memory at a time can be formed using the normal read/write to first read the word and then write it back shifted one bit position. By changing the address, all register file bits can be serially controlled and observed. Because no modification of the register cell array is needed, and normal read/write operations are used, the dynamic scan approach could be implemented by the memory user. The shared latch approach and reuse of the normal read/write access for scan keep the overhead limited and the word-level access mechanism is very suitable for debug (e.g., reading/writing one particular word). The disadvantage of the dynamic approach is that the address-driven scan operation is not supported by the currently available DFT and ATPG tools.

Regardless of the implementation details, scannable register files are modeled at the gate level and tested as part of the chip logic. However, they increase the number of scan cells as well as the size of the

ATPG netlist (the address decoding and read/write logic are modeled explicitly at the gate level), and thereby can possibly increase test time. Moreover, ATPG may have to be enhanced to recognize the register files and create intelligently structured tests for better compaction. Otherwise, longer than normal test sets can result.

Modern ATPG tools tend to have some sequential test generation capabilities and can handle nonscannable embedded memories to some extent. It should be noted in this context that the tools use highly abstract built-in memory models. Neither the memory cell array nor the decoding and access logic are explicitly modeled, meaning that no faults can be assigned to them for ATPG. Hence, some other means must be provided for testing the memory proper. Sequential test generation can take substantially longer and result in lower test coverage. It is recommended to make sure that the memory inputs can be controlled from and the memory outputs can be observed at scan cells or chip pins through combinational logic only. For logic BIST and other test methods that use signatures without masking, it must also be considered that nonscannable embedded memories are potential x-state sources until they are initialized. If memory access is desired as part of such a test (e.g., it is not unusual for performance-limiting paths to include embedded memories such that memory access may be required for accurate performance screening/characterization), then it may be necessary to provide some mechanism for initializing the memories. Multiple write ports can also be a source of x states if the result cannot be predicted when trying to write different data to the same word. To avoid multiport write conflicts it may be necessary to add some form of port-priority, for example, by adding logic that detects the address coincidence and degates the write-clock(s) for the port(s) without priority, or by delaying the write-clock for the port with enough priority to assure that its data will be written last.

For best predictability in terms of ATPG run-times and achievable logic test coverage, it may be desirable to remove entirely the burden of having to consider the memories for ATPG. That can be accomplished, for example, by providing a memory bypass (e.g., combinational connection between data inputs and data outputs) in conjunction with observe points for the address and control inputs. Rather than a combinational bypass, boundary scan can be used. With bypass or boundary scan, it may not be necessary to model the memory behavior for ATPG and a simple black-box model can be used instead. However, the bypass or boundary scan can introduce artificial timing paths or boundaries that potentially mask performance-limiting paths.

21.3.3.2.3 Testing Embedded Memories.

It is possible to model smaller nonscannable embedded memories at the gate level. In that case, each memory bit cell is represented by a latch or flip-flop, and all decoding and access logic is represented with logic gates. Such a gate-level model makes it possible to attach faults to the logic elements representing the memory internals and to use ATPG to generate tests for those faults. For large, dense memories this approach is not very practical because the typical logic fault models may not be sufficient to represent the memory failure modes and because the ATPG fault selection and cube compaction algorithms may not be suited for generating compact tests for regular structures like memories. As a consequence, it is customary to use special memory tests for the large, dense memories and, in many cases, also for smaller memories. These memory tests are typically specified to be applied to the memory interface pins. The complicating factor for embedded memories is that the memory interface is buried inside the chip design and some mechanism is needed to transport the memory tests to/from the embedded interface through the intervening logic.

21.3.3.2.3.1 Direct Access Testing. *Direct access testing* requires that all memory interface pins are individually accessible from chip pins through combinational access paths. For today's complex chips it is rare to have natural combinational paths to/from chip pins in the functional design. It is, hence, up to DFT to provide the access paths and also provide some chip-level controls to selectively enable the access paths for testing purposes and disable them for normal functional chip operation. With direct access testing, the embedded memory essentially can be tested as if it were a stand-alone memory. It requires that the memory test program is stored in the external test equipment. Memory tests for larger memories are many cycles long and could quickly exceed the test equipment limits if the patterns for each cycle have to be stored in vector memory. This memory problem does not arise if the test equipment contains dedicated algorithmic pattern generator (APG) hardware that can be programmed to generate a wide range of memory test sequences from a very compact code. In that case, direct access testing has the benefit of being able to take full advantage of the flexibility and high degree of programmability of the APG and other memory-specific hardware/software features of the equipment.

If the chip contains multiple embedded memories or, memories with a too many pins, and there are not enough chip pins available to accommodate access for all memory interface pins, then a multiplexing scheme with appropriate selection control has to be implemented. It should be noted that the need to connect the access paths to chip pins can possibly create considerable wiring overhead if too many paths have to be routed over long distances. It should also be noted that particularly for high-performance memories, it could be difficult to control the timing characteristics of the access paths accurately enough to meet stringent test timing requirements. Variants of direct access testing may permit the use of tightly timed pipeline flip-flops or latches in the data and nonclock access paths to overcome the effect of inaccurate access path timings. The pipelining and latency of such sequential access paths must be taken into account when translating the memory test program from the memory interface to chip interface.

Another potential method for reducing the chip pin footprint required for direct access testing is to serialize the access to some memory interface pins, e.g., the data pins. To that effect, scan chains are built to shift serially wide bit patterns for the data words from/to a small number of chip pins. It may take many scan clock cycles to serially load/unload an arbitrary new bit pattern and the memory may have to "wait" until that is done. The serial access, hence, would not be compatible with test sequences that depend on back-to-back memory accesses with arbitrary data changes. Test time also is a concern with serial access for larger memories.

21.3.3.2.3.2 Memory BIST. Although direct access testing is a viable approach in many cases, it is not always easy to implement, can lead to long test times if not enough bandwidth is available between the chip under test and the test equipment, and may consume excessive vector memory if no APG hardware is available. Moreover, it may not be easy or possible to design the access paths with sufficient timing accuracy for a thorough performance test. *Memory BIST* has become a widely used alternative. For memory BIST, one or more simplified small APG hardware macros, also known as *memory BIST controllers*, are added to the design and connected to the embedded macros using a multiplexing interface. The interface selectively connects the BIST hardware resources or the normal functional logic to the memory interface pins.

A variety of more or less sophisticated controller types are available. Some implementations use pseudo-random data patterns and signature analysis similar to logic BIST in conjunction with simple address stepping logic. Signature analysis has the advantage that the BIST controller does not have to generate expected responses for compare after read. This simplification comes at the expense of limited diagnostic resolution in case the test fails.

The memory test sequences used with APG hardware in the test equipment are constructed from heavily looped algorithms that repeatedly traverse the address space and write/read regular bit patterns into/from the memory cell array. Most memory BIST implementations used in practice follow the same scheme. *Hardwired controllers* use customized finite state machines to generate the data-in pattern sequences, expected data-out pattern sequences, as well address traversal sequences for some common memory test algorithms (e.g., march-type algorithms). The so-called *programmable controllers* generally contain several hardwired test programs that can be selected at run-time by loading programming register in the controller. *Microcoded controllers* offer additional flexibility by using a dedicated microengine with a memory-test-specific instruction set and an associated code memory that must be loaded at run-time to realize a range of user customizable algorithms.

The complexity of the controller not only depends on the level of programmability but also on the types of algorithms it can support. Many controllers are designed for so-called linear algorithms in which the address is counted up/down so that the full address space is traversed a certain number of times. In terms of hardware, the linear traversal of the address space can be accomplished by one address register with increment/decrement logic. Some more complex nonlinear algorithms jump back and forth between a test address and a disturb address, meaning that two address registers with increment/decrement and more complex control logic are required. For data-in generation, a data register with some logic manipulation features (invert, shift, rotate, mask, etc.) is needed. The register may not necessarily have the full data word width as long as the data patterns are regular and the missing bits can be created by replication. If bit-level compare of the data-outs is used for the BIST algorithm, then similar logic plus compare logic is required to generate the expected data-out patterns and perform the comparison. Clock and control signals for the embedded memory are generated by timing circuitry that, for example, generates memory clock and control waveforms from a single free-running reference clock. The complexity of the timing circuitry depends on how flexible it is in terms of generating different event sequences to accommodate different memory access modes and how much programmability it offers in terms of varying the relative edge offsets.

Certain memory tests, like those for retention and pattern-sensitive problems, require that specific bit values are set up in the memory cell array according to physical adjacency. The physical column/row arrangement is not always identical to the logical bit/word arrangement. Address/bit scrambling as well as cell layout details may have to be known and taken into account to generate physically meaningful patterns.

The impact of memory BIST on design complexity and design effort depends, among other things, on methodology and flow. In some ASIC flows, for example, the memory compiler returns the memories requested by the user complete with fully configured, hardwired, and preverified memory BIST hardware already connected to the memory. In other flows it is entirely up to the user to select, add, and connect memory BIST after the memories are instantiated in the design. Automation tools for configuring, inserting, and verifying memory BIST hardware according to memory type and configuration are available. The tools generate customized BIST RTL or gate-level controllers and memory interfaces from input information about memory size/configuration, number/type of ports, read/write timings, BIST/bypass interface specification, address/bit scrambling, and physical layout characteristics. Some flows may offer some relief through optional *shared controllers* where a single controller can drive multiple embedded memories. For shared controllers, the users have to decide and specify the sharing method (e.g., testing multiple memories in parallel or one after the other).

In any scenario, memory BIST can add quite a bit of logic to the design and this additional logic should be planned and taken into account early enough in the physical design planning stage to avoid unpleasant surprises late in the design cycle. In addition to the additional transistor count, wiring issues and BIST timing closure can cause conflicts with the demands of the normal function mode. Shared controllers make placement and wiring between the controller and the memory interface more complicated and require that the trade-off between wiring complexity and additional transistor count is well understood and the sharing strategy is planned appropriately (e.g., it may not make sense to share a controller for memories that are too far away from each other).

21.3.3.2.3.3 Complex Memories and Memory Substructures. In addition to embedded single-/multiport embedded SRAMs, some chips contain more complex memory architectures. Embedded DRAMs can have more complex addressing and timing than "simple" SRAMs. eDRAMs are derived from stand-alone DRAM architectures, and, for example, they may have time multiplexed addressing (i.e., the address is supplied in two cycles and assembled in the integrated memory controller) and programmable access modes (e.g., different latency, fast page/column, etc.). The BIST controller and algorithm design also has to accommodate the need for periodic refresh, all of which can make BIST controllers for eDRAM more complex.

Other fairly widely used memory types include logic capabilities. Content addressable memories, for example, contain logic to compare the memory contents with a supplied data word. BIST controllers for CAMs use enhanced algorithms to thoroughly test the compare logic in addition to the memory array. The algorithms depend on the compare capabilities and the details of the circuit design and layout.

High-performance processors tend to utilize tightly integrated custom memory subsystems where, for example, one dense memory is used to generate addresses for another dense memory. Performance considerations do not allow for separating the memories with scan. Likewise, compare logic and other high-performance logic may be included as well. Memories subsystems of this nature generally cannot be tested with standard memory test algorithms, and hence require detailed analysis of possible failure modes and custom development of suitable algorithms.

Some micro-controllers and many chips for Smartcards, for example, contain embedded nonvolatile flash memory. Flash memory is fairly slow and poses no challenge for designing a BIST controller with sufficient performance, but flash memory access and operation are different from SRAM/DRAM access. Embedded flash memory is commonly accessible via an embedded microprocessor that can, in many cases, be used for testing the embedded flash memory.

21.3.3.2.3.4 Performance Characterization. In addition to use in production test, a BIST approach may be desired for characterizing the embedded memory performance. The BIST controller, memory interface, and timing circuitry in this scenario have to be designed for higher speed and accuracy than what is required for production testing alone. It may also be necessary to add more algorithm variations or programmability to enable stressing particular areas of the embedded memory structure. The BIST designers in this case also may want to accommodate the use of special lab equipment used for characterization.

If BIST is to be used for characterization and speed binning, it should be designed with enough performance headroom to make sure that the BIST circuitry itself is not the performance-limiter that dominates the measurements. Designing the BIST circuitry to operate at "system cycle time," as advertised by some BIST tools, may not be good enough.

The quality of speed binning and performance characterization not only depends on the performance of BIST controller and memory interface, but also on the accuracy and programmability of the timing edges used for the test. The relative offset between timing edges supplied from external test equipment generally can be programmed individually and in very small increments. Simple timing circuitry for memory BIST, by contrast, tends to be aligned to the edges of a reference clock running at or near system speed. The relative edge offset can only be changed by changing the frequency, and affects all edge offsets similarly.

Some memory timing parameters can be measured with reasonable accuracy without implementing commensurate on-chip timing circuitry. For example, data-out bit values can be fed back to the address. The memory is loaded with appropriate values such that the feedback results in oscillation and the oscillation frequency is an indicator of read access time. By triggering a clock pulse from a data-out change, it may be similarly possible to obtain an indication of write access time.

21.3.3.2.3.5 Diagnosis, Redundancy, and Repair. Embedded memories can be yield limiters and useful sources of information for yield improvement. To that effect, the memory BIST implementation should permit the collection of memory fail-bit maps. The fail-bit maps are created by data logging the compare status or compare data and, optionally, BIST controller status register contents.

A simple data logging interface with limited diagnostic resolution can be implemented by issuing a BIST-start edge when the controller begins with the test and a cycle-by-cycle pass/fail signal that indicates whether a mismatch occurred during compare or not. Assuming the relationship between cycle count and BIST progress is known, the memory addresses associated with cycles indicating mis-compare can be derived. This simple method allows for creating an address-level fail log. However, the information is generally not sufficient for FA.

To create detailed fail-bit maps suitable for FA, the mis-compares must be logged at the bit level. This can, for example, be done by unloading the bit-level compare vectors to the test equipment through a fully parallel or a serial interface. The nature of the interface determines how long it takes to log a complete fail-bit map. It also has an impact on the design of the controller. If a fully parallel interface is used that is fast enough to keep up with the test, then the test can progress uninterrupted while the data are logged. If a serial or slower interface is used, then the test algorithm must be interrupted for data logging. This approach sometimes is referred to as *stop-on-nth-error*. Depending on the timing sensitivity of the tests, the test algorithm may have to be restarted from the beginning or from some suitable checkpoint each time the compare data have been logged for one fail. In other cases, the algorithm can be paused and can resume after logging.

Stop-on-*n*th-error with restart and serial data logging is relatively simple to implement, but it can be quite time consuming if mis-compares happen at many addresses (e.g., defective column). Although that is considered tolerable for FA applications in many cases, there have been efforts to reduce the data-logging time without needing a high-speed parallel interface by using on-chip data reduction techniques. For example, if several addresses fail with the same compare vector (e.g., defective column), then it would be sufficient to log the compare data once and only log the addresses for subsequent fails. To that effect, the on-chip data reduction hardware would have to remember one or more already encountered compare vectors and for each new compare vector, check whether it is different or not.

Large and dense embedded memories, for example, those used for processor cache memory, offer *redundancy* and *repair* features, meaning that the memory blocks include spare rows/columns and address manipulation features to replace the address/bit corresponding to the failing rows/columns with those corresponding to the spares. The amount and type of redundancy depend on the size of the memory blocks, their physical design, and the expected failure modes, trying to optimize postrepair memory density and yield. Memories that have either spare rows or spare columns, but not both, are said to have *one-dimensional* redundancy, and memories that have both spare rows and columns are said to have *two-dimensional* redundancy. It should be noted that large memories may be composed of smaller blocks and it is also possible to have block-level redundancy/repair.

Hard repair means that the repair information consisting of the failing row/column address(es) is stored in nonvolatile form. Common programming mechanisms for hard repair include laser-programmable fuses, electrically programmable fuses, or flash-type memory. The advantage of hard repair is that the failing rows/columns only need to be identified once upfront. However, hard repair often requires special equipment for programming, and one-time programmable fuses are not suitable for updating the repair information later in the chip's life cycle. *Soft repair*, by contrast, uses volatile memory, for example flip-flops, to store the repair information. This eliminates the need for special equipment and the repair information can be updated later. However, the repair information is not persistent and must be redetermined after each power-on.

All embedded memory blocks with redundancy/repair must be tested prior to repair, during manufacturing production test, with sufficient data logging to determine whether they can be repaired and if so, which rows/columns must be replaced. Chips with hard repair fuses may have to be brought to a special programming station for fuse-blow and then be returned to a tester for postrepair retest. Stop-on-*n*th-error data logging with a serial interface is much too slow for practical repair during production test. Some test equipment has special hardware/software features to determine memory repair information from raw bit-level fail data. In that scenario, no special features for repair may be required in the memory BIST engine, except for providing a data-logging interface that is fast enough for real-time logging.

If real-time data logging from memory BIST is not possible or not desired, then on-chip redundancy analysis (ORA) is required. Designing a compact ORA macro is relatively simple for one-dimensional redundancy. In essence all that is needed is memory to hold row/column addresses corresponding to the number of spares, and logic that determines whether a new failing row/column address already is in memory; if not, the memory uses the new memory if there still is room, and, if there is no more room, sets a flag indicating that the memory cannot be repaired. Dealing with two-dimensional redundancy is much more complicated, and ORA engines for two-dimensional redundancy are not commonly available. Embedded DRAM, for example, may use two-dimensional redundancy and come with custom ORA engines from the eDRAM provider. The ORA engines for two-dimensional redundancy are relatively large, and typically one such engine is shared by several eDRAM blocks, meaning that the blocks must be tested one after the other. To minimize test time, the eDRAM BIST controller should stagger the tests for the blocks such that ORA for one block is performed when the other blocks are idle for retention testing.

If hard repair is used, the repair data are serially logged out. Even if hard repair is used, the repair information may also be written into a repair data register. Then repair is turned on, and the memory is retested to verify its full postrepair functionality.

Large complex chips may contain many memory blocks with redundancy, creating a large amount of repair data. In that scenario, it can be useful to employ data compression techniques to reduce the amount of storage needed for repair or to keep the size of the fuse bay for repair small enough.

21.3.3.3 Embedded Digital Cores

The notion of embedded cores has become popular with the advent of SoC products. Embedded cores are predesigned and preverified blocks that are assembled with other cores and user-defined blocks into a complex chip.

21.3.3.3.1 Types of Embedded Digital Cores.

Embedded cores are generally classified into *hard cores*, which are already physically completed and delivered with layout data, and *soft cores*, which in many cases are delivered as synthesizable RTL code. In some cases the cores may already be synthesized to a gate-level netlist that still needs to be placed and wired (*firm cores*).

For hard cores, it is important whether or not a "test-ready" detailed gate-level netlist is made available (*white box*), or the block internals are not disclosed (*black box*). In some cases, partial information about some block features may be made available (*gray box*). Also, the details of what DFT features have been implemented in the core can be very important.

21.3.3.3.2 Merging Embedded Cores.

Soft cores and some white cores can be *merged* with other compatible cores and user-defined blocks into a single netlist for test generation.

For hard cores, the success of merging depends on whether the DFT features in the core are compatible with the test methodology chosen for the combined netlist. For example, if the chosen methodology is full-scan ATPG, then the core should be designed with full scan and the scan architecture as well as the netlist should be compatible with the ATPG tool. If logic BIST or test data compression is used, then the scan-chain segments in the core should be short enough and balanced to enable scan-chain balancing in the combined netlist.

21.3.3.3.3 Direct Access Test.

Direct access test generally requires the inputs of the embedded core to be individually and simultaneously controlled from, and the core outputs to be individually and simultaneously observed at, chip pins, through combinational access paths. This is only possible if enough chip pins are available and the test equipment has enough appropriately configured digital tester channels. Some direct access test guidelines may allow for multiplexing subsets of outputs onto common chip pins. In that case, the test must be repeated with a different output subset selected until all outputs have been observed. That is, the required chip output pin footprint for direct access can be somewhat reduced at the expense of test time. All inputs must however remain controllable at all times. For multiple identical cores, it may be possible to broadcast the input signals to like core inputs from a common set of chip input pins and test the cores in parallel, as long as enough chip output pins are available for observing the core outputs and testing the cores together fits does not exceed the power budget. Concurrent testing of nonidentical cores seems conceptually possible if all core inputs and outputs can be directly accessed simultaneously. However, even with parallel access it may not be possible to align the test waveforms for nonidentical cores well enough to fit within the capabilities of typical production test equipment (the waveforms for most equipment must fit into a common tester cycle length and limited set time set memory; the tester channels of some equipment, on the other hand, can be partitioned into multiple groups that can be programmed independently). If the chip contains multiple cores that cannot be tested in parallel, then some chip-level control scheme must be implemented to select one core (or a small enough group of compatible cores) at a time for direct access test.

In addition to implementing access paths for direct access testing, some additional DFT may be required to make sure that a complete chip test is possible and does not cause unwanted side effects. For example, extra care may be required to avoid potential three-state burn-out conditions resulting from a core with three-state outputs and some other driver on the same net trying to drive opposite values. In general, it must be expected that currently de-selected cores and other logic could be exposed to unusual input conditions during the test of the currently selected core(s). The integration of cores is easier if there is a simple control state to put each core into safe state that protects the core internals from being affected by unpredictable input conditions, asserts known values at the core outputs, and keeps power/noise for/from the core minimal. For cores that can be integrated into chips in which IDDq testing is possible, the safe state or another state should prevent static current paths in the core. The core test selection mechanism should activate the access paths for the currently selected core(s), while asserting the appropriate safe/IDDq state for nonselected cores.

The inputs of black-box cores cannot be observed and the black-box outputs cannot be fully controlled. Other observe/control means must be added to the logic feeding the core inputs and fed by the core outputs (*shadow logic*) to make that logic fully testable. Isolating and diagnosing core-internal defects can be very difficult for black-box cores or cores using proprietary test approaches that are not supported by the tools available and used for chip-level test generation.

There are some DFT synthesis and DRC tools that help with the generation, connection, and verification of direct access paths. Also, there are some test generation tools that help translate core test program pin references from the core pins to the respective chip pins, and integrate the tests into an overall chip-level test program.

21.3.3.3.4 Serializing the Embedded Core Test Interface.

There may not be enough chip pins available or the overhead for full parallel access to all core pins may be unacceptable. In that case, it may be possible and useful to serialize the access to some core pins. One common serialization method is to control core input pins and observe core output pins with scan cells. Updating the input pattern or comparing the output pattern in this scenario entails a scan load/unload operation. The core test program must be able to allow insertion of multiple tester cycles worth of wait time for the scan load/unload to complete. Moreover, the output of scan cells change state during scan. If the core input pin and core test program cannot tolerate multiple arbitrary state changes during scan load/unload, then a hold latch or flip-flop may have to be provided between the scan cell output and core input to retain the previous state required by the core test program until scan load/unload is done and the new value is available. For digital logic cores, data pins may be suitable for serialized access, assuming there is no asynchronous feedback and the internal memory cells are made immune to state changes (e.g., by turning off the clocks) during the core-external scan load/unload. Although it is conceptually possible to synthesize clock pulses and arbitrary control sequences via serialized access with hold, it generally is recommended or required to provide direct combinational control from chip pins for core clock and control pins.

Functional test programs for digital logic cores can be much shorter than memory test programs for large embedded memories. Serialized access for digital logic cores can, hence, be more practical and is more widely practiced for digital logic cores. In many cases, it is possible to find existing scan cells and to sensitize logic paths between those scan cells and the core pins with little or no need for new DFT logic. This minimizes the design impact and permits testing through functional paths rather than test-only paths.

However, it should be noted that with serialized access, it generally is not possible to create multiple arbitrary back-to-back input state changes and to observe multiple back-to-back responses. Hence, serialized access may not be compatible with all types of at-speed testing.

There are some DFT synthesis/DRC and test program translation tools that work with serialized access. Test program translation in this case is not as simple as changing pin references (and possibly polarity) and adjusting time sets as in the case of combinational direct access testing. Input value changes and output measures on core pins with serialized access entail inserting a core-external scan load/unload procedure. Additional scan load operations may be required to configure the access paths.

The capabilities of the pattern translation tool may have a significant impact on the efficiency of the translated test program. For example, if the embedded macro is or contains an embedded memory with BIST, then at least a portion of the test may consist of a loop. If the translation tool cannot preserve the loop, then the loop must be unrolled into parallel vectors prior to translation. That can result in excessive vector memory demand from the test equipment. Black-box cores that come with internal scan and with a scan-based test program are another special case. If the core scan-in/-out pins are identified and the associated scan load/unload procedures are appropriately defined and referenced in the core test program, then the pattern translation software may (or may not) be able to retain the scan load/unload procedure information in the translated test program, resulting in a more (or less) efficient test program.

21.3.3.3.5 Standardized Embedded Core Access and Isolation Architectures.

There have been several proprietary and industry-wide attempts to create an interoperable architecture for embedded core access and isolation. These architectures generally contain at least two elements. One element is the core-level test interface, often referred to as a core test *wrapper*, and the other one is a chip-level test access mechanism (TAM) to connect the interfaces among themselves and to chip pins, as well as some control infrastructure to selectively enable/disable the core-level test interfaces.

When selecting/designing a core-level test interface and TAM, it is important to have a strategy for both, i.e., for testing the core itself and for testing the rest of the chip. For testing the core itself, the core test interface should have an internal test mode in which the core input signals are controlled from the TAM and the core output signals are observed by the TAM without needing participation from other surrounding logic outside of the core. The issue for testing the surrounding logic is how to observe the signals connected to the core inputs and how to control the signals driven by the core outputs. If the core is a black-box core for example, the core inputs cannot be observed and the core outputs cannot be controlled. Even if the core is a white box, there may be no internal DFT that would make it easy. To simplify test generation for the surrounding logic, the core test interface should have an *external test mode* that provides alternate means for observing the core input signals and for controlling the core output signals, without needing participation from the core internals. Even if the core internals do not have to participate in the external test mode, it may be important to assure that the core does not go into some illegal state (e.g., three-state burn-out) or create undesired interference (e.g., noise, power, and drain) with the rest of the chip. It therefore may be recommended or required that the core test interface should have controls for a *safe mode* that protects the core itself from arbitrary external stimulation and vice versa. Last but not the least, there has to be a *normal mode* in which the core pins can be accessed for normal system function, without interference from the TAM and core test interface. Layout considerations may make it desirable to allow for routing access paths for one core "through" another core. The core test interface may offer a dedicated *TAM bypass mode* (combinational or registered) to facilitate the daisy chaining of access connections.

Overall, the TAM, so to speak, establishes switched connections between the core pins and the chip pins for the purpose of transporting test data between test equipment and embedded cores. It should be noted that some portions of the test equipment could be integrated on the chip (e.g., a BIST macro) such that some data sources/sinks for the TAM are internal macro pins and others are chip I/Os. The transport mechanism in general can be parallel or serialized. It is up to the core provider to inform the core user about any restrictions and constraints regarding the type and characteristics of the TAM connections as well as what the required/permissible data sources/sinks are for each core pin. Different core pins, e.g., clock pins and data pins, tend to have different restrictions and constraints. Depending on the restrictions and constraints, there probably will be some number of pins that must have parallel connections (with or without permitted pipe-lining), some number of pins that could be serialized (with some of those possibly requiring a hold latch or flip-flop). If certain core input pins need to be set to a specific state to initialize the core to some mode of operation, then it may be sufficient to assure that the respective state is asserted (e.g., decoded from some test mode control signals) without needing full blown independent access from chip pins. It further should be noted that certain cores may have different modes of operations, including different test modes (e.g., logic test and memory BIST), such that there could be different mode-specific access restrictions and constraints on any given pin. The TAM will have to accommodate that and if necessary, be dynamically adjustable.

Serialization of the core pin access interface reduces the number of chip pins and the bit-width (e.g., affecting the wiring overhead) of the TAM required for access, at the expense of test time. In addition to serializing some portion of the TAM for an individual core, there often is a choice to be made about testing multiple cores in parallel vs. serially. If the chip contains many cores, the trade-off becomes increasingly complex. For example, the TAM for all or subsets of cores could be arranged in a daisy-chain or star configuration, and decisions need to be made about the bit-width of each TAM. More decisions need to be made about testing groups of cores in parallel or in sequence. The decisions are influenced by the availability of pin, wiring and test equipment resources for parallel access, by the power/noise implications of testing several cores in parallel, by test time, and more.

Core test wrappers can be predesigned into a core by the core designer/provider (so-called *wrapped core*) or not (*unwrapped core*). In the latter case, the core user could wrap the core prior to, during, or after assembling it into a chip. In some cases, the wrapper overhead can be kept smaller by taking advantage of already existing core-internal design elements (e.g., flip-flops on the boundary of the core to implement a serial shift register for serialized access). If performance is critical, then it could be better to integrate the multiplexing function required for test access on the core input side with the first level of core-logic, or take advantage of hand-optimized custom design optimization that is possible in a custom core but not in synthesized logic. In any case, building a wrapper that requires modification of the core logic proper in hard cores in general can only be done by the core designer. Prewrapping a core can have the disadvantage of not being able to optimize the wrapper for a particular usage instance and of not being able to share wrapper components between cores. The once popular expectation that each third-party core assembled into a chip would be individually wrapped and tested using core-based testing has given way to a more pragmatic approach where un-wrapped cores are merged into larger partitions based on other design flow and design (team, location, schedule, etc.) management decisions, and only the larger partitions are wrapped and tested using core-based testing.

A chip-level control infrastructure is needed in addition to the wrapper and TAM components to create an architecture for core-based test. The control infrastructure is responsible for distributing and (locally) decoding instructions that configure the core/wrapper modes and the TAM according to the intended test objective (e.g., testing a particular core vs. testing the logic in between cores). The distribution mechanism of the control infrastructure is, in some architectures. combined with a mechanism for retrieving local core test status/result information. For example, local core-attached instruction and status/result registers can be configured into serial scan chains (dedicated chains for core test or part of "normal" chains) for core test instruction load and status/result unload.

21.3.3.4 Embedded Field Programmable Gate Arrays

Although eFPGAs are not very prevalent yet, they can have some unique properties that affect chip-level testing.

21.3.3.4.1 Embedded Field Programmable Gate Array Characteristics.

The term "field programmable" means that the eFPGA function can be programmed in the field, that is long after manufacturing test. At manufacturing test time, the final function generally is not known. FPGAs contain both functional resources and programming resources. Both types of resources tend to have their own I/O interface. The functional resources are configured by the programming resources prior to actual "use" to realize the intended function, and the rest of the chip logic communicates with the functional resources of an eFPGA through the functional I/O interface of the eFPGA. The normal chip functional logic in general does not see or interact directly with the programming resources or the programming I/O interface of the eFPGAs.

The customers of chips with eFPGAs expect that all functional and programming resources of the eFPGAs are working. Hence, it is necessary at chip manufacturing test time, to fully test all functional and programming resources (even if only a few of them will eventually be used).

21.3.3.4.2 Embedded Field Programmable Gate Array Test and Test Integration Issues.

The chip-level DFT for chips with eFPGAs will have to deal with the duality of the programming and functional resources. Many eFPGAs, for example, use an SRAM-like array of storage elements to hold the programming information. However, the outputs of the memory cells are not connected to data selection logic for the purpose of reading back a data word from some memory address. Instead, the individual memory cell outputs are intended to control the configuration of logic function blocks and of interconnect switches that constitute the functional resources of the eFPGA. The data-in side of the eFPGA programming memory array may also be different than in a normal SRAM that is designed for random access read/write. The design is optimized for loading a full eFPGA configuration program from some standardized programming interface. The programming interfaces tend to be relatively narrow even if the internal eFPGA programming memory is relatively wide. Hence, the address/data information may have to be brought in sequentially in several chunks.

Another idiosyncrasy of FPGAs is that the use of BIST techniques for testing does not necessarily mean that the test is (nearly) as autonomous as logic BIST, for example. Several BIST techniques have been proposed and are being used for testing the functional resources of the FPGA. To that effect, the reprogrammability of the FPGA is used to temporarily configure pattern generation and response compression logic from some functional resources and configure other functional resources as test target. Once configured, the BIST is indeed largely autonomous and needs only a little support from external test equipment. The difference is that potentially large amounts of programming data need to be downloaded from the external test equipment to create a suite of BIST configurations with sufficient test coverage. The programming data can consume large amounts of vector memory and downloading them costs test time. Non-BIST tests still need programming data and have the additional problem of needing test equipment support at the functional I/O interface during test application.

The chip-level DFT approach for chips with eFPGAs has to deal with the vector memory and test time issue as well as implement a suitable test interface for the I/O interface of the functional resources (even for eFPGAs with BIST there is a need to test the interface between the chip logic and the functional resources of the eFPGA). Also to be considered is that an unconfigured eFPGA is of little help in testing the surrounding logic, and a decision has to be made whether to exploit (and depend on) the programmability of the eFPGA's functional resources or to hardwire the DFT circuitry.

21.3.3.4.3 Embedded Field Programmable Gate Array Test Access Techniques.
To test fully an eFPGA, it is necessary to test both the programming resources and the functional resources, meaning that test access to the interfaces of the two different types of resources is needed. Of course, it is possible to treat the eFPGA like other embedded cores and implement an interface wrapper connected to the chip-level TAM. This approach however, does not take advantage of the fact that the chip already provides access to the eFPGA programming interface for functional configuration. The chip-level programming interface may be a good starting point for test access, especially if signature-based BIST is used to test the functional resources such that there is limited return traffic from the eFPGAs during test.

Regardless of whether the programming interface is adapted or another TAM is used, it may be desirable to test multiple sFPGA cores in parallel to reduce the demand on vector memory and test time. For example, if there are multiple identical core instances, then it makes sense to broadcast the same programming data to all instances simultaneously. The normal programming interface may have no need for and, therefore, not offer a broadcast load option for multiple eFPGA macros in parallel. Hence, such an option may have to be added to make the normal programming interface more useable for testing. Likewise, if another TAM is used, then the TAM should be configurable for the optional broadcast of programming data.

21.4 Conclusion

There are many other topics that could be covered in this chapter. These include the issues of embedded analog/mixed-signal DFT (which is covered in the chapter on "Analog Test" by Bozena Kaminska, in this handbook); and DFT and I/Os, both normal and high speed. In addition, there are issues of top-level DFT, including the integration of DFT elements, boundary scan for high-level assembly test, and test interface considerations (including IEEE 1149.1 protocol, BSDL, HSDL, COP/ESP, interrupts, burn-in, etc.). However, considerations of space and time do not let us go any further into these many interesting details at this point; perhaps in a future edition of this handbook, we will be able to cover them in some detail.

References

[1] K.J. Lee et al., Using a single input to support multiple scan chains, *Proceedings of ICCAD*, 1998, pp. 74–78 (broadcast scan).

[2] I.M. Ratiu and H.B. Bakoglu, Pseudorandom built-in self-test and methodology and implementation for the IBM RISC System/6000 processor, *IBM J. R&D*, Vol. 34, 78–84, 1990.

[3] B.L. Keller and D.A. Haynes, Design automation for the ES/9000 series processor, *Proceedings of ICCD*, 1991, pp. 550–553.

[4] A. Samad and M. Bell, Automating ASIC design-for-testability — the VLSI test assistant, *Proceedings of ITC*, 1989, pp. 819–828 (DFT automation including direct access macro test).

[5] V. Immaneni and S. Raman, Direct access test scheme — design of block and core cells for embedded ASICs, *Proceedings of ITC*, 1990, pp. 488–492.
[6] B.H. Seiss et al., Test point insertion for scan-based BIST, *Proceedings of the European Test Conference*, 1991, pp. 253–262.
[7] D. Kay and S. Mourad, Controllable LFSR for BIST, *Proceedings of the IEEE Instrument and Measurement Technology Conference*, 2000, pp. 223–229 (streaming LFSR-based decompressor).
[8] G.A. Sarrica and B.R. Kessler, Theory and implementation of LSSD scan ring & STUMPS channel test and diagnosis, *Proceedings of the International Electronics Manufacturing Technology Symposium*, 1992, pp. 195–201 (chain diagnosis for LBIST, lateral insertion concept).
[9] R. Rajski and J. Tyszer, Parallel Decompressor and Related Methods and Apparatuses, US Patent, US 5,991,909, 1999.
[10] I. Bayraktaroglu and A. Orailoglu, Test volume and application time reduction through scan chain concealment, *Proceedings of DAC*, 2001, pp. 151–155 (combinational linear decompressor).
[11] D.L. Fett, Current Mode Simultaneous Dual-Read/Write Memory Device, US Patent, US 4,070,657, 1978 (scannable register file).
[12] N.N. Tendolkar, Diagnosis of TCM failures in the IBM 3081 processor complex, *Proceedings of DAC*, 1983, pp. 196–200 (system-level ED/FI).
[13] J. Reilly et al., Processor controller for the IBM 3081, *IBM J. R&D*, Vol. 26, 22–29, January 1982 (system-level scan architecture).
[14] H.W. Miller, Design for test via standardized design and display techniques, *Elect. Test*, 108–116, 1983 (system-level scan architecture; scannable register arrays, control, data, and shadow chains; reset and inversion for chain diagnosis).
[15] A.M. Rincon et al., Core design and system-on-a-chip integration, *IEEE Des. Test*, Vol. 14, 26–35, 1997 (test access mechanism for core testing).
[16] E.K. Vida-Torku et al., Bipolar, CMOS and BiCMOS circuit technologies examined for testability, *Proceedings of the 34th Midwest Symposium on Circuits and Systems*, 1992, pp. 1015–1020 (circuit-level inductive fault analysis).
[17] J. Dreibelbis et al. Processor-based built-in self-test for embedded DRAM, *IEEE J. Solid-State Circ.*, Vol. 33, 1731–1740, November 1998 (microcoded BIST engine with 1-dimensional on-chip redundancy allocation).
[18] U. Diebold et al., Method and Apparatus for Testing a VLSI Device, European Patent EP 0 481 097 B1, 1995 (LFSR re-seeding with equation solving).
[19] D. Westcott, The self-assist test approach to embedded arrays, *Proceedings of ITC*, 1981, pp. 203–207 (hybrid memory BIST with external control interface).
[20] E.B. Eichelberger and T.W. Williams, A logic design structure for LSI testability, *Proceedings of DAC*, 1977, pp. 462–468.
[21] D.R. Resnick, Testability and maintainability with a new 6k gate array, *VLSI Design*, Vol. IV, 34–38, March/April 1983 (BILBO implementation).
[22] Zasio J.J., Shifting away from probes for wafer test, *COMPCON S'83*, 1983, pp. 317–320 (boundary scan and RPCT).
[23] P. Goel, PODEM-X: an automatic test generation system for VLSI logic structures, *Proceedings of DAC*, 1981, pp. 260–268.
[24] N. Benowitz et al., An advanced fault isolation system for digital logic, *IEEE Trans. Comput.*, Vol. C-24, 489–497, May 1975 (pseudo-random logic BIST).
[25] P. Goel and M.T. McMahon, Electronic chip-in-place test, *Proceedings of ITC*, 1982, pp. 83–90 (sort-of boundary scan).
[26] Y. Arzoumanian and J. Waicukauski, Fault diagnosis in an LSSD environment, *Proceedings of ITC*, 1981, pp. 86–88 (post-test simulation/dynamic fault dictionary; single location paradigm with exact matching).
[27] K.D. Wagner, Design for testability in the Amdahl 580, *COMPCON 83*, 1983, pp. 383–388 (random access scan).
[28] H-J. Wunderlich, PROTEST: a tool for probabilistic testability analysis, *Proceedings of DAC*, 1985, pp. 204–211 (testability analysis).
[29] H.C. Godoy et al., Automatic checking of logic design structures for compliance with testability ground rules, *Proceedings of DAC*, 1977, pp. 469–478 (early DFT DRC tool).

[30] B. Koenemann, J. Mucha, and G. Zwiehoff, Built-in test for complex digital integrated circuits, *IEEE J. Solid State Circ.*, Vol. 15, 315–319, June 1980 (hierarchical TM-Bus concept).
[31] B. Koenemann et al., Built-in logic block observation techniques, *Proceedings of ITC*, 1979, pp. 37–41.
[32] R.W. Berry et al., Method and Apparatus for Memory Dynamic Burn-in, US Patent, US 5,375,091, 1994 (design for burn-in).
[33] J.A. Waicukauski et al., Fault simulation for structured VLSI, *VLSI Syst. Des.*, 20–32, December 1985 (PPSFP).
[34] P.H. Bardell and W.H. McAnney, Self-testing of multi-chip logic modules, *Proceedings of ITC*, 1982, pp. 200–204 (STUMPS).
[35] K. Maling and E.L. Allen, A computer organization and programming system for automated maintenance, *IEEE Trans. Electr. Comput.*, 887–895, 1963 (early system-level scan methodology).
[36] R.D. Eldred, Test routines based on symbolic logic statements, *ACM J.*, Vol. 6, 33–36, January 1959.
[37] M. Nagamine, An automated method for designing logic circuit diagnostics programs, *Design Automation Workshop*, 1971, pp. 236–241 (parallel pattern compiled code fault simulation).
[38] P. Agrawal and V.D. Agrawal, On improving the efficiency of Monte Carlo test generation, *Proceedings of FTCS*, 1975, pp. 205–209 (weighted random patterns).
[39] V.S. Iyengar et al., On computing the sizes of detected delay faults, *IEEE Trans. CAD*, Vol. 9, 299–312, March 1990 (small delay fault simulator).
[40] Y. Aizenbud et al., AC test quality: beyond transition fault grading, *Proceedings of ITC*, 1992, pp. 568–577 (small delay fault simulator).
[41] R.C. Wong, An AC test structure for fast memory arrays, *IBM J. R&D*, Vol. 34, 314–324, March/May 1990 (SCAT-like memory test timing).
[42] M.H. McLeod, Test Circuitry for Delay Measurements on a LSI chip, US Patent No. 4,392,105, 1983 (on chip delay measurement method).
[43] W.S. Klara et al., Self-Contained Array Timing, US Patent, US 4,608,669, 1986.
[44] J.A. Monzel et al., AC BIST for a compilable ASIC embedded memory library, Digital Papers North Atlantic Test Workshop, 1996 (SCAT-like approach).
[45] L. Ternullo et al., Deterministic self-test of a high-speed embedded memory and logic processor subsystem, *Proceedings of ITC*, 1995, pp. 33–44 (BIST for complex memory subsystem).
[46] W.V. Huott et al., Advanced microprocessor test strategy and methodology, *IBM J. R&D*, Vol. 41, 611–627, 1997 (microcoded memory BIST).
[47] H. Koike et al., A BIST scheme using microprogram ROM for large capacity memories, *Proceedings of ITC*, 1990, pp. 815–822.
[48] S.B. Akers, The use of linear sums in exhaustive testing, *Comput. Math Appl.*, 13, 475–483, 1987 (combinational linear decompressor for locally exhaustive testing).
[49] D. Komonytsky, LSI self-test using level sensitive design and signature analysis, *Proceedings of ITC*, 1982, pp. 414–424 (scan BIST).
[50] S.K. Jain and V.D. Agrawal, STAFAN: an alternative to fault simulation, *Proceedings of DAC*, 1984, pp. 18–23.
[51] B. Nadeau-Dostie et al., A serial interfacing technique for built-in and external testing, *Proceedings of CICC*, 1989, pp. 22.2.1–22.2.5.
[52] T.M. Storey and J.W. Barry, Delay test simulation, *Proceedings of DAC*, 1977, pp. 492–494 (simulation with calculation of test slack).
[53] E.P. Hsieh et al., Delay test generation, *Proceedings of DAC*, 1977, pp. 486–491.
[54] K. Kishida et al., A delay test system for high speed logic LSI's, *Proceedings of DAC*, 1986, pp. 786–790 (Hitachi delay test generation system).
[55] A. Toth and C. Holt, Automated database-driven digital testing, *IEEE Comput.*, 13–19, January 1974 (scan design).
[56] A. Kobayashi et al., Flip-flop circuit with FLT capability, *Proceedings of IECEO Conference*, 1968, p. 962.
[57] M.J.Y. Williams and J.B. Angell, Enhancing testability of large scale circuits via test points and additional logic, *IEEE Trans. Comput.*, Vol. C-22, 46–60, 1973.
[58] R.R. Ramseyer et al., Strategy for testing VHSIC chips, *Proceedings of ITC*, 1982, pp. 515–518.

22

Automatic Test Pattern Generation

Kwang-Ting (Tim) Cheng
University of California
Santa Barbara, California

Li-C. Wang
University of California
Santa Barbara, California

22.1 Introduction

Test development for complex designs can be time-consuming, sometimes stretching over several months of tedious work. In the past three decades, various test development automation tools have attempted to address this problem and eliminate bottlenecks that hinder the product's time to market. These tools, which automate dozens of tasks essential for developing adequate tests, generally fall into four categories: design-for-testability (DFT), test pattern generation, pattern grading, and test program development and debugging. The focus of this chapter is on automatic test pattern generation (ATPG).

Because ATPG is one of the most difficult problems for electronic design automation, it has been researched for more than 30 years. Researchers, both theoreticians and industrial tool developers, have focused on issues such as scalability, ability to handle various fault models, and methods for extending the algorithms beyond Boolean domains to handle various abstraction levels.

Historically, ATPG has focused on a set of *faults* derived from a gate-level fault model. For a given target fault ATPG consists of two phases: *fault activation* and *fault propagation. Fault activation* establishes a signal value at the fault site opposite that produced by the fault. *Fault propagation* propagates the fault effect forward by sensitizing a path from the fault site to a primary output. The objective of ATPG is to find an input (or test) sequence that, when applied to the circuit, enables testers to distinguish between the correct circuit behavior and the faulty circuit behavior caused by a particular fault. Effectiveness of ATPG is measured by the fault coverage achieved for the fault model and the number of generated vectors, which should be directly proportional to test application time.

ATPG efficiency is another important consideration. It is influenced by the fault model under consideration, the type of circuit under test (full scan, synchronous sequential, or asynchronous sequential), the level of abstraction used to represent the circuit under test (gate, register transistor, switch), and the required test quality.

As design trends move toward nanometer technology, new ATPG problems are emerging. During design validation, engineers can no longer ignore the effects of crosstalk and power supply noise on reliability and performance. Current modeling and vector-generation techniques must give way to new techniques that consider timing information during test generation, that are scalable to larger designs, and that can capture extreme design conditions. For nanometer technology, many current design validation problems are becoming manufacturing test problems as well, so new fault-modeling and ATPG techniques will be needed.

This chapter is divided into five sections. Section 22.2 introduces gate-level fault models and concepts in traditional combinational ATPG. Section 22.3 discusses ATPG on gate-level sequential circuits. Section 22.4 describes circuit-based Boolean satistifiability (SAT) techniques for solving circuit-oriented problems. Section 22.5 illustrates ATPG for faults such as crosstalk and power supply noise, which involve timing and applications other than manufacturing testing. Section 22.6 presents sequential ATPG approaches that go beyond the traditional gate-level model.

22.2 Combinational ATPG

A fault model is a hypothesis of how the circuit may go wrong in the manufacturing process. In the past several decades, the most popular fault model used in practice is the *single stuck-at fault* model. In this model, one of the signal lines in a circuit is assumed to be stuck at a fixed logic value, regardless of what inputs are supplied to the circuit. Hence, if a circuit has n signal lines, there are potentially $2n$ stuck-at faults defined on the circuit, of which some can be viewed as being equivalent to others [1].

The stuck-at fault model is a *logical fault* model because no delay information is associated with the fault definition. It is also called a *permanent fault* model because the faulty effect is assumed to be permanent, in contrast to *intermittent* and *transient* faults that can appear randomly through time. The fault model is *structural* because it is defined based on a structural gate-level circuit model.

A stuck-at fault is said to be detected by a test pattern if, when applying the pattern to the circuit, different logic values can be observed, in at least one of the circuit's primary outputs, between the original circuit and the faulty circuit. A pattern set with 100% stuck-at fault coverage consists of tests to detect every possible stuck-at fault in a circuit.

Stuck-at fault coverage of 100% does not necessarily guarantee high quality. Earlier studies demonstrate that not all fault coverages are created equal [2,3] with respect to the quality levels they achieve. As fault coverage approaches 100%, additional stuck-at fault tests have diminishing chances to detect nontarget defects [4]. Experimental results have shown that in order to capture all nontarget defects, generating multiple tests for a fault may be required [5]. Generating tests to observe faulty sites multiple times may help to achieve higher quality [6,7]. Correlating fault coverages to test quality is a fruitful research area beyond the scope of this chapter. We use stuck-at fault as an example to illustrate the ATPG techniques.

A test pattern that detects a stuck-at fault satisfies two criteria simultaneously: *fault activation* and *fault propagation.* Consider Figure 22.1 as an example. In this example, input line a of the AND gate is assumed to be stuck-at 0. In order to activate this fault, a test pattern must produce logic value 1 at line a. Then, under the good-circuit assumption, line a has logic value 1 when the test pattern is applied. Under the faulty-circuit assumption, line a has logic value 0. The symbol $D = 1/0$ is used to denote the situation. D needs to be *propagated* through a sensitized path to one of the primary outputs. In order for D to be propagated from line a to line c, input line b has to be set at logic value 1. The logic value 1 is called the *non controlling value* for an AND gate. Once b is set at the noncontrolling value, line c will have whatever logic value that line a has.

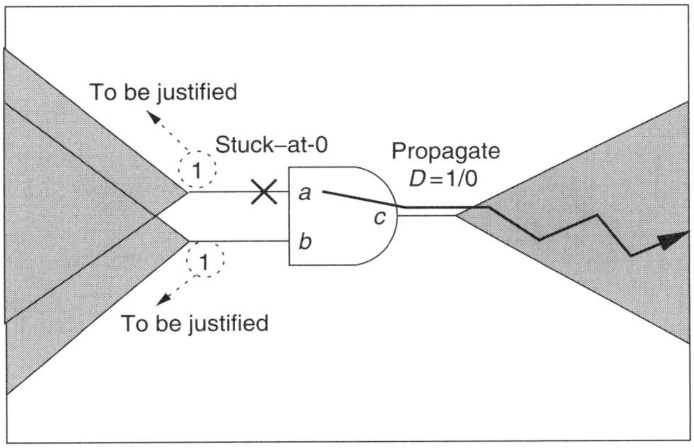

FIGURE 22.1 Fault activation and fault propagation for a stuck-at.

The ATPG process involves simultaneous justification of the logic value 1 at lines *a* and *b*, and propagation of the fault difference *D* to a primary output. In a typical circuit with reconvergent fanouts, the process involves a search for the right decisions to assign logic values at primary inputs and at internal signal lines in order to accomplish both justification and propagation. The ATPG problem is an NP-complete problem [8]. Hence, all known algorithms have an exponential worst-case run time.

Algorithm 22.1: BRANCH-AND-BOUND ATPG*(circuit, a fault)*

Solve()
if (*bound_by_implication()* = *FAILURE*)
  then return (*FAILURE*)
if (error difference at PO) **and** all lines are justified
  then return (*SUCCESS*)
while (there is an untried way to solve the problem)
  do { make a decision to select an untried way to propagate or to justify
    if (*Solve()* = *SUCCESS*)
    then return (*SUCCESS*) }
return (*FAILURE*)

Algorithm 22.1 illustrates a typical branch-and-bound approach to implement ATPG. The efficiency of this algorithm is affected by two things:

- The *bound_by_implication()* procedure determines whether there is a conflict in the current value assignments. This procedure helps the search to avoid making decisions in subspaces that contain no solution.
- The decision-making step inside the while loop determines how branches should be ordered in the search tree. This determines how quickly a solution can be reached. For example, one can make decisions to propagate the fault difference to a primary output (PO) before making decisions to justify value assignments.

22.2.1 Implication and Necessary Assignments

The ATPG process operates on at least a five-value logic defined over $\{0,1,D,\overline{D},X\}$; X denotes unassigned value [9]. D denotes that in the good circuit, the value should be logic 1, and in the faulty circuit the value should be logic 0. $\overline{D}$ is the complement of D. Logical AND, OR, and NOT can be defined based on these five values [1,9]. When a signal line is assigned with a value, it can be one of the four values $\{0,1,D,\overline{D}\}$.

After certain value assignments have been made, *necessary assignments* are those implied by the current assignments. For example, for an n-input AND gate, its output being assigned with logic 1 implies that all its inputs have to be assigned with logic 1. If its output is assigned with logic 0 and $n - 1$ inputs are assigned with logic 1, then the remaining input has to be assigned with logic 0. These necessary assignments derive from an analysis of circuit structure are called *implications*. If the analysis is done individually for each gate, the implications are *direct implications*. There are other situations where implications can be *indirect*.

Figure 22.2(a) shows a simple example of indirect implication. Suppose that a value 0 is being justified backward through line d where line b and line c have been already assigned with logic 1. To justify $d = 0$, there are three choices to set the values of the AND's inputs. Regardless of which way is used to justify $d = 0$, the line a must be assigned with logic value 0. Hence, in this case $d = 0$, $b = 1$, $c = 1$ together implies $a = 0$. Figure 22.2 (b) shows another example where $f = 1$ implies $e = 1$. This implication holds regardless of other assignments.

Figure 22.3 shows an example where the fault difference D is propagated through line a. Suppose that there are two possible paths to propagate D, one through line b and the other through line c. Suppose that both paths eventually converge at a 3-input AND gate. In this case, regardless of which path is chosen as the propagation path, line d must be assigned with logic value 1. Therefore, in this case $a = D$ implies $d = 1$.

The *bound_by_implication()* procedure in Algorithm 22.1 performs implications to derive all necessary assignments. A *conflict* occurs if a line is assigned with two different values after all the necessary assignments have been derived. In this case the procedure returns failure. It can be seen that the greater the number of necessary assignments that can be derived by implications, the more likely that a conflict can be detected. Because of this, efficiently finding necessary assignments through implications has been an important research topic for improving the performance of ATPG since the introduction of the first complete ATPG algorithm in Ref. [9].

In addition to the direct implications that can easily be derived based on the definitions of logic gates, indirect implications can be obtained by analyzing circuit structure. This analysis is called *learning* where

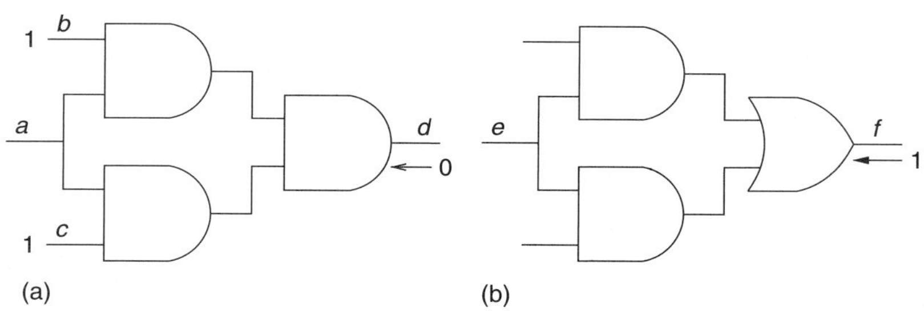

FIGURE 22.2 Implication examples [10]. (a) $d = 0$ implies $a = 0$ when $b = 1$, $c = 1$; (b) $f = 1$ implies $e = 1$.

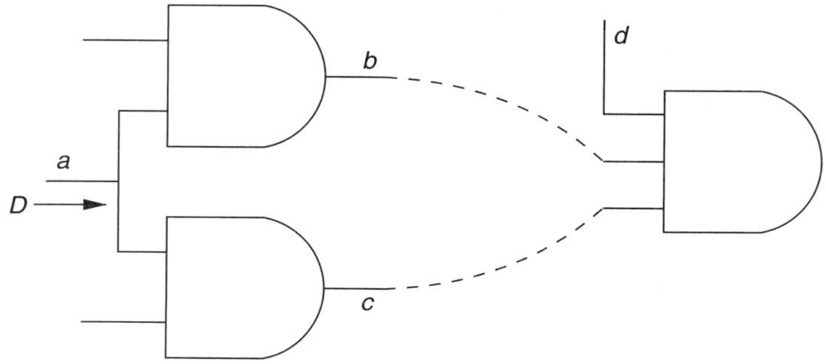

FIGURE 22.3 Implication $d = 1$ because of the unique propagation path [10].

correlations among signals have been established by simple and efficient methods. Learning methods should be simple and efficient enough so that the speedup in search by exploring the resulting implications should outweigh the cost of establishing the signal correlations. There are two types of learning approaches proposed in the literature:

Static learning Signal correlations are established before the search. For example, in Figure 22.2(b), forward logic simulation with $e = 0$ obtains $f = 0$. Because $e = 0$ implies $f = 0$, we obtain that $f = 1$ implies $e = 1$. For example, in Figure 22.3, the implication $a = D \Rightarrow d = 1$ is universal. This can be obtained by analyzing the *unique propagation path* from signal line a. These implications can be applied at any time during the search process.

Dynamic learning Signal correlations are established during the search. If the learned implications are conditioned on a set of value assignments, these implications can only be used to prune the search subspace based on those assignments. For example, in Figure 22.2(a), $d = 0$ implies $a = 0$ only when b and c have been assigned with 1. However, unconditional implications of the form $(x = v_1) \Rightarrow (y = v_2)$ can also be learned during dynamic learning.

The concepts of static and dynamic learning were suggested in [10]. A more sophisticated learning approach called *recursive learning* was presented in [11]. Recursive learning can be applied statically or dynamically. Because of its high cost, it is more efficient to apply recursive learning dynamically. Conflict-driven recursive learning and conflict-learning for ATPG were recently proposed in [12].

Knowing when to apply a particular learning technique is crucial in dynamic learning to ensure that the gain from learning outweighs the cost of learning. For example, it is more efficient to apply recursive learning on hard-to-detect faults where most of the subspaces during search contain no solution [11]. Because of this, search in these subspaces is inefficient. On the other hand, recursive learning can quickly prove that no solution exists in these subspaces. From this perspective, it appears that recursive learning implements a complementary strategy with respect to the decision-tree-based search strategy, since one is more efficient for proving the absence of a solution but the other is more efficient for finding a solution. Conflict-learning is another example in which learning is triggered by a conflict [12]. The conflict is analyzed and the cause of the conflict is recorded. The assumption is that during a search in a neighboring region, the same conflict might recur. By recording the cause of the conflict, the search subspace can be pruned more efficiently in the neighboring region. Conflict learning was first proposed in [13] with application in SAT. The authors in [12] implement the idea in their ATPG with circuit-based techniques.

22.2.2 ATPG Algorithms and Decision Ordering

One of the first complete ATPG algorithms is the D-algorithm [9]. Subsequently, other algorithms were proposed, including PODEM [15], FAN [16], and SOCRATES [10].

D-algorithm

The D-algorithm is based on the five-value logic defined on $\{0,1,D,\overline{D},X\}$. The search process makes decisions at primary inputs as well at internal signal lines. The D-algorithm is able to find a test even though a fault difference may necessitate propagation through multiple paths. Figure 22.4 illustrates such an example.

Suppose that fault difference D is propagated to line b. If the decision is made to propagate D through path d,g,j,l, we will require setting $a = 1$ and $k = 1$. Since $a = 1$ implies $i = 0$, $k = 1$ implies $h = 1$ which further implies $e = 1$. A conflict occurs between $e = 1$ and $b = D$. If the decision is made to propagate D through path e,h,k,l, we will require setting $i = 0$ and $j = 1$. $j = 1$ implies $g = 1$, which further implies $d = 1$. Again, $d = 1$ and $b = D$ cause a conflict. In this case, D has to be propagated through both paths. The required assignments are setting $a = 1$, $c = 1$, $f = 1$. This example illustrates a case when *multiple path sensitization* is required to detect a fault.

The D-algorithm is the first ATPG algorithm that can produce a test for a fault even though it requires multiple path sensitization. However, because the decisions are based on a five-value logic system, the search

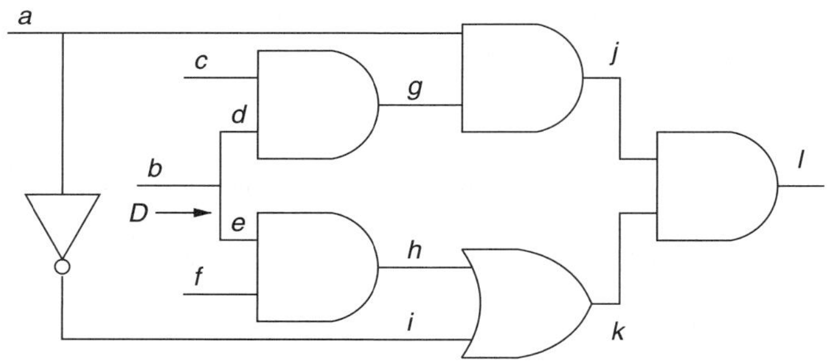

FIGURE 22.4 Fault propagation through multiple paths.

can be time-consuming. In practice, most faults may require only single-path sensitization and hence, explicit consideration of multiple path sensitization in search may become an overhead for ATPG [14].

PODEM

The D-algorithm can be characterized as an *indirect search* approach because the goal of an ATPG is to find a test at primary inputs, while the search decisions in D-algorithm are made on primary inputs and internal signal lines. PODEM implements a *direct search* approach where value assignments are made only on primary inputs, so that potentially the search tree is smaller. PODEM was proposed in [15]. A recent implementation of PODEM, called ATOM, is presented in [17].

FAN

FAN [16] adds two new concepts to PODEM. First, decisions can be made on internal *head lines* that are the end points of a tree logic cone. Therefore, a value assigned to a head line is guaranteed to be justifiable because of the tree circuit structure. Second, FAN uses a *multiple-backtrace* procedure so that a set of objectives can be satisfied simultaneously. In contrast, the original PODEM tries to satisfy one objective at a time.

SOCRATES

SOCRATES [10] is a FAN-based implementation with improvements in the implication and multiple-backtrace procedures. It also offers an improved procedure to identify a *unique sensitization path* [16].

The efficiency of an ATPG implementation depends primarily on the decision ordering it takes. There can be two approaches to influence the decision ordering: one by analyzing the fault-difference propagation paths and the other by measuring the testability of signal lines.

Analyzing the potential propagation paths can help to make decisions more likely to reach a solution. For example, a simple X-path check [15] can determine whether there exists a path from the point of a *D* line to a primary output, where no signal lines on the path have been assigned any value. *Unique sensitization* [16] can identify signal lines necessary for *D* propagation regardless of which path it takes. The *Dominator* approach [18] can identify necessary assignments for *D* propagation. Henftling et al. in [14] propose a single-path-oriented ATPG where fault propagation is explicitly made to have a higher priority than value justification.

Decision ordering can be guided through *testability measures* [1,19]. There are two types of testability measures: *controllability measures* and *observability measures.* Controllability measures indicate the relative difficulty of justifying a value assignment to a signal line. Observability measures indicate the relative difficulty of propagating the fault difference *D* from a line to a primary output. One popular strategy is to select a more difficult problem to solve before selecting the easier ones [1]. However, there can be two difficulties with a testability-guided search approach. First, the testability measures may not be sufficiently accurate. Second, always solving the hard problems first may bias the decisions too much in some cases. Wang et al. in [12] suggest a *dynamic decision ordering* approach in which failures in the justification process will trigger changes in decision ordering.

22.2.3 Boolean Satisfiability-Based ATPG

ATPG can also be viewed as solving a SAT problem. SAT-based ATPG was originally proposed in [20]. This approach duplicates part of the circuit which is influenced by the fault and constructs a satisfiability circuit instance by combining the good circuit with the faulty part. An input assignment that differentiates the faulty circuit from the good circuit is a test to detect the fault. Several other SAT-based ATPG approaches were developed later [21] [22]. Recently, a SAT-based ATPG called SPIRIT [23] was proposed, which includes almost all known ATPG techniques with improved heuristics for learning and search. ATPG and SAT will be discussed further in Section 22.4.

Practical implementation of an ATPG tool often involves a mixture of learning heuristics and search strategies. Popular commerical ATPG tools support full-scan designs where ATPG is mostly combinational. Although ATPG efficiency is important, other considerations such as test compression rate and diagnosability are also crucial for the success of an ATPG tool.

22.3 Sequential ATPG

The first ATPG algorithm for sequential circuits was reported in 1962 by Seshu and Freeman [24]. Since then, tremendous progress has been made in the development of algorithms and tools. One of the earliest commercial tools, LASAR [25], was reported in the early 1970s.

Due to the high complexity of the sequential ATPG, it remains a challenging task for large, highly sequential circuits that do not incorporate any design for testability (DfT) scheme. However, these test generators, combined with low-overhead DfT techniques such as partial scan, have shown a certain degree of success in testing large designs. For designs that are sensitive to area and performance overhead, the solution of using sequential-circuit ATPG and partial scan offers an attractive alternative to the popular full-scan solution, which is based on combinational-circuit ATPG.

It requires a sequence of vectors to detect a single stuck-at fault in a sequential circuit. Also, due to the presence of memory elements, the controllability and observability of the internal signals in a sequential circuit are in general much more difficult than those in a combinational circuit. These factors make the complexity of sequential ATPG much higher than that of combinational ATPG.

Sequential-circuit ATPG searches for *a sequence of vectors* to detect a particular fault through the space of all possible vector sequences. Various search strategies and heuristics have been devised to find a shorter sequence and to find a sequence faster. However, according to reported results, no single strategy/heuristic outperforms others for all applications/circuits. This observation implies that a test generator should include a comprehensive set of heuristics.

In this section, we will discuss the basics and give a survey of methods and techniques for sequential ATPG. We focus on the methods that are based on gate-level circuit models. Examples will be given to illustrate the basics of representative methods. The problem of sequential justification, sometimes referred to as *sequential SAT*, will be discussed in more detail in Section 22.4.

Figure 22.5 shows the taxonomy for sequential test generation approaches. Few approaches can directly deal with the timing issues present in highly asynchronous circuits. Most sequential-circuit test-generation approaches neglect the circuit delays during test generation. Such approaches primarily target synchronous or almost synchronous (i.e., with some asynchronous reset/clear and/or few asynchronous loops) sequential circuits, but they cannot properly handle highly asynchronous circuits whose functions are strongly related to the circuit delays and are sensitive to races and hazards. One engineering solution to using such approaches for asynchronous circuits is to divide the test-generation process into two phases. A potential test is first generated by ignoring the circuit delays. The potential test is then simulated using proper delay models in the second phase to check its validity. If the potential test is invalid due to race conditions, hazards, or oscillations, test generation is called again to produce a new potential test.

The approaches for (almost) synchronous circuits can be classified according to the level of abstraction at which the circuit is described. A class of approaches uses the state transition graph (STG) for test generation [26–29]. This class is suitable for pure controllers for which the STGS are either readily available or

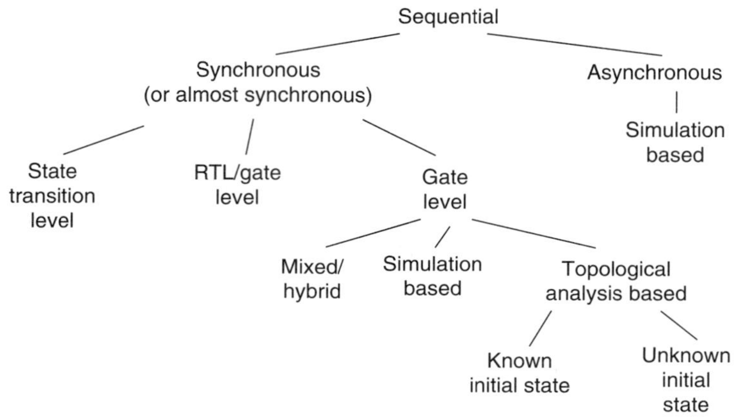

FIGURE 22.5 Sequential test generation: Taxonomy.

easily extractable from a lower level description. For data-dominated circuits, if both register transfer level (RTL) and gate-level descriptions are provided, several approaches can effectively use the RTL description for state justification and fault propagation [30–32].

Most of the commercial test generators are based on the gate-level description. Some of them employ the iterative array model [33,34] and use topological analysis algorithms [35–38] or they might be enhanced from a fault simulator [24,40–43]. Some use the mixed/hybrid methods that combine the topological-analysis-based methods and the simulation-based methods [44–46]. Most of these gate-level approaches assume an unknown initial state in the flip-flops, whereas some approaches assume a known initial state to avoid initialization of the state-holding elements [47–49]. The highlighted models and approaches in Figure 22.5 are those commonly adopted in most of today's sequential ATPG approaches.

22.3.1 Topological-Analysis-Based Approaches

Many sequential circuit test generators have been devised on the basis of fundamental combinational algorithms. Figure 22.6(a) shows the Huffman model of a sequential circuit. Figure 22.6(b) shows an array of combinational logic through *time-frame expansion*. In any time frame, logic values can be assigned only to the primary inputs (PIs). The values on the next state lines (NSs) depend on the values of the current state lines (PSs) at the end of the previous time frame. The iterative combinational model is used to approximate the timing behavior of the circuit. Topological analysis algorithms that activate faults and sensitize paths through these multiple copies of the combinational circuit are used to generate input assignments at the primary inputs. Note that a single stuck-at fault in a sequential circuit will correspond to a multiple stuck-at fault in the iterative array model where each time frame contains the stuck-at fault at the corresponding fault site.

The earliest algorithms extended the D-algorithm [9] based on the iterative array model [33,34]. It starts with one copy of the combinational logic and sets it to time frame 0. The D-algorithm is used for time frame 0 to generate a combinational test. When the fault effect is propagated to the next state lines, a new copy of the combinational logic is created as the next time frame, and the fault propagation continues. When there are values required at the present state lines, a new copy of the combinational logic is created as the previous time frame. The state justification is then performed backwards in the previous time frame. The process continues until there is no value requirement at the present state lines, and a fault effect appears at a primary output.

Muth [50] pointed out that the five-value logic based on $\{0,1,D,D,X\}$ used in the D-algorithm is not sufficient for sequential ATPG. A nine-value logic is suggested to take into account the possible repeated effects of the fault in the iterative array model. Each of the nine values is defined by an ordered pair of

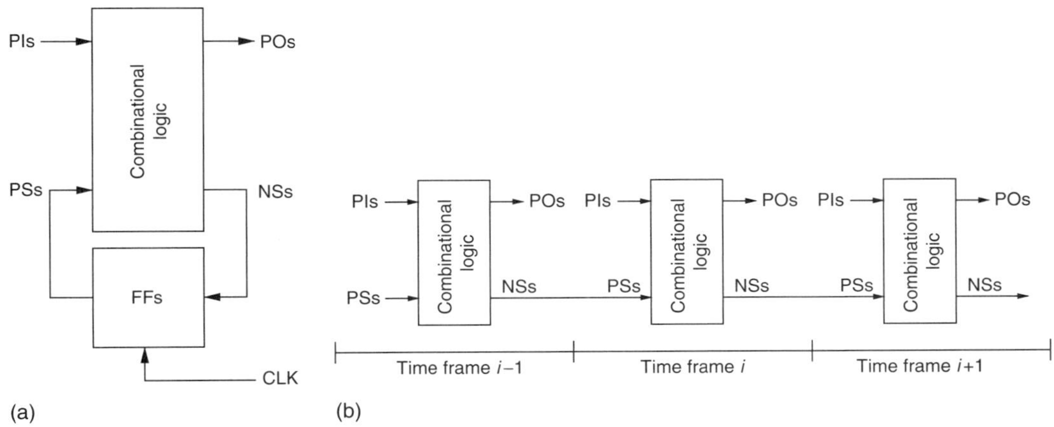

FIGURE 22.6 Synchronous sequential circuit model and time-frame expansion. (a) Huffman model; (b) Iterative array model.

binary values – the first value of the pair represents the ternary value (0,1,or X) of a single line in the fault-free circuit, and the second value represents the ternary value of the signal line in the faulty circuit. Hence, for a signal, there are possibly nine distinct ordered pairs (0/0, 0/1, 0/*X*, 1/0, 1/1, 1/*X*, *X*/0, *X*/1 and *X*/*X*).

The extended D-algorithm and the nine-value-based algorithm use *mixed forward and reverse time processing* techniques during test generation. The requirements created during the forward process (fault propagation) have to be justified by the backward process later. The mixed time processing techniques have some disadvantages. The test generator may need to maintain a large number of time frames during test generation because all time frames are partially processed, and the implementation is somewhat complicated.

The reverse time processing (RTP) technique used in the extended backtrace algorithm (EBT) [35] overcomes the problems caused by the mixed time processing technique. RTP works backwards in time from the last time frame to the first time frame. For a given fault, it preselects a path from the fault site to a primary output. This path may involve several time frames. The selected path is then sensitized backwards starting from the primary output. If the path is successfully sensitized, backward justification is performed for the required value at the fault site. If the sensitization process fails, another path is selected.

RTP has two main advantages: (1) At any time during the test-generation process, only two time frames need to be maintained: the current time frame and the previous one. For such a unidirectional algorithm, the backward justification process is done in a breadth-first manner. The value requirements in time frame n are completely justified before the justification of the requirements in time frame $n-1$. Therefore, the justified values at internal nodes of time-frame n can be discarded when the justification of time frame $n-1$ starts. As a result, the memory usage is low and the implementation is easier. Note that the decision points and their corresponding circuit status still need to be stacked for the purpose of backtracking. (2) It is easier to identify repetition of state requirements. A state requirement is defined as the state specified at the present state lines of a time frame during the backward-justification process. If a state requirement has been visited earlier during the current backward-justification process, the test generator has found a loop in the state-transition diagram. This situation is called *state repetition.* The backward-justification process should not continue to circle that loop, so backtracking should take place immediately. Since justification in time frame n is completed before the justification in time frame n - 1, state repetition can be easily identified by simply recording the state requirement after the completion of backward justification of each time frame, and then comparing each newly visited state requirement with the list of previously visited state requirements. Therefore, the search can be conducted more effectively. Similarly, the test generator can maintain a list of illegal states, i.e., the states that have been previously determined as unjustifiable. Each newly visited state requirement should also be compared against this list to determine whether the state requirement is an identified illegal state, in order to avoid repetitive and unnecessary searches.

There are two major problems with the EBT algorithm: (1) Only a single path is selected for sensitization. Faults that require multiple-path sensitization for detection may not be covered. (2) The number of possible paths from the fault site to the primary outputs can be very large; trying path-by-path may not be efficient.

After the EBT approach, several other sequential ATPGs were proposed, including the BACK algorithm [36], HITEC [37], and FASTEST [38].

BACK : The BACK algorithm [36] is an improvement over the EBT algorithm. It also employs the RTP technique. Instead of preselecting a path, the BACK algorithm preselects a primary output. It assigns a D or $\overline{\mathrm{D}}$ to the selected primary output and justifies the value backwards. A testability measure (called drivability) is used to guide the backward D-justification from the selected primary output to the fault site. Drivability is a measure associated with a signal that estimates the effort of propagating a D or $\overline{\mathrm{D}}$ from the fault site to the signal. The drivability measurement is derived based on the SCOAP [51] controllability measurement of both fault-free and faulty circuits. For a given fault, the drivability measure of each signal is computed before test generation starts.

HITEC : HITEC [37] employs several new techniques to improve the performance of test generation. Even though it uses both forward and reverse time processing, it clearly divides the test-generation process into two phases. The first is the forward time processing phase in which the fault is activated and propagated to a primary output. The second phase is the justification of the initial state determined in the first phase using the reverse time processing. Due to the use of the forward time processing for fault propagation, several efficient techniques (such as the use of dominators, unique sensitization, and mandatory assignments [10,16,18,39] used in combinational ATPG) can be extended and applied in phase one. In the reverse time processing algorithms, such techniques are of no use. Also, no drivability is needed for the fault-propagation phase which further saves some computing time.

FASTEST : FASTEST [38] uses only forward time processing and uses PODEM [15] as the underlying test-generation algorithm. For a given fault, FASTEST first attempts to estimate the total number of time frames required for detecting the fault and also to estimate at which time frame the fault is activated. The estimation is based on SCOAP[51]-like controllability and observability measures. An iterative array model with the estimated number of time frames is then constructed. The present state lines of the very first time frame have an unknown value and cannot be assigned to either binary value. A PODEM-like algorithm is employed where the initial objective is to activate the target fault at the estimated time frame. After an initial objective has been determined, it backtraces starting from the line of the initial objective until it reaches an unassigned primary input or a present state line in the first time frame. For the later case, backtracking is performed immediately. This process is very similar to the PODEM algorithm except that the process now works on a circuit model with multiple time frames. If the algorithm fails to find a test within the number of time frames currently in the iterative array, the number of time frames is increased, and test generation is attempted again based on the new iterative array.

Compared to the reverse time processing algorithms, the main advantage of the forward time processing algorithm is that it will not waste time in justifying unreachable states and will usually generate a shorter justification sequence for bringing the circuit to a hard-to-reach state. For circuits with a large number of unreachable states or hard-to-reach states, the reverse time processing algorithms may spend too much time in proving that unreachable states are unreachable or generating an unduly long sequence to bring the circuit to a hard-to-reach state. However, the forward time processing algorithm requires a good estimate of the total number of time frames and the time frame for activating each target fault. If that estimation is not accurate, the test generator may waste much effort in the smaller-than-necessary iterative array model.

22.3.2 Undetectability and Redundancy

For combinational circuits or full-scan sequential circuits, a fault is called *undetectable* if no input sequence can produce a fault effect at any primary output. A fault is called *redundant* if the presence of the fault does not change the input/output behavior of the circuit. The detectability is associated with a test-generation procedure whereas the redundancy is associated with the functional specification of a design.

A fault is combinationally redundant if it is reported as undetectable by a complete combinational test generator [52]. The definitions of detectability and redundancy for (nonscan) sequential circuits are much more complicated [1,53,54] and these two properties (the redundancy and undetectability of stuck-at faults) are no longer equivalent [1,53,54].

It is pointed out in [54] that undetectability could be precisely defined only if a test strategy is specified, and redundancy cannot be defined unless the operational mode of the circuit is known. The authors give formal and precise definitions of undetectability with respect to four different test strategies — namely fullscan, reset, multiple observation time, and single observation time. They also explain redundancies with respect to three different circuit operational modes — namely reset, synchronization, and nonsynchronization [54].

A fault is called undetectable under full scan if it is combinationally undetectable [55a]. In the case where hardware reset is available, a fault is said to be undetectable under the reset strategy if no input sequence exists such that the output response of the fault-free circuit is different from the response of the faulty circuit, both starting from their reset states. In the case where hardware reset is not available, there are two different test strategies: the multiple observation time (MOT) strategy and the single observation time (SOT) strategy.

Under the SOT strategy, a sequence detects a fault only if a fault effect appears at the same primary output *Oi* and at the same vector *vj* for all power-up initial state-pairs of the fault-free and faulty circuits (O_i could be any primary output, and vj could be any vector in the sequence). Most gate-level test generators and the sequential ATPG algorithms mentioned above assume the SOT test strategy.

Under the MOT strategy, a fault can be detected by multiple input sequences — each input sequence produces a fault effect at some primary output for a subset of power-up initial state-pairs and the union of the subsets covers all possible power-up initial state-pairs (for an n-flip-flop circuit, there are 2^{2n} power-up initial state-pairs). Under the MOT strategy, it is also possible to detect a fault using a single test sequence for which fault effects appear at different primary outputs or different vectors for different power-up initial state-pairs.

22.3.3 Approaches Assuming a Known Reset State

To avoid generating an initialization sequence, a class of ATPG approaches assumes the existence of a known initial state. For example, this assumption is valid for circuits like controllers that usually have a hardware reset (i.e., there is an external reset signal, and the memory elements are implemented by resettable flip-flops). Approaches like STALLION [47], STEED [48], and VERITAS [49] belong to this category.

STALLION

STALLION first extracts the STG for the fault-free circuit. For a given fault, it finds an activation state *S* and a fault-propagation sequence *T* that will propagate the fault effect to a primary output. This process is based on PODEM and the iterative array model. There is no backward state justification in this step. Using the STG, it then finds a state transfer sequence T0 from the initial state S0 to the activation state *S*. Because the derivation of the state transfer sequence is based on the state graph of the fault-free circuit, the sequence may be corrupted by the fault and hence, may not bring the faulty circuit into the required state *S*. Therefore, fault simulation for the concatenated sequence T0 T is required. If the concatenated sequence is not a valid test, an alternative transfer sequence or propagation sequence will be generated.

STALLION performs well for controllers for which the STG can be extracted easily. However, the extraction of STG is not feasible for large circuits. To overcome this deficiency, STALLION constructs a partial STG only. If the required transfer sequence cannot be derived from the partial STG, the partial STG is then dynamically augmented.

STEED

STEED is an improvement upon STALLION. Instead of extracting the complete or partial STG, it generates ON-set and OFF-set for each primary output and each next-state line for the fault-free circuit during the preprocessing phase. The ON-set (OFF-set) of a signal is the complete set of cubes (in terms of the primary inputs and the present state lines) that produces a logic 1 (logic 0) at a signal. The ON-sets and OFF-sets of the primary outputs and next state lines can be generated using a modified PODEM algorithm.

For a given fault, PODEM is used to generate one combinational test. The state transfer sequence and fault propagation sequence are constructed by intersecting the proper ON/ OFF-sets. In general, ON/OFF-set is a more compact representation than the STG. Therefore, STEED can handle larger circuits than STALLION. STEED shows good performance for circuits that have relatively small ON/OFF-sets. However, generating, storing, and intersecting the ON/ OFF-sets can be very expensive (in terms of both CPU time and memory) for certain functions such as parity trees. Therefore, STEED may have difficulties generating tests for circuits containing such function blocks. Also, like STALLION, the transfer and fault-propagation sequences derived from the ON/OFF-sets of the fault-free circuit may not be valid for the faulty circuit and therefore need to be verified by a fault simulator.

VERITAS

VERITAS is a BDD-based test generator which uses the binary decision diagram (BDD) [55b] to represent the state transition relations as well as sets of states. In the preprocessing phase, a state enumeration algorithm based on such BDD representations is used to find the set of states that are reachable from the reset state and the corresponding shortest transfer sequence for each of the reachable states. In the test-generation phase, as with STEED, a combinational test is first generated. The state transfer sequence to drive the machine into the activation state is readily available from the data derived from reachability analysis done in the preprocessing phase. Due to the advances in BDD representation, construction and manipulation, VERITAS in general achieves better performance than STEED.

In addition to the assumption of a known reset state, another common principle used by the above three approaches is to incorporate a preprocessing phase to (explicitly or implicitly) compute the state transition information. Such information could be used during test generation to save some repeated and unnecessary state justification effort. However, for large designs with huge state space, such preprocessing could be excessive. For example, complete reachability analysis used in the preprocessing phase of VERITAS typically fails (due to memory explosion) for designs with several hundreds of flip-flops. Either using partial reachability analysis, or simply performing state justification on demand during test generation, is a necessary modification to these approaches for large designs.

22.3.4 Summary

The presence of flip-flops and feedback loops substantially increases the complexity of the ATPG. Due to the inherent intractability of the problem, it remains infeasible to automatically derive high quality tests for large, nonscan sequential designs. However, because considerable progress has been made during the past few decades, and since robust commercial ATPG tools are now available, the partial-scan design methodology that relies on such tools for test generation might become a reasonable alternative to the full-scan design methodology [56].

As is the case with most other CAD tools, there are many engineering issues involved in building a test generator to handle large industrial designs. Industrial designs may contain tristate logic, bidirectional elements, gated clocks, I/O terminals, etc. Proper modeling is required for such elements, and the test-generation process would also benefit from some modifications. Many of these issues are similar to those present in the combinational ATPG problem that have been addressed (See e.g. [57–59]).

Developing special versions of ATPG algorithms/tools for circuits with special circuit structures and properties could be a good way to further improve the ATPG performance. For example, if the target circuit has a pipeline structure and is feedback-free, the algorithm described in [60pp. 98–101] is much more efficient than any algorithm surveyed in this section, which focuses on circuits with a more general structure. Many partial-scan circuits have unique circuit structures. A more detailed comparison of various sequential ATPG algorithms, practical implementation issues, and applications with partial-scan designs can be found in the survey [56].

22.4 ATPG and SAT

SAT has attracted tremendous research effort in recent years, resulting in the development of various efficient SAT solver packages. Popular SAT solvers [13,61–64] are designed based upon the conjunctive normal form (CNF).

Given a finite set of variables, V, over the set of Boolean values $\mathbf{B} \in \{0,1\}$, a *literal*, l or $\bar{l}$ is an instance of a variable v or its complement $\neg v$, where $v \in V$. A *clause* c_i, is a disjunction of literals $(l_1 \vee l_2 \vee \ldots \vee l_n)$. A formula f, is a conjunction of clauses $c_1 \wedge c_2 \wedge \ldots \wedge c_m$. Hence, a clause is considered as a set of literals, and a formula as a set of clauses. An assignment A *satisfies* a formula f if $f(A) = 1$. In a SAT problem, a formula f is given and the problem is to find an assignment A to satisfy f or prove that no such assignment exists.

22.4.1 Search in SAT

Modern SAT solvers are based on the search paradigm proposed in GRASP [13], which is an extension from the original DPLL [65] search algorithm. Algorithm 22.2 [13] describes the basic GRASP search procedure.

Algorithm 22.2: SAT()

comment: *B* is the backtracking decision level

comment: *d* is the current decision level

```
Search(d,B)
if (decided(d) = SUCCESS)
   then return (SUCCESS)
while (true)
     ⎧ if deduce(d) ≠ (CONFLICT)
     ⎪            ⎧ if (Search(d + 1 B) = SUCCESS)
     ⎪   then     ⎨  then return (SUCCESS)
  do ⎨            ⎪ if (B ≠ d)
     ⎪            ⎩  then erase(); return (CONFLICT)
     ⎪ if (diagnose(d,B) = CONFLICT)
     ⎪   then erase(); return (CONFLICT)
     ⎩ erase();
```

In the algorithm, function *decide()* selects an unassigned variable and assigns it with a logic value. This variable assignment is referred to as a *decision*. If no unassigned variable exists, *decide()* will return *SUCCESS* which means that a solution has been found. Otherwise *decide()* will return *CONFLICT* to invoke the *deduce()* procedure to check for conflict. A decision level *d* is associated with each decision. The first decision has decision level 1, and the decision level increases by one for each new decision.

The purpose of *deduce()* is to check for conflict by finding all necessary assignments induced by the current decisions. This step is similar to performing *implications* in ATPG. For example, in order to satisfy *f*, every clause of it must be satisfied. Therefore, if a clause has only one unassigned literal, and all the other literals are assigned with 0, then the unassigned literal must be assigned with value 1. A conflict occurs when a variable is assigned with both 1 and 0 or a clause becomes unsatisfiable.

The purpose of *diagnose()* is to analyze the reason that causes the conflict. The reason can be recorded as a *conflict clause*. The procedure can also determine a backtracking level other than backtracking to the previous decision level, a feature which can be used to implement *nonchronological backtracking* [66]. The *erase()* procedure deletes the value assignments at the current decision level.

In a modern SAT solver, one of the key concepts is *conflict-driven learning*. *Conflict-driven learning* is a method to analyze the causes of a conflict and then record the reason as a conflict clause to prevent the search from reentering the same search subspace. Since the introduction of conflict-driven learning, a hot research topic has been to find ways to derive conflict clauses that could efficiently prune the search space.

22.4.2 Comparison of ATPG and Circuit SAT

From all appearances, the problem formulation of ATPG is more complicated. ATPG involves fault activation and fault propagation, whereas circuit SAT concerns only justifying the value 1 at the single primary output of a circuit. However, as we have mentioned in Section 22.2.3, the ATPG problem can also be converted into a SAT problem [20].

Conflict-driven learning was originally proposed for SAT. One nice property of conflict-driven learning is that the reason for a conflict can be recorded as a *conflict clause* whose representation is consistent with that of the original problem. This simplifies the SAT solver implementation, in contrast to ATPGs where various learning heuristics are used, each of which may require a different data structure for efficient implementation. Silva and Sakallah in [22] argued that this simplification in implementation might provide benefits for runtime efficiency.

For circuit SAT, the conflict clauses can also be stored as gates. Figure 22.7 illustrates this. During the application of SAT search, constraints on the signal lines can be accumulated and added onto the circuit. These constraints are represented as OR gates where the outputs are set with logic value 1. These constraints encode the signal correlations that have to hold, due to the given circuit structure.

The idea of conflict-driven learning was implemented in the ATPG in [12]. For many other applications in computer-aided design (CAD) of integrated circuits, applying SAT to solve a circuit-oriented problem often requires transformation of the circuit gate-level netlist into its corresponding CNF format [67]. In a typical circuit-to-CNF transformation, the topological ordering among the internal signals is obscured in the CNF formula. In CNF format, all signals become (input) *variables*.

For solving circuit-oriented problems, circuit structural information has proved to be very useful. Tafertshofer et al. [68] developed a structural graph model called an *implication graph* for efficient implication and learning in SAT. Methods were also provided in [69,70] to utilize structural information in

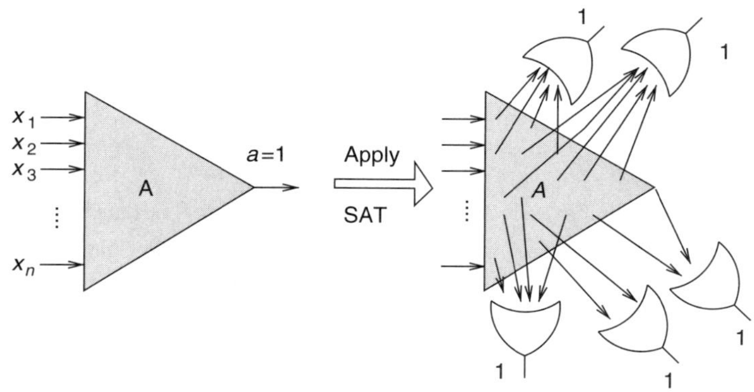

FIGURE 22.7 Learned gates accumulated by solving $a = 1$ in a circuit SAT problem.

SAT algorithms, which required minor modifications to the existing SAT algorithms. Gupta et al. [71] implemented a circuit-based SAT-solver that used structural information to identify unobservable gates and to remove the clauses for those gates. The work in [72] represented Boolean circuits in terms of 2-input AND gates and inverters. Based on this circuit model, a circuit SAT solver could be integrated with BDD sweeping [73].

Ganai et al. [74] developed a circuit-based SAT solver that adopted the techniques used in CNF-based SAT solver zChaff [62], e.g., the watched literal technique for efficient implication. Ostrowski et al. [75] tried to recover the structural information from CNF formulas, utilizing the structural information to eliminate clauses and variables. Theoretical results regarding circuit-based SAT algorithms were presented in[76]. Lu et al. [77] developed a SAT solver employing circuit-based implicit and explicit learning, applying the solver on industrial hard cases [78]. Lu et al. [79] developed a sequential circuit SAT solver for the sequential justification problem. Below, we summarize the ideas in [77,79] to illustrate how circuit information can be used in circuit SAT.

22.4.3 Combinational Circuit SAT

Consider the circuit in Figure 22.8 where shaded area B contains shaded area A, and shaded area C contains shaded area B. Suppose we want to solve a circuit SAT problem with the output objective $c = 1$. When we apply a circuit SAT solver to prove that $c = 0$ or to find an input assignment to make $c = 1$, potentially the search space for the solver is the entire circuit. Now suppose we identify, in advance, two internal signals a and b, such that $a = 1$ and $b = 0$ will be very unlikely outcomes when random inputs are supplied to the circuit. Then, we can divide the original problem into three subproblems: (1) solving $a = 1$, (2) solving $b = 0$, and then (3) solving $c = 1$.

Since $a = 1$ is unlikely to happen, when a circuit SAT solver makes decisions trying to satisfy $a = 1$, it is likely to encounter conflicts. As a result, much conflict-driven information can be learned and stored as *conflict gates* (as illustrated in Figure 22.7). If we assume that solving $a = 1$ is done only based upon the *cone of influence* headed by the signal a (the shaded area A in Figure 22.8), then the conflict gates will be based upon the signals contained in the area A only.

As the solver finishes solving $a = 1$ and starts solving $b = 0$, all the learned information regarding the circuit area A can be used to help in solving $b = 0$. In addition, if $a = 1$ is indeed unsatisfiable, then signal a can be assigned with 0 when the solver is solving $b = 0$. Similarly, learned information from solving $a = 1$ and $b = 0$ can be reused to help in solving $c = 1$.

Intuitively, we believe that solving the three subproblems following their topological order could be accomplished much faster than directly solving the original problem. This is because when solving $b = 0$, hopefully fewer (or no) decisions are required to go into area A. Hence, the search space is more restricted within the portion of area B that is not part of area A. Similarly, solving $c = 1$ requires most decisions to be made only within the portion of area C which is not part of area B. Moreover, the conflict gates accumulated by solving $a = 1$ could be smaller because they are based upon the signals in area A

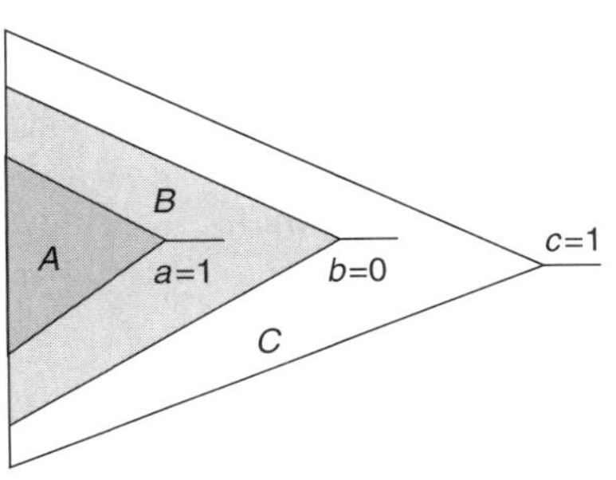

FIGURE 22.8 An example for incremental SAT solving.

only. Similarly, the conflict gates accumulated during the solving of $b = 0$ could be smaller. Conceptually, this strategy allows solving a complex problem incrementally.

Two observations can be made: (1) the incremental process suggests that we can guide the search process by solving a sequence of preselected subproblems following their topological order. (2) The selection of the sub-problems such as $a = 1$ and $b = 0$ should be those most likely to be unsatisfiable. Intuitively, we might expect that the search for a solution to a probably unsatisfiable subproblem will encounter more conflicts and can accumulate conflict-driven learning information more quickly.

If few or no conflicts arise in solving $a = 1$ and solving $b = 0$, then there may be not much information to be learned from solving these two problems. In this case, the time to spend on solving $a = 1$ may seem to be unnecessary overhead. Usually, this observation would suggest the ineffectiveness of the method used to determine that, both $a = 1$ and $b = 0$ are unlikely to be true. Moreover, if solving $c = 1$ does not depend much on signals a and b, then the above incremental strategy cannot be effective either. For example, if c is the output signal of a 2-input AND gate with two inputs as a function g and its complement g, then c = 1 can be directly proved to be unsatisfiable whether or not the actual function g is known.

Intuitively, we think the above incremental strategy would not be effective for solving a circuit SAT problem whose input is given in CNF form. This is because by treating the CNF form as a two-level OR–AND circuit structure, the topological ordering among the signals is lost. With a two-level structure, the incremental strategy has little room to proceed. In the example discussed above, both a and b become primary inputs in the 2-level OR–AND CNF circuit. Then, the ordering for solving the subproblems may become solving $b = 0$ followed by solving $a = 1$.

The authors in Ref. [77] implement a circuit SAT solver based on the idea just described. The decision ordering in the SAT solver is guided through *signal correlations* identified before the search process. A group of signals $s_1, s_2, \ldots, s_i$ (where $i > 1$) are said to be correlated if their values satisfy a certain Boolean function $f(s_1, s_2, \ldots, s_i)$ during random simulation, e.g., the values of s_1 and s_2 satisfy $s_1 = s_2$ during random simulation. Examples of signal correlations are *equivalence correlation*, *inverted equivalence correlation*, and *constant correlation.*

Two signals s_1 and s_2 have an equivalence correlation (inverted equivalence correlation) if and only if the values of the two signals satisfy $s_1 = s_2$ ($s_1 = s_2$) during random simulation. If s_2 is replaced by constant 0 or 1 in the notation, then that is a constant correlation. Note that the correlations are defined based on the results of random simulation. Signals may be correlated in one run of random simulation, yet not be correlated in another run of random simulation.

In [77], signal correlations can be used in two types of learning strategies: *explicit learning* and *implicit learning.* In explicit learning, the search is constrained by solving a sequence of subproblems constructed from the signal correlations to implement the incremental search as described above. In implicit learning, the decision-making in solving the original SAT objective is influenced by the signal correlations. No explicit solving of subproblems is performed. When the circuit SAT solver was applied to some industrial hard cases, it obtained promising performance results [78].

22.4.4 Sequential Circuit SAT

Recently, a sequential SAT solver was proposed [79]. The authors utilize combined ATPG and SAT techniques to implement a sequential SAT solver by retaining the efficiency of Boolean SAT and being complete in the search. Given a circuit following the Huffman synchronous sequential-circuit model, *Sequential SAT* (or *sequential justification)* is the problem of finding an *ordered input assignment sequence* such that a desired objective is satisfied, or proving that no such sequence exists. Under this model, a sequential SAT problem can fit into one of the following two categories:

- In a weak SAT problem, an initial-state value assignment is given. The problem is to find an ordered sequence of input assignments such that together with the initial state, the desired objective is satisfied or proved to be unsatisfiable.

- In a strong SAT problem, no initial state is given. Hence, it is necessary to identify an input sequence to satisfy the objective starting from the unknown state. To prove unsatisfiability, a SAT solver needs to prove that no input sequence can satisfy the given objective for *all* reachable initial states.

A strong SAT problem can be translated to a weak SAT problem by encoding technique [80]. In sequential SAT, a sequential circuit is conceptually unfolded into multiple copies of the combinational circuit through time frame expansion. In each time frame, the circuit becomes combinational and hence, a combinational SAT solver can be applied. In each time frame, a state element such as a flip-flop is translated into two corresponding signals: a pseudo primary input (PPI) and a pseudo-primary output (PPO). The initial state is specified with the PPIs in time frame 0. The objective is specified with the signals in time frame n (the last time frame, where n is unknown before solving the problem). During search in the intermediate time frames, intermediate solutions are produced at the PPIs, and they become the intermediate PPO objectives to be justified in the previous time frames.

Given a sequential circuit, a *state clause* is a clause consisting only of state variables. A state clause encodes a state combination where no solution can be found. Due to the usage of state clauses, the time frame expansion can be implemented implicitly by keeping only one copy of the combinational circuit.

22.4.4.1 Sequential SAT and State Clauses

To illustrate the usage of state clauses in sequential SAT, Figure 22.9 depicts a simple example circuit with three primary inputs a b c, one primary output f, and three state-holding elements (i.e., flip-flops) x,y,z. The initial state is $x = 1$, $y = 0$, $z = 1$. Suppose the SAT objective is to satisfy $f = 1$.

Starting from time frame n where n is unknown, the circuit is treated as a combinational circuit with state variables duplicated as PPOs and PPIs. This is illustrated as (1) in the figure. Since this represents a combinational SAT problem, a combinational circuit SAT solver can be applied.

Suppose after the combinational SAT solving, we can identify a solution $a = 1$, $b = 0$, c $= 0$, $PPI_x = 0$, $PPI_y = 1$ $PPI_z = 0$ to satisfy $f = 1$ (step (2)). The PPI assignment implies a state assignment $x = 0$, $y = 1$, $z = 0$. Since it is not the initial state, at this point, we may choose to continue the search by expanding into time frame $n - 1$ (this follows a depth-first search strategy).

Before solving in time frame $n - 1$, we need to examine the solution state $x = 0$, $y=1$, $z = 0$ more closely. This is because this solution may not represent the minimal assignment to satisfy the objective $f = 1$.

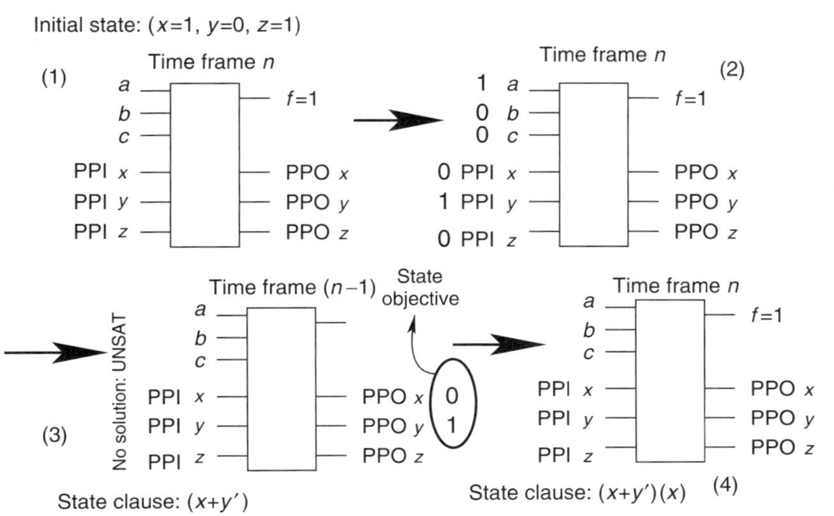

FIGURE 22.9 Sequential circuit SAT with state clauses.

Suppose that after analysis, we can determine that $z = 0$ is unnecessary. In other words, by keeping the assignment "$x = 0$, $y = 1$," we may discover that $f = 1$ can still be satisfied. This step is called *state minimization.*

After state minimization, backward time-frame expansion is achieved by adding a *state objective* $PPO_x=0$, $PPO_y = 1$ to the combinational copy of the circuit. Also, a state clause "$(x + y')$" is generated to prevent reaching those same state solutions defined by the state assignment "$x = 0$, $y = 1$." The new combinational SAT instance is then passed to the combinational circuit SAT solver for solving.

Suppose in time frame $n - 1$, no solution can be found. Then, we need to backtrack to time frame n to find another solution other than the state assignment "$x = 0$, $y = 1$." In a way, we have proved that from state "$x = 0$, $y = 1$," there exists no solution. This implies that there is no need to remove the state clause "$(x+y)$." However, at this point, we may want to perform further analysis to determine whether both $PPO_x = 0$ and $PPOy = 1$ are necessary to cause the unsatisfiability. Suppose that after *conflict analysis*, we discover that PPO_x 0 alone is sufficient to cause the conflict, then, for the sake of efficiency, we want to add another state clause "(x)." This is illustrated in (4) of the Figure 22.9. This step is called *state conflict analysis.*

The new combinational SAT instance now has the state clauses "$(x+y')(x)$" that record the non solution state subspaces previously identified. The solving continues until either one of the following two conditions is reached:

1. After state minimization, a solution is found with a state assignment *containing* the initial state. For example, a solution with the state assignment "$x=1$, $z = 1$" contains the initial state "$x=1$, $y = 0$, $z = 1$." In this case, a solution for the sequential SAT problem is found. We note that a solution without any state assignment and with only PI assignments is considered as one containing any initial state.
2. If in time frame n, the initial objective $f = 1$ and state clauses together cannot be satisfied, then the original problem is unsatisfiable. This is equivalent to saying that, without adding any PPO objective, if the initial objective $f = 1$ and state clauses together cannot be satisfied, then the original problem is unsatisfiable. Here, with the state clauses, the time-frame expansion is conducted implicitly rather than explicitly, i.e., only one copy of the combinational part of the circuit is required to be kept in the search process.

The above example illustrates several important concepts in the design of a sequential-circuit SAT solver:

- The *state minimization* involves finding the minimal state assignments for an intermediate PPI solution. The search can be more efficient with an intermediate PPI solution containing a smaller number of assigned states.
- The use of state clauses serves two purposes: (1) to record those state subspaces that have been explored, and (2) to record those state subspaces containing no solution. The first is to prevent the search from entering a *state justification loop.* The second follows the same spirit as that in the combinational SAT. When enough state clauses are accumulated, combinational SAT can determine that there is no solution to satisfy the initial objective and the state clauses. In this case, the problem is unsatisfiable. Note that while in sequential SAT, unsatisfiability is determined through combinational SAT, in combinational SAT, unsatisfiability is determined by implications based on the *conflict clauses.*
- Although conceptually, the search follows backward time frame expansion, the above example demonstrates that for implementation, explicit time frame expansion is not necessary. In other words, a sequential SAT solver needs only one copy of the combinational circuit. Moreover, there is no need to memorize the number of time frames being expanded.
- The above example demonstrates the use of a depth-first search strategy. As mentioned before, decision ordering can significantly affect the efficiency of a search process. In sequential circuit SAT, the issue is when to proceed with time-frame expansion. In depth-first fashion, time-frame

expansion is triggered by trying to solve the most recent state objective generated. In breadth-first fashion, the newly generated state objectives are resolved after solving all state objectives in the current time frame. A hybrid search strategy can be implemented by managing the state objective queue in various orders.

A *frame objective* is an objective to be satisfied which is passed to the combinational SAT solver. A frame objective can be either the initial objective or a state objective. *A frame solution* is an assignment on the PIs and PPIs, which satisfies a given frame objective without conflicting with the current state clauses.

When a frame objective is proved by the combinational SAT to be unsatisfiable, conflict analysis is applied to derive additional state clauses based on only the PPOs. In other words, conflict analysis traces back to PPOs to determine which of of them actually contribute to the conflict. Here, the conflict analysis can be similar to that in the combinational SAT [81], but the goal is to analyze the conflict sources up to the PPOs only.

As in the combinational SAT, where conflict clauses are accumulated through the solving process, in sequential SAT design, state clauses are accumulated through the solving process. The sequential solving process consists of a sequence of combinational solving tasks based on given frame objectives. At the beginning, the frame objective is the initial objective. As the solving proceeds, many state objectives become frame objectives. Hence, frame objectives also accumulate through the solving process.

A frame objective can be removed from the *objective array* only if it is proved to be unsatisfiable by the combinational SAT. If it is satisfiable, the frame objective stays in the *objective array.* The sequential SAT solver stops only when the *objective array* becomes empty. This means that it has exhausted all the state objectives and has also proved that the initial objective is unsatisfiable based on the accumulated state clauses.

During each step of combinational SAT, conflict clauses also accumulate through the combinational SAT solving process. When the sequential solving switches from one frame objective to another, these conflict clauses stay. Hence, in the sequential solving process, the conflict clauses generated by the combinational SAT are also accumulated. Although these conflict clauses can help to speed up the combinational SAT solving, experiences show that for sequential SAT, managing the state clauses dominates the overall sequential search efficiency [79].

Algorithm 22.3: SEQUENTIAL SAT(C, obj, s_0)

comment: C is the circuit with PPIs and PPOs expanded
comment: obj is the initial objective
comment: s_0 is the initial state
comment: FO is the objective array

```
FO ← {obj};
while (FO ≠ 0)
do { f_obj ← selecta_frame_objective(FO);
     f_sol ← combinational_solve_a_frame_objective(C,f_obj);
     if (f_sol = NULL)
       { clause ← PPO_state_conflict_analysis(C,f_obj);
         add_state_clause(C,clause);
         FO ← FO − {f_obj};
     else { stateassignment ← state_minimization(C,f_obj,f_sol);
            if (s_0 ∈ stateassignment)
            {return (SAT);
            else { clause ← convert_to_clause(stateassignment);
                   add_state_clause(C,clause);
                   FO ← FO + (stateassignment);
return (UNSAT)
```

The overall algorithm of the sequential circuit SAT solver is described in Algorithm 22.3. Note that in this algorithm, solving each frame objective produces only one solution. However, it is easy to extend this algorithm so that solving each frame objective produces many solutions at once. The search strategy is based on the selection of one frame objective at a time from the objective array *FO*, where different heuristics can be implemented. The efficiency of the sequential circuit SAT solver highly depends on the selection heuristic [79].

22.5 Applications of ATPG

In this section, we show that ATPG technology, in addition to generating high-quality tests for various fault models, also offers efficient techniques for analyzing designs during design verification and optimization. Already, ATPG has been used to generate tests not only to screen out chips with manufacturing defects but also to identify design errors and timing problems during design verification. It has also been used as a powerful logic-analysis engine for applications such as logic optimization, timing analysis, and design-property checking.

22.5.1 ATPG for Delay Faults and Noise Faults

The move toward nanometer technology is introducing new failure modes and a new set of design and test problems [82]. Device features continue to shrink as the number of interconnect layers and gate density increases. The result is increased current density and a higher voltage drop along the power nets as well as increased signal interference from coupling capacitance. All this gives rise to noise-induced failures, such as power supply noise or crosstalk. These faults may cause logic errors or excessive propagation delays which degrade circuit performance.

Demands for higher circuit operating frequencies, lower cost, and higher quality mean that testing must ascertain that the circuit's timing is correct. Timing defects can stay undetected after logic-fault testing such as testing of stuck-at faults, but they can be detected using delay tests. Unlike ATPG for stuck-at faults, ATPG for delay faults is closely tied to the test application strategy [83]. Before tests for delay faults are derived, the test application strategy has to be decided. The strategy depends on the circuit type as well as on the test equipment's speed.

In structural delay testing, detecting delay faults requires applying 2-vector patterns to the combinational part of the circuit at the circuit's intended operating speed. However, because high-speed testers require huge investments, most testers could be slower than the designs being tested. Testing high-speed designs on slower testers requires special test application and test-generation strategies — a topic that has been investigated for many years [83,84]. Because an arbitrary vector pair cannot be applied to the combinational part of a sequential circuit, ATPG for delay faults may be significantly more difficult for these than for full-scan circuits. Various testing strategies for sequential circuits have been proposed, but breakthroughs are needed [84]. Most researchers believe that some form of DFT is required to achieve high-quality delay testing for these circuits.

Noise faults must be detected during both design verification and manufacturing testing. When coupled with process variations, noise effects can exercise worst-case design corners that exceed operating conditions. These corners must be identified and checked as part of design validation. This task is extremely difficult, however, because noise effects are highly sensitive to the input pattern and to timing. Timing analysis that cannot consider how noise effects influence propagation delays will not provide an accurate estimation of performance, nor will it reliably identify problem areas in the design.

An efficient ATPG method must be able to generate validation vectors that can exercise worst-case design corners. To do this, it must integrate accurate timing information when the test vectors are derived. For manufacturing testing, ATPG techniques must be augmented and adapted to new failure conditions introduced by nanometer technology. Tests for conventional fault models, such as stuck-at and transition faults, obviously cannot detect these conditions. Thus, to check worst-case design corners, test vectors must sensitize the faults and propagate their effects to the primary outputs, as well as activate the conditions of worst-case noise effects. They must also scale to increasingly larger designs.

Power supply noise. For a highly integrated system-on-a-chip, more devices are switching simultaneously, which increases power supply noise. One component of this noise, inductive noise, results from sudden current changes on either the package lead or wire/substrate inductance. The other component, net IR voltage drop, is caused by current flowing through the resistive power and ground lines. The noise can cause a voltage glitch on these lines, resulting in timing or logic errors. Large voltage drops through the power supply lines can cause electromigration, which in turn can cause short or open circuits. To activate these defects and propagate them to the primary outputs, ATPG must carefully select test vectors.

Power supply noise can affect both reliability and performance. It reduces the actual voltage level that reaches a device, which in turn can increase cell and interconnection propagation delays. One way to detect these effects is to apply delay tests. Unfortunately, most existing delay techniques are based on simplified, logic-level models that cannot be directly used to model, and test timing defects in high-speed designs that use deep sub-micron technologies. New delay testing strategies are needed to close the gap between the logic-level delay fault models and physical defects. The tests must produce the worst-case power supply noise along the sensitized paths, and thus cause the worst-case propagation delays on these paths [85,86].

Crosstalk effects. The increased design density in deep-submicron designs leads to more significant interference between the signals because of capacitive coupling, or crosstalk. Crosstalk can induce both Boolean errors and delay faults. Therefore, ATPG for worst-case crosstalk effects must produce vectors that can create and propagate crosstalk pulses as well as crosstalk-induced delays [87–91].

Crosstalk-induced pulses are likely to cause errors on hazard-sensitive lines such as inputs to dynamic gates, clock, set/reset, and data inputs to flip-flops. Crosstalk pulses can result in logic errors or degraded voltage levels, which increase propagation delays. ATPG for worst-case crosstalk pulse aims to generate a pulse of maximum amplitude and width at the fault site and propagate its effects to primary outputs with minimal attenuation [92].

Studies show that increased coupling effects between signals can cause signal delay to increase (slowdown) or decrease (speedup) significantly. Both conditions can cause errors. Signal slowdown can cause delay faults if a transition is propagated along paths with small slacks. Signal speedup can cause race conditions if transitions are propagated along short paths. To guarantee design performance, ATPG techniques must consider how worst-case crosstalk affects propagation delays [88,89].

22.5.2 Design Applications

ATPG technology has been applied successfully in several areas of IC design automation, including logic optimization, logic equivalence checking, design property checking, and timing analysis.

22.5.2.1 Logic Optimization

To optimize logic, design aids can either remove redundancy or restructure the logic by adding and removing redundancy.

Redundancy Removal Redundancy is the main link between test and logic optimization. If there are untestable stuck-at faults, there is likely to be redundant logic. The reasoning is that, if a stuck-at fault does not have any test (the fault is untestable), the output responses of the faulty circuit (with this untestable fault) will be identical to the responses of the fault-free circuit for all possible input patterns applied to these two circuits. Thus, the faulty circuit (with an untestable stuck-at fault) is indeed a valid implementation of the fault-free circuit. Therefore, when ATPG identifies a stuck-at-1 (stuck-at-0) fault as untestable, one can simplify the circuit by setting the faulty net to logic

1(0) and thus effectively removing the faulty net from the circuit. This operation, called *redundancy removal,* also removes all the logic driving the faulty net.

Figure 22.10 illustrates an example. However, note that output Z in Figure 22.10(a) is hazard-free, but output Z in Figure 22.10(b) may have glitches. Testers must ensure that redundancy is removed only if having glitches is not a concern (as in synchronous design, for example).

Because this method only removes logic from the circuits, the circuit is smaller when the process ends; the topological delay of the longest paths will be shorter than or at most equal to that of the original circuit. The power dissipation of the optimized circuit will also be lower.

Logic Restructuring Removing a redundant fault can change the status of other faults. Those that were redundant might no longer be redundant, and *vice versa.* Although these changes complicate redundancy removal, they also pave the way for more rigorous optimization methods. Even for a circuit with no redundancies, designers can add redundancies to create new redundancies elsewhere in the circuit. By removing the created new redundancies, they may obtain an optimized circuit. This technique is called logic restructuring. For example, Figure 22.11 shows a circuit example that has no redundant logic. In Figure 22.12(a), a signal line is artificially added that does not change the function of the circuit but does create redundant logic. Figure 22.12(b) shows the resulting circuit after redundancy removal. This circuit is simpler than the one in Figure 22.11.

Efficient algorithms for finding effective logic restructuring [93] have been proposed in the past few years. By properly orienting the search for redundancy, these techniques can be adapted to target several optimizing goals

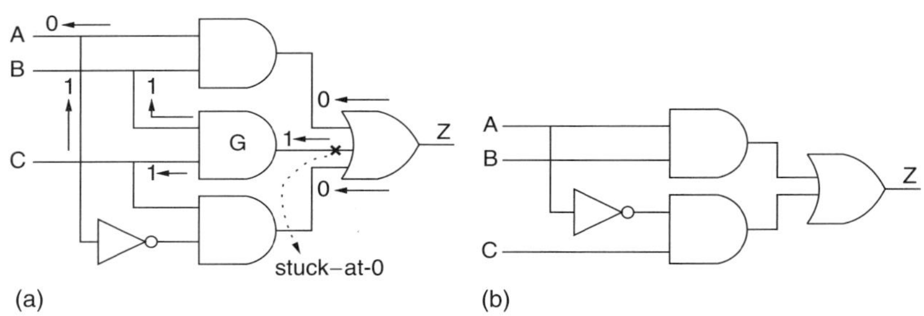

FIGURE 22.10 How ATPG works for redundancy removal: (a) the stuck-at 0 fault is untestable; (b) remove gate G and simply the logic.

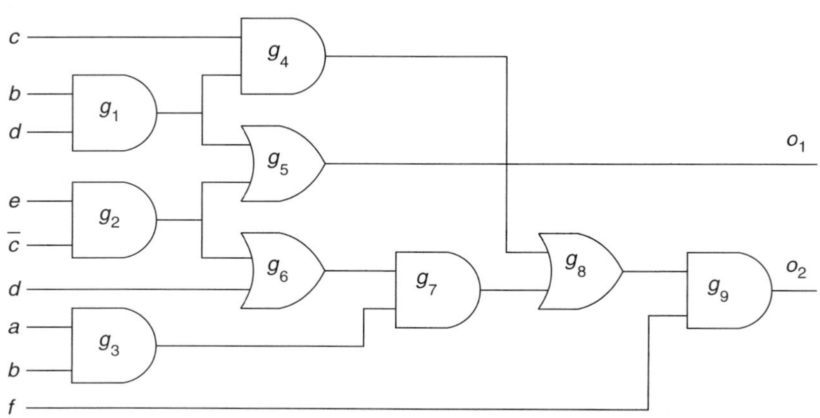

FIGURE 22.11 A circuit that is not redundant.

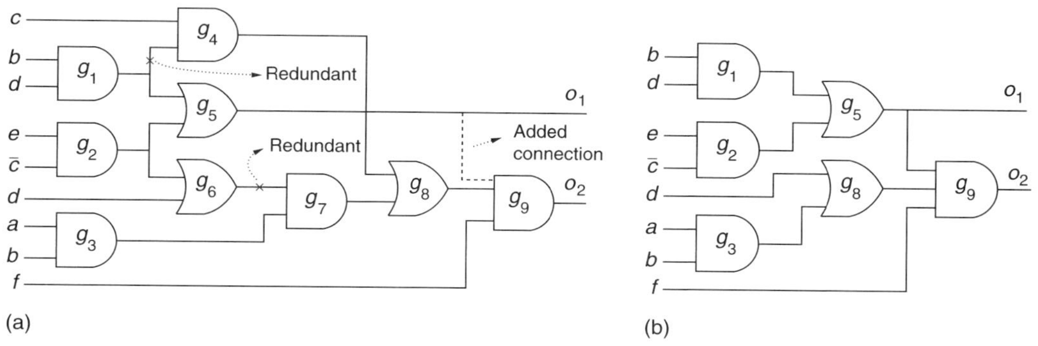

FIGURE 22.12 An example of logic restructuring by adding and removing redundancy: (a) adding a redundant connection; (b) optimized circuit after redundancy removal.

22.5.2.2 Design Verification

Techniques used to verify designs include checking logic equivalence and determining that a circuit does or does not violate certain properties.

Logic Equivalence Checking. It is important to check the equivalence of two designs described at the same or different levels of abstraction. Checking the functional equivalence of the optimized implementation against the RTL specification, for example, guarantees that no error is introduced during logic synthesis and optimization, especially if part of the process is manual. Checking the equivalence of the gate-level implementation and the gate-level model extracted from the layout assures that no error is made during physical design.

Traditionally, designers check the functional equivalence of two Boolean functions by constructing their canonical representations, as truth tables or BDDs, for example. Two circuits are equivalent if and only if their canonical representations are isomorphic.

Consider the comparison of two Boolean networks in Figure 22.13. A joint network can be formed by connecting the corresponding primary input pairs of the two networks and by connecting the corresponding primary output pairs to XOR gates. The outputs of these XOR gates become the new primary outputs of the joint network. The two networks are functionally equivalent if the primary output response of the joint network is 0 for any input vector. Therefore, to prove that two circuits are equivalent, designers must merely prove that no input vector produces 1 at this model's output signal *g*.

Another way to do equivalence checking is to formulate it as a problem that searches for a distinguishing vector, for which the two circuits under verification produce different output responses. If no distinguishing vector can be found after the entire space is searched, the two circuits are equivalent. Otherwise, a counterexample is generated to disprove equivalence. Because a distinguishing vector is also a test vector for the stuck-at-0 fault on the joint network's output *g*, equivalence checking becomes a test-generation process for *g's* stuck-at-0 fault. However, directly applying ATPG to check the output equivalence (finding a test for stuck-at-0 fault) could be CPU-intensive for large designs.

Figure 22.14 shows how complexity can be reduced substantially by finding an internal functional similarity between the two circuits being compared [94]. Designers first use naming information or structure analysis to identify a set of potentially equivalent internal signal pairs. They then build a model, as in Figure 22.14(a), where signals *a1* and *a2* are candidate internal equivalent signals. To check the equivalence between these signals, we run ATPG for a stuck-at-0 fault at signal line f. If ATPG concludes that no test exists for that fault, the joint network can be simplified to the one in Figure 22.14(b), where signal *a1* has been replaced with signal *a2*

With the simplified model, the complexity of ATPG for the output *g* stuck-at-0 fault will be reduced. The process identifies internal equivalent pairs sequentially from primary inputs to primary outputs. By the time it gets to the output of the joint network, the joint network could be substantially smaller, and

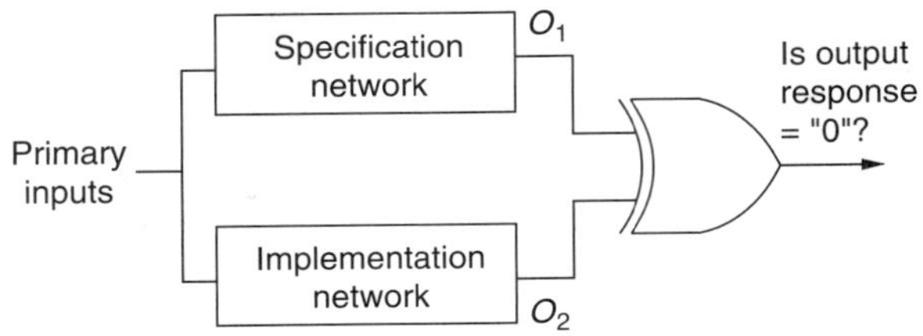

FIGURE 22.13 Circuit model for equivalence checking of two networks.

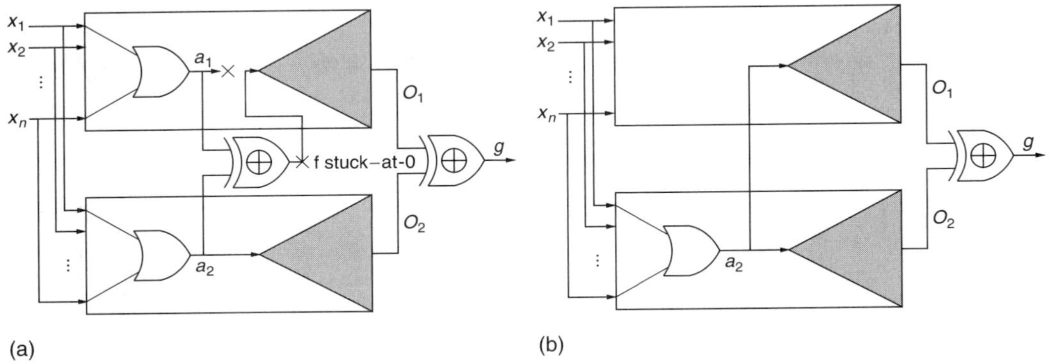

FIGURE 22.14 Pruning a joint network by finding internal equivalent pair. (a) Model for checking if signals *a1* and *a2* are equivalent; (b) Reducing complexity by replacing *a1* with *a2*.

ATPG for the *g* stuck-at-0 fault will be quite trivial. Various heuristics for enhancing this idea and combining it with BDD techniques have been developed in the past few years [95]. Commercial equivalence checking tools can now handle circuit modules of more than a million gates within tens of CPU minutes.

Property Checking An ATPG engine can find an example for proving that the circuit violates certain properties or, after exhausting the search space, can prove that no such example exists and thus that the circuit meets certain properties [96,97]. One example of this is checking for tristate bus contention, which occurs when multiple tristate bus drivers are enabled and their data is not consistent. Figure 22.15 shows a sample application. If the ATPG engine finds a test for the output stuck-at-0 fault, the test found will be the vector that causes bus contention. If no test exists, the bus can never have contention. Similarly, ATPG can check to see if a bus is floating — all tristate bus drivers are disabled — simply by checking for a vector that sets all enable lines to an inactive state.

ATPG can also identify races, which occur when data travels through two levels of latches in one clock cycle. Finally, an ATPG engine can check for effects (memory effect or an oscillation) from asynchronous feedback loops that might be in a pure synchronous circuit [96]. For each asynchronous loop starting and ending at a signal S, the ATPG engine simply checks to see whether there is a test to sensitize this loop. If such a test exists, the loop will cause either a memory effect (the parity from S to S is even) or an oscillation (the parity is odd).

Timing Verification and Analysis Test vectors that sensitize selected long paths are often used in simulations to verify circuit timing. In determining the circuit's clock period, designers look for the slowest true path. Various sensitization criteria have been developed for determining a true path. Such criteria set requirements (in terms of logic values and the arrival times of the logic values) at side inputs of the gates along the path. These requirements are somewhat similar to those for deriving tests for path delay faults. Thus, an ATPG engine can be used directly for this application.

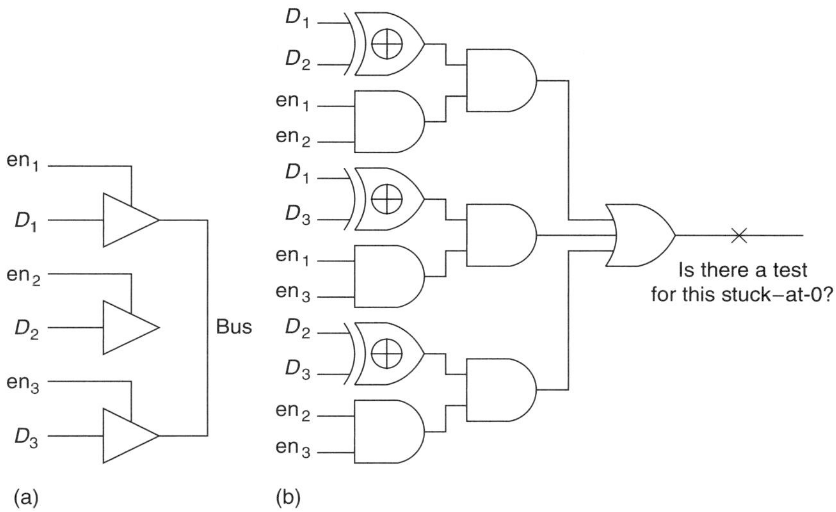

FIGURE 22.15 Application to bus contention checking. (a) A bus example; (b) An ATPG model for checking bus contention.

22.5.3 Summary

ATPG remains an active research area in both the computer-aided design and the test communities, but the new emphasis is on moving ATPG operations toward higher levels of abstraction and on targeting new types of faults in deep submicron devices. Test tools have evolved beyond merely gate-level test generation and fault simulation. Most design work now takes place at RTL and above, and test tools must support RTL handoff. New noise faults, including those from power supply noise, and crosstalk-induced noise, as well as substrate and thermal noise, will need models for manufacturing testing. The behaviors of these noise faults need to be modeled at levels of abstraction higher than the electrical, circuit, and transistor levels. Finding test vectors that can cover these faults is a challenge for ATPG.

22.6 High-Level ATPG

Test generation could be significantly speeded up if a circuit model at a higher level of abstraction is used. In this section, we discuss briefly the principles of approaches using RTL models and STGs. Here, we do not intend to give a detailed survey for such approaches, only a brief description of representative methods.

Approaches using RTL models have the potential to handle larger circuits because the number of primitives in an RTL description is much smaller than the gate count. Some methods in this class of approaches use only RTL description of the circuit [98–103], while others assume that both gate-level and RTL models are available [30–32]. Note that automatic extraction of the RTL description from a lower level of description is still not possible and therefore the RTL descriptions must be given by the designers. It is also generally assumed that data and control can be separated in the RTL model.

For approaches using both RTL and gate-level models [30–32], typically a combinational test is first generated using the gate-level model. The fault-free justification sequence and the fault propagation sequence are generated using the (fault-free) RTL description. Justification and fault propagation sequences generated in such a manner may not be valid and therefore need to be verified by a fault simulator. These approaches, in general, are suitable for data-dominated circuits but are not appropriate for control-dominated circuits.

For approaches using only RTL models [98–103], functional fault models at RTL are targeted, instead of the single-stuck-at fault model at the gate-level. The approaches in [98,99] target microprocessors and functional fault models are defined for various functions at the control-sequencing level. Because tests are generated for the functional fault models, a high coverage for gate-level stuck-at faults cannot be guaranteed. The methods suggested in [100,101] focus on minimizing the value conflicts during the value justification and fault propagation processes, using the high-level information. The technique in [103] intends to guarantee that the functional tests for their proposed functional faults achieve a complete gate-level stuck-at fault coverage. To do so, mappings from gate-level faults to functional operations of modules need to be established. This approach also uses an efficient method for resolving the value conflicts during propagation/justification at the RTL . A method of characterizing a design's functional information using a model extended from the traditional finite-state machine model, with the capability of modeling both the data-path operations and the control state transitions, is suggested in Ref.[102]. However, this method does not target any fault model and only generates functional vectors for design verification.

For finite-state machines (FSMs) for which the STGS are available, test sequences can be derived using the state transition information. In general, this class of approaches can handle only relatively small circuits due to the known state-explosion problem in representing a sequential circuit using its state table. However, successful applications of such approaches to protocol performance testing [104] and to testing the boundary-scan Test Access Port (TAP) controller [105] have been reported. The earliest method is the checking experiment [26] that is based on distinguishing sequences. The distinguishing sequence is defined as an input sequence that produces different output responses for each initial state of the FSM. This approach is concerned with the problem of determining whether or not a given state machine is distinguishable from all other possible machines with the same number of states or fewer. No explicit fault model is used. The distinguishing sequence may not exist, and the bound on length, if it exists, is proportional to the factorial of the number of states. This method is impractical because of the long test sequence. Improved checking experiments, based on either the simple I/O sequence [27] or the unique input/output (UIO) sequence [104] of the FSM, significantly reduce the test length.

A functional fault model in the state transition level has recently been used in a test generator FTG for FSMs [29]. In the single-state-transition (SST) fault model, a fault causes the destination state of a single state transition to be faulty. It has been shown [29] that the test sequence generated for the SST faults in the given STG achieves high fault coverages for the single stuck-at faults as well as the transistor faults in its multi-level logic implementation. As an approximation, FTG uses the fault-free STG to generate the fault-propagation sequence. AccuraTest [28] further improves the technique by using both fault-free and faulty circuits' STGs to generate accurate test sequences for the SST faults as well as for some multiple-state-transition faults.

More recent works in high-level ATPG can be found in [106–109]. The work in [106] utilizes *program slicing*, which was originally proposed as a static program analysis technique, for hierarchical test generation. Program slicing extracts environmental constraints for a given module, after which the module with the constraints can be synthesized into a gate-level model for test generation using a commercial sequential ATPG tool. Since program slicing extracts only relevant constraints for a module and ignores the rest of the design, it can significantly reduce the complexity of ATPG for each individual module. Huang and Cheng [107] propose a *word-level* ATPG combined with an arithmetic constraint solver in an ATPG application for property checking. The word-level ATPG involves a *word-level* implication engine. The arithmetic constraint solver is used as a powerful implication engine on the arithmetic datapath. Zhang et al. [109] propose a sequential ATPG guided by RTL information represented as an *assignment decision diagram* (ADD). STGs extracted from ADDs are used to guide the ATPG process. Iyer [108] developed a constraint solver for application in functional verification based on *random test program generation* (RTPG) methodology. The constraint solver utilizes word-level ATPG techniques to derive functional test sequences under user-supplied functional constraints.

References

[1] Abramovici, M., Breuer, M.A., and Friedman, A.D., *Digital Systems Testing and Testable Design*, Rev. Print Edition, Wiley-IEEE Press, 1994.

[2] Maxwell, P.C., Aitken, R.C., Johansen, V., and Chiang, I., The effect of different test sets on quality level prediction: When is 80% better than 90%? *Proceedings of the International Test Conference*, Washington, DC, 1991, pp. 26–30, 358–364.

[3] Maxwell, P.C. and Aitken, R.C., Test sets and reject rates: All fault coverages are not created equal, *IEEE Des. Test Comput.*, 10, 42–51, 1993.

[4] Wang, L.-C., Mercer, M.R., Kao, S.W., and Williams, T.W., On the decline of testing efficiency as fault coverage approaches 100%, *13th IEEE VLSI Test Symposium*, Princeton, NJ, 30 April–3 May 1995, pp. 74–83.

[5] Ma, S.C., Franco, P., and McCluskey, E.J., An experimental chip to evaluate test techniques experiment results, *Proceeding of the International Test Conference*, Washington, DC, 21–25 October 1995, pp. 663–672.

[6] Grimaila, M.R., Sooryong Lee., Dworak, J., Butler, K. M., Stewart, B., Balachandran, H., Houchins, B., Mathur, V., Jaehong Park, Wang, L.-C., and Mercer, M. R., REDO-random excitation and deterministic observation – first commercial experiment, *17th IEEE VLSI Test Symposium*, Dana Point, California, 25–29 April 1999, pp. 268–274.

[7] Dworak, J., Wicker, J.D., Lee, S., Grimaila, M.R., Mercer, M.R., Butler, K.M., Stewart, B., and Wang, L.-C., Defect-oriented testing and defective-part-level prediction, *IEEE Des. Test Comput.*, 18, 31–41, 2001.

[8] Fujiwara, H. and Toida, S., The complexity of fault detection problems for combinational logic circuits, *IEEE Trans. Comput.*, 31, 555–560, 1982.

[9] Roth, J.P., Diagnosis of automata failure: a calculus & a method, *IBM J. Res. Develop.*, 10, 278–291, 1966.

[10] Schulz, M.H., Trischler, E., and Sarfert, T.M., SOCRATES: a highly efficient automatic test pattern generation system, *IEEE Trans. Comput.-Aided Des. ICs*,. 7, 126–137, 1988.

[11] W. Kunz, and Pradhan, D.K., Recursive learning: an attractive alternative to the decision tree for test generation in digital circuits, *International Test Conference*, Washington, DC, 1992, pp. 816–825.

[12] Wang, C., Reddy, S.M., Pomeranz, I., Lin, X., and Rajski, J., Conflict driven techniques for improving deterministic test pattern generation, *International Conference on Computer-Aided Design*, San Jose, CA, 2002, pp. 87–93.

[13] Marques-Silva, J.P., and Sakallah, K.A., GRASP: a search algorithm for propositional satisfiability, *IEEE Trans. Comput.*, 48, 506–521, 1999.

[14] Henftling, M., Wittmann, H.C., and Antreich, K.J., A Single-Path-Oriented Fault-Effect Propagation in Digital Circuits Considering Multiple-Path Sensitization, *International Conference on Computer-Aided Design*, 1995, pp. 304–309.

[15] Goel, P., An implicit enumeration algorithm to generate tests for combinational logic circuits, *IEEE Trans. Comput.*, C–30, 215–222, 1981.

[16] Fujiwara, H., and Shimono, T., On the acceleration of test generation algorithms. *IEEE Trans. Comput.*, C-32, 1137–1144, 1983.

[17] Hamzaoglu, I., and Patel, J.H., New techniques for deterministic test pattern generation, *IEEE VLSI Test Symposium*, Monterey, CA, 1998, pp. 446–451.

[18] Kirkland, T., and Mercer, M.R., A topological search algorithm for ATPG, *24th ACM/IEEE Design Automation Conference*, Miami Beach, FL, 1987, pp. 502–508.

[19] Ivanov, A., and Agarwal, V.K., Dynamic testability measures for ATPG, *IEEE Trans. Comput.-Aided Des. ICs*, 7, 1988, pp. 598–608.

[20] Larrabee, T., Test pattern generation using Boolean Satisfiability. *IEEE Trans. Comput.-Aided Des. ICs*, 11, 4–15, 1992.

[21] Stephan, P., Brayton, R.K., and Sagiovanni-Vincentelli. A.L., Combinational test generation using satisfiability, *IEEE Trans. Comput.-Aided Des. ICs*, 15, 1167–1176, 1996.

[22] Marques-Silva, J.P., and Sakallah, K.A., Robust search algorithms for test pattern generation, *IEEE Fault-Tolerance Computing Symposium*, Seattle, WA, 1997, pp. 152–161.

[23] Gizdarski, E., and Fujiwara, H., SPIRIT: A highly robust combinational test generation algorithm. *IEEE VLSI Test Symposium*, Marina Del, Rey, CA, 2001, pp. 346–351.

[24] Seshu, S., and Freeman, D.N., The diagnosis of asynchronous sequential switching systems, *IRE Trans. Electron. Computing*, EC-11, 459–465, 1962.

[25] Thomas, J.J., Automated diagnostic test program for digital networks, *Comput. Des.*, 63–67, 1971.

[26] Hennie, F.C., Fault-detecting experiments for sequential circuits, *5th Annual Symposium on Switching Circuit Theory and Logical Design*, 1964, pp. 95–110.

[27] Hsieh, E.P., Checking experiments for sequential machines, *IEEE Trans. Comput.*, C-20, 1152–1166, 1971.

[28] Pomeranz, I., and Reddy, S.M., On achieving a complete fault coverage for sequential machines using the transition fault model, *28th Design Automation Conference*, San Francisco, CA, 1991, pp. 341–346.
[29] Cheng, K.-T., and Jou, J.-Y., A functional fault model for sequential circuits, *IEEE Trans. Comput.-Aided Des. ICs*, 11, 1992, pp. 1065–1073.
[30] Hill, F.J., and Huey, B., SCIRTSS: A search system for sequential circuit test sequences, *Transactions on Computers*, C-26, 490–502, 1977.
[31] Breuer, M.A., and Friedman, A.D., Functional level primitives in test generation, *IEEE Trans. Comput.*, C-29, 223–235, 1980.
[32] Ghosh, A., Devadas, S., and Newton, A.R., Sequential test generation at the register-transfer and logic levels, *27th ACM/IEEE Design Automation Conference*, Orlando, FL, June 1990, pp. 580–586.
[33] Kubo, H., A procedure for generating test sequences to detect sequential circuit failures, NEC *Res. Dev.*, 12, 69–78, 1968.
[34] Putzolu, G.R, and Roth, J.P., A heuristic algorithm for the testing of asynchronous circuits, *IEEE Trans. Comput.* C-20, 639–647, 1971.
[35] Marlett, R., EBT, A comprehensive test generation technique for highly sequential circuits, *15th Design Automation Conference*, Las Vegas, NV, 1978, pp. 332–339.
[36] Cheng, W.T., The BACK algorithm for sequential test generation, *International Conference on Computer Design*, Rye Brook, NY, 1988, pp. 66–69.
[37] Niermann, T., and Patel, J.H., HITEC: A test generation package for sequential circuits, *European Design Automation Conference*, Amsterdam, The Netherlands, 1991, pp. 214–218.
[38] Kelsey, T.P., Saluja, K.K., and Lee, S.Y., An efficient algorithm for sequential circuit test generation, *IEEE Trans. Comput.* 42–11, 1361–1371, 1993.
[39] Schulz, M.H., and Auth, E., Advanced automatic test pattern generation and redundancy identification techniques, *18th International Symposium on Fault-Tolerant Computing*, Tokyo, Japan, 1988, pp. 30–35.
[40] Cheng, K.-T., and Agrawal, V.D., *Unified methods for VLSI simulation and test generation*, Kluwer, Dordrecht, 1989.
[41] Saab, D.G., Saab, Y.G., and Abraham, J.A., CRIS: A test cultivation program for sequential VLSI circuits, *International Conference on Computer-Aided Design*, San Jose, CA, 1992, pp. 216–219.
[42] Rudnick, E.M., Patel, J.H., Greenstein, G.S., and Niermann, T.M., Sequential circuit test generation in a genetic algorithm framework, *ACM/IEEE Design Automation Conference*, San Diego, CA, 1994, pp. 698–704.
[43] Prinetto, P. Rebaudengo, M., and Sonza Reorda, M., An automatic test pattern generator for large sequential circuits based on genetic algorithm, *International Test Conference*, Baltimore, MD, 1994, pp. 240–249.
[44] Saab, D.G., Saab, Y.G., and Abraham, J.A., Iterative [simulation-based genetics + deterministic techniques] = complete ATPG, *International Conference on Computer-Aided Design*, San Jose, CA, 1994, pp. 40–43.
[45] Rudnick, E.M., and Patel, J.H., Combining deterministic and genetic approaches for sequential circuit test generation, *32nd Design Automation Conference*, San Francisco, CA, 1995, pp. 183–188.
[46] Hsiao, M.S., Rudnick, E.M., and Patel, J.H., Alternating strategy for sequential circuit ATPG, *European Design and Test Conference*, Paris, France, 1996, pp. 368–374.
[47] Ma, H-K. T., Devadas, S., Newton, A.R., and Sangiovanni-Vincentelli, A., Test generation for sequential circuit, *IEEE Trans. Comput.-Aided Design*, 7, 1081–1093, 1988.
[48] Ghosh, A., Devadas, S., and Newton, A.R., Test generation and verification for highly sequential circuits, *IEEE Trans. Comput.-Aided Design*, 10, 652–667, 1991.
[49] Cho, H., Hachtel, G. D., Somenzi, F., Redundancy identification/removal and test generation for sequential circuits using implicit state enumeration, *IEEE Trans. CAD*, 12, 935–945, 1993.
[50] Muth, P., A Nine-valued circuit model for test generation, *IEEE Trans. Comput.* C-25, 630–636, 1976.
[51] Goldstein, L.H., Controllability/observability analysis for digital circuits, *IEEE Trans. Circuits and Syst.* CAS-26, 685–693, 1979.
[52] Agrawal, V.D., and Chakradhar, S.T., Combinational ATPG theorems for identifying untestable faults in sequential circuits, *IEEE Trans. Comput.-Aided Des.* 14, 1155–1160, 1995.

[53] Cheng, K.-T., Redundancy removal for sequential circuits without reset states, *IEEE Trans.-Aided Des.*, 12, 13–24, 1993.
[54] Pomeranz, I., and Reddy, S.M., Classification of faults in sequential circuits, *IEEE Trans. Comput.* 42, 1066–1077, 1993.
[55a] Devadas, S., Ma, H.-K. T., and Newton, A.R., Redundancies and don't-cares in sequential logic synthesis, *J. Electron. Test. (JETTA)*, 1, 15–30, 1990.
[55b] Bryant, R.E., Graph-based algorithms for boolean function manipulation, *IEEE Trans. Comput.*, C-35, 677–691, 1986.
[56] Cheng, K.-T., Gate-level test generation for sequential circuits, *ACM Trans Des. Autom. Electron. Syst.*, 1, 405–442, 1996.
[57] Breuer, M.A., Test generation models for busses and tri-state drivers, *IEEE ATPG Workshop*, March 1983, pp. 53–58.
[58] Ogihara, T., Murai, S., Takamatsu, Y., Kinoshita, K., and Fujiwara, H., Test generation for scan design circuits with tri-state modules and bidirectional terminals, *Design Automation Conference*, Miami Beach, FL, June 1983, pp. 71–78.
[59] Chakradhar, S.T., Rothweiler, S., and Agrawal, V D., Redundancy removal and test generation for circuits with non-boolean primitives, *IEEE VLSI Test Symposium*, Princeton, NJ, 1995, pp. 12–19.
[60] Miczo, A., Digital logic testing and simulation, Harper & Row, New York, 1986.
[61] Zhang, H., SATO: an efficient propositional prover, *Proc. Int. Conf. Automated Deduction*, 1249, 272–275, 1997.
[62] Moskewicz, M., Madigan, C., Zhao, Y., Zhang, L., and Malik, S., Chaff: engineering an efficient SAT solver, *Proceedings of, Design Automation Conference*, Las Vegas, NV, 2001, pp. 530–535.
[63] Goldberg, E., and Novikov, Y., BerkMin: a fast and robust Sat-Solver, *Proceeding of, Design, Automation and Test in Europe*, Paris, France, 2002, pp. 142–149.
[64] Ryan, L., the siege satisfiability solver. http://www.cs.sfu.ca/ loryan/personal/.
[65] Davis, M., Longeman, G., and Loveland, D., A machine program for theorem proving, *Commn. ACM*, 5, 394–397, 1962.
[66] McAllester, D.A., An outlook on truth maintenance, AIMemo 551, MIT AI Laboratory, 1980.
[67] Tseitin, G.S., On the complexity of derivation in propositional calculus, in *studies in Constructive Mathematics and Mathematical Logic*, Part 2, 1968, pp. 115–125. Reprinted in Siekmann, J., and Wrightson, G., Eds., *Automation of Reasoning*, Vol. 2, Springer-Heiduberg., 1983, pp. 466–483.
[68] Tafertshofer, P., Ganz, A., and Henftling, M., A SAT-based implication engine for efficient ATPG, equivalence checking, and optimization of netlists, *Proceeding of, International Conference on Computer-Aided Design*, San Jose, CA, 1997, pp. 648–657.
[69] Silva, L., Silveira, L., and Marques-Silva, J.P., Algorithms for solving boolean satisfiability in combinational circuits, *Proceedinf of, Design, Automation and Test in Europe*, Munich, Germany, 1999, pp. 526–530.
[70] Silva, L., and Silva, J.M., Solving satisfiability in combinational circuits, *IEEE Design and Test of Computers*, 2003, pp. 16–21.
[71] Gupta, A., Yang, Z., and Ashar, P., Dynamic detection and removal of inactive clauses in SAT with application in image computation, *Proceeding of the, ACM/IEEE Design Automation Conference*, Las Vegas, NV, 2001, pp. 536–541.
[72] Kuehlmann, A., Ganai, M., and Paruthi, V., Circuit-based boolean reasoning, *Proceeding of, the ACM/IEEE Design Automation Conference*, Las Vegas, NV, 2001, pp. 232–237.
[73] Kuehlmann, A., and Krohm, F., Equivalence checking using cuts and heaps, *Proceedings Design Automation Conference*, Anahein, California, *1997*, pp. 263–268.
[74] Ganai, M.K., Zhang, L., Ashar, P., Gupta, A., and Malik, S., Combining strengths of circuit-based and CNF-based algorithms for a high-performance SAT solver, *Proceeding of the, ACM/IEEE Design Automation Conference*, New Orleans, Louisiana, 2002, pp. 747–750.
[75] Ostrowski, R., Grgoire, E., Mazure, B., and Sas, L., Recovering and exploiting structural knowledge from CNF formulas, *Principles and Practice of Constraint Programming (CP '02)*, Van Henten-ryck, P., Ed., LNCS 2470, Springer-Heidelberg, 2002, pp. 185–199.
[76] Broering, E., and Lokam, S.V., Width-based algorithms for SAT and CIRCUIT-SAT, *Proceeding of Theory and Applications of Satisfiability Testing*, Santa Margherita Ligure, Italy, 2003, pp. 162–171.
[77] Lu, F., Wang, L.C., Cheng, K.T., and Huang, R., A circuit SAT solver with signal correlation guided learning, *Proceeding of the Design, Automation and Test in Europe*, Munich, Germany, 2003, pp. 92–97.

[78] Lu, F., Wang, L.C., Cheng, K.-T., Moondanos, J., and Hanna. Z., A signal correlation guided ATPG solver and its applications for solving difficult industrial cases. *Proceeding of the IEEE/ACM Design Automation Conference*, Anaheim, CA, 2003, pp. 436–441.
[79] Lu, F., Iyer, M.K., Parthasarathy, G., Wang, L.-C., Cheng, K.-T., and Chen, K.C., An efficient sequential SAT solver with improved search strategies, *Proceeding of the European Design Automation and Test Conference*, Munich, Germany, 2005.
[80] Tafertshofer, P., Ganz, A., and Henftling, M., A SAT-based implication engine for efficient ATPG, equivalence checking, and optimization of netlists. *Proceeding of the IEEE/ACM International Conference on Computer-Aided Design*, San Jose, CA, 1997, pp. 648–655.
[81] Zhang, L., Madigan, C., Moskewicz, M., and Malik, S., Efficient conflict driven learning in a boolean satisfiability solver, *Proceeding of the International Conference on Computer-Aided Design*, San Jose, CA, 2001, pp. 279–285.
[82] Cheng, K.-T., Dey, S., Rodgers, M., and Roy, K., Test challenges for deep sub-micron technologies, *Design Automation Conference*, Los Angeles, CA, 2000, pp. 142–149.
[83] Special issue on speed test and speed binning for complex ICs, *IEEE D&T*, 20, 2003.
[84] Krstic, A., and Cheng, K.-T., *Delay Fault Testing for VLSI Circuits*, Kluwer, Boston, 1998.
[85] Jiang, Y.-M., Cheng, K.-T., and Deng, A.-C., Estimation of maximum power supply noise for deep submicron designs, *International Symposium on Low Power Electronics and Design*, ACM Order Dept., New York, NY, Monterey, CA, 1998, pp. 233–238.
[86] Chang, Y.-S., Gupta, S. K., and Breuer. M.A., Test generation for maximizing ground bounce considering circuit delay, *IEEE VLSI Test Symposium*, Napa Valley, CA, 2003, pp. 151–157.
[87] Lee, K.T., Nordquist, C., and Abraham, J., Automatic test pattern generation for crosstalk glitches in digital circuits, *IEEE VSLI Test Symposium*, Monterey, CA, 1998, pp. 34–39.
[88] Chen, L.H., and Marek-Sadowska, M., Aggressor alignment for worst-case coupling noise, *International Symposium Physical Design*, San Diego, CA, 2000, pp. 48–54.
[89] Chen, W.Y., Gupta, S.K., and Breuer, M.A., Test generation for crosstalk-induced delay in integrated circuits, *International Test Conference*, Atlantic City, NJ, 1999, pp. 191–200.
[90] Krstic, A., Liou, J.-J., Jiang, Y.-M., and Cheng, K.-T., Delay testing considering crosstalk-induced effects, *International Test Conference*, Baltimore, MD, 2001, pp. 558–567.
[91] Chen, L.-C., Mak, T.M., Breuer, M.A., and Gupta, S.A., Crosstalk test generation on pseudo industrial circuits: a case study, *International Test Conference*, Baltimore, MD, 2001, pp. 548–557.
[92] Chen, W., Gupta, S.K., and Breuer, M.A., Analytic models for crosstalk delay and pulse analysis under non-ideal inputs, *International Test Conference*, Washington, DC, 1997, pp. 809–818.
[93] Entrena, L.A., and Cheng, K.-T., Combinational and sequential logic optimization by redundancy addition and removal, *IEEE Trans. Comput.-Aided Des.* 14, 909–916, 1995.
[94] Brand, D., Verification of large synthesized designs, *International Conference on Computer-Aided Design*, 1993, pp. 534–537.
[95] Huang, S.-Y., and Cheng, K.-T., *Formal Equivalence Checking and Design Debugging*, Kluwer, Boston, 1998.
[96] Keller, B., McCauley, K., Swenton, J., and Youngs, J., ATPG in practical and non-traditional applications, *IEEE International Test Conference*, Washington, DC, 1998, pp. 632–640.
[97] Wohl, P., and Waicukauski, J., Using ATPG for clock rules checking in complex scan designs, *IEEE VLSI Test Symposium*, Monterey, CA, 1997, pp. 130–136.
[98] Thatte, S.M., and Abraham, J.A., Test generation for microprocessors, *IEEE Trans. Comput.* C-29, 429–441, 1980.
[99] Brahme, D., and Abraham, J.A., Functional testing of microprocessors, *IEEE Trans. Comput.* C-33, pp. 475–485, 1984.
[100] Lee, J., and Patel, J.H., A signal-driven discrete relaxation technique for architectural level test generation, *International Conference on Computer-Aided Design*, 1991, pp. 458–461.
[101] Lee, V., and Patel, J.H., Architectural level test generation for microprocessors, *IEEE Trans. Compu.-Aided Des.ICs*, 1310, 1288–1300, 1994.
[102] Cheng, K.-T., and Krishnakumar, A.S., Automatic generation of functional vectors using the extended finite state machine model, *ACM Trans. Des. Autom. Electron. Syst.*, 1, 57–79, 1996.
[103] Hansen, M.C., and Hayes, J.P., High-level test generation using symbolic scheduling, *International Test Conference*, Washington, DC, 1995, pp. 586–595.

[104] Sabnani, K., and Dahbura, A., A protocol test generation procedure, *Comput. Networks*, 15, 285–297, 1988.
[105] Dahbura, A.T., Uyar, M.U., and Yau, C.sW., An optimal test sequence for the JTAG/IEEE P1 149.1 test access port controller, *International Test Conference*, Washington, DC, 1989, pp. 55–62.
[106] Vedula, V. M., Abraham, A., and Bhadra., Program slicing for hierarchical test generation, *IEEE VLSI Test Symposium*, Monterey, CA, 2002, pp. 237-243.
[107] Huang, C., and Cheng, K.-T., Using word-level ATPG and modular arithmetic constraint-solving techniques for assertion property checking, *IEEE Trans. CAD*, 20, 381–391, 2001.
[108] Iyer, M.A., RACE: a word-level ATPG-based constraints solver system for smart random simulation, *International Test Conference*, Charlotte, NC, 2003, pp. 299–308.
[109] Zhang, L., Ghosh, I., and Hsiao, M., Efficient sequential ATPG for functional RTL Circuits, *International Test Conference*, Charlotte, NC, 2003, pp. 290–298.

23

Analog and Mixed Signal Test

Bozena Kaminska
Simon Fraser University and Pultronics Incorporated
Burnaby, British Columbia
Canada

23.1 Introduction

With the ever-increasing levels of integration of system-on-chip (SoC) designs, more and more of which include analog and mixed-signal (A/M-S) elements, test equipment, test development and test execution times, and costs are being increasingly impacted. The convergence of computer, telecommunication, and data communications technologies is driving the trend toward integrating more A/M-S circuit elements into larger deep submicron chips. The forces of "smaller, cheaper, faster" will only increase as we move into the future. But increased integrated functionality comes at a price when test is included in the equation.

This is especially true when the whole chip-synthesis model becomes untenable, as is the case when processor and other cores are designed by, or acquired from multiple sources, either in-house or from a third party. Adding the A/M-S elements, for which no truly effective synthesis and automatic test pattern generation (ATPG) techniques exist, aggravates the test issues considerably. Capital equipment cost increases to millions of dollars. Test development time can exceed functional circuit development time. It is thus necessary for design and test engineers to work together, early in the SoC architecture design phase, in order to keep the testing costs under control. Trade-offs between traditional automatic test equipment (ATE) and new integrated techniques need to be considered, and the test resources partitioned between internal and external test methods.

There is a trend [1–3] toward increasing use of design-for-test (DfT) methodologies that focus on testing the structure of a design rather than its macrofunctionality. Design for test with the purpose of testing the structure of the device is called structural DfT. This trend is being driven by several factors. The traditional driving forces for the use of DfT have been, and remain, observability into the design for diagnostics and system check-out, and achieving acceptable fault coverage levels, typically between 95 and 98%, in a predictable timeframe. As the designs have become larger with the advent of the SoC methodologies, DfT is being used to provide test portability for reusable intellectual-property (IP) blocks or cores. Additionally, DfT tools have advanced to permit a more comprehensive device test that, in some cases, has been proven to eliminate the need for a traditional functional test. As a consequence of these advances, DfT methods are seen as an enabling technology to break the cost trend. The reduced cost trend will cause a shift in the market acceptance of these nontraditional methods over the traditional ones. This rationale stems from the fact that much of the "tester" is now on-chip or in the form of partitioned test.

This also becomes a motivation for the focus and the detailed overview of the available test techniques for A/M-S circuits and, in particular, when they are part of an SoC. The fault modeling and test specification is reviewed, followed by practical DfT with emphasis on some specific mixed-signal blocks.

23.2 Analog Circuits and Analog Specifications

The following terms are used in discussing A/M-S testing:

- A *circuit* is a system containing a set of components (elements) connected together.
- *Parameters* are circuit characteristics obtained by measuring the output signals. A mathematical function in terms of some components describes each parameter. For example, the mathematical expression for the cut-off frequency is $f_c = 1/(2RC)$.
- A *nominal value* is the value of the component or the parameter of a good circuit.
- A *relative deviation* indicates the deviation of the element or the parameter from its nominal value, divided by its nominal value.

In designing electronic circuits, the designer should know the circuit's performance deviations due to changes in the value of its elements. The deviation of the elements' values from their nominal values depends on the manufacturing process and on the temperature of the element [4].

Various measurement techniques prove the diverse output parameters that characterize analog circuits and analog specifications. Four categories of commonly used measurements can be distinguished:

1. *Harmonic* techniques measure the frequency response of the circuit under test. From this measurement, we can extract diverse parameters — for example, gain at a known frequency, cut-off frequency, Q factor, and phase as a function of frequency. The input stimulus used in harmonic measurement is usually a sinusoidal waveform with a variable frequency.
2. *Time-domain* measurements use pulse signals, including a square wave, step, and pulse train, as the input stimuli of a circuit. We can then observe the circuit's transient response at the output. We can derive many parameters from this measurement and use them to predict the defective components in the analog circuit. Some of these parameters are rise, delay, and fall times.
3. *Static* measurements attempt to determine the parameters of the stable states of the circuit under test. This measurement includes the determination of the DC operating point, leakage currents, output resistance, transfer characteristics, and offset.
4. *Noise* measurements determine the variations in signal that appear at the circuit's output when the input is set to zero.

Tests involving sinusoidal signals for excitation are the most common among linear circuits, such as amplifiers, data converters, and filters. Among all waveforms, the sinusoid is unique in that its shape is not altered by its transmission through a linear circuit; only its magnitude and phase are changed. In contrast, a nonlinear circuit will alter the shape of a sinusoidal input. The more nonlinear the circuit is, the greater the change in the shape of the sinusoid. One means of quantifying the extent of the nonlinearity present in a circuit is by observing the power distributed in the frequency components contained in the output signal using a Fourier analysis. For example, a circuit can be excited by a sinusoid signal. At the output of a band pass filter that is connected to the output of a circuit under test, a power spectral density plot can be observed. In general, the fundamental component of the output signal is clearly visible, followed by several harmonics. The noise floor is also visible. By comparing the power contained in the harmonics to that in the fundamental signal, a measure of total harmonic distortion (TDH) is obtained. By comparing the fundamental power to the noise power over a specified bandwidth, one obtains the signal-to-noise ratio (SNR). By altering the frequency and amplitude of the input sinusoid signal, or by adding an additional tone with the input signal, other transmission parameters can be derived from the power spectral density plot [5].

Faults in analog circuits can be categorized as *catastrophic* (*hard*) *and parametric* (*soft*) (see Figure 23.1). Catastrophic faults are open and short circuits, caused by sudden and large variations of components. These usually induce a complete loss of correct functionality. Parametric faults are caused by an abnormal deviation of parameter values and result in altered performance [6–8]. In analog circuits, the concern is the selection of parameters to be tested, and the accuracy to which they should be tested, to detect the deviation of faulty components.

The most accepted test strategy for analog circuits depends on *functional testing* that is based on the verification of a circuit's functionality by applying stimuli signals at the input and verifying its outputs [5,9]. This type of test is convenient for complex analog circuits. Its major drawbacks, however, are (1) the difficulty of detecting and identifying the defective elements, (2) the complexity in writing and executing the test program, and (3) the access to the primary inputs and outputs of a circuit. The access problem is becoming especially critical as complexity and SoC integration increases. The test bus standards 1149.4 and 1149.1 (see below) are increasingly used to assure the access mechanism to the embedded A/M-S blocks [10,11]. This technique is known as design-for-functional-testing (DfFT).

The above-mentioned drawbacks have been a subject of intense research in recent years. Two main directions can be observed: one that deals with an extensive effort in fault modeling and test specification, and the second, which in contrast researches DfT techniques. Design for test includes built-in-self-test (BIST), and test access mechanisms such as test bus standard 1149.4.

The following observations can be made based on the current industry practices. A successful test strategy is domain-specific in digital circuits. For example, there are dedicated memory BIST solutions, random logic is well suited for full scan, and boundary scan is useful to tie the whole chip together. For mixed-signal testing, success is associated with the functional definition of a module and its related test specifications; for example, analog-to-digital converter (ADC), digital-to-analog converter (DAC), phase lack loop (PLL), filter, and transceiver. DfT and BIST are largely developed for each specific functional module and very often even for a particular design style or implementation. The trend of maintaining functional testing despite the recognized benefits of fault-based testing is much stronger for A/M-S modules. The standalone circuits as well as embedded modules (for example, in the case of SoC) are tested for the specification

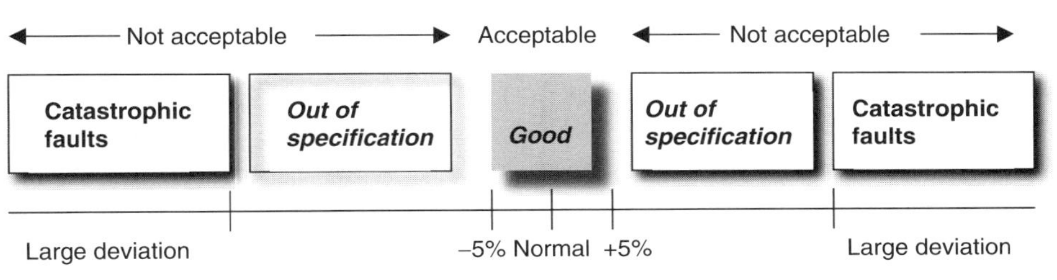

FIGURE 23.1 Fault classification: parametric and catastrophic faults.

verification and their related tolerances. A notion of DfFT becomes an SoC practice to deal with the access problem and specification verification.

In practice, functional testing can include the following tests:

1. *parametric*, which verifies the analog characteristics within a specified tolerance (for example, voltages, currents, impedances, and load conditions);
2. *dynamic*, which verifies the dynamic characteristics of the system under test — in particular, a transient analysis in the time domain;
3. *static*, which verifies the stable states of the system.

There is a common understanding that mixed signal represents the interface between the real world and digital processing. ADC or DAC are some examples. But it is not the only meaning of A/M-S. Otherwise, why are SPICE-like simulators developed for fully analog simulation used for digital-cell design? Each time a critical performance is considered, such as delay, jitter, or rise and fall times, an analog simulation takes place. The same is true with testing. Performance testing of digital cells is primarily an analog function. So, together, digital and analog design and test form the mixed-signal environment. Fault modeling and tolerance analysis relate to both digital and analog circuits. Performance testing requires analog techniques and instruments.

23.3 Testability Analysis

Testability analysis in analog circuits is an important task and a desirable approach for producing testable complex systems [12,70]. The densities of analog circuits continue to increase and the detection and isolation of faults in these circuits becomes more difficult owing to the nature of analog faults, the densities of today's analog circuits, and very often their embedded placement in SoC. Testability information is useful to designers who must know which nodes to make accessible for testing, and to test engineers who must plan test strategies. By analyzing testability, we can predict what components can be isolated by a given set of tests, and what kind of defects can be observed at a given test node. Analyzing testability in the frequency domain is an approach introduced for analog circuits to choose adequate test frequencies for increasing fault diagnosis [6]. It has also been observed that not all test points and frequencies are equally useful, and hence some selection criteria must be applied for a robust test selection.

The first published testability evaluation methods are based on calculating the rank-test algorithm and determining the solvability of a set of fault diagnosis equations describing the relation between measurements and parameters. Recently, a more practical approach [1,6,7] based on a test selection that allows maximization of the deviation between the output voltage of the good and faulty circuits, has been developed. The input stimulus that magnifies the difference between the response of the good circuit and that of the faulty one is derived. A more comprehensive study is presented in [4,13,14] and is based on the analog-fault observability concept and multifrequency testability analysis.

In [7], test nodes and test frequencies are chosen that increase the testability of a circuit in terms of searching for an effective way to observe defective components. A fault is judged to be easy to test if its effect can be observed at one of the outputs. The proposed multifrequency analysis consists of applying a sinusoidal input signal with a variable frequency and observing the amplitude of the output signals at the various test nodes. The influence of a defective component's deviation on the output is different from one interval of frequencies to another. This means that the observability of a fault depends on the frequencies.

In [6], it has been observed that for a parameter (such as gain) that reacts poorly to a component (for example, a resistance), deviation makes it impossible to observe a defect in this component. The defect *observability* has been defined as the sensitivity of the output parameter with respect to the variations of a component. As the absolute value of the sensitivity of an output parameter's variation with respect to a component's variation is high, the observability of a defect in a circuit is also high. We can see that the sensitivity gives information about component deviation observability and the observability depends also on the selected output parameters. A component may be highly observable with respect to one parameter, but may not be observable at all with respect to another parameter. A definition of sensitivity is provided in [15].

To analyze various symptoms of a fault, we can compute or measure the different observabilities in the frequency domain either from the sensitivity of the output signals' amplitudes with respect to each component of the circuit, or from the sensitivity of transfer functions relating the output test nodes to the input. Thus, it is possible to optimize the number of parameters necessary for testing and, at the same time, to achieve high fault coverage with high fault observability while reducing the test time.

To increase observability of the defective components in a circuit, a parameter that has a high sensitivity with respect to this component needs to be tested. Concepts such as fault masking, fault dominance, fault equivalence, and fault isolation for analog circuits have been defined in [6,16,17] and are based on a sensitivity concept. It can be noted that the testability analysis not only plays an important role in test set specification, but also in defining test access points for functional testing of embedded devices in SoC and for implementation of test bus standard 1149.4 for DfFT. Finally, testability analysis constitutes an important source of information about a circuit of interest for design, test, and application purposes.

23.4 Fault Modeling and Test Specification

Test generators have become an important part of functional verification of digital circuits. In the advent of the ever-increasing complexity of designs, decreasing design cycles, and cost-constrained projects resulting in an increased burden on verification engineers, design teams are becoming increasingly dependent on automatic test generators. Analog design teams do not have the same level of tool support. Automatic test specification is still an open research area.

The introduction of the *stuck-at* fault model for digital circuits enabled digital testing to cope with the exponential growth in the digital circuit size and complexity. Indeed, the stuck-at fault model enabled functional testing to be replaced by structural testing and acted as a measure to quantify the quality of the test plan, permitting test requirements and benchmarking of DfT strategies.

Hard and soft fault modeling and simulation, which address A/M-S circuits, have been the subject of many publications [2,8,16,18–23]. Millor et al. [24] reported on a test generation algorithm for detecting catastrophic faults under normal parameter variations. The later extension resulted in an approach based on a statistical process fluctuation model that was derived to select a subset of circuit specifications that detect parametric faults and minimize the test time. Test generation is formulated in [25] as a quadratic programming problem. This approach was developed for parametric faults and it determines an input stimulus $x(t)$ that maximizes the quadratic difference of response from the good and the faulty circuits, with all other parameters at their nominal values. The test generation approach for hard and parametric faults based on sensitivity analysis and tolerance computation was proposed in [26]. In this approach, the worst-case performance was expressed in terms of sensitivity and parameter tolerance; however, frequency analysis was not considered, and the model was a linearization obtained from first-order partial derivatives. Based on [26], Ben Hamida et al. [20] developed an automated sensitivity tool that allows adjunct network-based sensitivity analysis for designing fault-resistant circuits and for generating test vectors for parametric and catastrophic faults under normal parameter variations.

A method presented in [18] is founded on a fault model and sensitivity. For a given fault list, perturbation of sensitivity with respect to frequency is used to find the direction toward the best test frequency. In [17], Huynh et al. derived a multifrequency test generation technique based upon testability analysis and fault observability concepts. The test frequencies selected are those where the output performance sensitivity is a maximum with respect to the faulty component deviation. In these approaches [17,18], the masking effect due to variations of the fault-free components in their tolerance box are not considered and the test frequencies may be not optimal. A DC test generation technique for catastrophic faults was developed in [27]. It is formulated as an optimization problem and includes the effects of normal parameter variations.

23.5 Catastrophic Fault Modeling and Simulation

The majority of approaches presented for A/M-S fault simulation are based on *cause–effect* analysis and do not allow parallel fault simulation. Indeed, cause–effect analysis enumerates all the possible faults (causes) existing in a fault model and determines all their corresponding responses (effects) to a given applied test in a serial manner. The required simulation time can become impractically long, especially for large analog designs.

Fault simulation is used to construct a fault dictionary. Conceptually, a fault dictionary stores the signatures (effects) of the faults for all stimuli T. This approach requires computing the response of every possible fault before testing, which is impractical. In [22], a new type of fault dictionary that does not store the signatures (effects) of the faults f, but computes and stores the fault value R_{fault} (cause) that if added to the circuit, will drive the output parameter out of its tolerance box. This approach allows parallel fault simulation. The following steps are required for fault dictionary construction:

1. Generate the fault list; all possible shorts and opens in a circuit. Two fault-list extractors can be used: a layout-based fault list (standard inductive fault analysis) and a schematic-based fault-list extractor.
2. Compute the output sensitivities with respect to all hard faults in the fault list in parallel, for example, using the adjoint network method [28]. For example, the initial value for R_{fault} can be defined as zero value resistance for the opens and as zero value conductance for the shorts. The computed R_{fault} value (cause) is defined as the smallest resistance value that if added to the circuit, will only deviate the output parameter to the edge of the tolerance box.
3. From the fault-free circuit output tolerance $\eth_{\text{out}}$ (effect) and the fault-free output sensitivities S with respect to all hard faults in the fault list (obtained in step 2), the following equation is used to compute the fault value R_{fault} (cause) for all the defects in the fault list:

$$R_{\text{fault}} = \eth_{\text{out}}/S_{\text{out}}\,(R_{\text{fault}})$$

The adjoint network method for sensitivity computation in AC, DC, and transient domains has been described in [21–23]. Other parallel sensitivity computations are possible as well.

After creating the fault dictionary, the best stimulus (test vector) for detecting each fault needs to be found. The stimulus should be selected to maximize the fault observability. First, knowing the component tolerance, the output parameter distribution of the fault-free and faulty circuit can be estimated. Then the resistor value R_{fault} that will cause the output parameter to go out of the tolerance range is computed. Then, the dominance of faults is analyzed to minimize the number of test vectors. The details of the method, an example of implementation and practical results are in [22]. This is a structural approach and a parallel fault simulation and test generation.

23.6 Parametric Faults, Worst-Case Tolerance Analysis, and Test Generation

A robust test set should detect parametric faults under maximum masking effect due to the tolerances of the circuit parameters. Indeed, the worst masking effect of fault-free parameters may create a region around the tolerance box of a parameter where its faulty deviations cannot be detected. The size of the region usually depends on the applied test. A test set is robust if it minimizes the area of the regions associated with the parameters and if the boundaries of these regions are accurately determined. Only in this case, the quality of the test set may be considered and guaranteed. If a test set is not robust, then some faulty circuits may be classified as good. In [4,13,14], a series of optimization problems is formulated and two solutions proposed to guarantee the detectability of parametric variations under the worst-case tolerance analysis: by the nonlinear programming method SQP (sequential quadratic programming) available in MATLAB, and by constraint logic programming (CLP) using relational interval arithmetic. The multifrequency testing, which is much better suited for subtle parameter variations than DC testing, is adopted.

23.7 Design for Test — An Overview

The impetus for the DfT and BIST techniques is to facilitate the use of low-performance test instrumentation or perhaps to eliminate any need whatsoever by adding self-test capabilities and strategic control and observation points within the silicon. DfT is very often *ad hoc* and largely a matter of early engagement with the design community to specify design block architectures, signal routing and external control, and access points to ensure maximal visibility and provide bypass control to critical functional blocks.

Great strides have been made in BIST techniques [1–3,5,9] for A/M-S and RF devices, but robust, production deployment of these techniques is not yet widespread. This is due in part to the challenges of soft (parametric) fault behaviors and blocks being difficult to model or simulate accurately enough to generate alternative structural test stimuli. In addition, many of the proposed BIST techniques consume an unattractive amount of silicon die area and are intrusive to sensitive circuits and design methodology, or impede the post-silicon debug process.

In a traditional production test approach for testing of the A/M-S circuits, the functional specifications are measured using the appropriate tester resources and using the same kind of test stimuli and configuration with respect to which the specification is defined, e.g., multitone signal generator for measuring distortion, gain for codec, ramp generator for measuring integral nonlinearity (INL), differential nonlinearity (DNL) of ADCs and DACs. The measurement procedures are in agreement with the general intuition of how the module behaves and, hence, the results of the measurement are easy to interpret, in contrast to the fault-based testing.

In the BIST approaches, the external ATE functionality is designed inside the device for applying appropriate test stimuli and measuring the test response corresponding to the specification. In [29], adjustable delay generators and counters are implemented next to the feedback path of the PLL to measure the RMS jitter. Since the additional circuitry does not modify the operation of the PLL, the same BIST circuitry can be employed online. Reference [29] also discusses different ways of measuring properties like loop gain, capture range, and lock-in time by modifying the feedback path to implement dedicated phase delay circuitry. All these components are automatically synthesized using the digital libraries available in the manufacturing process. This kind of automation provides scalability and easy migration to different technologies. The approach of Seongwon and Soma [30] is similar to this approach in the sense that the extra tester circuitry is all-digital and can be easily integrated into an IEEE 1149.1 interface. In this work, the BIST approaches reuse the charge pump and the divide-by-N counter of the PLL in order to generate a defect-oriented test approach, which can structurally verify the PLL. While [29] can also be implemented on a tester board, Ref. [30] is limited to the BIST since a multiplexer must be inserted into the delay-sensitive path between the phase detector and the charge pump. Since both examples employ all-digital test circuitry, their application is limited to a few analog components like PLLs, where digital control is possible.

In [5], an attempt is made to implement simple on-chip signal generators and on-chip test response data capture techniques for testing the performance of high-frequency analog circuits. The communication between the BIST hardware and the external world takes place through a low-frequency digital channel.

It can be observed that there are the following groups of approaches: (1) a device under test is modified to improve testability (DfT), (2) a signal generator and signal analysis devices are incorporated on chip (BIST) or on a tester board, (3) test access is improved by implementing test bus standards 1144.1 and 1149.4, and (4) a device under test is reconfigured into an oscillator for test purposes. The typical solutions in these categories will be reviewed, followed by a dedicated discussion of the most popular mixed-signal devices and their respective test techniques.

23.8 Analog Test Bus Standard

The significant advancement [10,11,31] in mixed-signal DfT area is the IEEE P1149.4 bus standard. It is a new IEEE standard that aims at providing a complete solution for testing analog and digital I/O pins and the interconnection between mixed-signal ICs. The secondary objective is to provide access to internal cores based on the Test Access Bus concept. It includes the IEEE 1149.1 boundary scan Test Access Port (TAP) controller and therefore provides a support infrastructure for the BIST and test setup.

Figure 23.2 shows the IEEE P1149.4 architecture that includes the following elements:

- Test Access Port (TAP) comprising a set of four dedicated test pins: test data in (TDI), test data out (TDO), test mode select (TMS), test clock (TCLK), and one optional test pin: test reset (TRSTn).
- Analog test access port (ATAP) comprising two dedicated pins: Analog test stimulus (AT1) and analog test output (AT2), and two optional pins: inverse AT1 (AT1n) and inverse AT2 (AT2n) for differential signals.
- Test bus interface circuit (TBIC).
- An analog boundary module (ABM) on each analog I/O.
- A digital boundary module (DBM) on each digital I/O.
- A standard TAP controller and its associated registers.

23.9 Oscillation-Based DFT/BIST

Currently, oscillation-based approaches seem to be the most applicable in practical implementations for all types of A/M-S devices. From the time the oscillation methodology was introduced as OBIST [32 to 46], many researchers have successfully explored more areas of application, better performance, and other innovations [1,47,48]. The commercial products have been developed and industrial implementations have been

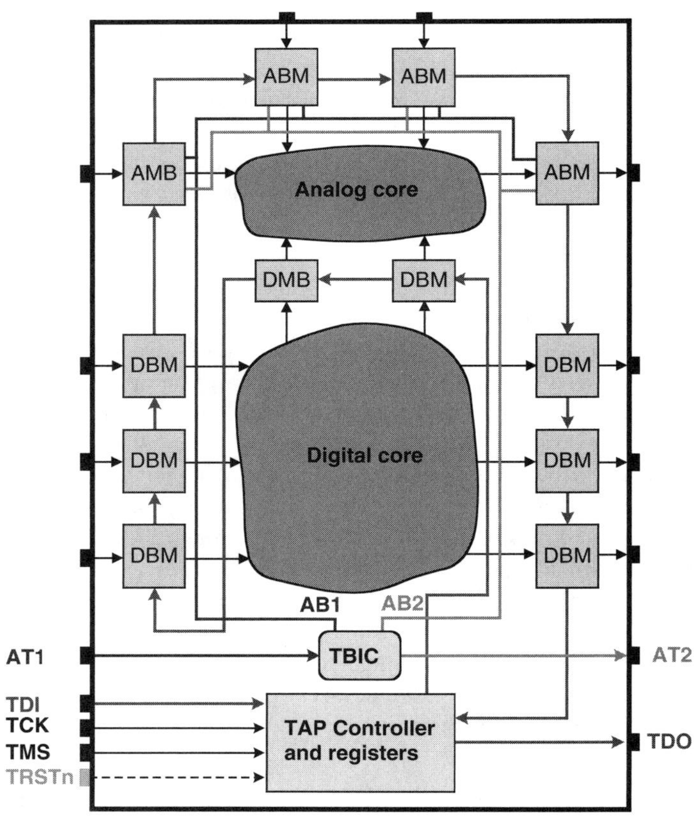

FIGURE 23.2 IEEE 1149.4 architecture.

successful as well [49]. The popularity of oscillation-based DfT/BIST techniques results from the following main characteristics:

- Adaptable for various functional modules: A/M-S, digital, and MEMS devices.
- Signal generator is not required.
- Immune to noise and technology-independent.
- Delivers very good fault coverage for parametric, hard, and functional errors.
- Easy to understand.

Oscillation DfT/BIST is a testing technique based upon converting a functional circuit to an oscillator and inferring something about the functionality of the circuit from the oscillation frequency (see Figure 23.3). To a first order, in any practical circuit, the oscillation frequency will be a function of several parameters of the circuit and also the elements that were added to convert the circuit to an oscillator. How then can oscillation BIST be used to definitively generate a go/no go decision in production test? To measure the parameters of the circuit under test? To characterize that circuit?

In a mathematical sense, the answer to these questions is very simple. If

$$\langle\text{Frequency of oscillation}\rangle = \text{Known\_function}\ [p1, p2, \ldots, p4, \ldots, pn]$$

and *pi* are all unknown circuit performance parameters, then *n* separate oscillation frequency measurements made under varying conditions can be used to calculate *pi*. So, by changing the feedback circuit topology, its frequency selection, its attenuation, etc., varying conditions may be imposed upon the circuit under test and varying frequencies of oscillation thereby determined. These *n* different equations with *n* unknowns can be solved for *pi*.

In a practical sense, not all of the *pi* may be of interest. Furthermore, the exact value of *pi* may be irrelevant. In most cases, what is needed is an assurance that each *pi* lies within a specific range of values, i.e., the criterion that was assumed when the circuit was designed and simulated — and if properly designed, the circuit will be functional over a known range of values for each of the *pi*. These are the values that can be used in a simulation of the circuit, with the BIST in place, to determine an acceptable range of frequencies of oscillation. Thus, the combination of a simulation and a single OBIST frequency measurement can be used to determine, with some level of certainty that the circuit under test is functioning properly. The level of certainty (including the feedback) relates to the parameters of its elements (the higher the sensitivity, the more certain that a single OBIST measurement will suffice). It is also a function of how tight the performance specifications of the circuit are. The tighter the specifications are, the more likely that additional measurements will be required to guarantee all the performance specifications.

The most extensive practical and commercial applications have been developed for PLL [49,50], general timing test, and characterization (including delay and 'stuck-at' fault testing in digital circuits [29,32,45,46]),

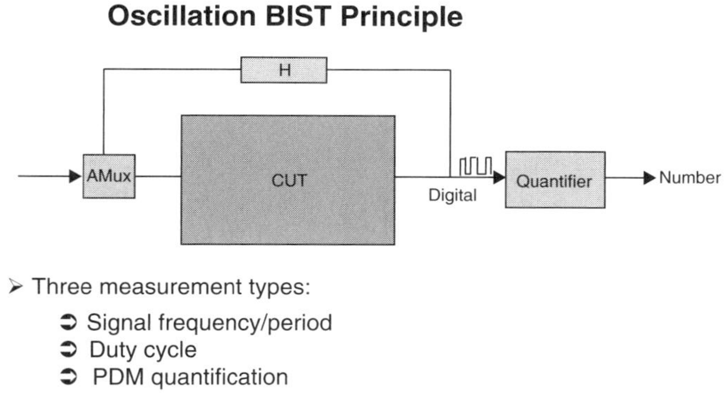

FIGURE 23.3 The principle of the oscillation technique- transforming a device into an oscillator.

operational amplifiers, and for a multitude of data converters. The differences in the implementations are mostly related to: (1) the configuration method of a device into an oscillator (choice of a transfer function, number of oscillators, control technique), (2) result processing techniques (from a custom signal processing to classical DSP to integrated method of histograms (HBIST): see below). In particular, each OBIST measurement can be targeted at a single parameter (or a small group of parameters) of a device under test by judicious selection of the feedback circuit. For example, a single application of OBIST can guarantee that the low-frequency gain of an operational amplifier is higher than some specified value. This is done by using a high attenuation factor in the feedback circuit, combined with reactive elements (one or more capacitors) which cause a low-frequency oscillation, provided the gain of the operational amplifier is greater than the specified value. Another example would be to guarantee operation at a certain input common mode voltage by feedback of the desired voltage, along with the feedback signal designed to make it oscillate into the input of the circuit to be tested. If the circuit oscillates, then the input range parameter requirement is fulfilled. A final example involves the measurement of the gain-bandwidth product of an operational amplifier. The operational amplifier is required to have a certain minimum gain in order to overcome the attenuation of the feedback loop — at a specific high frequency determined by frequency selective feedback.

In general, reconfiguration of the application circuit is accomplished by using digital circuits to control analog switches or multiplexers. The switches and multiplexers are designed to have a minimal impact on the measurement.

23.10 PLL, VCO, and Jitter Testing

PLL voltage-controlled oscillator (VCO) testing has generated significant interest recently due to the widespread integration of embedded, high-performance PLLs in mixed-signal communications, and data processing devices. In popular serial data communication standards such as Fiber Channel, FireWire, and GigaBit Ethernet, the data and clock are embedded within the signal codes. Because of these characteristics the receiver must have special clock recovery circuitry implemented with the PLL that extracts the data and clock received through the media. These clock recovery circuits are sensitive to input jitter (time distortion) and level distortion. To guarantee proper data reception in any network, the transmitter and receiver must meet a certain jitter budget.

Typical PLL applications include frequency synthesis, phase demodulation, clock distribution, and timing recovery — essential operations for systems like wireless phones, optical fiber links, and microcomputers, along with multimedia, space, and automotive applications. Serial data techniques and parameters also apply to hard disk drive read/write channels. Although the serial data and embedded clock information are stored on and retrieved from a magnetic medium rather than being transmitted over cable, the sensitivity to jitter is similar to datacomm and telecomm applications.

In recent years, the use of PLLs for clock generation has been widely popular for modern microprocessors. This is because PLL has the advantage of allowing multiplication of the reference clock frequency and allowing phase alignment between chips. It is impossible, however, to have these advantages if PLL has excessive jitter or variation in phase alignment. If the jitter is too large, then the cycle time for logic propagation is reduced and the error probability is higher.

Before discussing how to measure jitter, it is important to distinguish what jitter means for two application areas: noncommunication applications, for example, processors and serial communication applications. In the first case, jitter is defined as the variation of the clock period and is referred to as period jitter or cycle-to-cycle jitter. The measure of the phase variation between the PLL's input and output clocks is known as a long-term or tracking jitter. In serial data communication, on the other hand, jitter is defined as the short-term variation of a digital signal's significant instants,for example, rising edges with respect to their ideal position in time. Such a jitter is known as accumulative jitter and is described as a phase modulation of a clock signal. Jitter sources can come from power supply noise, thermal noise from the PLL components, and limited bandwidth of the transmitting media. There are some applications where the absolute jitter is important, for example, in clock synthesis circuits; there is a need for the use of a jitter-free or low-jitter reference signal. The difference between the position of corresponding edges of the signal of interest and the reference signal indicates the jitter. Although in a production environment the focus is on overall jitter, for characterization purposes, the following jitter components are distinguished: data-dependent, random, and duty-cycle distortion. Data-dependent or periodic jitter is caused by one or more sine waves and their harmonics. Random jitter has a probability distribution function — usually assumed to be Gaussian, but often it is not — and has a power spectral density that is a function of frequency. Duty-cycle distortion is caused by differing propagation delays for positive and negative data transitions. Based on the above definitions, a number of samples are collected to determine the jitter characteristics of interest. The most common characteristics include RMS, peak-to-peak, and frequency.

23.10.1 High-Speed Serial Links

In serial data communication, jitter plays a key role in clock extraction and network timing. The recovery of the clock from the data signal poses more stringent requirements on the jitter of the data signal than would exist when shipping a synchronous clock along with the data signal. The latter is typically done in very short links, where the effort of using a full-fledged clock-data-recovery receiver does not pay off in terms of silicon area and power consumption. Bit-error rate (BER) measurements or eye-diagram plots characterize the signal quality of serial links. BER measurements are subject to statistical uncertainties because of the tradeoff between test time and measurement accuracy.

The easiest way to measure jitter is to plot eye diagrams on an oscilloscope and apply histogram functions to the zero crossings, as Figure 23.10 shows. This procedure yields values for the horizontal or vertical eye opening as well for the total peak-to-peak jitter. However, this type of jitter measurement does not provide much insight into jitter properties. Some improvements come with the sampling scopes that add jitter measurement functionality based on the clever bookkeeping of sampling-point positions. Due to the sub-rate function of a sampling scope, it does not analyze every bit, which leads to accuracy problems at low error rates. A more adequate method of measuring jitter is based on the scan technique which measures the BER of every single bit within the data eye, and fits the resulting BER curve to a mathematical jitter model to obtain the required jitter properties. For data stems with several gigabits-per-second data rates, a BER scan tests many billions of bits each second and thus maintains accuracy for low BERs. In eye-diagram plots, distributed transitions of the threshold as data toggle between the logic states indicate jitter. The histograms measured at the zero crossing represent probability density function of the jitter, and statistically describe the data transitions' temporal locations. To derive a mathematical jitter model, jitter must be subdivided into different categories. In general, it is possible to split jitter into random and deterministic components. Random jitter has a Gaussian distribution and stems, for instance, from the phase noise of a VCO or power supply noise. Deterministic jitter can be divided into different subcategories of origin. Predominant types are sinusoidal and data-dependent jitter as well as jitter arising from duty-cycle distortion. Sinusoidal jitter can stem from slow variations of the supply voltage, the temperature, or the clock reference. The jitter components are not necessarily sinusoidal waveforms in reality, but we can model this jitter contribution with a sinusoidal representation. Data-dependent jitter of deterministic jitter comes from circuit asymmetries or duty-cycle distortion.

23.11 Review of Jitter Measurement Techniques

From the user perspective, there are three established ways of making jitter measurements: using ATE equipment, a real-time sampling oscilloscope, or a dedicated jitter measurement instrument. Because jitter is becoming such an important factor in the overall system performance, it is worthwhile to examine the pros and cons of using each of these techniques. A fourth technique based on BIST or just DFT has recently emerged (see Table 23.1).

23.11.1 Spectrum Analyzer Measurement

An analog spectrum analyzer can be used to measure the jitter of a signal in the frequency domain in terms of phase noise. For this measurement, the jitter is modeled as phase modulation. For example, a

TABLE 23.1 A Comparison of the Four Different Ways to Measure Jitter

	ATE Testers	Real-Time Sampling Oscilloscope	Dedicated Instrumentation	BIST/DFT
Performance	Limited accuracy, resolution, and throughput	Very high accuracy, resolution, and throughput	High accuracy, resolution, and moderate throughput	High accuracy, resolution, and throughput
Additional hardware	None or dedicated instrumentation	High-speed digitizer system per channel	Digital jitter measurement system per ATE	Modular, flexibility with different levels of integration
Additional software	Minimal	Significant signal processing and application-specific code	Insignificant signal processing and application-specific code	Insignificant signal processing and application-specific code
Cost	Nil or ATE-related	About $10,000 per channel for large channel counts, more for small channel counts	About $100,000 per ATE	Implementation dependence,low
Application	>30 psec jitter in CMOS production test	1 to 5 psec jitter on a few critical signals/clocks precise characterization of communications channels, optical fiber clocks, prediction of jitter	>1 psec jitter lock jitter, in ICs. serial data communications, carrier jitter in integrated RF subsystems	1 or more psec jitter Clock , data jitter, serial data communication, measurements in a noisy environment
Manufacturers	ATE companies	Tektronix, LeCroy, Agilent	Wavecrest, Guidetech	Emerging companies

sine wave signal is represented as a perfect sine wave with amplitude and phase modulation. With a spectrum analyzer, one can measure the power of the signal as a frequency with wide dynamic range; however, one cannot distinguish between the amplitude and phase-modulation components. A common assumption made when using a spectrum analyzer to measure jitter is that the amplitude modulation component of the signal is negligible. This assumption may be valid for signals internal to a pure digital system where undistorted square waves or pulses are the norm. This assumption is typically not true for a serial communication or data channel. Isolating the noise from the actual signal frequency components and translating that into jitter is nontrivial.

23.11.2 Real-Time Time Interval Analyzer Measurements

This technique measures the time interval between a reference voltage crossing of the transmitted signal. There is no need for an abstract model of the signal when using this technique because the time intervals are measured directly. The real-time interval analyzer gives complete knowledge of the nature of the jitter and the components of the jitter. With this measurement technique the position and time of every edge is measured, thus allowing statistical space and frequency-based models of the signal, as well as absolute peak-to-peak measurements. The clear advantages of this technique are that there are no skipped edges and measurement acquisition time is limited only by the signal itself. In practice, instrumentation that has the necessary acquisition rate and resolution to test gigabit data rates does not exist.

23.11.3 Repetitive Start/Stop Measurements

This is a common technique that gives high resolution and accuracy using a direct time measurement. Time measurements are made by starting a counter on the first occurrence of an edge, and stopping the counter on the next edge. Enhancements to this technique include skipping multiple edges for cumulative measurements, comparing two different signals, and time interpolation for achieving resolution greater than the counter clock period. Re-triggering of time interval measurement normally requires a significant dead time, particularly when time interpolation is used. After collecting many of these time interval measurements, postprocessing is applied to extract statistical parameters and jitter components. This technique has been used to good effect in ATE equipment to measure jitter of low-frequency clock (<100 MHz) or in some bench-top instrument implementations, such as Wavecrest, that include special software features for jitter component characterization.

23.11.4 ATE-Based Equipment

Using automatic tester equipment, a signal may be repeatedly acquired at slightly different time settings for the capture clock. The distribution of signal timing can be determined from the average number of successful acquisitions at each clock setting. Although using ATE equipment lowers the cost for those designers who have the test equipment, it does take some time to make multiple acquisitions of a given transition. What is more of a concern is that the accuracy and resolution are limited by the tester resolution, accuracy, and jitter. One plus point is that the processing load imposed on the tester's computer is minimal.

23.11.5 Real-Time Digital Sampling Oscilloscope

When using a real-time digital sampling oscilloscope to measure jitter, a signal is acquired by the oscilloscope's oversampling clock and transitions through a fixed voltage threshold that is determined by filtering and interpolation. The variation of the transitions with respect to a fixed clock is interpreted with special jitter software. Advanced real-time oscilloscopes are typically used to measure 1 to 5 psec jitter on a few critical signals/clocks for precise signal characterization in communications channels, optical fiber clocks, and predicting the BER. Making jitter measurements with an oscilloscope requires a high-speed digitizer on each of the instrument's channels, along with a high-speed memory system and possibly DSP hardware. The oscilloscope's system has the potential for very high throughput and accuracy because the signal is being continuously observed and the threshold transitions can be interpolated to very high precision, in part, because of the oscilloscope's multibit acquisition.

The drawback of this approach is that the processing load can be quite high and increases with the channel count, and so it may become impractical to realize a tolerable throughput when there are multiple channels involved. Moreover, the power dissipation in practical systems is of the order of 15 W per channel, so large channel counts may become problematic for this reason, too. Also with most oscilloscopes, when the full complement of channels is in use, the sampling acquisition is no longer continuous, and so the probability of capturing infrequent jitter faults drops off quickly.

23.11.6 Dedicated Jitter Instrumentation

Dedicated jitter instrument hardware is used to measure jitter directly and report the result as a number. A dedicated jitter measurement instrument is used, offloading the measurement burden from the ATE or oscilloscope. However, since jitter measurements are now made independent of general-purpose equipment, the overall test time is increased to accommodate these special measurements. Moreover, a method must be provided to switch the measurement system to the required channels. There are two main providers of this instrumentation: Wavecrest and Guidetech. Guidetech's hardware is based on time-interval analyzer (TIA) technology, whereas Wavecrest's hardware is based on counter–timer technology. TIAs create precise time steps of trigger events, and within limits (Guidetech's limit is 2 million triggers/sec) can measure the timing of every trigger event. At higher frequencies, however, TIAs can make precise measurements only on some fraction of all of the cycles received. Counter-based analyzers, on the other hand, make timing measurements by "stretching" cycles, for example, by ramping an analog integrator rapidly until a trigger event is detected, and then ramping the integrator slowly back down to zero, so that a relatively low-speed counter can measure the ramp-down time. This technique is allegedly slower than that used in TIAs, and, in the case of the Wavecrest boxes, limits the maximum frequency of signals that can be completely sampled (sampled in every cycle) to about 30,000 waveforms/sec.

23.11.7 BIST and DFT

All the methods discussed above rely on external measurement equipment to examine jitter. In contrast, the new BIST/DFT approaches can be used for SoC/ICs that exhibit jitter as low as 1 to 5 psec. Applications can include clock and data jitter, differential jitter, serial data communications, and measurements in a noisy environment.

23.11.7.1 DFT/ BIST Based on Oscillation Method

The PLL test circuit, DFT/BIST, which follows the oscillation methodology, is based on the Vernier principle transposed to the time domain. Due to a very high measurement resolution, it is practical to measure the frequency and the jitter of PLL circuits. As a Vernier caliper allows for a precise measurement of a linear distance between two points, the same principle applied in a time space allows for a precise measurement of time interval between two events. Instead of two linear scales having a very small difference in linear step, two oscillators with a very small difference in oscillation frequency are used for the PLL test circuit. The measurement resolution is that of the time difference between the periods of two oscillators. These oscillators have two essential characteristics:

1. When triggered, the oscillation will start virtually instantaneously and with a fixed phase relationship to the external trigger pulse.
2. When oscillating, the stability of the frequency is preserved.

The system uses phase-startable oscillators in both the start ($T1$) and stop ($T2$) channels. The stop oscillator has a slightly shorter period of oscillation than the reference (start) oscillator, such that once started, they will reach coincidence some number of cycles later, depending on the measured time interval. If the reference (start) oscillator period is $T0$, then the stop oscillator has period equal to $T0 - \Delta T$, where ΔT is the difference between the periods of both the oscillators. The measured time interval T is expressed as follows:

$$T = N1\,T1 - N2\;T2 = n(T1 - T2) = n\Delta T$$

where $N1$ is the number of start oscillator pulses to coincidence and equals n if the frequency difference is sufficiently small, and $N2$ the number of stop oscillator pulses to coincidence and equals n if the frequency difference is sufficiently small.

A practical limitation on the interpolation factor n is imposed by the inherent noise, and there is little advantage to be gained by having an interpolation factor capable of giving resolutions substantially better than this. The recently implemented [51] interpolation scheme provides better than 1 psec single-shot resolution. Other reported implementations (like that by Fluence [52]) give tens of picoseconds single-shot resolution.

The high-single-shot resolution allows the collection of meaningful statistical information. Important characteristics concerning time intervals like maximum, minimum, and standard deviation values are collected in addition to the mean value. With low-speed data collection, full histogram-building information is obtained. In many time interval measurement situations, these statistical properties are of prime importance. For example, in PLL jitter testing, the standard deviation gives an excellent indication of jitter.

In all practical implementations [48,53], the test circuit contains two matched phase-startable oscillators (see Figure 23.4). The coincidence is detected by a coincidence detector. The test circuit consists of two main parts: the functional circuit and the result processing circuit or software. The functional circuit is equivalent to about 2000 to 2500 gates, depending on the implementation. For result processing, different techniques can be adopted depending on the application requirements. It can be a hardware-implemented DSP technique (such as histogram building [54–56], used by Fluence) or custom software module [51]. The result processing is, in general, very fast, and can easily be executed in real time. The dedicated algorithms for fast-performance computation include: RMS and peak-to-peak jitter, instantaneous period, frequency, phase jitter, delay, and similar characteristics.

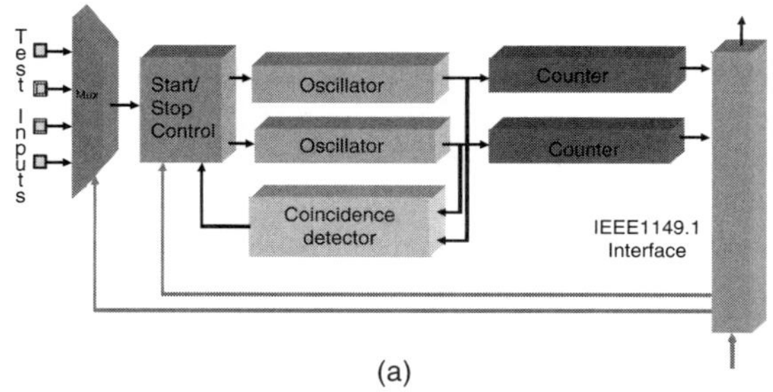

(a)

Vernier Period Measurement

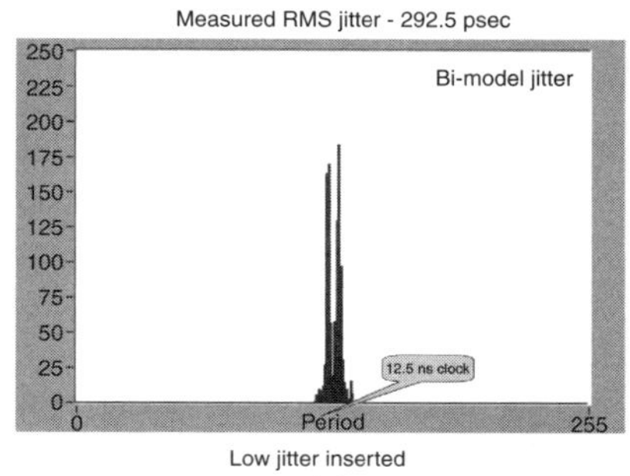

(d)

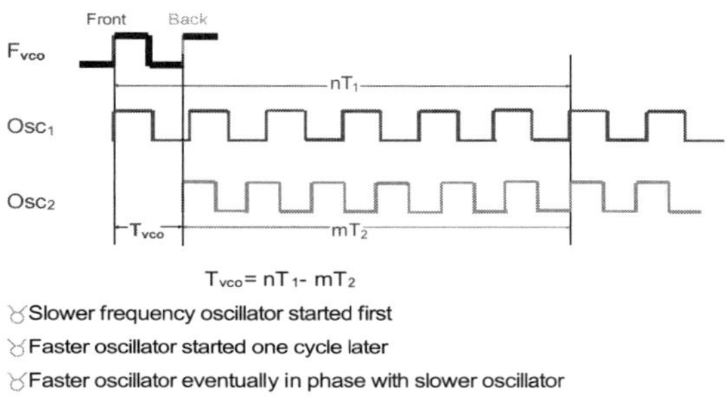

(b)

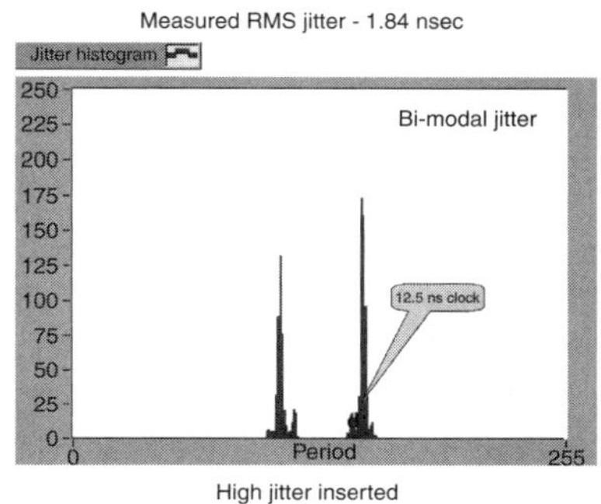

(e)

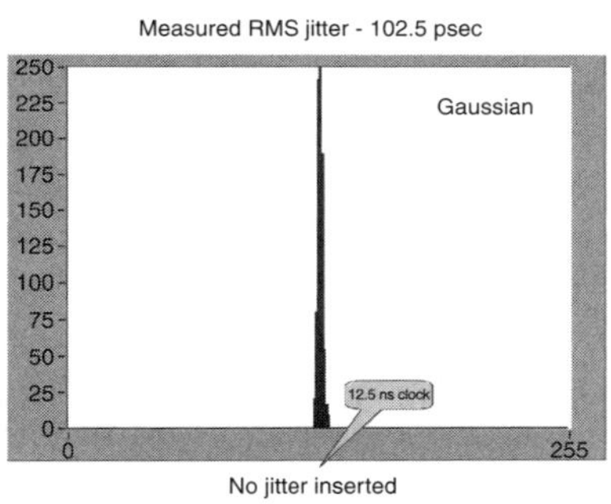

(c)

Specification	FPGA	0.25m CMOS	.18 SiGe
Test Strategy	**Off Chip**	**On Chip**	**Distributed**
Instantaneous period	100 ps	10 ps	1 ps
RMS Jitter	100 ps	10 ps	2 ps
On Chip System Clock	100 MHz	500 MHz	2.5 GHz
Phase jitter (rms)	100 ps	10 ps	2 ps
Long-term jitter *	200 ps	50 ps	5 ps
Delay	100 ps	10 ps	2 ps
Lock Indicator	50 ns	10 ns	2.5 ps

1024 Samples

(f)

On Chip System Clock	250 MHz	622 MHz	2.5 GHz
Accuracy	8 ps	3 ps	1 ps
Calibration Time	8.4 ms	6.7 ms	2.1 ms
Measurement Time	4 ms	1.7 ms	0.5 ms
Total Time	**12.4 ms**	**8.4 ms**	**2.6 ms**

Samples = 1024

(g)

FIGURE 23.4 Jitter measurements principle, Jitter measurements: a) principle of implementation; b) principle of operation; c, d, e) measurement examples, f) performance example for different implementations; g) example of final possible results as a function of implementation technology.

Besides its impressive jitter measurement capability, there are a multitude of benefits that flow from this DFT approach to PLL measurement. First, the dedicated test circuit based on the DFT approach requires only minor tester capabilities, which in turn lowers test cost development and equipment requirements. The test circuit can be placed directly on a low-cost tester [51] or can follow the integrated implementation. In the case of a fully integrated version, time-dependent process variations incurred during the fabrication of the SoC/IC can have an impact on the jitter of the oscillators used in the test circuitry. What should be noted is that the oscillator itself can inject jitter depending on process variations. Basically, there are two types of jitter that can occur in these particular situations. The first is the correlated jitter, where the two oscillation frequencies vary in time but always by the same amount. This jitter will only impact the absolute value of the frequencies but the difference will remain constant. The second type of jitter is the noncorrelated one and it impacts the difference between the two frequencies. The latest has the greatest impact on the DFT approach to measuring jitter and should be analyzed and minimized.

With a Monte-Carlo simulation, it is only possible to quantify the impact of the static variations—those variations that do not change with time. Injecting time-dependent variation involves an indirect use of the simulation, where the variations are generated externally and injected in the simulation as noise sources.

Other DFT/BIST techniques for PLL testing are reported in [5,29,53]. Each of them has, however, an important technical limitation and as a result, a limited applicability.

23.11.8 ADC and DAC Testing

Among frequently used mixed-signal circuits, data converters are typical mixed-signal devices that bridge the gap between the analog and digital world. They determine the overall precision and speed performances of the system and therefore, dedicated test techniques should not affect their specifications. For instance, it is difficult to test the analog and the digital portions of data converters separately using structural test methods and conclude that the whole device specifications are fully respected. Therefore, it is necessary to test data converters as an entity using at least some functional specifications. There is also a strict requirement in terms of precision related to hardware used to test data converters. Most of the efforts in on-chip testing data converters are devoted to ADC converters. In [57,58], conventional test techniques have been applied to the ADC under test, using a microcontroller available on the same chip. Oscillation-test strategy is a promising technique to test mixed-signal circuits and is very practical for designing effective BIST circuits [40]. Based on the oscillation-test method, in [40] the ADC under test is put into an oscillator and the system oscillates between two preestablished codes by the aid of some small additional circuits in the feedback loop. Functional specifications such as offset, DNL, INL, and gain error are then evaluated by measuring the oscillation frequency of the circuit under test. This technique has the advantage of delivering a digital signature that can be analyzed on-chip or by a conventional digital tester. Testing ADC–DAC pairs has been addressed in [59,60,72]. Such techniques use the DAC to apply analog stimuli to the ADC under test and use the ADC to convert DAC under test signatures into digital. Three problems have to be considered for these techniques. First, it is limited to applications, where one can find an ADC–DAC pair on the same IC. Second, the ADC (or DAC) used to test the DAC (or ADC) should have at least 2 bits of resolution more than the DAC (or ADC) under test. The third problem is fault masking in which a fault in DAC (or ADC) compensates another fault in ADC (or DAC). Therefore, it is very important to be able to test the DAC or ADC individually without using another data converter. The only BIST approach for solitary DACs has been proposed in [60], that verifies all static specifications using some additional analog circuitry and some control logic. The accuracy of the analog circuitry limits the test precision and the authors propose an auto-calibration scheme to overcome this limitation.

The main difficulty when dealing with BIST for DAC is the analog nature of its output signal that requires one to design high-resolution but still area-efficient analog signature analyzers. Oscillation-test strategy deals with this problem by establishing a closed-loop oscillation including the circuit under test, in which one does not have to apply analog input stimuli, and the test output is a pure digital signal.

Figure 23.5 shows the implementation of oscillation-test method to apply BIST to embedded data converters. Other interesting references include [47,61–64].

23.11.8.1 ADC Testing by Available ATE Systems

Testing A/M-S circuits often involves much more data and much more computation than is necessary for testing digital-only circuits. Traditional methods for A/M-S testing usually involve adding large suites of expensive analog instrumentation to the highest speed, highest pin count digital ATE system available. For testing an ADC, the most expensive instrument is usually the very precise, very low-noise floor stimulus source. In addition, there is the challenge of getting that stimulus signal to the input pin(s) of the

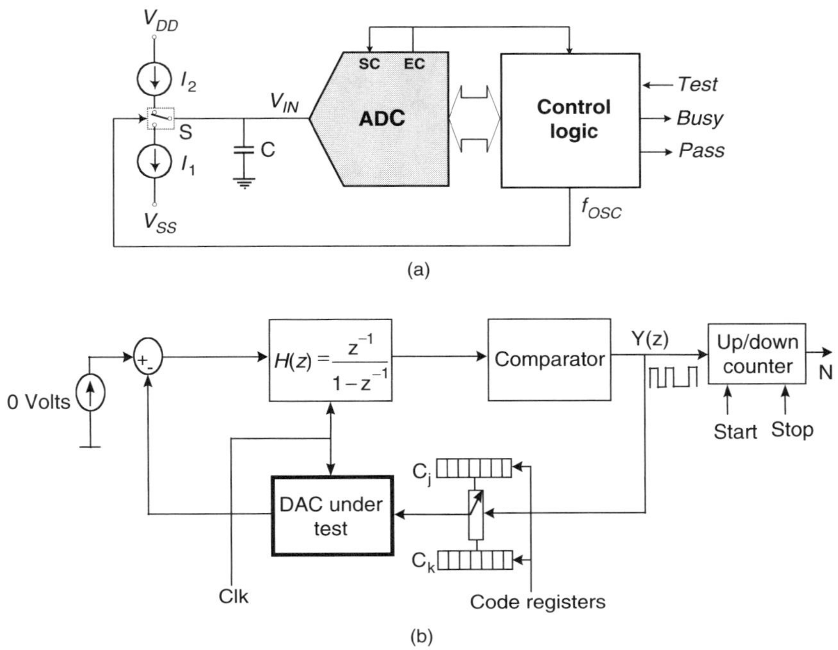

FIGURE 23.5 (a) Oscillation-test method used to apply BIST to ADCs; (b) Oscillation-test used to apply BIST to embedded DACs.

device under test, and the most time-consuming process is applying the requisite precision of input values to the ADC, while simultaneously capturing the digital output codes that must then be analyzed to determine the quality of the ADC portion of the design.

The input stimulus for ADC testing is most often either a ramp signal or a sine wave. The ramp signal is usually the preferred stimulus since it is easier to generate, using either a DAC or a specially designed counter driving analog voltage or current sources, rather than the sine wave. And if histogram-based techniques are used, the ramp signal usually results in a more even distribution of information on results. This has the advantage of reducing the data storage size and the resulting analysis time. The ramp signal stimulus is, however, unusable for AC-coupled designs. These designs require the constant movement typified by the sine wave signal, typically at a frequency higher than the ADC sampling rate, and need the use of either over-sampling or under-sampling techniques. Coherency between the analog and digital signals associated with the ADC is also required with a sine wave stimulus — a requirement that is not necessary when using a ramp signal.

Another issue associated with ADC testing is the type of testing that should be done. ADC test types are usually divided into two categories — static and dynamic. The so-called static tests include mainly INL and DNL, from which other test results such as gain, offset, and missing codes can be determined. The so-called dynamic tests include, for example, directly measured (as compared to computed) SNR, signal including noise and distortion (SINAD), and total harmonic distortion (THD). However, these parameters are only obtainable by computing results using fast Fourier transform (FFT) techniques.

It is also interesting to note that there is currently much debate over the advisability and fault coverage of "structural" testing techniques as compared to traditional functional performance of the testing methods. The static ADC tests — INL, DNL, offset, gain, and missing codes — will identify structural defects. Testing for "performance" defects requires using actual SNR, SINAD, and THD measurement techniques, or the use of sophisticated software routines to calculate these values from the static test results. So, in addition to the philosophical differences in the two testing strategies, there are also instrument and computation time trade-offs to be taken into consideration.

Linearity— INL, DNL: Integral and differential nonlinearity are the two basic building blocks for ADC testing. They can be used to detect structural defects in a chip.

Noise and distortion — SNR, SINAD, and THD: Noise and distortion can be measured directly using sophisticated ATE instrumentation or calculated from the INL and DNL figures while taking clock jitter into account.

Sophisticated instrumentation is used to supply the stimulus, and the ATE digital pin electronics are used to capture the resulting ADC output codes. These codes are then analyzed using either direct bit-by-bit, clock cycle-by-clock cycle comparisons, or by reading the acquired digital states and analyzing them with digital signal processing (DSP) techniques. The DSP approach is often much faster than the discrete comparison approach and can be used to yield qualitative information using tolerance values. The direct comparison approach will only provide go/no go information and is very unforgiving.

The histogram-based testing described below can also be applied to ADC testing using BIST. A specially designed linear feedback shift register (LFSR) is used to create the histogram generator whose results are stored in memory and accessed via the IEEE 1149.1 (JTAG) port that is often already present in the SoC design.

23.11.9 Histogram-Based DfT/BIST

Histogram-based methods have been used for analog testing for many years. In Tektronix PTS101 test system, a histogram is used to extract information about the code density in the output of A to D converter using a sine wave input. Reference [55] describes the use of the code density histogram-based method for distortion and gain tracking measurements on an A to D converter. Histograms have also been used to make quantitative measurements in digital systems, e.g., for performance analysis, a histogram built in real time gives an accurate picture of the way a digital computer spends its time [55].

In contrast to these informal test methods that somehow use histograms, an efficient integrated version based on histograms has been developed and commercialized [50,52,54],, as HBIST. The HBIST permits use of the low-impact applications including ADCs, DACs, and PLLs. Employing HBIST results in digital test results which can be analyzed for such parameters as INL and DNL, gain and offset errors, effective least-significant bit (LSB), and clipping and modulation distortion.

During an HBIST implementation, essential information about the signal under test is converted to a histogram, from which characteristics can be studied to gain valuable information about circuit performance. The sample-and-hold circuit and ADC perform the conversion from analog domain to the digital domain (Figure 23.6). Once the data from the ADC are fed to the histogram generator, the results can be downloaded and read by a digital ATE system. The technique uses under-sampling of the analog signal(s) under test to quantify how long the signal remains at each amplitude level, placing the values in various bins in a histogram (Figure 23.7). The histogram characterizes the waveform of the signal under test, capturing its essential elements.

Using software simulation tools, an ideal histogram for each signal under test can be created, as can histograms for signals due to certain defects, like stuck bits and various types of nonlinearity and distortion. These signatures for various types of faulty circuit behavior can be stored for use in determining the pass/fail status of analog circuits under test during production testing. Should the signal under test vary from the expected signal, the histogram normally undergoes significant changes (Figure 23.8). The clipped sine wave shown does not spend nearly as much time at the high and low boundaries. Therefore, the resulting histogram has fewer entries in the outside bins and many more entries in the bins adjacent to them. Subtracting the acquired histogram from the ideal histogram creates a difference histogram that can be analyzed to determine which defects are present in the circuit under test.

In addition, the histogram-based method can be deployed to test the ADC that is part of the circuit itself when a proper stimulus signal, usually a ramp, is applied to its input. The ramp signal can be supplied either by an external signal generator or by a DAC that might already be present in the design. Multiplexers at the DAC inputs and output allow it to be employed not only for on-chip functional purposes, but also

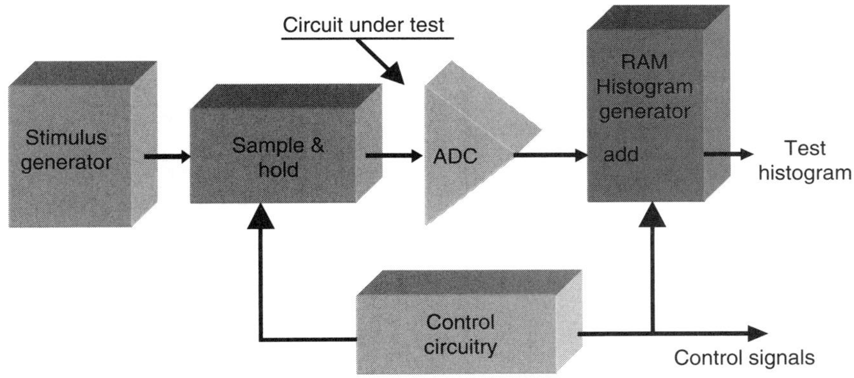

FIGURE 23.6 A typical HBIST configuration for analog circuit testing.

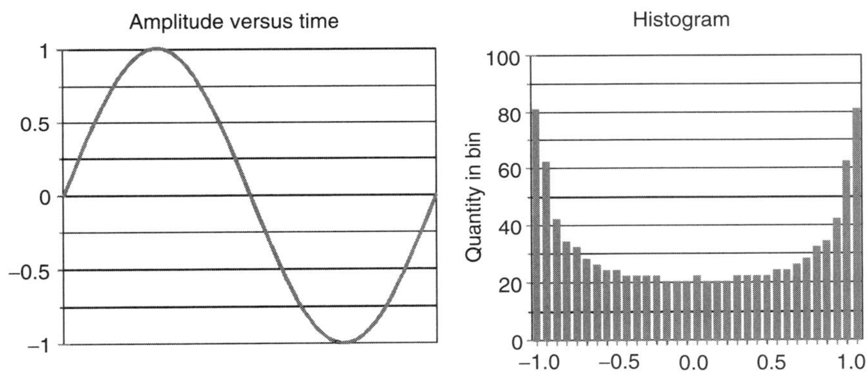

FIGURE 23.7 An example of histogram building from a sine wave.

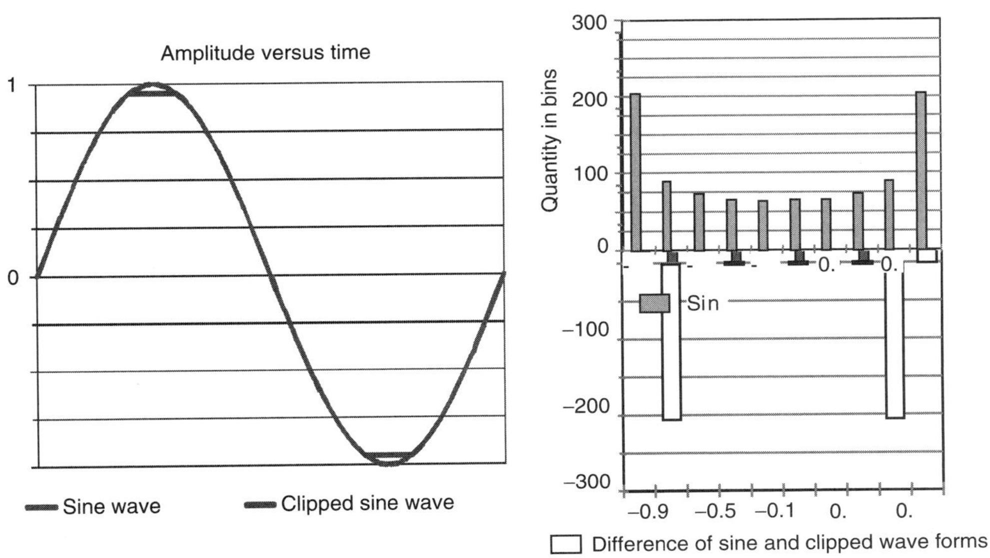

FIGURE 23.8 A difference histogram when clipping occurs.

as the stimulus generator for ADC testing. The results from the histogram-based BIST circuitry can be accessed by a digital-only ATE, offering significant savings in cost as well as complexity. The IEEE 1149.1 test bus interface can be used to access the HBIST results. The same histogram-based technology applies to DACs. It can solve the many problems that designers encounter when DAC is embedded in a complex SoC design. The PLL/VCO BIST technique based on oscillators, as described above, can also benefit from the histogram-based result analysis. The histogram infrastructure can be shared on an SoC for converters and PLL testing.

23.11.9.1 The Elements of HBIST Method

HBIST is highly adaptable to test requirements and to the realities of the device under test. A complete summary is given by the following items:

1. Generate a histogram of the expected signal at each test point under the desired conditions of stimulus. The histograms can be generated theoretically — e.g., from a SPICE simulation, or they can be experimentally derived, as from testing a "golden device." Stimulus can be supplied from external or built-in sources.
2. Determine the range of variance of offset, gain, noise, and distortion that are acceptable in the signature for each signal under test. This can be done by simulating limit cases with the limits of circuit parameters selected so that signals are marginally acceptable. Alternatively, these parameters can be derived empirically.
3. Provide a method for accessing each test point. Access can be achieved with: an oscilloscope probe or input cable — in cases where the test point is accessible; an electron beam probe such as a scanning electron microscope in voltage-contrast mode; an embedded analog buffer, probe, or sample and hold; built-in higher level test equipment, including a digitizer.
4. Generate a histogram of the signal at each test point under the desired conditions of stimulus. The histogram can be generated by built-in high-level test equipment, or by sending a replica of the signal at the test point to external facilities.
5. Process the acquired histogram using the HBIST algorithm and the expected histogram as a template. This process generates a signature that describes the difference between the histogram of the signal at the test point and its template.
6. Use the HBIST signature as both a go/no go test criterion and as a basis for the diagnosis of the cause of test failure. The part of the signature that results in test failure contains information that can be used by an expert system for diagnosis.

23.11.10 RF Test Practices

Until recently, RF functionality has been provided by single integrated circuits like mixers, PLLs, and transceivers. In fact, one has succeeded in reaching low cost, time to market, and low ppm, owing to the traditional test methods. Future RF circuits, however, are either in silicon or on substrate-integrated (SoC, SiP, and MCM), and represent new challenges and requirements. Since these RF IPs are embedded in the system, it is difficult to access all RF ports and as such, current test practices need to be revised. The test time needs to be reduced to acceptable limits within the digital testing time domains, and it also implies the incorporation of DfT, BIST, and DfFT techniques.

23.11.10.1 Testing Transceiver

The inclusion of wireless communication facilities in the SoCs and SiPs means the addition of a transceiver that can be seen as an IP. Typical transceiver architecture contains three predefined partitions, namely, the receiver path, the transmitter path, and the logic part. Without loss of generality, the following typical transceiver parameters can be specified for testing: (1) at the receiver path some of these parameters are the frequency bands of operation, the path gain, intermodulation distortion, noise figure, blocking, and harmonic distortion; (2) at the transmitter path parameters to consider are the frequency bands, transmitted power, automatic gain control, and the frequency spectrum; (3) the logic part deals with issues like VCO tuning, sensitivity, and leakage. The RF architecture is similar for different products, but small variations that are product-specific, imply a dedicated set of test parameters, which need to be included into test program. Very often, a tester is unable to measure some critical parameters.

The typical test strategies include block-by-block test and system-level test. Characterizing an isolated block in a whole transmitter is rather laborious since it involves proper loading conditions from one building block to another, and the requirement of special DfT to place isolation switches. In a system test strategy the transceiver is tested as used by end user, and also the product is specified in terms of communication standards. The preferred system test is based on BER measurements. A BER test compares a string of bits coming out of the device under test, against its corresponding golden sequence. BER is actually the ratio of the total number of errors to the total number of bits checked. Typically, a standard limit is one error for every 1000 demodulated bits. BER tests are applied as modulated and control signals to the device under test, such that it is possible to test parameters such as sensitivity, blocking signals, and other sorts of channel interference.

During the test development process, the important phase comprises the definition of test modes and failure conditions, followed by the required test parameter set. The appropriate DfT and the possible BIST can be decided based on the test requirements and test resource availability. External equipment often imposes the limitations on achievable performance of test. Reference [65] describes a number of the best RF test practices that are currently in use.

23.11.10.2 Transceiver Loop—Back Technique

Few methods in the literature address DfT and BIST for RF circuits. These techniques are based on the basic idea of loop-back in order to reuse the transmitter or receiver section [49,66–68,71,73] in a transceiver. The output of a system is routed back to its input directly, without using the wireless link.

In the case of integrated transceivers in the SoC/SiP environment, the loop-back test technique can reuse the DSP and memory resources already available in the system for the test response analyzer and signal generation. This approach may be able to reach the lowest test cost possible, if only switches and attenuators are needed. This may be a very important quality, mainly for systems operating at frequencies in the order of gigahertz. Another advantage is the lower effort needed in order to implement the test, and high flexibility, as the test is implemented in software and does not depend on the technology of the transceiver.

There are some disadvantages of the loop-back technique, mainly related to the fact that the complete transceiver is tested as a whole. This way, faults in the transmitter could be masked by the receiver, reducing fault coverage and ruling out fault diagnosis. There are some attempts to increase observability and controllability of the signal path [69]. The transmitter portion of the transceiver is used to create modulated signals, testing sensitivity, and several other RF tests. Further improvement can be found in [3] as shown in Figure 23.9. The masking effects are eliminated and high-test performance is achieved. A sample test result is shown in Figure 23.10.

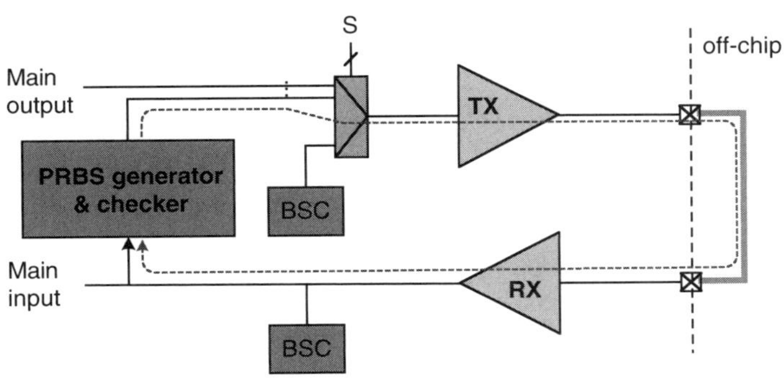

FIGURE 23.9 Enhanced loop-back technique (BSC - Boundary Scan Cell).

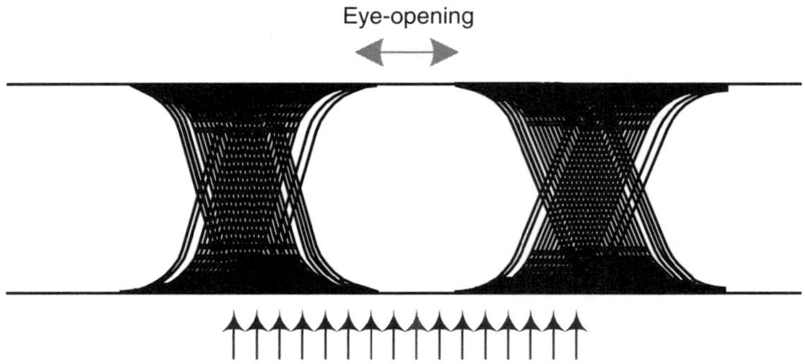

FIGURE 23.10 Sampling points in test mode to measure eye-opening.

It can be noted that the enhanced loop-back technique represents a functional BIST solution aimed at enabling current tester platforms to test high-speed interfaces at-speed, with increased functional test coverage. As presented in [3], the tests include: (1) supports external and internal loop-back from transmitter to receiver; (2) pseudo random bit sequence (PRBS) generator to generate test pattern; (3) ability to inject jitter and phase hit at the transmitter; (4) ability to check for correct incoming data at the receiver and count BER; (5) ability to measure eye-opening at the receiver; (6) ability to measure receiver sensitivity; (7) ability to uncover the fault masking effect between the receiver and the transmitter; and (8) offer embedded boundary scan cell (BSC) to reduce their impact on performance.

23.12 Summary

A short overview of A/M-S testing and practical DFT techniques has been presented. The practical aspects of such typical blocks as PLL, ADC, and DAC have been highlighted by reviewing testing by traditional and DfT approaches.

References

[1] *Proceedings IEEE Mixed Signal Testing Workshops*, 1994–2005.

[2] B. Vinnakota, Ed., *Analog and Mixed Signal Test*, Prentice-Hall, Englewood Cliffs, NJ, 1998.

[3] Kaminska, B. and Arabi, K., Mixed-signal testing — a concise overview, tutorial, *IEEE ICCAD 2003*, Santa Clara, November 2004.

[4] Abderrahman, A., Cerny, E., and Kaminska, B., Worst-case tolerance analysis and CLP-based multifrequency test generation for analog circuits, *IEEE Trans. CAD*, 18, 332–345, 1999.

[5] Burns, M. and Roberts, G., *An Introduction to Mixed Signal IC Test and Measurement*, Oxford University Press, Oxford, UK, 2001.

[6] Slamani, M. and Kaminska, B., Soft large deviation and hard fault multifrequency analysis in analog circuits, *IEEE Des. Test Comput.*, 12, 70–80, 1995.

[7] Slamani, M., Kaminska, B., Multifrequency testability analysis for analog circuits, *IEEE Trans. Circuits Syst., Part II*, 13, 134–139, 1996.

[8] Ben-Hamida, N., Saab, K., and Kaminska, B., A perturbation-based fault modeling and simulation for mixed-signal circuits, *Asian Test Symposium*, November 1997.

[9] Roberts, G., Metrics, techniques and recent developments in mixed-signal testing, *IEEE ICCAD*, November 1996.

[10] IEEE Standard Test Access Port and Boundary SCAN Architecture, IEEE Std. 1149.1-2001.

[11] IEEE Standard for Mixed-Signal Test Bus, IEEE Std. 1149.4-1999.

[12] Hemnek, G. J., Meijer, B. W., and Kerkhoff, H., G., Testability analysis of analog systems, *IEEE Trans. CAD*, 9, 573–583, 1990.

[13] Abderrahman, A., Cerny, E., Kaminska, B., Optimization-based multifrequency test generation for analog circuits, *J. Electron. Testing: Theory Appl.*, 13, 59–73, 1996.

[14] Abderrahman A., Cerny, E., and Kaminska, B., CLP-based multifrequency test generation for analog circuits, *IEEE VLSI Test Symposium*, April 1997.
[15] Slamani, M. and Kaminska, B., Fault diagnosis of analog circuits based on sensitivity computation and functional testing, *IEEE Des. Test Comput.*, 9, 30–39, 1992.
[16] Slamani, M., Kaminska, B., and Quesnel, G., An integrated approach for analog circuit testing with a minimum number of detected parameters, *IEEE International Test Conference*, Washington, DC, October 1994, pp. 631–640.
[17] Huynh, S. Dd. et al., Automatic analog test signal generation using multifrequency analysis, *IEEE Trans. Circuits Syst. II*, 46, 565–576, 1999.
[18] Balivada, A. Zheng, H., Nagi, N., Chatterjee, A., and Abraham, J., A unified approach for fault simulation of linear mixed-signal circuits, *J. Electron. Testing: Theory Appl.*, 9, 29–41, 1996.
[19] Harvey, J. A. et al., Analog fault simulation based on layout-dependent fault models, *IEEE ITC*, 1995, pp. 641–649.
[20] Ben Hamida, N., Saab, K., Marche, D., and Kaminska, B., LimSoft: automated tool for design and test integration of analog circuits, *IEEE International Test Conference*, Washington, DC, October 1996, pp. 571–580.
[21] Saab, K., Ben Hamida, N., Marche, D., and Kaminska, B., LIMSoft: automated tool for sensitivity analysis and test vector generation, *IEEE Proc. Circuits, Devices Syst.*, 143, 1998, pp. 118–124.
[22] Saab, K., Benhamida, N., Kaminska, B., Closing the gap between analog and digital testing, *IEEE Trans. CAD*, 20, 307–314, 2001.
[23] Saab, K. and Kaminska, B., Method for Parallel Analog and Digital Circuit Fault Simulation and Test Set Specification, U.S. Patent No. 09/380,386, June 2002.
[24] Millor, L. et al., Detection of catastrophic faults in analog IC, *IEEE Trans. CAD*, 8, 114–130, 1989.
[25] Tsai, S.J., Test vector generation for linear analog devices, *IEEE ITC*, 1990, pp. 592–597.
[26] Ben Hamida, N. and Kaminska, B., Analog circuits testing based on sensitivity computation and new circuit modeling, *IEEE International Test Conference*, 1993, pp. 652–661.
[27] Devarayandurg, G. and Soma, M., Analytical fault modeling and static test generation for analog IC, *IEEE ICCAD*, 1994, pp. 44–47.
[28] Director, S.W. and Rohrer, A., The generalized adjoint network and network sensitivities, *IEEE Trans. Circuit Theory*, CT-16, 318–323, 1969.
[29] Sunter, S. and Roy, A., BIST for PLL SIN digital applications, *IEEE International Test Conference*, September 1999, pp. 532–540.
[30] Seongwon, K., and Soma, M., An all-digital BIST for high-speed PLL, *IEEE Trans. Circuits Syst. II*, 48, 141–150, 2001.
[31] Sunter, S., The P1149.4 mixed-signal test bus: cost and benefits, *Proceedings of the IEEE ITC*, 1995, pp. 444–45.
[32] Arabi, K., and Kaminska, B., Method of Dynamic On-Chip Digital Integrated Circuit Testing, U.S. Patent No. 6,223,314, B1, April 2001.
[33] Arabi, K., Kaminska, B., Oscillation–Based Test Method for Testing an at least Partially Analog Circuit, U.S. Patent No. 6,005,407, December 1999.
[34] Arabi, K., and Kaminska, B., A new BIST scheme dedicated to digital-to-analog and analog-to-digital converters, *IEEE Des. Test Comput.*, 13, 37–42, 1996.
[35] Arabi, K., and Kaminska, B., Oscillation-test strategy for analog and mixed-signal circuits, *IEEE VLSI Test Symposium*, Princeton, 1996, pp. 476–482.
[36] Arabi, K., and Kaminska, B., Design for testability of integrated operational amplifiers using oscillation-test strategy, *IEEE International Conference on Computer Design (ICCD)*, Austin, October 1996, pp. 40–45.
[37] Arabi, K., and Kaminska, B., Testing analog and mixed-signal integrated circuits using oscillation-test method, *IEEE Trans. CAD IC&S*, 16, 745–753, 1997.
[38] Arabi, K. and Kaminska, B., Oscillation built-in self-test scheme for functional and structural testing of analog and mixed-signal integrated circuits, *IEEE International Test Conference*, November 1997.
[39] Arabi, K., and Kaminska, B., Parametric and catastrophic fault coverage of oscillation built-in self-test, *IEEE VLSI Test Symposium*, April 1997.
[40] Arabi, K., and Kaminska, B., Efficient and accurate testing of analog-to-digital converters using oscillation-test method, *European Design & Test Conference*, Paris, France, March 1997.

[41] Arabi, K., and Kaminska, B., Design for testability of embedded integrated operational amplifiers, *IEEE J. Solid-State Circuits*, , 33, 573–581, 1998.
[42] Arabi, K., and Kaminska, B., Integrated temperature sensors for on-line thermal monitoring of microelectronic structure, *J. Electron. Testing, JETTA*, 12,81–92, 1998.
[43] Arabi, K. and Kaminska, B., Oscillation built-in self-test of mixed-signal IC with temperature and current monitoring, *JETTA Special Issue on On-Line Testing*, 12, 93–100, 1998.
[44] Arabi, K. and Kaminska, B., Oscillation test methodology for low-cost testing of active analog filters, *IEEE Trans. Instrum. Meas.*, 48, 798–806, 1999.
[45] Arabi, B., His, H., and Kaminska, B., Dynamic digital integrated circuit testing using oscillation test method, *IEE Electron. Lett.*, 34, 762–764, 1998.
[46] Arabi, K., Hs, H., Dufaza, C., and Kaminska, B., Digital oscillation test method for delay and structural testing of digital circuits, *IEEE International Test Conference*, October 1998, pp. 91–100.
[47] Norworthy, R., Schreier, R., and Temes, G.C., *Delta–Sigma Data Converters: Theory, Design, and Simulation*, IEEE Press, Washington, DC, 1997.
[48] Tabatabaei, S. and Ivanov, A., An embedded core for subpicoseconds timing measurements, *IEEE ITC*, , October 2002, pp. 129–137.
[49] Kossel, M. and Schmatz, M.L., Jitter measurements of high-speed serial links, *IEEE Des. Test Comput.*, 536–543, 2004..
[50] Frisch, A. and Rinderknecht, T., Jitter Measurement System and Method, U.S. Patent No. 6,295,315, Fluence Technologies, September 2001.
[51] Kaminska, B. and Sokolowska, E., Floating infrastructure IP, *IEEE ITC Test Week*, 2003.
[52] OPMAXX, Fluence, Credence Corporation — Technical Notes, 1997–2002.
[53] Ong, C. et al., A scalable on-chip jitter extraction technique, *IEEE VTS*, , 2004, pp. 267–272.
[54] Frisch, A. and Almy, T., HBIST: histogram-based analog BIST, *IEEE ITC*, November 1997.
[55] Kerman et al., Hardware histogram processor, *COMPCON'1979*, p. 79.
[56] Crosby, P., Data Occurrence Frequency Analyzer, U.S. Patent No. 5,097,428, 1989.
[57] Browing, C., Testing A/D converter on microprocessors, *IEEE ITC*, 1985, pp. 818–824.
[58] Bobba, R. et al., Fast embedded A/D converter testing using the microcontrollers' resources, *IEEE ITC*, 1990, pp. 598–604.
[59] Arabi, K., Kaminska, B., and Rzeszut, J., BIST for DAC and ADC, *IEEE Des.Test Comput.s*, 13, 40–49, 1996.
[60] Arabi, K., Kaminska, B., and Rzeszut, J., A new BIST approach for medium to high-resolution DAC, *The 3rd Asian Test Symposium*, Nara, Japan, November 1994.
[61] Teraoca, E. et al., A built-in self test for ADC and DAC in a single-chip speech Codec, *IEEE ITC*, 1997, pp. 791–796.
[62] Aziz, P.M. et al., An overview of sigma–delta converters, *IEEE Signal Processing Magazine*, January 1996, pp. 61–84.
[63] Kenney, J.G. and Carley, L.R., Design of multi-bit noise-shaping data converters, *J. Analog Integrated Circuits Signal Process.*, 3, 259–272, 1993.
[64] Vazquez, D. et al., On-chip evaluation of oscillation-based test output signals for switched capacitor circuits, *Analog IC Signal Process.*, 33, 201–211, 2002.
[65] Pineda, J. et al., RF test best practices, *RF Test Workshop, DATE 2004*, Paris, 2004, pp. 43–49.
[66] Akbay, S. and Chatterjee, A., Feature extraction-based built-in alternate test of RF components using a noise reference, *IEEE VTS*, April 2004, pp. 273–278.
[67] Jarwala, M. et al., End-to-end strategy for wireless systems, *IEEE ITC*, 1995, pp. 940-946,
[68] Ferrario, J. et al., Architecting millisecond test solution for wireless phone RFIC's, *IEEE ITC*, 2003, pp.1325–1332.
[69] Force, C., Reducing the cost of high-frequency test, *RF Test Workshop, DATE 2004*, Paris, 2004, pp. 36–39.
[70] Kaminska, B. et al., Analog and mixed-signal benchmark circuits — First release, *IEEE International Test Conference*, November 1997.
[71] Kaminska, B., Multi-gigahertz electronic components: practicality of CAD tools, *DATE 2004 — W3*, Paris, March 2004.
[72] Arabi, K., Kaminska, B., and Rzeszut, J., A new built-in self-test approach for digital-to-analog and analog-to-digital converters, *IEEE/ACM International Conference on CAD*, San Jose, CA, November 1994, pp. 491–494.
[73] www.logicvision.com.

Index

A

D

E

I

J

K

L

M

N

O

P

Q

R

S

T

U

V

W

X

Z

EDA for IC Implementation, Circuit Design, and Process Technology

Electronic Design Automation for Integrated Circuits Handbook

Edited by
Louis Scheffer, Luciano Lavagno, and Grant Martin

EDA for IC System Design, Verification, and Testing

EDA for IC Implementation, Circuit Design, and Process Technology

EDA for IC Implementation, Circuit Design, and Process Technology

Edited by

Louis Scheffer
Cadence Design Systems
San Jose, California, U.S.A.

Luciano Lavagno
Cadence Berkeley Laboratories
Berkeley, California, U.S.A.

Grant Martin
Tensilica Inc.
Santa Clara, California, U.S.A.

CRC Taylor & Francis
Taylor & Francis Group
Boca Raton London New York

A CRC title, part of the Taylor & Francis imprint, a member of the Taylor & Francis Group, the academic division of T&F Informa plc.

Published in 2006 by
CRC Press
Taylor & Francis Group
6000 Broken Sound Parkway NW, Suite 300
Boca Raton, FL 33487-2742

Printed in the United States of America on acid-free paper
10 9 8 7 6 5 4 3 2 1

International Standard Book Number-10: 0-8493-7924-5 (Hardcover)
International Standard Book Number-13: 978-0-8493-7924-6 (Hardcover)
Library of Congress Card Number 2005052941

Library of Congress Cataloging-in-Publication Data

EDA for IC implementation, circuit design, and process technology / editors, Louis Scheffer, Luciano Lavagno, Grant Martin.
p. cm. -- (Electronic design automation for integrated circuits handbook)
Includes bibliographical references and index.
ISBN 0-8493-7924-5
1. Integrated circuits--Computer-aided design. 2. Integrated circuits--Design and construction. I. Title: Electronic design automation for integrated circuit implementation, circuit design, and process technology. II. Scheffer, Louis. III. Lavagno, Luciano, 1959- IV. Martin, Grant (Grant Edmund) V. Series.

TK7874.E257 2005
621.3815--dc22 2005052941

informa
Taylor & Francis Group
is the Academic Division of Informa plc.

Visit the Taylor & Francis Web site at
http://www.taylorandfrancis.com

and the CRC Press Web site at
http://www.crcpress.com

Acknowledgments and Dedication for the EDA Handbook

The editors would like to acknowledge the unsung heroes of EDA, those who work to advance the field in addition to their own personal, corporate, or academic agendas. These men and women serve in a variety of ways — they run the smaller conferences, they edit technical journals, and they serve on standards committees, just to name a few. These largely volunteer jobs won't make anyone rich or famous despite the time and effort that goes into them, but they do contribute mightily to the remarkable and sustained advancement of EDA. Our kudos to these folks, who don't get the credit they deserve.

On a more personal note, Louis Scheffer would like to acknowledge the love, support, encouragement, and help of his wife Lynde, his daughter Lucynda, and his son Loukos. Without them this project would not have been possible.

Luciano Lavagno would like to thank his wife Paola and his daughter Alessandra Chiara for making his life so wonderful.

Grant Martin would like to acknowledge, as always, the love and support of his wife, Margaret Steele, and his two daughters, Jennifer and Fiona.

Preface

Preface for Volume 2

Electronic Design Automation (EDA) is a spectacular success in the art of engineering. Over the last quarter of a century, improved tools have raised designers' productivity by a factor of more than a thousand. Without EDA, Moore's law would remain a useless curiosity. Not a single billion-transistor chip could be designed or debugged without these sophisticated tools, so without EDA we would have no laptops, cell phones, video games, or any of the other electronic devices we take for granted.

But spurred on by the ability to build bigger chips, EDA developers have largely kept pace, and these enormous chips can still be designed, debugged, and tested,-and in fact, with decreasing time to market.

The story of EDA is much more complex than the progression of integrated circuit (IC) manufacturing, which is based on simple physical scaling of critical dimensions. Instead, EDA evolves by a series of paradigm shifts. Every chapter in this book, all 49 of them, was just a gleam in some expert's eye just a few decades ago. Then it became a research topic, then an academic tool, and then the focus of a startup or two. Within a few years, it was supported by large commercial EDA vendors, and is now part of the conventional wisdom. Although users always complain that today's tools are not quite adequate for today's designs, the overall improvements in productivity have been remarkable. After all, in what other field do people complain of *only* a 21% compound annual growth in productivity, sustained over three decades, as did the International Technology Roadmap for Semiconductors in 1999?

And what is the future of EDA tools? As we look at the state of electronics and integrated circuit design in the 2005–2006 timeframe, we see that we may soon enter a major period of change in the discipline. The classical scaling approach to integrated circuits, spanning multiple orders of magnitude in the size of devices over the last 40+ years, looks set to last only a few more generations or process nodes (though this has been argued many times in the past, and has invariably been proved to be too pessimistic a projection). Conventional transistors and wiring may well be replaced by new nano and biologically-based technologies that we are currently only beginning to experiment with. This profound change will surely have a considerable impact on the tools and methodologies used to design integrated circuits. Should we be spending our efforts looking at CAD for these future technologies, or continue to improve the tools we currently use?

Upon further consideration, it is clear that the current EDA approaches have a lot of life left in them. With at least a decade remaining in the evolution of current design approaches, and hundreds of thousands or millions of designs left that must either craft new ICs or use programmable versions of them, it is far too soon to forget about today's EDA approaches. And even if the technology changes to radically new forms and structures, many of today's EDA concepts will be reused and evolved for design into technologies well beyond the current scope and thinking.

The field of EDA for ICs has grown well beyond the point where any single individual can master it all, or even be aware of the progress on all fronts. Therefore, there is a pressing need to create a snapshot of this extremely broad and diverse subject. Students need a way of learning about the many disciplines and topics involved in the design tools in widespread use today. As design grows multi-disciplinary, electronics designers and EDA tool developers need to broaden their scope. The methods used in one subtopic may well have applicability to new topics as they arise. All of electronics design can utilize a comprehensive reference work in this field.

With this in mind, we invited many experts from across all the disciplines involved in EDA to contribute chapters summarizing and giving a comprehensive overview of their particular topic or field. As might be appreciated, such chapters represent a snapshot of the state of the art, written in 2004–2005. However, as surveys and overviews, they retain a lasting educational and reference value that will be useful to students and practitioners for many years to come.

With a large number of topics to cover, we decided to split the Handbook into two volumes. Volume 1 covers system-level design, micro-architectural design, and verification and test. Volume 2 covers the classical "RTL to GDS II" design flow, incorporating synthesis, placement and routing, along with related topics; analog and mixed-signal design, physical verification, analysis and extraction, and technology CAD topics. These roughly correspond to the classical "front-end/back-end" split in IC design, where the front end (or logical design) focuses on making sure that the design does the right thing, assuming it can be implemented, and the back-end (or physical design) concentrates on generating the detailed tooling required, while taking the logical function as given. Despite limitations, this split has persisted through the years — a complete and correct logical design, independent of implementation, remains an excellent handoff point between the two major portions of an IC design flow. Since IC designers and EDA developers often concentrate on one side of this logical/physical split, this seemed to be a good place to divide the book as well.

Volume II opens with an overview of the classical RTL to GDS II design flows, and then steps immediately into the logic synthesis aspect of "synthesis, place and route." Power analysis and optimization methods recur at several stages in the flow. Recently, equivalence checking has increased the reliability and automation possible in the standard IC flows. We then see chapters on placement and routing and associated topics of static timing analysis and structured digital design. The standard back end flow relies on standard digital libraries and design databases, and must produce IC designs that fit well into packages and onto boards and hybrids. The relatively new emphasis on design closure knits many aspects of the flow together. Indeed, chapter 10, on design closure, is a good one to read right after Chapter 1, on design flows.

Before diving into the area of analog and mixed-signal design, the handbook looks at the special methods appropriate to FPGA design-this is a growing area for rapid IC design using underlying fixed but reprogrammable platforms. Then we turn to analog design, where we cover simulation methods, advanced modeling, and layout tools. Physical verification, analysis and extraction covers design rule checking, transformation of designs for manufacturability, analysis of power supply noise and other noise issues, and layout extraction. Finally, the handbook looks at process simulation and device modeling, and advanced parasitic extraction as aspects of technology CAD for ICs.

This handbook with its two constituent constitutes a valuable learning and reference work for everyone involved and interested in learning about electronic design and its associated tools and methods. We hope that all readers will find it of interest and a well-thumbed resource.

Louis Scheffer
Luciano Lavagno
Grant Martin
San Jose, Berkeley, and Santa Clara

Editors

Louis Scheffer

Louis Scheffer received the B.S. and M.S. degrees from Caltech in 1974 and 1975, and a Ph.D. from Stanford in 1984. He worked at Hewlett Packard from 1975 to 1981 as a chip designer and CAD tool developer. In 1981, he joined Valid Logic Systems, where he did hardware design, developed a schematic editor, and built an IC layout, routing, and verification system. In 1991, Valid merged with Cadence, and since then he has been working on place and route, floorplanning systems, and signal integrity issues.

His main interests are floorplanning and deep sub-micron effects. He has written many technical papers, tutorials, invited talks and panels, and has served the conferences DAC, ICCAD, ISPD, SLIP, and TAU as a technical committee member. He is currently the general chair of TAU and ISPD, and on the steering committee of SLIP. He holds five patents in the field of EDA, and has taught courses on CAD for electronics at Berkeley and Stanford. He is also interested in SETI, and serves on the technical advisory board for the Allen Telescope Array at the SETI institute, and is a co-author of the book SETI-2020, in addition to several technical articles in the field.

Luciano Lavagno

Luciano Lavagno received his Ph.D. in EECS from U.C. Berkeley in 1992 and from Politecnico di Torino in 1993. He is a co-author of two books on asynchronous circuit design, of a book on hardware/software co-design of embedded systems, and of over 160 scientific papers.

Between 1993 and 2000, he was the architect of the POLIS project, a cooperation between U.C. Berkeley, Cadence Design Systems, Magneti Marelli and Politecnico di Torino, which developed a complete hardware/software co-design environment for control-dominated embedded systems.

He is currently an Associate Professor with Politecnico di Torino, Italy and a research scientist with Cadence Berkeley Laboratories. He is serving on the technical committees of several international conferences in his field (e.g. DAC, DATE, ICCAD, ICCD) and of various workshops and symposia. He has been the technical program and tutorial chair of DAC, and the TPC and general chair of CODES. He has been associate and guest editor of IEEE Transactions on CAD, IEEE Transactions on VLSI and ACM Transactions on Embedded Computing Systems.

His research interests include the synthesis of asynchronous and low-power circuits, the concurrent design of mixed hardware and software embedded systems, and compilation tools and architectural design of dynamically reconfigurable processors.

Grant Martin

Grant Martin is a chief scientist at Tensilica, Inc. in Santa Clara, California. Before that, Grant worked for Burroughs in Scotland for 6 years; Nortel/BNR in Canada for 10 years; and Cadence Design Systems for 9 years, eventually becoming a Cadence Fellow in their Labs. He received his Bachelor's and Master's degrees in Mathematics (Combinatorics and Optimization) from the University of Waterloo, Canada, in 1977 and 1978.

Grant is a co-author of "Surviving the SOC Revolution: A Guide to Platform-Based Design", 1999, and "System Design with SystemC", 2002, and a co-editor of the books "Winning the SoC Revolution: Experiences in Real Design", and "UML for Real: Design of Embedded Real-Time Systems", June 2003, all published by Springer (originally by Kluwer). In 2004, he co-wrote with Vladimir Nemudrov the first book on SoC design published in Russian by Technosphera, Moscow. Recently he co-edited "Taxonomies for the Development and Verification of Digital Systems" (Springer, 2005), and "UML for SoC Design" (Springer, 2005).

He has also presented many papers, talks and tutorials, and participated in panels, at a number of major conferences. He co-chaired the VSI Alliance Embedded Systems study group in the summer of 2001, and is currently co-chair of the DAC Technical Programme Committee for Methods for 2005 and 2006. His particular areas of interest include system-level design, IP-based design of system-on-chip, platform-based design, and embedded software. Grant is a Senior Member of the Institute of Electrical and Electronics Engineers (IEEE).

Contributors

Mark Bales
Reshape, Inc.
San Jose, California

Mark Basel
Mentor Graphics, Inc.
Wilsonville, Oregon

Scott T. Becker
Tela Technologies, Inc.
Houston, Texas

Vaughn Betz
Altera Corp.
Toronto, Ontario, Canada

David Blaauw
University of Michigan
Ann Arbor, Michigan

Robert K. Brayton
University of California
Berkeley, California

Rajat Chaudhry
Freescale Semiconductor Inc.
Austin, Texas

David Chinnery
University of California
Berkeley, California

Chang-Hoon Choi
Stanford University
Palo Alto, California

John M. Cohn
IBM Systems and Technology Group
Essex Junction, Vermont

Nicola Dragone
PDF Solutions, Inc.
Brescia, Italy

Robert W. Dutton
Stanford University
Palo Alto, California

Katherine Fetty
Mentor Graphics, Inc.
Wilsonville, Oregon

Paul D. Franzon
North Carolina State University
Raleigh, North Carolina

Georges G.E. Gielen
Katholieke Universiteit Leuven
Leuven, Belgium

Laurence Grodd
Mentor Graphics, Inc.
Wilsonville, Oregon

Carlo Guardiani
PDF Solutions, Inc.
Brescia, Italy

David Hathaway
IBM Microelectronics
Essex Junction, Vermont

James Hogan
Telos Venture Partners, Inc.
Palo Alto, California

Mike Hutton
Altera Corp.
San Jose, California

Ralph Iverson
Magma Design Automation
Arlington, Massachusetts

Mark D. Johnson
Synopsys, Inc.
Mountain View, California

Andrew B. Kahng
University of California
San Diego, California

Mattan Kamon
Coventor, Inc.
Cambridge, Massachusetts

Edwin C. Kan
Stanford University
Palo Alto, California

William Kao
Cadence Design Systems, Inc.
San Jose, California

Vinod Kariat
Cadence Berkeley Laboratories
Berkeley, California

Kurt Keutzer
University of California
Berkeley, California

Sunil P. Khatri
Texas A&M University
College Station, Texas

Andreas Kuehlmann
Cadence Berkeley Laboratories
Berkeley, California

Chi-Yuan Lo
Cadence Design Systems, Inc.
New Providence, New Jersey

Alan Mantooth
University of Arkansas
Fayetteville, Arkansas

Fan Mo
Synplicity
Sunnyvale, California

Jose Monteiro
INESC-Instituto de Engenharia de Sistems e Computadores
Lisboa, Portugal

Makoto Nagata
Kobe University
Kobe, Japan

Peter J. Osler
IBM Systems and Technology Group
Essex Junction, Vermont

Rajendran Panda
Freescale Semiconductor Inc.
Austin, Texas

Sanjay Pant
University of Michigan
Ann Arbor, Michigan

Rakesh Patel
Intel Corp.
Chandler, Arizona

Joel R. Phillips
Cadence Berkeley Laboratories
Berkeley, California

Sherief Reda
University of California
San Diego, California

Jaijeet Roychowdhury
University of Minnesota
Minneapolis, Minnesota

Rob A. Rutenbar
Carnegie Mellon University
Pittsburgh, Pennsylvania

Sachin S. Sapatnekar
University of Minnesota
St. Paul, Minnesota

Louis Scheffer
Cadence Design Systems, Inc.
San Jose, California

Franklin M. Schellenberg
Mentor Graphics, Inc.
San Jose, California

Narendra V. Shenoy
Synopsys Inc.
Mountain View, California

Raminderpal Singh
IBM Corporation
Cartlandt Manor, New York

Fabio Somenzi
University of Colorado
Boulder, Colorado

Peter Spink
Cadence Design Systems, Inc.
Berkeley, California

Leon Stok
IBM Corporation
TJ Watson Research Center
Yorktown Heights, New York

Andrzej J. Strojwas
PDF Solutions, Inc.
San Jose, California

Vivek Tiwari
Intel Corp.
Santa Clara, California

Robert Todd
Mentor Graphics, Inc.
Wilsonville, Oregon

Nishath Verghese
Cadence Berkeley Laboratories
Berkeley, California

Contents

SECTION I RTL to GDS-II, or Synthesis, Place, and Route

SECTION III Physical Verification

SECTION IV Technology CAD

SECTION I

RTL TO GDS-II, OR SYNTHESIS, PLACE, AND ROUTE

1

Design Flows

Leon Stok
IBM Corporation
TJ Watson Research Center
Yorktown Heights, New York

David Hathaway
IBM Microelectronics
Essex Junction, Vermont

Kurt Keutzer
University of California
Berkeley, California

David Chinnery
University of California
Berkeley, California

1.1 Introduction

Scaling has driven the entire IC implementation RTL to GDSII design flow from one which uses primarily standalone synthesis, placement, and routing algorithms to an integrated construction and analysis flow for design closure. This chapter will address how the challenges of rising interconnect delay led to a new way of thinking about and integrating design closure tools. New scaling challenges such as leakage power, variability, and reliability will keep on challenging the current state of the art in design closure.

The RTL to GDSII flow has undergone significant changes in the last 25 years. The continued scaling of CMOS technologies significantly changed the objectives of the various design steps. The lack of good predictors for delay has led to significant changes in recent design flows. Challenges like leakage power, variability, and reliability will continue to require significant changes to the design-closure process in the future. In this chapter we will describe what drove the design flow from a set of separate design steps to a fully integrated approach, and what further changes we see coming to address the latest challenges.

In his keynote at the 40th Design Automation Conference entitled "The Tides of EDA" [1], Alberto Sangiovanni-Vincentelli distinguished three periods of EDA: The Age of the Gods, The Age of the Heroes, and The Age of the Men. These eras were characterized respectively by senses, imagination, and reason.

When we limit ourselves to the RTL to GDSII flow of the CAD area, we can distinguish three main eras in its development: the Age of Invention, the Age of Implementation, and the Age of Integration. During the invention era, routing, placement, static timing analysis and logic synthesis were invented. In the age of implementation they were drastically improved by designing sophisticated data structures and advanced algorithms. This allowed the tools in each of these design steps to keep pace with the rapidly increasing design sizes. However, due to the lack of good predictive cost functions, it became impossible to execute a design flow by a set of discrete steps, no matter how efficiently implemented each of the steps was. This led to the age of integration where most of the design steps are performed in an integrated environment, driven by a set of incremental cost analyzers.

Let us look at each of the eras in more detail and describe some of their characteristics.

1.2 Invention

In the early days, basic algorithms for routing, placement, timing analysis, and synthesis were *invented.* Most of the early invention in physical design algorithms was driven by package and board designs. Real estate was at a premium, only a few routing layers were available and pins were limited. Relatively few discrete components needed to be placed. Optimal algorithms of high complexity were not a problem since we were dealing with few components.

In this era, basic routing, partitioning, and placement algorithms were invented. Partitioning is one of the fundamental steps in the physical design flow. Kernighan and Lin [2] pioneered one of the basic partitioning techniques in 1970. Simulated annealing [3] algorithms were pioneered for placement, and allowed for a wide range of optimization criteria to be deployed. Basic algorithms for channel, switchbox, and maze routing [4] were invented. By taking advantage of restricted topologies and design sizes, optimal algorithms could be devised to deal with these particular situations.

1.3 Implementation

With the advent of integrated circuits, more and more focus shifted to design automation algorithms to deal with them, rather than boards. Traditional CMOS scaling allowed the sizes of these designs to grow very rapidly. As design sizes grew, design tool implementation became extremely important to keep up with the increasingly larger designs and to keep design time under control. New implementations and data structures were pioneered and algorithms that scaled most efficiently became the standard.

As design sizes started to pick up, new layers of abstraction were invented. The invention of standard cells allowed one to separate the detailed physical implementation of the cells from the footprint image that is seen by the placement and routing algorithms. Large-scale application of routing, placement, and later synthesis algorithms took off with the introduction of the concept of standard cells.

The invention of the standard cell can be compared to the invention of the printing press. While manual book writing was known before, it was a labor-intensive process. The concept of keeping the height of the letters fixed and letting the width of the lead base of each of the letters vary according to the letter's size, enabled significant automation in the development of printing. Similarly, in standard cell Application Specific Integrated Circuits (ASIC) design, one uses standard cells of common height but varying widths depending on the complexity of the single standard cell. These libraries (Chapter 9) created significant levels of standardization and enabled large degrees of automation. The invention of the first gate-arrays took the standardization to an even higher level.

This standardization allowed the creation of the ASICs business model, which created a huge market opportunity for automation tools and spawned a number of innovations. Logic synthesis [5] was invented to bridge the gap from language descriptions to these standard cell implementations.

In the implementation era, a design flow could be pasted together from a sequence of discrete steps (see Figure 1.1). High-level synthesis translated a Verilog or VHDL description into an RTL netlist. Logic synthesis optimized the netlist and mapped it into a netlist with technology gates. This was followed by placement to place the gates, and routing to connect them together. Finally, a timing simulator was used to verify the timing using a limited amount of extracted data.

Design sizes kept on increasing rapidly in this era. While in the discrete space, several tens of components needed to be placed and routed, logic synthesis allowed for rapid creation of netlists with millions of gates, growing from 40,000 gates in 1984 to 40,000,000 gate designs in 2000.

New data structures like Quadtrees [6] allowed very efficient searches in the geometric space. Applications of Boolean Decision Diagrams (BDDs) [7] enabled efficient Boolean reasoning on significantly larger logic partitions.

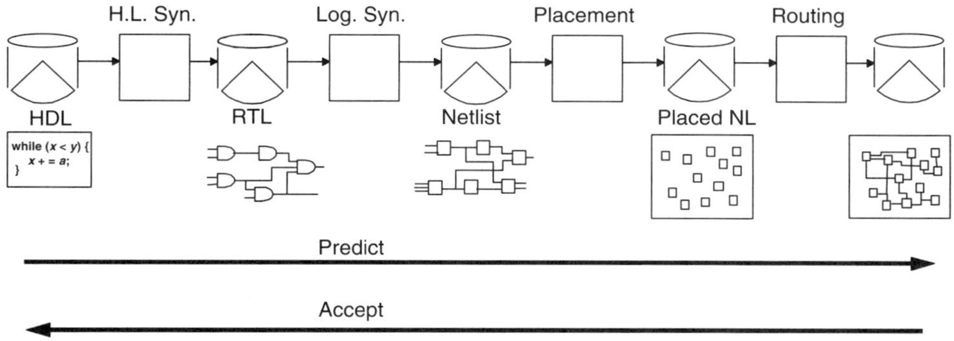

FIGURE 1.1 Sequential design flow.

Much progress was made in implementing partitioning algorithms. A more efficient version of Kernighan and Lin's partitioning algorithm was given by Fidducia and Mattheyses [8]. They used a specific algorithm for selecting vertices to move across the cut that saved runtime and allowed for the handling of unbalanced partitions and nonuniform edge weights. An implementation using spectral methods [9] proved to be very effective for certain problems. Yang [10] demonstrated results that outperformed the two earlier mentioned methods by applying a network flow algorithm iteratively.

Optimizing quadratic wire length became the holy grail in placement. Quadratic algorithms took full advantage of this by deploying efficient quadratic optimization algorithms, intermixed with various types of partitioning schemes [11].

Original techniques in logic synthesis, such as kernel and cube factoring, were applied to small partitions of the network at a time. More efficient algorithms like global flow [12] and redundancy removal [13] based on test generation could be applied to much larger designs. Complete coverage of all timing paths by timing simulation became too impractical due to its exponential dependence on design size, and static timing analysis [14] based on early work in [15] was invented.

With larger designs came more routing layers, allowing over-the-cell routing and sharing of intracell and intercell routing areas. Gridded routing abstractions matched the standard cell templates well and became the base for much improvement in routing speed. Hierarchical routing abstractions such as global routing, switch box, and area routing were pioneered as effective ways to decouple the routing problem.

Algorithms that are applicable to large-scale designs must have order of complexity less than $O(n^2)$ and preferably not more than $O(n \log n)$. These complexities were met because of the above-mentioned advances in data structures and algorithms. This allowed design tools to be applied to large real problems. However, it became increasingly difficult to find appropriate cost functions for these algorithms. Accurate prediction of the physical effects earlier in the design flow became more difficult.

Let us discuss how the prediction of important design metrics evolved over time during the implementation era. In the beginning of the implementation era, most of the important design metrics such as area and delay were quite easy to predict. The optimization algorithms in each of the discrete design steps were guided by objective functions that rely on these predictions. As long as the final values of these metrics could be predicted with good accuracy, the RTL to GDSII flow could indeed be executed in a sequence of fairly independent steps. However, the prediction of important design metrics was becoming increasingly difficult. As we will see in the following sections, this led to fundamental changes in the design closure flow. The simple sequencing of design steps was not sufficient anymore.

Let us look at one of the prediction functions, the measurement of delay, in more detail. In the early technologies, the delay along a path was dominated by the delay of the gates. In addition, the delay of most gates was very similar. As a result, as long as one knew how many gates there were on the critical path, one could reasonably predict the delay of the path, by counting the number of levels of gates on a path and multiplying that with a typical gate delay. The delay of a circuit was therefore known as soon as logic synthesis had determined the number of logic levels on each path. In fact, in early timing optimization, multiple gate sizes were used to keep delays reasonably constant across different loadings, rather than to actually improve the delay of a gate. Right after the mapping to technology dependent standard cells, the area could be pretty well predicted by adding up the cell areas. Neither subsequent placement nor routing steps would change these two quantities significantly. Power and noise were not of very much concern in these times.

In newer libraries, the delays of gates with different logic complexities started to vary significantly. Table 1.1 shows the relative delays of different types of gates. The logical effort indicates how the delay of the gate increases with load and the intrinsic delay is the load-independent contribution of the gate delay. The FO4 delay is the delay of one gate of the specified type driving four identical gates of the same size. The fourth column of the table shows that the delay of a more complex NOR4, driving four copies of itself, can be as much as three times the delay of a simple inverter. The logical effort-based calculation to compute this already assumes that the gates have been ideally sized to best match the load they are driving. Simple addition of logic levels is therefore becoming insufficient, and one needs to know what gates the logic is actually mapped to, to predict the delay of a design with reasonable accuracy. It became necessary to include a static timing analysis engine (Chapter 6) in the synthesis system to calculate these delays. The combination of timing and synthesis was the first step on the way to the era of integration. This trend started gradually in the 1980s; but by the beginning of the 1990s, integrated static timing analysis tools were essential to predict delays accurately. Once a netlist was mapped to a particular technology and the gate loads could be approximated, a pretty accurate prediction of the delay could be made by the timing analyzer.

At that time, approximating the gate load was relatively easy. The load was dominated by the input capacitances of the gates that were driven. The fact that the capacitance of the wire was assumed by a bad wire load model was hardly an issue. Therefore, as long as the netlist was not modified in the subsequent steps, the delay prediction was quite reliable.

Toward the middle of the 1990s, these predictions based on gate delays started to be much more inaccurate. Gate delays became increasingly dependent on the load they were driving, and on the rise and fall rates of the input signals to the gates. At the same time, the fraction of net load due to wires started to increase. Knowledge of the physical design became essential to predict reasonably the delay of a path. Initially, it was mostly just the placement of the cells that was needed. The placement affects the delay, but wiring does so much less, since any route close to the minimum length will have similar load.

In newer technologies, more and more of the delay started to shift toward the interconnect. Both gate and wire (RC) delay really began to matter. Figure 1.2 shows how the gate delay and interconnect delay compare over a series of technology generations. With a Steiner tree approximation of the global routing, the lengths of the nets could be reasonably predicted. Using these lengths, delays could be calculated for the longer nets. The loads from the short nets were not very significant and a guesstimate of these was still appropriate. Rapidly, it became clear that it was very important to buffer the long nets really well. In Figure 1.2, we see for example that around the 130 nm node, the difference between a net with repeaters inserted at the right places and an unbuffered net starts to have a significant impact on the delay. Buffering of long lines became an integral part of physical design, in addition to placing and routing [16].

Recently, we are seeing that a wire's environment is becoming more significant. The cross-coupling capacitance between wires has increased as the ratio between wire spacing and wire height decreases. The detailed routing is therefore becoming significant in predicting actual delays.

TABLE 1.1 Gate Delays

Gate	Logical Effort	Intrinsic Delay	FO4 Delay
INV	1.00	1.00	5.00
NAND2	1.18	1.34	6.06
NAND3	1.39	2.01	7.57
NAND4	1.62	2.36	8.84
NOR2	1.54	1.83	7.99
NOR3	2.08	2.78	11.10
NOR4	2.63	3.53	14.05

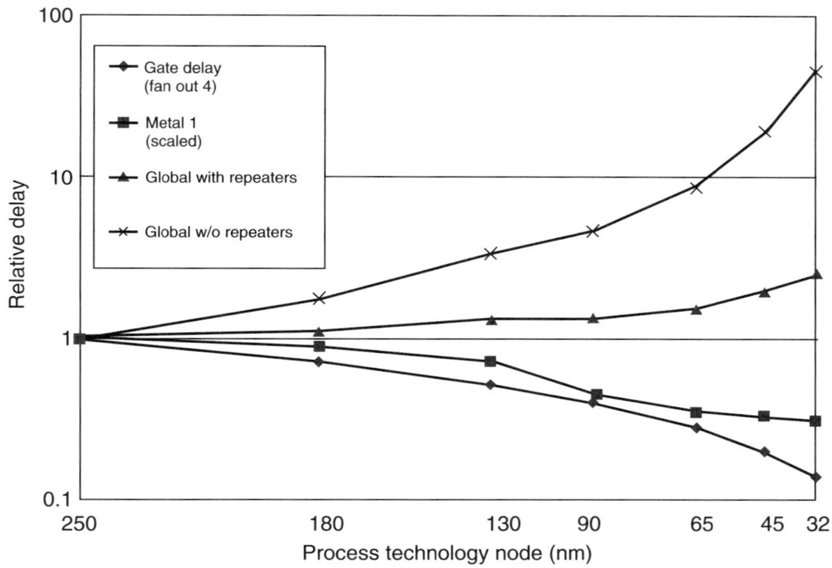

FIGURE 1.2 Gate and interconnect delay.

Net list alterations, traditionally done in logic synthesis, became an important part of place and route. Placement and routing systems that were designed to deal with static (not changing) netlists had to be reinvented.

1.4 Integration

This decrease in predictability continued and firmly planted us in the age of *integration*. The following are some of the characteristics of this era:

- The impact of later designs steps is increasing
- Prediction is difficult
- Larger designs allow less manual intervention
- New cost functions are becoming more important
- Design decisions interact
- Aggressive designs allow less guardbanding

The iterations between sequential design steps such as repeater insertion, gate sizing, and placement steps not only became cumbersome and slow, but also often did not even converge. People have explored several possible solutions to this convergence problem including:

- Insert controls for later design steps into the design source
- Fix problems at the end
- Improve prediction
- Concurrently design in different domains

The insertion of controls proved to be very difficult. As illustrated in Figure 1.3, the path through which source modifications influence the final design result can be very indirect, and it can be very hard to understand the effect of particular controls with respect to a specific design and tools methodology. Controls inserted early in the design flow might have a very different (side-)effect than desired or anticipated on the final result. The fix-up solution requires an enormous increase in manual design effort. Improving predictions has proven to be very difficult. Gain-based delay models trade off area predictability for significant delay predictability and gave some temporary relief, but in general it has been extremely hard to improve

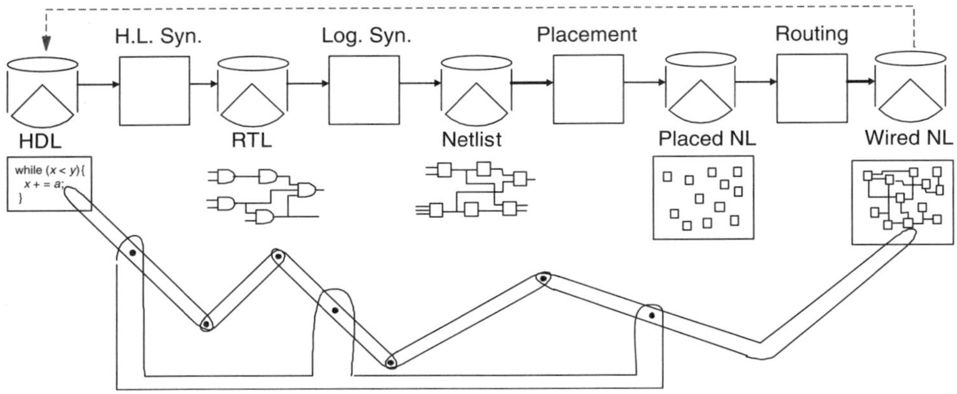

FIGURE 1.3 Controlled design flow.

predictions. The main lever seems to be concurrent design by integrating the synthesis, placement, and routing domains and coupling them with the appropriate design analyzers.

After timing/synthesis integration, placement driven (physical) synthesis was the next major step on the integration path. Placement algorithms were added to the integrated static timing analysis and logic synthesis environments. Well-integrated physical synthesis systems became essential to tackle the design closure problems of the increasingly larger chips.

This integration trend is continuing. Gate-to-gate delay depends on the wire length (unknown during synthesis), the layer of the wire (determined during routing), the configuration of the neighboring wires (e.g., distance — near/far, which is unknown before detailed routing), and the signal arrival times and slope of signals on the neighboring wires. Therefore, in the latest technologies, we see that most of the routing needs to be completed to have a good handle on the timing of a design. Local congestion issues might force a significant number of routing detours. This needs to be accounted for and requires a significantly larger number of nets to be routed earlier in the design closure flow. Coupling between nets affects both noise and delay in larger portions of the design. Therefore, knowledge of the neighbors of a particular net is essential to carry out the analysis to the required level of detail, and requires a significant portion of the local routing to be completed. In addition, power is rapidly becoming a very important design metric, and noise issues are starting to affect significantly delays and even make designs function incorrectly. In addition to integrated static timing analysis, power and noise analysis need to be included as well.

All these analysis and optimization algorithms need to work in an incremental fashion, because runtime prevents us from recalculating all the analysis results when frequent design changes are made by the other algorithms. In the age of integration, not only are the individual algorithms important, but the way they are integrated and can work together to reach closure on the objectives of the design has become the differentiating factor. A fine-grained interaction between these algorithms, guided by fast incremental timing, power, and area calculators has become essential.

In the era of integration, most progress has been made in the way the algorithms cooperate with each other. Most principles of the original synthesis, placement, and routing algorithms, and their efficient implementations are still applicable, but the way in which they are developed into EDA tools has changed significantly. In other cases, the required incrementality has led to interesting new algorithms. While focusing on the integration, we must retain our ability to focus on advanced problems in individual tool domains in order to address new problems posed by technology, and to continue to advance the capabilities of design algorithms.

To achieve this, we have seen a shift to the development of EDA tools guided by three basic interrelated principles:

- Tools are integrated
- Tools are modular
- Tools operate incrementally

Let us look at each of these in more detail in the following sections.

1.4.1 Integrated

Tool integration provides the ability for tools to directly communicate with each other. A superficial form of integration can be provided by initiating the execution of traditional point tools, which continue to operate to and from files, from a common user interface. When we discuss tool integration, however, we will mean a tighter form of integration, which allows tools to communicate while concurrently executing, rather than only through files. This tight integration is generally accomplished by building the tools on a common runtime model (see Chapter 12) or through a standardized message-passing protocol.

Tool integration enables the reuse of functions in different domains because the overhead of repeated file reading and writing is eliminated. This helps to reduce development resource requirements and improve consistency between applications.

Although tool integration enables incremental tool operation, it does not require it. For example, one could integrate placement and wiring programs, and still have the placement run to completion before starting wiring.

Careful design of an application can make it easier to integrate with other applications on a common runtime model, even if it was originally written as a standalone application.

Achieving tool integration requires an agreed upon set of semantic elements in the design representation in terms of which the tools can comunicate. In the design closure flow, these elements generally consist of blocks, pins, and nets and their connectivity, block placement locations, and wiring routes. Individual applications will augment this common model data with domain-specific information. For example, a static timing analysis tool will typically include delay and test edge data structures. In order for an integrated tool to accept queries in terms of the common data model elements, it must be able to find efficiently the components of the domain-specific data associated with these elements. Although this can be accomplished by name look-up, when integrated tools operate within a common memory space it is more efficient to use direct pointers. This in turn requires that the common data model provide methods for applications to attach private pointers to the common model elements.

1.4.2 Modular

Modular tools are developed in small, independent pieces. This gives several benefits. It simplifies incremental development because new algorithms can more easily be substituted for old ones if the original implementation was modular. It facilitates reuse. Smaller modular utilities are more likely to be able to be reused, since they are less likely to have side-effects, which are not wanted in the reuse environment. It simplifies code development and maintenance, since problems are likely to be easier to isolate, and modules are easier to test independently. Tool modularity should be made visible to and usable by application engineers and sophisticated users, allowing them to integrate modular utilities through an extension language.

In the past, some projects have failed in large part due to a lack of modularity [17]. In the interest of collecting all behavior associated with the common runtime model in one place, they also concentrated control of what would appropriately be application-specific data. This made the runtime model too large and complicated, and inhibited the tuning and reorganization of data structures by individual applications.

1.4.3 Incremental

Incrementally operating tools can update information about a design, or the design itself, without revisiting or reprocessing the entire design. This enables finer-grained interaction between integrated tools. For example, incremental processing capability in static timing analysis makes it possible for logic synthesis to change a design and see the effect of that change on timing, without requiring a complete retiming of the design to be performed.

Incremental processing can reduce the time required to make a "loop" between different analysis and optimization tools. As a result, it can make a higher frequency of tool interaction practical, allowing the complete consequences of each optimization decision to be more accurately understood.

The ordering of actions between a set of incremental applications is important. If a tool like synthesis is to make use of another incremental tool like timing analysis, it needs to be able to immediately see the effects of any actions it takes in the results reported by the tool being used. This means that the incremental application must behave as if every incremental update occurs immediately after the event which precipitates it.

There are four basic characteristics desirable in an incremental tool:

1.4.3.1 Autonomicity

Autonomicity means that applications initiating events which precipitate changes in the incremental tool do not need to notify explicitly the incremental tool of those events, and that applications using results from an incremental tool do not need to initiate explicitly or control the incremental processing in that tool. The first of these is important because it simplifies the process required for an application to make changes to a design, and eliminates the need to update the application when new incremental tools are added to the design tool environment. It is usually achieved by providing a facility for incremental applications to register callbacks, allowing them to be notified of events which are of interest.

The second is important because it keeps the application using the incremental tool from needing to understand the details of the incremental algorithm. This allows changes to the incremental tool algorithm without requiring changes in the applications that use it, and reduces the likelihood of errors in the control of the incremental processing, since that control is not scattered among many applications that use the incremental tool. It also makes the use of the incremental tool by other applications much simpler.

1.4.3.2 Lazy Evaluation (Full and Partial)

Lazy evaluation means that an incremental tool should try to the greatest extent possible to defer processing until the results are needed. This can save considerable processing time, which would otherwise be spent getting results that are never used. For example, consider a logic synthesis application making a series of individual disconnections and reconnections of pins to accomplish some logic transformation. If an incremental timing analyzer updates the timing results after each of these individual actions, it will end up recomputing time values many times for many or all of the same points in the design, while only the last values computed are actually used.

Lazy evaluation also simplifies the flow of control when recalculating values. If we have several incremental analysis functions with interdependencies (e.g., timing depends on extraction results), and all try to immediately update results when notified of a netlist change, some callback ordering mechanism would be needed to ensure that the updates are made in the correct order. If in the example given above, timing updates were made before extraction updates, the results would not be correct, as they would be using stale extraction results. However, if each application performs only invalidation when notified of a design change, the updates can be ordered correctly through demand-driven recomputation. In the above example, if the first result requested after a netlist change is from timing, it would in turn request updated extraction results it needed for delay computation. Since all invalidation in all applications would have been completed before any request was made for updated analysis results, the extraction tool would have received its callback and performed any necessary invalidation, so when a value was requested from it by timing, the new value would be computed and returned.

Lazy evaluation may be full if all pending updates are performed as soon as any information is requested, or partial if only those values needed to give the requested result are updated. Note that if partial lazy evaluation is used, the application must still retain enough information to be able to determine which information has not yet been updated, since subsequent requests may be made for some of this other information. Partial lazy evaluation is employed in some timing analyzers [18] by levelizing the design and limiting the propagation of arrival and required times based on this levelization, providing significant benefits in the runtime of the tool.

1.4.3.3 Change Notification (Callbacks and Undirected Queries)

Change notification means having the incremental tool notify other applications of changes that concern them. This is more than just providing a means to query specific pieces of information from the incremental tool, since the application needing the information may not be aware of all changes that have occurred. In the simplest situations, a tool initiating a design change can assume it knows where consequent changes to analysis results (e.g., timing) will result. In this case, no change notification is required. But in other situations, a tool may need to respond not only to direct changes in the semantic elements of the common runtime model, but also to secondary changes within specific application domains (e.g., changes in timing).

Change notification is important because the applications may not know where to find all the incremental changes that affect them. For example, consider an incremental placement tool used by logic synthesis. Logic synthesis might be able to determine the blocks which will be directly replaced as a consequence of some logic change. But if the replacement of these blocks has a ripple effect, which causes the replacement of other blocks (e.g., to open spaces for the first set of replaced blocks), it would be much more difficult for logic synthesis to determine which blocks are in this second set. Without change notification, the logic synthesis system would need to examine all blocks in the design before and after every transformation to ensure that it has accounted for all consequences of that transformation.

Change notification may occur immediately after the precipitating event, or may be deferred until requested. Immediate change notification can be provided through callback routines which applications can register with it, and which are called whenever certain design changes occur.

Deferred change notification requires the incremental tool to accumulate change information until a requesting application is ready for it. This requesting application will issue an undirected query to ask for all changes of a particular type that have occurred since some checkpoint (the same sort of checkpoint required for undoing design changes). It is particularly important for analysis results, since an evaluation routine may be interested only in the cumulative effect of a series of changes, and may neither need nor want to pay the price (in nonlazy evaluation) of immediate change notification.

A typical use of undirected queries is to get information about changes that have occurred in the design, in order to decide whether or not to reverse the actions that caused the changes. For this purpose, information is needed not only about the resultant state of the design, but also about the original state, so that the delta may be determined (did things get better or worse?). Thus, the application making an undirected query needs to specify the starting point from which changes will be measured. This should use the same checkpoint capability used to provide reversibility.

1.4.3.4 Reversibility (Save/Restore and Recompute)

We need to be able to use incremental applications in an "experimental mode," because of the many subtle effects any particular design change may cause, and because of the order of complexity of most design problems. An application makes a trial design change, examines the effects of that change as determined in part by the incremental tools with which it is integrated, and then decides whether to accept or reject the change. If such a change is rejected, we need to make sure that we can accurately recover the previous design state, which means that all incremental tool results must be reversible. It is appropriate for each incremental tool to handle the reversing of changes to the data for which it is responsible. In some cases such as timing analysis, the changed data (e.g., arrival and required times) can be deterministically derived from other design data, and it may be more efficient to recompute them rather than to store and recover them. In other cases such as incremental placement, changes involve heuristics that are not reversible, and previous state data must be saved. The incremental tool which handles the reversing of changes should be the one actually responsible for storing the data being reversed. Thus, changes to the netlist initiated by logic synthesis should be reversed by the runtime model (or an independent layer built on top of it), and not by logic synthesis itself.

All such model changes should be coordinated through a central undo facility. Such a facility allows an application to set checkpoints to which it could return, and applications which might have information to undo register callbacks with the facility to be called when such a checkpoint is established or when a request is made to revert to a previous checkpoint.

Ordering of callbacks to various applications to undo changes requires care. One approach is to examine the dependencies between incremental applications and decide on an overall ordering that would eliminate conflicts. A simpler solution is to ensure that each atomic change can be reversed, and to have the central undo facility call for the reversal of all changes in the opposite order from that in which they were originally made.

Applications can undo changes in their data in one of two ways. Save/restore applications (e.g., placement) may merely store the previous state and restore it without requiring calls to other applications. Recompute applications (e.g., timing analysis) may recompute previous state data that can be uniquely determined from the state of the model. Recompute applications generally do not have to register specific undo callbacks, but to allow them to operate, model change callbacks (and all other change notification callbacks) must be issued for the reversal of model changes just as they are for the original changes.

Such a central undo facility can also be used to help capture model change information, either to save to a file (e.g., an Engineering Change Order, or ECO, file) or to transmit to a concurrently executing parallel process, on the same machine or another one, with which the current applications need to synchronize.

Even when we choose to keep the results of a change, we may need to undo it and then redo it. A typical incremental processing environment might first identify a timing problem area, and then try and evaluate several alternative transformations. The original model state must be restored before each alternative is tried, and after all alternatives have been evaluated, the one that gives the best result is then redone.

To make sure that we get the same result when we redo a transformation that we got when we did it originally, we also need to be able to undo an undo. Even though the operation of a transformation is deterministic, the exact result may depend on the order in which certain objects are visited, and such orderings may not be semantic invariants and thus may not be preserved when a change is undone. Also, a considerable amount of analysis may be required by some transformations to determine the exact changes which are to be made, and we would like to avoid having to redo this analysis when we redo the transformation.

Because trial transformations to a network may be nested, we also need to be able to stack multiple checkpoints.

For these reasons, the central undo/ECO facility should be able to store and manage a tree of checkpoints, rather than just a single checkpoint.

It is important to remember that there is no magic bullet that will turn a nonincremental tool into an incremental one. In addition to having a common infrastructure, including such things as a common runtime model and a callback mechanism, appropriate incremental algorithms need to be developed.

Following some of these guidelines also encourages incremental design automation tool development. Rewriting tools is an expensive proposition. Few new ideas change all aspects of a tool. Incremental development allows more stability in the tool interface, which is particularly important for integrated applications. It also allows new ideas to be implemented and evaluated more cheaply, and it can make a required function available more quickly.

1.5 Future Scaling Challenges

In the previous sections, we mainly focused on how continued scaling changed the way we are able to predict delay throughout to the design flow. This has been one of the main drivers to the design flow changes over the last two decades. However, new challenges are arising that will again require us to rethink the way we automate the design flow. In the following sections, we will describe some of these in more detail and argue that they are leading to a design closure that requires an integrated, incremental analysis of power, noise and variability (with required incremental extraction tools) in addition to the well-established incremental timing analysis.

1.5.1 Leakage Power

Tools have traditionally focused on minimizing both critical path delay and circuit area. As technology dimensions have scaled down, the density of transistors has increased, mitigating the constraints on area, but increasing the power density (power dissipated per unit area). Heat dissipation limits the maximum chip power, which in turn limits switching frequency and hence how fast a chip can run. In 90 nm, even some high-end microprocessor designers have found power to be a more important constraint than delay.

The price of increasing circuit speed as per Moore's law has been an even faster increase in dynamic power. The dynamic power due to switching a capacitance C with supply voltage V_{dd} is $fCV_{dd}^2/2$. Increasing circuit speed increases switching frequency f proportionally, and capacitance per unit area also increases.

Transistor capacitance varies inversely to the transistor gate oxide thickness t_{ox} ($C_{gate} \propto WL/t_{ox}$, where W and L are transistor width and length). Gate oxide thickness has been scaled down linearly with transistor length, so as to limit short channel effects and maintain gate drive strength to increase circuit speed [19]. As device dimensions scale down linearly, the transistor density increases quadratically, and the capacitance per unit area *increases* linearly. Additionally, wire capacitance has increased relative to gate capacitance, as wires are spaced closer together with taller aspect ratios to limit wire resistance and corresponding wire RC delays.

Thus, if the supply voltage is constant, the dynamic power per unit area increases less than quadratically due to increasing switching frequency and increasing circuit capacitance. To reduce dynamic power, supply voltage has been scaled down. However, this reduces the drive current $I_{D,sat} \propto W(V_{dd} - V_{th})^{\alpha}/Lt_{ox}$, where V_{th} is the transistor threshold voltage, and α is between 1 and 2 [20], which reduces the circuit speed. To avoid reducing speed, threshold voltage has been scaled down with supply voltage.

Static power has increased with reductions in transistor threshold voltage and gate oxide thickness. The subthreshold leakage, current when the transistor gate-to-source voltage is below V_{th} and the transistor is nominally off, depends exponentially on V_{th} ($I_{subthreshold} \propto e^{-qV_{th}/nkT}$ where T is the temperature, and n, q, and k are constants). As the gate oxide becomes very thin, electrons have a nonzero probability of quantum tunneling through the thin gate oxide. While gate-tunneling current was magnitudes smaller than subthreshold leakage, it has been increasing much faster, and will be significant in future process technologies [21].

Synthesis tools have focused primarily on dynamic power when logic evaluates within the circuit, as static power was a minimal portion of the total power when a circuit was active. For example, automated clock-gating reduces unnecessary switching of logic. Managing standby static power was left for the designers to deal with, by using methods such as powering down modules that are not in use, or choosing standard cell libraries with lower leakage.

Standby power may be reduced by using high threshold voltage sleep transistors to connect to the power rails [22]. In standby, these transistors are switched off to stop the subthreshold leakage path from supply to ground rails. When the circuit is active these sleep transistors are on, and they must be sized sufficiently wide to cause only a small drop in voltage swing, but not so wide as to consume excessive power. To support this technique, place and route software must support connection to the "virtual" power rail provided by the sleep transistors, and clustering of cells which enter standby at the same time, so that they can share a common sleep transistor to reduce overheads.

Today, leakage can contribute a significant portion of the total power — when the circuit is active, in the range of 5 to 40% — There is a trade-off between dynamic power and leakage power. Reducing threshold voltage and gate oxide thickness allows the same drive current to be achieved with narrower transistors, with correspondingly lower capacitance and reduced dynamic power, but this increases leakage power. It is essential that tools treat leakage power on equal terms with dynamic power when trying to minimize the total power. Designers are now using standard cell libraries with multiple threshold voltages, using low threshold voltage transistors to maintain speed on the critical path and high threshold voltage transistors to reduce the leakage power of gates that have slack. Synthesis tools must support a choice between cells with smaller and larger leakage as appropriate for delay requirements on the gate within the circuit context.

1.5.2 Variability

To account for process and temperature variation, circuit delay is traditionally estimated from the (worst case) slow process and high temperature corner for the standard cell library. Hold-time violations and worst-case dynamic power may be estimated from the fast process (corresponding to lower threshold voltages and shorter channel lengths due to variability) and low temperature corner. Worst-case leakage power can be estimated at the fast process and high temperature corner. It is assumed that this simple analysis is sufficient to ensure that timing constraints are met. In practice, a mix of "slow" and "fast" process variations and differences in spot temperatures on a chip may lead to unforeseen failures not predicted by a single process corner. Secondly, the majority of chips fabricated may be substantially faster and have lower average power under normal circuit conditions. The design costs are significant for this overly conservative analysis.

Although some elements of a circuit are under tighter control in today's processes, the variation of other elements has increased. For example, a small reduction in transistor threshold voltage or channel length can cause a large increase in leakage. The layout of a logic cell's wires and transistors has become more complicated. Optical interference changes the resulting layout. Phase shift lithography attempts to correct this. Varying etch rates have a greater impact on narrower wires.

Certain cell layouts are more likely to reduce yield (whether due to increased gate delay or complete failure such as when a wire is not connected) due to these sorts of problems. Ideally, cell layouts with lower yield would not be used, but these cells may be of higher speed. It may be desirable to accept a small decrease in yield to meet delay requirements. To support yield trade-offs, foundries must give some yield information to customers along with corresponding price points.

Yield and variability data can be annotated to standard cell libraries, enabling software to perform statistical analysis of timing, power, and yield. This requires a detailed breakdown of the variability into systematic (e.g., spatially correlated) and random components. With this information, tools can estimate the yield of chips that will satisfy delay and power constraints. This is not straightforward, as timing paths are statistically correlated, and variability is also spatially correlated. However, these correlations can be accounted for in a conservative fashion that is still less conservative than worst-case corner analysis.

1.5.3 Reliability

Glitches can be caused by cross-coupling noise, and alpha particle and neutron bombardment. These glitches can cause a temporary error in a circuit's operation.

Cross-coupling noise has increased with higher circuit frequencies and greater coupling capacitance between wires. Wire cross-coupling capacitance has increased with wires being spaced closer together and with higher aspect ratios, which have been used to reduce wire RC delays as dimensions scale down. Wires can be laid out to reduce cross-coupling noise (e.g., by twizzling, or shielding with ground wires). There are tools for analyzing cross-coupling noise, but automated routing to reduce coupling is not supported in tools yet.

With gate capacitance decreasing with device dimensions, and due to reduced supply voltages, smaller amounts of charge are stored, which are more easily disrupted by an alpha particle or neutron strike. Soft error rates due to alpha particles increase by a large factor as device dimensions are scaled down [23]. Soft error rates can be reduced by using silicon on insulator and other manufacturing methods, or by using latches that are more tolerant to transient pulses [23].

For fault tolerance to glitches, circuits can have error detection and correction, or additional redundancy. Tool support for synthesis to such circuits will simplify a designer's task.

1.6 Conclusion

In this chapter, we gave a flavor of how the RTL to GDSII design flow has changed over time and will continue to evolve. As depicted in Figure 1.4, we foresee continuing integration of more analysis functions into the integrated environment to cope with design for robustness and design for power.

The other chapters of this book describe each of the algorithms, data structures, and their role in the RTL to GDSII flow in much more detail.

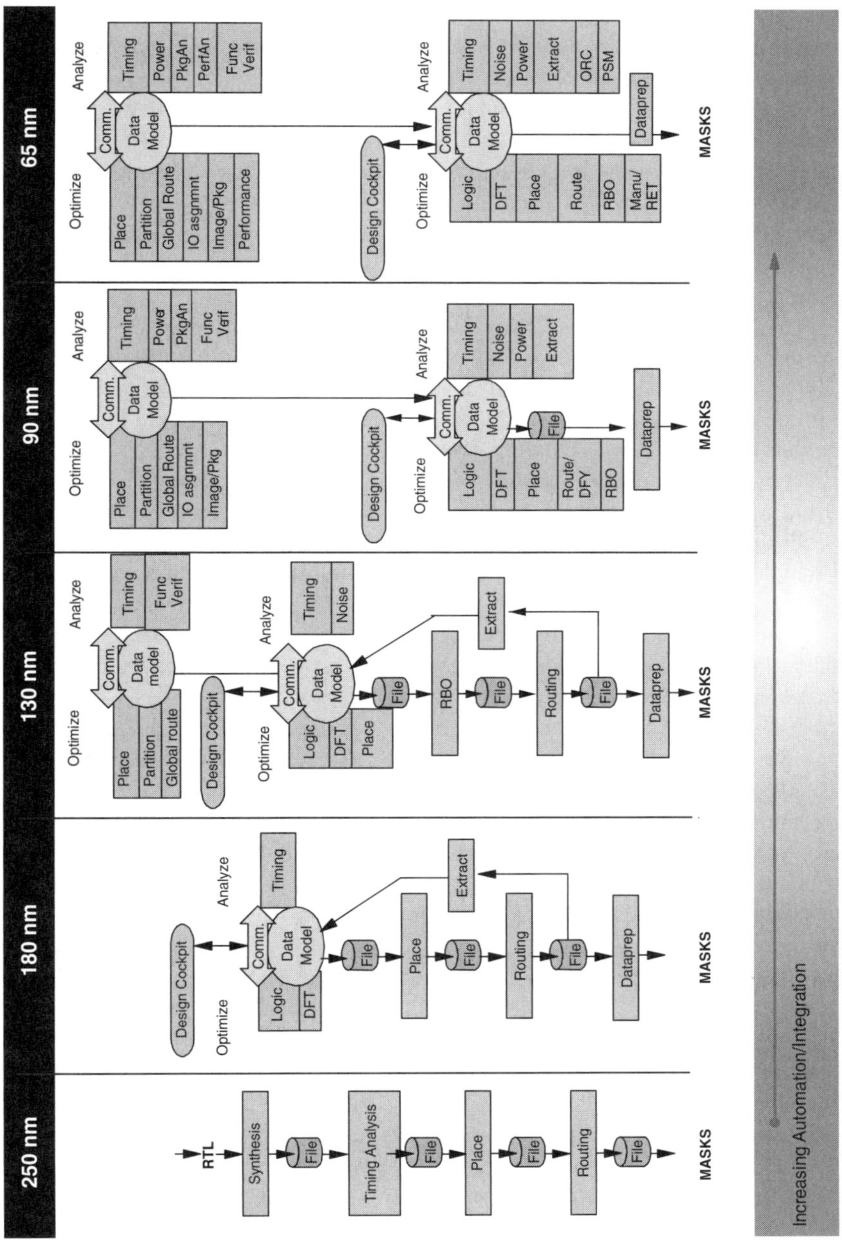

FIGURE 1.4 Integrated design flow.

References

[1] A. Sangiovanni-Vincentelli, The tides of EDA. *Des. Test Comput., IEEE*, 20, 59–75, 2003.

[2] B.W. Kernighan and S. Lin, An efficient heuristic procedure for partitioning graphs, *Bell Syst. Tech. J.*, 49, 291–308, 1970.

[3] S. Kirkpatrick, C.D. Gelatt Jr., and M.P. Vevvhi, Optimization by simulated annealing, *Science*, 220, 671–680, 1983.

[4] K.A. Chen, M. Feuer, K.H. Khokhani, N. Nan, and S. Schmidt, The chip layout problem: an automatic wiring procedure, *Proceedings of the 14th Design Automation Conference,* 1977, pp. 298–302.

[5] J.A. Darringer and W.H. Joyner, A new look at logic synthesis, *Proceedings of the 17th Design Automation Conference*, 1980, pp. 543-549.

[6] J.L. Bentley, Multidimensional binary search trees used for associative searching, *Commn. ACM*, 18, 509–517, 1975.

[7] R.E. Bryant, Graph-based algorithms for Boolean function manipulation, *IEEE Trans. Comput.*, C-35, 677-691, 1986.

[8] C.M. Fiduccia and R.M. Mattheyses, A linear time heuristics for improving network partitions, *Proceedings of the 19th Design Automation Conference*, 1982, pp. 175–181.

[9] L. Hagen and A.B. Kahng, Fast spectral methods for ratio cut partitioning and clustering, *Proceedings of the International Conference on Computer Aided Design*, 1991, pp. 10–13.

[10] H. Yang and D.F. Wong, Efficient network flow based min-cut balanced partitioning, *Proceedings of the International Conference on Computer Aided Design*, 1994, pp. 50–55.

[11] J.M. Kleinhaus, G. Sigl, and F.M. Johannes, Gordian: a new global optimization/rectangle dissection method for cell placement, *Proceedings of the International Conference on Computer Aided Design*, 1999, pp. 506–509.

[12] C.L. Berman, L. Trevillyan, and W.H. Joyner, Global flow analysis in automatic logic design, *IEEE Trans. Comput.*, C-35, 77–81, 1986.

[13] D. Brand, Redundancy and don't cares in logic synthesis, *IEEE Trans. Comput.*, C-32, 947–952, 1983.

[14] R.B. Hitchcock Sr., Timing verification and the timing analysis program, *Proceedings of the 19th Design Automation Conference*, 1982, pp. 594–604.

[15] T.I. Kirkpatrick and N.R. Clark, Pert as an aid to logic design, *IBM JRD*, 10, 135–141, 1966.

[16] L.P.P.P. van Ginneken, Buffer placement in distributed RC-tree networks for minimal Elmore delay, *International Symposium on Circuits and Systems*, 1990, pp. 865–868.

[17] J. Lakos, *Large-Scale C++ Software Design*, Addison-Wesley, Reading, MA, 1996.

[18] R.P. Abato, A.D. Drumm, D.J. Hathaway, and L.P.P.P. van Ginneken, Incremental Timing Analysis, U.S. patent 5 508 937, 1996.

[19] S. Thompson, P. Packan, and M. Bohr, MOS scaling: transistor challenges for the 21st century, *Intel Technol. J.*, Q3, 1998.

[20] T. Sakurai and R. Newton, Delay analysis of series-connected MOSFET circuits, *IEEE J. Solid-State Circuits*, 26, 122–131, 1991.

[21] D. Lee, W. Kwong, D. Blaauw, and D. Sylvester, Analysis and minimization techniques for total leakage considering gate oxide leakage, *Proceedings of the 40th Design Automation Conference*, 2003, pp. 175–180.

[22] S. Mutoh et al., 1-V power supply high-speed digital circuit technology with multithreshold-voltage CMOS, *IEEE J. Solid-State Circuits*, 30, 847–854, 1995.

[23] C. Constantinescu, Trends and challenges in VLSI circuit reliability, *IEEE Micro*, 23, 14–19, 2003.

2

Logic Synthesis

Sunil P. Khatri
Texas A&M University
College Station, Texas

Narendra V. Shenoy*
Synopsys Inc.
Mountain View, California

2.1 Introduction

The roots of logic synthesis can be traced to the treatment of logic by George Boole (1815 to 1865), in what is now termed Boolean algebra. Shannon's [1] discovery in 1938 showed that two-valued Boolean algebra can describe the operation of switching circuits. In the early days, logic design involved manipulating the truth table representations as Karnaugh maps [2,3]. The Karnaugh map-based minimization of logic is guided by a set of rules on how entries in the maps can be combined. A human designer can only work with Karnaugh maps containing four to six variables. The first step toward automation of logic minimization was the introduction of the Quine–McCluskey [4,5] procedure that could be implemented on a computer. This exact minimization technique presented the notion of prime implicants and minimum cost covers that would become the cornerstone of two-level minimization. Another area of early research was in state minimization and encoding of finite-state machines (FSMs)[6–8], a task that was the bane of designers. The applications for logic synthesis lay primarily in digital computer design. Hence, IBM and Bell Laboratories played a pivotal role in the early automation of logic synthesis. The evolution from discrete logic components to programmable logic arrays (PLAs) hastened the need for efficient two-level minimization, since minimizing terms in a two-level representation reduces the area of the corresponding PLA. MINI [9] was an early two-level minimizer based on heuristics. Espresso [10] is an improvement over MINI; it uses the unate recursive paradigm (URP) as a central theme for many optimization steps.

* Narendra Shenoy did not contribute to the material set forth in Section 2.4.3.2.

However, two-level logic circuits are of limited importance in very large-scale integrated (VLSI) design; most designs use multiple levels of logic. An early system that was used to design multilevel circuits was LSS [11] from IBM. It used local transformations to simplify logic. Work on LSS and the Yorktown Silicon Compiler [12] spurred rapid research progress in logic synthesis in the 1980s. Several universities contributed by making their research available to the public; most notably, MIS [13] from University of California, Berkeley and BOLD [14] from University of Colorado, Boulder. Within a decade, the technology migrated to commercial logic synthesis products offered by electronic design automation companies.

The last two decades have seen tremendous progress in the field of logic synthesis. It has provided a dramatic productivity boost to digital circuit design, enabling teams to fully utilize the large number of transistors made available by decreasing feature sizes. This chapter provides a brief survey of the rapid advances made in this area. A more comprehensive treatment can be found in books by De Micheli [15], Devadas et al. [16], Hassoun and Sasao [17], and Hachtel and Somenzi [18].

2.2 Behavioral and Register Transfer-Level Synthesis

To increase designer productivity, it is crucial to be able to specify and optimize designs at higher levels of abstraction. There has been a significant amount of research on synthesis of circuits, which are behaviorally specified using a hardware description language (HDL). The goal of behavioral synthesis is to transform a behavioral HDL specification into a register transfer level (RTL) specification, which can be used as input to a gate-level logic synthesis flow. A general overview of behavioral synthesis can be found in [19–22].

Behavioral optimization decisions are guided by cost functions that are based on the number of hardware resources and states required. These cost functions provide a coarse estimate of the combinational and sequential circuitry required to implement the design.

In a behavioral optimization tool, a front-end parser translates the behavioral HDL description of the design into a control and data flow graph (CDFG). Sometimes, separate control flow graphs (CFG) and data flow graphs (DFG) are created. This CDFG is subjected to high-level transformations, many of which are based on compiler optimizations [24], such as constant propagation, loop unrolling, dead code elimination, common subexpression elimination, code motion [25], and dataflow analysis. Some hardware-specific optimizations performed in this step include making hardware-specific transformations (shifting to perform multiplication by a power of 2), minimizing the number of logic levels to achieve speedup, and increasing parallelism. The tasks of *scheduling* and *resource allocation and sharing* generate the FSM and the datapath of the RTL description of the design.

Scheduling [26] assigns operations to points in time, while allocation assigns each operation or variable to a hardware resource. Scheduling identifies places in the CFG where states begin and end, yielding the FSM for the design. Scheduling usually precedes allocation, although they can be intertwined. Among the scheduling algorithms in use are as-soon-as-possible (ASAP) and its counterpart, as-late-as-possible (ALAP). Other algorithms include force-directed scheduling [27], list scheduling [28,29], path-based scheduling [30], and symbolic scheduling [31].

Given a schedule, the allocation operation optimizes the amount of hardware required to implement the design. Allocation consists of three parts: functional unit allocation, register allocation, and bus allocation. The goal during allocation is to share hardware units maximally. Allocation for low power has been studied in [32], while [33] reports joint scheduling and allocation for low power. Behavioral synthesis for low power has been studied in [34–36].

Behavioral synthesis typically ignores the size and delay of the required control logic. The CDFG representation is significantly different from the representation used in the RTL network. As a result, a behavioral network graph (BNG) is utilized in [37]. The BNG is an RTL network, with the ability to represent unscheduled behavioral descriptions. Since wiring delays are becoming increasingly important in VLSI design, early estimation of such delays is very helpful. The fast bus delay predictor for high-level synthesis [38] fulfills this need. In [39], the notion of don't cares has been exploited in behavioral optimization. Much research has been conducted in the area of high-level synthesis for testability. For details, we refer the interested reader to a survey paper [40] on this topic. Examples of behavioral synthesis systems include [41,42]. The Olympus [43] system combines behavioral and logic synthesis.

2.3 Two-Level Minimization

Two-level logic minimization is arguably the workhorse of logic synthesis. Its early and most direct application included logic minimization for PLA-based designs. Since then, it has been used extensively in multilevel technology-independent logic optimization as well, where it is utilized in performing node optimization.

Two-level logic minimization is an instance of the classical *unate covering problem* (UCP) [44–48]. For a logic function with n inputs and 1 output, the covering matrix has $O(2^n)$ rows (corresponding to minterms) and $O(3^n/n)$ columns (corresponding to primes of the function). In a typical solution to the UCP, we iterate the steps of row and column dominance, and extraction of row singletons (essential primes) until the cover matrix cannot be further reduced (referred to as the *cyclic core*). At this point, techniques such as branch-and-bound are used to solve the cyclic core. Early solutions include the Quine–McCluskey approach [4,5]. The maximum independent set of primes can be used to bound the solution cost. In the early 1990s, two new exact techniques based on the computation of *signature cubes* were developed. In one of these methods [49], the size of the covering matrix is reduced (both rows and columns) yielding more efficient solutions. The other method [50] is based on the use of reduced ordered binary decision diagrams (ROBDDS) [51]. Characteristic ROBDDs of all primes and minterms are created, and dominance steps are formulated in terms of ROBDD operations. Similarly, in [52], the authors implicitly create the cyclic core, with a significantly lower computational cost than previous approaches. In [44–46], improved bounding and pruning approaches are introduced, and implemented in SCHERZO. The approach of Goldberg et al. [47,48] is based on performing branch-and-bound with a goal of proving that a given subspace cannot yield a better solution (negative thinking) as opposed to trying to find a better solution via branching (positive thinking). In [53], the authors provide a technique that combines the use of zero-suppressed binary decision diagrams (BDDs) [54] (to represent the data) and Lagrangian relaxation to solve the integer formulation of the UCP, yielding significant improvements.

Unate covering can be an expensive (though exact) approach to solving the two-level minimization problem. Several heuristic approaches have been developed as well. MINI [9] was one of the early heuristic two-level minimizers. ESPRESSO [10] improved on MINI, utilizing the unate recursive paradigm (URP) at the core of the operations. In this heuristic approach, primes are never enumerated. Rather, operations are performed on a subset of primes. In ESPRESSO, the operations of *Reduce* (which reduces cubes in an ordered manner, such that the new set of cubes is still a cover), *Expand* (which expands cubes into primes, removing cubes that are covered by some other cubes in the cover), and *Irredundant* (which removes redundant cubes in the cover) are iterated until no further improvement is possible. These algorithms are based on cofactoring with respect to the most binate variable in the cover until unate leaves are obtained. The operation is efficiently performed on unate leaves, and then the results are recursively merged upward until the result of the operation on the original cover is obtained. When the reduce, expand, and irredundant iteration encounters a local minimum, a LASTGASP algorithm is called, which attempts to add more primes in a selective manner, in an attempt to escape the local minimum. After running LASTGASP, we again iterate on reduce, expand and irredundant until no improvement is possible. ESPRESSO requires the computation of the complement of a function, which can become unmanageable for functions such as the Achilles heel function. In [55], a reduced offset computation is proposed to avoid this potential cube explosion.

ESPRESSO yields close to optimum results, with significant speedup compared to exact techniques. To bridge further the gap between ESPRESSO and exact approaches, iterative applications of ESPRESSO [56] can be used. In this technique, after an ESPRESSO iteration, cubes are selectively extracted from the onset and treated as don't care cubes. Iterating ESPRESSO in this manner has been shown to result in bridging the (albeit small) optimality gap between ESPRESSO and exact techniques, with a modest run-time penalty.

ESPRESSO-MV [57] is the generalization of ESPRESSO to multivalued minimization. Two-level logic minimization has applications in minimizing Internet Protocol (IP) routing tables [58,59]. Hardware implementations of ESPRESSO tailored to this application [60,61] have reported significant speedups.

Boolean relations are a generalization of incompletely specified functions (ISFs), in that they are one-to-many multi-output Boolean mappings. Minimizing such relations involves finding the best two-level logic function which is *compatible* with the relation. The minimization of Boolean relations can be cast as a binate covering problem (BCP). In [62], a Quine–McCluskey-like procedure is provided to find an optimum two-level representation for a Boolean relation. The solution is based on a branch-and-bound covering method, applied to the BCP. Implicit techniques to solve the BCP are provided in [63], along with a comparison with explicit BCP solvers. In [64], a branch-and-bound BCP algorithm is provided, with the input specified as a conjunction of several ROBDDs. Heuristic minimization of multivalued relations, using the two-level logic minimization paradigm of Brayton et al. [10], is presented in [65]. This multivalued decision diagram [66] (MDD)-based approach is implemented in a tool called GYOCRO [65].

2.4 Multilevel Logic Minimization

2.4.1 Technology-Independent Optimization

Typical practical implementations of a logic function utilize a multilevel network of logic elements. A standard-cell-based logic netlist, for example, utilizes such a network. A multilevel logic network can be abstracted as a directed acyclic graph (DAG), with edges representing wires and nodes representing memory elements or combinational logic primitives. Typically, we consider each node to have a single output, although this can be generalized as well. The combinational logic of a node can be represented in several ways; however, a two-level cover is the most commonly utilized method.

From an RTL description of a design, we can construct a corresponding multilevel Boolean network. This network is optimized using several technology-independent techniques before technology-dependent optimizations are performed. The typical cost function during technology-independent optimizations is total literal† count of the factored representation of the logic function (which correlates quite well with circuit area).

Many technology-independent optimizations can be performed on a multilevel Boolean network. We present several such optimizations, with a brief discussion of the salient techniques for each.

2.4.1.1 Division

A function g is a *divisor* for f if $f = gh + r$, where r is a *remainder* function and h a *quotient* function. Note that h and r may not be unique. If $r = 0$, we refer to the process as *factoring*. Division can be of two types: Boolean [13,67,68] or algebraic [13,69,70]. In general, Boolean division explores all possible functions that divide f. It is therefore computationally expensive. Algebraic division is less flexible, but extremely fast since no Boolean operations are performed. Division can be performed recursively on g, h, and r, to obtain an initial multilevel network.

2.4.1.2 Kerneling

One important decision in division is the choice of divisor g. Kernels [69] are used for this purpose. Kernels are cube-free‡ primary divisors.§ Kernels can be computed efficiently using algebraic operations. The use of two-cube kernels [71] results in significant speedup in kerneling, with a minimal penalty in quality over the full-fledged kernel extraction.

The above division techniques can be utilized to optimize a multilevel network in several ways:

- If there exists a node g that can be divided into another node f in the network, we perform *substitution* of g into f.
- By *extracting* common subexpressions among one or more nodes, we may be able to reimplement a network with fewer literals. We find kernels of several nodes, choose a best

† A *literal* is a variable or its complement.

‡ A cube is a conjunction of literals. An expression is cube-free if no single cube can divide it evenly.

§ Primary divisors are expressions obtained by algebraically dividing f with some cube c.

kernel, and create a new node with the logic function of the kernel. Now we substitute the new node into all remaining nodes in the design.

- We may *eliminate* a node by *collapsing* it into its fanouts, to get out of a local minimum and enable further multilevel optimizations.

2.4.2 Multilevel Don't Cares

The use of multilevel don't cares can be very effective in reducing the literal count of a network. Multilevel don't care-based optimization is typically invoked at the end of the structural optimizations mentioned above. Multilevel don't cares are first computed as functions of primary input variables as well as internal node variables. Then, by performing an (usually ROBDD-based) image computation, we find the image of the multilevel don't cares of a node in terms of the fanin variables of that node. This allows us to perform two-level minimization on the node using the newly computed don't cares, thereby reducing the literal count.

Early multilevel don't care computation methods [72] were restricted to networks of NOR gates. The corresponding don't care computation technique was referred to as the *Transduction* method. Two sets of *permissible functions*¶ — the maximum set of permissible functions (MSPFs) and the compatible set of permissible functions (CSPFs) were defined. The downside of both is that they are defined to be global functions. Further, MSPFs are not compatible in the sense that if a function at some node j is changed, it may require the recomputation of the MSPFs of other nodes in the network. This limitation is removed for CSPFs.

More general multilevel don't care techniques [73,74] were developed shortly after the transduction method. For these methods, the multilevel don't cares were computed as a disjunction of external don't cares (XDCs), satisfiability don't cares (SDCs), and observability don't cares (ODCs). External don't cares for each output are typically specified as global functions, while SDCs for the circuit encode local logical conditions that cannot occur in the network. The network SDC is $\Sigma_i(y_i \oplus f_i)$, where f_i is the logic function of node y_i of the network. Observability don't cares are computed as $\prod_k (\partial z_k/\partial y_i)$, where $\partial z_k/\partial y_i$ is the Boolean difference of primary output z_k with respect to y_i, and encodes the minterms in the global space for which z_k is sensitive to changes in y_i.

The disjunction of the above don't cares is then imaged back to the local fanins of the node y_i, and the cover f_i is then minimized with respect to these local don't cares. Two-level minimization is used for this purpose.

If a node is minimized with respect to these don't cares (which include the ODC), it results in changes to the ODCs of other nodes in the design. This may be unimportant when the nodes are minimized one after another. However, if the optimization context is one in which the don't cares of all the nodes are utilized simultaneously, this does not guarantee correct results. For such applications, compatible output don't cares (CODCs) [75] can be used. Compatible output don't cares have the property that after they are computed for any design, a node can be optimized against its CODCs without the need to recompute CODCs of other nodes.

The don't care computations outlined above rely on an image computation, which requires computation of global ROBDDs of the circuit nodes. This limits the effectiveness of the don't care computation. Recently, approximate CODC [76] and ODC [77] computations were introduced. It was shown that by using a *window* of small topological depth in the forward and reverse directions from the node under consideration, a robust and effective don't care computation can be devised. In [76], a depth of $k = 4$ resulted in a $30 \times$ speedup and a $25 \times$ reduction in memory requirements, while 80% of the literal reduction of the traditional CODC technique was obtained.

Another recent development in this area is the introduction of Sets of pairs of Functions to be Distinguished (SPFDs) [78]. SPDFs were first introduced in the context of FPGA optimization. Their applicability to general logic network optimization was identified in [79]. The SPFD of a node is a set of

¶ A *permissible function* f_j at a node j is a function of the primary inputs, such that if the node j is replaced by f_j, the network functionality is unchanged.

Incompletely Specified Functions (ISFs).[**] As a result, the SPFD of a node encapsulates more information than the don't cares of that node. Sets of pairs of functions to be distinguished can be represented as bipartite graphs. The SPFD of a node encodes the set of pairs of points in the primary input space, which must be *distinguished* or assigned different functional values. These pairs are redistributed among the fanins of the node, resulting in new SPFDs at the fanins of the node. Coloring these graphs results in new implementations of the fanin functions. In [80], an ROBDD-based implementation of an SPFD-based network optimization package was demonstrated. Results for wire replacement and fanin minimization were provided, with about 10% improvement over CODC-based optimizations. It was also shown that the flexibility offered by SPFDs contains that of traditional don't care-based optimizations. Subsequently, SPFD-based optimizations have been demonstrated in other contexts as well, including rewiring [81], power, and delay minimization for FPGA synthesis [82], wire removal for PLA networks using multivalued SPFDs [83], and topologically constrained logic synthesis [84]. Sequential extensions to SPFDs were reported in [85].

2.4.3 Technology-Dependent Optimization

The multilevel technology-independent logic optimizations discussed so far in Section 2.4 have utilized a simple cost function, such as literal count. Technology-dependent optimization transforms an optimized technology-independent circuit into a network of gates in a given technology. The simple cost estimates are replaced by more concrete, implementation-driven estimates during and after technology mapping. The circuit structure created by mapping is crucial for final quality of optimization. Technology mapping and several local optimizations available after technology mapping are described in this section.

2.4.3.1 Technology Mapping

Mapping is constrained by factors such as the available gates (logic functions) in the technology library, the drive sizes for each gate, and the delay, power, and area characteristics of each gate. Further details on a technology library are given in Section 2.5.1. The initial work on technology mapping relied on rule-based heuristic transforms [86,87]. These were later replaced by rigorous algorithmic approaches. The process of technology mapping is conceptually best described by three generic steps. In the first step, called *subject graph construction*, an optimized technology-independent circuit is decomposed into a DAG, consisting of a set of primitive gates (such as nand gates and inverters). There are many logically equivalent decompositions for a given circuit and an appropriate choice is critical for good quality of mapping. The logic gates in the library are also decomposed into the same set of primitive gates. These are known as *patterns*. The second step, called *pattern matching*, generates matches of logic gates in the library to each primitive gate in the subject graph. The third step, called *covering*, constructs a tiling of the subject graph with a selection of a subset of patterns generated in the previous step. The first constraint of covering is that each primitive gate of the subject graph is contained in at least one selected pattern. The second constraint requires the inputs to a selected pattern to be outputs of another selected pattern. The objective of covering is to find a minimum cost covering. The cost function can be area, delay, power, or a combination of these.

Matching can be performed by a variety of methods: structural matching using graph isomorphism or Boolean matching using BDDs. In the case of structural matching for trees, efficient string matching techniques can be used [88]. Boolean matching [89,90] is more powerful as it can detect matches independent of the local decomposition of the subject graph, under arbitrary input/output phase assignments and input permutations (of the library gate).[*†] To avoid the combinatorial explosion due to permutations and phases, signatures [91,92] have been proposed.

The techniques utilized for the covering step depend on whether we are operating on a DAG or a tree. Directed acyclic graph covering can be formulated as a BCP [93]. It is not viable for very large circuits. Keutzer [94] describes an approach in which the subject graph is split into a set of fanout-free trees. Each

[**] An ISF partitions the set of points in B^n into onset, offset, and don't care set points.

[*†] This is often termed as NPN equivalence.

tree is mapped optimally (for area). Thus, DAG covering is approximated with a sequence of tree mapping steps. Tree covering is solved using dynamic programming, which leads to optimum coverings [95] under certain conditions on the cost function. Minimum area tree-based technology mapping is solvable in time polynomial in the number of nodes in the subject graph, and in the number of patterns in the library. Rudell [93] extends technology mapping to include timing using a binning technique. Touati [96] considers optimal delay mapping using a linear delay model. When the delay models are complex, the covering process requires an area–delay cost curve to be stored at each node in the subject graph. This can result in a high-memory consumption. Roy et al. [97] present a compression algorithm that yields a solution bounded from the optimal solution by a user-specified threshold. Minimum area mapping under delay constraints is more difficult and there is no known optimal algorithm [98]. The main disadvantages with applying tree covering are twofold. First, decomposing DAGs into trees leads to too many small trees, thereby losing any optimality with respect to the original DAG. The unpredictability of load estimation at the multifanout points, which are cut in order to convert DAGs into trees, causes errors in delay estimation. Directed acyclic graph covering is feasible if the delay models are modified to be load-independent [99]. Stok et al. [100] and Kukimoto et al. [101] present DAG covering approaches under gain-based and load-independent delay models. Since the quality of technology mapping is very sensitive to the initial decomposition of the subject graph, Lehman et al. [102] propose decomposition within the covering step. A simple way to improve the quality of circuits constructed by tree covering is to use the inverter-pair heuristic [16].

To illustrate tree-based technology mapping for minimum area, consider a library of ten combinational logic gates decomposed into patterns consisting of two-input nand gates and inverters as shown in Figure 2.1. Each gate may have multiple decompositions (only one is shown for each gate). Consider a tree of logic and its associated decomposition into a subject graph using two-input nand gates and inverters as shown in Figure 2.2. Observe that two inverters are added on a wire to enable detection of more patterns as suggested by the inverter-pair heuristic mentioned above. A subset of matches of library patterns on the subject graph is indicated in Figure 2.3. Note that multiple matches can occur at the output of a gate. Technology mapping selects the best match of a pattern at a gate from a set of candidate matches using dynamic programming. In this approach, the least area cost of implementing a pattern at a gate is computed from the inputs toward the output of the tree. The selection of a pattern implies that the inputs to the pattern must be selected as well in the final implementation. Hence, the least area cost of selecting a pattern can be computed as the sum of the area of the pattern and the least area costs of implementing the signals at the inputs to the pattern. Since the solution is constructed from inputs toward the output,

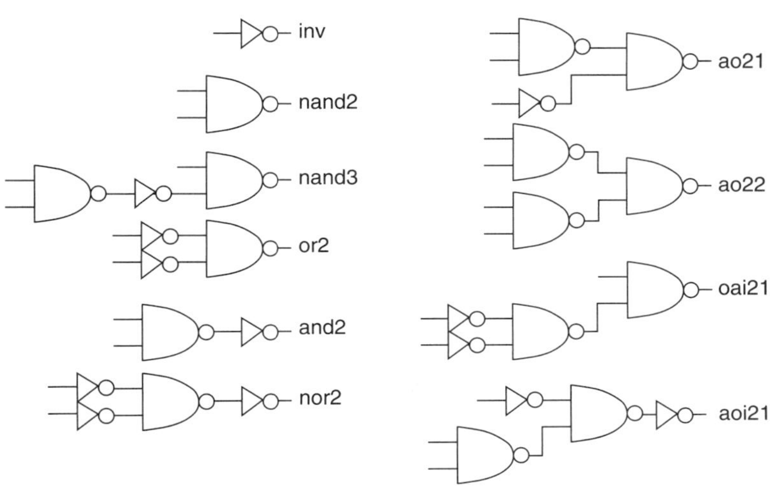

FIGURE 2.1 Decomposition of library gates into patterns.

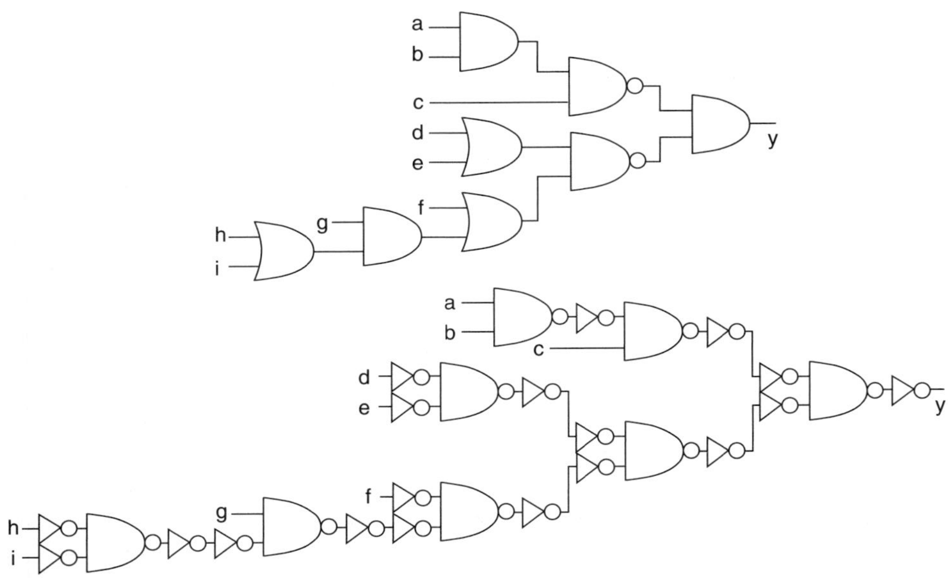

FIGURE 2.2 Logic undergoing mapping and its associated subject graph.

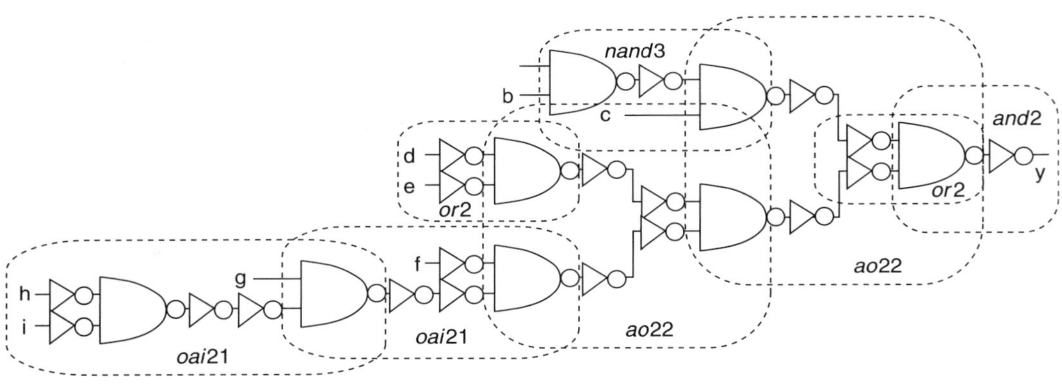

FIGURE 2.3 Subset of pattern matches on the subject graph.

the latter quantities are readily available. Figure 2.4 shows two options of selecting a gate to implement the logic at *y*. Selecting a *nor*2 gate (area of 3) requires the inputs to the pattern be implemented with area costs of 3 and 12, respectively, yielding a total area cost of 18. Selecting an *aoi*21 gate (area of three) requires the inputs to the pattern be implemented with area costs of 2 and 12 respectively; yielding a total area cost of 17.

Although area and delay have been traditional objective functions in mapping, power is now a central concern as well. The tree covering approach for low power under a delay constraint mimics the minimum area covering under a delay constraint. Tiwari et al. [103] and Tsui et al. [104] discuss technology mapping with a low-power focus.

2.4.3.2 Logical Effort-Based Optimizations

The delay of a circuit can be expressed using the notions of logical and electrical effort [105]. The delay of a gate can be expressed as

$$d = \tau(p + gh)$$

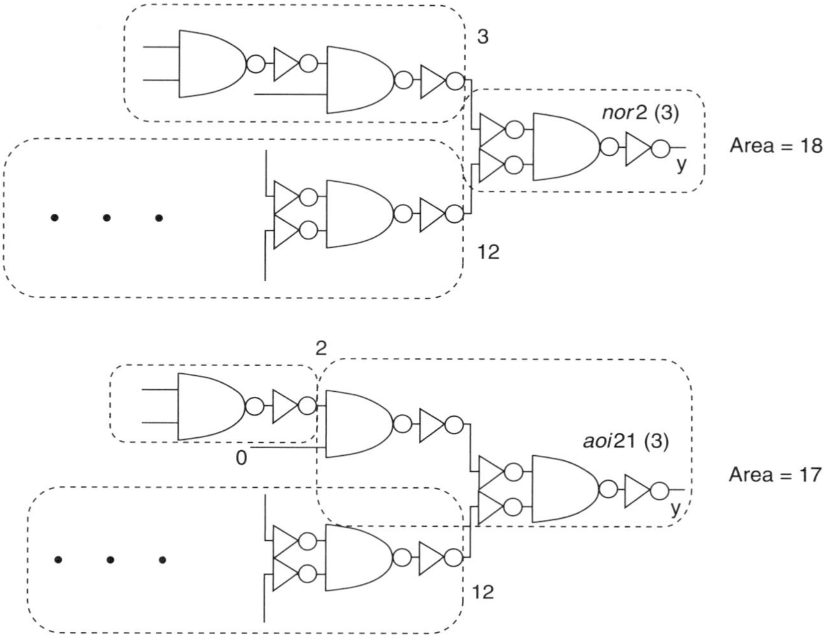

FIGURE 2.4 Dynamic programming during technology mapping for area.

Here τ is a process-dependent parameter, while p is the parasitic delay of the gate (which arises due to the source/drain capacitances of the gate output). The parameter g is the *logical effort* of the gate, and it characterizes the ability of the gate to drive an output current. We assume that an inverter has a logical effort of unity. The logical effort of other gates depends on their topology, and describes how much worse these gates are (in comparison with an inverter) in producing an output current, assuming that the input capacitances are identical to that of an inverter. The parameter h is the *electrical effort* (or gain) of the gate, and is the ratio of the output capacitance to the input capacitance of a pin of the gate. Note that p and g are independent of the size of the gate, while h is sizing-dependent.

In [105], the authors minimize delay along a circuit path by assigning an equal delay budget for each topological level along the circuit path. The resulting solution minimizes delay, with a solution that is not necessarily area minimal. In [106], the authors solve the fanout optimization problem to minimize the input capacitance of the source gate, subject to sink timing constraints, without considering area. In [107], the authors address the problem of minimizing buffer area subject to driver capacitance constraints. In [108], the idea of logical effort is applied to technology mapping, addressing the problem of load distribution at multifanout points. An analogous notion of interconnect effort is introduced in [109,110], combining the tasks of logic sizing and buffer insertion. Logical effort can also be applied to the design of regular structures such as adders [111,112].

2.4.3.3 Other Technology-Dependent Optimizations

After technology mapping, a set of technology-dependent optimizations is carried out. Gate sizing, gate replication, and de Morgan transformations make local changes to the circuit. The impact of fanout optimization is more widespread. Critical path restructuring can make significant non-local changes. These approaches have the following common strategy. The first phase identifies portions of the circuit that need to be modified for possible benefit. The second phase selects a subset of these portions for change, modifies the circuit, and accepts the change if it results in an improvement. These steps rely on accurate incremental timing analysis to make this decision (see Section 2.5.2.1).

The early work on sizing focused on transistors and used a simple delay model (such as RC trees). This makes the objective function convex and easy to optimize [113–115]. Nonlinear programming techniques have also been used [116–118]. However in the ASIC context, logic optimization can only size gates and not individual transistors. Hence the next wave of research focused on this aspect [119,120]. Also the simple delay models cease to be accurate in modern technologies. Coudert et al. [121] provide an excellent survey on gate sizing.

Fanout optimization seeks to distribute optimally a signal to various input pins. This is critical for gates driving a large number of fanouts. Berman et al. [122], Hoover et al. [123], and Singh and Sangiovanni-Vincentelli [124] describe solutions to the fanout problem. Fanout optimization for libraries rich in buffer sizes is addressed by Kung [125]. Using the logical effort model, Rezvani et al. [107] discuss a fanout optimization tool for area and delay.

Bartlett et al. [126] were the first to focus on minimizing the area of a circuit under a delay constraint in the SOCRATES system. Singh et al. [127] describe a strategy for identifying regions for resynthesis using a weighted min-cut heuristic. The weights are determined based on the potential for speedup and an estimate of the area penalty incurred during optimization. Fishburn [128] describes a heuristic for speeding up combinational logic. Further refinements have been made by Yoshikawa et al. [129].

2.5 Enabling Technologies for Logic Synthesis

Some technologies have played a key role in the rapid advances in logic synthesis. We discuss some of the enabling research in this section.

2.5.1 Library Modeling

A key reason for the commercial success of logic synthesis is the availability of technology library models for various semiconductor foundries. Two popular commercial formats for library description are Liberty from Synopsys [130] and Advanced Library Format (ALF) [131]. The library data can be categorized into two distinct sections.

Technology data. This includes information such as operating conditions (power supply and temperature) and wire load models. A wire load model is a statistical estimate of the capacitances observed on "typical" nets as a function of the number of pins and the size of the design (in terms of area). It is used as a proxy for net capacitance during synthesis when the physical data required for accurate estimation of net capacitance is unavailable (see discussion in Section 2.7 for more information).

Library cell data. Each cell in the library is accurately characterized for timing, power, and area. The delay is a function of circuit parameters (such as the input slew, output load, threshold voltage, and critical dimension) and operating conditions (such as local power and ground levels, temperature). For synthesis purposes, the delay is typically modeled as a nonlinear function of input slew and output load. It is specified as a two-dimensional table. The move to smaller geometries is forcing the use of polynomials to capture the dependence on many parameters. The table-based approach does not scale in terms of memory usage as the number of parameters increases. The timing information also includes constraints such as setup and hold constraints. The power data include internal switching power and leakage power.

An issue of debate in the last decade was the nature and number of different logic functions in a standard cell library [132,133]. We currently have libraries with hundreds of logic functions, with multiple sizes for each function. More recent approaches have modeled gate delays using the concept of logical effort [105,135]. This approach assumes that the area of a gate is a continuous function of its load. Thus, a gate can be sized to keep the delay constant. This was first proposed in [99] and its use in technology mapping was explored in [102]. For this methodology to work, each gate in the library needs to have sufficiently many sizes so that the error in discretization in the final step is minimal [136,137].

2.5.2 Timing Analysis

A good logic optimization system requires access to an efficient timing analysis engine. In this section, we briefly address some of the work in this area.

2.5.2.1 Incremental Static Timing Analysis

Static timing analysis is covered in detail in Chapter 6. A key requirement for efficient timing optimization in a logic synthesis system is to have a fast incremental static timing analyzer coupled with the netlist representation. Technology-dependent optimizations (Section 2.4.3) operate on mapped circuits. Typically, local changes are made to the circuit. After each change, the design is queried to check if the timing has improved. Invoking static timing analysis on the whole design is a waste of computational resources if the impact of the changes is local. For example, when a gate is sized up, all gates in the transitive fanout cone of the gate and the immediate fanin gates are candidates for a change in arrival times. The actual update in timing is triggered by a query for timing information and is also dictated by the location of the query. This is known as *level limiting* [138]. However, not all arrival times may change as other paths can dominate the arrival. This is known as *dominance limiting.* The objective of incremental static timing analysis is to perform the minimal computation required to update the timing information that is invalidated by logic optimization changes. Timing information includes net capacitances, arrival times, required times, slack, transition times, and pin-to-pin delays. One of the published works on incremental static timing analysis in the context of netlist changes during synthesis is [139].

2.5.2.2 False Paths

False paths are covered in detail in Chapter 6. Our interest is in logic optimization performed in the context of handling false paths. Keutzer et al. [140] demonstrate that redundancy is not necessary to reduce delay. They provide an algorithm (called the KMS algorithm after the authors) that derives an equivalent irredundant circuit with no increase in delay. Their approach converts the well-known carry-skip adder into a novel irredundant design. Saldanha et al. [141] study the relationship of circuit structure to redundancy and delay. Kukimoto and Brayton [142] extend the notion of false paths to safe replaceabilty under all input arrival considerations.

2.6 Sequential Optimization

Historically, the first logic synthesis approaches were combinational in nature. Then, sequential techniques were developed to manipulate and optimize single FSMs. Following this, sequential techniques to manipulate a hierarchy of FSMs were developed. Finite-state machines are similar to finite automata [143], except that FSMs produce output signals while automata produce no outputs and simply accept or reject input sequences. Moore machines [144] are FSMs whose output functions depend only on the present state, while Mealy machines [144] are FSMs whose outputs depend on the present state as well as the applied input.

A combinational circuit implements a Boolean function, which depends only on the inputs applied to the circuit. Acyclic circuits are necessarily combinational while cyclic circuits may be combinational [145–148]. However, such circuits are not commonly used in practice.

The rest of this section deals with synchronous sequential circuits. Such circuits typically implement state elements using clocked latches or flip-flops. Asynchronous sequential circuits, in contrast, are not clocked. Synchronous sequential behavior can be specified in the form of a netlist consisting of memory elements and combinational logic, as state transition graphs (STGs) or as transition relations. Transition relations are typically represented implicitly using ROBDDs. To avoid memory explosion, the ROBDDs of a transition relation can be represented in a partitioned manner [149]. Often, the FSM behavior is expressed *non-deterministically* by allowing transitions to different states (with different outputs) from a given input state and input combination. Nondeterminism allows for compact representations of the FSM behavior. Also, the transition and output functions may be *incompletely specified.* In such a case, any behavior is considered to be allowed. Of course, any implementation must be deterministic and completely specified; however, the relaxation of these restrictions during optimization allows us to express a multitude of allowable behaviors in a compact fashion.

2.6.1 State Minimization

Given an STG, with symbolic states and transitions, we perform state minimization to combine *equivalent states* (states that produce identical output sequences, given identical input sequences). This is cast as a fixed-point computation. Typically, for completely specified machines, at step i, we construct a set of sets of equivalent states that form a partition of the original state space, and refine the set by separating the states that can be distinguished by an additional step. This is continued until convergence, starting with an initial set that consists of all states in the state space. For completely specified machines, the problem has polynomial complexity.

For incompletely specified machines, we compute the set of *prime compatibles.* For incompletely specified machines, the problem is NP-hard [150]. One of the early approaches for deterministic incompletely specified FSMs was given in [151]. In [152], it was shown that a minimum state cover for an incompletely specified machine could be found using prime compatibles. In [153], an efficient minimization method was introduced, using the notion of compatibles. In [154], exact as well as heuristic minimization results were reported, and implemented in a tool called STAMINA. An implicit (using ROBDDs) computation of the compatibles was reported in [155], allowing the technique to handle extremely large numbers of compatibles. In [156], the authors address the minimization of nondeterministic FSMs based on the notion of generalized compatibles, which are computed implicitly.

One of the early works in the minimization of a network of FSMs based on input don't care sequences was by Kim and Newborn [157], which computed the input don't care sequences for the driven machine. Later, Rho and Somenzi [158] showed that for an incompletely specified machine, there exists an *analogous* machine, which has to be minimized for the optimization of two interacting completely specified machines. In [159], an implicit procedure to compute all input don't care sequences in a general FSM network was introduced. The work of Watanabe and Brayton [160] computes the maximum set of permissible behaviors for any FSM in a network of FSMs. The resulting nondeterministic machine is referred to as the *E-machine.* State minimization of this machine is reported in [161].

In [162], the problem of minimizing a network of interacting FSMs in the context of language containment-based formal property verification is addressed. The problem of synthesizing low-power FSMs is addressed in [163], while don't care-based minimization of extended FSMs is reported in [164].

The decomposition of an FSM is addressed in [165], where the objective is power reduction. A Kernighan-Lin-style [166] partitioning procedure is utilized. In [167], the goal of the FSM decomposition is I/O minimization, and it is achieved using a Fiduccia-Mattheyses-based [168] partitioning approach.

2.6.2 State Assignment

Given a minimized STG, we perform state assignment to assign binary codes to each symbolic state. The choice of binary codes has a great impact on the area and power of the final synthesized design. State assignment can be viewed as an encoding problem, using either input-encoding, output-encoding, or input–output-encoding.

The encoding problem is split into two steps. In the first step, a multivalued representation is minimized, along with a set of constraints on the codes used for the symbolic states. The next step involves finding an encoding that satisfies these constraints. The encoded representation has the same size as that of the minimized multivalued representation. Encoding can be performed for either two- or for multilevel implementations of the FSM. In two-level implementations, we attempt to minimize the number of cubes while in multilevel implementations, we attempt to minimize the number of literals in the design.

The input constraints are *face-embedding* constraints — they specify that any symbol can be assigned to a single face of the Boolean n cube, without that face being shared by other symbols. Output constraints are *dominance*[*‡] or *disjunctive*[*§] constraints. In [169], it was shown that finding a minimum-length code satisfying input constraints is NP-hard [150].

In [170,171], input-encoding-based state assignment for minimum two-level realizations is reported. A solution producing multilevel implementations is given in [172]. The output-encoding procedure of Devadas and Newton [173] generates disjunctive constraints with nested conjunctive terms. The NOVA algorithm of Villa and Sangiovanni-Vincentelli [174] exploits dominance constraints. In [169,175], both input-encoding and output-encoding constraints are handled simultaneously. Multilevel encoding is reported in [176–178]. State assignment with the goal of minimizing the size of the ROBDD of the FSM transition relation is reported in [179]. Parallel state assignment is explored in [180] while state assignment for low power is studied in [181–185]. State assignment for testability is studied in [185].

2.6.3 Retiming

Once a sequential netlist has been generated, further optimization can be achieved by *retiming*, which involves moving registers across logic elements, in a manner such that the sequential behavior of the circuit is maintained. The goal of retiming may be to minimize clock period (*min-period retiming*), or the number of registers (*min-area retiming*), or to minimize the number of registers subject to a maximum clock period constraint (*constrained min-area retiming*). Retiming may be coupled with resynthesis as well. In all retiming approaches, the circuit is represented as a graph, with edges representing wires and vertices representing gates. The weight of a vertex corresponds to the gate delay, and the weight of an edge corresponds to the number of registers on that edge.

Early retiming efforts [186] were based on mixed integer linear programming approaches. Later, relaxation-based [187] approaches were reported. In general, retimed circuits require a new initial state to be computed [188]. Resetable circuits [189] may be utilized to implement retimed designs so that an initializing sequence, which can be used to bring the machine into a known state after power-up, can be easily computed. In [190], techniques to minimize the effort of finding new equivalent initial states after retiming are integrated in the retiming approach. In [191], a min-area retiming methodology, which guarantees the existence of equivalent initial states, is described.

Retiming for level-sensitive designs was studied in [192–195]. Efficient implementations of min-period and constrained min-area retiming were studied in [196], while Maheshwari and Sapatnekar [197] demonstrated efficient min-area retiming.

In [198], the approach is to combine retiming and resynthesis by moving registers to the circuit boundary, then optimizing the combinational logic and moving registers back into the design. Peripheral retiming for pipelined logic [199] involves retiming such that internal edges have zero weight, although peripheral edges may have negative weights. Once the retiming is legalized (all weights are made greater than or equal to zero), we obtain a functionally equivalent circuit (equivalence is exhibited after an initializing sequence is applied). Another resynthesis approach [200] utilizes the retiming-induced register equivalence to optimize logic.

Retiming has been studied in the FPGA context [201,202], as well as in the context of testability preservation and enhancement [203,204]. Retiming for low power is described in [205]. Recent retiming approaches account for DSM delay constraints [206], interconnect and gate delay [207], and clock distribution [208,209]. Retiming is used to guide state assignment in [210].

2.7 Physical Synthesis

With the advance of technology in the late 1990s, it became evident that physical effects could not be ignored during logic synthesis. We briefly summarize some of the technology issues that led to this crisis. Decreasing transistor sizes translate to smaller and faster gates. Scaling down gate sizes implies that

[*‡] *Dominance constraints* express the condition in which the code for any symbol covers that of another.

[*§] *Disjunctive constraints* express the condition in which the code of any symbol is the bit-wise OR of other symbols.

capacitances of input pins of gates decrease. To incorporate more logic, the routing resources cannot be scaled in a commensurate manner. Consequently, the number of metal interconnect layers increased from three to a much larger number (8 to 10 is common in current technologies). In order to lower the resistance of interconnect, the cross-sectional area has to be increased. If the width of the wire is increased, then routability suffers as fewer wires can be packed in a given area. Consequently, the height of metal interconnect needs to be increased. This raises the lateral capacitance of an interconnect line to neighboring interconnect lines. Until the mid-1990s, synthesis and physical design (placement and routing) had been very weakly coupled. A simple notion of net weights to indicate timing criticality was sufficient for placement. The concept of net list sign-off served as a viable business model among the foundries, design teams, and tool vendors. The delay estimate of a gate during synthesis relies on an accurate estimate of the capacitance on a net driven by the gate. The net capacitance is the sum of the pin capacitance and the wire capacitance (the loading from the interconnect used to connect electrically all the pins on the net). Synthesis uses wire load models (Section 2.5.1) as a proxy for wire capacitances. As long as the postrouted wire capacitance does not account for a significant fraction of the total net capacitance, the inaccuracies in delay calculation during synthesis do not cause a problem. At 0.25 μm, wire capacitances began to dominate pin capacitances. Consequently, technology-dependent optimizations were less effective without a means of correctly estimating wire capacitances. This in turn required access to the physical placement of cells and pins. The placement algorithms had relatively weak timing models and analysis capabilities. Recall that static timing analysis was primarily developed in the context of synthesis. These factors led to non convergence of a design through the steps of synthesis, placement, and routing in the design flow. Moreover, effects such as signal integrity issues, which were ignored in previous technology generations, could no longer be ignored due to the strong lateral coupling capacitances.

Today, physical synthesis is a key step that has replaced placement and late timing correction in synthesis. The early work in this area focused on incorporating physical effects during logic optimization [211–214]. Keutzer et al. [215], in an invited paper, discussed the technical and business issues of this discontinuity. The second-generation efforts looked toward a closer integration of synthesis and placement [216–218]. The third-generation research [219,220] enabled commercial products to enter the market. Gosti et al. [221] improve the companion placement model proposed earlier [211] for better area and delay performance. The concept of logical effort [135] enables an alternative view of deferring technology-dependent optimizations until some placement information is available. Physical synthesis uses two sets of distinct optimizations to achieve convergence of the final design. Logic synthesis during placement is able to change the netlist structure undergoing placement. This enables operations such as effective buffering of long wires, sizing up of weak gates, etc. Placement will penalize cell congestion, cell overlaps, and nets with long wire lengths. Working in tandem, the two techniques produce much better results than either one alone.

Carragher et al. [222] discuss layout-driven logic optimization strategies. Logic restructuring [223] and technology mapping [224] with physical information have also been studied. Murgai [225] investigates area recovery using layout-driven buffer optimization. Kudva et al. [226] discuss metrics for measuring routability in the technology-independent optimization phase when the structure of the netlist is defined. Saxena and Halpin [227] discuss incorporating repeaters within the placement formulation.

2.8 Multivalued Logic Synthesis

Multivalued logic synthesis has received much attention in recent times. Although research on multivalued circuit [228–230] as well as memory [231] implementations has been reported, such circuits are hard to design and fabricate, as evidenced by the significantly greater attention given to binary-valued circuits.

The role of multivalued techniques is arguably during the early stages of synthesis. After multivalued logic optimizations have been performed, the design can be encoded into binary. At this point, the traditional binary-valued logic optimization flow can be applied.

Early multivalued techniques included the generalization of ESPRESSO to multiple values [57]. The reduced offset computation, which is used in ESPRESSO, also has a multivalued counterpart [232]. ROBDDs have also been generalized to multiple values [66]. Many other binary-valued logic optimization techniques have been generalized to multiple values, such as algebraic division [233], factorization [234], generalized cofactoring [235], don't cares [236], redundancy removal [237], wire removal [238], and satisfiability (SAT) [239]. In [240], nondeterministic multivalued network simplification is addressed. A nondeterministic multivalued relation is used to express the maximum flexibility of a node. In [241], a technique to reduce multivalued algebraic operations into binary is reported, yielding significant speedups in operations on multivalued networks. Multivalued, multilevel synthesis techniques are discussed in [242,243]. MVSIS [243] implements many of the above techniques, in addition to encoding.

Static [244] and dynamic [245] MDD variable reordering have been implemented, in addition to dynamic reencoding [246]. MDD-based synthesis [247] and functional decomposition [248] approaches have also been reported. The D-algorithm for ATPG has been generalized to multiple values [249]. In [250], the authors describe fault simulation for sequential multivalued networks.

2.9 Summary

In this chapter we have explored some of the key state-of-the-art logic synthesis techniques. Our focus has been on mainstream synthesis approaches, based on the CMOS standard cell design paradigm.

We covered high-level, two-level, and multilevel synthesis as well as sequential synthesis in this chapter. Related enabling technologies were covered, along with recent approaches that combine physical design and logic synthesis.

Starting with early logic synthesis techniques, there has been a steady push toward an increased level of abstraction. At the same time, technology considerations have forced logic synthesis to become physically aware. In the future, we see several challenges. Improved multivalued logic synthesis techniques will enable more effective algorithmic approaches to high-level synthesis. At the same time, the quality of the netlists produced by logic synthesis (from a physical design perspective) will continue to be important. The capacity and speed of optimization continue to be hurdles for flat multimillion gate synthesis.

A change in the underlying implementation style, such as a diversion from CMOS, can be a driver for new research in logic synthesis approaches. Currently there is much excitement about logic synthesis for quantum computing.

References

[1] C. Shannon, A symbolic analysis of relay and switching circuits, *Transactions of the American Institute of Electrical Engineers*, 1938, pp. 713–723.
[2] M. Karnaugh, The map method for synthesis of combinational logic circuits, *Transactions of the American Institute of Electrical Engineers*, 1953, pp. 593–599.
[3] E. Veitch, A chart method for simplifying truth functions, *Proceedings of the Association for Computing Machinery*, 1952, pp. 127–133.
[4] W. Quine, The problem of simplifying truth functions, *Am. Math. Monthly*, 59, 521–531, 1952.
[5] E. McCluskey, Minimization of Boolean functions, *Bell Syst. Tech. J.*, 35, 1417–1444, 1956.
[6] D. Huffman, The synthesis of sequential switching circuits, *J. Franklin Institute*, 1954, pp.161–190.
[7] G. Mealy, A method for synthesizing sequential circuits, *Bell Syst. Tech. J.*, 34, 1045–1079, 1955.
[8] E. Moore, Gedanken-experiments on sequential machines, *Automata Studies*, Princeton University Press, Vol. 34, 1956, pp. 129–153.
[9] S. Hong, R. Cain, and D. Ostapko, MINI: a heuristic approach for logic minimization, *IBM J. Res. Develop.*, 18, 443–458, 1974.
[10] R. Brayton, A. Sangiovanni-Vincentelli, C. McMullen, and G. Hachtel, *Logic Minimization Algorithms for VLSI Synthesis*, Kluwer Academic Publishers, Dordrecht, 1984.
[11] J. Darringer, D. Brand, J. Gerbi, W. Joyner, and L. Trevillyan, LSS: a system for production logic synthesis, *IBM J. Res. Develop.*, 28, 272–280, 1984.

[12] R. Brayton, R. Camposano, G. DeMicheli, R.H.J.M. Otten, and J.T.J. van Eijndhoven, The Yorktown silicon compiler system, in *Silicon Compilation*, D. Gajski, Ed., Addison-Wesley, Reading, MA, 1988.

[13] R. Brayton, R. Rudell, A. Sangiovanni-Vincentelli, and A. Wang, MIS: a multiple-level logic optimization system, *IEEE Trans. Comput.-Aided Des. Integrated Circuits Syst.*, 6, 1062–1081, 1987.

[14] D. Bostick, G.D. Hachtel, R. Jacoby, M.R. Lightner, P. Moceyunas, C.R. Morrison, and D. Ravenscroft, The Boulder optimal logic design system, *Proceedings of IEEE International Conference on Computer-Aided Design*, Santa Clara, CA, 1987, pp. 62–65.

[15] G. De Micheli, *Synthesis and Optimization of Digital Circuits*, McGraw-Hill, New York, 1994.

[16] S. Devadas, A. Ghosh, and K. Keutzer, *Logic Synthesis*, McGraw-Hill, New York, 1994.

[17] S. Hassoun and T. Sasao, Eds., *Logic Synthesis and Verification*, Kluwer Academic Publishers, Dordrecht, 2002.

[18] G. Hachtel and F. Somenzi, *Logic Synthesis and Verification Algorithms*, Kluwer Academic Publishers, Dordrecht, 2000.

[19] R. Camposano, From behavior to structure: high-level synthesis, *IEEE Design Test Comput.*, 7, 8–19, 1990.

[20] M. McFarland, A. Parker, and R. Camposano, The high-level synthesis of digital systems, *Proc. IEEE*, 78, 301–318, 1990.

[21] R. Camposano and W. Wolf, *High Level VLSI Synthesis*, Kluwer Academic Publishers, Boston, 1991.

[22] D. Gajski and L. Ramachandran, Introduction to high-level synthesis, *IEEE Design Test Comput.*, 11, 44–54, 1994.

[23] R. Camposano and R. Tebet, Design representation for the synthesis of behavioral VHDL models, *Proceedings of the International Symposium on Computer Hardware Description Languages and their Applications*, Washington, D.C., 1989, pp. 49–58.

[24] A. Aho, R. Sethi, and J. Ullman, *Compiler Principles, Techniques and Tools*, Addison-Wesley, Reading, MA, 1986.

[25] S. Gupta, N. Savoiu, N. Dutt, R. Gupta, and A. Nicolau, Using global code motions to improve the quality of results for high-level synthesis, *IEEE Trans. Comput.-Aided Des. Integrated Circuits Syst.*, 23, 302–312, 2004.

[26] S. Amellal and B. Kaminska, Scheduling of a control data flow graph, *Proceedings of the IEEE International Symposium on Circuits and Systems*, Portland, OR, Vol. 3, 1993, pp. 1666–1669.

[27] P. Paulin and J. Knight, Force-directed scheduling for the behavioral synthesis of ASICs, *IEEE Trans. Comput.-Aided Des. Integrated Circuits Syst.*, 8, 661–679, 1989.

[28] A. Parker, J. Pizarro, and M. Mlinar, MAHA: a program for datapath synthesis, *Proceedings of ACM/IEEE Design Automation Conference*, Las Vegas, NV, 1986, pp. 461–466.

[29] S. Davidson, D. Landskov, B. Shriver, and P. Mallet, Some experiments in local microcode compaction for horizontal machines, *IEEE Trans. Comput.*, C-30, 460–477, 1981.

[30] R. Camposano, Path-based scheduling for synthesis, *IEEE Trans. Comput.-Aided Des. Integrated Circuits Syst.*, 10, 85–93, 1991.

[31] I. Radivojevic and F. Brewer, Symbolic techniques for optimal scheduling, *Proceedings of SASIMI Workshop*, Nara, Japan, 1993, pp.145–154.

[32] A. Raghunathan and N. Jha, Behavioral synthesis for low power, *Proceedings of IEEE International Conference on Computer Design*, Cambridge, MA, 1994, pp. 318–322.

[33] Y. Fang and A. Albicki, Joint scheduling and allocation for low power, *Proceedings of the IEEE International Symposium on Circuits and Systems*, Atlanta, GA, Vol. 4, 1996, pp. 556–559.

[34] C. Gopalakrishnan and S. Katkoori, Behavioral synthesis of datapaths with low leakage power, *Proceedings of IEEE International Symposium on Circuits and Systems*, Scottsdale, AZ, Vol. 4, 2002, pp. 699–702.

[35] R. San Martin and J. Knight, Optimizing power in ASIC behavioral synthesis, *IEEE Design Test Comput.*, 13, 58–70, 1996.

[36] K. Khouri and N. Jha, Leakage power analysis and reduction during behavioral synthesis, *IEEE Trans. Very Large Scale Integration (VLSI) Syst.*, 10, 876–885, 2002.

[37] R. Bergamaschi, Bridging the domains of high-level and logic synthesis, *IEEE Trans. Comput.-Aided Des. Integrated Circuits Syst.*, Santa Clara, CA, 21, 582–596, 2002.

[38] R. Pomerleau, P. Franzon, and G. Bilbro, Improved delay prediction for on-chip buses, *Proceedings of ACM/IEEE Design Automation Conference*, New Orleans, LA, 1999, 497–501.

[39] R. Gupta and J. Li, Control optimizations using behavioral don't cares, *Proceedings of IEEE International Symposium on Circuits and Systems*, Atlanta, GA, Vol. 4, 1996, pp. 404–407.

[40] K. Wagner and S. Dey, High-level synthesis for testability: a survey and perspective, *Proceedings of ACM/IEEE Design Automation Conference*, Las Vegas, NV, 1996, pp. 131–136.

[41] W. Wolf, A. Takach, C.-Y. Huang, R. Manno, and E. Wu, The Princeton University behavioral synthesis system, *Proceedings of ACM/IEEE Design Automation Conference*, Anaheim, CA, 1992, pp. 182–187.

[42] G. DeMicheli and D. Ku, HERCULES: a system for high-level synthesis, *Proceedings of ACM/IEEE Design Automation Conference*, Atlantic City, NJ, 1988, pp. 483–488.

[43] G. DeMicheli, D. Ku, F. Mailhot, and T. Truong, The Olympus synthesis system, *IEEE Design Test Comput.*, 7, 37–53, Oct 1990.

[44] O. Coudert, Two-level minimization: an overview, *Integration*, 17, 97–140, 1994.

[45] O. Coudert and J.-C. Madre, New ideas for solving covering problems, *Proceedings of ACM/IEEE Design Automation Conference*, San Francisco, CA, 1995, pp. 641–646.

[46] O. Coudert, On solving covering problems, *Proceedings of ACM/IEEE Design Automation Conference*, Las Vegas, NV, 1996, pp. 197–202.

[47] E. Goldberg, L. Carloni, T. Villa, R. Brayton, and A. Sangiovanni-Vincentelli, Negative thinking by incremental problem solving: application to unate covering, *Proceedings of IEEE International Conference on Computer-Aided Design*, Santa Clara, CA, 1997, pp. 91–99.

[48] E. Goldberg, L. Carloni, T. Villa, R. Brayton, and A. Sangiovanni-Vincentelli, Negative thinking in branch-and-bound: the case of unate covering, *IEEE Trans. Comput.-Aided Des. Integrated Circuits and Syst.*, 19, 281–294, 2000.

[49] P. McGeer, J. Sanghavi, R. Brayton, and A. Sangiovanni-Vicentelli, ESPRESSO-SIGNATURE: a new exact minimizer for logic functions, *IEEE Trans. Very Large Scale Integration (VLSI) Syst.*, 1, 432–440, 1993.

[50] G. Swamy, R. Brayton, and P. McGeer, A fully implicit Quine-McCluskey procedure using BDDs, *Proceedings of the International Workshop on Logic Synthesis*, Tahoe City, CA, 1993.

[51] R. Bryant, Graph-based algorithms for Boolean function manipulation, *IEEE Trans. Comput.*, C-35, 677–691, 1986.

[52] O. Coudert, J.-C. Madre, and H. Fraisse, A new viewpoint on two-level logic minimization, *Proceedings of ACM/IEEE Design Automation Conference*, Dallas, TX, 1993, pp. 625–630.

[53] R. Cordone, F. Ferrandi, D. Sciuto, and R. Calvo, An efficient heuristic approach to solve the unate covering problem, *Proceedings, Design, Automation and Test in Europe Conference*, Paris, France, 2000, pp. 364–371.

[54] S. Minato, Implicit manipulation of polynomials using zero-suppressed BDDs, *Proceedings of European Design and Test Conference*, Paris, France, 1995, pp. 449–454.

[55] A. Malik, R. Brayton, A.R. Newton, and A. Sangiovanni-Vincentelli, Reduced offsets for two-level multi-valued logic minimization, *Proceedings of ACM/IEEE Design Automation Conference*, Orlando, FL, 1990, pp. 290–296.

[56] K. Shenoy, N. Saluja, and S. Khatri, An iterative technique for improved two-level logic minimization, *Proceedings of the International Workshop on Logic Synthesis*, Temecula, CA, 2004, pp. 119–126.

[57] R. Rudell and A. Sangiovanni-Vincentelli, Espresso-MV: algorithms for multiple-valued logic minimization, *Proceedings of IEEE Custom Integrated Circuit Conference*, Portland, OR, 1985, pp. 230–234.

[58] H. Liu, Reducing routing table size using ternary CAM, *Proceedings of Hot Interconnects*, Stanford, CA, 2001, pp. 69–73.

[59] J. Bian and S. Khatri, IP routing table compression using ESPRESSO-MV, *Proceedings of 11th IEEE International Conference on Networks*, Sydney, Australia, 2003, pp. 167–172.

[60] S. Ahmed and R. Mahapatra, m-Trie: an efficient approach to on-chip logic minimization, *Proceedings of IEEE International Conference on Computer-Aided Design*, Santa Clara, CA, 2004, pp. 428–435.

[61] R. Lysecky and F. Vahid, On-chip logic minimization, *Proceedings of ACM/IEEE Design Automation Conference*, 2003, Anaheim, CA, pp. 334–337.

[62] R. Brayton and F. Somenzi, An exact minimizer for Boolean relations, *Proceedings of IEEE International Conference on Computer-Aided Design*, Santa Clara, CA, 1989, pp. 316–319.

[63] T. Villa, T. Kam, R. Brayton, and A. Sangiovanni-Vincentelli, Explicit and implicit algorithms for binate covering problems, *IEEE Trans. Comput.-Aided Des. Integrated Circuits and Syst.*, 16, 671–691, 1997.

[64] S.-W. Jeong and F. Somenzi, A new algorithm for the binate covering problem and its application to the minimization of Boolean relations, *Proceedings of IEEE International Conference on Computer-Aided Design*, Santa Clara, CA, 1992, pp. 417–420.

[65] Y. Watanabe and R. Brayton, Heuristic minimization of multiple-valued relations, *IEEE Trans. Comput.-Aided Des. Integrated Circuits Syst.*, 12, 1458–1472, 1993.

[66] A. Srinivasan, T. Kam, S. Malik, and R. Brayton, Algorithms for discrete function manipulation, *Proceedings of IEEE International Conference on Computer-Aided Design*, Santa Clara, CA, 1990, pp. 92–95.

[67] R. Ashenhurst, The decomposition of switching functions, *Proceedings of International Symposium on the Theory of Switching*, Cambridge, MA, 1957, pp. 74–116.

[68] E. Lawler, An approach to multilevel Boolean minimization, *J. Assoc. Computing Mach.*, 11, 283–295, 1964.

[69] R. Brayton and C. McMullen, The decomposition and factorization of Boolean expressions, *Proceedings of IEEE International Symposium on Circuits and Systems*, Rome, Italy, 1982, pp. 49–54.

[70] R. Brayton and C. McMullen, Synthesis and optimization of multistage logic, *Proceedings of IEEE International Conference on Computer Design*, Port Chester, NY, 1984, pp. 23–38.

[71] J. Vasudevamurthy and J. Rajski, A Method for concurrent decomposition and factorization of boolean expressions, *Proceedings of IEEE International Conference on Computer-Aided Design*, 1990, Santa Clara, CA, pp. 510–513.

[72] S. Muroga, Y. Kambayashi, H. Lai, and J. Culliney, The transduction method-design of logic networks based on permissible functions, *IEEE Trans. Comput.*, 38, 1404–1424, 1989.

[73] H. Savoj and R. Brayton, The use of observability and external don't cares for the simplification of multilevel networks, *Proceedings of ACM/IEEE Design Automation Conference*, Orlando, FL 1990, pp. 297–301.

[74] H. Savoj, R. Brayton, and H. Touati, Extracting local don't cares for network optimization, *Proceedings of IEEE International Conference on Computer-Aided Design*, Santa Clara, CA, 1991, pp. 514–517.

[75] H. Savoj, Don't Cares in Multilevel Network Optimization, PhD thesis, University of California Berkeley, Electronics Research Laboratory, College of Engineering, University of California, Berkeley, CA 94720, 1992.

[76] N. Saluja and S. Khatri, A robust algorithm for approximate compatible observability don't care (CODC) computation, *Proceedings of ACM/IEEE Design Automation Conference*, San Diego, CA, 2004, pp. 422–427.

[77] A. Mishchenko and R. Brayton, SAT-based complete don't care computation for network optimization, *Proceedings of the International Workshop on Logic Synthesis*, Temecula, CA, 2004, pp. 353–360.

[78] S. Yamashita, H. Sawada, and A. Nagoya, A new method to express functional permissibilities for LUT based FPGAs and its applications, *Proceedings of IEEE International Conference on Computer-Aided Design*, 1996, pp. 254–261.

[79] R. Brayton, Understanding SPFDs: a new method for specifying flexibility, *Proceedings of the International Workshop on Logic Synthesis*, Tahoe City, CA, 1997.

[80] S. Sinha and R. Brayton, Implementation and use of SPFDs in optimizing boolean networks, *Proceedings of IEEE International Conference on Computer-Aided Design*, Santa Clara, CA, 1998, pp. 103–110.

[81] J. Cong, J. Lin, and W. Long, A new enhanced SPFD rewiring algorithm, *Proceedings of IEEE International Conference on Computer-Aided Design*, Santa Clara, CA, 2002, pp. 672–678.

[82] B. Kumthekar and F. Somenzi, Power and delay reduction via simultaneous logic and placement optimization in FPGAs, *Proceedings, Design, Automation and Test in Europe Conference*, Paris, France, 2000, pp. 202–207.

[83] S. Khatri, S. Sinha, R. Brayton, and A. Sangiovanni-Vincentelli, SPFD-based wire removal in standard-cell and network-of-PLA circuits, *IEEE Trans. Comput.-Aided Des. Integrated Circuits Syst.*, 23, 1020–1030, July 2004.

[84] S. Sinha, A. Mishchenko, and R. Brayton, Topologically constrained logic synthesis, *Proceedings of IEEE International Conference on Computer-Aided Design*, Santa Clara, CA, 2002, pp. 679–686.

[85] S. Sinha, A. Kuehlmann, and R. Brayton, Sequential SPFDs, *Proceedings of IEEE International Conference on Computer-Aided Design*, Santa Clara, CA, 2001, pp. 84–90.

[86] D. Gregory, K. Bartlett, A. de Geus, and G. Hachtel, SOCRATES: a system for automatically synthesizing and optimizing combinational logic, *Proceedings of ACM/IEEE Design Automation Conference*, Las Vegas, NV, 1986, pp. 79–85.

[87] W. Joyner, L. Trevillyan, D. Brand, T. Nix, and S. Gunderson, Technology adaptation in logic synthesis, *Proceedings of ACM/IEEE Design Automation Conference*, Las Vegas, NV, 1986, pp. 94–100.

[88] A. Aho and M. Corasick, Efficient string matching: an aid to bibliographic search, *Commn. Assoc. Computing Mach.*, 333–340, 1975.

[89] F. Mailhot and G. De Micheli, Algorithms for technology mapping based on binary decision diagrams and on Boolean operations, *IEEE Trans. Comput.-Aided Des. Integrated Circuits Syst.*, 559–620, 1993.

[90] J. Burch and D. Long, Efficient Boolean function matching, *Proceedings of IEEE International Conference on Computer-Aided Design*, Santa Clara, CA, 1992, pp. 408–411.

[91] J. Mohnke, P. Molitor, and S. Malik, Limits of using signatures for permutation independent Boolean comparison, *Proceedings of the Asia and South Pacific Design Automation Conference*, Makuhari, 1995.

[92] U. Schlichtman and F. Brglez, Efficient Boolean matching in technology mapping with very large cell libraries, *Proceedings of IEEE Custom Integrated Circuit Conference*, San Diego, CA, Japan 1993.

[93] R. Rudell, Logic Synthesis for VLSI Design, Ph.D. thesis, University of California, Berkeley, 1989.

[94] K. Keutzer, DAGON: technology binding and local optimization by DAG matching, *Proceedings of ACM/IEEE Design Automation Conference*, Miami Beach, 1987, pp. 341–347.

[95] A. Aho and S. Johnson, Optimal code generation for expression trees, *J. Assoc. Computing Mach.*, 1976.

[96] H. Touati, Performance Oriented Technology Mapping, Ph.D. thesis, University of California, Berkeley, 1990.

[97] S. Roy, K. Belkhale, and P. Banerjee, An α-approximate algorithm for delay-constraint technology mapping, *Proceedings of ACM/IEEE Design Automation Conference*, New Orleans, 1999, pp. 367–372.

[98] K. Chaudhary and M. Pedram, A near-optimal algorithm for technology mapping minimizing area under delay constraints, *Proceedings of ACM/IEEE Design Automation Conference*, Anaheim, CA, 1992, pp. 492–498.

[99] J. Grodstein, E. Lehman, H. Harkness, B. Grundmann, and Y. Watanabe, A delay model for logic synthesis of continuously sized networks, *Proceedings of IEEE International Conference on Computer-Aided Design*, San Jose, CA, 1995, pp. 458–462.

[100] L. Stok, M.A. Iyer, and A. Sullivan, Wavefront technology mapping, *Proceedings, Design, Automation and Test in Europe Conference*, Munich, 1999.

[101] Y. Kukimoto, R. Brayton, and P. Sawkar, Delay-optimal technology mapping by DAG covering, *Proceedings of ACM/IEEE Design Automation Conference*, San Francisco, CA, 1998.

[102] E. Lehman, E. Watanabe, J. Grodstein, and H. Harkness, Logic decomposition during technology mapping, *Proceedings of IEEE International Conference on Computer-Aided Design*, San Jose, CA, 1995, pp. 264–271.

[103] V. Tiwari, P. Ashar, and S. Malik, Technology mapping for low power in logic synthesis, *Integration, the VLSI J.*, 3, 243–268, 1996.

[104] C. Tsui, M. Pedram, and A. Despain, Technology decomposition and mapping targetting low power dissipation, *Proceedings of ACM/IEEE Design Automation Conference*, Dallas, TX, 1993, pp. 68–73.

[105] I. Sutherland and R. Sproull, Logical effort: designing for speed on the back of an envelope, *Advanced Research in VLSI*, Santa Cruz, CA, 1991, pp. 1–16.

[106] D. Kung, A fast fanout optimization algorithm for near-continuous buffer libraries, *Proceedings of ACM/IEEE Design Automation Conference*, San Francisco, CA, 1998, pp. 352–355.

[107] P. Rezvani, A. Ajami, M. Pedram, and H. Savoj, LEOPARD: a logical effort-based fanout optimizer for area and delay, *Proceedings of IEEE International Conference on Computer-Aided Design*, Santa Clara, CA, 1999, pp. 516–519.

[108] S. Karandikar and S. Sapatnekar, Logical effort based technology mapping, *Proceedings of IEEE International Conference on Computer-Aided Design*, Santa Clara, CA, 2004, pp. 419–422.

[109] S. Srinivasaraghavan and W. Burleson, Interconnect effort — a unification of repeater insertion and logical effort, *Proceedings of IEEE Computer Society Annual Symposium on VLSI*, Tampa, FL, 2003, pp. 55–61.

[110] K. Venkat, Generalized delay optimization of resistive interconnections through an extension of logical effort, *IEEE International Symposium on Circuits and Systems (ISCAS)*, Portland, OR, Vol. 3, 1993, pp. 2106–2109.

[111] D. Harris and I. Sutherland, Logical effort of carry propagate adders, *Conference Record of the Thirty-Seventh Asilomar Conference on Signals, Systems and Computers*, Pacific Grove, CA, Vol. 1, 2003, pp. 873–878.

[112] H. Dao and V. Oklobdzija, Application of logical effort techniques for speed optimization and analysis of representative adders, *Conference Record of the Thirty-Fifth Asilomar Conference on Signals, Systems and Computers*, Pacific Grove, CA, Vol. 2, 2001, pp. 1666–1669.

[113] J. Shyu, A. Sangiovanni-Vincentelli, J. Fishburn, and A. Dunlop, Optimization based transistor sizing, *IEEE J. Solid State Circuits*, 1988.

[114] J. Fishburn and A. Dunlop, TILOS: a posynomial programming approach to transistor sizing, *Proceedings of IEEE International Conference on Computer-Aided Design*, Santa Clara, CA, 1985, pp. 326–328.

[115] S. Sapatnekar, V. Rao, P. Vaidya, and S. Kang, An exact solution to the transistor sizing problem for CMOS circuits using convex optimization, *IEEE Trans. Comput.-Aided Des. Integrated Circuits Syst.*, 12, 1621–1634, 1993.

[116] M. Cirit, Transistor sizing in CMOS circuits, *Proceedings of ACM/IEEE Design Automation Conference*, Miami Beach, 1987, pp. 121–124.

[117] K. Hedlund, AESOP: a tool for automated transistor sizing, *Proceedings of ACM/IEEE Design Automation Conference*, Miami Beach, 1987, pp. 114–120.

[118] D. Marple, Transistor size optimization in the tailor layout system, *Proceedings of ACM/IEEE Design Automation Conference*, Las Vegas, NV, 1989, pp. 43–48.

[119] M. Berkelaar and J. Jess, Gate sizing in MOS digital circuits with linear programming, *Proceedings of European Design Automation Conference*, Glasgow, 1990.

[120] O. Coudert, Gate sizing: a general purpose optimization approach, *Proceedings of European Design and Test Conference*, Paris, 1996.

[121] O. Coudert, R. Haddad, and S. Manne, New algorithms for gate sizing: a comparative study, *Proceedings of ACM/IEEE Design Automation Conference*, Las Vegas, NV, 1996, pp. 734–739.

[122] C. Berman, J. Carter, and K. Day, The fanout problem: from theory to practice, in *Advanced Research in VLSI: Proceedings of the Decennial Caltech Conference*, C.L. Seitz, Ed., MIT press, Cambridge, MA, 1989, pp. 66–99.

[123] H. Hoover, M. Klawe, and N. Pippenger, Bounding fanout in logical networks, *J. Assoc. Computing Mach.*, 31, 13–18, 1984.
[124] K. Singh and A. Sangiovanni-Vincentelli, A heuristic algorithm for the fanout problem, *Proceedings of ACM/IEEE Design Automation Conference*, Orlando, FL, 1990, pp. 357–360.
[125] D. Kung, A fast fanout optimization algorithm for near-continuous buffer libraries, *Proceedings of ACM/IEEE Design Automation Conference*, San Francisco, CA, 1998, pp. 352–355.
[126] K. Bartlett, W. Cohen, A. De Geus, and G. Hachtel, Synthesis and optimization of multilevel logic under timing constraints, *IEEE Trans. Comput.-Aided Des. Integrated Circuits Syst.*, 5, 582–596, 1986.
[127] K. Singh, A. Wang, R. Brayton, and A. Sangiovanni-Vincentelli, Timing optimization of combinational logic, *Proceedings of IEEE International Conference on Computer-Aided Design*, Santa Clara, CA, 1988, pp. 282–285.
[128] J. Fishburn, A depth-decreasing heuristic for combinational logic, *Proceedings of ACM/IEEE Design Automation Conference*, Orlando, FL, 1990, pp. 361–364.
[129] K. Yoshikawa, H. Ichiryu, H. Tanishita, S. Suzuki, N. Nomizu, and A. Kondoh, Timing optimization on mapped circuits, *Proceedings of ACM/IEEE Design Automation Conference*, San Francisco, CA, 1991, pp. 112–117.
[130] http://www.synopsys.com/partners/tapin/lib_info.html. Liberty.
[131] http://www.eda.org/alf/. Advanced Library Format (alf) home page.
[132] K. Keutzer, K. Kolwicz, and M. Lega, Impact of library size on the quality of automated synthesis, *Proceedings of IEEE International Conference on Computer-Aided Design*, Santa Clara, CA, 1987, pp. 120–123.
[133] K. Scott and K. Keutzer, Improving cell libraries for synthesis, *Proceedings of IEEE Custom Integrated Circuit Conference*, San Diego, CA, 1994, pp. 128–131.
[134] I. Sutherland, R. Sproull, and D. Harris, *Logical Effort*, Morgan-Kaufmann, San Franciso, CA, 1999.
[135] R. Haddad, L.P.P.P. van Ginneken, and N. Shenoy, Discrete drive selection for continuous sizing, *Proceedings of IEEE International Conference on Computer Design*, Austin, TX, 1997, pp. 110–115.
[136] F. Beeftink, P. Kudva, D. Kung, and L. Stok, Gate-size selection for standard cell libraries, *Proceedings of IEEE International Conference on Computer-Aided Design*, San Jose, CA, 1998, pp. 545–550.
[137] R. Abato, A. Drumm, D. Hathaway, and L.P.P.P. van Ginneken, Incremental Timing Analysis, US Patent 5,508,937, 1993.
[138] A.R. Wang, Algorithms for Multilevel Logic Optimization, Ph.D. thesis, University of California, Berkeley, 1989.
[139] K. Keutzer, S. Malik, and A. Saldanha, Is redundancy necessary to reduce delay? *IEEE Trans. Comput.-Aided Des. Integrated Circuits Syst.*, 10, 427–435, 1991.
[140] A. Saldanha, R. Brayton, and A. Sangiovanni-Vincentelli, Circuit structure relations to redundancy and delay, *IEEE Trans. Comput.-Aided Des. Integrated Circuits Syst.*, 13, 875–883, 1994.
[141] Y. Kukimoto and R. Brayton, Timing-safe false path removal for combinational modules, *Proceedings of IEEE International Conference on Computer-Aided Design*, San Jose, CA, 1999.
[142] J. Hopcroft and J. Ullman, *Introduction to Automata Theory, Languages and Computation*, Addison-Wesley, Reading, MA, 1979.
[143] Z. Kohavi, *Switching and Finite Automata Theory*, Computer Science Series, McGraw-Hill, New York, 1970.
[144] W. Kautz, The necessity of closed circuit loops in minimal combinational circuits, *IEEE Trans. Comput.*, C-19, 162–166, 1971.
[145] S. Malik, Analysis of cyclic combinational circuits, *IEEE Trans. Comput.-Aided Des. Integrated Circuits Syst.*, 13, 950–956, 1994.
[146] T. Shiple, G. Berry, and H. Touati, Constructive analysis of cyclic circuits, *Proceedings of European Design and Test Conference*, Paris, France, 1996, pp. 328–333.
[147] M. Riedel and J. Bruck, The synthesis of cyclic combinational circuits, *Proceedings of ACM/IEEE Design Automation Conference*, Anaheim, CA, 2003, pp. 163–168.

[148] The VIS group, VIS: a system for verification and synthesis, in *Proceedings of the 8th International Conference on Computer Aided Verification*, number 1102, R. Alur and T. Henzinger, Eds., New Brunswick, NJ, 1996, pp. 428–432.

[149] M.R. Garey and D.S. Johnson, *Computers and Intractability: A Guide to the Theory of NP-Completeness*, W. H. Freeman and Company, New York, 1979.

[150] M. Paull and S. Unger, Minimizing the number of states in incompletely specified sequential switching functions, *IRE Trans. Electron. Comput.*, EC-8, 356–367, 1959.

[151] A. Grasselli and F. Luccio, A method for minimizing the number of internal states in incompletely specified sequential networks, *IEEE Trans. Electron. Comput.*, EC-14, 350–359, 1965.

[152] L. Kannan and D. Sarma, Fast heuristic algorithms for finite state machine minimization, *Proceedings of European Conference on Design Automation*, Amsterdam, 1991, pp. 192–196.

[153] J.-K. Rho, G. Hachtel, F. Somenzi, and R. Jacoby, Exact and heuristic algorithms for the minimization of incompletely specified state machines, *IEEE Trans. Comput.-Aided Des. Integrated Circuits Syst.*, 13, 167–177, 1994.

[154] T. Kam, T. Villa, R. Brayton, and A. Sangiovanni-Vincentelli, Implicit computation of compatible sets for state minimization of ISFSMs, *IEEE Trans. Comput.-Aided Des. Integrated Circuits Syst.*, 16, 657–676, 1997.

[155] T. Kam, T. Villa, R. Brayton, and A. Sangiovanni-Vincentelli, Theory and algorithms for state minimization of nondeterministic FSMs, *IEEE Trans. Comput.-Aided Des. Integrated Circuits Syst.*, 16, 1311–1322, 1997.

[156] J. Kim and M. Newborn, The simplification of sequential machines with input restrictions, *IEEE Trans. Comput.*, 1440–1443, Cambridge, MA, 1972.

[157] J.-K. Rho and F. Somenzi, The role of prime compatibles in the minimization of finite state machines, *Proceedings of IEEE International Conference on Computer Design*, Cambridge, MA, 1992, pp. 324–327.

[158] H.-Y. Wang and R. Brayton, Input don't care sequences in FSM networks, *Proceedings of IEEE International Conference on Computer-Aided Design*, Santa Clara, CA, 1993, pp. 321–328.

[159] Y. Watanabe and R. Brayton, The maximum set of permissible behaviors for FSM networks, *Proceedings of IEEE International Conference on Computer-Aided Design*, Santa Clara, CA, 1993, pp. 136–320.

[160] Y. Watanabe and R. Brayton, State minimization of pseudo non-deterministic FSMs, *Proceedings of European Design and Test Conference*, Paris, France, 1994, pp. 184–191.

[161] A. Aziz, V. Singhal, R. Brayton, and G. Swamy, Minimizing interacting finite state machines: a compositional approach to language containment, *Proceedings of IEEE International Conference on Computer Design*, Cambridge, MA, 1994, pp. 255–261.

[162] A. Dasgupta and S. Ganguly, Divide and conquer: a strategy for synthesis of low power finite state machines, *Proceedings of IEEE International Conference on Computer Design*, Austin, TX, 1997, pp. 740–745.

[163] Y. Jiang and R. Brayton, Don't cares in logic minimization of extended finite state machines, *Proceedings of the Asia and South Pacific Design Automation Conference*, Kitakyushu, Japan, 2003, pp. 809–815.

[164] J. Monteiro and A. Oliveira, Finite state machine decomposition for low power, *Proceedings of ACM/IEEE Design, Automation Conference*, San Francisco, CA, 1998, pp. 758–763.

[165] B.W. Kernighan and S. Lin, An efficient heuristic procedure for partitioning graphs, *Bell Syst. Tech. J.*, 49, 291–307, 1970.

[166] M.-T. Kuo, L.-T. Liu, and C.-K. Cheng, Finite state machine decomposition for I/O minimization, *Proceedings of IEEE International Symposium on Circuits and Systems*, Seattle, WA, Vol. 2, 1995, pp. 1061–1064.

[167] C.M. Fiduuccia and R.M. Mattheyses, A linear-time heuristic for improving network partitions, *Proceedings of ACM/IEEE Design, Automation Conference*, Las Vegas, NV, 1982, pp. 175–181.

[168] A. Saldanha, T. Villa, R. Brayton, and A. Sangiovanni-Vincentelli, Satisfaction of input and output encoding constraints, *IEEE Trans. Comput.-Aided Des. Integrated Circuits Syst.*, 13, 589–602, 1994.
[169] G. DeMicheli, R. Brayton, and A. Sangiovanni-Vincentelli, Optimal state assignment for finite state machines, *IEEE Trans. Comput.-Aided Des. Integrated Circuits Syst.*, 4, 269–285, 1985.
[170] G. DeMicheli, R. Brayton, and A. Sangiovanni-Vincentelli, Correction to "Optimal state assignment for finite state machines", *IEEE Trans. Comput.-Aided Des. Integrated Circuits Syst.*, 5, 239–239, 1986.
[171] L. Lavagno, S. Malik, R. Brayton, and A. Sangiovanni-Vincentelli, Symbolic minimization of multilevel logic and the input encoding problem, *IEEE Trans. Comput.-Aided Des. Integrated Circuits Syst.*, 11, 825–843, 1992.
[172] S. Devadas and A.R. Newton, Exact algorithms for output encoding, state assignment, and four-level Boolean minimization, *IEEE Trans. Comput.-Aided Des. Integrated Circuits Syst.*, 10, 13–27, 1991.
[173] T. Villa and A. Sangiovanni-Vincentelli, NOVA: state assignment of finite state machines for optimal two-level logic implementation, *IEEE Trans. Comput.-Aided Des. Integrated Circuits Syst.*, 9, 905–924, 1990.
[174] M. Ciesielski, J.-J. Shen, and M. Davio, A unified approach to input-output encoding for FSM state assignment, *Proceedings of ACM/IEEE Design Automation Conference*, San Francisco, CA, 1991, pp. 176–181.
[175] S. Devadas, H.-K. Ma, A.R. Newton, and A. Sangiovanni-Vincentelli, MUSTANG: state assignment of finite state machines targeting multilevel logic implementations, *IEEE Trans. Comput.-Aided Des. Integrated Circuits Syst.*, 7, 1290–1300, 1988.
[176] B. Lin and A.R. Newton, Synthesis of multiple level logic from symbolic high-level description languages, *Proceedings of the International Conference on VLSI*, Munich, Germany, 1989, pp. 187–196.
[177] X. Du, G. Hachtel, and P. Moceyunas, MUSE: a multilevel symbolic encoding algorithm for state assignment, *Proceedings of the 23rd Annual Hawaii International Conference on System Sciences*, Vol. 1, Hawaii, 1990, pp. 367–376.
[178] R. Forth and P. Molitor, An efficient heuristic for state encoding minimizing the BDD representations of the transition relations of finite state machines, *Proceedings of the Asia and South Pacific Design Automation Conference*, Yokohama, Japan, 2000, pp. 61–66.
[179] G. Hasteer and P. Banerjee, A parallel algorithm for state assignment of finite state machines, *IEEE Trans. Comput.*, 47, 242–246, 1998.
[180] L. Benini and G. DeMicheli, State assignment for low power dissipation, *IEEE J. Solid-State Circuits*, 30, 258–268, 1995.
[181] K.-H. Wang, W.-S. Wang, T. Hwang, A. Wu, and Y.-L. Lin, State assignment for power and area minimization, *Proceedings of IEEE International Conference on Computer Design*, Cambridge, MA, 1994, pp. 250–254.
[182] C.-Y. Tsui, M. Pedram, and A. Despain, Low-power state assignment targeting two- and multilevel logic implementations, *IEEE Trans. Comput.-Aided Des. Integrated Circuits Syst.*, 17, 1281–1291, 1998.
[183] X. Wu, M. Pedram, and L. Wang, Multi-code state assignment for low power design, *IEEE Proceedings G-Circuits, Devices and Systems*, 147, 271–275, 2000.
[184] S. Park, S. Cho, S. Yang, and M. Ciesielski, A new state assignment technique for testing and low power, *Proceedings of ACM/IEEE Design Automation Conference*, San Diego, CA, 2004, pp. 510–513.
[185] C. Leiserson and J. Saxe, Optimizing Synchronous Systems, *J. VLSI Comput. Syst.*, 1, 41–67, 1983.
[186] J. Saxe, Decomposable Searching Problems and Circuit Optimization by Retiming: Two Studies in General Transformations of Computational Structures, Ph.D. thesis, Carnegie Mellon University, 1985, CMU-CS-85-162.
[187] H. Touati and R. Brayton, Computing the initial states of retimed circuits, *IEEE Trans. Comput.-Aided Des. Integrated Circuits Syst.*, 12, 157–162, 1993.
[188] C. Pixley and G. Beihl, Calculating resetability and reset sequences, *Proceedings of IEEE International Conference on Computer-Aided Design*, Santa Clara, CA, 1991, pp. 376–379.

[189] G. Even, I. Spillinger, and L. Stok, Retiming revisited and reversed, *IEEE Trans. Comput.-Aided Des. Integrated Circuits Syst.*, 15, 348–357, 1996.

[190] N. Maheshwari and S. Sapatnekar, Minimum area retiming with equivalent initial states, *Proceedings of IEEE International Conference on Computer-Aided Design*, Santa Clara, CA, 1997, pp. 216–219.

[191] N. Shenoy, R. Brayton, and A. Sangiovanni-Vincentelli, Retiming of circuits with single phase transparent latches, *Proceedings of IEEE International Conference on Computer Design*, Cambridge, MA, 1991, pp. 86–89.

[192] A. Ishii, C. Leiserson, and M. Papaefthymiou, Optimizing two-phase, level-clocked circuitry, *Proceedings of the Brown/MIT Conference on Advanced Research in VLSI and Parallel Systems*, MIT Press, Cambridge, MA, 1992, pp. 245–264.

[193] B. Lockyear and C. Ebeling, Optimal retiming of multi-phase, level-clocked circuits, *Proceedings of the Brown/MIT Conference on Advanced Research in VLSI and Parallel Systems*, MIT Press, Cambridge, MA, 1992, pp. 265–280.

[194] N. Maheshwari and S. Sapatnekar, Efficient minarea retiming of large level-clocked circuits, *Proceedings, Design, Automation and Test in Europe Conference*, Paris, France, 1998, pp. 840–845.

[195] N. Shenoy and R. Rudell, Efficient implementation of retiming, *Proceedings of IEEE International Conference on Computer-Aided Design*, Santa Clara, CA, 1994, pp. 226–233.

[196] N. Maheshwari and S. Sapatnekar, Efficient retiming of large circuits, *IEEE Trans. Very Large Scale Integration (VLSI) Syst.*, 6, 74–83, 1998.

[197] S. Malik, E. Sentovich, R. Brayton, and A. Sangiovanni-Vincentelli, Retiming and resynthesis: optimizing sequential networks with combinational techniques, *IEEE Trans. Comput.-Aided Des. Integrated Circuits Syst.*, 10, 74–84, 1991.

[198] S. Malik, K. Singh, R. Brayton, and A. Sangiovanni-Vincentelli, Performance optimization of pipelined logic circuits using peripheral retiming and resynthesis, *IEEE Trans. Comput.-Aided Des. Integrated Circuits Syst.*, 12, 568–578, 1993.

[199] P. Kalla and M. Ciesielski, Performance driven resynthesis by exploiting retiming-induced state register equivalence, *Proceedings, Design, Automation and Test in Europe Conference*, Munich, Germany, 1999, pp. 638–642.

[200] J. Cong and C. Wu, Optimal FPGA mapping and retiming with efficient initial state computation, *IEEE Trans. Comput.-Aided Des. Integrated Circuits Syst.*, 18, 1595–1607, 1999.

[201] J. Cong and C. Wu, An efficient algorithm for performance-optimal FPGA technology mapping with retiming, *IEEE Trans. Comput.-Aided Des. Integrated Circuits Syst.*, 17, 738–748, 1998.

[202] A. El-Maleh, T. Marchok, J. Rajski, and W. Maly, Behavior and testability preservation under the retiming transformation, *IEEE Trans. Comput.-Aided Des. Integrated Circuits Syst.*, 16, 528–543, 1997.

[203] S. Dey and S. Chakradhar, Retiming sequential circuits to enhance testability, *Proceedings of IEEE VLSI Test Symposium*, Cherry Hill, NJ, 1994, pp. 28–33.

[204] J. Monteiro, S. Devadas, and A. Ghosh, Retiming sequential circuits for low power, *Proceedings of IEEE International Conference on Computer-Aided Design*, Santa Clara, CA, 1993, 398–402.

[205] A. Tabarra, R. Brayton, and A.R. Newton, Retiming for DSM with area-delay trade-offs and delay constraints, *Proceedings of ACM/IEEE Design Automation Conference*, New Orleans, LA, 1999, pp. 725–730.

[206] C. Chu, E. Young, D. Tong, and S. Dechu, Retiming with interconnect and gate delay, *Proceedings of IEEE International Conference on Computer-Aided Design*, Santa Clara, CA, 2003, pp. 221–226.

[207] T. Soyata, E. Friedman, and J. Mulligan, Incorporating interconnect, register, and clock distribution delays into the retiming process, *IEEE Trans. Comput.-Aided Des. Integrated Circuits Syst.*, 16, 105–120, 1997.

[208] X. Liu, M. Papaefthymiou, and E. Friedman, Retiming and clock scheduling for digital circuit optimization, *IEEE Trans. Comput.-Aided Des. Integrated Circuits Syst.*, 21, 184–203, 2002.

[209] B. Iyer and M. Ciesielski, Metamorphosis: state assignment by retiming and re-encoding, *Proceedings of IEEE International Conference on Computer-Aided Design*, Santa Clara, CA, 1996, pp. 614–617.

[210] M. Pedram and N. Bhat, Layout driven technology mapping, *Proceedings of ACM/IEEE Design Automation Conference*, San Francisco, CA, 1991, pp. 99–105.

[211] M. Pedram and N. Bhat, Layout driven logic restructuring/decomposition, *Proceedings of IEEE International Conference on Computer-Aided Design*, Santa Clara, CA, 1991, pp. 134–137.

[212] P. Abouzeid, K. Sakouti, G. Saucier, and F. Poirot, Multilevel synthesis minimizing the routing factor, *Proceedings of ACM/IEEE Design Automation Conference*, Orlando, FL, 1990, pp. 365–368.

[213] H. Vaishnav and M. Pedram, Routability driven fanout optimization, *Proceedings of ACM/IEEE Design Automation Conference*, Dallas, TX, 1993, pp. 230–235.

[214] K. Keutzer, A.R. Newton, and N. Shenoy, The future of logic synthesis and physical design in deep sub-micron process geometries, *Proceedings of the International Symposium on Physical Design*, Napa Valley, 1997.

[215] L. Kannan, P. Suaris, and H. Fang, A methodology and algorithms for post-placement delay optimization, *Proceedings of ACM/IEEE Design Automation Conference*, San Diego, CA, 1994.

[216] T. Ishioka, M. Murofushi, and M. Murakata, Layout driven delay optimization with logic resynthesis, *Proceedings of the International Workshop on Logic Synthesis*, Tahoe City, CA, 1997.

[217] G. Stenz, B. Riess, B. Rohfleisch, and F. Johannes, Timing driven placement in interaction with netlist transformations, *Proceedings of the International Symposium on Physical Design*, Napa Valley, 1997, pp. 36–41.

[218] S. Hojat and P. Villarubia, An integrated placement and synthesis approach for timing closure of PowerPC microprocessor, *Proceedings of IEEE International Conference on Computer Design*, Austin, TX, 1997, pp. 206–210.

[219] N. Shenoy, M. Iyer, R. Damiano, K. Harer, H.-K. Ma, and P. Thilking, A robust solution to the timing convergence problem in high performance design, *Proceedings of IEEE International Conference on Computer Design*, Austin, TX, 1999, pp. 250–257.

[220] W. Gosti, S. Khatri, and A. Sangiovanni-Vincentelli, Addressing the timing closure problem by integrating logic optimization and placement, *Proceedings of IEEE International Conference on Computer-Aided Design*, San Jose, CA, 2001, pp. 224–231.

[221] R. Carragher, R. Murgai, S. Chakraborty, M. Prasad, A. Srivastava, and N. Vemure, Layout driven logic synthesis, *Proceedings of the International Workshop on Logic Synthesis*, Dana Point, 2000.

[222] J. Lou, A. Salek, and M. Pedram, Concurrent logic restructuring and placement for timing closure, *Proceedings of IEEE International Conference on Computer-Aided Design*, San Jose, CA, 1999, pp. 31–35.

[223] J. Lou, W. Chen, and M. Pedram, An exact solution to simultaneous technology mapping and linear placement problem, *Proceedings of IEEE International Conference on Computer-Aided Design*, San Jose, CA, 1997, pp. 671–675.

[224] R. Murgai, Delay constrained area recovery via layout-driven buffer optimization, *Proceedings of the International Workshop on Logic Synthesis*, Tahoe City, CA, 1999, pp. 217–221.

[225] P. Kudva, A. Sullivan, and W. Dougherty, Metrics for structural logic synthesis, *Proceedings of IEEE International Conference on Computer-Aided Design*, San Jose, CA, 2002, pp. 551–556.

[226] P. Saxena and B. Halpin, Modeling repeaters explicitly within analytical placement, *Proceedings of ACM/IEEE Design Automation Conference*, San Diego, CA, 2004, pp. 699–704.

[227] Y.-J. Chang and C. Lee, Synthesis of multi-variable MVL functions using hybrid mode CMOS logic, *Proceedings of International Symposium on Multiple-Valued Logic*, Boston, MA, 1994, pp. 35–41.

[228] M. Syuto, J. Shen, K. Tanno, and O. Ishizuka, Multi-input variable-threshold circuits for multi-valued logic functions, *Proceedings of International Symposium on Multiple-Valued Logic*, Portland, OR, 2000, pp. 27–32.

[229] T. Temel and A. Morgul, Implementation of multi-valued logic, simultaneous literal operations with full CMOS current-mode threshold circuits, *Electron. Lett.*, 38, 160–161, 2002.

[230] E. Lee and P. Gulak, Dynamic current-mode multi-valued MOS memory with error correction, *Proceedings of International Symposium on Multiple-Valued Logic*, Sendai, Japan, 1992, pp. 208–215.

[231] A. Malik, R. Brayton, A.R. Newton, and A. Sangiovanni-Vincentelli, Reduced offsets for two-level multivalued logic minimization, *Proceedings of ACM/IEEE Design Automation Conference*, Orlando, FL, 1990, pp. 290–296.

[232] H. Wang, C. Lee, and J. Chen, Algebraic division for multilevel logic synthesis of multi-valued logic circuits, *Proceedings of International Symposium on Multiple-Valued Logic*, Boston, MA, 1994, pp. 44–51.

[233] H. Wang, C. Lee, and J. Chen, Factorization of multi-valued logic functions, *Proceedings of International Symposium on Multiple-Valued Logic*, Bloomington, IN, 1995, pp. 164–169.

[234] Y. Jiang, S. Matic, and R. Brayton, Generalized cofactoring for logic function evaluation, *Proceedings of ACM/IEEE Design Automation Conference*, Anahein, CA, 2003, pp. 155–158.

[235] Y. Jiang and R. Brayton, Don't cares and multi-valued logic network minimization, *Proceedings of IEEE International Conference on Computer-Aided Design*, Santa Clara, CA, 2000, pp. 520–525.

[236] S. Khatri, R. Brayton, and A. Sangiovanni-Vincentelli, Sequential multi-valued network simplification using redundancy removal, *Proceedings of International Conference on VLSI Design*, Goa, India, 1999, pp. 206–211.

[237] S. Sinha, S. Khatri, R. Brayton, and A. Sangiovanni-Vincentelli, Binary and multi-valued SPFD-based wire removal in PLA networks, *Proceedings of IEEE International Conference on Computer Design*, Austin, TX, 2000, pp. 494–503.

[238] C. Liu, A. Kuehlmann, and M. Moskewicz, CAMA: a multi-valued satisfiability solver, *Proceedings of IEEE International Conference on Computer-Aided Design*, Santa Clara, CA, 2003, pp. 326–333.

[239] A. Mishchenko and R. Brayton, Simplification of non-deterministic multi-valued networks, *Proceedings of IEEE International Conference on Computer-Aided Design*, Santa Clara, CA, 2002, pp. 557–562.

[240] J. Jiang, A. Mishchenko, and R. Brayton, Reducing multi-valued algebraic operations to binary, *Proceedings of Design, Automation and Test in Europe Conference*, Munich, Germany, 2003, pp. 752–757.

[241] R. Brayton and S. Khatri, Multi-valued logic synthesis, *Proceedings of International Conference on VLSI Design*, Goa, India, 1999, pp. 196–205.

[242] M. Gao, J. Jiang, Y. Jiang, Y. Li, A. Mishchenko, S. Sinha, T. Villa, and R. Brayton, Optimization of multivalued multi-level networks, *Proceedings of International Symposium on Multiple-Valued Logic*, Boston, MA, 2002, pp. 168–177.

[243] R. Drechsler, Evaluation of static variable ordering heuristics for MDD construction, *Proceedings of International Symposium on Multiple-Valued Logic*, Boston, MA, 2002, pp. 254–260.

[244] F. Schmiedle, W. Gunther, and R. Drechsler, Selection of efficient re-ordering heuristics for MDD construction, *Proceedings of International Symposium on Multiple-Valued Logic*, Warsaw, Poland, 2001, pp. 299–304.

[245] F. Schmiedle, W. Gunther, and R. Drechsler, Dynamic re-encoding during MDD minimization, *Proceedings of International Symposium on Multiple-Valued Logic*, Portland, OR, 2000, pp. 239–244.

[246] R. Drechsler, M. Thornton, and D. Wessels, MDD-based synthesis of multi-valued logic networks, *Proceedings of International Symposium on Multiple-Valued Logic*, Portland, OR, 2000, pp. 41–46.

[247] C. Files, R. Drechsler, and M. Perkowski, Functional decomposition of MVL functions using multi-valued decision diagrams, *Proceedings of International Symposium on Multiple-Valued Logic*, Antigonish, Canada, 1997, pp. 27–32.

[248] V. Shmerko, S. Yanushkevich, and V. Levashenko, Fault simulation in sequential multi-valued logic networks, *Proceedings of International Symposium on Multiple-Valued Logic*, Antigonish, Canada, 1997, pp. 139–144.

[249] R. Drechsler, M. Keim, and B. Becker, Fault simulation in sequential multi-valued logic networks, *Proceedings of International Symposium on Multiple-Valued Logic*, Antigonish, Canada,1997, pp. 145–150.

3

Power Analysis and Optimization from Circuit to Register-Transfer Levels

Jose Monteiro
INESC-Instituto de Engenharia de Sistems e Computadores
Lisboa, Portugal

Rakesh Patel
Intel Corp.
Chandler, Arizona

Vivek Tiwari
Intel Corp.
Santa Clara, California

3.1 Introduction

The increasing speed and complexity of today's designs implies a significant increase in the power consumption of very large-scale integration (VLSI) circuits in the near future. To meet this challenge, we need to develop design techniques to reduce power. The complexity of today's ICs with over 100 million transistors, clocked at over 1 GHz, demands computer-aided design (CAD) tools and methodologies for a successful design in time to market.

One of the key features that led to the success of complementary metal-oxide semiconductor (CMOS) technology was its intrinsic low-power consumption. This meant that circuit designers and electronic design automation (EDA) tools could afford to concentrate on maximizing circuit performance and minimizing circuit area. Another interesting feature of CMOS technology is its nice scaling properties, which has permitted a steady decrease in the feature size, allowing for more and more complex systems on a single chip, working at higher clock frequencies.

Power consumption concerns came into play with the appearance of the first portable electronic systems in the late 1980s. In this market, battery lifetime is a decisive factor for the commercial success of the product. Another fact that became apparent at about the same time was that the increasing integration of more active elements per die area would lead to prohibitively large-energy consumption of an integrated circuit. A high absolute level of power is not only undesirable for economic and environmental reasons, but it also creates the problem of heat dissipation. In order to keep the device working at acceptable temperature levels, excessive heat may require expensive heat removal systems.

These factors have contributed to the rise of power as a major design parameter on par with performance and die size. In fact, power consumption is regarded as the limiting factor in the continuing scaling of CMOS technology.

To respond to this challenge, in the last decade or so, intensive research has been put into developing computed-aided design (CAD) tools that address the problem of power optimization. Initial efforts were directed to circuit and logic-level tools because at this level CAD tools were more mature and there was a better handle on the issues. Today, most of the research for CAD tools targets system or architectural level optimization, which potentially have a higher overall impact, given the breadth of their application. Together with optimization tools, efficient techniques for power estimation are required, both as an absolute indicator that the circuit's consumption meets some target value and as a relative indicator of the power merits of different alternatives during design space exploration.

This chapter provides an overview of the most significant CAD techniques proposed for low power. We start in Section 3.2 by describing the issues and methods for power estimation at different levels of abstraction, thus defining the targets for the tools presented in the following sections. In Sections 3.3 and 3.4, we review power optimization techniques at the circuit and logic levels of abstraction respectively.

3.2 Power Analysis

Given the importance of the power consumption parameter in circuit design, EDA tools are required to provide power estimates for a design. When evaluating different design alternatives, we need estimates that indicate the most effective alternative in terms of power. Since estimates may be required for multiple alternatives, often iteratively, and relative accuracy suffices, absolute power accuracy is sometimes sacrificed for tool response speed. Second, an accurate absolute power consumption estimate is required before fabrication to guarantee that the circuit meets the allocated power budget.

Obtaining a power estimate is significantly more complex than circuit area estimates, or even delay estimates, because power is related not only to the circuit topology, but also to the activity of the signals.

Typically, design exploration is performed at each level of abstraction, thus creating the need for power estimation tools at the different levels. The higher the abstraction level, the less the information there is about the actual circuit implementation, implying less assurance about the power estimate accuracy.

In this section, we first discuss the components of power consumption in CMOS circuits. We then discuss how each of these components is estimated at the different design abstraction levels.

3.2.1 Power Components in CMOS Circuits

Power consumption of digital CMOS circuits is generally considered in terms of three components [1]:

- Dynamic power (P_{dyn})
- Short-circuit power (P_{short})
- Static power (P_{static})

The total power consumption is given by the sum of these components:

$$P = P_{\text{dyn}} + P_{\text{short}} + P_{\text{static}} \tag{3.1}$$

The dynamic power component, P_{dyn}, is related to the charging and discharging of the load capacitance at the gate output, C_{out}. This is a parasitic capacitance that can be lumped at the output of the gate. Today, this component is still the dominant source of power consumption in a CMOS gate.

As an illustrative example, consider the inverter circuit depicted in Figure 3.1 (to form a generic CMOS gate, the bottom transistor, nMOS, can be replaced by a network of nMOS transistors, and the top transistor, pMOS, by a complementary network of pMOS transistors). When the input goes low, the nMOS transistor is cutoff and the pMOS transistor conducts. This creates a direct path between the voltage supply and C_{out}. Current I_p flows from the supply to charge C_{out} up to the voltage level V_{dd}. The amount of charge drawn from the supply is $C_{out}V_{dd}$ and the energy drawn from the supply equals $C_{out}V_{dd}^2$. The energy actually stored in the capacitor, E_c, is only half of this, i.e., $E_c = 1/2C_{out}V_{dd}^2$. The other half is dissipated in the resistance represented by the pMOS transistor. During the subsequent low to high input transition, the pMOS transistor is cutoff and the nMOS transistor conducts. This connects the capacitor C_{out} to the ground, leading to the flow of current I_n. C_{out} discharges and its stored energy, E_c, is dissipated in the resistance represented by the nMOS transistor. Therefore, an amount of energy equal to E_c is dissipated (consumed) every time the output makes a transition. Given the number of transitions N that a gate makes in a period of time T, its dynamic power consumption during that time period is given by

$$P_{dyn} = E_c N/T = \tfrac{1}{2}C_{out}V_{dd}^2 N/T \tag{3.2}$$

In the case of synchronous circuits, an estimate, α, of the average number of transitions the gate makes per clock cycle, $T_{clk} = 1/f_{clk}$, can be used to compute average dynamic power

$$P_{dyn} = E_c\alpha f_{clk} = \tfrac{1}{2}C_{out}V_{dd}^2\alpha f_{clk} \tag{3.3}$$

C_{out} is the sum of three components, C_{int}, C_{wire}, and C_{load}. Of these, C_{int} represents the internal capacitance of the gate. This basically consists of the diffusion capacitance of the drain regions connected to the output. C_{load} represents the sum of gate capacitances of the transistors this logic gate is driving; and C_{wire} the parasitic capacitance of the wiring used to interconnect the gates, including the capacitance between the wire and the substrate, the capacitance between neighboring wires, and the capacitance due to the fringe effect of electric fields. The term αC_{out} is generally called the switched capacitance, which measures the amount of capacitance that is charged or discharged in one clock cycle.

The short-circuit power component, P_{short}, is also related to the switching of the output of the gate. During the transition of the input line from one voltage level to the other, there is a period of time when both the pMOS and the nMOS transistors are on, thus creating a path from V_{dd} to ground. Thus, each time a gate switches, some amount of energy is consumed by the current, indicated as I_{short} in Figure 3.1, that flows through both transistors during this period. The short-circuit power is determined by the time the input voltage V_{in} remains between V_{Tn} and $V_{dd} - V_{Tp}$, where V_{Tn} and V_{Tp} are the threshold voltages of the nMOS and the pMOS transistors, respectively. Careful design to minimize low slope input ramps, namely through the appropriate sizing of the transistors, can keep this component as a small fraction of the total power, and hence it is generally considered only as a second-order effect. Given an estimate of the average amount of charge, Q_{short}, which is carried by the short-circuit current per output transition, the short-circuit power is obtained with

$$P_{short} = Q_{short}V_{dd}\alpha f_{clk} \tag{3.4}$$

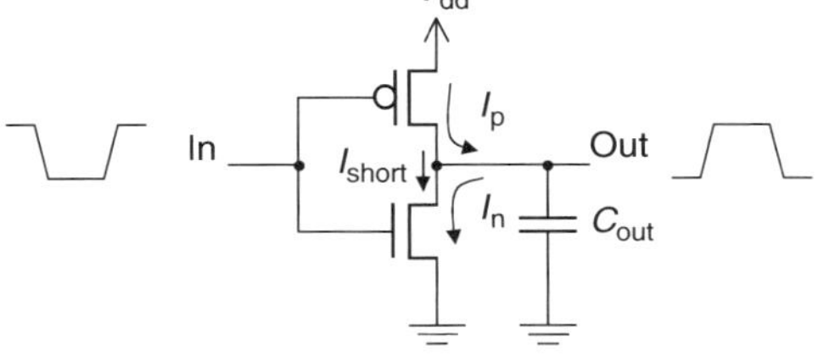

FIGURE 3.1 Illustration of the dynamic and short-circuit power components.

The static power component, P_{static}, is due to leakage currents in the MOS transistors. As the name indicates, this component is not related to the circuit activity and exists as long as the circuit is powered. The source and drain regions of a MOS transistor (MOSFET) can form reverse-biased parasitic diodes with the substrate. There is leakage current associated with these diodes. This current is very small and used to be negligible compared to dynamic power consumption. Another type of leakage current occurs due to the diffusion of carriers between the source and drain even when the MOSFET is in the cutoff region, i.e., when the magnitude of the gate-source voltage, V_{GS}, is below the threshold voltage, V_T. In this region, the MOSFET behaves like a bipolar transistor and the subthreshold current is exponentially dependent on $V_{GS} - V_T$.

Another situation that can lead to static power dissipation in CMOS is when a degraded voltage level (e.g., the "high" output level of an nMOS pass transistor) is applied to the inputs of a CMOS gate. A degraded voltage level may leave both the nMOS and pMOS transistors in a conducting state, leading to continuous flow of short-circuit current. This again is undesirable and care should be taken to avoid it in practice.

The above is true for pure CMOS design styles. In certain specialized circuits, namely for performance reasons, alternative design styles may be used. Some design styles produce a current when the output is constant at one voltage level, thus contributing to the increase in static power consumption. One example is the domino design style, where a precharged node needs to be recharged on every clock cycle if the output of the gate happens to be the opposite of the precharge value. Another example is the pseudo-nMOS logic family, where the pMOS network of a CMOS gate is replaced by a single pMOS transistor that always conducts. This logic style will have a constant current flowing at all times that the output is at logic 0, i.e., when there is a direct path to ground through the nMOS network.

3.2.2 Analysis at the Circuit Level

Power estimates at the circuit level are generally obtained using a circuit-level simulator such as Spice [2]. Given a user-specified representative sequence of input values, the simulator solves the circuit equations to compute voltage and current waveforms at all nodes in the electrical circuit. By averaging the current values drawn from the source, I_{avg}, the simulator can output the average power consumed by the circuit, $P = I_{avg}V_{dd}$ (if multiple power sources are used, the total average power will be the sum of the power drawn from the different power sources).

At this level, complex models for the circuit devices can be used. These models allow for the accurate computation of the three components of power — dynamic, short, and static power. Since the circuit is described at the transistor level, correct estimates can be computed not only for CMOS, but also for any logic design style and even analog modules. If the layout and routing of the circuit has been performed, and the simulation is made on the back-annotated circuit description, i.e., with realistic interconnect capacitive and resistive values, the power estimates obtained can be very close to those of the actual fabricated circuit.

The problem is that the simulation process with this level of detail implies the solution of complex systems of equations, hence creating severe limitations in terms of the size of the circuit that can be handled. Another limitation is that the length of the input sequences must necessarily be very short since simulation is very time consuming. The consequence of this is that power estimates may not reflect the real statistics of the inputs well.

For these reasons, full-fledged circuit-level power estimation is typically performed only for the accurate characterization of small circuit modules.

An approach to permit the applicability of the circuit-level simulation to larger designs is to resort to very simple models for the active devices. Naturally, this simplification comes at the cost of some accuracy loss.

Switch-level simulation is a limiting case, where transistor models are simply reduced to switches, which can be either opened or closed, with some associated parasitic resistive and capacitive values. This approach makes simulation run much more efficiently, allowing for the estimation of significantly larger circuit modules under much longer input sequences. Switch-level simulation can still model with fair accuracy the dynamic and short circuit components of power, but this is no longer true for leakage power. Although a few years ago designers were willing to ignore this power component since it would make for a negligible fraction of total power, its relative importance is increasing. Leakage-power estimations must then be performed independently using specifically tailored tools. Many different approaches for leakage-power estimation have been proposed recently, which are presented in the next section.

Intermediate complexity solutions exist. For example, PowerMill [3] is a circuit-level power estimation tool from Synopsys, Inc, which employs table lookup of current models for given transistor sizes. Furthermore, its algorithm uses circuit partitioning and solves the circuit equations independently on each partition. Although some error is introduced because interactions between different partitions are not accounted for, this technique greatly simplifies the problem to be solved, allowing for fast circuit-level simulation of large designs.

3.2.3 Static Power Estimation

Static power analysis is typically performed using the sub-threshold model to estimate leakage per micron of gate width of a minimum-length transistor. Then that model is extended to estimate leakage over the entire chip. Typically, the stacking factor (leakage reduction from stacking of devices) is a first-order component of this extension and serves to modify the total effective width of devices under analysis [4]. Analysis can be viewed as the modification of this total width by the stacking factor.

Most analytical works on leakage have used the BSIM2 sub-threshold current model [5],

$$I_{sub} = A\exp\left(\frac{(V_{GS} - V_T - \gamma' V_{SB} + \eta V_{DS})}{nV_{TH}}\right)\left(1 - e^{-\frac{V_{DS}}{V_{TH}}}\right) \tag{3.5}$$

where V_{GS}, V_{DS}, and V_{SB} are the gate-source, drain-source, and source-bulk voltages, respectively, V_T the zero bias threshold voltage, V_{TH} the thermal voltage (kT/q), γ' the linearized body-effect coefficient, η is the drain induced barrier lowering (DIBL) coefficient and

$$A = \mu_0 C_{ox}(W/L_{eff}) V_{TH}^2 e^{1.8}$$

The BSIM2 leakage model incorporates all the leakage behavior that we are presently concerned with. In summary, it accounts for the exponential increase in leakage with reduction in threshold voltage and gate-source voltage. It also accounts for the temperature dependence of leakage.

Calculating leakage current by applying Equation (3.5) to every single transistor in the chip can be very time-consuming. To overcome this barrier, empirical models for dealing with leakage at a higher level of abstraction have been studied [6,7]. For example, a simple empirical model is as follows [7]:

$$I_{leak} = I_{off} \frac{W_{tot}}{X_s} X_t \tag{3.6}$$

where I_{off} is the leakage current per micron of a single transistor measured from actual silicon at a given temperature, W_{tot} the total transistor width (sum of all N and P devices), X_s an empirical stacking factor based on the observation that transistor stacks leak less than single devices. X_t is the temperature factor and is used to scale I_{off} to the appropriate junction temperature of interest. The I_{off} value is typically specified at room temperature (therefore the need for a temperature factor to translate to the temperature of interest).

3.2.4 Logic-Level Power Estimation

A key observation made in Section 3.2.1 that makes power estimation at the logic level feasible is that if the input of the gate rises fast enough, the energy consumed by each output transition does not depend on the resistive characteristics of the transistors and is simply a function of the capacitive load, C_{out}, the gate is driving, i.e., $E_c = 1/2 C_{out} V_{dd}^2$. Given parasitic gate and wire capacitance models that allow the computation of C_{out_i} for each gate i in a gate-level description of the circuit, power estimation at the logic level reduces to computing the number of transitions that each gate makes in a given period of time, i.e., the switching activity of the gate. This corresponds to either parameter N or α, and we need only apply Equation (3.2) or Equation (3.3), respectively, to obtain power.

Naturally, this estimate refers only to the dynamic power component. Although this component used to make for over 90% of total power, current estimates for total power consumption must take leakage power into account, meaning that the methods described in the previous section must complement the logic-level estimate.

In many cases, power estimates at the logic level serve as indicators for guiding logic-level power optimization techniques, which typically target the dynamic power reduction, hence only an estimate for this component is required.

There are two classes of techniques for the switching activity computation — simulation-based and probabilistic analysis (also known as dynamic and static techniques, respectively).

3.2.4.1 Simulation-Based Techniques

In simulation-based switching activity estimation, highly optimized logic simulators are used, allowing for a fast simulation of a large number of input vectors. This approach raises two main issues: the number of input vectors to simulate and the delay model to use for the logic gates.

The simplest approach to model the gate delay is to assume zero delay for all the gates, meaning that all transitions in the circuit occur at the same time instant. As a consequence, each gate has at most one transition per input vector. In reality, logic gates have a nonzero transport delay, which may lead to different arrival times of transitions at the inputs of a logic gate due to different signal propagation paths. As a consequence, the output of the gate may switch multiple times in response to a single input vector. An illustrative example is shown in Figure 3.2.

Consider that initially signal x is set to 1 and signal y is set to 0, meaning that both signals w and z are set to 1. If y makes a transition to 1, then z will first respond to this transition by switching to 0. However, at about the same time, w is switching to 0, hence causing z to switch back to 1.

This spurious activity can make for a significant fraction of the overall switching activity, which in the case of circuits with a high degree of re-convergent signals, such as multipliers, may be above 50%. The modeling of gate delays in logic-level power estimation is thus of crucial significance. For an accurate switching activity estimate, the simulation must use a general delay model where gate delays are retrieved from a precharacterized library of gates.

The second issue is determining the number of input vectors to simulate. If the objective is to obtain a power estimate of a logic circuit under a user-specified, potentially long sequence of input vectors, then the switching activity can be easily obtained through logic simulation. When what is available are some statistics for the inputs, then this sequence of input vectors needs to be generated. One option is to generate a sequence of input vectors that approximates the given input statistics and simulate until the average power converges, that is, until this value stays within a margin ε during the last n input vectors, where ε and n are user-defined parameters.

An alternative is to compute beforehand the number of input vectors required for a given allowed percentage error ε and confidence level θ. Under a basic assumption that the power consumed by a circuit over a period of time T has a Normal distribution, the approach described in [8] uses the Central Limit Theorem to determine the number of input vectors with which to simulate the circuit

$$N \geq \left(\frac{z_{\theta/2} s}{\bar{p} \varepsilon} \right)^2$$

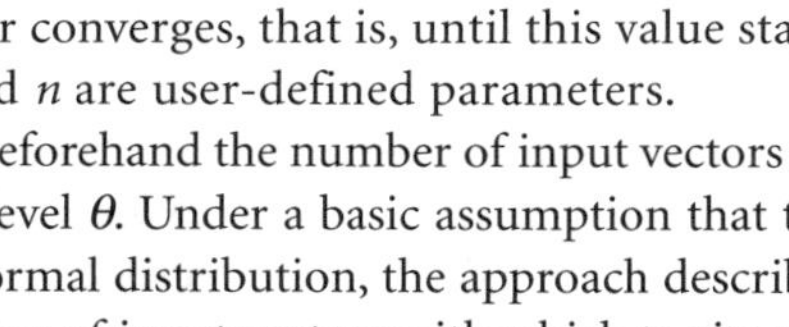

FIGURE 3.2 Example of a logic circuit with glitching and spatial correlation.

where N is the number of input vectors, $\bar{p}$ and s the measured average and standard deviation of the power, and $z_{\theta/2}$ is obtained from the Normal distribution.

In practice, for typical combinational circuits and reasonable error and confidence levels, the number of input vectors that need to be simulated to obtain the overall average switching activity is typically very small (of the order of thousands) even for complex logic circuits. However, in many situations, accurate average switching activity for each of the nodes in the circuit is required. A high level of accuracy for low-switching nodes may require a prohibitively large number of input vectors. The designer may need to relax the accuracy in these nodes, considering that these nodes have less impact in the dynamic power consumption of the circuit.

3.2.4.2 Probabilistic Techniques

The idea behind probabilistic techniques is to propagate directly the input statistics to obtain the switching probability of each node in the circuit. This approach is potentially very efficient, as only a single pass through the circuit is needed. However, it requires a new simulation engine with a set of rules for propagating the signal statistics. For example, the probability that the output of an AND gate evaluates to 1 is the intersection of the conditions that set each of its inputs to 1. If the inputs are independent, then this is just the multiplication of the probability that each input evaluates to 1. Similar rules can be derived for any logic gate and for different statistics, namely transition probabilities. Although all of these rules are simple, there is a new set of complex issues to be solved.

One of them is the delay model, as discussed above. Under a general delay model, each gate may switch at different time instants in response to a single-input change. Thus, we need to compute switching probabilities for each of these time instants. Assuming transport delays Δ_1 and Δ_2 for the gates in the circuit of Figure 3.2, this means that signal z will have some probability of making a transition at instant Δ_2 and some other probability of making a transition at instant $\Delta_1 + \Delta_2$. Naturally, the total switching activity of signal z will be the sum of these two probabilities.

Another issue is spatial correlation. When two logic signals considered together, they can only be assumed to be independent if they do not have any common input signal in their support. If there is one or more common input, we say that these signals are spatially correlated. To illustrate this point, consider again the logic circuit of Figure 3.2 and assume that both input signals x and y are independent and have a $p_x = p_y = 0.5$ probability of being at 1. Then p_w, the probability that w is 1, is simply $p_w = 1 - p_x p_y = 0.75$. However, it is not true that $p_z = 1 - p_w p_y = 0.625$, because signals w and y are not independent: $p_w p_y = (1 - p_x p_y) p_y = p_y - p_x p_y$ (note that $p_y p_y = p_y$), with $p_z = (1 - p_y + p_x p_y) = 0.75$. This indicates that not accounting for spatial correlation can lead to significant errors in the calculations.

Input signals may also be spatially correlated. Yet, in many practical cases, this correlation is ignored, either because it is simply not known or because of the difficulty in modeling this correlation. For a method that is able to account for all types of correlations among signals, see [9], but its complexity is such that it cannot be applied to very large designs.

A third important issue is temporal correlation. In probabilistic methods, the average switching activity is computed from the probability of a signal making a transition 0 to 1 or 1 to 0. Temporal correlation measures the probability that a signal is 0 or 1 in the next instant, given that its present value is 0 or 1. This means that computing the static probability of a signal being 1 is not sufficient; we need to calculate the transition probabilities directly so that temporal correlation is taken into account. Consider signals x and y in Figure 3.3, where the vertical lines indicate clock periods. The number of periods where these two signals are 0 or 1 is the same, hence the probability of the signals being at 1 is $p_x^1 = p_y^1 = 0.5$ (and the probability being at 0 is $p_x^0 = p_y^0 = 0.5$). If we only consider this parameter, thus ignoring temporal correlation, the transition probability for both signals is the same and can be computed as $\alpha = p^{01} + p^{10} = p^0 p^1 + p^1 p^0 = 0.5$. However, we can see that during the depicted time interval, signal x stays low for three clock cycles, stays high for another three cycles, and has a single clock cycle with a rising transition and another with a falling transition. Averaging over the number of clock periods, we have $p_x^{00} = 3/8 = 0.375$, $p_x^{01} = 1/8 = 0.125$, $p_x^{10} = 1/8 = 0.125$, and $p_x^{11} = 3/8 = 0.375$. Therefore,

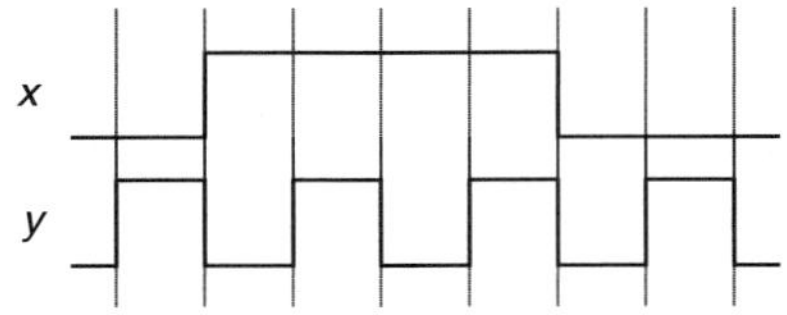

FIGURE 3.3 Example signal to illustrate the concept of temporal correlation.

the actual average switching activity of x is $\alpha_x = p_x^{01} + p_x^{10} = 0.25$. As for signal y, it never stays low or high, making a transition on every clock cycle. Hence, $p_y^{00} = p_y^{11} = 0$ and $p_y^{01} = p_y^{10} = 4/8 = 0.5$, and the actual average switching activity of y is $\alpha_y = p_y^{01} + p_y^{10} = 1.0$. This example illustrates the importance of modeling temporal correlation and indicates that probabilistic techniques need to work with transition probabilities for accurate switching activity estimates.

It has been shown that an exact modeling of these issues makes the computation of the average switching activity an NP-complete problem, meaning that exact methods are only applicable to small circuits, thus of little practical interest. Many different approximation schemes have been proposed [10].

3.2.4.3 Sequential Circuits

Computing the switching activity for sequential circuits is significantly more difficult because we need to ensure that the state space has been visited in a representative manner. For simulation-based methods, this requirement may imply too large an input sequence and, in practice, convergence is hard to guarantee.

Probabilistic methods can be effectively applied to sequential circuits as the statistics for the state lines can be derived from the circuit. The exact solution would require the computation of the transition probabilities between all pairs of states in the sequential circuit. In many cases, enumerating the states in the circuit is not possible, since these are exponential in the number of latches in the circuit. A common approximation is to compute the transition probabilities for each state line [11]. To recover to some degree the spatial correlation between state lines, a typical approach is to duplicate the subset of logic that generates the next state signals and append it to the present state lines, as illustrated in Figure 3.4. Then the method for combinatorial circuits is applied on this modified network, ignoring the switching activity in the duplicated next state logic block.

3.2.5 Analysis at the Register-Transfer Level

At the register-transfer level (RTL), the circuit is described in terms of a set of interconnected modules of varied complexity, from simple logic gates to full-blown multipliers. Power estimation at this level basically consists of determining the signal statistics at the input of each of these modules and then feeding these values to the module's power model to evaluate its dissipation. These models are normally available from the library of modules. One way to obtain these power models is to characterize the module using logic or circuit-level estimators; this process is known as macro-modeling. We refer to Chapters 7 and 13, vol.1 of this handbook, where this topic is discussed in more detail.

3.3 Circuit-Level Power Optimization

From the equations that model power consumption, it is easy to observe that reduction of the supply voltage has the largest impact on power reduction, given its quadratic dependency. This has been the largest source of power reductions and is widely applied across the VLSI industry. However, unless accompanied by the appropriate process scaling, reducing V_{dd} comes at the cost of increased propagation delays, necessitating the use of techniques to recover the lost performance.

Lowering the frequency of operation, f_{clk}, also reduces power consumption. This may be an attractive option in situations with low-performance requirements. Yet the power efficiency of the circuit is not improved, as the amount of energy per operation remains constant.

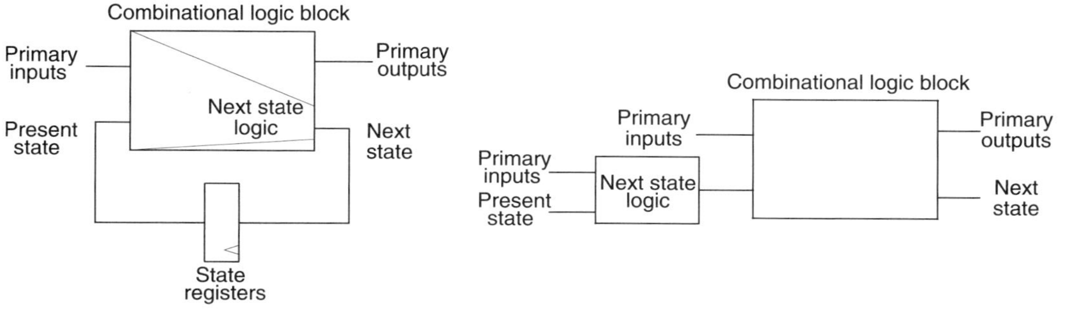

FIGURE 3.4 Creating temporal and spatial correlation among state lines.

A more interesting option is to address the switched capacitance term, αC_{out}, by redesigning the circuit such that the overall switching activity is reduced, or the overall circuit capacitance is reduced, or by a combination of both, where the switching activity in high-capacitance nodes is reduced possibly by exchanging this with higher switching in nodes with lower capacitance.

In the following, we discuss techniques at different levels of abstraction that have been developed to address each of these points.

3.3.1 Transistor Sizing

A large part of high-performance CPUs is typically custom-designed. These designs typically involve manual tweaking of transistors to upsize drivers in critical paths. If too many transistors are upsized unnecessarily, certain designs can operate on the steep part of a circuit's power-delay curve. In addition, the choice of logic family used (e.g., static vs. dynamic logic) can also greatly influence the circuit's power consumption. The traditional emphasis on performance often leads to over-design that is wasteful for power. An emphasis on lower power, however, motivates identification of such sources of power wastage. An example of this is the case where paths that are designed faster than they ultimately need to be. For synthesized blocks, the synthesis tool can automatically reduce power by downsizing devices in such paths. For manually designed blocks, on the other hand, downsizing may not always be done. Automated downsizing tools can thus have a big impact. The benefit of such tools is power savings as well as productivity enhancement over manual designs. Many custom designers are now exploring multiple-V_T techniques to take greater advantage of the transistor sizing. The main idea here is to use lower V_T transistors in critical paths rather than large high-V_T transistors. The main issue with this technique is the increase in subthreshold leakage due to low V_T. So it is very important to use low-V_T transistors selectively and to optimize their usage to achieve a good balance between capacitive current and leakage current in order to minimize total current.

3.3.2 Voltage Scaling, Voltage Islands, Variable V_{dd}

Power dissipation consists of a static component and a dynamic component. Since dynamic power is proportional to the square of the supply voltage (V_{dd}), the reduction in V_{dd} is the most effective way for reducing power. The industry has thus steadily moved to lower V_{dd}. Indeed, reducing the supply voltage is the best for low-power operation, even after taking into account the modifications to the system architecture, which are required to maintain the computational throughput. Another issue with voltage scaling is that to maintain performance, threshold voltage (V_T) also needs to be scaled down since circuit speed is roughly inversely proportional to ($V_{dd} - V_T$). Typically, V_{dd} should be higher than $4V_T$ if speed is not to suffer excessively. As the threshold voltage decreases, sub-threshold leakage current increases exponentially. At present, V_T is high enough that sub-threshold current is dominated by the dynamic component of the total active current, which dominates the total standby current (including gate and well leakage components). However, with every 0.1 V reduction in V_T, sub-threshold current increases ten times. In nanometer technologies, with further V_T reduction, sub-threshold current will become a significant portion, or even a dominant portion, of the overall chip current. At sub-0.1 μm feature sizes, the leakage power starts eating into the benefits of lower V_{dd}. In addition, design of dynamic circuits, caches, sense-amps, PLAs, etc., becomes difficult at higher sub-threshold leakage currents. Lower V_{dd} also exacerbates noise and reliability concerns. To combat sub-threshold current increase, various techniques have been developed.

Voltage islands and variable V_{dd} are variations of voltage scaling that can be used at the lower level. Voltage scaling is mainly technology-dependent and typically applied to the whole chip. Voltage islands are more suitable for system-on-chip (SOC) designs, which integrate different functional modules with various performance requirements on a single chip. We refer the reader to the RTL power analysis and optimization techniques chapter (chapter 13 of vol.1 of this handbook) for more details on the voltage island technique. The variable voltage and voltage-island techniques are complementary and can be implemented on the same block and used simultaneously. In the variable voltage technique, the supply voltage is varied based on throughput requirements. For higher throughput applications, the supply voltage is increased along with operating frequency, and vice versa for the lower throughput application. Sometimes this technique is also used to control power consumption and surface temperature. On-chip sensors sense temperature or current requirements and lower the supply voltage to reduce power consumption. Leakage-power mitigation can be achieved at the device level by applying multi-threshold voltage devices, multi-channel length devices, Stacking and Parking-states techniques. The following section gives details on these techniques.

3.3.3 Multiple-Threshold Voltages

Multiple-threshold voltages have been available on many, if not most, CMOS processes for a number of years. Since the standby power is so sensitive to the number of low-V_T transistors, their usage, of the order of 5–10% of total transistors, is generally limited to fixing critical timing paths, especially in the physical design stage where other design changes have very long turnaround time. For instance, if the low V_T is 110 mV less than the high V_T, 20% usage of the former will increase the chip standby power by nearly 500%. Low-V_T insertion does not impact the active power component or design size. Obvious candidate circuits are SRAMs, whose power is dominated by leakage; higher V_T generally also improves SRAM stability (as does a longer channel). The main drawbacks of low-V_T transistors are that variation due to doping is uncorrelated between the high- and low-threshold transistors and that extra mask steps are needed, which incur additional process cost.

3.3.4 Long-Channel Transistors

The use of transistors that have longer than nominal channel length is another method of reducing leakage power. For example, by drawing a transistor 10 nm longer (long-L) than a minimum sized one, the Drain Induced Barrier Lowering (DIBL) is attenuated and the leakage can be reduced by 7 to 10 times on a 90 nm process. With this one change, nearly 20% of the total SRAM leakage component can be alleviated while maintaining performance. The loss in drive current due to increased channel resistance, of the order of 10–20%, can be made up for by an increase in width or, since the impact is to a single-gate stage, can be ignored for most of the designs [12]. The use of long-L is especially useful for SRAMs, since their overall performance is relatively insensitive to transistor delay. It can also be employed in other circuits if used judiciously. Compared with multiple threshold voltages, long-channel insertion has similar or lower process cost — it manifests as size increase rather than mask cost. It allows lower process complexity, and the different channel lengths track over process variation. It can be applied opportunistically to an existing design to limit leakage. A potential penalty is the increase in gate capacitance. Overall active power does not increase significantly if the activity factor of the affected gates is low, so this should also be considered when choosing target gates.

The target gate selection is driven by two main criteria. First, transistors must lie on paths with sufficient timing margin. Second, the highest leakage transistors should be chosen first from the selected paths. The first criterion ensures that the performance goals are met. The second helps in maximizing leakage-power reduction. In order to use all of the available positive timing slack and avoid errors, long-L insertion is most advisable at the late design stages.

The long-L insertion can be done using standard cells designed using long-L transistors or by selecting individual transistors from the transistor-level design. Only the latter is applicable to full-custom design. There are advantages and disadvantages to both methods. For the cell level method, low-performance cells are designed with long-L transistors. For leakage reduction, high-performance cells on noncritical paths are replaced with lower performance cells with long-L. If the footprint and port locations are identical, then this method simplifies the physical convergence. Disadvantages of this method are that it requires a much larger cell library. It also requires a fine-tuned synthesis methodology to ensure long-L cell selection rather than lower performance nominal channel length cells. The transistor level flow has its own benefits. A unified flow can be used for custom- and auto-placed and routed blocks. Only a single nominal cell library is needed, albeit with space for long-L as mentioned.

3.3.5 Stacking and Parking States

There are two other design techniques for leakage-power reduction, Stacking and Parking states. In the Stacking technique, two or more transistors are placed in series. These transistor structures increase channel resistance of the leakage-current path in the OFF state. This technique gives ~2 to 3 times leakage-power reduction depending on the number of devices in the series. One needs to be careful about using this technique as it can increase switching power and gate leakage significantly. Stacking is beneficial in cases where a small number of transistors can add extra stack to a wide cone of logic or gate the power supply to it.

The main idea behind the Parking-states technique is to force the gates in the circuit to the low-leakage logic state when not in use [13]. There are three main pitfalls in this technique. First, additional logic is needed to generate the desirable state, which has area, switching power, as well as leakage-power cost. The second pitfall is that it is very specific to a design implementation. For example, a low-leakage state for a cone of logic may become high-leakage state if the logic implementation changes while keeping the same functionality. This technique is not advisable for random logic, but with careful implementation on structured data-path and memory arrays, it can save 2 to 5 times leakage power in the OFF state. The third pitfall is about making sure the design remains in the OFF state for long enough to make up for the dynamic power that is wasted during the forced transition to the low-leakage state.

Additional research on both of these techniques is needed with respect to the algorithmic aspects as well as the best way to incorporate them into synthesis flows.

3.3.6 Logic Styles

Dynamic circuits are generally regarded as more power dissipating than their static counterparts. While the power consumption of a static CMOS gates with constant inputs is limited to leakage power, dynamic gates may be continually precharging and discharging their output capacitance under certain input conditions.

For instance, if the inputs to the NAND gate in Figure 3.5(a) are stable, the output is stable. On the other hand, the dynamic NAND gate of Figure 3.5(b), under constant inputs A = B = 1, will keep rasing and lowering the output node, thus leading to a high level of energy consumption.

While this is true, it may not be such a relevant issue in high-performance designs, where most of the nodes will be switching most of the time. In fact, for several reasons, dynamic logic families are preferred in many high-speed, high-density designs (such as microprocessors). First, dynamic gates require fewer transistors, which means not only that they take up less space, but also that they exhibit a lower capacitance load, hence allowing for increased operation speed and for a reduction in the dynamic power dissipation component. Secondly, the evaluation of the output node can be performed solely through N-type MOSFET transistors, which further contributes to the improvement in performance. Thirdly, there is never a direct path from V_{dd} to ground, thus totally eliminating the short-circuit power component. Finally, dynamic circuits intrinsically do not exhibit any hazards, potentially resulting in a significant

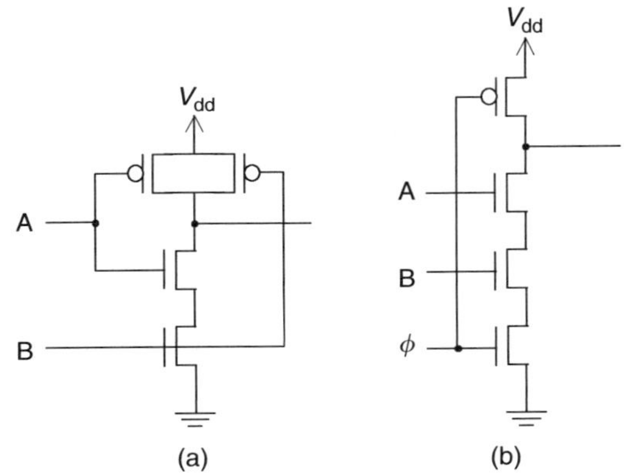

FIGURE 3.5 NAND gate: (a) static CMOS; (b) dynamic domino.

reduction in power consumption. However, the design of dynamic circuits presents several issues that have been addressed through different design families [14].

Pass-transistor logic is another design style whose merits for low power, mainly due to the lower capacitance load of the input signal path, have been pointed out. The problem is that this design style may imply a significantly larger circuit.

3.4 Logic Synthesis for Low Power

A significant amount of CAD research has been carried out in the area of low power logic synthesis. By adding power consumption as a parameter for the synthesis tools, it is possible to save power with no, or minimal, delay penalty.

3.4.1 Logic Factorization

A primary means of technology-independent optimization is the factoring of logical expressions. For example, the expression $xy \vee xz \vee wy \vee wz$ can be factored into $(x \vee w)(y \vee z)$, reducing transistor count considerably. Common subexpressions can be found across multiple functions and reused. For area optimization, the kernels of the given expressions are generated and kernels that maximally reduce literal count are selected. Even though minimizing transistor count may in general reduce power consumption, in some cases the total switched capacitance actually increases. When targeting power dissipation, the cost function must take into account switching activity. The algorithms proposed for low power kernel extraction compute the switching activity associated with the selection of each kernel. Kernel selection then is based on the reduction of both area and switching activity [15].

3.4.2 Don't Care Optimization

Multilevel circuits are optimized by repeated two-level minimization with appropriate don't-care sets. The structure of the logic circuit may imply that some combinations on the inputs of a given logic gate never occur. These combinations form the controllability or satisfiability Don't care set (SDC) of the gate. Similarly, there may be some input combinations for which the output value of the gate is not used in the computation of any of the outputs of the circuit. The set of these combinations is called the observability Don't care set (ODC). Although initially don't-care sets were used for area minimization, techniques have been proposed for the use of don't-cares to reduce the switching activity at the output of a logic gate [16]. The transition probability of a static CMOS gate is given by $\alpha_x = 2p_x^0 p_x^1 = 2p_x^1(1 - p_x^1)$ (ignoring temporal correlation). The maximum for this function occurs for $p_x^1 = 0.5$. Therefore, in order to minimize the switching activity, the strategy is to include don't care miniterms in the on-set of the function if $p_x^1 > 0.5$ or in the off-set if $p_x^1 < 0.5$.

3.4.3 Path Balancing

Spurious transitions account for a significant fraction of the switching activity power in typical combinational logic circuits. In order to reduce spurious switching activity, the delay of paths that converge at each gate in the circuit should be roughly equal, a problem known as path balancing. In the previous section, we discussed that transistor sizing can be tailored to minimize power basically at the cost of delaying signals not on the critical path. This approach has the additional feature of contributing to path balancing. Alternatively, path balancing can be achieved through the restructuring of the logic circuit, as illustrated in Figure 3.6.

3.4.4 Technology Mapping

Technology mapping is the process by which a logic circuit is implemented in terms of the logic elements available in a particular technology library. Associated with each logic element is the information about its area, delay, and internal and external capacitances. The optimization problem is to find the implementation that meets some delay constraint while minimizing cost that is a function of area and power consumption [17,18]. One of the main strategies to minimize power dissipation is to hide nodes with high switching activity within complex logic elements, as capacitances internal to gates are generally much smaller.

In many cases, the inputs of a logic gate are commutative in a Boolean sense. However, due to the way the gate is implemented, equivalent pins may present different input capacitance loads. In these cases, gate input assignment should be performed such that signals with high-switching activity map to the inputs that have lower input capacitance.

Additionally, most technology libraries include the same logic elements with different sizes (i.e., driving capability). Thus, in technology mapping for low power, the choice of the size of each logic element is made such that the delay constraints are met with minimum power consumption. This problem is the discrete counterpart of the transistor-sizing problem of the previous section.

3.4.5 State Encoding

The synthesis of sequential circuits offers new avenues of power optimization. State encoding is the process by which a unique binary code is assigned to each state in a finite-state machine (FSM). Although

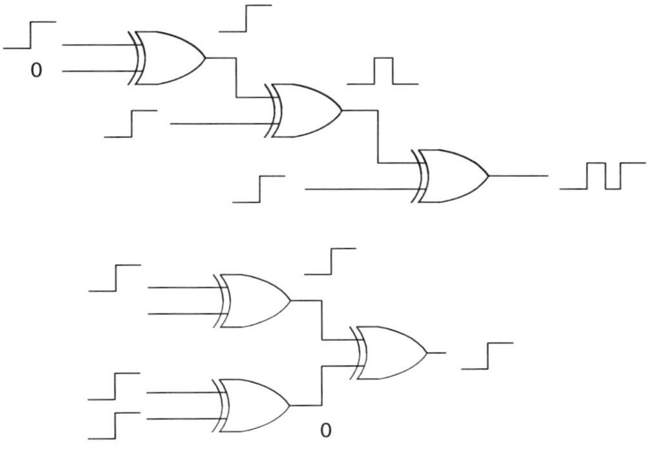

FIGURE 3.6 Path balancing through logic restructuring to reduce spurious transitions.

this assignment does not influence the functionality of the FSM, it determines the complexity of the combinational logic block in the FSM implementation. State encoding for low power uses heuristics that assigns minimum Hamming distance codes to states that are connected by edges that have larger probability of being traversed [19]. The probability that a given edge in the state transition graph (STG) is traversed is given by the steady-state probability of the STG being in the start state of the edge, multiplied by the static probability of the input combination associated with that edge. Whenever this edge is exercised, only a small number of state lines (ideally one) will change, leading to reduced overall switching activity in the combinational logic block.

3.4.6 Finite-State Machine Decomposition

Finite-state machine decomposition has been proposed as a means for a low-power implementation of an FSM. The basic idea is to decompose the STG of the original FSM into two coupled STGs that together have the same functionality as the original FSM. Except for transitions that involve going from one state in one sub-FSM to a state in the other, only one of the sub-FSMs needs to be clocked. The strategy for state selection is such that only a small number is selected for one of the sub-FSMs. This selection consists in searching for a small cluster of states, such that the total probability of transitions between states in the cluster is high and the probability of transition to and from states outside the cluster is very low. The aim is to have a small sub-FSM that is active most of the time, disabling the larger sub-FSM. The reason for requiring a small number of transitions to and from the other sub-FSM is that this corresponds to the worst situation when both sub-FSMs are active. Each sub-FSM has an extra output that disables the state registers of the other sub-FSM, as shown in Figure 3.7. This extra output is also used to stop transitions at the inputs of the large sub-FSM. An approach to perform this decomposition solely using circuit techniques, thus without obtaining the STG, explicitly or implicitly, was proposed in [20].

Other techniques also based on blocking input signal propagation and clock-gating, such as pre-computation and guarded-evaluation, are covered in some detail in Chapter 13 of the first volume of this handbook.

3.4.7 Retiming

Retiming was first proposed as a technique to improve throughput by moving the registers in a circuit while maintaining input–output functionality. The use of retiming to minimize switching activity is based on the observation that the register outputs have significantly fewer transitions than the register inputs. In particular, no glitching is present. Moving registers across nodes through retiming may change the switching activity at several nodes in the circuit. In the circuit of Figure 3.8(a), the switched capacitance is given by $N_0C_B + N_1C_{FF} + N_2C_C$ and the switched capacitance in its retimed version, shown in

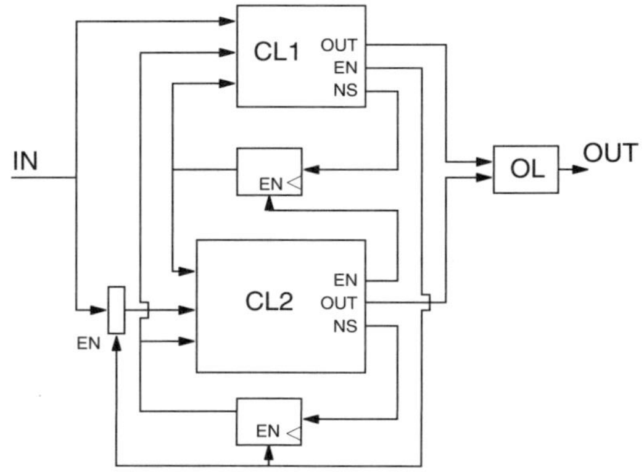

FIGURE 3.7 Implementation diagram of a decomposed FSM for low power.

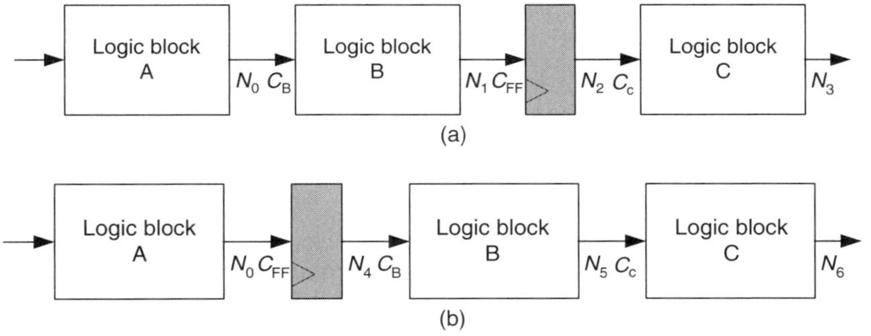

FIGURE 3.8 Retiming for low power.

Figure 3.8(b), is $N_0C_{FF} + N_4C_B + N_5C_C$. One of these two circuits may have significantly less switched capacitance. Heuristics to place registers such that nodes driving large capacitances have reduced switching activity with a given throughput constraint have been proposed [21].

3.5 Conclusion

This chapter has covered methodologies for the reduction of power dissipation of digital circuits at the lower levels of design abstraction. The reduction of supply voltage has a large impact on power. However, it has the undesired consequence of reducing performance. Some of the techniques described apply local voltage reduction and dynamic voltage changes to minimize the impact of lost performance.

For most designs, the principal component of power consumption is related to the switching activity of the circuit during normal operation (dynamic power). The main strategy is to reduce the overall average switched capacitance, that is, the average amount of capacitance that is charged or discharged during circuit operation. The techniques presented address this issue by selectively reducing the switching activity of high-capacitive nodes, possibly at the expense of increasing this activity in other less capacitive nodes. Design automation tools using these approaches can save from 10 to 50% in power consumption with little area and delay overhead.

The static power component has been rising in importance with the reduction of feature size due to increased leakage and subthreshold currents. The most relevant methods, mostly at the circuit level, to minimize this power component have been presented.

Also covered in this chapter are power analysis tools. The power estimates provided can be used not only to indicate the absolute level of power consumption of the circuit, but also to direct the optimization process by indicating the most power efficient design alternatives.

References

[1] N. Weste and K. Eshraghian, *Principles of CMOS VLSI Design: A Systems Perspective*, Addison-Wesley, Reading, MA, 1985.

[2] *The SPICE3 Implementation Guide*, University of California at Berkeley, ERL M89/44, 1989.

[3] PowerMill, Synopsys, http://www.synopsys.com/products/etg/powermill_ds.html.

[4] M. Johnson, D. Somasekhar, and K. Roy, Models and algorithms for bounds on leakage in CMOS circuits, *IEEE Trans. Comput.-Aided Design Integrated Circuits*, 18, 714–725, 1999.

[5] B. Sheu, D. Scharfetter, P. Ko, and M. Jeng, BSIM: Berkeley short-channel IGFET model for MOS transistors, *IEEE J. Solid-State Circuits*, 22, 558–566, 1987.

[6] J. Butts and G. Sohi, A static power model for architects, *Proceedings of MICRO-33*, Monterey, CA, 2000, pp. 191–201.

[7] W. Jiang, V. Tiwari, E. La Iglesia, and A. Sinha, Topological analysis for leakage prediction of digital circuits, *Proceedings of the International Conference on VLSI Design*, Bangalore, India, 2002, pp. 39–44.

[8] R. Burch, F. Najm, P. Yang, and T. Trick, A Monte Carlo approach to power estimation, *IEEE Trans. VLSI Systems*, 1, 63–71, 1993.

[9] A. Freitas and A.L. Oliveira, Implicit resolution of the Chapman–Kolmogorov equations for sequential circuits: an application in power estimation, in *Design, Automation and Test in Europe*, IEEE Computer Society, Los Alamitos, CA 2003, pp. 764–769.

[10] M. Pedram, Power minimization in IC design: principles and applications, *ACM Trans. Design Automation Electron. Syst.*, 1, 3–56, 1996.

[11] C-Y. Tsui, J. Monteiro, M. Pedram, S. Devadas, A. Despain, and B. Lin, Power estimation methods for sequential logic circuits, *IEEE Trans. VLSI Syst.*, 3, 404–416, 1995.

[12] L. Clark, R. Patel, and T. Beatty, Managing standby and active mode leakage power in deep submicron design, *International Symposium on Low Power Electronics and Design*, San Diego, CA, 2005.

[13] D. Lee and D. Blaauw, Static leakage reduction through simultaneous threshold voltage and state assignment, *Proceedings of the Design Automation Conference*, 2003, 2–6 June 2003, pp. 191–194.

[14] J. Yuan and C. Svensson, New single-clock CMOS latches and flipflops with improved speed and power savings, *IEEE J. Solid-State Circuits*, 32, 62–69, 1997.

[15] C-Y. Tsui, M. Pedram, and A. Despain, Power-efficient technology decomposition and mapping under an extended power consumption model, *IEEE Trans. CAD*, 13, 1110–1122, 1994.

[16] A. Shen, S. Devadas, A. Ghosh, and K. Keutzer, On average power dissipation and random pattern testability of combinational logic circuits, *Proceedings of the International Conference on Computer-Aided Design*, Santa Clara, CA, 1992, pp. 402–407.

[17] V. Tiwari, P. Ashar, and S. Malik, Technology mapping for low power in logic synthesis, *Integration, VLSI J.*, 20, pp.243–268, 1996.

[18] C. Tsui, M. Pedram, and A. Despain, Technology decomposition and mapping targeting low power dissipation, *Proceedings of the Design Automation Conference*, Dallas, USA, 1993, pp. 68–73.

[19] L. Benini and G. Micheli, State assignment for low power dissipation, *IEEE J. Solid-State Circuits*, 30, 258–268, 1995.

[20] J. Monteiro and A. Oliveira, Implicit FSM decomposition applied to low power design, *IEEE Trans. Very Large Scale Integration Syst.*, 10, 560–565, 2002.

[21] J. Monteiro, S. Devadas, and A. Ghosh, Retiming sequential circuits for low power, *Proceedings of the International Conference on Computer-Aided Design*, Santa Clara, CA, 1993, pp. 398–402.

4

Equivalence Checking

Andreas Kuehlmann
Cadence Berkeley Laboratories
Berkeley, California

Fabio Somenzi
University of Colorado
Boulder, Colorado

4.1 Introduction

This chapter covers the problem of formally checking whether two design specifications are functionally equivalent. In general, there is a wide range of possible definitions of functional equivalence covering comparisons between different levels of abstraction and varying granularity of timing details. For example, one common approach is to consider the problem of machine equivalence which defines two synchronous design specifications functionally equivalent if, clock by clock, they produce exactly the same sequence of output signals for any valid sequence of input signals. A more general notion of functional equivalence is of importance in the area of microprocessor design. Here, a crucial check is to compare functionally the specification of the instruction set architecture (ISA) with the register transfer level (RTL) implementation, ensuring that any program executed on both models will cause an identical update of the main memory content. A system design flow provides a third example for yet another notion of functional equivalence to be checked between a transaction level model (TLM), e.g., written in SystemC and its corresponding RTL specification. Such a check is becoming of increasing interest in a system-on-a-chip (SoC) design environment. The focus of this chapter will be on the aforementioned machine equivalence, which in practical design flows is currently the most established application of functionally comparing two design specifications. Following common terminology, we will use the term *equivalence checking* synonymously with this narrow view of the problem.

The key to the practical success of formal equivalence checking is the separation of function and timing concerns in contemporary design flows. By adopting a synchronous circuit style in combination with a combinational design and verification paradigm, it enables an efficient validation of the overall design correctness by statically and independently checking the timing and function. This approach is crucial for many steps in automated design optimization and analysis, for example, logic synthesis, static timing analysis, test pattern generation, and test synthesis. As outlined in later sections, this paradigm allows equivalence checking to reduce the problem of general machine comparison to the problem of checking pairs of combinational circuits for Boolean equivalence. This in turn can be performed efficiently for many practical designs by exploiting structural similarities between the two circuits under comparison.

The need for formal equivalence checking methods was triggered by the necessity to use multiple models in hardware design flows. Each of these models is tuned for a distinct purpose to address conflicting requirements. For example, in a typical case of using two models, the first would be designed to yield fast functional simulation whereas the second would describe in detail the envisioned circuit implementation, thus ensuring high performance of the final hardware. IBM was one of the first companies following such multimodel design flow. Beginning in the late 1970s, the design flow of IBM's mainframe computer chips included RTL modeling to facilitate fast and cycle-accurate functional simulation at the early design stages. In the initial methodology which predated automatic logic synthesis, the actual logic implementation was designed independently and verified against the RTL by formal equivalence checking [1]. The capability of performing equivalence checking on large, industrial-scale designs was key for the success of complex methodologies that use a multitude of models, each specifically tuned for a distinct purpose.

In the early 1980s, IBM's mainframe design flow introduced automatic logic synthesis which largely replaced the manual design step. This did not diminish the importance of equivalence checking which became instrumental in verifying the absence of functional discrepancies that are introduced by tool or methodology bugs, or through frequent manual intervention. In the early 1990s, IBM began to complement synthesis-based design flows with custom-designed circuit blocks. Here again, formal equivalence checking between the RTL specification and manually designed transistor-level implementation provided a critical component to ensure functional correctness [2]. A similar approach was followed in the design flow of later versions of the Alpha microprocessor [3] and Intel's microprocessors [4]. In the mid-1990s, after logic synthesis became broadly adopted in typical ASIC design flows, equivalence checking technology moved into the domain of CAD tool vendors. Today, in many ASIC and custom-design methodologies, formal equivalence checking is a key component of the overall chip validation procedure, similar to static timing analysis and design rule checking.

Equivalence checking has multiple important applications in hardware design flows. The most common use in ASIC flows utilizes equivalence checking as part of the *sign-off* verification step which compares the presynthesis RTL specification with the postsynthesis gate-level implementation. This check can catch errors introduced by *engineering changes* which are regularly applied to the gate-level circuit for implementing late updates of timing or functionality. It can further uncover errors introduced by bugs in the logic synthesis tool, test synthesis step, or the overall composition of the tool flow. As mentioned before, another important application can be found in custom-design styles typically utilized for high-speed designs such as microprocessors and video image processors. Here, equivalence checking is employed as a design aid, helping the circuit designer to implement a custom circuit at the transistor level according to its RTL specification. Other common usages of equivalence checking include verification of library cells or larger blocks implementing intellectual property (IP), validation of equivalent but more compact simulation models, or the validation of hardware emulation models.

The above-mentioned application spectrum of equivalence checking is reflected in differences in the use mode, the tool integration into an overall flow, and the tool interface in terms of debugging feedback and control options. For example, in an ASIC sign-off application, equivalence checking is typically applied in batch mode, verifying an entire chip design in a flat manner. In contrast, in a custom-design environment, equivalence checking can provide an interactive design aid for which a tight integration into the schematic editor is highly valuable. Interestingly, such a convenient setup has sometimes resulted in an evolution of the usage of equivalence checking from an application as *verification tool* that just confirms the correctness of an independently designed circuit, to a *design tool* which is used in a quick circuit manipulations loop in a "trial-and-error" manner. Clearly, the importance of debugging support is significantly different for a sign-off verification mode and a design aid mode. The less frequent occurrence of miscompares in the first case requires less sophisticated debugging support — a simple counterexample often suffices. In the latter cases, design miscompares are more common and advanced debugging can offer significant gains in productivity.

4.2 Equivalence Checking Problem

Informally, two designs are defined to be functionally equivalent in the above-mentioned narrow view, if they produce identical output sequences for all valid input sequences. For a more formal definition of machine equivalence, we assume that both designs under comparison are modeled as finite-state machines. In the following notation, we generally use uppercase and lowercase letters for sets and variables, respectively. Furthermore, we utilize underlined letters for vectors of variables and denote their individual members by superscripts. We will refer to the original and complemented value of a function as its positive and negative phase, respectively.

Let $M = (X, Y, S, S_0, \delta, \lambda)$ denote a (Mealy-type) finite-state machine in the common manner where X, Y, S, and S_0 are the sets of inputs, outputs, states, and initial states, respectively. $\delta : X \times S \rightarrow S$ denotes the next-state function and $\lambda : X \times S \rightarrow Y$ is the output function. Furthermore, let $s = (s^1,\ldots,s^u)$, $\underline{x} = (x^1,\ldots x^v)$, and $\underline{y} = (y^1,\ldots,y^w)$ be variable vectors holding the values for the encoded states S, inputs X, and outputs Y, respectively, of a finite-state machine M. We will refer to the encoding variables for the states as state bits or registers, and to the variables for the inputs and outputs as input and output signals, respectively. Furthermore, let $\underline{\delta}(\underline{x}, \underline{s}) = (\underline{\delta}^1(\underline{x}, \underline{s}),\ldots,\delta^u(\underline{x}, \underline{s}))$ denote the vector of next-state functions for the state bits. A state $\underline{s}_n$ is considered *reachable* if there exists a sequence of transitions from an initial state $\underline{s}_0$ to $\underline{s}_n$, i.e., $(\underline{s}_0, \underline{s}_1), (\underline{s}_0, \underline{s}_2),\ldots,(\underline{s}_{n-1}, \underline{s}_n)$, where $\exists \underline{x}.(\underline{s}_{i+1} = \delta(\underline{x}, \underline{s}_i))$ for $0 \leq i < n$.

In the sequel, we assume the following simplifications. First, we restrict comparison to machines that have exactly one initial state, i.e., $S_0 = \{\underline{s}_0\}$. Second, we require a one-to-one mapping between the inputs of the two machines and similarly a one-to-one mapping between their outputs. Furthermore, we assume that the timing of both machines is identical, i.e., they run at the same clock speed, consume inputs at the same rate, and similarly produce outputs at identical rates. All three assumptions can be lifted in a straightforward manner.

For checking functional equivalence of two machines M_1 and M_2 with pairs of mapped input and outputs, we construct a *product machine* $M = (X, Y, S, S_0, \delta, \lambda) = M_1 \times M_2$ in the following manner:

$$X = X_1 = X_2$$

$$S = S_1 \times S_2$$

$$S_0 = S_{0,1} \times S_{0,2}$$

$$Y = \{0, 1\}$$

$$\underline{\delta}(\underline{x},(\underline{s}_1, \underline{s}_2)) = (\underline{\delta}_1(\underline{x},\underline{s}_1), \underline{\delta}_2(\underline{x},\underline{s}_2))$$

$$\lambda(\underline{x},(\underline{s}_1, \underline{s}_2)) = \begin{cases} 1 & \text{if } \underline{\lambda}_1(\underline{x}, \underline{s}_1) = \underline{\lambda}_2(\underline{x},\underline{s}_2) \\ 0 & \text{otherwise} \end{cases}$$

A schematic view of the construction of the product machine is given in Figure 4.1.

The two machines, M_1 and M_2, are said to be functionally equivalent, iff the output function of M produces "1" for all of its reachable states and input vectors, i.e., $\lambda \leftrightarrow 1$.

In general, verifying that the product machine M produces $\lambda \leftrightarrow 1$ for all reachable states is achieved by identifying a characteristic function $\varrho(\underline{s})$ that relates the states of M_1 and M_2 and partitions the state space of M in two parts as depicted in Figure 4.2.

Informally, this characteristic function splits the state space of M into two halves such that:

- All initial states are in the first part ($\varrho = 1$).
- All "miscomparing states," i.e., $\lambda \leftrightarrow 0$, are in the second part ($\varrho = 0$).
- There is no transition from the first half to the second half.

If such a characteristic function exists, then one can use an inductive argument to reason that no miscomparing state can be reached from any of the initial states, i.e., both machines are functionally equivalent.

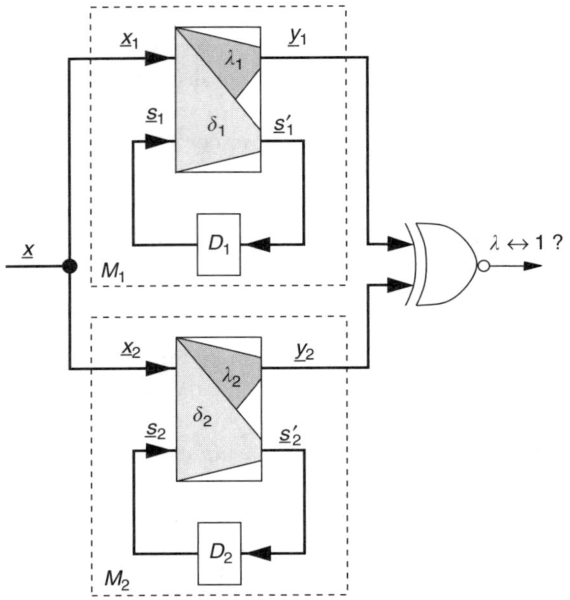

FIGURE 4.1 Product machine for comparing two finite-state machines.

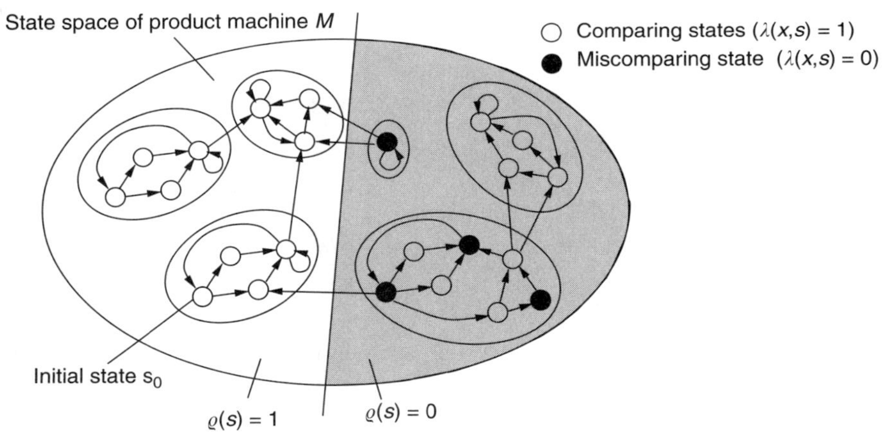

FIGURE 4.2 State-space structure of the product machine.

In fact, one can show that the inverse also holds true, namely, equivalence of M_1 and M_2 implies the existence of such a characteristic function.

Formally, the product machine M produces $\lambda \leftrightarrow 1$ for all reachable states iff there exists a function $\varrho: S \rightarrow \{0, 1\}$ such that

$$\begin{aligned} \varrho(\underline{s}_0) &= 1 \\ \varrho(\underline{s}) &= 1 \Rightarrow \forall\ \underline{x}.\varrho(\delta(\underline{x},\underline{s}) = 1 \\ \varrho(\underline{s}) &= 1 \Rightarrow \forall\ \underline{x}.\lambda(\underline{x},\underline{s}) = 1 \end{aligned} \quad (4.1)$$

The task of equivalence checking consists of two parts. First, a candidate for the characteristic function ϱ must be identified, and second, for the given ϱ, conditions Equation (4.1) must be checked for validity. The latter check requires efficient methods for Boolean reasoning which are described in Section 4.3.

Various methods for equivalence checking differ in the way candidates for ϱ are obtained. The most common approach assumes a combinational equivalence checking paradigm and uses a known or "guessed" correspondence of the registers of M_1 and M_2 to derive ϱ implicitly. In this case, the general check for functional equivalence can be reduced to verifying Boolean equivalence of a set of combinational circuits. More details on combinational equivalence checking can be found in Section 4.4, whereas Section 4.5 outlines methods for general sequential equivalence checking.

4.3 Boolean Reasoning

In this section, we review the techniques for Boolean reasoning that form the foundation for equivalence checking algorithms. We first present binary decision diagrams (BDDs) [5] — a popular data structure for Boolean functions, which are especially suited for the manipulation of the characteristic function ϱ. We then review algorithms for the propositional (or Boolean) satisfiability problem, which can be used to decide the validity of Equation (4.1) without explicitly building a representation of ϱ. In particular, we first restrict our attention to the satisfiability of conjunctive normal form (CNF) formulae and then briefly consider the general case.

4.3.1 Binary Decision Diagrams

Besides equivalence checking, many algorithms in synthesis and verification manipulate complex logic function. A data structure to support this task is therefore of great practical usefulness. Binary decision diagrams [5] have become highly popular because of their efficiency and versatility.

A BDD is an acyclic graph with two leaves representing the constant functions 0 and 1. Each internal node n is labeled with a variable v and has two children t (*then*) and e (*else*). Its function is inductively defined as

$$f(n) = (v \wedge f(t)) \vee (\neg v \wedge f(e)) \qquad (4.2)$$

Three restrictions are customarily imposed on BDDs:

1. There may not be isomorphic subgraphs.
2. For all internal nodes, $t \neq e$.
3. The variables are totally ordered. The variables labeling the nonconstant children of a node must follow in the order of the variable labeling the node itself.

Under these restrictions, BDDs provide a canonical representation of functions, in the sense that for a fixed variable order there is a bijection between logic functions and BDDs. Unless otherwise stated, BDDs will be assumed to be reduced and ordered. The BDD for $F = (x_1 \wedge x_3) \vee (x_2 \wedge x_3)$ with variable order $x_1 < x_2 < x_3$ is shown in Figure 4.3.

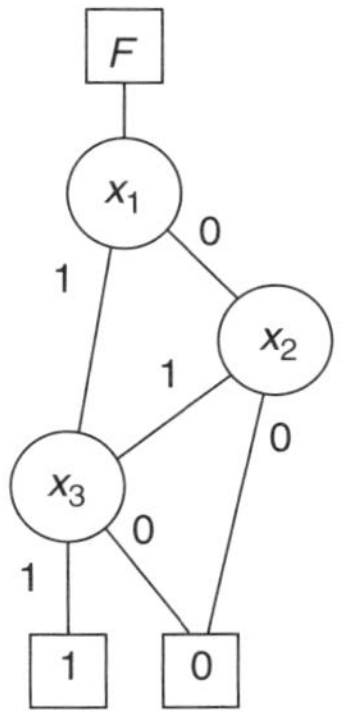

FIGURE 4.3 Binary decision diagram for $F = (x_1 \wedge x_3) \vee (x_1 \wedge x_3)$.

Canonicity is important in two main respects: it makes equivalence tests easy, and it increases the efficiency of *memoization* (the recording of results to avoid their recomputation). On the other hand, canonicity makes BDDs less concise than general circuits. The best-known case is that of multipliers, for which circuits are polynomial, and BDDs exponential [6]. Several variants of BDDs have been devised to address this limitation. Some of them have been quite successful for certain classes of problems. (For instance, Binary moment diagrams [7] for multipliers.) Other variants of BDDs have been motivated by the desire to represent functions that map $\{0, 1\}^n$ to some arbitrary set (e.g., the real numbers) [8].

For ordered BDDs, $f(t)$ and $f(e)$ do not depend on v; hence, comparison of Equation (4.2) to Boole's expansion theorem,

$$f = (x \wedge f_x) \vee (\neg x \wedge f_{\neg x}) \tag{4.3}$$

shows that $f(t) = f(n)_v$, the *positive cofactor* (or *restriction*) of $f(n)$ with respect to v and $f(e) = f(n)_{\neg v}$, the *negative cofactor*. This is the basis of most algorithms that manipulate BDDs, because for a generic Boolean connective $\langle op \rangle$,

$$f \langle op \rangle g = (x \wedge (f_x \langle op \rangle g_x)) \vee (\neg x \wedge (f_{\neg x} \langle op \rangle g_{\neg x})) \tag{4.4}$$

Equation (4.4) is applied with x chosen as the first variable in the order that appears in either f or g. This guarantees that the cofactors can be computed easily. If x does not appear in f, then $f_x = f_{\neg x} = f$; otherwise, f_x is the *then* child of f, and $f_{\neg x}$ is the *else* child. It is likewise for g. The terminal cases of the recursion depend on the specific operator. For instance, when computing the conjunction of two BDDs, the result is immediately known if either operand is constant or if the two operands are identical or complementary. All these conditions can be checked in constant time if the right provisions are made in the data structures [9].

Two tables are used by most BDD algorithms. The *unique table* allows the algorithm to access all nodes using the triple (v, t, e) as key. The unique table is consulted before creating a new node. If a node with the desired key is already in existence, it is reutilized. This approach guarantees that equivalent functions will share the same BDD, rather than having isomorphic BDDs; therefore, equivalence checks are performed in constant time.

The *computed table* stores recent operations and their results. Without the computed table, most operations on BDDs would take time exponential in the number of variables. With a lossless computed table (one that records the results of all previous computations) the time for most common operations is polynomial in the size of the operands. The details of the implementation of the unique and computed tables dramatically affect the performance of BDD manipulation algorithms, and have been the subject of careful study [10].

The order of variables may have a large impact on the size of the BDD for a given function. For adders, for instance, the optimal orders give BDDs of linear size, while bad orders lead to exponential BDDs. The optimal ordering problem for BDDs is hard [11]. Hence, various methods have been proposed to either derive a variable order from inspection of the circuit for which BDDs must be built, or by dynamically computing a good order while the BDDs are built [12].

An exhaustive list of applications of BDDs to problems in CAD is too long to be attempted here. Besides equivalence checking and symbolic model checking, BDD-based algorithms have been proposed for most synthesis tasks, including two-level minimization, local optimization, factorization, and technology mapping. In spite of their undeniable success, BDDs are not a panacea; their use is most profitable when the algorithms capitalize on their strengths [13], and avoid their weaknesses by combining BDDs with other representations [14–16]; for instance, with satisfiability solvers for CNF expressions.

4.3.2 Conjunctive Normal Form Satisfiability

A SAT solver returns an assignment to the variables of a propositional formula that satisfies it if such an assignments exists. A *literal* is either a variable or its negation, a *clause* is a disjunction of literals from distinct variables, and a CNF formula is a conjunction of clauses. We shall use the following simple CNF formula as a running example:

$$(\neg a \vee b \vee c) \wedge (a \vee \neg b \vee c) \wedge (\neg c \vee d) \wedge (\neg c \vee \neg d) \tag{4.5}$$

The variables are the letters a through d; a and $\neg a$ are the two literals (positive and negative) of variable a. The formula is the conjunction of four clauses, each consisting of two or three literals. The Boolean connectives $\vee$, $\wedge$, and $\neg$ stand for disjunction (OR), conjunction (AND), and negation (NOT), respectively.

The CNF satisfiability problem (CNF SAT for short) is the one of finding an assignment (a mapping of each variable to either 0 or 1) such that all clauses contain at least one true literal. For our example Equation (4.5), one such *satisfying assignment* is $a = b = c = d = 0$. A CNF formula with no clauses is trivially satisfiable, whereas a formula containing a clause with no literals (the *empty clause*) is not satisfiable.

The CNF SAT problem is hard, in the sense that no worst-case polynomial-time algorithm is known for it. It is in fact the archetypal NP-complete problem [17]. It can be solved in polynomial time by a nondeterministic machine which can guess a satisfying assignment and then verify it in linear time. However, if a deterministic polynomial-time algorithm were known for CNF SAT, then a polynomial time algorithm would also exist for *all* problems that a nondeterministic machine can solve in polynomial time.

Practical algorithms for CNF SAT rely on search and on the notion of *resolution*, which says that $(x \vee \neg y) \wedge (y \vee z)$ implies the *resolvent* of the two clauses, $(x \vee z)$. A CNF formula is not satisfiable iff the empty clause can be resolved from its clauses. A search algorithm for SAT finds a satisfying assignment to a given formula if it exists, and may produce a resolution of the empty clause otherwise. The approach was first described in work by Davis, Putnam, Logemann, and Loveland [18,19]. Hence, it is commonly called the DPLL procedure. However, important details have changed in the last 40 years. The GRASP solver [20], in particular has popularized the approach to search based on *clause recording* and *nonchronological backtracking* which we outline. Even if we present the algorithm in the context of CNF SAT, it is of wider applicability. In particular, it works on a generic Boolean circuit.

Figure 4.4 shows the pseudocode for the DPLL procedure, which works on three major data structures. The first is the *clause database*, which at an abstract level is just a list of clauses. The second data structure is the *assignment stack* in which all variable assignments currently in effect are kept together with their *levels* and *causes*, that is with information about when and why the assignment was made. The third data structure is the *assignment queue*, which stores the assignments that have been identified, but not yet effected.

Before they are handed to the DPLL procedure, the clauses are reprocessed to identify *unit clauses* and *pure literals*. A unit clause consists of one literal only, which consequently must be true in all satisfying assignments. The assignments for the unit literals are entered in the assignment queue. Throughout the algorithm, if all the literals of a clause except one are false and the remaining one is unassigned, then that

```
DPLL() {
    while (ChooseNextAssignment()) {
        while (Deduce() == CONFLICT) {
            blevel = AnalyzeConflict();
            if (blevel < 0) return UNSATISFIABLE;
            else Backtrack(blevel);
        }
    }
    return SATISFIABLE;
}
```

FIGURE 4.4 DPLL algorithm.

last literal is implied to 1 and the corresponding assignment is *implied*. The implication is added to the assignment queue. A *conflict* occurs when both literals of a variable are implied to true.

A pure literal is such that the opposite literal does not appear in any clause. If there is a satisfying assignment with the pure literal set to 0, then the assignment obtained from it by changing that literal to 1 is also satisfying. Hence, assignments that make pure literals true are also entered in the assignment queue. The handling of unit clauses and pure literals comprises preprocessing.

The procedure ChooseNextAssignment checks the assignment queue. If the queue is empty, the procedure makes a *decision*: it chooses one unassigned variable and a value for it, and adds the assignment to the queue. Every time a decision is made, the *decision level* is increased. The assignments made during preprocessing are at level 0 and each subsequent decision increases the decision level by 1. The decision level decreases when backtracking occurs. If no unassigned variable can be found, ChooseNextAssignment returns false. This causes DPLL to return an affirmative answer, because a complete assignment to the variables has been found that causes no conflict. Since it causes no conflict, it is satisfying.

If the queue was not empty or a decision has been added to it, one assignment is removed from it and added to the assignment stack for the current decision level. The stack also records the cause for the assignment: either a decision, or an implication. In the latter case, the clause that caused the implication is recorded.

When a new assignment is added to the stack, its implications are added by Deduce to the assignment queue. An implication occurs in a clause when all literals except one are false and the remaining literal is unassigned. Efficient computation of implications for clauses is discussed in [21].

If the implications yield a conflict (both literals of a variable are implied), AnalyzeConflict() is launched. Conflict analysis relies on the (implicit) construction of an *implication graph*. Each literal in the conflicting clause has been assigned at some level either by a decision, or by an implication. If there are multiple literals from the current decision level, at least one of them is implied. Conflict analysis locates the clause that is the source of that implication and extends the implication graph by adding arcs from the antecedents of the implication to the consequent one. This process is carried out until there is exactly one assignment for the current level among the leaves of the tree. The disjunction of the negation of the leaf assignments gives then the *conflict clause*.

For the example Equation (4.5), preprocessing produces no assignment. Suppose that the first decision is to set a to 0. This is the level-1 decision, and is denoted by $\neg a@1$. It is not the choice an SAT solver would likely make, but is convenient for illustrative purposes. No assignments are deduced from $\neg a@1$. Suppose the second decision is $b@2$; from it we deduce $c@2$ from the second clause. This, in turn, yields $d@2$ and $\neg d@2$ through the last two clauses. The two opposite assignments to d signal a conflict, which is therefore analyzed.

From Figure 4.5, one sees that $\neg a \wedge b$ is enough to cause the conflict. Hence, every satisfying assignment must satisfy the clause $(a \vee \neg b)$. One also sees that c is sufficient to cause the conflict and that consequently, $\neg c$ is a (unit) clause that can be added. Both $(a \vee \neg b)$ and c contain exactly one literal assigned at the current level (that is, at level 2).

The highest level of the assignments in the conflict clause, excluding the current one, is the backtracking level. Backtracking over some decisions that did not contribute to the conflict goes under the name of *nonchronological backtracking*. The single assignment at the current level is known as *unique implication point* (UIP). Conflict clauses based on the first UIP (the one closest to the conflict, like $\neg c$ in the example) have been empirically found to work well [22]. When the conflict clause contains a UIP — not necessarily the first UIP — the effect of backtracking is to make that clause *asserting*. That is, all literals in the clause are false except for the UIP literal. This causes that literal to be asserted to the opposite value

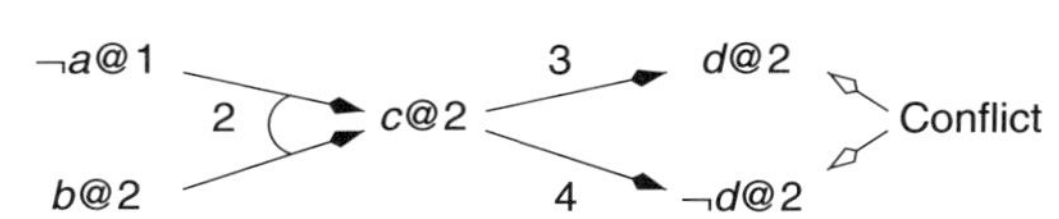

FIGURE 4.5 Implication graph. Each implication is annotated with the ordinal of the clause in (4.5) that caused it.

that it had when the conflict was detected. The procedure of Figure 4.4 relies on asserting conflict clauses to guarantee termination [23].

Returning to our example, if $(a \vee \neg b)$ is used as conflict clause, the backtracking level is 1, where the assignment $\neg b@1$ is deduced. If the next decisions are $\neg c@2$ and $d@3$, all variables are assigned, and the search terminates with a satisfying assignment. If, on the other hand, $\neg c$ is added as conflict clause, the backtracking level is 0 (since there is no literal in the clause except the UIP), and $\neg c@0$ is deduced. If the next decision is again $\neg a@1$, we get $\neg b@1$ by implication. At this point all clauses are satisfied, and, no matter what value is chosen for the only unassigned variable, d, we obtain a satisfying assignment at level 2.

In our example, the computation of the first UIP proceeds as follows. Suppose the conflict is detected when examining the fourth clause; that is, $(\neg c \vee \neg d)$ is the *conflicting clause*. Resolution is applied to this clause and to the clause that caused the other implication for d, $(\neg c \vee d)$. The result is $\neg c$. Since this clause contains exactly one literal at level 2, the first UIP has been found and the process stops. If one desires another UIP, one would continue by resolving $\neg c$ with the clause that caused the implication $c@2$, namely $(a \vee \neg b \vee c)$. The result, $(a \vee \neg b)$ is the other conflict clause that we have examined and contains the remaining UIP for this example.

The clauses that result from conflict analysis are used to steer the search away from a current partial assignment that cannot be extended to a complete assignment. They may also prevent the exploration of equally fruitless regions of the search space. For that reason, conflict clauses are *recorded* in the clause database. Periodically, those that are deemed less useful may be discarded to conserve space. Conflict clauses fulfill three important additional tasks: they are used to decide which decisions to make [21]; they are passed from one problem instance to the next to implement efficient *incremental* SAT solvers [24]; and they are recorded, together with the implication graphs from which they were derived, to produce the resolution proofs of unsatisfiable instances [23]. For the sake of conciseness, the details of these and other tasks are left to the reader to find in the referenced works. It would suffice to say that a skillful blend of the techniques we have briefly outlined has produced SAT solvers that can deal with many problems with hundreds of thousands of variables and millions of clauses. This kind of capacity is crucial in applications of SAT to equivalence checking.

4.3.3 Hybrid Satisfiability Methods

Given a generic combinational Boolean circuit, it is possible to translate it into an equisatisfiable CNF formula in linear time. For each logic gate, one generates a set of clauses relating the gate inputs to the gate output. For instance, the clauses that express the behavior of the AND gate $z = a \wedge b$ are $(a \vee \neg z)$, $(b \vee \neg z)$, and $(\neg a \vee \neg b \vee z)$. Thanks to this translation, one can, at least in theory, rely exclusively on a CNF SAT solver when reasoning about Boolean circuits. However, there are advantages to be had from dealing directly with the original circuit as well as from allowing mixed representations of the formulae whose satisfiability is investigated.

When trying to prove combinational circuits equivalent, reasoning on the circuits themselves allows one to detect structurally identical subcircuits, which can be merged to simplify the proof of equivalence of the remaining parts. This approach can be extended in two important ways. First, one can use a *semi-canonical* circuit representation to increase the chances that equivalent circuits will have the same structure. One example of this semicanonical representation is the *and-inverter graph* (AIG) [14], in which every node of the circuit is an AND gate whose inputs may be inverted. One can quickly identify isomorphic subcircuits in an AIG by hashing the gates. The second extension to structural comparison consists of computing signatures for the nodes of a circuit so that nodes with different signatures are known not to be equivalent. The few nodes with identical signatures can be subjected to further, more expensive, analysis to establish whether they are equivalent. Signature computation is usually based on simulation [25], while the more expensive analysis techniques can be based on either BDDs or CNF SAT.

In summary, direct manipulation of a circuit representation often leads to significant simplification in the task handed to the complete decision procedures. The combination of Boolean circuits, CNF clauses, and BDDs also finds application in general SAT solvers [26–28], which can leverage the strengths of the individual representations.

4.3.4 Related Applications of Boolean Reasoning in Computer-Aided Design

The application of Boolean reasoning in CAD is not limited to equivalence checking. Rather, Boolean reasoning is one of the fundamental tools for analysis and optimization in a wide range of problems in verification, synthesis, and testing. Many CAD problems can be modeled as discrete decision problem with a finite domain and therefore can be expressed as (a sequence of) satisfiability checks. The recent rapid improvements in size of the SAT problems that can be efficiently handled makes such an approach practical for surprisingly large instances.

Satisfiability checking was first introduced to the CAD domain for automatic test pattern generation (ATPG). ATPG computes a suitable set of test vectors that can demonstrate the absence of certain manufacturing faults [29]. Owing to its early use and algorithmic advances, the application of ATPG-style Boolean reasoning was adapted in many areas such as equivalence checking [30,31], logic synthesis [32–34], and verification [35,36]. Later, the introduction and improvement of BDDs, CNF SAT, circuit-based SAT, and hybrid methods substituted or complemented classical ATPG techniques in these domains.

The chapters on Assertion-Based Verification (Chapter 17, Vol. 1), Formal Property Verification (Chapter 19, Vol. 1), Automatic Test Pattern Generation (Chapter 21, Vol. 1), and Logic Synthesis (Chapter 2, Vol. 2) provide detailed summaries of corresponding CAD areas that make extensive use of Boolean reasoning.

4.4 Combinational Equivalence Checking

Most practical equivalence checking approaches assume a *combinational verification paradigm*, i.e., they require that the general problem of checking the product machine shown in Figure 4.1 can be reduced to comparing pairs of combinational circuits. In the simplest (and also most common case), they expect a one-to-one correspondence between the registers of both designs under comparison. Such requirements are enforced by an overall design methodology that strikes for a practical compromise which minimizes the restrictions on the designers, yet ensures an efficient verification flow. For example, by using only combinational logic optimization methods, a one-to-one register correspondence is automatically achieved in a synthesis-based ASIC design flow. Similarly, in a custom design flow, the circuit designer can be required to use the same state encoding for the implementation as given in the RTL specification. This can be further restricted by also requiring the use of identical register names, therefore facilitating an efficient computation of the correspondence.

4.4.1 Basic Approach

In case of a one-to-one register correspondence, equivalence of the two machines is shown if all corresponding next-state functions, and output functions are proved to be functionally equal. For this, all pairs of corresponding register inputs and primary inputs are connected and driven by a unique variable each. This setup corresponds to a specific characteristic function ϱ that is tested for validity. Figure 4.6 shows the corresponding setup and Figure 4.7 gives the derived *Miter** structure which was first introduced in [31].

Equivalence checking of two designs with corresponding registers is done in two steps:

- In the first step, the register correspondence is either guessed using simple heuristics or computed exactly. The spectrum of methods range from simple name-based methods to systematic methods (e.g., the ones outlined in Section 4.4.2) and often utilize combinations of both. Name-based methods simply match up instance names of registers, or names of signals connected to registers and design ports. This process is typically controlled by user-defined name-matching rules that allow the use of partial name matches, substitution of strings in names, etc.

* According to Brand [31] the Logic Constituting a Miter resembles two slabs of wood connected with a miter joint.

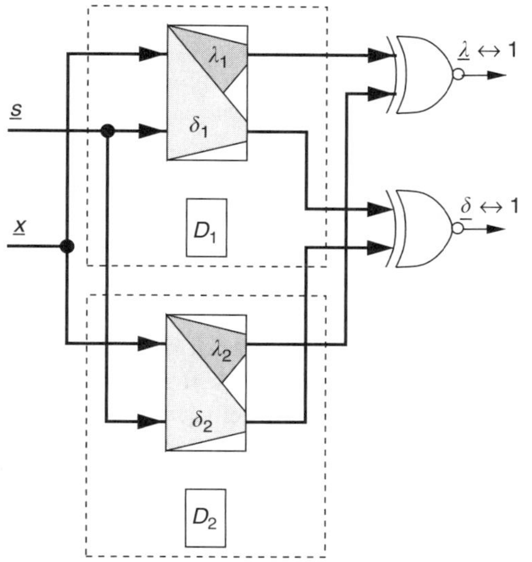

FIGURE 4.6 Combinational equivalence check based on register correspondence.

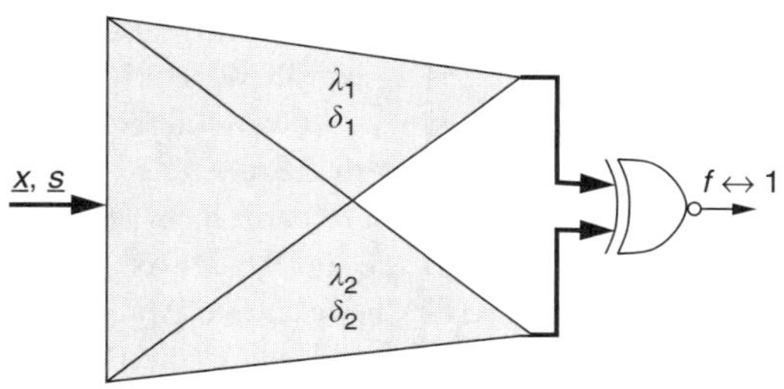

FIGURE 4.7 Miter circuit derived from configuration of Figure 4.6 for combinational equivalence checking.

- The second step involves the actual functional comparison of the individual combinational circuits. For this, a variety of methods for Boolean reasoning are applied, including the base methods outlined in Section 4.3.

It is important to note that an equivalence check based on register correspondence is *sound* but not *complete*. Soundness states that if all combinational checks pass, the two circuits are indeed functionally equivalent. The lack of completeness means that if any check fails, it cannot be decided whether the two machines are truly inequivalent or if only the given register equivalence is incorrect. In other words, referring to the concept of ϱ introduced in Section 4.2, the failing of check (1) cannot distinguish between an incorrectly chosen candidate for the characteristic function ϱ and true machine inequivalence.

In a practical design methodology, the incompleteness is addressed by rejecting all designs for which the comparison fails and demand a design update with an equivalent input/output behavior and a valid register correspondence. This effectively tightens the correctness criteria for design equivalence by requiring identical input/output behavior *and* a valid register correspondence that can prove equivalence by a combinational check.

4.4.2 Register Correspondence

In many practical design flows, a candidate register correspondence is derived from naming conventions. In the absence of naming rules, or if registers are only partially matched by names, the register correspondence can be computed automatically as a greatest fixed point using the algorithm outlined in Figure 4.8. The algorithm starts with one equivalence class (bucket) for the register correspondence, containing all registers (line 2). During each iteration, a unique variable is introduced for the outputs of all registers of each bucket (line 5) and all next-state functions are computed (line 8) based on these variables. Next, the buckets are partitioned into pieces that have identical next-state functions (line 11). This process is repeated until a fixed point is reached (line 10). The partitioning of the register buckets is monotonic, i.e., the number of buckets increases strictly from iteration to iteration until a fixed point is reached. Thus the algorithm will terminate after at most u iterations, where u is the total number of registers in the product machines.

If a canonical or semicanonical representation is used for the next-state functions, the comparison and refinement in lines 10 and 11 can simply be implemented by array sorting and splitting. For example, if BDDs (see Section 4.3.1) are applied, sorting the array of next-state functions by BDD pointer values moves identical functions to adjacent index positions. A single array sweep can then perform the partitioning step. Furthermore, if the representation allows efficient function complementation, this approach can also handle complemented registers: the construction of the next-state functions uses the complement of the bucket variable for these registers, and similarly, the resulting functions for them are complemented before comparison. Other examples of function representations that are applicable in this setting are the AND/INVERTER graph [14] or Boolean Expression Diagrams [37].

If a register correspondence between two machines exists, the algorithm of Figure 4.8 is guaranteed to find it. However, in case of discrepancies in any of the next-state functions, the algorithm generally fails. In the extreme case, the effect of a single incorrect next-state function may ripple through each iteration of the algorithm and split all buckets until they contain only one register.

A more relaxed criterion under which registers are considered equivalent can make a correspondence algorithm more robust. For example, an algorithm may consider two registers equivalent if the support sets of their next-state functions are equivalent with respect to the register correspondence. The algorithm given in Figure 4.8 can be modified to accommodate such relaxed criteria by adjusting the criteria in lines 10 and 11. Instead of performing a functional comparison, the support sets need to be simply compared. Clearly, this criterion may result in incorrect matching results. For example, two pairs of registers that represent two distinct state bits may have identical support sets and would get matched by such a criterion. Similarly, a register function may be functionally independent of particular inputs of its support set. If another functionally corresponding register omits any of these inputs, the matching based on input support would fail. Despite the possibility of producing incorrect results, a structural corresponding algorithm based on support sets works well in practice. It is significantly faster than the full functional correspondence algorithm

```
1   REGISTER_CORRESPONDENCE() {
2       put all registers r into bucket[0]
3       do {
4           forall buckets i do {
5               initialize output of all registers r ∈ i with variable v[i]
6           }
7           forall registers r do {
8               compute next state functions δ[r] based on inputs v
9           }
10          if ∀ buckets i: r1, r2 ∈ i ⇔ δ[r1] = δ[r2] return
11          split all buckets i into multiple buckets ij s.t.  r1, r2 ∈ ij ⇔ δ[r1] = δ[r2]
12      }
13  }
```

FIGURE 4.8 Algorithm for computing functional register correspondence.

and provides valuable information, even if next-state functions are erroneous. For the application in an equivalence checking tool, the structural correspondence algorithm can be used first, possibly in combination with complete or partial name matching. Only for register pairs where this algorithm produces incorrect matchings, the precise, but also more expensive functional correspondence is applied.

More details and improvements of register correspondence algorithms can be found in the literature [38,39]. The automated detection of a register correspondence can be extended to handle don't cares [40].

4.4.3 Equivalence Checking of Retimed Circuits

Sequential logic synthesis techniques have been researched for many years and there are a number of efficient approaches available that are applicable to practical designs. However, thus far their adoption in contemporary tool flows has been slow. The lack of an efficient equivalence checking flow is often cited as one of the reasons for this limited use. In the following, we focus the discussion on verifying circuits that have been optimized by retiming.

Retiming [41,42] is commonly referred to as a structural transformation of a sequential circuit that changes the positions of registers and latches without modifying the design's input–output behavior. The objective of retiming is to decrease the overall design cycle time by balancing the combinational path delays through a realignment of the synchronization grid. Figure 4.9 illustrates the impact of a retiming transformation on the finite-state machine structure of a circuit. The registers are either moved forward or backward, without changing the structure of the remaining logic. As shown in the figure, the register repositioning effectively moves the corresponding logic from one side of the combination logic block to the other.

As mentioned before, the most robust and scalable application of equivalence checking is based on a combinational verification paradigm. In the case of retiming, the next-state functions are not comparable and therefore, a straightforward application of equivalence checking methods mentioned in the previous subsections is not possible. However, by preserving the retime logic from the synthesis flow and applying it to make both designs comparable, the general sequential verification problem can also be reduced to a combinational case. Figure 4.10 illustrates this approach. In essence, both machines are "patched" with pieces of the retime logic to make the interfaces comparable. For this, the forward retime logic is added to the inputs of s_r^f to make the new inputs comparable to s^f. It is also added to the outputs of $s^{f'}$ to make them comparable with the outputs $s_r^{f'}$ of the retimed machine. Similarly, the backward retime logic is added to $s_r^{b'}$ and s^b. As a result, both machines have compatible current- and next-state

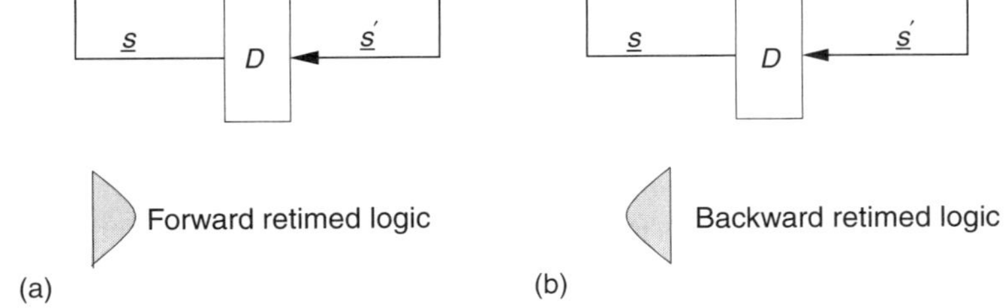

FIGURE 4.9 Effect of retiming on the FSM structure of a design (a) Forward retimed logic and (b) backward retimed logic.

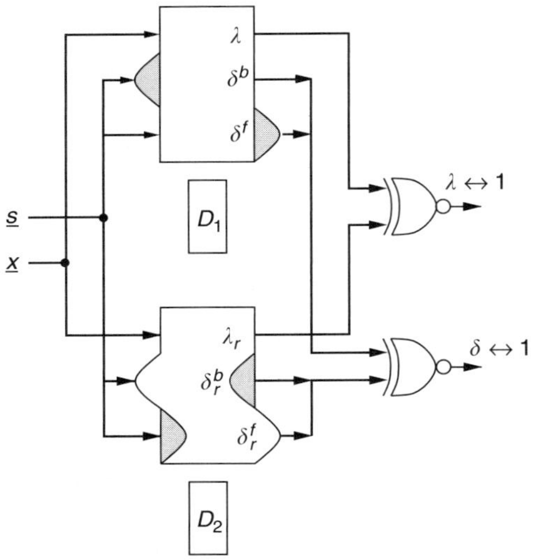

FIGURE 4.10 Application of the retime functions for equivalence checking based on a combinational verification paradigm.

variables which allow the application of a combinational verification paradigm. Note that this scheme is sound and cannot produce false-positive verification results.

Adding retime logic for equivalence checking can be done in multiple iterations, for example, to patch a sequence of retiming transformations interleaved with combinational optimization. At each step, the number of logic levels increases, which makes the combinational comparison more difficult. Further, efficient equivalence checking algorithms heavily exploit structural similarities of the two designs under comparison [43]. For too many retiming "shells" with little or no structural similarity, the equivalence checking problem may become intractable. However, a targeted and localized application of retiming for alleviating timing problems will typically result in small pieces of retime logic and therefore will not impact the performance of equivalence checking in the given setting.

Further generalizations to verify circuits which have been optimized by retiming can be found in [44–46].

4.5 Sequential Equivalence Checking

When the two sequential models to be tested for equivalence are substantially different, in particular when they have not been obtained one from the other through well-understood transformations like retiming or combinational optimizations, one has to resort to a fully general approach to verification that checks that the outputs of the two models are the same for the same inputs in all reachable states. In terms of the characteristic function introduced in Section 4.2, this corresponds to not relying on a guess for ϱ derived from the structure of the two machines.

Instead, one can take ϱ such that it holds for all the states where $\forall \underline{x}.\lambda(\underline{x}, \underline{s}) = 1$, that is, ϱ is the characteristic functions of all the states of the product machine M, where the outputs of the two machines M_1 and M_2 are identical. Equivalence is established if the initial states are in ϱ, and so are all the successors of the states in ϱ. In this case, equivalence is an *inductive invariant* for M. However, if the check on the successor states fails, the result is inconclusive, because the states that caused the failure may not be reachable from the initial states of M. In case of failure, the general approach is to *strengthen* ϱ, that is, to remove some states from it. The strongest ϱ for a given M is the characteristic functions of all its reachable states. The weakest ϱ, on the other hand, that guarantees completeness of the method is the characteristic functions of all the states of M from which no miscomparing state can be reached. Both these functions can be computed by reachability analysis as described in Section 4.5.1. It should be noted that this analysis can be quite expensive for large sequential circuits. Hence, it is sometimes useful to manually strengthen ϱ.

4.5.1 Reachability Analysis

Reachability analysis computes the states of a sequential machine M that either are reachable from a set of designated states, or from which the designated states are reachable. In equivalence verification, the states reachable from the initial states form the strongest possible ϱ: the outputs of M_1 and M_2 agree in all reachable states of $M = M_1 \times M_2$ if and only if M_1 is equivalent to M_2. Conversely, the complement of the states from which miscomparing states can be reached form the weakest ϱ for which the "if and only if" condition holds. The two cases corresponds to *forward* and *backward* reachability analysis, respectively. We first discuss forward analysis, and then briefly outline the differences.

Forward reachability analysis of a graph can be accomplished by the algorithm of Figure 4.11. The algorithm receives as input the graph and the set of initial states. It uses two main set variables: Q, to hold the states to be processed, and R to hold the states found to be reachable. At each iteration of the while loop, some unprocessed states are removed from Q, their successors are computed by function *Image*, and the states thus found that were not previously reached are added to both sets. The policy for choosing the states that are extracted from Q determines the type of search. If only the most recently added state is extracted, then one obtain *depth-first search* (DFS). Conversely, if only the least recently added states are extracted, the resulting search is *breadth-first* (BFS). Other policies are also possible. For hardware verification, the usefulness of DFS is limited by the state explosion problem: in most practical examples, there are too many states, and each state has too many successors for a search that considers each state and each transition individually.

If all the states in Q are extracted at once, one obtains *symbolic* BFS, in which states are reached in an order consistent with their distance from the initial states. In symbolic reachability analysis it is customary to represent the sets of states as BDDs. The transitions of the sequential machine are also represented by a BDD. Image computation, that is, the computation of the successors of the states in F, can then be performed by BDD operations that do not enumerate explicitly the states or the transitions. This approach can sometimes deal with huge numbers of states — in excess of 10^{100}. However, this ability depends on the BDDs remaining compact. This in turn depends on the state encoding of M, on the order of the state variables in the BDDs, and, of course, on the subsets of states that must be represented. Departure from strict BFS processing [13,47–49] may result in sets of states that have much smaller BDDs; this may considerably speed up the analysis.

```
FORWARD_REACHABILITY_ANALYSIS() {
        initialize Q and R to the initial states
        while   (Q ≠ ∅) {
              let F be nonempty subset of Q
              Q = Q \ F // remove F from Q
              S = Image(F) // compute successors
              N = S \ R // identity new states
              Q = Q ∪ N // add new states to Q ...
              R = R ∪ N // ... and to R
        }
     return R
}
```

FIGURE 4.11 Algorithm for computing forward reachable states.

Besides the order of traversal, the specifics of the computation of the successors of a set of states have profound influence on the performance of the symbolic reachability analysis algorithm. There are two main approaches, one based on recursive case analysis, which normally employs the transition *functions* [50], and the one based on conjunction and quantification, which employs a properly clustered transition *relation* [51,52]. The conjunction method is often faster, mostly because it is more efficient at caching intermediate computations, but neither is superior; their combination, which increases robustness, is described in [53]. The key problem in the image computation method that relies on conjunction and quantification is the order in which the components of the transition relation are combined to obtain the final result [54,55].

Backward reachability differs from forward reachability in two respects: the starting states are the miscomparing states instead of the initial states of *M*, and the computation of the successors (image computation) is replaced by the computation of the predecessors of a set of states (preimage computation). There are small differences in the algorithms used for image and preimage computations, due to the fact that the model of sequential circuits express the next states directly as a function of the present states and not vice versa.

Forward reachability analysis is more efficient than backward analysis on some models and less efficient on others. As an extreme case, consider checking the equivalence of two modulo-n counters. The algorithm of Figure 4.11 requires n iterations to converge. However, if the two counters are indeed equivalent, backward analysis converges after one iteration. In fact, in this example, equivalence is an inductive invariant, in which case backward analysis always converges in one iteration. Each step of backward reachability corresponds to a strengthening of the invariant, until it becomes inductive or equivalence is refuted.

4.5.2 Dependent Variables in Reachability Analysis

Suppose a model contains three state variables s_1, s_2, and s_3 such that

$$s'_1 = x_1, \quad s'_2 = s_1, \quad s'_3 = x_1 \wedge s_1$$

Suppose that in the initial states, $s_1 = s_2 = s_3$. Then it is easy to see that $s_3 = s_1 \wedge s_2$ in all reachable states. It is therefore possible to remove s_3 from the model and replace any usage of this variable with $s_1 \wedge s_2$. The resulting model is typically easier to deal with. We say that s_3 is a dependent variable in the given model. In this section, we study how we can identify dependent variables, and how we can use this knowledge to improve reachability analysis. The identification and extraction of *dependent variables* is based on the observation that for a Boolean function *f*,

$$f = (s \leftrightarrow f_s) \wedge (\exists s \cdot f) \text{ if and only if } \forall s \cdot f = 0$$

Suppose the set $R(s)$ of the reachable states of a model has been computed. If s_i is functionally dependent in R, then the state variable corresponding to s_i can be removed from the model, and replaced by a combinational function of other state variables. The resulting model satisfies the same properties as the original model. The advantage of removing dependent variables is that the BDDs are likely to be smaller [56].

For equivalence checking, however, the detection of functional dependencies would be more profitable if performed during fixpoint computation. In particular, one could extract dependencies from the set of states $P(s)$ whose image is being computed. The resulting functions can be used to eliminate the dependent variables from the transition relation used to compute the successors of the states in $P(s)$.

The problem with this approach is that the overhead required to detect the dependencies can be substantial. A more practical approach, albeit a less powerful one, concentrates on special forms of functional dependence, namely variable equivalence and complementarity. Variables a and b are equivalent in Boolean function f if $f \leq (a \leftrightarrow b)$. They are complementary in f if $f \leq (a \oplus b)$. This means that two variables are equivalent (complementary) if they have the same value (opposite values) in all assignments for which $f = 1$.

When computing the image of $P(s)$, if s_i is equivalent to s_j in P, we can use the same BDD variable for s_j and s_i in both P and in the transition relation T. If there is just one initial state, then for the first image computation in reachability analysis only one BDD variable is needed for the current-state variables. In successive image computations, the number of required BDD variables will likely increase, but for some circuits it will remain substantially lower than the number of state variables. In particular, corresponding register pairs in equivalent circuits will remain represented by the same variable.

The effectiveness of this approach relies on the ability to compute variable equivalence classes efficiently. A partition refinement algorithm for reachability analysis based on forward traversal was implemented in the TiGeR tool in the early 1990s [57], but not published. The main reason why equivalence detection can be made efficient is that it boils down to comparing Boolean functions for equivalence or complementarity — operations that are well supported by BDDs. Later, a method similar to that of TiGeR was discussed by van Eijk [58]; his thesis, in particular, discusses the identification of equivalent variables before reachability analysis, which was described in Figure 4.8.

4.6 Summary

In this chapter, we have examined the role of equivalence checking in the design cycle and we have reviewed the major approaches and techniques applied to this problem. Highly efficient algorithms have been developed for combinational equivalence checking; they rely on structural analysis and on sophisticated decision procedures that can establish the satisfiability of large propositional formulae. Since sequential equivalence checking is a much less tractable problem, every effort is made to reduce it to combinational equivalence by constraints in the design methodology and by the application of algorithms that deal with unknown register correspondence and retiming. Equivalence checking based on combinational techniques has become an integral part of most design flows and can deal with large-scale designs. In the general case however, one must analyze the so-called product machine to determine whether miscomparing states are reachable from the initial states. Techniques for this task are based on reachability analysis. Although capacity is a significant concern whenever such an analysis is undertaken, symbolic algorithms can deal with circuits with hundreds of state variables, and can therefore be applied to blocks of a design that have undergone deep transformations.

References

[1] G.L. Smith, R.J. Bahnsen, and H. Halliwell, Boolean comparison of hardware and flowcharts, *IBM J. Res. Dev.*, 26, 106–116, 1982.

[2] A. Kuehlmann, A. Srinivasan, and D.P. LaPotin, Verity — a formal verification program for custom CMOS circuits, *IBM J. Res. Dev.*, 39, 149–165, 1995.

[3] G.P. Bischoff, K.S. Brace, S. Jain, and R. Razdan, Formal implementation verification of the bus interface unit for the Alpha 21264 microprocessor, *Proceedings of the IEEE International Conference on Computer Design*, 1997, pp. 16–24.

[4] J. Moondanos, C.H. Seger, Z. Hanna, and D. Kaiss, CLEVER: divide and conquer combinational logic equivalence verification with false negative elimination, in *Computer Aided Verification (CAV'01)*, Paris, France, 2001, pp. 131–143.

[5] R.E. Bryant, Graph-based algorithms for Boolean function manipulation, *IEEE Trans. Comput.*, C-35, 677–691, 1986.

[6] R.E. Bryant, On the complexity of VLSI implementations and graph representations of Boolean functions with application to integer multiplication, *IEEE Trans. Comput.*, 40, 205–213, 1991.

[7] R. Bryant and Y.-A. Chen, Verification of arithmetic circuits with binary moment diagrams, *Proceedings of the Design Automation Conference*, San Francisco, CA, 1995, pp. 535–541.

[8] Y.-T. Lai and S. Sastry, Edge-valued binary decision diagrams for multi-level hierarchical verification, *Proceedings of the Design Automation Conference*, Anaheim, CA, 1992, pp. 608–613.

[9] F. Somenzi, Binary decision diagrams, in *Calculational System Design*, M. Broy and R. Steinbrüggen, Eds., IOS Press, Amsterdam, 1999, pp. 303–366.

[10] S. Manne, D.C. Grunwald, and F. Somenzi, Remembrance of things past: locality and memory in BDDs, *Proceedings of the Design Automation Conference*, Anaheim, CA, 1997, pp. 196–201.

[11] D. Sieling, The Nonapproximability of OBDD Minimization, Technical report 663, University of Dortmund, 1998.

[12] R. Rudell, Dynamic variable ordering for ordered binary decision diagrams, *Proceedings of the International Conference on Computer-Aided Design*, Santa Clara, CA, 1993, pp. 42–47.

[13] K. Ravi and F. Somenzi, Hints to accelerate symbolic traversal, in *Correct Hardware Design and Verification Methods (CHARME'99)*, Lecture Notes in Computer Science, Vol. 1703, Springer, Berlin, 1999, pp. 250–264.

[14] A. Kuehlmann and F. Krohm, Equivalence checking using cuts and heaps, *Proceedings of the 34th ACM/IEEE Design Automation Conference*, Anaheim, CA, 1997, pp. 263–268.

[15] J.R. Burch and V. Singhal, Tight integration of combinational verification methods, *Proceedings of the International Conference on Computer-Aided Design*, San Jose, CA, 1998, pp. 570–576.

[16] A. Biere, A. Cimatti, E. Clarke, and Y. Zhu, Symbolic model checking without BDDs, *Fifth International Conference on Tools and Algorithms for Construction and Analysis of Systems (TACAS'99)*, Lecture Notes in Computer Science, Vol. 1579, Amsterdam, The Netherlands, 1999, pp. 193–207.

[17] M.R. Garey and D.S. Johnson, *Computers and Intractability: A Guide to the Theory of NP-Completeness.* 2nd ed., W.H. Freeman and Company, New York, 1979.

[18] M. Davis and H. Putnam, A computing procedure for quantification theory, *J. Assoc. Computing Mach.*, 7, 201–215, 1960.

[19] M. Davis, G. Logemann, and D. Loveland, A machine program for theorem proving, *Commn. ACM*, 5, 394–397, 1962.

[20] J.P.M. Silva and K.A. Sakallah, Grasp — a new search algorithm for satisfiability, *Proceedings of the International Conference on Computer-Aided Design*, San Jose, CA, 1996, pp. 220–227.

[21] M. Moskewicz, C.F. Madigan, Y. Zhao, L. Zhang, and S. Malik, Chaff: engineering an efficient SAT solver, *Proceedings of the Design Automation Conference*, Las Vegas, NV, 2001, pp. 530–535.

[22] L. Zhang, C. Madigan, M. Moskewicz, and S. Malik, Efficient conflict driven learning in Boolean satisfiability solver, *Proceedings of the International Conference on Computer-Aided Design*, San Jose, CA, 2001, pp. 279–285.

[23] L. Zhang and S. Malik, Validating SAT solvers using an independent resolution-based checker: practical implementations and other applications, *Design, Automation and Test in Europe (DATE'03)*, Munich, Germany, 2003, pp. 880–885.

[24] H. Jin and F. Somenzi, An incremental algorithm to check satisfiability for bounded model checking, *Electronic Notes in Theoretical Computer Science*, 2004, Second International Workshop on Bounded Model Checking, http://www.elsevier.nl/locate/entcs/.

[25] C.L. Berman and L.H. Trevillyan, Functional comparison of logic designs for VLSI circuits, *Proceedings of the International Conference on Computer-Aided Design*, Santa Clara, CA, 1989, pp. 456–459.

[26] A. Kuehlmann, V. Paruthi, F. Krohm, and M.K. Ganai, Robust Boolean reasoning for equivalence checking and functional property verification, *IEEE Trans. Comput.-Aided Design*, 21, 1377–1394, 2002.

[27] M.K. Ganai, P. Ashar, A. Gupta, L. Zhang, and S. Malik, Combining strengths of circuit-based and CNF-based algorithms for a high-performance SAT solver, *Proceedings of the Design Automation Conference*, New Orleans, LA, 2002, pp. 747–750.

[28] H. Jin and F. Somenzi, CirCUs: a hybrid satisfiability solver, *International Conference on Theory and Applications of Satisfiability Testing (SAT 2004)*, Vancouver, Canada, 2004.

[29] J.P. Roth, Diagnosis of automata failures: a calculus & method, *IBM J. Res. Dev.*, 10, 278–291, 1966.

[30] J.P. Roth, Hardware verification, *IEEE Trans. Comput.*, C-26, 1292–1294, 1977.

[31] D. Brand, Verification of large synthesized designs, *Digest of Technical Papers of the IEEE/ACM International Conference on Computer-Aided Design*, Santa Clara, CA, 1993, pp. 534–537.

[32] M.H. Schulz and E. Auth, Improved deterministic test pattern generation with applications to redundancy identification, *IEEE Trans. Comput.-Aided Design*, 8, 811–816, 1989.

[33] D. Brand, Incremental synthesis, *Digest of Technical Papers of the IEEE/ACM International Conference on Computer-Aided Design*, San Jose, CA, 1994, pp. 14–18.

[34] L.A. Entrena and K.-T. Cheng, Combinational and sequential logic optimization by redundancy addition and removal, *IEEE Trans. Comput.-Aided Design*, 14, 909–916, 1995.

[35] H. Cho, G. Hachtel, S.-W. Jeong, B. Plessier, E. Schwarz, and F. Somenzi, ATPG aspects of FSM verification, *Digest of Technical Papers of the IEEE International Conference on Computer-Aided Design*, 1990, pp. 134–137.

[36] M.K. Ganai, A. Aziz, and A. Kuehlmann, Enhancing simulation with BDDs and ATPG, *Proceedings of the 36th ACM/IEEE Design Automation Conference*, New Orleans, Louisiana, 1999, pp. 385–390.

[37] H. Hulgaard, P.F. Williams, and H.R. Andersen, Equivalence checking of combinational circuits using Boolean expression diagrams, *IEEE Trans. Comput.-Aided Design*, 18, 903–917, 1999.

[38] T. Filkorn, *Symbolische Methoden für die Verifikation endlicher Zustandssysteme*, Dissertation, Technische Universität München, 1992.

[39] C.A.J. van Eijk and J.A.G. Jess, Detection of equivalent state variables in finite state machine verification, *1995 ACM/IEEE International Workshop on Logic Synthesis*, Tahoe City, CA, 1995, pp. 3–35–3–44.

[40] J.R. Burch and V. Singhal, Robust latch mapping for combinational equivalence checking, *Digest of Technical Papers of the IEEE/ACM International Conference on Computer-Aided Design*, San Jose, CA, 1998, pp. 563–569.

[41] C. Leiserson and J. Saxe, Optimizing synchronous systems, *J. VLSI Comput. Sys.*, 1, 41–67, 1983.

[42] C. Leiserson and J. Saxe, Retiming synchronous circuitry, *Algorithmica*, 6, 5–35, 1991.

[43] A. Kuehlmann and C.A. van Eijk, Combinational and sequential equivalence checking, in *Logic Synthesis and Verification*, S. Hassoun and T. Sasao, Eds., Kluwer Academic Publisher, Boston/Dordrecht/London, 2001, chap. 13.

[44] D. Stoffel and W. Kunz, A structural fixpoint iteration for sequential logic equivalence checking based on retiming, *International Workshop on Logic Synthesis*, Tahoe City, CA, 1997.

[45] C. van Eijk, Sequential equivalence checking based on structural similarities, *IEEE Trans. Comput.-Aided Design*, 19, 814–819, 2000.

[46] S.-Y. Huang, K.-T. Cheng, K.-C. Chen, C.-Y. Huang, and F. Brewer, AQUILA: an equivalence checking system for large sequential designs, *IEEE Trans. Comput.*, 49, 443–464, 2000.

[47] K. Ravi and F. Somenzi, High-density reachability analysis, *Proceedings of the International Conference on Computer-Aided Design*, San Jose, CA, 1995, pp. 154–158.

[48] G. Cabodi, P. Camurati, L. Lavagno, and S. Quer, Disjunctive partitionining and partial iterative squaring: an effective approach for symbolic traversal of large circuits, *Proceedings of the Design Automation Conference*, Anaheim, CA, 1997, pp. 728–733.

[49] R. Fraer, G. Kamhi, B. Ziv, M.Y. Vardi, and L. Fix, Prioritized traversal: efficient reachability analysis for verification and falsification, *Twelfth Conference on Computer Aided Verification (CAV'00)*, E.A. Emerson and A.P. Sistla, Eds., Lecture Notes in Computer science, Vol. 1855, Springer, Berlin, 2000, pp. 389–402.

[50] O. Coudert, C. Berthet, and J.C. Madre, Verification of sequential machines using Boolean functional vectors, *Proceedings IFIP International Workshop on Applied Formal Methods for Correct VLSI Design*, L. Claesen, Ed., Leuven, Belgium, 1989, pp. 111–128.

[51] H. Touati, H. Savoj, B. Lin, R.K. Brayton, and A. Sangiovanni-Vincentelli, Implicit enumeration of finite state machines using BDD's, *Proceedings of the International Conference on Computer-Aided Design*, 1990, pp. 130–133.

[52] J.R. Burch, E.M. Clarke, and D.E. Long, Representing circuits more efficiently in symbolic model checking, *Proceedings of the Design Automation Conference*, San Francisco, CA, 1991, pp. 403–407.

[53] I.-H. Moon, J.H. Kukula, K. Ravi, and F. Somenzi, To split or to conjoin: the question in image computation, *Proceedings of the Design Automation Conference*, Los Angeles, CA, 2000, pp. 23–28.

[54] D. Geist and I. Beer, Efficient model checking by automated ordering of transition relation partitions, *Sixth Conference on Computer Aided Verification (CAV'94)* D.L. Dill, Ed., Lecture Notes in Computer Science, Vol. 818, Springer, Berlin, 1994, pp. 299–310.

[55] I.-H. Moon, G.D. Hachtel, and F. Somenzi, Border-block triangular form and conjunction schedule in image computation, in *Formal Methods in Computer Aided Design*, W.A. Hunt, Jr. and S.D. Johnson, Eds., Lecture Notes in Computer Science, Vol. 1954, Springer, Austin, TX, 2000, pp. 73–90.

[56] A.J. Hu and D. Dill, Reducing BDD size by exploiting functional dependencies, *Proceedings of the Design Automation Conference*, Dallas, TX, 1993, pp. 266–271.

[57] J.C. Madre, Private communication, 1996.

[58] C.A.J. van Eijk, Formal Methods for the Verification of Digital Circuits, Ph.D. thesis, Eindhoven University of Technology, 1997.

5
Digital Layout — Placement

Andrew B. Kahng
University of California
San Diego, California

Sherief Reda
University of California
San Diego, California

5.1 Introduction: Placement Problem and Contexts

Placement is an essential step in the physical design flow since it assigns exact locations for various circuit components within the chip's core area. An inferior placement assignment will not only affect the chip's performance but might also make it nonmanufacturable by producing excessive wirelength, which is beyond available routing resources. Consequently, a placer must perform the assignment while optimizing a number of objectives to ensure that a circuit meets its performance demands. Typical placement objectives include total wirelength, timing, congestion, and power. In this chapter, we survey the main algorithmic methods used in state-of-the-art placers.

Figure 5.1 shows the position of placement within the EDA design flow. A placer takes a given synthesized circuit netlist together with a technology library and produces a valid placement layout. The layout is optimized according to the aforementioned objectives and ready for cell resizing and buffering — a step essential for timing and signal integrity satisfaction. Clock-tree synthesis and routing follow completing the physical design process. In many cases, parts of, or the entire, physical design flow are iterated a number of times until timing closure is achieved.

A circuit netlist is composed of a number of *components* and a number of *nets* representing the required electrical connectivity between the various components, where a net connects two or more components. In the case of application specific integrated circuits (ASIC), the chip's core layout area is comprised of a number of fixed height *rows*, with either some or no space between them. Each row consists of a number of *sites* which can be occupied by the circuit components. A *free site* is a site that is not occupied by any

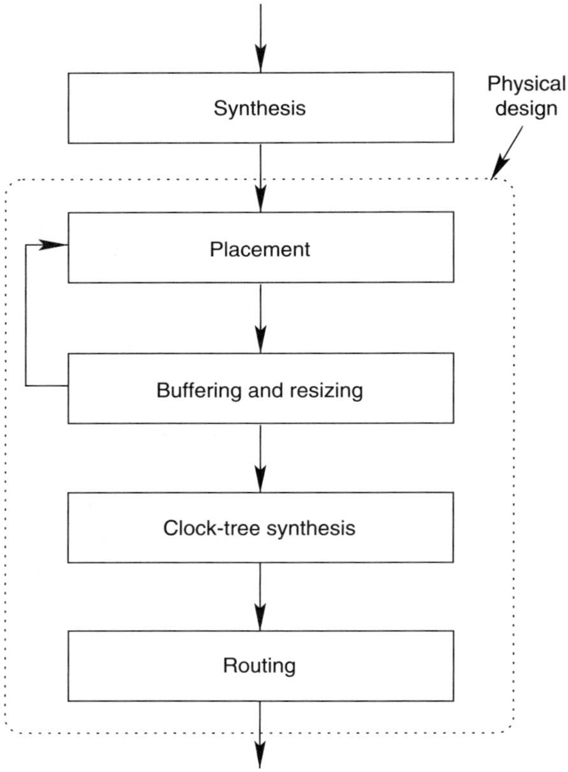

FIGURE 5.1 Placement position in a physical design flow.

component. Circuit components are either *standard cells, macro blocks,* or *I/O pads.** Standard cells have a fixed height equal to a row's height, but have variable widths. The width of a cell is an integral number of sites. On the other hand, blocks are typically larger than cells and have variable heights that can stretch a multiple number of rows. Figure 5.2 gives a view of a typical placement layout. Blocks can have preassigned locations — say from a previous floorplanning process — which limit the placer's task to assigning locations for just the cells. In this case, the blocks are typically referred to by *fixed blocks.* Alternatively, some or all of the blocks may not have preassigned locations. In this case, they have to be placed with the cells in what is commonly referred to as *mixed-mode* placement.

In addition to ASICs, placement retains its prime importance in gate array structures such as field programmable gate arrays (FPGAs). In FPGAs, placement maps the circuit's subcircuits into programmable FPGA logic blocks in a manner that guarantees the completion of the subsequent stage of routing.

The major placement objectives can be summarized as follows:

- *Wirelength.* Minimizing the total wirelength, or just wirelength, is the primary objective of most existing placers. This is no surprise given that the wirelength must be less than the limited total routing supply of the chip, and that power and delay are proportional to the wirelength and wirelength square, respectively. Consequently, minimizing the wirelength improves the chip's performance. Wirelength is measured by the sum of minimum Steiner tree costs of the various nets, where the Steiner tree cost of a given point set is the minimum cost, or length, of a tree that spans all the given points as well as any subset of additional points (Steiner points) [1]. Routed wirelength is typically slightly larger than the Steiner tree cost since contention on routing resources by different nets might

* We will refer to standard cells by just *cells* and macro blocks by just *blocks.*

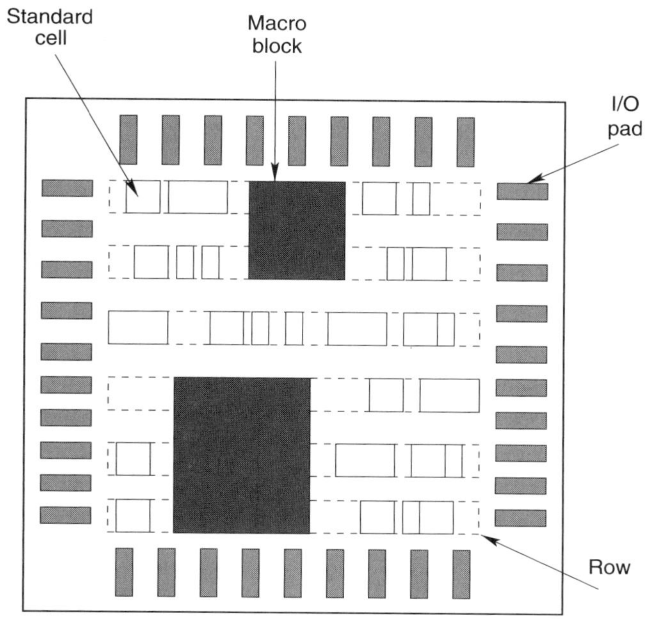

FIGURE 5.2 Placement layout area.

lead to detours, eventually increasing the wirelength. Given that the Steiner tree problem is NP-hard, placers typically minimize and report other metrics that are faster and easier to compute. These include half-perimeter wirelength (HPWL) and minimum spanning tree (MST). Half-perimeter wirelength is half the perimeter of the smallest bounding box enclosing a given net's components. Half-perimeter wirelength is equivalent to the Steiner minimum tree cost for two- and three-pin nets, and is well-correlated for multi-pin (≥ 4) nets [2]. Minimum spanning tree is also equivalent to the Steiner minimium tree cost for two- and three-pin nets, and within a constant factor for multi-pin (≥ 4) nets [1].

- *Timing.* The clock cycle of a chip is determined by the delay of its longest path, usually referred to as the *critical path.* Given a performance specification, a placer must ensure that no path exists with delay exceeding the maximum specified delay. Any delay in excess over such specified value is considered *negative slack*, and timing-driven placers must minimize the worst negative slack and the total negative slack to meet performance requirements.
- *Congestion.* While it is necessary to minimize the total wirelength to meet the total routing resources, it is also necessary to meet the routing resources within various local regions of the chip's core area. A congested region might lead to excessive routing detours. Such detours can ultimately lead to excessive increase in wirelength, adversely impacting both timing and power.
- *Power.* With increasing clock frequencies and demand for battery-powered mobile devices, minimizing power is becoming an increasingly important objective. Power minimization typically involves distributing the locations of cell components so as to reduce the overall power consumption, alleviate hot spots, and smooth temperature gradients.

Another secondary objective is placement *runtime* minimization. For a given netlist, placement is a one-time effort, and consequently it is usually tolerable to have increased runtimes if this has a positive impact on the placement quality. However, for state-of-the-art designs with multimillion components, placement can take a few days, which is deemed unacceptable for fast prototyping or in timing-closure iterations. For such cases, it is important to seek methods that minimize the total placement runtime with little or no impact to the placement quality.

Given that placement is one of the oldest and first problems in EDA, placement has a rich history of solutions. More than four decades ago, Steinberg [3] considered placement of circuit components on a back board such that the total wirelength is minimized. Steinberg solution starts with an initial placement, obtained say via random placement. Steinberg then selects an *independent set* of components, i.e., a set of components that do not share any connections, and optimally reembeds these components within their pool of locations via optimal linear assignment (OLA). The process of selection and optimal reembedding is iterated until no further improvement in solution quality is possible.

Analytical techniques approximate the wirelength objective using quadratic [4–8] or nonlinear formulations [9–12]. The first proposal to use such methods is due to Hall [4], where he suggested minimizing the squared length and devised an eigenvalue approach to solve the problem. Since that point, the central problems in analytical techniques are how to better approximate the wirelength objective, how to numerically solve the nonlinear objective and how to spread out the components that typically heavily overlap in analytical solutions. Approaches for solving the nonlinear objectives include sparse matrix and conjugate-gradient (CG) methods. Cell-spreading techniques include the use of partitioners in top-down frameworks [5,6,13], network flows [7], or additional repelling forces [8,14].

The advent of min-cut partitioners [15] paved the way to the introduction of min-cut placers [16]. The introduction of a linear-time min-cut partitioning heuristic by Fiduccia and Mattheyses [17] and terminal propagation mechanisms [18] further bolstered min-cut placement as an attractive solution. Finally, development of multilevel hypergraph partitioners [19] has initiated a revival in min-cut placement [20–22].

Another thread of placement techniques started with the proposal of simulated annealing as a general combinatorial optimization technique [23]. As a matter of fact, placement along with the Traveling Salesman Problem were the original two problems experimented by Kirkpatrick and Vecchi [23]. Simulated annealing was quickly adopted as a leading technique in placement [24,25], with methods such as clustering used to improve its execution time [26]. Simulated annealing can also be used with other placement methods such as min-cut to improve their performance [21].

Placement approaches typically differentiate between a *global placement* phase and a *detailed placement* phase. At the beginning of the global phase, all cells belong to one rectangular *bin* that spans the entire chip's core area. As global placement proceeds, cells are spread over the chip's core area into a number of smaller bins. By the end of this phase, each bin will typically contain few cells. In detailed placement, cells are assigned exact locations and all overlaps are removed.

In this chapter, we will a give a brief survey of the various placement approaches. We describe global placement algorithms in Section 5.2, detailed placement algorithms in Section 5.3, and recent placement trends in Section 5.4. Section 5.5 gives a brief of view of state-of-the-art academic and industrial placers, and Section 5.6 summarizes this chapter content.

5.2 Global Placement

The goal of global placement is to find a well-spread, ideally with no overlaps, placement for the given netlist that attains required objectives such as wirelength minimization or timing specifications. Formally, a circuit netlist is represented as a hypergraph $H = (V,E)$, where $V = \{v_1,v_2, \dots , v_n\}$ is the set of nodes with each node representing a circuit component, and $E = \{e_1, e_2, \dots ,e_m\}$ is the set of hyperedges corresponding to the nets, where a hyperedge $e_i \in E$ is a subset of nodes. Traditionally placers are HPWL driven with the main objective to minimize

$$z = \sum_{e_i \in E} (\max_{v_j \in e_i} x_j - \min_{v_j \in e_i} x_j + \max_{v_j \in e_i} y_j - \min_{v_j \in e_i} y_j) \tag{5.1}$$

where x_j and y_j are the vertical and horizontal coordinates of a node v_j, and such that all nodes are placed on sites with no overlaps. Fundamentally, z is the HPWL sum of all hyperedges. The placement problem is NP-hard [27], and solutions for the problem rely on heuristics to achieve this objective suboptimally. In this section we survey some of the algorithmic solutions to global placement. Specifically, we describe

(1) min-cut placers, (2) simulated-annealing placers, and (3) analytical placers. We also survey some of the main techniques to handle other objectives such as timing, congestion, and placement runtime.

5.2.1 Min-Cut Placers

Min-cut placers operate in a top-down hierarchical fashion by recursively partitioning a given netlist into 2^k, $k \geq 1$, partitions. When $k > 1$, partitioning is commonly referred to as *multiway* partitioning. In the case $k = 1$, partitioning is called *bisection*, and when $k = 2$, partitioning is called *quadrisection*. In addition to netlist partitioning, the placer also recursively divides the layout area into a number of *bins*, and assigns each of the netlist partitions into one of the bins. Min-cut placement is essentially a top-down refinement process, where each bin gets divided into two or four smaller bins with fewer number of cells. Bin dividing is achieved through either a horizontal or a vertical cut in case of bisection [16], or through simultaneous horizontal and vertical cuts in case of quadrisection [28]. Thus the outcome of recursive bin division is a slicing floorplan as illustrated in Figure 5.3. The process of partitioning and dividing all bins exactly once is called *placement level*. Placement levels created by simultaneous netlist partitioning and bin division continue until each bin has a few cells, beyond which detailed placers are used to assign final locations for all nodes within their corresponding bins [20,21,29].

The key concerns of min-cut placement are as follows. How to partition a netlist?, and into how many partitions, e.g., bisection or quadrisection? Given a bin, should it be divided horizontally or vertically in case of bisection? Finally, how to capture the global netlist connectivity information when partitioning local netlist instances inside the bins. We tackle each of the aforementioned concerns in the next subsections.

5.2.1.1 Min-Cut Partitioners

Given a number of netlist partitions, a hyperedge is considered *cut* if its nodes span or reside in more than one partition. The *k-way min-cut partitioning* problem is defined as follows. Given a hypergraph with an associated cost to each hyperedge, partition the nodes of the hypergraph into balanced 2^k subsets while minimizing the total cost of cut hyperedges. Subset balancing is achieved by imposing the constraint that no subset exceeds a given maximum size bound. The min-cut partitioning problem is NP-hard [1].

Kernighan and Lin [15] suggested the first heuristic solution (KL) to this problem within the context of graph partitioning. Assuming $k = 1$, the KL algorithm starts with a random initial balanced partitioning. For each pair of nodes — a pair is comprised of a node from each partition — the algorithm calculates the pair's *score*, or impact on cut size if the pair is swapped. The pair with the largest score or equivalently

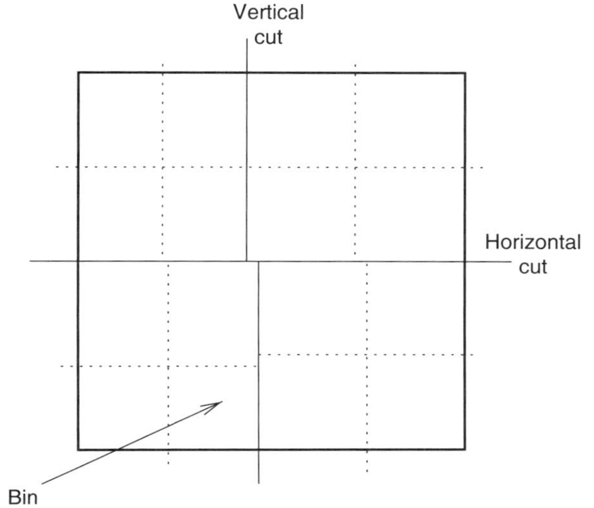

FIGURE 5.3 Slicing outline.

largest reduction in cut cost is identified, swapped, and *locked* to be prevented from any future swapping. Furthermore, their score is entered in a *record* and the swap score of their neighbors is updated. This process of identifying the best pair to swap, swapping the pair, locking, recording, and updating the neighbors' score is repeated until all pairs of cells are exchanged, essentially yielding the original partitioning. Using the record, the KL algorithm then executes the sequence of swaps that yields the overall minimum cut cost and unlocks all nodes. This constitutes a *KL iteration.* The KL algorithm keeps on iterating until an iteration causes no further reduction in cut size. At this point the algorithm stops.

The KL algorithm suffers from large runtimes due to the need to calculate and update the scores of all pairs of nodes. To overcome these limitations, Fiduccia and Mattheyses proposed a linear time heuristic (FM) based on the KL algorithm [17]. The key differences between the FM and KL algorithms are (1) moves instead of swaps are only considered, (2) hyperedge handling, and (3) a bucket data structure that allows a total update of move scores in linear time.

Despite the improvement in runtime of the FM heuristic in comparison to the KL heuristic, the performance of the FM heuristic, as measured by the cut cost, typically tends to degrade as the problem size increases. To address this problem, modern min-cut partitioners use the FM heuristic in a *multilevel* framework [19,30], where *clustering* or *coarsening* is used to reduce the size of a given flat hypergraph. A given hypergraph H_0 is coarsened k times until a hypergraph H_k of suitable size is attained as shown in Figure 5.4. The FM partitioner is then executed on H_k. Instead of projecting the partitioning results of H_k directly back to H_0, the unclustering is carried out in a controlled *refined fashion,* where the results of H_k are projected *to* H_{k-1} and followed by refinement using the FM partitioner. The process of uncoarsening and refinement is repeated until the original flat hypergraph H_0 is obtained.

In many cases a bin capacity is larger than the total cell area that belongs to it, thus creating an amount of *freespace.* Freespace can be distributed by the partitioner to ensure "smooth" partitioning by allowing tolerances in the specified partition sizes [20,31]. In addition, distributing freespace in a uniform hierarchical manner ensures smooth partitioning till the late placement levels, which results in a reduced amount of final overlaps. The issue of freespace distribution will be further investigated in Section 5.4.2.

5.2.1.2 Cut Sequences

Another important ingredient in min-cut placement is the cut direction. In the case of quadrisection, cut direction selection is trivial; each bin will be simultaneously divided with both a vertical and a horizontal cut [28]. In case of bisection, a cut sequence has to be determined. An *alternating sequence* that strictly alternates between horizontal and vertical cuts is a favorite choice [16]. However, recent approaches [20,32] suggest selecting a direction based on the bin's aspect ratio. For example, if the

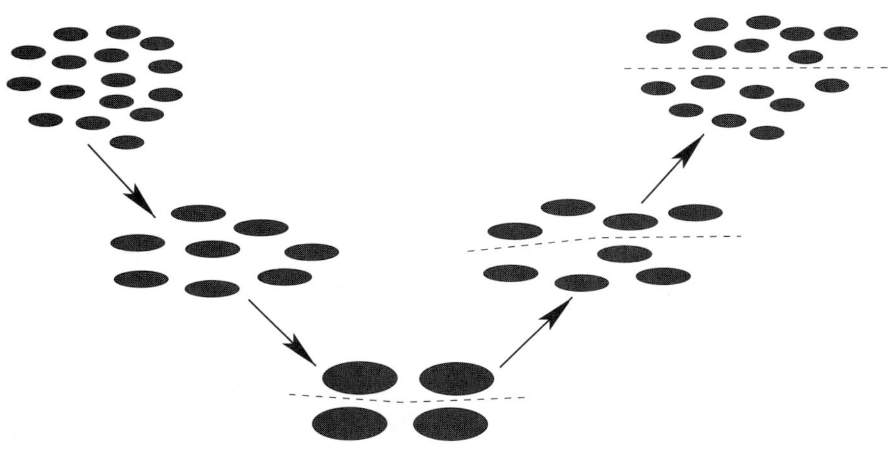

FIGURE 5.4 Coarsening and uncoarsening in a multilevel framework.

height of a bin exceeds its width by a certain ratio then division is carried out horizontally. Such selection is typically justified by wirelength estimates which exhibit an increase if the wrong division direction is chosen.

When comparing multiway partitioning, e.g., quadrisection, to bisection, we find out that (1) multiway partitioning fixes the shape of the partitions in contrast to bisection which allows more flexible outlines, and furthermore (2) the added computational complexity of multiway partitioning does not seem to translate to improvements in cut values in comparison to recursive bisection [33].

Another important consideration is whether to allow horizontal cuts that are only aligned with row boundaries or allow "fractional" horizontal cuts that run through the rows [34]. The latter approach might improve placement results but requires an extra effort from the detailed placer to snap the cells to the rows.

5.2.1.3 Capturing Global Connectivity

While a partitioner might deliver partitions with close to minimum cut costs, this result might not translate to total placement HPWL minimization if each bin is partitioned locally and in isolation of other bin results. Thus it is essential to capture global connectivity while partitioning a particular local bin instance. *Terminal propagation* [18,35] is a mechanism through which connectivity between cells residing in different bins is propagated into local partitioning instances. With terminal propagation, nodes external to a bin being partitioned are propagated as fixed terminals (nodes) to the bin. These terminals bias the partitioner toward placing movable nodes close to their terminals, reducing the overall placement wirelength. Given a bin being partitioned into two child bins and an external node connected to the cells of this bin, it is critical to determine which child bin the node will be propagated to. This is determined according to the distance between the node's position and the centers of the two new child bins as illustrated by the following example.

Example 5.1

Assume a bin B being partitioned into child bins U and L as shown in Figure 5.5. Given a number of external nodes a, b, c, and d that are connected to nodes in block B via a number of nets, it is necessary to determine to which block, U or L, where these nodes should propagate. The distances between each cell location and the centers of U and L are calculated, and each cell is propagated to the closest bin. For example, node a propagates to U; node b is not propagated, or propagated to both bins, since it is equally close to both U and L; node c is propagated to L; and node d is propagated to L.

Another possible terminal propagation case occurs when the nodes of a net are propagated as terminal to both child bins. In this case, the net will be cut anyway, and consequently it is labeled *inessential* and ignored during partitioning [35].

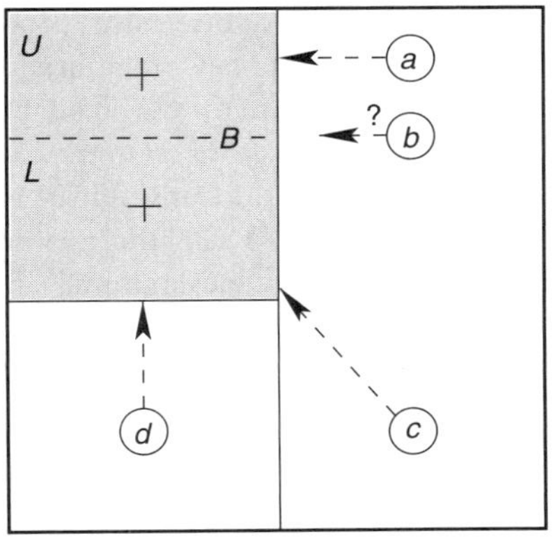

FIGURE 5.5 Terminal propagation. Bin B is being partioned into two bins U and L. Cell a is propagated to U. Cell b is not progated at all. Cell c is propagated to L. Cell d is propagated to L.

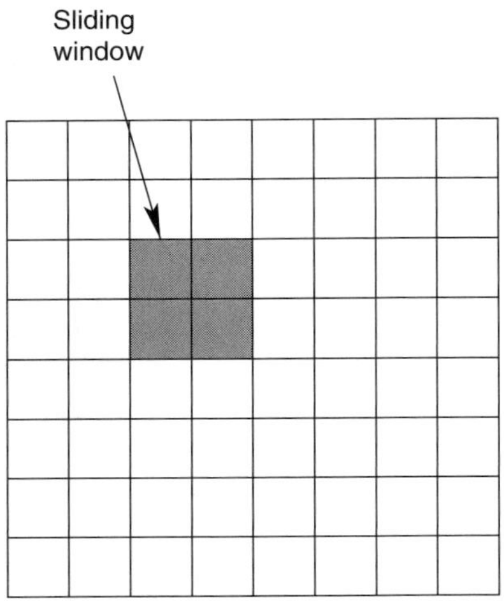

FIGURE 5.6 A 2 × 2 sliding window over a grid of 8 × 8.

Since the partitioning results of one bin affect the terminal propagation decisions of that bin cells to other bins, a placement level result depends on the order of bin processing. Furthermore, there might exist cyclic dependencies between the different bins [36,37]. For example, one bin B_i results might affect another bin B_j results, with B_j results affecting B_i results in turn. To improve placement results and alleviate these effects, a number of approaches suggest iterating the partitioning process of a placement level until a certain stopping criterion is satisfied [36,37]. It is also possible to use a *sliding window* as shown in Figure 5.6, where at each position of the sliding window, the bins that underlie the window are merged and repartitioned [33]. Since the window positions overlap in their movement, it is possible to move cells out of their assigned bins to improve wirelength.

It is also possible to capture global connectivity in a more direct manner by driving the partitioner to improve the global exact HPWL or MST objective [36,38]. In this case, the score of moving a node across a partitioning line is the exact amount of reduction in placement HPWL due to such a move. These exact approaches, however, incur an increased amount of runtime and are not typically used in practice.

5.2.2 Simulated Annealing-Based Placers

Simulated annealing is a successful generic combinatorial optimization technique that has its origins in statistical mechanics [23]. Given a combinatorial optimization problem, simulated annealing starts from an initial configuration, and iteratively moves to new configurations until some stopping criterion is satisfied. In each iteration, a random transition from the current configuration to a new configuration is generated. If the new configuration yields an improvement in the cost function or the objective being optimized then the transition is accepted; otherwise the transition is probabilistically rejected, where the rejection probability increases as iterations unfold. The ability to transit to new configurations with worse cost value allows simulated annealing to escape local minima, and is the essential ingredient to its success.

In placement, a new configuration can be generated by either (1) displacing a cell to a new location, (2) swapping two cells, or (3) changing a cell orientation [24,25]. If the change in cost function, ΔC, due to a transition from the current configuration to a new configuration is greater than or equal to zero, i.e., deteriorating a minimization objective, then the rejection probability is given by $1 - e^{-\frac{\Delta C}{T}}$, where T, commonly referred to as the *temperature*, is a parameter that controls the rejection probability. To control the value of T, a *cooling scheduling* is set up, where T is initially set to a high value and then gradually decreased so that transitions to configurations of worse cost value are less likely. At the end of simulated annealing,

$T = 0$, and the algorithm reduces to a greedy improvement iterator which stops when there is no possible further improvement in the solution.

To obtain solutions with good cost values, the cooling scheduling has to be slow. Consequently, simulated annealing typically suffers from excessive runtimes. To improve the runtime, it is possible to cluster or condense the netlist before the start of the simulated annealing process [26]. A three-stage approach to solve the placement problem can be as follows: (1) the netlist is condensed by clustering into a suitable size, (2) simulated annealing is applied on the condensed netlist, and (3) the netlist is unclustered and simulated annealing is applied on the flattened netlist for refinement.

Simulated annealing can be also used in combination with other placement techniques as an improvement operator. For example, the min-cut placer Dragon [21] uses simulated annealing to improve its results.

5.2.3 Analytical Placers

Since their introduction more than three decades ago [4], analytical methods have become one of the most successful techniques in solving the placement problem. This success can be attributed to a number of factors including: (1) the ability to capture mathematically the placement problem in a concise set of equations; (2) the availability of efficient numerical solvers that are able to solve the mathematical formulation in practical runtimes; and (3) the possibility to include analytical solvers within top-down frameworks to produce quality solutions. In this subsection, we survey the major analytical approaches suggested to solve the placement problem.

Most analytical placers typically start by converting a given hypergraph to a graph. Such conversion allows a more convenient mathematical formulation. A *k-pin* hyperedge can be converted into a k clique or a $k + 1$ star using weighted two-pin edges as shown in Figure 5.7. The problem of total HPWL minimization of a graph can be formulated as follows.

Given n objects, let c_{ij} denote the number, or weight, of connections between a pair of nodes v_i and v_j. The placement problem can be solved by minimizing the following equation:

$$z_1 = \frac{1}{2} \sum_{i=1} \sum_{j=1} c_{ij}(|x_i - x_j| + |y_i - y_j|) \tag{5.2}$$

under the constraint that no cells overlap. Since the absolute distance $|x_i - x_j| + |y_i - y_j|$ in Equation (5.2) is mathematically inconvenient, it is possible to approximate it by a number of methods. A possible approximation is to replace it by a quadratic term that leads to minimizing the following objective:

$$z_1 = \frac{1}{2} \sum_{i=1} \sum_{j=1} c_{ij}((x_i - x_j)^2 + (y_i - y_j)^2) \tag{5.3}$$

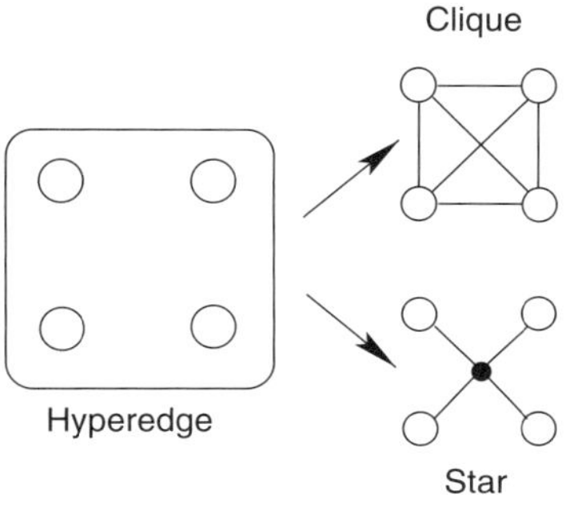

FIGURE 5.7 Net models are used to convert hyperedges into two-pin edges using either cliques or stars.

which minimizes the total squared edge length measured in Euclidean geometry. z_2 is a continuously differentiable convex function and thus can be minimized by equating its derivative to zero, which reduces to solving a system of linear equations. The approximation of the linear objective by a quadratic one will likely lead to an increase in the total wirelength. As a better approximation, it is possible to use the following objective:

$$z_G = \frac{1}{2}\sum_{i=1}\sum_{j=1} c_{ij}\left(\frac{(x_i - x_j)^2}{|x_i - x_j|} + \frac{(y_i - y_j)^2}{|y_i - y_j|}\right) = \frac{1}{2}\sum_{i=1}\sum_{j=1} c_{ij}\left(\frac{(x_i - x_j)^2}{\hat{x}_{ij}} + \frac{(y_i - y_j)^2}{\hat{y}_{ij}}\right) \tag{5.4}$$

where $\hat{x}_{ij} = |x_i - x_j|$, and $\hat{y}_{ij}\, |y_i - y_j|$ as proposed in the GORDIANL placer [39]. Since numerical solvers typically operate in an iterative manner, it is possible to incoporate $\hat{x}_{ij}$ and $\hat{y}_{ij}$ by calculating their values from the results of one iteration and using them as constants in the next iteration. A recent approximation proposed by Naylor [12] and advocated by Kahng and Wang [11] in their APlace placer, skips the hypergraph to graph conversion step and directly approximates Equation (5.1) as follows:

$$z_A = \alpha\sum_{e_i\in E}\left(\ln\Big(\sum_{v_j\in e_i} e^{x_j/\alpha}\Big) + \ln\Big(\sum_{v_j\in e_i} e^{-x_j/\alpha}\Big) + \ln\Big(\sum_{v_j\in e_i} e^{y_j/\alpha}\Big) + \ln\Big(\sum_{v_j\in e_i} e^{-y_j/\alpha}\Big)\right) \tag{5.5}$$

where α is defined as a "smoothing parameter." More accurate nonlinear formulations based on substituting the quadratic terms by higher-degree terms are also possible, but not as computationally efficient as Equation (5.3) or Equation (5.5) [10]. Note that the presence of fixed components, e.g., I/O pads, prevents the trivial solution $x_i = y_i = 0$ from being a possible minimum for the previous equations. Equation (5.3) can also be written in a matrix formulation as follows. Let C denote a connection matrix, where c_{ij} gives the number of connections between nodes v_i and v_j. Define a diagonal matrix D as follows: $d_{ij} = 0$ if $i \neq j$ else if $i = j$ then $d_{ii} = \Sigma_{j=1}^{n} c_{ij}$. In addition, define the matrix $B = D - C$, and the row vectors $X^{\mathrm{T}} = (x_1, x_2, \ldots, x_n)$ and $Y^{\mathrm{T}} = (y_1, y_2, \ldots, y_n)$. Using matrix B, Equation (5.3) can be rewritten as

$$z_2 = X^{\mathrm{T}} BX + Y^{\mathrm{T}} BY \tag{5.6}$$

For illustration of the possible outcome of just numerically solving the analytical objectives, we plot the placement results of a 400 cell netlist in a 20 × 20 grid layout in Figure 5.8 solved using the quadratic objective of Equation (5.3), or equivalently Equation (5.6). It is quite clear that just solving the previous

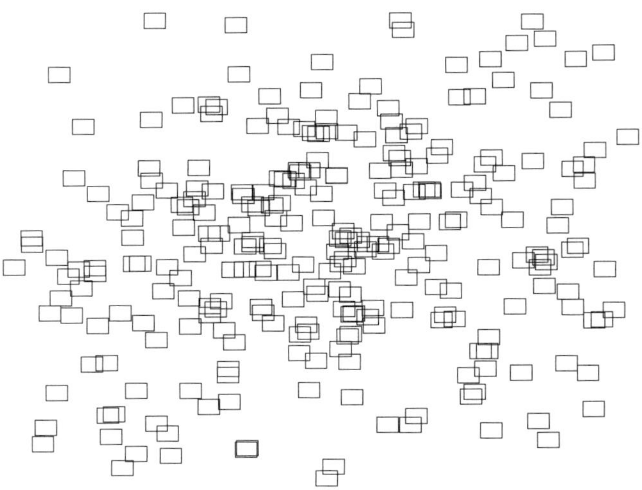

FIGURE 5.8 Results of analytical solvers are typically overlapping as shown by this 400 cell example solved using the quadratic objective.

equations yields quite an overlapping placement. Research in analytical techniques mainly focuses on how to numerically minimize either of the previous equations, while spreading out circuit components with minimum component overlap.

5.2.3.1 Spreading Out Circuit Components

There are two possible strategies in removing component overlap in analytical methods. In the first strategy, the overlap constraint is ignored while solving the previous objectives, leading to a highly overlapped placement. This overlap is then removed by another means. We refer to this strategy as "Remove after Solving." In the second strategy, a measure of overlap is incorporated into the previous objectives while minimizing them. This incorporation can be done via an additional set of terms to minimize or through a number of linear constraints. Thus the outcome solution is minimized with respect to both wirelength and overlaps. We refer to this strategy as "Remove while Solving." We next discuss a number of methods that can be used to implement both strategies.

5.2.3.1.1 Remove after Solving.

Perhaps one of the simplest methods to remove overlaps — assuming all cells have the same width — is to move the cells from their placement locations to legal placement sites with the minimum total displacement. This can be solved optimally in polynomial time using optimal linear assignment, where the cost of assigning a cell to a placement site is equal to the Manhattan distance between the cell location and the site position. Such an approach requires a runtime of $O(n^3)$ which is deemed expensive, and typically does not yield good solutions if the placement is highly overlapped.

A better approach for overlap removal is through the use of the top-down framework in a manner similar to min-cut placement. Recall that in top-down placement, the layout area is recursively divided by means of bisection or quadrisection, into finer and finer bins, where the min-cut partitioner takes the role of determining the cells that belong to each bin. This approach can be similarly used for analytical placement, where the solution of the analytical program rather than the min-cut partitioner is used to determine the cells that belong to each bin. We next explore a number of methods where analytical placement is used to partition the set of cells among the bin's child bins.

In setting up the analytical program of a subset of cells belonging to a particular bin, the placer is faced with the problem of handling the connections between those cells and the components that lie outside the bin, e.g., other cells or I/O pads. To solve this problem, cells outside the bin can be propagated as I/O pads to the periphery of the bin in a manner resembling that of terminal propagation in min-cut placement [6,7]. Such propagation, or *splitting*, of the connection mitigates the mismatch between quadratic and linear wirelength objectives since it leads to shorter connections from the fixed cells to the movable cells.

Once the analytical program is solved, the cells are partitioned among the child bins — produced from either bisection or quadrisection — using the spatial information provided from the analytical solution. In case of bisection, one possibility is to divide the bin into two equal child bins in a direction perpendicular to its longest side. The cell are then sorted along the direction of the longest side and the median cell is identified. All cells before and including the median cell are assigned to the closest child bin, while the other cells are assigned to the other child bin [6]. In case of quadrisection, Vygen [7] has devised an almost optimum linear-time method that reassigns cells to child bins to satisfy their capacities with the minimum total displacement from the analytical solution spatial results.

While the top-down framework offers a good method to remove overlap, it has the drawback of restricting cell movement since cells remain confined within their assigned bin boundaries. This restriction may adversely impact the wirelength. To allow cells to move from their bins, there are two possible solutions. First, it is possible to incorporate the top-down partitioning results "softly" as follows. Instead of solving the analytical placement of each bin separately, the analytical program of the whole netlist can be resolved, while imposing a number of additional linear constraints that coerce the center of gravity of a bin's cells to coincide with the center of the bin they belong to [13]. The additional constraints allow relative flexibility in cell movement while maintaining the global spatial cell distribution. Second, it is possible to use repartitioning via a sliding window moving over all bins in a manner similar to min-cut placement [7,36]. At each position of the sliding window, the bins that lie under the window are merged, analytically resolved, and repartitioned. The overlap in the sliding window positions allows cells to move out of their assigned bins to improve wirelength. While both previous solutions offer a good method to improve the wirelength, they incur an increase in placement runtime.

Another possible improvement to top-down analytical methods is to equip the analytical solver with a min-cut partitioner to improve its results [13]. In this latter scenario, the analytical solution can be thought of as a "seed" for the subsequent min-cut partitioner.

5.2.3.1.2 *Remove while Solving.*

In this approach, the wirelength objective is replaced by an objective that simultaneously minimizes wirelength and overlap. The analytical program is then solved in an iterative manner where every *iteration* yields a better spread placement than the previous iteration.

One possibility to achieve this objective is through the use of a set of *repelling forces* that push cells from high-density regions to low-density ones [8,14]. These forces are calculated and added to the wirelength minimization function in an iterative manner as follows. After solving Equation (5.3), the additional forces are calculated for each cell and inserted as linear terms in the quadratic program of Equation (5.3). The quadratic program is then resolved resulting in better-spread placement. This constitutes one *transformation iteration* that can be iterated a number of times until a desired even cell distribution is attained.

A recent approach that has yielded fast runtimes utilizes the concept of spreading forces in a different manner [40]. In each iteration, the utilization of all bins or regions is first calculated, and cells are shifted from bins of high utilization to new locations in bins of low utilization. To encourage the placer to keep the cells close to their newly assigned locations, spreading forces are applied to the shifted cells. These forces are realized via the addition of pseudo-nets connecting every shifted cell to the boundary of the layout region along the direction of its spreading force.

Recently, Naylor [12] proposed to divide the layout area into a grid and to minimize the uneven cell distribution in the grid to achieve uniform cell spreading. This can be achieved by adding a differentiable penalty function as a weighted term to the wirelength minimization function. The penalty function decreases in value when cells are spread more uniformly. Thus, minimizing the analytical objective function simultaneously minimizes both wirelength and overlaps.

5.2.3.2 Nonlinear Numerical Optimization

Minimization of Equation (5.3), Equation (5.4), or Equation (5.5) is carried out using numerical techniques [41]. Placement researchers have typically used numerical optimization methods as "black boxes" to provide the required solutions. We next give a glimpse of some of the possible techniques; we refer the interested reader to [41].

One of the first methods suggested to solve Equation (5.6) is through the use of eigenvalues [4]. However, eigenvalue calculation is typically slow. Instead, Cheng and Kuh [5] noticed that matrix B is sparse since a circuit component is typically connected to very few components, and thus sparse-matrix techniques represent a runtime-efficient method to solve the analytical program. Later, Tsay et al. [6] suggested the use of a generalized Gauss–Seidel method, called successive over-relaxation, to solve the set of linear equations.

Given that nonlinear programs described by Equation (5.3), Equation (5.4), or Equation (5.5) are convex — since matrix B of Equation (5.6) is positive- semidefinite — and typically sparse, gradient methods offer a fast method to find the local minimum. Given a convex function $f(\mathbf{x})$ that has a minimum at $\mathbf{x}^*$, gradient methods find a sequence of solutions $\mathbf{x}_0,\mathbf{x}_1,\mathbf{x}_2, \ldots ,\mathbf{x}_k = \mathbf{x}^*$, where each solution better approximates the minimum. One method is to proceed in the direction of *steepest descent*, which is in the direction of the gradient $\nabla f(\mathbf{x})$. Thus $\mathbf{x}_i = \mathbf{x}_{i-1}+\lambda_{i-1}\nabla f(\mathbf{x}_{i-1})$, where $\lambda_{i-1} < 0$ is the step that minimizes the function $f(x_{i-1}+\lambda\nabla f(\mathbf{x}_{i-1}))$ with respect to λ. The main disadvantage of steepest descent methods is that they might take an unnecessarily long time to converge to the optimal solution.

To improve the convergence rate, the Conjugate Gradient method is typically used [13,41]. Assume that $\mathbf{x}_0$ is the first approximation to $\mathbf{x}^*$, and define $\mathbf{v}_0=\nabla f(\mathbf{x}_0)$. Conjugate-gradient proceeds by executing the following n times:

1. Calculate $\mathbf{x}_i = \mathbf{x}_{i-1} + \lambda_{i-1}\mathbf{v}_{i-1}$, where again $\lambda_{i-1} < 0$ is the step that minimizes the function $f(x_{i-1} + \lambda\mathbf{v}_{i-1})$
2. Calculate $\mathbf{v}_i = -\nabla f(\mathbf{x}_i) + \dfrac{\|\nabla f(\mathbf{x}_i)\|^2}{\|\nabla f(\mathbf{x}_{i-1})\|^2}\,\mathbf{v}_{i-1}$
3. Increment i

It can be proven that the vectors $\mathbf{v}_i$ are mutually conjugate, i.e., $\mathbf{v}_i^{\mathrm{T}} B\mathbf{v} = 0$, and that the CG method offers a fast convergence rate to the optimal solution in comparison with the regular steepest descent method.

5.2.4 Handling Other Objectives

In addition to total wirelength minimization, placers also have to minimize a number of additional objectives such as the longest path delay, congestion and power. We discuss next how to handle these objectives within the different placement solvers discussed earlier.

5.2.4.1 Timing-Driven Placement

Circuit signal paths start at the input pads and flip–flip outputs, and end at the output pads and flip–flop inputs. The delay of a path is the sum of interconnect and gate delays that make up the path. *Critical paths* are all signal paths of delay larger than the specified maximum delay, which is typically the clock period. By controlling the proximity of interconnected components so that no signal path exceeds the specified timing constraint, timing-driven placement can achieve its goal of eliminating all critical paths if feasible.

Timing-driven placement methods depend on the underlying placement solver. Analytical methods can incorporate timing constraints in a number of ways. One way is to formulate the delay of critical paths as a number of linear constraints and include these constraints in the analytical placement program [42]. The linear constraints eliminate placement solutions that do not satisfy the timing constraints, and consequently solving the constrained analytical program automatically leads — if feasible — to solutions with no critical paths. Another way is to multiply the net connectivity coefficients, c_{ip}, in Equation (5.2) and Equation (5.3) by additional weights to reflect their timing criticality [8,43]. Nets included in critical paths receive higher weights than other nets, causing the analytical placer to reduce their length. Linear timing constraints can be indirectly incorporated in a recursive top-down paradigm — whether using a min-cut or analytical solver — as follows [44]. After each placement level, a linear program is constructed where the objective is to minimize the reassignment of cells to bins such that all timing delay constraints are satisfied. Solving the linear program leads to a minimum perturbation from an existing illegal-timing placement to a legal-timing placement.

One possibility to achieve timing-driven placement in min-cut placers is to perform static timing analysis at each placement level and then translate timing slacks to net weights, where nets participating in critical paths receive larger weights [45]. Another solution seeks to control the number of cuts experienced by a critical path, since the more cuts a path experiences, the longer the path tends to be [46]. To control the number of cuts, it is possible to give larger weights to critical nets and to impose an upper bound on the maximum number of times a path can be cut. These weights bias the min-cut partitioner to avoid cutting critical nets eventually reducing critical paths delay. Another recent method prioritizes cells that belong to critical paths [47]. Given a bin under partition, cells that belong to critical paths are preassigned and fixed to the child bins so as to reduce the negative slack of critical paths. After this cell preassignment, the hypergraph partitioner is invoked on the remaining cells.

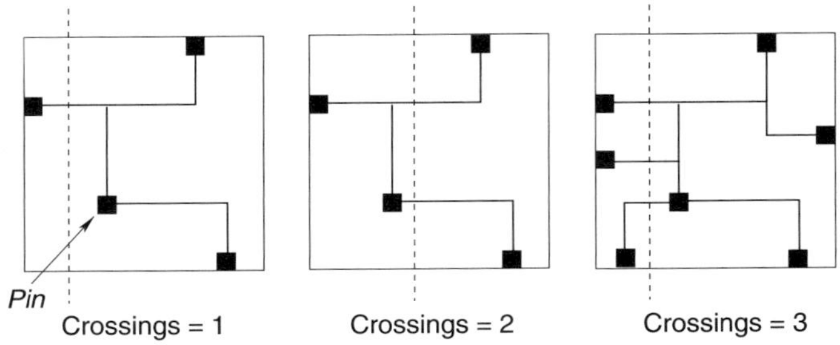

FIGURE 5.9 Number of wire crossings depends on cut location and number of pins [14].

5.2.4.2 Congestion-Driven Placement

Minimizing total wirelength ensures meeting the total routing supply. Nevertheless, congestion might occur in localized regions where local routing demand exceeds local routing supply. Congestion impacts routability in two ways. First, congestion might lead to the failure of routing some nets compromising the integrity of the complete physical design stage. Second, congestion forces routers to detour nets away from congested regions, leading to an increase in their wirelength. This increase in wirelength adversely impacts performance if the detoured nets are part of the critical paths.

Given a placement, congestion–reduction methods typically first divide the layout area into rectangular regions. A fast global routing method is then used to estimate the number of wires crossing each region boundary. This estimate is compared against each region's routing supply. If the demand exceeds the supply then the region is either expanded to increase the supply or cells are moved out of the region to reduce the demand. Congestion–reduction methods can be applied during placement or after placement as a postprocessing step. We next discuss briefly fast routing estimates and a number of methods to reduce the congestion.

Routing-demand estimation methods are numerous and there is no space to discuss them in detail. However, we discuss an attractive routing demand model that was developed from a placement perspective [2]. Cheng [2] observed that the number of wire crossings of a cut line through the bounding box of a net depends on the number of pins of the net as well as the cut line location as shown in Figure 5.9. Given a k-pin net, it is possible to calculate the average number of wire crossings at any location inside the net bounding box as follows: (1) place the k pins of the net randomly; (2) construct a Steiner minimum tree over the k pins and calculate the number of wire crossings at all possible locations; and (3) execute (1) and (2) a number of times and report the average results. The average results for different values of k are calculated once and stored in a lookup table. The table is then used for fast routing estimation for all future placement runs.

As for congestion reduction methods, one simple way to reduce the demand within a congested region is to *inflate* the cells within the region, and utilize repartitioning (see min-cut placers) — originally used to improve wirelength — to move cells out of the congested regions [48]. For example, using a sliding window of size 2×2 as shown in Figure 5.6 enables the cells to leave the regions they are currently placed in, thus reducing their congestion.

Alternatively, to increase the supply of a region, Yang et al. [49] consider two possible expansion areas for each region. Since expanded areas of different congested regions can overlap creating new congested regions, an integer linear program is constructed that finds an expansion of all regions that minimizes the maximum congested region. After all congested regions are expanded, cells within each region are placed within the region boundaries using a detailed placer. Another approach [50] expands regions during quadratic placement either vertically or horizontally depending on the demand.

5.2.4.3 Reducing Placement Runtime

Circuit clustering is an attractive solution to manage the runtime and quality of placement results on a large scale VLSI designs [26,33,51–54]. Clustering takes an input netlist and coarsens or condenses it by

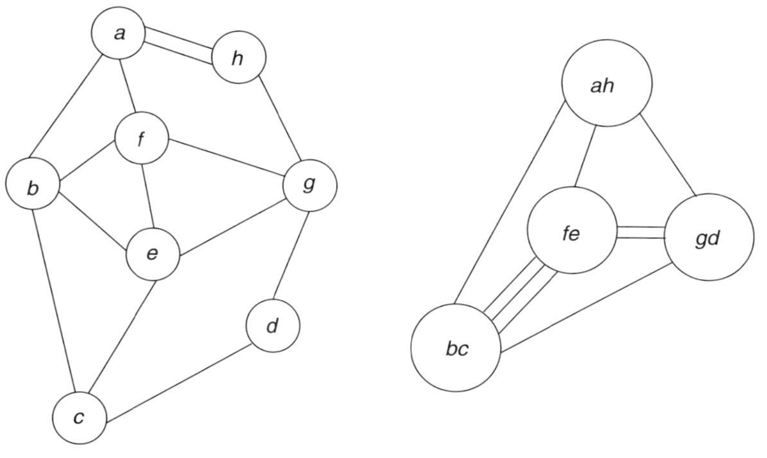

FIGURE 5.10 Clustering of netlist: (a) produces netlist; (b) reducing the number of nodes from 8 to 4 and the number of edges from 14 to 9.

merging nodes together to form larger nodes or clusters, and adjusts the nets or hyperedges accordingly as shown in Figure 5.10. Clustering methods have a long history and they are out of the scope of the discussion. We will only focus on the use of clustering in placement.

The interaction between clustering and placement can be classified into two categories. In the first category, clustering is used as a core part of the placement solver [20–22], such as clustering and unclustering within the multilevel partitioner of min-cut placers [19]. In this case, a cluster hierarchy is first generated, followed by a sequence of partitioning and unclustering. Partitioning results of prior clustered netlists are projected to the next level by unclustering, which become the seed for the subsequent partitioning. In the second category of clustering for VLSI placement, the cluster hierarchy is generated at the beginning of the placement as a preprocessing step in order to reduce the netlist size. The coarsened netlist is then passed on to the placer [26,53,54]. Usually, the clustered objects will be dissolved at or near the end of the placement process [54], with a "clean-up" operation applied to the fully uncoarsened netlist to improve the results. In some cases, unclustering and reclustering are executed during intermediate points in the placement flow to allow the placer escape any earlier bad clustering decisions [26,55].

5.3 Detailed Placement and Legalizers

A placement is illegal if cells/blocks overlap and/or occupy illegal sites, e.g., between placement rows. Illegal placements might be an outcome of global (twice) placement, or produced from *incremental* changes to legal placements. Such changes include cell resizing and buffer insertion, which are necessary for signal integrity. Figure 5.11 shows a possible outcome from global placement where cells are quite spread out, yet the placement is illegal since there is a small amount of overlap and some cells are not snapped to sites. A placement *legalizer* snaps cells to the sites of rows such that no cells overlap. This has to be done with minimum adverse impact on the quality of the placement. A *detailed* placer takes a legal placement and improves some placement objective such as wirelength or congestion, and produces a new legal placement. In many cases, a detailed placer includes a legalizer that is invoked to legalize any given illegal placement or to legalize the detailed placer's own placement. Detailed placers are also used in *incremental placement* to legalize the placement after netlist changes.

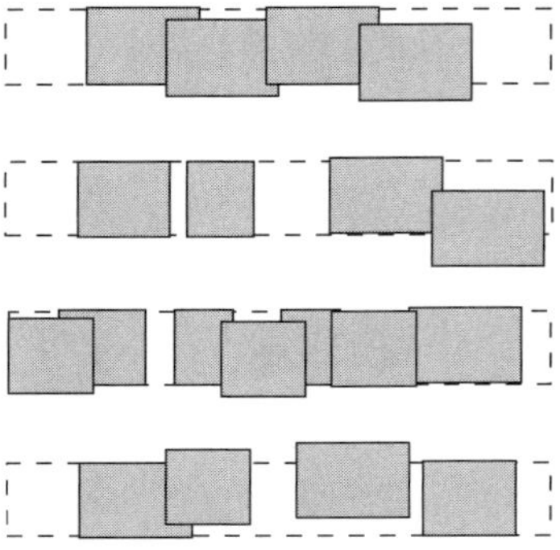

FIGURE 5.11 A possible outcome of global placement. A legalizer legalizes a placement by snapping cells to sites with no overlap.

In this case, they are called *Engineering Change Order* placers. We can summarize the goals of a detailed placer/legalizer as follows:

1. Remove all overlaps, and snap cells to sites with the minimum impact on wirelength, timing, and congestion.
2. Improve wirelength by reordering groups of cells. Such groups typically include spatially close cells.
3. Improve routability by carefully redistributing free sites.

We classify detailed placement methods as either heuristic or exact. Heuristic methods typically achieve good results in fast runtimes. Exact methods are only applicable for a few cases and usually take longer runtimes.

5.3.1 Heuristic Techniques

A possible legalizing approach starts by dividing the layout area into a number of regions [56]. Regions are then traversed from left to right, where the cells of each region are legalized and placed after preceding regions. During the course of legalization, a *borderline* can be envisioned where cells of regions to its left are legalized, but cells of regions to its right are yet to be legalized. A region is legalized by assigning its cells to new positions by solving a transportation problem, where cells are transported to their new legal positions with the minimum impact to wirelength.

A relatively recent approach that also divides the layout into regions and balances region utilization, is due to Hur and Lillis [57]. In their approach, an overutilized region, say *S*, and its nearest underutilized region, say *T*, are identified, and cells are ripple-moved from *S* to *T* along a *monotone path*, i.e., a sequence of adjacent bins. Since there are possibly many paths between *S* and *T* and many cells to move, the sequence of moves that minimizes the increase in wirelength is identified and executed. In addition, if row-size constraints are not met, then the utilized bin from the largest row, say *S*, and the most least utilized bin from the smallest row, say *T*, are identified, and cells are again ripple-moved along a monotone path from *S* to *T*.

A recent detailed placement approach divides the legalization effort into two stages [7,58,59]. In the first stage, the layout area is divided into *zones*. Since some of these zones might be over capacity and consequently produce overlaps, the detailed placer first balances these zones. Such balancing can be achieved while minimizing the total displacement using network flow programs. After all zones are balanced, cells are legalized within rows and free sites are distributed to further minimize the HPWL.

A simple greedy legalization heuristic that empirically proved useful is as follows [34,60]. Initially all cells are sorted by their horizontal coordinate. Cells are then visited in that order and assigned to the location which results in the minimal displacement from their current location, such that the row capacity is not exceeded. The heuristic can also be slightly modified to handle mixed cell and block legalization.

5.3.2 Exact Techniques

As for exact techniques in detailed placement, optimal end-case placers can use branch and bound algorithms to place optimally a set of few cells within a row [35]. The algorithm can be suboptimally scaled by sliding a rectangular *window* over the layout area, and placing cells within it optimally [20]. Sliding over the layout area can be iterated a number of times until improvement drops below a certain threshold. Branch and bound techniques are typically plagued with excessive runtime requirements, and window sizes are usually restricted to 7–8 cells to keep runtime manageable. However, if the subset of cells selected are *independent*, i.e., they do not share any common nets, and are of the same size, then it is possible to place them optimally in polynomial runtime using OLA [3,61].

Another detailed placement problem that is getting increasingly important due to the ubiquitous presence of free sites in most modern designs is as follows. Given a number of cells placed in a placement row, shift the cells to minimize the wirelength, such that the ordering of cells before and after the shifting is the same [62]. Kahng et al. [62] propose a solution that solves the problem optimally in polynomial time using dynamic programming. The dynamic programming algorithm uses the piecewise-linearity attribute of HPWL to distribute optimally the freespace for wirelength minimization purposes. A correction of the complexity bounds, and further speed up through efficient data structures are realized by Brenner and Vygen [63]. Recently, Kahng et al. [64] proposed a generic dynamic programming method to remove overlaps within a row given a fixed cell ordering while optimally minimizing either HPWL, total cell displacement, or maximum cell displacement.

5.4 Placement Trends

Technological advances introduce challenging extensions to the placement problem. For example, the recent proliferation of Intellectual Property cores transformed the placement problem from just cell placement to mixed cell and core, or block, placement. Additionally, once these blocks are placed, they represent a *blockage* where routing cannot be conducted on layers on top of them. Such blockage imposes additional challenges to congestion-driven placement methods.

Increasing system complexity of recent designs coupled with diminishing feature sizes created further demand for routing resources. This increased demand forced physical design engineers to expand layout areas to be more than the total components area creating *free space* for additional routing resources. Such *free* or *white* space must be distributed by the placer in a manner that improves the design's performance.

Another active research area in placement is that of placer *suboptimality evaluation.* Sub-optimality evaluation enables the quantification of existing placers' performance on arbitrary benchmarks. Essentially, suboptimality evaluation reveals how much room for improvement still exists, which guides the continued investment of attention to the placement problem. We next examine a number of recent solutions for the aforementioned placement challenges, starting with mixed-size placement.

5.4.1 Mixed-Size Placement

Rather than placing blocks before placement either manually or using automatic floorplanning, mixed-size placement simultaneously places cells and blocks, while considering a number of fixed blocks. A number of solutions have been recently proposed.

One solution clusters the cells in a bottom–up fashion condensing the netlist into blocks and clusters of cells [65]. Quadratic placement in a top–down partitioning framework is then used to globally place the netlist. After global placement, clusters are flattened and their cells are placed without overlap.

Instead of clustering cells, Adya and Markov [66] suggest "shredding" every block into a number of connected cells and then placing the shredded blocks and cells. Afterwards, the initial locations of the blocks are calculated by averaging the locations of the cells created during the shredding process, and the cells are clustered into soft blocks (soft blocks have adjustable aspect ratio) based on their spatial locations in the placement. A fixed outline floorplanner is then used to place both the blocks and the soft blocks without overlaps. Finally, blocks are fixed into place and detailed placers locate positions for the cells of each soft block.

A new simple approach that has demonstrated strong empirical results treats cells and blocks transparently in min-cut placement [67]. That is, the min-cut placer does not distinguish between cells and blocks. The output of min-cut placement in this case typically contains overlap and thus it is necessary to use a legalizer.

5.4.2 Whitespace Distribution

Whitespace or *freespace* is the percentage of placement sites not occupied by cells and blocks. Whitespace enlarges the core layout area more than necessary for placement, in order to provide larger routing area. Placement algorithms can allocate whitespace to improve performance in a number of ways including congestion reduction, overlap minimization, and timing improvement.

One of the first approaches sought whitespace allocation in a hierarchical uniform fashion to improve smooth operation of the min-cut partitioner [20,31]. With careful whitespace allocation, bins would have enough whitespace to avoid cell overlaps resulting from unbalanced partitioning. Such overlaps typically occur in min-cut placers due to lack of whitespace, variations in cell dimensions, as well as the choice of cut sequences [64].

Instead of allocating whitespace uniformly, it can be allocated according to congestion to improve the final routability of the placement. In a two-step approach [68], whitespace is first allocated to each placement row proportional to its degree of congestion, with the least congested row receiving a fixed minimum amount of whitespace. In the second step, rows are divided into bins, and bins are assigned whitespace proportional to their congestion, where a bin can be assigned zero whitespace if it is not congested. More recently, Li et al. [69] improved routability by simultaneously migrating cells connected to nets responsible for the overflow over routing demand, as well as by redistributing whitespace, allocating more whitespace to regions with a deficit in routing supplies.

When large amounts of whitespace is present, it can also be allocated to improve timing [70]. By observing that analytical placers better capture the global placement view than min-cut placers, it is possible to utilize analytical solvers in distributing the whitespace by determining the balance constraints for each partitioning step. Empirical results show that such methods lead to better timing results. In addition, whitespace can be compacted by inserting disconnected *free cells*, only to remove them after placement [71].

5.4.3 Placement Benchmarking

Placement suboptimality quantification is concerned with estimating the performance gap between reported results e.g., wirelength, of placers on a given benchmark and the optimal results of the benchmark. Since the placement problem is NP-hard, it is unlikely that there will exist optimal algorithms for arbitrary instances. Thus researchers attempt to indirectly estimate the performance gap. Existing approaches either use scaling [72] or synthetic constructive benchmarks [73,74] to quantify the suboptimality of existing placers.

Hagen et al. [72] scale a given placement instance by creating a number of identical instances and "patching" them into a single larger scaled instance. The performance of the placer is then compared against a value precalculated from the initial instance.

A recent paper by Chang et al. [73] uses an overlooked construction method by Hagen et al. [72] to construct optimally a number of benchmarks (PEKO) with known optimal HPWL. Such optimal construction is possible if each *k*-pin net is constructed with the least possible HPWL. For example, a 2-pin net takes a HPWL of 1 unit, a 3-pin net takes a HPWL of 2 units, a 4-pin net takes a HPWL of 2 units, and so forth. A sample construction is given in Figure 5.12. Empirical results on such artificially constructed benchmarks show that there exists a significant gap between the HPWL from the placer results and the optimal placement HPWL. However, these benchmarks are unrealistic since they only consider local hyperedges. To mitigate this drawback, Cong et al. [74] added global hyperedges to the PEKO benchmarks producing the PEKU benchmarks. The optimal placement of the PEKU benchmarks is unknown; however, it is possible to calculate an upper bound to their optimal placement. By using the optimal placement of the

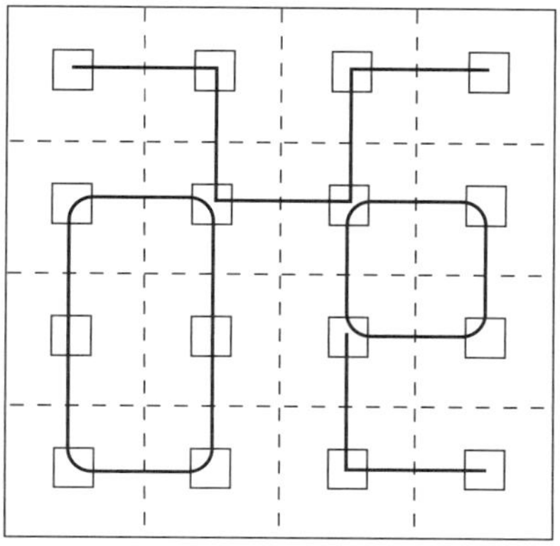

FIGURE 5.12 A benchmark constructed with known optimal HPWL. There are seven 2-pin edges, one 4-pin hyperedge, and one 6-pin hyperedge with a total HPWL of 12.

original PEKO benchmarks, an upper bound to the optimal HPWL of the added global hyperedges is readily calculated. Experimental results on the PEKU benchmarks show that the performance of available placers approaches the upper bounds as the percentage of global edges increases. It is not known whether the calculated upper bounds are tight or loose.

Another recent approach quantifies placer suboptimality using arbitrary netlists [75]. Given a netlist and its placement, a number of *zero-change* netlist transformations are applied to the netlist. These transformations do not alter the placement HPWL and can only increase the unknown optimal HPWL value. Executing the placer on the new transformed netlist produces a new suboptimal HPWL result. The difference between the new and old HPWL results is a lower bound on the suboptimality gap.

5.4.4 Other Trends and Practical Concerns

In addition to the previous trends, a number of practical issues arise in modern day designs that should be handled by placers. These include:

- Flat placement of 10 million components in mixed-mode designs. Modern designs typically require a number of placement iterations to achieve timing closure. These iterations routinely consume enormous amounts of time — a number of days — till completion. Thus it is necessary to have fast and efficient placement algorithms. Furthermore, components of modern designs include movable cells and movable and fixed blocks. All these need to be smoothly and efficiently handled by the placer.
- Given the ubiquitous presence of mobile devices, temperature- and power-driven placement where the objective is to reduce the total power consumption as well as to smooth temperature differences between spatially proximate regions, is getting increasingly important.
- Placement-driven synthesis, where placement is used to derive logic synthesis optimizations [76–78].

5.5 Academic and Industrial Placers

A number of competitive academic and industrial placers exist to satisfy academic research and commercial needs. We now give a brief overview of the main academic and industrial tools. The algorithmic techniques used by these placers have been already covered in Section 5.2. Academic placers include:

- APlace [11]: an analytical placer that is algorithmically similar to the Physical Compiler from Synopsys. It uses a nonlinear wirelength formulation (Equation [5.5]) while penalizing overlaps in a "remove while solving" fashion.
- BonnPlacer [7]: an analytical placer that uses a quadratic formulation in a top-down framework. It also uses repartitioning to improve its results.

- Capo [20]: a top-down min-cut placer that is based on recursive bisection.
- Dragon [21]: a top-down min-cut placer that is based on recursive quadrisection. The placer also uses simulated annealing to improve its results.
- FengShui [22]: a top-down min-cut placer that is based on recursive bisection.
- Kraftwerk [8]: an analytical placer that uses repelling forces in a "remove while solving" fashion.
- mPL [55]: an analytical placer that uses a nonlinear wirelength formulation, and uses clustering in a multilevel framework as a means to reduce instance sizes.
- Timberwolf [24]: a simulated annealing-based placer.

While details of academic placers are readily available in academic publication, published data on industrial placers are typically less detailed.

- IBM CPlace [70]: an analytical placer that uses the quadratic formulation in a top-down framework. It also distributes whitespace to improve timing and wirelength.
- Cadence QPlace: it is believed that QPlace uses the quadratic formulation in a top-down framework.
- Synopsys Physical Compiler [12]: similar to APlace.
- InternetCAD iTools: a commercial package based on Timberwolf.

5.6 Conclusions

Placement is — and will likely remain — one of the essential steps in the physical design part of any integrated circuit's life. In this chapter, we have surveyed the main algorithmic approaches to solve the placement problem. We have also explained how different placement objectives can be handled within those various approaches. Detailed placement as well as placement trends have also been covered. For the future, it seems that the core algorithmic approaches are likely to stay the same; however, they have to be extended to handle smoothly and efficiently multi-million gate mixed-size designs and to achieve timing closure in fewer turn-around cycles.

References

[1] V.V. Vazirani, *Approximation Algorithms*, 1st ed., Springer, Heidelberg, 2001.

[2] C.E. Cheng, RISA: accurate and efficient placement routability modeling, *Proceedings of IEEE International Conference on Computer Aided Design*, 1994, pp. 690–695.

[3] L. Steinberg, The backboard wiring problem: a placement algorithm, *SIAM Rev.*, 3, 37–50, 1961.

[4] K.M. Hall, An *r*-dimensional quadratic placement algorithm, *Manage. Sci.*, 17, 219–229, 1970.

[5] C.K. Cheng and E.S. Kuh, Module placement based on resistive network optimization, *IEEE Trans. Comput.-Aided Des. Integrated Circuits Syst.*, 4, 115–122, 1984.

[6] R.S. Tsay, E.S. Kuh, and C.P. Hsu, PROUD: a sea-of-gates placement algorithm, *International Conference on Computer Aided Design, IEEE Des. Test Comput.*, 5, 44–56, 1988.

[7] J. Vygen, Algorithms for large-scale flat placement, *Proceedings of ACM/IEEE Design Automation Conference*, 1997, pp. 746–751.

[8] H. Eisenmann and F.M. Johannes, Generic global placement and floorplanning, *Proceedings of ACM/IEEE Design Automation Conference*, 1998, pp. 269–274.

[9] B.X. Weis and D.A. Mlynski, A graph theoretical approach to the relative placement problem, *IEEE Trans. Circuits Syst.*, 35, 286–293, 1988.

[10] A.A. Kennings and I.L. Markov, Analytical minimization of half-perimeter wirelength, *Proceedings of IEEE Asia and South Pacific Design Automation Conference*, 2000, pp. 179–184.

[11] A. Kahng and Q. Wang, Implementation and extensibility of an analytical placer, *Proceedings of ACM/IEEE International Symposium on Physical Design*, 2004, pp. 18–25.

[12] W. Naylor, Non-Linear Optimization System and Method for Wirelength and Density Within an Automatic Electronic Circuit Placer, US Patent 6282693, 2001.

[13] J.M. Kleinhans, G. Sigl, F.M. Johannes, and K.J. Antreich, GORDIAN: VLSI placement by quadratic programming and slicing optimization, *IEEE Trans. Comput.-Aided Des. Integrated Circuits Syst.*, 10, 356–365, 1991.

[14] K. Vorwerk, A. Kennings, and A. Vannelli, Engineering details of a stable force-directed placer, *Proceedings of IEEE International Conference on Computer Aided Design*, 2004, pp. 573–580.

[15] B. Kernighan and S. Lin, An efficient heuristic procedure for partitioning graphs, *Bell Syst. Tech. J.*, Feb, 291–307, 1970.

[16] M.A. Breuer, Min-cut placement, *J. Des. Automat. Fault Tolerant Comput.*, 1, 343–362, 1977.

[17] C.M. Fiduccia and R.M. Mattheyses, A linear-time heuristic for improving network partitions, *Proceedings of ACM/IEEE Design Automation Conference*, 1982, pp. 175–181.

[18] A.E. Dunlop and B.W. Kernighan, A procedure for placement of standard-cell VLSI circuits, *IEEE Trans.Comput.-Aided Des. Integrated Circuits Syst.*, 4, 92–98, 1985.

[19] G. Karypis, R. Aggarwal, V. Kumar, and S. Shekhar, Multilevel hypergraph partitioning: application in VLSI domain, *Proceedings of ACM/IEEE Design Automation Conference*, 1997, pp. 526–529.

[20] A.E. Caldwell, A.B. Kahng, and I.L. Markov, Can recursive bisection alone produce routable placements? *Proceedings of ACM/IEEE Design Automation Conference*, 2000, pp. 477–482.

[21] M. Wang, X. Yang, and M. Sarrafzadeh, DRAGON2000: standard-cell placement tool for large industry circuits, *Proceedings of IEEE International Conference on Computer Aided Design*, 2001, pp. 260–263.

[22] M. Yildiz and P. Madden, Global objectives for standard-cell placement, *Proceedings of IEEE Great Lakes Symposium on VLSI*, 2001, pp. 68–72.

[23] S. Kirkpatrick, C.D. Gelatt, Jr., and M.P. Vecchi, Optimization by simulated annealing, *Science*, 220, 671–680, 1983.

[24] C. Sechen and A. Sangiovanni-Vincentelli, TimberWolf3.2: a new standard cell placement and global routing package, *Proceedings of ACM/IEEE Design Automation Conference*, 1986, pp. 432–439.

[25] C. Sechen and K.W. Lee, An improved simulated annealing algorithm for row-based placement, *Proceedings of IEEE International Conference on Computer Aided Design*, 1987, pp. 478–481.

[26] W.-J. Sun and C. Sechen, Efficient and effective placement for very large circuits, *IEEE Trans. Comput.-Aided Des. Integrated Circuits Syst.*, 14, 349–359, 1995.

[27] S. Sahni and T. Gonzalez, P-complete approximation problems, *J. ACM*, 23, 555–565, 1976.

[28] P.R. Suaris and G. Kedem, A quadrisection-based combined place and route scheme for standard cells, *IEEE Trans. Comput.-Aided Des. Integrated Circuits Syst.*, 8, 234–244, 1989.

[29] A. Caldwell, A. Kahng, and I. Markov, End-case placers for standard-cell layout, *Proceedings of ACM/IEEE International Symposium on Physical Design*, 1999, pp. 90–96.

[30] C.J. Alpert, J.H. Huang, and A.B. Kahng, Multilevel circuit partitioning, *Proceedings of ACM/IEEE Design Automation Conference*, 1997, pp. 530–533.

[31] A. Caldwell, I. Markov, and A. Kahng, Hierarchical whitespace allocation in top-down placement, *IEEE Trans. Comput.-Aided Des. Integrated Circuits Syst.*, 22, 1550–1556, 2003.

[32] M.C. Yildiz and P.H. Madden, Improved cut sequences for partitioning based placement, *Proceedings of ACM/IEEE Design Automation Conference*, 2001, pp. 776–779.

[33] G. Karypis and V. Kumar, Multilevel *k*-way hypergraph partitioning, *Proceedings of ACM/IEEE Design Automation Conference*, 1999, pp. 343–348.

[34] A. Agnihotri, M. Yildiz, A. Khatkhate, A. Mathur, S. Ono, and P. Madden, Fractional cut: improved recursive bisection placement, *Proceedings of IEEE International Conference on Computer Aided Design*, 2003.

[35] A.E. Caldwell, A.B. Kahng, and I.L. Markov, Optimal partitioners and end-case placers for standard-cell layout, *IEEE Trans. Comput.-Aided Des. Integrated Circuits Syst.*, 19, 1304–1313, 2000.

[36] D.J.-H. Huang and A.B. Kahng, Partitioning-based standard-cell global placement with an exact objective, *Proceedings of ACM/IEEE International Symposium on Physical Design*, 1997, pp. 18–25.

[37] A.B. Kahng and S. Reda, Placement feedback: a concept and method for better min-cut placement, *Proceedings of ACM/IEEE Design Automation Conference*, 2004, pp. 357–362.

[38] K. Zhong and S. Dutt, Effective partition-driven placement with simultaneous level processing and global net views, *Proceedings of IEEE International Conference on Computer Aided Design*, 2000, pp. 171–176.

[39] G. Sigl, K. Doll, and F.M. Johannes, Analytical placement: a linear or a quadratic objective function? *Proceedings of ACM/IEEE Design Automation Conference*, 1991, pp. 427–431.

[40] N. Viswanathan and C. Chu, FastPlace: efficient analytical placement using cell shifting, iterative local refinement and a hybrid net model, *Proceedings of ACM/IEEE International Symposium on Physical Design*, 2004, pp. 26–33.

[41] D. Wismer and R. Chattergy, *Introduction to Nonlinear Optimization: A Problem Solving Approach*, 1st ed., Elsevier, Amsterdam, 1978.

[42] A. Srinivasan, K. Chaudhary, and E.S. Kuh, RITUAL: A performance driven placement algorithm for small cell ICs, *Proceedings of IEEE International Conference on Computer Aided Design*, 1991, pp. 48–51.

[43] B. Riess and G. Ettelt, SPEED: fast and efficient timing driven placement, *IEEE International Symposium on Circuits and Systems*, 1995, pp. 377–380.

[44] B. Halpin, C.Y.R. Chen, and N. Sehgal, Timing driven placement using physical net constraints, *Proceedings of ACM/IEEE Design Automation Conference*, 2001, pp. 780–783.

[45] M. Marek-Sadowska and S. Lin, Timing driven placement, *Proceedings of IEEE International Conference on Computer Aided Design*, 1989, pp. 94–97.

[46] S.L. Ou and M. Pedram, Timing-driven placement based on partitioning with dynamic cut-net control, *Proceedings of ACM/IEEE Design Automation Conference*, 2000, pp. 472–476.

[47] A.B. Kahng, S. Mantik, and I.L. Markov, Min-max placement for large-scale timing optimization, *Proceedings of ACM/IEEE International Symposium on Physical Design*, 2002, pp. 143–148.

[48] U. Brenner and A. Rohe, An effective congestion driven placement framework, *Proceedings of ACM/IEEE International Symposium on Physical Design*, 2002, pp. 6–11.

[49] X. Yang, R. Kastner, and M. Sarrafzadeh, Congestion Reduction During Placement Based on Integer Programming, 2001, pp. 573–576.

[50] P.N. Parakh, R.B. Brown, and K.A. Sakallah, Congestion driven quadratic placement, *Proceedings of ACM/IEEE Design Automation Conference*, 1998, pp. 275–278.

[51] D.M. Schuler and E.G. Ulrich, Clustering and linear placement, *Proceedings of ACM/IEEE Design Automation Conference*, 1972, pp. 50–56.

[52] J. Cong and S.K. Lim, Edge separability-based circuit clustering with application to multilevel circuit partitioning, *IEEE Trans. Comput.-Aided Des. Integrated Circuits Syst.*, 23, 346–357, 2004.

[53] B. Hu and M.M. Sadowska, Fine granularity clustering-based placement, *IEEE Trans. Comput.-Aided Des. Integrated Circuits Syst.*, 23, 527–536, 2004.

[54] C. Alpert, A. Kahng, G.-J. Nam, S. Reda, and P. Villarrubia, A semi-persistent clustering technique for VLSI circuit placement, *Proceedings of ACM/IEEE International Symposium on Physical Design*, 2005, pp. 200–207.

[55] C.-C. Chang, J. Cong, D. Pan, and X. Yuan, Multilevel global placement with congestion control, *IEEE Trans. Comput.-Aided Des. Integrated Circuits Syst.*, 22, 395–409, 2003.

[56] K. Doll, F. Johannes, and K. Antreich, Iterative placement improvement by network flow methods, *IEEE Trans. Comput.-Aided Des. Integrated Circuits Syst.*, 13, 1189–1200, 1994.

[57] S.W. Hur and J. Lillis, Mongrel: hybrid techniques for standard cell placement, *Proceedings of IEEE International Conference on Computer Aided Design*, 2000, pp. 165–170.

[58] J. Vygen, Algorithms for detailed placement of standard cells, in *Design, Automation and Test in Europe*, 1998, pp. 321-324.

[59] U. Brenner, A. Pauli, and J. Vygen, Almost optimum placement legalization by minimum cost flow and dynamic programming, *Proceedings of ACM/IEEE International Symposium on Physical Design*, 2004, pp. 2–9.

[60] D. Hill, Method and System for High Speed Detailed Placement of Cells Within an Integrated Circuit Design, US Patent 6370673, 2001.

[61] S. Akers, On the use of the linear assignment algorithm in module placement, *Proceedings of ACM/IEEE Design Automation Conference*, 1981, pp. 13–144.

[62] A.B. Kahng, P. Tucker, and A. Zelikovsky, Optimization of linear placements for wirelength minimization with free sites, *Proceedings of IEEE Asia and South Pacific Design Automation Conference*, 1999, pp. 241–244.

[63] U. Brenner and J. Vygen, Faster optimal single-row placement with fixed ordering, in *Design, Automation and Test in Europe*, 2000, pp. 117–122.

[64] A.B. Kahng, I. Markov, and S. Reda, On legalization of row-based placements, *Proceedings of IEEE Great Lakes Symposium on VLSI*, 2004, pp. 214–219.

[65] H. Yu, X. Hing, and Y. Cai, MMP: a novel placement algorithm for combined macro block and standard cell layout design, *Proceedings of IEEE Asia and South Pacific Design Automation Conference*, 2000, pp. 271–276.

[66] S. Adya and I. Markov, Consistent placement of macro-blocks using floorplanning and standard cell placement, *Proceedings of ACM/IEEE International Symposium on Physical Design*, 2002, pp. 12–17.

[67] A. Khatkhate, C.Li, A.R. Agnihotri, M.C. Yildiz, S. Ono, C–K. Koh, P.H. Madden. Recursive bisection based mixed block placement, *Proceedings of ACM/IEEE International Symposium on Physical Design*, 2004, pp. 84–89.

[68] X. Yan, B.K. Choi, and M. Sarrafzadeh, Routability driven whitespace allocation for fixed-die standard-cell placement, *Proceedings of ACM/IEEE International Symposium on Physical Design*, 2002, pp. 42–47.

[69] C. Li, M. Xie, C–K. Koh, J. Cong, P.H. Madden. Routability-driven placement and whitespace allocation, *Proceedings of IEEE International Conference on Computer Aided Design*, 2004, pp. 394–401.

[70] C. Alpert, G.-J. Nam, and P. Villarrubia, Free space management for cut-based placement, *Proceedings of IEEE International Conference on Computer Aided Design*, 2002, pp. 746–751.

[71] S. Adya, I. Markov, and P. Villarrubia, On whitespace and stability in mixed-size placement, *Proceedings of IEEE International Conference on Computer Aided Design*, 2003, pp. 311–318.

[72] L.W. Hagen, D.J.H. Huang, and A.B. Kahng, Quantified suboptimality of VLSI layout heuristics, *Proceedings of ACM/IEEE Design Automation Conference*, 1995, pp. 216–221.

[73] C. Chang, J. Cong, and M. Xie, Optimality and scalability study of existing placement algorithms, *Proceedings of IEEE Asia and South Pacific Design Automation Conference*, 2003, pp. 621–627.

[74] J. Cong, M. Romesis, and M. Xie, Optimality and scalability study of partitioning and placement algorithms, *Proceedings of ACM/IEEE International Symposium on Physical Design*, 2003, pp. 88–94.

[75] A.B. Kahng and S. Reda, Evaluation of placer suboptimality via zero-change netlist transformations, *Proceedings of ACM/IEEE International Symposium on Physical Design*, 2005, pp. 208–215.

[76] J. Lou, W. Chen, and M. Pedram, Concurrent logic restructuring and placement for timing closure, *Proceedings of IEEE International Conference on Computer Aided Design*, 1999, pp. 31–36.

[77] W. Gosti, S. P. Khatri, and A. L. Sangiovanni-Vincentelli, Addressing the timing closure problem by intgerating logic optimization and placement, *Proceedings of IEEE International Conference on Computer Aided Design*, 2001, pp. 224–231.

[78] M. Hrkić, J. Lillis, and G. Beraudo, An approach to placement-coupled logic replication, *Proceedings of ACM/IEEE Design Automation Conference*, 2004, pp. 711–716.

6 Static Timing Analysis

Sachin S. Sapatnekar
University of Minnesota
St. Paul, Minnesota

6.1 Introduction

High-performance circuits have traditionally been characterized by the clock frequency at which they operate. Gauging the ability of a circuit to operate at the specified speed requires an ability to measure, during the design process, its delay at numerous steps. Moreover, delay calculation must be incorporated into the inner loop of timing optimizers at various phases of design, such as synthesis, layout (placement and routing), and in-place optimizations performed late in the design cycle. While timing measurements can theoretically be performed using a rigorous circuit simulation, such an approach is liable to be too slow to be practical. Static timing analysis plays a vital role in facilitating the fast and reasonably accurate measurement of circuit timing. The speedup appears due to the use of simplified delay models, and on account of the fact that its ability to consider the effects of logical interactions between signals is limited. Nevertheless, it has become a mainstay of design over the last few decades; one of the earliest descriptions of a static timing approach can be found in [1]. This chapter will first overview some of the most prominent techniques for Static Timing Analysis (STA). It will then outline issues related to Statistical Static Timing Analysis (SSTA), a procedure that is becoming increasingly necessary to handle the complexities of process and environmental variations in integrated circuits.

6.2 Representation of Combinational and Sequential Circuits

A combinational logic circuit may be represented as a timing graph $G = (V, E)$, where the elements of V, the vertex set, are the inputs and outputs of the logic gates in the circuit. The vertices are connected by two types of edges: one set of edges connects each input of a gate to its output, which represents the maximum delay paths from the input pin to the output pin, while another set of edges connects the output of each gate to the inputs of its fanout gates, and corresponds to the interconnect delays.

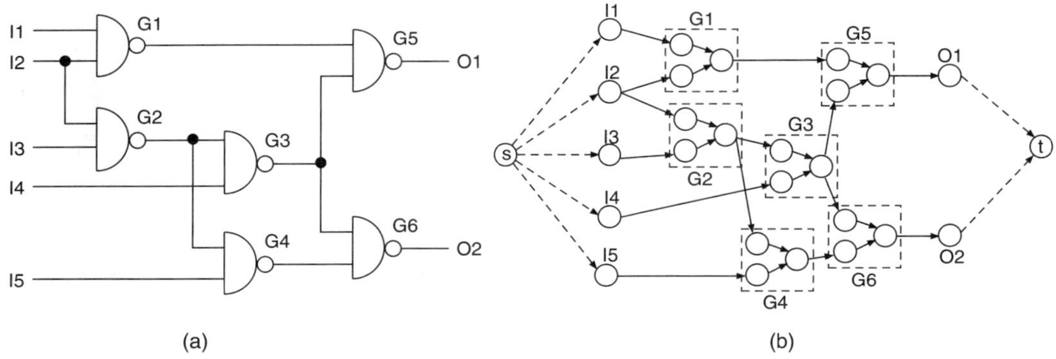

FIGURE 6.1 (a) Example of a combinational circuit and (b) its timing graph [2] (with permission from Springer).

An example logic circuit and its corresponding graph are illustrated in Figures 6.1(a) and (b), respectively. A simple transform is used to ensure that the graph has a single source *s* and a single sink *t*. In the event that all primary inputs are connected to flip-flops and transition at the same time, edges are added from the node *s* to each primary input, and from each primary output to the sink *t*. The case where the primary inputs arrive at different times $a_i \geq 0$ can be handled with a minor transformation to this graph,* adding a dummy node with a delay of a_i along each edge from *s* to the primary input *i*.

The above construction generally results in a combinational circuit that is represented as a directed acyclic graph (DAG). This is because combinational circuits, as conventionally described, do not have any cycles (though some implementations of combinational circuits are not acyclic). Many timing analyzers handle loops in combinational circuits by breaking them either in an *ad hoc* manner, or according to a heuristic, so that the circuit to be analyzed may be represented as a DAG.

A sequential circuit that consists both of combinational elements and sequential elements (flip-flops and latches) may be represented as a set of combinational blocks that lie between latches, and a timing graph may be constructed for each of these blocks. For any such block, the sequential elements or circuit inputs that fan out to a gate in the block constitute its primary inputs; similarly, the sequential elements or circuit outputs for which a fanin gate belongs to the block together represent its primary outputs. It is rather easy to find these blocks: to begin with, we construct a graph in which each vertex corresponds to a combinational element, an undirected edge is drawn between a combinational element and the combinational elements that it fans out to, and sequential elements are left unrepresented (this is substantially similar to the directed graph described above for a combinational circuit without the *s* and *t* nodes, except that it is undirected). The connected components of this graph correspond to the combinational blocks in the circuit.

The computation of the delay of such a combinational block is an important step in timing analysis. For an edge-triggered circuit, the signals at the primary inputs of such a block make a transition at exactly the same time, and the clock period must satisfy the constraint that it is no smaller than the maximum delay of any path through the logic block, plus the setup time for a flip-flop at the primary output at the end of that path. Therefore, finding the maximum delay through a combinational block is a vital subtask, and we will address this issue in this chapter. Alternatively, a combinational block that is to be treated as a blackbox must be characterized in terms of its delays from each input to each output; this is a vital problem, for example, when the analysis involves hierarchically defined blocks with such timing abstractions. For level-clocked circuits, the timing relations are more complex, but the machinery developed in this chapter will nevertheless be useful.

*If some arrival times are negative, the time variable can be shifted to ensure that $a_i \geq 0$ at each primary input.

6.3 Gate Delay Models

Consider a logic gate driving an interconnect net to one or more sinks. If the entire net can be considered to be fully capacitive, the problem of determining the delay of this logic stage reduces to that of finding the delay of the gate for a specific capacitive load. If the gate is a cell from a library, its delay is typically precharacterized under various loads C_L and input transition times τ_{in}.

Some commonly used delay characterization techniques under capacitive loads include the following (similar equations used to characterize the output transition time):

- Look-up table methods, with each entry in the table corresponding to the delay under different capacitive loads and input transition times. While this can achieve arbitrarily good accuracy provided the table has enough entries, it is seen to be memory-intensive.
- Traditional methods for characterizing a cell driving a load use an equation of the form $k_1C_L + k_2$, where k_1 is a characterized slope and k_2 an intrinsic delay. However, such an equation neglects the effect of the input transition time on the delay.
- The k-factor equations compact the table look-up by storing the delay D as a fitted function of the form $(k_1 + k_2C_L)\tau_{in} + k_3C_L^3 + k_4C_L + k_5$.
- The nonlinear delay model (NLDM) from Synopsys uses characterization equations of the form $\alpha\tau_{in} + \beta C_L + \gamma\tau_{in}C_L + \delta$.
- The scalable polynomial delay model (SPDM) from Synopsys uses a product of polynomials to fit the delay data. For example, for two parameters C_L and τ_{in}, a product of an mth-order polynomial in C_L with an nth-order polynomial in τ_{in}, of the form $(a_0 + a_1C_L + \cdots + a_mC_L^m) \cdot (b_0 + b_1\tau_{in} + \cdots + b_n\tau_{in}^n)$, may be used.

The lumped capacitance model for interconnects is only accurate when the driver resistance overwhelms the wire resistance; when the two are comparable, such a model could have significant errors. In particular, the phenomenon of "resistive shielding" causes the delay at the driver output (referred to as the *driving point*) to be equivalent to a situation where it drives a lumped load that is less than the total capacitance of the interconnect. In effect, the interconnect resistance shields a part of the total capacitance from the driving point. To overcome this, the concept of an *effective capacitance* at the driving point is widely used for delay calculation. For details, the reader is referred to [3,4].

6.4 Timing Analysis for Combinational Circuits

6.4.1 Delay Calculation for a Combinational Logic Block

In this section, we will present techniques that are used for the STA of digital combinational circuits. The word "static" alludes to the fact that this timing analysis is carried out in an input-independent manner, and purports to find the worst-case delay of the circuit over all possible input combinations. The computational efficiency (linear in the number of edges in the graph) of such an approach has resulted in its widespread use, even though it has some limitations; this will be described in Section 6.4.5.

A method that is commonly referred to as PERT (program evaluation and review technique) is popularly used in STA. In fact, PERT is a misnomer, and the so-called PERT method discussed in most of the literature on timing analysis refers to the CPM (critical path method) that is widely used in project management.

Before proceeding, it is worth pointing out that while the CPM-based methods are the dominant ones in use today, other methods for traversing circuit graphs have been used by various timing analyzers: for example, depth-first techniques have been presented in [5].

The algorithm applied to a timing graph $G = (V, E)$ can be summarized by the pseudocode shown in Figure 6.2. The procedure is best illustrated by means of a simple example. Consider the circuit in Figure 6.3, which shows an interconnection of blocks. Each of these blocks could be as simple as a logic gate or could be a more complex combinational block, and is characterized by the delay from each input pin to each output pin. For simplicity, this example will assume that for each block, the delay from any input to the output

```
Algorithm CRITICAL_PATH_METHOD
Q = ∅;
for all vertices i ∈ V
   n_visited_inputs[i] = 0;
/* Add a vertex to the tail of Q if all inputs are ready */
for all primary inputs i          /* Fanout gates of i */
   for all vertices j such that (i → j) ∈ E
      if (++n_visited_inputs[j] == n_inputs[j]) addQ(j,Q);
while (Q ≠ ∅) {
   g = top(Q); remove(g,Q); compute_delay[g]
   for all vertices k such that (g → k) ∈ E        /* Fanout gates of g */
      if (++n_visited_inputs[k] == n_inputs[k]) addQ(k,Q);
}
```

FIGURE 6.2 Pseudocode for the Critical Path Method.

is identical. Moreover, we will assume that each block is an inverting logic gate such as a NAND or a NOR, as shown by the "bubble" at the output. The two numbers, d_r/d_f, inside each gate represent the delay corresponding to the delay of the output rising transition, d_r, and that of the output fall transition, d_f, respectively. We assume that all primary inputs are available at time zero, so that the numbers "0/0" against each primary input represent the worst-case rise and fall arrival times, respectively, at each of these nodes. The CPM proceeds from the primary inputs to the primary outputs in topological order, computing the worst-case rise and fall arrival times at each intermediate node, and eventually at the output(s) of a circuit.

A block is said to be *ready for processing* when the signal arrival time information is available for all of its inputs; in other words, when the number of processed inputs of a gate *g*, `n_visited_inputs` [*g*], equals the number of inputs of the gate, `n_inputs` [*g*]. Notationally, we will refer to each block by the symbol for its output node. Initially, since the signal arrival times are known only at the primary inputs, only those blocks that are fed solely by primary inputs are ready for processing. In the example, these correspond to the gates *i*, *j*, *k*, and *l*. These are placed in a queue Q using the function `addQ`, and are processed in the order in which they appear in the queue.

In the iterative process, the block at the head of the queue Q is taken off the queue and scheduled for processing. Each processing step consists of:

- Finding the latest arriving input to the block that triggers the output transition (this involves finding the maximum of all the worst-case arrival times of inputs to the block), and then adding the delay of the block to the latest arriving input time to obtain the worst-case arrival time at the output. This is represented by `compute_delay` in the pseudocode.
- Checking the entire block that the current block fans out to, to find out whether theyare ready for processing. If so, the block is added to the tail of the queue using `addQ`.The iterations end when the queue is empty. In the example, the algorithm is executed by processing the gates in the sequence *i*, *j*, *k*, *l*, *m*, *n*, *p*, *o*, and the worst-case delay for the entire block is found to be max(7,11) = 11 units, as illustrated in Figure 6.3.

6.4.2 Critical Paths, Required Times, and Slacks

The *critical path*, defined as the path between an input and an output with the maximum delay, can now be easily found by using a traceback method. We begin with the block whose output is the primary output with the latest arrival time: this is the last block on the critical path. Next, the latest arriving input to this block is identified, as also the block that causes this transition is the preceding block on the critical path. The process is repeated recursively until a primary input is reached.

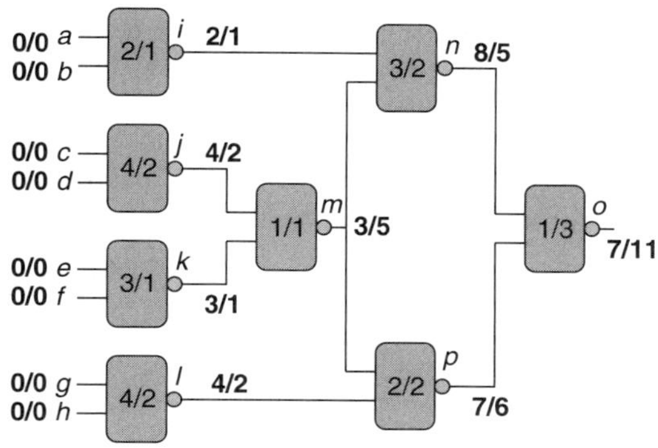

FIGURE 6.3 An example illustrating the application of the CPM on a circuit with inverting gates. The numbers within the gates correspond to the rise delay/fall delay of the block, and the bold numbers at each block output represent the rise/fall arrival times at that point. The primary inputs are assumed to have arrival times of zero, as shown [2] (with permission from Springer).

In the example, we begin with gate o at the output, whose falling transition corresponds to the maximum delay. This transition is caused by the rising transition at the output of gate n, which must therefore precede o on the critical path. Similarly, the transition at n is effected by the falling transition at the output of m, and so on. By continuing this process, the critical path from the input to the output is identified as being caused by a falling transition at either input c or d, and then progressing as follows: rising $j \rightarrow$ falling $m \rightarrow$ rising $n \rightarrow$ falling o.

A useful concept is the notion of the *required time*, R, at a node in the circuit. If a circuit is provided with a set of arrival time constraints at each output node, then on completion of the CPM traversal, one may check whether those constraints are met or not. If they are not met, it is useful to know which parts of the circuit should be sped up and how, and if they are met, it may be possible to save design resources by resynthesizing the circuit to slow it down. In either case, the required time at each node in the circuit provides useful information. At any node in the circuit, if the arrival time exceeds the required time, then the path that this node lies on must be sped up.

The computation of the required time proceeds as follows. At each primary output, the required times for rise/fall are set according to the specifications provided to the circuit. Next, a backward topological traversal is carried out, processing each gate when the required times at all of its fanouts are known. In essence, this is equivalent to performing a CPM traversal with the directions of the edges in $G(V, E)$ reversed, the block delays negated, and the required time at the outputs set to the timing specification.

The *slack* associated with each connection is then simply computed as the difference between the required time and the arrival time. A positive slack s at a node implies that the arrival time at that node may be increased by s without affecting the overall delay of the circuit. Such a delay increase will only eat into the slack, and may provide the potential to free up excessive design resources used to build the circuit.

6.4.3 Extensions to More Complex Cases

For ease of exposition, the example in the previous section contained a number of simplifying assumptions. The CPM can work under more general problem formulations, and a few of these that are commonly encountered are listed below:

- *Nonzero arrival times at the primary inputs*: If the combinational block is a part of a larger circuit, we may have nonzero rise/fall arrival times at the primary inputs. If so, the CPM traversal can be carried out by simply using these values instead of zero as the arrival times at the primary inputs.
- *Minimum delay calculations*: When the gate delays are fixed, the method described above can easily be adapted to find the minimum instead of the maximum delay from any input to any output. The only changes in the procedure is the manner in which an individual block is processed: the earliest arrival time over all inputs is now added to the delay of the block to find the earliest arrival time at its output.

- *Minmax delay calculations*: If the gate delay is specified in terms of an interval, $[d_{\min}, d_{\max}]$, then the minimum and maximum arrival time intervals can be propagated in a similar manner; again, these values may be maintained separately for the rising and falling output transitions. The values of $d_{\min}$ and $d_{\max}$ can be computed on the fly while processing a gate.
- *Generalized block delay models*: If the delay from input pin p to output pin q of a block is d_{pq}, and the values of d_{pq} are not all uniform for a block, then the arrival time at the output q can be computed as

$$\max_{\text{all inputs } p}(t_p + d_{pq})$$

where t_p is the arrival time (of the appropriate polarity) at input p. Note that if $d_{pq} = d \forall p$, then this simply reduces to the expression,

$$\max_{\text{all inputs } p}(t_p) + d$$

which was used in the example in Figure 6.3.
- *Incorporating input signal transition times*: In the example, the delay of each of the individual gates was prespecified as an input to the problem. However, in practice, the delay of a logic gate depends on factors such as the input transition time (i.e., the 10 to 90% rise or fall times), which are unknown until the predecessor gate has been processed. This problem is overcome by propagating not only the worst-case rise and fall arrival times, but also the input signal transition times corresponding to those arrival times. Note that this fits in well with the fact that the delay of a gate is only required at the time when it is marked as ready for processing during the CPM traversal. At this time, the input signal arrival times as well as the corresponding signal transition times are available, which implies that the delay of the gate may be computed on the fly, just before it is needed. More complex techniques for handling slope effects are presented in [6].

6.4.4 Incremental Timing Analysis

During the process of circuit design or optimization, it is frequently the case that a designer or a CAD tool may alter a small part of a circuit and consider the effects of this change on the timing behavior of the circuit. Since the circuit remains largely unchanged, performing a full STA entails needless computation, since many of the arrival times remain unchanged. The notion of incremental timing analysis provides a way of propagating the changes in circuit timing caused by this alteration only to those parts of the circuit that require it. Generally, incremental analysis is cheaper than carrying out a complete timing analysis.

An *event-driven* propagation scheme may be used: an event is said to occur when the timing information at the input to a gate is changed, and the gate is then processed so that the event is propagated to its outputs. Unlike CPM, in an event-driven approach, it is possible that a gate may be processed more than once. However, when only a small change is made to the circuit, and a small fraction of the gates have their arrival times altered, such a method is highly efficient.

Incremental approaches have been used in timing analysis for decades, and a good and relatively recent source describing an incremental timing analysis approach is [7].

6.4.5 False Paths

The true delay of a combinational circuit corresponds to the worst case over all possible logic values that may be applied at the primary inputs. Given that each input can take on one of four values (a steady 0, a steady 1, a $0 \rightarrow 1$ transition, and a $1 \rightarrow 0$ transition), the number of possible combinations for a circuit with m inputs is 4^m, which shows an exponential trend. However, it can be verified that the CPM for finding the delay of a combinational circuit can be completed in $O(|V| + |E|)$ time for a timing graph $G = (V, E)$, and in practice, this computation is considerably cheaper on large circuits than a full enumeration. The difference arises because of a specific assumption made by CPM: namely, that the logic function implemented by a gate is inconsequential and only its delay is relevant.

This may result in estimation errors that are pessimistic. As an example, consider the circuit shown in Figure 6.4, with three inputs *a*, *b*, and *c*, and one output, out. Assume, for simplicity, that the multiplexer

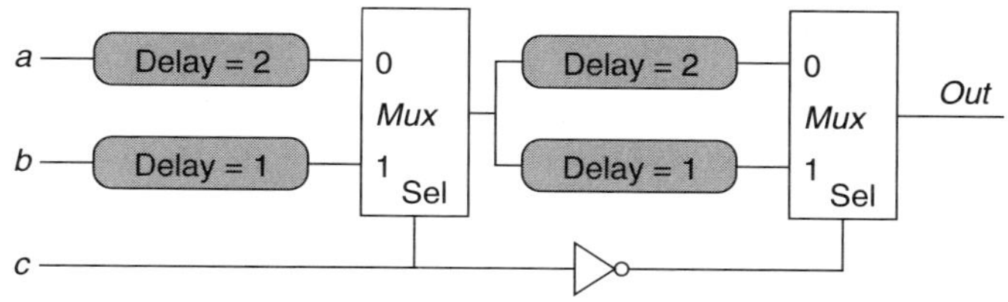

FIGURE 6.4 Illustration of a circuit with false paths.

and inverter have zero delays, and that the four blocks whose delays are shown are purely combinational. It can easily be verified that the worst-case delay for this circuit computed using the CPM is 4 units. However, by enumerating both possible logic values for the multiplexer, namely, $c = 0$ and $c = 1$, it can be seen that the delay in both cases is 3 units, implying that the circuit delay is 3 units. The reason for this discrepancy is simple: the path with a delay of 4 units can never be sensitized because of the restrictions placed by the Boolean dependencies among the inputs.

While many approaches to false path analysis have been proposed, most are rather too complex to be applied in practice. The identification of false paths includes numerous subtleties. Some paths may not be statically insensitizable when delays are not considered, but dynamically sensitizable. For instance, if the inverter has a delay of 3 units, then the path of delay of 4 units is indeed dynamically sensitizable. Various definitions have been proposed in the literature, and an excellent survey, written at a time when research on false paths was most active, is presented in [8].

However, most practical approaches use some designer input and case enumeration, as in the case of Figure 6.4, where the cases of $c = 0$ and $c = 1$ were enumerated. Approaches for pruning user-specified false paths from timing graphs have been presented in [9–12].

6.5 Timing Analysis for Sequential Circuits

A general sequential circuit is a network of computational nodes (gates) and memory elements (registers). The computational nodes may be conceptualized as being clustered together in an acyclic network of gates that forms a combinational logic circuit. A cyclic path in the direction of signal propagation is permitted in the sequential circuit only if it contains at least one register. In general, it is possible to represent any sequential circuit in terms of the schematic shown in Figure 6.5, which has I inputs, O outputs, and M registers. The registers outputs feed into the combinational logic which, in turn, feeds the register inputs. Thus, the combinational logic has $I + M$ inputs and $O + M$ outputs.

The functionality of registers plays a key role in determining the behavior of any sequential circuit, and there are several choices of register types that are available to a designer. The behavior of each register is controlled by the clock signal, and depending on the state of the clock, the data at the register input are either isolated from its output or transmitted to the output. The types of registers in a synchronous system are differentiated by the manner in which they use the clock to transmit data from the input to the output.

Level-clocked latches: These are commonly referred to merely as latches, and permit data to be transmitted from the input to the output whenever the clock is high. During this time, the latch is said to be "transparent."

Edge-triggered flip-flops: These are commonly called flip-flops (FFs), and use the clock edge to determine when the data are transmitted to the output. In a positive (negative) edge-triggered FF, data are transmitted from the input to the output when the clock makes a transition from 0 to 1 (1 to 0). Flip-flops can typically be constructed by connecting two level-clocked latches in a master–slave fashion.

A circuit consisting only of level-clocked latches and gates is referred to as a level-clocked circuit, and a circuit composed of edge-triggered FFs is called an edge-triggered circuit. In our discussions, we will primarily deal with edge-triggered D FFs and level-clocked D latches as representative memory elements.

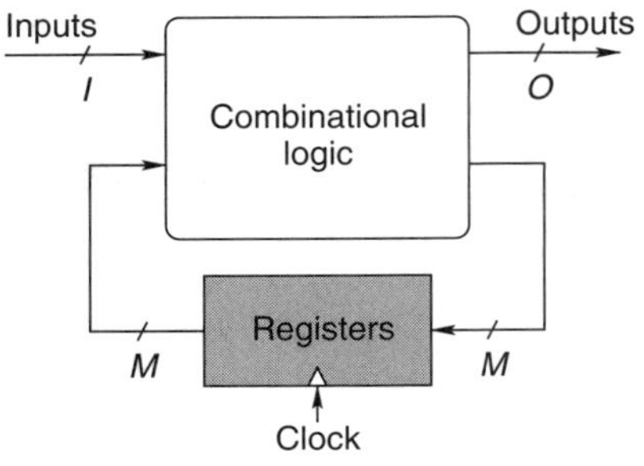

FIGURE 6.5 A general sequential circuit.

6.6 Clocking Disciplines: Edge-Triggered Circuits

We will now overview the timing requirements for edge-triggered sequential circuits, which consist of combinational blocks that lie between D FFs. The basic parameters associated with an FF can be summarized as follows:

- The data input of the register, commonly referred to as the D input, must receive incoming data at a time that is at least T_{setup} units before the onset of the latching edge of the clock. The data will then be available at the output node Q, after the latching edge. The quantity, T_{setup}, is referred to as the setup time of the FF.
- The input, D, must be kept stable for a time of T_{hold} units, where T_{hold} is called the hold time, so that the data are allowed to be stored correctly in the FF.
- Each latch has a delay between the time the data and clock are both available at the input, and the time when it is latched; this is referred to as the clock-to-Q delay, T_q.

In the edge-triggered scenario, let us consider two FFs, i and j, connected only by purely combinational paths. Over all such paths $i \rightsquigarrow j$, let the largest delay from FF i to FF j be $\overline{d}(i, j)$, and the smallest delay be $\underline{d}(i, j)$. Therefore, for any path $i \rightsquigarrow j$ with delay $d(i, j)$, it must be true that

$$\underline{d}(i, j) \leq d(i, j) \leq \overline{d}(i, j)$$

We will denote the setup time, hold time, and the maximum and minimum clock-to-Q delay of any arbitrary FF k as T_{s_k}, T_{h_k}, and Δ_k and δ_k, respectively. For a negative edge-triggered register, the setup and hold time requirements are illustrated in Figure 6.6. The clock is a periodic waveform that repeats after every P units of time, called the clock period or the cycle time.

The data are available at the launching FF, i, after the clock-to-Q delay, and will arrive at the latching FF, j, at a time no later than $\Delta_i + \overline{d}(i, j)$. For correct clocking, the data are required to arrive one setup time before the latching edge of the clock at FF j as shown in Figure 6.6, i.e., at a time no later than $P - T_{s_j}$. This leads to the following constraint:

$$\Delta_i + \overline{d}(i, j) \leq P - T_{s_j}$$

i.e.,

$$\overline{d}(i, j) \leq P - T_{s_j} - \Delta_i \tag{6.1}$$

For obvious reasons, this constraint is often referred to as the *setup time constraint.* Since this requirement places an upper bound on the delay of a combinational path, it is also called the *long path constraint.* A third name that can be used for this is the *zero clocking constraint*, because the data will not arrive in time to be latched at the next clock period if the combinational delay does not satisfy this constraint.

The data must be stable for an interval that is at least as long as the hold time after the clock edge if it is to be correctly captured by the FF. Hence, it is essential to ensure that the new data do not arrive at FF

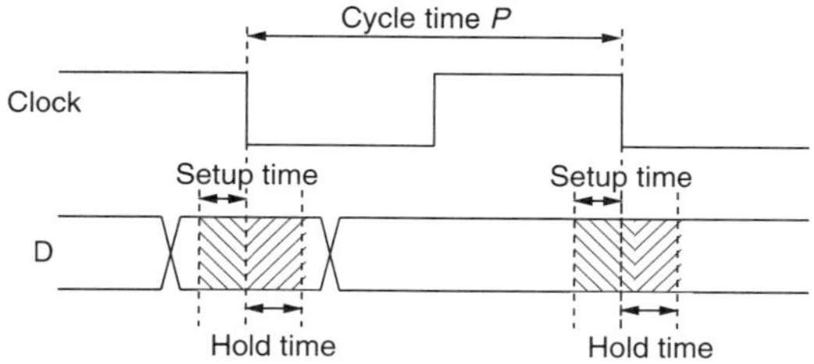

FIGURE 6.6 Illustration of the clocking parameters for a register [2] (with permission from Springer).

j before time T_{h_j}. Since the earliest time that the incoming data can arrive is $\delta_i + \underline{d}(i, j)$, this gives us the following *hold time constraint*:

$$\delta_i + \underline{d}(i, j) \geq T_{h_j}$$

i.e.,

$$\underline{d}(i, j) \geq T_{h_j} - \delta_i \tag{6.2}$$

Since this constraint puts a lower bound on the combinational delay on a path, it is referred to as a *short path constraint.* If this constraint is violated, then the data in the current clock cycle are corrupted by the data from the next clock cycle; as a result, data are latched twice instead of once in a clock cycle, and hence it is also called the *double clocking constraint.* Notice that if the minimum clock-to-Q delay of FF i is greater than the hold time of FF j, i.e., $\delta_i \geq T_{h_j}$ (this condition is not always true in practice), then the right-hand side of the constraint is negative. In this case, since $\underline{d}(i, j) \geq 0$, the short-path constraint is always satisfied.

An important observation is that both the long- and the short-path constraints refer to combinational paths that lie between the FFs. Therefore, for timing verification of edge-triggered circuits, it is possible to decompose the circuit into combinational blocks, and to verify the validity of the constraints on each such block independently. This is not so for level-clocked circuits, which present a greater complexity to the timing verifier.

Since a level-clocked latch is transparent during the active period of the clock, the analysis and design of level-clocked circuits is more complex than that of the edge-triggered ones, since combinational blocks are not insulated from each other by the memory elements, and multicycle paths are possible when the data are latched during the transparent phase of the clock. Methods for this purpose are described in [13–17].

6.7 Clocking and Clock-Skew Optimization

Conventional synchronous circuit design is predicated on the assumption that each clock signal of the same phase arrives at each memory element at exactly the same time. In a sequential VLSI circuit, due to differences in interconnect delays on the clock distribution network, this simultaneity is difficult to achieve and clock signals do not arrive at all of the registers at the same time. This is referred to as a skew in the clock. In a single-phase edge-triggered circuit, in the case where there is no clock skew, the designer must ensure that for correct operation, each input–output path of a combinational subcircuit has a delay that is less than the clock period. In the presence of skew, however, the relation grows more complex and the design of the combinational subcircuits becomes more involved.

The conventional approach to design builds the clock distribution network so as to ensure zero clock skew. An alternative approach views clock skews as manageable resources rather than liabilities, and uses them to advantage by intentionally introducing skews to improve the performance of the circuit. To illus-

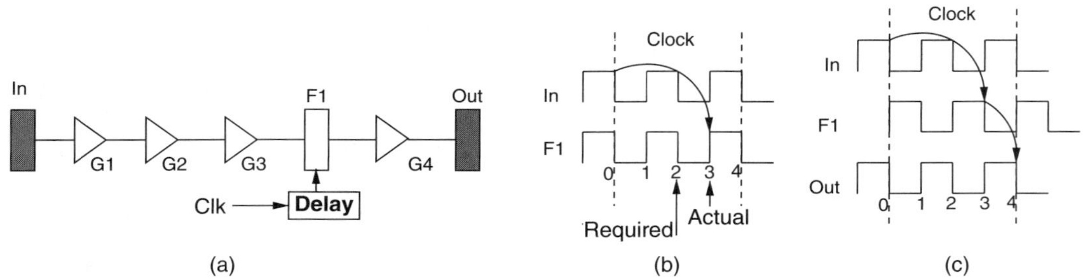

FIGURE 6.7 (a) Example of an edge-triggered circuit, where each gate has a unit delay and the setup time of each FF is zero. If a clock period of 2 units is applied under an edge-triggered discipline, then in (b), where delay = 0, the circuit functions incorrectly since the actual arrival time at F1 is later than the required time. In (c), where delay = 1, i.e., the clock to F1 is skewed by 1 unit, the circuit operates correctly [19] (with permission from IEEE).

trate how this may be done, consider the circuit shown in Figure 6.7(a), and assume each of the inverters to have a unit delay, and the delay in the clocking network to F1 to be 0. This circuit cannot be properly clocked to function at a period of 2 time units, because as shown in Figure 6.7(b), the required arrival time of the signal at register L1 is 2 units, while the data arrive after 3 time units. It is readily verifiable that the fastest allowable clock for this circuit has a period of 3 units. However, if a skew of +1 unit is applied to the clock line to register L1, the circuit can operate under a clock period of 2 units. This is possible because as shown in Figure 6.7(c), the application of a skew of +1 unit delays the clock arrival at register L1 by one unit, thus changing the required data arrival time to the new arrival time of the first clock tick, which is 3 units (i.e., the period of 2 units delayed by +1 unit). Under these circumstances, the actual data arrival time of 3 units does not cause a timing violation, and the circuit is correctly clocked. A formal method for determining the minimum clock period and the optimal skews was first presented in the work by Fishburn [18], where the clock-skew optimization problem was formulated as a linear program that was solved to find the optimal clock period.

Another application of deliberate skews relates to the fact that the simultaneous switching of FFs at the latching edge of the lock requires a large current on the power distribution network.

Consider a combinational block in a sequential circuit. Let FF_i and FF_j be a pair of FFs at the input and output of the combinational block, respectively, with skews of x_i and x_j, respectively. Let us denote the delay of the combinational block between them as $d(i, j)$, with $\underline{d}(i, j)$ being the minimum delay and $\overline{d}(i, j)$ being the maximum delay. If T_{hold} is the FF hold time, T_{setup} the FF setup time, and P the clock period, then in the presence of skews, the timing constraints take the following form:

Long-path constraints: The data from the current clock cycle should arrive at FF_j no later than a time T_{setup} before the next clock. The long-path constraint, shown in Figure 6.8(a), may be expressed as

$$x_i + \overline{d}(i, j) \leq x_j + P - T_{\text{setup}} \tag{6.3}$$

Short-path constraints: The data from the next clock cycle should arrive at FF_j no earlier than a time T_{hold} after the current clock. Thus, the short-path constraint illustrated in Figure 6.8(b) is

$$x_i + \underline{d}(i, j) \geq x_j + T_{\text{hold}} \tag{6.4}$$

The problem of minimizing the clock period P by controlling the clock skew at each FF, subject to correct timing, can now be formulated as the following linear program:

$$\begin{aligned} \text{minimize} \quad & P \\ \text{such that} \quad & x_i + \underline{d}(i, j) \geq x_j + T_{\text{hold}} \\ & x_i + \overline{d}(i, j) \leq x_j + P - T_{\text{setup}} \end{aligned} \tag{6.5}$$

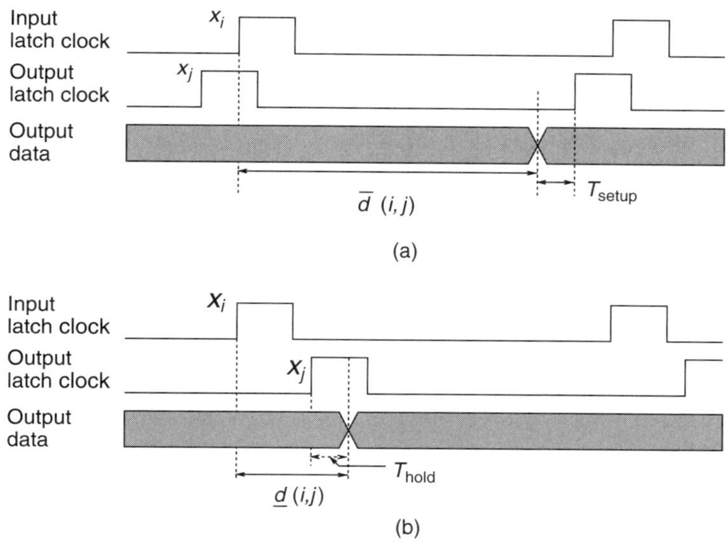

FIGURE 6.8 Pictorial depiction of (a) long-path constraints and (b) short-path constraints [2] (with permission from Springer).

The linear program above may be extended to model clock skew uncertainties [18].

The constraints of the linear program Equation (6.5) are rewritten as

$$
\begin{aligned}
x_i - x_j &\geq T_{\text{hold}} - \underline{d}(i, j) \\
x_j - x_i &\geq T_{\text{setup}} + \bar{d}(i, j) - P
\end{aligned}
\tag{6.6}
$$

Note that the skews at the primary inputs and the primary outputs may be set to zero under the assumption that they cannot be controlled or adjusted. For a constant value of P, the constraint matrix for this problem reduces to a system of difference constraints [20]. For such a system, it is possible to construct a constraint graph. The vertices of the constraint graph correspond to the x_i variables, and each constraint $x_a - x_b \geq c$ corresponds to a directed edge from node b to node a with a weight of c. If P is achievable, then a set of skews that satisfies P can be found by solving the longest path problem on this directed graph. The clock period P is feasible, provided the corresponding constraint graph contains no positive cycles.

This observation was utilized in [21] for efficient solution of the skew optimization problem. The optimal clock period is obtained by performing a binary search. At each step in the binary search (for clock period P), the Bellman–Ford algorithm [20] is applied to the corresponding constraint graph to check for positive cycles. The procedure continues until the smallest feasible clock period is found.

The optimization above, and indeed, many other procedures that work with clock skew, requires the detection of violations of the clocking constraints. This may be carried out by using a modification of the CPM for delay estimation, which is generalized to handle sequential circuits [22]. The procedure for timing analysis requires the determination of the optimal skews and the arrival times at all gate outputs. Since the constraint graph for an acyclic pipeline is acyclic, and since timing analysis involves the solution of the longest-path problem on this constraint graph, the complexity can be reduced substantially by processing the vertices in topological order [20]. Specifically, the timing analysis procedure is equivalent to applying the CPM described in Section 6.4.1 to the graph, with gates being represented in the normal way, and the FFs being represented by blocks with "delay" values of $T_{\text{setup}} - P - \delta$, where T_{setup} is the setup time for the FF, P the applied clock period, and δ the uncertainty in the clock skew.

It is noteworthy that the CPM is valid if the circuit is acyclic; if not, techniques such as those used in [23] will have to be applied for analysis. Alternatively, for optimization purposes, a cyclic graph can be forced to be acyclic using the techniques in [24], with additional constraints being imposed at points where the cycles are broken.

The process of calculating the optimal skews falls out as a natural consequence of this procedure: if the timing requirements are met, then the arrival time at the output of an FF, as calculated by CPM, is a valid value of the skew to be applied to that FF. It is also worth pointing out here that there is an intimate connection between clock-skew optimization and another sequential optimization technique called retiming [25]. For details, the reader is referred to [19,26].

6.8 Statistical Static Timing Analysis

Current-day integrated circuits are afflicted with a wide variety of variations that affect their performance. Essentially, under true operating conditions, the parameters chosen by the circuit designer are perturbed from their nominal values due to various types of variations. As a consequence, a single SPICE-level transistor or interconnect model (or an abstraction thereof) is seldom an adequate predictor of the exact behavior of a circuit. These sources of variation can broadly be categorized into two classes, both of which can result in changes in the timing characteristics of a circuit.

Process variations result from perturbations in the fabrication process that change the values of parameters such as the effective channel length (L_{eff}), the oxide thickness (t_{ox}), the dopant concentration (N_a), the transistor width (w), the interlayer dielectric (ILD) thickness (t_{ILD}), and the interconnect height and width (h_{int} and w_{int}, respectively).

Environmental variations arise due to changes in the operating environment of the circuit, such as the temperature or variations in the supply voltage (V_{dd} and ground) levels. There is a wide range of work on analysis techniques to determine environmental variations, both for thermal issues [27–29], and for supply net analysis [30].

Process variations can also be classified into the following categories:

Interdie variations are the variations from die to die, and affect all the devices on the same chip in the same way, e.g., they may cause all of the transistor gate lengths of devices on the same chip to be larger or all of them to be smaller.

Intradie variations correspond to variability within a single chip, and may affect different devices on the same chip differently, e.g., they may result in some devices having smaller oxide thicknesses than the nominal, while others may have larger oxide thicknesses.

Interdie variations have been a long-standing design issue, and for several decades, designers have striven to make their circuits robust under the unpredictability of such variations. This has typically been achieved by simulating the design at not just one design point, but at multiple "corners." These corners are chosen to encapsulate the behavior of the circuit under the worst-case variations, and have served designers well in the past. In nanometer technologies, intradie variations have become significant and can no longer be ignored. As a result, a process corner-based methodology, which would simulate the entire chip at a small number of design corners, is no longer sustainable, and such a procedure will be very conservative and pessimistic. For true accuracy, this can be overcome by using a larger number of process corners, but this number may be too large to permit computational efficiency.

Unlike interdie variations, whose effects can be captured by a small number of STA runs at the process corners, a more sophisticated approach is called for in dealing with intradie variations. This requires an extension of traditional STA techniques to move beyond their deterministic nature. An alternative approach that overcomes these problems is SSTA, which treats delays not as fixed numbers, but as probability density functions (PDFs), taking the statistical distribution of parametric variations into consideration while analyzing the circuit.

The sources of these variations may be used to create another taxonomy:

Random variations (as the name implies) depict random behavior that can be characterized in terms of a distribution. This distribution may either be explicit, in terms of a large number of samples provided from fabrication line measurements, or implicit, in terms of a known PDF (such as a Gaussian or a lognormal distribution) that has been fitted to the measurements. Random variations in some process or

environmental parameters (such as the temperature, supply voltage, or L_{eff}) can often show some degree of local *spatial correlation*, whereby variations in one transistor in a chip are remarkably similar in nature to those in spatially neighboring transistors, but may differ significantly from those that are far away. Other process parameters (such as t_{ox} and N_a) do not show much spatial correlation at all, so that for all practical purposes variations in neighboring transistors are uncorrelated.
Systematic variations show predictable variational trends across a chip, and are caused by known physical phenomena during manufacturing. Strictly speaking, environmental changes are entirely predictable, but practically, due to the fact that these may change under a large number (potentially exponential in the number of inputs and internal states) of operating modes of a circuit, it is easier to capture them in terms of random variations. Examples of systematic variations include those due to spatial intrachip gate length variability, which observes systematic changes in the value of L_{eff} across a reticle due to effects such as changes in the stepper-induced illumination and imaging nonuniformity due to lens aberrations [31], or ILD variations, due to the effects of chemical–mechanical polishing (CMP) on metal density patterns.

The existence of *correlations* between intradie variations complicates the task of statistical analysis. These correlations are of two types:

Spatial correlations: To model the intradie spatial correlations of parameters, the die region may be tessellated into n grids. Since devices or wires close to each other are more likely to have similar characteristics than those placed far away, it is reasonable to assume perfect correlations among the devices (wires) in the same grid, high correlations among those in close grids and low or zero correlations in far-away grids. Under this model, a parameter variation in a single grid at location (x, y) can be modeled using a single random variable $p(x, y)$. For each type of parameter, n random variables are needed, each representing the value of a parameter in one of the n grids.

Structural correlations: The structure of the circuit can also lead to correlations that mus be incorporated into SSTA. Consider the reconvergent fanout structure shown in Figure 6.9. The circuit has two paths, a-b-d and a-c-d. If, for example, we assume that each gate delay is a Gaussian random variable, then the PDF of the delay of each path is easy to compute, since it is the sum of Gaussians, which admits a closed form. However, the circuit delay is the maximum of the delays of these two paths, and these are correlated since the delays of a and d contribute to both paths. It is important to take into account such structural correlations, which arise due to reconvergences in the circuit, while performing an SSTA.

The geometrical parameters associated with the gate and interconnect can reasonably be modeled as normally distributed random variables. Before we introduce how the distributions of gate and interconnect delays will be modeled, let us first consider an arbitrary function $d = f(\mathbf{P})$ that is assumed to be a function of a set of parameters $\mathbf{P}$, where each $p_i \in \mathbf{P}$ is a random variable with a normal distribution given by $p_i \sim N(\mu_{p_i}, \sigma_{p_i})$. We can approximate d linearly using a first-order Taylor expansion:

$$d = d_0 + \sum_{\forall \text{parameters } p_i} \left[\frac{\partial f}{\partial p_i} \right]_0 \Delta p_i \qquad (6.7)$$

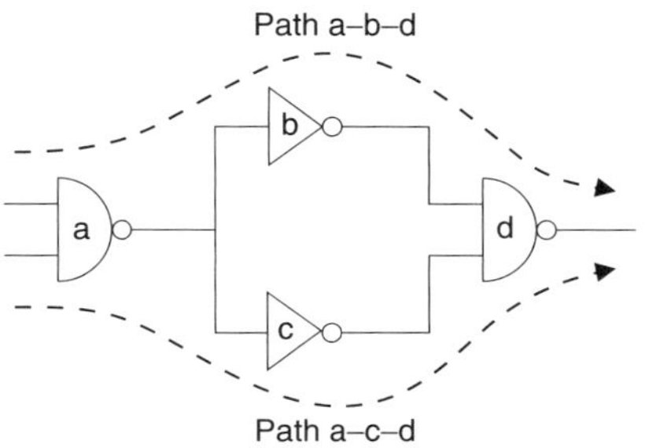

FIGURE 6.9 An example to illustrate structural correlations in a circuit.

where d_0 is tthe nominal value of d, calculated at the nominal values of parameters in the set **P**, $[\partial f/\partial_{p_i}]_0$ is computed at the nominal values of p_i, $\Delta p_i = p_i - \mu_{p_i}$ is a normally distributed random variable and $\Delta p_i \sim N(0, \sigma_{p_i})$. The delay function here is arbitrary, and may include, for example, the effects of the input transition time on the gate delay.

If all of the parameter variations can be modeled by Gaussian distributions, this approximation implies that d is a linear combination of Gaussians, and is therefore Gaussian. Its mean μ_d and variance σ_d^2 are

$$\mu_d = d_0 \tag{6.8}$$

$$\sigma_d^2 = \sum_{\forall i}\left[\frac{\partial f}{\partial p_i}\right]_0^2 \sigma_{p_i}^2 + 2\sum_{\forall i \neq j}\left[\frac{\partial f}{\partial p_i}\right]_0 \left[\frac{\partial f}{\partial p_j}\right]_0 \mathrm{cov}(p_i, p_j) \tag{6.9}$$

where $\mathrm{cov}(p_i, p_j)$ is the covariance of p_i and p_j.

This approximation is valid when Δp_i has relatively small variations, in which domain the first-order Taylor expansion is adequate and the approximation is acceptable with little loss of accuracy. This is generally true of the impact of intrachip variations on delay, where the process parameter variations are relatively small in comparison with the nominal values, and the function changes by a small amount under this perturbation. Hence, the delays, as functions of the process parameters, can be approximated as normal distributions when the parameter variations are assumed to be normal.

The existence of on-chip variations requires an extension of traditional STA techniques to move beyond their deterministic nature. The SSTA approach, which overcomes these problems, treats delays not as fixed numbers but as PDFs, taking the statistical distribution of parametric variations into consideration while analyzing the circuit. The simplest way to achieve this, in terms of the complexity of implementation, may be through Monte Carlo analysis. While such an analysis can handle arbitrarily complex variations, its major disadvantage is in its extremely large runtimes. Therefore, more efficient methods are called for.

The task of STA can be distilled into two types of operations:

- A gate is being processed in STA when the arrival times of all inputs are known, at which time the candidate delay values at the output are computed using the "sum" operation that adds the delay at each input with the input-to-output pin delay.
- Once these candidate delays have been found, the "max" operation is applied to determine the maximum arrival time at the output.

In SSTA, the operations are identical to STA; the difference is that the pin-to-pin delays and the arrival times are PDFs instead of single numbers. For simplicity, we consider variations only in gate delays here; these techniques can be extended to handle variations in clock skews.

The first method for SSTA to process successfully large benchmarks under probabilistic delay models was initially proposed in 1997, and published in [32]. Like STA, this approach was purely topological, and ignored the Boolean structure of the circuit. It assumed that each gate in the circuit has a delay distribution that is described by a Gaussian PDF, and assumed that all process variations were uncorrelated.

The approach maintains an invariant that expresses all arrival times as Gaussians. As a consequence, since the gate delays are Gaussian, the "sum" operation is merely an addition of Gaussians, which is well known to be a Gaussian. The computation of the max function, however, poses greater problems. The set of candidate delays are all Gaussian, so that this function must find the maximum of Gaussians. In general, the maximum of two Gaussians is *not* a Gaussian, but it may be reasonable to approximate this maximum using a Gaussian. Two approaches for this approximation have been used widely: one based on a classic paper by Clark [33], and another based on the notion of tightness probabilities [34].

Another class of methods includes the work in [35], which uses bounding techniques to arrive at the delay distribution of a circuit, but again, these ignore any spatial correlation effects, and it is important to take this into consideration.

Figure 6.10 shows a comparison of the PDF yielded by an SSTA technique that is unaware of spatial correlations, as compared with a Monte Carlo simulation that incorporates these spatial correlations.

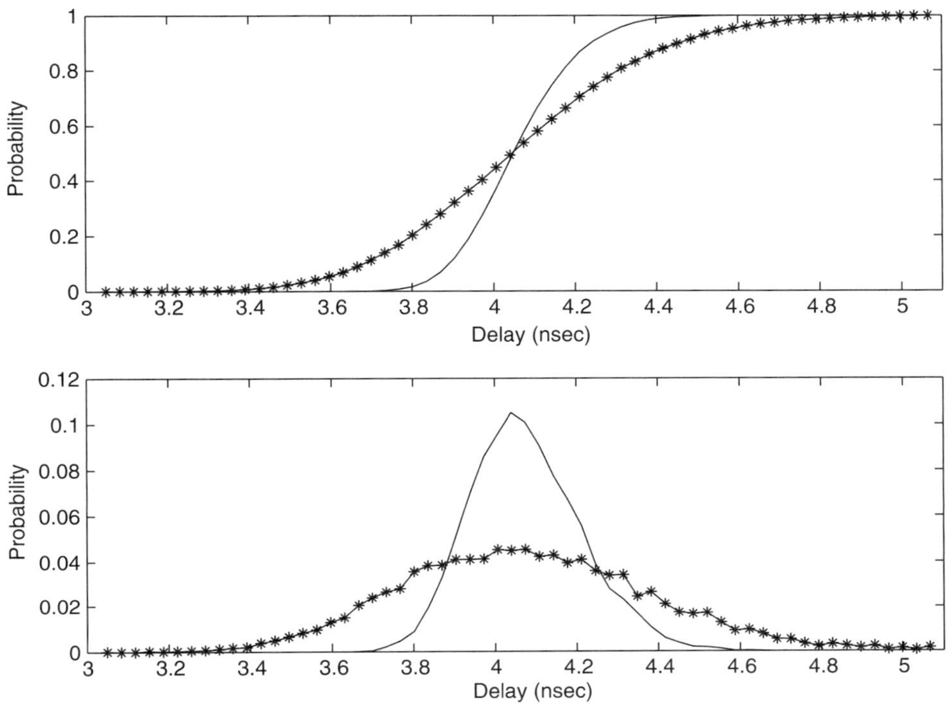

FIGURE 6.10 A comparison of the results of SSTA when the random variables are spatially correlated. The line on which points are marked with stars represents the accurate results obtained by a lengthy Monte Carlo simulation, and the the solid curve shows the results when spatial correlations are entirely ignored. The upper panel shows the CDFs, and the lower panel shows the PDFs [36] (With permission from IEEE.).

The clear difference between the curves demonstrates the need for developing methods that can handle these dependencies.

The approach in [36] presents a novel and simple method based on the application of principal component analysis (PCA) techniques [37] to convert a set of correlated random variables into a set of uncorrelated variables in a transformed space; this method has been widely adopted by numerous SSTA approaches that have appeared subsequently. The PCA step can be performed as a preprocessing step for a design and does not require extensive repeated computation. The use of PCA is effective in permitting rapid and efficient processing of spatial correlations. In cases where some parameters may be spatially correlated and others (such as T_{ox} and N_d) may be uncorrelated, a simple extension of [36] may be used. Once PCA has been performed, Chang and Sapatnekar [36] proceed in a manner that is similar to [32] in that it uses Gaussian models and CPM-like processing, and it develops a theory on how the sum and max operations may be performed under the principal components decomposition. An overview of the procedure is provided in Figure 6.11. The complexity of the method is $p \cdot n$ times the complexity of CPM, where n is the number of squares in the grid and p is the number of correlated parameters, plus the complexity of finding the principal components, which requires very low runtimes in practice. The overall CPU times for this method are low and the method yields accurate results.

6.9 Conclusion

The techniques described in the first six sections have broad applicability and are used widely in a number of commercial timing analyzers. Clock-skew scheduling, after an initial skeptical reception from

Input: Process parameter variations

Output: Distribution of circuit delay

1. Tesselate the chip into n grids, each modeled by spatially correlated variables.
2. For each type of parameter, determine the n jointly normally distributed random variables and the corresponding covariance matrix.
3. Perform an orthogonal transformation to represent each random variable with a set of principal components.
4. For each gate and net connection, model their delays as linear combinations of the principal components generated in step 3.
5. Using "sum" and "max" functions on Gaussian random variables, perform a CPM-like traversal on the graph to find the distribution of the statistical longest path.

FIGURE 6.11 Overall flow of the PCA-based SSTA method.

designers who once considered it "risky," is now being used more widely as there is a growing realization that the fear is unjustified, and that the method can be a very useful tool for speeding up the circuit. The methods in the previous section, on Statistical Static Timing Analysis, relate to a field that is currently an active area of research. A few techniques have been utilized in some design automation tools, and it is expected that the idea of performing STA under a statistical framework will evolve and mature over time, as new algorithms are developed, and as the parameters passed on by the fab to the designer move from a deterministic/process corner-based approach to a truly statistical paradigm.

References

[1] R.B. Hitchcock, Sr., G.L. Smith, and D.D. Cheng, Timing analysis of computer hardware, *IBM J. Res. Dev.*, 26, 100–105, 1982.

[2] S.S. Sapatnekar, *Timing*, Kluwer Academic Publishers, Boston, MA, 2004.

[3] J. Qian, S. Pullela, and L.T. Pillage, Modeling the "effective capacitance" for the RC interconnect of CMOS gates, *IEEE Trans. Comput.-Aided Des. Integrated Circuits Syst.*, 13, 1526–1535,1994.

[4] F. Dartu, N. Menezes, and L.T. Pileggi, Performance computation for precharacterized CMOS gates with RC loads, *IEEE Trans. Comput.-Aided Des. Integrated Circuits Syst.*, 15, 544–553, 1996.

[5] N.P. Jouppi, Timing analysis and performance improvement of MOS VLSI design, *IEEE Trans. Comput.-Aided Des. Integrated Circuits Syst.*, CAD-4, 650–665, 1987.

[6] D. Blaauw, V. Zolotov, and S. Sundareswaran, Slope propagation in static timing analysis, *IEEE Trans. Comput.-Aided Des. Integrated Circuits Syst.*, 21, 1180–1195, 2002.

[7] R.P. Abato, A.D. Drumm, D.J. Hathaway, and L.P.P.P. van Ginneken, Incremental Timing Analysis, U.S. Patent 5508937, 1996.

[8] P.C. McGeer and R.K. Brayton, *Integrating Functional and Temporal Domains in Logic Design*, Kluwer Academic Publishers, Boston, MA, 1991.

[9] K.P. Belkhale, Timing analysis with known false sub-graphs, *Proceedings of the IEEE/ACM International Conference on Computer-Aided Design*, 1995, pp. 736–740.

[10] D.J. Hathaway, J.P. Alvarez, and K.P. Belkhale, Network Timing Analysis Method which Eliminates Timing Variations between Signals Traversing a Common Circuit Path, U.S. Patent 5636372, 1997.

[11] E. Goldberg and A. Saldanha, Timing analysis with implicitly specified false paths, *Workshop Notes, International Workshop on Timing Issues in the Specification and Synthesis of Digital Systems*, 1999, pp. 157–164.

[12] D. Blaauw, R. Panda, and A. Das, Removing user-specified false paths from timing graphs, *Proceedings of the ACM/IEEE Design Automation Conference*, 2000, pp. 270–273.

[13] K.A. Sakallah, T.N. Mudge, and O.A. Olukotun, Analysis and design of latch-controlled synchronous digital circuits, *IEEE Trans. Comput.-Aided Des. Integrated Circuits Syst.*, 11, 322–333, 1992.
[14] T.M. Burks, K.A. Sakallah, and T.N. Mudge, Critical paths in circuits with level-sensitive latches, *IEEE Trans. VLSI Syst.*, 3, 273–291, 1995.
[15] T.G. Szymanski and N. Shenoy, Verifying clock schedules, *Proceedings of the IEEE/ACM International Conference on Computer-Aided Design*, 1992, pp. 124–131.
[16] T.G. Szymanski, Computing optimal clock schedules, *Proceedings of the ACM/IEEE Design Automation Conference*, 1992, pp. 399–404.
[17] J.-F. Lee, D.T. Tang, and C.K. Wong, A timing analysis algorithm for circuits with level-sensitive latches, *Proceedings of the IEEE/ACM International Conference on Computer-Aided Design*, 1994, pp. 743–748.
[18] J.P. Fishburn, Clock skew optimization, *IEEE Trans. Comput.*, 39, 945–951, 1990.
[19] N. Maheshwari and S.S. Sapatnekar, Optimizing large multi-phase level-clocked circuits, *IEEE Trans. Comput.-Aided Des. Integrated Circuits Syst.*, 18, 1249–1264, 1999.
[20] T.H. Cormen, C.E. Leiserson, and R.L. Rivest, *Introduction to Algorithms*, McGraw-Hill, New York, NY, 1990.
[21] R.B. Deokar and S.S. Sapatnekar, A graph-theoretic approach to clock skew optimization, *Proceedings of the IEEE International Symposium on Circuits and Systems*, 1994, pp. 1.407-1.410.
[22] H. Sathyamurthy, S.S. Sapatnekar, and J.R Fishburn, Speeding up pipelined circuits through a combination of gate sizing and clock skew optimization, *Proceedings of the IEEE/ACM International Conference on Computer-Aided Design*, 1995, pp. 467–470.
[23] Y.Z. Liao and C.K. Wong, An algorithm to compact a VLSI symbolic layout with mixed constraints, *IEEE Trans. Comput.-Aided Des. Integrated Circuits Syst.*, 2, 62–69, 1983.
[24] T.M. Burks, K.A. Sakallah, and T.N. Mudge, Optimization of critical paths in circuits with level-sensitive latches, *Proceedings of the IEEE/ACM International Conference on Computer-Aided Design*, 1994, pp. 468–473.
[25] C.E. Leiserson and J.B. Saxe, Retiming synchronous circuitry, *Algorithmica*, 6, 5–35, 1991.
[26] N. Maheshwari and S.S. Sapatnekar, Efficient retiming of large circuits, *IEEE Trans. VLSI Syst.*, 6, 74–83, 1998.
[27] Y. Cheng and S.M. Kang, A temperature-aware simulation environment for reliable ULSI chip design, *IEEE Trans. Comput.-Aided Des. Integrated Circuits Syst.*, 19, 1211–1220, 2000.
[28] T.-Y. Wang and C.C.-P. Chen, 3-D thermal-ADI: a linear-time chip level transient thermal simulator, *IEEE Trans. Comput.-Aided Des. Integrated Circuits Syst.*, 21, 1434–1445, 2002.
[29] B. Goplen and S.S. Sapatnekar, Efficient thermal placement of standard cells in 3d ICs using a force directed approach, *Proceedings of the IEEE/ACM International Conference on Comput.-Aided Design*, 2003, pp. 86–89.
[30] S.S. Sapatnekar and H. Su, Analysis and optimization of power grids, *IEEE Design Test*, 20, 7–15, 2002.
[31] M. Orshansky, L. Milor, P. Chen, K. Keutzer, and C. Hu, Impact of spatial intrachip gate length variability on the performance of high-speed digital circuits, *IEEE Trans. Comput.-Aided Des. Integrated Circuits Syst.*, 21, 544–553, 2002.
[32] E. Jacobs and M.R.C.M. Berkelaar, Gate sizing using a statistical delay model, *Proceedings of Designand Test in Europe*, 2000, pp. 283–290.
[33] C.E. Clark, The greatest of a finite set of random variables, *Oper. Res.*, 9, 85–91, 1961.
[34] C. Visweswariah, K. Ravindran, K. Kalafala, S.G. Walker, and S. Narayan, First-order incremental block-based statistical timing analysis, *Proceedings of the ACM/IEEE Design Automation Conference*, 2004, pp. 331–336.
[35] A. Agarwal, V. Zolotov, and D.T. Blaauw, Statistical timing analysis using bounds and selective enumeration, *IEEE Trans. Comput.-Aided Des. Integrated Circuits Syst.*, 22, 1243–1260, 2003.
[36] H. Chang and S.S. Sapatnekar, Statistical timing analysis considering spatial correlations using a single PERT-like traversal, *Proceedings of the IEEE/ACM International Conference on Comput.-Aided Design*, 2003, pp. 621–625.
[37] D.F. Morrison, *Multivariate Statistical Methods*, McGraw-Hill, New York, NY, 1976.

7

Structured Digital Design

Fan Mo
Synplicity
Sunnyvale, California

Robert K. Brayton
University of California
Berkeley, California

7.1 Introduction

Structured digital circuits, such as datapaths, programmable logic arrays (PLAs), and memories, are regular in both netlist and layout. Structured digital designs usually involve multibit signals and multibit operations, and hence regularity occurs naturally. Regularity in designs is desirable because regular circuit structures offer compact layout and good delay estimation.

One motivation for using regularity is the so-called "timing closure problem," which arises because design flows are sequential; early steps need to predict what the later steps will do. Inaccurate prediction leads to wrong decisions that can only be discovered later, resulting in many design iterations. In the deep submicron (DSM) domain, wiring delays are beginning to dominate gate delays, thereby increasing the importance of obtaining good estimates of wiring effects early in the flow. Structured circuits, because of their layout regularity, allow more accurate area and timing estimation. Also, regularity provides better guarantees that the layout designed by a CAD tool is faithfully replicated in the fabrication. This is because as the limits of the mask-making system are approached, the actual layout patterns on the wafer differ from those produced by a CAD tool. To compensate, techniques like optical proximity correction (OPC) are used. However, the number of layout patterns generated by a conventional design flow can lead to OPC requiring an unreasonable amount of time and generating an enormous data set. Regular structures can reduce these requirements as well as reduce the variations between the CAD tool and fabrication results by using fewer and simpler layout patterns.

Unfortunately, structured digital design is probably the area that involves the least automation today. Many circuits still have their structured parts manually designed, with computer aids limited to layout editing, design rule checking, and parasitic parameter extraction and simulation. Automatic structured digital design is challenging, because a complete structured design flow involves almost every step of a conventional flow, but each step differs from its conventional counterpart. This means that an entire new design flow and a whole set of new algorithms need to be developed to explore the uniqueness of the structural regularity. This subject area is still far from being mature and more or less requires manual intervention. In this chapter, we focus on automation techniques for structured design, and review some of the algorithms and methodologies being developed. These can be individually employed, supplementing existing conventional flows, or they can be organized in a new flow specific to structured digital design.

Use of regularity is not limited to structured circuit modules. Regular chip structures also exist. Gate arrays (GA), a regular array of transistors and routing channels, have been used for years. Recently, a few new regular chip architectures have emerged, including structured application-specific integrated circuits (SASICs) and via patterned gate arrays (VPGAs).

7.2 Datapaths

A datapath is a type of circuit structure that mainly performs multibit operations and has logical and physical regularity. Compared to control logic (sometimes called random logic), it is often more used in the design of arithmetic units. A datapath is composed of a set of datapath components, control logic, and their interconnections.

A datapath example is illustrated in Figure 7.1. The vertical data flow and the horizontal control flow are orthogonal. Along the data flow, which consists of multiple slices, lie multibit functional components including multiplexers, adders, multipliers, registers, etc., called the datapath components. Each datapath component forms a stage. The logic parts within a functional component that are separated by registers are also treated as stages. Due to pipelining some datapath components, like the top component in the figure, are partitioned into stages. Note that the data flow represents the main direction in which multiple data bits transit (vertical in the figure). Some components, like those of $stage_3$, can have internal data direction that is orthogonal to the main data flow. It is also possible that some data bits shift as they enter the next stage, like those at $stage_2$. The regular schematic can be easily translated to a regular layout with the arrangement of the stages and slices intact. In some cases, the data flow might be too long to implement in one single set of slices, resulting in a datapath floorplanning problem to be discussed in Section 7.2.2.3.

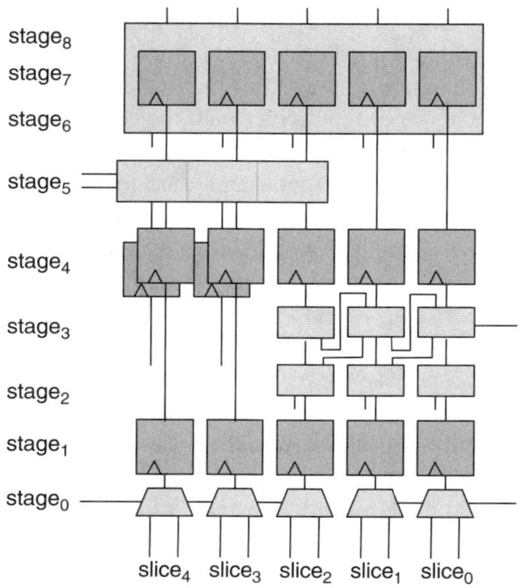

FIGURE 7.1 A datapath example.

7.2.1 Datapath Components

Datapath components are function blocks that realize certain multibit arithmetic or logic operations. There are many references about the details of the structures and functionalities of these components [1]. Here we briefly describe adders and multipliers with concentration on the trade-offs between area, delay, and regularity. Some complex datapath components are often pipelined to reduce the clock cycle. Pipelining is not described in this chapter.

7.2.1.1 Adders

An adder is one of the most widely used arithmetic elements in digital circuits [1]. Variations of adders include components implementing subtraction, comparison, increment, and decrement. Adders are also the building blocks of larger and more complicated arithmetic elements like multipliers.

Several types of adders are illustrated in Figure 7.2. A ripple carry adder (RCA), as shown in Figure 7.2(a), is one of the simplest types. It simulates the hand-calculating process of addition, i.e., computes one sum bit at a time, from the least significant bit (LSB) to the most significant bit (MSB). The carry signals are propagated from the LSB to the MSB. The RCA might be the most compact and regular parallel adder. However, the obvious drawback of RCA is its slower speed because of the long carry propagation path. Carry look-ahead adders (CLA), as shown in Figure 7.2(b), are much faster, making use of the following two facts. If at least one of the two data inputs of a bit is one, then the carry input can be "propagated" to the carry output. If both data inputs are one, then the carry output can be "generated" regardless of the carry input. The idea is generalized to the carry of a segment of bits. The disadvantages of a CLA are the larger area and more irregular structure. A better balance between area and delay is provided by the carry skip adder, as shown in Figure 7.2(c). It divides the set of data bits into several segments and each segment uses an RCA while their carries use the "propagate" logic. Another type is the carry select adder, as shown in Figure 7.2(d). It computes two versions of the sum and carry output of the higher bits, one assuming zero carry input and the other assuming one carry input. Which version becomes the final result is selected by the carry output of the lower bits. Hence, the additions of the higher bits and lower bits can be done simultaneously. Table 7.1 summarizes the delay, area, and regularity of the adders, where N is the bit width [1]. The delays of these "classical" adders are derived under the assumption that all the data inputs and the carry input arrive at the same time.

Modern logic synthesis tools can produce adders as if they are random logic. Area and delay can be somewhat continuously adjusted. This can be useful when the timing conditions for the adder are not normal, e.g., the input arrival times are different. For example, if the arrival times of the data bits are in ascending order, an RCA might be the best choice in terms of area, speed, and regularity. Another example is that certain sum bits, not necessarily including MSB, may be given tighter required times than the carry output, due to the delay characteristics of the surrounding circuits. In this case, the known fast adders may be inadequate. A logic synthesis tool could treat the adder as a normal combinational circuit and apply all the available optimization techniques to meet the timing requirement. A big disadvantage of the synthesis approach is that the synthesized adders are seldom regular. However, it is doubtful whether an adder with abnormal timing requirement can be made regular at all without sacrificing either area or speed. A good approach might be to choose a classical adder that roughly meets the timing requirement, and then apply incremental synthesis to improve it while maintaining its regularity. The combination is implemented in the module generation and resynthesis steps in the datapath design flow described in Section 7.2.2.

7.2.1.2 Multiplier

In the following discussion, an adder with three inputs (carry input as one input) is called a full adder (FA), and an adder with two inputs a half adder (HA). Both have two outputs, one is the *sum* and the other the *carry out*. Note that the *carry out* is one-bit higher in bit significance than the *sum*, because it is supposed to be an input to the adder on the next bit.

An array multiplier schematic is illustrated in Figure 7.3(a). The array multiplier has a speed of N and is quite regular. In each row of the multiplier, there is no carry propagation as in a normal multibit adder.

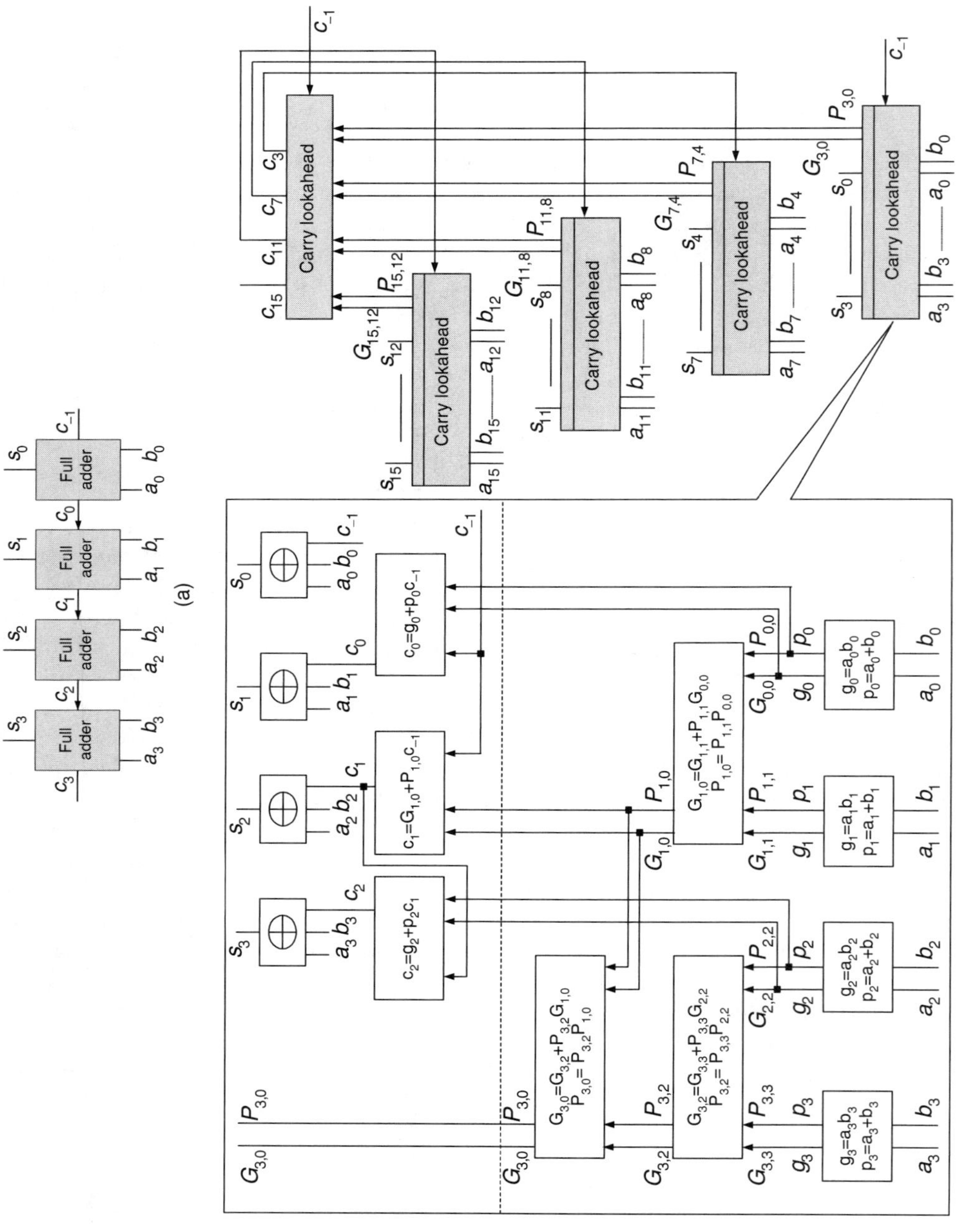

FIGURE 7.2 Several types of adder: (a) a 4-bit ripple carry adder; (b) a 16-bit carry look-ahead adder; (c) a 16-bit carry skip adder; (d) an 8-bit carry select adder.

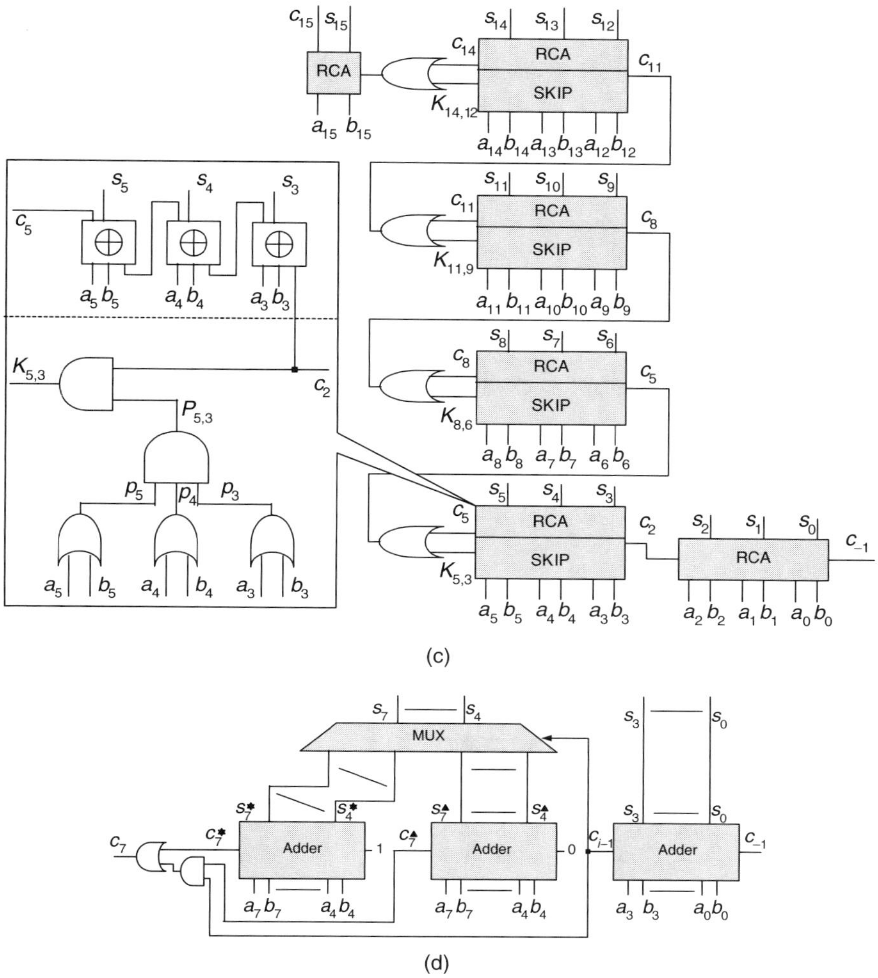

FIGURE 7.2 Continued.

TABLE 7.1

Adder Type	Delay	Area	Regularity
Ripple carry adder	N	N	High
Carry skip adder	$(N)^{1/2}$	N	Medium
Carry select adder	$(N)^{1/2}$	N	Medium
Carry look-ahead adder	$\log N$	$N \log N$	Medium–low

Instead, the carry signals are passed on to the next level. Such an adder organization is called a carry save adder (CSA). The final multibit adder is a conventional adder, or a carry propagate adder. To make the layout a rectangle, the schematic is reshaped, maintaining the array structure and the signal flow, as illustrated in Figure 7.3(b). The two input multiplicands enter the module from the left and bottom, and the product leaves from the right and top. The diagonal connections can be implemented with dogleg-shaped wires.

Improvement in speed can be achieved through rearranging the CSAs such that the longest path is reduced from N to $\log N$. An example is the so-called Wallace tree multiplier illustrated in Figure 7.4 [1].

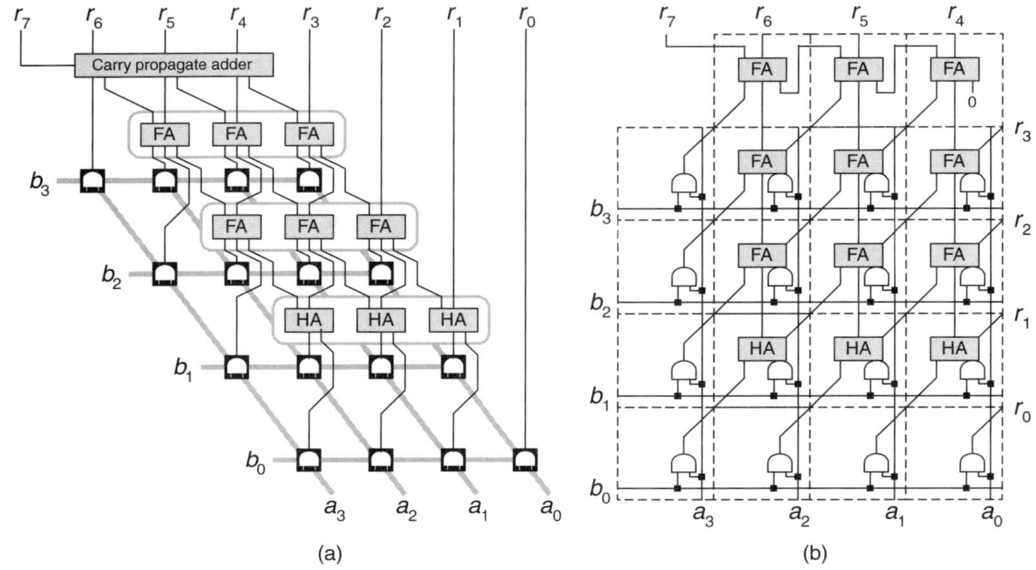

FIGURE 7.3 A 4-bit array multiplier.

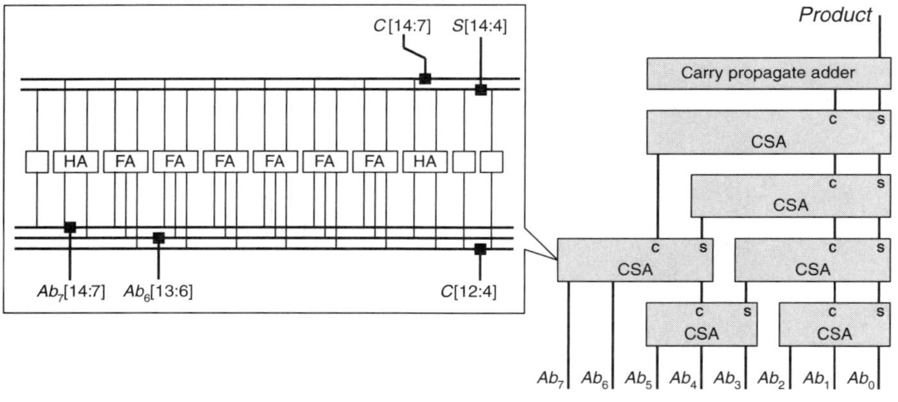

FIGURE 7.4 An 8-bit Wallace tree multiplier.

The notation Ab_i in the figure means a partial product vector. The big disadvantage of a Wallace tree multiplier is the loss of regularity in the layout. Other techniques like Booth encoding can also be used to improve the performance of the multiplier [1].

7.2.1.3 Other Datapath Components

Other datapath components, like the barrel shifter and bit-wise logic operators, are simpler than adders, multipliers, and their variations. They seldom have long combinational paths through all bits like the carry chain in the adders. Thus, most of these components contain a vector of identical elements, which are individually optimized. Area and delay trade-off for the entire component is often unnecessary, and regularity is well preserved.

7.2.2 Datapath Design Flow

A typical design flow for a datapath is illustrated in Figure 7.5. The design input is written in a hardware description language (HDL). Problems like resource scheduling, resource sharing, and pipelining are not

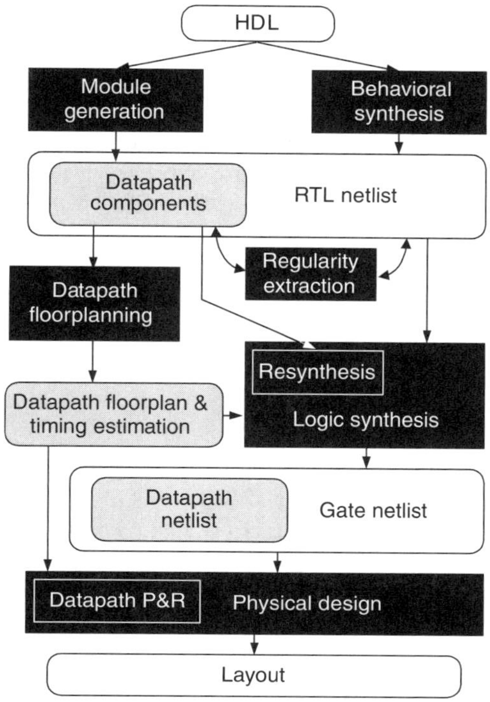

FIGURE 7.5 Datapath design flow.

described in this chapter, but are treated more in depth in chapter 11 of vol. 1 of this handbook. Datapath components can be identified from the register transfer level (RTL) description, and a module generator creates appropriate implementations. The rest of the circuit is compiled to an RTL netlist. When a module generator is not available, those parts of the circuit that have a certain level of regularity can be extracted from the RTL netlist. Regularity extraction can also be applied to a general netlist with or without specific datapath components. The datapath components are floorplanned such that their relative positions in the datapath are determined, and timing information can thus be estimated based on the floorplan. The RTL netlist is synthesized into a gate netlist, during which logic belonging to the datapath components is resynthesized to further improve area and timing. The synthesis process also takes as input the timing estimation based on the datapath floorplan. When the gate netlist is finalized, physical design begins. The datapath floorplan is used in the placement in order to maintain the layout regularity of the datapath and the fidelity of the timing estimation made earlier.

In the following module generation, regularity extraction, floorplanning, resynthesis, and datapath physical design are detailed. Some of these steps can be replaced or enhanced in practice by manual design. Design automation does not conflict with manual intervention, rather it can serve as a means of quick evaluation and fast design-rule checking while the user is making design changes. The flow may also contain loops, which are triggered when violations of design requirements occur. Some commercial synthesis tools label themselves "datapath aware" by integrating module generation, regularity extraction, and resynthesis into conventional flows.

7.2.2.1 Module Generation

Module generation is an automatic method of creating parameterized datapath components. The input is an operator, such as addition, multiplication, comparison, etc., with parameters like data width and speed requirements. The output is a gate netlist of the datapath component for cell-based design. Also, a simulation model is generally required. Generation of a hard datapath component, which also has its layout, requires a library of predesigned modules. This can happen if components laid out in previous designs are kept for reuse, provided that the fabrication technology is unchanged.

Module generation is based on a set of known structures. The generation of an adder, for instance, may involve a set of known structures such as RCA, CLA, and carry select adder. The choice of structure and its area and delay estimations are functions of the required bit width and timing, which can be implemented as a set of empirical equations or look-up tables. It is preferable to generate the smallest module that meets the timing requirement. Module generation is highly technology-dependent, and switching to a different technology may result in totally different modules under the same functional and timing requirements.

Besides the basic parameterized structure selection, there exist various optimization techniques that can be applied during module generation. These increase the complexity of a module generator but can achieve more economic implementations. The generation process becomes an iterative approach as a result, starting from a rough choice of structure and gradually improving it. In the following, we list several such optimization methods.

7.2.2.1.1 Arithmetic Transformation.

Applying arithmetic transformations to the expression may simplify the operations, resulting in smaller area and delay. For instance, an input expression $Z = A \times B + A \times C$ can be transformed to $Z = A \times (B + C)$, saving one multiplier. Additive operands can swap their positions to change the topology of the addition network without changing the number of adders and the functionality [2]. The transformation allows later inputs to feed adders closer to the end of the network, thus shortening the critical paths.

7.2.2.1.2 Operator Merge.

Some complex functions involve several operations, each of which corresponds to a known datapath component like an adder or multiplier. However, merging the operations may enable resource sharing among different components, reducing the area and the number of levels of logic required. For instance, multiplication and accumulation $Z = A \times B + C$ could be directly implemented as a multiplier followed by an adder. However, it is possible to let operand C merge into the CSA tree of the multiplier to reduce the area and delay. Such operator-merging possibilities are hard to recognize later in the resynthesis once separate multiplier and adder are generated.

7.2.2.1.3 Constant Propagation and Redundancy Removal.

Some operations involve constants. For example, $Z = A + 3$. Generating an adder with one input tied to constant "3" can be a starting point. Then the bits of the constant are injected into the adder component, and redundant logic gates are removed along the propagation. Similarly, unused outputs can trigger a backward redundancy removal.

7.2.2.1.4 Special Coefficients.

Operations with special coefficients can be replaced by other operations giving the same results but saving area and reducing delay. An example is $Z = 2^M \times X$. Building a multiplier and then using constant propagation can gain some optimization, but a simpler way is to build an M-shifter directly.

7.2.2.1.5 Restructuring through Multiplexing.

The complex operation $Z = A \times X$ if $(C = D)$ or $B \times X$ if $(C \neq D)$ can be implemented in two ways, as illustrated in Figures 7.6(a) and (b). Obviously implementation of (a) is faster but larger than (b). This could be very useful for datapath design, providing alternatives with different area and delay tradeoffs.

7.2.2.1.6 Restructuring through Topology Transformations.

As shown in Figure 7.7(a), a complex operation involves three sequential operations. Assume input A takes one of the three values A_1, A_2, or A_3. The operation chain can be triplicated, each copy of which has the A input fixed to one of its values. Through constant propagation, OP_1 can be optimized and sped up. The results of the three copies are selected through a multiplexer controlled by A. The carry select adder is an application of this kind of generalized select transformation. Another kind is the generalized bypass transformation [3], the application of which is the carry skip adder. Figure 7.7(c) shows an example. Inputs C and D are used to produce the "bypass" or "skip" signal, indicating whether Z can be solely represented by the output of OP_1.

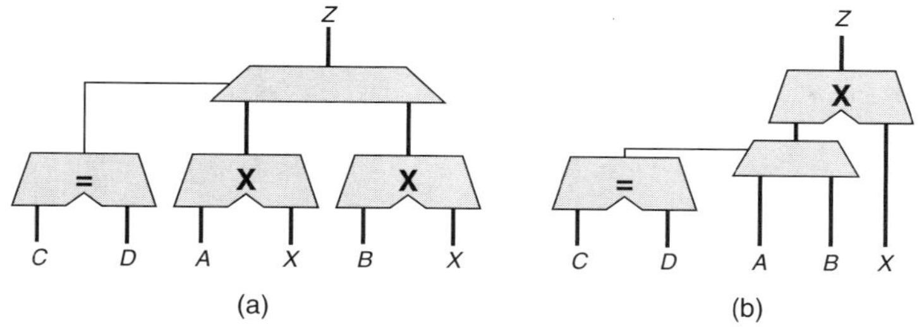

FIGURE 7.6 Restructuring through multiplexing.

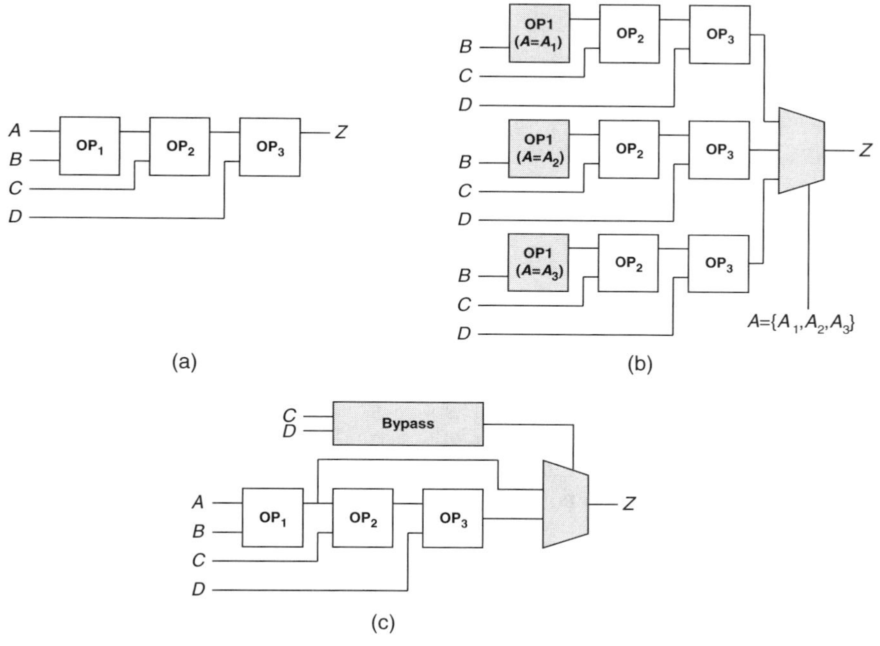

FIGURE 7.7 Restructuring through topology transformations.

7.2.2.2 Regularity Extraction

Regularity extraction can serve as a supplement to module generation. It can collect logic that is not described as datapath components but is regular in nature, e.g., doing multibit operations. Usually such regular logic appears in datapath components previously designed by hand. There are two main approaches to regularity extraction. One is to cover the circuit with templates. The other is to expand from "seeds" of multibit structures.

7.2.2.2.1 The Covering Approach.

The covering approach is composed of two steps, template generation and covering. A template is a subcircuit that can be instantiated multiple times in the circuit. In contrast to standard cells, templates can be as large as an entire column of the array multiplier. Templates can be given by the user or generated automatically from the circuit. The covering step is similar to the technology-mapping step in standard-cell designs.

The example shown in Figure 7.8(a) contains 14 instances where library cells are used as templates. Obviously, this is not what regularity extraction is meant to do. Figure 7.8(b) and Figure 7.8 (c) show two possible coverings of the circuit with larger templates. Even this simple example discloses two interesting

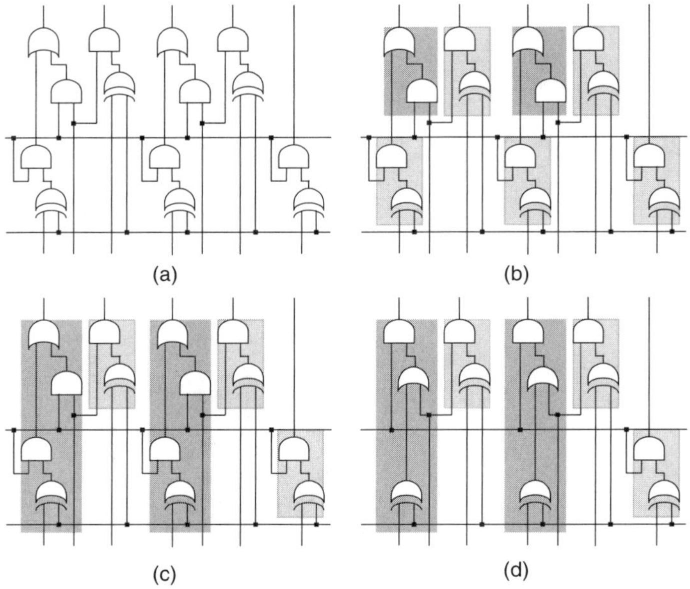

FIGURE 7.8 Regularity extraction.

aspects: first, the choice of a template set greatly affects the covering result; second, the size of a template and the number of its instantiations are contradictory. Assuming that only one instantiation of the same template is further optimized and the rest follows the optimization result, a larger number of instantiations is preferred. Smaller templates, which are easier to be instantiated more frequently, are harder to further optimize. In contrast, a large template has more possibilities of being optimized further. The large template in Figure 7.8(c), for example, can be optimized into a smaller and faster one as shown in Figure 7.8(d). Therefore, generating a good set of templates is essential to regularity extraction.

Of course, the template generation work can be done manually, especially when the designer knows what subcircuits really need to be used frequently in the circuit. An automatic template-generation algorithm was developed [4] in which a pair of gates in the circuit is compared. If identical, then the largest identical (functional and structural) fan-in trees are generated for each pair of their input pins. The gate and all its fan-in trees form a template. The authors also developed an algorithm that generates templates with multiple fanouts.

Given a set of templates, the next step is to cover the circuit. An iterative approach with greedy covering can be used, based on a combination of the "Largest-Fit-First" and "Most-Frequent-Fit-First" heuristics [4]. In each iteration, the template with the best covering effect is chosen, the subcircuits that are covered are removed, and the template set is updated. The iteration is terminated when either the whole circuit is covered or no more covering can take place.

7.2.2.2.2 The Expansion Approach

The expansion approach uses a template set. It starts with some "seeds" of regularity, which can be identified in various ways. For instance, the net names of a bus in the input netlist usually have the same prefix, so the driver gates of the bus can serve as "seeds." Signals like reset and preset span identical gates in multiple slices, and these gates can serve as "seeds."

When a set of seeds is collected, the expansion starts. A seed containing one gate per slice is chosen. The expansion in each slice must be exactly identical, with respect to gate functionality, gate pin order, and gate interconnection. The expansion is stopped if mismatches in the slices occur. At this point, the number of mismatched slices is compared against a threshold. If too many slices mismatch, then the expansion is terminated. Otherwise, the slices are partitioned into two groups, with all the slices still matching put in one group and this group is expanded further [5,6].

7.2.2.3 Floorplanning

As mentioned, components of a large datapath may not be able to be aligned in a single set of slices. Thus, the data flow must turn, and multiple sets of slices must be constructed. The datapath floorplanning problem is to determine the relative locations of the datapath components, and thus the data flow directions. The general floorplanning problem dealing with the placement of the blocks, has been well studied [7]. The objective is to achieve a nonoverlapping placement (locations and orientations) of all the blocks such that the total layout area and interconnection lengths are minimized. The datapath floorplanning problem is unique in that the majority of the nets in a datapath are buses and follow the data flow direction.

One method is to split one long set of slices into several sets and abut them side by side, as shown in Figure 7.9(a). With restriction on the total width or height, the goal is to find an optimal partitioning and pack the slice sets such that the total area is minimized. A dynamic programming method has been used to solve this problem [8]. Another method, as shown in Figure 7.9(b), is to use a general block-packing algorithm but to add in data flow constraints, i.e., adjacent stages should be abutted with the interconnections between them crossing no other stages. Sequence pair, a representation of nonoverlapping block placement [7], can be modified to take such constraints into account [9].

It is possible to regenerate some datapath components during the floorplanning based on more accurate wiring estimation [10,11]. Initially, the module generator produces implementations of the datapath components that just meet the timing requirements, with the wiring delays ignored. The floorplanning of the components makes available the wiring estimation, in which the datapath components can be replaced by faster implementations if necessary. Swapping implementations of datapath components is not the only way of improving timing. Other methods, especially when wire delays become significant, include component replication, reducing depths (number of stages) of critical paths, etc. [12].

7.2.2.4 Resynthesis

As mentioned, the module generation step generates datapath components based on rough or no wiring delay estimation. Although the module regeneration during floorplanning receives more realistic timing information, it still makes choices from a fixed set of classical implementations. It is often hard to generate a datapath component that meets timing requirements and has small area, because of the limited number of implementations created from a few classical structures. Therefore, after the floorplanning, the datapath components may still have room for further optimization. On the other hand, the regularity of the datapath components should be maintained somehow; complete structural change of the datapath components is not desirable. Another reason for datapath resynthesis is that a datapath component created by a module generator has a rigid boundary with other components and random logic. Breaking the boundary allows extra optimization. These two factors imply that the datapath resynthesis should focus on the boundary between the existing datapath components and the rest of the logic.

Common logic optimization techniques can be adopted [13,14]. However, the resynthesis is limited to a subset of the netlist and only small structural changes should take place. Peephole optimization is a useful method for this task [15]. It is a kind of pattern-driven improvement borrowed from compiler techniques, which searches through the netlist under optimization using a small "window." The subcircuit falling into the window is compared with a set of known patterns. Each identified pattern is replaced by another one that is better in area or delay.

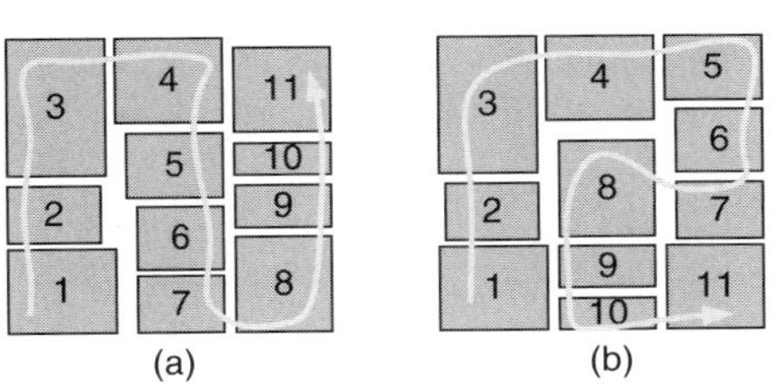

FIGURE 7.9 Dataflow-aware floorplanning.

The sizing of gates is also feasible [16]. Roughly, upsizing gates along the critical paths may reduce path delays, while downsizing along noncritical paths may save area. Buffer insertion can help shield heavy loads for driver gates and improve timing.

7.2.2.5 Physical Design

Once the datapath netlist is optimized, datapath physical design begins. The bit slice and stage information is either maintained from the very beginning of the flow or captured during regularity extraction. If the datapath has to be partitioned in the floorplanning into several sets of slices, the physical design is done for each set and the connections between the sets are made according to the floorplan. We will focus on the placement and routing problem of one set of slices. The perpendicular arrangement of the data flow and control flow allows us to cope with two placement problems independently — the ordering of the stages within one slice and the ordering of slices. The slices usually take their natural order, i.e., from the LSB to the MSB, so we focus on the ordering of stages.

The ordering of the stages for a set of slices might have been determined by the floorplanning. However, it can be adjusted because after resynthesis, the internal structure of the stages as well as the interconnections between the stages may have changed. Moreover, details like the wiring resources, omitted during the floorplanning, could result in inaccurate estimation of routability and timing satisfiability. The stage-ordering problem takes these aspects into account.

As illustrated in Figure 7.10(a), a slice contains four stages, labeled G_A to G_D. The black squares in the figure are the inputs of the stage and the white ones are the outputs. At the bottom of the slice are the external inputs and at the top are the external outputs. The geometric width of a bit of a stage is denoted by $w(stage)$. A stage may have internal routing that occupies $R(stage)$ vertical routing tracks, e.g., $R(G_C) = 1$ in the figure. A signal, denoted by l_i, may span several stages. The number of signals crossing a stage is denoted by $r(stage)$, which varies with the ordering of the stages. As mentioned, a signal can come from an adjacent slice in an orthogonal direction, such as the carry signal in an adder. Such signals are also counted, although this could be an overestimation for the leftmost and rightmost slices. The geometric width of the slice is thus

$$\max\{\max[w(stage)], P \times \max[r(stage) + R(stage)]\}$$

where P is the vertical wiring pitch. Timing constraints are normally transformed to signal length constraints. The goal of stage ordering is to minimize the slice width while meeting all signal–length constraints. Several algorithms can be used to solve this problem. Starting from an initial ordering, usually given by the floorplanning, the ordering of the stages can be altered through simulated annealing [17]; an analytical approach is also feasible [18].

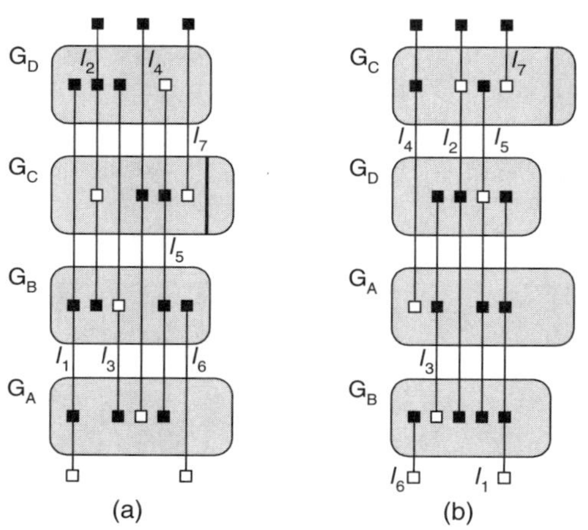

FIGURE 7.10 Ordering of the stages.

General-purpose routing algorithms can be employed to finish the interconnections within and between slices. The routing quality depends on the floorplanning and placement. A datapath-specific router [19] uses pin rails, which are small predefined horizontal segments crossing several vertical tracks. The routing of a pin pair within a slice seeks to find a path between the two, pin rails being used to shift the path from one track to another. The algorithm starts by collecting possible paths for each two-pin connection and splitting routing probability among them. Then the path probabilities of all the connections are added to the tracks. If congestion occurs on a certain segment of a track, some path choices (their probabilities) occupying it are removed. Finally, each path has one and only one choice left, realizing the final routing.

7.2.2.6 Summary of the Datapath Design Flow

The datapath design flow has been described. The input HDL description is translated into datapath components through a module generator and random logic. The latter can be processed for regularity extraction. A floorplanner determines the relative locations of the datapath components. Resynthesis is applied to explore optimization opportunities that lie across the boundary of the components. Physical design is the last step of the datapath flow. The algorithms involved in the flow take advantage of the regularity of the datapath structure.

7.3 Programmable Logic Arrays

7.3.1 Programmable Logic Array Structures and Their Design Methodology

A Programmable Logic Array (PLA) is a regular structure that is widely used to implement control logic. In the following, the PLA structure is briefly described, and then its design methodology is discussed.

As illustrated in Figure 7.11, the main elements of a PLA are two NOR arrays [20]. At each cross point of a NOR array is either an NMOS transistor or a void (an unconnected or deactivated transistor). The first

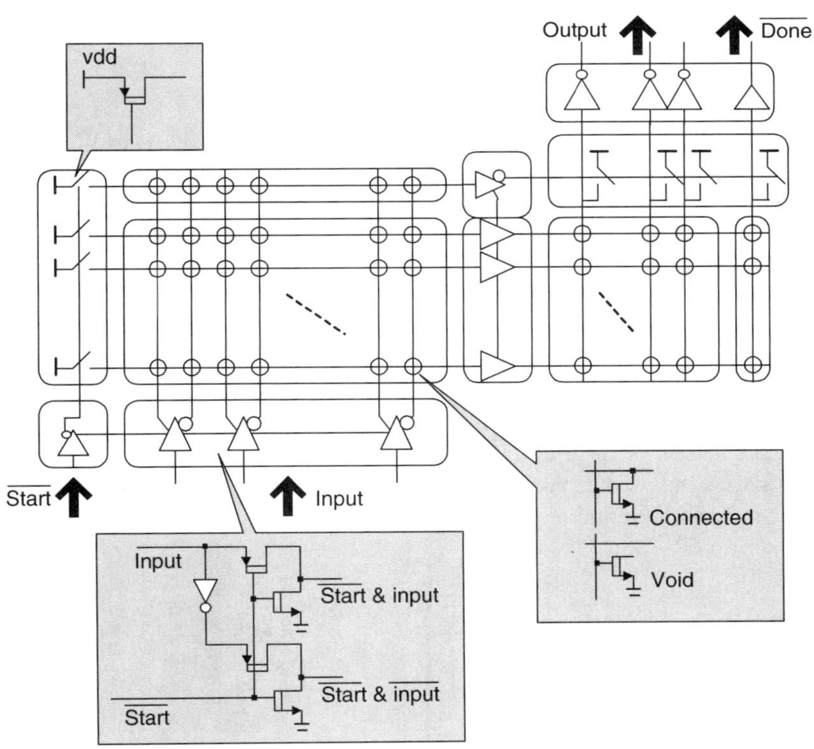

FIGURE 7.11 Programmable logic array structure (dynamic).

array is called the AND plane and its output lines called products. The second array is called the OR plane. A PLA represents the logic output functions in a sum-of-products (SOP) form. The "start"–"done" chain controls the timing of the dynamic PLAs. Owing to its structural regularity, the area and speed of a PLA can be precisely formulated from the SOP expressions that are implemented on it.

The layout generation of a PLA is straightforward. Since the horizontal (product) and vertical (input and output) lines are orthogonal, they can be independently permuted. This permutation acts as placement and routing of the transistors. Hence, the simple PLA structure shown in Figure 7.11 does not need physical design. A PLA layout generation program takes the transistor bit-maps of the two arrays and outputs a PLA layout. Note that although the PLA itself does not require physical design, a circuit being instantiated by multiple PLAs requires placement and routing for the PLA blocks. Synthesis of PLAs has been well studied, and efficient algorithms exist [21,13] for their optimization.

The ratio of the void transistors in the AND and OR planes to the actual transistors can be high in many designs, wasting area and slowing down the PLA because of longer wires. As shown in the simplified PLA diagram in Figure 7.12, the pair of inputs i_1 and i_3 can share the same column (a pair of complementary vertical lines). The same thing applies to outputs o_1 and o_2. Such horizontal compaction is called column folding of the PLA [22]. A restricted version of PLA column folding is discussed below. The two complementary lines of an input signal should remain together so that there is only one input pin per input signal. At most two signals can share one column; hence pins are always on the top and bottom boundaries of the PLA. The column folding of the AND and OR planes are separate, so the basic dual-plane structure of the PLA is preserved. With these restrictions, the PLA column-folding problem is reduced to separate interval-packing problems for the AND and OR planes [23]. In the literature there are also mentions of row-folding and simultaneous column-/row-folding methods [24,25], removing the above restrictions. However, in reality, it is hard for a dynamic PLA structure to support these more aggressive folding methods, because of the arrangement of the "start"–"done" chain and the precharging circuits. In addition, too much folding may cause the input and output pins to spread irregularly on the PLA, which in turn generates placement and routing problems when integrating the PLA onto the chip. These techniques were popular in earlier times when NMOS static PLAs were widely used.

7.3.2 Other Programmable Logic Array Structures

The regular array structure of PLAs is attractive. A single PLA implements several two-level logic functions. However, many logic functions cannot be efficiently implemented using only two levels. In general, multi-level logic synthesis techniques must be adopted to represent various logic functions [14]. A straightforward approach is to have a multilevel Boolean network, with each node in the network being an SOP. After minimization of the network, the nodes can be mapped to the PLAs. However, this "Network of PLAs" approach [26], requires block-level placement and routing, causing a loss of global regularity.

Newer PLA-based structures have been proposed. These structures not only maintain the internal regularity of the PLA, but also maintain a global regularity when putting multiple PLAs together. The "Whirlpool PLA" is a four-level cyclic structure [27]. The Whirlpool PLA can be synthesized with a four-level minimizer called "Doppio Espresso," which iteratively calls *Espresso* [21], a two-level minimizer, to minimize a pair of adjacent NOR arrays, i.e., an SOP. Another PLA-based structure is the "River PLA" [28], which is a stack of PLAs. Interconnections are only present between pairs of adjacent PLAs, and these are implemented with a simple river routing owing to the full permutability of the vertical lines in

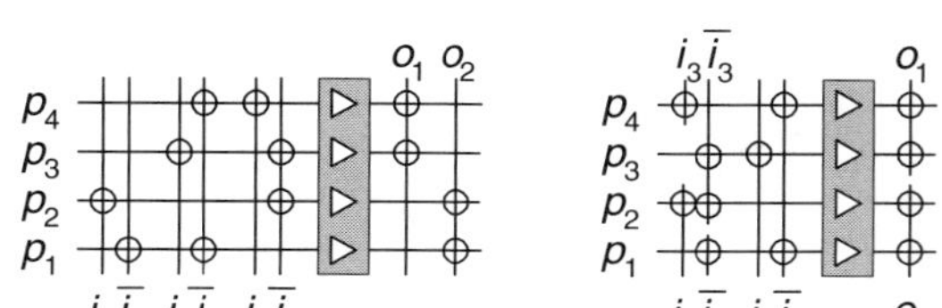

FIGURE 7.12 Programmable logic array folding.

the PLAs. The synthesis of a River PLA involves multilevel logic minimization, clustering nodes into PLAs in the stack, two-level minimization for each PLA, and river routing with vertical line permutation. A Checkerboard is an array of NOR planes [29]. The synthesis algorithms developed for the Checkerboard make use of its structural regularity and flexibility. The multilevel logic network is optimized and decomposed into a netlist of NOR gates. Then the gates are placed into the Checkerboard. Input pin sharing can occur between gates placed in the same NOR plane. Routing between the NOR planes uses a simple spine topology, which is fast enough to allow the placement and routing to be done at the same time. A detailed description and comparison of all these PLA-based structures can be found in [30,31].

7.4 Memory and Register Files

7.4.1 Memory

Memory components are usually generated by a memory compiler that comes with the technology library [32]. A static RAM (SRAM) compiler is discussed. Although other memory structures like dynamic RAM and read only memory have different memory units and surrounding logic, they use very similar compilation techniques.

Large SRAM blocks can be manually partitioned into smaller SRAM banks and floorplanned, or the compiler can be given some predefined partitioning schemes and floorplans [33]. Each bank has a small set of choices of implementations, which differ in area, performance, and power. When input parameters like the address space and the choice of area and speed are given, the memory compiler simply tiles up an array of predefined memory unit cells and places peripheral elements like the row decoder and sense amplifiers around the memory array. The output of a memory block compilation includes a layout (which could be an abstracted view of the actual layout), a behavioral model and a SPICE model. In the following discussion, we focus on how the memory compiler itself is created; in other words, how the memory unit cells and the peripheral elements are systematically, and, to some extent, automatically designed. Although the work can still be done manually, it is much more complicated than the design of a standard cell.

The problem can be stated as follows. Given the address space 2^N, data bit width 2^B, and speed requirement F, determine s_X and s_Y, the X/Y size of the memory unit, the number of rows (2^r), and the number of columns (2^c), such that the speed requirement is satisfied while the total area is minimized. In the statement, the speed requirement F is abstracted, which in reality could represent a set of timing requirements, such as read delay, write delay, and so on.

In Figure 7.13, a simplified 512 × 8 bit SRAM structure is shown. The main body is an array of 2^r-row-by-2^c-column memory units. Note that $2^c = 2^{N-r}B$. Figure 7.13 also shows the structure of a typical six-transistor SRAM unit. The high address $A[N-1{:}N-r]$ is decoded and one of the 2^r rows is activated. The activated row enables all the memory units in that row, which are connected to the complementary vertical bit-line pairs. If the SRAM is under the "read" mode, then the lower address $A[N{-}r{-}1{:}0]$ selects B data bits by the multiplexers and the signals are regenerated by the sense amplifiers. If the SRAM is under the "write" mode, then the data input D_{IN} is demuxed to the corresponding complementary bit-line pairs by $A[N{-}r{-}1{:}0]$.

Since most of the SRAM area is occupied by the memory units, it is reasonable to let s_X and s_Y, rather than the sizes of row decoder, sense amplifiers, etc., determine the column width and row height. The area of the SRAM block is approximately $2^{r+c}s_X\, s_Y$, and this is what we want to minimize. For simplicity, we will assume that s_X and s_Y are equal and are written as a single variable s. The sizes of the row decoder and column peripheral circuits are denoted by s_1 and s_2, respectively. The problem, with extensive abstraction, can be formulated as:

variables: s, s_1, s_2, r, c
minimizing s
subject to

$$s \geq S_0$$
$$\text{aspect\_ratio}(s,r,c) \leq ASP_0$$
$$f(s,r,c,h_1(s,s_1,c),\, h_2(s,s_2,r)) > F$$

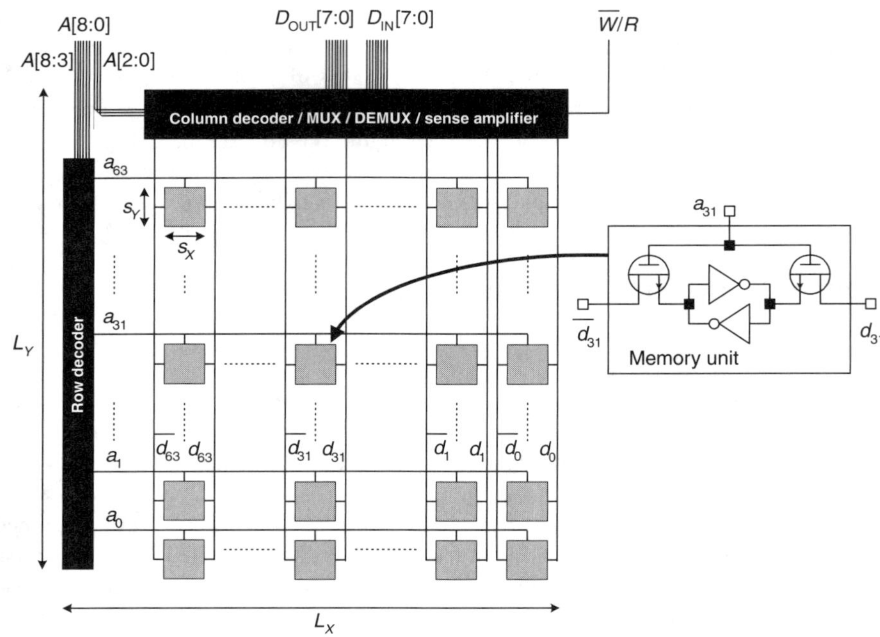

FIGURE 7.13 A 512 × 8-bit SRAM example.

The first constraint means that the memory unit cannot be so small that the sense amplifiers, etc., cannot be properly laid out. S_0 is the lower bound for *s*. The second constraint limits the aspect ratio of the block by ASP_0. The third constraint represents the performance requirement. Performance *f*, the set of various speeds, is a function of the memory unit size, row and column numbers, row decoder performance h_1, and column periphery performance h_2. Given memory unit size *s*, the word line capacitance can be extracted. Together with the load capacitance of a memory unit cell, which is also a function of *s*, the total load of a row on the row decoder can be computed. When the row decoder is sized to s_1, its performance h_1 can be calculated. Similarly, h_2, the performance of the column periphery, can be calculated. The overall performance *f*, depends on *r*, *c*, h_1, h_2, and the cell performance. Strictly speaking, memory unit performance is state-dependent. For example, when computing the write delay for a memory unit, the previous value of its memory state matters. As a result, *f* is actually the minimum performance among all possible state transitions (from 0 to 1, 1 to 1, and so on) [34]. Evaluation of function *f*, which involves the constructions of the unit cell, row decoder and column peripheral circuits, parameter extraction, and SPICE simulation, is time consuming.

The constrained optimization problem, although nonlinear or even discrete in nature, could be solved with various techniques. Run time spent on finding a solution is rarely a problem, because this only needs to be run once for a given technology when creating the memory compiler. Gradient search is a simple method to use [34]. Exhaustive search is also possible, because the number of variables (roughly the number of transistors that need to be sized) is not large, and their ranges are usually small. Even if the sizes of the transistors are manually adjusted, the evaluation of *f* can still be programmed and automated, which greatly eases the burden of complicated extraction and simulation.

7.4.2 Register Files

A register file is a special kind of SRAM, which is usually small in memory size, but has more than one read/write port. It is widely used in microprocessors and digital signal processors as a fast and versatile storage element. The circuit structure of the register file may differ from that of the SRAM in order to

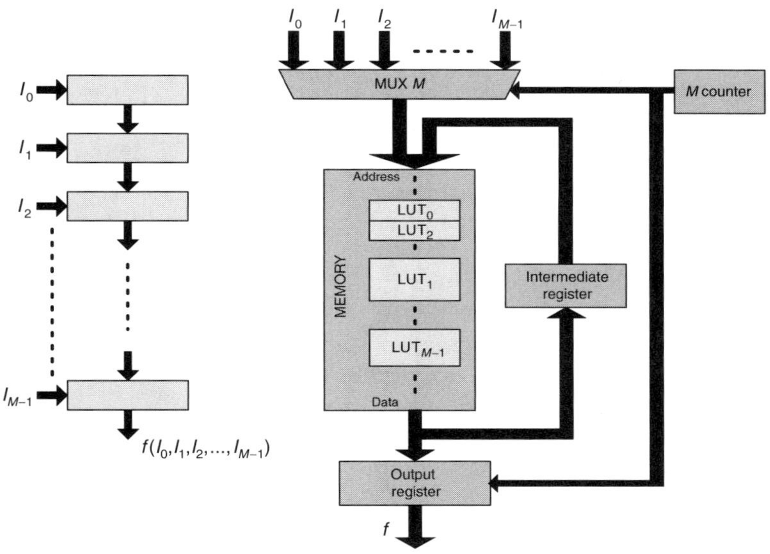

FIGURE 7.14 Cascaded look-up table structure.

achieve higher performance. However, the basic organization of the memory unit array and the peripheries remains the same. Although register files are usually designed by hand to achieve the highest performance, the automatic or semiautomatic optimization approach as done in the SRAM compilation can still be used. The optimization would now focus on the performance rather than the area.

7.4.3 Special Applications of Memory

Memory can be used not only to store data but also to implement logic evaluations. A Look-Up Table (LUT) is a memory block that stores a truth table in the memory cells. The input, feeding the address of the memory, gives the index of the memory cell that stores the output value.

A Field-Programmable Logic Array (FPGA) is a LUT-based structure, using SRAM cells to configure logic functions and the interconnections. The design flow for FPGAs is similar to that for standard cells, but the major difference is that the technology mapping targets LUTs. The output of the flow is not a layout but a bit stream that configures the LUTs.

Another application of memory is the cascaded realization of logic function evaluation [35]. As shown in Figure 7.14, the input vector I of logic function $f(I)$ is partitioned into M disjoint subsets labeled as I_0, I_1, I_2, ... , I_{M-1}, and the evaluation of f becomes a sequence of M evaluations, each involving only one subset of the input. The intermediate output of an evaluation is an input to the next evaluation, joined by another subset of I. A sequential version of this approach includes a memory containing M look-up tables, a multiplexer selecting a subset of the external inputs, a register storing the intermediate output, an output register storing the output, and a counter sequencing the whole operation. The LUTs, which can be of different sizes, need to be packed into the memory. The partitioning of the input and the realization of the function at each stage can be implemented by binary decision diagram (BDD) manipulation. The structure is regular and very flexible, but has low speed.

7.5 Structured Chip Design

In this section, we extend our view from regular circuit modules like datapaths and PLAs to regular chip structures. Several regular chip structures, GAs, SASICs, and VPGAs as well as their design automation are discussed. The guidelines for using these structures are also given.

7.5.1 Gate Arrays

Although these structures are called GAs, the name "transistor array" might be a more appropriate term. An example of a GA structure is illustrated in Figure 7.15(a). The core of a GA is composed of an array of basic cells, also called "bases" interleaved by routing channels. The most widely used base structure in CMOS technology is a four-transistor block, two NMOS transistors and two PMOS transistors. As shown in Figure 7.15(b), a base can be configured as a two-input NAND gate by connecting the terminals of the transistors. To configure more complex gates, such as the four-input NAND gate example in Figure 7.15(b), multiple adjacent bases can be used. All the metal layers of the GA structure are programmable. If channels exist, the metal layers on top of the bases are used to build the internal connections of the gates; the interconnections between the gates are built within the channels. When more metal layers are available, channels are used to construct larger modules, and the GA becomes a sea of gates with intra-gate routing taking place on lower metal layers and inter-gate routing on the upper ones. Masks like diffusion layers are unchanged from design to design if the same GA template is used. The patterns on the fixed layers are regular. Therefore, the GA structure, compared to standard cells, is more cost-effective and manufacturable.

The synthesis for GAs is almost identical to standard cells, except that the library cells become master cells like the ones shown in Figure 7.15(b). The area of a gate is measured by the number of bases it occupies. The physical design for the GA modules is also very similar to standard cells. If the GA structure contains channels, a channel router is needed; if the GA does not contain any channel, and routing takes place over the cell, then an area router is needed.

7.5.2 Structured Application-Specific Integrated Circuits

Cell-based chip structures that dominate today's IC market, provide high area utilization and performance. However, several disadvantages such as the timing closure problem, long turnaround time, and expensive mask costs jeopardize the usefulness of the standard-cell structure to meet the fast-growing design complexity and DSM challenges. The ultimate reason for these disadvantages is that standard cells are not regular enough, leading to timing unpredictability and manufacturability problems. Structured Application-Specific Integrated Circuit (Structured ASIC or SASIC) is a new type of chip structure [36,37] introduced to address some of these problems. In contrast to the transistor array structure in the GA, it embeds a cell array on the chip. The SASIC inherits from the GA the benefits of fewer design-dependent masks, enhancing manufacturability and saving mask costs. Power, clock, and scan chain networks are often optimized and fixed for a particular SASIC by the vendor. This greatly simplifies the design flow and reduces the turnaround time. For small to medium volume productions, SASIC could be a good alternative to standard cells.

The core area of a typical SASIC, as illustrated in Figure 7.16, consists of arrays of SRAM and gate sites. A site is a physical location on the die where a few types (usually only one) of logic primitives can be

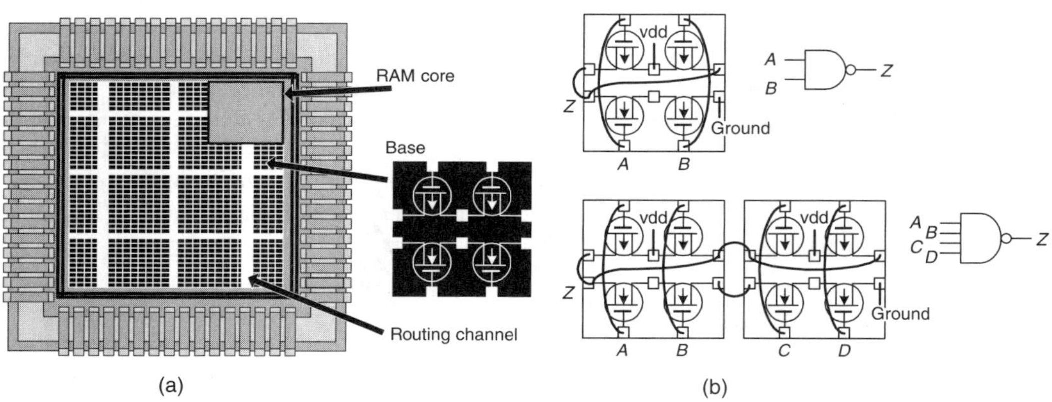

FIGURE 7.15 Gate array example.

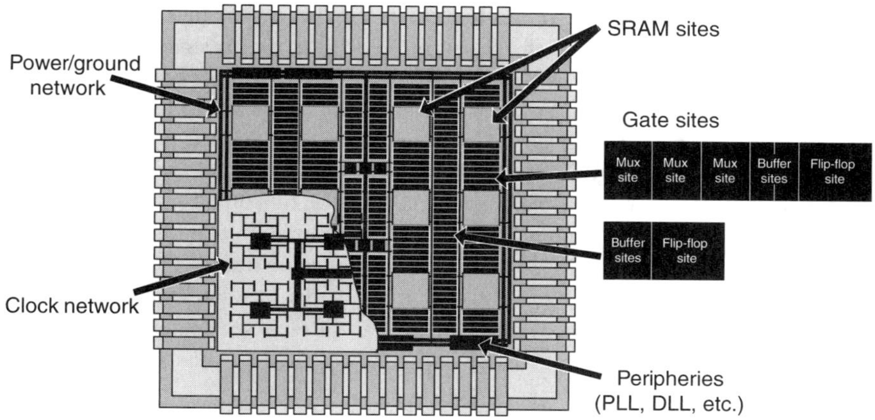

FIGURE 7.16 A SASIC example.

implemented. In some SASIC architectures, if an SRAM site is not used, it can be treated as a set of gate sites. The fixed site locations reduce the number of design-dependent masks; hence the placement problem of logic gates and SRAM modules becomes that of assigning objects to valid and vacant sites. The types of SASIC gate sites are limited as compared to the types of gates in a typical standard-cell library. The lower metal layers of a SASIC are mostly occupied by the internal connections and configurations of the gates and SRAM, as well as power, ground and clock networks. A few higher metal layers can be used for signal routing, but some regions on these layers might be occupied by prerouted nets, such as clocks, and some of the internal connections of the SRAMs.

The design methodology for SASICs is very similar to that for standard cells, from logic synthesis to physical design. The discussion in this chapter focuses on the unique features of the SASIC design flow. The technology mapping for SASICs deals with a relatively small gate library. Although this implies that some degree of optimization in technology mapping is lost, run time can be reduced because of the fewer gate types, or the algorithm can afford to use more CPU-intensive algorithms. Gate sizing and buffering for SASICs have the same limitations and benefits due to the restricted gate choices. The placements of the cells and macro-blocks have the site constraints, i.e., an object must be placed in an appropriate predefined site on the chip. This is somewhat similar to the GA placement, but the difference is that the sites of a SASIC are heterogeneous. A general-purpose cell placer, which is totally site unaware, would have difficulty in translating its placement result to an assignment of gates to legal sites. The placement algorithm needs to control site utilization during the placement. An analytical cell placer could be modified so that it alleviates not only overall row utilization, but also utilization of different site types. A partitioning-based placer could add similar criteria when swapping cells across the cuts. The power/ground routing for SASICs is normally unnecessary. Clock routing, if not fully predesigned by the SASIC vendor, is still needed but greatly simplified because of the regularly arranged register sites.

7.5.3 Via Patterned Gate Array

Both GAs and SASICs require some metal and via layers for configuring the interconnections. Making only via layers configurable leads to the regular chip structure called the VPGA [38,39]. VPGAs adopt the logic unit used in FPGAs, the configurable logic block (CLB). A CLB consists of an LUT or two, one or two flip-flops, and a few other elements like multiplexers. All the metal layers are fixed. The functionalities of the CLBs are configured using lower via layers. On top of each CLB, higher metal layers form a fixed crossbar structure, and signals are routed through vias between these layers. "Jumpers" between the CLBs or the crossbars relay the signal wires from one crossbar to another. A similar routing structure is used in the Checkerboard [29], a PLA-based regular structure described in Section 7.3.2. Using a crossbar is not the only choice for VPGAs; switchboxes or other topologies can also be employed [38]. The only design-dependent masks of a VPGA structure are the via layers, reducing mask costs significantly. Most other layers are laid out in a regular and optimal way, so the manufacturability is enhanced. An example of a VPGA is illustrated in Figure 7.17, with the internal structure of a CLB detailed. The three-input LUT is implemented as a symmetric tree rooted at the output terminal *Z*. The input signals *A*, *B*, and *C* and their complements control the conductive paths from *Z*, realizing any logic function of the three binary inputs. The via configuration of the LUT in the example implements the logic function $Z = AB\overline{C} + \overline{A}BC$.

Synthesis for VPGAs, including technology-independent optimization and LUT mapping, can be borrowed largely from FPGA methodologies. Routing for VPGAs is more flexible than for FPGAs, because of the "sea-of-wires" crossbars. Area-routing methods can be tailored to work with the crossbar structure. Modifications come from the fact that two signals cannot share the same segment in a crossbar of the VPGA.

7.5.4 General Guidelines for Structured Chip Design

Different methodologies can be compared with respect to a number of metrics such as area, performance, power, simpler design methodologies, lower mask costs, faster time-to-market, and improved manufacturability. The last is hard to quantify and needs more research. The last four metrics are becoming increasingly important as the technology advances. GA, SASIC, and VPGA are all competitors of standard cell-based design, but enjoy advantages of simpler design methodologies, lower mask costs, faster time-to-market, and improved manufacturability. On the other hand, area and performance disadvantages would prevent them from being used in high-volume and high-performance applications. If any one of area or performance or power is critical (e.g., in microprocessors or wireless applications), standard cells and manual techniques are still the best choice. Gate arrays require the most mask designs, but their placement and routing are flexible and free, potentially allowing higher area utilization and performance than the SASICs and VPGAs. VPGAs have the greatest area and performance penalties. SASICs sacrifice some placement flexibility but maintain some routing flexibility, and thus are placed between GAs and VPGAs in area and performance. FPGAs are the best for fast prototyping and low-volume applications, but are relatively slow and use a lot of area and power. Their regularity and programmability may increase

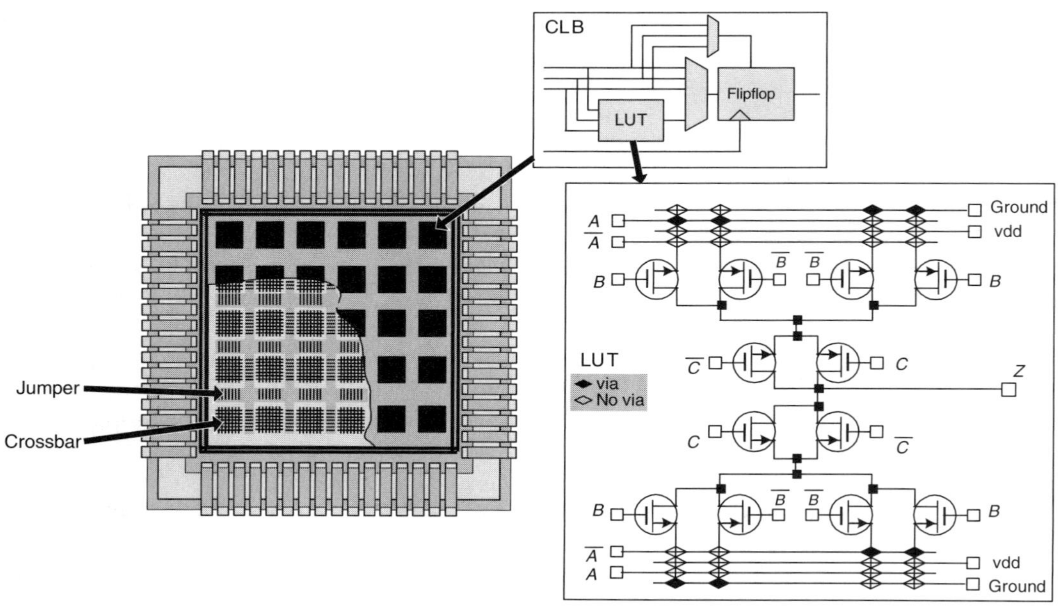

FIGURE 7.17 An example of the VPGA.

their usage in ultra-DSM technologies. Recent research suggests that PLA-based design may be competitive with SCs in terms of area, performance, and power while offering improved manufacturabilty and predictability.

In terms of availability of design automation tools, GAs have been on the market for a long time, so tools are readily available; Cadence and Synopsys both provide tools for various GA design stages. Synplicity already provides SASIC design solutions. Academic VPGA tools are available [38]. Most major EDA companies as well as FPGA vendors themselves provide FPGA design tools.

7.6 Summary

In this chapter, we discussed structured digital design, including datapaths, PLAs, and memories. Regular chip structures like GAs, SASICs, and VPGAs were also discussed. The design automation approaches for these structures were overviewed. The regularity of structured designs can offer high performance, compact layout, reduced mask cost, and improved manufacturability. Many structure-specific algorithms have been developed to identify, maintain, and make use of regularity. In practice, although many structured designs are still done manually, or with little automation, automatic approaches are becoming more attractive as both design complexity and design requirements increase.

References

[1] D.A. Patterson and J.L. Hennessy, *Computer Architecture, A Quantitative Approach*, Appendix A, Morgan Kaufmann, San Francisco, CA, 1996.

[2] I. Neumann, D. Stoffel, M. Berkelaar, and W. Kunz, Layout-driven synthesis of datapath circuits using arithmetic reasoning, *International Workshop on Logic and Synthesis*, 2003, Laguna Beach, CA, pp. 1–6.

[3] P.C. McGeer, R.K. Brayton, A.L. Sangiovanni-Vincentelli, and S. Sahni, Performance enhancement through the generalized bypass transform, *International Conference on Computer-Aided Design*, 1991, pp. 184–187.

[4] A. Chowdhary, S. Kale, P. Saripella, N. Sehgal, and R. Gupta, A general approach for regularity extraction in datapath circuits, *International Conference on Computer Aided Design*, 1998, San Jose, CA, pp. 332–338.

[5] T. Kutzschebauch, Efficient logic optimization using regularity extraction, *International Conference on Computer Design*, 2000, pp. 487–493.

[6] S.R. Arikati and R. Varadarajan, A signature-based approach to regularity extraction, *International Conference on Computer-Aided Design*, 1997, pp. 542–545.

[7] H. Murata, K. Fujiyoshi, S. Nakatake, and Y. Kajitani, VLSI module placement based on rectangle packing by the sequence-pair, *IEEE Trans. Comput.-Aided Design Integrated Circuits Syst.*, 15, 1518–1524, 1996.

[8] D. Paik and S. Sahni, Optimal folding of bit sliced stacks, *IEEE Trans. Comput.-Aided Design*, 12, 1679–1685, 1993.

[9] M. Moe and H. Schmit, Floorplanning of pipelined array modules using sequence pair, *International Symposium on Physical Design*, 2002, pp. 143–150.

[10] V.G. Moshnyaga et al., Layout-driven module selection for register-transfer synthesis of sub-micron ASICs, *International Conference on Computer-Aided Design*, 1993, pp. 100–103.

[11] F. Mo and R.K. Brayton, A timing-driven module-based chip design flow, *Design Automation Conference*, 2004, pp. 67–70.

[12] V.G. Moshnyaga and K. Tamaru, A floorplan-based methodology for data-path synthesis of sub-micron ASICs, *IEICE Trans. Inform. Syst.*, E79-D, 1389–1396, 1996.

[13] R. Brayton, G. Hachtel, C. McMullen, and A. Sangiovanni-Vincentelli, *Logic Minimization Algorithms for VLSI Synthesis*, Kluwer Academic, Dordrecht, 1984.

[14] R. Brayton, G. Hachtel, and A. Sangiovanni-Vincentelli, Multi-level logic synthesis, *Proc. IEEE*, v. 78, 1990, pp. 264–300.

[15] T. Chelcea and S.M. Nowick, Resynthesis and peephole transformations for the optimization of large-scale asynchronous systems, *Design Automation Conference*, 2002, pp. 405–410.

[16] A.J. Kim, C. Bamji, Y. Jiang, and S. Sapatnekar, Concurrent transistor sizing and buffer insertion by considering cost–delay tradeoffs, *International Symposium on Physical Design*, 1997, pp. 130–135.

[17] J.S. Yim and C.M. Kyung, Data path layout optimization using genetic algorithm and simulated annealing, *IEEE Proceedings of Computers and Digital Techniques*, 1998, pp. 135–141.

[18] T.T. Ye and G. De Micheli, Data path placement with regularity, *International Conference on Computer-Aided Design*, 2000, pp. 264–271.

[19] S. Raman, S.S. Sapatnekar, and C.J. Alpert, Datapath routing based on a decongestion metric, *International Symposium on Physical Design*, 2000, pp. 122–127.

[20] Y.B. Dhong and C.P. Tsang, High-speed CMOS POS PLA using predischarged OR array and charge sharing AND array, *IEEE Trans. Circuits Syst. II: Analog Digital Signal Process.*, 39, 557–564, 1992.

[21] R. Rudell and A. Sangiovanni-Vincentelli, Multiple-valued minimization for PLA optimization, *IEEE Trans. Comput.-Aided Design*, 6, 727–750, 1987.

[22] R.A. Wood, A high-density programmable logic array chip, *IEEE Trans. Comput.*, C-28, 602–608, 1979.

[23] N.A. Sherwani, *Algorithms for Physical Design Automation*, Kluwer Academic, Dordrecht, 1993.

[24] D.F. Wong, H.W. Leong, and C.L. Liu, PLA folding by simulated annealing, *IEEE J. Solid-State Circuits*, SC-22, 208–215, 1987.

[25] G. Hachtel, A.R. Newton, and A. Sangiovanni-Vincentelli, An algorithm for optimal PLA folding, *IEEE Trans. Comput.-Aided Des. Integrated Circuits Syst.*, 1, 63–77, 1982.

[26] S. Khatri, R. Brayton, and A. Sangiovanni-Vincentelli, Cross-talk immune VLSI design using a network of PLAs embedded in a regular layout fabric, *International Conference on Computer-Aided Design*, 2000, pp. 412–418.

[27] F. Mo and R.K. Brayton, Whirlpool PLAs: a regular logic structure and their synthesis, *International Conference on Computer-Aided Design*, 2002, pp. 543–550.

[28] F. Mo and R.K. Brayton, River PLA: a regular circuit structure, *Design Automation Conference*, 2002, pp. 201–206.

[29] F. Mo and R.K. Brayton, Checkerboard: a regular structure and its synthesis, *International Workshop on Logic and Synthesis*, 2003, pp. 7–13.

[30] F. Mo and R.K. Brayton, PLA-based regular structures and their synthesis, *IEEE Trans. Comput.-Aided Design Integrated Circuits Syst.*, vol. 22, no. 6, 2003, pp. 723–729.

[31] F. Mo and R.K. Brayton, *Regular Fabrics in Deep Sub-Micron Integrated-Circuit Design*, Kluwer Academic, Dordrecht, 2004.

[32] J. Tou, P. Gee, J. Duh, and R. Eesley, A submicrometer CMOS embedded SRAM compiler, *IEEE J. Solid-State Circuits*, 27, 417–424, 1992.

[33] H. Shinohara, N. Matsumoto, K. Fujimori, Y. Tsujihashi, H. Nakao, S. Kato, Y. Horiba, and A. Tada, A flexible multi-port RAM compiler for data path, *IEEE J. Solid-State Circuits,* vol. 26, 1991, pp. 343–349.

[34] A. Chandna, C. Kibler, R. Brown, M. Roberts, and K. Sakallah, The Aurora RAM compiler, *Design Automation Conference*, 1995, San Francisco, CA, pp. 261–266.

[35] T. Sasao, M. Matsuura, and Y. Iguchi, A cascade realization of multiple output function for reconfigurable hardware, *International Workshop on Logic Synthesis,* 2001, Lake Tahoe, CA, pp. 225–230.

[36] T. Okamoto, T. Kimoto, and N. Maeda, Design methodology and tools for NEC electronics' structured ASIC ISSP, *International Symposium on Physical Design*, 2004, pp. 90–96.

[37] K.-C. Wu and Y.-W. Tsai, Structured ASIC, evolution or revolution?, *International Symposium on Physical Design*, 2004, pp. 103–106.

[38] C. Patel, A. Cozzie, H. Schmit, and L. Pileggi, An architectural exploration of via patterned gate array, *International Symposium on Physical Design*, 2003, pp. 184–189.

[39] L. Pileggi, H. Schmit, A. Strojwas, P. Gopalakrishnan, V. Kheterpal, A. Koorapaty, C. Patel, V. Rovner, and K.Y. Tong, Exploring regular fabrics to optimize the performance-cost trade-off, *Design Automation Conference*, 2003, Anaheim, CA, pp. 782–787.

8 Routing

Louis Scheffer
Cadence Design Systems, Inc.
San Jose, California

8.1 Introduction

The placement step, as described in Chapter 5 *Digital Layout — Placement*, determines the location of each active element of an IC. However, routing remains a crucial step. Routing is the process of creating all the wires needed to properly connect all the placed components, while obeying the design rules of the process.

Routing is the blue-collar work of IC design. There are no conceptual difficulties and very little use of higher mathematics. Almost everything is based on a few basic techniques which have been known for a long time — the papers describing these were published before most readers of this work were born. The success of a router is determined by hard work, heuristics, and implementation — reducing memory usage, increasing speed, and dealing with a multitude of obscure but necessary design rules. Perhaps as a result, almost all state-of-the-art routers are built in industry, and relatively little is published about them. (however, see [1] for an academic router, and [2] for a description of an industrial router.) In contrast, the placement problem can be attacked using a wide variety of techniques, some quite sophisticated, and many academic groups develop their own placers. A placement contest at ISPD in 2005, using state-of-the-art examples, drew nine contestants from academia. A similar contest for routing would probably not find even one academic router that can solve a state-of-the-art routing problem.

Almost every problem associated with routing is known to be intractable. The simplest routing problem, finding the shortest route for *one* net in *one* layer with *no obstacles* and *no design rules*, is called the Steiner tree problem. It is NP-hard if all angles are allowed [3] and NP-complete using horizontal and vertical wires [4]. Variants of channel routing are shown to be NP-complete in [5,6]. Routing considering crosstalk is shown to be NP-complete in [7]; via minimization is shown to be NP-complete in [8], and so on.

Routers therefore, seldom attempt to find an optimum result. Instead, almost all routing is based on heuristics trying to find a solution that is "good enough."

A specific concern for IC routers is that the design rules vary considerably from layer to layer. The allowed width and spacing on the lower layers may be four or more times smaller than the legal widths and spacings on the upper layers. This introduces many additional complications not faced by routers for other applications such as printed wiring boards (PWBs) or multi-chip modules (MCMs). Particular difficulties ensue if the rules are not simple multiples of each other, and when vias must traverse between layers with different rules.

8.2 Types of Routers

The task of all routers is the same. They are given some pre-existing polygons consisting of *pins* (also called *terminals*) on cells, and (optionally) some pre-existing wiring called *preroutes*. Each of these polygons is associated with a net, usually by name or number. The primary task of the router is to create geometries such that all terminals assigned to the same net are connected, no terminals assigned to different nets are connected, and all design rules are obeyed. A router can fail by not connecting terminals that should be connected (an *open*), by mistakenly connecting two terminals that should not be connected (a *short*), or by creating a *design rule violation*. In addition to correctly connecting the nets, routers may also be expected to make sure the design meets timing, has no crosstalk problems, meets any metal density requirements, and so on. This long list of often conflicting objectives is what makes routing difficult.

8.2.1 Basic Router Types

Many different basic types of routers have evolved over the years. The three main forms are illustrated in Figure 8.1. A *channel* is a rectangular region with the terminals on the top and bottom. This is illustrated in Figure 8.1(a). Some signals may be required to connect to the left or right ends of the channel (or both ends). The end signals may have an order specified, but not an exact position. The output of the channel

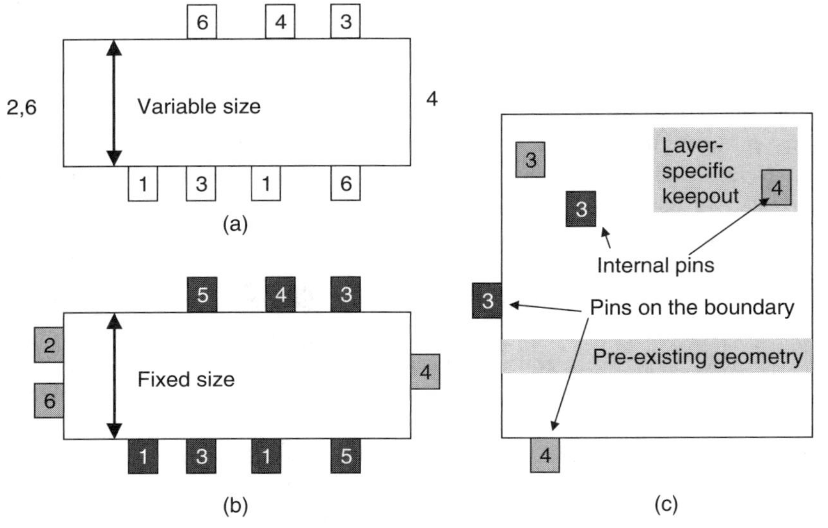

FIGURE 8.1 The three main routing problems: (a) a channel-routing problem; (b) a switchbox-routing problem; and (c) an area-routing problem. In the channel-routing problem, the width of the channel is variable and chosen to make all routes possible. In the switchbox and area problems, the size is fixed, and one or more routes will fail if the routes cannot be completed in the area allotted.

router is a set of geometries that make the required connections. Conceptually, a channel router can fail if there is not enough room for all nets to be routed. However, most channel routers do not do this; instead, they widen the channel until all the nets can be routed. However, to make use of this feature, the channels that make up all the routing on a chip must be routed in a carefully chosen order, often defined by a sliceable placement. Given such a placement, a chip can be routed by calling a channel router repeatedly until the whole chip is built [9,10].

A switchbox router also routes in a rectangular region. However, instead of pins on the top and bottom only, pins may be on all sides of the routing area. An example of a switchbox is shown in Figure 8.1(b). Normally, a switchbox is of defined size, and fails if the routes cannot be completed in the designated area.

An area router, as the name implies, routes in an arbitrary area. This is a much more general problem, and may include layer-specific obstacles and pins, as well as predefined geometries, at any location within the area as well as on the boundary. An example is illustrated in Figure 8.1(c).

8.2.2 Specialized Routers

In addition to these three general routers, there are many specialized routers designed for specific purposes.

8.2.2.1 Single-Layer Routers

Any routing that must be completed in a single layer will require a specialized algorithm, since routes cannot cross. Examples of this are pin-escape routers [11], which try to route a dense array of pins to the periphery of a region (this is normally a PCB and packaging problem), river routers, which change the pitch of a number of parallel wires in the smallest possible length [12–14], and some specialized power supply routers, such as in [15].

8.2.2.2 Power Supply Routers

Another specialty is power supply routers; these route either a highly redundant mesh in multiple layers or interdigitated trees when only a single layer is used. In addition to minimum area, other goals for power supply routers are achieving an acceptable IR drop and an adequate margin for electromigration. (See Chapter 20 *Design and Analysis of Power Supply Networks* for an overview of many of the problems that set the requirements for these routers. Examples are [15,16].)

8.2.2.3 Clock Routers

A third specialty is clock routers. Here the objective is to minimize the skew, or difference between net delays, rather than the usual goal of making every net as fast as possible. Most are variations on the basic idea of [17], which is that after routing between any two leaf nodes, there is some point along this route which will give zero skew. This procedure can be repeated, reducing the number of nodes by two each time, until a zero skew tree is complete (see, e.g., [18–22] and many more).

8.2.2.4 Analog Routers

Another specialty is routers for analog circuits. These can implement many additional routing types (such as single layer, shielded, differential, or bus) and can honor many more constraint types — minimum and maximum Rs and Cs, matching, odd widths, keepouts, and so on. They can often route at additional angles — at least 45°. Routers for analog are normally built on gridless routers for maximum flexibility in meeting the additional constraints. Normally, the capacity and speed will be less than that of a pure digital router (see [23,24] for examples).

8.2.2.5 FPGA Routers

These routers perform the same logical function as IC routers — they decide the exact connections between logical elements. However, since FPGAs are prefabricated, they are picking from a set of predetermined resources. This makes the problem much more combinatorial, and the routers share little except general ideas with conventional IC routers. More information on them can be found in Chapter 13 *FPGA Synthesis and Physical Design.*

8.2.3 Gridded and Gridless

Routers, in general, either generate results whose centerlines lie on a grid (roughly a design rule in size), or they allow the results to lie anywhere, subject only to the resolution of the representation. Not surprisingly, the first group of routers is called *gridded* and the second *gridless.* Internally, gridless routers usually explicitly represent each object (route, pin, or obstruction) whereas gridded routers represent the whole search space, used or not. Therefore gridless routers may offer a memory advantage if the design is sparse or contains objects of very different sizes. When the design is dense with small objects, typical of large flat digital designs, then the memory usage of the two router types tends to be comparable.

In theory, a gridless router should always give results at least as good as a gridded router, since the gridded output is but one of its possible solutions. In practice, however, gridded routers are often superior in terms of runtime, solution quality, or both. This is because the gridded form allows a much simpler representation of the search space, which can (sometimes) allow better global decision making. For example, the capacity of a region to accept new routes is easy to calculate if the routes are gridded, but quite difficult with arbitrary geometries. The result is that gridded routers are typically used on large flat digital designs, but gridless routers may be used for chip assembly, or on designs with analog or other special requirements.

8.3 A Brief History of Routing

Before ICs, there were PWBs. Routers for these [25–27] used maze routing (also called Lee's [28] algorithm) or line-probe routing (sometimes called Hightower's [29] algorithm). Maze routing was slow, but would find a route if one existed. Line-probe routing was faster, but would sometimes fail to find a route for a net, even when such a route existed. In general, the size of a PWB is fixed. Therefore, the general use model was to start with a placement, try to route it, and if it failed, either complete the remaining nets by hand, or add more layers to the board.

The first ICs had only a few layers available for routing, and so shared many similarities to routing a two-layer PWB. However, unlike a PWB, it is possible to change the size of a chip to allow room for all routing (of course, while it is possible to enlarge the chip to allow the routing to complete, it is still preferable to minimize the cost by making it as small as possible). This led to new form of routing, called channel routing. MP2D [30] was an example of a very early system using the channel routing style. Since channel routers (almost) always succeed, as they are allowed to set the size of the routing region, their figure of merit is the height of the required channel in tracks. Some thought revealed that there was a function called the *density*, which at every point along the channel is the number of nets that have pins on both the right and the left. Therefore at least one wire segment must be allocated to each of these nets at this point. The maximum density over the length of the channel (often just called the density) determines the minimum size attainable by any two-layer channel router. Research soon resulted in routers such as YACR2 [35] that could route almost any channel in density.

However, the number of interconnect layers available on ICs soon increased beyond two. Many cells did not require all these layers for their internal connections (for simple cells at least), and designers wanted to use the layers above the cells as routing resources. There were a few attempts to extend channel routers to multiple layers and over the cell routing, but in general they were not widely accepted. Area routers, on the other hand, could easily cope with this, and became the routers of choice.

To address both the memory and speed problems, the next step was the introduction of a "global routing" stage before the detailed route. In this stage, an approximate route for each wire is obtained by routing on a coarse grid, comprised of squares many routing tracks on a side. Global routing is much faster than pure maze routing — the smaller grid can fit entirely in the main memory, and search is much faster since fewer grid squares are involved. After the global route is complete, the area defined by each grid square must be detail-routed, but these are smaller problems and can be solved quickly. (See the section on global routing for more details.)

As the years passed, global routing itself grew more complex. Simple squares with capacities were replaced by more complex squares with capacities per layer, per direction, and via capacities per layer pair.

In the early 1990s, plasma etching replaced wet etching in the manufacture of ICs. This introduced a new problem, the *antenna effect*, that can best be fixed during routing.

In the late 1990s, crosstalk became a serious problem as well. Routers evolved to cope with this in several ways. One way was to assign a higher cost to the regions near sensitive nets. A more complex, but more effective approach is "track assignment." This assigns routes, especially those that go a long distance, to a particular track and layer. This has two main advantages — it reduces the number of jogs on long wires, and it allows certain tests and optimizations (such as timing analysis and crosstalk control) to be done before detail routing is complete.

In the early 2000s, features were still shrinking, though the exposing wavelengths had bottomed out. This made it very hard to get small features, particularly contacts, right, and the yield of contacts and vias became a serious problem. Routers evolved to use doubled vias, first on as "as fits" basis, then built in as a part of the routing process.

Also in the early 2000s, the use of chemical mechanical processing (CMP) introduced the need for more or less uniform metal densities on all layers and all locations. This can be done by postprocessing, but better results can be obtained if the router does it as part of the routing process.

An always existing problem has been engineering change orders (ECOs). These are netlist changes after routing is complete. The router is expected to remove the connections that are now wrong, and add the new connections needed, with minimal disturbance to the nets that need not be changed.

There have also been attempts to extend the two-level global/detailed routing hierarchy to full multilevel methods [31]. These look promising but are not yet in general use as of 2005.

8.4 Common Routing Algorithms

8.4.1 Maze Routing

This is also called Lee routing after an early paper describing the algorithm [28]. The basic algorithm is extremely simple. The main data structure is a priority queue, holding a list of all locations that can be reached from the starting point, but have not yet been examined. The queue is ordered by the cost to reach each point. The lowest cost point is pulled from the queue, all accessible neighbors that have not already been visited are added to the queue, and each visited point is marked with the direction the expansion came from. The algorithm terminates when either the destination is reached, or the queue is empty (in which case no route exists). If a route exists, the arrows are followed backward to the source. It is easy to show that this algorithm will always find a route if one exists, and is *admissable*, meaning it always finds the lowest cost route. Figure 8.2 shows the first steps of a maze router.

Maze routing, though, can be slow and memory-intensive. Imagine attempting to use a pure maze router to do the final routing on a ten-layer chip, 20×20 mm, with 100 nm rules. Even with a coarse grid of 1 pitch (200 nm in this case) the chip is 10^5 tracks across, so there are $10^5 \times 10^5 \times 10 = 10^{11}$ grid points. If each takes a single byte, that is a total of 100 GB, a rather expensive amount of memory (as of 2005). Worse, for a long and obstructed route, a significant number of these cells will need to be set and reset, even with good heuristics. This is already slow, and made worse by the bad cache behavior of such an algorithm. The overall result is that pure maze routing on big designs is not practical.

Although this example uses a particular set of technology rules, memory costs, memory speeds, and dates, it has been valid for many decades and promises to continue to be so. This is because computer memory is made of roughly constant numbers of chips, and each chip is made with last generation design rules, which are proportional to the current rules, given the more or less constant advance of Moore's law. Therefore, both the size of affordable main memory and the number of grid points needed to route a large chip scale in the same way. Both are proportional to the chip area, and the square of the inverse of

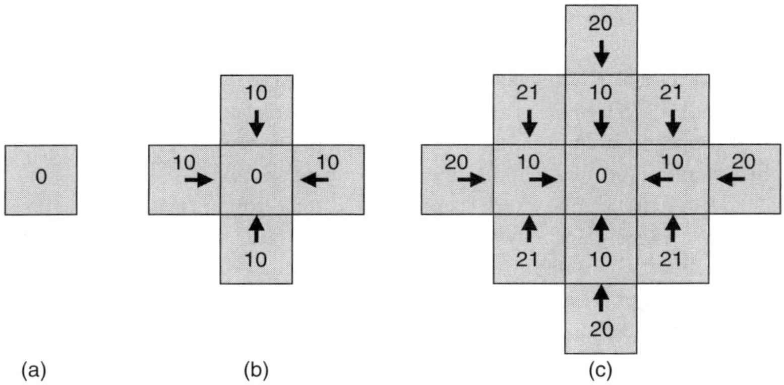

FIGURE 8.2 Maze routing, assuming a cost of 10 for a step and 1 for a turn: (a) the initial state; (b) the state after all the cells of cost 10 are expanded; and (c) the state after all the cells of cost 20 are expanded.

the design rules. Thus, pure-maze routing always has and always will require more main memory than is economical to purchase or fast to access. This drives the use of two-level (global and detailed) and the more general multilevel techniques.

Many improvements to maze routing algorithms are possible [32]. Some of the main improvements include bi-directional search, biasing the search in the direction of the target pin, and better computation of the costs.

8.4.2 The A* Algorithm

A* is a well-known algorithm used to speed up search without sacrificing optimality. It uses heuristic estimates of the remaining cost to bias the search in favor of examining promising solutions first. It is quite a general technique that can be used in many search applications, not just routing.

The basic idea is simple. Instead of ordering the search by the cost needed to get to a given node (as basic maze routing does), order it by the minimum possible cost of any route through that node. This cost is composed of two parts, the cost to get to the node (which is known exactly) and the cost to finish the route, which must be estimated [33]. (See Figure 8.3 for an example of how the algorithm works.)

As long as the estimation function is always optimistic (it never reports more the actual cost), A* is both *admissable* (always finds the lowest cost route, where one exists) and *optimal* (it looks at the minimum possible number of grid squares possible, given a particular estimator).

The optimality, however, is only with respect to a given estimator. The better the estimator, the better A* works. A remaining cost estimate of 0, for example, technically meets the A* requirement. However, performance in this case will be the same as maze routing. An accurate estimation function, on the other hand, can dramatically reduce the number of nodes that must be searched to find a solution.

In routing, a very good estimation function is the remaining Manhatten distance to the target. It is often correct, very fast to compute, and errs on the side of optimism as required. Usually, at least a small penalty for turns should be included. This will bias the algorithm in favor of long straight lines, which is good for both final quality and search time. This is illustrated in Figure 8.4, which shows the squares that are examined using different estimators.

8.4.3 Line-Probe Routing

Line-probe routing is also known as Hightower's algorithm after an early paper in this area [29]. The basic idea is simple — instead of searching in all directions, simply proceed as far as possible in one direction, until either an obstacle is encountered or the route can get no closer to the target. Then repeat the procedure. This can be done on a cellular representation (mainly speeding up the search phase) or on a representation

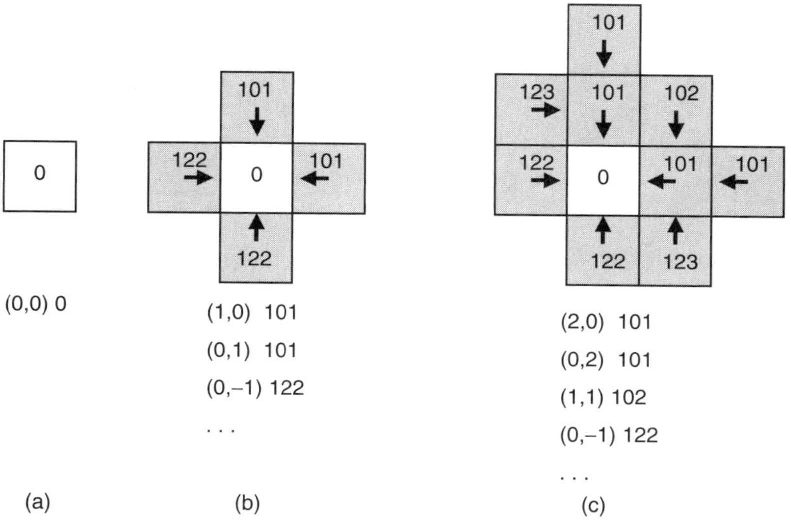

FIGURE 8.3 Example of the A* algorithm with target at (6, 4). (a) the initial state of the queue; (b) the state after the first square has been expanded; and (c) the state after all cells of cost 101 have been expanded.

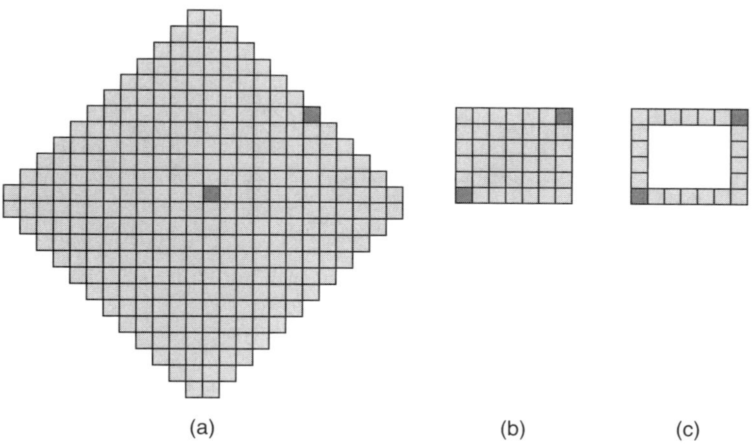

FIGURE 8.4 Illustration of the advantage of the A* algorithm: (a) the cells examined by a pure maze router, when connecting the two shaded cells in the absence of obstacles; (b) the cells examined by an A* router, where the estimated cost of completion is just the Manhatten distance to the destination; (c) shows still more efficient performance where the estimated cost of completion includes a term for turns as well as distance.

containing only line segments (which can save substantial memory, especially in sparse designs). See Figure 8.5 for an example of line-probe routing. Two of the most important properties are shown in this example — it succeeds with only a few points searched, but it does not find the shortest solution.

8.4.4 Channel Routing

The crux of channel routing is placing the required horizontal segments into tracks without creating any shorts. The *Left Edge First* algorithm placed horizontal segments into the first empty track in order of the their left edges. This will always place the wires in at most D tracks, where D is the density [34]. However, this may violate *vertical constraints*, which require some tracks to be routed above others, as shown in the

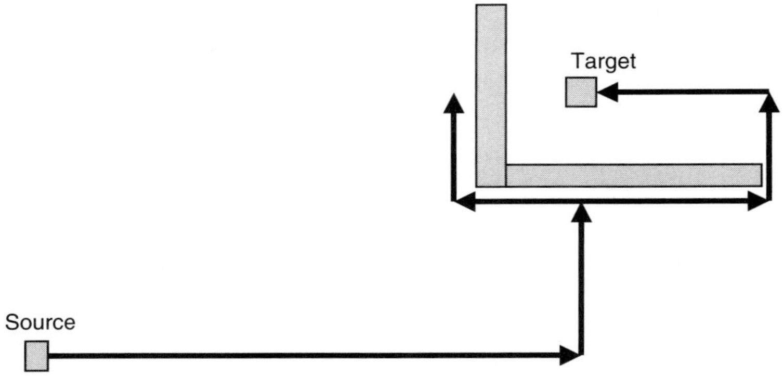

FIGURE 8.5 Line-probe routing. The solution is not optimal, but relatively few decisions were needed to find a solution.

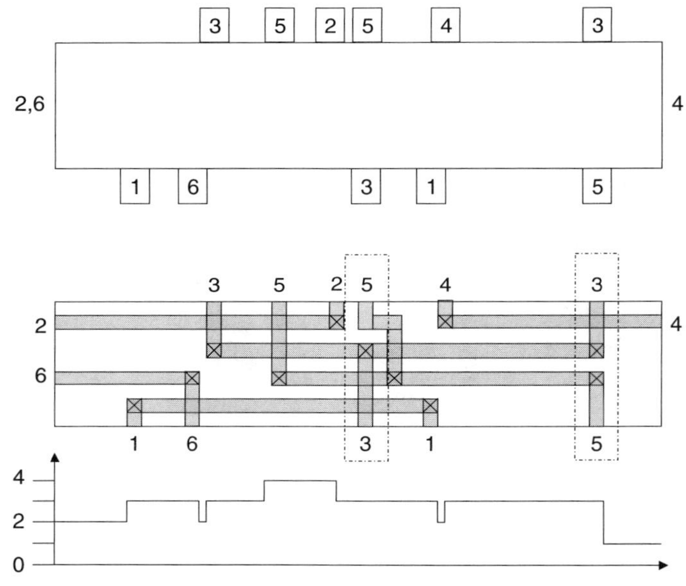

FIGURE 8.6 Channel routing. The upper panel shows the definition of the problem; the middle panel a possible solution; the bottom panel the *density* along the channel. The two pins in each dotted box define a vertical constraint; taken together they form a vertical constraint loop, which cannot be satisfied by any track assignment. The solution is the *dogleg* that connects pin 5 on the top of the channel to the track.

dashed boxes of Figure 8.6. More advanced channel routers use heuristics to attempt to assign segments to tracks in a way that minimizes the vertical constraints. The vertical constraints that cannot be removed by track assignment must be worked around by dogleg routing of the vertical segments, as shown in Figure 8.6. YACR2 is the best known of such routers, and can route large and dense channels in very little time [35].

Traditional channel routing was a two-layer problem. There have been attempts to extend channel routing to more routing layers [36,37]. Although these extensions technically worked, it turned out that area routing was a better solution to the overall problem.

8.4.5 Global Routing

As mentioned earlier, pure-maze routing alone was often impractical in terms of time and memory. The solution is *global routing*, which decides the rough route on the chip for each net, and handles almost all conflicts for resources. During global route, all wires are mapped to a coarse grid made of larger cells called *gcells* for global routing cells. Multiple wires are assigned to a gcell, subject to capacity constraints.

Global routing can be done with a maze router using rip-up and retry, similar to a detailed router, or an algorithm specific to global routing can be used. Several such techniques have been proposed. The most straight foward one builds several potential Steiner trees for each net, finds the per-gcell routing needed for each tree, and then defines a 0–1 variable that defines which tree is used for each net. Then global routing can be expressed as an integer linear programming problem, with the constraints that each net must pick exactly one tree, and that each gcell is assigned no more than its capacity. (See [38] for a survey of different attempts to reduce routing problems to integer linear programming.) Of course, integer linear programming is not only NP-complete in theory, but very hard in practice as well, so this problem cannot be solved exactly either. However, as a heuristic, it can be relaxed to linear programming to generate an approximate but nonintegral solution, which is then snapped to an integer case. This is one of the few uses of sophisticated mathematics in routing.

The representations of gcells has become more complex with time. Simple squares with capacities have largely been replaced by more complex representations with capacities per layer and per direction, and via capacities per layer pair.

Global routing is much faster than detailed routing — the smaller grid can fit entirely in main memory, and search (including rip-up and reroute) is much faster since fewer grid squares are involved. It also gives a very good idea of the difficulty of the detailed routing problem, producing *congestion maps* and reports that show the number of routes assigned to each grid cell, compared to the capacity. This is often expressed as a ratio (demand/supply) and presented to the user as a color-coded map. Cells that have more demand than supply are said to be *overcongested.* If only a few isolated squares are overcongested, and those by a small amount, the detailed router may complete anyway, using tricks such as detours through neighboring squares, or wrong-way routing that avoids the capacity limits. But if there is a high percentage of overcongested gcells and/or any gcells that are dramatically overcongested, then the detailed routing is likely to fail. Also, meeting other objectives such as timing and crosstalk control become dramatically more difficult when the congestion is high, since the detailed router has very little freedom.

Users quickly become adept at predicting design completion from the global routing map. If the design appears unroutable, they must return to the placement or floorplanning steps — (see chapter 1 – *Design Flows,* Chapter 10 – *Design Closure,* and vol, chapter 14 – *Design Planning.*) often incorporate some global routing approximation in an attempt to ensure their results are routable.

The size of the global route cell is critical. Larger cells reduce both memory and run-time requirements, but decrease the accuracy of the global routing step, leading to possible problems in the detailed routing that follows. Estimating how much capacity is consumed by routes contained entirely within a gcell is extremely difficult, and this problem rapidly becomes serious as the grid cell size is increased. Typically, a cell size of 10 to 20 tracks on a side is picked as a compromise between speed and accuracy.

After the global route is complete, the area defined by each of the many global grid squares must be detail-routed. Routing one of these areas takes much less memory than routing the whole chip, has much better cache behavior, and fails much more gracefully, when compared to routing the whole chip. In addition, these smaller routing jobs are mostly independent and can be easily parallelized. However, a good deal of care must be taken at the edges of each global grid square to make sure the individual detailed routes mesh together correctly at the boundaries.

Early works in global routing are [39,40]. A very detailed survey of the various techniques proposed, and their advantages and drawbacks, can be found in [41].

8.5 Additional Router Considerations

8.5.1 Constraints versus Costs

An interesting and almost philosophical question is whether router requests should be phrased as constraints or costs. If the design is feasible, constraints are preferable, since they guarantee that the results will meet the requirements. If the design is infeasible, however, constraints are less useful. In this case, ripup and reroute, to use one example, may loop forever looking for a solution. Furthermore, if a constraint cannot be met, there is often no distinction between slightly infeasible and completely impossible.

Costs behave the opposite way. In the feasible case, they may allow violations if enough improvements are obtained elsewhere, even if those improvements are not really required. In the infeasible case, however, their behavior is much more useful than that of constraints. Cost-driven algorithms simply reduce the total cost as much as possible, and terminate when they can reduce it no further. The resulting design will still contain errors, but may still be useful to determine the likelihood of fixing the problems.

Most routers use a combination of costs and constraints, driven by flow needs and the user's desires. Global routing is usually cost based, for two reasons. First, there is some chance a design can be routed even if it looks infeasible. Second, the congestion map, even if infeasible, gives good feedback to the designer about what needs to change to allow the design to succeed. Detailed routing, on the other hand, is often constraint based. If some nets cannot be routed, the router simply leaves them unconnected rather than violate the rules. Designers normally prefer this to a result that has all nets connected but contains design rule violations.

Combinations of the two approaches are often used as well. A router may well treat design rules as a constraint, but the desire for an empty track next to a timing-critical signal may be treated as a cost. Similarly, a constraint-driven algorithm may convert the constraints to costs to attempt to get closer to a solution. Once a solution is close, the cost can be turned back to a constraint in the hope that it can be satisfied. The best known example is placement, where overlap is treated as a cost early and a constraint later, but the same principle can be used in most routing problems.

8.5.2 Net Ordering

The order in which nets are routed may be significant, but the best order may not be obvious. (See, e.g., Figure 8.7.) Nets A and B look identical — they have the same length and same number of bends, if routed independently. However, if net A is routed first, it has two equally good choices, and may choose Route 1, which blocks net B. If B is routed first no such problem exists — net A can be routed in minimum length no matter which route is chosen for B.

This problem is normally attacked by a technique called "ripup and reroute." If a net cannot be routed, one or more of the blocking nets is ripped up or removed from the routing. Then the previously blocked net is routed, then the router tries to find a new route for the ripped-up net.

In general, this works well. However, there are cases where more intelligent planning helps. For example, routers using only rip-up and reroute will find it hard to route difficult two-layer examples in density.

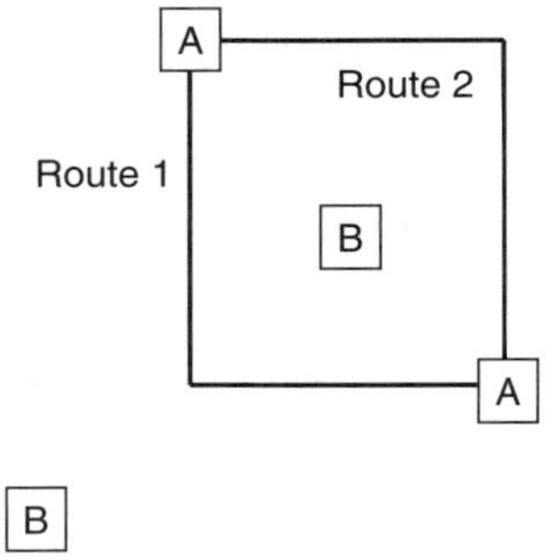

FIGURE 8.7 Example of net order dependence. If net A is routed first, and chooses route 1, then net B must detour. But if B is routed first, then both nets A and B can find a minimum length route, no matter which route is chosen by B.

8.5.3 Timing Driven

The placer (see Chapter 5 *Digital Layout — Placement*) does most of the work in timing-driven design, since it determines the minimum possible distance (and hence delay) for a net (see, e.g., [42]). So, in general the router's job is to make sure that critical or near-critical nets are routed in near-minimum distance. Usually this is done with heuristics:

- During layer assignment, make sure nets get assigned to layers consistent with their timing requirements.
- In both the global and detailed routing phases, route the nets in the order of increasing slack. This helps ensure that critical nets get the shortest (or at least adequately short) routes [43,7].
- Rip-up the timing-critical nets last, since replacing ripped-up nets often increases their length.
- If there is white space available, preferentially allocate it near timing-critical nets. This reduces the capacitance on such nets, and just as importantly reduces the timing pushout from coupling into the net.

8.5.4 Multiple Same-Name Pins on Cells

A purely practical routing problem comes up when a cell has more than one terminal of the same name. In most hardware design languages, only one terminal of a given name is permitted, so the physical meaning is usually clear — all of these terminals should carry the same signal. However, depending on how the cell was designed, the router can be faced with three very different cases:

- The two (or more) pins are not connected internally, so the router must connect them together, along with the other pins on the net, if any. This is called a *must-connect*, since the router must connect the pins.
- The pins are connected internally by a connection that is low in resistance. The router can hook up to only one of the pins if it wishes, or can use the cell to bridge otherwise separate parts of the net. This is called a *strong-connect.*
- The pins are connected internally, but by a connection of high resistance or low current-carrying capacity. In this case the router is free to pick whichever pin it prefers, but the other may not be used. This is called a *weak-connect.*
- Combinations of the above are possible. For example, a cell might have four pins of the same name. Two are strongly connected, one of these is more weakly connected to these two, and the remaining pin is a must-connect.

8.5.5 Multiterminal Nets

Many nets will have more than two pins. Routing such a net in minimum distance is a (usually rectilinear) Steiner tree problem. Since this problem is NP-complete, approximations are almost always employed. A common starting point is the *minimum spanning tree.* Like a Steiner tree, a spanning tree connects all the pins, but unlike the Steiner tree, it consists entirely of pin-to-pin connections — no new junctions are added. Minimum spanning trees are never worse than 50% longer than the optimal Steiner tree, but they are much cheaper to compute. For an N pin net, they can be computed in $O(N^2)$ time by a very simple algorithm, and in $O(N \log(N))$ time by a sophisticated algorithm (assuming the pins are in a plane — normally true for ICs).

One simple approach uses the regular routing process to approximate a Steiner tree. The tree is routed one terminal at a time, and the entire net routed so far is a target for the next terminal connection. Although simple, this approach sometimes generates results no better than a minimum spanning tree.

Other approaches try to construct near-optimal Steiner trees. Typical approaches start with a minimum spanning tree and then improve it using greedy optimizations. Other approaches are to generate a subset (not necessarily optimal) of Steiner trees, such as "single trunk" trees [44], or approximate optimal trees by using table lookup for small trees and composition of small trees for the rest [45].

8.5.6 Engineering Change Orders

Engineering change orders are another router function that is of little theoretical, but considerable practical, value. In theory, if the netlist changes then the router can be run again. In practice, however, there are at least two reasons why good ECO handling is crucial. First, considerable work may have been performed on the previous layout, and the designer wishes to take advantage of this.

Second, good ECO performance is critical to make flows work and achieve design closure (See Chapter 10 *Design Closure*). A router deals with millions of nets, and cannot afford to use the most accurate techniques for each routing decision. Therefore, it uses heuristics to estimate effects such as timing and crosstalk. Since these are approximations, a later more detailed sign-off analysis may reveal problems that the router missed. If these problems are fixed, perhaps by sizing and buffer insertion, and the router rerun from scratch, it is very likely that an entirely new set of very similar problems will crop up. Fixing these will cause yet another set of errors, and the process will converge slowly if at all.

It is crucial, therefore, that the router take an existing routing, and a slightly modified netlist and placement, and route this while preserving as much of the existing routing as possible. To do this, the router must:

- Examine the input netlist, the input placement, and the input routing, to see what is still correct and what must be removed. Nets may be connected to wrong pins, and new placements may obstruct already existing nets. A critical portion of ECO processing is deciding just what to delete. For example, if a net previously had five pins, and now has four, the connection to the pin no longer on the net must be removed. However, this can be done in may ways — just the very last segment to the pin can be deleted, or the branch of the net back to the last Steiner point or pin can be removed, or the whole net can be deleted. Likewise, if a new cell shorts one segment of an existing net, what should be torn up? Just the obstructed segment? The whole from-to? The whole net? There are cases where each of these is the best solution, and heuristics are used to choose between them.
- Once the no-longer needed nets are unrouted, the new connections must be added. This may involve a global routing step, or just detailed routing, depending on the router and the magnitude of the change. Global routing through an already completed design can be very difficult, since it is very difficult to estimate the capacity of each gcell when most nets are already detail-routed. (This is somewhat counterintuitive, but true. The capacity abstracts many shapes into just a few numbers, and is typically tuned to be conservative. Hence you can often rip one net out of a successfully completed gcell, and the global router will still estimate the remaining capacity as zero.)
- If the routes cannot be completed, the router must return to the rip-up-and-reroute step. There may be special cases here for use during ECOs. For example, if there are nets that were added to meet the metal-fill requirements, these should be the first to be ripped up. Otherwise the router will use its usual algorithm.
- If all the nets can be completed, the router must restore the metal fill and any other "finishing" steps it normally performs.

8.5.7 Crosstalk Control

Crosstalk is a very strong function of the routing, so crosstalk control is one of the main jobs of a router, after making sure the connections are correct. There are two parts in most flows — first attempting to reduce the number of problems, then removing any errors found with more sophisticated analysis tools.

The first part, prevention, involves any number of heuristics. These may be purely geometrical (don't run any two lines in parallel for more than distance X) or may be electrically based. A fully detailed crosstalk model is quite complex (see Chapter 21 *Noise Considerations in Digital ICs*), but a router often has enough information for a simple crosstalk model. This includes the approximate slew rates of the nets, and the output impedances and input capacitances of the cells. The router itself can estimate the resistance of each net and coupling to neighboring nets for any proposed routing. Armed with this information, the router can estimate the magnitude of the induced noise on each net. Finally the router, by judicious usage of white space, and the half-shielded tracks next to supply nets, attempts to keep the crosstalk on each net within bounds. Timing windows can be useful here, for if two nets have timing windows that do not overlap, each can act as a constant during the switching of the other. These heuristics work very well when the density is not too high, but in very congested portions of designs it may be very difficult to get satisfactory results for all nets. In this case, buffer insertion or driver sizing may be required, followed by ECO routing.

After the router completes, the next step usually involves invoking a more accurate extractor followed by a sophisticated crosstalk analysis tool (see Chapter 21 *Noise Considerations in Digital ICs* for details). This tool will use the full arsenal of detailed cell models, nonlinear analyses, timing windows, glitch propagation, the effect of noise on delay, and so on, to find out which nets have crosstalk (or crosstalk-induced delay) problems. Then the router is expected to fix these without hurting others too much. Normally, the router starts by computing a "crosstalk slack," or getting it from the crosstalk tool. Like any other slack, this is the amount by which each net could be made worse before it becomes unacceptable. Then the router tries to reduce the noise contributors to the nets with negative slack without hurting the other nets more than is allowable.

8.5.8 Metal Fill

Metal fill was originally done as a post-process. However, having the router do it has some major advantages. First, if the router is timing driven, as most are nowadays, it can be more intelligent about the placement of fill. It can keep areas around critical nets clear, while meeting fill targets by adding metal near nets with bigger margins. Second, since the fill is done during routing, the extractors have a detailed description of the fill, including whether each fill geometry is grounded or floating. This means the designer gets more accurate feedback, and fewer surprises when the detailed sign-off analysis is done. Finally, the fill generated by a router looks much more like regular nets than the checkerboard pattern used by most post-processing. This is very helpful for achieving lithographic uniformity.

Note that the fill must be marked as such in the database, even if the router does it. This is needed since the fill is the first thing to be ripped up during a rip-up-and-reroute process. (Removing all fill, and reinserting later, can lead to reconvergence problems as discussed in the section on ECOs.)

8.5.9 Antenna Effects

The *antenna effect* is a manufacturing problem that exists since ICs are built one layer at a time. Every net, when fully constructed, contains at least one connection to diffusion (at the driver if nowhere else). (See, e.g., Figure 8.8(a).) On the completed net, the diffusion junction protects the rest of the net by (harmlessly) breaking down before the voltage can rise high enough to damage the gate oxide of the inputs tied to the net. However, while the net is being constructed, there may be a significant area of metal attached

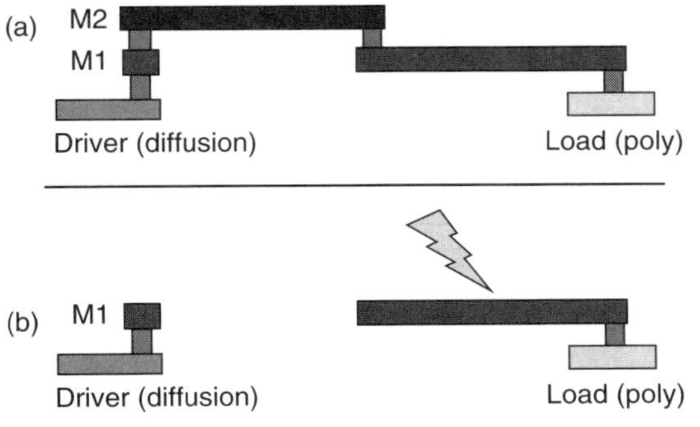

FIGURE 8.8 Diagram showing the antenna problem: (a) the final configuration of the net, which is OK; (b) the net as it is being built. Note the long segment on metal-1 that is connected to the gate, with no connected diffusion to drain any static charge that accumulates.

to inputs while not being connected to any diffusion at all, as shown in Figure 8.8(b), which shows the situation near the end of the metal-1 etch. At this time, the connection to the source–drain region not yet complete, so excess charge on the portion of the net connected to the gate (as may be caused by plasma etching, for example) cannot discharge through a diffusion. If the voltage on the isolated metal piece becomes too high, the gate oxide can break down, creating a permanent defect, either catastrophic or parametric. Fabs limit this problem by publishing *antenna rules*, which limit the ratio of the gate area and the metal area that is allowable at any stage of construction.

A router can fix an antenna problem in many different ways. If the highest layers used in the net routing are connected to the inputs, then the problem will be minimized. This fix is shown in Figure 8.9(a), and the resulting net during construction is shown in Figure 8.9(b), showing why the fix works. For a two-pin net this fix may have very little cost, but for nets with more than one input it may result in an increased number of vias.

A somewhat more expensive fix is shown in Figure 8.10(a). Here a "jumper" is added near the input, on one of the higher levels. This disconnects the input from the rest of the net during construction, as shown in Figure 8.10(b). Of course, this costs even more via congestion than the previous solution.

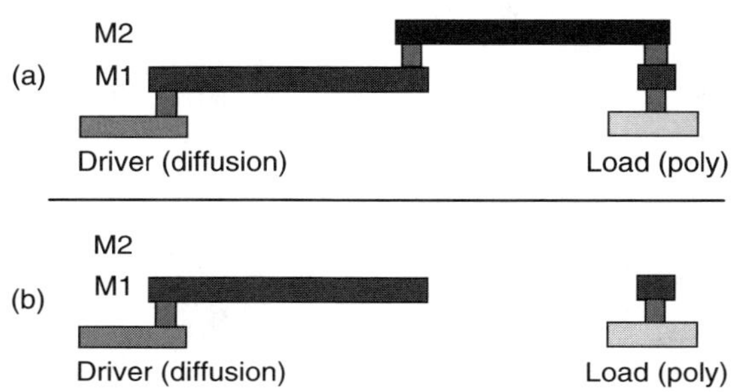

FIGURE 8.9 Diagram showing one way a router can avoid antenna problems: (a) the net with the layer assignments reversed; (b) the net as it is being built. As long as each input is connected directly to the highest layer used in a route, antenna problems are minimized.

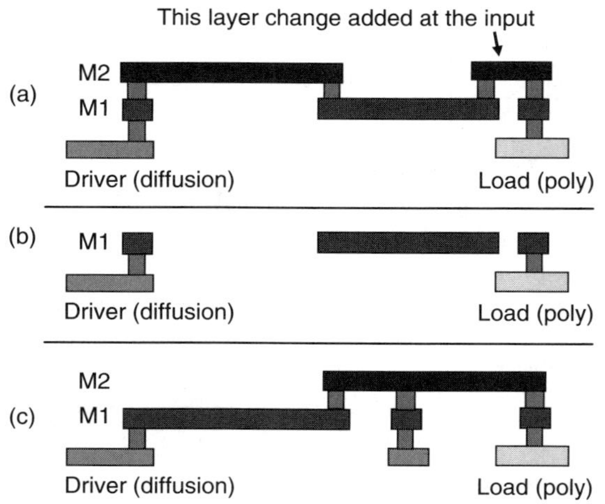

FIGURE 8.10 Diagram showing other ways a router can avoid antenna problems: (a) the net with a "jumper" inserted near the input; (b) why this works, showing the net after layer M1 is constructed, but before M2 is built; (c) yet another solution — inserting diodes on segments connected to inputs.

A third solution to the antenna problem is shown in Figure 8.10(c). Here the router adds a diode (a piece of diffusion) to a portion of the net that would otherwise have an antenna problem. This fix not only requires work from the router (to find a place to put the diode, and route to it) but also hurts performance by adding the diode, which always has at least some parasitic capacitance.

For completeness, it is also possible to avoid antenna problems by including protection diodes on every input. Such a methodology fix makes any effort by the router unnecessary. However, this is the worst solution for performance, since now every input has the parasitic diode, even those with no antenna problem. Leakage and standby current are also increased.

Although a serious practical problem, only a few papers have been published about solutions to antenna effects [46–48].

8.5.10 Track Assignment

The term *track assignment* is used to refer to two different applications. The first is carefully packing signals into tracks to minimize the number of tracks needed, and the number of conflicts in the perpendicular direction. This is crucial when the number of layers is small, and forms the heart of channel-routing algorithms. These algorithms still give the best known results for the classic channel-routing examples. However, they do not deal gracefully with many routing layers and multiple different constraint types, and so have been largely replaced by the "rip-up-and-reroute" algorithms.

The other meaning of track assignment is to assign tracks and layers to the long routes in large digital designs [49]. This step, sometimes called a "trial route," can be done after the global route and before the detailed route. Typically, it is done only for signals traversing at least two global routing squares, and the portion of the route within each global cell remains unassigned. The main advantage is that it is possible to get a very early picture of the routing status — whether it is likely that timing, noise, and capacity can all be satisfied simultaneously.

8.5.11 Angled Routing

Most routers assign adjacent layers to alternating horizontal and vertical directions, as in the channel-routing examples we have seen. For example, layer 1 might be mostly horizontal wires, layer 2 mostly vertical wires, and so on. It is possible that angled routes might be an improvement, since for a wire with equal horizontal and vertical runs the straight line diagonal is $\sqrt{2}$ shorter. In total, if four directions are used, this is called X routing, and layers are routed in a pattern such as horizontal, then vertical, then diagonal left, then diagonal right; Y routing uses three equally spaced directions (the first either vertical or horizontal, and the others 120° either way from this direction). See [50] for an overview of X routing, and [51] for an overview of Y. (There have also been routers that could create routes at any angle whatsoever [52], but these face severe practical problems and remain a curiosity.)

Angled routing offers possible reduced wire length, or even the use of fewer routing layers, but has some potential disadvantages as well.

- Most placers have an internal congestion model, to avoid generating placements too crowded to route. Conventional placements can be used with X or Y routing, but will not get the full benefit. To obtain the full gain, an X-(or Y-) specific placer must be used.
- Layers with mixtures of angles may be difficult to generate optical proximity correction (OPC) for, or even to build. Modern IC manufacturing increasing concentrates on just a few specific geometries (such as arrays of lines and spaces at minimum pitch) and all steps of mask data preparation, mask making, exposure, and etching are optimized for these cases. (See Chapter 18 on *Mask Data Preparation* for more details.) Depending on the degree of pattern-specific optimization, angled lines may not work well as horizontal or vertical lines. Even if they do work as well as lines of more conventional orientation, the fab may not have much experience with them, and may be reluctant to allow them.
- Even if they can be manufactured, manufacturing and mask making are not as well characterized for angled lines, or perhaps not characterized at all. Special process test vehicles may be required.

- The need for a manufacturing grid, and the need to keep vias on this grid, may interfere with the desire to pack angled lines at minimum pitch.
- Y routing in particular may not give good solutions for many data path routing problems. (See Chapter 7 on *Structured Digital Design.*)

Because of these limitations, diagonal routing has so far been used only in designs with very high-performance needs and high volumes, where the smaller numbers of layers needed can be a big advantage.

8.5.12 Wiring Space Prediction

An important related problem is predicting how much space to leave for routing [53], starting from the early stages of design. This is a major topic at conferences such as System Level Interconnect Prediction (SLIP).

References

[1] H. Shin and A. Sangiovanni-Vincentelli, A detailed router based on incremental routing modifications: mighty, *IEEE Trans. Comput.-Aided Design Integrated Circuits Syst.*, 66, 942–955, 1987.

[2] A. Hetzel, A sequential detailed router for huge grid graphs, *DATE '98: Proceedings of the Conference on Design, Automation and Test in Europe*, IEEE Computer Society, Washington, D.C., 1998, pp. 332–339.

[3] M.R. Garey, R.L. Graham, and D.S. Johnson, The complexity of computing Steiner minimal trees, *SIAM J. Appl. Math.*, 32, 835–859, 1977.

[4] M.R. Garey and D.S. Johnson, The rectilinear Steiner tree problem is NP-complete, *SIAM J. Appl. Math.*, 32, 826–834, 1977.

[5] T.G. Szymanski, Dogleg channel routing is NP-complete, *IEEE Trans. CAD*, 4, 31–41, 1985.

[6] M. Sarrafzadeh, Channel-routing problem in the knock-knee mode is NP-complete, *IEEE Trans. CAD*, 6, 503–506, 1987.

[7] H.-P. Tseng, L. Scheffer, and C. Sechen, Timing- and crosstalk-driven area routing, *IEEE Trans. CAD*, 20, 528–544, 2001.

[8] M. Marek-Sadowska, An unconstrained topological via minimization problem for two-layer routing, *IEEE Trans. CAD*, 3, 184–190, 1984.

[9] S. Kimura, N. Kubo, T. Chiba, and I. Nishioka, An automatic routing scheme for general cell LSI, *IEEE Trans. CAD*, 2, 285–292, 1983.

[10] Z. Syed, A.E. Gamal, and M.A. Breuer, On routing for custom integrated circuits, *DAC '82: Proceedings of the 19th Conference on Design Automation*, IEEE Press, Piscataway, NJ, 1982, pp. 887–893.

[11] M.-F. Yu and W.W.-M. Dai, Single-layer fanout routing and routability analysis for ball grid arrays, *ICCAD '95: Proceedings of the 1995 IEEE/ACM International Conference on Computer-Aided Design*, IEEE Computer Society, Washington, D.C., 1995, pp. 581–586.

[12] G. Dastghaibyfard, T.C. Tuan, and P.S. Chang, Multi-terminal-net river routing for VLSI layout design, *Proceedings of the 33rd Midwest Symposium on Circuits and Systems*, Vol. 1, 1990, pp. 211–214.

[13] T.C. Tuan and K.H. Teo, On river routing with minimum number of jogs, *IEEE Trans. CAD*, 10, 271–273, 1991.

[14] T.-C. Tuan, On optimal single jog river routing [VLSI layout], *IEEE Trans. Comput.*, 41, 366–369, 1992.

[15] H. Cai, Multi-pads, single layer power net routing in VLSI circuits, *Proceedings of the 25th ACM/IEEE Design Automation Conference*, 1988, pp. 183–188.

[16] Jaewon Oh and M. Pedram, Multi-pad power/ground network design for uniform distribution of ground bounce, *Proceedings of the 35th Design Automation Conference*, 1998, pp. 287–290.

[17] R.-S. Tsay, An exact zero-skew clock routing algorithm, *IEEE Trans. CAD*, 12, 242–249, 1993.

[18] M. Edahiro, A clustering-based optimization algorithm in zero-skew routings, *Design Automation Conference*, 1993, pp. 612–616.

[19] J. Cong and C.-K. Koh, Minimum-cost bounded-skew clock routing, *ISCAS*, 1995, pp. 215–218.

[20] M. Edahiro, Delay minimization for zero-skew routing, *Proceedings of IEEE International Conference on Computer-Aided Design*, 1993, pp. 563–566.

[21] J. Cong, B. Kahng, C.-K. Koh, and C.-W.A. Tsao, Bounded-skew clock and Steiner routing under elmore delay, *Proceedings of International Conference on Computer-Aided Design*, 1995, pp. 66–71.

[22] A.B. Kahng and C.-W.A. Tsao, Low-cost single-layer clock trees with exact zero elmore delay skew, *Proceedings of IEEE International Conference on Computer-Aided Design*, 1994, pp. 213–218.

[23] E. Malavasi, M. Chilanti, and R. Guerrieri, A general router for analog layout, *CompEuro'89, 'VLSI and Computer Peripherals. VLSI and Microelectronic Applications in Intelligent Peripherals and their Interconnection Networks', Proceedings*, 1989, pp. 5/49–5/51.

[24] K. Lampaert, G. Gielen, and W. Sansen, Analog routing for manufacturability, *Proceedings of the IEEE Custom Integrated Circuits Conference*, 1996, pp. 175–178.

[25] R.W. Kadis, K.L. Thompson, Jr., W.J. Volkman, W.L. Hill, and C.E. Gillette, Building block programs for the layout of printed circuit boards utilizing integrated circuit packs (dapsys v.2), *DAC'68: Proceedings of the 5th Annual Workshop on Design Automation*, ACM Press, New York, 1968, pp. 5.1–5.16.

[26] J.G. Harvey, Automated board layout, *DAC'72: Proceedings of the 9th Workshop on Design Automation*, ACM Press, New York, 1972, pp. 264–271.

[27] D.W. Hightower, The interconnection problem — a tutorial, *DAC'73: Proceedings of the 10th Workshop on Design Automation*, IEEE Press, Piscataway, NJ, 1973, pp. 1–21.

[28] C.Y. Lee, An algorithm for path connections and its applications, *IRE Trans. Electronic Comput.*, EC-10, 346–365, 1961.

[29] D.W. Hightower, A solution to line-routing problems on the continuous plane, *DAC'69: Proceedings of the 6th Annual Conference on Design Automation*, ACM Press, 1969, pp. 1–24.

[30] A. Feller, Automatic layout of low-cost quick-turnaround random-logic custom LSI devices, *DAC'76: Proceedings of the 13th Conference on Design Automation*, ACM Press, New York, 1976, pp. 79–85.

[31] J. Cong, J. Fang, and Y. Zhang, Multilevel approach to full-chip gridless routing, *ICCAD'01: Proceedings of the 2001 IEEE/ACM International Conference on Computer-Aided Design*, IEEE Press, Piscataway, NJ, 2001, pp. 396–403.

[32] J.H. Hoel, Some variations of Lee's algorithm, *IEEE Trans. Comput.*, 25, 19–24, 1976.

[33] P.E. Hart, N.J. Nilsson, and B. Raphael, A formal basis for the heuristic determination of minimum cost paths, *IEEE Trans. Syst. Sci. Cybernetics*, 4, 100–107, 1968. This is the original reference for the A^* algorithm.

[34] A. Hashimoto and J. Stevens, Wire routing by optimizing channel assignment within large apertures, *Proceedings of the 8th Design Automation Workshop*, 1971, pp. 155–163.

[35] J. Reed, A. Sangiovanni-Vincentelli, and M. Santamauro, A new symbolic channel router: YACR2, *IEEE Trans. CAD*, 4, 203–219, 1985. Probably the most widely known channel router.

[36] D. Braun, J.L. Burns, F. Romeo, A. Sangiovanni-Vincentelli, and K. Mayaram, Techniques for multilayer channel routing, *IEEE Trans. CAD*, 7, 698–712, 1988.

[37] B. Berger, M. Brady, D. Brown, and T. Leighton, Nearly optimal algorithms and bounds for multilayer channel routing, *J. ACM*, 42, 500–542, 1995.

[38] T. Lengauer and M. Lugering, Integer program formulations of global routing and placement problems, 1991, Reihe Informatik Nr. 95, Universit at- Gesamthochschule-Paderborn, Paderborn.

[39] R. Nair, S.J. Hong, S. Liles, and R. Villani, Global wiring on a wire-routing machine, *Proceedings of the 19th Design Automation Conference*, Las Vegas, NV, 1982, pp. 224–231.

[40] R. Nair, A simple yet effective technique for global wiring, *IEEE Trans. CAD*, 6, 165–172, 1987.

[41] J. Hu and S.S. Sapatnekar, A survey on multi-net global routing for integrated circuits, *Integration: VLSI J.*, 31, 1–49, 2001.

[42] W. Swartz and C. Sechen, Timing driven placement for large standard cell circuits, *DAC'95: Proceedings of the 32nd ACM/IEEE Conference on Design Automation*, ACM Press, New York, 1995, pp. 211–215.

[43] C.R. Selinger, T.A. Schell, and Y. Kleinman, Net ordering for timing and wiring optimization, *Proceedings of the 6th Annual IEEE International ASIC Conference and Exhibit*, 1993, pp. 550–554.

[44] H. Chen, C. Qiao, F. Zhou, and C.-K. Cheng, Refined single trunk tree: a rectilinear Steiner tree generator for interconnect prediction, *SLIP'02: Proceedings of the 2002 International Workshop on System-Level Interconnect Prediction*, ACM Press, New York, 2002, pp. 85–89.

[45] C. Chu, Fast and accurate rectilinear Steiner minimal tree algorithm for VLSI design, *ISPD'05: Proceedings of the 2005 International Symposium on Physical Design*, 2005, pp. 28–35.

[46] H. Shirota, T. Sadakane, M. Terai, and K. Okazaki, A new router for reducing "antenna effect" in ASIC design, *Proceedings of the IEEE 1998 Custom Integrated Circuits Conference*, 1998, pp. 601–604.

[47] T.-Y. Ho, Y.-W. Chang, and S.-J. Chen, Multilevel routing with antenna avoidance, *ISPD'04: Proceedings of the 2004 International Symposium on Physical Design*, ACM Press, New York, 2004, pp. 34–40.

[48] D. Wu, J. Hu, and R. Mahapatra, Coupling aware timing optimization and antenna avoidance in layer assignment, *ISPD'05: Proceedings of the 2005 International Symposium on Physical Design*, ACM Press, New York, 2005, pp. 20–27.

[49] S. Batterywala, N. Shenoy, W. Nicholls, and H. Zhou, Track assignment: a desirable intermediate step between global routing and detailed routing, *ICCAD'02: Proceedings of the 2002 IEEE/ACM International Conference on Computer-Aided Design*, ACM Press, New York, 2002, pp. 59–66.

[50] S.L. Teig, The X architecture: not your father's diagonal wiring, *SLIP'02: Proceedings of the 2002 International Workshop on System-Level Interconnect Prediction*, ACM Press, New York, 2002, pp. 33–37.

[51] H. Chen, C.-K. Cheng, A.B. Kahng, I. Mandoiu, Q. Wang, and B. Yao, The Y-architecture for on-chip interconnect: analysis and methodology, *ICCAD'03: Proceedings of the 2003 IEEE/ACM International Conference on Computer-Aided Design*, IEEE Computer Society, Washington, D.C., 2003, p. 13.

[52] D. Staepelaere, J. Jue, T. Dayan, and W. Dai, SURF: a rubberband routing system for multichip modules, *IEEE Design Test Comput.*, 10, 18–26, 1993.

[53] W.R. Heller, W.F. Michail, and W.E. Donath, Prediction of wiring space requirements for LSI, *DAC'77: Proceedings of the 14th Conference on Design Automation*, IEEE Press, Piscataway, NJ, 1977, pp. 32–42.

9

Exploring Challenges of Libraries for Electronic Design

James Hogan
Telos Venture Partners, Inc.
Palo Alto, California

Scott T. Becker
Tela Technologies, Inc.
Houston, Texas

9.1 Introduction

Explaining how to best design libraries is usually a difficult task. We all know that a library is a collection of design behavior models at specific points in the design process, but in order to understand fully what it means to design libraries, we have to explore the intricacies and challenges of designing libraries. This includes examining what it means to design libraries, understanding the background, exploring the design process, and perhaps, even analyzing the business models for libraries.

9.2 What Does It Mean to Design Libraries?

Good designers must optimize constraints to achieve market requirements in terms of a finite number of cost functions. For example, they have to consider costs, performance, features, power consumption, quality, and reliability. These considerations are pretty universal for any design — even if you were designing a car, cell phone, or toaster.

On top of traditional design constraints, the dramatic shortening of product life cycles also impacts design engineers. Often, this concern results from the dominance of consumer applications in the marketplace. The components of this trend are:

- Hardware continues to be commoditized.
  - Spiraling design costs lead to an increasing use of design platforms.
  - Original device manufacturers (ODMs) build private label hardware (e.g., Wal-Mart).
  - Original equipment manufacturers (OEMs) differentiate software and build brand value (e.g., Dell).
- Value is created through algorithms, system architecture, and software.
  - Partitioning hardware and software has become the key decision.
  - Reusing design platforms provides market leverage over multiple product cycles.

Thus, there are two opposing issues you must consider when designing libraries: time to market and costs. Industry economics (related partially to the complex nature of manufacturing small geometry silicon and short product lives driven by consumers) have little room for political or technical arguments. Instead, the answer for many system providers is reusable design platforms. Some key advantages of design platforms are:

- Allowing a provider to capture multiple market segments by amortizing large and growing design costs and reducing time to market.
- Reducing the number of core processor architectures while allowing more differentiation at the software application level.
- Increasing the percentage of mixed-signal designs as high speed and mobile applications are integrated into a single silicon system on chip (this does not need to be an SoC, — it could be a multi-die solution or even a small form factor by board implementation).
- Providing flexibility of outsourcing and integrating pre-verified intellectual property (IP) functional blocks.
- Increasing hardware and software programmability. For the system architects the trade-off is hardware or software. Software offers flexibility, but costs silicon real estate, degraded system performance, and increased power consumption.
- Enabling special emphasis or "special sauce" to be captured in custom blocks or in software.
- Including retargeted IP. Previously used IP substantially reduces functional risk.

9.3 How Did We Get Here, Anyway?

Electronic system design has evolved in the last 30 years into a hierarchical process, which generally can be separated into three groups: system design, hardware and software implementation, and manufacturing and test. Although each group has its own area of optimization, they each must maintain the design intent originally specified in the system requirements. Each level of the design hierarchy must preserve the design intent of the preceding level.

As the design progresses through the hierarchy, the details of design intent become more specific. Design intent at the system design level guarantees that the system performs the desired function under certain specifications such as power and speed. At the hardware implementation level, design intent is preserved by block and instance-level specifications. Finally, at the manufacturing level, design intent is preserved by the lowest level primitives such as transistors and metallization.

As a case of example, we can analyze the design of a cell phone.

The system supplier (e.g., Nokia) is typically the company that is familiar to the consumer. With ever increasing complexity in electronic systems, consumers have come to rely on a brand to make their purchase decisions. If all brands have equal qualities, the market becomes commoditized and the only factor then is cost. For example, brands that command a premium price are Sony and Apple Computers. To avoid commoditization, system suppliers must continue to increase their systems' features and performance.

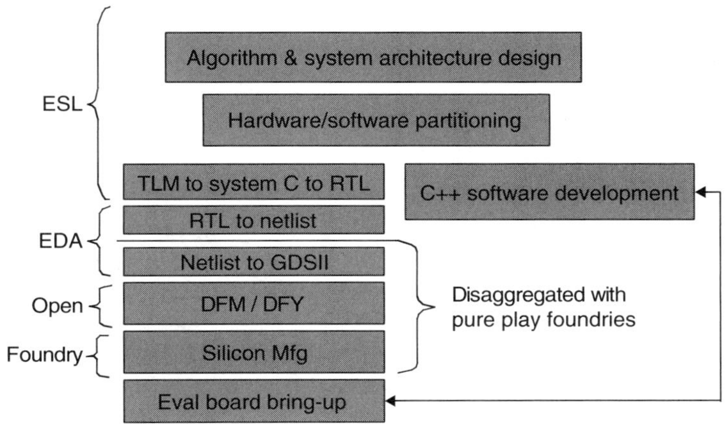

FIGURE 9.1 Design hierarchy.

Their tasks include market definition, product specification, brand identity, and distribution. The system designer in a company like Nokia works from a specification and describes the design in terms of behavior. The behavior can be expressed in a high-level language such as C, System C, or C++.

At the hardware and software implementation, the implementation company (e.g., Texas Instruments) will create a logical and physical description. Brands in this area are less of a factor with the consumer (the notable exception being the lingo "Intel Inside" which managed to commoditize everyone else on the motherboard). The main concerns are design closure for performance, cost, and power. In the implementation phase, there must be a convergence of the design or closure. The litmus test for closure is whether the design has met the performance specifications in the physical implementation.

At the manufacturing level, the manufacturing company (e.g., TSMC or Flextronics) provides manufacturing integration and test, process control, supply logistics, and capital utilization. In electronics design manufacturing, the measure of success is different from implementation. At the implementation phase, there is a binary decision (i.e., were the specifications met or not?). In manufacturing, it is statistical. Does the finished good fall in the statistical standard deviation that defines a good product?

In an increasingly disaggregated design chain, there must be efficient and accurate methods to communicate both vertically and horizontally to ensure the integrity of design intent. Figure 9.2 shows how horizontal and vertical solutions may appear for our cell phone example.

9.3.1 What Defines Horizontal Design Space?

At any given step in the vertical hierarchy of the design process, there is a need to explore the design space. Engineers at each level must guarantee that the design intent from preceding levels is preserved. This entails understanding the design constraints, optimization of the design, and validation of the implementation at the current level.

Generally the progression horizontally is the following:

- *Measure.* Using instruments automatic test equipment (ATE) or a computer program (timing analyzer) to measure the attributes of the current design.
- *Model.* A parameterized mathematical model of the behavior found through measurement, the accuracy of which is directly related to the level of statistical control. Typically there is a trade-off between model accuracy and complexity. Often developers look to develop the perfect model. This in turn drives more computation time and subsequently reduces the number of possible experiments due to limited resources. Generally speaking, there is more value in exploring a larger design space.

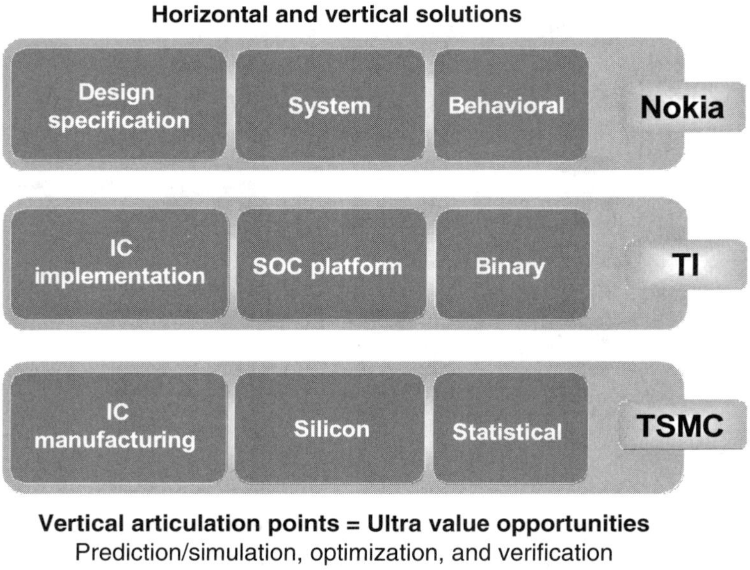

FIGURE 9.2 Horizontal and vertical design solutions for cell phone example.

- *Analyze.* Tools to view and edit the results of the designer's experiments in optimizing the design against constraints.
- *Simulate.* An engine using the models defined in the lower levels of the design hierarchy to explore "what if" scenarios. This can expose more widely the opportunities to optimize.
- *Verify.* Often the same engine as used in simulation, but with assertions on the design to verify that design intent was not changed during design optimization.
- *Optimize.* A process that uses weighted design constraints to perform experiments to best fit the design. The optimization process can be performed using an automated or manual approach depending upon the complexity of optimization and the number of experiments required. The drive is toward automation but this is not always feasible.

Libraries in the horizontal solution will contain models, preverified IP, and a set of rules (test benches, design rules, or lower level models) for verifying the design. A test bench implements an automated method for generating stimulus to apply to the model of the design, along with the means of comparing the expected results with the simulated results.

In terms of economic value, in the history of electronic design automation (EDA), more value has been awarded to simulation and optimization products that follow analysis in the horizontal flow. While modeling is a very difficult and necessary step, it has not been able to extract the same value.

9.3.2 What Is a Vertical Design Solution?

In the vertical axis of the design chain, the critical step is verification of the design as it moves through the vertical design chain flow.

An excellent example is at the physical verification step, where you verify whether an IC design has met the manufacturer's rules. The best execution of verification is to use the same physical verification engine and rules that the manufacturing semiconductor foundry will use to verify the design for manufacturing. In the past, verification tests were based solely on geometric rules, but they now rely increasingly on a model-based approach to ensure manufacturability. It has become an extremely difficult problem to ensure that design intent is maintained as the design moves from a binary to a statistical world. Several companies (e.g., Arm, Virage) have established an excellent business of developing standard physical library elements for pure play foundries (e.g., UMC, TSMC) that already take into account the manufacturing variability bridging graphic data system II (GDSII) to silicon.

In another example, at the system level of the design chain, the software developer relies on hardware behavior models to optimize application software. In this area, the software engineer has no concept of a clock as a hardware design would. The processor model only needs to be "cycle" accurate to ensure that the application can be verified effectively. Several companies (e.g., Virtutech, VAST, and Carbon) have offered their approach to modeling the processor behavior.

9.4 Commercial Efforts

These examples have illustrated that there is a need for IP or libraries in both the horizontal and vertical design chain. This need was recognized far back in the early 1970s. Early efforts such as in [1,2] concentrated on supplying the physical design of standard cells. Then new forms of modelling were added, typically behavioral and timing [3]. Integrated design and manufacturing houses explicitly built libraries to ease vertical integration [4], and vertically integrated systems implicitly included libraries [5,6]. As processes became more complex, libraries were also characterized for faults [7], signal integrity [8], and yield [9]. There are now companies, such as ARM [10] and web sites such as Design and Reuse [11], that are completely devoted to IP and libraries.

Economically, the commercial efforts that have done well have developed a *de facto* standard for communicating design intent across the design chain. The design chain adopts them as the standard protocol to communicate design data. They have enjoyed a proprietary position through a combination of business models and technologies, enabling them to establish broader product and service portfolios. Examples are Synopsys with their DesignWare and timing libraries, Rambus with memory controllers, Artisan with standard cells and memories, and Meta Software with SPICE models.

It is interesting that the market capitalization for the major IP companies has grown significantly (e.g., ARM, RAMBUS, Tessera, and Synopsys with DesignWare). These commercial IP companies have been able to capture significant value from the design chain as it has become disaggregated. What is different with regard to the traditional EDA companies is that once IP is established there is room for only one dominant supplier. As the design chain disaggregates, the requirements of preserving design intent create high value opportunities (i.e., preverified IP, EDA tools, or semiconductor manufacturing). In the physical IP space, the use of the IP can insulate the designer from the changing physics of manufacturing, so the designer can focus on creating value in the design, not on whether manufacturing can deliver the design intent. The details are embedded in the physical IP.

IP will continue to increase in value as the design chain continues to disaggregate and the difficulty of communicating design intent increases. For example, the complexities of manufacturing both in the physics (lithography, etch, and CMP) and logistics (global supply chains) require efficient and accurate virtual use of information by teams that know no political or time boundaries. The winners will not only solve the modeling problem technically, but also the business challenges of addressing this $24 \times 7 \times 365$ world.

There will be growth in the IP and libraries in the horizontal domain as current EDA models continue to mature. We can think of it in terms of an "evolution" that will identify more parameters (e.g., 48 term BSIM4 SPICE models that now include power), new analysis technologies (software and hardware) for dealing with the increased compute and data volumes, and new assertion cases for verification.

9.5 What Makes the Effort Easier?

IP and library standardization of information would be a great step forward. Standardization is usually the instrument of market expansion. Common interfaces between vertical domains allow the free and unencumbered exchange of information.

9.6 The Enemies of Progress

The design solutions that made winners in the past will not work with the speed of the market today. The following is a list of issues that have caused problems in the past, especially in IP and libraries. These should be avoided if possible:

- Warring tribes that miss the "Big Picture" of consumer demands (UNIX vs. DEC/VMS; VHSIC Hardware Description Language vs. Verilog; System Verilog vs. SystemC, etc.).
  - Unclear benefits such as the supplier controlling the agenda.
  - Companies holding onto proprietary formats until it becomes detrimental to their growth. Some enlightened suppliers have seen the future and have opened their standards.
- Political motivation that limits competition (DivX and IBM Microchannel).
- Customer edict without sufficient vendor input (e.g., OLA, another logical format (ALF), VHDL, Wal-Mart, and radio frequency identification (RFID)).
  - Difficult to adopt and no mapping of existing infrastructure. No one has the time or money to do something that does not have clear and efficient cost/benefit over existing solutions.
  - The installed base is incompatible and the cost to adopt is prohibitive.
- Standards that are ahead of their time and solutions looking for problems.
  - Perfection can be the enemy of a good solution. Good enough will win the day. Do not postpone for perfection.
  - Abstraction without effective automation. This is generally a gap between domains that are not interconnected. As a result the standard becomes useless.

9.7 Environments That Drive Progress

On the other hand, there are circumstances that make IP less of a limiting factor and more of an enabler. In order to increase system complexity and continue manufacturing that enables differing versions of IP, you need to consider the following attributes:

- Design is increasingly driven by cost, efficiency, and functionality over raw performance. Does the world need any more MIPS?
- Effective communication of design intent and use of models to the surrounding levels in the design chain.
- Cost pressures will cause industry consolidation around new market aggregation layers (IP and library companies have and will continue to capture value). Other companies like eSilicon and OpenASIC are successfully using design and logistics information to develop new business models.
- Reliance on standards will amortize the cost by members of standard conforming groups.
- Open IP standards are allowing for a new road map of requirements and the development of commercial offerings (PCI-X, DDR1/2). Systems companies will innovate in the applications, not in the protocols.
- Standards (products, architectures, interfaces, and abstractions) enable economies of scale and the reuse of platforms over multiple product cycles.

9.8 Libraries and What They Contain

The following lists are examples of libraries at various levels in the vertical design hierarchy (from low to high) as well as the methods for communication:

9.8.1 Low-Level Physical IP

- These include standard cells, memory, analog/mixed signal IP, and interface IP such as I/O cells and PHYs. (PHY is a generic term referring to a special electronic circuit or functional block of a circuit that provide physical access to a digital interconnect cable.)

- Examples: ARM Holdings, Virage, internal design groups, etc.
- Typical libraries enable variations to optimize area, speed power, and increase the yield. These are implemented via configuration or cell variants.
- Outputs to higher levels in the design hierarchy: cell layout (typically GDS-II), routing model (typically LEF), functional models (typically Verilog), timing models (typically .lib), circuit models (typically SPICE), and yield models (often PDF solutions format).
- Inputs from lower levels in the design hierarchy (manufacturing): geometric rules, transistor device models, and yield tradeoff rules.

9.8.2 High-Level Physical IP (Cores)

- Examples: ARM, MIPS, and OAK.
- Typical libraries offer time to market advantages with a functional and performance guarantee. Offers no aspect ratio control or functional flexibility.
- Outputs to higher levels in the design hierarchy: cell layout, routing model (abstract), functional model, timing model, and yield model.
- Inputs from lower levels in the design hierarchy: process models, lower level IP models, and system specification.

9.8.3 High-Level Soft IP

- Examples: serial bus controllers, processor cores.
- Advantages: flexible aspect ratio and usable with many processes.
- Disadvantages: may be difficult to communicate design intent to the physical implementer, complex soft IP may diminish due to complexities introduced in the manufacturing process. Hard to provide a performance guarantee since detailed implementation is unknown at design time.
- Inputs from higher levels in the design hierarchy: block-level specifications for performance and functionality.
- Outputs to lower levels in the design hierarchy: logic net list, performance parameters and placement information.

9.8.4 System Design/Implementation

- Inputs from higher levels in the design hierarchy: high-level C code and system specifications.
- Inputs from lower levels in the design hierarchy: low level models for function, timing, power, and area.
- Outputs to lower levels in the design hierarchy: block-level specifications, overall system specifications for performance, and cost.

9.8.5 System Architecture

- System specification — market-driven specification starts the design process.
- Outputs to lower levels in the design hierarchy: system functional code communicates system functionality along with the system performance requirements to the system design phase.

9.9 Summary

The EDA industry is grappling with many simultaneous technical challenges. They include:

- Increasing product complexity, the drive to improve design productivity, and shorten design cycles.

- The increased complexity of manufacturing processes forcing the knowledge of manufacturing physics up the design hierarchy to cause multiple nonconvergent optimization problems (Litho and DFM/DFY).
- The critical nature of verification and solving the equivalence problem of design intent.
- The increasing need for a higher level of design abstraction/handoff to foster more design starts.
- The optimization of multiple, simultaneous design objectives including power, timing, signal integrity, etc.

The market has matured — the EDA standards process must also rise to a new level. This will include IP as well as tools in their design solutions. IP and library products that address the stated challenges by bridging the horizontal and vertical design chain articulation points are and will be more valuable.

References

[1] R.L. Mattison, A high quality, low cost router for MOS/LSI, *Proceedings of 9th Design Automation Workshop*, 1972, pp. 94–103.
[2] A. Feller, Automatic layout of low-cost quick-turnaround random-logic custom LSI devices, *Proceedings 13th Design Automation Conference*, 1976, pp. 79–85.
[3] I. Jones, Characterization of standard cell libraries, *Proceedings CICC*, 1985, pp. 438–441.
[4] A. Martinez, S. Dholakia, and S. Bush, Compilation of standard cell libraries, *IEEE J. Solid–St. Circ.*, SC-22 190–197, 1987.
[5] J. Rosenberg, D. Boyer, J. Dallen, S. Daniel, C. Poirier, J. Poulton, D. Rogers, and N.Weste, A vertically integrated VLSI design environment, *Proceedings of the 20th Conference on Design Automation*, 1983.
[6] A.F. Hutchings, R.J. Bonneau, and W.M. Fisher, Integrated VLSI CAD systems at digital equipment corporation, *Proceedings of the 22nd ACM/IEEE Conference on Design Automation*, 1985.
[7] J. Khare, W. Maly, and N. Tiday, Fault characterization of standard cell libraries using inductive contamination analysis (ICA), *Proceedings of 14th VLSI Test Symposium*, Princeton, NJ, 1996, pp. 405–413.
[8] Artisan Inks Deals with Cadence, Synopsys, *Electronic News*, 11/19/2002, http://www.reed-electronics.com/electronicnews/article/CA260087.html.
[9] P. Clarke, PDF solutions, Virage to add DFM to cell libraries, *EETimes*, 6 October, 2004.
[10] ARM, see http://www.arm.com.
[11] Design and ReUse, http://www.us.design-reuse.com.

10
Design Closure

Peter J. Osler
IBM Systems and Technology Group
Essex Junction, Vermont

John M. Cohn
IBM Systems and Technology Group
Essex Junction, Vermont

10.1 Introduction

Design closure is the process by which a VLSI design is modified from its initial description to meet a growing list of design constraints and objectives. This chapter describes the common constraints in VLSI design, and how they are enforced through the steps of a design flow.

Every chip starts off as someone's idea of a good thing: "If we can make a part that performs function *X*, we will all be rich!" Once the concept is established, someone from marketing says "In order to make this chip and sell it profitably, it needs to cost $*C* and run at frequency *F*." Someone from manufacturing says "In order to make this chip's targets, it must have a yield of *Y*%." Someone from packaging says "And it has to fit in the *P* package and dissipate no more than *W* watts." Eventually, the team generates an extensive list of all the constraints and objectives that need to be met in order to manufacture a product that can be sold profitably. The management then forms a design team, consisting of chip architects, logic designers, functional verification engineers, physical designers, and timing engineers, and tasks them to create the chip to these specifications. Other chapters in this book have dealt with the details of each specific step in this design process (e.g., static timing analysis, placement, routing, etc.). This chapter looks at the overall *design closure* process, which takes a chip from its initial design state to the final form in which all of its design constraints are met.

We begin the chapter by briefly introducing a reference design flow. We then discuss the nature and evolution of design constraints. This is followed by a high-level overview of the dominant design closure constraints that currently face a VLSI designer. With this background we present a step-by-step walk-through of a typical design flow for *application-specific integrated circuits* (*ASICS*) and discuss the ways in which design constraints and objectives are handled at each stage. We will conclude with some thoughts on future design closure issues.

10.1.1 Evolution of the Design Closure Flow

Designing a chip used to be a much simpler task. In the early days of VLSI, a chip consisted of a few thousand logic circuits which performed a simple function at speeds of a few MHz. Design closure at this point was simple: if all of the necessary circuits and wires "fit," the chip would perform the desired function. Since that time, the problem of design closure has grown orders of magnitude more complex. Modern logic chips can have tens to hundreds of millions of logic elements switching at speeds of several GHz. This improvement has been driven by the Moore's law of scaling of technology which has introduced a whole host of new design considerations. As a result, a modern VLSI designer must simultaneously consider the performance of his/her chip against a list of dozens of design *constraints* and *objectives** including performance, power, signal integrity, reliability, and yield. We will discuss each of these design constraints in more detail in Section 10.1.2. In response to this growing list of constraints, the design closure flow has evolved from a simple linear list of tasks to a very complex, highly iterative flow such as the following simplified *ASICS* design flow:

1. *Concept phase*: The functional objectives and architecture of a chip are developed.
2. *Logic design*: [1] The architecture is implemented in a register transfer level (RTL) language, then simulated to verify that it performs the desired functions.
3. *Floorplanning*: The RTL of the chip is assigned to gross regions of the chip, input/output (I/O) pins are assigned and large objects (arrays, cores, etc.) are placed.
4. *Synthesis*: [2] The RTL is mapped into a gate-level netlist in the target technology of the chip.
5. *Placement*: [3] The gates in the netlist are assigned to nonoverlapping locations on the chip.
6. *Logic/placement refinement*: [4,5] Iterative logical and placement transformations to close performance and power constraints.
7. *Clock insertion*: Balanced buffered clock trees are introduced into the design.
8. *Routing (or wiring)*: [6] The wires that connect the gates in the netlist are added.
9. *Postwiring optimization*: [7–10] Remaining performance, noise, and yield violations are removed; final checking is done.

We will use this reference flow throughout the chapter to illustrate points about design closure.† The purpose of the flow is to take a design from concept phase to working chip. The complexity of the flow is a direct result of the addition and evolution of the list of design closure constraints. To understand this evolution it is important to understand the *life cycle* of a design constraint. In general, design constraints influence the design flow via the following five-stage evolution:

1. *Early warnings*: Before chip issues begin occurring, academics and industry visionaries make dire predictions about the future impact of some new technology effect.
2. *Hardware problems*: Sporadic hardware failures start showing up in the field due to the new effect. Postmanufacturing redesign and hardware re-spins are required to get the chip to function.

* The distinction between constraints and objectives is straightforward: a constraint is a design target that must be met in order for the design to be considered successful. For example, a chip may be required to run at a specific frequency in order to interface with other components in a system. In contrast, an objective is a design target where more (or less) is better. For example, yield is generally an objective, which is maximized to lower manufacturing cost. For the purposes of this chapter, the distinction between constraints and objectives is not all that important and we will use the words interchangeably.

†See Chapter 2, Volume 1 of this handbook for a good overview of the entire design flow.

3. *Trial and error*: Constraints on the effect are formulated and used to drive postdesign checking. Violations of the constraint are fixed manually.
4. *Find and repair*: Large number of violations of the constraint drives the creation of automatic postdesign analysis and repair flows.
5. *Predict and prevent*: Constraint checking moves earlier in the flow using predictive estimations of the effect. These drive optimizations to prevent violations of the constraint.

A good example of this evolution can be found in the coupling noise constraint. In the mid-1990s (180 nm node), industry visionaries were describing the impending dangers of coupling noise long before chips were failing [11]. By the mid-late 1990s, noise problems were cropping up in advanced microprocessor designs. By 2000, automated noise analysis tools were available and were used to guide manual fix-up [12]. The total number of noise problems identified by the analysis tools identified by the flow quickly became too many to correct manually. In response, CAD companies developed the noise avoidance flows that are currently in use in the industry [13].

At any point in time, the constraints in the design flow are at different stages of their life cycle. At the time of this writing, for example, performance optimization is the most mature and is well into the fifth phase with the widespread use of *timing-driven* design flows. Power- and defect-oriented yield optimization is well into the fourth phase; power supply integrity, a type of noise constraint, is in the third phase; circuit-limited yield optimization is in the second phase, etc. A list of the first-phase impending constraint crises can always by found in the *International Technology Roadmap for Semiconductors* (*ITRS*) [14] 15-year-outlook technology roadmaps.

As a constraint matures in the design flow, it tends to work its way from the end of the flow to the beginning. As it does this, it also tends to increase in complexity and in the degree that it contends with other constraints. Constraints tend to move up in the flow due to one of the basic paradoxes of design: *accuracy vs. influence*. Specifically, the earlier in a design flow a constraint is addressed, the more flexibility there is to address the constraint. Ironically, the earlier one is in a design flow, the more difficult it is to predict compliance. For example, an architectural decision to pipeline a logic function can have a far greater impact on total chip performance than any amount of postrouting fix-up. At the same time, accurately predicting the performance impact of such a change before the chip logic is synthesized, let alone placed or routed, is very difficult. This paradox has shaped the evolution of the design closure flow in several ways. First, it requires that the design flow is no longer composed of a linear set of discrete steps. In the early stages of VLSI it was sufficient to break the design into discrete stages, i.e., first do logic synthesis, then do placement, then do routing. As the number and complexity of design closure constraints has increased, the linear design flow has broken down. In the past, if there were too many timing constraint violations left after routing, it was necessary to loop back, modify the tool settings slightly, and reexecute the previous placement steps. If the constraints were still not met, it was necessary to reach further back in the flow and modify the chip logic and repeat the synthesis and placement steps. This type of looping is both time consuming and unable to guarantee convergence i.e., it is possible to loop back in the flow to correct one constraint violation only to find that the correction induced another unrelated violation.

To minimize the requirement for frequent iteration, the design flow has evolved to use the concept of *variable detail accuracy* in which estimates of downstream attributes, e.g., wire length and gate area, are used to drive upstream optimization of timing and power dissipation. As the design evolves, these approximations are refined and used to drive more precise optimizations. In the limit, the lines that separate two sequential steps such as synthesis and placement can be removed to further improve downstream constraint estimates. In today's most advanced design closure flows, the lines between once-discrete steps have been blurred to the point that steps such as synthesis, placement, and routing can be simultaneously co-optimized [4]. The evolution from discrete stand-alone design steps to integrated co-optimizations has had a profound influence on the software architecture of design closure systems. Modern design closure suites are composed of three major components: a central database, a set of optimization engines, i.e., logic optimization, placement, and wiring, and a set of analysis engines, i.e., *static timing analysis*, *power analysis*, *noise analysis*, etc. The central database manages the evolving state of the design. The optimization engines modify the database directly, while the analysis engines track incremental changes in the database and report back on the results. By allowing a certain degree of independence between optimization engines and analysis tools, this *data-driven* architecture greatly simplifies the addition of new design constraints or the refinement of existing ones. More detail on data-driven design closure architectures can be found in the chapter on design flows [15].

The evolution of chip timing constraints provides a good overall illustration of a constraint's movement up the flow. In the first integrated circuits, analyzing performance consisted of summing the number of blocks in each path. After a couple of technology generations, hardware measurements began to show that this simple calculation was becoming less accurate. In response, performance analysis tools were extended to take into consideration that not all gates have the same delay. This worked for a time, until measured hardware performance again began to deviate from prediction. It was clear that gate output loading was beginning to become a factor. The performance analysis flow was modified to include a factor that modified the delay of a gate based on the total input capacitances of the downstream gates it drove. As performance increased further, wire delay started to gain in importance. Wire-length measures were used to calculate total wire capacitive load, which was converted into a delay component. As performance increased further, wire resistance became a factor in all interconnect, so delays were approximated using a simple, single-pole *Elmore delay* model. In the early 1990s, it was observed that the Elmore approximation was a poor predictor of wire delay for multi-sink wires, so more complex *moment matching* [16] methods were introduced. As timing analysis matured further, it became increasingly difficult to correct all of the timing constraint problems found after routing was complete. As the flow matured from *find and repair* to *predict and prevent*, it became increasingly important to accurately predict total wire delay as part of a *timing-driven* design flow. Crude models of wire delay were added to placement, in order to shorten preferentially wires on critical paths [3]. The first placement-based wire delay models used gross estimates of total wire length such as calculating the bounding box of all pins on a wire. As interconnect delay grew in importance, such nonphysical wire-length estimates proved to be poor predictors of actual wire delay. Delay estimates based on Steiner-tree approximation of routed wire topologies were introduced to predict delay better. As interconnect delay increased further, it became necessary to push wire-length calculation even further back into preplacement logic synthesis. Initially, this was done using simple *wireload models*, which assigned average wire length based on estimates of placed block size. Eventually this too proved too inaccurate for high-performance logic. By the late-1990s, *placement-driven synthesis* [4] was introduced to bias logic synthesis based on predictions of critical wire length. As the handling of timing constraints matured, it also began to contend more with other constraints. During the early days of design closure, chip-timing problems could be mitigated by increasing the drive strength of all circuits on a slow path. Doing this, however, increases active power and increases the chance of coupling noise onto adjacent wires. These design trade-offs are discussed in the context of the actual design closure flow in the chapter on design flows [15].

We will now shift our focus to learn more about the specific design constraints that are facing chip designers. Armed with this, we will then begin a detailed walk-through of a typical design closure flow.

10.1.2 Introduction of Design Constraints

Chip designers face an ever-growing list of design constraints, which must be met for the design to be successful. The purpose of the design closure flow is to move the design from concept to completion while eliminating all constraint violations. This section will briefly introduce the current taxonomy of design closure constraints, objectives and considerations, and discuss the impact and future trends for each. For the sake of this discussion we will divide the constraint types into economic, realizability, performance, power, signal integrity, reliability, and yield.

10.1.2.1 Economic Constraints

While not explicitly technical, the economic constraints governing design closure are perhaps the most important constraints of all. Economic constraints pertain to the overall affordability and marketability of a chip. Economic constraints include *design cost, time to market,* and *unit cost.*

- *Design-cost constraints* govern the *nonrecurring expense* (*NRE*) associated with completing a design. This includes the cost of the skilled resources needed to run the design flow, the cost of any test hardware and additional design passes which occur due to errors. These costs also include the amortized cost of facilities, computer resources, design software licenses, etc., which the team needs to complete the task. Design cost can be traded off against other design constraints such as performance and power because increased effort and skill generally will yield better optimization. Missing a design cost estimate can be a very serious problem as cost overruns generally come out of projected profit.

- *Time-to-market (TTM) constraints* govern the schedule of a design project. This includes the time required for development, manufacturing, and any additional hardware re-spins necessary to yield a functional and manufacturable part in sufficient volumes to meet the customer's requirements. Time-to-market cost can be traded off against other design constraints such as performance and power because, like increased design effort, increased design time generally yields better optimization albeit at the expense of design cost. Missing a time-to-market constraint can imply missing a customer deadline or market window. In competitive market segments, being late to market can mean the difference between huge profits and huge losses.
- *Unit-cost constraints* govern the cost of each manufactured chip. This includes the cost of the chip itself accounting for any yield loss, the package, the cost of the module assembly, and the cost of all testing and reliability screens. Unit cost is a strong function of chip die size, chip yield, and package cost. This can be traded off against design cost and time to market by allowing additional effort to optimize density, power, and yield. Missing a unit-cost constraint can make a chip non-competitive in the marketplace.

10.1.2.2 Realizability Constraints

The most basic constraint of VLSI design is "does the chip fit and does it work?" These "realizability" constraints pertain to the basic logical correctness of the chip. Realizability constraints include *area constraints, routability constraints,* and *logical correctness constraints.*

- *Area constraints* are one of the most basic constraints of any VLSI design, i.e., does the chip "fit" in the desired *die size* and *package*? To fit, the total area required by the sum of the chip's circuitry plus additional area required for routing must be less than the total useable area of the die. The die size and package combination must also support the type and number of I/O pins required by the design. Because silicon area and package complexity are major components of chip cost, minimizing die size and package complexity is a major goal in cost-sensitive designs. The implication of mis-predicting capacity can be great. If the required area is over-estimated, die utilization is low and the chip costs more than it should. If the required area is underestimated, the design must move to a larger die size or more complex package, which implies more cost and additional design time.
- *Routability constraints* ensure that the resulting placement can be completely connected legally by a router. The impact of mis-predicting routability can impact average wire length, which can, in turn, affect timing. If routability is compromised enough, the designers must manually route the overflows that could not be routed automatically. In the worst case, mis-predicting routability can require that the chip be bumped up to a larger die size at great impact to cost and schedule. The challenges of this problem have been growing due to a number of factors such as increasing gate count, increased complexity of metallurgy, including complex via stack rules and multiple wire widths and heights, manufacturability constraints, and increased interrelations between performance and routing. However, the addition of extra routing layers and improvements to routing technology have alleviated this problem to some extent. Routability problems can be mitigated by adding additional area for routing, which contends with chip area constraints.
- *Logical correctness constraints* ensure that the design remains logically correct through all manipulations used by design closure. Modern design closure flows make significant use of local logical transformations such as buffering, inversions, logic cloning, and retiming to meet performance constraints. As these logical transformations have become more complex, there is an increased opportunity for introducing errors. To ensure that logical correctness is maintained, modern flows use equivalence checking to verify that the design function remains unchanged after each design closure step. Obviously, the impact of failing this constraint is that the chip no longer functions as intended.

10.1.2.3 Performance Constraints/Objectives

Once we ensure that the design will fit and is logically correct, the primary function of design closure is to ensure that the chip performance targets are met. Traditionally, chip timing has received the most focus as a design objective. Performance constraints can be divided into *late-* and *early-mode timing constraints.*

- *Late-mode timing* establishes the longest delay, or *critical path*, through all paths which sets the maximum speed a chip can run. Late-mode timing considerations can either be a constraint, i.e., *the chip* must run at least as fast as *x*, or they can be a design objective, i.e., the faster this chip runs the better. Late-mode timing constraints are enforced by comparing late-mode static timing [17] results against desired performance targets and clock cycle time. Late-mode timing constraints violations are removed by decreasing the gate or interconnect delay along critical paths. Gate delays can be reduced by decreasing logic depth, increasing gate drive strength, increasing supply voltage, or substituting gates with *low threshold* (low V_t) logic. Interconnect delay can be reduced by decreasing wiring length, adding buffers, or widening wires. Late-mode timing constraint violations can also be resolved by allowing more time for a path to evaluate through the addition of *useful clock skew.* Meeting the late-mode timing constraint for all paths is becoming increasingly difficult due to the combination of increasing chip complexity and clock speeds, combined with an increasingly pronounced "roll-off" of technology performance scaling. Both device performance and interconnect performance are failing to keep pace with the decade-long Moore's law improvement rate. More worrisome is that, as illustrated in Figure 10.1, interconnect performance is scaling even more slowly than gate performance. In fact, increases in wiring resistance are beginning to cause *reverse scaling* in which the relative interconnect delay actually increases with each new technology node. This implies more design closure effort in additional buffering and multi-cycle pipelining of long signals. Optimizations for late-mode timing constraints on one path may contend with late-mode timing constraints on other paths. For example, increasing the gate size to reduce delay on one critical path may increase gate load on another critical path. Late-mode timing optimizations also contend with power optimization. Most techniques used to optimize late-mode timing, e.g., gate sizing, buffering, and low-threshold logic substitution increase total chip power.

  The impact of missing a late-mode timing constraint is that the chip is slower than required. In most designs, missing the late-mode timing constraints implies increasing the cycle time of a machine. This may imply that the system specification must be renegotiated, or at worst case, redesigned at great cost of money and time. In some rare cases however, the final timing objectives may be negotiable, i.e., one may be able to sell a slower microprocessor for less money.

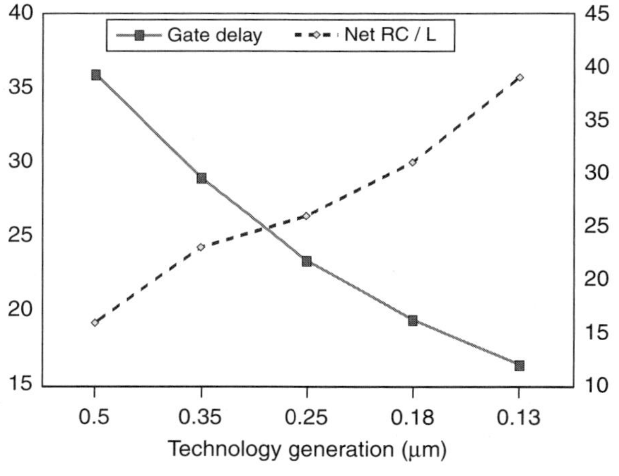

FIGURE 10.1 Interconnect and gate delay scaling trends.

- *Early-mode timing constraints* are designed to prevent a path from being too fast to be correctly captured by the capture clock. Unlike late-mode timing constraints, early-mode timing is always a constraint and never an objective. A single early-mode constraint violation means that the chip will not function at any speed. Early-mode timing issues are difficult to estimate early in the design flow. They generally must be analyzed and corrected after final clock routing is added to the design. At this point, static timing can be used to verify *hold* tests to ensure that the final logic state analyzed is stable before the capture clock fires. This testing must account for possible clock skew induced by manufacturing variations. It is necessary to run static timing over multiple *process corners*, e.g., fast wiring with slow logic, fast logic with slow wiring, to predict these cases correctly. Poor control of clock overlap can also give rise to a large number of early-mode timing violations. Early-mode timing violations are removed by adding delay elements such as buffers to fast paths. By adding circuitry, early-mode constraints contend with capacity, power, and late-mode timing constraints.

10.1.2.4 Power Constraints/Objectives

Power dissipation can either be a constraint, i.e., the chip must consume no more than x Watts, or it can be a design objective, i.e., the less power this chip uses, the longer the battery will last. As chip geometries have scaled, total chip power has become an increasingly important design closure constraint. Power can be divided into two types: *active* power and *static* power.

- *Active power* is dissipated through the charging and discharging of the capacitance of the switching nodes. Active power is proportional to the sum of $\frac{1}{2}FSCV^2$ of all switching signals on a chip, where F is the clock frequency, S the *switching factor*, i.e., the average fraction of clock cycles in which the signal switches, C the total capacitive load presented by logic fanout and interconnect, and V the supply voltage. Active power increases due to higher chip logic densities, increased switching speeds, and a slowdown in voltage scaling due to limits on scaling gate oxide thickness. Active power can be lowered by decreasing gate size, decreasing switched wire load, and decreasing switching frequency or duty cycle via clock gating. Reducing gate size to reduce active power contends directly with performance optimization, which gives rise to the essential power/performance trade-off.
- *Static* or *leakage power* is related to current that leaks through a device channel or gate even when it is turned off. Leakage power is increasing relative to active power due to the use of smaller active FET gate geometries and thinner FET gate. Leakage power is a function of supply voltage, temperature, device threshold voltage and logical state. Figure 10.2 shows the active and leakage power density trends by technology node. Static power can be mitigated by substituting in *high threshold* (*High* V_t) logic, by lowering supply voltage, or by removing power to inactive portions of the chip via *power gating*. Using High V_t logic and lowering supply voltage contends directly with performance optimization.

In high-performance applications such as server microprocessors, power constraints are generally imposed by the amount of heat that a particular chip, system, and package combination can remove before the chip temperature rises too high to allow proper function. In low-performance applications such as consumer products, factors such as battery life, packaging, and cooling expense are the major limits. Missing either an active or static power constraint can necessitate costly redesign of chip, package, or system. In the worst case, it can render a chip unusable in its intended application.

10.1.2.5 Signal Integrity Constraints

Signal Integrity constraints prevent chip function from being disrupted by electrical *noise*. Signal integrity constraints include *power integrity constraints* and *coupling constraints*.

- *Power integrity constraints* are used to ensure that the chip power supply is robust enough to limit unacceptable supply voltage variations. As chip logic elements switch, current is sourced through the chip power routing. The current must either come from off-chip or from the reserve capacity provided by on-chip capacitance provided by the diffusion structures and routing attached to the power bus and any *decoupling capacitors* (*DCaps*), which are structures added to provide small reservoirs of charge to smooth out switching transients. A switching event with net current I, which

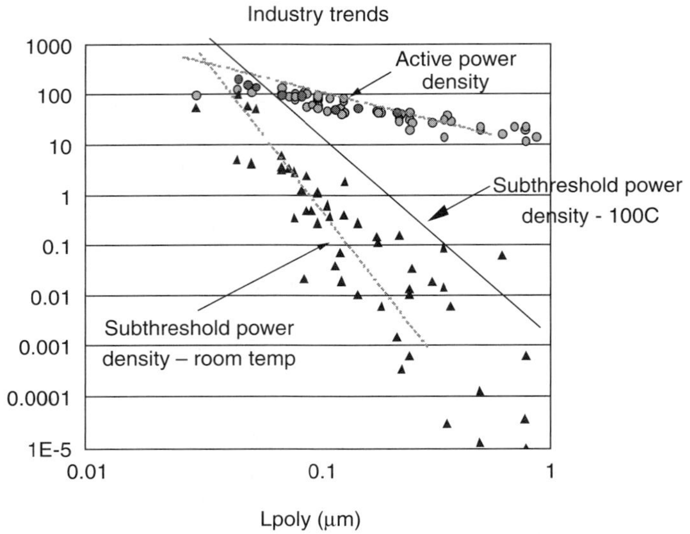

FIGURE 10.2 Active and static power trends.

cannot be supplied by local DCaps, induces a voltage drop $\Delta V = IR + L\,\mathrm{d}I/\mathrm{d}T$, as it flows through the resistance R and the inductance L of the power supply routing. These components are depicted in Figure 10.3. The voltage drop breaks down roughly into three components: *IR drop*, which is affected by power bus resistance and average current flow, *steady-state AC voltage drop*, which is affected primarily by intra-cycle current variation and local decoupling capacitance, and *switching response* which is affected primarily by the switching current variation and package inductance. When added together, the voltage drops induced by each switching event lead to potentially large variations in supply voltage both in space and time. Figure 10.4 shows a map of the spatial distribution of the average or steady-state voltage drop across a single, large ASIC calculated by IBM's ALSIM tool. The resulting voltage fluctuations cause time-varying delays across a chip. If large enough, these delay variations lead to late- or early-mode timing violations. Power voltage transients can also introduce disruptive electrical noise into sensitive analog circuitry. The relative magnitude of power voltage variations has increased as chip total power and operating frequency have increased, and supply voltages have decreased. Power supply integrity can be improved by increasing the wire width, package power/ground pin count, and via count used for power supply routing. Transient power integrity can be further improved by the judicious use of DCaps. These optimizations contend with capacity and routability constraints.

- *Coupling constraints* are used to ensure that inter-signal coupling noise does not disrupt chip timing or logical function. Noise is always a design constraint as even a single violation is sufficient to render a chip inoperable. Coupling noise occurs when a voltage transition on a noisy *aggressor* wire causes current to be injected into an adjacent sensitive victim wire through the mutual capacitance of the two wires. The injected current, ΔI, is proportional to $C\ \mathrm{d}V/\mathrm{d}T$, where C is the mutual capacitance and $\mathrm{d}V/\mathrm{d}T$ the time rate of change of the voltage transition. If the coupled signal exceeds the logic threshold on the victim wire, an incorrect logic value or *glitch* is induced in the victim circuit as shown in Figure 10.5. If the victim wire is transitioning during the coupling event, its delay is affected.

If not modeled correctly, this variation in delay can cause unexpected variations in chip performance. If these delay variations are large enough, they can lead to improper operation. Coupling has increased markedly with increased switching speeds, decreased supply voltage, decreased inter-wire spacing, and increased wire aspect ratios. Coupling can be reduced by segregating noisy and sensitive wiring, increasing inter-wire spacing, and, in extreme cases, by adding shielding wires. All of these optimizations contend with the routability constraint.

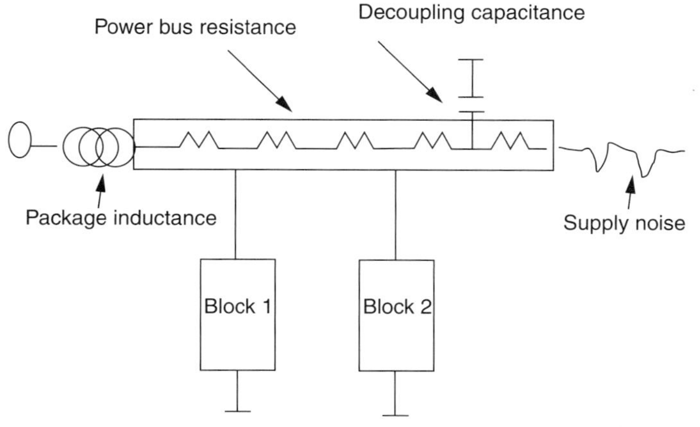

FIGURE 10.3 Power bus voltage drop components.

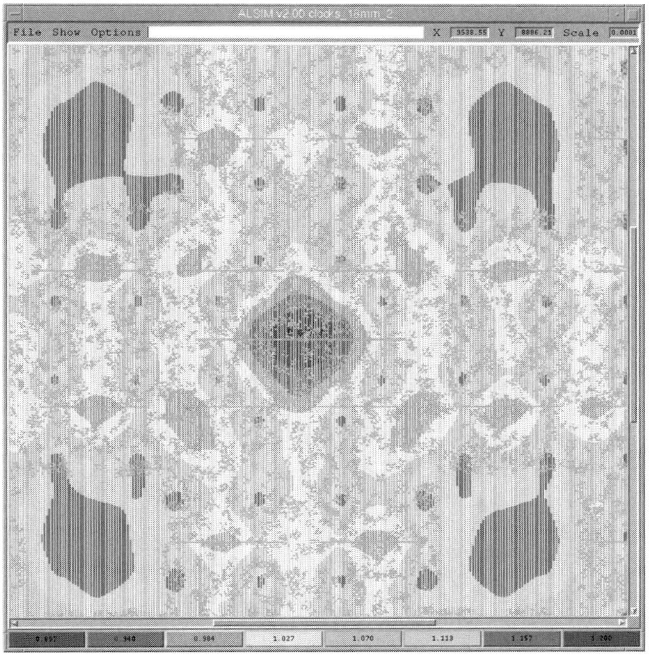

FIGURE 10.4 Power supply voltage drop map.

10.1.2.6 Reliability Constraints

Once the basic performance targets are met, we need to ensure that the chip will function properly through its required working life. Reliability concerns are related to processes that may allow a chip to function correctly immediately after manufacture but may cause the chip to malfunction at some later point in time. In the best case, this type of unexpected chip failure may present a costly inconvenience to the customer. In other, more mission-critical functions such as automotive, avionics, or security, the result of a malfunction can be a risk to life. There are many reliability factors that can affect the long-term operation of a chip. Most fit into one of the three categories: *device wear-out constraints*, *interconnect wear-out constraints*, and *transient disruption constraints*.

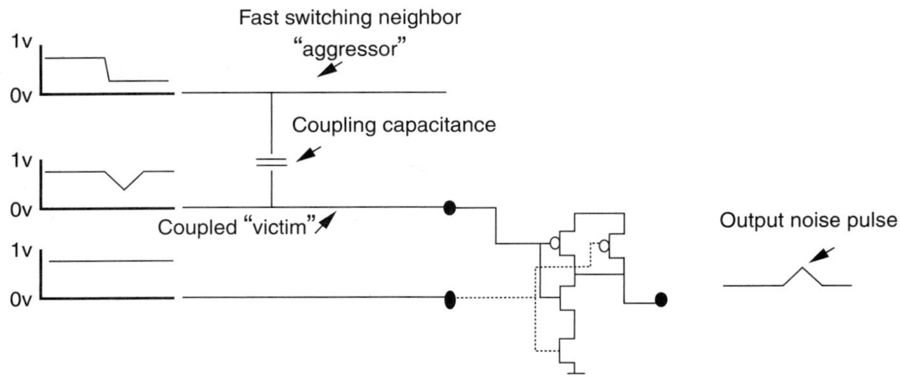

FIGURE 10.5 Coupling-induced logic glitch.

- *Device wear-out constraints* relate to mechanisms that can cause a gradual shift of electrical characteristics of a chip's transistors over time. Two common concerns are hot carrier injection (HCI) and negative bias threshold instability (NBTI). HCI occurs when electrons in the channel of a transistor are accelerated by the high electric field found near the drain of devices, which are on or switching. These highly energetic electrons are injected into the gate oxide where they create electron or hole traps. Over time these traps lead to charge build-up in the gate, which effectively causes a threshold voltage shift for the device. HCI is accelerated for devices, which have high applied gate voltages, high switching activity, and drive large loads. NBTI is similar in effect, but does not require high electric fields. It affects both switching and nonswitching devices. NBTI is accelerated by high operating temperatures and can be induced during burn-in test. The effect of these threshold shifting mechanisms is increasing over time due to the use of thinner gate oxides and lower threshold voltages required by scaling. Shifting the threshold of devices has a direct effect on the delay through a device. Delay can either increase or decrease with time depending on the nature of the injected charge and the type of device. In time, the performance shift may be large enough to cause the chip to malfunction. Device wear-out can be minimized by using the lowest switching voltage necessary and reducing load capacitance on the outputs of high-slew signals, both of which must be traded off against performance.
- *Interconnect wear-out constraints* relate to mechanisms that can cause a gradual shift in the electrical characteristics of chip interconnect with time. The principal mechanism is *electro-migration* (*EM*). EM occurs when ballistic collisions between energetic electrons and the metal atoms in the interconnect cause the interconnect atoms to creep away from their original position. This movement of metal can thin wires to the point that their resistance increases or that they fail completely. Like device wear-out, EM-induced wire wear-out can affect delay to the point that the chip begins to malfunction. In the limit, EM can cause wires to completely open, which clearly changes the function of the chip. Electromigration is accelerated by increased current densities. Additionally, the new low-permittivity dielectric materials introduced to help performance have inferior thermal characteristics. The result is an increase in wire self-heating, which further accelerates EM. EM problems can be mitigated by widening high-current wires and lowering the loads on high duty-cycle signals. These mitigations contend slightly with both routability and performance constraints.
- *Transient disruption constraints* relate to mechanisms which cause a sudden failure of devices or interconnect. The two most common mechanisms are *electro-static discharge* (*ESD*) and *soft error*

upset (*SEU*). ESD occurs when unacceptably high voltage is inadvertently presented to chip structures. This can occur either due to *induced charge build-up* during manufacturing or *a postmanufacture ESD event*. Induced charge build-up can be caused by certain manufacturing processes which involve high electric fields that can induce charge on electrically isolated structures such as transistor gates. If too much charge is induced, the resultant electric field can damage sensitive gate oxides. Induced charge ESD is mitigated by insuring that all gate inputs are tied to at least one diffusion connection. This allows a leakage path to ground, which prevents build-up of dangerously high fields. In some cases, this requires the addition of special *floating gate contacts*. Postmanufacture ESD events occur when high voltages are inadvertently applied to the chip I/O by improper handling, installation, or grounding of equipment and cables. When an ESD event occurs, the gate of any devices connected to the transient high voltage is destroyed due to the high field induced in its gate oxide. In some cases, the wiring that connects the pin to the device may also be destroyed due to the induced transient currents. Postmanufacture ESD events can be minimized by proper chip handling and by the addition of *ESD diodes* on all chip I/Os. These diodes protect chip I/O by shunting high-voltage transients to ground.

Soft error upsets are recoverable events caused by high-energy charged particles which either originate from outer space or from nearby radioactive materials. The carriers induced by the charged particle as it travels through the silicon substrate of the chip can disrupt the logic state of sensitive storage elements. The amount of charge required to upset a logic gate is dependent on its *critical charge* or Q_{crit}. As device structures shrink, the amount of charge required to cause logic upset is decreasing. SEU is currently a concern for memory arrays and dynamic logic, though some projections show that standard logic latch structures might also soon be susceptible to particle-induced soft errors. SEU is best addressed at the architecture level through the addition of logical redundancy or error-correction logic, which is not a design closure step *per se*.

10.1.2.7 Yield Constraints

For all its sophistication, semiconductor manufacturing remains an inexact science. Random defects and parametric variations can be introduced at almost any step of manufacturing which can cause a chip not to function as intended. The more the number of chips affected by manufacturing errors, the more the chips that must be manufactured to guarantee a sufficient number of working chips. Mis-predicting yield can require costly additional manufacturing, expensive delays, and possible product supply problems. In this way, yield might be considered a cost-oriented design constraint. In many cases, though, maximizing yield is considered an economic design objective. The two types of yield constraints that must be considered during design closure are *defect-limited yield* and *circuit-limited yield*:

- *Defect-limited yield constraints* relate to a product that is rendered faulty during manufacturing due to *foreign material defects* or *printability defects*. Foreign material defects result either when small bits of material accidentally fall on the chip surface or the mask reticule and interfere with the proper creation of a chip structure. Printability defects result when local geometry, chip topography, optical interference, or other manufacturing processes prevent correct printing of a desired width or spacing. These defects may take the form of unintended shorts between adjacent structures, unexpected holes in insulating materials such as device gates, or unintended opens in a conductor. Defect-limited yield can be improved by decreasing the amount of *critical area*, i.e., the inter-geometry spacing which is less than or equal to the size of likely defects. Critical area can be minimized by using *relaxed* or *recommended design rules* rather than *minimum design rules* wherever density will allow. This additional spacing contends with capacity and routability constraints.
- *Circuit-limited yield constraints* relate to yield loss due to the effect of manufacturing variations on chip performance. Despite advances in every phase of processing, there remain uncontrollable variations in the properties of the dimensions and materials of the finished product. These small variations cause identically designed devices or wires to have significantly different electrical characteristics. The most-studied variation is *across chip line-width variation* (*ACLV*), which creates

perturbations in the critical gate length of devices. ACLV has many causes including uneven etching due to local shape density variations as well as lithographic distortions caused by optical interactions between shape regions. As interconnect dimensions have shrunk, variations in wire delay are becoming equally significant. Recent data showing interconnect delay on long, unbuffered nets as high as ±60% have been reported [18]. Large amounts of this variation are introduced by both lithographic distortion and issues related to etch rate variations in processing steps such as *chemical mechanical polishing* (*CMP*). Many trends are contributing to the growing concern on parametric yield including increasingly deep *subresolution lithography* and increasingly complex manufacturing processes. In addition, decreased device and interconnect dimensions contribute to the relative impact of variation. For example, decreased device channel dimensions give rise to *micro-implant* dopant variation in which the distribution of a countable number of implanted dopant ions creates small differences in device thresholds. Similarly, gate oxides are approaching dimensions of only ten or so atomic layers. In such small configurations, a change of just one atomic layer can induce a quantized threshold voltage shift of nearly 10% as shown in Figure 10.6.

These parametric variations in turn give rise to statistical variations in design performance characteristics such as delay and leakage power. Figure 10.7 shows a typical manufacturing distribution of a large sample of *performance screen ring oscillator* (*PSRO*) circuits used to characterize the performance of a microprocessor. In the example, the PSRO circuits in the left-most tail of the distribution repre-

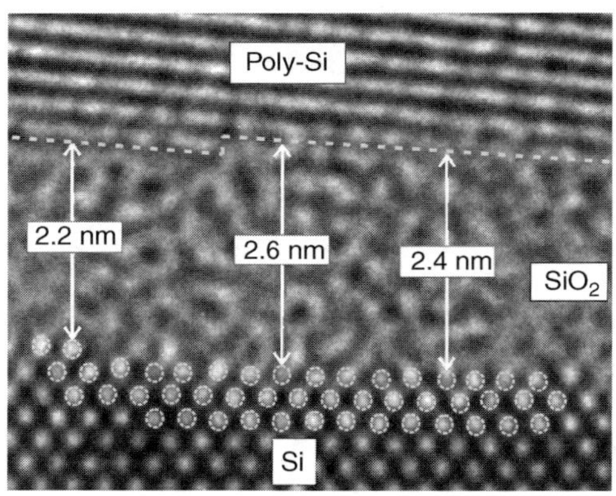

FIGURE 10.6 Gate oxide thickness variation.

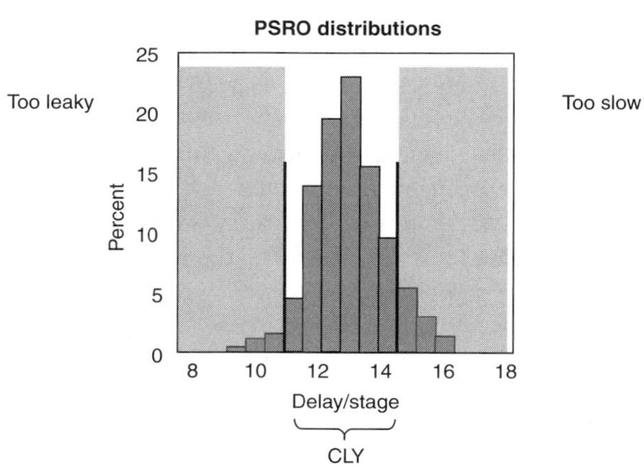

FIGURE 10.7 Ring oscillator performance distributions.

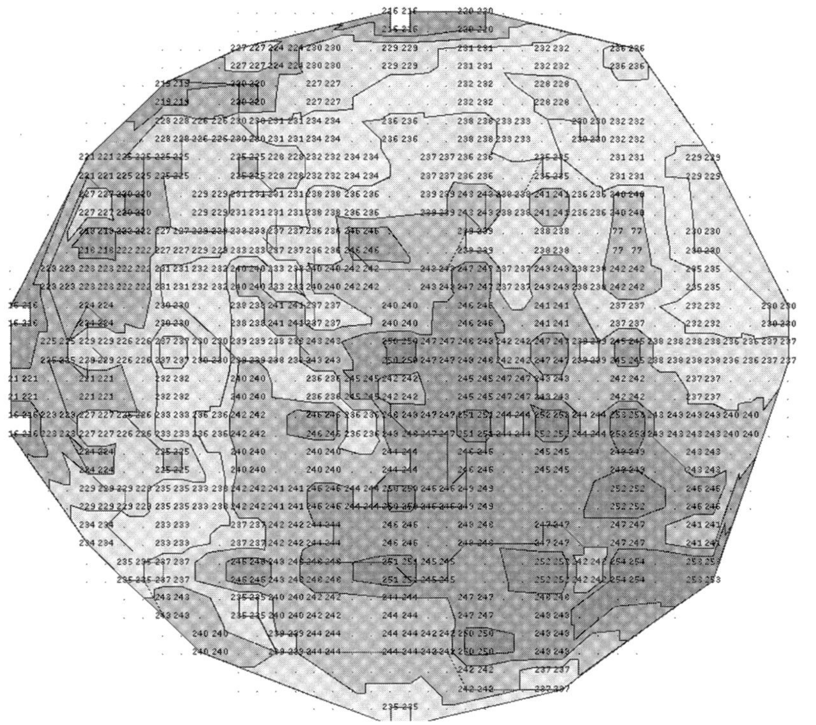

FIGURE 10.8 Wafer map of inter-die parametric variations.

sent the fastest circuits, but these same circuits violate the leakage power constraint and must be discarded. Similarly, the PSRO circuits on the right tail have low leakage, but they fail to meet the minimum late-mode timing constraint and must also be scrapped. The portion of the distribution between these two bounds is the circuit-limited yield.

These variations may be observable both when measuring the same device on different chips (*inter-die*) or between identical devices on the same chip (*intra-die*). Both inter-die and intra-die variations cause the actual performance of a given chip to deviate from its intended value. For example, Figure 10.8 illustrates inter-die variations as measured using identical ring oscillators placed on each die on a 200-mm silicon wafer. Areas of identical shading have identical frequency measurements. The total range of variation is 30%.

Parametric yield is emerging as a design closure constraint for structured logic. We will discuss how this will affect the design closure flow in our discussion of the future of design closure in Section 10.3.

10.2 Current Practice

In this section, we will examine the design closure implications of each phase of the ASIC design flow (Figure 10.9). At each phase we will examine the design constraints that are addressed, estimations and approximations that are made, and the trade-offs that can be made between constraints. We will also explore the interactions — both forward and backward — between the phases.

10.2.1 Concept Phase

The concept phase concerns itself with setting the overall scope of a chip project. During the concept phase, many aspects of the design are estimated: area, operating frequencies, voltages, power dissipation, I/O count, wiring uplift to area, yield percentages, and requirements for special processing (such as

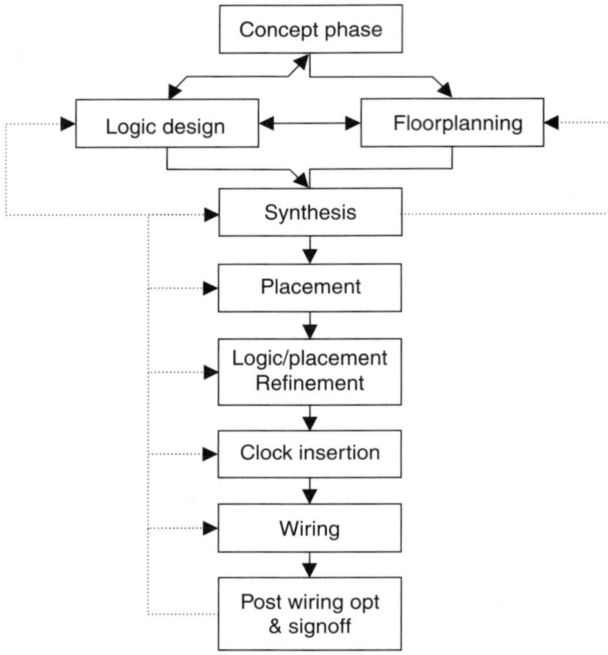

FIGURE 10.9 The design closure flow.

Embedded Dynamic RAM, analog circuits, or multiple threshold voltage [V_t] logic). All of these factors are estimated and combined into a business case: the chip cost to manufacture, its yield, its design time, its design cost, and its market price. Mis-estimating any of these factors can significantly impact the business case — sometimes to the point of jeopardizing the entire project.

Many trade-offs are made during this phase. The design is still at an abstract state and as a result changing significant aspects is easy to do: changing die sizes, making the power design more robust, adding or deleting RAMs or functional units to trade off power for speed, speed for area, and area for design time. The challenge is to estimate characteristics of the design as accurately as possible with only an inexact notion of what the design will eventually look like. The aspects of the design that need to be estimated are:

- *Amount of logic*: This is estimated based on a number of factors, including the size requirements of large reused components, e.g., memory, embedded processors, I/O circuitry, arithmetic functions, and other data-path functions. The amount of small gate-level *glue* or *dust logic* is estimated by comparing with past designs, experience, and technology insights. From the logic count we can accurately calculate required active logic area.
- *Die size*: This is derived from active logic by adding extra space for routing. First, a target placement density is chosen. Then an uplift factor based on empirical rules of achievable *wireability* is applied to the chosen placement density. A typical wireability curve is shown in Figure 10.10. This resulting value may be adjusted up based on the type of design, for example, a densely interconnected structure such as a cross-bar switch will require additional routing area. Finally, the density can be adjusted up or down by trading off design time. Higher densities can be achieved with extra effort in manual placement. Finally the limiting factor in die size may be the number of I/Os, for example, the die size of an extremely high pin-count design with simple logic functionality will be defined by the I/O count.
- *Defect-limited yield*: This factor is calculated from logic count and die size. Yield prediction is based on empirical tables as in Figure 10.11, which take into consideration logic density, die size, and technology information.

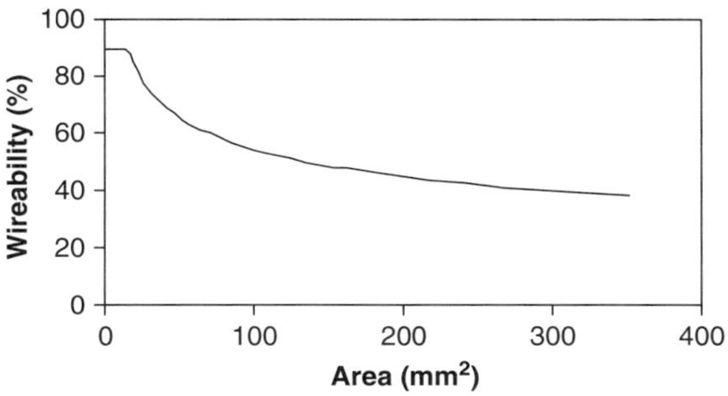

FIGURE 10.10 Chip wireability curve.

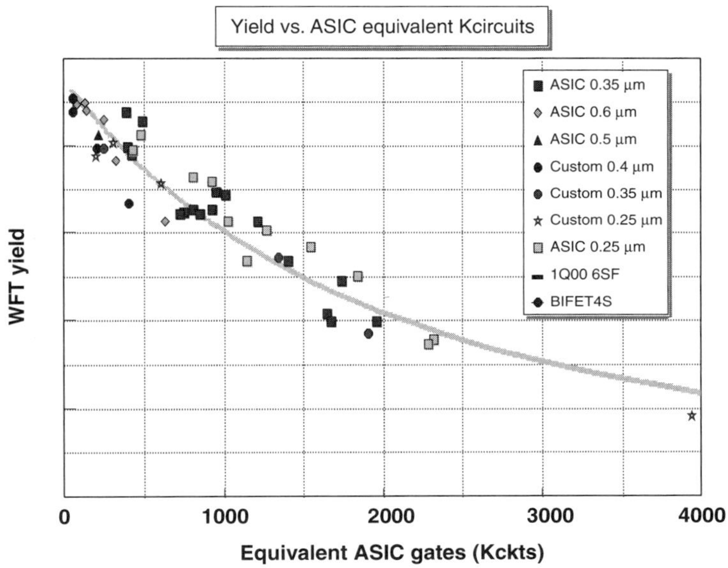

FIGURE 10.11 Yield vs. gate count plot.

- *Performance*: This is estimated by examining factors such as maximum logical path length, which is in itself derived in the absence of an actual design by experience with similar designs, information about technology parameters, *a priori* knowledge of the performance of embedded IPs, required I/O rates, estimated interconnect delays based on projected die size, voltage, and power limitations. Performance is a strong function of voltage, and as such the voltage is set to achieve performance targets defined by system or marketing constraints.
- *Power*: Active power is estimated based on performance, voltage, technology parameters, and empirical information about switching factors. Leakage power is estimated based on logic count, technology parameters, and voltage. There are significant architectural levers that can affect the power, for example, voltage or frequency scaling can be specified and significant sub-systems can be identified as candidates for clock gating.
- *Package*: Once the power is known, the package can be determined — generally the cheapest package that supports the frequency of operation, power dissipation, I/O requirements, and mechanical constraints.
- *Unit cost*: This is derived from yield, die size, and package calculations. Projected volumes from marketing are also a factor in unit-cost calculations.

In reality, the calculations are more complicated than this linear list implies. There are many trade-offs that are made, for example, architectural accommodations that favor performance at the expense of power and die size, such as adding additional arithmetic units to improve throughput. There are trade-offs balancing areas for yield by adding redundancy, such as extra word lines on array structures or area for reliability trade-offs by adding error-correction logic. Design effort is balanced against unit cost, or time to market, or performance, or power. The calculations outlined above are made and revisited as different trade-off decisions are made.

Platform-based design, where a design is made up of one or more reused large components including processors, integer, and floating-point arithmetic units, digital signal processors, memories, etc., allows for a much higher degree of accuracy of these early estimates.

After all the trade-offs have been made, the product of the concept phase is an architectural specification of the chip and a set of design constraints. These are fed forward to the floorplan and logic design phases.

10.2.2 Logic Design

The logic design phase involves implementing the register transfer logic (RTL) description of the chip based on the concept phase architectural spec. First, the specification is mapped into a hierarchical RTL structure in as clear and concise a fashion as possible and the details are filled in. Then the RTL is simulated against a set of test cases to ensure compliance with the specs. Because logic design and floorplanning are so intimately linked, the two steps generally proceed in parallel with each other.

There are three main design closure aspects that are dealt with during the logic design phase — performance, power, and routability. Because the design is so easy to change at this phase, mitigations of problems in these areas are easy to implement. However, it is difficult to measure any of these parameters directly from the RTL, which has not yet been mapped to gates, and is not yet placed or routed. There are some virtual prototyping tools that can provide quick but low-accuracy measurements based on either table-based analysis of the RTL or a quick-and-dirty encapsulated pass through synthesis, placement, and routing. In general, insight into a design's performance, power consumption, and routability is fed back to this phase from subsequent phases such as logic synthesis and placement.

The basic performance-improving RTL change involves identifying a set of critical paths, and modifying the RTL in some way to reduce logic depth. Reducing the amount of logic in a path is done by analyzing the logical content of the path and identifying portions that are not required. Owing to the hierarchical nature of design, there are often logical redundancies that can be removed by restructuring the RTL hierarchy. It is also possible to improve performance via *path balancing* or *retiming*; moving logic from one side of a long path's source or capture *latches* or *flip-flops* to the other.

There is also significant leverage for mitigation of power issues at the RTL . Chip logic can be modified to use *clock gating* which saves active power by shutting off clock switching to idle portions of logic or *power gating*, which saves both active and static power by switching off power to unused portions of the design. Both clock and power gating require careful attention to ensure that gating signals are calculated correctly and arrive in time to allow logic to stabilize as it comes out of its idle state.

In addition to these logical transformations, power/performance trade-offs can be made by using *frequency scaling* or *voltage scaling*. In frequency scaling, portions of the design that can run more slowly are segregated into more power-efficient, lower-frequency clock domains while more performance-critical logic is assigned to higher-frequency, and therefore higher power, domains. Voltage islands allow a similar power/performance trade-off by assigning less performance-critical logic to a lower voltage "island." Using a lower supply voltage saves both active and static power at the cost of additional delay. Voltage islands also require the addition of *level shifting logic* which must be added to allow logic level translation between circuits running at different voltages. The granularity of voltage islands need to be chosen carefully to ensure that the benefits of their implementation outweigh their performance, area, and power overhead.

Routability can also be optimized during the logic design phase by identifying congested regions and restructuring the RTL hierarchy such that either the congested regions are all in the same hierarchically designed unit allowing for better logical optimization and placement, or in extreme cases the RTL can be hand-instantiated and hand-placed. This is particularly useful in regular dataflow or "bit-stacked" logic.

Owing to the enormous amount of simulation time required to ensure logical correctness, logic design is the most time-consuming phase of the design closure flow, and it happens in parallel with the rest of the flow. At regular intervals, the RTL is brought into a consistent state and the rest of the design closure flow is launched. The purpose of these trial runs is to give engineers responsible for subsequent steps opportunities to tune their recipes, and provide feedback to the logic designers and floorplanners about factors such as late paths and poor structure.

The product of this phase is a hierarchical RTL design and amended constraints. These are fed forward to the final stages of floorplanning and the logic synthesis phase.

10.2.3 Floorplanning

The floorplanning phase prepares a design for the placement of the standard cell logic. Owing to their tight linkage, this phase generally proceeds in parallel with the logic design phase. Design work at this phase includes placing large objects, creating power grids and some portions of the clock distribution logic, placing I/O cells and pads, and wiring the cells to the pads, and creating circuit rows in the remaining areas for the placement of the "dust logic." Large objects consist of I/O cells, Random Access Memory (RAMs), Content Addressable Memory (CAMs), register arrays, large clock buffers, DCaps, and analog circuits such as phase lock loops. In a hierarchical design, subcell pin locations are assigned as part of the floorplan, as are restricted placement areas. This step also calculates rough load values for global nets, which can be used to guide logic synthesis. Figure 10.12 shows the floorplan of a large ASIC design.

By establishing large object and I/O placement, floorplanning has a large impact on interconnect delay on critical paths. As interconnect scaling continues to worsen, the importance of floorplanning is increasing. The major design closure aspects that are treated in this phase are wireability, performance, power-supply integrity, and power. The initial large-object placement is guided by insights about their interconnectivity and expected participation in critical paths. The floorplan is refined based on feedback from subsequent steps, for

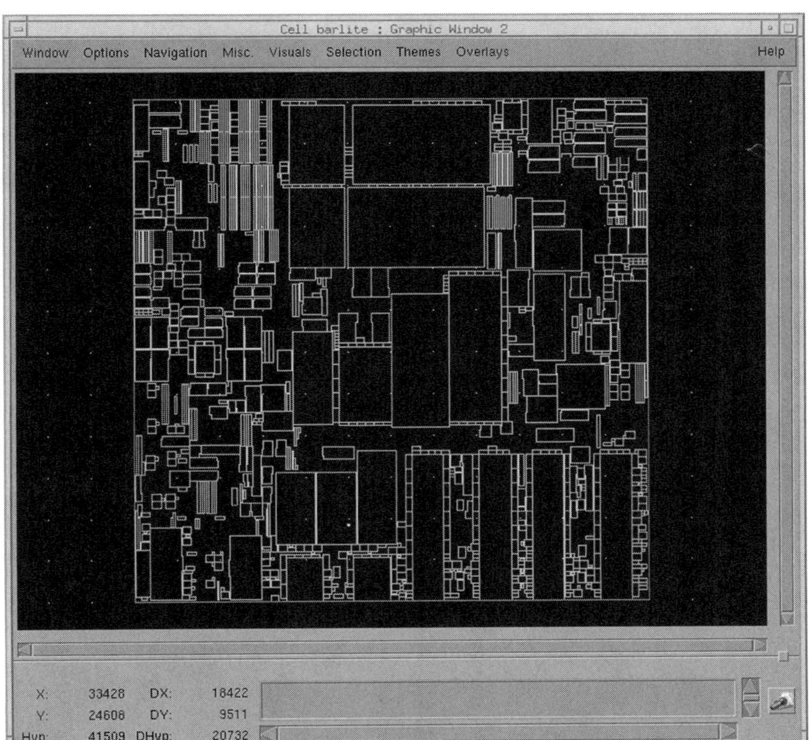

FIGURE 10.12 ASIC floorplan showing large objects.

example, insight into critical performance and congestion issues garnered from postplacement-phase global wiring and timing runs.

Wireability is the primary design closure consideration dealt with during the floorplanning phase. The placement of the large objects significantly affects the eventual congestion of the design. Sets of large objects with a high degree of interconnection are placed close together, with central areas of reduced placement density defined to accommodate the high wiring load — the classic example of this is a cross-bar switch. Initially, the large objects are placed based on *a priori* knowledge of design interconnectivity, or from insights about connectivity garnered from floorplanning tools. There are a number of such tools that can provide assistance, ranging from the simplest that give an abstract view of the large objects, the dust logic, and their interconnectivity as in Figure 10.13, to virtual prototyping tools that do quick low-accuracy synthesis, placement, and routing and give almost real-time feedback. Later in the design closure flow, feedback from the actual placement and routing steps is used to adjust the location of the large objects to reduce congestion.

Performance problems are also addressed during the floorplanning phase. The key action is to place large objects that have timing-critical connections close together. Initially this is done based on *a priori* knowledge of the timing paths in the design. The floorplan is adjusted as either low-accuracy timing feedback from virtual prototyping tools, or higher accuracy feedback from timing runs performed after placement and global wiring become available. Floorplanning insights from these sources are used to drive restructuring of the RTL back into the logic design phase to keep clock domains and critical logic closer together.

Power supply integrity can also be addressed during the floorplanning phase. DCaps are in general placed to provide power supply isolation or specifically around particularly noisy elements, such as CAMs and large clock drivers. Early power supply design is based on factors such as the location of chip power pins, power requirements of large fixed objects, expected dust-logic densities, and locations and sizes of voltage islands. Postplacement feedback is used to augment the power-supply design if necessary. If power density is estimated to be too high in certain areas, the placement density in these areas may be reduced.

Power is addressed indirectly during floorplanning. The most important power optimization at this stage is the planning of the voltage islands introduced in the logic design section. Each voltage island requires the design and analysis of its own separate power supply and power pin routing.

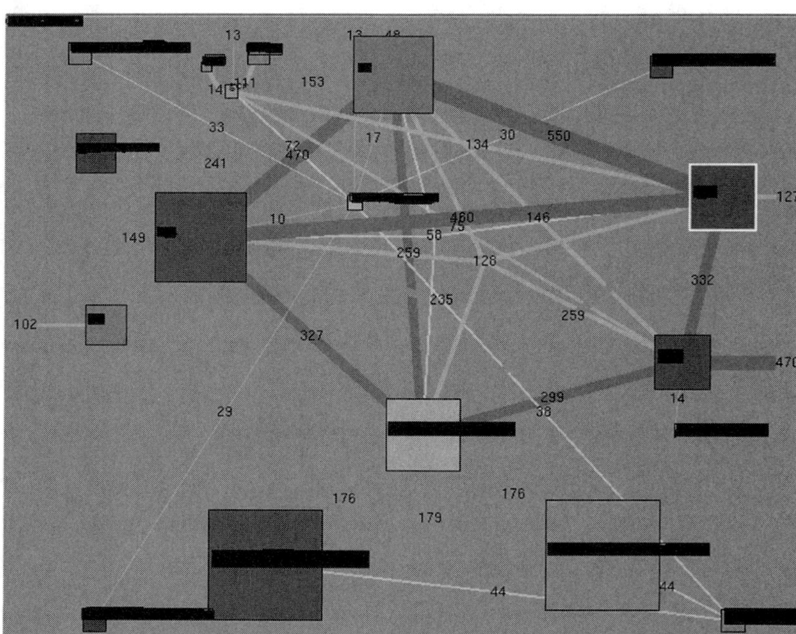

FIGURE 10.13 Interconnectivity and logic size visualization tool.

During the floorplanning phase, there are a number of trade-offs that are made. Addition of DCaps or lowered placement density regions can cause extra wire length, which impacts performance. Overdesign in the power distribution can significantly affect routability, which again can impact performance. Finding the right design points among all these factors can often take a number of iterations between floorplanning, logic design, and the subsequent design closure flow steps.

Handoff from the floorplanning phase is a detailed floorplan, including large object placements, power grids, move bounds, pin assignments, circuit rows, I/O wiring, and amended constraints. Handoff from here goes to placement.

10.2.4 Logic Synthesis

The purpose of the logic synthesis step is to map the RTL description of the design to a netlist rendered in library elements of the chosen technology meeting the performance targets with the fewest number of gates.[‡] The main design closure issues addressed in the phase are performance, power, and area. First, the RTL is compiled into a technology independent netlist format. Then, this netlist is subjected to logical analysis to identify and remove redundancies and balance cones of logic. Next, the optimized technology independent netlist is mapped into technology-dependent gates, and finally, the technology dependent netlist is timed and critical timing paths are corrected.

Because of the relative ease of running logic synthesis with different optimization targets, designers use it to explore the design space and make the best performance, power, and area trade-offs for their design. In order to achieve the area, power, and performance goals, logic synthesis applies a number of techniques to the technology-mapped netlist. In order to do this, these factors need to be measured: logic area is measured by adding up the sizes of the various gates; power is assumed to be a function of gate size — reducing gate sizes reduces power. Measuring performance is more complicated, and the logic synthesis is the first phase to rely heavily on static timing analysis. Since the placement of the dust logic has yet to be defined, *wire-load models*, usually a function of fanout, chip size, and technology, are used to estimate parasitic effects. There are subtle interactions between the wire-load-based parasitic estimation of the logic synthesis phase and subsequent placement and logic/placement refinement phases. If the wire-load models overestimate loading, then the power levels in the gates in the resulting design passed to placement will be excessively large, with no real way to recover the over-design. However, if the parasitic estimation is optimistic, then optimization in this phase will not focus on the correct problems. In general, erring on the side of optimism produces better results, especially with the advent of truly effective timing driven placement flows, to the point that many chips are now synthesized with zero-wire-load models for local signals, and load estimates for global signals derived from the floorplanning step. Since the clock distribution circuitry has yet to be added, idealized clock arrival times are applied to launch and capture clocks at latches. This step must also anticipate the impact of buffer insertion that will be performed during and after placement to prevent interconnect delay from being over estimated.

There are a number of environmental factors that need to be set in synthesis to guide static timing, such as voltage and temperature, based on information from the concept phase, guard banded for manufacturing variation and reliability factors such as HCI and NBTI. Performance is measured using incremental static timing analysis, and transforms that trade-off area, performance, and power applied. The basic approach to easing power and area problems is reducing gate size via a global repowering step, where all the gate sizes in the design are determined simultaneously. Then, critical paths are individually timing-corrected using a slack-take-down approach — an ordered list of the critical paths is created, the top path has one or more timing optimization transforms applied to it which moves the path "toward the good" in the critical path list, and the process loops back to creating a new ordered list of bad paths, etc. (see Figure 10.14).

This slack-take-down is repeated until all paths meet the performance constraint or there are no more optimizations that can be applied to the top critical path. The slack-take-down optimization scenario is

[‡] For a full description of logic synthesis techniques see Chapter 2, Volume 2 of this handbook.

```
                                                     Worst slack =  -0.98
                                      Number of negative slack =     10
                                                    Total tests = 11132
  |
-1.20 |                   Freq   Cumu1     %      Cum%
-1.00 | *                 ----   -----    ---     ----
-0.80 |                      0       0    0.00    0.00
-0.60 | *                    2       2    0.02    0.02
-0.40 | *                    0       2    0.00    0.02
-0.20 |                      4       6    0.04    0.05
 0.00 |                      4      10    0.04    0.09
 0.20 | *****                0      10    0.00    0.09
 0.40 | ******               0      10    0.00    0.09
 0.60 | *********          479     489    4.30    4.39
 0.80 | ***                607    1096    5.45    9.85
      ----+---+-----+      814    1910    7.31   17.16
          4   8   100      292    2202    2.62   19.78
```

FIGURE 10.14 Timing-slack histogram.

used extensively in the subsequent design closure phases. A short list of some of these optimizations used to improve timing on the critical path includes:

- *retiming* where logic is moved across latch boundaries to balance the amount of logic between latches;
- *rewiring* where timing critical signals are moved "later," i.e., further toward the sinks in a cone of logic to reduce the overall path delay;
- *refactoring* where a group of logic is mapped back into technology independent form and then resynthesized to preferentially shorten the logic length of a critical path;
- *cloning* where a section of logic is replicated to allow logical fan-out to be divided over more drivers; and
- *repowering* where a logic function is replaced by a similar function with higher drive strength (repowering used the other way, to reduce gate sizes, is the work-horse transform for both area reduction and power mitigation in the logic synthesis phase).

The first time that accurate gate count and timing are available in the logic synthesis phase is after technology mapping. Previous phases have all relied on gate-count and timing estimates. Surprises encountered when these numbers become available cause looping back to the floorplanning phase to reallocate space due to excess logic, or to reposition floorplanned elements for performance reasons, or go back to the logic design phase for various area and performance mitigations available in that phase.

The output of the logic synthesis phase is an area, performance, and power optimized technology-mapped netlist. This is combined with the floorplan and the most recent updated list of constraints and passed on to the placement phase.

10.2.5 Placement

The purpose of the placement phase is to assign locations to the logic that shortens timing-critical wires and minimizes wiring congestion.[§] Figure 10.15 shows the placement of a small portion of a larger ASIC design.

Until recently, placers solved congestion and performance problems by creating a minimum-wire-length placement — both min-cut or quadrisection placement techniques provide good results. As interconnect delay has become more dominant, it has become necessary to make the placement flow more timing driven. An effective timing-driven placement flow involves two placement passes: *global placement* and *detail*

[§]For a full description of placement techniques see [3].

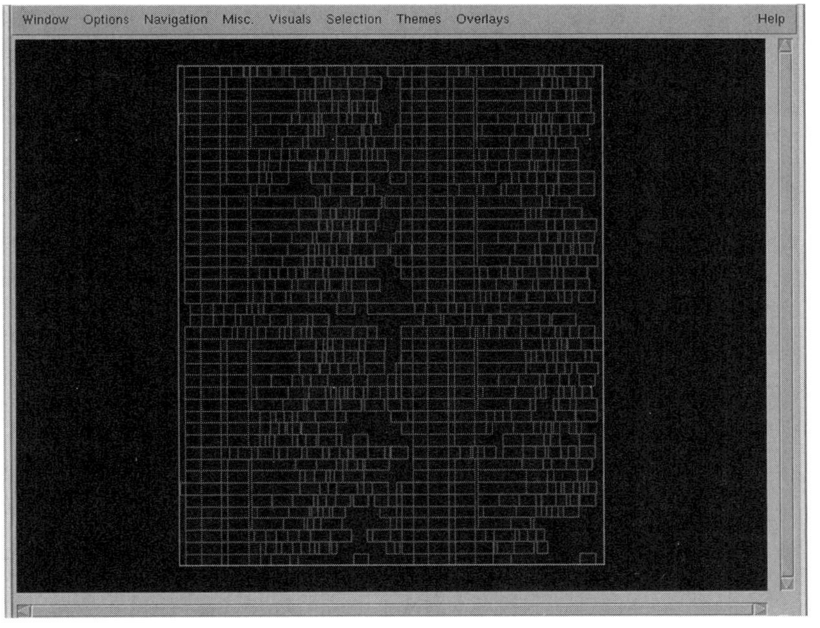

FIGURE 10.15 Logic placement.

placement. Before the global placement, the design is preprocessed to remove artifacts such as buffers on large-fanout nets, repeaters on long nets, scan and clock connections, which may bias the placement improperly. Then, a timing-independent congestion-mitigation placement is run. Next, a series of timing optimizations including gate sizing, buffer tree insertion and long wire buffering is performed. These optimizations require only analysis of slew and capacitance violations. Resized gates and new buffers and repeaters are added to the design without regard to legal placement constraints. The design is then timed using ideal clocks and interconnect delay calculated from Steiner estimates of wire topology. The timing problems that are identified are the "hard" problems that need to be fixed by the timing-driven detail placement. A set of attraction factors or *net weights* are then calculated, that bias the placement engine to move timing-critical objects closer together. These net weights are used to guide a second placement step on the optimized design. Finally, scans are reconnected and *N*-well contacts are added.

A key point in this timing-driven flow is the stability of the placement algorithm — a small change in the input to the placement engine, such as adding attractions between a small percentage of the objects being placed, must generate a small change in the result. If this is not the case, although the timing-driven placement will have pulled the timing-critical objects from the first placement closer together, a completely new set of timing problems will manifest. This same stability can help limit the disruption caused by late *engineering changes* (*EC*) to chip logic.

Another key consideration in placement is design *hierarchy*. All of the previous design steps, i.e., concept, logic design, floorplanning, and logic synthesis, rely on hierarchy to limit problem complexity. At the placement stage it is possible to either keep the design hierarchy set at logic design, or *flatten* it. If the hierarchy is kept, each hierarchical unit is placed separately, and then the units are combined to form the chip. In flat design the borders between some or all logical hierarchies are dissolved and the flattened logic is placed together. Generally, flattening a design allows significantly better optimization of performance and power during placement and subsequent design steps and requires less manual effort. On the other hand, retaining all or some of the hierarchy allows better parallelization of design effort, easier reuse of design, and faster incorporation of design changes. In addition, hierarchy is a natural way to keep the highest frequency portions of a design physically compact, which can help reduce clock skew. It should be noted that some flat placement flows provide some form of *move bounds* mechanism to manage the proximity of the most performance-critical circuits. The debate on advantages and disadvantages of flat vs. hierarchy continues in the industry and is worthy of its own chapter.

The output from the placement phase is a placed, sized netlist including buffers, repeaters, and N-well contacts which is passed to the logic/placement refinement phase.

10.2.6 Logic/Placement Refinement

The purpose of the logic/placement refinement phase is to apply placement, timing and congestion aware transformations to the logic to correct performance and power problems left over from the placement phase. The timing-driven flow outlined in the placement phase is excellent at localizing large numbers of gates to solve general performance and routability problems. However, upon the exit from that phase, there are performance-critical paths that need to be fixed individually. In addition, local power and routability issues are considered while making these optimizations. This phase has two steps. First, electrical problems such as slew limit exceptions at signal sink pins and capacitance limit exceptions at output pins are corrected. These problems are fixed by gate sizing, buffering of large fan-out nets, and repeater insertion. Second, the design is timed and optimized using the slack-take-down approach.

The timing environment for the second step includes using ideal clocks, worst-case timing rules, and parasitics extracted from Steiner wires. Also, this is the first place in the design closure flow where useful information about spatially dependent power-supply *IR* drop is available — this information is applied to the timing model where it is used to adjust individual gate delays. Feedback from this analysis can be used to guide redesign of the power supply routing.

The logic optimizations used in the second step are similar to the timing correction transforms mentioned in the logic synthesis phase; here they have been augmented to also consider assignment of locations to changed gates, impact to placement density, and wiring congestion. When logic is modified at this step, its placement must be *legalized* to a legitimate nonoverlapping placement site. This generally involves moving nearby logic as well, which can induce new timing or congestion constraint violations. These new violations are queued to be optimized. The step is complete when no more resolvable violations exist. An important part of this phase is the tools infrastructure that allows for the incremental analysis of timing and congestion caused by simultaneous changes to the logical netlist and the placement [15]. As the placement changes, previously calculated interconnects delays are invalidated, and when new timing values are requested, these values are recalculated using new Steiner estimates of routing topology. To measure congestion, an incrementally maintained probabilistic congestion map is used.

Routability can be affected in a number of different ways in this phase. Congestion information is used to assign lower placement densities in overly congested regions. As transforms are applied and logic moves, it avoids these low-density regions, mitigating congestion. This can impact timing in that gates that cannot be placed in their optimal locations due to placement density constraints, are placed further away. Another method of congestion mitigation is via congestion avoidance buffer placement, guiding the routes that long nets take by placing the repeaters along those nets in less-congested regions. Figure 10.16 shows a postplacement chip routability map.

Power and area can be recovered at this stage with the application of gate resizing. Despite the fact that the bulk placement is already done, reducing the gate sizes wherever possible provides for additional space for subsequent changes and additions, and reduced power. Also, at this point there is enough accuracy in the timing to do *mixed threshold logic* optimization. Mixed threshold logic gates have the same logical function and footprint of their standard threshold counterparts. They differ in their use of high or low threshold transistors. High threshold logic can be used to reduce static power at the cost of increased delay — off-critical-path standard-threshold-voltage gates can be replaced with their high-threshold-voltage equivalents. These substitutions are done in a "least impact to timing" fashion, where the set of possible swaps is generated, and the swap that degrades the overall timing of the chip the least is chosen. Then the set of possible swaps is regenerated based on the new design and the function is repeated until there are no more swaps that meet the criticality-plus-guardband specification. In contrast, low threshold logic can be used to decrease path delay on critical paths at the cost of increased power. Figure 10.17 shows a delay

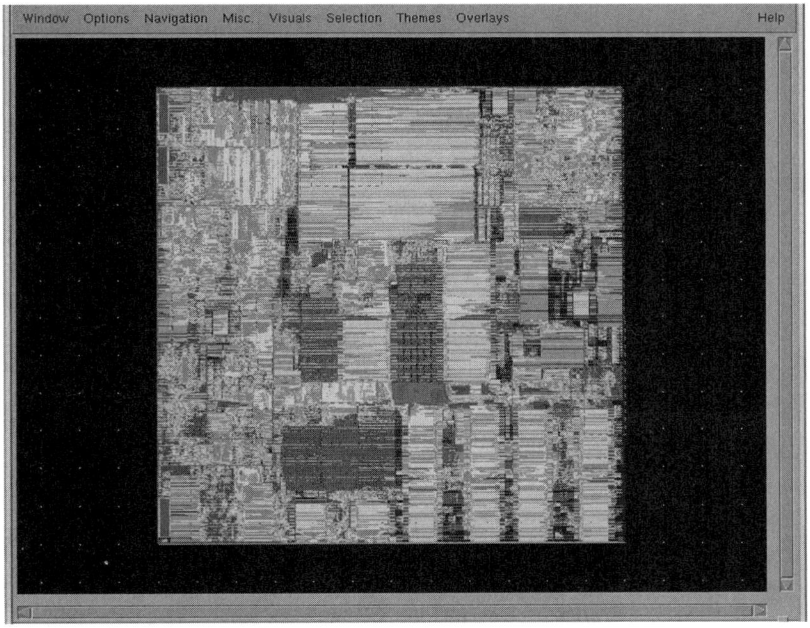

FIGURE 10.16 Chip routability map.

comparison between low and standard threshold logic. These substitutions are done generally by sequentially substituting logic on the most critical paths until all timing violations are resolved or until the amount of low threshold logic exceeds some static power-limited threshold. Low-V_t substitutions can be done at this phase or after clocking when more timing accuracy is available. Note that the introduction of multiple threshold voltages has two main costs: the first is an additional mask per threshold voltage, and the second is timing complexity associated with threshold-voltage mis-track.

The product of this phase is a timing-closed legally placed layout, which is passed to the *introduction-of-clocks* phase.

10.2.7 Introduction of Clocks

This phase inserts clock buffers and wires to implement the clock distribution logic. Figure 10.18 shows the clock distribution logic of a large ASIC design. By this phase, the amount of design optimization that can be accomplished is growing more limited. The main design closure issues dealt with in this phase are performance, signal integrity, manufacturability, and power supply integrity. The first step in this phase is the clustering of latches and the placement of the first-stage clock buffers to drive those clusters. Then, the first-stage buffers are clustered and the process repeats recursively. The clock buffers are placed with priority in the dust-logic regions, which causes dust logic to be moved out from under the buffers. After the clock buffers have been placed, the clock wires are inserted into the design.

Managing skew in the clock trees is critically important. Any unplanned skew is deducted directly from the cycle time. The buffers and wires in the clock distribution logic are carefully designed to provide as little unplanned skew as possible. To provide low skew, a number of different clock topologies can be used, including spines, H-trees, and grids. For many years *zero skew* was the optimization goal of clock design. In a zero-skew clock, the clock signal arrives at every latch in a clock domain at precisely the same moment. Recently, flows have begun using *intentional* or *useful skew* to further improve performance. By advancing or delaying clocks to latches on the critical path where there is a significant difference between the slack of the data-input and data-output signal, the cycle time of the design can be improved. Flow-based algorithms are used to recalculate the useful skew targets for all the critical latches in the design. Implementation of a

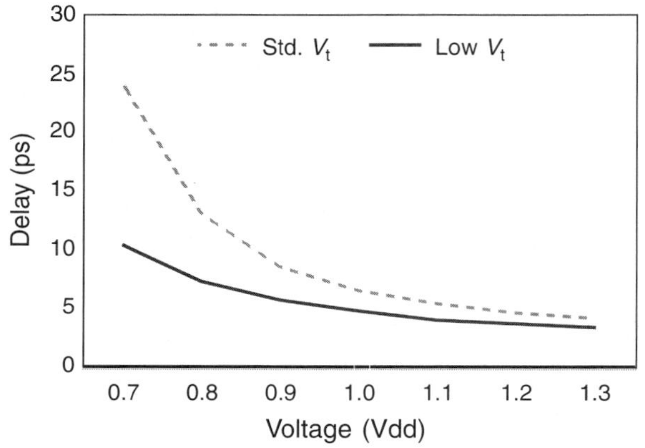

FIGURE 10.17 Effect of low threshold logic on delay.

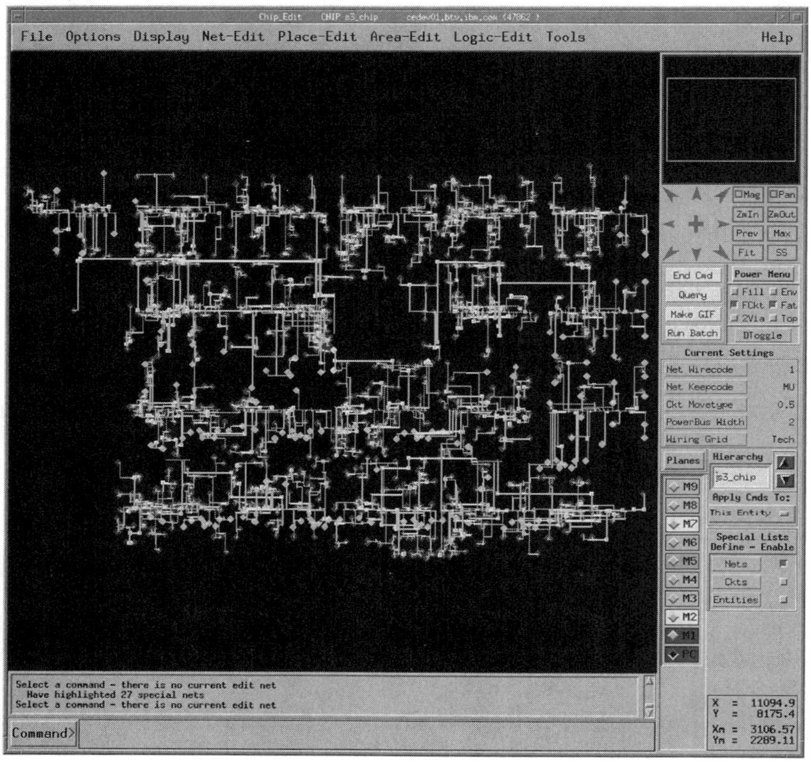

FIGURE 10.18 ASIC clock plan.

skew schedule is done by clustering latches with similar skew targets, and delaying clock arrival times to them by the addition of delay gates or extra wires. Calculating the skew schedule and then implementing it in this phase can yield a significant cycle-time improvement. However, if this skew-schedule is fed back to the logic-synthesis/placement/refinement phases, further gains in both performance and area can be realized.

Optimizing the clock for power is an important task, because a significant portion of a chip's active power is dissipated in the clock logic and routing. Clock buffers are tuned to ensure that the minimum gate size is used to provide for clock slew and latency requirements. Since most clock power is dissipated in the last stage or *leaves* of the clock tree, particular care is applied to minimize leaf level clock wire length. There are also interesting power trade-offs to be made around the way clock gating is handled in a clock tree. Moving clock-gates as close to the root of the clock trees as possible allows for the greatest amount of the clock tree to be turned off. However, there are generally both gated and nongated versions of the same clock required. The farther up the clock tree the gates are moved, the greater the amount of the design that must be spanned by both trees, which increases the total capacitance of the clock tree. At some point, the added power savings of moving the clock further toward the root is offset by the extra power burned in the additional capacitance of the second clock tree.

Transitions on clock wires have crisp edge rates, and as a result can cause coupling noise to adjacent wires. In addition, sequential elements can be susceptible to noise on their clock inputs. As a result, clock wires are sometimes shielded with parallel wires or given nonminimum spacing from adjacent wires to protect against coupling noise and to reduce signal dependent *clock jitter*. The switching of high-power clock buffers can also introduce significant power supply noise. To combat this, it is a good practice to surround main clock buffers with a large number of DCaps.

As manufacturing variation has become worse, it has become increasingly important to do *variation aware clocking*. Variation aware clocking techniques include the use of matched clock buffers, the use of wider metal lines, and preferential use of thicker wiring layers. In sensitive cases, it is necessary to perfectly match the order of the layer and via usage of all paths to skew-sensitive latches. It is also helpful to group latches which share critical timing paths on to the same branches of a clock tree. Timing can then use *common path pessimism removal* (*CPPR*) to account for the improved skew tolerance derived from the shared portion of the clock tree.

The output of this phase is a fully placed, presumably routable layout with fully instantiated and routed clock trees. This is passed to the postclocking optimizations phase.

10.2.8 Postclocking Optimizations

The purpose of this phase is to clean up the performance problems caused by the introduction of the clocks. There are two factors which drive the performance to degrade. First, the clock distribution logic can now be timed fully using 2.5D or 3D extractions of the real clock wires. Up until this point, all timing has been done with idealized clocks. If the idealized clocks were assigned properly with guardbands for skew, the introduction of the real clocks is fairly painless — only a few problems surface. The second factor that drives the performance to change is the replacement of the logic under the clock buffers — when the clock buffers are placed in the design, other logic is moved out from underneath them. The performance problems introduced by these two factors are fixed by applying the same timing and placement aware transforms that are applied in the logic/placement refinement phase.

In addition to fixing the performance problems caused by the introduction of the clocks, the added accuracy of timing the real clock distribution logic in this phase allows for the correction of hold-time problems. The workhorse hold-time fix is the introduction of delay gates between the launch and capture latches. Sometimes the launch and capture clocks can be de-overlapped to eliminate hold-time violations. Because this involves reworking the clock distribution logic, this is usually the option of choice only when a large number of hold-time violations can be eliminated with a single change.

To analyze properly timing problems at this stage, and to ensure that any fixes inserted do not break other timing paths, timing must be run on multiple timing "corners" simultaneously. The four standard corners are: slow-process/worst-case-voltage-and-temperature, slow-process/best-case, fast-process/worst-case, and fast-process/best-case. This requirement for multiple simultaneous timing runs further complicates the infrastructure requirements of the design-closure tool flow. As variation effects become more pronounced, this set of corners must be extended to include the possibility of *process mis-tracking* between device types and interconnect layer characteristics. Recent work in *variation-aware* and *statistical static timing analysis* provides much of the benefit of exhaustive corner analysis at much lower cost in tool run time [18].

Finally, in preparation for inevitable engineering changes, all empty spaces in the dust-logic regions of the chip are filled with gate array-style "filler cells." These cells provide unused transistors, which can be configured into logic gates by customizing one or more wiring layers. This technique can be used to implement emergency fixes for design problems found late in the flow or after hardware has been built. It is often possible to provide a patch to a design by rebuilding only one or two mask levels.

The output of the postclocking optimizations phase is a timing-closed clock instantiated presumably routable netlist. This is passed to the routing phase.

10.2.9 Routing

The purpose of the routing phase is to add wires to the design.[¶] By this phase, all of the main timing objectives of the design should be met. It is very difficult for timing problems to be fixed in routing; rather it is routing's job to deliver on the "promises" made by the wire length and congestions estimates used in synthesis, placement, and so on. There are three major design closure issues dealt with in this phase: performance, signal integrity, and manufacturability. Routing is generally broken up into two steps: *global* routing and *detail* routing. The goal of global routing is to localize all wires on the chip in such a way that all the routing capacity and demands of the chip are roughly balanced. To do this, the chip is partitioned into horizontal and vertical wiring tracks, where each track has some fixed capacity. Steiner wires are generated for all the nets, and these are laid into the tracks. Routability is optimized during global routing by permuting the track assignments to flow excess capacity out of highly utilized tracks. At this step, the wiring *porosity* of fixed objects is also analyzed to ensure correct calculation of routing capacity. As interconnect scaling worsens, it is becoming increasingly important to assign layer usage as well as track assignments during global routing. Overestimating track and layer capacity can have large impacts on the actual routed net length. The important performance-related design closure mitigation available during global routing is the preferential routing of timing-critical nets, where preidentified timing-critical nets are routed first, and to the extent possible will follow a direct path with the minimum number of jogs between source and sinks. Coupling noise can also be addressed at this phase by forcing the global router to segregate noisy and sensitive nets into different global routing tracks [19].

After the global routing step comes detail routing. In this step, the global routes are mapped into the real wire resources on the chip. In this phase, the real wire topology is determined and all wire widths, layer, via, and contact choices are made. As wires are added, wiring congestion increases and average wire length goes up. To ensure that timing constraints are met, timing-critical signals are wired first. This gives them access to the shortest wire length and preferred wiring layers. In addition, certain signals may be assigned to be wired on specific layers or specific widths for electrical reasons. For example, long timing-critical wires may be designed as wide wires on *thick* upper wiring layers to minimize resistance. Particularly skew-sensitive situations such as busses or differential signals may require *balanced routing* in which the lengths, topology, and layer usage of two or more skew-sensitive wires are matched. Wide busses are often prewired to ensure that they have balanced delay.

In addition to managing performance, wiring must optimize for coupling noise. The detail router can be guided to segregate noisy and sensitive nets [19], and to insert additional empty space between them to further reduce coupling if needed. In some cases it may be necessary to route adjacent grounded shielding nets to protect particularly sensitive signals, or to prevent interference by a particularly noisy signal. Coupling issues on busses can be improved by using *random Z-shaped routing* to prevent overly long parallel wires. Inductance effects in large busses can be mitigated somewhat by the addition of interspaced power or ground wires which serve as low-resistance current return paths.

Manufacturability can also be optimized during the routing phase. Routing can add redundant vias and contacts to improve both reliability and yield. As interconnect variability becomes more important, routing can also add wide wires and matched vias and layers to manage back-end-of-the-line variability on sensitive wires. Overall manufacturability can be further improved by adding extra *fill shapes* during routing to ensure more uniform shapes density, which improves dimensional control during lithography and CMP.

[¶]For a full description of routing techniques see [6].

If congestion estimates were inaccurate or other constraints such as coupling decreased routability, routing may not be able to embed all wires. The remaining wiring *overflows* must be manually embedded. Figure 10.19 shows a section of a fully routed ASIC.

The output of the wiring phase is a wired legally placed netlist. This is passed to the postrouting optimization phase.

10.2.10 Postrouting Optimization and Final Signoff

This phase deals with optimizing out the last problems using the most accurate analyses. The main design closure issues dealt with in this phase are performance, signal integrity, yield, and reliability. At this point, since the design is nearly finished, any changes that are made must be very local in nature.

Timing is now performed using fully extracted wires with real clocks. Any timing problems that occur at this point are fixed with gate sizing, buffering, automatic or manual rerouting, and wire-widening.

Now that real wires are available, mutual capacitances can be extracted to drive noise analysis. Noise analysis uses a simple model of coupling combined with analysis of possible logic switching windows derived from timing to determine if adjacent wire switching will create timing violations or logic glitches. If errors are found they may be fixed by reducing common run length, spreading wires, adding shields, adding buffers, resizing gates to modify slew rates, and rerouting to segregate wires. Yield issues may also be analyzed and corrected in a similar manner; wire-limited yield can be addressed at this stage by increasing wire width, increasing inter-wire spacing or decreasing common run length between wires. Yield can also be optimized via wire spreading (see Figure 10.20), though this processing can complicate optical proximity processing and can interfere with wire uniformity.

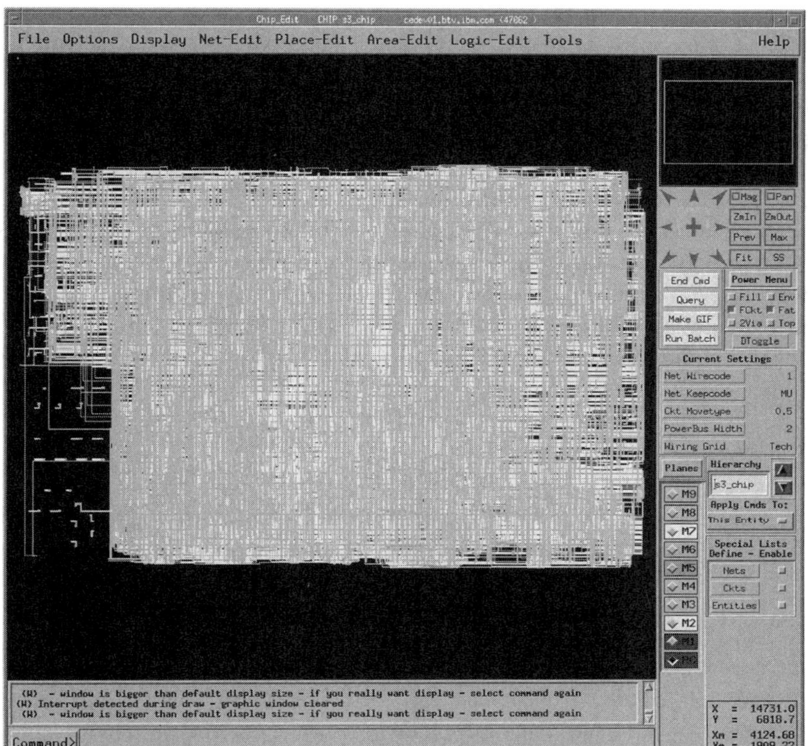

FIGURE 10.19 ASIC chip routing.

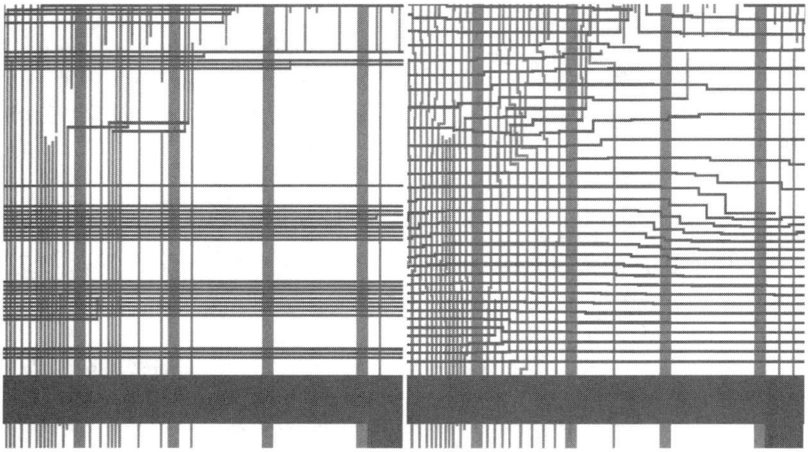

FIGURE 10.20 Before and after wire spreading.

Any timing problems, coupling or yield constraint violations found at this point can be fixed manually or by automatic routing-based optimization, which uses the same logic, placement, and timing optimization framework described in logic/placement refinement. In such an automated flow, wires with constraint violations are deleted and queued for rerouting. Final closure is performed using transformations that adjust gate sizes, do minimal logic modification, and minimally adjust placement density on the offending wire's logical and physical neighbors.

In addition to these final optimizations, there are a number of final checks and optimizations that are run on the design. One of these is ESD *checking and optimization.* The first of these is the *floating gate detection* and the addition of any necessary floating *gate contacts* to ensure that all gates in the design electrically connect to at least one diffusion. The second is *wiring antenna* detection and correction, which remove any unterminated wiring segments that might be left in the design by previous editing. Antennas can also cause ESD problems. Reliability checks are performed to ensure that no wire violates its electromigration limit and any wire that does is modified via wire-widening and load adjustment. Finally, design rule checking and logical-to-physical verification are performed.

Final timing sign-off is performed using *variation aware timing* by checking the design against an exhaustive set of process corners. This exhaustive checking tests the design for robustness to possible mistracking between layers, device strengths, multiple supply voltages, etc. This guarantees conservatively that the design will be manufacturable over the entire process window.

The product of this phase is a fully placed, routed, and manufacturable design, ready to be sent to the mask house.**

10.3 The Future of Design Closure

The next big challenges in design closure will be dealing with the increased importance of *power-limited performance optimization* and the increased need to do *design for variability.*

10.3.1 Power-Limited Performance Optimization

Current design closure flows treat performance as the primary optimization objective, with constraints such as power and area handled as secondary concerns. As both active and static power continue to increase, the design flow will have to be modified to treat the power/performance trade-off as the primary

** It's time to go home and have a beer!

optimization objective. Because power is best addressed early in the flow, the most leverage will come from advances in the concept and logic design phases. The most important design closure advances in this area will need to be the creation of power-oriented design exploration and optimization flows. To enable these flows, the industry will need to develop more accurate early power prediction techniques.

In the logic synthesis, placement, routing, and refinement stages, the flow will have to be modified to manage the power impact of every optimization rather than sequentially optimizing for performance, and then assessing the power impact. The flow will also have to be modified for more aggressive power management design techniques including wider use of dynamic voltage scaling, and wider use of power gating. Design closure flows will also need to be extended to handle emerging circuit-oriented power management technique. These techniques are likely to include the use of new circuit families and dynamic *body bias*, a technique that lowers static power by dynamically adjusting the gate to body bias.

10.3.2 Design for Variability

In Section 10.1.2.7, we mentioned the trend of increasing parametric variability for both devices and wires as chip geometry continues to shrink. As this has occurred, the relative amount of performance guardbanding has had to increase. This forced conservatism effectively reduces the amount of performance gain extractable from new technology nodes. As we move into 65-nm design and below, we are now entering a phase in design closure where parametric variability is becoming a first-order optimization objective. Design closure for variability can be addressed in two distinct ways. The first is *design for manufacturability* (*DFM*); the second is by doing *statistically driven optimization*.

Design-for-manufacturability concerns have been slowly working their way into the design closure flow. DFM optimizations involve constructively modifying the design to minimize sources of variability in order to make the design more robust to manufacture variations. There are many constructive DFM techniques, which have been or are currently being automated as part of the design closure flow. These include:

Matching devices. Identical circuits tend to track each other better than dissimilar circuits in the face of manufacturing variation. Design closure flows can exploit this by using identical buffering in skew-sensitive circuits such as clocks.

Variation aware routing. Automatic routers are being modified to create more variation-tolerant routing. Optimizations include: the use of wide wires on variation-sensitive routing, the use of geometrically balanced (i.e., matching topology, layer, and via usage) routing for critically matched signals, and the design of maximum common subtrees to reduce process-induced delay variation in skew-sensitive routes such as clocks.

Geometric regularity. Designs with a high degree of device and wiring regularity tend to have lower levels of dimensional variation. This is because both optical lithography and etch processes such as *CMP* respond better with uniform shapes densities. Regularity can be imposed at many levels of the design. At the cell level, regularity can be achieved by placing all critical geometry such as transistor polysilicon gates on a uniform grid. At a placement level, the flow can ensure that all gates see essentially uniform density on critical polysilicon and contact layers. At a routing level, the flow can enforce a fixed routing grid and uniform routing density through careful route planning and the selective addition of routing fill. New regular design styles such as *structured ASIC* [20–22] are being introduced to reduce sensitivity to manufacturing variation.

Adaptive circuit techniques. Chips will include greater number of *adaptive variability management* circuits. These circuits will either use automatic feedback or digital controls to "dial out" variability. For example, adaptive control could be used to cancel a manufacturing-induced offset in a differential receiver, or could be used to adjust out delay variation in a critical signals timing.

In addition to these constructive DFM techniques, a new paradigm for managing variability based on the availability of new *statistical static timing* (*SST*) [18] is emerging. Previous static timing tools characterized gate and wire delays based on specific parametric assumptions: i.e., *best, worst, nominal*; statistical timing tools, in contrast, characterize delay using statistical delay distributions derived from measured hardware. Rather than calculating path delay as the sum of all the worst-case (or best-case) delays, statistical timing calculates a probability distribution for delays. This distribution indicates the likelihood of a given path having a specific delay. By calculating this measure for all paths, one can determine the percentage of manufactured chips that will run at a given speed. Calculating delay in this way avoids the compounding conservatism of using the worst-case delay of all elements for the circuit. Combining statistical timing with an automatic design closure flow will allow designers to make yield vs. performance trade-offs. By optimizing a design to an acceptable circuit-limited yield value rather than an exhaustive set of worst/best-case parameters, a designer can greatly reduce the performance lost to over-design. Though these techniques are still at the early stages of deployment, we are certain that DFM and statistical design will be the dominant themes in design closure for the next several years.

10.4 Conclusion

We have defined the design closure problem, and the current design closure constraints and how they have evolved. We then explored how these constraints are addressed throughout the design flow. New constraints will continue to evolve and the closure problem will continue to grow more complex and more interesting.

Acknowledgments

The authors would like to acknowledge Dr. Leon Stok, Dr. Ruchir Puri, Dr. Juergen Koehl, and David Hathaway for their helpful input and feedback on this chapter.

References

[1] Design and Verification Languages, chap. 14, this handbook, Vol. 1.
[2] Logic Synthesis, chap. 2, this handbook, vol.2.
[3] Digital Layout Placement, chap. 5, this handbook.
[4] Trevillyan, L., Kung, D., Puri, R., Reddy, L., and Kazda, M., An integrated design environment for technology closure of deep-submicron IC designs, *IEEE Design Test*, 21, 14–22, 2004.
[5] Shenoy, N., Iyer, M., Damiano, R., Harer, K., Ma, H.-K., and Thilking, P., A Robust Solution to the Timing Convergence Problem in High-Performance Design, *International Conference on Computer, Design*, San Jose, CA, May, 1999, pp. 250–254, this handbook, vol.2.
[6] Routing, chap. 8, this handbook, Vol. 2.
[7] Design and Analysis of Power Supply Networks, chap. 20, this handbook, Vol. 2.
[8] Noise Considerations in Digital IC's, chap. 21, this handbook.
[9] Design for Manufacturability in Nanometer Era, chap. 19, this handbook.
[10] Design Rule Checking, chap. 17, this handbook, Vol. 2.
[11] Shepard, K., and Narayanan, V., Noise in deep submicron digital design, *ICCAD 1996*, pp. 524–531.
[12] Shepard, K., Narayanan, V., and Rose R., Harmony: static noise analysis of deep submicron digital integrated circuits, *IEEE TCAD*, pp. 1132–1150, August 1999.
[13] Shepard, K., and Narayanan, V., Conquering noise in deep-submicron digital ICs, *IEEE Des. Test Comput.*, 15, 51–62, 1998.
[14] *International Technology Roadmap for Semiconductors, 2004 Update for Design.* http://www.itrs.net/Common/2004Update/2004_01_Design.pdf
[15] Design Flows, chap. 1, this handbook, Vol. 2.
[16] Pillage L.T., and R.A. Rohrer, Asymptotic waveform evaluation, *IEEE Trans. Comput. Aided Des.* (1991 IEEE Best Paper Award), 352–366, 1990.
[17] Static Timing Analysis, chap. 6, this handbook, Vol. 2.
[18] Visweswariah C., Death, Taxes and Failing Chips, *IEEE/ACM Design Automation Conference*, Anaheim, CA, June 2003, pp. 343–347.

[19] Stohr, T., Alt, H., Hetzel, A., and Koehl, J., Analysis, Reduction and Avoidance of Cross Talk on VLSI Chips, *International Symposium on Physical Design*, Monterey, CA, 1998 (ISPD98).
[20] AMIs XpressArray structured ASIC: http://www.amis.com/asics/structured_asics/XPressArray.html
[21] eASIC http://www.easic.com/
[22] Pileggi, L., Schmit, H., Strojwas, A.J., Gopalakrishnan, P., Kheterpal, V., Koorapaty, A., Patel, C., Rovner, V., and Tong, K.Y., Exploring Regular Fabrics to Optimize the Performance-Cost Trade-off, *Proceedings of IEEE Design Automation Conference*, Anaheim, CA, June 2003.
[23] Power Analysis and Optimization from circuit to Register Levels, chap. 3, this hand book.

11

Tools for Chip-Package Codesign

Paul D. Franzon
North Carolina State University
Raleigh, North Carolina

11.1 Introduction

Chip-package codesign refers to design scenarios in which the design of the chip impacts the package design or vice versa. Computer aided tools are needed for codesign in situations where simple bookkeeping is insufficient. The most classical and most used chip-package codesign tool is the I/O buffer interface standard (IBIS) macromodeling tools for conversion of integrated circuit (IC) I/O buffer information into a format suited for rapid cosimulation. Tool issues of IBIS will be discussed toward the end of this chapter.

However, the need for more sophisticated tools for chip-package codesign is rapidly appearing. High-frequency designs require more accurate modeling of the chip and package components. Reduced order of macromodeling is often needed, especially to speed up the IC portion of a simulation. Chip-package cosynthesis is starting to emerge as a new area, especially for simultaneous floorplanning, pin assignment, and routability analysis for high pin count packages and three dimensional ICs.

11.2 Drivers for Chip-Package Codesign

Until recently, there was relatively little need for chip-package codesign. The package, and subsequently the board, could be designed after the fact, mainly with a sole focus of obtaining connectivity. The only information that needed to be communicated between the chip and package design was pin functionality.

As operating frequencies and system density increased, this "over the wall" sequential design strategy no longer worked. Table 11.1 summarizes the issues that have become important over the last decade, together with the solutions available and challenges still open. The next two sections discuss this table in detail, first looking at digital codesign and then mixed-signal codesign. The following section provides a brief overview of the most successful codesign tool to date, the IBIS macromodeling language, before presenting conclusions and an annotated bibliography.

TABLE 11.1 Summary of Some of the Main Issues, Solutions, and Open Challenges in Chip-Package Codesign

Type of System	System Issue	Technological Solutions	Design Solutions	Open Challenges
Digital	Board/package level timing and noise		Simulation with accurate timing and accurate I/O macromodels	
	SSN management	Package-embedded decoupling capacitors	Chip-package cosimulation	SSN-accurate macromodels for core and I/O
	High-speed transceiver design	Flip-chip solder bump; low loss materials	Equalization	Continued scaling beyond 10 Gbps; increased high-speed cost pin counts; effectiveness
	High pin counts	High-density laminates; 3D ICs	System floorplanning and pin assignment; coverification	Effective floorplanning and pin assignment tools
	Thermal dissipation and stress management	Heat spreaders, advanced air cooling, spray cooling, etc.	Thermal design	Thermal macromodeling
Mixed signal	Cost effective passive integration (Ls, Cs, baluns, antennae, etc.) for miniatu-rization and cost reduction	System on a package (SOP) technologies (embedded passives; 3D-IC)	Cosimulation and coextraction	Simulation models
	Noise management e.g., for voltage controlled oscillators (VCOs) and low noise amplifiers (LNAs)		Cosimulation and coextraction noise reduction	Accurate SSN prediction

Figure 11.1 presents a summary flow of the major design activities involved in chip-package codesign. Early planning requires determining design size plan, anticipated I/O counts, and selection of the basic package type. Often package selection might require detailed evaluation of some of the later activities in the flow. Although thermal design is shown next, it is usually conducted concurrently with the other activities. For more power hungry chips, it has become important to pay attention to "hot spots" and thermal transients when specific chip functions turn on or off.

Pin assignment and I/O electrical designs are major codesign activities that, especially for higher pin counts and mixed-signal systems, require simultaneous evaluation of chip, package, and board. For example, poor pin assignment will lead to larger packages and higher levels of simultaneous switching noise (SSN). Delay planning must include budgets for chip, package, and board.The SSN management will require macromodeling of on-chip drivers and other sources of power/ground transients (e.g., clock distribution circuits) as well as detailed simulation of the package, including any embedded capacitors or the board. For a large digital system, SSN modeling and simulation are often the most complex task in this flow. For mixed-signal systems, RF and analog circuits must be carefully separated from digital circuits, e.g., excess noise from the latter can prevent a phase locked loop (PLL) from even locking.

11.3 Digital System Codesign Issues

Table 11.1 lists digital issues roughly in their chronological emergence as codesign drivers. When system clock speeds started exceeding a few tens of megahertz, it became necessary to design the traces on a printed circuit board (PCB) to meet timing (delay) and reflection noise requirements. Two pieces of information had to be

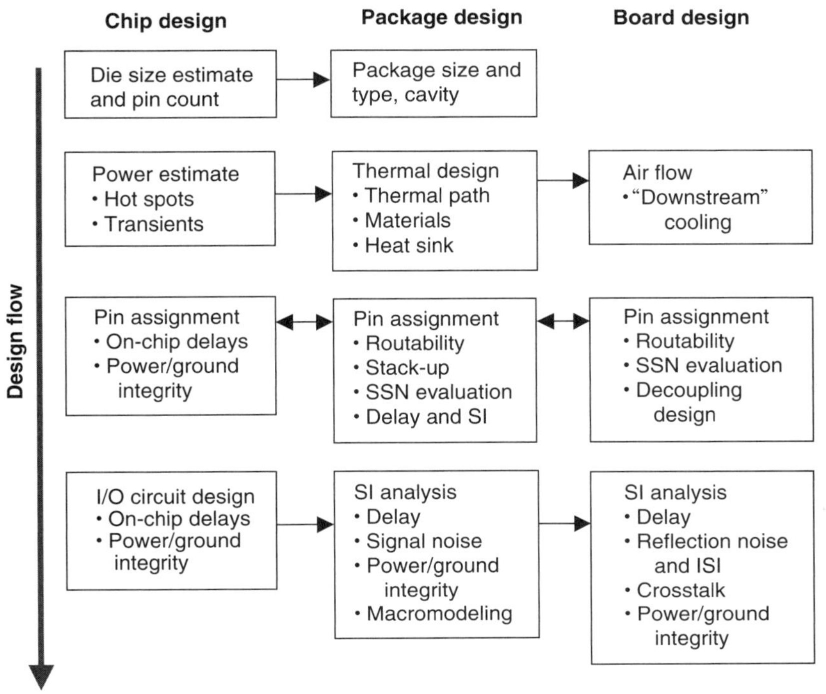

FIGURE 11.1 Major flows and design activities used in chip-package codesign.

passed from the system/IC design to the board designer. The first was a list of timing slacks or what board delays would be acceptable. Timing slacks can easily be conveyed via a spreadsheet, or similar simple mechanism. The second was a reasonably accurate simulatable model of the chip I/O, including package parasitics. A transistor-level SPICE (or other circuit-level simulator — see chapter 14 on *Simulation of Analog and RF Circuits and Systems*) model is rarely a desirable solution, mainly because it conveys proprietary data about the chip design and fabrication process, and also because chip transistor models are often proprietary to a specific simulator. Instead a portable macromodel is needed. Since the mid-1990s, the highly successful IBIS macromodel has fulfilled this need, but new, more accurate, techniques are emerging (see the next section).

As IC transistor counts increase, the amount of SSN increases. Simultaneous switching noise occurs when on-chip transients lead to current pulses (di/dt spikes) in the power and ground system. Since any practical power and ground system has inductance associated with it, noise is induced (as $V = L\, di/dt$) on the power and ground rails. High noise levels can lead to transient errors in the chip. While most of the di/dt pulsing is caused by the output switching, on-chip logic blocks also contribute to it. Techniques to control SSN include the provision of decoupling capacitors between power and ground, and also design methods to reduce L and di/dt. Accurate SSN simulation requires a combination of fairly detailed macromodels for on-chip di/dt sources, along with high fidelity models of the package and the PCB. There is no widely accepted macromodeling procedure for the di/dt models and, thus, this is an important open issue. While there are existent proofs of accurate SSN cosimulation, these are generally confined to sophisticated teams working in vertically integrated companies. No general methodology is available, though one is sorely needed.

The SSN management has started getting into cosimulation as well as codesign issues. An example would be codesign of on-chip and off-chip decoupling capacitors, together with optimized pin assignment so as to minimize the effective inductance. While sophisticated teams have performed such codesign, the methods for doing it effectively when the chip design team and system design team are vendor and customer are yet to be established. Many signal integrity textbooks document the impact of pin assignment on SSN. In general, it is important to intersperse power and ground pins among signal pins in order to control SSN. The ratio and degree of interspersion depends on the pin bit rate and overall SSN issues.

With increased interchip bandwidths also comes the requirement for multi-Gbps I/O and high pin counts. True codesign is needed to build successfully multi-Gbps I/O. Since the channel characteristics are strongly frequency-dependent, on-chip signal processing or other circuit techniques, are used to compensate (equalize) for the poor channel frequency response. Macromodels of IBIS are not used here due to the use of circuit and DSP techniques and the high degree of modeling fidelity required at these speeds. Transistor-level design is required, along with accurate PCB and package characteristics. However, some emerging macromodel techniques are under investigation.

Chips with over 1000 pins are available today and multithousand pin packages are anticipated in the future. With such high I/O pin counts, the details of chip I/O location are important. Doing a poor job at pin assignment could greatly increase the layer count and cost of the package and PCB, or even make the system unbuildable! Quality I/O assignment requires simultaneous consideration of chip floorplan and package/PCB routability. Relatively little work has been done in this area of integrated cosynthesis of chip and package. Anecdotal evidence shows that good I/O pin assignment can reduce the amount of package-level wiring by more than 10%.

Three dimensional ICs, in which chips are stacked and integrated with internal vias, is likely to be the epitome of a high pin count system. Anticipated via technologies will be able to support tens of thousands of "pins" between adjacent chips in the 3D stack. Since the chips are fabricated separately and then joined, the 3D IC is more properly viewed as a packaging technology than as an IC technology. Given the high pin count, automatic codesign will be needed for any but the most regular system. At the least, a 3D floorplanner with interlayer pin assignment will be needed.

After design, coverification is important for these high pin count systems, but is fortunately more straightforward. It is relatively easy to import the package design into the IC design tools (Figure 11.2) and

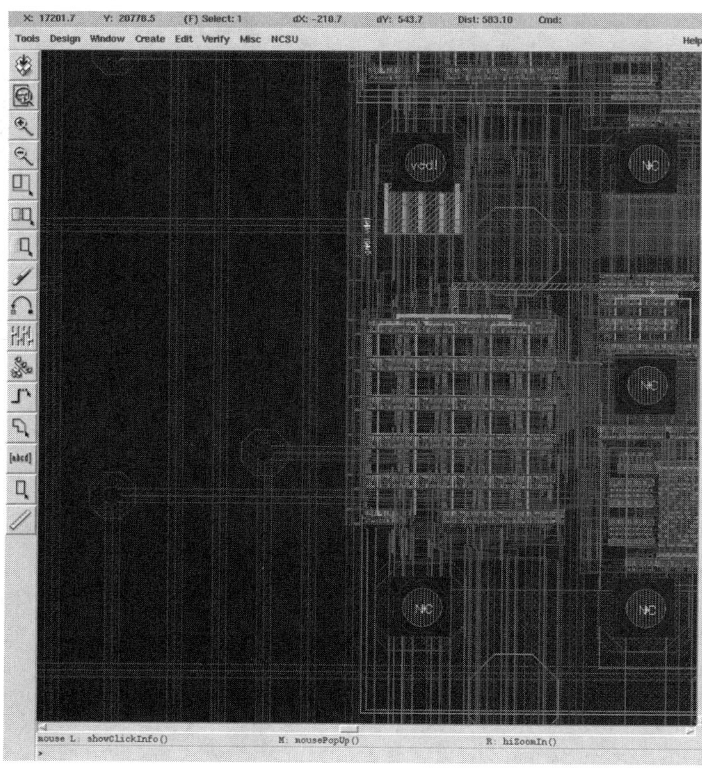

FIGURE 11.2 Result of a flip-chip package after importation into an IC design environment for the purposes of coverification.

conduct verification tasks such as extraction and comparison of the extracted view with the initial schematic.

The final issue in Table 11.1 is thermal dissipation. Today, thermal chip-package codesign is not practiced, since the package can be safely designed after the chip. However, in future, the spatial and temporal rates of change of heat flux will increase. Chip-package codesign will be needed to match these rates of change to the capability of the package. For example, a high spatial variation of heat flux could lead to unacceptable temperature gradients and hot spots, and might best be fixed through an adjustment in the chip floorplan. Similarly, the temporal rate of change might have to be reduced for the same reason. This issue is likely to be particularly important in 3D chip stacks, wherein heat removal is particularly difficult.

11.4 Mixed-Signal Codesign Issues

For mixed-signal systems, the combination of increasing operating frequencies, together with a strong drive towards miniaturization, leads to an increasing need for coverification and codesign. This need is very important for the correct execution of a SOP design, in which digital and analog RF IC portions are separately or together combined with on-package passives, such as inductors, capacitors, baluns, and antennae. This combination allows separate optimization of the analog/RF and digital IC processes, tightly integrated with passives that perform better than their on-chip alternatives, and are smaller than discrete components.

For example, an inductor placed on a package structure can be built with a higher Q factor (lower losses) than its on-chip equivalent. It has been shown that SOP-based filters and antennae can have lower insertion losses than their discrete equivalents. A chip package with co-implemented radio is likely to have lower power consumption and decreased size, compared to its conventional equivalent. One drawback of the SOP approach is the increasing stress it places on the need for good parametric models to enable cosimulation of the transistors and SOP passives.

Accurate analog transistor models are needed together with accurate models for the passives integrated onto the package. Such models must include the impact of processing variations. Collecting such models is a tedious and time-consuming task, especially for the first run on a new technology. Good parametric models are needed to size correctly the passives, while sizing the transistors at the same time. When such models are lacking, multiple design-build-test cycles are often needed to achieve a working design.

In some circumstances, use of an SOP might increase the modeling and design effort required. For example, if an RF chip is placed just above a package-integrated inductor or antenna, the designer should model and simulate the current induced in the on-chip circuits due to the integrated passive and the impact of the chip substrate on the performance of the passive. For the former, it might be necessary to move a highly sensitive circuit, such as a low noise amplifier (LNA) away from the peak fields induced by the passive. If shield layers are used it is important to model the amount of digital SSN introduced onto the "grounds," especially floating grounds. Remember, there is no such thing as a perfect voltage reference — common mode noise is always a concern.

Even if integrated passives are not used, accurate cosimulation is important to achieve a quality design. Often the most critical issue is the impact of the digitally-sourced SSN on the jitter of VCOs and noise figure in LNAs. Phase noise and noise figure are generally hard to simulate accurately. However, to come close, the power and ground noise needs to be included. To predict accurately the power and ground noise requires good di/dt modeling and the inclusion of accurate package and PCB parameters for the power and ground system. True cosimulation is also needed. (See also Chapter 23, *Mixed Signal Noise Coupling in System-on-Chip Design,* which is faced with very similar problems.)

11.5 I/O Buffer Interface Standard and Other Macromodels

The IBIS has been a highly successful technique to macromodel I/O buffers. Although an accurate count of the number of users is not available, a web search shows over 800 companies using IBIS and over 1000 users have downloaded the newest version of the SPICE to IBIS tool. The reasons for its success are as follows:

- No proprietary information is conveyed in an IBIS model.
- It is sufficiently accurate to lead to design success for timing and noise prediction in board level designs operating at clock rates of around or just above 100 MHz.
- It can be automatically produced from a SPICE netlist, using the SPICE to IBIS utility.
- IBIS-compatible simulators can be very fast, as IBIS IV curves are continuous and analytic (continuous in the first derivative).

The circuit elements in an IBIS-compatible output model are shown in Figure 11.3. The pull-up and pull-down components are used in the output models only. They include an IV data table and ramp rates (to give accurate rise and fall times). The power_clamp and gnd_clamp curves are used in input models and output models to enable outputs. . They are intended to model the clamp diodes that turn on when the voltage goes too far outside the Gnd–Vcc range. They are also captured as IV tables. All IV tables are interpolated during simulation. L_pkg, R_pkg, and C_pkg are a simple attempt to capture package parasitics. Note that there is a lot of ambiguity here, as to how to account for mutual inductance and capacitance between neighboring leads in the package. It is up to the IBIS vendor whether to leave them out or to simply add them in; neither is entirely accurate. C_comp is used to capture the on-chip pin capacitances, including the pad capacitance and the drain and sources capacitances of the final drive transistors.

While highly successful, IBIS does have a number of limitations. It tends to overpredict SSN (Figure 11.4). The main reason for this overprediction is that the pull-up and pull-down components do not lose current drive capability during the power ground noise collapse event, as they do in the full fidelity simulation (i.e., normally *i*_out and d*i*/d*t* is reduced during the voltage collapse and this is not captured in IBIS).

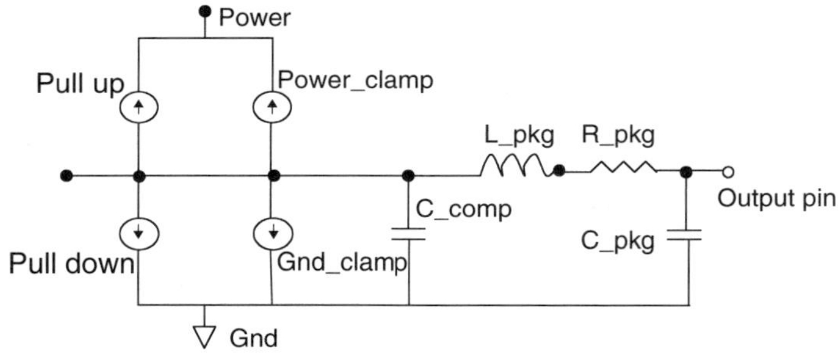

FIGURE 11.3 IBIS output behavioral model.

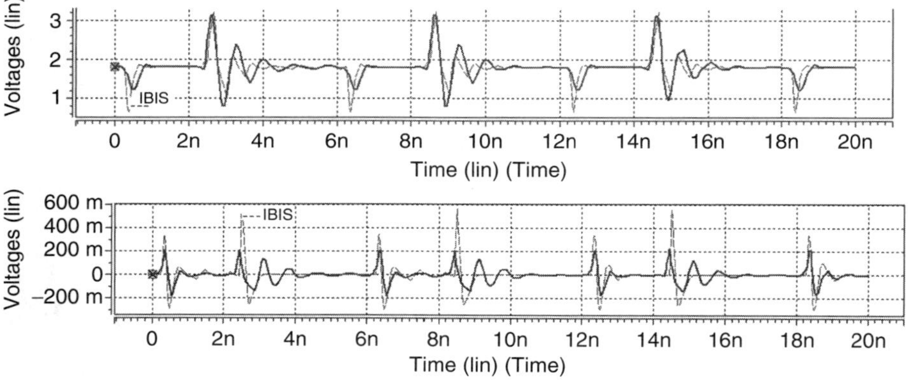

FIGURE 11.4 Power (top) and ground (bottom) signals for IBIS (broken line) and SPICE (solid line).

The simplicity of IBIS models also leads to limitations. The model can be incorrect if used in a different loading environment than it was produced for. In particular, the ramp times for the pull up and pull down are very loading specific. If package parasitics or the characteristics of the transmission line being driven change, then the IV tables should be produced for the new load, and not just pulled from a previous model. Also, the IBIS model is not sufficiently accurate for multi-Gbps simulations.

Despite these limitations there is no ready successor for IBIS. However, at the time of writing this chapter, several alternatives are under investigation, including "template" modeling and "black box" modeling (see the references below). An important challenge is in achieving the same level of automation that was achievable with IBIS.

11.6 Conclusions

Traditionally when one refers to EDA tools for chip package codesign, modeling tools are usually implied. The most classic and most successful tool examples are the reduced complexity macromodeling tools available to support IBIS I/O models. However codesign is not just about modeling. In future, physical cosynthesis will be needed to support the design of high pin-count systems.

Annotated Bibliography

System on a package technologies as it impacts codesign:

- R. Tummala, M. Swaminathan, M.M. Tentzeris, J. Laskar, Gee-Kung Chang, S. Sitaraman, D. Keezer, D. Guidotti, Zhaoran Huang, Kyutoe Lim, Lixi Wan, S.K. Bhattacharya, V. Sundaram, Fuhan Liu, and P.M. Raj, The SOP for miniaturized, mixed-signal computing, communication, and consumer systems of the next decade, *IEEE Trans. Adv. Pack.*, 27, 250–267, 2004.

Digital codesign and coverification:

- J.T. Schaffer, A. Glaser, S. Lipa, and P. Franzon, Chip package codesign of a triple DES processor, *IEEE Trans. Adv. Pack.*, 27, 194–202, 2004.
- J. Wang, K. K. Muchherla, and J.G. Kumar, A clustering based area I/O planning for flip-chip technology, *Proceedings of the 5th International Symposium on Quality Electronic Design*, 2004, 2, pp. 196–201.
- A.K. Varma, A.W. Glaser, and P.D. Franzon, CAD flows for chip-package coverification, *IEEE Trans. Adv. Pack.*, 28, 96–101, 2005.
- M. Shen, L. Zheng, and H. Etenhunen, Robustness enhancement through chip-package co-design for high-speed electronics, *Proceedings of the 5th International Symposium on Quality Electronic Design*, 2004, 3, pp. 184–189.

Mixed-signal codesign examples:

- S. Donnay, P. Pieters, K. Vasen, W. Diels, P. Wambacq, W. De Raedt, E. Beyne, M. Engles, and I. Bolsens, Chip-package codesign of a low-power 5-GHz RF front end, *Proc. IEEE*, 88, 1583–1597, 2000.
- S. Brebels, J. Ryckaert, B. Come, S. Donnay, W. De Raedt, E. Beyne, and R. Mertens, SOP integration and codesign of antennas, *IEEE Trans. Adv. Pack.*, 27, 341–351, 2004.
- G. Nayak and P.R. Mukund, Chip package codesign of a heterogeneously integrated 2.45 GHz, CMOS VCO using embedded passives in a silicon package, *Proceedings of the 17th International Conference on VLSI Design*, 2004, 4, pp. 627–630.

IBIS and other macromodeling:

- A. Varma, A. Glaser, S. Lipa, M. Steer, and P. Franzon, The development of a macromodeling tool to develop IBIS models, *Proc. IEEE Electric. Perform. Electron. Pack.*, 177–280, October 2003.

- B. Mutnury, M. Swaminathan, and J. Libous, Macromodeling of nonlinear I/O drivers using Spline functions and finite time difference approximation, *12th Tropical Meeting on Electrical Performance of Electronic Packaging (EPEP 2003)*, Princeton, NJ, 2003.
- I. Stievano, I. Maio, F. Canavero, and C. Siviero, Parametric macromodels of differential drivers and receivers, in Advanced packaging, *IEEE Trans. Adv. Pack.*, 28, 189–196, 2005.

3D-ICs:

- J. Cong, J. Wei, and Y. Zhang, A thermal-driven floorplanning algorithm for 3D ICs, *ICCAD* 2004, pp. 306–313.
- V. Chan, P. Chan, and M. Chan, Three-dimensional CMOS SOI integrated circuit using high-temperature metal-induced lateral crystallization, *Electron Devices, IEEE Trans.*, 48, 1394–1399, 2001.

Thermal codesign:

- K. Shadron, The importance of computer architecture in microprocessor thermal design, in thermal and thermomechanical phenomena in electronic systems, 2004, *ITHERM 2004. The 9th Intersociety Conference*, 2, 2004, pp. 729, 730.

12

Design Databases

Mark Bales
Reshape, Inc.
San Jose, California

12.1 Introduction

When Chevrolet introduced the C5 Corvette in 1997, the chassis was 4 1/2 times sturdier than the chassis of the C4. This is the main reason that the C5 Corvette runs so well on the skid pad, and its handling has improved to the point where it is competing with even Porsche sports cars. The new C6 Corvette has improved even further on this. I can state with a high degree of certainty that *not a single person* bought a C5 Corvette because of the chassis. It is a necessary feature but it always takes a back seat to the flash, so it is with EDA design databases. The design database is at the core of any EDA system. It is expected and to perform flawlessly, with a very high level of performance, and be as miserly with memory as possible. End users tend to ignore the database unless something goes wrong. A file is corrupted; the design gets too large to fit in the memory; a design with 10,000,000 cells all superimposed at 0,0 performs poorly. Of course, in each of these cases, the database is at fault. Application developers have their own requirements of design databases. The API must be so intuitive that it can be used without a manual; it must be completely tolerant of any bad data given to it and fail in a graceful way; it must maintain a high level of performance even when the database is used for purposes for which it was never designed. But a design database *is* the heart of the system. While it is possible to build a bad EDA tool or flow on *any* database, it is *impossible* to build a good EDA tool or flow on a *bad* database.

This chapter describes the place of a design database in an integrated system. It documents some historical databases and design systems by way of reference, for those who are interested in how we got to today's state of the art. We then focus on the current best practices in the EDA design database area, starting with a generic conceptual schema, and building in complexity on that. Real-world examples of commercially available design databases are presented, with special emphasis placed on the OpenAccess Coalition and their community-open-sourced EDA database. Finally, a series of historic and current references are given as a resource for those who would like to learn more. As there is entirely too much information to present in these pages, readers are encouraged to seek out the current references.

12.2 History

12.2.1 Design Databases as Part of Integrated System

In examining EDA design databases, it is useful to look at hypothetical tool architecture, to determine which parts are to be considered part of the design database, and which parts are the application levels. In Figure 12.1, the portion within the dotted line represents the design database components. On the left is the language system (which, although not directly part of the database, will be used by *parameterized cells* described below). On top of the database are built the algorithmic engines within the tool (such as timing, floor-planning, place and route, or simulation engines), and the highest level represents the applications built from these component blocks. Note that the scope of the design database includes the actual design, library information, technology information, and the set of translators to and from external formats. All of these components will be discussed in this chapter.

12.2.2 Mature Design Databases

Many instances of mature design databases exist in the EDA industry, both as a basis for commercial EDA tools as well as proprietary EDA tools developed by the CAD groups of major electronics companies. Systems from IBM [3], Hewlett-Packard [10,24], SDA Systems (Cadence), ECAD (Cadence), High Level Design Systems [16] and many other companies have been developed over the last 20 years and continue to be the basis of IC-design systems today. Many of these systems took ideas from university research [6,9,11,15] and successfully productized them.

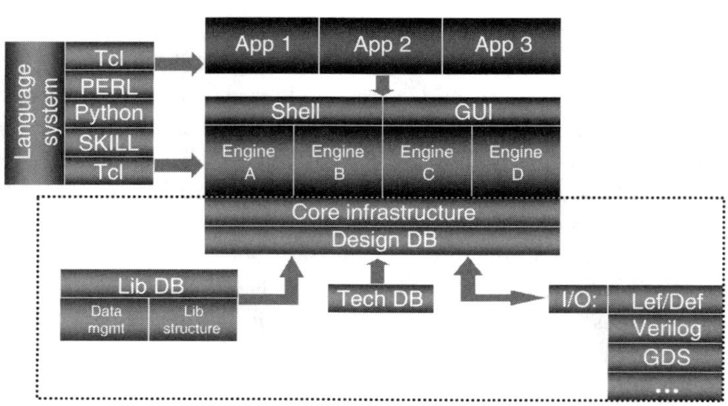

FIGURE 12.1 High-level architecture of an EDA system.

Most of the mature design databases have evolved to the point where they can represent netlist data, layout data, and the ties between the two. They are hierarchical to allow for reuse and smaller designs. They can support styles of layout from digital through pure analog and many styles of mixed-signal design. Most of these systems are having performance and/or capacity problems as designs grow larger and larger. Although it is possible to increase capacity by using 64-bit versions of these systems, their performance falls off even more. A new generation of databases is needed that has higher capacity while maintaining performance. Those systems are arriving now.

12.3 Modern Database Examples

12.3.1 The OpenAccess Design Database

Given the importance of a common design database in the EDA industry, the OpenAccess Coalition has been formed to develop, deploy, and support an open-sourced EDA design database with shared control [37]. The data model presented in the OA DB provides a unified model that currently extends from structural RTL through GDSII-level mask data, and now into the reticle and wafer space. It provides a rich enough capability to support digital, analog, and mixed-signal design data. It provides technology data that can express foundry process design rules through 90 nm, contains the definitions of the layers and purposes used in the design, definitions of VIAs and routing rules, definitions of operating points used for analysis, and so on. OA makes extensive use of IC-specific data compression techniques to reduce the memory footprint, to address the size, capacity, and performance problems of previous DBs.

The details of the data model and the OA DB implementation have been described previously [42]. As of 2005, OA is the only modern IC database where the implementation is publically available, so the examples of this chapter primarily refer to the OA implementation.

12.3.2 Open Milkyway

Perhaps in reaction to the OpenAccess Coalition, in early 2003 Synopsys decided to open up their Milkyway physical design database, and started the MAP-In (Milkyway Access Program). This has been described in [28]. The similarities to OpenAccess outweigh the differences. Milkyway is a more mature database, and has many more chips taped out using it. At the same time, its data model is a subset of what is available in OA, with about 70% of the OA model available in Milkyway. It is written in C, and was never built to be open-sourced or a standard. As a result, its API is not as clean as OpenAccess. Its internal implementation is not available even for MAP-In members, so no comments may be made about the implementation, other than to stress that this is a tried-and-true production-level EDA database. One observation that can be made is that the capacity of the Milkyway database seems to be that of other databases of the same age (for example, DF-II from Cadence). In that regard, it is probable that OpenAccess is as much as an order of magnitude smaller than the Milkyway data.

12.3.3 Magma, Mentor, and Others

Other significant design databases have been built by Mentor Graphics (their Falcon database, one of the first in the industry written in C++) and Magma Design Automation [34]. Neither of these databases is available publicly, so little can be said about their features or performance relative to other industry standards. Like Milkyway is for Synopsys, Falcon seems to be a stable and mature platform for Mentor's IC products. Falcon seems to be similar in capability to Milkyway. Magma's database is not just a disk format with an API, but is an entire system built around their DB as a central data structure. This is very advanced, and is more advanced than the levels of integration currently taking place using OpenAccess. Again, since the details of the system are not publicly available, a direct comparison of features or performance is not possible. Looking at the capabilities of the Magma tools would indicate that this DB has a similar functionality to OpenAccess, and may be capable of representing behavioral (synthesis input) information.

12.4 Fundamental Features

For the following descriptions, refer to Figure 12.2, the general-purpose physical/logical database schema diagram. In this diagram, a shaded bubble represents a database object. Unshaded bubbles represent classes that are base classes for some of the objects. The dashed lines represent the class hierarchy relationships. The solid lines represent relationships among the objects, and their end point arrows represent the ordinality of the relationship. A single solid arrow represents a 1 to 1 relationship. An example of this is the Term to Net relationship. A Term may not exist without a corresponding Net. A double-hollow arrow represents a 1 to 0 or more relationship. An example of this is a Net to Term relationship. A net may have one or more associated terminals, but it does not necessarily have even one.

12.4.1 The Design as the Basic Unit

The fundamental basic unit of a design block is called a *Design.* It has also been called a *cell, view, block, cellview, representation,* and other names in other databases. In essence, a Design is a container that holds the contents of what a designer thinks of as a component within the design. Most designs are hierarchical, and they are composed of many Design objects. In a flat design, there is only one Design object. It then contains every piece of physical geometry, all of the connectivity, all of the timing and P&R objects, and in general everything about the design with the exception of the technology information and the directory information for the storage of the Design in a library. Each of these classes of information is kept elsewhere.

The Design can be viewed as being the file descriptor for the information in this design component. In addition to containing general information about the component, it is the starting point from which all of the other information contained in this Design may be found.

12.4.2 Shapes and Physical Geometry

The simplest and longest-lived information in a Design (from a historical viewpoint) are *shapes.* These represent the physical shapes used to construct the mask patterns for implementation of the integrated-circuit masks. While some shapes are used purely for annotations or other forms of graphical display, most shapes are actually part of the design, and become part of the ultimate masks that are built at the tail-end of the IC design flow. Shapes include the *Dot,* the *Path,* the *Rect,* the *Polygon,* the *Ellipse,* the *Text,*

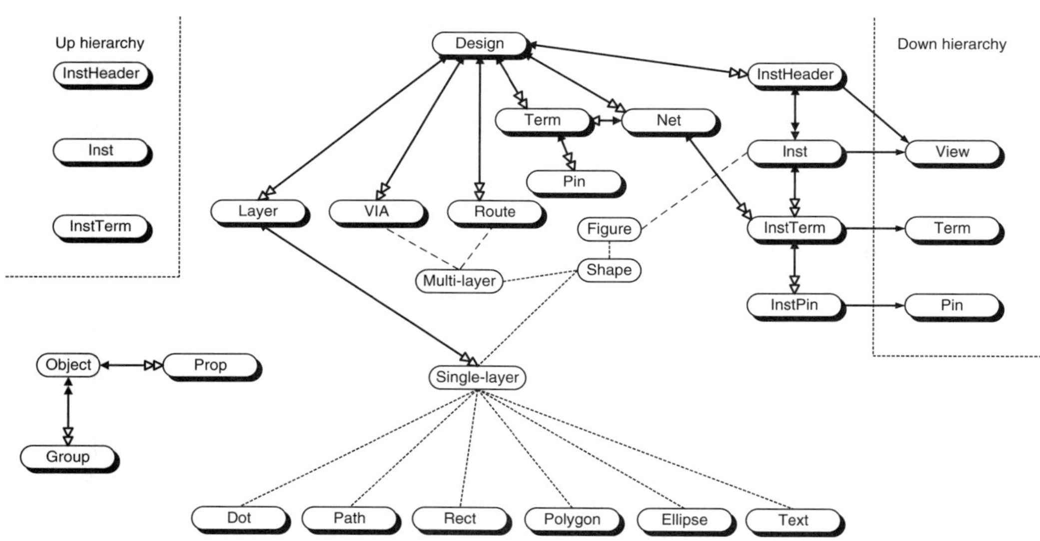

FIGURE 12.2 General EDA logical/physical database schema diagram.

the *VIA*, and the *Route*. They are most often assigned to a *Layer*, which is an object that links together all of the shapes in a given Design on the same mask layer.

The *Dot* (also called a *Point* or *Keystone* in other databases) represents a zero-dimensional object, which just has a position in the *X–Y* plane. Some implementations may assign a "size" to the dot, which is used for display purposes. Some implementations may assign a layer to the dot as well. Dots are used to represent the center of a pin; to represent an anchor point for some sort of constraint or annotation; possibly to represent the origin of a cell. Since they are a zero-dimensional object, their use is semantic, and they do not become part of the mask, even if they are on a mask layer.

The *Path* (also called a *Line* or *Segment* in other databases) represents a one-dimensional object, which is defined by two (or more) points in the *X–Y* plane. Since it has a width, which may be greater than 0, it is in reality a 2D object, but it is more useful to think of it as a line plus a width. Note that some databases consider a zero-width path and a path with nonzero width as different objects. Some implementations restrict a Path to a single segment (defined by its two endpoints). Most implementations allow a Path to have a series of points. Paths are assigned to a layer. If their width is greater than zero and they are assigned to a mask layer, they become part of the ultimate mask for that layer. Paths may be kept as *Manhattan*, meaning their segments are always parallel to one of the *X* or *Y* coordinate axes, they may be *Manhattan plus 45*, meaning their angle is one of 0, 45, 90, 135, or 180, or they may be *All-angle*, meaning there is no restriction on the slope of the path segments. The least-restrictive segment of a path defines the type of path it is. In other words, if every segment but one in a path is Manhattan, and the last is All-angle, then the entire path is considered All-angle. Paths have an end-style, which describe a possible extension of the path beyond its end points by an amount in a direction given by the slope of the end segment. The path end-style may be *flush* (with no extension), *half-width* (where the extension is half the width of the wire), *variable* (where the amount of extension is explicitly specified), *octagonal* (where the end forms an octagon of the same width as the path centered around the end point of the path segment), or (in rare cases today) *round* (with the diameter of the half-circle end matching the width of the segment).

The *Rect* (sometimes called a *Box* or by its full name of *Rectangle*) is the simplest and historically most-used shape. It is a simple rectangle whose sides are parallel to the *X–Y* coordinate axes. It can be represented by two points, typically the *lower-left* and *upper-right* (although some systems use *upper-left* and *lower-right* points, and still others use a single point plus a *width* and *height*). Each Rect is assigned to a layer, and becomes part of the mask for that layer.

The *Polygon* (sometimes called a *Poly*) is a shape defined by a closed set of points in the *X–Y* plane. Similar to a Path, it may be Manhattan, Manhattan plus 45, or All-angle. The type of polygon is defined by the least restrictive of the segments that make up its boundary. Most systems assume that the closing edge of the polygon goes from the last point to the first point, although some databases explicitly represent the first and last points as the same point. Some databases allow holes to be explicitly represented in polygons, but most databases have the hole split at some point so the polygon touches itself but does not overlap. Polygons are not allowed to be self-intersecting in any modern system. As with the other shapes to this point, polygons are assigned to a layer and become part of the mask.

The *Ellipse* (sometimes simplified as a *Circle*) may sometimes be used in Analog IC design, or when the database is used to represent a printed circuit board. Their greatest use is in representing connection points, both in IC layout and in schematic diagrams. The ellipse has axes that are parallel with the coordinate axes. An ellipse where the defining points are coincident forms a circle — some databases only represent circles and do not allow the full freedom of the ellipse. Unless used in Analog design, the ellipses are almost never part of the final IC mask. They are most often used as annotation shapes in nonmask layers.

The *Text* (sometimes called a *String* or *Message*) is a nonmask shape that is used for labels and other forms of annotation. While there is often a need for "mask text," this is almost always done with polygons. Modern text objects may allow for different fonts, styles, and sizes. Their size may be represented using a word-processing "point" value, or it may be given in units that match the ones being used in the Design. Text strings are assigned to a layer, which controls their display color.

The *VIA* is the first shape introduced that represents a multilayer object. As such, it is not assigned to a layer, but rather is assigned directly to the Design, or carried along with a *route* (described next). It is defined by two conducting layers and a contact-cut layer that will join the two. VIAs may be simple, including only a single cut, or they may be quite complex, having an array of cuts, possibly with some cuts missing. Their definition is usually described in the technology database in modern systems. In older databases, they may be represented as anything from a single-contact cut (with the connecting metal layers implicitly derived from the cut), a series of shapes joined together (which can be horrendously inefficient), an instance of a separate cell (again, very inefficient), or some form of parameterized cell (which is defined later).

The *Route* (shown in Figure 12.3) is the other shape that represents a multilayer (and in this case, multi-object) shape. It is like a Path, in that it goes from one point to another, may include a single segment or many segments, and may be Manhattan, Manhattan plus 45, or All-angle. It differs from the Path in that it may change width, layer (through a VIA), or both during the course of its travels. It also has an explicit record of what its end points are connected to. It is the equivalent of many paths and VIAs, all in a single object. Its use is one of the main reasons for the space efficiency of the OpenAccess database.

12.4.3 Hierarchy

Most integrated-circuit designs are hierarchical. Even for designs that are small enough to be implemented as a single Design, the leaf library cells are kept separate from the Design, for reasons of capacity, update, and just because it makes sense. This means that a database should be hierarchical, and every database of the modern era is. Hierarchy is established through an *instance/master* relationship, in which a Design may be *instantiated* or placed within another design. In the simplest case described above, library leaf cells are placed and then connected using additional shapes to implement the circuit under construction. The Inst object refers to a Design (called the master). The Inst has a placement location and an orientation used for placement. The location is just an offset of the origin (the 0, 0 point) of the Design being instantiated from the origin of the higher-level Design containing the Inst. The orientation is most generally one of eight Manhattan-based orientations that are achieved by giving a rotation of 0, 90, 180, or 270°, and mirror-inverting each of these rotated instantiations. Some databases support fewer of these eight possible orientations, and a few databases support all-angle orientations, so that a Design may be, for example, rotated by 38°. Most modern databases are limited to the eight Manhattan-based orientations. Note that the hierarchy is most often *folded*, in that each Inst of the same master Design points to that single master Design (see Figure 12.4).

The introduction of hierarchy and physical instantiation prompts a discussion of two additional concepts, those of units and coordinate spaces. In every physical design database, there is some notion of units. Most databases keep coordinate data as integers, because rounding errors that might occur in floating-point numbers would be unacceptable if, for example, they resulted in layout shapes not touching. There is then a mapping from these integer database units (often abbreviated as *DBUs*, for Data Base Units) into physical dimensions. OpenAccess represents this mapping in the technology data, and uses a *single* such mapping for every Design in the entire integrated-circuit implementation. This makes for a simpler, more robust system as it becomes impossible to have mismatch of a unit between Designs across the instance/master boundary. Other systems have used a single mapping throughout, for example, considering each DBU as a nanometer. Other systems allow a different mapping in each Design. This can make for an error-prone system; since the databases present coordinates as the raw integers it becomes

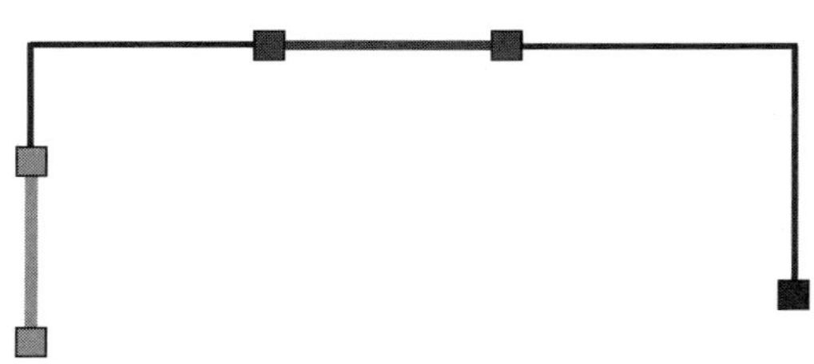

FIGURE 12.3 An example of a route.

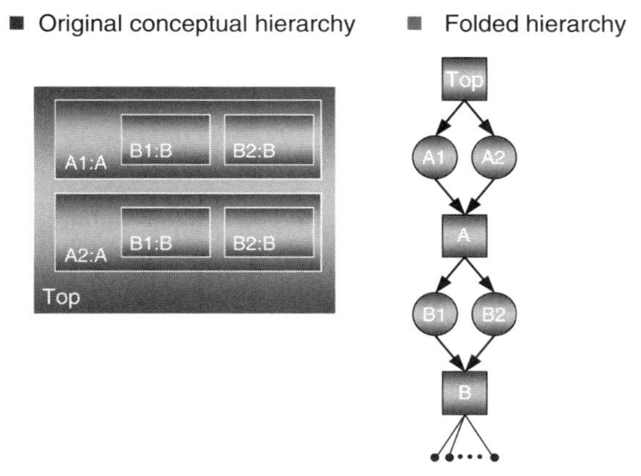

FIGURE 12.4 Example of a folded hierarchy.

the responsibility of each application to understand when hierarchy boundaries have been crossed and to make the necessary adjustments in units. A single application in a flow that does not properly treat differing units can corrupt data or generate incorrect analyses. From experience, using different units for different parts of a design is confusing to both man and machine, and should be avoided wherever possible, even if the database supports it.

Each Design has its own coordinate space. In a hierarchical physical design, the coordinates of shapes within an instantiated master are different inside the master and from the point of view of the Inst. The coordinates are transformed by the translation and orientation of the Inst in the instantiating Design. In a multilevel hierarchy, these transformations must be concatenated. Most modern databases provide support for this, so that an ultimate flattening of the physical design data that must be done in order to create the IC masks is simple.

12.4.4 Connectivity and Hierarchical Connectivity

One of the advantages of design databases of the modern era is the unification of physical layout with connectivity. Being able to represent simultaneously a netlist along with the shapes that implement the design and being able to relate specific shapes to specific nets within the netlist provided a quantum step in capability within the EDA tools. The typical connectivity structures include a *Net*, which represents a logical or physical node within the netlist, a *Term*, which represents an available connection point on a Design, and an *InstTerm*, which represents a connection between a Net and the Term on the master of a given Inst. These constructs form the basic scalar connectivity. In addition, these constructs can be combined into a *Bus*, which is an aggregation of nets that are all related and have the same root name, or a *Bundle*, which is an aggregation of nets that are not related and have different root names. This is true for both the *Net* and *Term* (and in some databases the *InstTerm*), so that we have the *BusNet* and *BusTerm* (and sometimes the *BusInstTerm*).

The *Net* is the core construct for connectivity. It represents a logical or physical node within the netlist. It has a name. It provides a linkage of zero (or more) *Terms* that "export" the net so that it becomes visible when the containing Design is instantiated. It provides a linkage of zero (or more) *InstTerms* that represent connections to Terms within Designs that have Insts at this level of hierarchy. It is the InstTerm that provides the linkage of connectivity across hierarchy.

The *Term* is a net that is exported, or made visible outside the Design to which it belongs. In some databases it may have a name that is different from the Net to which it is connected, while other systems have a single name for the net and terminal. The Term usually contains a linkage of the *Pin* shapes that define the physical connection points and layers for the physical implementation. Most modern databases allow multiple Terms for a single Net (sometimes called *shorted terminals*) while some restrict the relationship to a single Term per Net. These shorted terminals have become necessary in the modern era mostly due to postsynthesis optimizations that remove one or more gates, having the side effect of shorting two or more terminals together. Systems that cannot support shorted terminals have problems with modern-day netlists that come from synthesis tools.

12.4.5 General Constructs

Two fundamental constructs that can be used with any object in a modern design database are the *Prop* and the *Group*. Props (also called Property or Attribute) form a simple extension mechanism that is intended for use in adding data that is sparsely populated. For example, if one needs to add a property "GridShape" that identifies shapes that are part of a power grid, one can do this with a property and not incur the overhead of storing this information on each and every shape, even though the properties are used only on a small fraction of the overall shapes. Props are of simple basic types, such as boolean, integer, float, and string. Some databases extend these basic types with specialized variations, such as file-name, mapped-name, etc. Some databases provide for default values for the properties; some restrict the legal set of property names per object type (to control unwarranted proliferation and abuse of properties); and some add range-values and/or enumerations to indicate legal values for the properties.

The *Group* is used to extend the database relationships among objects. Most databases allow groups to be either a collection or a set (where each member must be unique). Many implementations allow groups to be declared as having a unique name or a common name. In this way, groups can be used to establish specific relationships among objects, or a class of relationships in the nonspecific name case. An example of the former is to have a group called "GridShapes" that represents the set of shapes that are used in a power grid. An example of the latter would be a series of groups called "macroArray," where each such group represents a set of macro cells that should be placed in a physical array structure.

Note that the use of both properties and groups is discouraged in general. They are very expensive in terms of space, and moderately expensive in terms of performance. In a typical implementation, if the percentage of objects in a given class that can have a specific property or group grows above about 15%, it is better to implement the extension as a "native" object within the database for superior space and speed performance. OpenAccess provides an extension mechanism which makes this possible without change to the core database, and this preferred mechanism is described in a subsequent section.

12.4.6 API Forms

Along with the general evolution of API forms in Computer Science, the form of Design Database APIs has evolved. Most modern database implementations use C++ classes to represent the objects in the database. In most implementations, the constructors and destructors for the objects are special methods and the default constructors and destructors are disabled. This is because the memory management of the objects must be closely controlled to allocate the objects within their associated Design, and the techniques that are used in the internal database implementation to represent the data in an efficient manner require the locality of the reference that these customized constructors supply. In addition, it is a good idea for the API that the developers use to make it perfectly clear when they are creating and deleting objects.

In most modern implementations, the conceptual schema represented by the public classes is very different from the internal implementation of the database. This is done to allow ongoing trade-offs between speed and space performance, as well as to allow implementation-level improvements to be made without needing to revise the interface of the public API. In this way the lifetime of the public APIs is longer and the structure of the APIs may remain unchanged through several different implementations of that API.

Because of this separation, most implementations provide access to all members through methods rather than a direct form of access. This has the added advantage of making it possible to provide a very robust form of *observer* (also called a callback, trigger, or daemon) that can be used to detect any change of interest in the data. The methods are generally broken up into "set()" and "get()" methods, to make the code more readable and less error prone.

In addition to the data that are accessed through member functions, special functions called iterators (also called traversal or generation methods) are created to allow traversal of the relationships among the data. For example, an Iterator is created from a Design that will allow iteration through all Nets in the Design. Some older implementations use macros to implement these traversal functions as a kind of loop, but most modern implementations create an object that contains the state of the iteration and which is used in a pure control loop within the application program. Iterators require special care to deal with situations where the design is modified during iteration.

12.4.7 Utility Layer

The set of classes and methods that would allow a database to be fully created, modified, queried, traversed, and deleted constitutes the core set of database functions. There are many common functions that are done on top of this core set, and it makes sense to unify them into a "utility layer" of classes and methods that come along with the code database. Most modern databases implement some such functions. Here are descriptions of a few of the possible utility areas.

12.4.7.1 Region Query

One operation that is almost universal throughout physical design tools is the ability to search quickly for shapes contained in an area of interest. Most modern databases provide a capability of doing this efficiently, and this capability is used by everything from graphical display to DRC utilities to P&R utilities. Most often, some form of hashed tree organization (K-D tree, quad-tree, etc. [25]) is used. Some implementations consider the region-query structures to be part of the database and persistently store them. Others re-compute this data when it is needed, and focus on high-performance computation to make this practical.

12.4.7.2 Error Handling

There are two main styles of error handling used in modern design databases — *return codes* and *exceptions*. In the return-code style, each method returns an error code when it fails, or it might return a known bad value and then provide a way of determining the error code. Exception-based methods will "throw" an exception when an error occurs, and it must be "caught" by the application-level code. This area is a somewhat controversial one. People who prefer the return-code style note that uncaught exceptions can make for system crashes that can enrage end-users. People who prefer the exception style note that unchecked return codes can make for programs that fail in obscure ways and possibly corrupt data. OpenAccess uses the exception style of error handling, and this method seems to result in more robust code over time.

12.4.7.3 Boolean Mask Operations

Another oft-used utility is the ability to perform Boolean mask operations (AND, OR, NOT, NAND, XOR, and GROW) on mask shapes. These can be used as the basis of a DRC capability, to create blockages, to cut power-grid structures, and for many other physical purposes. The algorithms used for this capability are complex, and often have a high value placed on them (see chapter 17, *Design Rule Checking*, and chapter 22, *Layout Extraction*). Most publicly supported design databases either charge extra for this capability, or, as is the case in OpenAccess, provide a way to hook in customer-supplied Boolean mask operations. This provides for a standard API while giving the end user the flexibility of applying a mask op tool set of appropriate performance, capacity, and price.

12.5 Advanced Features

12.5.1 Parameterized Designs

Most of the *Designs* that exist are just aggregations of shapes and/or connectivity that represent a particular design. In many cases however, a library or technology requires many cells that share a common design but have different sizes, etc. The simplest example of this is a VIA, which can easily be imagined as a more complex object with a set of parameters. In fact, this is what the VIA object is. There is the notion of a *Parameterized Design* that abstracts this concept, allowing a user to write a small piece of code that, when coupled with a set of the parameters, will evaluate to a Design with layout and/or connectivity. This requires that the database supports one or more extension languages. Historically, the first extension languages were proprietary and specific to the database, but recent efforts use conventional programming languages and APIs. For example, OpenAccess supports C++ and now TCL. Unofficial additions to OA include Python and PERL, and the proprietary language SKILL is supported by Cadence. In order to be successful, the system must be architected so that any user of the given database can at least *read* the Parameterized Designs. OpenAccess provides the capability of shipping an interpreter along with the parameterized library designs, making the use of this capability portable. OA also provides for the management of *variants* (a Design evaluated with a specific set of parameters) in order to make the system as efficient as possible. Several other systems support this capability, but usually in a single-language-specific way.

12.5.2 Namespaces and Name Mapping

A concept that every design database must deal with at some point is *name mapping*. Every database, every file format, and every hardware description language (such as Verilog or VHDL) has a legal name syntax, and may (or may not) have some characters that convey semantic meaning. Examples of this are hierarchy delimiters, bus-name constructs, and path-terminator separating constructs. If a design database supports characters with semantic meaning, there needs to be some form of escaping to be able to treat these characters in names *without* their semantic meaning.

Add to this complexity the fact that of the many formats that might be used to drive a third-party tool, every one has a slightly different version of its own name-mapping and namespace characteristics. The problem is simple. When you input from one format and then output to the same format, the names need to remain as they originally were. When names are input in one format and then output in another, the transformations that occur must be consistent, so another tool that uses these files can be presented with a consistent group of files. As an example, consider a third-party tool that takes the library data as a LEF file, the netlist as a Verilog file, and the parasitic data as a SPEF file. Each of these file formats has a different namespace, and it is tricky to represent names in any design database so that they will be consistent when they meet up in a single tool.

The most modern approach is to use some variation of algorithmic conversion of names, so that the transformations are invertible. This was the method used in the Pillar database [16] and is the method used in OpenAccess. A namespace is defined for each different format, and a native namespace is used within the design database. Other systems can deal with the problem by building maps to map names that are not representable in both namespaces. This can be problematic, since if the map is lost or if the names are changed out of context of the map, then it may no longer be possible to correctly map the names.

12.5.3 Place-and-Route Constructs

Many place-and-route-specific constructs may be added to an EDA design database. (see the chapters on "Placement" and "Routing" for a more detailed description of these operations.) *Blockages* and *Halos* describe areas where operations such as placement or routing are prohibited. A *cluster* defines a group of instances that are clustered together and possibly an area where they should be placed. A *GCell* (Global-routing Cell) divides the chip area up into a grid, and is used to store global routing and congestion information. A *Steiner* point is a virtual object that acts as a junction object for more-than-two-way route junctions. These objects may or may not be implemented in some design databases. OpenAccess implements all of these objects and you can find a detailed description of these items in [35].

12.5.4 Timing and Parasitic Constructs

Modern-day design databases contain constructs for dealing with timing-driven design and analysis. *Parasitic* elements represent the results of parasitic extraction, storing a network representing the full parasitic model, including resistor, capacitor, and inductor elements. It is possible to store a reduced form of the network, either as a pole-residue model or as a simpler *Elmore* model. It is necessary that each net be represented at a different level of abstraction, which reflects on the fact that the routing is fully complete for some nets but only partially complete for others. *Timing* models for Designs must be possible, and can serve both as the de facto model for leaf cells and as an abstraction for an intermediate-level Design. Constraints must be stored, both in the technology area (as definitions of operating conditions) and within the Design (as design-specific constraints). Timing arcs should be able to be stored, to allow for long-term lazy evaluation of timing information. Few modern databases contain much of this information. The Magma database contains most of it. OpenAccess contains everything but timing arcs and specific timing constraints, though a publically available extension includes these objects [44].

12.5.5 Occurrence Models and Logical/Physical Mapping

An advanced feature beyond the scope of this chapter is the addition of Occurrence Models to the database. Remember that most modern design databases use *folded* hierarchy, in that every instance of a given master Design points to the same Design. This has the advantage of making it easy to fix a problem in a cell and have it reflected everywhere instantly. However, there is no place in a folded database to attach data that will be specific to each unique *occurrence* of a master within a chip design (see Figure 12.5). Most systems today deal with this problem by using auxiliary data structures to represent the *unfolded* (occurrence) hierarchy, or they proceed to make copies of every intermediate-level cell that is used more than once, a process called *uniquification.*

OpenAccess contains an Occurrence Model called Embedded Module Hierarchy (EMH). It includes a logical view (called the *Module* view), a physical view (called the *Block* view), and an unfolded occurrence view (called the *Occurrence* view) that relates the two. Figure 12.6 shows a logical hierarchy with its corresponding flat physical hierarchy, and Figure 12.7 shows the EMH representation for this conceptual design, illustrating the Module, Occurrence, and Block hierarchies side-by-side. This is very useful as you can refer to an object by using its logical name or its physical name, and this correlation persists even through some re-partitioning. One limitation of the OA model is that it is not permitted to move a logical Module past a physical Block boundary. This means that some common forms of re-partitioning are not yet possible with the current version of OpenAccess. See the OpenAccess manual [35] for more details.

12.5.6 Extensibility

One key feature of a design database is extensibility. We have already discussed Props and Groups, and know that while they are suitable for simple extensions, they are too costly to use for anything substantial. A modern design database must have some way of quickly adding new objects and relationships. If the database is proprietary, the only issue with changing the database is the effects on customer

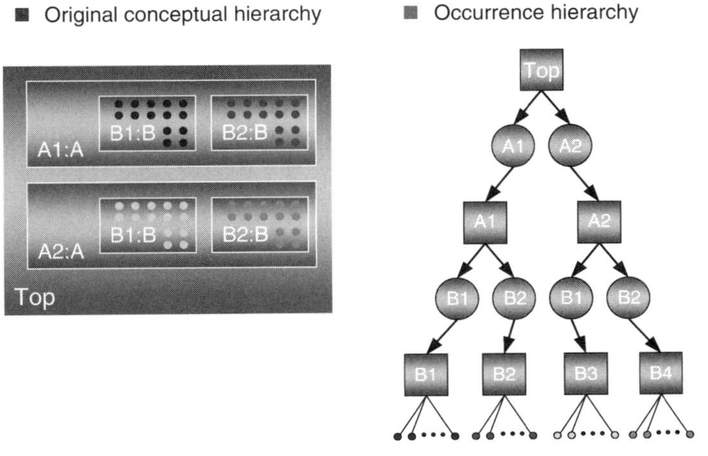

FIGURE 12.5 Different occurrences of the same master.

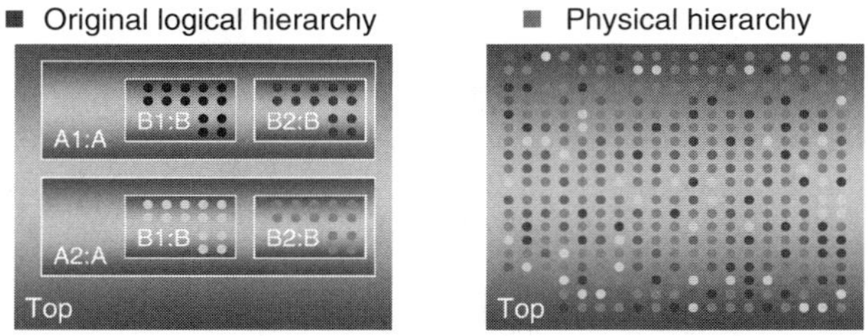

FIGURE 12.6 A logical hierarchy with its corresponding flat physical hierarchy.

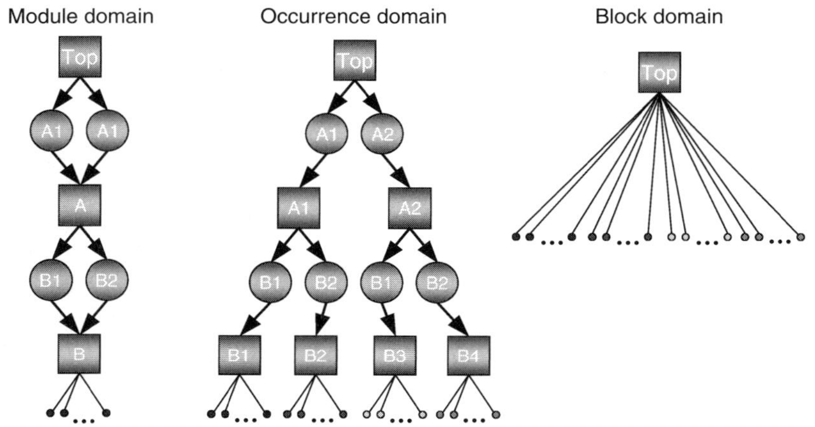

FIGURE 12.7 Three components of the EMH of OpenAccess.

migration. Most customers will not move to a new version of EDA tools that requires a database change. They will wait until the current chip is finished. This can lead to excessive time spent supporting two or more versions of the database and hence the tools. If, on the other hand, a more portable extension mechanism is available, the core information may remain unchanged while the extensions are added.

This is the approach taken by OpenAccess. Extension mechanisms to add attributes to existing objects, create new objects, and establish new relationships among new and/or existing objects are available in OA. While they take some work to set up, they are nearly identical in performance and capacity to similar additions that would be added in a "native" way.

As a last comment, interoperability among a set of tools that originates from many different sources is greatly improved by using the OA approach. As long as the core data are understood by all tools, the additions may be kept and communicated only as necessary. This makes for more robust multivendor tool flows.

12.6 Technology Data

Any design database must be able to represent process-level design rules, such as spacing and width requirements as well as rules for the construction of VIAs. The rules must be able to represent technology nodes down to at least 90 nm. This means that, for example, opposition- and width-based spacing rules are possible. Also, there should be a way of recording data to be used for *constructing* routing as well as for *analyzing* shapes. OpenAccess uses *Constraint* and *ConstraintGroup* objects to implement these capabilities.

12.7 Library Data and Structures: Design-Data Management

12.7.1 Library Organization: From Designs to Disk Files

Because most IC implementation teams these days are comprised of many members, and sometimes as many as a hundred members, the Design database must allow for the individual Design databases that comprise a chip hierarchy to be worked on simultaneously. Each Design may be worked on by a single person. A finer-grained level of concurrent access is not warranted. This makes it natural to divide up the data so that each individual Design is contained in one or more files that may be put under design-data management control (it also makes it possible to use the native operating system file locking and access permissions, which reduces implementation effort and may already be familiar to the designer). In general, a library is a directory, and in it are contained a number of Designs, each of which is identified by a unique cell and view name. Beyond that, the only requirements are to store things in such a way as to make the performance of searching for and reading Designs as high as is humanly possible. Most end-users would like a library organization scheme that makes it easy to locate a file in the file system, given a library directory and the cell and view names. Historically, this has been done by using directories for cells (and in some cases, views). While this worked well in the past, modern designs are seeing an explosion in the number of unique Designs, and directory searches are very poor when the number of cells in a given library might reach 50,000 or more. Also, in such cases the actual data might be relatively small, and the overhead for a directory might be as large as or larger than the Design file. This means that the traditional lib/cell/view structures found on disk will have poor speed and space performance.

OpenAccess provides two different mechanisms for mapping Designs to disk files. The first is an upward-compatible system, which represents the data in the traditional file system-based way. The second is a server-based method, which has significantly higher performance than the older method. The effects of this are only visible in designs with more than about 5000 Cellviews.

12.7.2 Design-Data Management

Design-data management (DDM) is a controversial area in which it is virtually impossible to achieve a consensus. The best that can be hoped for is to build a flexible enough system that it will allow incorporation of *any* DDM system, either commercial or home-grown. OpenAccess has taken this approach starting with its 2.2 release. Instances of interfacing this to RCS, ClearCase, Synchronicity, and IC Manage exist, as also many OA users that have adapted their own proprietary DDM systems.

12.8 Interoperability Models

It is useful to look at what interoperability requirements are from an application-developer's point of view. We can look at data interoperability levels ranging from file or interchange format through a common in-memory model, and discuss which mechanism is appropriate under which circumstances. In addition, we can look at the control-interface level for components within a part of an inter-operable system. Given an underlying framework that supports the interoperability required at the data level, we will discuss what is necessary at the algorithmic and application levels to insure interoperability. File-level interoperability is the simplest kind.

It has the benefit that it decouples the tools involved from each other and probably from the exact version of the interchange format used. It provides a readable format (in most cases) that can be used for purposes of debugging and testing. Using files presents many challenges and issues, however. The time needed to parse a readable format is usually much greater than that required to read a specific database file. The size of the readable files is usually much greater than the design data in a database format. Interchange formats act as "lossy filters" when compared to design databases. If a format does not contain a particular type of information, then a "round trip" through that format will lose possibly critical information. File-level interchange is most useful for applications that create a different form of output and is not responsible for round-trip fidelity of the data. Using a database, but with a different in-memory model, treats it almost as an interchange format, but without several of the problems.

It shares the file-format advantage of allowing the tools to remain uncoupled. It does not directly provide a readable version of the data, but databases often provide a readable archiving or versioning format that can be used if need arises. Databases mitigate almost all of the problems found in file-based formats. They require little or no parsing, and are hence much faster to read. Applications only need be concerned with the relevant data and the database system manages the rest of the data. In this sense they are not "lossy" like the file-based formats. Round-trip fidelity is achieved by updating the relevant information in the database system. The problems unique to database-use arise when significant mapping must be done between the database information model and the in-memory model used by the application. This can make opening and saving a design too slow, and can even cause problems in the fidelity of the data mapping in the worst case. In addition, it is very difficult to share components among applications that use the same database but have different in-memory models. If this is a requirement you are much better off developing a common in-memory model.

Once you have a common in-memory model, you have the best of all worlds.

If you take the additional step of making the system components incremental and based upon the core in-memory model wherever possible, then you can collect an entire set of application components; technology and algorithmic engines that can be combined and reused in many different products. This lends consistency to the different tool suites within a design flow, reduces development and maintenance costs over time, makes for a modular system that can remain vital by the possibility of easily upgrading modules one at a time, and provides a rich base for more rapid development of new products and technologies.

Some would think (with more than a bit of justification) that having a common in-memory model can slow progress. It is true that any sweeping change must be well planned and deployed. It is less true that additions must be well planned; they often benefit from a developmental deployment where an initial design is tested with actual products and refined before being more broadly deployed. The need for rapid progress can be met through use of the extension mechanisms mentioned in the previous section. This allows rapid prototyping, the ability to keep proprietary extensions private, and makes sure that the progress of the standard does not limit the ability of the tools to keep up with changes in technology. See [44] for an example of the use of this mechanism.

Even though the common in-memory model provides the greatest gains, it is important to choose an integration method that is well matched to the task at hand. We continue to dismiss file-based formats for the problems previously mentioned. However, using a database as an exchange format is appropriate if:

- The subsystem is not incremental
- The I/O time as a fraction of the total run time is small
- The components within the subsystem are built on a different in-memory model and switching would be expensive, or would cause degradation in performance.

It is possible to build a highly productive design flow that uses tools plugged into a common database as a backplane that interfaces each tool in the flow in the most appropriate manner. When done in a careful and elegant way, the integration seems completely seamless.

One advantage the common in-memory model gives over the use of the database as an interchange format is the ability to reuse system components. The biggest impact this makes on the overall design flow is the level of consistency and convergence it brings. Results of timing analysis right after synthesis correlate properly with the answers given after physical implementation and extraction, even if the analysis is done in using different tools.

References

The references listed below represent many of the important papers on design databases. They are broken up into two sections — historical and current references. In many of the papers (especially the historical ones), the main thrust is not on the design database, so you will need to read through these papers to find the description of the data structures, models, and actual database technology used.

Historical References

[1] T. Barnes, D. Harrison, and A.R. Newton, *Rick Spickelmier, Electronic CAD Frameworks*, Kluwer Academic Publishers, Boston, MA, 1992.

[2] P. Bingley, O. ten Bosch, and P. van der Wolf, Incorporating design flow management in a framework based CAD system, *Proceedings of the 1992 IEEE/ACM International Conference on Computer-Aided Design*, 1992.

[3] R.G. Bushroe, S. DasGupta, A. Dengi, P. Fisher, S. Grout, G. Ledenbach, N.S. Nagaraj, and R. Steele, Chip hierarchical design system (CHDS): a foundation for timing-driven physical design into the 21st century, *International Symposium on Physical Design*, 1997.

[4] F.L. Chan, M.D. Spiller, and A.R. Newton, WELD – an environment for web-based electronic design, *Proceedings of the 35th Conference on Design Automation*, 1998.

[5] G. Chin, W. Dietrich, Jr., D. Boning, A. Wong, A. Neureuther, and R. Dutton, Linking TCAD to EDA — benefits and issues, *Proceedings of the 28th Conference on Design Automation*, 1991.

[6] J. Crawford, An electronic design interchange format, *Proceedings of the 21st Conference on Design Automation*, 1984.

[7] J. Eurich, A tutorial introduction to the electronic design interchange format, *Proceedings of the 23rd Conference on Design Automation*, 1986.

[8] N. Filer, M. Brown, and Z. Moosa, Integrating CAD tools into a framework environment using a flexible and adaptable procedural interface, *Proceedings of the Conference on European Design Automation*, 1994.

[9] D. Harrison, P. Moore, R. Spickelmier, and A.R. Newton, Data management and graphics editing in the Berkeley design environment, *Proceedings of the 1986 IEEE/ACM International Conference on Computer-Aided Design*, 1986.

[10] B. Infante, D. Bracken, B. McCalla, S. Yamakoshi, and E. Cohen, An interactive graphics system for the design of integrated circuits, *Proceedings of the 15th Conference on Design Automation*, 1978.

[11] K.H. Keller, A.R. Newton, and S. Ellis, A symbolic design system for integrated circuits, *Proceedings of the 19th Conference on Design Automation*, 1982.

[12] L.-C. Liu, P.-C. Wu, and C.-H. Wu, Design data management in a CAD framework environment, *Proceedings of the 27th Conference on Design Automation*, 1990.

[13] T. Mitsuhashi, T. Chiba, M. Takashima, and K. Yoshida, Integrated mask artwork analysis system, *Proceedings of the 17th Conference on Design Automation*, 1980.

[14] D. Nash, Topics in design automation data bases, *Proceedings of the 15th Conference on Design Automation*, 1978.

[15] J. Ousterhout, G. Hamachi, R. Mayo, W. Scott, and G. Taylor, Magic: a VLSI layout system, *Proceedings of the 21st Conference on Design Automation*, 1984.

[16] High Level Design Systems, *Pillar Language Manual*, Version 3.2, 1995.

[17] K. Roberts, T. Baker, and D. Jerome, A vertically organized computer-aided design database, *Proceedings of the 18th Conference on Design Automation*, 1981.

[18] B. Schürmann and J. Altmeyer, Modeling design tasks and tools – the link between product and flow model, *Proceedings of the 34th Conference on Design Automation*, 1997.

[19] M. Silva, D. Gedye, R. Katz, and R. Newton, Protection and versioning for OCT, *Proceedings of the 26th Conference on Design Automation*, 1989.

[20] W.D. Smith, D. Duff, M. Dragomirecky, J. Caldwell, M. Hartman, J. Jasica, and M.A. d'Abreu, FACE core environment: the model and its application in CAE/CAD tool development, *Proceedings of the 26th Conference on Design Automation*, 1989.

[21] J. Soukup, Organized C: a unified method of handling data in CAD algorithms and databases, *Proceedings of the 27th Conference on Design Automation*, 1990.

[22] M. Spiller and A.R. Newton, EDA and the network, *Proceedings of the 1997 IEEE/ACM International Conference on Computer-Aided Design*, 1997.

[23] P. van der Wolf and T.G.R. van Leuken, Object type oriented data modeling for VLSI data management, *Proceedings of the 25th Conference on Design Automation*, 1988.

[24] J. Wilmore, The design of an efficient data base to support an interactive LSI layout system, *Proceedings of the 16th Conference on Design Automation*, 1979.
[25] J.B. Rosenberg, Geographical data structures compared: a study of data structures supporting region queries, *IEEE Trans. Comput.-Aided Des.*, 4, 53–67, 1985.

Current References

[26] M. Bales, Facilitating EDA flow interoperability with the openaccess design database, *Electronic Design Processes 2003 Workshop*, 2003.
[27] T. Blanchard, Assessment of the OpenAccess standard: insights on the new EDA industry standard from Hewlett-Packard, a beta partner and contributing developer, *International Symposium on Quality Electronic Design*, 2003.
[28] L. Brevard, Introduction to Milkyway, *Electronic Design Processes Workshop 2003*, 2003.
[29] D. Cottrell and T. Grebinski, Interoperability beyond design: sharing knowledge between design and manufacturing, *International Symposium on Quality Electronic Design*, 2003.
[30] J. Darringer and J. Morrell, Design systems evolution and the need for a standard data model, *Electronic Design Processes 2003 Workshop*, 2003.
[31] W. Grobman, R. Boone, C. Philbin, and B. Jarvis, Reticle enhancement technology trends: resource and manufacturability implications for the implementation of physical designs, *International Symposium on Physical Design*, 2001.
[32] W. Grobman, M. Thompson, R. Wang, C. Yuan, R. Tian, and E. Demircan, Reticle enhancement technology: implications and challenges for physical design, *Proceedings of the 38th Conference on Design Automation*, 2001.
[33] A. Khang and I. Markov, Impact of interoperability on CADIP reuse: an academic viewpoint, *International Symposium on Quality Electronic Design*, 2003.
[34] Magma Design Automation White Paper, *Gain-Based Synthesis: Speeding RTL to Silicon*, http://www.magma-da.com, 2002.
[35] D. Mallis and E. Leavitt, *OpenAccess: The Standard API for Rapid EDA Tool Integration*, Silicon Integration Initiative, Austin, TX, 2003.
[36] S.N. Adya, M.C. Yildiz, I.L. Markov, P.G. Villarrubia, P.N. Parakh, and P.H. Madden, Benchmarking for large-scale placement and beyond, *International Symposium on Physical Design*, 2002.
[37] Silicon Integration Initiative, *OpenAccess: Goals and Status*, http://www.si2.org/openaccess/Presentations/OAC_Intro_092802.pdf, 2003.
[38] R.H.J.M. Otten, R. Camposano, and P.R. Groeneveld, Design automation for deepsubmicron: present and future, *Proceedings of the 2002 Design, Automation and Test in Europe Conference and Exhibition*, 2002.
[39] M.A. Riepe, Interoperability, datamodels, and databases, *Electronic Design Processes 2003 Workshop*, 2003.
[40] P. Rodman, R. Collett, L. Lev, P. Groeneveld, N. Nettleton, and L. van den Hoven, Tools or users: which is the bigger bottleneck? *Proceedings of the 39th Conference on Design Automation*, 2002.
[41] J. Santos, OpenAccess architecture and design philosophy, *OpenAccess 2002 Conference*, 2002.
[42] J. Santos, Overview of OpenAccess: the next-generation database for IC design, *OpenAccess 2003 Conference*, 2003.
[43] N.V. Shenoy and W. Nicholls, An efficient routing database, *Proceedings of the 39th Conference on Design Automation*, 2002.
[44] Z. Xiu, D. Papa, P. Chong, C. Albrecht, A. Kuehlmann, R.A. Rutenbar, and I.L. Markov, Early research experience with OpenAccess gear: an open source development environment for physical design, *ACM International Symposium on Physical Design (ISPD'05)*, 2005, pp. 94–100.

13

FPGA Synthesis and Physical Design

Mike Hutton
Altera Corp.
San Jose, California

Vaughn Betz
Altera Corp.
Toronto, Ontario, Canada

13.1 Introduction

Since their introduction in the early 1980s, *programmable logic devices* (PLDs) have evolved from implementing small glue-logic designs to large, complete systems. PLDs can be divided into two categories: *complex programmable logic devices* (CPLDs) and *field-programmable gate arrays* (FPGAs). CPLDs are lower-density devices employing nonvolatile programming — in other words, the CPLD programming is not lost when the device is powered down. FPGAs, the topic of this chapter, are typically based on *look-up table* (LUT) cells, and reminiscent of ASIC gate arrays in structure. FPGAs are typically static-RAM programmed and thus require power to maintain their configuration. Today, the majority of all design starts (though typically not the largest ones) target PLDs, with the higher-density designs on FPGAs and smaller designs and designs that require nonvolatility targeting CPLDs. The increasing use of PLD devices has resulted in significant research in computer-aided design (CAD) algorithms and tools targeting programmable logic.

There are two branches of FPGA CAD-tool research: one concerned with developing algorithms for a given FPGA and the second parallel branch developing the tools required to design FPGA architectures. This emphasizes the interdependence of FPGA CAD tools and architectures: unlike ASIC flows where CAD is an implementation of a design in silicon, the CAD flow for FPGAs is an embedding of the design into a device architecture that is fixed in terms of both cells and routing. Although some algorithms for FPGAs continue to overlap with ASIC-targeted tools, notably language and technology-independent synthesis, and to some extent placement, technology mapping and routing are notably different.

In addition to the core synthesis, placement, and routing algorithms, emerging tools for FPGAs now concentrate on physical synthesis, power and timing analysis and optimization, and the growing area of system-level design. Going forward, power optimizations will be a combination of semiconductor industry-wide techniques and ideas targeting the programmable nature of FPGAs, closely tuned to the evolving FPGA architectures. System-level design tools will attempt to exploit two of the key benefits of FPGAs — fast design and verification time, and a high degree of programmable flexibility.

FPGA tools differ from ASIC tools in that the core portions of the tool flow are owned by the silicon vendors. Third-party EDA tools are often used for synthesis and verification, but with very few exceptions physical design tools are supplied by the FPGA vendor, and support only that vendor's products. The two largest FPGA vendors (Altera and Xilinx) offer complete CAD flows from language extraction and synthesis through placement and routing to power and timing analysis. One of our goals in this chapter is to describe CAD for programmable logic not just from an individual algorithm perspective, but in the context of the end-to-end flow that a designer using a commercial FPGA would experience.

After some introductory description of FPGA architectures and CAD flows, we will describe research in the key areas of CAD for FPGAs, roughly in flow order.

13.1.1 Architecture of Field-Programmable Gate Arrays

To appreciate CAD algorithms for FPGAs, it is necessary to have some understanding of FPGA architectures and how they are specified.

Figure 13.1 shows a high-level block diagram of a modern FPGA. The FPGA in Figure 13.1 is a Stratix FPGA, but other modern FPGAs have fairly similar features. Components of the floorplan are logic blocks, which in Stratix are called *logic array blocks* (LABs), digital signal processing (DSP) blocks containing multiplier/accumulator functionality, and various sizes of embedded RAM blocks (here 512 bit, 4 kbit and the 576 kbit M-RAM). Horizontal rows and vertical columns of routing interconnect are shown behind the functional blocks.

Figure 13.2 details the internals of a Stratix logic array block. A LAB is composed of several logic elements (LEs) and local routing to interconnect them. Each LE is composed of an LUT, a register, and dedicated circuitry for arithmetic functions. Figure 13.2(a) shows a 4-input LUT. The 16 bits of programmable LUT mask specify the truth table for a 4-input function, which is then controlled by the A, B, C, and D inputs to the LUT. Figure 13.2(b) shows a complete LE, consisting of a 4-input LUT and a DFF with selectable control signals. Arithmetic in the Stratix LE is accomplished by using the two 3-input LUTs independently to generate sum and ripple-carry functions. The carry-out becomes a dedicated fifth connection into the neighboring LE, and optionally replaces the C input. By "chaining" the carry-out of

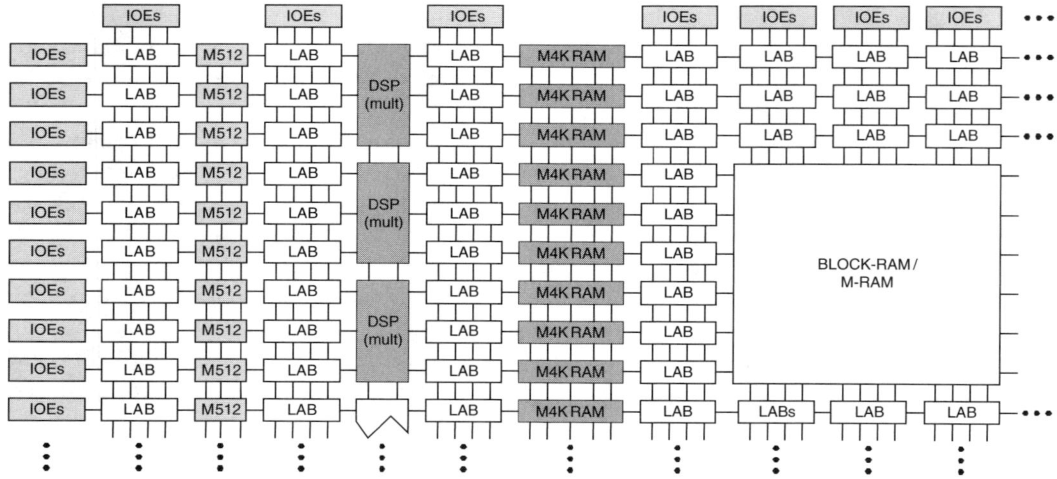

FIGURE 13.1 Stratix FPGA high-level block diagram.

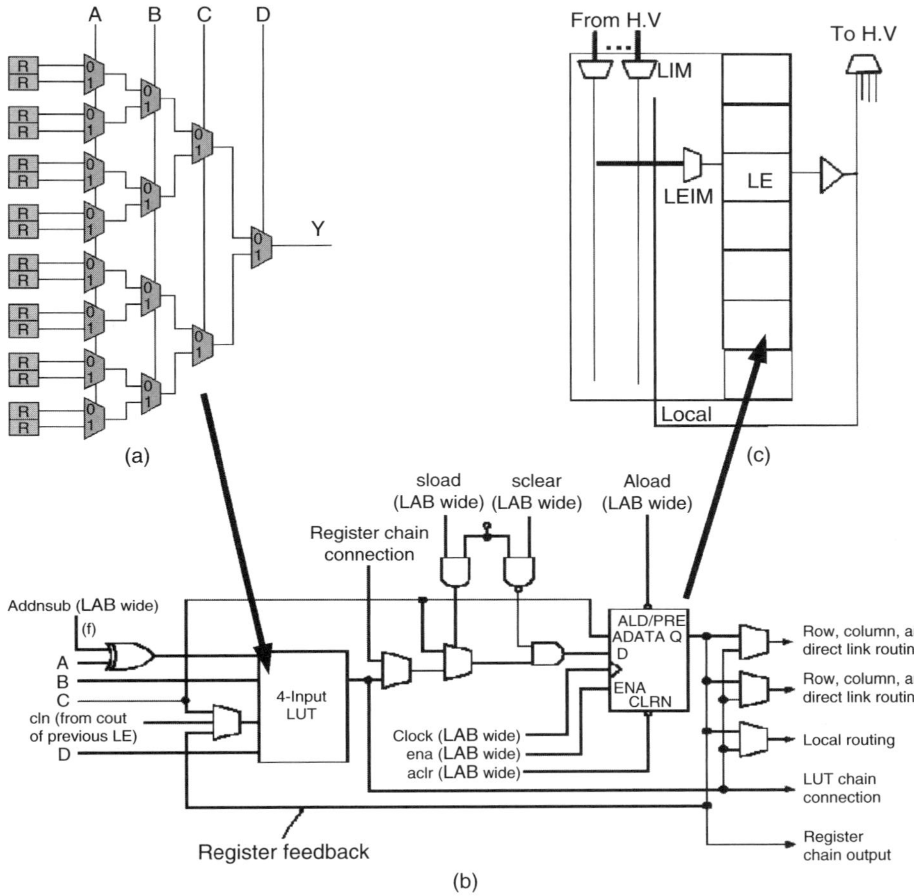

FIGURE 13.2 Inside an FPGA logic block: (a) 4-LUT showing LUT mask; (b) complete Stratix LE, and (c) LAB cluster showing routing interface to inter-logic-block H and V routing wires.

several LEs together, a fast multi-bit adder can be formed. Figure 13.2(c) shows how several LEs (ten in Stratix) are grouped together to form a LAB. A subset of the global horizontal (H) and vertical (V) lines that form the FPGA interconnect can drive into a number of *LAB input muxes* (LIMs). The signals generated by the LIMs (along with feedback from other LEs in the same LAB) can drive the inputs to the LE through LEIM (*logic-element input muxes*). On the output side of the LE, *routing multiplexers* allow signals from one or more LEs to drive H and V routing wires in order to reach other LABs. Each of the LAB inputs, LE inputs, and routing muxes are controlled by SRAM configuration bits. Thus, the total of the SRAM bits programming the LUT, DFF, modes of the LE, and the various muxes comprise the configuration of the FPGA to perform a specific logical function.

Depending on the architectural choices, some of the blocks in Figure 13.1 may not be present, e.g., the device may have no DSP blocks, the routing could be organized differently, and control signals on the DFF could be added or removed. Virtex FPGAs, for example, have no small RAMs, but have additional hardware to allow the LUT mask in the normal LE to be used as a 16-bit RAM.

There is a large amount of literature on good choices of FPGA architecture. Rose et al. [1] originally showed that the most area-efficient LUT is one whose number of inputs (k) is 4. Betz and Rose [3] investigated M, the number of LEs a logic block should contain, and found that cluster sizes from 1 to 8 LEs resulted in FPGAs with the best area. Ahmed and Rose [2] later updated these studies by examining the interaction of LUT size and logic cluster size, and found that a LUT size of $k = 4$–to 6 and logic cluster size of $M = 8$ to 10 is best when considering both area and delay. Rose and Brown [4] evaluated the flexibility required of the routing structure, meaning the size of programmable muxes (LIMs, LEIMs, and routing muxes in the earlier terminology), and Lemieux and Lewis [5] describe circuit-level issues. Cong and Hwang [6] evaluated FPGAs, such as the Xilinx 4000, with some hard-wired (rather than programmable) connections between LEs,. For a complete description of this early work on FPGA architecture see Brown et al. [7] or Betz et al. [8].

The VPR toolset [9] introduced what is now a standard paradigm for empirical architecture evaluation. An architecture specification file controls a parameterized CAD flow capable of targeting a wide variety of FPGA architectures, and individual architecture parameters are swept across different values to determine their impact on FPGA area, delay, and recently power. Figure 13.3 illustrates a commercial version of this flow called the "FPGA Modeling Toolkit" used at Altera. Yan et al. [10] validated the importance of the combined hardware and software exploration methodology by showing the sensitivity of architecture results to the tool flow.

Li et al. [11] expanded the VPR toolset to perform power exploration on FPGA architectures, using it to examine voltage islands and other recent FPGA architectural issues. Wilton [12] evaluated the architecture of embedded memory blocks in FPGAs.

In terms of commercial FPGA architectures, descriptions have been published by Lewis for Altera's Stratix [14] and Stratix II [15] families and by Trimberger et al. [16] for the Xilinx 4000. For the Virtex I–IV FPGA families, see the Xilinx website [17], as there is no academic publication. Ahanin [18] described the architecture of the MAX 7000 CPLD.

13.1.2 CAD Flow for FPGAs

The core CAD flow seen by an FPGA user is shown in Figure 13.4. After design entry, the design is synthesized at the register-transfer level (RTL) into operators (adders, muxes, and multipliers), state machines, and memory blocks, followed by gate-level logic synthesis, then technology mapping to LUTs

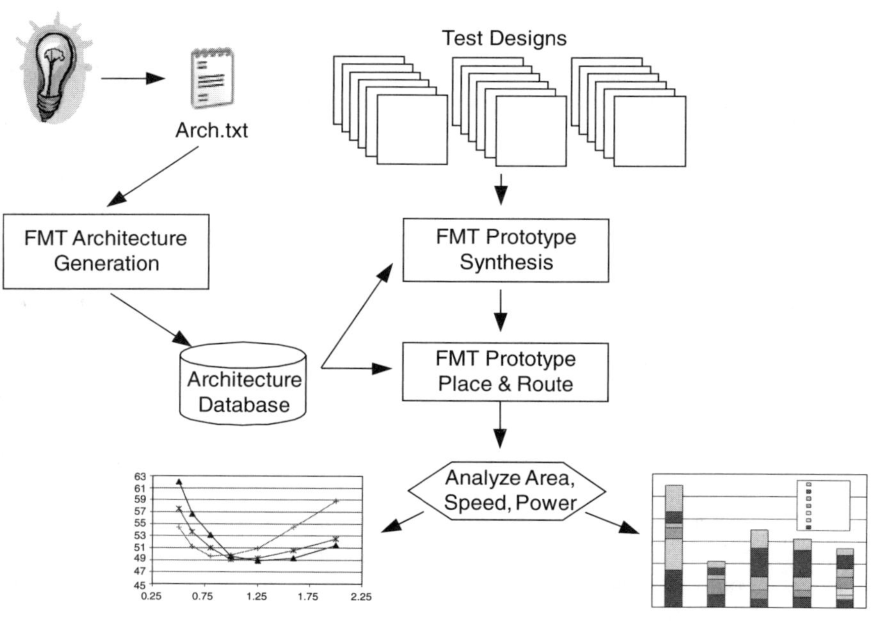

FIGURE 13.3 FPGA modeling toolkit flow based on VPR.

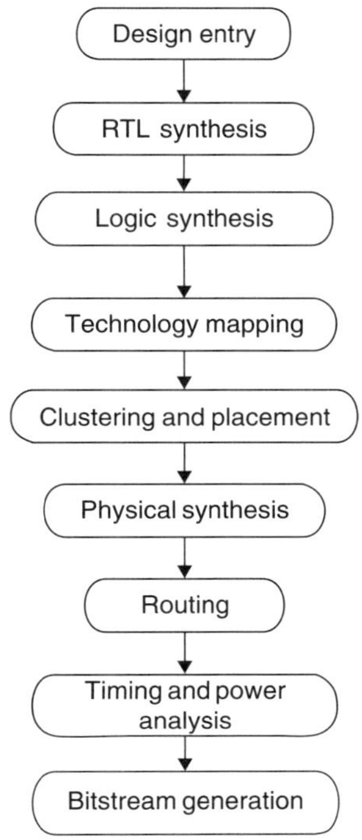

FIGURE 13.4 FPGA CAD flow.

and registers. Clustering groups LEs into logic blocks (LABs), and placement determines a fixed location for each logic or other block. Routing selects which programmable switches to turn on in order to connect all the terminals of each signal net, essentially determining the configuration settings for each of the LIM, LEIM, and routing muxes described in Section 13.1.1. Physical synthesis, shown between placement and routing in this flow, is an optional step, which re-synthesizes the netlist to improve area or delay, now that preliminary timing is known. Timing analysis [19] and power analysis [20] for FPGAs are approximated at many points in the flow to guide the optimization tools, but are performed for final analysis and reporting at the end of the flow. The last step is bit-stream generation, which determines the sequence of 1's and 0's, which will be serially loaded to configure the device. Notice that some ASIC CAD steps are absent: sizing and compaction, buffer insertion, clock-tree, and power grid synthesis are not relevant to a prefabricated architecture.

Research on FPGA algorithms can take place at all parts of this flow. However, academic literature has generally been dominated by synthesis (particularly LUT mapping) at the front end and place and route algorithms at the back end, with most other aspects left to commercial tools and the FPGA vendors. Literature on FPGA architecture and CAD can be found in the major FPGA conferences (FPGA, FPL, FPT, and FCCM) as well as the general CAD conferences (IWLS, ISPD, DATE, ICCAD, and DAC). Device architecture is often presented at CICC, ISSCC, ISLPED, or HotChips.

In this chapter, we will survey algorithms, tools, and techniques for both the end-user and architecture-development FPGA CAD flows. The references we have chosen to include are representative of a large body of research in FPGAs, but the list cannot be comprehensive. For more complete discussion, the reader is referred to the chapter on Logic Synthesis and texts such as [21] on synthesis, and [22,23] on FPGA synthesis and mapping. Commercial manufacturers [17] of LUT-based FPGAs include Altera, Xilinx, Actel, and Lattice. Dominant third-party EDA vendors for FPGAs include Synplicity, Mentor Graphics and Synopsys for synthesis, Synopsys for timing analysis, Mentor Graphics for simulation, and Cadence and Synopsys for formal verification.

13.2 System-Level Tools

Although the vast majority of current FPGA designs are entered directly in VHDL or Verilog, there have been a number of attempts to raise the level of abstraction to the behavioral or block-integration level with system-level tools.

Berkeley Design Technologies Inc. [26] found that FPGAs have better price/performance than DSP processors for many DSP applications. However, HDL-based flows are not natural for most DSP designers, so higher-level DSP design flows are an important research area. Altera DSPBuilder™ and Xilinx SystemGenerator™ link the MatLab™ and Simulink™ algorithm exploration and simulation environments popular with DSP designers to VHDL and Verilog descriptions targeting FPGAs. Accelchip [17] and Catalytic also offer Matlab-based flows. Hwang et al. [24] give an overview of SystemGenerator, and Stroomer [25] describes *jg*, a tool for providing Java specifications generating Matlab code which targets SystemGenerator.

The FPX system of Lockwood et al. [27] developed system-level and modular design methodologies for the network processing domain. The modular nature of FPX also allows for exploration of hardware and software implementations.

Altera's System on a Programmable Chip (SOPC) Builder™ (see Kempa [28]) and Xilinx's Embedded Development Kit (EDK)™ are examples of system integration tools. These tools allow the designer to integrate each company's soft-core (synthesized from the FPGA fabric) or hard-core (prebuilt on the FPGA) processors, and other intellectual property cores and peripherals together easily. The details of bus interfacing, bus mastering, and peripheral memory-map creation are implemented automatically from a high-level description. These tools also provide some ability for exploration of the hardware–software co-design space. In the case of SOPC Builder, this includes mechanisms for adding custom instructions to the Nios II™ soft processor, which are executed with custom-created hardware units.

Work has also proceeded on synthesizing high-level design languages to FPGAs. Celoxica [17], for example, provides commercial synthesis of Handel-C to RTL. In a different direction, Hutchings et al. [29] describe JBits which uses Java for both high-level design specification and low-level bit manipulation for FPGAs. Brandolese et al. [30] gave a methodology for estimating required FPGA LE counts from SystemC design descriptions.

13.3 Logic Synthesis

FPGA logic synthesis can be seen as four high-level steps: language synthesis, RTL synthesis, technology-independent logic synthesis, and technology mapping. Language synthesis, meaning VHDL/Verilog extraction, analysis and elaboration, overlaps closely with ASIC tools, and is rarely discussed in a purely FPGA context. The other steps are discussed in the next sections.

13.3.1 RTL Synthesis

Register-transfer-level synthesis includes the inference and processing of high-level structures — adders, multipliers, muxes, busses, shifters, crossbars, RAMs, shift registers, and finite-state machines — prior to decomposition into generic gate-level logic. This is an important area of work for FPGA and CAD-tool vendors, but has been given very little attention in the published literature. One reason for this is that the implementation of operators can be architecture-specific. However, a more pedantic issue is simply that there are very few public-domain VHDL/Verilog front-end tools or high-level designs with these structures and both are necessary for research in the area. Approximately 20–30% of logic elements eventually seen by placement are mapped directly from RTL and are not processed by gate-level logic synthesis.

In commercial FPGA architectures, dedicated hardware is provided for multipliers, addition, clock enables, clear, preset, RAM, and shift registers. Arithmetic was discussed briefly earlier; all major FPGAs either convert 4-LUTs into 3-LUT-based sum and carry computations or provide dedicated arithmetic hardware separate from the LUT [15]. The most important effect of arithmetic is the restriction it imposes on placement, since cells in a carry chain must be placed in fixed relative positions.

One of the goals of RTL-level synthesis is to take better advantage of the hardware provided. This will often result in different synthesis than would take place in an ASIC flow because the goal is to minimize LEs rather than gates. A 4:1 mux, for example, uses three 2-input gates and cannot fit in a single 4-LUT. However, transformations such as shown in Figure 13.5 allow that mux to be implemented optimally in two 4-LUTs, even though these LUTs contain five gates. RTL synthesis typically produces these premapped cells and then protects them from processing by logic synthesis, particularly when they occur in a bus and there is a timing advantage from a symmetric implementation. RTL synthesis will recognize barrel shifters ("$y<=(x>>s)$" in Verilog) and convert these into shifting networks as shown in Figure 13.6, again protecting them from gate-level manipulation.

Recognition of register control signals such as clock enable, clear/preset, and sync/asynch load signals is also interesting in FPGAs. Since these preexist in the logic cell hardware (see Figure 13.2), there is a strong incentive to synthesize them even when it would not make sense in ASIC synthesis. For example, a 4:1 mux with one constant input does not fit in a 4-LUT, but when it occurs in a datapath it can be synthesized for most commercial LEs by using the LAB-wide (i.e., shared by all LEs in a LAB) synchronous load signal as a fifth input. Similarly, a clock enable already exists in the hardware, so register feedback can be converted into an alternative structure with a clock enable to hold the current value, but no routed register feedback. For example, if $f = z$ in Figure 13.7, we can re-express the cone of logic with a clock-enable signal

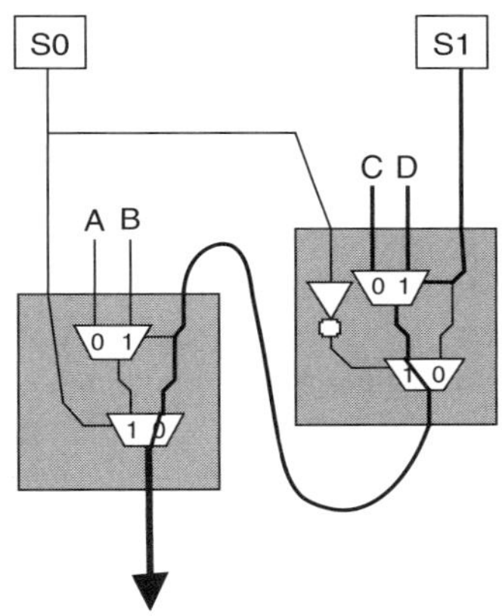

FIGURE 13.5 Implementing a 4:1 mux in two 4-LUTs.

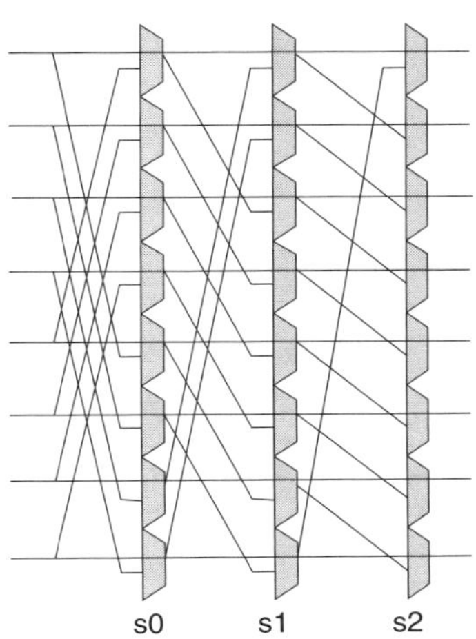

FIGURE 13.6 Eight-bit barrel shifter network.

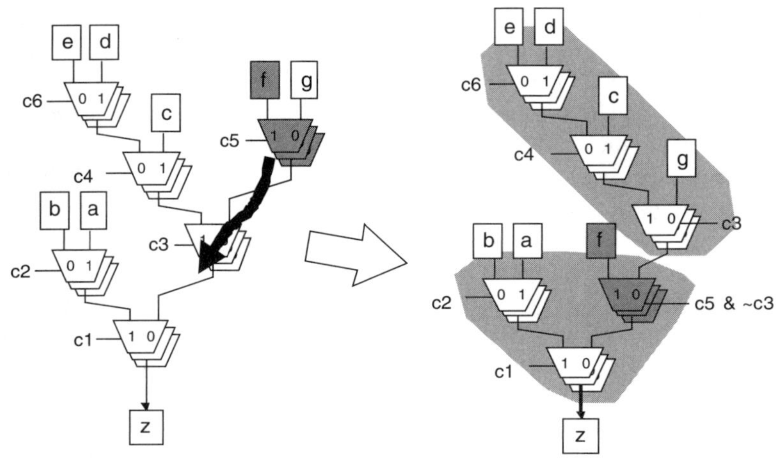

FIGURE 13.7 Multiplexor bus restructuring for LUT-packing.

CE = c5*c3*c1′ with the *f,g* mux replaced by a wire from *g* — this is a win for bus widths of 2 or more. This transformation can be computed with a BDD or other functional techniques.

Several algorithms have recently been published in the RTL-synthesis area. Metzgen and Nancekievill [32,33], for example, showed algorithms for optimization of multiplexer-based busses which would otherwise be inefficiently decomposed into gates. Most modern FPGA architectures do not provide on-chip tri-state buses, so muxes are the only choice for busses, and are heavily used in designs. Multiplexors are very interesting structures for FPGAs, because the LUT cell yields a relatively inefficient implementation of a mux, and hence special-purpose hardware for handling muxes is common. Figure 13.7 shows an example taken from Ref. [32] that re-structures busses of muxes for better technology mapping (covering) into 4-LUTs. The structure on the left requires three LUTs per bit to implement in 4-LUTs while the structure on the right requires only 2 LUTs per bit.

Also to address muxes, newer FPGA architectures have added clever hardware to aid in the synthesis of mux structures such as crossbars and barrel shifters. The Xilinx Virtex family of FPGAs [17] provides additional "stitching" multiplexors for adjacent LEs, which can be combined to efficiently build efficient larger multiplexors; an abstraction of this composable LUT is shown in Figure 13.8(a). These are also used for stitching RAM bits together when the LUT is used as a 16-bit RAM (discussed earlier). Altera's Stratix II adaptive logic module [31] shown abstractly in Figure 13.8(b) allows a 6-LUT to be fractured to implement a 6-LUT, two independent 4-LUTs, two 5-LUTs which share two input signals, and also two 6-LUTs that have four common signals and two different signals (a total of 8). This latter feature allows two 4:1 muxes with common data and different select signals to be implemented in a single LE, which means crossbars and barrel shifters built out of 4:1 muxes use half the area they would otherwise require.

13.3.2 Logic Optimization

Technology-independent logic synthesis for FPGAs has followed the same methodology of point algorithms in a script flow, as popularized by SIS [34]. The general topic of logic synthesis is described in the chapter on Logic Synthesis of this handbook, and in textbooks such as Ref. [21]. Synthesis tools for FPGAs contain basically the same two-level minimization algorithms, and algebraic and Boolean algorithms for multi-level synthesis. Here we will generally restrict our discussion to the differences from ASIC synthesis.

One major difference between standard and FPGA synthesis is in cost metrics. The target technology in a standard cell ASIC library is a more finely grained cell (e.g., a two-input NAND gate) while a typical FPGA cell is a generic *k*-input LUT. A 4-LUT is a 16-bit SRAM LUT mask driving a 4-level tree of 2:1 muxes controlled by the inputs A,B,C, and D (Figure 13.2(a)). Thus A + B + C + D and AB + CD + AB′D′ + A′B′C′

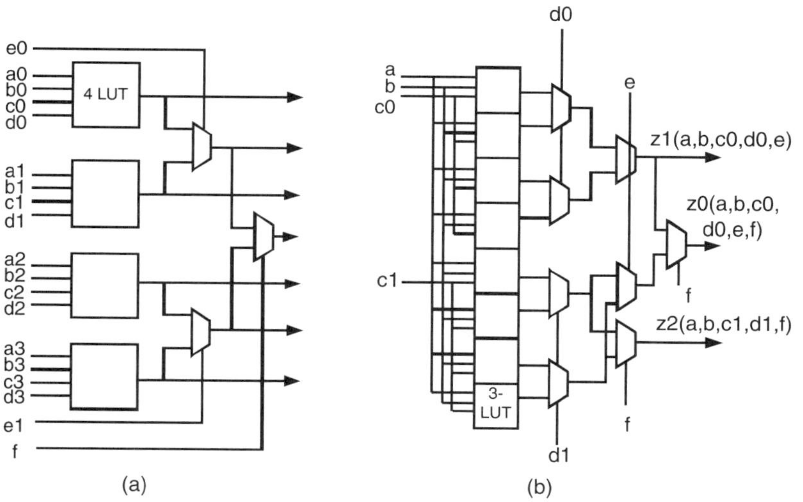

FIGURE 13.8 Composable and fracturable LEs.

have identical costs in LUTs even though the former has 4 literals and the latter 10. In general, the count of 2-input gates correlates much better to 4-LUT implementation cost than the literal-count cost often used in these algorithms, but this is not always the case, as described for the 4:1 mux in the preceding section. A related difference is that inverters are free in FPGAs because (1) the LUT mask can always be reprogrammed to remove an inverter feeding or fed by a LUT, and (2) programmable inversion at the inputs to RAM, IO and DSP blocks is available in most FPGA architectures. In general, registers are also free because all LEs have a built-in DFF. This changes cost functions for re-timing and state-machine encoding as well as designer preference for pipelining.

Subfactor extraction algorithms are much more important for FPGA synthesis than commonly reported in academic literature where ASIC gates are assumed. It is not clear whether this arises from the much larger and more datapath-oriented designs seen in industrial flows (vs. MCNC gate-level circuits), from the more structured synthesis, from a complete HDL to gates flow, or due to the larger cell granularity. In contrast, algorithms in the class of "speed_up" [34] do not have significant effects on circuit performance for commercial FPGA designs. Again, this may be due either to the flow and reference circuits, or to differing area/depth trade-offs. Commercial tools carefully balance area and depth during multilevel synthesis.

Synthesis of arithmetic functions is typically performed separately in commercial FPGA tools, although most academic tools synthesize arithmetic into LUTs. This can result in a dramatic difference in the efficacy of synthesis algorithms which perform well on arithmetic circuits compared to random or multiplexor/selector-based logic. Typical industrial designs contain 10–25% of logic cells in arithmetic mode, in which the dedicated carry circuitry is used with or instead of the LUT.

Retiming algorithms from general logic synthesis [35] have been adapted specifically for FPGAs [36], taking into account practical restrictions such as meta-stability, input/output (I/O) vs. core timing trade-offs, power-up conditions, and the abundance of registers.

There are a number of resynthesis algorithms that are of particular interest to FPGAs, specifically structural decomposition, functional decomposition, and postoptimization using SPFD-based rewiring. These will be discussed in Sections 13.3.3.1 and 13.3.3.2.

An alternative, more FPGA-specific, approach to synthesis was taken by Vemuri [37] using BDS [38] building on BDD-based decomposition [39]. These authors argued that the separation of tech-independent synthesis from tech-mapping disadvantaged FPGAs, which need to optimize LUTs rather than literals due to their greater flexibility and larger granularity. The BDS system integrated technology-independent optimization using BDDs with LUT-based logic restructuring, and used functional decomposition to target decompositions of k-feasible LUTs for mapping. The standard sweep, eliminate, decomposition, and factoring algorithms from SIS were implemented in a BDD framework. The end result uses a technology mapping step, but on a netlist more amenable to LUT mapping. Comparisons between SIS and BDS-pga using the same technology mapper showed area and delay benefits to the BDD-based algorithms.

A new area for CAD optimization in FPGAs involves power optimization, particularly leakage. Anderson et al. [40] propose some interesting ideas on modifying the LUT mask during synthesis to place LUTs into a state that will reduce leakage power in the FPGA routing. Commercial tools synthesize clock-enable circuitry to reduce dynamic power consumption on blocks. These are likely the beginning of many future treatments for power management in FPGAs.

13.3.3 Technology Mapping

Technology mapping for FPGAs is the process of turning a network of primitive gates into a network of LUTs of size at most k. The constant k is historically 4 [1], although recent commercial architectures have used fracturable logic cells with $k = 6$ [31]. LUT-based tech-mapping is best seen as a covering problem, since it is both common and necessary to cover some gates by multiple LUTs for an efficient solution. Figure 13.9 (taken from Ref. [41]), illustrates this concept. Technology mapping aims for the least unit depth combined with the least number of cells in the mapped network.

FPGA technology mapping differs from library-based mapping for cell-based ASICs (e.g., [42]). Technology mapping into k-input LUTs is a well-studied problem, with at least 100 related papers on the topic. The most successful attempts divide into two paradigms: dynamic programming approaches such as Chortle [43], and network flow based approaches branching from FlowMap [44]. Many of these technology-mapping algorithms are implemented in the RASP system from UCLA [45]. Technology mapping is usually preceded by decomposition of the netlist into 2-input gates, but we will defer that topic to Section 13.3.3.1 because it draws on techniques from mapping.

In the Chortle algorithm [43] by Francis, the netlist is decomposed to 2-input gates, and then divided into a forest of trees, a starting point used by Keutzer [42] and most library-based mappers. For each node in topological order, a set of k-feasible mappings for children nodes is known (by induction). To compute an optimum set of k-feasible mappings for the current node, Chortle combines solutions for children within reach of one LUT implemented at the current node, following the dynamic programming paradigm.

Improvements on Chortle considered not trees, but maximum fanout-free cones, which allowed for mapping with duplication. Area mapping with no duplication was later shown to be optimally solvable in polynomial time for MFFCs [46]. But, perhaps contrary to intuition, duplication is important in

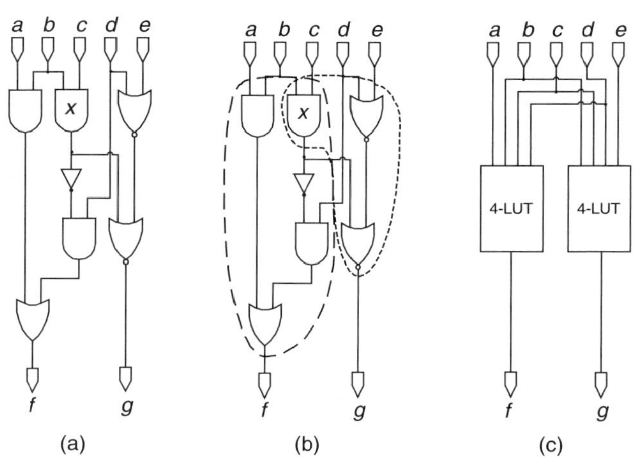

FIGURE 13.9 Technology mapping as a covering problem. (From Ling, A., Singh, D.P., and Brown, S.D., FPGA technology mapping: a study of optimality, *Proceedings of the Design Automation Conference*, 2005.)

improving results for LUTs because it allows nodes with fanout greater than 1 to be implemented as internal nodes of the cover.This is required to obtain improved delay and also can contribute to improved area [47]. Figure 13.9(b) shows a mapping to illustrate this point. Chortle-crf [48] deals more efficiently with re-convergent fanout. Chortle-d [49] adds a new bin-packing step on nodes, which considers all decompositions into 2-input gates as part of the dynamic programming step, and is shown to be depth optimal for mapping on trees.

FlowMap is a two-step algorithm proposed by Cong and Ding [44]. In the labeling phase, a topological traversal of the network assigns each node a Lawler label $L(v)$ [50], which is constructed to be the worst-case depth of node v in a depth-optimal mapping solution. Primary inputs have $L(v) = 0$. For other nodes, $L(v)$ is either D or $D + 1$, where D is the max $L(w)$ over all fan-in w of v. To determine this, the fan-in with $L(w) = D$ are collapsed into v (to simulate a LUT implemented at v), and a feasible cut computation is made to determine if $L(w)$ can be D. Otherwise it is $D + 1$. A k-feasible cut is a partition of the network into A and $\bar{A}$ such that the output of no more than k nodes is cut. The size of a cut is the number of cut-edges, the height is the largest label of the nodes cut, and the volume is the number of nodes in $\bar{A}$. The key aspect to the FlowMap algorithm is to reduce the minimum-height k-feasible cut computation to the well-known network flow problem [51]. Figure 13.10 (reproduced from Ref. [44]) illustrates a network of nodes with depth labeling, the auxiliary S (source) and T (sink) nodes required for network flows, and a 3-feasible cut with size 10, volume 9, and height 2.

The result of FlowMap is provably depth-optimal for arbitrary k-bounded networks, meaning that for a fixed decomposition into 2-input gates, it always generates a unit-delay-minimal solution. Later enhancements allow for finding a minimum-height maximal-volume cut as an area reduction heuristic. Since most problems in tech-mapping are NP-hard (e.g., area minimization [52]), this also makes FlowMap theoretically interesting as well as practical. FlowMap can be extended to use more general models of delay [53]. Cong and Ding [46] added a duplication-free re-mapping step to FlowMap to improve area, and also explored these trade-offs.

CutMap by Cong and Hwang [54] is an area improvement on FlowMap which maintains the property of optimal delay. The key feature of CutMap is the computation of both min-cost min-height cuts for nodes on the critical path and min-cost k-feasible cuts for other nodes for the implementation phase. This allows for an area/delay tradeoff directly in the cost function of the mapping phase, unlike FlowMap which addressed area only in a postprocessing step.

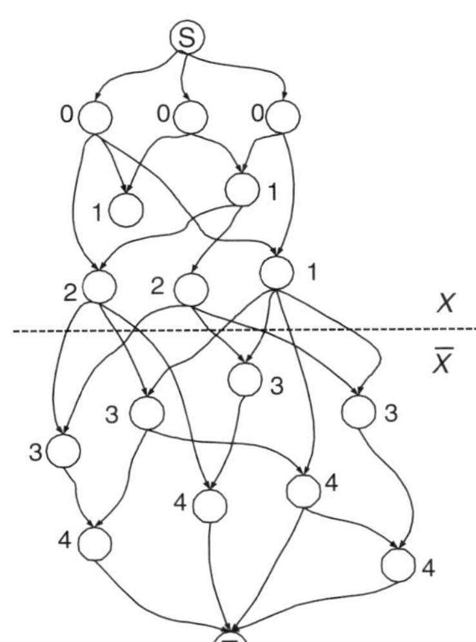

FIGURE 13.10 A FlowMap 3-feasible cut with cut-size 10, volume 9, height 2. (Reproduced from Cong, J. and Ding, E., An optimal technology mapping algorithm for delay optimization in lookup table based FPGA designs, *IEEE Trans. CAD*, 13, 1–12, 1994.)

In the first step, CutMap computes arrival labels using the FlowMap labeling phase and marks each node also with its required implementation time, so that unit-delay slack can be computed. In the implementation phase, nodes with zero-slack follow the FlowMap calculation — predecessor nodes u with label(u) = label(v) are collapsed into v for a min-cost k-feasible cut with minimum height. However, for nodes v with positive slack, the nodes are not collapsed, and only the min-cost k-feasible cut calculation is made. In this way, implementation of all noncritical nodes favors area minimization. Empirical results for CutMap show 15% better area than FlowMap with the same unit delay, however, at the cost of longer run-time.

Manohararajah et al. [55] gave a mapping algorithm IMAP (iterative map), based on dynamic programming that simultaneously addresses depth and area. This paper also introduces new metrics of area flow and depth bounds, and uses an edge-delay model based on Ref. [56]. IMAP generates k-feasible cones for nodes as in [47] and then iteratively traverses the graph forward and backward. The forward traversal identifies covering cones for each node (depth-optimal for critical and area-optimal for noncritical nodes), and the backward traversal then selects the covering set and updates the heights of remaining nodes. The benefit of iteration is to relax the need for delay-optimal implementation once updated heights mark nodes as no longer critical, allowing greater area improvement while maintaining depth optimality.

DAOmap from Chen and Cong [57] uses enumeration and iteration/relaxation techniques similar in spirit to IMAP, but additionally considers the potential node duplications during the cut enumeration procedure. There are, however, numerous differences in the details of the cost functions and heuristics. For example, DAOmap has a "cut probing" or lookahead step during the local cost adjustment phase on the backward traversal. DAOmap and IMAP would be the current best-known solutions for delay-minimum, area-minimal LUT technology mapping; however, there is no published comparison of the two showing equivalent netlist starting points.

Pan [58] and Pan and Lin [59] integrated technology mapping and re-timing. This not only provides the performance benefit of re-timing but, since registers are obstacles to efficient covering, can also contribute to area gains.

The Stratix II "adaptable LE" or ALM introduced by Hutton et al. [31] and shown in Figure 13.8(b) poses interesting new problems for technology mapping. The ALM structure can implement one 6-LUT, two 5-LUTs with 2 common inputs, or two independent 4-LUTs (among other combinations). The 6-LUTs are useful for depth reduction, but the most efficient use of the structure for area is the two 5-LUTs with sharing. This makes the area-cost function no longer a simple count of the number of covering LUTs, because the sharing needs to be accounted for. Dynamic programming approaches such as IMAP are more amenable to these modified LEs.

Recently technology-mapping algorithms have begun to address power [60]. Anderson and Najm [61] proposed modification to technology-mapping algorithms to minimize node duplication and to minimize the number of wires between LUTs (as a proxy for dynamic power required to charge up the global routing wires). EMAP by Lamoureaux and Wilton [62] modifies CutMAP with an additional cost-function component to favor cuts that reuse nodes already cut and those which have a high-activity factor (using probabilistic activity factors). Chen et al. [63] extend this type of mapping to a heterogeneous FPGA architecture with dual voltage supplies, where high-activity nodes additionally need to be routed to low-Vdd (low power) LEs and critical nodes to high-Vdd (high performance) LEs.

13.3.3.1 Gate Decomposition

An important preprocessing step for technology mapping is to decompose the network into a k-bounded (most often 2) network in preparation for LUT covering. The previously mentioned covering algorithms are only depth optimal for k-bounded networks: Cong and Hwang [64] showed that depth-optimal tech-mapping is NP-hard for unbounded networks when $k = 3$, and further for bounded networks when $k>4$.

Different decompositions will give very different mapping results [64]. There are multiple approaches to structural decomposition; tech_decomp [34] and DMIG [65] as part of SIS were used originally and Chortle contains a modified decomposition step, but the improved DOGMA algorithm [64,66] is currently the best-known structural decomposition technique for tech-mapping.

DOGMA combines the bin-packing step of Chortle-d with FlowMap's min-height k-feasible cut. The goal is to produce a 2-bounded network that will minimize depth in the final mapping solution, so at each step bin-nodes representing a level boundary node are used to store information. For each node v in topological order, labels are computed as follows: The sets of fan-in nodes previously labeled with value q (the stratum of depth q) are computed. Groups of ascending stratum q are packed into a minimum number of bins such that a k-feasible cut of height q–1 exists (using network flows), and then stored in bin-nodes. The bin-nodes are then packed in the $q + 1$ step into a minimal number of minimal height k-feasible bins. This continues until all nodes are packed into one bin corresponding to the node v. At the conclusion of the algorithm, the packing gives the decomposition (and additionally the labeling for the first step of FlowMap or CutMap) and the bin-nodes are removed.

Legl et al. [67] gave a method for technology mapping which combines functional decomposition and mapping, using Boolean techniques. This showed better results than DMIG + FlowMap, but has practical issues with size and computation time for the BDDs. Cong and Hwang [68] also applied partially dependent functional decomposition with technology mapping to target Xilinx 4000 CLBs, which are a block containing two 4-LUTs hard-wired into a 3:1 mux.

As a final note on technology mapping, some commercial FPGAs such as Altera's APEX family allow embedded RAM to be used as product-term logic or as combinational ROMs implementing large LUTs with multiple outputs. Wilton [12] and Lin [13] proposed pMapster as an algorithm for dynamically mapping to both LUT and pterm logic, and Cong and Xu [69] gave HeteroMap which covers the netlist using both LUTs and ROMs.

13.3.3.2 SPFD-Based Rewiring

An interesting recent development in FPGA synthesis is the use of SPFDs for exploiting the inherent flexibility in LUT-based netlists. *Sets of pairs of functions to be distinguished* (SPFDs) were proposed by Yamashita et al. [70]. SPFDs are a generalization of *observability don't care* (ODC) functions (see Chapter 2 on Logic Synthesis), wherein the on/off/dc set of functions is represented abstractly as a bipartite graph denoting "distinguishing" edges between min-terms, and a coloring of the graph gives an alternative implementation of the function. An inherent flexibility of LUTs in FPGAs is that they do not need to represent inverters, because these can always be absorbed by changing the destination node's LUT mask. By storing distinctions rather than functions, SPFDs generalize this to allow for more efficient expressions of logic.

Cong et al. [71,72] applied SPFD calculations to the problem of rewiring a previously tech-mapped netlist. The algorithm consists of precomputing the SPFDs for each node in the network, identifying a target wire (e.g., a delay-critical input of a LUT after tech-mapping), and then trying to replace that wire with another LUT output that satisfies its SPFD. The don't-care-sets in the SPFDs occur in the internal nodes of the network where flexibility exists in the LUT implementation after synthesis and tech mapping. Rewiring was shown to have benefits both for delay and area. Hwang et al. [73] and Kumthekar and Somenzi [74] also applied SPFD-based techniques for power reduction.

13.4 Physical Design

The physical design aspect of FPGA tools consists of clustering, placement, physical re-synthesis and routing. Commercial tools have additional preprocessing steps to allocate clock and reset signals to special low-skew "global networks," to place phase-locked loops, and to place transceiver blocks and I/O pins to meet the many electrical restrictions imposed on them by the FPGA device and the package. With the exception of some work on placing FPGA I/Os to respect electrical restrictions [75,76], however, these preprocessing steps are typically not seen in any literature.

FPGA physical design can broadly be divided into *routability* and *timing-driven* algorithms. Routability-driven algorithms seek primarily to find a legal placement and routing of the design by optimizing for reduced routing demand. In addition to optimizing for routability, timing-driven algorithms also use timing analysis to identify critical paths or connections, and attempt to optimize the delay of those connections. Since most of the delay in an FPGA is contributed by the programmable interconnect, timing-driven placement and routing can achieve a large circuit speed-up as compared to routability-driven approaches. For example, a recent commercial CAD system achieves an average of 50% higher design performance with full effort timing-driven placement and routing vs. routability-only placement and routing, at a cost of 5× run-time [77].

In addition to optimizing timing and routability, some recent FPGA physical design algorithms also implement circuits such that power is minimized.

13.4.1 Placement and Clustering

13.4.1.1 Problem Formulation

The placement problem for FPGAs differs from the placement problem for ASICs in several important ways. First, placement for FPGAs is a slot assignment problem — each circuit element in the technology-mapped netlist must be assigned to a discrete location, or slot, on the FPGA device which can accommodate it. Figure 13.1 showed the floorplan of a typical modern FPGA. An LE, for example, must be assigned to a location on the FPGA where an LE has been fabricated, while an I/O block or RAM block must be placed in a location where the appropriate resource exists on the FPGA. Second, there are usually a large number of constraints that must be satisfied by a legal FPGA placement. For example, groups of LEs that are placed in the same logic block have limits on the maximum number of distinct input signals and the number of distinct clocks they can use [14], and cells in carry chains must be placed together as a macro. Finally, all routing in FPGAs consists of prefabricated wires and transistor-based switches to interconnect them. Hence, the amount of routing required to connect two circuit elements, and the delay between them, is a function not just of the distance between the circuit elements, but also of the FPGA routing architecture. It also means that the amount of routing is strictly limited, and a placement that requires more routing in some region of the FPGA than exists there cannot be routed.

13.4.1.2 Clustering

A common adjunct to FPGA placement algorithms is a bottom-up clustering step that runs before the main placement algorithm in order to group related circuit elements together into clusters (LABs in Figure 13.2). Clustering reduces the number of elements to place, improving run-time of the main placement algorithm. In addition, the clustering algorithm usually deals with many of the complex FPGA legality constraints by grouping LEs into legal logic blocks, simplifying legality checking for the main placement algorithm.

The RASP system [45] includes one of the first logic block clustering algorithms. It performs maximum weighted matching on a graph where edge weights between LEs reflect the desirability of clustering them together. However, it has a high computational complexity of $O(n^3)$, where n is the number of LEs in the circuit, and this prevents it from scaling to very large problems. The VPack algorithm [78] clusters LEs into logic blocks by choosing a *seed* LE for a new cluster, and then greedily packing the LE with the highest *attraction* to the current cluster until no further LEs can be legally added to the cluster. The attraction function is the number of nets in common between an LE and the current cluster. VPack has computational complexity of $O(k_{max}n)$, where k_{max} is the maximum fanout of any net in the design. The T-VPack algorithm from Marquardt et al. [80] is a timing-driven enhancement of Vpack, where the attraction function for an LE, L, to cluster C becomes

$$\text{Attraction}(L) = 0.75\text{Crit}(L,C) + 0.25\frac{|\text{Nets}(L)\cap\text{Nets}(C)|}{\text{MaxNets}} \tag{13.1}$$

The first term gives higher attraction to LEs which are connected to the current cluster by timing-critical connections, while the second term is taken from VPack and favors grouping LEs with many common signals together. Surprisingly, T-VPack improves not only circuit speed vs. VPack but also routability, by absorbing more connections within clusters. The iRAC [81] clustering algorithm achieves further reductions in the amount of routing necessary to interconnect the logic blocks by using attraction functions that favor the absorption of small nets within a cluster.

Lamoureaux and Wilton [62] developed a power-aware modification of T-VPack that adds a term to the attraction function of Equation (13.1), such that LEs connected to the current cluster by connections with a high rate of switching have a larger attraction to the cluster. This favors the absorption of nets that frequently switch logic states, resulting in lower capacitance for these nets, and lower dynamic power. Chen and Cong [82] developed a clustering algorithm that reduces power in an FPGA architecture where each logic block can be run at either a high or a low voltage. Their algorithm groups non-timing-critical LEs into different clusters from timing-critical LEs, enabling many logic blocks to run at reduced voltage and hence reduced power.

13.4.1.3 Placement

Although the literature includes numerous techniques, simulated annealing is the most widely used placement algorithm for FPGAs. Figure 13.11 shows the basic flow of simulated annealing. An initial placement is generated, and a placement perturbation is proposed by a *move generator*, generally by moving a small number of circuit elements to new locations. A *cost function* is used to evaluate the impact of each proposed move. Moves that reduce cost are always accepted, or applied to the placement, while those that increase cost are accepted with probability $e^{-\Delta \mathrm{Cost}/T}$, where T is the current *temperature*. Temperature starts at a high level, and gradually decreases throughout the anneal, according to the *annealing schedule*. The annealing schedule also controls how many moves are performed between temperature updates, and when the *ExitCriterion* that terminates the anneal is met.

Two key strengths of simulated annealing that many other approaches lack are:

1. It is possible to enforce all the legality constraints imposed by the FPGA architecture in a fairly direct manner. The two basic techniques are to forbid the creation of illegal placements in the move generator, or to add a penalty cost to illegal placements.
2. By creating an appropriate cost function, it is possible to directly model the impact of the FPGA routing architecture on circuit delay and routing congestion.

VPR [8,79,80] contains a timing-driven simulated-annealing placement algorithm as well as timing-driven routing. The VPR placement algorithm is usually used in conjunction with T-VPack, which preclusters the LEs into legal logic blocks. The placement annealing schedule is based on monitoring statistics generated during the anneal, such as the fraction of proposed moves that are accepted. This adaptive annealing schedule allows VPR to automatically adjust to different FPGA architectures. VPR's cost function also automatically adapts to different FPGA architectures [80]:

$$\mathrm{Cost} = (1-\lambda) \sum_{i \in \mathrm{AllNets}} q(i) \left[\frac{bb_x(i)}{C_{\mathrm{av},x}(i)} + \frac{bb_y(i)}{C_{\mathrm{av},y}(i)} \right] + \lambda \sum_{j \in \mathrm{AllConnection}} \mathrm{Criticality}(j) \cdot \mathrm{Delay}(j) \quad (13.2)$$

```
P = InitialPlacement ();
T = InitialTemperature ();

while (ExitCriterion () == False) {
    while (InnerLoopCriterion () == False) {  /* "Inner Loop" */
        P_new = PerturbPlacementViaMove (P);
        ΔCost = Cost (P_new) – Cost (P);
        r = random (0,1);
        if (r < e^(-ΔCost/T)) {
            P = P_new;      /* Move Accepted */
        }
    } /* End "Inner Loop" */
    T = UpdateTemp (T);
}
```

FIGURE 13.11 Pseudo-code of a generic simulated-annealing placement algorithm.

The first term in Equation (13.2) causes the placement algorithm to optimize an estimate of the routed wirelength, normalized to the average wiring supply in each region of the FPGA. The wirelength needed to route each net i is estimated as the bounding box span (bb_x and bb_y) in each direction, multiplied by a fanout-based correction factor, $q(i)$. In FPGAs with differing amounts of routing available in different regions or channels, it is beneficial to move wiring demand to the more routing-rich regions, so the estimated wiring required is divided by the average routing capacity over the bounding box in the appropriate direction.

The second term in Equation (13.2) optimizes timing by favoring placements in which timing-critical connections have the potential to be routed with low delay. To evaluate the second term quickly, VPR needs to be able to estimate quickly the delay of a connection. To accomplish this, VPR assumes that the delay is a function only of the difference in the coordinates of a connection's endpoints, $(\Delta x, \Delta y)$, and invokes the VPR router with each possible $(\Delta x, \Delta y)$ to determine a table of delays vs. $(\Delta x, \Delta y)$ for the current FPGA architecture before the simulated-annealing algorithm begins. The criticality of connections is determined via periodic timing analysis using delays computed from the current placement.

Many enhancements to the original VPR algorithm have been made and published. The PATH algorithm from Kong [84] uses a new timing-criticality formulation in which the timing criticality of a connection is a function of the slacks of all paths passing through it, rather than just a function of the worst-case (smallest) slack of any path through that connection. This technique significantly improves timing optimization, and results in circuits with 15% smaller critical path delay, on average. The SCPlace algorithm [83] enhances VPR so that a portion of the moves are *fragment moves* in which a single logic element is moved, instead of an entire logic block. This allows the placement algorithm to modify the initial clustering, and improves both circuit timing and wirelength. Lamoureaux and Wilton [62] modified VPR's cost function by adding a third term, PowerCost, to Equation (13.2):

$$\text{PowerCost} = \sum_{i \varepsilon \text{AllNets}} q(i)\left[bb_x(i) + bb_y(i)\right]\text{Activity}(i) \tag{13.3}$$

where Activity(i) represents the average number of times net i transitions per second. This additional cost function term reduces circuit power, although the gains are less than those obtained by power-aware clustering.

PROXI [85] uses simulated annealing for placement, but its cost function is based not on fast heuristics to estimate placement routability and timing, but instead on maintaining at least a partially routed design at all times. The PROXI cost function is a weighted sum of the number of unrouted nets and the delay of the circuit critical path. After each placement perturbation, all nets connected to moved cells are ripped-up and rerouted via a fast, directed-search maze router. To keep the CPU time tolerable, PROXI allows the maze router to explore only a small portion of the routing fabric at high temperatures — if no unblocked routing path is found quickly, the net is marked as unrouted. At lower temperatures, the placement is of higher quality and the router is allowed to explore a larger portion of the graph. After each net is re-routed, the critical path is recomputed incrementally. PROXI produces high quality results, but requires relatively high CPU time.

Sankar and Rose [86] seek the opposite tradeoff of reduced result quality for extremely low placement run-times. They create a hierarchical annealer that employs greedy clustering with a net-absorption-based attraction function to reduce the size of the placement problem. The best run-time quality tradeoff occurs when they cluster the circuit logic blocks twice — first clustering into level 1 clusters of approximately 64 logic blocks, and then clustering four of these level 1 clusters into each level 2 cluster. The level 2 clusters are placed with a greedy (temperature = 0) anneal seeded by a fast constructive initial placement. Next each level 1 cluster is initially placed within the boundary of the level 2 cluster that contained it, and another temperature = 0 anneal is performed. Finally, the placement of each logic block is refined with a low-temperature anneal. For very fast CPU times, this algorithm significantly outperforms VPR in terms of achieved wirelength, while for longer permissible CPU times, it lags VPR.

Another popular placement approach for FPGAs is recursive partitioning. ALTOR [87] was originally developed for standard cell circuits, but was adapted to FPGAs and widely used in FPGA research. ALTOR employs a recursive min-cut bi-partitioning technique with terminal propagation [88] to gradually partition the design into small portions of the FPGA floorplan, at which point a complete placement is obtained. Figure 13.12 shows the sequence of cut lines used by ALTOR to partition the FPGA area. This sequence of cut lines means ALTOR assumes that terminals that have a small Manhattan distance between them can be connected efficiently by the FPGA routing fabric. This assumption matches the capabilities of the segmented routing architectures used by most modern commercial FPGAs.

An approach by Maidee et al. [89] is also based on recursive bi-partitioning, but it adds timing-driven features. Before partitioning begins, the VPR routing algorithm is used to generate a table of net delay vs. distance spanned by the net that takes into account the FPGA routing architecture. As partitioning proceeds, the algorithm records the minimum length each net could achieve, given the current number of portioning boundaries it crosses. The delay corresponding to each net's span is retrieved from the precalculated table, and a timing analysis is performed to identify critical connections. Timing-critical connections to terminals outside of the region being partitioned act as anchor points during each partitioning. This forces the other end of the connection to be allocated to the partition that allows the critical connection to be short. Once partitioning has proceeded to the point that each region contains only a few cells, any overfilled regions are legalized with a greedy movement heuristic. Finally, the placement is further optimized by using VPR to perform a low-temperature anneal. The technique achieves wirelength and speed results comparable to VPR, with significantly reduced CPU time.

A commercial recursive partitioning placement algorithm for the Altera Apex 20K family is described in [90]. Apex has a hierarchical routing architecture, making it well suited to partitioning-based placement. Recursive partitioning is conducted along the natural cut lines formed by the various hierarchy levels of the routing architecture, as shown in Figure 13.13. Notice that the sequence of partitions in this algorithm is significantly different than that of ALTOR, showing the large impact an FPGA's routing architecture has on placement algorithms. This algorithm is made timing driven by weighting connections with low slack highly during each partitioning phase, to encourage partitioning solutions in which such connections can be routed using only fast, lower-hierarchy-level routing. To improve the prediction of the critical path, the delay estimate for each connection is a function both of the known

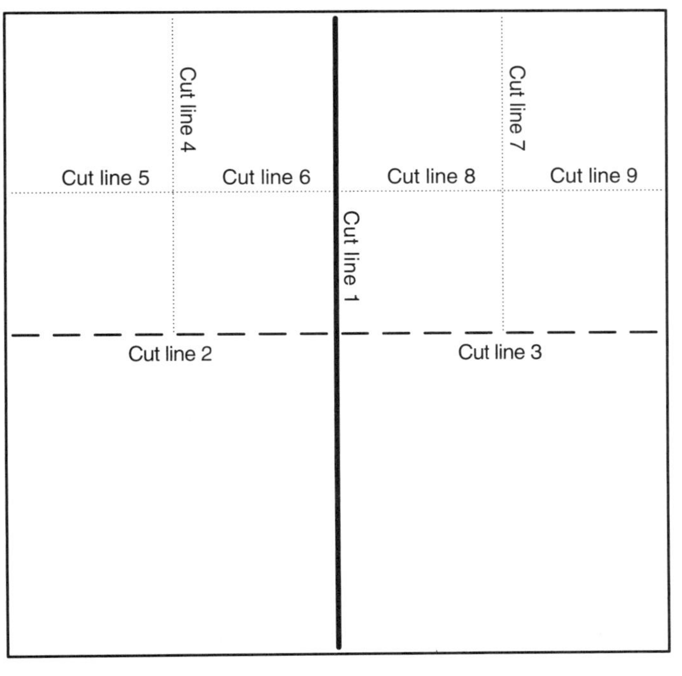

FIGURE 13.12 ALTOR partitioning sequence.

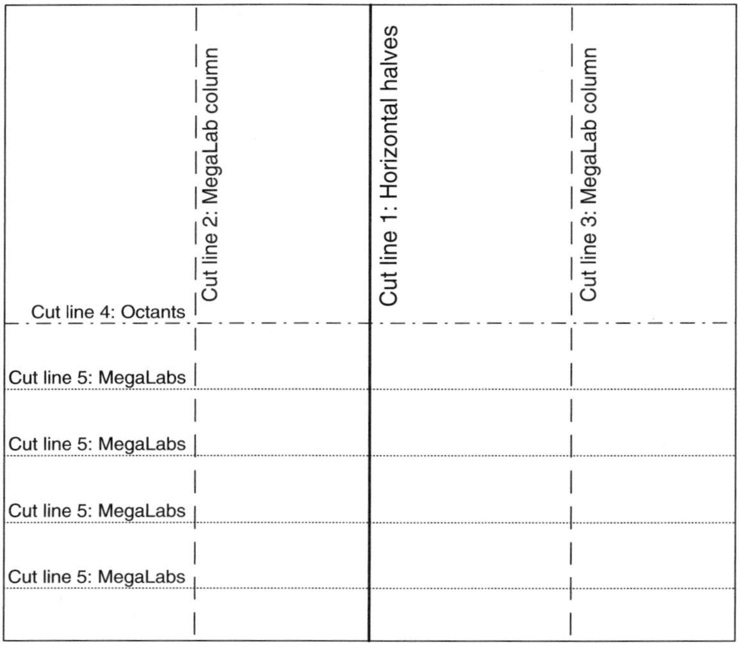

FIGURE 13.13 Sequence of cut lines for APEX architecture recursive partitioning placement.

number of hierarchy boundaries the net must traverse due to partitionings at the higher levels of the routing hierarchy, and statistical estimates of how many hierarchy boundaries the connection will cross at future partitioning steps.

Analytic algorithms are the third major approach to the placement problem. Analytic algorithms are based on creating a convex function that approximates wirelength. Finding the global minimum of this function yields a placement which has good wirelength if the function approximated routed wirelength well. However, this global minimum is usually an illegal placement solution, so constraints and heuristics must be applied to guide the algorithm to a legal solution.

While analytic placement approaches are popular for ASICs, there are few analytic FPGA algorithms, probably due to the more difficult legality constraints in FPGA placement. The Negotiated Analytic Placement (NAP) algorithm from Chan and Schlag [91] combines global analytic placement [92] to determine wirelength optimized, but overlapping cell locations with a negotiated congestion algorithm that gradually reduces overuse of LE locations until a legal placement is achieved. NAP also includes features that make it suitable for parallelization across multiple processors. First, the circuit netlist is covered by a set of sub-netlists, each of which is a tree. The placement of each tree can then proceed in parallel, which results in those cells contained in multiple trees being placed in multiple locations. Extra edges between copies of the same cell to the center of gravity of the cell are added to the edge-weight matrix to gradually pull copies of the cells together. After every iteration of analytic placement, a negotiated congestion algorithm is invoked to spread out the cells within each region of the placement. After a sufficient number of analytic placement/negotiated congestion movement iterations, an overlap-free placement is obtained.

13.4.2 Physical Synthesis Optimizations

Timing visibility can be poor during FPGA synthesis. What appears to be a noncritical path can turn out to be critical after placement and routing. This is true for ASICs as well, but the problem is especially acute for FPGAs. Unlike an ASIC implementation, FPGAs have a predefined logic and routing fabric, and hence cannot use drive-strength selection, wire sizing or buffer insertion to increase the speed of long routes.

Recently, physical synthesis techniques have arisen both in the literature and in commercial tools. Physical synthesis techniques for FPGAs generally refer either to resynthesis of the netlist once some approximate placement has occurred and thus some visibility of timing exists, or to local modifications to the netlist during placement itself. Figure 13.14 highlights the difference between the two styles of physical synthesis flow. The "iterative" flow of Figure 13.14(a) iterates between synthesis and physical design. A positive of this flow is that the synthesis tool is free to make large-scale changes to the circuit implementation, but a negative is that the placement and routing of this new design may not match the synthesis tool expectations, and hence the loop may not converge well. The "incremental" flow of Figure 13.14(b) instead makes only more localized changes to the circuit netlist, such that it can integrate these changes into the current placement with only minor perturbations. This flow has the advantage that convergence is easier, since a legal or near-legal placement is maintained at all times, but it has the disadvantage that it is more difficult to make large-scale changes to the circuit structure.

Commercial tools from Synplicity and Mentor Graphics [17] largely follow the "iterative" flow, and re-synthesize a netlist given output from the FPGA vendor place and route tool. However, these tools can also provide constraints to the place and route tool in subsequent iterations to assist convergence. Lin et al. [93] described a similar academic flow in which re-mapping is performed after either a placement estimate or after actual placement delays are known. Suaris et al. [94] used timing budgets for resynthesis, where the budget is calculated using a quick layout of the design. This work also makes modifications to the netlist to facilitate re-timing in the re-synthesis step. In a later improvement [95] this flow was altered to modify incrementally the placement after each netlist transform, assisting convergence.

There are commercial and academic examples of the "incremental" physical synthesis flow as well. Altera's Quartus CAD system tightly integrates the physical synthesis and placement engines. Schabas and Brown [96] used logic duplication as a postprocessing step at the end of placement, with an algorithm that simultaneously duplicates logic and finds legal and optimized locations for the duplicates. Logic

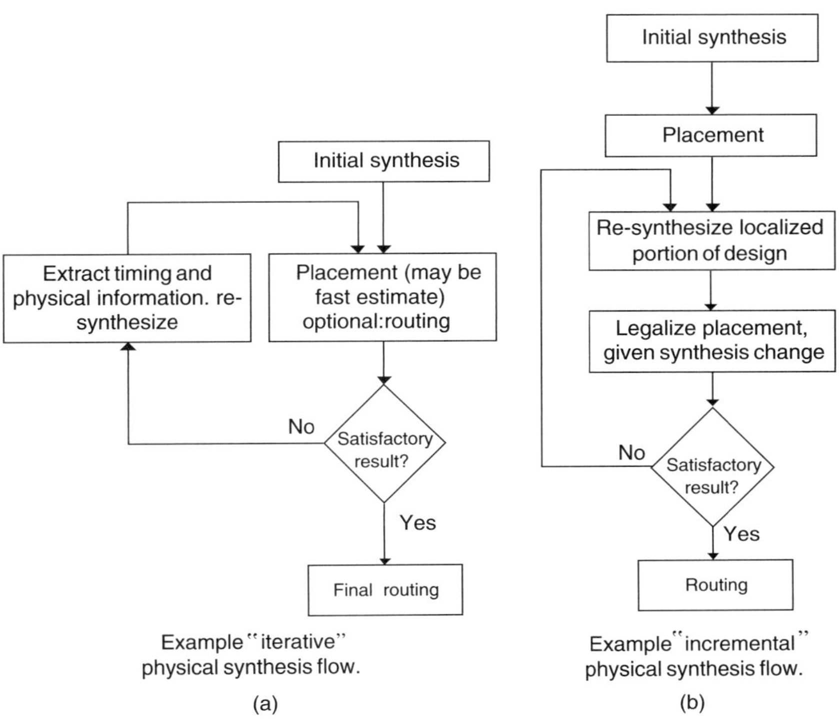

FIGURE 13.14 Physical synthesis flows: (a) example of "iterative" physical synthesis flow; (b) example of "incremental" physical synthesis flow.

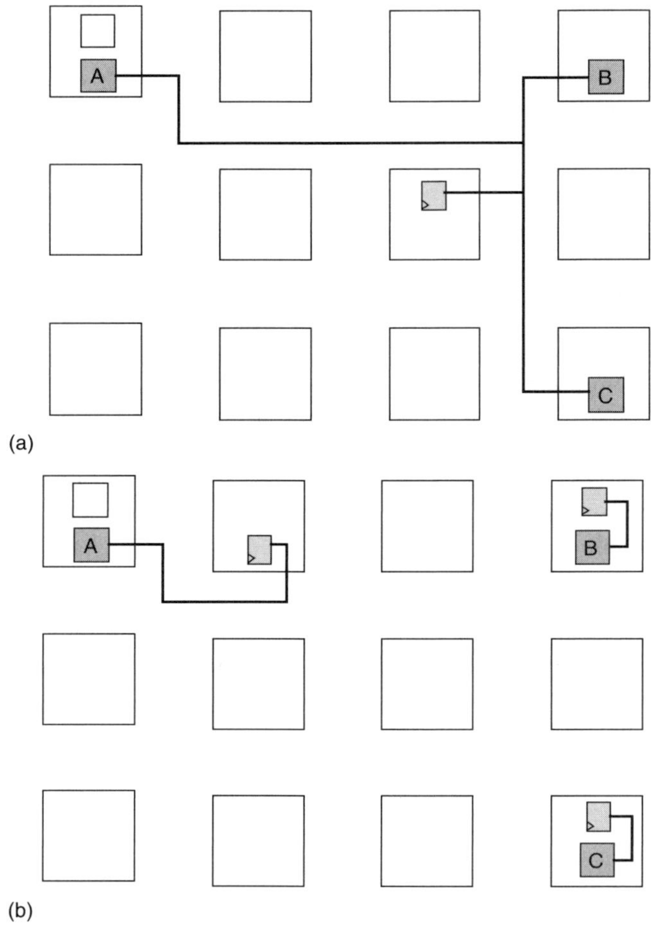

FIGURE 13.15 Duplicating registers to optimize timing in physical synthesis: (a) register with three time-critical output connections; (b) three register duplicates created and legally placed to optimize timing.

duplication, particularly on high-fanout registers, allows significant relaxation on placement critical paths because it is common for a multi-fanout register to be "pulled" in multiple directions by its fanouts, as shown in Figure 13.15. Chen and Cong [97] integrated duplication throughout a simulated-annealing-based placement algorithm. Before each temperature update, logic duplicates are created and placed if it will assist timing, and previously duplicated logic may be "unduplicated" if the duplicates are no longer necessary.

Manohararajah et al. [98,99] performed local restructuring of timing-critical logic to shift delay from the critical path to less critical paths. A LUT and a timing-critical portion of its fanin are considered for functional decomposition or BDD-based resynthesis. An incremental placement algorithm then integrates any changed or added LUTs into a legal placement. Ding et al. [100] gave an algorithm for postplacement pin permutation in LUTs. This algorithm re-orders LUT inputs to take advantage of the fact that each input typically has a different delay. This algorithm also swaps inputs amongst several LUTs that form a logic cone in which inputs can be legally swapped , such as an and-tree or xor-tree. An advantage of this algorithm is that no placement change is required, since only the ordering of inputs is affected, and no new LUTs are created.

Singh and Brown [103] present a postplacement retiming algorithm. This algorithm initially places added registers and duplicated logic at the same location as the original logic. It then invokes an incremental placement algorithm to legalize the placement. This incremental placement algorithm is similar to an annealing algorithm, but it includes costs for various types of resource overuse as well as timing and wiring costs, and it accepts only moves that reduce cost (i.e., temperature = 0). A later improvement [104] altered the retiming algorithm so that it incrementally modifies the design using local re-timing operations, each of which is separately legalized before moving onto the next operation. This simplifies the legalization of each modification, and saves compile time.

In an alternative to re-timing, Singh and Brown [101] employed unused PLLs and global clocking networks to create several shifted versions of a clock, and developed a postplacement algorithm which selected whichever time-shifted clock resulted in the best timing performance for each register. This approach is conceptually similar to re-timing after placement, but involves shifting clock edges at registers rather than moving registers across combinational logic. Chao-Yang and Marek-Sadowska [102] extended this beneficial clock skew timing optimization to a proposed FPGA architecture where clocks can be delayed via programmable delay elements on the global clock distribution networks.

13.4.3 Routing

13.4.3.1 Problem Formulation

All FPGA routing consists of prefabricated metal wires and programmable switches to connect the wires to each other and to the circuit element input and output pins. Figure 13.16 shows an example of FPGA routing architecture. In this example, each routing channel contains four wires of length 4 — wires that span four logic blocks before terminating — and one wire of length 1. In the example of Figure 13.16, the

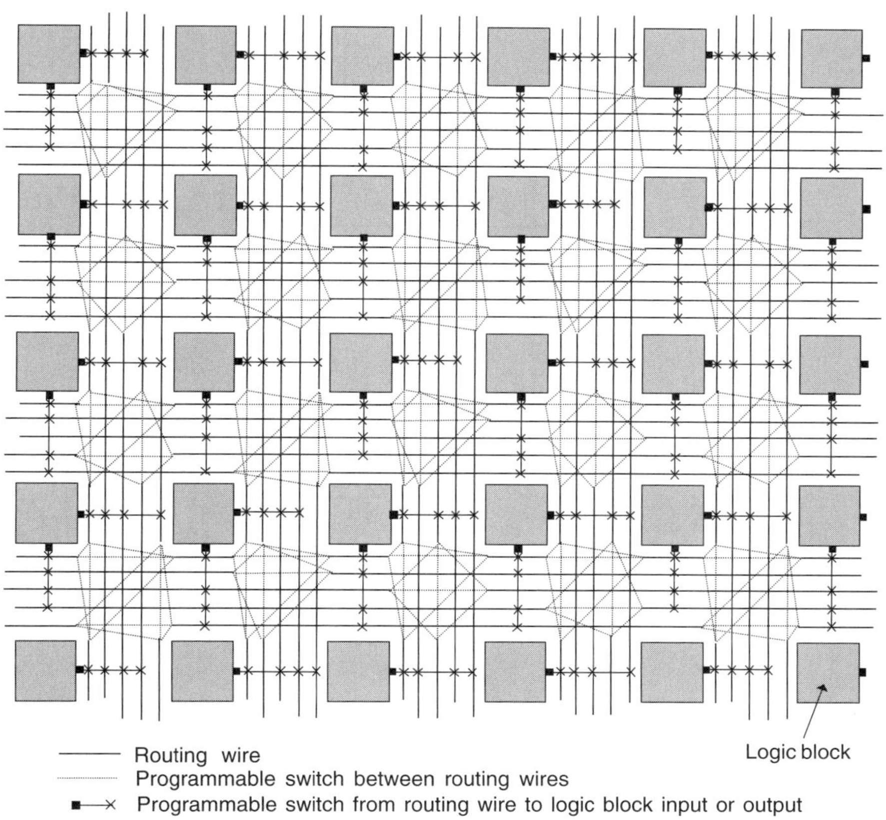

FIGURE 13.16 Example of FPGA routing architecture.

programmable switches allow wires to connect only at their endpoints, but many FPGA architectures also allow programmable connections from interior points of long wires as well.

Usually the wires and the circuit element input and output pins are represented as nodes in a *routing-resource graph*, while programmable switches that allow connections to be made between the wires and pins become directed edges. Programmable switches can be fabricated as pass transistors, tri-state buffers, or multiplexers. Multiplexers are the dominant forms of programmable interconnect in recent FPGAs such as the Altera Stratix [14,17] and Xilinx Virtex [17] families, since multiplexer-based routing produces FPGAs with a superior area-delay product [14,106]. Figure 13.17 shows how a small portion of an FPGA's routing is transformed into a routing-resource graph. This graph can also efficiently store information on which pins are logically equivalent and hence may be swapped by the router, by including *source* and *sink* nodes that connect to all the pins which can perform a desired function. It is common to have many logically equivalent pins in commercial FPGAs — for example, all the inputs to a LUT are logically equivalent, and may be swapped by the router. A legal routing of a design consists of a tree of routing-resource nodes for each net in the design such that (1) each tree electrically connects the net source to all the net sinks, and (2) no two trees contain the same node, as that would imply a short between two signal nets.

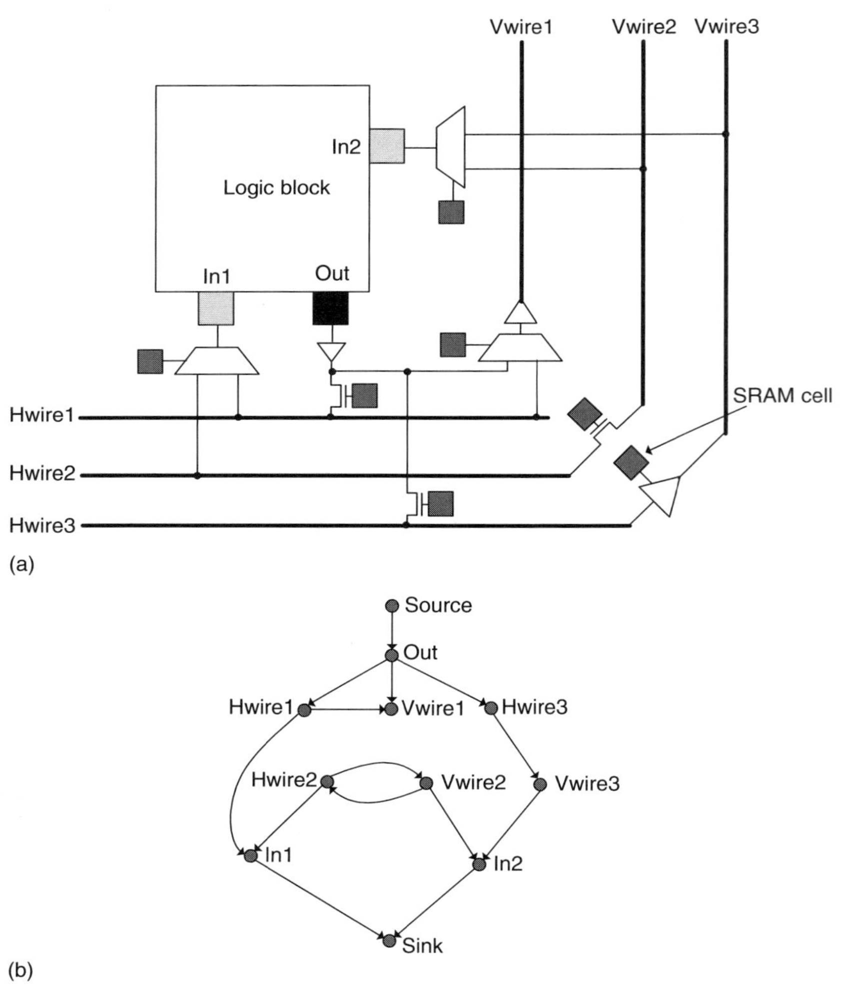

FIGURE 13.17 Transforming FPGA routing circuitry to a routing-resource graph: (a) example of FPGA routing circuitry; (b) equivalent routing-resource graph.

Since the number of routing wires in an FPGA is limited, and the limited number of programmable switches also creates many constraints on which wires can be connected to each other, congestion detection and avoidance is a key feature of FPGA routers. Since most delay in FPGAs is due to the programmable routing, timing-driven routing is important to obtain the best speed as well.

13.4.3.2 Two-Step Routing

Some FPGA routers operate in two sequential phases as shown in Figure 13.18. First, a global route for each net in the design is determined using channeled global routing algorithms that are essentially the same as those for ASICs [107]. The output of this stage is the series of channel segments through which each connection should pass. Next a detailed router is invoked to determine exactly which wire segment should be used within each channel segment. The CGE [108] and SEGA [109] algorithms find detailed routes by employing different levels of effort in searching the routing graph. A search of only a few routing options is conducted first in order to quickly find detailed routes for nets that are in un-congested regions, while a more exhaustive search is employed for nets experiencing routing difficulty. An alternative approach by

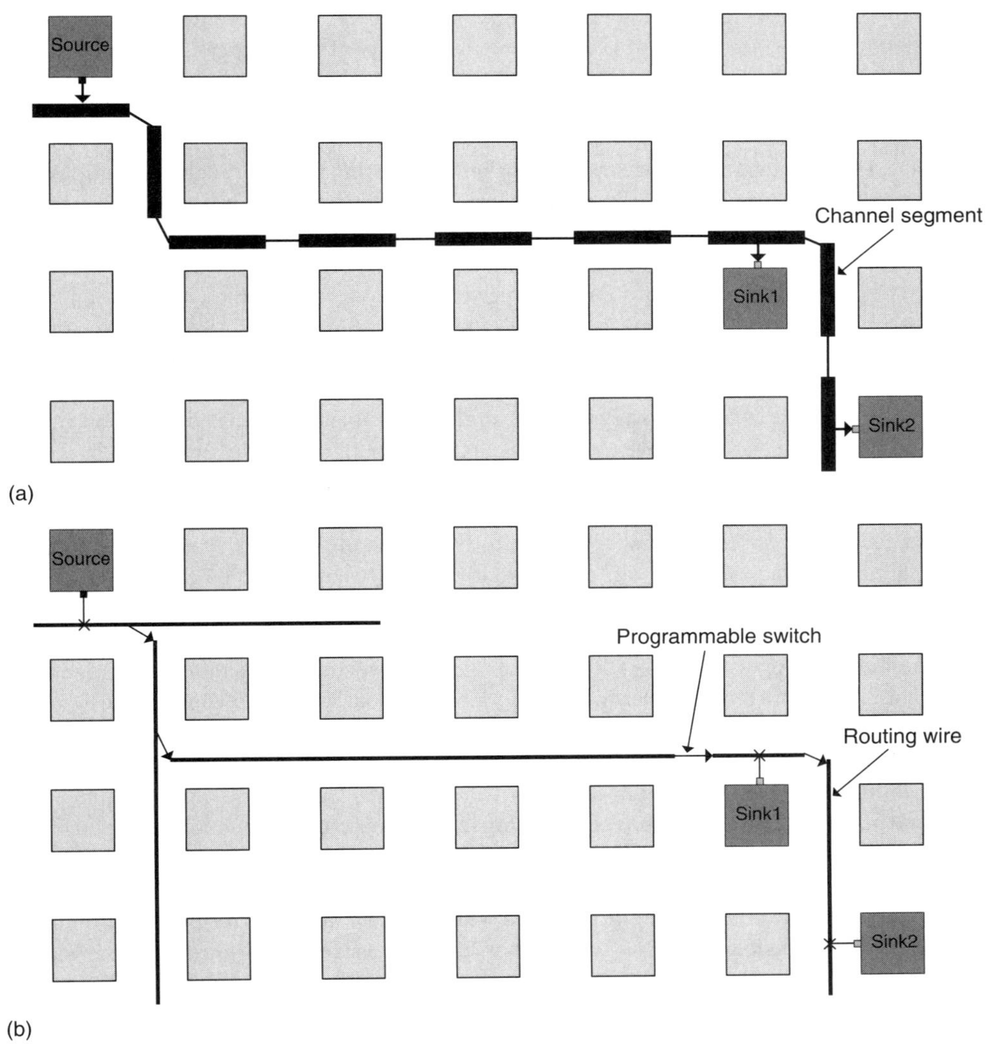

FIGURE 13.18 Two-step FPGA routing flow. (a) Step 1: global routing chooses a set of channel segment for a net; (b) step 2: detailed routing wires within each channel segment and switches to connect them.

Nam et al. [110] formulates the FPGA detailed routing problem as a Boolean satisfiability problem. This approach guarantees that a legal detailed routing (that obeys the current global routing) will be found if one exists. However, the CPU time can be high for large problems. None of these detailed routers use timing analysis to determine timing-critical nets and optimize their delay. SEGA attempts to minimize the delay of all nets equally while the other algorithms are purely routability driven.

The divide-and-conquer approach of two-step routing reduces the problem space for both the global and detailed routers, helping to keep their CPU times down. However, the flexibility loss of dividing the routing problem into two phases in this way can result in significantly reduced result quality. The global router optimizes only the wirelength of each route, and attempts to control congestion by trying to keep the number of nets assigned to a channel segment comfortably below the number of routing wires in that channel segment. However, the fact that FPGA wiring is prefabricated and can be interconnected only in limited patterns makes the global router's view of both wirelength and congestion inaccurate. For example, a global route one-logic-block long may require the detailed router to use a wire that is four logic-blocks long to complete actually the connection, wasting wire, and increasing delay. Figure 13.18 highlights this behavior; the global route requires nine units of wire, but the final wires used in the detailed routing of the net are 13 wiring units long in total. Similarly, a global route where the number of nets assigned to each channel segment is well below the capacity of each segment may still fail detailed routing because the wiring patterns may not permit this pattern of global routes.

13.4.3.3 Single-Step Routers

Most modern FPGA routers are *single-step* routers that find routing paths through the routing-resource graph in a single, unified search algorithm. These routers differ primarily in their costing of various routing alternatives, their search technique through the routing-resource graph, and their congestion-resolution techniques.

Lee and Wu [111] introduced a tracer algorithm, which routes each net using a multiple-component growth algorithm that is an extension of a traditional maze router [112]. This multiple-component growth algorithm tries to find a minimum wirelength routing tree for each net. Multiple nets are allowed to use the same routing-resource node, but the cost of such an overused node is ten times the normal cost. If some routing-resource nodes are overused once all nets are routed, a simulated evolution rip up and retry scheme is invoked. Nets that have routing lengths longer than minimum, or that contain overused nodes, are more likely to be ripped up — however, every net has some chance of being ripped-up and re-routed, since the routing of any net may indirectly cause routing congestion. Once a legal route is achieved, further rip-up and retry iterations are performed to improve timing. Part of the rip-up criteria considers connection slack — nets with either negative slack (which could be routed faster) or large positive slack (which could take a more circuitous route to free up resources for other nets) are more likely to be selected for rip-up.

The Limit Bumping Algorithm (LBA) of Frankle [113] considers timing in a more direct manner. First, a timing analysis is performed to determine a worst-case path slack for each connection. Next, a slack allocator is invoked to convert these path slacks into an upper delay limit, $U(c)$, on the permitted delay for each connection, c. The LBA ensures that each of these $U(c)$ is larger than a lower bound, $L(c)$, on the achievable routing delay for each connection given the placement and FPGA routing architecture, so there is some possibility of routing success. A routing in which each connection is routed with delay less than its upper delay limit will meet all maximum delay-timing constraints on paths. The LBA routes connections in decreasing order of $L(c)/U(c)$, so connections with tighter delay limits compared to what is achievable go first. Each connection must be routed with delay less than $U(c)$ or it is left unrouted. Once an attempt has been made to route all connections, the $U(c)$ values of all unrouted connections are increased by 20%, and routing is retried.

The PathFinder algorithm by McMurchie and Ebeling [105] introduced the concept of *negotiated congestion routing*. The negotiated congestion technique now underlies many FPGA routers, including those with the best routability results on a set of standard academic FPGA benchmarks. In a negotiated congestion router, each connection is initially routed to minimize some metric such as delay or wirelength, with little regard to congestion or overuse of routing resources. After each *routing iteration*, in which every net in the circuit is ripped-up and re-routed, the cost of congestion is increased such that it is less likely that overused nodes will occur in the next routing iteration. Over the course of many routing iterations, the increasing cost of congestion gradually forces some nets to accept suboptimal routing in order to resolve congestion and achieve a legal routing.

The congestion cost of a node is

$$\text{CongestionCost}(n) = [b(n) + h(n)]p(n) \tag{13.4}$$

where $b(n)$ is the base cost of the node, $p(n)$ the present congestion of the node, and $h(n)$ the historical cost of the node. The base cost of a node could be its intrinsic delay, its length, or simply 1 for all nodes. The present congestion cost of a node is a function of the overuse of the node and the routing iteration. For nodes that are not currently overused, $p(n)$ is 1. In early routing iterations, $p(n)$ will be only slightly higher than 1 for nodes that are overused, while in later routing iterations, to ensure congestion is resolved, $p(n)$ becomes very large for overused nodes. $h(n)$ maintains a congestion history for each node. $h(n)$ is initially 0 for all nodes, but is increased by the amount of overuse on node n at the end of each routing iteration. The incorporation of not just the present congestion, but also the entire history of congestion of a node, into the cost of that node is a key innovation of negotiated congestion. Historical congestion ensures that nets which are "trapped" in a situation where all their routing choices have present congestion can see which choices have been overused the most in the past. Exploring the least historically congested choices ensures that new portions of the solution space are being looked at, and resolves many cases of congestion that the present congestion cost term alone cannot resolve.

In the PathFinder algorithm, the complete cost of using a routing-resource node n in the routing of a connection c is

$$\text{Cost}(n) = [1\text{-Crit}(c)]\text{CongestionCost}(n) + \text{Crit}(c)\text{Delay}(n) \tag{13.5}$$

The criticality is the ratio of the connection slack to the longest delay in the circuit

$$\text{Crit}(c) = \frac{\text{Slack}(c)}{D_{\text{max}}} \tag{13.6}$$

The total cost of a routing-resource node is therefore a weighted sum of its congestion cost and its delay, with the weighting being determined by the timing criticality of the connection being routed. This formulation results in the most timing-critical connections receiving delay-optimized routes, with nontiming-critical connections using routes optimized for minimal wirelength and congestion. Since timing-critical connections see less cost from congestion, these connections are also less likely to be forced off their optimal routing paths due to congestion — instead, nontiming-critical connections will be moved out of the way of timing-critical connections.

The VPR router [8,79] is based on the PathFinder algorithm, but introduces several enhancements. First, instead of assuming that each routing-resource node has a constant delay, the VPR router models the delay of a route with the more accurate Elmore delay [114], and directly optimizes this delay. The original Elmore delay is only capable of modeling linear networks of resistors and capacitors, but by using linearized RC models of FPGA routing switches, it can be applied to FPGA routing and yields good accuracy.

Second, instead of using a breadth-first search, or an A^* search through the routing-resource graph to determine good routes, VPR uses a more aggressive directed search technique. This directed search sorts each routing-resource node, n, found during graph search toward a sink, j, by a total cost given by

$$\text{TotalCost}(n) = \text{PathCost}(n) + \alpha \cdot \text{ExpectedCost}(n,j) \tag{13.7}$$

Here PathCost(n) is the known cost of the routing path from the connection source to node n, while ExpectedCost(n,j) is a prediction of the remaining cost that will be incurred in completing the route from node n to the target sink. The "directedness" of the search is controlled by α. An $\alpha = 0$ results in a breadth-first search of the graph, while $\alpha > 1$ makes the search more efficient, but may result in suboptimal routes. An $\alpha = 1.2$ leads to improved CPU time without a noticeable reduction in result quality.

The VPR router also uses an alternative form of Equation (13.4) that more easily adapts to different FPGA architectures (see [8,79] for details, and for a discussion of how best to set $b(n)$, $h(n)$, and $p(n)$ to achieve good results in reasonable CPU times).

An FPGA router based on negotiated congestion, but designed for very low CPU times, is presented by Swartz et al. [115]. This router achieves very fast run-times through the use of an aggressive directed search during routing graph exploration, and by using a "binning" technique to speed the routing of high-fanout nets. When routing the kth terminal of a net, most algorithms examine every routing-resource node used in routing the previous k–1 terminals. For a k-terminal net, this results in an $O(k^2)$ algorithm which becomes slow for large k. By examining only the portion of the routing of the previous terminals that is in a "bin" near the sink for connection k, the algorithm achieves a significant CPU reduction.

Wilton [116] developed a crosstalk-aware FPGA routing algorithm. This algorithm enhances the VPR router by adding an additional term to the routing cost function that penalizes routes in proportion to the amount of delay they will add to neighboring routes due to crosstalk, weighted by the timing criticality of those neighboring routes. Hence, this router achieves a circuit speed-up by leaving routing tracks near those used by critical connections vacant.

Lamoureaux and Wilton [62] enhanced the VPR router to optimize power by adding a term to the routing node cost, Equation (13.5), that includes the capacitance of a routing node multiplied by the switching activity of the net being routed. This drives the router to achieve low-energy routes for rapidly toggling nets.

The Routing Cost Valleys (RCV) algorithm [117] combines negotiated congestion with a new routing cost function and an enhanced slack allocation algorithm. The RCV is the first FPGA routing algorithm that optimizes not only long-path timing constraints which specify the delay on a path must be less than some value, but also addresses the increasing importance of short-path timing constraints, which specify that the delay on a path must be greater than some value. Short-path timing constraints arise in FPGA designs as a consequence of hold-time constraints within the FPGA, or of system-level hold-time constraints on FPGA input pins and system-level minimum clock-to-output constraints on FPGA output pins. To meet short-path timing constraints, RCV will intentionally use slow or circuitous routes to increase the delay of a connection.

The RCV algorithm allocates both short- and long-path slack to determine a pair of delay budgets, $D_{\text{Budget,Min}}(c)$ and $D_{\text{Budget,Max}}(c)$, for each connection, c, in the circuit. A routing of the circuit in which every connection has delay between $D_{\text{Budget,Min}}(c)$ and $D_{\text{Budget,Max}}(c)$ will satisfy all the long- and short-path timing constraints. Such a routing may not exist however, so it is advantageous for connections to seek not simply to achieve a delay in the window between the two delay budgets, but instead to try to achieve a target delay $D_{\text{Target}}(c)$ near the middle of the window. The extra timing margin achieved by this connection may allow another connection to have a delay outside its delay budget window, without violating any of the path-based timing constraints. Figure 13.19 shows the form of the RCV routing cost function compared to that of the original PathFinder algorithm. The RCV alogrithm strongly penalizes routes that have delays outside the delay budget window, and weakly guides routes to achieve D_{Target}. RCV achieves superior results on short-path timing constraints and also outperforms PathFinder in optimizing traditional long-path timing constraints.

13.5 Looking Forward

In this chapter we have surveyed the current algorithms for FPGA synthesis, placement, and routing. Some of the more recent publications in this area point to the growth areas in CAD tools for FPGAs. The relatively few papers on power modeling and optimization algorithms are likely just the beginning, as lower process geometries force tools to be more and more aware of power effects. Timing modeling for FPGA interconnect will need to take into account variation, min–max analysis, crosstalk, and other physical effects that have been largely ignored to date, and incorporate these into the optimization algorithms. Timing estimation and physical synthesis approaches will likely contribute to improve performance in the

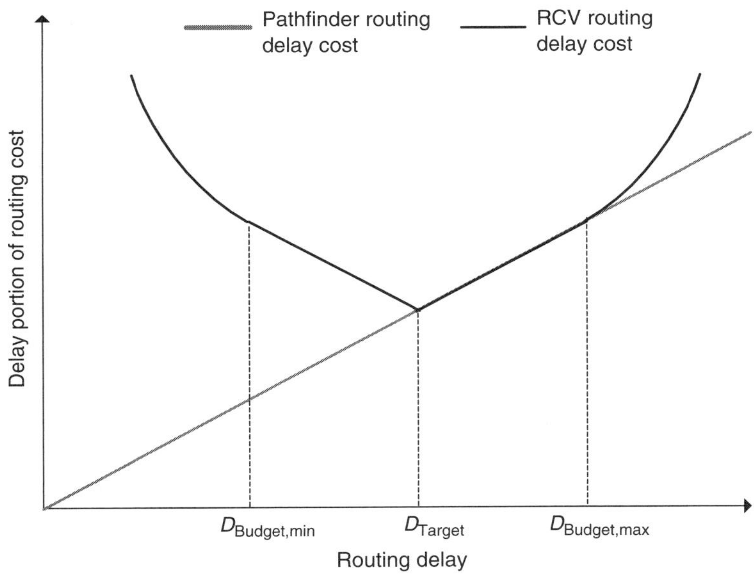

FIGURE 13.19 RCV routing delay cost compared to PathFinder routing delay cost.

future. Finally, as more and more of the lower-end ASIC market migrates to FPGAs, the system-level tools to which these designers are accustomed will be required for FPGA flows.

Acknowledgments

We thank Babette van Antwerpen for her help with the sections on synthesis and technology mapping, and to Paul Metzgen, Valavan Manohararajah, and Andrew Ling for providing several figures.

References

[1] J. Rose, R. Francis, D. Lewis, and P. Chow, Architecture of programmable gate arrays: the effect of logic block functionality on area efficiency, *IEEE J. Solid State Circuits*, 25, 1217–1225, 1990.

[2] E. Ahmed and J. Rose, The effect of LUT and cluster size on deep-submicron FPGA performance and density, *Proceedings of the Eighth International Symposium on FPGAs*, 2000, pp. 3–12.

[3] V. Betz and J. Rose, How much logic should go in an FPGA logic block? *IEEE Des. Test*, 15, 10–15, 1998.

[4] J. Rose and S. Brown, Flexibility of interconnection structures for FPGAs, *JSSC*, 26, 277–282, 1992.

[5] G. Lemieux and D. Lewis, Circuit design of routing switches, *Proceedings of the Tenth International Symposium on FPGAs*, 2002, pp. 19–28.

[6] J. Cong and Y. Hwang, Boolean matching for LUT-based logic blocks with applications to architecture evaluation and technology mapping, *IEEE Trans. CAD*, 20, 1077–1090, 2001.

[7] S. Brown, R. Francis, J. Rose, and Z. Vranesic, *Field-Programmable Gate Arrays*, Kluwer, Dordrecht, 1992.

[8] V. Betz, J. Rose, and A. Marquardt, *Architecture and CAD for Deep-Submicron FPGAs*, Kluwer, Dordrecht, 1999.

[9] V. Betz and J. Rose, Automatic generation of FPGA routing architectures from high-level descriptions, *Proceedings of the Seventh International Symposium on FPGAs*, 2000, pp. 175–184.

[10] A. Yan, R. Cheng, and S. Wilton, On the sensitivity of FPGA architectural conclusions to experimental assumptions, tools and techniques, *Proceedings of the Tenth International Symposium on FPGAs*, 2003, pp. 147–156.

[11] F. Li, D. Chen, L. He, and J. Cong, Architecture evaluation for power-efficient FPGAs, *Proceedings of the 11th International Symposium on FPGAs*, 2003, pp. 175–184.

[12] S. Wilton, Heterogeneous technology mapping for area reduction in FPGAs with embedded memory arrays, *IEEE Trans.* CAD, 19, 56–68, 2000.

[13] E. Lin, and S. Wilton, Macrocell architectures for product term embedded memory arrays, *Proceedings of the 11th International Conference on Field-Programmable Logic and Applications,* 2001, pp. 48–58.

[14] D. Lewis, V. Betz, D. Jefferson, A. Lee, C. Lane, P. Leventis, S. Marquardt, C. McClintock, B. Pedersen, G. Powell, S. Reddy, C. Wysocki, R. Cliff, and J. Rose, The Stratix routing and logic architecture, *Proceedings of the 11th ACM International Symposium on FPGAs*, 2003, pp. 12–20.

[15] D. Lewis, E. Ahmed, G. Baeckler, V. Betz, M. Bourgeault, D. Cashman, D. Galloway, M. Hutton, C. Lane, A. Lee, P. Leventis, S. Marquardt, C. McClintock, K. Padalia, B. Pedersen, G. Powell, B. Ratchev, S. Reddy, J. Schleicher, K. Stevens, R. Yuan, R. Cliff, and J. Rose, The Stratix II routing and logic architecture, *Proceedings of the 13th ACM International Symposium on FPGAs*, 2005, pp. 14–20.

[16] S. Trimberger, K. Duong, and B. Conn, Architecture issues and solutions for a high-capacity FPGA, *Proceedings of the Fifth ACM International Symposium on FPGAs*, 1997, pp. 3–9.

[17] See www.<companyname>.com for commercial tools and architecture information.

[18] B. Ahanin, and S.S. Vij, A high density, high-speed, array-based erasable programmable logic device with programmable speed/power optimization, *Proceedings of the First ACM International Symposium on FPGAs*, 1992, pp. 29–32.

[19] M. Hutton, D. Karchmer, B. Archell, and J. Govig, Efficient static timing analysis and applications using edge masks, *Proceedings of the 13th International Symposium on FPGAs*, 2005, pp. 174–183.

[20] K. Poon, A. Yan, and S.J.E. Wilton, A flexible power model for FPGAs, *ACM Trans. Des. Automat. Digital Syst.*, 10, 279–302, 2005.

[21] G. De Micheli, *Synthesis and Optimization of Digital Circuits*, McGraw-Hill, New York, 1994.

[22] J. Cong and Y. Ding, Combinational logic synthesis for LUT-based FPGAs, *ACM Trans. Des. Automat. Digital Syst.*, 1, 145–204, 1996.

[23] Murgai, R. Brayton, R. and Sangiovanni-Vincentelli, A. *Logic Synthesis for Field-Programmable Gate Arrays*, Kluwer, Norwell, MA, 2000.

[24] J. Hwang, B. Milne, N. Shirazi, and J. Stroomer, System-level tools for DSP in FPGAs, *Proceedings of the 14th Symposium on Field-Programmable Logic (FPL)*, 2001.

[25] J. Stroomer, J. Ballagh, H. Ma, B. Milne, J. Hwang, and N. Shirazi, Creating system generator design using jg, *Proceedings of the IEEE Symposium on Field-Programmable Custom Computing Machines (FCCM)*, 2003.

[26] Berkeley Design Technology Inc., Evaluating FPGAs for Communication Infrastructure Applications, *Proceedings of the Communications Design Conference*, 2003.

[27] J. Lockwood, N. Naufel, J. Turner, and D. Taylor, Reprogrammable network packet processing on the Field-Programamble Port Extender (FPX), *Proceedings of the Ninth International Symposium on FPGAs*, 2001, pp. 87–93.

[28] J. Kempa, S.Y. Lim, C. Robinson, and J. Seely, SOPC builder: performance by design, in *Winning the SOPC Revolution*, Chapter 8, G. Martin, and H. Chang, Eds., Springer, Heidelberg, 2003.

[29] B. Hutchings, P. Bellows, J. Hawkins, S. Hemmert, B. Nelson, and Rytting, M. A CAD suite for high-performance FPGA design, *Proceedings of the Ninth International Workshop on Field-Programmable Logic*, 1999.

[30] C. Brandolese, W. Fornaciari, and F. Salice, An area estimation methodology for FPGA-based designs at SystemC-level, *Proceedings of the Design Automation Conference*, 2004, pp. 129–132.

[31] M. Hutton, J. Schleicher, D. Lewis, B. Pedersen, R. Yuan, S. Kaptanoglu, G. Baeckler, B. Ratchev, K. Padalia, M. Bourgeault, A. Lee, H. Kim, and R. Saini, Improving FPGA performance and area using an adaptive logic module, *Proceedings of the 14th International Symposium on Field-Programmable Logic*, 2004, pp. 134–144.

[32] P. Metzgen and D. Nancekievill, Multiplexor restructuring for FPGA implementation cost reduction, *Proceedings of the Design Automation Conference*, 2005.

[33] D. Nancekievill and P. Metzgen, Factorizing multiplexors in the datapath to reduce cost in FPGAs, *Proceedings of the International Workshop on Logic Synthesis*, 2005.

[34] E. M. Sentovich, K.J. Singh, L. Lavagno, C. Moon, R. Murgai, A. Saldanha, H. Savoj, P.R. Stephan, R.K. Brayton, and A.L. Sangiovanni-Vincentelli, SIS: A System for Sequential Circuit Analysis, Technical Report No. UCB/ERL M92/41, UC Berkeley, 1992.

[35] N. Shenoy and R. Rudell, Efficient implementation of retiming, *Proceedings of the International Conference on CAD (ICCAD)*, 1994, pp. 226–233.

[36] B. van Antwerpen, M. Hutton, G. Baeckler, and R. A Yuan, safe and complete gate-level register retiming algorithm, *Proceedings of the IWLS*, 2003.

[37] N. Vemuri, P. Kalla and R. Tessier, BDD-based logic synthesis for LUT-based FPGAs, *ACM Trans. Des. Automat. Electr. Syst.*, Vol. 7, 501–525, 2002.

[38] C. Yang, M. Ciesielski, and V. Singhal, BDS a BDD-based logic optimization system, *Proceedings of the Design Automation Conference*, 2000, pp. 92–97.

[39] Y. Lai, M. Pedram and S. Vrudhala, BDD-based decomposition of logic functions with application to FPGA synthesis, *Proceedings of the Design Automation Conference*, 1992, pp. 448–451.

[40] J. Anderson, F. Najm, and T. Tuan, Active leakage power estimation for FPGAs, *Proceedings of the 12th International Symposium on FPGAs*, 2004, pp. 33–41.

[41] A. Ling, D.P. Singh, and S.D. Brown, FPGA technology mapping: a study of optimality, *Proceedings of the Design Automation Conference*, 2005.

[42] K. Keutzer, DAGON: technology binding and local optimization by DAG matching, *Proceedings of 24th Design Automation Conference*, 1987, pp. 341–347.

[43] R.J. Francis, J. Rose, and K. Chung, Chortle: a technology mapping program for lookup table-based field-programmable gate arrays, *Proceeding of the Design Automation Conference*, 1990, pp. 613–619.

[44] J. Cong and E. Ding, An optimal technology mapping algorithm for delay optimization in lookup table based FPGA designs, *IEEE Trans. CAD*, 13, 1–12, 1994.

[45] J. Cong and J. Peck, RASP: a general logic synthesis system for SRAM-based FPGAs, *Proceedings of Fifth International Symposium on FPGAs*, 1996, pp. 137–143.

[46] J. Cong and Y. Ding, On area/depth trade-off in LUT-based FPGA technology mapping, *IEEE Trans. VLSI*, 2, 137–148, 1994.

[47] J. Cong, C. Wu, and Y. Ding, Cut ranking and pruning: enabling a general and efficient FPGA mapping solution, *Proceedings of the Ninth International Symposium on FPGAs*, 1999, pp. 29–35.

[48] R.J. Francis, J. Rose, and Vranesic, Z. Chortle-crf: fast technology mapping for lookup table-based FPGAs, *Proceedings of the Design Automation Conference*, 1991, pp. 227–233.

[49] R.J. Francis, J. Rose, and Vranesic, Z. Technology mapping of lookup table-based FPGAs for performance, *Proceedings of the International Conference on CAD (ICCAD)*, 1991.

[50] E.L. Lawler, K.N. Levitt, and J. Turner, Module clustering to minimize delay in digital networks, *IEEE Trans. Comput.*, C-18, 47–57, 1969.

[51] Tarjan, R.E. *Data Structures and Network Algorithms*, SIAM, Philadelphia, PA, 1983.

[52] A. Farrahi and M. Sarrafzadeh, Complexity of the lookup-table minimization problem for FPGA technology mapping, *IEEE Trans. CAD*, 13, 1319–1332, 1994.

[53] J. Cong, Y. Ding, T. Gao, and K.C. Chen, An optimal performance-driven technology mapping algorithm for LUT-based FPGAs under arbitrary net-delay models, *Proceedings of the International Conference on CAD (ICCAD)*, 1993, pp. 599–603.

[54] J. Cong and Y. Hwang, Simultaneous depth and area minimization in LUT-based FPGA mapping, *Proceedings of the Fourth International Symposium on FPGAs*, 1995, pp. 68–74.

[55] V. Manohararajah, S. Brown, and Z. Vranesic, Heuristics for area minimization in LUT-based FPGA technology mapping, *Proceedings of the International Workshop on Logic Synthesis*, 2004, pp. 14–21.

[56] H. Yang and D.F. Wong, Edge-map: optimal performance driven technology mapping for iterative LUT based FPGA designs, *Proceedings of the IEEE International Conference on CAD (ICCAD)*, 1994, pp. 150–155.

[57] D. Chen and D. Cong, DAOmap: a depth-optimal area optimization mapping algorithm for FPGA designs, *Proceedings of the International Conference on CAD (ICCAD)*, Nov. 2004.

[58] P. Pan, Performance-driven integration of retiming and resynthesis, *Proceedings of the Design Automation Conference*, 1999, pp. 243–246.

[59] P. Pan and Lin, C.-C. A new retiming-based technology mapping algorithm for LUT-based FPGAs, *Proceedings of the Sixth International Symposium on FPGAs*, pp. 35–42, 1998.

[60] A.H. Farrahi and M. Sarrafzadeh, FPGA technology mapping for power minimization, *Proceedings of the International Workshop on Field-Programmable Logic and Applications*, 1994.

[61] J. Anderson and F.N. Najm, Power-aware technology mapping for LUT-based FPGAs, *Proceedings of the International Conference on Field-Programmable Technology*, 2002.

[62] J. Lamoureaux and S. Wilton, On the interaction between power-aware FPGA CAD algorithms, *Proceedings of the International Symposium on CAD (ICCAD)*, 2003, pp. 701–708.

[63] D. Chen, J. Cong, F. Li, and He, Low-power technology mapping for FPGA architectures with dual supply voltages, *Proceedings of the 12th International Symposium on FPGAs*, 2004, pp. 109–117.

[64] J. Cong and Y. Hwang, Structural gate decomposition for depth-optimal technology mapping in LUT-based FPGA design, *Proceedings of the Design Automation Conference*, 1996, pp. 726–729.

[65] A. Wang, *Algorithms for Multilevel Logic Optimization*, Ph.D. Dissertation, UC Berkeley Computer Science Department, 1991.

[66] J. Cong and Y. Hwang, Structural gate decomposition for depth-optimal technology mapping in LUT-based FPGA designs, *ACM Trans. Design Automat. Digital Syst.*, 5, 193–225, 2000.

[67] C. Legl, B. Wurth, and K. Eckl, A Boolean approach to performance-directed technology mapping for LUT-based FPGA designs, *Proceedings of the Design Automation Conference*, 1996.

[68] J. Cong and Y. Hwang, Partially-dependent functional decomposition with applications in FPGA synthesis and mapping, *Proceedings of the Sixth International Symposium on FPGAs*, 1997, pp. 35–42.

[69] J. Cong and S. Xu, Performance-driven technology mapping for heterogeneous FPGSs, *IEEE Trans. CAD*, 19, 1268–1281, 2000.

[70] S. Yamashita, H. Sawada, and A. Nagoya, A new method to express functional permissibilities for LUT-based FPGAs and its applications, *Proceedings of the International Conference on CAD (ICCAD)*, 1996, pp. 254–261.

[71] J. Cong, Y. Lin, and W. Long, SPFD-based global re-wiring, *Proceedings of the 10th International Symposium on FPGAs*, 2002, pp. 77–84.

[72] J. Cong, Y. Lin, and W. Long, A new enhanced SPFD rewiring algorithm, *Proceedings of the International Conference on CAD (ICCAD)*, 2002, pp. 672–678.

[73] J.M. Hwang, F.Y. Chiang, and T.T. Hwang, A re-engineering approach to low power FPGA design using SPFD, *Proceedings of the 35th ACM/IEEE Design Automation Conference*, 1998, pp. 722–725.

[74] B. Kumthekar and F. Somenzi, Power and delay reduction via simultaneous logic and placement optimization in FPGAs, *Proceedings of the Design and Test in Europe (DATE)*, 2000, pp. 202–207.

[75] J. Anderson, J. Saunders, S. Nag, C. Madabhushi, and R. Jayarman, A placement algorithm for FPGA designs with multiple I/O standards, *Proceedings of the International Conference on Field Programmable Logic and Applications*, 2000, pp. 211–220.

[76] W. Mak, I/O placement for FPGA with multiple I/O standards, *Proceedings of the 11th International Symposium on FPGAs*, 2003, pp. 51–57.

[77] J. Anderson, S. Nag, K. Chaudhary, S. Kalman, C. Madabhushi, and P. Cheng, Run-time conscious automatic timing-driven FPGA layout synthesis, *Proceedings of the 14th International Conference on Field-Programmable Logic and Applications*, 2004, pp. 168–178.

[78] A. Marquardt, V. Betz, and J. Rose, Using cluster-based logic blocks and timing-driven packing to improve FPGA speed and density, *Proceedings of the Seventh International Symposium on FPGAs*, 1999, pp. 37–46.

[79] V. Betz and J. Rose, VPR: a new packing, placement and routing tool for FPGA research, *Proceedings of the Seventh International Conference on Field-Programmable Logic and Applications*, 1997, pp. 213–222.

[80] A. Marquardt, V. Betz, and J. Rose, Timing-driven placement for FPGAs, *Proceedings of the International Symposium on FPGAs*, 2000, pp. 203–213.
[81] A. Singh and M. Marek-Sadowska, Efficient circuit clustering for area and power reduction in FPGAs, *Proceedings of the International Symposium on FPGAs*, 2002, pp. 59–66.
[82] D. Chen and J. Cong, Delay optimal low-power circuit clustering for FPGAs with dual supply voltages, *Proceedings of the International Symposium on Low Power Electronics and Design (ISLPED)*, 2004, pp. 70–73.
[83] G. Chen and J. Cong, Simultaneous timing driven clustering and placement for FPGAs, *Proceedings of the International Conference on Field Programmable Logic and Applications*, 2004, pp. 158–167.
[84] T. Kong, A novel net weighting algorithm for timing-driven placement, *Proceedings of the International Conference on CAD (ICCAD)*, 2002, pp. 172–176.
[85] S.K. Nag and R.A. Rutenbar, Performance-driven simultaneous placement and routing for FPGAs, *IEEE Trans. CAD*, 1998, pp. 499–518.
[86] Y. Sankar and J. Rose, Trading quality for compile time: ultra-fast placement for FPGAs, *Proceedings of the International Symposium on FPGAs*, 1999, pp. 157–166.
[87] J. Rose, W. Snelgrove, and Z. Vranesic, ALTOR: an automatic standard cell layout program, *Proceedings of the Canadian Conference VLSI*, 1985, pp. 169–173.
[88] A. Dunlop and B. Kernighan, A procedure for placement of standard-cell VLSI circuits, *IEEE Trans. CAD*, 1985, pp. 92–98.
[89] M. Maidee, C. Ababei, and K. Bazargan, Fast timing-driven partitioning-based placement for island style FPGAs, *Proceedings of the Design Automation Conference (DAC)*, 2003, pp. 598–603.
[90] M. Hutton, K. Adibsamii, and A. Leaver, Adaptive delay estimation for partitioning-driven PLD placement, *IEEE Trans. VLSI*, 11, 60–63, 2003.
[91] P. Chan and M. Schlag, Parallel placement for field-programmable gate arrays, *Proceedings of the 11th International Symposium on FPGAs*, 2003, pp. 43–50.
[92] J. Kleinhans, G. Sigl, F. Johannes, and Antreich, K. Gordian: VLSI placement by quadratic programming and slicing optimization, *IEEE Trans. CAD*, 10, 356–365, 1991.
[93] J. Lin, A. Jagannathan, and J. Cong, Placement-driven technology mapping for LUT-based FPGAs, *Proceedings of the 11th International Symposium FPGAs*, 2003, pp. 121–126.
[94] P. Suaris, D. Wang, and N. Chou, Smart move: a placement-aware retiming and replication method for field-programmable gate arrays, *Proceedings of the Fifth International Conference on ASICs*, 2003.
[95] P. Suaris, L. Liu, Y. Ding, and N. Chou, Incremental physical re-synthesis for timing optimization, *Proceedings of the 12th International Symposium on FPGAs*, 2004, pp. 99–108.
[96] K. Schabas and S. Brown, Using logic duplication to improve performance in FPGAs, *Proceedings of the 11th International Symposium on FPGAs*, 2003, pp. 136–142.
[97] G. Chen and J. Cong, Simultaneous timing-driven placement and duplication, *Proceedings of the 13th International Symposium on FPGAs*, 2005, pp. 51–61.
[98] V. Manohararajah, D. Singh, S. Brown, and Z. Vranesic, Post-placement functional decomposition for FPGAs, *Proceedings of the International Workshop on Logic Synthesis*, 2004, pp. 114–118.
[99] V. Manohararajah, D.P. Singh, and S. Brown, Timing-driven functional decomposition for FPGAs, *Proceedings of the International Workshop on Logic and Synthesis*, 2005.
[100] Y. Ding, P. Suaris, and N. Chou, The effect of post-layout pin permutation on timing, *Proceedings of the 13th International Symposium on FPGAs*, 2005, pp. 41–50.
[101] D. Singh and S. Brown, Constrained clock shifting for field programmable gate arrays, *Proceedings of the Tenth International Symposium on FPGAs*, 2002, pp. 121–126.
[102] Y. Chao-Yang, M. Marek-Sadowska, Skew-programmable clock design for FPGA and skew-aware placement, *Proceedings of the International Symposium on FPGAs*, 2005, pp. 33–40.
[103] D. Singh and S. Brown, Integrated retiming and placement for FPGAs, *Proceedings of the Tenth International Symposium on FPGAs*, 2002, pp. 67–76.
[104] D.P. Singh, V. Manohararajah, and Brown, S.D. Incremental retiming for FPGA physical synthesis, *Proceedings of the Design Automation Conference*, 2005.

[105] L. McMurchie and C. Ebeling, PathFinder: a negotiation-based performance-driven router for FPGAs, *Proceedings of the Fifth International Symposium on FPGAs*, 1995, pp. 111–117.
[106] G. Lemieux and D. Lewis, *Design of Interconnection Networks for Programmable Logic*, Kluwer, Dordrecht, 2004.
[107] J. Rose, Parallel global routing for standard cells, *IEEE Trans. CAD*, 9, pp. 1085–1095, 1990.
[108] S. Brown, J. Rose, and Z. Vranesic, A detailed router for field-programmable gate arrays, *IEEE Trans. CAD*, 11, 620–628, 1992.
[109] G. Lemieux and S. Brown, A detailed router for allocating wire segments in FPGAs, *Proceedings of the Physical Design Workshop*, 1993, pp. 215–226.
[110] G.-J. Nam, F. Aloul, K. Sakallah, and R. Rutenbar, A comparative study of two Boolean formulations of FPGA detailed routing constraints, *Proceedings of the International Symposium on Physical Designs*, 2001, pp. 222–227.
[111] Y.S. Lee and A. Wu, A performance and routability-driven router for FPGAs considering path delays, *Proceedings of the Design Automation Conference*, 1995, pp. 557–561.
[112] C.Y. Lee, An algorithm for path connections and applications, *IRE Trans. Electron. Comput.*, Vol. EC-10, 346–365, 1961.
[113] J. Frankle, Iterative and adaptive slack allocation for performance-driven layout and FPGA routing, *Proceedings of the Design Automation Conference*, 1992, pp. 536–542.
[114] W. Elmore, The transient response of damped linear networks with particular regard to wideband amplifiers, *J. Appl. Phys.*, 19, 55–63, 1948.
[115] J. Swartz, V. Betz, and J. Rose, A fast routability-driven router for FPGAs, *Proceedings of the Sixth International Symposium on FPGAs*, 1998, pp. 140–151.
[116] S. Wilton, A crosstalk-aware timing-driven router for FPGAs, *Proceedings of the Ninth ACM International Symposium on FPGAs*, 2001, pp. 21–28.
[117] R. Fung, V. Betz, and W. Chow, Simultaneous short-path and long-path timing optimization for FPGAs, *Proceedings of the International Conference on CAD (ICCAD)*, 2004.

SECTION II

ANALOG AND MIXED-SIGNAL DESIGN

14

Simulation of Analog and RF Circuits and Systems

Jaijeet Roychowdhury
University of Minnesota
Minneapolis, Minnesota

Alan Mantooth
University of Arkansas
Fayetteville, Arkansas

14.1 Introduction

Circuit simulation has always been a crucial component of analog system design and is becoming even more so today. Ever-shrinking DSM technologies are resulting in both analog and digital designs becoming less ideal. Indeed, the separation between digital and analog design is becoming blurred; from the digital standpoint, analog effects, often undesired, are becoming pervasive. Integrated radio frequency (RF) and communication system design involving both analog and digital components on the same substrate, now constitutes an important part of the semiconductor industry's growth, while traditionally digital circuits (such as microprocessors) are now critically limited by analog effects such as delays, the need to synchronize internal busses, and so on. In short, advances in analog, RF, digital, and mixed-signal design over the last few years, combined with the effects of shrinking technologies, have spurred a renaissance in simulation. Old simulation problems have assumed renewed significance and new simulation challenges — in some cases already addressed by novel and elegant algorithmic solutions — have arisen.

Every circuit designer is familiar with the program SPICE, the original circuit simulation program released into the public domain in the mid-1970s by the University of California at Berkeley. The basic algorithms and capabilities within SPICE engendered an entire industry of circuit simulation tools, with noted commercial offerings such as HSPICE, PSPICE, and SPECTRE enjoying widespread adoption. Several of these programs have evolved from originally providing roughly the same capabilities as Berkeley SPICE, into tools with significantly more advanced simulation capabilities. Furthermore, the aforementioned resurgence of interest and research in advanced circuit simulation algorithms promise more powerful capabilities, which will enable designers to make higher-performance circuits and systems faster, better, and with fewer mistakes.

In this chapter, we provide a quick tour of modern circuit simulation. The style we adopt attempts to combine mathematical rigor with physical intuition, in order to provide a simple, clean exposition that links the various facets of the subject logically. We start with showing how one can write circuit equations as nonlinear differential equations and comment on why it is a good idea to do so. Next, we delve into "device models" — i.e., the equations that describe current–voltage and charge–voltage relationships of the semiconductor devices that pervade most modern circuits. We then show how the basic circuit analyses can be simply derived as specific ways of solving the circuit's differential equations. We touch on an important feature typical of circuit equations — sparsity — that makes these analyses computationally feasible for large circuits, and hence practically useful.

Next, motivated by recent interest in RF circuit simulation, we review more advanced analysis techniques for the circuit's differential equations, starting with the computation of periodic steady states. We explain two alternative methods in the frequency and time domains — harmonic balance (HB) and shooting — for computing periodic steady states that have complementary numerical properties. We show how the equations involved in these analyses lose sparsity, and hence can be much more difficult to solve computationally than those for the basic analyses above. We then explain how methods based on preconditioned iterative solution of matrix equations can be used to alleviate the computational problem to a large extent.

A recurring theme in circuit simulation is the simultaneous presence of fast and slow rates of time variation, which presents challenges for basic analyses such as transient simulation. We show how the use of multiple artificial time scales can be used to separate fast/slow behavior in a fundamental manner at the circuit equation level, leading to a special circuit equation form called multitime partial differential equations (MPDEs). We then outline how MPDEs can be solved numerically using different techniques to solve fast/slow problems efficiently. We touch on special MPDE forms for oscillators and also outline how MPDEs can be used as a link to enable automatic macromodeling of time-varying systems.

Finally, we touch on the important issue of (statistical) noise analysis of circuits. We first outline the fundamental concepts of basic stationary noise analysis and explain the kinds of circuits it applies to. Then we show how stationary noise analysis is insufficient for noise analysis of RF, switching, and other nonlinear circuits. We explain the fundamental concepts of cyclostationary noise and outline the use of these concepts for the noise analysis of RF circuitry, including computational aspects. Finally, we touch on the important problem of noise analysis of oscillators — in particular, calculating phase noise and jitter in oscillators.

14.2 Differential-Algebraic Equations for Circuits via Modified Nodal Analysis

The fundamental premise behind virtually all types of computer simulation is to first abstract a physical system's behavior using appropriate mathematical representations and then to use algorithms to solve these representations by computer. For most kinds of circuits, the appropriate mathematical abstraction turns out to be systems of differential equations, or more precisely, systems of nonlinear, differential-algebraic equations (DAEs). In the circuit context, the following special DAE form can be shown to be appropriate:

$$q(x) + f(x) = b(t). \tag{14.1}$$

In Equation 14.1, all quantities (except *t*, the time variable) are vectors of size *n*, which are (roughly) the size of the circuit represented by the DAE. The nonlinear vector functions $q(x)$ and $f(x)$ represent the charge/flux and resistive parts of the circuit respectively, while $b(t)$ is a forcing term representing external inputs from independent current and voltage sources. This DAE form is obtained by writing out the fundamental charge and potential conservation equations of the circuit — Kirchoff's current laws (KCL) and Kirchoff's voltage law (KVL) — together with the branch constitutive relationships (BCR) of the individual elements. We refer the reader to e.g., [1–3], for a systematic exposition of how these equations can be obtained for a circuit topology. It is not difficult, however, to understand the basic procedure of defining the quantities in Equation (14.1) and to apply it to any circuit of interest, "by hand" or via a computer procedure. Consider the simple circuit shown in Figure 14.1, consisting of a voltage source driving a parallel-RC load through a nonlinear diode. The electrical quantities of interest in this circuit are the two node voltages $v_1\ (t)$ and $v_2(t)$, as well as the current through the voltage source, $i(t)$.

The resistor's action on the circuit is defined by its current–voltage relationship (Ohm's law), $i_R(t) = v_2(t)/R$; similarly, the capacitor's current–voltage relationship is given by $i_C(t) = (\mathrm{d}/\mathrm{d}t)(Cv_2(t))$. The current–voltage relationship of the diode element, which is nonlinear, is given by (see Section 14.3.2)

$$i = \mathrm{d}(v_1 - v_2) = I_s\left(\mathrm{e}^{\frac{v_1 - v_2}{V_T}} - 1\right) \tag{14.2}$$

where I_s and V_T are constant numbers. Finally, the voltage source is captured simply by its setting the voltage of node 1 to its value: $v_1\ (t) = E$. These four relationships, one for each of the four circuit elements, are the BCRs relevant to this circuit.

By summing currents at each node of interest, the circuit equations are obtained. At node 1, we obtain

$$i(t) + \mathrm{d}(v_1 - v_2) = 0 \tag{14.3}$$

while at node 2, we have

$$\frac{\mathrm{d}}{\mathrm{d}t}(Cv_2) + \frac{v_2}{R} - \mathrm{d}(v_1 - v_2) = 0 \tag{14.4}$$

Finally, we also have (for the voltage source)

$$v_1 = E. \tag{14.5}$$

These are known as the modified nodal analysis equations (MNA equations) of the circuit.

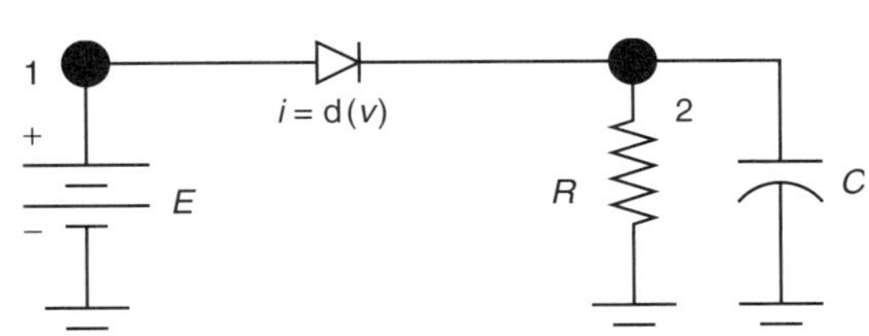

FIGURE 14.1 Simple diode–resistor–capacitor circuit.

By writing Equation 14.3, Equation 14.4, and Equation 14.5 in vector form, the form of Equation (14.1) can be obtained. First, the voltage/current unknowns are written in vector form as

$$x = \begin{bmatrix} v_1(t) \\ v_2(t) \\ i(t) \end{bmatrix}. \tag{14.6}$$

Define the vector function $q(x)$ in Equation (14.1) to be

$$q(x) \equiv \begin{bmatrix} 0 & & \\ & C & \\ & & 0 \end{bmatrix} x = \begin{bmatrix} 0 \\ Cv_2(t) \\ 0 \end{bmatrix}. \tag{14.7}$$

$f(x)$ to be

$$f(x) \equiv \begin{bmatrix} i(t) + d(v_1 - v_2) \\ \frac{v_2}{R} - d(v_1 - v_2) \\ v_1 \end{bmatrix}, \tag{14.8}$$

and the vector $b(t)$ to be

$$b(t) = \begin{bmatrix} 0 \\ 0 \\ E(t) \end{bmatrix}. \tag{14.9}$$

With these definitions, Equation (14.1) simply represents the circuit Equations (14.3, 14.4, and 14.5), but in vector DAE form. The utility of writing circuit equations in this canonical form is that a variety of mathematical and computational techniques may be brought to bear on Equation (14.1), without having to worry about the details of $f(x)$, $q(x)$, and $b(t)$, so long as they can be realistically assumed to have some simple properties like continuity, smoothness, etc. Indeed, the DAE form Equation (14.1) is not restricted to circuits; it subsumes virtually all physical systems, including those from mechanical, optical, chemical, etc., applications.

It will be readily understood, of course, that DAEs for typical industrial circuits are usually much larger and more complex than the one for the simple example of Figure 14.1. Much of the complexity of these equations stems from the complicated equations describing the *semiconductor devices,* such as metal oxide semiconductor (MOS) and bipolar transistors, that populate circuits in large numbers. These equations, called *device models,* are described further in Section 14.3.

14.3 Device Models

Many books that describe the details of various device models available in circuit simulators [4–8] have been written. In these few pages, it is not possible to delve into the derivation of device physics that represent any specific device models. Rather, the focus of this section will be on the role the models play and constraints placed on models by simulators. This section first describes models in the context of traditional circuit simulation and then touches on two key device models, the diode and MOSFET, which are illustrative of issues relevant to this chapter. In each case, some remarks are made with respect to versions of these models suitable for power electronic applications as well.

14.3.1 Role of Models in Circuit Simulation

In the context of this chapter, models are the representation of the properties of devices or a group of interconnected devices by means of mathematical equations, circuit representations (i.e., macromodels), or tables. Care has been taken to avoid defining models strictly as electrical devices. In the era of hardware description languages (HDLs), circuit simulation and modeling have expanded beyond the boundaries of passive and active electrical devices [9–12]. However, for the purpose of this section, our focus will remain on electrical device models.

Semiconductor device models such as the MOSFET involve many complicated equations. Often these equations are defined piecewise according to the different regions of operation observed in the device. Timing studies performed on circuit simulations indicate that most of the computational effort in network analysis is spent on evaluation of these complicated model equations. This is not hard to imagine when "compact" model implementations of advanced MOSFET technologies are over 15,000 lines of C code in SPICE3 [13]. While much of this code involves data structures in the simulator, several thousand lines of code are devoted to describing model behavior alone. As such, models and modeling techniques are as important to addressing simulation efficiency and accuracy as the solution algorithms employed. In fact, the fidelity of simulation results will be dictated by the weakest link between the two, so both are vital to the simulation process.

Figure 14.2 below depicts a very high level view of the simulation process, which specifically pulls out the role the models play in the solution of the system. This flow diagram begins with the processing of the input netlist to the simulator and proceeds to the formulation of the system of equations. The system formulation step involves both the interconnectivity of the specific models in the circuit as well as the matrix stamps of the models. The interconnectivity defines what models contribute current to which nodes and thus leads to the matrix statement of KCL. The matrix stamps of the models, which will be described further in Section 14.4, indicate the specific contributions a model makes to various rows and columns of the matrix. Once the system formulation is complete, the process proceeds to the numerical solution based on the user-specified analysis. In any event, one thing that is common to all analyses (e.g., DC, AC, transient) and solution methods (e.g., Newton–Raphson [NR]) is the need to evaluate the models' constitutive relationships at each iteration, subject to any bypass techniques that may be employed in a given algorithm. This is the model evaluation time referred to above. The independent variables indicated in Figure 14.2 are those variables being solved for by the simulator and are thus given to the models as known values for an iteration. Then, the model is evaluated and provides the dependent variable values back to the simulator for the purpose of determining if the simulator's guess for the independent variable values was the correct solution to the system.

14.3.2 Diode Model

For the model of a basic p–n junction diode, the constitutive relations are derived in [14] and also shown in [4, 15]. Figure 14.3 shows the basic large-signal model of the diode. The series resistance is a straightforward application of Ohm's law. The current source is represented by the BCR

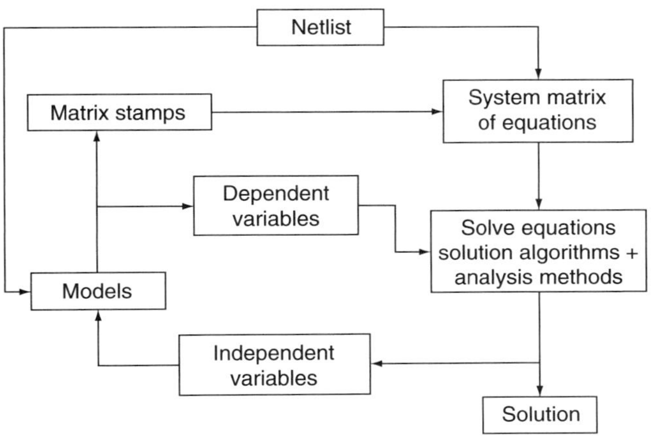

FIGURE 14.2 Depiction of the role of models in the traditional circuit simulation process.

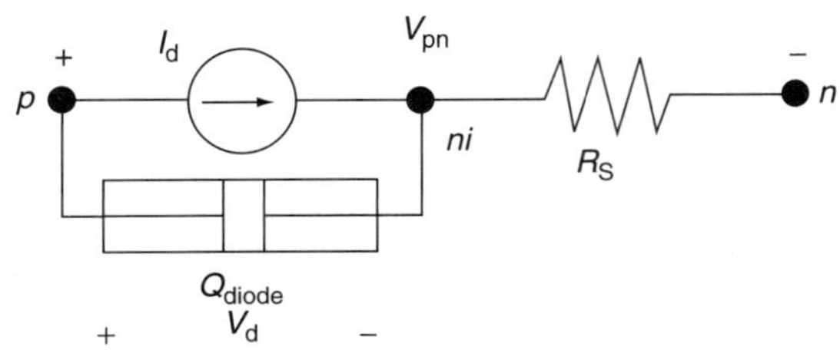

FIGURE 14.3 Large-signal model topology of a p–n junction diode.

$$I_d = I_S\left(e^{\frac{V_d}{nV_T}} - 1\right), \tag{14.10}$$

where I_S is the saturation current, V_T the thermal voltage, V_d the diode junction voltage, and n the emission coefficient of the diode. The saturation current is a function of the geometry of the device and is given by

$$I_S = Aqn_i^2\left(\frac{D_P}{L_P N_D} + \frac{D_n}{L_n N_A}\right), \tag{14.11}$$

where A is the cross-sectional area of the junction, q the electronic charge, n_i the intrinsic carrier concentration, N_A and N_D the acceptor and donor concentrations, $L_n(D_n)$ and $L_p(D_p)$ are the diffusion lengths (diffusion constants) for electrons in p-type material and holes in n-type material, respectively. The thermal voltage is equal to

$$V_T = \frac{kT}{q}, \tag{14.12}$$

where $k = 1.3806 \times 10^{-23}$ J/K is Boltzmann's constant, T the absolute temperature in K, and $q = 1.6022 \times 10^{-19}$ C the electronic charge.

The well-known relationships in Equation (14.10), Equation (14.11), and Equation (14.12) can be used to illustrate some of the options that a model developer has when creating a model for circuit simulation. In this case, a single continuous (and continuously differentiable) equation has been used to describe the BCR. This is the most desirable scenario for the solution algorithms in traditional circuit simulators. Most solvers require continuity in the first derivative of the BCRs in order to avoid convergence problems associated with the Jacobian. Saber [16] is an exception to this requirement, but even in that case models that possess continuous first derivatives simulate faster and more robustly than those with only continuous BCRs. From a pragmatic perspective, fewer convergence problems are introduced with BCRs that are not piecewise defined [8]. A second option that faces the model developer is the degree of physics to represent in the model. Most diode models simply specify the saturation current as a parameter and avoid calculating it from Equation (14.11). This allows diodes to be effectively modeled without the need to know doping concentrations and other low-level process information about how the device was constructed. It still remains possible to make the saturation current scalable with area by simply providing a scale factor for I_S in the model equations. Then, since I_S varies so significantly with temperature, primarily due to the n_i^2 factor in Equation (14.11), an empirical relationship is commonly used to model the variation of I_S with temperature, as observed from data sheets or measurements.

The governing equation for the charge in the diode in Figure 14.3 is

$$Q_{diode} = Q_{diff} + Q_{depl} \tag{14.13}$$

where Q_{diode} is the total charge, Q_{diff} the charge stored due to diffusion in the diode, and Q_{depl} the charge stored in the depletion region of the junction. The relationships [4] for each of these components of charge are given by

$$Q_{diff} = \tau_T I_d \tag{14.14}$$

where τ_T is known as the mean transit time of the diode and is typically a model parameter. This charge-storage effect is only significant in forward bias and thus a piecewise definition for the charge is typically created at this point by setting $Q_{diff} = 0$ for reverse bias voltages. This charge equation is continuous as is its derivative.

The depletion charge is also defined piecewise as follows:

$$Q_{depl} = \begin{cases} V_{j0}C_{j0}\dfrac{\left(1-\dfrac{V_d}{V_{j0}}\right)^{1-m}}{m-1} & V_d < f_c V_{j0}, \\ V_{j0}C_{j0}\dfrac{(1-f_c)^{1-m}}{m-1} + \dfrac{C_{j0}}{(1-f_c)^{1+m}}\left[1-f_c(1+m)(V_d - f_c V_{j0}) + m\dfrac{V_d^2-(f_c V_{j0})^2}{2V_{j0}}\right], & V_d \geq f_c V_{j0}. \end{cases} \tag{14.15}$$

f_c is a fitting parameter (typically = 0.5), m the grading coefficient of the junction (= 0.5 for a step junction), C_{j0} the zero-bias junction capacitance, and V_{j0} the built-in junction potential. In this case, a great deal of care is taken to produce a continuous charge and a continuous capacitance (i.e., first derivative) for Equation (14.15). The diode model possesses nonlinear behavior in the current–voltage relationships as well as in the charge relationships. Details of how such behavior is "stamped" into the circuit matrices and thus used in the flow of Figure 14.2 will be covered in the next section.

The modeling of power diodes is conceptually the same as that for low-voltage diodes. The primary differences arise from the different conditions the device is subjected to and the physical structure of the device. These differences lead to a different set of assumptions for deriving the model. In the end, this does have an impact on the nature of the equations that must be implemented. In the case of power diode models, the levels of injection are significantly higher, requiring the model developer to account for ambipolar diffusion effects and pay particular attention to conductivity modulation within lightly doped regions. Some power diodes are manufactured with a lightly doped "base" region between the p and n regions, making a p–ν–n or a p–π–n diode. Such diodes are generically referred to as p–i–n diodes, where the *i* refers to an almost intrinsic region. Figure 14.4 is a cross-sectional diagram of a p–i–n diode. In these devices, holes are injected from the p region and electrons are injected from the n region into the base. The normally high-resistance region of the base becomes saturated with carriers and the on-state resistance of the region becomes quite low. When the device is switched off, the carriers must be removed from the base region before the device can effectively turn off.

Through mechanisms such as recombination and carrier sweep-out effects, the device has a significant reverse recovery where the current through the device changes from positive to negative and then recovers back to near zero (Figure 14.5). The depletion region that forms during this recovery process changes width, producing a nonquasi-static effect that must also be accounted for to model reverse recovery accurately. The diode model in [18] possesses all of this behavior which typical SPICE diode models do not.

Figure 14.6 shows the large-signal topology of the power diode model in [18]. The DC characteristics are far more complicated than those described above for the low-voltage diode. Due to the injection levels, there are four separate exponential relationships that are used to model low-level recombination, normal injection, high-level injection, and emitter (i.e., end region) recombination effects. All of this DC behavior is represented by the current I_d represented by the diode symbol in Figure 14.6. An additional series resistance effect to R_S is R_{MOD}, which represents the conductivity modulation of the base region. The nonlinear charge effects are represented by four different charges in parallel with the diode current in Figure 14.6. The capacitance C_j is the voltage-dependent diode junction capacitance, Q_{SW} and Q_R represent the two-time constant response of the base charge, and C_r is the moving boundary redistribution capacitance. The total diode voltage v_d is the sum of the contact voltage v_r, the junction voltage v_j, and the midregion voltage v_m.

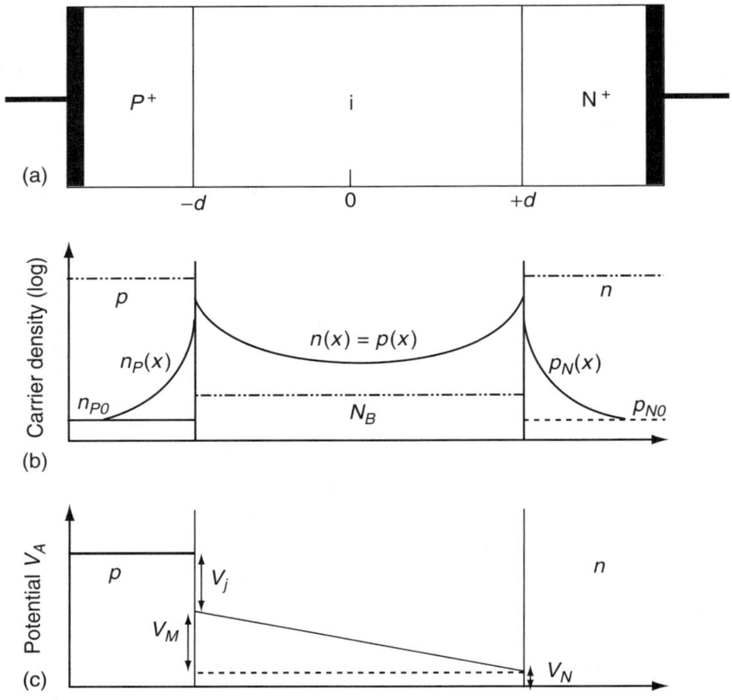

FIGURE 14.4 (a) Cross-sectional diagram of a p–i–n power diode; (b) corresponding carrier distribution profile; and (c) potential distribution for high-level injection conditions.

14.3.3 MOSFET Models

The most commonly used device model in circuit simulation is the MOSFET model, given the volume of integrated circuits that are designed using this workhorse device. MOS technology has undergone rapid advances that have enabled the semiconductor industry to keep pace with Moore's Law. As a result of these advances, MOSFET models have been an ongoing subject of research for the past two decades. Five versions of the BSIM model have been developed by a research team from UC Berkeley [19]. Many semiconductor companies have made impressive MOSFET modeling efforts including Texas Instruments, Motorola, Philips, and IBM to name a few. Other research efforts including the EKV model [20] and the latest efforts on the surface potential (SP) model [21] have made impacts on device modeling. It is well beyond the scope of this section to delve into any significant details of these or any other leading-edge MOSFET model. Our intent here is merely to provide another example of how linear and nonlinear BCRs arise in device models and segue into how such nonlinearities are dealt with in simulators. For this purpose, we will illustrate a simple MOSFET model with nonlinear drain–source current and charges inspired by the work in [22].

The MOSFET equations for the three regions of drain–source current i_{DS} are given below:

$$i_{DS} = \begin{cases} 0, & v_{GS} - V_t \leq 0 \quad \text{(cut-off)} \\ \dfrac{\beta}{2}(v_{GS} - V_t)^2, & v_{DS} \geq V_{GS} - V_t > 0 \quad \text{(saturation)} \\ \beta\left[(v_{GS} - V_t)v_{DS} - \dfrac{1}{2}v_{DS}^2\right], & v_{GS} - V_t > V_{DS} \text{ and } v_{GS} - V_t > 0 \quad \text{(triode)} \end{cases} \tag{14.16}$$

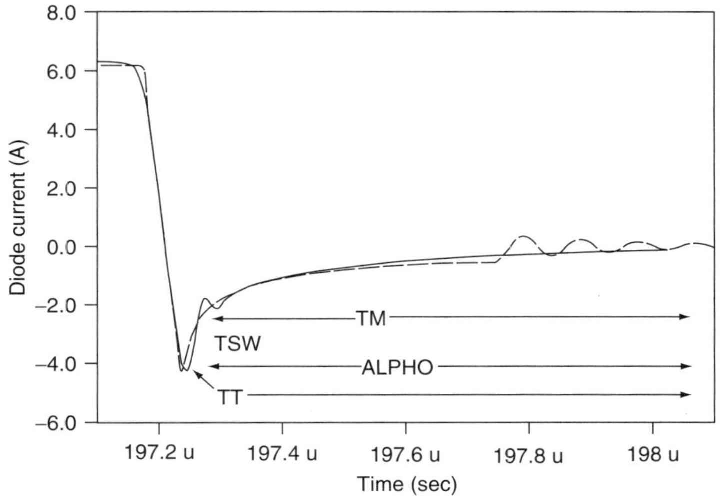

FIGURE 14.5 Reverse recovery waveform for a silicon carbide p–i–n power diode. (For further reading on SiC diode modeling, please see T.R. McNutt et al., *IEEE Trans. Power Electronics*, 19, 573–581, 2004.)

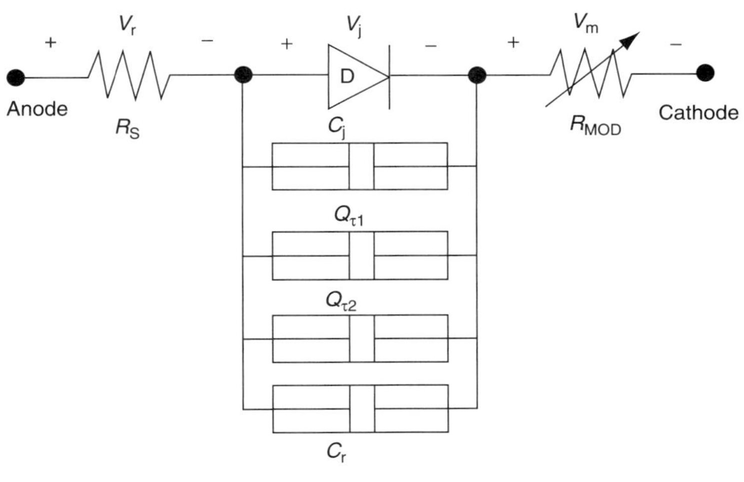

FIGURE 14.6 Large-signal model topology of power diode.

where V_t is the effective threshold voltage and $\beta = k'(W/L)(1+\lambda v_{DS})$.

The governing equations for the bulk–drain and bulk–source junctions are similar to those provided in the previous section, but typically do not possess series resistance effects. The governing equations for the intrinsic MOSFET charge are taken from [22] over four regions of operation as shown below:

Below flatband; $v_{GB} \leq V_{fb}$:

$$\begin{aligned} Q_{gate} &= C_{ox}\,(V_{GB} - V_{fb}) \\ Q_{bulk} &= -Q_{gate} \\ Q_{drain} &= 0 \\ Q_{source} &= 0 \end{aligned} \tag{14.17}$$

Below threshold; $v_{GDt} \leq v_{GSt} \leq 0$:

$$
\begin{aligned}
Q_{gate} &= C_{ox}\,\gamma\left(\sqrt{\frac{\gamma^2}{2} + v_{GB} - v_{fb}} - \frac{\gamma}{2}\right)\\
Q_{bulk} &= -Q_{gate}\\
Q_{drain} &= 0\\
Q_{source} &= 0
\end{aligned}
\tag{14.18}
$$

Saturation; $v_{GDt} \leq 0$ and $v_{GSt} > 0$:

$$
\begin{aligned}
Q_{gate} &= C_{ox}\left(\frac{2}{3}\,v_{GSt} + (V_t - V_{bi})\right)\\
Q_{bulk} &= -C_{ox}\,(V_t - V_{bi})\\
Q_{drain} &= -\frac{4}{15}\,C_{ox}\,V_{GSt}\\
Q_{source} &= -\frac{2}{5}\,C_{ox}v_{GSt}
\end{aligned}
\tag{14.19}
$$

Triode; $v_{GDt} \geq 0$ and $v_{GSt} > v_{GDt}$:

$$
\begin{aligned}
Q_{gate} &= C_{ox}\left[V_t - V_{bi} + \frac{2}{3}\left(v_{GSDt} + v_{GSt} - \frac{v_{GSt}v_{GDt}}{v_{GSt} + v_{GDt}}\right)\right]\\
Q_{bulk} &= -\,C_{ox}(V_t - V_{bi})\\
Q_{drain} &= -\,\frac{C_{ox}}{3}\left[\frac{1}{5}v_{GDt} + \frac{4}{5}\,v_{GSt} + v_{GDt}\left(\frac{v_{GDt}}{v_{GSDt}}\right) + \frac{1}{5}\,\frac{v_{GSt}v_{GDt}\,(v_{GDt} - v_{GSt})}{v_{GSDt}^2}\right]\\
Q_{source} &= -\,\frac{C_{ox}}{3}\left[\frac{1}{5}v_{GSt} + \frac{4}{5}\,v_{GDt} + v_{GSt}\left(\frac{v_{GSt}}{v_{GSDt}}\right) + \frac{1}{5}\,\frac{v_{GSt}v_{GDt}\,(v_{GSt} - v_{GDt})}{v_{GSDt}^2}\right]
\end{aligned}
\tag{14.20}
$$

In the above, $v_{GDt} = v_{GD} - V_t$, $v_{GSDt} = v_{GSt} - v_{GDt}$, $V_{bi} = V_{fb} + 2\phi_f$, and $C_{ox} = \varepsilon_{Si} WL/t_{ox}$.

The large-signal model topology of this simple MOSFET model is given in Figure 14.7. In addition to the drain-source current, body diodes, and nonlinear charges, some linear capacitances are shown as well as series resistances on the drain and source.

As in the case of the power diode, several additional effects must be taken into account when dealing with power MOSFET models. The device structure is very different in that power MOSFETs are typically vertical conduction devices, where the source and gate are on the top surface and the drain is on the back surface of the die or wafer. Many of the same issues arise as they did in the case of the power diode. Higher levels of injection, conductivity modulation of a drift layer, and moving-boundary conditions are just some of the major effects that must be modeled.

14.4 Basic Circuit Simulation: DC Analysis

In Section 14.2, we noted how any circuit could be written as a set of nonlinear differential equations (Equation 14.1), while in Section 14.3, we looked into the details of some of the more complex equations involved, notably those from semiconductor models. We now look at how to solve Equation (14.1)

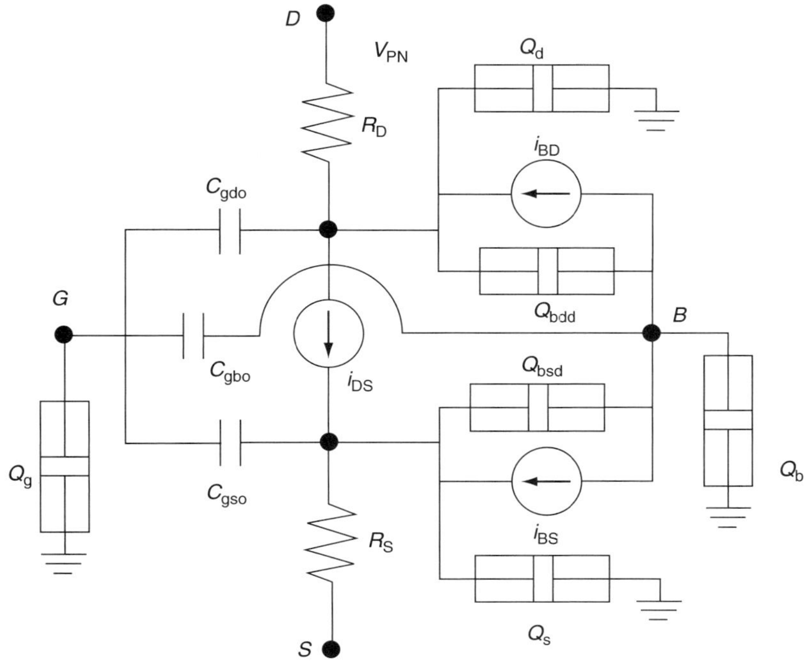

FIGURE 14.7 Large-signal model topology of a MOSFET.

numerically for a basic design problem: finding the DC operating point of the circuit, i.e., solving the circuit when all voltages/currents within it remain constant with time.

Finding the DC operating point is a fundamental and basic requirement in the design of virtually every circuit. The voltages and currents in a circuit at its DC operating point provide important information about correct functioning of the circuit. The DC operating point is essential not only as a first basic check of circuit operation, but is also a prerequisite for further analyses. Small-signal AC analysis, transient analysis, noise analysis, etc., all rely on a prior DC operating point having been calculated; furthermore, the operating point is also useful for steady-state and envelope analyses.

When no quantity in Equation (14.1) is changing with time, the input $b(t)$ must be a constant vector b_{DC}; and the circuit variables $x(t)$ must similarly be unchanging, i.e., $x(t) \equiv x_{DC}$. When nothing changes with time, all derivatives become zero, hence Equation (14.1) becomes

$$f(x_{DC}) = b_{DC}, \text{ or } g(x_{DC}) \equiv f(x_{DC}) - b_{DC} = 0. \tag{14.21}$$

This is a *nonlinear equation system,* where the unknown to be solved for is the vector x_{DC}; Solving such a system of equations is generally a nontrivial numerical task. In circuit simulation applications, a numerical method known as the Newton–Raphson algorithm is most often used to solve this system.

14.4.1 The Newton–Raphson Algorithm for Nonlinear Equation Solution

The Newton–Raphson (NR) algorithm (e.g., [23]) is a technique that is widely used in many disciplines for finding solutions to nonlinear equations. It is an *iterative* method — that is to say, it starts from a guess regarding the solution x_0 of Equation (14.21), and keeps refining the guess until it finds a solution. It is important to note that there is no guarantee that NR will succeed in finding a solution, although it often does so in practice. Therefore, the algorithm is typically implemented to announce failure after some number of iterations (e.g., after 100 tries).

NR relies on having available to it two functionalities related to the nonlinear function $g(x)$ in Equation (14.21):

1. It needs a means to evaluate $g(x)$, given any x
2. It needs a means to evaluate the derivative of $g(x)$, i.e., $J(x) = \mathrm{d}g(x)/\mathrm{d}x$, given any x. Note that since both x and $g(x)$ are vectors of size n, J is an $n \times n$ matrix. The matrix J is known as the *Jacobian matrix* of $g(x)$.

The following is an outline of the basic NR algorithm, in MATLAB-like pseudo-code:

```
1     function solution = NR(xguess)
2     x = xguess;
3     finished = 0;
4     epsilon = 1e-10;
5     while (~finished) % iteration loop
6           gx = g(x); % evaluate g(x)
7           if (norm(gx) < epsilon)
8                 finished = 1; % solution to accuracy epsilon found
9                 solution = x;
10                break;
11          end % of if
12          Jx = dgdx(x); % find the Jacobian matrix J(x)
13          delta_x = - inverse(Jx)*gx; % calculate update delta_x
14          %to current guess by solving J(x) delta_x = - g(x)
15          x = x + delta_x; % update current guess; re-do loop
16    end % of while
```

Observe that the key step is line 13, where the current guess for the solution is updated by an amount Δx, calculated by solving

$$J(x)\Delta x = -g(x) \quad \text{or} \quad \Delta x = -J^{-1}(x)g(x). \tag{14.22}$$

The above involves solving a *linear matrix equation*, with the Jacobian matrix $J(x)$, at every iteration of the NR algorithm. The computational ease with which this can be done depends on the structure of the Jacobian matrix, which we will discuss in more detail below.

14.4.2 Derivative ("Jacobian") Matrices and Device "Stamps"

To illustrate the structure of the Jacobian matrix, consider again the example of Figure 14.1, together with its $f(x)$ defined in Equation (14.8). The Jacobian matrix of $f(x)$ in Equation (14.8) is

$$J(x) = \begin{pmatrix} \mathrm{d}'(v_1 - v_2) & -\mathrm{d}'(v_1 - v_2) & 1 \\ -\mathrm{d}'(v_1 - v_2) & \frac{1}{R} + \mathrm{d}'(v_1 - v_2) & 0 \\ 1 & 0 & 0 \end{pmatrix} \tag{14.23}$$

where $\mathrm{d}'(\cdot)$ is the derivative of the diode's current–voltage relationship, defined in Equation (14.2). Observe, first, that each element in the circuit has a characteristic pattern of entries that it contributes to the Jacobian matrix. For example, the resistor R from node 2 to ground contributes the pattern

$$\begin{pmatrix} \cdot & \cdot & \cdot \\ \cdot & 1/R & \cdot \\ \cdot & \cdot & \cdot \end{pmatrix}$$

while the diode between nodes 1 and 2 contributes the pattern

$$\begin{pmatrix} d'(v_1 - v_2) & d'(v_1 - v_2) & \cdot \\ -d'(v_1 - v_2) & -d'(v_1 - v_2) & \cdot \\ \cdot & \cdot & \cdot \end{pmatrix} \tag{14.25}$$

The pattern of Jacobian entries contributed by each element is called its *matrix stamp*. The Jacobian matrix $J(x)$ thus consists of the addition of the stamps of all the elements in the circuit.

14.4.3 Jacobian Sparsity and Its Importance

Observe also that several entries in the Jacobian matrix in Equation (14.23) are zero. As the size of the circuit grows larger, the number of zero entries in the matrix predominates. The reason for this is easy to see — if the circuit has roughly n nodes and each node is connected to only 2 or 3 circuit elements (as is typical), the total number of nonzero matrix stamps is of the order of 2n or 3n. As a result, the remaining entries of the matrix (which has a total of n^2 entries) must be zero or "unstamped". Such matrices, where there are relatively few nonzero entries and many zero entries, are called *sparse matrices*. When a matrix has no zero entries, it is called *dense*. We have just noted that because each node in typical circuits is connected only to a few elements, Jacobian matrices of circuits tend to be sparse.

The fact that circuit matrices are usually sparse is of enormous importance for circuit simulation. The reason is that, while the Newton update step in Equation (14.22) is in general very computationally expensive for arbitrary matrices J (especially if J is dense), efficient techniques, collectively dubbed "sparse matrix technology," exist for solving for Δx in Equation (14.22) when J is sparse. More specifically, the computational complexity of general linear equation solution (e.g., when J is dense) is $O(n^3)$, while that for typical sparse J is $O(n)$. When the circuit size n reaches the thousands or tens of thousands, this difference in computational complexity is extremely significant from a practical standpoint. Thus, sparsity of circuit matrices is critical for enabling efficient and practically effective linear circuit solution so far we have noted the importance of this linear solution only for DC analysis (as part of the NR loop in line 13), we will see that exploiting sparsity is critical for virtually all other analyses related to circuits.

14.5 Steady-State Analysis

It is often important in RF design to find the periodic steady state of a circuit driven by one or more periodic inputs. For example, a power amplifier driven to saturation by a large single-tone input is operating in a periodic steady state. A variant is the quasiperiodic steady state, i.e., when the circuit is driven by more than one signal tone; for example, an amplifier driven by two closely spaced sinusoidal tones at 1 GHz and 990 MHz. Such excitations are closer approximations of real-life signals than pure tones and are useful for estimating intermodulation distortion.

The workhorse of analog verification, SPICE (and its derivatives), can of course be applied to find the (quasi) periodic steady state of a circuit, simply by performing a time-stepping integration of the circuit's differential equations ("transient simulation") long enough for the transients to subside and the circuit's response to become (quasi)periodic. This approach has several disadvantages, however. In typical RF circuits, the transients take thousands of periods to die out, and hence the procedure can be very inefficient. Further, harmonics are typically orders of magnitude smaller than the fundamental, hence long transient simulations are not well suited for their accurate capture, because numerical errors from time-stepping integration can mask them. These issues are exacerbated in the presence of quasiperiodic excitations, because simulations need to be much longer — e.g., for excitations of 1 GHz and 990 MHz, the system needs to be simulated for thousands of multiples of the common period, 1/10 MHz, yet the simulation time-steps must be much smaller than 1 nsec, the period of the fastest tone. For these reasons, more efficient and accurate specialized techniques have been developed. We will focus on two different methods with complementary properties, Harmonic Balance (HB) and shooting.

14.5.1 Harmonic Balance and Shooting

In the well-known method of HB (e.g., [24–32]), $x(t)$ and $b(t)$ of Equation (14.1) are expanded in a Fourier series. The Fourier series can be one-tone for periodic excitations (e.g., 1 GHz and its harmonics) or multitone in the case of quasiperiodic excitations (e.g., 1 GHz and 990 MHz, their harmonics, and intermodulation mixes). The DAE is rewritten directly in terms of the Fourier coefficients of $x(t)$ (which are unknown) and of $b(t)$; the resulting system of nonlinear equations is larger by a factor of the number of harmonic/mix components used, but are algebraic (i.e., there are no differential components). Hence they can be solved numerically using, e.g., the well-known NR method [33].

Example 14. 5.1

We illustrate the one-tone procedure with the following scalar DAE:

$$\dot{x} + x - \varepsilon x^2 - \cos(2\pi 1000t) = 0 \tag{14.26}$$

First, we expand all time variations in a Fourier series of (say) M = 3 terms, i.e., the DC component, fundamental, and second harmonic components:

$$x(t) = \sum_{i=-2}^{2} X_i e^{j2\pi i 10^3 t}$$

Here, X_i are the unknown Fourier coefficients of $x(t)$. For notational convenience, we express them as the vector $X = [X_2, \ldots, X_{-2}]^T$.

Similarly, $x^2(t)$ is also expanded in a Fourier series in t; the Fourier coefficients of this expansion are functions of the elements of X, which we denote by $F_i(X)$.

$$x^2(t) = \sum_{i=-2}^{2}\sum_{k=-2}^{2} X_i X_k e^{j2\pi(i+k)10^3 t} = \sum_{i=-2}^{2} F_i(X) e^{j2\pi i 10^3 t} + \text{higher terms}$$

In this case, where the nonlinearity is a simple quadratic, F_i can be obtained analytically; but in general, numerical techniques like those used in HB need to be employed for computing these functions. For convenience, we write $F(X) = [F_2(X), \ldots, F_{-2}(X)]^T$.

We also write the Fourier coefficients of the excitation $\cos(2\pi 1000t)$ as a vector $B = [0, 1/2, 0, 1/2, 0]^T$. Finally, we write the differential term x also as a vector of Fourier coefficients. Because the differentiation operator is diagonal in the Fourier basis, this becomes simply ΩX, where $\Omega = j2\pi 1000\ \mathrm{diag}(2, 1, 0, -1\ -2)$ is the diagonal frequency-domain differentiation matrix.

Invoking the orthogonality of the Fourier basis, we now obtain the HB equations for our DAE:

$$H(X) \equiv \Omega X + X - \varepsilon F(X) - B = 0$$

This is a set of nonlinear algebraic equations in five unknowns, and can be solved by numerical techniques such as NR.

The above example illustrates that the size of the HB equations is larger than that of the underlying DAE, by a factor of the number of harmonic/mix components used for the analysis. In fact, the HB equations are not only larger in size than the DAE, but also considerably more difficult to solve using standard numerical techniques. The reason for this is the dense structure of the derivative, or Jacobian matrix, of the HB equations. If the size of the DAE is n, and a total of N harmonics and mix components are used for the HB analysis, the Jacobian matrix has $Nn \times Nn$. Just the storage for the nonzero entries can become prohibitive for relatively moderate values of n and N; for example, $n = 1000$ (for a medium-sized circuit) and $N = 100$ (e.g., for a two-tone problem with about 10 harmonics each) require 10 GB of storage for the matrix alone. Further, inverting the matrix, or solving linear systems with it, requires $O(N^3n^3)$ operations, which is usually infeasible for moderate to large-sized problems. Such linear solutions are typically required as steps in solving the HB equations, for example by the NR method. Despite this disadvantage, HB is a useful tool for small circuits and few harmonics, especially for microwave circuit design. Moreover, as we will see later, new algorithms have been developed for HB that make it much faster for larger problems.

Another technique for finding periodic solutions is the shooting method (e.g., [34–37]). Shooting works by finding an initial condition for the DAE that also satisfies the periodicity constraint. A guess is made for the initial condition, the system is simulated for one period of the excitation using time-stepping DAE solution methods, and the error from periodicity used to update the initial condition guess, often using an NR scheme.

More precisely, shooting computes the *state transition function* $\Phi(t,x_0)$ of Equation (14.1). $\Phi(t,x_0)$ represents the solution of the system at time t, given initial condition x_0 at time 0. Shooting finds an initial condition x^* that leads to the same state after one period T of the excitation $b(t)$; in other words, shooting solves the equation $H(x) \equiv \Phi(t,x) - x = 0$.

The shooting equation is typically solved numerically using the NR method, which requires evaluations of $H(x)$ and its derivative (or Jacobian) matrix. Evaluation of $H(x)$ is straightforward using time-stepping, i.e., transient simulation, of Equation (14.1). However, evaluating its Jacobian is more involved. The Jacobian matrix is of the same size n as the number of circuit equations, but it is dense, hence storage of its elements and linear solutions with it is prohibitive in cost for large problems. In this respect, shooting suffers from size limitations similar to HB. In other respects though, shooting has properties complementary to HB. The following list compares and contrasts the main properties of HB and shooting:

- *Problem* size: The problem size is limited for both HB and shooting, due to the density of their Jacobian matrices. However, since the HB system is larger by a factor of the number of harmonics used, shooting can handle somewhat larger problems given the same resources. Roughly speaking, sizes of about 40 circuit elements for HB and 400 for shooting represent practical limits.
- *Accuracy/dynamic range:* Because HB uses orthogonal Fourier bases to represent the waveform, it is capable of very high dynamic range — a good implementation can deliver 120 dB of overall numerical accuracy. Shooting, being based on time-stepping solution of the DAE with time-steps of different sizes, is considerably poorer in this regard.
- *Handling of nonlinearities:* HB is not well suited for problems that contain strongly nonlinear elements. The main reason for this is that strong nonlinearities (e.g., clipping elements) generate sharp waveforms that do not represent compactly in a Fourier series basis. Hence many harmonics/mix components need to be considered for an accurate simulation, which raises the overall problem size.

  Shooting, on the other hand, is well suited for strong nonlinearities. By approaching the problem as a series of initial value problems for which it uses time-stepping DAE methods, shooting is able to handle the sharp waveform features caused by strong nonlinearities quite effectively.
- *Multitone problems:* A big attraction of HB is its ability to handle multitone or quasiperiodic problems as a straightforward extension of the one-tone case, by using multitone Fourier bases to represent quasiperiodic signals. Shooting, on the other hand, is limited in this regard. Since it uses time-stepping DAE solution, shooting requires an excessive number of timepoints when the waveforms involved have widely separated rates of change; hence it is not well suited for such problems.

14.5.2 Fast Methods

A disadvantage of both HB and shooting is their limitation to circuits of relatively small size. This was not a serious problem as long as microwave/RF circuits contained only a few nonlinear devices. Since the mid-1990s however, economic and technological developments have changed this situation. The market for cheap, portable wireless communication devices has expanded greatly, leading to increased competition and consequent cost pressures. This has spurred on-chip integration of RF communication circuits and the reduction of discrete (off-chip) components. On-chip design techniques favor the use of many integrated nonlinear transistors over even a few linear external components. Hence the need has arisen to apply HB and shooting to large circuits in practical times.

To address this issue, so-called *fast* algorithms have arisen to enable both HB and shooting to handle large circuits. The key property of these methods is that computation/memory usage grows approximately linearly with problem size. The enabling idea behind the improved speed is to express the dense Jacobian matrices as sums and products of simpler matrices that are either sparse, or have very regular structure and so can be applied/inverted efficiently. Using these expansions for the Jacobian, special solution algorithms called *preconditioned iterative linear solvers* are applied to solve linear equations involving the Jacobian, without forming it explicitly.

A detailed description of fast techniques is beyond the scope of this chapter; the interested reader is referred to [29, 31,32,37,38] for further information. Here we outline the main ideas behind these methods in a simplified form using Example 14.5.1 for illustration, and summarize their main properties.

From Example 14.5.1, the Jacobian matrix of the HB system is

$$J = \frac{\partial H(X)}{\partial X} = \Omega + I - \varepsilon \frac{\partial F(X)}{\partial X}$$

Now, $F(X)$ in this case represents the vector of Fourier coefficients of the nonlinear term $f(x) = x^2$. One way in which these can be computed numerically is (1) to use the inverse Fast Fourier Transform (FFT) to convert the Fourier coefficients X into samples of the time-domain waveform $x(t)$, then (2) evaluate the nonlinear function $f(x) = x^2(t)$ at each of these samples in the time domain, and finally, (3) use the FFT to reconvert the time-domain samples of $f(x)$ back to the frequency domain, to obtain $F(X)$. The derivative of these operations can be expressed as

$$\frac{\partial F(X)}{\partial X} = DGD^*$$

where D is a block-diagonal matrix with each block equal to the discrete Fourier transform (DFT) matrix, D^* is its inverse, and G is a diagonal matrix with entries $f'()$ evaluated at the time-domain samples of $x(t)$. Hence the overall Jacobian matrix can be represented as

$$J = \Omega + I - \varepsilon DGD^*$$

Observe that each of the matrices in this expansion is either sparse or consists of DFT matrices. Hence multiplication of J with a vector is efficient, since the sparse matrices can be applied in approximately linear time, and the DFT matrix and its inverse can be applied in $N \log N$ time using the FFT, where N is the number of harmonics. It is this key property, that multiplications of J with a vector can be performed in almost-linear computation despite its dense structure, that enables the use of preconditioned iterative linear methods for this problem.

Preconditioned iterative linear methods (e.g., [39–41]) are a set of numerical techniques for solving linear systems of the form $Jc = d$. Modern iterative solvers like QMR [40] and GMRES [39] use Krylov-subspace techniques for superior performance. The key feature of these solvers is that the only way in which J is used is in matrix–vector products Jz. This constrasts with traditional methods for linear solution that use Gaussian elimination or variants like LU factorizations directly on elements of J. Due to this property of iterative linear solvers, it is not necessary to even form J explicitly in order to solve linear systems with it, so long as a means is available for computing matrix–vector products with it.

As we have seen above, products with the HB Jacobian can be conveniently computed in almost-linear time without having to build the matrix explicitly. Hence preconditioned linear iterative techniques are well suited to solving the linear systems that arise when solving the nonlinear HB equations using the NR method. If the iterative linear methods use only a few matrix–vector products with the Jacobian to compute the linear system's solution, and the NR is well behaved, the overall cost of solving the HB equations remains almost linear in problem size.

An important issue with preconditioned iterative linear solvers, especially those based on Krylov subspace methods, is that they require a good *preconditioner* to converge reliably in a few iterations. The convergence of the iterative linear method is accelerated by applying a preconditioner $\tilde{J}$, replacing the original system $Jc = d$ with *the preconditioned* system $\tilde{J}^{-1}\tilde{J}c = \tilde{J}^{-1}d$, which has the same solution. For robust and efficient convergence, the preconditioner matrix $\tilde{J}$ should be in some sense a good approximation of J, and also "easy" to invert, usually with a direct method such as LU factorization. Finding good preconditioners that work well for a wide variety of circuits is a challenging task, especially when the nonlinearities become strong.

The ideas behind the fast techniques outlined above are applicable not just to HB but also to shooting [37]. Jacobian matrices in to from shooting can be decomposed into products and sums of the sparse circuit Jacobians matrices. Preconditioned linear iterative techniques can then be applied to invert the Jacobian efficiently.

As an example of the application of the fast methods, consider the HB simulation of an RFIC quadrature modulator reported in [32]. The circuit of about 9500 devices was simulated by fast HB a three-tone excitation, a baseband signal at 80 kHz, and local oscillators at 178 MHz and 1.62 GHz. The size of the circuit's DAE was $n = 4800$; the three tones, their harmonics, and mixes totalled $N = 4320$ components. Simulating a circuit with these specifications is completely infeasible using traditional HB techniques.

Using fast HB, the simulation required only 350 MB of memory, and took 5 days of computation on an SGI 150 MHz R4400 machine. The results of the the simulation are shown in Figure 14.8.

14.6 Multitime Analysis

In the previous section, we noted that HB and shooting have complementary strengths and weaknesses, stemming from their use of Fourier and time-domain bases respectively. While HB is best suited for multitone problems that are only mildly nonlinear, shooting is best for single-tone problems that can be strongly nonlinear. Neither method is suitable for circuits that have both multitone signals (i.e., widely separated time scales of variation) *and* strongly nonlinear components. With greater RF integration, tools that can analyze precisely this combination of circuit characteristics effectively are required. In this section, we review a promising family of techniques, based on partial differential equations (PDEs) using multiple artificial time scales [30, 42–46].

Consider the waveform $y(t)$ shown in Figure 14.9, a simple two-tone signal given by

$$y(t) = \sin\left(\frac{2\pi}{T_1}t\right)\sin\left(\frac{2\pi}{T_2}t\right),\ T_1 = 0.02 \text{ sec},\ T_2 = 1 \text{ sec} \tag{14.27}$$

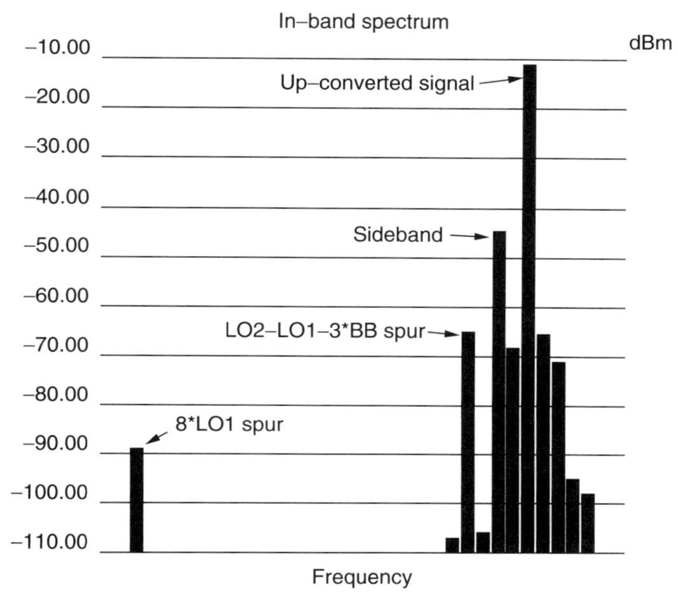

FIGURE 14.8 Quadrature modulator spectrum.

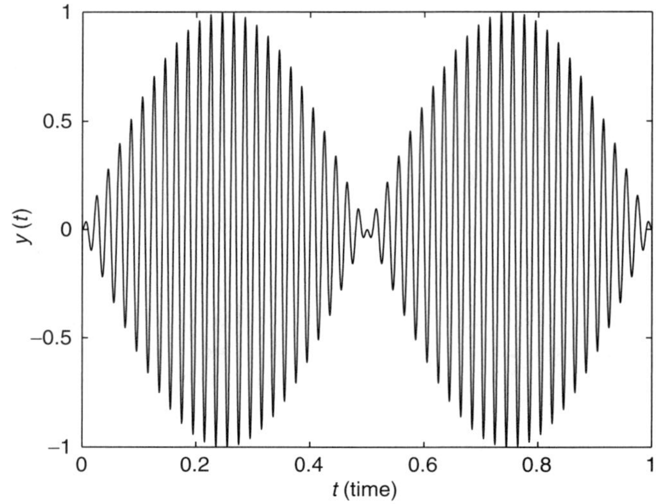

FIGURE 14.9 Example of a two-tone quasiperiodic signal $y(t)$.

The two tones are at frequencies $f_1 = 1/T_1 = 50$ Hz and $f_2 = 1/T_2 = 1$ Hz, i.e., there are 50 fast-varying cycles of period $T_1 = 0.02$ sec modulated by a slowly varying sinusoid of period $T_2 = 1$ sec. If each fast cycle is sampled at n points, the total number of time steps needed for one period of the slow modulation is nT_2/T_1 To generate Figure 14.9, 15 points were used per cycle, hence the total number of samples was 750. This number can be much larger in applications where the rates are more widely separated, e.g., separation factors of 1000 or more are common in electronic circuits. Now consider a multivariate representation of $y(t)$ using two artificial timescales, as follows: for the 'fast-varying' parts of $y(t)$, t is replaced by a new variable t_1; for the 'slowly varying' parts, by t_2. The resulting function of two variables is denoted by

$$\hat{y}(t_1, t_2) = \sin\left(\frac{2\pi}{T_1}t_1\right)\sin\left(\frac{2\pi}{T_2}t_2\right). \tag{14.28}$$

The plot of $\hat{y}(t_1,t_2)$ on the rectangle $0 \le t_1 \le T_1$, $0 \le t_2 \le T_2$ is shown in Figure 14.10. Observe that $\hat{y}(t_1,t_2)$ does not have many undulations, unlike $y(t)$ in Figure 14.9. *Hence it can be represented by relatively few points, which, moreover, do not depend on the relative values of* T_1 *and* T_2. Figure 14.10 was plotted with 225 samples on a uniform 15 × 15 grid — three times fewer than for Figure 14.9. This saving increases with increasing separation of the periods T_1 and T_2.

Further, note that $\hat{y}\,(t_1,t_2)$ is periodic with respect to both t_1 and t_2, i.e., $\hat{y}(t_1 + T_1, t_2 + T_2) = \hat{y}(t_1,t_2)$. This makes it easy to recover $y(t)$ from $\hat{y}(t_1,t_2)$, simply by setting $t_1 = t_2 = t$, and using the fact that $\hat{y}$ is biperiodic. It is easy, from direct inspection of the three-dimensional plot of $\hat{y}(t_1,t_2)$, to visualize what $y(t)$ looks like. As t increases from 0, the path given by $\{t_i = t \bmod T_i\}$ traces the sawtooth path shown in Figure 14.11. By noting how $\hat{y}$ changes as this path is traced in the t_1–t_2 plane, $y(t)$ can be traced.

When the time scales are widely separated, inspection of the bivariate waveform directly provides information about the slow and fast variations of $y(t)$ more naturally and conveniently than $y(t)$ itself.

We observe that the bivariate form may require far fewer points to represent numerically than the original quasiperiodic signal, yet it contains all the information needed to recover the original signal completely. This observation is the basis of the partial differential formulation to be introduced shortly. The waveforms in a circuit are represented in their bivariate forms (or multivariate forms if there are more than two timescales). The key to efficiency is to solve for these waveforms directly, without involving the numerically inefficient one-dimensional forms at any point. To do this, it is necessary to first describe the circuit's equations using the multivariate functions. If the circuit is described by the differential Equation (14.1),

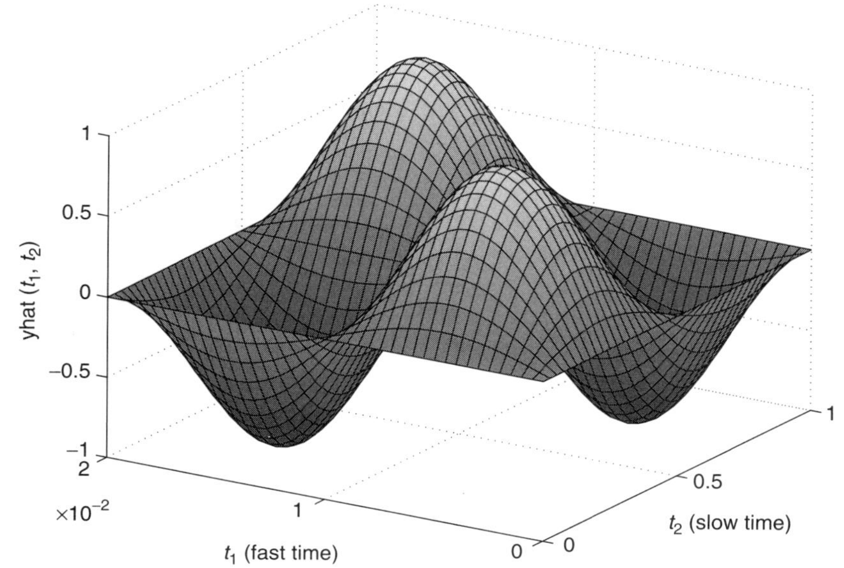

FIGURE 14.10 Corresponding two-periodic bivariate form $\hat{y}(t_1, t_2)$.

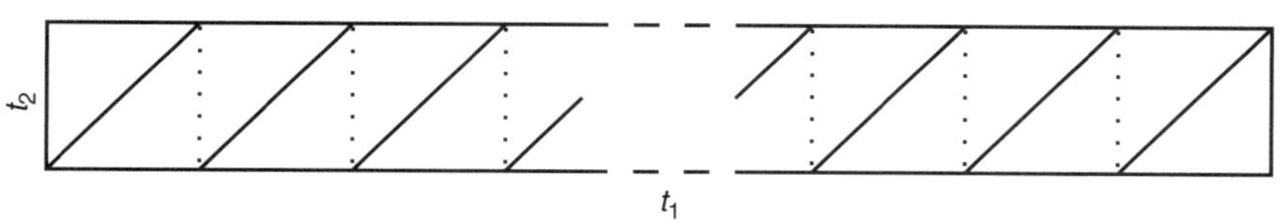

FIGURE 14.11 Path in the t_1–t_2 plane.

then it can be shown that if $\hat{x}(t_1,t_2)$ and $\hat{b}(t_1,t_2)$ denote the bivariate forms of the circuit unknowns and excitations, then the following MPDE is the correct generalization of Equation (14.1) to the bivariate case:

$$\frac{\partial q(\hat{x})}{\partial t_1} + \frac{\partial q(\hat{x})}{\partial t_2} + f(\hat{x}) = \hat{b}(t_1,t_2) \tag{14.29}$$

More precisely, if $\hat{b}$ is chosen to satisfy $b(t) = \hat{b}(t,t)$, and x satisfies Equation (14.29), then it can be shown that $x(t) = \hat{x}(t,t)$ satisfies Equation (14.1). Also, if Equation (14.1) has a quasiperiodic solution, then Equation (14.29) can be shown to have a corresponding bivariate solution.

By solving the MPDE numerically in the time domain, strong nonlinearities can be handled efficiently. Several numerical methods are possible, including discretization of the MPDE on a grid in the t_1–t_2 plane, or using a mixed time–frequency method in which the variation along one of the timescales is expressed in a short Fourier series. Quasiperiodic and envelope solutions can both be generated, by appropriate selection of boundary conditions for the MPDE. Sparse matrix and iterative linear methods are used to keep the numerical algorithms efficient even for large systems.

As an example, Figure 14.12 depicts the output voltage of a switched-capacitor integrator block, obtained from a multitime simulation based on the above concepts. The cross-section parallel to the signal timescale represents the envelope of the signal riding on the switching variations. By moving these cross-sections to different points along the clock timescale, the signal envelope at different points of the clock waveform can be seen.

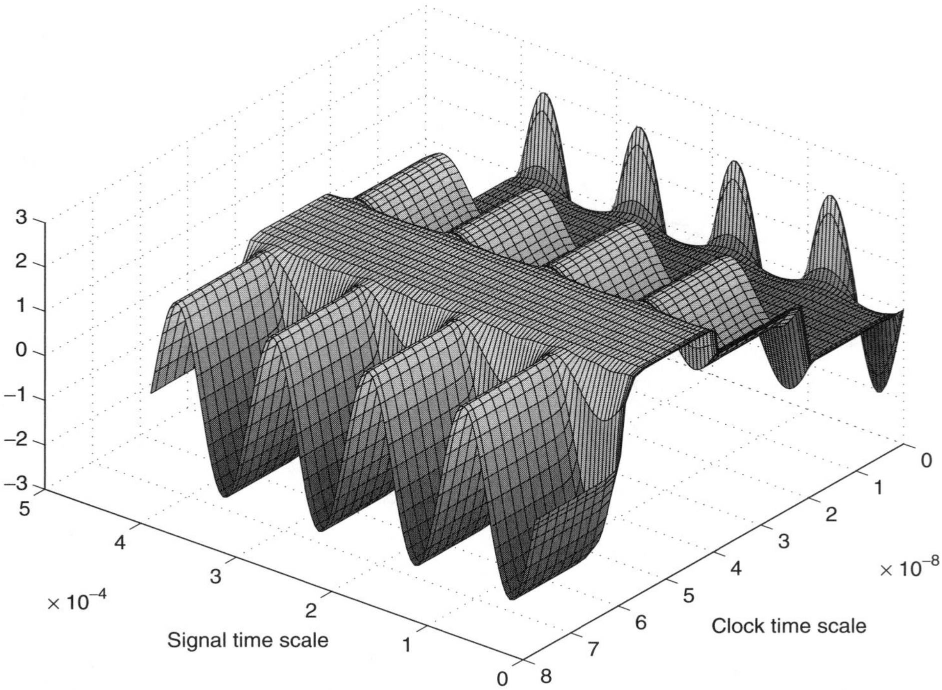

FIGURE 14.12 Multitime output waveform of SC integrator.

14.6.1 Autonomous Systems: The Warped MPDE

When the DAEs under consideration are oscillatory, frequency modulation (FM) can be generated. Unfortunately, FM cannot be represented compactly using multiple timescales as easily as the waveform in Figure 14.10. We illustrate the difficulty with an example. Consider the following prototypical FM signal:

$$x(t) = \cos\,(2\pi f_0 t + k\cos(2\pi f_2 t)), \quad f_0 \gg f_2 \tag{14.30}$$

with instantaneous frequency

$$f(t) = f_0 - kf_2 \sin(2\pi f_2 t) \tag{14.31}$$

$x(t)$ is plotted in Figure 14.13 for $f_0 = 1$ MHz, $f_2 = 20$ kHz, and modulation index $k = 8\pi$. Following the same approach as for Equation (14.27), a bivariate form can be defined to be

$$\hat{x}(t_1,t_2) = \cos(2\pi f_0 t_1) + k\cos(2\pi f_2 t_2)), \quad \text{with } x(t) = \hat{x}_1(t,t). \tag{14.32}$$

Note that $\hat{x}_1$ is periodic in t_1 and t_2, hence $x(t)$ is quasiperiodic with frequencies f_0 and f_2. Unfortunately, $\hat{x}_1(t_1,t_2)$, illustrated in Figure 14.14, is not a simple surface with only a few undulations like in Figure 14.10. When $k \gg 2\pi$, i.e., $k \approx 2\pi m$ for some large integer m, then $\hat{x}(t_1,t_2)$ will undergo about m oscillations as a function of t_2 over one period T_2. In practice, k is often of the order of $f_0/f_2 \gg 2\pi$, and hence this number of undulations can be very large. Therefore, it becomes difficult to represent $\hat{x}_1$ efficiently by sampling on a two-dimensional grid. It turns out that resolving this problem requires the stretching, or warping of one of the timescales. We illustrate this by returning to Equation (14.30). Consider the following new multivariate representation:

$$\hat{x}_2(\tau_1,\tau_2) = \cos(2\pi\tau_1), \tag{14.33}$$

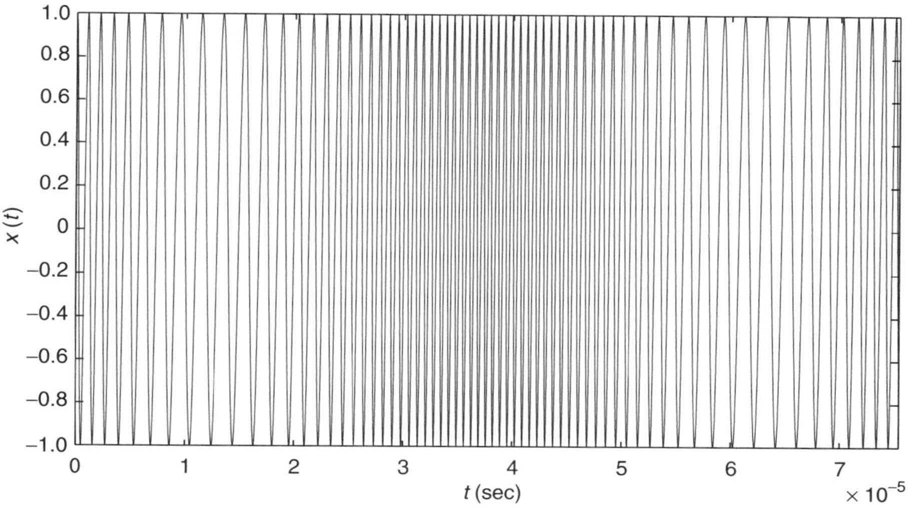

FIGURE 14.13 FM signal.

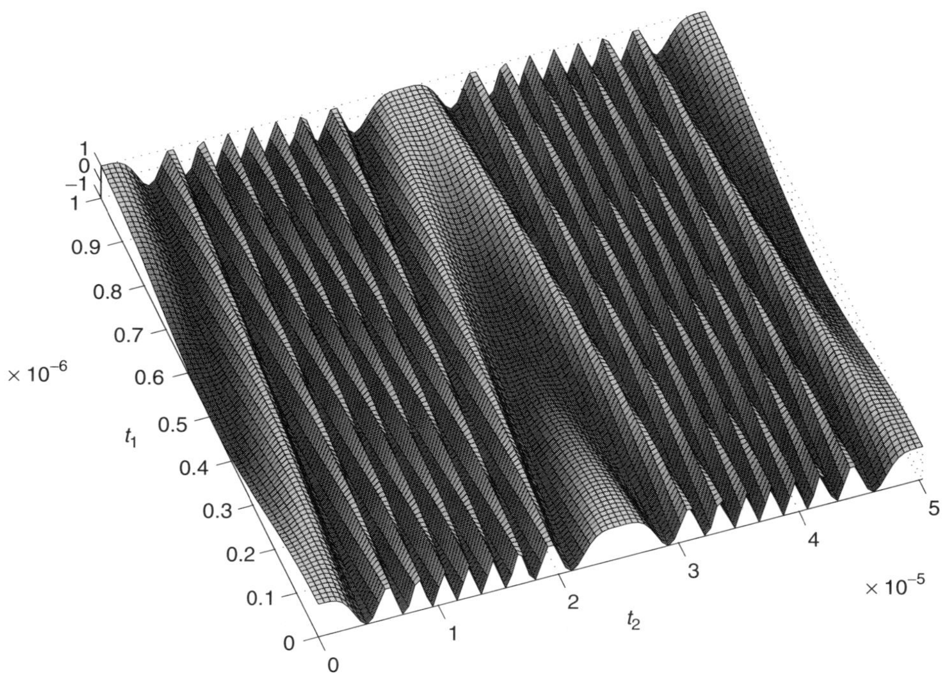

FIGURE 14.14 $\hat{x}_1$ unwarped bivariate representation of FM signal.

together with the warping function

$$\phi(\tau_2) = f_0\tau_2 + \frac{k}{2\pi}\cos(2\pi f_2\tau_2). \tag{14.34}$$

We now retrieve our one-dimensional FM signal i.e., Equation (14.30) as

$$x(t) = \hat{x}_2(\phi(t),t). \tag{14.35}$$

Note that both $\hat{x}_2$ and ϕ, given in Equation (14.33) and Equation (14.34), can easily be represented with relatively few samples, unlike $\hat{x}_1$ in Equation (14.32). What we have achieved with Equation (14.34) is simply a stretching of the time axis differently at different times, to even out the period of the fast undulations in Figure 14.13. The extent of the stretching or the derivative of $\phi(\tau_2)$ at a given point is simply the local frequency $\omega(\tau_2)$, which modifies the original MPDE to result in the warped multirate partial differential equation (WaMPDE)

$$\omega(\tau_2)\,\frac{\partial q(\hat{x})}{\partial \tau_1} + \frac{\partial q(\hat{x})}{\partial \tau_2} + f(\hat{x}(\tau_1,\tau_2)) = b(\tau_2). \tag{14.36}$$

The usefulness of Equation (14.36) lies in that specifying

$$x(t) = \hat{x}(\phi(t),t),\ \phi(t) = \int_0^t \omega(t_2)\,\mathrm{d}\tau_2 \tag{14.37}$$

results in $x(t)$ being a solution to Equation (14.1). Furthermore, when Equation (14.36) is solved numerically, the local frequency $\omega(\tau_2)$ is also obtained, which is desirable for applications such as VCOs and also difficult to obtain by any other means.

As an example, Figure 14.15 shows the changing local frequency in an LC tank VCO simulated with WaMPDE-based numerical techniques. The controlling input to the VCO was about 30 times slower than its nominal frequency. Figure 14.16 depicts the bivariate waveform of the capacitor voltage. It is seen that the controlling voltage changes not only the local frequency, but also the amplitude and shape of the oscillator waveform.

The circuit was also simulated by traditional numerical ODE methods ("transient simulation"). The waveform from this simulation, together with the one-dimensional waveform obtained by applying Equation (14.37) to Figure 14.16, are shown in Figure 14.17. Frequency modulation can be observed in the varying density of the undulations.

14.6.2 Macromodeling Time-Varying Systems

Another useful application of multiple time scales is in macromodeling linear time-varying (LTV) systems Equation [47, 48]. These approximations are adequate for many apparently nonlinear systems, like mixers and switched-capacitor filters, where the signal path is designed to be linear, even though other inputs (e.g., local oscillators, clocks) cause "nonlinear" parametric changes to the system. LTV approximations of large systems with few inputs and outputs are particularly useful, because it is possible to automatically generate *macromodels* or *reduced-order models* of such systems. The macromodels are much smaller dynamical systems than the originals, but retain similar input–output behavior within a

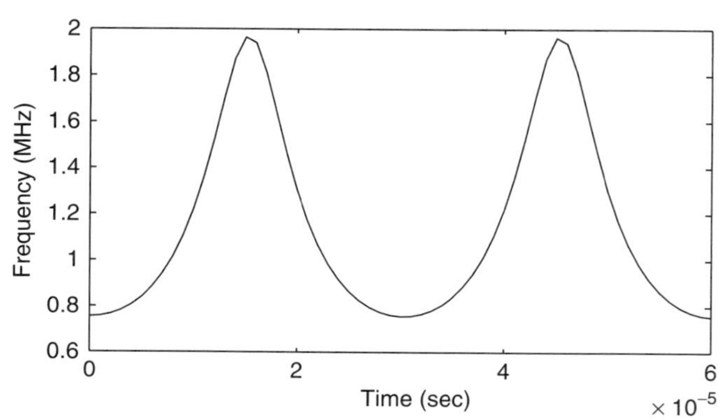

FIGURE 14.15 VCO: frequency modulation.

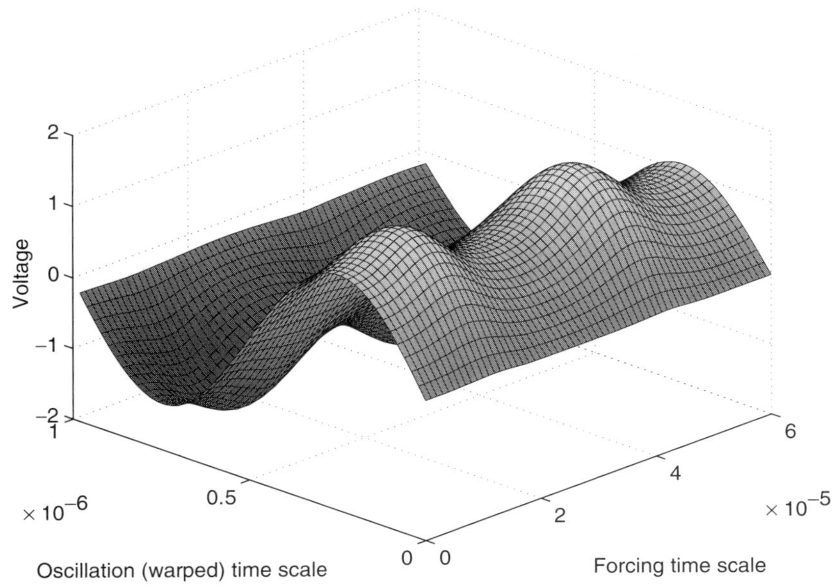

FIGURE 14.16 VCO: bivariate representation of capacitor voltage.

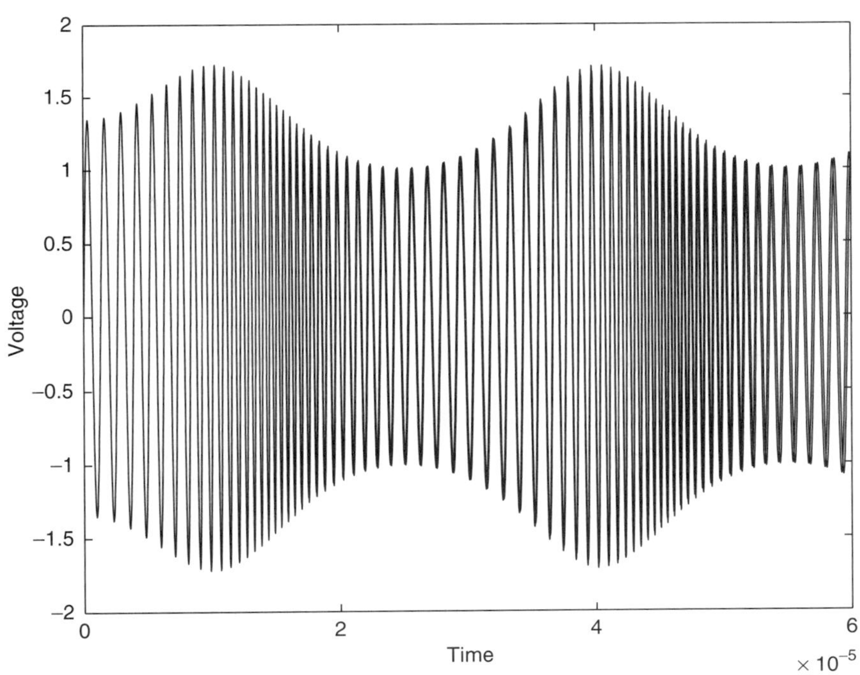

FIGURE 14.17 VCO: WaMPDE vs. transient simulation.

given accuracy. Such macromodels are useful in verifying systems hierarchically at different levels of abstraction, an important task in communication system design.

While mature techniques are available for the simpler task of reduced-order modeling of linear time-invariant (LTI) systems (e.g., [49–55]), a difficulty in extending them to handle LTV systems has been the interference of the time variations of the system and the input. By separating the two with artificial time variables, the MPDE provides an elegant solution to this problem.

The time-varying small-signal equations obtained by linearizing Equation (14.1) around a steady-state solution are given by

$$C(t)\dot{x}(t) + G(t)x(t) = ru(t)$$
$$y(t) = d^{\mathrm{T}}x(t) \tag{14.38}$$

In Equation (14.38), the input to the system is the scalar $u(t)$ while the output is $y(t)$. If the above equation is Laplace-transformed (following the LTI procedure), the system time variation in $C(t)$ and $G(t)$ interferes with the I/O time variation through a convolution. The LTV transfer function $H(t,s)$ is therefore hard to obtain; this is the difficulty alluded to earlier. The problem can be avoided by casting Equation (14.38) as an MPDE:

$$C(t_1)\left[\frac{\partial \hat{x}}{\partial t_1}(t_1,t_2) + \frac{\partial \hat{x}}{\partial t_2}(t_1,t_2)\right] + G(t_1)\hat{x}(t_1,t_2) = ru(t_2)$$
$$\hat{y}(t_1,t_2) = d^{\mathrm{T}}\hat{x}(t_1,t_2), \quad y(t) = \hat{y}(t,t) \tag{14.39}$$

Notice that the input and system time variables are now separated. By taking Laplace transforms in t_2 and eliminating x, the time-varying transfer function $H(t_1,s)$ is obtained:

$$Y(t_1,s) = \underbrace{\left\{d^{\mathrm{T}}\left[C(t_1)\left\{\frac{\partial}{\partial t_1} + s\right\} + G(t_1)\right]^{-1}[r]\right\}}_{H(t_1,s)}U(s) \tag{14.40}$$

Observe that $H(t_1, s)$ in Equation (14.40) is periodic in t_1; hence, discretizing the t_1 axis, it can also be represented as *several* time-invariant transfer functions $H_i(s) = H(t_{1i},s)$. Or, a frequency-domain discretization using harmonics of the t_1 -variation can be used. Once an equivalent system of LTI transfer functions has been obtained, existing reduced-order modeling techniques for LTI systems can be used to find a smaller system of equations, in the same form as Equation (14.39), that has the same input–output relationship to within a given accuracy.

The reduced-order modeling technique (dubbed *time-varying Padé*, or TVP) was run on an RFIC I-channel mixer circuit of size about $n = 360$ nodes, excited by a local oscillator at 178 Mhz [48]. A frequency-domain discretization of the t_1 axis in Equation (14.39) was employed in the model reduction process. Figure 14.18 shows frequency plots of $H_1(s)$, the up-conversion transfer function (the first harmonic w.r.t t_1 of $H(t_1, s)$). The points marked '+' were obtained by direct computation of the full system, while the lines were computed using the reduced models of size $q = 2$ and 10, respectively.* Even with $q = 2$, a size reduction of two orders of magnitude, the reduced model provides a good match up to the LO frequency. When the order of approximation is increased to 10, the reduced model is identical up to well beyond the LO frequency. The reduced models were more than three orders of magnitude faster to evaluate than the original system, and hence are useful for system-level verification.

The poles of the reduced models for $H_1(s)$, easily calculated on account of their small size, are shown in Table 14.1. These are useful in design because they constitute an excellent approximations of the full system's poles, which are difficult to determine otherwise.

* The order q of the reduced model is the number of state variables in its differential equation description.

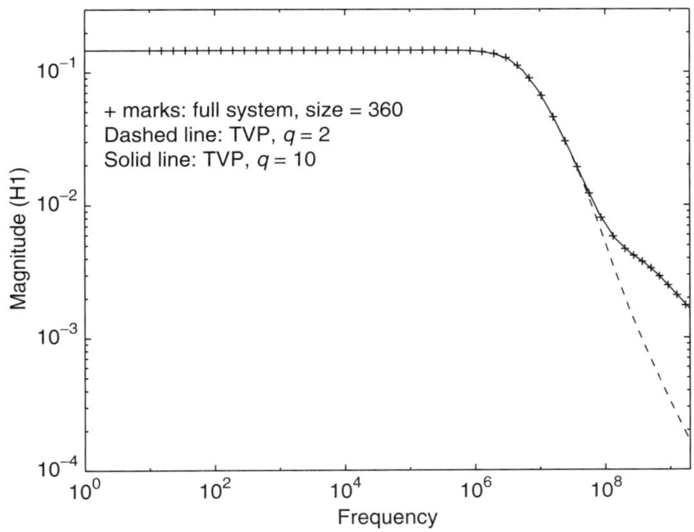

FIGURE 14.18 I-channel mixer H_1 (s): reduced vs. full system.

TABLE 14.1 Poles of H_1 (s) for the I-Channel Buffer/Mixer

TVP, $q = 2$	TVP, $q = 10$
− 5.3951 e + 06	− 5.3951 e + 06
− 6.9196e + 07 − *j*3.0085e + 05	− 9.4175e + 06
	− 1.5588e + 07− *j*2.5296e + 07
	− 1.5588e + 07 + *j*2.5296e + 07
	− 6.2659e + 08 − *j*1.6898e + 06
	− 1.0741e + 09− *j*2.2011e + 09
	− 1.0856e + 09 + *j*2.3771 e + 09
	− 7.5073e + 07− *j*1.4271e + 04
	− 5.0365e + 07 + *j*1.8329e + 02
	− 5.2000e + 07 + *j*7.8679e + 05

14.7 Noise in RF Design

Predicting noise correctly in order to minimize its impact is central to RF design. Traditional circuit noise analysis is based on three assumptions: that noise sources and their effects are *small enough* not to change the operating point; that all noise sources are *stationary;* and that the small-signal linearization of the circuit is *time invariant.* These assumptions break down when there are large signal variations, as is typical in RF circuits. Because of changing operating points, small-signal linearizations do not remain constant but become *time-varying.* In addition, noise sources that depend on operating point parameters (such as shot noise and flicker noise) also vary with time and no longer remain stationary. Finally, even though noise sources remain small, their impact upon circuit operation may or may not. In nonautonomous (driven) circuits, circuit effects of small noise remain small, allowing the use of linearized *mixing noise* analysis. In autonomous circuits (oscillators), however, noise creates frequency changes that lead to large deviations in waveforms over time — this phenomenon is called *phase noise.* Because of this, analysis based on linearization is not correct, and nonlinear analysis is required.

Figure 14.19 illustrates mixing noise. A periodic noiseless waveform in a circuit is shown as a function of time. The presence of small noise corrupts the waveform, as indicated. The extent of corruption at any

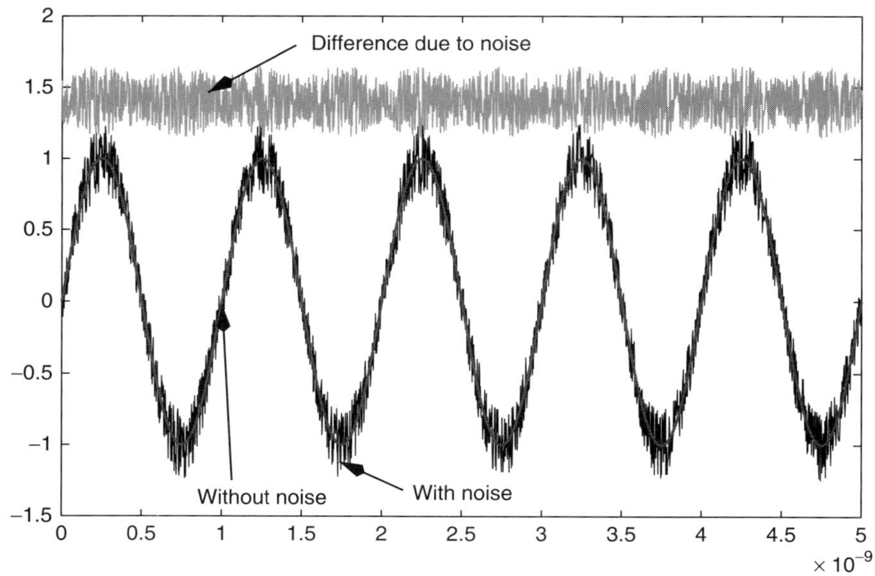

FIGURE 14.19 Time-domain view of mixing noise.

given time remains small, as shown by the third trace, which depicts the difference between the noiseless and noisy waveforms. The noise power can, however, vary depending on the large signal swing, as indicated by the roughly periodic appearance of the difference trace — depicting *cyclostationary* noise, where the statistics of the noise are periodic.

Figure 14.20 illustrates oscillator phase noise. Note that the noisy waveform's frequency is now slightly different from that of that of the noise-free one leading to increasing deviations between the two with the progress of time. As a result, the difference between the two does not remain small, but reaches magnitudes of the order of the large signal itself. Small additive corruptions remain here also just as in the mixing noise case, but the main distinguishing characteristic of oscillator noise is the frequency deviation.

The difference between mixing and phase noise is also apparent in the frequency domain, shown in Figure 14.21. Noise-free periodic waveforms appear as the impulse in the upper graph. If this is corrupted by small mixing noise, the impulse is not modified, but a small possibly broadband noise floor appears. In the case of free-running oscillators, the impulse disappears in the presence of any noise, no matter how small. It is replaced by a continuous spectrum that peaks at the oscillation frequency, and retains the power of the noise-free signal. The width and shape of this *phase noise spectrum* (i.e., the spread of power over neighboring frequencies) is related to the amount and nature of noise in the circuit.

14.7.1 Mixing Noise

Correct calculation of noise in nonlinear circuits with large signal swings (e.g., mixers and gain-compressed amplifiers) requires a sufficiently powerful stochastic process model. In the following, we use cyclostationary time-domain processes (e.g., [56–59]), although a different but equivalent formulation, i.e., that of correlated processes in the frequency domain (e.g., [27, 60]), is often used. The statistics of cyclostationary processes (in particular, the second-order statistics) are periodic or quasiperiodic, and hence can be expressed in Fourier series. The coefficients of the Fourier series, termed *cyclostationary components,* capture the variations of noise power over time. The DC term of the Fourier series, or the *stationary* component, is typically the most relevant for design, since it captures the average noise power over a long time. It is important to realize, though, that calculating the correct value of the stationary component of noise over a circuit does require *all* the Fourier components to be properly accounted for.

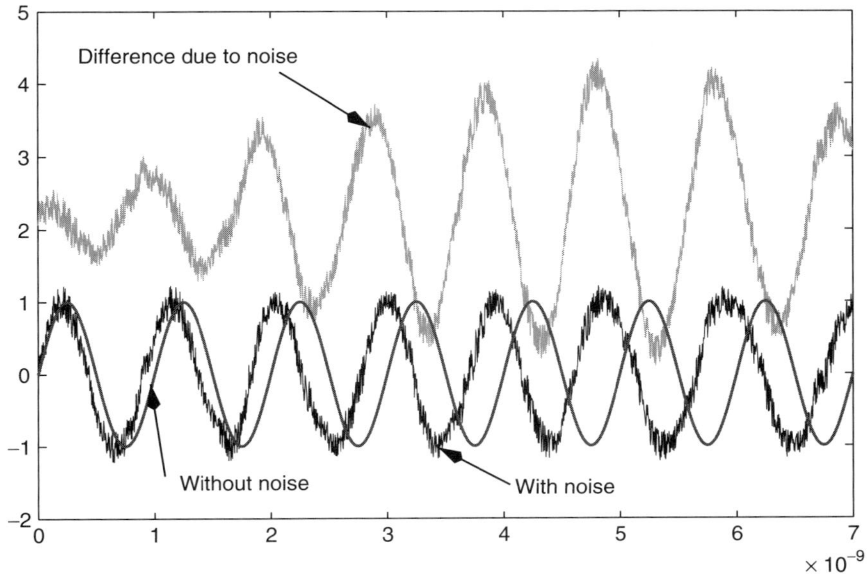

FIGURE 14.20 Time-domain view of phase noise.

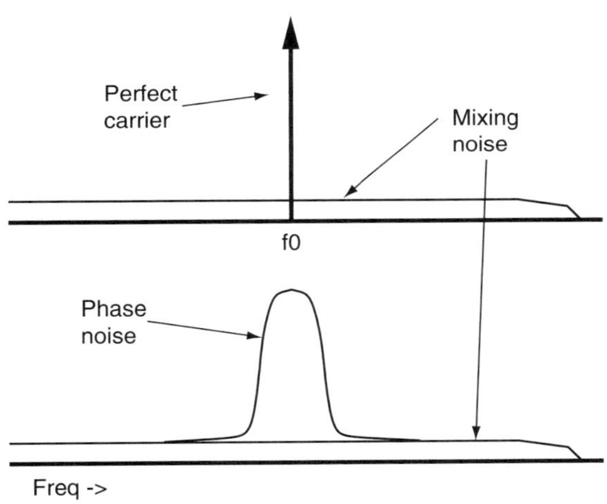

FIGURE 14.21 Frequency-domain view of mixing and phase noise.

Basing calculations only on the stationary component at each node or branch current in the circuit will, in general, produce wrong results. This is analogous to computing the DC term of the product of two sinusoidal waveforms by simply multiplying the DC terms of each.

We highlight the need for cyclostationary analysis with an example. The circuit of Figure 14.22 consists of a mixer, followed by a bandpass filter, followed by another mixer. This is a simplification of, for example, the bias-dependent noise generation mechanism in semiconductor devices [61]. Both mixers multiply their inputs by a local oscillator of frequency f_0, i.e., by $\cos(2\pi f_0 t)$. The bandpass filter is centered around f_0 and has a bandwidth of $B \ll f_0$. The circuit is noiseless, but the input to the first mixer is stationary band-limited noise with two-sided bandwidth B.

A naive attempt to determine the output noise power would consist of the following analysis, illustrated in Figure 14.22. The first mixer shifts the input noise spectrum by $\pm f_0$ and scales it by 1/4. The

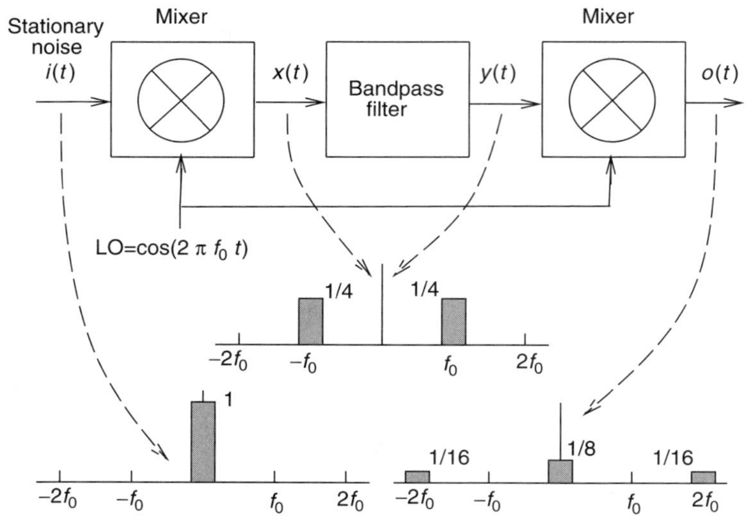

FIGURE 14.22 Mixer–filter–mixer circuit: naïve analysis.

resulting spectrum is multiplied by the squared magnitude of the filter's transfer function. Since this spectrum falls within the pass-band of the filter, it is not modified. Finally, the second mixer shifts the spectrum again by $\pm f_0$ and scales it by 1/4, resulting in the spectrum with three components shown in the figure. The total noise power at the output, i.e., the area under the spectrum, is 1/4th that at the input.

This common but simplistic analysis is inconsistent with the following alternative argument. Note that the bandpass filter, which does not modify the spectrum of its input, can be ignored. The input then passes through only the two successive mixers, resulting in the output noise voltage $o(t) = i(t)\cos^2(2\pi f_0 t)$. The output power is

$$o^2(t) = i^2(t)\left[\frac{3}{8} + \frac{\cos(2\pi 2f_0t) + \cos(2\pi 4f_0t)}{2}\right]$$

The average output power consists of only the 3/8 $i^2(t)$ term, since the cosine terms time-average to zero. Hence the average output power is 3/8th of the input power, 50% more than that predicted by the previous naive analysis. This is, however, the correct result.

The contradiction between the arguments above underscores the need for cyclostationary analysis. The autocorrelation function of any cyclostationary process $z(t)$ (defined as $R_{zz}(t,\tau) = E\,[z(t)z(t+\tau)]$, $E[\cdot]$ denoting expectation) can be expanded in a Fourier series in t:

$$R_{zz}(t,\tau) = \sum_{i=-\infty}^{\infty} R_{Zi}(\tau)e^{ji2\pi f_0 t} \tag{14.41}$$

$R_{zi}(\tau)$ are termed *harmonic autocorrelation functions.* The periodical time-varying power of $z(t)$ is its autocorrelation function evaluated at $\tau = 0$, i.e., $R_{zz}(t,0)$. The quantities $R_{zi}(0)$ represent the harmonic components of the periodically varying power. The average power is simply the value of the DC or *stationary component,* $R_{z0}(0)$.† The frequency-domain representation of the harmonic autocorrelations are termed *harmonic power spectral densities* (HPSDs) $S_{zi}(f)$ of $z(t)$, defined as the Fourier transforms

$$S_{Zi}(f) = \int_{-\infty}^{\infty} R_{Zi}(\tau)e^{-j\pi f\tau}\,d\tau \tag{14.42}$$

† *Stationary* processes are a special case of cyclostationary processes, where the autocorrelation function (hence the power) is independent of the time t; it follows that $R_{zi}(\tau) \equiv 0$ if $i \neq 0$.

Equations can be derived that relate the HPSDs at the inputs and outputs of various circuit blocks. By solving these equations, any HPSD in the circuit can be determined.

Consider, for example, the circuit in Figure 14.22. The input and output HPSDs of a perfect cosine mixer with unit amplitude can be shown [62] to be related by

$$S_{vk}(f) = \frac{S_{u_{k-2}}(f-f_0)}{4} + \frac{S_{u_k}(f-f_0) + S_{u_k}(f+f_0)}{4} + \frac{S_{u_{k+2}}(f+f_0)}{2} \tag{14.43}$$

where u and v denote the input and output, respectively. The HPSD relation for a filter with transfer function $H(f)$ is [62]

$$S_{v_k}(f) = H(-f)H(f+kf_0)\, S_{u_k}(f) \tag{14.44}$$

The HPSDs of the circuit are illustrated in Figure 14.23. Since the input noise $i(t)$ is stationary, its only nonzero HPSD is the stationary component $S_{i_0}(f)$, assumed to be unity in the frequency band $[-B/2, B/2]$, as shown. From Equation (14.43) applied to the first mixer, *three* nonzero HPSDs (S_{x_0}, S_{x_2}, and S_{x-2}, shown in the figure) are obtained for $x(t)$. These are generated by shifting the input PSD by $\pm f_0$ and scaling by 1/4; in contrast to the naive analysis, the stationary HPSD is not the only spectrum used to describe the up-converted noise. From Equation 14.44, it is seen that the ideal bandpass filter propagates the three HPSDs of $x(t)$ unchanged to $y(t)$. Through Equation 14.43, the second mixer generates five nonzero HPSDs, of which only the stationary component $S_{o0}(f)$ is shown in the figure. This is obtained by scaling and shifting not only the stationary HPSD of $y(t)$, but also the cyclostationary HPSDs, which in fact contribute an extra 1/4 to the lobe centered at zero. The average output noise (the shaded area under $S_{o0}(f)$) equals 3/8 of the input noise.

We now sketch the general procedure for analyzing mixing noise in circuits. The noise sources within a circuit can be represented by a small additive term $Au(t)$ to Equation (14.1), where $u(t)$ is a vector of noise sources, and A an incidence matrix capturing their connections to the circuit. Equation 14.1 is first solved for a (quasi)periodic steady state in the absence of noise, and then linearized as in Equation (14.38), to obtain

$$C(t)\dot{x} + G(t)x + Au(t) = 0 \tag{14.45}$$

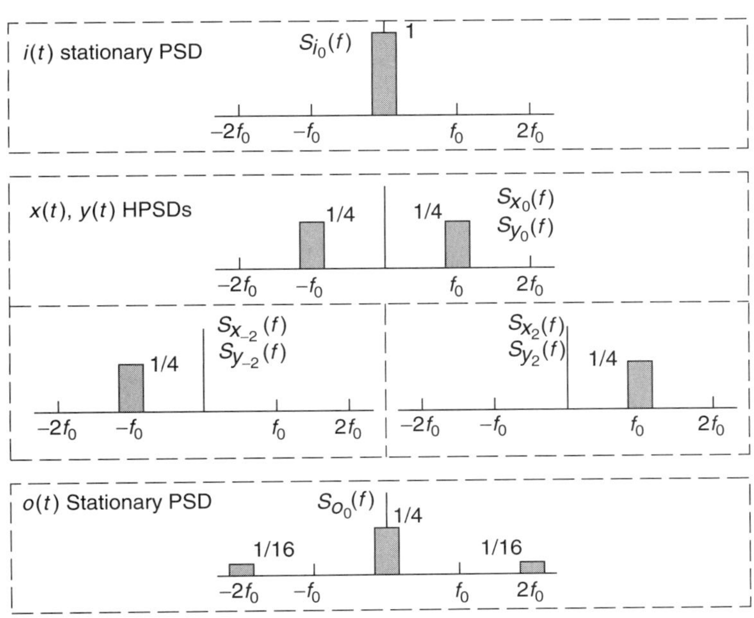

FIGURE 14.23 HPSDs of mixer–filter–mixer circuit.

where $x(t)$ represents the small-signal deviations due to noise. Equation 14.45 describes a linear periodically time-varying (LPTV) system with input $u(t)$ and output $x(t)$. The system can be characterized by its time-varying transfer function $H(t,f)$. $H(t,f)$ is periodic in t and can be expanded in a Fourier series similar to Equation (14.41). Denote the Fourier components *(harmonic transfer functions)* by $H_i(f)$,.

Since $u(t)$ and $x(t)$ are vectors, their autocorrelation functions are *matrices* $R_{zz}(t,\tau) = E[z(t)z^{\mathrm{T}}(t+\tau)]$, consisting of auto-and cross-correlations. Similarly, the HPSDs $S_{zi}(f)$ are also matrices. It can be shown [59] that the HPSD matrices of y and u are related by:

$$S_{xx}(f) = \mathcal{H}(f)\, S_{uu}(f)\, \mathcal{H}^{*}(f) \tag{14.46}$$

$\mathcal{H}(f)$ (the *conversion matrix)* is the following block-structured matrix *(*f^k denotes $f+kf_0$*)*:

$$\mathcal{H}(f) = \begin{pmatrix} & \vdots & \vdots & \vdots & \\ \cdots & H_0(f^1) & H_1(f^0) & H_2(f^{-1}) & \cdots \\ \cdots & H_{-1}(f^1) & H_0(f^0) & H_1(f^{-1}) & \cdots \\ \cdots & H_{-2}(f^1) & H_{-1}(f^0) & H_0(f^{-1}) & \cdots \\ & \vdots & \vdots & \vdots & \end{pmatrix} \tag{14.47}$$

$S_{uu}(f)$ and $S_{xx}(f)$ are similar to $\mathcal{H}(f)$: their transposes $S_{zz}^{\mathrm{T}}(f)$ have the same structure, but with $H_i(f^k)$ replaced by $S_{zi}^{\mathrm{T}}(f^k)$.

Equation (14.46) expresses the output HPSDs contained in $S_{xx}(f)$, in terms of the input HPSDs contained in $S_{uu}(f)$, and the harmonic transfer functions of the circuit contained in $\mathcal{H}(f)$. The HPSDs of a single output variable $x_p(t)$ (both auto- and cross-terms with all other output variables) are available in the *pth* column of the central block-column of $S_{xx}^{\mathrm{T}}(f)$. To pick this column, $S_{xx}^{\mathrm{T}}(f)$ is applied to a unit vector E_{0p}, as follows ($\bar{\ }$ denotes the conjugate):

$$S_{xx}^{\mathrm{T}}(f)E_{0p} = \bar{\mathcal{H}}(f)S_{uu}^{\mathrm{T}}(f)H^{\mathrm{T}}(f)E_{0p} \tag{14.48}$$

Evaluating Equation (14.48) involves two kinds of matrix–vector products, $\mathcal{H}(f)\,z$ and $S_{uu}(f)\,z$ for some vectors z. Consider the latter product first. If the inputs $u(t)$ are stationary, as can be assumed without loss of generality [62], then $S_{uu}(f)$ is block-diagonal. In practical circuits, the inputs $u(t)$ are either uncorrelated or sparsely correlated. This results in each diagonal block of $S_{uu}(f)$ being either diagonal or sparse. In both cases, the matrix–vector product can be performed efficiently.

The product with $\mathcal{H}(f)$ can also be performed efficiently by exploiting the relation $\mathcal{H}(f) = J^{-1}(f)\,\mathcal{A}$ [26]. $\mathcal{A}$ is a sparse incidence matrix of the device noise generators, hence its product with a vector can be computed efficiently. $J(0)$ is the HB Jacobian matrix [31] at the large-signal solution $x^*(t)$. $J(f)$ is obtained by replacing kf_0 by $kf_0 + f$ in the expression for the Jacobian. The product $J^{-1}z$ can therefore be computed efficiently using the fast techniques outlined in Section 14.5.2. As a result, Equation (14.48) can be computed efficiently for large circuits to provide the auto- and cross-HPSDs of any output of interest.

For example, a portion of the Lucent W2013 RFIC, consisting of an I-channel buffer feeding a mixer, was simulated using Equation (14.48). The circuit consisted of about 360 nodes, and was excited by two tones — a local oscillator at 178 MHz driving the mixer, and a strong RF signal tone at 80 kHz feeding into the I-channel buffer. Two noise analyses were performed. The first analysis included both LO and RF tones (sometimes called a three-tone noise analysis). The circuit was also analyzed with only the LO tone to determine if the RF signal affects the noise significantly. The two-tone noise simulation, using a total of 525 large-signal mix components, required 300 MB of memory, and for each frequency point, took 40 min on an SGI machine (200 MHz, R10000 CPU). The one-tone noise simulation, using 45 harmonics, needed 70 MB of memory and took 2 min per point.

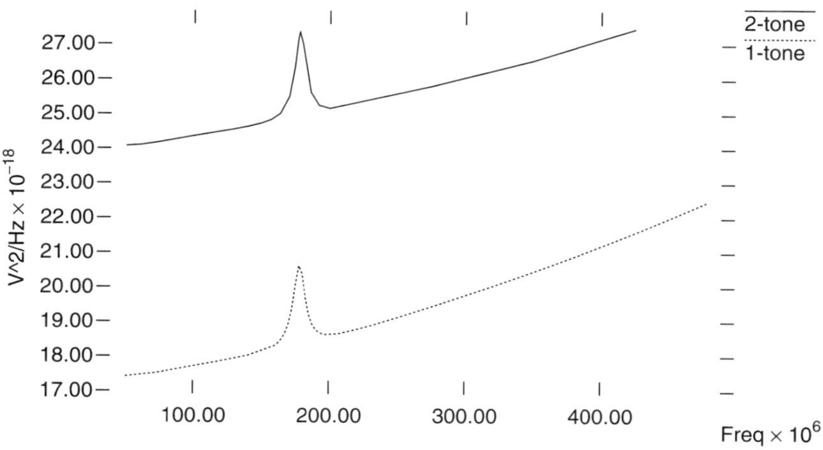

FIGURE 14.24 Stationary PSDs for the I–Q mixer/buffer circuit.

The stationary PSDs of the mixer output noise for the two simulations are shown in Figure 14.24. It can be seen that the presence of the large RF signal increases the noise by about 1/3. This is due to *noise folding*, the result of devices being driven into nonlinear regions by the strong RF input tone. The peaks in the two waveforms located at the LO frequency are due to up and down conversion of noise from other frequencies.

14.7.2 Phase Noise

Even small noise in an oscillator leads to dramatic changes in its frequency spectrum and timing properties, i.e., to phase noise. This effect can lead to interchannel interference and increased bit-error rates (BER) in RF communication systems. Another manifestation of the same phenomenon, jitter, is important in clocked and sampled-data systems: uncertainties in switching instants caused by noise can affect synchronization.

Although a large body of literature is available on phase noise,[‡] treatments of the phenomenon from the design perspective have typically been phenomenological, e.g., the well-known treatment of Leeson [63]. Most analyses have been based on linear time-invariant or time-varying approaches, which though providing useful design guidelines, contain qualitative inaccuracies — e.g., they can predict infinite noise power. Recently, however, the work of Kärtner [64] and Demir et al. [65] have provided a more correct understanding of phase noise. Here, we sketch the approach in [65].

The starting point for phase noise analysis is Equation (14.1), reproduced here for oscillators with no external forcing:

$$\dot{q}(x) + f(x) = 0 \tag{14.49}$$

We assume Equation (14.49) to be the equation for an oscillator with an orbitally stable,[§] nontrivial periodic solution, i.e., an oscillation waveform $x_s(t)$. With small noise generators in the circuit, possibly dependent on circuit state, the equation becomes

$$\dot{q}(x) + f(x) = B(x)b(t) \tag{14.50}$$

where $b(t)$ now represents small perturbations.

[‡] BSIM homepage [65] contains a list of references.

[§] See, e.g., [66] for a precise definition; roughly speaking, an orbitally stable oscillator is one that eventually reaches a unique, periodic waveform with a definite magnitude.

When $b(t)$ is small, it can be shown [67] that the originally periodic oscillation $x_s(t)$ changes to

$$x(t) = x_s(t + \alpha(t)) + y(t) \tag{14.51}$$

where $y(t)$ remains small, but $\alpha(t)$ (a time/phase deviation) can grow unboundedly with time, no matter how small the perturbation $b(t)$ is (see Figure 14.25). For driven circuits (the mixing noise case) $\alpha(t)$ remains bounded and small and its effects can therefore be lumped into the $y(t)$ term. This is the difference illustrated in Figure 14.19, Figure 14.20, and Figure 14.21. The underlying reason for this difference is that oscillators by their very definition are phase unstable, and hence phase errors build up indefinitely.

Furthermore, it can be shown that $\alpha(t)$ is given by a nonlinear scalar differential equation

$$\dot{\alpha} = v_1^{\mathrm{T}}(t + \alpha(t))B(x_s(t + \alpha(t)))b(t) \tag{14.52}$$

where v_1 (t) is a periodic vector function dubbed the perturbation projection vector (PPV). The PPV, which is characteristic of an oscillator in steady state and does not depend on noise parameters, is an important quantity for phase noise calculation. Roughly speaking, it is a "transfer function" that relates perturbations to resulting time or phase jitter in the oscillator. The PPV can be found only through a linear time-varying analysis of the oscillator around its oscillatory solution, and simple techniques to calculate it using HB or shooting are available [68].

In general, Equation (14.52) can be difficult to solve analytically. When the perturbation $b(t)$ is white noise however, it can be shown that $\alpha(t)$ becomes a Gaussian random walk process with linearly increasing variance ct, where c is a scalar constant given by

$$c = \frac{1}{T}\int_0^{\mathrm{T}} v_1^{\mathrm{T}}(t)B(x_s(t))B^{\mathrm{T}}(x_s(t))v_1(t)\mathrm{d}t \tag{14.53}$$

with T being the period of the unperturbed oscillation.

This random-walk stochastic characterization of the phase error $\alpha(t)$ implies that:

1. The average spread of the jitter (mean-square jitter) increases *linearly* with time, with cT being the jitter per cycle.

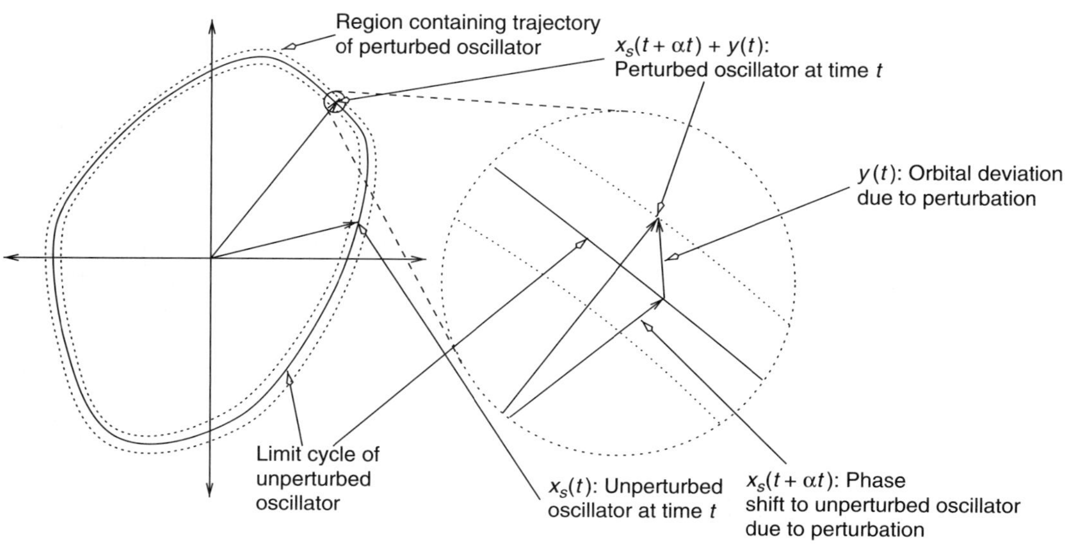

FIGURE 14.25 Oscillator trajectories.

2. The spectrum of the oscillator's output, i.e., the power spectrum of $x_s(t + \alpha(t))$, is *Lorenzian*[¶] about each harmonic. For example, around the fundamental (with angular frequency $\omega_0 = 2\pi/T$ and power P_{fund}), the spectrum is

$$S_p(f) = P_{\text{fund}} \frac{\omega_0^2 c}{w_0^4 c^2/4 + (2\pi f - \omega_0)^2} \tag{14.54}$$

   This means that the spectrum decays as $1/f^2$ beyond a certain knee distance away from the original oscillation frequency and its harmonics, as is well known for white noise in oscillators [63]. The $1/f^2$ dependence does not, however, continue as $f \rightarrow 0$, i.e., close to and at the oscillation frequency; instead, the spectrum reaches a finite maximum value.
3. The oscillator's output is a *stationary* stochastic process.

The Lorenzian shape of the spectrum also implies that the power spectral density at the carrier frequency and its harmonics have a finite value, and that the total carrier power is preserved despite spectral spreading due to noise. Equation (14.52) can also be solved for colored noise perturbations *b(t)* [67], and it can be shown that if $S(f)$ is the spectrum of the colored noise, then the phase noise spectrum generated falls as $S(f)/(f-f_0)^2$, away from f_0.

Numerical methods based on the above insights are available to calculate phase noise. The main effort is calculating the PPV; once it is known, c can be calculated easily using Equation (14.53) and the spectrum obtained directly from Equation (14.54). The PPV can be found from the time-varying linearization of the oscillator around its steady state. Two numerical methods can be used to find the PPV. The first calculates the time-domain monodromy (or state-transition) matrix of the linearized oscillator explicitly, and obtains the PPV by eigendecomposing this matrix [65]. A more recent method [68] relies on simple post-processing of internal matrices generated during the solution of the operator's steady state using HB or shooting, and as such, can take advantage of the fast techniques of Section 14.5.2. The separate contributions of noise sources, and the sensitivity of phase noise to individual circuit devices and nodes, can be obtained easily.

As an example, the oscillator in Figure 14.26 consists of a Tow–Thomas second-order bandpass filter and a comparator [69]. If the OpAmps are considered to be ideal, it can be shown that this oscillator is equivalent (in the sense of the differential equations that describe it) to a parallel RLC circuit in parallel with a nonlinear voltage-controlled current source (or equivalently, a series RLC circuit in series with a nonlinear

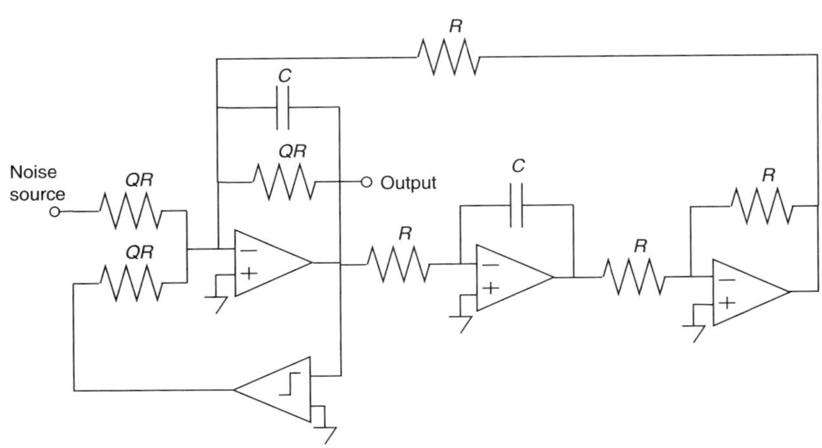

FIGURE 14.26 Oscillator with a band-pass filter and a comparator. (For further information on such oscillators, please see [69].)

[¶] A Lorenzian is the shape of the squared magnitude of a one-pole lowpass filter transfer function.

current-controlled voltage source). In [69], authors breadboarded this circuit with an external white noise source (intensity of which was chosen such that its effect is much larger than the other internal noise sources), and measured the PSD of the output with a spectrum analyzer. For $Q = 1$ and $f_o = 6.66$ kHz, a phase noise characterization of this oscillator was performed to yield the periodic oscillation waveform $x_s(t)$ for the output and $c = 7.56 \times 10^{-8} \text{sec}^2$ Hz. Figure 14.27(a) shows the PSD of the oscillator output and Figure 14.27(b) shows the spectrum analyzer measurement.[||] Figure 14.27(c) shows a blown-up version of the PSD around the first harmonic. The single-sideband phase noise spectrum is shown in Figure 14.27(d). The oscillator model that was simulated has two state variables and a single stationary noise source. Figure 14.27(e) shows a plot of the periodic nonnegative scalar (essentially the squared magnitude of the PPV)

$$v_1^{\mathrm{T}}(t)B(x_s(t))B^{\mathrm{T}}(x_s(t))v_1(t) = (v_1^{\mathrm{T}}(t)B)^2$$

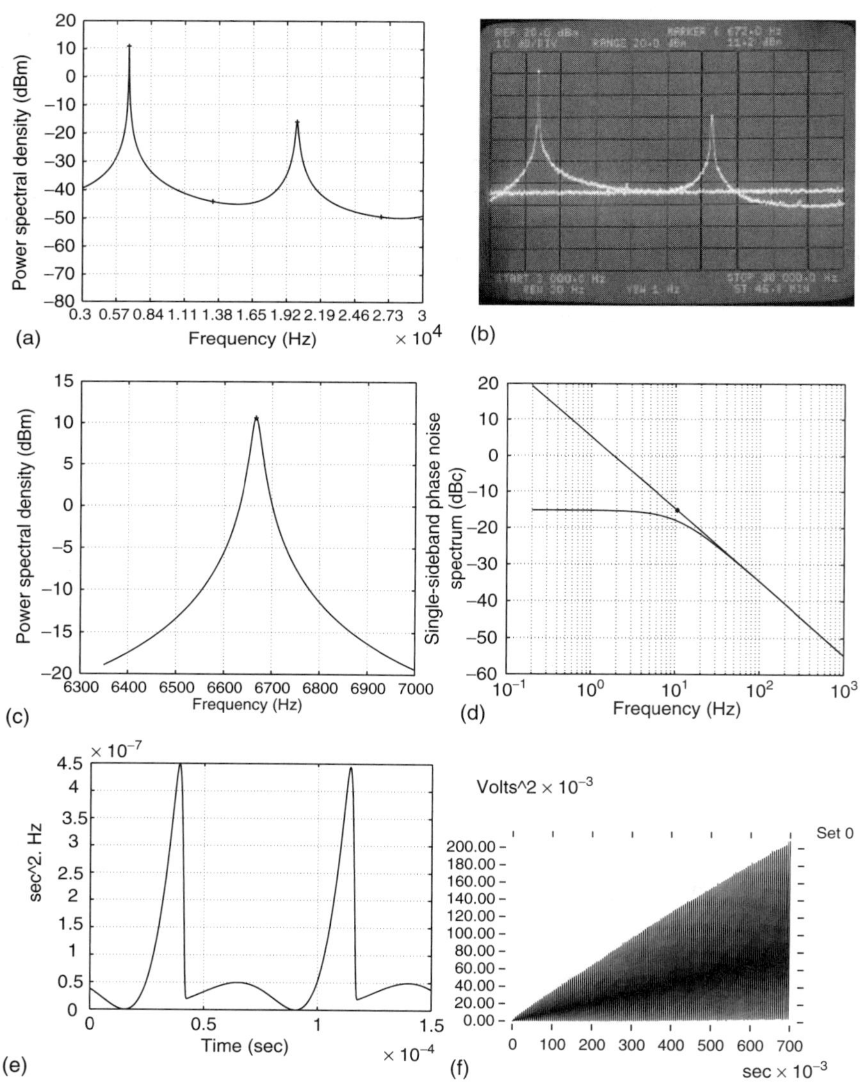

FIGURE 14.27 Phase noise characterisation for the oscillator in Figure 26.

[||] The PSDs are plotted in units of dBm.

where B is independent of t since the noise source is stationary. Recall that c is the time average of this scalar that is periodic in time.

c can also be obtained relatively accurately in this case using Monte-Carlo analysis (in general, however, Monte-Carlo based on transient simulations can be extremely time-consuming and also inaccurate, and should be avoided except as a sanity check). The circuit was simulated with 10,000 random excitations and the results averaged to obtain the mean-square difference between the perturbed and unperturbed systems as a function of time. Figure 14.27(f) illustrates the result, in which the slope of the envelope determines c. The Monte-Carlo simulations required small time-steps to produce accurate results, since numerical integration methods easily lose accuracy for autonomous circuits.

14.8 Conclusions

In this chapter, we have taken the reader through a quick tour of both basic and advanced topics in analog simulation — from writing circuit equations, to solving them numerically for simple analyses (like DC), to steady-state, envelope, multitime, and noise simulation. We have also looked into special challenges for oscillatory circuits. We hope that this tour has served to provide the reader an appreciation of concepts in simulation, in some depth as well as breadth.

Acknowledgments

Roychowdhury takes pleasure in acknowledging Alper Demir and Amit Mehrotra for joint work on which much of Section 14.7.2 is based. He would also like to acknowledge his other colleagues in the Design Principles Research Department of Bell Laboratories during the period 1995–2000, notably Peter Feldmann, David Long, Bob Melville, and Al Dunlop. Discussions with these colleagues have influenced the style and content of much of this chapter, in particular Sections 14.2, 14.4, 14.5, 14.6, and 14.7.

References

[1] L.O. Chua and P-M. Lin, *Computer-Aided Analysis of Electronic Circuits: Algorithms and Computational Techniques*, Prentice-Hall, Englewood Cliffs, NJ, 1975.

[2] Kenneth S. Kundert and Alberto Sangiovanni-Vincentelli, Simulation of nonlinear circuits in the frequency domain, *IEEE Trans. Comput.-Aided Design of Integrated Circuits Syst.*, *CAD-5*, 521–535, October 1986.

[3] Kenneth S. Kundert, Gregory B. Sorkin, and Alberto Sangiovanni-Vincentelli, Applying harmonic balance to almost-periodic circuits, *IEEE Trans. Microwave Theory and Techniques*, *MTT-36*, 366–378, 1988.

[4] P. Antognetti and G. Massobrio, *Semiconductor Device Modeling with SPICE*, McGraw-Hill, New Year, NY, 1988, Chap. 1.

[5] W. Liu, *MOSFET Models for SPICE Simulation Including BSIM3v3 and BSIM4*, Wiley, New York, NY, 2001.

[6] Y.P. Tsividis, *Operation and Modeling of the MOS Transistor*, McGraw-Hill, New York, 1987.

[7] I.E. Getreu, *Modeling the Bipolar Transistor*, Tektronix, Inc., Beaverton, Oregon, 1976.

[8] D. Foty, *MOSFET Modeling with SPICE: Principles and Practice*, Prentice Hall, Upper Saddle River, NJ, 1997.

[9] K. Kundert and O. Zinke, *The Designer's Guide to Verilog-AMS*, Kluwer Academic Publishers, Norwell, MA, 2004.

[10] D. Fitzpatrick and I. Miller, *Analog Behavioral Modeling with the Verilog-A Language*, Kluwer Academic Publishers, Norwell, MA, 1998.

[11] H.A. Mantooth and M. Fiegenbaum, *Modeling with an Analog Hardware Description Language*, Kluwer Academic Publishers, Norwell, MA, 1995.

[12] P. Ashenden, G. Peterson and D. Teegarden, *System Designer's Guide to VHDL-AMS*, Morgan Kaufman, SanFrancisco, CA, 2003.

[13] P. Su, S.K.H. Fung, S. Tang, F. Assaderaghi and C. Hu, BSIMPD: A partial-depletion SOI MOSFET model for deep-submicron CMOS designs, *IEEE Proceedings of Custom Integrated Circuits Conference*, 2000, pp. 197–200.

[14] B.G. Streetman, *Solid State Electronic Devices*, Prentice-Hall, New York, 1980, pp. 172–173.

[15] A.S. Sedra and K.C. Smith, *Microelectronic Circuits*, 5th Ed., Oxford University Publishing, New York, NY, 2004, Chap. 3.

[16] H.A. Mantooth and M. Vlach, Beyond SPICE with Saber and MAST, *IEEE Proceedings of International Symposium on Circuits Syst.*, San Diego, California, Vol. 1, May 1992, pp. 77–80.

[17] T.R. McNutt, A.R. Hefner, H.A. Mantooth, J.L. Duliere, D. Berning, and R. Singh, Silicon carbide PiN and merged PiN Schottky power diode models implemented in the Saber circuit simulator, *IEEE Trans. Power Electron.*, *19*, 573–581, 2004.

[18] H.A. Mantooth and J.L. Duliere, A unified diode model for circuit simulation, *IEEE Trans. Power Electron.*, 12, 816–823, 1997.

[19] BSIM Homepage, http://www-device.eecs.berkeley.edu/bsim3/.

[20] C. Enz, F. Krummenacher, and E. Vittoz, An analytical MOS transistor model valid in all regions of operation and dedicated to low-voltage and low-current applications, *J. Analog Integrated Circuits Signal Process.*, 83–114, 1995.

[21] G. Gildenblat, H. Wang, T.-L Chen, X. Gu, and X. Cai, SP: An advanced surface-potential-based compact MOSFET model, *IEEE J. Solid-State Circuits*, 39, 1394–1406, 2004.

[22] K.A. Sakallah, Y.T. Yen, and S.S. Greenberg, First-order charge conserving MOS capacitance model, *IEEE Trans. Comput.-Aided Design*, 9, pp. 99–108, 1990.

[23] W.H. Press, S.A. Teukolsky, W.T. Vetterling, and B.P. Flannery, *Numerical Recipes — The Art of Scientific Computing*. Cambridge University Press, Cambridge, 1989.

[24] K.S. Kundert, J.K. White, and A. Sangiovanni-Vincentelli, *Steady-State Methods for Simulating Analog and Microwave Circuits*, Kluwer Academic Publishers, Norwell, MA, 1990.

[25] M.S. Nakhla and J. Vlach. A piecewise harmonic balance technique for determination of periodic responses of nonlinear systems, *IEEE Trans. Circuit. Syst.*, CAS-23, 85, 1976.

[26] S.A. Haas, *Nonlinear Microwave Circuits*, Artech House, Norwood, MA, 1988.

[27] V. Rizzoli and A. Neri, State of the art and present trends in nonlinear microwave CAD techniques, *IEEE Trans. MTT*, *36*, 343–365, 1988.

[28] R.J. Gilmore and M.B. Steer, Nonlinear circuit analysis using the method of harmonic balance — a review of the art. Part I. Introductory concepts. *Int. J. Microwave Millimeter Wave CAE*, *1*, 22–37, 1991.

[29] M. Rösch, Schnell simulation des stationären Verhaltens nichtlinearer Schaltungen. Ph.D. thesis, Technischen Universität München, 1992.

[30] R. Mickens, *Oscillations in Planar Dynamic Systems*, World Scientific, Singapore, 1995.

[31] R.C. Melville, P. Feldmann, and J. Roychowdhury, Efficient multi-tone distortion analysis of analog integrated circuits, *Proc. IEEE CICC*, May 1995, pp. 241–244.

[32] D. Long, R.C. Melville, K. Ashby, and B. Horton, Full chip harmonic balance, *Proc. IEEE CICC*, 379–382, May 1997.

[33] W.H. Press, S.A. Teukolsky, W.T. Vetterling, and B.P. Flannery, *Numerical Recipes — The Art of Scientific Computing*, Cambridge University Press, Cambridge, 1989.

[34] T.J. Aprille and T.N. Trick, Steady-state analysis of nonlinear circuits with periodic inputs, *Proc. IEEE*, 60, 108–114, 1972.

[35] S. Skelboe, Computation of the periodic steady-state response of nonlinear networks by extrapolation methods, *IEEE Trans. Circuit Syst.*, *CAS-27*, 161–175, 1980.

[36] A. Nayfeh and B. Balachandran, *Applied Nonlinear Dynamics*, Wiley, New York, 1995.

[37] R. Telichevesky, K. Kundert, and J. White, Efficient steady-state analysis based on matrix-free Krylov subspace methods, *Proceedings of the IEEE DAC*, 1995, pp. 480–484.

[38] P. Feldmann, R.C. Melville, and D. Long, Efficient frequency domain analysis of large nonlinear analog circuits, *Proc. IEEE CICC*, 461–464, May 1996.
[39] Y. Saad, *Iterative Methods for Sparse Linear Systems*, PWS, Boston, 1996.
[40] R.W Freund, Reduced-Order Modeling Techniques Based on Krylov Subspaces and Their Use in Circuit Simulation, Technical report 11273-980217-02TM, Bell Laboratories, 1998.
[41] R.W. Freund, Reduced-order modeling techniques based on Krylov subspaces and their use in circuit simulation, *Applied Comput. Control, Signals Circuits, 1*, 435–498, 1999.
[42] J. Kevorkian and J.D. Cole, *Perturbation Methods in Applied Mathematics.* Springer, Berlin, 1981.
[43] E. Ngoya and R. Larcheveque, Envelop transient analysis: a new method for the transient and steady state analysis of microwave communication circuits and systems, *Proceedings of the IEEE MTT Symposium*, 1996.
[44] H.G. Brachtendorf, G. Welsch, R. Laur, and A. Bunse-Gerstner, Numerical steady-state analysis of electronic circuits driven by multi-tone signals, *Electrical Engineering, 79*, 103–112,1996.
[45] J. Roychowdhury, Efficient methods for simulating highly nonlinear multi-rate circuits, *Proceedings of the IEEE DAC*, 1997.
[46] O. Narayan and J. Roychowdhury, Multi-time simulation of voltage-controlled oscillators, *Proceedings of the IEEE DAC*, New Orleans, LA, June 1999.
[47] J. Phillips, Model reduction of time-varying linear systems using approximate multipoint Krylov-subspace projectors, *Proceedings of the ICCAD*, November 1998.
[48] J. Roychowdhury, Reduced-order modelling of time-varying systems, *IEEE Trans. Circuit Syst. — II: Sig. Proc., 46*, 1273–1288, November 1999.
[49] L.T. Pillage and R.A. Rohrer, Asymptotic waveform evaluation for timing analysis, *IEEE Trans. CAD, 9*, 352–366, 1990.
[50] X. Huang, V. Raghavan, and R.A. Rohrer, AWEsim: A program for the efficient analysis of linear(ized) circuits, *Proceedings of the ICCAD*, November 1990, pp. 534–537.
[51] E. Chiprout and M.S. Nakhla, *Asymptotic Waveform Evaluation*, Kluwer, Norwell, MA, 1994.
[52] P. Feldmann and R.W. Freund, Efficient linear circuit analysis by Padé approximation via the Lanczos process, *IEEE Trans. CAD, 14*, 639–649, 1995.
[53] P. Feldmann and R.W. Freund, Reduced-order modeling of large linear subcircuits via a block Lanczos algorithm, *Proceedings of the IEEE DAC*, 1995, pp. 474–479.
[54] P. Feldmann and R.W. Freund, Circuit noise evaluation by Padé approximation based model-reduction techniques, *Proceedings of the ICCAD*, November 1997, pp. 132–138.
[55] A. Odabasioglu, M. Celik, and L.T. Pileggi, PRIMA: passive reduced-order interconnect macromodelling algorithm, *Proceedings of the ICCAD*, November 1997, pp. 58–65.
[56] W. Gardner, *Introduction to Random Processes*, McGraw-Hill, New York, 1986.
[57] T. Ström and S. Signell, Analysis of periodically switched linear circuits. *IEEE Trans. Circuits Syst., CAS-24*, 531–541, 1977.
[58] M. Okumura, H. Tanimoto, T. Itakura, and T. Sugawara, Numerical noise analysis for nonlinear circuits with a periodic large signal excitation including cyclostationary noise sources, *IEEE Trans. Circuits Syst. — I: Fund. Th. Appl., 40*, 581–590, 1993.
[59] J. Roychowdhury, D. Long, and P. Feldmann, Cyclostationary noise analysis of large RF circuits with multitone excitations, *IEEE J. Solid-State Circuits, 33*, 324–336, 1998.
[60] V. Rizzoli, F. Mastri, and D. Masotti, General noise analysis of nonlinear microwave circuits by the piecewise harmonic balance technique, *IEEE Trans, MTT, 42*, 807–819, 1994.
[61] A.R. Kerr, Noise and loss in balanced and subharmonically pumped mixers: Part 1 — Theory, *IEEE Trans. MTT, MTT-27*, 938–943, 1979.
[62] J. Roychowdhury and P. Feldmann, A new linear-time harmonic balance algorithm for cyclostationary noise analysis in RF circuits, *Proceedings of the ASP-DAC*, 1997, pp. 483–492.
[63] D.B. Leeson, A simple model of feedback oscillator noise spectrum, *Proc. IEEE, 54*, 329, 1966.
[64] F. Kärtner, Analysis of white and $f^{-\alpha}$ noise in oscillators, *Int. J. Circuit Theory Appl.*, 18, 485–519, 1990.

[65] A. Demir, A. Mehrotra, and J. Roychowdhury, Phase noise in oscillators: a unifying theory and numerical methods for characterization, *IEEE Trans. Circuits Syst.—I: Fund. Th. Appl.*, *47*, 655–674, 2000.

[66] M. Farkas, *Periodic Motions*, Springer, Berlin, 1994.

[67] A. Demir, Phase noise in oscillators: DAEs and colored noise sources, *Proceedings of the ICCAD*, 1998, pp. 170–177.

[68] A. Demir, D. Long, and J. Roychowdhury, Computing phase noise eigenfunctions directly from steady-state Jacobian matrices, *In Proceedings of the ICCAD*, November 2000.

[69] A. Dec, L. Toth, and K. Suyama, Noise analysis of a class of oscillators, *IEEE Trans. Circuits Syst.*, *45*, 757–760, 1998.

15

Simulation and Modeling for Analog and Mixed-Signal Integrated Circuits

Georges G.E. Gielen
Katholieke Universiteit Leuven
Leuven, Belgium

Joel R. Phillips
Cadence Berkeley Laboratories
Berkeley, California

15.1 Introduction

This chapter presents an overview of the modeling and simulation methods that are needed to design and embed analog and RF blocks in mixed-signal integrated systems (ASICs, SoCs, and Systems in Package. The design of these integrated systems is characterized by growing design complexities and shortening time-to-market constraints. Handling these requires mixed-signal design methodologies and flows that include system-level architectural explorations and hierarchical design refinements with behavioral models in the top-down design path, and detailed behavioral model extraction and efficient mixed-signal behavioral simulation in the bottom-up verification path. Mixed-signal simulation methods at different hierarchical levels are reviewed, and techniques to generate analog behavioral models, including regression-based methods as well as model-order reduction techniques, are described in detail. Also the generation of performance models for analog circuit synthesis and of symbolic models that provide designers with insight into the relationships governing the performance behavior of a circuit are described.

With the evolution toward ultra-deep-submicron and nanometer CMOS technologies [1], the design of complex integrated systems, be they ASICs, SoCs, or SiPs, is emerging not only in consumer-market applications such as telecom and multimedia, but also in more traditional application domains like automotive or instrumentation. Driven by cost reduction, these markets demand for low-cost optimized and highly integrated solutions with very demanding performance specifications. These integrated systems are increasingly mixed-signal designs, embedding on a single die both high-performance analog or mixed-signal blocks and possibly sensitive RF front-ends, together with complex digital circuitry (multiple processors, some logic blocks, and several large memory blocks) that form the core of most electronic systems today. In addition to the technical challenges related to the increasing design complexity and the problems posed by analog–digital integration, shortening time-to-market constraints put pressure on the design methodology and tools used to design these systems.

Hence, the design of today's integrated systems calls for mixed-signal design methodologies and flows that include system-level architectural explorations and hierarchical design refinements with behavioral models in the top-down design path to reduce the chance of design iterations and to improve the overall optimality of the design solution [2]. In addition, to avoid design errors before tape-out, detailed behavioral model extraction and efficient mixed-signal behavioral simulation are needed in the bottom-up verification path. This chapter presents an overview of the modeling and simulation methods used in this context.

The chapter is organized as follows. Section 15.2 addresses mixed-signal design methodologies and describes techniques and examples for architectural exploration and top-down hierarchical design refinement. Section 15.3 discusses mixed-signal simulation techniques. Section 15.4 describes analog and mixed-signal behavioral simulation and the corresponding hardware description languages. It also gives an overview of techniques to generate automatically analog behavioral models, including regression-based methods as well as model-order reduction techniques. Also the generation of performance models for analog circuit synthesis is described. Section 15.5 then presents methods to generate symbolic models that provide designers with insight into the relationships governing the performance behavior of a circuit. Conclusions are drawn in Section 15.6, followed by an extensive list of references.

15.2 Top-Down Mixed-Signal Design Methodology

The growing complexity of the systems that can be integrated on a single die today, in combination with the tightening time-to-market constraints, results in a growing design productivity gap. That is why new design methodologies are being developed that allow designers to shift to a higher level of design abstraction, such as the use of platform-based design, object-oriented system-level hierarchical design refinement flows, hardware–software co-design, and IP reuse, on top of the already established use of CAD tools for logic synthesis and digital place and route. However, these flows have to be extended to incorporate the embedded analog/RF blocks.

A typical top-down design flow for mixed-signal integrated systems may appear as shown in Figure 15.1, where the following distinct phases can be identified: system specification, architectural design, cell design, cell layout, and system-layout assembly [2,3]. The advantages of adopting a top-down design methodology are:

- the possibility to perform system-architectural exploration and a better overall system optimization (e.g., finding an architecture that consumes less power) at a high level before starting detailed circuit implementations;
- the elimination of problems that often cause overall design iterations, such as the anticipation of problems related to interfacing different blocks;
- the possibility to do early test development in parallel to the actual block design, etc.

The ultimate advantage of top-down design therefore is to catch problems early in the design flow, and as a result have a higher chance of first-time success with fewer or no overall design iterations, hence

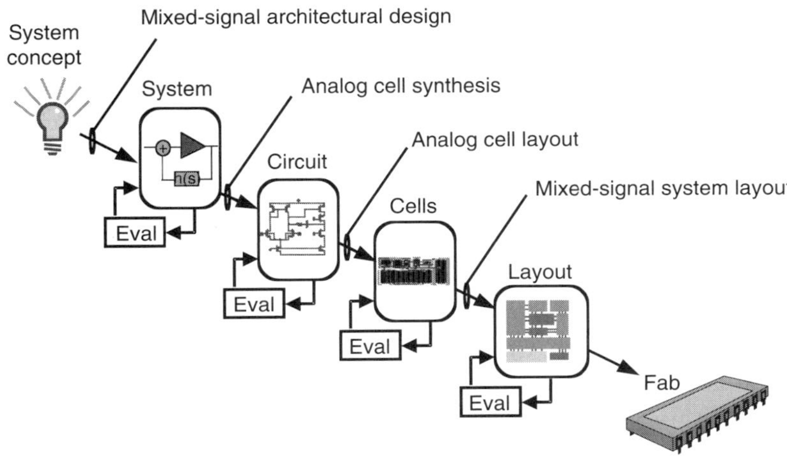

FIGURE 15.1 Top-down view of the mixed-signal IC design process.

shortening design time, while at the same time obtaining a better overall system design. A top-down design example will be presented later on. The methodology, however, does not come for free, and requires some investment from the design team, especially in terms of high-level modeling and setting up a sufficient model library for the targeted application. Even then, there remains the risk that at higher levels in the design hierarchy also, low-level details (e.g., matching limitations, circuit nonidealities, and layout effects) may be important to determine the feasibility or optimality of an analog solution. The high-level models used must therefore include such effects to the extent possible, but it remains difficult in practice to anticipate or model everything accurately at higher levels. Besides the models, efficient simulation methods are also needed at the architectural level in order to allow efficient interactive explorations. The subjects of system exploration and simulation as well as behavioral modeling will now be discussed in more detail.

15.2.1 System-Level Architectural Exploration

The general objective of analog architectural system exploration is twofold [4,5]. First of all, a proper (and preferably optimal) architecture for the system has to be decided upon. Secondly, the required specifications for each of the blocks in the chosen architecture must be determined, so that the overall system meets its requirements at minimum implementation cost (power, chip area, etc.). The aim of a system-exploration environment is to provide the system designer with the platform and the supporting tool-set to explore different architectural alternatives in a short time and to take the above decisions based on quantified rather than heuristic information.

Consider, for instance, the digital telecommunication link of Figure 15.2. It is clear that digital bits are going into the link to be transmitted over the channel, and that the received signals are being converted again into digital bits. One of the major considerations in digital telecom system design is the bit error rate, which characterizes the reliability of the link. This bit error rate is not only impacted by the characteristics of the transmission channel itself, but also by the architecture chosen for the transmitter and receiver front-end, and by the performances achieved and the non-idealities exhibited by the analog/RF blocks in this front-end. For example, the noise figure and nonlinear distortion of the input low-noise amplifier (LNA) are key parameters. Similarly, the resolution and sampling speed of the analog-to-digital converter (ADC) used may have a large influence on the bit error rate, but they also determine the requirements for the other analog subblocks: a higher ADC resolution may relax the filtering requirements in the transceiver, resulting in simpler filter structures, though it will also consume more power

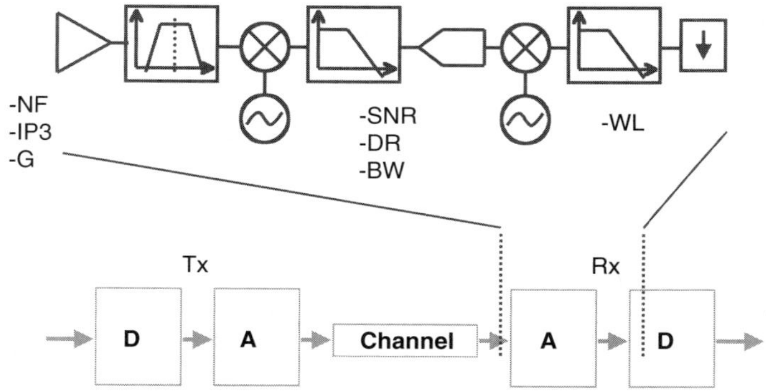

FIGURE 15.2 Digital telecommunication link, indicating a possible receiver front-end architecture with some building block specifications to be determined during front-end architectural exploration.

and chip area than a lower-resolution converter. At the same time, the best trade-off solution, i.e., the minimum required ADC resolution and therefore also the minimum power and area, depends on the architecture chosen for the transceiver front-end.

Clearly, there is a large interaction between system-level architectural decisions and the performance requirements for the different subblocks, which in turn are bounded by technological limits that shift with every new technology process being employed. Hence, it is important to offer designers an exploration environment where they can define different front-end architectures and analyze and compare their performance quantitatively and derive the necessary building block specifications. Today, the alternative architectures that are explored are still to be provided by the system designer, but future tools might also derive or synthesize these architectures automatically from a high-level language description [6].

The important ingredients that are needed to set up such an architectural exploration environment are [4,5]:

- a fast high-level simulation method that allows the evaluation of the performance (e.g., SNR or BER) of the front-end;
- a library of high-level (behavioral) models for the building blocks used in the targeted application domain, including a correct modeling of the important building block non-idealities (offset, noise, distortion, mirror signals, phase noise, etc.);
- power and area estimation models that, starting from the block specifications, allow estimation of the power consumption and chip area that would be consumed by a real implementation of the block, without really designing the block.

The above ingredients allow a system designer to explore interactively front-end architectures. Combining this with an optimization engine would additionally allow optimization of the selected front-end architecture in determining the optimal building block requirements so as to meet the system requirements at minimum implementation cost (power/area). Repeating this optimization for different architectures then makes a quantitative comparison between these architectures possible before they are implemented down to the transistor level. In addition, the high-level exploration environment would also help in deciding on other important system-level decisions, such as determining the optimal partitioning between analog and digital implementations in a mixed-signal system [7], or deciding on the frequency planning of the system, all based on quantitative data rather than ad hoc heuristics or past experiences.

To some extent, this approach can be implemented in existing commercial tools such as COSSAP, PTOLEMY, Matlab/Simulink, ADS, and SPW. However, not all desired aspects for system-level

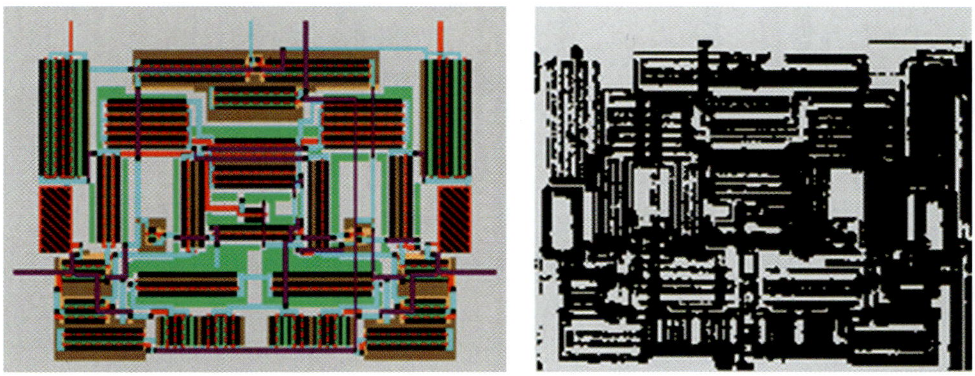

COLOR FIGURE 16.1 Automatic analog cell layout for a CMOS amplifier, layout (left) and die shot (right) done using KOAN/ANAGRAM tools. (From Cohn et al., *IEEE JSSC*, 26, 330–342, March 1991. With permission.)

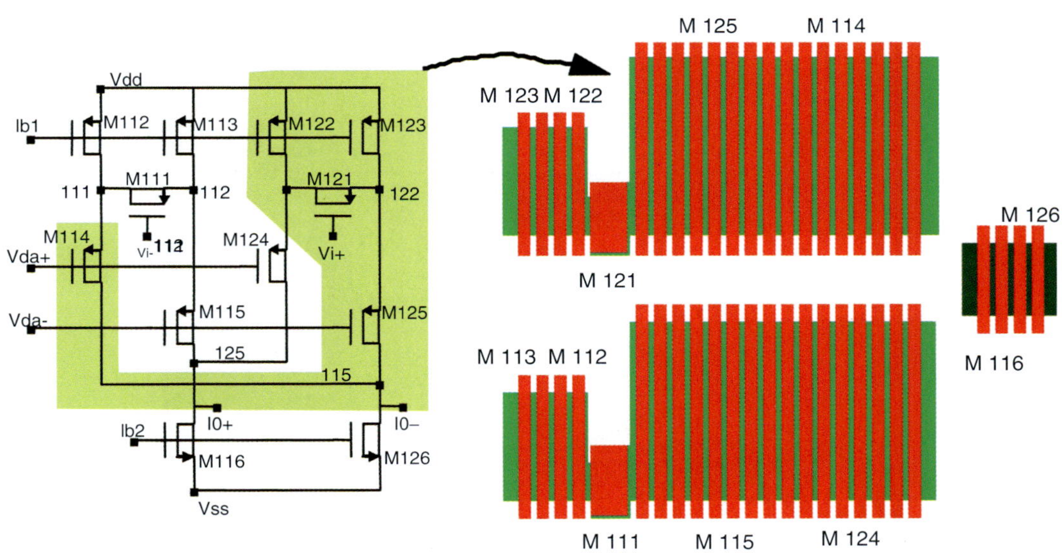

COLOR FIGURE 16.2 Example of automatic symmetric stack extraction from a CMOS amplifier circuit, using Basaran's algorithm. (From Basaran and Rutenbar, *Proceedings of ACM/IEEE DAC*, Las Vegas, NV, 1996, pp. 221–226. With permission.)

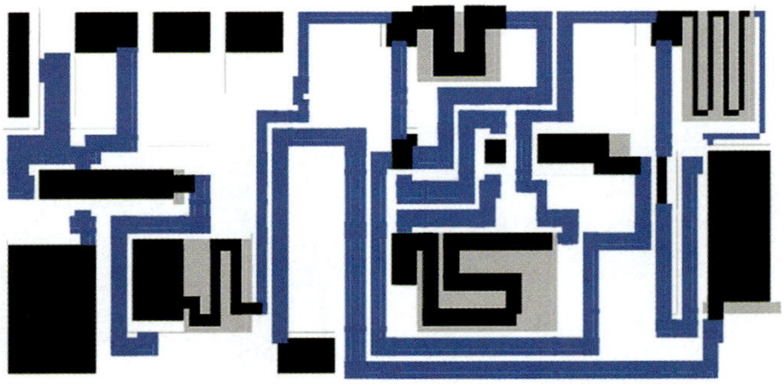

COLOR FIGURE 16.3 Microwave frequency (~60 GHz) device-level floorplan, shows devices, planar wiring on a single layer, and several instances of "detours" which are inserted to match lengths and electrical properties of wires in this small design. (From Aktuna et al., *IEEE Trans. CAD*, 18, 375–388, 1999. With permission.)

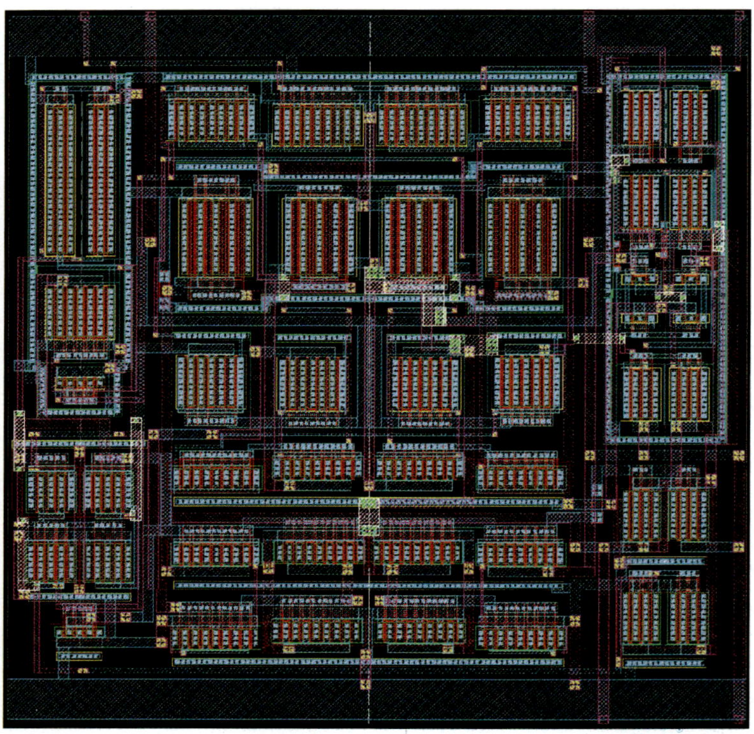

COLOR FIGURE 16.4 Commercial amplifier circuit, with custom device generators, custom wells, symmetric placement and routing, all synthesized automatically. (Courtesy Cadence Design Systems.)

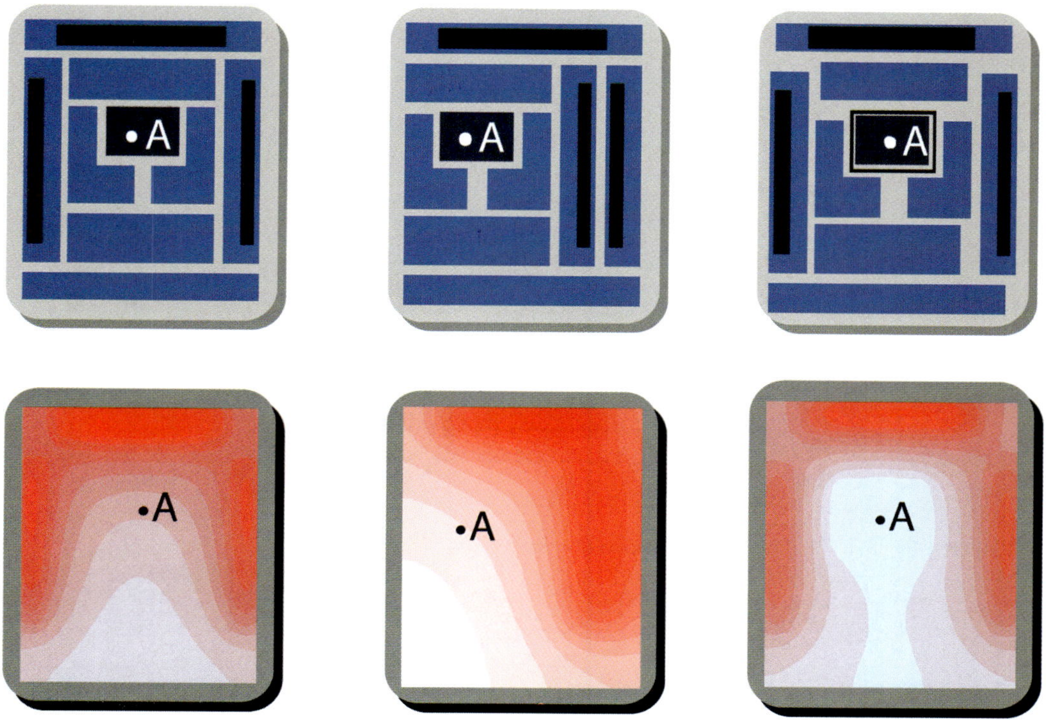

COLOR FIGURE 16.5 Example of a small, synthetic floorplan optimized by WRIGHT [95,96] under substrate noise constraints. Dark bars show noise sources; the module labeled "A" is sensitive to the overall noise level. From left to right, we see the result of floorplans generated under increasingly tight constraints on the allowed noise seen at "A". Bottom figures show iso-voltage contours in the substrate. At the far right, we see "A" has had a guard ring added to meet the tight noise spec.

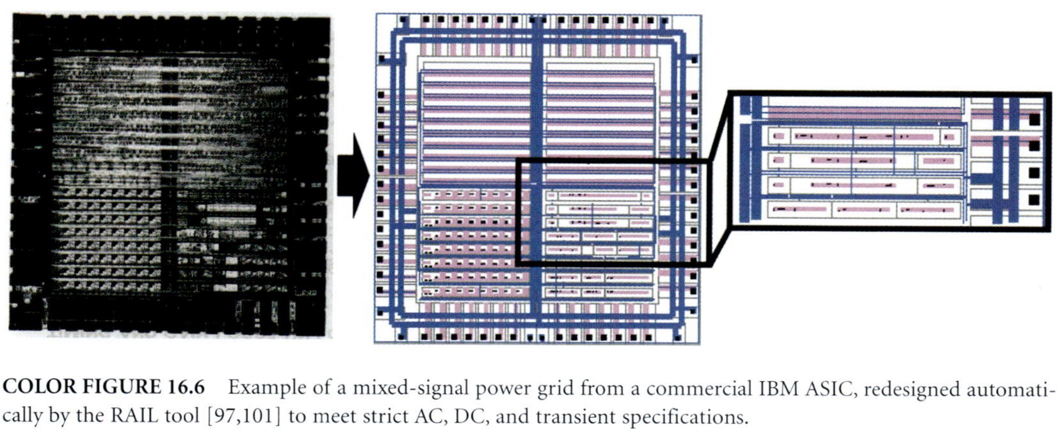

COLOR FIGURE 16.6 Example of a mixed-signal power grid from a commercial IBM ASIC, redesigned automatically by the RAIL tool [97,101] to meet strict AC, DC, and transient specifications.

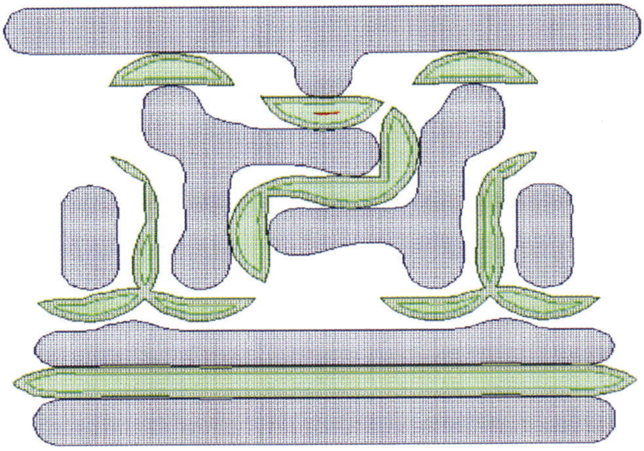

COLOR FIGURE 19.4 Graphical representation of critical area.

COLOR FIGURE 19.10 Via chain structure for characterizing failure rate dependency on the environment.

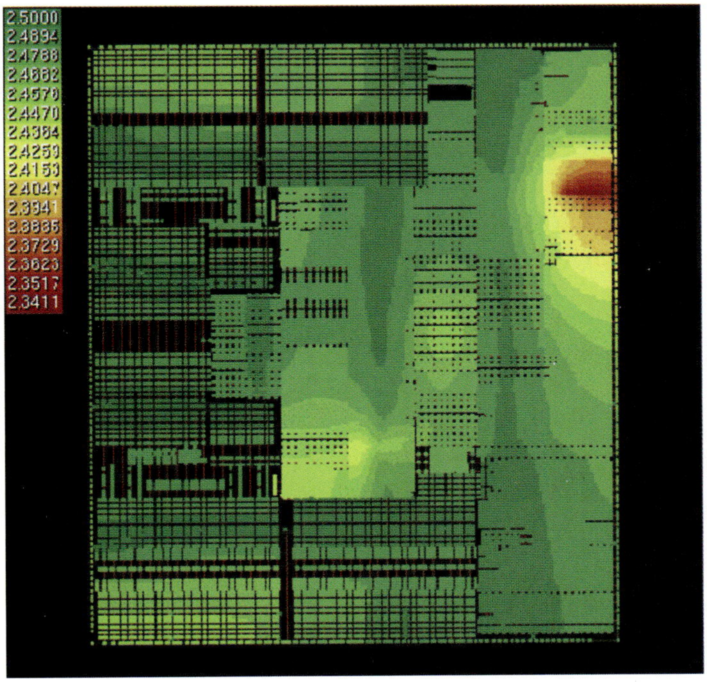

COLOR FIGURE 20.2 Postlayout IR-drop map of PowerPC™ 750 microprocessor.

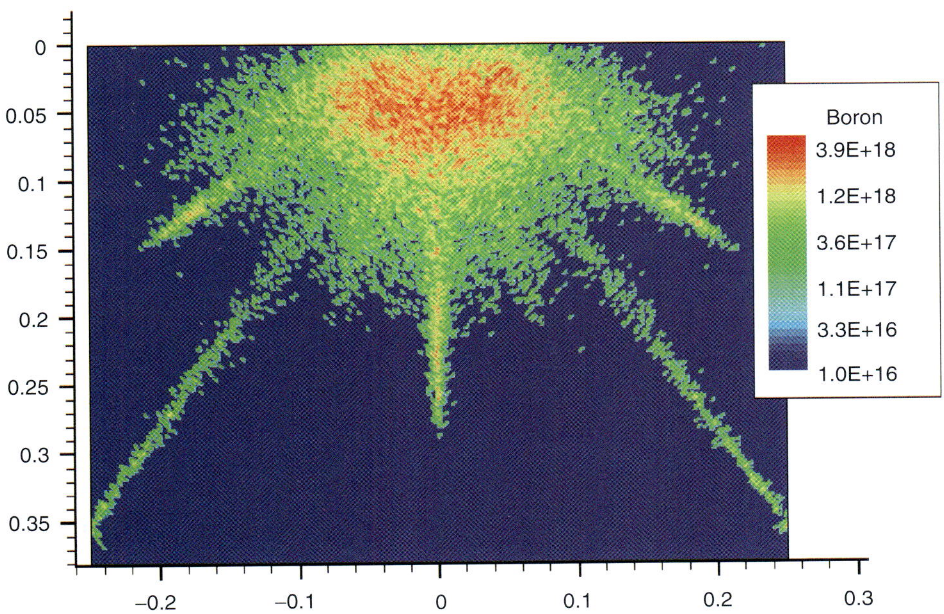

COLOR FIGURE 24.1 Figure created using Monte–Carlo implant by varying the tilt from 0 to 75° to probe the channeling behavior in silicon, dramatically illustrating the channeling tails in silicon.

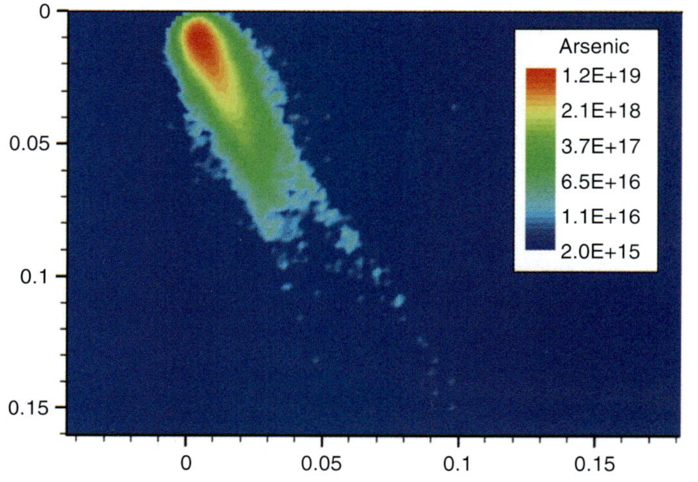

COLOR FIGURE 24.5 Point response Monte–Carlo implantation simulation for a 10^{14} cm^{-2}, 15 keV arsenic implant tilt 25°.

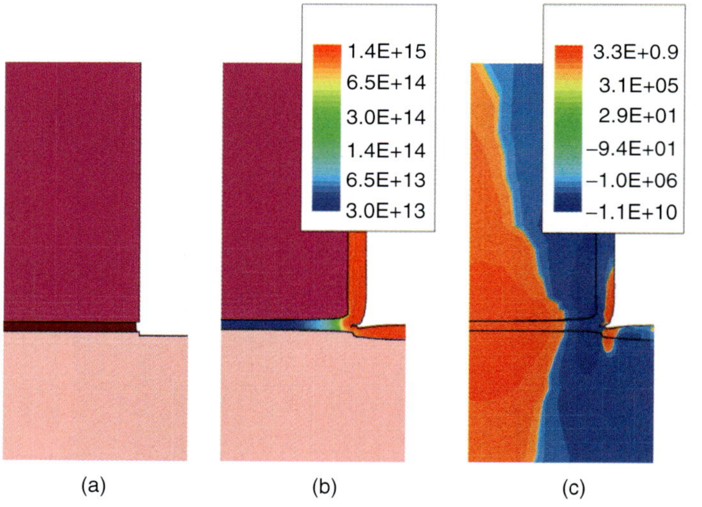

COLOR FIGURE 24.6 An oxidation simulation showing oxidant diffusion, Si consumption, and SiO_2 expansion. (a) Light pink is silicon, brown is oxide, and magenta is polysilicon; (b) Same simulation as in (a), but showing oxidant concentration in the oxide. Because the oxidant concentration does not reach far underneath the gate, oxidation occurs mostly at the edge of the poly gate; (c) Same simulation as in (a), (b) but the shading shows the component of stress in the vertical direction. Stress, which is concentrated in the region where oxidation happens, is a by-product of the consumption of silicon by the oxidant-forming less-dense SiO_2.

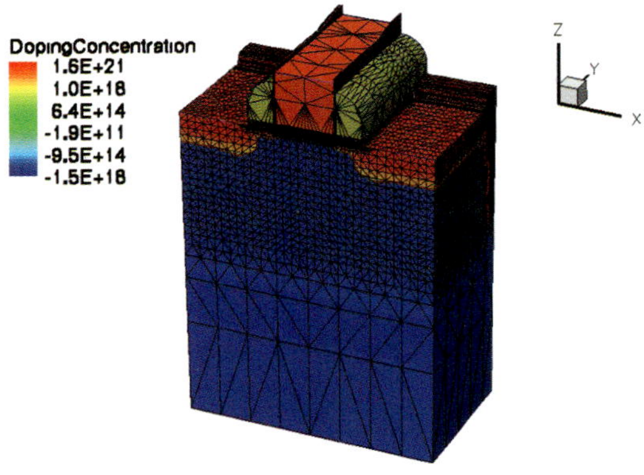

COLOR FIGURE 24.16 Final 3-D structure ready for device simulation. The half structure is reflected and contacts are added (outlined in purple on the top). The mesh lines are shown, and the doping concentration is shown in a rainbow, shaded red for n type and blue for p type.

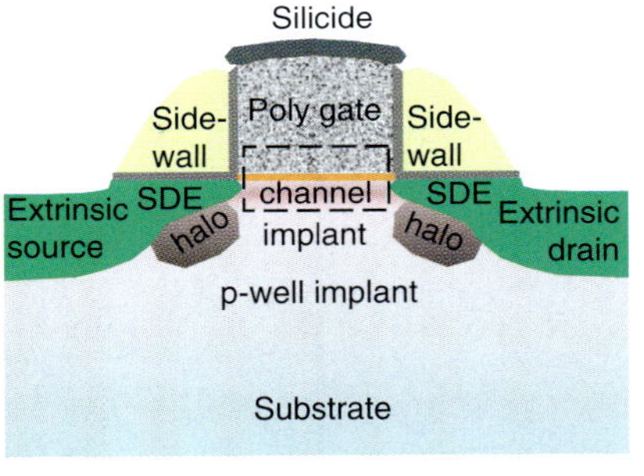

COLOR FIGURE 25.3b Cross-section of the n-channel MOS transistor, including details of gate, sidewalls and substrate doping profiles (SDE, channel/well implants, halo doping, etc.).

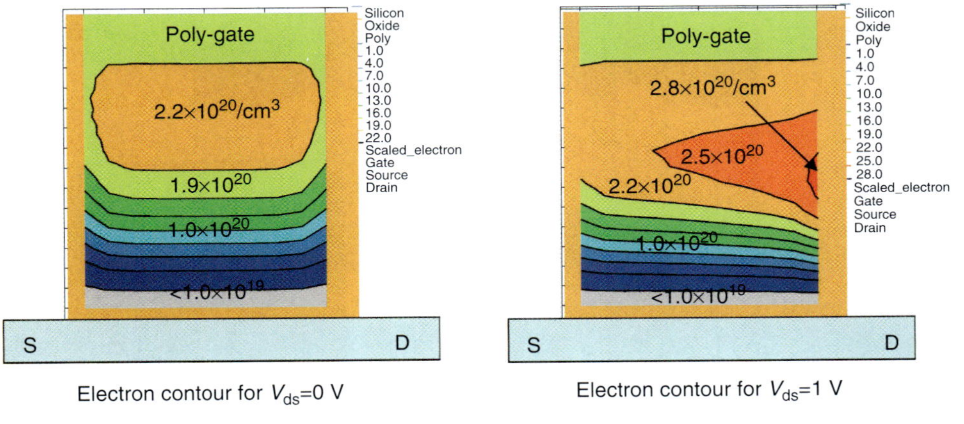

COLOR FIGURE 25.12b Simulations of QM poly depletion effects for two drain bias conditions (same device and I-V as Figure 25.12a). Results show significant gate depletion (left) and lateral effects that influence drain-induced barrier lowering (right).

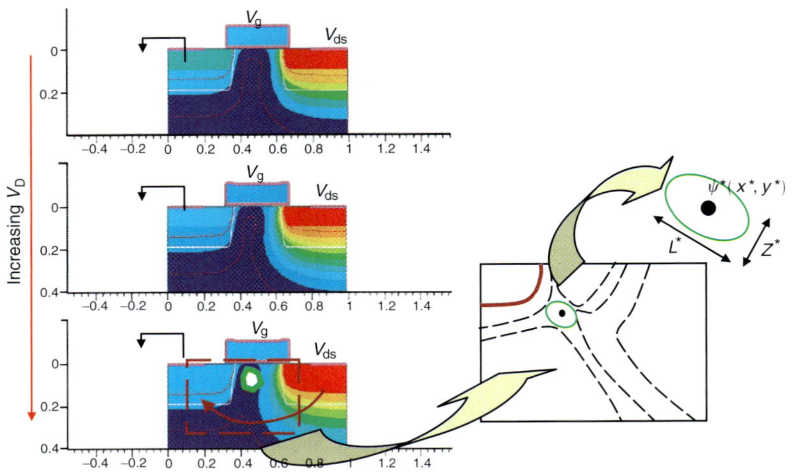

COLOR FIGURE 25.14c Two-dimensional plot of electrostatic potential in MOS device for three drain bias (Vds) conditions. The results show typical operation as well as drain-induced barrier lowering (DIBL) for sufficiently high drain bias (see insert).

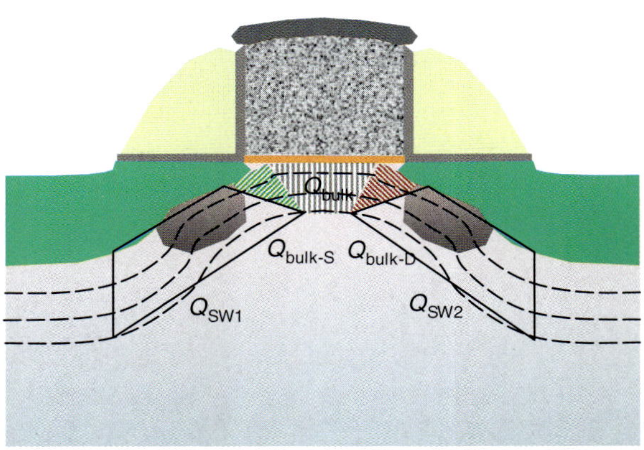

COLOR FIGURE 25.16a Cross-section on NMOS (same as Figure 25.3b) showing electrostatic potential contours (broken lines) and trapezoids and triangles of charge, representing influence of gate, source and drain potential as well as side-wall (SW) effects due to doping.

exploration are readily available in the present commercial system-level simulators, asking for more effective and more efficient solutions to be developed. To make system-level exploration really fast and interactive, dedicated algorithms can be developed that speed up the calculations by maximally exploiting the properties of the system under investigation and using proper approximations where possible. ORCA, for instance, is targeted toward telecom applications and uses dedicated signal spectral manipulations to gain efficiency [8]. A more recent development is the FAST tool which performs a time-domain dataflow type of simulation without iterations [9], and which easily allows dataflow co-simulation with digital blocks. Compared to commercial simulators such as COSSAP, PTOLEMY, or SPW, this simulator is more efficient because it uses block processing instead of point-by-point calculations for the different time points in circuits without feedback. In addition, the signals are represented as complex equivalent baseband signals with multiple carriers. The signal representation is local and fully optimized, as the signal at each node in the circuit can have a set of multiple carriers and each corresponding equivalent baseband component can be sampled with a different time-step depending on its bandwidth. Large feedback loops, especially when they contain nonlinearities, are however more difficult to handle with this approach. A method to simulate efficiently bit error rates with this simulator has been presented in Ref. [10].

Example

As an example [4,5], consider a front-end for a cable TV modem receiver, based on the MCNS standard. The MCNS frequency band for upstream communication on the CATV network is from 5 to 42 MHz (extended subsplit band). Two architectures are shown in Figure 15.3: (a) an all-digital architecture where both the channel selection and the down-conversion are done in the digital domain, and (b) the classical architecture where the channel selection is performed in the analog domain.

A typical input spectrum is shown in Figure 15.4. For this example we have used 12 QAM-16 channels with a 3 MHz bandwidth. We assume a signal variation of the different channels of maximally ±5 dB around the average level. The average channel noise is 30 dB below this level. Figure 15.5 and Figure 15.6 show the spectrum simulated by ORCA [8] for the all-digital architecture of Figure 15.3a. Figure 15.5 shows the spectrum of all eight channels after initial filtering at the output of the ADC, whereas Figure 15.6 shows the spectrum of the desired channel at the receiver output after digital channel selection and quadrature down-conversion. The desired channel signal and the effects of the channel noise, the ADC quantization noise, and the second- and third-order distortion are generated separately, providing useful feedback to the system designer. The resulting SNDR is equal to 22.7 dB in this case, which corresponds to a symbol error rate of less than 10^{-10} for QAM-16.

By performing the same analysis for different architectures, and by linking the required subblock specifications to the power or chip area required to implement the subblocks, a quantitative comparison of different alternative architectures becomes possible with respect to (1) their suitability to implement the system specifications, and (2) the corresponding implementation cost in power consumption or silicon

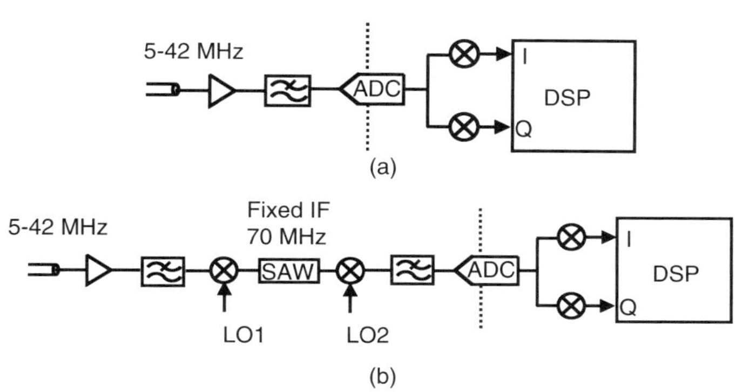

FIGURE 15.3 Two possible architectures for a cable TV application: (a) all-digital architecture, (b) classical architecture.

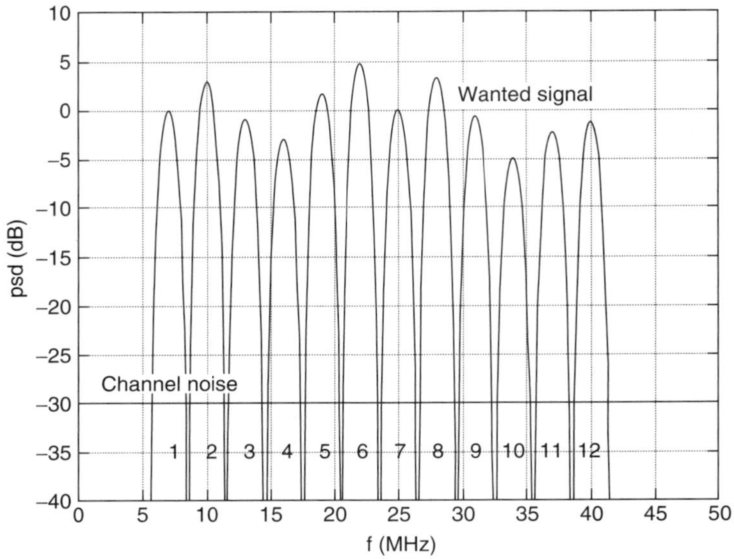

FIGURE 15.4 Typical input spectrum for a CATV front-end architecture using 12 QAM-16 channels.

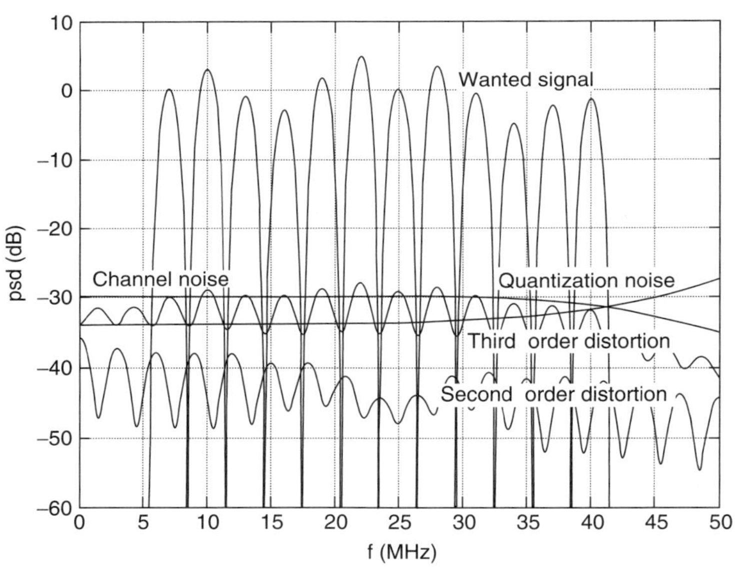

FIGURE 15.5 Simulated spectrum of the eight input channels at the output of the ADC in the all-digital CATV architecture.

real estate. To assess the latter, high-level power or area estimators must be used to quantify the implementation cost. In this way, the system designer can choose the most promising architecture for the application at hand.

Figure 15.7 shows a comparison between the estimated total power consumption required by the all-digital and by the classical CATV receiver architectures of Figure 15.3, as a function of the required SNR [11]. These results were obtained with the simulator FAST [9]. Clearly, for the technology used in the experiment, the classical architecture still required much less power than the all-digital solution.

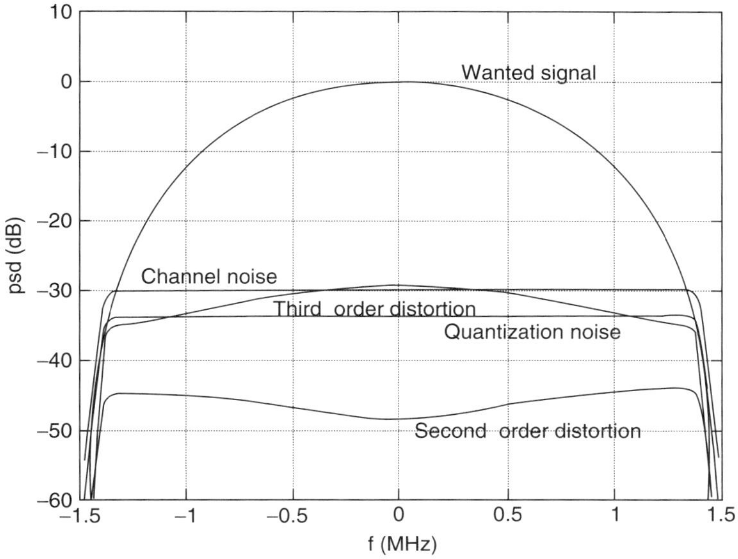

FIGURE 15.6 Simulated spectrum for the desired channel at the receiver output in the all-digital CATV architecture, indicating both the signal as well as noise and distortion added in the receiver.

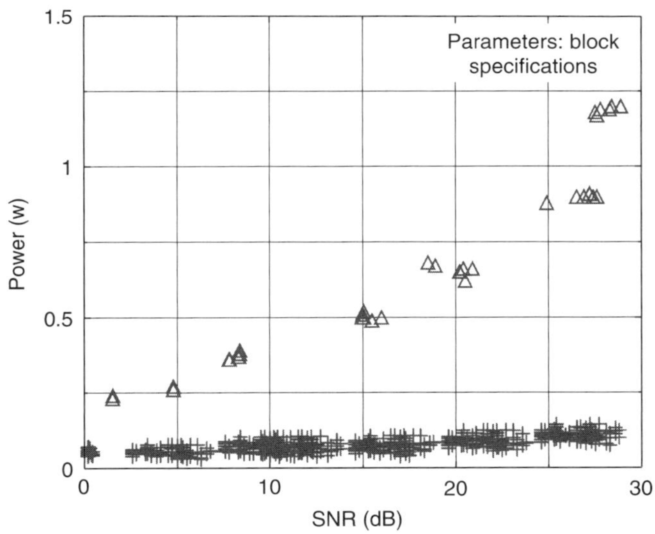

FIGURE 15.7 Power consumption comparison between the all-digital CATV architecture (triangles) and the classical architecture (crosses), as a function of the required SNR. (From P. Wambacq, G. Vandersteen, S. Donnay, M. Engels, I. Bolsens, E. Lauwers, P. Vanassche, and G. Gielen, High-level simulation and power modeling of mixed-signal front-ends for digital telecommunications, Proceedings of the International Conference on Electronics, Circuits and Systems (ICECS), September 1999, pp. 525–528. With permission.)

Finally, Figure 15.8 shows the result of a BER simulation with the FAST tool, for a 5-GHz 802.11 WLAN architecture [9]. The straight curve shows the result without taking into account nonlinear distortion caused by the building blocks; the dashed curve takes this distortion into account. Clearly, the BER considerably worsens in the presence of nonlinear distortion. Note that the whole BER analysis was

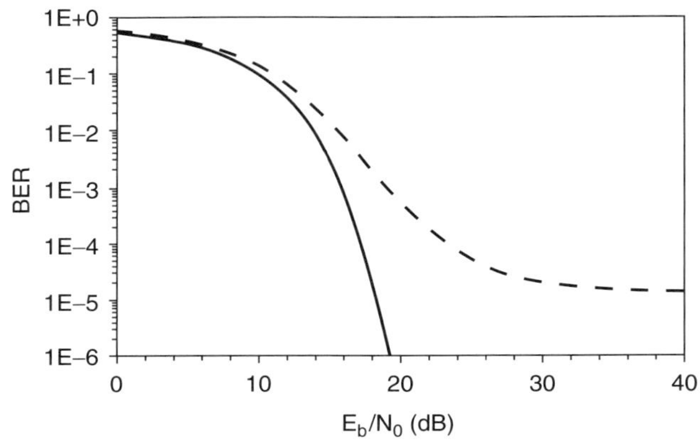

FIGURE 15.8 Simulated BER analysis result for a 5-GHz 802.11 WLAN architecture with (dashed), and without (straight), nonlinear distortion included. (From G. Vandersteen et al., Efficient bit-error-rate estimation of multicarrier transceivers, Proceedings of the Conference on Design, Automation and Test in Europe (DATE), 2001, pp. 164–168. With permission.)

performed in a simulation time that is two orders of magnitude faster than traditional Monte-Carlo analysis performed on a large number of OFDM symbols.

15.2.2 Example of Top-Down Design

Top-down design is already heavily used in industry today for the design of complex analog blocks, such as Delta–Sigma converters or phase-locked loops (PLL). In these cases, first a high-level design of the block is done with the block represented as an architecture of subblocks, each modeled with a behavioral model that includes the major non-idealities as parameters, rather than a transistor schematic. This step is often done using Matlab/Simulink and it allows the determination of the optimal architecture of the block at this level, together with the minimum requirements for the subblocks (e.g., integrators, quantizers, VCO, etc.), so that the entire block meets its requirements in some optimal sense. This is then followed by a detailed device-level (SPICE) design step for each of the chosen architecture's subblocks, targeted to the derived subblock specifications. This is now illustrated for a PLL.

Example

The basic block diagram of a PLL is shown in Figure 15.9. If all subblocks like the phase-frequency detector or the voltage-controlled oscillator (VCO) are represented by behavioral models instead of device-level circuits, then enormous time savings with regard to simulation time can be obtained during the design and verification phase of the PLL. For example, for requirements arising from a GSM-1800 design example (frequency range around 1.8 GHz, phase noise –121 dB/Hz at 600 kHz frequency offset, settling time of the loop for channel frequency changes below 1 ms within 1×10^{-6} accuracy), the following characteristics can be derived for the PLL subblocks using behavioral simulations with generic behavioral models for the subblocks [12]: $A_{LPF}=1$, $K_{VCO}=1 \times 10^6$ Hz/V, $N_{div}=64$, $f_{LPF}=100$ kHz. These specifications are then the starting point for the device-level design of each of the subblocks.

15.3 Mixed-Signal and Behavioral Simulation

We will now describe the state of the art in mixed-signal and behavioral simulation and modeling.

15.3.1 Analog and Mixed-Signal Simulation

Circuit simulation began with the early development of the SPICE simulator [13,14] which even today remains the cornerstone of many circuit designs. This simulator numerically solves the system of

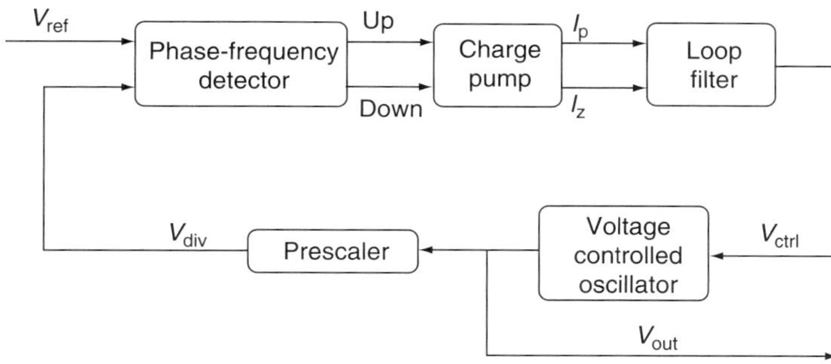

FIGURE 15.9 Basic block diagram of a phase-locked loop analog block.

nonlinear differential-algebraic equations (DAE) that characterize the circuit by using traditional techniques of numerical analysis. For example, to compute the evolution of the circuit's response in time, the time-derivative operator is first discretized by an implicit integration scheme such as the backward-Euler or trapezoidal rule. At each time point, the resulting set of nonlinear algebraic equations is solved iteratively via the Newton–Raphson method, with the matrix solution required to update the approximate solution computed using sparse matrix techniques. For details, refer to the companion chapter on Circuit simulation in this volume [15]. Advances in mathematics and the evolutionary improvement of the core numerical algorithms (e.g., adaptive time-step control, sparse matrix factorization, and improved convergence) have over the years contributed to a vast number of commercial CAD tools. Many descendents of the SPICE simulator are now marketed by a number of CAD vendors and many IC manufacturers have in-house versions of the SPICE simulator that have been adapted to their own proprietary processes and designs. A few examples of the many commercial SPICE-class simulators are HSPICE (Synopsys), Spectre (Cadence Design Systems), and Eldo (Mentor Graphics). SPICE or its many derivatives have evolved into an established designer utility that is being used both during the design phase (often in a designer-guided trial-and-error fashion) and for extensive postlayout design verification.

Although SPICE is a general-purpose circuit simulator, its adaptive time-stepping approach is very slow for circuits with widely separated time constants. This is why for certain circuit classes, more dedicated, faster solutions have been developed. For instance, simulators for switched-capacitor [16] and switched-current circuits, as well as discrete-time $\Delta\Sigma$ modulators [17] take advantage of the switched timing of the circuits to cut down on simulation time. Another important domain is RF simulation, where shooting and harmonic balance techniques have been developed to directly simulate the steady-state behavior of these circuits without having to follow slow initial transients or many carrier cycles in a slow signal envelope [18].

With the explosion of mixed-signal designs, the need has also arisen for simulation tools that allow not only simulation of analog or digital circuits separately, but also simulation of truly mixed analog–digital designs [19]. Simulating the large digital parts with full SPICE accuracy results in very long overall simulation times, whereas efficient event-driven techniques exist to simulate digital circuits at higher abstraction levels than the transistor level. Therefore, mixed-mode simulators were developed that glue together an accurate SPICE-like analog simulator to an efficient digital simulator. The general scheme of these so-called glued mixed-signal simulators is shown in Figure 15.10. These simulators partition the netlist into digital and analog parts that are each simulated by a separate kernel that interfaces through a simulation "backplane": an analog kernel with adaptive time-step control, and a digital kernel with its event-driven mechanism. The integration of two disparate simulation technologies requires reconciliation of the different computation models. Conversion of the signals between analog and digital signal representations and the appropriate driving/loading effects are addressed by inserting interface elements at the boundaries between the analog and digital circuitry. Effective treatment of the synchronization between the analog simulation kernel with its nonuniform time-steps due to its adaptive time-step control and the digital simulation kernel with its event-driven mechanism determines the efficiency of the overall simulation.

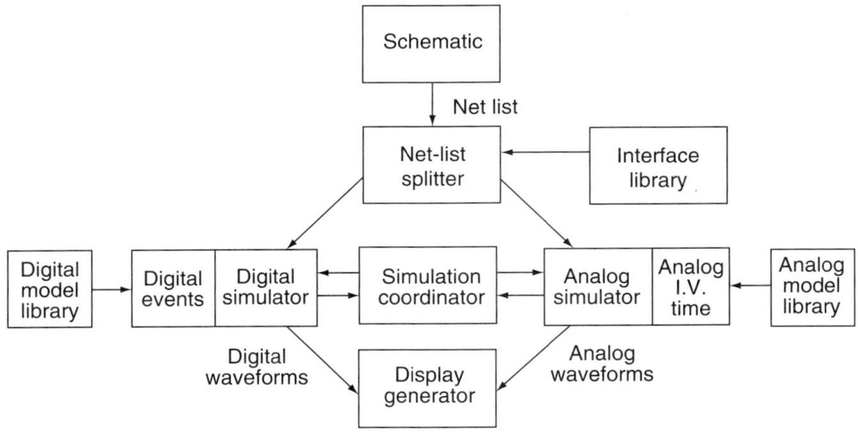

FIGURE 15.10 General scheme of a mixed-signal simulator, gluing an analog and a digital simulation kernel together.

Such synchronization is needed at each time point when there is a signal event at the boundary between the analog and digital parts of the system: whenever an analog signal at a boundary node passes the logic's thresholds, an event is created on the digital simulator side, and alternatively a digital event at a boundary node is translated into the corresponding analog signal transition on the analog simulator side.

Problems arising due to the weak integration mechanism in these glued simulators, such as poor simulation performance, difficulty in describing modules with mixed semantics, and difficulty in debugging across the mixed-signal boundary, have recently led to more integrated simulators that combine the analog and digital capabilities in a single kernel. Such simulators that feature the ability to seamlessly mix Verilog-AMS, VHDL-AMS (see below for details), and SPICE descriptions are today available in the commercial marketplace. As representative examples of modern mixed-signal simulators, we would mention the AdvanceMS (Mentor Graphics) and AMS Designer tools (Cadence Design Systems).

Besides the eternal problem of the limited accuracy of the device models used in SPICE, the main problems with the standard SPICE simulator are that it is essentially a structural circuit simulator, and that its CPU time increases rapidly with the size of the circuit, making the simulation of really large designs infeasible. This is why in the past years the need has arisen for higher levels of abstraction to describe and simulate analog circuits more efficiently, at the expense of a (little) loss in accuracy.

15.3.2 Analog Behavioral Simulation

There are (at least) four reasons for using higher-level analog modeling (functional, behavioral, or macromodeling) for describing and simulating mixed-signal systems [2]:

- The simulation time of circuits with widely spaced time constants (e.g., oversampling converters, PLLs, etc.) is quite large since the time-step control mechanism of the analog solver follows the fastest signals in the circuit. Use of higher-level modeling for the blocks will accelerate the simulation of these systems, particularly if the "fast" time-scale behavior can be "abstracted away", e.g., by replacing transistor-level descriptions of RF blocks by baseband-equivalent behavioral models.
- In a top-down design methodology based on hierarchical design refinement (as in Figure 15.1 and the example of Section 15.2.2) at higher levels of the design hierarchy, there is a need for higher-level models describing the pin-to-pin behavior of the circuits in a mathematical format, rather than representing it as an internal structural netlist of components. This is unavoidable during top-down design, since at higher levels in the design hierarchy, the details of the underlying circuit implementation are simply not yet known and hence only generic mathematical models can be used.

- A third use of behavioral models is during bottom-up system verification when these models are needed to reduce the CPU time required to simulate the block as part of a larger system. The difference is that in this case the underlying implementation is known in detail, and that peculiarities of the block's implementation can be incorporated in the model as far as possible without slowing down the simulation too much.
- Fourthly, when providing or using analog IP macrocells in a system-on-a-chip context, the virtual component has to be accompanied by an executable model that efficiently models the pin-to-pin behavior of the virtual component. This model can then be used in system-level design and verification by the SoC integrating company, even without knowing the detailed circuit implementation of the macrocell [20].

For all these reasons, analog/mixed-signal (AMS) behavioral simulation models are needed, that describe analog circuits at a higher level than the circuit level, i.e., that describe the input–output behavior of the circuit in a mathematical model rather than as a structural network of basic components. These higher-level models must describe the desired behavior of the block (e.g., amplification, filtering, mixing, or quantization) and simulate efficiently, while still including the major non-idealities of real implementations with sufficient accuracy.

In addition, high-level models also form a key part of several analog circuit synthesis and optimization systems, both at the circuit level as well as for the hierarchical synthesis of more complex blocks. Most of the basic techniques in both circuit and layout synthesis today rely on powerful numerical optimization engines coupled to "evaluation engines" that qualify the merit of some evolving analog circuit or layout candidate [2]. This basic scheme of optimization-based analog circuit sizing is shown in Figure 15.11. The most general but also by far the slowest solution is to call the transistor-level simulator as evaluation engine during each iteration of the optimization of a circuit. The CPU time can be reduced significantly by replacing the simulations by model evaluations. These models can be behavioral simulation models as described above, or they can be what are termed *performance models* [21]. Rather than modeling the input–output behavior, performance models directly relate the achievable performances of a circuit (e.g., gain, bandwidth, or slew rate) to the design variables (e.g., device sizes and biasing). In the synthesis procedure, calls to the transistor-level simulation are then replaced by performance model evaluations, resulting in substantial speed-ups of the overall synthesis, once the performance models have been created and calibrated. For the synthesis of more complex analog blocks, a hierarchical approach is needed, in which higher-level models are indispensable to bridge between the different levels [22,17].

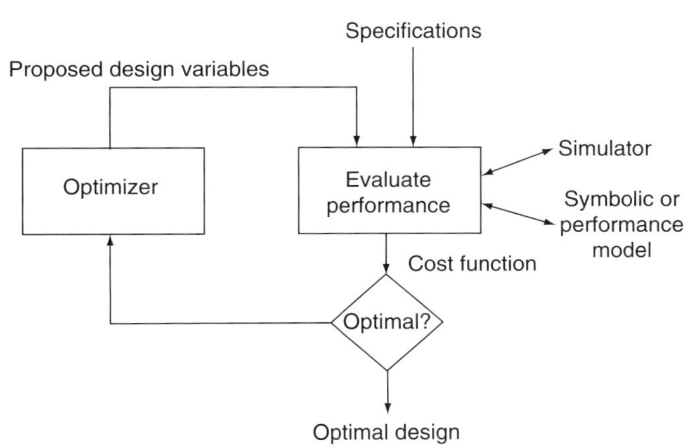

FIGURE 15.11 Basic flow of optimization-based analog circuit sizing.

Example

For example, the realistic dynamic behavior of a (current-steering) digital-to-analog converter shows finite settling and ringing (glitch), as shown in Figure 15.12. Such behavior can mathematically be approximated by superposition of an exponentially damped sine for the glitch and a hyperbolic tangent [23]:

$$i_{out}=A_{gl}\sin\left(\frac{2\pi}{t_{gl}}(t-t_0)\right)\exp\left(-\sin(t-t_0)\frac{2\pi}{t_{gl}}(t-t_0)\right)+\frac{\text{level}_{i+1}-\text{level}_i}{2}\tanh\left(\frac{2\pi}{t_{gl}}(t-t_0)\right)+\frac{\text{level}_{i+1}+\text{level}_i}{2} \tag{15.1}$$

where level_i and level_{i+1} are the DAC output levels before and after the considered transition, and where the parameters such as A_{gl}, t_0, and t_{gl} need to be determined, for instance, by regression fitting to simulation results of a real circuit. Figure 15.13 compares the response of the behavioral model (with parameter values extracted from SPICE simulations) with SPICE simulation results of the original circuit. The speedup in CPU time is almost three orders of magnitude, while the error is below 1% [23].

Table 15.1 gives an overview of the different analog hardware description levels considered in design practice today, and the implications of those abstraction levels on the modeling [24]. The *circuit level* is the traditional level where a circuit is simulated as a network of physical components. In a *macromodel,* an equivalent but computationally cheaper circuit representation is used that has approximately the same behavior as the original circuit. Equivalent sources combine the effect of several other elements that are eliminated from the netlist. The simulation speedup is roughly proportional to the number of nonlinear devices that can be eliminated. In a *behavioral or functional model,* a purely mathematical description of the input–output behavior of the block is used. This typically will be in the form of a set of DAE or transfer functions. At the behavioral level, conservation laws still have to be satisfied; at the functional level this is no longer the case, and the simulated system turns into a kind of signal-flow diagram.

The industrial use of analog higher-level (functional, behavioral, macro-) modeling is today enabled by the availability of standardized mixed-signal hardware description languages such as VHDL-AMS [25,26] and Verilog-AMS [27,28], both of which are extensions of the corresponding digital hardware description languages, and both of which are supported by commercial simulators today. These languages allow description and simulation of separate analog circuits, separate digital circuits, and mixed analog–digital circuits, at the different abstraction levels mentioned above. In general, they also allow

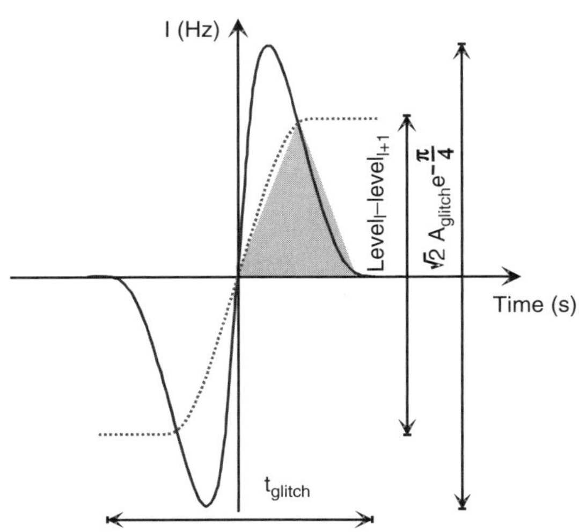

FIGURE 15.12 Typical dynamic behavior of a current-steering digital-to-analog converter output showing both finite settling and glitch behavior when switching the digital input code.

TABLE 15.1 Different Analog Hardware Description Levels to Describe Analog Circuits

Level	Modeling Primitives	Implications
Functional	Mathematical signal flow description per block, connected in signal flow diagram	No internal block structure; conservation laws need not be satisfied on pins
Behavioral	Mathematical description (equations, procedures) per block	No internal block structure; conservation laws must be satisfied on pins
Macromodel	Simplified circuit with controlled sources	Spatially unrelated to actual circuit; conservation laws must be satisfied
Circuit	Connection of SPICE primitives	Spatially one-to-one related to actual circuit; conservation laws must be satisfied

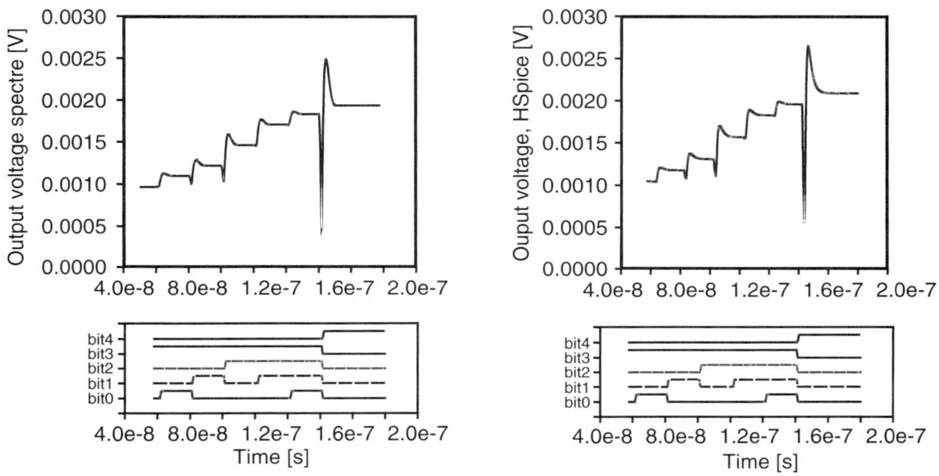

FIGURE 15.13 Comparison between the device-level simulation results (on the right) and the response of the behavioral model (on the left). (From J. Vandenbussche et al., Systematic design of high-accuracy current-steering D/A converter macrocells for integrated VLSI systems, *IEEE Trans. Circuit Syst.*, Part II, 48, 300–309, 2001. With permission.)

description and simulation of both electrical and nonelectrical systems, as long as they can be modeled by a set of (nonlinear) DAE. Note that while originally restricted to low-to-medium frequencies with circuits having lumped elements only, efforts to standardize the extension of these languages, e.g., VDHL-AMS, toward RF and microwave circuits with distributed elements, have been started in recent years. Likewise, effort is being made to leverage language-based modeling to modernize descriptions of the transistor models themselves by means of language extensions that support automatic compilation of semiconductor device compact models [29,30].

Example

Below is an illustrative example of a VHDL-AMS model for a circuit combining a VCO and a frequency divider (DIVN):

```
entity vcodivn is
generic (vco_gain : real := 1.0e6;
            f0 : real := 1.0e9;
         rin : real := 1.0e6;
         propdelay : time : = 5.0 ns;
         pn : integer := 64);
port (terminal input, ref : electrical;
      signal sdived : out std_ulogic);
```

```
end entity vcodivn;
architecture behavioral of vcodivn is
quantity phase : real := 0.0;
quantity freq : real := 0.0;
quantity vin across iin through input to ref;
begin
   break phase => 0.0 when
     phase'above(real(pn/2));
   freq == f0 + vco_gain*vin;
   phase == freq'integ;
   iin == vin/rin;
   triggerOutput : process is
         variable ostate : std_ulogic := '0';
         begin
         wait on phase'above(real(pn/2));
         ostate := not ostate;
         sdived <= ostate after propdelay;
   end process triggerOutput;
end architecture behavioral;
```

The model contains an entity part, describing the circuit as a box, i.e., the parameters and the pins of the model, and an architecture part, describing the behavior of the circuit. In the example, the circuit has five parameters (after the keyword "generic") and two electrical terminals and one signal port (after the keyword "port"). The architecture description first defines the analog variables ("quantities") that are being used and then describes the equations that govern the behavior of the circuit. Both algebraic and an integral equation (freq'integ means applying the integral operator to the quantity freq) are used. The model is a mixed analog–digital model as it also contains a standard VHDL process description as part of the model.

15.3.3 Entry and Analysis Environments

While transistor- and behavioral-level simulation tools provide the core analysis capability for analog design, practical usability of simulation tools for design and verification of complex circuits has been enabled by advances in user interface technology: schematic entry systems, graphical interfaces, waveform displays, and scripting languages. The bulk of analog design at present takes place within such integrated design environments that tie analysis capabilities together with physical layout, verification, and parasitic extraction tools. While emerging analog synthesis tools [2] are beginning to appear in commercial use, the more traditional environments are centered around the capture–simulate–analyze task loop. In this, loop circuit topology and properties are described via a graphical schematic entry tool. From the database description thus created, a circuit netlist or other elaborated structural circuit description is generated, which is read and processed by the simulator to produce simulation results data (i.e., waveforms). Simulation waveforms are either processed in-line to obtain circuit performance measurements (delay, rise time, distortion, etc.) or stored in a waveform database for later analysis by a waveform display and calculation tool. As commercial examples of such environments, we would mention the Virtuoso Analog Design Environment (aka "Artist") from Cadence Design Systems, and the Cosmos system marketed by Synopsys.

15.4 Analog Behavioral and Power Model Generation Techniques

One of the biggest problems today is the lack of systematic methods to create good analog behavioral or performance models — a skill not yet mastered by the majority of analog designers — as well as the lack of any tools to automate this process. Fortunately, in recent years, research has started to develop methods that can automatically create models for analog circuits, both behavioral models for behavioral simulation and performance models for circuit sizing. Techniques used here can roughly be divided into fitting or regression approaches, constructive approaches, and model-order reduction methods.

15.4.1 Fitting or Regression Methods

In the *fitting or regression approaches*, a parameterized mathematical model is proposed or constructed by the model developer, and the values of the parameters p are selected as to best approximate the known (simulated) circuit behavior, as schematically depicted in Figure 15.14. A systematic approach to regression-based model construction consists of several steps:

1. Selection of an appropriate model structure or template, often done ad hoc by the designer based on knowledge of the circuit behavior. The possible choices of model are vast. Some of the more common include polynomials, rational functions [31], and artificial neural networks. Recently, EDA researchers have begun to develop more constructive methods to build models from underlying circuits by utilizing results from statistical inference [32] and data mining [33], and we expect to see regression tree, k-nearest neighbor, and kernel forms such as support vector machines [34–36] to become more prominent in the future. Posynomial forms have attracted particular interest for optimization applications [37,21], as optimization problems involving models of this form can be recast as convex programs, leading to very efficient sizing of analog circuits.
2. Creation or selection of the simulation data to which to fit the model via an appropriate design-of-experiments scheme [21].
3. Selection of a model fidelity criterion. For example, the model can be fit by a least-squares error optimization where the model response matches the simulated (or measured) time-domain response of the real circuit as closely as possible in an average sense [25]. This is schematically depicted in Figure 15.14. The error could, for instance, be calculated as

$$error = \int_0^T ||v_{\text{out,real}}(t) - v_{\text{out,model}}(t)||^2 \mathrm{d}t \tag{15.2}$$

4. Selection of the optimization procedure to select the parameters (in some cases this step and the previous one are combined), such as by gradient descent or other gradient-based optimization, "boosting" [38], or stochastic optimization such as Markov-chain Monte-Carlo or simulated annealing.
5. Validation of the final model. Without specific attention paid to model validation, it is quite common to find "overfit" models. Such models may have small error for the simulation data on which they were "trained", but very poor accuracy when slightly different circuit excitations are introduced when the model is put into use. Regularization may be introduced in step 3 to attempt to suppress such behavior, e.g., by modifying the model fidelity criterion to penalize large coefficients in a least-squares fit.

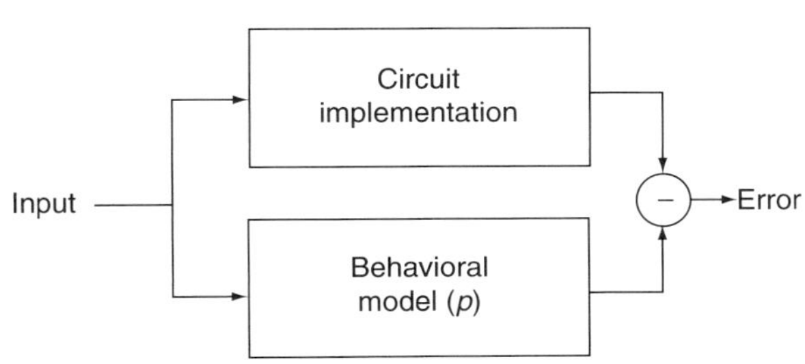

FIGURE 15.14 Basic flow of fitting or regression approach for analog behavioral model generation.

It should be clear that these fitting approaches can, in principle, be very generic as they consider the block as a blackbox and only look at the input–output behavior of the block which can easily be simulated (or measured). Once the model is generated, it becomes an implicit model of the circuit. However, hiding in each of the steps outlined above are daunting practical challenges. Chief among these comes the first step: for any hope of success, first a good model template must be proposed, which is not always trivial to do in an accurate way without knowing the details of the circuit. Even when good choices are possible, it may happen that the resulting model is specific for one particular implementation of the circuit. Likewise, the training set must exercise all possible operating modes of the circuit, but these can be hard to predict in advance.

To address these challenges, progress made in other areas such as in research on time-series prediction (e.g., support vector machines [35,36]) and data-mining techniques [33] are being pursued.

15.4.2 Symbolic Model Generation Methods

The second class of methods, the *constructive approaches*, try to generate or build a model from the underlying circuit description. Inherently, these are therefore white-box methods, as the resulting model is specific for the particular circuit, but there is a higher guarantee than with the fitting methods that it tracks the real circuit behavior well in a wider range. One approach, for instance, uses symbolic analysis techniques to generate first the exact set of algebraic/differential equations describing the circuit, which are then simplified within a given error bound of the exact response using both global and local simplifications [39]. The resulting simplified set of equations then constitutes the behavioral model of the circuit and tracks the behavior of the circuit nicely. The biggest drawback, however, is that the error estimation is difficult, and for nonlinear circuits heavily depends on the targeted response. Until now, the gains in CPU time obtained in this way are not high enough for practical circuits. More research in this area is definitely needed.

15.4.3 Model-Order Reduction Methods

The third group of methods, the *model-order reduction methods*, are mathematical techniques that generate a model with lower order for a given circuit by direct analysis and manipulation of its detailed, low-level description; for example the nonlinear differential equations in a SPICE simulator, or the resistor–capacitor model describing extracted interconnect. Classical model-order reduction algorithms take as input a linear, time-invariant set of differential equations describing a state-space model, for example,

$$\frac{dx}{dt} = Ax + Bu, \; y = Cx + Du \tag{15.3}$$

where x represents the circuit state, u the circuit inputs, y the circuit outputs, and the matrices A, B, C, and D determine the circuit properties. As output model-order reduction methods produce a similar state-space model $\tilde{A}$, $\tilde{B}$, $\tilde{C}$, $\tilde{D}$, but with a state vector $\tilde{x}$ (thus matrix description) of lower dimensionality, i.e., of lower order:

$$\frac{d\tilde{x}}{dt} = \tilde{A}\tilde{x} + \tilde{B}u, \; \tilde{y} = \tilde{C}\tilde{x} + \tilde{D}u \tag{15.4}$$

These reduced-order models simulate much more efficiently, while closely approximating the exact response; for example matching the original model closely up to some specified frequency.

Originally developed to reduce the complexity of interconnect networks for timing analysis, techniques such as asymptotic waveform evaluation (AWE) [40] used Padé approximation to generate a lower-order model for the response of the linear interconnect network. The early AWE efforts used explicit moment-matching techniques which were not numerically stable, and thus could not produce higher-order models that were needed to model circuits more complicated than resistor–capacitor networks, and Padé approximations often generate unstable or nonpassive reduced-order models. Subsequent developments using Krylov-subspace-based methods [41,42] resulted in methods like PVL that overcome many of the deficiencies of the earlier AWE efforts, and passive model construction is now guaranteed via projection-via-congruence such as used in PRIMA [43].

In recent years, similar techniques have also been extended in an effort to create reduced-order macromodels for analog/RF circuits. Techniques have been developed for time-varying models, particularly periodically time-varying circuits [44,45], and for weakly nonlinear circuits via polynomial-type methods that have a strong relation to Volterra series [45–47]. Current research focuses on methods to model more strongly nonlinear circuits (e.g., using trajectory piecewise-linear [48] or piecewise-polynomial approximations [49]), and is starting to overlap with the construction of performance models, through the mutual connection to the regression and data-mining ideas [33,35,36].

Despite the progress made so far, more research in the area of automatic or systematic behavioral model generation or model-order reduction is certainly needed.

15.4.4 Power/Area Estimation Methods

Besides behavioral models, the other crucial element to compare different architectural alternatives and to explore trade-offs during system-level exploration and optimization are accurate and efficient power and area estimators [50]. They allow the assessment and comparison of the optimality of different design alternatives. Such estimators are functions that predict the power or area that is going to be consumed by a circuit implementation of an analog block (e.g., an ADC) with given specification values (e.g., resolution and speed). Since the implementation of the block is not yet known during high-level system design, and considering the large number of different possible implementations for a block, it is very difficult to generate these estimators with high absolute accuracy. However, for the purpose of comparing different design alternatives, the tracking accuracy of estimators with varying block specifications is of much more importance.

Such functions can be obtained in two ways. A first possibility is the derivation of analytic functions or procedures that given the block's specifications, return the power or area estimate. An example of a general yet relatively accurate power estimator that was derived based on the underlying operating principles for the whole class of CMOS high-speed Nyquist-rate ADCs (such as flash, two-step, pipelined, etc., architectures) is given by [50]

$$\text{power} = \frac{V_{dd}^2 L_{min}(F_{sample} + F_{signal})}{10^{(-0.1525 \cdot \text{ENOB} + 4.8381)}} \tag{15.5}$$

where F_{sample} and F_{signal} are the clock and signal frequency respectively, and where ENOB is the effective number of bits at the signal frequency. The estimator is technology-scalable (V_{dd} and L_{min} are parameters of the model), and has been fitted with published data of real converters, and for more than 85% of the designs checked, the estimator has an accuracy better than 2.2 ×. Similar functions are developed for other blocks, but of course often a more elaborate procedure than a simple formula is needed. For example, for the case of high-speed continuous-time filters [50], a crude filter synthesis procedure in combination with operational transconductor amplifier behavioral models had to be developed to generate accurate results. The reason is that the filter implementation details, and hence the power and chip area, vary quite significantly with the specifications, requiring a rough filter synthesis to be performed to gather sufficient implementation detail knowledge to allow reliable estimates of power and area.

A second possibility to develop power/area estimators is to extract them from a whole set of data samples from available or generated designs through interpolation or fitting of a predefined function or an implicit function, e.g., a neural network. As these methods do not rely on underlying operating principles, extrapolations of the models have no guaranteed accuracy.

In addition to power and area estimators, *feasibility functions* are also needed that limit the high-level optimization to realizable values of the building block specifications. These can be implemented under the form of functions (e.g., a trained neural network or a support vector machine [51]) that return whether a block is feasible or not, or of the geometrically calculated feasible performance space of a circuit (e.g., using polytopes [52] or using radial base functions [53]).

15.5 Symbolic Analysis of Analog Circuits

Analog design is a very complex and knowledge-intensive process, which heavily relies on circuit understanding and related design heuristics. Symbolic circuit analysis techniques have been developed to help designers gain a better understanding of a circuit's behavior. A symbolic simulator is a computer tool that takes as input an ordinary (SPICE type) netlist, and returns as output (simplified) analytic expressions for the requested circuit network functions in terms of the symbolic representations of the frequency variable and (some or all of) the circuit elements [54,55]. These simulators perform the same function that designers traditionally do by hand analysis (even the simplification). The difference is that the analysis is now done by the computer, which is much faster, can handle more complex circuits, and does not make as many errors. An example of a complicated BiCMOS opamp is shown in Figure 15.15. The (simplified) analytic expression for the differential small-signal gain of this opamp in terms of the symbolic small-signal parameters of the opamp's devices, has been analyzed with the SYMBA tool [56], and is shown below:

$$A_{V0} = \frac{g_{m,M2}}{g_{m,M1}} \frac{g_{m,M4}}{(g_{o,M4}g_{o,M5}/g_{m,M5} + g_{mb,M5}) + (G_a + g_{o,M9} + g_{o,Q2}/\beta_{Q2})} \tag{15.6}$$

The symbolic expression gives a better insight into which small-signal circuit parameters predominantly determine the gain in this opamp and how the user has to design the circuit to meet a certain gain constraint. In this way, symbolic circuit analysis is complementary to numerical (SPICE) circuit simulation that was described in the previous section. Symbolic analysis provides a different perspective that is more suited for obtaining insight into a circuit's behavior and for circuit explorations, whereas numerical simulation is more appropriate for detailed design validation once a design point has been decided upon. In addition, the generated symbolic design equations also constitute a model of the circuit's behavior that can be used in CAD tasks, such as analog synthesis, statistical analysis, behavioral model generation, or formal verification [54].

At this moment, only symbolic analysis of linear or small-signal linearized circuits in the frequency domain is possible, both for continuous- and discrete-time (switched) analog circuits [54,55,57]. In this way, symbolic expressions can be generated for transfer functions, impedances, noise functions, etc. In addition to understanding the first-order functional behavior of an analog circuit, a good understanding of the second-order effects in a circuit is equally important for the correct functioning of the design in its system application later on. Typical examples are the Power Supply Rejection Ratio (PSRR) and Common Mode Rejection Ratio (CMRR) of a circuit, which are limited by the mismatches between circuit elements. These mismatches are represented symbolically in the formulas. Another example is the distortion or intermodulation behavior,

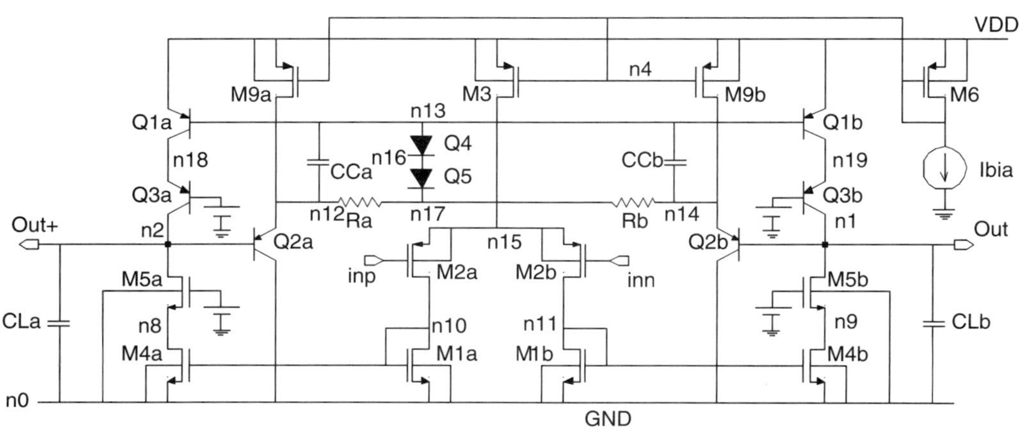

FIGURE 15.15 BiCMOS operational amplifier to illustrate symbolic analysis.

which is critical in telecom applications. To this end, the technique of symbolic simulation has been extended to the symbolic analysis of distortion and intermodulation in weakly nonlinear analog circuits where the nonlinearity coefficients of the device small-signal elements appear in the expressions [46].

Exact symbolic solutions for network functions are however too complex for linear(ized) circuits of practical size, and even impossible to calculate for many nonlinear effects. Even rather small circuits lead to an astronomically high number of terms in the expressions, which can neither be handled by the computer nor interpreted by the circuit designer. Therefore, since the late 1980s, and in principle similar to what designers do during hand calculations, dedicated symbolic analysis tools have been developed that use heuristic simplification and pruning algorithms based on the relative importance of the different circuit elements to reduce the complexity of the resulting expressions and retain only the dominant contributions within user-controlled error tolerances. Examples of such tools are ISAAC [57], SYNAP [58], and ASAP [59] among many others. Although successful for relatively small circuits, the fast increase of the CPU time with the circuit size restricted their applicability to circuits between 10 and 15 transistors only, which was too small for many practical applications.

In the past years, however, an algorithmic breakthrough in the field of symbolic circuit analysis has been realized. The techniques of simplification before and during the symbolic expression generation, as implemented in tools such as SYMBA [56] and RAINIER [60], highly reduce the computation time and therefore enable the symbolic analysis of large analog circuits of practical size (like the entire 741 opamp or the example of Figure 15.15). In simplification before generation (SBG), the circuit schematic or some associated matrix or graph(s), are simplified before the symbolic analysis starts [61,62]. In simplification during generation (SDG), instead of generating the exact symbolic expression followed by pruning the unimportant contributions, the desired simplified expression is built up directly by generating the contributing dominant terms one by one in decreasing order of magnitude, until the expression has been generated with the desired accuracy [56,60]. In addition, the technique of determinant decision diagrams (DDDs) has been developed as a very efficient canonical representation of symbolic determinants in a compact nested format [63]. The advantage is that all operations on these DDDs are linear with the size of the DDD, but the DDD itself is not always linear with the size of the circuit. Very efficient methods have been developed using these DDDs [63,64].

All these techniques however, still result in large, expanded expressions, which restricts their usefulness for larger circuits. Therefore, for really large circuits, the technique of hierarchical decomposition has been developed [65,66]. The circuit is recursively decomposed into loosely connected subcircuits. The lowest-level subcircuits are analyzed separately, and the resulting symbolic expressions are combined according to the decomposition hierarchy. This results in the global nested expression for the complete circuit, which is much more compact than the expanded expression. The CPU time increases about linearly with the circuit size, provided that the coupling between the different subcircuits is not too strong. Also the DDD technique has been combined with hierarchical analysis in Ref. [67].

Another recent extension is toward the symbolic analysis of linear periodically time-varying circuits, such as mixers [68]. The approach generalizes the concept of transfer functions to harmonic transfer matrices, generating symbolic expressions for the transfer function from any frequency band from the circuit's input signal to any frequency band from the circuit's output signal.

In addition, recently, approaches have also started to appear that generate symbolic expressions for large-signal and transient characteristics, for instance, using piecewise-linear approximations or using regression methods that fit simulation data to predefined symbolic templates. A recent example of such a fitting approach is the automatic generation of symbolic posynomial performance models for analog circuits, which are created by fitting a pre-assumed posynomial equation template to simulation data created according to some design-of-experiments scheme [21]. These kinds of methods are very promising, since they are no longer limited to simple device models nor to small-signal characteristics only, but they still need some further research.

Based on the many research results in this area over the last decade, it can be expected that symbolic analysis techniques might soon be part of the standard tool suite of every analog designer, as an add-on to numerical simulation.

15.6 Conclusions

The last few years have seen significant advances in both design methodology and CAD tool support for analog, mixed-signal, and RF designs. Mixed-signal simulators allow simulation of combined analog and digital circuits. The emergence of commercial AMS simulators supporting multiple analog abstraction levels (functional, behavioral, macromodel, and circuit level) enables top-down design flows in many industrial scenarios. In addition, there is increasing progress in system-level modeling and analysis allowing architectural exploration of entire systems, as well as in mixed-signal verification to anticipate problems related to embedding the analog blocks in a digital environment. A crucial element to enable this is the development of techniques to generate efficient behavioral models. An overview of model generation techniques has been given, including regression-based methods as well as model-order reduction techniques. Also the generation of performance models for analog circuit synthesis and of symbolic models that provide designers with insight into the relationships governing the performance behavior of a circuit have been described. Finally, the progress in symbolic analysis of both linear(ized) and weakly nonlinear analog circuits has been presented.

Despite this enormous progress in research and commercial EDA offerings that today enable the efficient design of analog blocks for embedding in mixed-signal integrated systems (ASICs, SoCs or SiPs), several problems still remain to be solved. Especially, behavioral model generation remains a difficult art that needs more research work toward automatic model generation techniques. Also, more advanced mixed-signal verification tools for signal integrity analysis (crosstalk, electromagnetic interference, etc.) need to be developed.

Acknowledgments

The authors acknowledge the Ph.D students who have contributed to the reported results, as well as the IEEE and IEE for their kind permission to allow to reuse some parts of earlier publications in their overview chapter.

References

[1] International Technology Roadmap for Semiconductors, 2004, http://public.itrs.net.

[2] G. Gielen and R. Rutenbar, Computer-aided design of analog and mixed-signal integrated circuits, *Proc. IEEE*, 88, 1825–1854, 2000.

[3] H. Chang, E. Charbon, U. Choudhury, A. Demir, E. Felt, E. Liu, E. Malavasi, A. Sangiovanni-Vincentelli, and I. Vassiliou, *A Top-down Constraint-driven Design Methodology for Analog Integrated Circuits*, Kluwer Academic Publishers, Dordrecht, 1997.

[4] G. Gielen, Modeling and analysis techniques for system-level architectural design of telecom frontends, *IEEE Trans. Microwave Theory Tech.*, 50, 360–368, 2002.

[5] G. Gielen, Top-down design of mixed-mode systems: challenges and solutions, in *Advances in Analog Circuit Design*, Chapter 18, J. Huijsing, W. Sansen, R. Van der Plassche Eds., Kluwer Academic Publishers, Dordrecht, 1999.

[6] The MEDEA+ Design Automation Roadmap, 2003, http://www.medea.org.

[7] S. Donnay, G. Gielen, and W. Sansen, High-level analog/digital partitioning in low-power signal processing applications, *Proceedings of the Seventh International Workshop on Power and Timing Modeling, Optimization and Simulation (PATMOS)*, 1997, pp. 47–56.

[8] J. Crols, S. Donnay, M. Steyaert, and G. Gielen, A high-level design and optimization tool for analog RF receiver front-ends, *Proceedings of the International Conference on Computer-Aided Design (ICCAD)*, 1995, pp. 550–553.

[9] P. Wambacq, G. Vandersteen, Y. Rolain, P. Dobrovolny, M. Goffioul, and S. Donnay, Dataflow simulation of mixed-signal communication circuits using a local multirate, multicarrier signal representation, *IEEE Trans. Circuits Syst., Part I*, 49, 1554–1562, 2002.

[10] G. Vandersteen et al., Efficient bit-error-rate estimation of multicarrier transceivers, *Proceedings of the Conference on Design, Automation and Test in Europe (DATE)*, 2001, pp. 164–168.
[11] P. Wambacq, G. Vandersteen, S. Donnay, M. Engels, I. Bolsens, E. Lauwers, P. Vanassche, and G. Gielen, High-level simulation and power modeling of mixed-signal front-ends for digital telecommunications, *Proceedings of the International Conference on Electronics, Circuits and Systems (ICECS)*, September 1999, pp. 525–528.
[12] B. De Smedt and G. Gielen, Models for systematic design and verification of frequency synthesizers, *IEEE Trans. Circuit Syst., Part II: Analog Digit. Signal Process.*, 46, 1301–1308, 1999.
[13] L. Nagle and R. Rohrer, Computer analysis of nonlinear circuits, excluding radiation (CANCER), *IEEE J. Solid-State Circuit*, 6, 166–182, 1971.
[14] A. Vladimirescu, *The SPICE Book*, Wiley, New York, 1994.
[15] J. Roychowdhury, A. Mantooth, chap. 14, this volume.
[16] S. Fang, Y. Tsividis, and O. Wing, SWITCAP: a switched-capacitor network analysis program, *IEEE Circuit Syst. Mag.*, 5, 4–10, 1983.
[17] K. Francken and G. Gielen, A high-level simulation and synthesis environment for Delta–Sigma modulators, *IEEE Trans. Comput. Aided Des.*, 22, 1049–1061, 2003.
[18] K. Kundert, Introduction to RF simulation and its applications, *IEEE J. Solid-State Circuit*, 34, 1298–1319, 1999.
[19] R. Saleh, B. Antao, and J. Singh, Multi-level and mixed-domain simulation of analog circuits and systems, *IEEE Trans. Comput. Aided Des.*, 15, 68–82, 1996.
[20] Virtual Socket Interface Alliance, Several documents including VSIA Architecture Document and Analog/Mixed-Signal Extension Document, http://www.vsia.org.
[21] W. Daems, G. Gielen, and W. Sansen, Simulation-based generation of posynomial performance models for the sizing of analog integrated circuits, *IEEE Trans. Comput. Aided Des.*, 22, 517–534, 2003.
[22] R. Phelps, M. Krasnicki, R. Rutenbar, L.R. Carley, and J. Hellums, A case study of synthesis for industrial-scale analog IP: redesign of the equalizer/filter frontend for an ADSL CODEC, *Proceedings of the ACM/IEEE Design Automation Conference (DAC)*, 2000, pp. 1–6.
[23] J. Vandenbussche et al., Systematic design of high-accuracy current-steering D/A converter macrocells for integrated VLSI systems, *IEEE Trans. Circuit Syst.*, Part II, 48, 300–309, 2001.
[24] A. Vachoux, J.-M. Bergé, O. Levia, and J. Rouillard, Eds., *Analog and Mixed-signal Hardware Description Languages*, Kluwer Academic Publishers, Dordrecht, 1997.
[25] E. Christen and K. Bakalar, VHDL-AMS — a hardware description language for analog and mixed-signal applications, IEEE Trans. Circuit Syst., Part II: Analog Digit. Signal Process., 46, 1999, pp. 1263–1272.
[26] IEEE 1076.1 Working Group, *Analog and Mixed-signal Extensions to VHDL*, http://www.eda.org/vhdl-ams/.
[27] K. Kundert and O. Zinke, *The Designer's Guide to Verilog-AMS*, Kluwer Academic Publishers, Dordrecht, 2004.
[28] *Verilog-AMS Language Reference Manual*, version 2.2, November 2004, http://www.eda.org/verilog-ams/.
[29] B. Wan, B.P. Hu, L. Zhou, and C.-J.R. Shi, MCAST: an abstract-syntax-tree based model compiler for circuit simulation, *Proceedings of the Custom Integrated Circuits Conference (CICC)*, September 2003, pp. 249–252.
[30] L. Lemaitre, G. Coram, C. McAndrew, and K. Kundert, Extensions to Verilog-A to support compact device modeling, *Proceedings of the Workshop on Behavioral Modeling and Simulation (BMAS)*, October 2003, pp. 134–138.
[31] B. Antao and F. El-Turky, Automatic analog model generation for behavioral simulation, *Proceedings of the IEEE Custom Integrated Circuits Conference (CICC)*, May 1992, pp. 12.2.1–12.2.4.
[32] T. Hastie, R. Tibshirani, and J. Friedman, *The Elements of Statistical Learning*, Springer, Heidelberg, 2001.

[33] H. Liu, A. Singhee, R. Rutenbar, and L.R. Carley, Remembrance of circuits past: macromodeling by data mining in large analog design spaces, Proceedings of the Design Automation Conference (DAC), June 2002, pp. 437–442.

[34] B. Scholkopf and A. Smola, *Learning with Kernels*, MIT Press, Cambridge, MA, 2001.

[35] J. Phillips, J. Afonso, A. Oliveira, and L.M. Silveira, Analog macromodeling using kernel methods, *Proceedings of the IEEE/ACM International Conference on Computer-Aided Design (ICCAD)*, November 2003.

[36] Th. Kiely and G. Gielen, Performance modeling of analog integrated circuits using least-squares support vector machines, *Proceedings of the Design, Automation and Test in Europe (DATE) Conference*, February 2004, pp. 448–453.

[37] M. Hershenson, S. Boyd, and T. Lee, Optimal design of a CMOS op-amp via geometric programming, *IEEE Trans. Comput. Aided Des. Int. Circuit Syst.*, 20, 1–21, 2001.

[38] Y. Freund and R. Schapire, A short introduction to boosting, *J. Jap. Soc. Artif. Intell.*, 14, 771–780, 1999.

[39] C. Borchers, L. Hedrich, and E. Barke, Equation-based behavioral model generation for nonlinear analog circuits, *Proceedings of the IEEE/ACM Design Automation Conference (DAC)*, 1996, pp. 236–239.

[40] L. Pillage and R. Rohrer, Asymptotic waveform evaluation for timing analysis, *IEEE Trans. Comput. Aided Des.*, 9, 352–366, 1990.

[41] L.M. Silveira et al., A coordinate-transformed Arnoldi algorithm for generating guaranteed stable reduced-order models of RLC circuits, *Proceedings of the IEEE/ACM International Conference on Computer-Aided Design (ICCAD)*, 1996, pp. 288–294.

[42] P. Feldmann and R. Freund, Efficient linear circuit analysis by Padé approximation via the Lanczos process, *IEEE Trans. Comput. Aided Des.*, 14, 639–649, 1995.

[43] A. Odabasioglu, M. Celik, and L. Pileggi, PRIMA: passive reduced-order interconnect macromodeling algorithm, *IEEE Trans. Comput. Aided Des.*, 17, 645–654, 1998.

[44] J. Roychowdhury, Reduced-order modeling of time-varying systems, *IEEE Trans. Circuit Syst., Part II*, 46, 1273–1288, 1999.

[45] J. Phillips, Projection-based approaches for model reduction of weakly nonlinear, time-varying systems, *IEEE Trans. Comput. Aided Des.*, 22, 171–187, 2003.

[46] P. Wambacq, G. Gielen, P. Kinget, and W. Sansen, High-frequency distortion analysis of analog integrated circuits, *IEEE Trans. Circuit Syst., Part II*, 46, 335–345, 1999.

[47] L. Peng and L. Pileggi, NORM: compact model order reduction of weakly nonlinear systems, *Proceedings of the Design Automation Conference*, June 2003, pp. 472–477.

[48] M. Rewienski and J. White, A trajectory piecewise-linear approach to model order reduction and fast simulation of nonlinear circuits and micromachined devices, *IEEE Trans. Comput. Aided Des.*, 22 , 155–170, 2003.

[49] N. Dong and J. Roychowdhury, Piecewise polynomial nonlinear model reduction, *Proceedings of the Design Automation Conference*, June 2003, pp. 484–489.

[50] E. Lauwers and G. Gielen, Power estimation methods for analog circuits for architectural exploration of integrated systems, *IEEE Trans. Very Large Scale Integration (VLSI) Syst.*, 10, 155–162, 2002.

[51] F. De Bernardinis, M. Jordan, and A. Sangiovanni-Vincentelli, Support vector machines for analog circuit performance representation, *Proceedings of the Design Automation Conference (DAC)*, June 2003, pp. 964–969.

[52] P. Veselinovic et al., A flexible topology selection program as part of an analog synthesis system, *Proceedings of the IEEE European Design & Test Conference (ED&TC)*, 1995, pp. 119–123.

[53] R. Harjani and J. Shao, Feasibility and performance region modeling of analog and digital circuits, *Kluwer Int. J. Analog Integr. Circuit Signal Process.*, 10, 23–43, 1996.

[54] G. Gielen, P. Wambacq, and W. Sansen, Symbolic analysis methods and applications for analog circuits: a tutorial overview, *Proc. IEEE*, 82, 287–304, 1994.

[55] F. Fernández, A. Rodríguez-Vázquez, J. Huertas, and G. Gielen, *Symbolic Analysis Techniques — Applications to Analog Design Automation*, IEEE Press, Washington, DC, 1998.
[56] P. Wambacq, F. Fernández, G. Gielen, W. Sansen, and A. Rodríguez-Vázquez, Efficient symbolic computation of approximated small-signal characteristics, *IEEE J. Solid-State Circuit*, 30, 327–330, 1995.
[57] G. Gielen, H. Walscharts, and W. Sansen, ISAAC: a symbolic simulator for analog integrated circuits, *IEEE J. Solid-State Circuit*, 24, 1587–1596, 1989.
[58] S. Seda, M. Degrauwe, and W. Fichtner, A symbolic analysis tool for analog circuit design automation, *Proceedings of the IEEE/ACM International Conference on Computer-Aided Design (ICCAD)*, 1988, pp. 488–491.
[59] F. Fernández, A. Rodríguez-Vázquez, and J. Huertas, Interactive AC modeling and characterization of analog circuits via symbolic analysis, *Kluwer Int. J. Analog Integr. Circuit Signal Process.*, 1, 183–208, 1991.
[60] Q. Yu and C. Sechen, A unified approach to the approximate symbolic analysis of large analog integrated circuits, *IEEE Trans. Circuit Syst., Part I*, 43, 656–669, 1996.
[61] W. Daems, G. Gielen, and W. Sansen, Circuit simplification for the symbolic analysis of analog integrated circuits, *IEEE Trans. Comput. Aided Des.*, 21, 395–407, 2002.
[62] J. Hsu and C. Sechen, DC small signal symbolic analysis of large analog integrated circuits, *IEEE Trans. Circuit Syst., Part I*, 41 (12), 817–828, 1994.
[63] C.-J. Shi and X.-D. Tan, Canonical symbolic analysis of large analog circuits with determinant decision diagrams, *IEEE Trans. Comput. Aided Des.*, 19, 1–18, 2000.
[64] W. Verhaegen and G. Gielen, Efficient DDD-based symbolic analysis of linear analog circuits, *IEEE Trans. Circuit Syst., Part II: Analog Digit. Signal Process.*, 49, 474–487, 2002.
[65] J. Starzyk and A. Konczykowska, Flowgraph analysis of large electronic networks, *IEEE Trans. Circuit Syst.*, 33, 302–315, 1986.
[66] O. Guerra, E. Roca, F. Fernández, and A. Rodríguez-Vázquez, A hierarchical approach for the symbolic analysis of large analog integrated circuits, *Proceedings of the IEEE Design Automation and Test in Europe Conference (DATE)*, 2000, pp. 48–52.
[67] X.-D. Tan and C.-J. Shi, Hierarchical symbolic analysis of analog integrated circuits via determinant decision diagrams, *IEEE Trans. Comput. Aided Des.*, 19, 401–412, 2000.
[68] P. Vanassche, G. Gielen, and W. Sansen, Symbolic modeling of periodically time-varying systems using harmonic transfer matrices, *IEEE Trans. Comput. Aided Des.*, 21, 1011–1024, 2002.

16

Layout Tools for Analog Integrated Circuits and Mixed-Signal Systems-on-Chip: A Survey

Rob A. Rutenbar
Carnegie Mellon University
Pittsburgh, Pennsylvania

John M. Cohn
IBM Systems and Technology Group
Essex Junction, Vermont

Abstract

Layout for analog circuits has historically been a time-consuming, manual, trial-and-error task. The problem is not so much the size (in terms of the number of active devices) of these designs, but rather the plethora of possible circuit and device interactions: from the chip substrate, from the devices and interconnects themselves, and from the chip package. In this short survey, we enumerate briefly the basic problems faced by those who need to do layout for analog and mixed-signal designs, and survey the evolution of the design tools and geometric/electrical optimization algorithms that have been directed at these problems.

16.1 Introduction

Layout for digital integrated circuits (ICs) is usually regarded as a difficult task because of the *scale* of the problem: millions of gates, kilometers of routed wires, complex delay, and timing interactions. Analog designs and the analog portions of mixed-signal systems-on-chip (SoCs) are usually much smaller — up to 100 devices in a cell, usually less than 20,000 devices in a complete subsystem — and yet they are nothing if not *more* difficult to lay out. Why is this? The answer is that the complexity of analog circuits is not so much due to the number of devices, as to the complex *interactions* among the devices, among the various continuous-valued performance specifications, and with the fabrication process and the operating environment.

This would be less of a problem if analog circuits and subsystems were rare or exotic commodities, or if they were sufficiently generic that a few stable designs could be easily re-targeted to each new application. Unfortunately, neither is true. The markets for application-specific ICs (ASICs), application-specific standard parts (ASSPs), and high-volume commodity ICs are characterized by an increasing level of integration. In recent years, complete systems that previously occupied separate chips, are being integrated on a single chip. Examples of such "systems on a chip" include telecommunications ICs such as modems, wireless designs such as components in radio frequency (RF) receivers and transmitters, and networking interfaces such as local area network ICs. Although most functions in such integrated systems are implemented with digital (or especially digital signal processing) circuitry, the analog circuits needed at the interface between the electronic system and the "real" world are now being integrated on the same die for reasons of cost and performance.

The booming market share of mixed-signal ASICs in complex systems for telecommunications, consumer, computing, and automotive applications is one direct result of this. But along with this increase in achievable complexity has come a significant increase in design complexity. And at the same time, many present ASIC application markets are characterized by shortening product life cycles and time-to-market constraints. This has put severe pressure on the designers of these analog circuits, and especially on those who lay out these designs. If cell-based library methodologies were workable for analog (as they are for semicustom digital designs), layout issues would be greatly mitigated. Unfortunately, such methodologies fare poorly here. Most analog circuits are one-of-a-kind, or at best few-of-a-kind on any given IC. Today, they are usually designed by hand and laid out by hand. The problem recurs at the system level: the discipline of row-based layout as the basis for large function blocks that is so successful in digital designs is not (yet) as rigorously applied in analog designs.

Despite the problems here, there is a thriving research community working to make custom analog layout tools a practical reality. This brief survey attempts to describe the history and evolution of these tools; it extends two earlier reviews [1,2]. Our survey is organized as follows. We begin with a brief taxonomy of problems and strategies for analog layout in Section 16.2. Then, in Section 16.3, we review attacks on the cell-level layout problem. In Section 16.4, we review mixed-signal system-level layout problems. In Section 16.5, we review recent work on field-programmable analog arrays (FPAAs). Section 16.6 offers some concluding remarks. We end with an extensive, annotated bibliography.

16.2 Analog Layout Problems and Approaches

Before trying to categorize the various geometric strategies that have been proposed for analog layout, it is essential first to understand what the electrical problems that affect analog design are. We enumerate here briefly the salient effects that layout can have on circuit performance. References [25,76,101] together comprise a fairly complete treatment of these issues. There are really three core problems, which we first describe. We then briefly survey solution strategies here.

16.2.1 Loading Problems

The nonideal nature of interdevice wiring introduces capacitive and resistive effects, which can degrade circuit performance. At sufficiently high frequencies, inductive effects arise as well. There are also parasitic RLC elements associated with the geometry of the devices themselves, e.g., the various capacitances associated with MOS diffusions. All these effects are remarkably sensitive to detailed (polygon-level) layout. For example, in MOS circuits, layout designers have a useful degree of freedom in that they can fold large devices (i.e., devices with large width-to-length ratios), thus altering both their overall geometric shape and detailed parasitics. Folding transforms a large device (large channel width) into a parallel connection of smaller devices. The smaller devices are *merged*: parallel side-by-side alignment allows a single source or drain diffusion to be shared between adjacent gate regions, thus minimizing overall capacitance. Every diffused structure also has an associated parasitic resistance which varies with its shape. These resistances can be reduced by minimizing the aspect ratio of all diffusions (reducing the width of the device), merging diffusions when possible, and strapping diffusion with low-resistance layers such as metal where possible. Of course, this "strapping" may interfere with signal routing. Cohn et al. [35] offer a careful treatment of the layout issues here for MOS devices.

One of the distinguishing characteristics of analog layout, in comparison to digital layout, is the amount of effort needed to create correct layouts for atomic devices. MOS devices with large width-to-length ratios, bipolar devices with large emitter areas, etc., are common in analog designs and require careful attention to detail. Passive components that implement resistors or capacitors or inductors are also more frequent in analog design, and require careful layout (see, e.g., Bruce et al. [15] as an example of procedural generation of complex MOS devices). Extremely low-level geometric details of the layout of individual devices and passives can have a significant circuit-level impact.

16.2.2 Coupling Problems

Layout can also introduce unexpected signal coupling into a circuit, which may inject unwanted electrical noise or even destroy its stability through unintended feedback. At lower frequencies, coupling may be introduced by a combination of capacitive, resistive, or thermal effects. At higher frequencies inductive coupling becomes an issue. Especially in the modern deep submicron digital processes in which analog systems are being integrated, coupling is an increasing problem. Metal conductors couple capacitively when two metal surfaces are sufficiently close, e.g., if wires run in parallel on a single layer or cross on adjacent layers. If a parallel run between incompatible signals is unavoidable, a neutral wire such as a ground or reference line can be placed between them as a coupling shield.

Current flowing through a conductor also gives rise to a fluctuation in the voltage drop across the conductor's finite resistance. This fluctuation is then coupled into all the devices attached to the conductor. This effect is particularly problematic in power supply routing for analog cells on digital ICs. Sensitive analog performance often depends on an assumption of moderate stability in the power rails — this may not be true if the power distribution network is improperly laid out (see Refs. 99,101 for a detailed treatment of the issues in mixed-signal power distribution).

Signals can also be coupled through the silicon substrate or bulk, either through capacitive, resistive, or thermal effects. Because all devices share the same substrate, noise injected into the substrate is capacitively or resistively coupled into every node of the circuit. This is particularly problematic when analog circuits must share a substrate with inherently noisy high-speed digital logic. On mixed-signal ICs, conventional solutions focus on *isolation* either by locating the sensitive analog far away from the noise-injecting source, or surrounding the noise-sensitive circuits with a low-impedance diffusion guard ring to reduce the substrate noise level in a particular area. Unfortunately, the structure of the substrate and details of the layout of the power supply network that bias the substrate greatly affect even the qualitative behavior of this coupling. For example, epitaxial substrates have such low resistivity that injected noise can "reappear" far from its origin on the chip surface; simple isolation schemes do not always work well here. Bulk substrates are somewhat more resistive and noise may remain more local. Evolving silicon-on-insulator (SOI) processes may dramatically reduce this problem, but these are not yet in widespread use for commodity mixed-signal designs. References [68,85,101] are a good starting point for an analysis of the substrate-coupling problem in mixed-signal design.

In addition, since silicon is an excellent conductor of heat, local temperature variations due to current changes in a device can also cause signal coupling in nearby thermally sensitive devices. This phenomenon is most prevalent in Bipolar or Bi–CMOS processes. Placing thermally sensitive devices far away from high-power thermally dissipating devices can reduce this effect. Placing matching devices symmetrically about thermally "noisy" sources can also be effective in reducing the effects of thermal coupling.

16.2.3 Matching Problems

Unavoidable variations which are present in all fabrication processes lead to small mismatches in electrical characteristics of identical devices. If these mismatches are large enough, they can affect circuit performance by introducing electrical problems such as offsets. Four major layout factors can affect the matching of identical devices: *area*, *shape*, *orientation*, and *separation*.

Device area is a factor because semiconductor processing introduces unavoidable distortions in the geometry which make up devices. Creating devices using identical geometry (identical shape) improves matching by insuring that both devices are subject to the same (or at least *similar*) geometric distortions. Similarly, since the proportional effect of these variations tends to decrease as the size of the device increases, matching devices are usually made as large as the circuit performance and area constraints will allow. Since many processing effects, e.g., ion-implantation, introduce anisotropic geometric differences, devices which must match should also be placed in the same orientation. Finally, spatial variations in process parameters tend to degrade the matching characteristics of devices as their separation increases. This is largely due to process-induced gradients in parameters such as mobility or oxide thickness. Sensitivity to these effects can be reduced by placing devices which must match well in close proximity. Devices which must be *extremely* well matched may be spatially interdigitated in an attempt to cancel out the effects of global process gradients.

Device matching, particularly of bipolar devices, may also be degraded by thermal gradients. Two identical devices at different points on a thermal gradient will have slight differences in VBE for a given collector current. To combat this, it is common practice to arrange thermally sensitive matching devices symmetrically around thermally generating noise sources. Parasitic capacitive and resistive components of interconnect can also introduce problems of matching in differential circuits, i.e., those circuits which are comprised of two matching halves. A mismatch in the parasitic capacitance and resistance between the two matching halves of the circuit can give rise to offsets and other electrical problems. The most powerful technique used to improve interconnect parasitic matching is layout symmetry, in which the placement and wiring of matching circuits are forced to be identical, or in the case of differential circuits, mirror-symmetric.

16.2.4 Layout Solution Strategies

We mentioned a variety of solution techniques in the above enumeration of layout problems: careful attention to atomic device layout, MOS merging, substrate noise isolation, symmetric layout, etc. However, these are really low-level *tactics* for dealing with specific problems for specific circuits in specific operating or fabrication environments. The more general question we wish to address next is how the analog portion of a large mixed-signal IC is laid out — what are the overall geometric *strategies* here?

We note first that, like digital designs, analog designs are attacked hierarchically. However, analog systems are usually significantly smaller than their digital counterparts: 10,000 to 20,000 analog devices vs. 1,00,000 to 1,000,000 digital gates. Thus, analog layout hierarchies are usually not as deep as their digital counterparts. The need for low-level attention to coupling and interaction issues is another force that tends to flatten these hierarchies. The typical analog layout hierarchy comprises two fundamentally different types of layout tasks:

- *Cell-level layout.* The focus here is really device-level layout, placement, and routing of individual active and passive devices. At this level, many of the low-level matching, symmetry, merging, reshaping, and proximity management tactics for polygon-level optimization are applied. The goal is to create cells that are suitably insulated not only from fluctuations in fabrication and operating environment, but from probable coupling with neighboring pieces of layout.
- *System-level layout.* The focus here is cell composition, arranging, and interconnecting the individual cells to complete an analog subsystem. At this level, isolation is one major concern: from nearby noisy digital circuits coupled in via substrate, power grid, or package. The other concern is signal integrity. Some number of digital signals from neighboring digital blocks need to penetrate

into any analog regions, and these signals may be fundamentally incompatible with some sensitive analog signals.

- *Programmable layout.* The focus here is applying field-programmable gate array (FPGA; see, e.g., Rose et al. [114]) ideas to analog designs. The idea is to bypass completely the need to do custom cells. Rather, a set of programmable active and passive elements is connected with programmable wiring to achieve low-performance analog functions. This is a relatively recent implementation strategy, but we expect to see it grow.

We visit the ideas behind each of these layout strategies in the following three sections.

16.3 Analog Cell Layout Strategies

For our purposes a "cell" is a small analog circuit, usually comprising not more than about 100 active and passive devices, which is designed and laid out as a single atomic unit. Common examples include operational amplifiers, comparators, voltage references, analog switches, oscillators, and mixers.

The earliest approaches to custom analog cell layout relied on procedural module generation. These approaches are a workable strategy when the analog cells to be laid out are relatively static, i.e., necessary changes in device sizing or biasing result in little need for global alterations in device layout, orientation, reshaping, etc. Procedural generation schemes usually start with a basic geometric template (sometimes called a *topology* for the circuit), which specifies all necessary device-to-device and device-to-wiring spatial relationships. Generation completes the template by correctly sizing the devices and wires, re-spacing them as necessary. References [6,11] are examples dedicated mainly to opamps. The mechanics for capturing the basic design specifications can often be as familiar as a common spreadsheet interface (see, e.g., [12]). Owen et al.'s work [14] is a more recent example focused on opamp generation, and describes both languages for specifying these layouts and several optimized layout results. The system at Philips [13] is another good example of practical application of these ideas on complex circuits. Bruce et al. [15] show an example of module generation useful for atomic MOS devices (Figure 16.1).

Often, however, changes in circuit design require full custom layout, which can be handled with a *macro-cell-style* strategy. The terminology is borrowed from digital floorplanning algorithms, which manipulate flexible layout blocks, arrange them topologically, and then route them. For analog cells, we regard the flexible blocks as devices to be reshaped and reoriented as necessary. Module generation techniques are used to generate the layouts of the individual devices. A placer then arranges these devices, and a router interconnects them — all while attending to the numerous parasitics and couplings to which analog circuits are sensitive.

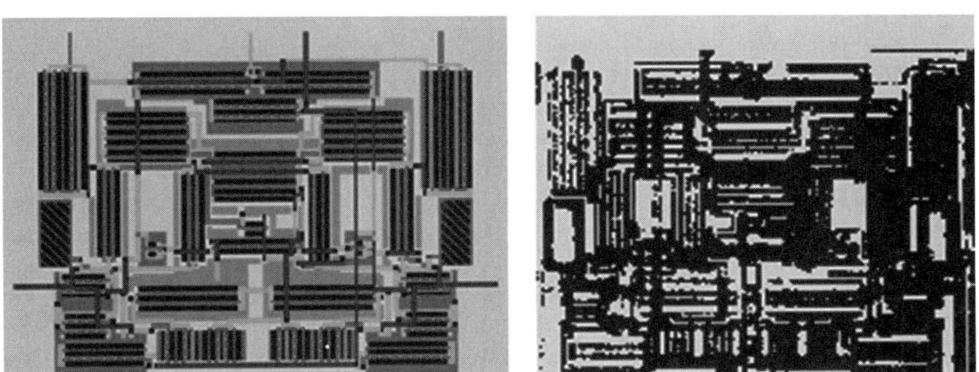

FIGURE 16.1 **(See color insert following page 15–4.)** Automatic analog cell layout for a CMOS amplifier, layout (left) and die shot (right) done using KOAN/ANAGRAM tools. (From Cohn et al., *IEEE JSSC*, 26, 330–342, March 1991. With permission.)

The earliest techniques used a mix of knowledge-based and constructive techniques for placement, with routers usually adapted from common semicustom digital applications [18,20–22]. For example, a common constructive placement heuristic is to use the spatial relationships in a drawn circuit schematic as an initial basis from mask-level device placement [23]. Unfortunately, these techniques tended to be rather narrow in terms of which circuits could be laid out effectively.

The ILAC tool from CSEM was an important early attempt in this style [17,22]. It borrowed heavily from the best ideas in digital layout: efficient slicing tree floorplanning with flexible blocks, global routing via maze routing, detailed routing via channel routing, and area optimization by compaction. The problem with the approach was that it was difficult to extend these primarily digital algorithms to handle all the low-level geometric optimizations that characterize expert manual design. Instead, ILAC relied on a large, very sophisticated library of device generators.

ANAGRAM and its successor KOAN/ANAGRAM II from CMU kept the macrocell style, but reinvented the necessary algorithms from the bottom-up, incorporating many manual design optimizations [19,25,27]. For example, the device placer KOAN relied on a very small library of device generators, and migrated important layout optimizations into the placer itself. KOAN could dynamically fold, merge, and abut MOS devices, and thus discover desirable optimizations to minimize parasitic capacitance during placement. KOAN was based on an efficient simulated annealing algorithm [3]. (Recent placers have also extended ideas from sequence-pair module-packing representations [5] to handle analog layout tasks [38].) KOAN's companion, ANAGRAM II, was a maze-style detailed area router capable of supporting several forms of symmetric differential routing, mechanisms for tagging compatible and incompatible classes of wires (e.g., noisy and sensitive wires), parasitic crosstalk avoidance, and over-the-device routing. Other device placers and routers operating in the *macrocell-style* have appeared (see, e.g., Refs. [26,31–34]), confirming its utility.

In the next generation of cell-level tools, the focus shifted to quantitative optimization of performance goals. For example, KOAN maximized MOS drain-source merging during layout, and ANAGRAM II minimized crosstalk, but without any specific, quantitative performance targets. The routers ROAD [29] and ANAGRAM III [30] use improved cost-based schemes that route instead to minimize the deviation from acceptable parasitic bounds derived from designers or sensitivity analysis. The router in Ref. [39] can manage not just parasitic sensitivities, but also basic yield and testability concerns. Similarly, the placers in Refs. [28,36,37] augment a KOAN-style model with sensitivity analysis so that performance degradations due to layout parasitics can be accurately controlled. Other tools in this style include Ref. [40].

In the newest generation of CMOS analog cell research, the device placement task has been separated into two distinct phases: device *stacking*, followed by *stack placement*. By rendering the circuit as an appropriate graph of connected drains and sources, it is possible to identify natural clusters of MOS devices that ought to be merged — called *stacks* — to minimize parasitic capacitance. Malavasi et al. [43,44] and Malavasi and Pandini [47] gave an exact algorithm to extract all the optimal stacks, and the placer in Refs. [45,46] extends a KOAN-style algorithm to choose dynamically the right stacking and the right placement of each stack. Basaran and Rutenbar [48] offer another variant of this idea: instead of extracting all the stacks (which can be time-consuming since the underlying algorithm is exponential), this technique extracts one optimal set of stacks very fast (in linear time). The technique is useful in either the inner loop of a layout algorithm (to evaluate quickly a merging opportunity) or in an interactive layout editor (to stack quickly a set of devices optimally) (Figure 16.2).

The notion of using sensitivity analysis to quantify the impact on final circuit performance of low-level layout decisions (e.g., device merging, symmetric placement/routing, parasitic coupling due to specific proximities, etc.) has emerged as a strategy to link the various approaches being taken for cell-level layout and system assembly. Several systems from U.C. Berkeley are notable here. The influential early formulation of the sensitivity analysis problem were by Choudhury and Sangiovanni-Vincentelli [59], who not only quantified layout impacts on circuit performance, but also showed how to use nonlinear-programming techniques to map these sensitivities into constraints on various portions of the layout task. Charbon et al. [61] extended these ideas to handle constraints involving nondeterminninistic parasitics of the type that arise from statistical fluctuations in the fabrication process. In related work, Charbon et al. [60] also showed how to extract critical constraints on symmetry and matching directly from a device schematic.

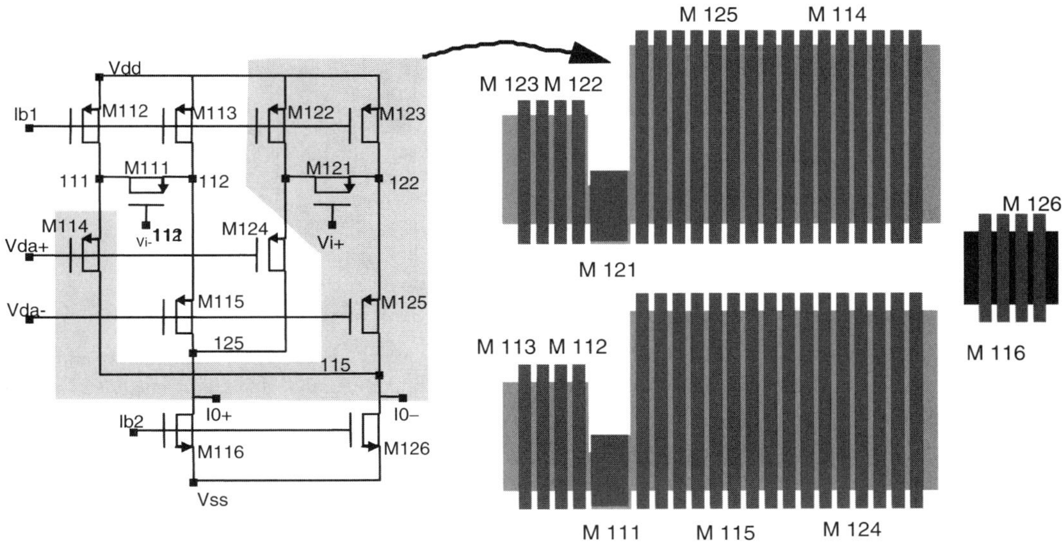

FIGURE 16.2 **(See color insert following page 15–4.)** Example of automatic symmetric stack extraction from a CMOS amplifier circuit, using Basaran's algorithm. (From Basaran and Rutenbar, *Proceedings of ACM/IEEE DAC*, Las Vegas, NV, 1996, pp. 221–226. With permission.)

One final problem in the macrocell style is the separation of the placement and routing steps. In manual cell layout, there is no effective difference between a rectangle representing a wire and one representing part of a device: they can each be manipulated simultaneously. In a place-then-route strategy, one problem is estimating how much space to leave around each device for the wires. One solution strategy is *analog compaction* (see, e.g., [49, 51, 52]), in which we leave extra space during device placement and then compact. A more radical alternative is *simultaneous device place-and-route.* An experimental version of KOAN [61] supported this by iteratively perturbing both the wires and the devices, with the goal of optimally planning polygon-level device and wire layout interactions.

As wireless and mobile design have proliferated, more RF designs have appeared. These offer yet another set of challenges at the cell level. RF circuits and higher frequency microwave circuits have unique properties which make their automated layout impossible with the techniques developed for lower frequency analog cells. Because every geometric property of the layout of an individual wire — its length, bends, proximity to other wires or devices — may play a key role in the electrical performance of the overall circuit, most RF layouts are optimized for performance first and density second.

Most layout work targeting RF circuits comprises interactive tools that aid the designer to speed manual design iterations [54,55]. Other work in the area includes semiautomated approaches, which rely on knowledge of the relative position of all cells [53]. However, these template-based approaches with predefined cells do limit the design alternatives possible. Charbon et al. [56] introduced a performance-driven router for RF circuits . In their approach, sensitivity analysis is employed to compute upper bounds for critical parasitics in the circuit, which the router then attempts to respect. Aktuna et al. [57,58] introduced the idea of device-level floorplanning for these circuits; using a genetic algorithm, the tool simultaneously evolves placement and detailed routing under constraints on length, bends, phase, proximity, and planarity (Figure 16.3).

There are several open problems in cell-level layout. For example, the optimal way to couple the various phases of cell layout — stacking, placement, routing, compaction — to each other and back to circuit design (or redesign) remains a challenging problem. However, there is now a maturing base of workable transistor-level layout techniques to build on. We note that commercial tools offering device-level analog layout synthesis have emerged and now been applied in a range of industrial applications, and are now in

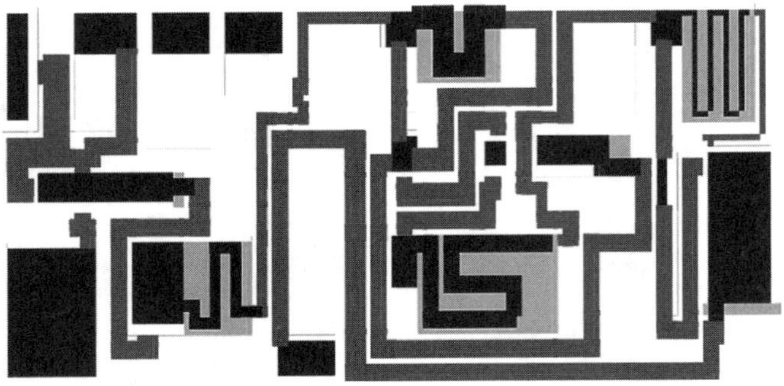

FIGURE 16.3 (**See color insert following page 15–4.**) Microwave frequency (~60 GHz) device-level floorplan, shows devices, planar wiring on a single layer, and several instances of "detours" which are inserted to match lengths and electrical properties of wires in this small design. (From Aktuna et al., *IEEE Trans. CAD*, 18, 375–388, 1999. With permission.)

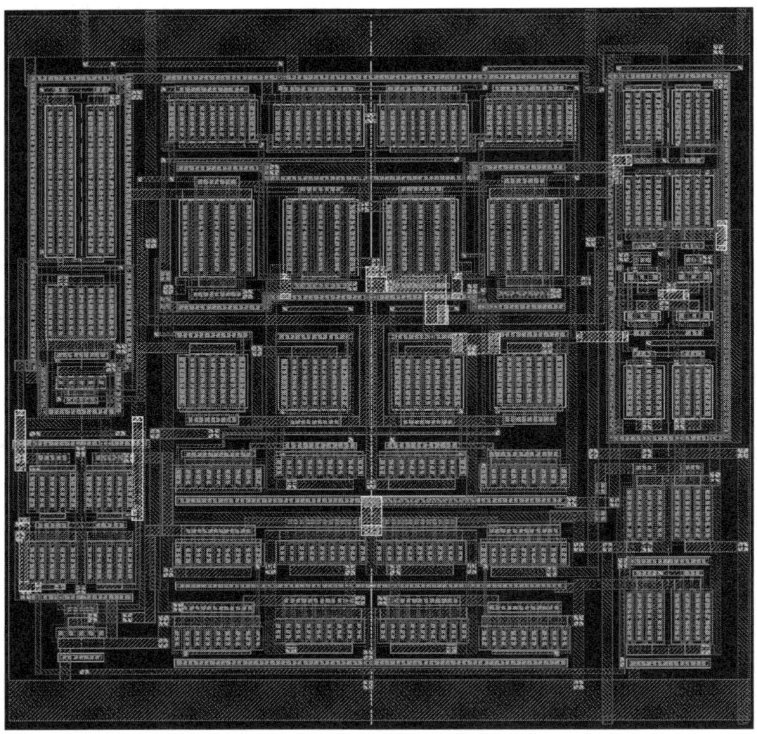

FIGURE 16.4 (**See color insert following page 15–4.**) Commercial amplifier circuit, with custom device generators, custom wells, symmetric placement and routing, all synthesized automatically. (Courtesy Cadence Design Systems.)

production use. Figure 16.4, an industrial analog cell produced by a commercial analog layout synthesis tool, is one such automatic layout example [41,42].

16.4 Mixed-Signal System Layout

A mixed-signal *system* is a set of custom analog and digital functional blocks. At the system level, the problem is really an *assembly* problem [2]. Assembly means block floorplanning, placement, global, and detailed routing (including the power grid). As well as parasitic sensitivities, the two new problem at the chip level are *coupling* between noisy signals and sensitive analog signals, and *isolation* from digital switching noise that couples through the substrate, power grid, or package.

Just as at the cell level, procedural generation remains a viable alternative for well-understood designs with substantial regularity. Many signal-processing applications have the necessary highly stylized, regular layout structure. Procedural generation has been successful for many switched capacitor filters and data converters [13,62,66], and especially regular, array-style blocks [16].

More generally though, work has focused on custom placement and routing at the block level, with layout optimization aimed at not only area, but also signal coupling and isolation. Trnka et al. [89] offered one very early attempt aimed at bipolar array-style (i.e., fixed device image) layout, adapting semicustom digital layout tools to this analog layout image. For row-based analog standard cell layout, an early elegant solution to the coupling problem was the *segregated channels* idea of authors in Refs. [86,87] to alternate noisy digital and sensitive analog wiring channels in a row-based cell layout. The strategy constrains digital and analog signals never to be in the same channel and remains a practical solution when the size of the layout is not too large. However, in modern multilevel interconnect technologies, this rigorous segregation can be overly expensive in area.

For large designs, analog channel routers were developed. In Gyurscik and Jeen [90], it was observed that a well-known digital channel routing algorithm [88] could be easily extended to handle critical analog problems that involve varying wire widths and wire separations needed to isolate interacting signals. Work at Berkeley substantially extended this strategy to handle complex analog symmetries and the insertion of shields between incompatible signals [91,92,94]. This work also introduced the idea of *constraint mapping*, which begins with parasitic sensitivities available from analysis of the system (or the cell) to be laid out, and transforms these into hard bounds on the allowable parasitics of each wire in each channel. The mapping process is itself a nonlinear-programming problem, in this case a quadratic programming formulation. These tools are particularly effective for stylized row-based layouts such as switched capacitor filters, where complex routing symmetries are necessary to balance subtle parasitics and adjoint simulation methods can yield the necessary sensitivities.

The WREN [93] and WRIGHT [95,96] systems from CMU generalized these ideas to the case of arbitrary layouts of mixed functional blocks. WREN comprises both a mixed-signal global router and channel router. WREN introduced the notion of *SNR-style* (signal-to-noise ratio) constraints for incompatible signals, and both the global and detailed routers strive to comply with designer-specified noise rejection limits on critical signals. WREN incorporates a constraint mapper (influenced by Choudhury and Sangiovanni-Vincentelli [59]) that transforms input noise rejection constraints from across-the-whole-chip form used by the global router into the per-channel per-segment form necessary for the channel router (as in [94]). WRIGHT uses a KOAN-style annealer to floorplan the blocks, but with a fast substrate noise-coupling evaluator so that a simplified view of the substrate noise influences the floorplan. WRIGHT used a coarse-resistive mesh model with numerical pruning to capture substrate coupling; the approach in Charbon et al. [83] use semianalytical substrate modeling techniques, which allow fast update when blocks move, and can also support efficient noise sensitivity analysis (Figure 16.5).

The substrate coupling problem is an increasingly difficult one as more and faster digital logic is placed side by side with sensitive analog parts. One avenue of relevant work here seeks to model the substrate accurately, efficiently extract tractable electrical models of its conduction of overall chip noise, and understand qualitatively how various isolation mechanisms (e.g., separation, guard rings) will work. This has been an active area of late. References [67,69,71,73,80] address basic computational electromagnetics attacks on modeling and analysis of substrate coupling. The approaches vary in their discretization of the substrate, their numerical technique to solve for the point-to-point resistance between two devices in the substrate, and their model-order reduction techniques to reduce potentially large, extracted circuit-level substrate models to smaller, more efficient circuit models. Su et al. [68] offer experimental data from test circuits on the mechanisms of substrate noise conduction for CMOS mixed-signal designs in epitaxial substrates. Charbon et al. [83] and Miliozzi et al. [82] address substrate coupling in the context of linking substrate modeling with the generation of constraints on allowable noise in the synthesis and layout

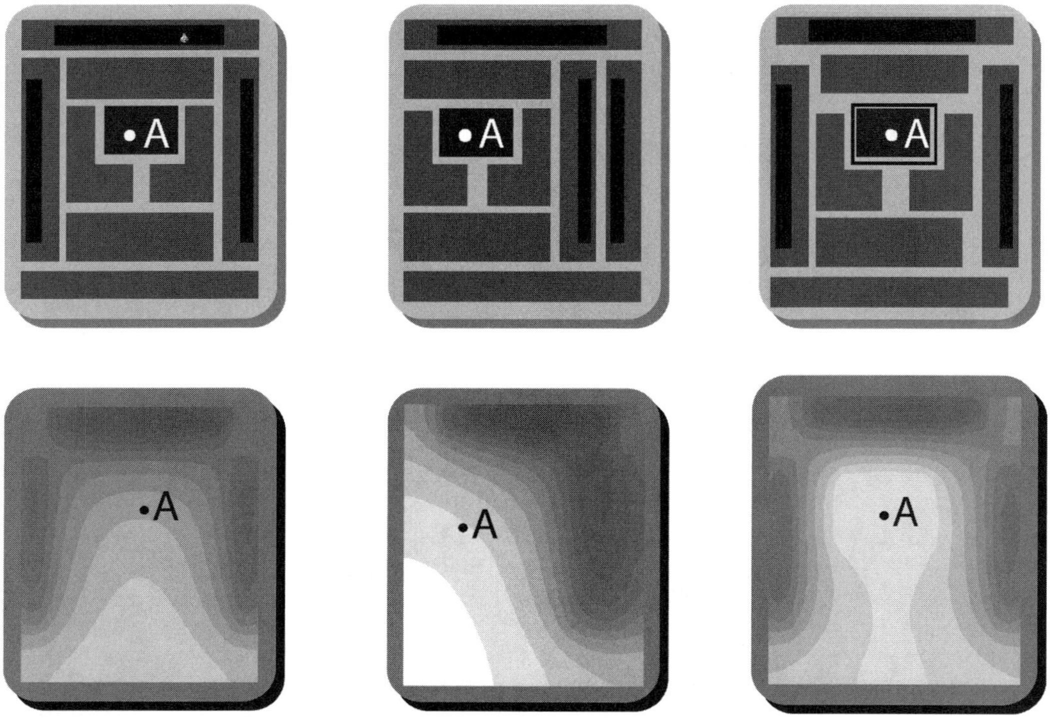

FIGURE 16.5 **(See color insert following page 15–4.)** Example of a small, synthetic floorplan optimized by WRIGHT [95,96] under substrate noise constraints. Dark bars show noise sources; the module labeled "A" is sensitive to the overall noise level. From left to right, we see the result of floorplans generated under increasingly tight constraints on the allowed noise seen at "A". Bottom figures show iso-voltage contours in the substrate. At the far right, we see "A" has had a guard ring added to meet the tight noise spec.

process. Mitra et al. [72] and Miliozzi et al. [81] address the problem of estimating substrate current injection; Mitra et al. [72]] use a circuit-level switching model and circuit simulation, and transforms simulation results into an equivalent single-tone sinusoid with the same total energy as the original random switching waveform. Miliozzi et al. [81] use a digital simulator to capture simple digital switching waveforms, which are then combined with precharacterized circuit-level injection models to estimate block-level injection. Tsukada and Makie-Fukuda [84] suggest an active guard ring structure to mitigate substrate noise, based on some of these modeling ideas. Verghese and Allstot [85] offer a recent survey of substrate modeling, extraction, reduction, and injection work, along with a review of how substrate issues are dealt with in current mixed-signal design methodologies.

Another important task in mixed-signal system layout is power grid design. Digital power grid layout schemes usually focus on connectivity, pad-to-pin ohmic drop, and electromigration effects. But these are only a small subset of the problems in high-performance mixed-signal chips which feature fast-switching digital systems next to sensitive analog parts. The need to mitigate unwanted substrate interactions, the need to handle arbitrary (nontree) grid topologies, and the need to design for transient effects such as current spikes are serious problems in mixed-signal power grids. The RAIL system [97,101] addresses these concerns by casting mixed-signal power grid synthesis as a routing problem that uses fast AWE-based [4] linear system evaluation to model electrically the entire power grid, package, and substrate during layout. By allowing changes in both grid topology (where segments are located, where power pins are located, where substrate contacts are located) and grid segment sizing, the tool can find power grid layouts to optimize AC, DC, and transient noise constraints. Techniques such as [72,81] are useful to estimate the digital

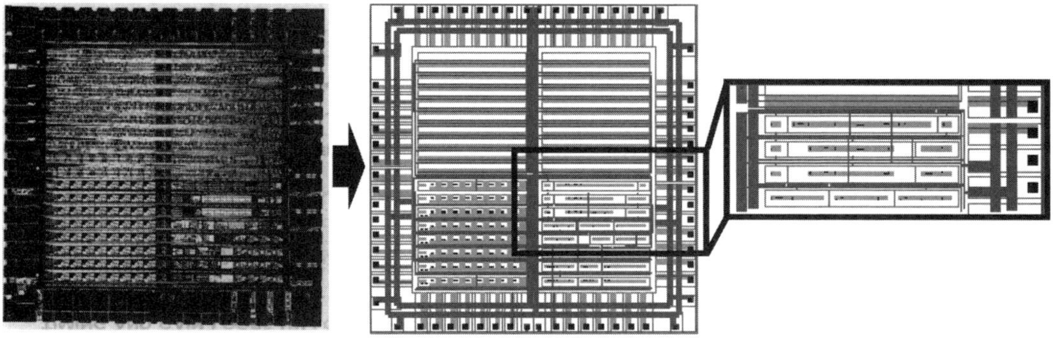

FIGURE 16.6 (**See color insert following page 15–4.**) Example of a mixed-signal power grid from a commercial IBM ASIC, redesigned automatically by the RAIL tool [97,101] to meet strict AC, DC, and transient specifications.

switching currents needed here for power grid optimization. Chen and Ling [102] discusses a similar power distribution formulation applied to digital circuits (Figure 16.6).

Most of these system layout tools are fairly recent, but because they often rely on mature core algorithms from similar digital layout problems, many have been prototyped both successfully and quickly. Several full, top-to-bottom prototypes have recently emerged (see, e.g. [83,103,106]).

There are many open problems here. Signal coupling and substrate/package isolation are overall still addressed via rather *ad hoc* means. There is still much work to be done to enhance existing constraint-mapping strategies and constraint-based layout tools to handle the full range of industrial concerns, and to be practical for practicing designers.

16.5 Field-Programmable Analog Arrays

Finally we mention one very recent analog layout style which is radically different from those mentioned above. In the digital realm, FPGAs have revolutionized digital prototyping and rapid time-to-market designs [114]. Integrating programmable logic elements and programmable interconnect, these allow rapid customization with no fabrication steps. An obvious question is: can a similar technology be adapted for analog and mixed-signal designs? The apparent answer is a qualified "yes".

Early work such as [107,109] directly mimicked the FPGA style of small primitive functions connectable by programmable interconnect. However, the loading which is already problematic in digital designs proved even more deleterious here. Later designs such as [110,113] moved up to higher-level building blocks (e.g., opamps, switches, capacitor arrays) and also focused new energy on sensitive analog design of the programmable interconnect. FPAAs are just beginning to be commercially available. They are currently so small however, that layout is not really a problem. It will be interesting to see if these designs become larger (e.g., larger digital blocks with smaller analog blocks), and if so, if automatic layout becomes a requirement.

16.6 Conclusions

There has been substantial progress on tools for custom analog and mixed-signal circuit layout. Cast mostly in the form of numerical and combinatorial optimization tasks, linked by various forms of sensitivity analysis and constraint mapping, leveraged by ever faster workstations, these tools are beginning to have practical — even commercial — application. There remain many open problems here, and some newly created problems as analog circuits are increasingly embedded in unfriendly digital deep submicron designs. Given the demand for mixed-signal ICs, we expect no reduction in the interest in various layout tools to speed the design of these important circuits.

Acknowledgments

We are grateful to our colleagues at Carnegie Mellon for their assistance with the preparation of this chapter. Rick Carley, Mehmet Aktuna, and Brian Bernberg in particular helped with early drafts. Sachin Sapatnekar of University of Minnesota also offered helpful comments on the final version of the paper. The writing of this survey was supported in part by the NSF under contract 9901164, and by the SRC.

References

For clarity, we have grouped the references by topic, in roughly the order in which each topic area is visited in our review.

General References

[1] R.A. Rutenbar, Analog design automation: Where are we? Where are we going? *Proceedings of IEEE Custom IC Conference*, San Diego, CA, 1993, pp. 13.1.1–13.1.7.
[2] L.R. Carley, G.G.E. Gielen, R.A. Rutenbar, and W.M.C. Sansen, Synthesis tools for mixed-signal ICs: Progress on frontend and backend strategies, *Proceedings of the ACM/IEEE Design Automation Conference*, Las Vegas, NV, 1996, pp. 298–303.
[3] S. Kirkpatrick, C.D. Gelatt, and M.P. Vecchi, Optimization by simulated annealing, *Science*, 220, 671–680, 1983.
[4] L.T. Pillage and R.A. Rohrer, Asymptotic waveform evaluation for timing analysis, *IEEE Trans. CAD*, 9, 352–366, 1990.
[5] H. Murata, K. Fujiyoshi, S. Nakatake, and Y. Kajitani, VLSI module placement based on rectangle-packing by the sequence-pair method, *IEEE Trans. CAD*, 15, 1518–1524, 1996.

Cell-Level Module Generators

[6] J. Kuhn, Analog module generators for silicon compilation, *VLSI Syst. Des.*, 75–80, 1987.
[7] E. Berkcan, M. d'Abreu, and W. Laughton, Analog compilation based on successive decompositions, *Proceedings of ACM/IEEE Design Automation Conference*, Atlantic City, NJ, 1988, pp. 369–375.
[8] H. Koh, C. Séquin, and P. Gray, OPASYN: a compiler for CMOS operational amplifiers, *IEEE Trans. CAD*, 9, 113–125, 1990.
[9] H. Onodera, et al., Operational amplifier compilation with performance optimization, *IEEE JSSC*, SC-25, 460–473, 1990.
[10] J.D. Conway and G.G. Schrooten, An automatic layout generator for analog circuits, *Proceedings of EDAC*, Brussels, 1992, pp. 466–473.
[11] J.P. Harvey, et al., STAIC: an interactive framework for synthesizing CMOS and BiCMOS analog circuits, *IEEE Trans. CAD*, 1402–1416, 1992.
[12] R. Henderson, et al., A spreadsheet interface for analog design knowledge capture and re-use, *Proceedings of IEEE CICC*, San Diego, CA, 13.3, 13.1.1–13.3.4, 1993.
[13] G. Beenker, J. Conway, G. Schrooten, and A. Slenter, Analog CAD for consumer ICs, in *Analog Circuit Design*, J. Huijsing, R. van der Plassche, and W. Sansen, Eds., Kluwer Academic Publishers, Dordrecht, 1993, pp. 347–367, chap. 15.
[14] B.R. Owen, R. Duncan, S. Jantzi, C. Ouslis, S. Rezania, and K. Martin, BALLISTIC: an analog layout language, *Proceedings of IEEE Custom IC Conference*, Santa Clara, CA, 1995, pp. 41–44.
[15] J.D. Bruce, H.W. Li, M.J. Dallabetta, and R.J. Baker, Analog layout using ALAS!, *IEEE JSSC*, 31, 271–274, 1996.
[16] G. van der Plas, J. Vandenbussche, G. Gielen, and W. Sansen, Mondriaan: a tool for automated layout of array-type analog blocks, *Proceedings of IEEE Custom IC Conference*, Santa Clara, CA, 1998, pp. 485–488.

Device-Level Placement and Routing

[17] J. Rijmenants, T.R. Schwarz, J.B. Litsios, and R. Zinszner, ILAC: an automated layout tool for CMOS circuits, *Proceedings of the IEEE Custom IC Conference*, Rochester, NY, 1988, pp. 7.6/1–7.6/4.

[18] M. Kayal, S. Piguet, M. Declerq, and B. Hochet, SALIM: a layout generation tool for analog ICs, *Proceedings of the IEEE Custom IC Conference*, Rochester, NY, 1988, pp. 7.5/1–7.5/4.

[19] D.J. Garrod, R.A. Rutenbar, and L.R. Carley, Automatic layout of custom analog cells in ANANGRAM, *Proceedings of ICCAD*, Santa Clara, CA, 1988, pp. 544-547.

[20] M. Kayal, S. Piguet, M. Declerq, and B. Hochet, An interactive layout generation tool for CMOS ICs, *Proceedings of IEEE ISCAS*, Espoo, Finland, 3, 2431–2434, 1988.

[21] M. Mogaki, et al., LADIES: an automatic layout system for analog LSI's, *Proceedings of ACM/IEEE ICCAD*, Santa Clara, CA, 1989, pp. 450–453.

[22] J. Rijmenants, J.B. Litsios, T.R. Schwarz, and M.G.R. Degrauwe, ILAC: an automated layout tool for analog CMOS circuits, *IEEE JSSC*, 24, 417–425, 1989.

[23] S.W. Mehranfar, STAT — A schematic to artwork translator for custom analog cells, *Proceedings of IEEE Custom IC Conference*, Boston, MA, 1990, pp. 30.2/1–30.2/4.

[24] S. Piguet, F. Rahali, M. Kayal, E. Zysman, and M. Declerq, A new routing method for full-custom analog ICs, *Proceedings of IEEE Custom IC Conference*, Boston, MA, 1990, pp. 27.7/1–27.7/4.

[25] J.M. Cohn, D.J. Garrod, R.A. Rutenbar, and L.R. Carley, New algorithms for placement and routing of custom analog cells in ACACIA, *Proceedings of IEEE Custom IC Conference*, Boston, MA, 1990, pp. 27.6/1–27.6/4.

[26] E. Malavasi, U. Choudhury, and A. Sangiovanni-Vincentelli, A routing methodology for analog integrated circuits, *Proceedings of ACM/IEEE ICCAD*, Santa Clara, CA, 1990, pp. 202–205.

[27] J.M. Cohn, D.J. Garrod, R.A. Rutenbar, and L.R. Carley, KOAN/ANAGRAM II: new tools for device-level analog placement and routing, *IEEE JSSC*, 26, 330–342, 1991.

[28] E. Charbon, E. Malavasi, U. Choudhury, A. Casotto, and A. Sangiovanni-Vincentelli, A constraint-driven placement methodology for analog integrated circuits, *Proceedings of IEEE CICC*, Boston, MA, 1992, pp. 28.2.1–28.2.4.

[29] E. Malavasi and A. Sangiovanni-Vincentelli, Area routing for analog layout, *IEEE Trans. CAD*, 12, 1186–1197, 1993.

[30] B.Basaran, R.A. Rutenbar, and L.R. Carley, Latchup-aware placement and parasitic-bounded routing of custom analog cells, *Proceedings of ACM/IEEE ICCAD*, Santa Clara, CA, 1993, pp. 415–421.

[31] M. Pillan and D. Sciuto, Constraint generation and placement for automatic layout design of analog integrated circuits, *Proceedings of IEEE ISCAS*, London, 1994, pp. 355–358.

[32] G.J. Gad El Karim, R.S. Gyurcsik, and G.L. Bilbro, Sensitivity driven placement of analog modules, *Proceedings of IEEE ISCAS*, London, 1994, pp. 363–366.

[33] J.A. Prieto, J.M. Quintana, A. Rueda, and J.L. Huertas, An algorithm for the place-and-route problem in the layout of analog circuits, *Proceedings of IEEE ISCAS*, London, 1994, pp. 491–494.

[34] E. Malavasi, J.L. Ganley, and E. Charbon, Quick placement with geometric constraints, *Proceedings of IEEE Custom IC Conference*, Santa Clara, CA, 1997, pp. 561–564.

[35] J.M Cohn, D.J. Garrod, R.A. Rutenbar, and L.R. Carley, *Analog Device-Level Layout Automation*, Kluwer, Dordrecht, 1994.

[36] K. Lampaert, G. Gielen, and W. Sansen, Direct performance-driven placement of mismatch-sensitive analog circuits, *Proceedings of the European Design and Test Conference*, Paris, 1995, p. 597.

[37] K. Lampaert, G. Gielen, and W.M. Sansen, A performance-driven placement tool for analog integrated circuits, *IEEE JSSC*, 30, 773–780, 1995.

[38] F. Balasa and K. Lampaert, Module placement for analog layout using the sequence pair representation, *Proceedings of ACM/IEE Design Automation Conference*, New Orleans, LA, 1999, pp. 274–279.

[39] K. Lampaert, G. Gielen, and W. Sansen, Analog routing for manufacturability, *Proceedings of IEEE CICC*, San Diego, CA, 1996, pp. 175–178.

[40] C. Brandolese, M. Pillan, F. Salice, and D. Sciuto, Analog circuits placement: a constraint driven methodology, *Proceedings of IEEE ISCAS*, Atlanta, GA, 4, 635–638, 1996.

[41] Stephan Ohr, Electronica: cell-builder tool anticipates analog synthesis, *EE Times*, 9, 1998.
[42] A.H. Shah, S. Dugalleix, and F. Lemery, High-performance CMOS-amplifier design uses front-to-back analog flow. EDN magazine, Reed Electronics Group, 2002.

Optimal MOS Device Stacking

[43] E. Malavasi, D. Pandini, and V. Liberali, Optimum stacked layout for analog CMOS ICs, *Proceedings of IEEE Custom IC Conference*, San Diego, CA, 1993, p. 17.1.1.
[44] V. Liberali, E. Malavasi, and D. Pandini, Automatic generation of transistor stacks for CMOS analog layout, *Proceedings of IEEE ISCAS*, Chicago, IL, 1993, pp. 2098–2101.
[45] E. Charbon, E. Malavasi, D. Pandini, and A. Sangiovanni-Vincentelli, Simultaneous placement and module optimization of analog ICs, *Proceedings of ACM/IEEE Design Automation Conference*, San Diego, CA, 1994, pp. 31–35.
[46] E. Charbon, E. Malavasi, D. Pandini, and A. Sangiovanni-Vincentelli, Imposing tight specifications on analog ICs through simultaneous placement and module optimization, *Proceedings of the IEEE Custom Integrated Circuits Conference* (CICC), San Diego, CA, 1994, pp. 525–528.
[47] E. Malavasi and D. Pandini, Optimum CMOS stack generation with analog constraints, *IEEE Trans. CAD*, 14, 107–112, 1995.
[48] B. Basaran and R.A. Rutenbar, An $O(n)$ algorithm for transistor stacking with performance constraints, *Proceedings of ACM/IEEE DAC*, Las Vegas, NV, 1996, pp. 221–226.

Device-Level Compaction and Layout Optimization

[49] R. Okuda, T. Sato, H. Onodera, and K. Tamuru, An efficient algorithm for layout compaction problem with symmetry constraints, *Proceedings of IEEE ICCAD*, Santa Clara, CA, 1989, pp. 148–151.
[50] J. Cohn, D. Garrod, R. Rutenbar, and L.R. Carley, Techniques for simultaneous placement and routing of custom analog cells in KOAN/ANAGRAM II, *Proceedings of ACM/IEEE ICCAD*, Santa Clara, CA, 1991, pp. 394–397.
[51] E. Felt, E. Malavasi, E. Charbon, R. Totaro, and A. Sangiovanni-Vincentelli, Performance-driven compaction for analog integrated circuits, *Proceedings of IEEE Custom IC Conference*, San Diego, CA, 1993, pp. 17.3.1–17.3.5.
[52] E. Malavasi, E. Felt, E. Charbon, and A. Sangiovanni-Vincentelli, Symbolic compaction with analog constraints, *Int. J. Circuit Theory Appl.*, 23, 433–452, 1995.

Radio Frequency Cell Layout

[53] J.F. Zurcher, MICROS 3 — A CAD/CAM program for fast realization of microstrip masks, 1985 *IEEE MTT-S International Microwave Symposium Digest*, St. Louis, Mo, June 1985.
[54] R.H. Jansen, LINMIC: a CAD package for the layout-oriented design of single- and multi-layer MICs/MMICs up to mm-wave frequencies, *Microwave J.*, 29, 151–161, 1986.
[55] R.H. Jansen, R.G. Arnonld, and I.G. Eddison, A comprehensive CAD approach to design of MMICs up to MM-wave frequencies, *IEEE J. MTT-T*, 36, 208–219, 1988.
[56] E. Charbon, G. Holmlund, B. Donecker, and A. Sangiovanni-Vincentelli, A Performance-Driven Router for RF and Microwave Analog Circuit Design, *Proceedings of IEEE Custom Integrated Circuits Conference*, Santa Clara, CA, 1995, pp. 383–386.
[57] M. Aktuna, R.A. Rutenbar, and L.R. Carley, Device level early floorplanning for RF circuits, *Proceedings of ACM International Symposium on Physical Design*, Monterey, CA, 1998, pp. 57–64.
[58] M. Aktuna, R.A. Rutenbar, and L.R. Carley, Device level early floorplanning for RF circuits, *IEEE Trans. CAD*, 18, 1999, pp. 375–388.

Constraint Generation and Mapping to Physical Design

[59] U. Choudhury and A. Sangiovanni-Vincentelli, Automatic generation of parasitic constraints for performance-constrained physical design of analog circuits, *IEEE Trans. CAD*, 12, 208–224, 1993.

[60] E. Charbon, E. Malavasi, and A. Sangiovanni-Vincentelli, Generalized constraint generation for analog circuit design, *Proceedings of IEEE/ACM ICCAD*, Santa Clara, CA, 1993, pp. 408–414.

[61] E. Charbon, P. Miliozzi, E. Malavasi, and A. Sangiovanni-Vincentelli, Generalized constraint generation in the presence of non-deterministic parasitics, *Proceedings of ACM/IEEE International Conference on CAD*, Santa Clara, CA, 1996, pp. 187–192.

System-Level Mixed-Signal Module Generators

[62] W.J. Helms and K.C. Russel, Switched capacitor filter compiler, *Proceedings of IEEE CICC*, Rochester, NY, 1986, pp. 125–128.

[63] H. Yaghutiel, A. Sangiovanni-Vincentelli, and P.R. Gray, A methodology for automated layout of switched capacitor filters, *Proceedings of ACM/IEEE ICCAD*, Santa Clara, CA, 1986, pp. 444–447.

[64] G. Jusef, P.R. Gray, and A. Sangiovanni-Vincentelli, CADICS — Cyclic analog-to-digital converter synthesis, *Proceedings of IEEE ICCAD*, Santa Clara, CA, 1990, pp. 286–289.

[65] H. Chang, A. Sangiovanni-Vincentelli, F. Balarin, E. Charbon, U. Choudhury, G. Jusuf, E. Liu, E. Malavasi, R. Neff, and P. Gray, A top-down, constraint-driven methodology for analog integrated circuits, *Proceedings of IEEE Custom IC Conference*, Boston, MA, 1992, pp. 8.4.1–8.4.6.

[66] R. Neff, P. Gray, and A. Sangiovanni-Vincentelli, A module generator for high speed CMOS current output digital/analog converters, *Proceedings of IEEE Custom IC Conference*, Santa Clara, CA, 1995, pp. 481–484.

Substrate Modeling, Extraction, and Coupling Analysis

[67] T.A. Johnson, R.W. Knepper, V. Marcello, and W. Wang, Chip substrate resistance modeling technique for integrated circuit design, *IEEE Trans. CAD*, CAD-3, 3, 126–134, 1984.

[68] D.K. Su, M. Loinaz, S. Masui, and B. Wooley, Experimental results and modeling techniques for substrate noise in mixed-signal integrated circuits, *IEEE JSSC*, 28, 420–430. 1993.

[69] N. Verghese, D. Allstot, and S. Masui, Rapid simulation of substrate coupling effects in mixed-mode ICs, *Proceedings of IEEE Custom IC Conference*, San Diego, CA, 1993, pp. 18.3.1–18.3.4.

[70] F. Clement, E. Zysman, M. Kayal, and M. Declerq, LAYIN: toward a global solution for parasitic coupling modeling and visualization, *Proceedings of IEEE Custom IC Conference*, San Diego, CA, 1994, pp. 537–540.

[71] R. Gharpurey and R.G. Meyer, Modeling and analysis of substrate coupling in ICs, *Proceedings of IEEE Custom IC Conference*, Santa Clara, CA, 1995, pp. 125–128.

[72] S. Mitra, R.A. Rutenbar, L.R. Carley, and D.J. Allstot, A methodology for rapid estimation of substrate-coupled switching noise, *Proceedings of IEEE Custom IC Conference*, Santa Clara, CA, 1995, pp. 129–132.

[73] N.K. Verghese, D.J. Allstot, and M.A. Wolfe, Fast parasitic extraction for substrate coupling in mixed-signal ICs, *Proceedings of IEEE Custom IC Conference*, Santa Clara, CA, 1995, pp. 121–124.

[74] I.L. Wemple and A.T. Yang, Mixed signal switching noise analysis using Voronoi-tessellated substrate macromodels, *Proceedings of ACM/IEEE Design Automation Conference*, San Francisco, CA, 1995, pp. 439–444.

[75] N. Verghese and D. Allstot, SUBTRACT: a program for efficient evaluation of substrate parasitics in integrated circuits, *Proceedings of ACM/IEEE ICCAD*, Santa Clara, CA, 1995, pp. 194–198.

[76] N. Verghese, T. Schmerbeck, and D. Allstot, *Simulation Techniques and Solutions for Mixed-Signal Coupling in Integrated Circuits*, Kluwer Academic Publishers, Norwell, MA, 1995.

[77] T. Smedes, N.P. van der Meijs, and A.J. van Genderen, Extraction of circuit models for substrate cross-talk, *Proceedings of ACM/IEEE ICCAD*, Santa Clara, CA, 1995, pp. 199–206.
[78] K.J. Kerns, I.L. Wemple, and A.T. Yang, Stable and efficient reduction of substrate model networks using congruence transforms, *Proceedings of ACM/IEEE ICCAD*, Santa Clara, CA, 1995, pp. 207–214.
[79] N.K. Verghese, D.J. Allstot, and M.A. Wolfe, Verification techniques for substrate coupling and their application to mixed signal IC design, *IEEE JSSC*, 31, 354–365, 1996.
[80] R. Gharpurey and R.G. Meyer, Modeling and analysis of substrate coupling in integrated circuits, *IEEE JSSC*, 31, 344–352, 1996.
[81] P. Miliozzi, L. Carloni, E. Charbon, and A.L. Sangiovanni-Vincentelli, SUBWAVE: a methodology for modeling digital substrate noise injection in mixed-signal ICs, *Proceedings of IEEE Custom IC Conference*, San Diego, CA, 1996, pp. 385–388.
[82] P. Miliozzi, I. Vassiliou, E. Charbon, E. Malavasi, and A. Sangiovanni-Vincentelli, Use of sensitivities and generalized substrate models in mixed-signal IC design, *Proceedings of ACM/IEEE Design Automation Conference*, Las Vegas, CA, 1996, pp. 227–232.
[83] E. Charbon, R. Gharpurey, R.G. Meyer, and A. Sangiovanni-Vincentelli, Semi-analytical techniques for substrate characterization in the design of mixed-signal ICs, *Proceedings of ACM/IEEE ICCAD*, Santa Clara, CA, 1996, pp. 455–462.
[84] T. Tsukada and K.M. Makie-Fukuda, Approaches to reducing digital noise coupling in CMOS mixed signal LSIs, *IEICE Trans. Fundam. Electron. Commun.Comput. Sci.*, E80-A, 2, 263–275, 1997.
[85] N.K. Verghese and D.J. Allstot, Verification of RF and mixed-signal integrated circuits for substrate coupling effects, *Proceedings of IEEE Custom IC Conference*, Santa Clara, CA, 1997, pp. 363–370.

System-Level Mixed-Signal Placement and Routing

[86] C.D. Kimble, A.E. Dunlop, G.F. Gross, V.L. Hein, M.Y. Luong, K.J. Stern, and E.J. Swanson, Autorouted Analog VLSI, *Proceedings of IEEE Custom Integrated Circuits Conference*, Portland, OR, 1985, pp. 72–78.
[87] A.E. Dunlop, G.F. Gross, C.D. Kimble, M.Y. Luong, K.J. Stern, and E.J. Swanson, Features in the LTX2 for analog layout, *Proceedings of IEEE ISCAS*, Kyoto, Japan, 1985, pp. 21–23.
[88] H.H. Chen and E. Kuh, Glitter: a gridless variable width channel router, *IEEE Trans. CAD*, CAD-5, 5, 459–465, 1986.
[89] J. Trnka, R. Hedman, G. Koehler, and K. Lading, A device level auto place and wire methodology for analog and digital masterslices, *IEEE ISSCC Dig. Tech. Pap.*, San Francisco, CA, 1988, pp. 260–261.
[90] R.S. Gyurcsik and J.C. Jeen, A generalized approach to routing mixed analog and digital signal nets in a channel, *IEEE JSSC*, 24, 436–442, 1989.
[91] U. Choudhury and A. Sangiovanni-Vincentelli, Use of performance sensitivities in routing analog circuits, *Proceedings of IEEE ISCAS*, New Orleans, LA, 1990, pp. 348–351.
[92] U. Choudhury and A. Sangiovanni-Vincentelli, Constraint generation for routing analog circuits, *Proceedings of ACM/IEEE DAC*, Orlando, FL, 1990, pp. 561–566.
[93] S. Mitra, S. Nag, R.A. Rutenbar, and L.R. Carley, System-level routing of mixed-signal ASICs in WREN, *Proceedings of ACM/IEEE ICCAD*, Santa Clara, CA, 1992, pp. 394–399.
[94] U. Choudhury and A. Sangiovanni-Vincentelli, Constraint-based channel routing for analog and mixed analog/digital circuits, *IEEE Trans. CAD*, 12, 497–510, 1993.
[95] S. Mitra, R.A. Rutenbar, L.R. Carley, and D.J. Allstot, Substrate-aware mixed-signal macrocell placement in WRIGHT, *Proceedings of IEEE Custom IC Conference*, San Diego, CA, 1994, pp. 529–532.
[96] S. Mitra, R.A. Rutenbar, L.R. Carley, and D.J. Allstot, Substrate-aware mixed-signal macrocell placement in WRIGHT, *IEEE JSSC*, 30, 269–278, 1995.

Mixed-Signal Power Distribution Layout

[97] B.R. Stanisic, R.A. Rutenbar, and L.R. Carley, Power distribution synthesis for analog and mixed signal ASICs in RAIL, *Proceedings of IEEE Custom IC Conference*, San Diego, CA, 1993, pp. 17.4.1–17.4.5.

[98] B.R. Stanisic, R.A. Rutenbar, and L.R. Carley, Mixed-signal noise decoupling via simultaneous power distribution design and cell customization in RAIL, *Proceedings of IEEE Custom IC Conference*, San Diego, CA, 1994, pp. 533–536.

[99] B.R. Stanisic, N.K. Verghese, R.A. Rutenbar, L.R. Carley, and D.J. Allstot, Addressing substrate coupling in mixed-mode IC's: simulation and power distribution synthesis, *IEEE JSSC*, 29, 226–238, 1994.

[100] B.R. Stanisic, R.A. Rutenbar, and L.R. Carley, Addressing noise decoupling in mixed-signal ICs: power distribution design and cell customization, *IEEE JSSC*, 30, 321–326, 1995.

[101] B.R. Stanisic, R.A. Rutenbar, and L.R. Carley, *Synthesis of Power Distribution to Manage Signal Integrity in Mixed-Signal ICs*, Kluwer Academic Publishers, Norwell, MA, 1996.

[102] H.H. Chen and D.D. Ling, Power supply noise analysis methodology for deep submicron VLSI chip design, *Proceedings of ACM/IEEE Design Automation Conference*, Anaheim, CA, 1997, pp. 638–643.

Examples of Complete Analog Layout Flows

[103] R. Rutenbar et al,, Synthesis and layout for mixed-signal ICs in the ACACIA system, in *Analog Circuit Design*, J.H. Huijsing, R.J. van de Plassche, and W.M.C Sansen, Eds., Kluwer Academic Publishers, Norwell, MA, 1996, pp. 127–146.

[104] I. Vassiliou, H. Chang, A. Demir, E. Charbon, P. Miliozzi, and A. Sangiovanni-Vincentelli, A video driver system designed using a top-down constraint-driven methodology, *Proceedings of ACM/IEEE ICCAD*, San Jose, CA, 1996, pp. 463–468.

[105] E. Malavasi, E. Felt, E. Charbon, and A. Sangiovanni-Vincentelli, Automation of IC layout with analog constraints, *IEEE Trans. CAD*, 15, 923–942, 1996.

[106] H. Chang, E. Charbon, U. Choudhury, A. Demir, E. Felt, E. Liu, E. Malavasi, A. Sangiovanni-Vincentelli, and I. Vassiliou, *A Top-Down, Constraint-Driven Design Methodology for Analog Integrated Circuits*, Kluwer Academic Publishers, Norwell, MA, 1997.

Field Programmable Analog Arrays

[107] M. Sivilotti, A dynamically configurable architecture for prototyping analog circuits, *Proceedings of the fifth MIT Conference on Advanced research in VLSI*, Cambridge, MA, 1988, pp. 237–258.

[108] E.K.F. Lee and P.G. Gulak, A field programmable analog array based on MOSFET transconductors, *Electron. Lett.*, 28, 28–29, 1992.

[109] E.K.F. Lee and P.G. Gulak, A transconductor-based field programmable analog array, *IEEE ISSCC Dig.Tech. Pap.*, San Francisco, 1995, pp. 198–199.

[110] H.W. Klein, Circuit development using EPAC technology: an analog FPGA, *Proceedings of the SPIE, International Society for Optical Engineering*, Philadelphia, PA, 2607, 136–144, 1995.

[111] A. Bratt and I. Macbeth, Design and implementation of a field programmable analogue array, *Proceedings of ACM International Symposium on FPGAs*, Monterey, CA, 1996, pp. 88–93.

[112] P. Chow and P.G. Gulak, A field programmable mixed analog digital array, *Proceedings of ACM International Symposium on FPGAs*, Monterey, CA, 1995, pp. 104–109.

[113] C.A. Looby and C. Lyden, A CMOS continuous time field programmable analog array, *Proceedings of ACM International Symposium on FPGAs*, Monterey, CA, 1997, pp. 137–141.

[114] S.D. Brown, R.J. Francis, J. Rose, and Z.G. Vranesic, *Field Programmable Gate Arrays*, Kluwer Academic Publishers, Norwell, MA, 1992.

SECTION III

PHYSICAL VERIFICATION

17

Design Rule Checking

Robert Todd
Mentor Graphics, Inc.
Wilsonville, Oregon

Laurence Grodd
Mentor Graphics, Inc.
Wilsonville, Oregon

Katherine Fetty
Mentor Graphics, Inc.
Wilsonville, Oregon

17.1 Introduction

After the physical mask layout is created for a circuit for a specific design process, the layout is measured by a set of geometric constraints, or rules, for that process. The main objective of design rule checking (DRC) is to achieve a high overall yield and reliability for the design. To meet this goal of improving die yields, DRC has evolved from simple measurement and Boolean checks, to more involved rules that modify existing features, insert new features, and check the entire design for process limitations such as layer density. A completed layout consists not only of the geometric representation of the design, but also data that provide support for manufacture of the design.

While design rule checks do not validate that the design will operate correctly, they are constructed to verify that the structure meets the process constraints for a given design type and process technology. Before discussing how DRC accomplishes this verification, an understanding of the concepts of DRC is useful.

17.1.1 Concepts

The physical mask layout consists of shapes on drawn layers that are grouped into one or more cells. A cell may contain a placement of another cell. If the entire design is represented in one cell, it is a flat design; otherwise it is a hierarchical design. Figure 17.1 shows an example of a hierarchical design with a top cell that contains other instances of cells and primitive objects; in the lower left, a flat view of one cell is magnified to show its content.

Early in design verification stages, the layout may contain text objects or other attributes that further define the purpose of an object in the layout or provide information that ties an object or set of objects to the logical representation of the design.

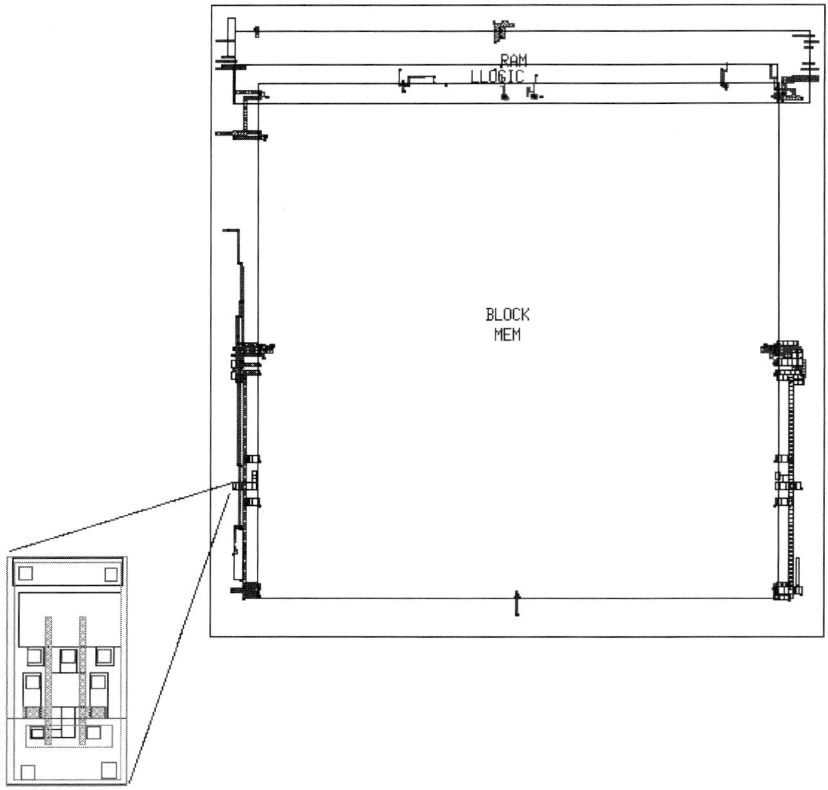

FIGURE 17.1 Example of design hierarchy for mask data.

17.1.2 Layers

Geometries in a verification flow are grouped into layers. Most verification systems provide unlimited numbers of layers.

The following four types of layers are shown in Figure 17.2:

- Drawn layers represent the original layout data. They are merged on input to the verification system to remove any overlap or abutment of geometry on the same layer.
- A polygon layer is the output of a layer creation operation such as a Boolean operation, a topological polygon operation, or area function. Examples of operations that generate a polygon layer are

```
Gate = poly AND diff
Big_island = AREA active > 100
Floating_met1 = met1 outside contact
```

- An edge layer represents the edges of merged polygons as output by length, angle, or other measurement operations. Example of operations that return edge layers are

```
Long_met1_edge = LENGTH met1 > 10
```

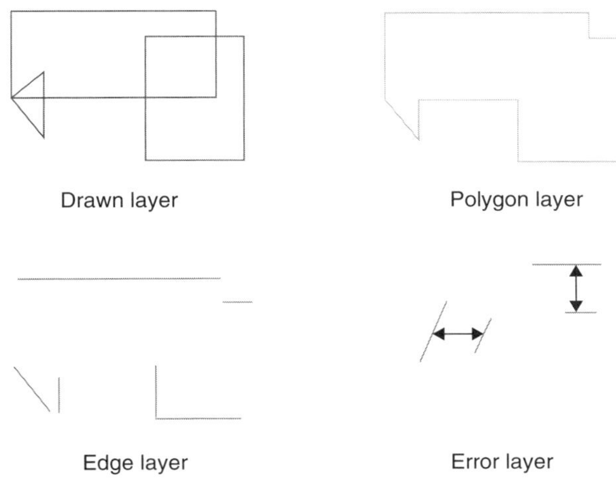

FIGURE 17.2 Types of layers defined for DRC.

```
Poly_gate_edges = poly INSIDE EDGE diff
```

- An error layer is a cluster of one to four edges from a DRC spatial measurement for use in graphical result representation. An example of an operation that returns an error layer is

```
Contact_in_met1_error = ENCLOSURE contact met1 < 0.6
```

17.1.3 Design Rule Checking Operations

To perform DRC, you need to be able to select data, access a rich set of operations to perform on the data, and choose how to report output. Output from a DRC operation is reported as an error or is provided as input to another DRC operation. Some operations have more information to return than an object or edge alone may convey, so they create a separate report with the results information. For example, a DENSITY operation may create rectangles that overlay regions of the design and create a separate ASCII report providing the same density and gradient values of each rectangle.

The following design rule combines two DRC operations and returns edge clusters representing the errors:

```
Rule1 {
  X = ENCLOSURE [CP] ME > 2.24 < 2.26
  ENCLOSURE X POLY < 1.25
}
```

Figure 17.3 shows the initial layers required for Rule 1, where the ME layer is previously filtered to show only a region immediately surrounding the CP layer. The first DRC operation creates layer X to represent all the CP edges that are enclosed by the ME layer between 2.24 and 2.26 units. Figure 17.4 shows what layer X looks like in this region of the design, and also circles the results of the second ENCLOSURE operation between X and POLY. This second DRC ENCLOSURE operation returns the three separate errors that are represented as paired edges.

17.1.3.1 Language-Based Design Rule Checking

A complete DRC system provides a common language to describe all phases of layout verification. The language will allow the same rules to apply to checks for flat, cell/block-based, and hierarchical full-chip modes. This DRC language may also define the device definitions, parasitic parameters, layers, and design rules for a process technology.

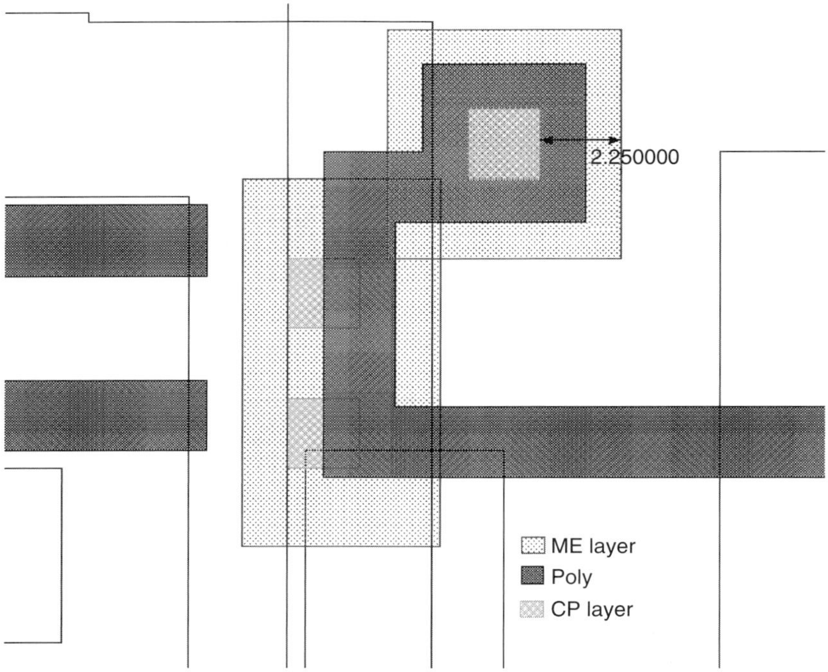

FIGURE 17.3 Geometries used in a design rule.

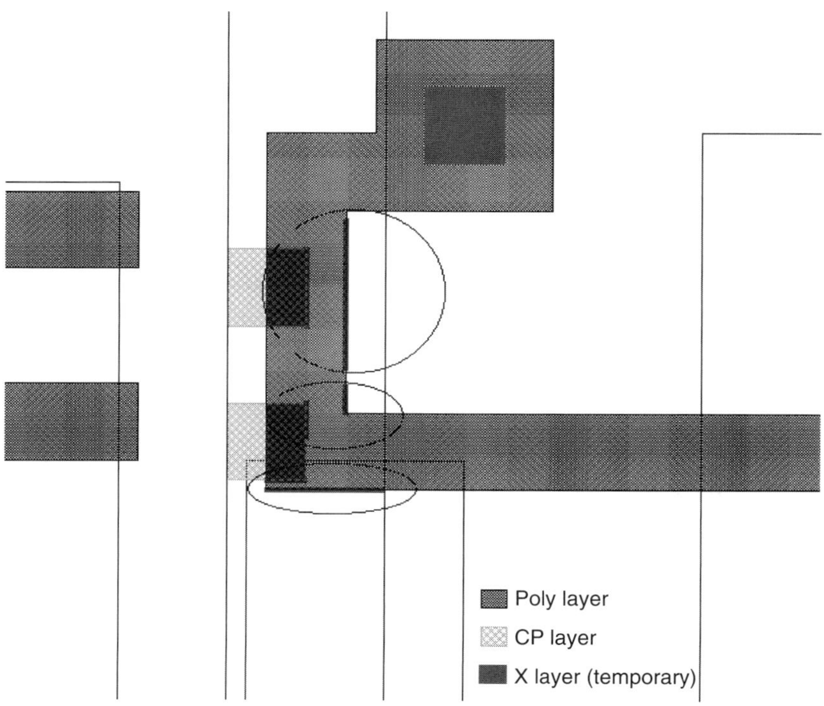

FIGURE 17.4 Three error layer results from a design rule.

The following example shows how DRC operations in a language-based system may be combined to modify layout by adding slots on wide metal lines and adding fill shapes to low-density regions of the design.

```
LAYER MET1 16
//smooth any small metal jogs
size_met1 = SIZE MET1 UNDEROVER BY 3

//undersize the wide metal for proper slot enclosure
fat_met1 = SIZE size_met1 BY -0.6

met1_slots {@ metal1 square slots for Cu
        RECTANGLES 0.5 0.5 0.6 INSIDE OF LAYER fat_met1
}

met1_density = DENSITY MET1 < 0.1 WINDOW 100 STEP 50
met1_fill = RECTANGLES 0.3 0.3 0.25 INSIDE OF LAYER met1_den-
sity
met1_size = SIZE MET1 BY 0.25
met1_mask { met1_fill OUTSIDE met1_size }
```

The results of this rule are shown in Figure 17.5, in which the dark squares are metal slots along the bottom of the frame, and the metal fill is in the upper right. First, the MET1 layer is sized in two DRC operations to smooth the contour of the metal route and to select only metal regions that may validly contain a slot. Then RECTANGLES operation creates squares of the specified dimension only inside the regions of fat_met1 previously selected. These squares are output as met1_slots. The second part of this example calculates the DENSITY of MET1 in a region of the layout, selecting regions of the layout with unacceptably small amounts of MET1. Then, the RECTANGLES command fills these regions with fill

FIGURE 17.5 Automatically generated metal slots and metal fill in a layout.

shapes. The RECTANGLES are contoured to be placed a minimum distance away from the original MET1 shapes.

17.1.3.2 When To Do Design Rule Checking

Traditionally, DRC was primarily considered at the cell creation and block assembly levels of design, and physical layout was done by hand [1]. As the complexity in the layout increased, DRC became a requirement at more stages in the manufacturing process.

Interactive checks verify small cells or cell libraries, often from within the layout editor. As blocks are assembled, the integrated blocks also need to be verified. Areas between blocks and transitions from the block to the routing areas are verified for layout accuracy. After the full chip is assembled, DRCs may also create fill objects, insert slots in wide metal routes, create and analyze features for Optical Proximity Correction, or insert Sub-Resolution Assist Features (SRAF or scatter bars). Systems-on-chip (SoC) designs require checks at both the interactive and batch phases of the design. At the full-chip phase, DRC is used as a sign-off tool.

17.1.3.3 Flat Design Rule Checking

In older verification systems, the cell-based hierarchical input for a design was flattened and the entire design was internally represented as one large cell. Overlaps between shapes on a layer were removed or merged, and the design rule checks performed on the merged data.

Errors in the design were reported from the top-level view of the design. Since the input hierarchy no longer existed, any output from the system was also represented as flat data.

17.1.3.4 Hierarchical Design Rule Checking

As designs become more complex, verification of a design in flat mode rapidly becomes impractical, consuming too many compute and storage resources and taking too long to complete. Modern verification systems take advantage of the input design hierarchy and other repetitions found in a physical layout design to identify blocks of the design that are analyzed once and then reused, to significantly reduce verification time. The DRC results are reported at the lowest practical level of the design hierarchy. Graphical output from a hierarchical verification tool retains or enhances the original design hierarchy.

17.2 Geometric Algorithms for Physical Verification

All DRC programs, whether hierarchical or flat, require low-level algorithms that analyze geometric relationships between primitive data objects such as polygons and edges. Many computational geometry algorithms are applied to perform this analysis, which seek to minimize the time and space resources required when the number of objects is large.

17.2.1 Scan-Line-Based Analysis

Scan-line-based sweep algorithms [11,2,3] have become the predominant form of low-level geometric analysis. A scan-line sweep analyzes relationships between objects that intersect a virtual line, either vertical or horizontal, as that line is swept across the layout extent. Figure 17.6 shows a scan line moving across the extent of the layout, analyzing geometric data represented as edges. The edges provided as input to the scan line are ordered — first by increasing X, then by increasing Y.

17.2.1.1 Time Complexity of a Scan-Line Algorithm

In practice, if an IC design contains a total of N objects, then the number of objects intersecting the scan line is of $O(\text{sqrt}(N))$, since the objects are roughly the same size and cover the surface of the chip. This has advantages in both space and time. For space, only $O(\text{sqrt}(N))$ objects need to be in active memory for analysis. For time, each of the N layout objects must first be inserted, then removed, from the scan

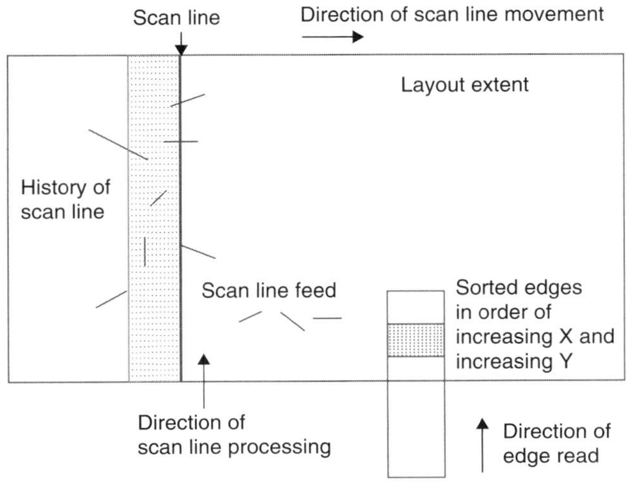

FIGURE 17.6 Components of an edge-based scan line algorithm.

line. Since the scan line is kept sorted, each such operation takes $O(\log(\text{sqrt}(N)) = O(\log N)$ time. Therefore, the time complexity [13] is approximately

$$N * O(\log\ (N) = O(N \log\ N)$$

In practice, this may be improved somewhat by indexing into an array of sorted objects [4], but can also be optimistic since the scan line operation is often more complex than a simple insert and delete. As a result, most implementations are between $O(N \log N)$ and $O(N ** 1.5)$. Any type of object can be placed in a scan-line sweep, and as a result, this approach is very flexible.

Typically, either edges or trapezoids are used for most implementations [4]. Edge representations have the advantage of directly representing the geometries being analyzed.

Trapezoids, which are formed by fracturing input polygons along either axis, provide several performance optimizations but require additional book-keeping complexity from the false boundaries created from fracturing.

By expanding the scan line to have some width, separation or distance relationships can be readily handled.

Another issue that must be addressed by the low-level algorithms that support all angle geometries arises from the fact that all layout objects have vertices that lie on a finite x–y grid. Orthogonal geometries intersect each other only at grid points. Nonorthogonal geometry can intersect off-grid and must be rounded in order to be represented in any output. The rounding perturbations can cause many issues in existing algorithms and must be addressed in order to provide robust implementations [5].

17.3 Hierarchical Data Structures

Introducing design hierarchy to the DRC process, first proposed in [12], requires additional structures for processing data. Hierarchical DRC operations determine the subset of data that can be acted upon on a per-cell basis. Each cell is processed once and the results applied to multiple areas of the original layout. Data outside of subset are promoted up the hierarchy to the lowest point at which it can be accurately acted upon by the DRC operation on a cell-specific basis. Hierarchical layout verification incorporates these concepts:

- *Intrinsic geometries.* Given a layer L and a cell C, an intrinsic geometry of C on L is a geometry on layer L common to every instance of cell C.
- *Interaction geometries.* Given a layer L and a cell C, an interaction geometry of C on layer L is a region where intrinsic geometries on layer L intersect C at some point higher up the hierarchy.

- *Promotion.* The process of moving intrinsic geometries up the hierarchy, as necessary, in order to achieve sufficient cell-specific context to execute accurately an operation.

Intrinsic geometries are promoted based on their proximity to the interaction geometries in the cell on related layers. Promotion is also dependent on the algorithmic intricacies of the type of DRC operation being executed. When promotion ceases, the intrinsic geometry may normally be analyzed or manipulated on a per-cell basis.

17.3.1 Area Interaction and Promotion

To show how the concepts of intrinsic geometries, interaction geometries, and object promotion work in a hierarchical DRC operation, consider the operation Z = X AND Y as operated on the objects in Figure 17.7.

In this design, there are two cells. Cell B is placed (or instantiated) in cell Top. Object 1 is an intrinsic geometry of cell Top and objects 2 to 4 are intrinsic geometries to cell B.

The overlap of object 1 and cell B has created an interaction geometry in cell B on layer X. Figure 17.8 shows how the interaction geometry affects execution of the AND operation. The AND operation is first performed in cell B. Objects 3 and 4 are sufficiently remote from the interaction geometry and may be processed with the AND operation on a per-cell basis. Object 2 and the interaction geometry overlap, and so object 2 must be promoted up the hierarchy (to cell Top). The AND operation is then completed in cell Top because no further promotion is required.

Layer Z contains two intrinsic geometries, one in cell Top and one in cell B. An interaction geometry on layer Z is also created in cell B, representing the overlap of the geometry in Top and cell B.

17.4 Time Complexity of Hierarchical Analysis

Hierarchical analysis is a two-edged sword in terms of efficiency. Flat analysis, as we saw earlier, has an asymptotic time complexity of $O(n \log n)$. It is based on sorting, inserting, and deleting. These are very predictable operations, and since the density of features across a chip is roughly constant (for economic reasons), flat analysis depends very little on design style or chip content.

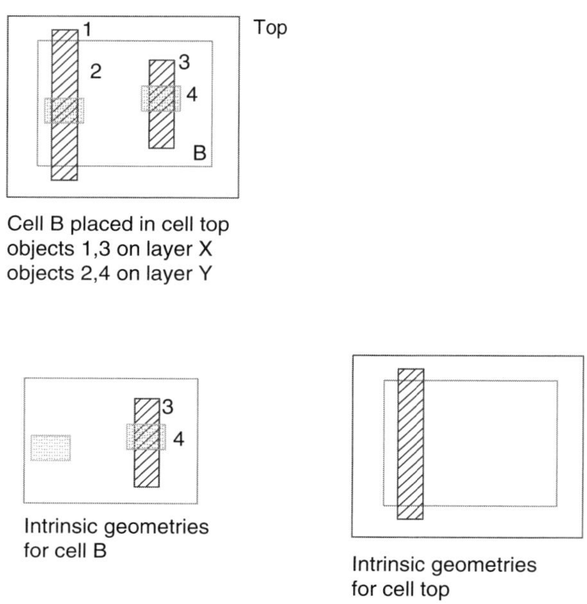

FIGURE 17.7 Initial intrinsic geometries.

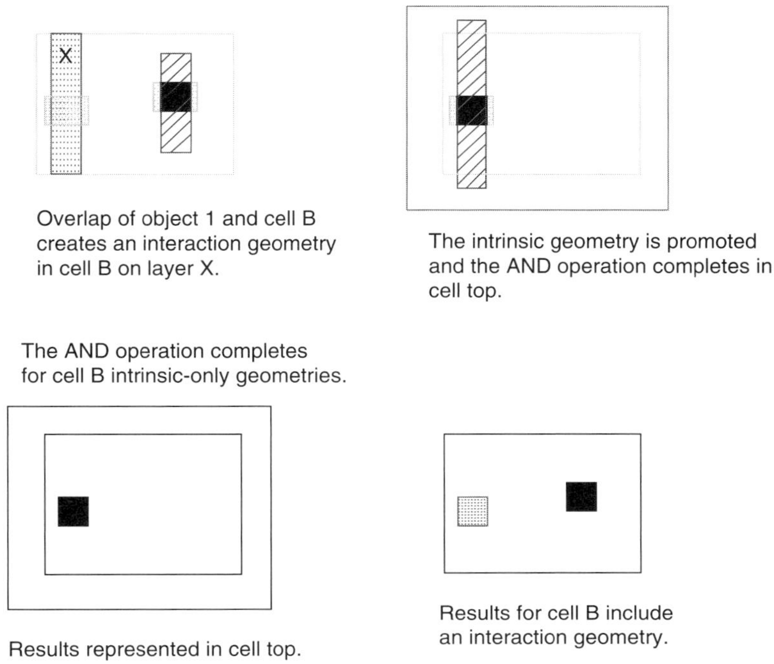

FIGURE 17.8 Interaction geometries and promotion.

Hierachical analysis is a different beast entirely. If the hierarchy is geometrically well organized so that the time spent checking cell interactions is small, then the checking time can improve to $O(N)$, as it is more efficient to analyze many small cells than one big one [15]. On the other hand, if the interactions are complex, then analyzing the hierarchy itself can be NP complete (in the size of the hierarchical design) [14]. This means that even in theory, there may be little advantage in expanding the hierarchy out, even though such an expansion might involve an exponential increase in size. In practice, the situation is even worse, and analyzing a hierarchy ill suited for verification can be worse, much worse, than analyzing it flat.

These distinctions are not merely theoretical. In practice, "good" hierarchies can be analyzed much more quickly than the same design treated flat, and "bad" hierarchies actually take longer if hierarchical analysis is attempted. Unfortunately, the DRC program has no control of the hierarchy used in its input, which is often chosen by the designer for completely unrelated reasons (ease of logical understanding, for example). A great deal of the work of a commercial DRC package is related to distinguishing "good" from "bad" hierarchy, and rebuilding the design (if needed) in a form more suited to hierarchical verification.

17.5 Connectivity Models

Another useful data structure for verification is a connectivity model, which encapsulates geometric interactions within a layer or between several layers, into a logical structure associated with the geometries.

Connectivity models in flat verification can be easily implemented by encapsulating the interaction sets as a single unique number. Hierarchical connectivity models require a more complex encapsulation using the concepts of pins and nets.

Within any given cell, a net organizes the geometric interactions with regard to connectivity. A net may also be an external pin, an internal pin, or both. An internal pin forms a connection to a net within the hierarchical subtree of the cell, while an external pin forms a connection to a net outside the hierarchical subtree of the cell. Pins allow hierarchical algorithms to traverse a net in the hierarchy both up, via an external pin, and down, via an internal pin. Examples of internal and external pins are shown in Figure 17.9. A net is considered complete at the topmost cell in which that net is not an external pin.

Hierarchical algorithms using connectivity models must work hand in hand with the geometrical promotion techniques described earlier. Logical promotion, via dynamic creation of new logical pins, must accompany geometric promotion.

17.5.1 Polygonal Connectivity

The polygon connectivity model determines which geometries interact with each other on a particular layer. The polygon connectivity model is useful for those topological operations (such as INSIDE, OUTSIDE, or OVERLAP) that require information about polygonal relationships.

Consider the operation Z = X INTERACT Y. The INTERACT operation selects all layer X polygons that overlap Y or have a coincident edge with a layer Y polygon. For flat verification, this operation is comparatively simple since full polygons in both X and Y are present at the same (single) hierarchical level. In hierarchical verification, a single polygon may be broken across hierarchical boundaries and exist at multiple locations in the hierarchy. Flattening this data in order to get full polygons at a single hierarchical level may be prohibitively expensive and can destroy the hierarchical notion of the design. Fortunately, selective geometric and logical promotion, along with careful traversal of the nets and pins of the polygon connectivity models in X and Y, can produce hierarchical output in Z without excessive flattening.

17.5.2 Nodal Connectivity

The nodal connectivity model, if specified, exists for the entire design and determines the geometries that interact with each other on layers predefined as having an electrical connectivity. This electrical connectivity is defined by the user with an operation such as CONNECT between two or more layers.

This connectivity is essential for device recognition and connectivity-based spacing checks. The nodal connectivity model is also useful for an operation such as NET AREA RATIO, which performs charge accumulation design rule checks by computing ratios of areas between layers on the same electrical node.

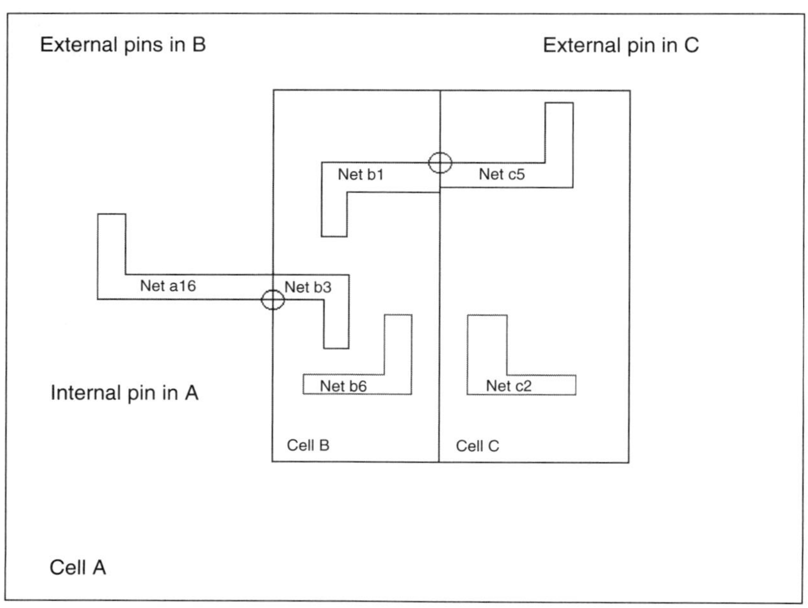

FIGURE 17.9 Internal and external pins for a hierarchical design.

17.6 Parallel Computing

The continuing and growing high computational costs of DRC creates the need to pursue parallel computing techniques to reduce the turn-around time for these tasks. Parallel computing requires decomposing the verification problem into subtasks that can be performed independently. A simple approach to this decomposition is to build a dependency graph of operations or commands in the rule deck, then execute the operations in parallel where the graph allows. The major problem with this approach is that scalability is constrained by the structure of the rule deck, which in current practice severely limits its usefulness. For instance, in a regular DRC rule deck, the scalability is typically limited at the total number of metal layers. For a scattering bar generation rule deck the scalability is limited to 1, since each operation feeds the next.

Another approach is to decompose the data into distinct spatial areas. This is natural for hierarchical algorithms, because the design hierarchy itself provides much of the spatial distinction [6]. The scalability is limited here by the hierarchy, which again in practice can be severely limited based on design style. A relatively flat design with many large custom blocks will scale well with the number of blocks. A standard cell design will scale poorly because the standard cells take very little computation time and the top-level interconnect will dominate. However, spatial areas can also be created internally by partitioning the flattened input data [7]. Note that spatial partitioning is just another form of hierarchy. The scalability is limited here only by the number of partitions [8,9].

17.7 Future Roles for Verification

As the physical size of layout features shrink, the steps required for manufacture become more complex and costly. This complexity translates to increasing amounts of layout data to verify, new types of process information represented as layout geometries, expanded postproduction analysis of chips, and many discussions centered about improving chip manufacture from design to completion [10].

To cope with the changes necessary for sub-100 nm manufacture processes, a successful DRC application will also grow in feature depth and runtime capacity. DRC then becomes a small portion of what a design verification tool can accomplish.

References

[1] C. Mead and L. Conway, Introduction to VLSI Systems, Addison-Wesley Publishers, 1980, pp. 91–111.

[2] M.I. Shamos and D.J. Hoey, Geometric intersection problems, *Proceedings of the 17th Annual Conference on Foundations of Computer Science*, Houston, TX, October 1976, pp. 208–215.

[3] J.L. Bently and T.A. Ottmann, Algorithms for reporting and counting geometric intersections, *IEEE Trans. Comp.*, C-28, 643–647, 1979.

[4] K.-W. Chiang, S. Nahar, and C.-Y. Lo, Time-efficient VLSI artwork analysis algorithms in GOALIE2, *IEEE Trans. Computer-Aided Des.*, 8, 640–648, 1989.

[5] D.H. Greene and F.F. Yao, Finite-resolution computational geometry. *Proceedings of the 27th Annual Symposium on Foundations of Computer Science*, Toronto, Ontario, Canada, 27–29 1986, pp. 143–152.

[6] F. Gregoretti and Z. Segall, Analysis and evaluation of VLSI design rule checking implementation in a multiprocessor, *Proceedings of International Conference on Parallel Processing*, August 1984, pp. 7–14.

[7] G.E. Bier and A.R. Pleszkun, An algorithm for design rule checking on a multiprocessor, *Proceedings of the 22nd Design Automation Conference*, 1985, pp. 299–304.

[8] L. Grodd, Placement Based Design Cells Injection Into an Integrated Circuit, U.S. Patent 6381731, 2002.

[9] Z. Bozkus and L. Grodd, Cell Based Parallel Verification on an Integrated Circuit Design, U.S. Patent 6397372, 1999.

[10] M.E. Mason, The rising cost and complexity of RETs, *Proc. SPIE*, 5379, 0–19, 2004.

[11] H.S. Baird, Fast algorithms for LSI artwork analysis, *DAC'77: Proceedings of the 14th Conference on Design Automation*, IEEE Press, Piscataway, NJ, USA, 1977, pp. 303–311.

[12] P. Losleben, Design validation in hierarchical systems, *DAC'75: Proceedings of the 12th Conference on Design Automation*, IEEE Press, Piscataway, NJ, 1975, pp. 431–438.

[13] U. Lauther, An O (N log N) algorithm for Boolean mask operations, *DAC'81: Proceedings of the 18th Conference on Design Automation*, Nashville, TN, IEEE Press, Piscataway, NJ, 1981, pp. 555–562.

[14] J.L. Bentley and T. Ottman, The complexity of manipulating hierarchically defined sets of rectangles, in *Proceedings of the 10th Symposium on Mathematical Foundations of Computer Science*, J. Gruska and M. Chytil, Eds., Springer, Strbske Pleso, Czechoslovakia, 1981, pp. 1–15.

[15] L.K. Scheffer, Computer-aided design of integrated circuits and systems, *Some conditions under which hierarchical verification is O(N)*, *IEEE Trans.*, 22, 643–646, 2003.

18

Resolution Enhancement Techniques and Mask Data Preparation

Franklin M. Schellenberg

Mentor Graphics, Inc.

San Jose, California

18.1 Introduction

Traditionally, after an IC design has been converted into a physical layout, the timing verified, and the polygons certified to be DRC-clean, the IC was ready for fabrication. The data files representing the various layers were shipped to a mask shop, which used mask-writing equipment to convert each data layer into a corresponding mask, and the masks were shipped to the fab where they were used to repeatedly manufacture the designs in silicon.

In the past, the creation of the layout was the end of EDA's involvement in this flow. However, as Moore's Law has driven features to ever-smaller dimensions, new physical effects that could be effectively ignored in the past are now having an impact on the features that are formed on the silicon wafer. So even though the final layout may represent what is desired in silicon, the layout can still undergo dramatic alteration through several EDA tools before the masks are fabricated and shipped.

These alterations are required not to make any change in the device as designed, but to simply allow the manufacturing equipment, often purchased and optimized for making ICs one or two generations behind, to deliver the new devices. The intent of these alterations is to precompensate for known manufacturing distortions that are inherent in the manufacturing process. These distortions can arise in almost any processing step: lithography, etching, planarization, and deposition, all of which introduce distortions of some kind. Fortunately, when these distortions are measured and quantified, an algorithm for their compensation can also be determined.

The first part of this chapter is concerned with these compensation schemes, particularly for the compensations required for lithographic processes. These lithographic compensations are usually grouped under the heading resolution enhancement techniques (RET), and we will go into these techniques and the consequences of their implementation on the IC design in some detail. They are also sometimes listed under the more general category of design for manufacturability (DFM), but we will not attempt to give an exhaustive treatment to all the possible manufacturing effects that may be treated in this same way, but instead direct the reader to the references for further information.

The second part of this chapter is concerned with the final step, that of converting the final compensated layout into a mask-writer data format. Although previously a simple translation task, the changes required for RET can create huge data volume problems for the mask writer, so some care must be applied to mitigate these negative effects.

18.2 Lithographic Effects

Although various processes[1,2] are used to create the physical structures of circuit elements in various layers of an IC, the physical dimensions of those structures are defined using lithography.[3–5] In a lithographic process, the wafer is coated with a sensitized resist material, and that material is exposed and processed to selectively remove the resist (see Figure 18.1). The subsequent manufacturing step, whether it be implantation, etching, deposition, or some other step, occurs on the silicon only where the resist is not protecting the surface.

The most common process for patterning wafers is optical lithography, using ultraviolet (UV) light as the source. In this case, a photomask (also called a reticle), with the layout pattern for a given layer mapped on it as a pattern of transparent or absorbing regions, is created for each layer.[6,6a] A reduced image of the photomask is formed on the resist-coated wafer using a highly specialized exposure tool called a stepper or scanner. This process therefore resembles a standard photographic printing process, with the photomask corresponding to the negative, the stepper to the enlarger, and the wafer to the photographic paper. However, the tolerances on mechanical motion and lens quality are far tighter than for any photographic system.

The elements of a typical lithographic stepper are shown in Figure 18.2. A UV light source (usually an excimer laser) is collimated, and the illumination shaped to illuminate the photomask from behind. The light passing through the mask is collected and focused as a reduced image (typically 4× smaller) on the resist-coated wafer using a complex lens, designed to be as aberration-free as possible. The resolution of the image is the primary figure of merit for a lithographic system. The photomask image only corresponds to one field, typically 26 mm × 32 mm in size (as of 2005). On a 300-mm-diameter silicon wafer, there is room for hundreds of fields. Therefore, once one field is exposed by the photomask, the wafer stage steps to the next field, where the exposure is repeated (hence the term "stepper"). Clearly, alignment of the photomask to the pre-existing layers on the wafer must be exact, making overlay the second critical figure of merit for lithography.

For some recent tool configurations, only part of the photomask is illuminated at one time, and the mask and wafer are scanned to expose the entire field. These systems are called "scanners" or "step-and-scan" systems.

When light illuminates the photomask, the transmitted light diffracts, with light from regions with higher spatial frequencies diffracting at higher angles. The relationship governing resolution in imaging can be understood by recognizing that the mathematical description of the diffraction of light in an imaging system is equivalent to a 2-D Fourier transform operation[7], with the wavefront at the lens pupil plane corresponding to the transform. This transform pattern then diffracts and is recollected by the remaining lens elements to form the image. The formation of the image is mathematically equivalent to taking the inverse transform of the pupil plane.[7,8]

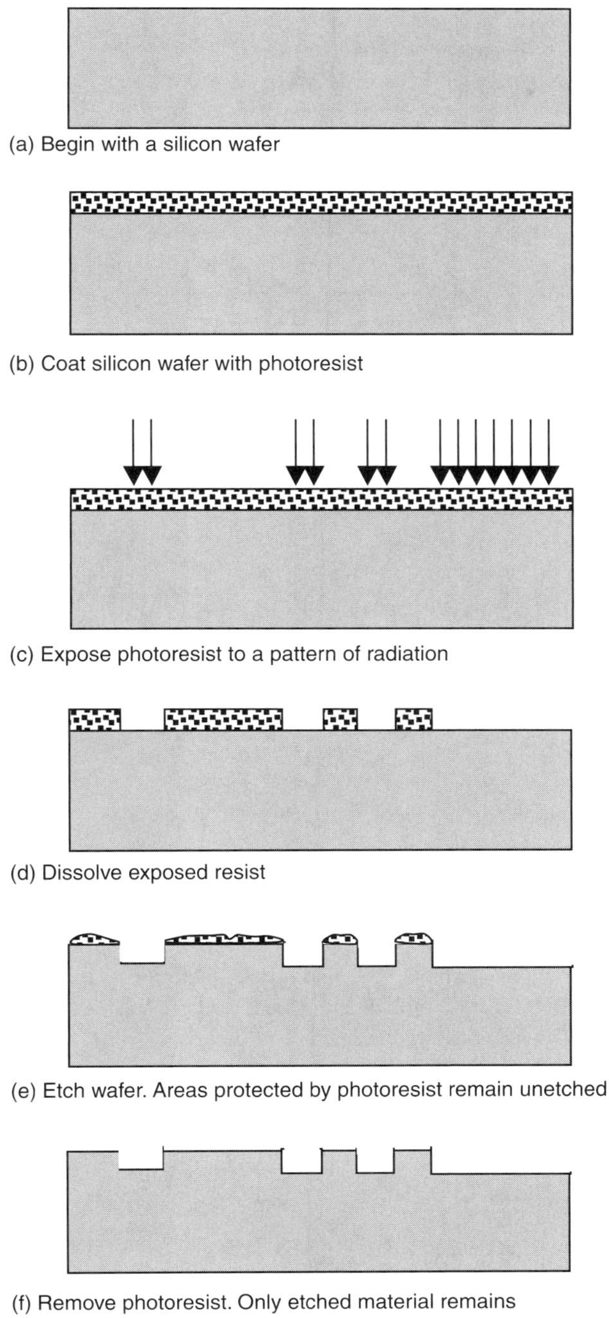

FIGURE 18.1 The photolithography process flow: exposure,development, and etching.

Because the lens pupil has a finite aperture, the light diffracted at the highest angles (corresponding to the finest gratings) is truncated at the pupil. The imaging system therefore effectively acts as a low-pass filter, removing the highest spatial frequencies from the 2-D layout pattern. The measure of this aperture, defined as the sine of the largest angle as light falls on the wafer, is called the numerical aperture (NA) of the system. This is one of the key factors limiting resolution.[9]

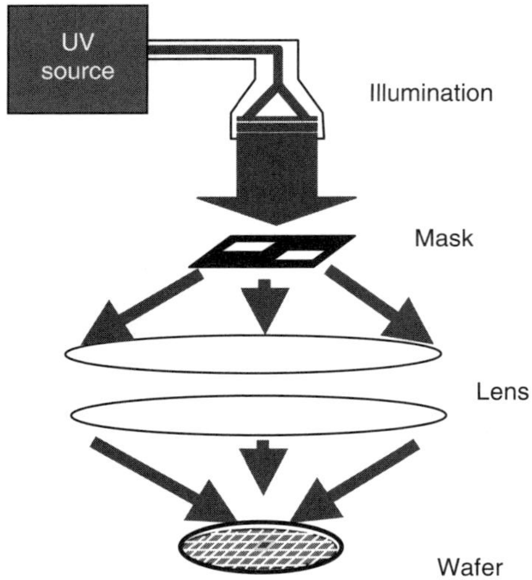

FIGURE 18.2 The elements of a typical UV stepper.

The other factor limiting resolution is the wavelength of the light source itself. With smaller wavelengths, the diffraction angles are smaller, and more light can be collected and passed through the pupil. These factors come together in what is often called the "Rayleigh equation" relating the smallest linewidth (or technically, the smallest half pitch dimension for equal lines and spaces) with the wavelength λ, refractive index n of the imaging medium, and NA[10–12]:

$$L_w = k_1 \frac{\lambda}{n\mathrm{NA}}$$

The factor k_1 represents the degree of difficulty of a process. The traditional resolution limit of an optical system would give $k_1 = 0.61$; anything larger can be easily resolved. The absolute limit for single-exposure imaging is $k_1 = 0.25$; below this the modulation transfer function of any lens system is effectively zero. It is this mid-range of $0.25 < k_1 < 0.61$ that represents the challenge for optical lithography.

Therefore there are two obvious paths to improving the capabilities of optical lithography. Historically, mercury lamps with emission lines at 436 and 365 nm were readily available, and so improvements in resolution came from better lens design with larger values of NA, increasing from NA = 0.3 in 1982 to NA = 0.85 in 2004. However, NA is the sine of an angle, and has an absolute maximum of 1.0. Recent tools have NA values approaching this, indicating that this is not a path open for development in the future.

"Effective" values of NA > 1, sometimes called "hyper-NA" lithography, can be achieved if n is increased.[13,14] For exposure in air, $n \approx 1$, but for immersion in water, $n \approx 1.4$. For a system designed with $\lambda = 193$ nm and NA = 0.95, this is equivalent to having an NA = 1.33 (or $\lambda = 134$ nm) in an $n = 1$ system. Recent models of lithographic tools are now being developed in which the lens–wafer interface is filled with flowing de-gassed water, effectively exploiting this effect.[15]

The second obvious path to improvement is to reduce the wavelength λ. This is shown in Table 18.1.[3–5,13–21] Lithographic exposure has followed a steady progression from the use of the mercury lamp atomic lines at 436 to 365 nm (near-UV), and more recently excimer lasers have been used at 248 and 193 nm in the deep UV (DUV).[16] Fluorine excimers, emitting at 157 nm in the vacuum UV (VUV), have also been explored, but technical problems have placed that development on hold.[17]

However, as shown in Figure 18.3, even with this decrease in wavelength, the ever-shrinking linewidths required to stay on the cost-per-function curve of Moore's Law[22,23] dictate that the feature sizes are now smaller than the wavelength of the light.

Other lithographic approaches exist. These are also shown in Table 18.1. Extreme ultraviolet (EUV) sources at 13.4 nm are being developed as a possible alternative,[18] but questions remain about whether

TABLE 18.1 Lithography Sources

Source	Wavelength (nm)	Tool NA	Comments	Reference
Hg g-line	436	0.3		[3–5]
Hg i-line	365	0.45–0.62		[3–5]
KrF excimer	248	0.63–0.70		[4,5,16]
ArF excimer	193	0.63–0.95	Hyper NA = 1.33	[4,13–16]
F_2 excimer	157	0.95	Development on hold	[17]
Xe or Sn plasma	13.4	0.3	EUV project — under development	[18]
Synchrotron	1.2	Contact/proximity printing	X-ray lithography project discontinued	[19]
Electron Projection System (EPL)	0.0039	4× reduction electron optics	100 keV electrons depth of focus 4 μm	[20]
E-beam direct write	0.0055	Beam technology	50 keV electrons	[21]

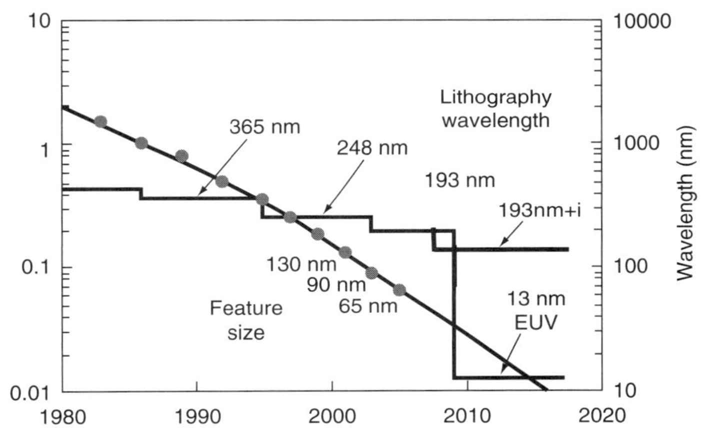

FIGURE 18.3 Ever-shrinking linewidths in accordance with Moore's Law became subwavelength in 1998. 193 nm + i represents 193 nm water immersion lithography. Figure reproduced courtesy of Intel Corporation.

the technology can achieve its goals with a reasonable cost of ownership. X-ray lithography using synchrotron radiation to generate photons with $\lambda = 1.2$ nm have been built and demonstrated,[19] but again the cost of ownership for these systems proved to be impractical.

High-energy electron beams ($\lambda = 0.0055$ nm for 50 keV electrons) have been successfully deployed in beam systems to write directly extremely small features, but these address the image serially, and are consequently extremely slow when used to write patterns over large areas.[20] These are therefore only commonly used to write the master photomasks at high resolution. Projection electron beam systems have been introduced,[21] but these have additional problems compensating for the interactions of the charged particles.

The push to stay on Moore's Law, given the limitations of NA and wavelength, have dictated the development of low-k_1 lithography solutions to perform lithography in this subwavelength regime. The general category these come under is RET.[24] We now turn our attention to these.

18.3 RET for Smaller k_1

An electromagnetic wave has four independent variables that define it: an amplitude, a phase, the direction of propagation, and the direction of the electric field (polarization). The first three variables have been exploited to provide resolution enhancement, while polarization is currently an active topic for exploration.

18.3.1 Amplitude

The photomasks typically used in lithography are coated with an opaque mixture of chrome and chrome oxide compounds, and are patterned by removing the opaque layer. This makes the masks effectively binary — transmitting ($T = 1.0$) or opaque ($T = 0$). It would therefore seem there is little to control for amplitude.

However, although the high resolution of the electron beam writing systems used to make photomasks can produce sharp corners and well-resolved structures, the resolution limits of the stepper lens make the lens act effectively as a low-pass filter for the various spatial frequencies in the 2-D layout. This manifests itself in three distinct effects:

1. A bias between isolated and dense structures (iso-dense bias)
2. A pullback of line-ends from their desired position (line-end pullback)
3. Corner rounding

These come under the general heading of "optical proximity effects,"[25,26] and are illustrated in Figure 18.4.[27]

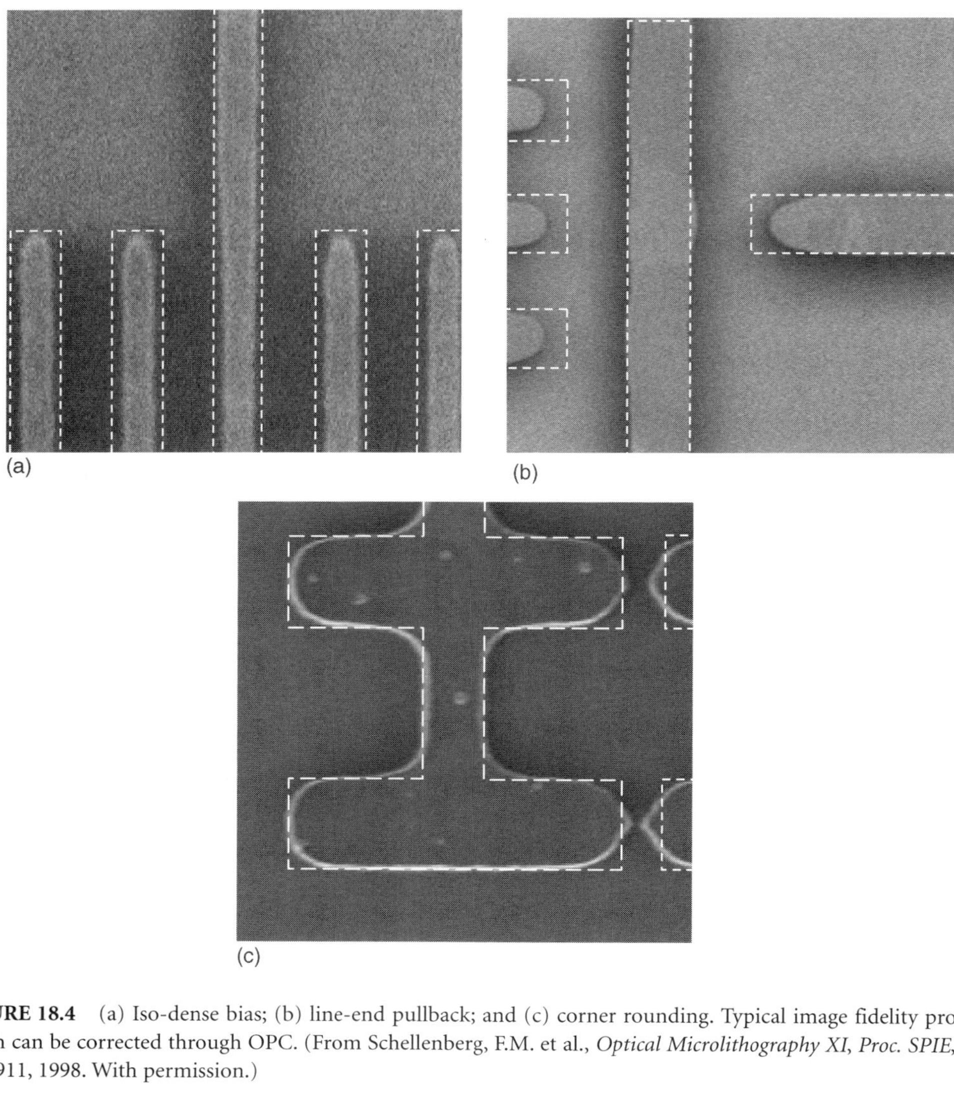

FIGURE 18.4 (a) Iso-dense bias; (b) line-end pullback; and (c) corner rounding. Typical image fidelity problems which can be corrected through OPC. (From Schellenberg, F.M. et al., *Optical Microlithography XI*, *Proc. SPIE*, 3334, 892–911, 1998. With permission.)

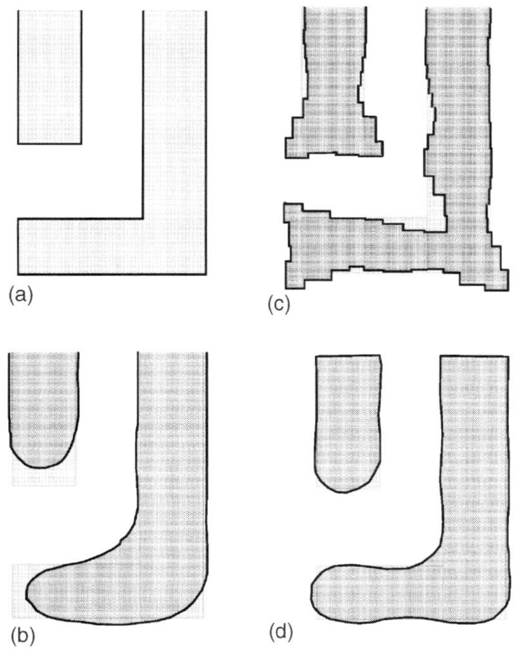

FIGURE 18.5 (a) Original drawn layout; (b) corresponding image of the layout of (a) on a wafer showing rounding and line-end pullback; (c) layout modified with OPC (original layout is shown faintly underneath); and (d) corresponding image of the layout of (c) showing that the image of the modified layout is much closer to the designer's intent (i.e., the original layout).

To compensate for these effects, the layout can be adjusted so that the image matches the desired pattern better. The actions, described in more detail below, generally serve to add higher spatial frequency content to the layout to adjust for the spatial frequency components that are attenuated by the lens system. This is illustrated in Figure 18.5.

These corrections have traditionally been given the name "optical proximity correction" (OPC). As other effects besides these proximity effects have been introduced to the solutions, the meaning of the acronym has broadened to "optical and process correction."

18.3.2 Phase

Interference phenomena produce fringes of dark and light that can be exploited to enhance the contrast of an image. A single-phase transition from 0 to 180° becomes an interference fringe that, in imaging, becomes a very thin dark line, as illustrated in Figure 18.6. With special exposure conditions, extremely thin individual lines can be imaged.[11]

The use of phase requires that light passing through different regions has different path lengths at various refractive indices. This is achieved by etching the photomask to different depths in different regions. The etch depth is given by

$$d_{\mathrm{etch}} = \frac{1}{2}\frac{\lambda}{(n-1)}$$

A mask with such phase-shifting structures is typically called a phase-shifting mask (PSM). For quartz at $\lambda = 193$ nm and $n = 1.56$, the etch depth becomes 172 nm.

Various names exist for certain types of PSMs. These are illustrated in Table 18.2.[28–38] The most straightforward version, in which the phase of alternating apertures are phase-shifted, is called

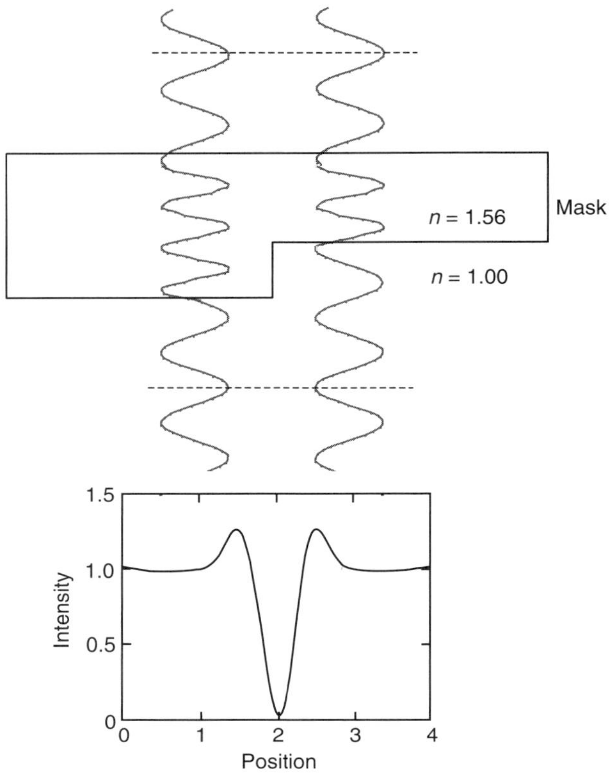

FIGURE 18.6 Cross-section of a phase-shifting mask, and an intensity plot of the interference fringe formed in the image of the phase edge.

"alternating PSM" (or altPSM).[28–31] Features with phase-shifted subresolution assist features (SRAF), sometimes nicknamed "outriggers," have been explored,[31] as have features with edges augmented with phase-shifters, also called "rim-shifters."[32] Use of a single-phase edge to form a pattern is a phase-edge PSM, also sometimes called "chromeless PSM."[33,34] Another "chromeless" technique merges the dark fringes of two phase edges to form a larger dark line.[35,36] This is commonly used in combination with off-axis illumination (OAI) in a technique called "phase-shifted chromeless and off-axis" (PCO) or "chromeless phase lithography" (CPL).[37,38]

18.3.3 Amplitude and Phase

The previous techniques produce clear regions somewhere on the mask that shift the phase of the light 180°. A variation of phase-shifting also shown in Table 18.2, called "attenuated PSM," is constructed with a choice of material composition and thickness such that the transmission is weak (typically 6 to 9%, well below the exposure threshold of the photo-resist) and 180° out of phase with the completely transparent regions. This has the effect of increasing contrast at the edges, which arises from the zero crossing for the electric field at the transition from dark to light.[39] This is illustrated in Figure 18.7.

Although the overall layout of an altPSM mask resembles a binary mask, the additional contrast given by having the phase shift can significantly improve the image quality. OPC can also be applied, assuming that the effects of the phase are also anticipated in the correction algorithms. These photomasks are fabricated from a mixture of material, usually a combination of molybdenum (Mo) and silicon (Si), deposited so that the film has the desired transmission and 180° phase shift. For more information on attenuated materials, see Ref. [40].

TABLE 18.2 Types of Phase-Shifting Masks (PSMs)

Mask Type and Layout	Mask Topography (Cross-section)	Comments	Reference
Alternating 	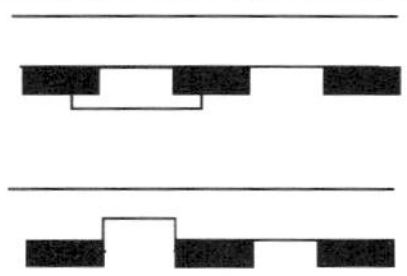	Created by adding a layer Created by etching the substrate	[28 – 30]
Outrigger 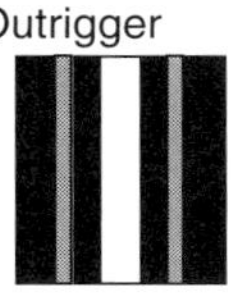		Phase-shifted SRAF	[31]
Rim-shifter 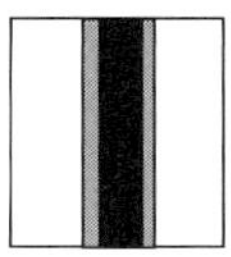	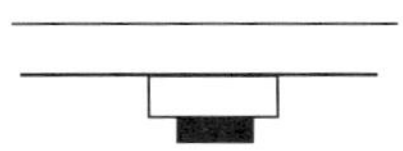	Rarely used	[32]
Chromeless 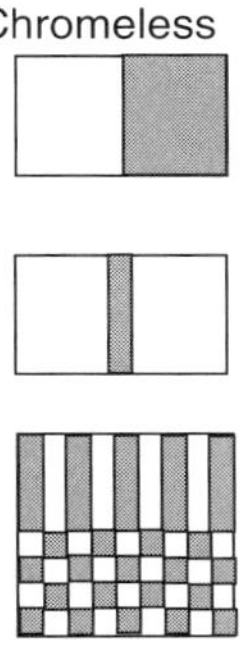	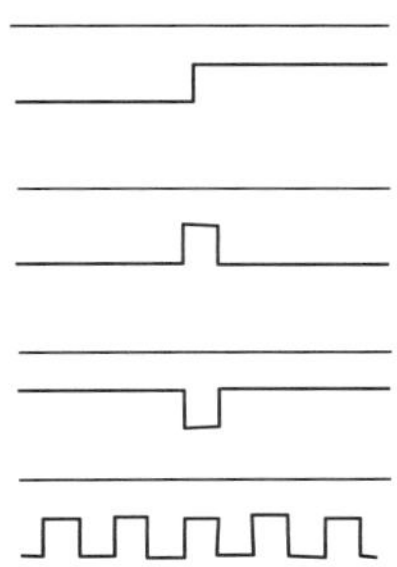	Single-phase edge Trench-type twin edges Mesa-type twin edges "Shifter shutter" opaque pattern	[33,34] [35,36]
Attenuated 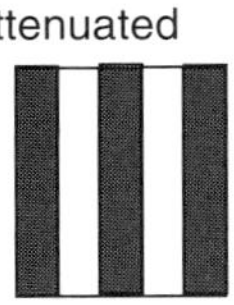	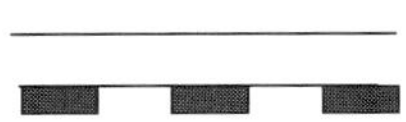	Material composition and thickness selected to transmit a small amount of phase-shifted light. Also called "half-tone" masks	[39,40]

18.3.4 Off-Axis Illumination (OAI)

A traditional lithography system uses uniform illumination falling perpendicular onto the photomask. However, if light falling at an angle is used, the diffraction from certain high spatial frequencies can be enhanced, effectively increasing the resolution.[41] This is illustrated in Figure 18.8.

Various patterns of OAI have been demonstrated. These are shown in Figure 18.9.[25,42–49] Annular illumination is the most common,[42] but does not offer the greatest possible benefit. Quadrupole illumination can emphasize certain pitches very well, and can work for layouts with Manhattan geometries, but diagonal lines will fail to print using such a system.[44,45,47,48] With dipole illumination, the greatest emphasis of certain specific pitches is achieved, but only for features aligned with the dipole.[49] Features of the orthogonal variation do not print. To print a Manhattan layout, a double exposure with horizontal and vertical dipoles must be done. For an arbitrary layout, more exposures may be necessary to produce the desired features.

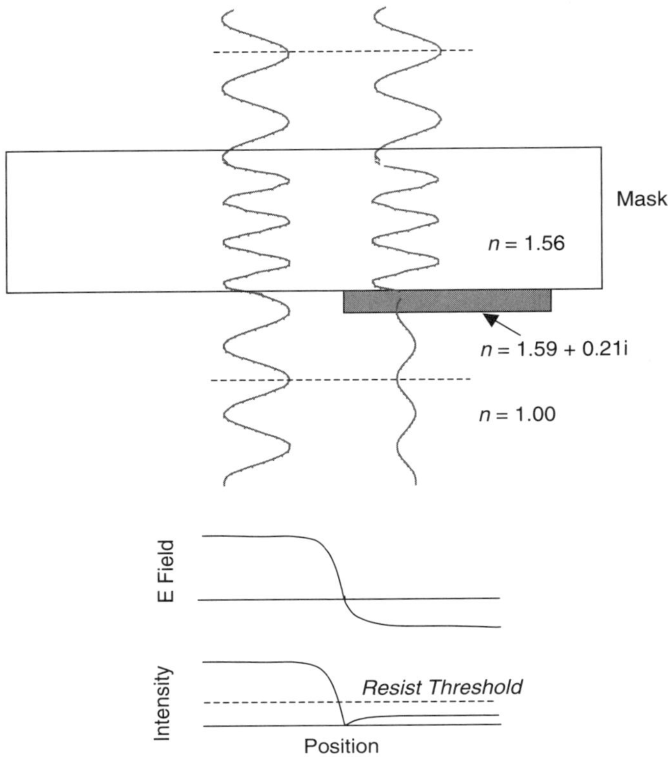

FIGURE 18.7 Cross-section of an attenuated phase-shifting mask, and plots of the corresponding image amplitude (*E* field) and intensity.

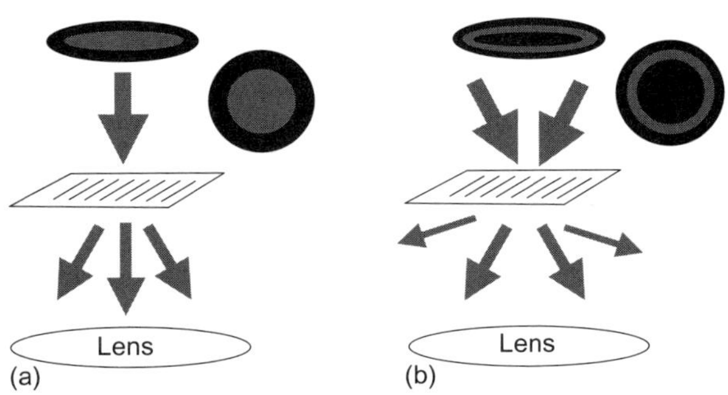

FIGURE 18.8 (a) Convetional circular illumination; and (b) off-axis (annular) illumination.

18.3.5 Polarization

The fourth independent variable of the electromagnetic wave is polarization. The illumination in a typical lithography system is unpolarized, or more accurately, randomly polarized. Until now, there has been relatively little need to explore this further. However, in the subwavelength domain, this may be impossible to ignore.

As feature sizes on the photomask approach the wavelength of light, they begin to act like wire grid polarizing structures,[50] depending on the conductivity of the opaque material on the mask at optical frequencies. They may therefore begin to polarize actively the transmitted light, but only in certain geometry-dependent regions.[51] Surface plasmons may also be excited at the surface, changing the local transmission factors.[52]

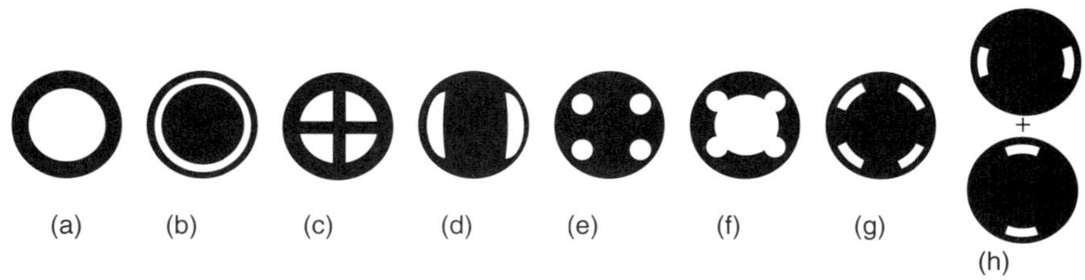

FIGURE 18.9 Off-axis pupil maps: (a) conventional illumination, (b) annular illumination [42], (c) four-fold source [25], (d) "separated" source [43],(e) quadrupole illumination [44,45], (f) "CQUEST" illumination [36], (g) QUASAR illumination [47,48], and (h) horizontal and vertical dipole illumination [49].

Since only light of the same polarization can form interference patterns, the polarization properties of the light can have an important impact on the image formed, especially when PSMs are used. The impact on the software flow for a layout compensating for these effects will be minimal. The impact instead will be felt through the accuracy of the models available to the RET software tools, with inaccurate models degrading the correction quality. We will therefore not discuss further these polarization and plasmon models here.

18.4 Software Implementations of RET Solutions

The main RET solutions can all be implemented in software that alters the layouts of integrated circuit. As such, it belongs to a part of an EDA design flow.

The correct insertion point for RET solutions has been a matter of some debate. A simplified flow diagram[53] of the process steps in the design of a typical IC, without explicitly accommodating RET is shown in Figure 18.10. Given this flow, insertion of RET was initially carried out after the layout had been generated, as an augmentation of the "data massaging" step after the layout has been completed and verified. However, insertion after verification runs the risk that the layout changes may introduce some unforeseen effect. An alternative insertion point for RET can be found as part of the layout creation/verification steps. This is illustrated in Figure 18.11.[54] Insertion in earlier parts of the physical verification flow means that, in principle, the results can be verified again after corrections have been made, reducing the risk to the functionality of the overall product. Modification of the design rules, using what are sometimes called "restrictive design rules (RDRs)," has been proposed as a way of accomplishing this insertion.[55,56]

Although it is clear that the insertion of RET solutions before there is even a layout would be counterproductive, there are certain places, for example, a place and route tool, where some awareness of the lithographic limitations might help. After we review the implementations of the various RET solutions, we will return to this question.

18.4.1 Optical Proximity Correction (OPC)

The geometric manipulations needed to implement OPC might appear to be fairly straightforward, and initial attempts to implement OPC followed what is now called *rules-based OPC.*

In rules-based OPC, the proximity effects as discussed in Section 18.3 are characterized, and specific solutions devised for specific geometric configurations. Then, the layout is searched using a DRC tool or a similar geometry-based software engine to find these geometric configurations. Once they are found, the problem areas are then replaced by the specific solution. The first OPC software products introduced were rules-based software systems.[57]

For iso-dense biasing, the "rules" simply represent a set of biases that are added to geometries, with different biases used when the nearest neighbors have certain distances. This can be easy to conceptualize, and can be implemented in a compact look-up table format.

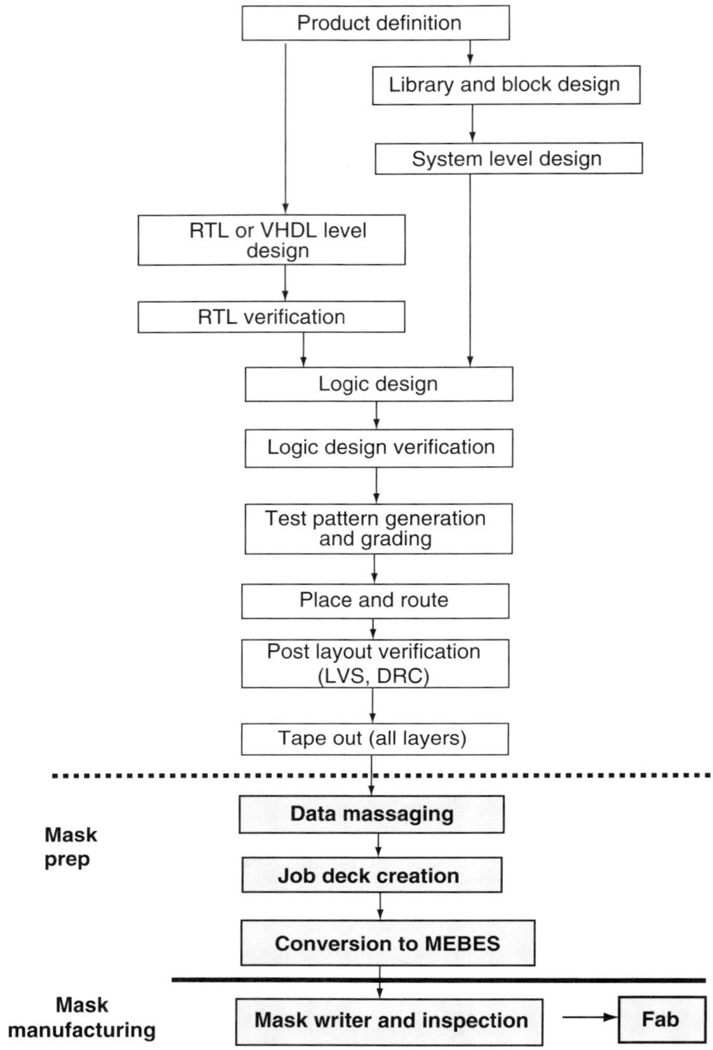

FIGURE 18.10 Data flows as identified in the SEMATECH Litho/Design Workshops. (From Schellenberg, F.M., *Proceedings of the 12th International Conference on VSLI Design*, IEEE Computer Society Press, Los Alamitos, CA, 1999, pp.111–119. With permission.)

For line-end pullback, a solution that extends a line with, for example, a hammerhead structure, can be substituted for every line end. When printed, the extended line-end falls much closer to the desired location, instead of being pulled back. Look-up tables can also be used to implement these solutions, with different sizes and shapes for line ends inserted for different original linewidths and nearest-neighbor conditions.

Corner rounding is generally addressed using a serif (or antiserif for the case of an inside corner). Again, the size and shape of the serif can be predetermined, and a look-up table for various feature sizes and nearest-neighbor conditions can be created.

An example of a rule-based implementation using Calibre SVRF script is shown below. The rules can be encompassed in only a few lines of code.

The results of running this script are shown in Figure 18.12.

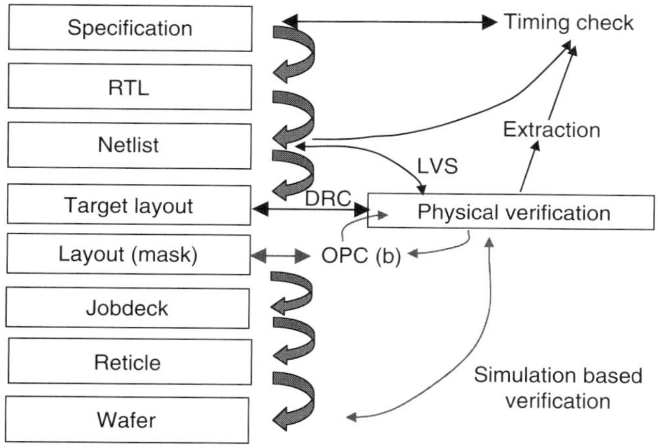

FIGURE 18.11 Simplified flow showing the insertion of OPC integrated with the steps of physical verification. (From Schellenberg, F.M., *Photomask and Next-Generation Lithography Mask Technology IX, Proc. SPIE*, 4754, 54–65, 2002. With permission.)

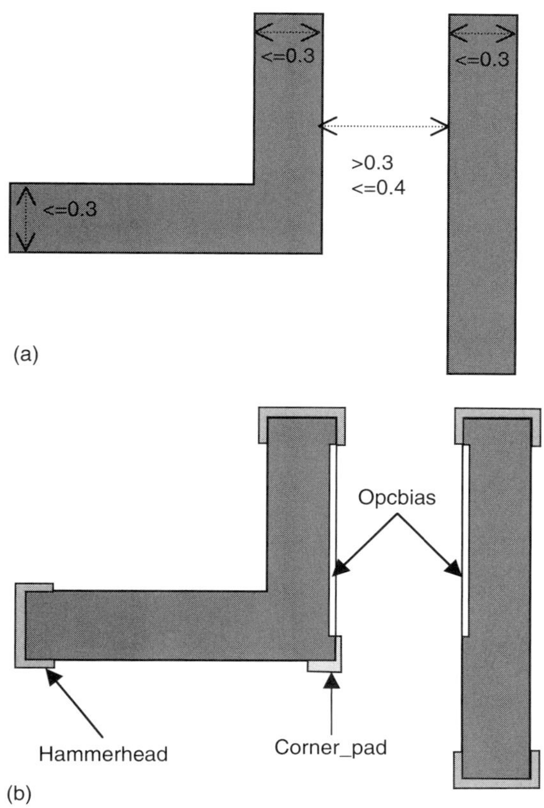

FIGURE 18.12 (a) original layout (target layout); (b) layout modified with a rule-based OPC script using biasing to correct proximity effects, hammerheads for line-end shortening, and a corner serif for corner rounding.

```
HAMMERHEAD = OPCLINEEND MET1
      WIDTH < =0.3 HEIGHT > 0.1 END 0.02 SERIF 0.02 0.01
   // OPCLINEEND to explained elsewhere

MET1_EDGE1 = MET1 OUTSIDE EDGE HAMMERHEAD

CORNER_OUT = INT [MET1_EDGE1] < =0.08 ABUT==90 INTERSECTING ONLY

CORNER_PAD = (OPCBIAS CORNER_OUT MET1
      SPACE > 0.3 <= 0.4 MOVE 0.01
      SPACE > 0.4 MOVE 0.02) NOT MET1

MET1_EDGE2 = MET1_EDGE1 NOT COINCIDENT EDGE CORNER_OUT
OPC { OPCBIAS MET1_EDGE2 MET1
 xxxxxSPACE>0.3 < = 0.4 WIDTH < =0.3 OPPOSITE EXTENDED 0.3 MOVE
0.01
  }
```

18.4.2 Subresolution Assist Features (SRAF)

A special case of rules-based OPC involves the insertion of SRAF, sometimes also known as "scattering bars."[58,59] These are typically additional opaque features, introduced to address the iso-dense bias problem that are themselves too small to be resolved by the imaging system. A simple case of SRAF insertion is shown in Figure 18.13. When SRAF are inserted into the layout, they provide a dense-like environment for the isolated feature. The isolated features therefore print more like the dense features, reducing the problem.

There are certainly special cases for introducing SRAF, which depend on nearest-neighbor proximity. The size and placement of the SRAF are also open to some optimization — SRAF that are too large may print, but those that are too small will not have the desired effect. Placement can also be important, depending on nearest-neighbor distance.

Additional SRAF can also be placed to further enhance the effect. This is illustrated in Figure 18.14. However, as is clear from the illustration, adding more SRAF to the layout can depend highly on nearest-neighbor geometry, and also depends on the size of the isolated feature and the room available in the immediate neighborhood. This again can be made much easier by the preparation of look-up tables based on feature size and nearest-neighbor spacing.

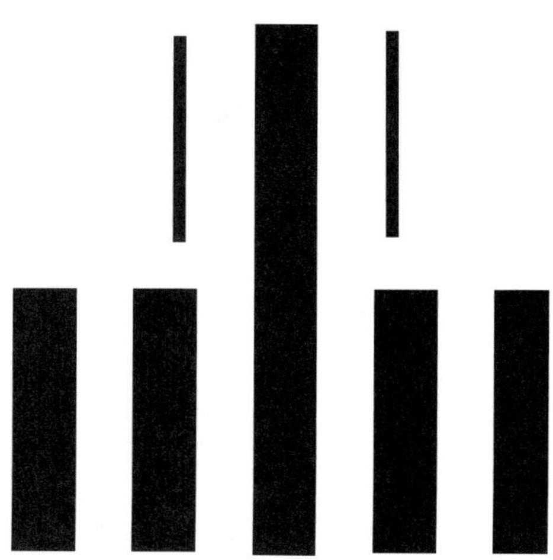

FIGURE 18.13 Layout with simple SRAF to compensate for iso-dense bias.

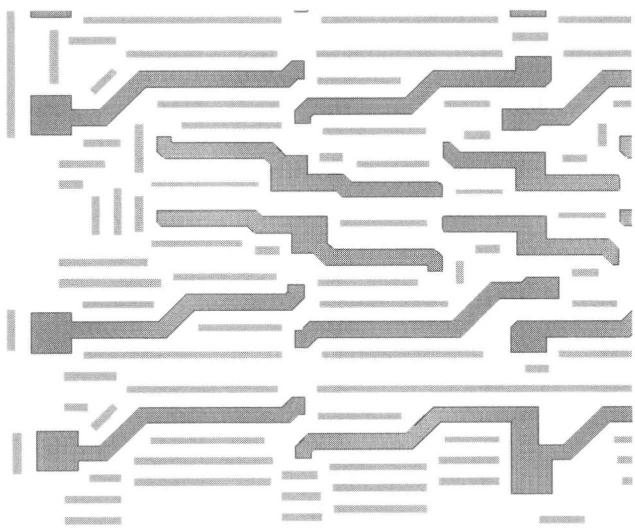

FIGURE 18.14 Layout with printing features (dark) and SRAF (light grey).

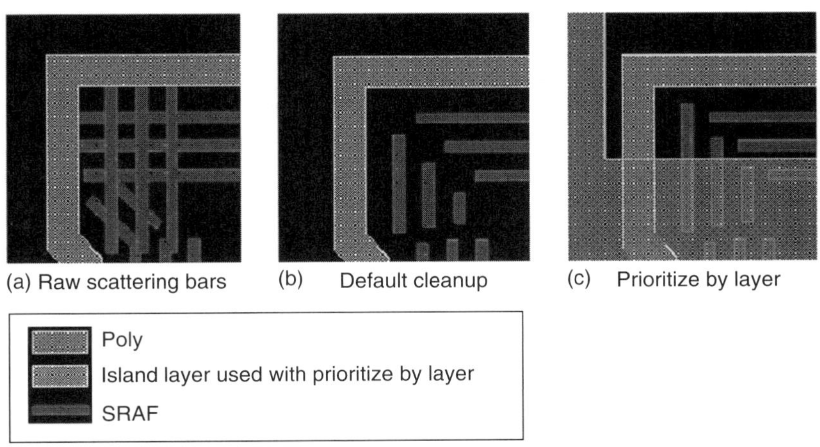

FIGURE 18.15 SRAF prioritization for a poly layer. (a) Raw output after the insertion of SRAF (b) layout after initial SRAF cleanup and (c) final layout, showing an extension (priority) given to SRAFs that overlap the island layer (also shown).

SRAF are typically added along 1-D structures (e.g., lines). Conflicts can arise, however, when SRAF are added for two orthogonal features that intersect. This is illustrated in Figure 18.15. For cases like this, prioritization schemes need to be developed that allow certain SRAF to be retained or extended while others are truncated or eliminated.

Rules-based approaches work well for simple cases. However, for complex layouts, the number of feature sizes and geometric environments can be huge, and it is not possible to encompass the behavior in a managable set of rules. For this, *model-based OPC* has been developed.

The original proposal for model-based OPC was made in the early 1980s, as an academic exercise in image processing.[60–63] In that work, small layouts were digitized into individual pixels, and the lithography system modeled as a low-pass filter followed by a high-contrast square law photosensor (the photo-resist). Iterative processing of the layout with several proposed correction functions was compared. The imaging system was modeled rather simply using a fast Fourier transform (FFT), low-pass filter, and inverse FFT to create the image. Although a simple model, this actually mimics what a real imaging system ideally does in controlling the wavefront with refractive surfaces. The intensity and phase of the image at the ideal pupil plane of the lens system will be mathematically equivalent to the Fourier transform of the object itself. Since the pupil has a finite extent, the low-pass filter has a physical meaning as characterized by the NA of the imaging system, which forms the image by taking the inverse transform of the pupil plane.

After the image is computed, the transfer of the gray-scale image into resist is modeled as a simple constant-threshold function to define the photo-resist boundaries. For image pixels that exceeded the desired tolerance, several correction functions were introduced, and the computation run again. As many as 500 iterations were necessary to reach convergence in such a system.

The results, as illustrated in Figure 18.16, are quite different than the results a rule-based approach would generate.[61] Here, additional "ripples" are generated in the edges of the lines, and irregular SRAF appear naturally out of the algorithm, and not at predetermined sizes and distances as a rule-based implementation would suggest. Although producing counterintuitive mask layouts, the corresponding wafer images are far closer to the desired patterns.

A problem with such a pixel-based approach was that large amounts of computation are used on regions where computation is not necessary. For example, in a large dark region, an FFT-based approach must still compute the transform and inverse transform for all the dark pixels, including those in the center that will certainly print as dark in any circumstance.

To streamline the process, contemporary model-based OPC differs from the pixel-based approach in two ways.

First, most contemporary model-based OPC software systems are edge-based. Here, the polygons of the layout are divided into edge fragments. Traditional GDSII layouts already contain these edge-based definitions in that the polygons are defined by vertex points defining polygon edges. For model-based OPC, additional edge vertices may be introduced to allow the desired fine motion of edge fragments.[64] This is illustrated in Figure 18.17. Once these edge fragments are defined, only certain points on or near the edges that are expected to require adjustment are selected in advance for simulation. This approach reduces the total amount of computation, and eliminates computation of image values in regions where there is no ambiguity in the outcome.

The second refinement is that the image simulation itself is based on the Hopkins method, not on FFTs.[65] The Hopkins simulation integrates the image over all the source points, then adds the various contributions and produces an aerial image of the intensity of the electric field at the wafer. This can be significantly more efficient computationally, especially if the series expansion of the computation is reasonably truncated after a certain number of kernels are used. However, most lithographic patterns are further distorted by the exposure in the photo-resist and by the subsequent processing. This means that the most accurate models will not be a constant-threshold model, as was used in the original Saleh work, but will require a more complex evaluation of the threshold conditions to be used.

Models, such as those of Rieger and Stirniman[66] or the variable-threshold resist models as proposed by Cobb,[67] typically require a test pattern to be fabricated on a photomask and printed using the wafer process in question. A set of measurements from that test pattern characterizes the signature for the effects of the process, including the proximity effects mentioned above (iso-dense bias, line-end pullback, and corner rounding). An example of a pitch curve from such a test pattern is shown in Figure 18.18. From a comparison of the actual line placement with the predicted line placement, variable-threshold models for edge placement can be generated. These empirical adjustments to the aerial image model allow the rapid computation of the wafer as processed, without additionally using a full physics-based process simulator for the various steps (e.g., post-exposure baking for the photo-resist, diffusion phenomena, and plasma-etch biasing).

Using such a model-based approach, convergence for the final edge placement is generally achieved in a few iterations.

18.4.3 Phase-Shifting Mask (PSM)

The geometric operations needed to implement phase-shifting in a photomask layout may seem to be straightforward, especially if alternating aperture/dark-field masks are used (see Table 18.2). However,

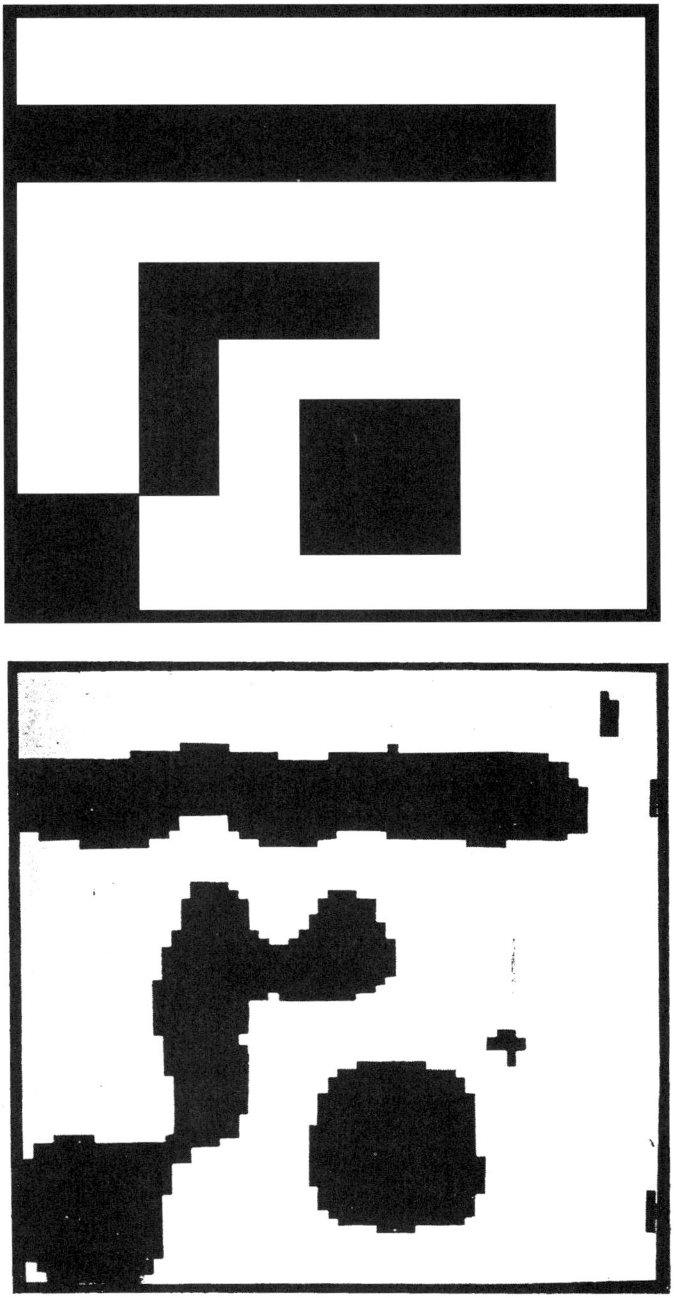

(a)

(b)

FIGURE 18.16 (a) Original layout (target) and (b) results of Saleh's model-based OPC calculation to achieve this target. (From Nashold, K.M. and Saleh, B.E.A., *J. Opt. Soc. Am. A*, 2, 635–643, 1985.)

certain topological structures, such as "T"-shaped junctions, may lead to phase conflicts that cannot be easily solved.[68,69] An example of conflict structures is shown in Figure 18.19.

18.4.3.1 Dark-Field Phase Shifting

When fine pitches are required in a dark-field mask, for example, for the lower metal layers of an IC, a novel approach to solve the problem has been presented by Ooi et al.[70,71] Here, the phases are

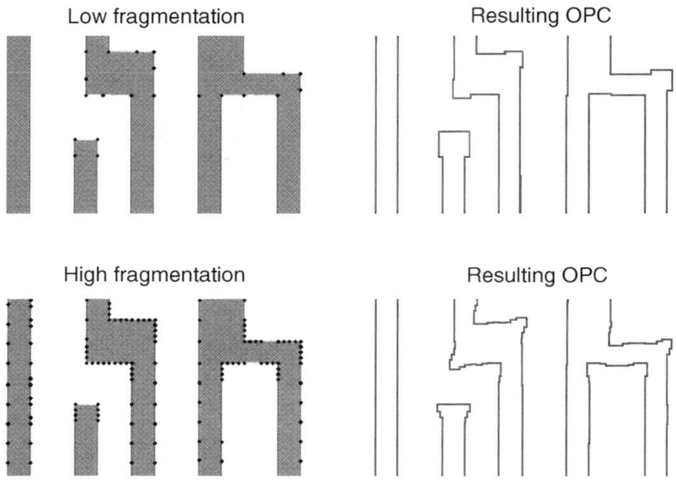

FIGURE 18.17 Effect of fragmentation settings on model-based OPC results.

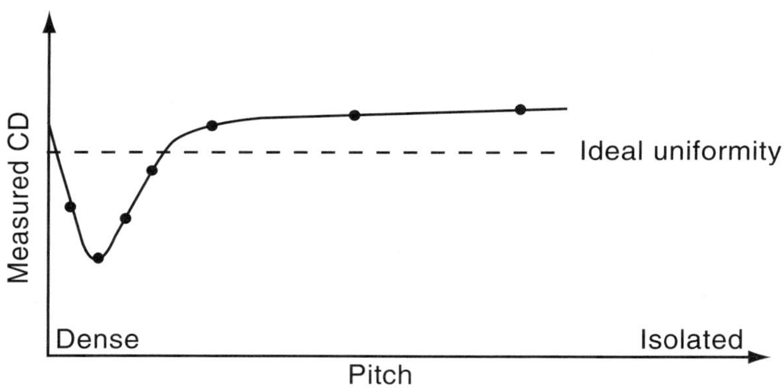

FIGURE 18.18 Pitch curve characterizing 1-D iso-dense bias for model-based OPC calibration.(After Cobb, N.B., Fast Optical and Process Proximity Correction Algorithms for Integrated Circuit Manufacturing, Ph.D. Dissertation, University of California at Berkeley,1998.)

assigned prior to the finalization of the layout. Then, using compactor software, the final layout is only created at the fine pitch using phase-shifting where there are no phase conflicts. In other situations, the feature sizes and spacings are kept larger, so that phase-shifting is not needed and no conflict occurs.

Fabrication of the mask now becomes relatively straightforward, in that two writing steps are required, one to clear the chrome and the other to etch the phase-shifted structures.

18.4.3.2 Bright-Field Phase-Shifting

When dark fringes are required for thin lines, such as for very small transistor gates, a bright-field approach to phase-shifting is called for. This assigns 0° and 180° on the opposite sides of the features desired to be thin, and has been proven extremely effective at shrinking transistor gates.[72] This is illustrated in Figure 18.20.

However, since both sides of the gate are not infinitely large, the phase polygons must somehow be closed. This represents the main problem of bright-field phase-shifting.

One solution to the problem is to create "partial shifters" of either 90° or 60°–120° combinations.[73] These partial phase steps scatter some light, but do not create the strong phase interference that a 0° to

FIGURE 18.19 Layout structures that represent topologies leading to "phase conflicts."

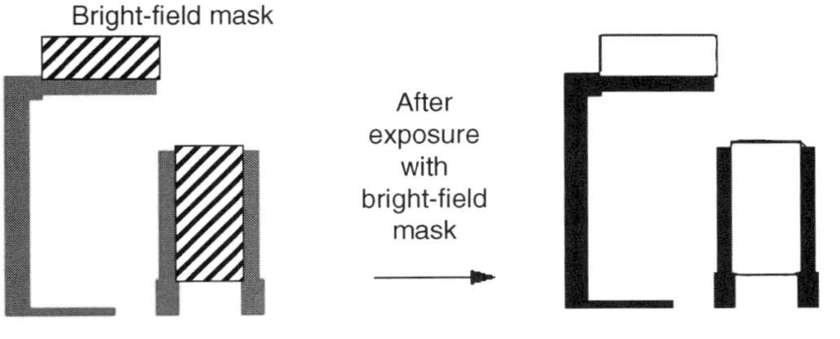

FIGURE 18.20 Bright-field shifiting. Left: original layout pattern, with clear 0° areas (white), chrome opaque areas (dark), and clear regions with a 180° phase shift. ▨ Right: the image on the wafer after this mask is used for exposure. Not only do the opaque features print, but the phase edges print as well.

180° transition does, and the dim fringe they create ideally never goes below the exposure threshold of the resist. This is illustrated in Figure 18.21.

Using simple routing algorithms or, in some cases, simple design rules, allows the boundaries of these partial shifters to be defined.[74] Although easily able to make fine lines and even dense groups of fine lines when alternating apertures are used, the drawback to this relatively straightforward technique is that the additional partial shifters use up space that could otherwise be dedicated to additional circuit elements, decreasing the overall possible device density. In addition, partial shifter masks are expensive, since four distinct writing steps are now required to fabricate the different etch depths of the photomask, each requiring precise alignment. Defects in such masks are often impossible to repair, and only about 1 in 10 masks are typically defect-free.

An additional solution to this problem is to leave the boundaries of the phase-shifters on the PSM, and to use a second exposure to "trim" or remove these unwanted features.[75] This is illustrated in Figure 18.22. The advantage here is that the benefits of the phase-edge fringe can be achieved using masks that are cheaper than the multi-write, defect-prone partial-shifter masks, but the disadvantage is that the multiple exposures

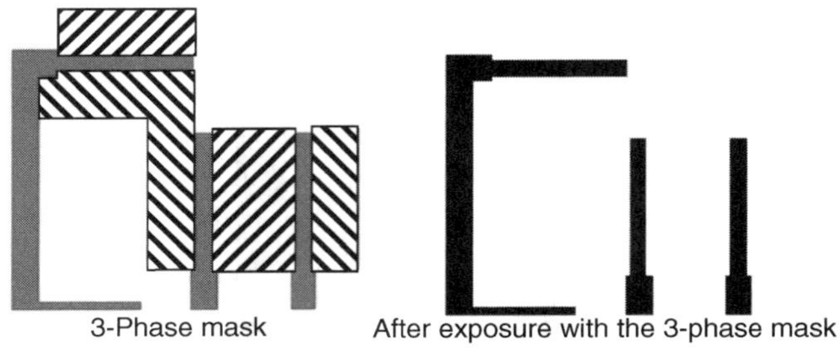

FIGURE 18.21 Partial-shifter approach. Left: original layout pattern, with clear 0° areas (white), chrome opaque areas (dark), and clear regions with a −90° phase shift or a +90° phase shift Right: the image on the wafer after this mask is used for exposure. Regions bounded by both partial shifters have a relative 180° phase shift, and print as dark fringes, while those with only a net 90° phase shift do not.

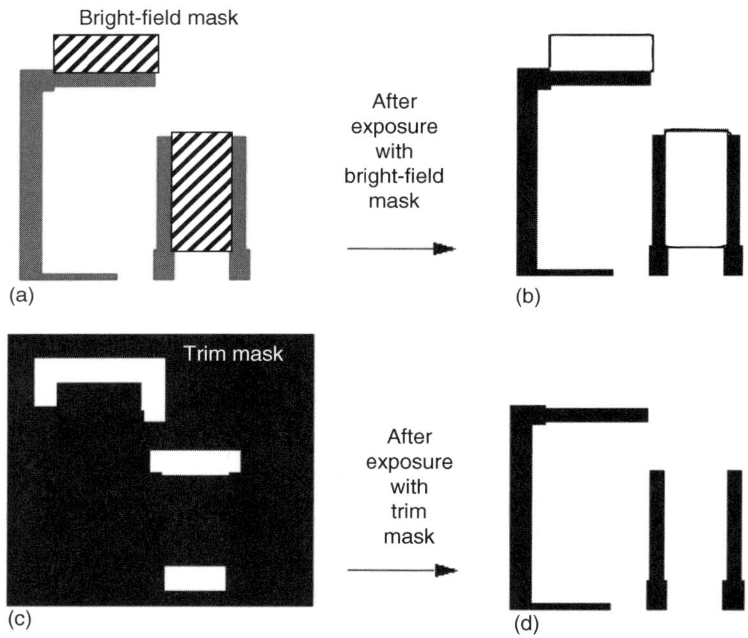

FIGURE 18.22 Trim-mask approach. After initial exposure to the mask in (a), the image shown in (b) stilll has unwanted bridging structures. With the additional exposure using a second trim mask (c), the unwanted features are also exposed and the final double-exposed image (d) is correct.

reduce the wafer stepper throughput. An interesting variation on this approach is one in which the phase-shifters are designed on the dark-field second-exposure mask, leaving the other circuit features and larger "protection" features on the first exposure.[76] This is illustrated in Figure 18.23. This allows the polygons of other layers, e.g., active for the gate layer, to serve as the starting point for the phase-mask polygons, eliminating the need for routing algorithms in the definition of the phase features. Furthermore, by limiting the area that is cleared in the phase mask, this reduces the susceptibility to defects, potentially making masks less expensive. However, the throughput disadvantages of the "trim"-mask approach are not mitigated in this approach. This means that these approaches to phase-shifting will only find application in large-volume products such as microprocessors, in which the higher mask and processing costs can be recovered.

We should also note that the problems OPC has been designed to fix are not eliminated with the adoption of phase-shifting. Although contrast may be improved, isolated lines can still behave differently than

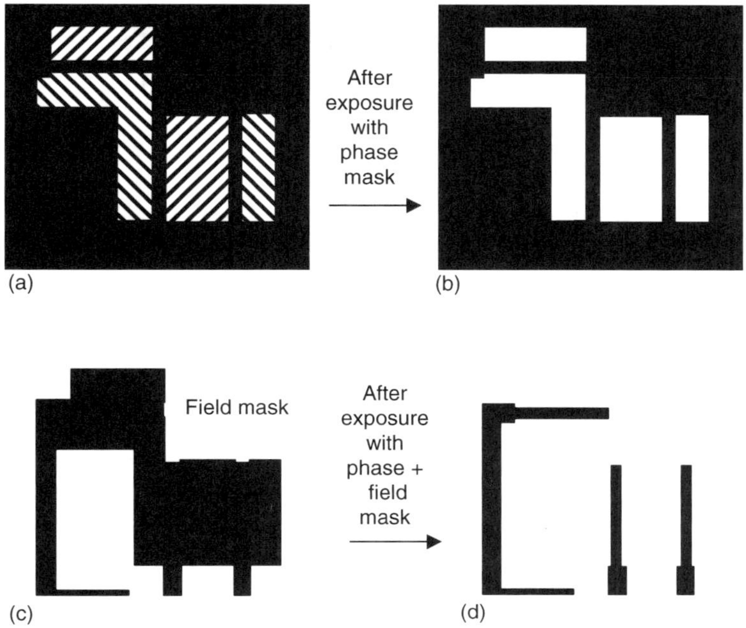

FIGURE 18.23 Dark-field phase-shifting trim-mask approach. After initial exposure to the mask in (a), the image shown in (b) has printed the fine features that benefit from phase-shifting, but everything else remains unexposed. Using a second exposure with a "field mask" shown in (c), the remaining larger dimensions of the layout are defined, leading to the desired image in (d).

dense lines, line ends can still pull back, and corners can still be rounded. For this reason, OPC is still typically required after conversion of a layout to phase-shifting has been completed.[77]

18.4.4 Off-Axis Illumination (OAI)

The third approach to RET is OAI. Although one would think that the choice of illumination angle would not be a factor in the creation of IC layouts, in fact there are significant ramifications to the layout that must be discussed.

As mentioned above, OAI functions by emphasizing the diffraction from certain pitches at the expense of others. This leads to the concept of "forbidden pitches": lines with certain spacings that are far more difficult to print (have smaller focus and exposure latitudes) than lines of other spacings.[47,48] Although related to the angle of the illumination and the NA of the lens, the exact pitches where performance degrades can only be found by simulation using the desired OAI pattern. The results can be surprising: dense pitches may print very well under a particular kind of OAI, while contrast is very poor for certain larger pitches that would traditionally print very well. This is illustrated in Figure 18.24.[78]

To mitigate the problems caused by forbidden pitches, combinations of RET are used. This is where SRAF are most commonly applied. The insertion of SRAF transforms a region with a "forbidden" pitch into one where the pitch between polygons is smaller and no longer "forbidden." This is also illustrated in Figure 18.24.

The orientation of the pitches is also a factor. For symmetric annular illumination, all orientations are treated equally, but for quadrupole illumination (see Figure 18.9), only vertical and horizontal features diffract light well. Diagonal structures will fail to print, allowing only Manhattan layouts to be used with this illumination system (see Figure 18.25).

Likewise, dipole illumination allows exceptionally high contrast for dense features oriented orthogonally to the dipole, but features parallel to the dipole will not print. For 2-D layouts, dipole illumination

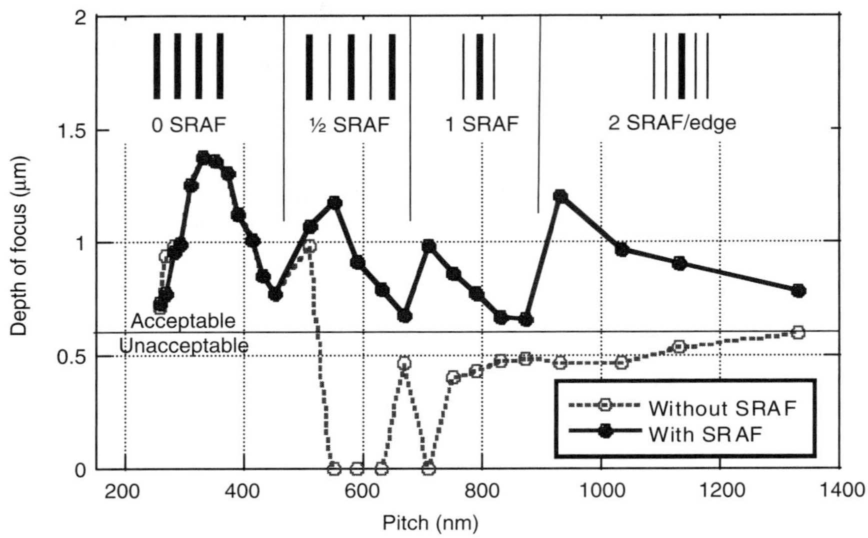

FIGURE 18.24 Off-axis illumination (OAI) and forbidden pitches, mitigated using SRAF. (From Schellenberg, F.M. et al., *Proceedings of the 38th Design Automation Conference*, ACM, New York, 2001, pp 89–92. With permission.)

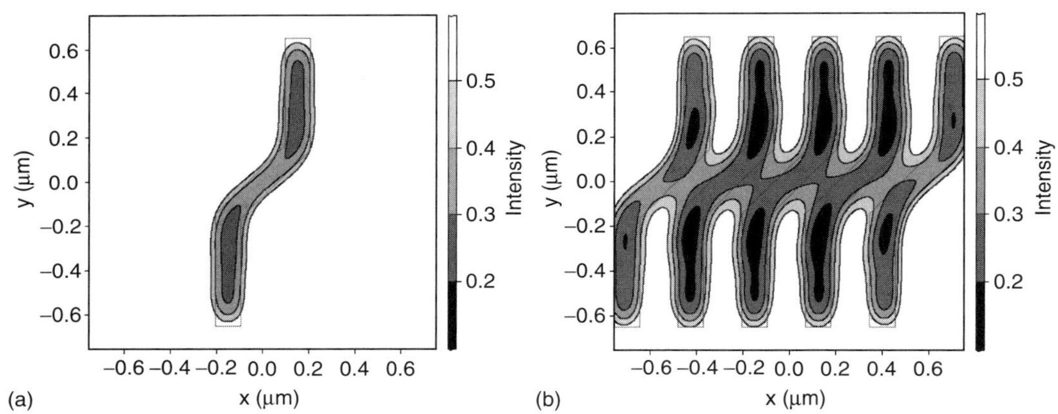

FIGURE 18.25 (a) Simulated image of an isolated line, and (b) dense lines with 45 nm portions under QUASAR illumination. The diagonal portions are significantly brighter, and will not be faithfully reproduced. (From Schellenberg, F.M. and Capodieci, L., *Proceedings of the 2001 ISPD*, ACM, New York, 2001. With permission.)

therefore requires a double exposure. Since not all features are either horizontal or vertical, the parsing of data into these two masks, and the correct treatment of the points of intersection of horizontal and vertical lines can be complicated. A number of algorithms have been explored that allow the automatic conversion of these layouts.

An area of active development to improve simulation accuracy for OAI relies on collecting actual pupil maps of the illumination source and using these in the simulation algorithms. For generations at 90 nm and below, this can make a significant difference in the accuracy of the model. An ideal source map for QUASAR illumination and the experimental results of a measurement of the source map for a particular stepper is shown in Figure 18.26.[79] Use of the actual map in the simulation software engine for the computation of OPC removed a 19nm error in the model that had remained when the idealized source map had been used. For a 90-nm process, 19nm represents a significant error.

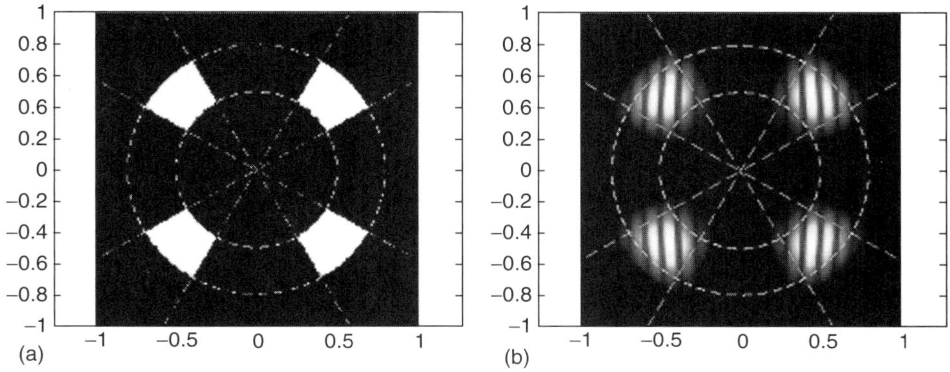

FIGURE 18.26 Pupil map of (a) an ideal QUASAR source, and (b) measured intensity map of an actual stepper illumination pattern. (From Granik, Y. and Cobb, N., *Optical Microlithography XVI, Proc. SPIE*, 5040, 1166–1175, 2003. With permission.)

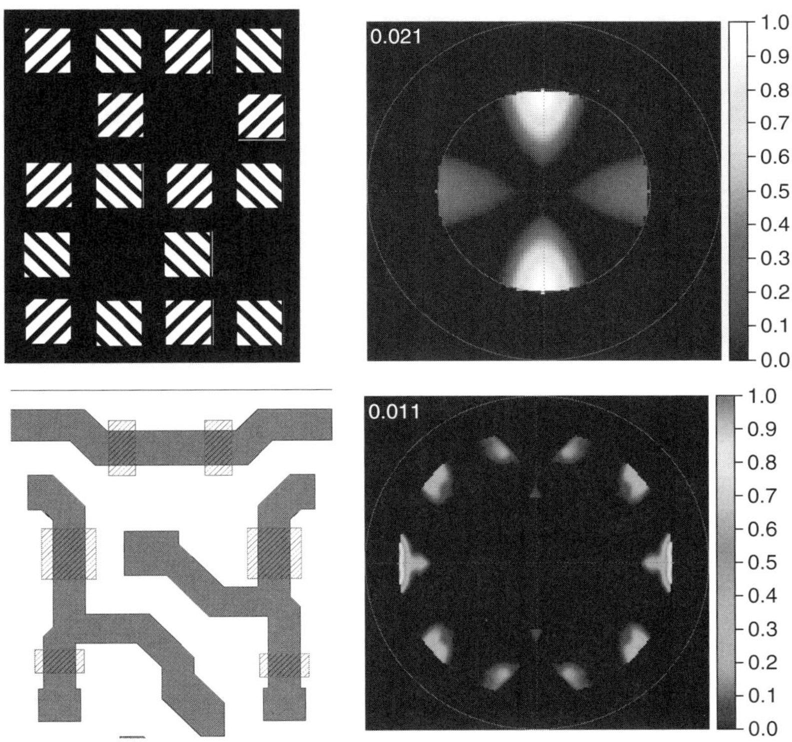

FIGURE 18.27 Mask/source optimization. On the left are representative layout patterns for (upper) a contact layer with alternating phase shifts, and (lower) the gate layer of an SRAM cell without phase-shifting. On the right are the corresponding optimized source illumination patterns that optimize lithographic performance. (From Granik, Y., *J. Microlith. Microfab. Microsyst.*, 3, 509–522, 2004. With permission.)

Another area of improvement is in the customization of the illumination pattern for the particular layout. This can be very effective for ICs that will see large volume production, such as DRAMs.[80,81] Two examples of layouts and a source map that produced the optimal lithographic performance for each are shown in Figure 18.27. It is clear that these are far from the simple patterns for illuminators shown in Figure 18.9. However, using customized diffractive optical elements, these can be produced and inserted into a stepper, further improving the potential lithographic performance.

TABLE 18.3 Optical Properties for the Complex Refractive Index *n* and *k* at Three Different Wavelengths for Metals Silver and Gold, Materials Commonly Found in the Photomask Industry, and an Insulator (Glass)

	564 nm		248 nm		193 nm	
	n	*k*	*n*	*k*	*n*	*k*
Silver (Ag)[a]	0.120	3.45	1.298	1.35	1.028	1.18
Gold (Au)[a]	0.306	2.88	1.484	1.636	1.426	1.156
Chromium (Cr)[a]	3.215	4.40	0.85	2.01	0.84	1.65
CrN[b]	2.466	0.041	1.600	1.007	1.292	0.899
Cr_2O_3[b]	2.185	0.229	1.863	0.708	1.616	0.021
CrO_3[b]	1.874	0.070	1.722	0.392	1.634	0.6504
Silicon (Si)[a]	4.042	0.032	1.730	3.222	0.883	2.78
Glass (SiO_2)[a]	1.459	$0(<10^{-6})$	1.508	$0(<10^{-6})$	1.563	$0(<10^{-6})$

Note: Electrically conductive behavior is typically indicated by small values of *n* and large values of *k*; an insulator by extremely small values of *k*.

Source:

[a]Palik, E.D., Ed., *Handbook of Optical Constants of Solids*, Vol. I, II, and III, Academic Press, San Diego, 1998.

[b]http://www.rit.edu/%7E635dept5/thinfilms/thinfilms.htm.

18.4.5 Polarization

As discussed above, polarization is an emerging variable that may need to be considered more rigorously as IC dimensions continue to shrink. Certain polarization effects causing contrast changes as SRAF become approximations to wire grid polarizers have been observed. Modeling these phenomena requires a full electromagnetic simulation of the fields around the photomask, and knowledge of the actual optical properties *n* and *k* for the mask firm as fabricated.

This can be complicated. Although nominally *n* and *k*, the real and imaginary parts of the complex refractive index of the materials of a photomask, can be easily estimated assuming that the opaque mask material is chrome, masks in fact are not only chrome, but a mixture of chrome, chrome oxides, and other compounds designed to make the mask opaque and antireflective. Table 18.3 shows the values for *n* and *k* at optical and lithography wavelengths.[82,83] The difference can be significant, with materials that behave optically like metals (forming polarizers, etc.) becoming less metallic at shorter wavelengths while the material properties of others become more metallic.

This is important because the various polarization effects (and also certain plasmon effects, which can lead to additional transmission through subwavelength apertures) depend on the metallic properties of the material. Without an accurate way to determine these material properties, there is no way to determine the effect they will have on imaging.

Simulation tools such as TEMPEST have been developed to carry out a finite-difference time domain (FDTD) simulation of the electromagnetic field at the mask, and can predict these effects.[84] However, the amount of time that is needed to compute these for each aperture and polygon for an IC with millions of components would be prohibitive.

Techniques have been developed to form a compromise between accuracy and speed by precomputing certain images with a full FDTD solution, storing these primitive image components, and then adding these precomputed images as required for specific image layouts. This "domain decomposition method" (DDM) has been proven for masks where the topography effects must be precomputed with a 3-D solver, and are currently showing promise when used for polarization and other E-M effects.[85]

18.5 Mask Data Preparation

Once the mask data layout has been created and modified to accommodate various RET algorithms, the final photomask still needs to be written. Although in principle a straightforward process, there are several practical issues that require further manipulation of the layout data. Usually, the data must be flattened to some degree, and the polygons must be reduced to a simple set of structures (typically rectangles and trapezoids) that the machine can use to write the patterns directly. This process of data conversion is called *fracturing*.[86]

18.5.1 Mask Writers: Physics

Mask writers are machines that modulate an exposure beam to expose photo-resist coated onto a mask blank in selected locations. Typically, either electron beams or laser beams are used. In one common architecture, the beam constantly scans in a predetermined pattern (a raster scan) and is turned on and off to write the pattern, as illustrated in Figure 18.28. This has the advantage of making the motion of the stage quite predictable, and allows the task of writing to be divided into several exposure beams that handle different portions of the total exposure field. However, the tool scans (with the beam off) over areas that require no exposure as well, reducing throughput for sparse layouts. As many as 24 beams are used to write simultaneously masks in some tools.

Other tools write with a "variable-shaped beam" (VSB), as illustrated in Figure 18.29. Here, a larger beam is shaped by an aperture into a primitive shape (usually a rectangle or a trapezoid), and the image of the aperture projected in individual "flashes" at appropriate locations. With VSB tools, many small and large features can written using only single flashes, while no flashes are used for large areas that require no exposure. This can be more efficient for many types of layouts, particularly sparse ones such as contact hole layers. Additionally, some "vector scan" tools combine aspects of VSB tools (vectoring only to

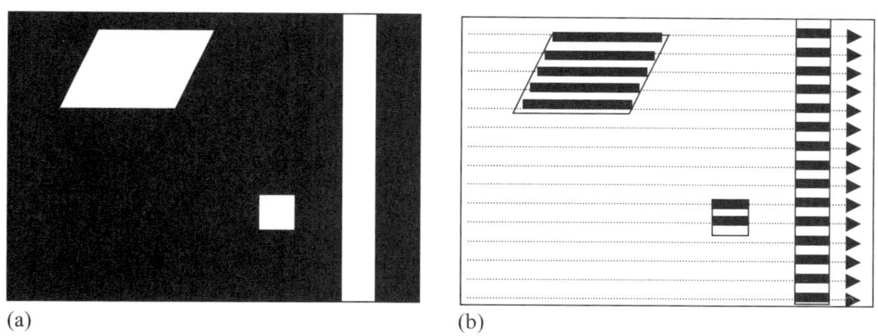

FIGURE 18.28 Exposure of (a) a representative layout geometry using (b) a raster scan. The beam coordinates for exposure are systematically swept over the area to be exposed, and the beam is turned on only as it passes over the areas to be cleared on the mask.

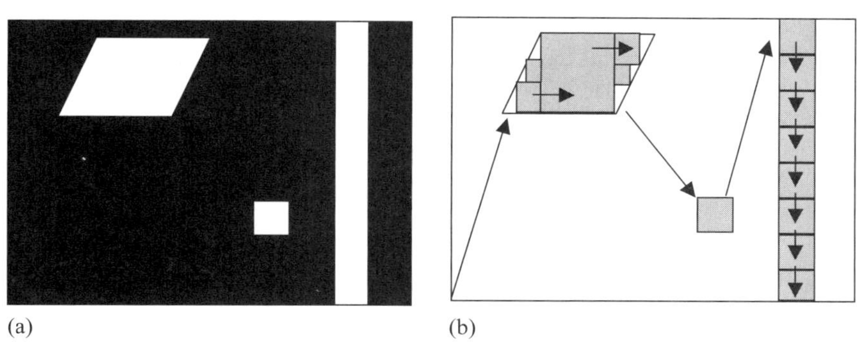

FIGURE 18.29 Exposure of (a) a representative layout geometry using (b) a variable shaped beam (VSB). The beam is 'vectored' to particular locations to be exposed, and the beam size and shape are adjusted to expose the areas to be cleared on the mask.

regions requiring exposure) and raster scanning (using raster scans to expose those regions once there). More information on different exposure strategies can be found in Ref. [6] and [6a].

18.5.2 Mask Data Preparation

Clearly, the exact machine instructions that will be required to write a mask will be different for these two tool architectures. For raster scan machines, the data are converted into primitive shapes (rectangles and trapezoids), and then converted by the tool itself into a pattern of scan lines to be used with each of the exposing beams. A common format used with raster scan machines is MEBES.[87] This format represents all the figures as composites of either rectangles and trapezoids, and partitions the data into individual stripes. The dimensions of the stripes are a multiple of the minimum address unit specified in the particular data file. The mask is written stripe by stripe. The MEBES format has no hierarchy — all polygons occur uniquely within their stripe location. However, MEBES also uses a jobdeck language that controls the placement and writing sequence of MEBES files. In this regard, the jobdeck can be regarded as a second layer of hierarchy for an otherwise flat format.

VSB machines require that the layout be fractured into shots of acceptable size, and the appropriate stage motion instructions be generated to create the pattern. Examples of a layout broken down into shots are shown in Figure 18.30. Because VSB format data is a collection of shot sizes and locations unrelated to a specific writing stripe on the mask, they can have a limited hierarchy structure. How this is utilized is a proprietary function of the different mask-writing tools themselves.

The introduction of RET can significantly change the mask-writing characteristics of an IC layout, especially with VSB. A layout after edges have been moved with the application of OPC, and the corresponding shot layout is shown in Figure 18.30. Clearly, the number of shots has increased dramatically, and the writing time would be expected to correspondingly increase. For the worst cases of aggressive OPC, mask-writing times over 24 hours have been reported, as illustrated in Figure 18.31.

With a judicious choice of OPC parameters, this can be significantly improved. An example of this is shown in Figures 18.32–18.34. Here, OPC has been run on a gate layer. The typical result, as shown in Figure

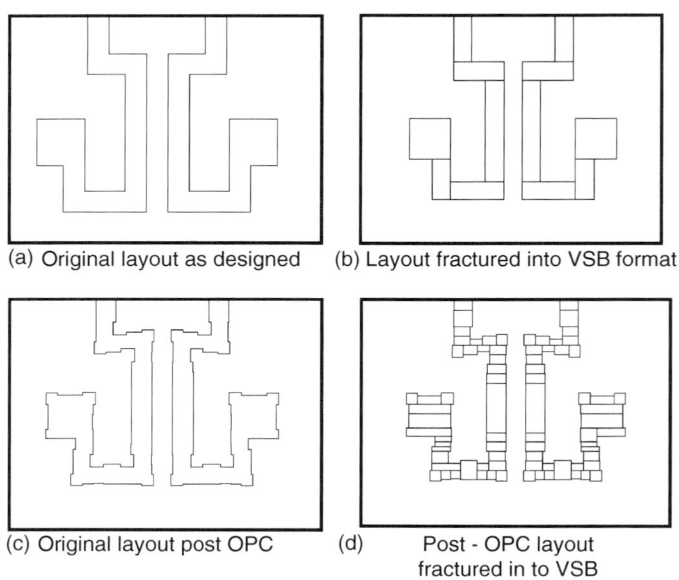

FIGURE 18.30 Visualization of the impact of the application OPC on the fractured result in VSB11 format. The more complex polygons require a more detailed trapezoiding, which results in a much larger number of trapezoids in the fractured output.

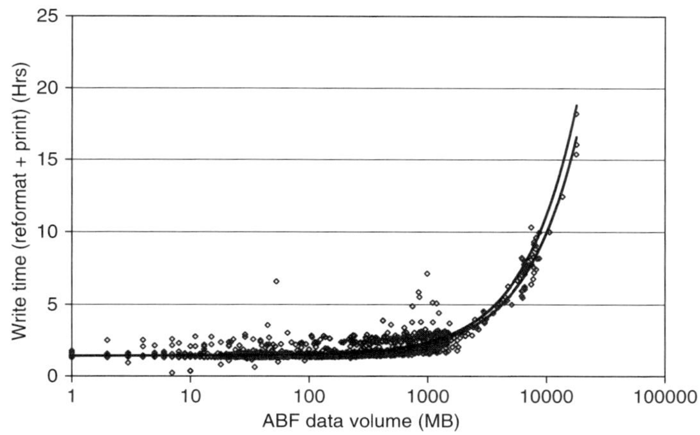

FIGURE 18.31 Measured writing times for an ALTA 3500 raster scan mask writer. Above 1 GB, write times increase in a dramatic and nonlinear manner. (Data courtesy Dupont Photomask. From Schellenberg F.M., *Photomask and Next–Generation Lithography Mask Technology IX, Proc. SPIE*, 4754, 54–65, 2002.)

18.33, is a significant increase in shot count and hence writing time. However, the portions of the gate layout that are not overlapping the underlying active area can be interpreted as poly interconnects, which can be fabricated with a more relaxed tolerance than the more critical gate regions themselves. Performing a logical AND operation of the gate layer and the underlying active area allows the fragmentation for the OPC run to be set dynamically, with finer fragmentation in the gate regions over active and more relaxed fragmentation in the other areas. The resulting VSB layout, with significantly smaller shot count, is shown in Figure 18.33.

In the long run, as long as feature sizes continue to shrink while IC density increases, the volume of data required to describe an IC will continue to grow exponentially. To mitigate this, in June 2003, a new data format called Open Artwork System Interchange Standard (OASIS) was approved by SEMI after 2 years of discussion and testing.[88] The goals achieved by this standard were:

- greater than 10× file size reduction for the same data when compared to GDSII;
- removal of 16- and 32-bit restrictions, allowing integers to extend to 64 bits and beyond;
- efficient handling of flat geometric data;
- improvement in the "richness" (e.g., annotation) that can be communicated along with the formatted files; and
- no significant increase in operation/fracture run times.

OASIS achieves its primary goal by defining the coordinates of polygons by difference functions rather than absolute coordinates for all vertices. In other words, once the initial coordinates for the first vertex have been set, the following coordinates can be expressed in length differences Δx and Δy from the initial coordinate. Since the distance between vertices is usually very small (especially after refragmentation for OPC has occurred), a single byte is often enough to represent the additional vertex. The success of the format is represented by the comparisons for the files of Table 18.4.[89] We expect to see an increasing use of OASIS to represent layout data as Moore's Law continues to progress and RET continues to grow in complexity.

18.6 Summary

The final layout verification step in IC design used to be a simple matter of checking design rules. With the more complex process distortions and manufacturing limitations being placed on designs by lithographic processes, RET has grown to be an important part of the tape-out process. Various insertions points have been proposed, but in general the adoption of RET has been migrating up the design flow, finding its most common application integrated with layout generation and verification.

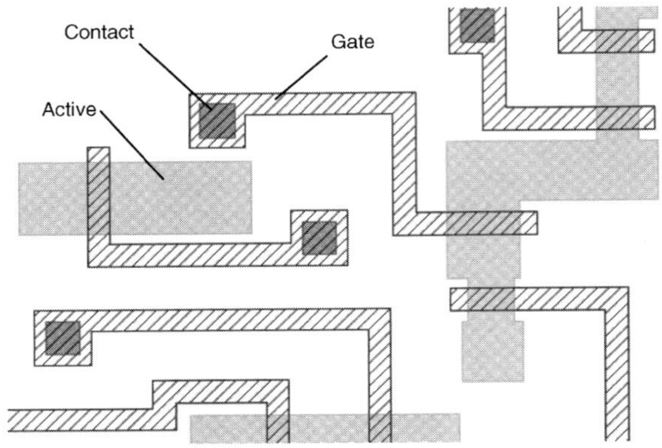

FIGURE 18.32 Representative portion of a microprocessor layout.

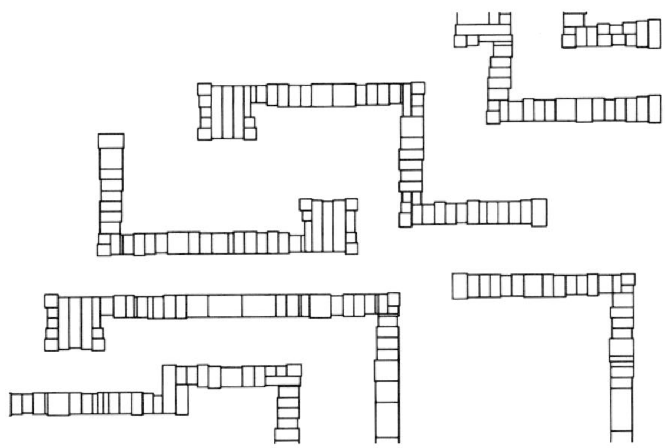

FIGURE 18.33 Fractured layout of the gate layer from FIGURE 18.32 in VSB11 format with standard fragmentation-uniform aggressiveness for the entire gate layer.

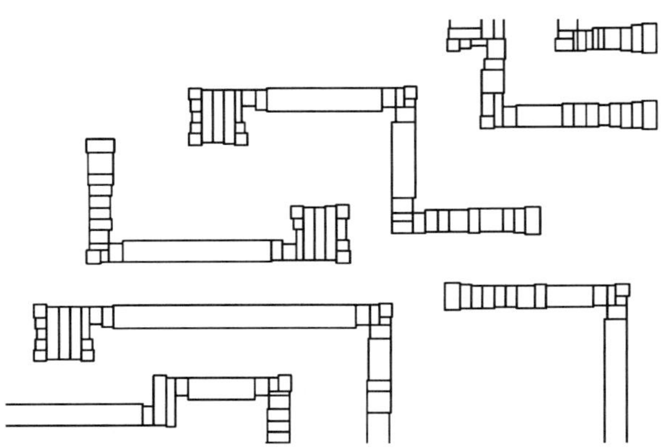

FIGURE 18.34 Fractured layout of the gate layer from FIGURE 18.32 in VSB11 format with interlayer-aware fragmentation. The layout is separated into the critical and noncritical regions. The aggressiveness of the fragmentation is retained for the critical areas while it is reduced for the noncritical areas. This significantly decreases the total data volume with no loss in performance.

TABLE 18.4 File Size Comparison for GDSII and OASIS Representation of Design Data

Test Case	GDSII	OASIS	Ratio (GDSII/OASIS)
1	5316341760	26438663	201.1
2	6433212416	417658458	15.4
3	8750811136	515677720	17.0
4	995145728	19078143	52.2
5	2111688704	122104830	17.3
6	2344345600	109435610	21.4
7	8191951330	749401369	10.9
8	1077782528	38230183	28.2
9	2329663888	136189555	17.1
10	1150431232	92419711	12.4
11	2379108352	91049913	26.1
12	2046351360	111240392	18.4

Note: The test cases represent a variety of design styles and device types.
Source: S. Schulze, P. LaCour and L. Grodd, OASIS-based data preparation flows: progress report on containing data size explosion, *Design and Process Integration for Microelectronic Manufacturing II, Proc SPIE*, 5379, pp. 149–157, 2004.

The various specific techniques involved with the layout changes for any of the RET approaches, whether they be OPC, phase-shifting, or OAI, all require that efficient models be provided that can efficiently describe the expected wafer behavior. Once the prescriptive action has been determined, polygon manipulation itself is fairly straightforward, and can be automated efficiently with new formats such as OASIS. However, with the new physical processes required for 65 and 45 nm IC generations, the ability to calibrate models and predict wafer behavior will determine whether the full promise of RET can be achieved.

References

[1] For a general reference on microfabrication, see M. Madou, *Fundamentals of Microfabrication: The Science of Miniaturization*, 2nd ed., CRC Press, Boca Raton, FL, 2002.

[2] For a general reference on IC manufacturing processes, see Y. Nishi and R. Doering, Eds., *Handbook of Semiconductor Manufacturing Technology*, Marcel Dekker, New York, 2000.

[3] For a general reference on lithography, see H. Levinson, *Principles of Lithography*, 2nd ed., SPIE Press, Bellingham, WA, 2005.

[4] For a general reference on lithography, see L. Thompson, C.G. Willson, and M. Bowden, Eds., *Introduction to Microlithography*, 2nd ed., American Chemical Society, Washington, DC, 1994.

[5] For a general reference on lithography, see J.R. Sheats and B. Smith, Eds., *Microlithography: Science and Technology*, Marcel Dekker, New York, 1998.

[6] For a general reference on photomask technology, see S. Rizvi, Ed., *Handbook of Photomask Manufacturing Technology*, CRC Press, Boca Raton, FL, 2005.

[6a] For a general reference on photomask technology, see B. Eynon, Jr. and B. Wu, *Photomask Fabrication Technology*, McGraw Hill, New York, NY, 2005.

[7] J.W. Goodman, *Introduction to Fourier Optics*, McGraw-Hill, San Francisco, 1968.

[8] M. Bowden, The lithographic process: the physics, chap. 2 of Ref. [4]; and B. Smith, Optics for photolithography, chap. 3 of Ref. [5].

[9] References on lithographic resolution and resolution enhancement are found in F.M. Schellenberg, Ed., *Selected Papers on Resolution Enhancement Techniques in Optical Lithography*, SPIE Press, Bellingham, WA, 2004.

[10] E. Abbe, Beiträge zur Theorie des Mikroskops und der mikroskopischen Wahrnehmung, *Archiv Mikroskopische Anatomie*, 9, 413–468, 1873; also reproduced in Ref. [9].

[11] Lord Rayleigh, On the theory of optical instruments, with special reference to the microscope, *Philos. Mag.*, 42, 167–195, 1896; also reproduced in Ref. [9].

[12] B.J. Lin, Where is the lost resolution? *Optical Microlithography V, Proc. SPIE*, 633, 1986, pp. 44–50; also reproduced in Ref. [9].

[13] B. Smith, A. Bourov, H. Kang, F. Cropanese, Y. Fan, N. Lafferty, and L. Zavyalova, Water immersion optical lithography at 193 nm, *J. Microlith. Microfab. Microsyst.*, 3, 44–51, 2004.

[14] B.J. Lin, Immersion lithography and its impact on semiconductor manufacturing, *J. Microlith. Microfab. Microsyst.*, 3, 377–395, 2004.

[15] S. Owa, H. Nagasaka, Y. Ishii, O. Hirakawa, and T. Yamamoto, Advantage and feasibility of immersion lithography, *J. Microlith. Microfab. Microsyst.*, 3, 97–103, 2004.

[16] K. Jain, *Excimer Laser Lithography*, SPIE Press, Bellingham, WA, 1990.

[17] N. Shiraishi, S. Owa, Y. Ohmura, T. Aoki, Y. Matsumoto, M. Hatasawa, T. Mori, and I. Tanaka, Current status of Nikon's F2 exposure tool development, *Optical Microlithography XIV, Proc. SPIE*, 4346, 2001, pp. 81–88.

[18] C.W. Gwyn, R. Stulen, D. Sweeney, and D. Attwood, Extreme ultraviolet lithography, *J. Vac. Sci. Technol B*, 16, 3142–3149, 1998.

[19] F. Cerrina, X-Ray lithography, in *Handbook of Microlithography, Micromachining and Microfabrication, Vol. 1: Microlithography*, SPIE Press, Bellingham, WA, 1997, chap. 3.

[20] M. McCord and M. Rooks, Electron beam lithography, in *Handbook of Microlithography, Micromachining and Microfabrication, Vol. 1: Microlithography*, SPIE Press, Bellingham, WA, 1997, chap. 2.

[21] K. Suzuki and T. Fujiwara, K. Hada, N. Hirayanagi, S. Kawata, Kenji Morita, K. Okamoto, T. Okinuo and S. Shimizu, Nikon EB Stepper: system concept and countermeasures for critical issues, *Emerging Lithographic Technologies IV, Proc. SPIE*, 3997, 2000, pp. 214–224.

[22] G. Moore, Cramming more components onto integrated circuits, *Electronics*, 38, 114–117, 1965; also reproduced in Ref. [9].

[23] G.E. Moore, Progress in digital integrated electronics, *IEDM 1975 Technical Digest*, 1975; also reproduced in Ref. [9].

[24] A.K.K. Wong, *Resolution Enhancement Techniques in Optical Lithography*, SPIE Press, Bellingham, WA, 2001.

[25] A. Rosenbluth, D. Goodman, and B.J. Lin, A critical examination of submicron optical lithography using simulated projection images, *J. Vac. Sci. Technol. B*, 1, 1190–1195, 1983.

[26] P. Chien and M. Chen, Proximity effects in submicron optical lithography, *Optical Microlithography VI, Proc. SPIE*, 772, 1987, pp. 35–40.

[27] F.M. Schellenberg, H. Zhang, and J. Morrow, SEMATECH J111 Project: OPC validation, *Optical Microlithography XI, Proc. SPIE*, 3334, 1998, pp. 892–911.

[28] Masato Shibuya, 透過照明用被投影原板 [Projection master for use with transmitted illumination], 公開特許公報 (A) 昭 57–62052, 特許公報 (B) 昭 62–50811 [Japan Patent Office Laid-Open Patent Publication (A) Showa 57-62052, Patent Publication (B) Showa 62– 50811] (filed Sept. 30, 1980; published April 14, 1982, issued October 27, 1987).

[29] M.D. Levenson, N.S. Viswanathan, and R.A. Simpson, Improving resolution in photolithography with a phase-shifting mask, *IEEE Trans. Electron Dev.*, ED-29, 1828–1836, 1982.

[30] M.D. Levenson, D.S. Goodman, S. Lindsey, P.W. Bayer, and H.A.E. Santini, The phase-shifting mask II: imaging simulations and submicrometer resist exposures, *IEEE Trans. Electron Dev.*, ED-31, 753–763, 1984.

[31] T. Terasawa, N. Hasegawa, T. Kurosaki, and T. Tanaka, 0.3-micron optical lithography using a phase-shifting mask, *Optical/Laser Microlithography II, Proc. SPIE*, 1088, 1989, pp. 25–33.

[32] A. Nitayama, T. Sato, K. Hashimoto, F. Shigemitsu, and M. Nakase, New phase shifting mask with self-aligned phase shifters for a quarter micron lithography, *IEDM 1989 Technical Digest*, 57–60, 1989.

[33] T. Tanaka, S. Uchino, N. Nasegawa, T. Yamanaka, T. Terasawa, and S. Okazaki, A novel optical lithography technique using the phase-shifter fringe, *Jpn. J. Appl. Phys.*, 30, 1131–1136, 1991.

[34] H. Jinbo and Y. Yamashita, 0.2 mm or less i-line lithography by phase-shifting-mask technology, *IEDM 1990 Technical Digest*, 825–828, 1990.

[35] K.K.H. Toh, G. Dao, R. Singh, and H. Gaw, Chromeless phase-shifted masks: a new approach to phase-shifting masks, *10th Annual Symposium on Microlithography, Proc. SPIE*, 1496, 1990, pp. 27–53.

[36] H. Watanabe, Y. Todokoro, Y. Hirai, and M. Inoue, Transparent phase shifting mask with multistage phase shifter and comb-shaped shifter, *Optical/Laser Microlithography IV, Proc. SPIE*, 1463, 1991, pp. 101–110.

[37] J.F. Chen, J.S. Petersen, R. Socha, T. Laidig, K.E. Wampler, K. Nakagawa, G. Hughes, S. MacDonald, and W. Ng, Binary halftone chromeless PSM technology for λ/4 optical lithography, *Optical Microlithography XIV, Proc. SPIE*, 4346, 2001, pp. 515–533.

[38] D.J. Van Den Broeke, J.F. Chen, T.L. Laidig, S. Hsu, K.E. Wampler, R.J. Socha, and J.S. Petersen, Complex two dimensional pattern lithography using chromeless phase lithography (CPL), *J. Microlith. Microfab. Microsyst.*, 1, 229–242, 2002.

[39] Y.-C. Ku, E.H. Anderson, M.L. Schattenburg, and H.I. Smith, Use of a pi-phase shifting x-ray mask to increase the intensity slope at feature edges, *J. Vac. Sci. Technol., B* 6, 150–153, 1988.

[40] F.D. Kalk, R.H. French, H.U. Alpay, and G. Hughes, Attenuated phase shifting photomasks fabricated from Cr-based embedded shifter blanks, *Photomask and X-Ray Mask Technology, Proc. SPIE*, 2254, 1994, pp. 64–70.

[41] Ernst Abbe's lecture notes on off-axis resolution for microscopy are mentioned in Ref. [10], and also are presented by O. Lummer and F. Reiche, *Die Lehre von der Bildentstehung im Mikroskop von Ernst Abbe*, Vieweg und Sohn, Braunschweig, Germany, 1910.

[42] D.L. Fehrs, H.B. Lovering, and R.T. Scruton, Illuminator Modification of an Optical Aligner, *Proceedings of KTI Microelectronics Seminar INTERFACE'89*, 1989, pp. 217–230.

[43] S. Asai, I. Hanyu, and K. Hikosaka, High performance optical lithography using a separated light source, *J. Vac. Sci. Technol. B*, 10, 3023–3026, 1992.

[44] M. Noguchi, M. Muraki, Y. Iwasaki, and A. Suzuki, Subhalf micron lithography system with phase-shifting effect, *Optical/Laser Microlithography V, Proc. SPIE*, 1674, 1992, pp. 92–104.

[45] N. Shiraishi, S. Hirukawa, Y. Takeuchi, and N. Magome, New imaging technique for 64M-DRAM, *Optical/Laser Microlithography V, Proc. SPIE*, 1674, 1992, pp. 741–752.

[46] T. Ogawa, M. Uematsu, T. Ishimaru, M. Kimura, and T. Tsumori, The effective light source optimization with the modified beam for the depth-of-focus enhancements, *Optical/Laser Microlithography VII, Proc. SPIE*, 2197, 1994, pp. 19–30.

[47] R.J. Socha, M.V. Dusa, L. Capodieci, J. Finders, J.F. Chen, D.G. Flagello, and K.D. Cummings, Forbidden pitches for 130 nm lithography and below, in *Optical Microlithography XIII, Proc. SPIE*, Vol. 4000, pp. 1140–1155, 2000.

[48] F.M. Schellenberg and L. Capodieci, Impact of RET on Physical Layouts, *Proceedings of the 2001 ISPD*, ACM, New York, 2001, pp. 52–55.

[49] M. Eurlings, E. van Setten, J.A. Torres, M.V. Dusa, R.J. Socha, L. Capodieci, and J. Finders, 0.11-μm imaging in KrF lithography using dipole illumination, *Lithography for Semiconductor Manufacturing II, Proc. SPIE*, 4404, 2001, pp. 266–278.

[50] Lord Rayleigh, On the remarkable case of diffraction spectra described by Prof. Wood, *Philos. Mag.*, 14, 60–65, 1907.

[51] A. Estroff, Y. Fan, A. Bourov, F.C. Cropanese, N.V. Lafferty, L.V. Zavyalova, and B.W. Smith, Mask-induced polarization, *Optical Microlithography XVII, Proc. SPIE*, 5377, 2004, pp. 1069–1080.

[52] T.W. Ebbesen, H.J. Lezec, H.F. Ghaemi, T. Thio, and P.A. Wolff, Extraordinary optical transmission through sub-wavelength hole arrays, *Nature*, 391, 667–669, 1998.

[53] F.M. Schellenberg, Design for Manufacturing in the Semiconductor Industry: The Litho/Design Workshops, *Proceedings of the 12th International Conference on VLSI Design*, IEEE Computer Society Press, Los Alamitos, CA, 1999, pp. 111–119.

[54] F.M. Schellenberg, Advanced data preparation and design automation, *Photomask and Next-Generation Lithography Mask Technology IX, Proc. SPIE*, 4754, 2002, pp. 54–65.

[55] L.W. Liebmann, G.A. Northrop, J. Culp, L. Sigal, A. Barish, and C.A. Fonseca, Layout optimization at the pinnacle of optical lithography, *Design and Process Integration for Microelectronic Manufacturing, Proc. SPIE*, 5042, 2003, pp. 1–14.

[56] L. Capodieci, P. Gupta, A.B. Kahng, D. Sylvester, and J. Yang, Toward a Methodology for a Manufacturability-Driven Design Rule Exploration, *Proceedings of the 41st Design Automation Conference*, ACM, New York, 2004, pp. 311–316.

[57] O.W. Otto, J.G. Garofalo, K.K. Low, C.-M. Yuan, R.C. Henderson, C. Pierrat, R.L. Kostelak, S. Vaidya, and P.K. Vasudev, Automated optical proximity correction: a rules-based approach, *Optical/Laser Microlithography VII, Proc. SPIE*, Vol. 2197, pp.278–293, 1994.

[58] J. Garofalo, C. Biddick, R.L. Kostelak, and S. Vaidya, Mask assisted off-axis illumination technique for random logic, *J. Vac. Sci. Technol. B*, 11, 2651–2658, 1993.

[59] J.F. Chen and J.A. Matthews, Mask for Photolithography, U.S. Patent 5,242,770, 1993.

[60] B.E.A. Saleh and S.I. Sayegh, Reduction of errors of microphotographic reproductions by optimal correction of original masks, *Opt. Eng.*, 20, 781–784, 1981.

[61] K.M. Nashold and B.E.A. Saleh, Image construction through diffraction-limited high-contrast imaging systems: an iterative approach, *J. Opt. Soc. Am. A*, 2, 635–643, 1985.

[62] B.E.A. Saleh and K. Nashold, Image construction: optimum amplitude and phase masks in photolithography, *Appl. Opt.*, 24, 1432–1437, 1985.

[63] B.E.A. Saleh, Image synthesis: discovery instead of recovery, in *Image Recovery: Theory and Application*, H. Stark, Ed., Academic Press, Orlando, FL, 1987, chap. 12, pp. 463–498.

[64] N.B. Cobb and A. Zakhor, Fast sparse aerial-image calculation for OPC, *15th Annual BACUS Symposium on Photomask Technology and Management, Proc. SPIE*, 2621, 1995, pp. 534–545.

[65] H.H. Hopkins, On the diffraction theory of optical images, *Proc. Royal Soc. London Series A*, 217, 408–432, 1953.

[66] M. Rieger and J. Stirniman, Using behavior modeling for proximity correction, *Optical/Laser Microlithography VII, Proc. SPIE*, Vol. 2197, pp. 371–376, 1994.

[67] N.B. Cobb, Fast Optical and Process Proximity Correction Algorithms for Integrated Circuit Manufacturing, Ph.D. Dissertation, University of California at Berkeley, 1998.

[68] L. Liebmann, J. Lund, F.L. Heng, and I. Graur, Enabling Alternating Phase Shifted Mask Designs for a Full Logic Gate Level: Design Rules and Design Rule Checking, *Proceedings of the 38th Design Automation Conference*, ACM, New York, 2001, pp. 79–84.

[69] L. Liebmann, J. Lund, and F.L. Heng, Enabling alternating phase shifted mask designs for a full logic gate level, *J. Microlith. Microfab. Microsyst.*, 1, 31–42, 2002.

[70] K. Ooi, S. Hara, and K. Kojima, Computer aided design software for designing phase-shifting masks, *Jpn. J. Appl. Phys.*, 32, 5887–5891, 1993.

[71] K. Ooi, K. Koyama, and M. Kiryu, Method of designing phase-shifting masks utilizing a compactor, *Jpn. J. Appl. Phys.*, 33, 6774–6778, 1994.

[72] M. Fritze, J.M. Burns, P.W. Wyatt, D.K. Astolfi, T. Forte, D. Yost, P. Davis, A. Curtis, D.M. Preble, S. Cann, S. Denault, H.Y. Liu, J.C. Shaw, N.T. Sullivan, R. Brandom, and M. Mastovich, Application of chromeless phase-shift masks to sub-100-nm SOI CMOS transistor fabrication, *Optical Microlithography XIII, Proc. SPIE*, 4000, 2000, pp. 388–407.

[73] T. Terasawa, N. Hasegawa, H. Fukuda, and S. Katagiri, Imaging characteristics of multi-phase-shifting and halftone phase-shifting masks, *Jpn. J. Appl. Phys.*, 30, 2991–2997, 1991.

[74] G. Galan, F. Lalanne, P. Schiavone, and J.M. Temerson, Application of alternating type phase shift mask to polysilicon level for random logic circuits, *Jpn. J. Appl. Phys.*, 33, 6779–6784, 1994.

[75] H. Jinbo and Y. Yamashita, Improvement of phase-shifter edge line mask method, *Jpn. J. Appl. Phys.*, 30, 2998–3003, 1991.

[76] H.-Y. Liu, L. Karklin, Y.-T. Wang, and Y.C. Pati, Application of alternating phase-shifting masks to 140-nm gate patterning: II. Mask design and manufacturing tolerances, *Optical Microlithography XI, Proc. SPIE*, 3334, 1998, pp. 2–14.
[77] C. Spence, M. Plat, E. Sahouria, N. Cobb, and F. Schellenberg, Integration of optical proximity correction strategies in strong phase shifter design for poly-gate layer, *19th Annual Symposium on Photomask Technology, Proc. SPIE*, 3873, 1999, pp. 277–287.
[78] F.M. Schellenberg, L. Capodieci, and R. Socha, Adoption of OPC and the Impact on Design and Layout, *Proceedings of the 38th Design Automation Conference*, ACM, New York, 2001, pp. 89–92.
[79] Y. Granik and N. Cobb, New process models for OPC at sub-90nm nodes, *Optical Microlithography XVI, Proc. SPIE*, 5040, 2003, pp. 1166–1175.
[80] A.E. Rosenbluth, S. Bukofsky, M. Hibbs, K. Lai, R.N. Singh, and A.K. Wong, Optimum mask and source patterns for printing a given shape, *J. Microlith. Microfab. Microsyst.*, 1, 13–30, 2002.
[81] Y. Granik, Source optimization for image fidelity and throughput, *J. Microlith. Microfab. Microsyst.*, 3, 509–522, 2004.
[82] E.D. Palik, Ed., *Handbook of Optical Constants of Solids*, Vol. I, II, and III, Academic Press, San Diego, 1998.
[83] B. Smith at Rochester Institute of Technology (RIT) maintains a website that allows the computation of optical properties for semiconductor materials at http://www.rit.edu/%7E635dept5/thinfilms/thinfilms.htm
[84] A.K.K. Wong and A.R. Neureuther, Rigorous three dimensional time-domain finite difference electromagnetic simulation, *IEEE Trans. Semicond. Manuf.*, 8, 419–431, 1995.
[85] K. Adam and A.R. Neureuther, Domain decomposition methods for the rapid electromagnetic simulation of photomask scattering, *J. Microlith. Microfab. Microsyst.*, 1, 253–269, 2002.
[86] P. van Adrichern and C. Kalus, Data preparation, chap. 2 of Ref. [6]; and P. DePesa, D. Kay, and G. Meyers, Data preparation, chap. 2 of Ref. [6A].
[87] MEBES is a format owned by ETEC Systems, an Applied Materials Company (www.amat.com).
[88] *SEMI P39-0304E2 - OASIS — Open Artwork System Interchange Standard* can be downloaded from the SEMI Standards Organization, www.semi.org/
[89] S. Schulze, P. LaCour, and L. Grodd, OASIS-based data preparation flows: progress report on containing data size explosion, *Design and Process Integration for Microelectronic Manufacturing II, Proc SPIE*, 5379, 2004, pp. 149–157.

19

Design for Manufacturability in the Nanometer Era

Nicola Dragone
PDF Solutions, Inc.
Brescia, Italy

Carlo Guardiani
PDF Solutions, Inc.
Brescia, Italy

Andrzej J. Strojwas
PDF Solutions, Inc.
San Jose, California

19.1 Introduction

Achieving high-yielding designs in the state of the art, Ultra Large Scale Integration (ULSI) technology has become an extremely challenging task due to the miniaturization as well as the complexity of leading-edge products. The design methodology called design for manufacturing (DFM) includes a set of techniques to modify the design of ICs in order to make them more manufacturable, i.e., to improve their functional yield, parametric yield, or their reliability. Traditionally, in the pre-nanometer era, DFM consisted of a set of different methodologies trying to enforce some soft (recommended) design rules (DRs) regarding the shapes and polygons of the physical layout of an IC product. These DFM methodologies worked primarily at the full chip level. Additionally, worst-case simulations at different levels of abstraction were applied to minimize the impact of process variations on performance and other types of parametric yield loss. All these different types of worst-case simulations were essentially based on a set of worst-case (or corner) SPICE device parameter files that were intended to represent the variability of transistor performance over the full range of variation in a fabrication process.

With the advent of nanometer technologies, namely at 130 nm and below, process and design systematic yield loss mechanisms (YLMs) have started to cross over random effects in the yield Pareto [1] (see Figure 19.1).

Random YLMs are caused by random contaminants or defects (e.g., particles) and are therefore characterized by the absence of strong spatial, temporal, or other kinds of correlations. On the contrary, process systematic YLMs are usually spatially correlated (e.g., their failure rate shows strong radial wafer patterns) or temporally correlated (e.g., lot-to-lot variations as a function of equipment). Design systematic YLMs are strongly correlated to specific local physical layout patterns, and therefore their failure rate may vary by orders of magnitude as a function of the layout characteristics.

The same physical design feature (e.g., a contact or a via) may yield completely differently depending upon neighboring features (short correlation distance effects) such as the presence or lack of certain shapes (e.g., the presence of a minimum distance, small area, metal island, and via) next to it . Other types of design systematic YLMs are characterized by medium or even relatively big correlation distances, for example, YLM due to CMP effects. Any design systematic YLM is, however, characterized by an intrachip correlation pattern. In other words, they depend on local physical design properties.

This design locality property of design systematic YLMs considerably complicates the effectiveness of the traditional DR approach because DRs are applied globally to every physical design feature in a chip.

In order to cope with the design locality property of design systematic YLMs, the total number of DRs exploded and the rules have become increasingly complex. The DRs for subnanometer technologies have been split into two sets: a fundamental set of DRs, i.e., rules that if violated, would cause a product's yield to drop to zero; and a set of "recommended DRs" (RDRs), i.e., rules that the design tools should try to implement as much as possible based upon specific layout patterns in the design. Unfortunately, it would be basically impossible or extremely impractical to include every such complex DR in the design tools to generate maximally manufacturable designs by construction.

For this reason, a large number of rule-driven reactive DFM methodologies have been introduced recently, which have the drawback of modifying the design patterns after timing closure and verification have been completed, hence introducing additional risk and delay in the design process completion.

Moreover, several of the RDRs create some conflicts among other rules. For example, adding via redundancy may create a large number of small dielectric regions that are difficult to manufacture and may cause systematic failures.

Model-based, proactive DFM philosophy consists in the development of accurate, silicon-verified YLM models that can evaluate the relative impact of each YLM and assess trade-offs. These models are then integrated in the design tool's cost function along with other design objectives such as speed, power, and signal integrity functions. Design tools are thus able to exploit the design locality property of design

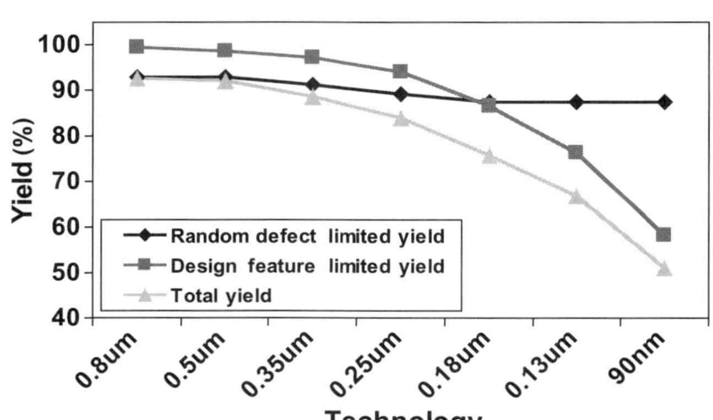

FIGURE 19.1 Yield limiters by technology node.

systematic YLMs as well as global random YLMs to achieve the optimal manufacturability compatible with the actual IC design specifications.

19.2 Taxonomy of Yield Loss Mechanisms

The most important YLMs for VLSI ICs can be classified into several categories based on their nature. Functional yield loss is still the dominant factor and is caused by mechanisms such as misprocessing (e.g., equipment-related problems), systematic effects such as printability or planarization problems, and purely random defects. High-performance products may exhibit parametric design marginalities caused by either process fluctuations or environmental factors (such as supply voltage or temperature). The test-related yield losses, which are caused by incorrect testing, can also play a significant role. In this section, we present a detailed description of how YLMs have evolved in nanometer technologies, and some of the standard yield modeling approaches.

19.2.1 Random Yield Loss Modeling

Random defects and contaminants are possibly the best-known and studied YLMs [1]. Typical random failure mechanisms involve active, poly, and metal shorts and opens due to particle defects, as well as contact and via opens due to formation defectivity. Some examples of open/short failures caused by random particles are shown in Figure 19.2.

Before deep submicron technologies, the yield loss was dominated by random defects, and yield was typically modeled as a function of the defect density and size distribution (Figure 19.3) and of the amount of the chip area that is susceptible to defects (also known as critical area). The concept of critical area is shown graphically in Figure 19.4.

A number of different models [1] have been developed over the years starting from the simple Poisson to negative binomial to more complicated critical-area-based models, also taking into account the defect size distribution (DSD) as shown in Figure 19.5.

19.2.2 Design Systematic Yield Loss

The contribution of systematic YLMs at the time when critical-area-based models were applied was typically less than 5% of the total, and often they were totally negligible for relatively mature processes. Critical-area-based yield models are not sufficient for state-of-the-art technologies that are affected by significant systematic YLMs. For example, it is often the case that two products characterized by similar critical area values have very different yields. A more comprehensive set of models had to be developed that included a parameterization of the yield loss of individual design components as a function of their design attributes in order to predict yield adequately, especially in the early production stages. Typical

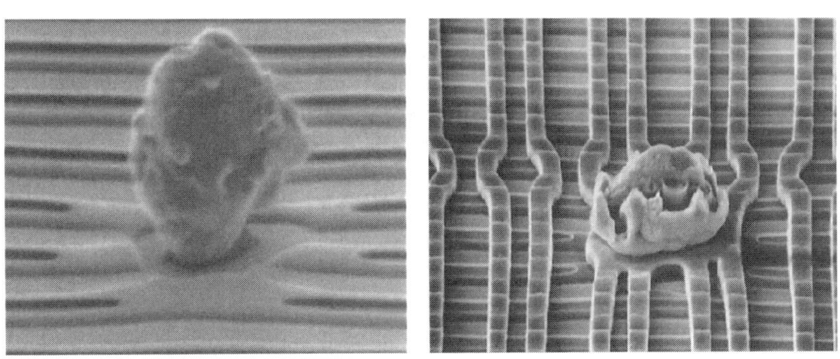

FIGURE 19.2 Random defects causing metal opens and shorts.

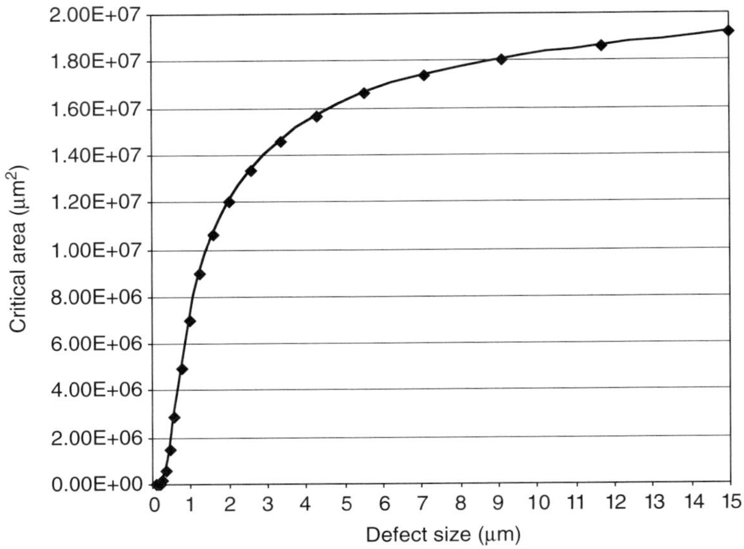

FIGURE 19.3 Critical area as a function of defect size distribution.

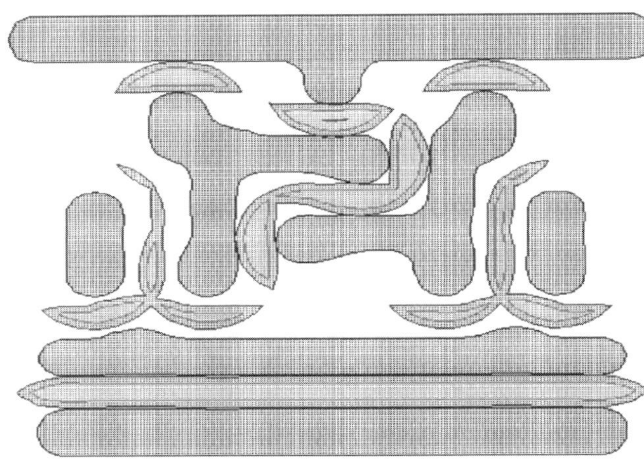

FIGURE 19.4 **(See color insert following page 15–4.)** Graphical representation of critical area.

design systematic failure mechanisms are due to mask misalignment, line-end and border effects, oxide stress, and more generally due to different DR marginalities sensitized by the impact of micro- and macro- loading. Figure 19.6 demonstrates the drastic profile difference between sparse and dense vias. Sparse vias or stacks are much more likely to exhibit the Cu pull-up-type failures. These profile differences result in much higher fail rates for sparse vias than for vias placed in a dense environment.

19.2.3 Printability Yield Loss

Printability failure mechanisms are often caused by errors in the optical proximity correction (OPC) algorithm and limited process window. Although it is sometimes difficult to separate potential flaws of the OPC algorithm from intrinsically "hard to print" features, DFM should only address the latter, as OPC/RET errors are, by definition, independent of the drawn layout features, and hence of the design.

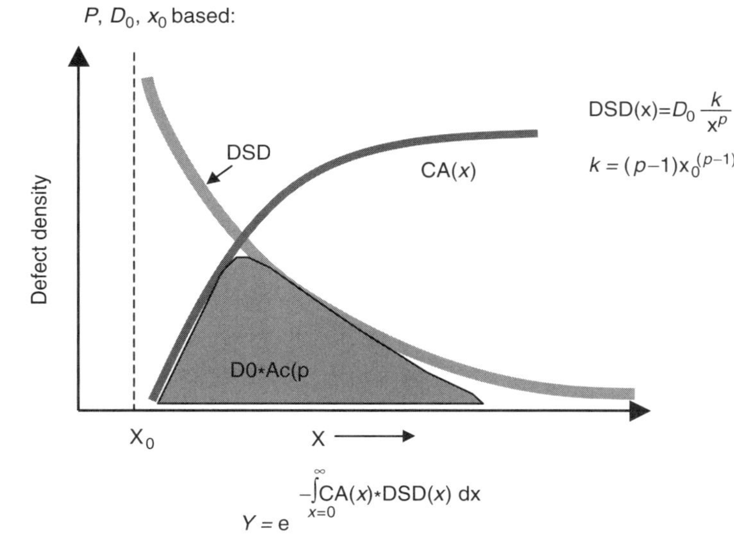

FIGURE 19.5 Random defect limited yield model.

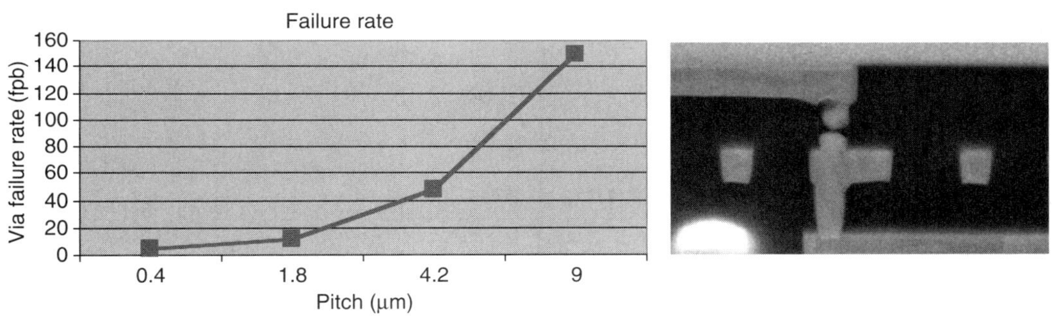

FIGURE 19.6 Pitch dependence of via failure rate in a sparse environment.

Therefore, in a DFM methodology, we are interested in identifying, characterizing, and fixing process window issues in the design features that are overly sensitive to lithography parameters variation. Common failures of this type are due to metal pull-back and consequent poor via coverage, metal necking (open), and metal bridging (short). Printability yield loss modeling must be based on physically accurate simulation of the printed shapes for the entire process window. Figure 19.7 illustrates printability simulation results of a layout pattern with high sensitivity to contact coverage.

19.2.4 Parametric Yield Loss

The electrical performances of IC devices are subject to natural random variation. Because of this, every IC performance parameter can be described as a random variable with a specific associated probability distribution. The tails of the probability distribution may fall outside the acceptable device specs and cause parametric yield loss. Parametric yield loss is characterized by inter- and intra-die level correlations, with the latter also known as device mismatch. There are two major sources of process variations that affect the electrical behavior of digital or analog circuits: electrical parameter variations, for example related to dopant fluctuations or oxide thickness variations; and lithography parameter variations, for example related to gate length or width variations. One more source of discrepancy between predicted

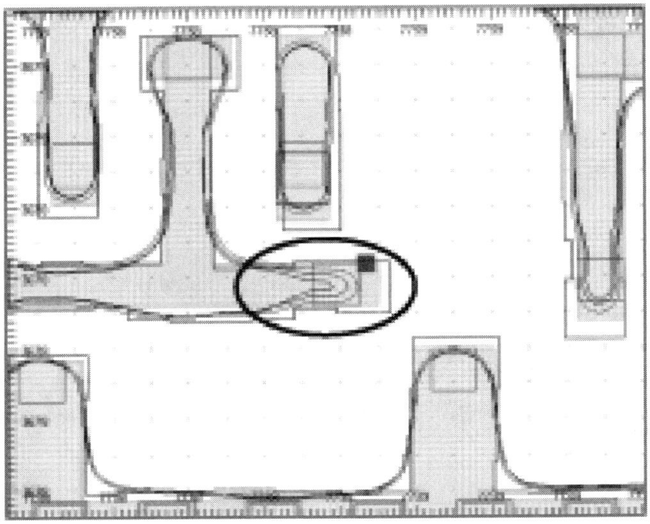

FIGURE 19.7 Simulation of contact coverage for different process corners.

and actual circuit performance is due to layout pattern-dependent device characteristics, such as stress-related drain extension-dependent I_{DSS} in MOSFET devices.

Figure 19.8 illustrates the dispersion of the I_{OFF} vs. I_{ON} data caused by poly CD variations for three different layout environments.

19.3 Logic Design for Manufacturing

In the previous sections we illustrated the major YLMs and how they relate to the design attributes of devices and interconnections. We have shown that the design-dependent and printability systematic YLMs are becoming increasingly critical and that they dominate the design-dependent yield loss, i.e., the fraction of yield loss that is recoverable by adopting different design configurations, which ultimately defines the scope of DFM changes. The process sensitivity to such systematic YLMs is hard to predict and varies considerably as a function of the particular process flow implementation. Therefore, although with some experience it is possible to predict the list of the potential YLMs including systematic effects that may impact a given technology, it is often hard, if not impossible, to know which of these potential YLMs are the most critical, and the Pareto of the dominant effects. As a consequence, the key enabler for a DFM methodology is an accurate characterization and modeling of the process technology to determine process sensitivities of the leading YLMs and to determine design trade-offs. In the following section, we will focus on bulk and SOI (silicon on insuator) CMOS technologies facing the integration challenges of new materials, advanced next generation lithography, and yield limitations above and beyond random defectivity.

Generation of the process–design mechanism database requires several major components:

1. Development and implementation of the required characterization structures to span all the relevant process–design interaction effects and to generate the required physical data for accurate yield modeling.
2. A test and analysis infrastructure with sufficient throughput, including both cost effectiveness and time efficiency.
3. Development and calibration of yield models as a function of design attributes and based on the data from test structures.
4. Time-based scaling of process maturity to enable predictability of the yield impact of YLMs.

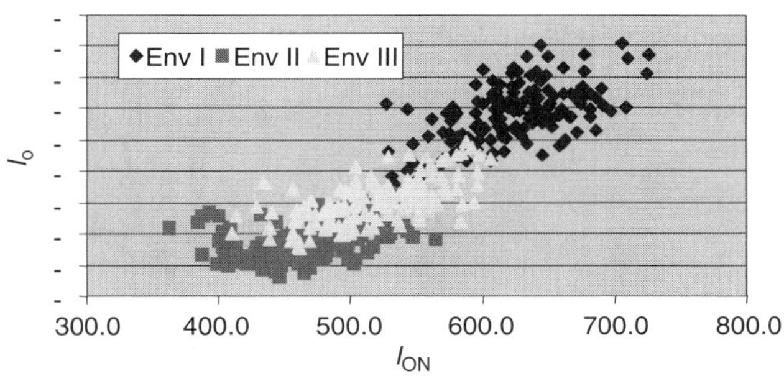

FIGURE 19.8 Effects of environment-dependent poly-CD variations on I_{ON}/I_{OFF}.

19.3.1 Characterization Structures

The ability to exploit fully the advantages of a DFM methodology requires a set of comprehensive characterization test structures. A comprehensive set of structures includes those required to characterize fully random, systematic, and parametric effects on the front-end-of-line (FEOL) and back-end-of-line (BEOL) process module yields. In addition, the structures must incorporate design of experiments (DOE) and response surface methodology [2,3] to enable product layout attribute dependence to take into account local and global loading, proximity, and density effects — as well as to optimize the amount of silicon area required to characterize each effect within the required confidence interval [1].

A large number of structures must be designed to capture defect densities on the order of parts per billion (ppb) and, at the same time, make efficient use of the available silicon area. Among those we can mention are harp, nest/comb, checkerboard, chains, and passive multipexers [4,5]. Figure 19.9 illustrates several structures used to characterize the DSD and open/short defectivity.

A rich DOE must be implemented in test structure design to identify process sensitivities and process–design interactions. Typical experiments include failure dependency on material density, both local and global, associated with particular layout patterns. Figure 19.10 illustrates a via experiment with variable chain and hole pitch, and metal density.

Characterization vehicles must include both calibration and verification structures for printability. Calibration structures are needed for tuning litho simulators for hotspot detection. Verification structures must be included to assess the quality of calibration and characterize process robustness against the variation in the parameters. Typical litho parameters that could play a critical role are exposure level, mask error, defocus, misalignment, and resist. Figure 19.11 illustrates structures designed to verify the quality of printability in environments, which can be considered *hard-to-print* for any OPC algorithm.

A heterogeneous set of structures should also cover the most relevant effects that have an impact on performance variability and performance predictability. Typical effects are CD line variations, poly and active corner rounding, and STI stress. Figure 19.12 illustrates a couple of structures that can be implemented to characterize poly CD variations with different local and global density levels, and the impact of layout styles on transistor performance.

19.3.2 Test and Analysis Infrastructure

The feasibility of fully utilizing a characterization test structure set and associated DOE, such as those presented in the previous section, must take into account the overall throughput capability as a function of testing and analysis. Novel structures that optimize the statistical sample size per test structure are available and reduce the test time and silicon area required for accurate characterization [3]. Nonetheless, the amount of data generated by a comprehensive test structure set remains massive.

Reducing the time and cost of test as a function of automated test equipment (ATE)-related costs [6,7] should be complemented with the reduction of analysis time and costs when time-to-market pressures

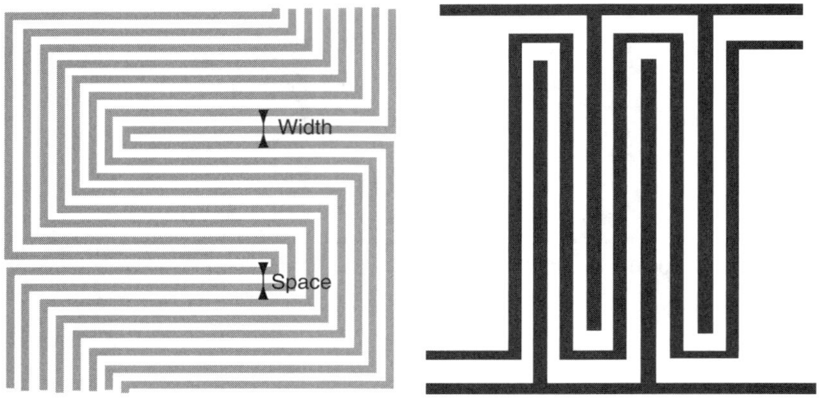

FIGURE 19.9 Nest and snake/comb structures for random defectivity characterization.

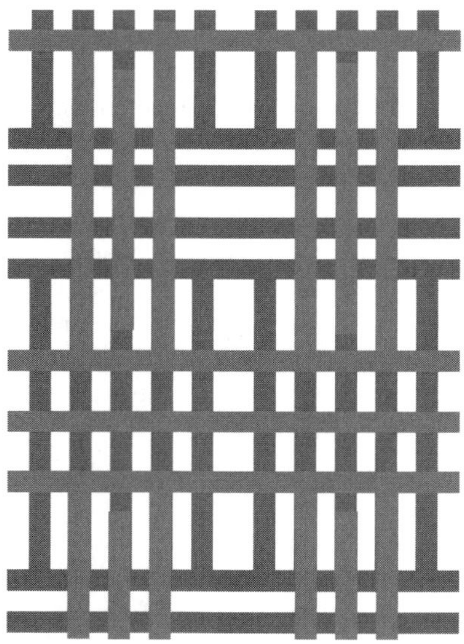

FIGURE 19.10 **(See color insert following page 15–4.)** Via chain structure for characterizing failure rate dependency on the environment.

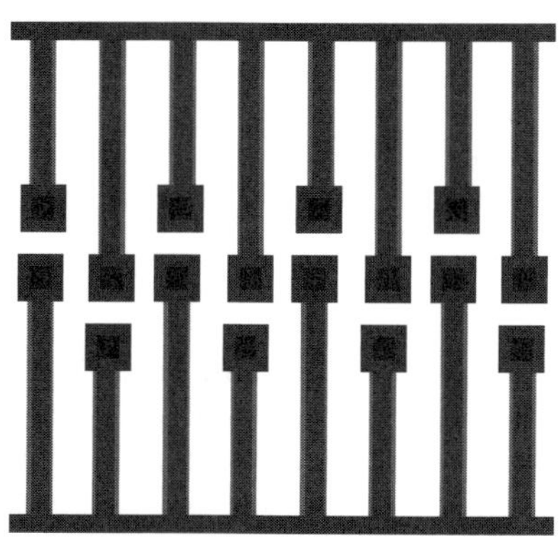

FIGURE 19.11 Litho verification structure with difficult-to-print OPC patterns.

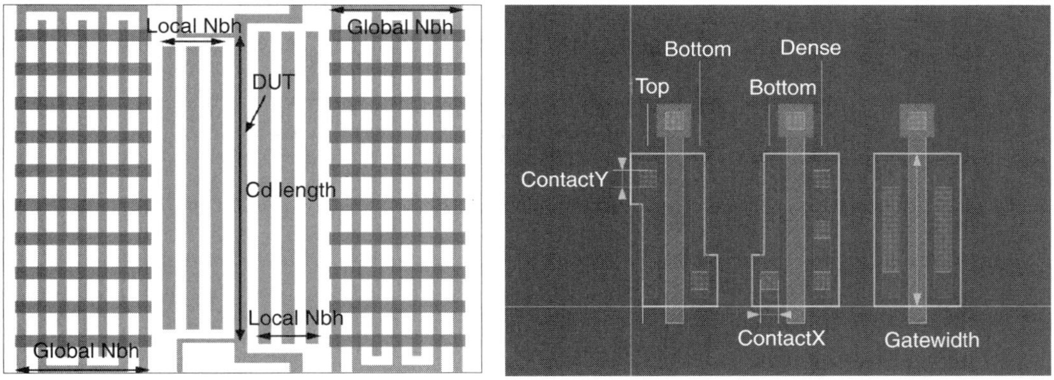

FIGURE 19.12 Structures for Idrive characterization: poly-CD variation and layout styles.

commonly demand early design tape-outs for test sample debugging, engineering samples for customers, product ramp in nonmature processes, and finally, long-term volume manufacturing. Therefore, the desired solution integrates and optimizes the ATE and analysis infrastructure [8] as much as possible to enable fast turnaround time such that the process characterization is used for calibrating yield models that can be utilized early in the product design cycle.

19.3.3 Yield Impact Matrix

Any type of DFM solution applied to an arbitrary circuit must consider two basic concepts. The first is the likelihood that a certain layout configuration will cause a fail. The second is the information on the number of occurrences of the layout configuration in the product design. The former must be derived from process characterization, whereas the latter must be computed through product analysis and extraction of design attributes. By combining these two pieces of information, we obtain yield models, as presented in previous section. A very effective way to summarize the yield numbers associated with a particular design is represented by the yield impact matrix (YIMP) [9,10], which identifies the most critical contributors to yield loss in terms of circuit components and process modules. Figure 19.13 shows a very simple YIMP in which columns represent design blocks and rows represent process modules. The relevance of such a representation is twofold: on the one hand it allows the identification of critical blocks and, on the other, it makes it possible to plan adequate DFM or process fixes to alleviate yield losses.

19.3.4 IP Optimization

19.3.4.1 Standard Cell Optimization

A substantial fraction of IC yield loss occurs in standard cell logic. Logic-limited yield associated with standard cells may become the most critical factor in designs such as graphic applications where logic content is much larger than the analog or memory content.

Composition of standard cell libraries, both in terms of their logic functionality and capability of steering current, often referred to as cell driving strength, is commonly being recognized as one of the most critical aspects to achieve high-quality synthesis [11,12]. A number of actions can be taken to improve standard cell manufacturability, such as adding feature redundancy, over-sizing, spacing, increasing uniformity, or density. However, the application of such techniques often conflicts with the optimization of other design parameters such as area, speed, and power. Therefore, the ability to quantify the contribution of different effects, and to rank them in order of priority is of utmost importance in a yield-aware design flow [13]. In addition to the above considerations, general-purpose standard cell libraries should also support differentiated portfolios of IC products that are characterized by different performance

YIMP Matrix				Limited yield		
	Failure mode			Full chip	SRAM	LOGIC
Random defects		Active	Random	98%	99%	99%
			Pattern dependent	99%	99%	100%
			Total	97%	98%	99%
		Poly	Random	97%	98%	99%
			Pattern dependent	94%	95%	99%
			Total	91%	95%	96%
		Metal	Random	97%	98%	99%
			Pattern dependent	97%	98%	99%
			Total	94%	95%	99%
		Holes		97%	98%	99%
	Total			81%	82%	99%
Systematic		Metal islands		87%	89%	90%
		Pattern density		91%	93%	94%
		Narrow space wide neighbor		97%	99%	100%
		Via induced metal shorts		77%	78%	79%
	Total			62%	64%	64%

FIGURE 19.13 Yield impact matrix.

specifications and volume production requirements. Such requirements can be translated into design objectives for the standard cell library as follows:

- Maximize average yield, which is typically the most important manufacturing objective for high-volume parts.
- Minimize lot-to-lot, wafer-to-wafer, and within-wafer yield excursions that are often a major concern for less mature processes and are a key objective for relatively low-volume products and obviously for engineering samples, where an excursion may mean no parts to test.
- Minimize performance variability, hence provide tighter speed bins and minimize parametric yield loss, which is obviously a main objective for high-performance products such as micro processors.

These high-level manufacturability optimization objectives are in general characterized by different and often conflicting process–layout interaction mechanisms. A manufacturing-aware standard cell library should support all these objectives; therefore, it should contain cell modules that are optimized for each of the key process–design interaction mechanisms that are relevant for a given class. A nonexhaustive example set of effects that are addressed in each class can consist of the following:

- Maximize average yield: hole (contact and via) opens, critical area for opens, shorts, etc.
- Minimize yield variability: hole coverage by the upper/lower layer, metal islands, field transitions, active leakage paths, etc.
- Minimize performance variability: poly flaring, STI stress, etc.

Figure 19.14 shows the high-level architecture of the proposed DFM standard cell library. A core library including a rich function set of high-density cells with all their required drive strength variants is complemented by additional cells that address one or more manufacturing objectives. Within each class and for each logic function/drive strength combination represented in the class, multiple yield strength variants are designed that implement different levels of yield vs. cell size and other design costs trade-off optimizations.

Some example statistics based on the application of a high-yield standard cell library design for a 130 nm process are reported in Table 19.1. The Limited Yield (LY) improvements have been obtained on two very similar 130nm processes but at a different stage of maturity. On a million-gate design, the average functional yield improvement is 2% and 5%, respectively.

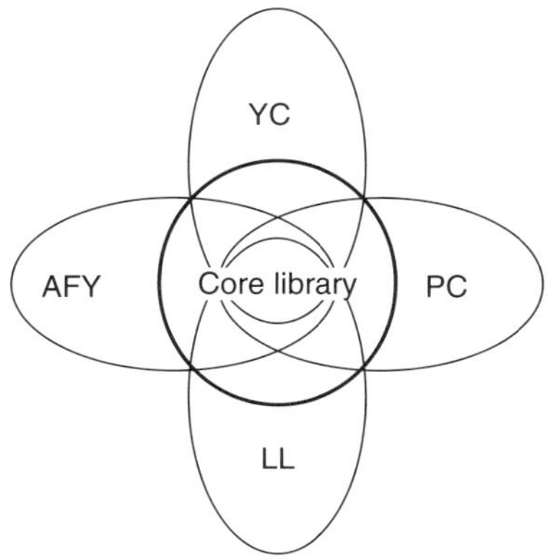

FIGURE 19.14 DFM standard cell library modules.

TABLE 19.1 Yield Improvements for High-Yield Library

	Equivalent Gates (million)	LY Improvement (%)
Process 1 (mature)	1	2.08
	2	4.03
	3	5.84
Process 2 (ramp)	1	4.96
	2	9.27
	3	12.99

Figure 19.15 illustrates the distribution of performance degradation for the same library cells. Most of the cells show performance degradation between 1% and 10%, but we also have standard cells with worst-case timing arcs exceeding 20% degradation.

Figure 19.16 shows the high-yield cell area increase distribution. The average standard cell area increase for this implementation is 15%, but also reaches peaks of up to 50%. This clearly demonstrates the need of accurate characterization of the yield benefit and of the separate contribution of all the important effects.

19.3.5 Memory Optimization

Along with logic, memories represent another significant source of yield loss for a large class of ICs, and systems-on-chip (SoCs) in particular. Yield loss in memories can be due to failures occurring in the core or the periphery, with the former being usually most relevant except for very small arrays.

The memory bit-cell still plays a critical role during technology development because it is often used to tune and optimize the process flow and OPC. The need for larger embedded memories drives the implementation of very high-density bit-cell layouts and of aggressive ad hoc DRs along with custom manually optimized OPC.

The challenge of designing highly manufacturable bit-cells is therefore often overwhelming, given the fact that multiple issues need to be addressed at the same time. As already mentioned, the first goal is to design bit-cell layouts that are as compact as possible to drive down bit-cell size and ultimately memory size, speed, and power consumption. Dense, sub-DR bit-cells often pose severe printability and lithography process window problems. The integration between process and design is exacerbated in memories where stress can have severe impact on the electrical behavior of the circuitry. On top of this, performance,

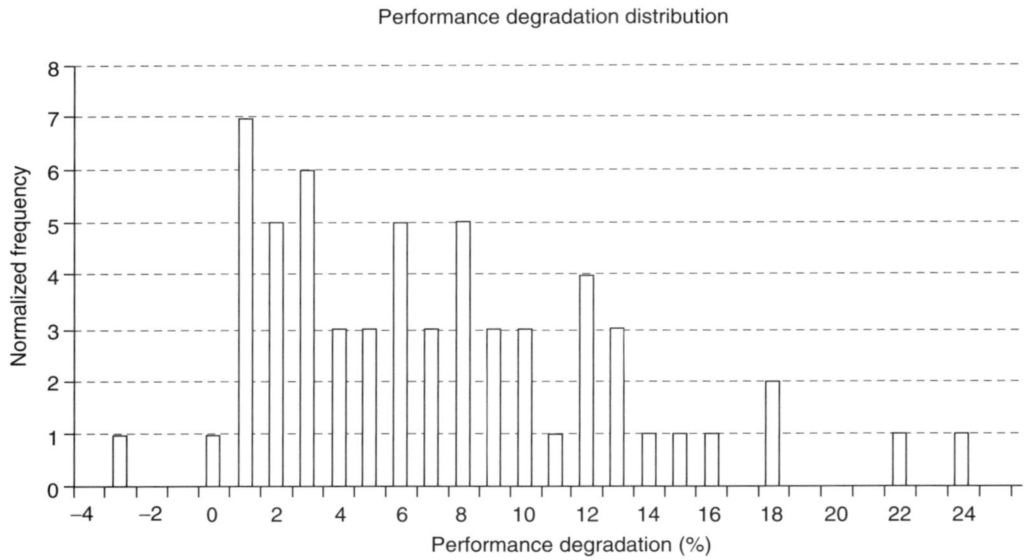

FIGURE 19.15 DFM performance degradation of a DFM library.

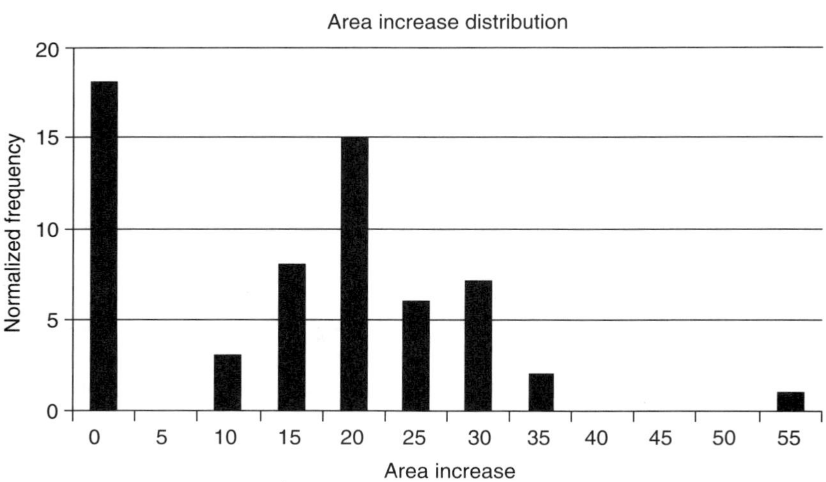

FIGURE 19.16 Area degradation of a DFM library.

power, and leakage obviously must all be optimized. All these issues must be addressed during technology development and they often require a significant amount of resources, silicon, and time in order to achieve satisfactory results.

Embedded as well as commodity memories, such as SRAMs, ROMs, FLASH, and DRAMs, make use of different fault recovery methods in order to achieve reasonable yields. The simplest and most popular method is memory redundancy that provides spare memory elements such as spare rows or columns of bit-cells, which are fuse-selectable at electrical test or by embedded self-test logic.

Many semiconductor companies started developing multiple bit-cell variants to more flexibly address different application needs and to improve manufacturability. It is in fact often the case that multiple bit-cells are employed within a single SoC product; smaller bit-cells are used for large memory arrays that usually include a significant fraction of spare resources (as the overhead in terms of block size is still tolerable), whereas relaxed or design-rule compliant bit-cells are used in smaller blocks with little or often no redundancy. A critical DFM decision has thus to occur at the memory partitioning and allocation level, where yield should be included along with speed and power considerations when allocating logic memory units to physical arrays as this may have a significant impact on the floorplan design, on the chip performance, and on its yield. Of course, this requires a somewhat accurate model of the memory yield, as well as of its repairability as a function of the available redundancy. The effort required to design and validate multiple bit-cell memory variants is relatively small if planned properly, especially when taking into account that this effort is already dominated by the high-density bit-cell designs that push DRs the hardest.

For a sound DFM strategy, it is therefore recommended that high-density memory bit-cells are complemented by at least two or three variants (depending on critical YLMs) of high-yield layouts; for example, a design-rule compliant bit-cell and an intermediate density variant with feature redundancy and litho-friendly layout. The area penalty for alternative bit-cell variants may range from 30% to 70%. The timing penalty depends on the design styles and the memory array architecture. The availability of high-yield, fast bit-cells for small embedded memories is critical to achieve better SoC yields. In fact unrepairable memory errors are often a significant factor in SoC yield loss.

As discussed previously, yield modeling is a critical DFM factor as it enables both bit-cell selection and optimized redundancy allocation.

A key enabler methodology for memory yield modeling is based on a technique called micro-event analysis [1]. The main goal of micro-event analysis is to propagate the probability of failure from the microscopic or individual feature level (such as the probability of contact opens or critical area for metal shorts) to the electrical level (such as the probability of a bit-line short event or single vs. paired bit fail). By feeding process characterization data into a micro-event probability model, it is possible to compute the frequency of certain faults and thus predict the effectiveness of any given type and amount of memory redundancy, and ultimately its yield. For example, a micro-event associated with M2 shorts may generate a paired bit-line (columns) shorts signature, which can be repaired if, and only if, at least two redundant columns per physical memory bank are available. On the other hand, an open on the access transistor gate will produce a single bit failure, which is often a repairable event up to high fail rate densities.

As discussed previously, functional failures in the logic of the memory periphery are usually a much less likely event unless there are also some significant design systematic effects such as increased fail rates in the transition region from the dense array to the surrounding logic. Parametric faults in the memory sensing circuitry are often more significant. The DFM methods for analysis and optimization of parametric yield loss are described in the next section.

19.4 Parametric Design for Manufacturing Methodologies

As discussed in the previous sections, the occurrence of random and systematic defects that modify the functionality of a circuit may cause functional yield losses. A defect-free circuit that has been tested for correct functionality can still fail to meet its performance specifications, such as speed, power consumption, and leakage. Yield loss caused by performance specs test failure is called parametric yield loss.

Analog circuits do not normally involve very large area or transistor counts, and they often adopt relaxed DRs, and hence their functional yield is often very high. However, both technology and voltage scaling increase the sensitivity of analog circuit performance to manufacturing variations, and thus high-performance, high-precision analog components are sensitive to die-to-die variation and mismatch. The design for manufacturability of analog and mixed-signal IP involves the verification and optimization of parametric variations over the measured or expected range of fabrication process perturbations.

Parametric yield loss in digital circuits often involves speed test failures — in other words the circuit meets its functional specs at low clock frequencies, but it starts failing once the internal clock speed is increased above a certain frequency value. Typically, this can happen because some of the logic gates on a critical path become too slow, thus causing an incorrect signal to be latched at the register outputs. On the contrary, it may also happen that the internal circuit synchronization is lost when some gates become too fast, thus causing hold-time violations or excessive clock skew.

Parametric/speed yield optimization of logic circuits involves accurate verification of critical path delays over all process corners by using static timing analysis (STA) or statistical static timing analysis (SSTA).

A simple way to detect parametric yield loss in products is to compare the product yield learning curve (equivalent product D_0 expressed in DEF/cm^2) to the learning curve of random defect monitors resulting, for example, from periodic measurements of random defect limited yield (RLY) structures of a characterization vehicle [14] as shown in Figure 19.17. Large deviations (slow down) from the ideal product yield learning curve with respect to RLY are an indication of the presence of parametric yield loss.

Correlation of product test bins with process control monitor (PCMs) measurements in scribe lines is also often used in product engineering to track the presence of parametric yield loss and to take suitable corrective actions.

19.4.1 Source of Parametric Yield Loss

In general, parametric yield loss can be caused by both random and systematic effects. Random sources of parametric yield loss are primarily represented by random variation of the electrical properties of the layers forming active and passive devices. For example, random fluctuations of channel dopants, oxide thickness, and effective length in MOSFETS cause I_{DSS} variation across identically designed transistors. Random variation of electrical device parameters can be extremely local, and can cause two identically drawn devices placed close to each other to have different electrical behavior. This phenomenon is called device mismatch, or intra-die variation, as opposed to inter-die variation, i.e., such that device parameters are quasi-constant for devices on the same die but change from one die to another.

Systematic sources of parametric yield loss are generally caused by electrical device parameter variation that is a function of the device layout attributes. Examples of systematic sources of parametric yield loss are:

- Device layout pattern distortion caused by printability effects
- Context-dependent poly/active flaring causing W_{EFF}/L_{EFF} variation in MOSFETs
- Stress-related mobility variation in the MOSFET channel
- Contact position and contact resistance
- Narrow salicide formation issues causing output device impedance to vary as a function of the contact to gate spacing

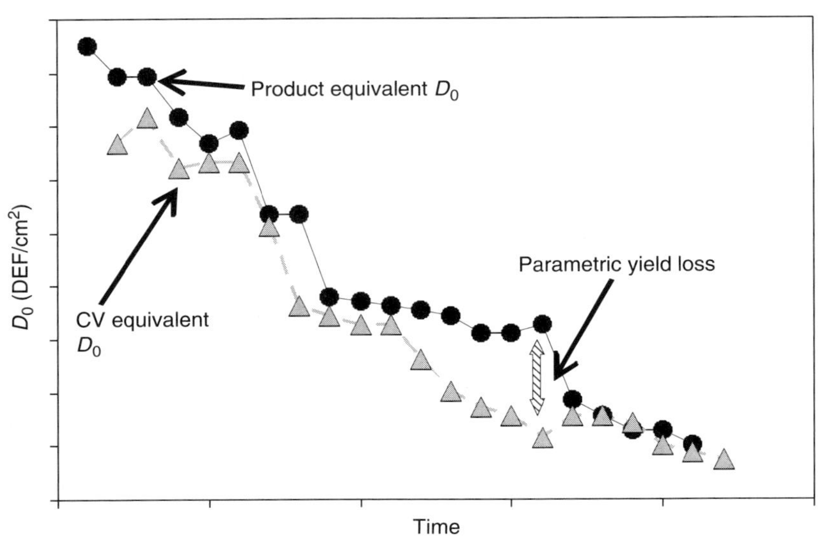

FIGURE 19.17 Identification of parametric yield loss by comparison with defect yield limited characterization vehicles.

Systematic sources of parametric yield loss, such as random sources, can have intra-die as well as inter-die components.

19.4.2 Corner Analysis

A typical approach to account for process variation is to simulate circuit performance at the so-called worst-case corners. These corners are estimated from the expected worst-case variation in key device parameters, such as I_{DSAT} and I_{OFF} of MOSFET transistors. The corners of the expected statistical distribution of the target performance parameters are translated into worst-case corner SPICE models, i.e., one or more combinations of SPICE model parameters, extracted from electrical device test data, or e-tests. The procedure of extracting corner model parameters from electrical test data often involves the identification of a set of first-order independent model parameters that can be uniquely extracted from specific regions of the I–V characteristics of certain transistors (such as TOX and NCH from I_{DSAT} and V_{TSAT} of long-channel MOSFETs). Availability of physical device models, such as BSIM4 [15], helps to determine and extract (i.e., analytically inverting the model to obtain parameter values corresponding to specific e-test data) a suitable set of such first-order physical parameters.

The set of device model parameters that corresponds to a predefined percentile point (e.g., $\pm 3\sigma$) of the e-test data statistical distribution is then used to define the SPICE models that are used for circuit corner analysis.

Corner analysis is powerful because it allows for capture of the impact of process variation on circuit performance with a small number of electrical or timing simulations; however, it presents a number of significant limitations, such as:

- Special attention is required to account properly for model parameter correlation, including different parameters of the same device (e.g., VT0 and TOX) as well as the same parameters of different devices and device mismatch. Lack of proper handling of device parameter correlation leads to unphysical corner models and often to unnecessary pessimism.
- Corner models are defined as $\pm 3\sigma$ (or $\pm N\sigma$.) percentiles of certain key device parameters that are assumed to map to the corresponding percentile of the target circuit performance parameters. Owing to the nonlinear mapping between device and circuit performance, this assumption is often inaccurate and also leads to excessive pessimism.
- Worst-case corners are typically derived by considering combinations of model parameter values that will cause slow or fast digital switching events. These do not necessarily result in the worst-case performance of analog circuits, where a different combination of parameter values could potentially be the worst case for many performances of interest (e.g., gain, offset voltage, and unity-gain bandwidth).

Statistical circuit simulation can be used to address these limitations.

19.4.3 Statistical SPICE Models

The active and passive SPICE device parameters can be modeled statistically in order to capture the effect of the electrical variation of the device parameters and ultimately to be able to simulate parametric yield loss.

As mentioned in the previous section, statistical circuit simulation represents a more accurate alternative to the worst-case corner approach. Statistical circuit simulation requires statistical and mismatch SPICE models. These models represent the change in SPICE model parameters caused by manufacturing variation. Statistical SPICE models are derived by estimating the joint probability density function (JPDF) of a certain number of model parameters. By carefully maintaining the correlation between model parameters, these models capture the correlated variation in device characteristics. For analog design, statistical and mismatch SPICE models can be used for parametric yield assessment and robust design [16]. For digital designs, these models can be utilized for estimating realistic and application-specific worst-case corners that take into account the observed device characteristics [17].

The problem of extracting statistical SPICE models is more challenging than it may seem. In fact, the most popular SPICE models such as BSIM4 employ a large number of different parameters, many of which are actually empirical fitting coefficients [15]. These parameters are practically obtained by using nonlinear, multiobjective, numeric optimization methods that are often trapped in local minima.

Therefore, the approach of estimating model parameter statistics by fitting a separate model for each individual sample in a large collection of measurement data sets (*I–V* and *C–V* curves), besides being impractical, is also subject to noise and variance inflation caused by numerical optimization noise.

Statistical modeling methods that address these issues generally fall into one of the following two categories:

1. *Monte Carlo device and process simulation (technology CAD or TCAD)*. These methods have the advantage that physical model parameters, such as TOX or LEFF, can be directly obtained from the TCAD simulators' outputs, thus reducing the need for aggressive numerical fitting [18,19]. Although TCAD-based Monte Carlo methods are very useful during the early manufacturing phase, when a large and stable set of statistical device measurements is not available, the main drawback of these approaches is the time and effort required to calibrate TCAD tools adequately.
2. *Direct extraction*. These approaches apply direct inversion to estimate the value of a reduced subset (core or first-order independent set of physical parameters) of the device model parameters [20]. Direct extraction methods are definitely more practical although they often require availability of a large number of different types of measurements, which are often difficult to obtain in a production environment.

19.4.4 Statistical Circuit Simulation

19.4.4.1 Monte Carlo Analysis

Monte Carlo analysis is the most typical and straightforward method of statistical circuit simulation. A random number generator is used to sample the value of the parameters of a set of statistical SPICE models from their probability density function (PDF). For each random sample, the circuit is simulated and the output circuit performance statistics of interest are collected.

The random Monte Carlo samples need to maintain the same statistical correlation of the original population of device parameters that are used in the target circuit. There are different ways to do this. Correlated random Monte Carlo samples can be extracted from the JPDF of the original population if this is known. Alternatively, if an estimated value of the covariance or correlation coefficient matrix is known, a linear transformation known as principal component analysis (PCA) or principal factor analysis (PFA) can be applied to the original device parameters. In this case, SPICE device parameters are defined as linear combinations of a set of new variables, which can be normalized to be independent $N(0,1)$ random variables.

The typical output of Monte Carlo simulations is the sample distribution and statistics (such as sample mean and variance) of the circuit performance specifications. An example of such sample distribution for the noise figure (NF) of an RF LNA circuit is shown in Figure 19.18.

The main drawback of Monte Carlo analysis is that a number of circuit simulations are required in order to estimate the circuit output parameter statistics with reasonable accuracy.

In fact, the Monte Carlo output statistics, such as mean and standard deviation of circuit parameters, are sample estimates of the statistical parameters of the underlying population. The variance of sample estimates is inversely proportional to the sample size, i.e., to the number of Monte Carlo circuit simulations. Therefore, in order to estimate the true underlying population distribution parameters with a reasonable confidence level (e.g., >90%), the number of simulations required is easily on the order of a few thousands, for typical circuit parameter variability.

The plots in Figure 19.19 show an example of the confidence intervals at 95% and 99% confidence levels as a function of the sample size, showing that several hundreds of simulations are required in order to get within ±5% of the true average value.

19.4.4.2 Response Surface Modeling

The method known as response surface modeling (RSM) is a statistical technique that can be applied to statistical circuit analysis in order to circumvent the runtime complexity of Monte Carlo simulations [21,22]. The principle of RSM is to determine an analytical regression model of the circuit response (e.g., VCO phase noise, amplifier BW, PM and GM, LNA gain, and NF and IP3) to SPICE device parameters variation. This regression model is typically obtained as the least square model fit of a set of circuit simulation results obtained by perturbing the SPICE device parameters according to a particular DOE plan. Figure 19.20 shows an example of a full factorial DOE in three variables.

The type of DOE plan is determined based on the regression model, the number of parameters, and the model accuracy requirements. The typical number of simulations that are required to build a quadratic regression model is less than hundred, with 10 to 12 input statistical device parameters, which is typical for the state-of-the-art MOSFET technology.

After the circuit output response parameter has been modeled by RSM, the input device parameters can be perturbed according to a Monte Carlo sampling plan, and statistical circuit behavior can be estimated using the RSM model instead of circuit simulations.

In this way very tight error bounds on the statistical parameters of the output circuit response distribution can be achieved with a much smaller computational effort.

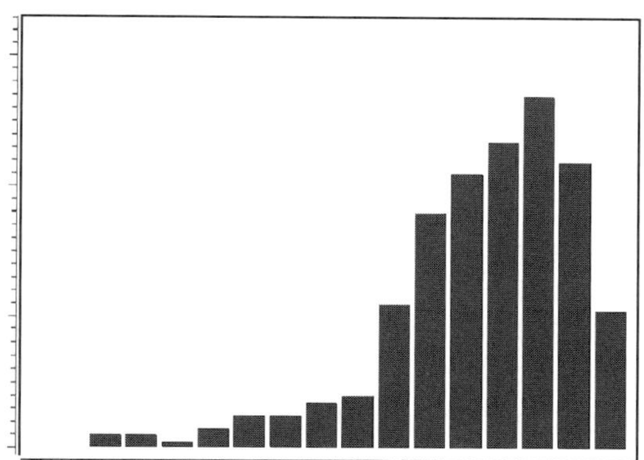

FIGURE 19.18 Example of Monte Carlo simulations results: frequency distribution of LNA noise figure.

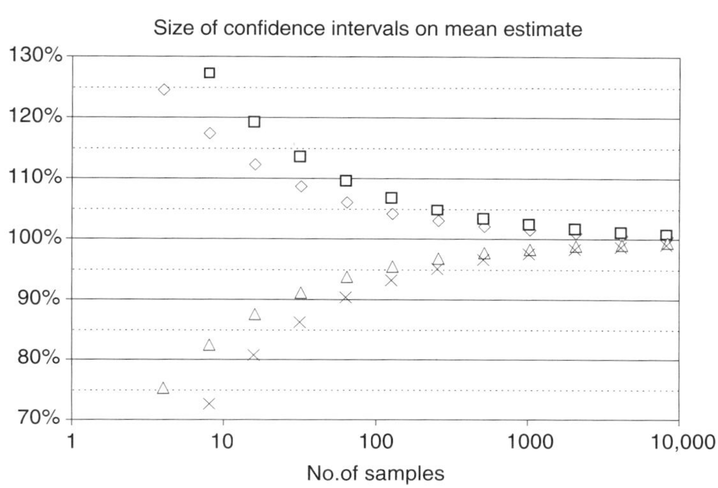

FIGURE 19.19 Confidence intervals vs. sample size.

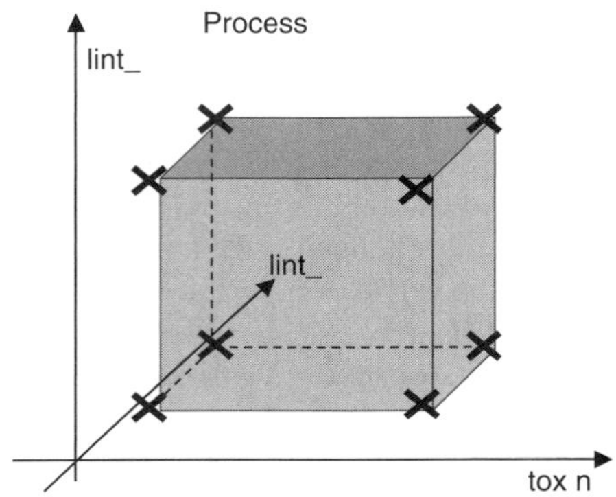

FIGURE 19.20 Full factorial DOE example in three variables.

The RSM methodology has been integrated in a commercial EDA tool [22] and is routinely used in IC industry.

19.4.4.3 Mismatch Simulation

Not only analog circuits and SRAM design, but also the clocking network of digital designs, are particularly sensitive to intra-die variation or mismatch. The reason is that in order to reduce the effect of die-to-die variability as well as to compensate for global environment parameter variation (e.g., temperature or supply voltage), circuit designers use differential circuit configurations that rely upon good matching properties of the semiconductor devices. Because of this reason, the problem of characterizing and modeling the intra-die component of device variability has been extensively studied by researchers in both academia and industry [23,24].

Nonetheless, including device mismatch in statistical circuit simulations is challenging because the number of statistical device parameters to be considered increases by a factor that is proportional to the number of transistors in the circuits. This inflation of parameters is a serious problem for RSM-based tools, as the number of DOE experiments (hence the number of simulations) grows exponentially with the number of input parameters.

This apparent roadblock can be circumvented by assuming that the parameters of identically drawn devices placed in close proximity, although not perfectly matched, are strongly correlated. Therefore, by decomposing the covariance matrix of a system of correlated random variables that describes the set of device parameters of a large circuit, it is possible to identify a small number of independent factors (linear combinations of the original correlated random variables) that can properly explain most of the system variance and correlation. This method [25], based on a two-step eigenvalue decomposition of the system covariance matrix, has been successfully applied to both analog and digital statistical circuit analysis [26,27].

19.5 Design for Manufacturing Integration in the Design Flow: Yield–Aware Physical Synthesis

19.5.1 Early Years: Reactive Design for Manufacturing

In the early years, DFM optimizations were applied just before final layout finishing and mask data preparation (MDP). In fact, some of the early types of DFM can be seen as a special kind of layout finishing or MDP operation.

For example, the introduction of feature redundancy, such as contact and via redundancy, wire spreading for critical area reduction, line biasing (over- or under-sizing), cheesing and dummy fill insertion, and to some extent OPC can be considered as the most basic form of DFM, and were all introduced after final tape-out of the mask layout — in other words after timing closure, and physical and signal integrity verification.

A typical design flow, which consists of these types of DFM operations, will be referred to as "reactive DFM" (for reasons that will become clear in the following discussion) is shown in Figure 19.21.

Despite its apparent simplicity and widespread popularity, reactive DFM presents some significant drawbacks. In fact, because reactive DFM manipulations occur after main timing closure and physical verification loops, they are supposed to leave the circuit's electrical (such as Rs and Cs) timing, and connectivity properties unchanged. This is of course not strictly possible, therefore, the typical approach is to limit the scope and location of DFM changes to transformations and places in the layout that have minimum impact on the circuit characteristics. Examples of such opportunistic optimizations are shown in Figure 19.22.

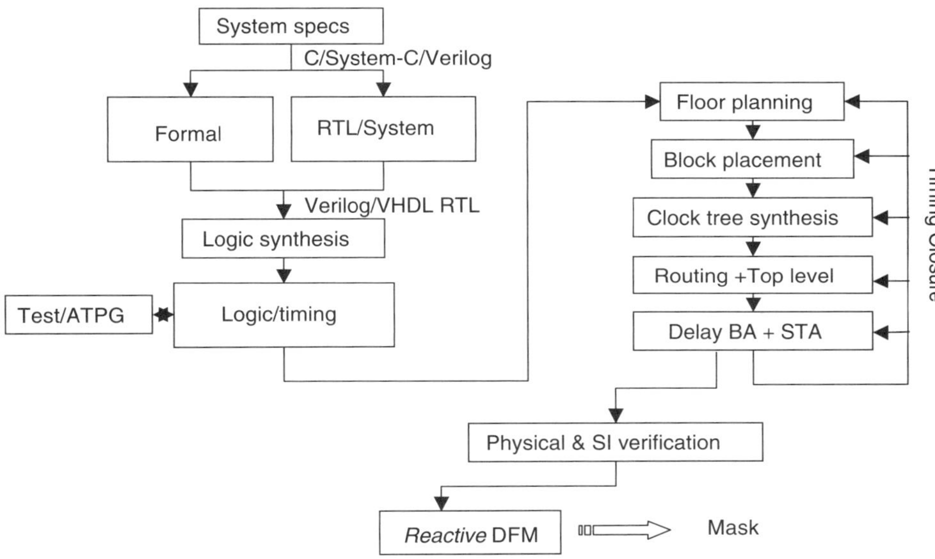

FIGURE 19.21 "Reactive DFM" design flow.

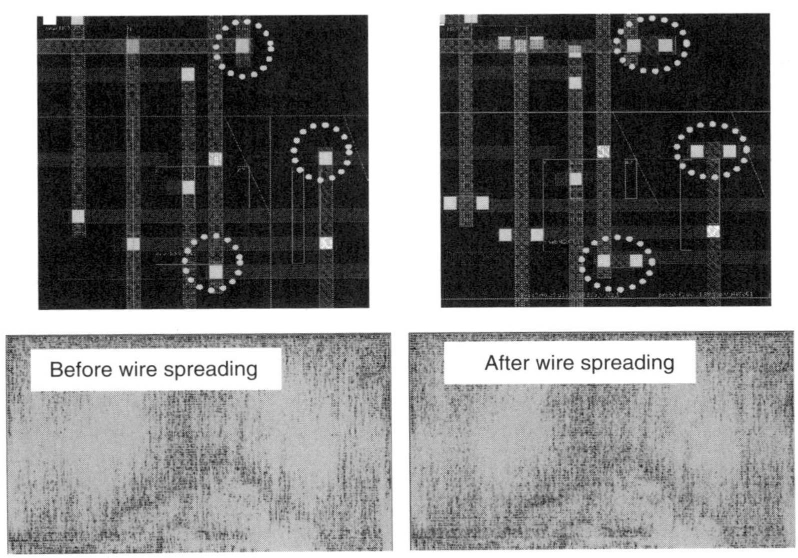

FIGURE 19.22 Opportunistic optimization examples: via redundancy (above) and wire spreading for CA optimization (below).

Although most of these simple modifications have minimum impact on the electrical characteristics, it is impossible to guarantee with absolute confidence that the impact is completely negligible, so reactive DFM methods may expose the designers to some risk and uncertainty. For example, redundant vias and contacts with their larger landing pads have a small but nonzero impact on the node resistance and capacitance; similarly slotting/cheesing (i.e., creating "holes" in fat Cu lines to improve CMP and achieve better planarity), and dummy fill insertion may change the interconnect coupling and grounded capacitance values and thus potentially invalidate timing and signal integrity simulation results.

Another drawback of reactive DFM techniques is the additional lag time that they introduce between design tape-out and mask making. In fact, most of these techniques can only be effective at full chip level and may create some additional hazards when used hierarchically, as some of the connectivity checks may fail to catch issues at the interface between blocks.

For large VLSI designs with the final flat GDS database size of an order of several Gbytes, the pure processing times of any of the above-described DFM steps may take a couple of days. Assuming a few days of physical verification, and even without taking into account the potential time to fix eventual problems and reloop, the total time for reactive DFM may easily total from a few weeks up to more than a month.

19.5.2 Proactive Design for Manufacturing

The limitations of reactive DFM techniques stem from the fact that they are applied at the end of the design cycle instead of being integrated into the design flow.

In order to address the limitation of reactive DFM, it is thus necessary to develop DFM methods that are integrated upstream in the design flow and that are capable of synthesizing DFM-friendly physical layouts, while meeting all remaining design constraints (such as timing, power, and signal integrity) before timing closure and physical verification.

A key enabler of DFM integration in the design flow is the capability of precisely quantifying the yield and manufacturability impact of every different design choice. In fact, in this way it is possible to expand the cost function of the various design optimization steps in the design flow to include a yield metric. Design flow tools are thus made aware of the manufacturability cost of each different choice that they take and trade-off yield for existing slacks in the remaining design dimensions.

An example of such a proactive approach of DFM has been presented in [28], where the cost function of logic synthesis has been modified in order to take into account different types of yield metrics. Other proactive DFM approaches have been developed introducing yield cost and metrics at different stages in the design flow such as floorplanning, physical synthesis, and routing [29,30].

The key feature of any of these methods is the silicon characterization of accurate yield models of the circuit IP blocks, such as standard cells, interconnects, memories, and hard-macros, and the integration of such models into the cost function of the design tools.

The implementation described in [29], for example, makes use of accurate standard cell yield models to optimize the tech-mapping and placement steps of logic synthesis by trading off some of the available timing slack of logic cells placed off the critical path for increased design manufacturability.

Planning for manufacturability at earlier stages of the design flow, such as floorplanning and global routing, leaves increased margins for DFM optimizations when compared to a methodology that performs DFM optimizations after detailed routing and timing ECOs, when the physical design hierarchy is basically frozen.

This allows substantial gains in terms of design manufacturability which may translate into 10% or more yield and GDPW (good die per wafer) improvement as per reported, for example in [30].

19.6 Summary

Until the deep-submicron era, worst-case simulation and DR compliance were the only methods that design engineers had to worry about in order to ensure consistent product yields. However, with each new generation of process technology, DRs become more complicated trying to provide tolerance to increasingly complex layout systematic effects, while intra-die variability and pattern dependency of MOSFET parameters make worst-case simulation less and less manageable. In this chapter, we have focused on the DFM requirements for technologies in the nanometer range where the spectrum of physical phenomena that may impact manufacturability of products is mind-boggling. We have shown that the systematic characterization of these phenomena and their impact on IC yield and performance is of crucial importance for true DFM.

We have stated that the most accurate way of accounting for these effects is to provide accurate physical models and then simulate the actual IC layout to estimate the impact on yield and performance. Such a simulator requires calibration to the actual manufacturing process by specially designed test structures that cover all the possible layout patterns found in real products. To provide observability in the range of a few failures per billion as a function of layout attributes, such test chips must contain many specially designed layout patterns that require large die size up to the full reticle area. In return, the necessary DFM characterization is achievable with just a few wafers, which actually reduces the cost and (equally importantly) the turnaround time.

We have shown that such a modeling-based approach can then serve as the basis for generating guaranteed-to-yield IP blocks and as a yield sign-off tool for the entire IC layout. Moreover, such a simulation-based model can also be used to generate guidelines for the physical-design tools (for place and route), without creating any extra effort for IC designers or drastically changing the design flow. We have shown that such an approach enables truly proactive DFM in which all the design modifications take place before verification and after tape-out.

As process technology moves into the sub-50nm era, new approaches are being developed to maximize layout regularity that will guarantee manufacturability of giga-scale ICs designed with these technologies. The extreme version of such regularity will be required where guaranteed-to-yield layout patterns are created in this DFM environment and layout synthesis becomes equivalent to pattern assembly [31].

References

[1] W. Maly, Computer-aided design for vlsi circuit manufacturability, *Proc. IEEE*, 78, 356–390, 1990.

[2] R. Myers and D. Montgomery, *Response Surface Methodology — Process and Product Optimization Using Designed Experiments*, Wiley, New York, 2002.

[3] S. Saxena et al., Tests structures and analysis techniques for estimation of the impact of layout on MOSFET performance and variability, *ICMTS 2004*, 2004.

[4] C. Hess, D. Stashower, B. Stine, G. Verma, and L. Weiland, Fast extraction of killer defect density and size distribution using a single layer short flow NEST structure, *IEEE 2000 International Conference on Microelectronic Test Structures*, Vol. 13, March 2000.

[5] C. Hess, B.E. Stine, L.H. Weiland, T. Mitchell, M. Karnett, and K. Gardner, Passive multiplexer test structure for fast and accurate contact and via fail rate evaluation, *ICMTS 2002*, 2002.

[6] J. Horgan, *TEST & ATE — Cost of Test*, http://www.edacafe.com/magazine/

[7] D. Hamling, ATE solutions for the global semiconductor industry, *Fabless Forum*, Vol. 11, Fabless Semiconductor Association, March 2004.

[8] J. Orbon et al., Integrated electrical and SEM based defect characterization for rapid yield ramp, *SPIE 2004*, 2004.

[9] D. Ciplickas, A. Joshi, S. Lee, and A.J. Strojwas, *Designing for High Product Yield*, Semiconductor International, October 2002.

[10] PDF Solutions, Inc., *Yield Ramp Simulator User Manual v3.1*, Copyright 2001-2003.

[11] K. Keutzer, K. Kolwicz, and M. Lega, Impact of library size on the quality of automated synthesis, *Proc. ICCAD*, 1987, pp. 120–123.

[12] K. Scott and K. Keutzer, Improving cell libraries for synthesis, *Proceedings of CICC*, 1994, pp. 128–131.

[13] N. Dragone et al., High yield standard cell libraries: optimization and modeling, *PATMOS 2004*, 2004.

[14] K. Miyamoto, K. Inoue, I. Tamura, N. Kondo, H. Inoto, I. Ito, K. Kasahara, and Y. Oshikiri, Yield management for SoC vertical yield ramp, *IEDM 2000*, 2000.

[15] *BSIM4v4 Manual*, Department of Electrical and Computer Science, University of California, Berkeley, CA, 2004, http://www.device.eecs.berkeley.edu/~bsim3/bsim4.html

[16] C. Guardiani, S. Saxena, P. Schumaker, P. McNamara, and D. Coder, An asymptotically constant, linearly bounded methodology for the statistical simulation of analog circuits including component mismatch effects, *Proceedings IEEE/ACM Design Automation Conference*, 1999, pp. 15–18.

[17] A.D. Fabbro, F. Franzini, L. Corce, and C. Guardiani, An assigned probability technique to derive realistic worst-case timing models of digital standard-cells, *Proceedings IEEE/ACM Design Automation Conference*, 1995, pp. 702–706.

[18] pdfab Manual, PDF Solutions, 333 W. San Carlos Street, Suite 700, San Jose, CA, http://www.pdf.com

[19] D.A. Hanson, R.J.G. Gossens, M. Redford, J. McGinty, J.K. Kibarian, and K.W. Micheals, Analysis of mixed-signal manufacturability with statistical TCAD, *IEEE Transactions on Semiconductor Manufacturing*, 1996.

[20] J.C. Chen, C. Hu, C.-P. Wan, P. Benedix, and A. Kapoor, E-T based statistical modeling and compact statistical circuit simulation methodologies, *Proceedings International Electron Device Meeting*, 1996, pp. 635–638.

[21] G. Nicollini and C. Guardiani, A 3.3-V 800-nV^{rms} noise, gain programmable CMOS microphone preamplifier design using yield modeling technique, *IEEE J. Solid State Circuits*, 28, 915–921, 1993.

[22] Circuit Surfer Manual, PDF Solutions, 333 W. San Carlos Street, Suite 700, San Jose, CA, http://www.pdf.com

[23] C. Guardiani, A. Tomasini, J. Benkoski, M. Quarantelli, and P. Gubian, Applying a submicron mismatch model to practical IC design, *Proceedings of IEEE Custom Integrated Circuits Conference*, San Diego (CA), May 1994.

[24] M. Quarantelli et al., Characterization and modeling of MOSFET mismatch of a deep submicron technology, *International Conference on Microelectronic Test Structures*, 17–20 March 2003, pp. 238–243.

[25] C. Guardiani, S. Saxena, P. McNamara, P. Schumaker, and D. Coder, An asymptotically constant, linearly bounded methodology for the statistical simulation of analog circuits including component mismatch effects, *Proceedings IEEE/ACM 37th Design Automation Conference*, Los Angeles, CA, June 2000.

[26] P. McNamara, S. Saxena, C. Guardiani, H. Taguchi, E. Yoshida, N. Takahashi, K. Miyamoto, K. Sugawara, and T. Matsunaga, Design for manufacturability characterization and optimization of mixed-signal IP", *Proceedings of IEEE Custom Integrated Circuits Conference*, San Diego, CA, May 2001.

[27] E. Malavasi, S. Zanella, J. Uschersohn, M. Misheloff, M. Cao, and C. Guardiani, Impact analysis of process variability on digital circuits with performance limited yield, *Proceedings IEEE International Workshop on Statistical Metrology*, Kyoto, JP, June 2001.

[28] A. Nardi and A.L. Sangiovanni-Vincentelli, Synthesis for manufacturability: a sanity check, *Design, Automation and Test in Europe Conference and Exhibition. Proceedings*, Vol. 2, 16–20 February 2004, pp. 796–801.

[29] C. Guardiani, N. Dragone, and P. McNamara, "Proactive design for manufacturing (DFM) for nanometer SoC designs" custom integrated circuits conference, 2004, *Proc. IEEE 2004*, pp. 309–316, 2004.

[30] J. Kibarian, C. Guardiani, and A.J. Strojwas, Design for manufacturability in nanometer era: system implementation and silicon results, *International Solid State Circuits Conference (ISSCC), Proceedings of the IEEE*, 6–10 February 2005, pp. 18–19.

[31] V. Kheterpal, V. Rovner, T.G. Hersan, D. Motiani, Y. Takegawa, A.J. Strojwas, and L. Pileggi, Design methodology for IC manufacturability based on regular logic-bricks, *DAC 2005*, 2005.

[32] D. Ciplickas, S.F. Lee, and A.I. Strojwas, A new paradigm for evaluating IC yield loss, *Solid State Technol.*, 47–52, October 2001.
[33] D.J. Ciplickas, X.Li, and A.J. Strojwas, *"Predictive Yield Modeling of VLSIC's"2000 Symposium on VLSI Technology*, Statistical Metrology Workshop, Honolulu, HI, June 2000.
[34] C. Hess and L. Weiland, Determination of defect size distributions based on electrical measurements at a novel harp test structure, *Proceedings IEEE 1997 International Conference on Microelectronic Test Stuctures*, Vol. 10, March 1997.
[35] M. Quarantelli et al., Characterization and modeling of MOSFET mismatch of a deep submicron technology, *IEEE International Conference on Microelectronic Test Structures*, 2003.
[36] E.P. Box and N.R. Draper, *Empirical Model-Building and Response Surfaces*, Wiley, New York, 1987.

20

Design and Analysis of Power Supply Networks

David Blaauw
University of Michigan
Ann Arbor, Michigan

Sanjay Pant
University of Michigan
Ann Arbor, Michigan

Rajat Chaudhry
Freescale Semiconductor Inc.
Austin, Texas

Rajendran Panda
Freescale Semiconductor Inc.
Austin, Texas

20.1 Introduction

Power distribution networks distribute power and ground voltages from pad locations to all devices in a design. Shrinking device dimensions, faster switching frequencies, and increasing power consumption in deep submicron technologies cause large switching currents to flow in the power and ground networks, which degrade performance and reliability. A robust power distribution network is essential to ensure reliable operation of circuits on a chip. Power supply integrity verification is a critical concern in high-performance designs. Due to the resistance of the interconnects constituting the network, there is a voltage drop across the network, commonly referred to as the *IR drop*. The package supplies currents to the pads of the power grid either by means of package leads in wire-bond chips or through C4 bump arrays [1] in flip-chip technology. Although the resistance of the package is quite small, the inductance of the package leads is significant, which causes a voltage drop at the pad locations due to the time-varying current drawn by the devices on the die. This voltage drop is referred to as the d*i*/d*t drop*. Therefore, the voltage seen at the devices is the supply voltage minus the *IR* drop and d*i*/d*t* drop.

Excessive voltage drops in the power grid reduce switching speeds [2–4] and noise margins of circuits, and inject noise which may lead to functional failures. High average current densities lead to the undesirable wearing out of metal wires due to electromigration (EM) [5]. Therefore, the challenge in the design of a power distribution network is in achieving excellent voltage regulation at the consumption points notwithstanding the wide fluctuations in power demand across the chip, and to build such a network using minimum area of the metal layers. These issues are prominent in high-performance chips such as microprocessors, since large amounts of power have to be distributed through a hierarchy of many metal layers. A robust power distribution network is vital in meeting performance guarantees and ensuring reliable operation.

Capacitance between power and ground distribution networks, referred to as the decoupling capacitance or *decap*, acts as local charge storage and is helpful in mitigating the voltage drop at supply points. Parasitic capacitance between metal wires of supply lines, device capacitance of the nonswitching devices, and capacitance between the N-well and substrate occur as implicit decoupling capacitance in a power distribution network. Unfortunately, this implicit decoupling capacitance is not enough to constrain the voltage drop within safe bounds and designers have to often add intentional explicit decoupling capacitance structures on the die at strategic locations [6]. These explicitly added decoupling capacitances are not free, and increase the area and leakage power consumption of the chip. Parasitic interconnect resistance, decoupling capacitance, and package/interconnect inductance form a complex RLC network which has its own resonant frequency [7, 8]. If the resonance frequency lies close to the operating frequency of the design, large voltage drops can develop in the grid.

The crux of the problem in designing a power grid is that there are many unknowns until the very end of the design cycle. Nevertheless, decisions about the structure, size, and layout of the power grid have to be made at very early stages when a large part of the chip design has not even begun. Unfortunately, most commercial tools focus on postlayout verification of the power grid when the entire chip design is complete and detailed information about the parasitics of the power and ground lines and the currents drawn by the transistors are known. Power grid problems revealed at this stage are usually very difficult or expensive to fix. A methodology has been presented in [9, 10] that helps to design an initial power grid and refine it progressively at various design stages. In this chapter, we show how such a methodology can be extended to include the effects of package inductance in the analysis. Due to the growth in power consumption and switching speeds of modern high-performance microprocessors, the di/dt effects are becoming a growing concern in these designs. Clock-gating (see the chapter on Power Optimization), which is a preferred scheme for power management of high-performance designs, can cause rapid surges in current demands of macroblocks and increase di/dt effects. Designers rely on the on-chip parasitic capacitances and intentionally added decoupling capacitors to counteract the di/dt variations in the voltage. But it is necessary to model accurately the inductance and capacitance of the package and chip and analyze the grid with such models, as otherwise the amount of decoupling to be added might be underestimated or overestimated. Also it is necessary to maintain the efficiency of the analysis even when including these detailed models.

A critical issue in the analysis of power grids is the large size of the network (typically millions of nodes in a state-of-the-art microprocessor). Simulating all the nonlinear devices in the chip together with the power grid is computationally infeasible. To make the size manageable, the simulation is done in two steps. First, the nonlinear devices are simulated assuming perfect supply voltages and the currents drawn by the devices are measured. Next, these devices are modeled as independent time-varying current sources for simulating the power grid, and the voltage drops at the transistors are measured. Since voltage drops are typically less than 10% of the power supply voltage, the error incurred by ignoring the interaction between the device currents and the supply voltage is small. By doing these two steps, the power grid analysis problem reduces to solving a linear network, which is still quite large. To reduce further the network size, we can exploit the hierarchy in the power distribution models.

Note that the circuit currents are not independent due to signal correlations between blocks. This is addressed by deriving the inputs for individual blocks of the chip from the results of logic simulation using a common set of chip-wide input patterns. An important issue in power grid analysis is to determine what these input patterns should be. For IR-drop analysis, patterns that produce maximum instantaneous currents are required, whereas for electromigration purposes, patterns producing large sustained (average) currents are of interest.

Power grid analysis can be classified into *input vector-dependent* methods and *vectorless* methods. The input vector pattern-dependent methods employ search techniques to find a set of input patterns that cause the worst drop in the grid. A number of methods have been proposed in the literature [11–13] which use genetic algorithms or other search techniques to find vectors or a pattern of vectors that maximize the total current drawn from the supply network. Input vector-pattern-dependent approaches are computationally intensive and are limited to circuit blocks rather than full-chip analysis. Furthermore, these approaches are inherently optimistic, underestimating the voltage drop and thus letting some of the supply noise problems go unnoticed. The vectorless approaches [14–16], on the other hand, aim to compute an upper bound on the worst-case drop in an efficient manner. These approaches have the advantage of being fast but may sometimes be too conservative, leading to over-design.

In this chapter, we present a methodology for the design and analysis of power distribution networks across different stages of the design process. We consider a microprocessor design flow since it includes all aspects of power distribution design and analysis, but the flow applies equally well to any high-performance design. We assume that suitable input patterns are available for simulation.

This chapter concentrates on the issue of computing the worst voltage drops in the power network. Electromigration is an equally serious concern, but is attacked with almost identical methods. Instead of the voltage at each node, EM analysis solves for current in each branch, and instead of a voltage limit, there is a current limit per wire, depending on its layer and width.

Other IC applications may use only a portion of the flows specified here. A gate array or FPGA designer, for example, will be used only in the design stages, since the detailed usage of these parts is not known when the power supply must be designed. Likewise, a user of FPGAs or gate arrays will only use the analysis portion, as the design is already fixed.

20.2 Voltage-Drop Analysis Modes

To apply the voltage-drop analysis methodology across all stages of the design of a complex microprocessor, we define several modes of operation of the voltage-drop analysis tool. These modes are distinguished by different models of the power distribution network and the currents being drawn by the functional blocks. The *modes* of operation can be broadly classified as *early* or *prefloorplan* mode, *postfloorplan* mode, and *postlayout* mode. As the design proceeds, voltage-drop analysis is run in these different modes using more and more accurate models of the power grid and the block currents.

20.2.1 Early Mode Analysis

At the very early stages of the design of a microprocessor, there are a number of issues related to the power distribution network that have to be addressed. These include locations of the clean Vdd/Gnd pads, nominal pitches and widths of metal layers, via styles (point or bar vias) and parameters of the chip package. Since at this early stage of the design, the power network has not yet been synthesized and the location and logic content of the blocks are not known, IR-drop analysis is performed using very simplistic models of grid topology and block currents. Following are the steps involved in early mode analysis:

1. A mock power grid down to the lowest metal layer is constructed using a simple uniform grid topology, where the metal lines in each layer have a user-specified pitch (separation) and width. At the areas where the metal lines of adjacent layers cross over, vias are placed according to user-specified via geometries and via styles. Other topologies such as rings can also be modeled. The clean Vdd/Gnd pads can be placed at the periphery of the chip or on the surface of the chip using C4 pads (for flip-chip packages).
2. To model the currents drawn by the devices, a simple area-based DC estimate of the current is used. This is obtained by taking the current estimate of a previous chip and scaling it by the power supply voltage, operating frequency, complexity, size, and technology variables. This estimate is inflated three to seven times to account for differences between the average and maximum instantaneous currents and to obtain a robust grid. The current sinks are placed on the lines of the lowest metal layer at points midway between adjacent vias that connect the lowest layer to the upper layer. The value of the current is obtained by multiplying the per-unit area current by the product of the pitches of the two metal layers.

Using simple length-based resistance formulae, a resistive electrical network is constructed from the mock grid topology. Direct current analysis of this network yields the IR drops at various locations of the chip. This analysis is very fast and allows the designer to evaluate a large number of different topologies and to trade off robustness and metal utilization in the power grid. This analysis is used to design the locations of the C4 pads and nominal pitches and widths of the metal layers. Moreover, if the processing technology allows different width and thickness combinations for some of the top metal layers, the user can determine the best values of these in terms of the IR drop. Even though the real power grid will not be as regular as the mock grid, and all the devices will not be drawing the estimated current simultaneously, important design decisions are made from the results of this simple analysis and an early picture of the robustness of the grid is obtained.

20.2.2 Postfloorplan Analysis

In this mode, the global power distribution network has been designed and the blocks have been placed. The locations and geometries of the power lines and the blocks are read from the design database. Even though the blocks are placed, the power grids within them have not yet been wired. The *power service terminals* (PSTs) of a block are the wires in the topmost metal layer within the block that connect the global and intrablock power networks. In this mode, the PSTs for a block may or may not be known — if they are not known, mock PSTs are constructed. Next, the block *ports* are determined by the intersection or overlays between the global lines and the block PSTs. In a typical design hierarchy, *blocks* are custom datapath components, synthesized random logic macros (RLM), and off-the-shelf (OTS) components (custom components that can be reused). Custom components are small, but RLMs can be large. Off-the-shelf components can range from small blocks (nands, nors, muxes, etc.) to large blocks (adders, comparators, etc.). A *functional block* (e.g., floating point unit, memory management unit) consists of several instances of custom, synthesized and OTS components.

Each block current can be independently described in one of the following ways, thus allowing a mixture of them:

1. If the logic content of a block is not yet defined, the current model is a DC estimate based on the block area. The total block current is divided equally among all the ports. Since the area-based numbers are calculated such that they reflect expected peak currents, the analysis where every block has area-based currents is likely to be pessimistic since it assumes that each block draws this current simultaneously.
2. The next more accurate block current model is derived from a full-chip gate-level power estimation tool. Given a set of chipwide input vectors, this tool computes the average power consumed by each block over a cycle. From the average power consumed by a block, an average block current is computed and distributed equally among all of its ports. Hence in this mode, a multicycle DC current signature is used for a block. Since chipwide vectors are used for the simulation, correlation among the blocks is preserved.
3. The most accurate current model comes from a detailed transistor-level simulation of a functional block using a fast transistor-level simulator. The input vectors for the functional block are derived from the chipwide vectors through logic simulation. This ensures that correlation across *functional blocks* is maintained. The transistor-level netlist of the block is available and capacitances are extracted for the signal nets. However, since the power grid within the block has not been designed, it is considered to be ideal. The transient current waveform drawn by each custom, RLM, or OTS block within the functional block is obtained from a power simulator and is divided equally among all of their individual ports. Since the blocks are not very large and the power grid within them have not yet been wired, this block current model is quite accurate for this stage of the design.

If all block current models are derived from methods 1 and 2 above, then a resistance-only electrical network is extracted from the geometrical information using length-based resistance formulae. Direct current analysis is then performed to yield the IR-drop values at each of the block ports (multiple DC analyses if multicycle DC current signatures are used). If transient current signatures are used for some of the blocks, then an RC network is extracted from the global grid using length-based resistance extraction and statistical rule-based capacitance extraction (since global routing is not done, these statistical rules account for coupling

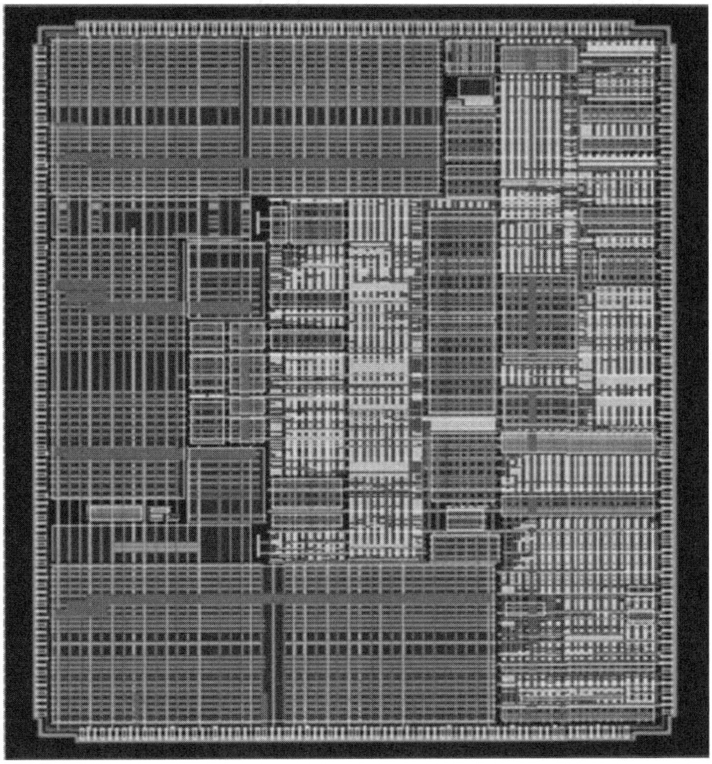

FIGURE 20.1 Power grid of PowerPC™ 750 microprocessor.

between the power and signal lines). Figure 20.1 shows the post-floorplan power distribution network for the PowerPCTM750 microprocessor. The worst IR drop was obtained to be 170mV after this analysis.

20.2.3 Postlayout Mode

This model is used when the global and block-level grids have been completely designed. Using the layout of the power grid, an accurate RC model of the power grid is extracted using a commercial extraction tool. In addition to the resistance of the network, the device capacitances and the inductance of the package can also be modeled. The models for inductance and capacitance are discussed in more detail in Section 20.4. The tap points for each device are determined during extraction. The current profile for each gate is determined by using a fast transistor-level simulator.

In this mode, the power network analysis produces time-varying voltages and currents at all points in the power grid. Figure 20.2 shows an example of the postlayout-based analysis for the PowerPC™ 750 microprocessor. The worst drop in this case is 170 mV, which corresponds well with the floorplan analysis mode.

20.3 Linear System Solution Techniques

With the increasing number of devices on a chip, the size of power networks has grown so large as to make the verification task very challenging. Power grid simulation involves solution of a system of differential equations using circuit simulation approaches such as the *modified nodal analysis* (MNA) [17]:

$$Gx(t) + Cx'(t) = b(t) \tag{20.1}$$

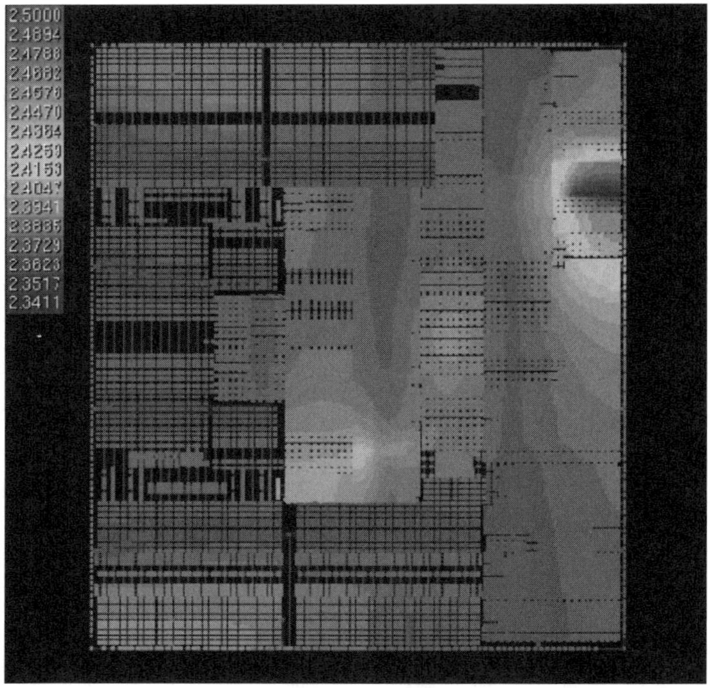

FIGURE 20.2 **(See color insert following page 15–4.)** Postlayout IR-drop map of PowerPC™ 750 microprocessor.

where G is the conductance matrix; C is the matrix resulting from capacitive and inductive elements; $b(t)$ is the vector of ideal voltage sources and time-varying current sources; and $x(t)$ is the vector of node-voltages, inductor currents and currents drawn from ideal voltage sources.

This differential system is very efficiently solved in the time domain by reducing it to a linear algebraic system:

$$\left(G + \frac{C}{h}\right)x(t) = b(t) + \frac{C}{h}x(t-h) \tag{20.2}$$

using the Backward Euler (BE) technique with a small fixed time-step, h. The BE reduction with a fixed time-step is advantageous for transient simulation since the left-hand side (LHS) matrix $(G + C/h)$, referred to as the coefficient matrix, does not change during simulation, allowing for preprocessing or factoring of the matrix for a one-time cost and for efficiently reusing to solve the system at successive time points.

Several *direct* [18] and *iterative* [19] approaches are available to solve the linear system of equations as in Equation (20.2). Direct techniques rely on factorizing the LHS matrix once and then using the LU factors repeatedly in a simple backward and forward substitution procedure [18] to solve the system at every time-step. Iterative methods, on the other hand, rely on efficient convergence techniques to steer the iterations from an initial guess to the final solution. In this section, we analyze the relative merits and limitations of these methods as applied to solving large power networks. The size and structure of the conductance matrix of the power grid is important in determining the type of linear solution technique that should be used. Typically, the power grid contains millions of nodes, but the conductance matrix is sparse (typically, less than five entries per row/column). This matrix is also symmetric positive-definite, but for a purely resistive network, it may be ill-conditioned.

Sparsity favors the use of iterative methods, but convergence is slowed down by ill conditioning and can be mitigated to some extent by preconditioning [20]. Iterative methods do not suffer from size limitations as long as the (sparse) matrix and some iteration vectors can fit into the memory. The single biggest problem with direct methods is the need for large amounts of memory to store the factors of the matrix. The number of fill-ins is of the order of $O(N^2)$, where N is the number of rows/columns in the matrix. However, if fixed time steps are used for transient analysis, then the initial factorization can be reused with subsequent current vectors, thus amortizing the large decomposition time. Iterative methods do not have this feature of reusability. Iterative methods are best suited for solving large systems using limited memory resources.

When the vector $x(t)$ in Equation 20.2 consists only of node voltages (power grid network of only R, C elements, and current sources), the coefficient matrix, $(G + C/h)$ can be shown to be symmetric and positive-definite. The symmetric positive definiteness of the coefficient matrix, which is also very sparse, is especially attractive as the system can now be solved very efficiently using specialized linear system solution techniques, such as Cholesky factorization (direct method) and conjugate gradient (iterative method) techniques. The MNA circuit formulation is no longer guaranteed to be positive-definite when inductance is included in the power grid model. However, using simple nodal or mesh current formulation, the RLC model of the power distribution network can also be converted into a symmetric positive-definite system and the above techniques can be effectively used. To speed up the simulation further by exploiting the hierarchy in the supply distribution network, a hierarchical macromodeling-based approach has been presented in [21], where the power grid is divided into a global grid and multiple local grids.

Several other approaches have also been proposed to compute the voltage drop. Kozhaya et al. [22] proposed a PDE-like multigrid method [23] for the simulation of large power grids. This method, which is particularly attractive for regular meshes, reduces the size complexity by solving several coarser meshes and interpolating the results to the original fine mesh. In [24], a frequency-based analysis has been proposed which, though efficient for repetitive signals, does not perform well for irregular simulation vectors. The approach in [16] formulates the voltage-drop maximization problem as a linear optimization formulation with constraints on block currents, while [15] formulates the voltage-drop computation as an integer linear-programming (ILP) problem to estimate the worst-case switching activity based on working modes of macro blocks. Recently, a statistical approach [25] based on random walks [26] has been proposed that exploits the localizing property of power grids.

20.4 Models for Power Distribution Networks

Modeling of power distribution networks depends on the analysis required. Although a purely resistive model may be sufficient for IR-drop computation, a more detailed model including the network capacitances, device parasitic capacitances, explicit decoupling capacitors, and the inductance of the package is often used when analyzing the network with time-varying currents. Due to higher switching speeds, the *di/dt drop* is becoming a significant part of the total voltage drop. Therefore, it is becoming very important to model the effect of package inductance. Analyzing the inductance effects without accurately modeling the capacitance will give meaningless results since the high fluctuation of currents at the pads will cause very large voltage swings. In the following subsections, we present techniques to model the resistance, capacitance, and inductance of a supply network.

20.4.1 Resistance and Capacitance Models

Three sources of capacitances affect the voltages in the power grid: (1) parasitic wire capacitances between power wires and ground wires, substrate, or signal nets, (2) parasitic capacitance of transistors, and (3) explicitly placed decoupling capacitors.

Parasitic wire capacitances can be extracted using either approximate Chern equations that use the width and spacing between wires, or by using a commercial extraction tool (See chapter 22, *Layout Extraction*). A difficult issue is the treatment of the coupling capacitors to signal nets. The effect of these capacitors on the voltage in the power grid depends on the state of the signal net. For example, coupling from a power network to a signal net that is high simply couples the power network to itself, with little or no effect on the voltage drop. Unfortunately, it is infeasible to model the signal nets and power grid simultaneously. Therefore, a statistical approach needs to be taken while modeling the coupling capacitance between the power grids and signal nets. The switching activity is determined by calculating the average number of signal nets that switch in a clock cycle. Since the low and high switching probability of a signal is equal, it is reasonable to assume that their effects cancel each other out. Therefore, switching nets can be ignored. Of the remaining nonswitching nets, half are considered to be in stable high state and half in stable low state. Each coupling capacitance is replaced by an effective capacitance to ground, in series with a resistor. The effective value of the coupling capacitor is

$$C_{\text{eff}} = \frac{1}{2} C_{\text{coupling}}(1 - P_{\text{active}}) \tag{20.3}$$

where P_{active} is the average switching activity. The resistor models the effective resistance of the gate holding the signal net in its stable state, and can be an average value over all gates or a specific value for each particular gate.

Device capacitances attenuate the voltage drop in the power grid and have a larger effect on the voltage in the power grid as they are much larger than the wire capacitances. The effect of device capacitances also depends on the state of the signal. Again, we propose a statistical approach to model these capacitances. Each transistor has five device capacitances, C_{sb} (source to bulk), C_{db} (drain to bulk), C_{gs} (gate to source), C_{gd} (gate to drain), and C_{gb} (gate to bulk). The C_{sb} can be ignored since the source and bulk for both the pmos and nmos are always at the same potential. Figure 20.3(a) shows the remaining four device capacitances for an inverter. In Figure 20.3(b), the capacitances are arranged across three inverters to make the analysis more convenient. The capacitances of switching devices contribute to the current drawn from the grid, which is already modeled by the time-varying current source in the power grid analysis. Therefore, in power grid analysis, we need to consider the device capacitances of only those gates which do not switch.

We first look at the case where net N in Figure 20.3(a) is in a low state. The device capacitances shown in Figure 20.3(a) can be modeled with the equivalent RC circuit shown in Figure 20.3(b). The resistance R_{p} corresponds to the effective pull-up resistance of inverter 1, the resistance R_{n} corresponds to the effective pull-down resistance of inverter 2, and the resistances R_{wp} and R_{wn} correspond to the P and N well resistances, respectively. Since net N is low, capacitances C_{dbn}, C_{gsn} and C_{gbn} are discharged and do not contribute to the decoupling between the power and ground grid. Furthermore, since R_{wp} is a relatively high resistance and since C_{dbp} and C_{gbp} are small, they can be ignored without significant loss in accuracy. An analogous

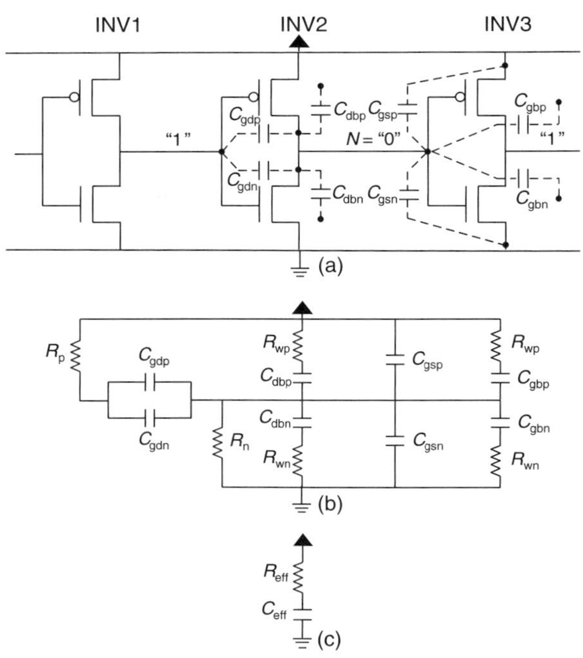

FIGURE 20.3 Capacitance model for device capacitance in power grid analysis.

analysis can be made for the decoupling capacitances when the state of signal *N* is high. This state is assumed to have equal probability of being high or low when the gate is not switching, although a different ratio of high to low signal states could easily be incorporated in the analysis. An approximate model of the device capacitances for power grid analysis is shown in Figure 20.3(c). The effective decoupling capacitance in this simplified model is the sum of the effective high and low decoupling capacitances weighted by the probability of the gate being in either state:

$$C_{\text{eff}} = (1 - P_{\text{active}})(C_{\text{gdp}} + C_{\text{gdn}}) + \frac{1}{2}(1 - P_{\text{active}})(C_{\text{gsn}} + C_{\text{gsp}}) \quad (20.4)$$

Similarly, the effective resistance is approximated by the sum of the high and low resistance of the gates:

$$R_{\text{eff}} = R_{\text{p}} + R_{\text{n}} \quad (20.5)$$

Next, we describe two other approximate methods [27], which can be useful for the quick estimation of the intrinsic decoupling capacitance in the design with reasonable accuracy. The first approach, which has been used extensively in traditional analysis, estimates the amount of decoupling (nonswitching) capacitance C_{eff} from the average total power using the expression

$$C_{\text{eff}} = \frac{P}{V^2 f} \frac{1-\alpha}{\alpha} \quad (20.6)$$

where *P* is the average power, *f* the switching frequency, *V* the supply voltage, and α the average switching factor. Since $(1-\alpha)$ is much larger than α, the error in this estimate is highly sensitive to any uncertainty in the estimated α factor. For instance, when α ranges from 0.1 to 0.3, as is the case in a typical design, the capacitance estimate varies by 285%.

The second approach, proposed in [27], estimates the intrinsic decoupling capacitance by using circuit simulation of several representative circuit blocks.

The procedure is shown in Figure 20.4. A representative circuit block is put in an arbitrary nonswitching state by applying a DC operating voltage (V_{dc}) to the power terminals and 0s and 1s randomly to input signals. Then a small AC (sinusoidal) perturbation, typically 5 to 15% of the V_{DC} representing the supply noise, is applied to the power rails to determine the decoupling action of the circuit in that state. Figure 20.4(a) shows the simulation setup and Figure 20.4(b) depicts the equivalent RC circuit of the underlying decoupling circuit. Figure 20.4(c) shows the voltage and current at the power terminals in phasor notation. Since the device capacitances are voltage-dependent in the circuit model, the resultant I_{ac} is not exactly sinusoidal, but the deviation is unnoticeably small due to the small change in voltages. The C_{eff} and R_{eff} elements are then calculated as follows:

$$C_{\text{eff}} = \left(\frac{1}{2\pi f}\right)\left(\frac{V_{\text{ac}}}{I_{\text{ac}}}\right)\sin\phi \quad (20.7)$$

$$zR_{\text{eff}} = \left(\frac{V_{\text{ac}}}{I_{\text{ac}}}\right)\cos\phi \quad (20.8)$$

For this method, a small number of representative circuit blocks are chosen and are simulated along with all the interconnect parasitics. To account for the loss in decoupling action due to the switching of some devices, the C_{eff} as determined above is scaled down by the factor $(1-\alpha)$. This procedure is less sensitive to any error in the estimation of α since α is typically much smaller than 1. For a variation from 0.1 to 0.3 in α, the estimated capacitance will vary by only 28%.

The intrinsic N-well capacitance is also modeled as a series RC whose time constant and capacitance per unit well area are characterized using a process simulator. The intrinsic as well as the explicit decoupling capacitances are distributed either according to the layout or when a layout is not available (as during the early design stage), uniformly across the power rails.

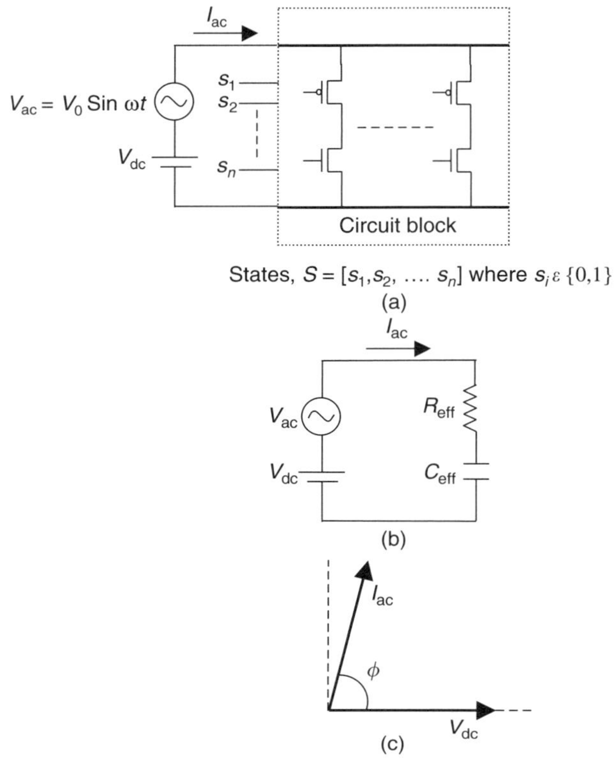

FIGURE 20.4 Intrinsic decoupling capacitance estimation.

It is important that the resistance and capacitance network that is extracted for the power grid is not reduced using standard reduction techniques, since these techniques are designed for reducing signal nets. Such techniques preserve the delay characteristics of a signal net with high fidelity, but may not preserve the voltage drop in a power grid correctly.

The intrinsic decoupling capacitance is usually not enough to confine the voltage drop within safe bounds and designers have to add specific decoupling capacitance structures on the chip. Several works [28–30] have been published which formulate the decoupling capacitance allocation as an optimization problem with the objective of decoupling capacitance area minimization and constraints on the worst voltage drop.

20.4.2 Inductance Models

Package leads have resistance, self-inductance, and mutual inductance to other package leads. The package model forms a dense symmetric partial inductance matrix [31,32], where there is mutual inductance among each of the entries. To simplify the computation, the package leads are grouped into equivalent leads, which supply power to a group of pads in close proximity. Figure 20.5 shows the reduced model of the package. If the package includes decoupling capacitances they should also be modeled appropriately.

Since there is mutual inductance among all the package leads, the current through both Vdd and Gnd leads need to be modeled simultaneously. This can be done by simultaneously analyzing both the Vdd and Gnd grids, but this doubles the size of the network to be analyzed. The alternative is to make a simplifying assumption that the current through each power lead is equal to an associated ground lead. This is a good assumption if the power and ground leads are in close proximity. With this assumption, we can take into account the mutual inductance of the ground leads on the power leads or vice versa, without modeling both Gnd and Vdd networks separately.

As explained in Section 20.3, certain computational advantages are lost when the RC network model is expanded to include the model for the package and off-chip effects. Note that an RC model of the power network produces a symmetric positive-definite LHS matrix for an MNA of the system $Ax = b$. However, when the package model is included in the analysis, the self and mutual inductors of the package now require the currents through the inductors to be declared as variables as well, resulting in an LHS matrix whose positive definiteness is not guaranteed. One could use general solution techniques, such as the LU decomposition, to solve the RLC model, but the large size of the network makes such a general solution infeasible. This difficulty is overcome using the following partitioning approach.

The network is partitioned at the interface between the package network and the power grid network. Then the power grid network (consisting only of Rs, Cs, and the current sources) is reduced to equivalent admittance and currents at the interface points. The package network can now be solved including the reduced representation of the power grid. This approach works well since the reduction of the power grid network involves solving a symmetric positive-definite system. The package network with the reduced power grid model is much smaller and hence can be solved using a general solution technique, such as the LU decomposition. The idea behind this approach is illustrated in Figure 20.6.

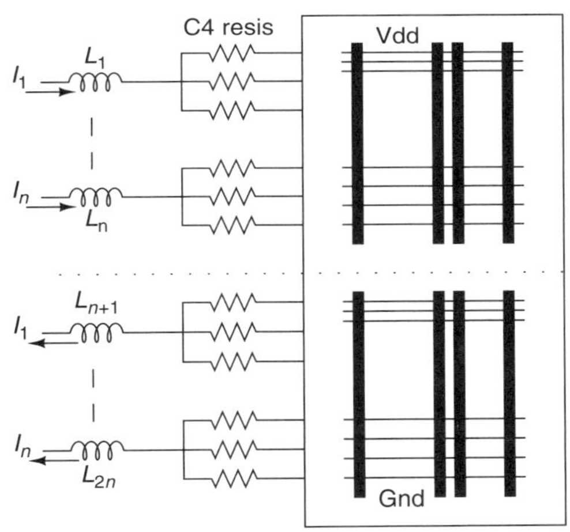

FIGURE 20.5 A simplified package model.

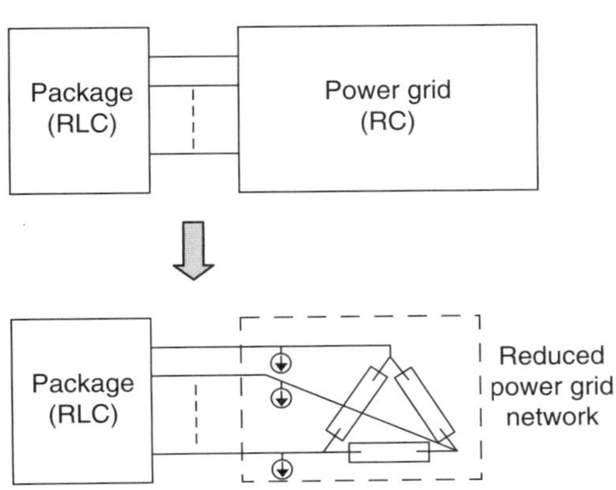

FIGURE 20.6 Power grid network reduction.

Suppose $Ax = b$ is the system representing the combined network

$$A = \begin{bmatrix} A_{11} & A_{12} \\ A_{12}^T & A_{22} \end{bmatrix}, x = \begin{bmatrix} x_1 \\ x_2 \end{bmatrix}, b = \begin{bmatrix} b_1 \\ b_2 \end{bmatrix} \tag{20.9}$$

where x_1 is the vector of voltages at the nodes in the power grid, and x_2 the vector of voltages and inductor currents in the package network.

The above linear system of equations is solved in two steps as follows:

$$x_2 = (A_{22} - A_{12}^T A_{11}^{-1} A_{12})^{-1}(b_2 - A_{12}^T A_{11}^{-1} b_1) \tag{20.10}$$

$$x_1 = A_{11}^{-1}(A_{22} - A_{12}x_2) \tag{20.11}$$

Note that the premultiplications with the inverse of A_{11} are efficiently done indirectly using the Cholesky factors or the conjugate gradient method. The model reduction requires solving the power network $p + 1$ times, where p is the number of interface nodes.

20.4.3 Comparison of the *R*, *RC*, and *RLC* Analyses

Figure 20.7 shows the analysis of the power grid of the PowerPC™ 750 processor with three types of models. The *y*-axis in Figure 20.7 represents the voltage in volts and the *x*-axis is time in nanoseconds. The supply voltage is 2.5 V and the cycle time is 5 nsec. The first curve (R) shows the analysis when only the resistance of the power grid is modeled. In this case the worst drop is 199 mV. The second curve (RC) shows the analysis when the power grid resistance and the intrinsic decoupling cap of the grid were modeled. In this case the worst drop improved to 170 from 199 mV for the R-only case. The total intrinsic decoupling cap for the whole chip is 30 nF. The third curve shows the analysis with power grid resistance, device decoupling caps, and package inductance. Due to the package inductance the worst drop has increased to 233 from 170 mV for the RC case.

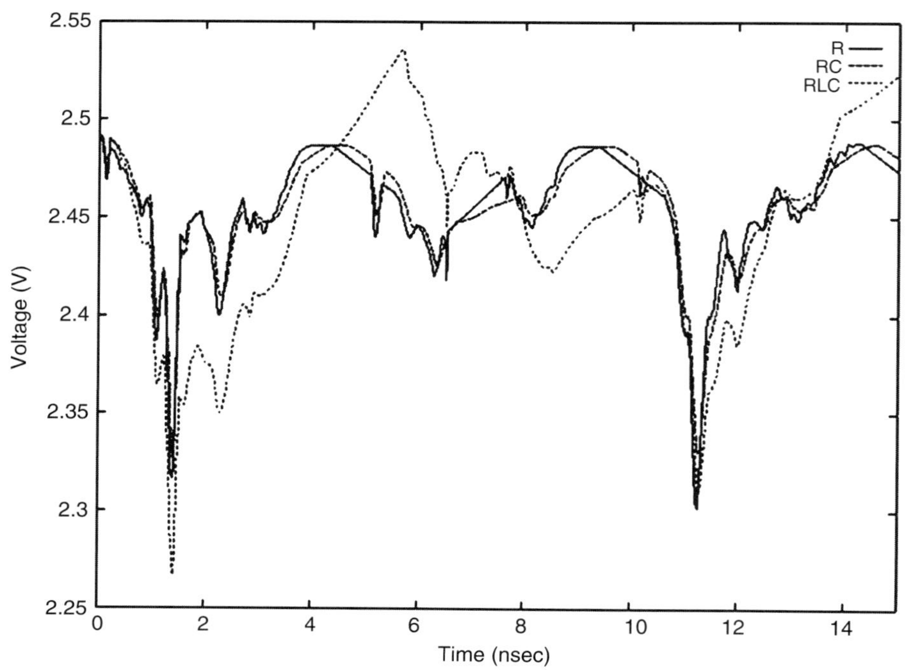

FIGURE 20.7 Supply drop waveform with R, RC, and RLC Models.

20.5 Conclusions

Power grid verification has become a major challenge for high-performance microprocessors. In this chapter, we presented a design and analysis flow for designing the power grids of large, high-performance processors. Linear system solution techniques for power grid simulation were discussed. Also, models for device decoupling capacitance and package inductance were presented. The usefulness of the multimode analysis capability in progressively refining the design of a power grid design is demonstrated through actual processor designs. The enormous size of a power supply network poses difficulty in accurate modeling, simulation, and optimization of power grid. The continued process scaling is likely to make power grid analysis an even more challenging task.

References

[1] L. Miller, Controlled collapse reflow chip joining, *IBM J. Res. Dev.*, 13, 239–250, 1969.

[2] S.R. Vemuru, Effects of simultaneous switching noise on the tapered buffer design, *IEEE Trans. Very Large Integration (VLSI) Syst.*, 5, 290–300, 1997.

[3] R. Saleh, S.Z. Hussain, S. Rochel, and D. Overhauser, Clock skew verification in the presence of IR-drop in the power distribution network, *IEEE Trans. Comput.-Aided Des. Integrated Circuits Syst.*, 19, 635–644, 2000.

[4] L.H. Chen, M. Marek-Sadowska, and F. Brewer, Buffer delay change in the presence of power and ground noise, *IEEE Trans. Very Large Integration (VLSI) Syst.*, 11, 461–473, 2003.

[5] J.R. Black, Electromigration failure modes in aluminum metallization for semiconductor devices, *Proc. IEEE*, 57, 1587–1594, 1969.

[6] H. Chen and D. Ling, Power supply noise analysis methodology for deep-submicron VLSI chip design, *Proceedings of the 34th ACM/IEEE Design Automation Conference (DAC'97)*, Anaheim, CA, 1997, pp. 638–643.

[7] P. Larsson, Resonance and damping in CMOS circuits with on-chip decoupling capacitance, *IEEE Trans. Circuits Syst. I: Fundam. Theory Applic.*, 45, 849–858, 1998.

[8] M.D. Pant, P. Pant, and D.S. Wills, On-chip decoupling capacitor optimization using architecture-level prediction, *IEEE Trans. Very Large Integration (VLSI) Syst.*, 10, 319–326, 2002.

[9] A. Dharchoudhury, R. Panda, D. Blaauw, R. Vaidyanathan, B. Tutuianu, and D. Bearden, Design and analysis of power distribution networks in PowerPC microprocessors, *Proceedings of the 35th ACM/IEEE Design Automation Conference (DAC'98)*, San Francisco, CA, 1998, pp. 738–743.

[10] D. Blaauw, Industrial perspectives on emerging CAD tools for low-power processor design, *Proceedings of the International Symposium on Low Power Electronics and Design (ISLPED'98)*, Monterey, CA, 1998.

[11] S. Chowdhry and J.S. Barkatullah, Estimation of maximum currents in MOS IC logic circuits, *IEEE Trans. Comput.-Aided Des. Integrated Circuits Syst.*, 9, 642–654, 1990.

[12] Y.M. Jiang, T.K. Young, and K.T. Cheng, VIP-an input pattern generator for identifying critical voltage drop for deep sub-micron designs, *Proceedings of the International Symposium, on Low-Power Electronics and Design (ISLPED'97)*, San Diego, CA, 1999, pp. 156–161.

[13] A. Krstic and K.T. Cheng, Vector generation for maximum instantaneous current through supply lines for CMOS circuits, *Proceedings of the 34th ACM/IEEE Design Automation Conference (DAC'97)*, Anaheim, CA, 1997, pp. 383–388.

[14] H. Kriplani, F.N. Najm, and I.N. Hajj, Pattern independent maximum current estimation in power and ground buses of CMOS VLSI circuits: algorithms, signal correlations, and their resolution, *IEEE Trans. Comput.-Aided Des. Integrated Circuits Syst.*, 14, 998–1012, 1995.

[15] H. Qian, S.R. Nassif, and S.S. Sapatnekar, Early-stage power grid analysis for uncertain working modes, *IEEE Trans. Comput.-Aided Des. Integrated Circuits Syst.*, 24, 676–682, 2005.

[16] D. Kuoroussis and F.N. Najm, A static pattern-independent technique for power grid voltage verification, *Proceedings of the ACM/IEEE Design Automation Conference (DAC'03)*, Anaheim, CA, 2003, pp. 99–104.

[17] C. Ho, A.E. Ruehli, and P. Brennan, The modified nodal approach to network analysis, *IEEE Trans. Comput.-Aided Des. Integrated Circuits Syst.*, CAS-22, 504–509, 1975.

[18] L. Pillage, R. Rohrer, and C. Visweswariah, *Electronic Circuit and System Simulation Methods*, McGraw-Hill, New York, 1994.

[19] G. Golub and C. Van Loan, *Matrix Computations*, The John Hopkins University Press, Baltimore, 1989.

[20] T. Chen and C.C. Chen, Efficient large-scale power grid analysis based on preconditioned Krylov-subspace iterative methods, *Proceedings of the Design Automation Conference (DAC'01)*, Las Vegas, NV, 2001, pp. 559–562.

[21] M. Zhao, R.V. Panda, S.S. Sapatnekar, and D. Blaauw, Hierarchical analysis of power distribution networks, *IEEE Trans. Comput.-Aided Des. Integrated Circuits Syst.*, 21, 159–168, 2002.

[22] J. Kozhaya, S.R. Nassif, and F.N. Najm, A multigrid-like technique for power grid analysis, *IEEE Trans. Comput.-Aided Des. Integrated Circuits Syst.*, 21, 1148–1160, 2002.

[23] W.L. Briggs, *A Multigrid Tutorial*, SIAM, Philadelphia, PA, 1987.

[24] S. Zhao, K. Roy, and Cheng-Kok Koh, Frequency domain analysis of switching noise on power supply network, *Proceedings of the ACM/IEEE International Conference on Computer-Aided Design (ICCAD'00)*, San Jose, CA, 2000, pp. 487–492.

[25] H. Qian, S.R. Nassif, and S.S. Sapatnekar, Power grid analysis using random walks, *IEEE Trans. Comput.-Aided Des. Integrated Circuits Syst.*, 24, 1204–1224, 2005.

[26] P.G. Doyle and J.L. Snell, *Random Walks and Electric Networks*, Mathematical Association of America, Washington, DC, 1984.

[27] R. Panda, D. Blaauw, R. Chaudhry, V. Zolotov, B. Young, and R. Ramaraju, Model and analysis for combined package and on-chip power grid simulation, *Proceedings of the Internation Symposium on Low-Power Electronics and Design (ISLPED'00)*, Rapallo, Italy, 2000, pp. 179–184.

[28] H. Su, K.H. Gala, and S.S. Sapatnekar, Analysis and optimization of structured power/ground networks, *IEEE Trans. Comput.-Aided Design of Integrated Circuits Syst.*, 22, 1533–1544, 2003.

[29] H. Su, S.S. Sapatnekar, and S.R. Nassif, Optimal decoupling capacitor sizing and placement for standard-cell layout designs, *IEEE Trans. Comput.-Aided Des. Integrated Circuits Syst.*, 22, 428–436, 2003.

[30] S. Zhao, K. Roy, and C.-K. Koh, Decoupling capacitance allocation and its application to power-supply noise-aware floorplanning, *IEEE Trans. Comput.-Aided Des. Integrated Circuits Syst.*, 21, 81–92, 2002.

[31] F.W. Grover, *Inductance Calculations*, Dover, New York, 1954.

[32] A.E. Ruehli, Inductance calculations in a complex integrated circuit environment, *IBM J. Res. Dev.*, 16, 470–481, 1972.

21 Noise Considerations in Digital ICs

Vinod Kariat
Cadence Berkeley Laboratories
Berkeley, California

21.1 Introduction

In the early days of VLSI design, digital chip circuit design and layout were manual processes. The use of abstraction and the application of automatic synthesis techniques have since allowed designers to express their designs using high-level languages and apply an automated design process to create very complex designs, ignoring the electrical characteristics of the underlying circuits to a large degree. However, scaling trends have again brought electrical effects to the forefront in recent technology nodes. At 0.25 μm and below, wire delays needed to be considered to achieve timing closure. In nanometer technologies at 0.13 μm and below, unintended interactions between signals (or *noise*) has become an important consideration for digital design. At these technology nodes, the performance and correctness of a design cannot be assured without considering noise effects.

Noise can have many drastic consequences for digital designs: (1) it can make the design slower, (2) it can make the design work incorrectly or even fail completely, and (3) it can create yield problems. For a chip designer, the cost of such a failure is very high, and includes mask cost, engineering cost, and opportunity cost due to delayed product introduction. For a high-volume chip, it is estimated that a 3-month product launch delay can lead to $500 million loss in product revenues [1,2].

The rest of this chapter is organized as follows. We first discuss the technology and design factors that have made noise a significant problem at nanometer technology nodes. We then discuss how noise affects the operation of a digital circuit. We will then review noise analysis techniques and algorithms in detail, and finally discuss how noise can be handled during the design process.

21.2 Why Has Noise Become a Problem for Digital Chips?

In analog circuits, designers are concerned with noise that arise from physical sources, such as thermal noise, flicker noise, and shot noise. These noise sources on the one hand present a lower limit to the smallest signal that can be amplified, and on the other, define an upper limit to useful amplification.

In digital circuits, noise arises not from fundamental physical sources, but from the operation of the circuit itself, primarily the switching of other signals. Due to the very nonlinear nature of the CMOS gate transfer characteristic, CMOS circuits are very noise-immune and noise suppressing. The noise immunity of a CMOS circuit is typically represented by the *DC noise margin* [3] of the circuit. For simplicity, consider the DC transfer characteristic of an inverter as shown in Figure 21.1. As is seen from the figure, when the input voltage is between 0 and V_{IL}, the output remains at V_{DD}, or logic 1. The output changes to logic 0 when the input voltage reaches V_{IH}. The CMOS gate is insensitive to voltage changes from 0 to V_{IL} when the input is at logic 0, and voltage changes from V_{DD} to V_{IH} when the input is at logic 1. However, in nanometer geometries, several factors have coalesced to make noise a major issue for digital designs:

- Higher interconnect density has led to each net having neighbors that are closer, thus leading to increased coupling capacitance between neighboring nets.
- With the introduction of copper, wire geometries changed in such a way as to increase the sidewall capacitance at the expense of ground capacitance. Wires have become longer, thinner, and much closer (see Figure 21.2). As a result , a much higher percentage of the total capacitance of a net comes from coupling with its neighbors, thus increasing the coupling cross-talk problem.
- Technology scaling [4] has lead to lower threshold voltages, and has also reduced the headroom between threshold and supply voltage, thus reducing noise margin.
- Clock speeds have increased significantly, thus leading to faster transition times. Faster transition times are closely linked to higher capacitive cross talk, as will be explained in the next section.

These effects have increased the interactions between signals and decreased the noise immunity of digital CMOS circuits. This has led to noise being a significant problem for digital ICs that must be considered by every digital chip designer prior to tapeout. A later section discusses how noise considerations affect the design methodology. A fair question to consider is whether cross-talk problems can be completely eliminated by design methodology constraints. For example, can we completely eliminate the

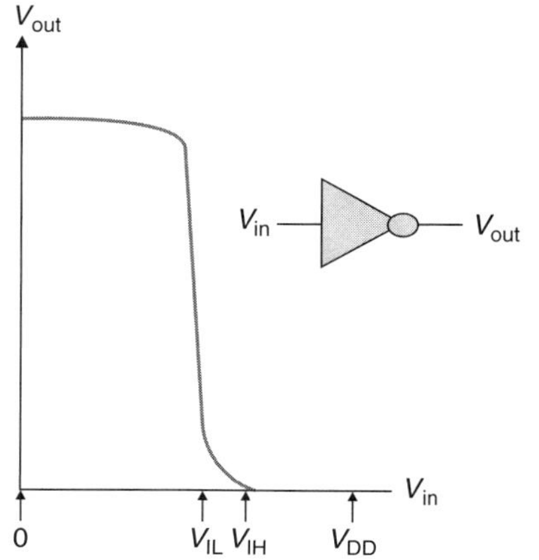

FIGURE 21.1 Inverter transfer characteristic and DC noise margins.

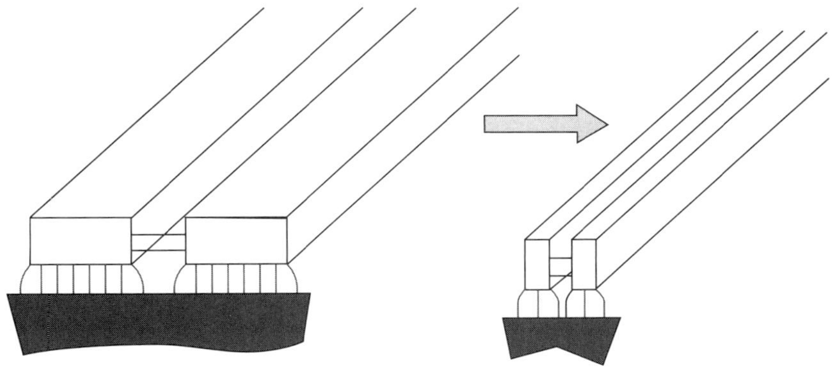

FIGURE 21.2 Wire cross-section scaling.

need to analyze a design for noise problems by restricting the number of neighbors any given net is allowed to have, and by inserting buffers often? In practice, if we restrict the amount of neighboring wiring to completely safe levels, there is a significant area increase and associated cost. Similarly, adding sufficient buffers to achieve safe levels leads to significant degradation in performance. Furthermore, as we will see in later sections, the same amount of noise may be harmful at one place in the design and not harmful at another place in the design. Hence, a designer must analyze a design for noise issues and fix noise problems that can lead to chip failures while allowing other, less significant noise effects to still exist in the design.

21.3 Noise Effects in Digital Designs

Noise can affect a digital design in three primary ways: (1) it can cause a signal to have the wrong value, leading to a functional failure, (2) it can cause a signal to arrive too late, thus causing the chip to run at a different frequency than intended, or (3) it can cause a signal to arrive too early, thus causing the chip to fail.

21.3.1 Capacitive Cross Talk

The dominant source of noise in most digital circuits is capacitive cross talk. Capacitive cross talk occurs due to unintended interaction of two signal lines due to parasitic coupling capacitance between them. The signal net that causes the cross talk is usually called an *attacker* or *aggressor* net, and the signal net that is affected by the cross talk is usually called the *victim* net. Note that the same net can be a victim in one situation and an attacker in another.

First, let us see how cross-talk noise can create a functional noise problem; this is often called a "cross-talk glitch". Consider a situation where a signal net A is being held at a steady logic 0 state; the nominal voltage on this net will then be 0. Now, suppose a neighboring net B switches from 0 to 1. Owing to the capacitive coupling between the nets, a voltage pulse will be induced on net A. This voltage will start from 0 V to some voltage usually less than V_{DD}, and then discharge back to 0 V, as shown in Figure 21.3. We call this a V_L noise pulse [5]. A similar situation can occur when net A is held at a logic 1 state, and net B switches from 1 to 0. In this case, the noise pulse that starts at an initial value of V_{DD} moves toward 0, and then returns to V_{DD}. We call this a V_H noise pulse [5, 27]. It is also important to consider a noise pulse which starts at logic 1 and is directed upward (sometimes called overshoot pulse), and similarly, a pulse having base level 0 and directed downward (undershoot pulse); these are called V_H^* and V_L^* noise, respectively.

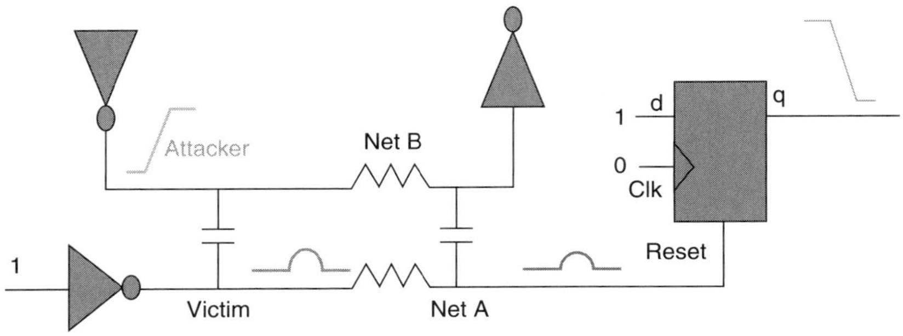

FIGURE 21.3 Cross-talk glitch.

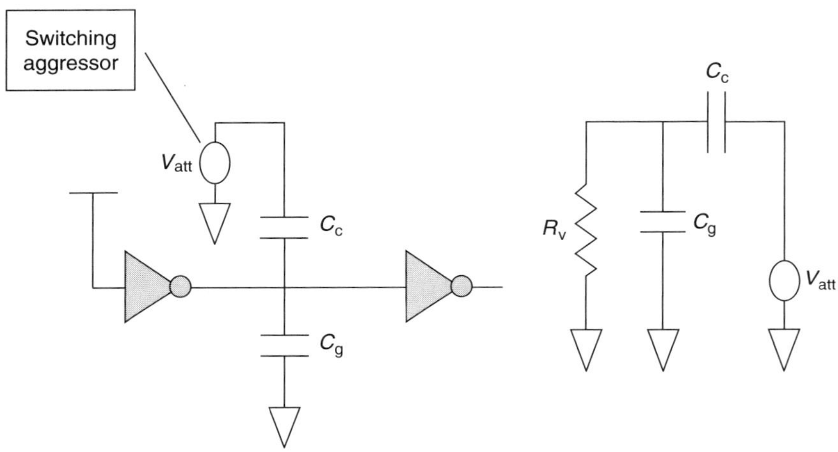

FIGURE 21.4 Simple noise glitch computation model.

In order to understand the different factors that affect cross talk, it is helpful to understand a simple model for computing cross-talk magnitude. Figure 21.4 shows such a model [28]. In order to simplify the computation, the following assumptions are made in this model:

- The parasitic networks for the attacker and victim net are represented by simple lumped capacitances, rather than with distributed RC networks.
- The driver which holds the victim net at its steady-state value is represented by a single holding resistance R_V, rather than using a nonlinear driver model that is closer to the behavior of the transistors.
- The attacker net is represented using an idealized saturated-ramp waveform with a transition time t_r. In reality, the attacker behaves nonlinearly.

Under this model, the maximum cross-talk voltage induced on the victim is given by the following equation:

$$V_{\text{pulse}} = \frac{1}{t_r} V_{\text{DD}} R_v C_c \left(1 - \exp\left(\frac{-t_r}{R_v(C_x + C_g)}\right)\right) \tag{21.1}$$

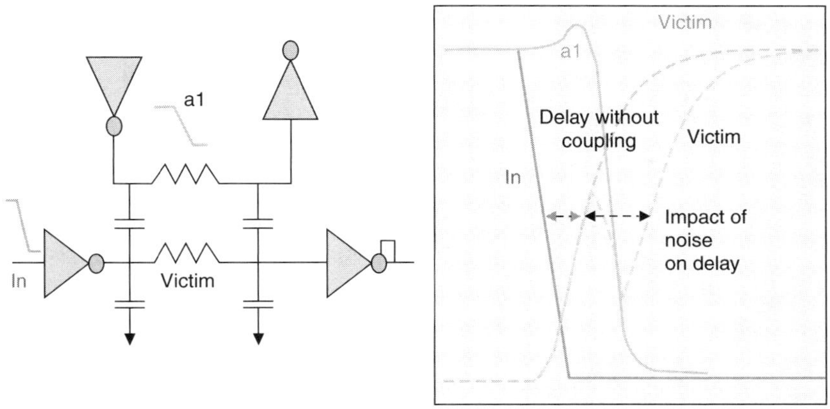

FIGURE 21.5 Effect of noise on delay.

As can be seem from Equation (21.1), the magnitude of the cross-talk glitch is directly proportional to the holding resistance of the driver (R_V) and the coupling capacitance (C_x), and inversely proportional to the transition time (t_r) of the attacker. This gives us a good indication of where we will see cross-talk problems. Nets with higher driver holding resistance (i.e., lower drive strength) and higher coupling capacitance are most susceptible to cross talk. If the attackers switch faster (i.e., have smaller transition times), the cross talk glitch will be worse. Equation (21.1) is often used in cross-talk analysis programs as a simple filter to eliminate low-risk nets from further analysis.

In addition to functional problems, noise can also create timing problems. Consider the case where a given victim net A is transitioning from 0 to V_{DD}. Now, if the neighboring net B switches at the same time from V_{DD} to 0, this will slow down the transition of the victim net A. This situation is shown in Figure 21.5. We call this phenomenon *delay slowdown*. If the neighboring net B were to switch in the same direction as the victim net (i.e., from 0 to V_{DD}), then this will result in speeding up the transition of the victim net A. This is called *delay speedup*. Both of these effects are influenced by the same factors as cross-talk glitch: namely, C_x and t_r of the attacker, and the drive strength of the victim. In this case, the victim parameter of interest is its switching waveform, which is modified by the noise effect. Collectively, we call delay speedup and delay slowdown as *noise-on-delay* effect.

In the future, inductive coupling could also become a serious noise issue for digital designs [6]. Many of the techniques described in this chapter can be extended to cover inductive coupling; however, existing simulation engines will need to be enhanced to cover inductive noise.

21.3.2 Propagated Noise

As we saw earlier in this section, capacitive interaction between signals can create unwanted cross-talk glitches on a signal line. The presence of a cross-talk glitch at the input of a gate can create a few different effects:

- It can weaken the driving power of the gate; this effect is called *driver weakening.*
- Under the right conditions, the noise pulse can actually appear at the output of the gate; then, the noise pulse is called *propagated noise.*

This propagated noise can combine with the noise on the output net, and continue downstream to other gates. Propagated noise must be considered along with cross-talk glitches to understand whether the circuit will function correctly in the presence of noise, since each gate is affected by the total noise it sees.

21.3.3 Charge-Sharing Noise

Standard-cell* based designs typically have to deal only with capacitive coupling noise and propagated noise. However, full-custom designs often implement dynamic circuit design techniques such as domino logic. These dynamic circuits are sensitive to an effect known as *charge-sharing* [7]. Here, a noise glitch can be induced on an otherwise quiet signal due to two nets coupling through a transistor, rather than through interconnect capacitance.

Figure 21.6(a) shows a domino circuit, and some waveforms under different charge-sharing conditions are shown in Figure 21.6(b) [5]. The circuit is first pre-charged by setting the clock signal (clk) to 0. This turns off the NFET at the bottom of the stack and turns on the PFET at the top of the stack. This will store charge on the output node of the circuit; this is called the precharge phase. When the clock signal changes its value to 1, the circuit is said to be in its evaluation phase. At this time, the inputs from the previous stages become valid. Consider the case where inputs A1, A2, A3, and A4 switch from 0 to 1 while inputs B1, B2, B3, and B4 remain at 0. This should not change the value of the output node. However, the output waveforms for this case are shown in Figure 21.6(b). The solid line represents the case where there is no *keeper* transistor (PFET shown as dotted line on the upper right side of the circuit). The dotted waveforms represent the output node voltage with different strengths for the keeper transistor. As can be seen from the figure, the redistribution of charge within different nodes of this dynamic circuit creates a significant noise problem for this particular circuit.

In the rest of this chapter, we will discuss noise analysis in terms of capacitive cross talk; however, charge-sharing analysis can be seamlessly integrated into a noise analysis tool using similar principles.

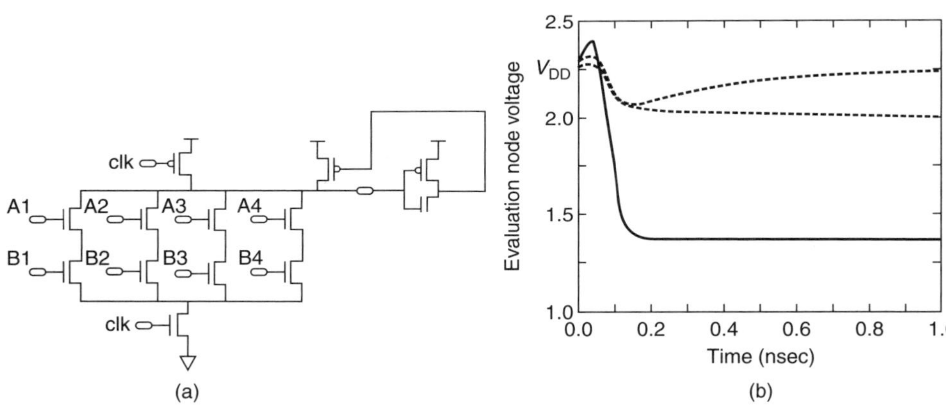

FIGURE 21.6 Charge-sharing noise: (a) circuit; (b) waveforms.

* For the purpose of this discussion, digital designs can be broadly classified into two categories or design styles: full-custom designs and standard-cell designs.

- *Full-custom* is typically used to implement very high-performance chips, such as microprocessors and high-end DSP cores. The design is done at the transistor level, and the layout is usually created manually, one transistor at a time.
- *Standard-cell* designs are done using a standard library of components or cells. The cells themselves are created using full-custom styles, but the design is created using these cells and laid out using Place & Route tools. Because standard cells must be used in a variety of unknown environments, they are more conservatively designed to avoid issues such as charge-sharing noise.

Although both styles of designs are affected by noise issues, the full-custom designs need to deal with more noise issues than a standard-cell design.

21.3.4 Multiple Attacker Considerations

In our discussion so far, we have considered the case where a victim is disturbed by the transition on a single neighboring attacker net. In reality, every net will have many neighbors, and we need to consider the effect of multiple attackers on the same victim net. We need to combine the noise pulses induced by different attackers on the victim net to create a noise pulse that realistically represents the combined effect of all the attackers on the victim net. The worst-case scenario is to assume that all the neighbors will contribute to the noise on the victim. However, this is usually not the case in reality; several factors contribute to making the situation less severe:

- Different nets are expected to switch at different times; we say that each net has a different *timing window*. The timing window represents the interval between the earliest time that a particular signal can switch and the latest time that a signal can switch. We can reduce the number of active aggressors by considering the overlap of the timing windows of different aggressors. We call this *timing window filtering*.
- Different nets may switch in opposite directions, or there might be logical relationships between different nets that prevent a net from switching when another net switches. The number of active aggressors can be reduced by using this information as well; we call this *logic filtering*.

In practice, it is very easy to obtain timing window information from static timing analysis [8,9], which already tracks the earliest and latest arrival times at every pin; the transition time of attackers and victims can also be obtained from static timing analysis. All commercially available noise analysis tools provide timing window filtering capabilities, and these capabilities are applied in the noise analysis of the vast majority of digital ICs. (However, some designers prefer to use a very conservative design methodology, and may not use timing windows.) Timing window considerations are described in more detail in the next section.

Logic filtering has also been implemented within the framework of noise analysis tools [10]; however, it has proven much more elusive to implement in practice, since it is computationally very expensive, and the improvements in analysis so far do not justify the cost. While the general case is difficult to handle, some special cases can be handled with low complexity. For example, many cross-talk analysis programs can recognize the fact that two attackers are separated by an inverter, and therefore will always switch in opposite directions.

21.4 Static Noise Analysis

As can be seen from the previous section, there are multiple noise effects that need to be analyzed in a digital IC. Prior to the introduction of static noise analysis [5], digital designers dealt with noise using a variety of *ad hoc* techniques based on circuit simulation. This was practical only for full-custom design styles, where a designer works on a very small circuit block — and still very cumbersome even for that case. *Static noise analysis* is a systematic way to deal with all the noise effects in a digital design.

In order to understand the motivation and roots of static noise analysis, let us first briefly try to understand how a circuit designer would go about analyzing a design for noise effects using circuit simulation. Typically, a designer would take the following steps for his verification:

- Create a list of expected noise events: this can include different types of noise events, such as coupling and charge-sharing. For example, if a sensitive victim net is routed close to two other nets, the designer might plan for either one noise event (both attacker nets switching in a particular direction at one logic value), or many noise events (each attacker switching in each directions, both attackers switching in same direction or different directions etc.)
- For each noise event, develop a way to excite the circuit so that the expected noise event will occur.
- Create a SPICE netlist that represents the desired excitation; run SPICE† and record the results.
- Analyze the simulation results manually and decide whether any re-design is required.

† We generically refer to all circuit simulators as SPICE for simplicity.

- Divide the circuit into small individually analyzable partitions
- Sort the partitions in order from source nodes to sink nodes (topological ordering)
- In topological sort order, for each partition:
  - Create noise events
  - Create simulation condition for each event
  - Simulate each noise event and calculate amount of noise
  - Determine worst case combination of noise events and relative timing of noise events
  - Calculate worst-case combined noise
- Propagate worst case noise to output of partition and perform appropriate tests.

FIGURE 21.7 High-level description of static noise analysis.

Although easy to describe, this procedure is error-prone and unwieldy in practice. Some of the difficulties are: the number of noise events can explode; creating excitations either requires tracing logic or cutting the SPICE netlist to insert appropriate voltage sources at the right places; it is difficult to determine the right transition and delay values; and there is no clear criteria to decide whether a noise problem requires re-design.

Static noise analysis overcomes these difficulties through a systematic procedure. A high-level procedural description of static noise analysis is shown in Figure 21.7. The steps outlined there are general enough to apply to both the analysis of functional noise effects and effect of noise on timing; they are also general enough to apply to both full-custom and standard-cell designs. In the rest of this section, we will discuss each of those steps in more detail, and describe how the steps can be implemented differently for different analysis scenarios.

21.4.1 Circuit Partitioning and Ordering

For ease and speed of analysis, each partition must be as small as possible. However, it is important to partition in a way that preserves circuit behavior; this can be accomplished by keeping some CMOS behavior in mind.

Let us first consider the case where we are dealing with a full-custom, transistor-level netlist. We want to make sure that any transistors that are strongly connected are analyzed together; for CMOS transistors, this then suggests that the most appropriate cut-point is the gate terminal of a transistor. So we create each partition with all the transistors that are connected through source or drain terminal connections, i.e., through the channel. Such a partition is called a channel connected component (CCC) [11]. In addition to the transistors, each CCC will contain all the nets and pins associated with the transistors in the set.

When we analyze a CCC, we also need to include all the nets that have significant coupling capacitances to any of the nets in the CCC (some nets that have insignificant coupling can be eliminated through *simple electrical filtering*). These nets are the attackers to the nets that belong to the CCC, and may be modeled in simpler form than the nets within the CCC itself. For example, one could simply model them as being driven by ideal voltage sources and not their real, highly nonlinear drivers. We define a *net complex* as a net with its parasitic elements, and all other nets that have significant coupling to it, including their parasitic elements. For each CCC, we need to include the net complexes of the nets within the CCC. In general, most CCCs consist of only a few transistors, and hence lead to quick analysis. However, certain types of circuits such as barrel shifters and multipliers, can lead to very large CCCs. In these cases, rigorous analysis of a CCC can become intractable and special heuristics may be required.

We can now extend this concept easily to standard-cell designs as well. Since the CCCs are contained within a cell, we can replace the CCC concept above by the idea of a *stage*. A stage is a cell driving a net; we include the appropriate electrical model for the driving cell, the driven net and its parasitic elements, and the net complex associated with that net. The receiving cell can usually be modeled in a simple fashion, for example, as a single fixed capacitance (more sophisticated models may be required for highly accurate analysis).

Once the circuit has been partitioned, we need to analyze the partitions in the right order. Before we can analyze a partition, we must have all the analysis results from any partitions that precede it, and provide input values to it. Let us use some graph terminology to describe this. Let each partition be a node n in a circuit graph G; also, let there be an edge E from any node $n1$ to a node $n2$ if the output of node corresponding to a gate $n1$ connects to an input of a gate corresponding to node $n2$; in other words, the edges represent the connectivity represented by each net. We now need to do a *topological sort* [12] on the graph G to decide the order in which the partitions are analyzed. If the circuit was fully combinational and contained no loops, this would be sufficient (i.e., if the circuit graph was a directed acyclic graph). A real circuit requires some modifications. First of all, the graph can be cut at latch boundaries. Also, combinational loops must be cut. In the simplest case, a combinational loop can be cut at any arbitrary net. However, more sophisticated algorithms are often employed for preserving important information about the loop, but these are beyond the scope of this chapter.

In general, static noise analysis may be categorized as a "breadth-first-search" of the circuit under analysis, and is very similar to the approach taken in static timing analysis [8].

21.4.2 Analyzing Partitions

Once the circuit has been partitioned appropriately and the partitions have been ordered for analysis, we can analyze each partition in detail. The fundamental analysis approach is similar for both full-custom and cell-level designs, although the actual engines employed are different. Figure 21.8 sketches the different steps involved in analyzing a partition at a high level. We will describe each of these steps in more detail below.

1. *Identify noise events:* In the case of coupling noise analysis, this includes deciding which attackers must be included. As discussed in an earlier section, both electrical filtering and timing window filtering are employed to reduce the set of relevant attackers.
2. *Analyze noise event*: Once we have identified the noise events, we need to analyze each noise event. In order to do this, we need to carry out the following steps.
   - *Create an electrical view of the noise event.* Conceptually, this step is equivalent to what a designer will do in the *ad hoc* technique described earlier, where he creates a SPICE netlist for simulation. The main difference is that we will be replacing many of the devices with appropriate models. The driver will be represented by an appropriate model, each of the attackers will be represented by a simplified model, and the receiver is also represented by a simplified model. The interconnect parasitics are often represented by a reduced RC or RLC or RLCK model. We will discuss some of the models in a later section; for a detailed treatment of models, see [13]. In order to select the right models, we also need to create an appropriate *sensitization condition* for the

- Select *noise events* to analyze
- For each noise event
  - Create sensitization condition
  - Create electrical analysis view of noise event
  - Simulate and calculate response
- Pick combination of noise events and relative alignment
- Create combined response

FIGURE 21.8 Steps to analyze each partition.

driver. For example, consider the case where the driver is a NAND gate and the output is being held at a logic value of "1". For this case, we need to have both inputs at "0". We need to choose an appropriate model for this input condition: in the case of a transistor model, it means setting up the input vector for the SPICE simulation to the right value, while for a cell model it may mean selecting the appropriate holding resistance or the set of I–V characteristics from a set of precharacterized library models. The end result of this step is an electrical view that can be simulated or "solved" for the resulting noise response on the output pins of the partition.

- *Simulate noise event*: During this step, the electrical view created in the previous step is evaluated to calculate the noise contribution of this particular event. Typically, the noise glitch contribution due to this event is calculated, and the effect of noise on delay is calculated in a separate step. Depending on the type of electrical model, different types of engines are used to solve the electrical view. The simplest one to explain and understand is a SPICE model. In this case, usually a SPICE engine with an API is employed. The API is used to create the circuit elements which constitute the netlist, and to instruct the SPICE engine to run for a given amount of time. The appropriate output nodes will be instrumented to measure the voltages — either the entire waveform or the peak voltage. An example of the required electrical model for a crosstalk noise event is shown in Figure 21.9. In this case, the driver is represented using transistor models, the interconnect is represented using resistors and capacitors, the attacker is represented using a saturated-ramp waveform, and the receivers are represented by equivalent passive capacitive loads.

3. *Select combination of noise events and relative alignment*: This is a crucial step where many pragmatic trade-offs must be considered. The objective of this step is to select a realistic subset of all possible noise events that can occur together to create the worst possible noise on an output node of this partition. Some of the factors to consider are the following:
   - Which combination of noise events can occur together? There will be several different sets of noise events that can occur together based on the timing windows of the different signals. When considering the effect of noise on delay, we must also look at the timing window of the victim signals.
   - For a given set of noise events, what is the relative timing of each of the attackers? Different relative timings can produce very different combined noise on the output of the partition. The step of finding the relative timing of the different attackers is usually called *alignment*, and is discussed in detail in [14].
   - Should one consider noise events which are only partially overlapping? For example, the last time a particular signal can switch may be 10 psec before the earliest time another signal can switch, but the effects of these two signals can overlap at the receiver.

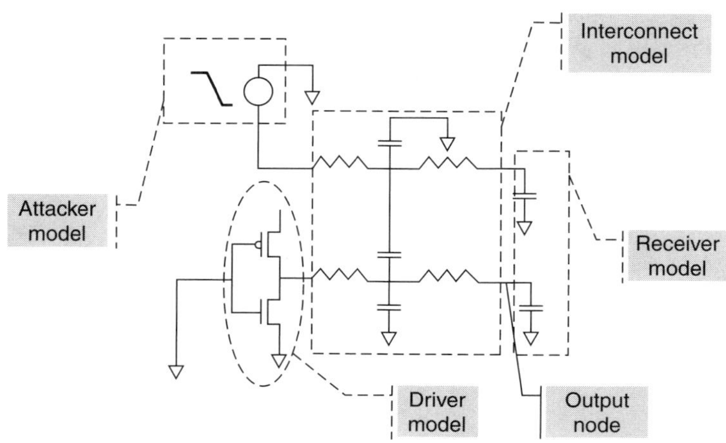

FIGURE 21.9 SPICE electrical model for simulation of a coupling noise event.

  - Given a set of signals which have overlapping timing windows, what logical relationships can one consider to eliminate some combinations? How does one consider logical relationships with timing information [10]? Can one consider a small subset of logical relationships without exploding the complexity of analysis?

   Each commercial noise analysis tool makes slightly different choices; there is no single definitive answer for any of the above questions.

4. *Create combined noise response*: Once Step 3 has been completed, this step is relatively straightforward. If the electrical model used in Steps 2 and 3 is linear (i.e., it contains no nonlinear circuit models like transistors), then linear superposition principles can be applied to calculate the combined noise response of the circuit. In the case of glitch noise, the glitches can be added together to create the final glitch. For calculating the effect of noise on delay, the noise glitch waveform can be added to the victim switching waveform to create the new switching waveform, from which the new delay and slew may be measured. It is also possible to calculate analytically the delay slowdown based on the magnitude of the glitch peak. Since the circuit behaves very nonlinearly, we must use nonlinear models to obtain high accuracy. When nonlinear models are used, a final *combined simulation* is usually required to get the final response. In this final simulation, all the attacker noise sources are instantiated with appropriate relative timings, and the output noise is measured. For noise-on-delay simulation, the victim net is also set up to switch at the appropriate time.

Once the final combined simulation is completed, the analysis of the partition is now complete. Once a partition has been analyzed, the results are available for the next stage to use. The result that is saved for the next stage depends on the type of analysis. For glitch noise, the final noise calculated at the output of this stage is usually required. In addition, if we are using a stage-based metric, then we also need to test whether the total noise computed at this stage is going to cause a failure. If we are using noise propagation, then the test is typically done only at storage nodes such as latches and flip-flops. For noise on delay, the delay change is available and usually back-annotated to the timing tool in order to update the timing information.

21.4.3 Noise Metrics

In the previous section, we have seen how to calculate the noise effect at the output of a stage or partition. In order to determine whether the calculated noise will cause incorrect operation of the circuit, some metrics are required; different metrics are required for glitch noise and noise-on-delay effects.

First, let us consider glitch noise; the following metrics are typically employed for determining the severity of a glitch noise problem. These different glitch noise metrics are shown in Figure 21.10.

- *Glitch peak*: This is the simplest of all noise metrics. We simply define a maximum absolute value for any allowed noise glitch peak. Typically, glitch peak limits are defined with respect to V_{DD}, so we may say that the glitch peak limit is 40% of V_{DD}. With a power supply of 1.2 V, only glitches less than 0.48 V will be allowed; any higher noise glitches will be labeled as a noise error, and must be fixed by changing the design. This metric ignores the shape of the noise pulse; in reality, the shape of the noise pulse significantly alters its impact on downstream logic. A sharp, narrow pulse is much less dangerous than a wide pulse of the same height. Usually, DC noise margins [3] are used as the glitch peak limits, which makes this metric very conservative. DC noise margins are illustrated pictorially in Figure 21.10(a). This metric can have a single value for the entire design, in which case the DC noise margin of the most noise-sensitive gate in the entire design must be applied throughout the design. In order to reduce pessimism, receiver-specific DC noise margins can be applied.
- *Noise rejection curve*: Noise rejection curves modify the glitch peak metric to also take the glitch shape into account. The glitch width/glitch peak space is divided into safe and unsafe regions. For a given glitch peak, a glitch with a smaller width would be safe, but one with larger width would be unsafe. This is pictorially shown in Figure 21.10(c).
- *Receiver output sensitivity*: The idea behind this metric is very simple—if a noise is strong enough to push a receiver into its amplification region, then it is a problem. Otherwise, the noise is being

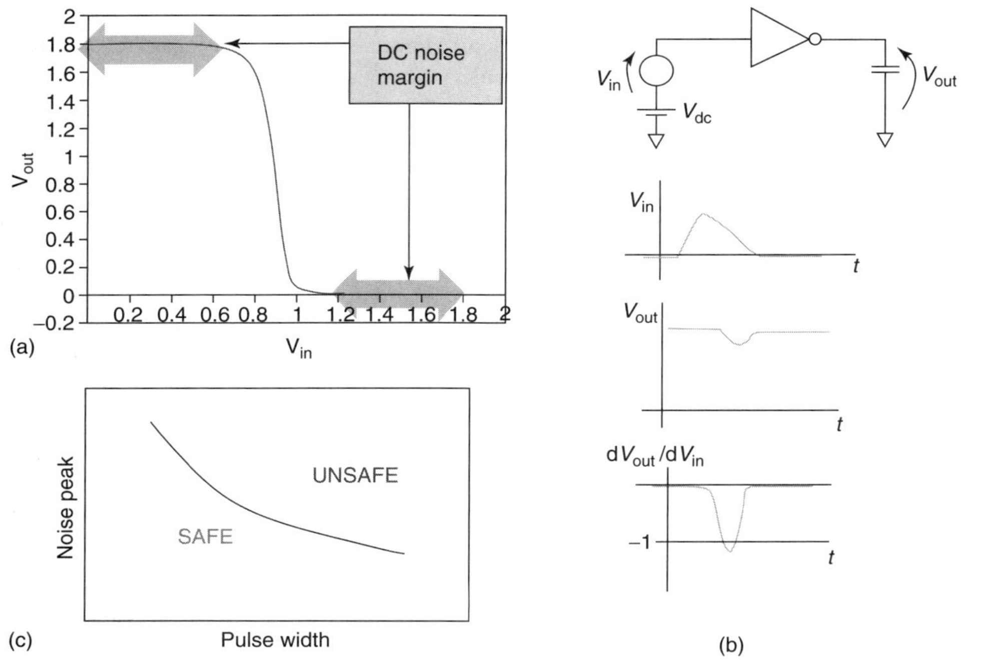

FIGURE 21.10 Glitch noise metrics: (a) glitch peak: DC noise margin; (b) noise sensitivity; (c) noise rejection curve.

suppressed by the receiver and will not have a significant impact on downstream logic. Receiver output sensitivity is shown pictorially in Figure 21.10(b). In practice, this metric must be implemented as a time-domain sensitivity calculation in the simulation engine.

- *Receiver output peak*: This is a modification of receiver output sensitivity. Instead of checking whether the receiver enters its amplification region, we check the propagated noise peak at the output of the receiver. Given the transfer characteristics of CMOS gates, usually there will be no propagated noise until the gate is well into its amplification region. This metric is more aggressive than sensitivity; fewer noise failures will be reported using this metric.
- *Noise propagation to latches*: All of the previous metrics are stage-based. They evaluate the noise at the output of a stage, and then evaluate its effect relative to its receiver. However, a glitch noise is not a problem unless it propagates all the way to a storage element. Noise propagation moves the tests away from all the combinational stages to the storage element stages. This is the most aggressive, least pessimistic, and most realistic noise metric for glitch noise.

Now, let us consider the effect of noise on delay. The metric of significance for a delay change is of course its effect on timing. However, we need to consider where the delay change is measured, and there are two approaches to measuring the delay change:

- *Receiver input measurement*: In this case, the delay change is measured at the input of the receiver. First, a nominal delay for the stage is calculated without any coupling effects. Then a "noisy delay" is calculated which includes coupling, and the difference between the two is the effect of noise on delay. All of these measurements are made at the input of the receiver. This change in delay at the input of each stage is essentially added to the path delay.
- *Receiver output measurement*: In this case, the delay change is measured at the output of the receiver. This method is more realistic, because some of the delay change seen at the input of the receiver may never be seen at the output; the receiver is able to absorb the change in the waveform shape at the input with only minor change in the output waveform.

Both of the above approaches are shown in Figure 21.11.

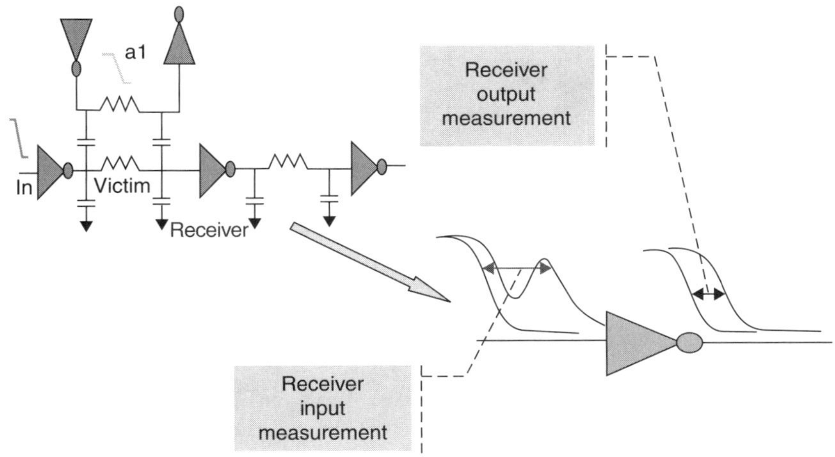

FIGURE 21.11 Delay measurement at receiver input and output.

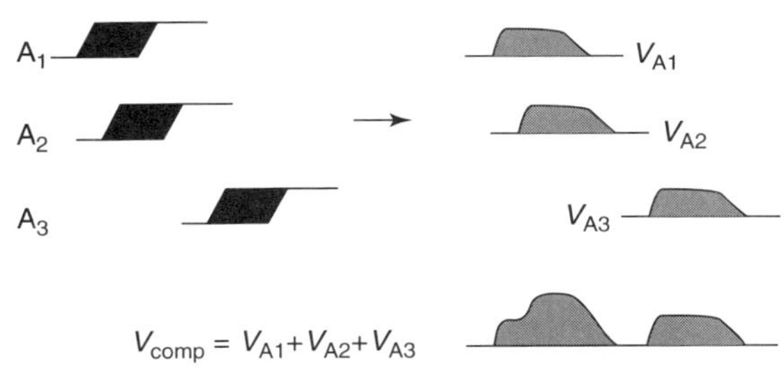

FIGURE 21.12 Timing window filtering.

21.4.4 Timing Window Considerations

Timing window filtering provides significant reduction in the number of noise problems that must be considered and analyzed, and hence noise analysis tools pay careful attention to how timing windows are computed and used. Figure 21.12 illustrates how timing windows are used to filter glitch noise events. The main factors to consider in computing timing windows are (1) the relationship between timing and noise, (2) the order of computing timing windows, and (3) how to determine timing window overlap. We will discuss each of these topics in this section.

First, let us look at the interaction between timing window computation and noise-on-delay analysis. Essentially, for a given victim net, we look at the earliest and latest arrival time of each potential attacker (its timing window) to decide whether the attacker can switch during the time that the current victim net can switch. Depending on the set of attackers that are selected, we compute both a noise-induced slowdown (or push-out) and a speedup (or pull-in) for the given net. Now, these values change the delay through this particular net, and affect the earliest and latest arrival times of all the other downstream nets. Some of these nets can be attackers to the current net; the timing windows of the attackers can be changed by noise computations on nets that feed them, and the timing window of the victim net itself can be changed by noise computations on nets that are upstream from it. To solve this interdependency, we need to do multiple rounds of timing window and noise-on-delay computations.This process is usually called *timing-window iteration* and is discussed extensively in [15], including convergence criteria for the iterations. The average number of iterations required to converge is often debated, but can be controlled by user settings.

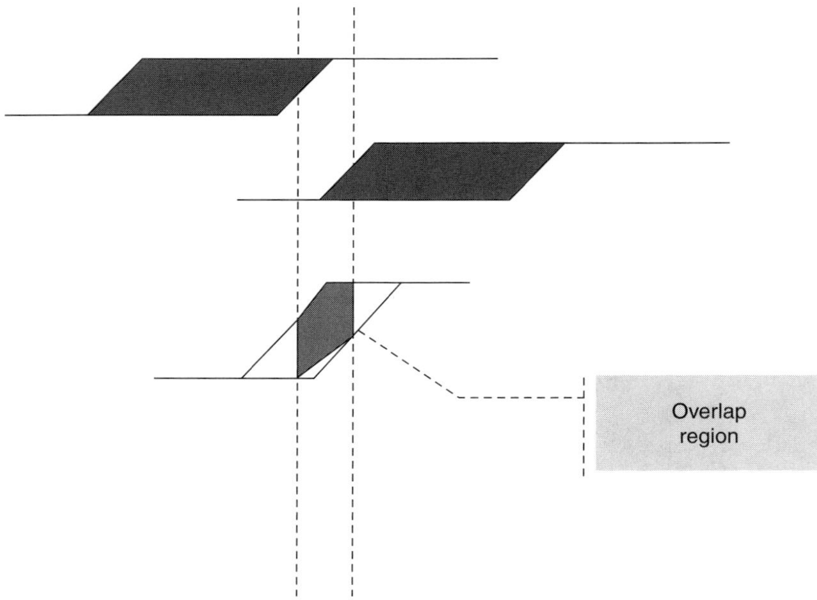

FIGURE 21.13 Timing window padding.

For the first pass of analysis, the initial timing windows can be infinite or nominal. In the infinite case, all the noise-induced delay changes are calculated for the worst-case timing, and then successively reduced as timing windows appear starting with the second pass of analysis. When starting with nominal timing windows, the first-pass analysis assumes no noise-induced delay changes, and then the timing windows expand as the early and late arrival times are modified for noise effects.

Early and late arrival times are usually measured based on the switching waveform crossing a specific threshold voltage, usually set to 50% of V_{DD}. Consider the case where the latest arrival time of a signal is slightly before the earliest arrival time of another signal. The waveforms resulting from this case are shown in Figure 21.13. The timing windows do not overlap, yet the waveforms definitely have some overlap and can affect each other. Typically, timing windows are padded by half of the slew for each signal to account for this case.

Even though a timing window constructed by using the earliest and latest arrival times covers the entire range of time for which a signal can switch, there are usually times during this interval when the signal will not transit; these can be considered as *holes* in the timing window[15]. With some extra overhead in timing analysis, it is possible to keep approximate track of timing window holes. Some noise analysis tools use this approach to add an extra level of filtering.

Another type of timing window information may be used to add further filtering to cross-talk glitch analysis. A noise glitch can create a functional failure only if it arrives at a latch during the time that the latch is sensitive to the value of the data input; this is a narrow window of time around a clock edge. We can calculate a possible timing window for the glitch (a *noise window*) based on the timing of the aggressors, and calculate the propagation delay for the glitch from the victim net to the latch. We can eliminate many glitches as harmless if they arrive at the latch during a time when the noise glitch cannot affect the state of the latch. This type of noise window filtering is treated in detail in [16].

21.5 Electrical Analysis

In this section, we will delve deeper into the models and solvers that are used in noise analysis. The speed and accuracy of any noise analysis tool depends on the models and solvers used for the analysis.

Furthermore, there is no perfect model or solver which delivers both best accuracy and best performance; every model and solver is a trade-off between accuracy and performance.

There are different models required to cover different aspects of the circuit being modeled. The most important models to consider are the following:

- *Driver models*: These models are used to represent the drivers for each stage or partition, and are probably the most important type of model for noise analysis. Similar but slightly different models are employed for glitch analysis and delay analysis.
- *Interconnect models*: These are usually *reduced-order* models used to represent the interconnect in a simpler form, so that it can be analyzed much faster.
- *Receiver models*: These models represent some characteristics of the receiving gate of the stage being analyzed.
- *Attacker models*: These models represent the attacking nets.

Depending on the type of model used and the computational requirements, different electrical solvers may be used for computing the response of the electrical network of a partition. The solvers range from closed-form analytical equations to fully SPICE-like solvers. In the rest of this section, we will discuss specific models and solvers in some more detail.

21.5.1 Driver Models

The early static noise analysis tools were implemented at the transistor level [17], and the driver model was obviously composed of transistors. It consisted of a CCC driving a coupled interconnect; often, the inactive (i.e., OFF) transistors of the CCC were dropped. An example of a transistor-based driver model is shown in Figure 21.14(a). A transistor-based driver model preserves the nonlinear behavior completely, and hence is very flexible. However, it requires a SPICE simulation to evaluate the model, which can be very expensive.

At the other extreme, early cell-based noise analysis tools [18] used a single-resistor model for representing the driver; this is shown in Figure 21.14(b). This model is very easy to characterize, and also very easy to simulate — it can be simulated very efficiently by a linear network simulator. This model can provide reasonable accuracy for glitch peak, but cannot accurately model the noise waveform. Also, since the actual behavior of the driver is nonlinear, a single resistance value can only accurately represent the behavior of the driver in some linear subregion. The model must be characterized in the subregion where the driver has the highest resistance, and then that value must be used to represent the driver across its entire region of operation. This will overestimate the noise in all other subregions. Furthermore, the

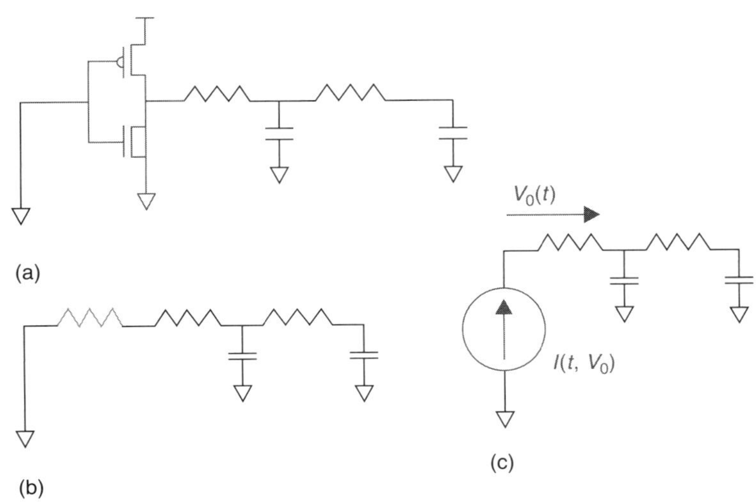

FIGURE 21.14 Driver models: (a) transistor driver model; (b) resistor driver model; (c) *I–V* driver model.

resistance used to model the driver for glitch noise (when the driver is holding state) is different from the resistance used to model the driver for noise on delay, since the switching resistance of the driver is different from the holding resistance.

A refinement of the resistor-based model is to use different resistance values for different regions. The selection of the right resistance value for a particular victim driver is still a difficult problem, since it depends on the amount of noise — we can iterate on the selection of the resistance and the calculation of the noise, but this will be computationally expensive and takes away from the simplicity and efficiency of the model.

A compromise between transistor models and resistor models is a specialized noise model based on storing the current–voltage (*I–V*) characteristics of the driver cell [13], as shown in Figure 21.14(c). Instead of modeling the individual transistors in the driving cell, we model the nonlinear electrical behavior of the entire cell through its *I–V* characteristics. These models may be based on DC (or static) *I–V* characteristics or on transient *I–V* characteristics. One significant advantage of the *I–V* model over the resistor model is that it can effectively model the behavior of the driver for both glitch noise and delay push-out. More details of this model are beyond the scope of this paper, but details can be found in [13].

Most commercial tools use a combination of models. Typically, they may use a fixed resistance for electrical filtering, an *I–V* characteristic-based model for the bulk of the computation, and transistor models selectively in cases where the highest accuracy is required.

21.5.2 Receiver Models

In most cases, receiver models tend to be simpler than driver models, and are used for fewer calculations. Typically, when calculating the noise response of a stage or partition, a very simple receiver model is used — the receiver is modeled by a simple capacitor. In the simplest case, the same capacitance is used for all the different calculations: base delay (without any noise), noise glitch, and as noise on delay. A slightly more sophisticated approach is to use different capacitances for each of these cases; there may even be a separate capacitance for each switching direction.

More information is required for propagating cross-talk glitches. The most accurate model for this purpose is again a transistor-based model. Since many glitches can be filtered from further propagation by using thresholds, it is possible to limit the number of places where noise must be propagated, and hence it is quite feasible to use transistor models for noise propagation. Another alternative is to build noise propagation tables, where the height and width of the output noise pulse is characterized and stored based on the input noise pulse width, input noise pulse height and the capacitance seen by the gate. This characterization tends to be expensive, and cannot easily capture the effect of the shape of the noise waveform. For noise on delay, a different model is required in the case of receiver output measurement. This can be accomplished by an *I–V* characteristic-based cell model [13].

21.5.3 Interconnect Models

In the simplest case, the interconnect can be modeled through a lumped RC network. A lumped model is shown in Figure 21.15(a). Here, the entire RC network is represented by a few lumped parasitic elements. In the model shown, all the resistances on the victim are summed up into a single resistor, all the capacitance to ground is lumped into a single capacitance, and the entire coupling capacitance to the attacker is lumped into a single coupling capacitance. Refinements of this scheme may include resistance for the attacker or a PI model for the victim, which consists of two ground capacitors separated by a resistor.

A more sophisticated interconnect modeling scheme is to use an RC reduction technique [6,19–26], as shown in Figure 21.15(c). The objective of RC reduction is to replace the detailed RC network with an equivalent mathematical model, which , at the *ports* and *taps* of the RC interconnect network,[‡] provides

[‡] A *port* is an input terminal, while a *tap* is an output terminal. Figure 21.15(a) shows the ports and taps of the network.

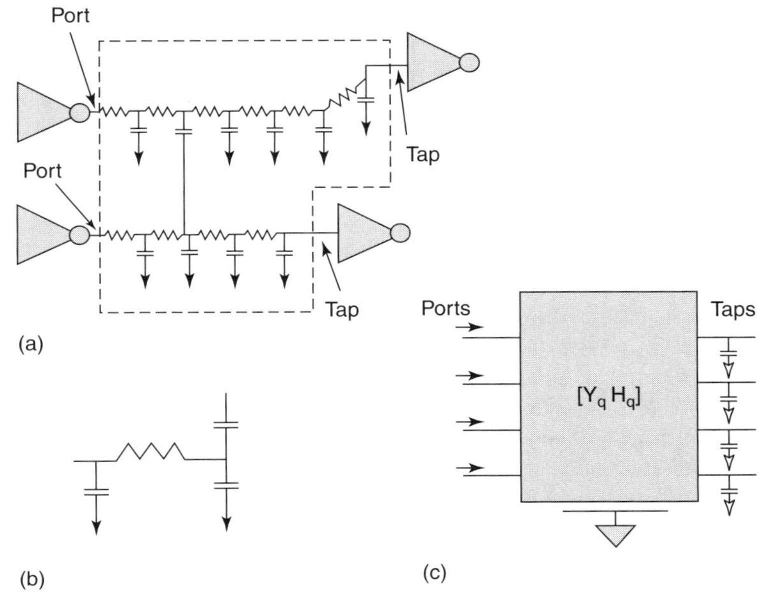

FIGURE 21.15 Interconnect models: (a) original interconnect; (b) lumped model; (c) reduced order model.

a response similar to the original unreduced network. The electrical characteristics of the intermediate nodes in the RC network are not preserved. Different RC reduction techniques are available with differing accuracy, performance characteristics, and features. Some RC reduction techniques can generate a reduced network which can be represented in terms of other R and C values, while other techniques may create a reduced network which can only be represented as a state-space matrix. Some reduction techniques can handle mutual- and self-inductances (and the reduction process is then called RLCK reduction). A detailed mathematical treatment of interconnect reduction is beyond the scope of this chapter. Interested readers can consult references [6,19–26] for more details.

21.5.4 Attacker Models

Attacker models are usually quite simple as well. The most commonly used attacker model is a so-called *saturated-ramp* voltage source. This is a three-piece voltage waveform as shown in Figure 21.16(a). The waveform starts with an initial value, and then switches to a final value with a strictly linear shape with a given transition time. The advantage of this model is that it is very simple to model and to treat analytically. In reality, the attacker waveform is much smoother. This can be modeled to some degree by adding a driver resistance in front of the saturated ramp waveform to create a softer waveform. This is illustrated in Figure 21.16(b). It is also possible to use a piece wise linear waveform that represents different parts of the attacker waveform better. When utmost accuracy is required, the attacking driver can be represented using transistors; this is computationally very expensive and should be avoided.

21.5.5 Solver Considerations

In order to achieve any given accuracy and performance requirements, a noise analysis tool must use both the right models and the right solvers. While simpler models can be handled by a more complex solver albeit at a lower performance, a given solver will limit the type of models that can be handled.

The simplest solver is an analytic closed form solution. This can usually be applied only with a lumped interconnect model and a resistor-based driver cell model. This is the approach usually used for electrical

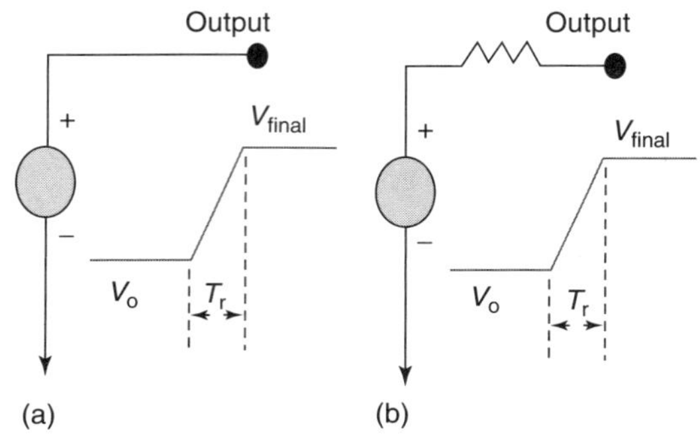

FIGURE 21.16 Attacker models: (a) saturate-ramp model; (b) modified saturate-ramp model.

filtering. A more complex analytic solver, similar to those discussed in [19], may be employed when distributed RC interconnect models are employed along with resistor-based cell models.

At the other end of the spectrum, the most complex and flexible solver used in noise analysis is a SPICE engine. Typically, this SPICE engine is accessed through a programming interface or API. SPICE is able to handle all the different types of driver models; special device models may need to be created for handling *I–V* cell models, and some modifications are necessary for handling reduced RC networks. Since noise analysis involves many small simulations rather than one large simulation, the SPICE engine may be tuned in different ways to improve its performance. (For more details, see chapter 14, *Simulation of Analog and RF Systems*, and chapter 15, *Simulation and Modeling for Analog and Mixed Signal Integrated Circuits.*)

A special-purpose numerical nonlinear solver is usually used for many of the simulations during the detailed analysis. Such a solver is able to handle the specialized cell *I–V* models, but not able to handle transistors. These solvers will make several simplifications to improve their performance. An example of such a simplification is to fix the time-steps used in simulation in a SPICE-like engine. Often, a hybrid linear/nonlinear approach is employed. The driver can drive a simplified lumped interconnect model, and this circuit is solved using a nonlinear numerical solver. Then, the time-domain response of the interconnect is computed by exciting the reduced network using the voltage computed by the numerical solver, and computing the output voltages using an analytic techniques like recursive convolution. An example of such a hybrid scheme is given in [13].

21.6 Fixing Noise Problems

It is not sufficient to know that different noise problems exist; we need to be able to create a design that works correctly. In this section, we will consider some prevention and correction techniques that may be applied in the design process. However, a caveat is in order: many of these techniques are heuristics employed in the design methodology or in particular design tools, and are hence difficult to discuss exhaustively. The effectiveness of any methodology or tool heuristics is in the quality of the design that is produced, and hence the discussion in this section should be considered only as examples for some possible approaches.

First of all, in order to understand noise prevention and correction, we must consider Equation 21.1. There, we have seen that the noise induced on a victim from a specific attacker is:

1. Directly proportional to the ratio of the coupling capacitance to the total capacitance
2. Inversely proportional to the transition time of the attacker
3. Directly proportional to the holding (or drive) resistance of the victim driver.

These relationships tell us the parameters that can control noise in the design. By making different design changes, we can alter one or more of these parameters to either reduce potential noise problems or to fix a particular noise problem. First, let us see what correction steps can be taken to fix a noise problem, and then we will see how to extend it to noise prevention. The typical correction steps are driver upsizing, attacker downsizing, buffer insertion, and routing changes. We will discuss each of these in more detail below:

- *Driver upsizing.* The victim driving cell is made stronger by upsizing. This will reduce the victim holding resistance, hence leading to a smaller noise. The same victim net may also be functioning as an attacker to other nets. In general, this step also improves the timing. We must be careful to either ensure that the upsizing does not create other noise problems, or iterate to make sure that no further problems are detected as a result of this step.
- *Buffer insertion.* In this approach, instead of upsizing the victim driver, a buffer is inserted at an appropriate point in the victim net. This will either require a complete re-routing of the victim net or a partial re-routing to connect up the inserted buffer. This helps reduce the cross talk in multiple ways. For the part of the original net that is in front of the inserted buffer, the coupling cap is decreased, and hence the effective strength of the driving cell is increased. In general, this step also improves the timing, since the RC delay will be reduced even if the gate delay is increased. For the part of the net that is after the inserted buffer, a better driver is provided. Multiple buffers may be inserted on the same net; inverters can be inserted if care is taken to ensure that pairs of inverters are inserted.
- *Attacker downsizing.* This works by increasing the transition time of the attacking net by reducing the strength of its driver. This action can cause the attacker to now become more susceptible to noise, or for it to not meet timing. Again, an iterative approach is usually required to ensure that the design is clean.
- *Routing changes.* Routing changes can be very effective in fixing noise problems. The basic cause of cross talk is coupling between wires, and routing changes the total coupling and the ratio of coupling capacitance to total capacitance. Many different routing heuristics can be employed to reduce the total amount of noise. The simplest technique is to add extra space around critical nets by leaving vacant tracks; however, this is rather expensive. Often, the router can selectively add extra space whenever it is able to do so without sacrificing its ability to route the design; this technique is usually called *soft spacing*. It is also possible to change the order of nets and to change routing layers to prevent nets from running parallel to each other. In addition, timing window information may be used to decide which nets can run parallel to each other. Commercial routers typically use a combination of these techniques to achieve good results.

After this overview of the basic correction techniques, let us see how to put them together into a flow. This is shown in Figure 21.17. After the initial routing is completed, a full analysis of the block (or chip) is performed. If no errors are found, the process can stop. If there are noise errors or noise-induced timing

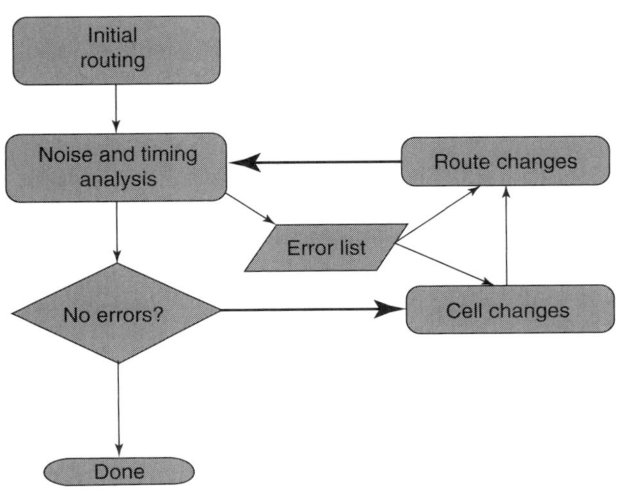

FIGURE 21.17 Noise correction flow.

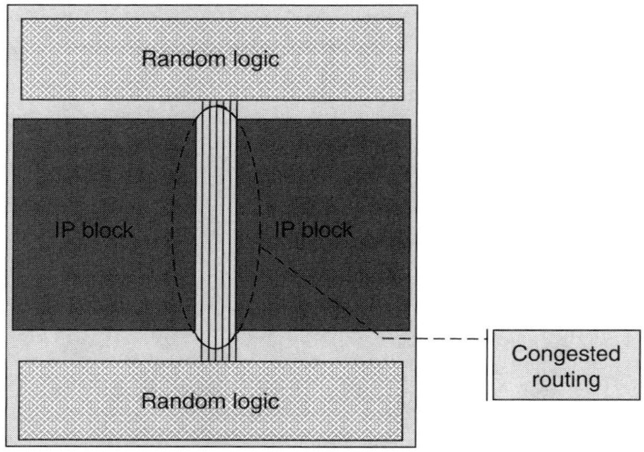

FIGURE 21.18 Floorplan-generated noise issues.

errors, then the list of errors is usually fed to a correction program, which heuristically determines the changes to be made. These may include both cell changes and routing changes; even if only cell changes are made, corrective routing is typically required to hook up inserted buffers. The correction program typically uses quick noise estimates to ensure that its corrections are effective, but the final effectiveness of the changes is verified by another round of noise analysis. The process iterates until there are no errors left; in practice, only a few automatic iterations are permitted before the designer evaluates the results and makes some changes to the settings and tolerances.

While it is good to have a correction flow, it is still important to practice noise prevention strategies throughout the design flow. However, there are certain design implementation steps where critical decisions are made, and it is important to focus on these design steps. (The various design steps are discussed in Chapter 1, *Design Flows*, and Chapter 10, *Design Closure*.)

Particular attention must be paid to buses and other very dense wiring structures during the floorplanning stage, especially if there are large inflexible IP blocks in the design. It is possible to create a situation as in Figure 21.18. Here, the random logic made up of standard cells is separated at the two ends of the chip, with many wires running between them. However, because of the placement of the two hard IP blocks, all the wires have to travel in a narrow area between the two blocks. This will cause many wires to run in parallel for long lengths, leading to cross-talk problems. Given the narrow area of the channel, it will not be possible to insert buffers to solve this problem.

Another important prevention step is to ensure that good transition time limits are set during the physical optimization step. If some nets have very long transition times, they will be very susceptible to cross-talk noise. Nets that are not timing critical can become so due to cross-talk effects. It is also very important to have a noise avoidance strategy in the initial routing. This can be in the form of limiting parallel lengths of wire whenever possible. However, more sophisticated routers usually employ simple heuristic noise calculators within the inner loop of the router to predict which nets are sensitive to noise. These nets can then be routed with special consideration to reduce coupling.

21.7 Summary and Conclusions

Noise has become a design metric of significance equal to timing and noise; it is not feasible to finish a digital design without considering noise. Noise must be considered throughout the design flow, and a comprehensive noise analysis and correction strategy must be applied in the design implementation tool suite. In this chapter, we have discussed the basic noise considerations, noise analysis strategies and some techniques for noise prevention. The references provide more in-depth treatment for any reader wishing to pursue this subject in further depth.

Acknowledgments

I would like to acknowledge the many valuable discussions with my colleagues and professional acquaintances which have contributed significantly to the material in this chapter. In particular, I would like to acknowledge Nishath Verghese, Igor Keller, Louis Scheffer, Ken Shepard, Ken Tseng, Joel Philips, Aurangzeb Khan, and Venkat Thanvantri. Many thanks are due to Louis also for reviewing the initial drafts of this chapter.

References

[1] A. Khan, An enhanced, system-centric, fully hierarchical design method to enable nanometer system-on-chip IC development, *International Conference on Solid-State and Integrated-Circuit-Technology*, 2004.

[2] A.Khan, Recent developments in high-performance system-on-chip IC design, *International Conference on IC Design and Technology*, 2004.

[3] J.P. Uyemura, *Circuit Design for CMOS VLSI*, Kulwer, Dordrecht, 1992, pp. 80–83.

[4] ITRS Roadmap, http://public.itrs.net

[5] K.L. Shepard, V. Narayanan, and R. Rose, Harmony: static noise analysis of deep submicron digital integrated circuits, *IEEE Trans. Comput.-Aided Des. Integrated Circuits Systems*, 18, 1132–1150, 1999.

[6] S.C. Chan and K.L. Shepard, Practical considerations in RLCK crosstalk analysis for digital integrated circuits, *ICCAD*, 2001, 598–604.

[7] J. Pretorius, A. Shubat, and C. Salama, Charge redistribution and noise margins in domino CMOS logic, *Circuits Systems, IEEE Trans.*, 33, 786–793, 1986.

[8] R.B. Hitchcock, Sr., G.L. Smith, and D.D. Cheng, Timing analysis of computer hardware, *IBM J. Res. Develop.*, 26, 100–105, 1982.

[9] K.A. Sakallah, T.N. Mudge, and O.A. Olukotun, CheckTc and minTc: timing verification and optimal clocking of synchronous digital circuits, *Computer-Aided Design*, 1990, *ICCAD-90, Digest of Technical Papers, IEEE International Conference*, 11–15 Nov., 1990, 552–555.

[10] D. Chai, A. Kondratyev, Y. Ran, K.H. Tseng, Y. Watanabe, and M. Marek-Sadowska, Temporo-functional crosstalk noise analysis, *Proceedings of the Design Automation Conference*, 2–6 June, 2003, pp. 860–863.

[11] R.E. Bryant, Boolean analysis of MOS circuits, *IEEE Trans. Comput. Aided Design*, CAD-6, 634–649, 1987.

[12] T.H. Cormen, C.E. Leiserson, and R.L. Rivest, *Introduction to Algorithms*, MIT Press, Cambridge, MA, 1996, 484–487.

[13] I. Keller, K. Tseng, and N. Verghese, A robust cell-level crosstalk delay change analysis, *ICCAD*, 2004, pp. 147–154.

[14] L.H. Chen and M. Marek-Sadowska, Aggressor alignment for worst case noise, *IEEE Trans. CAD*, 20, 612–620, 2001.

[15] C. Pinhong, Static Crosstalk Analysis for Deep Sub-Micron Digital Designs, Ph.D. thesis, University of California, Berkeley, 2003.

[16] K. Tseng and V. Kariat, Static noise analysis with noise windows, *Design Automation Conference*, 2003, pp. 864–868.

[17] K.L. Shepard and V. Narayanan, Noise in deep submicron design, *ICCAD-96*, 1996, pp. 524–531.

[18] L. Scheffer, A roadmap of CAD tool changes for sub-micron interconnect Problems, *ISPD-97*, 1997, pp. 104–109.

[19] L.T. Pillage and R.A. Rohrer, Asymptotic waveform evaluation for timing analysis, *IEEE Trans. Comput. Aided Design*, 352–366, 1990.

[20] A. Odabasioglu, M. Celik, and L. Pileggi, PRIMA: passive reduced-order interconnect macromodeling algorithm, *IEEE Trans. Comput.-Aided Design*, 645–654, 1998.

[21] K.J. Kerns and A.T. Yang, Stable and efficient reduction of large multiport RC networks by pole analysis via congruence transformation, *IEEE Trans. Comput.-Aided Design*, pp.734–744, 1997.

[22] H. Levy, W. Scott, D. MacMillen, and J. White, A rank-one update method for efficient processing of interconnect parasitics in timing analysis, *Proceedings of the Design Automation Conference*, June 2000, pp. 75–79.

[23] C. Ratzlaff, N. Gopal, and L. Pillage, TRICE: rapid interconnect evaluator, *Proceedings of 28th DAC*, June 1991, pp. 555–560.

[24] D. Anastasakis, N. Gopal, S.Y. Kim, and L.T. Pillage, On the stability of approximations in Asymptotic Waveform Evaluation, *Proceedings of 29th DAC*, June 1992.

[25] L.M. Silveira, M. Kamon, and J. White, Efficient reduced-order modelling of frequency-dependent coupling inductances associated with 3-D interconnect structures, *EDTC*, 1995, 534–538.

[26] J.R. Phillips and L.M. Silveira, Poor Mans TBR: a simple model reduction scheme, *IEEE Trans. CAD*, 43–55, 2005.

[27] K.L. Shepard and V. Narayanan, Conquering noise in deep-submicron digital ICs, *IEEE Design and Test of Computers*, 1998, pp. 51–62.

[28] A. Rubio, et al., An approach to the analysis and detection of crosstalk faults in digital VLSI circuits, *IEEE Trans. Comput. -Aided Design*, 387–395, 1994.

22

Layout Extraction

William Kao
Cadence Design Systems, Inc.
San Jose, California

Chi-Yuan Lo
Cadence Design Systems, Inc.
New Providence, New Jersey

Mark Basel
Mentor Graphics, Inc.
Wilsonville, Oregon

Raminderpal Singh
IBM Corporation
Cortlandt Manor, New York

Peter Spink
Cadence Design Systems, Inc.
Berkeley, California

Louis Scheffer
Cadence Design Systems, Inc.
San Jose, California

22.1 Introduction

Layout extraction is the translation of the topological layout back into the electrical circuit it is intended to represent. This extracted circuit is needed for various purposes including simulation, timing analysis, and logic to layout comparison (see the Chapters on Digital and Analog Simulation, Timing Analysis, and Formal Verification). Each of these functions requires a slightly different representation of the circuit, resulting in the need for multiple layout extractions. In addition, there may be a postprocessing step of converting the device-level circuit into a gate-level circuit [97–101], but this is not considered part of the extraction process.

The detailed functionality of an extraction process will depend on its system environment. The simplest form of extracted circuit may be in the form of a netlist, which is formatted for a particular simulator or analysis program. A more complex extraction may involve writing the extracted circuit back into the original database containing the physical layout and the logic diagram. In this case, by associating the extracted circuit with the layout and the logic network, the user can cross-reference any point in the circuit to its equivalent points in the logic and layout (cross-probing). For simulation or analysis, various formats of netlist can then be generated using programs that read the database and generate the appropriate text information.

In this chapter, we will make a (informal) distinction between *designed devices*, which are devices that are deliberately created by the designer, and *parasitic devices*, which were not explicitly intended by the designer but are inherent in the layout of the circuit.

Primarily there are three different parts to the extraction process. These are designed device extraction, interconnect extraction, and parasitic device extraction. These parts are inter-related since various device extractions can change the connectivity of the circuit, e.g., resistors (whether designed or parasitic) convert single nets into multiple nodes. Usually one level of interconnect extraction is used with designed device extraction to provide a circuit for simulation or gate-level reduction, and a second level of interconnect extraction is used with parasitic device extraction to provide a circuit for timing analysis.

22.2 Early History

Early integrated circuits (ICs) were small, and designers manually determined the electrical model corresponding to the geometry, including devices, their interconnections, and estimates of parasitics. However, even in a small layout it was easy to miss a width or spacing error, so design rule checking (DRC) programs emerged to find these errors (See chapter 17, *Design Rule Checking* in this volume). DRC programs require a machine-readable layout, and designers soon realized such a representation could be used to derive automatically a circuit representation from the geometry. Extraction was first implemented as an afterthought to these early DRC programs — for example, the first mention of a computer program that processes layouts into netlists is a single paragraph at the end of a DRC paper [83]. As chips grew larger, and processes more sophisticated, extraction became an important task in its own right, and emerged as a specific task distinct from DRC, and soon there were programs (and papers) concentrating entirely on extraction [87]. Despite this divergence, the two tasks share considerable basic technology. The now-universal scan line approach, used in both DRC and extraction, appears in [84], and is improved to near current form in [88]. Early consideration of the use of hierarchy can be found in [85].

22.3 Problem Analysis

For layout extraction, there are eight main areas to be evaluated and analyzed. These are overall system capabilities, designed device extraction, connectivity extraction, parasitic in-line device (resistance) extraction, parasitic cross-coupled device (capacitance and inductance) extraction, and network reduction. These issues are summarized briefly here and covered in detail in later sections.

- *Overall system.* What is the intended application — analog, digital, radio frequency (RF), or other? What are the input and output formats and databases? Is extraction flat or hierarchical? Batch or incremental?
- *Creating the as-built silicon geometries from the drawn geometries.* The designer draws an idealized version of the overhead view of the layout. Extraction needs the as-built geometries of the devices and interconnect, including thicknesses, and this must be derived.
- *Designed device extraction.* Find the designed devices and determine their parameters. Remove them from the underlying artwork, if needed, so that what remains is interconnect.
- *Connectivity extraction.* Find which combinations of the remaining geometry form nets. Assign a name to each net. Determine which nets are connected to which pins of the devices, and which are externally visible.
- *Resistance extraction.* Optionally, each net may be divided into one or more sub-nodes to account for the parasitic resistance of each net. This is a 2D problem. Along similar lines, the substrate may need to be subdivided as well. This is a 3D problem over a large area, and different techniques are used.
- *Capacitance and inductance estimation.* Each piece of each net must have its capacitance and inductance estimated. Both capacitance and inductance may require both self and mutual terms.
- *Reduction.* Straightforward extraction gives netlists that are far too big for most practical uses. Reduction generates smaller netlists with very similar properties.

- *Process variation.* Because the fabrication process can vary significantly, even from place to place on the same die, the user may be interested in not only the nominal component values, but also in how they change with changes in physical dimensions.

We examine each of these steps in turn.

22.4 System Capabilities

The overall design of the extraction system is dictated by the design of the overall system in which it resides. Usually layout extraction is a single component in a much larger system involving a complex database, layout editor, simulator, design rule checker, etc. The two major impacts this has on the layout extraction are in the areas of hierarchical processing and results processing.

There are three main possibilities for hierarchical processing:

- First, the complete circuit being processed can be flattened to a single level of hierarchy. This is not usually done since it generates extremely large data sets and it eliminates the optimization of the cell-level repetitions. It does, however, facilitate the extraction of all designed and parasitic devices without imposing any design or layout restrictions.
- Second, the layout can be processed on a cell-by-cell basis, ignoring the hierarchical relationships between cells other than for the determination of inter-cell connectivity. This is a popular processing methodology since layout cells are normally designed to match the logic cells, which are normally complete mini-circuits in themselves. This method does, however, require some form of hierarchical processing to form connections between cells, and it cannot handle intra-cell parasitic capacitances or devices formed across the hierarchy.
- Third, a true hierarchical system can be designed which allows for the extraction of inter-cell parasitic devices and permits optimizations beyond the cell-level repetitions. This theoretically allows both fast processing and unrestricted layouts, but involves a complex special-purpose database and complex algorithms. Also, even if the extraction program allows it, it is often a good idea to prohibit designed device formation between cells as they violate the hierarchy and eliminate repetition optimizations.

For the purposes of defining extraction techniques, this document will look only at the flat- or cell-level extraction, without reference to the hierarchy. Extension to fully hierarchical extraction is a fascinating problem but beyond the scope of this chapter.

The ultimate result of layout extraction will be a circuit description in a form that can be used by a simulator or analysis program. This can be (and originally was) achieved by writing the extracted results out directly as a text file in the simulator language. Today however, it is more likely that the extraction program is part of a larger system. The overall system will enable such features as viewing the extracted circuit on top of the layout; cross-probing the extracted circuit relative to the layout and the logic diagram; and translating the extracted circuit into different text formats for different analysis programs. Each of these functions will require different information from the extraction process stored in the database. What the content and format of this output is will depend on the overall system and database design, but it will probably include physical shapes representing the extracted circuit and devices, in addition to circuit interconnect information to represent the logical extracted circuit.

For the purposes of defining extraction techniques, these detailed output requirements will be ignored. The focus will be on the detection of the circuit elements as logical entities.

A large system for the design of ICs will already have many tools available for the manipulation of layout data (polygons). Many of these can be utilized in the development of an extraction program to save development time and to minimize new buggy code. These tools may be low-level tools such as scan lines, sort programs, hash tables, etc., or they may be higher-level tools such as programs to generate logical operations like AND and OR between polygons. Choosing to use these tools will push the developer in specific directions when deciding on the techniques to use.

22.5 Converting Drawn Geometries to Actual Geometries

The designer of an IC draws an idealized overhead view of all the layers to be created. Accurate parasitic extraction, however, depends on accurate knowledge of the dimensions of structures as they are actually built on the wafer. For many years this difference was small and could be ignored, and extraction was based on the drawn geometries. However, with sub-wavelength lithography (See chapter 18, *Resolution Enhancement Techniques and Mask Data Preparation*) and modern planarization techniques (such as chemical mechanical polishing [CMP]) this difference is no longer small, and must be accounted for. There are two main effects. First, the size and shape of the geometries may be different than drawn, primarily because of limitations in mask making, exposure, resist, and etching processes. Second, the thickness of the various layers is not specified at all by the designer, and must be determined (at least implicitly) by the extraction program.

The first step is to take into account the limitations of mask making, lithography, exposure, and etching. The designers, of course, may draw whatever they wish, but the as-built IC geometries will be limited by the resolution of light and lenses. Specialized software, often called optical proximity correction (OPC) or resolution enhancement technology (RET) is used to preprocess the user-defined geometries before making masks, so that the result on the wafer will best match the designer's intent. This process is not perfect, however. In some cases full correction is simply not possible (the physics of exposure prevents a truly square corner, for example) and in other cases a very accurate correction is deemed too expensive. Therefore, to best match the silicon result, the layout may need to be preprocessed to accurately represent what will really be built. For optical effects, the radius is small (perhaps a micron or so), but the edge positions can depend in complex ways upon the surrounding geometries and the OPC process.

Next, the sizes of geometries in the vertical direction (perpendicular to the wafer) are not specified at all by the designer, and are determined entirely by the process. Originally, these dimensions were either specified by a technology file, or assumed implicitly by the extraction coefficients (amount of capacitance per unit area, for example) and were assumed to be the same for all geometries on a given layer. However, with modern processes, the vertical dimensions of shapes are a function of the surrounding neighborhood. For example, CMP makes it possible to stack many metal layers by grinding each layer flat as it is processed. However, metal and oxide grind at different rates, leading to phenomena such as "erosion" and dishing, as shown in Figure 22.1. These must be accounted for during the extraction process. In general, these effects depend both upon the details of features (such as the width and spacing to neighbors) and the overall density over the surrounding area (typically averaged over a few hundred microns) [89–91].

There are tools and techniques that can analyze these effects accurately, but most are simulation based and too slow for extraction of large circuits. Therefore, extractors usually use simple empirical models. For OPC, the most common model is an "edge bias," or displacement of the edge from the drawn position, which depends on the width and spacing to the neighbors. For CMP, the extractor may define the layer thickness as a function of wire width, spacing to neighbors, and local density integrated over some region.

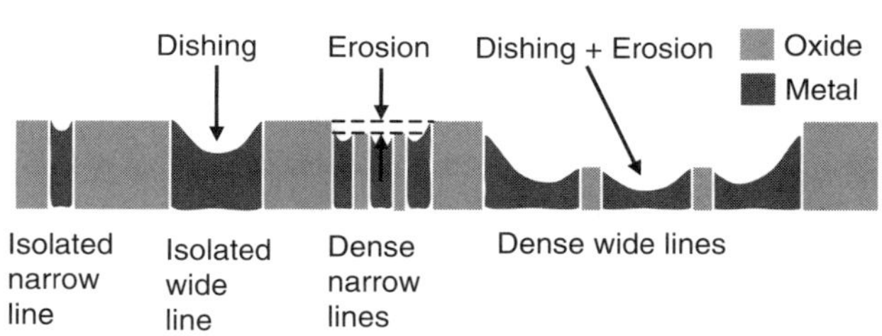

FIGURE 22.1 Layout dependent variation of metal thickness depending on local layout properties. This effect is caused by the chemical–mechanical polishing (CMP) process, which grinds down the metal faster than the oxide, since the metal is softer.

22.6 Designed Device Extraction

A designed device is any device required in the circuit that is explicitly created in the layout. These devices can be created by the inter-relationship of the active circuit layers as in the case of a metal-oxide-semiconductor (MOS) transistor or capacitor occurring anywhere when polysilicon overlaps diffusion (Figure 22.2a), or by the application of specific device definition layers as in the case of a designed bipolar transistor (Figure 22.2b), or by inclusion of specific cells implementing a device (Figure 22.2c), as is typical in RF design.

There are number of steps involved in extracting a device. Firstly, the existence of the device must be recognized. The exact form of the device must be extracted (and verified for correctness) and an instance of the device added to the circuit database. Each device must be given a unique identification and its terminal connections identified. If the existence of a device impacts the flow of connectivity in an interconnect layer, like an MOS transistor that breaks the continuity of diffusion, then that device area must be removed from the connectivity layer. Finally, if required, the parameters of the device must be measured.

The recognition of devices is linked directly to the circuit connectivity extraction. The connectivity extraction cannot be done until the device areas are removed from the interconnect layers, yet the devices cannot be fully recognized until the interconnect nets are extracted so the device terminal connections can be identified. (For example, the device identity is sometimes different depending on the number of distinct nets connected.) This means the two functions of device recognition and interconnect extraction must be done in parallel or interwoven.

There are five stages involved in designed device extraction:

1. Firstly, before any interconnect analysis is performed the device must be recognized, a device definition shape created, and the device given a unique identifier. Devices must be recognized by the specific combination of design layers, by the presence of a specially created device definition layer, or by the presence of designated cells (refer again to Figure 22.2). For each device that is recognized, a series of polygonal logical operations must be performed to create a single shape (polygon) that represents the physical location and extent of that device.
2. Second, if the existence of the device breaks the continuity of an interconnect layer, the area of that device must be removed from the interconnect layer prior to interconnect analysis being performed.
3. After interconnect analysis, the device definition shape must be related back to the associated interconnect layer to determine the existence and extent of terminals. Each terminal found must be identified as a physical and a nodal connection to the device. The layer involved in each connection will sometimes define the type of terminal (e.g., for MOS transistors, diffusion terminals will form sources and drains, and polysilicon will form gates).

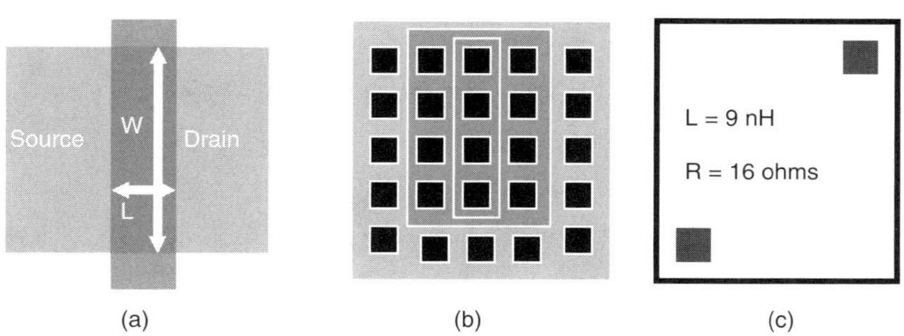

FIGURE 22.2 Three styles of defined devices. (a) Typical for MOS design — the transistor is defined by poly overlap diffusion, and the extractor should measure *W*, *L*, and the area of source and drain. (b) Typical bipolar design — transistor is defined by emitter inside base inside collector. Extractor should measure number of contacts, length of perimeter, and area of each polygon. (c) Device inside a cell commonly used for RF or analog design. Extractor should just make connections, but take device values from cell name or properties.

4. Next, given the terminals and nodal connections to the device, it must be validated. Each device must have the correct number of terminals of each type. A different number may mean the device is badly formed, or it may mean it is a different device type. In addition, the nodes connecting to the device may affect the validity or type of device. For example, a MOS transistor having three physical source drains may be a badly formed device unless that is a valid configuration for the technology involved. However, if two of those terminals are connected to the same electrical node, the device has only two logical source drains and that may be acceptable. If the transistor has two physical source-drain terminals which are connected to the same electrical node, then functionally it is an MOS capacitor and may be redefined as that device type.
5. Finally, if the measurement of device parameters is required, this operation must be performed. The measurement process itself is fairly simple. For a MOS transistor where the parameters of gate length, gate width, and source and drain area are required, measuring the appropriate edge lengths of the gate polygon and the areas of the source and drain diffusions gives the required answers. For bipolar devices, the area and perimeter of each of the constituent polygons is required. Add to these the association of the correct terminal node number to each measurement and the parameters are complete. When and how these measurements are made will depend on the algorithms being used for general polygon manipulation and device recognition. It may be possible to make these measurements during the process of recognizing and extracting the device recognition shapes (gates) or it may be necessary to perform secondary measurements. For devices contained in explicit cells, the device parameters can be obtained from the name or the properties of the master cell instance (or the occurrence, in the case of programmable cells, or pcells). For unusual devices, user-defined code may be needed to establish the device parameters. (See chapter 12, *Design Databases*, for more information about pcells and extension languages, which are used for these tasks.)

A very real problem for device parameter measurement is the definition of the required parameters and their relationship to interconnect parasitic measurements. The layout is inherently a distributed system, but the desired output is almost always in terms of lumped elements. This conversion is inherently ambiguous. Consider Figure 22.3, and the questions it raises:

- If transistor source and drain area are required as parameters, how should the area of diffusion between gates A and B be considered? Is the total area applied to each gate or is it divided between them?
- If the source and drain area are used to define the inherent capacitance in a device, should the areas measured for device parameters be excluded from the interconnect parasitic capacitance measurement?

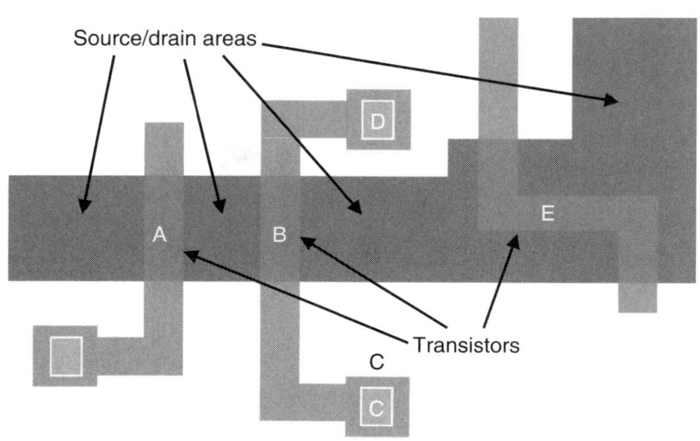

FIGURE 22.3 Diagram illustrating the inherent ambiguity of converting a distributed layout model into a lumped model.

- For the extension of the diffusion to the right of gate B, how much should be allotted to the source/drain area and how much is just an interconnect path to the next device?
- The resistive polysilicon path between points C and D has a single logical connection point to gate B so the polysilicon area over the gate is still part of the interconnect resistance. However, the gate area itself is part of the device and can be included as part of the device characteristics or measurements. How should this be resolved?
- Not all devices are constructed as simple rectangles, as shown by transistor E, but SPICE (for example) wants a single value for the width W and the length L. A transistor gate can be shaped like an L, U, or Z, or even a more complex shape with any number of bends. Some technologies even allow Y shaped or matrix transistors. How is *W* and *L* for these devices defined?

Only the electrical engineer responsible for the technology can answer these questions and a major component of designing a system will be the interaction with that engineer to completely clarify what is required. The parameter definitions provided by the electrical engineer must cover all device types allowed in the technology for which the program is being written, and the measurement techniques developed for their extraction must be adapted to suit.

All the extracted information (device shapes and location, terminal locations and types, and nodal connections and parameters) must be stored back into a database accessible by other functions of the extraction process and ultimately by the programs that will generate the netlists required for the simulation and analysis programs.

22.7 Connectivity Extraction

A circuit consists of devices, device terminals, I/O terminals and the network that interconnects them. The network is made up of conductive elements and the contacts that join them together. For the circuit to be useful in subsequent analysis programs, certain terminals or nodes in the circuit must be identified by their function name.

For clarification, in this context, a net is a physical collection of shapes in a layout that connect together to form a single logical entity, which ideally shares a single voltage value. A node is a point in the physical net that forms a terminal or branch. It can also be used to define the single logical entity in the circuit that is represented by a net.

For connectivity extraction the following issues must be addressed:

- In-line designed device identification and removal. The first step in interconnect extraction must be the removal of any designed device that changes the connectivity of the circuit. This is addressed in Section 22.3.3. Figure 22.4 shows the interconnect layers for the circuit from Figure 22.2 after the designed devices have been removed.
- Interconnect continuity recognition through contacts and vias. There are two main forms of connection in IC layouts. Layer to layer connections are defined by contacts and vias. Connections within the same layer are formed by abutment or overlap. Within a cell, connection by abutment is not needed, since two shapes on the same layer that touch are in general made part of the same polygon. Connection by abutment or overlap is more commonly used to connect cells together, or to connect interconnect to cells.
- Circuit terminal identification. There are three forms of terminals to be considered during extraction. When a designed device is identified, the body of the device may form a single terminal as in the case of the gate of MOS transistor. This terminal can be recognized by the overlap of the device body and the terminal layer.

The body of a designed device may form a terminal by butting up against the edge of a shape on the interconnect layer, as in the case of the source and drain of an MOS transistor.

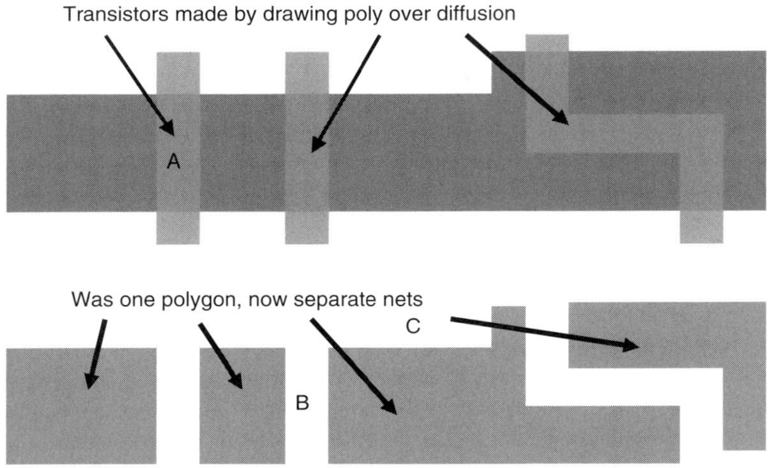

FIGURE 22.4 Why designed devices (transistors) must be removed from drawn layer diffusion before interconnect extraction. The upper panel shows a portion of a layout as drawn; the bottom panel shows diffusion as used for extraction. Note that poly gates do not need to be removed from the poly layer, but we need to make sure the gate capacitance is not double counted.

The input and output terminals of a circuit must also be identified, and this is normally done through a terminal definition layer such as a bonding pad with the addition of a text label, or in some cases by the use of a text label alone without the identification layer. Additional issues are:

- Terminal naming and net labeling. In addition to the identification of I/O terminals, internal nets in the circuit may need specific identification to help the application programs that are going to be applied to the extracted circuit. These nets are normally identified simply with text labels.
- Internal net numbering. Every net extracted must be identified in some way. Some will be defined by net and terminal labels. All others must be given a unique identification, usually in the form of a number.
- Net subdivision through parasitic resistance identification. When parasitic resistance is required, the nets must be extracted twice. First the nonresistive interconnect must be extracted. Then for each physical net that contains shapes on resistive layers, a resistive sub-network must be extracted. The issues of resistance extraction will be dealt with later, but for the actual network extraction there is one additional issue that must be addressed.

For nodes in the circuit that form inputs and outputs, the exact point of the I/O terminal must be identified. For normal connectivity extraction, a label anywhere on the net suffices to identify the function of the net; however, if a resistive network is to be extracted, that is not sufficient. The resistance network must form a path between the I/O and the device terminals. Either an identifier shape must be used to exactly position the I/O terminal, or a text label must be placed at the exact location of the terminal.

22.8 Parasitic Resistance Extraction

Once the connectivity extraction is complete, nets with resistive interconnect layers can be converted into parasitic resistance networks. This is best done one resistive layer at a time. Multiple layers in one net can be extracted separately but the connection between the layers, whether it is vias or sections of a nonresistive layer, must be identified as additional nodes in the network. There are a number of steps involved:

- Terminal location. Polygons on each resistive net must be related to contacts, I/O terminal definitions, and to device terminals to determine the exact physical position of the resistive network terminals.

- Breaking the nets into shapes. The basic idea is to break a complex shape into a collection of simpler shapes. The *R* of each simple shape can be estimated, then the results combined to create a resistive network representing the net. How the net is broken up determines the accuracy and efficiency of the process (see [23] for details).
- Resistance calculation. For each shape extracted from the layout, a resistance value must be calculated using the dimensions of the shape, its function (linear resistor, bend, junction, etc.), and a resistance coefficient value. In addition, the resistance values of the contacts must be calculated if these are required.

In modern copper processes, several additional effects complicate resistance extraction. Unlike aluminum interconnect, the entire conductor cross-section is not one homogeneous material. Instead, there is a thin layer of cladding, intended to keep the core of copper in place. The cladding is a much poorer conductor than the copper, and this must be taken into account, particularly for microwave frequencies where the skin effect concentrates current in this part of the wire. Also, for thin wires at deep sub-micron dimensions, a significant fraction of electrons scatter off the walls of the conductor. This leads to an increase in resistance of narrow lines that depend on the surface roughness of the conductor.

The resistance of contacts can be significant, and for accurate extractions it must be considered in more detail. There is the vertical resistance of the contact material itself which for a given technology is inversely proportional to the contact area, and there is the contact edge resistance (where the resistive material thins out as it dips over the edge of a via) that is inversely proportional to the length of periphery of a contact. Complicating the issue even further is the inclusion in the layout of contact arrays where the array itself forms a very complex resistive network or where the multiple resistive layers are in parallel. Two other issues are associated with resistance extraction:

- Network reduction. In general, an accurate *R* reduction requires splitting a net into far more pieces than the user desires. Finding a smaller network with the same response is the problem of "reduction." The reduction that can be achieved depends strongly on the application and the need for accurate capacitance and inductance. This will be covered in more detail in Section 22.3.8.
- Substrate coupling. Extracting the equivalent circuit for the substrate is similar to resistance calculations in that it requires dividing the network (here the substrate) into smaller pieces, then extracting each piece and computing a reduced model. The coupling mechanisms are shown in Figure 22.5, and the desired result is shown in Figure 22.6, where the entire substrate is condensed into a relatively simple electrical model. Substrate extraction differs from interconnect extraction since the problem is 3D, not 2D, and the networks are much larger. Therefore specialized techniques have been developed for both extraction and reduction of substrate models — (see the Chapter 23 of this volume). A wide variety of methods, including finite element, finite difference, and boundary element methods have been used for this task.

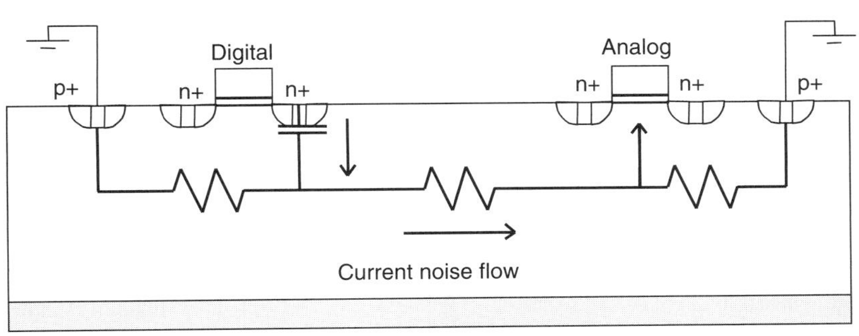

FIGURE 22.5 Current is injected into the substrate and flows to other parts of the circuit.

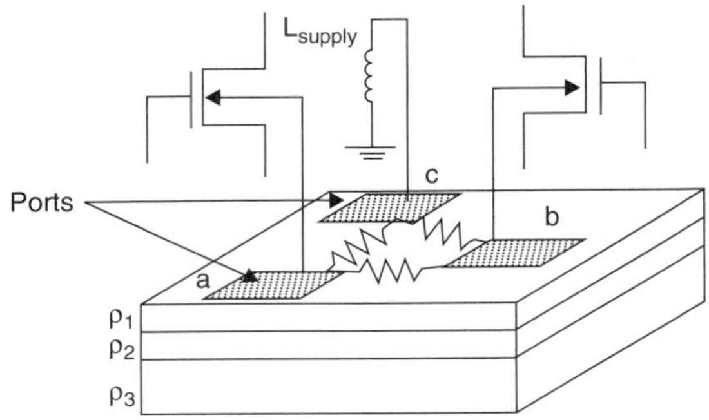

FIGURE 22.6 Substrate ports (nodes) created to connect extracted substrate resistive network to the circuit.

22.9 Capacitance Extraction Techniques

Capacitance and resistance are the dominant interconnect parasitics in the very deep sub-micron regime. Even though the recent processing technology advancements of copper interconnect reduces the effect of R, and the low k (<3) dielectric material reduces the effect of C, the continued scaling down of the feature size keeps the parasitic effects dominant, and makes accurate capacitance extraction a necessity. Capacitance extraction has advanced from 1D, 2D, 2.5D to 3D effects to meet the required accuracy. Here, the extraction definitions of 1D, 2D,..., etc. are given roughly to mark the evolutions.

In 1D capacitance extraction, the area and perimeter parameters of interconnect geometries are obtained. A fine-tuned set of area and perimeter weights per routing layer can be used to calculate capacitance values as an inner product.

$$\text{Cap} = (\text{area\_amount}, \text{perimeter\_amount}) \cdot (\text{area\_weight}, \text{perimeter\_weight})$$

Such area and perimeter weights can be obtained by pre-characterization of an "average" environment of a wire. The area can be single layer or a combination of layer overlaps.

In 2D capacitance extraction, the parameter sets are extended to closely account for the same layer lateral capacitive effect. In other words, not only is the conductor's overlap of interest, but also the capacitive effect across emptiness (free space). In 2D mode, all capacitance effect are modeled as long routing structures with per unit length values. The pure 2D capacitance model does not address the effect of 3D structures (such as two wires in layers metal 2 (m2) and metal 3 (m3) crossing).

An extension to address 3D effects is the 2.5D method [39], where the 3D effect is modeled as a combination of two orthogonal 2D structures. This is illustrated in Figure 22.7.

In Figure 22.7, an m2 wire crosses an m1 wire. Along a vertical cut line (looking down), a 2D cross-sectional view is shown in the middle. Along a horizontal cut line (looking left-right), the other 2D cross-section view is shown to the right. By carefully composing a 3D solution from the two orthogonal 2D ones, most 3D effects are captured.

Obviously, true 3D extractors would go straight to a 3D pattern for a solution in the above crossing structure. However, there are numerous variations in 3D structures. Three dimensional capacitance extractors are not trivial extensions of 2D ones.

22.9.1 How Modern 3D Capacitance Extractors Work

Modern capacitance extraction tools divide the full-chip extraction tasks into three major steps. The first step is called "technology pre-characterization." Given a description of the process cross-sections, tens of thousands of test structures are enumerated and simulated with 2D or 3D field solvers [40–43]. The resulting data are collected either to fit empirical formulas or to build look-up tables (call either one a pattern library). We cite [44] as a reference for model fitting using analytical equations. A good fit would require fewer simulation points. The number of patterns can be reduced by pattern reduction techniques.

We cite [45] as a reference for pattern compression technique. Pattern compression reduces the total number of pre-characterization patterns. Specifically, consider the following layout pattern and two precharacterization patterns in Figure 22.8. In the layout pattern, c12 (m2 to m2 lateral capacitance) is to be obtained over some fixed separation distance. The bottom is fully covered by m1 plate while the top is partially covered by m3 plate and partially covered by m4 plate. A vertically disjoint partition is made to yield two volumes, A and B. There are two corresponding precharacterization patterns A and B, each producing a distinct c12 value at the fixed spacing for m2 to m2 lateral cap. The actual c12 can be composed from c12a and c12b.

Capacitance field solvers employ different numerical algorithms (See Chapter 26 of this volume, *High Accuracy Parasitic Extraction*). Even though they often give similar answers for dense layout structures, they may give very different answers for sparse layout structures, depending on the problem setup and boundary conditions. Therefore, the precharacterization software should have the flexibility to incorporate any third party field solvers. The first step needs to be performed only once per process technology. The challenge in this area includes the handling of ever-more-complex processing technology: low *k*, air bubble dielectric, nonrectangular conductor cross-sections, conformal dielectrics, shallow trench isolations, etc. An effort of standardization on the process file format is sponsored by SII (Silicon Integration Initiative). (For more details please visit the website www.si2.org.)

The second major step in capacitance extraction is the geometric parameter extraction. The geometric parameters are an integral part of precharacterization. If a geometric pattern requires ten parameters to describe, each with 5 possible values, there is a corresponding pre-characterization of 5^{10} ($\sim$10,000,000) patterns to simulate. This is assuming that all sample points are taken in each of the ten parameters, resulting in a 10-dimensional table of the above size. This is clearly not feasible. On the other hand, if a geometric pattern can be described by very few parameters, then it is difficult for it to be accurate. In a full-chip situation, the run time of geometric parameter extraction can be very time/space consuming,

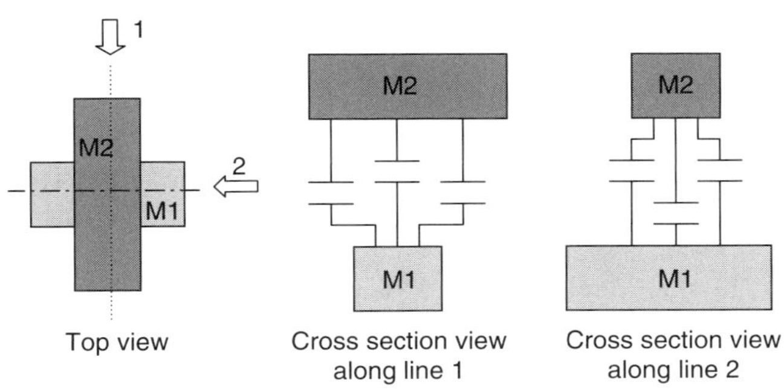

FIGURE 22.7 Three-dimensional effect modeling using 2D structures.

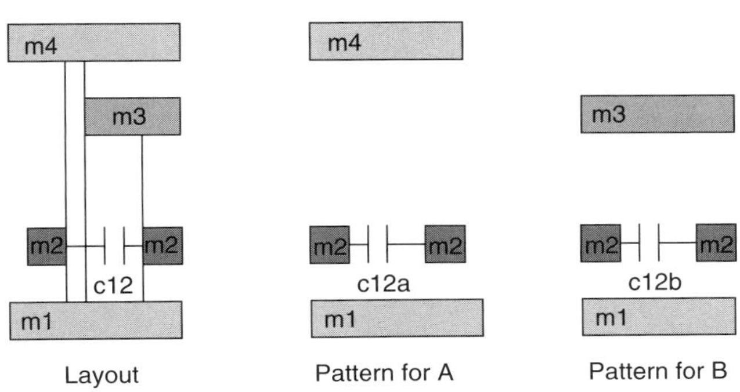

FIGURE 22.8 Layout pattern and two precharacterization patterns.

with millions of interconnect polygons to analyze. Time- and space-efficient geometric processing algorithms are very important.

A geometric parameter reduction technique can be found in [46], where geometric parameters can be dramatically reduced by taking advantage of the shielding effect. Conductors two layers away from the main conductor of interest do not require a precise description. This is particularly useful for the very deep sub-micron geometry, where a far away conductor mesh behaves like a big plate.

The last step in capacitance extraction is to calculate capacitance from geometric parameters. Here, the geometric parameters are matched to entries in the pattern library. One major source of error is called the "pattern mismatch," where extracted geometry parameters do not have an exact match in the pattern library. At this time, there are two remedies to perform the capacitance calculations. One method is to enhance the pattern library by running field solvers at the full-chip extraction time. The other method is to employ heuristics to synthesize a solution from closely matched precharacterization patterns. Even if all the geometric patterns match the library completely, there could still be discontinuities in the layout pattern decomposition, which is another source of error. If the sources of errors can be identified and characterized, it is possible to estimate the error committed due to each error source. One such error bound can be obtained by the "empty" and "full" boundary conditions [47,50].

22.9.2 Critical Net Extraction

Given that full-chip pattern matching capacitance extractors are subject to errors, they should allow a third party field solver to be plugged in to reextract critical nets. A SEMATECH critical net application program interface, or API, is a possible standard. In this API, coupling capacitances are supported. The standardization effort of this API is headed by SII.

22.9.3 Area Fill

Modern chemical mechanical polishing (CMP) process requires area-fill patterns to help planarization. The area-fill patterns becomes a floating net in a chip. The presence of floating nets can make traditional delay calculation or signal integrity verification unreliable, or cause to fail completely. A typical solution is to ground the floating nets (either physically by wire routing or artificially by the extraction tool). In either approach, the resulting cap values are larger (more pessimistic) than reality. In Figure 22.9, a simple demonstration with three nets is shown. Net f is the floating net, which does not connect to anything.

Let us assume c1=3ff and c2=4ff. Using a grounding method, net a would have 3ff to ground, while net b, 4ff to ground; there is no coupling between a and b. Using the well-known series capacitor merging formula, the equivalent capacitance between net a and b becomes 1.7ff, which is much smaller than either one. Once the equivalent cap is obtained, both net a and net b will have 1.7ff to ground, and net a–b coupling is also 1.7ff.

In certain 3D capacitance field solvers (e.g., BEM), floating nets can be modeled with a zero total charge boundary condition on them (with equal electric potential on their surfaces). With such modeling, the

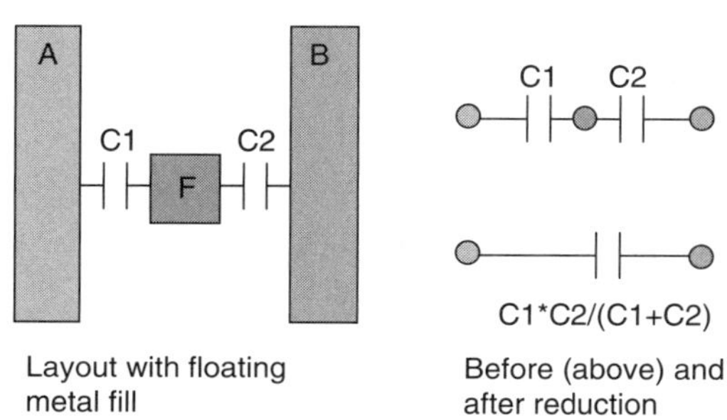

FIGURE 22.9 Area-fill pattern becomes a floating net.

same results can be obtained as above. However, the 3D solvers have limited capacity and cannot be used in the full-chip environment.

In reality, the area-fill patterns form an array of polygons, which means many floating nets need to be eliminated. A sparse capacitance matrix solver is required to eliminate them for aggressive designs.

22.10 Inductance Extraction Techniques

For the idealized case of a lossless, homogeneous dielectric with an array of conductors, the inductance matrix $[L]$ can be derived directly from the capacitance matrix $[C]$ by

$$[L] = \frac{1}{v_0^2}\,[C]^{-1} \tag{22.1}$$

where v_0 is the phase velocity of the medium [51]. Unfortunately, in the IC domain, these assumptions do not hold and more complicated methods are necessary.

Inductance is much more complicated to extract than resistance or capacitance, because of the loop current definition of inductance. In other words, the calculation of inductance for a particular structure relies on knowledge of the return current paths, in addition to the current flow through the wire itself.

22.10.1 The Return Current Problem

To complicate things further, not all of the return currents follow a simple DC path. Instead, they follow the path of least impedance, so the return path may change with frequency or even with the circuit state. Some return currents are in the form of displacement currents through the interconnect capacitances (Figure 22.10) or even the substrate. If these displacement currents are not taken into account and only the DC path is used, the loop inductance will be overestimated.

Capacitance is a function of the electric field density (i.e., flux) between conductors. By and large electric field lines conveniently terminate on adjacent conductors, physically limiting the scope of the problem. Furthermore, it is possible to limit artificially the scope of the problem by only taking into account nearest-neighbors. Doing so will only affect the accuracy of the capacitance calculation but would not result in an unstable RC interconnection model. This is not the case for inductance.

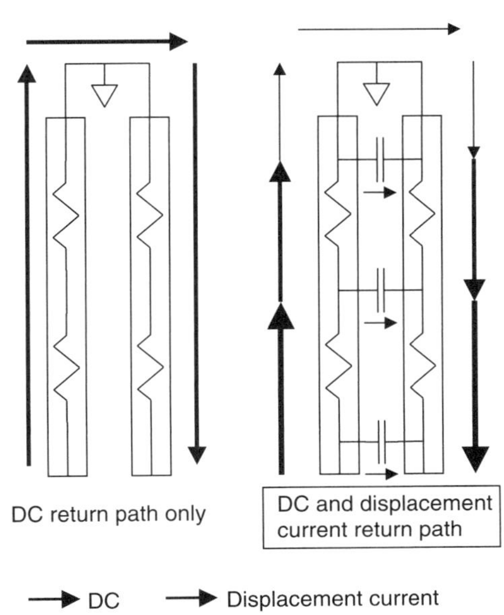

FIGURE 22.10 Inductance return current loop with and without considering displacement currents.

Artificially restricting the physical problem during inductance extraction by using a nearest-neighbor approximation leads not only to obvious inaccuracies but more insidiously, it can result in unstable interconnect models [52].

The magnetic field lines due to a current in a wire do not terminate nicely on adjacent conductors but will spread out to encompass as many wires as necessary to provide a sufficient return current path. Since arbitrarily eliminating some of these wires from consideration could result in ill-behaved models, the inductance matrix produced for a given structure will be dense (i.e., no wire interactions can be eliminated and therefore all off-diagonal elements will be nonzero).

It is therefore highly desirable to use some form of circuit reduction technique when including inductance in the interconnect models. Otherwise the size of the interconnect models produced could overwhelm the analysis tools. Note that not all circuit reduction methods can handle inductance, especially mutual inductance.

Since a simplistic nearest-neighbor strategy cannot be applied to the inductance problem, the simple pattern matching (i.e., rule based) methods used in capacitance extraction would not work (at least not for the general case of arbitrarily shaped and placed conductors).

22.10.2 Numerical Techniques

Most inductance extraction methods in use today are numerically intensive and involve breaking the 3–D conductors into filaments or panels (Figure 22.11) [53–55]. Unfortunately, this method requires that the return current path be known, and may require meshing adjacent conductors, including the substrate. The rapid growth in computational resources required by such methods limit them to small problems on the order of a few dozen shapes. In reality, even a few lines above a nonideal substrate may overwhelm purely numerical methods.

The partial inductance [56,57] method avoids the problem of determining return current paths by assuming that the loops are closed at infinity.

As seen in Figure 22.12, the return loop of line 2 is closed at infinity. The partial inductance of a segment is defined by the flux due to the current in another segment captured inside this semi-infinite loop. The total loop inductance is defined as the sum of the partial self- and mutual inductances along the closed path.

Since the area of each partial inductance loop is much larger than it needs to be, their values tend to be much larger than the actual inductance. The trade-off then in not having to know that the return path is both deviations (errors) from the actual values and potentially ill-conditioned matrices in the interconnect model due to these overly large values.

Various methods have been proposed [52] to sparsify a dense partial inductance matrix in order to reduce the complexity and simulation time of the interconnect model. These turn out to be impractical for a number of reasons. First, they do not reduce the computationally intensive extraction time significantly (at least compared to a rule based capacitance extraction tool). Secondly, the window around the target conductor for sub-micron designs is still fairly large, requiring the inclusion of a large number of wires in order to maintain stability.

Another method to produce a sparse inductance matrix is by use of equipotential shells [58,59]. Rather than defining the return current path at infinity, an equipotential shell is placed around the conductor segment. Instead of an electrical charge as one would use in capacitance calculations, this shell has a uniform current distribution (Figure 22.13). This shell then negates the magnetic vector potential created by the current in the segment for regions outside of the shell. This forces the mutual partial inductance to segments outside the shell to 0.

As long as the shell is a sphere or other equipotential surface, the only effect on the partial inductances inside the shell is to shift them by a constant. It turns out that in many cases this does not produce a corresponding change in the circuit solution [59].

Unfortunately, there are a couple of problems with this method. Firstly, creating the shells can be nontrivial when dealing with complex interconnect structures. If only part of a segment lies within another's shell, the partial inductance calculations become complex. Secondly, even though this technique may

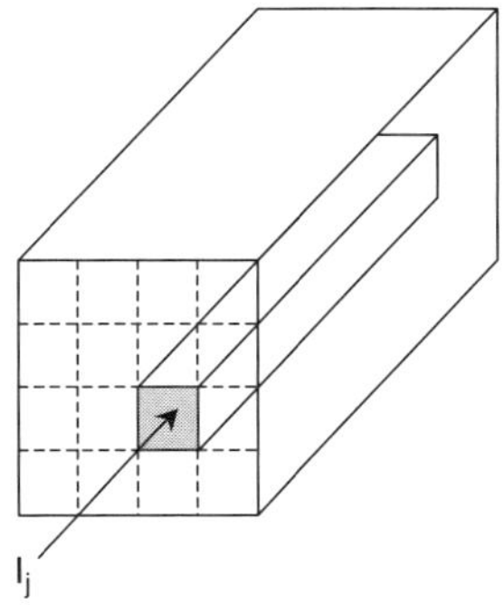

FIGURE 22.11 Example showing one of the current filaments created as part of the discretization of a rectangular conductor.

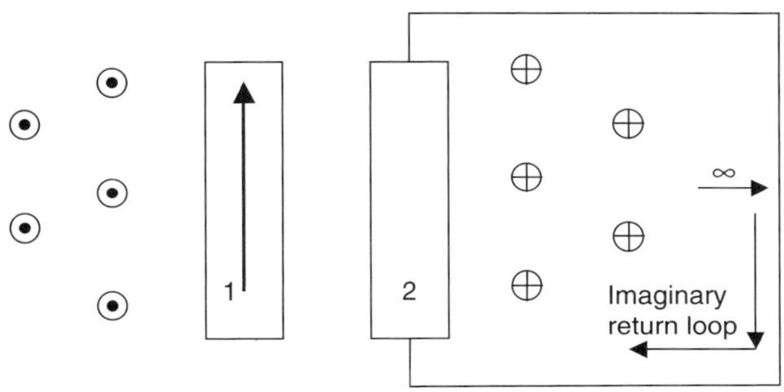

FIGURE 22.12 Computation of partial inductance using an imaginary return path at infinity.

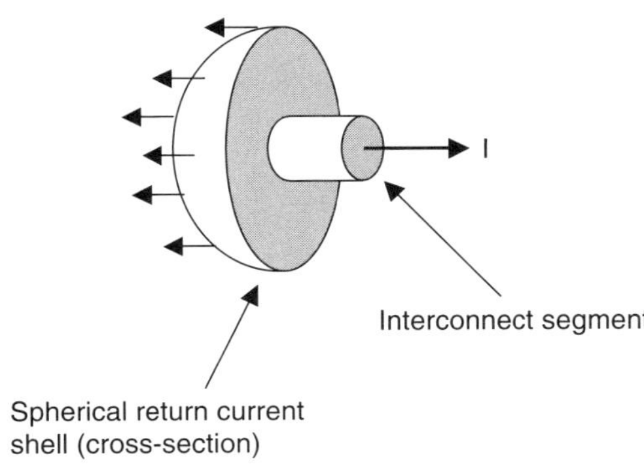

FIGURE 22.13 Example of an equipotential shell of return currents used to decrease the density of the partial inductance matrix.

greatly enhance the sparsity of the resulting matrix, it is still computationally expensive and not well suited to large-scale (particularly full chip) problems.

22.10.3 Rule-Based Methods

Due to the return current problem, it is difficult to create a rule-based scheme such as that used in capacitance and resistance extraction. For the special case of long parallel lines, Krauter and Mehrotra [60] have proposed a quasi-rule-based scheme that uses closed-form expressions instead of numerically intensive iterative methods. By exploiting the fact that inductive issues are a greater problem in the long, wide lines found in upper-level routings, they were able to reduce the scope of the problem. In general, these upper-level routes are often composed of long stretches of parallel lines resembling a traditional microwave coplanar waveguide [51]. (A related approach [71] assumes return currents flow in the nearest power supply line.) This assumption allows the use of well-established, closed-form expressions for the inductance of such structures [56]. Unfortunately, there are closed-form expressions only for the simplest of structures.

22.10.3.1 Only Extract Inductance When You Really Need To

Since inductance is difficult to extract and reduce, it makes sense to extract it only when necessary. Several approaches have been proposed to determine exactly where inductance extraction is necessary.

One technique looks at the bandwidth of the signal to be transmitted. The faster the edge rate of a signal, the greater the bandwidth requirements for the interconnect structures. Insufficient bandwidth will result in severe distortion of the signal, affecting timing, power consumption, and other performance parameters. This need for high bandwidth implies that more complex interconnect models are necessary. Simple RC models are insufficient to simulate accurately the interconnection wiring in extreme conditions. A multistage RLC model would be more appropriate.

Modeling the interconnect as a true frequency-dependent transmission line captures the behavior of the line more accurately by including phenomenon such as retardation (i.e., time of flight). Note that it is impossible to generate a constant signal delay at all frequencies (due to the finite speed of light) with models using only discrete passive elements.

The critical length of a line (the length above which inductance must be included) can be determined from knowledge of the desired edge rate of the signal, along with the speed of propagation of the interconnect structure. Several rules-of-thumb can be used to determine what this critical length is. In general, an interconnect structure should be considered as a transmission line when its physical length approaches 1/4 and 1/10 the wavelength of the highest frequency component.

For the case of a simple microstrip line, it can be shown [64] that the wavelength at a given frequency is

$$\lambda_g = \frac{300}{F\sqrt{\varepsilon_{eef}}}\ \text{mm} \tag{22.2}$$

where ε_{eff} is the effective dielectric constant given by

$$\varepsilon_{eff} = \frac{1}{2}(\varepsilon_r + 1) \tag{22.3}$$

and F is the frequency in GHz.

Consider for example, the case of a 500 MHz signal (square wave). Assuming that the highest frequency component we wish to pass is the sixth harmonic, then the line should be able to pass a signal at 3 GHz. Also assuming a relative dielectric constant of 4, the wavelength for the highest component is then 63.25 mm. Assuming the 1/4 λ rule, the line length for which transmission line properties become important is about 16 mm. In other words, any line longer than about 16 mm should probably be considered for inductance modeling.

The choice of using an RC, RLC, or transmission line model is also dependent on what analysis will be performed with it. For short lines (a few mm), it may be possible to get away with a distributed RC model if timing is the sole criteria. Even though the waveform shapes will be significantly different than those obtained with an RLC model, the timing at the 50% points would not be drastically different.

The case is different for medium to long lines, however. An RC-only representation can underestimate delay and cross-talk values by upwards of 20% or more [65,66].

Another approach takes into account the relatively high resistance of IC interconnect. In a theoretical study using a distributed RLC transmission line with a CMOS driver, Ismail et al. [67] have proposed metrics for determining what line lengths may experience inductive effects. For the case where the driver and line impedances match, it can be shown that line lengths that fall within the inequalities given in Equation (22.4) have significant inductive effects.

$$\frac{t_r}{2\sqrt{LC}} < l < \frac{2}{R}\sqrt{\frac{L}{C}} \tag{22.4}$$

Here R, L, and C are the per unit length parameters of the line, and t_r is the rise time of the signal. If the line is shorter than the lower limit, then inductance is not important because the rise time is longer than the propagation time, so the line behaves as an equipotential. If the line is greater than the upper limit, inductance is not important because the resistive component is so large it overwhelms any inductive contribution. For narrow lines on most modern IC processes, these two regions overlap, and there is no length for which inductance is important. It may need to be considered only for wide wires on the upper metal layers.

22.10.3.2 Simplifying the Inductance Problem by Using Specific Design Methodologies

Another way to reduce the inductance extraction problem is to simplify the determination of the return current path by imposing a strict interconnection structure for all global routing. In other words, follow the path already established by high-speed printed circuit board (PCB) technologies.

For future large-scale, high-speed, DSM designs to meet performance goals, interconnect lines must have predictable delay and noise characteristics.

As shown in several studies [61–63], the most predictable and well-behaved interconnect lines are those with either ground planes above or below them, or neighboring ground/power lines (i.e., coplanar waveguides), as shown in Figure 22.14. As mentioned previously, the key to the behavior of an interconnect wire is its electrical length, not its physical dimensions.

In today's random interconnect environment, it is impractical to perform inductance extraction on even a modest scale except for a few special cases (e.g., coplanar structures). If sufficient care is not exercised when determining the return current paths, not only can significant errors arise but also the resulting models can be unstable. Restricting the interconnect structures to those with well-defined return paths (either power and ground planes or coplanar lines) will simplify the extraction of not only inductance but capacitance and resistance as well [61,68].

While resorting to these techniques will limit the density of the signal lines, the result will be simpler and faster CAD tools along with more predictable interconnect wiring. On the other hand the need for accuracy will require the use of 3D inductance extraction methods, such as explained in Chapter 26, *High Accuracy Parasitic Extraction*, and references [69,70,72].

22.11 Network Reduction

The techniques of the previous sections can generate very large netlists. This is particularly true if accurate resistances are attempted, if substrate analysis is included, or if mutual inductances are wanted. Each of these can generate huge numbers of internal nodes (for R and substrate) or mutual components (for inductance). This causes several problems — the storage alone is prohibitive, the downstream tools often

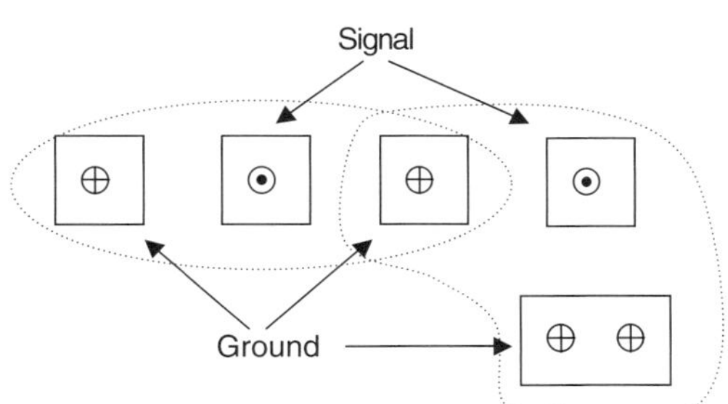

FIGURE 22.14 Example of a coplanar interconnect structure with alternating signal and ground return lines.

run slowly (or fail entirely) when faced with huge networks, and the wide range of time constants in the raw networks causes numerical and stability problems.

In most cases, the intended use of the netlist does not strictly require all these internal nodes and components. In a digital circuit the user is primarily interested in the delay from an output to the inputs on the net. In an analog circuit, the user may be interested in the frequency response, or the transient response, at one particular output.

This defines the job of reduction — take a (usually large) netlist, and generate a smaller netlist (or a mathematical expression) that accurately represents the desired response. There are at least three variants in common use:

- Take an *RC*(*L*) netlist, and generate a smaller *RC*(*L*) netlist. Reducers like these are generally built into extraction tools. The methods tend to be ad hoc, though they try to preserve basic properties like total *C* and total *R* [81,82].
- Take an *RC*(*L*) netlist, and generate an approximation to the desired result directly. The desired result is typically a delay, or a waveform from which delay and slope information can be derived. Elmore delay [12], and asymptoptic waveform evaluation [13–16] are the best known examples here. However, because of the increasing interconnect series resistance, these methods are no longer sufficiently accurate, thus requiring the use of higher order moment matching [17,18] or Krylov subspace methods [19–21].
- Take an RC(L) netlist, and generate a mathematical model (poles and zeros, a state space model, or rational polynomial expressions). This is most commonly used in RF designs, and there is a large and specialized literature on this topic. (See chapter 15, *Simulation and Modeling for Analog and Mixed-Signal Integrated Circuits* for more details.)

22.12 Process Variation

Unfortunately, even if CMP and OPC/RET are taken into account, the final result of fabrication is not always what was intended, due to variations both systematic and random. For systematic effects, as in the previous section, it is best to model them if possible. If they cannot be modeled, or if they are in fact truly random, then we are in the realm of *statistical analysis.* This is a complex area due to the wide variety of correlations that may exist and the difficulty of treating them accurately. For example, if the metal is thicker at a given point, it is not clear without detailed analysis whether the delay goes up (due to increased *C*), down (due to decreased *R*), or stays the same (due to cancellation of these effects).

Since the extractor does not know how the extracted data will be used, it cannot solve the correlation problem. What it *can* do is express how the extracted values will vary with changes in the physical structure. For example, if the local change in metal thickness is $\Delta t1$, the local change of width is $\Delta w1$, and the local change in interconnect thickness is $\Delta i1$, then the *R* of a wire might be expressed as

$$R = \mathrm{R}_0 + k_1\Delta t1 + k_2\Delta w1$$

The capacitance *C* of the same wire might be

$$C = C_0 + k_3\,\Delta t1 + k_4\,\Delta w1 + k_5\,\Delta i1$$

In this case, we would expect k_1, k_2, and k_5 to be negative, and k_3 and k_4 to be positive. Note that delays, slacks, moments, and other circuit metrics can be expressed in the same form, which is very useful during statistical timing and optimization [92,93].

The coefficients k_N are called "sensitivities," and the expression of values in this form is called an "affine representation." As a linear form it is not completely accurate, but for interconnect modeling in particular the accuracy is quite good [94]. The big advantage of this form is that it preserves the correlation information. See reference [95] and chapter 6, *Static Timing Analysis*, for a more detailed discussion of this. A practical advantage of this form is that a single extraction run can be used instead of multiple interconnect corner run. Of course, such parasitic data needs to be stored somehow, but this can be done by extensions to existing databases (See chapter 12, *Design Databases*).

22.13 Conclusions

Extracting a circuit from a physical description is conceptually simple, but quite complex in practice. The requirements, the required capacities, and extraction programs themselves, keep changing as new technologies are introduced. The best way to keep abreast of developments is to monitor the conferences, ISPD, DAC, ICCAD, and DATE, the IEEE journals, and the commercial trade publications that cover IC design.

References

These include articles, not mentioned in the text, which the authors have found helpful.

[1] M. Kamon, S. McCormick, and K. Shepard, Interconnect parasitic extraction in the digital IC design methodology, *Proceedings of the ICCAD*, 1999, pp. 223–230.

[2] W.H. Kao, L. Chi-Yuan, M. Basel, and R. Singh, Parasitic extraction: current state of the art and future trends, *IEEE Proceedings*, pp. 729–739.

[3] D. Bailey and B. Benschneider, Clocking design and analysis for a 600 MHz alpha microprocessor, *IEEE J. Solid State Circ.*, 33, 1627–1633, 1998.

[4] D. Edelstein et al., Full copper wiring in a sub-0.25 um CMOS ULSI technology, *Proceedings of the IEDM*, 1997, pp. 773–776.

Finite Difference Methods

[5] E. Dengi and R. Rohrer, Hierarchical 2-D field solution for capacitance extraction for VLSI interconnect modeling, *Proceedings of the 34th ACM/IEEE DAC*, Anaheim, CA, June 1997, pp. 127–132.

Finite Element Methods

[6] G. Costache, Finite element method applied to skin effect problems in strip transmission lines, *IEEE Trans. Microw. Theory Tech.*, MTT-35, 1009–1013, 1987.

Method of Moments

[7] W. Cao, R. Harrington, J. Mantz, and T. Sarkar, Multiconductor transmission lines in multilayered dielectric media, *IEEE Trans. Microw. Theory Tech.*, MTT-32, 439–450, 1984.

Boundary Element Methods

[8] S. Rao, T. Sarkar, and R. Harrington, The electrostatic field of conducting bodies in multiple dielectric media, *IEEE Trans. Microw. Theory Tech.*, 32, 1441–1448, 1984.

[9] E. Dengi and R. Rohrer, Boundary element method macromodels for 2-D hierarchical capacitance extraction, *Proceedings of the 35th ACM/IEE DAC*, June 1998, pp. 218–223.

3D Interconnects

[10] W. Sun, W.M. Dai, and W. Hong, Fast parameters extraction of general 3-D interconnects using geometry independent measured equation of invariance, *Proceedings of the 33rd ACM/IEEE DAC*, 1996, pp. 105–110.

[11] M. Kamon, N. Marques, and J. White, FastPep: a fast parasitic extraction program for complex three dimensional geometries, *Proceedings of the ICCAD*, San Jose, CA, 1997, pp. 456–460.

Elmore Delay, Asymptoptic Waveform Evaluation

[12] W.C. Elmore, The transient response of damped linear networks with particular regard to wideband amplifiers, *J. Appl. Phys.* 19, 55–63, 1948.

[13] J. Rubinstein, J.P. Penfield, and M.A Horowitz, Signal delay in RC tree networks, *IEEE Trans. CAD*, CAD-2, 202–211, 1983.

[14] L.T. Pillage and R.A Rohrer, Asymptotic waveform evaluation for timing analysis, *IEEE Trans. CAD*, 9, 352–366, 1990.

[15] C. Ratzlaff and L.T. Pillage, RICE: rapid interconnect circuit evaluator using asymptotic waveform evaluator, *IEEE Trans. CAD*, 13(6), pp. 763–776, June 1994.

[16] J.E. Bracken, V. Raghavan, and R. Rohrer, Interconnect simulation with asymptotic waveform evaluation, *IEEE Trans. Circ. Syst.*, 39(11), pp. 869–878, Nov. 1992.

Moment Matching Methods

[17] E. Chiprout and M. Nakhla, Generalized moment-matching methods for transient analysis of interconnect networks, *Proceedings of the 29th ACM/IEEE DAC*, Anaheim, CA, June 1992, pp. 201–206.

[18] J.R. Phillips, E. Chiprout, and D. Ling, Efficient full-wave electromagnetic analysis via model order reduction of fast integral transforms, *Proceedings of the 33rd ACM/IEEE DAC*, Las Vegas, NV, June 1996, pp. 377–382.

Krylov Subspace Methods

[19] P. Feldmann and R. Freund, Efficient linear circuit analysis by Pade approximation via the Lanczos process, *IEEE Trans. CAD*, 14, 639–649, 1995.

[20] K. Gallivan, E. Grimme, and P. Van Dooren, Asymptotic waveform evaluation via a Lanczos method, *Appl. Math. Lett.*, 7, 75–80, 1994.

[21] A. Odabasioglu, M. Celik, and L. Pileggi, PRIMA: passive reduced-order interconnect macromodeling algorithm, *Proceedings of the ICCAD*, San Jose, CA, August 1997, pp. 58–65.

Resistance Extraction

[22] R. Singh, A review of substrate coupling issues and modeling strategies, *Proceedings of the IEEE Custom Integrated Circuits Conference*, 1999, pp. 491–499.

[23] M. Horowitz and R.W. Dutton, Resistance Extraction from Mask Layout Data, *IEEE Trans. Comput. Aided Des. Integr. Circ. Syst.*, 2, 1983, 145–150.

[24] D. Su, M.J. Loinaz, S. Masui, and B.A. Wooley, Experimental results and modeling techniques for substrate noise and mixed-signal integrated circuits, *IEEE J. Solid-State Circ.*, 28, 420–430, 1993.

[25] N. Verghese, T. Schmerbeck, and D. Allstot, *Simulation Techniques and Solutions for Mixed-Signal Coupling in Integrated Circuits*, 2nd ed., Kluwer Academic, Dordrecht, 1995.

[26] X. Aragones, J. Luis Gonzalez, and A. Rubio, *Analysis and Solutions for Switching Noise Coupling in Mixed Signal ICs*, Kluwer Academic Publishers, Dordrecht, 1999.

[27] F.J. Clement, E. Zysman, M. Kayal, and M. Declercq, LAYIN: toward a global solution for parasitic coupling modeling and visualization, *Proceedings of the IEEE Custom Integrated Circuits Conference*, 1994, pp. 537–540.

[28] K. Nabors and J. White, Multipole-accelerated 3-D capacitance extraction algorithms for structures with conformal dielectrics, *Proceedings of the 29th ACM/IEEE Design Automation Conference*, 1992, pp. 710–715.

[29] A.E. Ruehli and P.A. Brennan, Efficient capacitance calculations for three dimensional multiconductor systems, *IEEE Trans. Microw. Theory Tech.*, MTT-21, pp. 76–82, Feb. 1973.
[30] N.K. Verghese and D. Allstot, SUBTRACT: a program for the efficient evaluation of substrate parasitics in integrated circuits, *Proceedings in IEEE/ACM International Conference on Computer-Aided Design*, 1995, pp. 194–198.
[31] N. Verghese, D. Allstot, and M. Wolfe, Verification techniques for substrate coupling and their application to mixed-signal IC design, *IEEE J. Solid-State Circ.*, 31, 354–365, 1996.
[32] R. Gharpurey and R. Meyer, Modelling and analysis of substrate coupling in integrated circuits, *IEEE J. Solid-State Circ.*, 31, 344–353, 1996.
[33] T. Smedes, N.P. van der Meijs, and A.J. van Genderen, Boundary element methods for capacitance and substrate resistance calculations in a VLSI layout verification package, *Proceedings of the ELECTROSOFT'93*, July 1993, pp. 337–344.
[34] R. Gharpurey, Hosur transform domain techniques for efficient extraction of substrate parasitics, *Proceedings of the IEEE/ACM International Conference on Computer-Aided Design*, 1997, pp. 461–467.
[35] M. Chou and J. White, Multilevel integral equation methods for the extraction of substrate coupling parameters in mixed-signal IC's, *Proceedings of the 35th Design Automation Conference*, 1998, pp. 20–25.
[36] J. Zhao, W.M. Dai, S. Kadur, and D.E. Long, Efficient three-dimensional extraction based on static and full-wave layered Green's functions, *Proceedings of the 35th Design Automation Conference*, 1998, pp. 224-229.
[37] S. Kapur and D.E. Long, IES3: a fast integral equation solver for efficient 3-dimensional extraction, *Proceedings of the IEEE/ACM International Conference on Computer-Aided Design*, November 1997, pp. 448–455.
[38] J. Zhao, W. Dai., R.C. Frye, and K.L. Tai, Green function via moment matching for rapid and accurate substrate parasitics evaluation, *Proceedings of the IEEE Custom Integrated Circuits Conference*, 1997, pp. 371–374.

Capacitance Extraction

[39] S. Napper, Technical white paper on RC extraction, Technical Report, *EPIC Design Technology*, 1995.
[40] K. Nabors and J. White, FastCap: a multipole-accelerated 3-D capacitance extraction program, *IEEE Trans. CAD*, 10, 1447–1459, 1991.
[41] K. Nabors and J. White, Multipole-accelerated 3-D capacitance extraction algorithms for structures with conformal dielectrics, *Proceedings of the 29th ACM/IEEE DAC*, June 1993, pp. 710–715.
[42] W. Hong, W.K. Sun, Z.H. Zhu, H. Ji, B. Song, and W. Dai, A novel dimension reduction technique for the capacitance extraction of 3-D VLSI interconnects, *MTT*, 46, 1037–1044, 1996.
[43] W. Shi, J. Liu, N. Kakani, and T. Yu, A fast hierarchical algorithm for 3-D capacitance extraction, *Proceedings of the 35th ACM/IEEE DAC*, San Francisco, CA, June 1998, pp. 212–217.
[44] U. Choudhury and A. Sangiovanni-Vicentelli, Automatic generation of analytical models for interconnect capacitances, *IEEE Trans. Comput. Aided Des. Integr. Circ. Syst.*, 14, 470–480, 1995.
[45] N.D. Arora and K.V. Roal, R. Schumann, and L.M. Richardson, Modeling and extraction of interconnect capacitances for multilayer VLSI circuits, *IEEE Trans. Comput. Aided Des. Integr. Circ. Syst.*, 15, 58–67, 1996.
[46] P.A Habitz and I.L. Wemple, A simpler, faster method of parasitic capacitance extraction, *Electr. J.* 11–15, Oct. 1997.
[47] E. Dengi, A parasitic capacitance extraction method for VLSI interconnect modeling, Research Report: CMUCAD-97-29, Carnegie Mellon University.
[48] E. You, S. Yew Choe, C. Kim, L. Varadadesikan, and K. Aingaran, J. MacDonald, Parasitic extraction for multimillion-transistor integrated circuits: methodology and design experience, *Proceedings of the IEEE 2000 Custom Integrated Circuits Conference, CICC*, 21–24 May 2000, pp. 491–494 (Hierarchical).

[49] A. Kurokawa, T. Kanamoto, A. Kasebe, Y. Inoue, and H. Masuda, Efficient capacitance extraction method for interconnects with dummy fills, *Proceedings of the IEEE 2004 Custom Integrated Circuits Conference*, 3–6 October 2004, pp. 485–488.

[50] M.W. Beattie and L.T. Pileggi, Error bounds for capacitance extraction via window techniques, *IEEE Trans. CAD*, 18, 1999, pp. 311–321.

Inductance Extraction

[51] K.C. Gupta, R. Garg, and R. Chadha, *Computer Aided Design of Microwave Circuits*, Artech House, Bedham, MA, 1981.

[52] M. Beattie and L. Pileggi, IC Analyses including extracted inductance models, *Proceedings of 36th International Conference on Design Automation*, June 1999, pp. 915–920.

[53] J. Wang, J. Tausch, and J. White, A wide frequency range surface integral formulation for 3-D inductance and resistance extraction, *Technical Proceedings of the 1999 International Conference on Modeling and Simulation of Microsystems*, 1999.

[54] M. Kamon, N. Marques, Y. Massoud, L. Silveira, and J. White, Interconnect analysis: from 3-D structures to circuit models, *Proceedings of 36th International Conference on Design Automation*, June 1999, pp. 910–914.

[55] Z. He, M. Celik, and L. Pileggi, SPIE: sparse partial inductance extraction, *Proceedings of 34th International Conference on Design Automation*, June 1997, pp. 137–140.

[56] E. Rosa, The self and mutual inductance of linear conductors, *Bull. Nat. Bur. Stand.*, No. 4, 301–344, 1908.

[57] A. Ruehli, Inductance calculations in a complex integrated circuit environment, *IBM J. Res. Dev.*, 16, 470–481, 1972.

[58] B. Krauter and L. Pileggi, Generating sparse partial inductance matrices with guaranteed stability, *International Conference on Computer Aided Design*, 1995, pp. 45–52.

[59] M. Beattie, L. Alatan, and L. Pileggi, Equipotential shells for efficient partial inductance extraction, *Proceedings of the 1998 IEDM*, December 1998.

[60] B. Krauter and S. Mehrotra, Layout based frequency dependent inductance and resistance extraction for on-chip interconnect and timing analysis, *Proceedings of 35th International Conference on Design Automation*, June 1998, pp. 303–308.

[61] P. Restle, A. Ruehli, and S. Walker, Dealing with inductance in high-speed chip design, *Proceedings of 36th International Conference on Design Automation*, June 1999, pp. 904–910.

[62] S. Morton, On-chip inductance issues in multiconductor systems, *Proceedings of 36th International Conference on Design Automation*, 1999, pp. 921–926.

[63] A. Deutsch, G.V. Kopcsay, C.W. Surovic, B.J. Rubin, L.M. Terman, R.P. Dunne, T.A. Gallo, and R.H. Dennard, Modeling and characterization of long on-chip interconnections for high-performance microprocessors, *IBM J. Res. Dev.*, 39, 547–567, 1995.

[64] T.C. Edwards, *Foundations for Microstrip Circuit Design*, Wiley, New York, NY, USA, 1981.

[65] A. Deutsch, G. Kopcsay, P. Restle, H. Smith, G. Katopis, et al., When are transmission line effects important for on-chip interconnections? *IEEE Trans. Microw. Theory Tech.*, 45, 1836–1846, 1997.

[66] A. Deutsch, H. Smith, C. Surovic, et al., Frequency dependent crosstalk simulation for on-chip interconnections, *IEEE Trans. Adv. Pack.*, 22, 292–308, 1999.

[67] Y. Ismail, E. Friedman, and J. Neves, Figures of merit to characterize the importance of on-chip inductance, *Proceedings of 35th International Conference on Design Automation*, June 1998, pp. 560–565.

[68] N. van der Meijs and T. Smedes, Accurate interconnect modeling: towards multi-million transistor chips as microwave circuits, *International Conference on Computer Aided Design*, 1996, pp. 244–251.

[69] M. Kamon, M. Tsuk, C. Smithhisler, and J. White, Efficient techniques for inductance extraction of complex 3-D geometries, *Proceedings of the ICCAD*, Santa Clara, 1992, pp. 438–442.

[70] M. Kamon, M.J. Tsuk, and J. White, FASTHENRY: a multipole-accelerated 3-D inductance program, *IEEE Trans Microw. Theory Tech.*, 42, 1750–1758, 1994.

[71] K.L. Shepard and Z. Tian, Return-limited inductances: a practical approach to on-chip inductance extraction, *IEEE Trans. CAD*, 19, 425–436, 2000.

[72] K.L. Shepard, D. Sitaram, and Y. Zheng, Full-chip, three-dimensional, shapes-based RLC extraction, *Proceedings of the International Conference on Computer-Aided Design*, 2000, pp. 142–149.
[73] M. Beattie and L. Pileggi, Efficient inductance extraction via windowing (2001), *Proceedings of the Design Automation and Test in Europe* (DATE), March 2001.

Interconnect Modeling, Model Order Reduction, and Delay Calculation

[74] B. Tutuianu, F. Dartu, and L. Pileggi, An explicit RC-circuit delay approximation based on the first three moments of the impulse response, *Proceedings of the 33rd ACM/IEEE DAC*, June 1996, pp. 611–616.
[75] Y. Ismail, E. Friedman, and J. Neves, *Equivalent Elmore delay for RLC trees, IEEE Trans. CAD*, 19, No. 1, pp. 83–97, Jan. 2000.
[76] L.M. Miguel Silveira, M. Kamon, I. Elfaldel, and J. White, A coordinate-transformed Arnoldi algorithm for generating guaranteed stable reduced-order models of RLC circuits, *Proceedings of the International Conference on CAD*, San Jose, CA, November 1996, pp. 288–294.
[77] I.M. Elfaldel and D. Ling, A block rational Arnoldi algorithms for multipoint passive model-order reduction of multiport RLC networks, *Proceedings of the ICCAD*, San Jose, CA, November 1997, pp. 66–71.
[78] Y. Liu, L. Pileggi, and A. Strojwas, Model order reduction of RCL interconnect including variational analysis, *Proceedings of the 36th ACM/IEEE DAC*, June 1999, pp. 201–206.
[79] Y. Ismail, E. Friedman, and J. Neves, Repeater insertion in tree structured inductive interconnect, *Proceedings of the 36th DAC*, June 1999, pp. 420–424.
[80] J. Kanapka, J. Phillips, and J. White, Fast methods of extraction and sparsification of substrate coupling, *Proceedings of 37th Design Automation Conference*, June 2000, pp. 738–743.
[81] B.N. Sheehan, TICER: realizable reduction of extracted RC circuits, *ICCAD'99: Proceedings of the 1999 IEEE/ACM International Conference on Computer-Aided Design*, San Jose, CA, 1999, pp. 200–203.
[82] V.B. Rao, J.P. Soreff, R. Ledalla, and F.L. Yang, Aggressive crunching of extracted RC netlists, *TAU'02: Proceedings of the Eighth ACM/IEEE International Workshop on Timing Issues in the Specification and Synthesis of Digital Systems*, Monterey, CA, 2002, pp. 70–77.

Historical

[83] Yamin, M., XYTOLR — A computer program for integrated circuit mask design checkout, *Bell Syst. Tech. J.*, 51, 1581–1593, 1972.
[84] H.S. Baird, Fast algorithms for LSI artwork analysis, *DAC'77: Proceedings of the 14th Conference on Design Automation*, 1977, pp. 303–311.
[85] P. Losleben, Design validation in hierarchical systems, *DAC'75: Proceedings of the 12th Conference on Design Automation*, 1975, pp. 431–438.
[86] R.M. Allgair and D.S. Evans, A comprehensive approach to a connectivity audit, or a fruitful comparison of apples and oranges, *DAC'77: Proceedings of the 14th Conference on Design Automation*, 1977, pp. 312–321.
[87] B.T. Preas, B.W. Lindsay, and C.W. Gwyn, Automatic circuit analysis based on mask information, *DAC'76: Proceedings of the 13th Conference on Design Automation*, 1976, pp. 309–317.
[88] U. Lauther, An O (N log N) algorithm for Boolean mask operations, *DAC'81: Proceedings of the 18th Conference on Design Automation*, 1981, pp. 555–562.

OPC, RET, and CMP Modeling (CMP modeling alone is a huge area. This is just an introduction)

[89] C. Heitzinger, A. Sheikholeslami, F. Badrieh, H. Puchner, and S. Selberherr, Feature-scale process Simulation and accurate capacitance extraction for the backend of a 100-nm aluminum/ TEOS process, *IEEE Trans. Electron Devices*, 51, 1129–1134, 2004.

[90] D.O. Ouma, D.S. Boning, J.E. Chung, W.G. Easter, V. Saxena, S. Misra, and A. Crevasse, Characterization and modeling of oxide chemical–mechanical polishing using planarization length and pattern density concepts, *IEEE Trans. Semiconduct. Manuf.*, 15, 232–244, 2002.

[91] T.E. Tugbawa, T.H. Park, and D.S. Boning, Integrated chip-scale simulation of pattern dependencies in copper electroplating and copper chemical mechanical polishing processes. *Proceedings of the IEEE 2002 International Interconnect Technology Conference*, 3–5 June 2002, pp. 167–169.

Statistical Extraction

The paper by Scheffer has some extraction experiments showing the linear form is quite accurate. The papers by Sapatnekar and Visweswariah show how to use results in this form in timing analysis. The paper by Rutenbar is about the mechanics of manipulating the affine forms. The paper by Liu and Pillagi [78], referenced in Section 22.3.8, discusses how this form can be reduced.

[92] H. Chang and S.S. Sapatnekar, Statistical timing analysis considering spatial correlations using a single pert-like traversal, *ICCAD'03: Proceedings of theIEEE/ACM International Conference on Computer-Aided Design*, 2003, pp. 621.

[93] C. Visweswariah, K. Ravindran, K. Kalafala, S.G. Walker, and S. Narayan, First-order incremental block-based statistical timing analysis, *DAC'4: Proceedings of the 41st Annual Conference on Design Automation*, 2004, pp. 331–336.

[94] L. Scheffer, Explicit computation of performance as a function of process variation, *TAU '02: Proceedings of the 8th ACM/IEEE International Workshop on Timing Issues in the Specification and Synthesis of Digital Systems*, 2002, pp. 1–8.

[95] J.D. Ma and R.A. Rutenbar, Interval-valued reduced order statistical interconnect modeling, *Proceedings of the IEEE/ACM International Conference on Computer Aided Design, ICCAD-2004*, 7–11 November 2004, pp. 460–467.

[96] A. Labun, Rapid method to account for process variation in full-chip capacitance extraction, *IEEE Trans. CAD*, 23, 941–951, 2004.

Converting Transistor Level to Gate Level

[97] G. Ditlow, W. Donath, and A. Ruehli, Logic equations for MOSFET circuits, *Proceeding of the IEEE Int. Symp. Circ. Syst.*, 752–755, May 1983.

[98] Z. Barzilai, L. Huisman, G.M. Silberman, D.T. Tang, and L.S. Woo, Simulating pass transistor circuits using logic simulation machines, *Design Automation Conference*, 1983, pp. 157–163.

[99] R.E. Bryant, Boolean analysis of MOS circuits, *IEEE Trans. CAD*, 6, 634–649, 1987.

[100] D.T. Blaauw, D.G. Saab, P. Banerjee, and J. Abraham, Functional abstraction of logic gates for switch level simulation, *European Conference on Design Automation*, 1991, pp. 329-333.

[101] R.E. Bryant, Extraction of gate level models from transistor circuits by four valued symbolic analysis, *International Conference in Computer-Aided Design*, 1991, pp. 350–353.

23

Mixed-Signal Noise Coupling in System-on-Chip Design: Modeling, Analysis, and Validation

Nishath Verghese
Cadence Berkeley Laboratories
Berkeley, California

Makoto Nagata
Kobe University
Kobe, Japan

Abstract

The impact of noise coupling in mixed-signal integrated circuits is described and the techniques to model, analyze, and validate it are reviewed. The physical phenomena responsible for the creation of noise as well as its transmission mechanisms and media are described, and the parameters affecting the strength of the noise coupling are discussed. The different modeling approaches and computer simulation methods used in quantifying noise coupling phenomena are presented. Measurement techniques for substrate noise and subsequent validation of noise analysis results are also presented. Application of substrate noise analysis to placement and power distribution synthesis are also reviewed.

23.1 Introduction

The push for reduced cost, more compact circuit boards, and added customer features has provided incentives for the inclusion of analog functions on primarily digital MOS integrated circuits (ICs) forming mixed-signal ICs. In these systems, the speed of digital circuits is constantly increasing, chips are becoming more densely packed, interconnect layers are added, and analog resolution is increased. In addition, recent increase in wireless applications and its growing market are introducing a new set of aggressive design goals for realizing mixed-signal systems. Here, the designer integrates radio-frequency (RF) analog and base band digital circuitry on a single chip. The goal is to make single-chip radio frequency integrated circuits (RFICs) on silicon, where all the blocks are fabricated on the same chip. One of the advantages of this integration is low power dissipation for portability due to a reduction in the number of package pins and associated bond wire capacitance. Another reason that an integrated solution offers lower power consumption is that routing high-frequency signals off-chip often requires a 50 Ω impedance match, which can result in higher power dissipation. Other advantages include improved high-frequency performance due to reduced package interconnect parasitics, higher system reliability, smaller package count, smaller package interconnect parasitics, and higher integration of RF components with VLSI-compatible digital circuits. In fact, the single-chip transceiver is now a reality [1].

The design of such systems, however, is a complicated task. There are two main challenges in realizing mixed-signal ICs. The first challenging task, specific to RFICs, is to fabricate good on-chip passive elements such as high-Q inductors [2]. The second challenging task, applicable to any mixed-signal IC and the subject of this chapter, is to minimize noise coupling between various parts of the system to avoid any malfunctioning of the system [3,4]. In other words, for successful system-on-chip integration of mixed-signal systems, the noise coupling caused by nonideal isolation must be minimized so that sensitive analog circuits and noisy digital circuits can effectively coexist, and the system operates correctly. To elaborate, note that in mixed-signal circuits, both sensitive analog circuits and high-swing high-frequency noise injector digital circuits may be present on the same chip, leading to undesired signal coupling between these two types of circuit via the conductive substrate. The reduced distance between these circuits, which is the result of constant technology scaling, exacerbates the coupling. The problem is severe, since signals of different nature and strength interfere, thus affecting the overall performance, which demands higher clock rates and greater analog precisions.

The primary mixed-signal noise coupling problem from fast-changing digital signals, coupling to sensitive analog nodes, is shown schematically in Figure 23.1. Another significant cause of undesired signal coupling is the cross talk between analog nodes themselves owing to the high-frequency/high-power analog signals. One of the media through which mixed-signal noise coupling occurs is the substrate. Digital operations cause fluctuations in the underlying substrate voltage, which spreads through the common substrate causing variations in the substrate potential of sensitive devices in the analog section. Similarly, in the case of cross talk between analog nodes, a signal can couple from one node to another via the substrate. This phenomenon is referred to as "substrate coupling" or "substrate noise coupling".

In this chapter, we discuss the mixed-signal coupling problem and review various techniques to model, analyze, and validate it. In Section 23.2, the random noise inherent to electronic devices and the deterministic noise generated by circuits are differentiated. Then, we discuss the physical phenomena responsible for the creation of undesired signals in a digital circuit and the mechanisms of their transport to other parts of the system, mainly via the substrate. Various modeling approaches and simulation techniques of digital noise generation as well as substrate impedance network determining noise coupling are reviewed in Section 23.3. Section 23.4 describes an application of the analysis techniques to placement and power distribution synthesis.

23.2 Mechanisms and Effects of Mixed-Signal Noise Coupling

23.2.1 Differentiation between Random Noise and Deterministic Noise

All undesired phenomena, behaviors, or influences that degrade performance are regarded as noise [5]. Noise in a mixed-signal IC can be classified into two types. One is the intrinsic noise of active and passive

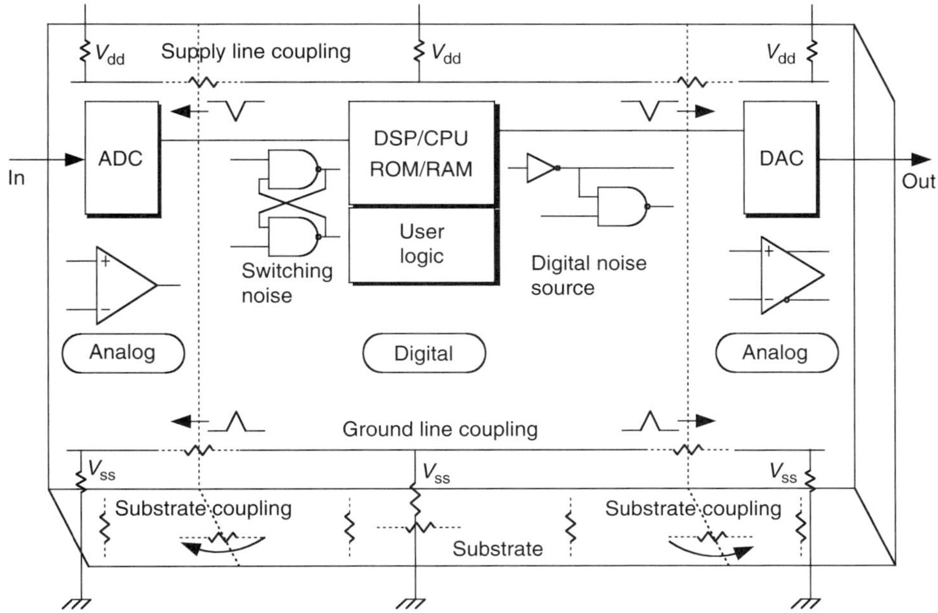

FIGURE 23.1 Digital noise coupling in a mixed-signal LSI circuit. For simplicity, some of the parasitic elements causing the coupling are not shown. (From T. Tsukada and K. Makie-Fukuda, Approaches to reducing digital-noise coupling in CMOS mixed-signal LSI's, *IEICE-T. Fund. Electr.*, E80-A, 263–275, 1997. With permission.)

devices in the circuit and the other is the undesired signal (e.g., switching noise of digital circuits) coupled from other parts of the circuit. Intrinsic noise originates from various physical phenomena within the device itself; some examples are thermal noise, shot noise, and flicker noise. This type of noise, which has a non-deterministic nature, is an important consideration in the design of sensitive analog circuits such as receiver RF circuitry where signal levels could be very small. The level of this type of noise represents a minimum level of noise in a system, and its control is accomplished through optimal circuit design, topology selection, bandwidth limiting of signals, and semiconductor process control. The parameters that are used to quantify this type of noise are noise figure (NF) and signal-to-noise ratio (SNR). Having the circuit topology of a block as well as its circuit component's noise characteristics, one can determine the NF or SNR for the block. The noise is usually represented by an input referred noise source and has its own frequency spectrum.

Contrary to the first noise type, the second type, which is an undesired signal, has a deterministic nature and, theoretically, can be quantified both in the frequency domain and in the time domain. An example of the second type of noise is digital switching noise, which is the major source of undesired signals in mixed-signal ICs [6] and can be very destructive because it can be broadcast over great distances, acting on all transistors by modulating threshold voltage and gain, and directly coupling to the signal node. It can also increase the average delay of digital blocks [7]. In addition to digital switching noise, any high-frequency signal of analog circuits can behave as the source of an undesired signal for other circuits. Therefore, not only digital circuits but also analog circuits could play the role of aggressor blocks for other parts of the system. However, we will intensively focus on deterministic noise from digital ICs in the rest of this chapter.

23.2.2 Coupling from Digital Integrated Circuits

Noise can couple to the substrate through various mechanisms during the operation of digital ICs, for instance, through the impact-ionization phenomenon in device-level operation, capacitive coupling from parasitic passive components in primitive circuit-level operation, and resistive coupling with ground wirings in an entire IC operation. Designers need to recognize the relative importance of these possible noise-generation mechanisms to tackle undesired coupling during design. In recent years, substrate noise measurements have provided unvarnished waveforms and distinctive physical properties of substrate noise within a chip, which are quite helpful for intuitive understanding of the noise injection mechanisms.

CMOS digital circuits can inject dynamic noise into a substrate through junction capacitances at source and drain terminals of MOSFETs. Also, every well and interconnect on an IC couples capacitively to the substrate through a reverse biased bulk/well and an overlap capacitance to substrate, respectively. Only high-frequency components of a signal can couple to substrate via capacitive coupling. This capacitively coupled substrate current is important in mixed-signal circuits due to the presence of both a large number of switching digital nodes and high-impedance analog nodes. With decreasing technology feature sizes, the interconnect capacitances to substrate are becoming increasingly important.

In contrast to the parasitic resistive (ohmic) and capacitive coupling described above, noise current in a substrate can be induced by the impact-ionization phenomenon, where electron–hole pairs are generated and cause a current flow to the substrate when the electric field in the depleted region of the drain bulk of a MOS transistor becomes large. This can be transported to other parts of the system via the substrate, even under DC operating conditions. Although the growth of impact-ionization current cannot be ignored with the increasing speed of operation and the decreasing feature sizes of technology [8,9], the leakage of ground bounce and the size of capacitive coupling also significantly increases for larger scale of integration. Therefore, it can be said that the contribution of impact-ionization phenomenon to substrate noise is insignificant from a macroscopic view of the entire circuit operation, which Briaire has pointed out in [10,11], with detailed comparisons made by device and circuit simulations.

23.2.3 Effects of Power/Ground Grid, Package/Bondwire Parasitics on Noise Generation

Since substrate noise is dominated by the leakage of power supply and ground bounce as discussed in Section 23.2.2, the interaction of power supply and transient currents with parasitic RLC impedance networks strongly affect the noise.

Interconnect layers are used to set the voltage of different terminals of a circuit and to transport the signals from one point to another point of the circuit. Interconnect layers have distributed parasitics such as inductance and resistance on each layer, and capacitance and mutual inductance between two interconnect layers and between an interconnect layer and the substrate. These parasitic elements can create or couple noise. When a digital gate switches, current transients pass through power supply interconnect layers and lead to transient voltage drops across the parasitic inductance and resistance of the power supply/ground (V_{dd}/G_{nd}) interconnect lines [12]. The current spikes can be proportional to the load capacitance. The voltage drops in the power grid cause power supply/ground line bounce, known as 'V_{dd} bounce' or '*ground bounce*', and can be obtained from the following expression:

$$V_{drop} = Ri + L\frac{di}{dt} \tag{23.1}$$

where R and L are the resistance and inductance parasitic to the interconnect layers and more dominantly to the package and bondwires. The second term on the right-hand side of Equation (23.1) is also referred to as 'inductive noise', 'Ldi/dt noise', or 'delta-I noise'. This equation implies that the ground and supply voltages bounce as a function of the switching current and its derivative. The former determines the resistive voltage drop, while the latter represents the inductive voltage drop across the supply lines. Another undesired effect of di/dt term is that it induces ringing due to parasitic reactance, which consists mainly of capacitance at pads and inductance of bonding [13,14]. Given the above discussion, a reference ground set by these layers may not be at real zero potential as it should be, and the same is true for the reference supply voltage. Through interconnect layers, the noise is distributed to other parts of the circuit, which are tied to the same power/ground network. In addition to direct coupling of the G_{nd}/V_{dd} bounce, if the same power lines are used for substrate bias, voltage changes in the power interconnect layers propagate to the substrate directly through substrate contacts or by means of the depletion capacitances associated with source and drain regions. This leakage is the primary cause of substrate noise as was observed in

Figure 23.2, where ringing dominated in the substrate noise waveform when large d*i*/d*t* was induced for the shorter interval (T_s) of inverter switching.

In reality, as in the equivalent model of a CMOS inverter of Figure 23.3, external ports are connected to lumped equivalent inductances and resistances representing links (e.g., bond wires) to the board. Although not shown, equivalent capacitances due to package cavities also exist. The nonzero package/bondwire parasitic elements cause signal coupling between the blocks directly (pin to pin and bondwire to bondwire mutual inductance and capacitance), or indirectly by initially coupling to the substrate and from the substrate to other devices. They can have a major impact on the amount of coupled switching noise in an IC design [15]. This is due to the fact that many cheaper packages have large bondwires and package pin inductances associated with the supplies carrying fast current variations, caused by digital gates and nondifferential analog

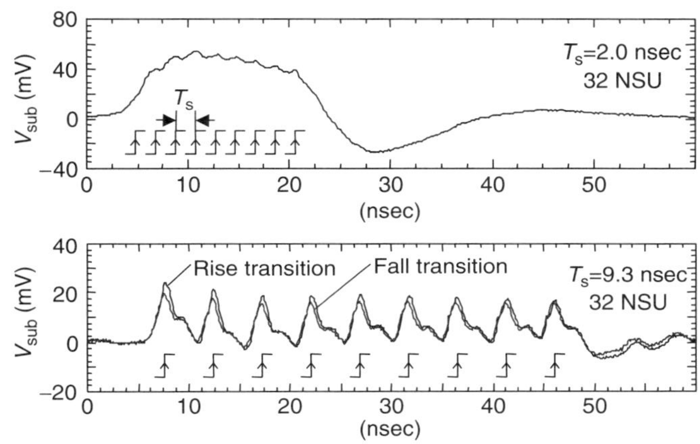

FIGURE 23.2 Measured substrate noise waveforms from transition-controlled noise source units (NSU) with stage delay T_s = 2.0 nsec (upper) and 9.3 nsec (lower).

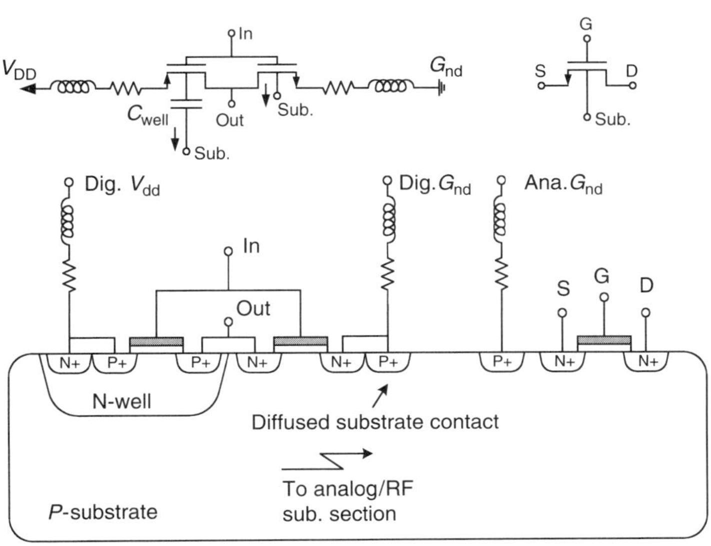

FIGURE 23.3 Noise coupling from the switching portion of the IC (CMOS-not gate) to the RF/analog portion via substrate. Some of the parasitic elements involved in the undesired signal coupling are shown. (From N.K. Verghese and D.J. Allstot, Computer-aided design considerations for mixed-signal coupling in RF integrated circuits, *IEEE J. Solid-State Circ.*, 33, 314–323, 1998. With permission.)

circuits, resulting in sizeable voltage drops [16]. One should also note that since package/bondwire parasitics along with substrate contacts can behave as RLC circuits, switching currents induced by logic circuits cause ringing in the power supply rails and in the output drive circuitry [6,17,18]. The oscillations (ringing), which modify the substrate voltage with a peak voltage and a settling time, are attributed to this ringing.

On the one hand, a major contribution of parasitic capacitance, *C*, is found in the determination of noise amplitude. Figure 23.4 shows the reduction of peak substrate noise amplitude, V_p, vs. the volume of inactive logic cells (gray area) on the same cell row as the active logic cells fixedly located at one end (black area), where an on-chip noise probing technique is applied [19]. The noise decreases as the volume increases, since an inactive logic cell serves as a local charge reservoir, namely, decoupling capacitance. On the other hand, the effect of noise reduction deteriorates as the inactive logic cells are located away from the active logic cells due to screening by resistive impedance on power supply and ground rails.

23.2.4 Coupling to Analog Integrated Circuits

The chip substrate can act as a collector, integrator, and distributor of switching currents, causing noise coupling to locations across the chip [15,20]. The injection of currents into the substrate causes substrate biases to vary, which in turn cause variations in MOS threshold voltages, depletion capacitances, and other circuit bias and performance quantities. The MOS threshold voltage modulation is due to the substrate bias fluctuation, which is called the body effect. The MOS drain current I_D is a function of both the gate voltage V_{GS} and the substrate voltage V_{BS}, where dependencies are modeled by two transconductances, g_m and g_{mb} respectively. While the former results in normal MOS operation, the latter increases the body effect, which in turn increases the circuit noise in analog circuits. This is because g_{mb} is a function of the substrate voltage and changing the substrate voltage due to any undesired current leads to the fluctuation of the drain current. Hence, a substrate voltage disturbance affects the neighboring transistors. Note that substrate coupling is a problem both in pure analog circuits and in mixed-signal circuits [15]. Although the mechanism of coupling in the substrate is identical in both cases, the effect of the parasitic cross talk tends to be slightly different in these two classes of circuits. In purely analog circuits, the substrate acts as a signal feedback path and the substrate coupling can lead to changes in small signal performance functions like amplifier gain and bandwidth due to feedback properties. In mixed-signal circuits, the switching noise injected into the substrate is picked up by sensitive analog devices on the same substrate, through both their junction capacitances to substrate and through

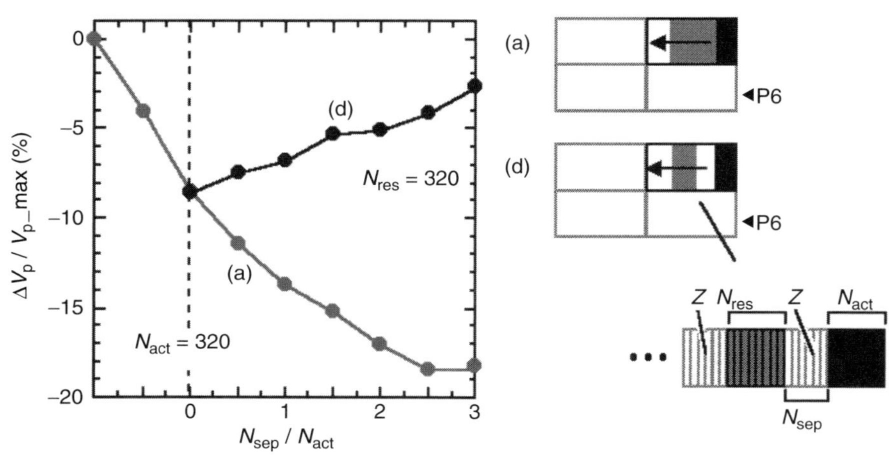

FIGURE 23.4 Effects of parasitic capacitances in the determination of noise amplitude. Number of inactive logic cells increase in curve (a), while the position of fixed number of inactive logic cells moves away from active logic cells in curve (d). (From M. Nagata et al., Effects of power-supply parasitic components on substrate noise generation in large-scale digital circuits, *Symposium on VLSI Circuits Digest of Technical Papers*, 2001, pp. 159–162. With permission.)

the MOS device body effect as stated above. This results in induced spikes of noise in both device currents and node voltages. It can be shown that the signal coupling via body effect becomes less important relative to depletion capacitance coupling as the substrate doping is increased [21]. It is noteworthy to mention that in addition to transistors, diffused resistors and capacitors may be affected capacitively by substrate noise [9].

23.3 Modeling of Mixed-Signal Noise Coupling

A complete analysis of substrate noise requires both extraction of the substrate as well as simulation. Any substrate noise analysis technique has to include some form of circuit simulation to assess the impact of substrate noise on a particular parameter of interest for a specific analog function. Extraction is the process by which the electrical equivalent model of the substrate, possibly including resistance, capacitance, or inductance, is determined. To extract a substrate accurately, the complex geometries of wells, contacts, well taps, diffusions, trenches, etc. must be extracted. Once extraction has been completed, simulation can be performed on a circuit including the extracted RC network for the substrate. To predict simulation of substrate noise requires some knowledge of the equivalent extracted network, as well as the nature and location of noise injectors that cause the noise. If a SPICE simulation is performed with devices and substrate parasitics present, the computational time required explodes very quickly; hence this approach is tractable only for analyzing small components of the order of a few hundred devices. Alternatives to circuit simulation of substrate noise are discussed below.

23.3.1 Modeling Coupling to Substrate

23.3.1.1 Capturing Noise Current

To model the effect of coupling, there are expressions or equivalent circuit elements that must be incorporated in a circuit simulator to calculate the current injected into the substrate and hence the signal received by sensitive analog circuits. Among them, device capacitances are included in the transistor models for SPICE-like circuit simulators. In the case of interconnect to substrate capacitances, layout extraction tools can easily extract these parasitic elements. Since these values are layout-dependent, they are extracted after the layout of the circuit is completed and are typically incorporated into the circuit simulator for postlayout simulation. A straightforward approach to capture noise current on power supply and ground wirings in an entire large scale digital circuit is to apply SPICE-like enhanced simulators to the extracted netlist. Those simulators provide the capability of hundred-thousands of components, through the use of macromodel devices based on a look-up table of current–voltage characteristics as well as hierarchical circuit macromodels relying on the similarity of circuit operation among primitive circuits with the same set of input signals. The contribution of impact-ionization phenomenon can be additionally included to the noise current, where analytical models representing *I–V* characteristics of the impact ionization current are built-in with the transistor model employed by SPICE. In one of these models, for example, the hot-electron-induced substrate current can be expressed in semianalytical form as [22]

$$I_{\text{sub}} = C_1(V_{\text{ds}} - V_{\text{dsat}})I_{\text{d}} \exp\left(-\frac{C_2 t_{\text{ox}}^{1/3} \cdot x_{\text{j}}^{1/2}}{V_{\text{ds}} - V_{\text{dsat}}}\right) \tag{23.2}$$

where C_1 and C_2 are process-related, empirically determined parameters, t_{ox} is the oxide thickness, x_{j} the junction depth, V_{ds} the drain–to-source voltage, and V_{dsat} the saturation voltage. Using the results obtained from device simulations or measurements, it is possible to determine the empirical coefficients C_1 and C_2, and incorporate impact ionization induced substrate currents into existing device models for circuit simulation. This straightforward approach can provide good accuracy in the noise current; however, one of the major drawbacks is the limitation of circuit scale to analyze due to the explosion of memory usage and CPU time for larger digital circuits.

23.3.1.2 Noise Current Macromodels

The use of simulators with higher levels of abstraction than transistor-level descriptions in SPICE can accelerate as well as improve the capacity of capturing noise current. There are many possible ways of macromodeling large scale digital operations.

A traditional idea of macromodeling assumes that a logic gate draws current with triangular waveshape when it toggles; the height and slope of the triangle depend on input–output logic function as well as fan-out load capacitance and are characterized in advance. Then, power supply current of a digital circuit with a given vector can be approximated by way of superimposing the triangular currents. Noise analyses use the captured noise current directly in terms of electromagnetic radiation such as EMI [23] and ground bounce where it interacts with impedance networks parasitic to power supply and ground wirings as well as to a substrate. This allows the substitution of gate-level logic simulation for transistor-level circuit simulation, which greatly improves the capability of noise analysis.

Another approach to modeling switching current sources utilizes the concept of substrate noise signature for each digital gate [24,25]. This methodology exploits the fact that any given logic gate injects a particular signal into the substrate through capacitive coupling and impact ionization. Such a signal, known as substrate injection pattern, is a unique fingerprint of the gate, input transition, circuit implementation, and technology. It can be accurately calculated using standard device modeling and circuit simulation, and is better than the simplest signature of triangular waveform approximation described above. The substrate noise signature of the entire circuit is then evaluated using the substrate injection patterns and a precise analysis of the switching activity in the circuit's internal nodes. Switching activities are computed from user-specified input vector sequences by a gate-level simulation. Since the input vector is not known *a priori*, the user should simulate a realistic load or perform a worst/bestcase analysis. Further improvements applied to this approach are discussed in [26], where the power-supply current flowing from V_{dd} to G_{nd}, and noise current directly injected to a substrate, are separately characterized for every digital gate with every possible input combination and stored in a standard cell library as auxiliary information. The pairs of currents are combined in synthesizing substrate noise current accordingly to the switching activities.

In contrast to the previous signature-based techniques that approximate a noise current by the superposition of predetermined waveforms, time series divided parasitic capacitance (TSDPC) substitutes a single capacitor charging process for a mass of logic toggles that occur in a time interval [27]. A train of switched capacitors, such as shown in Figure 23.5, represents a large-scale digital circuit for simulating power-supply currents. The size of capacitance can be derived from toggle records by gate-level logic simulation. A circuit-level simulator solves power-supply current through real charge transfer process within

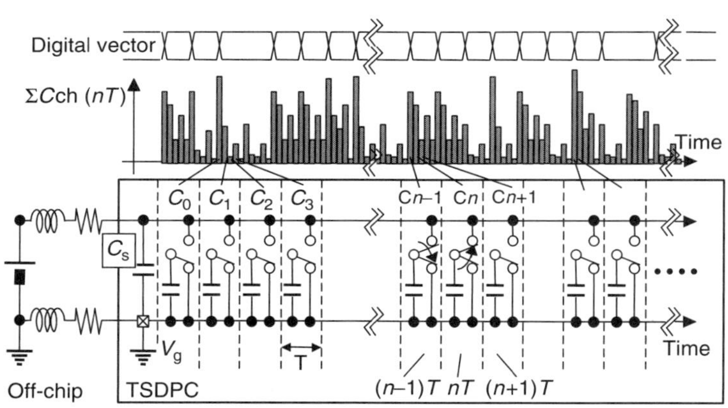

FIGURE 23.5 Time series divided parasitic capacitance (TSDPC) model. (From M. Nagata et al., Modeling substrate noise generation in CMOS digital integrated circuits, *Proceedings of IEEE Custom Integrated Circuits Conference*, 2002, pp. 501–504.)

digital circuits along with the parasitic impedances. This improves the accuracy of the noise current, and the simultaneous ground-bounce waveform that couples to the substrate.

23.3.2 Modeling Substrate Parasitics

Due to its distributed nature, the substrate cannot be translated into a compact analytical model accounting for the entire chip area whose global effects are felt everywhere in the chip. In general, models for substrate coupling can be derived by one of three methods: from the full 3-D numerical simulation (Maxwell's equation), or using suitable discretization of a simplified form of Maxwell's equation, or using lumped element models. Common techniques to model the substrate include the finite difference method (FDM) followed by network reduction of the resulting mesh, or the boundary element method (BEM) followed by fast BEM matrix solution techniques.

23.3.2.1 Box Integration Formulation

To obtain a distributed RC network, the box integration technique may be utilized [28]. Using this technique, a 3-D rectangular RC mesh network as the equivalent circuit representation of the modeled substrate is constructed. The mesh topology could be correlated to the circuit's physical design by distributing grid points according to the layout features on relevant fabrication photomasks. The substrate is treated as a 3-D mesh where each mesh edge is a parallel combination of a resistor and a capacitor. In this approach, the edge surfaces (boundaries) are assumed to be Neumann (reflective) boundaries for voltages, while the diffusion/active areas and contact areas are treated as Dirichlet (fixed) boundaries for voltages (equipotential regions) in the resulting 3-D RC mesh. These areas are represented as ports in the multiport network and connected to corresponding nodes in the electrical circuit [29]. Outside the diffusion/active areas which are called ports here, the substrate can be approximated as layers of uniformly doped semiconductor of varying doping density. In these regions, a box integration method [30] can be applied to spatially discretize simplified Maxwell's equations. Ignoring magnetic fields and using the identity $\nabla\cdot(\nabla \times \mathrm{A}) = 0$, Maxwell's equations can be written as

$$\nabla\cdot \mathrm{J} + \nabla\cdot\frac{\partial \mathrm{D}}{\partial t} = 0 \tag{23.3}$$

where J is the current density, D the displacement vector, and t the time. Using the relations, $\mathrm{D} = \varepsilon \mathrm{E}$ and $\mathrm{J} = \sigma \mathrm{E}$, this reduces to

$$\sigma\nabla\cdot\mathrm{E} + \varepsilon\frac{\partial}{\partial t}(\nabla\cdot\mathrm{E}) = 0 \tag{23.4}$$

where E is the electric field, σ the conductivity, and ε the permitivity. The above equation can be discretized on the substrate volume either in differential form using FDM, or in integral form using the BEM, which is explained later in this section.

In the FDM technique, the substrate is expressed as a collection of square cubes as shown in Figure 23.6 [18]. An electric field normal to a contact plane of two adjacent cubes (i,j) with distance h_{ij} is given in Equation (23.5), and thus Equation (23.4) is rewritten as Equation (23.6) by Gauss' law under the assumption of uniform impurity concentration in the cube and by applying the box integration method [31,30]:

$$E_{ij} = \frac{\mathrm{V}_i - \mathrm{V}_j}{h_{ij}} \tag{23.5}$$

$$\sum_j \left[G_{ij}(V_i - V_j) + C_{ij}\left(\frac{\partial}{\partial t}V_i - \frac{\partial}{\partial t}V_j\right)\right] = 0 \tag{23.6}$$

where

$$G_{ij} = \sigma\left(\frac{W_{ij} \times d_{ij}}{h_{ij}}\right) \tag{23.7}$$

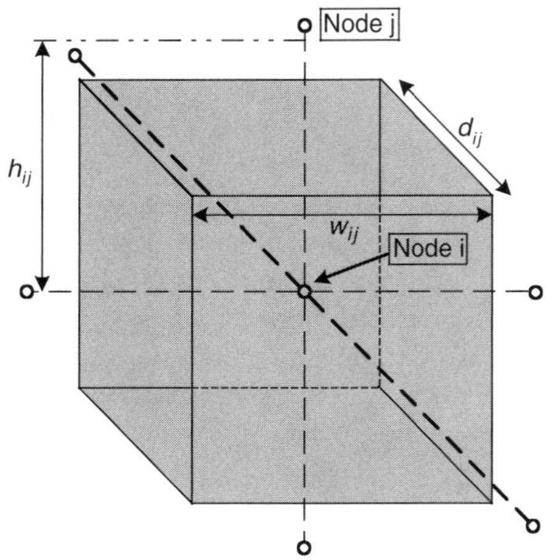

FIGURE 23.6 Substrate is expressed as a collection of square cubes. (From N.K. Verghese et al., *Simulation Techniques and Solutions for Mixed-signal Coupling in Integrated Circuits*, Boston, MA: Kluwer, 1995. With permission.)

and

$$C_{ij} = \varepsilon\left(\frac{W_{ij} \times d_{ij}}{h_{ij}}\right) \tag{23.8}$$

as shown in Figure 23.7.

This approach is not efficient because large 3-D meshes are created, which would be prohibitive to simulate using variable time-step trapezoidal integration techniques. For instance, consider the case of a lightly doped substrate where a lot of empty space (between ports) must be discretized to get an accurate estimate of port-to-port substrate admittance [20].

23.3.2.2 Network Reduction

To address this problem, the generated linear RC network should be approximated by a smaller circuit that exhibits similar electrical properties. A small percentage of the network nodes, called ports, are physically connected to the external circuit (at the top surface of the substrate). In theory, an "equivalent" multiport network (similar to a Thevenin equivalent circuit for a one-port network) can be formulated by eliminating a substantial fraction of the internal nodes. This technique is generally referred to as "*network reduction*." For network reduction using congruence transformations [28], full-network conductance and susceptance matrices are transformed to reduced equivalents, which can be directly realized with resistors and capacitors. This algorithm utilizes the well-conditioned symmetric Lanczos process, which exploits the specialized structure of the extracted substrate networks to formulate Pade approximations of the network port admittance. Congruence transformations are employed to ensure stability and to create reduced networks that are easily realizable with RC elements. The approximated networks are guaranteed to be passive, and are thus well behaved in subsequent simulations. The reduced models retain the accuracy of original models, but contain orders of magnitude of fewer circuit nodes.

The network reduction problem can be simplified if the capacitances in the RC mesh can be ignored. Neglecting the intrinsic substrate capacitance is a reasonable assumption for operating speeds of up to a few GHz and switching times of the order of 0.1 nsec. This is due to the fact that the relaxation time of the substrate (outside of active areas and well diffusions) which is given by $\tau = \rho\varepsilon$, is of the order of 15 ps (with $\rho = 15$ Ω-cm and $\varepsilon_r = 11.9$). The capacitive behavior of the substrate outside the active area is negligible for frequencies below one tenth of τ. As we see here, if the capacitance to substrate, which is introduced by the depletion regions of well diffusions and interconnect overlying field oxide, can be accurately modeled as lumped circuit elements outside the mesh located near the chip surface, the substrate can be modeled as a purely resistive mesh. If the substrate is multilayered, then the

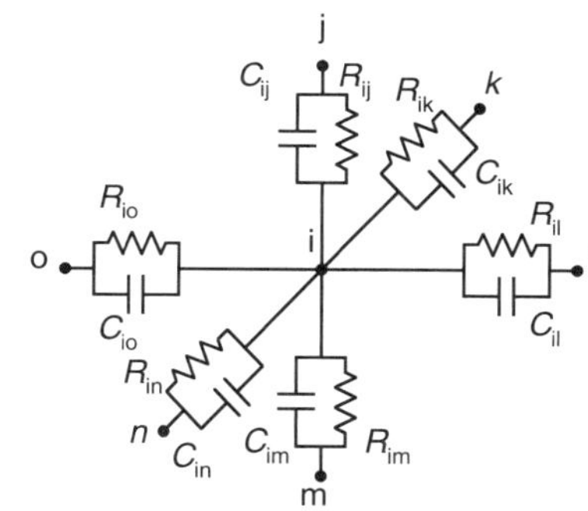

FIGURE 23.7 Resistances and capacitances around a mesh node in the electrical substrate mesh. (From B.R. Stanisic et al., Addressing substrate coupling in mixed-mode ICs: simulation and power distribution systems, *IEEE J. Solid-State Circ.*, 29, 226–237, 1994. With permission.)

network may not be approximated by a resistive network even at lower frequencies [15]. Note that in circuit simulators, such as SPICE, junction capacitances of active devices are modeled outside the mesh as lumped capacitances.

The substrate may be modeled as a resistive grid, which is determined through box integration or the Delaunay tessellation (for greater accuracy in and around the more interesting regions of the substrate) [12]. Using a simple DC macromodeling approach, the 3-D resistive mesh can be reduced to an equivalent set of $n(n + 1)/2$ resistances interconnecting the n ports [31,32]. In this approach, wells are considered ports and connected to lumped capacitances outside the mesh. Since the substrate is modeled as a purely resistive mesh, the macromodel consists of only the steady-state/DC values of the admittance parameters, with the higher-order mesh moment being zero. The computational complexity of the DC macromodel is much lower than that of the congruence transform-based method, and the resulting macromodel is more compact.

23.3.2.3 Boundary Element Method Using Green's Function

An alternative approach to solving the simplified Maxwell's equations is the BEM, which can be used for parasitic and substrate extraction [33]. In this method only the ports that connect the substrate to the devices/contacts/wells are discretized. BEM is more appealing compared to FDM, since in this method instead of discretizing the whole structure, only the relevant boundary features, the 2-D substrate contacts (port areas) are discretized, and hence the resulting matrix to be inverted in the network reduction process is much smaller albeit fully dense [34]. The extraction can be combined with a model-reduction technique to obtain simpler models for cross talk [35]. Another major advantage of the BEM is that it is not very discretization-dependent (unlike FDM) [20]. For example, by discretizing a port into a single panel, i.e., assuming a constant current density across the port, the results are within 10% of the actual answer. By a proper choice of the Green's function of the BEM, only those parts of the substrate boundary (called 'contacts') have to be discretized that directly interact with the designed circuit.

The Green's function is used to determine point-to-point impedance between each pair of discretized ports as shown schematically in Figure 23.8. The resulting impedance matrix is then inverted to yield the required substrate admittances [36]. The Green's function is the potential at any point in a medium due to a current injected at any point in the medium, and can be determined for the substrate in quasianalytical form. The areas of the substrate that connect to the external world (device/contact areas) are discretized into a collection of n panels, and the contribution to the potential at each panel, due to currents injected at every panel is stenciled into an $n \times n$ matrix of impedances, which is then inverted to determine the substrate admittances. This technique effectively reduces a 3-D problem into a 2-D one.

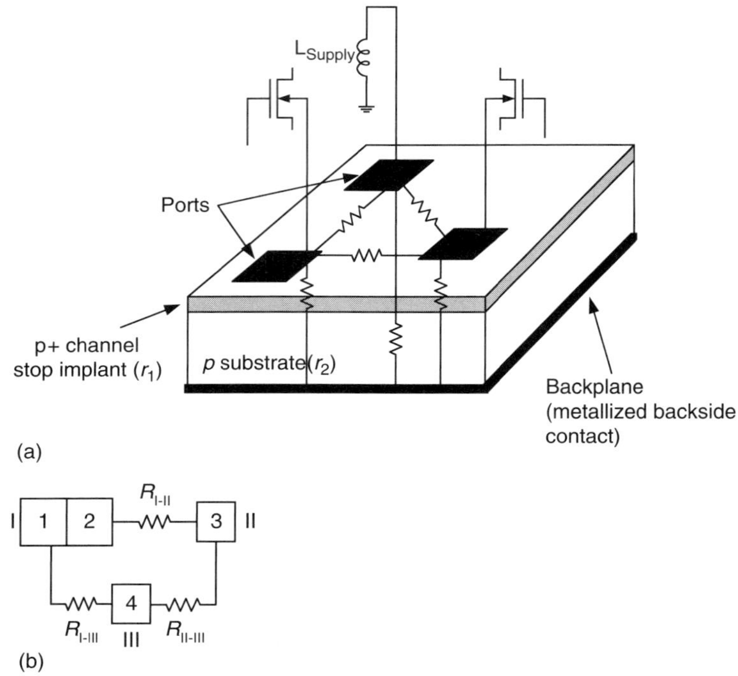

FIGURE 23.8 (a) Ports on a substrate resistively connected to one another; (b) Determining resistive coupling between ports using the BEM. (From N.K. Verghese and D.J. Allstot, Computer-aided design considerations for mixed-signal coupling in RF integrated circuits, *IEEE J. Solid-State Circ.*, 33, 314–323, 1998. With permission.)

23.3.2.4 Green's Function Formulation

For the resistive substrate case ($\varepsilon_r = 0$), the Maxwell's equations reduce to the well-known Laplace's equation [20]

$$\nabla^2\phi = 0 \tag{23.10}$$

where ϕ is the electrostatic potential. Applying Green's theorem to the above equation gives the electrostatic potential at an observation point, r, due to a unit current injected at a source point, r', as

$$\phi(r) = \int_v J(r')G(r,r')\mathrm{d}^3r \tag{23.11}$$

where $G(r,r')$ is the substrate Green's function satisfying the boundary conditions of the substrate, and $J(r')$ is the source current density. Since all the sources and observation points are at the defined ports on the substrate and these are planar and practically 2-D, the above volume integral reduces to a surface integral.

$$\phi(r) = \int_s J(r')G(r,r')\mathrm{d}a \tag{23.12}$$

Essentially, this reduces a 3-D problem into a 2-D problem. In addition, since the Green's function implicitly takes the substrate boundaries into account, there is no need to consider them explicitly when solving the above equation where only the port areas (that actually connect to the substrate) need to be discretized to solve the equation.

The Green's function of the substrate can be determined analytically using classical mathematical techniques [36,37] and has been reported in the literature [33,38]. The substrate Green's function $G(x,x',y,y')$, with (x,y) and (x',y') being the coordinate locations of the observation and source points on the substrate surface is:

$$G(x,x',y,y') = \sum_{m=0}^{M}\sum_{n=0}^{N} f_{\mathrm{mn}} \cos\left(\frac{m\pi x}{a}\right)\cos\left(\frac{m\pi x'}{a}\right)\sin\left(\frac{n\pi y}{b}\right)\sin\left(\frac{n\pi y'}{b}\right) \tag{23.13}$$

where f_{mn} for a homogeneously doped substrate is given by

$$f_{\mathrm{mn}} = \frac{C_{\mathrm{mn}}}{ab\sigma}\tanh\left(\sqrt{\frac{m^2\pi^2}{a^2}+\frac{n^2\pi^2}{b^2}}\cdot c\right) \tag{23.14}$$

Here, C_{mn} is a constant, σ the substrate conductivity, and (a,b,c) are the (X,Y,Z) substrate dimensions. For a multilayered substrate profile (of uniform sheet resistivities) a more complicated expression is obtained for f_{mn}. Once the Green's function is determined, Equation (23.12) remains to be solved [20]. The solution is obtained using a suitable discretization technique which discretizes each port on the substrate into a set of panels. A system of equations can be formulated that relate the currents and potentials at all panels in the system. In matrix form this is represented as

$$\phi = Z_i \tag{23.15}$$

where each entry in the impedance matrix, z_{ij}, is given as

$$z_{ij} = \int_{S_i}\int_{S_j} J(r')G(r,r')\mathrm{d}a'\mathrm{d}a \tag{23.16}$$

In this equation, S_i and S_j are the surface areas of panel i and j, respectively. The impedance, given by Equation (23.16), can be analytically determined for rectangular panels once the Green's function is derived. The matrix, Z, is then inverted to obtain the substrate admittance matrix, Y. We can determine the substrate resistance between any two ports as the reciprocal of the sum of corresponding admittance matrix entries.

From a computational point of view, the direct evaluation of the quasianalytical Green's function given above involves several million floating-point multiplications and additions and since it must be repeated for every pair of panels, the formulation of the impedance matrix becomes an expensive task for large problems [36]. As an alternative technique, after discretizing the entire substrate surface into a uniform grid of panels, a 2-D discrete cosine transformation (DCT) can be utilized to pre-compute all the panel-to-panel impedances on the substrate in $O(N \log N)$ time [38]. Another problem with the BEM approach in general is that inversion of the dense $n \times n$ matrix is a cumbersome task. Direct LU factorization requires $O(n^3)$ operations, which is clearly infeasible for a reasonably sized problem.

23.3.2.5 Fast Solution of BEM Matrices

To improve the efficiency of the BEM , multilevel (multigrid) methods, which are efficient iterative techniques can be developed for first-kind integral equations defined over complicated geometries [35]. Based on the method, a multigrid iterative solver integrated with sparsification algorithms specially tuned to account accurately for substrate edge effect allows for almost an order of magnitude improvement in the speed of the BEM solution process.

Alternatively, a fast eigendecomposition technique that accelerates operator application in BEMs and avoids the dense matrix storage while taking all of the substrate boundary effects into account has been presented explicitly [34,39]. For efficient extraction of the substrate coupling model in a BEM formulation, the authors use the eigendecomposition-based technique in a Krylov subspace solver. The model can be incorporated into a circuit simulator such as SPICE to perform coupled circuit–substrate simulation. To speed up the model-computation process at the cost of a slight decrease in accuracy, the use of pre-corrected-DCT (PcDCT) algorithm was also proposed [40,41]. The main idea behind the PcDCT algorithm is to realize that the effect of an injected current in a panel on the potential of another far-away panel can be considered the same for small variations in the distance between panels. A heavily doped bulk substrate was used in the experiment that the authors reported. The results indicated better accuracy as well a speed up of around 180 times compared to the vanilla Green's function method, and 12 times speed up compared to the eigendecomposition method. The memory requirements for the PcDCT algorithm were seen to be considerably smaller (20 times) than those of the vanilla Green's function method. In the example, some savings in memory (three times) were also obtained with respect to the unaccelerated eigendecomposition algorithm.

Several other problems plague the BEM-based substrate modeling and subsequent simulation problem [42]. First, the density of the extracted coupling matrix makes later circuit simulation prohibitively costly, because the now dense circuit matrix must be factored hundreds or thousands of times in each simulation. Second, most methods of obtaining the n columns of the coupling matrix require n matrix solutions, which is computationally quite costly, making it impractical to solve problems with n larger than a few hundred. To address these problems on a multiscale, wavelet-like basis for fast integral equation solutions has been proposed [42]. The wavelet basis efficiently represents coarse grain information of the IC geometry. Using such a basis, many entries would become small and could simply be dropped with only a small loss of accuracy. The wavelet basis has a multiresolution property. It was shown, for example, that reducing the number of non-zero elements by 90% led to 1% accuracy loss. The results presented showed that for a problem with a few thousand contacts, this method was almost ten times faster in constructing the matrix.

23.3.2.6 Chip-Level Substrate Network Extraction

SeismIC™ performs a mixed-signal noise simulation without the presence of devices (using equivalent noise sources) in order to compute the time- or frequency-domain substrate noise waveforms at the bulk nodes of interest and can be utilized to analyze chips with 1 million or more devices [43]. After the substrate noise waveforms have been computed, a circuit simulation with devices and noise sources attached can be performed to assess the impact of the noise on the subcircuits of interest. A typical flow for substrate noise analysis for verification using SeismIC™ is shown in Figure 23.9.

In order to efficiently model and analyze the substrate of large designs, an adaptive substrate modeling approach is used. This is achieved through the use of sensitivity analysis to determine which areas of the chip need high model accuracy and where the model accuracy can be relaxed without impacting the accuracy of the overall analysis. Noise sensitivity analysis is also used to measure the impact on substrate

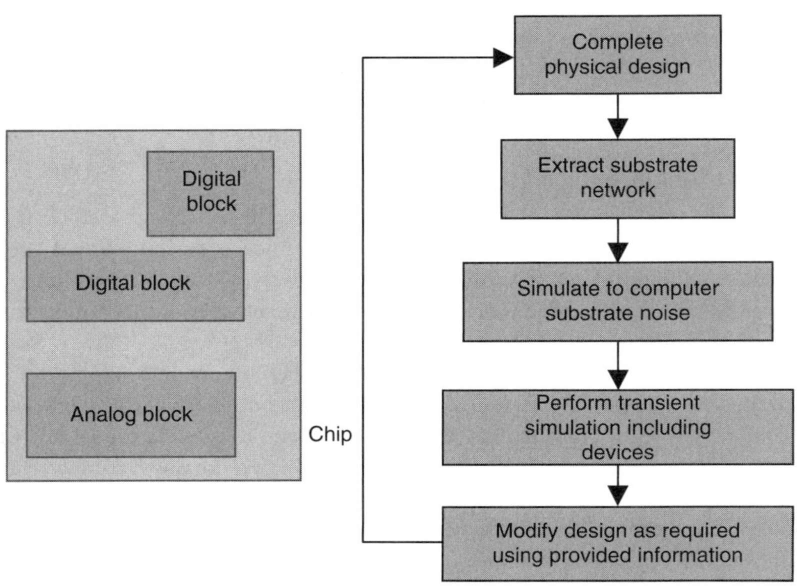

FIGURE 23.9 Verification flow for full chip level substrate noise analysis.

noise of a change in any given parameter. By calculating the sensitivity to various layout, process, and package parameters, the appropriate measures to minimize substrate noise are determined.

23.3.3 Chip-Level Mixed-Signal Substrate Coupling Analysis

Once an accurate substrate extraction has been performed, the location and magnitude of noise injectors need to be determined to facilitate simulation of the substrate noise waveforms. The location of noise injectors can be determined from the layout and schematic netlist information. To determine the magnitude and phase of injected currents, some form of simulation input is required under assumed switching activity. Once this has been ascertained, the problem is reduced to solving a very large RC network with active current sources, as shown in Figure 23.10. The number of current sources can be extremely large, for example, a million transistor mixed-signal design may have a million current sources. To see how a large RC network driven by active current sources is analyzed, consider that the voltage response at a bulk node of interest, *vb*, is desired. The voltage response can be written as follows:

$$vb(s) = z1(s)\cdot i1(s) + z2(s)\cdot i2(s) + z3(s)\cdot i3(s)+\ldots. \tag{23.17}$$

where *i*1, *i*2, *i*3, … are the current sources at various locations on the substrate and *z*1, *z*2, *z*3, … are their corresponding impedances to the bulk node of interest. The current source values, *i*1, *i*2, *i*3 … can be determined from a simulation of the original circuit (without parasitics) by observing the currents flowing in the power/ground nodes and the device bulk terminals. This can be accomplished either with a transistor-level circuit simulator or a gate-level event driven simulator in conjunction with precharacterized cell libraries, as discussed in Section 23.3.1. The currents can be either time-domain waveforms or a composition of spectral values at every frequency ($s = j\omega$) of interest. The impedances, *z*1, *z*2, … can be obtained by inverting the admittance matrix formed by the RC substrate network and package inductances (Figure 23.10) at every frequency ($s = j\omega$) of interest. The frequency-domain response of *vb* can be obtained by solving Equation (23.17) at every frequency of interest. Applying the inverse Laplace transform to this response results in the corresponding time-domain waveform.

One advantage of using Equation (23.17) to calculate the noise response of a bulk node of interest is that each individual noise contributor can be calculated independently. Hence, from Equation (23.17), the noise contribution from injector 1 at the bulk node of interest is $z1(s)i1(s)$. Similarly, $z2(s)i2(s)$ is the contribution from injector 2, $z3(s)i3(s)$ from injector 3 and so on. Thus, the most significant noise contributors can be identified and appropriate measures can be taken to minimize their impact.

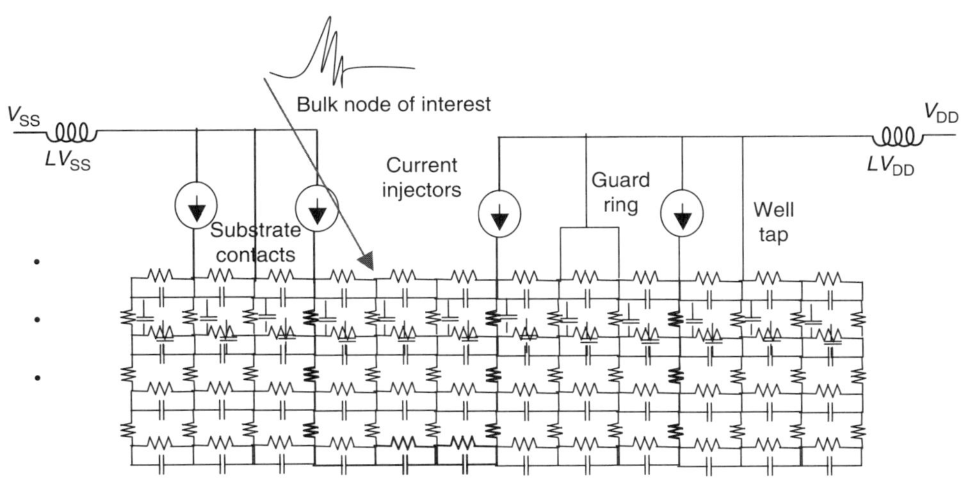

FIGURE 23.10 Simulation model for full chip substrate with a large number of noise injectors.

23.3.3.1 Macroscopic Substrate Noise Analysis Using High-Level Simulation

A macroscopic substrate noise model that expresses coupling noise as a function of logic state transition frequencies among digital blocks has been proposed [13,44]. The coupled noise is defined as one of the state variables in behavioral description of victim circuits such as analog circuits. The noise can be typically expressed by the superimposition of voltage changes arising from digital state transitions in unit time; thus a function of state transition frequencies is evolved in the digital block, which can be easily extracted from digital logic simulation. This results in the introduction of behavioral noise modeling using a hardware description language (HDL)-based system-level design [44–46]. A simulation system based on the model was implemented in a mixed-signal simulation environment as shown in Figure 23.11, where performance degradation of a second order Delta Sigma Analog-to-Digital Converter (ADC) coupled to digital noise sources was simulated. The computation of the noise proceeded according to a noise waveform function $F(f_{\text{eff}}, t)$ in parallel with a transient analysis of the mixed-signal circuit under design. The calculated noise waveform was injected into the analog circuit. Here, $f_{\text{eff}}(n)$ is the effective transition frequency, a global state transition count per unit time, T at the nth sampling interval defined by

$$f_{\text{eff}}(n) = \frac{1}{T}\sum_{i=1}^{m}[W_i^+ N_i^+ + W_i^- N_i^-] \tag{23.18}$$

where $N_i(n)$ is the local state transition count of the ith digital sub-block and W_i the weight coefficient corresponding to a substrate coupling intensity of the sub-block to the sensitive analog circuit, and is a relative quantity among the digital sub-blocks. Superscripts '+' and '–' stand for rising and falling transitions, respectively. T is the noise sampling period introduced to discretize the noise-generation process and F_{nclk} $(=1/T)$ is synchronous to the system clock (F_{sclk}). $F(f_{\text{eff}}, t)$ must be continuous and reflect the nature of its transient behavior. The author adopted a successive function system $\{\phi_n(t)\}$ of Equation (23.19) as $F(f_{\text{eff}}, t)$, where $\phi_n(t)$ is defined in $t \in [0,T]$ of the nth sampling period, and α, β are model parameters.

$$\phi_n(t) = \phi_{n-1}(T) + \alpha[f_{\text{eff}}(n) - f_{\text{eff}}(n-1)]\left[1-\exp\left(-\frac{t}{\beta}\right)\right] \tag{23.19}$$

The weight coefficient W_i models the attenuation of the noise amplitude by distance and guardbanding, and the ratio of the noise amplitude for rising to falling transitions. To determine the coefficients beforehand, evaluating the substrate noise transmission by circuit simulation is required. Hence, one of the modeling methods for substrate equivalent circuits explained in Section 23.3.2 should be used. One of the shortcomings of this technique is that the coefficients are functions of technology, circuit, layout, etc., and therefore the circuit simulator must be run each time any of these parameters change. In addition, α and β are parameters that determine the amplitudes and the widths of the generated noise waveforms, respectively. These are dominated by the substrate structure and thus should be evaluated from experimental results with dedicated test chips consisting of simple noise sources like inverter arrays and wide bandwidth substrate

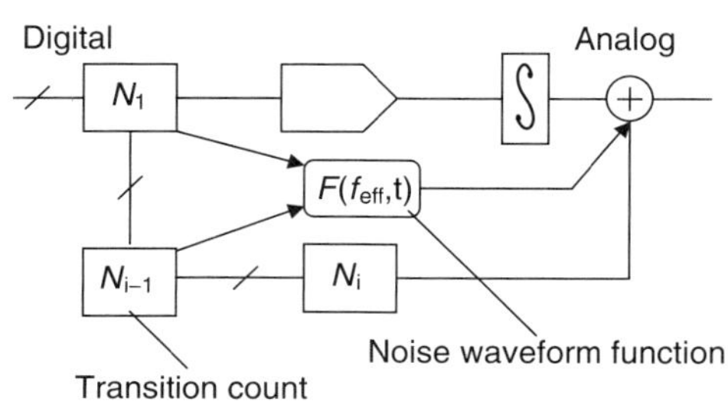

FIGURE 23.11 Proposed macroscopic substrate noise model. (From M. Nagata and A. Iwata, Substrate noise simulation techniques for analog-digital mixed LSI design, *IEICE Trans. Fund. Electr.*, E82-A, 271–278, 1999.)

noise sensors. However, the macro-modeling approach is appreciated for capturing the sensitivity, and measuring the response of mixed analog and digital circuits to coupling noise in terms of performance metrics such as SNR and BER, for which the designer cannot apply circuit-level simulation. For instance, the observed degradation of THD performance shown in Figure 23.12 agrees with the reported experiments [47], and therefore the model successfully expresses the interaction of delta-sigma modulation loop dynamics with transient voltage noises injected into analog signal paths mainly through integrators.

23.3.3.2 Periodic Analysis of Mixed-Signal Noise in Radio Frequency Circuits

For RF circuits, specialized simulation techniques for the analysis of periodic circuits can be used to calculate quickly the response of such circuits to mixed-signal noise [6,48]. Following a periodic steady-state operating point analysis, a transfer function analysis computes the transfer functions to the RF circuit output at a single frequency and from every noise source in the circuit at every input frequency (i.e., output frequency and all frequencies offset from it by a harmonic of the periodic signal). Using this approach, it is possible to compute transfer functions from the bulk node of every device in the (RF) circuit to a specific output at a given frequency of interest. Note that for a given frequency of interest, this results in a set of transfer functions for each bulk node. Once the transfer functions have been computed for the (RF) circuits, it no longer needs to be represented at the transistor level. A transient simulation can be performed on the digital (and analog) circuits to obtain the transient substrate noise signals. A postprocessing of the signal (Fourier transform) determines the equivalent noise spectra at the device bulk nodes. These noise frequency components multiplied with the transfer functions obtained above calculate the coupling of substrate noise to the RF circuit output. To calculate the periodic transfer functions, efficient matrix-free iterative methods for the periodic analysis of RF circuits are used [49–52]. The methodology was reported to have been applied to the verification of the transmit section of a portable radio front-end IC [6,48]. Measured results on the fabricated IC indicated an RF spur at the output of an up-conversion mixer (modulator) in the transmit section of the circuit, which was adequately predicted after analyzing the circuit using this method. With roughly 1900 devices, 717 nodes, and 3234 equations to solve, a transient analysis of the modulator and reference frequency generator would have required 2 days of computation to simulate 20 periods which are required for the modulator to attain a steady state. Using periodic analysis instead, a macromodel of the modulator (containing 982 devices, 438 nodes, and 1445 equations) was obtained in less than an hour of CPU time. Another hour was required to simulate the transient noise coupling from the reference generator.

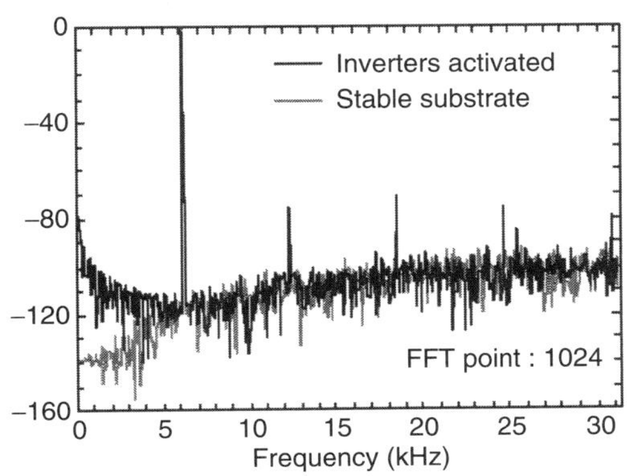

FIGURE 23.12 Simulated in-band power spectrum of second order delta sigma modulator coupled to substrate noise generator. (From M. Nagata and A. Iwata, Substrate noise simulation techniques for analog-digital mixed LSI design, *IEICE Trans. Fund. Electr.*, E82-A, 271–278, 1999. With permission.)

23.4 Mixed-Signal Noise Measurement and Validation

Nagata et al. [53] proposed a direct sampling technique for substrate noise measurement, where a detector named SFLC consists of a P-channel source follower (SF) with an input probe located around P+ area on the surface of a P-type substrate and a latch comparator (LC) connecting to the SF output, as shown in Figure 23.13. The SF picks up substrate potential around the probe and the LC discretizes the SF's level-shifted output voltage through successive comparisons, with stepwise reference voltage externally provided and with sampling occurring at every latch operation. The SF provides good linearity with the input voltage range of the order of 1 V, along with gain of slightly less than unity and bandwidth of a few GHz, even when followed by the LC. The authors demonstrated waveform-accurate substrate noise measurements with voltage and time resolution of 100 uV and 100 ps, respectively, and also showed the consistency between substrate noise waveforms acquired by direct measurements and comparator-based indirect measurements described in the previous section [53]. Here, only the former can achieve absolute-voltage quantitative evaluation of substrate noise while the latter is restricted to relative evaluation.

A transition-controllable noise source (TCNS) shown in Figure 23.14 includes a multiphase clock ($C_k[0{:}8]$) generator consisting of nine delay elements and a matrix of noise source unit (NSU) in the form of 9 rows × 12 columns, where the number of NSUs to activate by each edge of C_k can be set from 0 to 12. The delay element has bias voltages V_n and V_p for regulating rise and fall delay, respectively. In addition, inverse or noninverse transitions among adjacent noise source blocks are selected by the signal "Sel". The NSU has 30 inverters operating in parallel, where each inverter has a 50 fF load capacitor to the substrate that corresponds to typical parasitic capacitance of 2 fanout gates and local wiring. Minimum gate length is used and widths are chosen to have a switching time of roughly 200 ps for both rise and fall transitions when driving the load capacitors. The TCNS can generate substrate noise with controlled transitions in size, inter-stage delay, and direction [53].

Multiple-point measurements on a single substrate were achieved by arraying the SFLC detectors shown in Figure 23.15 [19]. A combination of TCNS and arrayed SF+LC can be a reference structure for assessing substrate noise generation and substrate coupling for a given CMOS mixed-signal technology. An example of such a test chip, which was fabricated in a commercial 0.3-μm 3.3-V CMOS process with P-type bulk substrate is shown in Figure 23.16 [54]. The chip includes two TCNS blocks on the top-right and bottom-left quadrants, a victim PLL circuit in the top-left quadrant, and 12 SF+LC placed along four different axes at the periphery of the noise source and inside the PLL.

The substrate noise waveforms shown in Figure 23.2 were measured by SF+LC for TCNS running with the shortest delay among $C_k[0{:}8]$ (T_s) on the top of the figure, and with a larger delay in the bottom of the figure. The shortest delay causes single large peak noise due to large di/dt coupling to inductance parasitic

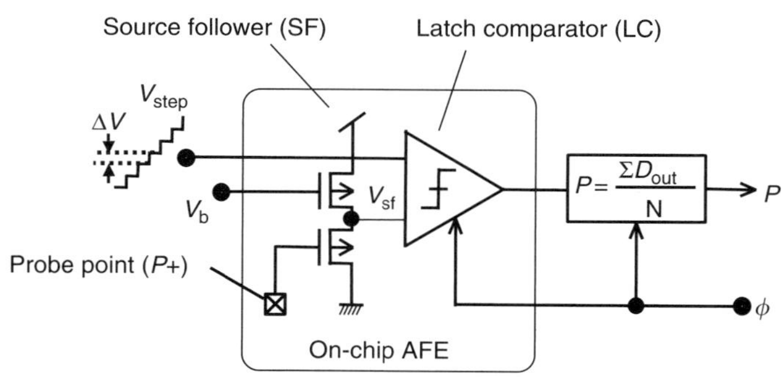

FIGURE 23.13 On-chip noise probing circuit consisting of source follower (SF) and latched comparator (LC). (From M. Nagata et al., Measurements and analyses of substrate noise waveform in mixed-signal IC environment, *IEEE Trans. Comput. Aid. Des.*, 19, 671–678, 2000. With permission.)

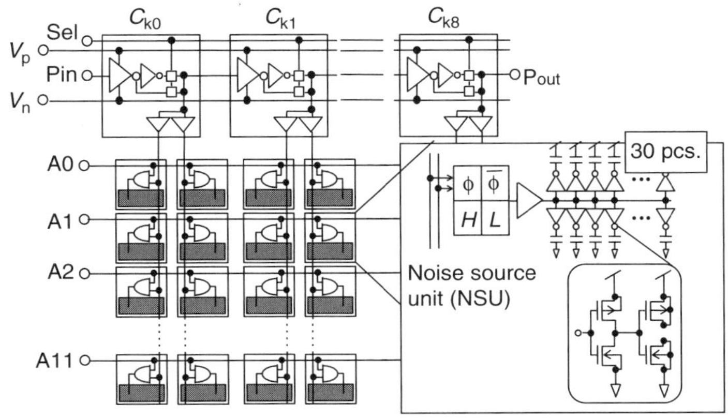

FIGURE 23.14 Transition controllable noise source (TCNS) circuit. (From M. Nagata et al., Measurements and analyses of substrate noise waveform in mixed-signal IC environment, *IEEE Trans. Comput. Aid. Des.*, 19, 671–678, 2000. With permission.)

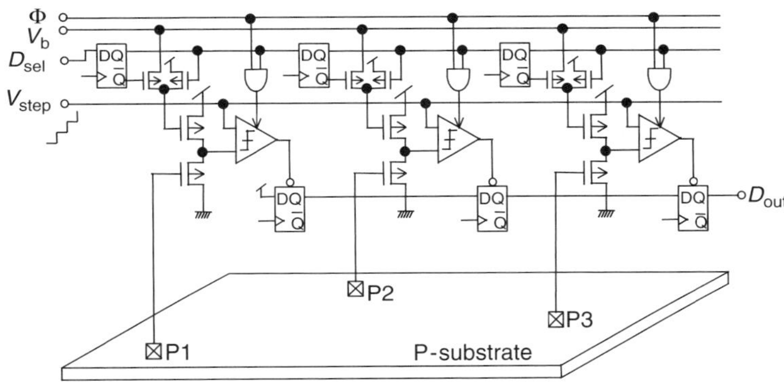

FIGURE 23.15 Arrayed SFLC detectors for multiple-point substrate noise measurement. (From M. Nagata et al., Effects of power-supply parasitic components on substrate noise generation in large-scale digital circuits, *Symposium on VLSI Circuits Digest of Technical Papers*, 2001, pp. 159–162. With permission.)

to assembly, while positive peaks corresponding to each edge of C_k with rise and fall transitions appear for the larger T_s. The observed difference in substrate noise waveform from the identical digital noise circuit also results from the fact that the major cause of substrate noise is the leakage of ground bounce.

The location dependence of peak-to-peak substrate noise amplitude obtained by measurements with arrayed SF+LC detectors and by simulation with chip-level substrate network extraction and noise analysis is shown in Figure 23.17. The distance of each detector from the first one on the axis is listed on the x-axis of each graph. The average absolute error between these simulated and measured results is 4.5 dBV [54].

23.5 Application to Placement and Power Distribution Synthesis

Simulation of mixed-signal switching noise has been integrated into several automatic layout tools including a power distribution synthesis program (RAIL) that automates the design of the power distribution network [31,55] and a substrate aware placement tool (WRIGHT) [56,57]. In RAIL, the topology of the power grid, the sizing of individual segments, and the choice of I/O pad number and location are simultaneously optimized. The optimization, which employs combinational optimization techniques, is performed under tight DC, AC, and transient electrical constraints arising from the interaction of the power grid with the rest of the IC — notably via substrate coupling. Coupling effects are included in the

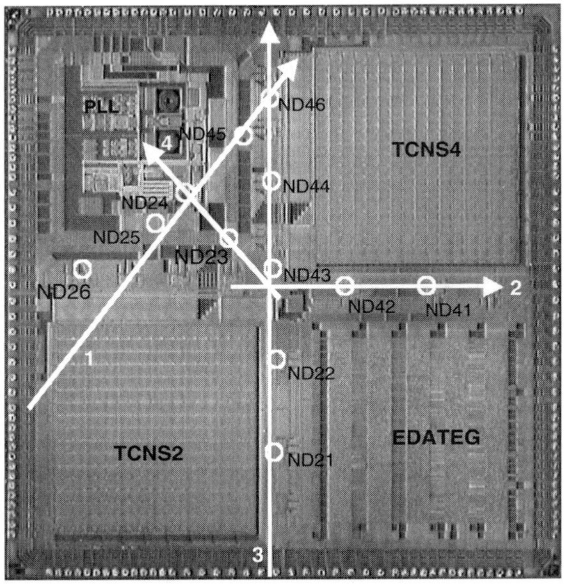

FIGURE 23.16 Micrograph of test chip carrying two TCNS circuits and twelve SF+LC detectors. (From W.K. Chu et al., A substrate noise analysis methodology for large-scale mixed-signal ICs, *Proceedings of IEEE Custom Integrated Circuits Conference*, 2003, pp. 369–372. With permission.)

cost function of a simulated annealing (SA) based power distribution synthesis system. In this work, linear macromodels for the digital switching logic circuits are created, and the capacitive and resistive coupling to the power rails are modeled as shown in Figure 23.18. Each logic circuit is replaced by its linear macromodel. To design a power grid, the tool begins with an initial "state" for the power bus geometry and power I/O pad configuration and then this geometry is perturbed to create a new candidate power grid or pad configuration and update the electrical models for the buses and I/O pads. In the next step, these models are combined with designer-supplied circuit macro-models for blocks being supplied by the power grid, and for the substrate. With this complete electrical model — power grid, blocks, pads, and substrate — the resulting electrical performance is evaluated and compared against designer constraints. For example, one might evaluate the coupled noise waveform at a sensitive node against a designer-supplied peak-to-peak noise amplitude constraint. Finally, the optimizer accepts or rejects the perturbation based on the result. The iterative improvement loop is continued until the optimizer determines no further improvement is possible. The main objective is to ensure that the power distribution as a whole (buses, power I/O cell assignment, and internal cell decoupling) is designed to meet DC voltage drop and current density constraints, while keeping transient voltage below user-specified targets.

In placement tools, traditionally area and the wire length have been the most important concerns, but other factors that deal with the interaction between analog and digital sections through the common substrate have added a new dimension to this problem. A set of algorithms for handling substrate-coupled switching noise in an iterative placement framework implemented in a substrate-aware mixed-signal placement tool called WRIGHT has been described [56,57]. The focus here is on physical design, in particular chip-level macro-cell placement and the approach incorporates simplified switching noise estimation into an SA placement algorithm. A coarse-resistive grid method analyzing the coupling of digital switching noise into the analog macros on the chip is used as shown in Figure 23.19. The tool includes models for the chip substrate, noise sources, and receivers on the macrocells. In addition, mitigation measures such as guard rings were incorporated into the inner loop of the placer by low impedance ties from the substrate to the reference potential. The accuracy available in design tools to analyze the substrate noise is not needed and is unaffordable within a placement framework since such a tool must visit thousands of candidate placement solutions.

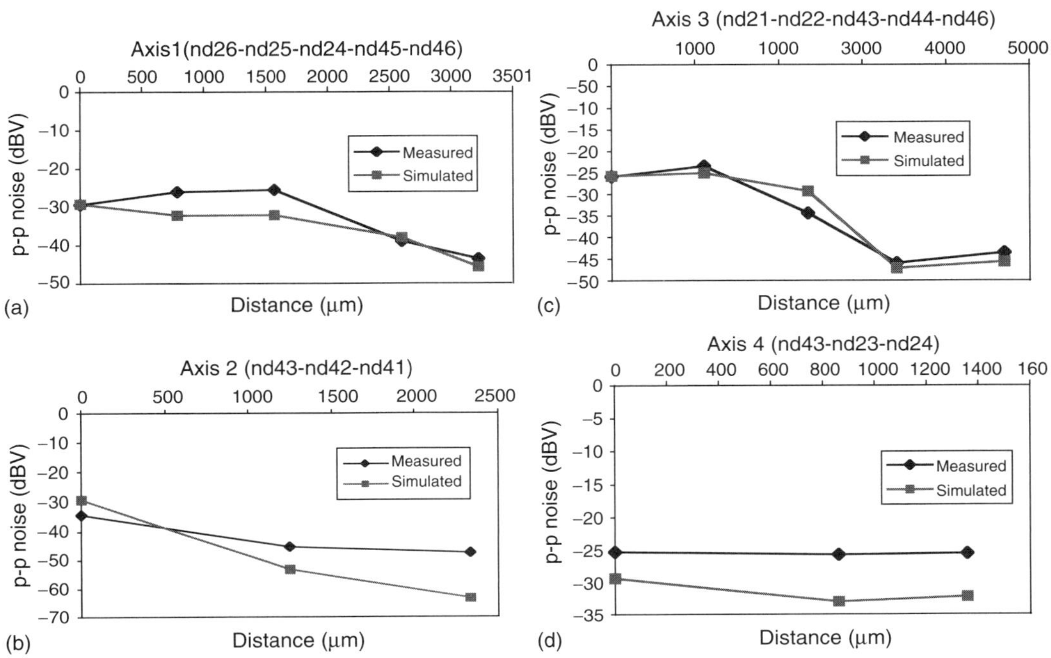

FIGURE 23.17 Simulated and measured peak-to-peak noise amplitude at various points. (From W.K. Chu et al., A substrate noise analysis methodology for large-scale mixed-signal ICs, Proceedings of *IEEE Custom Integrated Circuits Conference*, 2003, pp. 369–372. With permission.)

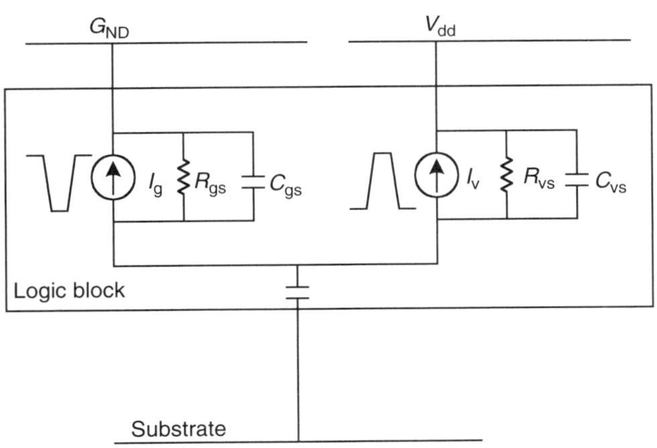

FIGURE 23.18 Simple linear macromodel of logic block for switching noise in power distribution synthesis. (From B.R Stanisic et al., Addressing noise decoupling in mixed-signal IC's: power distribution design and cell customization, *Proc. IEEE J. Solid-State Circ.*, 321–326, 1995. With permission.)

23.6 Summary

To understand and address the problem of noise coupling in mixed-signal ICs many modeling methods and computer simulation techniques for mixed-signal noise coupling have been proposed. These efforts have been reviewed in this chapter. The physical phenomena responsible for the generation of the undesired signal have been discussed, and the media transporting the signal from the source to the destination described. In addition, different approaches for modeling the source and coupling media, and subsequent computer methods to simulate the coupling have been discussed. Measurement techniques for substrate

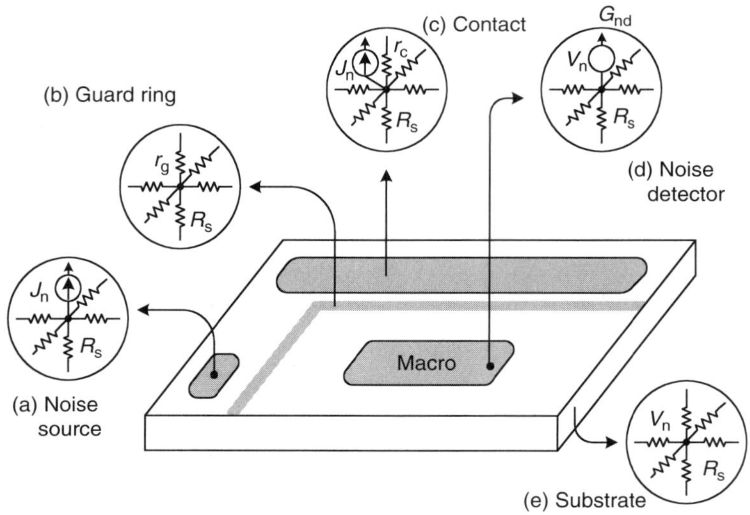

FIGURE 23.19 Substrate-resistive macromodel for placement. (From S. Mitra et al., Substrate-aware mixed-signal macro-cell placement in WRIGHT, *Proceedings of IEEE Custom Integrated Circuits Conference*, 1994, pp. 529–532. With permission.)

noise and validation of computer simulation approaches have also been discussed. Finally, an application of substrate noise analysis to placement and power distribution synthesis has been reviewed.

References

[1] A. Rofougaran, G. Chang, J.J. Rael, J.Y.-C. Chang, M. Rofougaran, P.J. Chang, M. Djafari, M.-K. Ku, E.W. Roth, A.A. Abidi, and H. Samueli, A single-chip 900-MHz spread-spectrum wireless transceiver in 1-mm CMOS — Part I Architecture and transmitter design, *IEEE J. Solid-State Circ.*, 33, 515–534, 1998.

[2] D.R. Pehlke, A. Burstein, and M.F. Chang, Extremely high-Q tunable inductor for Si-based RF integrated circuit applications, *IEEE International Electronic Devices Meeting Digest*, 1997, pp. 63–66.

[3] J.A. Olmstead and S. Vulih, Noise problems in mixed analog-digital integrated circuits, *Proceedings of IEEE Custom Integrated Circuit Conference*, Portland, OR, May 4–7, 1987, pp. 659–662.

[4] S. Masui, Simulation of substrate-coupling in mixed-signal MOS circuits, *Proceedings of VLSI Circuit Symposium*, Seattle, WA, June 4–6, 1992, pp. 42–43.

[5] T. Tsukada and K. Makie-Fukuda, Approaches to reducing digital-noise coupling in CMOS mixed-signal LSI's, *IEICE-T. Fund. Electr.*, E80-A, 263–275, 1997.

[6] N.K. Verghese and D.J. Allstot, Computer-aided design considerations for mixed-signal coupling in RF integrated circuits, *IEEE J. Solid-State Circ.*, 33, 314–323, 1998.

[7] E. Charbon, R. Gharpurey, R.G. Meyer, and A. Sangiovanni-Vincentelli, Substrate optimization based on semi-analytical techniques, *IEEE Trans. Comput. Aid. Des.*, 18, 172–190, 1999.

[8] R.B. Merrill, W.M. Young, and K. Brehmer, Effect of substrate material on crosstalk in mixed analog/digital integrated circuit, *International Electronic Devices Meeting Technical Digest*, San Francisco, CA, Dec. 11–14, 1994, pp. 433–436.

[9] X. Aragones and A. Rubio, Experimental comparison of substrate noise coupling using different wafer types, *IEEE J. Solid-State Circ.*, 34, 1405–1409, 1999.

[10] J. Briaire and K.S. Krisch, Substrate injection and crosstalk in CMOS circuits, *Proceedings of IEEE Custom Integrated Circuits Conference*, San Diego, CA, May 16–19, 1999, pp. 483–486.

[11] J. Briaire and KS. Krisch, Principles of substrate crosstalk generation in CMOS circuits, *IEEE Trans. Comput. Aid. Des.*, 19, 645–653, 2000.

[12] S. Mitra, R.A. Rutenbar, L.R. Carley, and D.J. Allstot, A methodology for rapid estimation of substrate-coupled switching noise, *Proceedings of IEEE Custom Integrated Circuits Conference*, Santa Clara, CA, May 1–4, 1995, pp. 129–132.

[13] M. Nagata and A. Iwata, Substrate noise simulation techniques for analog-digital mixed LSI design, *IEICE Trans. Fund. Electr.*, E82-A, 271–278, 1999.

[14] M. Nagata and A. Iwata, Substrate crosstalk analysis in mixed signal CMOS integrated circuits, *Proceedings of IEEE Asia and South Pacific Design Automation Conference 2000 with EDA TechnoFair*, Yokohama, Japan, Jan. 25–28, 2000, pp. 623–629.

[15] S. Kiaei, D.J. Allstot, K Hansen, and N.K. Verghese, Noise consideration for mixed-signal RF IC transceivers, *Wirel. Netw.*, 4, 41–53, 1998.

[16] A.J. Rainal, Eliminating inductive noise of external chip interconnections, *IEEE J. Solid-State Circ.*, 29, 126–129, 1994.

[17] D.K. Su, M.J. Loinaz, S.Masui, and B.A. Wooley, Experimental results and modeling techniques for substrate noise in mixed-signal integrated circuits, *IEEE J. Solid-State Circ.*, 28, 420–430, 1993.

[18] N.K. Verghese, T.J. Schmerbeck, and D.J. Allstot, *Simulation Techniques and Solutions for Mixed-signal Coupling in Integrated Circuits*, Kluwer, Boston, MA, 1995.

[19] M. Nagata, T. Ohmoto, Y. Murasaka, T. Morie, and A. Iwata, Effects of power-supply parasitic components on substrate noise generation in large-scale digital circuits, *Symposium on VLSI Circuits Digest of Technical Papers*, Kyoto, Japan, June 14–16, 2001, pp. 159–162.

[20] N.K. Verghese, D.J. Allstot, and M.A. Wolfe, Verification techniques for substrate coupling and their application to mixed-signal IC design, *IEEE J. Solid-State Circ.*, 31, 354–365, 1996.

[21] J.M. Casalta, X. Aragones, and A. Rubio, Substrate coupling evaluation in BiCMOS technology, *IEEE J. Solid-State Circ.*, 32, 598–603, 1997.

[22] C. Hu, *VLSI Electronics: Microstructure Science*, Vol. 18, Academic Press, New York, 1981.

[23] K. Shimazaki, H. Tsujikawa, S. Kojima, and S. Hirano, LEMINGS: LSI's EMI-noise analysis with gate level simulator, *Proceedings of IEEE International Symposium on Quality Electronic Design*, San Jose, CA, Mar. 24–26, 2003, pp. 129–136.

[24] E. Charbon, P. Miliozzi, L. Carloni, A. Ferrari, and A. Sangiovanni-Vincentelli, Modeling digital substrate noise injection in mixed-signal IC's, *IEEE Trans. Comput. Aid. Des.*, 18, 301–310, 1999.

[25] P. Miliozzi, L. Carloni, E. Charbon, and A. Sangiovanni-Vincentelli, SUBWAVE: a methodology for modeling digital substrate noise injection in mixed-signal IC's, *Proceedings on IEEE Custom Integrated Circuits Conference*, San Diego, CA, May 5–8, 1996, pp. 385–388.

[26] M. van Heijningen, M. Badaroglu, S. Donnay, M. Engels, and I. Bolsens, High-level simulation of substrate noise generation including power-supply noise coupling, *Proceedings of 37th Design Automation Conference*, Los Angeles, CA, June 5–9, 2000, pp. 446–451.

[27] M. Nagata, T. Morie, and A. Iwata, Modeling substrate noise generation in CMOS digital integrated circuits, *Proceedings of IEEE Custom Integrated Circuits Conference*, Orlando, FL, May 12–15, 2002, pp. 501–504.

[28] K.J. Kerns, I.L. Wemple, and A.T. Yang, Stable and efficient reduction of substrate model networks using congruence transforms, *Proceedings of IEEE/ACM International Conference on Computer-Aided Design*, San Jose, CA, Nov. 5–9, 1995, pp. 207–214.

[29] N.K. Verghese and D.J. Allstot, Rapid simulation of substrate coupling effects in mixed-mode IC's, *Proceedings of IEEE Custom Integrated Circuits Conference*, San Diego, CA, May 9–12, 1993, pp. 422–426.

[30] S. Kumasiro, R.A. Rohrer, and A. Strowas, A new efficient method for transient simulation of three-dimensional interconnect structures, San Francisco, CA, Dec. 9–12, *Digest of the IEEE International Electron Devices Meeting*, San Francisco, CA, Dec. 9–12, 1990, pp. 193–196.

[31] B.R. Stanisic, N.K. Verghese, R.A. Rutenbar, L.R. Carley, and D.J. Allstot, Addressing substrate coupling in mixed-mode ICs: simulation and power distribution systems, *IEEE J. Solid-State Circ.*, 29, 226–237, 1994.

[32] F.J.R. Clement, E. Zysman, M. Kayal, and M. Declercq, LAYIN: toward a global solution for parasitic coupling modeling and visualization, *Procedings IEEE Custom Integrated Circuits Conference*, San Diego, CA, May 1–4, 1994, pp. 537–540.

[33] T. Smedes, Substrate resistance extraction for physics-based layout verification, *IEEE/PRORISC Workshop on Circuits Systems and Signal Processing*, Mierlo, The Netherlands, 1993, pp. 101–106.

[34] J.P. Costa, M. Chou, and L.M. Silveria, Efficient techniques for accurate modeling and simulation of substrate coupling in mixed-signal IC's, *IEEE Trans. Comput. Aid. Des.*, 18, 597–607, 1999.

[35] M. Chou and J. White, Multilevel integral equation methods for the extraction of substrate coupling parameters in mixed-signal IC's, *Proceedings on ACM/IEEE Design Automation Conference*, San Francisco, CA, June 15–19, 1998, pp. 20–25.

[36] N.K. Verghese, D.J. Allstot, and M.A. Wolfe, Fast parasitic extraction for substrate coupling in mixed-signal ICs, *Proceedings of IEEE Custom Integrated Circuits Conference*, Santa Clara, CA, May 1–4, 1995, pp. 121–124.

[37] J.D. Jackson, *Classical Electrodynamics*, Wiley, New York, 1962.

[38] R. Gharpurey and R.G. Meyer, Modeling and analysis of substrate coupling in integrated circuits, *Proceedings of IEEE Custom Integrated Circuits Conference*, Santa Clara, CA, May 1–4, 1995, pp. 125–128.

[39] J.P. Costa, M. Chou, and K.M. Silveira, Efficient techniques for accurate modeling and simulation of substrate coupling in mixed-signal ICs, *Proceedings of Design, Automation and Test in Europe*, Paris, Feb. 23–26, 1998, pp. 892–898.

[40] J.P. Costa, M. Chou, and K.M. Silveira, Precorrected-DCT techniques for modeling and simulation of substrate coupling in mixed-signal IC's, *Proceedings of IEEE International Symposium on Circuits and Systems, 1998; ISCAS 1998*, Monterey, CA, May 31–June 3, Vol. 6, 1998, pp. 358–362.

[41] J.P. Costa, M. Chou, and K.M. Silveira, Efficient techniques for accurate extraction and modeling of substrate coupling in mixed-signal IC's, *Proceedings of Design, Automation and Test in Europe, 1999; Conference and Exhibition*, Munich, Mar. 9–12, 1999, pp. 396–400.

[42] J. Kanapka, J. Phillips, and J. White, Fast methods for extraction and sparsification of substrate coupling, *Proceedings of ACM/IEEE Design Automation Conference*, San Francisco, CA, 2000, pp. 738–743.

[43] S. Ponnapalli, N. Verghese, W. Chu, and G. Coram, Preventing a noisequake, *IEEE Circ. Dev.*, 17, 19–28, 2001.

[44] M. Nagata and A. Iwata, A macroscopic substrate noise model for full chip mixed-signal verification, *Symposium on VLSI Circuits Digest Technical Papers*, Kyoto Japan, June 12–14, 1997, pp. 37–38.

[45] M.K. Mayes and S.W. Chin, All Verilog mixed-signal simulator with analog behavioral and noise models, *Symposium on VLSI Circuits Digest Technical Papers*, Honolulu, HI, June 13–15, 1996, pp. 186–187.

[46] M.K. Mayes and S.W. Chin, All Verilog mixed-signal simulator with analog behavioral and noise models, *IEEE-CAS Region 8 Workshop on Analog and Mixed IC Design Proceedings*, 1996, pp. 50–54.

[47] T. Blalack and B.A. Wooley, The effects of switching noise on an oversampling A/D converter, *IEEE International Solid-State Circuits Conference on Digest of Technical Papers*, San Francisco, CA, 1995, pp. 200–201.

[48] N.K. Verghese and D.J. Allstot, Verification of RF and mixed-signal integrated circuits for substrate coupling effects, *Proceedings of IEEE Custom Integrated Circuits Conf*erence, Santa Clara, CA, May 5–8, 1997, pp. 363–370.

[49] R. Telichevesky, K. Kundert, I. Elfadel, and J. White, Fast simulation algorithms for RF circuits, *Proeedings of IEEE Custom Integrated Circuits Conference*, San Diego, CA, May 5–8, 1996, pp. 437–444.

[50] R. Telichevesky, K. Kundert, and J. White, Receiver characterization using periodic small-signal analysis, *Proceedings of IEEE Custom Integrated Circuits Conference*, San Diego, CA, May 5–8, 1996, pp. 449–452.

[51] R. Telichevesky, K. Kundert, and J. White, Efficient steady-state analysis based on matrix-free Krylov-subspace methods, *Proceedings on ACM/IEEE Design Automation Conf*erence, San Francisco, CA, June 12–16, 1995, pp. 480–484.

[52] K.H. Kwan, I.L. Wemple, and A.T. Yang, Simulation and analysis of substrate coupling in realistically-large mixed-A/D circuits, *Symposium on VLSI Circuits Digest of Technical Papers*, Honolulu, HI, June 13–15, 1996, pp. 184–185.

[53] M. Nagata, J. Nagai, T. Morie, and A. Iwata, Measurements and analyses of substrate noise waveform in mixed-signal IC environment, *IEEE Trans. Comput. Aid. Des.*, 19, 671–678, 2000.

[54] W.K. Chu, N. Verghese, H.J. Cho, K. Shimazaki, H. Tsujikawa, S. Hirano, S. Doushoh, M. Nagata, A. Iwata, and T. Ohmoto, A substrate noise analysis methodology for large-scale mixed-signal ICs, *Proceedings of IEEE Custom Integrated Circuits Conference*, San Jose, CA, Sep. 21–24, 2003, pp. 369–372.

[55] B.R. Stanisic, R.A. Rutenbar, and L.R. Carley, Mixed-signal noise decoupling via simultaneous power distribution and cell customization in RAIL, *Proceedings of IEEE Custom Integrated Circuits Conference*, San Diego, CA, May, 1994, pp. 533–536.

[56] S. Mitra, R.A. Rutenbar, L.R. Carley, and D.J. Allstot, Substrate-aware mixed-signal macro-cell placement in WRIGHT, *Proceedings of IEEE Custom Integrated Circuits Conference*, San Diego, CA, May, 1994, pp. 529–532.

[57] S. Mitra, R.A. Rutenbar, L.R. Carley, and D.J. Allstot, Substrate-aware mixed-signal macro-cell placement in WRIGHT, *IEEE J. Solid-State Circ.*, 30, 269–278, 1995.

SECTION IV

TECHNOLOGY CAD

24

Process Simulation

Mark D. Johnson
Synopsys, Inc.
Mountain View, California

24.1 Introduction

Process simulation is the modeling of the fabrication of semiconductor devices such as transistors. The ultimate goal of process simulation is an accurate prediction of the active dopant distribution, the stress distribution, and the device geometry. Process simulation is typically used as an input for device simulation, the modeling of device electrical characteristics. Collectively, process and device simulation form the core tools for the design phase known as technology computer-aided design (TCAD). Considering the design process as a series of steps with decreasing levels of abstraction, synthesis would be at the highest level and TCAD, being closest to fabrication, would be the phase with the least amount of abstraction. Because of the detailed physical modeling involved, process simulation is almost exclusively used to aid in the development of single devices — whether discrete or as a part of an integrated circuit.

The fabrication of integrated circuit devices requires a series of processing steps called a process flow. Process simulation involves modeling all essential steps in the process flow in order to obtain dopant and stress profiles and, to a lesser extent, device geometry. The input for process simulation is the process flow and a layout. The layout is selected as a linear cut in a full layout for a 2-D simulation or a rectangular cut from the layout for a 3-D simulation.

Technology computer-aided design has traditionally focused mainly on the transistor fabrication part of the process flow, ending with the formation of source and drain contacts — also known as front-end of line manufacturing. The back-end of line manufacturing, e.g., interconnect and dielectric layers are not considered. One reason for delineation is the availability of powerful analysis tools such as electron microscopy techniques, scanning electron microscopy (SEM) and transmission electron microscopy (TEM), which allow for accurate measurement of device geometry. There are no similar tools available for accurate high-resolution measurement of dopant or stress profiles.

Nevertheless, there is a growing interest to investigate the interaction between front-end and back-end manufacturing steps. For example, back-end manufacturing may cause stress in the transistor region, changing device performance. These interactions will stimulate the need for better interfaces to back-end simulation tools or lead to integration of some of those capabilities into TCAD tools.

In addition to the recent expanding scope of process simulation, there has always been a desire to have more accurate simulations. However, simplified physical models have been most commonly used in order to minimize computation time. But shrinking device dimensions put increasing demands on the accuracy of dopant and stress profiles, so new process models are added for each generation of devices to match new accuracy demands. Many of the models were conceived by researchers long before they were needed, but sometimes new effects are only recognized and understood once process engineers discover a problem and experiments are performed. In any case, the trend of adding more physical models and considering more detailed physical effects will continue and may accelerate.

The history of commercial process simulators began with the development of the Stanford University Process Modeling (SUPREM) program.[1] Building upon this beginning with improved models, SUPREM II and SUPREM III were developed. Technology Modeling Associates, Inc. (TMA), which was formed in 1979, was the first company to commercialize SUPREM III. Later, Silvaco also commercialized SUPREM and named the product ATHENA. TMA commercialized SUPREM-IV (2-D version) and called it TSUPREM4. In 1992, Integrated Systems Engineering (ISE) came out with the 1-D process simulator TESIM and the 2-D process simulator DIOS. At about the same time, development of a new 3-D process and device simulator began at TMA, and after TMA was acquired by Avant!, Corp. The product was released in 1998 as Taurus. Around 1994, the first version of the Florida Object Oriented Process Simulator (FLOOPS) was completed. FLOOPS was later commercialized by ISE in 2002. Another process simulator PROPHET was created around 1994 at Bell Labs, which later became Agere, but has not been sold commercially. In 2002, Synopsys acquired Avant!, Corp. and in 2004, it acquired ISE. Synopsys released Sentaurus Process in 2005 which it says combines the best features of FLOOPS, TSuprem4 and Taurus using FLOOPS as a platform. Besides these simulators, there are numerous other university and commercial simulators such as PROMIS, PREDICT, PROSIM, ICECREAM, DADOS, TITAN, MicroTec, DOPDEES, and ALAMODE.

This chapter describes the fabrication steps most often modeled with process simulation tools including both the important physical effects, and models and techniques used to simulate them. The process steps most often associated with process simulation are ion implantation, annealing (diffusion and dopant activation), etch, deposition, oxidation, and epitaxy. Other common steps include CMP, silicidation, and reflow. In addition to the current state of the art, some attempt will be made to look forward to what new physical effects will be required to develop the future devices described by the ITRS roadmap. The last section discusses some practical aspects commonly used in process simulation, and gives a description of the steps required to create a structure suitable for device simulation.

24.2 Process Simulation Methods

Since all commercial process simulators use a combination of the finite-element (FE) and finite-volume (FV) methods, we begin with a brief introduction to the topic, including common techniques and strategies. A complete description of FE/FV methods is out of the scope of this chapter but there are many fine books[2] which describe the topic thoroughly. However, it is important to discuss requirements for process simulation for achieving accurate results. These requirements are based on the same requirements which are generic to FE/FV techniques, with an additional difficulty coming from the changes in the geometry during the simulated fabrication of the device. Process simulation uses an FE/FV mesh to compute and store the dopant and stress profiles. Each geometrical change in the simulation domain requires a new mesh which fits to the new boundaries. As will be described below, the large number of geometry-modifying steps involved and the nature of process simulation where each step depends on the cumulative results of all previous steps, make process simulation an especially challenging application of the FE/FV technique.

One of the most important results of process simulation is the dopant profile after processing. The accuracy of the profile strongly depends on maintaining a proper density of mesh points at any time during the simulation. The density of points should be just enough to resolve all dopant and defect profiles but not more, because the computational expense of solving the diffusion equations increases with the number of mesh points. A typical full-flow CMOS process simulation can have more than 50 mesh changes, and the number of mesh changes can increase dramatically if adaptive meshing is performed. For each mesh change, interpolation is used to obtain data values on the new mesh. It is important to manage the mesh changes in such a way as to avoid accuracy degradation due to interpolation error. The easiest way to do this is to always keep points once they are introduced into the mesh, but this has the drawback of producing too many mesh points, which can be computationally expensive. Maintaining a balance between interpolation error, computational expense, and minimization of required user input is important for obtaining accurate results with a minimum of computational expense. This is especially true when simulating devices in 3-D. Without careful placement of mesh, either the accuracy will suffer unacceptably, or the computational expense will be too great to be useful. Process simulation tools so far have had limited success in completely automating mesh adaptation such that no user intervention is required. This places a requirement on the user to understand meshing and how it affects simulation accuracy and run time, and burdens the user with the task of tracking mesh changes during the simulation to ensure that proper mesh is maintained.

One of the most important uses of TCAD tools is to explore new device technology where many exploratory simulations are performed in order to give the device designer a better understanding of possible benefits as well as drawbacks of a given technology. This use case demands sequential simulations with some analysis in between. In order to be useful, many simulation cycles must be run within the time allotted for exploration, putting a high priority on minimization of simulation run-time. Currently, full-flow standard CMOS simulations are most often accomplished with a combination of 1-D and 2-D simulation and take less than a few hours on a 2.6 GHz Pentium 4. To perform these simulations (from gate formation onward) in 3-D would take a minimum of 24 h for minimum-accuracy simulation. Most of the information desired from TCAD simulations can be extracted from the simplification that the device can be treated uniformly in depth (i.e., as a 2-D simulation). To include the effects device shape along the depth or to investigate implant shadowing, 3-D simulations must be performed.

24.3 Ion Implantation

Ion implantation introduces dopant atoms into a wafer by ionizing the atoms or molecules containing the atoms, and these ionized species are then subjected to an electric field driving the ionized species into the wafer. The resulting depth profile for each type of implanted species depends mainly on the energy of the species and the dose. The energy is set by the process engineer by adjusting the potential difference between the ion source and the wafer, and the dose is determined by ion current and exposure time. Other factors which affect the profile are wafer orientation, and to a lesser extent, dose rate (ion current) and implantation temperature.

Physically, ionized dopants incident on the wafer will scatter if they come close to nuclei in the solid. In crystalline materials such as silicon, there are many "channeling" directions in which the nuclei are lined up in columns one behind the other, providing "channels" between the columns for the ions to travel long distances without scattering. Channeling is greatly enhanced if the incident ion beam is very nearly lined up with a channeling direction, but any orientation will give some channeling depending on the energy, dose, and species. In any case, dopants end up in exposed areas of the device, and some fraction are scattered uniformly in all directions, giving "lateral straggle" under a mask (such as resist). Figure 24.1 illustrates channeling by probing ion penetration depth vs. tilt angle. The inherent randomness of amorphous materials

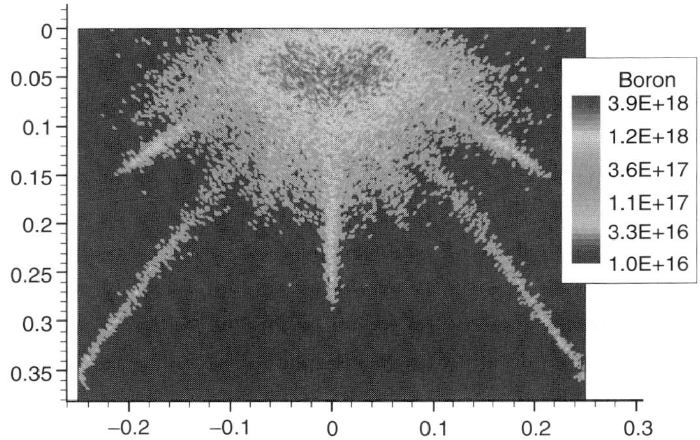

FIGURE 24.1 **(See Color insert following page 15–4.)** Figure created using Monte-Carlo implant by varying the tilt from 0 to 75° to probe the channeling behavior in silicon, dramatically illustrating the channeling tails in silicon.

such as SiO_2 scatters incoming ions more often, as there are essentially no channels for ions to travel. Process engineers often use a sacrificial oxide layer over a silicon region where a shallow implanted profile is desired. When an incoming ion scatters from a nucleus, energy can be transferred to the nucleus, creating a recoil. The recoil is typically a silicon atom, which has been removed from its lattice site in the crystal (creating a vacancy in the lattice). The recoil atom is scattered elsewhere — typically to an interstitial location. Interstitials and vacancies are known as point defects and when a pair is formed as part of a recoil collision, they are called a Frenkel pair. Figure 24.2 shows the atomic configuration of two interstitial types, and a vacancy compared with the defect-free silicon crystal lattice.

After an implant, the concentration of Frenkel pairs is very large compared to the concentration of ions/dopants, because the recoiling nuclei can themselves cause secondary damage in addition to the primary damage caused by ion/dopant scattering. As the implant dose is increased, fewer and fewer silicon nuclei are in their crystalline positions. Eventually the damage accumulation can cause amorphization of the material in those areas where the Frenkel pair concentration exceeds about 20% of the lattice density. After amorphization, channeling is essentially completely suppressed. This effect is sometimes exploited by process engineers by implanting an impurity such as Ge to create an amorphous layer to suppress channeling in a subsequent dopant implant. Figure 24.3 shows the effect of increasing damage on channeling. As the dose of an implant is increased, the proportion of newly arriving ions encountering a pristine lattice is reduced, hence reducing the proportion of ions which end up in channeling trajectories.

There are two common techniques for computing ion-implantation profiles available in commercial process simulators — analytic and Monte-Carlo. Analytic implantation uses analytic functions to compute the dopant and defect profile resulting from an implant. The analytic functions are normally expressed in two directions: along the ion trajectory, and perpendicular to the trajectory which models scattering of ions. Because of this, it is necessary to perform a convolution integral to compute the contribution from all surface points to the concentration at every point in the bulk. Figure 24.4 shows how this is done. Because of the rapid roll-off in both directions, the integration can be limited to surface points nearby. Alternatively, Monte-Carlo implantation computes the individual ion trajectories starting from a randomly chosen position in space. This technique will be described in more detail below. Because of the detailed nature of the calculation, Monte-Carlo implantation can be between ten and hundred times slower, depending on the desired concentration resolution. There are a number of techniques which are typically employed to speed up the Monte-Carlo implantation, but in general the analytic technique is much less computationally expensive than Monte-Carlo. The trade-offs are that analytic implantation can be quite inaccurate in complicated geometries where the ion beam passes through multiple regions, and analytic implantation does not handle damage properly for multiple sequential implants with no annealing step in between.

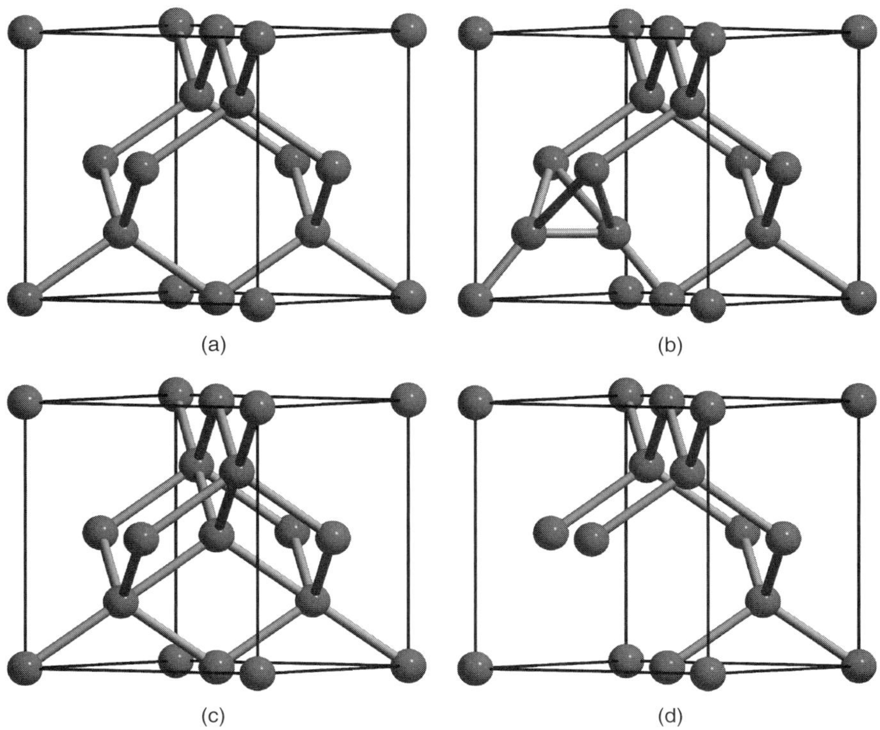

FIGURE 24.2 Atomic configuration of (a) perfect silicon lattice; (b) a split ⟨110⟩ interstitial; (c) a tetrahedral interstitial; and (d) a vacancy in the silicon crystal lattice. These configurations are not the lowest energy configuration so it is expected that there would be some adjustment of the atomic locations in the real material.

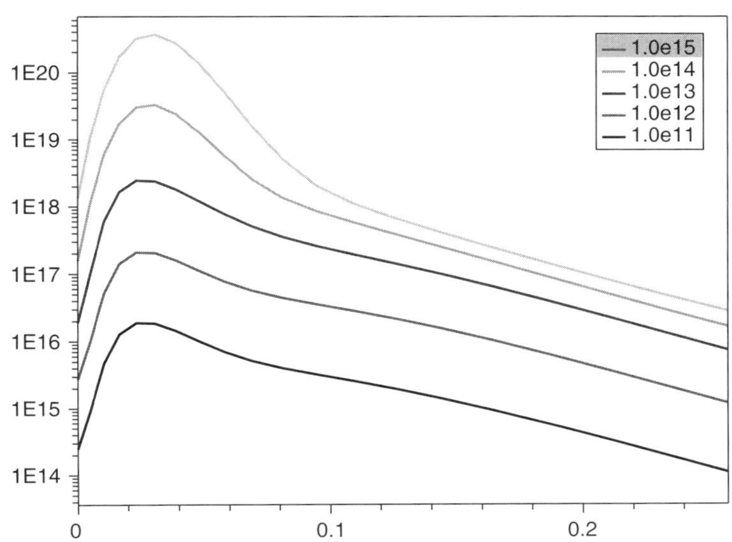

FIGURE 24.3 A series of arsenic implants for different doses, all 40 keV energy. Larger dose implants have a larger fraction of de-channeling due to increased damage accumulation. This is apparent from the increase in the peak height relative to the channeling (i.e., shoulder height).

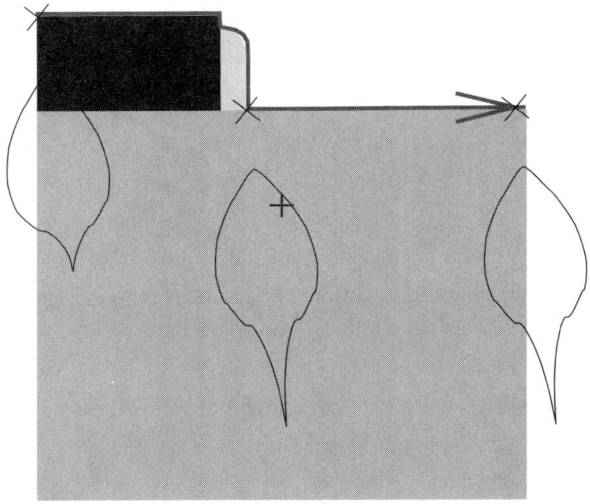

FIGURE 24.4 Schematic figure showing implantation integration scheme. The concentration distribution which results from ions impinging at one point on the surface is called a point response (constant concentration profiles in black). Analytic implantation uses functions to define the point responses. In a structure, the concentration of a point (cross) is computed by integrating the contribution from each point on the surface. However, there is no practical solution to handle multiple layers correctly with the analytic method. It would require a huge database of combinations of materials of varying thicknesses.

24.3.1 Analytic Implantation

Analytic implantation relies on a set of look-up tables which supply parameters for the profile of the implanted species, and a separate table which gives the profile for the implant damage, i.e., Frenkel pair distribution. The implant parameters are a function of energy, species, dose, rotation, and tilt angles, and in addition may be a function of an overlayer thickness. Typically, Pearson IV functions are used for the primary profile and complementary error functions are used for the lateral roll-off due to scattering.

Many specific techniques have been developed to enhance the accuracy of analytic implantation for technology simulation. One common technique to reduce channeling as mentioned before is to use a sacrificial oxide. As the oxide thickness has a strong influence on scattering in the silicon layer below, special tables have been created which give the implant parameters as a function of oxide overlayer thickness. Similarly, the effect of damage accumulation on subsequent implants requires special handling. It is possible to employ some heuristics to estimate the effect of all implants since the last anneal cycle. The most important effect which must be mimicked is tracking the effect of amorphizing implant on subsequent implants. An amorphous layer has a similar impact on implant profiles as oxide, so implant parameter tables which have oxide thickness dependence can be used to estimate the resulting profiles. Amorphization is typically estimated by a threshold for Frenkel pair formation — above the threshold, the material is considered amorphous. However, experimental implant tables are normally only obtained for implants into crystalline silicon. Another difficulty comes from the fact that multiple implants may involve multiple species, and so far, because of the number of combinations possible and the expense, no experimental multiple implant tables with or without amorphization are commercially available.

In addition to the difficulty of modeling multiple implants, the geometry of devices also present challenges for analytic implantation. High-angle implants such as source drain engineering implants, are especially problematic. These implants are targeted at a corner near the gate edge and because of the high angle, the thickness of polysilicon and oxide overlayers that the ion beam passes through before hitting silicon varies dramatically there. Since the implant parameters depend on the thickness of the material covering silicon, these implants cause great difficulty in the analytic technique. In addition, modern

CMOS devices have very thin layers, so a large fraction of ions can pass through layers with perhaps only a few scattering events. Modeling ions, which scatter from more than one type of material, present an enormous challenge for any type of analytic technique.

One model which has been developed to overcome these difficulties can be found in [3]. In this model, amorphization and damage accumulation are approximated by adjusting implant parameters to account for reduced channeling. Another technique based on the same idea was employed to adjust parameters to account for the dechanneling effect of SiO_2 over layers when there are no implant parameters available for such conditions.

24.3.2 Monte-Carlo Implantation

Monte-Carlo implantation, as its name implies, explicitly computes individual ion, and optionally, recoil trajectories starting from a random location in space. Each scattering event will reduce the ion's kinetic energy, change the direction of the ion, and possibly produce secondary recoils, which may also be followed in a manner similar to the ion. The implantation damage is obtained naturally from the number of recoils produced per unit volume. Also, ions and even recoils can be tracked from one material into another so that multiple material layers can be handled accurately. A central strength of the technique is the wealth of experimental data and theoretical understanding of nuclear scattering. The ions start with a sufficiently large energy that nuclear scattering dominates the process. Only after many collisions with target atoms is the incident ion slow enough for the electronic stopping effects to become important. These effects are also fairly well understood, but there are functions used which require fitting parameters to simplify the computation. The least well-understood aspect of the technique, and an area of some ongoing research, is in differences in the detailed nature of the damage produced by different ions. In addition to the more accurate ion and damage profiles produced by the Monte-Carlo method, dose loss due to backscattering events near the surface is obtained naturally.

Regarding performance, computing the scattering events is the most expensive part of the Monte-Carlo implant calculation. Look-up tables can be used to compute these events, but since the number of ions incident upon a single device for a single implant can be very large (of the order of 10^7 μm^{-2}) and there may be 10 to 30 implants for a process flow, Monte-Carlo implant computations can be the most expensive step in a process simulation if they are used to simulate every implant. For typical conditions, analytic implant is of the order of ten times faster than Monte-Carlo in 2-D, but for 3-D the times are somewhat closer. There are a number of techniques which can be employed to speed up the computation. One such technique is trajectory replication, where in a homogeneous part of a structure, a trajectory may be reused instead of being re-computed. Another technique commonly used is rare-event enhancement. This technique keeps track of the statistical significance of all the elements of the structure. When ions enter a region of the simulation where no data exists, they are split into multiple "pseudo-particles" with reduced weight, so that more of the simulation time is spent on the rare events instead of those events which already have good statistics.

There are a couple of things which can be done to combine the accuracy of Monte-Carlo with the speed of analytic implantation. The most straightforward technique which is commonly employed by TCAD vendors and users is to calibrate the Monte-Carlo implantation to a limited technology-specific set of implantation conditions. Then, it is possible to create specialized implant tables for each implant by using Monte-Carlo implantation in 1-D and extracting Pearson IV parameters for the analytic tables. For a small subset of implantation conditions, this is not an enormous task and can have great benefits in terms of accuracy. This technique effectively creates process-specific ion implant tables. Another technique used to enhance analytic implantation is to create 2-D or 3-D point response functions using Monte-Carlo. Instead of two perpendicular analytic functions, a 2-D or 3-D matrix is computed using the Monte-Carlo technique by introducing numerous ions at a single location in a block of silicon (see Figure 24.5). The resulting intensity distribution is used as a look-up table to perform the integration over nearby surface points. This technique is used to account for the anisotropic nature of implants into crystalline materials, specifically the off-axis channeling.

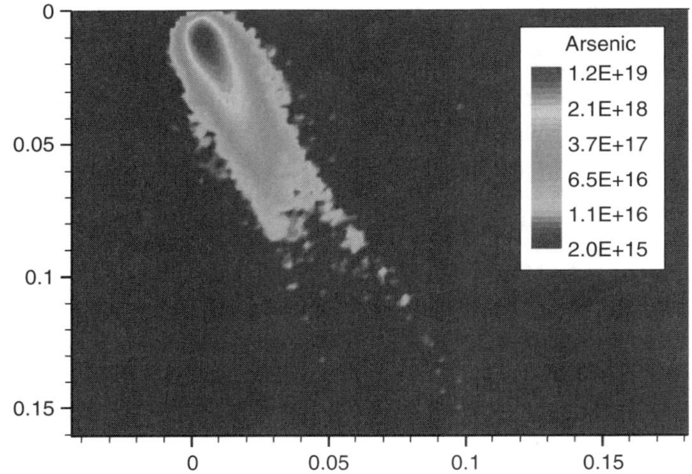

FIGURE 24.5 (**See color insert following page 15–4.**) Point response Monte-Carlo implantation simulation for a 10^{14} cm^{-2}, 15 keV arsenic implant tilt 25°.

24.4 Diffusion

One of the most important goals of process simulation is the accurate modeling of the active dopant profile evolution. In the early years of semiconductor devices, the device dimensions were huge compared to today's standards, and dopants were typically introduced through in-diffusion. This process occurs in near-equilibrium conditions and results in smooth and deep profiles by today's standards. The push toward smaller devices has fostered the development of process conditions which are further away from equilibrium. Specifically, device designers require high peak active concentrations and steep dopant profile gradients. With near-equilibrium conditions, the kinetic mechanisms of diffusion are unimportant because they are either constant or nearly constant through the whole step. As conditions are pushed further from equilibrium, the detailed kinetics of diffusion becomes more important. This can be seen in the development of physical models where the first process simulators used one equation per dopant and assumed equilibrium concentration of defects; later two defect equations were added to simulate defect kinetics but the dopant–defect pairs were still considered to be in local equilibrium; and now sometimes three equations per dopant in addition to two defect equations are solved to simulate the full dopant, defect, and dopant–defect kinetics. This effect is similarly seen in the evolution of dopant and defect cluster equations, which were not even used at first, but now are relied upon for day-to-day use. Research groups are exploring simulations which include tens to thousands of clustering species, most of which are transient as they are not typically present in significant concentrations by the end of the anneal step. The most detailed calculations are now accomplished with kinetic Monte-Carlo diffusion simulations where the number of species types is essentially unlimited.

In addition to accurate models for dopants, accurate diffusion modeling strongly depends on models of point and extended defects which come mostly from ion implantation, but are also introduced during oxidation as well as a small, thermally activated concentration. After ion implantation, there is a tremendous amount of disorder or "damage" in the implanted regions. Annealing of the sample is necessary to reduce damage and activate the dopants. Reducing damage reduces electron scattering, improves electron mobility and reduces leakage current. Dopant activation will be discussed in more detail below, but involves the dopants occupying substitutional locations in the silicon lattice. The side effect of annealing is that the ion/dopant profiles redistribute or diffuse and this diffusion can strongly depend on the nature of the implanted damage.

Understanding the kinetics of diffusion is important to understanding the trends in thermal processing. But before discussing kinetics, it is first necessary to discuss the states and structures that dopants and defects can be in. Dopants can occupy substitutional silicon lattice sites, in which case they are normally considered active.

Alternatively, dopants can be paired with an interstitial, vacancy, combination of interstitials and vacancies, or simply be in an interstitial location in which case they are inactive because in this state they do not affect the electron or hole concentrations. The substitutional sites are stable sites for the dopants so they are considered immobile species. On the other hand, the interstitial sites are not energetically favorable, so the dopants tend to migrate if they are in interstitial locations. When a point defect such as an interstitial or vacancy comes close to a dopant, it may change the bonding or local structure in a way to "kick out" the dopant from its substitutional site and hence allow the dopant to migrate. The reverse process whereby a dopant, paired with a point defect or simply in an interstitial location, may exchange with a silicon atom in the lattice creating an interstitial and leaving the dopant in a substitutional location. Consequently, point defects play an important role in dopant diffusion. Additionally, dopants can form immobile clusters consisting of both defects and dopants.

Interstitials play an especially important role in diffusion of ion-implanted wafers. Not only is there a large amount of damage in the form of Frenkel pairs as discussed above, but there is also an additional interstitial for every dopant atom implanted. Because of the conservation of lattice sites, an implanted ion adds an extra atom into the lattice — eventually most of the dopant atoms end up in substitutional sites, displacing one silicon atom per dopant.

The evolution of interstitial and vacancy profiles is governed by a number of physical processes. Interstitials can recombine with vacancies to reduce damage, but they might also combine with other interstitials to form interstitial clusters. Interstitial clusters form in part because of a reduction in the total strain energy as the two particles come close together. For a given size of interstitial cluster, it is possible that many different structures with different corresponding formation energies exist. Normally, models only consider the most stable form of the cluster. Additionally, the clusters can grow one interstitial at a time as other interstitials come close by and attach. The clusters can grow large enough to form extended defects visible in TEM and are generally called 311s because they are linear in shape and grow along the ⟨311⟩ direction in Si. The 311s can also grow larger and will at some point unfault into dislocation loops with hundreds to tens of thousands of interstitials each. Conversely, interstitial clusters can shrink and dissolve away as well. It is generally accepted that clusters are constantly releasing interstitials at a temperature-dependent rate and are acquiring interstitials at a rate dependent on the free interstitial concentration. Typically, right after implantation, the free interstitial concentration is very large, which favors the growth of interstitial clusters. After a while, free interstitials have recombined with vacancies, recombined with the surface, or clustered with a dopant, and the population is reduced. Once the free interstitial population is sufficiently low, the rate of release of interstitials is greater than the acquisition rate and the clusters begin shrinking. So the clusters are both sinks and sources for interstitials. The evolution of the free interstitials is responsible for an effect observed long ago called transient enhanced diffusion (TED).[4] What was observed was an initial burst of dopant migration after ion implantation, followed by much slower migration — close to what would be expected from measured bulk boron diffusivity. This effect is seen to a limited extent for all ion-implanted dopants but is most important for boron. TED was attributed to kinetics of interstitial clustering and dissolution[5] and continues to be a very challenging effect to simulate accurately.

Similar to defects, dopants can also cluster. Dopant clustering is assumed to freeze migration by trapping the dopants in immobile clusters, as is the case with defects, but activation is also reduced. Dopant clusters have been assumed to be inactive in the past, but the charge state of dopant clusters is now a topic of ongoing research.

Both boron and arsenic are known to cluster with point defects; boron with interstitials, and arsenic with vacancies. These species further complicate the TED picture because these clusters compete with 311s for interstitials and dissolve at different rates than do 311s. As for the other dopants, there is less certainty. There are some indications of phosphorus clustering with interstitials[6] and antimony with vacancies.[9] There appears to be no significant concentration of dopant–defect clusters of indium for typical process conditions.

When the concentration of dopants becomes very large, as is the case in the highly doped source and drain regions, precipitates usually in the form of silicides may form. The temperature-dependent concentration for dopants dissolved in Si (i.e., in substitutional sites) to form precipitates is called the solubility limit. All dopants are expected to have this behavior, which limits the maximum active concentration of dopants. Dopant clustering competes with precipitation and may be more important depending on the conditions. Some extremely fast anneal steps including laser annealing have shown that activation above solid solubility can be achieved, but this is a metastable state which can relax (forming clusters and precipitates) upon further annealing.

There are two main methods used to solve diffusion: methods which solve partial differential equations such as FE/FV which will be called "continuum methods" and kinetic Monte-Carlo techniques. The continuum method for diffusion can handle most of today's important effects to a high degree of accuracy, and therefore has been and continues to be the method of choice. However, trends in MOS scaling have prompted the use of extremely rapid anneal steps with anneal times of the order of a few seconds and ramp rates in excess of 1000°C/sec. It was found that faster thermal ramp rates at the same peak temperature can reduce dopant redistribution while still activating the dopants and reducing damage. As was mentioned before, the trend away from equilibrium makes it important to solve the detailed diffusion kinetics, which makes the continuum method for diffusion increasingly more expensive. Meanwhile, as the device dimensions shrink, kinetic Monte-Carlo has become increasingly less expensive because the simulation time strongly depends on the number of diffusing species, which in turn depends on the size of the simulation domain. In addition, kinetic Monte-Carlo can solve for an essentially unlimited number of species without paying any performance penalty. So it is likely that the use of kinetic Monte-Carlo will some day overtake continuum diffusion.

24.4.1 Continuum Methods for Diffusion

Continuum methods for solving a diffusion/reaction system use partial differential equations which can be solved using the finite difference or finite element (FD/FE) technique. The number of equations which are solved depends on the desired accuracy. As few as one equation per dopant can be used for long-time high-temperature furnace anneals, and to simulate more detailed kinetics, as many as three equations per dopant and dozens of defect and dopant cluster equations could be used for an rapid thermal annealing (RTA) step where ramp rates exceed 250°C/sec and total annealing time may only be a few seconds. The extra equations cost both memory and execution time and the effect is especially detrimental in 3-D where the simulation time is dominated by linear solvers for which simulation times scale faster than linearly with the number of equations. To reduce the number of partial differential equation (PDE's), the complete physical picture is simplified by assuming some species can be approximated using an expression instead of solving a separate PDE. Normally, this is done by making steady-state approximations. The closer the system is to equilibrium, the better these steady-state assumptions are. As computing power increases, it is expected that increasingly complete physical models will be practical and better results can be expected.

The most commonly used model in the industry today relies on one equation per dopant and two defect equations. Dopant–defect pairing is assumed to be in a steady state such that the concentration of dopant–defects is proportional to the product of the dopant and defect concentrations. Without going into all the details, the model can be formulated as follows[7]:

$$\frac{\partial C_{\mathrm{A}}}{\partial t} = -\nabla \bullet J_{\mathrm{A}} - R_{\mathrm{A}}^{\mathrm{clus}} \tag{24.1}$$

$$\frac{\partial C_{\mathrm{X}}^{\mathrm{total}}}{\partial t} = -\nabla \bullet J_{\mathrm{X}} - \nabla \bullet J_{\mathrm{A}} - R_{\mathrm{IV}} - R_{\mathrm{X}}^{\mathrm{clus}} \tag{24.2}$$

$$C_{\mathrm{X}}^{\mathrm{total}} = C_{\mathrm{X}} + C_{\mathrm{AX}} \tag{24.3}$$

$$J_{\mathrm{A}} \propto \sum_{\mathrm{c,X}} D_{\mathrm{AX}^{\mathrm{c}}} \left(\frac{n}{n_{\mathrm{i}}}\right)^{-\mathrm{c}-\mathrm{z}} \nabla\left(C_{\mathrm{A}}^{+}\left(\frac{n}{n_{\mathrm{i}}}\right)^{-\mathrm{z}}\right) \tag{24.4}$$

$$J_{\mathrm{X}} \propto \sum_{\mathrm{c}} D_{\mathrm{X}^{\mathrm{c}}} \left(\frac{n}{n_{\mathrm{i}}}\right)^{-\mathrm{c}} \nabla\left(\frac{C_{\mathrm{X}}}{C_{\mathrm{X}}^{*}}\right) \tag{24.5}$$

where C_A is the concentration of dopant A, C_X^{total} the total concentration of defect X (X is either interstitial or vacancy) and is the sum of the free defect X concentration and the pair concentration, C_{AX}. J_A and J_X are the particle current of dopant A and defect X, respectively. $D_{AX}{}^c$ is the dopant–defect X diffusivity which depends on the charge state c, $D_X{}^c$ the diffusivity of defect X at charge state c, n the electron concentration, and n_i the intrinsic electron concentration. R_{IV} is the bulk interstitial vacancy recombination. Finally, the various terms labeled R^{clus} are recombination/clustering terms, which depend upon what clustering models are selected. One thing to notice about the model is the strong coupling between Equation (24.1) and Equation (24.2) which reflects the dopant diffusion dependence on defects. The extension of this model to include nonequilibrium defect pairing can be found in [8]. A nice comprehensive review of diffusion modeling can be found in [9].

The performance of continuum diffusion solvers is affected by many factors, but the most important ones are number of mesh nodes, number of PDEs to be solved, and the "nonlinearity" of the PDE equations. Roughly speaking, the linearity of an equation refers to the time rate of change of the solution variable, given a perturbation of one of the variables being solved. If a small perturbation causes a large nonlinear change in the solution variable, then the solver must take shorter steps in order to accurately capture the time evolution of the system. Because diffusion equations are nonlinear time-dependent PDEs, the normal procedure for solving the equations is using an implicit time-stepping scheme where each time-step is solved using the Newton method. Assembly of the Jacobian (a critical part of solving nonlinear PDEs) has become a significant portion of the total time because of the complexity of the equations which are solved. Once the Jacobian is assembled, a system of linear equations must be solved. The total number of unknowns in the linear system can be estimated by the number of nodes multiplied by the number of solution variables. Because the computational expense of linear solvers in general scale faster than linearly, for very big problems with a number of nodes and PDEs, linear solve times dominate the total time, and for small problems, the assembly dominates. Another consideration for large problems is memory consumption, though with the adoption of 64-bit CPUs, this is less of a problem than it used to be.

24.4.2 Kinetic Monte-Carlo Diffusion

Kinetic Monte-Carlo diffusion, unlike the continuum method, does not place restrictions on the complexity of the physical model. Kinetic Monte-Carlo tracks the positions defects, dopants and clusters (collectively referred to as "particles") of all particles within the simulation domain. Diffusion is simulated by particles hopping from lattice site to lattice site in random directions. If diffusing particles come close to dopants or clusters, they can react to form a new species or increase the size of a cluster. Clusters can break up by losing an interstitial, and interstitials can recombine at the surface or with vacancies. All of the possible events are put into a large event table weighted with their relative event rates, and the next event is chosen with a random number. Keeping track of all available species, their locations, and their possible reactions to form new species is simply a book-keeping exercise and costs essentially nothing in terms of performance but can require a significant amount of memory. The most expensive step in kinetic Monte-Carlo is determining the neighbors after a diffusing species hops. In order to speed up the computation, a regular grid is usually created which stores the list of species in each grid cell. This reduces the list of neighbor candidates considerably, but can be very costly in terms of memory usage.

The main cause of the poor performance of kinetic Monte-Carlo with respect to continuum diffusion is the large number of fast-diffusing species. Right after an implant, there is an enormous amount of damage in the form of point defects. The average time step per event is extremely small, perhaps of the order of picoseconds. The total reaction rate of the system is given by $R = \sum_i r_i n_i$ where r_i is the rate of event type i and n_i is current multiplicity of event i. The Monte-Carlo time step is proportional to $1/R$, which and R is very large just after an implant because of the very large number of possible diffusion events. By the end of a simulation, there are far fewer diffusing species because of various re-combination events, so the time step increases dramatically. However, once the number of diffusing species becomes very small, one can also run into finite size effects in the simulation. One important finite size effect comes from the handling of the interstitials, which in a very large simulation box would form a very smooth, deep profile, and in the active part of the device would result in a slowly decreasing concentration. This effect is not easy to reproduce accurately with a small simulation domain.

As the size of devices shrink, the cost of kinetic Monte-Carlo is reduced because there are fewer species and the number of cells required for neighbor look-up is reduced. Considering the shrinking device dimension trend, the complexity of kinetic reactions which can be included using kinetic Monte-Carlo, and the fact that kinetic Monte-Carlo does not depend on a high-quality mesh, it is likely that at some point and for some problems, kinetic Monte-Carlo will become the diffusion module of choice.

One way to take advantage of both methods even today is to use the full kinetic Monte-Carlo for a very short initial time period. Essentially, the kinetic Monte-Carlo would serve to compute the starting conditions for the continuum method for diffusion. Without going into too many details, there is not a good understanding of the precise starting conditions after an implant. So, for example, the proportion of dopants which end up in interstitial as opposed to substitutional sites after an implant is not known. Similarly, dopant, dopant–defect, and interstitial clusters form very quickly after diffusion starts and without solving for hundreds of species, it may be difficult to know which are the dominant species at the beginning of the anneal. So one way to get the best of both techniques may be to start with kinetic Monte-Carlo and then finish the simulation using FD/FE method with a simpler physical model.

24.5 Oxidation

Thermal oxidation of silicon forms a very high-quality dielectric, SiO_2, and is perhaps the most compelling reason that Si is the dominant material used for electronics. Very thin defect-free layers of SiO_2 are used to insulate the gate from the channel in CMOS devices. Recently, small amounts of nitrogen have been added to SiO_2-forming oxynitrides to improve reliability and reduce unwanted diffusion of boron from the gate to the channel. In addition to its use as a gate dielectric, oxidation is also used to line isolation trenches between devices, which are subsequently filled with a deposited oxide such as tetraethyl orthosilicate (TEOS).

The simulation of oxidation has two functions — one is the formation and growth of SiO_2 when Si is exposed to an oxidizing ambient, and the other is the re-distribution of dopants due to diffusion and due to the growth and flow of the oxide layer. After an initial native layer of oxide is formed on the silicon surface, the subsequent growth of the oxide layer occurs at the Si–SiO_2 interface and is caused by oxidant(s) diffusing through SiO_2 and reacting at the Si surface. Oxidants can be O_2, H_2O, O, or compounds involving N, to create oxynitrides. Figure 24.6 shows the simulation of a poly re-oxidation step. The growth rate of the SiO_2 is known to be affected by the type(s) and concentration(s) of oxidant(s), stress, and dopant concentration. In addition, the oxidation rate depends on the orientation of the oxidizing surface as is shown in Figure 24.7. One important aspect of oxidation is that SiO_2 is not naturally as tightly packed as Si, which causes compression in SiO_2 laterally and tension in Si near the interface. The lateral compression of the SiO_2 in turn causes expansion in the vertical direction through Poisson's ratio. This expansion causes motion of the entire SiO_2 layer so that both the Si–SiO_2 and the SiO_2–gas surfaces are in motion. If there are any exposed polysilicon layers, those will behave similarly to silicon. Because oxidation is always solved using FD/FE methods and moving boundaries, this boundary movement creates meshing challenges which are particularly difficult to overcome in a full 3-D simulation.

In addition to re-distributing dopants during inert thermal ramps, diffusion modules must also take into account several important effects during oxidation. The most important of those effects is oxidation-enhanced diffusion (OED). The conversion of Si to SiO_2 introduces interstitials at the Si–SiO_2 interface which then contribute to enhanced diffusion of interstitial diffusing dopants. The conversion of silicon to oxide is not by itself considered to affect the dopants, so for each step part of the dopant profile in silicon must be transferred to oxide, but at the same time silicon/oxide segregation must also be solved. In general, dopants favor either silicon or oxide — most likely as a result of chemical effects. In addition, it is critical to maintain the dose of the dopants locally. Therefore global remeshing, which is normally used to produce a clean mesh but introduces interpolation errors, must be minimized as much as possible.

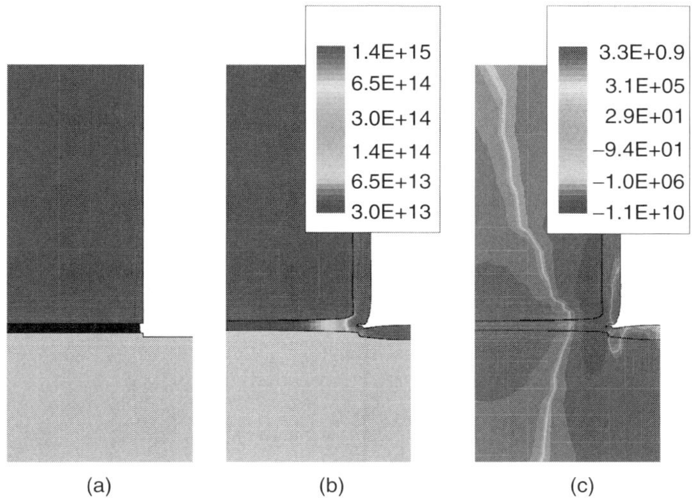

FIGURE 24.6 **(See color insert following page 15–4.)** An oxidation simulation showing oxidant diffusion, Si consumption, and SiO_2 expansion. (a) Light pink is silicon, brown is oxide, and magenta is polysilicon; (b) Same simulation as in (a), but showing oxidant concentration in the oxide. Because the oxidant concentration does not reach far underneath the gate, oxidation occurs mostly at the edge of the poly gate; (c) Same simulation as in (a), (b) but the shading shows the component of stress in the vertical direction. Stress, which is concentrated in the region where oxidation happens, is a by-product of the consumption of silicon by the oxidant-forming less-dense SiO_2.

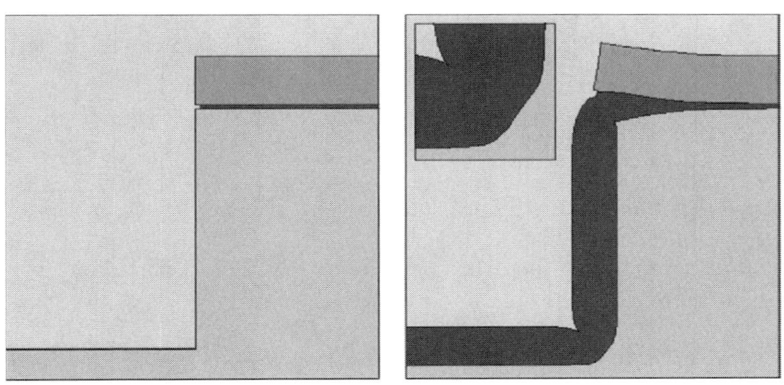

FIGURE 24.7 Simulation of trench oxidation showing orientation-dependent oxidation of a trench. Oxidation rate is assumed to be slowest along the ⟨111⟩ direction in silicon, resulting in a facet at the bottom of the trench. The wafer orientation is ⟨100⟩ and devices are normally oriented along ⟨110⟩.

24.6 Etch and Deposition

Traditionally, TCAD has focused on front-end fabrication simulation ending with source/drain implant and anneal steps. So it has not been necessary to include a full physical etch and deposition module. The main goal of TCAD simulations is ultimately the device electrical characteristics and there are only a few device dimensions which have a significant impact on the electrical characteristics including, for example, gate oxide thickness, gate width and length, shape of the poly gate at the bottom, and spacer width. For these critical parts of the device, simple geometrical etch and deposition models have been sufficient. Additionally, process engineers developed very good etch and deposition techniques, creating fairly simple shapes. As device dimensions shrink and more complex techniques are required to realize structures, it is becoming increasingly important especially in the gate and spacer formation to take into account physical effects in etching, deposition, and even resist topography. While this has traditionally been outside the scope of process simulation, it is useful in the context of this chapter to point out the mechanisms which lead to nonidealistic device structures and techniques which have been developed to simulate those effects.

24.6.1 Etch

For the purposes of this chapter it is sufficient to focus on two main etch types which are dominant in the industry, dry etching or plasma related etching (including reactive ion etching), and wet chemical etching. The two types of etch produce different shapes. Dry etching tends to be highly directional and is used to produce nearly square bottom holes, whereas wet chemical etching tends to be more isotropic producing rounded bottom holes and under etching (see Figure 24.8).

Reactive ion etching involves creation of a plasma of reactants and biasing of a sample in a way that ions from the plasma are incident on the sample surface. A reaction occurs at the wafer surface and the waste products are pumped away. Since the incident ions are driven by a mostly vertical electric field, the etch rate is mostly vertical and a vertical directional etching results. Material selectivity is enhanced by the use of reactive ions, which preferentially react with one material over another. The make-up of the plasma can contain reactive species and inert species. Reactive species give higher etch rates and better selectivity but tend to produce isotropic profiles, whereas inert species have lower etch-rate selectivity and produce more directional profiles. It is also possible to combine inert with reactive species to obtain a compromise.

Accurate physical simulation of these processes can be very challenging. A full simulation would include the following: ionization reactions in the plasma, the interaction of the plasma with the wafer, trajectories and energies of incoming species, surface reactions for the reactive species, surface diffusion of adsorbed reactive species, re-emission of inert, reacted, or unreacted species back into the gas (which can land at other surface sites if there is direct visibility), and finally desorption of reacted species back into the gas. So in addition to many chemical reactions, a set of coupled integral equations must be solved which compute the re-emitted flux coming from other surface elements, which are in the line of sight of the current surface element (see Figure 24.9). After these computations are performed, the etch rate as a function of position on the surface is computed, and finally the boundary is moved.

24.6.2 Deposition

Depending on the material to be deposited, there are several deposition techniques currently used which are mostly variants of chemical vapor deposition (CVD) or sputtering. These two classes of deposition

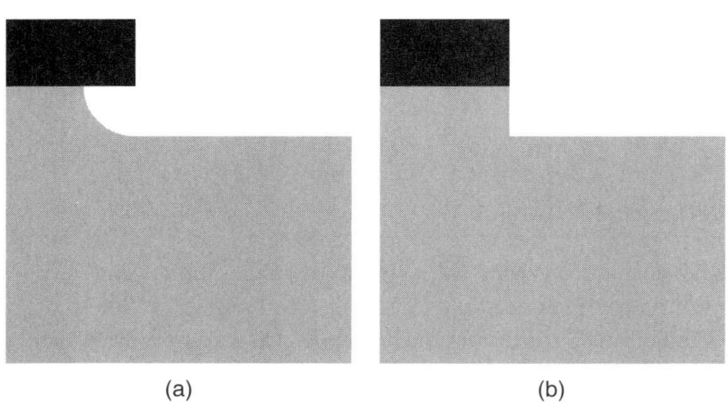

FIGURE 24.8 Basic etch types; (a) "Wet" etch normally is simulated with isotropic etching which causes material to be removed under masks. The dark material is assumed to be an etch mask; (b) "Dry" etching (reactive ion etching or plasma etching) is normally approximated with what is labeled anisotropic etch or sometimes called directional etch.

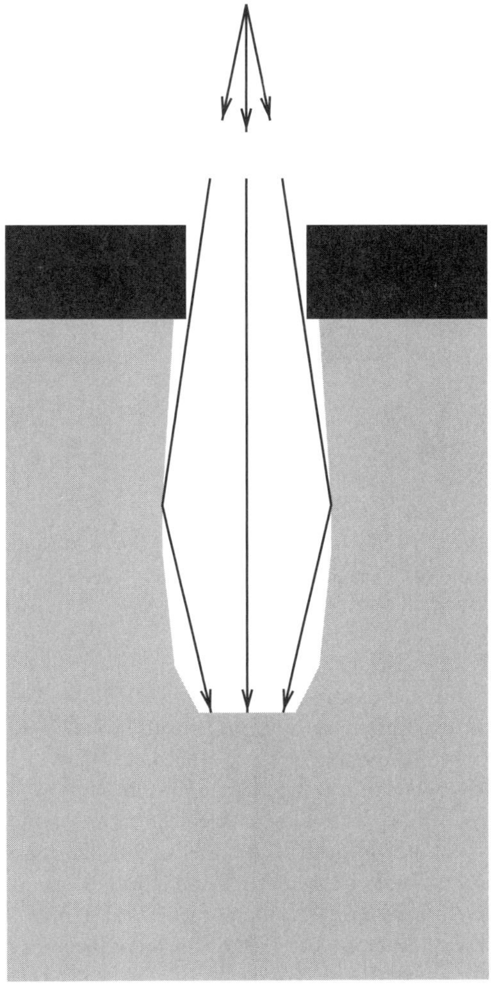

FIGURE 24.9 Schematic figure illustrating geometrical effects leading to more realistic shape. In a real dry-etch process, the incoming beam has a slight spread (exaggerated here for clarity). The beam-spread can create bowed side walls and rounded trench bottoms. The sides of the hole are only exposed to a part of the incoming beam, thus slowing the etch rate on the sides compared with the middle of the hole. In addition, the etchant flux is incident at a glancing angle on the side walls, increasing the chance of reflection re-directing part of the flux to the bottom of the hole.

processes, isotropic and sputter deposition, generate different layer topography, and most deposit steps are handled with one of these two types of deposition modules. Figure 24.10 shows a schematic of the two deposition types. A description of some of the physical effects responsible for the deposited geometries is given below.

For deposition of oxide, nitride, polysilicon, or silicon, it is common to use low-pressure CVD (LPCVD) or plasma-enhanced CVD (PECVD). In general, the following set of steps describes CVD: adsorption of a gas molecule, possible surface diffusion of the adsorbed species to find a favorable site (such as kinks in atomic layer steps on the surface), a reaction taking place which is either temperature- or plasma-assisted, the reaction depositing the material and releasing reacted species, and finally desorption of reacted species. Depending on the conditions in the reactor, the deposition can occur isotropically or anisotropically. In addition to the possible directionality of the incoming beam, many other processes can affect the resulting topography including re-deposition, surface diffusion, and changes in deposition

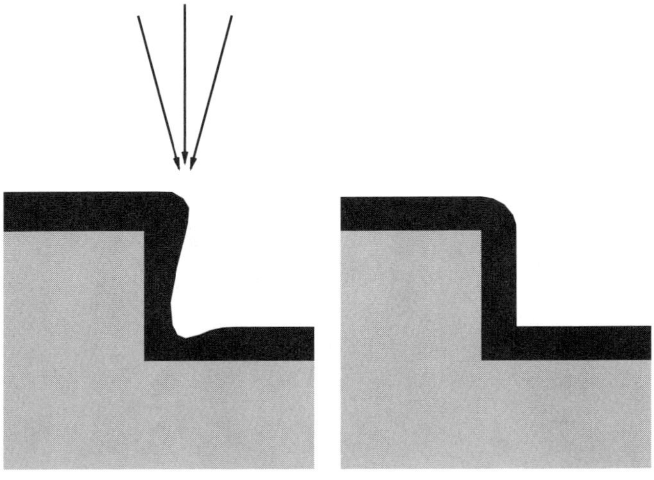

FIGURE 24.10 Schematic figure of (a) sputter deposition vs. (b) isotropic deposition. In sputter deposition, the directionality of the incoming flux leads to irregular shapes near steep surface features. The bottom of such features can be shadowed by an overhang created at the top.

conditions (partial pressures of mixed gases) inside deep trenches. TCAD simulators have traditionally stopped short of modeling gas or plasma reactions but instead have offered a set of geometrically based models which can effectively model directional, re-deposition, and surface diffusion.

For metal over-layers, sputter deposition is the most common method. With this technique, electrons incident on a target of the desired deposition metal kick out atoms, which are then deposited on the wafer. This technique results in an atomic beam which has an angular distribution. The directionality of the atomic beam can cause an instability in the deposited shape. For example, in Figure 24.10(a), deposition takes place near a step in the surface. At the top corner of the step, deposition can occur on the side wall which is exposed to the incoming flux, whereas the bottom of the step is first partially shadowed and then fully shadowed as deposition progresses. This results in a shape like the dashed line shown. For comparison, the right side of Figure 24.10 shows isotropic deposition, which can result from CVD-type deposition where all areas of the surface have a uniform growth rate because the deposition rate depends on the gas concentration which is uniform everywhere along the surface.

24.6.3 Reflow and Surface Diffusion

Doped oxides such as TEOS or silane-oxygen doped with phosphorus (PSG), boron (BSG), or a combination of both (BPSG), at moderate to high temperatures become soft and flow like a viscous liquid. This process is most often associated with back-end processing and can be used to fill in trenches and smooth deposited features (see Figure 24.11 for a schematic illustration). PSG re-flow occurs at a fairly high temperature (>950°C) and can cause significant dopant re-distribution, whereas for BPSG the re-flow temperatures are in the range of 400 to 450°C, which would not cause dopant re-distribution. Normally, the shape of the deposited doped oxide layer is not important because it is followed by a chemical mechanical polishing (CMP) step which flattens the top surface. So most often, an inert anneal step following a trench fill would be sufficient to model doped oxide deposition with re-flow.

High-temperature deposition can be used to smooth out instabilities caused by a directional beam because of surface diffusion effects. Topography changes due to surface diffusion and re-flow, to a certain extent, are driven by a minimization of surface energy. Surface energy is proportional to the surface curvature: convex corners have a high surface energy and concave corners have a low surface energy. So, when

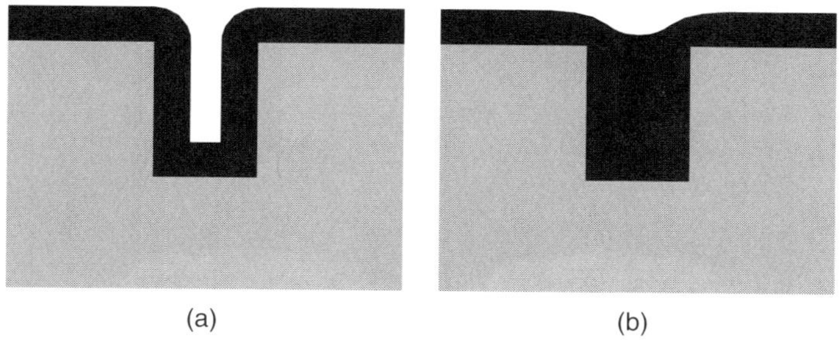

FIGURE 24.11 Schematic figure of (a) re-flow and (b) surface diffusion. Both re-flow and surface diffusion are driven by the surface free energy (flow is from high-energy to low-energy regions). Convex corners have a high energy and concave corners have a low energy.

doped oxides are heated, re-flow will cause material to flow from convex to concave regions, which tend to smooth films and fill vias and holes. Surface diffusion normally only occurs on semiconductor and metal surfaces at higher temperatures than re-flow of TEOS and only in high vacuum, but the simulation technique is pretty much the same. Simulating re-flow or surface diffusion can be accomplished by computing and applying a normal speed function where the growth/removal rate (assumed to be normal to the surface) is a function of the local surface curvature.

24.6.4 Re-Deposition

Re-deposition occurs when a deposition molecule scatters from a part of a surface and is adsorbed or deposited somewhere else (see Figure 24.12). Re-deposition can be modeled in a geometric way by using the concept of a sticking coefficient. Each location on the growing surface is exposed to an incoming flux, which depends on the visibility if the deposition is directional. The rate at which each individual surface element moves depends on the direct incoming flux plus the indirect flux coming from other parts of the surface, multiplied by the sticking coefficient. In order to account for multiple reflections, an integral equation must be solved which couples every surface element to every other surface element (some coupling may be discarded if the pair of surface elements do not see each other).

24.6.5 Epitaxy

The growth of silicon epitaxial films is similar to CVD except that the temperature is higher, and the growth rate lower. The low deposition rates allow incoming species to find favorable growth sites such as kink sites, leading to atomic layer-by-layer growth. Epitaxy is often done to prepare the starting wafer with a very flat contaminant-free surface, as well as to control the wafer doping. There are various other applications as well. Recently, SiGe epitaxy has been used in the source and drain regions to introduce stress in the channel, thereby increasing electron mobility. Epitaxy most often occurs at relatively high temperature compared with CVD of dielectric films. Often, dopants are introduced during epitaxy or it may be that dopants are already present in the structure when epitaxy is performed. For those cases, it is necessary to compute the simultaneous diffusion of dopants during the deposition process. There are two possible techniques to accomplish this. One technique involves breaking up the deposition into smaller alternating diffusion and deposition steps; the other solves a moving boundary problem like oxidation where a dopant flux is introduced into the growing boundary nodes and provisions are made to account for the change in neighboring element size and the introduction of a dopant flux.

24.6.6 Selective Deposition

Because deposition occurs as a reaction with a gas species, it is possible to tailor the gas in such a way that it reacts with only one of the materials present on the surface of the wafer. For example, silicon tetrachloride

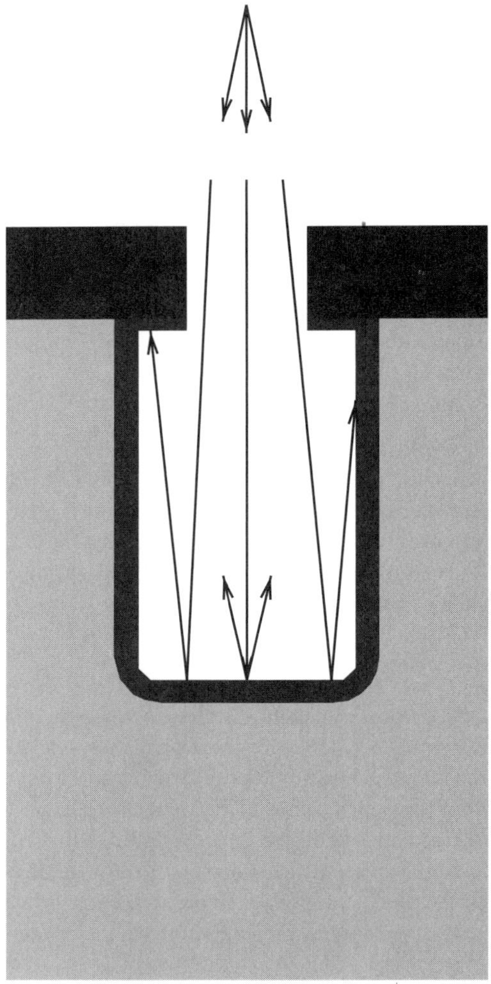

FIGURE 24.12 Schematic showing re-deposition. Some deposition flux may be reflected to be re-deposited elsewhere in the structure.

($SiCl_4$) will deposit polysilicon on silicon at temperatures above 800°C and will grow a layer of Si on Si without depositing on SiO_2 or Si_3N_4 at temperatures above 1100°C (HCl may be added to improve selectivity). Other gas mixtures have also been used to reduce the required deposition temperature. This type of deposition can be used to explore novel devices or other specific uses. Simulation of this type of deposition is fairly easy. The only concern is how to handle the edges where the active material meets an inert material. Typically, it is handled as an isotropic material-specific deposition. And similarly to standard epitaxy, if dopants are involved, diffusion must be solved simultaneously with deposition.

24.6.7 Techniques for Modeling Etch and Deposition

We begin by discussing the two main methods which have been employed to compute the layer topography and topology.

24.6.7.1 Levelset Method

Levelset methods[10] solve equations that describe the evolution of surface motion given a normal speed function. Instead of solving potentially complex discrete-topology evolution problems, the levelset equation resolves the motion of a front in a continuous manner. Discretized levelset equations can be solved using FE/FV methods. Because of this, levelset methods handle changes in topology naturally and reliably. The normal speed function of the evolving surface can be quite general. It can be a function of the local surface orientation, curvature or visibility, or can be computed with an integral along the front. Another possibility is to import a speed function from an external simulator. This allows a straightforward way to couple physical etch and deposition tools to process simulators. The current simulation boundaries can be passed to an etch and deposit module which would compute the normal speed on the boundary points. This information can be passed back to the process simulation tool which would use the levelset method to move the boundaries, handle topology changes, update the mesh, and interpolate the previous values onto the current mesh.

The main drawback of the levelset methods is the difficulty in handling thin layers and sharp corners. Also, for a nontrivial normal speed function, accuracy of the evolving front may be a concern making it necessary to maintain a proper mesh density to solve the levelset function. In this case there is a trade-off between accuracy and etch, and deposit computation cost both in terms of memory use and execution speed. Discontinuities in the speed function and sharp corners have to be handled with care. Discontinuities in the speed function come from the visibility — meaning that a surface segment is either exposed to the beam or is shadowed by some other part of the surface. Sharp corners in general cannot be handled exactly — normally a specified geometrical tolerance is used to maintain the accuracy of the boundary evolution. The fundamental reason for these limitations is that the levelset function which defines the boundary location is assumed to be a continuous function (see Figure 24.13). The other main difficulty with levelset methods is in handling thin layers for reasons similar to the corner problem. In the case of thin regions, mesh edges which cross the region entirely without a node inside the region will miss the two levelset crossing inside, causing a hole in the thin layer (see Figure 24.13). Even if the layer is recognized, an accurate solution needs at least 4 to 5 points across the layer, which can be very computationally expensive.

24.6.7.2 String Algorithms

There are many variations of string algorithms and other analytic techniques to move boundaries. They all involve breaking up a surface into a set of edges; the positions of the points are moved analytically and collisions must be detected and resolved. Therein lies one of the difficulties with these methods — the resolution of colliding boundaries can be very complicated. This method has not been very successful for 3-D simulation. However, for simple geometries and for 2-D simulation, string algorithms can perform better than levelset-based methods, because string algorithms are extremely accurate, can easily handle discontinuities in both the speed function and the surface itself, and are computationally inexpensive.

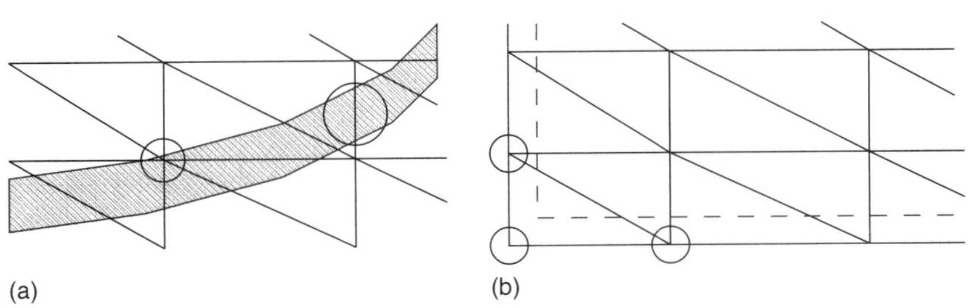

FIGURE 24.13 Difficulties with the levelset method; (a) A thin blue shaded region is formed during a deposition step. The levelset method defines the location of a surface as the contour of constant values of the levelset function which is a field defined on the mesh at each nodal point. This implies that there can only be one level crossing for each edge or the topology of the region will be incorrect. Inside the large circle, the double crossing will be missed whereas if there is mesh point inside the region, as with the small circle, then the topology will be correct; In (b), deposition is simulated starting with the solid blue thin line. However, all the edges between the circled mesh points will either have no level crossing or will have an undetectable double crossing, so the corner will be lost. Adaptive meshing can be used to solve both problems.

24.7 Lithography and Photoresist Modeling

Lithography/photoresist modeling is out of the scope of this chapter, but it is mentioned here for completeness. The shape of the resist will affect the final etch shape, and not just the width; but it may be important to the final structure if the resist shape has sloped and/or rough walls. A full litho/etch simulation would include photo-resist exposure, postexposure bake, and development — essentially all steps responsible for the final resist topography. In the exposure step, it is important to include diffraction effects, and allow for phase-shifting masks (PSM) and masks with optical proximity corrections (OPC). There has been some work done recently to transfer resist shapes from the lithography and resist simulators to the process simulators to get more accurate gate shapes.

24.8 Silicidation

The formation of silicides is simulated in a manner similar to oxidation. A metal is deposited on Si, then a native layer of the silicide is created between the metal and the Si, and finally the moving boundary problem is solved to move the Si–silicide and metal–silicide boundaries. In the silicide, two species can be present — silicon diffusing from the silicon side reacting at the metal, and metal atoms diffusing through the silicide and reacting at the Si–silicide boundary. Each reaction can either introduce compressive or tensile stresses in silicon, depending on the relative density of the silicide compared to silicon.

24.9 Mechanics Modeling

Stress and strain modeling have become an evermore important goal of process modeling. Not only is the stress state itself important as an estimate of the likelihood of dislocations which have a deleterious effect on yield, but also the stress and strain can affect dopant diffusion and be used to improve device performance[11] through changes to the silicon bandgap, and electron mobility. In fact, stress is now used as a tool to improve electrical characteristics in the latest devices produced by many major manufacturers.[12–14] Known sources of strain come from many sources: thermal mismatch, deposited stress, oxidation, lattice mismatch, densification, and high dose implantation. A more detailed discussion of various stress sources except oxidation and silicidation, which have already been discussed separately in Sections 24.5 and 24.8, respectively, will be covered below.

Thermal mismatch is modeled as a natural by-product of thermal expansion, which varies from material to material. This is normally accomplished by referencing all expansion coefficients to the silicon substrate. The over layers expand and contract relative to the substrate, introducing stress.

The deposition of Si_3N_4 is known to introduce stress. Thermal mismatch is at least partially responsible, and chemical reactions involving Si–H and N–H bonds are thought to account for the rest. Stress increases with film thickness up to a certain level, leading to the conclusion that the dominant mechanism may be similar to lattice mismatch stress. In addition, subsequent annealing of the film can cause further stress. Studies indicate a maximum stress which can be attained during deposition depending on the temperature which has been attributed to a visco-elastic behavior of the films.[15] Because of the lack of a good model for the stress state of the deposited film, normally an experimentally determined (user-defined) constant stress is applied to the deposited film and a subsequent relaxation of the whole system is performed to give an estimate for the deposited stress state. It is quite likely that because of the renewed interest in stress and strain modeling, commercial simulators will implement more sophisticated stress models in the future.

The source of strain in strained silicon and SiGe is lattice mismatch. Typically, both of these technologies start with a Si wafer on top of which is grown a relaxed SiGe buffer layer. The buffer layer growth creates dislocations due to the build-up of stress from the difference in lattice constants between SiGe and pure Si. Typically, the Ge mole fraction is varied in the buffer layer to control the density of dislocations and ensure that they are predominantly located near the bottom of the layer. At the top of the buffer layer, the stress is minimal, so the top of the buffer layer can be thought of as a starting substrate with a variable lattice constant. A pure Si or SiGe layer with a mole fraction different from the relaxed buffer layer is grown on top of the buffer layer, so a tensile layer of Si or a tensile or compressive layer of SiGe can be produced in this manner. Because there are no good analytic models which can accurately reproduce the stress state during the buffer layer growth, and because the top part of the buffer layer is mostly or completely relaxed, it is normally sufficient to treat the lattice mismatch by assigning the lattice constant to be fixed at some point near the top of the buffer layer, and apply an external strain proportional to the concentration of Ge (see Figure 24.14).

Densification is the process where the density of a material can increase during annealing. The deposited dielectric TEOS undergoes densification due to the evaporation of residual organics from the deposition and due to chemical changes in the material induced by annealing. These reactions cause the natural density of the material to increase, normally creating a tensile stress in the material because the region boundaries are bonded to neighboring regions.

Implantation has been shown to introduce compressive stress in Si of the order of 10^8 (N/m^2).[16] Good models for the evolution of stress coming from implantation are not available currently. It seems likely even without a good model that the stress is in large part due to the Frenkel pair concentration generated during implantation, and in amorphous regions volume increase of approximately 6% have been observed leading to very large stresses. The recombination of the Frenkel pairs happens very quickly compared to dopant diffusion, so it could be that stress due to implantation is greatly reduced before significant dopant diffusion takes place. Despite this, it may still be important for extremely fast ramp rates and short anneal times which are being increasingly utilized. Stress coming from high concentrations of dopants could be significant, especially if coupled with stress-dependent diffusion models.

A final note about material properties: silicon is normally modeled as an elastic material, but this is not necessarily a good approximation if dislocations are to be considered. Dislocations can relax stresses in silicon but they allow leakage current to flow and can potentially short the transistor. However, if they are carefully controlled, as is the case with SiGe-relaxed substrates, their existence can be tolerated. SiO_2 is modeled as a visco-elastic material which can flow at high temperatures but remains elastic at low temperatures. Si_3N_4 is normally modeled as a stiff elastic material, and polysilicon is elastic like silicon. These properties influence the evolution of stress and strain during the simulation.

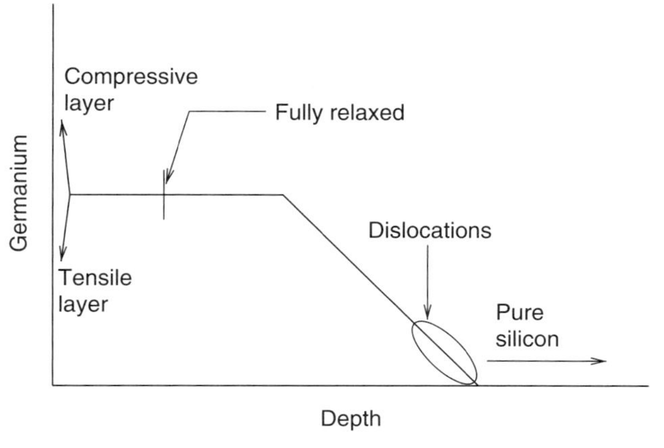

FIGURE 24.14 Schematic figure of graded buffer layer. SiGe buffer layers allow for a buffer layer with a customized lattice constant between that of Si and that of SiGe (approximately 4% larger than Si lattice constant). The formation of the buffer layer uses a grading of Ge concentration, which helps to trap dislocations deep in the graded region. Because there are no known analytic models for handling dislocations and stress relaxation, it is generally assumed in process simulation that a relaxed buffer layer of given Ge concentration can be produced. The user chooses a relaxed depth, where the lattice constant is fixed. Above the relaxed depth, no dislocations are assumed to be formed, so the strain is a simple linear function of Ge concentration.

24.10 Putting It All Together

There are a number of simulation strategies which are commonly used to improve simulation accuracy and performance. A very important capability for a simulator is to be able to easily change the number of spatial dimensions of the simulation (see Figure 24.15). Many initial process steps such as epitaxy, well implants, and gate implants and subsequent anneal steps can be performed in 1-D. Some shallow trench isolation (STI) trench formation and liner oxide steps can be performed in 2-D. Each increase in dimension will cost ten times or more in CPU and memory, so staying in the lowest possible spatial dimension for as long as possible will reduce simulation time without hurting accuracy. Many structures such as CMOS devices have one- or two-fold symmetry. Simulating all process steps with either half structures in 2-D or one-fourth structures in 3-D and only building the full structure as a last step not only can cut CPU and memory by their respective fractions, but will also help improve device simulation accuracy because the mesh will reflect the symmetry of the device.

Another technique which has been employed to assist in difficulties of producing 3-D meshes is to produce one composite structure at the beginning of the simulation which contains all intermediate boundaries.[7] Once this structure is meshed, etch and deposition steps can be mimicked by changing the material of a given region (deposition would be changing a gas region to a material, and etching the reverse). This technique eliminates interpolation error from changing meshes and reduces the simulation time by reducing many mesh operations, but it cannot be used to solve any moving boundary problems such as oxidation or silicidation.

Transferring information from process simulation to device simulation requires a structure and associated data fields for active dopants and stress. Often the mesh must be recreated so it can be tailored to device simulation, and hence the process data are then interpolated onto the new mesh. For device simulation, it is necessary to create contacts in the mesh where potentials will be applied during device simulation. The contacts are simply a collection of surface elements and a name is designated for each contact.

More sophisticated device modeling has led to more information transfer. At first, all that was needed was net doping (donors–acceptors) and total doping (sum of all dopants) for mobility degradation due to scattering. If partial ionization of dopants is to be considered, then the concentration of each dopant must be sent separately. In addition, for 90-nm node and below, the deliberate introduction of stress is used to increase drive currents. Advances in mobility models and band structure models depending on

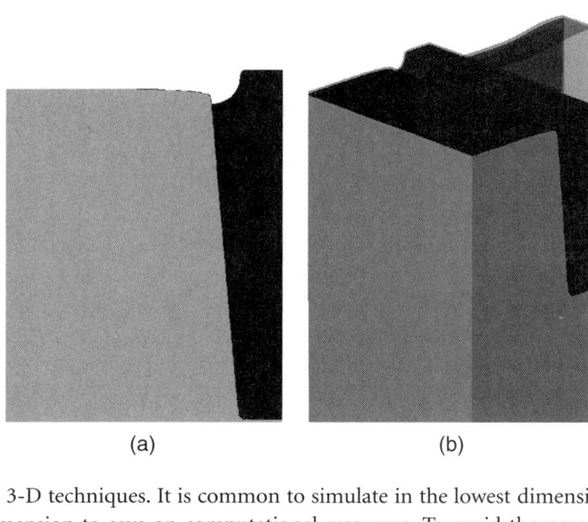

FIGURE 24.15 2-D and 3-D techniques. It is common to simulate in the lowest dimensionality until it is necessary to extrude to a higher dimension to save on computational resources. To avoid the necessity of creating many 3-D meshes, it is highly beneficial to create a structure containing all intermediate boundaries mesh the structure once. After this the structure that is meshed during the simulation regions only need to be turned "on" and "off" to reproduce all intermediate structures. This technique minimizes 3-D meshing difficulties and interpolation, but cannot simulate reaction/diffusion situations such as oxidation where the boundaries move at the same time the dopants are diffusing.

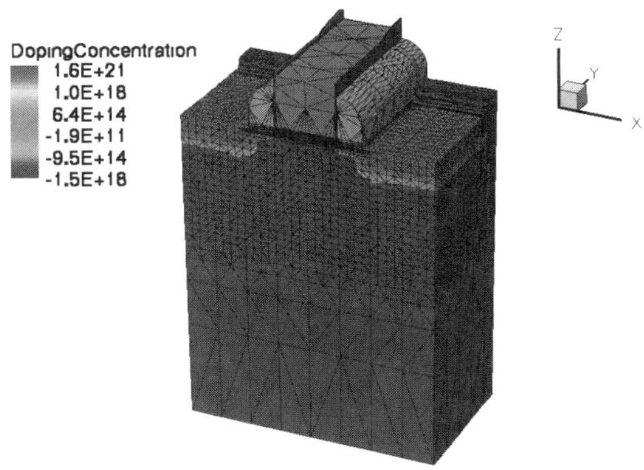

FIGURE 24.16 **(See color figure following page 15–4.)** Final 3-D structure ready for device simulation. The half structure is reflected and contacts are added (outlined in purple on the top). The mesh lines are shown, and the doping concentration is shown in a rainbow, shaded red for n type and blue for p type.

the full strain/stress tensor and is now transferred from process simulators directly to device simulators. Figure 24.16 shows an example of a structure created by combination of Synopsys process simulation and mesh generation tools which is ready for device simulation.

24.11 Conclusions

Process simulation is a very powerful tool to investigate next-generation device manufacturing strategies and problems. In combination with device simulation, investigations of the benefits and trade-off with novel device structures can be explored at relatively low cost. In addition, parametric yield issues can also be probed at a fraction of the time and cost of a purely experimental approach. These trends will likely give rise to increased use and reliance on TCAD in future process development and yield improvement.

Acknowledgments

I would like to thank Synopsys for support in preparing this document, and Victor Moroz and Vinay Rao for helping me review the contents. I would also like to thank Beat Sahli for contributing Figure 24.2.

References

[1] D.A. Antoniadis, S.E. Hansen, R.W. Dutton, and A.G. Gonzalez, SUPREM I-A Program for IC Process Modeling and Simulation, Stanford Technical report, No. 5019-2, 1978.

[2] S. Selberher, *Analysis and Simulation of Semiconductor Devices*, Springer, New York, 1984 and O.C. Zienkiewicz and R.L. Taylor, *The Finite Element Method*, McGraw-Hill, London, 1991.

[3] Synopsys, *Taurus-Process Reference Manual Release 2004.09*, Mountain View, CA, 2004.

[4] W.K. Hofker, H.W. Werner, D.P. Oosthoek, and N.J. Koeman, *Taurus-Process Reference Manual Release 2004.09*, *Appl. Phys.*, 4, 125, 1974.

[5] A nice review of the topic can be found in P.A. Stolk et al., Physical mechanisms of transient enhanced dopant diffusion in ion-implanted silicon, *J. Appl. Phys.*, 81, 6031, 1997.

[6] M. Uematsu, Simulation of high concentration phosphorus diffusion in silicon taking into account Ostwald Repening of defects, *Jpn. J. Appl. Phys.*, 38, 6188, 1999; and P.H. Keys, R. Brindos, V. Krishnamoorthy, M. Puga-Lambers, K.S. Jones, and M.E. Law, Phosphorus/silicon interstitial annealing after ion implantation, *Mat. Res. Soc. Symp. Proc.*, 610, B6.6.1-6, 2000.

[7] Synopsys, *FLOOPS-ISE Release10.0 Manual*, Mountain View, CA, 2004.

[8] S.T. Dunham, A quantitative model for the coupled diffusion of phosphorus and point defects in silicon, *J. Electrochem. Soc.*, 139, 2628, 1992.

[9] P. Pichler, *Intrinsic Point Defects, Impurities, and Their Diffusion in Silicon*, Springer, Wien, 2004 and references therein.

[10] A review of the topic can be found in J.A. Sethian, *Level Set Methods and Fast Marching Methods*, Cambridge University Press, New York, 2002.

[11] Rim, J.L. Hoyt, and J.F. Gibbons, Fabrication and analysis of deep submicron strained-Si N-MOSFET's, *IEEE Trans. Electron Devices*, 47, 1406–1415, 2000.

[12] T. Ghani, M. Armstrong, C. Auth et al., A 90 nm high volume manufacturing logic technology featuring novel 45 nm gate length strained silicon CMOS transistors, *IEDM Technical Digest*, Washington, D.C., USA, December 7–10, 2003.

[13] H. Shang, J. Chu, S. Bedell, E.P. Gusev, P. Jamison, Y. Zhang, J. Ott, M. Copel, D. Sadana, K. Guarini, and M. Ieong, Selectively formed high mobility strained Ge PMOSFETs for high performance CMOS, *IEDM Technical Digest*, San Francisco, CA, December 13–15, 2004.

[14] S. Takagi, T. Mizuno, T. Tezuka et al., Channel structure design, fabrication and carrier transport properties of strained-Si/SiGe-on-insulator (Strained-SOI) MOSFETs, *IEDM Technical Digest*, Washington, D.C., USA, December 7–10, 2003.

[15] A.G. Noskov, E.B. Gorokhov, G.A. Sokolova, E.M. Trukhanov, and S.I. Stenin, Correlation between stress and structure in chemically vapour deposited silicon nitride films, *Thin Solid Films*, 162, 129, 1988.

[16] C.A. Volkert, *Stress and Plastic Flow in Silicon during Amorphization by Ion Bombardment, J. Appl. Phys.*, 70, 3521, 1991.

25

Device Modeling — From Physics to Electrical Parameter Extraction

Robert W. Dutton
Stanford University
Palo Alto, California

Chang-Hoon Choi
Stanford University
Palo Alto, California

Edwin C. Kan
Sanford University
Palo Alto, California

25.1 Introduction

Technology files and design rules are essential building blocks of the IC design process. Their accuracy and robustness over process technology, its variability and the operating conditions of the IC — environmental, parasitic interactions and testing, including adverse conditions such as electrostatic discharge (ESD) — are critical in determining performance, yield and reliability. Development of these technology and design rule files involves an iterative process that spans across the boundaries of technology and device development, product design, and quality assurance. Modeling and simulation play a critical role in support of many aspects of this evolution process.

The goals of this chapter are to start from the physical description of integrated circuit devices, considering both the physical configuration and related device properties, and then to build the links between the broad range of physics and electrical behavior models that support circuit design. Physics-based modeling of devices, in distributed and lumped forms, is an essential part of the IC process development. It seeks to quantify the underlying understanding of the technology, and abstract that knowledge to the device design level, including extraction of the key parameters that support circuit design and statistical metrology. Although the emphasis of this chapter will be on metal-oxide-semiconductor (MOS) transistors — the workhorse of the IC industry — it is useful to overview briefly the development history of the modeling tools and methodology that has set the stage for the present state of the art.

The evolution of technology computer-aided design (TCAD) — the synergistic combination of process, device and circuit simulation, and modeling tools — has its roots in bipolar technology starting in the late 1960s, and the challenges of junction isolated, double- and triple-diffused transistors. These devices and technology were the basis of the first integrated circuits; nonetheless, many of the scaling issues and underlying physical effects are still integral to IC design, even after four decades of IC development. With these early generations of IC, process variability and parametric yield were an issue — a theme that will re-emerge as a controlling factor in future IC technology as well.

Process control issues — both for the intrinsic devices and all the associated parasitics — presented formidable challenges and mandated the development of a range of advanced physical models for process and device simulation. Starting in the late 1960s and into the 1970s, the modeling approaches exploited were dominantly one- and two-dimensional (2-D) simulators. While TCAD in these early generations showed exciting promise in addressing the physics-oriented challenges of bipolar technology, the superior scalability and power consumption of MOS technology revolutionized the IC industry. By the mid-1980s, CMOS became the dominant driver for integrated electronics. Nonetheless, these early TCAD developments [1,2] set the stage for their growth and broad deployment as an essential toolset that has leveraged technology development through the very-large and ultra-large-scale-integration (VLSI and ULSI) eras, which are now the mainstream.

IC development for more than a quarter century has been dominated by the MOS technology. In the 1970s and 1980s n-channel MOS (NMOS) was favored owing to speed and area advantages, coupled with technology limitations and concerns related to isolation, parasitic effects, and process complexity. During that era of NMOS-dominated LSI and the emergence of VLSI, the fundamental scaling laws of MOS technology were codified and broadly applied [3]. It was also during this period that TCAD reached maturity in terms of realizing robust process modeling (primarily one-dimensional) which then became an integral technology design tool, used universally across the industry [4]. At the same time device simulation, dominantly 2-D owing to the nature of MOS devices, became the workhorse of technologists in the design and scaling of devices [5,6]. The transition from NMOS to CMOS technology resulted in the necessity of tightly coupled and fully 2-D simulators for process and device simulations. This third generation of TCAD tools became critical to address the full complexity of twin-well CMOS technology (see Figure 25.3a), including issues of design rules and parasitic effects such as latch-up [7,8]. An abbreviated but prospective view of this period, through the mid-1980s, is given in [9], from the point of view of how TCAD tools were used in the design process [10].

Today the requirements for and use of TCAD cut across a very broad landscape of design automation issues, including many fundamental physical limits. At the core are still a host of process and device modeling challenges that support intrinsic device scaling and parasitic extraction. These applications include technology and design rule development, extraction of compact models and more generally design for manufacturability (DFM) [11]. The dominance of interconnects for giga-scale integration (transistor counts in O[billion]) and clocking frequencies in O(10 GHz) has mandated the development of tools and methodologies that embrace patterning by electromagnetic simulations — both for optical patterns, and electronic and optical interconnect performance modeling — as well as circuit-level modeling. This broad range of issues at the device and interconnect levels, including links to underlying patterning and processing technologies, is summarized in Figure 25.1, and provides a conceptual framework for the discussion that now follows.

Figure 25.1a depicts a hierarchy of process, device, and circuit levels of simulation tools. On each side of the boxes indicating modeling level are icons that schematically depict representative applications for TCAD. The left side gives emphasis to DFM issues such as shallow-trench isolation (STI), extra features required for phase-shift masking (PSM), and challenges for multilevel interconnects that include processing issues of chemical–mechanical polishing (CMP) and the need to consider electromagnetic effects using field solvers. The icons on the right side show the more traditional hierarchy of expected TCAD results and applications: complete process simulations of the intrinsic devices, predictions of drive current scaling, and extraction of technology files for the complete set of devices and parasitics.

Figure 25.2 again looks at TCAD capabilities but this time more in the context of design flow information and how this relates to the physical layers and modeling of the electronic design automation (EDA) world. Here the simulation levels of process and device modeling are considered as integral

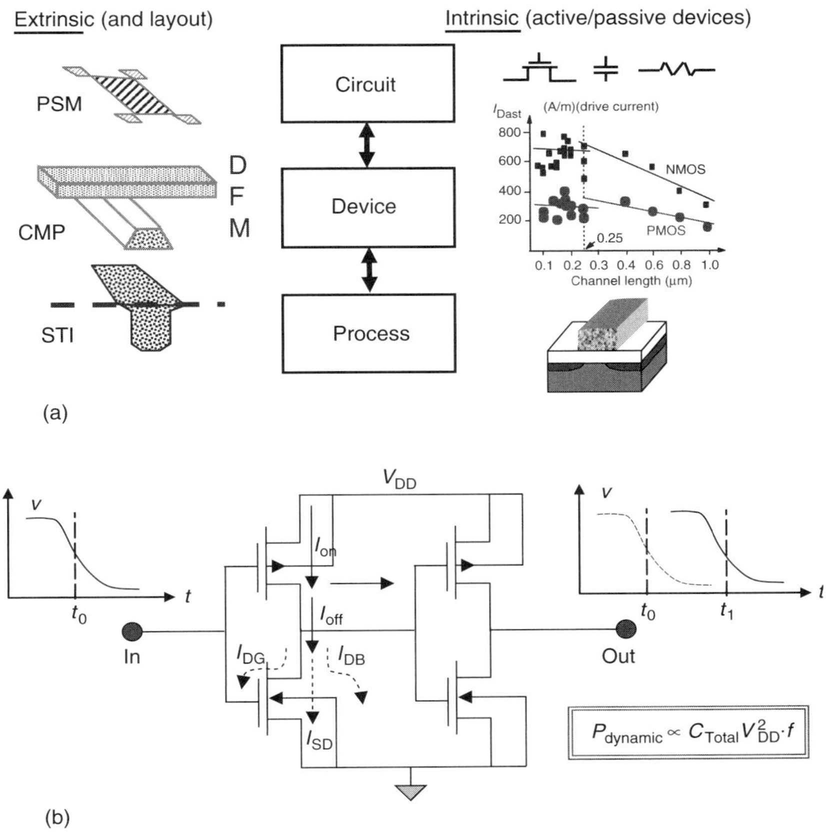

FIGURE 25.1 (a) Hierarchy of technology CAD tools building from the process level to circuits. Icons on the left show typical manufacturing issues; icons on the right reflect MOS scaling results based on TCAD. (b) Schematic view of circuit-oriented aspects of modeling that illustrate voltage-time behavior of inverters, components of "on" and "off" currents and dynamic power constraints.

capabilities (within TCAD) that together provide the "mapping" from mask-level information to the functional capabilities needed at the EDA level, such as compact models ("technology files") and even higher-level behavioral models. Also shown is the extraction and electrical rule checking (ERC) this indicates that many of the details that to date have been embedded in analytical formulations, may in fact also be linked to the deeper TCAD level in order to support the growing complexity of technology scaling.

25.2 MOS Technology and Intrinsic Device Modeling

The previous section has motivated the scope of device- and technology-level modeling with emphasis on TCAD as the toolset that provides details ranging from deep physical models of fabrication processes and device physics to the higher-level EDA interfaces, to electrical extraction — both for technology files and design rules in support of behavior-level models. Figure 25.1a points out the modeling hierarchy going from process technology to circuit modeling; the hierarchy has a dual "ladder" in terms of intrinsic and extrinsic devices. Chapter 24 considers the process modeling level of the hierarchy; Chapter 19 considers the statistical effects associated with the design process that have a profound effect on overall performance. The extrinsic devices and modeling of ICs will be considered in later sections of this chapter. This section will focus on considerations of the intrinsic MOS device, including supporting details of the variety of p–n junction effects that are inherent to the bulk CMOS processing technology.

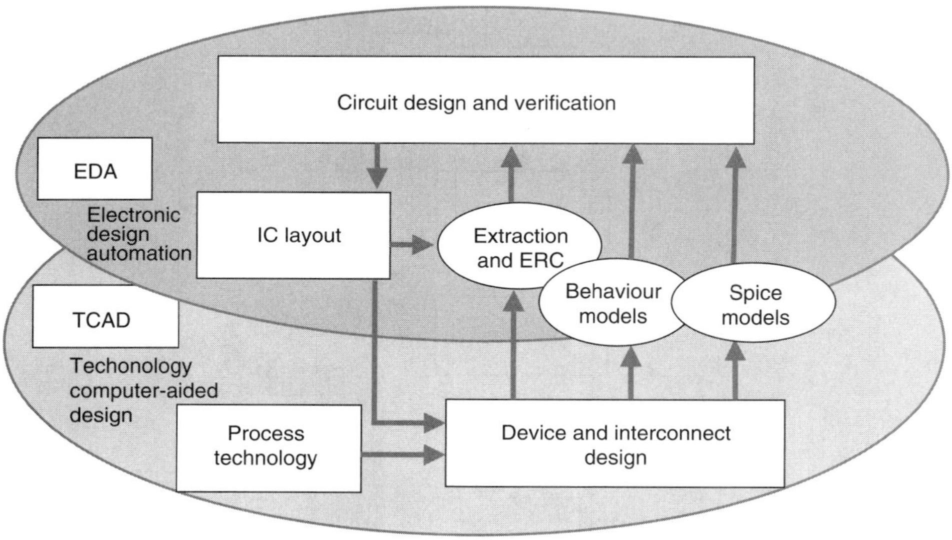

FIGURE 25.2 Information links between EDA and TCAD toolsets. Layout and process technology files support the TCAD flow which in turn outputs device (and interconnect) electrical data used for EDA models.

It is useful to briefly consider the objectives of device design, both from the system and technology perspectives, prior to going into greater detail at the physical modeling level. Figure 25.1b provides a simplified conceptual view of "the big picture." This figure shows two inverter stages and the resulting input–output voltage–time plot of the circuit. Clearly, from the digital systems point of view, the key parameters of interest are timing delays, switching power, leakage current, and cross-coupling ("crosstalk") with other blocks. The voltage levels and transition speed are also of concern. The figure also shows schematically the importance of I_{on} vs. I_{off}, which in turn is related to drive current (and mobility) for the "on" device and several leakage paths for the "off" devices. Not shown explicitly in Figure 25.1b are the capacitances — both intrinsic and parasitic — that affect dynamic performance. Moreover, the power scaling which is now a major driving force in the industry is reflected in the simplified equation shown in Figure 25.1b — critical parameters are capacitance, power supply, and clocking frequency. Key parameters that relate device behavior to system performance include the threshold voltage, driving current, and sub-threshold characteristics. It is the confluence of system performance issues with the underlying technology and device design variables that results in the ongoing scaling laws that we now codify as "Moore's law" [42]. With this backdrop of system performance metrics that relate to the technology perspective, we now begin the discussion of the physical (and physics-based) considerations of the MOS transistor.

The physics and modeling of integrated circuits can roughly be divided between the surface effects associated with MOS devices and a range of bulk junction effects that involve bipolar and associated leakage and parasitic effects. As mentioned in the introduction, CMOS transistors for ULSI now claim the lion's share of practical IC devices. In CMOS technology the bulk junction device effects are virtually all parasitic. In Section 25.4 some of the bipolar options, for example the use of BiCMOS for some power and RF applications will be considered. There are of course also issues of reliability engineering — for example, ESD protection circuits and devices — where substrate and parasitic devices are of pivotal importance. These effects and modeling are not considered here; the reader is referred to several excellent monographs in the area of ESD and I/O modeling [12–14]. The following discussion will use CMOS devices as the driver for discussing the physical modeling issues. Figure 25.3a shows a low-resolution cross-sectional version of a basic CMOS technology. Figure 25.3b and the subsequent discussions will focus on the n-channel MOS technology using a twin-well process and STI on a lightly doped substrate.

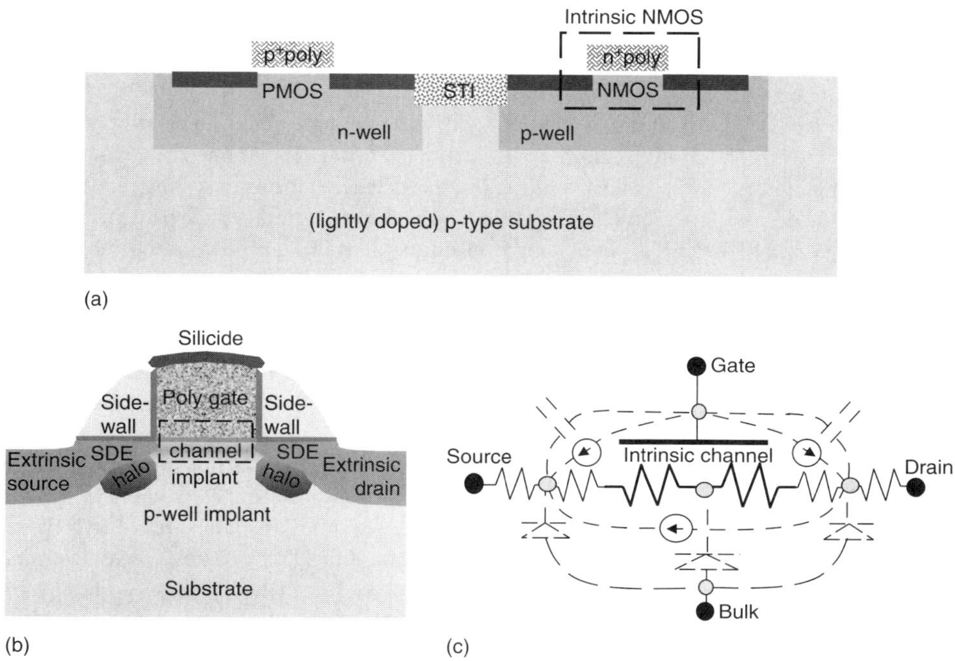

FIGURE 25.3 (**See color insert following page 15–4.**) (a) Simplified cross-section of a twin-well CMOS process using a lightly doped p-type substrate p and n-channel MOS transistors are shown along with STI isolation. (b) Cross-section of the n-channel MOS transistor, including details of gate, sidewalls and substrate doping profiles (SDE, channel/well implants, halo doping, etc.). (c) Schematic view of MOS transistor model that reflects intrinsic channel field effects, bulk node interactions with the channel ("back gate"), S/D diodes and various R,C, and leakage paths.

25.2.1 Intrinsic MOS Transistor

The intrinsic MOS device relies on gate control to capacitively induce inversion charge in the channel between source and drain regions. Historically, polycrystalline silicon gates have been used, where they are highly doped, in concert with the source-drain regions — using the so-called self-aligned process — to control both the threshold work function of the gate and reduce gate resistance. With ever-shrinking channel lengths, a range of substrate parasitics associated with the channel region have become first-order effects in fabrication and device modeling. The shallowness and lateral abruptness of the source-drain extension (SDE) regions (see Figure 25.3b) are critical to controlling parasitic resistances and overlap capacitance. They also have a direct impact on the on-to-off current ratios — an important figure of merit. Source-drain leakage is another factor limited by various implanted doping profiles such as "halo" doping (i.e., HALO), which is introduced to prevent parasitic substrate leakage and electrostatic punch-through between source and drain. Yet as doping in these regions below the channel as well as in the source-drain regions (along with the SDE) increase, consistent with scaling laws [3], the band-to-band tunneling (BTBT) leakage and breakdown phenomena become more severe. BTBT leakage and junction breakdown have to be kept in check by accurate doping placement to form a sandwich low-doping area, where nanometer precision is often necessary. Moreover, ultra-thin gate dielectrics now also have significant leakage current. The following subsections summarize the physics and modeling needed to capture properly the essential features of these diverse physical effects.

Figure 25.3b shows the technology cross-section of a typical scaled (i.e., sub-90 nm) MOS, including highlighted information that indicates 2-D doping effects related to threshold, sub-threshold and source-drain "engineering" (all the details of SDE, extrinsic resistance, capacitance, and substrate coupling). In Figure 25.3c, a simplified equivalent circuit of various intrinsic, extrinsic, and distributed substrate components is indicated.

The key points to understand, both from the physical cross-section of the device and the simplified lumped equivalent circuit, are now briefly discussed. This is followed by the formulation of the modeling equations that are used for simulation and in creating technology files for circuit design.

The MOS transistor is inherently a 2-D field-effect device with gate electrode creating a vertical field that induces carriers from the source region that are extracted toward the drain. The technological complexity shown in Figure 25.3b reflects the evolution over several generations of scaling that has been necessary to achieve the desired inversion charge density and at the same time reduce parasitics that include resistance, capacitance, and leakage — both source-to-drain and bulk leakage from the substrate diodes. The intrinsic part of the MOS transistor (denoted by the dashed region in Figure 25.3b) continues to be scaled using electrostatics that tries to maintain greater vertical electric field (from the gate) compared to the lateral field down the channel. However, with each generation of scaling this intrinsic portion of the device becomes smaller and contributions of the extrinsic parasitics, including source-drain contact regions, are not scaling as rapidly. The implications can be seen by considering the equivalent circuit shown in Figure 25.3c.

The simplified schematic of the MOS transistor shown in Figure 25.3c is not specifically targeted to represent the equivalent circuit used in compact modeling. It is to point out how many of the physical device effects (coming from the technology cross-section shown in Figure 25.3b) relate to intrinsic and parasitic effects. The key points to be emphasized here are as follows: (1) the channel region couples electrostatically to both gate and bulk and is a distributed effect (i.e., the channel charge can be shared with both source and drain); (2) there are several leakage paths that couple gate-to-channel (i.e., tunneling of carriers through the gate) and source-to-drain (i.e., thermionic emission and diffusion as well as quantum mechanical (QM) tunneling); (3) there are both resistive and capacitive parasitics associated with source and drain regions. As noted above, these passive parasitics are increasingly coming to dominate the device and only serve to degrade the intrinsic transistor action.

The following discussion will first address the intrinsic behavior of the MOS transistor and associated simulation and modeling issues, followed by a discussion of considerations of the extrinsic and parasitic effects. Figure 25.4a gives a first-order transistor equation that describes how the drain current depends on electrostatic effects of gate voltage and source-to-drain electric field. This simplified formulation indicates the importance of both the quantity of channel charge (capacitively induced by the gate) and the value of carrier velocity (as reflected by the mobility). The supporting equations show the voltage dependence of C_g and how the various scattering mechanisms affect the overall channel mobility. From a scaling perspective the technology approaches now being considered (and implemented in many research demonstrations) involve increasing the dielectric constant of the gate and the implementation of materials and processing that increase the mobility. Figure 25.4b shows how drive current is scaled with technology generation. The reduced slope for bulk silicon devices over the past several generations of scaling has motivated serious research efforts and industrial demonstrations of materials with increased mobility. Which options might become the mainstream are still under investigation. Models that embrace the necessary physical effects — both for conventional silicon and for advanced material options — are considered as part of the following discussion.

25.2.1.1 Inversion-Layer Mobility Modeling

The modeling of mobility and its dependence on physical parameters, ambient and operating conditions, is arguably one of the most important dependencies both for TCAD physical models and for circuit-level compact models. For mobility modeling at the TCAD level the electrical variables are the local electric field at each point in the device (i.e., components parallel and perpendicular to the planar device surface) and the various scattering mechanisms, including their technology and ambient dependencies. By contrast, the circuit-level models represent a macroscopic approximation of the physical phenomena and parameterize the effects only in terms of terminal voltages vis-à-vis internal (and microscopic) electric fields. Through the electrical extraction process the two representations can be compared and calibrated. The following discussion will consider two typical and well-known TCAD-based mobility models, show comparisons with representative measured data, and then finally contrast these physical models with a typical compact model representation for mobility.

$$I_{\text{channel}} \propto \overbrace{\text{charge}} \;*\; \overbrace{\text{carrier velocity}}$$

$$\propto C_{\text{ox}}\,(V_{\text{gs}} - V_{\text{th}}) \;*\; E_{\text{lateral}}\;\mu_{\text{channel}}$$

$$C_{\text{ox}} = \frac{\varepsilon_{\text{gate}}}{t_{\text{ox}}}$$

$$\frac{1}{C_{\text{g}}(V_{\text{gs}})} = \frac{1}{C_{\text{ox}}} + \frac{t_{\text{depl}}(V_{\text{gs}})}{\varepsilon_{\text{silicon}}} = \frac{1}{C_{\text{ox}}} + \frac{1}{C_{\text{depl}}(V_{\text{gs}})}$$

$$\frac{1}{\mu_{\text{channel}}} = \sum_{i_{\text{th}}\text{scattering mechanisms}} \frac{1}{\mu_i}$$

(a)

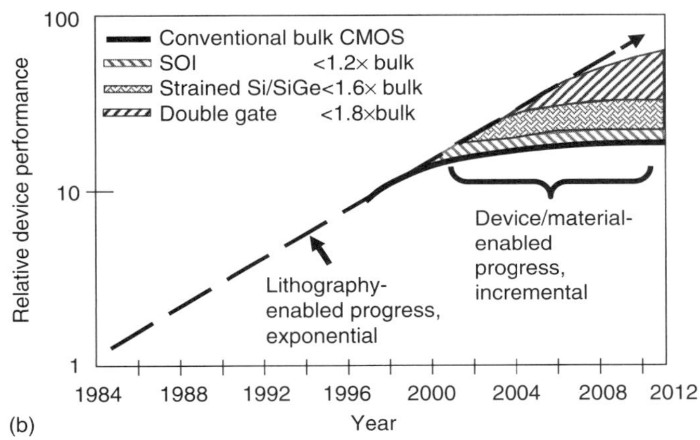

(b)

FIGURE 25.4 (a) First-order model of how MOS inversion layer channel current (Ichannel) depends on gate-included charge and carrier velocity. The physical parameters are Cg and μ_{channel}. (b) Relative MOS device performance occurring for conventional bulk CMOS and projects incremental improvements with new materials.

The Lombardi surface mobility (LSM) model was first proposed in the late 1980s as an empirical model that captures the vertical field dependence and accounts for the doping dependence of bulk mobility (μ_{b}) as well as acoustic phonon (μ_{ac}) and surface roughness (μ_{sr}) scattering effects [15]. The overall formulation exploits the classical Matthiessen's summation rule, shown as the first equation in Figure 25.5, i.e., the inverse of the total surface mobility (μ_{s}) is the sum of the inverse of all constituent mobility terms. The set of relationships shown in Figure 25.5 give details of one popular implementation of the LSM model in the MEDICI simulator. It is beyond the scope of this work to discuss in detail the parameters of the model, other than to point out that the key physical quantities are total doping (N_{total}), maximum mobility for the respective material ($\mu_{\text{max},n}$ here for electrons) and perpendicular local electric field ($E_{\perp,n}$). All model parameters are indicated with capital letters and a ".LSM" suffix.

The LSM model was developed with emphasis on the vertical electric field effects, hence additional modeling terms are needed for the lateral field effects. The work of Shin et al. [16,17] is an example of a model that considers both vertical and lateral field effects, especially since it is able to match the so-called "universal mobility" curves [19] and was suitable for implementation as a local field model — a key requirement for TCAD simulator implementation. Another alternative model developed by Darwish et al. [18] considers lateral field effects and allows the model to be localized in space ($\vec{r}$). Figure 25.6 summarizes the key equations used by Darwish, again using Matthiessen's summation rule, where the subscripts refer to: b, bulk; ac, acoustic phonon; sr, surface roughness; L, lattice, and I, ionized impurity scattering events, respectively. The term involving acoustic phonons has been refined and, while still similar in form to that in the Lombardi model, is based on improved physical modeling of the velocity dependence. The perpendicular field terms have the same meaning as considered above and the form of

$$\mu_S = \left[\frac{1}{\mu_{ac}} + \frac{1}{\mu_b} + \frac{1}{\mu_{sr}}\right]^{-1}$$

$$\mu_{ac} = \frac{BN.LSM}{E_{\perp,n}} + \frac{CN.LSM.N_{total}^{EXN4.LSM}}{T\sqrt[3]{E_{\perp,n}}}$$

$$\mu_b = MUN0.LSM + \frac{\mu_{max,n} - MUN0.LSM}{1+\left(\dfrac{N_{total}}{CRN.LSM}\right)^{EXN1.LSM}} - \frac{MUN1.LSM}{1+\left(\dfrac{CSN.LSM}{N_{total}}\right)^{EXN2.LSM}}$$

$$\mu_{max,n} = MUN2.LSM\left(\frac{T}{300}\right)^{-EXN3.LSM}$$

$$\mu_{sr} = \left(\frac{DN.LSM}{E_{\perp}^{EXN8.LSM}}\right)$$

FIGURE 25.5 Lombardi surface mobility (LSM) model formulation [15], including parameterization as implemented in MEDICI. Matthiessin's used in summation of various scatting terms.

$$\frac{1}{\mu_0(\vec{r})} = \frac{1}{\mu_b(\vec{r})} + \frac{1}{\mu_{ac}(\vec{r})} + \frac{1}{\mu_{sr}(\vec{r})}$$

$$\frac{1}{\mu_b(\vec{r})} = \frac{1}{\mu_L} + \frac{1}{\mu_I(\vec{r})} \qquad \mu_{sr}(\vec{r}) = \frac{\delta}{E_{\perp}^{\gamma}(\vec{r})}$$

$$\mu_{ac}(\vec{r}) = \left(\frac{BT'}{E_{\perp}(\vec{r})} + \frac{CN_I^{T}}{E_{\perp}^{1/3}(\vec{r})} - \frac{1}{T'}\right)$$

FIGURE 25.6 Darwish mobility model expression [18] that considers local position dependence ($\vec{r}$). Comparisons of this model and data from Takagi et al [19] are given in Figure 25.7.

the surface roughness term is also unchanged. Figure 25.7 shows a plot of mobility vs. normal electric field with doping as a parameter; the curves compare the Darwish model with measured data of Takagi et al. [19]. There is generally good agreement between both models and the data; the Darwish model shows good agreement for larger effective electric field (E_{eff}). The notion of a "universal curve" for mobility simply refers to the fact that the channel mobility can be modeled as a function of E_{eff} over several orders of magnitude, where the substrate doping effect is represented in terms of only μ_b and E_{eff}, and the $\mu(E_{eff})$ is independent of the other device parameters such as oxide thickness and interface states.

The above discussion is representative of physical models for mobility used in TCAD simulators. These models utilize detailed information about the behavior of the inversion-layer channel in MOS transistors and parameterize it in a form suitable for implementation in numerical device simulators. There a variety of models available in TCAD tools and the interested reader should refer to the documentation provided by the respective tool developers. It is useful to look briefly at the mobility modeling used in circuit simulation and to contrast it with that used at the TCAD level. Figure 25.8 shows a typical mobility formulation used in a circuit-level compact model (i.e., BSIM3 used in HSPICE). In contrast to the TCAD model that has physical parameters such as doping level and local electric fields occurring inside the device, the

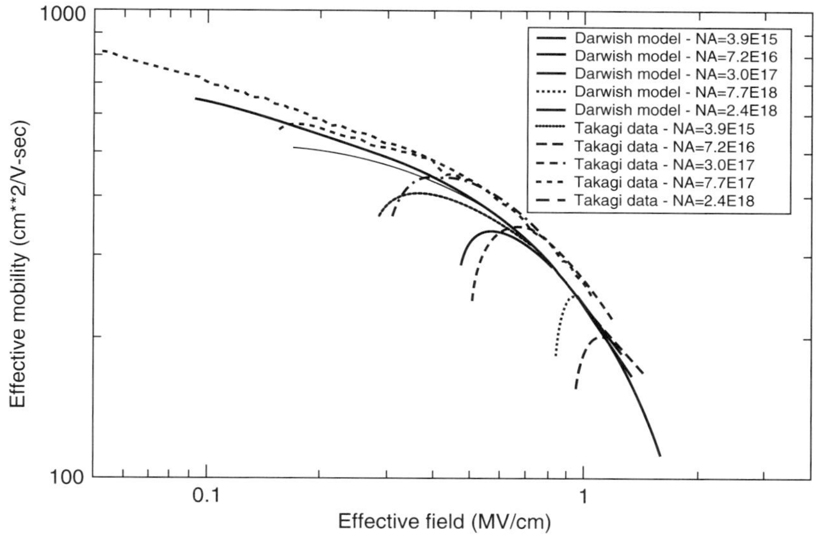

FIGURE 25.7 Effective mobility vs. effective electric field showing experimental results of Takagi et al. [19] and Darwish TCAD model formulation [18] over range of doping levels with good agreement.

$$\mu_{\text{eff}} = \frac{\mu_0}{1 + U_a \cdot \left(\frac{V_{gs}+V_{th}}{T_{ox}}\right) + U_b \cdot \left(\frac{V_{gs}+V_{th}}{T_{ox}}\right)^2 + U_c \cdot V_{bs}}$$

$$E_{\text{sat}} = 2 \cdot \frac{v_{\text{sat}}}{\mu_{\text{eff}}}$$

$$\mu_{\text{total}}\left(V_{\text{sat}}, V_{ds}, V_{bs}\right) = \frac{\mu_{\text{eff}}}{\left[1 + \left(V_{ds,\text{eff}} \big/ E_{\text{sat}} \cdot L\right)^{\beta}\right]^{1/\beta}}$$

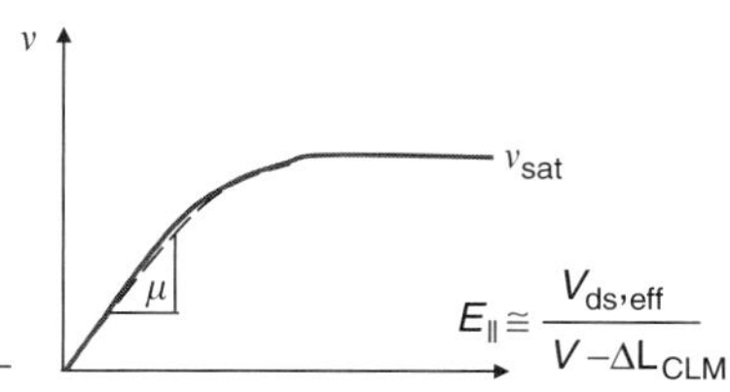

$$V_{ds,\text{eff}} = V_{ds},\ V_{ds} < V_{dsat}$$

$$V_{ds,\text{eff}} = V_{dsa},\ V_{ds} > V_{dsat}$$

FIGURE 25.8 Compact (circuit-level) model for effective carrier mobility showing gate (Vgs) and bulk (Vbs) voltage dependencies, along with definition of critical field term, where velocity saturation occurs and simple "scaling" term used in linear region.

compact model for mobility only considers terminal voltages and fitting parameters, including expressions used to take into account effects of velocity saturation. The inserted figure shows the carrier velocity vs. the parallel electric field plot, indicating the region where carriers become velocity-saturated. The dashed curve in Figure 25.8 shows the approximation used to smooth the transition between the low and high parallel field mobility regions as a function of the drain voltage; the last equation shows one form used to achieve that smoothing. The main point to emphasize about compact models is the fact that they require curve fitting to terminal *I–V* data for transistors. This can be done either with experimental data or TCAD-generated *I–V* curves, which in turn exploit the physical mobility models discussed above.

25.2.1.2 Channel Charge Modeling

The channel charge and all the effects of gate and dielectric materials are as important as the mobility as indicated by the formula shown in Figure 25.4a. The basic formula indicates a simple relationship between gate capacitance and gate-to-source voltage based on the series combination of oxide capacitance (C_{ox}) and the bulk depletion layer capacitance (determined by t_{depl}). In fact, the dependence for (of?) C_g is much more complex as reflected in Figure 25.9a. This figure shows typical measured and simulated *C–V* characteristics for a 2-nm oxide, along with the TCAD-simulated characteristics. The idealized formula given in Figure 25.4a is shown with a broken line. For gate bias both above and below threshold (~+0 V) there are possibly poly-depletion QM effects in the channel region (bulk) and in the poly-silicon gate region [20].

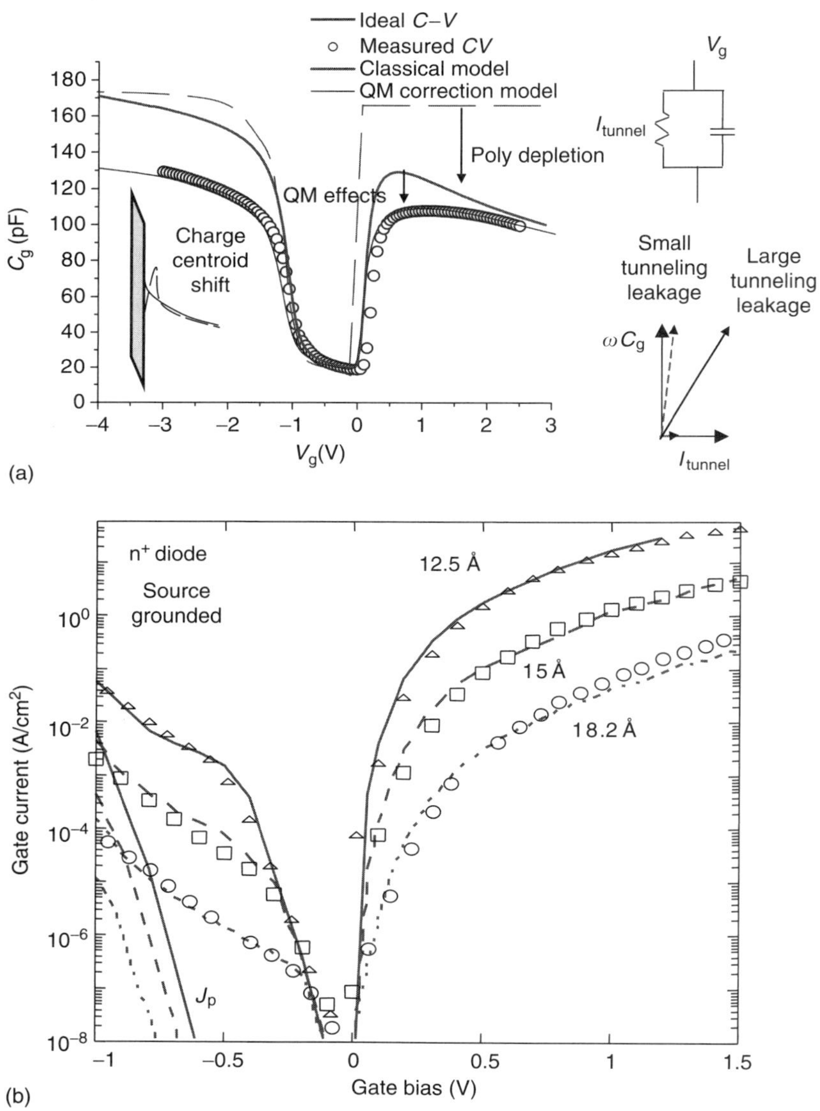

FIGURE 25.9 (a) Measured and simulated curves of MOS gate capacitance vs. gate voltage for an n-channel, polysilicon gate with 2 nm-oxide thickness. Simulated curves compare impact of QM effects. (b) Simulated and measured data for gate current in an MOS capacitor as a function of gate oxide thickness. Simulations were performed using NEMO; data provided by Hewlett-Packard.

Moreover, as the gate thickness decreases further, tunneling of carriers through the gate region becomes distinguishable. Quantum mechanical effects in *C–V* include the quantum repulsion (i.e., the peak of carrier concentration has to be away from the interface in inversion and accumulation) and tunneling which corrects the impedance phase during capacitance extraction. Figure 25.9b shows a semilog plot of simulated and measured gate currents for much thinner oxides as a function of bias. The simulations were performed using the NEMO simulator [21]. Tunneling currents, along with electrostatic effects of depletion in the poly-gate regions and QM confinement effects, play a key role in determining the observed *C–V* characteristics as well as *I–V* characteristics in scaled MOS transistors.

From both the technology scaling and circuit design perspectives, it is important to separate the effect of gate capacitance (which controls channel charge) from the gate leakage current which is a parasitic effect. In order to extract properly and model the channel charge component separately from the effects of tunneling current on the measured *C–V* curves, an equivalent circuit extraction approach has been used as reflected in Figure 25.10a [20]. This circuit represents a large area MOS device (100 μm × 100 μm)

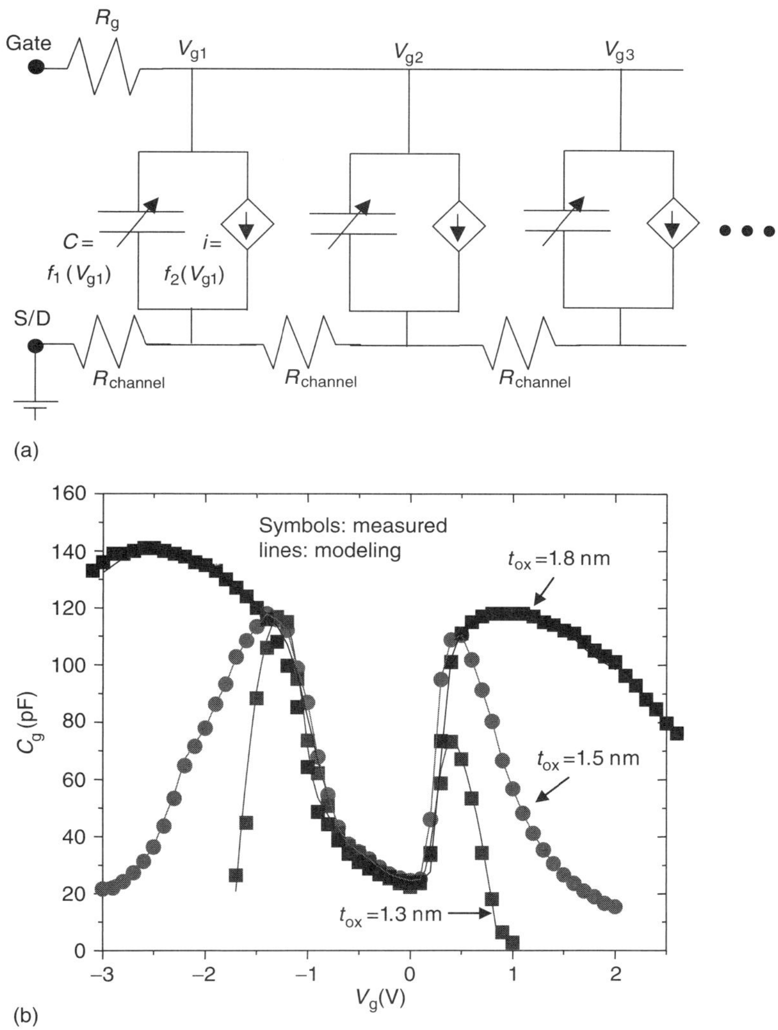

FIGURE 25.10 (a) Equivalent circuit (lumped) model for large area (100 mm 3100 mm) MOs capacitor, S/D grounded showing distributed R-C effects well as dependent generators representing tunneling current. (b) Results comparing measure and equivalent circuit simulations of imaginary component of input admittance for large area MOS capacitors structure with oxide thickness as a parameter.

that has been modeled as a distributed RC network, including voltage-dependent tunneling current generators extracted from the data shown in Figure 25.9b. Circuit simulation and optimization are used to fit the capacitive terms such that the simulated and measured data for the imaginary component of ac input admittance agree, as shown in Figure 25.10b. For an oxide thickness of 1.8 nm, the C_g plot shown in Figure 25.10b is not unlike that shown in Figure 25.9a. However, as the oxide thickness decreases further (i.e., 1.5 and 1.3 nm), the curves are dominated by the tunneling current component. That is, as direct (tunneling) current becomes a larger fraction of the total current (AC+DC), there is a fall-off in the apparent capacitance component. This spurious effect needs to be corrected for in order to provide the correct information to the IC designer. These results illustrate an important link between TCAD-based and compact modeling. The equivalent circuit used for extraction of the actual gate capacitance (from measured data) was parameterized using TCAD, where each component can be more independently extracted that is not separable in actual measurement. In turn, the resulting model (including tunneling current generators) is essential for circuit-level modeling to fit a wider range of technologies. Implications of tunneling current on circuit performance, using this overall modeling approach, are discussed elsewhere [22].

As stated above, there are important QM effects that impact the confinement of inversion-layer charge in the channel and gate of MOS devices. Figure 25.11 shows the comparison of electron distributions in both the bulk silicon (at the surface inversion layer) and in the poly-silicon gate regions. Figure 25.11a contrasts the classical solution for the inversion layer with that of the QM solution where there are discrete energy levels in the channel region, resulting in a solution that has zero carriers at $x=0$. The energy levels also impose constraints in terms of the number of available states (density of states [DOS]) that can be occupied by electrons. This in turn places limits on the amount of charge that can be induced in each level and ensemble.

In Figure 25.11b, the electrons in the poly-gate have a spread-out distribution for the QM solution vs. a sharply peaked distribution at the interface for the classical solution. These QM effects have a very significant impact on both *C–V* and *I–V* characteristics of scaled devices. The impact on the *C–V* curves, as demonstrated in Figure 25.9a, is a reduction in capacitance. This can be understood readily by looking at the simplified expression for C_g given in Figure 25.4a; an increase in t_{depl} in the bulk region causes the total gate capacitance to decrease. Similarly, by adding a dipole layer in the poly-silicon gate there is an additional series capacitance that again reduces C_g. As alternative silicon-on-insulator (SOI) and double-gated (DG) devices are considered in future scaling nodes, there are coupled QM effects that can potentially impact performance. For example, in ultra-thin layers, resonant (QM) tunneling between the gates becomes possible [23]. Even without considering such unusual effects, the 2-D electrostatic (QM) effects can have a major impact on scaled SOI devices.

It is beyond the scope of this chapter to delve seriously into future trends of MOS scaling. The current projections of the International Technology Roadmap for Semiconductors (ITRS) [42] outlines the ongoing needs to consider alternative device geometries and materials for channel lengths below the 45 nm technology node. As discussed above for bulk silicon devices, many of the issues related to carrier mobility and channel charge need to be revisited based on these new structures. The beauty and genius of the poly-silicon, self-aligned gate technology has been its scalability — the dimensional and electrical control of key parameters. The above discussions of gate leakage and QM effects in the gate now brings to the foreground the motivation for changing both dielectric materials of the gate and the gate material itself. Some of these limits are now illustrated for ultra-scaled, DG SOI devices, one of the contenders for future device scaling.

Figure 25.12 shows the impact of QM effects on both threshold and subthreshold characteristics of an idealized 20 nm device structure. Figure 25.12a shows the simulated electrical characteristics — *C*− *V* and *I*− *V* curves. For $V_{ds}=0$ V, the main effect of the poly-QM is the shift in threshold voltage from subband formation and charge centroid repulsion at the interface. Notice that the sub-threshold slope hardly degrades by including the poly-QM effect. Especially as the drain voltage increases to $V_{ds}=1$ V, the I_{ds} curve shows drain-induced coupling effects that occur through the poly-gate region, which increases the drive current quite significantly. Figure 25.12b shows the 2-D contours of charge in the poly-gate region for $V_{ds}=0$ and 1 V when $V_{gs}=0.8$ V; the poly-depletion effects are first-order in importance for such ultra-scaled devices [24]. The take-home message from these simulations is that even with improvements of channel scaling based on SOI,

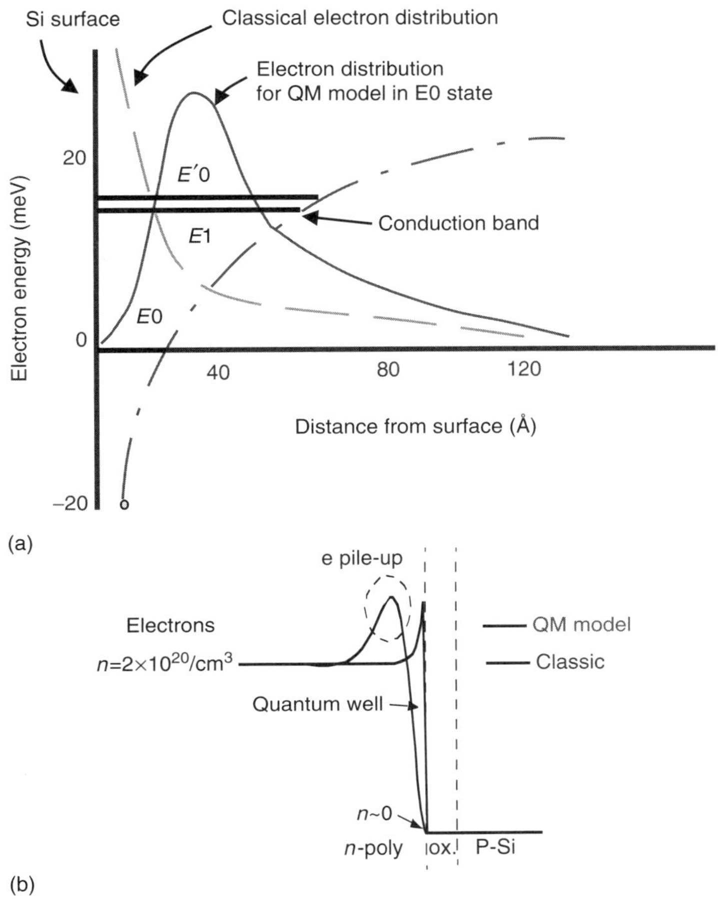

FIGURE 25.11 (a) Surface region of bulk MOS device that shows well potential created by the conduction band energy (-); discrete energy levels imposed by QM (E0,E1); electron distribution based on classical theory (- - -). (b) Electron distribution in the n-type poly-silicon gate of an NMOS transistor, comparing charge buildup based on classical theory vs. that for a quantum-based (QM) solution for charge. The peak region shows a "QM depletion" resulting in Vth and C-V shifts.

the scaling of the poly-silicon gates is rapidly reaching electrostatic limits. The issues of threshold voltage, its control — both electrical and with process variations — and technology options are now briefly explored.

25.2.1.3 Threshold Voltage Modeling

Threshold voltage is the third key parameter that affects performance of MOS devices, namely it controls the channel charge as given by the equation in Figure 25.4a. This equation for channel charge involves both the gate capacitance and "overdrive" voltage (the amount of voltage greater than the threshold voltage, V_{th}). There are many factors that affect V_{th}, including choice of gate electrode materials, device topology (bulk vs. SOI) and bias conditions of all four terminals (see Figure 25.3c). A brief discussion of technology options complimentary MOS, silicon-on-insulator, Bipolar-CMOS (i.e., bulk CMOS, SOI, BiCMOS, etc.) will be covered in Section 25.4. The next section will consider many of the threshold voltage and other parasitic effects associated with bulk CMOS. The emphasis of this subsection is to consider first-order, vertical field only, threshold behavior of bulk CMOS, and the basic silicon-on-insulator NMOS, where the "back gate" is the electrically floating bulk wafer. Alternative SOI substrate configurations, including the most advanced three-dimensional (3-D) structures, are summarized elsewhere [25].

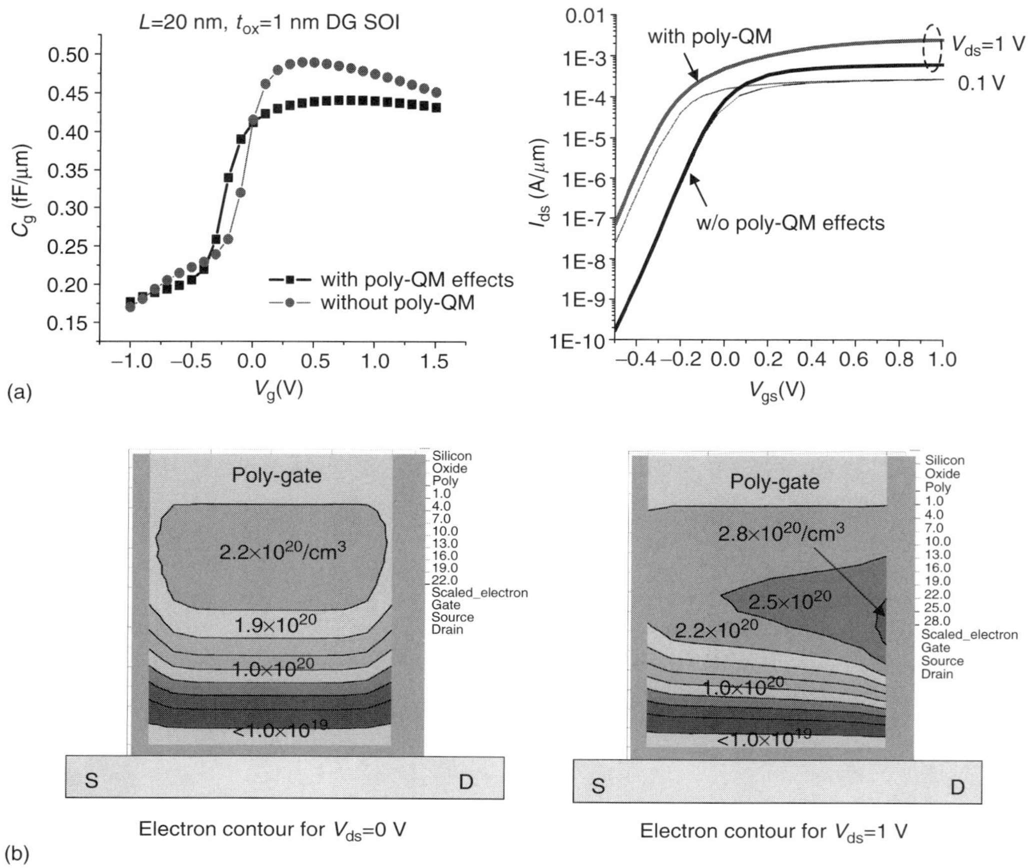

FIGURE 25.12 **(See color insert following page 15–4.)** (a) C-V and I-V characteristic for a 20 nm channel length DG SOI transistor, comparing effects of QM effects in the poly-gate region with idealized (metal-like) gate. Results show shifts in threshold as well as serious drain-induced current effects. (b) Simulations of QM poly depletion effects for two drain bias conditions (same device and I-V as Figure 25.12a). Results show significant gate depletion (left) and lateral effects that influence drain-induced barrier lowering (right).

Figure 25.13 summarizes two typical expressions for threshold voltage, along with details of the supporting equations and definitions. Threshold (V_{th}) is typically defined in terms of an electrostatic condition that results in a specified amount of inversion charge or drain current in the device. The two equations shown in Figure 25.13 represent typical formulations for bulk (Figure 25.13a) and SOI (Figure 25.13b) MOS devices, respectively; the threshold condition is defined in terms of a specific surface potential. The threshold expressions consist of several terms; these are contributions to the overall gate potential coming from effects of the gate(s) and bulk/SOI material properties, as will be discussed now. From a physical modeling point of view, process and device simulations are used to characterize the multidimensional electrostatic effects in each device structure. The threshold expressions as shown in Figure 25.13 seek to capture the key dependencies and parameterize them based on analytical approximations of the numerical simulations. The following discussion represents the long-channel results; the effects of source, drain, and substrate are discussed further in the next section. In device simulation, since the 2-D Poisson equation is explicitly solved, all electrostatic effects are included, provided that the correct DOS information is available. Lumping the effects to different parts of the device is useful for device design and compact model construction.

$$V_{th,N} = V_{FB} + 2\phi_{F,sub} + \frac{1}{C_{ox}}\sqrt{2q\varepsilon_{si}N_{A_{bulk}}\left[2\phi_{F,sub} + V_{SB} - \frac{qx_c D_I}{\varepsilon_{si}}\right]} + \frac{qD_I}{C_{ox}}$$

$$\phi_{F,sub} = \frac{kT}{q}\ln\left(\frac{N_{A_{bulk}}}{n_i}\right)$$

$$C_{ox} = \frac{\varepsilon_{ox}}{t_{ox}}$$

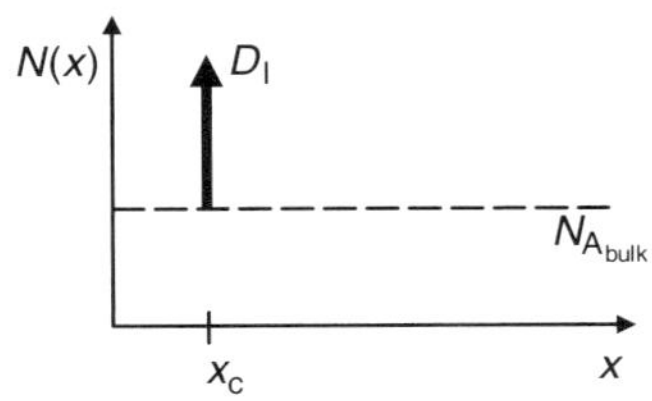

(a)

$$V_{th} = \phi_{F,poly} + \phi_{F,SOI} + \frac{\varepsilon_{si}}{C_{ox1}}\left(\alpha + \sqrt{\beta + \frac{2qN_{A_{bulk}}}{\varepsilon_{si}}\phi_{F,SOI}}\right)$$

$$\alpha = \frac{qN_{A_{SOI}}}{C_S} - qN_{A_{bulk}}\left(\frac{1}{C_S} + \frac{1}{C_{ox2}}\right)$$

$$\beta = (qN_{A_{bulk}})^2\left(\frac{1}{C_S} + \frac{1}{C_{ox2}}\right)^2 - \left(\frac{q}{C_S}\right)^2 N_{A_{bulk}}N_{A_{SOI}} + \frac{2qN_{A_{bulk}}}{\varepsilon_{si}}(-\Psi_{sub})$$

$$C_{ox1} = \frac{\varepsilon_{ox1}}{t_{ox1}},\quad C_{ox2} = \frac{\varepsilon_{ox2}}{t_{ox2}},\quad C_S = \frac{\varepsilon_{si}}{t_{SOI}},\quad \phi_{F,SOI} = \frac{kT}{q}\ln\left(\frac{N_{A_{SOI}}}{n_i}\right)$$

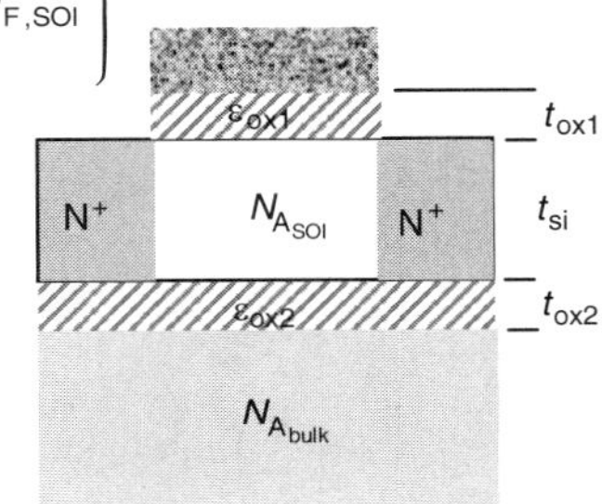

(b)

FIGURE 25.13 (a) Threshold voltage expression of a traditional bulk CMOS device (NMOS) with an equivalent threshold adjustment implant dose (DI) centered at Xc. The notation follows that given in [26]. (b) Threshold voltage expression for an SOI (fully depleted) NMOS, where the channel is induced at the top surface of the SOI layer (tsi). The definition of terms follows the notation and derivation in [27].

The material-related effects reflect electrical differences between the bulk/SOI regions and the gate(s). The V_{FB} and $\phi_{F,poly}$ terms in Figures 25.13a and b are the specific terms that capture the gate-related dependencies. The expression given in Figure 25.13a (for bulk CMOS) considers both gate-related and fixed interface charge effects, which are lumped into V_{fb}. The gate voltage needs to have the semiconductor to be in the "flat-band" condition (i.e., no vertical electric field at the interface). The expression given in Figure 25.13b considers the gate material to be heavily (n^+) doped poly-silicon; the term $\phi_{F,poly}$ represents the flat-band offset due to only that contribution (i.e., interface or other fixed charge has been neglected). As threshold voltage continues to be scaled down with channel length, these gate-related contributions pose scaling limits; changes in gate materials with a wider selection of work functions are becoming necessary to coordinate the scaling of these terms with the other threshold contributions discussed next.

The remaining terms in both Figures 25.13a and b represent capacitive coupling terms, which generally have the form Q/C_{ox}. The important trends to observe related to these terms include interdependence of doping ($N_{A_{bulk}}$ or $N_{A,SOI}$) with gates(s) (C_{ox} or dual gates if SOI) and substrate bias (V_{sb} for bulk CMOS or capacitively through the substrate potential as a "second gate" via Ψ_{sub}). In Figure 25.13a there are two such terms, one representing the bulk depletion layer (i.e., the term involving $N_{A_{bulk}}$) and another involving an implanted dose (D_I) of charge. For bulk CMOS, as shown in Figure 25.3b, the substrate doping profiles can be complex, and 2-D effects such as SDE and HALO profiles cannot be ignored. The simple threshold model shown in Figure 25.13a lumps all vertical nonuniform doping effects into an equivalent dose of dopant charge. In the case of Figure 25.13b for an SOI device, the bracketed term involving α and

β terms as well as $N_{A,SOI}$ also demonstrates the general form for threshold voltage, controlled by the top gate via C_{ox1}. The α and β terms reflect the effects of the substrate ($N_{A_{bulk}}$) and bottom gate via C_{ox2}. Key points to note here are the multielectrode effects (i.e., dependence on C_{ox1}, C_{ox2}, and Ψ_{sub}) as well as the silicon layer thickness. The expression shown in Figure 25.13b is for a fully-depleted (FD) SOI in all V_{gs} operating conditions; here the thickness as well as silicon layer doping are inter-related in determining the depletion charge contribution to V_{th}. For ultra-thin SOI, the layer doping becomes less important. Details of these two threshold formulations can be found elsewhere [27,28].

The complexity of nonuniformly doped, single-gate structures and DG SOI structures leads to increasingly complex threshold expressions; the two shown in Figure 25.13 are only representative. For bulk CMOS the trade-off between flat-band voltage, surface potential and associated bulk charge pose ongoing challenges in the design of the gate stack and choices in substrate doping profiles. Silicon-on-insulator provides additional parameters to consider, particularly the semiconductor layer thickness (t_{si}) and the additional control of a DG structure. With FD, DG SOI the control of layer thickness becomes a first-order concern and layer doping becomes much less of an important controlling factor. Looking for "the optimal" device structure can be very challenging; the optimization process is largely one of economics (fabrication costs) and performance objectives. However, it is worth pointing out that threshold voltage (and its relationship to power supply level) will be a key factor in the trade-off between power and speed for large digital circuits.

25.2.2 Substrate Effects on MOS Transistors

The drain current expression given in Figure 25.4a is useful in understanding the basic issues of the MOS as an electrostatically induced conducting channel "resistor," controlled by the gate voltage with respect to the threshold voltage. For bulk CMOS the substrate terminal (the *p*- or *n*-well for the *p*- or *n*-channel device, respectively, as in Figure 25.3a) can directly influence the threshold voltage. That is, for the most basic formulation of threshold voltage used in circuit simulation, it is modulated by the substrate bias (V_{sb}) as shown in the following equation:

$$V_{th}(V_{sb}) = V_{th0} + \frac{\sqrt{2q\varepsilon_{si}N_{A_{bulk}}}}{C_{ox}}\left[(2\phi_{F,sub} + V_{sb})^{1/2} - (2\phi_{F,sub})^{1/2}\right]$$

where the first term represents the threshold voltage with zero V_{sb} applied and the second term represents the correction of the "VTO" term (using SPICE terminology) for the additional bulk charge (Q/C_{ox}) that needs to be included. Note that the threshold expression given in Figure 25.13a does include the first term from the square-bracketed expression given above. The second term represents the component of bulk charge term that must be removed from V_{th0}, that is, it must be replaced by the first term (only) to be correct (and not "double count" the Q/C_{ox} contributions). The term in front of the square-bracketed terms is the so-called body coefficient, GAMMA $= \gamma \equiv \sqrt{2q\varepsilon_{si}N_{A_{bulk}}}/C_{ox}$ (again using SPICE notation). This threshold expression and body coefficient are a simplified form of that given in Figure 25.13a and only serve to link the first-order SPICE terminology to the discussion given in the previous section. Additionally, this form clearly points out the fact that threshold voltage shifts as a result of applying "body bias" (V_{sb}). The assumption that no current flows below threshold is discussed next, along with other parasitic substrate effects.

In reality, drain current flows for gate voltages below "threshold"; this is the so-called subthreshold region that is important both for low-power design and in terms of understanding parasitic leakage effects. Prior to strong inversion at the MOS interface that creates the channel, which in turn pins the surface potential, there is an exponential buildup of charge at the surface due primarily to the gate-induced (V_{gs}) effects. This is the expected (and desirable) subthreshold effect. There are also electrostatic effects of V_{ds} and V_{bs} which are generally of a parasitic nature in that they shift the *I–V* curves and can induce unwanted leakage which is not controlled by the primary (front) gate. Figure 25.14a gives representative expressions for the drain current dependencies on these potentials. Figure 25.14b shows a semilog plot of I_{ds} vs. V_{gs} with V_{ds} as a parameter for three channel length devices where the channel doping profiles have purposefully not been scaled to adjust for the shorter channel effects. That is, electrostatically the long-channel device is properly scaled in terms of dopant profiles; the shorter channel-length devices allow too much coupling of drain potential with the source region.

$$I_{ds} = \mu_{eff} C_{ox} \frac{W}{L}(m-1)\left(\frac{kT}{q}\right)^2 e^{q(V_g - V_{th})/mkT}\left(1 - e^{-qV_{ds}/kT}\right)$$

$$V_g = V_{th} + m(\psi_s - 2\phi_{F,sub})$$

$$m = 1 + \frac{\sqrt{\varepsilon_{si} q N_{A_{bulk}} / 4\phi_{F,sub}}}{C_{ox}} = 1 + \frac{C_{depl}}{C_{ox}}$$

$$C_{depl} \equiv \frac{\varepsilon_{si}}{t_{depl}(\psi_s = 2\phi_{F,sub})}$$

$$I_{ds} = qD_n \frac{W \cdot Z^*}{L^*} \frac{n_i^2}{N_{A_{sub}}} e^{\left(\frac{q(\psi^* - V_s)}{kT}\right)}\left[1 - e^{\left(\frac{-qV_{ds}}{kT}\right)}\right] \quad \psi^*(x^*, y^*) = \psi^*(V_{gs}, V_{ds}, V_{sb})$$

(a)

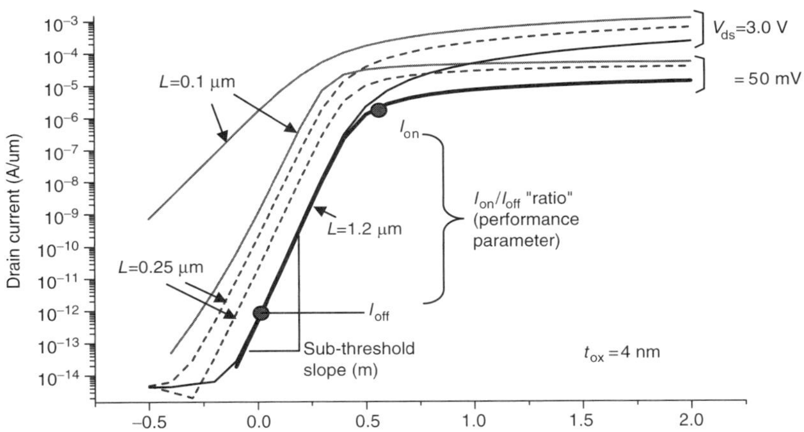

(b)

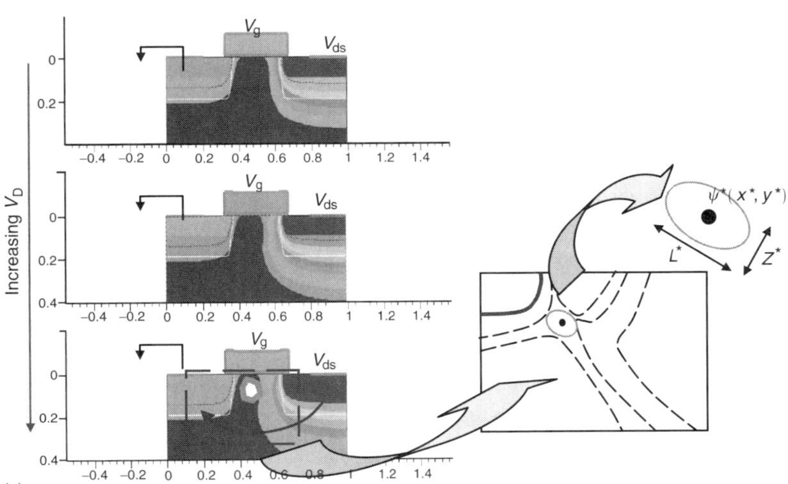

(c)

FIGURE 25.14 **(See color insert following page 15–4.)** (a) Drain current vs. gate, source, drain-source, and bulk (barrier) potential for two regimes of sub threshold operation: (top) gate-controlled (bottom) bulk-controlled. (b) Semilog plot of drain voltage (Vds) as a parameter for three channel length NMOS devices. (c) Two-dimensional plot of electrostatic potential in MOS device for three drain bias (Vds) conditions. The results show typical operation as well as drain-induced barrier lowering (DIBL) for sufficiently high drain bias.

25.2.2.1 Subthreshold and Drain-Induced-Barrier-Lowering (DIBL) Currents

The first set of expressions in Figure 25.14a are for the subthreshold current regime [29] and demonstrate several key features of device performance in this region of operation. The exponential dependence of drain current on V_{gs} reflects the buildup of channel charge before the surface potential becomes pinned; the current flow is a result of diffusion rather than due to the lateral drift field (i.e., Figure 25.4a) [27]. The "m" factor in the exponential is determined by the ratio of the substrate depletion capacitance to the oxide capacitance, which reflects the division of potential between gate and bulk regions; if the ratio is 0 (m=1) this means that the gate potential is fully coupled to the surface. The impact of this m factor on slope of the drain current is shown in Figure 25.14b. The theoretical limit is m=1 but for practical doping levels m> 1 and increases with the square root of the doping. Careful readers may notice that the prefactor of I_D contains $(m-1)$ which makes the ideal case without sub-threshold leakage, which is nonphysical but a common problem in deriving equations by artificial separation of two operating regions. In SPICE model implementation, there will often be a fitting factor to join the equations for above threshold and sub-threshold regions with a current at the threshold voltage defined, so the $(m-1)$ prefactor does not matter even for ideal sub-threshold leakage case. This m factor determines the slope of the semilog I_{ds} vs. V_{gs} in the subthreshold region. From a circuit perspective the ratio of "on current" to "off current" is critical; this ratio is directly impacted by the m factor. This determines how large a voltage difference is required to achieve the desired on–off current ratio.

The first set of equations in Figure 25.14a does not include the drain bias effect on the threshold voltage or the sub-threshold slope, because when V_{ds}>$3kT$, it has almost no influence on the drain current. The second equation shown in Figure 25.14a represents the drain current that flows as the drain voltage overpowers the gate control — the so-called drain-induced barrier-lowering (DIBL). Notice that the most leaky path (MLP) for the sub-threshold drain current can only be determined from 2-D potential contours, and can be away from the surface channels depending on the doping and SOI design parameters. The form of the expression is rather similar to the previous subthreshold equation, but this apparent similarity is superficial. The DIBL expression is a bulk conduction that is controlled by the potential (Ψ^*) which occurs at a "saddle point" as indicated in Figure 25.14c. The current is controlled by the width/length ratio, Z^*/L^*, which are the TCAD-extracted parameters of the saddle point shown in the figure. The product of WZ^* (a cross-sectional area factor) in the numerator indicates that this can be a bulk conduction term in the MLP, and the difference between Ψ^* and V_s indicates the "barrier control" at the saddle point. Details of the method for extraction of these saddle-point parameters are given in [30].

The expressions given in Figures 25.14a and b represent electrostatically controlled current conditions — the first where the gate is the primary controlling factor and the second expression results from the MLP induced by the drain voltage. As stated above, the subthreshold conduction regime with strong gate control is expected and is useful in low-power design. The condition that results in DIBL-degraded subthreshold leakage is one of several parasitic substrate conduction effects. In order to reduce this effect, the substrate doping, the HALO implant dose, or the body thickness of SOI can be scaled [43]. However, this in turn can result in parasitic conduction at the drain end of the channel. For heavily doped junctions there can be band-to-band (Zener) tunneling as well as trap-assisted tunneling due to damage introduced during the HALO implant. A simulation-based model for extraction of these tunneling currents has been proposed [31] and further refined based on empirical data from experiments using different HALO dose conditions [32]. This again provides evidence that the doping distribution details (and their statistics) are of critical importance in device design and modeling.

25.2.2.2 Band-to-Band Tunneling (Zener) Current

Figure 25.15a shows an expression for the BTBT given in [31] based on pioneering work in the 1950s by Kane and Keldysh on Zener tunneling in junctions. The rate equation (R_{btbt}) is expressed in terms of tunneling current density per unit of energy, parameterized in terms of electric field ($\vec{E}$) and energy band parameters (which are denoted with superscript "~"). The supporting equations give further details for R_{btbt} including an expression for D that is suitable for TCAD-based calculations and a modified form for E'_0 [32] that accounts for the doping dependence (that can also be extracted based on TCAD process simulations).

Figure 25.15b shows schematically the MOS device, the HALO (ion implanted) region and the energy bands between the p^+ and n^+ regions where the BTBT occurs. The plot to the right in Figure 25.15b shows the experimental data and simulation results based on the formulation shown in Figure 25.15a for different HALO implantation doses [32]. There is a strong doping (dose) dependence; changing the angle of the HALO implantation also has an effect on how defects influence the leakage currents.

$$R_{btbt} - \frac{dJ_{btbt}}{d\tilde{E}}\vec{E}$$

$$R_{btbt} = -B|\vec{E}|^{5/2} D(\vec{E}, \tilde{E}, \tilde{E}_{fn}, \tilde{E}_{fp}) e^{\left(\frac{E_0'}{|\vec{E}|}\right)}$$

$$D(\psi, \tilde{E}_{fn}, \tilde{E}_{fp}) \quad \frac{1}{\exp\left[(-\tilde{E}_{fp} - q\psi)/kT\right]+1} - \frac{1}{\exp\left[(-\tilde{E}_{fn} - q\psi)/kT\right]+1}$$

$$E_0' = E_0 - a(N_A - N_0)$$

(a)

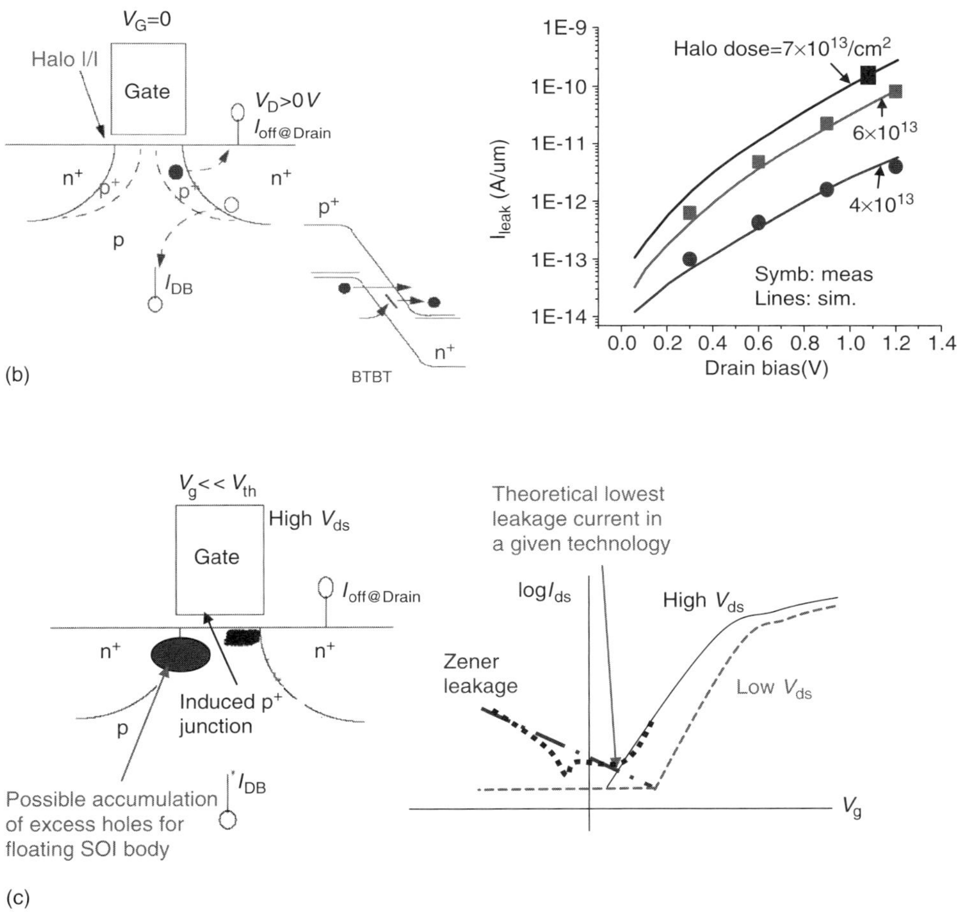

FIGURE 25.15 (a) Rate equation for band-to-band tunneling (Rbtbt) expressed in terms of the tunneling current density per unit of energy (dJ/dE) with supporting equations for the rate equation for the rate equation [31], including a modified term for the doping dependence [32].(b) Schematic view of the band-to-band tunneling, including trap-assisted tunneling, and plots of the modeling results (Figure 25.15a) along with experimental data for different HALO implant doses [32].(c) Schematic view of the band-to-band tunneling in the gate-induced drain leakage (GIDL) region. The leakage current has a negative transconductance and determines the lowest possible leakage current in high Vds

Band-to-band tunneling is also an important concern for drain and substrate currents in the deep subthreshold and high V_D, where the channel region close to the drain side is driven to accumulation by the gate bias as shown in Figure 25.15c. In high V_D, this induced Zener junction can break down and carries a significant current, which has a negative transconductance usually called gate-induced drain leakage (GIDL). Notice that GIDL can happen even for low-doped substrates and can inject significant amounts of carriers to the floating substrate in SOI structures, similar to those created by impact ionization, until the body potential is raised high enough by the excess hole concentration. This can potentially turn on the parasitic bipolar junction transistors (BJT) in the substrate and cannot be shut off by the gate bias. Detailed models including the self-consistent body potential for modeling BTBT can again be obtained from TCAD simulation, and appropriate substrate and drain-doping engineering needs to be employed to control this parasitic effect.

Drain-induced barrier lowering, together with channel length modulation (CLM) by velocity saturation or velocity overshoot also sets the output resistance in the saturation region when $V_{ds} > V_{dsat} \sim V_g - V_{th}$. The product of small-signal output resistance and transconductance is referred to as the self-limited gain in that operating point, which is a critical parameter in analog and RF circuit designs. In digital circuits, the self-limited gain is also important for sense amplifiers and phase-lock-loop (PLL) designs. Since DIBL and CLM are quite complicated in their multidimensional potential profile nature, detailed TCAD simulation is usually necessary to guide the design process [44].

The above discussion has considered subthreshold conduction as well as parasitic drain-to-source (DIBL) and drain junction (BTBT) leakage currents. The discussion also addressed the first-order body-bias effect in terms of threshold shift with V_{sb}. There are also 2-D effects on threshold voltage that result from the influence of source and drain potential on the depletion charge in the channel region. The next following discussion considers generally the inhomogeneous substrate depletion effects, including influence on threshold voltage and parasitic junction capacitance.

25.3 Parasitic Junction and Inhomogeneous Substrate Effects

The previous discussions of threshold voltage have focused on the vertical field effects and have generally ignored the two-dimension (2-D) effects, other than the DIBL and BTBT leakage effects. From the perspective of achieving an electrostatically "well-tempered" MOS device where the vertical field dominates over lateral field effects, this is the desirable condition. However, as scaling continues for bulk CMOS there are an increasing number of doping profiles — HALO, SDE, etc. — that influence the 2-D electrostatics. Figure 25.3b shows a fairly realistic cross-section of such a scaled device (i.e., 90 nm technology node). In this section, we will briefly look at some of the other substrate effects that impact circuit performance.

Figure 25.16 considers that basic NMOS device, now showing electrostatic potential contours under that gate and junction regions (broken lines). Also shown in Figure 25.16 are several trapezoidal and triangular regions that schematically indicate regions of influence in terms of electrostatic potential and dopant distributions. The trapezoid directly under the gate represents the bulk charge dominantly controlled by the gate — the term Q_{bulk}/C_{ox} in the threshold expression. The two neighboring triangles ($Q_{bulk\text{-}S}$ and $Q_{bulk\text{-}D}$) represent the portions of bulk charge that are strongly influenced by source and drain potentials. That is, electrostatically, as these potentials change they have a greater influence on the bulk charge than does the gate potential, resulting in 2-D effects on V_{th}. Also shown in Figure 25.16a are the sidewall (SW) contributions of charge of the source and drain junctions. The portion of charge within the L_{eff} region but not controlled by the gate bias has been used as the initial estimation of the short-channel effect (SCE) [27]. It can be readily seen that the source/drain charge effect has more influence in short-channel devices than in long-channel devices. While these charges may only weakly influence the threshold voltage in the "well-tempered" device design, they do have a major impact on SW capacitance as discussed below. We will not specifically consider how these 2-D bulk charge effects impact threshold voltage or the circuit-level models. Suffice it to say that TCAD simulations and electrostatic results such as those shown in Figure 25.16a are the basis for developing appropriate threshold expressions that account for these 2-D effects.

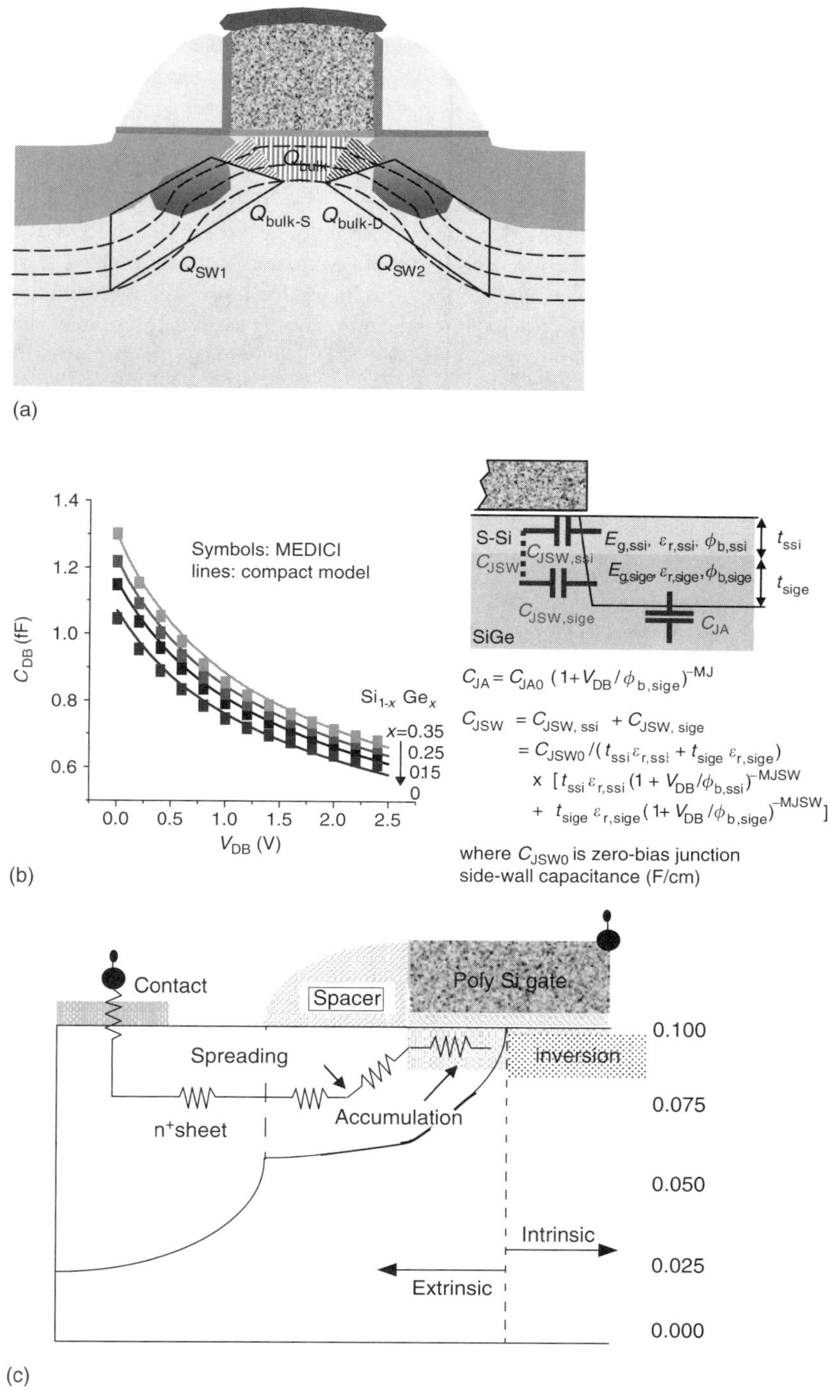

FIGURE 25.16 (See color insert following page 15–4.) (a) Cross-section on NMOS (same as Figure 25.3b) showing electrostatic potential contours (broken lines) and trapezoids and triangles of charge, representing influence of gate, source and drain potential as well as side-wall (SW) effects due to doping. (b) Junction capacitance (CDB) as a function of bias (VDB) for junctions with varying degress of strain, induced by different germanium fraction (x). The components of capacitance area (CJA) and side-wall (CJSW) are indicated, along with compact model forms. (c) Cross-section of contact region (source end) of a scaled NMOS, with emphasis on the several components of resistance: SDE region, spreading resistance, bulk (n+) sheet resistance and contact resistance.

As noted above, the source-bulk and drain-bulk junctions exhibit 2-D effects as a result of various doping and other isolation-related effects along the edges. Figure 25.16a shows one example of these edge-effects resulting from the HALO implant doping profiles. Recently, technologists have begun using "strain engineering" to improve mobility in transistors [33]. There are several ways to induce this strain, for example using epitaxial layers with silicon-germanium (SiGe) or alternatively localized (i.e., ion-implanted) material changes . These material changes have localized effects on the 2-D junction parameters both in terms of dielectric constants and energy band parameters. Figure 25.16b shows the results of one set of comparisons for junction capacitance in an epitaxial SiGe process. Results of 2-D (MEDICI) device-simulated drain junction capacitance are compared with parameterized compact (circuit) models for the area and SW components of the capacitance, including different components for the SiGe and strained-silicon (S-Si) regions. Such parameterization of the various components is essential for accurate circuit modeling; TCAD simulations provide an excellent means to perform the necessary extraction. From a technology point of view the capacitance increases with increased germanium fraction (x), leading to a design trade-off for this particular example between increased drive current (due to higher mobility) and increased capacitance as indicated in Figure 25.16b. Part of the drain junction capacitance will be reflected also as the capacitance between drain and gate, especially when $V_{ds} < V_{dsat}$. The increased C_{dg} is a serious concern in RF and analog CMOS circuit design, and has to be accounted for carefully in determining technology alternatives.

There are other substrate-related parasitic effects that influence bulk CMOS performance including edge-effects in the isolation (i.e., see STI region in Figure 25.3a), source and drain resistance (i.e., the SDE and extrinsic regions in Figure 25.3b), and bulk resistance of the substrate. Some of these issues are considered in Chapter 24 from the technology perspective. The device implications range from leakage current concerns for the isolation to system-on-chip (SoC) noise coupling resulting from substrate coupling as is discussed in Chapter 21. Parasitic source and drain resistance deserves some further discussion since it has major implications on limits of scaling and also issues related to statistics of the devices, namely, the potential drop across the source-side resistance directly subtracts from the gate over-drive that in turn reduces the drive current.

Figure 25.16c shows in detail the various components that contribute to the parasitic source resistance. These include the SDE "link up" (or overlap) region between the channel and extrinsic region, spreading resistance that results as the current goes from the confinement of the SDE into the deeper junction region, and finally the bulk junction and contact regions. As junctions become shallower, consistent with the scaling laws, the sheet resistance of these regions also increases and it becomes more difficult to fully activate the high-concentration doping profiles (see Chapter 24). In addition, the SDE "tip regions" are influenced by surface effects and the lateral diffusion and abruptness of this profile is critical in controlling devices performance — both statistically and in terms of the "on–off ratio." The determination of the parameters associated with source/drain parasitic resistance are very important; Taur has developed the "shift-and-ratio" method [27] which is broadly used in this extraction process. There are also many details related to the profile scaling that can be obtained using Taur's approach in combination with TCAD-based modeling [34].

At this point it is useful to specifically point out the importance of process modeling and its relationship to statistical effects (see Chapters 24 and 19). Namely, in the regime of sub-100 nm scaling, the issues of dopant activation and lateral diffusion are critical. That is, they increasingly tighten the constraints on thermal budget for diffusion (the product of diffusivity and time in the Gaussian solution of a diffusion equation — "Dt") which in turn directly affects junction depth. The shrinking device dimension also results in smaller volumes for a given dopant density, which in turn bring statistics into play in the modeling. This is also a good vantage point to return to the "closure" between system-level and technology-driven perspectives of device scaling as initially introduced in discussing Figure 25.1b. Namely, many of the system-level care-about issues such as leakage current, parasitic R/C effects, and generally I_{on}/I_{off} ratio come down to the complete array of process technology and device scaling issues representative of the cross-section shown in Figure 25.16c. These dopant profile and dopant activation issues, constrained by physical constants such as dopant solid solubility and other parameters of dopant kinetics, directly impact the electrical device properties, including their statistical distributions.

In this section and the previous one on intrinsic MOS device effects, we have considered the use of TCAD to understand parameter dependencies and the relationship between these physical effects and models that bridge to the circuit world. Figure 25.17 summarizes schematically the set of physical modeling effects considered to this point and how they relate to the world of compact modeling. We have purposefully used the NMOS device as the primary vehicle and have emphasized bulk CMOS technology. While there have been limited discussions of technology shifts to SOI and the use of other materials such as SiGe to induce S-Si, we have primarily discussed the ways to address known "roadblocks" that are becoming performance limitations with bulk CMOS. The next section presents a slightly broader view of other device technology concerns and points the reader toward supplemental reading that can go beyond this very cursory discussion.

25.4 Device Technology Alternatives

The above discussion has emphasized CMOS technology that is indeed the engine of the integrated electronics economy and will continue to be so for the indefinite future. The coordinated scaling of analog, digital and a variety of memory technologies using the same basic infrastructure is the driving force. Over the years there have been a variety of contenders seeking to change this "all MOS" landscape. As noted in the introduction, BJT preceded MOS and for some applications, especially in mixed-signals, it still has competitive advantages. For about a decade ca. the 1980s bulk BiCMOS (BJT devices integrated with CMOS) had been predicted as mainstream; however, due to limitations of power, "on voltage" and generally issues related to scalability, "all CMOS" regained its dominant position.

In certain analog areas there have been alternative devices that held out promise based on unique performance, for example the charge-coupled device (CCD) technology used for imaging. The CCD while generally compatible with bulk CMOS is finding itself squeezed for market space by the growing prevalence of CMOS-based digital cameras, due largely to the digitally dominated infrastructure for information processing. This digital signal processing (DSP) approach also has impacted wireless applications as well as data conversion and other interface circuits. What this means is that chips are dominantly digital with add-on analog or other interface capabilities only to the degree that they can be justified *economically* instead of just technically [35]. This economics-driven perspective (which is in fact the true message of Moore's law) also poses a very real challenge for SoC vs. system-in-package (SiP). The following few sub-sections will consider device options that have had some success over the past decade in terms of integration with mainstream CMOS technology into single-chip solutions, primarily from the perspective of physical modeling.

25.4.1 Silicon-on-Insulator Technology

There is growing interest in 3-D MOS structures including SOI-like devices [25]. A DG SOI device example was discussed above (Figure 25.12), primarily to illustrate the severity of scaling limits when using

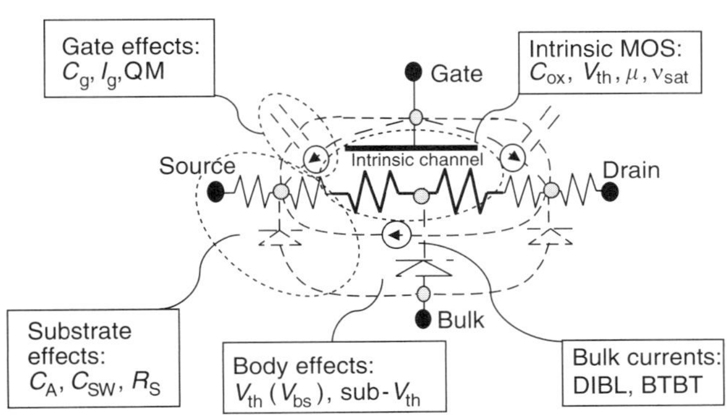

FIGURE 25.17 Summary of modeling considerations presented in Sections 25.2 and 25.3. Intrinsic MOS, gate, body/bulk effects (Section 25.2) and parasitic substrate capacitance/resistance effects (Section 25.3).

poly-silicon gates. The major advantages of SOI technology come from the elimination of the underlying bulk silicon layers, associated dopant distributions needed to avoid parasitic currents, and capacitive effects. For example, having an insulator rather than silicon bulk region dramatically alters — mainly reduces — junction and body effects for the MOS devices. Additionally, there are advantages to DG SOI structures, including more complete surrounding of the channel region, that are manifest in increased gate control and greater effective channel charge. This in turn can give tighter SCE control for V_{th} parametric yield with a given gate oxide thickness. The infrastructure for SOI technology is generally compatible with the mainstream. For applications where the cost is warranted in return for performance advantages, this technology will continue to be used. An excellent discussion of the technology and scaling issues for SOI can be found in [26].

In spite of the benefits and promise of SOI-like technologies, including commercial availability, there are also some limitations. For example, the so-called FD SOI technology, which embraces a range of device topologies [25], relies on the silicon body thickness and gate work function to set threshold voltages in contrast to the bulk and gate doping levels traditionally introduced by ion implantation. This shifts the design burden of scaling, which was dominated by selective introduction of dopants into the substrate using ion implantation, to other process control issues such as deposition or growth of thin layers and control of gate work functions. Additionally, heat generation and its removal is of growing concern, not only for SOI but even for bulk CMOS; thermal management has become a fundamental limitation to scaling across the landscape of IC technologies [36]. This growing concern with thermal management, especially for SOI technologies, also leads to new metrics to be considered in technology design; one example related to use of germanium on insulator (GOI) will be considered later in this section.

25.4.2 Silicon-Germanium Heterojunction Bipolar Transistor Technology

In spite of previous scaling challenges encountered by bulk BiCMOS, the introduction of SiGe bipolar transistors into the same technology with bulk CMOS has found an economic window, particularly in the telecommunications sector, both wireline and wireless applications. In contrast to earlier BiCMOS technologies that tried to exploit diffused junctions to create bipolar transistors within the fabric of a CMOS technology, the heterojunction bipolar transistor (HBT) technology uses selective epitaxial addition of SiGe layers that allow a more loosely coupled process flow. Details of the process flow and device design trade-offs can be found elsewhere [37]. This discussion will briefly contrast the physical modeling issues for bipolar vis à vis MOS devices.

In contrast to the MOS devices that are voltage-controlled and resistive in nature, bipolar transistors are current-controlled so that current gain ($\beta \equiv I_C / I_B$) is a critical parameter. With increased interest in low-power design, the use of the subthreshold regime with MOS devices results in an exponential dependence of I_d on V_{gs} with subthreshold slope (semilog plot) given by the "*m* factor" as discussed above. This *m* factor, similar in concept to the ideality factor in bipolar designs, is greater than unity due to the voltage sharing between the gate capacitance and bulk depletion capacitance as discussed above. The bipolar transistor does not suffer from this limitation so that the transconductance is simply $g_m = I_C / V_{thermal}$, where $V_{thermal} = kT/q$. Hence, the bipolar transistor offers a larger transconductance compared to the MOS device for a given current level and, for circuit configurations with low impedance and capacitance, the bipolar transistors can deliver excellent high-frequency performance. Due to the many telecommunications applications for such high-frequency devices, there have been intense scaling efforts to reduce unwanted parasitics in the HBT technology, particularly junction capacitance and resistance in the base and collector leads.

It is important to emphasize two fundamental differences between bipolar and MOS devices, besides those mentioned above. First, the bipolar transistor conducts current vertically in bulk material, under control of the base-emitter voltage. Hence, the current density tends to be uniformly spread over the emitter contact, if we for the moment ignore current crowding due to base resistance, compared to spreading resistance effects in the MOS SDE region, leading into the channel region. The second point relates to the current control vs. voltage control nature of the bipolar device. Namely, the flow of base current is necessary for the HBT operation, yet at the same time the controlling base current is an unavoidable consequence of bipolar current flow in junction diodes and it is strongly process-dependent. Interestingly enough, with the ongoing scaling of MOS devices the importance of gate current has been discussed above. Nevertheless, this gate current is truly a parasitic effect whereas for the HBT the flow of base current is unavoidable and required for bipolar transistor operation. Alternatively, the control base current can be isolated using a BiCMOS configuration, but this will inevitably reduce the layout efficiency and increase the technology difficulty.

Circuit design techniques that fully leverage the capabilities of bipolar transistors are well established. In the golden era of LSI, bipolar devices were used for both digital as well as analog design [38,39]. With ongoing scaling, the bipolar technology has been almost completely replaced by MOS in digital applications, especially due to power limitations. One important limitation of the bipolar device is that in order to have it turned on, the base-emitter voltage must be sufficiently large (i.e., $V_{BE}(\text{on}) = kT / q \ln(I_C/I_S)$, where I_S represents the saturation current parameter). By contrast, the MOS threshold voltage can be much smaller than this; the primary limitation becomes the I_{on}/I_{off} ratio as mentioned earlier. With the ongoing reduction of the supply voltage, the bipolar "on voltage" poses a nearly insurmountable and fundamental obstacle for low-voltage design with bipolar technology.

25.4.3 Future Technology Trends

The next few generations of technology scaling pose a range of challenges as well as opportunities. This chapter has focused primarily on bulk CMOS as the ongoing mainstream, with previews of scaling issues that will come into play for SOI type devices. The ongoing challenges related to leakage currents — both in the substrate and through the gate — bring forward the need for rethinking the materials and device architectures used for future technologies. As pointed out in earlier discussions, going to SOI and FD, DG structures shifts the challenges of substrate doping to thickness control and further concerns about heat generation and removal. The shift to SOI will continue to be evolutionary and driven by specific application requirements — speed, reduced substrate coupling (i.e., radiation effects) or other performance related issues. By contrast to evolution toward SOI-like structures, the re-engineering of the gate-stack materials is an imminent concern.

The challenges of increased gate leakage and limitations to further scale of threshold voltages using polycrystalline silicon gates have brought the issues of alternative gate-stack materials into the foreground. The two related changes — gate dielectric and gate metal — have a profound impact on scaling and device performance. The introduction of materials with increased dielectric constant (so-called "high-K" materials) offers the possibility to increase gate capacitance (and the vertical electric field) while using thicker layers compared to the ~1.5–2 nm oxide limit discussed above. This increased dielectric capacitance improves the sub-threshold slope ("*m* factor") and also can, depending on processing conditions, reduce the gate tunneling current. The changes required to a metal gate, replacing the classic silicon gate technology, are mandated to achieve desired threshold voltages and to overcome gate depletion effects as discussed earlier. The complexity of these changes — moving away from the traditional silicon-dioxide and poly-silicon materials — represents one of the potentially most "disruptive" technology changes over the most recent generations of scaling. This is an area of intense ongoing technology development; one interesting benchmark study is presented from the perspective of a foundry [40].

There are a variety of more radical, speculative technologies that are on the horizon, each seeking to find a niche in providing value. The elegance and genius of the monolithic silicon scaling that has supported Moore's law has been the ability to pattern layers at ever smaller dimensions and to integrate new materials while maintaining backward compatibility with devices and circuit fabrics from earlier technology nodes. It is beyond the scope of this chapter and premature to speculate on the impact of nanotechnology options, ranging from carbon nano-tubes (CNT) and nano-wires to self-assembly of layers of other materials and functional capabilities. New "fabrics" for memory and sensor technologies seem to be the most promising opportunities for nano-technology. In the context of the well-established monolithic process, the overarching challenge is how to structure these new layers to be aligned with underlying layers. This challenge equally applies to alternatives that fall into the class of new 3-D IC options.

As discussed briefly in considering SOI, various alternative structures, including geometry innovations and new semiconductor materials such as germanium, are now being seriously considered. Germanium has already been mentioned in the context of SiGe HBT technology; its benefits as an elemental semiconductor in field-effect transistor (FET) technology are also being reconsidered. Namely, germanium preceded silicon as a leading contender for making junction transistors in the 1940s and 1950s. However, the stability of silicon dioxide compared to water-soluble germanium oxide was a leading factor that secured the future for silicon. At the same time, the higher mobility for both holes and electrons in germanium, compared to silicon, has continued to be a very attractive feature. Moreover, there has been impressive progress in the fabrication of nearly single-crystal germanium-on-insulator (GOI) devices at low processing temperatures. There are still a host of technology challenges including the ongoing issues of gate-stack engineering as well as dopant properties to mention only two. Nonetheless, the ongoing push in scaling and increased search for new materials with improved electrical performance, have considerably broadened the playing field for innovations compared to a decade ago. An interesting point to emphasize about this quest for new materials and revisiting materials such as germanium is the fact that substantial data about the properties, including their process dependencies and statistical variations, need to be gathered in order to validate their suitability in production technologies.

To illustrate the point about how these new materials and reconsideration of device design trade-offs can lead in interesting directions, a recent simulation-based study has compared SOI and GOI devices in terms of performance [41]. This study has leveraged a detailed thermal modeling of performance, coupled with the electrical analysis. Figure 25.18 shows a plot of gate delay for the two technologies, parameterized in terms of thickness of the S/D extension regions. For the GOI device it is assumed the germanium layer thickness is $t_{Ge}=3/4t_{Si}$, where $t_{Si}=L_g/4$ with L_g being the device gate length. While the silicon has higher thermal conductivity, there is greater spatial falloff due to larger phonon mean free path. One key reason for the improved GOI performance compared to the SOI case, is the fact that germanium has a two-fold mobility advantage over silicon, allowing a 40% lower V_{dd} to be used and thus also reducing the power dissipation. The parameterized sets of curves show performance as a function of increasing the thickness of the raised S/D extension regions; it is shown that beyond $\sim 3t_{film}$ increase provide no additional benefit in terms of increased speed. These results suggest that further scaling of SOI, GOI, and possibly other thin-film semiconductor materials will need to understand carefully the electro-thermal trade-offs. In a sense, this kind of detailed physics-based study is the prelude to the ongoing challenges of 3-D integration of active devices.

25.5 Conclusions

This chapter concludes with comments very similar to those with which it began; physics-based modeling of devices is an essential part of the development process for IC electronics. By means of examples for bulk CMOS we have attempted to quantify much of the underlying device and technology understanding, and

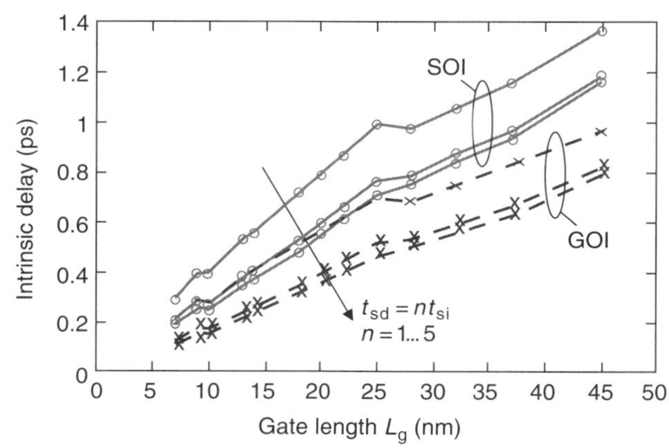

FIGURE 25.18 Self-consistent electro-thermal computational result of intrinsic delay for SOI and GOI devices. The S/D extension region thickness t_{sd} is varied, ranging from 1X to 5X the t_{si} thickness.

abstract that knowledge to levels that link with circuit design and compact modeling. Although the emphasis of the chapter has been bulk MOS transistors — the workhorse of the IC industry — there have been suitable previews of evolution toward SOI-like technologies as well as limited consideration of future technology trends. Even more so than over the past three decades, future technologies will exploit new materials in concert with detailed physical understanding and modeling. As device dimensions continue to shrink and packets of charge and current reach fundamental and quantized limits, the importance of these physics-based models is critical. Hence, many of the examples used in this chapter have explored these limits and demonstrated their impact on scaling — quantized channel/gate charge, gate tunneling current and thermal limits for SOI/GOI are illustrative. Future scaling will continue to be leveraged by such modeling that will in turn provide better devices and models for design.

References

[1] H.J. DeMan and R. Mertens, SITCAP — a simulator for bipolar transistors for computer-aided circuit analysis programs, *International Solid-State Circuits Conference (ISSCC), Technical Digest*, February, 1973, pp. 104–105.

[2] R.W. Dutton and D.A. Antoniadis, Process simulation for device design and control, *International Solid-State Circuits Conference (ISSCC), Technical Digest*, February, 1979, pp. 244–245.

[3] R.H. Dennard, F.H. Gaensslen, H.N. Yu, V.L. Rodeout, E. Bassous, and A.R. LeBlanc, Design of ion-implanted MOSFETs with very small physical dimensions, *IEEE J. Solid State Circ.*, SC-9, 256–268, 1974.

[4] R.W. Dutton and S.E. Hansen, Process modeling of integrated circuit device technology, *Proc. IEEE*, 69, 1305–1320, 1981.

[5] P.E. Cottrell and E.M. Buturla, Two-dimensional static and transient simulation of mobile carrier transport in a semiconductor, *Proceedings NASECODE I (Numerical Analysis of Semiconductor Devices)*, Boole Press, Dublin, 1979, pp. 31–64.

[6] S. Selberherr, W. Fichtner, and H.W. Potzl, Miminos — a program package to facilitate MOS device design and analysis, *Proceedings NASECODE I (Numerical Analysis of Semiconductor Devices)*, Boole Press, Ireland, 1979, pp. 275–279.

[7] C.S. Rafferty, M.R. Pinto, and R.W. Dutton, Iterative methods in semiconductor device simulation, *IEEE Trans. Electr. Dev.*, ED-32 (10), 2018–2027, 1985.

[8] M.R. Pinto and R.W. Dutton, Accurate trigger condition analysis for CMOS latchup, *IEEE Electr. Dev. Lett.*, EDL-6 (2), pp. 100–102, 1985.

[9] R.W. Dutton, Modeling and simulation for VLSI, *International Electron Devices Meeting (IEDM), Technical Digest*, December 1986, pp. 2–7.

[10] K.M. Cham, S.-Y. Oh, D. Chin, and J.L. Moll, *Computer-Aided Design and VLSI Device Development*, Kluwer, Dordrecht, 1986.

[11] R.W. Dutton and A.J. Strojwas, Perspectives on technology and technology-driven CAD, *IEEE Trans. CAD-ICAS*, 19 (12), 2000, 1544–1560.

[12] C. Duvvury and A. Amerasekera, ESD: a pervasive reliability concern for IC technologies, *Proc. IEEE*, 81, 690–702, 1993.

[13] A. Amerasekera and C. Duvvury, *ESD in Silicon Integrated Circuits*, 2nd ed., Wiley, New York, 2002.

[14] S. Dabral and T.J. Maloney, *Basic ESD and I/O Design*, Wiley, New York, 1998.

[15] C. Lombardi, S. Manzini, A. Saporito, and M. Vanzi, A physically based mobility model for numerical simulation of nonplanar devices, *IEEE Trans. CAD*, 7 (11), 1164–1171, 1988.

[16] H. Shin, A.F. Tasch, C.M. Mazier and S.K. Banerjee, A new approach to verify and derive a transverse-field-dependent mobility model for electrons in MOS inversion layers, *IEEE Trans. Electr. Dev.*, TED-36 (6), 1117–1124, 1989.

[17] H. Shin, G.M. Yeric, A.F. Tasch and C.M. Mazier, Physcially-based models for effective mobility and local-field mobility of electrons in MOS inversion layers, *Solid State Electr.*, 34 (6), 545–552, 1991.

[18] M.N. Darwish, J.L. Lentz, M.R. Pinto, P.M. Zeitzoff, T.J. Krutsick, and J.J. Vuong, An improved electron and hole mobility model for general purpose device simulation, *IEEE Trans. Electr. Dev.*, 44(9), pp. 1529–1538, September 1997.

[19] S. Takagi, M. Iwase, and A. Toriumi, On universality of inversion-layer mobility in *n*- and *p*-channel MOSFETs, *International Electron Devices Meeting (IEDM), Technical Digest*, December, 1988, pp. 398–401.

[20] C.-H. Choi, J.-S. Goo, T.-Y. Oh, Z. Yu, R.W. Dutton, A. Bayoumi, M. Cao, P. Vande Voorde, D. Vook, and C.H. Diaz, MOS *C–V* characterization of ultrathin gate oxide thickness (1.3–1.8 nm), *IEEE Elec. Dev. Lett.*, EDL-20 (6), 292–294, 1999.
[21] C. Bowen, C.L. Fernando, G. Klimeck, A. Chatterjee, D. Blanks, R. Lake, J. Hu, J. Davis, M. Kulkarni, S. Hattangady, and I.-C. Chen, Physical oxide thickness extraction and verification using quantum mechanical simulation, *International Electron Devices Meeting (IEDM), Technical Digest*, December, 1997, pp. 869–872.
[22] C.-H. Choi, K.-Y. Nam, Z. Yu, and R.W. Dutton, Impact of gate direct tunneling current on circuit performance: a simulation study, *IEEE Trans. Electr. Dev.*, ED-48 (12), 2001.
[23] C.H. Choi, Z. Yu, and R.W. Dutton, Resonant gate tunneling current in double-gate SOI: a simulation study, *IEEE Trans. Electr. Dev.*, 50 (12), 2579–2581, 2003.
[24] C.-H. Choi, Z. Yu, and R.W. Dutton, Two-dimensional polysilicon quantum-mechanical effects in double-gate SOI, *International Electron Devices Meeting (IEDM), Technical Digest*, December, 2002, pp. 723–726.
[25] H.-S. P. Wong, D.J. Frank, P.M. Solomon, H.-J. Wann, and J. Welser, Nanoscale CMOS, *Proc. IEEE*, 87, 537–570, 1999.
[26] J.P. Colinge, *Silicon-on-Insulator Technology*, 3rd ed., Kluwer, Dordrecht, 2004.
[27] Y. Taur and T.H. Ning, *Fundamentals of Modern VLSI Devices*, Cambridge University Press, Cambridge, UK, 1998.
[28] J.B. Kuo and K.-W. Su, *CMOS VLSI Engineering Silicon-on-Insulator (SOI)*, Kluwer, Boston, 1998.
[29] R.M. Swanson and J.D. Meindl, Ion-implanted complementary MOS transistors in low-voltage circuits, *IEEE J. Solid State Circ.*, SC-7, 146, 1972.
[30] J.A. Greenfield and R.W. Dutton, Nonplanar VLSI device analysis using the solution of Poisson's equation, *IEEE Trans. Electr. Dev.*, ED-27 (8), 1520–1532, 1980.
[31] G.A.M. Hurkx, D.B.M. Klaassen, and M.P.G. Knuvers, A new recombination model for device simulation including tunneling, *IEEE Trans. Electr. Dev.*, 39 (2), 331–338, 1992.
[32] C.-H. Choi, S.-H. Yang, G. Pollack, S. Ekbote, P.R. Chidambaram, S. Johnson, C. Machala, and R.W. Dutton, Characterization of Zener-tunneling drain leakage current in high-dose halo implants, *2003 IEEE International Conference on Simulation of Semiconductor Processes and Devices (SISPAD), Technical Digest*, Boston, September 2003, pp. 133–136.
[33] J.L. Hoyt, H.M. Nayfeh, S. Eguchi, I Aberg, G. Xia, T. Drake, E.A. Fitzgerald, and D.A. Antoniadis, "Strained silicon MOSFET technology, *International Electron Devices Meeting (IEDM), Technical Digest*, December 2002, pp. 23–26.
[34] M.Y. Kwong, R. Kasnavi, P. Griffin, J.D. Plummer, and R.W. Dutton, Impact of lateral source/drain abruptness on device performance, *IEEE Trans. Electr. Dev.*, 49 (11), 1882–1890, 2002.
[35] D. Buss, B.L. Evans, J. Bellay, W. Krenik, B. Haroun, D. Leipold, K. Maggio, J.-Y. Yang, and T. Moise, SOC CMOS technology for personal Internet products, *IEEE Trans. Electr. Dev.*, 50 (3), 546–556, 2003.
[36] E. Pop, R. Dutton, and K. Goodson, Thermal analysis of ultra-thin body device scaling, *International Electron Devices Meeting (IEDM), Technical Digest*, December 2003, pp. 883–886.
[37] G. Freeman, B. Jagannathan, S.-J. Jeng, J.-S. Rieh, A.D. Stricker, D.C. Ahlgren, and S. Subbanna, Transistor design and application considerations for>200-GHz SiGe HBTs, *IEEE Trans. Electr. Dev.*, 50 (3), 645–655, 2003.
[38] P.R. Gray, P.J. Hurst, S.H. Lewis, and R.G. Meyer, *Analysis and Design of Analog Integrated Circuits*, 4th ed., Wiley, New York, 2001.
[39] M.I. Elmasry, *Digital Bipolar Integrated Circuits*, Wiley Interscience, New York, 1983.
[40] H.C.-H. Wang, S.-J. Chen, M.-F. Wang, P.-Y. Tsai, C.-W. Tsai, T.-W. Wang, S. M. Ting, T.-H. Hou, P.-S. Lim, H.-J. Lin, Y. Jin, H.-J. Tao, S.-C. Chen, C. H. Diaz, M.-S. Liang, and C. Hu, Low power device technology with SiGe channel, HfSiON, and poly-Si gate, *IEDM Tech. Dig.*, 2004.
[41] E. Pop. C.O. Chui, S. Sinha, R. Dutton, and K. Goodson, Electro-thermal performance comparison and optimization of thin-body SOI and GOI MOSFETs, *International Electron Devices Meeting (IEDM), Technical Digest*, December, 2004.
[42] International Technology Roadmap for Semiconductors (available from http://public.itrs.net) ITRS (2003).
[43] R.-H. Yan, A. Ourmazd, and K. F. Lee, Scaling the Si MOSFET: from bulk to SOI to bulk, *IEEE Trans. Electr. Dev.*, 39 (7), 1704–1710, 1992.
[44] R.W. Dutton, B. Troyanovsky, Z. Yu, E.C. Kan, K. Wang, and T. Chen, Advance analog circuit modeling with virtual device and instrument, *Proceedings of the ISSCC*, San Francisco, CA, February 1996, p. 78.

26

High-Accuracy Parasitic Extraction

Mattan Kamon
Coventor, Inc.
Cambridge, Massachusetts

Ralph Iverson
Magma Design Automation
Arlington, Massachusetts

Abstract

In this chapter, we describe high-accuracy parasitic extraction by both fast integral equation methods as well as random-walk-based methods. For any extraction application, the appropriate simplification of Maxwell's equations must be chosen and then solved to generate an appropriate model suitable for circuit simulation. We will briefly describe these various domains of electromagnetics, and then discuss integral equation approaches as they are used for capacitance, inductance, and full-wave analysis. We complete the chapter with a discussion of random-walk methods as used for capacitance extraciton.

26.1 Introduction

The extraction of parasitic circuit models is important for various aspects of physical verification such as timing, signal noise, substrate noise, and power grid analysis. As circuit speeds and densities have increased, the need has grown to account accurately for parasitic effects for larger and more complicated interconnect structures. In addition, the electromagnetic complexity has grown as well, from resistance and capacitance, to inductance, and now even full electromagnetic wave propagation. This increase in complexity has also grown for the analysis of passive devices such as integrated inductors.

Electromagnetic behavior is governed by Maxwell's equations, and all parasitic extraction requires solving some form of Maxwell's equations. That form may be a simple analytic parallel plate capacitance equation, or may involve a full numerical solution for a complicated 3D geometry with wave propagation. Analytic formulas for simple or simplified geometry can be used where accuracy is less important than speed, but when the geometric configuration is not simple and accuracy demands do not allow simplification, numerical solution of the appropriate form of Maxwell's equations must be employed (see Chapter 22, *Layout Extraction*, for details).

The appropriate form of Maxwell's equations is typically solved by one of two classes of methods. The first uses a differential form of the governing equations and requires the discretization (meshing) of the entire domain in which the electromagnetic fields reside. Two of the most common approaches in this first class are the finite difference (FD) and finite element method (FEM). The resultant linear algebraic system (matrix) that must be solved is large but *sparse* (contains very few nonzero entries). Sparse linear solution methods, such as sparse factorization, conjugate-gradient, or multigrid methods can be used to solve these systems, the best of which require CPU time and memory of $O(N)$ time, where N is the number of elements in the discretization. However most problems in electronic design automation (EDA) are *open* problems, also called exterior problems, and since the fields decrease slowly toward infinity, these methods can require extremely large N.

The second class of methods are integral equation methods which instead require a discretization of only the *sources* of electromagnetic field. Those sources can be physical quantities, such as the surface charge density for the capacitance problem, or mathematical abstractions resulting from the application of Green's theorem. When the sources exist only on two-dimensional surfaces for three-dimensional problems, the method is often called a boundary-element method (BEM). For open problems, the sources of the field exist in a much smaller domain than the fields themselves, and thus the size of linear systems generated by integral equations methods are much smaller than FD or FEM, as illustrated for a small portion of two signal lines in Figure 26.1.

Integral equation methods, however, generate *dense* (all entries are nonzero) linear systems that makes such methods preferable to FD or FEM only for small problems. Such systems require $O(N^2)$ memory to

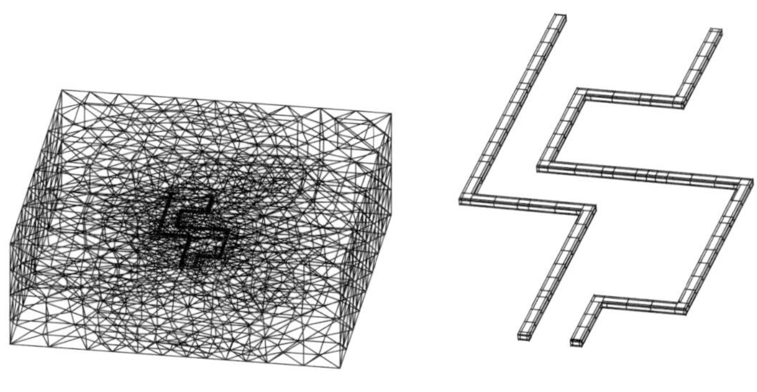

FIGURE 26.1 The FEM mesh on the left discretizes the space (infinite) surrounding the conductors while the BEM mesh on the right only discretizes the conductor surfaces.

store and $O(N^3)$ to solve via direct Gaussian elimination or at best $O(N^2)$ if solved iteratively. Increasing circuit speeds and densities require the solution of increasingly complicated interconnect, making dense integral equation approaches unsuitable due to these high growth rates of computational cost with increasing problem size.

In the past two decades, much work has gone into improving both the differential and integral equation approaches as well as new approaches based on random-walk methods [1]. Methods of truncating the discretization required by the FD and FEM approaches has greatly reduced the number of elements required [2,3]. Integral equation approaches have become particularly popular for interconnect extraction due to *sparsification* techniques, also sometimes called matrix compression, acceleration, or matrix-free techniques, which have brought nearly $O(N)$ growth in storage and solution time to integral equation methods [4–10].

The focus of this chapter will be the sparsified integral equation approaches and the random-walk methods. The sparsified integral equation techniques are typically used to solve capacitance and inductance-extraction problems. The random-walk methods have become quite mature for capacitance extraction. For problems requiring the solution of the full Maxwell's equations (full-wave), both differential and integral equation approaches are common.

Part I of this chapter discusses the sparsification methods in relation to the various integral forms of Maxwell's equations used in EDA. Part II discusses the random-walk approach and compares it to one of the deterministic approaches of Part I.

Part I: Extraction via Fast Integral Equation Methods

26.2 Introduction

In this part, we describe the popular detailed parasitic extraction approaches by integral-equation solution of the governing electromagnetics. For any extraction application, the appropriate simplification of Maxwell's equations must be chosen and then solved to generate an appropriate model suitable for circuit simulation. We will briefly describe these various domains of electromagnetics. *Sparsified* integral equation approaches have become quite popular for their solution. We will briefly describe the sparsification techniques in the context of each of the relevant electromagnetic domains as well as methods of generating models appropriate for circuit simulation.

In the next section we explore the various forms of Maxwell's equations, then in Section 26.4 we describe sparsification methods for capacitance extraction. These fast methods can also be applied to inductance extraction as we see in Section 26.5, as well as the distributed RLC and full-wave solutions in Sections 26.6.1 and 26.6.2.

26.3 Forms of Maxwell's Equations

The term *field solver* commonly refers to any tool that numerically solves some form of Maxwell's equations. Most problems related to EDA do not require solving the full set of Maxwell's equations. In this section, we will describe the simplified forms of Maxwell's equations for various applications in extraction. Table 26.1 summarizes the rest of this section.

While any parasitic behavior is described by Maxwell's equations, it would be far too inefficient to request multiple field solver calls during the course of a higher level simulation, such as in a circuit simulation. Instead, the simulator typically takes some responsibility for the electromagnetic behavior by assuming some network topology of parasitic elements and then the field solver is used to *extract* parameters for the elements of the topology.

For example, parasitic extraction for timing analysis where inductance is not important would require solution of the electroquasistatic (distributed RC) equations. However, using a field solver for the electroquasistatic equations is typically not done. Instead, as described in Chapter 22, static resistance and static

TABLE 26.1 Forms of Maxwell's Equations

Form of Maxwell's Equations	Common Circuit Description	Section	Example Application
Electrostatic	C	Sec 26.4	Charge-dominated, $L << \lambda$
Conduction	R	See Chapter 23, Sec 23.3.2	Line resistance IR drop
Electroquasistatic	RC	See Chapter 23, Sec 23.3.2	SRAM cell Substrate noise
Magnetoquasistatic	RL	Sec 26.5	Current-dominated $L << \lambda$
Electromagneto-quasistatic	RLC	Sec 26.6.1	$L < \lambda, L \approx \lambda$ On-chip inductors
Full Maxwell (Full-wave)	*S*-parameters	Sec 26.6.2	Interconnect, $L > \lambda$ RF/microwave circuits passive devices

Note: L is a characteristic length of interconnect; λ is the wavelength of an electromagnetic wave at the frequency of interest.

capacitance are computed separately on small sections of interconnect and the higher-level simulator is responsible for the dynamic RC behavior. The simulator accomplishes this by connecting the R and C together in an appropriate ladder network. This permits the use of a field solver that solves the much simpler electrostatic (C) problem. Solution of the electrostatic problem is discussed in Section 26.4.

The delegation of the dynamics to the simulator for RC extraction is only possible because a circuit topology (a ladder network) is available *a priori* which meets the accuracy needs of the application. In contrast, consider analyzing the dynamic behavior of an SRAM cell that is very sensitive to amount and distribution of coupling capacitances and resistance. Here, attempting to associate automatically a circuit topology would require heuristics that would likely incur intolerable error. Instead, the full electroquasistatic equations would be the best approach [11, 12].

In contrast to parasitic RC extraction, field solvers for parasitic inductance extraction must solve the magnetoquasistatic (RL), not the magnetostatic (L) problem. The resistance cannot be separated from the inductance because it directly affects the current distribution in a frequency-dependent manner, and the inductance is very dependent on this current distribution. Two examples are the *skin effect* which causes currents to travel more toward the surface of conductors as the frequency rises, and the *return path* of signals in a ground plane or grid arrangement which is closer to the signal as the frequency rises. (In fact, the challenge for full-chip inductance extraction centers around heuristics for the return path. See Chapter 22 or [13]). Solution of the inductance problem is described in Section 26.5.

Historically, many extraction problems could be *lumped* into inductance-dominated portions, such as the leads or wire bonds of lead-frame packages, and capacitance-dominated portions, such as sections of on-chip wiring. However, as circuit speeds have increased, more interconnect problems require a *distributed* representation of not only resistance and capacitance, but also inductance. A good example is the on-chip inductor, which has parasitic capacitance to the substrate as well as turn-to-turn capacitance. Accurately determining the self-resonance cannot be accomplished with a single inductance extraction separate from capacitance. Solving this distributed RLC problem has sometimes been called the electromagnetoquasistatic form of Maxwell's equations [14] and will be described in Section 26.6.1.

Modeling a system as distributed is also important for capturing the finite propagation speed of electromagnetic phenomena. When such effects are important, the full Maxwell equations (full-wave) problem must be solved as briefly described in Section 26.6.2.

A common method for determining whether a system is distributed or lumped is to compare the characteristic length of interconnect to the wavelength of electromagnetic waves at the maximum frequency of signals in the system. A rule-of-thumb is that interconnect longer than one-tenth the wavelength is distributed. To determine the maximum frequency for digital systems, one possible metric described in [15] is

$$F = \frac{0.5}{T_r} \tag{26.1}$$

where F is in Hertz and T_r the 10–90% pulse rise time in seconds. For instance, a 2 mm wire bond in an encapsulant with a relative electrical permittivity of 4 should use a distributed model for rise times faster than 67 ps.

For distributed problems, extraction might be the creation of an RLC network [16]. At today's circuit densities, such methods are impractical due to the density of mutual couplings, and a field solver that directly generates a reduced-order model in the form of a few linear ordinary-differential equations is typically required, as will be discussed in Section 26.6.

The next sections briefly derive the various classes of equations described above and discuss recent sparsification approaches to solve for the quantities of interest. For detailed derivations and descriptions of each of the regimes in the above table, see [17]. A good summary of the relationship between Maxwell's equations and circuit equations can be found in [18].

26.4 Fast Field Solvers: Capacitance Solution

Describing sparsification for integral equation solution is best done through the capacitance problem. Methods of fast inductance and full-wave solution build on the fast techniques for capacitance.

The capacitance problem is to find the relation between net charge and voltage for a set of k-conducting bodies: $Q = CV$, where C is the $k \times k$ capacitance matrix, V the vector of k conductor voltages, and Q the vector of net charge on each of the k conductors. A column i of this matrix can be determined by computing the net charge on all k conductors when conductor i is set to 1 V and the rest to zero.

To compute the net charge from the voltage, we assume magnetic effects are negligible and solve the Laplace equation for the potential ϕ

$$\nabla^2\phi(r) = 0, \quad r \in \Omega \tag{26.2}$$

where $\phi(r)$ is the potential in the space Ω surrounding all the conductors, $\phi(r) = V_i$ on the surface of each conductor i, and V_i the voltage of conductor i. Given a solution to Equation (26.2) in Ω, it can be shown that the surface charge density, ρ_s, is

$$\rho_s = -\varepsilon\frac{d\phi}{dn} \tag{26.3}$$

where n is the surface normal pointing out of a conductor and ε the permittivity of the space around the conductor. The net charge on any conductor i is then

$$Q_i = \int_{S_i}\rho_s\, da \tag{26.4}$$

where S_i is the surface of conductor i.

The Laplace Equation (26.2) is the differential form for electrostatics as would be solved by an FD [19, 20] or FEM [21, 22] method. Many integral forms exist [23–25]. The most familiar is the first-kind formulation also known as charge superposition,

$$\phi(\boldsymbol{r}) = \int_S \rho_s(\boldsymbol{r}')\frac{1}{4\pi\varepsilon|\boldsymbol{r}-\boldsymbol{r}'|}da' \tag{26.5}$$

where S is the union of all conductor surfaces, $|\boldsymbol{r} - \boldsymbol{r}'|$ the distance between charge point $\boldsymbol{r}'$ and evaluation point $\boldsymbol{r}$. Equation (26.5) may be more familiar from introductory electromagnetics as the integral form of the potential due to a set of point charges, q_i: $\phi(\boldsymbol{r}) = \Sigma_i\, q_i/4\pi\varepsilon|\boldsymbol{r} - \boldsymbol{r}_i|$.

Equation (26.5) directly relates the potential at a point on a conductor surface, which is known, to the unknown charge distribution on the surface. After solving this integral equation for ρ_s, one can use Equation (26.4) to compute the net charge.

The presentation above assumes the space around the conductors is a single, infinite, homogenous dielectric. Methods that solve the diffential form can naturally handle arbitrary dielectrics and methods that solve the integral forms can be augmented for piecewise constant regions of isotropic dielectrics [24,26,27] as are common in integrated circuit and package geometries.

A common approach to solve Equation (26.5) numerically is to discretize the surface of the conductors into panels as shown on the right in Figure 26.1 and assume the charge density is uniformly distributed on each panel. That is, ρ_s is approximated as a sum of basis functions, w_i,

$$\rho_s(\boldsymbol{r}) \approx \sum_{i=1}^{N} q_i w_i(\boldsymbol{r}) \tag{26.6}$$

where $w_i(\boldsymbol{r}) = 1$, for $\boldsymbol{r}$ on panel i and 0 elsewhere. The goal is then to solve for the unknown magnitude of the charge density, q_i, on each panel.

To turn this discretization into a system of linear equations, one can require that the integral equation be satisfied at the center of each panel. Such an approach is formally called a weighted residuals [28] approach based on collocation [29]. For a problem with n panels, inserting Equation (26.6) into Equation (26.5) enforced at the n panel centers results in a linear system

$$Pq = v \tag{26.7}$$

where $q \in \mathbb{R}^n$ are the unknown charges on each panel, $v \in \mathbb{R}^n$ the known panel voltages, and $P \in \mathbb{R}^{n\times n}$. The entries of the matrix, P_{ij}, can be computed from the integral relation

$$P_{ij} = \frac{1}{a_j}\int_{p_j} \frac{1}{4\pi\varepsilon|\boldsymbol{r}_i - \boldsymbol{r}'|}\,\mathrm{d}a' \tag{26.8}$$

where $\boldsymbol{r}_i$ is the center of panel i, p_j the surface of panel j, and a_j the area of panel j. The integral can be computed analytically for quadrilateral panels [30]. Note that because the potential at any point has a contribution from the charge on *every* panel, $P_{ij} \neq 0$ for all i,j, and the matrix P is dense.

Computing each column of the capacitance matrix, C, is now a matter of solving Equation (26.7) for the charge vector, q, and summing up the charge for each conductor to arrive at the net charge vector Q. Solving Equation (26.7) by a direct matrix solution method such as Gaussian elimination requires $O(n^3)$ computation, which is intractable for complicated interconnect geometries requiring hundreds of thousands of panels. To avoid the $O(n^3)$ cost, an iterative technique, such as GMRES [31], can be employed. The dominant cost of applying an iterative scheme is the $O(n^2)$ cost at each iteration of computing a matrix–vector product, Px, where x is a vector computed at each iteration.

Sparsification approaches turn the $O(n^2)$ matrix–vector product cost into $O(n)$ or $O(n \log n)$ by exploiting the fact that $y = Px$ can be interpreted as the evaluation of the potentials, y, due to a set of known charges, x. Most of these approaches hinge on the fact that the potential contribution from far away charges is much smaller than those nearby. These far-field interactions are treated approximately without significantly impacting overall accuracy.

One of the first sparsification techniques applied to capacitance extraction was the fast multipole method (FMM) [4–6] with $O(n)$ complexity in time and memory. The idea, illustrated in Figure 26.2, is to approximate a group of distant charges within some radius R as a single charge, and then use that approximation for evaluation of potentials at points a distance r away, where $r >> R$. Using a single charge to represent this group is a monopole expansion, and using a series of higher-order representations, such as a monopole, dipole, quadrapole, etc., is a multipole expansion. Similarly, the potential due to many multipole expansions can be expressed with a local expansion centered at a point around a cluster of evaluation points. By using multipole and local expansions and by keeping the ratio r/R constant for all expansions, Px can be computed in $O(n)$ time and memory for a given error tolerance. The details

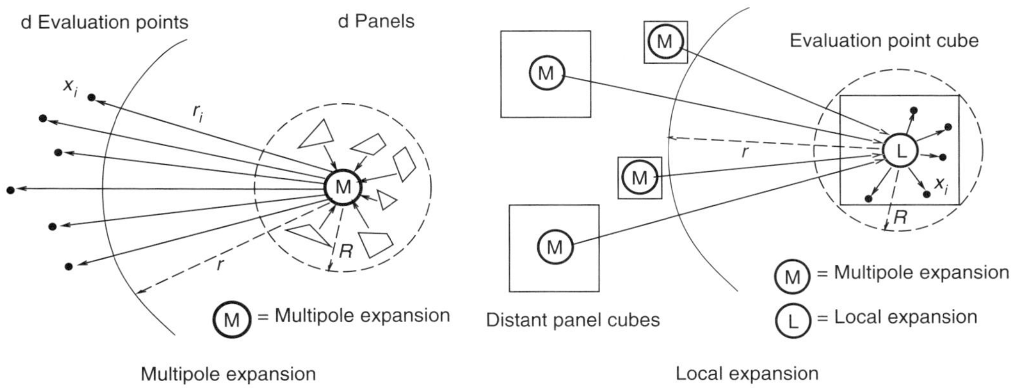

FIGURE 26.2 Multipole expansions represent the potential due to a cluster of charges. Local expansions represent the potential at a cluster of evaluation points due to many multipole expansions.

of applying an iterative solution algorithm and FMM sparsification to the capacitance problem can be found in [32], with a proof of $O(n)$ given in [33].

FMM methods reduce the computation time and memory by orders of magnitude for even modest capacitance problems. It was shown in [6] that even a small problem of roughly 6000 panels could be solved 10 times faster than a dense matrix–vector product iterative scheme, and 100 times faster than direct factorization.

The drive to solve even larger problems motivates continued work on multipole-based sparsification for interconnect extraction [34, 35]. In addition, other methods of sparsification such as the precorrected FFT [8] discussed later and SVD-based approaches [9] further improve performance.

26.5 Fast Inductance Solution

Given an interconnect network with some user-defined ports, inductance extraction is the process of computing the complex frequency-dependent port admittance matrix, Y_p, under the magnetoquasistatic approximation. In this section, we give a brief overview of fast inductance extraction.

Because of the strong dependence of the port inductance on the current distribution, the inductance problem is best understood when the governing integral equation is cast into an equivalent circuit. The resistive and inductive part of the partial element equivalent circuit (PEEC) method [36,37] is the best known method for generating an equivalent circuit. Originally, this equivalent circuit was then used directly within a circuit simulator to model the interconnect. Unfortunately, present day interconnect problems generate equivalent circuits too large to insert directly into a circuit simulator. Instead, the equivalent circuit must either be solved to extract the port admittances at a particular frequency of interest, or the equivalent circuit model must be reduced to something tractable for insertion into a circuit simulator. For either approach, sparsification methods greatly increase the complexity of interconnect these methods can analyze.

In this section, we describe the governing integral equation, derive the equivalent circuit, and then describe the circuit solution methods necessary to apply sparsification methods. The model reduction approach will be described briefly later in the chapter.

26.5.1 Integral Equation and Equivalent Circuit

Several integral equation-based approaches have been used to derive the Y_t associated with a given package or interconnect structure [37–40]. These integral formulations start with Maxwell's equations in the frequency domain, and then apply the magnetoquasistatic assumption that the displacement current is negligible everywhere.

This leaves the current density, J, and scalar potential, Φ, as the unknown quantities for the integral equation

$$\frac{J(r)}{\sigma} + \frac{s\mu}{4\pi}\int_V \frac{J(r')}{|r-r'|}\mathrm{d}v' = -\nabla\Phi(r), \quad r \in V \tag{26.9}$$

where s is the Laplace frequency, σ the conductivity, μ the permeability, and V the union of the volumes of all conductors. The unknown quantities can be computed by combining Equation (26.9) with conservation of current

$$\nabla \cdot J = 0 \tag{26.10}$$

To create a linear system from Equation (26.9), note that current travels in the *volume* of the conductors, so the interior of all conductors must be discretized. Current within a long thin conductor can be assumed to flow parallel to its surface since the magnetoquasistatic assumption implies there is no charge accumulation on the surface. Thus, for long thin structures such as pins of a package or signal lines on an integrated circuit, the conductor can be divided along its length into segments. In order to properly capture skin and proximity effects in these long, thin conductors, the cross-section of the segments can be divided into a bundle of parallel *filaments*. It is also possible to use the filament approach for planar structures, such as ground planes, where the current distribution is two-dimensional. In such cases, a grid of filaments must be used.

Since the current density inside each filament is assumed to be constant, the unknown current distribution can be approximated as a sum of basis functions,

$$J(r) \approx \sum_{i=1}^{b} I_i w_i(r)\mathbf{l}_i \tag{26.11}$$

where b is the number of filaments, I_i the current inside filament i, $\mathbf{l}_i$ a unit vector along the length of the filament, and $w_i(r)$ has a value of zero outside filament i, and $1/a_i$ inside, where a_i is the cross-sectional area. Note that unlike capacitance extraction, these basis functions are vector quantities.

By following an approach similar to that for capacitance extraction to generate a linear system, we can arrive at

$$(R + sL)I_b = \tilde{\Phi}_A - \tilde{\Phi}_B \tag{26.12}$$

where $I_b \in \mathbb{C}^b$ is the vector of b filament currents,

$$R_{ii} = \frac{l_i}{\sigma a_i} \tag{26.13}$$

is the $b \times b$ diagonal matrix of filament DC resistances,

$$L_{ij} = \frac{\mu}{4\pi a_i a_j}\int_{V_i}\int_{V_j'} \frac{\mathbf{l}_i \cdot \mathbf{l}_j}{|r-r'|}\mathrm{d}V'\,\mathrm{d}V \tag{26.14}$$

is the $b \times b$ dense, symmetric positive-definite matrix of partial inductances, V_i and V_j the volumes of filaments i and j, respectively, and $\tilde{\Phi}_A$ and $\tilde{\Phi}_B$ the averages of the potentials over the cross-sections of the filament end faces. Analytic methods for computing the terms of this matrix are given in [37,41,42]. Equation (26.12) can also be written as

$$ZI_b = V_b \tag{26.15}$$

where $Z = R + sL \in \mathbb{C}^{b\times b}$ is called the branch impedance matrix and $V_b = \tilde{\Phi}_A - \tilde{\Phi}_B$ the vector of branch voltages.

Note that intuitively one can view filament i as a resistor with resistance R_{ii} in series with an inductor with self-inductance L_{ii} and b mutual inductances L_{ij} each magnetically coupling filament i to one other filament j. To enforce Equation (26.10), the interconnection of the current filaments can be represented with a planar graph, where the n nodes in the graph are associated with connection points between filaments, and the b branches in the graph represent the filaments into which each conductor segment is discretized. These ideas are illustrated in Figure 26.3.

The circuit obtained from the graph described above is the resistive and inductive part of the well-known PEEC method [36, 37].

The description above of filaments as inductors is not precise since inductance is a closed-loop quantity but each filament is a straight section. A filament's inductance represents only part of a loop and is thus termed the "partial" inductance. It can be shown that the correct loop inductance will be extracted if (1) all the filaments that make up the loop of current are included in the PEEC circuit, and (2) all the mutual inductances between the filaments are included, that is, the full L matrix described above is used for extraction. A full description of the connection between the partial inductance concept and the definition of loop inductance is given in [43]. More detail is also given in Chapter 22.

The practical application of the partial inductance concept is that creating a PEEC circuit of just an on-chip signal line or just a single bondwire is not adequate to model the inductive behavior of signals carried on that conductor. Instead, all the lines which represent the returning current and all the mutual inductances must also be included in the model to capture the true loop inductance.

26.5.2 Extraction of Port Admittances via Sparsification

The creation of the PEEC circuit and the R and L values could be the end of the extraction step. The PEEC circuit network could be inserted into a circuit simulator to represent the interconnect. However, most interconnect problems generate thousands to hundreds of thousands of filaments, and thus insertion in a circuit simulator would be computationally intractable because of the dense mutual couplings between filaments. Instead, it is best to solve the PEEC circuit for the port admittances, Y_t, and use those for circuit simulation. Since the L matrix is dense, the computational cost of solving will be at least $O(b^3)$ for direct methods. Instead, we wish to apply an iterative method with some sparsification approach.

A standard approach to form a linear system for the circuit would be to apply nodal analysis and generate a sparse tableau form [44]. Such a system includes both the constitutive relations Equation (26.15) as well as the circuit topology relations implied by Equation (26.10). Unfortunately, applying an iterative

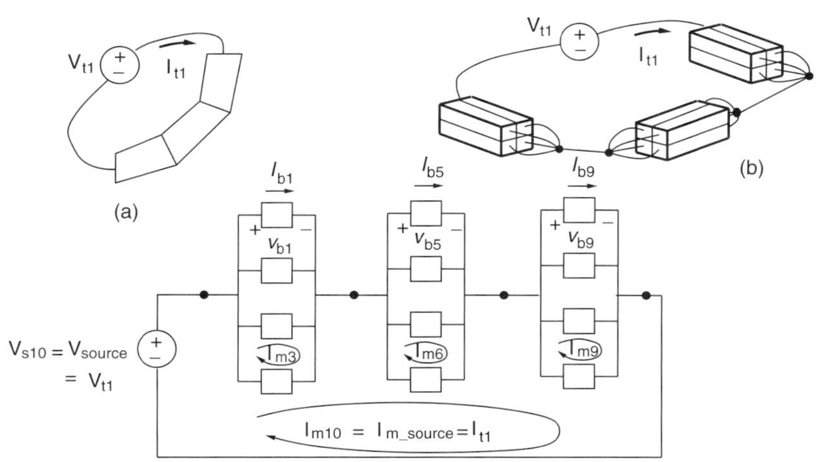

FIGURE 26.3 One conductor, (a) as piecewise-straight segments, (b) discretized into filaments, (c) modeled as a circuit with each box element representing a filament as a resistor in series with an inductor. Note that each filament has mutual inductance to every other inductor. Mesh analysis quantities indicated with m suffix.

technique to the sparse tableau form requires a very large number of iterations to achieve a solution [14]. A common method of reducing the iteration count is to apply a *preconditioner* (see [45]); however, developing a preconditioner for this problem is difficult because the sparse tableau approach includes unknowns of different types.

Instead, we can reformulate the equations using mesh analysis, and then apply an iterative method. Here "mesh" refers to the circuit concept rather than the discretization mesh described previously. In mesh analysis [46], a mesh is any loop of branches in the circuit graph that does not enclose any other branches. Kirchoff's voltage law, which implies that the sum of branch voltages around each mesh in the network must be zero, is represented by

$$MV_{\mathrm{b}} = V_{\mathrm{s}} \tag{26.16}$$

where V_{b} is the vector of voltages across each branch except for the source branches, V_{s} mostly the zero vector of source branch voltages, and $M \in \mathbb{R}^{m\times b}$ the mesh matrix, where $m = b' - n + 1$ is the number of meshes and b' the number of filaments branches plus the number of source branches.

By defining $I_m \in \mathbb{C}^m$ as the vector of currents that circulate around each mesh as shown in Figure 26.3, it can be shown that

$$M\,Z\,M^{\mathrm{T}}I_m = V_{\mathrm{s}} \tag{26.17}$$

This system has a single type of unknown quantity, I_m, and preconditioning this system for $O(m^2)$, iterative solution is possible [40].

The system is dense due to the dense mutual couplings of L from Equation (26.14). Note that Equation (26.14) shares the same *kernel*, $1/|\mathbf{r} - \mathbf{r}'|$, as the capacitance problem, and thus similar spasification techniques can also be applied to inductance.

Applying sparsification techniques such as FMM for the inductance problem is not as effective as doing so for the capacitance problem for two reasons. First, the mutual inductance depends on the dot product between the vector directions of the two filaments of the coupling as shown in Equation (26.14). Since sparsification techniques represent many different filaments as a group, this representation must be done component-wise, requiring three sparsified evaluations to perform a matrix–vector product for L, compared to the single one for capacitance. Secondly, capturing skin effect can require the cross-sections of wires to be divided into bundles of 25 to 100 filaments. Since the current density typically does not change along the length of a straight wire, the best mesh is often a very long bundle of densely packed filaments. Unfortunately, the long length of the wires limits the multipole radius, R, to be large, and since $r >> R$, much less of the matrix-vector product can be sparsified.

Nonetheless, applying a mesh-analysis-based iterative scheme sparsified by the fast multipole method can result in significant speedup as shown on the left in Figure 26.4. The data are for two long traces with return paths through a finite-conductivity ground plane detailed in [40]. Using a mesh-analysis-based iterative scheme was nearly two orders of magnitude faster than direct factorization for $m = 12,000$. Usage of FMM sparsification cut the time further by nearly a factor of 4. The memory savings usage of FMM was also over an order of magnitude for this modest problem size as shown on the right in Figure 26.5.

Some of the shortcomings of applying sparsification for inductance extraction under skin effect can be overcome by using basis functions whose shape more closely match the actual exponentially varying current distribution instead of using the piecewise constant functions of Equation (26.11). Such techniques include those which use an analytic form for the basis functions [47,48], as well as numerically determined ones [49,50]. Alternately, if an integral equation could be derived that requires only a discretization of the surface yet correctly captures the interior current distribution for both low and high frequency, then the many interior filaments could be avoided. Such a pure surface formulation for Maxwell's equations is pursued in [51,52].

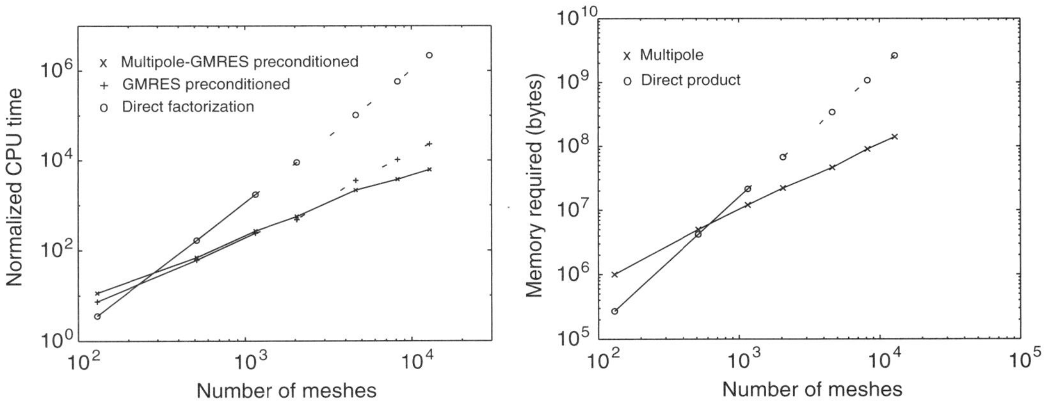

FIGURE 26.4 Comparison of the CPU time (left) and memory (right) to compute the reduced inductance matrix for two traces over a solid plane.

26.6 Distributed RLC and Full Wave

When an application requires a distributed interconnect model, the full Maxwell equations must be considered. While both finite element and integral equation-based approaches are common, we restrict the discussion to integral equation techniques. The integral equation commonly used extends that from Equation (26.9):

$$\frac{J(r)}{\sigma} + \frac{s\mu}{4\pi}\int_V \frac{J(r')e^{(s/c)|r-r'|}}{|r-r'|}\mathrm{d}v' = -\nabla\Phi(r), \quad r \in V \tag{26.18}$$

$$\Phi(r) = \frac{1}{4\pi\varepsilon}\int_S \frac{\rho_s(r')e^{(s/c)|r-r'|}}{|r-r'|}\mathrm{d}v', \quad r \in \mathbb{R}^3 \tag{26.19}$$

where s is the Laplace frequency, c the speed of light, V the interior of all conductors, J the current density in V, S the surface of all conductors and ρ_s the charge density on S. Additionally, the currents and charge obey the conservation equation,

$$\nabla \cdot J(r) = 0, \quad r \in V \tag{26.20}$$

$$\boldsymbol{n} \cdot J(r) = -s\rho_s(r), \quad r \in S \tag{26.21}$$

where $\boldsymbol{n}$ is the inward normal on S. These equations form the basis of many integral equation approaches. By adding capacitance-like circuit elements to the circuit in Figure 26.3 to represent Equation (26.19), we have the rPEEC method [53]. By assuming the propagation delay is negligible ($e^{(s/c)|r-r'|} \approx 1$), we arrive at the original PEEC method [16].

Other similar integral equations form the basis of codes in other fields. If all current is assumed to have a direction in the *xy* plane and conductors are thin in the *z* direction compared to their width, we arrive at the $2\frac{1}{2}$ D approaches common in the RF and microwave community. In 3D, if the frequency is high enough to assume current is confined to an infinitesimally thin layer on the surface of conductors, then we arrive at the high-frequency surface approaches, such as [54] often used for electromagnetic scattering and radiation calculations.

For continuity from previous sections and to highlight recent work, we will describe the PEEC-like approaches in the rest of this section.

The integral operators of Equation (26.18) and Equation (26.19) are similar to those described for capacitance Equation (26.5) and inductance Equation (26.9) extraction, but with the addition of a propagation delay term $e^{(s/c)|r-r'|}$. By modifying the L and P matrices appropriately and by considering the charge panels as PEEC circuit elements as well, we can follow a mesh-based circuit approach to generate a linear system for extracting the port admittances, Y_p,

$$M Z_{\mathrm{EM}} M^{\mathrm{T}} I_m = V_s \tag{26.22}$$

where

$$Z_{\mathrm{EM}} = \begin{bmatrix} R + sL(s) & 0 \\ 0 & P(s)/s \end{bmatrix} \tag{26.23}$$

Note the matrices L and P are now frequency-dependent. In this context, the mesh analysis approach has similarities to the "divergence-free basis functions" of Ref. [55] used in scattering analysis.

26.6.1 Distributed RLC

If the frequency is low enough that $e^{(s/c)|r-r'|} \approx 1$, then L and P are no longer frequency-dependent, and we arrive at the distributed RLC problem. Again, one would not want to insert the PEEC circuit of tens to hundreds of thousands of elements into a circuit simulator, but instead reduce Equation (26.22) to something manageable for circuit simulation.

One approach is to solve Equation (26.22) for either the admittance, Y_t, or the scattering parameters at a discrete set of frequency points. Making this solution computationally tractable again requires the application of a sparsified iterative solution technique. Because L and P are identical to the matrices from the previous sections, the fast multipole method can be applied, provided a good preconditioner is available to keep the iteration count small.

However, the frequency dependence of distributed RLC problems is much stronger than the RL problem due to resonant behavior, and thus solving Equation (26.22) at a small set of discrete frequency points can incur significant error. For instance, this approach can completely miss a resonant peak which falls between two adjacent frequency points. In addition, using frequency-domain discrete data for time-domain simulation requires an additional data fitting approximation step for inclusion in circuit simulation (for a discussion of this issue, see [56]).

Instead, one can apply model order reduction methods to Equation (26.22) to directly derive a model for circuit simulation. These methods require that the original system be written in a state space form such as the nth-order system

$$\begin{aligned} s\mathcal{L}x &= -\mathcal{R}x + BV_t \\ I_t &= C^{\mathrm{T}}x \end{aligned} \tag{26.24}$$

where $x \in \mathbb{R}^n$ is the vector of states with dimension comparable to the number of discretization elements (panels and filaments), s is the Laplace frequency, $V_t, I_t \in \mathbb{R}^t$ are the input and output vectors, respectively, $\mathcal{L},\mathcal{R} \in \mathbb{R}^{n\times n}$, and $B,C \in \mathbb{R}^{n\times t}$. The idea of model reduction is to derive a much smaller qth-order system of form similar to Equation (26.24) with $q \ll n$ but which still accurately models the system behavior. For circuit simulation, a qth-order system of this form can be written directly in an analog hardware description language such as Verilog-A or VHDL-AMS.

Similar to applying an iterative technique, the cost of model generation is dominated by the dense matrix–vector products, and thus sparsification techniques are just as important.

The challenge of these model reduction methods is threefold: (1) to be numerically robust and efficient for the large systems of EDA, (2) preserve the passivity of the original system, and (3) generate an optimally compact model that is accurate over the desired frequency range. A brief history of these methods can be found in Chapter 15, Section 15.4.3. The PRIMA model reduction algorithm described in [57] satisfies (1) and passivity (2) is preserved if the original system meets certain structural criteria.

In [58], it was shown that Equation (26.22) could be rewritten in a form suitable for the application of a multipoint version of PRIMA [59], however generating an optimally compact model required manual exploration. In [60–62], methods are developed to satisfy (1) and (3), and in [63] a numerically robust method is presented satisfying (2) and (3). Finally, the two-step procedure in [64] suggests that all three requirements could be satisfied by combining, for instance, a first step of [59] followed by a second step of [63]. Such a procedure is not fully automatic since a model order must be chosen for the first step, which accurately models the desired system behavior but is not too large for the more computationally expensive second step.

26.6.2 Full Wave

When the propagation delay of mutual couplings is important, the full-wave problem must be solved, and $L = L(s)$ and $P = P(s)$ now have the $e^{(s/c)|r-r'|}$ oscillatory term. This frequency dependence changes the strategies necessary for both sparsification and model reduction.

Sparsification techniques, such as the FMM, are well tuned for the quasistatic cases described earlier, but to date are not as efficient for surface integral equation solutions to the full Maxwell equations [10]. Sparsification techniques such as the precorrected FFT [8], which are nearly kernel-independent, can be applied instead. Instead of treating the far-field interaction as groups of multipole expansions which must be translated in a kernel-dependent manner, the precorrected FFT approximates the distribution of sources as point sources lying on a uniform grid as shown in Figure 26.5. Since the charges lie on a uniform grid, the potential they produce on the grid can be computed with an FFT for any translation-invariant kernel. Such an approach has been shown to have $O(N \log N)$ complexity when the geometry is reasonably homogenously distributed over a domain.

The model reduction problem for full wave becomes more difficult because the $e^{(s/c)|r-r'|}$ terms in $L(s)$ and $P(s)$ prevent Equation (26.17) from being written as the first-order linear system of Equation (26.24). Approaches for model reduction can be found in [65, 66], however automatically generating a passive, optimally compact model is still an open problem. Under suitable approximation, the matrices generated by a FEM are not frequency-dependent and some of the recent methods described in the previous section could be applied. An approach that uses a Lanczos method is described in [56].

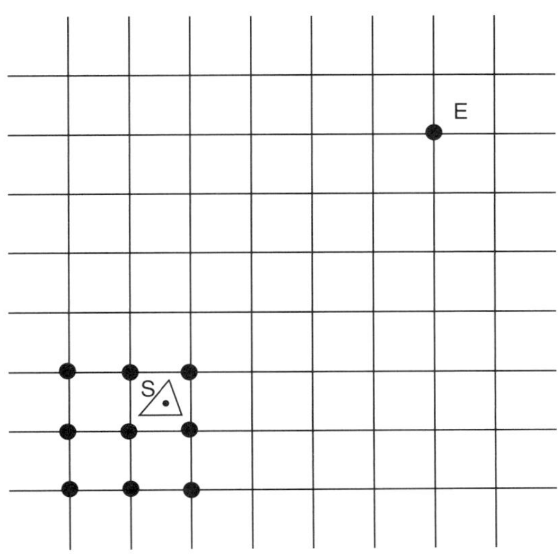

FIGURE 26.5 The precorrected-FFT algorithm treats the far-field by projecting panel charges onto a grid (panel S projected onto nine grid points) and then uses an FFT to compute the potential everywhere on the grid. An interpolation step uses these grid potentials to compute panel potentials (not shown).

Similar to the inductance problem, capturing skin effect in a wide-band sense for a single formulation can lead to many current filaments. Here again, the exponentially shaped basis functions and surface formulations described at the end of Section 26.5 are advantageous for the efficient computation of the full wave frequency response. However, such techniques typically trade size for added complexity in the frequency-dependent behavior of matrices $L(s)$ and $P(s)$. Hence, their full potential could be realized only when used in combination with a robust and efficient model order reduction procedure for frequency-dependent matrices.

26.7 Conclusions

In this chapter, various regimes of electromagnetics and their governing equations as they relate to interconnect extraction and simulation were discussed. The use of integral equation techniques to solve these systems results in dense linear systems whose $O(N^3)$ computational and $O(N^2)$ memory complexity make it impractical to solve for complicated interconnect. To remedy this, we discussed sparsification techniques to bring this complexity down to $O(N)$ or $O(N \log N)$ improving the computation time and memory by orders of magnitude for capacitance, inductance, as well as distributed RLC and full-wave solution.

Acknowledgments

The author would like to thank Tom Korsmeyer and Luca Daniel for many suggestions regarding this chapter, as well as Mike Chou, Keith Nabors, Joel Phillips, and Zhenhai Zhu for diagrams and other contributions used in this paper.

Part II: Statistical Capacitance Extraction

26.8 Introduction

The floating random-walk technique, a Monte Carlo method, was used in the 1950s to calculate voltage[67]. 3D Monte Carlo capacitance extraction using the floating random-walk technique was developed in the early 1990s[68,69]. It has been also used for a similar problem, which is thermal analysis[70]. QuickCap, a commercial capacitance extractor incorporating this approach, has been available since the mid-1990s. Other capacitance extractors have also been developed using similar approaches[71,72]. Monte Carlo capacitance extraction has been extensively used to extract capacitance for

- all nets in test structures and critical cells,
- critical nets in full layouts, and
- layout patterns (used to drive library-lookup methods).

It is especially valuable for accurate modeling of 3D physical features such as conformal and planar dielectrics, bias remaining after optical proximity correction (OPC), thickness variation due to chemical-mechanical planarization (CMP), and trapezoidal cross sections in small test structures and in large IC layouts.

Deterministic field solvers, such as those in the first part of this chapter, approximate voltage, charge, or other state variables and introduce error that is difficult to precisely quantize. In contrast, the floating random-walk technique essentially samples an exact integral for capacitance. The statistical error can be calculated and reported. Depending on the implementation of the floating random-walk method, the statistical error could be the only significant source of error.

Due to its statistical nature, Monte Carlo extraction has several desirable characteristics that are not available to deterministic methods

- User-specified accuracy can be relaxed where possible, decreasing run time.
- Memory usage, orders of magnitude below that of deterministic field solvers, is not impacted when the accuracy goal is increased.

- Run time is only weakly dependent on the size of the problem.
- Capacitance values that matter collectively but not individually can be calculated to low accuracy, saving run time. A circuit simulation can be significantly more accurate than the individual capacitance values because of statistical cancellation.

Because this last point is not widely recognized in the EDA industry, the accuracy goal for statistical capacitance extraction might be specified stricter than it needs to be, resulting in much slower run times that are actually needed.

The following sections describe the theory underlying 3D Monte Carlo capacitance extraction, accuracy benchmarks, characteristics, and statistical cancellation.

26.9 Theory

The floating random-walk method is equivalent to Monte Carlo integration of an integral formulation for capacitance. Because the integral is exact, the error is limited only by the number of samples evaluated.

26.9.1 Integral Formulation for Capacitance

The integral formulation for the capacitance associated with net i is based on the integral form of Gauss' law to find the charge on net i.

$$Q_i = \oiint d\mathbf{A}_1 \, \{\varepsilon\}[(\hat{\mathbf{n}} \cdot \vec{\mathbf{E}}(\mathbf{r}_1))] \tag{26.25}$$

To develop this into an equation for capacitance, we define an operator $\vec{\nabla}_V$ that changes its scalar operand into a vector by taking the derivative with respect to the voltage on each net in turn. Applying this operator to the both sides, Equation 26.25 becomes a vector equation determining all capacitance values associated with net i. In addition, $\vec{\mathbf{E}}(\mathbf{r}_1)$ can be expressed as a surface integral of voltage weighted by a Green's function, and voltage can be similarly expressed. This yields an exact infinite-dimensional integral representation for column $\vec{\mathbf{C}}_i$ of the capacitance matrix:

$$\vec{\mathbf{C}}_i = \oiint d\mathbf{A}_1 \, \{\varepsilon\}[\vec{\nabla}_V(\hat{\mathbf{n}} \cdot \vec{\mathbf{E}}(\mathbf{r}_1))] \tag{26.26a}$$

$$\vec{\nabla}_V(\hat{\mathbf{n}} \cdot \vec{\mathbf{E}}(\mathbf{r}_1)) = \oiint d\mathbf{A}_2 \, \{\hat{\mathbf{n}} \cdot \vec{\mathbf{G}}_E(\mathbf{r}_2|\mathbf{r}_1)\}[\vec{\nabla}_V\phi(\mathbf{r}_2)] \tag{26.26b}$$

$$\vec{\nabla}_V\phi(\mathbf{r}_k) = \oiint d\mathbf{A}_{k+1} \, \{G_\phi(\mathbf{r}_{k+1}|\mathbf{r}_k)\}[\vec{\nabla}_V\phi(\mathbf{r}_{k+1})], \quad k = 2, \ldots, \infty \tag{26.26c}$$

$\vec{\nabla}_V(\hat{\mathbf{n}} \cdot \vec{\mathbf{E}}(\mathbf{r}_1))$ in Equation 26.26a is expanded in Equation 26.26b. $\vec{\nabla}_V\phi(\mathbf{r}_2)$ in Equation 26.26b is expanded in Equation 26.26c. $\vec{\nabla}_V\phi(\mathbf{r}_{k+1})$ in Equation 26.26c is expanded recursively. When $\mathbf{r}$ is on net j, the term $\vec{\nabla}_V\phi(\mathbf{r})$ is an incidence vector which is zero everywhere, except for the jth element, which is unity. Green's functions have been developed in 2D and 3D[67,73].

The integration surface in Equation 26.26b encloses $\mathbf{r}_1$, contains no nets, and is a shape for which the electric field $\vec{\mathbf{E}}(\mathbf{r}_1)$ can be written as a surface Green's function of voltage. $\vec{\mathbf{G}}_E(\mathbf{r}_2|\mathbf{r}_1)$ is the electric field (a vector) at $\mathbf{r}_1$ when the voltage on the surface is an impulse function at $\mathbf{r}_2$. For a sphere radius R centered at $\mathbf{r}_1$, $\vec{\mathbf{G}}_E(\mathbf{r}_2|\mathbf{r}_1)$ is $3(\mathbf{r}_1-\mathbf{r}_2)/4\pi R^4$, as can be derived from the Poisson integral for a sphere[74].

Similarly, the integration surface in Equation 26.26c encloses $\mathbf{r}_k$, contains no nets, and is a shape for which the voltage $\phi(\mathbf{r}_{k+1})$ can be written as a surface Green's function. $G_\phi(\mathbf{r}_{k+1}|\mathbf{r}_k)$ is the voltage at $\mathbf{r}_k$ when the voltage on the surface is an impulse function at $\mathbf{r}_k+_1$. For a spherical integration surface radius R centered at $\mathbf{r}_k$, $G_\phi(\mathbf{r}_{k+1}|\mathbf{r}_k)$ is a constant, $1/4\pi R^2$. In other words, the voltage at the center of a sphere is the average of the voltage on the surface of the sphere.

26.9.2 Monte Carlo Integration

In Monte Carlo integration, the integral of a function $f()$ is evaluated by repeatedly sampling $f()$ randomly to find the average, which is then scaled by the domain size.

A single sample of Equation 26.26 is equivalent to a floating random walk, as is illustrated in Figure 26.6. Here, a 2D sample of the integral using circles as the integration surfaces for the Green's functions starts on a surface around net i and ends on net j. Each sample is a scaled incidence vector. For the walk pictured here, only the jth value of the vector is nonzero.

- To sample Equation 26.26a for net i, $\mathbf{r}_1$ is selected at random on the closed integration surface around net i. The final incidence vector (generated when the walk hits a net) will need to be scaled by ε and the perimeter of the integration surface.
- To sample Equation 26.26b at $\mathbf{r}_1$, $\mathbf{r}_2$ is selected at random on the largest circle around $\mathbf{r}_1$ that contains no nets. The incidence vector will need to be additionally scaled by $\hat{\mathbf{n}} \cdot \vec{\mathbf{G}}_E(\mathbf{r}_2|\mathbf{r}_1)$ (a known value) and the circumference of the circle.
- To sample Equation 26.26c at $\mathbf{r}_2$ and successive points, $\mathbf{r}_3$, $\mathbf{r}_4$, and $\mathbf{r}_5$ are similarly selected at random, each on the largest circle around the previous point. The incidence vector will need to be scaled by $G_\phi(\mathbf{r}_3|\mathbf{r}_2)$, $G_\phi(\mathbf{r}_4|\mathbf{r}_3)$, and $G_\phi(\mathbf{r}_5|\mathbf{r}_4)$ (known values) and the circumference of each circle.
- The function $\vec{\nabla}_V\phi(\mathbf{r}_5)$, here, is taken to be an incidence vector that is zero except for the jth element, which is unity. Since a random point selected on the surface of a circle will never lie exactly on a prespecified flat tangential surface, using circular Green's functions requires some level of approximation in order to complete the Monte Carlo sample. The commercial capacitance extractor QuickCap, referenced in the next section, uses Green's functions based on cubes and does not require such an approximation.

The series of points used to sample Equation 26.26 is equivalent to a floating random walk, where each point is selected on the perimeter of the largest homogeneous circle centered at the previous point. This example describes a simple implementation. Advanced implementations might use importance sampling or other techniques[75].

The scaled incidence vectors from many samples (or walks) are averaged together. An averaged capacitance value $\langle C_{ij}\rangle$ has a 1-σ standard error δ_{ij} given by

$$\delta_{ij} = \sqrt{(\langle C_{ij}^2\rangle - \langle C_{ij}\rangle^2)/n} \tag{26.27}$$

$\langle C_{ij}^2\rangle$ is the average of the square of the capacitance, and n the number of samples. As n increases, the probability distribution associated with the statistical error conforms to a normal distribution (bell curve), allowing a confidence level to be evaluated.

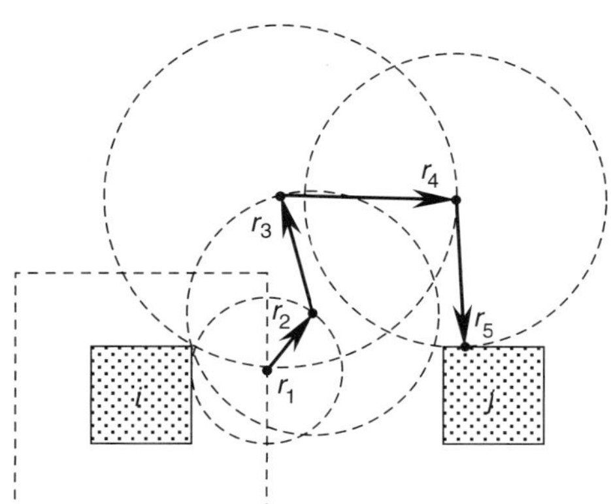

FIGURE 26.6 A 2D floating random walk for capacitance, using circles.

26.10 Characteristics

Characteristics for a Monte Carlo extractor that follow from theory are listed below and illustrated in Sections 26.10.1–26.10.4. An example in Section 26.10.5 highlights a cautionary note concerning the applicability of a "gold extractor." Results are shown using QuickCap, a commercial capacitance extractor developed by then Random Logic Corporation, now acquired by Magma Design Automation:

- *Reportable error.* The statistical error can be calculated (Equation 26.27) and reported. Section 26.10.1 shows how this can be used to calculate confidence levels.
- *Little or no bias.* No significant approximations are introduced when evaluating the capacitance integral. Benchmarks are shown in Section 26.10.2.
- *Memory efficient.* The bulk of required memory is to represent the geometric structure. The method uses no mesh. Examples are shown in Section 26.10.3.
- *Known convergence rate.* The statistical error should decrease with run time t as $t^{-1/2}$. Twice the accuracy requires four times the run time.
- *Convergence rate is a weak function of problem size.* Expanding the problem size and adding many structures should have little effect on the speed of a single walk. Examples are shown in Section 26.10.3.
- *Statistically independent errors.* Statistical cancellation occurs whenever capacitance values are effectively added, whether before or during simulation. This applies to tiled capacitance extraction, RC analysis, delay times involving multiple nets in a critical path, and crosstalk involving multiple aggressor nets. This is further described in Section 26.10.4.

Because these characteristics are markedly different from those of deterministic methods, some care is required when comparing the two types of extractors.

26.10.1 Standard Error

The standard error δ of a result is a measure of uncertainty of that result due to statistical considerations. The standard error of a single statistical result should agree with the standard deviation of a population of results.

QuickCap reports the standard error (1-σ error) associated with each capacitance value it extracts. Figure 26.7 shows a histogram of 10,000 independent Monte Carlo calculations of a single capacitance value. These 10,000 runs reported standard errors between 4.0 and 4.3 aF, agreeing with the standard deviation of the entire population, 4.01 aF.

When a result is based on many Monte Carlo samples, the probability distribution associated with the statistical error due to statistics can generally be expected to be a *normal distribution* (a bell curve), which is fully described by an average (the statistical result) and a width, δ. In such a case, the standard error can be used to calculate a confidence level. Table 26.2 lists some confidence levels.

When a statistical result is based on few samples, the standard error can be a reasonable estimate of the standard deviation of a population of results, even though the distribution of the population is not normal. The examples in Section 26.10.4 apply even when the probability distribution associated with the statistical error is not a normal distribution.

26.10.2 Bias

Bias is the amount of error that is not due to statistics. Bias can arise from discretization of equations (does not occur in the floating random-walk method), from round-off error, from boundary error, from limitations of the random-number generator, etc. Bias can be measured on a case-by-case basis when the correct answer is known.

Table 26.3 lists QuickCap bias for three problems with analytic solutions [76]: the capacitance between infinite-area parallel plates (1D electric field), the capacitance between infinite-length parallel circular

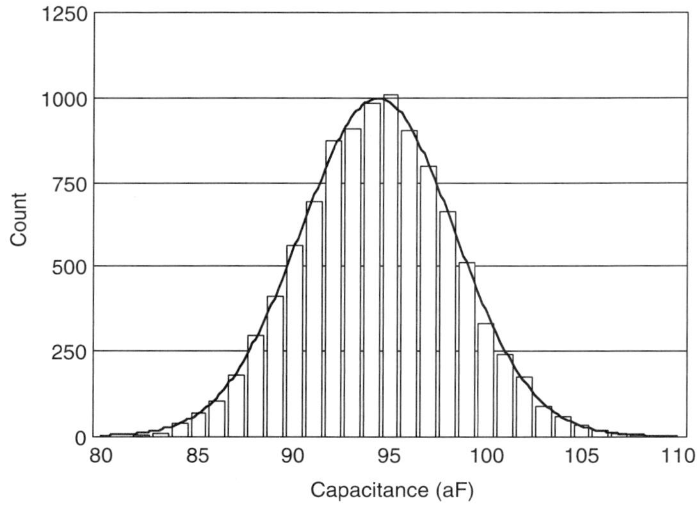

FIGURE 26.7 Histogram of 10,000 independent Monte Carlo samples. The solid line is a Gaussian characterized by the average (95.39 aF) and standard deviation (4.01 aF) of the results. (From data provided by Magma Design Automation, 1994.)

TABLE 26.2 Confidence Levels when Statistical Error has a Normal Distribution

n	Confidence that Result is Within $\pm n\delta$	Confidence that Result is Not More Than $n\delta$ Overestimated (Underestimated)
1	68% (2/3)	84% (5/6)
2	95% (19/20)	98% (49/50)
3	99.7% (349/350)	99.87% (749/750)
4	99.994% (1–1/15,000)	99.997% (1–1/30,000)
5	99.99994% (1–1/1,500,000)	99.99997% (1–1/3,000,000)

TABLE 26.3 QuickCap Bias for Known Problems (From Data provided by Magma Design Automation, 1994.)

Problem Description	Analytic Solution	Normalized QuickCap Bias
1D: Infinite parallel plates, spaced d apart	ε/d	−0.004% ± 0.001%
2D: Two infinite wires radius R, spaced D (center to center)	$\pi\varepsilon/\ln\left[\frac{D}{2R} + \sqrt{\left(\frac{D}{2R}\right)^2 - 1}\right]$	−0.001% ± 0.002%
3D: Sphere radius R in free space	$4\pi\varepsilon R$	−0.002% ± 0.003%

wires (2D electric field), and the capacitance of a sphere in free space (3D electric field). The bias is quite small compared to the accuracy required for the analysis of IC layouts. Because the random-walk method is local (it has no mesh that depends on the problem size, and the size of each hop is limited by the *nearest* object), these bias values are expected to apply to self-(total) capacitance in an environment with uniform dielectric. QuickCap's coupling–capacitance bias has also been checked and has been found to be small whether in a uniform dielectric, or in an environment with multiple dielectrics. Naturally, this behavior is implementation-dependent.

Because the floating random-walk method for capacitance extraction consists of Monte Carlo samples of an exact integral representation of capacitance, it can have negligible bias. The error in a method that attempts to solve for charge, voltage, or electric field by discretization is more difficult to quantify.

26.10.3 Accuracy, Memory, Problem Size, and Run Time

Any capacitance extractor has tradeoffs between accuracy, memory, problem size, and run time. To add perspective, Monte Carlo capacitance extraction and deterministic approaches are compared to each other, here. "Deterministic" approaches include finite element methods, boundary-element methods, and transform methods. For fairness, methods should be compared at the same accuracy levels. Also, a range of problems should be tested since each method has its own strengths and weaknesses.

The parameters that control the level of error are quite different for Monte Carlo and deterministic capacitance extraction. For the floating random-walk approach, the error is basically controlled by the number of samples, proportional to the run time for a given problem. Since this method can report its own error, error control is straightforward.

For mesh-based approaches, the error is controlled by discretization which in turn affects both memory requirements and run time. The relationship between discretization and error is not straightforward. By varying the amount of discretization, however, the effect of discretization on the capacitance value can be observed, allowing an estimate of the error to be extrapolated.

Figure 26.8 shows how run time and memory requirements vary with problem "complexity," the size, or volume of a problem. The boundary-element results were generated with an academic version of a multipole-accelerated capacitance tool. QuickCap runtime for a given level of accuracy has a weak dependence complexity, while the boundary-element method used has an approximately linear dependency. The memory required to achieve the same level of accuracy increases linearly in both cases, though QuickCap memory requirements are many orders of magnitude below those of the boundary-element method. Note that the error of the deterministic method, here, is that due to spatial discretization and does not include boundary error. Also note that the important feature in the run-time plot of Figure 26.8 is the slope. Not only can a commercial implementation of boundary-element extractor be significantly faster than the academic version used here, the relative speed can change with the type of problem analyzed and the convergence goal.

One characteristic not shown in Figure 26.8 is the dependence of accuracy on memory. For QuickCap, the accuracy does not depend on memory. For the boundary-element method used here, the error is inversely proportional to the memory used. For other method tested in Ref. [77], the error was difficult to control. It was inversely proportional to the fourth root of memory.

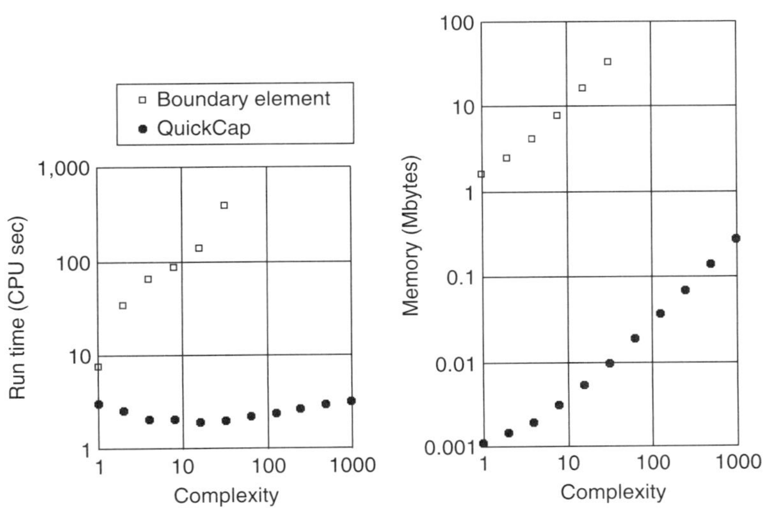

FIGURE 26.8 Characteristics of QuickCap and a boundary-element approach. The results, normalized to 2% error, are based on published data. (From Le Coz, Y.L. et al., *Solid State Electron.*, 42, 581–588, 1998.) Complexity, arbitrary units, refers to the number of elements and to the calculation volume.

Sample characteristics for statistical and deterministic capacitance extractors are summarized in Table 26.4. For QuickCap, where the run time is proportional to the number of samples, the error drops slower than for the boundary-element method of Ref. [77]. However, of primary importance is the run time in the 1 to 10% range required for accurate circuit analysis. For the boundary- and finite-element methods, discretization error is related to memory. This is quite different from QuickCap's characteristics. Because QuickCap's run time is only weakly dependent on problem size, QuickCap will be faster than deterministic methods on large problems (for the equivalent levels of accuracy). Finally, QuickCap's memory requirements are low.

Boundary error, not addressed above, is the error due to neglecting objects outside of some "calculation region." Whether one applies periodic boundary conditions, reflective boundary conditions, some "mixed" condition, or simply ignored objects outside the calculation region, the error can be calculated by finding the range of capacitance when varying the environment beyond the calculation region. Because the floating random-walk method is memory efficient, expanding the calculation region to include additional objects is straightforward.

26.10.4 Statistical Cancellation

When the implementation of the floating random-walk approach has negligible bias, many advantages can be realized, including efficient tiled capacitance extraction, and error reduction during circuit simulation with respect to accumulative delay time, crosstalk, and the behavior of RC networks.

Statistical error cancellation occurs when adding are averaging statistical results. The sum of two statistical values, $A \pm \delta_A$ and $B \pm \delta_B$, is $(A + B) \pm \sqrt{\delta_A^2 + \delta_B^2}$. While the total error increases, the relative error decreases.

Tiled capacitance extraction can be performed with nearly 100% efficiency. In other words, the total run time and memory used can be about the same as without tiling. A tiled capacitance run can be implemented in parallel, allowing huge layouts to be analyzed in reasonable time. Consider, for example, the capacitance of the net shown in Figure 26.9. The layout is divided into two tiles, each including some of the physical data from the adjacent tile. The total capacitance is the integral (from Equation 26.6) over the dashed line. Whether this is sampled in two pieces (left) or as a single untiled run (right) makes no difference to the capacitance value or the statistical error. The only source of error is stitching error due to walks that terminate outside the valid (shaded) region. Furthermore, performing 64,000 walks on the upper part and 36,000 walks on the lower part is approximately equivalent to performing 100,000 walks on the nontiled net in terms of the total numeric result and statistical error. Note that because of statistical cancellation, even though the relative errors associated with 64 and 36fF, here, are ±1.25 and ±1.67%, the relative error of the sum is ±1%.

Whenever a critical path involves multiple nets, the total delay time will exhibit statistical error cancellation as long as the errors are independent, even when the analysis does not consider statistical error. In Figure 26.10, even though the individual delay times have approximately ±2% statistical error, the statistical error of the total delay time is ±1%. In order for a deterministic capacitance extractor to match this, it would need to generate ±1% results on each of the four nets in the critical path.

While small coupling capacitances might be individually negligible, they can be collectively important when many aggressor nets switch simultaneously. In Figure 26.11, each coupling capacitance is known to ±10%. Simultaneous switching results in 25% crosstalk (±2% relative error or ±0.5% crosstalk error). For deterministic results to match this level of accuracy, each result would need to be ±2%. Using statistical

TABLE 26.4 Sample Characteristics of Statistical and Deterministic Capacitance Extractors

	Floating Random Walk	Boundary Element	Finite Element
Accuracy vs. run time	$\varepsilon \propto 1/t^{1/2}$	$\varepsilon \propto 1/t$	$\varepsilon \propto 1/t^{1/6}$
Accuracy vs. memory	Independent	$\varepsilon \propto 1/m$	$\varepsilon \propto 1/m^{1/4}$
Run time vs. complexity	Weak dependence	$t \propto C$	$t \propto C$
Memory vs. complexity	$m \propto C$	$m \propto C$	$m \propto C$

Note: This summary is based on published data. Error of the deterministic capacitance does not include error introduced by boundary approximations. (From Le Coz, Y.L. et. al., *Solid State Electron.*, 42, 581–588, 1998.)

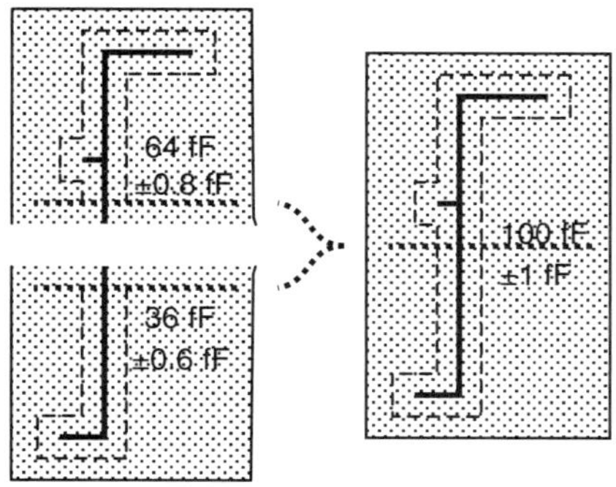

FIGURE 26.9 Tiled capacitance extraction can be almost 100% efficient while introducing negligible stitching error.

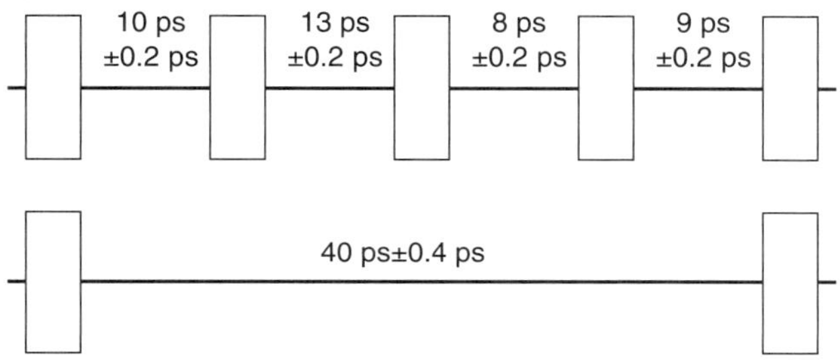

FIGURE 26.10 Statistical error is canceled in multinet critical paths even when a timing analyzer is not aware of the error.

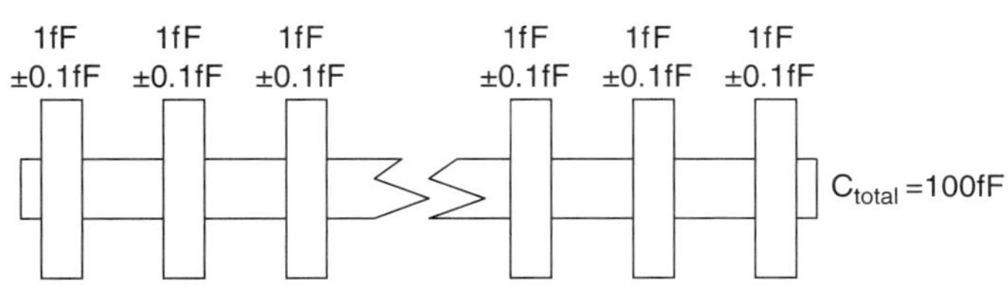

FIGURE 26.11 Statistical error cancellation occurs when aggressor nets switch simultaneously. The total coupling capacitance for 25 crosswires, here, is 25±0.5 fF.

methods, the extraction time for the capacitance associated with a given victim net need not be increased when it has lots of coupling capacitance. While the individual coupling capacitance may have significant relative error, the effect of any single capacitance on circuit behavior is small, and the collective effect incorporates statistical error cancellation.

Similar statistical cancellation occurs for waveforms of RC networks. In Figure 26.12, each of 25 capacitance values is known to be ±10%. The statistical error associated with the Elmore delay time (the first-order response time) is ±2.3%, comparable to the statistical error associated with the sum (±2%). For deterministic results to match this level of accuracy, each result would need to be ±2.3%. This effect is more pronounced for more complex RC networks.

26.10.5 Technology Modeling

The accuracy of a capacitance extractor is compromised when it is given bad input. To address this issue, details of a given IC fabrication technology should be examined to decide what is important. For example, the physical line widths may not match layout line widths due to bias from OPC, CMP effects can

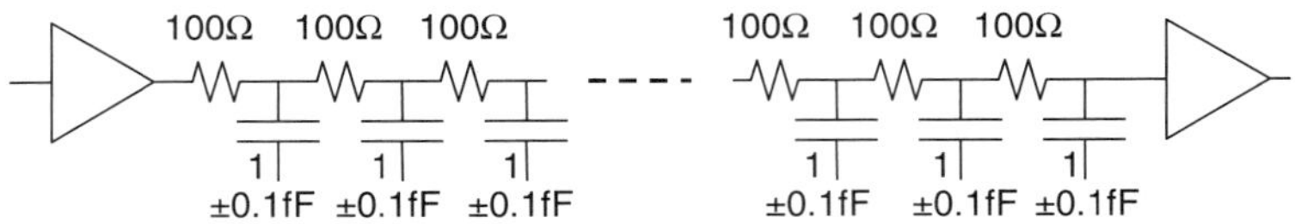

FIGURE 26.12 Statistical error cancellation occurs in RC networks. The total Elmore delay time due to 25 RC stages here is 32.5±7.4 ps (±2.3%).

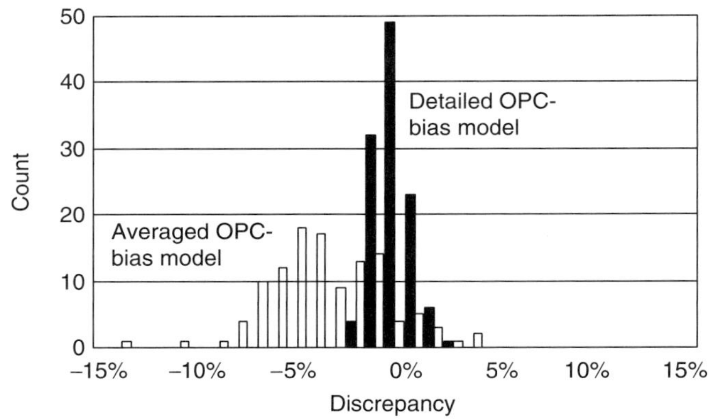

FIGURE 26.13 Comparison between silicon measurements and capacitance values extracted using two technology models. (From Data provided by Magma Design Automation, 2004.)

cause thickness variations, and sidewalls may not be vertical. Care must be taken when using statistical capacitance extraction or a 3D field solution as a gold standard. It can only be "gold" if given the correct input. Claims that a "fast" capacitance calculator matches a gold standard capacitance extractor are meaningless, when the gold standard extractor is given an approximated geometry.

Figure 26.13 shows a comparison between QuickCap and silicon measurements for test structures in a 90 nm process. For the detailed OPC-bias model, most of the discrepancy can be attributed to imprecise representation of the silicon (wire width, wire spacing, dielectric values, etc.) rather than any QuickCap bias. In fact, discrepancies of almost 15% result if the technology model uses an average OPC correction rather than one that is a function of line width and spacing. Fast methods to calculate capacitance should be based on comparisons to full 3D results that use the detailed-OPC bias model, here, not an averaged OPC-bias model.

26.11 Summary

The floating random-walk method for capacitance extraction has many characteristics not found in deterministic methods. In addition to reporting the statistical error (the only significant source of error), it is memory efficient and capable of handling large layouts. Statistical error cancellation results in circuit simulations that are more accurate than one might expect looking at the accuracy of individual capacitance values.

References

[1] Y.L. Le Coz and R.B. Iverson, A stochastic algorithm for high speed capacitance extraction in integrated circuits, *Solid State Electron.*, 35, 1005–1012, 1992.

[2] O.M. Ramahi and B. Archambeault, Adaptive absorbing boundary conditions in finite-difference time domain applications for emc simulations, *IEEE Trans. Electromagn. Comp.* 37, 580–583, 1995.

[3] J.C. Veihl and R. Mittra, An efficient implementation of berenger's perfectly matched layer (pml) for finite-difference time-domain mesh truncation, *IEEE Microwave Guided Wave Lett.*, 6, 94, 1996.

[4] L. Greengard, *The Rapid Evaluation of Potential Fields in Particle Systems*, MIT. Press, Cambridge, MA, 1988.

5] V. Rokhlin, Rapid solution of integral equations of classical potential theory. *J. Comput. Phys.*, 60, 187–207, 1985.

[6] K. Nabors and J. White, Fastcap: a multipole accelerated 3-D capacitance extraction program, *IEEE Trans Comput.-Aided Des. Integrated Circuits Systems*, 10, 1447–1459, 1991.

[7] A. Brandt, Multilevel computations of integral transforms and particle interactions with oscillatory kernls, *Comput. Phys. Commun.*, 65, 24–38, 1991.

[8] J.R. Phillips and J.K. White, A precorrected-fft method for electrostatic analysis of complicated 3-d structures. *IEEE Trans. Comput.-Aided Des. Integrated Circuits Systems*, 16, 1059–1072, 1997.

[9] S. Kapur and D.E. Long, Ies3: Efficient electrostatic and electromagnetic simulation, *IEEE Comput. Sci. Eng.*, 5, 60–67, 1998.

[10] J.M. Song, C.C. Lu, W.C. Chew, and S.W. Lee. Fast illinois solver code (fisc). *IEEE Antennas Propagation Magazine*, 40, 27–34, 1998.

[11] S. Kumashiro, R. Rohrer, and A. Strojwas, A new efficient method for the transient simulation of 3-d interconnect structures, *Proccedings of Inernational, Electron Devices Meeting*, 1990, pp. 193–196.

[12] M. Chou and J.K. White, Efficient formulation and model-order reduction for the transient simulation of three-dimensional vlsi interconnect, *IEEE Trans. Computer-Aided Des. Integrated Circuits Systems*, 16, 1454–1476, 1997.

[13] K.L. Shepard and T. Zhong, Return-limited inductances: a practical approach to on-chip inductance extraction, *IEEE Trans. Comput.-Aided Des. Integrated Circuits Systems*, 19, 425–436, 2000.

[14] M. Kamon, *Fast Parasitic Extraction and Simulation of Three-dimensional Interconnect via Quasistatic Analysis*, Ph.D. dissertation, Massachusetts Institute of Technology, Electrical Engineering and Computer Science, 1998.

[15] H. Johnson and M. Graham, *High-Speed Digital Design: A Handbook of Black Magic*, Prentice-Hall, Upper Saddle River, NJ, 1993.

[16] A.E. Ruehli, Equivalent circuit models for three-dimensional multiconductor systems, *IEEE Trans. Microwave Theory Techniques*, MTT-22, 216–221, 1974.

[17] H.A. Haus and J.R. Melcher, *Electromagnetic Fields and Energy*, Prentice-Hall, Englewood Cliffs, NJ, 1989.

[18] S. Ramo, J.R. Whinnery, and T.V Duzer, *Fields and Waves in Communication Electronics*, Wiley, New York, 1994.

[19] A.H. Zemanian, R.P. Tewarson, C.P. Ju, and J.F. Jen. Three-dimensional capacitance computations for VLSI/ULSI interconnections, *IEEE Trans. Comput.-Aided Des.*, 8, 1319–1326, 1989.

[20] A. Seidl, H. Klose, M. Svoboda, J. Oberndorfer, and W. Rösner. CAPCAL — a 3-d capacitance solver for support of CAD systems, *IEEE Trans, Comput.-Aided Des. Integr. Circuits Syst.*, 7, 549–556, 1988.

[21] P.E. Cottrell and E.M. Buturla, VLSI wiring capacitance. *IBM J. Res. Dev.*, 29, 277–287, 1985.

[22] T. Chou and Z.J. Cendes, Capacitance calculation of IC packages using the finite element method and planes of symmetry, *IEEE Trans. Comput.-Aided Des.*, 13, 1159–1166, 1994.

[23] I. Stakgold, *Boundary Value Problems of Mathematical Physics*, Vol. 2, Macmillan, New York, 1968.

[24] M.A. Jaswon and G.T. Symm, *Integral Equation Methods in Potential Theory and Elastostatics*, Academic Press, London, 1977.

[25] J. Tausch and J.K. White, Capacitance extraction of 3-d conductor systems in dielectric media with high-permittivity ratios, *IEEE Trans. Microwave Theory Tech.*, 47, 18–26, 1999.

[26] S.M. Rao, T.K. Sarkar, and R.F. Harrington, The electrostatic field of conducting bodies in multiple di-electric media, *IEEE Trans. Microwave Theory Tech.*, MTT-32, 1441–1448, 1984.

[27] K. Nabors and J. White, Multipole-accelerated capacitance extraction algorithms for 3-d structures with multiple dielectrics, *IEEE Trans. Circuits Syst. I: Fundam. Theory Appl.*, 39, 946–954, 1992.

[28] S.H. Crandall, *Engineering Analysis*, McGraw-Hill, New York, 1956.

[29] R.F. Harrington, *Field Computation by Moment Methods*, MacMillan, New York, 1968.
[30] J.N. Newman, Distributions of sources and normal dipoles over a quadrilateral panel, *J. Eng. Math.*, 20, 113–126, 1986.
[31] Y. Saad and M.H. Schultz, GMRES: A generalized minimal residual algorithm for solving nonsymmetric linear systems, *SIAM J. Sci. Stat. Comp.*, 7,856–869, 1986.
[32] K. Nabors and J. White, Fast capacitance extraction of general three-dimensional structures, *IEEE Trans. Microwave Theory Tech.*, 40, 7, pp. 1496–1506, 1992.
[33] K. Nabors, F.T. Korsmeyer, F.T. Leighton, and J. White, Multipole accelerated preconditioned iterative methods for three-dimensional potential integral equations of the first kind, *SIAM J. Sci. Stat. Comp.*, 15, 713–735, 1994.
[34] M. Bachtold, M. Spasojevic, C. Lage, and P.B. Ljung, A system for full-chip and critical net parasitic extraction for ulsi interconnects using a fast 3-d field solver,*IEEE Trans. Comput.-Aided Des. Integr. Circuits Syst.*, 19, 325–338, 2000.
[35] M.W. Beattie and L.T. Pileggi, Parasitics extraction with multipole refinement, *IEEE Trans. Comput.-Aided Des. Integrated Circuits Syst.*, 23, 288–292, 2004.
[36] W.T. Weeks, L.L. Wu, M.F. McAllister, and A. Singh, Resistive and inductive skin effect in rectangular conductors, *IBM J. Res. Dev.*, 23, 652–660, 1979.
[37] A.E. Ruehli, Inductance calculations in a complex integrated circuit environment, *IBM J. Res. Develop.*, 16, 470–481, 1972.
[38] A.C. Cangellaris, J.L. Prince, and L.P. Vakanas, Frequency-dependent inductance and resistance calculation for three-dimensional structures in high-speed interconnect systems, *IEEE Trans. Compon., Hybrids and Manufacturing Tech.*, 13, 154–159, 1990.
[39] M.J. Tsuk and J.A. Kong, A hybrid method for the calculation of the resistance and inductance of transmission lines with arbitrary cross sections, *IEEE Trans. Microwave Theory Tech.*,39, 1338–1347, 1991.
[40] M. Kamon, M.J. Tsuk, and J. White, Fasthenry: a multipole-accelerated 3-d inductance extraction program, *IEEE Trans. Microwave Theory Tech.*, 42, 1750–1758, 1994.
[41] F.W. Grover, *Inductance Calculations, Working Formulas and Tables*, Dover, New York, 1962.
[42] C. Hoer and C. Love, Exact inductance equations for rectangular conductors with applications to more complicated geometries, *J. Res. Natl. Bureau Standards*, 69C, 127–137, 1965.
[43] D. Ling and A.E. Ruehli, Interconnect modeling, in *Circuit Analysis, Simulation and Design, 2*, A. E. Ruehli, Ed., Elsevier Scienc, North-Holland, 1987, chap. 11, pp. 211–332.
[44] G.D. Hachtel, R.K. Brayton, and F.G. Gustavson, The sparse tableau approach to network analysis and design, *IEEE Trans. Circuit Theory*, 18, 101–113, 1971.
[45] G.H. Golub and C.F. Van Loan, *Matrix Computations*, 2nd Ed., The Johns Hopkins University Press, Baltimore, 1989.
[46] C. Desoer and E. Kuh, *Basic Circuit Theory*, McGraw-Hill, New York, 1969.
[47] E. Tuncer, B.-T. Lee, and D.P. Neikirk, Interconnect series impedance determination using a surface ribbon method, *Proceedings of the IEEE Topical Meeting on Electrical Performance of Electronic Packaging*, 1994, pp. 249–252.
[48] L. Daniel, A. Sangiovanni-Vincentelli, and J. White, Using conduction modes basis functions for efficient electromagneticanalysis of on-chip and off-chip interconnect, *Proceedings of the Design Automation Conference*, 2001, pp. 563–566.
[49] L. Daniel, A. Sangiovanni-Vincentelli, and J. White, Proximity templates for modeling of skin and proximity effects on packages and high frequency interconnect, *Proceedings of the IEEE/ACM International Conference on Computer Aided-Design*, 2002, pp. 326–333.
[50] K.M. Coperich, A.E. Ruehli, and A. Cangellaris, Enhanced skin effect for partial-element equivalent-circuit (peec) models, *IEEE Trans. Microwave Theory Techniques*, 48, 1435–1442, 2000.
[51] J. Wang and J.K. White, A wide frequency range surface integral formualtion for 3d rlc extraction, *Proceedings of the International Conference on Computer Aided-Design*, November 1999, pp. 453–457.

[52] Z. Zhu, B. Song, and J. White, Algorithms in fastimp: a fast and wideband impedance extraction, program for complicated 3-d geometries, *Proceedings of the Design Automation Conference*, 2003, pp. 712–717.
[53] Hansruedi Heeb and Albert E. Ruehli, Three-dimensional interconnect analysis using partial element equivalent circuits, *IEEE Trans. Circuits Syst. I:Fundam. Theory Appl.*, 39, 974–982, 1992.
[54] S.M. Rao, D.R. Wilton, and A.W. Glisson, Electromagnetic scattering by surfaces of arbitrary shape, *IEEE Trans. Antennas Propagation*, 30, 409–418, 1982.
[55] E. Arvas and R.F. Harrington, Computation of the magnetic polarizability of conducting disks and the electric polarizability of apertures, *IEEE Trans. Antennas Propag.*, 31, 719–725, 1983.
[56] B. Anderson, J.E. Bracken, J.B. Manges, G. Peng, and Z. Cendes, Full-wave analysis in spice via model-order reduction. *IEEE Trans. Microwave Theory Tech.*, 52, 2314–2320, 2004.
[57] A. Odabasioglu, M. Celik, and L. Pileggi, Prima: Passive Reduced-Order Interconnect Macro-Modeling Algorithm, *CMU Report*, 1997.
[58] M. Kamon, N. Marques, L.M. Silveira, and J. White, Automatic generation of accurate circuit models of 3-d interconnect. *IEEE Trans. Compon. Packaging Manuf. Tech.-Part B*, 21, 225–240, 1998.
[59] Ibrahim M. Elfadel and D.D. Ling. A block rational Arnoldi algorithm for multipoint passive model-order reduction of multiport RLC networks, *Proceedings of IEEE/ACM International Conference on Computer Aided-Design*, San Jose, CA, 1997.
[60] J.-R. Li, F. Wang, and J. White, Efficient model reduction of interconnect via approximate system grammians, *Proceedings of International conference Computer Aided-Design*, pp. 380–383, 1999.
[61] P. Rabiei and M. Pedram, Model order reduction of large circuits using balanced truncation, *Proceedings of Asia and South Pacific Design Automation Conf*, pp. 237–240, Hong Kong, 1999.
[62] I.M. Jaimoukha and E.M. Kasenally, Krylov subspsace methods for solving large lyapunov equations, *SIAM J. Numer. Anal.*, 31, 227–251, 1994.
[63] J.R. Phillips, L. Daniel, and L.M. Silveira, Guaranteed passive balancing transformations for model order reduction, *IEEE Trans. Comput.-Aid. Des. Integr. Circuits Syst.* 22, 1027–1041, 2003.
[64] M. Kamon, F. Wang, and J.K. White, Generating nearly optimally compact models from krylov-subspace based reduced-order models, *IEEE Trans. Circuits Syst.— II: Analog Digital Signal Process*, 47(4), 239–248, 2000.
[65] J.R. Phillips, E. Chiprout, and D.D. Ling, Efficient full-wave electromagnetic analysis via model-order reduction of fast integral transforms, *Proceedings of the 33rd Design Automation Conference*, Las Vegas, Nevada, 1996.
[66] Eli Chiprout, Hansruedi Heeb, Michael S. Nakhla, and Albert E. Ruehli, Simulating 3-D retarded interconnect models using complex frequency hopping (CFH). *International Conference on Computer Aided-Design*, Santa Clara, CA November 1993, pp. 6–72.
[67] G.M. Brown, *Modern Mathematics for Engineers*, E.F. Beckenbach, (Ed.), McGraw-Hill, New York, 1956.
[68] Y.L. Le Coz and R.B. Iverson, A stochastic algorithm for high-speed capacitance extraction in integrated circuits, *Solid-State Electron.*, 35, 1005–1012, 1992.
[69] J. Jere and Y.L. Le Coz, An improved method for the multi-dielectric dirichlet problem, *IEEE Trans. Microwave Theory Tech.*, 41, 325–329, 1993.
[70] Y.L. Le Coz , R.B. Iverson, T.-L. Sham, H.F. Tiersten, and M.S. Shephard, Theory of a floating random-walk algorithm for solving the steady-state heat equation in complex, materially inhomogeneous rectilinear domains, *Numerical Heat Transfer, Part B: Fundam.*, 26, pp. 353–366, 1994.
[71] M. Mascagni and N.A. Simonov, The random walk on the boundary method for calculating capacitance, *J. Compu. Phys.*, 195, 465–473, 2004.
[72] M.P. Desai, CAPEM: The Capacitance Extraction Tool, http://www.ee.iitb.ac.in/~microel/download
[73] R.B. Iverson and Y.L. Le Coz, A floating random-walk algorithm for extracting electrical capacitance, *J. Math. Comput. Simulation*, 55, 59–66, 2001.

[74] J.D. Jackson, *Classical Electrodynamics*, 3rd Ed., Wiley, New York, 1998, p. 65, Eq. (2.19) ff.

[75] W.H. Press et al., *Numerical Recipes in C*, 2nd Ed., Cambridge University Press, Cambridge, pp. 316–328.

[76] Data provided by Magma Design Automation, 2004.

[77] Y.L. Le Coz, H.J. Greub, and R.B. Iverson, Performance of random-walk capacitance extractors for IC interconnects: a numerical study, *Solid-State Electron.*, 42, pp. 581–588, 1998.

Index

C

D

E

F

G

H

M

Q

R

S

T

U

V